Essential Reagents for Organic Synthesis

Essential Reagents for Organic Synthesis

Edited by

Philip L. Fuchs
Purdue University, West Lafayette, IN, USA

André B. Charette
Université de Montréal, Montréal, Québec, Canada

Tomislav Rovis
Colorado State University, Fort Collins, CO, USA

Jeffrey W. Bode
ETH Zürich, Switzerland

WILEY

Library of Congress Cataloging-in-Publication Data

Names: Fuchs, Philip L., 1945- editor. | Charette, A. B. (André B.), 1961-
    editor. | Rovis, Tomislav, 1968- editor. | Bode, J. W. (Jeffrey W.),
    editor.
Title: Essential reagents for organic synthesis / edited by Philip L. Fuchs,
    Andre B. Charette, Tomislav Rovis, Jeffrey W. Bode.
Description: Chichester, UK ; Hoboken, NJ : John Wiley & Sons, 2016. |
    Includes index.
Identifiers: LCCN 2016021468| ISBN 9781119278306 (paperback) | ISBN 9781119279877
    (epub)
Subjects: LCSH: Chemical tests and reagents. | Organic compounds–Synthesis.
Classification: LCC QD77 .E77 2016 | DDC 547/.2–dc23 LC record available at
    https://lccn.loc.gov/2016021468

A catalogue record for this book is available from the British Library.

ISBN 13: 978-1-119-27830-6

Set in 9½/11½ pt Times Roman by Thomson Press (India) Ltd., New Delhi.
Printed and bound in Singapore by Markono Print Media Pte Ltd.

Contents

Preface

This handbook is a subset of the *Encyclopedia of Reagents for Organic Synthesis* (EROS), a knowledge base with detailed information on organic-chemical reagents and catalysts. As of mid-2016, the online collection offers reviews on 4959 different reagents and catalysts that are regularly updated. To keep up with the continual evolution in the field, about 200 new or updated reagent articles are added per year to the online database.

In addition to the complete collection that is available only online (see *http://wileyonlinelibrary.com/ref/eros*), a number of highly focused single-volume handbooks in print and electronic format on editor-selected subjects have been published (*Handbook for Reagents in Organic Synthesis*, HROS). Recent titles in the HROS series include:

- *Reagents for Organocatalysis*
  Edited by Tomislav Rovis (2016)

- *Reagents for Heteroarene Functionalization*
  Edited by André B. Charette (2015)

- *Catalytic Oxidation Reagents*
  Edited by Philip L. Fuchs (2013)

- *Reagents for Silicon-Mediated Organic Synthesis*
  Edited by Philip L. Fuchs (2011)

- *Sulfur-Containing Reagents*
  Edited by Leo A. Paquette (2010)

- *Reagents for Radical and Radical Ion Chemistry*
  Edited by David Crich (2009)

Data mining of EROS user downloads guided by editorial adjudication has yielded the present collection of 52 often-used reagents that will facilitate the daily laboratory endeavor of every organic chemist. The collection contains oxidants (15), reductants (10), metal and organic catalysts (11), Brønsted and Lewis acids (8), and bases (6), to cite a few general mechanistic categories.

We hope that this handbook will prove to be an invaluable primary resource for both beginning graduate students and experienced Ph.D. researchers.

Philip L. Fuchs
Purdue University, West Lafayette, IN, USA

André B. Charette
Université de Montréal, Montréal, Québec, Canada

Tomislav Rovis
Colorado State University, Fort Collins, CO, USA

Jeffrey W. Bode
ETH Zürich, Switzerland

Short Note on InChIs and InChIKeys

The IUPAC International Chemical Identifier (InChI^{TM}) and its compressed form, the InChIKey, are strings of letters representing organic chemical structures that allow for structure searching with a wide range of online search engines and databases such as Google and PubChem. While they are obviously an important development for online reference works, such as *Encyclopedia of Reagents for Organic Synthesis (e-EROS)*, readers of this volume may be surprised to find printed InChI and InChIKey information for each of the reagents.

We introduced InChI and InChIKey to e-EROS in autumn 2009, including the strings in all HTML and PDF files. While we wanted to make sure that all users of *e-EROS*, the second print edition of *EROS*, and all derivative handbooks would find the same information, we appreciate that the strings will be of little use to the readers of the print editions, unless they treat them simply as reminders that *e-EROS* now offers the convenience of InChIs and InChIKeys, allowing the online users to make best use of their browsers and perform searches in a wide range of media.

If you would like to know more about InChIs and InChIKeys, please go to the *e-EROS* website: www.wileyonlinelibrary.com/ref/eros and click on the InChI and InChIKey link.

General Abbreviations

Ac	acetyl
acac	acetylacetonate
AIBN	2,2′-azobisisobutyronitrile
Ar	aryl
BBN	borabicyclo[3.3.1]nonane
BCME	dis(chloromethyl)ether
BHT	butylated hydroxytoluene (2,6-di-*t*-butyl-*p*-cresol)
BINAL-H	2,2′-dihydroxy-1,1′-binaphthyl-lithium aluminum hydride
BINAP	2,2′-bis(diphenylphosphino)-1,1′-binaphthyl
BINOL	1,1′-bi-2,2′-naphthol
bipy	2,2′-bipyridyl
BMS	borane–dimethyl sulfide
Bn	benzyl
Boc	*t*-butoxycarbonyl
BOM	benzyloxymethyl
bp	boiling point
Bs	brosyl (4-bromobenzenesulfonyl)
BSA	*N,O*-bis(trimethylsilyl)acetamide
Bu	*n*-butyl
Bz	benzoyl
CAN	cerium(IV) ammonium nitrate
Cbz	benzyloxycarbonyl
CDI	*N,N′*-carbonyldiimidazole
CHIRAPHOS	2,3-bis(diphenylphosphino)butane
Chx	=Cy
cod	cyclooctadiene
cot	cyclooctatetraene
Cp	cyclopentadienyl
CRA	complex reducing agent
CSA	10-camphorsulfonic acid
CSI	chlorosulfonyl isocyanate
Cy	cyclohexyl
d	density
DABCO	1,4-diazabicyclo[2.2.2]octane
DAST	*N,N′*-diethylaminosulfur trifluoride
dba	dibenzylideneacetone
DBAD	di-*t*-butyl azodicarboxylate
DBN	1,5-diazabicyclo[4.3.0]non-5-ene
DBU	1,8-diazabicyclo[5.4.0]undec-7-ene
DCC	*N,N′*-dicyclohexylcarbodiimide
DCME	dichloromethyl methyl ether
DDO	dimethyldioxirane
DDQ	2,3-dichloro-5,6-dicyano-1,4-benzoquinone
de	diastereomeric excess

DEAD	diethyl azodicarboxylate
DET	diethyl tartrate
DIBAL	diisobutylaluminum hydride
DIEA	=DIPEA
DIOP	2,3-*O*-isopropylidene-2,3-dihydroxy-1,4-bis-(diphenylphosphino)butane
DIPEA	diisopropylethylamine
diphos	=dppe
DIPT	diisopropyl tartrate
DMA	dimethylacetamide
DMAD	dimethyl acetylenedicarboxylate
DMAP	4-(dimethylamino)pyridine
DME	1,2-dimethoxyethane
DMF	dimethylformamide
dmg	dimethylglyoximato
DMPU	*N,N′*-dimethylpropyleneurea
DMS	dimethyl sulfide
DMSO	dimethyl sulfoxide
DMTSF	dimethyl(methylthio) sulfonium tetrafluoroborate
dppb	1,4-bis(diphenylphosphino)butane
dppe	1,2-bis(diphenylphosphino)ethane
dppf	1,1′-bis(diphenylphosphino)ferrocene
dppp	1,3-bis(diphenylphosphino)propane
DTBP	di-*t*-butyl peroxide
EDA	ethyl diazoacetate
EDC	1-ethyl-3-(3-dimethylaminopropyl)-carbodiimide
EDCI	=EDC
ee	enantiomeric excess
EE	1-ethoxyethyl
Et	ethyl
ETSA	ethyl trimethylsilylacetate
EWG	electron withdrawing group
Fc	ferrocenyl
Fmoc	9-fluorenylmethoxycarbonyl
fp	flash point
Hex	*n*-hexyl
HMDS	hexamethyldisilazane
HMPA	hexamethylphosphoric triamide
HOBt	l-hydroxybenzotriazole
HOBT	=HOBt
HOSu	*N*-hydroxysuccinimide
Im	imidazole (imidazolyl)
Ipc	isopinocampheyl
IR	infrared

KHDMS   potassium hexamethyldisilazide

LAH   lithium aluminum hydride
LD$_{50}$ dose that is lethal to 50% of test subjects
LDA   lithium diisopropylamide
LDMAN   lithium 1-(dimethylamino)naphthalenide
LHMDS   =LiHMDS
LICA   lithium isopropylcyclohexylamide
LiHMDS   lithium hexamethyldisilazide
LiTMP   lithium 2,2,6,6-tetramethylpiperidide
LTMP   =LiTMP
LTA   lead tetraacetate
lut   lutidine

m-CPBA   *m*-chloroperbenzoic acid
MA   maleic anhydride
MAD   methylaluminum bis(2,6-di-*t*-butyl-4-
             methylphenoxide)
MAT   methylaluminum bis(2,4,6-tri-*t*-
             butylphenoxide)
Me   methyl
MEK   methyl ethyl ketone
MEM   (2-methoxyethoxy)methyl
MIC   methyl isocyanate
MMPP   magnesium monoperoxyphthalate
MOM   methoxymethyl
MoOPH   oxodiperoxomolybdenum(pyridine)-
             (hexamethylphosphoric triamide)
mp   melting point
MPM   =PMB
Ms   mesyl (methanesulfonyl)
MS   mass spectrometry; molecular sieves
MTBE   methyl *t*-butyl ether
MTM   methylthiomethyl
MVK   methyl vinyl ketone

n   refractive index
NaHDMS   sodium hexamethyldisilazide
Naph   naphthyl
NBA   *N*-bromoacetamide
nbd   norbornadiene (bicyclo[2.2.1]hepta-
             2,5-diene)
NBS   *N*-bromosuccinimide
NCS   *N*-chlorosuccinimide
NIS   *N*-iodosuccinimide
NMO   *N*-methylmorpholine *N*-oxide
NMP   *N*-methyl-2-pyrrolidinone
NMR   nuclear magnetic resonance
NORPHOS   bis(diphenylphosphino)bicyclo[2.2.1]-hept-
             5-ene
Np   =Naph

PCC   pyridinium chlorochromate
PDC   pyridinium dichromate
Pent   *n*-pentyl
Ph   phenyl
phen   1,10-phenanthroline
Phth   phthaloyl

Piv   pivaloyl
PMB   *p*-methoxybenzyl
PMDTA   *N,N,N′,N″,N″*-pentamethyldiethylene-
             triamine
PPA   polyphosphoric acid
PPE   polyphosphate ester
PPTS   pyridinium *p*-toluenesulfonate
Pr   *n*-propyl
PTC   phase transfer catalyst/catalysis
PTSA   *p*-toluenesulfonic acid
py   pyridine

RAMP   (R)-1-amino-2-(methoxymethyl)pyrrolidine
rt   room temperature

salen   bis(salicylidene)ethylenediamine
SAMP   (S)-1-amino-2-(methoxymethyl)pyrrolidine
SET   single electron transfer
Sia   siamyl (3-methyl-2-butyl)

TASF   tris(diethylamino)sulfonium
             difluorotrimethylsilicate
TBAB   tetrabutylammonium bromide
TBAF   tetrabutylammonium fluoride
TBAD   =DBAD
TBAI   tetrabutylammonium iodide
TBAP   tetrabutylammonium perruthenate
TBDMS   *t*-butyldimethylsilyl
TBDPS   *t*-butyldiphenylsilyl
TBHP   *t*-butyl hydroperoxide
TBS   =TBDMS
TCNE   tetracyanoethylene
TCNQ   7,7,8,8-tetracyanoquinodimethane
TEA   triethylamine
TEBA   triethylbenzylammonium chloride
TEBAC   =TEBA
TEMPO   2,2,6,6-tetramethylpiperidinoxyl
TES   triethylsilyl
Tf   triflyl (trifluoromethanesulfonyl)
TFA   trifluoroacetic acid
TFAA   trifluoroacetic anhydride
THF   tetrahydrofuran
THP   tetrahydropyran; tetrahydropyranyl
Thx   thexyl (2,3-dimethyl-2-butyl)
TIPS   triisopropylsilyl
TMANO   trimethylamine *N*-oxide
TMEDA   *N,N,N′,N′*-tetramethylethylenediamine
TMG   1,1,3,3-tetramethylguanidine
TMS   trimethylsilyl
Tol   *p*-tolyl
TPAP   tetrapropylammonium perruthenate
TBHP   *t*-butyl hydroperoxide
TPP   tetraphenylporphyrin
Tr   trityl (triphenylmethyl)
Ts   tosyl (*p*-toluenesulfonyl)
TTN   thallium(III) nitrate

UHP   urea–hydrogen peroxide complex

Z   =Cbz

B

Essential Reagents for Organic Synthesis, Edited by Philip L. Fuchs, André B. Charette, Tomislav Rovis, and Jeffrey W. Bode.
©2016 John Wiley & Sons, Ltd. Published 2016 by John Wiley & Sons, Ltd.

Bis(dibenzylideneacetone)palladium(0)

(PhCH=CHCOCH=CHPh)₂Pd

[32005-36-0]   C₃₄H₂₈O₂Pd   (MW 575.01)
InChI = 1S/2C17H14O.Pd/c2*18-17(13-11-15-7-3-1-4-8-15)14-
    12-16-9-5-2-6-10-16;/h2*1-14H;/b2*13-11+,14-12+;
InChIKey = UKSZBOKPHAQOMP-SVLSSHOZSA-N

(catalyst for allylation of stabilized anions,[1] cross coupling of
allyl, alkenyl, and aryl halides with organostannanes,[2] cross
coupling of vinyl halides with alkenyl zinc species,[3] cyclization
reactions,[4] and carbonylation of alkenyl and aryl halides,[5] air stable Pd⁰ complex used as a homogeneous Pd⁰-precatalyst in the
presence of additional external ligands)

Alternative Name: palladium(0) bis(dibenzylideneacetone).
Physical Data: mp 135 °C (dec).
Solubility: insoluble in H₂O, soluble in organic solvents
(dichloromethane, chloroform, 1,2-dichloroethane, acetone,
acetonitrile, benzene, and others)
Form Supplied in: black solid; commercially available.
Preparative Method: prepared by the addition of sodium acetate
to a hot methanolic solution of dibenzylideneacetone (dba)
and Na₂[Pd₂Cl₆] (from **Palladium(II) Chloride** and NaCl),
cooling, filtering, washing with MeOH, and air drying, gives
Pd(dba)₂; which is formulated more correctly as [Pd₂(dba)₃]-
dba.[6,12a] Alternatively, palladium(II) chloride, sodium acetate,
and dba can be added to a 40 °C methanolic solution, cooled,
filtered, and washed copiously with H₂O and acetone, in suc-
cession, and dried in vacuo.[112a]
Handling, Storage, and Precautions: moderately air stable in the
solid state; slowly decomposes in solution to metallic palladium
and dibenzylideneacetone.

Tris(dibenzylideneacetone)dipalladium(0)-Chloroform

[Ph—CH=CH—C(O)—CH=CH—Ph]₂ Pd

[52522-40-4]   C₅₂H₄₃Cl₃O₃Pd₂   (MW 1035.14)
InChI = 1/3C17H14O.CHCl3.2Pd/c3*18-17(13-11-15-7-3-1-4-
    8-15)14-12-16-9-5-2-6-10-16;2-1(3)4;;/h3*1-14H;1H;;/
    b3*13-11+,14-12+;;;
InChIKey = LNAMMBFJMYMQTO-FNEBRGMMBW

Alternative Name: dipalladium-tris(dibenzylideneacetone)-
chloroform complex.
Physical Data: mp 131–135 °C,[111] 122–124 °C (dec).[6,112]
Solubility: insoluble in H₂O; soluble in chloroform, dichloromethane, and benzene.
Form Supplied in: purple solid; commercially available.
Preparative Method: the reaction of Na₂[Pd₂Cl₆] and dba give
Bis(dibenzylideneacetone)palladium(0), [Pd(dba)₂dba], recrystallization from chloroform displaces the uncoordinated
dba with chloroform to gives the title reagent as deep purple
needles.[112a]

Tris(dibenzylideneacetone)dipalladium

[Ph—CH=CH—C(O)—CH=CH—Ph]₃ Pd₂·CHCl₃

[51364-51-3]   C₅₁H₄₂O₃Pd₂   (MW 915.72)
InChI = 1/3C17H14O.2Pd/c3*18-17(13-11-15-7-3-1-4-8-15)14-
    12-16-9-5-2-6-10-16;;h3*1-14H;;/b3*13-11+,14-12+;;
InChIKey = CYPYTURSJDMMP-WVCUSYJEBM

Alternative Name: dipalladium-tris(dibenzylideneacetone).
Physical Data: mp 152–155 °C.[6a]
Form Supplied in: dark purple to black solid; commercially
available.
Preparative Method: prepared from dba and sodium tetra-
chloropalladate.[6a]

Original Commentary

John R. Stille
Michigan State University, East Lansing, MI, USA

Allylation of Stabilized Anions. Pd(dba)₂ is an effective
catalyst for the coupling of electrophiles and nucleophiles, and has
found extensive use in organic synthesis (for a similar complex
with distinctive reactivities, see also ***Tris(dibenzylideneacetone)-
dipalladium***). Addition of a catalytic amount of Pd(dba)₂ activates
allylic species, such as allylic acetates or carbonate derivatives,
toward nucleophilic attack.[1] The intermediate organometallic
complex, a π-allylpalladium species, is formed by backside
displacement of the allylic leaving group, and stereochemical
inversion of the original allylic position results. Subsequent nucleophilic attack on the external face of the allyl ligand displaces the
palladium in this double inversion process to regenerate the original stereochemical orientation (eq 1).[7] The allylpalladium intermediate can also be generated from a variety of other substrates,
such as allyl sulfones,[8] allenes,[9] vinyl epoxides,[10] or α-allenic
phosphates.[11] In general, the efficiency of Pd(dba)₂ catalysis is
optimized through the addition of either **Triphenylphosphine** or
1,2-Bis(diphenylphosphino)ethane (dppe).

$$\text{(1)}$$
>95% cis

The anions of malonate esters,[12] cyclopentadiene,[12] β-
keto esters,[13] ketones,[13,14] aldehydes,[14] α-nitroacetate esters,[15]
Meldrum's acid,[15] diethylaminophosphonate Schiff bases,[16] β-
diketones,[17] β-sulfonyl ketones and esters,[17] and polyketides[18,19]
represent the wide variety of carbon nucleophiles effective in
this reaction. Generation of the stabilized anions normally is

Essential Reactions for Organic Synthesis, First Edition. Edited by Philip L. Fuchs.
© 2016 John Wiley & Sons, Ltd. Published 2016 by John Wiley & Sons, Ltd.

accomplished by addition of **Sodium Hydride**, **Potassium Hydride**, or basic **Alumina**[15] However, when allyl substrates such as allylisoureas,[14] allyl oxime carbonates,[17] or allyl imidates[20] are used, the allylation reaction proceeds without added base. Nitrogen nucleophiles, such as azide[10] and nucleotide[21] anions, are useful as well.

The coupling reaction generally proceeds regioselectively with attack by the nucleophile at the least hindered terminus of the allyl moiety,[22] accompanied by retention of alkene geometry (eq 2). Even electron-rich enol ethers can be used as the allylic moiety when an allylic trifluoroacetyl leaving group is present.[23] When steric constraints of substrates are equivalent, attack will occur at the more electron rich site.[19] Although this reaction is usually performed in THF, higher yields and greater selectivity are observed for some systems with the use of DME, DMF, or DMSO.[14,16,20] Alternatively, Pd(dba)$_2$ can promote efficient substitution of allylic substrates in a two-phase aqueous–organic medium through the use of P(C$_6$H$_4$-m-SO$_3$Na)$_3$ as a phase transfer ligand.[24]

$$\text{(2)}$$

(E) isomer	47%	95 (E only):5
(Z) isomer	62%	76 (Z only):23

Intramolecular reaction of a β-dicarbonyl functionality with a π-allyl species can selectively produce three-,[25] five-,[25] or six-membered[26] rings (eq 3).

$$\text{(3)}$$

E = CO$_2$Me   >98% *cis*

Asymmetric Allylation Reactions. Employing chiral bidentate phosphine ligands in conjunction with Pd(dba)$_2$ promotes allylation reactions with moderate to good enantioselectivities, which are dependent upon the solvent,[27] counterion,[28] and nature of the allylic leaving group.[27] Chiral phosphine ligands have been used for the asymmetric allylation of α-hydroxy acids (5–30% ee),[29] the preparation of optically active methylenecyclopropane derivatives (52% ee),[22] and chiral 3-alkylidenebicyclo[3.3.0]octane and 1-alkylidenecyclohexane systems (49–90% ee).[27] Allylation of a glycine derivative provides a route to optically active α-amino acid esters (eq 4).[28] The intramolecular reaction can produce up to 69% ee when vicinal stereocenters are generated during bond formation (eq 5).[30]

$$\text{(4)}$$

60%
57% ee

$$\text{(5)}$$

Cross-coupling Reactions. Allylic halides,[5,31] aryl diazonium salts,[32] allylic acetates,[33] and vinyl epoxides[34] are excellent substrates for Pd(dba)$_2$ catalyzed selective cross-coupling reactions with alkenyl-, aryl-, and allylstannanes. The reaction of an allylic halide or acetate proceeds through a π-allyl intermediate with inversion of sp^3 stereochemistry, and transmetalation with the organostannane followed by reductive elimination results in coupling from the palladium face of the allyl ligand. Coupling produces overall inversion of allylic stereochemistry, a preference for reaction at the least substituted carbon of the allyl framework, and retention of allylic alkene geometry. In addition, the alkene geometry of alkenylstannane reagents is conserved (eq 6). Functional group compatibility is extensive, and includes the presence of CO$_2$Bn, OH, OR, CHO, OTHP, β-lactams, and CN functionality.

$$\text{(6)}$$

Similar methodology is used for the coupling of alkenyl halides and triflates with 1) alkenyl-, aryl-, or alkynylstannanes,[35] 2) alkenylzinc species,[3,36] or 3) arylboron species.[37] This methodology is applied in the synthesis of cephalosporin derivatives (eq 7),[35] and can be used for the introduction of acyl[3,36] and vinylogous acyl[3] equivalents (eq 8).

$$\text{(7)}$$

$$\text{(8)}$$

Intramolecular Reaction with Alkenes. Palladium π-allyl complexes can undergo intramolecular insertion reactions with alkenes to produce five- and six-membered rings through a 'metallo-ene-type' cyclization.[4] This reaction produces good stereoselectivity when resident chirality is vicinal to a newly formed stereogenic center (eq 9), and can be used to form tricyclic and tetracyclic ring systems through tandem insertion reactions.[38] In the presence of Pd(dba)$_2$ and triisopropyl phosphate, α,β-alkynic esters and α,β-unsaturated enones undergo intramolecular [3 + 2] cycloaddition reactions when tethered to methylenecyclopropane to give a bicyclo[3.3.0]octane ring system (eq 10).[39]

$$(9)$$

$$trans:cis = 93:7$$

$$(10)$$

Carbonylation Reactions. In the presence of CO and Pd(dba)$_2$, unsaturated carbonyl derivatives can also be prepared through carbonylative coupling reactions. Variations of this reaction include the initial coupling of allyl halides with carbon monoxide, followed by a second coupling with either alkenyl- or arylstannanes (eq 11).[5] This reaction proceeds with overall inversion of allylic sp^3 stereochemistry, and retains the alkene geometry of both the allyl species and the stannyl group. Similarly, aryl and alkenyl halides will undergo carbonylative coupling to generate intermediate acylpalladium complexes. Intermolecular reaction of these acyl complexes with HSnBu$_3$ produces aldehydes,[35,40] while reaction with MeOH or amines generates the corresponding carboxylic acid methyl ester[41] or amides, respectively.[42]

$$(11)$$

Palladium acyl species can also undergo intramolecular acylpalladation with alkenes to form five- and six-membered ring γ-keto esters through exocyclic alkene insertion (eq 12).[43] The carbonylative coupling of o-iodoaryl alkenyl ketones is also promoted by Pd(dba)$_2$ to give bicyclic and polycyclic quinones through endocyclization followed by β-H elimination.[44] Sequential carbonylation and intramolecular insertion of propargylic and allylic alcohols provides a route to γ-butyrolactones (eq 13).[45]

$$(12)$$

$$(13)$$

First Update

F. Christopher Pigge
University of Iowa, Iowa City, IA, USA

Bis(dibenzylideneacetone)palladium(0) or Pd(dba)$_2$ continues to be a popular source of Pd(0), used extensively in transition metal-catalyzed reactions. The reagent is widely available from commercial sources and exhibits greater air stability than Pd(PPh$_3$)$_4$. The dba ligands are generally viewed as weakly co-ordinated and so are readily displaced by added ligands (usually mono- or bidentate phosphines) to generate active catalysts. Detailed mechanistic studies, however, have revealed that dba ligands are not as innocent as originally thought and exert a profound influence upon catalyst activity through formation of mixed ligand species of the type (dba)PdL$_2$ (L = phosphine).[46] The reagent is also a convenient source of phosphine-free Pd(0). Synthetic applications of Pd(dba)$_2$ include catalysis of allylic alkylation reactions, various cross-coupling reactions, Heck-type reactions, and multi-component couplings.

Allylation of Stabilized Anions. Electrophilic π-allyl Pd(0) complexes can be generated from Pd(dba)$_2$ and functionalized allylic acetates, carbonates, halides, etc. These complexes are susceptible to reaction with a range of stabilized nucleophiles, such as malonate anions. Alkylation usually occurs at the less-substituted allylic terminus. Silyl-substituted π-allyl complexes undergo regioselective alkylation at the allyl terminus farthest removed from the silyl group (eq 14).[47]

$$(14)$$

$$67\%$$

Allylic alkylation catalyzed by Pd(dba)$_2$ and (^{i}PrO)$_3$P has been used for incorporation of nucleobases (pyrimidines and purines) into carbocyclic nucleoside analogs.[48] In certain cases, unstabilized nucleophiles have been found to participate in allylic alkylation reactions. For example, an allenic double bond is sufficiently nucleophilic to react with the π-allyl complex generated upon heating Pd(dba)$_2$ and **1** in toluene (eq 15).[49] Formation of the *trans*-fused product (**2**) was interpreted to arise via the double inversion pathway commonly observed in conventional Pd-catalyzed allylic alkylation reactions. Interestingly, changing to a coordinating solvent (CH$_3$CN) resulted in allene insertion into the π-allyl complex to form the isomeric *cis*-fused product (**3**).

Asymmetric Allylation Reactions. Enantioselective allylic alkylation is used extensively in asymmetric synthesis with chiral nonracemic phosphines often serving as the source of enantiodiscrimination.[50] A monodentate phosphabicyclononane derivative in conjunction with Pd(dba)$_2$ was found to be effective in promoting the asymmetric allylation of 2-substituted cyclopentenyl and cyclohexenyl carbonates with malonate and sulfonamide nucleophiles with ee's ranging from 50 to 95% (eq 16).[51]

$$E = CO_2Me \tag{15}$$

$$\tag{16}$$

97%
95% ee

Catalysts generated from aminophosphine phosphinite chelates and Pd(dba)$_2$ were found to be effective at promoting alkylation of 1,3-diphenylpropenyl acetate with low to moderate enantiomeric excess.[52] An unusual monoylide monophosphine ligand (Yliphos) structurally related to BINAP also has been used to generate an active asymmetric allylic alkylation catalyst from Pd(dba)$_2$.[53] Axially chiral allenes have been prepared via asymmetric alkylation of in situ-generated alkylidene π-allyl palladium complexes. The reaction proceeds with reasonable levels of stereocontrol in the presence of BINAP (eq 17)[54] or a modified bis(silyl)-substituted BINAP derivative.[55] Interestingly, higher levels of enantioselectivity were observed in reactions using catalysts generated from Pd(dba)$_2$ and BINAP than in reactions performed using preformed Pd(BINAP)$_2$. It is believed that the presence of dba in the reaction mixture promotes equilibration of two diastereomeric (π-allyl)Pd(BINAP) intermediates.

$$\tag{17}$$

41–89% ee

Nuc = C(NHAc)(CO$_2$Et)$_2$; CMe(CO$_2$Me)$_2$

Cross-coupling Reactions. Metal-mediated C–C and C–X bond formation via various cross-coupling reactions has emerged as a powerful tool in organic synthesis. Palladium-catalyzed processes are ubiquitous and Pd(dba)$_2$ is frequently employed as a catalyst precursor. Cross-coupling sequences involving π-allyl palladium complexes generally proceed with overall inversion of stereochemistry with respect to the allylic leaving group and so are stereocomplementary to allylic alkylation reactions. Stereo- and regioselectivities of alkylation and cross-coupling reactions involving substituted cyclic (π-allyl)Pd intermediates have been investigated. Tetrabutylammonium triphenyldifluorosilicate (TBAT) was found to be a better transmetallation agent than an organostannane (eq 18).[56]

$$\tag{18}$$

10:1

Readily available functionalized aryl siloxanes are also viable cross-coupling partners for Pd(dba)$_2$-catalyzed allylic arylations.[57] A mixture of 5% Pd(dba)$_2$, allylic halide, and in situ-generated aryl zinc reagent produces allylated arenes in high yield.[58] Aryl boronic acids have been converted to allylated arenes as well.[59] Diastereoselective intramolecular Stille-type coupling of two allylic moieties (allylic acetate and allylic stannane) has been performed in high yield to produce the key intermediate in the synthesis of racemic 10-*epi*-elemol (eq 19).[60]

$$\tag{19}$$

91%

E = SO$_2$Ph

Catalysts derived from Pd(dba)$_2$ readily participate in oxidative addition reactions with aryl and alkenyl substrates and this forms the basis for a range of C–C couplings. The displacement of dba groups by added ligands provides a means to easily alter the electronic and steric environment around the metal center. For example, aryl bromides and iodides undergo Stille cross-coupling reactions with organostannanes using a catalyst prepared from Pd(dba)$_2$ and dicyclohexyl diazabutadiene with turnover numbers approaching one million.[61] Suzuki-type couplings between aryl halides and aryl boronic acids have been reported using Pd(dba)$_2$ in combination with mixed phosphine/sulfur[62] and phosphine/oxygen donor ligands.[63] Biaryl couplings with aryl chlorides are readily facilitated by the combination of Pd(dba)$_2$ and an *N*-heterocyclic carbene ligand generated via in situ deprotonation of an imidazolium salt (eq 20).[64] The addition of tetrabutylammonium bromide was found to be crucial for successful coupling.

Heterocyclic aryl chlorides can be coupled with aryl magnesium chlorides using a Pd(dba)$_2$–dppf catalyst system.[65] Even unactivated aryl tosylates have been successfully coupled with aryl Grignard reagents in the presence of as little as 0.1% of a catalyst prepared from Pd(dba)$_2$ and chelating phosphines of the Josiphos-type.[66] Symmetrical biaryls can be prepared from the direct homocoupling of aryl iodides and bromides using a combination of phosphine-free Pd(dba)$_2$ and TBAF in DMF.[67]

$$\text{(20)}$$

$$83\%$$

IPr-H$^+$ =

(Ar = 2,6-diisopropylphenyl)

$$\text{(22)}$$

L =

Although known for some time, the ability of organosilanes to participate in metal-mediated cross-coupling reactions has received considerable attention in recent years.[68] While several palladium sources have been employed in such reactions, Pd(dba)$_2$ often gives the best results. Aryl and alkenyl halides undergo Pd-catalyzed cross-coupling with vinyl and aryl siletanes,[69] organosiloxanes,[57,70] organosilanols,[71] and silyl ethers[72] under slightly different reaction conditions (i.e., with or without fluoride ion additives). This feature has resulted in development of a sequential cross-coupling approach for the synthesis of unsymmetrical 1,4-dienes (eq 21).[73] Hypervalent silicates have been found to give cross-coupled products with aryl bromides under microwave irradiation.[74]

Enolate Arylation Reactions. The direct coupling of aryl halides with enolates (or enolate equivalents) of ketones, esters, and amides is now well established. Malonic esters, cyanoacetates, and malononitrile can be arylated upon treatment with aryl halides in the presence of Pd(dba)$_2$ and electron-rich phosphines[81] or N-heterocyclic carbenes.[82] Carbene ligands have also proven effective in promoting the α-arylation of protected amino acids.[83] As a caveat to the use of Pd(dba)$_2$, the arylation of azlactones in the presence of this palladium source and phosphines was less efficient than that with Pd(OAc)$_2$. The dba ligands were found to react with azlactone substrates to form catalytically inactive palladium complexes.[84] Diastereoselective enolate arylation has been achieved through the use of chiral auxiliaries appended to preformed enol silyl ethers (eq 23).[85] The role of the zinc additive is not clear, however, it appears that discrete zinc enolates are not involved.

$$\text{(23)}$$

$$70\%, 82\% \text{ de}$$

In contrast, lactams such as 2-piperidinone have been α-arylated via the zinc enolate.[86] Intramolecular ketone arylation has been used to construct 4-arylisoquinoline derivatives that have been subsequently converted to the naturally occurring alkaloids cherylline and latifine.[87]

Heck Reactions. The Heck reaction is a Pd-catalyzed olefination usually performed between an aryl halide or triflate and an acrylate ester. While phosphines are traditionally used as ancillary ligands, new Pd(dba)$_2$-mediated reactions have been performed with a variety of other ligand types. These include chelating N-heterocyclic carbene/phosphine ligands,[88,89] benzimidazoles,[90] and quinolinyl oxazolines.[91] Air stable catalysts have been prepared from Pd(dba)$_2$ and sterically hindered thiourea ligands (eq 24).[92] An effective immobilized catalyst has been prepared from Pd(dba)$_2$ and a dendritic phosphine-containing polymer.[93]

Multicomponent Coupling Reactions. Tandem one-pot Pd-catalyzed processes have been developed that permit the coupling of three (or more) reactants in a single step. For example, allenes, aryl halides, and aryl boronic acids react in the presence of Pd(dba)$_2$ and CsF to afford functionalized olefins (eq 25).[94] In related transformations, in situ-generated benzynes have been

$$\text{(21)}$$

$$\sim 80\%$$

Cross-coupling reactions leading to the formation of C–X (X = heteroatom) bonds catalyzed by Pd(dba)$_2$ have been reported. Aniline derivatives have been prepared via reaction of amine nucleophiles with aryl halides in the presence of Pd(dba)$_2$ and phosphines, especially P(^{t}Bu)$_3$.[75,76] Likewise, diaryl and aryl alkyl ethers are produced from aryl halides (Cl, Br, I) and sodium aryloxides and alkoxides under similar conditions.[77] Conditions effective for the coupling of aryl chlorides with amines, boronic acids, and ketone enolates using an easily prepared phosphine chloride as a ligand have recently been uncovered (eq 22).[78] The preparation of aryl siloxanes[79] and allyl boronates[80] via Pd(dba)$_2$-catalyzed C–Si and C–B coupling have been reported as well.

coupled with allylic halides and alkynyl stannanes[95] or aryl metal reagents.[96]

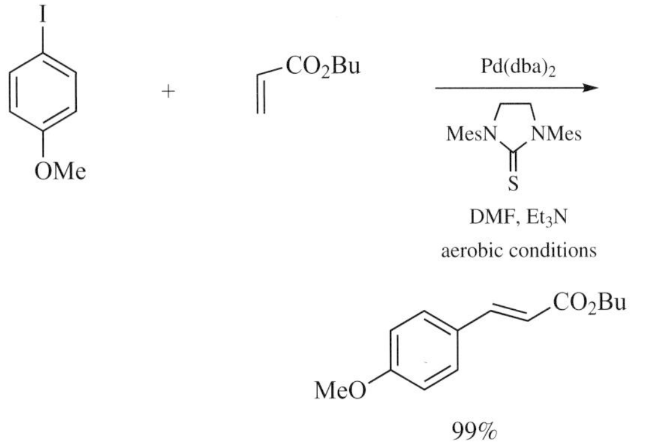

(24)

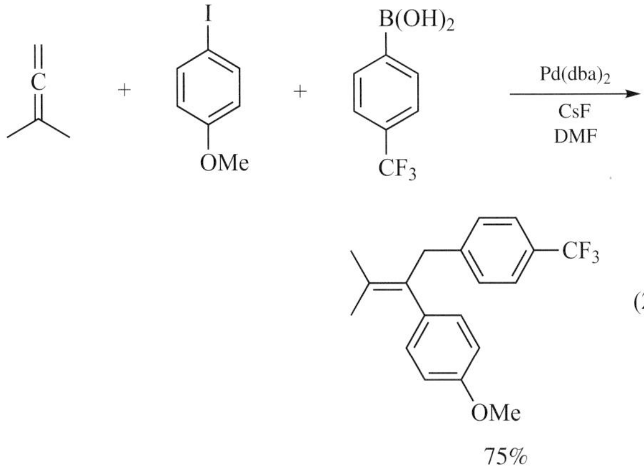

99%

(25)

75%

A four-component coupling between benzyl halides and alkynyl stannanes has been developed for the preparation of functionalized enynes.[97] Activated olefins participate in a regioselective Pd(dba)$_2$-catalyzed three-component coupling with allylic acetates and Bu$_3$SnH.[98] Allylic amines have been prepared via reaction of vinyl halides, alkenes, and amines in the presence of Pd(dba)$_2$ and Bu$_4$NCl.[99] Organogermanes and silanes have been constructed via multicomponent carbogermanylation[100] and carbosilylation[101] sequences.

Miscellaneous Reactions. Palladium dba has been employed as a catalyst for effecting various annulation reactions. Medium-sized nitrogen heterocycles have been prepared from allenes and amino alkenyl halides in the presence of a Pd(dba)$_2$/PPh$_3$ catalyst system.[102] 1,3-Dienes can be converted to benzofuran derivatives upon reaction with o-iodoacetoxy arenes and this reaction has been applied in the synthesis of new coumarins.[103,104] Dihydroquinoxalines and quinoxalinones have been obtained via reductive annulation of nitro enamines (eq 26).[105]

Cyclobutylidene derivatives have been regio- and stereoselectively reduced to substituted vinyl cyclobutanes with Pd(dba)$_2$ and sodium formate.[106] Heteroaryl benzylic acetates (including 2° acetates) undergo Pd-catalyzed benzylic nucleophilic substitution with malonate nucleophiles.[107] Cyclobutanone O-benzoyloximes have been converted to a variety of nitrile derivatives using Pd(dba)$_2$ in combination with chelating phosphines (eq 27).[108]

The ratio of cyclic to acyclic product was found to be a function of added phosphine.

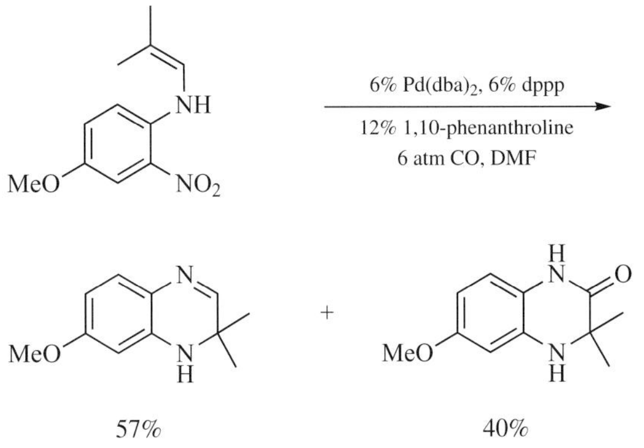

(26)

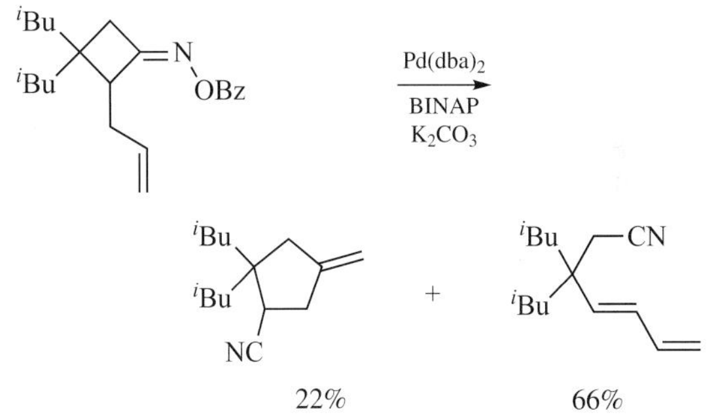

57%   40%

(27)

22%   66%

A novel route to biaryls has been reported starting from 1,4-epoxy-1,4-dihydroarenes. These substrates participate in a symmetrical coupling reaction in the presence of Pd(dba)$_2$, Zn, and HSiCl$_3$ (eq 28).[109] Finally, a heterogeneous catalyst prepared from Pd(dba)$_2$ and a phosphine-containing polymer resin has been found to facilitate the cycloisomerization of enynes in water.[110]

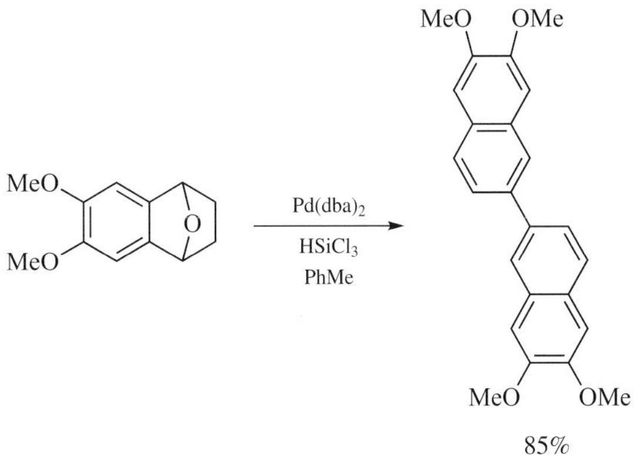

(28)

85%

Second Update

Christopher S. Regens, Ke Chen, Adrian Ortiz & Martin D. Eastgate
Bristol-Myers Squibb, New Brunswick, NJ, USA

Introduction. Bis(dibenzylideneacetone)palladium(0) and its derivatives are the most widely employed commercial sources of air stable "ligand-free" Pd(0). In its most typical applications the weakly coordinating dba ligand of Pd(dba)$_2$ are exchanged in situ with better and more donating ligands (e.g., phosphines, N-heterocyclic carbenes, amines, etc.), enabling the formation of a more competent catalytic Pd-species with a range of activities. However, Pd(dba)$_2$ and its derivatives are homogenous catalysts on their own merits and show activity in numerous synthetic transformations such as cross-coupling, Heck-coupling, C–X cross-coupling, asymmetric allylation, α-arylation, ene, carbonylation, and so on. While nonoxidative Pd(0)-mediated transformations are the usual applications of Pd(dba)$_2$, the focus of this update will give a "snapshot" of the use of Pd(dba)$_2$ and derivatives in oxidation reactions. This type of "oxidase" reaction employs molecular oxygen (or a stoichiometric oxidant) as an electron/proton acceptor in the substrate oxidation and does not involve direct oxygen atom transfer. In this strategy palladium must mediate both key steps in the catalytic cycle: (1) selective oxidation of the organic substrate by an active oxidized palladium intermediate, and (2) reoxidation of reduced palladium (i.e., oxidation of Pd(0) to Pd(II)).[113]

Oxidation of Primary and Secondary Alcohols. The conversion of alcohols to aldehydes and ketones using palladium-catalyzed aerobic oxidation offers significant advantage over more traditional approaches; those that utilize toxic metals (e.g., chromium, osmium, etc.) or super-/stoichiometric amounts of reagents, for example, MnO$_2$, SO$_3$·pyridine, and Dess-Martin periodinane. Such reactions are Pd(II) mediated, where Pd(0) is the initial palladium product of the alcohol oxidation. Reoxidation of Pd(0) to Pd(II) is therefore crucial to establishing a catalytic cycle; this is generally accomplished by adding co-oxidants such as molecular oxygen, benzoquinone, or copper(II) salts. For example, Pd(dba)$_2$ in conjunction with a cyclic thiourea ligand is an active catalyst for the aerobic oxidation of primary and secondary alcohols to aldehydes and ketones. It is postulated that a thiourea-Pd(dba)$_2$ complex is formed, which sufficiently stabilizes palladium in the presence of molecular oxygen to maintain a soluble Pd-complex by suppressing the formation of palladium black (eqs 29 and 30).[114]

$$R = Me, F, H, i\text{-}Pr$$
$$R' = Me, Et, H$$

$$(29)$$

70–98%

In another example, Pd$_2$(dba)$_3$ was used in conjunction with allyl diethyl phosphate, an unusual stoichiometric hydrogen acceptor in the oxidation of simple alcohols. Oxidative addition of Pd(0) into the allylic phosphate (generating a π-allyl-Pd(II) complex), is followed by an alcohol/phosphate displacement and subsequent β-hydride elimination giving the oxidized alcohol (aldehyde/ketone) and an (allyl)(Pd(II))-H intermediate. Reductive elimination of this intermediate affords propylene gas and regenerates Pd(0), completing the catalytic cycle. Under these conditions a wide range of secondary alcohols were oxidized to the corresponding ketones in good yields with primary alcohols mainly producing the corresponding aldehydes; however, in some examples esters were obtained (from over-oxidation) (eq 31).[115]

$$(30)$$

87%

Palladium catalysts have found application in the oxidative kinetic resolution of secondary alcohols such as 1-phenylethanol. (−)-Sparteine, was used to obtain high levels of enantioselection; however, it was found that the nature of the palladium source was critical in obtaining a high chemical selectivity factor; Pd$_2$(dba)$_3$ proved superior to Pd(OAc)$_2$ but not as effective as Pd(nbd)Cl$_2$. The observed difference in reactivity, for various palladium catalysts, was attributed to subtle differences in the solubility of the palladium-precatalysts in toluene; as well as their ability to complex with (−)-sparteine (eq 32).[116]

Wacker-type Oxidation. The palladium-catalyzed oxidation of terminal olefins to ketones (Wacker oxidation) is an important chemical process both in the laboratory and in industrial settings. Pd(dba)$_2$ has shown useful activity in this area, for example, 1-dodecene was readily oxidized to the corresponding methyl ketone, using a mixed catalytic system comprising Pd(dba)$_2$/PPh$_3$ and AgNO$_2$/HNO$_3$. Kinetic and mechanistic studies indicate that a Wacker-type mechanism, where the transferred oxygen is coming from the nucleophilic addition of H$_2$O to a Pd-olefin complex (not from O$_2$). When a nonoxidizable alcohol such as *tert*-butanol was used in combination with AgNO$_2$/HNO$_3$, *tert*-butyl nitrite was observed. From this finding, the authors suggest that the alkyl nitrite by-product is responsible for the regeneration of the Pd(II) active complex. The authors also note that the presence of silver ions is necessary for higher yields and conversions; however, the exact role of silver is unclear (eq 33).[117]

6 h, 93% conversion (31)

KOP(O)(OEt)$_2$

RCH$_2$OH

L$_n$Pd —]$^+$ K$_2$CO$_3$

–OP(O)(OEt)$_2$ HOP(O)(OEt)$_2$

OP(O)(OEt)$_2$

L$_n$Pd(0)   L$_n$Pd — >   L$_n$Pd —]$^+$

OCH$_2$R   RCH$_2$O$^-$

L$_n$Pd

CH$_3$CH=CH$_2$ H   RCH=O

(–)-sparteine (20 mol %)
Pd-precat (5 mol %)

MS3Å, O$_2$, toluene, 80 °C

(32)

Pd Source	Conversion, %	ee ROH, %	s Note
Pd(OAc)$_2$	15.1	13.7	8.8
Pd$_2$(dba)$_3$	66.2	81.5	5.7
Pd(nbd)Cl$_2$	59.9	98.7	23.1

(–)-sparteine

Note: The selectivity factor (s) was determined using the equation:
s = k_{rel} = ln[(1 - C)(1 - ee)]/ln[(1 - C)(1 + ee)], where C = conversion.

Pd(dba)$_2$ (1.1 mol %), PPh$_3$ (1.1 mol %)

HNO$_3$/AgNO$_2$, 10% t-BuOH/water
80 °C, 1 atm O$_2$

(33)

81%

The next section will switch from "classic" oxidation processes to formal oxidation state changes at the carbon center. These types of processes parallel nonoxidative cross-coupling processes; however, in these examples the carbon atom is fully reduced and a directing group is typically employed to guide the C–H palladium insertion event. Thus after reductive elimination the carbon atom has been formally oxidized/functionalized.

C–H Functionalization. A mild method for the perfluoroalkylation of simple arenes has been developed using Pd$_2$(dba)$_3$ in the presence of BINAP ligand. A variety of aromatic substrates undergo selective perfluoroalkylation in the presence of perfluoroalkyl iodides, giving the desired substituted arene in moderate to excellent yields. Arenes containing electron-donating methyl and alkoxy substituents generally reacted smoothly to provide the desired perfluoroalkylated products in good to excellent yields (in some instances the low yield is attributed to the volatility of the product). However, low reactivity was observed with aromatic substrates containing electron-withdrawing substituents. Substrates containing benzylic sp^2 C–H sites were highly selective for functionalization of the aromatic C–H bond in preference to the benzylic center. Preliminary mechanistic studies did not support a purely free radical pathway and suggested the formation of either caged radical or organometallic intermediates, rather than a traditional palladium cross-coupling mechanism (eqs. 34 & 35).[118]

$$R{-}\langle\rangle \ + \ C_{10}F_{21}{-}I \xrightarrow[\substack{Cs_2CO_3\ (2\ equiv),\ 40{-}100\,°C \\ 15{-}24\ h}]{\substack{Pd_2(dba)_3\ (5\ mol\ \%) \\ BINAP\ (20\ mol\ \%)}} R{-}\langle\rangle{-}C_{10}F_{21} \quad (34)$$

70–98%

$$\langle\rangle \ + \ F_3C{-}I \xrightarrow[\substack{Cs_2CO_3\ (2\ equiv),\ 100\,°C,\ 15\ h}]{\substack{Pd_2(dba)_3\ (5\ mol\ \%) \\ BINAP\ (20\ mol\ \%)}} \langle\rangle{-}CF_3 \quad (35)$$

26%

In addition to the perfluoroalkylation of arenes Pd(dba)$_2$ competently facilities the *ortho*-C–H amination of *N*-aryl benzamides with electrophilic *O*-benzoyl hydroxylamines. Pd(dba)$_2$ showed equal reactivity to other precatalysts, such as Pd(OAc)$_2$. Although mechanistically unclear, the addition of AgOAc significantly improved conversion, while the addition of external phosphine ligands was detrimental. Finally, the 4-CF$_3$(C$_6$F$_4$) directing group was optimal because acidic N–H bonds are essential for this transformation (eq 36).[119]

The extension of this concept to Pd-catalyzed *ortho* C–H borylation (using the same *N*-arylbenzamide fluorinated directing group [Ar = (4-CF$_3$)C$_6$F$_4$]) was developed. Amongst the various palladium sources surveyed, Pd$_2$(dba)$_3$ was found to be the precatalyst of choice, affording the *ortho*-borylated product in 51% yield. Further optimization of the dba ligand showed that 4,4′-Cl-dibenzylideneacetone (**L**) was found to be superior to the parent dba ligand, improving the yield to 78% (eq 37).[120]

Site selective arylation of the sp^3 benzylic position of 2-methyl pyridine *N*-oxide and 2,3-dimethyldiazine *N*-oxide has been demonstrated. In this process, Pd$_2$(dba)$_3$ was employed as a precatalyst along with XPhos, representing a unique strategy in the funtionalization of heterocycles (eqs 38 and 39).[121]

cat. [Pd] (10 mol %)
AgOAc (1 equiv)
CsF (2 equiv), DCE, 130 °C
18 h

2.0 equiv

Pd Source	Yield, %[a]
Pd(OAc)$_2$	98 (74)
Pd(dba)$_2$	96 (70)
Pd(OAc)$_2$/Pt-BuMe·HBF$_4$	32

(36)

[a] The yields in parentheses are obtained in the absence of AgOAc.

$$\xrightarrow[\substack{TsONa,\ K_2S_2O_8,\ MeCN \\ 80\,°C,\ 12\ h}]{\substack{Pd(0)\ (10\ mol\ \%)}}$$

Ar = (4-CF$_3$)C$_6$F$_4$

(37)

with Pd$_2$(dba)$_3$: 51%
with Pd(OAc)$_2$ and **L**: 78%

L =

During the course of an investigation into the Suzuki–Miyaura cross-coupling of 1-bromo-2,4,6-tri-*tert*-butylbenzene with phenylboronic acid, α,α-dimethyl-β-phenyl dihydrostyrene by-product was isolated in excellent yield, while the desired biaryl product was not observed. This unexpected transformation likely proceeded via a pathway involving a tandem C–H activation/ Suzuki–Miyaura cross-coupling sequence (eq 40).[122]

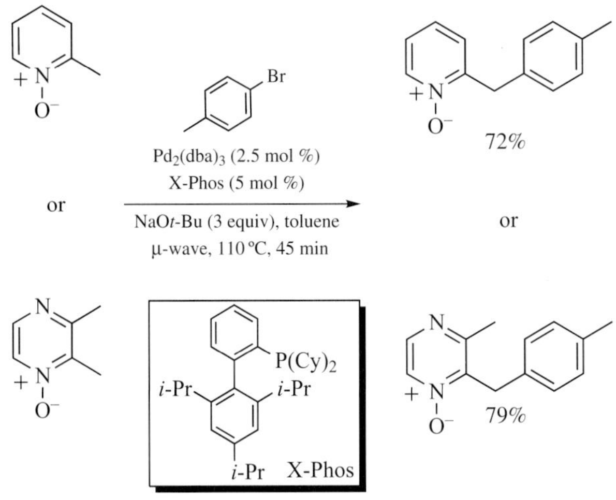

$$(38)$$

72%

or

$$(39)$$

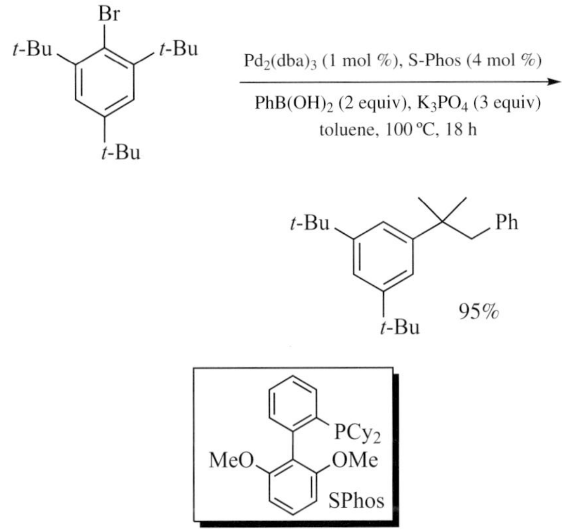

79%

$$(40)$$

95%

Building upon this methodology, the C–H activation/C–N couplings of related aromatic bromides with anilines were developed. It was observed that an alternative ligand class, the *N*-heterocyclic carbene (SIPr·HBF$_4$) was superior to SPhos, enabling the reaction of both electron-rich and electron-deficient anilines. Heteroaryl amines also provide the desired product in good yields; however, *N*-substituted anilines, along with alkyl amines were not suitable as coupling partners. Interestingly, extending the aromatic bromide by one carbon led to the formation of the desired C–H functionalized product, along with the dehydrogenated olefin by-product. It is postulated that this by-product arises via C–H activation of the ethyl group followed by β-hydride elimination (eqs 41 and 42).[123]

In some settings the presence of the dba ligand significantly impedes the use of this reagent; an example is the palladium-catalyzed halogenation of sp^2 C–H bonds. This type of oxidative C–H functionalization has potential advantages over the classic methods for the halogenation of the *ortho*-postion of arenes, such as electrophilic aromatic substitution, or directed *ortho*-lithiation (D*o*L) followed by a halogen quench. For example, phenylacetic acid and its derivatives belong to a substrate class that are not applicable to D*o*L protocol, because: (1) the chelating functional group is too remote from the C–H bond to be activated, and (2)

the acid protons at the benzylic position and the acid proton of the carboxcyclic acid moiety would quench the organolithium reagent. However, these substrates smoothly undergo selective *ortho*-halogenation under Pd-catalyzed conditions using IOAc as oxidant. From an extensive screen of palladium catalysts it was determined that Pd(II) precatalysts were effective, while common Pd(0) precatalysts, such as Pd(PPh$_3$)$_4$ and Pd$_2$(dba)$_3$, gave unsatisfactory yields of the desired iodoarene. The authors speculate that the ligands used to stabilize Pd(0) may be inhibitory to the reaction (eq 43).[124]

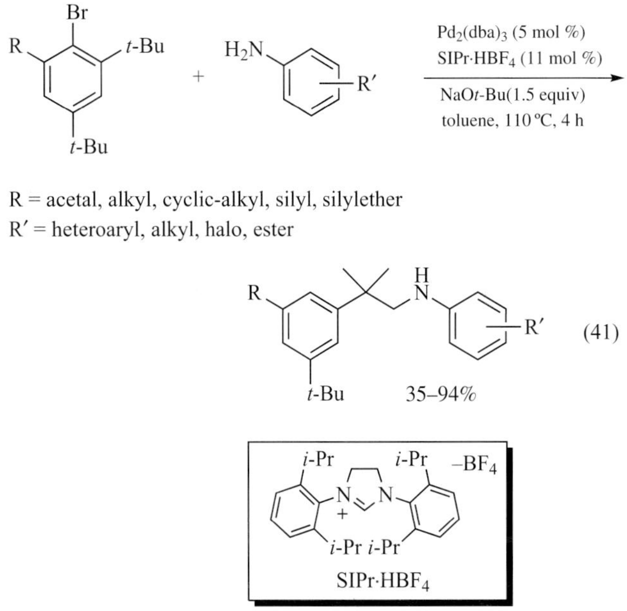

R = acetal, alkyl, cyclic-alkyl, silyl, silylether
R′ = heteroaryl, alkyl, halo, ester

$$(41)$$

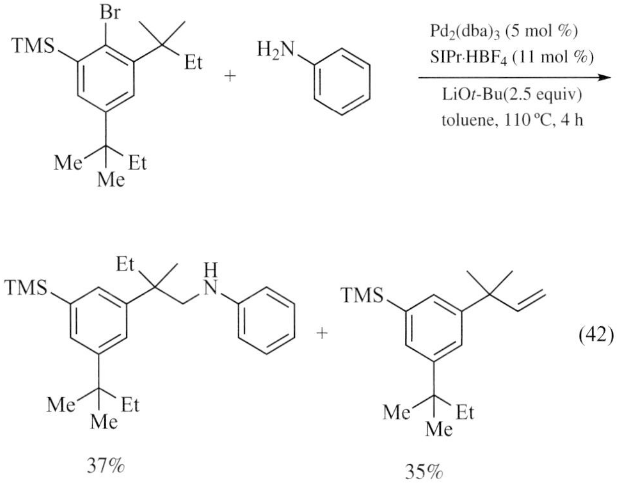

35–94%

$$(42)$$

37%   35%

Olefin Activation. Mechanistic studies on the 1,4-oxidation of 1,3-dienes led to the discovery of a new palladium catalyst [Pd(DA)$_2$], which was readily prepared from the reaction of Pd$_2$(dba)$_3$ and the Diels–Alder adduct derived from 1,3-cyclohexadiene and *p*-benzoquinone. With *p*-benzoquinone as the stoichiometric oxidant, this palladium complex proves more reactive and selective than the Pd(II) carboxylate, typically used in classic 1,4-oxidation of cyclohexadiene. The use of $Pd_2(dba)_3$ directly in the oxidation (in the presence of the diene ligand **A**), enables in situ formation of this more active complex from the $Pd_2(dba)_3$ precatalyst (eq 44).[125]

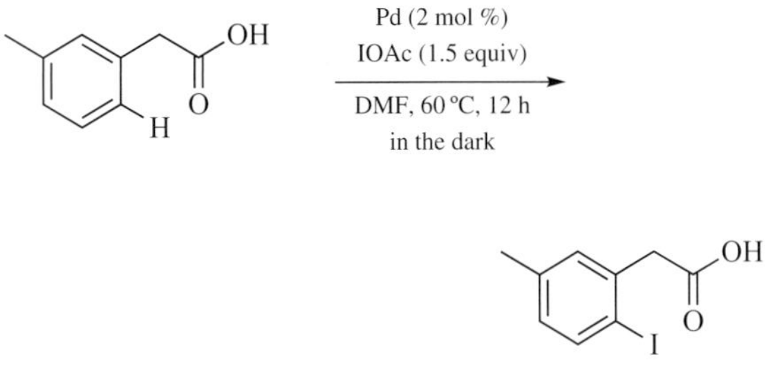

(43)

Pd Source	Yield, %
$Pd(OAc)_2$	92
PdI_2	90
$Pd(OTFA)_2$	85
$Pd(PPh_3)_4$	20
$Pd_2(dba)_3$	30

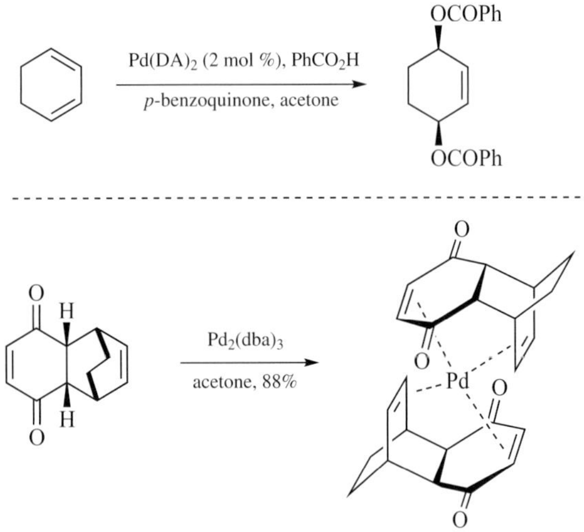

Pd(DA)₂

(44)

Pd(DA)₂

The direct enantioselective diamination at the allylic and homoallylic carbons of terminal olefins has been demonstrated in the presence of $Pd_2(dba)_3$. This formal C–H diamination required a catalyst system that could effectively convert the terminal olefin to a conjugated diene in situ, while inducing the enantioselective addition of di-*tert*-butyl diaziridinone. This was achieved using $Pd_2(dba)_3$, in the presence of phosphoramidite ligand (**L1**), giving the desired diaminated products in good yields (50–85%) and with high enantioselectivities (89–94%). It was found through empirical and mechanistic studies that the Pd/ligand ratio significantly influenced reaction conversion. From these studies it was demonstrated that a 2:1 ligand/Pd ratio was necessary to achieve

complete conversion of the terminal olefin. In addition, the authors demonstrated that the Z,E-olefin geometry was preserved during the course of the reaction. Furthermore, the chiral catalyst primarily determined the stereochemistry of the diaminated products, while the stereochemical information within terminal olefin has minimal effect on the overall diastereoselectivity. Finally, the authors were able to extend this methodology to the synthesis of (+)-CP-99994 (eq 45).[126]

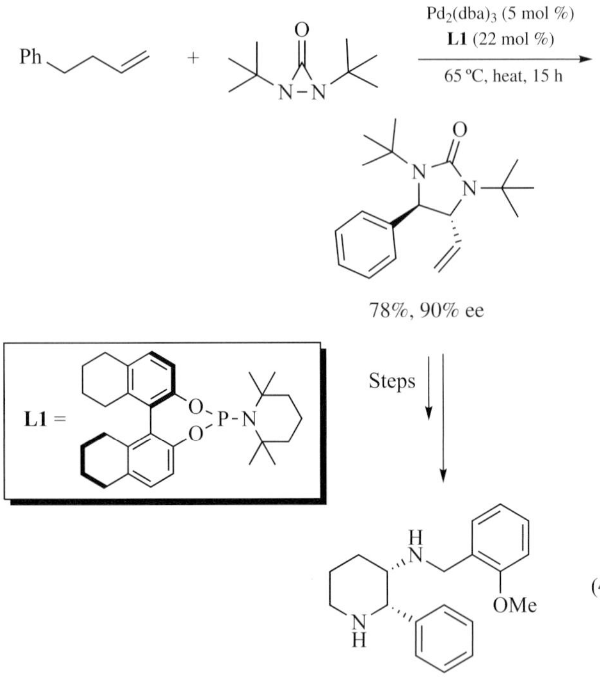

78%, 90% ee

Steps

L1 =

(45)

(+)-CP-99,994

Finally, the asymmetric elementometalation[127] across unsaturated olefins enables the design of multicomponent reactions for the development of one-pot enantioselective carbon-carbon bond forming sequences. For instance, diborylated olefins can be engaged in the allylation of both aldehydes and imines producing products in a highly selective fashion. One such example of asymmetric elementometalation is the diboration of prochiral allenes. This transformation, catalyzed by a $Pd_2(dba)_3$/(R,R)-(TADDOL)PNMe$_2$ complex, is operative for a wide range of monosubstituted allenes, affording the desired 1,2-bis(boronate)-ester products in good yields and with high enantioselectivites (eq 18).[128]

Oxidative Cyclization and Ring Contraction. An intramolecular cyclization/carboalkoxylation reaction takes place when 2-(penten-1-yl)-indole is treated with $Pd_2(dba)_3$ and a stoichiometric amount of copper (II) chloride in methanol under CO (1 atm). The polycyclic indole derivative was obtained in 68% yield, along with the minor C(3)-chlorinated by-product. A heterobimetallic Pd/Cu complex was proposed as the active catalyst in this transformation (eq 47).[129]

The ability of palladium to serve as both nucleophile (Pd(0)) and electrophile (Pd(II)) has led to the development of oxidase-type reactions that exploit the electrophilic nature of Pd(II). One such example is an extension of the above kinetic resolution of secondary alcohols catalyzed by Pd(nbd)Cl$_2$ in the presence of $(-)$-sparteine (described earlier), for the oxidative cyclization of substituted phenols. This racemic aerobic cyclization utilizes a Pd(II) salt in the presence of pyridine, O$_2$, and 3Å molecular sieves. Numerous palladium precatalysts were screened, Pd(TFA)$_2$ was optimal, yielding the desired cyclized product in 87% yield. Pd$_2$(dba)$_3$ also enabled the cyclization but was less effective, providing the desired cyclized product in significantly reduced yield (25%) (eq 48).[130]

(46)

52–85%
90–98% ee

R = alkyl, cyclic-alkyl, EW-arene,
      ED-arene, distial silyl, and benzyl ethers

(R,R)-**L**

1-methyl-2-(pent-4-enyl)-1*H*-indole

(47)

68%   10%

Contrary to the above results, it has been reported that Pd(dba)$_2$ was effective for the cyclization of *ortho*-allylic phenols. In the presence of Pd(dba)$_2$, the 6-membered 2-*H*-1-benzopyran was the sole product; while Pd(OAc)$_2$ predominately gave mixture of 5-, and 6-membered adducts. In addition, the choice of base was crucial for the formation of the desired benzopyran. It was observed that carbonate bases gave exclusively the 6-membered

ether, while acetate bases gave a mixture of 5- and 6-membered ethers (eq 49).[131]

(48)

with Pd$_2$(dba)$_3$:   25%
with Pd(TFA)$_2$:   87%

(49)

80%

Finally, it has been demonstrated that *tert*-cyclobutanols could be transformed, via Pd$_2$(dba)$_3$ catalyzed oxidative ring contraction, to acyl cyclopropane adducts in good yields. In the proposed mechanism, the first step is a β-elimination of the palladium alkoxide, producing an alkyl palladium intermediate. Enol formation of this intermediate is believed to be in equilibrium with a four membered palladacycle. Reductive elimination of this intermediate delivered the cyclopropane product and concurrently liberated the Pd(0) species. It was found that acetic acid accelerated the reaction significantly, due to the increased rate in the formation of the divalent AcOPdOOH. This Pd (II) species is proposed to be the active catalyst, responsible for producing the palladium alkoxide and hydrogen peroxide, completing the catalytic cycle (eq 50).[132]

Oxidative Cross-coupling. Molecular oxygen has been employed as the bulk oxidant in numerous Pd(0) and Pd(II)-catalyzed oxidation processes. Another example is the cross-coupling of organoboranes with olefins, in an oxidative Heck-type reaction. Both Pd(OAc)$_2$ and Pd$_2$(dba)$_3$ were productive in this process affording the desired product in 85–87% yield. It was found that O$_2$ was crucial for this reaction as very little product was formed under anaerobic conditions. The authors infer that these results suggest that O$_2$ plays a pivotal role in the Pd(II)-catalyzed reaction though the reoxidation of Pd(0) species to Pd(II) (eq 51).[133]

Another example of an oxidative Heck-type process is the cross-coupling of imidazo[1,2]pyridines with alkenes, using the combination of catalytic Pd(II) or Pd(0) source and stoichiometric copper(II) acetate. Several palladium sources worked well in promoting this oxidative coupling, including Pd(OAc)$_2$, Pd$_2$(dba)$_3$, and Pd(PPh$_3$)$_4$. However, reactions using PdCl$_2$ encountered low conversion, even with prolonged reaction times. This transformation proved highly regioselective at the C-(3) position, offering a direct route to 3-alkenylimidazo[1,2]pyridine derivatives in good to excellent yields (eq 52).[134]

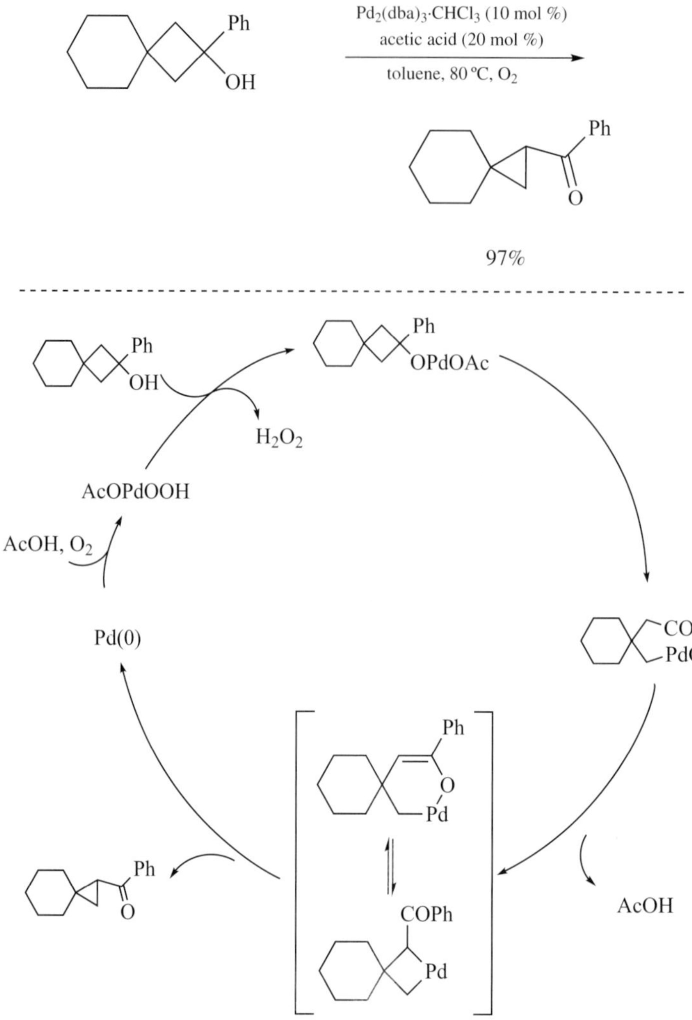

97%

(50)

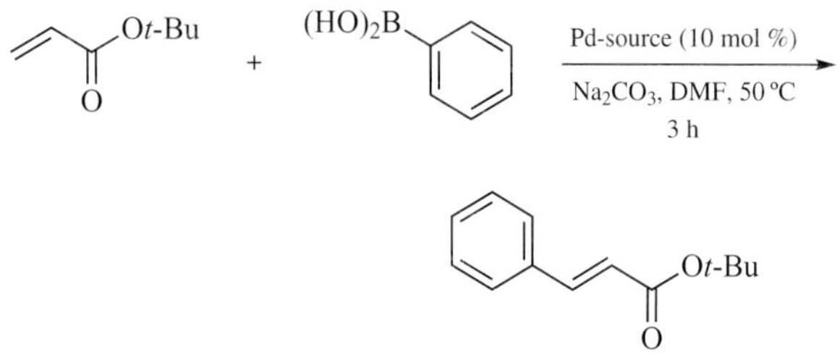

$$\text{(51)}$$

Pd Source	Oxidant	Yield, %
Pd(OAc)$_2$	air	44
Pd(OAc)$_2$	O$_2$	87
Pd$_2$(dba)$_3$	O$_2$	85
Pd(OAc)$_2$	none	12
Pd$_2$(dba)$_3$	none	0

Finally, Pd$_2$(dba)$_3$ has been utilized as a Pd(0) precatalyst for the palladium-catalyzed carbonylation of triarylstibines using ceric(IV) ammonium nitrate (CAN) as the bulk oxidant. Numerous oxidants were surveyed and CAN gave optimal results. It is believed that CAN serves the same role as molecular oxygen, in

previous examples, to reoxidize the Pd(0) species to Pd(II), regenerating the active Pd(II) catalyst. However, the authors note that the real role of CAN may be complex and requires further investigation (eq 53).[135]

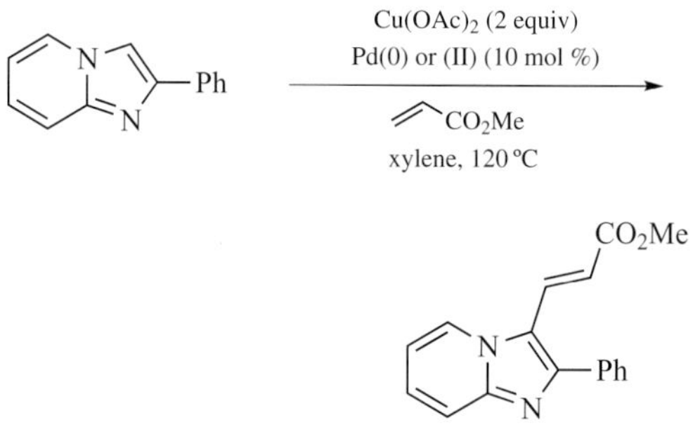

$$\text{(52)}$$

Pd Source	Time, h	Yield, %
Pd(OAc)$_2$	3	85
Pd$_2$(dba)$_3$	3	77
Pd(PPh$_3$)$_4$	3	78
PdCl$_2$	30	35

$$(53)$$

82%   16%

In summary, Pd(dba)$_2$ is an unusual precatalyst in palladium-related oxidation reactions. In several settings Pd(dba)$_2$ and its related complexes, in the presence/absence of a bulk oxidant, facilitate a multitude of synthetically useful oxidation reactions. However, the presence of the dba ligand can be a complicating factor, reducing reactivity in certain cases.

1. Trost, B. M., *Angew. Chem., Int. Ed. Engl.* **1989**, *28*, 1173.

2. Stille, J. K., *Angew. Chem., Int. Ed. Engl.* **1986**, *25*, 508.

3. (a) Rao, C. J.; Knochel, P., *J. Org. Chem.* **1991**, *56*, 4593. (b) Wass, J. R.; Sidduri, A.; Knochel, P., *Tetrahedron Lett.* **1992**, *33*, 3717. (c) Knochel, P.; Rao, C. J., *Tetrahedron* **1993**, *49*, 29.

4. (a) Oppolzer, W.; Gaudin, J.-M., *Helv. Chim. Acta* **1987**, *70*, 1477. (b) Oppolzer, W.; Swenson, R. E.; Gaudin, J.-M., *Tetrahedron Lett.*, **1988**, *29*, 5529. (c) Oppolzer, W.; Keller, T. H.; Kuo, D. L.; Pachinger, W., *Tetrahedron Lett.* **1990**, *31*, 1265.

5. (a) Sheffy, F. K.; Stille, J. K., *J. Am. Chem. Soc.* **1983**, *105*, 7173. (b) Sheffy, F. K.; Godschalx, J. P.; Stille, J. K., *J. Am. Chem. Soc.* **1984**, *106*, 4833.

6. (a) Takahashi, Y.; Ito, T.; Sakai, S.; Ishii, Y., *J. Chem. Soc., Chem. Commun.* **1970**, 1065. (b) Rettig, M. F.; Maitlis, P. M., *Inorg. Synth.* **1990**, *28*, 110.

7. Fiaud, J.-C.; Legros, J.-Y., *J. Org. Chem.* **1987**, *52*, 1907.

8. Backväll, J.-E.; Juntunen, S. K., *J. Am. Chem. Soc.* **1987**, *109*, 6396.

9. (a) Ahmar, M.; Barieux, J.-J.; Cazes, B.; Goré, J., *Tetrahedron* **1987**, *43*, 513. (b) Chaptal, N.; Colovray-Gotteland, V.; Grandjean, C.; Cazes, B.; Goré, J., *Tetrahedron Lett.* **1991**, *32*, 1795.

10. Tenaglia, A.; Waegell, B., *Tetrahedron Lett.* **1988**, *29*, 4851.

11. Cazes, B.; Djahanbini, D.; Goré, J.; Genêt, J.-P.; Gaudin, J.-M., *Synthesis* **1988**, 983.

12. Fiaud, J. C.; Malleron, J. L., *Tetrahedron Lett.* **1980**, *21*, 4437.

13. Fiaud, J.-C.; Malleron, J.-L., *J. Chem. Soc., Chem. Commun.* **1981**, 1159.

14. Inoue, Y.; Toyofuku, M.; Taguchi, M.; Okada, S.; Hashimoto, H., *Bull. Chem. Soc. Jpn.* **1986**, *59*, 885.

15. Ferroud, D.; Genet, J. P.; Muzart, J., *Tetrahedron Lett.* **1984**, *25*, 4379.

16. Genet, J. P.; Uziel, J.; Juge, S., *Tetrahedron Lett.* **1988**, *29*, 4559.

17. Suzuki, O.; Hashiguchi, Y.; Inoue, S.; Sato, K., *Chem. Lett.* **1988**, 291.

18. Marquet, J.; Moreno-Mañas, M.; Prat, M., *Tetrahedron Lett.* **1989**, *30*, 3105.

19. Prat, M.; Ribas, J.; Moreno-Mañas, M., *Tetrahedron* **1992**, *48*, 1695.

20. Suzuki, O.; Inoue, S.; Sato, K., *Bull. Chem. Soc. Jpn.* **1989**, *62*, 239.

21. Liotta, F.; Unelius, R.; Kozak, J.; Norin, T., *Acta Chem. Scand.* **1992**, *46*, 686.

22. Stolle, A.; Ollivier, J.; Piras, P. P.; Salaün, J.; de Meijere, A., *J. Am. Chem. Soc.* **1992**, *114*, 4051.

23. RajanBabu, T. V., *J. Org. Chem.* **1985**, *50*, 3642.

24. Safi, M.; Sinou, D., *Tetrahedron Lett.* **1991**, *32*, 2025.

25. (a) Ahmar, M.; Cazes, B.; Goré, J., *Tetrahedron Lett.* **1985**, *26*, 3795. (b) Ahmar, M.; Cazes, B.; Goré, J., *Tetrahedron* **1987**, *43*, 3453. (c) Fournet, G.; Balme, G.; Barieux, J. J.; Goré, J., *Tetrahedron* **1988**, *44*, 5821. (d) Geng, L.; Lu, X., *J. Chem. Soc., Perkin Trans. 1* **1992**, 17.

26. Bäckvall, J.-E.; Vågberg, J.-O.; Granberg, K. L., *Tetrahedron Lett.* **1989**, *30*, 617.

27. Fiaud, J.-C.; Legros, J.-Y., *J. Org. Chem.* **1990**, *55*, 4840.

28. (a) Genet, J. P.; Ferroud, D.; Juge, S.; Montes, J. R., *Tetrahedron Lett.* **1986**, *27*, 4573. (b) Genêt, J.-P.; Jugé, S.; Montès, J. R.; Gaudin, J.-M., *J. Chem. Soc., Chem. Commun.* **1988**, 718. (c) Genet, J.-P.; Juge, S.; Achi, S.; Mallart, S.; Montes, J. R.; Levif, G., *Tetrahedron* **1988**, *44*, 5263.

29. Moorlag, H.; de Vries, J. G.; Kaptein, B.; Schoemaker, H. E.; Kamphuis, J.; Kellogg, R. M., *Recl. Trav. Chim. Pays-Bas* **1992**, *111*, 129.

30. Genet, J. P.; Grisoni, S., *Tetrahedron Lett.* **1988**, *29*, 4543.

31. Farina, V.; Baker, S. R.; Benigni, D. A.; Sapino, Jr., C., *Tetrahedron Lett.* **1988**, *29*, 5739.

32. Kikukawa, K.; Kono, K.; Wada, F.; Matsuda, T., *J. Org. Chem.* **1983**, *48*, 1333.

33. Del Valle, L.; Stille, J. K.; Hegedus, L. S., *J. Org. Chem.* **1990**, *55*, 3019.

34. Tueting, D. R.; Echavarren, A. M.; Stille, J. K., *Tetrahedron* **1989**, *45*, 979.

35. Farina, V.; Baker, S. R.; Sapino, C. Jr., *Tetrahedron Lett.* **1988**, *29*, 6043.

36. Russell, C. E.; Hegedus, L. S., *J. Am. Chem. Soc.* **1983**, *105*, 943.

37. (a) Legros, J.-Y.; Fiaud, J.-C., *Tetrahedron Lett.* **1990**, *31*, 7453. (b) Tour, J. M.; Lamba, J. J. S., *J. Am. Chem. Soc.* **1993**, *115*, 4935.

38. Oppolzer, W.; DeVita, R. J., *J. Org. Chem.* **1991**, *56*, 6256.

39. Lewis, R. T.; Motherwell, W. B.; Shipman, M., *J. Chem. Soc., Chem. Commun.* **1988**, 948.

40. Baillargeon, V. P.; Stille, J. K., *J. Am. Chem. Soc.* **1986**, *108*, 452.

41. Takeuchi, R.; Suzuki, K.; Sato, N., *Synthesis* **1990**, 923.

42. Meyers, A. I.; Robichaud, A. J.; McKennon, M. J., *Tetrahedron Lett.* **1992**, *33*, 1181.

43. Tour, J. M.; Negishi, E., *J. Am. Chem. Soc.* **1985**, *107*, 8289.

44. Negishi, E.; Tour, J. M., *Tetrahedron Lett.* **1986**, *27*, 4869.

45. Ali, B. E.; Alper, H., *J. Org. Chem.* **1991**, *56*, 5357.

46. Amatore, C.; Jutand, A., *Coord. Chem. Rev.* **1998**, *178*, 511.

47. Macsári, I.; Hupe, E.; Szabó, K. J., *J. Org. Chem.* **1999**, *64*, 9547.

48. Velcicky, J.; Lanver, A.; Lex, J.; Prokop, A.; Wieder, T.; Schmalz, H. G., *Chem. Eur. J.* **2004**, *10*, 5087.

49. Franzén, J.; Löfstedt, J.; Falk, J.; Bäckvall, J. E., *J. Am. Chem. Soc.* **2003**, *125*, 14140.

50. Trost, B. M.; Van Vranken, D. L., *Chem. Rev.* **1996**, *96*, 395.

51. Hamada, Y.; Sakaguchi, K.; Hatano, K.; Hara, O., *Tetrahedron Lett.* **2001**, *42*, 1297.

52. Gong, L.; Chen, G.; Mi, A.; Jiang, Y.; Fu, F.; Cui, X.; Chan, A. S. C., *Tetrahedron: Asymmetry* **2000**, *11*, 4297.

53. Ohta, T.; Sasayama, H.; Nakajima, O.; Kurahashi, N.; Fujii, T.; Furukawa, I., *Tetrahedron: Asymmetry* **2003**, *14*, 537.

54. Ogasawara, M.; Ikeda, H.; Nagano, T.; Hayashi, T., *J. Am. Chem. Soc.* **2001**, *123*, 2089.

55. Ogasawara, M.; Ngo, H. L.; Sakamoto, T.; Takahashi, T.; Lin, W., *Org. Lett.* **2005**, *7*, 2881.

56. Hoke, M. E.; Brescia, M. R.; Bogaczyk, S.; DeShong, P.; King, B. W.; Crimmins, M. T., *J. Org. Chem.* **2002**, *67*, 327.

57. Correia, R.; DeShong, P., *J. Org. Chem.* **2001**, *66*, 7159.

58. Ikegami, R.; Koresawa, A.; Shibata, T.; Takagi, K., *J. Org. Chem.* **2003**, *68*, 2195.

59. Moreno-Mañas, M.; Pajuelo, F.; Plexats, R., *J. Org. Chem.* **1995**, *60*, 2396.

60. Cuerva, J. M.; Gómez-Bengoa, E.; Méndez, M.; Echavarran, A. M., *J. Org. Chem.* **1997**, *62*, 7540.

61. Li, J. H.; Liang, Y.; Xie, Y. X., *Tetrahedron* **2005**, *61*, 7289.

62. Zhang, W.; Shi, M., *Tetrahedron Lett.* **2004**, *45*, 8921.

63. Bei, X.; Turner, H. W.; Weinberg, W. H.; Guram, A. S.; Peterson, J. L., *J. Org. Chem.* **1999**, *64*, 6797.

64. Arentsen, K.; Caddick, S.; Cloke, F. G. N.; Herring, A. P.; Hitchcock, P. B., *Tetrahedron Lett.* **2004**, *45*, 3511.

65. Bonnet, V.; Mongin, F.; Trécourt, F.; Quéguiner, G.; Knochel, P., *Tetrahedron* **2002**, *58*, 4429.

66. Roy, A. H.; Hartwig, J. F., *J. Am. Chem. Soc.* **2003**, *125*, 8704.

67. Seganish, W. M.; Mowery, M. E.; Riggleman, S.; DeShong, P., *Tetrahedron* **2005**, *61*, 2117.

68. Denmark, S. E.; Sweis, R. F., *Acc. Chem. Res.* **2002**, *35*, 835.

69. Denmark, S. E.; Choi, J. Y., *J. Am. Chem. Soc.* **1999**, *121*, 5821.

70. Denmark, S. E.; Wang, Z., *J. Organomet. Chem.* **2001**, *624*, 372.

71. Denmark, S. E.; Sweis, R. F.; Wehrli, D., *J. Am. Chem. Soc.* **2004**, *126*, 4865.

72. Denmark, S. E.; Pan, W., *Org. Lett.* **2001**, *3*, 61.

73. Denmark, S. E.; Tymonko, S. A., *J. Am. Chem. Soc.* **2005**, *127*, 8004.

74. Seganish, W. M.; DeShong, P., *Org. Lett.* **2004**, *6*, 4379.

75. Lee, S.; Jørgensen, M.; Hartwig, J. F., *Org. Lett.* **2001**, *3*, 2729.

76. Hooper, M. W.; Utsunomiya, M.; Hartwig, J. F., *J. Org. Chem.* **2003**, *68*, 2861.

77. Mann, G.; Incarvito, C.; Rheingold, A. L.; Hartwig, J. F., *J. Am. Chem. Soc.* **1999**, *121*, 3224.

78. Ackermann, L.; Born, R., *Angew. Chem., Int. Ed.* **2005**, *44*, 2444.

79. Manoso, A. S.; DeShong, P., *J. Org. Chem.* **2001**, *66*, 7449.

80. Ishiyama, T.; Ahiko, T.; Miyaura, N., *Tetrahedron Lett.* **1996**, *37*, 6889.

81. Beare, N. A.; Hartwig, J. F., *J. Org. Chem.* **2002**, *67*, 541.

82. Gao, C.; Tao, X.; Qian, Y.; Huang, J., *Chem. Commun.* **2003**, 1444.

83. Lee, S.; Beare, N. A.; Hartwig, J. F., *J. Am. Chem. Soc.* **2001**, *123*, 8410.

84. Liu, X.; Hartwig, J. F., *Org. Lett.* **2003**, *5*, 1915.

85. Liu, X.; Hartwig, J. F., *J. Am. Chem. Soc.* **2004**, *126*, 5182.

86. de Filippis, A.; Pardo, D. G.; Cossy, J., *Tetrahedron* **2004**, *60*, 9757.

87. Honda, T.; Namiki, H.; Satoh, F., *Org. Lett.* **2001**, *3*, 631.

88. Yang, C.; Lee, H. M.; Nolan, P., *Org. Lett.* **2001**, *3*, 1511.

89. Wang, A. E.; Xie, J. H.; Wang, L. X.; Zhou, Q. L., *Tetrahedron* **2005**, *61*, 259.

90. Reddy, K. R.; Krishna, G. G., *Tetrahedron Lett.* **2005**, *46*, 661.

91. Wu, X. Y.; Xu, H. D.; Zhou, Q. L.; Chan, A. S. C., *Tetrahedron: Asymmetry* **2000**, *11*, 1255.

92. Yang, D.; Chen, Y. C.; Zhu, N. Y., *Org. Lett.* **2004**, *6*, 1577.

93. Dahan, A.; Portnoy, M., *Org. Lett.* **2003**, *5*, 1197.

94. Huang, T. H.; Chang, H. M.; Wu, M. Y.; Cheng, C. H., *J. Org. Chem.* **2002**, *67*, 99.

95. Jeganmohan, M.; Cheng, C. H., *Org. Lett.* **2004**, *6*, 2821.

96. Jayanth, T. T.; Jeganmohan, M.; Cheng, C. H., *Org. Lett.* **2005**, *7*, 2921.

97. Pottier, L. R.; Peyrat, J. F.; Alami, M.; Brion, J. D., *Tetrahedron Lett.* **2004**, *45*, 4035.

98. Shim, J. G.; Park, J. C.; Cho, C. S.; Shim, S. C.; Yamamoto, Y., *Chem. Commun.* **2002**, 852.

99. Larock, R. C.; Tu, C., *Tetrahedron* **1995**, *51*, 6635.

100. Jeganmohan, M.; Shanmugasundaram, M.; Cheng, C. H., *Chem. Commun.* **2003**, 1746.

101. Obora, Y.; Tsuji, Y.; Kawamura, T., *J. Am. Chem. Soc.* **1995**, *117*, 9814.

102. Larock, R. C.; Tu, C.; Pace, P., *J. Org. Chem.* **1998**, *63*, 6859.

103. Rozhkov, R. V.; Larock, R. C., *J. Org. Chem.* **2003**, *68*, 6314.

104. Rozhkov, R. V.; Larock, R. C., *Tetrahedron Lett.* **2004**, *45*, 911.

105. Söderberg, B. C. G.; Wallace, J. M.; Tamariz, J., *Org. Lett.* **2002**, *4*, 1339.

106. Bernard, A. M.; Frongia, A.; Secci, F.; Delogu, G.; Ollivier, J.; Piras, P. P.; Salaün, J., *Tetrahedron* **2003**, *59*, 9433.

107. Legros, J. Y.; Primault, G.; Toffano, M.; Rivière, M. A.; Fiaud, J. C., *Org. Lett.* **2000**, *2*, 433.

108. Nishimura, T.; Nishiguchi, Y.; Maeda, Y.; Uemura, S., *J. Org. Chem.* **2004**, *69*, 5342.

109. Shih, H. T.; Shih, H. H.; Cheng, C. H., *Org. Lett.* **2001**, *3*, 811.

110. Nakai, Y.; Uozumi, Y., *Org. Lett.* **2005**, *7*, 291.

111. *Sigma-Aldrich Catalog Web.* http://www. sigmaaldrich. com/united-states. html (accessed January 2012).

112. (a) Ukai, T.; Kawazura, H.; Ishii, Y.; Bonnett, J.; Ibers, J. A., *J. Organomet. Chem.* **1974**, *65*, 253. (b) Maitlis, P. M.; Russell, M. J. H. In *Comprehensive Organometallic Chemistry;* Wilkinson, G., Ed.; Pergamon: Oxford, **1982**; Vol. 6, p 259.

113. (a) Gligorich, K. M.; Sigman, M. S., *Chem. Commun.* **2009**, 3854. (b) Stahl, S. S., *Angew. Chem., Int. Ed.* **2004**, *43*, 3400.

114. Yang, M.; Yip, K.-T.; Pan, J.-H.; Chen, Y.-C.; Zhu, N.-Y.; Yang, D., *Synlett.* **2006**, 3057.

115. Shvo, Y.; Goldman-Lev, V., *J. Organomet. Chem.* **2002**, *650*, 151.

116. Ferreira, E. M.; Stoltz, B. M., *J. Am. Chem. Soc.* **2001**, *123*, 7725.

117. Sage, J.-M.; Gore, J.; Guilmet, E., *Tetrahedron Lett.* **1989**, *30*, 6319.

118. Loy, R. N.; Sanford, M. S., *Org. Lett.* **2011**, *13*, 2548.

119. Yoo, E.-J.; Ma, S.; Mei, T.-S.; Chan, K. S. L.; Yu, J.-Q., *J. Am. Chem. Soc.* **2011**, *133*, 7652.

120. Dai, H. X.; Yu, J.-Q., *J. Am. Chem. Soc.* **2012**, *134*, 134.

121. Campeau, L.-C.; Schipper, D. J.; Fagnou, K., *J. Am. Chem. Soc.* **2008**, *130*, 3266.

122. Barder, T. E.; Walker, S. D.; Martinelli, J. R.; Buchwald, S. L., *J. Am. Chem. Soc.* **2005**, *127*, 4685.

123. Pan, J.; Su, M.; Buchwald, S. L., *Angew. Chem., Int. Ed.* **2011**, *50*, 8647.

124. Mei, T.-S.; Wang, D.-H.; Yu, J.-Q., *Org. Lett.* **2010**, *12*, 3140.

125. Eastgate, M. D.; Buono, F. G., *Angew. Chem., Int. Ed.* **2009**, *48*, 5958.

126. (a) Du, H.; Zhao, B.; Shi, Y., *J. Am. Chem. Soc.* **2008**, *130*, 8590. (b) Fu, R.; Zhao, B.; Shi, Y., *J. Org. Chem.* **2009**, *74*, 7577.

127. Elementometalation is defined as: reactions that include the functionalization of alkynes, for example, hydrometalation (B, Al, Zr, etc.), carbometalation (Cu, Al–Zr, etc.), and haloboration (BX3 where X is Cl, Br, and I), see: Negishi, E.; Wang, G.; Rao, H.; Xu, Z., *J. Org. Chem.* **2010**, *75*, 3151.

128. Burks, H. E.; Liu, S.; Morken, J. P., *J. Am. Chem. Soc.* **2007**, *129*, 8766.

129. Liu, C.; Widenhoefer, R. A., *Chem. Eur. J.* **2006**, *12*, 2371.

130. Trend, R. M.; Ramtohul, Y. K.; Stolz, B. M., *J. Am. Chem. Soc.* **2005**, *127*, 17778.

131. Larock, R. C.; Wei, L.; Hightower, T. R., *Synlett* **1998**, 522.

132. Nishimura, T.; Ohe, K.; Uemura, S., *J. Org. Chem.* **2001**, *66*, 1455.

133. Jung, Y. C.; Mishra, R. K.; Yoon, C. H.; Jung, K. W., *Org. Lett.* **2003**, *5*, 2231.

134. Koubachi, J.; Berteina-Raboin, S.; Mouaddib, A.; Guillaumet, G., *Synthesis* **2009**, 0271.

135. Cho, C. S.; Tanabe, K.; Itoh, O.; Uemura, S., *J. Org. Chem.* **1995**, *60*, 274.

9-Borabicyclo[3.3.1]nonane Dimer[1]

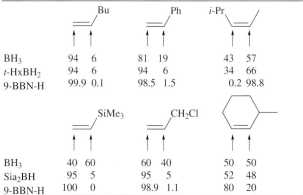

[70658-61-6] $C_{16}H_{30}B_2$ (MW 244.03)

InChI = 1/2C8H15B/c2*1-3-7-5-2-6-8(4-1)9-7/h2*7-9H,1-6H2/t2*7-,8+

InChIKey = FPEQNDQUWPJCKZ-MOGCCKQLBC

(highly selective, stable hydroborating agent;[1,3] anti-Markovnikov hydration of alkenes and alkynes;[1d] effective ligation for alkyl-, aryl-, allyl-, allenyl-, alkenyl- and alkynylboranes;[1a,4,5] forms stable dialkylboryl derivatives, borinate esters, and haloboranes;[1f] organoboranes from hydroboration and organometallic reagents;[1,5] precursor to boracycles;[1,11] can selectively reduce conjugated enones to allylic alcohols;[1a,30] organoborane derivatives for α-alkylation and arylation of α-halo ketones, nitriles, and esters;[1b] vinylation and alkynylation of carbonyl compounds;[1a,46] conjugate addition to enones;[1a,47] homologation; asymmetric reduction;[1,8] Diels–Alder reactions;[1a,18,50] enolboranes for crossed aldol condensations;[1a,20,52] Suzuki–Miyaura coupling[1a,54–57])

Alternate Name: 9-BBN-H.

Physical Data: mp 153–155 °C (sealed capillary); bp 195 °C/12 mmHg.[1,3]

Solubility: sparingly sol cyclohexane, dimethoxyethane, diglyme, dioxane (<0.1 M at 25 °C); sol THF, ether, hexane, benzene, toluene, CCl$_4$, CHCl$_3$, CH$_2$Cl$_2$, SMe$_2$ (ca. 0.2–0.6 M at 25 °C); reacts with alcohols, acetals, aldehydes, and ketones.[1,3]

Form Supplied in: colorless, stable crystalline solid; 0.5 M solution in THF or hexanes.

Analysis of Reagent Purity: the melting point of 9-BBN-H dimer is very sensitive to trace amounts of impurities. Recrystallization from dimethoxyethane is recommended for samples melting below 146 °C. The dimer exhibits a single ^{11}B NMR (C$_6$D$_6$) resonance at δ 28 ppm and ^{13}C NMR signals at 20.2 (br), 24.3 (t), and 33.6 (t) ppm.[1,3]

Handling, Storage, and Precaution: the crystalline 9-BBN-H dimer can be handled in the atmosphere for brief periods without significant decomposition. However, the reagent should be stored under an inert atmosphere, preferably below 0 °C. Under these conditions the reagent is indefinitely stable. In solution, 9-BBN is more susceptible both to hydrolysis and oxidation, and contact with the open atmosphere should be rigorously avoided. Many 9-BBN derivatives are pyrophoric and/or susceptible to hydrolysis so that individuals planning to use 9-BBN-H dimer should thoroughly familiarize themselves with the special techniques required for the safe handling of such reagents prior to their use.[1b] The reagent should be used in a well-ventilated hood.

Original Commentary

John A. Soderquist
University of Puerto Rico, Rio Piedras, Puerto Rico

Essential Reactions for Organic Synthesis, First Edition. Edited by Philip L. Fuchs.
© 2016 John Wiley & Sons, Ltd. Published 2016 by John Wiley & Sons, Ltd.

Organoboranes from 9-BBN-H. First identified by Köster,[2] 9-BBN-H dimer is prepared from the cyclic hydroboration of 1,5-cyclooctadiene (eq 1).[3] As a dialkylborane, 9-BBN-H exhibits extraordinary steric- and electronic-based regioselectivities which distinguish these derivatives from the less useful polyhydridic reagents (Table 1).[4]

However, in contrast to other dialkylborane reagents (e.g. ***Diisoamylborane*** or ***Dicyclohexylborane***) which must be freshly prepared immediately prior to their use, 9-BBN-H dimer is a stable crystalline reagent[1,3] which is commercially available in high purity. This feature of the reagent facilitates the control of reaction stoichiometry at a level unattainable with most borane reagents. The remarkable thermal stability of 9-BBN derivatives permits hydroborations to be conducted over a broad range of temperatures (from 0 °C to above 100 °C) either neat or in a variety of solvents.[4] The B-alkyl-9-BBN products can frequently be isolated by distillation without decomposition and fully characterized spectroscopically.[1,4,5] The integrity of the 9-BBN ring is retained even at elevated temperatures (200 °C), but positional isomerization in the B-alkyl portion can take place at ca. 160 °C.[6]

Table 1 Boron Atom Placement in the Hydroboration of Simple Alkenes

	Bu		Ph		i-Pr	
BH$_3$	94	6	81	19	43	57
t-HxBH$_2$	94	6	94	6	34	66
9-BBN-H	99.9	0.1	98.5	1.5	0.2	98.8

	SiMe$_3$		CH$_2$Cl			
BH$_3$	40	60	60	40	50	50
Sia$_2$BH	95	5	95	5	52	48
9-BBN-H	100	0	98.9	1.1	80	20

Like most other dialkylboranes, 9-BBN-H exists as a dimer, but hydroborates as a monomer (eq 2).[7] In general, the rates of hydroboration follow the order R$_2$C=CH$_2$ > RHC=CH$_2$ > *cis*-RHC=CHR > *trans*-RHC=CHR > RHC=CR$_2$ > R$_2$C=CR$_2$.[1,4] For relatively unsubstituted alkenes the dissociation of the 9-BBN-H dimer is rate-limiting ($T_{1/2}$ at 25 °C ≈ 20 min) so that the hydroborations of typical 1-alkenes are normally complete in less than 3 h at room temperature. Competitive rate studies have revealed that electron-donating groups enhance the rates within these groups, e.g. for p-XC$_6$H$_4$CH=CH$_2$, $k_{rel} = 1$ (X = CF$_3$), 5 (X = H), 70 (X = OMe).[4c]

$$[9\text{-BBN-H}]_2 \underset{k_{-1}}{\overset{k_1}{\rightleftarrows}} 2\,[9\text{-BBN-H}]$$

$$9\text{-BBN-H} \xrightarrow[k_2]{\text{alkene}} \text{R-9-BBN} \qquad (2)$$

Hydroborations of more substituted alkenes such as α-pinene[8] or 2,3-dimethyl-2-butene[4b] with 9-BBN-H are slower (k_2 is rate-limiting) and require heating at reflux temperature in THF for 2 and 8 h, respectively, for complete reaction to occur. However, the enantioselective reducing agent[8a] Alpine-borane® (see **B-3-Pinanyl-9-borabicyclo[3.3.1]nonane**) is formed quantitatively as a single enantiomer, the process taking place with complete Markovnikov regiochemistry, exclusively through *syn* addition from the least hindered face of the alkene (eq 3).

$$(3)$$

While the monohydroboration of symmetrical nonconjugated dienes with 9-BBN-H is thwarted by competing dihydroboration because these remote functionalities act as essentially independent entities, with nonequivalent sites the chemoselectivity of the reagent can be excellent (eq 4).[4] Also, whereas the monohydroboration of conjugated dienes is not always a useful process because of competitive dihydroboration, highly substituted dienes and 1,3-cyclohexadiene produce allylborane products efficiently. In contrast to the monohydroboration of allene itself, which gives a 1,3-diboryl adduct, 9-BBN-H is an effective reagent for the preparation of allylboranes from substituted allenes.[4g] For example, excellent selectivity has been observed for silylated allenes where hydroboration occurs *anti* to the silyl group on the allene and at the terminal position (eq 5).[9] It is also important to note that the diastereofacial selectivities of 9-BBN-H can be complementary to those obtained with Rh-catalyzed hydroborations (eq 6).[10]

$$(4)$$

$$(5)$$

$$(6)$$

9-BBN-H 11:89
CatBH, ClRh(PPh$_3$)$_3$ 96:4

Medium-ring boracycles are efficiently prepared by the dihydroboration of α,ω-dienes with 9-BBN-H followed by exchange with borane.[11] In this process 9-BBN-H is particularly useful because it not only fixes the key 1,5-diboryl relationship, but also the 9-BBN ligands do not participate in the exchange process (eq 7).

$$(7)$$

Unlike most dialkylboranes, 9-BBN-H hydroborates alkenes faster than its does the corresponding alkynes, a feature which leads to the competitive formation of 1,1-diboryl adducts in the hydroboration of 1-alkynes with 9-BBN-H employing a 1:1 stoichiometry (eq 8).[1,12,13] In some cases, the (E)-1-alkenyl-9-BBN derivative can be efficiently prepared by employing either a large excess of the alkyne[12,13] or through the use of 1-trimethylsilyl derivatives.[6] However, these vinylboranes are now perhaps best prepared through the dehydroborylation of their 1,1-diboryl adducts with aromatic aldehydes (eq 8).[12]

$$(8)$$

By contrast, 9-BBN-H effectively monohydroborates internal alkynes to produce the corresponding vinylboranes in >90% yields.[1,13] Compared to 1-alkynes, their 1-silyl counterparts also produce better yields of vinylboranes but, in contrast to normal internal alkynes, produce *vicinal* rather than *geminal* diboryl adducts with dihydroboration.[6,13,14] Larger silyl groups can effectively be used to redirect the boron to the internal position producing the 'silyl-Markovnikov' vinylborane, exclusively, without competitive dihydroboration.[6] For 1-halo-1-alkynes, hydroboration with 9-BBN is slow, but the (Z)-1-halovinylboranes (eq 9) are produced cleanly and these are protonolyzed to provide *cis*-vinyl halides.[15] The isomeric (Z)-2-bromovinyl-9-BBN derivatives are available from the bromoboration of 1-alkynes with B-Br-9-BBN[16] (see **9-Bromo-9-borabicyclo[3.3.1]nonane**). It is important to point out that the preference of 9-BBN-H to hydroborate alkenes in the presence of alkynes can have useful synthetic applications (eq 10).[17]

(9)

(10)

Organometallic reagents can provide very useful entries to many *B*-substituted 9-BBN derivatives. These are particularly important for organoboranes which cannot be prepared by hydroboration.[5] Both *B*-alkoxy and *B*-halo derivatives serve as useful precursors to *B*-alkyl, -allyl, -aryl, -vinyl or -alkynyl-9-BBN derivatives (eqs 11–16). *B*-Halo-9-BBN derivatives are effectively vinylated with organotin reagents.[18] However, *B*-MeO-9-BBN is superior to its *B*-halo counterparts for secondary and tertiary alkyllithium reagents where the latter undergo some reduction to 9-BBN-H through β-hydride transfer from the organolithium.

(11)

(12)

(13)

(14)

(15)

(16)

Hydroboration of the byproduct alkene gives isomeric *B*-alkyl-9-BBN products.[5b] Generally, hydrocarbon solvents are preferable to ether or THF for this process because the greater stability of the intermediate methoxyborate complexes (i.e. Li[R(MeO)-9-BBN]) at −78 °C in these solvents prevents the product from being formed and competing with *B*-MeO-9-BBN for the alkyllithium reagent prior to its complete consumption.[5b] The complex is stable for alkenyl and alkynyl derivatives which require

BF$_3$·Et$_2$O to remove the methoxy moiety. The procedure has also been used for the preparation of *cis*-vinyl-9-BBN derivatives[19] since the normal route to such derivatives based upon the hydroboration of 1-haloalkynes, followed by hydride-induced rearrangement gives ring expansion products competitively with (*Z*)-1-halovinyl-9-BBNs.[20] Similar behavior has been observed for the reaction of α-methoxyvinyllithium with *B*-alkyl-9-BBNs (see *1-Methoxyvinyllithium*).[21]

Derivatives of 9-BBN. Like other boron hydrides, a variety of proton sources (ROH, RCO$_2$H, RSO$_3$H, HX (X = Cl, Br, OH, SH, O$_2$P(OH)$_2$, NHR)) as well as boron halides can be effectively employed to prepare useful derivatives from 9-BBN-H.[1d,5b,22] The synthetic value of *B*-MeO-9-BBN lies principally in the preparation of *B*-alkyl derivatives through organometallic reagents as described above. As a byproduct in other processes, it is also easily converted to 9-BBN-H with BMS (eq 17).[22]

(17)

Efficient procedures have been developed for the preparation of *B*-Cl-9-BBN from 9-BBN-H (HCl in Et$_2$O)[5b] and *B*-Br-9-BBN (BBr$_3$ in CH$_2$Cl$_2$),[23] the latter being a useful reagent for ether cleavage, the bromoboration of 1-alkynes, and for conjugative additions to enones (see *9-Bromo-9-borabicyclo[3.3.1]nonane*). 9-BBN triflate is highly useful in formation of enolboranes for stereoselective crossed aldol reactions[24] (see *9-Borabicyclononyl Trifluoromethanesulfonate*). The *B*-acyloxy-9-BBN derivatives have been employed in conjuction with borohydrides for the reduction of carboxylic acids to aldehydes[25] (see *Lithium 9-boratabicyclo[3.3.1]nonane*). Amine complexes of 9-BBN-H and borohydride derivatives are easily prepared from the addition of amines or metal hydrides to 9-BBN-H.[26] *B*-Alkyl-9-BBNs and their borohydrides are very selective reducing agents and, with chiral terpenoid or sugar appendages, can also effectively function as enantioselective reagents (eq 18)[8,27] (see *B-3-Pinanyl-9-borabicyclo-[3.3.1]nonane* and *Potassium 9-Siamyl-9-boratabicyclo[3.3.1]-nonane*). 9,9-Dialkylborate derivatives of 9-BBN are also highly selective reducing agents[28] (see *Lithium 9,9-Dibutyl-9-borabicyclo[3.3.1]nonanate*), transferring a bridgehead hydride with rearrangement to bicyclo[3.3.0]octylboranes. This process is best accomplished with *Acetyl Chloride* and provides a highly versatile entry to these organoboranes for subsequent conversions (eq 19).[29]

(18)

R = *n*-C$_8$H$_{17}$, 97% ee

(19)

Functional Group Conversions with 9-BBN-H. 9-BBN-H selectively reduces acid chlorides, aldehydes, ketones, lactones, and sulfoxides at 25 °C.[1e] Alcohol rather than amine products are produced as the major products from tertiary amides, while primary derivatives are not reduced effectively. Reduction is slow with esters, carboxylic acids, nitriles, and epoxides, and does not occur with nitro compounds, nor with alkyl or aryl halides. At 65 °C, carboxylic acids and esters are cleanly reduced to alcohols, the former being significantly slower (18 vs. 4 h). Moreover, 9-BBN-H is a highly selective reducing agent for the reduction of enones to allylic alcohols (eq 20).[30]

As noted earlier, *B*-substituted-9-BBN derivatives are available from a variety of sources and organoboranes serve as a versatile entry to other functionalities. Their oxidative conversion to alcohols with alkaline **Hydrogen Peroxide** or **Sodium Perborate**[31] is quantitative and occurs with complete retention of configuration, making the process highly useful.[1] The 9-BBN moiety is oxidized to *cis*-1,5-cyclooctanediol (eq 21), a compound which can be removed from less polar products through extraction with water, selective crystallization, or by chromatography.[1] The monooxidation of 9-BBN derivatives with anhydrous **Trimethylamine N-Oxide** (TMANO) produces 9-oxa-10-borabicyclo[3.3.2]decanes (eq 22),[14] many of which are air-stable and undergo useful coupling reactions.

Anomalous oxidation products are observed from the oxidation of tetraalkylborate salts (i.e. Li(R$_2$-9-BBN)), which produces bicyclo[3.3.0]octan-l-ol as a co-product through a skeletal rearrangement which occurs during the oxidation process.[32] Moreover, the alkaline hydrogen peroxide oxidation of 1,1-di-9-BBN derivatives gives primary alcohols rather than aldehydes because of their solvolysis prior to oxidation.[6,13]

While the protonolysis of *B*-alkyl-9-BBNs, like other trialkylboranes, with carboxylic acids takes place only at temperatures above 100 °C, *B*-vinyl derivatives are readily cleaved by HOAc at 0 °C with complete retention of configuration.[1,13] This can be combined with other 9-BBN processes (e.g. thermal isomerization or dehydroborylation) to give remarkable overall conversions (eq 23).[6] The hydrolysis of allylic and alkynic 9-BBN derivatives is more facile, occurring even with water or alcohols.[5c,32]

In the absence of light, molecular bromine readily cleaves *B*-(*s*-alkyl)-9-BBN derivatives through a hydrogen abstraction process, to give excellent yields of the corresponding alkyl bromides, the 9-BBN moiety being converted to 9-Br-9-BBN (eq 24).[33] However, bicyclo[3.3.0]octylborinic and -boronic acids are produced from

B-Me- and *B*-MeO-9-BBN through this radical bromination under hydrolytic conditions where the facile 1,2-alkyl migration of a ring B–C bond occurs (eq 25).[1b,34] This latter compound serves as a convenient source of 9-oxabicyclo[3.3.1]nonane through base-induced iodination via an S$_E$2-type inversion.[35]

Mechanistically similar to the oxidation of 9-BBN derivatives with TMANO, the amination of *B*-alkyl-9-BBN proceeds through ring B–C migration rather than through *B*-alkyl migration (eq 26).[36] Similar behavior is observed for the thermal reaction of organic azides with these derivatives. Dichloroboryl derivatives have proved to be more versatile and general for the synthesis of amines, including optically active derivatives.[37]

Carbon–Carbon Bond Formation via 9-BBN Derivatives. Consistent with the versatility of organoboranes in synthetically useful chemical transformations, most conversions with 9-BBN derivatives are very efficient and occur with strict stereochemical control.[1] Of particular importance are those which involve the formation of new carbon–carbon bonds because valuable R groups can often be selectively transferred to the substrates without competition from the 9-BBN ring. For example, whereas only one of the alkyl, vinyl or aryl groups can be transferred from BR$_3$ to the anions derived from α-halo ketones, esters and nitriles, these reactions are ideally suited to *B*-R-9-BBN derivatives which transfer the *B*-R group selectively (eq 27).[1a,38] The vinylogous γ-alkylation of γ-bromo-α,β-unsaturated esters efficiently leads to γ-substituted-β,γ-unsaturated esters, the double bond transposition being commonly observed in the kinetic protonation of enolates with extended conjugation. Both sulfur ylides and α-bromo sulfones undergo related alkylations.[1a]

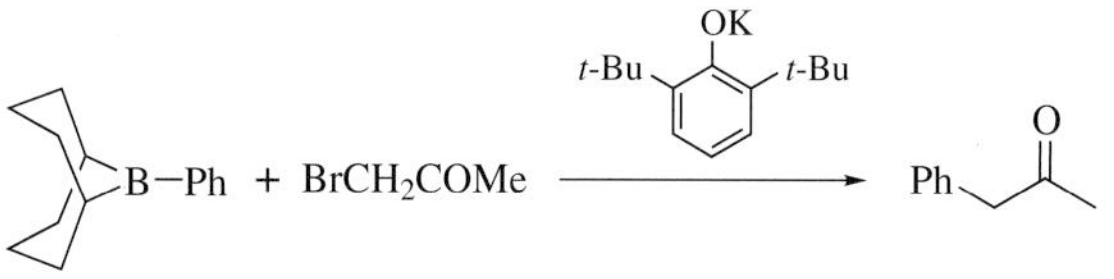

$$Ph-B + BrCH_2COMe \xrightarrow{\quad} Ph \overset{O}{\frown} \qquad (27)$$

Base-induced eliminations of γ-haloalkyl-9-BBN derivatives give cycloalkanes (C_3 to C_6)[1b] with inversion of configuration at both carbon centers.[1,39] 1,1-Diboryl adducts from the dihydroboration of 1-alkynes with 9-BBN-H serve as useful precursors to *B*-cyclopropyl-9-BBN derivatives by a similar process (eq 28).

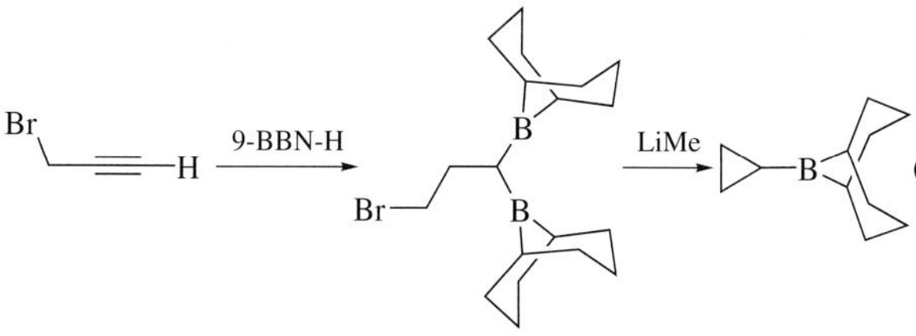

$$(28)$$

The carbonylation of *B*-alkyl-9-BBNs at 70 atm, 150 °C in the presence of ethylene glycol produces intermediate boronate esters which are oxidized with alkaline hydrogen peroxide to give high yields of the corresponding carbinols (Scheme 1).[40] In the presence of hydride-reducing agents (e.g. LiHAl(OMe)$_3$ or *Potassium Triisopropoxyborohydride*), the carbonylation of *B*-R-9-BBN derivatives can be carried out at atmospheric pressures at 0 °C, producing an intermediate α-alkoxyalkylborane which can be further reduced with *Lithium Aluminum Hydride*. This results in homologated organoboranes and, after oxidation, alcohols. Alternatively, the intermediate α-alkoxyalkylborane can be directly oxidized to produce aldehydes (Scheme 1).[1a,38c,41]

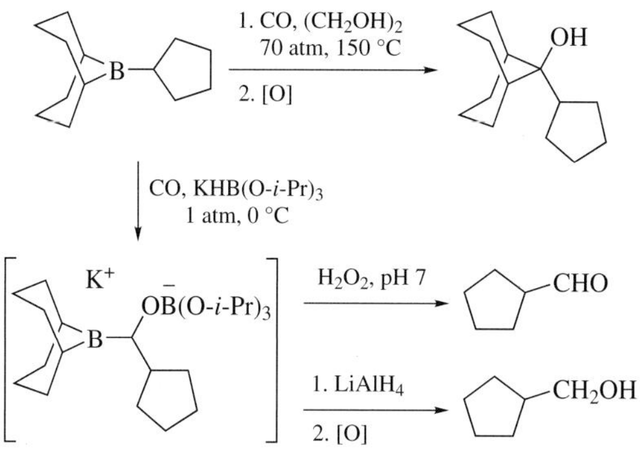

Scheme 1

As a useful alternative to carbonylation, the Brown dichloromethyl methyl ether (DCME) process has been effectively used for the synthesis of 9-alkylbicyclo[3.3.1]nonan-9-ols.[42] The ketone bicyclo[3.3.1]nonan-9-one (eq 29)[42b] has also been prepared from a hindered *B*-aryloxy-9-BBN derivative, with simple *B*-alkoxy-9-BBN derivatives failing to undergo this process. However, most borinate esters are smoothly converted to ketones through this process, including germa- and silaborinanes (eq 30).[11e,f] In these cases, 9-BBN-H provides the essential 1,5-

diboryl relationship which allows the formation of borinane by the exchange reaction described earlier.

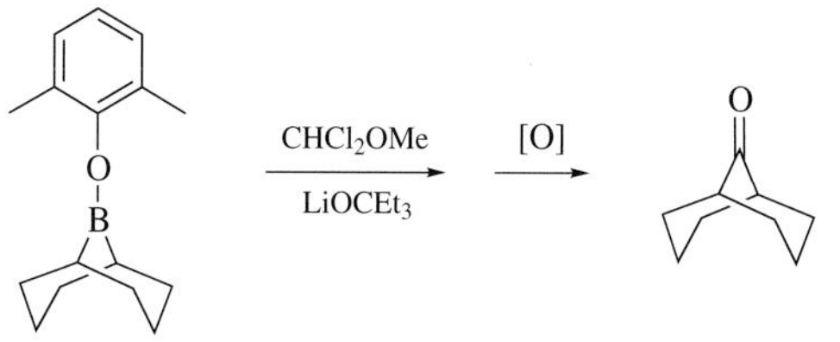

$$(29)$$

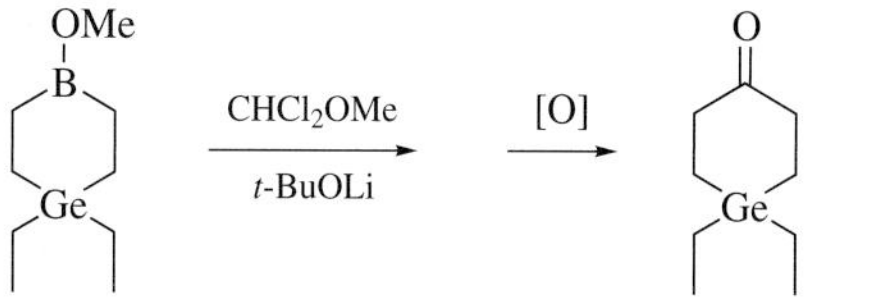

$$(30)$$

Allylboration with 9-BBN derivatives (see *B-Allyl-9-borabicyclo[3.3.1]nonane*) is an efficient process, resulting in the smooth formation of homoallylic alcohols (eq 31).[43] Alkynylboranes also undergo 1,2-addition to both aldehydes and ketones.[44] As with other reactions producing *B*-alkoxy-9-BBN byproducts, the conversion of these to alcohols with *Ethanolamine* also results in the formation of an alkane-insoluble 9-BBN complex which is conveniently removed, thereby greatly simplifying the workup procedure.

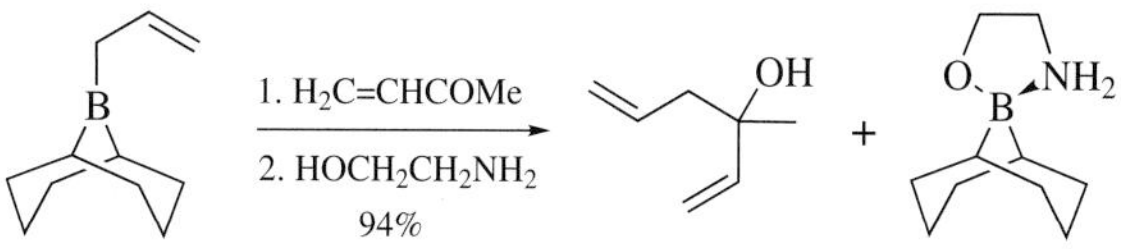

$$(31)$$

Vinyl derivatives of 9-BBN uniquely undergo 'Grignard-like' 1,2-additions to aldehydes to produce stereodefined allylic alcohols (eq 32).[12,45] The thermal stability of the 9-BBN ring system, as well as its resistance to serve as a β-hydride source, facilitates this highly effective process. These vinylboranes are also the borane reagents of choice for the conjugate additions to enones which can adopt a *cisoid* conformation, providing a convenient entry to γ,δ-unsaturated ketones from enones (eq 33).[46] Alkynylboranes undergo a related addition–elimination process with β-methoxyenones, giving enynones (eq 34).[47]

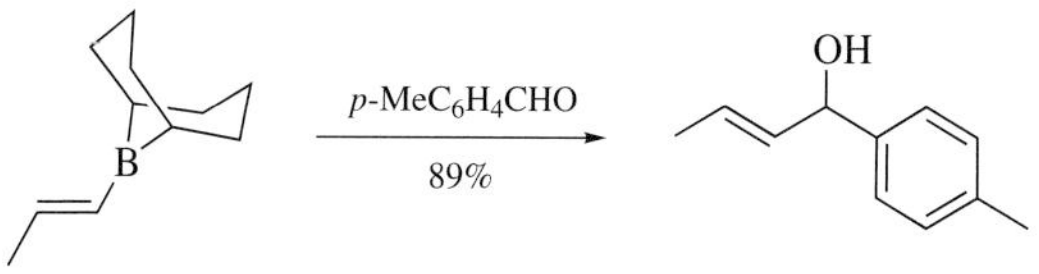

$$(32)$$

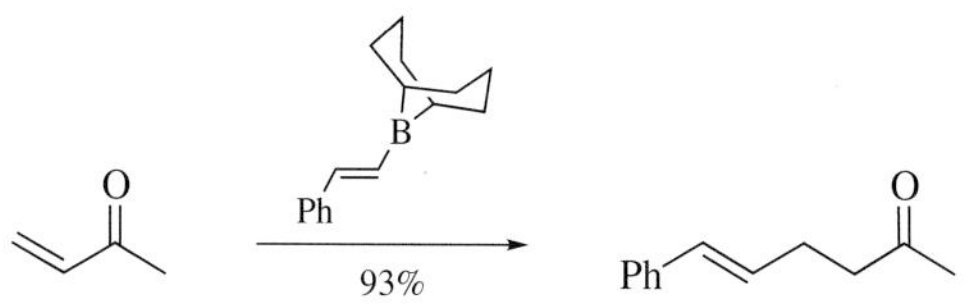

$$(33)$$

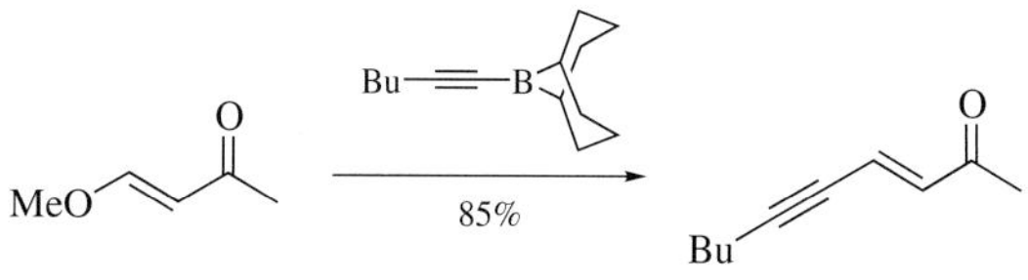

$$(34)$$

Vinyl(methoxy)-9-BBN 'ate' complexes undergo an unusual homocoupling reaction when treated with *Zinc Chloride* (0.5

equiv) (eq 35).[48] Related intermediates, formed through the Cu[I]-catalyzed addition of stannylborate complexes to 1-alkynes, can be coupled either through catalytic palladium or stoichiometric copper chemistry to produce stereodefined vinylstannanes (eq 36).[49]

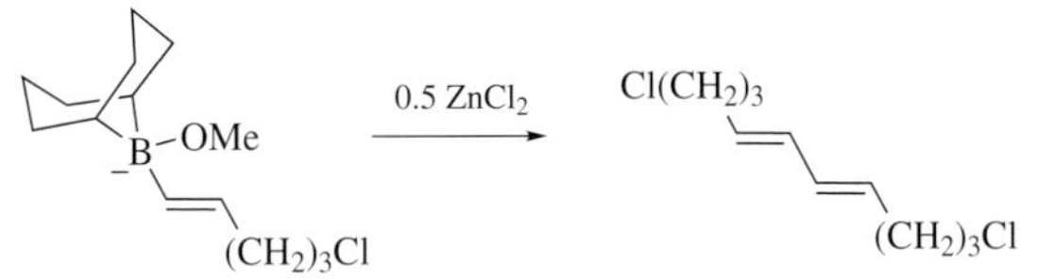

(35)

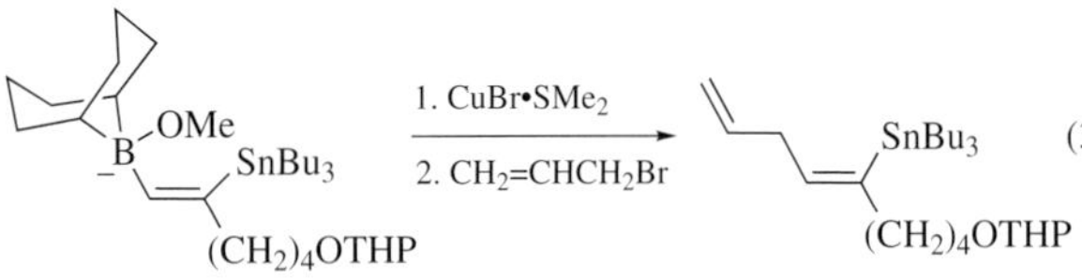

(36)

Both vinyl- and alkynyl-9-BBN derivatives are effective dienophiles in Diels–Alder cycloadditions, leading to boron-functionalized cyclohexenes in a selective manner (eqs 37 and 38).[18,50] Silylated allenylboranes add selectively as allylboranes to aldehydes, a reaction which has been effectively used to prepare the steroid nucleus through a Hudrlik elimination followed by a Bergman rearrangement (eq 39).[51]

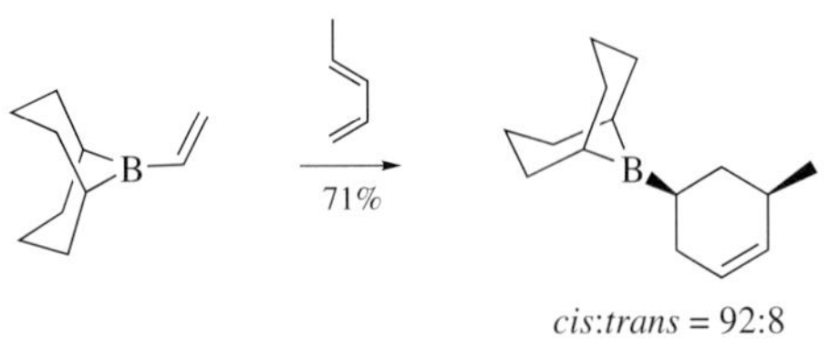

(37)

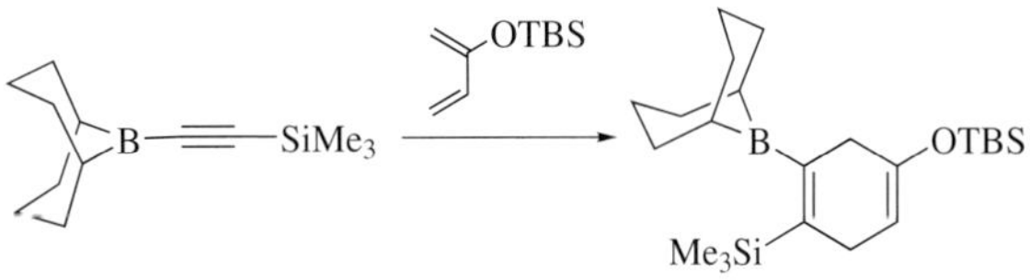

cis:trans = 92:8

(38)

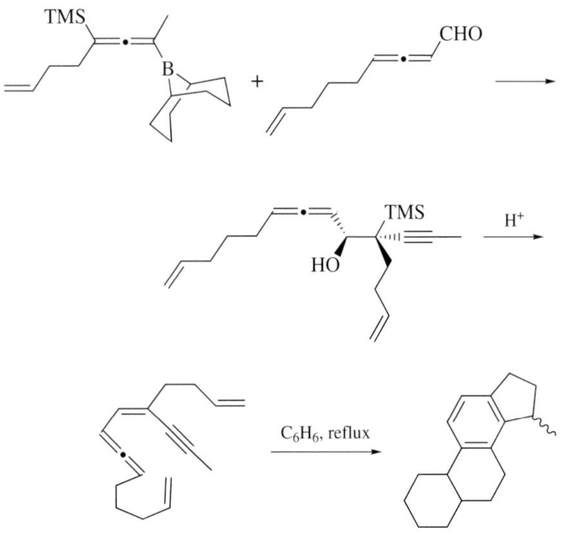

(39)

Stereodefined 9-BBN enolboranes which contain a directing chiral auxiliary undergo highly selective crossed aldol condensations as do other dialkylboryl systems (eq 40).[20,52] The conjugate addition of B-Br-9-BBN also produces enolboranes which condense with aldehydes to produce, after the elimination of the elements of B-HO-9-BBN, α-bromomethyl enones stereoselectively (eq 41).[53]

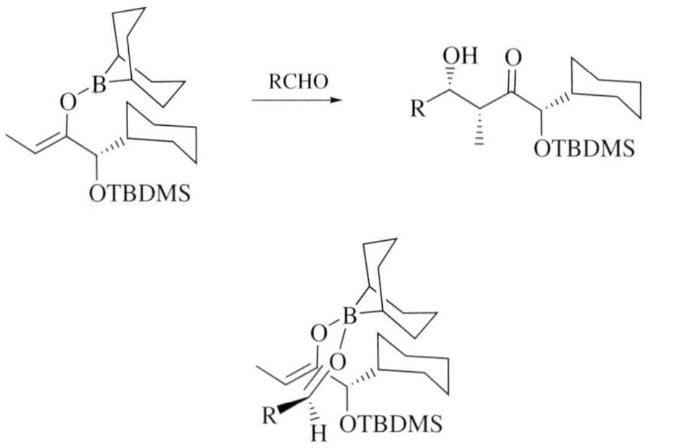

(40)

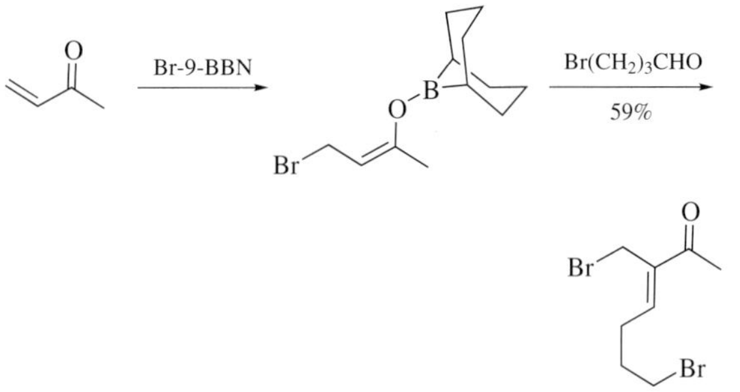

(41)

While catechol- and disiamylborane derivatives were originally employed in the Pd-catalyzed cross coupling of organoboranes to unsaturated organic halides under basic conditions (Suzuki–Miyaura coupling), 9-BBN has recently found an important place in this process. Initially, B-(primary alkyl)-9-BBNs, with added bases (NaOH, TlOH, NaOMe, or K_3PO_4), were found to undergo efficient coupling with iodobenzene using **Dichloro[1,1′-bis(diphenylphosphino)ferrocene]-palladium(II)** as the catalyst.[54] However, while secondary alkylboranes may require this catalyst, **Tetrakis(triphenylphosphine)palladium(0)** is perfectly satisfactory for the coupling of n- or i-alkyl-9-BBN derivatives to unsaturated bromides, iodides, or triflates under basic conditions (eqs 42–45).

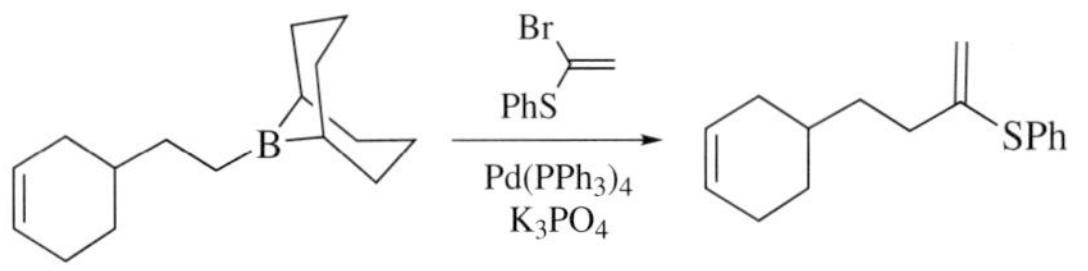

(42)

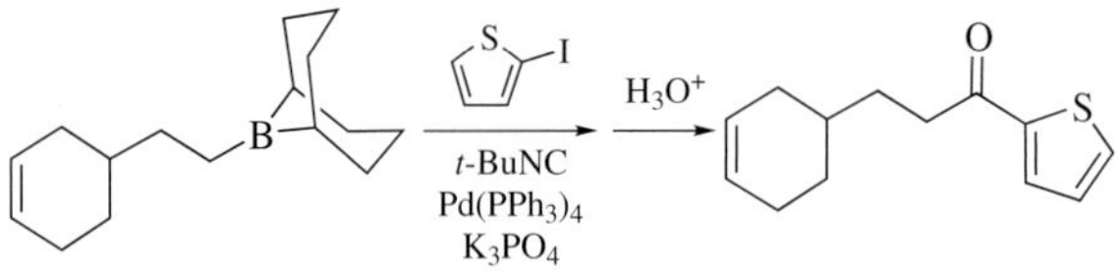

(43)

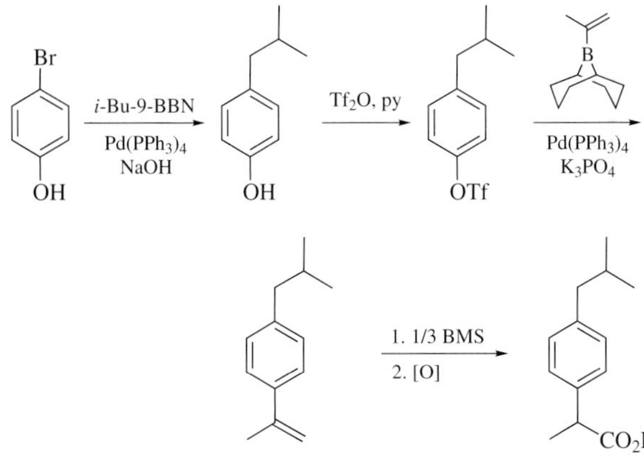

(44)

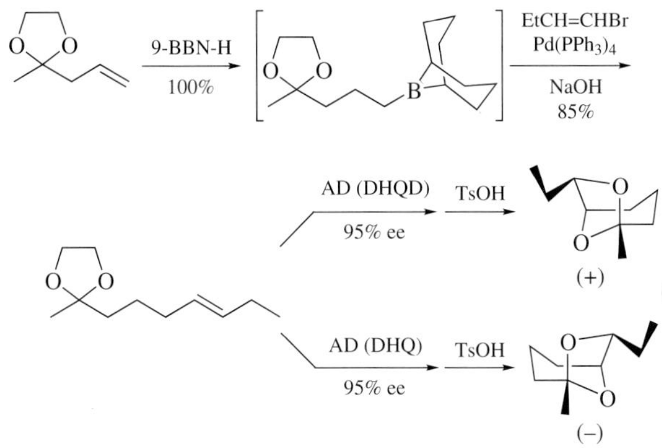

(45)

Several recent applications include the syntheses of pharmaceuticals, pheromones, and prostaglandins, with complete retention of configuration being observed with alkenyl substrates.[54b,55] Either carbon monoxide or ***tert*-Butyl Isocyanide** can be used to prepare ketones through the sequential formation of two new carbon–carbon bonds with this reaction.[56] Moreover, vinyl-9-BBNs (eqs 46–48) are also smoothly cross-coupled with retention of configuration to these substrates and, with these now being readily available, their expanded use in this process should flourish.[12,57] It is important to mention that vinyl vs. primary alkyl group transfer is favored by oxygenated ligation on the borane.[52b,58]

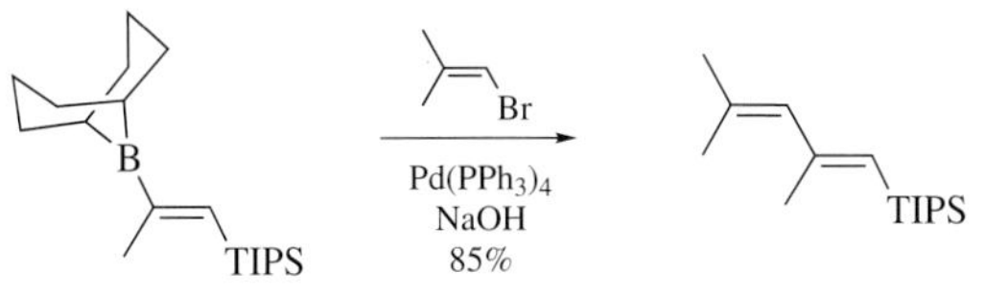

(46)

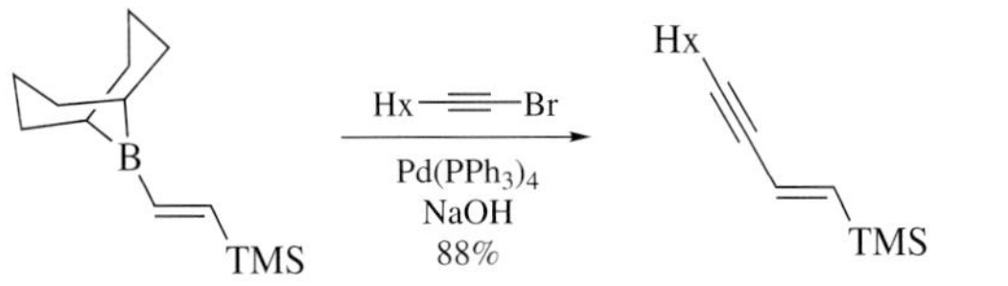

(47)

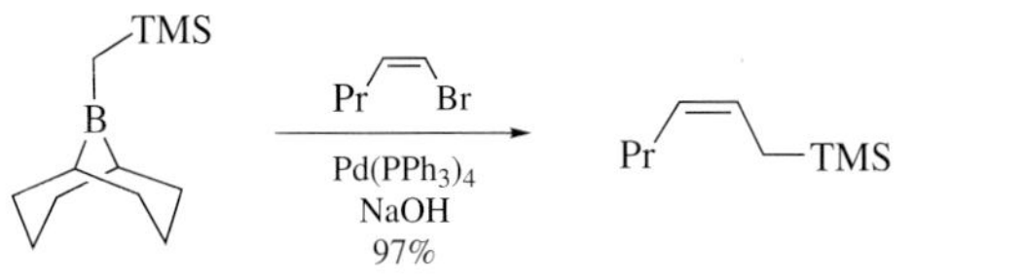

(48)

First Update

William R. Roush
University of Michigan, Ann Arbor, MI, USA

Jung-Nyoung Heo
Korea Research Institute for Chemical Technology, Daejeon, Korea

Suzuki–Miyaura Coupling. The Suzuki–Miyaura coupling reaction is very useful for construction of carbon-carbon bonds under mild conditions. Because it is tolerant of many functional groups, it has been employed widely in the synthesis of many natural products and related compounds.[59] The stereochemistry of such coupling reactions proceeds with retention of configuration at both the C-X and C-(9-BBN) centers, confirmed recently by independent studies from the groups of Woerpel[60] and Soderquist.[61]

Based on the general catalytic cycle (oxidative addition, transmetalation, reductive elimination), alkyl halides historically have not been good substrates for the Suzuki–Miyaura coupling due to their slow rate of oxidative addition and facile β-hydride elimination of the alkylpalladium intermediate. However, Suzuki and co-workers carried out pioneering work in which the catalytic coupling of alkyl iodides with alkylboranes succeeded to form a $C(sp^3)$-$C(sp^3)$ bond in moderate to good yields.[55i] Since then, the scope of the alkyl-alkyl Suzuki coupling reaction has been successfully expanded to include alkyl bromides[62] and chlorides[63] (eqs 49 and 50).

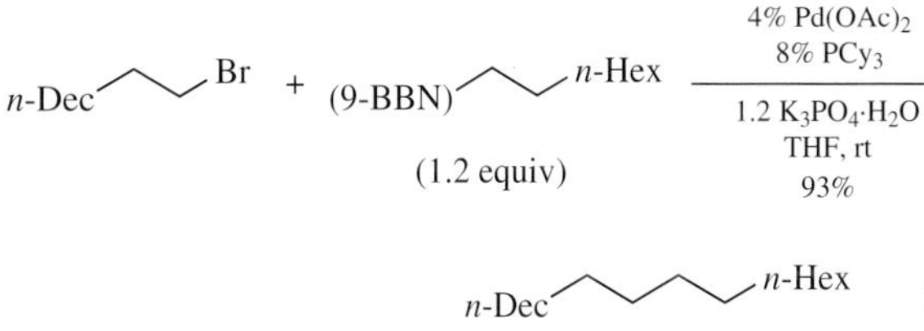

(49)

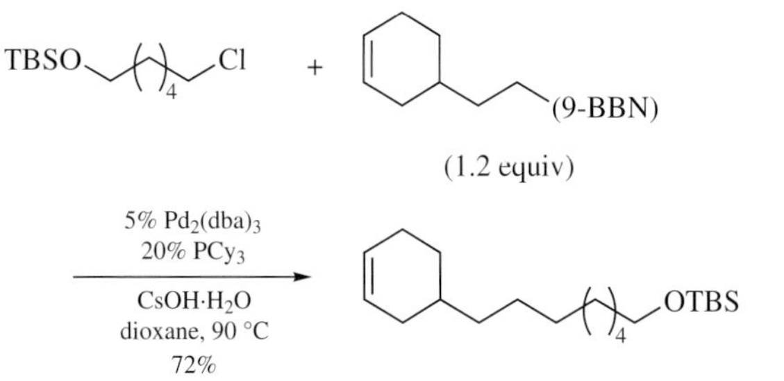

(50)

Protecting Group. Boroxazolidinone complexes prepared by the reaction of free amino acids with 9-BBN dimer can serve as protected amino acids (eq 51).[64] The 9-BBN protection of amino acids improves the solubility of these compounds in organic solvents and is stable to a wide range of reaction conditions. Removal of this protecting group is easily accomplished in good yield under mild conditions either by exchange with ethylenediamine or by treatment with dilute methanolic HCl without epimerization of the amino acid (Table 2). For example, the 9-BBN complex of 3-iodo-L-tyrosine is utilized for a side chain transformation via a palladium-catalyzed coupling reaction (eq 52).[65]

Table 2 Formation and cleavage of 9-BBN complexes

		Formation	Cleavage	
Entry	R (Amino acid)	Yield (%)	aq HCl	Ethylenediamine
1	(CH$_2$)$_4$NH$_2$ (lysine)	90	78	91
2	(CH$_2$)$_3$NH$_3$Cl (ornithine HCl)	96	75	94
3	C$_6$H$_6$OH (tyrosine)	92	89	94
4	CH$_2$SH (cysteine)	50	75	83
5	CH$_2$CO$_2$H (aspartic acid)	98	80	76
6	(CH$_2$)$_3$NHC(=NH)NH$_2$ (arginine)	70	78	62

$$(51)$$

$$(52)$$

***N*-Alkylation Reactions.** The 9-BBN complexes of diamines such as 1,8-diaminonaphthalene and 2-aminobenzylamine serve as substrates for selective *N*-alkylation.[66] *N*-chelated borane complexes undergo regioselective mono-*N*-alkylation with an alkyl halide under basic conditions without further alkylation. Hydrolysis of the corresponding intermediate provides the mono-*N*-alkylated product in excellent yields (eq 53).[66c] In this case, the 9-BBN chelation of the diamine both protects the aromatic amine and activates the benzylic amine.

$$(53)$$

Reduction of Lactams to Cyclic Amines. 9-BBN-H is a well-known reagent for the reduction of aldehydes, ketones, acid chlorides, and alkenes. Recently, 9-BBN-H has been employed for the reduction of lactams to the corresponding cyclic amines in good yield (eq 54).[67] Although BH$_3$·THF selectively reduces lactams to amines in the presence of an ester group, an excess of borane is required due to the strong coordination of borane with the amine products.[68] In contrast, 9-BBN-H generally does not form a complex with the tertiary amines. Consequently, only a slight excess of 9-BBN-H is sufficient to reduce lactams to amines.

$$(54)$$

9-BBN-H is also effective for reductive cleavage of the oxazolidine C-O bond concomitant with the reduction of the lactam carbonyl (eq 55).[69] In this case, excess amounts of 9-BBN-H were necessary, and inversion of stereochemistry at the angular benzyloxy group occurred.

$$(55)$$

Radical Initiator. 9-BBN-H effectively initiates the reduction of common radical precursors by tributyltin hydride at low temperatures (eq 56).[70] Importantly, the radical reaction proceeds without the need for oxygen, whereas similar initiation by trialkylboranes requires at least some oxygen.

$$(56)$$

Related Reagents. Borane–Dimethyl Sulfide; Borane–Tetrahydrofuran.

---

1. (a) Pelter, A.; Smith, K.; Brown, H. C., *Borane Reagents*; Academic: London, 1988. (b) Brown, H. C.; Midland, M. M.; Levy, A. B.; Kramer, G. W., *Organic Synthesis via Boranes*; Wiley: New York, 1975. (c) Brown, H. C.; Lane, C. F., *Heterocyles* **1977**, *7*, 453. (d) Zaidlewicz, M., In *Comprehensive Organometallic Chemistry*; Wilkinson, G.; Stone, F. G. A.; Abel, E. W., Eds.; Pergamon: Oxford, 1982; Vol 7, p 199. (e) Negishi, E.-I., In *Comprehensive Organometallic Chemistry*; Wilkinson, G.; Stone, F. G. A.; Abel, E. W., Eds.; Pergamon: Oxford, 1982; Vol 7, p 255. (f) Köster, R.; Yalpani, M., *Pure. Appl. Chem.* **1991**, *63*, 387.

2. Köster, R., *Angew. Chem.* **1960**, *72*, 626.

3. (a) Knights, E. F.; Brown, H. C., *J. Am. Chem. Soc.* **1968**, *90*, 5281. (b) Soderquist, J. A.; Brown, H. C., *J. Org. Chem.* **1981**, *46*, 4599. (c) Soderquist, J. A.; Negron, A., *Org. Synth.* **1991**, *70*, 169. (d) Brauer, D. J.; Kruger, C., *Acta Crystallogr. B* **1973**, *29*, 1684.

4. (a) Scouten, C. G.; Brown, H. C., *J. Org. Chem.* **1973**, *38*, 4092. (b) Brown, H. C.; Knights, E. F. Scouten, C. G., *J. Am. Chem. Soc.* **1974**, *96*, 7765. (c) Brown, H. C.; Liotta, R.; Scouten, C. G., *J. Am. Chem. Soc.* **1976**, *98*, 5297. (d) Liotta, R.; Brown, H. C., *J. Org. Chem.* **1977**, *42*, 2836. (e) Brener, L.; Brown, H. C., *J. Org. Chem.* **1977**, *42*, 2702. (f) Brown, H. C.; Liotta, R.; Brener, L., *J. Am. Chem. Soc.* **1977**, *99*, 3427. (g) Brown, H. C.; Liotta, R.; Kramer, G. W., *J. Org. Chem.* **1978**, *43*, 1058. (h) Soderquist, J. A.; Hassner, A., *J. Organomet. Chem.* **1978**, *156*, C12. (i) Brown, H. C.; Liotta, R.; Kramer, G. W., *J. Am. Chem. Soc.* **1979**, *101*, 2966. (j) Brown, H. C.; Vara Prasad, J. V. N.; Zee, S.-H., *J. Org. Chem.* **1985**, *50*, 1582. (k) Brown, H. C.; Vara Prasad, J. V. N., *J. Org. Chem.* **1985**, *50*, 3002. (l) Brown, H. C.; Ramachandran, P. V.; Vara Prasad, J. V. N., *J. Org. Chem.* **1985**, *50*, 5583. (m) Soderquist, J. A.; Anderson, C. L., *Tetrahedron Lett.* **1986**, *27*, 3961. (n) Fleming, I., *Pure. Appl. Chem.* **1988**, *60*, 71.

5. (a) Brown, H. C.; Rogić, M. M., *J. Am. Chem. Soc.* **1969**, *91*, 4304. (b) Kramer, G. W.; Brown, H. C., *J. Organomet. Chem.* **1974**, *73*, 1. (c) ibid., *J. Organomet. Chem.* **1977**, *132*, 9. (d) Soderquist, J. A.; Brown, H. C., *J. Org. Chem.* **1980**, *45*, 3571. (e) Soderquist, J. A.; Rivera, I.; Negron, A., *J. Org. Chem.* **1989**, *54*, 4051.

6. (a) Soderquist, J. A.; Colberg, J. C.; Del Valle, L., *J. Am. Chem. Soc.* **1989**, *111*, 4873. See also: (b) Negishi, E.-I., In *Comprehensive Organometallic Chemistry*; Wilkinson, G.; Stone, F. G. A.; Abel, E. W., Eds.; Pergamon: Oxford, 1982; Vol 7, p 265.

7. (a) Brown, H. C.; Scouten, C. G.; Wang, K. K., *J. Org. Chem.* **1979**, *44*, 2589. (b) Brown, H. C.; Wang, K. K.; Scouten, C. G., *Proc. Natl. Acad. Sci. USA* **1980**, *77*, 698. (c) Wang, K. K.; Brown, H. C., *J. Org. Chem.* **1980**, *45*, 5303. (d) Nelson, D. J.; Cooper, P. J., *Tetrahedron Lett.* **1986**, *27*, 4693. (e) Brown, H. C.; Chandrasekharan, J.; Nelson, D. J., *J. Am. Chem. Soc.* **1984**, *106*, 3768. (f) Chandrasekharan, J.; Brown, H. C., *J. Org. Chem.* **1985**, *50*, 518.

8. (a) Midland, M. M., *C. R. Hebd. Seances Acad. Sci.* **1989**, *89*, 1553. (b) Brown, H. C.; Ramachandran, P. V., *Pure. Appl. Chem.* **1991**, *63*, 307; ibid., *Acc. Chem. (Res.)* **1992**, *25*, 16. (c) Srebnik, M.; Ramachandran, P. V., *Aldrichim. Acta* **1987**, *20*, 9. (d) Brown, H. C.; Srebnik, M.; Ramachandran, P. V., *J. Org. Chem.* **1989**, *54*, 1577.

9. Liu, C.; Wang, K. K., *J. Org. Chem.* **1986**, *51*, 4733.

10. (a) Evans, D. A.; Fu, G. C.; Hoveyda, A. H., *J. Am. Chem. Soc.* **1988**, *110*, 6917. (b) Burgess, K.; van der Donk, W. A.; Jarstfer, M. B.; Ohlmeyer, M., *J. Am. Chem. Soc.* **1991**, *113*, 6139.

11. (a) Negishi, E.-I.; Burke, P. L.; Brown, H. C., *J. Am. Chem. Soc.* **1972**, *94*, 7431. (b) Burke, P. L.; Negishi, E.-I.; Brown, H. C., *J. Am. Chem. Soc.* **1973**, *95*, 3654. (c) Brown, H. C.; Pai, G. G., *Heterocyles* **1982**, *17*, 77. (d) ibid., *J. Organomet. Chem.* **1983**, *250*, 13. (e) Soderquist, J. A.; Shiau, F.-Y.; Lemesh, R. A., *J. Org. Chem.* **1984**, *49*, 2565. (f) Soderquist, J. A.; Negron, A., *J. Org. Chem.* **1989**, *54*, 2462. However,

for the unusual behavior of the 9-BBN systems with alkynyltins, see: (g) Bihlmayer, C.; Kerschl, S.; Wrackmeyer, B., *Z. Naturforsch., Teil B* **1987**, *42*, 715. (h) Wrackmeyer, B.; Abu-Orabi, S. T., *Ber. Dtsch. Chem. Ges.* **1987**, *120*, 1603. (i) Bihlmayer, C.; Abu-Orabi, S. T.; Wrackmeyer, B., *J. Organomet. Chem.* **1987**, *322*, 25.

12. Colberg, J. C.; Rane, A.; Vaquer, J.; Soderquist, J. A., *J. Am. Chem. Soc.* **1993**, *115*, 6065.

13. (a) Brown, H. C., Scouten, C. G.; Liotta, R., *J. Am. Chem. Soc.* **1979**, *101*, 96. (b) Wang, K. K.; Scouten, C. G.; Brown, H. C., *J. Am. Chem. Soc.* **1982**, *104*, 531. (c) Blue, C. D.; Nelson, D. J., *J. Org. Chem.* **1983**, *48*, 4538.

14. Soderquist, J. A.; Najafi, M. R., *J. Org. Chem.* **1986**, *51*, 1330.

15. Nelson, D. J.; Blue, C. D.; Brown, H. C., *J. Am. Chem. Soc.* **1982**, *104*, 4913.

16. Hara, S.; Dojo, H.; Takinami, S.; Suzuki, A., *Tetrahedron Lett.* **1983**, *24*, 731.

17. Brown, C. A.; Coleman, R. A., *J. Org. Chem.* **1979**, *44*, 2328.

18. Singleton, D. A.; Martinez, J. P., *J. Am. Chem. Soc.* **1990**, *112*, 7423.

19. Brown, H. C.; Bhat, N. G.; Rajagopalan, S., *Organometallics* **1986**, *5*, 816.

20. Campbell, Jr., J. B.; Molander, G. A., *J. Organomet. Chem.* **1978**, *156*, 71.

21. Soderquist, J. A.; Rivera, I., *Tetrahedron Lett.* **1989**, *30*, 3919.

22. Soderquist, J. A.; Negron, A., *J. Org. Chem.* **1987**, *52*, 3441.

23. (a) Bhatt, M. V., *J. Organomet. Chem.* **1978**, *156*, 221. See also: (b) Köster, R.; Grassberger, M. A., *Justus Liebigs Ann. Chem.* **1968**, *719*, 169. (c) Brown, H. C.; Kulkarni, S. U., *J. Org. Chem.* **1979**, *44*, 281. (d) ibid., *J. Org. Chem.* **1979**, *44*, 2422.

24. (a) Inoue, T.; Uchimaru, T.; Mukaiyama, T., *Chem. Lett.* **1977**, 153. (b) Inoue, T.; Mukaiyama, T., *Bull. Chem. Soc. Jpn.* **1980**, *53*, 174. (c) Masamune, S.; Choi, W.; Kerdesky, F. A. J.; Imperiali, B., *J. Am. Chem. Soc.* **1981**, *103*, 1566. (d) Masamune, S.; Choi, W.; Peterson, S. S.; Sita, L. R., *Angew. Chem., Int. Ed. Engl.* **1985**, *24*, 1.

25. Cha, J. S.; Kim, J. E.; Oh, S. Y.; Kim, J. D., *Tetrahedron Lett.* **1987**, *28*, 4575.

26. (a) Brown, H. C.; Kulkarni, S. U., *Inorg. Chem.* **1977**, *16*, 3090. (b) ibid., *J. Org. Chem.* **1977**, *42*, 4169. (c) Brown, H. C.; Soderquist, J. A., *J. Org. Chem.* **1980**, *45*, 846. (d) Brown, H. C.; Singaram, B.; Mathew, C. P., *J. Org. Chem.* **1981**, *46*, 2712. (e) Brown, H. C.; Mathew, C. P.; Pyun, C.; Son, J. C.; Yoon, Y. M., *J. Org. Chem.* **1984**, *49*, 3091. (f) Soderquist, J. A.; Rivera, I., *Tetrahedron Lett.* **1988**, *29*, 3195. (g) Hubbard, J. L., *Tetrahedron Lett.* **1988**, *29*, 3197.

27. (a) Brown, H. C.; Krishnamurthy, S., *Tetrahedron* **1979**, *35*, 567. (b) Brown, H. C.; Park, W. S.; Cho, B. T., *J. Org. Chem.* **1986**, *51*, 1934. (c) Brown, H. C.; Cho, B. T.; Park, W. S., *J. Org. Chem.* **1988**, *53*, 1231. (d) Narasimhan, S., *Indian J. Chem., Sect. B* **1986**, *25B*, 847 (e) Cha, J. S.; Yoon, M. S.; Kim, Y. S.; Lee, K. W., *Tetrahedron Lett.* **1988**, *29*, 1069. (f) Soderquist, J. A.; Rivera, I., *Tetrahedron Lett.* **1988**, *29*, 3195. (g) Cha, J. S.; Lee, K. W.; Yoon, M. S.; Lee, J. C.; Yoon, N. M., *Heterocyles* **1988**, *27*, 1713. (h) Hutchins, R. O.; Abdel-Magid, A.; Stercho, Y. P.; Wambsgams, A., *J. Org. Chem.* **1987**, *52*, 702.

28. Toi, H.; Yamamoto, Y.; Sonoda, A.; Murahashi, S.-I., *Tetrahedron* **1981**, *37*, 2261.

29. (a) Kramer, G. W.; Brown, H. C., *J. Organomet. Chem.* **1975**, *90*, Cl. (b) ibid., *J. Am. Chem. Soc.*, **1976**, *98*, 1964. (c) ibid., *J. Org. Chem.* **1977**, *42*, 2832.

30. (a) Krishnamurthy, S.; Brown, H. C., *J. Org. Chem.* **1975**, *40*, 1864 (b) ibid., *J. Org. Chem.* **1977**, *42*, 1197. (c) Molin, H.; Pring, B. G., *Tetrahedron Lett.* **1985**, *26*, 677.

31. Kabalka, G. A.; Shoup, T. M.; Goudgaon, N. M., *J. Org. Chem.* **1989**, *54*, 5930.

32. Negishi, E.-I., In *Comprehensive Organometallic Chemistry*; Wilkinson, G.; Stone, F. G. A.; Abel, E. W., Eds.; Pergamon: Oxford, 1982; Vol 7, p 337.

33. (a) Lane, C. L.; Brown, H. C., *J. Organomet. Chem.* **1971**, *26*, C51. (b) Brown, H. C.; DeLue, N. R., *J. Am. Chem. Soc.* **1974**, *96*, 311.

34. (a) Yamamoto, Y.; Brown, H. C., *Chem. Commun.* **1973**, 801. (b) *ibid.*, *J. Org. Chem.* **1974**, *39*, 861.

35. Brown, H. C.; DeLue, N. R.; Kabalka, G. W.; Hedgecock, Jr, H. C., *J. Am. Chem. Soc.* **1976**, *98*, 1290.

36. Lane, C. L., PhD Thesis, Purdue University, 1972.

37. (a) Brown, H. C.; Midland, M. M.; Levy, A. B.; Suzuki, A.; Sano, S.; Itoh, M., *Tetrahedron* **1987**, *43*, 4079. (b) Brown, H. C.; Salunkhe, A. M.; Singaram, B., *J. Org. Chem.* **1991**, *56*, 1170. (c) For related behavior with α-diazo ketones, see: Hooz, J.; Gunn, D. M., *Tetrahedron Lett.* **1969**, 3455.

38. (a) Brown, H. C.; Rogić, M. M., *J. Am. Chem. Soc.* **1969**, *91*, 2146. (b) Brown, H. C.; Rogić, M. M.; Nambu, H.; Rathke, M. W., *J. Am. Chem. Soc.* **1969**, *91*, 2147. (c) Brown, H. C.; Rogić, M. M.; Rathke, M. W.; Kabalka, G. W., *J. Am. Chem. Soc.* **1969**, *91*, 2150.

39. (a) Brown, H. C.; Rhodes, S. P., *J. Am. Chem. Soc.* **1969**, *91*, 2149. (b) *ibid.*, *J. Am. Chem. Soc.* **1969**, *91*, 4306.

40. Brown, H. C.; Knights, E. F., *J. Am. Chem. Soc.* **1968**, *90*, 5283.

41. (a) Brown, H. C.; Hubbard, J. L.; Smith, K., *Synthesis* **1979**, 701. (b) Hubbard, J. L.; Smith, K., *J. Organomet. Chem.* **1984**, *276*, C41. (c) Kabalka, G. W.; Delgado, M. C.; Kunda, U. S.; Kunda, S. A., *J. Org. Chem.* **1984**, *49*, 174.

42. (a) Carlson, B. A.; Katz, J.-J.; Brown, H. C., *J. Org. Chem.* **1973**, *38*, 3968. (b) Carlson, B. A.; Brown, H. C., *Org. Synth., Coll. Vol.* **1988**, *6*, 137.

43. (a) Kramer, G. W.; Brown, H. C., *J. Org. Chem.* **1977**, *42*, 2292. (b) Yamamoto, Y.; Yatagai, H.; Maruyama, K., *J. Am. Chem. Soc.* **1981**, *103*, 3229.

44. Brown, H. C.; Molander, G. A.; Singh, S. M.; Racherla, U. S., *J. Org. Chem.* **1985**, *50*, 1577.

45. (a) Jacob, III, P.; Brown, H. C., *J. Org. Chem.* **1977**, *42*, 579. (b) Soderquist, J. A.; Vaquer, J., *Tetrahedron Lett.* **1990**, *31*, 4545.

46. (a) Jacob, III, P.; Brown, H. C., *J. Am. Chem. Soc.* **1976**, *98*, 7832. (b) Sinclair, J. A.; Molander, G. A.; Brown, H. C., *J. Am. Chem. Soc.* **1977**, *99*, 954.

47. Molander, G. A.; Brown, H. C., *J. Org. Chem.* **1977**, *42*, 3106.

48. Molander, G. A.; Zinke, P. W., *Organometallics* **1986**, *5*, 2161.

49. (a) Sharma, S.; Oehlschlager, A. C., *Tetrahedron Lett.* **1988**, *29*, 261. See also: (b) *ibid.*, *J. Org. Chem.* **1989**, *54*, 5064. (c) Hutzinger, M. W.; Singer, R. D.; Oehlschlager, A. C., *J. Am. Chem. Soc.* **1990**, *112*, 9397.

50. (a) Singleton, D. A.; Martinez, J. P., *Tetrahedron Lett.* **1991**, *32*, 7365. (b) Singleton, D. A.; Leung, S.-W., *J. Org. Chem.* **1992**, *57*, 4796. (c) Singleton, D. A.; Redman, A. M., *Tetrahedron Lett.* **1994**, *35*, 509.

51. Andemichael, Y. W.; Huang, Y.; Wang, K. K., *J. Org. Chem.* **1993**, *58*, 1651.

52. (a) Evans, D. A.; Bartroli, J.; Shih, T. L., *J. Am. Chem. Soc.* 1981, **103**, 2127. (b) Gage, J. R.; Evans, D. A., *Org. Synth.* **1989**, *68, 77*, 83. (c) Evans, D. A.; Dow, R. L.; Shih, T. I.; Takacs, J. M.; Zahler, R., *J. Am. Chem. Soc.* **1990**, *112*, 5290. (d) Heathcock, C. H., *Aldrichim. Acta* **1990**, *23*, 99.

53. Shimizu, H.; Hara, S.; Suzuki, A., *Synth. Commun.* **1990**, *20*, 549.

54. (a) Miyaura, N.; Ishiyama, T.; Ishikawa, M.; Suzuki, A., *Tetrahedron Lett.* **1986**, *27*, 6369. (b) Miyaura, N.; Ishiyama, T.; Sasaki, H.; Ishikawa, M.; Satoh, M.; Suzuki, A., *J. Am. Chem. Soc.* **1989**, *111*, 314.

55. (a) Hoshino, Y.; Ishiyama, T.; Miyaura, N.; Suzuki, A., *Tetrahedron Lett.* **1988**, *29*, 3983. (b) Oh-e, T.; Miyaura, N.; Suzuki, A., *Synlett* **1990**, 221. (c) Soderquist, J. A.; Santiago, B.; Rivera, I., *Tetrahedron Lett.* **1990**, *31*, 4981. (d) Soderquist, J. A.; Santiago, B., *Tetrahedron Lett.* **1990**, *31*, 5113. (e) Suzuki, A., *Pure. Appl. Chem.* **1991**, *63*, 419. (f) Ishiyama, T.; Miyaura, N.; Suzuki, A., *Synlett* **1991**, 687. (g) Ishiyama, T.; Miyaura, N.; Suzuki, A., *Org. Synth.* **1992**, *71*, 89. (h) Miyaura, N.; Ishikawa, M.; Suzuki, A., *Tetrahedron Lett.* **1992**, *33*, 2571. (i) Ishiyama, T.; Abe, S.; Miyaura, N.; Suzuki, A., *Chem. Lett.* **1992**, 691. (j) Santiago, B.; Soderquist, J. A., *J. Org. Chem.* **1992**, *57*, 5844. (k) Rivera, I.; Colberg, J. C.; Soderquist, J. A., *Tetrahedron Lett.* **1992**, *33*, 6919. (l) Nomoto, Y.; Miyaura, N.; Suzuki, A., *Synlett* **1992**, 727. (m) Oh-e, T.; Miyaura, N.; Suzuki, A., *J. Org. Chem.* **1993**, *58*, 2201. (n) Soderquist, J. A.; Rane, A. M., *Tetrahedron Lett.* **1993**, *34*, 5031. (o) Johnson, C. R.; Braun, M. P., *J. Am. Chem. Soc.* **1993**, *115*, 11014.

56. (a) Wakita, Y.; Yasunaga, T.; Akita, M.; Kojima, M., *J. Organomet. Chem.* **1986**, *301*, C17. (b) Ishiyama, T.; Miyaura, N.; Suzuki, A., *Bull. Chem. Soc. Jpn.* **1991**, *64*, 1999. (c) *ibid.*, *Tetrahedron Lett.* **1991**, *32*, 6923. (d) Ishiyama, T.; Oh-e, T.; Miyaura, N.; Suzuki, A., *Tetrahedron Lett.* **1992**, *33*, 4465.

57. (a) Soderquist, J. A.; Colberg, J. C., *Synlett* **1989**, 25. (b) Soderquist, J. A.; Colberg, J. C., *Tetrahedron Lett.* **1994**, *35*, 27.

58. Rivera, I.; Soderquist, J. A., *Tetrahedron Lett.* **1991**, *32*, 2311.

59. (a) Miyaura, N.; Suzuki, A., *Chem. Rev.* **1995**, *95*, 2457. (b) Suzuki, A., *J. Organomet. Chem.* **1999**, *576*, 147. (c) Chemler, S. R.; Trauner, D.; Danishefsky, S. J., *Angew. Chem., Int. Ed.* **2001**, *40*, 4544.

60. Ridgway, D. H.; Woerpel, K. A., *J. Org. Chem.* **1998**, *63*, 458.

61. Matos, K.; Soderquist, J. A., *J. Org. Chem.* **1998**, *63*, 461.

62. Netherton, M. R.; Dai, C.; Neuschutz, K.; Fu, G. C., *J. Am. Chem. Soc.* **2001**, *123*, 10099.

63. Kirchhoff, J. H.; Dai, C.; Fu, G. C., *Angew. Chem., Int. Ed.* **2002**, *41*, 1945.

64. Dent, W. H.; Erickson, W. R.; Fields, S. C.; Parker, M. H.; Tromiczak, E. G., *Org. Lett.* **2002**, *4*, 1249.

65. Walker, W. H.; Rokita, S. E., *J. Org. Chem.* **2003**, *68*, 1563.

66. (a) Bar-Haim, G.; Kol, M., *J. Org. Chem.* **1997**, *62*, 6682. (b) Bar-Haim, G.; Shach, R.; Kol, M., *Synth. Commun.* **1997**, 229. (c) Bar-Haim, G.; Kol, M., *Tetrahedron Lett.* **1998**, *39*, 2643.

67. Collins, C. J.; Lanz, M.; Singaram, B., *Tetrahedron Lett.* **1999**, *40*, 3673.

68. Brown, H. C.; Heim, P., *J. Org. Chem.* **1973**, *38*, 912.

69. Meyers, A. I.; Andres, C. J.; Resek, J. E.; McLaughlin, M. A.; Woodall, C. C.; Lee, P. H., *J. Org. Chem.* **1996**, *61*, 2586.

70. Perchyonok, V. T.; Schiesser, C. H., *Tetrahedron Lett.* **1998**, *39*, 5437.

Boron Trifluoride Etherate

$$BF_3 \cdot OEt_2$$

(BF$_3$·OEt$_2$)

[109-63-7] C$_4$H$_{10}$BF$_3$O (MW 141.94)

InChI = 1/C4H10O.BF3/c1-3-5-4-2;2-1(3)4/h3-4H2,1-2H3;

InChIKey = KZMGYPLQYOPHEL-UHFFFAOYAN

(BF$_3$·MeOH)

[373-57-9] CH$_4$BF$_3$O (MW 99.85)

InChI = 1/CH4O.BF3/c1-2;2-1(3)4/h2H,1H3;

InChIKey = JBXYCUKPDAAYAS-UHFFFAOYAA

(BF$_3$·OEt$_2$: easy-to-handle and convenient source of BF$_3$; Lewis acid catalyst; promotes epoxide cleavage and rearrangement, control of stereoselectivity; BF$_3$·MeOH: esterification of aliphatic and aromatic acids; cleavage of trityl ethers)

Alternate Names: boron trifluoride diethyl etherate; boron trifluoride ethyl etherate; boron trifluoride ethyl ether complex; trifluoroboron diethyl etherate.

Physical Data: BF$_3$·OEt$_2$: bp 126 °C; *d* 1.15 g cm^{-3}; BF$_3$·MeOH: bp 59 °C/4 mmHg; *d* 1.203 g cm^{-3} for 50 wt % BF$_3$, 0.868 g cm^{-3} for 12 wt % BF$_3$.

Solubility: sol benzene, chloromethanes, dioxane, ether, methanol, THF, toluene.

Form Supplied in: BF$_3$·OEt$_2$: light yellow liquid, packaged under nitrogen or argon; BF$_3$·MeOH is available in solutions of 10–50% BF$_3$ in MeOH.

Preparative Methods: BF$_3$·OEt$_2$ is prepared by passing BF$_3$ through anhydrous ether;[1a] the BF$_3$·MeOH complex is formed from BF$_3$·OEt$_2$ and methanol.

Purification: oxidation in air darkens commercial boron trifluoride etherate; therefore the reagent should be redistilled prior to use. An excess of the etherate in ether should be distilled in an all-glass apparatus with calcium hydroxide to remove volatile acids and to reduce bumping.[1b]

Handling, Storage, and Precautions: keep away from moisture and oxidants; avoid skin contact and work in a well-ventilated fume hood.

Original Commentary

Veronica Cornel
Emory University, Atlanta, GA, USA

Addition Reactions. BF$_3$·OEt$_2$ facilitates the addition of moderately basic nucleophiles like alkyl-, alkenyl-, and aryl-lithium, imines, Grignard reagents, and enolates to a variety of electrophiles.[2]

Organolithiums undergo addition reactions with 2-isoxazolines to afford *N*-unsubstituted isoxazolidines, and to the carbon–nitrogen double bond of oxime *O*-ethers to give *O*-alkylhydroxylamines.[3] Aliphatic esters react with lithium acetylides in the presence of BF$_3$·OEt$_2$ in THF at −78 °C to form alkynyl ketones in 40–80% yields.[4] Alkynylboranes, generated in situ from lithium acetylides and BF$_3$·OEt$_2$, were found to react with oxiranes[5] and oxetanes[6] under mild conditions to afford β-hydroxyalkynes and

γ-alkoxyalkynes, respectively. (1-Alkenyl)dialkoxyboranes react stereoselectively with α,β-unsaturated ketones[7] and esters[8] in the presence of BF$_3$·OEt$_2$ to give γ,δ-unsaturated ketones and α-acyl-γ,δ-unsaturated esters, respectively.

The reaction of imines activated by BF$_3$·OEt$_2$ with 4-(phenyl-sulfonyl)butanoic acid dianion leads to 2-piperidones in high yields.[9] (Perfluoroalkyl)lithiums, generated in situ, add to imines in the presence of BF$_3$·OEt$_2$ to give perfluoroalkylated amines.[10] Enolate esters add to 3-thiazolines under mild conditions to form thiazolidines if these imines are first activated with BF$_3$·OEt$_2$.[11] The carbon–nitrogen double bond of imines can be alkylated with various organometallic reagents to produce amines.[12] A solution of benzalaniline in acetone treated with BF$_3$·OEt$_2$ results in the formation of β-phenyl-β-anilinoethyl methyl ketone.[13] Anilinobenzylphosphonates are synthesized in one pot using aniline, benzaldehyde, dialkyl phosphite, and BF$_3$·OEt$_2$;[14] the reagent accelerates imine generation and dialkyl phosphite addition. Similarly, BF$_3$·OEt$_2$ activates the nitrile group of cyanocuprates, thereby accelerating Michael reactions.[15]

The reagent activates iodobenzene for the allylation of aromatics, alcohols, and acids.[16] Allylstannanes are likewise activated for the allylation of *p*-benzoquinones, e.g. in the formation of coenzyme Q$_n$ using polyprenylalkylstannane.[17]

Nucleophilic silanes undergo stereospecific addition to electrophilic glycols activated by Lewis acids. The glycosidation is highly stereoselective with respect to the glycosidic linkage in some cases using BF$_3$·OEt$_2$. Protected pyranosides undergo stereospecific *C*-glycosidation with C-1-oxygenated allylsilanes to form α-glycosides.[18,19] α-Methoxyglycine esters react with allylsilanes and silyl enol ethers in the presence of BF$_3$·OEt$_2$ to give racemic γ,δ-unsaturated α-amino acids and γ-oxo-α-amino acids, respectively.[20] β-Glucopyranosides are synthesized from an aglycon and 2,3,4,6-tetra-*O*-acetyl-β-D-glucopyranose.[21] Alcohols and silyl ethers also undergo stereoselective glycosylation with protected glycosyl fluorides to form β-glycosides.[22]

BF$_3$·OEt$_2$ reverses the usual *anti* selectivity observed in the reaction of crotyl organometallic compounds (based on Cu, Cd, Hg, Sn, Tl, Ti, Zr, and V, but not on Mg, Zn, or B) with aldehydes (eq 1a) and imines (eq 1b), so that homoallyl alcohols and homoallylamines are formed, respectively.[23–28] The products show mainly *syn* diastereoselectivity. BF$_3$·OEt$_2$ is the only Lewis acid which produces hydroxy- rather than halo-tetrahydropyrans from the reaction of allylstannanes with pyranosides.[29] The BF$_3$·OEt$_2$ mediated condensations of γ-oxygenated allylstannanes with aldehydes (eq 1c) and with 'activated' imines (eq 1d) affords vicinal diol derivatives and 1,2-amino alcohols, respectively, with *syn* diastereoselectivity.[30,31] The 'activated' imines are obtained from

(a) X = O, Y = Me
(b) X = NR^2, Y = Me
(c) X = O, Y = OMe, OTBDMS
(d) X = NR^2, Y = OMe, OTBDMS
  or OCH$_2$OMe

(a) X = OH, Y = Me
(b) X = NHR^2, Y = Me
(c) X = OH, Y = OMe, OTBDMS
(d) X = NHR^2, Y = OH or derivative

Essential Reactions for Organic Synthesis, First Edition. Edited by Philip L. Fuchs.

aromatic amines, aliphatic aldehydes, and α-ethoxycarbamates. The reaction of aldehydes with α-(alkoxy)-β-methylallylstannanes with aldehydes in the presence of BF$_3$·OEt$_2$ gives almost exclusively *syn-(E)*-isomers.[31]

The reaction of α-diketones with allyltrimethylstannane in the presence of BF$_3$·OEt$_2$ yields a mixture of homoallylic alcohols, with the less hindered carbonyl group being allylated predominantly.[32] The reaction between aldehydes and allylic silanes with an asymmetric ethereal functionality produces *syn*-homoallyl alcohols when **Titanium(IV) Chloride** is coordinated with the allylic silane and *anti* isomers with BF$_3$·OEt$_2$.[33]

Chiral oxetanes can be synthesized by the BF$_3$·OEt$_2$ catalyzed [2 + 2] cycloaddition reactions of 2,3-*O*-isopropylidenealdehyde-D-aldose derivatives with allylsilanes, vinyl ethers, or vinyl sulfides.[34] The regiospecificity and stereoselectivity is greater than in the photochemical reaction; *trans*-2-alkoxy- and *trans*-2-phenylthiooxetanes are the resulting products.

2-Alkylthioethyl acetates can be formed from vinyl acetates by the addition of thiols with BF$_3$·OEt$_2$ as the catalyst.[35] The yield is 79%, compared to 75% when BF$_3$·OEt$_2$ is used in conjunction with **Mercury(II) Sulfate** or **Mercury(II) Oxide**.

α-Alkoxycarbonylallylsilanes react with acetals in the presence of BF$_3$·OEt$_2$ (eq 2).[36] The products can be converted into α-methylene-γ-butyrolactones by dealkylation with **Iodotrimethylsilane**.

$$(2)$$

The cuprate 1,4-conjugate addition step in the synthesis of (+)-modhephene is difficult due to the neopentyl environment of C-4 in the enone, but it can occur in the presence of BF$_3$·OEt$_2$ (eq 3).[37]

$$(3)$$

The reagent is used as a Lewis acid catalyst for the intramolecular addition of diazo ketones to alkenes.[38] The direct synthesis of bicyclo[3.2.1]octenones from the appropriate diazo ketones using BF$_3$·OEt$_2$ (eq 4) is superior to the copper-catalyzed thermal decomposition of the diazo ketone to a cyclopropyl ketone and subsequent acid-catalyzed cleavage.[38]

$$(4)$$

BF$_3$·OEt$_2$ reacts with fluorinated amines to form salts which are analogous to Vilsmeier reagents, Arnold reagents, or phosgene–immonium salts (eq 5).[39] These salts are used to acylate electron-rich aromatic compounds, introducing a fluorinated carbonyl group (eq 6).

$$(5)$$

R = Et; X = Cl, F, CF$_3$ (**1**)

$$(6)$$

Xenon(II) Fluoride and methanol react to form **Methyl Hypofluorite**, which reacts as a positive oxygen electrophile in the presence of BF$_3$ (etherate or methanol complex) to yield anti-Markovnikov fluoromethoxy products from alkenes.[40,41]

Aldol Reactions. Although **Titanium(IV) Chloride** is a better Lewis acid in effecting aldol reactions of aldehydes, acetals, and silyl enol ethers, BF$_3$·OEt$_2$ is more effective for aldol reactions with anions generated from transition metal carbenes and with tetrasubstituted enol ethers such as (Z)- and (E)-3-methyl-2-(trimethylsilyloxy)-2-pentene.[42,43] One exception involves the preparation of substituted cyclopentanediones from acetals by the aldol condensation of protected four-membered acyloin derivatives with BF$_3$·OEt$_2$ rather than TiCl$_4$ (eq 7).[44] The latter catalyst causes some loss of the silyl protecting group. The pinacol rearrangement is driven by the release of ring strain in the four-membered ring and controlled by an acyl group adjacent to the diol moiety.

$$(7)$$

The reagent is the best promoter of the aldol reaction of 2-(trimethylsilyloxy)acrylate esters, prepared by the silylation of pyruvate esters, to afford γ-alkoxy-α-keto esters (eq 8).[45] These esters occur in a variety of important natural products.

$$(8)$$

BF$_3$·OEt$_2$ can improve or reverse the aldehyde diastereofacial selectivity in the aldol reaction of silyl enol ethers with aldehydes, forming the *syn* adducts. For example, the reaction of the silyl enol ether of pinacolone with 2-phenylpropanal using BF$_3$·OEt$_2$ gives enhanced levels of Felkin selectivity relative to the addition of the corresponding lithium enolate.[46,47] In the reaction of silyl enol ethers with 3-formyl-Δ^2-isoxazolines, BF$_3$·OEt$_2$ gives predominantly *anti* aldol adducts, whereas other Lewis acids give *syn* aldol adducts.[48] The reagent can give high diastereofacial selectivity in the addition of silyl enol ethers or silyl ketones to chiral aldehydes.[49] In the addition of a nonstereogenic silylketene acetal to chiral, racemic α-thioaldehydes, BF$_3$·OEt$_2$ leads exclusively to the *anti* product.[49]

1,5-Dicarbonyl compounds are formed from the reaction of silyl enol ethers with methyl vinyl ketones in the presence of BF$_3$·OEt$_2$ and an alcohol (eq 9).[50] α-Methoxy ketones are formed from α-diazo ketones with BF$_3$·OEt$_2$ and methanol, or directly from silyl enol ethers using iodobenzene/BF$_3$·OEt$_2$ in methanol.[51]

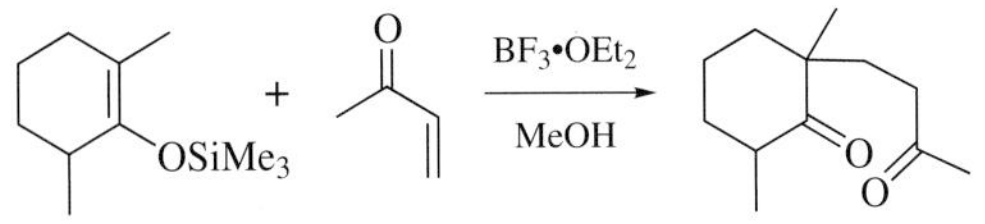

$$(9)$$

α-Mercurio ketones condense with aldehydes in the presence of BF$_3$·OEt$_2$ with predominant *erythro* selectivity (eq 10).[52] Enaminosilanes derived from acylic and cyclic ketones undergo *syn* selective aldol condensations in the presence of BF$_3$·OEt$_2$.[53]

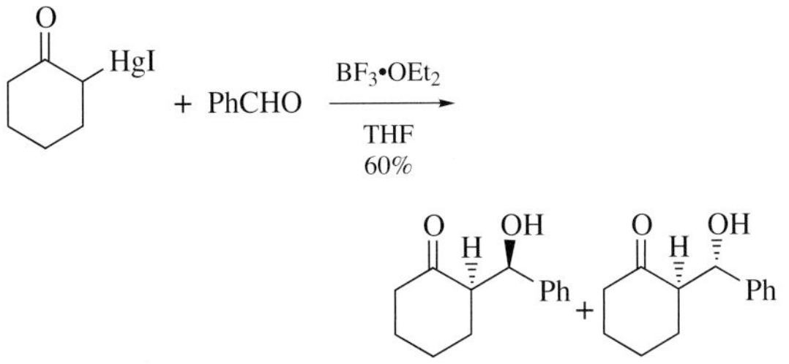

$$(10)$$

erythro 90:10 *threo*

Cyclizations. Arylamines can undergo photocyclization in the presence of BF$_3$·OEt$_2$ to give tricyclic products, e.g. 9-azaphenanthrene derivatives (eq 11).[54]

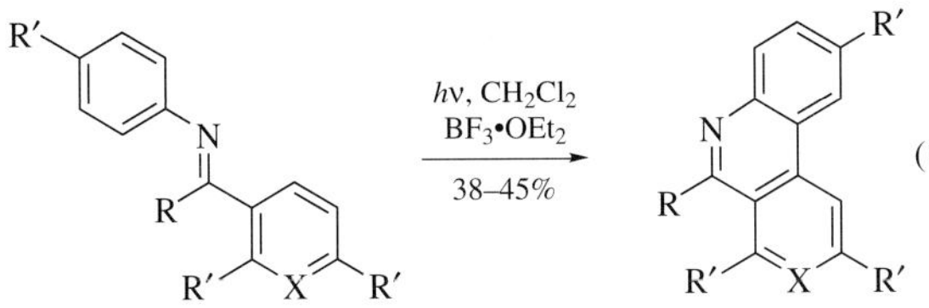

$$(11)$$

R = H, Me; R′ = H, OMe; X = CH, N

Substituted phenethyl isocyanates undergo cyclization to lactams when treated with BF$_3$·OEt$_2$.[55] Vinyl ether epoxides (eq 12),[56] vinyl aldehydes,[57] and epoxy β-keto esters[58] all undergo cyclization with BF$_3$·OEt$_2$.

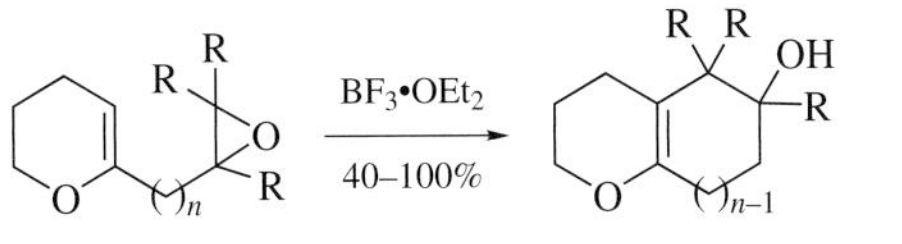

$$(12)$$

R = H, Me

β-Silyl divinyl ketones (Nazarov reagents) in the presence of BF$_3$·OEt$_2$ cyclize to give cyclopentenones, generally with retention of the silyl group.[59] BF$_3$·OEt$_2$ is used for the key step in the synthesis of the sesquiterpene trichodiene, which has adjacent quaternary centers, by catalyzing the cyclization of the dienone

to the tricyclic ketone (eq 13).[60] Trifluoroacetic acid and trifluoroacetic anhydride do not catalyze this cyclization.

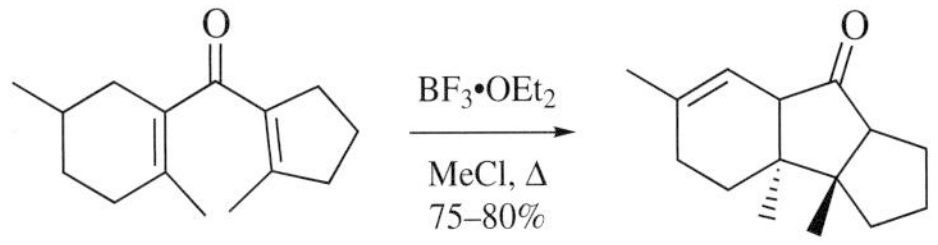

$$(13)$$

Costunolide, treated with BF$_3$·OEt$_2$, produces the cyclocostunolide (**2**) and a C-4 oxygenated sesquiterpene lactone (**3**), 4α-hydroxycyclocostunolide (eq 14).[61]

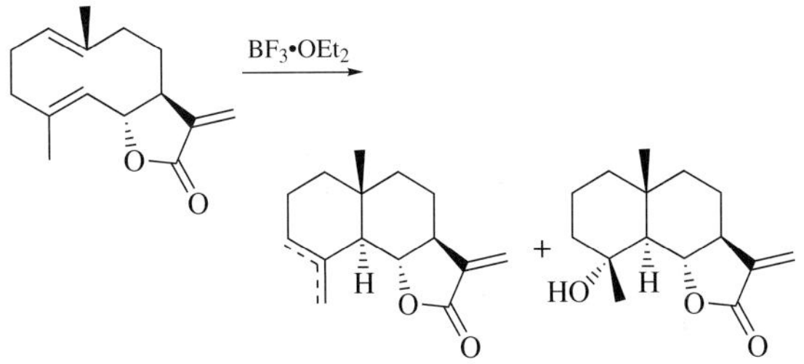

$$(14)$$

(**2**)   (**3**)

Other Condensation Reactions. BF$_3$·MeOH and BF$_3$·OEt$_2$ with ethanol are widely used in the esterification of various kinds of aliphatic, aromatic, and carboxylic acids;[62] the reaction is mild, and no rearrangement of double bonds occurs. This esterification is used routinely for stable acids prior to GLC analysis. Heterocyclic carboxylic acids,[63] unsaturated organic acids,[64] biphenyl-4,4′-dicarboxylic acid,[65] 4-aminobenzoic acid,[63] and the very sensitive 1,4-dihydrobenzoic acid[65] are esterified directly.

The dianion of acetoacetate undergoes Claisen condensations with tetramethyldiamide derivatives of dicarboxylic acids to produce polyketides in the presence of BF$_3$·OEt$_2$ (eq 15).[66] Similarly, 3,5-dioxoalkanoates are synthesized from tertiary amides or esters with the acetoacetate dianion in the presence of BF$_3$·OEt$_2$ (eq 16).[66]

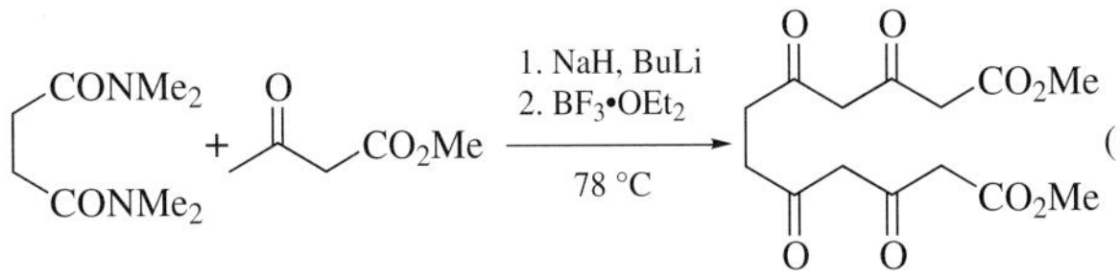

$$(15)$$

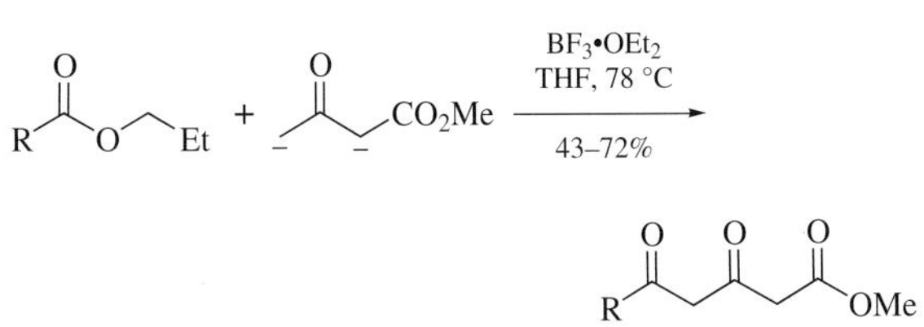

$$(16)$$

R = n-C$_9$H$_{19}$, ClCH$_2$, Ph

Aldehydes and siloxydienes undergo cyclocondensation with BF$_3$·OEt$_2$ to form pyrones (eq 17).[67] The stereoselectivity is influenced by the solvent.

$$\text{(17)}$$

solvent: CH_2Cl_2 1:2.3
       PhMe   7:1

$BF_3 \cdot OEt_2$ is effective in the direct amidation of carboxylic acids to form carboxamides (eq 18).[68] The reaction is accelerated by bases and by azeotropic removal of water.

$$\text{(18)}$$

Carbamates of secondary alcohols can be prepared by a condensation reaction with the isocyanate and $BF_3 \cdot OEt_2$ or *Aluminum Chloride*.[69] These catalysts are superior to basic catalysts such as pyridine and triethylamine. Some phenylsulfonylureas have been prepared from phenylsulfonamides and isocyanates using $BF_3 \cdot OEt_2$ as a catalyst; for example, 1-butyl-3-(p-tolylsulfonyl) urea is prepared from p-toluenesulfonamide and butyl isocyanate.[70] $BF_3 \cdot OEt_2$ is an excellent catalyst for the condensation of amines to form azomethines (eq 19).[71] The temperatures required are much lower than with *Zinc Chloride*.

$$\text{(19)}$$

Acyltetrahydrofurans can be obtained by $BF_3 \cdot OEt_2$ catalyzed condensation of (Z)-4-hydroxy-1-alkenylcarbamates with aldehydes, with high diastereo- and enantioselectivity.[72] Pentasubstituted hydrofurans are obtained by the use of ketones.

Isobornyl ethers are obtained in high yields by the condensation of camphene with phenols at low temperatures using $BF_3 \cdot OEt_2$ as catalyst.[73] Thus camphene and 2,4-dimethylphenol react to give isobornyl 2,4-dimethylphenyl ether, which can undergo further rearrangement with $BF_3 \cdot OEt_2$ to give 2,4-dimethyl-6-isobornylphenol.[73]

The title reagent is also useful for the condensation of allylic alcohols with enols. A classic example is the reaction of phytol in dioxane with 2-methyl-1,4-naphthohydroquinone 1-monoacetate to form the dihydro monoacetate of vitamin K_1 (eq 20), which can be easily oxidized to the quinone.[74]

$$R = H \text{ or } COMe \qquad \text{(20)}$$

$BF_3 \cdot OEt_2$ promotes fast, mild, clean regioselective dehydration of tertiary alcohols to the thermodynamically most stable alkenes.[75] 11β-Hydroxysteroids are dehydrated by $BF_3 \cdot OEt_2$ to give $\Delta^{9(11)}$-enes (eq 21).[76,77]

$$\text{(21)}$$

Epoxide Cleavage and Rearrangements. The treatment of epoxides with $BF_3 \cdot OEt_2$ results in rearrangements to form aldehydes and ketones (eq 22).[78] The carbon α to the carbonyl group of an epoxy ketone migrates to give the dicarbonyl product.[79] The acyl migration in acyclic α,β-epoxy ketones proceeds through a highly concerted process, with inversion of configuration at the migration terminus.[80] With 5-substituted 2,3-epoxycyclohexanes the stereochemistry of the quaternary carbon center of the cyclopentanecarbaldehyde product is directed by the chirality of the 5-position.[81] Diketones are formed if the β-position of the α,β-epoxy ketone is unsubstituted. The 1,2-carbonyl migration of an α,β-epoxy ketone, 2-cycloheptylidenecyclopentanone oxide, occurs with $BF_3 \cdot OEt_2$ at $25 \,^\circ C$ to form the cyclic spiro -1,3-diketone in 1 min (eq 23).[82]

$$\text{(22)}$$

$$R^1 = Me, H; \; R^2 = Me, Ph$$

$$\text{(23)}$$

The migration of the carbonyl during epoxide cleavage is used to produce hydroxy lactones from epoxides of carboxylic acids (eq 24).[83] α-Acyl-2-indanones,[84] furans,[85] and Δ^2-oxazolines[86] (eq 25) can also be synthesized by the cleavage and rearrangement of epoxides with $BF_3 \cdot OEt_2$. The last reaction has been conducted with sulfuric acid and with tin chloride, but the yields were lower. γ,δ-Epoxy tin compounds react with $BF_3 \cdot OEt_2$ to give the corresponding cyclopropylcarbinyl alcohols (eq 26).[87]

$$(24)$$

$$(25)$$

$$(26)$$

Remotely unsaturated epoxy acids undergo fission rearrangement when treated with $BF_3 \cdot OEt_2$. Hence, *cis* and *trans* ketocyclopropane esters are produced from the unsaturated epoxy ester methyl vernolate (eq 27).[88]

$$(27)$$

Epoxy sulfones undergo rearrangement with $BF_3 \cdot OEt_2$ to give the corresponding aldehydes.[89] α-Epoxy sulfoxides, like other negatively substituted epoxides, undergo rearrangement in which the sulfinyl group migrates and not the hydrogen, alkyl, or aryl groups (eq 28).[89]

$$(28)$$

α,β-Epoxy alcohols undergo cleavage and rearrangement with $BF_3 \cdot OEt_2$ to form β-hydroxy ketones.[90] The rearrangement is stereospecific with respect to the epoxide and generally results in *anti* migration. The rearrangement of epoxy alcohols with β-substituents leads to α,α-disubstituted carbonyl compounds.[91]

The $BF_3 \cdot OEt_2$-induced opening of epoxides with alcohols is regioselective, but the regioselectivity varies with the nature of the substituents on the oxirane ring.[92] If the substituent provides charge stabilization (as with a phenyl ring), the internal position is attacked exclusively. On the other hand, terminal ethers are formed by the regioselective cleavage of the epoxide ring of glycidyl tosylate.[92]

A combination of cyanoborohydride and $BF_3 \cdot OEt_2$ is used for the regio- and stereoselective cleavage of most epoxides to the less substituted alcohols resulting from *anti* ring opening.[93] The

reaction rate of organocopper and cuprate reagents with slightly reactive epoxides, e.g. cyclohexene oxide, is dramatically enhanced by $BF_3 \cdot OEt_2$.[94] The Lewis acid and nucleophile work in a concerted manner so that *anti* products are formed.

Azanaphthalene *N*-oxides undergo photochemical deoxygenation reactions in benzene containing $BF_3 \cdot OEt_2$, resulting in amines in 70–80% yield;[95] these amines are important in the synthesis of heterocyclic compounds. *Azidotrimethylsilane* reacts with *trans*-1,2-epoxyalkylsilanes in the presence of $BF_3 \cdot OEt_2$ to produce (*Z*)-1-alkenyl azides.[96] The *cis*-1,2-epoxyalkylsilanes undergo rapid polymerization in the presence of Lewis acids.

Other Rearrangements. $BF_3 \cdot OEt_2$ is used for the regioselective rearrangement of polyprenyl aryl ethers to yield polyprenyl substituted phenols, e.g. coenzyme Q_n.[97] The reagent is used in the Fries rearrangement; for example, 5-acetyl-6-hydroxycoumaran is obtained in 96% yield from 6-acetoxycoumaran using this reagent (eq 29).[98]

$$(29)$$

Formyl bicyclo[2.2.2]octane undergoes the retro-Claisen rearrangement to a vinyl ether in the presence of $BF_3 \cdot OEt_2$ at $0\,^{\circ}C$ (eq 30), rather than with HOAc at $110\,^{\circ}C$.[99]

$$(30)$$

$BF_3 \cdot OEt_2$ is used for a stereospecific 1,3-alkyl migration to form *trans*-2-alkyltetrahydrofuran-3-carbaldehydes from 4,5-dihydrodioxepins (eq 31), which are obtained by the isomerization of 4,7-dihydro-1,3-dioxepins.[100] Similarly, α-alkyl-β-alkoxyaldehydes can be prepared from 1-alkenyl alkyl acetals by a 1,3-migration using $BF_3 \cdot OEt_2$ as catalyst.[101] *Syn* products are obtained from (*E*)-1-alkenyl alkyl acetals and *anti* products from the (*Z*)-acetals.

$$(31)$$

The methyl substituent, and not the cyano group, of 4-methyl-4-cyanocyclohexadienone migrates in the presence of $BF_3 \cdot OEt_2$ to give 3-methyl-4-cyanocyclohexadienone.[102] $BF_3 \cdot OEt_2$-promoted regioselective rearrangements of polyprenyl aryl ethers provide a convenient route for the preparation of polyprenyl-substituted hydroquinones (eq 32), which can be oxidized to polyprenyl-quinones.[103]

$$(32)$$

The (E)–(Z) photoisomerization of α,β-unsaturated esters,[104] cinnamic esters,[105] butenoic esters,[106] and dienoic esters[106] is catalyzed by BF$_3$·OEt$_2$ or **Ethylaluminum Dichloride**. The latter two reactions also involve the photodeconjugation of α,β-unsaturated esters to β,γ-unsaturated esters. The BF$_3$·MeOH complex is used for the isomerization of 1- and 2-butenes to form equal quantities of cis- and trans-but-2-enes;[107] the BF$_3$·OEt$_2$–acetic acid complex is not as effective.

The complex formed with BF$_3$·OEt$_2$ and **Epichlorohydrin** in DMF acts as a catalyst for the Beckmann rearrangement of oximes.[108] Cyclohexanone, acetaldehyde, and syn-benzaldehyde oximes are converted into ε-caprolactam, a mixture of N-methylformamide and acetamide, and N-phenylacetamide, respectively.

The addition of BF$_3$·OEt$_2$ to an α-phosphorylated imine results in the 1,3-transfer of a diphenylphosphinoyl group, with resultant migration of the C–N=C triad.[109] This method is less destructive than the thermal rearrangement. The decomposition of dimethyldioxirane in acetone to methyl acetate is accelerated with BF$_3$·OEt$_2$, but acetol is also formed.[110] Propene oxide undergoes polymerization with BF$_3$·OEt$_2$ in most solvents, but isomerizes to propionaldehyde and acetone in dioxane.[111]

Hydrolysis. BF$_3$·OEt$_2$ is used for stereospecific hydrolysis of methyl ethers, e.g. in the synthesis of ($\pm$)-aklavone.[112] The reagent is also used for the mild hydrolysis of dimethylhydrazones.[113] The precipitate formed by the addition of BF$_3$·OEt$_2$ to a dimethylhydrazone in ether is readily hydrolyzed by water to the ketone; the reaction is fast and does not affect enol acetate functionality.

Cleavage of Ethers. In aprotic, anhydrous solvents, BF$_3$·MeOH is useful for the cleavage of trityl ethers at rt.[114] Under these conditions, O- and N-acyl groups, O-sulfonyl, N-alkoxycarbonyl, O-methyl, O-benzyl, and acetal groups are not cleaved.

BF$_3$·OEt$_2$ and iodide ion are extremely useful for the mild and regioselective cleavage of aliphatic ethers and for the removal of the acetal protecting group of carbonyl compounds.[115,116] Aromatic ethers are not cleaved, in contrast to other boron reagents. BF$_3$·OEt$_2$, in chloroform or dichloromethane, can be used for the removal of the t-butyldimethylsilyl (TBDMS) protecting group of hydroxyls, at 0–25 °C in 85–90% yield.[117] This is an alternative to ether cleavage with **Tetrabutylammonium Fluoride** or hydrolysis with aqueous **Acetic Acid**.

In the presence of BF$_3$·OEt$_2$, dithio-substituted allylic anions react exclusively at the α-carbons of cyclic ethers, to give high yields of the corresponding alcohol products (eq 33).[118] The dithiane moiety is readily hydrolyzed with **Mercury(II) Chloride** to give the keto derivatives.

$$(33)$$

Inexpensive di-, tri-, and tetramethoxyanthraquinones can be selectively dealkylated to hydroxymethoxyanthraquinones by the formation of difluoroboron chelates with BF$_3$·OEt$_2$ in benzene and subsequent hydrolysis with methanol.[119] These unsymmetrically functionalized anthraquinone derivatives are useful intermediates for the synthesis of adriamycin, an antitumor agent. 2,4,6-Trimethoxytoluene reacts with cinnamic acid and BF$_3$·OEt$_2$, with selective demethylation, to form a boron heterocycle which can be hydrolyzed to the chalcone aurentiacin (eq 34).[120]

$$(34)$$

Reductions. In contrast to hydrosilylation reactions catalyzed by metal chlorides, aldehydes and ketones are rapidly reduced at rt by **Triethylsilane** and BF$_3$·OEt$_2$, primarily to symmetrical ethers and borate esters, respectively.[121] Aryl ketones like acetophenone and benzophenone are converted to ethylbenzene and diphenylmethane, respectively. Friedel–Crafts acylation–silane reduction reactions can also occur in one step using these reagents; thus **Benzoyl Chloride** reacts with benzene, triethylsilane, and BF$_3$·OEt$_2$ to give diphenylmethane in 30% yield.[121]

BF$_3$·OEt$_2$ followed by **Diisobutylaluminum Hydride** is used for the 1,2-reduction of γ-amino-α,β-unsaturated esters to give unsaturated amino alcohols, which are chiral building blocks for α-amino acids.[122] α,β-Unsaturated nitroalkenes can be reduced to hydroxylamines by **Sodium Borohydride** and BF$_3$·OEt$_2$ in THF;[123,124] extended reaction times result in the reduction of the hydroxylamines to alkylamines. Diphenylamine–borane is prepared from sodium borohydride, BF$_3$·OEt$_2$, and diphenylamine in THF at 0 °C.[125] This solid is more stable in air than BF$_3$·THF and is almost as reactive in the reduction of aldehydes, ketones, carboxylic acids, esters, and anhydrides, as well as in the hydroboration of alkenes.

Bromination. BF$_3$·OEt$_2$ can catalyze the bromination of steroids that cannot be brominated in the presence of HBr or sodium acetate. Hence, 11α-bromoketones are obtained in high yields from methyl 3α,7α-diacetoxy-12-ketocholanate.[126] Bromination (at the 6α-position) and dibromination (at the 6α- and 11α-positions) of methyl 3α-acetoxy-7,12-dioxocholanate can occur, depending on the concentration of bromine.[127]

A combination of BF$_3$·OEt$_2$ and a halide ion (tetraethylammonium bromide or iodide in dichloromethane or chloroform, or sodium bromide or iodide in acetonitrile) is useful for the conversion of allyl, benzyl, and tertiary alcohols to the corresponding halides.[128,129]

Diels–Alder Reactions. $BF_3 \cdot OEt_2$ is used to catalyze and reverse the regiospecificity of some Diels–Alder reactions, e.g. with *peri*-hydroxylated naphthoquinones,[130] sulfur-containing compounds,[131] the reaction of 1-substituted *trans*-1,3-dienes with 2,6-dimethylbenzoquinones,[132] and the reaction of 6-methoxy-1-vinyl-3,4-dihydronaphthalene with *p*-quinones.[133] $BF_3 \cdot OEt_2$ has a drastic effect on the regioselectivity of the Diels–Alder reaction of quinoline- and isoquinoline-5,8-dione with piperylene, which produces substituted azaanthraquinones.[134] This Lewis acid is the most effective catalyst for the Diels–Alder reaction of furan with methyl acrylate, giving high *endo* selectivity in the 7-oxabicyclo[2.2.1]heptene product (eq 35).[135]

(35)

7:3

α-Vinylidenecycloalkanones, obtained by the reaction of *Lithium Acetylide* with epoxides and subsequent oxidation, undergo a Diels–Alder reaction at low temperature with $BF_3 \cdot OEt_2$ to form spirocyclic dienones (eq 36).[136]

(36)

Other Reactions. The 17-hydroxy group of steroids can be protected by forming the THP (*O*-tetrahydropyran-2-yl) derivative with 2,3-dihydropyran, using $BF_3 \cdot OEt_2$ as catalyst;[137] the yields are higher and the reaction times shorter than with *p*-toluenesulfonic acid monohydrate.

$BF_3 \cdot OEt_2$ catalyzes the decomposition of β,γ-unsaturated diazomethyl ketones to cyclopentenone derivatives (eq 37).[138,139] Similarly, γ,δ-unsaturated diazo ketones are decomposed to β,γ-unsaturated cyclohexenones, but in lower yields.[140]

(37)

$BF_3 \cdot OEt_2$ is an effective reagent for debenzyloxycarbonylations of methionine-containing peptides.[141] Substituted $6H$-1,3-thiazines can be prepared in high yields from $BF_3 \cdot OEt_2$-catalyzed reactions between α,β-unsaturated aldehydes, ketones, or acetals with thioamides, thioureas, and dithiocarbamates (eq 38).[142]

(38)

α-Alkoxy ketones can be prepared from α-diazo ketones and primary, secondary, and tertiary alcohols using $BF_3 \cdot OEt_2$ in

ethanol.[143] Nitrogen is released from a solution of **α-Diazoacetophenone** and $BF_3 \cdot OEt_2$ in ethanol to give α-ethoxyacetophenone.[143]

Anti-diols can be formed from β-hydroxy ketones using **Tin(IV) Chloride** or $BF_3 \cdot OEt_2$.[144] The hydroxy ketones are silylated, treated with the Lewis acid, and then desilylated with **Hydrogen Fluoride**. *Syn*-diols are formed if **Zinc Chloride** is used as the catalyst.

$BF_3 \cdot OEt_2$ activates the formal substitution reaction of the hydroxyl group of γ- or δ-lactols with some organometallic reagents (M = Al, Zn, Sn), so that 2,5-disubstituted tetrahydrofurans or 2,6-disubstituted tetrahydropyrans are formed.[145]

A new method of nitrile synthesis from aldehydes has been discovered using *O*-(2-aminobenzoyl)hydroxylamine and $BF_3 \cdot OEt_2$, achieving 78–94% yields (eq 39).[146]

(39)

Carbonyl compounds react predominantly at the α site of dithiocinnamyllithium if $BF_3 \cdot OEt_2$ is present, as the hardness of the carbonyl compound is increased (eq 40).[147] The products can be hydrolyzed to α-hydroxyenones.

(40)

Optically active sulfinates can be synthesized from sulfinamides and alcohols using $BF_3 \cdot OEt_2$.[148] The reaction proceeds stereospecifically with inversion of sulfinyl configuration; the mild conditions ensure that the reaction will proceed even with alcohols with acid-labile functionality.

First Update

Carl J. Lovely
University of Texas at Arlington, Arlington, TX, USA

Addition Reactions. $BF_3 \cdot Et_2O$ is often among the first Lewis acids evaluated for the activation of electrophiles toward the addition of less reactive nucleophiles, and equilibrium constants between this Lewis acid and several carbonyl compounds have been determined.[149] TMSCN undergoes addition to *N*-sulfonyl and *N*-alkyl imines in high yields (82–85%) in the presence of $BF_3 \cdot Et_2O$. Non-racemic amines engage in this transformation, providing efficient access to non-racemic amino acid derivatives after nitrile hydrolysis.[150] Vinylogous formamides can be converted into the corresponding trienes in moderate to good yields through the addition of in situ generated allyl indium reagents and subsequent elimination.[151] $BF_3 \cdot Et_2O$ catalyzes the vinylogous aldol reaction between aldehydes and 3-siloxyfuran derivatives, in good yields (42–100%) and with moderate to high diastereoselectivity (1:1–19), favoring the syn adduct.[152] Potassium aryl and vinyl trifluoroborate salts react with in situ generated iminium ions catalyzed by $BF_3 \cdot Et_2O$, providing the Mannich-type adducts in 21–95% yield.[153] N-Sulfonylimines and N-sulfinylimines react with potassium allyl- and crotyltrifluoroborates in high yield (54-99%) and generally high diastereoselectivity (91–98:9–2) (eq 41).[154] In addition, asymmetric variants with Ellman-type N-sulfinylimines were investigated, leading to the formation of the expected addition product in excellent yield and diastereoselectivity.[154]

(41)

This reagent activates imines toward vinylogous Mannich-type reactions. For example, trifluoromethyl aldimines react in both high yield and with high anti diastereoselectivities (>98% de) with 2-trialkylsiloxyfurans, affording the corresponding substituted butyrolactones.[155] β-Amino-α,α-difluoroketones can be generated through an imino aldol reaction from acylsilanes, trifluoromethyltrimethylsilane, and $BF_3 \cdot Et_2O$ in good yields (53–83%) and reasonable levels of diastereoselectivity.[156] In the presence of $BF_3 \cdot Et_2O$, vinyl epoxides rearrange to the isomeric β,γ-unsaturated aldehyde, which further tautomerizes to the vinylogous enol, and then engages in a Mannich reaction with N-aryl aldimines (eq 42).[157] Reformatsky reactions of aldehydes have been reported to be catalyzed by $BF_3 \cdot Et_2O$ leading to the formation of β-hydroxyl esters in good to excellent yield (66–92%).[158]

(42)

Methylene cyclopropene derivatives undergo reaction with aldehydes and imines that are activated by Lewis acid to generate the indene derivative as the major product in most cases, although occasionally dihydropyrans are obtained.[159] Alternatively, if the reaction is conducted at −25 °C, rather than room temperature, the corresponding furan or pyrrolidine derivative is formed (eq 43). A mechanistic rationale for the divergent reaction pathways was provided.[159]

(43)

An unusual addition reaction of alcohols to olefins involving the net formation of a C–C bond between two sp^3-hybridized carbon atoms through the action of Wilkinson's catalyst and $BF_3 \cdot Et_2O$ has been reported (eq 44). The reaction appears to be reasonably general and is interpreted in terms of a radical-based process.[160]

(44)

1,2-Bis(trimethylsilyloxy)cyclobutene derivatives react with ketones and their dialkyl acetals to afford the corresponding spiro-fused 1,3-cyclopentadiones. Depending on the structure of the cyclobutene, the corresponding aldehyde-derived acetals give 2-substituted 3-hydroxycylopentenones, or 3-substituted 2-hydroxy pentenones.[161] $BF_3 \cdot Et_2O$ has been used in the Baylis–Hillman reaction in conjuction with sulfides, including a non-racemic sulfide, leading to a moderately effective asymmetric variant.[162] It was proposed that the Lewis acid plays a dual role of activating both carbonyl components: first to activate the donor towards Michael addition of the sulfide, and second, to activate the electrophile to nucleophilic attack.[162]

Non-racemic oxime ethers undergo addition with Grignard or organolithium reagents upon activation with $BF_3 \cdot Et_2O$ at low temperature, providing the 1,2-adducts in high chemical yield and generally high diastereoselectivity. The resulting adducts have utility in the asymmetric construction of nitrogen heterocycles.[163]

$BF_3 \cdot Et_2O$ has been employed to activate vinyl oxazolidinones towards S_N2' reactions with organocuprates. These reactions proceed primarily via an anti mode of addition, providing dipeptide isosteres in good yield.[164] Similar results were obtained using allylic mesylates, although in this case access to both the cis and trans product was possible by appropriate choice of solvent (Et_2O = trans; THF = cis).[165]

2,3,4-Trisubstituted tetrahydrofurans can be prepared in good yields with moderate to good levels of diastereocontrol through the reaction of cylic allylsiloxane derivatives (eq 45). Plausible mechanisms were provided to account for both products formed and the stereochemical control observed.[166]

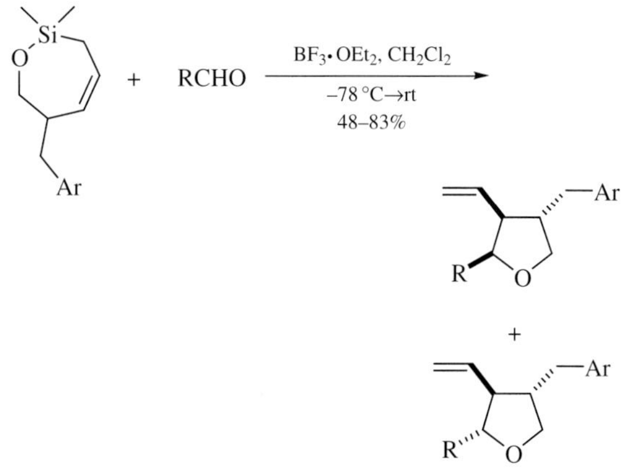

(45)

Lewis acid assisted ring-opening reactions of strained rings continue to be a prevalent application of $BF_3 \cdot Et_2O$. For example, the reagent has been used to activate ethylene oxide towards nucleophilic addition with an organolithium derivative generated from the sparteine-mediated asymmetric deprotonative lithiation of BOC-protected pyrrolidine.[167] A related approach has been employed to desymmetrize meso epoxides (eq 46).[168]

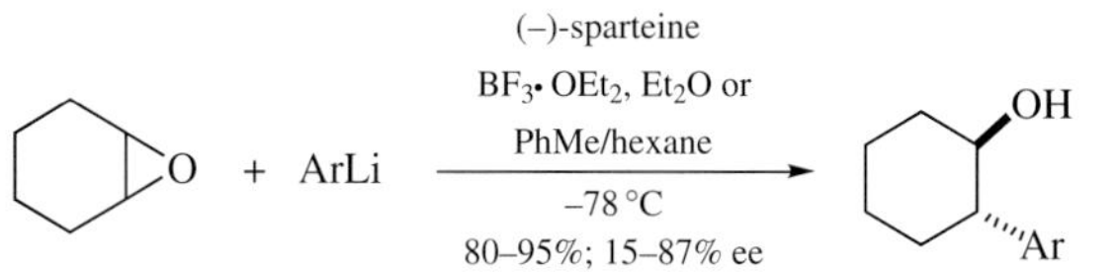

(46)

Amino epoxides react with ketones in the presence of the reagent to afford the dioxolane in a stereospecific fashion with respect to the epoxide (eq 47). Subsequent treatment of the dioxolane with dilute HCl provides the corresponding diols in high yield and stereochemical purity.[169] Similarly in the presence of nitriles, amino epoxides undergo a tandem ring-opening Ritter-type of process to provide amido hydroxy amine derivatives (eq 47).[170] N-Tosyl aziridines have been shown to undergo ring-opening and intramolecular nucleophilic trapping with a variety of π-nucleophiles, including allyl- and vinylsilanes,[171,172] aromatic rings, and simple olefins.[173]

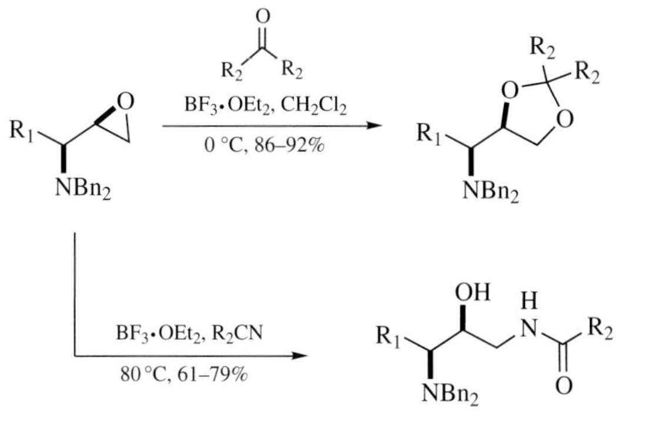

(47)

Substitution. A broad range of substitution reactions catalyzed or mediated by $BF_3 \cdot Et_2O$ are known. For example, acetals undergo substitution with organomanganese reagents, its presence leading to the formation of ethers.[174] The use of this reagent for the in situ generation of iminium ions has been reported, for example, α-methoxy amides react with $BF_3 \cdot Et_2O$ in the presence of various nucleophiles leading to substitution.[175] Similar

transformations can be accomplished intramolecularly from α-methoxy carbamates.[176] Friedel–Crafts reactions have been initiated by treatment with the reagent. For instance, tertiary propargylic silyl ethers react in an intramolecular sense with electron-rich aromatics to provide the allene, which upon exposure to acid affords the corresponding aromatized adduct (eq 48).[177]

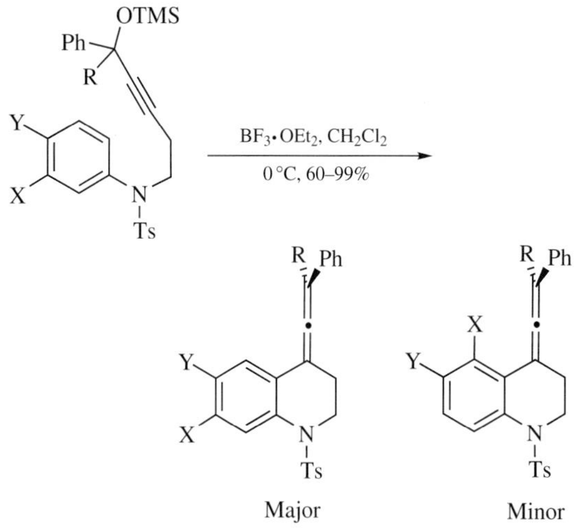

(48)

Major　　　　Minor

An intramolecular Friedel–Crafts reaction of a diene and an electron-rich aromatic moiety, promoted by $BF_3 \cdot Et_2O$, was used in the late stage construction of cyclohexyl ring in an approach to the marine natural product pseudopterosin aglycone.[178] In a somewhat related transformation, it has been demonstrated that aromatics can be converted into the corresponding diarylmethane derivative on exposure to α-methoxyacetic acid in the presence of trifluoroacetic anhydride (TFAA) and $BF_3 \cdot Et_2O$ (eq 49).[179]

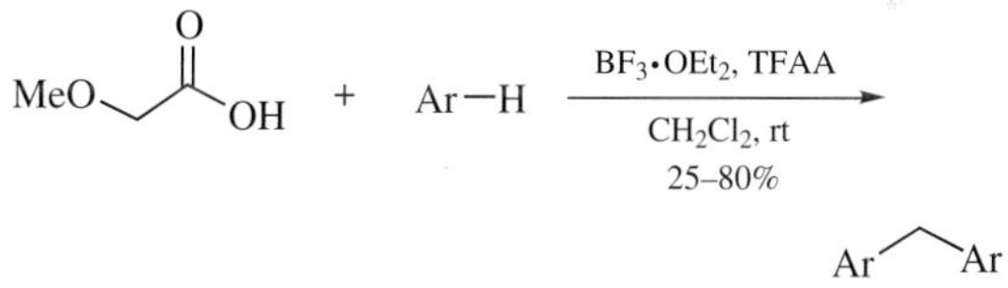

(49)

A regiocomplementary method to the acid-catalyzed synthesis of aryl benzothiophenes from 2-arylthio acetophenone derivatives is possible using $BF_3 \cdot Et_2O$. If polyphosphoric acid is used, the 2-isomer is formed, whereas, in the presence of the Lewis acid, the isomeric 3-substituted product is obtained.[180] An unusual substitution reaction was observed on exposure of aryl α-bromoketones to allyl tributyltin and $BF_3 \cdot Et_2O$. Allyl ketones were obtained, resulting from the net addition of the allyl moiety to the carbonyl group and migration of the aryl moiety. This outcome was rationalized in terms of the intermediacy of the epoxide and its subsequent Lewis acid-catalyzed rearrangement (eq 50).[181]

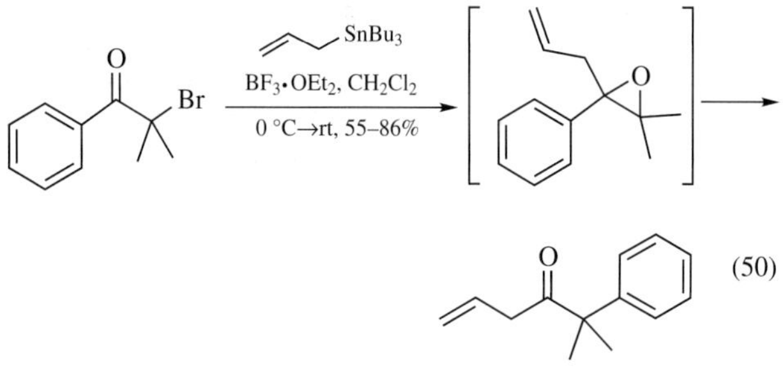

(50)

Cyclic ethers can be obtained in good yields and high stereoselectivities from 1,2,n-triols on treatment with an ortho ester and $BF_3 \cdot Et_2O$. The intermediacy of an acetoxonium species was postulated which undergoes intramolecular nucleophilic attack to provide the observed products (eq 51).[182]

$$R = \text{alkyl, c-alkyl, aryl,} \quad (51)$$
$$n = 1\text{-}4$$

A Pummerer-type of α-functionalization of arylselenyl acetates with allyl trimethylsilanes or allyl tri-n-butylstannanes occurs in the presence of $BF_3 \cdot Et_2O$ or TFAA, providing homo allylic ester derivatives (eq 52).[183]

$$63\% \quad (52)$$

$BF_3 \cdot Et_2O$ has been shown to mediate and catalyze the substitution of *tert*-alkyl fluorides with various nucleophiles including silyl enol ethers, allyl silanes, hydrosilanes and phosphines (eq 53).[184,185]

$$(53)$$

An elegant in situ protection-activation strategy using $BF_3 \cdot Et_2O$ has been described for the functionalization of amino acids, leading to the facile preparation of amides (including dipeptides) of the carboxylic acid, circumventing the need to protect–deprotect the amino moiety (eq 54).[186]

$$(54)$$

Several transition metal mediated processes rely on the participation of Lewis acids, among these, the Nicholas reaction figures prominently. After treatment with $BF_3 \cdot Et_2O$, $Co_2(CO)_6$ complexed propargylic alcohols provide the carbocation, which can be trapped intramolecularly with various nucleophiles including epoxides (eq 55),[187] or via hydride transfer from an appropriately poised benzyl moiety (eq 56)[188] or aromatic rings.[189] In the latter case, access to the relatively small [7] metacyclophane derivatives was possible.

$$(55)$$

$$(56)$$

Addition of the Schwartz' reagent ($Cp_2Zr(H)Cl$) to N-allyl oxazolines, followed by $BF_3 \cdot Et_2O$ leads to the formation of trans 1,5-disubstituted pyrrolidines in moderate yields (48–52%), but with high levels of diastereocontrol.[190] It has been reported that aryl triazines will function as coupling partners in Suzuki cross-coupling reactions using an aryl boronic acid, Pd_2dba_3, $P(Bu\text{-}t)_3$, and $BF_3 \cdot Et_2O$. Generally fast reaction times were observed and high yields were obtained.[191] This represents a significant improvement over the previous conditions which relied on protic acids to catalyze the transformation.

Rearrangements. The epoxide to aldehyde/ketone rearrangement is a classic reaction mediated by $BF_3 \cdot Et_2O$. Recent reports have disclosed the utility of this rearrangement in concert with a second process. For example, terminal epoxides rearrange in the presence of BF$_3$·Et$_2$O, to provide the corresponding aldehyde, which undergoes ene reactions with a pendant allyl silane, providing either an cyclic allyl silane or vinyl silane (eq 57).[192] A related reaction was observed with optically active α-aminoallenylstannanes and epoxides. Rather than the expected direct addition to the epoxide, rearrangement to the aldehyde occurred prior to addition, leading to the formation of syn homo propargylic alcohols in good yield and excellent diastereoselectivity (eq 58).[193]

2-Aminomethylazetidines have been shown to undergo ring expansion to 3-aminopyrrolidines upon exposure to BF$_3$·Et$_2$O in good yields, and in high diastereoselectivity, other Lewis acids or protic acids did not promote the rearrangement (eq 59).[194]

The rearrangement of divinyl ketones to cyclopentenones (the Nazarov rearrangement) can be promoted with BF$_3$·Et$_2$O. This transformation has recently been applied to good effect in the construction of the hydroazulene core found in guanacastepene.[195] Other variants of the Nazarov reaction have been developed and are usually coupled with a second event, such as the reductive

Nazarov cyclization (in presence of Et$_3$SiH)[196] and the interrupted Nazarov reaction (in presence of pendant olefin, eq 60).[197,198]

This reagent has been shown to be particularly effective for the net O to C rearrangement of alkynyl tributylstannane derivatives from mixed acetals derived from furanyl and pyranyl lactols (eq 61).[199] Attempts to accomplish the same transformation with other Lewis acids (TiCl$_4$, SnCl$_4$, EtAlCl$_2$, Me$_2$AlCl) were not successful.

Cycloadditions. A novel BF$_3$·Et$_2$O catalyzed ring-opening reaction of methylene aziridines has been reported leading to the in situ formation of 2-aminoallyl cations, which undergo [4 + 3] cycloaddition with pendant dienes (eq 62). These reactions proceed in moderate to good yield affording polycyclic ring systems that appear to have substantial potential for use in approaches to a variety of natural products. Similar reactions with Sc(OTf)$_3$ were less efficient.[200]

Methylenecyclopropenes react with a pendant imine moiety through the action of BF$_3$·Et$_2$O to provide the indolizidine in moderate yields (eq 63). BF$_3$·Et$_2$O was the only Lewis acid out of a range of reagents that led to the observed cycloaddition.[201]

Oximes can participate in cycloaddition reactions after tautomerizing to the isomeric nitrone derivative generating the isoxazoline. In early reports this tautomerization was achieved at high temperatures, or in the presence of Pd(II) salts, although this latter approach has some structural limitations. It has been demonstrated that $BF_3 \cdot Et_2O$ can be used to accomplish the same transformation with a wider array of substrates and at substantially lower temperatures via the O-silyloxime derivatives (eq 64). For comparative purposes, the parent oxime requires a week at reflux in toluene to provide 54% yield.[202]

$$ \text{(64)} $$

cis/trans = 1:1

Through a net [2 + 2 + 1] cycloaddition, iminocyclopentadienes can be obtained through the initial formation of a zirconacyclopentadiene derivative (from two alkynes and a low valent zirconocene component), and subsequent reaction with an aryl isocyanate and $BF_3 \cdot Et_2O$ (eq 65). Quite remarkably, no other Lewis acid investigated provided this cycloaddition; only $EtAlCl_2$ led to a productive reaction, although different adducts were obtained.[203]

$$ \text{(65)} $$

51–91%

Pyrano[3,2-c]benzothiopyrans can be constructed through the reaction of thiosalicylaldehyde derivatives, unsaturated alcohols and trimethyl orthoformate in the presence of $BF_3 \cdot Et_2O$. This procedure is an improvement on the Bronsted acid-catalyzed process.[204]

Oxidative or Reductive Processes. Polymer supported IBX (2-iodoxybenzoic acid) in the presence of $BF_3 \cdot Et_2O$ leads to the smooth and rapid conversion of various alcohols to the corresponding aldehyde or ketone.[205] A novel and direct oxidative coupling reaction to form biaryls has been reported which avoids the use of transition metals and aryl halides. This transformation can be accomplished by using $BF_3 \cdot Et_2O$, bis(trifluoroacetoxy) iodosobenzene (PIFA), and the arene (eq 66).[206,207] The reaction has reasonable scope and provides the biaryl derivative in generally high yield. A mechanistic pathway involving radical cations was posited to explain the coupling process.[208] α-Oxidation of ketones can be accomplished through the use of this Lewis acid, a peroxyacid (MCPBA) and bis(acetoxy) iodosobenzene, providing the corresponding α-acetoxy ketones in moderate yield (43–63%).[209] A combination of MCPBA and $BF_3 \cdot Et_2O$ reacts with various aldimines to provide the corresponding amide, via the intermediacy of the oxaziridine in most cases (eq 67).[210,211] Similar chemistry has been reported with ketimines.[212]

$$ \text{(66)} $$

$$ \text{(67)} $$

Imines can be reduced through the action of tributyltin hydride and $BF_3 \cdot Et_2O$. Essentially any class of imine can be reduced under these conditions, including oximes, hydrazones and nitrones (eq 68).[213] O-Silyl ketoximes on treatment with $BF_3 \cdot Et_2O$ and $BH_3 \cdot THF$ or $BH_3 \cdot DMS$ provide rearranged secondary amines in good yields.[214] $BF_3 \cdot Et_2O$ has also been used in conjunction with $NaBH_4$ to reduce a wide variety of carboxylic acid derivatives, including acid chlorides, esters, amides, nitriles and carboxylic acids.[215]

$$ \text{(68)} $$

R_1 = Alkyl, aryl

R_2 = H, alkyl

X = OH, OBn, NPh_2, $N^+(O^-)Bn$, NTs

In an interesting redox process mediated by $BF_3 \cdot Et_2O$, it has been demonstrated that ketals or aminals can be generated from the corresponding heterocycle (tetrahydrofuran, pyran, and pyrrolidine) and a pendant aldehyde moiety via a net 1,5-hydride shift (eq 69).[216] In a related transformation, electron deficient double bonds engage in a similar redox process to form a C–C bond, although in this case, it appears that other Lewis acids ($Sc(OTf)_3$ and $PtCl_4$) are more general (eq 70).[217]

$$ \text{(69)} $$

$$ \text{(70)} $$

1.5:1

Fluorinations. Although not widespread, there are limited examples of $BF_3 \cdot Et_2O$ functioning as a fluorinating agent. For example, reaction with phenylglycidyl ethers provides a convenient synthesis of enantiopure, stereodefined fluorohydrins (eq 71).[218] Similarly, aziridines can be fluorinated on exposure to $BF_3 \cdot Et_2O$.[219] Aryl fluorides can be prepared in good yields from aryl lead triacetates on treatment with $BF_3 \cdot Et_2O$ at room temperature (eq 72).[220]

$$(71)$$

$$(72)$$

$$(73)$$

Miscellaneous. There are several uses of $BF_3 \cdot Et_2O$ that do not fall into the categories described above. For example, it has been shown in the presence of $BF_3 \cdot Et_2O$ that regiocontrol in the formation of non-racemic enamines can be exercised, leading to the formation of the exocyclic enamine (eq 73). Whereas formation of the enamine via an acid-catalyzed pathway leads to the endocyclic enamine. These enamines exhibit enantiodivergent behavior in a Cu(II)-catalyzed Michael reaction with methyl vinyl ketone eq 73.[221] Alkynes via a net metathesis process ([2 + 2] cycloaddition then [2 + 2] cycloreversion) with aldehydes generate the corresponding enone (eq 74).[222] The overall process is equivalent to a Wittig process without the use of phosphines or strong base. The reaction can be performed with $BF_3 \cdot Et_2O$, or HBF_4 or $AgSbF_6$ as catalysts with essentially similar results, only in a couple of instances is the silver-catalyzed process superior.

$$(74)$$

Related Reagents. See entries for other Lewis Acids, e.g. *Zinc Chloride, Aluminum Chloride, Titanium(IV) Chloride*; also see entries for *Boron Trifluoride* (and combination reagents), and combination reagents employing Boron Trifluoride Etherate, e.g. *Butyllithium–Boron Trifluoride Etherate, Cerium(III) Acetate–Boron Trifluoride Etherate, Lithium Aluminum Hydride–Boron Trifluoride Etherate, Methylcopper–Boron Trifluoride Etherate.*

---

1. (a) Hennion, G. F.; Hinton, H. D.; Nieuwland, J. A., *J. Am. Chem. Soc.* **1933**, *55*, 2857. (b) Zweifel, G.; Brown, H. C., *Org. React.* **1963**, *13*, 28.

2. Eis, M. J.; Wrobel, J. E.; Ganem, B., *J. Am. Chem. Soc.* **1984**, *106*, 3693.

3. (a) Uno, H.; Terakawa, T.; Suzuki, H., *Chem. Lett.* **1989**, 1079. (b) *Synlett* **1991**, 559.

4. Yamaguchi, M.; Shibato, K.; Fujiwara, S.; Hirao, I., *Synthesis* **1986**, 421.

5. Yamaguchi, M.; Hirao, I., *Tetrahedron Lett.* **1983**, *24*, 391.

6. Yamaguchi, M.; Nobayashi, N.; Hirao, I., *Tetrahedron* **1984**, *40*, 4261.

7. Hara, S.; Hyuga, S.; Aoyama, M.; Sato, M.; Suzuki, A., *Tetrahedron Lett.* **1990**, *31*, 247.

8. Aoyama, M.; Hara, S.; Suzuki, A., *Synth. Commun.* **1992**, *22*, 2563.

9. Thompson, C. M.; Green, D. L. C.; Kubas, R., *J. Org. Chem.* **1988**, *53*, 5389.

10. Uno, H.; Okada, S.; Ono, T.; Shiraishi, Y.; Suzuki, H., *J. Org. Chem.* **1992**, *57*, 1504.

11. Volkmann, R. A.; Davies, J. T.; Meltz, C. N., *J. Am. Chem. Soc.* **1983**, *105*, 5946.

12. Kawate, T.; Nakagawa, M.; Yamazaki, H.; Hirayama, M.; Hino, T., *Chem. Pharm. Bull.* **1993**, *41*, 287.

13. Snyder, H. R.; Kornberg, H. A.; Romig, J. R., *J. Am. Chem. Soc.* **1939**, *61*, 3556.

14. Ha, H. J.; Nam, G. S., *Synth. Commun.* **1992**, *22*, 1143.

15. (a) Lipshutz, B. H.; Ellsworth, E. L.; Siahaan, T. J., *J. Am. Chem. Soc.* **1988**, *110*, 4834. (b) *J. Am. Chem. Soc.* **1989**, *111*, 1351.

16. Ochiai, M.; Fujita, E.; Arimoto, M.; Yamaguchi, H., *Chem. Pharm. Bull.* **1985**, *33*, 41.

17. Maruyama, K.; Naruta, Y., *J. Org. Chem.* **1978**, *43*, 3796.

18. Panek, J. S.; Sparks, M. A., *J. Org. Chem.* **1989**, *54*, 2034.

19. Giannis, A.; Sandhoff, K., *Tetrahedron Lett.* **1985**, *26*, 1479.

20. Roos, E. C.; Hiemstra, H.; Speckamp, W. N.; Kaptein, B.; Kamphuis, J.; Schoemaker, H. E., *Recl. Trav. Chim. Pays-Bas* **1992**, *111*, 360.

21. Kuhn, M.; von Wartburg, A., *Helv. Chim. Acta* **1968**, *51*, 1631.

22. Kunz, H.; Sager, W., *Helv. Chim. Acta* **1985**, *68*, 283.

23. Yamamoto, Y.; Schmid, M., *J. Chem. Soc., Chem. Commun.* **1989**, 1310.

24. Yamamoto, Y.; Maruyama, K., *J. Organomet. Chem.* **1985**, *284*, C45.

25. (a) Keck, G. E.; Abbott, D. E., *Tetrahedron Lett.* **1984**, *25*, 1883. (b) Keck, G. E.; Boden, E. P., *Tetrahedron Lett.* **1984**, *25*, 265.

26. Keck, G. E.; Enholm, E. J., *J. Org. Chem.* **1985**, *50*, 146.

27. Trost, B. M.; Bonk, P. J., *J. Am. Chem. Soc.* **1985**, *107*, 1778.

28. Marshall, J. A.; DeHoff, B. S.; Crooks, S. L., *Tetrahedron Lett.* **1987**, *28*, 527.

29. Marton, D.; Tagliavini, G.; Zordan, M.; Wardell, J. L., *J. Organomet. Chem.* **1990**, *390*, 127.

30. Ciufolini, M. A.; Spencer, G. O., *J. Org. Chem.* **1989**, *54*, 4739.

31. Gung, B. W.; Smith, D. T.; Wolf, M. A., *Tetrahedron Lett.* **1991**, *32*, 13.

32. Takuwa, A.; Nishigaichi, Y.; Yamashita, K.; Iwamoto, H., *Chem. Lett.* **1990**, 1761.

33. Nishigaichi, Y.; Takuwa, A.; Jodai, A., *Tetrahedron Lett.* **1991**, *32*, 2383.

34. Sugimura, H.; Osumi, K., *Tetrahedron Lett.* **1989**, *30*, 1571.

35. Croxall, W. J.; Glavis, F. J.; Neher, H. T., *J. Am. Chem. Soc.* **1948**, *70*, 2805.

36. Hosomi, A.; Hashimoto, H.; Sakurai, H., *Tetrahedron Lett.* **1980**, *21*, 951.

37. Smith, A. B. III; Jerris, P. J., *J. Am. Chem. Soc.* **1981**, *103*, 194.

38. Erman, W. F.; Stone, L. C., *J. Am. Chem. Soc.* **1971**, *93*, 2821.

39. Wakselman, C.; Tordeux, M., *J. Chem. Soc., Chem. Commun.* **1975**, 956.

40. Shellhamer, D. F.; Curtis, C. M.; Hollingsworth, D. R.; Ragains, M. L.; Richardson, R. E.; Heasley, V. L.; Shakelford, S. A.; Heasley, G. E., *J. Org. Chem.* **1985**, *50*, 2751.

41. Shellhamer, D. F.; Curtis, C. M.; Hollingsworth, D. R.; Ragains, M. L.; Richardson, R. E.; Heasley, V. L.; Heasley, G. E., *Tetrahedron Lett.* **1982**, *23*, 2157.

42. Wulff, W. D.; Gilbertson, S. R., *J. Am. Chem. Soc.* **1985**, *107*, 503.

43. Yamago, S.; Machii, D.; Nakamura, E., *J. Org. Chem.* **1991**, *56*, 2098.

44. Nakamura, E.; Kuwajima, I., *J. Am. Chem. Soc.* **1977**, *99*, 961.

45. Sugimura, H.; Shigekawa, Y.; Uematsu, M., *Synlett* **1991**, 153.

46. Heathcock, C. H.; Flippin, L. A., *J. Am. Chem. Soc.* **1983**, *105*, 1667.

47. Evans, D. A.; Gage, J. R., *Tetrahedron Lett.* **1990**, *31*, 6129.

48. Kamimura, A.; Marumo, S., *Tetrahedron Lett.* **1990**, *31*, 5053.

49. Annunziata, R.; Cinquini, M.; Cozzi, F.; Cozzi, P. G., *Tetrahedron Lett.* **1990**, *31*, 6733.

50. Duhamel, P.; Hennequin, L.; Poirier, N.; Poirier, J.-M., *Tetrahedron Lett.* **1985**, *26*, 6201.

51. Moriarty, R. M.; Prakash, O.; Duncan, M. P.; Vaid, R. K., *J. Org. Chem.* **1987**, *52*, 150.

52. Yamamoto, Y.; Maruyama, K., *J. Am. Chem. Soc.* **1982**, *104*, 2323.

53. Ando, W.; Tsumaki, H., *Chem. Lett.* **1983**, 1409.

54. Thompson, C. M.; Docter, S., *Tetrahedron Lett.* **1988**, *29*, 5213.

55. Ohta, S.; Kimoto, S., *Tetrahedron Lett.* **1975**, 2279.

56. Boeckman, R. K. Jr.; Bruza, K. J.; Heinrich, G. R., *J. Am. Chem. Soc.* **1978**, *100*, 7101.

57. Rigby, J. H., *Tetrahedron Lett.* **1982**, *23*, 1863.

58. Sum, P.-E.; Weiler, L., *Synlett* **1979**, *57*, 1475.

59. Chenard, B. L.; Van Zyl, C. M.; Sanderson, D. R., *Tetrahedron Lett.* **1986**, *27*, 2801.

60. Harding, K. E.; Clement, K. S., *J. Org. Chem.* **1984**, *49*, 3870.

61. Jain, T. C.; McCloskey, J. E., *Tetrahedron Lett.* **1971**, 1415.

62. (a) Hinton, H. D.; Nieuwland, J. A., *J. Am. Chem. Soc.* **1932**, *54*, 2017. (b) Sowa, F. J.; Nieuwland, J. A., *J. Am. Chem. Soc.* **1936**, *58*, 271. (c) Hallas, G., *J. Chem. Soc* **1965**, 5770.

63. Kadaba, P. K., *Synthesis* **1972**, 628.

64. Kadaba, P. K., *Synthesis* **1971**, 316.

65. Marshall, J. L.; Erikson, K. C.; Folsom, T. K., *Tetrahedron Lett.* **1970**, 4011.

66. Yamaguchi, M.; Shibato, K.; Nakashima, H.; Minami, T., *Tetrahedron* **1988**, *44*, 4767.

67. Danishefsky, S.; Chao, K.-H.; Schulte, G., *J. Org. Chem.* **1985**, *50*, 4650.

68. Tani, J.; Oine, T.; Inoue, I., *Synthesis* **1975**, 714.

69. Ibuka, T.; Chu, G.-N.; Aoyagi, T.; Kitada, K.; Tsukida, T.; Yoneda, F., *Chem. Pharm. Bull.* **1985**, *33*, 451.

70. Irie, H.; Nishimura, M.; Yoshida, M.; Ibuka, T., *J. Chem. Soc., Perkin Trans. 1* **1989**, 1209.

71. Taylor, M. E.; Fletcher, T. L., *J. Org. Chem.* **1961**, *26*, 940.

72. Hoppe, D.; Krämer, T.; Erdbrügger, C. F.; Egert, E., *Tetrahedron Lett.* **1989**, *30*, 1233.

73. Kitchen, L. J., *J. Am. Chem. Soc.* **1948**, *70*, 3608.

74. Hirschmann, R.; Miller, R.; Wendler, N. L., *J. Am. Chem. Soc.* **1954**, *76*, 4592.

75. Posner, G. H.; Shulman-Roskes, E. M.; Oh, C. H.; Carry, J.-C.; Green, J. V.; Clark, A. B.; Dai, H.; Anjeh, T. E. N., *Tetrahedron Lett.* **1991**, *32*, 6489.

76. Heymann, H.; Fieser, L. F., *J. Am. Chem. Soc.* **1952**, *74*, 5938.

77. Clinton, R. O.; Christiansen, R. G.; Neumann, H. C.; Laskowski, S. C., *J. Am. Chem. Soc.* **1957**, *79*, 6475.

78. House, H. O.; Wasson, R. L., *J. Am. Chem. Soc.* **1957**, *79*, 1488.

79. Bird, C. W.; Yeong, Y. C.; Hudec, J., *Synthesis* **1974**, 27.

80. Domagala, J. M.; Bach, R. D., *J. Am. Chem. Soc.* **1978**, *100*, 1605.

81. Obuchi, K.; Hayashibe, S.; Asaoka, M.; Takei, H., *Bull. Chem. Soc. Jpn.* **1992**, *65*, 3206.

82. Bach, R. D.; Klix, R. C., *J. Org. Chem.* **1985**, *50*, 5438.

83. Hancock, W. S.; Mander, L. N.; Massy-Westropp, R. A., *J. Org. Chem.* **1973**, *38*, 4090.

84. French, L. G.; Fenlon, E. E.; Charlton, T. P., *Tetrahedron Lett.* **1991**, *32*, 851.

85. Loubinoux, B.; Viriot-Villaume, M. L.; Chanot, J. J.; Caubere, P., *Tetrahedron Lett.* **1975**, 843.

86. Smith, J. R. L.; Norman, R. O. C.; Stillings, M. R., *J. Chem. Soc., Perkin Trans. 1* **1975**, 1200.

87. Sato, T.; Watanabe, M.; Murayama, E., *Synth. Commun.* **1987**, *17*, 781.

88. Conacher, H. B. S.; Gunstone, F. D., *J. Chem. Soc., Chem. Commun.* **1967**, 984.

89. Durst, T.; Tin, K.-C., *Tetrahedron Lett.* **1970**, 2369.

90. Maruoka, K.; Hasegawa, M.; Yamamoto, H.; Suzuki, K.; Shimazaki, M.; Tsuchihashi, G., *J. Am. Chem. Soc.* **1986**, *108*, 3827.

91. Shimazaki, M.; Hara, H.; Suzuki, K.; Tsuchihashi, G., *Tetrahedron Lett.* **1987**, *28*, 5891.

92. Liu, Y.; Chu, T.; Engel, R., *Synth. Commun.* **1992**, *22*, 2367.

93. Hutchins, R. O.; Taffer, I. M.; Burgoyne, W., *J. Org. Chem.* **1981**, *46*, 5214.

94. Alexakis, A.; Jachiet, D.; Normant, J. F., *Tetrahedron* **1986**, *42*, 5607.

95. Hata, N.; Ono, I.; Kawasaki, M., *Chem. Lett.* **1975**, 25.

96. Tomoda, S.; Matsumoto, Y.; Takeuchi, Y.; Nomura, Y., *Bull. Chem. Soc. Jpn.* **1986**, *59*, 3283.

97. Yoshizawa, T.; Toyofuku, H.; Tachibana, K.; Kuroda, T., *Chem. Lett.* **1982**, 1131.

98. Davies, J. S. H.; McCrea, P. A.; Norris, W. L.; Ramage, G. R., *J. Chem. Soc.* **1950**, 3206.

99. Boeckman, R. K. Jr.; Flann, C. J.; Poss, K. M., *J. Am. Chem. Soc.* **1985**, *107*, 4359.

100. Suzuki, H.; Yashima, H.; Hirose, T.; Takahashi, M., Moro-Oka, Y.; Ikawa, T., *Tetrahedron Lett.* **1980**, *21*, 4927.

101. Takahashi, M.; Suzuki, H.; Moro-Oka, Y.; Ikawa, T., *Tetrahedron Lett.* **1982**, *23*, 4031.

102. Marx, J. N.; Zuerker, J.; Hahn, Y. P., *Tetrahedron Lett.* **1991**, *32*, 1921.

103. Yoshizawa, T.; Toyofuku, H.; Tachibana, K.; Kuroda, T., *Chem. Lett.* **1982**, 1131.

104. Lewis, F. D.; Oxman, J. D., *J. Am. Chem. Soc.* **1981**, *103*, 7345.

105. Lewis, F. D.; Oxman, J. D.; Gibson, L. L.; Hampsch, H. L.; Quillen, S. L., *J. Am. Chem. Soc.* **1986**, *108*, 3005.

106. Lewis, F. D.; Howard, D. K.; Barancyk, S. V.; Oxman, J. D., *J. Am. Chem. Soc.* **1986**, *108*, 3016.

107. Roberts, J. M.; Katovic, Z.; Eastham, A. M., *J. Polym. Sci. A1* **1970**, *8*, 3503.

108. Izumi, Y., *Chem. Lett.* **1990**, 2171.

109. Onys'ko, P. P.; Kim, T. V.; Kiseleva, E. I.; Sinitsa, A. D., *Tetrahedron Lett.* **1992**, *33*, 691.

110. Singh, M.; Murray, R. W., *J. Org. Chem.* **1992**, *57*, 4263.

111. Sugiyama, S.; Ohigashi, S.; Sato, K.; Fukunaga, S.; Hayashi, H., *Bull. Chem. Soc. Jpn.* **1989**, *62*, 3757.

112. Pearlman, B. A.; McNamara, J. M.; Hasan, I.; Hatakeyama, S.; Sekizaki, H.; Kishi, Y., *J. Am. Chem. Soc.* **1981**, *103*, 4248.

113. Gawley, R. E.; Termine, E. J., *Synth. Commun.* **1982**, *12*, 15.

114. Mandal, A. K.; Soni, N. R.; Ratnam, K. R., *Synthesis* **1985**, 274.

115. Mandal, A. K.; Shrotri, P. Y.; Ghogare, A. D., *Synthesis* **1986**, 221.

116. Pelter, A.; Ward, R. S.; Venkateswarlu, R.; Kamakshi, C., *Tetrahedron* **1992**, *48*, 7209.

117. Kelly, D. R.; Roberts, S. M.; Newton, R. F., *Synth. Commun.* **1979**, *9*, 295.

118. Fang, J.-M.; Chen, M.-Y., *Tetrahedron Lett.* **1988**, *29*, 5939.

119. Preston, P. N.; Winwick, T.; Morley, J. O., *J. Chem. Soc., Perkin Trans. 1* **1983**, 1439.

120. Schiemenz, G. P.; Schmidt, U., *Liebigs Ann. Chem.* **1982**, 1509.

121. Doyle, M. P.; West, C. T.; Donnelly, S. J.; McOsker, C. C., *J. Organomet. Chem.* **1976**, *117*, 129.

122. Moriwake, T.; Hamano, S.; Miki, D.; Saito, S.; Torii, S., *Chem. Lett.* **1986**, 815.

123. Varma, R. S.; Kabalka, G. W., *Org. Prep. Proced. Int.* **1985**, *17*, 254.

124. Varma, R. S.; Kabalka, G. W., *Synth. Commun.* **1985**, *15*, 843.

125. Camacho, C.; Uribe, G.; Contreras, R., *Synthesis* **1982**, 1027.

126. Yanuka, Y.; Halperin, G., *J. Org. Chem.* **1973**, *38*, 2587.

127. Takeda, K.; Komeno, T.; Igarashi, K., *Chem. Pharm. Bull.* **1956**, *4*, 343.

128. Mandal, A. K.; Mahajan, S. W., *Tetrahedron Lett.* **1985**, *26*, 3863.

129. Vankar, Y. D.; Rao, C. T., *Tetrahedron Lett.* **1985**, *26*, 2717.

130. Trost, B. M.; Ippen, J.; Vladuchick, W. C., *J. Am. Chem. Soc.* **1977**, *99*, 8116.

131. Kelly, T. R.; Montury, M., *Tetrahedron Lett.* **1978**, 4311.

132. Stojanać, Z.; Dickinson, R. A.; Stojanác, N.; Woznow, R. J.; Valenta, Z., *Can. J. Chem.* **1975**, *53*, 616.

133. Das, J.; Kubela, R.; MacAlpine, G. A.; Stojanac, Z.; Valenta, Z., *Can. J. Chem.* **1979**, *57*, 3308.

134. Ohgaki, E.; Motoyoshiya, J.; Narita, S.; Kakurai, T.; Hayashi, S.; Hirakawa, K., *J. Chem. Soc., Perkin Trans. 1* **1990**, 3109.

135. Kotsuki, H.; Asao, K.; Ohnishi, H., *Bull. Chem. Soc. Jpn.* **1984**, *57*, 3339.

136. Gras, J.-L.; Guerin, A., *Tetrahedron Lett.* **1985**, *26*, 1781.

137. Alper, H.; Dinkes, L., *Synthesis* **1972**, 81.

138. Smith, A. B., III; Branca, S. J.; Toder, B. H., *Tetrahedron Lett.* **1975**, 4225.

139. Smith, A. B., III., *J. Chem. Soc., Chem. Commun.* **1975**, 274.

140. Smith, A. B., III, Toder, B. H.; Branca, S. J.; Dieter, R. K., *J. Am. Chem. Soc.* **1981**, *103*, 1996.

141. Okamoto, M.; Kimoto, S.; Oshima, T.; Kinomura, Y.; Kawasaki, K.; Yajima, H., *Chem. Pharm. Bull.* **1967**, *15*, 1618.

142. Hoff, S.; Blok, A. P., *Recl. Trav. Chim. Pays-Bas* **1973**, *92*, 631.

143. Newman, M. S.; Beal, P. F., III., *J. Am. Chem. Soc.* **1950**, *72*, 5161.

144. Anwar, S.; Davis, A. P., *Tetrahedron* **1988**, *44*, 3761.

145. Tomooka, K.; Matsuzawa, K.; Suzuki, K.; Tsuchihashi, G., *Tetrahedron Lett.* **1987**, *28*, 6339.

146. Reddy, P. S. N.; Reddy, P. P., *Synth. Commun.* **1988**, *18*, 2179.

147. Fang, J.-M.; Chen, M.-Y.; Yang, W.-J., *Tetrahedron Lett.* **1988**, *29*, 5937.

148. Hiroi, K.; Kitayama, R.; Sato, S., *Synthesis* **1983**, 1040.

149. Gajeweski, J. J.; Ngernmeesri, P., *Org. Lett.* **2000**, *2*, 2813.

150. Prasad, B. A. B.; Bisai, A.; Singh, V. K., *Tetrahedron Lett.* **2004**, *45*, 9565.

151. Kumar, V.; Chimni, S. S.; Kumar, S., *Tetrahedron Lett.* **2004**, *45*, 3409.

152. Winkler, J. D.; Oh, K.; Asselin, S. M., *Org. Lett.* **2005**, *7*, 387.

153. Tremblay-Morin, J.-P.; Raeppel, S.; Gaudette, F., *Tetrahedron Lett.* **2004**, *45*, 3471.

154. Li, S.-L.; Batey, R. A., *Chem. Commun.* **2004**, 1382.

155. Spanedda, M. V.; Ourévitch, M.; Crousse, B.; Béguéa, J.-P.; Bonnet-Delpon, D., *Tetrahedron Lett.* **2004**, *45*, 5023.

156. Jonet, S.; Cherouvrier, F.; Brigaud, T.; Portella, C., *Eur. J. Org. Chem.* **2005**, 4304.

157. Lautens, M.; Tayama, E.; Nguyen, D., *Tetrahedron Lett.* **2004**, *45*, 5131.

158. Chattopadhyay, A.; Salaskar, A., *Synthesis* **2000**, 561.

159. Shi, M.; Xu, B.; Huang, J.-W., *Org. Lett.* **2004**, *6*, 1175.

160. Shi, L.; Tu, Y.-Q.; Wang, M.; Zhang, F.-M.; Fan, C.-A.; Zhao, Y.-M.; Xia, W. J., *J. Am. Chem. Soc.* **2005**, *127*, 10836.

161. Gao, F.; Burnell, D. J., *J. Org. Chem.* **2006**, *71*, 356.

162. Walsh, L. M.; Winn, C. L.; Goodman, J. M., *Tetrahedron Lett.* **2002**, *43*, 8219.

163. Cooper, T. S.; Larigo, A. S.; Laurent, P.; Moody, C. J.; Takle, A. K., *Org. Biomol. Chem.* **2005**, *3*, 1252.

164. Oishi, S.; Niida, A.; Kamano, T.; Odagaki, Y.; Tamamura, H.; Otaka, A.; Hamanaka, N.; Fujii, N., *Org. Lett.* **2002**, *4*, 1055.

165. Oishi, S.; Kamano, T.; Niida, A.; Odagaki, Y.; Tamamura, H.; Otaka, A.; Hamanaka, N.; Fujii, N., *Org. Lett.* **2002**, *4*, 1051.

166. Miles, S. M.; Marsden, S. P.; Leatherbarrow, R. J.; Coates, W. J., *J. Org. Chem.* **2004**, *69*, 6874.

167. Deng, X.; Mani, N. S., *Tetrahedron: Asymmetry* **2005**, *16*, 661.

168. Vrancken, E.; Alexakis, A.; Mangeney, P., *Eur. J. Org. Chem.* **2005**, 1354.

169. Concellón, J. M.; Suárez, J. R.; García-Granda, S.; Díaz, M. R., *Org. Lett.* **2005**, *7*, 247.

170. Concellón, J. M.; Suárez, J. R.; del Solar, V., *J. Org. Chem.* **2005**, *70*, 7447.

171. Bergmeier, S. C.; Seth, P. P., *Tetrahedron Lett.* **1995**, *36*, 3793.

172. Bergmeier, S. C.; Seth, P. P., *J. Org. Chem.* **1999**, *64*, 3237.

173. Bergmeier, S. C.; Katz, S. J.; Huang, J.; McPherson, H.; Donoghue, P. J.; Reed, D. D., *Tetrahedron Lett.* **2004**, *45*, 5011.

174. Hojo, M.; Ushioda, N.; Hosomi, A., *Tetrahedron Lett.* **2004**, *45*, 4499.

175. Chen, B.-F.; Tasi, M.-R.; Yang, C.-Y.; Chang, J.-K.; Chang, N.-C., *Tetrahedron* **2004**, *60*, 10223.

176. Hioki, H.; Okuda, M.; Miyagi, W.; Itôa, S., *Tetrahedron Lett.* **1993**, *34*, 6131.

177. Ishikawa, T.; Manabe, S.; Aikawa, T.; Kudo, T.; Saito, S., *Org. Lett.* **2004**, *6*, 2361.

178. Harrowven, D. C.; Tyte, M. J., *Tetrahedron Lett.* **2004**, *45*, 2089.

179. Jobashi, T.; Hino, T.; Maeyama, K.; Ozaki, H.; Ogino, K.; Yonezawa, N., *Chem. Lett.* **2005**, *34*, 860.

180. Kim, S.; Yang, J.; DiNinno, F., *Tetrahedron Lett.* **1999**, *40*, 2909.

181. Miyake, H.; Hirai, R.; Nakajima, Y.; Sasaki, M., *Chem. Lett.* **2003**, *32*, 164.

182. Zheng, T.; Narayan, R. S.; Schomaker, J. M.; Borhan, B., *J. Am. Chem. Soc.* **2005**, *127*, 6946.

183. Shimada, K.; Kikuta, Y.; Koganebuchi, H.; Yonezawa, F.; Aoyagi, S.; Takikawa, Y., *Tetrahedron Lett.* **2000**, *41*, 4637.

184. Hirano, K.; Fujita, K.; Yorimitsu, H.; Shinokubo, H.; Oshima, K., *Tetrahedron Lett.* **2004**, *45*, 2555.

185. Hirano, K.; Yorimitsu, H.; Oshima, K., *Org. Lett.* **2004**, *6*, 4873.

186. Van Leeuwen, S. H.; Quaedflieg, P. J. L. M.; Broxterman, Q. B.; Milhajlovic, Y.; Liskamp, R. M. J., *Tetrahedron Lett.* **2005**, *46*, 653.

187. Crisóstomo, F. R. P.; Martín, T.; Martín, V. S., *Org. Lett.* **2004**, *6*, 565.

188. Díaz, D.; Martín, V. S., *Org. Lett.* **2000**, *2*, 335.

189. Guo, R.; Green, J. R., *Chem. Commun.* **1999**, 2503.

190. Vasse, J. L.; Joosten, A.; Denhez, C.; Szymoniak, J., *Org. Lett.* **2005**, *7*, 4887.

191. Saeki, T.; Son, E.-C.; Tamao, K., *Org. Lett.* **2004**, *6*, 617.

192. Barbero, A.; Castreño, P.; Fernández, G.; Pulido, F. J., *J. Org. Chem.* **2005**, *70*, 10747.

193. de los Rios, C.; Hegedus, L. S., *J. Org. Chem.* **2005**, *70*, 6541.

194. Vargas-Sanchez, M.; Couty, F.; Evano, G.; Prim, D.; Marrot, J., *Org. Lett.* **2005**, *7*, 5861.

195. Chiu, P.; Li, S., *Org. Lett.* **2004**, *6*, 613.

196. Giese, S.; West, F. G., *Tetrahedron Lett.* **1998**, *39*, 8393.

197. Bender, J. A.; Blize, A. E.; Browder, C. C.; Giese, S.; West, F. G., *J. Org. Chem.* **1998**, *63*, 2430.

198. Giese, S.; Mazzola Jr., R. D.; Amann, C. M.; Arif, A. M.; West, F. G., *Angew. Chem. Int. Ed.* **2005**, *44*, 6546.

199. Buffet, M. F.; Dixon, D. J.; Ley, S. V.; Reynolds, D. J.; Storer, R. I., *Org. Biomol. Chem.* **2004**, *2*, 1145.

200. Priéa, G.; Prévost, N.; Twin, H.; Fernandes, S. A.; Hayes, J. F.; Shipman, M., *Angew. Chem. Int. Ed.* **2004**, *43*, 6517.

201. Rajamaki, S.; Kilburn, J. D., *Chem. Commun.* **2005**, 1637.

202. Tamura, O.; Mitsuya, T.; Ishibashi, H., *Chem. Commun.* **2002**, 1128.

203. Lu, J.; Mao, G.; Zhang, W.; Xi, Z., *Chem. Commun.* **2005**, 4848.

204. Inoue, S.; Wang, P.; Nagao, M.; Hoshino, Y.; Honda, K., *Synlett* **2005**, 469.

205. Chung, W.-J.; Kim, D.-K.; Lee, Y.-S., *Synlett* **2005**, 2175.

206. Tohma, H.; Morioka, H.; Takizawa, S.; Arisawa, M.; Kita, Y., *Tetrahedron* **2001**, *57*, 345.

207. Tohma, H.; Iwata, M.; Maegawa, T.; Kiyono, Y.; Maruyama, A.; Kita, Y., *Org. Biomol. Chem.* **2003**, *1*, 1647.

208. Tohma, H.; Iwata, M.; Maegawa, T.; Kita, Y., *Tetrahedron Lett.* **2002**, *43*, 9241.

209. Ochiai, M.; Takeuchi, Y.; Katayama, T.; Sueda, T.; Miyamoto, K., *J. Am. Chem. Soc.* **2005**, *127*, 12244.

210. An, G.-I.; Kim, M.; Kim, J. Y.; Rhee, H., *Tetrahedron Lett.* **2003**, *44*, 2183.

211. An, G. -I.; Rhee, H., *Synlett* **2003**, 876.

212. Kim, S. Y.; An, G.-I.; Rhee, H., *Synlett* **2003**, 112.

213. Ueda, M.; Miyabe, H.; Namba, M.; Nakabayashi, T.; Naito, T., *Tetrahedron Lett.* **2002**, *43*, 4371.

214. Ortiz-Marciales, M.; Rivera, L. D.; De Jesus, M.; Espinosa, S.; Benjamin, J. A.; Casanova, O. E.; Figueroa, I. G.; Rodriguez, S.; Correa, W., *J. Org. Chem.* **2005**, *70*, 10132.

215. Cho, S.-D.; Park, Y.-D.; Kim, J.-J.; Falck, J. R.; Yoonm, Y.-J., *Bull. Kor. Chem. Soc.* **2004**, *25*, 407.

216. Pastine, S. J.; McQuaid, K. M.; Sames, D., *J. Am. Chem. Soc.* **2005**, *127*, 12180.

217. Pastine, S. J.; Sames, D., *Org. Lett.* **2005**, *7*, 5429.

218. Islas-González, G.; Puigjaner, C.; Vidal-Ferran, A.; Moyano, A.; Riera, A.; Pericàs, M. A., *Tetrahedron Lett.* **2004**, *45*, 6337.

219. Ding, C.-H.; Dai, L.-X.; Hou, X.-L., *Synlett* **2004**, 2218.

220. De Melo, G.; Morgan, J.; Pinhey, J. T., *Tetrahedron* **1993**, *49*, 8129.

221. Kreidler, B.; Baro, A.; Frey, W.; Christoffers, J., *Chem. Eur. J.* **2005**, *11*, 2660.

222. Rhee, J. U.; Krische, M. J., *Org. Lett.* **2005**, *7*, 2493.

N-Bromosuccinimide[1]

[128-08-5]　　　　　C$_4$H$_4$BrNO$_2$　　　　(MW 177.99)
InChI = 1/C4H4BrNO2/c5-6-3(7)1-2-4(6)8/h1-2H2
InChIKey = PCLIMKBDDGJMGD-UHFFFAOYAS

(radical bromination of allylic and benzylic positions; electrophilic bromination of ketones, aromatic and hetero-cyclic compounds; bromohydration, bromoetherification, and bromolactonization of alkenes)

Alternative Names: NBS; 1-bromo-2,5-pyrrolidinedione.
Physical Data: mp 173–175 °C (dec); *d* 2.098 g cm^{-3}.
Solubility: sol acetone, THF, DMF, DMSO, MeCN; slightly sol H$_2$O, AcOH; insol ether, hexane, CCl$_4$ (at 25 °C).
Form Supplied in: white powder or crystals having a faint odor of bromine when pure; widely available.
Purification: in many applications the use of unrecrystallized material has led to erratic results. Material stored for extended periods often contains significant amounts of molecular bromine and is easily purified by recrystallization from H$_2$O (AcOH has also been used). In an efficient fume hood (caution: bromine evolution), an impure sample of NBS (200 g) is dissolved as quickly as possible in 2.5 L of preheated water at 90–95 °C. As filtration is usually unnecessary, the solution is then chilled well in an ice bath to effect crystallization. After most of the aqueous portion has been decanted, the white crystals are collected by filtration through a bed of ice and washed well with water. The crystals are dried on the filter and then in vacuo. The purity of NBS may be determined by the standard iodide–thiosulfate titration method.
Handling, Storage, and Precautions: should be stored in a refrigerator and protected from moisture to avoid decomposition. One of the advantages of using NBS is that it is easier and safer to handle than bromine; however, the solid is an irritant and bromine may be released during some operations. Therefore, precautions should be taken to avoid inhalation of the powder and contact with skin. All operations with this reagent are best conducted in an efficient fume hood. In addition, since reactions involving NBS are generally quite exothermic, large-scale operations (>0.1 mol) should be approached with particular caution.

Original Commentary

Scott C. Virgil
Massachusetts Institute of Technology, Cambridge, MA, USA

Introduction. *N*-Bromosuccinimide is a convenient source of bromine for both radical substitution and electrophilic addition reactions. For radical substitution reactions, NBS has several advantages over the use of molecular **Bromine**, while **1,3-Dibromo-5,5-dimethylhydantoin** is another reagent of use. **N-Chlorosuccinimide** and **N-Iodosuccinimide** generally do not facilitate analogous substitution reactions. For electrophilic substitutions, **Bromine**, **N-Bromoacetamide**, **Bromonium Di-sym-collidine Perchlorate**, **1,3-Dibromoisocyanuric Acid**, and **2,4,4,6-Tetrabromo-2,5-cyclohexadienone** also have applicability and the analogous halogenation reactions are generally possible using NCS, NIS, and I$_2$. Possible impurities generated during NBS brominations include conjugates of succinimide and, if basic conditions are employed, β-alanine (formed by the Hofmann reaction) and its derivatives may be isolated.

Allylic Bromination of Alkenes.[2] Standard conditions for allylic bromination involve refluxing of a solution of the alkene and recrystallized NBS in anhydrous CCl$_4$ using **Dibenzoyl Peroxide**, irradiation with visible light (ordinary 100 W light bulb or sunlamp[3]), or both to effect initiation. Both NBS and the co-product succinimide are insoluble in CCl$_4$ and succinimide collects at the surface of the reaction mixture as the reaction proceeds.[4] High levels of regioselectivity operate during the hydrogen-abstraction step of the chain mechanism, such that allylic methylene groups are attacked much more rapidly than allylic methyl groups.[5] However, a thermodynamic mixture of allylic bromides is generally isolated since both the allylic radical and the allylic bromide are subject to isomerization under the reaction conditions.[6] High levels of functional group selectivity are characteristic of this reaction, for example alkenic esters may be converted to allylic bromides prior to intramolecular cyclization (eq 1).[7] Brominations of α,β-unsaturated esters (eq 2)[8] and lactones (eq 3) are also successful.[9]

$$\text{(1)}$$

$$\text{(2)}$$

$$\text{(3)}$$

Benzylic Bromination of Aromatic Compounds. Using the conditions described above, NBS also effects the bromination of benzylic positions.[10] Bromine is also regularly used for benzylic bromination (eq 4);[11] however, many functional groups are sensitive to the generation of HBr during the reaction, including carbonyl groups which suffer competing acid-catalyzed bromination. These considerations render NBS as the reagent of choice for bromination of polyfunctional aromatic compounds. Selectivity can be anticipated with polyfunctional molecules based on the predicted stabilities of the radical intermediates (eq 5).[12] Accordingly, the use of NBS allows the bromination of alkyl

Essential Reactions for Organic Synthesis, First Edition. Edited by Philip L. Fuchs.
© 2016 John Wiley & Sons, Ltd. Published 2016 by John Wiley & Sons, Ltd.

groups attached to sensitive heterocyclic compounds (eq 6).[13] Complications which may arise from this method include *gem*-dibromination (eq 7)[14] of methyl substituents as well as in situ elimination of the product benzylic bromide (see also *1,3-Dibromo-5,5-dimethylhydantoin*).

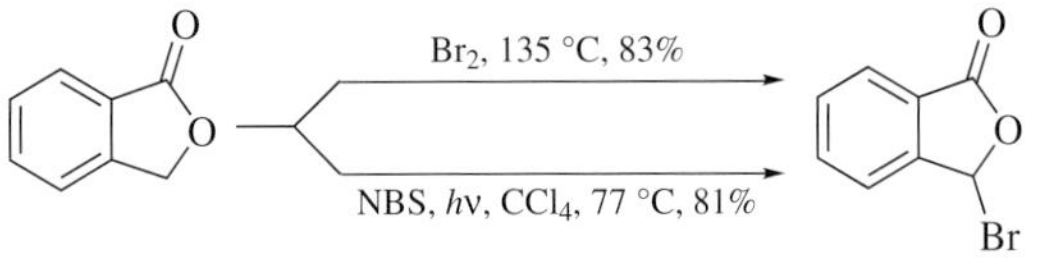

$$\text{(4)}$$

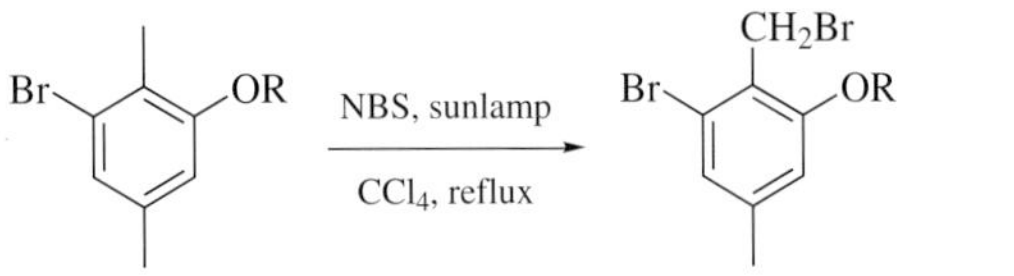

$$\text{(5)}$$

R = Me, 82%
R = TBDMS, 60%

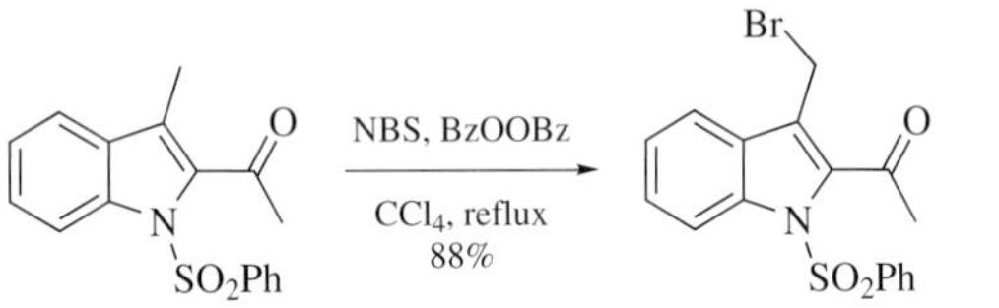

$$\text{(6)}$$

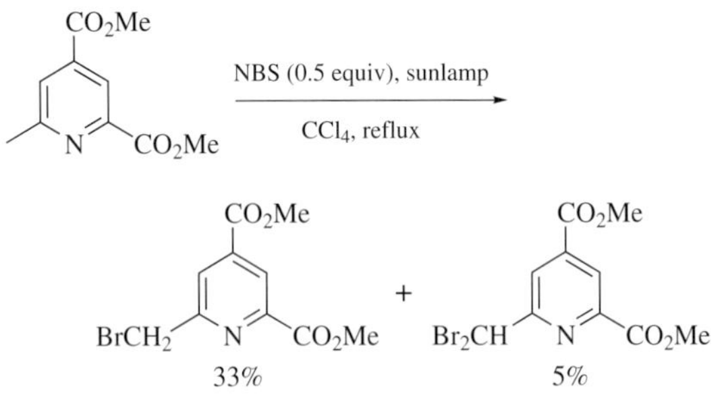

$$\text{(7)}$$

BrCH$_2$ 33%   +   Br$_2$CH 5%

The regioselective cleavage of benzylidene acetals using NBS has been used widely in the synthesis of natural products from carbohydrates (eq 8)[15] and other chiral materials (eq 9).[16] It is rather important that the reaction be conducted in anhydrous CCl$_4$ (passage through activated alumina is sufficient), since in the presence of water the hydroxy benzoate is formed.[17] Barium carbonate is generally added to maintain anhydrous and acid-free conditions, and the addition of Cl$_2$CHCHCl$_2$ often improves solubility of the substrate. Selectivity is usually very high in cases in which a primary bromide can be produced, but may also be obtained in systems such as shown in eq 10.[18] As alkoxy substituents serve to further stabilize the adjacent radicals, these reactions proceed with high selectivity in the presence of other functional groups. Other applications in the carbohydrate field include the cleavage of benzyl ethers and benzyl glycosides (to the corresponding glycosyl bromides) and the bromination of pyranoses in the 5-position.[19]

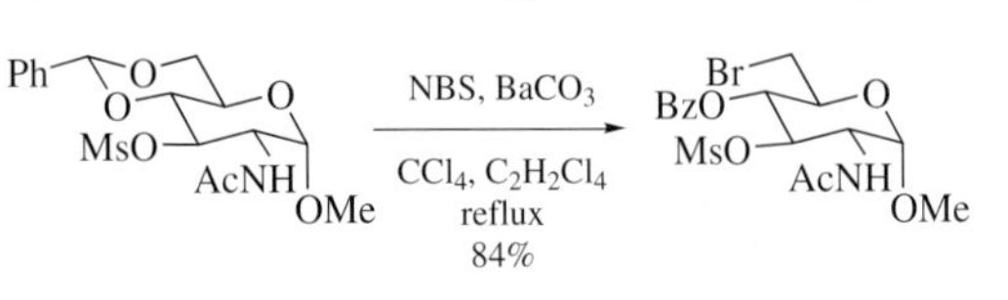

$$\text{(8)}$$

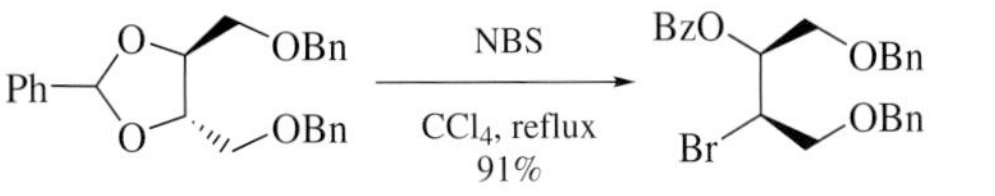

$$\text{(9)}$$

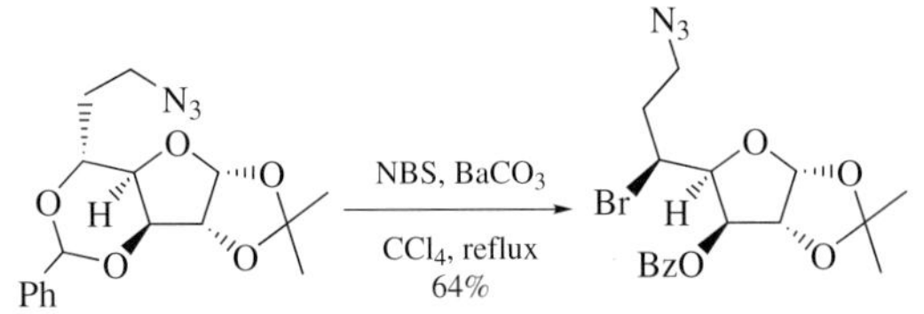

$$\text{(10)}$$

Unsaturation and Aromatization Reactions.[20] Unsaturated aldehydes, esters, and lactones can be accessed via strategies involving radical bromination and subsequent elimination. The allylic bromination of unsaturated lactones may be followed by elimination with base to obtain dienoic and trienoic lactones (eqs 11 and 12).[21] Conversion of an aldehyde to the enol acetate allows the radical bromination at the C$_\beta$ position to proceed smoothly and, upon ester hydrolysis, the α,β-unsaturated aldehyde is obtained (eq 13).[22]

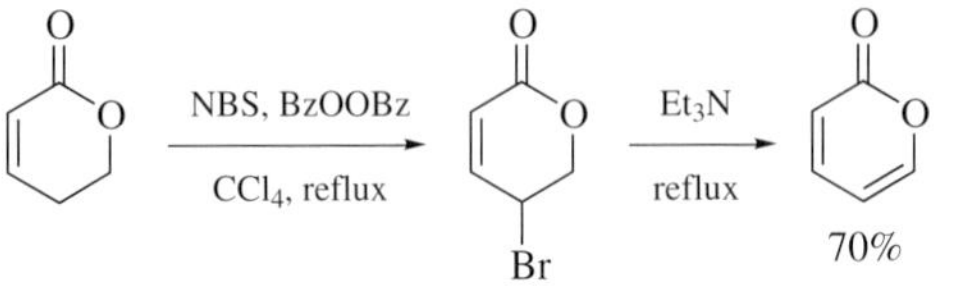

$$\text{(11)}$$

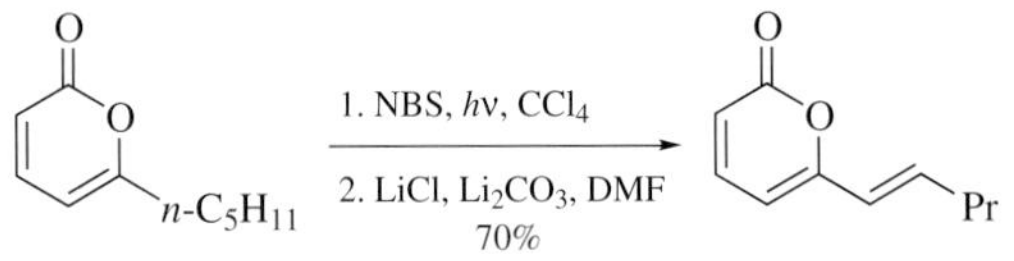

$$\text{(12)}$$

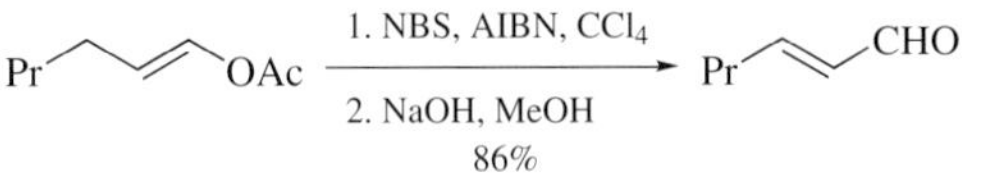

$$\text{(13)}$$

The direct bromination of β-alkoxylactones at the β position initially generates the α,β-unsaturated lactones (eq 14); however, the required radical abstraction is not so facile and further bromination of the α,β-unsaturated lactone proceeds competitively to afford the mono- and dibrominated products.[23] NBS is also used for the oxidative aromatization of polycyclic compounds, including steroids and anthraquinone precursors (eq 15).[24]

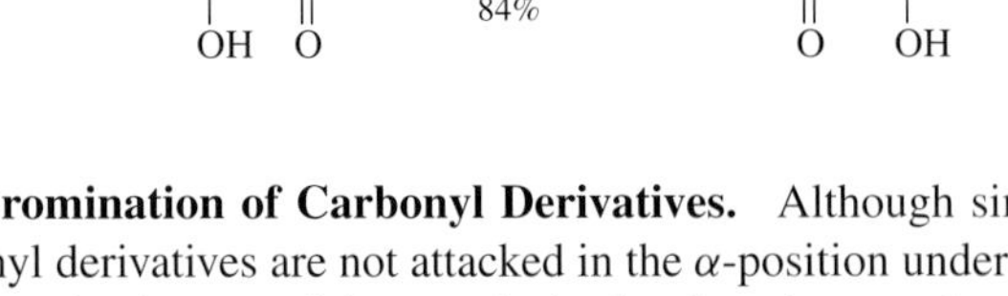

$$\text{(14)}$$

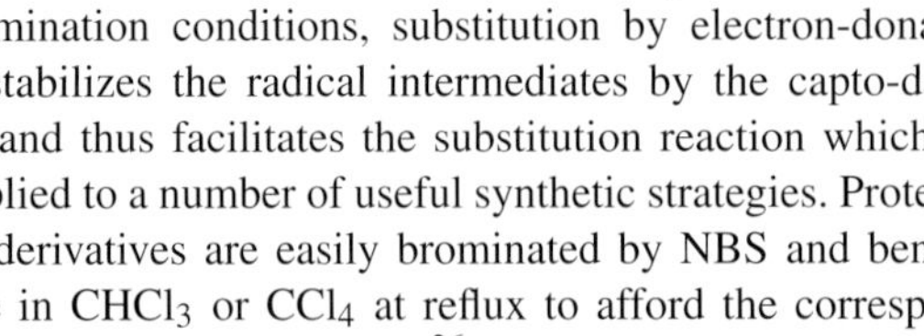

X = H and X = Br

$$\text{(15)}$$

α-Bromination of Carbonyl Derivatives. Although simple carbonyl derivatives are not attacked in the α-position under radical bromination conditions, substitution by electron-donating groups stabilizes the radical intermediates by the capto-dative effect[25] and thus facilitates the substitution reaction which has been applied to a number of useful synthetic strategies. Protected glycine derivatives are easily brominated by NBS and benzoyl peroxide in CHCl$_3$ or CCl$_4$ at reflux to afford the corresponding α-bromoglycine derivatives.[26] These compounds are stable precursors of *N*-acyliminoacetates, which may be alkylated by silyl enol ethers in the presence of Lewis acids, organometallic reagents, and other nucleophiles to afford novel α-amino acids (eq 16).[27] Diketopiperazines and related heterocycles are also substituted in good yields (eq 17).[28] Furthermore, in contrast to aldehydes which undergo abstraction of the aldehydic hydrogen (see below), *O*-trimethylsilylaldoximes are readily brominated at the α-position under radical bromination conditions and can be converted to substituted nitrile oxides (*O*-trimethylsilylketoximes react similarly).[29]

(16)

(17)

The use of NBS in the presence of catalytic **Hydrogen Bromide** has proven to be more convenient than Br_2 for the conversion of acid chlorides to α-bromo acid chlorides.[30] The reaction of the corresponding enolates, enol ethers, or enol acetates with NBS (and other halogenating agents) offers considerable advantages over direct acid-catalyzed halogenation of ketones and esters.[31] Although both reagents may afford the α-brominated products in high yields, NBS is more compatible than is bromine with sensitive functional groups and has been used in the asymmetric synthesis of α-amino acids.[32] The bromination of cyanoacetic acid proceeds rapidly with NBS to afford dibromoacetonitrile[33] and, similarly, β-keto esters, β-diketones, and β-sulfonyl ketones may be reacted with NBS in the presence of base to afford the products of bromination and in situ deacylation (see **N-Chlorosuccinimide**).[34] (5*E*)-Bromovinyluridine derivatives are readily prepared by bromodecarboxylation of the corresponding α,β-unsaturated acids with NBS (eq 18).[35]

(18)

R = 2-deoxyribosyl

Reaction with Vinylic and Alkynic Derivatives. NBS is a suitable source of bromine for the conversion of vinylcopper and other organometallic derivatives to the corresponding vinyl bromides.[36] Vinylsilanes, prepared from the corresponding 1-trimethylsilylalkyne by reduction with **Diisobutylaluminum Hydride**, can be isomerized from the (*Z*) to the (*E*) geometry by irradiation with NBS and **Pyridine**, thus making (*E*)-vinylsilanes readily available stereoselectively in three steps from the corresponding alkyne (eq 19).[37] Allylsilane can be brominated by NBS under radical conditions, whereas more reactive allylsilanes are bromodesilated by NBS in CH_2Cl_2 at $-78\,°C$.[38] 1-Bromoalkynes can be prepared under mild conditions by reaction with NBS in acetone in the presence of catalytic **Silver(I) Nitrate**.[39]

(19)

Bromination of Aromatic Compounds. Phenols, anilines, and other electron-rich aromatic compounds can be monobrominated using NBS in DMF with higher yields and higher levels of *para* selectivity than with Br_2.[40] *N*-Trimethylsilylanilines and aromatic ethers are also selectively brominated by NBS in $CHCl_3$ or CCl_4.[41] *N*-Substituted pyrroles are brominated with NBS in THF to afford 2-bromopyrroles (1 equiv) or 2,5-dibromopyrroles (2 equiv) with high selectivity, whereas bromination with Br_2 affords the thermodynamically more stable 3-bromopyrroles.[42] The use of NBS in DMF also achieves the controlled bromination of imidazole and nitroimidazole.[43] Thiophenes are also selectively brominated in the 2-position using NBS in acetic acid–chloroform.[44]

Bromohydration, Bromolactonization, and Other Additions to C=C.[45] The preferred conditions for the bromohydration of alkenes involves the portionwise addition of solid or predissolved NBS (recrystallized) to a solution of the alkene in 50–75% aqueous DME, THF, or *t*-butanol at $0\,°C$. The formation of dibromide and α-bromo ketone byproducts can be minimized by using recrystallized NBS. High selectivity for Markovnikov addition and *anti* stereochemistry results from attack of the bromonium ion intermediate by water. Aqueous DMSO can also be used as the solvent; however, since DMSO is readily oxidized under the reaction conditions, significant amounts of the dibromide byproduct may be produced.[46,47] In the bromohydration of polyalkenic compounds, high selectivity is regularly achieved for attack of the most electron-rich double bond (eq 20).[48] With farnesol acetate, squalene, and other polyisoprenes, choice of the optimum proportion of water is used to effect the selective bromohydration at the terminal double bond (eq 21),[49] and the two-step sequence shown is often the method of choice for the preparation of the corresponding epoxides.[50]

(20)

(21)

Bromoetherification of alkenes can be achieved using NBS in the desired alcohol as the solvent. The reaction of 1,3-dichloropropene with NBS in methanol yields an α-bromo dimethyl acetal in the first step in a convenient synthesis of cyclopropenone.[51] Using propargyl alcohol the reaction depicted in eq 22 has been extended to an annulation method for the synthesis of α-methylene-γ-butyrolactones.[52] Intramolecular bromoetherification and bromoamination reactions are generally very facile (eq 23).[53] In natural products synthesis, bromoetherification has been used for the synthesis of cyclic ethers (by subsequent debromination, see

Tributylstannane) and for the protection of alkene appendages as cyclic bromoethers (regenerated by reaction with zinc).[54]

(22)

(23)

Ar = p-MeOC$_6$H$_4$

NBS is also an effective reagent for bromolactonization of unsaturated acids and acid derivatives with the same high stereo- and Markovnikov selectivity (see also *Iodine*). Dienes, such as the cycloheptadiene derivative shown, may react exclusively via *syn*-1,4-addition (eq 24).[55] Alkynic acids are converted to the (E)-bromo enol lactones by NBS in a biphasic medium, whereas the combination of bromine and silver nitrate afford the (Z)-bromo enol lactones (eq 25).[56] α,β-Unsaturated acylprolines react with NBS in anhydrous DMF to afford the corresponding bromolactones having diastereomeric excesses up to 93%, which can be converted to chiral α-hydroxy acids by debromination followed by acidic hydrolysis (eq 26).[57] In contrast to alkenic amides, which generally react with NBS to afford bromolactones (via the cyclic iminoether derivatives), alkenic sulfonamides readily undergo cyclization on nitrogen when reacted with NBS to afford the bromosulfonamides in high yields.[58] *N*-Methoxyamides have also proven effective for bromolactamization, leading to diketopiperazines (eq 27)[59] (see also *Bromonium Di-sym-collidine Perchlorate*).

(24)

(25)

(26)

89% de

(27)

Addition of NBS to an alkene in the presence of aqueous *Sodium Azide* affords fair yields of the corresponding β-bromo-

azides, which can be converted by *Lithium Aluminum Hydride* reduction to aziridines.[60] Intermolecular reactions of alkenes with NBS and weaker nucleophiles can be achieved if conducted under anhydrous conditions to avoid the facile bromohydration reaction. In this manner, bromofluorination of alkenes has been extensively studied using *Pyridinium Poly(hydrogen fluoride)*, triethylammonium dihydrogentrifluoride or tetrabutylammonium hydrogendifluoride as the fluoride ion source.[61]

Oxidation and Bromination of Other Functional Groups. Conjugate bases of other functional groups can be α-brominated with NBS. Nitronate anions of aliphatic nitro compounds react with NBS to afford the *gem*-bromonitro compounds in high yield.[62] The α-bromination of sulfoxides can be performed in the presence of pyridine and proceeds more satisfactorily using NBS in the presence of catalytic Br$_2$ than with either reagent alone.[63] NBS also reacts with sulfides to afford sulfoxides when methanol is used as a solvent, or to form α-bromo sulfides in anhydrous solvents.[64] NBS is a favored reagent for the deprotection of dithianes and dithioacetals to regenerate carbonyl groups (eq 28)[65] (see also *N-Chlorosuccinimide* and *1,3-Diiodo-5,5-dimethylhydantoin*).

(28)

In polar media, NBS effectively oxidizes primary and secondary alcohols to carbonyl compounds via hypobromite or alkoxysuccinimide intermediates. Although this transformation is more commonly effected by the use of chromium reagents or activated *Dimethyl Sulfoxide*, the most notable application of NBS and related reagents lies in its selectivity for the oxidation of axial vs. equatorial hydroxy groups in steroid systems (see *N-Bromoacetamide*).[66] Often, a single secondary alcohol may be converted to the ketone in the presence of many other alcohol groups.

Under radical conditions, aldehydes are readily oxidized by NBS to acid bromides.[67] The oxidation of aldoximes to nitrile oxides using NBS and *Triethylamine* in DMF is superior to the use of aqueous hypochlorite.[68] Tosylhydrazones are cleaved by reaction with NBS in methanol,[69] and hydrazines and hydrazides are oxidized to azo compounds.[70]

First Update

P. R. Jenkins, A. J. Wilson & M. D. García Romero
University of Leicester, Leicester, UK

Allylic Bromination of Alkenes. The alcohols (**1**) were converted into the rearranged primary allylic bromides (**2**) via S$_N$2′ displacement by treatment with NBS/Me$_2$S (eq 29).[71] A well researched procedure for the allylic bromination of 1,5-cyclooctadiene has also appeared.[72] NBS and water react with allylic ethers to regenerate alcohols.[73]

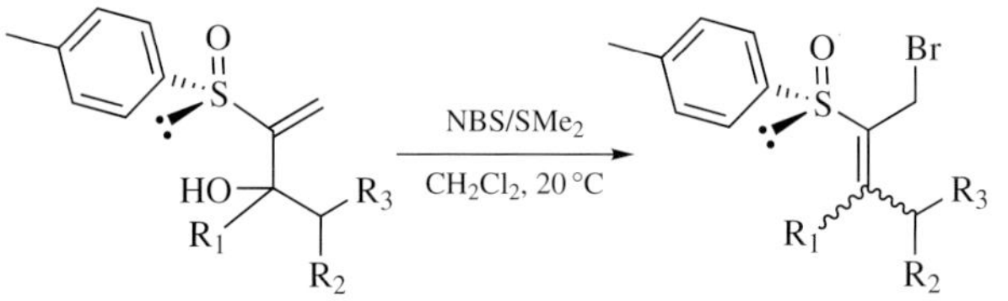

(29)

1

R_1 = Me, *i*-Pr, Ph   2
R_2 = H, Me
R_3 = Me, H

Bromination of Cyclopropanes. NBS gives bromination of donor-acceptor cyclopropanes by an electron-transfer (ET) mechanism (eq 30).[74]

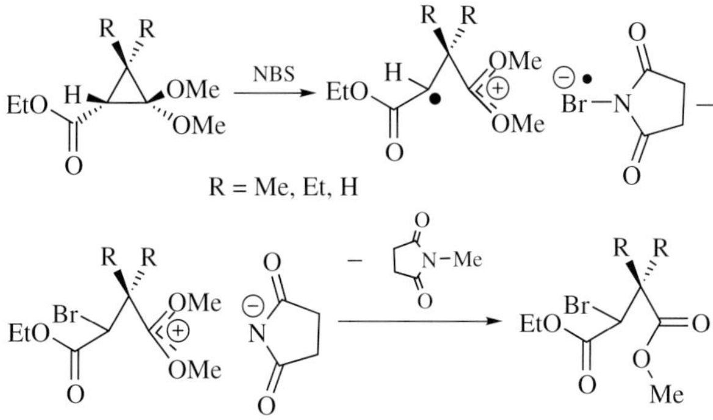

R = Me, Et, H

(30)

Benzylic Bromination of Aromatic Compounds. An efficient and fast microwave-assisted method for the preparation of benzylic bromides has appeared.[75] The 2-trimethylsilylethyl substituent on the benzenoid ring of **3** undergoes benzylic bromination followed by elimination of Me₃SiBr and addition of bromine to produce the dibromo compound (**4**) (eq 31). The ketone (**5**) is also observed from the hydrolysis of (**4**).[76]

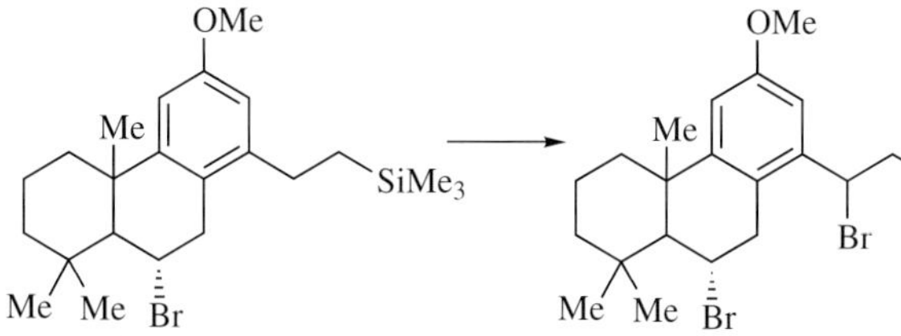

(31)

3   4

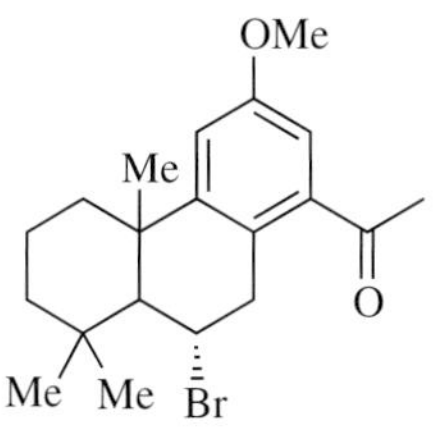

5

α-Bromination of Carbonyl Derivatives. Reaction of a complex silyl enol ether with NBS leads to an α-bromo ketone in the Ogasawara synthesis of (−)-morphine.[77] Amberlyst-15® promotes the bromination of 1,3-keto esters and cyclic ketones with NBS.[78] α-Bromination of carbonyl compounds has been achieved using NBS in the presence of silica-supported sodium hydrogen

sulfate as a heterogeneous catalyst.[79] *C*-Alkylation of Meldrum's acid is possible using triphenylphosphine and NBS (eq 32).[80]

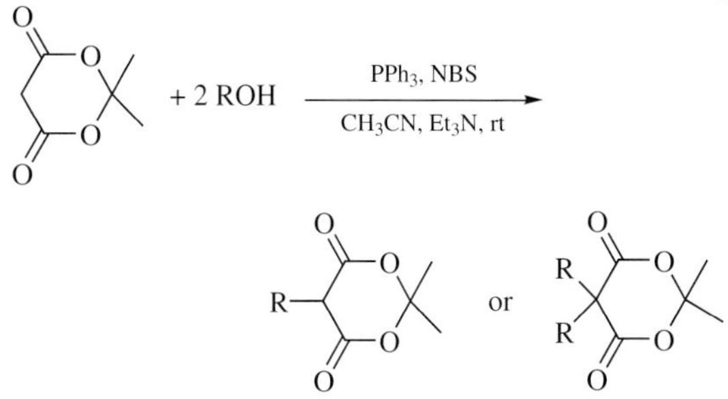

or   (32)

A process of selenocatalytic α-halogenation using NBS has been reported.[81] A catalytic enantioselective bromination of β-keto esters has been achieved using a combination of NBS and TiCl₂(TADDOLato) complexes as enantioselective catalyst;[82] modest enantiomeric excesses were obtained.

Decarboxylation. Bromodecarboxylation (Hunsdiecker reaction) of α,β-unsaturated carboxylic acids was achieved employing IBD or IBDA as catalysis.[83] Manganese(II) acetate[84] and lithium acetate[85] (eq 33) can also catalyze this kind of reaction.

On the other hand, a slight modification of the latter reaction protocol can be employed for the synthesis of α-bromo-β-lactams when the starting material is a α,β-unsaturated aromatic amide, with catalysis by NaOAc instead of LiOAc (eq 33).[86]

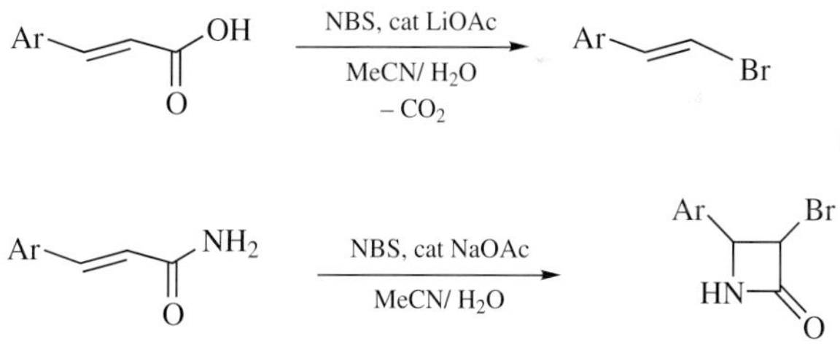

(33)

Reaction with Vinylic and Alkynic Derivatives. Vinylic boronic acids are converted with good yields to alkenyl bromides, keeping the same geometry, by treatment with NBS (eq 34).[87]

(34)

Propiolates can be brominated with or without decarboxylation.[88,89]

Bromination of Aromatic Compounds. Studies on the bromination of monocyclic and polycyclic aromatic compounds with NBS have continued[90,91] and in particular the bromination of phenols and naphthols has received attention,[92,93] e.g., the conversion of **6** into **7** (eq 35).

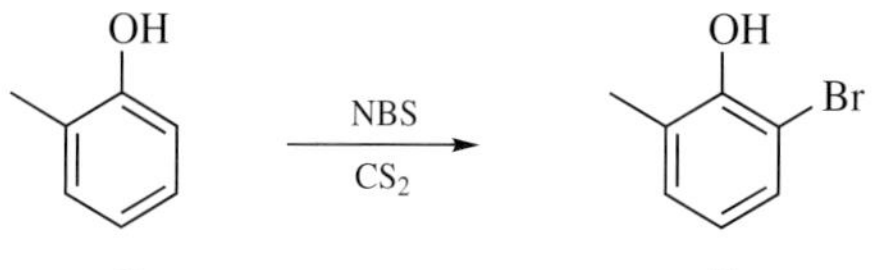

(35)

6   7

Aromatic bromination is also achieved using NBS,[94] in some cases using strong acids as catalysts.[95] Deactivated aromatic compounds are brominated by NBS in trifluoroacetic acid and sulphuric acid.[96] NBS and aqueous sodium hydroxide is used to brominate activated benzoic acid derivatives.[97] An intriguing effect of lithium perchlorate dispersed on silica gel on the bromination of aromatic compounds with NBS has been reported.[98] Finally a method for the *ipso*-substitution of phenyl boronic acids (**8**) with NBS leading to the aromatic bromides (**9**) has appeared (eq 36).[99]

$$(36)$$

Heterocyclic Bromination. Pyridines with electron-donating groups undergo regioselective bromination with NBS under mild acidic conditions as shown by the conversion of **10** into **11** (eq 37).[100]

$$(37)$$

A range of "pyridine-type" hydroxyl heterocycles are brominated effectively with NBS/PPh₃[101] while NBS is used as a synthesis of pyridines.[102] Polysubstituted pyrroles,[103] furans,[104] pyrrolidin-2-ones,[105] thiophenes,[106] and 3,4-disubstituted indoles[107] have also been prepared using NBS as a key reagent. 3-Methyl indole derivatives of general structure **12** are brominated in the methyl group to **13** with NBS under radical conditions. Under ionic conditions bromination occurs at the 2-position of the indole structure (**12**) to give products with general structure **14** (eq 38).[108]

$$(38)$$

NBS is used as a reagent for phenylselenyl activation in a route to aziridines and oxazolidin-2-ones.[109] The synthesis of 5-bromoisoquinoline and 5-bromo-8-nitroisoquinoline has been achieved using NBS.[110] 3-Bromo-*N*-methylpyrrole can be obtained from *N*-methylpyrrole by the use of NBS and a catalytic amount of PBr₃.[111] A new synthetic route to indoloquinones has appeared in which 2-methoxy-2*H*-azepine derivatives react with NBS to form 3*H*-azepines.[112] Convenient methods for the bromination of 3,5-diarylisoxazoles[113] and for the synthesis of

3-halogeno-1-methylpyridazino[3,4-*b*]quinoxalin-4(1*H*)-ones[114] using NBS have appeared.

Purine derivative (**15**) undegoes regioselective bromination with NBS in DMF to give the brominated product (**16**) (eq 39).[115]

$$(39)$$

N-Thiosuccinimide Formation. The reagent **17** is prepared from NBS (eq 40) and is very useful in the synthesis of cyanoethyl-protected nucleotides due to its solubility in pyridine. It is also used in the selective reactions of *H*-phosphonate derivatives.[116]

$$(40)$$

Acetal Bromination and Formation. The bromination of an acetal by NBS under radical conditions does not require the presence of an aromatic group (eq 41).[117]

$$(41)$$

NBS can also be used to make acetals: the reaction of *para*-chlorobenzaldehyde, NBS and PPh₃ produces a reagent which forms an acetal with 1,2-*O*-isopropylidene-α-D-xylofuranose (eq 42).[118]

$$(42)$$

1,2,4-Trioxones are produced by reaction of aldehydes with allylic peroxide (**18**) (eq 43); yields are in the range 25–35%, when R = Me, Et and Pr.[119]

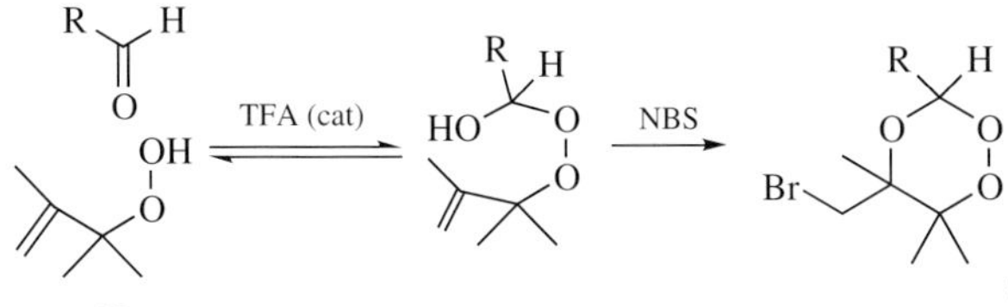

(43)

In the carbohydrate area, two important uses of the reagent have appeared: one uses NBS-Me₃SiOTf as the promoter for the glycosidic bond formation and simultaneous bromination of an activated aryl aglycon.[120] In the second, the synthesis of branched polysaccharides by polymerization of 6-*O*-*t*-butyldimethylsilyl-D-glucal through stereoregular bromoglycosylation was achieved by the use of NBS.[121]

NBS is a chemoselective catalyst for the acetalization of carbonyl compounds using triethyl orthoformate under almost neutral conditions (eq 44).[122,123]

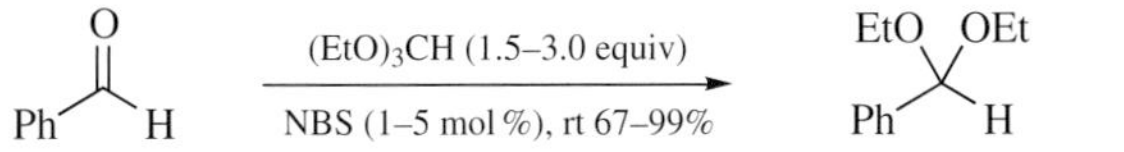

(44)

NBS is an effective catalyst for the acetalation of alcohols under mild conditions:[124] aldehydes are converted to 1,1-diacetates by reaction of acetic anhydride with NBS as a catalyst.[125]

Reactions of Thioacetals.   The ring expansion of aromatic thioacetals can be achieved using NBS: initial bromination α to the thioacetal is followed by ring expansion and proton transfer (eq 45).[126]

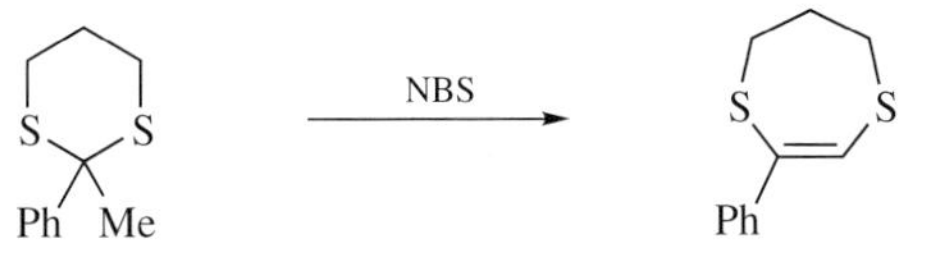

(45)

The use of NBS as an alternative for HgCl₂ in the deprotection of 2-silyl-1,3-dithianes into the corresponding acylsilanes has been investigated;[127] trithioorthoesters are converted to α-oxo thiolcarboxylates.[128] Sulfoxides are reduced to sulfides by the reaction of a thioacetal and NBS (eq 46),[129] and 1,3-oxathioacetals and dithioacetals are converted into acetals using NBS.[130]

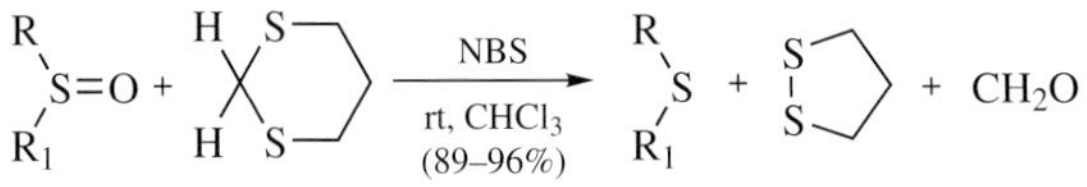

(46)

1,3-Oxathiolanes may be synthesised from aldehydes and mercaptoethanol using NBS as a catalyst;[131] the reverse reaction is also possible in aqueous acetone.[132] Glycosidation can be achieved using thioglycosides activated by NBS and a catalytic amount of strong acid salts.[133]

Bromination of Olefins.   In the Corey synthesis of epibatidine[134] the cyclohexene (**19**) reacts with NBS to give bromination with neighboring group participation, producing **20** (eq 47). This reaction has been studied in detail by Vasella.[135]

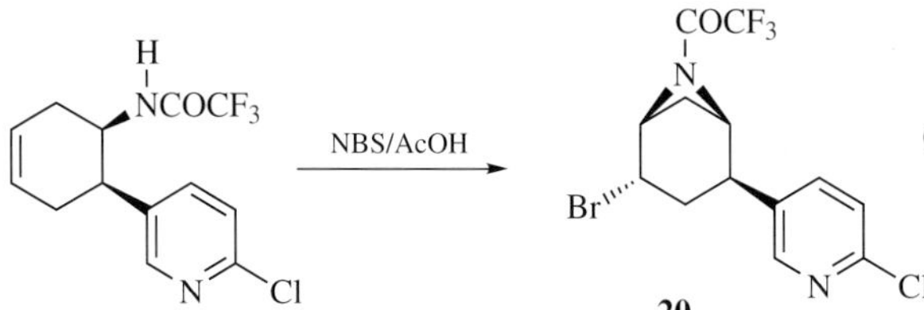

(47)

NBS and diphenylacetic acid add regiospecifically to olefins, e.g., the conversion of **21** to **22** (eq 48).[136]

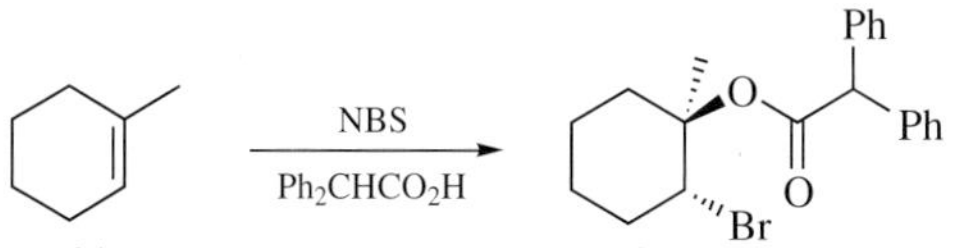

(48)

ω-Alkenyl glycoside (**23**) reacts with aq NBS to give bromo alcohols (**24** and **25**) (eq 49).[137] The observed selectivity is explained by the formation of a cyclic bromonium ion intermediate.

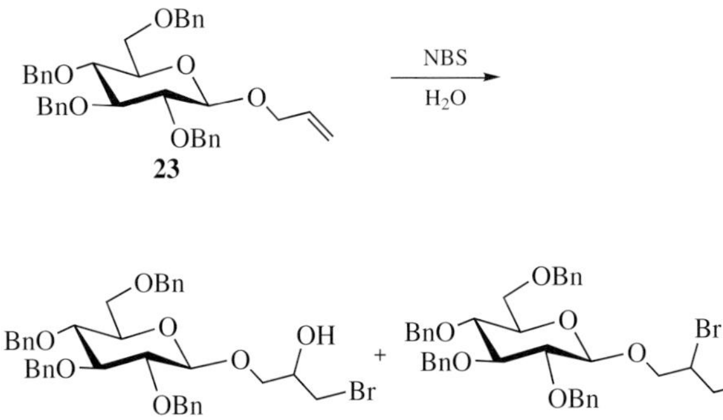

(49)

Transition metal-catalyzed regio- and stereoselective aminobromination of olefins with TsNH₂ and NBS as nitrogen and bromine sources.[138] Studies have appeared on the use of NBS in additions to alkenes[139] and in the isomerization of alkenes.[140]

Bromination of Amides and Amines.   Although yields are low, radical bromination α to nitrogen is possible (eq 50) and indicates a novel use of NBS.[105]

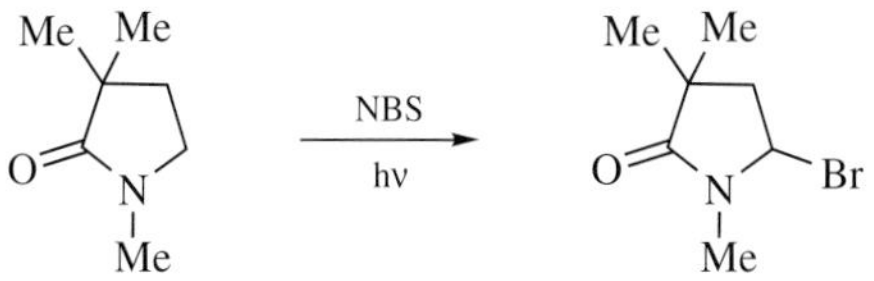

(50)

Secondary or tertiary amides are prepared in good yield from amines and alcohols using an in situ generated *N*-bromophosphonium salt from the reaction of NBS and PPh₃.[141] Benzylamines are debenzylated by NBS and AIBN[142] and the conversion of amides into carbamates was achieved in a Hofmann rearrangement using NBS/NaOMe,[143] or NBS/DBU/MeOH.[144]

Bromohydration, Bromolactonization and Other Additions to C=C. The first catalytic method for the halolactonization of olefins has appeared.[145] The selenium-catalyzed method using NBS leads to a mixture of regioisomers depending on the reaction conditions (eq 51).

No catalyst	2	1	0
5 mol % PhSeSePh	17	1	1

Oxidations.

Oxidation and Bromination of Other Functional Groups. Selective oxidation of alcohols may be achieved using a 1:1 complex of NBS and tetrabutylammonium iodide,[146] whereas 1,2-diols are converted into 1,2-diketones using *N*-bromosuccinimide.[147] An efficient and mild procedure has been reported for the preparation of benzoic acids via oxidation of aromatic carbonyl compounds by employing NBS and mercuric acetate.[148] Selective and efficient oxidation of sulfides to sulfoxides has been achieved with NBS in the presence of β-cyclodextrin in water.[149] Epoxides and aziridines are conveniently oxidized to the corresponding α-hydroxy or α-amino ketones using cerium(IV) ammonium nitrate and NBS.[150]

New Reaction Techniques Involving NBS. Several new reaction techniques have been applied to NBS reactions to develop potentially useful new synthetic methods, a selection of these are outlined below.

Solid State and Related Reactions. The area of solid/solid organic reactions has been explored.[151–153] Results on the solid state nuclear bromination of aromatic compounds with NBS as well as some theoretical insights into the mechanism of the reaction have been reported. NBS on a solid support has been used to sythesize benzylic bromides under neutral conditions[154] and for the functionalization of α-oxoaldehyde-supported silicas.[155]

Microwave Reactions. Side chain bromination of mono and dimethyl heteroaromatic and aromatic compounds by a solid phase *N*-bromosuccinimide reaction without radical initiator under microwave conditions was developed.[156] The stereoselective synthesis of (*E*)-β-arylvinyl bromides by microwave-induced Hunsdiecker-type reaction has also appeared.[157]

Reactions in Ionic Liquids. NBS in an ionic liquid has been used to oxidize benzylic alcohols to carbonyl compounds[158] to convert olefins to *vic*-bromohydrins[159] and for the regioselective monobromination of aromatic substrates.[160]

NBS as a Ligand in Organometallic Chemistry. Bromobis(triphenylphosphine)(*N*-succinimide)palladium(II) has been reported as a novel catalyst for Stille cross-coupling reactions.[161]

NBS in Water with Cyclodextrin. NBS in water with cyclodextrin has been used as a deprotecting agent for silyl ethers[162] and THP ethers[163] in the conversion of oxiranes to α-hydroxylmethyl aryl ketones,[164] in the conversion of aryl aziradines to α-tosyl amino ketones[165] and in the conversion of oximes into a carbonyl compounds.[166]

Related Reagents. *N*-Bromosuccinimide–Dimethylformamide; *N*-Bromosuccinimide–dimethyl sulfide; *N*-Bromosuccinimide–hydrogen fluoride; *N*-Bromosuccinimide–sodium azide; Triphenylphosphine–*N*-Bromosuccinimide.

1. Pizey, J. S. *Synthetic Reagents*; Wiley: New York, 1974; Vol. 2, p 1.

2. (a) Djerassi, C., *Chem. Rev.* **1948**, *43*, 271. (b) Horner, L.; Winkelmann, E. H., *Angew. Chem.* **1959**, *71*, 349.

3. UV irradiation through Pyrex (λ > 313 nm) can lead to Cl- and Cl_3C-substituted products from the solvent CCl_4. Futamura, S.; Zong, Z.-M., *Bull. Chem. Soc. Jpn.* **1992**, *65*, 345.

4. Greenwood, F. L.; Kellert, M. D.; Sedlak, J., *Org. Synth., Coll. Vol.* **1963**, *4*, 108.

5. (a) Ziegler, K.; Spaeth, A.; Schaaf, E.; Schumann, W.; Winkelmann, E., *Justus Liebigs Ann. Chem.* **1942**, *551*, 80. (b) Using the solvents $CHCl_3$ and MeCN, different selectivities are observed. Day, J. C.; Lindstrom, M. J.; Skell, P. S., *J. Am. Chem. Soc.* **1974**, *96*, 5616.

6. Accordingly, the product obtained in Ref. 4 is almost certainly a mixture of isomers.

7. Inokuchi, T.; Asanuma, G.; Torii, S., *J. Org. Chem.* **1982**, *47*, 4622.

8. (a) Franck-Neumann, M.; Martina, D.; Heitz, M.-P., *Tetrahedron Lett.* **1989**, *30*, 6679. (b) Martin, R.; Chapleo, C. B.; Svanholt, K. L.; Dreiding, A. S., *Helv. Chim. Acta* **1976**, *59*, 2724.

9. Yoda, H.; Shirakawa, K.; Takabe, K., *Chem. Lett.* **1989**, 1391.

10. (a) Corbin, T. F.; Hahn, R. C.; Shechter, H., *Org. Synth., Coll. Vol.* **1973**, *5*, 328. (b) Kalir, A., *Org. Synth., Coll. Vol.* **1973**, *5*, 825.

11. (a) Koten, I. A.; Sauer, R. J., *Org. Synth., Coll. Vol.* **1973**, *5*, 145. (b) Shriner, R. L.; Wolf, F. J., *Org. Synth., Coll. Vol.* **1955**, *3*, 737.

12. (a) Leed, A. R.; Boettger, S. D.; Ganem, B., *J. Org. Chem.* 1980, *45*, 1098. (b) Goldberg, Y.; Bensimon, C.; Alper, H., *J. Org. Chem.* 1992, *57*, 6374.

13. (a) Gribble, G. W.; Keavy, D. J.; Davis, D. A.; Saulnier, M. G.; Pelcman, B.; Barden, T. C.; Sibi, M. P.; Olson, E. R.; BelBruno, J. J., *J. Org. Chem.* **1992**, *57*, 5878. (b) Campaigne, E.; Tullar, B. F., *Org. Synth., Coll. Vol.* **1963**, *4*, 921.

14. Hendrickson, J. B.; de Vries, J. G., *J. Org. Chem.* **1985**, *50*, 1688.

15. (a) Hanessian, S., *Org. Synth.* **1987**, *65*, 243; *Org. Synth., Coll. Vol.* **1993**, *8*, 363. (b) Hanessian, S., *Methods Carbohydr. Chem.* 1972, *6*, 183. (c) Hanessian, S.; Plessas, N. R., *J. Org. Chem.* **1969**, *34*, 1035. (d) Hanessian, S.; Plessas, N. R., *J. Org. Chem.* **1969**, *34*, 1045.

16. (a) Wenger, R. M., *Helv. Chim. Acta* **1983**, *66*, 2308. (b) Machinaga, N.; Kibayashi, C., *J. Org. Chem.* **1992**, *57*, 5178.

17. Binkley, R. W.; Goewey, G. S.; Johnston, J. C., *J. Org. Chem.* **1984**, *49*, 992.

18. Hendry, D.; Hough, L.; Richardson, A. C., *Tetrahedron Lett.* **1987**, *28*, 4597.

19. (a) Binkley, R. W.; Hehemann, D. G. *J. Org. Chem.* **1990**, *55*, 378. (b) Hashimoto, H.; Kawa, M.; Saito, Y.; Date, T.; Horito, S.; Yoshimura, J., *Tetrahedron Lett.* **1987**, *28*, 3505. (c) Giese, B.; Linker, T., *Synthesis* **1992**, 46. (d) Ferrier, R. J.; Tyler, P. C., *J. Chem. Soc., Perkin Trans. 1* **1980**, 2767.

20. Filler, R., *Chem. Rev.* **1963**, *63*, 21.

21. (a) Nakagawa, M.; Saegusa, J.; Tonozuka, M.; Obi, M.; Kiuchi, M.; Hino, T.; Ban, Y., *Org. Synth., Coll. Vol.* **1988**, *6*, 462. (b) Jones, T. H.; Fales, H. M., *Tetrahedron Lett.* **1983**, *24*, 5439.

22. Jung, F.; Ladjama, D.; Riehl, J. J., *Synthesis* **1979**, 507.

23. (a) Zimmermann, J.; Seebach, D., *Helv. Chim. Acta* **1987**, *70*, 1104. (b) Lange, G. L.; Organ, M. G.; Roche, M. R., *J. Org. Chem.* **1992**, *57*, 6000. (c) Seebach, D.; Gysel, U.; Job, K.; Beck, A. K., *Synthesis* **1992**, 39.

24. Hauser, F. M.; Prasanna, S., *J. Org. Chem.* **1982**, *47*, 383.

25. Viehe, H. G.; Merényi, R.; Stella, L.; Janousek, Z., *Angew. Chem., Int. Ed. Engl.* **1979**, *18*, 917.

26. (a) Yamaura, M.; Suzuki, T.; Hashimoto, H.; Yoshimura, J.; Shin, C., *Bull. Chem. Soc. Jpn.* **1985**, *58*, 2812. (b) Lidert, Z.; Gronowitz, S., *Synthesis* **1980**, 322.

27. (a) Bretschneider, T.; Miltz, W.; Münster, P.; Steglich, W., *Tetrahedron* **1988**, *44*, 5403. (b) Mühlemann, C.; Hartmann, P.; Odrecht, J.-P, *Org. Synth.* **1992**, *71*, 200. (c) Allmendinger, T.; Rihs, G.; Wetter, H., *Helv. Chim. Acta* **1988**, *71*, 395. (d) Ermert, P.; Meyer, J.; Stucki, C.; Schneebeli, J.; Obrecht, J.-P., *Tetrahedron Lett.* **1988**, *29*, 1265.

28. (a) Kishi, Y.; Fukuyama, T.; Nakatsuka, S.; Havel, M., *J. Am. Chem. Soc.* **1973**, *95*, 6493. (b) Zimmermann, J.; Seebach, D., *Helv. Chim. Acta* **1987**, *70*, 1104.

29. Hassner, A.; Murthy, K., *Tetrahedron Lett.* **1987**, *28*, 683.

30. (a) Harpp, D. N.; Bao, L. Q.; Coyle, C.; Gleason, J. G.; Horovitch, S., *Org. Synth., Coll. Vol.* **1988**, *6*, 190. (b) Harpp, D. N.; Bao, L. Q.; Black, C. J.; Gleason, J. G.; Smith, R. A., *J. Org. Chem.* **1975**, *40*, 3420.

31. (a) Stotter, P. L.; Hill, K. A., *J. Org. Chem.* **1973**, *38*, 2576. (b) Blanco, L.; Amice, P.; Conia, J. M., *Synthesis* **1976**, 194. (c) Hooz, J.; Bridson, J. N., *Can. J. Chem.* **1972**, *50*, 2387. (d) Lichtenthaler, F. W.; Kläres, U.; Lergenmüller, M.; Schwidetzky, S., *Synthesis* **1992**, 179.

32. (a) Evans, D. A.; Ellman, J. A.; Dorow, R. L., *Tetrahedron Lett.* **1987**, *28*, 1123. (b) Oppolzer, W.; Dudfield, P., *Tetrahedron Lett.* **1985**, *26*, 5037.

33. Wilt, J. W.; Diebold, J. L., *Org. Synth., Coll. Vol.* **1963**, *4*, 254.

34. Mignani, G.; Morel, D.; Grass, F., *Tetrahedron Lett.* **1987**, *28*, 5505.

35. (a) Izawa, T.; Nishiyama, S.; Yamamura, S.; Kato, K.; Takita, T., *J. Chem. Soc., Perkin Trans. 1* **1992**, 2519. (b) Jones, A. S.; Verhelst, G.; Walker, R. T., *Tetrahedron Lett.* **1979**, 4415.

36. Levy, A. B.; Talley, P.; Dunford, J. A., *Tetrahedron Lett.* **1977**, 3545.

37. (a) Zweifel, G.; On, H. P., *Synthesis* **1980**, 803. (b) Camps, F.; Chamorro, E.; Gasol, V.; Guerrero, A., *Synth. Commun.* **1989**, *19*, 3211.

38. (a) Fleming, I.; Dunogues, J.; Smithers, R., *Org. React.* **1989**, *37*, 57. (b) Angell, R.; Parsons, P. J.; Naylor, A., *Synlett* **1993**, 189. (c) Weng, W.-W.; Luh, T.-Y., *J. Org. Chem.* **1992**, *57*, 2760.

39. Hofmeister, H.; Annen, K.; Laurent, H.; Wiechert, R., *Angew. Chem., Int. Ed. Engl.* **1984**, *23*, 727.

40. Mitchell, R. H.; Lai, Y.-H.; Williams, R. V., *J. Org. Chem.* **1979**, *44*, 4733.

41. (a) Ando, W.; Tsumaki, H., *Synthesis* **1982**, 263. (b) Townsend, C. A.; Davis, S. G.; Christensen, S. B.; Link, J. C.; Lewis, C. P., *J. Am. Chem. Soc.* **1981**, *103*, 6885.

42. (a) Gilow, H. M.; Burton, D. E., *J. Org. Chem.* **1981**, *46*, 2221. (b) Martina, S.; Enkelmann, V.; Wegner, G.; Schlüter, A.-D., *Synthesis* **1991**, 613.

43. Palmer, B. D.; Denny, W. A., *J. Chem. Soc., Perkin Trans. 1* **1989**, 95.

44. (a) Kellogg, R. M.; Schaap, A. P.; Harper, E. T.; Wynberg, H., *J. Org. Chem.* **1968**, *33*, 2902. (b) Goldberg, Y.; Alper, H., *J. Org. Chem.* **1993**, *58*, 3072.

45. (a) Bartlett, P. A. *Asymmetric Synthesis*; Morrison, J. D., Ed.; Academic: New York: 1984; Vol. 3, Chapter 6. (b) Beger, J., *J. Prakt. Chem.* **1991**, *333*, 677.

46. (a) Dalton, D. R.; Dutta, V. P.; Jones, D. C., *J. Am. Chem. Soc.* **1968**, *90*, 5498. (b) Langman, A. W.; Dalton, D. R., *Org. Synth., Coll. Vol.* **1988**, *6*, 184.

47. NBS in anhydrous DMSO converts dihydropyrans to α-bromolactones. Berkowitz, W. F.; Sasson, I.; Sampathkumar, P. S.; Hrabie, J.; Choudhry, S.; Pierce, D., *Tetrahedron Lett.* **1979**, 1641.

48. Kutney, J. P.; Singh, A. K., *Synlett* **1982**, *60*, 1842.

49. (a) van Tamelen, E. E.; Curphey, T. J., *Tetrahedron Lett.* **1962**, 121. (b) van Tamalen, E. E.; Sharpless, K. B., *Tetrahedron Lett.* **1967**, 2655. (c) Hanzlik, R. P., *Org. Synth., Coll. Vol.* **1988**, *6*, 560. (d) Nadeau, R.; Hanzlik, R., *Methods Enzymol.* **1969**, *15*, 346.

50. (a) Jennings, R. C.; Ottridge, A. P., *J. Chem. Soc., Chem. Commun.* **1979**, 920. (b) Gold, A.; Brewster, J.; Eisenstadt, E., *J. Chem. Soc., Chem. Commun.* **1979**, 903.

51. Breslow, R.; Pecoraro, J.; Sugimoto, T., *Org. Synth., Coll. Vol.* **1988**, *6*, 361.

52. Dulcere, J. P.; Mihoubi, M. N.; Rodriguez, J., *J. Chem. Soc., Chem. Commun.* **1988**, 237.

53. (a) Demole, E.; Enggist, P., *Helv. Chim. Acta* **1971**, *54*, 456. (b) Hart, D. J.; Leroy, V.; Merriman, G. H.; Young, D. G. J., *J. Org. Chem.* **1992**, *57*, 5670. (c) Michael, J. P.; Ting, P. C.; Bartlett, P. A., *J. Org. Chem.* **1985**, *50*, 2416. (d) Baskaran, S.; Islam, I.; Chandrasekaran, S., *J. Org. Chem.* **1990**, *55*, 891.

54. (a) Corey, E. J.; Pearce, H. L., *J. Am. Chem. Soc.* **1979**, *101*, 5841. (b) Schlessinger, R. H.; Nugent, R. A., *J. Am. Chem. Soc.* **1982**, *104*, 1116.

55. Pearson, A. J.; Ray, T., *Tetrahedron Lett.* **1986**, *27*, 3111.

56. Dai, W.; Katzenellenbogen, J. A., *J. Org. Chem.* **1991**, *56*, 6893.

57. (a) Jew, S-s.; Terashima, S.; Koga, K., *Tetrahedron* **1979**, *35*, 2337. (b) Hayashi, M.; Terashima, S.; Koga, K., *Tetrahedron* **1981**, *37*, 2797.

58. (a) Tamaru, Y.; Kawamura, S.; Tanaka, K.; Yoshida, Z., *Tetrahedron Lett.* **1984**, *25*, 1063. (b) Balko, T. W.; Brinkmeyer, R. S.; Terando, N. H., *Tetrahedron Lett.* **1989**, *30*, 2045.

59. Miknis, G. F.; Williams, R. M., *J. Am. Chem. Soc.* **1993**, *115*, 536.

60. (a) Van Ende, D.; Krief, A., *Angew. Chem., Int. Ed. Engl.* **1974**, *13*, 279. (b) Nagorski, R. W.; Brown, R. S., *J. Am. Chem. Soc.* **1992**, *114*, 7773.

61. (a) Olah, G. A.; Welch, J. T.; Vankar, Y. D.; Nojima, M.; Kerekes, I.; Olah, J. A., *J. Org. Chem.* **1979**, *44*, 3872. (b) Alvernhe, G.; Laurent, A.; Haufe, G., *Synthesis* **1987**, 562. (c) Camps, F.; Chamorro, E.; Gasol, V.; Guerrero, A., *J. Org. Chem.* **1989**, *54*, 4294. (d) Kuroboshi, M.; Hiyama, T., *Tetrahedron Lett.* **1991**, *32*, 1215.

62. Amrollah-Madjdabadi, A.; Beugelmans, R.; Lechevallier, A., *Synthesis* **1986**, 828.

63. (a) Iriuchijima, S.; Tsuchihashi, G., *Synthesis* **1970**, 588. (b) Drabowicz, J., *Synthesis* **1986**, 831.

64. Harville, R.; Reed, Jr., S. F., *J. Org. Chem.* **1968**, *33*, 3976.

65. (a) Corey, E. J.; Erickson, B. W., *J. Org. Chem.* **1971**, *36*, 3553. (b) Bari, S. S.; Trehan, I. R.; Sharma, A. K.; Manhas, M. S., *Synthesis* **1992**, 439.

66. Filler, R., *Chem. Rev.* **1963**, *63*, 21.

67. Cheung, Y.-F., *Tetrahedron Lett.* **1979**, 3809.

68. Grundmann, C.; Richter, R., *J. Org. Chem.* **1968**, *33*, 476.

69. Rosini, G., *J. Org. Chem.* **1974**, *39*, 3504.

70. (a) Carpino, L. A.; Crowley, P. J., *Org. Synth., Coll. Vol.* **1973**, *5*, 160. (b) Bock, H.; Rudolph, G.; Baltin, E., *Chem. Ber.* **1965**, *98*, 2054.

71. Bonfand, E.; Gosselin, P.; Maignan, C., *Tetrahedron: Asymmetry* **1993**, *4*, 1667.

72. Oda, M.; Kawase, T.; Kurata, H., *Org. Synth. Col. Vol.* **1998**, *9*, 19.

73. Diaz, R. R.; Melgarejo, C. R.; Lopez-Espinosa, M. T. P.; Cubero, II, *J. Org. Chem.* **1994**, *59*, 7928.

74. Piccialli, V.; Graziano, M. L.; Iesce, M. R.; Cermola, F., *Tetrahedron Lett.* **2002**, *43*, 45, 8067.

75. Lee, J. C.; Hwang, E. Y., *Synth. Commun.* **2004**, *34*, 16, 2959.

76. Harris, P. W. R.; Rickard, C. E. F.; Woodgate, P. D., *J. Organomet. Chem.* **2000**, *601*, 172.

77. Hagata, H.; Miyazawa, N.; Ogasawara, K., *Chem. Commun.* **2001**, 1094.

78. Meshram, H. M.; Reddy, P. N.; Sadashiv, K.; Yadav, J. S., *Tetrahedron Lett.* **2005**, *46*, 623.

79. Das, B.; Venkateswarlu, K.; Mahender, G.; Mahender, I., *Tetrahedron Lett.* **2005**, *46*, 3041.

80. Dhuru, S. P.; Mohe, N. U.; Salunkhe, M. M., *Synth. Commun.* **2001**, *31*, 3653.

81. Wang, C.; Tunge, J., *Chem. Commun.* **2004**, *23*, 2694.

82. Hintermann, L.; Togni, A., *Helv. Chim. Acta* **2000**, *83*, 2425.

83. Graven, A.; Jorgensen, K. A.; Dahl, S.; Stanczak, A., *J. Org. Chem.* **1994**, *59*, 3543.

84. Chowdhury, S.; Roy, S., *Tetrahedron Lett.* **1996**, *37*, 2623.

85. Chowdhury, S.; Roy, S., *J. Org. Chem.* **1997**, *62*, 199.

86. Naskar, D.; Roy, S., *J. Chem. Soc., Perkin Trans. 1* **1999**, 2435.

87. Petasis, N. A.; Zavialov, I. A., *Tetrahedron Lett.* **1996**, *37*, 567.

88. Leroy, J., *Org. Synth. Coll. Vol.* **1998**, *9*, 129.

89. Naskar, D.; Roy, S., *J. Org. Chem.* **1999**, *64*, 6896.

90. Tanemura, K.; Suzuki, T.; Nishida, Y.; Satsumabayashi, K.; Horaguchi, T., *Chem. Lett.* **2003**, *32*, 932.

91. Andersh, B.; Murphy, D. L.; Olson, R. J., *Synth. Commun.* **2000**, *30*, 2091.

92. Carreño, M. C.; García Ruano, J. L.; Toledo, M. A.; Urbano, A., *Synlett* **1997**, 1241.

93. Roush, W. R.; Madar, D. J.; Coffey, D. S., *Can. J. Chem.* **2001**, *79*, 1711.

94. Goldberg, Y.; Alper, H., *J. Org. Chem.* **1993**, *58*, 3072.

95. Duan, S.; Turk, J.; Speigle, J.; Corbin, J.; Masnovi, J.; Baker, R. J., *J. Org. Chem.* **2000**, *65*, 3005.

96. Duan, J.; Zang, L. H.; Dolbier, W. R., Jr., *Synlett* **1999**, *8*, 1245.

97. Auerbach, J.; Weissman, S. A.; Blacklock, T. J.; Angeles, M. R.; Hoogsteen, K., *Tetrahedron Lett.* **1993**, *34*, 931.

98. Bagheri, M.; Azizi, N.; Saidi, M. R., *Can. J. Chem.* **2005**, *83*, 146.

99. Thiebes, C.; Parkash, G. K. S.; Petasis, N. A.; Olah, G. A., *Synlett* **1998**, 141.

100. Canibano, V.; Rodriguez, J. F.; Santos, M.; Sanz-Tejedor, A.; Carreno, M. C.; Gonzalez, G.; Garcia-Ruano, J. L., *Synthesis* **2001**, 2175.

101. Sugimoto, O.; Mori, M.; Tanji, K., *Tetrahedron Lett.* **1999**, *40*, 7477.

102. Bagley, M. C.; Glover, C.; Merritt, E. A.; Xiong, X., *Synlett* **2004**, *5*, 811.

103. Agami, C.; Dechoux, L.; Hamon, L.; Hebbe, S., *Synthesis* **2003**, *6*, 859.

104. Dvornikova, E.; Kamienska-Trela, K., *Synlett* **2002**, *7*, 1152.

105. Easton, C. J.; Pitt, M. J.; Ward, C. M., *Tetrahedron* **1995**, *51*, 46, 12781.

106. Turbiez, M.; Frere, P.; Roncali, J., *J. Org. Chem.* **2003**, *68*, 5357.

107. Amat, M.; Hadida, S.; Sathyanarayana, S.; Bosch, J., *Org. Synth. Coll. Vol.* **1998**, *9*, 417.

108. Zhang, P.; Liu, R.; Cook, J. M., *Tetrahedron Lett.* **1995**, *36*, 3103.

109. Miniejew, C.; Outurquin, F.; Pannecoucke, X., *Org. Biomol. Chem.* **2004**, *2*, 1575.

110. Brown, W. D.; Gouliaev, A. H., *Org. Synth.* **2005**, *81*, 98.

111. Kamal, A.; Chouhan, G., *Synlett* **2002**, *3*, 474.

112. Satake, K.; Cordonier, C.; Kubota, Y.; Jin, Y.; Kimura, M., *Heterocycles* **2003**, *60*, 2211.

113. Day, R. A.; Blake, J. A.; Stephens, C. E., *Synthesis* **2003**, 1586.

114. Kurasawa, Y.; Satoh, W.; Matsuzaki, I.; Maesaki, Y.; Okamoto, Y.; Kim, H. S., *J. Het. Chem.* **2003**, *40*, 837.

115. Ramzaeva, N.; Mittelbach, C.; Seela, F., *Helv. Chim. Acta* **1999**, *82*, 12.

116. Reese, C. B.; Yan, H., *J. Chem. Soc., Perkin Trans 1.* **2002**, 2619.

117. Hon, Y. S.; Yan, J. L., *Tetrahedron* **1998**, *54*, 8525.

118. Hodosi, G., *Tetrahedron Lett.* **1994**, *35*, 6129.

119. Bloodworth, A. J.; Shah, A., *Tetrahedron Lett.* **1993**, *34*, 6643.

120. Qin, Z. H.; Li, H.; Cai, M. S.; Li, Z. J., *Carbohydr. Res.* **2002**, *337*, 31.

121. Kadokawa, J. I.; Yamamoto, M.; Tagaya, H.; Chiba, K., *Carbohydr. Lett.* **2001**, *4*, 97.

122. Karimi, B.; Ebrahimian, G. R.; Seradj, H., *Org. Lett.* **1999**, *1*, 1737.

123. Karimi, B.; Seradj, H.; Ebrahimian, G. R., *Synlett* **1999**, 1456.

124. Karimi, B.; Seradj, H., *Synlett* **2001**, 519.

125. Karimi, B.; Seradj, H.; Ebrahimian, G. R., *Synlett* **2000**, 623.

126. Firouzabadi, H.; Iranpoor, N.; Garzan, A.; Shaterian, H. R.; Ebrahimzadeh, F., *Eur. J. Org. Chem.* **2005**, 416.

127. Patrocínio, A. F.; Moran, P. J. S., *J. Organomet. Chem.* **2000**, *603*, 220.

128. Degani, J.; Dughera, S.; Fochi, R.; Gatti, A., *Synthesis* **1996**, 467.

129. Iranpoor, N.; Firouzabadi, H.; Shaterian, H. R., *J. Org. Chem.* **2002**, *67*, 2826.

130. Karimi, B.; Seradj, H.; Maleki, J., *Tetrahedron* **2002**, *58*, 22, 4513.

131. Kamal, A.; Chouhan, G.; Ahmed, K., *Tetrahedron Lett.* **2002**, *43*, 6947.

132. Karimi, B.; Seradj, H.; Tabaei, M. H., *Synlett* **2000**, *12*, 1798.

133. Fukase, K.; Hasuoka, A.; Kinoshita, I.; Aoki, Y.; Kusumoto, S., *Tetrahedron* **1995**, *51*, 4923.

134. Corey, E. J.; Loh, T.-P.; AchyuthaRao, S.; Daley, D. C.; Sarshar, S., *J. Org. Chem.* **1993**, *58*, 5600.

135. Kapferer, P.; Vasella, A., *Helv. Chim. Acta* **2004**, *87*, 2764.

136. Dulcère, J. P.; Rodriguez, J., *Synlett* **1992**, 347.

137. Rodebaugh, R.; Fraser-Reid, B., *Tetrahedron* **1996**, *52*, 7663.

138. Thakur, V. V.; Talluri, S. K.; Sudalai, A., *Org. Lett.* **2003**, *5*, 861.

139. Dulcere, J. P.; Agati, V.; Faure, R., *Chem. Commun.* **1993**, 270.

140. Baag, M. M.; Kar, A.; Argade, N. P., *Tetrahedron* **2003**, *59*, 34, 6489.

141. Frøyen, P.; Juvvik, P., *Tetrahedron Lett.* **1995**, *36*, 9555.

142. Baker, S. R.; Parsons, A. F.; Wilson, M., *Tetrahedron Lett.* **1998**, *39*, 331.

143. Huang, X.; Keillor, J. W., *Tetrahedron Lett.* **1997**, *38*, 313.

144. Huang, X.; Seid, M.; Keillor, J. W., *J. Org. Chem.* **1997**, *62*, 7495.

145. Mellegaard, S. R.; Tunge, J. A., *J. Org. Chem.* **2004**, *69*, 8979.

146. Beebe, T. R.; Boyd, L.; Fonkeng, S. B.; Horn, J.; Money, T. M.; Saderholm, M. J.; Skidmore, M. V., *J. Org. Chem.* **1995**, *60*, 6602.

147. Khurana, J. M.; Kandpal, B. M., *Tetrahedron Lett.* **2003**, *44*, 4909.

148. Anjum, A.; Srinivas, P., *Chem. Lett.* **2001**, 900.

149. Surendra, K.; Krishnaveni, N. S.; Kumar, V. P.; Sridhar, R.; Rao, K. R., *Tetrahedron Lett.* **2005**, *46*, 4581.

150. Surendra, K.; Krishnaveni, N. S.; Rama Rao, K., *Tetrahedron Lett.* **2005**, *46*, 4111.

151. Rothenberg, G.; Downie, A. P.; Raston, C. L.; Scott, J. L., *J. Am. Chem. Soc.* **2001**, *123*, 8701.

152. Sarma, J. A. R. P.; Nagaraju, A., *J. Chem. Soc., Perkin Trans 2* **2000**, *6*, 1113.

153. Sarma, J. A. R. P.; Nagaraju, A.; Majumdar, K. K.; Samuel, P. M.; Das, I.; Roy, S.; McGhie, A. J., *J. Chem. Soc., Perkin Trans 2* **2000**, *6*, 1119.

154. Zoller, T.; Ducep, J. B.; Hibert, M., *Tetrahedron Lett.* **2000**, *41*, 9985.

155. Kar, S.; Joly, P.; Granier, M.; Melnyk, O.; Durand, J.-O., *Eur. J. Org. Chem.* **2003**, 4132.

156. Goswami, S.; Dey, S.; Jana, S.; Adak, A. K., *Chem. Lett.* **2004**, *33*, 916.

157. Kuang, C.; Yang, Q.; Senboku, H.; Tokuda, M., *Synthesis* **2005**, *8*, 1319.

158. Lee, J. C.; Lee, J. Y.; Lee, J. M., *Synth. Commun.* **2005**, *35*, 1911.

159. Yadav, J. S.; Reddy, B. V. S.; Baishya, G.; Harshavardhan, S. J.; Chary, C. J.; Gupta, M. K., *Tetrahedron Lett.* **2005**, *46*, 3569.

160. Rajagopal, R.; Jarikote, D. V.; Lahoti, R. J.; Daniel, T.; Srinivasan, K. V., *Tetrahedron Lett.* **2003**, *44*, 1815.

161. Crawforth, C. M.; Burling, S.; Fairlamb, I. J. S.; Taylor, R. J. K.; Whitwood, A. C., *Chem. Commun.* **2003**, 2194.

162. Somi-Reddy, M.; Narender, M.; Nageswar, Y. V. D.; Rama Rao, K., *Synthesis* **2005**, 714.

163. Narender, M.; Somi-Reddy, M.; Rama Rao, K., *Synthesis* **2004**, 1741.

164. Arjun Reddy, M.; Bhanumathi, N.; Rama Rao, K., *Tetrahedron Lett.* **2002**, *43*, 3237.

165. Somi-Reddy, M.; Narender, M.; Rama Rao, K., *Tetrahedron Lett.* **2005**, *46*, 1299.

166. Somi-Reddy, M.; Narender, M.; Rama Rao, K., *Synth. Commun.* **2004**, *34*, 3875.

n-Butyllithium[1]

[109-72-8]   C$_4$H$_9$Li   (MW 64.05)

InChI = 1/C4H9.Li/c1-3-4-2;/h1,3-4H2,2H3;/rC4H9Li/c1-2-3-
    4-5/h2-4H2,1H3

InChIKey = MZRVEZGGRBJDDB-NESCHKHYAE

(strong base capable of lithiating carbon acids;[1] useful for
heteroatom-facilitated lithiations;[2,3] useful for lithium–halogen
exchange;[1,4] reagent of choice for lithium–metal transmetalation
reactions[1a,c,5])

Physical Data: colorless liquid; stable at rt; eliminates LiH on
    heating; d^{25} 0.765; mp −76 °C; bp 80–90 °C/0.0001 mmHg;
    dipole moment 0.97 D.[6] ^{13}C NMR, ^{1}H NMR, ^{6}Li NMR[8] and
    MS studies have been reported.[7–9]
Solubility: sol hydrocarbon and ethereal solvents, but should be
    used at low temperature in the latter solvent type: half-lives in
    diethyl ether and THF have been reported;[10] reacts violently
    with H$_2$O and other protic solvents.
Form Supplied in: commercially available as approximately
    1.6 M, 2.5 M, and 10.0 M solution in hexanes and in cyclo-
    hexane, approximately 2.0 M solution in pentane, and approx-
    imately 1.7 M, and 2.7 M solution in *n*-heptane. Hexameric in
    hydrocarbons;[1c] tetrameric in diethyl ether;[1c] dimer–tetramer
    equilibrium mixture in THF;[11] when used in combination with
    tertiary polyamines such as TMEDA and DABCO, reactivity is
    usually increased.[1,12]
Analysis of Reagent Purity: since the concentration of com-
    mercial solutions may vary appreciably it is necessary to
    standardize solutions of the reagent prior to use. A recom-
    mended method for routine analyses involves titration of the
    reagent with *s*-butyl alcohol using 1,10-phenanthroline or 2,2′-
    biquinoline as indicator.[15] Several other methods have been
    described.[16]
Preparative Methods: may be prepared in high yield from *n*-butyl
    chloride[13] or *n*-butyl bromide[14] and **Lithium** metal in ether or
    hydrocarbon solvents.
Handling, Storage, and Precautions: solutions of the reagent are
    pyrophoric and the reagent may catch fire if exposed to air
    or moisture. Handling of the reagent should be done behind
    a shield in a chemical fume hood. Safety goggles, chemical
    resistant gloves, and other protective clothing should be worn.
    In case of fire, a dry-powder extinguisher should be used: in no
    case should an extinguisher containing water or halogenated
    hydrocarbons be used to fight an alkyllithium fire. Bottles and
    reaction flasks containing the reagent should be flushed with
    N$_2$ or preferably Ar and kept tightly sealed to preclude contact
    with oxygen or moisture. Standard syringe/cannula techniques
    for air and moisture sensitive chemicals should be applied when
    transferring the reagent. For detailed handling techniques see
    Wakefield.[1b]

Original Commentary

Timo V. Ovaska
Connecticut College, New London, CT, USA

Lithiations. *n*-Butyllithium is a commonly used reagent
for deprotonation of a variety of nitrogen (see **Lithium
Diisopropylamide**), oxygen, phosphorus (see **Diphenylphosphine**), and carbon acids to form lithium salts. Compared to its *s*-
and *t*-butyl analogs, *n*-BuLi is less basic[17] and less reactive but it is
usually the reagent of choice for deprotonation of relatively strong
carbon acids. These lithiations are most favorable when the conjugated bases are stabilized by resonance or when the carbanion
forms at the sp hybridized carbon of a triple bond. Thus indene,[18]
triphenylmethane,[19] allylbenzene,[20] and methyl heteroaromatics
(e.g. **Pyridine**, quinoline, and isoquinoline derivatives)[21] are readily lithiated with *n*-BuLi at the benzylic position. Allenes are lithiated at C-1 or C-3 depending on the number and size of the alkyl
groups at these positions,[22] and terminal alkynes react with *n*-
BuLi to give lithium acetylides.[4] Propargylic hydrogens can also
be removed,[23] and treatment of terminal alkynes with 2 equiv of
n-BuLi results in the lithiation of both propargylic and acetylenic
positions (eq 1).[23a]

$$\text{(1)}$$

82%

The metalating ability of *n*-BuLi (and other organolithiums)
is greater in electron-donating solvents than in hydrocarbons.
Electron-donating solvents such as diethyl ether or THF provide
coordination sites for the electron deficient lithium and promote
the formation of lower-order organolithium aggregates.[1a,c] These
exhibit significantly higher levels of reactivity than do the higher-
order oligomers present in hydrocarbon solvents. *n*-Butyllithium
is often used in the presence of added lithium complexing ligands,
such as ***N,N,N′,N′-Tetramethylethylenediamine*** (TMEDA) and
1,4-Diazabicyclo[2.2.2]octane (DABCO), which further enhance
the reactivity of this reagent.[12] Many compounds, normally unreactive toward *n*-BuLi alone (e.g. benzene), are readily lithiated[24]
and even polylithiated[25] by a combination of *n*-BuLi and one of
these additives. Allylic[26,27] and benzylic sites bearing no additional activating groups at the α-position (e.g. toluene)[12c] are also
lithiated in the presence of lithium complexing donors (eq 2).[26]

$$\text{(2)}$$

70%

Essential Reactions for Organic Synthesis, First Edition. Edited by Philip L. Fuchs.
© 2016 John Wiley & Sons, Ltd. Published 2016 by John Wiley & Sons, Ltd.

Metal alkoxides are often employed as additives to enhance the metalating ability of *n*-BuLi. Particular use has been made of the ***n-Butyllithium–Potassium t-Butoxide*** mixture,[28] which is capable of effecting rapid metalation of benzylic[29] and allylic[30] systems as well as aromatic rings.[28] Although the products of these metalations are not organolithium compounds they do, nevertheless, readily react with electrophiles. Alternatively, they may be converted into the corresponding organolithium derivatives by the addition of ***Lithium Bromide***.[30a, 31] Using *t*-BuOK/*n*-BuLi for deprotonation, the one-pot synthesis of 2-(4-isobutyl-phenyl)propanoic acid (ibuprofen) from *p*-xylene, through a sequence of metalations and alkylations, has been achieved in 52% overall yield.[29] Dimetalation of arylalkynes with *n*-BuLi/*t*-BuOK followed by addition of electrophiles has been used as a route to *ortho*-substituted arylalkynes (eq 3).[31b]

(3)

88%

Facile and regioselective α-deprotonation is often effected by treatment of heteroatom-containing (e.g. oxygen, sulfur, nitrogen, and the like) compounds with *n*-BuLi, and these reactions have been extensively reviewed.[2,3,32,33] Thus sulfones,[34] certain sulfides,[35] and sulfoxides[36] can be lithiated adjacent to sulfur under various conditions, and α-heterosubstituted vinyllithium compounds are often available from the corresponding ethers,[37] thioethers,[38] chlorides,[39] and fluorides[40] via lithiation using *n*-BuLi. Isocyanides[41] and nitro compounds[42] have been lithiated adjacent to nitrogen with *n*-BuLi in THF at low temperatures (less than $-60\,°C$) and the formation of various α-phosphorus alkyllithiums has been reported.[43] Simple ethers are also susceptible to lithiation at the α site by *n*-BuLi, especially at elevated temperatures.[44] The initial proton abstraction in these systems is generally followed by various cleavage reactions resulting from α,β- and α′,β′-eliminations as well as the Wittig rearrangement.[45] Tetrahydrofuran, for example, is rapidly lithiated by *n*-BuLi at $35\,°C$ ($t_{1/2} = 10$ min) to give ethylene and the lithium enolate of acetaldehyde.[46]

Due to the stabilizing effect of two sulfur atoms, 1,3-dithianes are easily lithiated at the α-position on treatment with *n*-BuLi (see ***2-Lithio-1,3-dithiane***).[47] The 2-lithio-1,3-dithianes constitute an important class of acyl anion equivalents, permitting electrophilic substitution to occur at the masked carbonyl carbon.[47] Hydrolysis of the 1,3-dithiane functionality into a carbonyl group is effected in the presence of mercury(II) ion (eq 4).[48]

When conducted at sufficiently low temperatures (less than $-78\,°C$), α-deprotonation can occur faster than nucleophilic addition to an electrophilic center present in the same molecule. This strategy, which involves initial lithiation followed by intramolecular nucleophilic addition of the newly generated C–Li bond to an electrophilic moiety, is useful for the construction of carbocycles including medium and large ring systems (eqs 5 and 6).[49,50]

(4)

(5)

91%

(6)

62%

A large number of heteroaromatic compounds,[2] such as furans (see ***Furan***),[51] thiophenes (eq 7),[52] oxazoles,[53] and *N*-alkyl- and *N*-aryl substituted pyrroles (see ***2-Lithio-N-phenylsulfonyl-indole***),[54] pyrazoles,[55] imidazoles,[56] triazoles,[57] and tetrazoles,[58] are lithiated under various conditions α to the ring heteroatom using *n*-BuLi. However, pyridine and other nitrogen heteroaromatics bearing the pyridine, pyrimidine, or pyrazine nucleus are generally not lithiated. Indeed, they have a tendency to undergo nucleophilic addition reactions with this reagent[1b] (see ***2-Lithiopyridine***).

(7)

79%

When the α-position is benzylic, propargylic, or allylic, deprotonation takes place more readily and *n*-BuLi is generally the reagent of choice for these reactions.[59,60] However, the more basic ***s-Butyllithium*** is a better reagent for deprotonation of alkyl allyl ethers[61] and certain alkyl allyl thioethers[62] which react slowly (if at all) with *n*-BuLi in THF at low temperatures (less than $-65\,°C$).

Proton removal adjacent to a heteroatom is further facilitated if the lithium can be coordinated to proximate electron donors, such as a carbonyl oxygen, permitting the formation of 'dipole-stabilized' carbanions.[33] Thus various 2-alkenyl *N,N*-dialkyl-carbamates undergo rapid α-deprotonation adjacent to oxygen on treatment with *n*-BuLi/TMEDA at $-78\,°C$.[63,64] The resulting dipole-stabilized lithium carbanions react with ketones and aldehydes in a highly regioselective fashion providing γ-hydroxyalkylated enol esters (eq 8) which, following cleavage of the carbamoyl moiety (***Titanium(IV) Chloride***/H$_2$O or MeOH), afford δ-hydroxy carbonyl compounds (homoaldols) as lactols or lactol ethers.[65] Similarly, aliphatic or aromatic amides[66,67] (eq 9),[67a] phosphoramides,[68] and some formamidine derivatives (e.g. 1,2,3,

6-tetrahydropyridine,[69] thiazolidine,[69] 1,3-thiazine,[69] and tetra-hydroisoquinolines[70]) are selectively lithiated α to the nitrogen at the activated position. Electrophilic substitution of the intermediate organolithiums followed by hydrolytic cleavage of the amide or formamidine group provides a synthetically valuable route to α-substituted (or γ-substituted[69]) secondary amines.[33] Use of the more basic s-BuLi (or **t-Butyllithium**) is generally required for deprotonation of the analogous nonbenzylic or nonallylic systems.

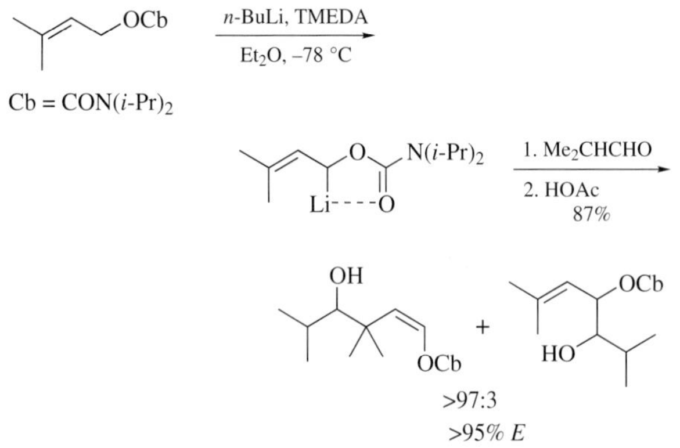

$$(8)$$

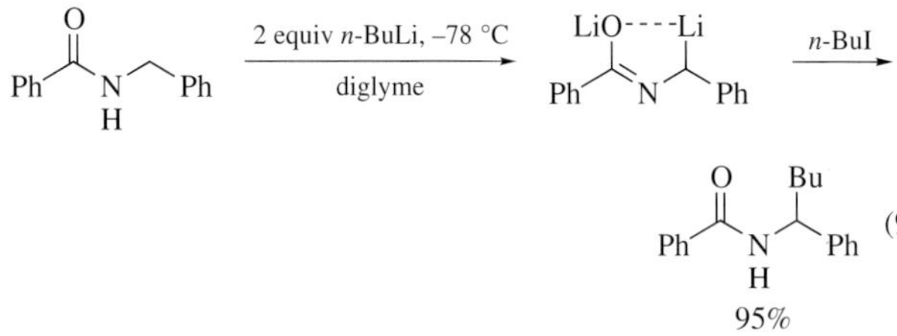

>97:3
>95% E

$$(9)$$

95%

n-Butyllithium is also used for the stereoselective α-lithiation of chiral sulfonyl compounds,[71] chiral 2-alkenyl carbamates,[64,72] various heterocyclic amine derivatives with chiral auxiliaries appended on the nitrogen (e.g. oxazoline[73] or formamidine[74,75] groups) (eq 10),[75] and chiral oxazolidinones derived from benzyl-amines (eq 11).[76] These elegant reactions have been applied to the asymmetric syntheses of a number of natural products.[77,78]

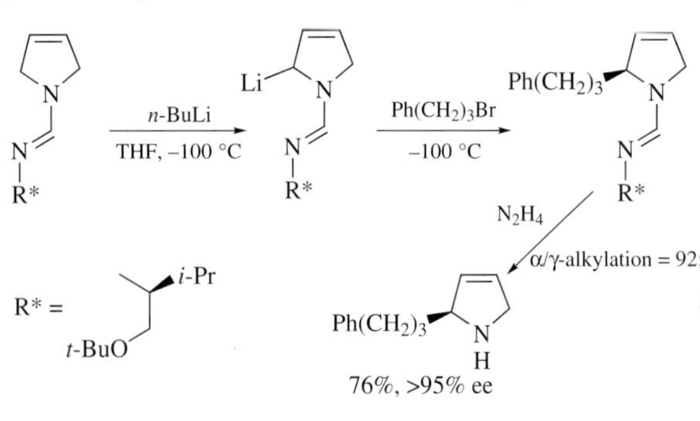

α/γ-alkylation = 92:8

$$(10)$$

76%, >95% ee

Ortho Lithiations. Heteroatom-containing substituents on aromatic rings facilitate metalation by organolithium reagents and direct the metal almost exclusively to the *ortho* position. This effect, usually referred to as *ortho* lithiation, is of considerable synthetic importance and the topic has been extensively reviewed.[2,4,79] Although n-BuLi (typically in THF or Et$_2$O with added TMEDA)[80] is capable of effecting a large number of *ortho* lithiations, the use of this reagent is somewhat limited by its tendency to undergo nucleophilic carbonyl additions with some

of the most potent and useful *ortho* directors, particularly tertiary amide and carbamate functionalities (e.g. CONEt$_2$ and OCONEt$_2$).[79,81] For example, N,N-dimethyl- and -diethyl-benzamides afford primarily aryl butyl ketones upon treatment with n-BuLi (see **2-Lithio-N,N-diethyl-benzamide**).[82] The reagent of choice for these lithiations is **s-Butyllithium**, which is less nucleophilic than n-BuLi and hence more tolerant of electrophilic functional groups.[79] There are, however, a variety of *ortho*-directing groups that are well suited for n-BuLi-promoted lithiations. These include NR$_2$,[80] CH$_2$NR$_2$,[80] CH$_2$CH$_2$NR$_2$,[80] OMe (see **o-Lithioanisole**),[83] OCH$_2$OMe,[84] SO$_2$NR$_2$,[85] C=NR,[86] 2-oxazolinyl,[85,87] F,[88] CF$_3$,[80] and groups that contain acidic hydrogens and themselves undergo deprotonation prior to lithiation of the aromatic ring (thus requiring the use of 2 equiv of n-BuLi), e.g. CONHR,[89] CH$_2$OH,[80] NHCO-t-Bu,[90] and SO$_2$NHR[80] (eq 12).[91] n-Butyllithium is also used frequently for the *ortho* lithiation of heterocyclic aromatic rings,[2] including those that contain the pyridine nucleus (eq 13).[92]

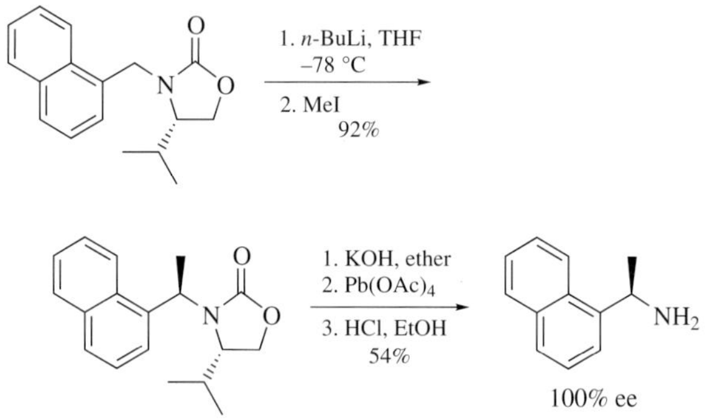

$$(11)$$

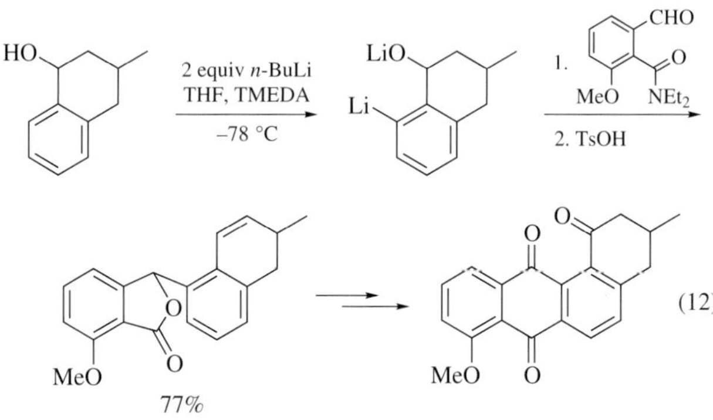

100% ee

$$(12)$$

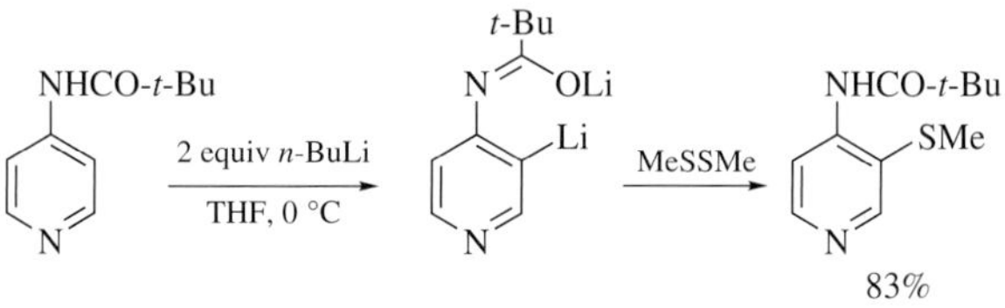

77%

$$(13)$$

83%

Formation of Enolate Anions and Enolate Equivalents. Owing to its tendency to undergo nucleophilic addition with carbonyl groups and other electrophilic carbon–heteroatom multiple bonds (C=NR, C≡N, C=S),[1] n-BuLi is usually not the reagent of choice for the generation of enolate anions or enolate equivalents from active hydrogen compounds. This is done most conveniently using the less nucleophilic lithium dialkylamides (e.g. **Lithium Diisopropylamide** (LDA), **Lithium 2,2,6,6-Tetra-**

methylpiperidide (LiTMP), and **Lithium Hexamethyldisilazide** (LTSA) prepared (often in situ) from sterically hindered secondary amines, typically by treatment with *n*-BuLi.[93] However, less reactive carbonyl compounds such as amides (eq 14)[94] and carboxylic acids[95] as well as those containing carbon–nitrogen or carbon–sulfur multiple bonds, e.g. imines,[96] oxazines,[97] nitriles,[98] some hydrazones,[99] and thioamides,[100] can be lithiated α to the electrophilic carbon with *n*-BuLi under various conditions. β-Keto esters can be alkylated at the α′-carbon using **Sodium Hydride** for the first deprotonation and *n*-BuLi for abstraction of the less acidic α′-proton followed by addition of alkyl halides.[101] Lithiation of unsymmetrical imines using *n*-BuLi takes place regioselectively at the most substituted α-carbon (eq 15).[96] In contrast, LDA directs metalation and subsequent alkylation predominantly to the less substituted α-position in similar systems.[102] Lithium enolates of camphor imine esters, prepared by addition of *n*-BuLi, undergo highly diastereoselective Michael additions with α,β-unsaturated esters (eq 16).[103] The tightly chelated structures of the intermediate enolates permit selective *re* face approach of the Michael acceptors, giving rise to the high degree of distereoselectivity observed.

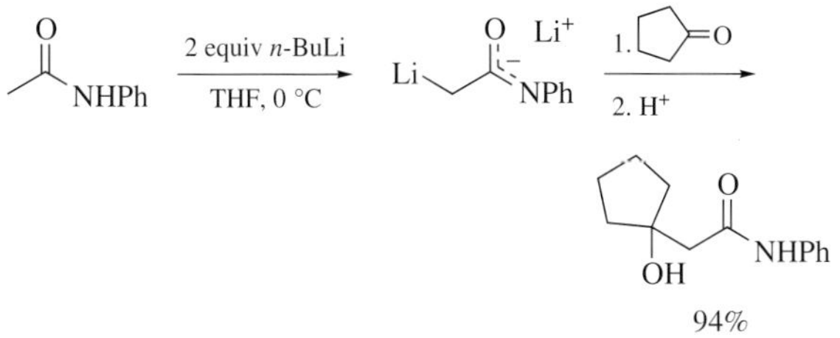

$$(14)$$

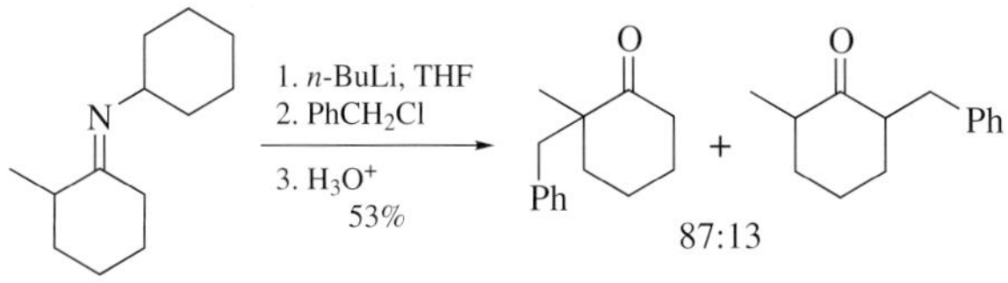

94%

$$(15)$$

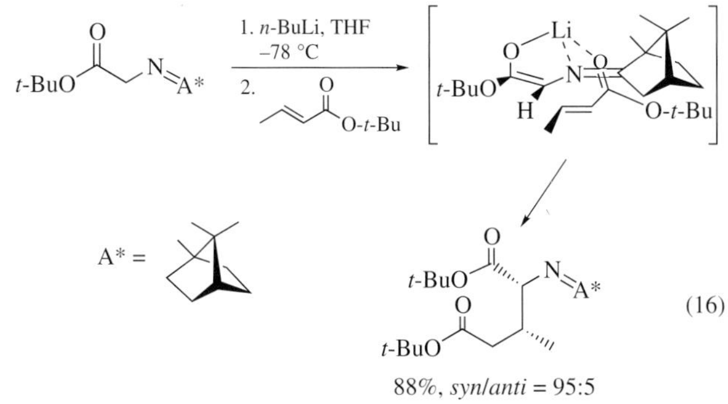

53%   87:13

$$(16)$$

88%, *syn/anti* = 95:5

Metal–Halogen Interchange and Transmetalation Reactions. The metal–halogen interchange reaction which involves the exchange of halogen and lithium atoms is an important method for the preparation of organolithium compounds not readily accessible through metalation. In particular, the generation of aryl-, vinyl-, and cyclopropyllithium derivatives from the corresponding bromides or iodides on treatment with *n*-BuLi (usually in ethereal solvents at or below −78 °C) is of considerable synthetic utility (eqs 17–19).[104–106] The relative rates of exchange depend on the halide and decrease in the order I > Br > Cl > F.

Fluorides and chlorides, with the exception of some polychlorinated aliphatic[107] and aromatic compounds,[108] are quite resistant to lithium–halogen interchange; instead, they tend to promote *ortho* and α lithiations.[79]

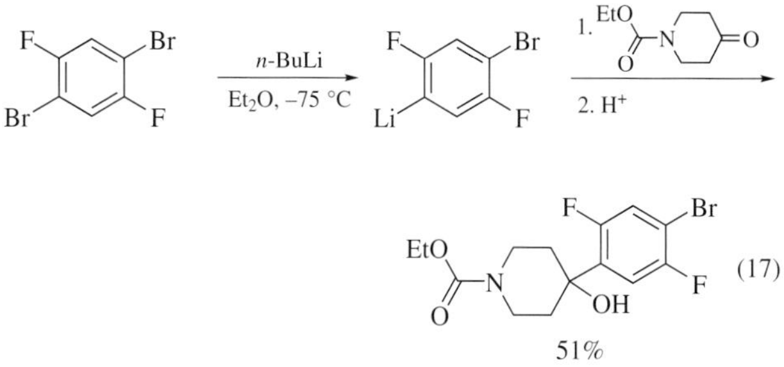

$$(17)$$

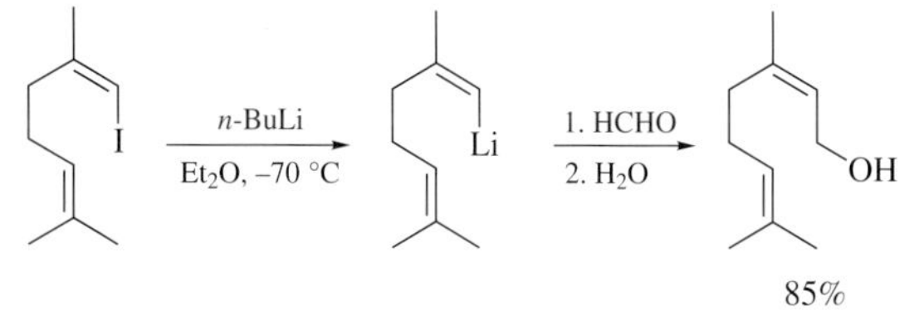

51%

$$(18)$$

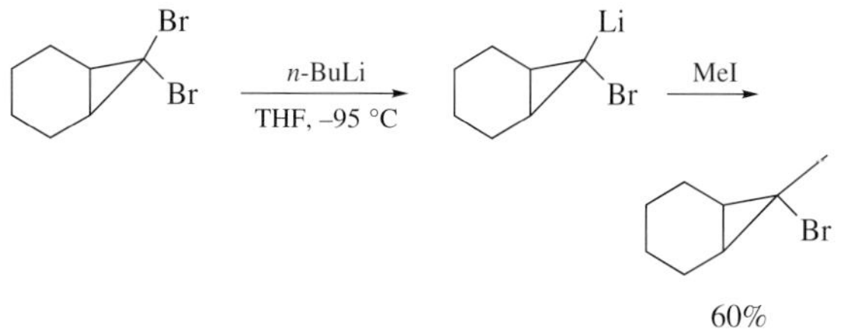

85%

$$(19)$$

60%

gem-Dihalocyclopropanes react with *n*-BuLi in a highly stereoselective fashion (eq 19), although subsequent isomerization can take place.[109] Alkyl-substituted vinyllithiums can be prepared with retention of configuration[105,110] and even aryl-substituted vinyllithium compounds retain their configuration under controlled conditions (<−78 °C in THF or at rt in hydrocarbon solvents).[111] Simple alkyllithiums are generally not accessible by this route because of the unfavorable interchange equilibrium that ensues when primary or secondary halides are treated with *n*-BuLi, except in cases where the initially formed organolithium is rapidly consumed in a subsequent, irreversible reaction (eq 20).[112] Primary alkyllithiums may be prepared, however, from the corresponding alkyl iodides (but not from the bromides)[113] on treatment with the more reactive *t***-Butyllithium** which renders the exchange operationally irreversible.[114]

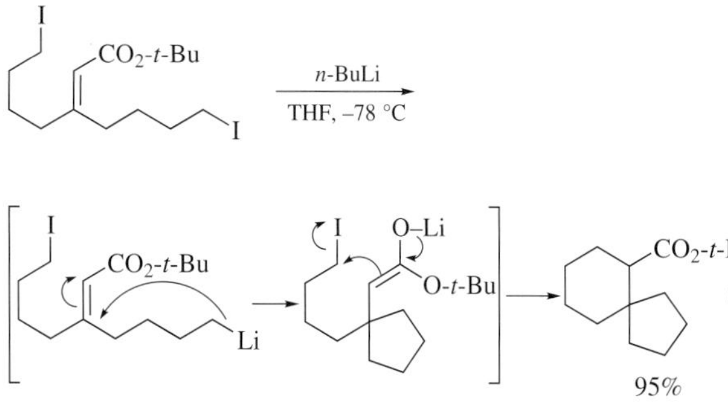

$$(20)$$

95%

n-Butyllithium is the reagent of choice for effecting a number of transmetalation reactions involving the replacement of tin, selenium, tellurium, or mercury by lithium.[1a,1c,5] These reactions are typically conducted at low temperatures (less than $-60\,°C$) in ethereal solvents, with THF being the most commonly employed reaction medium. The tin–lithium exchange is a particularly important operation, providing a convenient route to aryl- and vinyllithiums,[115] as well as functionalized lithio derivatives which are not readily accessible by other means, such as α-alkoxylithium compounds,[116] oxiranyllithiums,[117] and amino-substituted organolithiums.[118] The Sn–Li exchange in these systems and subsequent trapping of the intermediate lithio derivatives proceed stereoselectively with retention of configuration at the tin-bearing carbon (eqs 21 and 22).[116c, 118c] Consequently, enantiomerically enriched functionalized products are available through this methodology from homochiral α-hetero-substituted stannanes.[119]

(21)

(22)

Appropriately substituted alkenic α-alkoxylithiums, derived from the corresponding tri-n-butyltin compounds on treatment with n-BuLi, have also been used to initiate 5-exo-trig cyclization reactions to give tetrahydrofuran derivatives (eq 23).[120] Other related ring-forming reactions have been reported.[121]

(23)

cis:trans = 10:1

Organoselenium compounds, particularly selenoacetals or selenothioacetals, undergo facile lithium–selenium exchange reactions on treatment with n-BuLi, typically in THF at $-78\,°C$.[122–124] The intermediate α-seleno organolithium derivatives formed in these reactions are readily trapped with electrophiles to afford a variety of synthetically useful α-functionalized selenide products[3,125] (eq 24).[126] Studies on cyclohexyl selenoacetals have shown that Li–Se exchange takes place almost exclusively at the axial position[123] and that equatorial α-lithio sulfides, derived from mixed Se,S-acetals by Li–Se exchange, epimerize within minutes at $-78\,°C$ to give the more stable axial lithio isomers.[124] n-Butyllithium may be used for the majority of Li–Se exchanges although selenoacetals derived from sterically

hindered ketones react quite slowly with this reagent, and in these cases the use of the more reactive **s-Butyllithium** is warranted.[123]

(24)

A variety of organolithiums, including benzylic, vinylic, alkynic, and 1-alkoxylithium compounds, are also accessible through the lithium–tellurium exchange, which involves the treatment of diorganotellurides with n-BuLi at $-78\,°C$ in THF (eq 25);[127] vinyllithium derivatives have been prepared from the corresponding organomercurials by essentially the same methodology.[128]

(25)

Rearrangements. n-Butyllithium is used frequently to promote various anionic rearrangement reactions. 2,3-Wittig rearrangements are commonly effected by treatment of allylic and propargylic ethers with n-BuLi,[129] and 1,2-, 1,4-, as well as 3,4-rearrangements with similar systems are also known.[130] These reactions are most often initiated at low temperatures either by α-deprotonation adjacent to oxygen or via Li–Sn exchange, and the rearrangement occurs upon warming of the reaction mixture to $0\,°C$. Additional benzylic, propargylic, or allylic stabilization is necessary for deprotonation by n-BuLi; however, the Li–Sn method does not suffer from this limitation.[131] 2,3-Wittig rearrangements involving (Z)-allylic ethers proceed generally with high syn stereoselectivity (eq 26),[131] whereas the opposite tendency is observed with (E)-allylic ethers.[132] The regiochemistry of unsymmetrical bis-allylic ethers depends on the degree of substitution at the α- and γ-carbons, with the less substituted allylic position being the preferred site of deprotonation (eq 27).[133] The cyclic variant of the 2,3-Wittig rearrangement can be used for ring contraction reactions and it provides a useful method for the construction of macrocyclic ring systems (eq 28).[134]

(26)

>120:1 syn:anti

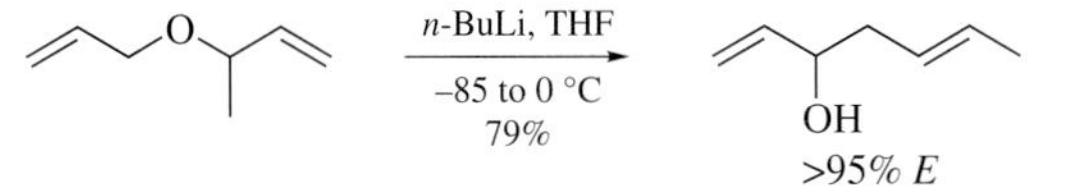

(27)

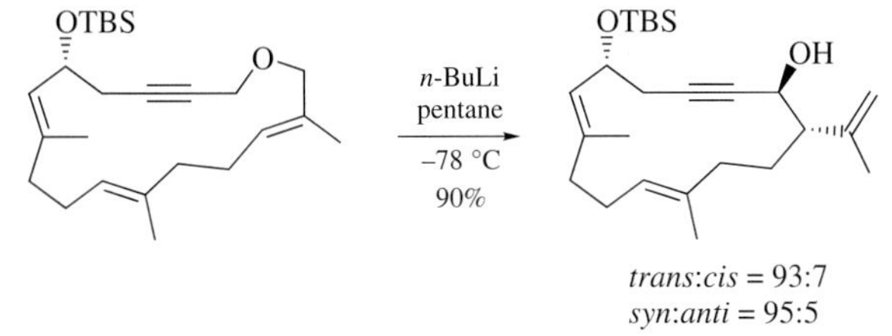

>95% *E*

(28)

trans:*cis* = 93:7
syn:*anti* = 95:5

(32)

80%

Secondary 1,2-epoxy alcohols may be prepared through the Payne rearrangement in aprotic media by treatment of primary 2,3-epoxy alcohols with *n*-BuLi in THF with catalytic amounts of **Lithium Chloride**,[135] and α,β-unsaturated ketones are obtained from trimethylsilyl substituted propargylic alcohols through the Brook rearrangement followed by alkylation of the intermediate allenyllithium products and acidic workup (eq 30).[136] Quaternary ammonium salts, such as benzyltrimethylammonium ion, react with *n*-BuLi to deliver nitrogen ylides which can undergo either the Stevens rearrangement to give tertiary amines or the Sommelet–Hauser rearrangement to afford *ortho* alkyl-substituted nitrogen-containing aromatics, depending on the reaction conditions (eq 31).[137] Dibenzyl thioether has been reported to undergo a Sommelet–Hauser type rearrangement on treatment with *n*-BuLi/TMEDA in HMPA,[138] and appropriately substituted sulfonic ylides, prepared from sulfonium salts on treatment with *n*-BuLi, undergo 2,3-sigmatropic rearrangements (eq 32).[139]

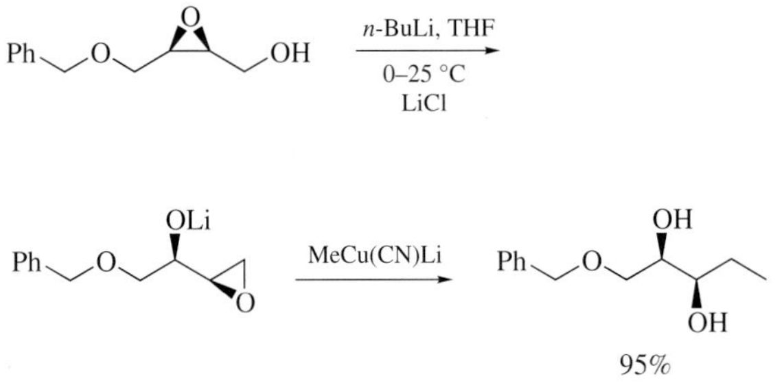

Elimination Reactions. A number of eliminations can be effected using *n*-BuLi. The formation of phosphorus, nitrogen, and sulfur ylides[1a] and the generation of benzyne intermediates from aromatic halides[140] are well established processes, and α-eliminations resulting in the formation of carbenes have been used to prepare cyclopropanes,[141] oxiranes,[142] and other products via subsequent carbenoid rearrangements (eq 33).[143] Vinylidene dihalides undergo dehalogenation with concomitant formation of a triple bond on treatment with 2 equiv of *n*-BuLi.[144] The initial product of this reaction is a lithium acetylide which can be quenched with methanol to give a terminal alkyne or alkylated in situ by addition of alkyl halides to afford internal alkynes (eq 34).[145] The decomposition of (arylsulfonyl)hydrazones of aldehydes or ketones upon treatment with at least 2 equiv of *n*-BuLi (Shapiro reaction) is a useful method for the generation of vinyllithium compounds (eq 35).[146] In cases where two regioisomers can be produced, *n*-BuLi appears to promote the formation of the less substituted vinyllithium via deprotonation of the kinetically more acidic proton, whereas the use of *s*-BuLi reportedly leads to the more substituted vinyllithium product.[147]

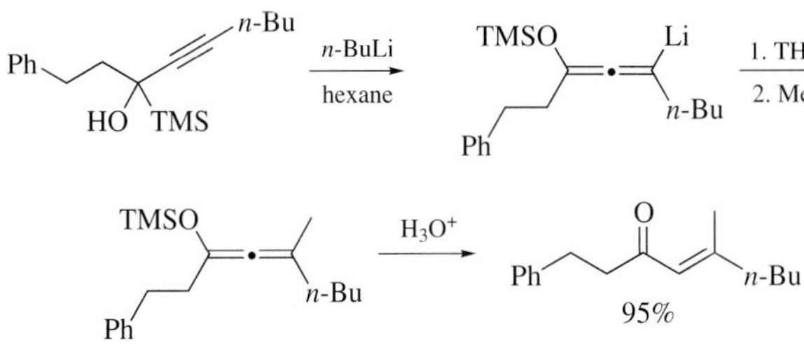

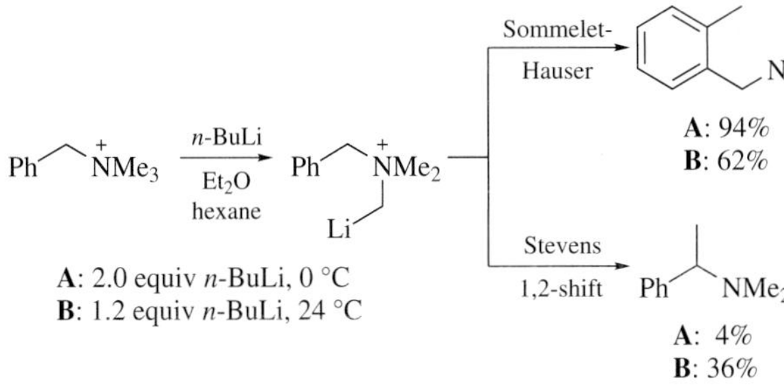

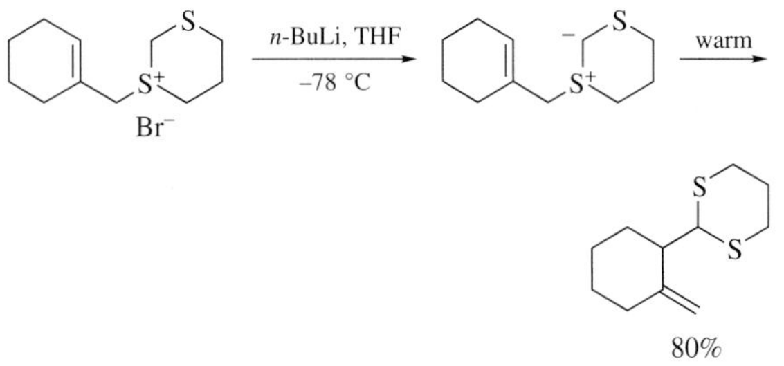

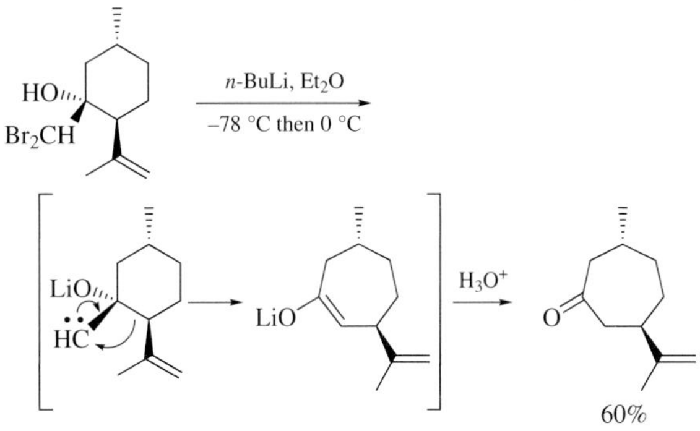

(29)

95%

(30)

95%

A: 2.0 equiv *n*-BuLi, 0 °C
B: 1.2 equiv *n*-BuLi, 24 °C

A: 94%
B: 62% (31)

A: 4%
B: 36%

(33)

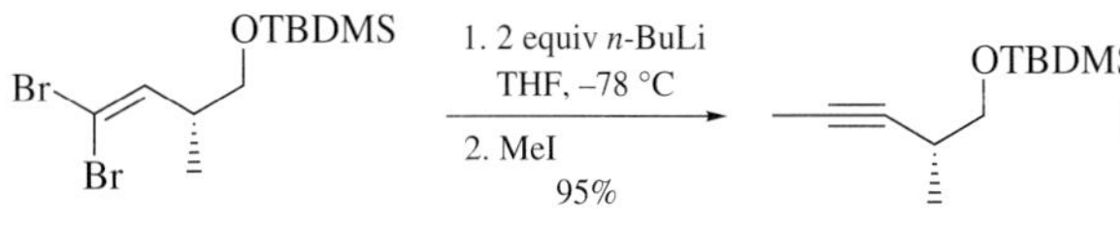

60%

(34)

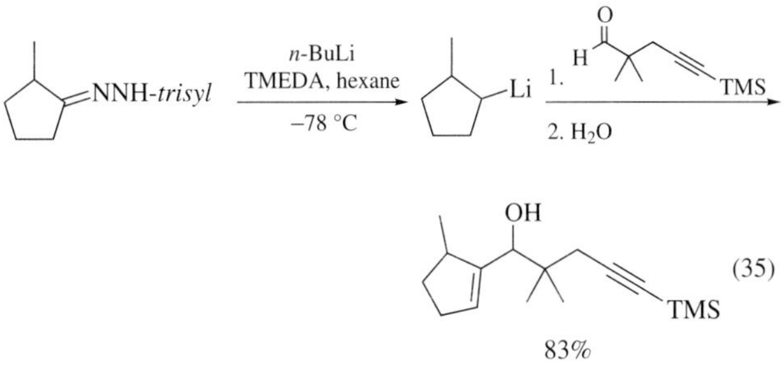

95%

(35)

83%

First Update

Victor Snieckus
Queen's University, Kingston, Ontario, Canada

Bert Nolte
Alantos Pharmaceuticals AG, Heidelberg, Germany

Rearrangements. Stereoselective synthesis of highly functionalized *C*-glycosides based on acetal [1,2]- and [1,4]-Wittig rearrangements is reported.[148] Treatment of the *O*-glycoside with BuLi at low temperature results in a [1,2]-Wittig rearrangement to yield the *C*-glycoside in a highly diastereoselective manner (eq 36).

$$77\% \qquad (36)$$

Another protocol represents the first example of a *C*-glycoside synthesis based on a [1,4]-Wittig rearrangement (eq 37). Moreover, interception of the intermediate lithium enolate with an external electrophile is achieved. In addition, the application of this synthetic strategy for the synthesis of zaragozic acid A is demonstrated.[149]

$$60\text{--}73\% \qquad \text{(E = CH}_2\text{OH: 60\% dr)} \qquad (37)$$

In the first asymmetric total synthesis of (+)-astrophylline, a [2,3]-Wittig–Still rearrangement is used to establish a 1,2-*trans* substituent relationship of the key cyclopentene intermediate (eq 38).[150] Careful optimization of the reaction conditions leads to the development of a multigram-scale protocol.

$$(38)$$

In an instructive application for the construction of hindered benzophenones, an initial metal–halogen exchange of a bromobenzyloxy ester with BuLi sets the stage for anionic homo-Fries rearrangement of the intermediate aryllithium species (eq 39).[151] The resulting tetra *ortho*-substituted benzophenone serves as an intermediate for the synthesis of the potent protein kinase C inhibitor (−)-balanol.

$$(39)$$

A new type of anionic oxygen to α- and β-vinyl carbamoyl migration reaction is reported (eq 40). In the case of the carbamoyl migration to the α-vinyllithium site, mechanistic evidence is secured by generation of the respective anion by metal–halogen exchange using BuLi. As expected, warming of the reaction mixture gives the product of an *O*-carbamoyl migration in high yield and complete stereochemical conservation of the (*Z*)-stilbene.[152]

$$(40)$$

The complete transfer of axial chirality to central chirality during the Stevens rearrangement of binaphthyl compounds is demonstrated (eq 41).[153] Treatment with an excess of BuLi results first into the expected Stevens [1,2]-benzyl shift, which is followed by base-induced 1,2-elimination to give rise, in a one-pot procedure, to pentahelicene with excellent enantioselectivity (>99%).

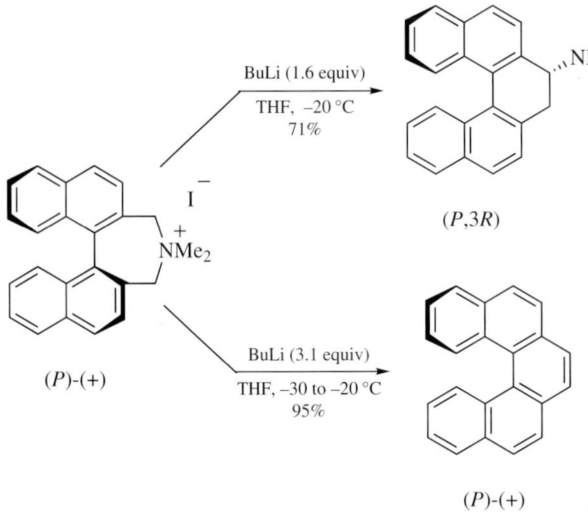

(41)

Carbolithiations. Carbolithiation of styrene derivatives constitutes an evolving methodology for the generation of benzylic carbanions.[154–157] This type of strategy, which has the potential for effecting two C–C bond forming processes in one pot, has found utility in the development of versatile new routes for heterocyclic synthesis (eq 42).[158] Best results for the (−)-sparteine-mediated asymmetric carbolithiation with BuLi are observed under coordinating solvent conditions. Thus, the synthesis of optically active isoquinolines is achieved by electrophilic quench of the lithiated intermediate with DMF, following acidification, subsequent deprotection, and oxidative aromatization. Similarly, starting from an *o*-OMe substrate, benzofurans are obtained. Further extensions are also shown for isoquinolinones and isobenzofurans. It is worth noting that the described carbolithiation protocol provides access also to enantioenriched 2-alkylanilines, substituted indoles, and indol-2-ones by reacting the chiral lithiated species with various electrophiles.[159,160]

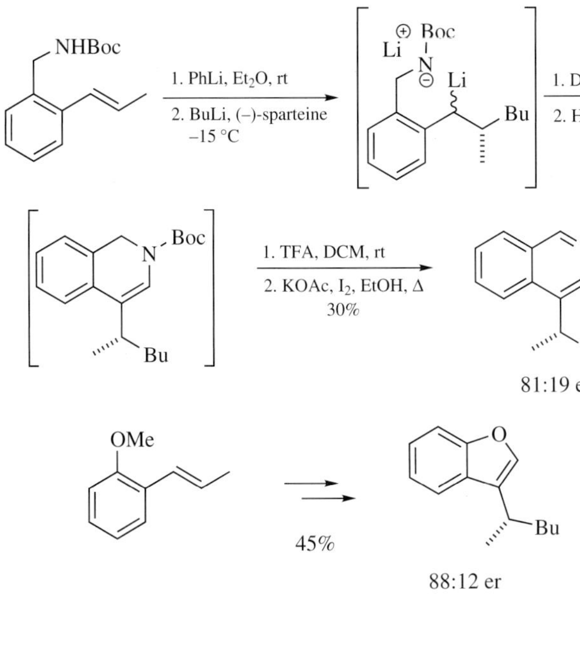

(42)

Lithiation α to Heteroatoms and at Benzylic Sites. The Ph₂P(=O)-assisted asymmetric deprotonation of benzylic amines using BuLi/(−)-sparteine allows the synthesis of a series of phosphinamides with good to excellent enantioinduction (eq 43).[161] The cleavage of these derivatives to the corresponding amines is effected by conc HCl in THF or under reductive conditions, for example, LiAlH₄ in THF.[162]

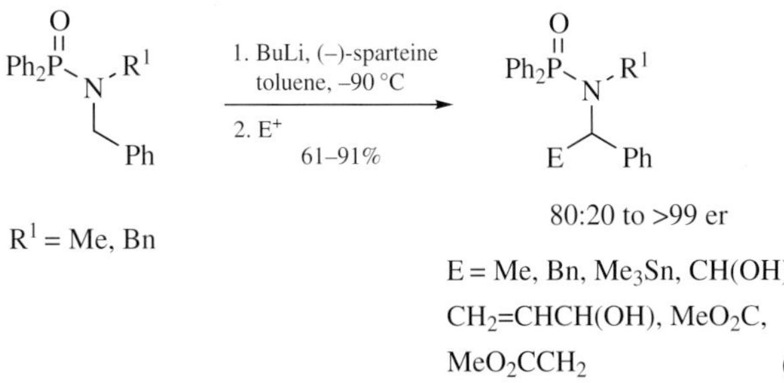

$R^1 = Me, Bn$

80:20 to >99 er

$E = Me, Bn, Me_3Sn, CH(OH)Ph,$
$CH_2=CHCH(OH), MeO_2C,$
MeO_2CCH_2 (43)

The Horner–Wadsworth–Emmons (HWE) reaction of a lithiated (diphenylphosphinoyl)methyl amine with a biaryl aldehyde furnishes the corresponding enecarbamates, which serve as key intermediates for the synthesis of the Amaryllidaceae alkaloid buflavine (eq 44).[163]

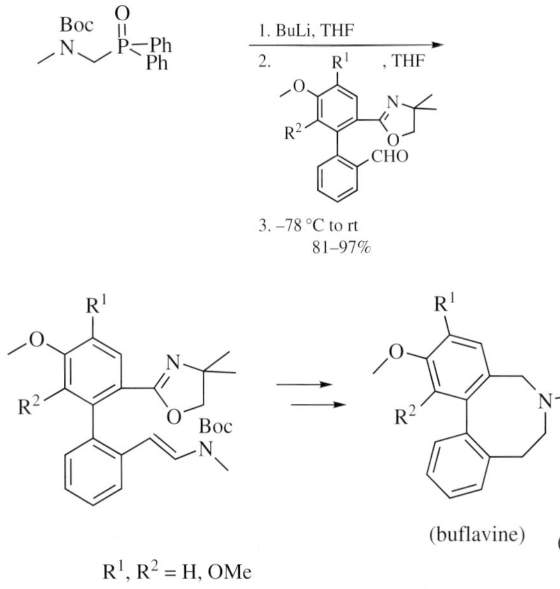

(buflavine) (44)

$R^1, R^2 = H, OMe$

The synthesis of modular pyridine-type *P,N*-ligands from monoterpenes is described (eq 45). An α-picolyl-type lithiation with BuLi followed by a stereoselective methylation using MeI gives new *P,N*-ligands.[164]

Lateral Lithiation. An enantioselective synthesis of tetrahydroisoquinolin-1-ones via a BuLi/(−)-sparteine-mediated lateral metalation–imine addition sequence proceeds with highest yields and induction in toluene/ether mixtures (eq 46).[165]

Selective *ortho* and benzylic functionalization of secondary and tertiary *p*-tolylsulfonamides is investigated (eq 47). For both R = H and R = Et, kinetic *ortho* metalation is achieved using BuLi, while thermodynamic conditions lead to *ortho* and benzylic deprotonation, respectively. Regioselective metalation of secondary sulfonamides, R = H, is achieved by using BuLi/KO*t*Bu superbase.[166]

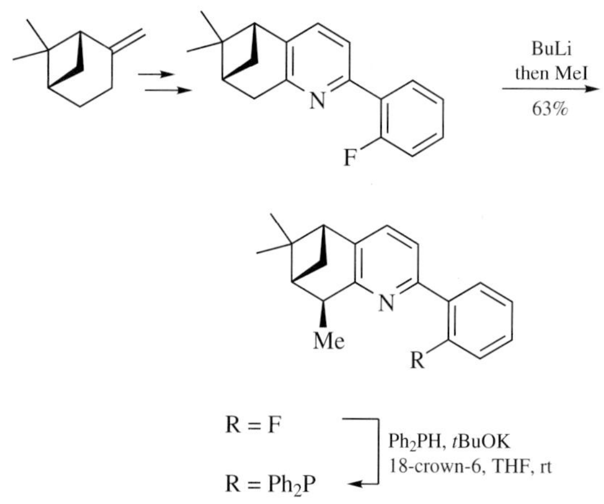

(45)

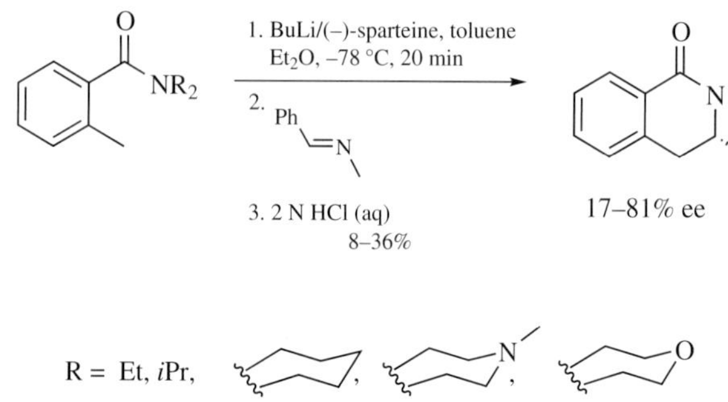

R = F
R = Ph₂P ← Ph₂PH, *t*BuOK
18-crown-6, THF, rt

17–81% ee (46) 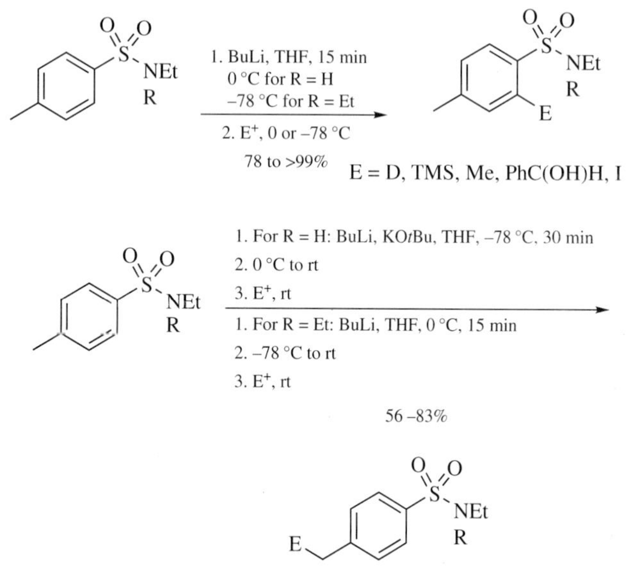
8–36%

R = Et, *i*Pr,

E = D, TMS, Me, PhC(OH)H, I

56–83%

E = D, TMS, Me, PhC(OH)H, Br, I (47)

Directed *ortho* Metalation (D*o*M). The synthesis of the alkaloid schumanniophytine requires a silicon protection of the more reactive *ortho* *O*-carbamate site to set the stage for the key step, a remote anionic Fries rearrangement that results in a pyridine ring carbamoyl translocation (eq 48).[167]

A general approach to 7-substituted saccharins (eq 49) based on two successive D*o*M reactions starting from a secondary sulfonamide suffers from a potential drawback, the formation of a *N*-methyl saccharin that precludes further functionalization.[168]

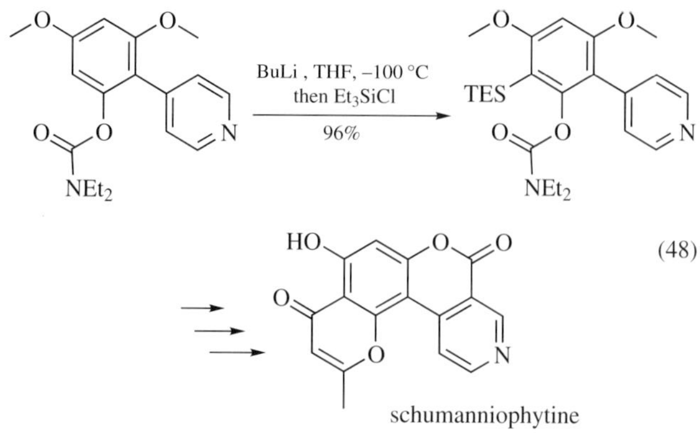

(48)

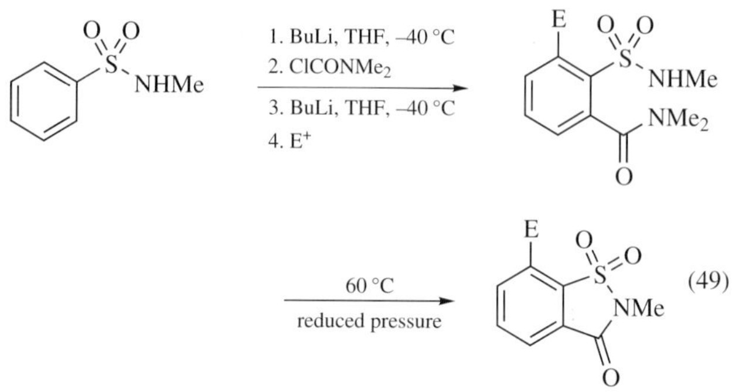

schumanniophytine

(49)

The discovery of the *N*-cumyl directed metalation group (DMG) for amides, *O*-carbamates, and sulfonamides helps to broaden the scope of new D*o*M strategies.[169] Synthetic utility is demonstrated by establishing viable routes to 4- and 4,7-substituted saccharin derivatives (eq 50). Thus, BuLi/TMEDA metalation of selected biaryl *N*-cumylsulfonamides followed by *N,N*-diethylcarbamoyl chloride quench results in a regioselective functionalization (eq 51).

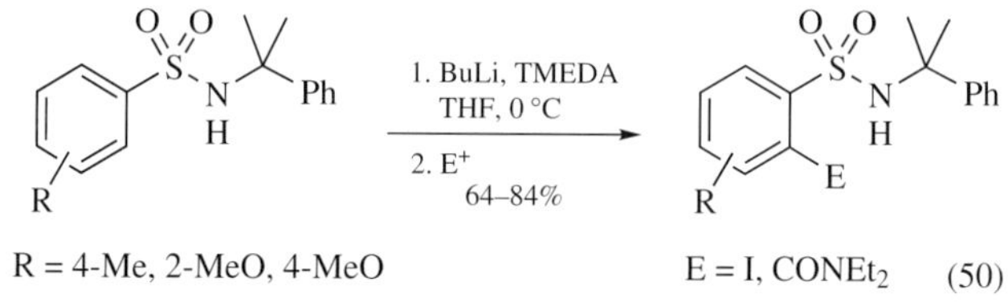

R = 4-Me, 2-MeO, 4-MeO   E = I, CONEt₂ (50)

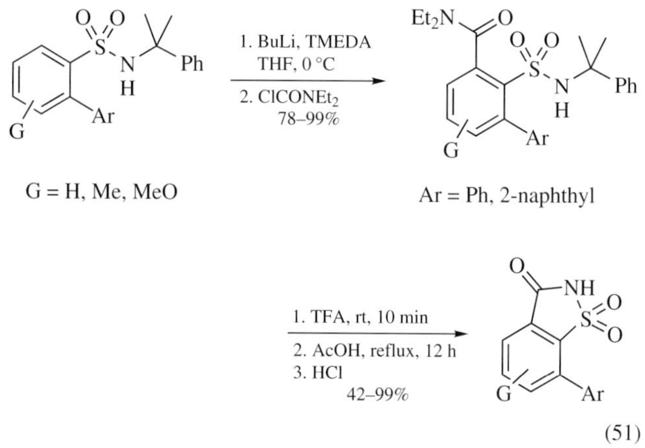

G = H, Me, MeO   Ar = Ph, 2-naphthyl

(51)

A mild TFA decumylation initiates the concluding transformations to obtain the *N*-substituted saccharins. The sequence leading to 4,7-disubstituted saccharins is initiated by a BuLi/iodination protocol to give the required intermediate for a Suzuki cross-coupling reaction toward the final compound (eq 52).[170]

$$(52)$$

$$(54)$$

The *N*-isopropyl-*N*-trimethylsilylcarbamoyl-DMG may be conveniently installed via treatment of a phenol with isopropyl isocyanate followed by a one-flask operation that begins the reaction with TMS triflate, lithiation with BuLi/TMEDA, and concludes with electrophile quench (eq 53).[171] The application of this methodology was demonstrated for the synthesis of hydroxybiaryls.[172] The regiospecific construction of a diverse set of isomeric chlorodihydroxybiphenyls and polychlorinated diphenyls is reported by applying BuLi-mediated DoM protocols and Suzuki cross-coupling tactics on commercially available starting materials.[173]

E = TMS, SnBu$_3$, I, Me, CH(OH)Ph, C(OH)Ph$_2$

$$(53)$$

The generality of a DoM strategy is demonstrated for the functionalization of deep-cavity cavitands by treatment with varying equivalents of butyllithiums, for example, BuLi, and trapping of the resulting carbanions with different electrophiles.[174]

In the course of the synthesis of potential D1 agonists, DoM of pyridine 2-carboxamide derivatives X = CR^2, Y = N, R^1 = H, Me, F, OMe, *n*-Pr, and R^2 = H, Me, *n*-Bu with 2 equiv of BuLi followed by Michael addition of nitroolefin and a triethylamine-mediated equilibration gives the *trans*-tetralin products (eq 54).[175] In contrast, if X = N, Y = CR^2, R^1 = H, and R^2 = Me is metalated with BuLi and the resulting dianion is added to the nitroolefin, the product of a benzylic lithiation–Michael addition is isolated.

An alternative route takes advantage of a halogen–metal exchange of the corresponding 3-bromo-5-methyl-pyridine derivative to yield the alternative Michael adduct in good yield.

Halogen–Lithium Exchange and Transmetalation. The generation of propynyllithium is achieved from inexpensive commercially available (*E*/*Z*)-1-bromopropene by treatment with BuLi in THF at 20 °C (eq 55). Consequent reaction with aldehydes, ketones, amides, and acid chlorides yields the corresponding propargylic alcohols or ketones.[176]

R^1: H, alkyl, Cl, NR$_2$ R^1: H, alkyl
R^2: alkyl, aryl, vinyl   R^2: alkyl, aryl, vinyl

$$(55)$$

In studies whose thrust is the development of the new *O*-sulfamate DMG, *ortho*-bromo aryl *O*-sulfamates are subjected to metal–halogen exchange in the presence of furan (eq 56) to give, following treatment with catalytic HCl, naphthol derivatives. The TMS substituent is cleaved by protodesilylation under these conditions.[177]

R = H, TMS, OMe   R = H, OMe

$$(56)$$

During studies toward the synthesis of isotopically labeled aryl triflates, metal–halogen exchange of a bromo aryl triflate with BuLi at low temperatures results in quantitative aryne generation as evidenced by in situ trapping with 1,3-diphenylisobenzofuran to give the corresponding cycloaddition product (eq 57).[178] As perhaps expected, the temperature difference for the generation of the benzyne intermediate from the O-sulfamate and triflate suggests the faster rate for the latter functionality.

$$(57)$$

The synthesis of a new core structure of ferroelectric liquid crystals takes advantage of a transmetalation protocol (eq 58). The aryl bromide to phenol conversion is achieved via lithium–halogen exchange followed by trapping with B(OiProp)$_3$ and subsequent peroxide oxidation in one pot.[179]

$$(58)$$

Compared to the arylboronic acids as borane source, significant higher yields are observed for the corresponding in situ generated lithium trimethylphenylborates in the rhodium-catalyzed 1,4-addition to α,β-unsaturated ketones (eq 59).[180] The borates are formed by a metal–halogen exchange–trimethoxyborane quench procedure. This constitutes an important new enantioselective Michael reaction type.[181,182]

R = H, 4-OMe, 4-Me, 4-CF$_3$, 3,5-di-Me-4-OMe

$$(59)$$

Rh(I)/L* = Rh(acac)(C$_2$H$_4$)$_2$/   98–99% ee (S)
(S)-binap (3 mol %)

In studies concerned with optimization of the Suzuki cross-coupling of 2-pyridyl nucleophiles, lithium triisopropyl 2-pyridylborates are shown to be the most suitable boron coupling partners (eq 60). The borates are isolated and subjected to coupling with various aryl and hetaryl halides (X = Cl, Br) to give the corresponding azabiaryls.[183]

$$(60)$$

Eliminations. An improved synthetic protocol based on a Wittig olefination strategy allows the preparation of carbon-linked pentasaccharides by a BuLi/HMPA-mediated generation of the ylide from a sugar phosphonium salt precursor and Wittig reaction with formyl mannopyranosides.[184]

During the enantioselective formal synthesis of 4-demethoxy-daunomycin, BuLi proves to be the base of choice for a Hofmann elimination process of the quaternary ammonium salt of a trans-β-amino alcohol to obtain the corresponding allylic alcohol with a slight erosion of ee (eq 61).[185] Other bases such as KHMDS and KOtBu afford 1,4-dimethoxy-naphthalene as a result of an undesired consequent dehydration.

$$(61)$$

The C22–C26 central tetrahydropyran ring of phorboxazole, a macrocyclic marine antifungal natural product, is prepared using a Petasis–Ferrier rearrangement (eq 62).[186] Unfortunately, the required Z-enol acetate is not selectively obtained from a type-II Julia olefination using BuLi deprotonation and exposure of the resulting anion to chloroiodoethane and i-PrMgBr (as reducing agent). But finally, treatment of the E/Z mixture with Me$_2$AlCl furnishes only the desired tetrahydropyran.

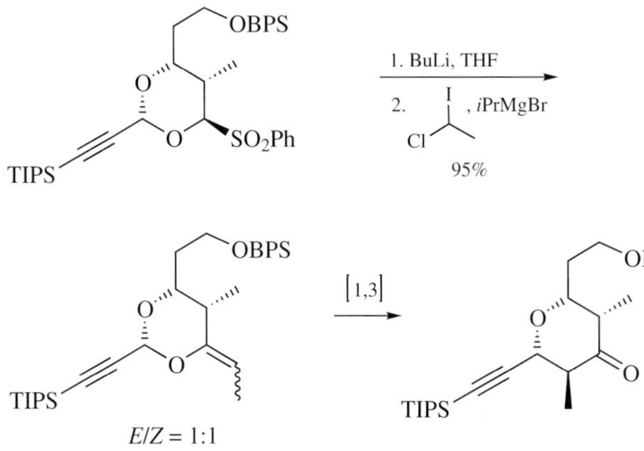

$E/Z = 1:1$ (62)

The synthesis of pyranoid and furanoid glycols from glycosyl sulfoxides by treatment with organolithium reagents (eq 63) involves a sulfoxide/metal exchange to generate a glycosyllithium derivative that undergoes β-elimination of its C-2 substituent followed by fragmentation.[187]

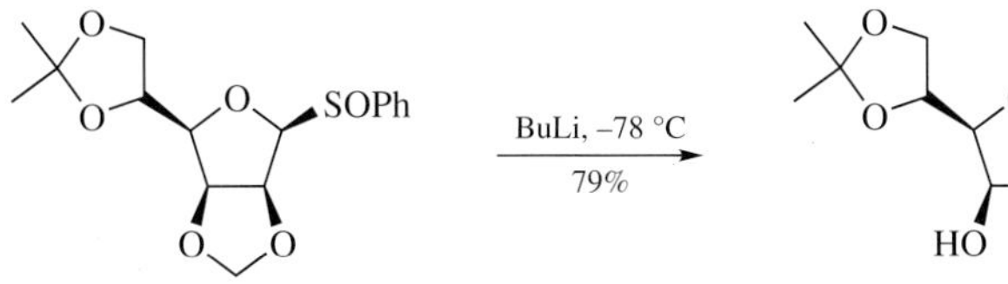

(63)

A general one-pot procedure for the synthesis of alkynes from aldehydes (Corey–Fuchs reaction) that involves the synthesis of the triphenylphosphonium dibromomethane reagent has been explored by Michel and coworkers.[188] In general, the base of choice is *t*-BuOK, but in case of the *N*-Boc piperidine, use of BuLi in order to avoid unwanted side reactions between the carbamate and the acetylide function provides a cleaner reaction at low temperature (eq 64).

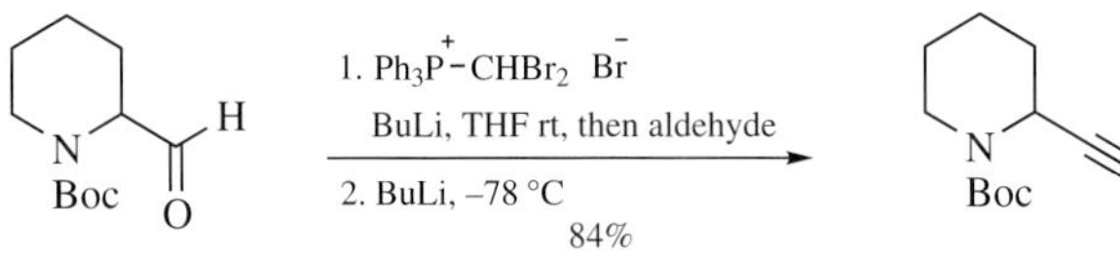

(64)

One of the key steps involved in the total synthesis of the marine natural product (+)-brasilenyne is a Peterson olefination to introduce the (*Z*)-enyne side chain (eq 65).[189] BuLi treatment gives the lithiated 1,3-bis(triisopropylsilyl)propyne, which is treated with the aldehyde to afford the desired enyne with a 6:1 (*Z:E*) ratio.

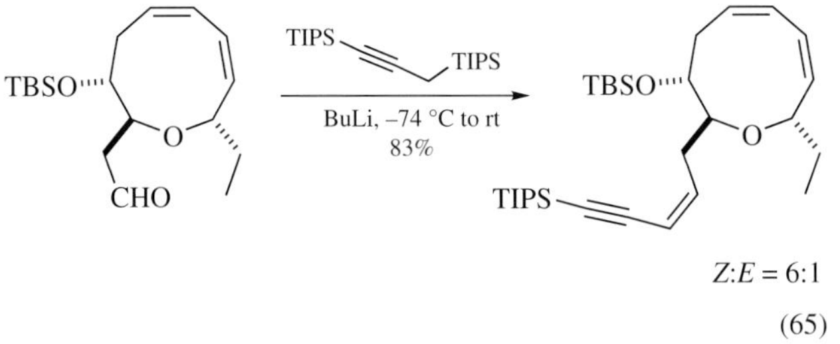

$Z:E = 6:1$ (65)

The Peterson olefination is known for its better performance compared to the corresponding Wittig process for hindered substrates. This is demonstrated in the first asymmetric synthesis of

(+)-maritimol, a member of the stemodane diterpenoids (eq 66). Thus, the key step, a Thorpe–Ziegler annulation, requires a 1,5-dinitrile motif. This is achieved by the generation of an α-silyl boronate, obtained by BuLi deprotonation of trimethylsilylacetonitrile and subsequent transmetalation with triisopropyl borate, which is then condensed with the tricyclic aldehyde.[190]

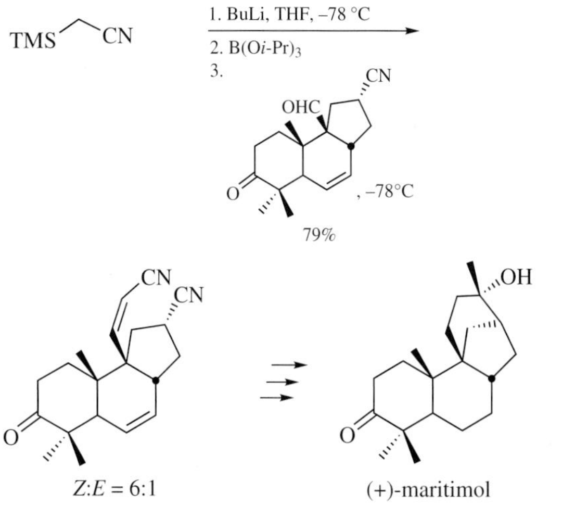

$Z:E = 6:1$ (+)-maritimol (66)

Ferrocenyls. Direct nonauxiliary-mediated enantioselective synthesis of ferrocenyl carboxamide derivatives with planar chirality using BuLi/(−)-sparteine-mediated D*o*M is demonstrated (eq 67).[191]

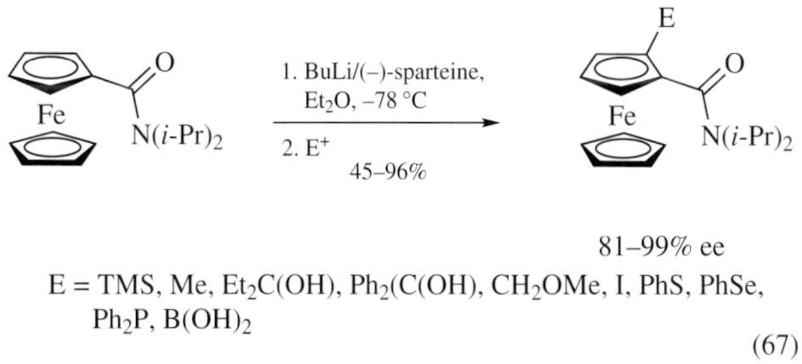

81–99% ee

E = TMS, Me, Et₂C(OH), Ph₂(C(OH), CH₂OMe, I, PhS, PhSe, Ph₂P, B(OH)₂

(67)

The electrophiles introduced provide diverse carbon- and heteroatom-based chiral ferrocenes, some of which are related to popular ligands currently used in small- and large-scale enantioselective catalysis. In extension of the methodology, a direct and highly efficient enantioselective synthesis of C_2-symmetric ferrocene diamides from a achiral ferrocenyldicarboxamide by BuLi/ (−)-sparteine-mediated D*o*M is described (eq 68).[192] In the same context, the first case of double asymmetric induction of ferrocene planar chirality is reported. Amplification of an otherwise low diastereoselectivity by induction of planar chirality via a matched chiral (DMG)/(−)-sparteine interaction is shown for a menthylferrocenylsulfonate derivative.[193] In another application, BuLi-mediated deprotonation of ferrocenyl amides and electrophilic quench with aldehydes and ketones yields carbinols from which, upon acid catalyzed fragmentation (Nesmeyanov ferrocenylcarbenium ion reaction), unusual 6-aryl and 6,6-diaryl ring functionalized fulvenes are prepared.[194]

$$E = TMS, Me, Et_2C(OH), Ph_2(C(OH) I, PhS, PhSe, Ph_2P \tag{68}$$

NH Deprotonations. During a modified Madelung indole synthesis, the convergent high-yielding condensation of the two main fragments of (−)-penitrem D, N-silylation (1.1 equiv BuLi) and lateral metalation (s-BuLi) of a complex o-toluidine substrate followed by condensation with a lactone fragment is carried out (eq 69).[195]

Treatment of a (2.2.1)-bicyclic amine with BuLi gives a chiral lithium amide that undergoes highly stereoselective Michael additions to α,β-unsaturated esters (eq 70).[196] Treatment with N-iodosuccinimide releases the chiral auxiliary as the bornylalde-hyde and furnishes the optically active β-amino esters.

Enolate Formation. Molander and coworkers report a synthetic route toward the marine metabolite eunicellin diterpenes

(eq 71).[197] Treatment of a tricyclic β-keto ester intermediate with BuLi and subsequent γ-methylation occur with complete diastereoselectivity; the concluding Krapcho reaction not only removes the methoxycarbonyl group, but also effects epimerization of the newly formed stereogenic center. A subsequent epoxide intermediate is reduced using Sharpless' reagent (WCl_6/BuLi) to provide the required olefin.[198,199]

Palacios and coworkers describe a simple *de novo* construction of fluorinated pyridines via a stereoselective synthesis of fluorine-containing β-aminophosphonate intermediates (eq 72).[200]

Thus, treatment of enamines with BuLi yields the lithiated species, which was selectively C-methylated to give the corresponding α-methyl-substituted enamines in good yields. The reaction of the enamines with α,β-unsaturated ketones is also explored (eq 73). Furthermore, the lithio-enamines undergo 1,2-addition to unsaturated ketones to give intermediates, which, upon cyclization and prototropic rearrangement, provide a regioselective de novo synthesis of fluorinated pyridines.

Miscellaneous. Lithium amides of chiral pyrrolidines, tetrahydrofurans, and tetrahydrothiophenes are evaluated as chiral ligands in the condensation of BuLi to o-tolualdehyde (eq 74).[201]

R^1, R^2 = H, alkyl, aryl   46–80% ee   (74)
Y = NMe, S, O

Related Reagents. *t*-Butoxide; Methyllithium; Tungsten(VI) Chloride–*n*-Butyllithium; *n*-Butyllithium–Lithium Dimethylaminoethanol; *s*-Butyllithium and *t*-Butyllithium.

1. (a) Wakefield, B. J. *The Chemistry of Organolithium Compounds*; Pergamon: Oxford, 1974. (b) Wakefield, B. J. *Organolithium Methods*; Academic: San Diego, 1990. (c) Wardell, J. L. In *Comprehensive Organometallic Chemistry*; Wilkinson, G., Ed.; Pergamon: Oxford, 1982; Chapter 2.

2. Gschwend, H. W.; Rodriguez, H. R., *Org. React.* **1979**, *26*, 1.

3. Krief, A., *Tetrahedron* **1980**, *36*, 2531.

4. Wardell, J. L. In *Inorganic Reactions and Methods*; Zuckerman, J. J., Ed.; VCH: New York, 1988; Vol. 11, p 107.

5. Wardell, J. L. In *Inorganic Reactions and Methods*; Zuckerman, J. J., Ed.; VCH: New York, 1988; Vol. 11, p 31.

6. *Dictionary of Organometallic Compounds*; Buckingham, J., Ed.; Chapman & Hall: London, 1984; Vol. 1, p 1213.

7. (a) Seebach, D.; Hässig, R.; Gabriel, J., *Helv. Chim. Acta* **1983**, *66*, 308. (b) Heinzer, J.; Oth, J. F. M.; Seebach, D., *Helv. Chim. Acta* **1985**, *68*, 1848.

8. McGarrity, J. F., *J. Am. Chem. Soc.* **1985**, *107*, 1805, 1810.

9. Plavsik, D.; Srzic, D.; Klasinc, L., *J. Phys. Chem.* **1986**, *90*, 2075.

10. Bates, R. B.; Kroposki, L. M.; Potter, P. E., *J. Org. Chem.* **1972**, *37*, 560.

11. Bauer, W.; Clark, T.; Schleyer, P. v. R., *J. Am. Chem. Soc.* **1987**, *109*, 970.

12. (a) Langer, A. W., *Adv. Chem. Ser.* **1974**, *130*, 1. (b) Smith, W. N., *Adv. Chem. Ser.* **1974**, *130*, 23. (c) West, R., *Adv. Chem. Ser.* **1974**, *130*, 211.

13. (a) Bryce-Smith, D.; Turner, E. E., *J. Chem. Soc.* **1953**, 861. (b) Amonoo-Neizer, E. H.; Shaw, R. A.; Skovlin, D. O.; Smith, B. C., *Inorg. Synth.* **1966**, *8*, 19.

14. Jones, R. G.; Gilman, H., *Org. React.* **1951**, *6*, 339.

15. Watson, S. C.; Eastham, J. F., *J. Organomet. Chem.* **1967**, *9*, 165.

16. See e. g. (a) Gilman, H.; Haubein, A. H., *J. Am. Chem. Soc.* **1944**, *66*, 1515. (b) Gilman, H.; Cartledge, F. K., *J. Organomet. Chem.* **1964**, *2*, 447. (c) Eppley, R. L.; Dixon, J. A., *J. Organomet. Chem.* **1967**, *8*, 176. (d) Collins, P. F.; Kamienski, C. W.; Esmay, D. L.; Ellestad, R. B., *Appl. Catal.* **1961**, *33*, 468. (e) Lipton, M. F.; Sorensen, C. M.; Sadler, A. C.; Shapiro, R. H., *J. Organomet. Chem.* **1980**, *186*, 155. (f) Bergbreiter, D. E.; Pendergrass, E., *J. Org. Chem.* **1981**, *46*, 219.

17. Arnett, E. M.; Moe, K. D., *J. Am. Chem. Soc.* **1991**, *113*, 7068.

18. Meth-Cohn, O.; Gronowitz, S., *J. Chem. Soc., Chem. Commun.* **1966**, 81.

19. Eisch, J. J., *Org. Synth.* **1981**, *2*, 98.

20. Herbrandson, H. F.; Mooney, D. S., *J. Am. Chem. Soc.* **1957**, *79*, 5809.

21. (a) Takashi, K.; Konishi, K.; Ushio, M.; Takaki, M.; Asami, R., *J. Organomet. Chem.* **1973**, *50*, 1. (b) Kaiser, E. M.; McLure, J. R., *J. Organomet. Chem.* **1979**, *175*, 11.

22. Michelot, D.; Clinet, J.-C.; Linstrumelle, G., *Synth. Commun.* **1982**, *12*, 739.

23. (a) Bailey, W. F.; Ovaska, T. V., *J. Am. Chem. Soc.* **1993**, *115*, 3080. (b) Quillinan, A. J.; Scheinmann, F., *Org. Synth.* **1978**, *58*, 1.

24. Rausch, M. D.; Ciappenelli, D. J., *J. Organomet. Chem.* **1967**, *10*, 127.

25. Klein, J.; Medlik-Balan, A., *J. Am. Chem. Soc.* **1977**, *99*, 1473.

26. Akiyama, S.; Hooz, J., *Tetrahedron Lett.* **1973**, 4115.

27. Cardillo, G.; Contento, M.; Sandri, S., *Tetrahedron Lett.* **1974**, 2215.

28. Schlosser, M.; Strunk, S., *Tetrahedron Lett.* **1984**, *25*, 741.

29. Faigl, F.; Schlosser, M., *Tetrahedron Lett.* **1991**, *32*, 3369.

30. (a) Heus-Kloos, Y. A.; de Jong, R. L. P.; Verkruisse, H. D.; Brandsma, L.; Julia, S., *Synthesis* **1985**, 958. (b) Mordini, A.; Palio, G.; Ricci, A.; Taddei, M., *Tetrahedron Lett.* **1988**, *29*, 4991.

31. (a) Ahlbrecht, H.; Dollinger, H., *Tetrahedron Lett.* **1984**, *25*, 1353. (b) Hommes, H.; Verkruijsse, H. D.; Brandsma, L., *Tetrahedron Lett.* **1981**, *22*, 2495.

32. (a) Ahlbrecht, H., *Chimia* **1977**, *31*, 391. (b) Seebach, D.; Geiss, K.-H. In *New Applications of Organometallic Reagents in Organic Synthesis*; Seyferth, D., Ed.; Elsevier: Amsterdam, 1976; p 1.

33. Beak, P.; Zajdel, W. J.; Reitz, D. B., *C. R. Hebd. Seances Acad. Sci.* **1984**, *84*, 471.

34. Magnus, P., *Tetrahedron* **1977**, *33*, 2019.

35. Corey, E. J.; Seebach, D., *J. Org. Chem.* **1966**, *31*, 4097.

36. Durst, T. In *Comprehensive Organic Chemistry*; Barton, D. H. R.; Ollis, W. D., Eds.; Pergamon: Oxford, 1979; Vol. 3, p 171.

37. Schlosser, M.; Schaub, B.; Spahic, B.; Sleiter, G., *Helv. Chim. Acta* **1973**, *56*, 2166.

38. Schoufs, M.; Meyer, J.; Vermeer, P.; Brandsma, L., *Recl. Trav. Chim. Pays-Bas* **1977**, *96*, 259.

39. Schlosser, M.; Ladenberger, V., *Chem. Ber.* **1967**, *100*, 3893.

40. Drakesmith, F. G.; Richardson, R. D. Stewart, O. J.; Tarrant, P., *J. Org. Chem.* **1968**, *33*, 286.

41. Schöllkopf, U.; Stafforst, D.; Jentsch, R., *Liebigs Ann. Chem.* **1977**, 1167.

42. Lehr, F.; Gonnermann, J.; Seebach, D., *Helv. Chim. Acta* **1979**, *62*, 2258.

43. (a) Patterson, D. J., *J. Organomet. Chem.* **1967**, *8*, 199. (b) Appel, R.; Wander, M.; Knoll, F., *Chem. Ber.* **1979**, *112*, 1093. (c) Issleib, K.; Abicht, H. P., *J. Prakt. Chem.* **1970**, *312*, 456.

44. Maercker, A., *Angew. Chem., Int. Ed. Engl.* **1987**, *26*, 972.

45. Maercker, A.; Demuth, W., *Liebigs Ann. Chem.* **1977**, 1909.

46. Bates, R. B.; Kroposki, L. M.; Potter, D. E., *J. Org. Chem.* **1972**, *37*, 560.

47. Gröbel, B. T.; Seebach, D., *Synthesis* **1977**, 357.

48. Seebach, D.; Beck, A. K., *Org. Synth., Coll. Vol.* **1988**, *6*, 316.

49. Cere, V.; Paolucci, C.; Pollicino, S.; Sandri, E.; Fava, A., *J. Org. Chem.* **1991**, *56*, 4513.

50. Kodama, M.; Matsuki, Y.; Ito, S., *Tetrahedron Lett.* **1975**, 3065.

51. Ramanathan, V.; Levine, R., *J. Org. Chem.* **1962**, *27*, 1216.

52. Labaudiniere, R.; Hilboll, G.; Leon-Lomeli, A.; Lautenschläger, H.-H.; Parnham, M.; Kuhl, P.; Dereu, N., *J. Med. Chem.* **1992**, *35*, 3156.

53. Schroeder, R.; Schöllkopf, U.; Blume, E.; Hoppe, I., *Liebigs Ann. Chem.* **1975**, 533.

54. Chadwick, D. J.; Cliffe, I. A.; *J. Chem. Soc., Perkin Trans. 1* **1979**, 2845.

55. Butler, D. E.; Alexander, S. M., *J. Org. Chem.* **1972**, *37*, 215.

56. Noyce, D. S.; Stowe, G. T., *J. Org. Chem.* **1973**, *38*, 3762.

57. Behringer, H. Ramert, R., *Liebigs Ann. Chem.* **1975**, 1264.

58. Raap, R., *Can. J. Chem.* **1971**, *49*, 2139.

59. Biellmann, J. F.; Ducep, J.-B., *Org. React.* **1982**, *27*, 1.

60. Yamamoto, Y., *Comprehensive Organic Synthesis* **1991**, *1*, 55.

61. Evans, D. A.; Andrews, G. C.; Buckwalter, B., *J. Am. Chem. Soc.* **1974**, *96*, 5560.

62. Stotter, P. L.; Hornish, R. E., *J. Am. Chem. Soc.* **1973**, *95*, 4444.

63. Hoppe, D.; Hanko, R.; Brönneke, A.; Lichtenberg, F.; Hülsen, E., *Chem. Ber.* **1985**, *118*, 2822.

64. Hoppe, D., *Angew. Chem., Int. Ed. Engl.* **1984**, *23*, 932.

65. Hoppe, D.; Hanko, R.; Brönneke, A.; Lichtenberg, F., *Angew. Chem., Int. Ed. Engl.* **1981**, *20*, 1024.

66. Beak, P.; Zajdel, W. J., *J. Am. Chem. Soc.* **1984**, *106*, 1010.

67. (a) Tischler, A. N.; Tischler, M. H., *Tetrahedron Lett.* **1978**, 3. (b) Tischler, A. N.; Tischler, M. H., *Tetrahedron Lett.* **1978**, 3407.

68. Savignac, P.; Leroux, Y.; Normant, H., *Tetrahedron* **1975**, *31*, 877.

69. Meyers, A. I.; Edwards, P. D.; Rieker, W. F.; Bailey, T. R., *J. Am. Chem. Soc.* **1984**, *106*, 3270.

70. (a) Gonzalez, M. A.; Meyers, A. I., *Tetrahedron Lett.* **1989**, *30*, 47. (b) Gonzalez, M. A.; Meyers, A. I., *Tetrahedron Lett.* **1989**, *30*, 43.

71. Gais, H. J.; Hellmann, G., *J. Am. Chem. Soc.* **1992**, *114*, 4439.

72. Hoppe, D.; Krämer, T., *Angew. Chem., Int. Ed. Engl.* **1986**, *25*, 160.

73. Rein, K.; Goicoechea-Pappas, M.; Anklekar, T. V.; Hart, G. C.; Smith, G. A.; Gawley, R. E., *J. Am. Chem. Soc.* **1989**, *111*, 2211.

74. Meyers, A. I.; Elworthy, T. R., *J. Org. Chem.* **1992**, *57*, 4732.

75. Meyers, A. I.; Dickman, D. I.; Bailey, T. R., *J. Am. Chem. Soc.* **1985**, *107*, 7974.

76. Gawley, R. E.; Rein, K.; Chemburkar, S. R., *J. Org. Chem.* **1989**, *54*, 3002.

77. Meyers, A. I., *Tetrahedron* **1992**, *48*, 2589.

78. Meyers, A. I., *Aldrichim. Acta* **1985**, *18*, 59.

79. Snieckus, V., *C. R. Hebd. Seances Acad. Sci.* **1990**, *90*, 879.

80. Slocum, D. W.; Jennings, C. A., *J. Org. Chem.* **1976**, *41*, 3653.

81. Beak, P.; Snieckus, V., *Acc. Chem. Res.* **1982**, *15*, 306.

82. (a) Ludt, R. E.; Griffiths, T. S.; McGrath, K. N.; Hauser, C. R., *J. Org. Chem.* **1973**, *38*, 1668. (b) Beak, P.; Brown, R. A., *J. Org. Chem.* **1982**, *47*, 34.

83. Newman, M. S.; Kanakarajan, J., *J. Org. Chem.* **1980**, *45*, 2301.

84. Winkle, M. R.; Ronald, R. C., *J. Org. Chem.* **1982**, *47*, 2101.

85. Meyers, A. I.; Lutomski, K., *J. Org. Chem.* **1979**, *44*, 4464.

86. Ziegler, F. E.; Fowler, K. W., *J. Org. Chem.* **1976**, *41*, 1564.

87. Reuman, M.; Meyers, A. I., *Tetrahedron* **1985**, *41*, 837.

88. Furlano, D. C.; Calderon, S. N.; Chen, G.; Kirk, K. L., *J. Org. Chem.* **1988**, *53*, 3145.

89. Baldwin, J. E.; Bair, K. W., *Tetrahedron Lett.* **1978**, 2559.

90. Fuhrer, W.; Gschwend, H. W., *J. Org. Chem.* **1979**, *44*, 1133.

91. Katsuura, K.; Snieckus, V., *Can. J. Chem.* **1987**, *65*, 3165.

92. Turner, J. A., *J. Org. Chem.* **1983**, *48*, 3401.

93. Moorhoff, C. M.; Paquette, L. A., *J. Org. Chem.* **1991**, *56*, 703.

94. Gay, R. L.; Hauser, C. R., *J. Am. Chem. Soc.* **1967**, *89*, 1647.

95. Adam, W.; Cueto, O., *J. Org. Chem.* **1977**, *42*, 38.

96. Hosomi, A.; Araki, Y.; Sakurai, H., *J. Am. Chem. Soc.* **1982**, *104*, 2081.

97. Meyers, A. I.; Malone, G. R.; Adickes, H. W., *Tetrahedron Lett.* **1970**, 3715.

98. (a) Sauvetre, R.; Roux-Schmitt, M.-C.; Seyden-Penne, J., *Tetrahedron Lett.* **1978**, *34*, 2135. (b) Sauvetre, R.; Seyden-Penne, J., *Tetrahedron Lett.* **1976**, 3949.

99. Takano, S.; Shimazaki, Y.; Takahashi, M.; Ogasawara, K., *J. Chem. Soc., Chem. Commun.* **1988**, 1004.

100. Tamaru, Y.; Kagotini, M.; Furukawa, Y.; Amino, Y.; Yoshida, Z., *Tetrahedron Lett.* **1981**, *22*, 3413.

101. Hucklin, S. N.; Weiler, L., *J. Am. Chem. Soc.* **1974**, *96*, 4691.

102. Fraser, R. R.; Banville, J.; Dhawan, K. C., *J. Am. Chem. Soc.* **1978**, *100*, 7999.

103. Kanemasa, S.; Tatsukawa, A.; Wada, E., *J. Org. Chem.* **1991**, *56*, 2875.

104. Laborde, E.; Kiely, J. S.; Culbertson, T. P.; Leheski, L. E., *J. Med. Chem.* **1993**, *36*, 1964.

105. Cahiez, G.; Bernard, D.; Normant, J. F., *Synthesis* **1976**, 245.

106. Kitatani, K.; Hiyama, T.; Nozaki, H., *J. Am. Chem. Soc.* **1975**, *97*, 949.

107. Hoeg, D. F.; Lusk, D. I.; Crumbliss, A. L., *J. Am. Chem. Soc.* **1965**, *87*, 4147.

108. Foulger, N. J.; Wakefield, B. J., *Organomet. Synth.* **1986**, *3*, 369.

109. Kitatani, K.; Hiyama, T.; Nozaki, H., *Bull. Chem. Soc. Jpn.* **1977**, *50*, 3288.

110. Georgoulis, C.; Meyet, J.; Smajda, W., *J. Organomet. Chem.* **1976**, *121*, 271.

111. Panek, E. J.; Neff, B. L.; Chu, H.; Panek, M. G., *J. Am. Chem. Soc.* **1975**, *97*, 3996.

112. Cooke, Jr., M. P., *J. Org. Chem.* **1993**, *58*, 2910.

113. Bailey, W. F.; Nurmi, T. T.; Patricia, J. J.; Wang, W., *J. Am. Chem. Soc.* **1987**, *109*, 2442.

114. (a) Bailey, W. F.; Punzalan, E. P., *J. Org. Chem.* **1990**, *55*, 5404. (b) Negishi, E.; Swanson, D. R.; Rousset, C. J., *J. Org. Chem.* **1990**, *55*, 5406.

115. (a) Peterson, D. J.; Ward, J. F., *J. Organomet. Chem.* **1974**, *66*, 209. (b) Seyferth, D.; Mammarella, R. E., *J. Organomet. Chem.* **1979**, *77*, 53.

116. (a) Still, W. C.; Sreekumar, C., *J. Am. Chem. Soc.* **1980**, *102*, 1201. (b) Sawyer, J. S.; Kucerovy, A.; Macdonald, T. L.; McGarvey, G. J., *J. Am. Chem. Soc.* **1988**, *110*, 842. (c) Sawyer, J. S.; Macdonald, T. L.; McGarvey, G. J., *J. Am. Chem. Soc.* **1984**, *106*, 3376.

117. Lohse, P.; Loner, H.; Acklin, P.; Sternfeld, F.; Pfaltz, A., *Tetrahedron Lett.* **1991**, *32*, 615.

118. (a) Chong, J. M.; Park, S. B., *J. Org. Chem.* **1992**, *57*, 2220. (b) Pearson, W. H.; Lindbeck, A. C., *J. Org. Chem.* **1989**, *54*, 5651. (c) Pearson, W. H.; Lindbeck, A. C., *J. Am. Chem. Soc.* **1991**, *113*, 8546. (d) N-stannylmethanimines: Pearson, W. H.; Szura, D. P.; Postich, M. J., *J. Am. Chem. Soc.* **1992**, *114*, 1329.

119. Chan, P. C.-M.; Chong, J. M., *Tetrahedron Lett.* **1990**, *31*, 1985.

120. Broka, C. A.; Shen, T., *J. Am. Chem. Soc.* **1989**, *111*, 2981.

121. (a) McGarvey, G. J.; Kimura, M., *J. Org. Chem.* **1985**, *50*, 4652. (b) Krief, A.; Hobe, M., *Tetrahedron Lett.* **1992**, *33*, 6527.

122. Krief, A.; Evrard, G.; Badaoui, E.; DeBeys, V.; Dieden, R., *Tetrahedron Lett.* **1989**, *30*, 5635.

123. Krief, A.; Dumont, W.; Clarembeau, M.; Bernard, G.; Badaoui, E., *Tetrahedron* **1989**, *45*, 2005.

124. Reich, H. J.; Bowe, M. D., *J. Am. Chem. Soc.* **1990**, *112*, 8994.

125. (a) Reich, H. J., *Acc. Chem. Res.* **1979**, *12*, 22. (b) Liotta, D., *Acc. Chem. Res.* **1984**, *17*, 28. (c) Clive, D. L. J., *Tetrahedron* **1978**, *34*, 1049. (d) Davis, F. A.; Reddy, R. T., *J. Org. Chem.* **1992**, 2599.

126. Seebach, D.; Beck, A. K., *Angew. Chem., Int. Ed. Engl.* **1974**, *13*, 806.

127. Tomoki, H.; Kambe, N.; Ogawa, A.; Miyoshi, N.; Murai, S.; Sonoda, N., *Angew. Chem., Int. Ed. Engl.* **1987**, *26*, 1187.

128. Curtin, D. Y.; Koehl, Jr., W. J., *J. Am. Chem. Soc.* **1962**, *84*, 1967.

129. Marshall, J. A., *Comprehensive Organic Synthesis* **1991**, *3*, 975.

130. Crombie, L.; Darnbrough, G.; Pattenden, G., *J. Chem. Soc., Chem. Commun.* **1976**, 684.

131. Midland, M. M.; Kwon, Y. C., *Tetrahedron Lett.* **1985**, *26*, 5021.

132. Tsai, D. J.-S.; Midland, M. M., *J. Org. Chem.* **1984**, *49*, 1842.

133. Nakai, T.; Mikami, K.; Taya, S.; Fujita, Y., *J. Am. Chem. Soc.* **1981**, *103*, 6492.

134. (a) Marshall, J. A.; Robinson, E. D.; Lebreton, J., *J. Org. Chem.* **1990**, *55*, 227. (b) Takashi, T.; Nemoto, H.; Kanda, Y.; Tsuji, J.; Fujise, Y., *J. Org. Chem.* **1986**, *51*, 4315.

135. Bulman Page, P. C.; Rayner, C. M.; Sutherland, I. O., *J. Chem. Soc., Chem. Commun.* **1988**, 356.

136. Kuwajima, I.; Kato, M., *Tetrahedron Lett.* **1980**, *21*, 623.

137. Klein, K. P.; Van Eenam, D. N.; Hauser, C. R., *J. Org. Chem.* **1967**, *32*, 1155.

138. Harvey, R. G.; Cho, H., *J. Am. Chem. Soc.* **1974**, *96*, 2434.

139. Hunt, E.; Lythgoe, B., *J. Chem. Soc., Chem. Commun.* **1972**, 757.

140. Gilman, H.; Gorsich, R. D., *J. Am. Chem. Soc.* **1957**, *79*, 2625.

141. Fischer, P.; Schaefer, G., *Angew. Chem., Int. Ed. Engl.* **1981**, *20*, 863.

142. Cainelli, G.; Ronchi, A. U.; Bertini, F.; Grasselli, P.; Zubiani, G., *Tetrahedron* **1971**, *27*, 6109.

143. Bandouy, R.; Gore, J.; Ruest, L., *Tetrahedron* **1987**, *43*, 1099.

144. Corey, E. J.; Fuchs, P. L., *Tetrahedron Lett.* **1972**, 3769.

145. Romo, D.; Johnson, D. D.; Plamondon, L.; Miwa, T.; Schreiber, S. L., *J. Org. Chem.* **1992**, *57*, 5060.

146. Schone, N. E.; Knudsen, M. J., *J. Org. Chem.* **1987**, *52*, 569.

147. Chamberlin, A. R.; Bond, F. T., *Synthesis* **1979**, 44.

148. Tomooka, K.; Yamamoto, H.; Nakai, T., *Angew. Chem. Int. Ed.* **2000**, *39*, 4500.

149. Tomooka, K.; Kikuchi, M.; Igawa, K.; Suzuki, M.; Keong, P.-H.; Nakai, T., *Angew. Chem., Int. Ed.* **2000**, *39*, 4502.

150. Schaudt, M.; Blechert, S., *J. Org. Chem.* **2003**, *68*, 2913.

151. Lampe, J. W.; Hughes, P. F.; Biggers, C. K.; Smith, S. H.; Hu, H., *J. Org. Chem.* **1996**, *61*, 4572.

152. Reed, M. A.; Chang, M. T.; Snieckus, V., *Org. Lett.* **2004**, *6*, 2297.

153. Stará, I. G.; Starý, I.; Tichý, M.; Závada, J.; Hanus, V., *J. Am. Chem. Soc.* **1994**, *116*, 5084.

154. Norsikian, S.; Marek, I.; Normant, J. F., *Tetrahedron Lett.* **1997**, *38*, 7523.

155. Norsikian, S.; Marek, I.; Klein, S.; Poisson, J. F.; Normant, J. F., *Chem. Eur. J.* **1995**, *5*, 2055.

156. Hogan, L. A.-M.; O'Shea, D. F., *J. Am. Chem. Soc.* **2006**, *128*, 10360.

157. Hogan, L. A.-M.; O'Shea, D. F., *J. Org. Chem.* **2008**, *73*, 2503.

158. Hogan, L. A.-M.; Tricotet, T.; Meek, A.; Khokhar, S. S.; O'Shea, D. F., *J. Org. Chem.* **2008**, *73*, 6041.

159. Hogan, A.-M. L.; O'Shea, D. F., *J. Am. Chem. Soc.* **2006**, *128*, 10360.

160. Hogan, A.-M. L.; O'Shea, D. F., *J. Org. Chem.* **2008**, *73*, 2503.

161. Ona-Burgos, P., Fernández, I.; Roces, L.; Torre-Fernández, L.; Gracía-Granada, S.; López-Ortiz, F., *Org. Lett.* **2008**, *10*, 3195.

162. Ona-Burgos, P., Fernández, I.; Iglesias, M. J.; Gracía-Granada, S.; López-Ortiz, F., *Org. Lett.* **2008**, *10*, 537.

163. Hoarau, C.; Couture, A.; Deniau, E.; Grandclaudon, P., *J. Org. Chem.* **2002**, *67*, 5846.

164. Malkov, A. V.; Bella, M.; Stará, I. G.; Kocovský, P., *Tetrahedron Lett.* **2001**, *42*, 3045.

165. Derdau, V.; Snieckus, V., *J. Org. Chem.* **2001**, *66*, 1992.

166. MacNeil, S. L.; Familoni, O. B.; Snieckus, S., *J. Org. Chem.* **2001**, *66*, 3662.

167. Macklin, T. K.; Reed, M. A.; Snieckus, V., *Eur. J. Org. Chem.* **2008**, *9*, 1507.

168. Proudfoot, J. R.; Patel, U. R.; Dyatkin, A. B., *J. Org. Chem.* **1997**, *62*, 1851.

169. Metallinos, C.; Nerdinger, S.; Snieckus, V., *Org. Lett.* **1999**, *1*, 1183.

170. Blanchet, J.; Macklin, T.; Ang, P.; Metallinos, C.; Snieckus, V., *J. Org. Chem.* **2007**, *72*, 3199.

171. Kauch, M.; Hoppe, D., *Can. J. Chem.* **2001**, *79*, 1736.

172. Kauch, M.; Snieckus, V.; Hoppe, D., *J. Org. Chem.* **2005**, *70*, 7149.

173. Nerdinger, S.; Kendall, C.; Cai, X.; Marchart, R.; Riebel, P.; Johnson, M. R.; Yin, C.-F.; Hénaff, N.; Eltis, L. D.; Snieckus, V., *J. Org. Chem.* **2007**, *72*, 5960.

174. Srinivasan, K.; Laughrey, Z. R.; Gibb, B. C., *Eur. J. Org. Chem.* **2008**, 3265.

175. Gu, Y. G.; Bayburt, E. K.; Michaelides, M. R.; Lin, C. W.; Shiosaki, K., *Bioorg. Med. Chem. Lett.* **1999**, *9*, 1341.

176. Suffert, J.; Toussaint, D., *J. Org. Chem.* **1995**, *60*, 3550.

177. Macklin, T. K.; Snieckus, V., *Org. Lett.* **2005**, *7*, 2519.

178. Dyke, A. M.; Gill, D. M.; Harvey, J. N.; Hester, A. J.; Lloyd-Jones, G. C.; Munoz, M. P.; Shepperson, I. R., *Angew. Chem., Int. Ed.* **2008**, *47*, 5067.

179. McCubbin, J. A.; Tong, X.; Wang, R.; Zhao, Y.; Snieckus, V.; Lemieux, R. P., *J. Am. Chem. Soc.* **2004**, *126*, 1161.

180. Takaya, Y.; Ogasawara, M.; Hayashi, T., *Tetrahedron Lett.* **1999**, *40*, 6957.

181. Skai, M.; Hayashi, H.; Miyaura, N., *Organometallics* **1997**, *16*, 4229.

182. Hayashi, T.; Yamasaki, K., *Chem. Rev.* **2003**, *103*, 2829.

183. Billingsley, K. L.; Buchwald, S. L., *Angew. Chem., Int. Ed.* **2008**, *47*, 4695.

184. Dondoni, A.; Marra, A.; Mizuno, M.; Giovannini, P. P., *J. Org. Chem.* **2002**, *67*, 4186.

185. Sekine, A.; Ohshima, T.; Shibasaki, M., *Tetrahedron* **2002**, *58*, 75.

186. Smith, A. B., III; Minbiole, K. P.; Verhoest, P. R.; Schelhaas, M., *J. Am. Chem. Soc.* **2001**, *123*, 10942.

187. Gómez, A. M.; Casillas, M.; Barrio, A.; Gawel, A.; López, J. C., *Eur. J. Org. Chem.* **2008**, 3933.

188. Michel, P.; Gennet, D.; Rassat, A., *Tetrahedron Lett.* **1999**, *40*, 8575.

189. Denmark, S. E.; Yang, S.-M., *J. Am. Chem. Soc.* **2002**, *124*, 15196.

190. Toro, A.; Nowak, P.; Deslongchamps, P., *J. Am. Chem. Soc.* **2000**, *122*, 4526.

191. Tsukazaki, M.; Tinkl, M.; Roglans, A.; Chapell, B. J.; Taylor, N. J.; Snieckus, V., *J. Am. Chem. Soc.* **1996**, *118*, 685.

192. Laufer, R. S.; Veith, U.; Taylor, N. J.; Snieckus, V., *Org. Lett.* **2000**, *2*, 629.

193. Metallinos, C.; Snieckus, V., *Org. Lett.* **2004**, *4*, 1935.

194. Miao, B.; Tinkl, M.; Snieckus, V., *Tetrahedron Lett.* **1999**, *40*, 2449.

195. Smith, A. B., III; Kanoh, N.; Ishiyama, H.; Minakawa, N.; Rainier, J. D.; Hartz, R. A.; Cho, Y. S.; Cui, H.; Moser, W. H., *J. Am. Chem. Soc.* **2003**, *125*, 8228.

196. Node, M.; Hashimoto, D.; Katoh, T.; Ochi, S.; Ozeki, M.; Watanabe, T.; Kajimoto, T., *Org. Lett.* **2008**, *10*, 2653.

197. Molander, G. A.; St. Jean, D. J., Jr.; Haas, J., *J. Am. Chem. Soc.* **2004**, *126*, 1642.

198. Sharpless, K. B.; Umbreit, M. A.; Nieth, M. T.; Flood, T. C., *J. Am. Chem. Soc.* **1972**, *94*, 6538.

199. Johnson, J.; Kim, S.-H.; Bifano, M.; DiMarco, J.; Fairchild, C.; Gougoutas, J.; Lee, F.; Long, B.; Tokarski, J.; Vite, G., *Org. Lett.* **2000**, *2*, 1537.

200. Palacios, F.; Ochoa de Retana, A.; Oyarzabal, J.; Pascual, S.; Fernández de Trocóniz, G., *J. Org. Chem.* **2008**, *73*, 4568.

201. Duguet, N.; Petit, S. M.; Marchand, P.; Harrison-Marchand, A.; Maddaluno, J., *J. Org. Chem.* **2008**, *73*, 5397.

N,N'-Carbonyl Diimidazole[1]

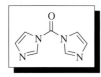

[530-62-1]   C$_7$H$_6$N$_4$O   (MW 162.15)

InChI = 1/C7H6N4O/c12-7(10-3-1-8-5-10)11-4-2-9-6-11/h1-6H

InChIKey = PFKFTWBEEFSNDU-UHFFFAOYAX

(reagent for activation of carboxylic acids;[1] synthesis of esters,[2] amides,[3] peptides,[4] aldehydes,[5] ketones,[6] β-keto thioesters,[7] tetronic acids;[8] ureas,[9] isocyanates,[10] carbonates;[9] halides from alcohols;[11] glycosidation;[12] dehydration[13−16])

Physical Data: mp 116–118 °C.

Solubility: no quantitative data available. Inert solvents such as THF, benzene, CHCl$_3$, DMF are commonly used for reactions.

Form Supplied in: commercially available white solid.

Analysis of Reagent Purity: purity can be determined by measuring the amount of CO$_2$ evolved on hydrolysis.

Preparative Method: prepared by mixing phosgene with four equivalents of imidazole in benzene/THF.[17]

Purification: may be purified by recrystallization from hot, anhydrous THF with careful exclusion of moisture.[17]

Handling, Storage, and Precautions: moisture sensitive; reacts readily with water with evolution of carbon dioxide. May be kept for long periods either in a sealed tube or in a desiccator over P$_2$O$_5$.

Original Commentary

Alan Armstrong

University of Bath, Bath, UK

Activation of Carboxylic Acids: Synthesis of Acyl Imidazoles. *N,N'* Carbonyldiimidazole (**1**) converts carboxylic acids into the corresponding acylimidazoles (**2**) (eq 1).[1] The method can be applied to a wide range of aliphatic, aromatic, and heterocyclic carboxylic acids, including some examples (such as formic acid and vitamin A acid) where acid chloride formation is difficult. The reactivity of (**2**) is similar to that of acid chlorides, but the former have the advantage that they are generally crystalline and easily handled. Isolation of (**2**) is simple, but often unnecessary; further reaction with nucleophiles is usually performed in the same reaction vessel. Conversion of (**2**) into acid chlorides (via reaction with HCl),[18] hydrazides,[3] hydroxamic acids,[3] and peroxy esters[19] have all been described. Preparation of the more important carboxylic acid derivatives is described below.

Esters from Carboxylic Acids. Reaction of equimolar amounts of carboxylic acid, alcohol, and (**1**) in an inert solvent (e.g. THF, benzene, or chloroform) results in ester formation (eq 2). Since alcoholysis of the intermediate acylimidazole is relatively slow, the reaction mixture must be heated at 60–70 °C for some time. However, addition of a catalytic amount of a base such as **Sodium Amide** to convert the alcohol to the alkoxide, or a

catalytic amount of the alkoxide itself, allows rapid and complete formation of the ester at room temperature.[2] The base catalyst must of course be added after formation of (**2**) from the acid is complete, as indicated by cessation of evolution of carbon dioxide.

$$RCO_2H + (1) \longrightarrow (2) + CO_2 + \text{imidazole} \quad (1)$$

$$RCO_2H + R^1OH \xrightarrow[\substack{2.\ cat.\ R^1ONa \\ or\ NaNH_2}]{1.\ (1)} RCO_2R^1 \quad (2)$$

Esters of tertiary alcohols may not be prepared from carboxylic acids containing acidic α-protons using this modified procedure, since deprotonation and subsequent condensation, competes. However, the use of stoichiometric **1,8-Diazabicyclo[5.4.0]-undec-7-ene** as base has been shown to provide good yields of *t*-butyl esters even for acids with acidic α-protons (eq 3).[20] This procedure was unsuccessful for pivalic acid or for *N*-acyl-α-amino acids.

$$Ph(CH_2)_3CO_2H \xrightarrow[\substack{2.\ 1\ equiv\ t\text{-BuOH} \\ 1\ equiv\ DBU \\ 40\ °C,\ 24\ h}]{\substack{1.\ 1\ equiv\ (1),\ DMF \\ 40\ °C,\ 1\ h}} Ph(CH_2)_3CO_2\text{-}t\text{-Bu} \quad (3)$$

An alternative approach to increasing the rate of esterification is to activate further the intermediate (**2**). **N-Bromosuccinimide** has been used for this purpose,[21] but unsaturation in the carboxylic acid or alcohol is not tolerated. More generally useful is the addition of an activated halide, usually **Allyl Bromide**, to a chloroform solution of (**1**) and a carboxylic acid, resulting in formation of the acylimidazolium salt (**3**) (eq 4).[22] Addition of the alcohol and stirring for 1–10 h at room temperature or at reflux affords good yields of ester in a one-pot procedure. These conditions work well for the formation of methyl, ethyl, and *t*-butyl esters of aliphatic, aromatic, and α,β-unsaturated acids. Hindered esters such as *t*-butyl pivalate can be prepared cleanly (90% yield). The only limitation is that substrates must not contain functionality that can be alkylated by the excess of the reactive halide.

$$RCO_2H \xrightarrow[\substack{(1),\ CHCl_3 \\ 4–5\ equiv\ allyl\ bromide}]{} (3) \xrightarrow{R^1OH} RCO_2R^1 \quad (4)$$

Since the purity of commercial (**1**) may be variable due to its water sensitivity, it is common to employ an excess in order to

Essential Reactions for Organic Synthesis, First Edition. Edited by Philip L. Fuchs.

© 2016 John Wiley & Sons, Ltd. Published 2016 by John Wiley & Sons, Ltd.

ensure complete conversion of the carboxylic acid to the acylimidazole. It has been suggested that alcohols react faster with residual (**1**) than with the acylimidazole (**2**), thus reducing the yield of ester. A procedure has been developed for removal of excess (**1**) before addition of the alcohol.[23]

Macrolactonization has been accomplished using (**1**).[24] Thiol and selenol esters can also be prepared in one pot from carboxylic acids using (**1**);[25] reaction of the intermediate (**2**) with aromatic thiols or selenols is complete within a few minutes in DMF, while aliphatic thiols require a few hours. Formation of the phenylthiol ester of *N*-Cbz-L-phenylalanine was accompanied by only slight racemization. *N,N'*-**Carbonyldi-sym-triazine** can be used in place of (**1**), with similar results.

Amides from Carboxylic Acids: Peptide Synthesis. Analogous to ester formation, reaction of equimolar amounts of a carboxylic acid and (**1**) in THF, DMF, or chloroform, followed by addition of an amine, allows amide bond formation.[3] The method has been applied to peptide synthesis (eq 5).[4] One equivalent of (**1**) is added to a 1 M solution of an acylamino acid in THF, followed after 1 h by the desired amino acid or peptide ester. The amino acid ester hydrochloride may be used directly instead of the free amino acid ester. An aqueous solution of the amino acid salt can even be used, but yields are lower.

$$\text{CbzNH} \begin{array}{c} R \\ | \\ \end{array} \text{OH} \xrightarrow[\text{2. EtO}_2\text{C}\begin{array}{c}|\\R^1\end{array}\text{NH}_2]{\text{1. (1)}} \text{CbzNH}\begin{array}{c}R\\|\end{array}\begin{array}{c}H\\N\end{array}\begin{array}{c}|\\R^1\end{array}\text{CO}_2\text{Et} \quad (5)$$

As is the case in esterification reactions, the presence of unreacted (**1**) can cause problems since the amine reacts with this as quickly as it does with the acylimidazole, forming urea byproducts that can be difficult to separate. Use of exactly one equivalent of (**1**) is difficult due to its moisture sensitivity, and also because of the tendency of some peptides or amino acids to form hydrates. Paul and Anderson solved this problem by use of an excess of (**1**) to form the acylimidazole, then cooling to −5 °C and adding a small amount of water to destroy the unreacted (**1**) before addition of the amine.[26]

For the sensitive coupling of Cbz-glycyl-L-phenylalanine and ethyl glycinate (the Anderson test), Paul and Anderson reported the level of racemization as 5% using THF as solvent at −10 °C, but as <0.5% in DMF.[4] Performing the same coupling reaction at room temperature, Beyerman and van den Brink later claimed that the degree of racemization in DMF was in fact as high as 17%, and reported better results (no detectable racemization) using the related reagent *N,N'*-carbonyl di-*sym*-triazine in place of (**1**).[27] In a comparative study of several reagents, Weygand and co-workers also observed extensive racemization using (**1**).[28] In the formation of tyrosine esters, Paul reported that the mixed anhydride method is to be preferred to use of (**1**), since *O*-acylation is a major side reaction with the latter.[29]

Aldehydes and Ketones from Carboxylic Acids. Reduction of the derived acylimidazole (**2**) with **Lithium Aluminum Hydride** achieves conversion of an aliphatic or aromatic carboxylic acid to an aldehyde (eq 6).[5] **Diisobutylaluminum Hydride** has also been used, allowing preparation of α-acylamino aldehydes from *N*-protected amino acids.[30] Similarly, reaction of

(**2**) with Grignard reagents affords ketones,[6] with little evidence for formation of tertiary alcohol.

$$(\mathbf{2}) \xrightarrow{\begin{array}{c}\text{LiAlH}_4\\\\\text{EtMgBr}\end{array}} \begin{array}{c}\text{RCHO}\\\\R\overset{O}{\underset{||}{C}}\text{Et}\end{array} \quad (6)$$

Reaction of acylimidazoles with the appropriate carbon nucleophile has also been used for the preparation of α-nitro ketones[31] and β-keto sulfoxides.[32]

C-Acylation of Active Methylene Compounds. Treatment of an acylimidazole, derived from a carboxylic acid and (**1**), with the magnesium salt of a malonic or methylmalonic half thiol ester results in *C*-acylation under neutral conditions (eq 7).[7] The presence of secondary hydroxyl functionality in the carboxylic acid is tolerated, but primary alcohols require protection. Magnesium salts of malonic esters may be used equally effectively. Intramolecular *C*-acylation of ketones has also been reported.[33]

$$\text{Ph(CH}_2)_2\text{CO}_2\text{H} \xrightarrow[\substack{\text{18 h, 25 °C}\\100\%}]{\substack{\text{1. (1)}\\\text{2.}\left[^-\text{O}_2\text{C}\overset{O}{\underset{||}{}}\text{SEt}\right]_2\text{Mg}^{2+}}} \text{Ph(CH}_2)_2\overset{O\quad O}{\underset{||\quad||}{}}\text{SEt} \quad (7)$$

Tetronic Acids. A synthesis of tetronic acids reported by Smith and co-workers relies on the reaction between (**1**) and the dianion derived from an α-hydroxy ketone (eq 8).[8] The reaction proceeds in moderate yield (31–57%).

$$\underset{\text{OH}}{\overset{O}{\underset{||}{}}} \xrightarrow[48\%]{\substack{\text{1. 2 equiv LDA}\\\text{2. (1)}}} \underset{O}{\overset{HO}{}}\overset{O}{\underset{||}{}} \quad (8)$$

Ureas and Carbonates. Reagent (**1**) may be used as a direct replacement for the highly toxic **Phosgene** in reactions with alcohols and amines. Reaction of (**1**) with two equivalents of a primary aliphatic or aromatic amine at room temperature rapidly yields a symmetrical urea (eq 9).[9] If only one equivalent of a primary amine is added to (**1**), then the imidazole-*N*-carboxamide (**4**) is formed (eq 10). These compounds can dissociate into isocyanates and imidazole, even at room temperature, and distillation from the reaction mixture provides a useful synthesis of isocyanates (eq 10).[10] Secondary amines react only at one side of (**1**) at room temperature, again giving the imidazole-*N*-carboxamide of type (**4**).

$$2\text{ RNH}_2 \xrightarrow{\text{(1)}} \text{RHN}\overset{O}{\underset{||}{C}}\text{NHR} \quad (9)$$

$$\text{R = alkyl, aryl}$$

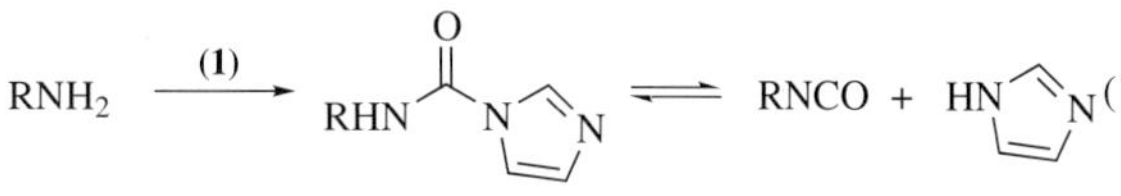

R = alkyl, aryl

(4)

Reaction of (**1**) with one equivalent of an alcohol provides the imidazole-*N*-carboxylic ester (**5**) (eq 11).[34] Further treatment with another alcohol or phenol yields an unsymmetrical carbonate; alternatively, reaction with an amine affords a carbamate (eq 12).[34]

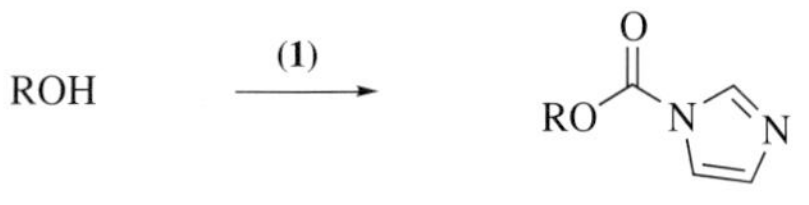

(5)

X = O, NH

Heating (**1**) with an excess of an alcohol or phenol gives the symmetrical carbonate.[9] This reaction can be accelerated dramatically by the presence of catalytic base (e.g. **Sodium Ethoxide**). Reaction under these conditions is so exothermic even at room temperature that only *t*-butanol stops at the imidazole-*N*-carboxylic ester stage.

1,2-Diamines, 1,2-diols, or 1,2-amino alcohols react with (**1**) to form cyclic ureas,[35] carbonates,[36] or oxazolidinones,[35] respectively. In the case of cyclohexane-1,2-diols, the *cis*-diol reacts much more rapidly than the *trans*, as would be expected.[36] Thiazolidinones can also be prepared using (**1**).[37]

Halides from Alcohols. Treatment of an alcohol with (**1**) and an excess (at least three equivalents) of an activated halide results in its conversion to the corresponding halide (eq 13).[11] Any halide more reactive than the product halide may be used, but in practice **Allyl Bromide** or **Iodomethane** give best results as they are effective and readily removed after the reaction. Acetonitrile is the best solvent and yields are generally high (>80%). Bromide or iodide formation work well, but not chlorination. Optically active alcohols are racemized.

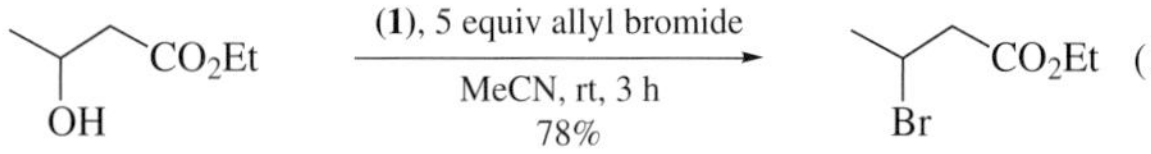

Glycosidation. A mild glycosidation procedure involving (**1**) has been reported by Ford and Ley.[12] A carbohydrate derivative in ether or dichloromethane reacts with (**1**) through the anomeric C-1-hydroxyl to give the (1-imidazolylcarbonyl) glycoside (IMG) (**6**) (eq 14). Isolation of (**6**) is not usually necessary; treatment with one equivalent of an alcohol and two equivalents of **Zinc Bromide** in ether at reflux gives the glycoside. Generally, higher $\alpha{:}\beta$ ratios are obtained for more hindered alcohols and when ether is used as solvent rather than the less polar dichloromethane. Along with the fact that the $\alpha{:}\beta$ ratio is independent of the configuration of (**6**), this suggests an S_N1-type mechanism. In contrast, treatment

of (**6**) with **Acetyl Chloride** provides the anomeric chloride with essentially exclusive inversion.

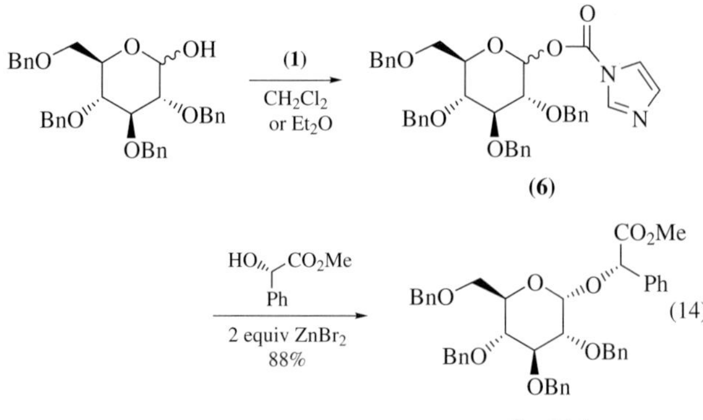

(6)

$\alpha{:}\beta = 10{:}1$

Dehydration. Reagent (**1**) has been used for the dehydration of various substrates, including aldoximes (to give nitriles),[13] β-hydroxy amino acids,[14] and β-hydroxy sulfones.[15] 3-Aryl-2-hydroxyiminopropionic acids undergo dehydration and decarboxylation, to give 2-aryl acetonitriles, upon reaction with (**1**).[16]

First Update

Wenju Li
University of Illinois at Chicago, Chicago, IL, USA

Preparation of β-Enamino Acid Derivatives. The reaction of ketimines (**7**) with *N,N'*-carbonyldiimidazole (CDI) (**1**) in the presence of $BF_3 \cdot OEt_2$ as catalyst affords β-enamino carbonylimidazole derivatives (**8**) in good to excellent yields (eq 15).[38,39]

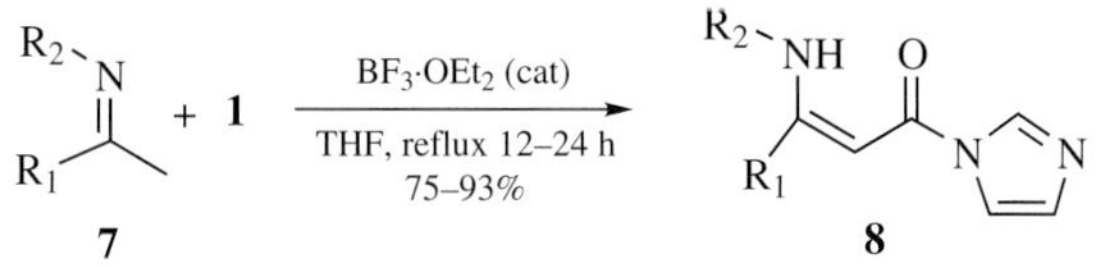

The derivatives of **8** can react with alcohols and thiols to produce β-enamino esters (**9**) and thioesters (**10**) (eq 16). Hydrolysis of **9** under mild acidic conditions achieves the production of the corresponding β-keto esters (**11**) (eq 17).[39]

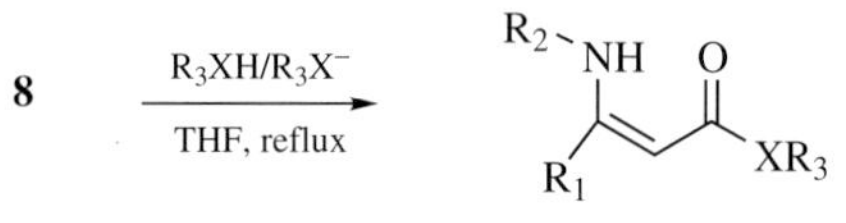

9: X = O; **10**: X = S

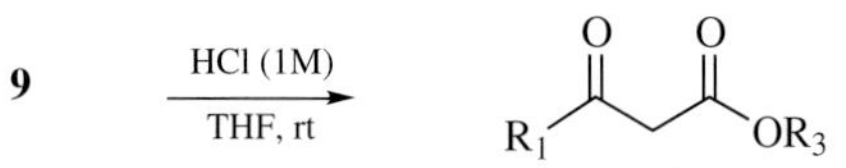

11

Synthesis of the Imidazo[1,2-*a*]pyridine Ring System. A simple one-pot procedure for the synthesis of imidazo-[1,2-*a*]pyridines (**13**) involves a novel condensation reaction of 1-(arylacetyl)imidazoles (**12**) with dimethyl acetylenedicarboxylate (eq 18).[40–42] Functionalization of **13** with hydrazine hydrate affords the imidazo[1',2':1,6]pyrido[2,3-*d*]pyridazine ring system (**14**).[42]

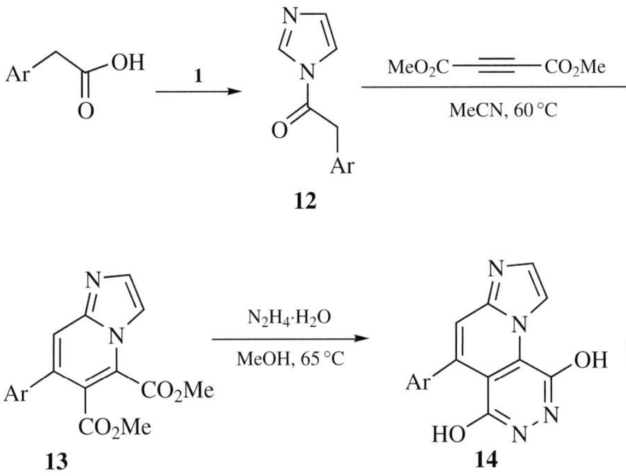

(18)

Formation of 1-[(*E*) 3-(1-Imidazolyl)-2-alkenoyl] Imidazoles. The reaction of (**1**) with propynoic acid was found to afford **16**, the *cis*-adduct of imidazole to the C–C triple bond of 1-propynoylimidazole (**15**) (eq 19).[43] In the first step, the reaction of (**1**) with propynoic acid generates **15**, CO$_2$, and imidazole. The by-product imidazole subsequently attacks the electron-deficient alkyne moiety of **15** to give **16**.

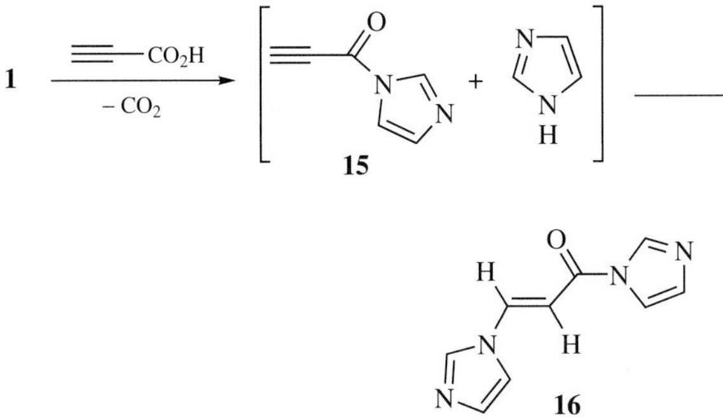

(19)

The reaction of 3-butynoic acid with **1** gives a similar result most likely through an imidazole-catalyzed isomerization of the initially formed 1-(3-butynoyl)imidazole (**17**) to 1-(2-butynoyl)-imidazole (**18**) (eq 20).

Synthesis of α-Diketones from α,β-Dihydroxyketones. Treatment of *syn*-α,β-dihydroxyketones (**19**) with **1** in CH$_2$Cl$_2$ at room temperature for 2 h affords the α-diketones (**21**) as the only isolated products in good yield (eq 21).[44] It was established that the reaction is based on the elimination of the cyclic carbonates (**20**) formed in situ. The *anti*-α,β-dihydroxyketones can also gave the corresponding α-diketones in good yields, albeit at higher temperature and with longer reaction times.

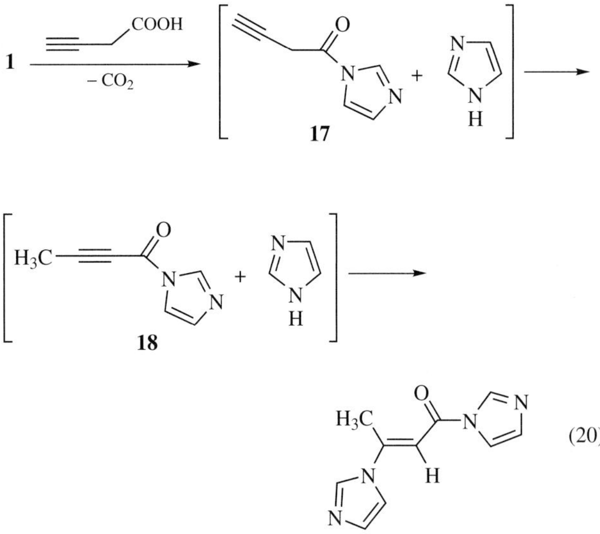

(20)

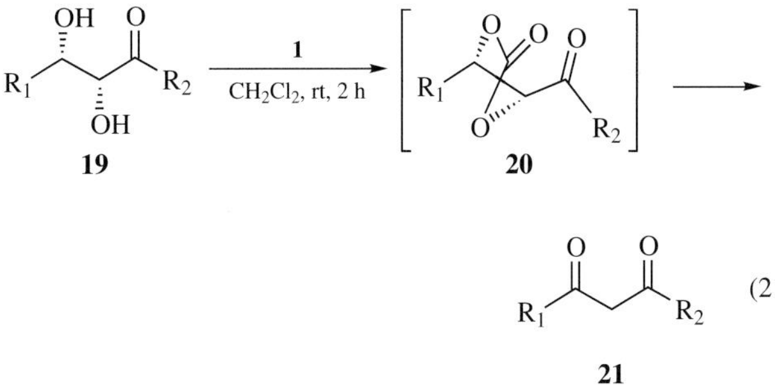

(21)

Synthesis of Bridged Dipyrrinones. Treatment of dipyrrinones (**22**) in anhydrous dichloromethane with 5 equiv of **1** in the presence of DBU afforded *N,N'*-carbonyl-bridged dipyrriones (**23**) in almost quantitative yield (eq 22).[45,46]

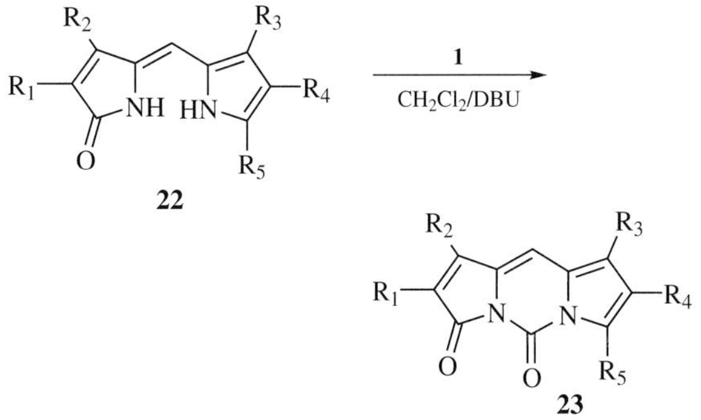

(22)

Asymmetric Synthesis of Tetramic Acid Derivatives. An asymmetric synthesis of tetramic acid derivatives was reported by Fustero et al.[47] The key step is a carbonyl transfer from **1** to α-diimines (**24**) to form *N*-alkyl-4-alkylamino-5-methylenepyrrol-2-ones (**25**). These can be transformed into tetramic acid derivatives (**26**) by asymmetric reduction of the exocyclic double bond induced by the chiral R* group and subsequent hydrolysis of the *N*-protected enamine (eq 23).

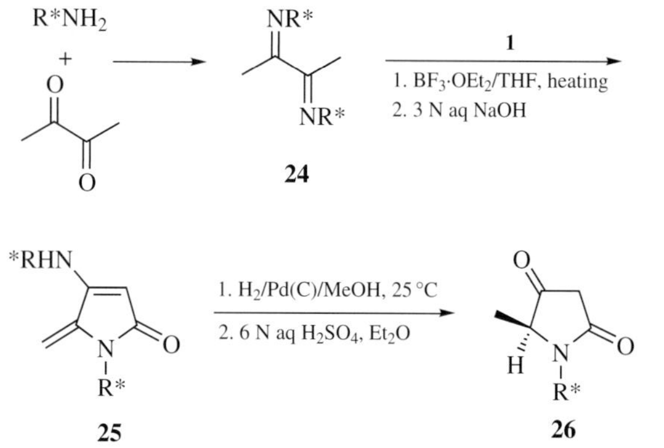

(23)

Ferrocenyl(alkyl)imidazole. A new method for the introduction of imidazolyl groups into ferrocenylalkyl compounds by the ferrocenylalkylation reaction of **1** was reported.[48] A mixture of equimolar amounts of ferrocenylcarbinols (**27**) and **1** in methylene chloride was boiled for 1 h, and *N*-ferrocenyl(alkyl)imidazoles (**28**) were synthesized in high yields (eq 24).

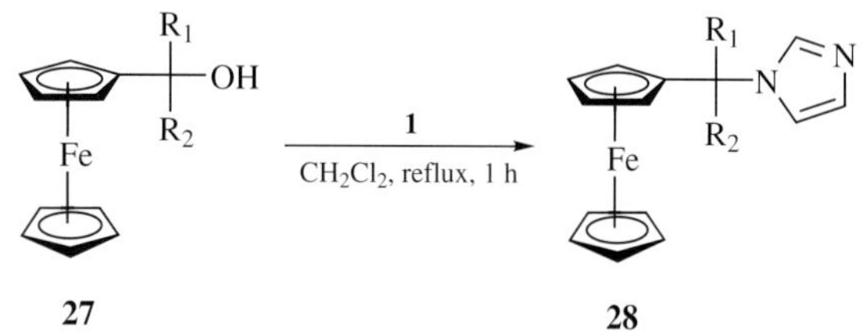

(24)

Further, ferrocenyl(alkyl)imidazole compound (**30**) was made by a similar method.[49] Reaction of compound **29** with **1** in CH₃CN affords imidazole (**30**) in excellent yield. After methylation with methyl iodide followed by deprotonation, **30** was transformed into a planar-chiral stable carbene (**31**) (eq 25).

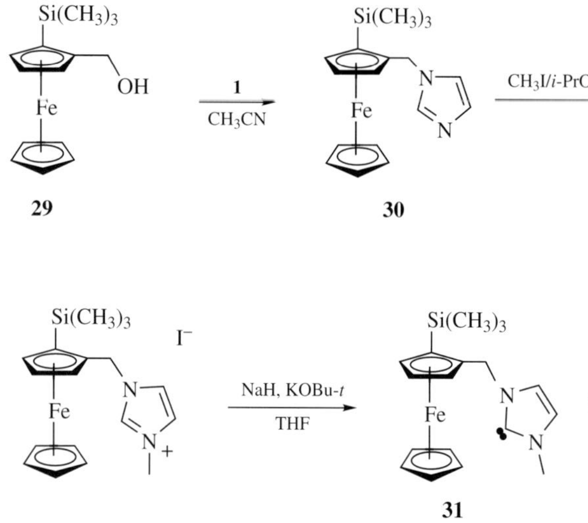

(25)

Bisimidazolylferrocenylmethane (**32**) can be prepared from the reaction of **1** and ferrocene carboxaldehyde at 80 °C in the presence of CoCl₂·6H₂O (eq 26).[50] Methylation of (**32**) with CH₃I in the presence of NH₄PF₆ affords the bis(methylimidazolium) salt (**33**), which is a precursor in the synthesis of some *N*-heterocyclic carbenes (NHCs).

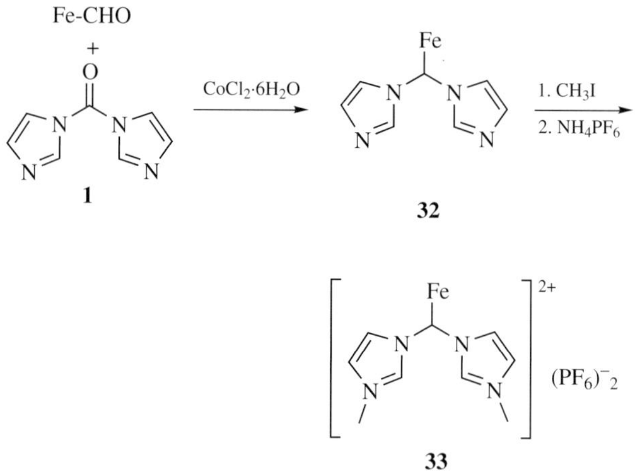

(26)

Asymmetric Synthesis of Carbonates. Reaction of **1** with 1 equiv of an alcohol provides the imidazole-*N*-carboxylic ester (**5**) (eq 11).[34] It was reported that imidazole-*N*-carboxylic esters (**34**, R = H, CH₃) formed from primary or secondary benzyl alcohols have high selectivity when reacted with the mixture of primary and secondary alcohols (eq 27).[51]

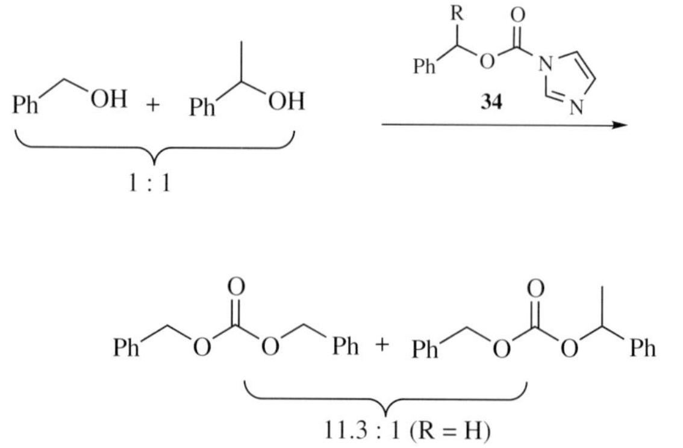

(27)

Using this simple method, the benzyl carbonate of the primary hydroxyl of methyl α-D-glucopyranoside was synthesized with high selectivity and moderate yield (eq 28).

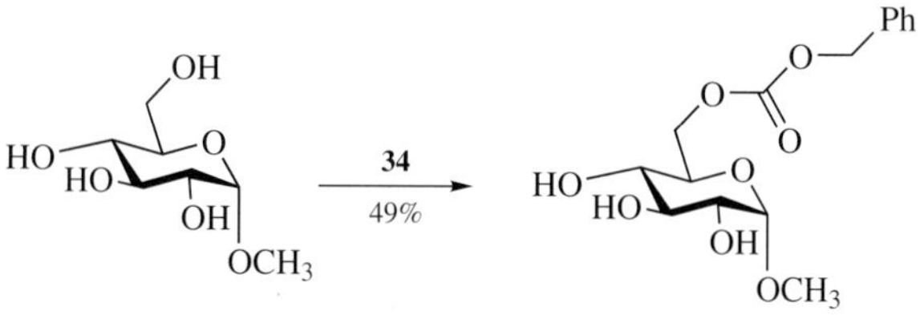

(28)

Later, it was found that the imidazole-*N*-carboxylate esters (**35** and **36**) formed by the reaction of **1** and either secondary or tertiary alcohols react selectively with primary hydroxyls in polyols containing mixtures of primary, secondary, and tertiary hydroxyls, without the need for protecting group chemistry (eq 29).[52] With this method, selective *t*-Boc protection of primary hydroxyl groups could be easily achieved through a single step.

35   **36**

$$HO\!\!\diagup\!\!\diagdown\!\!OH \xrightarrow[97\%]{\textbf{35 or 36}} R-O\diagup\!\!O\diagup\!\!\diagdown\!\!OH \quad (29)$$

When **35** or **36** were reacted with polyols consisting of 1,2-diol substitution, cyclic carbonates were formed without unwanted side reactions or reaction with secondary amines (eqs 30 and 31). The cyclic carbonate formation most possibly proceeds via the selective reaction of **35** or **36** at the primary hydroxyl followed by cyclization involving the neighboring secondary hydroxyl.

$$HO\!\!\diagup\!\!\diagdown\!\!OH \xrightarrow[85\%]{\textbf{35 or 36}} \text{(cyclic carbonate)} \quad (30)$$

$$HO\!\!\diagup\!\!\diagdown\!\!N \xrightarrow[85\%]{\textbf{35 or 36}} \text{(cyclic carbonate)} \quad (31)$$

3,5-Dimethoxybenzoin (DMB) carbonates (**39**) were found to be photochemically removable alcohol protecting groups.[53] DMB carbonates can be installed by the carbamate product (**38**) of DMB-OH (**37**) and *N,N′*-carbonyldiimidazole activated by methylation (eqs 32 and 33).[54]

$$\textbf{1} \xrightarrow{\text{MeOTf}} \text{(imidazolium triflate)} \; + \text{DMB-OH, 37}$$

DMB-OH, **37**

38 (32)

$$ROH \xrightarrow[\text{MeNO}_2, \text{ pyridine}]{\textbf{38}} RO\diagup\!\!O-DMB \xrightarrow[\text{THF}]{\text{light, 350 nm}}$$

39

$$ROH \; + \; \text{(benzofuran)} \quad (33)$$

Formation of Carbamates and *N*-Alkylimidazoles. The conversion of a variety of primary and secondary alcohols and phenols into their corresponding *N*-alkylimidazoles (**40**) and triazoles by high yield reactions with **1** or carbonyl ditriazole at room temperature, or in refluxing MeCN, EtOAc, and CH_2Cl_2 has been reported (eq 34).[55] While this method was subsequently applied with success in some published cases,[49,50,56,57] it was found that the treatment of 1,5-diisopropylphenol with **1** under these conditions, or at higher temperatures, gave only the corresponding carbamate (**41**) (eq 35).[58,59]

$$ROH \xrightarrow{\textbf{1}} \text{(N-alkylimidazole)} \quad (34)$$

40

$$\text{(diisopropylphenol)} \xrightarrow[\substack{CH_2Cl_2, \text{ refulx (5 h)} \\ 97\%}]{\textbf{1}} \text{(carbamate)} \quad (35)$$

41

This series of reactions was investigated further, leading to the conclusion that the reactions of nonbenzylic primary and secondary aliphatic alcohols with **1** afford the corresponding carbamates but not *N*-alkylimidazoles.[60] For benzylic primary alcohols, formation of *N*-alkylimidazoles occurs at 170 °C in several different solvents by way of the initially formed carbamates. Under these forcing conditions, or even at lower temperatures, elimination is a significant side reaction for benzylic secondary alcohols with β-hydrogen atoms. It was also found that reactions of *N,N*-disubstituted β-amino alcohols with **1** proceed under relatively mild conditions to form *N*-alkylimidazoles possibly by way of an aziridinium intermediate.

Synthesis of 1,2-Imidazolylpropylamines. *syn*-1,2-Imidazolylpropylamines and *anti*-1,2-Imidazolylpropylamines (**45** and **47**) can be synthesized regio- and stereo-specifically by the reaction of the commercially available *syn*- and *anti*-isomers of 2-dimethylamino-1-phenylpropane (**42** and **46**) with **1** (eqs 36 and 37).[61] The reaction of *syn*-isomer (+)-(**42**) with **1** in acetonitrile at room temperature afforded imidazolyl-derived carbamate (**43**) as the sole product (quenching the reaction with methanol afforded methyl carbonate (**44**). When the same reaction was heated to 80 °C, *syn*-imidazolylpropylamine (**44**) was formed exclusively in 76% yield. Heating carbamate (**43**) in acetonitrile to 80 °C also afforded (**45**) In contrast, the reaction of *anti*-isomer (+)-(**46**) with **1** in acetonitrile at room temperature afforded *anti*-imidazolylpropylamine (**47**) exclusively. Based on the net retention of stereochemistry in each reaction, a mechanism involving double-inversion of configuration through an aziridinium intermediate was proposed.

$$(36)$$

reaction and a mechanism was proposed involving the intermediate alkylammonium *N*-alkyl carbamate (**48**) formed from the reaction of amine with CO_2 (eq 39).

$$2\ RNH_2\ +\ CO_2\ \rightleftharpoons\ \underset{48}{R\diagdown\underset{H}{N}\diagup\overset{O}{C}\diagdown\bar{O}\ \overset{+}{N}H_3R}\qquad(39)$$

Other Applications. **1** was applied in the synthesis of some important heterocyclic compounds such as oxadiazole,[67] triazolone,[68] quinazoline-2,4-diones,[69] tetrahydroisoquinoline-hydantoins,[70] and pyrazole.[71] The activation of phosphates to phosphorimidazolides, phosphate-phosphate coupling and nucleotide formation can also rely on **1**.[72–75]

$$(37)$$

C-Acylation. *C*-Acylation can be realized by treatment of acylimidazoles, derived from carboxylic acids and **1**, with the magnesium salts of malonic esters prepared by using Bu_2Mg.[62,63] Acyl imidazoles can also undergo Claisen-type condensation with the dianion of β-keto esters (eq 38).[64,65]

$$(38)$$

Catalytic Effect of Carbon Dioxide in Amidation Reactions using CDI. **1** is widely used in the formation of amides from carboxylic acids and amines.[3] It was found that when the CO_2 released in the acylimidazole formation step was thoroughly removed from the reaction vessel, by means of better agitation and ventilation, the amidation became substantially slower.[66] Further investigation confirmed the catalytic effect of CO_2 in amidation

1. Staab, H. A., *Angew. Chem., Int. Ed. Engl.* **1962**, *1*, 351.

2. Staab, H. A.; Mannschreck, A., *Chem. Ber.* **1962**, *95*, 1284 (*Chem. Abstr.* **1962**, *57*, 5846d).

3. Staab, H. A.; Lüking, M.; Dürr, F. H., *Chem. Ber.* **1962**, *95*, 1275 (*Chem. Abstr.* **1962**, *57*, 5908a).

4. Paul, R.; Anderson, G. W., *J. Am. Chem. Soc.* **1960**, *82*, 4596.

5. Staab, H. A.; Bräunling, H., *Justus Liebigs Ann. Chem.* **1962**, *654*, 119 (*Chem. Abstr.* **1962**, *57*, 5906c).

6. Staab, H. A.; Jost, E., *Justus Liebigs Ann. Chem.* **1962**, *655*, 90 (*Chem. Abstr.* **1962**, *57*, 15 090g).

7. Brooks, D. W.; Lu, L. D.-L.; Masamune, S., *Angew. Chem., Int. Ed. Engl.* **1979**, *18*, 72.

8. Jerris, P. J.; Wovkulich, P. M.; Smith, A. B., III, *Tetrahedron Lett.* **1979**, 4517.

9. Staab, H. A., *Justus Liebigs Ann. Chem.* **1957**, *609*, 75 (*Chem. Abstr.* **1958**, *52*, 7332e).

10. Staab, H. A.; Benz, W., *Justus Liebigs Ann. Chem.* **1961**, *648*, 72 (*Chem. Abstr.* **1962**, *57*, 4649g).

11. Kamijo, T.; Harada, H.; Iizuka, K., *Chem. Pharm. Bull.* **1983**, *31*, 4189.

12. Ford, M. J.; Ley, S. V., *Synlett* **1990**, 255.

13. Foley, H. G.; Dalton, D. R., *J. Chem. Soc., Chem. Commun.* **1973**, 628.

14. Andruszkiewicz, R.; Czerwinski, A., *Synthesis* **1982**, 968.

15. Kang, S.-K.; Park, Y.-W.; Park, Y.-W.; Kim, S.-G.; Jeon, J.-H., *J. Chem. Soc., Perkin Trans. 1* **1992**, 405.

16. Kitagawa, T.; Kawaguchi, M.; Inoue, S.; Katayama, S., *Chem. Pharm. Bull.* **1991**, *39*, 3030.

17. Staab, H. A.; Wendel, K., *Q. Rev., Chem. Soc.* **1968**, *48*, 44.

18. Staab, H. A.; Datta, A. P., *Angew. Chem., Int. Ed. Engl.* **1964**, *3*, 132.

19. Hecht, R.; Rüchardt, C., *Chem. Ber.* **1963**, *96*, 1281 (*Chem. Abstr.* **1963**, *59*, 1523h).

20. Ohta, S.; Shimabayashi, A.; Aono, M.; Okamoto, M., *Synthesis* **1982**, 833.

21. Katsuki, T., *Bull. Chem. Soc. Jpn.* **1976**, *49*, 2019.

22. Kamijo, T.; Harada, H.; Iizuka, K., *Chem. Pharm. Bull.* **1984**, *32*, 5044.

23. Morton, R. C.; Mangroo, D.; Gerber, G. E., *Can. J. Chem.* **1988**, *66*, 1701.

24. (a) White, J. D.; Lodwig, S. N.; Trammell, G. L.; Fleming, M. P., *Tetrahedron Lett.* **1974**, 3263. (b) Colvin, E. W.; Purcell, T. A.; Raphael, R. A., *J. Chem. Soc., Chem. Commun.* **1972**, 1031.

25. Gais, H. -J., *Angew. Chem., Int. Ed. Engl.* **1977**, *16*, 244.

26. Paul, R.; Anderson, G. W., *J. Org. Chem.* **1962**, *27*, 2094.

27. Beyerman, H. C.; Van Den Brink, W. M., *Recl. Trav. Chim. Pays-Bas* **1961**, *80*, 1372.

28. Weygand, F.; Prox, A.; Schmidhammer, L.; König, W., *Angew. Chem., Int. Ed. Engl.* **1963**, *2*, 183.

29. Paul, R., *J. Org. Chem.* **1963**, *28*, 236.

30. Khatri, H.; Stammer, C. H., *J. Chem. Commun.* **1979**, 79.

31. Baker, D. C.; Putt, S. R., *Synthesis* **1978**, 478.

32. Ibarra, C. A.; Rodríguez, R. C.; Monreal, M. C. F.; Navarro, F. J. G.; Tesorero, J. M., *J. Org. Chem.* **1989**, *54*, 5620.

33. Garigipati, R. S.; Tschaen, D. M.; Weinreb, S. M., *J. Am. Chem. Soc.* **1985**, *107*, 7790.

34. Staab, H. A., *Justus Liebigs Ann. Chem.* **1957**, *609*, 83 (*Chem. Abstr.* **1957**, *52*, 16341e).

35. Wright, W. B., Jr., *J. Heterocycl. Chem.* **1965**, *2*, 41.

36. Kutney, J. P.; Ratcliffe, A. H., *Synth. Commun.* **1975**, *5*, 47.

37. D'Ischia, M.; Prota, G.; Rotteveel, R. C.; Westerhof, W., *Synth. Commun.* **1987**, *17*, 1577.

38. Fustero, S.; Díaz, M. D.; Carlón, P. R., *Tetrahedron Lett.* **1993**, *34*, 725.

39. Fustero, S.; García de la Torre, M.; Jofré, V.; Carlón, R. P.; Navarro, A.; Fuentes, A. S.; Carrión, J. S., *J. Org. Chem.* **1998**, *63*, 8825.

40. Knölker, H.-J.; Boese, R., *J. Chem. Soc., Chem. Commun.* **1988**, 1151.

41. Knölker, H.-J.; Boese, R.; Hitzemann, R., *Chem. Ber.* **1990**, *123*, 327.

42. Knölker, H.-J.; Boese, R.; Hitzemann, R., *Heterocycles* **1990**, *31*, 1435.

43. Knölker, H.-J.; El-Ahl, A.-A., *Heterocycles* **1993**, *36*, 1381.

44. Kang, S.-K.; Park, D.-C.; Rho, H.-S.; Han, S.-M., *Synth. Commun.* **1993**, *23*, 2219.

45. Brower, J. O.; Lightner, D. A., *J. Org. Chem.* **2002**, *67*, 2713.

46. Boiadjiev, S. E.; Lightner, D. A., *J. Heterocyclic Chem.* **2004**, *41*, 1033.

47. Fustero, S.; Torre, G. M.; Sanz-Cervera, J. F.; Ramírez de Arellano, C.; Piera, J.; Simón, A., *Org. Lett.* **2002**, *4*, 3651.

48. Simenel, A. A.; Morozova, E. A.; Kuzmenko, Y. V.; Snegur, L. V., *J. Organomet. Chem.* **2003**, *665*, 13.

49. Bolm, C.; Kesselgruber, M.; Raabe, G., *Organometallics* **2002**, *21*, 707.

50. Viciano, M.; Mas-Marzá, E.; Poyatos, M.; Sanaú, M.; Crabtree, R. H.; Peris, E., *Angew. Chem., Int. Ed.* **2005**, *44*, 444.

51. Bertolini, G.; Pavich, G.; Vergani, B., *J. Org. Chem.* **1998**, *63*, 6031.

52. Rannard, S. P.; Davis, N. J., *Org. Lett.* **1999**, *1*, 933.

53. Pirrung, M. C.; Bradley, J.-C., *J. Org. Chem.* **1995**, *60*, 1116.

54. Saha, A. K.; Rapoport, H. Schultz, P., *J. Am. Chem. Soc.* **1989**, *111*, 4856.

55. Njar, V. C. O., *Synthesis* **2000**, 2019.

56. Patel, J. B.; Huynh, C. K.; Handratta, V. D.; Gediya, L. K.; Brodie, A. M. H.; Goloubeva, O. G.; Clement, O. O.; Nanne, I. P.; Soprano, D. R.; Njar, V. C. O., *J. Med. Chem.* **2004**, *47*, 6716.

57. Vatèle, J.-M., *Tetrahedron* **2004**, *60*, 4251.

58. Palencia, H.; Garcia-Jimenez, F.; Takacs, J. M., *Tetrahedron Lett.* **2004**, *45*, 3849.

59. Fischer, W., *Synthesis* **2002**, 29.

60. Tang, Y.; Dong, Y.; Vennerstrom, J. L., *Synthesis* **2004**, 2540.

61. Mulvihill, M. J.; Cesario, C.; Smith, V.; Beck, P.; Nigro, A., *J. Org. Chem.* **2004**, *69*, 5124.

62. Hodgson, M. D.; Stupple, P. A.; Pierard, F. Y. T. M.; Labande, A. H.; Johnstone, C., *Chem. Eur. J.* **2001**, *7*, 4465.

63. Hodgson, M. D.; Labande, A. H.; Pierard, Y. T. M. F., *Synlett* **2003**, 59.

64. Parker, A. K.; Lim, Y.-H., *J. Am. Chem. Soc.* **2004**, *126*, 15968.

65. Liang, G.; Seiple, I. B.; Trauner, D., *Org. Lett.* **2005**, *7*, 2837.

66. Vaidyanathan, R.; Kalthod, V. G.; Ngo, D. P.; Manley, J. M.; Lapekas, S. P., *J. Org. Chem.* **2004**, *69*, 2565.

67. Liang, G.-B.; Feng, D. D., *Tetrahedron Lett.* **1996**, *37*, 6627.

68. Romine, J. L.; Martin, S. W.; Gribkoff, V. K.; Boissard, C. G.; Dworetzky, S.; Natale, J.; Li, Y.; Gao, Q.; Meanwell, N. A.; Starrett, J. E., Jr., *J. Med. Chem.* **2002**, *45*, 2942.

69. Okuzumi, T.; Nakanishi, E.; Tsuji, T.; Makino, S., *Tetrahedron* **2003**, *59*, 5603.

70. Charton, J.; Gassiot, A. C.; Melnyk, P.; Girault-Mizzi, S.; Sergheraert, C., *Tetrahedron Lett.* **2004**, *45*, 7081.

71. Armstrong, A.; Jones, L. H.; Knight, J. D.; Kelsey, R. D., *Org. Lett.* **2005**, *7*, 713.

72. Sawai, H.; Hirano, A.; Mori, H.; Shinozuka, K.; Dong, B.; Silverman, R. H., *J. Med. Chem.* **2003**, *46*, 4926.

73. Ye, X.-Y.; Lo, M.-C.; Brunner, L.; Walker, D.; Kahne, D.; Walker, S., *J. Am. Chem. Soc.* **2001**, *123*, 3155.

74. Wahler, D.; Reymond, J.-L., *Can. J. Chem.* **2002**, *80*, 665.

75. Sawai, H.; Kuroda, K.; Hojo, T., *Bull. Chem. Soc. Jpn.* **1989**, *62*, 2018.

Cerium(IV) Ammonium Nitrate[1]

$$(NH_4)_2Ce(NO_3)_6$$

[16774-21-3] $H_8CeN_8O_{18}$ (MW 548.26)

InChI = 1/Ce.6NO3.2H3N/c;6*2-1(3)4;;/h;;;;;;;2*1H3/q+4;6*-1;;/p+2/fCe.6NO3.2H4N/h;;;;;;;2*1H/q7m;2*+1

InChIKey = XMPZTFVPEKAKFH-XPZOHPFVCX

(volumetric standard oxidant;[2] oxidant for many functional groups;[1] can promote oxidative halogenation[3])

Alternate Name: ammonium cerium(IV) nitrate; ceric ammonium nitrate; CAN.

Solubility: sol water (1.41 g mL^{-1} at 25 °C, 2.27 g mL^{-1} at 80 °C); sol nitric acid.

Form Supplied in: orange crystals; widely available.

Handling, Storage, and Precautions: solid used as supplied. No toxicity data available, but cerium is reputed to be of low toxicity.

Original Commentary

Tse-Lok Ho

National Chiao-Tung University, Hsinchu, Taiwan, Republic of China

Functional Group Oxidation. Ce^{IV} in acidic media is a stronger oxidant than elemental chlorine and is exceeded in oxidizing power only by a few reagents (F$_2$, XeO$_3$, Ag^{2+}, O$_3$, HN$_3$). The thermodynamically unstable solutions can be kept for days because of kinetic stability. CAN is a one-electron oxidant soluble in water and to a smaller extent in polar solvents such as acetic acid. Its consumption can be judged by the fading of an orange color to pale yellow, if the substrate or product is not strongly colored. Because of its extremely limited solubility in common organic solvents, oxidations are often carried out in mixed solvents such as aqueous acetonitrile. There are advantages in using dual oxidant systems in which Ce^{IV} is present in catalytic amounts. Cooxidants such as **Sodium Bromate**,[4] **tert-Butyl Hydroperoxide**,[5] and **Oxygen**[6] have been employed. Electrolytic recycling[7] of Ce^{IV} species is also possible.

Cerium(IV) sulfate and a few other ligand-modified CAN reagents have been used for the oxidation. The differences in their oxidation patterns are small, and consequently it is quite safe to replace one particular oxidizing system with another. More rarely employed is cerium(IV) perchlorate.

Oxidation of Alkenes and Arenes. The outcome of the oxidation of alkenes is solvent dependent, but dinitroxylation (eq 1)[8] has been achieved. Certain arylcyclopropanes are converted into the 1,3-diol dinitrates.[9]

$$(1)$$

CAN promotes benzylic oxidation of arenes,[10] e.g. methyl groups are converted into formyl groups but less efficiently when an electron-withdrawing group is present in the aromatic ring. A very interesting molecule, hexaoxo[1$_6$]orthocyclophane in an internal acetal form (eq 2),[11] has been generated via CAN oxidation. Regioselective oxidation is observed with certain substrates, e.g. 2,4-dimethylanisole gives 3-methyl-*p*-anisaldehyde. Oxidation may be diverted into formation of non-aldehyde products by using different media: benzylic acetates[12] are formed in glacial acetic acid, ethers[13] in alcohol solvents, and nitrates[14] in acetonitrile under photolytic conditions.

$$(2)$$

hexaoxo[1$_6$]orthocyclophane

Polynuclear aromatic systems can be oxidized to quinones,[15] but unsymmetrical substrates will often give a mixture of products. It has been reported that mononitro derivatives were formed by the oxidation of polynuclear arenes with CAN adsorbed in silica,[16] whereas dinitro compounds and quinones were obtained from oxidation in solution.

Oxidation of Alcohols, Phenols, and Ethers. A primary alcohol can be retained when a secondary alcohol is oxidized to a ketone.[17] Tetrahydrofuran formation (eq 3)[18] predominates in molecules with rigid frameworks, which are favorable to δ-hydrogen abstraction by an alkoxyl radical.

$$(3)$$

Tertiary alcohols are prone to fragmentation;[19] this process is facilitated by a β-trimethylsilyl group (eq 4).[20] Other alcohols prone to fragmentation are cyclobutanols,[21] strained bicyclo [*x.y.z*]alkan-2-ols,[22] and homoallylic alcohols.[18d]

$$(4)$$

CAN converts benzylic alcohols into carbonyl compounds.[23] Even *p*-nitrobenzyl alcohol gives *p*-nitrobenzaldehyde in the catalytic oxidation system.[23c] Oxygen can be used[6] as the stoichiometric oxidant.

Catechols, hydroquinones, and their methyl ethers readily afford quinones on Ce^{IV} oxidation.[4,24] Partial demethylative oxidation is feasible, as shown in the preparation of several intramolecular quinhydrones (eq 5)[25] and a precursor of daunomycinone.[26] Sometimes the dual oxidant system of CAN–$NaBrO_3$ is useful. In a synthesis of methoxatin (eq 6)[27] the *o*-quinone moiety was generated from an aryl methyl ether.

$$n = 1\text{--}4 \qquad (5)$$

$$(6)$$

Oxidative regeneration of the carboxylic acid from its 2,6-di-*t*-butyl-4-methoxyphenyl ester[28] is the basis for the use of this auxiliary in a stereoselective α-hydroxyalkylation of carboxylic acids. The smooth removal of *p*-anisyl (eq 7)[29] and *p*-anisylmethyl[30] groups from an amidic nitrogen atom by Ce^{IV} oxidation makes these protective groups valuable in synthesis.

$$(7)$$

Simple ethers are oxidized[31] to carbonyl products and the intermediate from tetrahydrofuran oxidation can be trapped by alcohols.[32]

Vicinal diols undergo oxidative cleavage.[33] There is no apparent steric limitation as both *cis*- and *trans*-cycloalkane-1,2-diols are susceptible to cleavage. However, under certain conditions α-hydroxy ketones may be oxidized without breaking the C–C bond.[6]

Oxidation of Carbonyl Compounds. The Ce^{IV} oxidation of aldehydes and ketones is of much less synthetic significance than methods using other reagents. However, cage ketones often provide lactones (eq 8)[34,35] in good yield. Tetracyclones furnish α-pyrones.[36]

$$(8)$$

Concerning carboxylic acids and their derivatives, transformations of practical value are restricted to oxidative hydrolysis such as the conversion of hydrazides[37] back to carboxylic acids, transamidation of *N*-acyl-5,6-dihydrophenanthridines,[38] and decarboxylative processes, especially the degradation of α-hydroxymalonic acids (eq 9).[39] In some cases the Ce^{IV} oxidation is much superior to periodate cleavage. A related reaction is involved in a route to lactones.[40]

$$(9)$$

Nitrogenous derivatives of carbonyl compounds such as oximes and semicarbazones are oxidatively cleaved by Ce^{IV},[41] but only a few synthetic applications have been reported.[42]

Oxidation of Nitroalkanes. Ce^{IV} oxidation provides an alternative to the Nef reaction.[43] At least in the case of a ketomacrolide synthesis (eq 10),[44] complications arising from side reactions caused by other reagents are avoided.

$$(10)$$

Oxidation of Organosulfur Compounds. Thiols are converted into disulfides using reagents such as *Bis[trinitratocerium (IV)] Chromate*.[45] Chemoselective oxidation of sulfides by Ce^{IV} reagents to sulfoxides[4,46] is easily accomplished. Stoichiometric oxidation under phase transfer conditions[46b] and the dual oxidant[4] protocols permit oxidation of a variety of sulfides.

The reaction of dithioacetals including 1,3-dithiolanes and 1,3-dithianes with CAN provides a convenient procedure for the generation of the corresponding carbonyl group.[47] The rapid reaction is serviceable in many systems and superior to other methods, e.g. in the synthesis of acylsilanes.[48] In a series of compounds in which the dithiolane group is sterically hindered, the reaction led to enones, i.e. dehydrogenation accompanied the deprotection (eq 11).[49]

$$(11)$$

R = H, Me

Oxidative Cleavage of Organometallic Compounds. Oxidative deligation of both σ- and π-complexes by treatment with CAN is common practice. Ligands including cyclobutadiene and derivatives (eq 12)[50] and α-methylene-γ-butyrolactone (eq 13)[51] have been liberated successfully and applied to achieving the intended research goals. In the recovery of organic products from a Dötz reaction, CAN is often employed to cleave off the metallic species.[52]

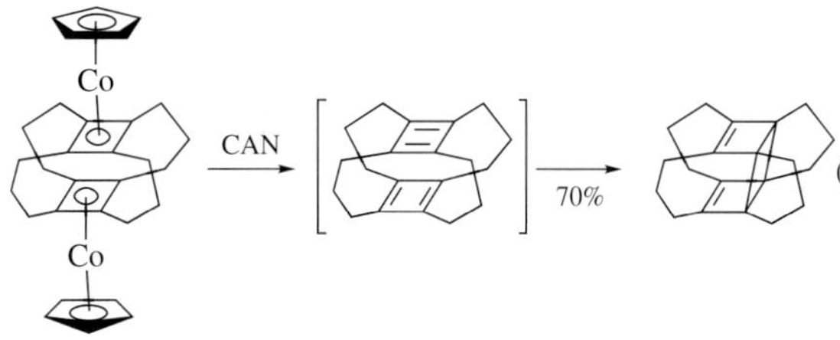

(12)

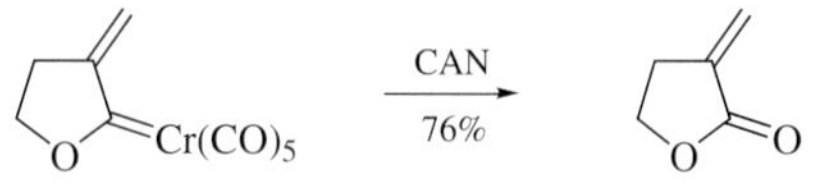

(13)

Generation of α-Acyl Radicals. As a one-electron oxidant, Ce^{IV} can promote the formation of radicals from carbonyl compounds. In the presence of interceptors such as butadiene and alkenyl acetates, the α-acyl radicals undergo addition.[53] The carbonyl compounds may be introduced as enol silyl ethers, and the oxidative coupling of two such ethers may be accomplished.[54] Some differences in the efficiency for oxidative cyclization of δ,ε-, and ε,ζ-unsaturated enol silyl ethers using CAN and other oxidants have been noted (eq 14).[55]

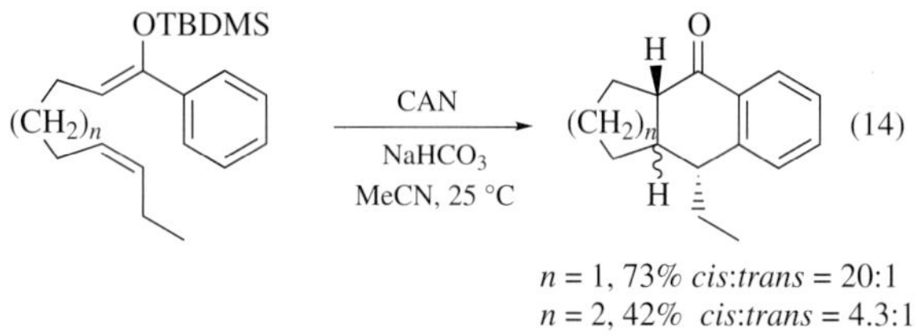

(14)

$n = 1$, 73% $cis{:}trans = 20{:}1$
$n = 2$, 42% $cis{:}trans = 4.3{:}1$

Oxidative Halogenation. Benzylic bromination[56] and α-iodination of ketones[3a] and uracil derivatives[3b] can be achieved with CAN as in situ oxidant.

First Update

Junhua Wang & Chaozhong Li
Shanghai Institute of Organic Chemistry, Shanghai, China

Carbon–Carbon Bond Formation. The CAN-mediated oxidative generation of carbon-centered radicals has been extensively investigated.[57] The radicals add to a C=C double bond resulting in the formation of a new carbon–carbon bond. The adduct radical can be further oxidized by another CAN molecule to give the carbocation, which is then trapped by a suitable nucleophile to give the final product. Active methylene compounds such as 1,3-dicarbonyls are among the typical substrates.[58] For example, the CAN-mediated oxidative addition of dimedone to 1-phenylcyclohexene affords the corresponding 2,3-dihydrofuran

compound in high yield under mild conditions (eq 15).[58c] This protocol has been found to be applicable to a variety of 1,3-dicarbonyls and alkenes. The addition can also be carried out intramolecularly leading to the formation of cyclized products such as β-lactams (eq 16).[59] In the absence of a suitable radical acceptor, the radicals are susceptible to dimerization. For instance, Nicolaou and Gray have reported the CAN-mediated dimerization of naphthazirin in the synthesis of racemic hybocarpone (eq 17).[60]

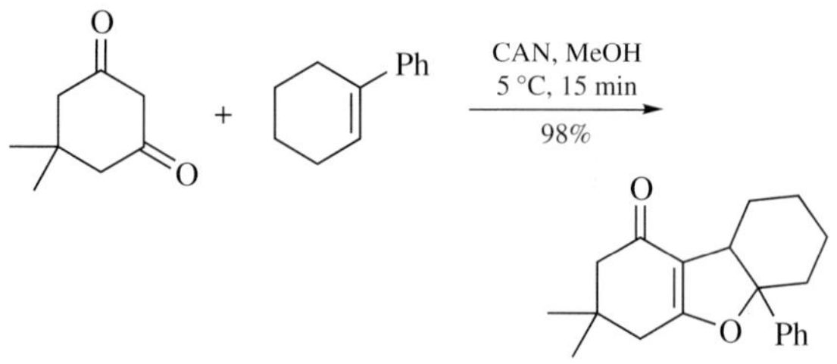

(15)

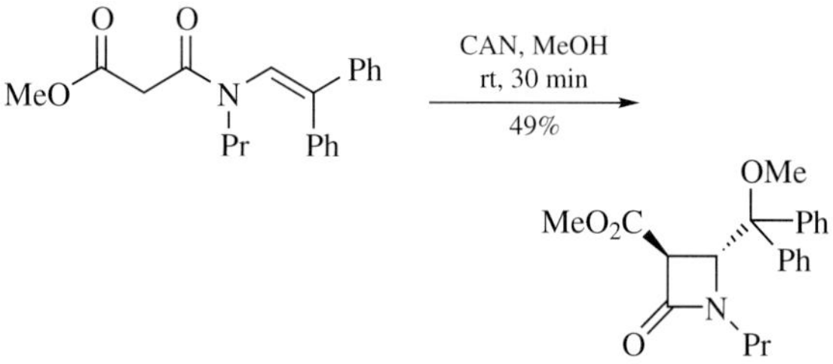

(16)

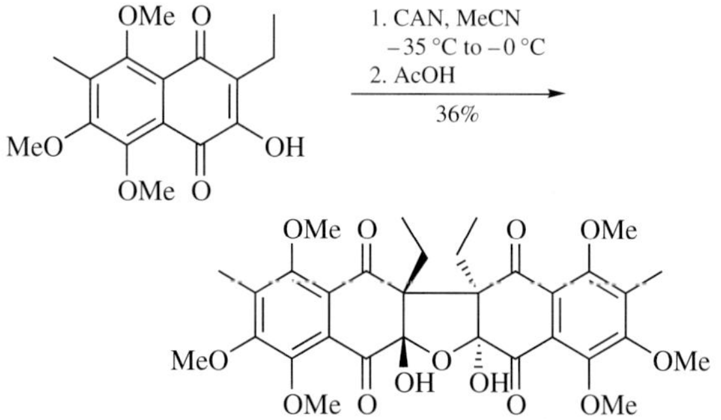

(17)

Vinyl (or cyclopropyl) silyl ethers have also been used to generate carbon-centered radicals by treatment with CAN.[61] 3,6-Dihydroxyphthalate esters are produced by dimerization when bisenolsilylated 1,3-diketones are treated with CAN.[61c] An elegant example is the three-component condensation of cyclopropyl silyl ether, cyclopentenone, and methyl vinyl ether (eq 18).[61b] Oxidation of cyclopropyl silyl ether gives the β-ester radical, which undergoes tandem radical addition processes apparently controlled by electronic effects. Subsequent oxidation and trapping affords the 2,3-disubstituted cyclopentanone in an excellent stereoselectivity. Other substrates include tertiary aminocyclopropanes,[62] N-(silylmethyl)amides,[63] and N,N-dialkylanilines.[64] For example, CAN-mediated oxidation of N,N-dialkylanilines in water affords the coupling products N,N,N′,N′-tetraalkylbenzidines.[64]

$$(18)$$

The CAN-mediated oxidation of electron-rich alkenes provides another facile entry to the construction of C–C bonds.[65] Nair et al. have uncovered that substituted styrenes can undergo dimerization to give 1-amino-4-aryltetralin derivatives in a one-pot procedure (eq 19).[66] A mechanistic rationale has been proposed for the formation of tetralin derivatives. The styrene undergoes oxidative electron transfer to afford the radical cation, which adds to another styrene molecule to generate a distonic radical cation. The radical cation undergoes 1,6-addition to the phenyl ring followed by the loss of a proton and an electron to give the corresponding carbocation, which is then trapped by the solvent acetonitrile in a manner analogous to the Ritter reaction to afford the final product.[66] An intramolecular version of this dimerization using dicinnamyl ethers as the substrates produces 3,4-*trans*-disubstituted tetrahydrofuran derivatives.[67a] Similarly, 3,4-*trans*-disubstituted pyrrolidines and cyclopentanes can be achieved by CAN-mediated oxidative cyclization of bis(cinnamyl)tosylamides and bis(cinnamyl)malonates.[67b] The reaction is also applicable to epoxypropyl cinnamyl amines, and 3,4,5-trisubstituted piperidines can be achieve with good stereoselectivity (eq 20).[67c]

$$(19)$$

$$(20)$$

Carbon–Nitrogen Bond Formation. Apart from the CAN-mediated reactions in which solvent (e.g., acetonitrile) incorporation results in carbon-heteroatom bond formation, the oxidative generation and subsequent addition of heteroatom-centered radicals to alkenes or alkynes provide means of direct construction of carbon–hetereoatom bonds.[68]

The introduction of an azide functionality with CAN/NaN$_3$ as the reagents has been shown to be a useful transformation in organic synthesis. It also offers a convenient protocol for the bis-functionalization of a variety of alkenes.[69–73] The treatment of silyl enol ethers with sodium azide and CAN gives the α-azido ketones.[69] The reactions of (substituted)styrenes with sodium azide and CAN in methanol under oxygenated conditions also furnish α-azido ketones (eq 21).[70] With the aid of sodium iodide,

azidoiodination can be accomplished for various alkenes in moderate to good yields.[71] The reaction of CAN/NaN$_3$ with triacetyl galactal provides a facile entry to aminosugars and glycopeptides (eq 22).[72b] This strategy has also been applied in the synthesis of α-amino acids.[72c] A one-pot synthesis of α-azido cinnamates can be achieved by treatment of cinnamates with CAN/NaN$_3$ in acetonitrile followed by the elimination of nitric acid with the use of sodium acetate as the base.[73] Under similar conditions, cinnamic acids can be converted to β-azido styrenes.[73]

$$(21)$$

$$(22)$$

CAN-mediated nitration provides a convenient route for the introduction of a nitro group into a variety of substrates. Alkenes on treatment with an excess of sodium nitrite and CAN in chloroform under sonication afford nitroalkenes.[74] When acetonitrile is used as the solvent, nitroacetamidation occurs in a Ritter-type fashion.[75] However, the attempted nitroacetamidation of cyclopentene-1-carboxaldehyde under similar conditions resulted in the formation of an unexpected dinitro-oxime compound.[76] A one-pot synthesis of 3-acetyl- or 3-benzoylisoxazole derivatives by reaction of alkenes (or alkynes) with CAN in acetone or acetophenone has been reported.[77] The proposed mechanism involves α-nitration of the solvent acetone, oxidation to generate the nitrile oxide, and subsequent 1,3-dipolar cycloaddition with alkenes or alkynes. The nitration of aromatic compounds[78] such as carbozole,[78a] naphthalene,[78b] and coumarins[78c] by CAN has also been investigated. As an example, coumarin on treatment with 1 equiv of CAN in acetic acid gives 6-nitrocoumarin in 92% yield.[78c]

Several other reactions involving C–N bond formation have been reported. A Ritter-type reaction of alkylbenzenes with nitriles has been achieved.[79] Thus, the treatment of ethylbenzene with CAN in the presence of a catalytic amount of N-hydroxyphthalimide (NHPI) in EtCN produces the corresponding amide in good selectivity (eq 23).[79a] The reaction is also applicable to a number of unactivated hydrocarbons. As a comparison, the photolysis of admantane with CAN gives a mixture of products.[80] In another case, the oxidation of monoterpenes such as pinene with CAN in acetonitrile affords the corresponding bisamides in good yields (eq 24).[81]

$$(23)$$

$$(24)$$

Carbon–Oxygen Bond Formation. CAN is an efficient reagent for the conversion of epoxides into β-nitrato alcohols.[82] 1,2-*cis*-Diols can be prepared from alkenes by reaction with CAN/I_2 followed by hydrolysis with KOH.[83] Of particular interest is the high-yield synthesis of various α-hydroxy ketones and α-amino ketones from oxiranes and aziridines, respectively.[84] The reactions are operated under mild conditions with the use of NBS and a catalytic amount of CAN as the reagents (eq 25). In another case, *N*-(silylmethyl)amides can be converted to *N*-(methoxymethyl)amides by CAN in methanol (eq 26).[85] This chemistry has found application in the removal of electroauxiliaries from peptide substrates. Other CAN-mediated C–O bond-forming reactions include the oxidative rearrangement of aryl cyclobutanes and oxetanes,[86] the conversion of allylic and tertiary benzylic alcohols into their corresponding ethers,[87] and the alkoxylation of cephem sulfoxides at the position α to the ester moiety.[88]

The condensation of arenenitriles with arenethiols in the presence of CAN furnishes 2-arylbenzothiazoles.[95]

Carbon–Halogen Bond Formation. The combination of CAN with a metal bromide offers a convenient generation of bromine radicals, which can be intercepted by a C=C bond leading to the formation of C–Br bonds. As a result, a variety of alkenes can be converted to 1,2-dibromides by reaction with CAN and KBr.[96] Similarly, acetylenes and arylcyclopropanes afford the corresponding vicinal dibromoalkenes and 1,3-dibromides, respectively. Cinnamyl esters or ketones on reaction with CAN, LiBr, and propargyl alcohol give the corresponding 2-alkoxy-1-bromoesters or ketones (eq 28).[97] The reaction, however, is only effective when there is an electron-donating group such as a methoxy group in the aromatic ring.

(28)

$$Ar = p\text{-MeOC}_6\text{H}_4$$

(25)

(26)

CAN in combination with iodine or an iodide has been demonstrated to be a powerful iodination reagent. Stereoselective iodo-acetoxylation of glycals using sodium iodide and CAN in a mixture of MeCN and acetic acid has been achieved.[98] In a similar fashion, the reaction of α,β-unsaturated ketones or esters with iodine and CAN in alcohol affords the corresponding β-alkoxy-α-iodoketones or esters in good yields.[99] The regioselective iodination of pyrazoles[100] and an alkoxybenzene[101] mediated by CAN has been reported.

Miscellaneous Reactions. A silyl-containing alcohol derived from cyclohexene oxide can be converted to a nine-membered lactone on treatment with CAN, presumably via the oxidative generation and subsequent transformations of an alkoxy radical (eq 29).[102]

(29)

Carbon–Sulfur Bond Formation. The oxidation of sulfinates by CAN provides an easy entry to sulfonyl radicals, which can be trapped by various alkenes, especially electron-rich ones, to afford sulfones. For example, the reaction of sodium 2-naphthalenesulfinate with 1-vinylcyclobutanol in the presence of CAN furnishes the ring-enlarged product (eq 27).[89] With the aid of sodium iodide, the CAN-mediated oxidative addition of sulfinates to styrene affords vinyl sulfones and the addition to alkynes leads to β-iodo vinyl sulfones.[90]

A mild protocol for the conversion of β-ketoesters and β-diketones to carboxylic acids with the use of CAN in acetonitrile is reported (eq 30).[103]

(27)

(30)

The treatment of styrenes with ammonium thiocyanate[91] and CAN in MeCN results in the formation of dithiocyanates.[91a] Under an oxygen atmosphere, phenacyl thiocyanates can be the major products.[91c] The thiocyanation of indoles also proceeds under similar conditions.[91b] Chemoselective thioacetalization of aldehydes[92] and the conversion of epoxides to their corresponding thiiranes[93] can be operated under mild conditions with the catalysis of CAN. As an extension, selenocyanation can be conducted in a similar fashion with CAN/KSeCN.[94]

CAN-mediated dehydrogenation leads to a variety of aromatic compounds such as quinoline,[104] pyridazine,[105] thiadiazole,[106] pyrido[4,3,2-kl]acridin-4-one,[107] azobenzene,[108] and tetrazole[109] derivatives.

CAN is often used to detach a metal ion from its complex. CAN-mediated decomplexation of the complexes of Co,[110] Mn,[111] Fe,[112] Os,[113] Ru,[114] and Mo[115] has also been reported.

Related Reagents. Cerium(IV) Ammonium Nitrate–Sodium Bromate; Iodine–Cerium(IV) Ammonium Nitrate.

1. (a) Richardson, W. H. In *Oxidation in Organic Chemistry*, Wiberg, K. B., Ed.; Academic: New York, 1965; Part A, Chapter IV. (b) Ho, T.-L., *Synthesis* **1973**, 347. (c) Ho, T.-L. In *Organic Syntheses by Oxidation with Metal Compounds*, Mijs, W. J.; de Jonge, C. R. H. I., Eds., Plenum: New York, 1986, Chapter 11.

2. Smith, G. F. *Cerate Oxidimetry*, G. Frederick Smith Chemical Co.: Columbus, OH, 1942.

3. (a) Horiuchi, C. A.; Kiji, S., *Chem. Lett.* **1988**, 31. (b) Asakura, J.; Robins, M. J., *J. Org. Chem.* **1990**, *55*, 4928.

4. Ho, T.-L., *Synth. Commun.* **1979**, *9*, 237.

5. Kanemoto, S.; Saimoto, H.; Oshima, K.; Nozaki, H., *Tetrahedron Lett.* **1984**, *25*, 3317.

6. Hatanaka, Y.; Imamoto, T.; Yokoyama, M., *Tetrahedron Lett.* **1983**, *24*, 2399.

7. Kreh, R. P.; Spotnitz, R. M.; Lundquist, J. T., *J. Org. Chem.* **1989**, *54*, 1526.

8. Baciocchi, E.; Giacco, T. D.; Murgia, S. M.; Sebastiani, G. V., *Tetrahedron* **1988**, *44*, 6651.

9. Young, L. B., *Tetrahedron Lett.* **1968**, 5105.

10. (a) Syper, L., *Tetrahedron Lett.* **1966**, 4493. (b) Laing, S. B., *J. Chem. Soc. (C)* **1968**, 2915.

11. Lee, W. Y.; Park, C. H.; Kim, S., *J. Am. Chem. Soc.* **1993**, *115*, 1184.

12. Baciocchi, E.; Della Cort, A.; Eberson, L.; Mandolini, L.; Rol, C., *J. Org. Chem.* **1986**, *51*, 4544.

13. Della Cort, A.; Barbera, A. L.; Mandolini, L., *J. Chem. Res. (S)* **1983**, 44.

14. Baciocchi, E.; Rol, C.; Sebastiani, G. V.; Serena, B., *Tetrahedron Lett.* **1984**, *25*, 1945.

15. (a) Ho, T.-L.; Hall, T.-W.; Wong, C. M., *Synthesis* **1973**, 206. (b) Periasamy, M.; Bhatt, M. V., *Tetrahedron Lett.* **1977**, 2357.

16. Chawla, H. M.; Mittal, R. S., *Synthesis* **1985**, 70.

17. Kanemoto, S.; Tomioka, H.; Oshima, K.; Nozaki, H., *Bull. Chem. Soc. Jpn.* **1986**, *58*, 105.

18. (a) Trahanovsky, W. S.; Young, M. G.; Nave, P. M., *Tetrahedron Lett.* **1969**, 2501. (b) Doyle, M. P.; Zuidema, L. J.; Bade, T. R., *J. Org. Chem.* **1975**, *40*, 1454. (c) Fujise, Y.; Kobayashi, E.; Tsuchida, H.; Ito, S., *Heterocycles* **1978**, *11*, 351. (d) Balasubramanian, V.; Robinson, C. H., *Tetrahedron Lett.* **1981**, 501.

19. Trahanovsky, W. S.; Macaulay, D. B., *J. Org. Chem.* **1973**, *38*, 1497.

20. Wilson, S. R.; Zucker, P. A.; Kim, C.; Villa, C. A., *Tetrahedron Lett.* **1985**, *26*, 1969.

21. (a) Meyer, K.; Rocek, J., *J. Am. Chem. Soc.* **1972**, *94*, 1209. (b) Hunter, N. R.; MacAlpine, G. A.; Liu, H.-J.; Valenta, Z., *Can. J. Chem.* **1970**, *48*, 1436.

22. Trahanovsky, W. S.; Flash, P. J.; Smith, L. M., *J. Am. Chem. Soc.* **1969**, *91*, 5068.

23. (a) Trahanovsky, W. S.; Cramer, J., *J. Org. Chem.* **1971**, *36*, 1890. (b) Trahanovsky, W. S.; Fox, N. S., *J. Am. Chem. Soc.* **1974**, *96*, 7968. (c) Ho, T.-L., *Synthesis* **1978**, 936.

24. (a) Ho, T.-L.; Hall, T.-W.; Wong, C. M., *Chem. Ind. (London)* **1972**, 729. (b) Jacob, P., III; Callery, P. S.; Shulgin, A. T.; Castagnoli, N., Jr., *J. Org. Chem.* **1976**, *41*, 3627. (c) Syper, L.; Kloc, K.; Mlochowski, J., *Synthesis* **1979**, 521.

25. Bauer, H.; Briaire, J.; Staab, H. A., *Angew. Chem., Int. Ed. Engl.* **1983**, *22*, 334.

26. Hauser, F. M.; Prasanna, S., *J. Am. Chem. Soc.* **1981**, *103*, 6378.

27. Corey, E. J.; Tramontano, A., *J. Am. Chem. Soc.* **1981**, *103*, 5599.

28. Heathcock, C. H.; Pirrung, M. C.; Montgomery, S. H.; Lampe, J., *Tetrahedron* **1981**, *37*, 4087.

29. (a) Fukuyama, T.; Frank, R. K.; Jewell, C. F., *J. Am. Chem. Soc.* **1980**, *102*, 2122. (b) Kronenthal, D. R.; Han, C. Y.; Taylor, M. K., *J. Org. Chem.* **1982**, *47*, 2765.

30. Yamaura, M.; Suzuki, T.; Hashimoto, H.; Yoshimura, J.; Okamoto, T.; Shin, C., *Bull. Chem. Soc. Jpn.* **1985**, *58*, 1413.

31. Olah, G. A.; Gupta, B. G. B.; Fung, A. P., *Synthesis* **1980**, 897.

32. Maione, A. M.; Romeo, A., *Synthesis* **1987**, 250.

33. (a) Hintz, H. L.; Johnson, D. C., *J. Org. Chem.* **1967**, *32*, 556. (b) Trahanovsky, W. S.; Young, L. H.; Bierman, M. H., *J. Org. Chem.* **1969**, *34*, 869.

34. Soucy, P.; Ho, T.-L.; Deslongchamps, P., *Can. J. Chem.* **1972**, *50*, 2047.

35. Mehta, G.; Pandey, P. N.; Ho, T.-L., *J. Org. Chem.* **1976**, *41*, 953.

36. Ho, T.-L.; Hall, T.-W.; Wong, C. M., *Synth. Commun.* **1973**, *3*, 79.

37. Ho, T.-L.; Ho, H. C.; Wong, C. M., *Synthesis* **1972**, 562.

38. (a) Uchimaru, T.; Narasaka, K.; Mukaiyama, T., *Chem. Lett.* **1981**, 1551. (b) Narasaka, K.; Hirose, T.; Uchimaru, T.; Mukaiyama, T., *Chem. Lett.* **1982**, 991.

39. Salomon, M. F.; Pardo, S. N.; Salomon, R. G., *J. Am. Chem. Soc.* **1980**, *102*, 2473.

40. Salomon, R. G.; Roy, S.; Salomon, M. F., *Tetrahedron Lett.* **1988**, *29*, 769.

41. Bird, J. W.; Diaper, D. G. M., *Synlett* **1969**, *47*, 145.

42. (a) Oppolzer, W.; Petrzilka, M.; Bättig, K., *Helv. Chim. Acta* **1977**, *60*, 2964. (b) Oppolzer, W.; Bättig, K.; Hudlicky, T., *Tetrahedron* **1981**, *37*, 4359.

43. Olah, G. A.; Gupta, B. G. B., *Synthesis* **1980**, 44.

44. Cookson, R. C.; Ray, P. S., *Tetrahedron Lett.* **1982**, *23*, 3521.

45. (a) Firouzabadi, H.; Iranpoor, N.; Parham, H.; Sardarian, A.; Toofan, J., *Synth. Commun.* **1984**, *14*, 717. (b) Firouzabadi, H.; Iranpoor, N.; Parham, H.; Toofan, J., *Synth. Commun.* **1984**, *14*, 631.

46. (a) Ho, T.-L.; Wong, C. M., *Synthesis* **1972**, 561. (b) Baciocchi, E.; Piermattei, A.; Ruzziconi, R., *Synth. Commun.* **1988**, *18*, 2167.

47. Ho, T.-L.; Ho, H. C.; Wong, C. M., *J. Chem. Soc., Chem. Commun.* **1972**, 791.

48. Tsai, Y.-M.; Nieh, H.-C.; Cherng, C.-D., *J. Org. Chem.* **1992**, *57*, 7010.

49. Lansbury, P. T.; Zhi, B., *Tetrahedron Lett.* **1988**, *29*, 179.

50. (a) Watts, L.; Fitzpatrick, J. D.; Pettit, R., *J. Am. Chem. Soc.* **1965**, *87*, 3253. (b) Gleiter, R.; Karcher, M., *Angew. Chem., Int. Ed. Engl.* **1988**, *27*, 840.

51. Casey, C. P.; Brunsvold, W. R., *J. Organomet. Chem.* **1975**, *102*, 175.

52. Wulff, W. D.; Tang, P. C.; McCallum, J. S., *J. Am. Chem. Soc.* **1981**, *103*, 7677.

53. (a) Baciocchi, E.; Ruzziconi, R., *J. Org. Chem.* **1986**, *51*, 1645. (b) Baciocchi, E.; Civitarese, G.; Ruzziconi, R., *Tetrahedron Lett.* **1987**, *28*, 5357. (c) Baciocchi, E.; Ruzziconi, R., *Synth. Commun.* **1988**, *28*, 1841.

54. Baciocchi, E.; Casu, A.; Ruzziconi, R., *Tetrahedron Lett.* **1989**, *30*, 3707.

55. Snider, B. B.; Kwon, T., *J. Org. Chem.* **1990**, *55*, 4786.

56. Maknon'kov, D. I.; Cheprakov, A. V.; Rodkin, M. A.; Mil'chenko, A. Y.; Beletskaya, I. P., *Zh. Org. Chem.* **1986**, *22*, 30.

57. Nair, V.; Balagopal, L.; Rajan, R.; Mathew, J., *Acc. Chem. Res.* **2004**, *37*, 21.

58. (a) Nair, V.; Mathew, J., *J. Chem. Soc., Perkin Trans. 1* **1995**, 187. (b) Nair, V.; Mathew, J.; Alexander, S., *Synth. Commun.* **1995**, *25*, 3981. (c) Nair, V.; Mathew, J.; Radhakrishnan, K. V., *J. Chem. Soc., Perkin Trans. 1* **1996**, 1487. (d) Kobayashi, K.; Mori, M.; Uneda, T.; Morikawa, O.; Konishi, H., *Chem. Lett.* **1996**, 451. (e) Cravotto, G.;

Nano, G. M.; Palmisano, G.; Tagliapietra, S., *Synthesis* **2003**, 1286. (f) Kobayashi, K.; Nagase, K.; Morikawa, O.; Konishi, H., *Heterocycles* **2003**, *60*, 939. (g) Waizumi, N.; Stankovic, A. R.; Rawal, V. H., *J. Am. Chem. Soc.* **2003**, *125*, 13022. (h) Kobayashi, K.; Umakoshi, H.; Hayashi, K.; Morikawa, O.; Konishi, H., *Chem. Lett.* **2004**, *33*, 1588. (i) Sommermann, T.; Kim, B. G.; Peters, K.; Peters, E.-M.; Linker, T., *Chem. Commun.* **2004**, 2624. (j) Chuang, C.-P.; Wu, Y.-L., *Tetrahedron* **2004**, *60*, 1841.

59. D'Annibale, A.; Pesce, A.; Resta, S.; Trogolo, C., *Tetrahedron Lett.* **1997**, *38*, 1829.

60. Nicolaou, K. C.; Gray, D., *Angew. Chem., Int. Ed.* **2001**, *40*, 761.

61. (a) Paolobelli, A. B.; Ceccherelli, P.; Pizzo, F.; Ruzziconi, R., *J. Org. Chem.* **1995**, *60*, 4954. (b) Paolobelli, A. B.; Ruzziconi, R., *J. Org. Chem.* **1996**, *61*, 6434. (c) Langer, P.; Kohler, V., *Chem. Commun.* **2000**, 1653.

62. Takemoto, Y.; Yamagata, S.; Furuse, S.; Hayase, H.; Echigo, T.; Iwata, C., *Chem. Commun.* **1998**, 651.

63. Kim, H. J.; Yoon, U. C.; Jung, Y.-S.; Park, N. S.; Cederstrom, E. M.; Mariano, P. S., *J. Org. Chem.* **1998**, *63*, 860.

64. Xi, C.; Jiang, Y.; Yang, X., *Tetrahedron Lett.* **2005**, *46*, 3909.

65. Nair, V.; Sheeba, V.; Panicker, S. B.; George, T. G.; Rajan, R.; Balagopal, L.; Vairamani, M.; Prabhakar, S., *Tetrahedron* **2000**, *56*, 2461.

66. Nair, V.; Rajan, R.; Rath, N. P., *Org. Lett.* **2002**, *4*, 1575.

67. (a) Nair, V.; Balagopal, L.; Sheeba, V.; Panicker, S. B.; Rath, N. P., *Chem. Commun.* **2001**, 1682. (b) Nair, V.; Mohanan, K.; Suja, T. D.; Suresh, E., *Tetrahedron Lett.* **2006**, *47*, 2803. (c) Nair, V.; Mohanan, K.; Suja, T. D.; Suresh, E., *Tetrahedron Lett.* **2006**, *47*, 705.

68. Nair, V.; Panicker, S. B.; Nair, L. G.; George, T. G.; Augustine, A., *Synlett* **2003**, 156.

69. (a) Magnus, P.; Barth, L., *Tetrahedron* **1995**, *51*, 11075. (b) Battaglia, A.; Baldelli, E.; Bombardelli, E.; Carenzi, G.; Fontana, G.; Gelmi, M. L.; Guerrini, A.; Pocar, D., *Tetrahedron* **2005**, *61*, 7727.

70. Nair, V.; Nair, L. G.; George, T. G.; Augustine, A., *Tetrahedron* **2000**, *56*, 7607.

71. Nair, V.; George, T. G.; Sheeba, V.; Augustine, A.; Balagopal, L.; Nair, L. G., *Synlett* **2000**, 1597.

72. (a) Matos, M. N.; Afonso, C. A. M.; Batey, R. A., *Tetrahedron* **2005**, *61*, 1221. (b) Renaudet, O.; Dumy, P., *Tetrahedron Lett.* **2004**, *45*, 65. (c) Clive, D. L. J.; Etkin, N., *Tetrahedron Lett.* **1994**, *35*, 2459.

73. Nair, V.; George, T. G., *Tetrahedron Lett.* **2000**, *41*, 3199.

74. (a) Hwu, J. R.; Chen, K. L.; Ananthan, S., *J. Chem. Soc., Chem. Commun.* **1994**, 1425. (b) Hwu, J. R.; Chen, K. L.; Ananthan, S.; Patel, H. V., *Organometallics* **1996**, *15*, 499.

75. Reddy, M. V. R.; Mehrotra, B.; Vankar, Y. D., *Tetrahedron Lett.* **1995**, *36*, 4861.

76. Smith, C. C.; Jacyno, J. M.; Zeiter, K. K.; Parkanzky, P. D.; Paxson, C. E.; Pekelnicky, P.; Harwood, J. S.; Hunter, A. D.; Lucarelli, V. G.; Lufaso, M. W.; Cutler, H. G., *Tetrahedron Lett.* **1998**, *39*, 6617.

77. Itoh, K.; Takahashi, S.; Ueki, T.; Sugiyama, T.; Takahashi, T. T.; Horiuchi, C. A., *Tetrahedron Lett.* **2002**, *43*, 7035.

78. (a) Chakrabarty, M.; Batabyal, A., *Synth. Commun.* **1994**, *24*, 1. (b) Mellor, J. M.; Parkes, R.; Millar, R. W., *Tetrahedron Lett.* **1997**, *38*, 8739. (c) Ganguly, N.; Sukai, A. K.; De, S., *Synth. Commun.* **2001**, *31*, 301. (d) Asghedom, H.; LaLonde, R. T.; Ramdayal, F., *Tetrahedron Lett.* **2002**, *43*, 3989.

79. (a) Sakaguchi, S.; Hirabayashi, T.; Ishii, Y., *Chem. Commun.* **2002**, 516. (b) Nair, V.; Suja, T. D.; Mohanan, K., *Tetrahedron Lett.* **2005**, *46*, 3217.

80. Mella, M.; Freccero, M.; Soldi, T.; Fasani, E.; Albini, A., *J. Org. Chem.* **1996**, *61*, 1413.

81. Nair, V.; Rajan, R.; Balagopal, L.; Thomas, S.; Narasimlu, K. A., *Tetrahedron Lett.* **2002**, *43*, 8971.

82. Iranpoor, N.; Salehi, P., *Tetrahedron* **1995**, *51*, 909.

83. Horiuchi, C. A.; Dan, G.; Sakamoto, M.; Suda, K.; Usui, S.; Sakamoto, O.; Kitoh, S.; Watanabe, S.; Utsukihara, T.; Nozaki, S., *Synthesis* **2005**, 2861.

84. Surendra, K.; Krishnaveni, N. S.; Rao, K. R., *Tetrahedron Lett.* **2005**, *46*, 4111.

85. Sun, H.; Moeller, K. D., *Org. Lett.* **2003**, *5*, 3189.

86. Nair, V.; Rajan, R.; Mohanan, K.; Sheeba, V., *Tetrahedron Lett.* **2003**, *44*, 4585.

87. Iranpoor, N.; Mothaghineghad, E., *Tetrahedron* **1994**, *50*, 1859.

88. Alpegiani, M.; Bissolino, P.; Borghi, D.; Perrone, E., *Synlett* **1994**, 233.

89. (a) Narasaka, K.; Mochizuki, T.; Hayakawa, S., *Chem. Lett.* **1994**, 1705. (b) Mochizuki, T.; Hayakawa, S.; Narasaka, K., *Bull. Chem. Soc. Jpn.* **1996**, *69*, 2317.

90. Nair, V.; Augustine, A.; George, T. G.; Nair, L. G., *Tetrahedron Lett.* **2001**, *42*, 6763.

91. (a) Nair, V.; Nair, L. G., *Tetrahedron Lett.* **1998**, *39*, 4585. (b) Nair, V.; George, T. G.; Nair, L. G.; Panicker, S. B., *Tetrahedron Lett.* **1999**, *40*, 1195. (c) Nair, V.; Nair, L. G.; George, T. G.; Augustine, A., *Tetrahedron* **2000**, *56*, 7607.

92. Mandal, P. K.; Roy, S. C., *Tetrahedron* **1995**, *51*, 7823.

93. Iranpoor, N.; Kazemi, F., *Synthesis* **1996**, 821.

94. Nair, V.; Augustine, A.; George, T. G., *Eur. J. Org. Chem.* **2002**, 2363.

95. Tale, R. H., *Org. Lett.* **2002**, *4*, 1641.

96. Nair, V.; Panicker, S. B.; Augustine, A.; George, T. G.; Thomas, S.; Vairamani, M., *Tetrahedron* **2001**, *57*, 7417.

97. Roy, S. C.; Guin, C.; Rana, K. K.; Maiti, G., *Synlett* **2001**, 226.

98. Roush, W. R.; Narayan, S.; Bennett, C. E.; Briner, K., *Org. Lett.* **1999**, *1*, 895.

99. Horiuchi, C. A.; Ochiai, K.; Fukunishi, H., *Chem. Lett.* **1994**, 185.

100. (a) Rodriguez-Franko, M. I.; Dorronsoro, I.; Hernandez-Higueras, A. I.; Antequera, G., *Tetrahedron Lett.* **2001**, *42*, 863. (b) Sammelson, R. E.; Casida, J. E., *J. Org. Chem.* **2003**, *68*, 8075.

101. Nishiyama, T.; Isobe, M.; Ichikawa, Y., *Angew. Chem., Int. Ed.* **2005**, *44*, 4372.

102. Hatcher, M. A.; Borstnik, K.; Posner, G. H., *Tetrahedron Lett.* **2003**, *44*, 5407.

103. Zhang, Y.; Jiao, J.; Flowers, R. A., II, *J. Org. Chem.* **2006**, *71*, 4516.

104. (a) Jimenez, O.; de la Rosa, G.; Lavilla, R., *Angew. Chem. Int. Ed.* **2005**, *44*, 6521. (b) Wolf, C.; Lerebours, R., *J. Org. Chem.* **2003**, *68*, 7077.

105. Požgan, F.; Polanc, S.; Kočevar, M., *Synthesis* **2003**, 2349.

106. Somogyi, L., *Heterocycles* **2004**, *63*, 2243.

107. Bouffier, L.; Demeunynck, M.; Milet, A.; Dumy, P., *J. Org. Chem.* **2004**, *69*, 8144.

108. Carreno, M. C.; Mudarra, G. F.; Merino, E.; Ribagorda, M., *J. Org. Chem.* **2004**, *69*, 3413.

109. De Lombaert, S.; Blanchard, L.; Stamford, L. B.; Tan, J.; Wallace, E. M.; Satoh, Y.; Fitt, J.; Hoyer, D.; Simonsbergen, D.; Moliterni, J.; Marcopoulos, N.; Savage, P.; Chou, M.; Trapani, A. J.; Jeng, A. Y., *J. Med. Chem.* **2000**, *43*, 488.

110. (a) Tanino, K.; Shimizu, T.; Miyama, M.; Kuwajima, I., *J. Am. Chem. Soc.* **2000**, *122*, 6116. (b) Crisostomo, F. R. P.; Carrillo, R.; Martin, T.; Martin, V. S., *Tetrahedron Lett.* **2005**, *46*, 2829.

111. Lepore, S. D.; Khoran, A.; Bromfield, D. C.; Cohn, P.; Jairaj, V.; Silvestri, M. A., *J. Org. Chem.* **2005**, *70*, 7443.

112. (a) Lukesh, J. M.; Donaldson, W. A., *Chem. Commun.* **2005**, 110. (b) Limanto, J.; Khuong, K. S.; Houk, K. N.; Snapper, M. L., *J. Am. Chem. Soc.* **2003**, *125*, 16310. (c) Yun, Y. K.; Godula, K.; Cao, Y.; Donaldson, W. A., *J. Org. Chem.* **2003**, *68*, 901.

113. Stokes, S. M., Jr; Ding, F.; Smith, P. L.; Keane, J. M.; Kopach, M. E.; Jervis, R.; Sabat, M.; Harman, W. D., *Organometallics* **2003**, *22*, 4170.

114. Pigge, F. C.; Coniglio, J. J.; Rath, N. P., *J. Org. Chem.* **2004**, *69*, 1161.

115. Kocienski, P. J.; Christopher, J. A.; Bell, R.; Otto, B., *Synthesis* **2005**, 75.

m-Chloroperbenzoic Acid[1]

[937-14-4]   C₇H₅ClO₃   (MW 172.57)

InChI = 1/C7H5ClO3/c8-6-3-1-2-5(4-6)7(9)11-10/h1-4,10H

InChIKey = NHQDETIJWKXCTC-UHFFFAOYAC

(electrophilic reagent capable of reacting with many functional groups; delivers oxygen to alkenes, sulfides, selenides, and amines)

Alternate Names: *m*-CPBA; MCPBA.

Physical Data: mp 92–94 °C.

Solubility: soluble in CH_2Cl_2, $CHCl_3$, 1,2-dichloroethane, ethyl acetate, benzene, and ether; slightly soluble in hexane; insoluble in H_2O.

Form Supplied in: white powder, available with purity of 50%, 85%, and 98% (the rest is 3-chlorobenzoic acid and water).

Analysis of Reagent Purity: iodometry.[2]

Purity: commercial material (purity 85%) is washed with a phosphate buffer of pH 7.5 and dried under reduced pressure to furnish reagent with purity >99%.[3]

Handling, Storage, and Precaution: pure *m*-CPBA is shock sensitive and can deflagrate;[4] potentially explosive, and care is required while carrying out the reactions and during workup.[5] Store in polyethylene containers under refrigeration.

Original Commentary

A. Somasekar Rao & H. Rama Mohan

Indian Institute of Chemical Technology, Hyderabad, India

Functional Group Oxidations. The weak O–O bond of *m*-CPBA undergoes attack by electron-rich substrates such as simple alkenes, alkenes carrying a variety of functional groups (such as ethers, alcohols, esters, ketones, and amides which are inert to this reagent), some aromatic compounds,[6] sulfides, selenides, amines, and N-heterocycles; the result is that an oxygen atom is transferred to the substrate. Ketones and aldehydes undergo oxygen insertion reactions (Baeyer–Villiger oxidation).

Organic peroxy acids (**1**) readily epoxidize alkenes (eq 1).[1b] This reaction is *syn* stereospecific;[7] the groups (R^1 and R^3) which are *cis* related in the alkene (**2**) are *cis* in the epoxidation product (**3**). The reaction is believed to take place via the transition state (**4**).[8] The reaction rate is high if the group R in (**1**) is electron withdrawing, and the groups R^1, R^2, R^3, and R^4 in (**2**) are electron releasing.

Epoxidations of alkenes with *m*-CPBA are usually carried out by mixing the reactants in CH_2Cl_2 or $CHCl_3$ at 0–25 °C.[9] After the reaction is complete the reaction mixture is cooled in an ice bath and the precipitated *m*-chlorobenzoic acid is removed by filtration. The organic layer is washed with sodium bisulfite solution, $NaHCO_3$ solution, and brine.[10] The organic layer is dried and concentrated under reduced pressure. Many epoxides have been purified chromatographically; however, some epoxides decompose during chromatography.[11] If distillation (*caution:* check for peroxides[12]) is employed to isolate volatile epoxides, a trace of alkali should be added to avoid acid-catalyzed rearrangement.

Alkenes having low reactivity (due to steric or electronic factors) can be epoxidized at high temperatures and by increasing the reaction time.[13] The weakly nucleophilic α,β-unsaturated ester (**5**) thus furnishes the epoxide (**6**) (eq 2).[13b] When alkenes are epoxidized at 90 °C, best results are obtained if radical inhibitor is added.[13a] For preparing acid-sensitive epoxides (benzyloxiranes, allyloxiranes) the pH of the reaction medium has to be controlled using $NaHCO_3$ (as solid or as aqueous solution),[14] Na_2HPO_4, or by using the *m*-CPBA–KF[9a] reagent.

Regioselective Epoxidations. In the epoxidation of simple alkenes (2) (eq 1), due to the electron-releasing effect of alkyl groups the reactivity rates are tetra- and trisubstituted alkenes > disubstituted alkenes > monosubstituted alkenes.[1a] High regioselectivity is observed in the epoxidation of diene hydrocarbons (e.g. **7**) having double bonds differing in degree of substitution (eq 3).[15] Epoxidation takes place selectively at the more electron-rich C-3–C-4 double bond in the dienes (**8**)[16] and (**9**).[17]

Essential Reactions for Organic Synthesis, First Edition. Edited by Philip L. Fuchs.
© 2016 John Wiley & Sons, Ltd. Published 2016 by John Wiley & Sons, Ltd.

Diasterereoselective Epoxidation of Cyclic Alkenes. π-Facial stereoselectivity (75% *anti*) is observed in the epoxidation of the allyl ether (**10a**) since reagent approach from the α-face is blocked by the allylic substituent; a higher diastereoselectivity (90% *anti* epoxidation) is observed when the bulkier O-*t*-Bu is located on the allylic carbon (eq 4).[18] Due to steric and other factors, the norbornene (**11**) undergoes selective (99%) epoxidation from the *exo* face.[19] In 7,7-dimethylnorbornene (**12**), approach to the *exo* face is effectively blocked by the methyl substituent at C-7, and (**12**) is epoxidized from the unfavored *endo* face, although much more slowly (1% of the rate of epoxidation of **11**).[19] The geminal methyl group at C-7 is able to block the approach of the peroxy acid even when the double bond is exocyclic to the norbornane ring system (for example, epoxidation of (**13**) proceeds with 86% *exo* attack, while (**14**) is oxidized with 84% *endo* attack). Folded molecules are epoxidized selectively from the less hindered convex side; *m*-CPBA epoxidation of the triene lactone (**15**) takes place from the α-face with 97% stereoselectivity.[20] The triepoxide (**16**) has been obtained in 74% yield by epoxidizing the corresponding triene;[21] in the epoxidation step, six new chiral centers are introduced stereoselectively as a result of steric effects.

idation of (**17**) the ratio of axial:equatorial attack is 86:14;[22] for the alkene (**18**) the ratio is 75:25.[23]

Epoxidation of Cyclic Alkenes Having Directing Groups. Henbest showed that in the absence of severe steric interference, allylic cyclohexenols are epoxidized stereoselectively by organic peroxy acids to furnish *cis*-epoxy alcohols;[24a] a large number of *cis*-epoxy alcohols have been prepared by epoxidizing allylic cyclohexenols.[7] A mixture (5:1) of labile bisallylic alcohols (**19**) and (**20**) was reacted with *m*-CPBA (eq 5); from the reaction mixture diepoxide (**21**) was isolated as a single isomer.[25] Epoxidation of (Z)-cyclooct-2-en-1-ol (**22**) furnishes exclusively (99.8%) the *trans*-epoxide (**23**) (eq 6).[24b] Similar observations have been made subsequently.[26] This result, as well as the stereoselectivity observed during the epoxidation of other allylic alcohols, both cyclic and acyclic, has been rationalized on the basis of transition state models.[24,27]

Stereoselectivity has been observed during the peroxy acid epoxidation of some homoallylic and bishomoallylic alcohols,[28] and the epoxidation of the allylic carbamate (**24**) is *syn* stereoselective (eq 7).[28]

Unhindered methylenecyclohexanes and related compounds show a moderate preference for axial epoxidation. In the epox-

Epoxidations of Acyclic Alkenes. Since acyclic systems normally are not rigid, high stereoselectivity has been observed only when special structural features are present. The presence of functional groups (OH, NH, CO, and ether) which form hydrogen bonds with the peroxy acid can facilitate stereoselective epoxidations by imparting rigidity to the system. High *anti* selectivity (>95%) has been observed in the epoxidation of both (**25**) and (**26**) each of which has a branched substituent adjacent to the carbon carrying the silicon group.[29] High *anti* selectivities have been noted during the epoxidation of (**27**) (95%),[30] (**28**) (96%),[31] (**29**) (95%),[32] and (**30**) (96%).[33] High *syn* selectivity has been observed in the reactions of (**31**) (98%)[33] and (**32**) (93%).[34] When the allyl alcohol (**28**) reacts with *m*-CPBA, in the transition state the reagent is hydrogen-bonded to the ether oxygen as well as to allylic hydroxyl. The high selectivity is due to the cooperative effect of the hydroxyl group and the ether oxygen.[31]

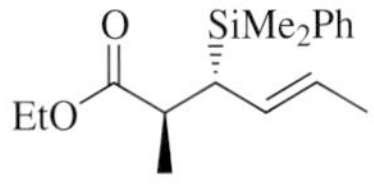

(**25**) 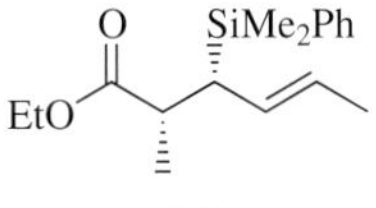   (**26**)

(**27**) 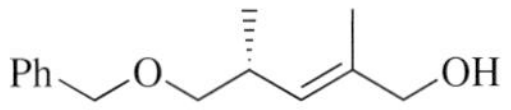   (**28**)

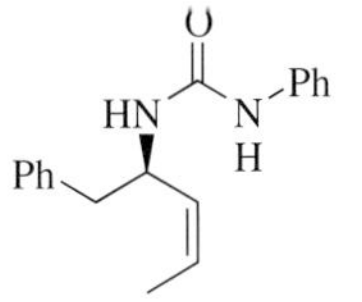

(**29**) 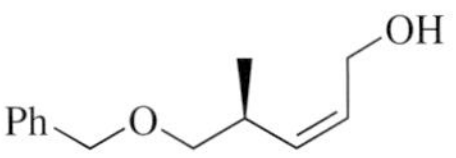   (**30**)

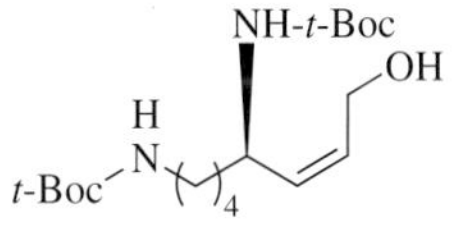

(**31**) 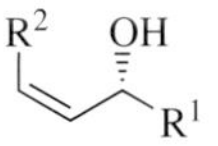   (**32**) $R^1 = (CH_2)_6CO_2Me$
                   $R^2 = (CH_2)_5Me$

High stereoselectivity has also been observed in the epoxidation of some acyclic homoallylic alcohols.[35]

Oxidation of Enol Silyl Ethers and Furans. Epoxides of enol silyl ethers undergo facile ring opening and only in rare cases have stable epoxides been isolated.[36] α-Hydroxy enones have been prepared in two steps from α,β-unsaturated ketones; the enol silyl ether (**33**) prepared from the corresponding enone is treated with *m*-CPBA and the resulting product reacts with triethylammonium fluoride to furnish an α-hydroxy enone (**34**) (eq 8).[37] This method has also been used for the preparation of α-hydroxy ketones,[38] α-hydroxy acids,[39] and α-hydroxy esters. As illustrated in (eq 9), aldehydes have been converted to protected α-hydroxy aldehydes in a similar fashion.[40] Epoxidation of enol silyl ethers according to eq 10 has been used in synthesizing α,α′-dihydroxy ketones from methyl secondary alkyl ketones; the silyl ether (**36**) furnishes the corresponding dihydroxy ketone quantitatively upon brief acidic treatment.[41] Peroxy acid oxidation of furfuryl alcohols yields pyranones according to eq 11.[42,43] Furfurylamides also react similarly.[44]

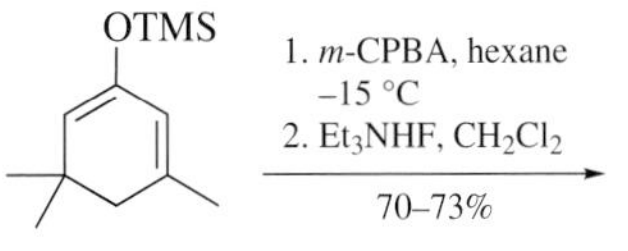

(**33**)   (**34**)   (8)

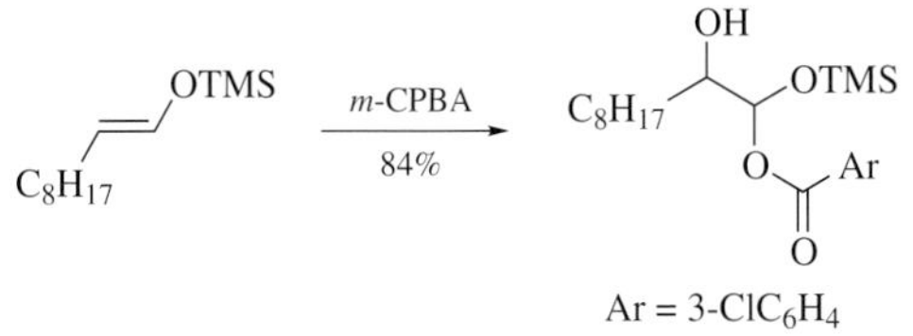

Ar = 3-ClC$_6$H$_4$ (9)

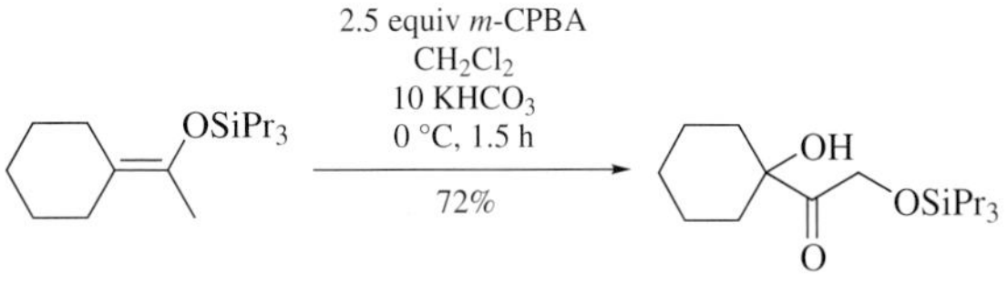

(**35**)   (**36**)   (10)

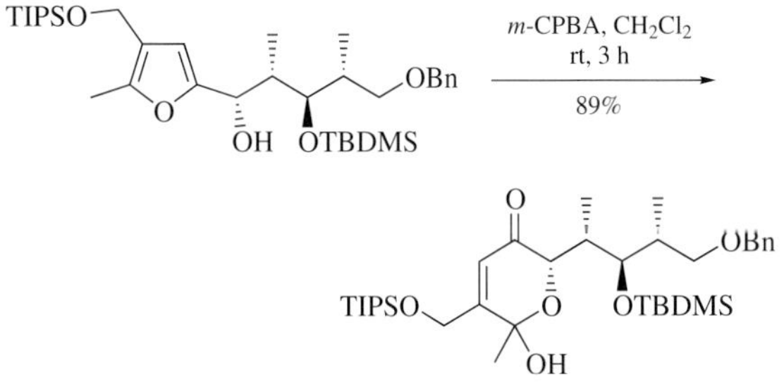

(11)

Baeyer–Villiger Rearrangement. Reaction of a ketone (**37**) with peroxy acid results in oxygen insertion to furnish the esters (**38**) and (**39**). This reaction, known as the Baeyer–Villiger rearrangement, has been reviewed recently.[45] Cyclobutanones undergo very facile rearrangement with peroxy acids, as well as with **Hydrogen Peroxide** in presence of base. The cyclobutanone (**40**) reacted readily with *m*-CPBA to furnish regio-, stereo-, and chemoselectively the lactone (**41**) (eq 12),[38b] which was elaborated to gingkolide. Baeyer–Villiger reaction of (**40**) with H$_2$O$_2$ /base furnished a γ-lactone which was the regioisomer of (**41**). When 1,2,3,8,9,9a-hexahydro-1-methyl-3a,8-methano-3a*H*-cyclopentacycloocten-10-one, which has double bonds as well as a keto group, was treated with *m*-CPBA, exclusive alkene epoxidation was observed.[46] Ketones having stannyl groups on the β-carbon undergo a tin-directed Baeyer–Villiger reaction.[47]

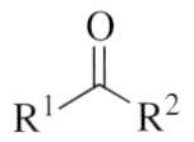

(**37**) 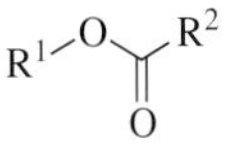   (**38**) 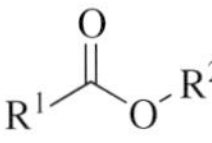   (**39**)

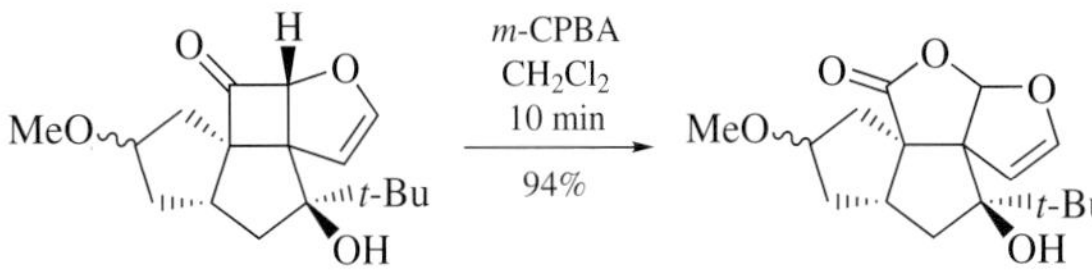

(**40**)   (**41**)   (12)

Oxidation of Nitrogen-containing Compounds. Primary amines are oxidized by *m*-CPBA to the corresponding nitro compounds. One of the intermediates formed in this reaction is the corresponding nitroso compound, which reacts sluggishly with the reagent. High yields are obtained by carrying out the reaction at a high temperature (~83 °C) and increasing the reaction time (3 hours). For example, *n*-hexylamine is oxidized to 1-nitrohexane in 66% yield.[48] When a substrate having the amino group at a chiral center was oxidized, the nitro compound was formed with substantial (~95%) retention of configuration.[49]

m-CPBA oxidation of the sulfilimine (**42**) prepared from 2-aminopyridine, furnished 2-nitrosopyridine (**43**) (eq 13).[50]

Secondary amines have been oxidized to hydroxylamines with *m*-CPBA.[26b] In this reaction, substantial amounts of nitrone as byproduct are expected. (The best method for the preparation of hydroxylamines is to oxidize the secondary amine with 2-(phenylsulfonyl)-3-aryloxaziridine (see e.g. (±) *trans-2-(Phenylsulfonyl)-3-phenyloxaziridine*) to the nitrone, and then to reduce the nitrone with **Sodium Cyanoborohydride**).[51]

m-CPBA oxidation of N-heterocycles furnishes in high yields the corresponding *N*-oxides.[52] Several tertiary *N*-oxides have been prepared by the reaction of tertiary amines with *m*-CPBA in CHCl$_3$ at 0–25 °C and employing chromatography on alkaline alumina; for example, trimethylamine *N*-oxide was obtained in 96% yield.[53] When the optically pure tertiary amine (**44**) is oxidized with *m*-CPBA, the initially formed amine oxide rearranges to the hydroxylamine (**45**) with complete 1,3-transfer of chirality (eq 14).[54]

Reaction of *m*-CPBA with the isoxazole (**46**) furnishes the nitrone (**47**) (eq 15).[55] *m*-CPBA oxidation of (−)-isoxazole (**48**) and subsequent workup results in the formation of the (−)-cyclopentanone (**49**) (eq 16);[56] the initially formed nitrone is hydrolyzed during workup. The oxaziridine (**51**) has been prepared by epoxidizing the sulfonimine (**50**) (eq 17).[57]

The cleavage of the *N,N*-dimethylhydrazone (**52**) proceeds rapidly in the presence of *m*-CPBA, even at low temperatures, to furnish the ketone (**53**), without isomerization to the more stable *cis* isomer (eq 18).[58]

Oxidation of Phosphorus-containing Compounds. *m*-CPBA oxidation of the phosphite (**54**) is stereospecific; it furnishes the phosphate (**55**) (eq 19).[59a] However, aqueous **Iodine** is the reagent of choice for the oxidation of nucleotidic phosphite triesters.[59b] *m*-CPBA oxidation of thiophosphate triesters furnishes the corresponding phosphate esters with retention of configuration.[60]

Oxidation of Sulfur-containing Compounds. *n*-Butanethiol is oxidized by *m*-CPBA in CH$_2$Cl$_2$ at −30 °C to furnish in 82% yield *n*-butanesulfinic acid (*n*-BuSO$_2$H); other thiols react similarly.[61] Sulfides are oxidized chemoselectively to sulfoxides by *m*-CPBA; the reaction is fast even at −70 °C, and the product is free from sulfone.[62] Three reagents (*m*-CPBA, **Sodium Periodate**, and **Iodosylbenzene**) are regarded as ideal for the oxidation of sulfides to sulfoxides.[63] Good diastereoselectivity has been observed in the oxidation of the sulfide (**56**) (eq 20).[64] Sulfides carrying suitably located hydroxyl groups are oxidized diastereoselectively, due to the directing influence of the hydroxyl group.[65] A phenyl sulfide carrying a variety of functional groups (epoxide, hydroxyl, ether, carbamate, and enediyne) has been chemoselectively oxidized in 99% yield to the corresponding sulfone.[52]

(56) R = Ph   68%   28%   (20)

Allenyl chloromethyl sulfoxide (**57**) reacts with *m*-CPBA to furnish allenyl chloromethyl sulfone (**58**) (eq 21).[66] The enethiolizable thioketone (**59**) has been oxidized to the (*E*)-sulfine (**60**) (eq 22).[67] The 2′-deoxy-4-pyrimidinone (**62**) has been prepared by reacting the 2-thiopyrimidine nucleoside (**61**) with *m*-CPBA (eq 23).[68] Thioamides have been transformed to the amides in high yields.[69]

(57) → *m*-CPBA → **(58)**   (21)

(59) → *m*-CPBA, CH₂Cl₂, 0 °C, 1–5 min, 100% → **(60)**   (22)

(61) R = DMTrO → *m*-CPBA, 10 min CH₂Cl₂, py (9:1), 98% → **(62)**   (23)

Oxidation of Selenides. Phenyl selenides react rapidly with *m*-CPBA at −10 °C to form phenyl selenoxides,[70a] which on warming to 0 °C or at rt undergo facile *cis* elimination. This procedure for introducing unsaturation under mild conditions has been used in the synthesis of thermally sensitive compounds; for an example see eq 24.[70b] The selenonyl moiety is a good leaving group, and its generation in the substrate can lead to the formation of cyclic compounds. The oxazoline (**64**) has been synthesized through oxidation of the selenide (**63**) and treatment of the oxidized material with base (eq 25).[71]

(63) → *m*-CPBA, CH₂Cl₂, 0 °C, 40 min, 93% →   (24)

(63) → 1. *m*-CPBA 2. KOH, 88% → **(64)**   (25)

Oxidation of Allylic Iodides. *m*-CPBA oxidation of the primary allylic iodide (**65**) furnishes the secondary allylic alcohol (**66**) (eq 26);[72] this involves rearrangement of the iodoxy compound formed initially.

(65) → *m*-CPBA, EtOAc CH₂Cl₂, Na₂CO₃ (aq) 5–20 °C, 1 h, 65–67% → **(66)**   (26)

Comparison with Other Reagents. To effect epoxidation, the most commonly used reagents are *m*-CPBA, **Peracetic Acid** (PAA), and **Trifluoroperacetic Acid** (TFPAA). TFPAA is not commercially available. *m*-CPBA is more reactive than PAA and is the reagent of choice for laboratory-scale reactions. For large-scale epoxidations the cheaper PAA is preferred. The highly reactive TFPAA is used for unreactive and heat-sensitive substrates; its reactivity permits the use of low reaction temperatures. The recently introduced reagent magnesium monoperphthalate (MMPP) (see **Monoperoxyphthalic Acid**) is more stable than *m*-CPBA and has many applications.[4]

Epoxidations of hydroxyalkenes have been carried out with **Butyl Hyroperoxide**/vanadium (TBHP/V). *m*-CPBA epoxidation of (*Z*)-cyclooct-2-en-1-ol is *anti* selective; with TBHP/V it is *cis* selective.[24b] Similar differences have been noticed in some acyclic systems.[27c] Since the directing effect of the hydroxyl group is larger in the TBHP/V system it is a better reagent for hydroxyl-directed regioselective epoxidations of polyunsaturated alcohols;[73] the TBHP/V system also exhibits higher hydroxyl-directed selectivity in highly hindered allylic alcohols.[74]

m-CPBA epoxidation of hindered alkenes takes place selectively from the less hindered side; the epoxide of opposite stereochemistry can be prepared by a two-step procedure involving initial preparation of bromohydrin, followed by base treatment.[28]

For the epoxidation of extremely unreactive alkenes[38b] and for the preparation of epoxides which are highly susceptible to nucleophilic attack, **Dimethyldioxirane** is the reagent of choice.[75] Electron-deficient alkenes such as α,β-unsaturated ketones are usually oxidized with **Hydrogen Peroxide**/base.

First Update

André Charette
Universite de Montreal, Montreal, Quebec, Canada

Oxidative Cleavage of Polymer-bound Sulfoximines. Oxidative cleavage of a Merrifield resin-bound sulfoximine afforded the corresponding sulfone in high yield (eq 27).[76] The reaction proceeds by the formation of the *N*-oxide that reacts with a second equivalent of the peracid to lead to the sulfone and the polymer-bound nitroso derivative.

→ *m*-CPBA (4.7 equiv) HCl (0.1 M), THF, rt → 81%   (27)

Oxidative Destannylation of β-Stannylsilylenol Ethers.
m-CPBA was used as the reagent of choice to functionalize the [3+4] annulation product of ketone enolates with
β-stannylacryloylsilanes leading to enediones (eq 28).[77]

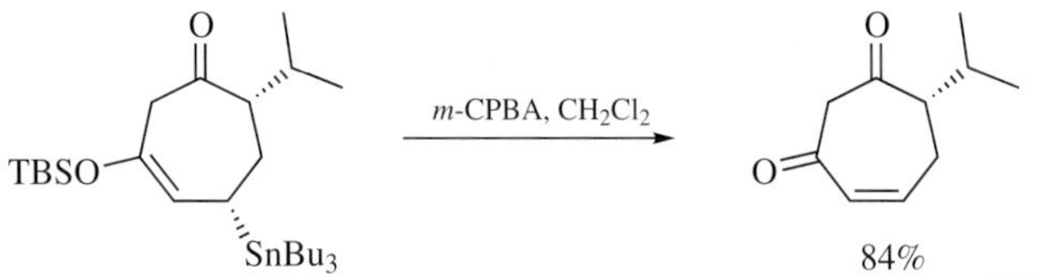

84%
(28)

Chiral Auxiliary-based Epoxidation of Substituted Alkenes.
High diastereoselectivities were found for the *m*-CPBA or
dimethyldioxirane epoxidation of chiral oxazolidine-substituted
alkenes bearing a strongly basic urea group (eq 29).[78] However, in
most cases, the diastereoselectivities were superior with dimethyldioxirane.

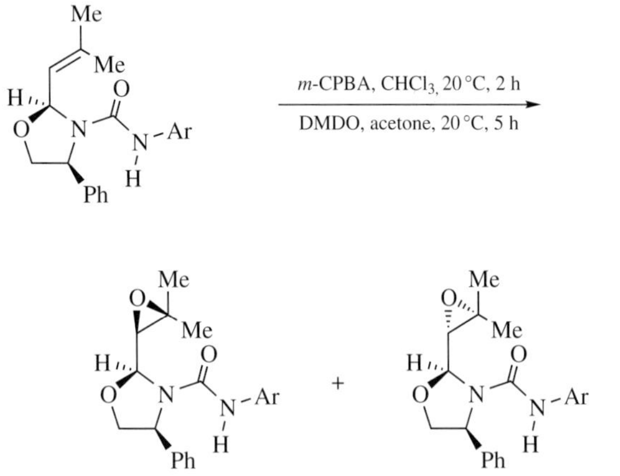

81:19 (94%)

>98:2 (>95%)

Chiral (*E*)-enamides derived from (*S*)-4-phenyl-5,5-dimethyl-
oxazolidin-2-one undergo highly diastereoselective epoxidation
upon treatment with *m*-CPBA followed by in situ epoxide opening
and trapping with *m*-chlorobenzoic acid (eq 30).[79]

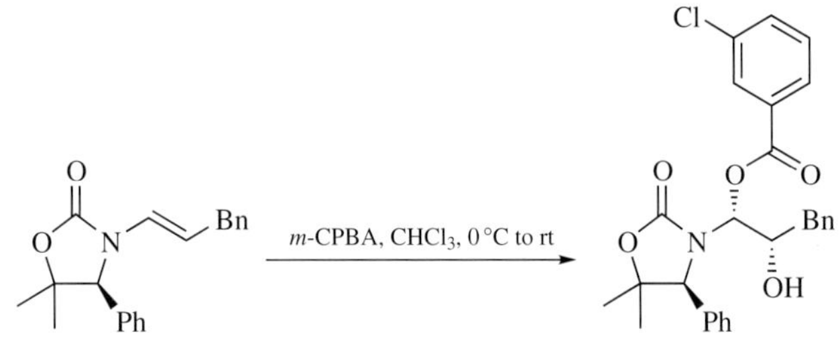

71%, >98% de
(30)

The epoxidation of chiral β-alkoxy-α,β-unsaturated esters bearing a chiral auxiliary with *m*-CPBA led to α-hydroxy esters with
high diastereoselectivities (eq 31).[80] Both enantiomers are readily accessible simply by protecting the free hydroxy group on the
auxiliary by a trimethylsilyl group (eq 32).

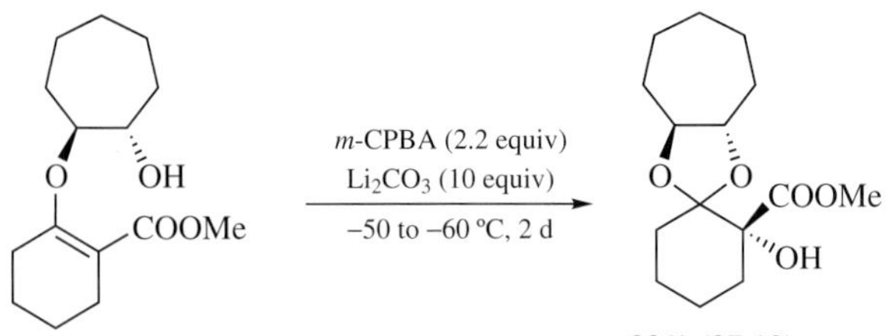

93% (87:13)

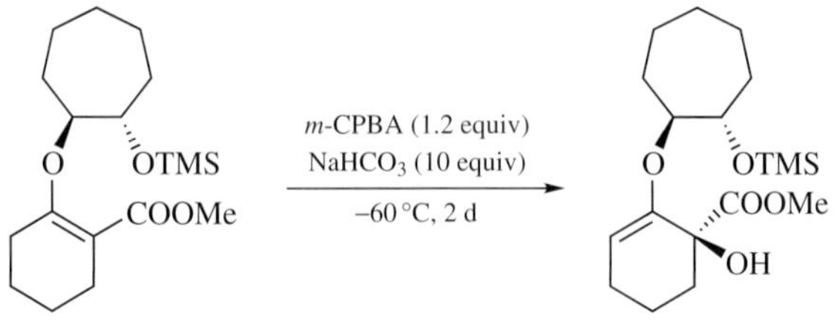

63% (>99:1)

Diastereoselective Formation of Glycosyl Sulfoxides. The
systematic oxidation of several *S*-alkyl and *S*-aryl D-thiopyrano-
sides was reported. The reaction is highly stereoselective with
α-glycosides leading to a single diastereomeric sulfoxide
(eq 33).[81] Although there are a few exceptions, this is a general
trend.

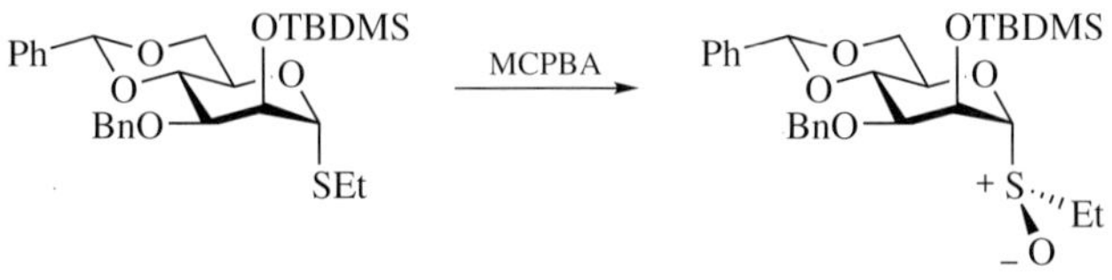

(33)

Epoxidation of Acyclic Alkenes. Stereoselective epoxidation
of a series of allylic carbamate methyl esters (eq 34), homoallylic
alcohols (eq 35), and acetates (eq 36) could be performed with
good to excellent stereocontrol.[82] It is believed that the directing
effect of the carbamate protecting group plays an important role
in dictating the level of stereocontrol.

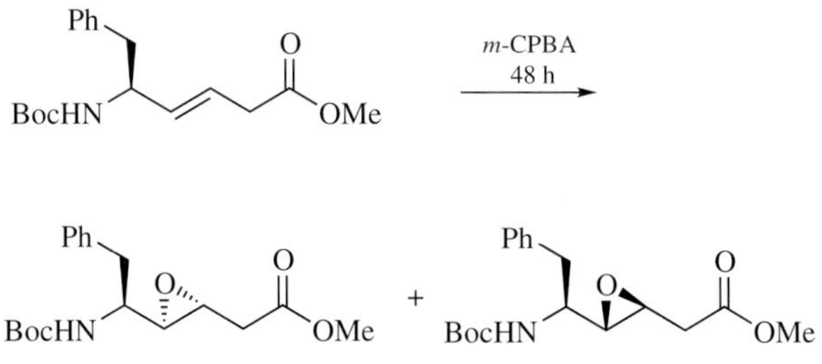

75% (8:1)

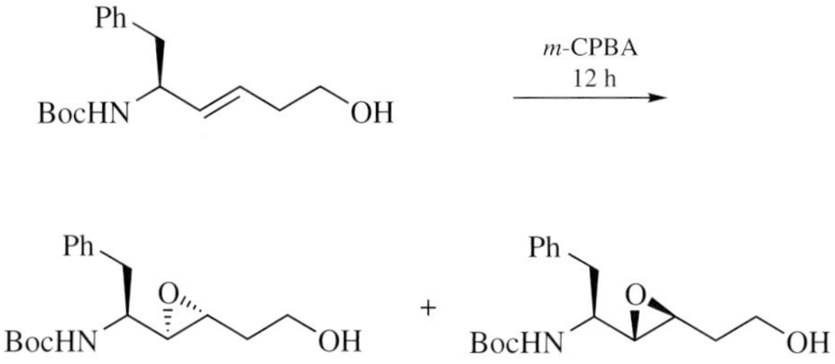

$$>95\%,\ 9{:}1\ (trans{:}cis)$$ (39)

(35)

68% (7:1)

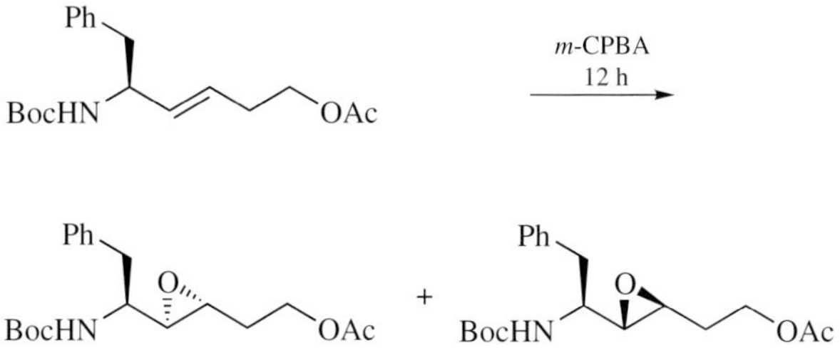

(36)

87% (5:1)

The diastereoselectivity of the epoxidation of allylic alcohols substituted by polypropionate units is highly dependant upon the nature of the protecting group and on the stereochemistry of the stereogenic centers (eqs 37 and 38).[83]

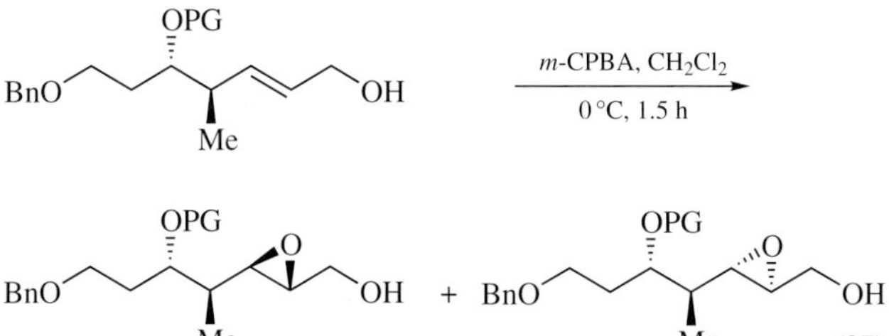

(37)

PG = MOM   82% (59:41)
PG = TES   82% (95:5)

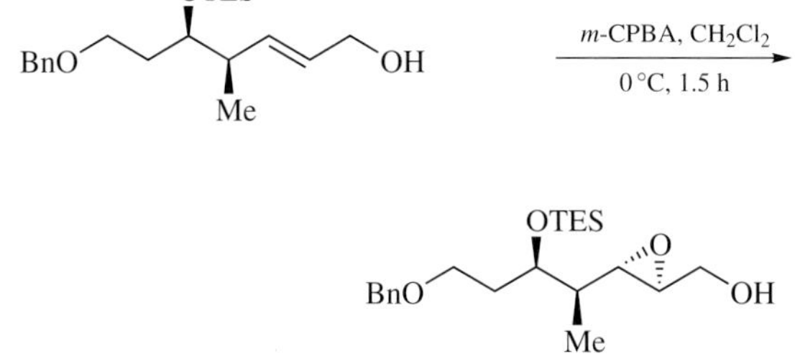

(38)

95% (>99:1)

Epoxidation of Glycals. The epoxidation of glycals has been traditionally accomplished using dimethyldioxirane. However, it was found that the Camps reagent that consists of a mixture of *m*-CPBA and KF allows the conversion of glycals to the corresponding epoxides in high yields (eq 39).[84] The presence of KF is necessary to avoid the concomitant epoxide opening by the nucleophilic by-product of the reaction (*m*-chlorobenzoic acid).

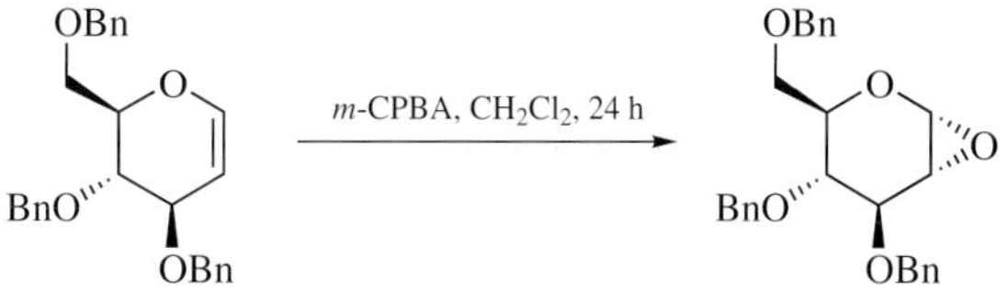

Monoxidation of Thiophene. Thiophenes can be oxidized by *m*-CPBA to produce thiophene-*S,S*-dioxides in high yields. However, when this reaction is carried out in the presence of a catalytic amount of $BF_3 \cdot OEt_2$ the oxidation affords good yields of thiophene *S*-monoxides that can be isolated in solution.[85,86] These products can also be trapped directly by dienophiles to lead to bicyclic compounds (eq 40).

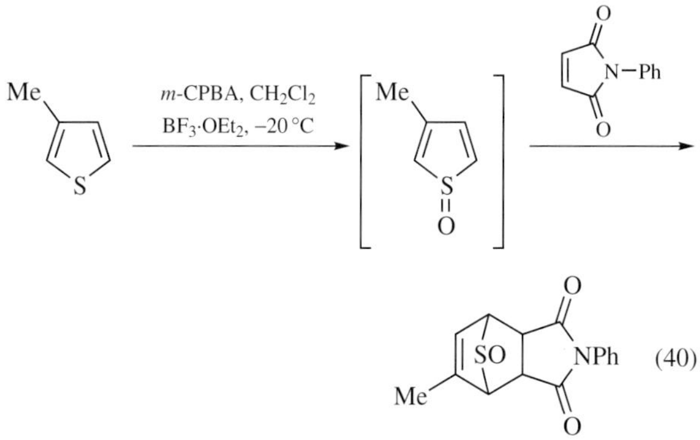

(40)

Epoxidation of 1,4-Dioxene. Oxidation of 2,3-disubstituted-1,4-dioxenes with *m*-CPBA in methanol gave predominantly only one acetal (eq 41).[87] The product can be used as a precursor to generate oxabicyclo[4.2.1]nonene systems.

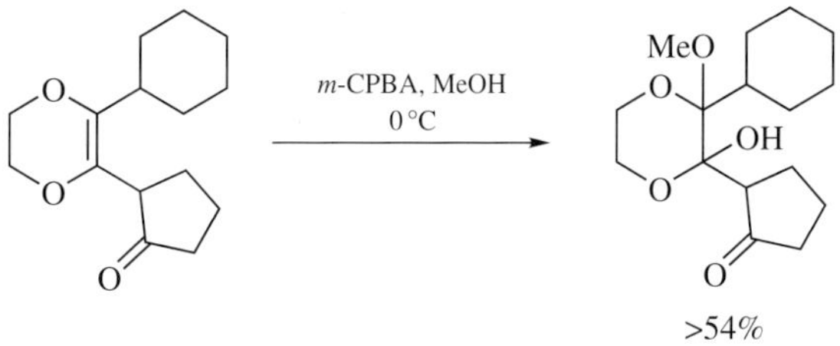

(41)

>54%

Baeyer–Villiger Reaction of Fluoroketones. The Baeyer–Villiger reaction of ketones in the presence of peracids is an important methodology to access esters and lactones via a skeletal rearrangement of the peracid/ketone adduct. The stereoselectronic requirement for correct antiperiplanar alignment of the migrating group and the O–O bond of the leaving group (RCOO⁻) are well established. However, the presence of strongly electron-withdrawing α-substituents may influence the regioselectivity of the C–C bond migration. For example, substituted 2-fluorocyclohexanones can give the two regioisomers depending of the relative configuration of the stereogenic centers and on the solvent used (eqs 42–44).[88]

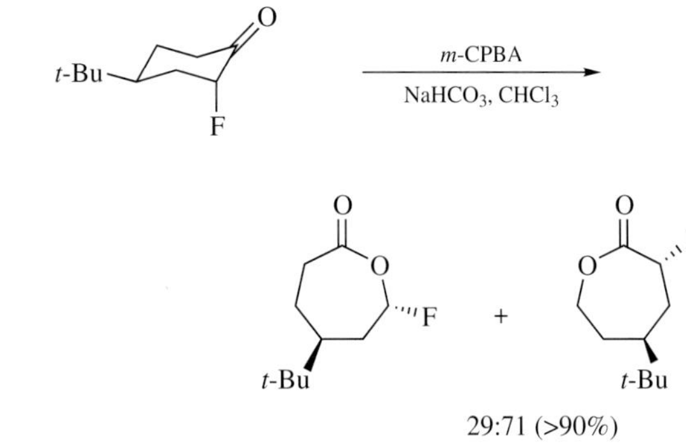

t-Bu

$$\xrightarrow[\text{NaHCO}_3, \text{CHCl}_3]{m\text{-CPBA}}$$

t-Bu   +   *t*-Bu   (42)

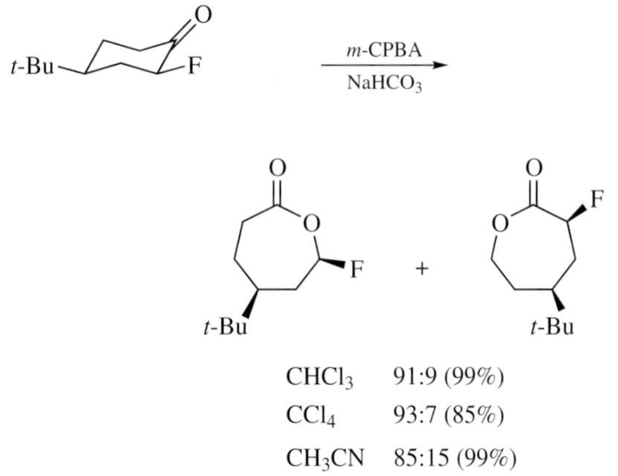

29:71 (>90%)

t-Bu

$$\xrightarrow[\text{NaHCO}_3]{m\text{-CPBA}}$$

t-Bu   +   *t*-Bu   (43)

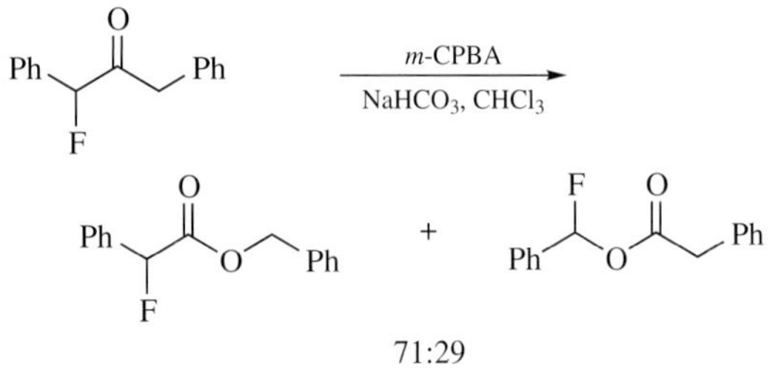

CHCl$_3$ 91:9 (99%)

CCl$_4$ 93:7 (85%)

CH$_3$CN   85:15 (99%)

Ph   Ph

$$\xrightarrow[\text{NaHCO}_3, \text{CHCl}_3]{m\text{-CPBA}}$$

Ph   Ph   +   Ph   Ph   (44)

71:29

However, the Baeyer–Villiger reaction of α-trifluoromethyl-cycloalkylketones show the reversal in regioselectivity (eq 45).[89]

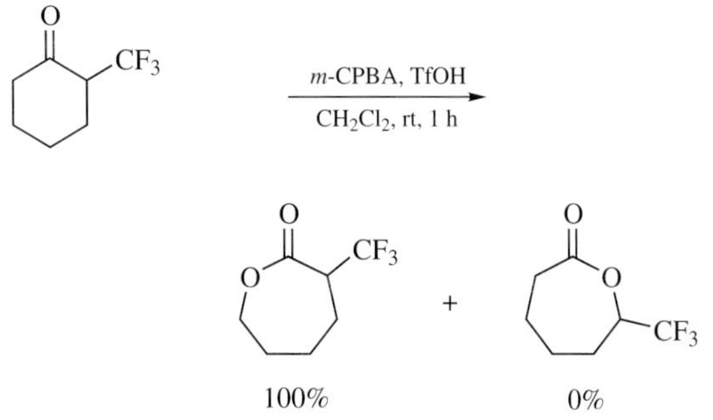

CF$_3$

$$\xrightarrow[\text{CH}_2\text{Cl}_2, \text{rt, 1 h}]{m\text{-CPBA, TfOH}}$$

CF$_3$ +   CF$_3$ (45)

100%   0%

The Baeyer–Villiger oxidation of acyclic 4-methoxyphenyl substituted fluoroketones provides an expedient access to α-fluoroesters (eq 46).[90] The regioselectivity of the carbon-carbon bond migration is highly dependent not only on the presence of an aryl substituent but also on the presence of the *p*-methoxy group.

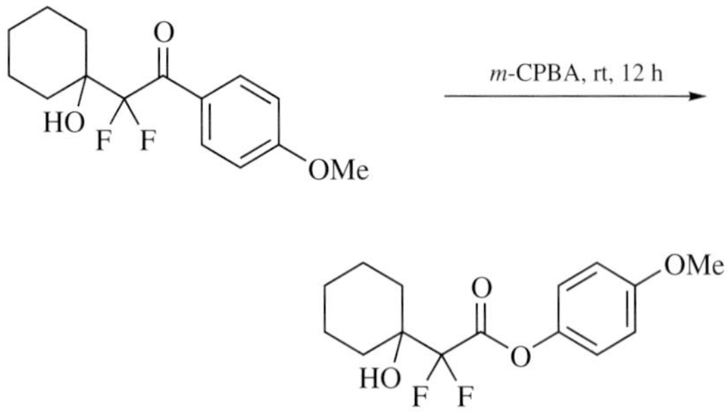

OMe

$$\xrightarrow[]{m\text{-CPBA, rt, 12 h}}$$

OMe   (46)

Baeyer–Villiger Reaction of Hemiacetals. Hemiacetals have been shown to react with *m*-CPBA to generate lactones with excellent regiocontrol (eq 47). The reaction is often much cleaner than that performed on the cyclic ketone and it has been shown to involve the formation of a cyclic oxocarbenium ion intermediate.[91]

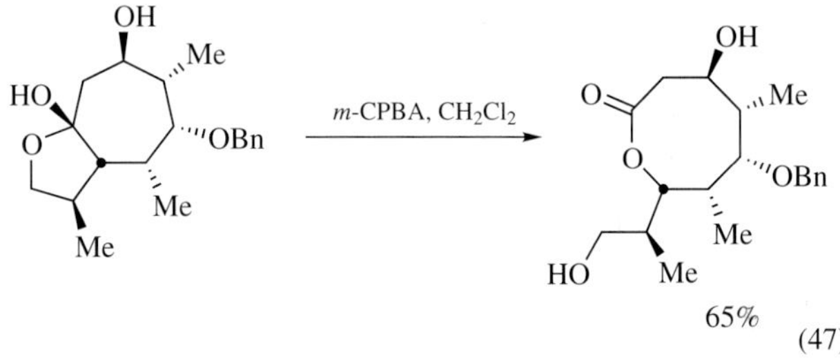

$$\xrightarrow[]{m\text{-CPBA, CH}_2\text{Cl}_2}$$

65%   (47)

Lewis Acid Catalyzed Baeyer–Villiger Oxidation. Although *m*-CPBA is one of the most commonly used reagents for carrying out the Baeyer–Villiger oxidation of ketones, its reactivity can be increased by adding Lewis acids. Recently, it has been shown that the addition of catalytic amounts of scandium(III) triflate,[92] triflic acid, or bismuth(III) triflate[93] significantly increases the yield and rate of the Baeyer–Villiger oxidation of ketones.

Oxidative Cleavage of Cyclic Acetals. Benzylidene acetals are cleaved to hydroxyesters upon treatment with a mixture of 2,2′-bipyridinium chlorochromate and *m*-CPBA (eq 48).[94]

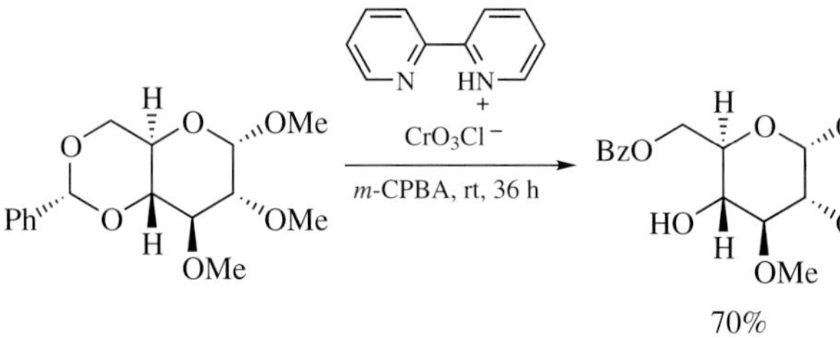

$$\xrightarrow[m\text{-CPBA, rt, 36 h}]{\text{CrO}_3\text{Cl}^-}$$

70%   (48)

Chlorination of Dichlorovinyl Ethers. A powerful chlorinating agent that successfully converts dichloroalkenes to tetrachloroalkanes is generated with *m*-CPBA in dichloromethane (eq 49).[95] The reaction presumably proceeds by the in situ formation of chlorine via a radical mechanism. This process occurs only when the alkene is not sufficiently nucleophile to react directly with *m*-CPBA.

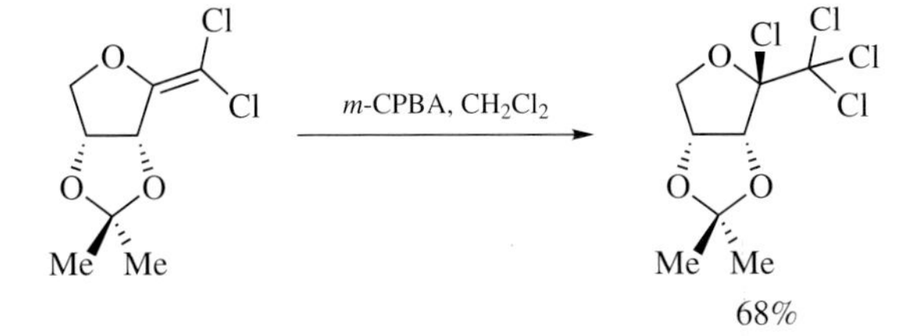

$$(49)$$

68%

However, if the reaction is run in the presence of methanol and of a radical scavenger, the α-chloro methyl ester is formed in good yield (eq 50).[96]

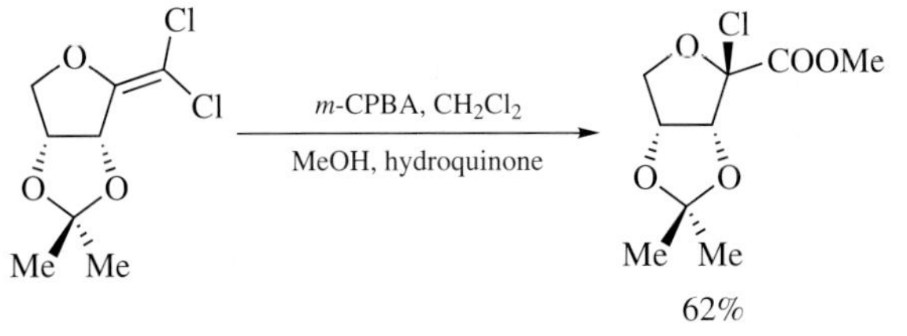

$$(50)$$

62%

Oxidation of Imines. The oxidative rearrangement of *N*-arylaldimines leads to amides upon treatment with *m*-CPBA and BF$_3$·OEt$_2$ (eq 51).[97] The reaction presumably proceeds via the Lewis acid mediated peracid imine adduct which then loses *m*-chlorobenzoic acid.

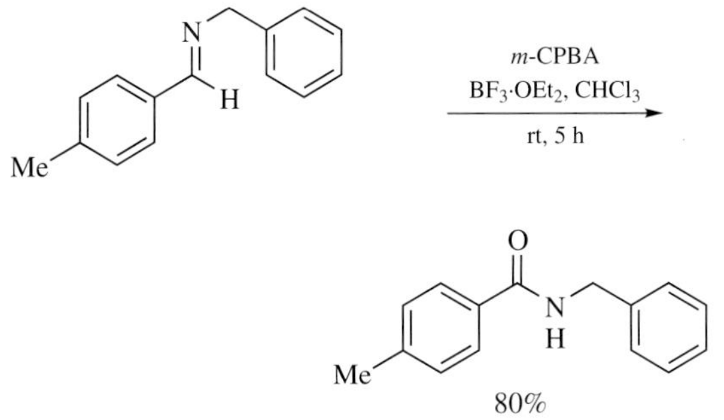

$$(51)$$

80%

Ketimines derived from either diarylketones or arylalkylketones and arylamines react in an analogous fashion to lead to amides (eq 52).[98] The reaction involves the initial formation of the oxaziridine which undergoes a Lewis acid mediated 1,2-migration of the aryl group.

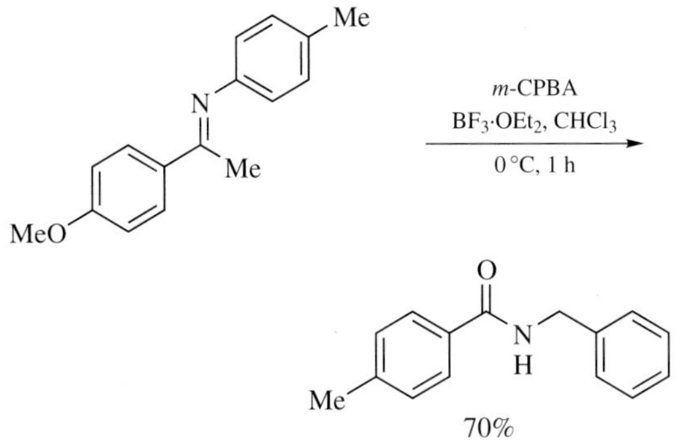

$$(52)$$

70%

Terminal Oxidant for the Low Temperature Catalyzed Epoxidation of Olefins. The enantioselective epoxidation of

olefins catalyzed by (salen)Mn(III) complexes[99] can be effected under anhydrous conditions and at $-78\,°C$ with a combination of *m*-CPBA and NMO (eq 53).[100,101] A variety of unfunctionalized alkenes undergo epoxidation with a significant increase in enantioselectivity relative to reaction using aqueous sodium hypochlorite.

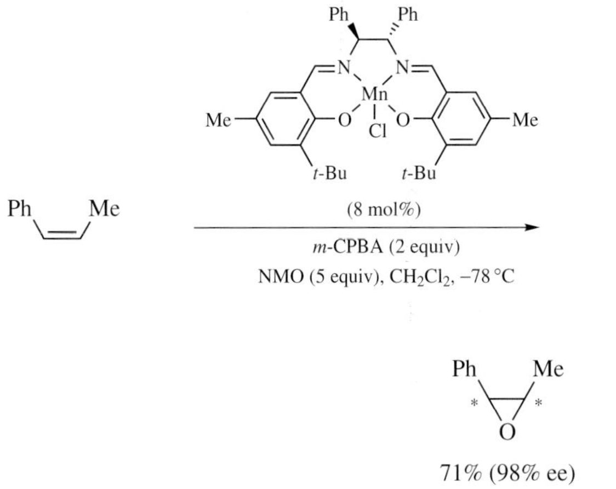

$$(53)$$

71% (98% ee)

m-CPBA was also used as the terminal oxidant in the low temperature Cu(CH$_3$CN)$_4$PF$_6$ catalyzed epoxidation of alkenes.[102]

Hydroxylamine Synthesis by Oxidation-cope Elimination. The oxidation of a range of β-cyanoethyl tertiary amines with *m*-CPBA gives the corresponding *N*-oxides which can undergo Cope-elimination to give secondary hydroxylamines in high yields (eq 54).[103]

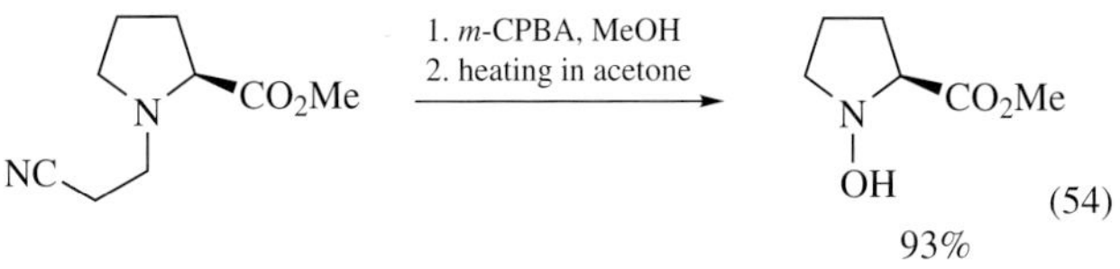

$$(54)$$

93%

The reaction has also been successfully used for the solid-phase synthesis of hydroxylamines (eq 55).[104] The resin used was a polymer-bound benzyl acrylate.

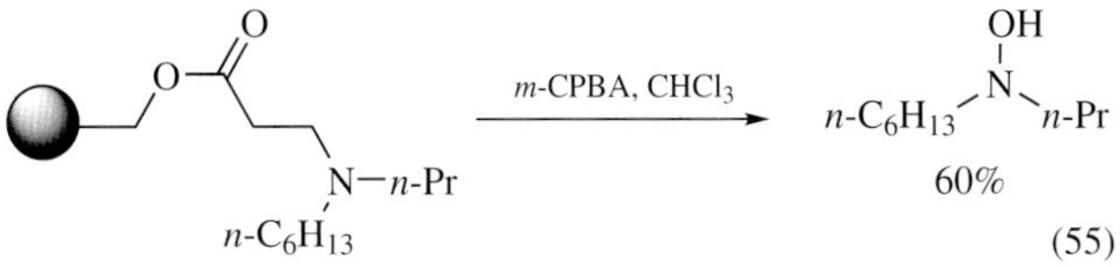

$$(55)$$

60%

Oxidation of Benzylic Methylene to Aryl Ketones. Benzylic positions are oxidized under treatment with *m*-CPBA, air, and NaHCO$_3$ to provide the corresponding ketones (eq 56).[105] The reaction proceeds very well with some substrates but poorly with those possessing electron-withdrawing substituents.

(56)

Benzylic oxidation of *N*-substituted phthalimidines by treatment with 2,2′-bipyridinium chlorochromate/*m*-CPBA affords phthalimides in good yields.[106]

Related Reagents. *m*-Chloroperbenzoic Acid–2,2,6,6-Tetramethylpiperidine Hydrochloride.

1. (a) Swern, D. *Organic Peroxides*; Wiley: New York, 1971; Vol 2, p 355. (b) Plesnicar, B. *Organic Chemistry*; Academic: New York, 1978; Vol. 5 C, p 211. (c) Rao, A. S., *Comprehensive Organic Synthesis* **1991**, *7*, Chapter 3.1.

2. McDonald, R. N.; Steppel, R. N.; Dorsey, J. E., *Org. Synth.* **1970**, *50*, 15.

3. (a) Schwartz, N. N.; Blumbergs, J. H., *J. Org. Chem.* **1964**, *29*, 1976. (b) Nakayama, J.; Kamiyama, H., *Tetrahedron Lett.* **1992**, *33*, 7539.

4. Brougham, P.; Cooper, M. S.; Cummerson, D. A.; Heany, H.; Thompson, N., *Synthesis* **1987**, 1015.

5. *Hazards in the Chemical Laboratory*; Luxon, S. G., Ed.; Royal Society of Chemistry: Cambridge, 1992.

6. (a) Ishikawa, K.; Charles, H. C.; Griffin, G. W., *Tetrahedron Lett.* **1977**, 427. (b) Srebnik, M.; Mechoulam, R.; *Synthesis* **1983**, 1046.

7. Berti, G., *Top. Stereochem.* **1973**, *7*, 93.

8. (a) Woods, K. W.; Beak, P., *J. Am. Chem. Soc.* **1991**, *113*, 6281. (b) Bartlett, P. D., *Rec. Chem. Prog.* **1950**, *11*, 47 (cited in Ref. 1a).

9. (a) Camps, F.; Coll, J.; Messeguer, A.; Pujol, F., *J. Org. Chem.* **1982**, *47*, 5402. (b) Chai, K.-B.; Sampson, P., *Tetrahedron Lett.* **1992**, *33*, 585.

10. Paquette, L. A.; Barrett, J. H., *Org. Synth.* **1969**, *49*, 62.

11. (a) Wender, P. A.; Zercher, C. K., *J. Am. Chem. Soc.* **1991**, *113*, 2311. (b) Philippo, C. M. G.; Vo, N. H.; Paquette, L. A., *J. Am. Chem. Soc.* **1991**, *113*, 2762.

12. Bach, R. D.; Knight, J. W., *Org. Synth.* **1981**, *60*, 63.

13. (a) Kishi, Y.; Aratani, M.; Tanino, H.; Fukuyama, T.; Goto, T., *J. Chem. Soc., Chem. Commun.* **1972**, 64. (b) Valente, V. R.; Wolfhagen, J. L., *J. Org. Chem.* **1966**, *31*, 2509.

14. Anderson, W. K.; Veysoglu, T., *J. Org. Chem.* **1973**, *38*, 2267. (b) Imuta, M.; Ziffer, H., *J. Org. Chem.* **1979**, *44*, 1351.

15. Urones, J. G.; Marcos, I. S.; Basabe, P.; Alonso, C.; Oliva, I. M.; Garrido, N. M.; Martin, D. D.; Lithgow, A. M., *Tetrahedron* **1993**, *49*, 4051.

16. Bäckvall, J.-E.; Juntunen, S. K., *J. Org. Chem.* **1988**, *53*, 2398.

17. Hudlicky, T.; Price, J. D.; Rulin, F.; Tsunoda, T., *J. Am. Chem. Soc.* **1990**, *112*, 9439.

18. Marsh, E. A., *Synlett* **1991**, 529.

19. Brown, H. C.; Kawakami, J. H.; Ikegami, S., *J. Am. Chem. Soc.* **1970**, *92*, 6914.

20. Devreese, A. A.; Demuynck, M.; De Clercq, P. J.; Vandewalle, M., *Tetrahedron* **1983**, *39*, 3049.

21. Still, W. C.; Romero, A. G., *J. Am. Chem. Soc.* **1986**, *108*, 2105.

22. Schneider, A.; Séquin, U., *Tetrahedron* **1985**, *41*, 949.

23. Johnson, C. R.; Tait, B. D.; Cieplak, A. S., *J. Am. Chem. Soc.* **1987**, *109*, 5875.

24. (a) Henbest, H. B.; Wilson, R. A. L., *J. Chem. Soc.* **1957**, 1958. (b) Itoh, T.; Jitsukawa, K.; Kaneda, K.; Teranishi, S., *J. Am. Chem. Soc.* **1979**, *101*, 159.

25. Wipf, P.; Kim, Y., *J. Org. Chem.* **1993**, *58*, 1649.

26. (a) Kim, G.; Chu-Moyer, M. Y.; Danishefsky, S. J.; Schulte, G. K., *J. Am. Chem. Soc.* **1993**, *115*, 30. (b) Fukuyama, T.; Xu, L.; Goto, S., *J. Am. Chem. Soc.* **1992**, *114*, 383.

27. (a) Sharpless, K. B.; Verhoeven, T. R., *Aldrichim. Acta* **1979**, *12*, 63. (b) Rossiter, B. E.; Verhoeven, T. R.; Sharpless, K. B., *Tetrahedron Lett.* **1979**, 4733. (c) Adam, W.; Nestler, B., *J. Am. Chem. Soc.* **1993**, *115*, 5041.

28. Kočovský, P.; Starý, I., *J. Org. Chem.* **1990**, *55*, 3236.

29. Murphy, P. J.; Russel, A. T.; Procter, G., *Tetrahedron Lett.* **1990**, *31*, 1055.

30. Fleming, I.; Sarkar, A. K.; Thomas, A. P., *J. Chem. Soc., Chem. Commun.* **1987**, 157.

31. Johnson, M. R.; Kishi, Y., *Tetrahedron Lett.* **1979**, 4347.

32. Roush, W. R.; Straub, J. A.; Brown, R. J., *J. Org. Chem.* **1987**, *52*, 5127.

33. Sakai, N.; Ohfune, Y., *J. Am. Chem. Soc.* **1992**, *114*, 998.

34. Lewis, M. D.; Menes, R., *Tetrahedron Lett.* **1987**, *28*, 5129.

35. Fukuyama, T.; Wang, C.-L. J.; Kishi, Y., *J. Am. Chem. Soc.* **1979**, *101*, 260.

36. Paquette, L. A.; Lin, H.-S.; Gallucci, J. C., *Tetrahedron Lett.* **1987**, *28*, 1363.

37. (a) Rubottom, G. M.; Gruber, J. M., *J. Org. Chem.* **1978**, *43*, 1599. (b) Rubottom, G. M.; Gruber, J. M.; Juve, H. D.; Charleson, D. A., *Org. Synth.* **1986**, *64*, 118.

38. (a) Herlem, D.; Kervagoret, J.; Yu, D.; Khuong-Huu, F.; Kende, A. S., *Tetrahedron* **1993**, *49*, 607. (b) Crimmins, M. T.; Jung, D. K.; Gray, J. L., *J. Am. Chem. Soc.* **1993**, *115*, 3146.

39. Rubottom, G. M.; Marrero, R., *J. Org. Chem.* **1975**, *40*, 3783.

40. Hassner, A.; Reuss, R. H.; Pinnick, H. W., *J. Org. Chem.* **1975**, *40*, 3427.

41. Horiguchi, Y.; Nakamura, E.; Kuwajima, I., *Tetrahedron Lett.* **1989**, *30*, 3323.

42. Shimshock, S. J.; Waltermire, R. E.; DeShong, P., *J. Am. Chem. Soc.* **1991**, *113*, 8791.

43. Honda, T.; Kobayashi, Y.; Tsubuki, M., *Tetrahedron* **1993**, *49*, 1211.

44. Zhou, W.-S.; Lu, Z.-H.; Wang, Z.-M., *Tetrahedron* **1993**, *49*, 2641.

45. Krow, G. R., *Comprehensive Organic Synthesis* **1991**, *7*, Chapter 5, 1.

46. Feldman, K. S.; Wu, M.-J.; Rotella, D. P., *J. Am. Chem. Soc.* **1990**, *112*, 8490.

47. Bakale, R. P.; Scialdone, M. A.; Johnson, C. R., *J. Am. Chem. Soc.* **1990**, *112*, 6729.

48. Gilbert, K. E.; Borden, W. T., *J. Org. Chem.* **1979**, *44*, 659.

49. Robinson, C. H.; Milewich, L.; Hofer, P., *J. Org. Chem.* **1966**, *31*, 524.

50. Taylor, E. C.; Tseng, C.-P.; Rampal, J. B., *J. Org. Chem.* **1982**, *47*, 552.

51. Jasys, V. J.; Kelbaugh, P. R.; Nason, D. M.; Phillips, D.; Rosnack, K. J.; Saccomano, N. A.; Stroh, J. G.; Volkmann, R. A., *J. Am. Chem. Soc.* **1990**, *112*, 6696.

52. Nicolaou, K. C.; Maligres, P.; Suzuki, T.; Wendeborn, S. V.; Dai, W.-M.; Chadha, R. K., *J. Am. Chem. Soc.* **1992**, *114*, 8890.

53. Craig, J. C.; Purushothaman, K. K., *J. Org. Chem.* **1970**, *35*, 1721.

54. Reetz, M. T.; Lauterbach, E. H., *Tetrahedron Lett.* **1991**, *32*, 4481.

55. Ali, Sk. A.; Wazeer, M. I. M., *Tetrahedron* **1993**, *49*, 4339.

56. Hwu, J. R.; Robl, J. A.; Gilbert, B. A., *J. Am. Chem. Soc.* **1992**, *114*, 3125.

57. Vishwakarma, L. C.; Stringer, O. D.; Davis, F. A., *Org. Synth.* **1988**, *66*, 203.

58. Duraisamy, M.; Walborsky, H. M., *J. Org. Chem.* **1984**, *49*, 3410.

59. (a) Sekine, M.; Iimura, S.; Nakanishi, T., *Tetrahedron Lett.* **1991**, *32*, 395. (b) Beaucage, S. L.; Iyer, R. P., *Tetrahedron* **1992**, *48*, 2223.

60. Cullis, P. M., *J. Chem. Soc., Chem. Commun.* **1984**, 1510.

61. Filby, W. G.; Günther, K.; Penzhorn, R. D., *J. Org. Chem.* **1973**, *38*, 4070.

62. Trost, B. M.; Salzmann, T. N.; Hiroi, K., *J. Am. Chem. Soc.* **1976**, *98*, 4887.

63. Madesclaire, M., *Tetrahedron* **1986**, *42*, 5459.

64. Evans, D. A.; Faul, M. M.; Colombo, L.; Bisaha, J. J.; Clardy, J.; Cherry, D., *J. Am. Chem. Soc.* **1992**, *114*, 5977.

65. (a) Wang, X.; Ni, Z.; Lu, X.; Smith, T. Y.; Rodriguez, A.; Padwa, A., *Tetrahedron Lett.* **1992**, *33*, 5917. (b) DeLucchi, O.; Fabri, D., *Synlett* **1990**, 287.

66. Block, E.; Putman, D., *J. Am. Chem. Soc.* **1990**, *112*, 4072.

67. Le Nocher, A. M.; Metzner, P., *Tetrahedron Lett.* **1991**, *32*, 747.

68. Kuimelis, R. G.; Nambiar, K. P., *Tetrahedron Lett.* **1993**, *34*, 3813.

69. Kochhar, K. S.; Cottrell, D. A.; Pinnick, H. W., *Tetrahedron Lett.* **1983**, *24*, 1323.

70. (a) Reich, H. J.; Shah, S. K., *J. Am. Chem. Soc.* **1975**, *97*, 3250. (b) Danheiser, R. L.; Choi, Y. M.; Menichincheri, M.; Stoner, E. J., *J. Org. Chem.* **1993**, *58*, 322.

71. Toshimitsu, A.; Hirosawa, C.; Tanimoto, S.; Uemura, S., *Tetrahedron Lett.* **1992**, *33*, 4017.

72. Yamamoto, S.; Itani, H.; Tsuji, T.; Nagata, W., *J. Am. Chem. Soc.* **1983**, *105*, 2908.

73. Boeckman, R. K. Jr.; Thomas, E. W., *J. Am. Chem. Soc.* **1979**, *101*, 987.

74. Sanghvi, Y. S.; Rao, A. S., *J. Heterocycl. Chem.* **1984**, *21*, 317.

75. (a) Murray, R. W., *C. R. Hebd. Seances Acad. Sci.* **1989**, *89*, 1187. (b) Halcomb, R. L.; Danishefsky, S. J., *J. Am. Chem. Soc.* **1989**, *111*, 6661. (c) Adam, W.; Hadjiarapoglou, L.; Wang, X., *Tetrahedron Lett.* **1989**, *30*, 6497. (d) Adam, W.; Curci, R.; Edwards, J. O., *Ann. Chim. (Rome)* **1989**, *22*, 205.

76. Hachtel, J.; Gais, H. J., *Eur. J. Org. Chem.* **2000**, 1457.

77. Takeda, K.; Nakajima, A.; Takeda, M.; Okamoto, Y.; Sato, T.; Yoshii, E.; Koizumi, T.; Shiro, M., *J. Am. Chem. Soc.* **1998**, *120*, 4947.

78. (a) Adam, W.; Peters, K.; Peters, E. M.; Schambony, S. B., *J. Am. Chem. Soc.* **2001**, *123*, 7228. (b) Adam, W.; Schambony, S. B., *Org. Lett.* **2001**, *3*, 79. (c) Adam, W.; Bosio, S. G.; Wolff, B. T., *Org. Lett.* **2003**, *5*, 819.

79. Davies, S. G.; Key, M. S.; Rodriguez-Solla, H.; Sanganee, H. J.; Savory, E. D.; Smith, A. D., *Synlett* **2003**, 1659.

80. Kato, K.; Suemune, H.; Sakai, K., *Tetrahedron Lett.* **1994**, *35*, 3103.

81. (a) Crich, D.; Mataka, J.; Zakharov, L. N.; Rheingold, A. L.; Wink, D. J., *J. Am. Chem. Soc.* **2002**, *124*, 6028.

82. (a) Jenmalm, A.; Berts, W.; Li, Y. L.; Luthman, K.; Csoregh, I.; Hacksell, U., *J. Org. Chem.* **1994**, *59*, 1139. (b) Jenmalm, A.; Berts, W.; Luthman, K.; Csoregh, I.; Hacksell, U., *J. Org. Chem.* **1995**, *60*, 1026.

83. Maruyama, K.; Ueda, M.; Sasaki, S.; Iwata, Y.; Miyazawa, M.; Miyashita, M., *Tetrahedron Lett.* **1998**, *39*, 4517.

84. Bellucci, G.; Catelani, G.; Chiappe, C.; Dandrea, F., *Tetrahedron Lett.* **1994**, *35*, 8433.

85. Li, Y. Q.; Thiemann, T.; Sawada, T.; Mataka, S.; Tashiro, M., *J. Org. Chem.* **1997**, *62*, 7926.

86. Furukawa, N.; Zhang, S. Z.; Sato, S.; Higaki, M., *Heterocycles* **1997**, *44*, 61.

87. Hanna, I.; Michaut, V., *Org. Lett.* **2000**, *2*, 1141.

88. Crudden, C. M.; Chen, A. C.; Calhoun, L. A., *Angew. Chem., Int. Ed.* **2000**, *39*, 2852.

89. Itoh, Y.; Yamanaka, M.; Mikami, K., *Org. Lett.* **2003**, *5*, 4803.

90. Kobayashi, S.; Tanaka, H.; Amii, H.; Uneyama, K., *Tetrahedron* **2003**, *59*, 1547.

91. Hunt, K. W.; Grieco, P. A., *Org. Lett.* **2000**, *2*, 1717.

92. Kotsuki, H.; Arimura, K.; Araki, T.; Shinohara, T., *Synlett* **1999**, 462.

93. Alam, M. M.; Varala, R.; Adapa, S. R., *Syn. Comm.* **2003**, *33*, 3035.

94. Luzzio, F. A.; Bobb, R. A., *Tetrahedron Lett.* **1997**, *38*, 1733.

95. Lakhrissi, M.; Carchon, G.; Schlama, T.; Mioskowski, C.; Chapleur, Y., *Tetrahedron Lett.* **1998**, *39*, 6453.

96. Lakhrissi, M.; Chapleur, Y., *Tetrahedron Lett.* **1998**, *39*, 4659.

97. An, G. I.; Rhee, H., *Synlett* **2003**, 876.

98. Kim, S. Y.; An, G. I.; Rhee, H., *Synlett* **2003**, 112.

99. Palucki, M.; Pospisil, P. J.; Zhang, W.; Jacobsen, E. N., *J. Am. Chem. Soc.* **1994**, *116*, 9333.

100. Palucki, M.; McCormick, G. J.; Jacobsen, E. N., *Tetrahedron Lett.* **1995**, *36*, 5457.

101. Prasad, J. S.; Vu, T.; Totleben, M. J.; Crispino, G. A.; Kacsur, D. J.; Swaminathan, S.; Thornton, J. E.; Fritz, A.; Singh, A. K., *Org. Process. Res. Dev.* **2003**, *7*, 821.

102. Andrus, M. B.; Poehlein, B. W., *Tetrahedron Lett.* **2000**, *41*, 1013.

103. O'Neil, I. A.; Cleator, E.; Tapolczay, D. J., *Tetrahedron Lett.* **2001**, *42*, 8247.

104. Sammelson, R. E.; Kurth, M. J., *Tetrahedron Lett.* **2001**, *42*, 3419.

105. Ma, D. W.; Xia, C. F.; Tian, H. Q., *Tetrahedron Lett.* **1999**, *40*, 8915.

106. Luzzio, F. A.; Zacherl, D. P.; Figg, W. D., *Tetrahedron Lett.* **1999**, *40*, 2087.

N-Chlorosuccinimide

[128-09-6]   C₄H₄ClNO₂   (MW 133.53)

$[128\text{-}09\text{-}6]$ $\text{C}_4\text{H}_4\text{ClNO}_2$ (MW 133.53)

InChI = 1S/C4H4ClNO2/c5-6-3(7)1-2-4(6)8/h1-2H2

InChIKey = JRNVZBWKYDBUCA-UHFFFAOYSA-N

(electrophilic α-chlorination of sulfides, sulfoxides, and ketones; preparation of *N*-chloroamines)

Alternative Names: 1-chloro-2,5-pyrrolidinedione; NCS.

Physical Data: mp 144–146 °C.

Solubility: sol H_2O; sl sol CCl_4, benzene, toluene, AcOH; insol ether.

Form Supplied in: white powder or crystals having a weak odor of chlorine when pure; widely available.

Purification: the commercial reagent acquires a light yellow color and a rather strong odor of chlorine after long storage but is easily recrystallized from acetic acid: rapidly dissolve 200 g of impure sample in 1 L preheated glacial AcOH at 65–70 °C (3–5 min); cool to 15–20 °C to effect crystallization; filter through a Buchner funnel and wash the white crystals once with glacial AcOH and twice with hexane; dry in vacuo (>85% recovery).

Analysis of Reagent Purity: the standard iodide–thiosulfate titration method is suitable.

Handling, Storage, and Precautions: store under refrigeration and protect from moisture; acutely irritating solid, with toxic effects similar to those of the free halogens; avoid inhalation; use an efficient fume hood; perform all operations as rapidly as possible to avoid extensive decomposition of the reagent.

Original Commentary

Scott C. Virgil

Massachusetts Institute of Technology, Cambridge, MA, USA

N-Chlorosuccinimide is a convenient reagent for the electrophilic substitution and addition of chlorine to organic compounds. Other chlorinating agents of use include **Chlorine**, **Sulfuryl Chloride**, **Chloramine-T**, **tert-Butyl Hypochlorite**, and **Trichloroisocyanuric Acid**. The primary advantages of using NCS include the ease in handling, the mild conditions under which chlorination proceeds, and the ease of removal of the inoffensive byproduct succinimide.

α-Chlorination of Carbonyl Derivatives. Carbonyl compounds can be chlorinated in the α-position by addition of NCS directly to the lithium enolates, enoxyborinates, or more commonly to the silyl enol ether derivatives.[1] In combination with methods for the regiospecific generation of enolates and silyl enol ethers, α-chloroketones of desired structure can be produced. For example, β-ionone can be chlorinated selectively in the α'-position by addition of NCS to the kinetic enolate (eq 1).[2] With the appropriate chiral auxiliary, NCS chlorinates silyl ketene acetals with high levels of diastereoselectivity (eq 2).[3] α-Chloro

ketones, α-chloro esters, and α-chloro sulfones may also be prepared by reaction of NCS with the β-keto derivatives and in situ deacylation in the presence of base (eq 3).[4] NCS is also an effective reagent for the α-chlorination of acid chlorides.[5]

(1)

(2)

(3)

Chlorination of Sulfides and Sulfoxides.[6] The reaction of alkyl sulfides with NCS has been used extensively for the preparation of α-chloro sulfides, and NCS is generally regarded as the reagent of choice for the preparation of these useful synthetic intermediates (see also **Trichloroisocyanuric Acid**). Since the mechanism of chlorination involves initial formation of an *S*-chlorosulfonium salt followed by a Pummerer-like rearrangement, monochlorination proceeds smoothly in CCl_4 or benzene in the absence of added acid or base.[7] The most straightforward procedure involves the addition of NCS to a solution of the sulfide in CCl_4 at rt or reflux, followed by removal of insoluble succinimide by filtration. The resulting α-chloro sulfides are easily hydrolyzed and, as this is usually undesirable, α-chloro sulfides must be prepared under strictly anhydrous conditions and are often used without further purification. A method has been developed for the conversion of benzylic halides to aromatic aldehydes (eq 4);[8] however, this transformation is more conveniently effected in one operation with other reagents (see **Hexamethylenetetramine**). Many advantages have led to the preferred use of NCS in the Ramberg–Bäcklund rearrangement sequence (eq 5), which has been recently reviewed.[9]

(4)

(5)

The chlorination of trimethylsilylmethyl sulfides with NCS and trifluoroacetic acid affords the product of chlorodesilation in high yield.[10] The degradation of carboxylic acids to ketones can be achieved by α-sulfenation followed by reaction with NCS in the presence of $NaHCO_3$ (eq 6).[11] The *S*-chlorosulfonium ion intermediate undergoes a decarboxylative Pummerer-like rearrangement to afford the ketone upon hydrolysis. α-Phenylthio esters

Essential Reactions for Organic Synthesis, First Edition. Edited by Philip L. Fuchs.
© 2016 John Wiley & Sons, Ltd. Published 2016 by John Wiley & Sons, Ltd.

and amides can be successfully α-chlorinated using NCS in CCl$_4$ at 0 °C (eq 7).[12] 1,3-Dithianes are deprotected to afford ketones by reaction with NCS alone or in combination with *Silver(I) Nitrate* in aqueous acetonitrile (see also *N-Bromosuccinimide, Mercury(II) Chloride, 1,3-Diiodo-5,5-dimethylhydantoin*).[13]

(6)

64%

(7)

100%

Sulfides can be oxidized to sulfoxides by reaction with NCS in methanol (0 °C, 1 h).[14] Similarly, selenides couple with amines when activated by NCS to form selenimide species. These have been generated from allylic selenides in order to prepare allylic amines and chiral secondary allylic carbamates by [2,3]-sigmatropic rearrangement (eq 8).[15]

(8)

69%

The α-chlorination of sulfoxides is generally performed in dichloromethane in the presence of a base (either K$_2$CO$_3$ or pyridine) and proceeds more slowly than the reactions with sulfides.[16] α-Chloro sulfoxides bearing high optical purity at sulfur are especially useful in asymmetric synthesis, but unfortunately the chlorination of optically active sulfoxides is generally accompanied by significant racemization at sulfur. Alternate procedures are available for achieving chlorination with predominant retention or inversion.[17] Using NCS and *Potassium Carbonate* the degree of racemization is minimized and chloromethyl *p*-tolyl sulfoxide can be prepared in 87% ee and 91% chemical yield (eq 9).[18]

(9)

91%, 87% ee

Reaction with Vinylic and Acetylenic Derivatives. NCS is a suitable source of chlorine for the conversion of vinylcopper and other organometallic derivatives to the corresponding vinyl chlorides.[19] (*E*)-(1-Chloro-1-alkenyl)silanes are available from the appropriate 1-trimethylsilylalkynes by hydroalumination with *Diisobutylaluminum Hydride* followed by direct treatment of the vinylaluminum intermediate with NCS in ether at −20 °C (eq 10) (the corresponding (*Z*)-isomer is obtained by NBS-catalyzed isomerization of the (*E*)-isomer).[20] 1-Chloroalkynes can be prepared

by reaction of the corresponding lithium acetylides with NCS in THF.[21]

(10)

84%

Chlorination of Aromatic Compounds. NCS has also been used for the chlorination of pyrroles and indoles; however, the reaction is less straightforward than when NBS and *N-Iodosuccinimide* are used.[22] In the chlorination of 1-methylpyrrole, it has been demonstrated that basic conditions (NaHCO$_3$, CHCl$_3$) lead to the formation of 1-methyl-2-succinimidylpyrrole (eq 11).[23] In the presence of catalytic amounts of perchloric acid, thiophenes and other electron-rich aromatic compounds have been chlorinated with NCS.[24] (*N-Chlorosuccinimide–Dimethyl Sulfide* is used for the selective *o*-substitution of phenols.)

(11)

| THF | 89% | 3% |
| CHCl$_3$, NaHCO$_3$ | – | 76% |

Synthesis of *N*-Chloroamines. The conversion of secondary amines to *N*-chloroamines by reaction with NCS in ether or dichloromethane has many advantages over the use of aqueous hypochlorite, including ease of isolation. This method has been used repeatedly in the preparation of *N*-chloroamines for alkene amination (eqs 12 and 13)[25] and other reactions.[26]

(12)

59%

(13)

83%

Other Oxidation and Chlorination Reactions.[27] *gem*-Chloronitro compounds are prepared by treating nitronate anions with NCS in aqueous dioxane, or alternatively by reaction of ketoximes with NCS (eq 14).[28] Oxidative decarboxylation of carboxylic acids with *Lead(IV) Acetate* and NCS has been used effectively for the synthesis of tertiary alkyl chlorides (eq 15).[29]

(14)

100%

$$\text{(15)}$$

NCS is also regularly used for the direct oxidation of alcohols to ketones. The presence of **Triethylamine** serves to activate the reagent for rapid quantitative oxidation of catechols and hydroquinones to o- and p-quinones, respectively, and for the oxidation of benzophenone hydrazone to diphenyldiazomethane.[30] **N-Chlorosuccinimide–Dimethyl Sulfide** is also used in the mild oxidation of alcohols, as well as in the conversion of allylic alcohols to allylic chlorides.

First Update

Terry V. Hughes
J&JPRD, Raritan, NJ, USA

α-Chlorination of Carbonyl Derivatives. The direct chlorination of β-keto esters and cyclic ketones by NCS proceeds readily at room temperature under acid catalysis by Amberlyst-15©. The reaction is general and works for acyclic, cyclic, and heterocyclic β-keto esters. For example, 3-oxo-3-pyridin-2-yl-propionic acid ethyl ester was α-chlorinated in excellent yield (eq 16).[31]

$$\text{(16)}$$

The enantioselective α-chlorination of β-keto esters was achieved with up to 88% ee using NCS with a commercially available TADDOL ligand.[32] The chiral bisoxazoline copper(II) complexes have also been reported to induce the asymmetric α-chlorination of β-keto esters when reacted with NCS.[33] The asymmetric α-chlorination of aldehydes has been achieved using NCS and (2R,5R)-diphenylpyrrolidine as a chiral catalyst. For example, the enantioselective chlorination of 3-methylbutanal with NCS proceeds in 95% yield and 94% ee (eq 17).[34]

$$\text{(17)}$$

The enantioselective α-chlorination reaction was also reported to proceed for β-keto phosphonates using NCS and bisoxazoline zinc(II) complexes in 70–91% ee.[35] Phenylselenyl chloride has been shown to enhance the electrophilicity of NCS in chlorination reactions. Allylic chlorination of olefins with NCS catalyzed by PhSeCl was reported to occur with ene regiochemistry in high yields at room temperature. For example, methyl oct-3-enoate was

smoothly converted to methyl 4-chloro-oct-2-enoate in excellent yield with no α-chlorination to the carbonyl detected (eq 18).[36]

$$\text{(18)}$$

Interestingly, NCS catalyzed by phenylselenyl chloride selectively α-chlorinates β-keto esters in the presence of olefins with no allylic chlorination observed (eq 19).[37]

$$\text{(19)}$$

Chlorination of Sulfides. The treatment of 1,3-oxathioacetals or dithioacetals with NCS in the presence of MeOH, EtOH, 1,2-ethanediol, or 1,3-propanediol results in the clean conversion to the corresponding acetal or cyclic acetal. The protecting group conversion occurs quickly and in excellent yield. For example, the reaction of 2-phenyl-1,3-dithiolane with 1 equiv of NCS and 3 equiv of 1,2-ethanediol in dichloromethane proceeds readily to afford 2-phenyl-1,3-dioxolane in almost quantitative yield (eq 20).[38]

$$\text{(20)}$$

In a similar reaction, 1,3-oxathioacetals or dithioacetals can be deprotected with 10 mol % NCS in chloroform with 5 equiv of DMSO to yield the corresponding carbonyl compound in excellent yield. The reaction is chemoselective and works in the presence of O,O-acetals.[39]

Conversion of Alcohols and Thiols to Chlorides. Primary and secondary alcohols are converted to the corresponding alkyl chlorides with the inversion of configuration when reacted with NCS and triphenylphosphine under Mitsunobu-type conditions.[40] The NCS and triphenylphosphine combination also transforms certain hydroxyheterocycles to the corresponding chloroheterocycle. The structural requirement for this transformation is that the hydroxyl needs to be *ortho* to a nitrogen atom in the heterocycle. For example, quinoxalin-2-ol is converted to 2-chloroquinoxaline in good yield when treated with NCS and triphenylphosphine in refluxing dioxane (eq 21).[41]

$$\text{(21)}$$

Benzylic, primary, and secondary thiols are readily converted to the corresponding alkyl chlorides when treated with NCS and triphenylphosphine in dichloromethane. The reaction for benzylic thiols is immediate and occurs within 24 h for secondary thiols. For example, α-toluenethiol is immediately converted to benzyl chloride in 90% yield when treated with NCS and triphenylphosphine at room temperature (eq 22).[42]

$$\text{(22)}$$

Hunsdiecker Reactions. The Hunsdiecker reaction is the decarboxylative halogenation of metal carboxylate salts. The reaction of α,β-unsaturated carboxylic acids with NCS and catalytic lithium acetate in acetonitrile–water provides the corresponding β-halostyrenes in moderate yields under mild conditions. The reaction proceeds with a good degree of stereospecificity. For example, the reaction of 3-(4-methoxy-phenyl)-acrylic acid with NCS with a catalytic amount of lithium acetate at room temperature provides 1-(2-chloro-vinyl)-4-methoxybenzene in good yield (eq 23).[43]

$$\text{(23)}$$

A modification of the Hunsdiecker reaction uses NCS catalyzed with tetrabutylammonium trifluoroacetate (TBATFA) and gives β-chlorostyrenes in excellent yields.[44] The use of the NCS/TBATFA-catalyzed Hunsdiecker reaction has been extended to various heterocyclic α,β-unsaturated carboxylic acids.[45]

Aromatic Chlorination. Many aromatic and heteroaromatic chlorinations using NCS are catalyzed by acetic acid.[46,47] Ferric chloride and ammonium nitrite have also been used to catalyze the chlorination of various heterocycles with NCS.[48] Although NCS has been used for halogenation of electron-rich aromatics, the halogenation of electron-poor aromatic systems with NCS has been difficult to achieve. However, the chlorination of various deactivated aromatic systems can be achieved when NCS is acid catalyzed with boron trifluoride monohydrate. The reaction is impressive in that even the deactivated 1-fluoro-2-nitrobenzene is chlorinated to afford 4-chloro-1-fluoro-2-nitrobenzene in 81% yield after 18 h at 100 °C (eq 24).[49]

$$\text{(24)}$$

Oxidation of Alcohols. The oxidation of primary, benzylic, and allylic alcohols to aldehydes can be selectively achieved when the alcohol is treated with NCS catalyzed by TEMPO. Reaction conditions are mild and do not chlorinate olefins or allylic positions. The reaction is run under typical phase-transfer conditions using a dichloromethane–water mixture and TBACl as the phase-transfer agent. The aqueous layer for the biphasic reaction is buffered at pH 8.6 with $NaHCO_3$–K_2CO_3. Primary alcohols were selectively oxidized to aldehydes in the presence of secondary alcohols and only 0–5% of the ketone resulting from oxidation of the secondary alcohol was observed.[50] Alternatively the oxidation of alcohols with NCS to the corresponding carbonyl compounds can be catalyzed with N-tert-butylbenzenesulfenamide. This reaction presumably proceeds via an initial oxidation of the sulfur atom of the catalyst. The N-tert-butylbenzenesulfenamide-catalyzed oxidation is selective for primary alcohols over secondary alcohols, works on a variety of substrates, and has the advantage that it can be performed without using phase-transfer conditions.[51] An interesting variant to the oxidation of alcohols to carbonyl compounds with NCS is the oxidation of diols to lactones. The reaction of 1,4-butanediol and 1,5-pentanediol with NCS in dichloromethane at room temperature provided the corresponding five- and six-membered-ring lactones in excellent yield (eq 25).[52]

$$\text{(25)}$$

Miscellaneous Uses. NCS catalyzes the transesterification of β-keto ethyl esters with substrate alcohols under neutral conditions in refluxing toluene in excellent yields. The ethanol formed during the reaction is removed by distillation. Surprisingly, the reaction conditions are selective and the chlorination of allylic positions or olefins is not observed. Additionally, the reaction proceeds with only 1 equiv of the substrate alcohol allowing for complex esters to be readily formed (eq 26).[53]

$$\text{(26)}$$

Oximes are converted to the corresponding carbonyl compound when treated with NCS in CCl_4 at room temperature in excellent yields. The workup of these deoximation reactions is especially simple with the removal of insoluble succinimide and concentration of the solvent to afford the product carbonyl compound in high purity. For example, 4-methoxyacetophenone oxime was readily converted to the corresponding ketone in 4 h at room temperature (eq 27).[54]

$$\text{(27)}$$

Alkenyl boronic acids are converted to the corresponding alkyl chlorides when treated with NCS and TEA in good to excellent yields. The reaction proceeds with retention of configuration at room temperature in 30 min. For example, (*E*)-β-styryl boronic acid is readily converted to (*E*)-β-chlorostyrene in 85% yield after 30 min at room temperature (eq 28).[55]

$$\text{(28)}$$

The conversion of a primary amine to the corresponding alkyl chloride can be achieved through NCS chemistry. *N*-Substituted-*N*-tosylhydrazines are readily available from the reaction of primary amines with tosyl chloride followed by subsequent amination *O*-(2,4-dinitrophenyl)hydroxylamine. Treatment of *N*-substituted-*N*-tosylhydrazines with NCS at room temperature affords the corresponding alkyl chlorides in good yields. Solvent choice for the reaction is critical with THF giving optimum results. It is presumed that the chlorodeamination reaction proceeds via a radical mechanism with the loss of nitrogen. The overall reaction sequence for conversion of a primary amine to the corresponding primary chloride is shown in (eq 29).[56]

$$\text{(29)}$$

A new synthesis of 5-chloro-1-phenyltetrazole, a useful activating group for the hydrogenolysis of phenols, was reported using NCS-mediated chemistry. The phase-transfer reaction of NCS with sodium azide in chloroform generates chloroazide in situ. The transient chloroazide reacts with phenyl isocyanide via a 1,3-dipolar cycloaddition at 0 °C to afford 5-chloro-1-phenyltetrazole in 69% yield (eq 30).[57]

$$\text{(30)}$$

An interesting rearrangement of cyclic dithiane alcohols to the corresponding one-carbon ring expanded 1,2-diketones is catalyzed by NCS. The reaction appears to be quite general and provides 1,2-diketones in high yields in a two-step sequence from cyclic ketones. The two-step reaction sequence from a cyclic ketone to a 1,2-diketone is high yielding and uses readily available reagents (eq 31).[58]

$$\text{(31)}$$

Chlorination of dialkylphosphites with NCS affords the corresponding dialkylchlorophosphate. The dialkylchlorophosphates generated react with alcohols to give phosphonate esters. The direct chlorination of dibenzylphosphite with NCS was used in the synthesis of phosphate prodrugs of the anti-HIV drug 3′-azido-2′,3′-dideoxythymidine (AZT) (eq 32).[59]

$$\text{(32)}$$

Second Update

Di Qiu & Jianbo Wang
Peking University, Beijing, China

α-Chlorination of Carbonyl Derivatives. NCS has been demonstrated to be a good chlorinating reagent, especially for α-chlorination of carbonyl compounds.[60] Direct chlorination of 1,3-diketones, β-ketoesters, and cyclic ketones can be carried out in ionic liquids (ILs) at room temperature, which has been used as a green recyclable reaction media. The recovered ionic liquid has been reused five to six times with consistent activity (eq 33).[61]

$$\text{(33)}$$

Organoselenide-catalyzed oxidative chlorination reactions included halolactonization, α-chlorination of ketones, and allylic chlorination. The ability of selenium to undergo reversible 2e⁻ oxidation–reduction chemistry facilitates halogenation through selenium-bound halogen intermediates.[62]

Enantioselective chlorination of β-oxoesters with NCS was catalyzed by chiral sulfoximine–copper complex. Chlorination, bromination, and fluorination reactions all proceeded well. An example is shown in eq 34.[63]

$$(34)$$

Enantioselective α-chlorination of cyclic β-oxoesters was promoted by chiral amino diol derivatives. Optimization of the catalyst structure and the reaction conditions has allowed the synthesis of optically active products with high enantioselectivities (up to 96% ee) using inexpensive NCS as the chlorine source under mild conditions (eq 35).[64]

$$(35)$$

Another efficient N,N'-dioxide organocatalyst system was also developed for the asymmetric α-chlorination of cyclic β-ketoesters using NCS to provide a series of optically active α-chloro-β-ketoesters in excellent yields with 90–98% ee.[65]

A Lewis acid-catalyzed one-pot sequential transformation of β-ketoesters, aromatic aldehydes, and NCS was reported. The reaction proceeds by way of Knoevenagel condensation/Nazarov cyclization/halogenations to give α-chloro-β-ketoesters in moderate yields with high diastereoselectivities.[66]

The chlorination of ketone derivatives was also explored. The Antilla group reported a high yielding and enantioselective chiral calcium phosphate-catalyzed chlorination of 3-substituted oxindoles with NCS (eq 36).[67]

$$(36)$$

Chlorination or Oxidation of Sulfides and Sulfoxides. The oxidation of aryl and alkyl thioacetates, as well as a variety of thiols and disulfides, by a combination of NCS and hydrochloric acid affords the corresponding sulfonyl chlorides in good yields. The reaction involves the rapid generation of reactive molecular chlorine.[68] The reaction between NCS and a thioether has been studied to provide additional insight into the Corey–Kim reaction.[69]

Allylic selenides react with NCS to form a Se–N bond (eq 37). The product, *N*-(2-nitrophenylselenenyl) succinimide, is a useful selenoetherification reagent.[70]

$$(37)$$

Chlorination of Alkenes and Alkynes. The haloamidation of olefins was carried out by reaction of *N*-haloimides (X = Cl, Br) with nitriles in the presence of Lewis acids. It was presumed that the reaction involves nucleophilic attack of nitrile on the halonium ion followed by hydrolysis of the products (eq 38).[71]

$$(38)$$

Thiourea-catalyzed chlorination of olefins with NCS in the presence of water afforded chlorohydrins in high yield (eq 39).[72] When using alcohol as solvent, the reaction gave β-chloroethers as the products (eq 40).[73]

$$(39)$$

$$(40)$$

For other types of conjugated alkenes, a variety of vinyl halides were prepared from the reaction of 1,1-diarylalkenes with NCS in good yields under mild and base-free conditions. The reaction proceeds well with electron-rich diarylalkenes, such as those with methoxy substituents on the aromatic rings.[74]

The halohydroxylation of methylenecyclopropane derivatives using NXS provides a variety of halocyclopropylmethanol. Solvent plays an important role in this halogenation (eq 41).[75]

$$(41)$$

The aminohalogenation of electron-deficient olefins with NCS can be promoted by hypervalent iodine reagents. First, the NXS may react with TsNH$_2$ to generate TsNHCl, which can be oxidized by PhI(OAc)$_2$ to form I–N bond. The chloro anion can dissociate and the generated nitrenium intermediate can immediately react with the double bond of olefins to form an aziridinium ion. A series of substrates are tolerable under the reaction conditions and are aminochlorinated or brominated in good yields with high diastereoselectivities (eq 42).[76]

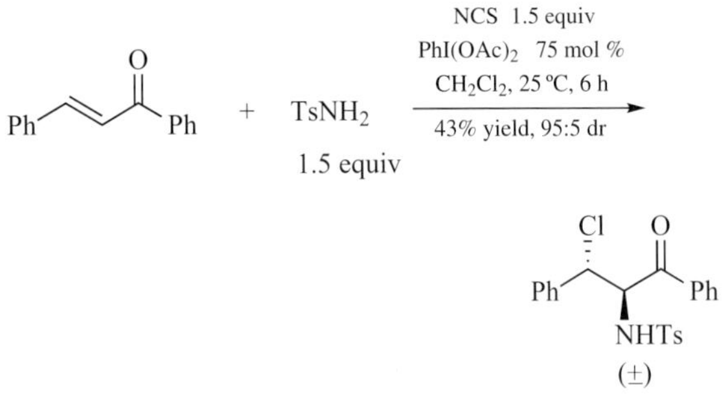

(42)

Compared to alkenes, alkynes are more difficult to react with NCS. For example, Frontier's group has reported the chlorination of conjugated alkyne in the presence of strong base (eq 43).[77]

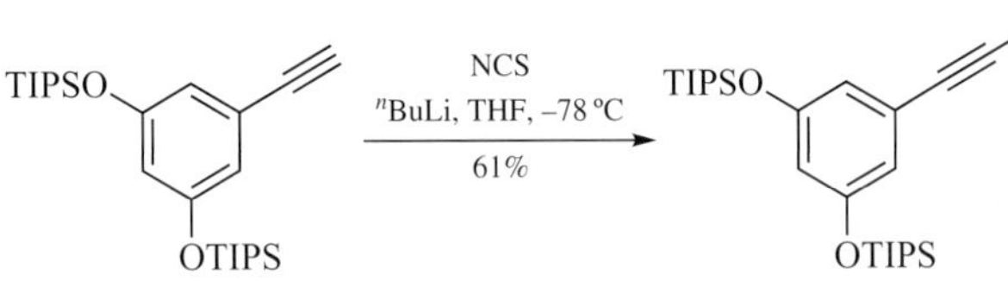

(43)

The difunctional additions of electrophiles and nucleophiles to 1-cyclopropylallenes have been investigated. In the presence of NCS and NaCl, 2,6-dichloro-1,3-hexadiene was formed in good yield (eq 44).[78]

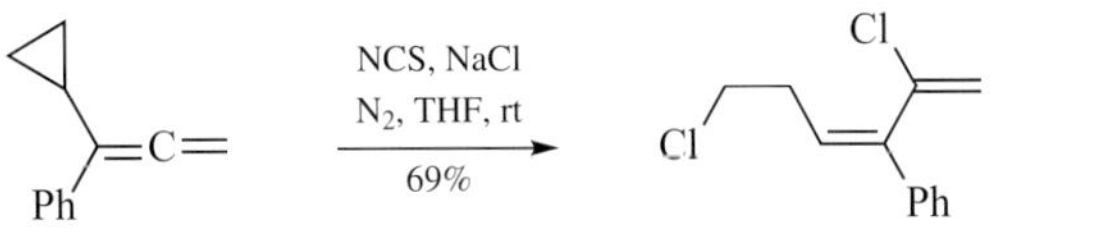

(44)

Conversion of Alcohols to Chlorides. A variety of trans-β-substituted cyclic secondary alcohols have been stereoselectively chlorinated to either the corresponding cis-chloride or trans-chloride in good to excellent yields; the stereochemical outcome is determined by the size of the ring and the nature of the β-substituents, especially the electronegativity of the substituent atom. When intermolecular S$_N$2 reaction occurs, the reaction provides cis-chloride product; when intramolecular S$_N$2 reaction occurs, a three-member ring is formed as intermediate, followed by nucleophilic attack to give trans-chloride product (eq 45).[79]

Chlorination of Epoxides. Epoxides are regioselectively converted into vic-haloalcohols with 1.2 equiv of NXS and triphenylphosphine in MeCN at room temperature. On the other hand, treatment with 2.5 equiv of NXS and PPh$_3$ at reflux temperature afforded symmetrical vic-dihalides. When different N-haloimides are used successively, unsymmetrical vic-dihalides can be obtained in high yield (eq 46).[80]

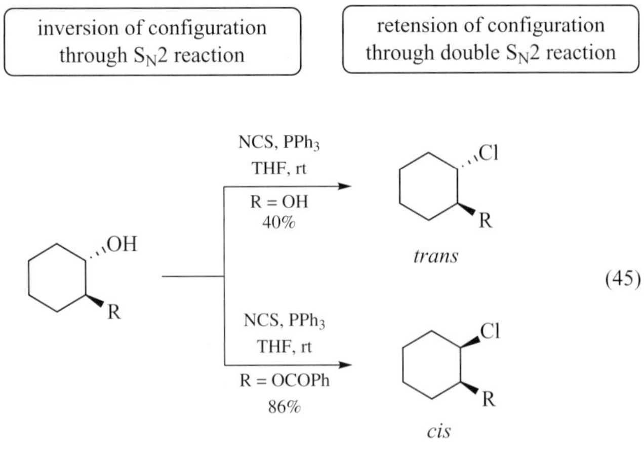

(45)

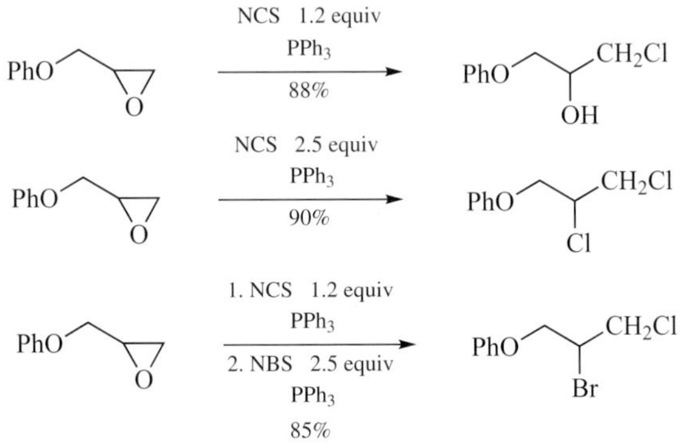

(46)

The chiral polychlorides are accessible stereospecifically by nucleophilic multiple chlorination reactions of internal epoxides using NCS and an organophosphine. PPh$_3$ may coordinate with the oxygen atom of epoxides, followed by double nucleophilic attack of chlorides.[81] This process has been used in asymmetric total synthesis.[82]

Chlorination of Nitroalkyl Compounds. Conversion of —CH$_2$NO$_2$ group attached to an alkyl or aryl moiety into a dichloronitromethyl (CCl$_2$NO$_2$) group has been achieved by using NCS and DBU in dichloromethane. The proton on the α-carbon of nitro group is readily deprotonated by base, followed by the electrophilic attack of NCS (eq 47).[83]

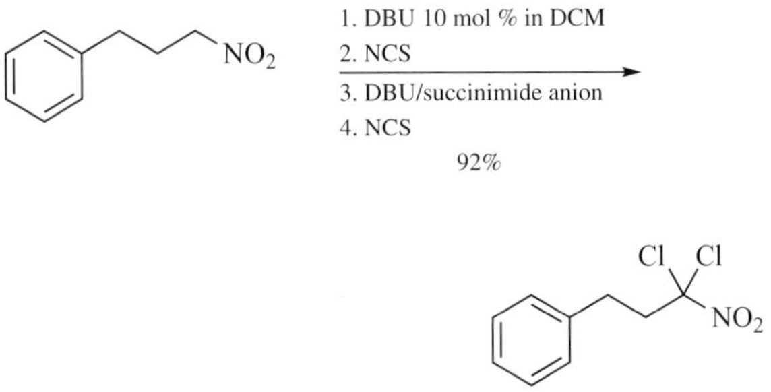

(47)

Chlorination of Aromatic Compounds. NCS can be used for the chlorination of electron-rich aromatic compounds, such as phenol[84] and xylene derivatives under the microwave conditions.[85] Palladium-catalyzed directing group-assisted chlorination of arenes has also been reported (eq 48).[86]

$$(48)$$

Gold-catalyzed direct chlorination of benzene derivatives has also been reported. The $AuCl_3$ may activate both the arenes and the carbonyl group of NCS.[87] Furthermore, gold-catalyzed direct chlorination of arylboronic esters occurs in high yield and with good regioselectivity (eq 49).[88] The regioselectivity is controlled by both electronic effect and steric effect of the substituents.

$$(49)$$

Oxidation Reaction. An efficient and regioselective method for iodination of electron-rich aromatic compounds has been reported by using NCS and sodium iodide in AcOH. This method is also applicable to nonbenzenoid aromatic or heteroaromatic compounds. NCS reacts with NaI to form the key intermediate ICl (eq 50).[89]

$$(50)$$

NCS can also be employed as an oxidant to promote the formation of heterocycles from diamine derivatives and aldehyde. The aldehyde moiety is oxidized by NCS, and finally imidazoline is obtained as the product. This method has been used in the late-stage of total synthesis (eq 51).[90]

$$(51)$$

The primary alcohols are oxidized to aldehydes in the presence of radical oxidant with NCS as the terminal oxidant. This reaction usually occurs in a two-phase system $CH_2Cl_2–H_2O$ (eq 52).[91]

$$(52)$$

$$R = CH_3(CH_2)_{12}$$

Miscellaneous Reactions. Many organometallic intermediates react with NCS. The Zr–C bond of zirconacyclopentadiene can be cleaved with NCS. For example, halogenation of the Zr–C intermediate with CuCl gives a linear triene (eq 53).[92]

$$(53)$$

Håkansson's group has synthesized both enantiomers of 1-chloroindene from achiral starting materials with high selectivity, and the reaction is free of optically active catalysts or auxiliaries. This process involves the reaction between diindenylzinc reagents with NCS.[93]

1. (a) Hambly, G. F.; Chan, T. H., *Tetrahedron Lett.* **1986**, *27*, 2563. (b) Hooz, J.; Bridson, J. N., *Can. J. Chem.* **1972**, *50*, 2387. (c) Ohkata, K.; Mase, M.; Akiba, K., *J. Chem. Soc., Chem. Commun.* **1987**, 1727.

2. Vaz, A. D. N.; Schoellmann, G., *J. Org. Chem.* **1984**, *49*, 1286.

3. Oppolzer, W.; Dudfield, P., *Tetrahedron Lett.* **1985**, *26*, 5037.

4. Mignani, G.; Morel, D.; Grass, F., *Tetrahedron Lett.* **1987**, *28*, 5505.

5. Harpp, D. N.; Bao, L. Q.; Black, C. J.; Gleason, J. G.; Smith, R. A., *J. Org. Chem.* **1975**, *40*, 3420.

6. Dilworth, B. M.; McKervey, M. A., *Tetrahedron* **1986**, *42*, 3731.

7. Tuleen, D. L.; Stephens, T. B., *J. Org. Chem.* **1969**, *34*, 31.

8. Paquette, L. A.; Klobucar, W. D.; Snow, R. A., *Synth. Commun.* **1976**, *6*, 575.

9. Paquette, L. A., *Org. React.* **1977**, *25*, 1.

10. Ishibashi, H.; Nakatani, H.; Maruyama, K.; Minami, K.; Ikeda, M., *J. Chem. Soc., Chem. Commun.* **1987**, 1443.

11. (a) Trost, B. M.; Tamaru, Y., *J. Am. Chem. Soc.* **1977**, *99*, 3101. (b) Trost, B. M.; Crimmin, M. J.; Butler, D., *J. Org. Chem.* **1978**, *43*, 4549.

12. Ishibashi, H.; Uemura, N.; Nakatani, H.; Okazaki, M.; Sato, T.; Nakamura, N.; Ikeda, M., *J. Org. Chem.* **1993**, *58*, 2360.

13. Corey, E. J.; Erickson, B. W., *J. Org. Chem.* **1971**, *36*, 3553.

14. Harville, R.; Reed, S. F., Jr., *J. Org. Chem.* **1968**, *33*, 3976.

15. (a) Fitzner, J. N.; Shea, R. G.; Fankhauser, J. E.; Hopkins, P. B., *J. Org. Chem.* **1985**, *50*, 418. (b) Spaltenstein, A.; Carpino, P. A.; Hopkins, P. B., *Tetrahedron Lett.* **1986**, *27*, 147.

16. (a) Tsuchihashi, G.; Ogura, K., *Bull. Chem. Soc. Jpn.* **1971**, *44*, 1726. (b) Ogura, K.; Imaizumi, J.; Iida, H.; Tsuchihashi, G., *Chem. Lett.* **1980**, 1587.

17. Calzavara, P.; Cinquini, M.; Colonna, S.; Fornasier, R.; Montanari, F., *J. Am. Chem. Soc.* **1973**, *95*, 7431.

18. (a) Satoh, T.; Oohara, T.; Ueda, Y.; Yamakawa, K., *Tetrahedron Lett.* **1988**, *29*, 313. (b) Drabowicz, J., *Synthesis* **1986**, 831.

19. Levy, A. B.; Talley, P.; Dunford, J. A., *Tetrahedron Lett.* **1977**, 3545.

20. Zweifel, G.; Lewis, W., *J. Org. Chem.* **1978**, *43*, 2739.

21. (a) Murray, R. E., *Synth. Commun.* **1980**, *10*, 345. (b) Verboom, W.; Westmijze, H.; De Noten, L. J.; Vermeer, P.; Bos, H. J. T., *Synthesis* **1979**, 296.

22. (a) Gilow, H. M.; Burton, D. E., *J. Org. Chem.* **1981**, *46*, 2221. (b) Powers, J. C., *J. Org. Chem.* **1966**, *31*, 2627.

23. De Rosa, M.; Nieto, G. C., *Tetrahedron Lett.* **1988**, *29*, 2405.

24. Goldberg, Y.; Alper, H., *J. Org. Chem.* **1993**, *58*, 3072.

25. (a) Kametani, T.; Suzuki, Y.; Ban, C.; Honda, T., *Heterocycles* **1987**, *26*, 1491. (b) Honda, T.; Yamamoto, A.; Cui, Y.; Tsubuki, M., *J. Chem. Soc., Perkin Trans. 1* **1992**, 531.

26. (a) Wolff, M. E., *Chem. Rev.* **1963**, *63*, 55. (b) Stella, L., *Angew. Chem., Int. Ed. Engl.* **1983**, *22*, 337.

27. Filler, R., *Chem. Rev.* **1963**, *63*, 21.

28. (a) Amrollah-Madjdabadi, A.; Beugelmans, R.; Lechevallier, A., *Synthesis* **1986**, 828. (b) Amrollah-Madjdabadi, A.; Beugelmans, R.; Lechevallier, A., *Synthesis* **1986**, 826. (c) Corey, E. J.; Estreicher, H., *Tetrahedron Lett.* **1980**, *21*, 1117.

29. Becker, K. B.; Geisel, M.; Grob, C. A.; Kuhnen, F., *Synthesis* **1973**, 493.

30. Durst, H. D.; Mack, M. P.; Wudl, F., *J. Org. Chem.* **1975**, *40*, 268.

31. Meshram, H. M.; Reddy, P. N.; Sadashiv, K.; Yadav, J. S., *Tetrahedron Lett.* **2005**, *46*, 623.

32. Hintermann, L.; Togni, A., *Helv. Chim. Acta* **2000**, *83*, 2425.

33. Marigo, M.; Kumaragurubaran, N.; Jørgensen, K. A., *Chem. Eur. J.* **2004**, *10*, 2133.

34. Halland, N.; Braunton, A.; Bachmann, S.; Marigo, M.; Jørgensen, K. A., *J. Am. Chem. Soc.* **2004**, *126*, 4790.

35. Bernardi, L.; Jørgensen, K. A., *Chem. Commun.* **2005**, 1324.

36. Tunge, J. A.; Mellegaard, S. R., *Org. Lett.* **2004**, *6*, 1205.

37. Wang, C.; Tunge, J., *Chem. Commun.* **2004**, 2694.

38. Karimi, B.; Seradj, H.; Maleki, J., *Tetrahedron* **2002**, *58*, 4513.

39. Iranpoor, N.; Firouzabadı, H.; Shaterian, H. R., *Tetrahedron Lett.* **2003**, *44*, 4769.

40. Mihovilovic, M. D.; Rudroff, F.; Grötzl, B.; Stanetty, P., *Eur. J. Org. Chem.* **2005**, *5*, 809.

41. Sugimoto, O.; Mori, M.; Tanji, K., *Tetrahedron Lett.* **1999**, *40*, 7477.

42. Iranpoor, N.; Firouzabadi, H.; Aghapour, G., *Synlett* **2001**, 1176.

43. Chowdhury, S.; Roy, S., *J. Org. Chem.* **1997**, *62*, 199.

44. Naskar, D.; Chowdhury, S.; Roy, S., *Tetrahedron Lett.* **1998**, *39*, 699.

45. Naskar, D.; Roy, S., *Tetrahedron* **2000**, *56*, 1369.

46. Day, R. A.; Blake, J. A.; Stephens, C. E., *Synthesis* **2003**, 1586.

47. Menichincheri, M.; Ballinari, D.; Bargiotti, A.; Bonomini, L.; Ceccarelli, W.; D'Alessio, R.; Fretta, A.; Moll, J.; Polucci, P.; Soncini, C.; Tibolla, M.; Trosset, J.-Y.; Vanotti, E., *J. Med. Chem.* **2004**, *47*, 6466.

48. Tanemura, K.; Suzuki, T.; Nishida, Y.; Satsumabayashi, K.; Horaguchi, T., *Chem. Lett.* **2003**, *32*, 932.

49. Prakash, G. K. S.; Mathew, T.; Hoole, D.; Esteves, P. M.; Wang, Q.; Rasul, G.; Olah, G. A., *J. Am. Chem. Soc.* **2004**, *126*, 15770.

50. Einhorn, J.; Einhorn, C.; Ratajczak, F.; Pierre, J.-L., *J. Org. Chem.* **1996**, *61*, 7452.

51. Matsuo, J.; Iida, D.; Yamanaka, H.; Mukaiyama, T., *Tetrahedron* **2003**, *59*, 6739.

52. Kondo, S.; Kawasoe, S.; Kunisada, H.; Yuli, Y., *Synth. Commun.* **1995**, *25*, 719.

53. Bandgar, B. P.; Uppalla, L. S.; Sadavarte, V. S., *Synlett* **2001**, 1715.

54. Bandgar, B. P.; Kunde, L. B.; Thote, J. L., *Synth. Commun.* **1997**, *27*, 1149.

55. Petasis, N. A.; Zavialov, I. A., *Tetrahedron Lett.* **1996**, *37*, 567.

56. Collazo, L. R.; Guziec, F. S.; Hu, W.; Pankayatselvan, R., *Tetrahedron Lett.* **1994**, *35*, 7911.

57. Collibee, W. L.; Nakajima, M.; Anselme, J.-P., *J. Org. Chem.* **1995**, *60*, 468.

58. Ranu, B. C.; Jana, U., *J. Org. Chem.* **1999**, *64*, 6380.

59. Cardona, V. M. F.; Ayi, A. I.; Aubertin, A.-M.; Guedj, R., *Antiviral Res.* **1999**, *42*, 189.

60. Gołębiewski, W. M.; Gucma, M., *Synthesis* **2007**, 3599.

61. Meshram, H. M.; Reddy, P. N.; Vishnu, P.; Sadashiv, K.; Yadav, J. S., *Tetrahedron Lett.* **2006**, *47*, 991.

62. Mellegaard-Waetzig, S. R.; Wang, C.; Tunge, J. A., *Tetrahedron* **2006**, *62*, 7191.

63. Frings, M.; Bolm, C., *Eur. J. Org. Chem.* **2009**, 4085.

64. Etayo, P.; Badorrey, R.; Díaz-de-Villegas, M. D.; Gálvez, J. A., *Adv. Synth. Catal.* **2010**, *352*, 3329.

65. Cai, Y.; Wang, W.; Shen, K.; Wang, J.; Hu, X.; Lin, L.; Liu, X.; Feng, X., *Chem. Commun.* **2010**, *46*, 1250.

66. Cui, H.; Dong, K.; Nie, J.; Zheng, Y.; Ma, J., *Tetrahedron Lett.* **2010**, *51*, 2374.

67. Zheng, W.; Zhang, Z.; Kaplan, M. J.; Antilla, J. C., *J. Am. Chem. Soc.* **2011**, *133*, 3339.

68. Nishiguchi, A.; Maeda, K.; Miki, S., *Synthesis* **2006**, 4131.

69. Cink, R. D.; Chambournier, G.; Surjono, H.; Xiao, Z.; Richter, S.; Naris, M.; Bhatia, A. V., *Org. Process Res. Dev.* **2007**, *11*, 270.

70. Denmark, S. E.; Kalyani, D.; Collins, W. R., *J. Am. Chem. Soc.* **2010**, *132*, 15752.

71. Yeung, Y.; Gao, X.; Corey, E. J., *J. Am. Chem. Soc.* **2006**, *128*, 9644.

72. Bentley, P. A.; Mei, Y.; Du, J., *Tetrahedron Lett.* **2008**, *49*, 1425.

73. Bentley, P. A.; Mei, Y.; Du, J., *Tetrahedron Lett.* **2008**, *49*, 2653.

74. Chang, M.; Lee, M.; Lin, C.; Lee, N., *Tetrahedron Lett.* **2011**, *52*, 826.

75. Yang, Y.; Su, C.; Huang, X.; Liu, Q., *Tetrahedron Lett.* **2009**, *50*, 5754.

76. Wu, X.; Wang, G., *Eur. J. Org. Chem.* **2008**, 6239.

77. Canterbury, D. P.; Herrick, I. R.; Um, J.; Houk, K. N.; Frontier, A. J., *Tetrahedron* **2009**, *65*, 3165.

78. Meng, B.; Yu, L.; Huang, X., *Tetrahedron Lett.* **2009**, *50*, 1947.

79. Jaseer, E. A.; Naidu, A. B.; Kumar, S. S.; Rao, R. K.; Thakur, K. G.; Sekar, G., *Chem. Commun.* **2007**, 867.

80. Iranpoor, N.; Firouzabadi, H.; Azadi, R.; Ebrahimzadeh, F., *Can. J. Chem.* **2006**, *84*, 69.

81. Yoshimitsu, T.; Fukumoto, N.; Tanaka, T., *J. Org. Chem.* **2009**, *74*, 696.

82. Yoshimitsu, T.; Fukumoto, N.; Nakatani, R.; Kojima, N.; Tanaka, T., *J. Org. Chem.* **2010**, *75*, 5425.

83. Butler, P.; Golding, B. T.; Laval, G.; Loghmani-Khouzani, H.; Ranjbar-Karimi, R.; Sadeghi, M. M., *Tetrahedron* **2007**, *63*, 11160.

84. Bovonsombat, P.; Ali, R.; Khan, C.; Leykajarakul, J.; Pla-on, K.; Aphimanchindakul, S.; Pungcharoenpong, N.; Timsuea, N.; Arunrat, A.; Punpongjareorn, N., *Tetrahedron* **2010**, *66*, 6928.

85. Bucos, M.; Villalonga-Barber, C.; Micha-Screttas, M.; Steele, B. R.; Screttas, C. G.; Heropoulos, G. A., *Tetrahedron* **2010**, *66*, 2061.

86. Kalyani, D.; Dick, A. R.; Anani, W. Q.; Sanford, M. S., *Org. Lett.* **2006**, *8*, 2523.

87. Mo, F.; Yan, J. M.; Qiu, D.; Li, F.; Zhang, Y.; Wang, J., *Angew. Chem., Int. Ed.* **2010**, *49*, 2028.

88. Qiu, D.; Mo, F.; Zheng, Z.; Zhang, Y.; Wang, J., *Org. Lett.* **2010**, *12*, 5474.

89. Yamamoto, T.; Toyota, K.; Morita, N., *Tetrahedron Lett.* **2010**, *51*, 1364.

90. Murai, K.; Morishita, M.; Nakatani, R.; Kubo, O.; Fujioka, H.; Kita, Y., *J. Org. Chem.* **2007**, *72*, 8947.

91. Ichikawa, Y.; Matsunaga, K.; Masuda, T.; Kotsuki, H.; Nakano, K., *Tetrahedron* **2008**, *64*, 11313.

92. Kanno, K.; Igarashi, E.; Zhou, L.; Nakajima, K.; Takahashi, T., *J. Am. Chem. Soc.* **2008**, *130*, 5624.

93. Lennartson, A.; Olsson, S.; Sundberg, J.; Håkansson, M., *Angew. Chem., Int. Ed.* **2009**, *48*, 3137.

Chlorotrimethylsilane[1,2]

[75-77-4]   C$_3$H$_9$ClSi   (MW 108.64)

InChI = 1/C3H9ClSi/c1-5(2,3)4/h1-3H3

InChIKey = IJOOHPMOJXWVHK-UHFFFAOYAM

(protection of silyl ethers,[3] transients,[5–7] and silylalkynes;[8] synthesis of silyl esters,[4] silyl enol ethers,[9,10] vinylsilanes,[13] and silylvinylallenes;[15] Boc deprotection;[11] TMSI generation;[12] epoxide cleavage;[14] conjugate addition reactions catalyst[16–18])

Alternate Names: trimethylsilyl chloride; TMSCl.

Physical Data: bp 57 °C; *d* 0.856 g cm^{-3}.

Solubility: sol THF, DMF, CH$_2$Cl$_2$, HMPA.

Form Supplied in: clear, colorless liquid; 98% purity; commercially available.

Analysis of Reagent Purity: bp, NMR.

Purification: distillation over calcium hydride with exclusion of moisture.

Handling, Storage, and Precautions: moisture sensitive and corrosive; store under an inert atmosphere; use in a fume hood.

Original Commentary

Ellen M. Leahy

Affymax Research Institute, Palo Alto, CA, USA

Protection of Alcohols as TMS Ethers. The most common method of forming a silyl ether involves the use of TMSCl and a base (eqs 1–3).[3,19–22] Mixtures of TMSCl and *Hexamethyldisilazane* (HMDS) have also been used to form TMS ethers. Primary, secondary, and tertiary alcohols can be silylated in this manner, depending on the relative amounts of TMS and HMDS (eqs 4–6).[23]

Trimethysilyl ethers can be easily removed under a variety of conditions,[19] including the use of *Tetrabutylammonium Fluoride* (TBAF) (eq 7),[20] citric acid (eq 8),[24] or *Potassium Carbonate* in methanol (eq 9).[25] Recently, resins (OH$^-$ and H$^+$ form) have been used to remove phenolic or alcoholic TMS ethers selectively (eq 10).[26]

Transient Protection. Silyl ethers can be used for the transient protection of alcohols (eq 11).[27] In this example the hydroxyl groups were silylated to allow tritylation with concomitant desilylation during aqueous workup. The ease of introduction and removal of TMS groups make them well suited for temporary protection.

Essential Reactions for Organic Synthesis, First Edition. Edited by Philip L. Fuchs.
© 2016 John Wiley & Sons, Ltd. Published 2016 by John Wiley & Sons, Ltd.

$$(11)$$

Trimethylsilyl derivatives of amino acids and peptides have been used to improve solubility, protect carboxyl groups, and improve acylation reactions. TMSCl has been used to prepare protected amino acids by forming the O,N-bis-trimethylsilylated amino acid, formed in situ, followed by addition of the acylating agent (eq 12).[5] This is a general method which obviates the production of oligomers normally formed using Schotten–Baumann conditions, and which can be applied to a variety of protecting groups.[5]

$$(12)$$

Transient hydroxylamine oxygen protection has been successfully used for the synthesis of N-hydroxamides.[6] Hydroxylamines can be silylated with TMSCl in pyridine to yield the N-substituted O-TMS derivative. Acylation with a mixed anhydride of a protected amino acid followed by workup affords the N-substituted hydroxamide (eq 13).[6]

$$(13)$$

Formation of Silyl Esters. TMS esters can be prepared in good yields by reacting the carboxylic acid with TMSCl in 1,2-dichloroethane (eq 14).[4] This method of carboxyl group protection has been used during hydroboration reactions. The organoborane can be transformed into a variety of different carboxylic acid derivatives (eqs 15 and 16).[7] TMS esters can also be reduced with metal hydrides to form alcohols and aldehydes or hydrolyzed to the starting acid, depending on the reducing agent and reaction conditions.[28]

$$(14)$$

$$(15)$$

$$(16)$$

Protection of Terminal Alkynes. Terminal alkynes can be protected as TMS alkynes by reaction with **Butyllithium** in THF followed by TMSCl (eq 17).[8] A one-pot β-elimination–silylation process (eq 18) can also yield the protected alkyne.

$$(17)$$

$$(18)$$

Silyl Enol Ethers. TMS enol ethers of aldehydes and symmetrical ketones are usually formed by reaction of the carbonyl compound with **Triethylamine** and TMSCl in DMF (eq 19), but other bases have been used, including **Sodium Hydride**[29] and **Potassium Hydride**.[30]

$$(19)$$

Under the conditions used for the generation of silyl enol ethers of symmetrical ketones, unsymmetrical ketones give mixtures of structurally isomeric enol ethers, with the predominant product being the more substituted enol ether (eq 20).[10] Highly hindered bases, such as **Lithium Diisopropylamide** (LDA),[31] favor formation of the kinetic, less substituted silyl enol ether, whereas **Bromomagnesium Diisopropylamide** (BMDA)[10] generates the more substituted, thermodynamic silyl enol ether. A combination of TMSCl/**Sodium Iodide** has also been used to form silyl enol ethers of simple aldehydes and ketones[32] as well as from α,β-unsaturated aldehydes and ketones.[33] Additionally, treatment of α-halo ketones with **Zinc**, TMSCl, and TMEDA in ether provides a regiospecific method for the preparation of the more substituted enol ether (eq 21).[34]

$$(20)$$

Reagents	Ratio (**A**):(**B**)
LDA, DME; TMSCl	1:99
NaH, DME; TMSCl	73:27
Et$_3$N, TMSCl, DMF	78:22
KH, THF; TMSCl	67:33
TMSCl, NaI, MeCN, Et$_3$N	90:10
BMDA, TMSCl, Et$_3$N	97:3

$$\text{(21)}$$

Reaction eq 21: cyclohexanone with Cl and methyl substituents + Zn, TMSCl / TMEDA, Et$_2$O, 85% → OTMS enol ether product.

Mild Deprotection of Boc Protecting Group. The Boc protecting group is used throughout peptide chemistry. Common ways of removing it include the use of 50% **Trifluoroacetic Acid** in CH$_2$Cl$_2$, **Trimethylsilyl Perchlorate**, or **Iodotrimethylsilane** (TMSI).[19] A new method has been developed, using TMSCl–phenol, which enables removal of the Boc group in less than one hour (eq 22).[11] The selectivity between Boc and benzyl groups is high enough to allow for selective deprotection.

$$\text{Boc-Val-OCH}_2\text{-resin} \xrightarrow[\substack{20 \text{ min} \\ 100\%}]{\text{TMSCl, phenol}} \text{Val-OCH}_2\text{-resin} \quad \text{(22)}$$

In Situ Generation of Iodotrimethylsilane. Of the published methods used to form TMSI in situ, the most convenient involves the use of TMSCl with NaI in acetonitrile.[12] This method has been used for a variety of synthetic transformations, including cleavage of phosphonate esters (eq 23),[35] conversion of vicinal diols to alkenes (eq 24),[36] and reductive removal of epoxides (eq 25).[37]

$$\text{Bn}-\overset{\overset{\displaystyle O}{\|}}{\underset{\underset{\displaystyle OMe}{|}}{P}}-\text{OMe} \xrightarrow[\substack{\text{MeCN} \\ 78\%}]{\text{TMSCl, NaI}} \text{Bn}-\overset{\overset{\displaystyle O}{\|}}{\underset{\underset{\displaystyle OH}{|}}{P}}-\text{OH} \quad \text{(23)}$$

Reaction eq 24: polycyclic diol with isobutyrate ester and OMe groups + TMSCl, NaI / MeCN, 90% → alkene product.

$$\text{(24)}$$

Reaction eq 25: cyclohexene oxide + TMSCl, NaI / MeCN, 94% → cyclohexene.

$$\text{(25)}$$

Conversion of Ketones to Vinylsilanes. Ketones can be transformed into vinylsilanes via intermediate trapping of the vinyl anion from a Shapiro reaction with TMSCl. Formation of either the tosylhydrazone[38] or benzenesulfonylhydrazone (eq 26)[13,39] followed by reaction with n-butyllithium in TMEDA and TMSCl gives the desired product.

Reaction eq 26: benzosuberone + 1. PhSO$_2$NHNH$_2$, p-TsOH, EtOH; 2. n-BuLi, TMEDA, TMSCl, 88% → vinylsilane (TMS) product.

$$\text{(26)}$$

Epoxide Cleavage. Epoxides open by reaction with TMSCl in the presence of **Triphenylphosphine** or tetra-n-butylammonium chloride to afford O-protected vicinal chlorohydrins (eq 27).[14]

Reaction eq 27: cyclohexene oxide + TMSCl, PPh$_3$ / CHCl$_3$, 99% → trans chlorohydrin with OTMS and Cl.

$$\text{(27)}$$

Formation of Silylvinylallenes. Enynes couple with TMSCl in the presence of Li/ether or Mg/**Hexamethylphosphoric Triamide** to afford silyl-substituted vinylallenes. The vinylallene can be subsequently oxidized to give the silylated cyclopentanone (eq 28).[15]

Reaction eq 28: chloro-enyne + 1. Li, Et$_2$O; 2. TMSCl, 80% → TMS-vinylallene → [O] → TMS silylated cyclopentenone.

$$\text{(28)}$$

Conjugate Addition Reactions. In the presence of TMSCl, cuprates undergo 1,2-addition to aldehydes and ketones to afford silyl enol ethers (eq 29).[16] In the case of a chiral aldehyde, addition of TMSCl follows typical Cram diastereofacial selectivity (eq 30).[16,40]

Reaction eq 29: cyclohexanone + Bu$_2$CuLi, TMSCl / THF, 75% → TMSO Bu product.

$$\text{(29)}$$

Reaction eq 30: Ph$_2$CHCHO + Bu$_2$CuLi, TMSCl / THF, 96% → two diastereomers Ph...OTMS/Bu, 9:1.

$$\text{(30)}$$

Conjugate addition of organocuprates to α,β-unsaturated carbonyl compounds, including ketones, esters, and amides, are accelerated by addition of TMSCl to provide good yields of the 1,4-addition products (eq 31).[17,41,42] The effect of additives such as HMPA, DMAP, and TMEDA have also been examined.[18,43] The role of the TMSCl on 1,2- and 1,4-addition has been explored by several groups, and a recent report has been published by Lipshutz.[40] His results appear to provide evidence that there is an interaction between the cuprate and TMSCl which influences the stereochemical outcome of these reactions.

Reaction eq 31: cyclohexenone + 1. (EtCH=CH)$_2$CuLi, TMSCl, Et$_2$O; 2. H$_2$O, 86% → 1,4-addition product.

$$\text{(31)}$$

The addition of TMSCl has made 1,4-conjugate addition reactions to α-(nitroalkyl)enones possible despite the presence of the acidic α-nitro protons (eq 32).[44] Copper-catalyzed addition of Grignard reagents proceeds in high yield in the presence of TMSCl and HMPA (eq 33).[45] In some instances the reaction gives dramatically improved ratios of 1,4-addition to 1,2-addition.

(32)

(33)

First Update

Wenming Zhang
Dupont Crop Protection, Newark, DE, USA

Protection of Alcohols as TMS Ethers. Several new methods have been developed for the protection of alcohols as TMS ethers. For example, TMS silyl ethers of alcohols and phenols can be prepared efficiently by treatment of the alcohol or phenol with TMSCl and catalytic amount of imidazole or iodine under the solvent-free and microwave irradiation conditions.[46] This transformation proved to be reversible. Under the same microwave conditions, treatment of the silyl ether in methanol and in the presence of catalytic amount of iodine releases the parent alcohol in quantitative yield.

In another new method, treatment of aliphatic alcohols with TMSCl in DMF containing magnesium turnings at rt produces the corresponding TMS silyl ethers in good to excellent yields. This protocol provides a viable way to protect sterically hindered alcohols as their TMS ethers. For example, *O*-silylation of ethyl 2-methyl-2-hydroxypropanate generates the desired TMS ether in 91% yield (eq 34).[47]

(34)

Mercaptans have been protected as trimethylsilylated sulfides by treating the mercaptan first with a strong base such as *n*-hexyllithium and then capturing the sulfide anion with TMSCl.[48] Aldehydes have been converted into *N*-trimethylsilylimines by sequential treatment with LiHMDS and TMSCl.[49]

Silyl Enol Ethers. α,α-Difluorotrimethylsilyl enol ethers can be prepared through a Mg/TMSCl-promoted selective C–F bond cleavage of the corresponding trifluoromethyl ketones. An α,α-difluorinated analog of Danishefsky's diene was prepared in good

yield when the ketone was exposed to excessive Mg/TMSCl in DMF at 50 °C for 3 min (eq 35).[50]

(35)

Treatment of a vinyl ketone with TMSCl in THF provides the corresponding 2-chloroenol TMS ether, which can be used to generate the Machenzie complex through reaction with Ni(cod)$_2$, and which can be further applied in several multicomponent assembly reactions.[51]

Epoxide Cleavage. Chlorotrimethylsilane has been employed in combination with various Lewis acids, such as SnCl$_2$,[52] BF$_3 \cdot$ Et$_2$O,[53] and concentrated LiClO$_4$ in ether,[54] to open epoxide rings and generate chlorohydrins or their derivatives. For example, when glycidyl phenyl ether was reacted with TMSCl in the presence of SnCl$_2$ as catalyst, the oxirane ring was cleaved to generate two chlorohydrin regioisomers that were further transformed into corresponding acetates (eq 36). In this particular case, the C$_3$–O bond cleavage is preferred due to stabilizing chelation of the tin(II) cation by the 1-phenoxy oxygen. However, when glycidyl benzoate was treated under the identical conditions, the C$_3$–O bond cleavage product now was obtained as the minor product (eq 37). The major product came from the C$_2$–O bond cleavage, due to the neighboring benzoate group participation and rearrangement.

(36)

(37)

It was demonstrated that the ring opening of expoxides with TMSCl can also be facilitated by nucleophilic catalysts, such as 1,2-ferrocenediylazaphosphinines,[55] phosphaferrocenes, and phosphazirconcenes.[56] For example, in the presence of 5% of 1,2-ferrocenediylazaphosphinine, 1-hexene oxide was converted into 1-chloro-2-hexanol in 97% yield with 100% regioselectivity (eq 38). Various toluenesulfonyl aziridines also undergo a similar ring opening reaction to produce the corresponding chloramine derivatives under mild conditions in DMF.[57] As solvent, DMF also serves as an activator.

$$n\text{-Bu} \xrightarrow[\text{rt, 70 min. 97\%}]{\text{TMSCl, cat}}$$

$$n\text{-Bu} \overset{\text{OH}}{\underset{\text{Cl}}{\diagdown}} + n\text{-Bu} \overset{\text{Cl}}{\underset{\text{OH}}{\diagdown}} \quad (38)$$

$$100 \quad : \quad 0$$

$$cat =$$

Related to the epoxide ring opening, TMSCl also mediates some cyclopropane ring opening reactions. For example, treatment of 1-aceto-2,2-dimethylcyclopropane with TMSCl and sodium chloride in acetonitrile at 55 °C for 24 h generated 5-chloro-5-methyl-2-hexanone in 84% yield (eq 39).[58] When sodium iodide was employed to replace sodium chloride, iodotrimethylsilane generated in situ, and the reaction completed under more facile conditions (rt and 12 h). Interestingly, the dominant product is 5-iodo-4,4-dimethyl-2-pentanone (eq 40), arising from iodide attacking at the less hindered secondary methylene carbon, instead of the quaternary dimethylmethylene carbon.

$$\xrightarrow[\text{CH}_3\text{CN, 55 °C, 24 h}]{\text{TMSCl, NaCl}}$$

$$\text{Cl}\diagup\diagdown \overset{O}{\diagdown} + \text{Cl}\diagup\diagdown \overset{O}{\diagdown} \quad (39)$$

$$84\% \quad : \quad 0\%$$

$$\xrightarrow[\text{CH}_3\text{CN, rt, 12 h}]{\text{TMSCl, NaI}}$$

$$\text{I}\diagup\diagdown \overset{O}{\diagdown} + \text{I}\diagup\diagdown \overset{O}{\diagdown} \quad (40)$$

$$33\% \quad : \quad 47\%$$

Conjugate Addition Reactions. Besides copper salts, a number of other catalyst systems, such as Ni(acac)$_2$/DIBAL,[59] ZnEt$_2$,[60] InCl$_3$,[61] and Pd(PPh$_3$)$_4$/LiCl,[62] catalyze Michael addition of various enones at the β-carbon to form new C–C bonds. In addition to the formation of C–C bonds, FeCl$_3 \cdot$ 6H$_2$O, when combined with a stoichiometric amount of TMSCl, can also catalyze the aza-Michael addition to form new C–N bonds.[63] For example, in the presence of 5% of FeCl$_3 \cdot$ 6H$_2$O and 1.1 equiv of TMSCl, stirring a mixture of 2-cyclohexenone and ethyl carbamate in dichloromethane provides an 89% yield of the protected β-aminocyclohexanone (eq 41).

$$\overset{O}{\diagdown} + \text{NH}_2\text{CO}_2\text{Et} \xrightarrow[\text{rt, 12 h, 89\%}]{\substack{\text{TMSCl, (1.1 equiv)} \\ \text{FeCl}_3\cdot\text{H}_2\text{O (5\%)}}} \overset{O}{\underset{\text{NHCO}_2\text{Et}}{\diagdown}} \quad (41)$$

Cohen and Liu have examined the yield and selectivity enhancement effects of TMSCl and TMSCl/HMPA on conjugate addition of some stabilized organolithium reagents onto, particularly, easily polymerized α,β-unsaturated carbonyl compounds. The authors proposed that the beneficial effect is due to the prevention of 1,2-addition and polymerization of the α,β-unsaturated carbonyl compounds.[64]

The regiochemistry of 1,2- versus 1,4-addition of lithiated N-Boc-N-(p-methoxyphenyl)benzylamine with 2-cyclohexenone can be ligand controlled.[65] In the absence of a ligand, the reaction provides a 77% yield of the 1,2-addition product allylic alcohol (eq 42). When $(-)$-sparteine is premixed with n-BuLi before lithiation, the reaction furnishes a 70% yield of the corresponding 1,4-addition product. Furthermore, in the presence of TMSCl, the same reaction now affords 82% yield of the Michael adduct, with excellent diastereoselectivity ($>$99:1 dr) and enantioselectivity (96:4 er) favoring the (S,S)-cyclohexanone.

$$(42)$$

Ligand			
None	77%	5%	
(-)-sparteine	<5%	70%	-
(-)-sparteine + TMSCl	-	-	82%, >99:1 dr, 94:6 er

Related to the copper-catalyzed Michael addition reaction, allyl phosphates furnish substitution products via the *anti* S$_N$2' mechanism when treated with an appropriate Grignard reagent in the presence of catalytic amount of CuCN and 1 equiv of TMSCl (eq 43).[66]

$$\xrightarrow[\text{CuCN, Et}_2\text{O, 0 °C, 73\%}]{\text{MeMgBr, TMSCl}}$$

$$86 \quad : \quad 4 \quad : \quad 10 \quad (43)$$

Anion Trap. Besides formation of silyl enol ethers, TMSCl has also been applied to trap other oxide anions to form the desired trimethylsilyl derivatives. For example, reaction of ethyl 4-phenylbutanoate with 1,1-dichloroethyllithium, which was generated from 1,1-dichloroethane and LDA, produced exclusively the mixed acetal as expected in 86% yield in the presence of TMSCl (eq 44).[67] Without TMSCl, the reaction gave the corresponding ketone as the final product. However, the yield is rather low and the ketone was obtained in only 16% (eq 45). When the same reaction run with methoxymethyl 4-phenylbutanoate, it always afforded the ketone as the final product, regardless of the presence of TMSCl, although, the yield was improved dramatically from 20% to 73% in the presence of the reagent (eq 46).

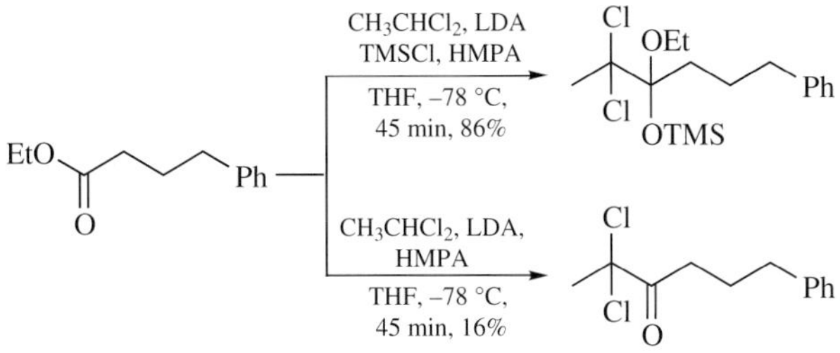

(44)

(45)

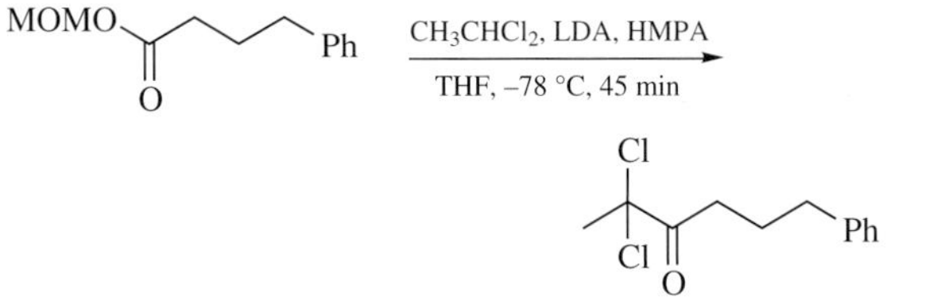

(46)

with TMSCl	73%
without TMSCl	20%

Dehydrating Agent. TMSCl is also a suitable dehydrating agent for scavenging water generated in various reactions. For example, in the esterification of 5-methylpyrazinoic acid with *n*-nonanol, a 44% yield of the desired ester can be distilled out directly from the reaction mixture of the acid, alcohol, and TMSCl (eq 47).[68]

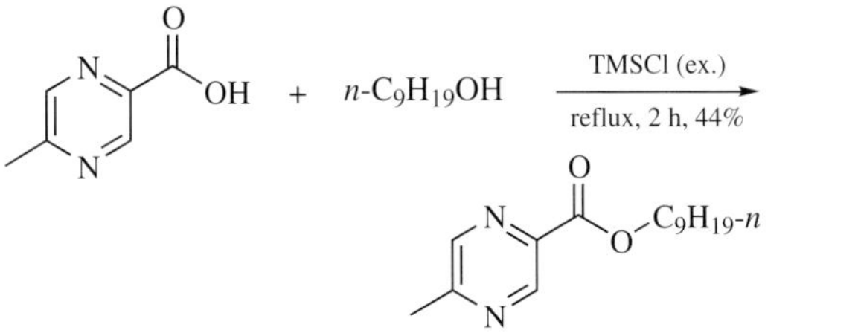

(47)

TMSCl also effects some dehydration cyclization reactions for the formation of heterocyclic compounds. For example, in the presence of TMSCl, boiling of *N*-benzoyl-α,β-dehydrophenylalanine anilide in DMF leads to isolation of the cyclization product imidazol-5-one in 80% yield (eq 48).[69] Interestingly, the same compound can be prepared by directly heating the corresponding oxazolone and aniline, also in the presence of TMSCl, although in somewhat lower yield (eq 49).

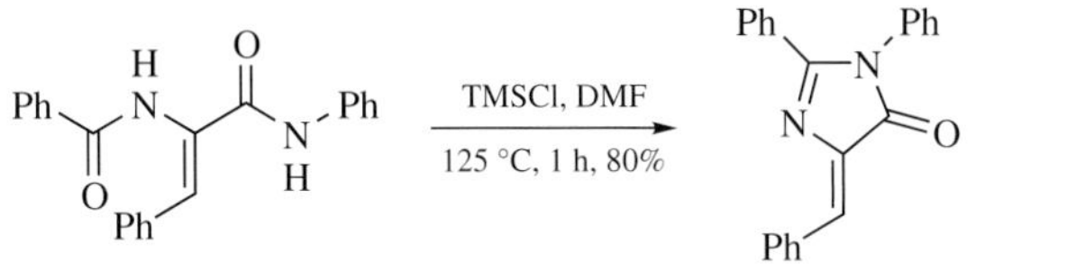

(48)

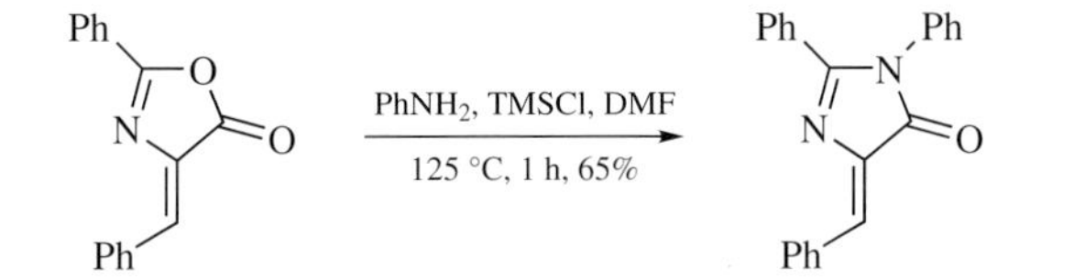

(49)

The combination of TMSCl and triethylamine provides another choice of dehydrating agent. For example, stirring a mixture of 1-nitronaphthyl tolyl sulfone, TMSCl, and triethylamine in DMF at rt for 4 days provides the corresponding anthranil in 72% yield (eq 50).[70] In a similar transformation, treatment of substituted *ortho*-nitroarylethanes with TMSCl and triethylamine in DMF results in formation of 1-hydroxyindoles as the dehydration/cyclization product.[71]

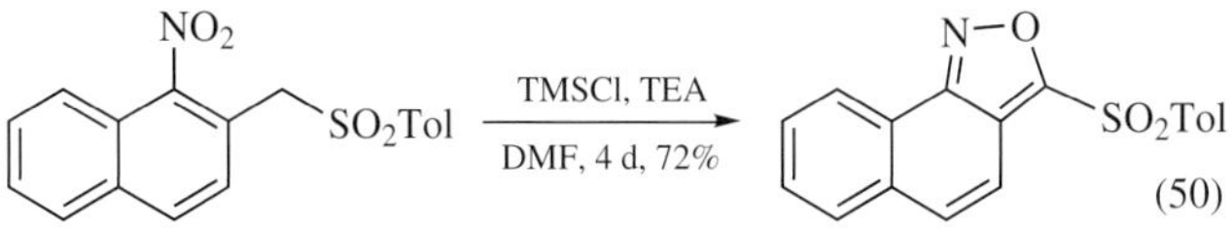

(50)

Chlorination. TMSCl has been used as the chloride source in a variety of substitution reactions. For example, when 1-bromoundecane is heated with stoichiometric amount of TMSCl in DMF at 90 °C and in the presence of imidazole, a Finkelstein-like reaction provides the corresponding 1-chloroundecane in quantitative yield (eq 51).[72]

$$n\text{-}C_{11}H_{23}Br \xrightarrow[\text{90 °C, 1 h, 99%}]{\text{TMSCl, Im, DMF}} n\text{-}C_{11}H_{23}Cl \qquad (51)$$

In another example, treatment of an oxazoline ester with TMSCl in THF at reflux leads to oxazoline ring opening and furnishes the corresponding β-chloro-amino ester in quantitative yield (eq 52).[73] Treatment with TMSBr and TMSI at rt affords the corresponding β-halo-amino esters in quantitative yield.

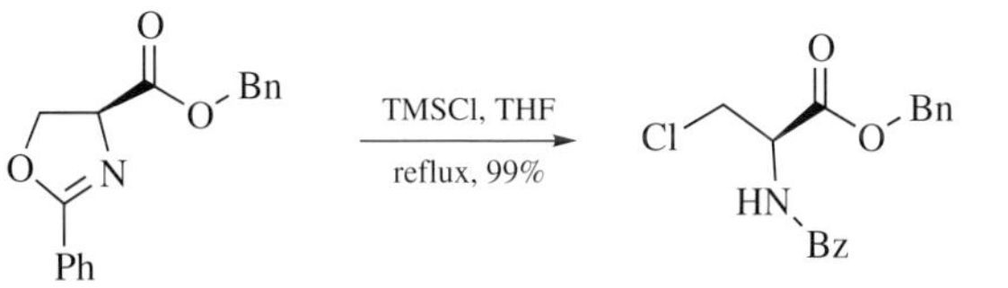

(52)

In a reaction generating β-halo-amino carboxylic acid derivatives similar to the ones above, stirring a mixture of a serine amide and TMSCl in acetonitrile at reflux for 8 h furnishes the β-chloro-amino amide in 72% yield (eq 53).[74] Treatment with TMSI results in the same transformation, but provides only 20% yield of the desired iodide.

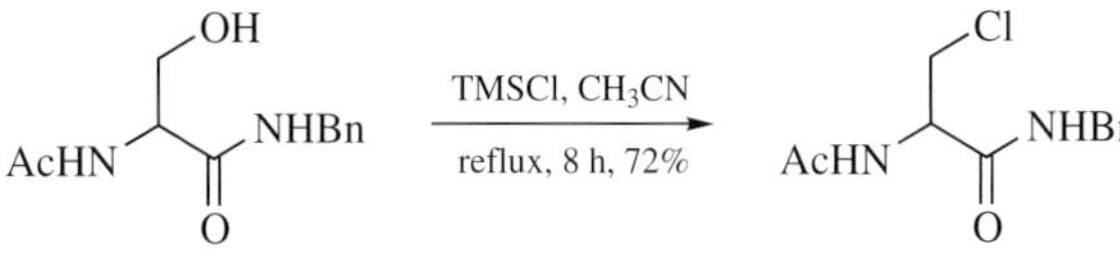

(53)

The above transformation of alcohols into chlorides does not occur for simple alcohols unless a catalytic amount of DMSO is added. Thus, addition of the catalyst into a mixture of 1-propanol and TMSCl (2 equiv) at rt provides 1-chloropropane in 93% yield in only 10 min (eq 54).[75] The reaction also works for other primary and tertiary alcohols, but not for secondary alcohols. The reaction of secondary alcohols with TMSCl and a stoichiometric amount of DMSO follows a procedure similar to that of the Swern oxidation and the corresponding ketones are produced.[76]

$$\text{(54)}$$

The TMSCl and DMSO combination can also be utilized to convert thiols into the corresponding disulfides if a stoichiometric amount of DMSO and a catalytic amount of TMSCl are used. For example, addition of 0.05 equiv of TMSCl into a mixture of thiophenol and DMSO in dichloromethane affords the diphenyl disulfide product in 90% yield (eq 55).[77]

$$PhSH \xrightarrow[\text{rt, 30 min, 90\%}]{\text{DMSO, TMSCl (cat)}} PhSSPh \qquad \text{(55)}$$

TMSCl is also a suitable chlorine source for the electrophilic chlorination reactions. For example, a combination of TMSCl, DMSO, and a catalytic amount of Bu$_4$NBr can be used for the chlorination of isoxazolin-5(4H)-ones (eq 56).[78] In another report, the combination of TMSCl and iodobenzene diacetate effectively converts various flavone derivatives into the corresponding 3-chloroflavones under mild conditions.[79]

$$\text{(56)}$$

TMSCl can be used as chlorine source for hydrochlorination of olefins or acetylenes. For example, treatment of a mixture of $\Delta^{9,10}$-octaline and a substoichiometric amount of water with TMSCl furnishes the *trans*-hydrochlorination product in quantitative yield (eq 57).[80]

$$\text{(57)}$$

In another reaction, in the presence of TMSCl and FeCl$_2$ catalyst, decomposition of propargyloxycarbonyl azide results in the formation of an intramolecular *syn*-aminochlorination product under mild conditions (eq 58).[81]

$$\text{(58)}$$

H$^+$ Surrogate. TMSCl can function similarly to a Brønsted acid or can generate a Brønsted acid in situ in various reactions. For example, addition of 2 equiv of TMSCl and 2 equiv of water to various nitriles furnishes the corresponding amides in good yield under ambient conditions (eq 59).[82] Furthermore, heating a

mixture of 2 equiv of TMSCl and an amide in an alcohol solvent at 40 °C provides the corresponding ester in moderate to good yield (eq 60).[83] Finally, one-pot direct conversion of a nitrile into the corresponding ester can be achieved by heating a mixture of the nitrile and 2 equiv of TMSCl in the desired alcohol solvent at 50 °C (eq 61).[84]

$$\text{(59)}$$

$$\text{(60)}$$

$$\text{(61)}$$

TMSCl has also been used as an efficient catalyst in acetalization reactions to protect ketones as dioxalanes[85] or thioacetals.[86] Treatment of 4-methylidene-3,4-dihydro-2H-pyrrole by TMSCl affords the corresponding pyrrole as the sole product (eq 62).[87]

$$\text{(62)}$$

A combination of TMSCl and triethylamine has allowed the successful cyclization of several deoxybenzoins to form isoflavones (eq 63), while a number of other protocols under either basic or acidic conditions failed to provide the desired products.[88]

$$\text{(63)}$$

Lewis Acid. In addition to the numerous examples of the type throughout this article, several specific examples of TMSCl as a Lewis acid are given here. Addition of diethylzinc to various imines is promoted by different Lewis acids, such as TMSCl, BF$_3 \cdot$ Et$_2$O, and ZnCl$_2$, to offer a variety of secondary amines (eq 64).[89] The reaction is simple and effective and, unlike the comparable additions to carbonyl compounds, no amino alcohol is needed.

$$ (64) $$

TMSCl also promotes the Pummerer reaction. For example, sequential addition LDA and TMSCl into a THF solution of 2-ethylbenzene sulfoxide produces 2-(1-hydroxy)-ethylbenzene thioether in reasonable yield and excellent stereoselectivity (eq 65).[90] Interestingly, treatment of 2-methylbenzene sulfoxide under the identical conditions gives 2-trimethylsilylmethylbenzene sulfoxide (eq 66).

$$ (65) $$

$$ (66) $$

In another example, addition of phenyl vinyl selenoxide into a mixture of an indanedione, hexamethyldisilazane, and TMSCl in dichloromethane affords the α-trimethylsiloxyselenide in 70% yield (eq 67).[91] The reaction proceeds through a sequence of in situ formation of TMS silyl enol ether, Michael addition onto the phenyl vinyl selenoxide, and seleno Pummerer rearrangement of the resulting selenoxide. Trifluoroacetic anhydride and various other trialkylchlorosilanes give the same product for this reaction, but in much lower yields.

$$ (67) $$

The reactivity of nitrones is enhanced by addition of TMSCl as well as TESCl and TMSOTf. Thus, stirring a mixture of TMSCl, the benzylnitrone of propanal (1 equiv), and indole (2 equiv) in dichloromethane at rt provides the bis-indole in good yields (eq 68).[92] In contrast, when equimolar amounts of TMSCl, benzylnitrone, and indole are mixed, and pyridine is added into the reaction mixture to trap hydrogen chloride liberated during the reaction, the reaction gives the indole hydroxylamine as the major product (eq 69).

$$ (68) $$

$$ (69) $$

NaBH4/TMSCl and NaBH3CN/TMSCl. TMSCl has been combined with NaBH4 to alter its reducing ability and applied to the reduction of a number of different functional groups. For example, when this combination was mixed with a succimide in ethanol at 0 °C, the corresponding mono- aminal was isolated in good yield, with no further reduction product detected (eq 70).[93]

$$ (70) $$

With the addition of a catalytic amount of (S)-α,α-diphenyl-pyrrolidinemethanol, this reagent combination of NaBH4/TMSCl has been successfully utilized in the enantioselective reduction of various ketones. The chiral alcohols were produced in excellent yields and very high enantiomeric excess (eq 71).[94]

$$ ee = 96\% \qquad (71) $$

In the presence of acetic acid, the NaBH4/TMSCl combination is the reagent of choice for reductive amination of urea and aromatic aldehydes. Either the mono-alkylated urea or the dialkylated urea can be obtained in good yields, depending on the ratio of urea and the aldehyde (eq 72).[95]

$$\text{(72)}$$

molar ratio				
0.5	:	1		75%
20	:	1	94%	6%

NaBH$_3$CN/TMSCl is another reducing agent combination that has been studied. With this reagent system, aldehydes, ketones, and acetals attached to benzo[*b*]furans or activated aromatic rings, such as methoxyphenyl rings, are completely deoxygenated to produce the corresponding alkyl arenes (eq 73).[96] When attached to nonactivated or deactivated aromatic rings, such as chlorophenyl and nitrophenyl rings, those aldehydes, ketones, and acetals groups are reduced to the alcohol or ether stage upon treatment with NaBH$_3$CN/TMSCl (eq 74). Ethyl and methyl benzofuroate esters are completely inactive under the same reaction conditions.

$$\text{(73)}$$

$$\text{(74)}$$

This reduction system also provides an alternative choice for the selective regeneration of alcohols from the corresponding allylic ethers, while nitro, ester, carbamate, and acetal groups maintain intaction under the reaction conditions (eq 75).[97]

$$\text{(75)}$$

Additive in Various Reduction Systems. TMSCl has been employed in many different reduction systems to improve the reactivity and/or increase the selectivity. The reagent has been used to activate zinc dust[98] or as additive for Reformatskii reactions.[99] While less reactive than Mg/TMSCl combination, the Zn/TMSCl combination proved to be the preferred reagent system for large scale and high concentrated preparation of diamines through the pinacol coupling of two molecules of imines (eq 76).[100]

$$\text{(76)}$$

A catalytic amount of TMSCl can substitute for toxic HgCl$_2$ as an excellent activating agent in the preparation of SmI$_2$, and also in Sm-promoted cyclopropanation of both allylic and α-allenic alcohols.[101] The Sm/TMSCl reduction system has been explored in the pinacol coupling reaction of aromatic carbonyl compounds, reductive dimerization/cyclization of 1,1-dicyanoalkenes,[102] debromination of vicinal dibromides to produce (*E*)-alkenes, and reductive coupling of sodium thiosulfates to generate disulfides.[103]

TMSCl has been combined with various metals and used in a variety of reactions, such as in combination with Na in acyloin condensation,[104] with Ti or Ti/Zn in the McMurry reaction and deoxygenation of epoxides,[105] and with Mn/PbCl$_2$ in three component coupling of alkyl iodides, electron-deficient olefins, and carbonyl compounds (eq 77).[106] The reagent has also been used in reductive ring opening of sugar-rings,[107] with La/I$_2$/CuI in deoxygenative dimerization of benzylic and allylic alcohols, ethers, and esters (eq 78),[108] with In to selectively reduce β-nitrostyrenes to α-phenyl-α-methoxy oximes,[109] with CHI$_3$/Mn/CrCl$_2$ to convert aldehydes into the corresponding homologated (*E*)-1-alkenyltrimethylsilanes,[110] and with CrCl$_2$/H$_2$O to transform terminal ynones and aldehydes into 2,5-disubstituted furans through the Baylis-Hillman type adduct intermediate and its subsequent cyclization (eq 79).[111]

$$\text{(77)}$$

$$\text{(78)}$$

$$\text{(79)}$$

TMSCl, in combination with Mn and a catalytic amount of CrCl$_2$ can mediate the addition of various organic halides and alkenyl triflates to aldehydes to form the corresponding secondary alcohols.[112] In particular, regardless of the configuration of the crotyl bromide, addition of this bromide to various aldehydes always generates *anti*-configured homoallyl alcohols with excellent diastereomeric excess and in good yields (eq 80). It was demonstrated that, as catalyst, $CrCl_3$ is equally efficient and is preferred for practical reasons. Chromocene (Cp_2Cr) or $CpCrCl_2 \cdot THF$ further upgrades the number of turnovers at chromium. This reduction system has also been used to produce 3-substituted furans in good yields via reductive annelation of 1,1,1-trichloroethyl propargyl ethers (eq 81).[113]

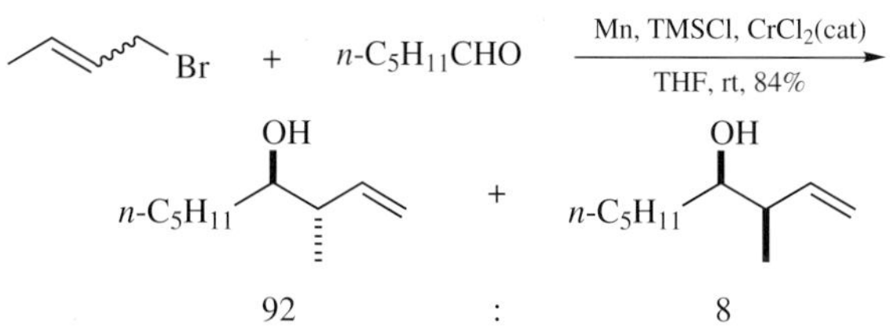

(80)

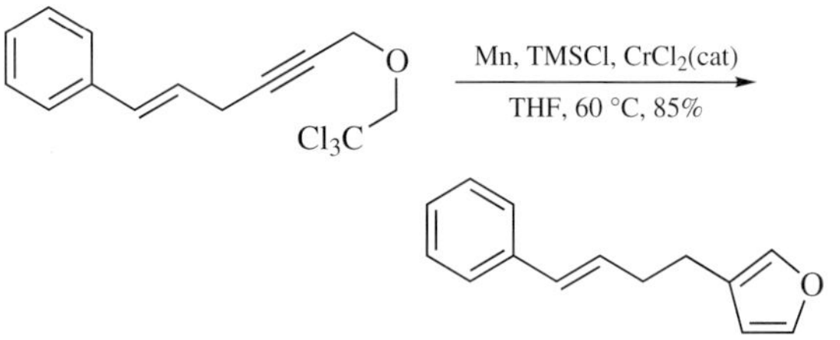

(81)

The Mg/TMSCl combination in DMF is another reducing system to be thoroughly studied. When treated with this reducing system, aromatic carbonyl compounds are converted into α-trimethylsilylalkyl trimethylsilyl ethers, generally, in reasonable to good yield.[114] When bis(chlorodimethylsilyl)ethane or 1,5-dichlorohexamethyltrisiloxane is used instead, the corresponding cyclic silylalkyl silyl ether is formed in moderate to good yield (eq 82).[115]

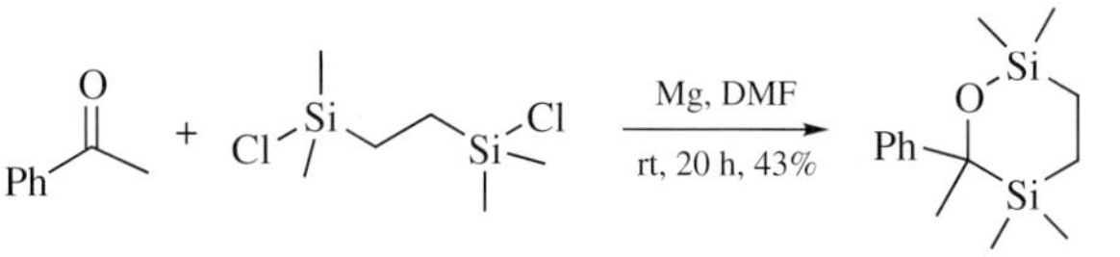

(82)

Reactions of various activated alkenes with Mg/TMSCl in DMF system have been explored as well. When treated with this reduction mixture, α-substituted arylvinyl sulfones affords (*E*)-β-substituted styrenes in high stereoselectivity and reasonable yields, through the desulfonation reaction (eq 83).[116] Under the reaction conditions, α,β-unsaturated ketones undergo facile and regioselective reductive dimerization to afford the corresponding bis(silyl enol ethers), that is, 1,6-bis(trimethylsilyloxy)-1,5-dienes (eq 84).[117] In contrast, α,β-unsaturated esters or α,β-epoxy esters were converted into β-trimethylsilyl esters when Mg/TMSCl in HMPA was used instead.[118] In the presence of an electrophile, such as an acid anhydride or acid chloride, α,β-unsaturated carbonyl compounds,[119] vinylphosphorus compounds,[120] and stilbenes and acenaphthylenes all experience a reductive cross-coupling reaction to give β-C-acylation products (eq 85).[121] Under these conditions, the reaction between ethyl β-arylacrylates and an aldehyde furnishes the corresponding γ-hydroxyester as the intermediate product, which is further converted into a β-aryl-γ-lactone as final product (eq 86).[122] In general, the *trans* γ-lactone is obtained as the preferred product.

(83)

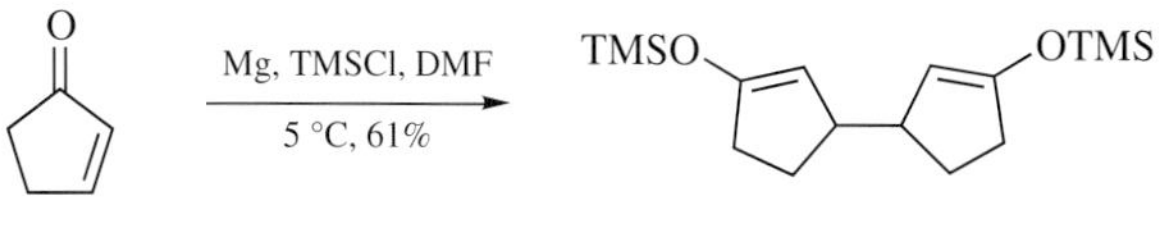

(84)

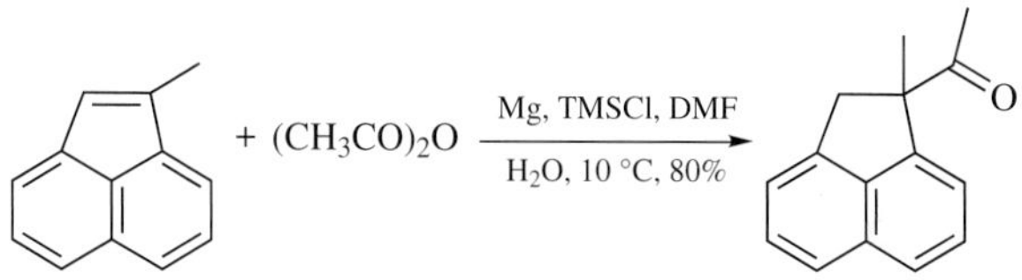

(85)

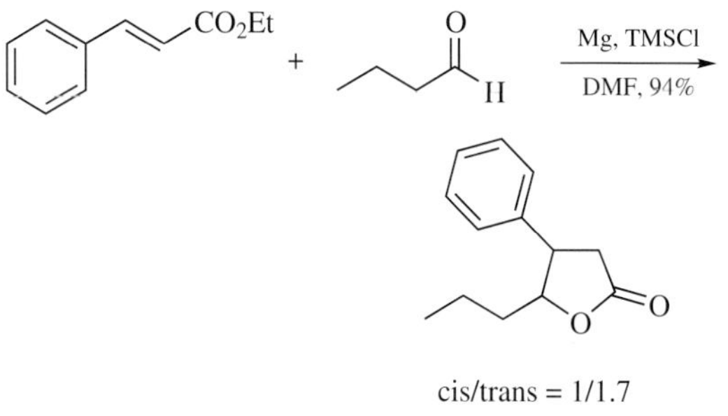

(86)

cis/trans = 1/1.7

Mg/TMSCl has also been applied to reduce $Ti(O-i-Pr)_4$ and generate a low-valent titanium species, which can mediate allyl and propargyl ether cleavage reactions to produce the corresponding aliphatic or aromatic alcohols (eq 87).[123] Selective cleavage of a propargyl ether in the presence of an allyl ether was made possible through addition of 2 equiv of ethyl acetate (eq 88).

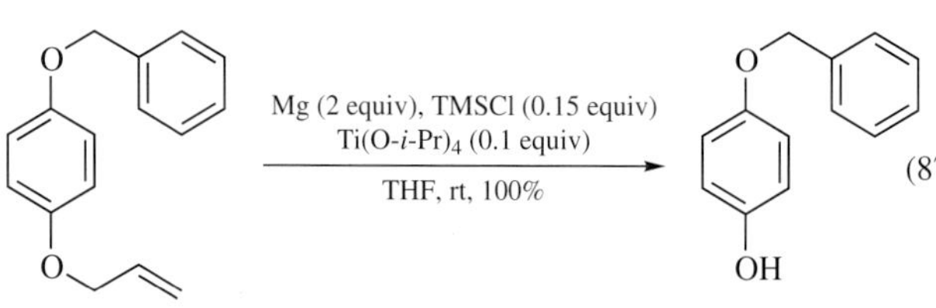

(87)

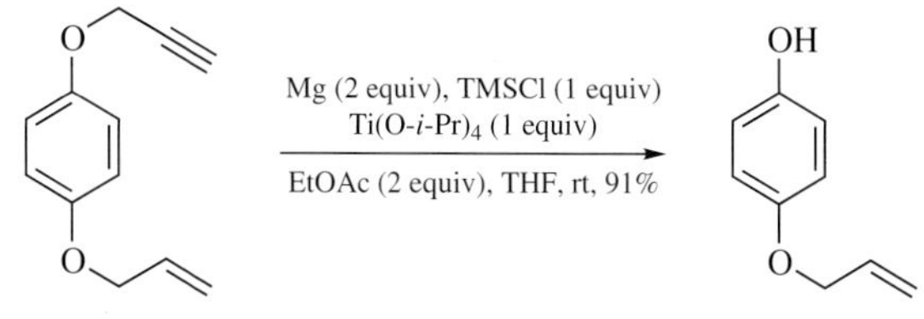

(88)

Additive in Miscellaneous Reactions. Concurrent addition of TMSCl and an enol triflate into a preformed Boc-protected α-aminoalkyl cuprate results in formation of Boc-protected secondary allylamine in good yield (eq 89).[124]

$$\text{(89)}$$

Treatment of terminal acetylenes with hydrogen iodide, generated in situ from TMSCl, sodium iodide, and water, provides a highly regioselective synthesis of 2-iodo-1-alkenes. Followed by addition of cuprous cyanide, the protocol offers a convenient one-pot preparation of 2-substituted acrylonitriles in fair to good yield (eq 90).[125] In contrast, when the reaction with arylacetylene proceeds in DMSO and with catalytic amount of sodium iodide, the 3-arylpropynenitrile is obtained as the preferred product.[126]

$$\text{(90)}$$

In the presence of TMSCl and $ZnCl_2$, Zn(Hg) can convert ortho formates, or acetals into organozinc carbenoid species, which then undergo a variety of reactions, such as cyclopropanation.[127] N-Diethoxymethyl amides function similarly and give amidocyclopropanation reactions,[128] or diastereoselective amidocyclopropanations if a chiral auxiliary is incorporated into the original amide (eq 91).[129]

$$\text{(91)}$$

TMSCl is a suitable additive for some Rh-[130] and $FeCl_2$-[131] catalyzed diazo decomposition reactions. TMSCl itself can also convert β-trimethylsiloxy α-diazocarbonyl compounds into a mixture of α-chloro-β,γ-unsaturated and γ-chloro-α,β-unsaturated carbonyl compounds.[132]

In a number of multicomponent condensation reactions, TMSCl has also been utilized to improve the yield or efficacy of the desired product, or is directly incorporated into the final molecules. Examples include the synthesis of N-aryl-3-arylamino acids from a three-component reaction of phenols, glyoxylates, and anilines,[133] preparation of 2,4,5-trisubstituted oxazoles,[134] or 4-cyanooxazoles,[135] and the three-component Biginelli reaction

(eq 92) and Biginelli-like Mannich reaction of carbamates, aldehydes, and ketones.[136]

$$\text{(92)}$$

In the presence of TMSCl, various sulfides are oxidized by KO_2 to afford sulfoxides in excellent yields, with little further oxidation to the sulfones (eq 93).[137] A trimethylsilylperoxy radical species is proposed to mediate the reaction. When promoted by a Lewis acid such as $SnCl_4$, $(SnO)_n$, or $Zr(O\text{-}i\text{-}Pr)_4$, bis(trimethylsilyl) peroxide (BTSP) and TMSCl convert olefins into chlorohydrins (eq 94).[138] Replacing TMSCl with TMS acetate furnishes the corresponding acetoxy alcohols as the final product. Stirring a mixture of allenic zinc reagents, TMSCl, and $ZnCl_2$ under an oxygen atmosphere produces propargyl hydroperoxides regioselectively; further transformation to the corresponding propargyl alcohols is achieved with Zn and hydrochloric acid.[139]

$$\text{(93)}$$

$$\text{(94)}$$

A mixture of an inorganic nitrate salt and TMSCl promotes the *ipso*-nitration of arylboronic acids to the corresponding nitroarenes in moderate to excellent yields, with high regioselectivity (eq 95).[140] Treatment of various secondary nitro compounds subsequently with 1 equiv of KH and then a catalytic amount of TMSCl furnishes the corresponding ketones in good to excellent yields.[141] In the Lewis acid mediated carboxylation of aromatic compounds with CO_2, addition of a large excess of TMSCl significantly improves the yields of the resulting aromatic carboxylic acids.[142]

$$\text{(95)}$$

Upon addition of TMSCl, the yield dramatically increased from 21% to 91% for the palladium-catalyzed intramolecular conversion of methyl 3-oxo-6-heptenonate to 2-carbomethoxycyclohexanone (eq 96).[143] In this particular case, however, hydrogen chloride, generated from hydrolysis of TMSCl with adventitious moisture, was identified as the active promoter. Thus, in a similar cyclization reaction, addition of hydrogen chloride, instead of TMSCl, provided the desired product with comparable yield (eq 97).

$$\text{(96)}$$

TMSCl	
-	21%
2–3 equiv	91%

$$\text{(97)}$$

CuCl$_2$ (1 equiv), TMSCl (2–3 equiv)	78%
CuCl$_2$ (0.3 equiv), HCl (0.1 equiv)	79%

1. Colvin, E. *Silicon in Organic Synthesis*; Butterworths: Boston, 1981.

2. Weber, W. P. *Silicon Reagents for Organic Synthesis*; Springer: New York, 1983.

3. Langer, S. H.; Connell, S.; Wender, I., *J. Org. Chem.* **1958**, *23*, 50.

4. Hergott, H. H.; Simchen, G., *Synthesis* **1980**, 626.

5. Bolin, D. R.; Sytwu, I.-I; Humiec, F.; Meinenhofer, J., *Int. J. Peptide Protein Res.* **1989**, *33*, 353.

6. Nakonieczna, L.; Chimiak, A., *Synthesis* **1987**, 418.

7. Kabalka, G. W.; Bierer, D. E., *Synth. Commun.* **1989**, *19*, 2783.

8. Valenti, E.; Pericàs, M. A.; Serratosa, F., *J. Org. Chem.* **1990**, *55*, 395.

9. House, H. O.; Czuba, L. J.; Gall, M.; Olmstead, H. D., *J. Org. Chem.* **1969**, *34*, 2324.

10. Krafft, M. E.; Holton, R. A., *Tetrahedron Lett.* **1983**, *24*, 1345.

11. Kaiser, E.; Tam, J. P.; Kubiak, T. M.; Merrifield, R. B., *Tetrahedron Lett.* **1988**, *29*, 303.

12. Olah, G. A.; Narang, S. C.; Gupta, B. G. B.; Malhotra, R., *J. Org. Chem.* **1979**, *44*, 1247.

13. Paquette, L. A.; Fristad, W. E.; Dime, D. S.; Bailey, T. R., *J. Org. Chem.* **1980**, *45*, 3017.

14. Andrews, G. C.; Crawford, T. C.; Contillo, L. G., *Tetrahedron Lett.* **1981**, *22*, 3803.

15. Dulcere, J.-P; Grimaldi, J.; Santelli, M., *Tetrahedron Lett.* **1981**, *22*, 3179.

16. Matsuzawa, S.; Isaka, M.; Nakamura, E.; Kuwajima, I., *Tetrahedron Lett.* **1989**, *30*, 1975.

17. Alexakis, A.; Berlan, J.; Besace, Y., *Tetrahedron Lett.* **1986**, *27*, 1047.

18. Horiguchi, Y.; Matsuzawa, S.; Nakamura, E.; Kuwajima, I., *Tetrahedron Lett.* **1986**, *27*, 4025.

19. Green, T. W.; Wuts, P. G. M. *Protective Groups in Organic Synthesis*; Wiley: New York, 1991.

20. Allevi, P.; Anastasia, M.; Ciufereda, P., *Tetrahedron Lett.* **1993**, *34*, 7313.

21. Olah, G. A.; Gupta, B. G. B.; Narang, S. C.; Malhotra, R., *J. Org. Chem.* **1979**, *44*, 4272.

22. Lissel, M.; Weiffen, J., *Synth. Commun.* **1981**, *11*, 545.

23. Cossy, J.; Pale, P., *Tetrahedron Lett.* **1987**, *28*, 6039.

24. Bundy, G. L.; Peterson, D. C., *Tetrahedron Lett.* **1978**, 41.

25. Hurst, D. T.; McInnes, A. G., *Synlett* **1965**, *43*, 2004.

26. Kawazoe, Y.; Nomura, M.; Kondo, Y.; Kohda, K., *Tetrahedron Lett.* **1987**, *28*, 4307.

27. Sekine, M.; Masuda, N.; Hata, T., *Tetrahedron* **1985**, *41*, 5445.

28. Larson, G. L.; Ortiz, M.; Rodrigues de Roca, M., *Synth. Commun.* **1981**, 583.

29. Stork, G.; Hudrlik, P. F., *J. Am. Chem. Soc.* **1968**, *90*, 4462.

30. Negishi, E.; Chatterjee, S., *Tetrahedron Lett.* **1983**, *24*, 1341.

31. Corey, E. J.; Gross, A. W., *Tetrahedron Lett.* **1984**, *25*, 495.

32. Cazeau, P.; Duboudin, F.; Moulines, F.; Babot, O.; Dunogues, J., *Tetrahedron* **1987**, *43*, 2075.

33. Cazeau, P.; Duboudin, F.; Moulines, F.; Babot, O.; Dunogues, J., *Tetrahedron* **1987**, *43*, 2089.

34. Rubottom, G. M.; Mott, R. C.; Krueger, D. S., *Synth. Commun.* **1977**, *7*, 327.

35. Morita, T.; Okamoto, Y.; Sakurai, H., *Tetrahedron Lett.* **1978**, *28*, 2523.

36. Barua, N. C.; Sharma, R. P., *Tetrahedron Lett.* **1982**, *23*, 1365.

37. Caputo, R.; Mangoni, L.; Neri, O.; Palumbo, G., *Tetrahedron Lett.* **1981**, *22*, 3551.

38. Taylor, R. T.; Degenhardt, C. R.; Melega, W. P.; Paquette, L. A., *Tetrahedron Lett.* **1977**, 159.

39. Fristad, W. E.; Bailey, T. R.; Paquette, L. A., *J. Org. Chem.* **1980**, *45*, 3028.

40. Lipschutz, B. H.; Dimock, S. H.; James, B., *J. Am. Chem. Soc.* **1993**, *115*, 9283.

41. Nakamura, E.; Matsuzawa, S.; Horiguchi, Y.; Kuwajima, I., *Tetrahedron Lett.* **1986**, *27*, 4029.

42. Corey, E. J.; Boaz, N. W., *Tetrahedron Lett.* **1985**, *26*, 6015.

43. Johnson, C. R.; Marren, T. J., *Tetrahedron Lett.* **1987**, *28*, 27.

44. Tamura, R.; Tamai, S.; Katayama, H.; Suzuki, H., *Tetrahedron Lett.* **1989**, *30*, 3685.

45. Booker-Milburn, K. I.; Thompson, D. F., *Tetrahedron Lett.* **1993**, *34*, 7291.

46. (a) Saxena, I.; Deka, N.; Sarma, J. C.; Tsuboi, S., *Synth. Commun.* **2003**, *33*, 4005. (b) Bastos, E. L.; Ciscato, L. F. M. L.; Baader, W. J., *Synth. Commun.* **2005**, *35*, 1501.

47. Nishiguchi, I.; Kita, Y.; Watanabe, M.; Ishino, Y.; Ohno, T.; Maekawa, H., *Synlett* **2000**, 1025.

48. Pesti, J. A.; Yin, J.; Chung, J., *Synth. Commun.* **1999**, *29*, 3811.

49. Barluenga, J.; del Pozo, C.; Olano, B., *Synthesis* **1995**, 1529.

50. Amii, H.; Kobayashi, T.; Terasawa, H.; Uneyama, K., *Org. Lett.* **2001**, *3*, 3103.

51. (a) García-Gómez, G.; Moretó, J. M., *Chem. Eur. J.* **2001**, *7*, 1503. (b) García-Gómez, G.; Moretó, J. M., *J. Am. Chem. Soc.* **1999**, *121*, 878.

52. Oriyama, T.; Ishiwata, A.; Hori, Y.; Yatabe, T.; Hasumi, N.; Koga, G., *Synlett* **1995**, 1004.

53. Concellon, J. M.; Suarez, J. R.; del Solar, V.; Llavona, R., *J. Org. Chem.* **2005**, *70*, 10348.

54. Azizi, N.; Saidi, M. R., *Tetrahedron Lett.* **2003**, *44*, 7933.

55. Paek, S. K.; Shim, S. C.; Cho, C. S.; Kim, T.-J., *Synlett* **2003**, 849.

56. Wang, L-S.; Hollis, T. K., *Org. Lett.* **2003**, *5*, 2543.

57. Wu, J.; Sun, X.; Xia, H-G., *Eur. J. Org. Chem.* **2005**, 4769.

58. Huang, H.; Forsyth, C. J., *Tetrahedron* **1997**, *53*, 16341.

59. Ikeda, S.; Kondo, K.; Sato, Y., *J. Org. Chem.* **1996**, *61*, 8248.

60. Reddy, C. K.; Devasagayaraj, A.; Knochel, P., *Tetrahedron Lett.* **1996**, *37*, 4495.

61. Lee, P. H.; Lee, K.; Sung, S.; Chang, S., *J. Org. Chem.* **2001**, *66*, 8646.

62. Yuguchi, M.; Tokuda, M.; Orito, K., *J. Org. Chem.* **2004**, *69*, 908.

63. Xu, L.-W.; Xia, C.-G.; Hu, X.-X., *Chem. Commun.* **2003**, 2570.

64. Liu, H.; Cohen, T., *Tetrahedron Lett.* **1995**, *36*, 8925.

65. Park, Y. S.; Weisenburger, G. A.; Beak, P., *J. Am. Chem. Soc.* **1997**, *119*, 10537.

66. Kimura, M.; Yamazaki, T.; Kitazume, T.; Kubota, T., *Org. Lett.* **2004**, *6*, 4651.

67. Shiina, I.; Imai, Y.; Suzuki, M.; Yanagisawa, M.; Mukaiyama, T., *Chem. Lett.* **2000**, 1062.

68. Cynamon, M. H.; Gimi, R.; Gyenes, F.; Sharpe, C. A.; Bergmann, K. E.; Han, H. J.; Gregor, L. B.; Rapolu, R.; Luciano, G.; Welch, J. T., *J. Med. Chem.* **1995**, *38*, 3902.

69. Topuzyan, V. O.; Oganesyan, A. A.; Panosyan, G. A., *Russ. J. Org. Chem.* **2004**, *40*, 1644.

70. Wrobel, Z., *Synthesis* **1997**, 753.

71. Wrobel, Z., *Tetrahedron* **1997**, *53*, 5501.

72. Peyrat, J-F.; Figadere, B.; Cave, A., *Synth. Commun.* **1996**, *26*, 4563.

73. Meyer, F.; Laaziri, A.; Papini, A. M.; Uziel, J.; Juge, S., *Tetrahedron: Asymmetry.* **2003**, *14*, 2229.

74. Choi, D.; Kohn, H., *Tetrahedron Lett.* **1995**, *36*, 7011.

75. Snyder, D. C., *J. Org. Chem.* **1995**, *60*, 2638.

76. Roa-Gutierrez, F.; Liu, H-J., *Bull. Inst. Chem., Acad. Sin.* **2000**, *47*, 19.

77. Karimi, B.; Hazarkhani, H.; Zareyee, D., *Synthesis* **2002**, 2513.

78. Huppe, S.; Rezaei, H.; Zard, S. Z., *Chem. Commun.* **2001**, 1894.

79. Rho, H. S.; Ko, B-S.; Ju, Y.-S., *Synth. Commun.* **2001**, *31*, 2101.

80. Boudjouk, P.; Kim, B-K.; Han, B-H., *Synth. Commun.* **1996**, *26*, 3479.

81. (a) Bach, T.; Schlummer, B.; Harms, K., *Synlett* **2000**, 1330. (b) Danielec, H.; Kluegge, J.; Schlummer, B.; Bach, T., *Synthesis* **2006**, 551.

82. Basu, M. K.; Luo, F-T., *Tetrahedron Lett.* **1998**, *39*, 3005.

83. Xue, C.; Luo, F-T., *J. Chin. Chem. Soc. (Taipei)* **2004**, *51*, 359.

84. (a) Luo, F-T.; Jeevanandam, A., *Tetrahedron Lett.* **1998**, *39*, 9455. (b) Cravotto, G.; Giovenzana, G. B.; Pilati, T.; Sisti, M.; Palmisano, G., *J. Org. Chem.* **2001**, *66*, 8447.

85. Su, X.; Gao, H.; Hang, L.; Li, Z., *Synth. Commun.* **1995**, *25*, 2807.

86. Papernaya, L. K.; Levanova, E. P.; Sukhomazova, E. N.; Albanov, A. I.; Deryagina, E. N., *Russ. J. Org. Chem.* **2003**, *39*, 1533.

87. Tsutsui, H.; Narasaka, K., *Chem. Lett.* **1999**, 45.

88. Pelter, A.; Ward, R. S.; Whalley, J. L., *Synthesis* **1998**, 1793.

89. Hou, X. L.; Zheng, X. L.; Dai, L. X., *Tetrahedron Lett.* **1998**, *39*, 6949.

90. Garcia Ruano, J. L.; Aleman, J.; Aranda, M. T.; Arevalo, M. J.; Padwa, A., *Org. Lett.* **2005**, *7*, 19.

91. Hagiwara, H.; Kafuku, K.; Sakai, H.; Kirita, M.; Hoshi, T.; Suzuki, T., *J. Chem. Soc., Perkin Trans.1* **2000**, 2577.

92. Chalaye-Mauger, H.; Denis, J.-L.; Averbuch-Pouchot, M.-T.; Vallee, Y., *Tetrahedron* **2000**, *56*, 791.

93. Romero, A. G.; Leiby, J. A.; Mizsak, S. A., *J. Org. Chem.* **1996**, *61*, 6974.

94. Jiang, B.; Feng, Y.; Zheng, J., *Tetrahedron Lett.* **2000**, *41*, 10281.

95. Xu, D.; Ciszewski, L.; Li, T.; Repic, O.; Blacklock, T. J., *Tetrahedron Lett.* **1998**, *39*, 1107.

96. Box, V. G. S.; Meleties, P. C., *Tetrahedron Lett.* **1998**, *39*, 7059.

97. Rao, G.; Venkat, R. D. S.; Mohan, G. H.; Iyengar, D. S., *Synth. Commun.* **2000**, *30*, 3565.

98. (a) Deboves, H. J. C.; Montalbetti, C. A. G.; Jackson, R. F. W., *J. Chem. Soc., Perkin Trans. 1* **2001**, 1876. (b) Kim, Y.; Choi, E. T.; Lee, M. H.; Park, Y. S., *Tetrahedron Lett.* **2007**, *48*, 2833.

99. Dartiguelongue, C.; Payan, S.; Duval, O.; Gomes, L. M.; Waigh, R. D., *Bull. Soc. Chem. Fr.* **1997**, *134*, 769.

100. Alexakis, A.; Aujard, I.; Mangeney, P., *Synlett* **1998**, 873.

101. Lautens, M.; Ren, Y., *J. Org. Chem.* **1996**, *61*, 2210.

102. Wang, L.; Zhang, Y., *Tetrahedron* **1998**, *54*, 11129.

103. Xu, X.; Lu, P.; Zhang, Y., *Synth. Commun.* **2000**, *30*, 1917.

104. Matzeit, A.; Schaefer, H. J.; Amatore, C., *Synthesis* **1995**, 1432.

105. Fürstner, A.; Hupperts, A., *J. Am. Chem. Soc.* **1995**, *117*, 4468.

106. Takai, K.; Ueda, T.; Ikeda, N.; Moriwake, T., *J. Org. Chem.* **1996**, *61*, 7990.

107. Tanaka, K.; Yamano, S.; Mitsunobu, O., *Synlett* **2001**, 1620.

108. Nishino, T.; Nishiyama, Y.; Sonoda, N., *Bull. Chem. Soc. Jpn.* **2003**, *76*, 635.

109. Yadav, J. S.; Reddy, B. V. S.; Srinivas, R.; Ramalingam, T., *Synlett* **2000**, 1447.

110. Takai, K.; Hikasa, S.; Ichiguchi, T.; Sumino, N., *Synlett* **1999**, 1769.

111. Takai, K.; Morita, R.; Sakamoto, S., *Synlett* **2001**, 1614.

112. (a) Fürstner, A.; Shi, N., *J. Am. Chem. Soc.* **1996**, *118*, 2533. (b) Fürstner, A.; Shi, N., *J. Am. Chem. Soc.* **1996**, *118*, 12349.

113. Barma, D. K.; Kundu, A.; Baati, R.; Mioskowski, C.; Falck, J. R., *Org. Lett.* **2002**, *4*, 1387.

114. Ishino, Y.; Maekawa, H.; Takeuchi, H.; Sukata, K.; Nishiguchi, I., *Chem. Lett.* **1995**, 829.

115. Uchida, T.; Kita, Y.; Maekawa, H.; Nishiguchi, I., *Tetrahedron* **2006**, *62*, 3103.

116. Nishiguchi, I.; Matsumoto, T.; Kuwahara, T.; Kyoda, M.; Maekawa, H., *Chem. Lett.* **2002**, 478.

117. Maekawa, H.; Sakai, M.; Uchida, T.; Kita, Y.; Nishiguchi, I., *Tetrahedron Lett.* **2004**, *45*, 607.

118. Bolourtchian, M.; Mamaghani, M.; Badrian, A., *Phos., Sulfur, Silicon, Rel. Elem.* **2003**, *178*, 2545.

119. Ohno, T.; Sakai, M.; Ishino, Y.; Shibata, T.; Maekawa, H.; Nishiguchi, I., *Org. Lett.* **2001**, *3*, 3439.

120. Kyoda, M.; Yokoyama, T.; Maekawa, H.; Ohno, T.; Nishiguchi, I., *Synlett* **2001**, 1535.

121. (a) Nishiguchi, I.; Yamamoto, Y.; Sakai, M.; Ohno, T.; Ishino, Y.; Maekawa, H., *Synlett* **2002**, 759. (b) Yamamoto, Y.; Kawano, S.; Maekawa, H.; Nishiguchi, I., *Synlett* **2004**, 30.

122. Ohno, T.; Ishino, Y.; Tsumagari, Y.; Nishiguchi, I., *J. Org. Chem.* **1995**, *60*, 458.

123. Ohkubo, M.; Mochizuki, S.; Sano, T.; Kawaguchi, Y.; Okamoto, S., *Org. Lett.* **2007**, *9*, 773.

124. Dieter, R. K.; Dieter, J. W.; Alexander, C. W.; Bhinderwala, N. S., *J. Org. Chem.* **1996**, *61*, 2930.

125. Luo, F-T.; Ko, S-L.; Chao, D-Y., *Tetrahedron Lett.* **1997**, *38*, 8061.

126. Cheng, Z-Y.; Li, W-J.; He, F.; Zhou, J-M.; Zhu, X-F., *Bioorg. Med. Chem.* **2007**, *15*, 1533.

127. Fletcher, R. J.; Motherwell, W. B.; Popkin, M. E., *Chem. Commun.* **1998**, 2191.

128. Begis, G.; Cladingboel, D.; Motherwell, W. B., *Chem. Commun.* **2003**, 2656.

129. Begis, G.; Sheppard, T. D.; Cladingboel, D. E.; Motherwell, W. B.; Tocher, D. A., *Synthesis* **2005**, 3186.

130. (a) Mori, T.; Oku, A., *Chem. Commun.* **1999**, 1339. (b) Sawada, Y.; Mori, T.; Oku, A., *Chem. Commun.* **2001**, 1086.

131. Bach, T.; Schlummer, B.; Harms, K., *Synlett* **2000**, 1330.

132. Xiao, F.; Zhang, Z.; Zhang, J.; Wang, J., *Tetrahedron Lett.* **2005**, *46*, 8873.

133. Huang, T.; Li, C-J., *Tetrahedron Lett.* **2000**, *41*, 6715.

134. Wang, Q.; Ganem, B., *Tetrahedron Lett.* **2003**, *44*, 6829.

135. Xia, Q.; Ganem, B., *Synthesis* **2002**, 1969.

136. Xu, L.-W.; Wang, Z.-T.; Xia, C.-G.; Li, L.; Zhao, P.-Q., *Helv. Chim. Acta* **2004**, *87*, 2608.

137. Chen, Y.-J.; Huang, Y.-P., *Tetrahedron Lett.* **2000**, *41*, 5233.

138. Sakurada, I.; Yamasaki, S.; Goettlich, R.; Iida, T.; Kanai, M.; Shibasaki, M., *J. Am. Chem. Soc.* **2000**, *122*, 1245.

139. Harada, T.; Kutsuwa, E., *J. Org. Chem.* **2003**, *68*, 6716.

140. Prakash, G. K. S.; Panja, C.; Mathew, T.; Surampudi, V.; Petasis, N. A.; Olah, G. A., *Org. Lett.* **2004**, *6*, 2205.

141. Hwu, J. R.; Josephrajan, T.; Tsay, S-C., *Synthesis* **2006**, 3305.

142. Nemoto, K.; Yoshida, H.; Suzuki, Y.; Morohashi, N.; Hattori, T., *Chem. Lett.* **2006**, 820.

143. Wang, X.; Pei, T.; Han, X.; Widenhoefer, R. A., *Org. Lett.* **2003**, *5*, 2699.

Chlorotris(triphenylphosphine)-rhodium(I)

$$\boxed{RhCl(PPh_3)_3}$$

[14694-95-2] $C_{54}H_{45}ClRhP_3$ (MW 925.24)

InChI = 1/3C18H15P.ClH.Rh/c3*1-4-10-16(11-5-1)19(17-12-6-2-7-13-17)18-14-8-3-9-15-18;;/h3*1-15H;1H;/q;;;;+1/p-1/f3C18H15P.Cl.Rh/h;;;1h;/q;;;-1;/m/r3C18H15P.Cl Rh/c3*1-4-10-16(11-5-1)19(17-12-6-2-7-13-17)18-14-8-3-9-15-18;1-2/h3*1-15H;

InChIKey = IXAYKDDZKIZSPV-KIDOFKJHCK

(catalyst precursor for many reactions involving alkenes, alkynes, halogenated organics, and organometallic reagents; notably hydrogenations, hydrosilylations, hydroformylations, hydroborations, isomerizations, oxidations, and cross-coupling processes)

Alternate Name: Wilkinson's catalyst.

Physical Data: mp 157 °C. It exists in burgundy-red and orange polymeric forms, which have identical chemical properties (as far as is known).

Solubility: about $20\,g\,L^{-1}$ in $CHCl_3$ or CH_2Cl_2, about $2\,g\,L^{-1}$ in benzene or toluene; much less in acetic acid, acetone, methanol, and other aliphatic alcohols. Virtually insol in alkanes and cyclohexane. Reacts with donor solvents like DMSO, pyridine, and acetonitrile.

Form Supplied in: burgundy-red powder, possibly containing excess triphenylphosphine, triphenylphosphine oxide, and traces of rhodium(II) and -(III) complexes.

Analysis of Reagent Purity: ^{31}P NMR displays resonances for the complex in equilibrium with dissociated triphenylphosphine (CH_2Cl_2, approximate δ ppm: 31.5 and 48.0 {*J* values: Rh–P^1 −142 Hz; Rh–P^2 −189 Hz; P^1–P^2 −38 Hz} shifted in the presence of excess PPh_3).[3] Triphenylphosphine oxide contaminant can also be observed (CH_2Cl_2, δ ppm: 29.2) but paramagnetic impurities are generally not evident. In rhodium NMR a signal is observed at −1291 ppm.

Preparative Methods: good quality material can be obtained using the latest *Inorganic Syntheses* procedure,[4] with careful exclusion of air. Recrystallization is *not* recommended.

Handling, Storage, and Precautions: the complex should be stored at reduced temperature under dinitrogen or argon. It oxidizes slowly when exposed to air in the solid state, and faster in solution. Such partial oxidation can influence the catalytic efficacy. Consequently, the necessary precautions are governed by the reaction in question. For mechanistic and kinetic studies, reproducible results may only be obtained if the catalyst is freshly prepared and manipulated in an inert atmosphere; even the substrate should be treated to remove peroxides. For hydrogenations of alkenes on a preparative scale, complex that has been handled in the air for very brief periods should be active, but competing isomerization processes may be enhanced as a result of partial oxidation of the catalyst. At the other extreme, exposure to air just before use is clearly acceptable for oxidations in the presence of O_2 and *t*-BuOOH.

Original Commentary

Kevin Burgess & Wilfred A. van der Donk
Texas A&M University, College Station, TX, USA

Background.[1] In solution, Wilkinson's catalyst is in equilibrium with the 14e species $RhCl(PPh_3)_2$ (**1**) and triphenylphosphine. The 14e complex is far more reactive than the parent material; consequently it is the reactive entity most likely to coordinate with the substrate and/or the reagents. Generally, the catalytic cycles involving this material then proceed via a cascade of oxidative addition, migratory insertion, and reductive elimination reactions. The postulated mechanism for the hydrogenation of alkenes illustrates these features (Scheme 1), and is typical of the rationales frequently applied to comprehend the reactivity of $RhCl(PPh_3)_3$. Other types of transformations may be important (e.g. transmetalations), and the actual mechanisms are certainly more complicated in many cases; nevertheless, the underlying concepts are similar.

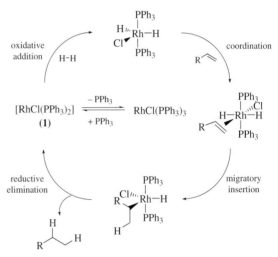

Scheme 1 Simplified mechanism for alkene hydrogenations mediated by $RhCl(PPh_3)_3$

Two important conclusions emerge from these mechanistic considerations. First, $RhCl(PPh_3)_3$ is not a catalyst in the most rigorous sense, but a catalyst precursor. This distinction is critical to the experimentalist because it implies that there are other ways to generate catalytically active rhodium(I) phosphine complexes in solution. Wilkinson's 'catalyst' is a convenient source of homogeneous rhodium(I); it has been extensively investigated because it is easily obtained, and because it was discovered early in the development of homogeneous transition metal catalysts. However, for any transformation there always may be better catalyst precursors than $RhCl(PPh_3)_3$. Secondly, reactions involving a catalytic cycle such as the one shown in Scheme 1 are inherently more complicated than most in organic chemistry. Equilibria and rates for each of the steps involved can be influenced by solvent, temperature, additives, and functional groups on the substrate. Competing reactions are likely to be involved and, if they are, the performance of the catalytic systems therefore is likely to be sensitive

Essential Reactions for Organic Synthesis, First Edition. Edited by Philip L. Fuchs.
© 2016 John Wiley & Sons, Ltd. Published 2016 by John Wiley & Sons, Ltd.

to these parameters. Consequently, the purity of the Wilkinson's catalyst used is an important factor. Indeed, less pure catalyst occasionally gives superior results because removal of a fraction of the triphenylphosphine in solution by oxidation to triphenylphosphine oxide gives more of the dissociation product (**1**).

In summary, practitioners of organometallic catalysis should consider the possible mechanistic pathways for the desired transformation, then screen likely catalyst systems and conditions until satisfactory results are obtained. Wilkinson's catalyst is one of the many possible sources of homogeneous rhodium(I).

Hydrogenations. Wilkinson's catalyst is highly active for hydrogenations of unconjugated alkenes at ambient temperatures and pressures. Steric effects are important insofar as less hindered alkenes react relatively quickly, whereas highly encumbered ones are not reduced (eq 1).[5] Hydrogen in the presence of RhCl(PPh$_3$)$_3$ under mild conditions does not reduce aromatic compounds, ketones, carboxylic acids, amides, or esters, nitriles, or nitro (eq 2) functionalities. Moreover, hydrogenations mediated by Wilkinson's catalyst are stereospecifically *cis* (eq 1). These characteristics have been successfully exploited to effect chemo-, regio-, and stereoselective alkene reductions in many organic syntheses (eqs 1–3). For instance, steric effects force delivery of dihydrogen to the least hindered face of the alkene in (eq 3).[6] eq 4 illustrates that 1,4-cyclohexadienes can be reduced with little competing isomerization/aromatization,[7] unlike many other common hydrogenation catalysts.[5]

(1)

(2)

(3)

(4)

Strongly coordinating ligands can suppress or completely inhibit hydrogenations mediated by Wilkinson's catalyst; examples

include 1,3-butadiene, many phosphorus(III) compounds, sulfides, pyridine, and acetonitrile. Similarly, strongly coordinating substrates are not hydrogenated in the presence of Wilkinson's catalyst, presumably because they bind too well. Compounds in this category include maleic anhydride, ethylene, some 1,3-dienes, and some alkynes. Conversely, transient coordination of functional groups on the substrate can be useful with respect to directing RhCl(PPh$_3$)$_3$ to particular regions of the molecule for stereoselective reactions. However, in directed hydrogenations Wilkinson's catalyst is generally inferior to more Lewis acidic cationic rhodium(I) and iridium(I) complexes.[8] The activity of Wilkinson's catalyst towards hydrogenation of alkenes has been reported to be enhanced by trace quantities of oxygen.[9]

Hydrogenations of alkynes mediated by Wilkinson's catalyst generally give alkanes. *Cis*-alkene intermediates formed in such reactions tend to be more reactive than the alkyne substrate, so this is usually not a viable route to alkenes. Some alkynes suppress the catalytic reactions of RhCl(PPh$_3$)$_3$ by coordination. Nevertheless, hydrogenation of alkynes mediated by RhCl(PPh$_3$)$_3$ can be useful in some cases, as in eq 5 in which the catalyst tolerates sulfoxide functionalities and gives significantly higher yields than the corresponding reduction catalyzed by *Palladium on Barium Sulfate*.[10]

(5)

Wilkinson's catalyst can mediate the hydrogenation of allenes to isolated alkenes via reduction of the least hindered bond.[11] Di-*t*-butyl hydroperoxide is 'hydrogenated' to *t*-BuOH in the presence of RhCl(PPh$_3$)$_3$, though this transformation could occur via a radical process.[12]

Hydrogen Transfer Reactions. Wilkinson's catalyst should lower the energy barrier for dehydrogenations of alkanes to alkenes since it catalyzes the reverse process, but no useful transformation of this kind have been discovered. Presumably, the activation energy for this reaction is too great since alkanes have no coordinating groups. Alcohols and amines, however, do have ligating centers, and can dehydrogenate in the presence of Wilkinson's catalyst. These reactions have been used quite often, mostly from the perspective of hydrogen transfer from an alcohol or amine to an alkene substrate, although occasionally to dehydrogenate alcohols or amines.

2-Propanol solvent under basic conditions has been extensively used to transfer hydrogen to alkenes and other substrates. Elevated temperatures are usually required and under these conditions RhCl(PPh$_3$)$_3$ may be extensively modified prior to the catalysis. Ketones, alkenes (eq 6), aldimines (eq 7),[13] nitrobenzene, and some quinones are reduced in this way.

(6)

Wilkinson's catalyst mediates a Cannizzaro-like process with benzaldehyde in ethanol; the aldehyde serves as a dihydrogen source to reduce itself, and the benzoic acid formed is esterified by the solvent (eq 8).[14] Pyrrolidine is *N*-methylated by methanol in the presence of RhCl(PPh$_3$)$_3$, a reaction that presumably occurs via hydrogen transfer from methanol, condensation of the formaldehyde formed with pyrrolidine, then hydrogen transfer to the iminium intermediate (eq 9).[15]

Hydrosilylations.[16] Wilkinson's catalyst is one of several complexes which promote hydrosilylation reactions, and it often seems to be among the best identified.[17] However, hydrosilylations with RhCl(PPh$_3$)$_3$ tend to be slower than those mediated by H$_2$PtCl$_6$. Good turnover numbers are observed, the catalyst eventually being inactivated by P–C bond cleavage reactions at the phosphine,[18] and other unidentified processes. Catalysts without phosphine ligands may be even more robust than RhCl(PPh$_3$)$_3$ because they are unable to decompose via P–C bond cleavage.[19] Wilkinson's catalyst is relatively efficient with respect to converting silanes to disilanes.[20] The latter reaction could be useful in its own right but in the context of hydrosilylation processes it means that the product yields based on the silane are less than quantitative.

For hydrosilylation of alkenes, the reaction rate increases with temperature and hence many of these reactions have been performed at 100 °C. Higher reaction rates are obtained for silanes with very electronegative substituents and low steric requirements (e.g. HSi(OEt)$_3$ > HSi(*i*-Pr)$_3$). Terminal alkenes usually are hydrosilylated in an anti-Markovnikov sense to give terminal silanes. Internal alkenes tend not to react (e.g. cyclohexene), or isomerize to the terminal alkene which is then hydrosilylated (eq 10). Conversely, terminal alkenes may be partially isomerized to unreactive internal alkenes before the addition of silane can occur. 1,4-Additions to dienes are frequently observed, and the product distributions are extremely sensitive to the silane used (eq 11).

α,β-Unsaturated nitriles are hydrosilylated, even γ-substituted ones, to give 2-silyl nitriles with good regioselectivity (eq 12).[21]

Secondary alkyl silanes are also formed in the hydrosilylation of phenylethylene. In fact, the latter reaction has been studied in some detail, and primary alkyl silanes, hydrogenation product (i.e. ethylbenzene), and *E*-2-silylphenylethylenes are also formed (eq 13).[22] Equimolar amounts of ethylbenzene (**2**) and *E*-2-silylphenylethylene (**4**) are produced, implying these products arise from the same reaction pathway. It has been suggested that this involves dimeric rhodium species because the relative amounts of these products increase with the rhodium:silane ratio; however, competing radical pathways cannot be ruled out. Certainly, product distributions are governed by the proportions of all the components in the reaction (i.e. catalyst, silane, and alkene), and the reaction temperature. Side products in the hydrosilylation of 1-octene include vinylsilanes and allylsilanes (eq 14).[23,24]

Hydrosilylation of alkynes gives both *trans* products (i.e. formally from *cis* addition), and *cis* products (from either isomerization or *trans* addition); H$_2$PtCl$_6$, however, gives almost completely *cis* addition to *trans* products.[25] Moreover, CC–H to CC–SiR$_3$ exchange processes can occur for terminal alkynes giving **6** (eq 15).[25–27] The product distribution in these reactions is temperature dependent, and other factors may be equally important. Nonstereospecific transition metal catalyzed hydrosilylations of alkynes are not confined to Wilkinson's catalyst, and the origin of the *trans* addition product has been investigated in detail for other homogeneous rhodium and iridium complexes.[19]

Hydrosilylation of terminal alkenes has been used in a polymerization process to form new polymeric organic materials.[28]

Hydrosilylation of α,β-unsaturated aldehydes and ketones gives silylenol ethers via 1,4-addition, even when the 4-position is relatively hindered.[29] Hydrolysis of the silyl enol ethers so formed gives saturated aldehydes. Combination of these reduction and hydrolysis steps gives overall reduction of alkenes conjugated to aldehydes, in selectivities which are generally superior to those obtained using hydridic reducing agents (eqs 16 and 17). Dihydrosilanes tend to reduce α,β-unsaturated carbonyl compounds to the corresponding alcohols, also with good regioselectivity (eq 18).

$$(16)$$

$$(17)$$

$$(18)$$

Similar hydrosilylations of α,β-unsaturated esters are useful for obtaining silyl ketene acetals with over 98:2 (Z) selectivity (eq 19);[30] this transformation is complementary to the reaction of α-bromo esters with zinc and chlorotrialkylsilanes, which favors the formation of the corresponding (E) products.[30] In cases where (E):(Z) stereoselectivity is not an issue, *Rhodium(III) Chloride* (RhCl$_3$·6H$_2$O) may be superior to Wilkinson's catalyst.[31] Unconjugated aldehydes and ketones are reduced by silanes in the presence of RhCl(PPh$_3$)$_3$; trihydrosilanes react quicker than di- than monohydrosilanes.[32,33]

$$(19)$$

Alcohols (eq 20)[34] and amines (eq 21)[35] react with silanes in the presence of Wilkinson's catalyst to give the silylated compounds and, presumably, hydrogen. These reactions are useful in protecting group strategies.

$$Ph_2SiH_2 + MeOH \xrightarrow{\text{cat RhCl(PPh}_3)_3} Ph_2SiH(OMe) + H_2 \quad (20)$$

$$(21)$$

N,N-Dimethylacrylamide and triethylsilane combine in the presence of Wilkinson's catalyst (50 °C) to give a *O,N*-silylketene acetal as the pure (Z) isomer after distillation; this reaction can be conveniently performed on a gram scale (eq 22). The products have been used in new aldol methodology.[36]

$$(22)$$

Hydrostannylations. Hydrostannanes add to alkynes in uncatalyzed reactions at 60 °C. Phenylacetylene, for instance gives a mixture of (E)- and (Z)-vinylstannanes, wherein the tin atom has added to the terminal carbons. In the presence of Wilkinson's catalyst, however, the hydrostannylation proceeds at 0 °C to give mostly the regioisomeric vinylstannanes (eq 23).[37] Terminal stannanes in the latter process seem to result from competing free radical additions. This may not be a complication with some other catalysts; the complexes PdCl$_2$(PPh$_3$)$_2$ and Mo(η^3-allyl)(CO)$_2$(NCMe)$_2$ also mediate hydrostannylations of alkynes, and they are reported to be 100% *cis* selective.[38] Hydrostannanes and thiols react in a similar way to silanes and alcohols (eq 24).[39]

$$(23)$$

$$(24)$$

Hydroacylations. Alkenes with aldehyde functionality in the same molecule, but displaced by two carbon atoms, can cyclize via intramolecular hydroacylation reactions. Substituent effects can have a profound influence on these transformations. For instance, 3,4-disubstituted 4-pentenals cyclize to cyclopentanones without serious complications,[40] but 2,3-disubstituted 4-pentenals give a cyclopropane as a competing product (eqs 25 and 26).[41] Formation of the latter material illustrates two features which restrict the applicability of this type of reaction. First decarbonylation of the aldehyde can occur, in this case presumably giving a rhodium alkyl complex which then inserts the pendant alkene functionality. Secondly, decarbonylation reactions convert the catalyst into RhCl(CO)(PPh$_3$)$_2$, which tends to be inactive. Moreover, the reaction is only generally applicable to the formation of five-membered rings, and it is apparently necessary to use quite large amounts of Wilkinson's catalyst to ensure good yields (eq 27).[42] Rhodium(I) complexes other than RhCl(PPh$_3$)$_3$ can give better results in some cases.[43]

$$(25)$$

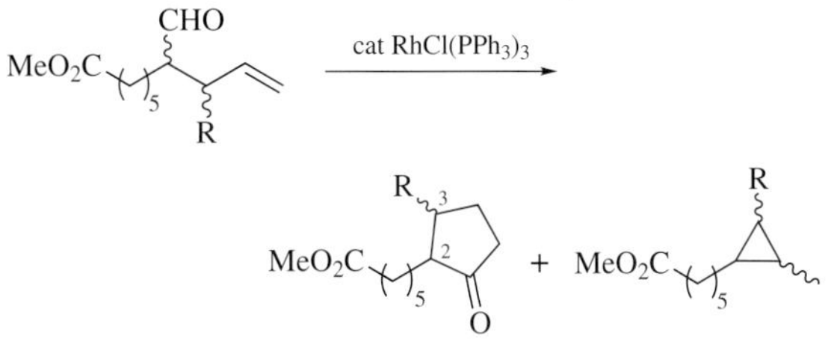

(26)

R = (CH₂)₇Me   1:1

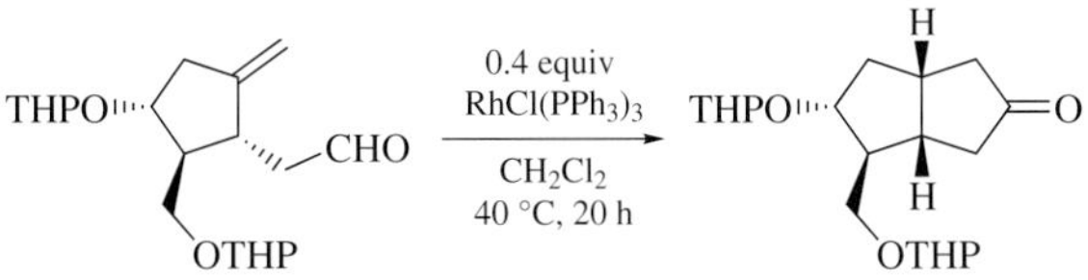

(27)

Lactols can be cyclized under the typical hydroacylation conditions (eq 28), presumably via equilibrium amounts of the corresponding aldehyde.[40] Finally, intermolecular hydroacylation has been formally achieved in the reaction of a pyridyl aldimine with ethylene under pressure at 160 °C; here the pyridine functionality anchors the aldimine to the rhodium, and decarbonylation is impossible (eq 29).

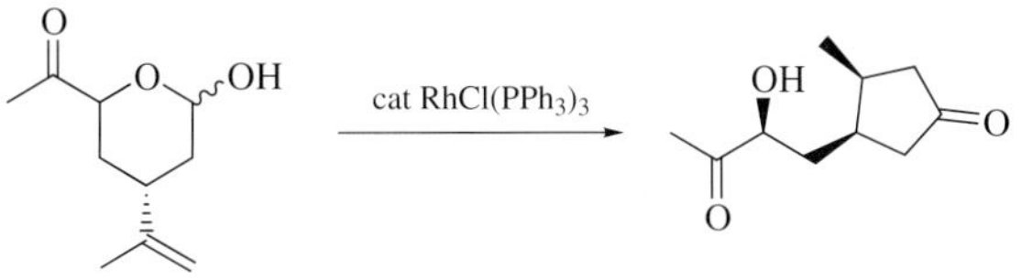

(28)

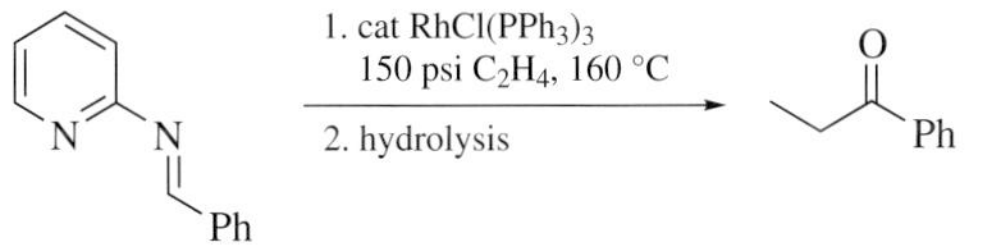

(29)

Decarbonylations. Wilkinson's catalyst has been known for some time to decarbonylate aldehydes, even heavily functionalized ones, to the corresponding hydrocarbons.[44] Some examples are shown in eqs 30–33, illustrating high stereochemical retention in the decarbonylation of chiral, cyclopropyl, and unsaturated aldehydes.[45,46] Acid chlorides are also decarbonylated by RhCl(PPh₃)₃.

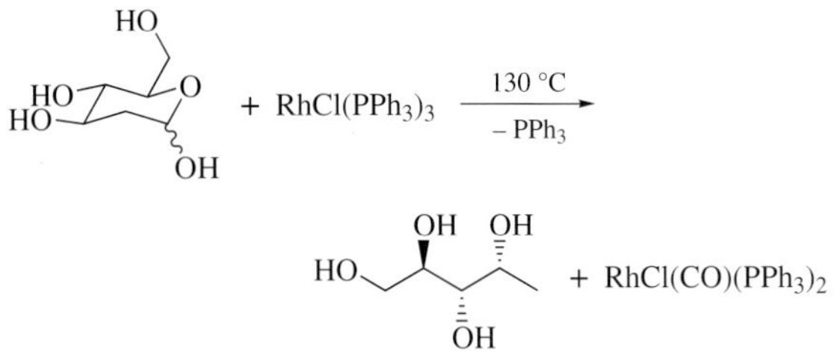

(30)

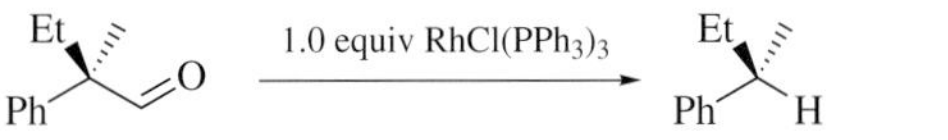

(31)

93% retention

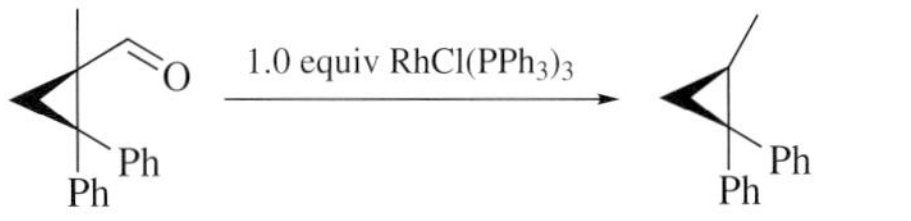

(32)

94% retention

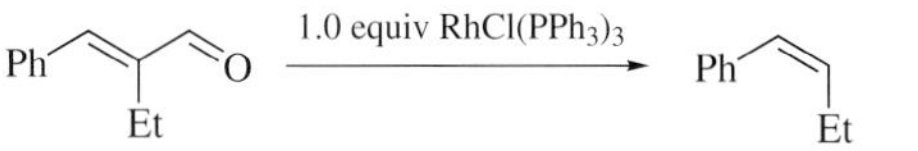

(33)

100% retention

The problem with all these reactions is that stoichiometric amounts of the catalyst are required, and the process is inordinately expensive. Consequently, it has only been used by those wishing to illustrate a decarbonylation occurs for some special reason, or in the closing stages of small scale syntheses of complex organic molecules. Very recently, however, it has been shown that the reaction can be made catalytic by adding **Diphenyl Phosphorazidate**.[47] The role of the latter is to decarbonylate the catalytically inactive RhCl(CO)(PPh₃)₂, regenerating rhodium(I) without carbonyl ligands. Examples of this catalytic process are shown in eqs 34 and 35. The path is now clear for extensive use of RhCl(PPh₃)₃ for catalytic decarbonylation reactions in organic synthesis.

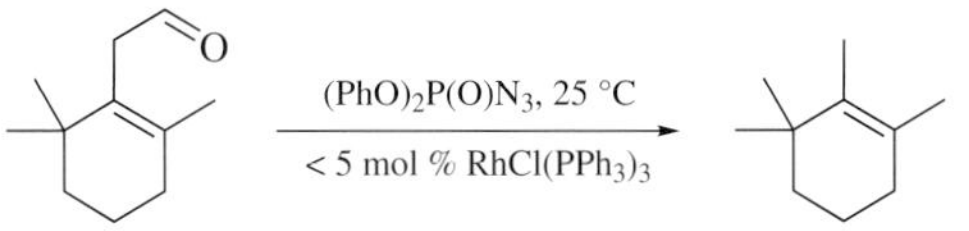

(34)

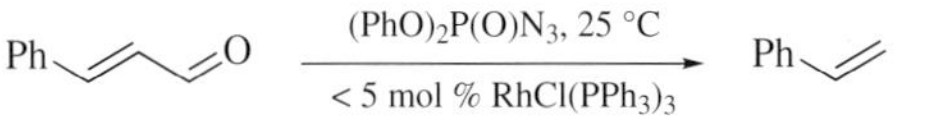

(35)

Catalytic decarbonylations of a few substrates other than aldehydes have been known for some time, e.g. conversion of benzoic anhydrides to fluorenones at high temperatures (ca. 225 °C).[48]

Hydroformylations.[49] Carbon monoxide reacts rapidly with RhCl(PPh₃)₃ to give RhCl(CO)(PPh₃)₂. With hydrogen, in the presence of triphenylphosphine, the latter carbonyl complex affords some **Carbonylhydridotris(triphenylphosphine)rhodium-(I)**, and this very actively mediates hydroformylations.[50] Reactions wherein RhCl(PPh₃)₃ is used as a hydroformylation catalyst probably proceed via this route. A more direct means of hydroformylation is to use RhH(CO)(PPh₃)₃. Nevertheless, Wilkinson's catalyst (an unfortunate term here because Wilkinson also pioneered hydroformylations using RhH(CO)-(PPh₃)₃) has been used to effect hydroformylations of some substrates. Eq 36 is one example and illustrates that transient coordination of the acyl group with rhodium apparently leads to predominant formation of a 'branched chain' aldehyde, whereas straight chain aldehydes are usually formed in these reactions.[51] Other hydroformylation catalysts that have been studied include cobalt and iridium based systems.[49]

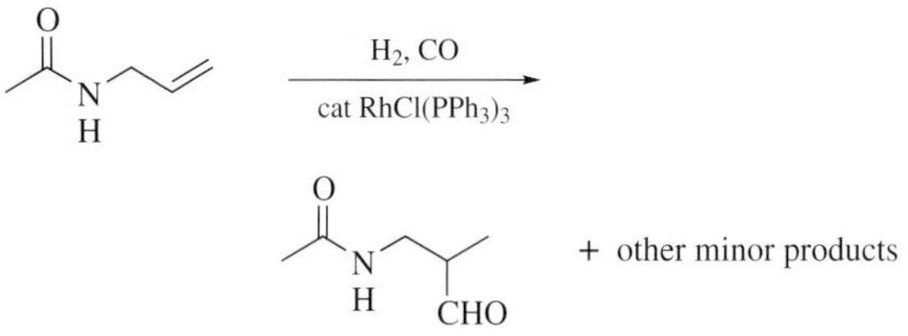

+ other minor products   (36)

Hydroborations.[52] Addition of *Catecholborane* to alkenes is accelerated by Wilkinson's catalyst, and other sources of rhodium-(I) complexes.[53] Unfortunately, the reaction of Wilkinson's catalyst with catecholborane is complex; hence if the conditions for these reactions are not carefully controlled, competing processes result. In the hydroboration of styrene, for instance, the secondary alcohol is formed almost exclusively (after oxidation of the intermediate boronate ester, eq 37); however, the primary alcohol also is formed if the catalyst is partially oxidized and this can be the major product in extreme cases.[54,55] Conversely, hydroboration of the allylic ether (**12**) catalyzed by pure Wilkinson's catalyst gives the expected alcohol (**13**), hydrogenation product (**14**), and aldehyde (**15**), but alcohol (**13**) is the exclusive (>95%) product if the RhCl(PPh$_3$)$_3$ is briefly exposed to air before use.[54] The *syn*-alcohol is generally the favored diastereomer in these and related reactions (eq 38), and the catalyzed reaction is therefore stereocomplementary to uncatalyzed hydroborations of allylic ether derivatives.[56-58]

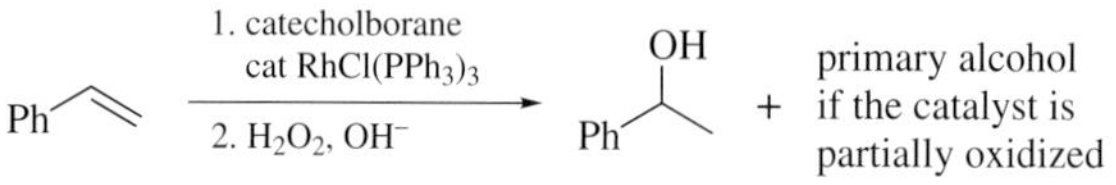

$$(37)$$

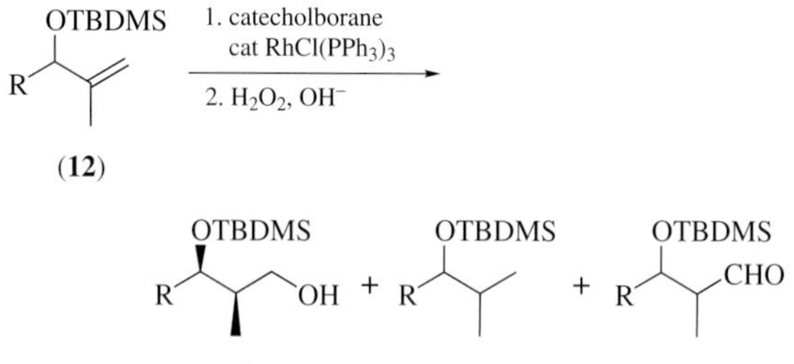

$$(38)$$

syn product favored

Other sources of rhodium(I) are equally viable catalysts for hydroborations, notably Rh(η^3-CH$_2$CMeCH$_2$)(i-Pr$_2$PCH$_2$CH$_2$P-i-Pr$_2$) which gives a much cleaner reaction with catecholborane than Wilkinson's catalyst.[59] Other catalysts for hydroborations are also emerging.[60-62]

Catecholborane hydroborations of carbonyl and related functionalities are also accelerated by RhCl(PPh$_3$)$_3$ (eqs 39–41); however, several related reactions proceed with similar selectivities in the absence of rhodium.[63-65]

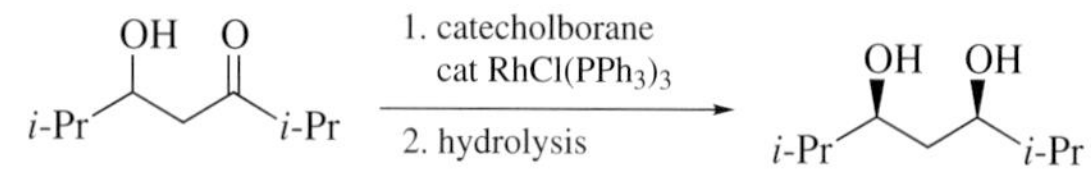

$$(39)$$

syn:anti = 12:1

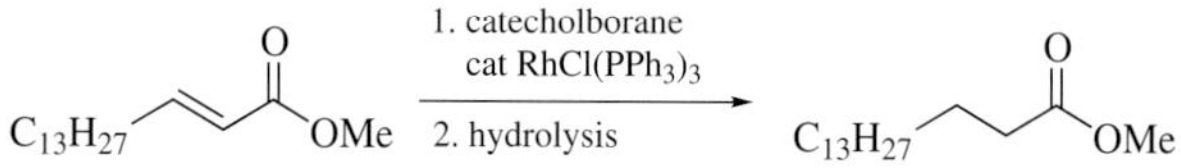

$$(40)$$

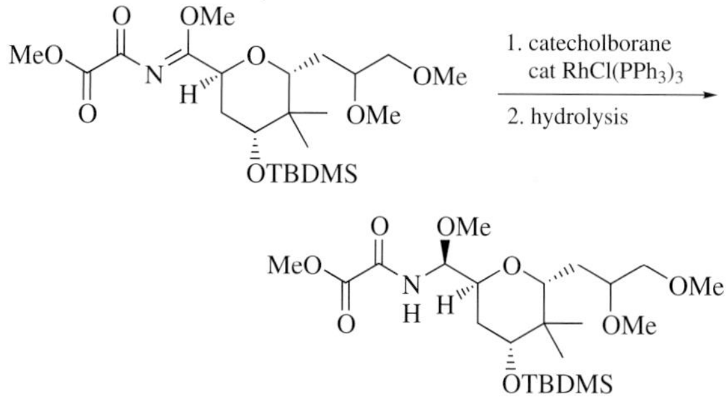

$$(41)$$

de 10:1

Cyclization, Isomerization, and Coupling Reactions. Inter-(eq 42)[66] and intramolecular (eq 43)[67] cyclotrimerizations of alkynes are mediated by Wilkinson's catalyst. This is an extremely efficient route to ring fused systems. Similarly, Diels–Alder-like [4 + 2] cyclization processes are promoted by RhCl(PPh$_3$)$_3$;[68] 'dienophile' components in these reactions need not be electron deficient, and they can be an alkene or alkyne (eqs 44 and 45). Allenes oligomerize in pathways determined by their substituents. For instance, four molecules of allene combine to give a spiro-cyclic system (eq 46), but tetraphenylallene isomerizes to give an indene (eq 47).[69]

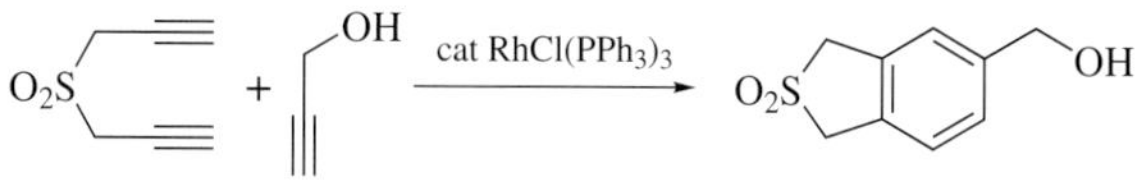

$$(42)$$

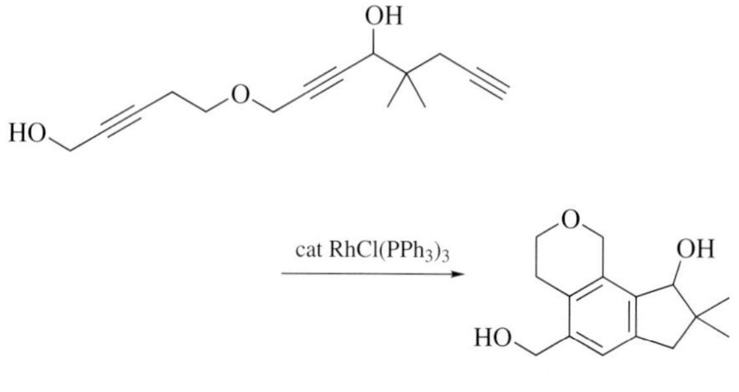

$$(43)$$

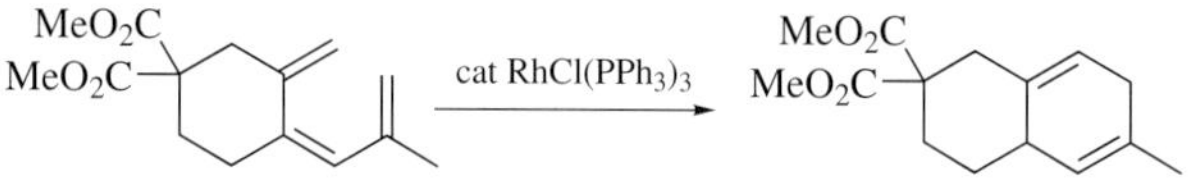

$$(44)$$

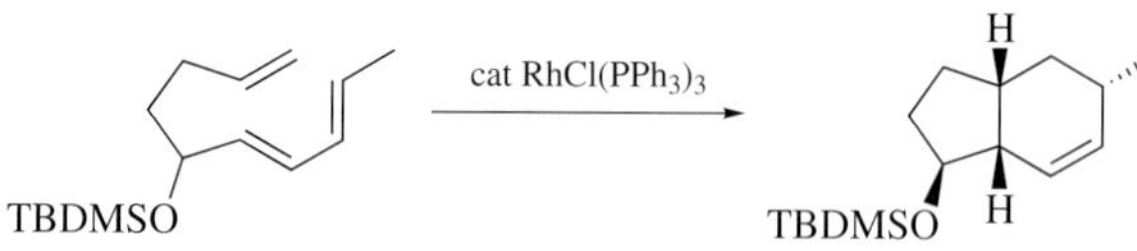

$$(45)$$

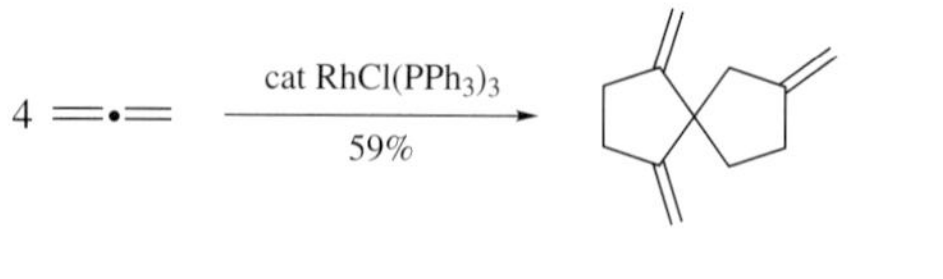

$$(46)$$

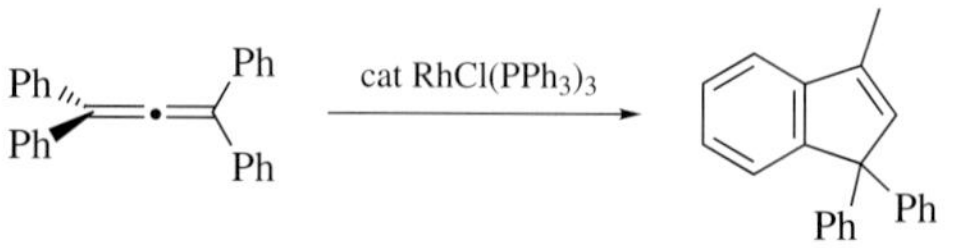

$$(47)$$

Wilkinson's catalyst is also capable of mediating the formation of C–C bonds in reactions which apparently proceed via oxidative addition of an unsaturated organohalide across the metal (eq 48),[70] or via transmetalation from an organometallic (eq 49).[71] These two transformation types are very similar to couplings developed by Heck so, predictably, some palladium complexes also mediate these reactions (see *Tetrakis(triphenylphosphine)palladium(0)* and *Palladium(II) Acetate*).

$$ (48) $$

$$ (49) $$

Intermolecular reactions of dienes, allenes, and methylenecyclopropanes with alkenes are mediated by RhCl(PPh$_3$)$_3$, although mixtures of products are usually formed (eqs 50–51).[72–75]

$$ (50) $$

$$ (51) $$

Wilkinson's catalyst mediates hydrogenation of 1,4-cyclohexadienes without double bond isomerization (see above), but at elevated temperatures in the absence of hydrogen it promotes isomerization to conjugated dienes (eq 52).[76] Isomerization of allylamines to imines followed by hydrolysis has also been performed using RhCl(PPh$_3$)$_3$ (eq 53),[77] although RhH(PPh$_3$)$_4$ and other catalysts are more frequently used for this reaction type.[78]

$$ (52) $$

$$ (53) $$

Oxidations. Cleavage of alkenes to aldehydes and ketones is promoted by Wilkinson's catalyst under pressures of air or oxygen,[79] but these reactions are inferior to ozonolysis because they tend to form a mixture of products. More useful are the oxidations of anthracene derivatives to anthraquinones in the presence of oxygen/*tert-Butyl Hydroperoxide* and catalytic RhCl(PPh$_3$)$_3$ (eq 54).[80,81] Wilkinson's catalyst reacts with oxygen to form an adduct so RhCl(PPh$_3$)$_3$ is clearly quite different from the true catalyst in all the reactions mentioned in this section.

$$ (54) $$

Other Transformations. At high temperatures ($>200\,°C$) aromatic sulfonyl chlorides are desulfonated to the corresponding aryl halides in the presence of Wilkinson's catalyst (eq 55).[82] Benzamides and malonamide also decompose under similar conditions, giving benzonitrile and acetamide, respectively.[83]

$$ (55) $$

Diazonium fluoroborates are reduced to the corresponding unsubstituted aryl compounds by Wilkinson's catalyst in DMF; the solvent is apparently the hydride source in this reaction (eq 56).[84]

$$ (56) $$

Finally, aryl group interchange between triarylphosphines is mediated by Wilkinson's catalyst at 120 °C, but a near statistical mixture of the exchanged materials is formed along with some byproducts.[85]

First Update

Chul-Ho Jun & Young Jun Park
Yonsei University, Seoul, Korea

Hydrogenations. Wilkinson's catalyst does not promote hydrogenation of aromatic compounds under mild reaction conditions. However, in special cases, certain aromatic compounds such as benzophenone can be partially reduced by molecular hydrogen in the presence of Wilkinson catalyst via an indirect activation process involving a germylene (eq 57).[86] Further reduction of the two remaining double bonds is most likely inhibited by steric hindrance of the germylene.

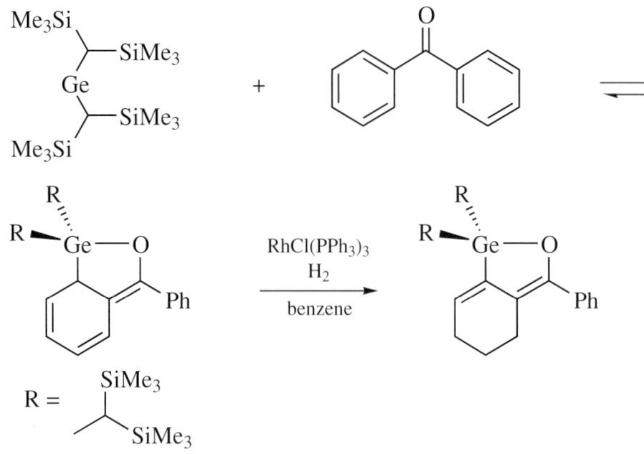

(57)

Examples of the regioselective hydrogenation of dienes by Wilkinson's catalyst to give allylic and homoallylic alcohols have been reported (eq 58).[87]

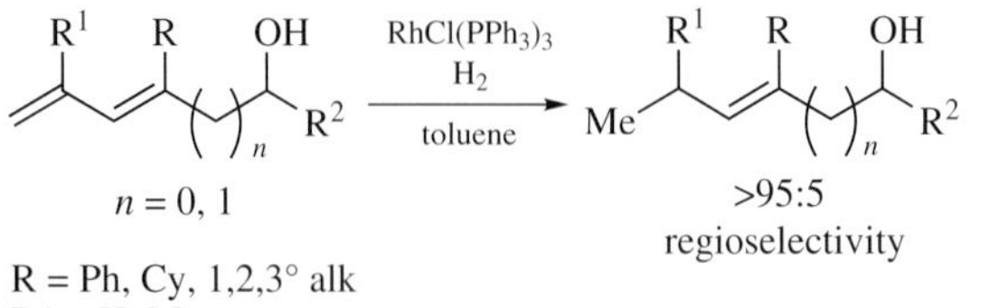

(58)

$n = 0, 1$

>95:5 regioselectivity

R = Ph, Cy, 1,2,3° alk
R1 = H, Me
R2 = 1,2 ° alk

Transfer Hydrogenations. Transition metal-catalyzed transfer hydrogenation is thought to occur via two different intermediates, monohydrides or dihydrides, depending on the transition metal species employed. For Wilkinson's catalyst the monohydridic pathway was shown to be operative with the aid of deuterium labeling experiments (eq 59).[88]

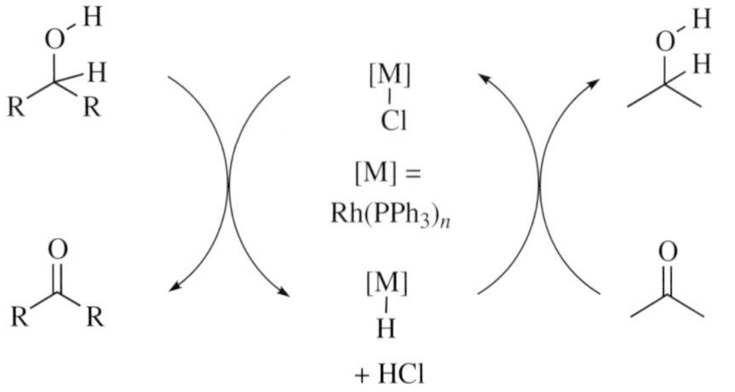

$[M] = Rh(PPh_3)_n$

(59)

+ HCl

In certain cases, additives are observed to have an accelerating effect in the RhCl(PPh$_3$)$_3$-catalyzed hydrogen transfer reaction. By adding Yb(OTf)$_3$, propionophenone is reduced to the alcohol under milder reaction conditions than with the conventional RhCl(PPh$_3$)$_3$-catalyzed hydrogen transfer reaction (eq 60).[89]

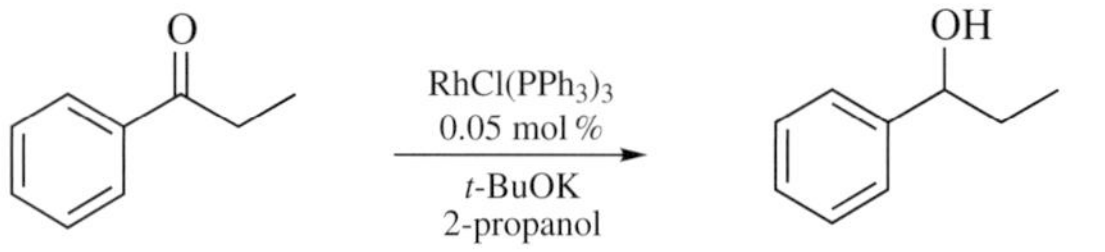

(60)

	Yield (%)	
	With Yb(OTf)$_3$	Without Yb(OTf)$_3$
	85	65

Hydrosilylations. Hydrosilylation of vinylcyclopropane in the presence of Wilkinson catalyst is accompanied by ring cleavage of vinylcyclopropane leading to the formation of terminal silyl-substituted regioisomeric alkenes in 80% yield (eq 61).[90] In the hydrosilylation of vinylcyclopropane analogs, the ring opening of the cyclopropyl group dominates over the simple addition of the Si–H bond to the vinyl group. This process provides an alternative synthesis of silyl-substituted alkenes to the hydrosilylation of dienes.

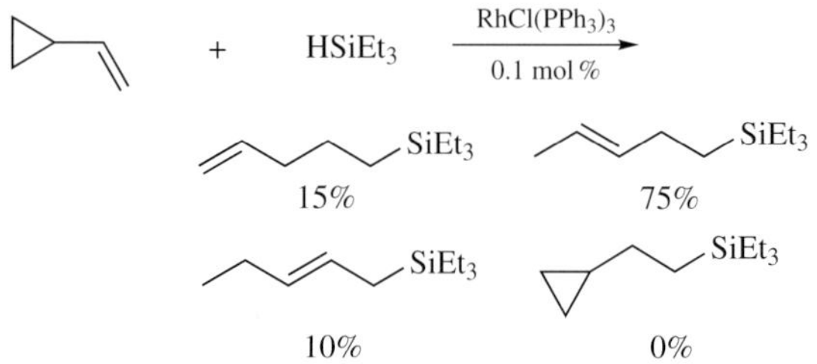

(61)

Hydrophosphorylations. Among the transition metal-catalyzed reactions for constructing carbon-hetero atom bonds, strategies for forming carbon-phosphorous bonds are relatively limited.[91] Most of the successful metal-catalyzed reactions of phosphorous(V) compounds have been conducted in the presence of a Pd catalyst at elevated temperatures.[91] However, with a highly reactive five-membered ring hydrogen phosphonate, Wilkinson's catalyst is capable of the hydrophosphorylation of alkynes to give the corresponding (E)-alkenylphosphonates with excellent regio- and stereoselectivities (eq 62).[92] Microwave-assisted RhCl(PPh$_3$)$_3$-catalyzed hydrophosphorylation of alkynes can also give the corresponding alkenyl phosphine oxide very efficiently under solvent-free conditions.[93]

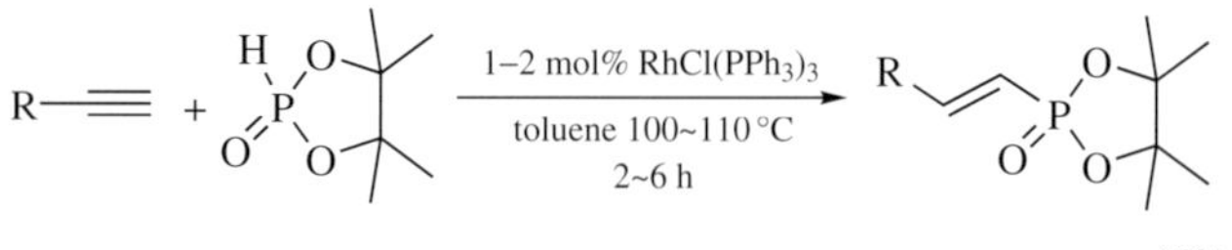

(62)

Hydroacylations. Wilkinson's catalyst is an extremely powerful catalyst for intermolecular hydroacylation when combined with several organococatalysts such as 2-amino-3-picoline, aniline, and benzoic acid (for details, see *2-amino-3-picoline*).[94] Equation 63 illustrates how benzaldehyde undergoes intermolecular hydroacylation very efficiently with terminal olefins by the chelation-assistance of 2-amino-3-picoline by a process involving C–H bond activation.

$$(63)$$

Hydroformylations. Wilkinson's catalyst can tolerate highly oxygenated functionality in hydroformylation reactions as illustrated by the cyclopentene of eq 64. This strategy has been applied to the synthesis of monosaccharide analogs such as a carba-D-fructofuranose.[95]

$$(64)$$

Hydroborations. Although the reaction of Wilkinson's catalyst with catecholborane is often complex, superior selectivity over the uncatalyzed reaction can be observed in RhCl(PPh$_3$)$_3$-catalyzed hydroborations (eq 65).[96]

$$(65)$$

5-endo + 5-exo

Conditions	*endo/exo*	Combined yield (%)
BH$_3$	89:11	50
RhCl(PPh$_3$)$_3$/CatBH	95:5	83

In the case of the hydroboration of perfluoroalkylolefins, the choice of Rh-complex and borane species can influence regioselectivity dramatically. For instance, while the reaction of perfluoroalkylolefins with catecholborane in the presence of a cationic Rh-catalyst produces the secondary alcohols predominantly after the oxidative work-up process, a Wilkinson's complex-catalyzed reaction with pinacolborane affords the primary alcohols (eq 66).[97]

C–H Functionalization by Ortho Alkylation. Among the several complexes that promote orthoalkylation (originally discovered by Murai using a ruthenium catalyst in 1993),[98] Wilkinson's catalyst appears to be the most successful, in terms of its broad scope of substrates.[99] For instance, while allylic protons in most olefins, especially terminal ones, are not tolerated in Murai's ruthenium-catalyzed orthoalkylation, terminal olefins with or without allylic protons can be used successfully with Wilkinson's catalyst in the orthoalkylation of benzyl imines. Remarkably, even dienes and internal olefins are also substrates for this reaction (eq 67).

$$(66)$$

The functional group tolerance of this reaction is illustrated by olefins containing ester, amide, sulfone, and nitrile groups which can be applied to RhCl(PPh$_3$)$_3$-catalyzed orthoalkylation with remarkable efficiency (eq 68).[100] These functionalized olefins are much more reactive than "nonfunctionalized" olefins. Nevertheless, when rhodium cationic species are employed as a catalyst, much higher yields of orthoalkylated products can be obtained under mild reaction conditions.

$$(67)$$

R$_1$ = Me, Et, *n*-Pent
R$_2$ = *t*-C$_4$H$_9$, C$_6$F$_5$, Cy, *n*-C$_4$H$_9$, *n*-C$_6$H$_{13}$, *n*-C$_{10}$H$_{25}$, (CH$_3$)$_3$Si
R$_3$ = CF$_3$, H, CH$_3$O

$$(68)$$

Functionalized olefins

This chelation-assisted cyclometallation using Wilkinson's catalyst can be extended to β-alkylation through aliphatic sp^2 C–H bond activation (eq 69).[101] When an enone is allowed to react with excess olefin in the presence of RhCl(PPh$_3$)$_3$, benzoic acid, and secondary amine at 130 °C for 12 h, β-alkylated products can be obtained in good yields.

$$R = \text{alkyl, } Si(CH_3)_3 \tag{69}$$

The Wilkinson's catalyst/benzyl imine system for orthoalkylation can be applied to RhCl(PPh$_3$)$_3$-catalyzed orthoalkenylation of aromatic benzyl imine with both terminal and internal alkynes.[102] Equation 70 illustrates reaction of the benzyl imine of acetophenone with several terminal alkynes giving mono- and doubly-alkenylated products, depending on the substituents in the alkyne and aromatic imine.

$$\tag{70}$$

$R_1 = H, CF_3, OMe \qquad R_3 = H, Ph$
$R_2 = Me, Et, {}^nPent \qquad R_4 = {}^nBu, {}^nHex, {}^tBu, Ph$

Under more vigorous reaction conditions, two isoquinoline derivatives are formed from the reaction of benzyl imine with diphenylacetylene (eq 71).

$$\tag{71}$$

$R = H, CF_3, OMe$

Carbonyl Methylenations. In an alternative to the classical Wittig reaction,[103] Wilkinson's catalyst mediates the olefination of carbonyl compounds in the presence of a diazo compound and triphenylphosphine.[104] This transformation is quite attractive because several drawbacks of the Wittig reaction, including the use of stoichiometric amounts of phosphonium salts, can be avoided. In the presence of 2-propanol, trimethylsilyldiazomethane, triphenylphosphine, and Wilkinson's catalyst can convert the ketone and aldehyde groups in various organic compounds into the corresponding methylene group (eqs 72 and 73).[104] Various other transition metals have been employed in the catalytic methylenation of carbonyl compounds in organic synthesis.[105-109]

$$\tag{72}$$

$$\tag{73}$$

Cyclization, Isomerization, and Coupling Reactions. The cyclization of diynes is an efficient route to the formation of 1,2-dialkylidenecycloalkanes. When silanes are included in Wilkinson's complex-catalyzed reaction of 1,6-diynes, silylative cyclization occurs (eq 74).[110]

$$\tag{74}$$

The cyclization of 1,6-enynes by RhCl(PPh$_3$)$_3$ can generate functionalized 1,3- or 1,4-diene cyclic compounds. For example, treatment of 1,6-enynes containing a haloalkenyl group with Wilkinson's catalyst in dichloromethane at reflux produces cyclization product which incorporates an intramolecular halogen shift: a wide spectrum of enynes can be applied in this transformation (eq 75).[111]

$$\tag{75}$$

α-Arylpropargyl alcohols can be isomerized to indenones in the presence of Wilkinson's catalyst under mild conditions (eq 76).[112] This isomerization, which includes a 1,4-hydrogen shift, is regioselective for the less hindered position of the aromatic ring.

$$(76)$$

Wilkinson's catalyst can also catalyze the formation of C–O bonds via a reductive coupling reaction of epoxides with aldehydes in the presence of Et_3B as a reductant (eq 77).[113]

$$(77)$$

>95:5
regioselectivity

Treatment of α,β-unsaturated ketones with CF_3I in the presence of Et_2Zn and $RhCl(PPh_3)_3$ gives an α-trifluoromethylated product, thereby providing alternative to previous electrophilic reaction[114] using chalcogenium reagents or photochemical reactions[115] of enamine with CF_3I. Eq. 78 illustrates α-trifluoromethylation of α,β-unsaturated ketones by Wilkinson's catalyst.[116]

$$(78)$$

Wilkinson's catalyst mediates stoichiometric intramolecular C–C bond forming reactions with certain substrates containing acidic C–H bonds via an intramolecular hydride migration yielding a 1,3-diketone (eq 79).[117]

$$(79)$$

The reaction of benzaldehyde and 4-pentynoic acid in the presence of Wilkinson's catalyst and 2-amino-3-picoline exhibits the exclusive formation of (E)-3-benzylidene-3H-furan-2-one, instead of the usual enone hydroacylation product (eq 80).[118,119]

$$(80)$$

While enynes are common products in the reaction of 1-alkynes under Wilkinson's catalyst,[120] hydrative dimerization products of 1-alkyne with H_2O are obtained in the presence of an additional cocatalyst, 2-amino-3-picoline. For instance, when the reaction of terminal alkynes and H_2O is carried out using the catalytic system of $RhCl(PPh_3)_3$, 2-amino-3-picoline, and benzoic acid in THF, a mixture of branched α,β-enone and linear enone can be obtained in a 4:1 ratio (eq 81).[121]

$$(81)$$

Other Transformations. Aldoxime groups can be converted to amide groups in the presence of Wilkinson's catalyst with high selectivity and efficiency (eq 82),[122] with no requirement for additives.

$$(82)$$

The synthesis of formaldehyde dithioacetals may be achieved through a reaction with thiols and dichloromethane in the presence of Wilkinson's catalyst and triethylamine (eq 83).[123] The reaction is simple and takes place under very mild reaction conditions.

$$(83)$$

Wilkinson's catalyst mediates the Reformatsky-type reaction of ethyl bromodifluoroacetate with various carbonyl compounds (eq 84).[124]

$$(84)$$

Related Reagents. Bis(bicyclo[2.2.1]hepta-2,5-diene)rhodium Perchlorate; [1,4-Bis(diphenylphosphino)butane](norboradiene)rhodium Tetrafluoroborate Catecholborane; (1,5-Cyclooctadiene)[1,4-Bis(diphenylphosphino)butane]iridium(I) Tetrafluoroborate; (1,5-Cyclooctadiene)(tricyclohexylphosphine)(pyridine)iridium(I) Hexafluorophosphate Octacarbonyldicobalt; Palladium(II) Chloride; Tetrakis(triphenylphosphine)palladium(0); 2-Amino-3-picoline; Benzylamine.

1. Jardine, F. H., *Prog. Inorg. Chem.* **1981**, *28*, 63.
2. Osborn, J. A.; Jardine, F. H.; Young, J. F.; Wilkinson, G., *J. Chem. Soc. (A)* **1966**, 1711.
3. Brown, T. H.; Green, P. J., *J. Am. Chem. Soc.* **1970**, *92*, 2359.
4. Osborn, J. A.; Wilkinson, G., *Inorg. Synth.* **1990**, *28*, 77.

5. Birch, A. J.; Williamson, D. H., *Org. React.* **1976**, *24*, 1.

6. Sum, P.-E.; Weiler, L., *Can. J. Chem.* **1978**, *56*, 2700.

7. Birch, A. J.; Walker, K. A. M., *J. Chem. Soc. (C)* **1966**, 1894.

8. Brown, J. M., *Angew. Chem., Int. Ed. Engl.* **1987**, *26*, 190.

9. van Bekkum, H.; van Rantwijk, F.; van de Putte, T., *Tetrahedron Lett.* **1969**, 1.

10. Kosugi, H.; Kitaoka, M.; Tagami, K.; Takahashi, A.; Uda, H., *J. Org. Chem.* **1987**, *52*, 1078.

11. Bhagwat, M. M.; Devaprabhakara, D., *Tetrahedron Lett.* **1972**, 1391.

12. Kim, L.; Dewhirst, K. C., *J. Org. Chem.* **1973**, *38*, 2722.

13. Grigg, R.; Mitchell, T. R. B.; Tongpenyai, N., *Synthesis* **1981**, 442.

14. Grigg, R.; Mitchell, T. R. B.; Sutthivaiyakit, S., *Tetrahedron* **1981**, *37*, 4313.

15. Grigg, R.; Mitchell, T. R. B.; Sutthivaivakit, S.; Tongpenyai, N., *J. Chem. Soc., Chem. Commun.* **1981**, 611.

16. Speier, J. L., *Adv. Organomet. Chem.* **1979**, *17*, 407.

17. Ojima, I. In *The Chemistry of Organic Silicon Compounds*; Patai, S.; Rappoport, Z., Eds.; Wiley: New York, 1989; Vol. 2, p 1479.

18. Garrou, P. E., *Chem. Rev.* **1985**, *85*, 171.

19. Tanke, R. S.; Crabtree, R. H., *J. Am. Chem. Soc.* **1990**, *112*, 7984.

20. Brown-Wensley, K. A., *Organometallics* **1987**, *6*, 1590.

21. Ojima, I.; Kumagai, M.; Nagai, Y., *J. Organomet. Chem.* **1976**, *111*, 43.

22. Onopchenko, A.; Sabourin, E. T.; Beach, D. L., *J. Org. Chem.* **1983**, *48*, 5101.

23. Onopchenko, A.; Sabourin, E. T.; Beach, D. L., *J. Org. Chem.* **1984**, *49*, 3389.

24. Millan, A.; Towns, E.; Maitlis, P. M., *J. Chem. Soc., Chem. Commun.* **1981**, 673.

25. Ojima, I.; Kumagai, M.; Nagai, Y., *J. Organomet. Chem.* **1974**, *66*, C14.

26. Dickers, H. M.; Haszeldine, R. N.; Mather, A. P.; Parish, R. V., *J. Organomet. Chem.* **1978**, *161*, 91.

27. Brady, K. A.; Nile, T. A., *J. Organomet. Chem.* **1981**, *206*, 299.

28. Crivello, J. V.; Fan, M., *J. Polym. Sci., Part A: Polym. Chem.* **1992**, *30*, 1.

29. Ojima, I.; Kogure, T., *Organometallics* **1982**, *1*, 1390.

30. Slougui, N.; Rousseau, G., *Synth. Commun.* **1987**, *17*, 1.

31. Revis, A.; Hilty, T. K., *J. Org. Chem.* **1990**, *55*, 2972.

32. Ojima, I.; Kogure, T.; Nihonyanagi, M.; Nagai, Y., *Bull. Chem. Soc. Jpn.* **1972**, *45*, 3506.

33. Ojima, I.; Nihonyanagi, M.; Kogure, T.; Kumagai, M.; Horiuchi, S.; Nakatsugawa, K.; Nagai, Y., *J. Organomet. Chem.* **1975**, *94*, 449.

34. Corriu, R. J. P.; Moreau, J. J. E., *J. Chem. Soc., Chem. Commun.* **1973**, 38.

35. Bonar-Law, R. P.; Davis, A. P.; Dorgan, B. J., *Tetrahedron Lett.* **1990**, *31*, 6721.

36. Myers, A. G.; Widdowson, K. L., *J. Am. Chem. Soc.* **1990**, *112*, 9672.

37. Kikukawa, K.; Umekawa, H.; Wada, F.; Matsuda, T., *Chem. Lett.* **1988**, 881.

38. Zhang, H. X.; Guibé, F.; Balavoine, G., *J. Org. Chem.* **1990**, *55*, 1857.

39. Talley, J. J.; Colley, A. M., *J. Organomet. Chem.* **1981**, *215*, C38.

40. Sakai, K.; Ishiguro, Y.; Funakoshi, K.; Ueno, K.; Suemune, H., *Tetrahedron Lett.* **1984**, *25*, 961.

41. Sakai, K.; Ide, J.; Oda, O.; Nakamura, N., *Tetrahedron Lett.* **1972**, 1287.

42. Ueno, K.; Suemune, H.; Sakai, K., *Chem. Pharm. Bull.* **1984**, *32*, 3768.

43. Larock, R. C.; Oertle, K.; Potter, G. F., *J. Am. Chem. Soc.* **1980**, *102*, 190.

44. Andrews, M. A.; Gould, G. L.; Klaeren, S. A., *J. Org. Chem.* **1989**, *54*, 5257.

45. Walborsky, H. M.; Allen, L. A., *Tetrahedron Lett.* **1970**, 823.

46. Walborsky, H. M.; Allen, L. E., *J. Am. Chem. Soc.* **1971**, *93*, 5465.

47. O'Connor, J. M.; Ma, J., *J. Org. Chem.* **1992**, *57*, 5075.

48. Blum, J.; Lipshes, Z., *J. Org. Chem.* **1969**, *34*, 3076.

49. Pruett, R. L., *Adv. Organomet. Chem.* **1979**, *17*, 1.

50. Jardine, F. H., *Polyhedron* **1982**, *1*, 569.

51. Ojima, I.; Zhang, Z., *J. Organomet. Chem.* **1991**, *417*, 253.

52. Burgess, K.; Ohlmeyer, M. J., *Chem. Rev.* **1991**, *91*, 1179.

53. Männig, D.; Nöth, H., *Angew. Chem., Int. Ed. Engl.* **1985**, *24*, 878.

54. Burgess, K.; vander Donk, W. A.; Westcott, S. A.; Marder, T. B.; Baker, R. T.; Calabrese, J. C., *J. Am. Chem. Soc.* **1992**, *114*, 9350.

55. Evans, D. A.; Fu, G. C.; Anderson, B. A., *J. Am. Chem. Soc.* **1992**, *114*, 6679.

56. Evans, D. A.; Fu, G. C.; Hoveyda, A. H., *J. Am. Chem. Soc.* **1988**, *110*, 6917.

57. Burgess, K.; Cassidy, J.; Ohlmeyer, M. J., *J. Org. Chem.* **1991**, *56*, 1020.

58. Burgess, K.; Ohlmeyer, M. J., *J. Org. Chem.* **1991**, *56*, 1027.

59. Westcott, S. A.; Blom, H. P.; Marder, T. B.; Baker, R. T., *J. Am. Chem. Soc.* **1992**, *114*, 8863.

60. Evans, D. A.; Fu, G. C., *J. Am. Chem. Soc.* **1991**, *113*, 4042.

61. Harrison, K. N.; Marks, T. J., *J. Am. Chem. Soc.* **1992**, *114*, 9220.

62. Burgess, K.; Jaspars, M., *Organometallics* **1993**, *12*, 4197.

63. Evans, D. A.; Hoveyda, A. H., *J. Org. Chem.* **1990**, *55*, 5190.

64. Evans, D. A.; Fu, G. C., *J. Org. Chem.* **1990**, *55*, 5678.

65. Kocieński, P.; Jarowicki, K.; Marczak, S., *Synthesis* **1991**, 1191.

66. Grigg, R.; Scott, R.; Stevenson, P., *Tetrahedron Lett.* **1982**, *23*, 2691.

67. Neeson, S. J.; Stevenson, P. J., *Tetrahedron* **1989**, *45*, 6239.

68. Jolly, R. S.; Luedtke, G.; Sheehan, D.; Livinghouse, T., *J. Am. Chem. Soc.* **1990**, *112*, 4965.

69. Jones, F. N.; Lindsey, R. V., Jr., *J. Org. Chem.* **1968**, *33*, 3838.

70. Grigg, R.; Stevenson, P.; Worakun, T., *J. Chem. Soc., Chem. Commun.* **1984**, 1073.

71. Larock, R. C.; Narayanan, K.; Hershberger, S. S., *J. Org. Chem.* **1983**, *48*, 4377.

72. Salerno, G.; Gigliotti, F.; Chiusoli, G. P., *J. Organomet. Chem.* **1986**, *314*, 231.

73. Salerno, G.; Gallo, C.; Chiusoli, G. P.; Costa, M., *J. Organomet. Chem.* **1986**, *317*, 373.

74. Chiusoli, G. P.; Costa, M.; Schianchi, P.; Salerno, G., *J. Organomet. Chem.* **1986**, *315*, C45.

75. Chiusoli, G. P.; Costa, M.; Pivetti, F., *J. Organomet. Chem.* **1989**, *373*, 377.

76. Harland, P. A.; Hodge, P., *Synthesis* **1983**, 419.

77. Laguzza, B. C.; Ganem, B., *Tetrahedron Lett.* **1981**, *22*, 1483.

78. Stille, J. K.; Becker, Y., *J. Org. Chem.* **1980**, *45*, 2139.

79. Bönnemann, H.; Nunez, W.; Rohe, D. M. M., *Helv. Chim. Acta* **1983**, *66*, 177.

80. Müller, P.; Bobillier, C., *Tetrahedron Lett.* **1981**, *22*, 5157.

81. Müller, P.; Bobillier, C., *Tetrahedron Lett.* **1983**, *24*, 5499.

82. Blum, J.; Scharf, G., *J. Org. Chem.* **1970**, *35*, 1895.

83. Blum, J.; Fisher, A.; Greener, E., *Tetrahedron* **1973**, *29*, 1073.

84. Marx, G. S., *J. Org. Chem.* **1971**, *36*, 1725.

85. Abatjoglou, A. G.; Bryant, D. R., *Organometallics* **1984**, *3*, 932.

86. Sweeder, R. D.; Cygan, Z. T.; Banaszak Holl, M. M.; Kampf, J. W., *Organometallics* **2003**, *22*, 4613.

87. Miller, K. M.; Luanphaisarnnont, T.; Molinaro, C.; Jamison, T. F., *J. Am. Chem. Soc.* **2004**, *126*, 4130.

88. Bäckvall, J.-E., *J. Organomet. Chem.* **2002**, *652*, 105.

89. Matsunaga, H.; Yoshioka, N.; Kunieda, T., *Tetrahedron Lett.* **2001**, *42*, 8857.

90. Bessmertnykh, A. G.; Blinov, K. A.; Grishin, Y. K.; Donskaya, N. A.; Beletskaya, I. P., *Tetrahedron Lett.* **1995**, *36*, 7901.

91. Han, L.-B.; Tanaka, M., *Chem. Commun.* **1999**, 395.

92. Zhao, C.-Q.; Han, L.-B.; Goto, M.; Tanaka, M., *Angew. Chem. Int. Ed.* **2001**, *40*, 1929.

93. Stone, J. J.; Stockland, Jr., R. A.; Reyes, Jr., J. M.; Kovach, J.; Goodman, C. C.; Tillman, E. S., *J. Mol. Catal. A: Chem.* **2005**, *226*, 11.

94. (a) Jun, C.-H.; Lee, D.-Y.; Lee, H.; Hong, J.-B., *Angew. Chem. Int. Ed.* **2000**, *39*, 3070. (b) Jun, C.-H.; Moon, C. W.; Lee, D.-Y., *Chem. Eur. J.* **2002**, *8*, 2423. (c) This volume, p 14.

95. Seepersaud, M.; Kettunen, M.; Abu-Surrah, A. S.; Repo, T.; Voelter, W.; Al-Abed, Y., *Tetrahedron Lett.* **2002**, *43*, 1793.

96. Bunch, L.; Liljefors, T.; Greenwood, J. R.; Frydenvang, K.; Bräuner-Osborne, H.; Krogsgaard-Larsen, P.; Madsen, U., *J. Org. Chem.* **2003**, *68*, 1489.

97. Ramachandran, P. V.; Jennings, M. P.; Brown, H. C., *Org. Lett.* **1999**, *1*, 1399.

98. Murai, S.; Kakiuchi, F.; Sekine, S.; Tanaka, Y.; Kamatani, A.; Sonoda, M.; Chatani, N., *Nature* **1993**, *366*, 529.

99. (a) Jun, C.-H.; Hong, J.-B.; Kim, Y.-H.; Chung, K.-W., *Angew. Chem. Int. Ed.* **2000**, *39*, 3440. (b) Jun, C.-H.; Moon, C. W.; Hong, J.-B.; Lim, S.-G.; Chung, K.-Y.; Kim, Y.-H., *Chem. Eur. J.* **2002**, *8*, 485.

100. Lim, S.-G.; Ahn, J.-A.; Jun, C.-H., *Org. Lett.* **2004**, *6*, 4687.

101. Jun, C.-H.; Moon, C. W.; Kim, Y.-M.; Lee, H.; Lee, J. H., *Tetrahedron Lett.* **2002**, *43*, 4233.

102. Lim, S.-G.; Lee, J. H.; Moon, C. W.; Hong, J.-B.; Jun, C.-H., *Org. Lett.* **2003**, *5*, 2759.

103. (a) Wittig, G.; Geissler, G., *Liebigs Ann. Chem.* **1953**, *580*, 44. (b) Wittig, G.; Schöllkopf, U., *Chem. Ber.* **1954**, *87*, 1318.

104. (a) Lebel, H.; Paquet, V.; Proulx, C., *Angew. Chem. Int. Ed.* **2001**, *40*, 2887. (b) Lebel, H.; Paquet, V., *J. Am. Chem. Soc.* **2004**, *126*, 320. (c) Lebel, H.; Guay, D.; Paquet, V.; Huard, K., *Org. Lett.* **2004**, *6*, 3047.

105. Kelly, S. E. Alkene Synthesis. In *Comprehensive Organic Synthesis*; Trost, B. M.; Fleming, I., Eds.; Pergamon: Oxford, 1991; Vol. 1, p 729.

106. Lu, X. Y.; Fang, H.; Ni, Z. J., *J. Organomet. Chem.* **1989**, *373*, 77.

107. (a) Herrmann, W. A.; Wang, M., *Angew. Chem., Int. Ed. Engl.* **1991**, *30*, 1641. (b) Herrmann, W. A.; Roesky, P. W.; Wang, M.; Scherer, W., *Organometallics* **1994**, *13*, 4531. (c) Carreira, E. M.; Ledford, B. E., *Tetrahedron Lett.* **1997**, *38*, 8125. (d) Herrmann, W. A., *Appl. Homogeneous Catal. Organomet. Compd* 2nd ed. **2002**, *3*, 1078. (e) Santos, A. M.; Romao, C. C.; Kuhn, F. E., *J. Am. Chem. Soc.* **2003**, *125*, 2414. (f) Zhang, X. Y.; Chen, P., *Chem. Eur. J.* **2003**, *9*, 1852.

108. (a) Mirafzal, G. A.; Cheng, G. L.; Woo, L. K., *J. Am. Chem. Soc.* **2002**, *124*, 176. (b) Cheng, G. L.; Mirafzal, G. A.; Woo, L. K., *Organometallics* **2003**, *22*, 1468. (c) Chen, Y.; Huang, L.; Ranade, M. A.; Zhang, X. P., *J. Org. Chem.* **2003**, *68*, 3714. (d) Chen, Y.; Huang, L.; Zhang, X. P., *J. Org. Chem.* **2003**, *68*, 5925. (e) Chen, Y.; Huang, L.; Zhang, X. P., *Org. Lett.* **2003**, *5*, 2493. (f) Aggarwal, V. K.; Fulton, J. R.; Sheldon, C. G.; de Vicente, J., *J. Am. Chem. Soc.* **2003**, *125*, 6034.

109. (a) Fujimura, O.; Honma, T., *Tetrahedron Lett.* **1998**, *39*, 625. (b) Graban, E.; Lemke, F. R., *Organometallics* **2002**, *21*, 3823.

110. Muraoka, T.; Matsuda, I.; Itoh, K., *Tetrahedron Lett.* **1998**, *39*, 7325.

111. Tong, X.; Li, D.; Zhang, Z.; Zhang, X., *J. Am. Chem. Soc.* **2004**, *126*, 7601.

112. Shintani, R.; Okamoto, K.; Hayashi, T., *J. Am. Chem. Soc.* **2005**, *127*, 2872.

113. Molinaro, C.; Jamison, T. F., *Angew. Chem. Int. Ed.* **2005**, *44*, 129.

114. (a) Ma, J.-A.; Cahard, D., *J. Org. Chem.* **2003**, *68*, 8726. (b) Umemoto, T.; Adachi, K., *J. Org. Chem.* **1994**, *59*, 5692.

115. (a) Crusiani, G.; Margaretha, P., *J. Fluorine Chem.* **1987**, *37*, 95. (b) Kitazume, T.; Ishikawa, N., *J. Am. Chem. Soc.* **1985**, *107*, 5186.

116. Sato, K.; Omote, M.; Ando, A.; Kumadaki, I., *Org. Lett.* **2004**, *6*, 4359.

117. Biju, P. J.; Rao, G. S. R. S., *Chem. Commun.* **1999**, 2225.

118. Jun, C.-H.; Lee, H.; Hong, J. B.; Kwon, B.-I., *Angew. Chem. Int. Ed.* **2002**, *41*, 2146.

119. Lim, S.-G.; Kwon, B.-I.; Choi, M.-G.; Jun, C.-H., *Synlett* **2005**, 1113.

120. (a) Oishita, J.; Furumori, K.; Matsuguchi, A.; Ishikawa, M., *J. Org. Chem.* **1990**, *55*, 3277. (b) Schmit, H. J.; Singer, H., *J. Organomet. Chem.* **1978**, *153*, 165. (c) Boese, W. T.; Goldman, A. S., *Organometallics* **1991**, *10*, 782. (d) Schaefer, M.; Mahr, N.; Wolf, J.; Werner, H., *Angew. Chem., Int. Ed. Engl.* **1993**, *32*, 1315. (e) Tanaka, K.; Shirasaka, K., *Org. Lett.* **2003**, *5*, 4697.

121. Park, Y. J.; Kwon, B.-I.; Ahn, J.-A.; Lee, H.; Jun, C.-H., *J. Am. Chem. Soc.* **2004**, *126*, 13892.

122. Park, S.; Choi, Y.-a.; Han, H.; Yangm, S. H.; Chang, S., *Chem. Commun.* **2003**, 1936.

123. Tanaka, K.; Ajiki, K., *Org. Lett.* **2005**, *7*, 1537.

124. Sato, K.; Tarui, A.; Kita, T.; Ishida, Y.; Tamura, H.; Omote, M.; Ando, A.; Kumadaki, I., *Tetrahedron Lett.* **2004**, *45*, 5735.

D

Essential Reagents for Organic Synthesis, Edited by Philip L. Fuchs, André B. Charette, Tomislav Rovis, and Jeffrey W. Bode.
©2016 John Wiley & Sons, Ltd. Published 2016 by John Wiley & Sons, Ltd.

(Diacetoxyiodo)benzene[1–3]

[3240-34-4] C$_{10}$H$_{11}$IO$_4$ (MW 322.10)

InChI = 1/C10H11IO4/c1-8(12)14-11(15-9(2)13)10-6-4-3-5-7-
10/h3-7H,1-2H3

InChIKey = ZBIKORITPGTTGI-UHFFFAOYAA

(transannular carbocyclization,[6] *vic*-diazide formation,[7] α-hydroxy dimethyl acetal formation,[8,10,11,13] oxetane formation,[9] chromone, flavone, chalcone oxidation,[11,12] arene–Cr(CO)$_3$ functionalization,[14] phenolic oxidation[16] and coupling,[17,18] lactol fragmentation,[19] iodonium ylides and intramolecular cyclopropanation,[20] oxidation of amines[24–28] and indoles,[30,31] hydrazine derivatives (diimide[32] and azodicarbonyls[33]) and radical type intramolecular oxide formation[44–46])

Alternate Names: phenyliodine(III) diacetate; DIB; iodobenzene diacetate; IBD.

Physical Data: mp 163–165 °C.

Solubility: sol AcOH, MeCN, CH$_2$Cl$_2$; in KOH or NaHCO$_3$/MeOH it is equivalent to PhI(OH)$_2$.

Form Supplied in: commercially available as a white solid.

Preparative Method: by reaction of iodobenzene with **Peracetic Acid**.[4,5]

Purification: recrystallization from 5 M acetic acid.[4]

Handling, Storage, and Precautions: a stable compound which can be stored indefinitely.

Original Commentary

Robert M. Moriarty, Calvin J. Chany II, & Jerome W. Kosmeder II
University of Illinois at Chicago, Chicago, IL, USA

Reactions with Alkenes. Reactions of simple alkenes with PhI(OAc)$_2$ are not synthetically useful because of formation of multiple products.

Transannular carbocyclization in the reaction of *cis,cis*-1,5-cyclooctadiene yields a mixture of three diastereomers of 2,6-diacetoxy-*cis*-bicyclo[3.3.0]octane, a useful precursor of *cis*-bicyclo[3.3.0]octane-2,6-dione (eq 1).[6]

$$\text{(1)}$$

PhI(OAc)$_2$/NaN$_3$/AcOH yields vicinal diazides (eq 2).[7]

$$\text{(2)}$$

Oxidation of Ketones to α-Hydroxyl Dimethyl Acetals. Ketones are converted to the α-hydroxy dimethyl acetal upon reaction with PhI(OAc)$_2$ in methanolic potassium hydroxide (eqs 3–5).[8]

$$\text{ArCOMe} \xrightarrow[\text{MeOH, KOH}]{\text{PhI(OAc)}_2} \text{ArC(OMe)}_2\text{CH}_2\text{OH} \quad (3)$$

$$\text{(4)}$$

$$\text{(5)}$$

Several potentially oxidizable groups are unaffected in this reaction (eq 6).[13]

$$\text{R}^1\text{COCH}_2\text{R}^2 \xrightarrow[\text{MeOH, KOH}]{\text{PhI(OAc)}_2} \text{R}^1\underset{\text{OMe}}{\overset{\text{OMe}}{-}}\text{CH(OH)R}^2 \quad (6)$$

In the case of a 17α-hydroxy steroid the hydroxy group acts as an intramolecular nucleophile to yield the 17-spirooxetan-20-one. It is noteworthy that the 3β-hydroxy-Δ^5-system is unaffected (eq 7).[9]

R = H, NHAc, Me

$$\text{(7)}$$

cis-3-Hydroxyflavonone is obtained via acid-catalyzed hydrolysis of *cis*-3-hydroxyflavone dimethyl acetal, which is formed upon treatment of flavanone with PhI(OAc)$_2$ (eq 8).[10,11]

R = H, Ac

$$\text{(8)}$$

α,β-Unsaturated ketones, such as chromone, flavone, chalcone, and flavanone, yield α-hydroxy-β-methoxy dimethyl acetal products (eqs 9–11).[12a]

Intramolecular participation by the *ortho* hydroxy group occurs in the reaction of substituted *o*-hydroxyacetophenones, yielding the corresponding coumaran-3-ones (eq 12).[12b]

Formation of the α-hydroxy dimethyl acetal occurs without reaction of the $Cr(CO)_3$ complex of η^6-benzo-cycloalkanones (eqs 13–15).[14]

Carbon–Carbon Bond Cleavage with PhI(OAc)$_2$/TMSN$_3$. PhI(OAc)$_2$/*Azidotrimethylsilane* reacts with unsaturated compounds even at $-53\,^\circ\text{C}$ to yield keto nitriles (eq 16).[15]

Oxidation of Phenols. Phenols are oxidized using PhI(OAc)$_2$ with nucleophilic attack by solvent (eqs 17 and 18),[16] or with intramolecular nucleophilic addition amounting to an overall oxidative coupling as with the bisnaphthol (eq 19)[17] and also with the conversion of reticuline to salutaridine (eq 20).[18]

Fragmentation of Lactols to Unsaturated Medium-ring Lactones. Ring cleavage to form a medium-sized ring lactone with a transannular double bond has been observed (eq 21).[19]

Reactions with β-Dicarbonyl Systems; Formation of Iodonium Ylides; Intramolecular Cyclopropanation. β-Dicarbonyl compounds upon reaction with PhI(OAc)$_2$ and KOH/MeOH at $0\,^\circ\text{C}$ yield isolable iodonium ylides (eq 22).[20] This is a general reaction which requires two stabilizing groups flanking the carbon of the C=I group, such as NO$_2$ and SO$_2$Ph.[21] Decomposition of unsaturated analogs in the presence of **Copper(I) Chloride** proceeds with intramolecular cyclopropanation (Table 1).[20]

An asymmetric synthesis of a vitamin D ring A synthon employed this intramolecular cyclopropanation reaction (eq 23).[22]

Oxidation of Amines. Aromatic amines are oxidized with PhI(OAc)$_2$ to azo compounds in variable yield. PhI(OAc)$_2$ in

Table 1 Intramolecular cyclopropanation of iodonium ylides

Reactants	Iodonium ylide	Product	Yield (%)
			76
			81
			85
			82

benzene oxidizes aniline in excellent yield (eq 24);[23] however, substituted anilines give substantially lower yields.

$$(24)$$

Intramolecular azo group formation is a useful reaction for the formation of dibenzo[c,f]diazepine (eq 25).[24,25] Other *ortho* groups may react intramolecularly to yield the benzotriazole (eq 26), benzofuroxan (eq 27), or anthranil (eq 28) derivatives.[24-28]

$$(25)$$

$$(26)$$

$$(27)$$

$$(28)$$

A number of examples of oxidative cyclization of 2-(2′-pyridyl-amino)imidazole[1,2-*a*]pyridines to dipyrido[1,2-*a*:2′,1′-*f*]-1,3,4,6-tetraazapentalenes with $PhI(OAc)_2/CF_3CH_2OH$ have been reported (eq 29).[29]

$$(29)$$

$$R =$$

In the case of the oxidation of indole derivatives, nucleophilic attack by solvent may occur (eq 30).[30] Reserpine undergoes an analogous alkoxylation.[30] In the absence of a nucleophilic solvent, intramolecular cyclization occurs, an example of which is illustrated in the total synthesis of sporidesmin A (eq 31).[31]

$$(30)$$

$$(31)$$

Hydrazine is oxidized by $PhI(OAc)_2$ to diimide, which may be used to reduce alkenes and alkynes under mild conditions (Table 2).[32]

Table 2 Diimide reduction of various compounds

Compound	Product	Yield (%)
$PhSCH=CH_2$	$PhSCH_2Me$	85
cis-$EtO_2CCH=CHCO_2Et$	$EtO_2CCH_2CH_2CO_2Et$	94
$EtO_2C-N=N-CO_2Et$	$EtO_2CNH-NHCO_2Et$	90
Maleic anhydride	$(MeCO)_2O$	83
$PhC\equiv CPh$	*cis*-$PhCH=CHPh$	80
$PhCH=CHCO_2Et$	$PhCH_2CH_2CO_2Et$	96
$CH_2=CHCN$	$MeCH_2CN$	97

The hydrazodicarbonyl group is smoothly oxidized by $PhI(OAc)_2$ to the azodicarbonyl group (eqs 32 and 33).[33]

$$(32)$$

$$(33)$$

An intramolecular application of this reaction was used in a tandem sequence with $PhI(OAc)_2$ oxidation and a Diels–Alder reaction in the synthesis of nonpeptide β-turn mimetics (eq 34).[34,35]

$$(34)$$

Oxidation of 5-Substituted Pyrazol-3(2H)-ones; Formation of Alkynyl Esters. Oxidation of various 5-substituted pyrazol-3(2H)-ones proceeded with fragmentative loss of molecular nitrogen to yield methyl-2-alkynoates (eq 35).[36] An analogous fragmentation process with pyrazol-3(2H)-ones occurs with **Thallium(III) Nitrate Trihydrate**[37,38] and **Lead(IV) Acetate**.[39]

$$(35)$$

R = Me, 60%   R = Ph, 59%
R = Et, 63%   R = p-ClC$_6$H$_4$, 61%
R = CO$_2$Me, 59%   R = p-MeC$_6$H$_4$, 59%
R = CH$_2$CO$_2$Me, 62% R = p-MeOC$_6$H$_4$, 64%

Oxidation of Hydrazones, Alkylhydrazones, N-Amino Heterocycles, N-Aminophthalimidates, and Aldazines. The oxidation of hydrazones to diazo compounds is not a generally useful reaction but it was uniquely effective in the oxidation of a triazole derivative (eq 36).[40]

$$(36)$$

Oxidation of arylhydrazones proceeds with intramolecular cyclizations (eqs 37 and 38)[41] and aziridines may be formed via nitrene additions (eq 39).[42]

$$(37)$$

$$(38)$$

$$(39)$$

A linear tetrazane is formed in the oxidation of N-aminophthalimide (eq 40).[43]

$$(40)$$

(Diacetoxyiodo)benzene/**Iodine** is reported to be a more efficient and convenient reagent for the generation of alkoxyl radicals than Pb^{IV}, Hg^{II}, or Ag^{I}, and this system is useful for intramolecular oxide formation (eqs 41 and 42).[44]

$$(41)$$

R = H, CH$_2$CO$_2$Et

$$(42)$$

Fragmentation processes of carbohydrate anomeric alkoxyl radicals[45] and steroidal lactols[46] using $PhI(OAc)_2$/I$_2$ have been reported.

First Update

Justin Du Bois
Stanford University, Stanford, CA, USA

Preparation. Methods for the synthesis of $PhI(OAc)_2$ and aryl-substituted derivatives involve oxidation of iodoarenes with CH_3CO_3H, $NaBO_3$, $NaIO_4$, and CrO_3.[47–49] Peracid oxidation is typically employed to prepare the title reagent as well as solid-supported forms;[50] reactions with $NaBO_3$ and CrO_3, however, appear to be more versatile and have been used to synthesize both arene and heteroarene iodine(III) reagents. Alternatively, ligand exchange between $PhI(OAc)_2$ and a variety of carboxylic acids is quite facile and proceeds in high yield (eq 43).[51] Derivatives of stronger acids (i.e., p-TsOH) are also accessible starting from $PhI(OAc)_2$.[52,53]

$$PhI(OAc)_2 + 2 RCO_2H \xrightarrow[\substack{50-55\,°C \\ \text{reduced pressure} \\ >90\%}]{C_6H_5Cl} PhI(O_2CR)_2 \quad (43)$$

$$R = C_6H_5,\, p\text{-}NO_2C_6H_4,$$
$$^tBu, Bn, CCl_3$$

Reactions with Alkenes. The combination of $PhI(OAc)_2$ with nucleophilic reagents including KSCN, Me_3SiSCN, $(PhSe)_2$, and $Et_4N^+Br^-$ provides for the *trans*-selective functionalization of alkene derivatives (eq 44).[54,55] Good levels of regiocontrol are often observed in such reactions employing unsymmetrical olefins.[56,57] Treatment of $PhI(OAc)_2$ with halogen salts is known to generate $(AcO)_2X^-$, thought to be the active oxidant in reactions with glycals and other unsaturated materials (eq 45).[58] Alternatively, activation of $PhI(OAc)_2$ with catalytic $BF_3 \cdot OEt_2$ enables the conversion of protected glycals to *trans*-1,2-bis(acetoxy)-glycosides.[59]

$$83:17 \quad (44)$$

$$X = Br \text{ or } I \quad (45)$$

Ligand exchange of $PhI(OAc)_2$ with $Mg(ClO_4)_2$ and subsequent introduction of terminal or cyclic alkenes has been reported to give vicinal-bis(perchlorato)alkanes.[60] Reaction with cyclohexene selectively affords the *cis*-product.

Alkene derivatives such as alkenylboronic acids and alkenylzirconanes react with $PhI(OAc)_2$ to furnish alkenyliodonium salts (eq 46).[61] These transformations proceed with retention of olefin configuration.[62] Similarly, alkenylboron species add to $PhI(OAc)_2$ in the presence of NaI to give vinyl acetate products (eq 47). In these examples, (E)-alkenylboronates give stereochemically pure (Z)-configured enol acetates.[63]

$$MeO \diagdown\diagup IPh^+ BF_4^- \quad (46)$$

$$NC \diagdown\diagup OAc \quad (47)$$

stereospecific

Rearrangement and Fragmentation Reactions. Hoffmann-type rearrangements of 1° amides were described originally using $PhI(OAc)_2$ in methanolic KOH solution.[64] Under such conditions, benzo-fused azolones are conveniently prepared (eq 48).[65] The need for strong base, however, does not appear essential for conducting such oxidations, as highlighted in the reaction of *N*-Boc asparagine (eq 49).[66] The mildness of these conditions for effecting the Hoffmann rearrangement and the inexpensive cost of $PhI(OAc)_2$ facilitate the large scale preparation of important amine derivatives. Similar transformation of *N*-substituted amidines with $PhI(OAc)_2$ leads to urea products via the corresponding carbodiimide intermediate.[67]

$$(48)$$

$$(49)$$

Rearrangement reactions of three- and four-membered cycloalkanols have been demonstrated in a number of notable contexts.[48] In the former case, fragmentation occurs with $PhI(OAc)_2$ in MeOH to yield the unsaturated ester product (eq 50).[68] Other combinations of hypervalent iodine reagents with Brønsted acids have also proven effective for this transformation.

$$(50)$$

Synthesis of substituted furanones is made possible starting from squaric ester derivatives (eq 51).[69] Ring-expansion occurs through a putative acylium species, which may be trapped with either AcOH or MeOH depending on the choice of reaction solvent.

$$(51)$$

Alcohol Oxidation. Catalytic 2,2,6,6-tetramethyl-1-piperidinyloxyl (TEMPO), employed in combination with PhI(OAc)$_2$, will effect the oxidation of 1° and 2° alcohols to aldehydes and ketones, respectively (eq 52).[70,71] Reactions are generally high yielding and complete in a few hours. A related protocol uses polymer-supported PhI(OAc)$_2$ with KBr in H$_2$O to generate carboxylic acids from 1° alcohols.[50,72] Ketones may also be prepared in excellent yield with this latter method.

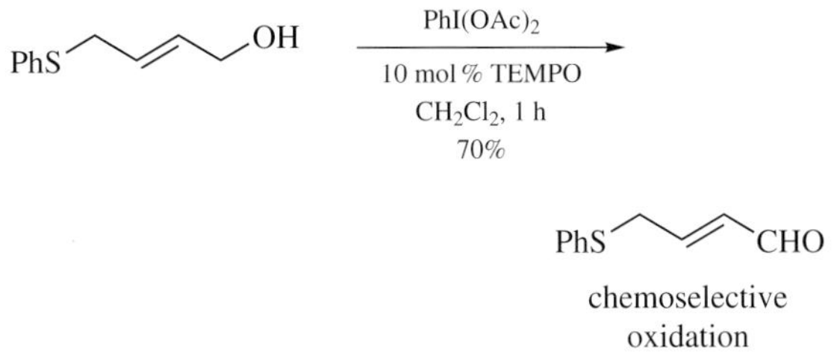

$$\text{(52)}$$

chemoselective
oxidation

Alcohol oxidation has been described using (salen)CrCl (10 mol %) and PhI(OAc)$_2$.[73] Over-oxidation of 1° alcohols to carboxylic acids is not observed under these conditions. Chemoselective synthesis of α,β-unsaturated enones from allylic alcohols is also possible. Optically active (salen)MnPF$_6$ catalysts perform with PhI(OAc)$_2$ for the kinetic resolution of a small collection of chiral 2° alcohols (eq 53).[74]

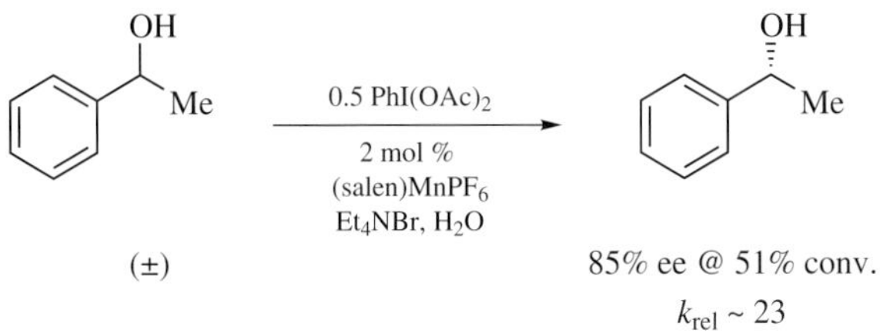

$$\text{(53)}$$

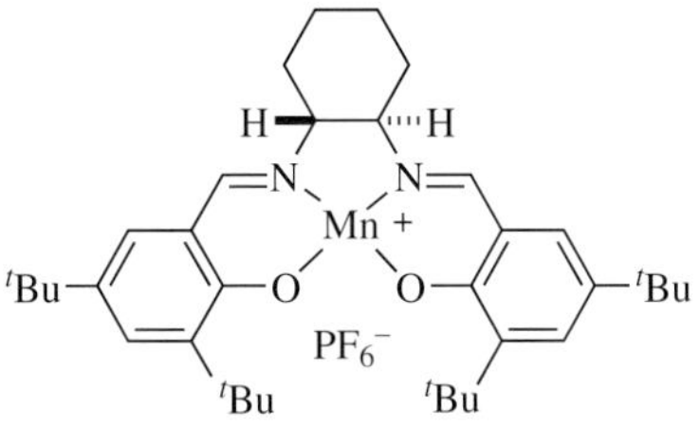

85% ee @ 51% conv.
$k_{rel} \sim 23$

(salen)MnPF$_6$

In the absence of TEMPO or a metal catalyst, benzylic alcohols can be oxidized with alumina-supported PhI(OAc)$_2$ using microwave irradiation.[75] The reaction is conducted without solvent and is completed in 1–3 min.

Decarboxylation and Related Radical Processes. Oxidation of alcohols and sulfonamides with PhI(OAc)$_2$ and I$_2$ under irradiation from a Hg- or W-lamp results in the formation of oxygen- and nitrogen-centered radicals, respectively.[76] Chroman derivatives may be synthesized in this manner from simple 3-arylpropan-1-ol starting materials (eq 54).[77] N-Alkylsaccharin products have been assembled in a similar fashion.[76] The PhI(OAc)$_2$/I$_2$ conditions also make available benzo-fused lactones from *ortho*-substituted benzoic acids.

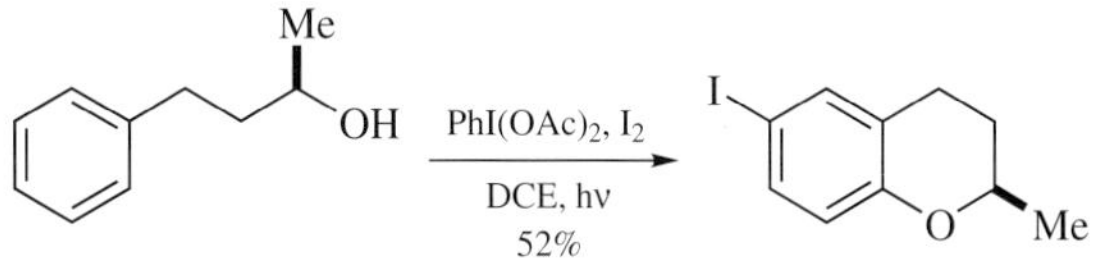

$$\text{(54)}$$

Decarboxylation of α-heteroatom-functionalized carboxylic acids occurs smoothly using PhI(OAc)$_2$ and I$_2$ without the requirement for photolysis.[78] When proline derivatives are employed for this reaction, the intermediate N,O-acetal may be treated with nucleophilic agents to give 2-substituted pyrrolidine products (eq 55).

Treatment of both aliphatic and electron-withdrawn aromatic carboxylic acids with PhI(OAc)$_2$, Br$_2$, and light affords alkyl- and aryl bromides, respectively.[79]

Alkene Epoxidation/Aziridination. Olefin epoxidation, principally of styrene derivatives, using a chiral Ru(II)-bisoxazoline catalyst and PhI(OAc)$_2$ takes place with modest enantiomeric induction (eq 56).[80,81] The rate of this reaction is enhanced by the presence of H$_2$O. An electron-deficient Fe-porphyrin complex also serves as a catalyst for O-atom transfer using PhI(OAc)$_2$ as the terminal oxidant.[82]

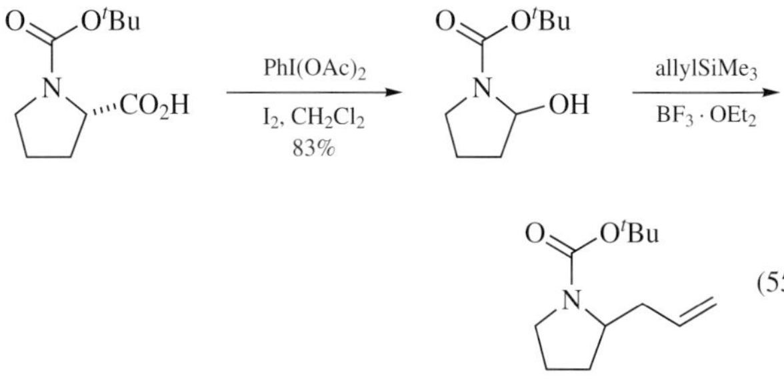

$$\text{(55)}$$

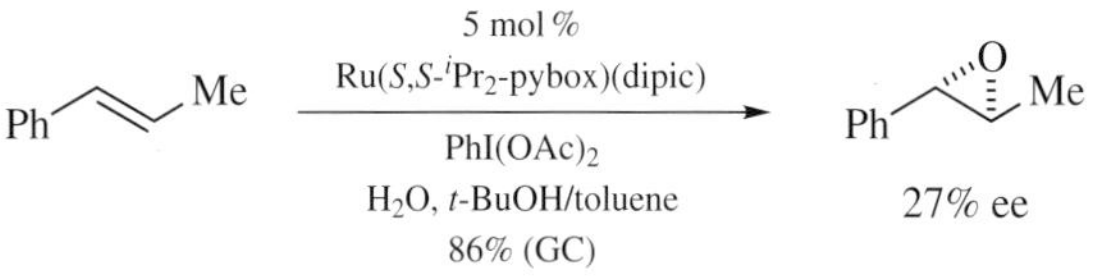

$$\text{(56)}$$

27% ee

Both intra- and intermolecular aziridination of alkenes can be accomplished with PhI(OAc)$_2$ and an appropriate nitrogen source (eqs 57 and 58).[83,84] The former reactions have been described using carbamate, sulfonamide, and sulfamate substrates. Typical catalysts utilized for these processes include Ru, Rh, and Cu complexes.[85–88] By employing chiral transition metal catalysts, asymmetric induction has been realized in both intra- and intermolecular reactions.

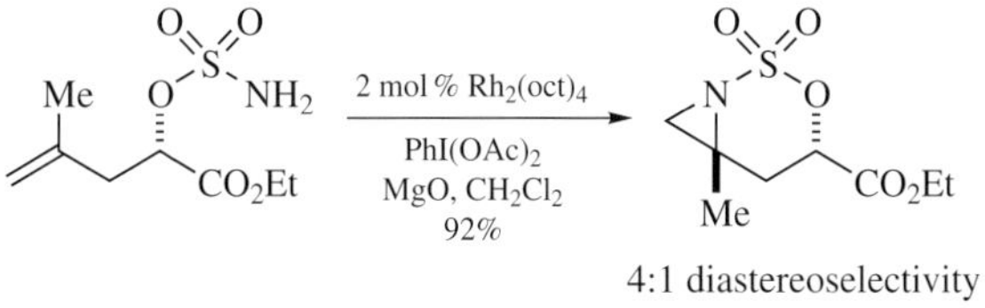

$$\text{(57)}$$

4:1 diastereoselectivity

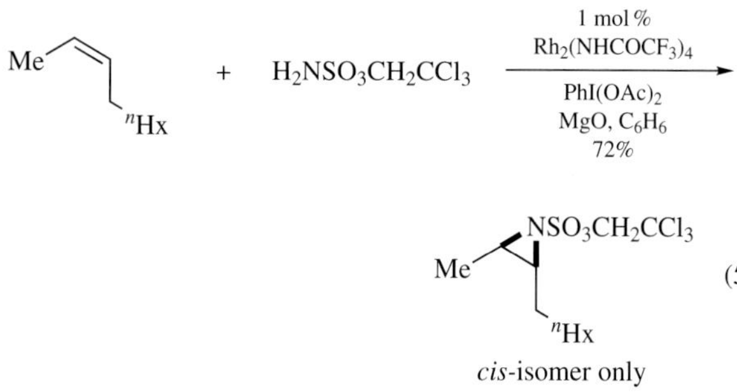

$$(58)$$

cis-isomer only

Alkene Cyclopropanation. Olefin cyclopropanation is possible in selected cases through in situ generation of stabilized phenyliodonium ylides using PhI(OAc)$_2$ and enolate equivalents such as Meldrum's acid or methyl nitroacetate (eq 59).[89] The intermediate ylide decomposes in the presence of Rh$_2$(O$_2$CR)$_4$ catalysts to generate a putative metallo-carbene species as the reactive cyclopropanating agent. With methyl nitroacetate, substituted cyclopropanes are formed with good levels of cis/trans stereocontrol.[90] PhI=O has also been employed for related reactions using both Rh and Cu complexes.[91,92]

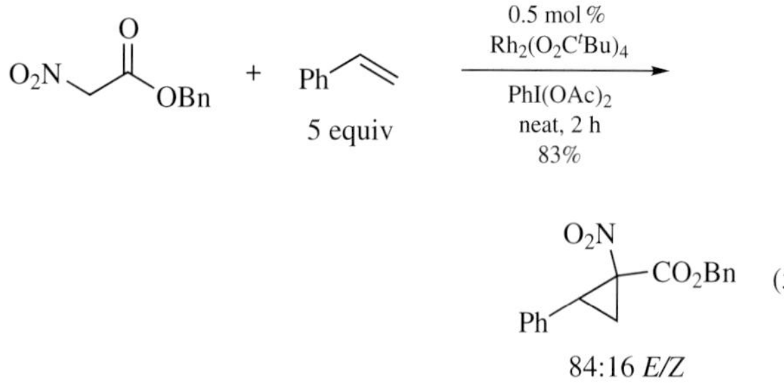

$$(59)$$

84:16 E/Z

C–H Amination. A number of amine-based starting materials will react with PhI(OAc)$_2$ and a transition metal catalyst to promote selective C–H bond amination.[84] Intramolecular oxidation of substrates such as carbamates, ureas, sulfamates, sulfonamides, and sulfamides affords the corresponding heterocycles in high yields and, in many cases, with excellent diastereocontrol (eqs 60 and 61).[93–95] Insertion into optically active 3° C–H centers is reported to be stereospecific (eq 62).[96,97] Chiral Ru, Mn, and Rh catalysts have all been utilized for asymmetric C–H amination, though product enantiomeric induction is variable. Many of the heterocyclic structures furnished from these reactions function as versatile precursors to 1,2- and 1,3-amine derivatives.

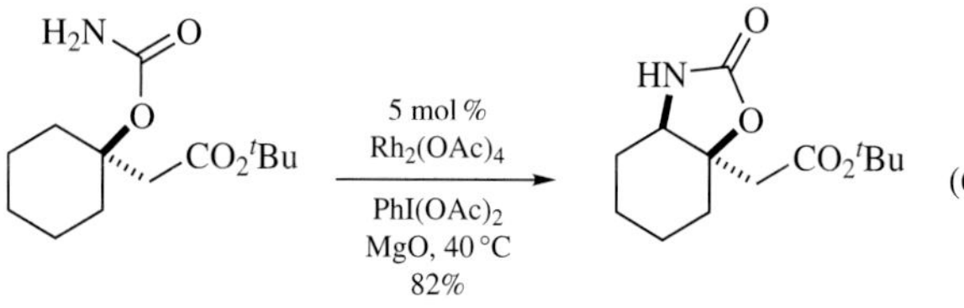

$$(60)$$

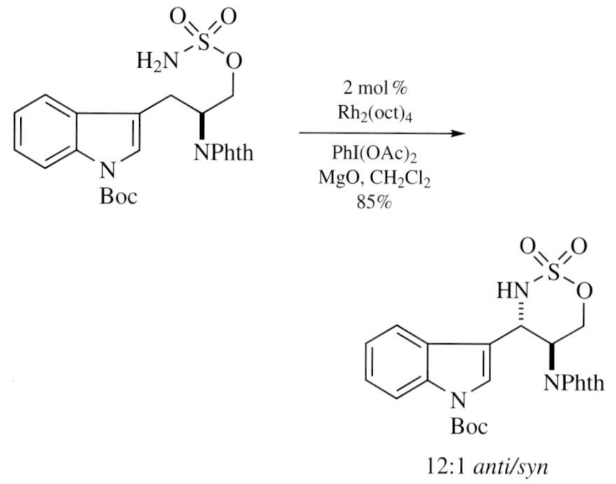

$$(61)$$

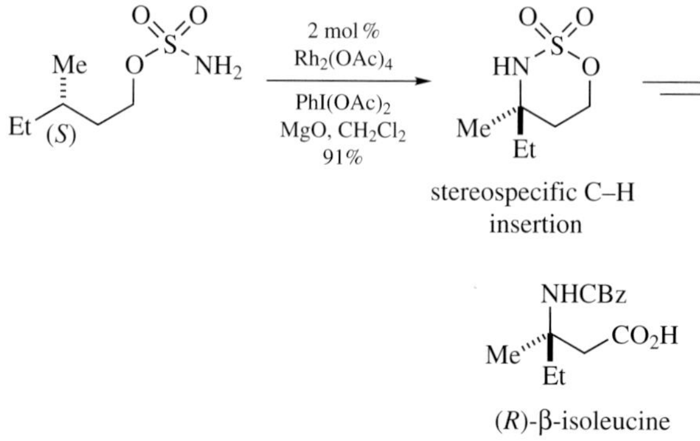

12:1 anti/syn

stereospecific C–H insertion

$$(62)$$

(R)-β-isoleucine

Intermolecular C–H amination has been demonstrated using primarily p-TsNH$_2$ or 2,2,2-trichloroethoxysulfonamide as the nitrogen source (eq 63).[93,95,98] Reactions operate most effectively with benzylic hydrocarbons, although C–H amination of 3° and 2° C–H bonds is possible. Other sulfonamides as well as certain acyl amides can be employed in this unique oxidation reaction.[99] These methods generally function with limiting amounts of the starting hydrocarbon and a slight excess of both the nitrogen source and PhI(OAc)$_2$.

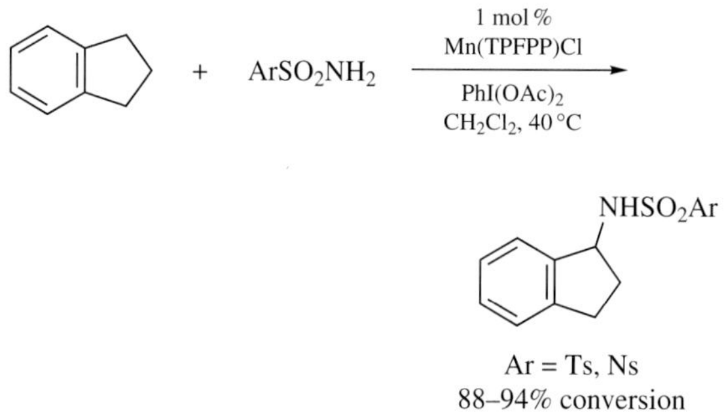

$$(63)$$

Ar = Ts, Ns
88–94% conversion

C–H Oxygenation. The combination of PhI(OAc)$_2$ and catalytic Pd(OAc)$_2$ can be used with functionalized aromatic and aliphatic hydrocarbons for the directed oxygenation of C–H bonds (eq 64).[100,101] Substrates containing pyridine, azole, imine, and oxime groups fuction under these reaction conditions to afford acetoxylated or methoxylated compounds. Oxidation of 1° methyl centers is strongly preferred over 2° and 3° C–H sites.

(64)

Heteroatom Oxidation. Oxidation of organosulfur compounds with PhI(OAc)$_2$ is noted in a number of different contexts. A particularly useful method for deprotection of dithiane-derived aldehydes and ketones employs PhI(OAc)$_2$ in aq acetone (eq 65).[102] Related to this process, organosulfides may be selectively oxidized to sulfoxides by employing the supported reagent PhI(OAc)$_2$/Al$_2$O$_3$ or with a combination of PhI(OAc)$_2$ and Ac$_2$O.[103]

(65)

R^1 = aryl, alkyl, vinyl
R^2 = H, alkyl

Synthesis of sulfilimines and sulfoximines from sulfides and sulfoxides, respectively, may be accomplished with PhI(OAc)$_2$ and o-NsNH$_2$ or CF$_3$CONH$_2$ as the nitrogen source.[104] Rh$_2$(OAc)$_4$ was found to be an optimal catalyst for this process (eq 66). Removal of the CF$_3$CO-protecting group from the products is facilitated with methanolic base, thereby affording the N-unsubstituted products. Under these conditions, reaction of a chiral sulfoxide is stereospecific. N-Sulfonyl sulfilimine formation has also been reported using arenesulfonamides and PhI(OAc)$_2$ in the absence of a metal catalyst.[105]

(66)

Diacetoxyiodobenzene has been utilized for the oxidation of organic derivatives of both bismuth and antimony.[103] Reactions of triaryl species proceed under mild, neutral conditions to yield the corresponding pentavalent diacetates (eq 67).[106]

(67)

Ar = C$_6$H$_5$, 2-MeC$_6$H$_4$,
4-MeOC$_6$H$_4$, 4-FC$_6$H$_4$,
3-F$_3$CC$_6$H$_4$, 1-naphthyl

1. Moriarty, R. M.; Prakash, O., *Acc. Chem. Res.* **1986**, *19*, 244.
2. Moriarty, R. M.; Vaid, R. K., *Synthesis* **1990**, 431.
3. Varvoglis, A. *The Organic Chemistry of Polycoordinated Iodine*; VCH: New York, 1992; p 131.
4. Sharefkin, J. G.; Saltzman, H., *Org. Synth., Coll. Vol.* **1973**, *5*, 660.
5. Lucas, H. J.; Kennedy, E. R.; Formo, M. W., *Org. Synth., Coll. Vol.* **1955**, *3*, 483.
6. Moriarty, R. M.; Duncan, M. P.; Vaid, R. K.; Prakash, O., *Org. Synth., Coll. Vol.* **1992**, *8*, 43.
7. Moriarty, R. M.; Kamernitskii, J. S., *Tetrahedron Lett.* **1986**, *27*, 2809.
8. Moriarty, R. M.; Hu, H.; Gupta, S. C., *Tetrahedron Lett.* **1981**, *22*, 1283.
9. Turuta, A. M.; Kamernitzky, A. V.; Fadeeva, T. M.; Zhulin, A. V., *Synthesis* **1985**, 1129.
10. Moriarty, R. M.; Prakash, O., *J. Org. Chem.* **1985**, *50*, 151.
11. Moriarty, R. M.; Prakash, O.; Musallam, H. A., *J. Heterocycl. Chem.* **1985**, *22*, 583.
12. (a) Moriarty, R. M.; Prakash, O.; Freeman, W. A., *J. Chem. Soc., Chem. Commun.* **1984**, 927. (b) Moriarty, R. M.; Prakash, O.; Prakash, I.; Musallam, H. A., *J. Chem. Soc., Chem. Commun.* **1984**, 1342.
13. Moriarty, R. M.; Prakash, O.; Thachet, C. T.; Musallam, H. A., *Heterocycles* **1985**, *23*, 633.
14. Moriarty, R. M.; Engerer, S. C.; Prakash, O.; Prakash, I.; Gill, U. S.; Freeman, W. A., *J. Org. Chem.* **1987**, *52*, 153.
15. Zbiral, E.; Nestler, G., *Tetrahedron* **1970**, *26*, 2945.
16. Pelter, A.; Elgendy, S., *Tetrahedron Lett.* **1988**, *29*, 677.
17. Bennett, D.; Dean, F. M.; Herbin, G. A.; Matkin, D. A.; Price, A. W., *J. Chem. Soc., Perkin Trans.* **1980**, 2, 1978,.
18. Szántay, C.; Blaskó, G.; Bárczai-Beke, M.; Pechy, P.; Dörnyei, G., *Tetrahedron Lett.* **1980**, *21*, 3509.
19. Ochiai, M.; Iwaki, S.; Ukita, T.; Nagao, Y., *Chem. Lett.* **1987**, 133.
20. Moriarty, R. M.; Prakash, O.; Vaid, R. K.; Zhao, L., *J. Am. Chem. Soc.* **1989**, *111*, 6443.
21. Koser, G. F. *The Chemistry of Functional Groups, Supplement D*; S., Patai andZ., Rappoport, Eds.; Wiley: New York, 1983; p 721.
22. Moriarty, R. M.; Kim, J.; Guo, L., *Tetrahedron Lett.* **1993**, *34*, 4129.
23. Pausacker, K. H., *J. Chem. Soc.* **1953**, 1989.
24. Szmant, H. H.; Lapinski, R. L., *J. Am. Chem. Soc.* **1956**, *78*, 458.
25. Szmant, H. H.; Infante, R., *J. Org. Chem.* **1961**, *26*, 4173.
26. Dyall, L. K., *Aust. J. Chem.* **1973**, *26*, 2665.
27. Dyall, L. K.; Kemp, J. E., *Aust. J. Chem.* **1973**, *26*, 1969.
28. Pausacker, K. H.; Scroggie, J. G., *J. Chem. Soc.* **1954**, 4499.
29. Devadas, B.; Leonard, N. J., *J. Am. Chem. Soc.* **1990**, *112*, 3125.
30. Awang, D. V. C.; Vincent, A., *Can. J. Chem.* **1980**, *58*, 1589.
31. Kishi, V.; Nakatsura, S.; Fukuyama, T.; Havel, M., *J. Am. Chem. Soc.* **1973**, *95*, 6493.
32. Moriarty, R. M.; Vald, R. K.; Duncan, M. P., *Synth. Commun.* **1987**, *17*, 703.
33. Moriarty, R. M.; Prakash, I.; Penmasta, R., *Synth. Commun.* **1987**, *17*, 409.
34. Kahn, M.; Bertenshaw, S., *Tetrahedron Lett.* **1989**, *30*, 2317.
35. Kahn, M.; Wilke, S.; Chen, B.; Fujita, K., *J. Am. Chem. Soc.* **1988**, *110*, 1638.
36. Moriarty, R. M.; Vaid, R. K.; Ravikumar, V. T.; Hopkins, T. E.; Farid, P., *Tetrahedron* **1989**, *45*, 1605.
37. Taylor, E. C.; Robey, R. L.; McKillop, A., *Angew. Chem., Int. Ed. Engl.* **1972**, *11*, 48.
38. Myrboh, B.; Ile, H.; Junjappa, H., *Synthesis* **1992**, 1101.
39. Smith, P. A. S.; Bruckmann, E. M., *J. Org. Chem.* **1974**, *39*, 1047.
40. Boulton, A. J.; Devi, P.; Henderson, N.; Jarrar, A. A.; Kiss, M., *J. Chem. Soc., Perkin Trans. 1* **1989**, *1*, 543.
41. Baumgarten, H. E.; Hwang, D.-R.; Rao, T. N., *J. Heterocycl. Chem.* **1986**, *23*, 945.
42. Schröppel, F.; Sauer, J., *Tetrahedron Lett.* **1974**, 2945.
43. Anderson, D. J.; Gilchrist, T. L.; Rees, C. W., *J. Chem. Soc., Chem. Commun.* **1971**, 800.
44. Dorta, R. L.; Francisco, C. G.; Hernández, R.; Salazar, J. A.; Suárez, E., *J. Chem. Res. (S)* **1990**, 240.
45. deArmas, P.; Francisco, C. G.; Suárez, E., *Angew. Chem., Int. Ed. Engl.* **1992**, *31*, 772.
46. Freire, R.; Marrero, J. J.; Rodríguez, M. S.; Suárez, E., *Tetrahedron Lett.* **1986**, *27*, 383.

47. Varvoglis, A., *Top. Curr. Chem.* **2003**, *224*, 69.

48. Zhdankin, V. V.; Stang, P. J., *Chem. Rev.* **2002**, *102*, 2523.

49. Varvoglis, A. *Hypervalent Iodine in Organic Synthesis*; Academic Press: London, 1997.

50. Togo, H.; Sakuratani, K., *Synlett* **2002**, 1966.

51. Stang, P. J.; Boehshar, M.; Wingert, H.; Kitamura, T., *J. Am. Chem. Soc.* **1988**, *110*, 3272.

52. Yusubov, M. S.; Wirth, T., *Org. Lett.* **2005**, *7*, 519.

53. Moriarty, R. M.; Prakash, O., *Org. React.* **1999**, *54*, 273.

54. Koser, G. F., *Top. Curr. Chem.* **2003**, *224*, 137.

55. Kirschning, A., *J. Prak. Chem.* **1998**, *340*, 184.

56. De Mico, A.; Margarita, R.; Mariani, A.; Piancatelli, G., *Chem. Commun.* **1997**, 1237.

57. Hashem, M. A.; Jung, A.; Ries, M.; Kirschning, A., *Synlett* **1998**, 195.

58. Kirschning, A.; Plumeier, C.; Rose, L., *Chem. Commun.* **1998**, 33.

59. Shi, L.; Kim, Y.-J.; Gin, D. Y., *J. Am. Chem. Soc.* **2001**, *123*, 6939.

60. De Mico, A.; Margarita, R.; Parlanti, L.; Piancatelli, G.; Vescovi, A., *Tetrahedron* **1997**, *53*, 16877.

61. Stang, P. J., *J. Org. Chem.* **2003**, *68*, 2997.

62. Huang, X.; Xu, X.-H., *J. Chem. Soc., Perkin Trans. 1* **1998**, 3321.

63. Murata, M.; Satoh, K.; Watanabe, S.; Masuda, Y., *J. Chem. Soc., Perkin Trans. 1* **1998**, 1465.

64. Moriarty, R. M.; ChanylI, C. J.; Vaid, R. K.; Prakash, O.; Tuladhar, S. M., *J. Org. Chem.* **1993**, *58*, 2478.

65. Prakash, O.; Batra, H.; Kaur, H.; Sharma, P. K.; Sharma, V.; Singh, S. P.; Moriarty, R. M., *Synthesis* **2001**, 541.

66. Zhang, L.-h.; Kauffman, G. S.; Pesti, J. A.; Yin, J., *J. Org. Chem.* **1997**, *62*, 6918.

67. Ramsden, C. A.; Rose, H. L., *J. Chem. Soc., Perkin Trans. 1* **1997**, 2319.

68. Kirihara, M.; Nishio, T.; Yokoyama, S.; Kakuda, H.; Momose, T., *Tetrahedron* **1999**, *55*, 2911.

69. Ohno, M.; Oguri, I.; Eguchi, S., *J. Org. Chem.* **1999**, *64*, 8995.

70. Tohma, H.; Kita, Y., *Adv. Synth. Catal.* **2004**, *346*, 111.

71. De Mico, A.; Margarita, R.; Parlanti, L.; Vescovi, A.; Piancatelli, G., *J. Org. Chem.* **1997**, *62*, 6974.

72. Tohma, H.; Takizawa, S.; Maegawa, T.; Kita, Y., *Angew. Chem., Int. Ed.* **2000**, *39*, 1306.

73. Adam, W.; Hajra, S.; Herderich, M.; Saha-Möller, *Org. Lett.* **2000**, *2*, 2773.

74. Sun, W.; Wang, H.; Xia, C.; Li, J.; Zhao, P., *Angew. Chem., Int. Ed.* **2003**, *42*, 1042.

75. Varma, R. S.; Saini, R. K.; Dahiya, R., *J. Chem. Res. (S)* **1998**, 120.

76. Togo, H.; Katohgi, M., *Synlett* **2001**, 565.

77. Muraki, T.; Togo, H.; Yokoyama, M., *Tetrahedron Lett.* **1996**, *37*, 2441.

78. Boto, A.; Hernández, R.; Suárez, E., *J. Org. Chem.* **2000**, *65*, 4930.

79. Camps, P.; Lukach, A. E.; Pujol, X.; Vázquez, S., *Tetrahedron* **2000**, *56*, 2703.

80. Nishiyama, H.; Shimada, T.; Itoh, H.; Sugiyama, H.; Motoyama, Y., *Chem. Commun.* **1997**, 1863.

81. Tse, M. K.; Bhor, S.; Klawonn, M.; Döbler, C.; Beller, M., *Tetrahedron Lett.* **2003**, *44*, 7479.

82. Park, S. E.; Song, R.; Nam, W., *Inorg. Chim. Acta* **2003**, *343*, 373.

83. Dauban, P.; Dodd, R. H., *Synlett* **2003**, 1571.

84. Müller, P.; Fruit, C., *Chem. Rev.* **2003**, *103*, 2905.

85. Liang, J.-L.; Yuan, S.-X.; Huang, J.-S.; Che, C.-M., *J. Org. Chem.* **2004**, *69*, 3610.

86. Guthikonda, K.; Du Bois, J., *J. Am. Chem. Soc.* **2002**, *124*, 13672.

87. Kwong, H.-L.; Liu, D.; Chan, K.-Y.; Lee, C.-S.; Huang, K.-H.; Che, C.-M., *Tetrahedron Lett.* **2004**, *45*, 3965.

88. Han, H.; Bae, I.; Yoo, E. J.; Lee, J.; Do, Y.; Chang, S., *Org. Lett.* **2004**, *6*, 4109.

89. Müller, P., *Acc. Chem. Res.* **2004**, *37*, 243.

90. Wurz, R. P.; Charette, A. B., *Org. Lett.* **2003**, *5*, 2327.

91. Müller, P.; Ghanem, A., *Org. Lett.* **2004**, *6*, 4347.

92. Dauban, P.; Sanière, L.; Tarrade, A.; Dodd, R. H., *J. Am. Chem. Soc.* **2001**, *123*, 7707.

93. Espino, C. G.; Fiori, K. W.; Kim, M.; Du Bois, J., *J. Am. Chem. Soc.* **2004**, *116*, 15378.

94. Cui, Y.; He, C., *Angew. Chem., Int. Ed.* **2004**, *43*, 4210.

95. Liang, J.-L.; Huang, J.-S.; Yu, X.-Q.; Zhu, N.; Che, C.-M., *Chem. Eur. J.* **2002**, *8*, 1563.

96. Espino, C. G.; Du Bois, J., *Angew. Chem., Int. Ed.* **2001**, *40*, 598.

97. Espino, C. G.; Wehn, P. M.; Chow, J.; Du Bois, J., *J. Am. Chem. Soc.* **2001**, *123*, 6935.

98. Au, S.-M.; Huang, J.-S.; Che, C.-M.; Yu, W.-Y., *J. Org. Chem.* **2000**, *65*, 7858.

99. Yu, X.-Q.; Huang, J.-S.; Zhou, X.-G.; Che, C.-M., *Org. Lett.* **2000**, *2*, 2233.

100. Dick, A. R.; Hull, K. L.; Sanford, M. S., *J. Am. Chem. Soc.* **2004**, *126*, 2300.

101. Desai, L. V.; Hull, K. L.; Sanford, M. S., *J. Am. Chem. Soc.* **2004**, *126*, 9542.

102. Shi, X.-X.; Wu, Q.-Q., *Synth. Commun.* **2000**, *30*, 4081.

103. Koser, G. F., *Top. Curr. Chem.* **2003**, *224*, 173.

104. Okamura, H.; Bolm, C., *Org. Lett.* **2004**, *6*, 1305.

105. Ou, W.; Chen, Z.-C., *Synth. Commun.* **1999**, *29*, 4443.

106. Combes, S.; Finet, J.-P., *Tetrahedron* **1998**, *54*, 4313.

Diazomethane[1]

$$H_2C=\overset{+}{N}=\overset{-}{N}$$

[334-88-3]   CH$_2$N$_2$ (MW 42.04)

InChI = 1/CH2N2/c1-3-2/h1H2

InChIKey = YXHKONLOYHBTNS-UHFFFAOYAZ

(methylating agent for various functional groups including carboxylic acids, alcohols, phenols, and amides; reagent for the synthesis of α-diazo ketones from acid chlorides, and the cyclopropanation of alkenes[1])

Physical Data: mp −145 °C; bp −23 °C.

Solubility: diazomethane is most often used as prepared in ether, or in ether containing a small amount of ethanol. It is less frequently prepared and used in other solvents such as dichloromethane.

Analysis of Reagent Purity: diazomethane is titrated[2] by adding a known quantity of benzoic acid to an aliquot of the solution such that the solution is colorless and excess benzoic acid remains. Water is then added, and the amount of benzoic acid remaining is back-titrated with NaOH solution. The difference between the amount of acid added and the amount remaining reveals the amount of active diazomethane present in the aliquot.

Preparative Methods: diazomethane is usually prepared by the decomposition of various derivatives of *N*-methyl-*N*-nitrosoamines. Numerous methods of preparation have been described,[3] but the most common and most frequently employed are those which utilize ***N-Methyl-N-nitroso-p-toluenesulfonamide*** (Diazald®; **1**),[4] ***1-Methyl-3-nitro-1-nitrosoguanidine*** (MNNG, **2**),[5] or *N*-methyl-*N*-nitrosourea (**3**)[2]

(1)   **(2)**   **(3)**

The various reagents each have their advantages and disadvantages, as discussed below. The original procedure[6] for the synthesis of diazomethane involved the use of *N*-methyl-*N*-nitrosourea, and similar procedures are still in use today. An advantage of using this reagent is that solutions of diazomethane can be prepared without distillation,[7] thus avoiding the most dangerous operation in other preparations of diazomethane. For small scale preparations (1 mmol or less) which do not contain any alcohol, a kit is available utilizing MNNG which produces distilled diazomethane in a closed environment. Furthermore, MNNG is a stable compound and has a shelf life of many years. For larger scale preparations, kits are available for the synthesis of up to 300 mmol of diazomethane using Diazald as the precursor. The shelf life of Diazald (about 1–2 years), however, is shorter than that of MNNG. Furthermore, the common procedure using Diazald produces an ethereal solution of diazomethane which contains ethanol; however, it can be modified to produce an alcohol-free solution. Typical preparations of diazomethane involve the slow addition of base to a heterogeneous aqueous ether mixture containing the precursor. The precursor reacts with the base to liberate diazomethane which partitions into the ether layer and is concomitantly distilled with the ether to provide an ethereal solution of diazomethane. Due to the potentially explosive nature of diazomethane, the chemist is advised to carefully follow the exact procedure given for a particular preparation. Furthermore, since diazomethane has been reported to explode upon contact with ground glass, apparatus which do not contain ground glass should be used. All of the kits previously mentioned avoid the use of ground glass.

Handling, Storage, and Precautions: diazomethane as well as the precursors for its synthesis can present several safety hazards, and must be used with great care.[8] The reagent itself is highly toxic and irritating. It is a sensitizer, and long term exposure can lead to symptoms similar to asthma. It can also detonate unexpectedly, especially when in contact with rough surfaces, or on crystallization. It is therefore essential that any glassware used in handling diazomethane be fire polished and not contain any scratches or ground glass joints. Furthermore, contact with certain metal ions can also cause explosions. Therefore metal salts such as calcium chloride, sodium sulfate, or magnesium sulfate must not be used to dry solutions of the reagent. The recommended drying agent is potassium hydroxide. Strong light is also known to initiate detonation. The reagent is usually generated immediately prior to use and is not stored for extended periods of time. Of course, the reagent must be prepared and used in a well-ventilated hood, preferably behind a blast shield. The precursors used to generate diazomethane are irritants and in some cases mutagens and suspected carcinogens, and care should be exercised in their handling as well.

Methylation of Heteroatoms. The most widely used feature of the chemistry of diazomethane is the methylation of carboxylic acids. Carboxylic acids are good substrates for reaction with diazomethane because the acid is capable of protonating the diazomethane on carbon to form a diazonium carboxylate. The carboxylate can then attack the diazonium salt in what is most likely an S$_N$2 reaction to provide the ester. Species which are not acidic enough to protonate diazomethane, such as alcohols, require an additional catalyst, such as ***Boron Trifluoride Etherate***, to increase their acidity and facilitate the reaction. The methylation reaction proceeds under mild conditions and is highly reliable and very selective for carboxylic acids. A typical procedure is to add a yellow solution of diazomethane to the carboxylic acid in portions. When the yellow color persists and no more gas is evolved, the reaction is deemed complete. Excess reagent can be destroyed by the addition of a few drops of acetic acid and the entire solution concentrated to provide the methyl ester.

Esterification of Carboxylic Acids and Other Acidic Functional Groups. A variety of functional groups will tolerate the esterification of acids with diazomethane. Thus α,β-unsaturated carboxylic acids and alcohols survive the reaction (eq 1),[9] as do ketones (eq 2),[10] isolated alkenes (eq 3),[11] and amines (eq 4).[12]

$$ \text{(1)} $$

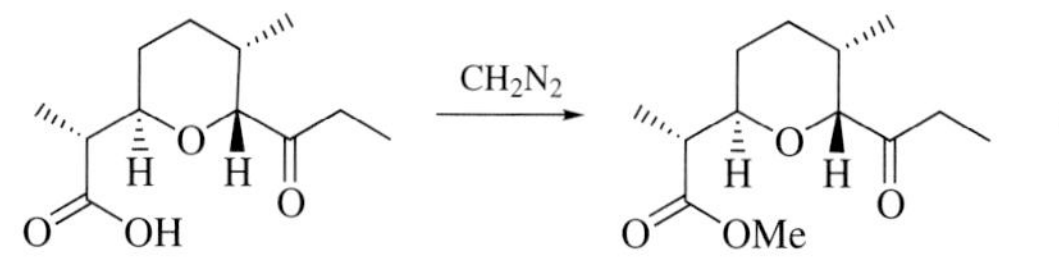

(2)

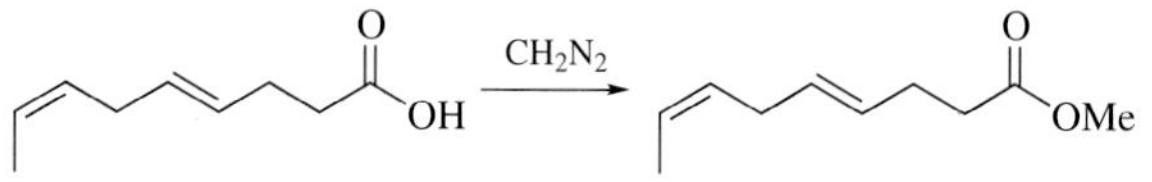

(3)

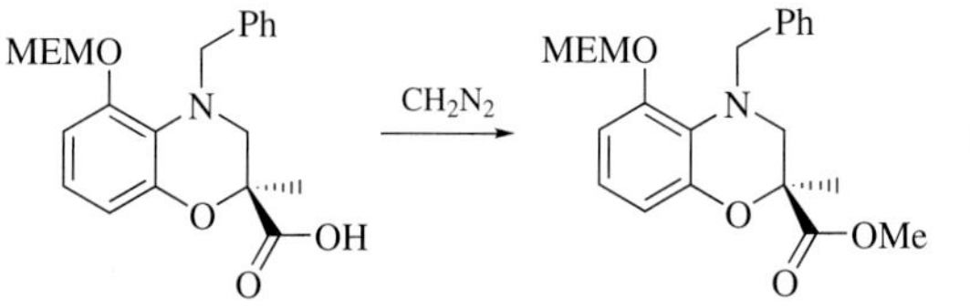

(4)

Other acidic functional groups will also undergo reaction with diazomethane. Thus phosphonic acids (eq 5)[13] and phenols (eq 6)[14] are methylated in high yields, as are hydroxytropolones (eq 7)[15] and vinylogous carboxylic acids (eq 8).[16] The origin of the selectivity in eq 6 is due to the greater acidity of the A-ring phenol.

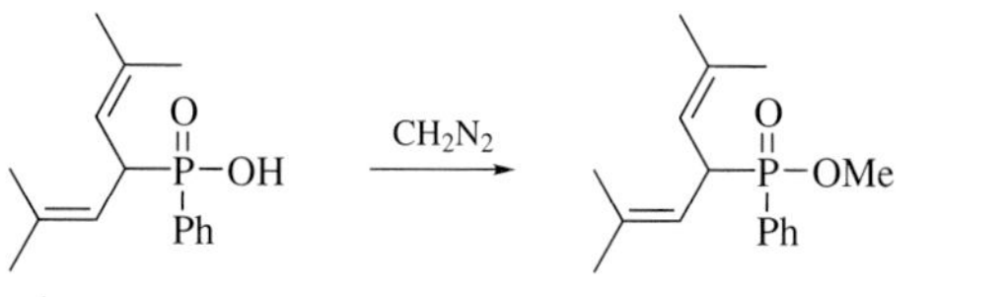

(5)

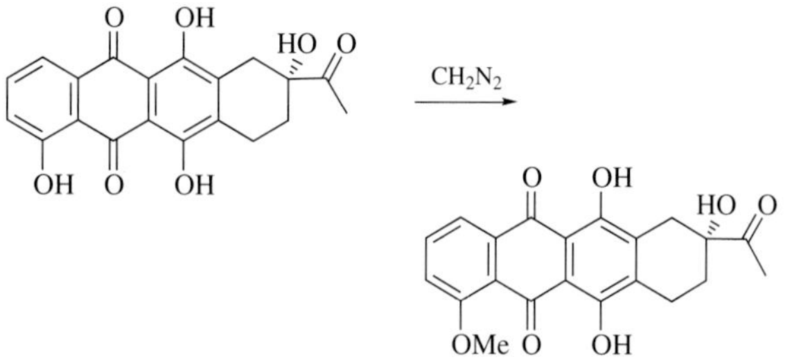

(6)

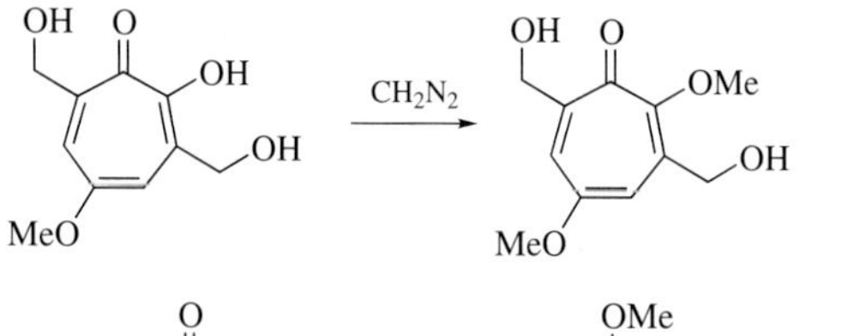

(7)

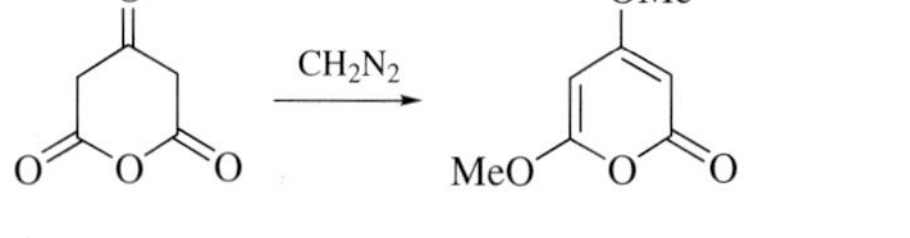

(8)

Selective monomethylation of dicarboxylic acids has been reported using **Alumina** as an additive (eq 9).[17] It is thought that one of the two carboxylic acid groups is bound to the surface of the alumina and is therefore not available for reaction. Carboxylic acids that are engaged as lactols will also undergo methylation with diazomethane to provide the methyl ester and aldehyde (eq 10).[18]

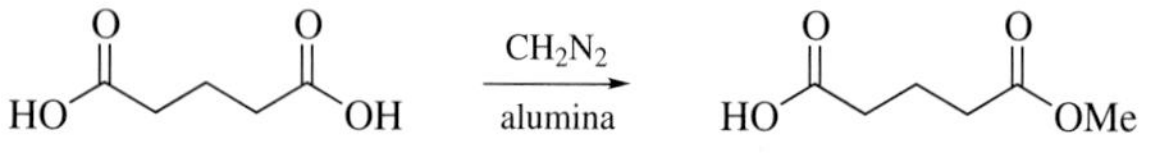

(9)

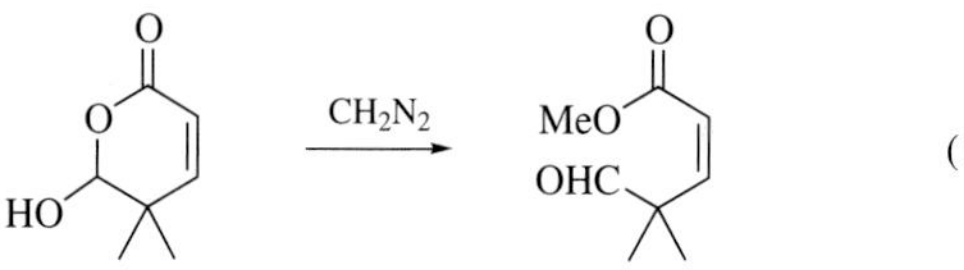

(10)

Methylation of Alcohols and Other Less Acidic Functional Groups. As previously mentioned, alcohols require the addition of a catalyst in order to react with diazomethane. The most commonly used is boron trifluoride etherate (eq 11),[19] but **Tetrafluoroboric Acid** has been used as well (eq 12).[20] Mineral acids are not effective since they rapidly react with diazomethane to provide the corresponding methyl halides. Acids as mild as silica gel have also been found to be effective (eq 13).[21] Monomethylation of 1,2-diols with diazomethane has been reported using various Lewis acids as promoters, the most effective of which is **Tin(II) Chloride** (eq 14).[22]

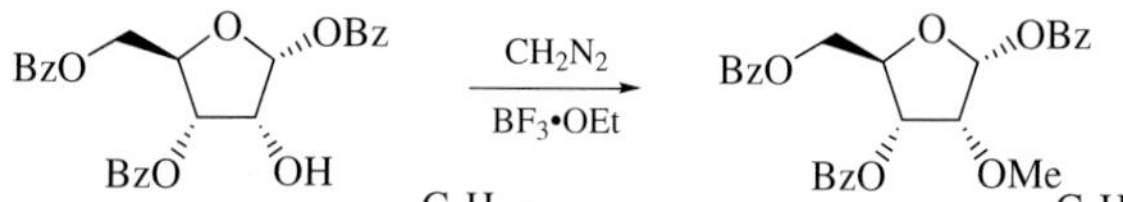

(11)

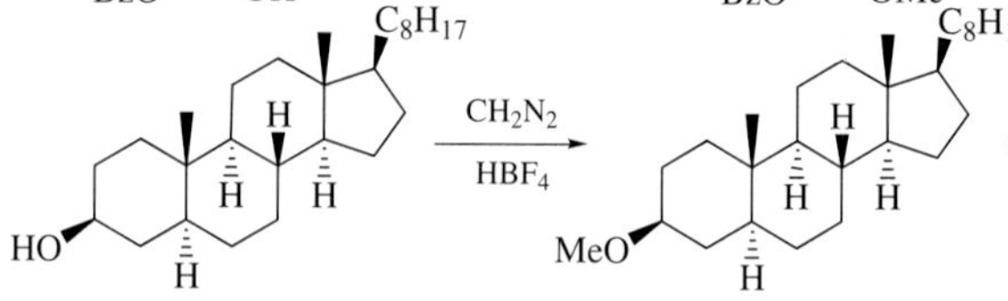

(12)

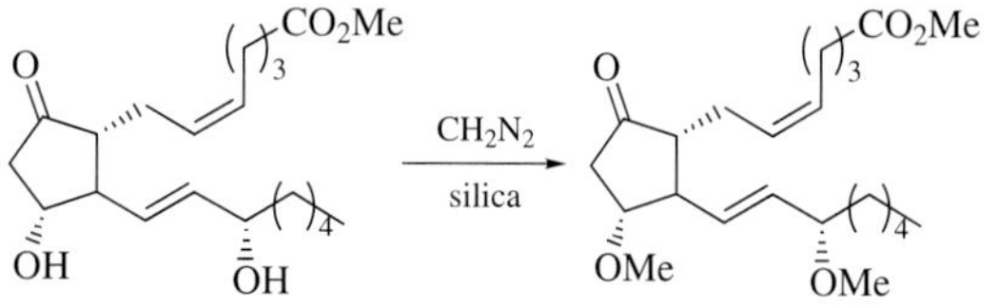

(13)

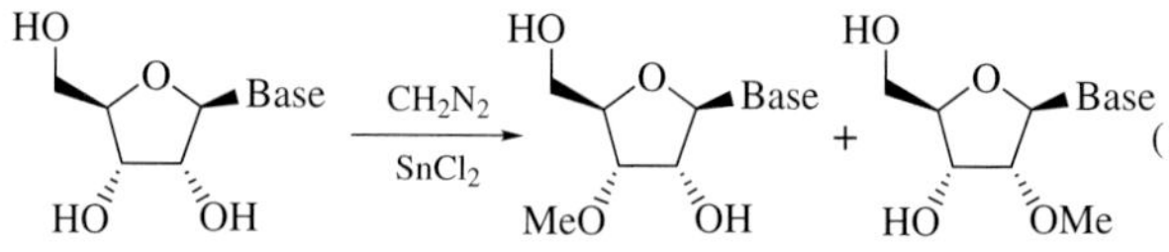

(14)

An interesting case of an alcohol reacting with diazomethane at a rate competitive with a carboxylic acid has been reported (eq 15).[23] In this case, the tertiary structure of the molecule is thought to place the alcohol and the carboxylic acid in proximity to each other. Protonation of the diazomethane by the carboxylic acid leads to a diazonium ion in proximity to the alcohol as well as the carboxylate. These species then attack the diazonium ion at competitive rates to provide the methyl ether and ester. No reaction is observed upon treatment of the corresponding hydroxy ester with diazomethane, indicating that the acid is required to activate the diazomethane.

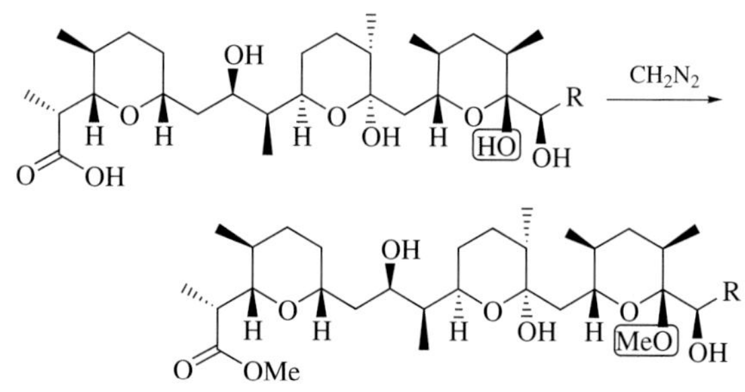

(15)

Amides can also be methylated with diazomethane in the presence of silica gel; however, the reaction requires a large excess of diazomethane (25–60 equiv, eq 16).[24] The reaction primarily provides *O*-methylated material; however, in one case a mixture of *O*- and *N*-methylation was reported. Thioamides are also effectively methylated with this procedure to provide *S*-methylated compounds. Finally, amines have been methylated with diazomethane in the presence of BF$_3$ etherate, fluoroboric acid,[25] or copper(I) salts;[26] however, the yields are low to moderate, and the method is not widely used.

$$\text{(16)}$$

The Arndt–Eistert Synthesis. Diazomethane is a useful reagent for the one-carbon homologation of acid chlorides via a sequence of reactions known as the Arndt–Eistert synthesis. The first step of this sequence takes advantage of the nucleophilicity of diazomethane in its addition to an active ester, typically an acid chloride,[27] to give an isolable α-diazo ketone and HCl. The HCl that is liberated from this step can react with diazomethane to produce methyl chloride and nitrogen, and therefore at least 2 equiv of diazomethane are typically used. The α-diazo ketone is then induced to undergo loss of the diazo group and insertion into the adjacent carbon–carbon bond of the ketone to provide a ketene. The ketene is finally attacked by water or an alcohol (or some other nucleophile) to provide the homologated carboxylic acid or ester. This insertion step of the sequence is known as the Wolff rearrangement[28] and can be accomplished either thermally (eq 17)[29] or, more commonly, by treatment with a metal ion (usually silver salts, eq 18),[30] or photochemically (eq 19).[31] It has been suggested that the photochemical method is the most efficient of the three.[32] As eqs 18 and 19 illustrate, retention of stereochemistry is observed in the migrating group. The obvious limitations of this reaction are that there must not be functional groups present in the molecule which will react with diazomethane more rapidly than it will attack the acid chloride. Thus carboxylic acids will be methylated under these conditions. Furthermore, electron-deficient alkenes will undergo [2,3] dipolar cycloaddition with diazomethane more rapidly than addition to the acid chloride. Thus when the Arndt–Eistert synthesis is attempted on α,β-unsaturated acid chlorides, cycloaddition to the alkene is observed in the product. In order to prevent this, the alkene must first be protected by addition of HBr and then the reaction carried out in the normal way (eq 20).[33] Cycloaddition to isolated alkenes, however, is not competitive with addition to acid chlorides.

$$\text{(17)}$$

$$\text{(18)}$$

$$\text{(19)}$$

$$\text{(20)}$$

Other Reactions of α-Diazo Ketones Derived from Diazomethane. Depending on the conditions employed, the Wolff rearrangement may proceed via a carbene or carbenoid intermediate, or it may proceed by a concerted mechanism where the insertion is concomitant with loss of N$_2$ and no intermediate is formed. In the case where a carbene or carbenoid is involved, other reactions which are characteristic of these species can occur, such as intramolecular cyclopropanation of alkenes. In fact, the reaction conditions can be adjusted to favor cyclopropanation or homologation depending on which is desired. Thus treatment of the dienoic acid chloride shown in eq 21 with diazomethane followed by decomposition of the α-diazo ketone with silver benzoate in the presence of methanol and base provides the homologated methyl ester. However, treatment of the same diazoketone intermediate with Cu^{II} salts provides the cyclopropanation products selectively.[34] This trend is generally observed; that is, silver salts as well as photochemical conditions (eqs 18 and 19) favor the homologation pathway while copper or rhodium salts favor cyclopropanation.[35] Using copper salts to decompose the diazo compounds, hindered alkenes as well as electron-rich aromatics can be cyclopropanated as illustrated in eqs 22 and 23,[36,37] respectively.

$$\text{(21)}$$

$$\text{(22)}$$

$$\text{(23)}$$

In addition to these reactions, α-diazo ketones will undergo protonation on carbon in the presence of protic acids[38] to provide the corresponding α-diazonium ketone. These species are highly electrophilic and can undergo nucleophilic attack. Thus if the proton source contains a nucleophile such as a halogen then the corresponding α-halo ketone is isolated (eq 24).[39] However, if the proton source does not contain a nucleophilic counterion then the diazonium species may react with other nucleophiles that are present in the molecule, such as alkenes (eq 25)[40] or aromatic rings (eq 26).[41] Note the similarity between the transformations in eqs 26 and 23 which occur using different catalysts and by different pathways. Also, eq 26 illustrates the fact that other active esters will undergo nucleophilic attack by diazomethane.

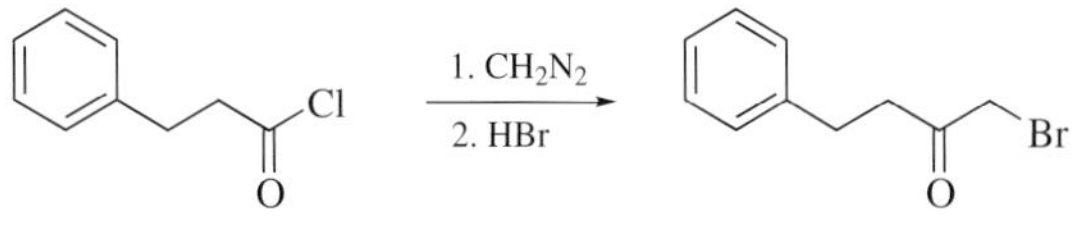

(24)

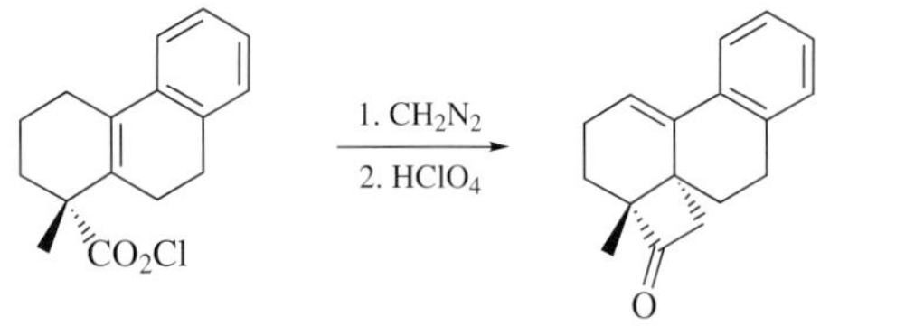

(25)

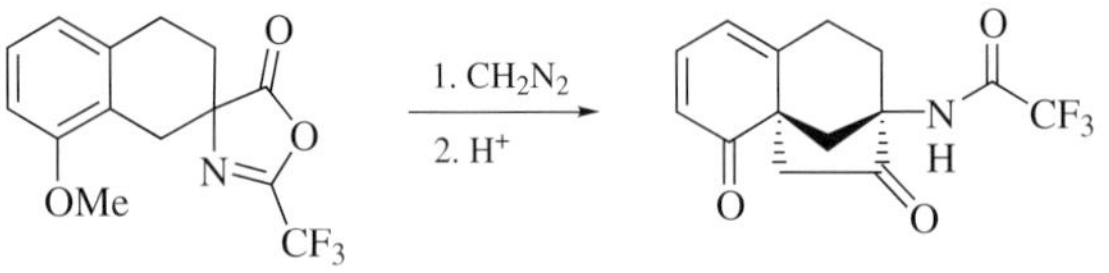

(26)

Lewis acids are also effective in activating α-diazo ketones towards intramolecular nucleophilic attack by alkenes and arenes.[42] The reaction has been used effectively for the synthesis of cyclopentenones (eq 27) starting with β,γ-unsaturated diazo ketones derived from the corresponding acid chloride and diazomethane. It has also been used to initiate polyalkene cyclizations (eq 28). Typically, boron trifluoride etherate is used as the Lewis acid, and electron-rich alkenes are most effective providing the best yields of annulation products.

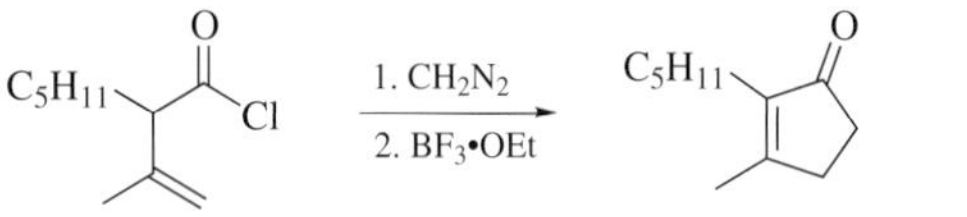

(27)

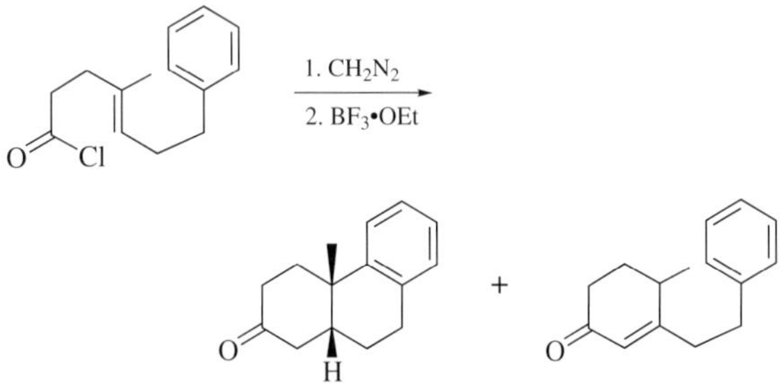

(28)

The Vinylogous Wolff Rearrangement. The vinylogous Wolff rearrangement[43] is a reaction that occurs when the Arndt–Eistert synthesis is attempted on β,γ-unsaturated acid chlorides using copper catalysis. Rather than the usual homologation products, the reaction proceeds to give what is formally the product of a [2,3]-sigmatropic shift, but is mechanistically not derived by this pathway.[44] The mechanism is thought to proceed by an initial cyclopropanation of the alkene by the α-diazo ketone to give a bicyclo[2.1.0]pentanone derivative. This compound then undergoes a fragmentation to a ketene alkene before being trapped by the solvent (eq 29). Inspection of the products reveals that they are identical with those derived from the Claisen rearrangement of the corresponding allylic alcohols, and as such this method can be thought of as an alternative to the Claisen procedure. However, the stereoselectivity of the alkene that is formed is not as high as is typically observed in the Claisen rearrangement (eq 30), and in some substrates the reaction proceeds with no selectivity (eq 31).

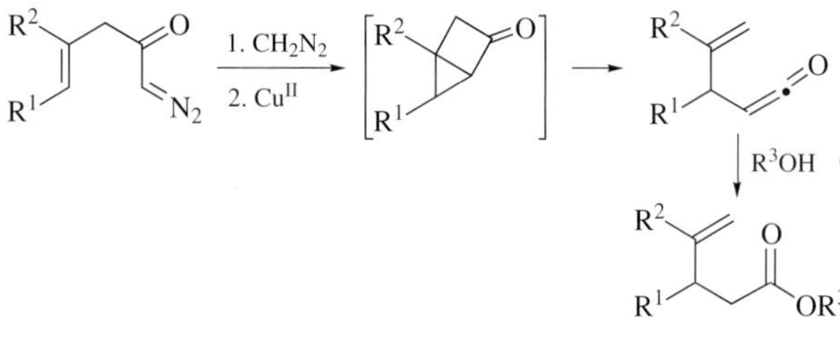

(29)

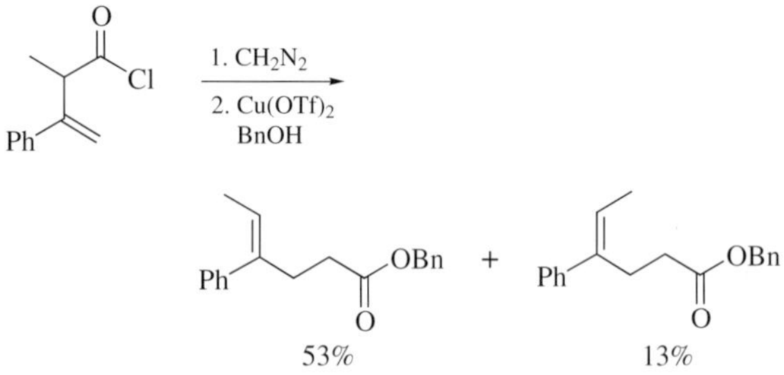

(30)

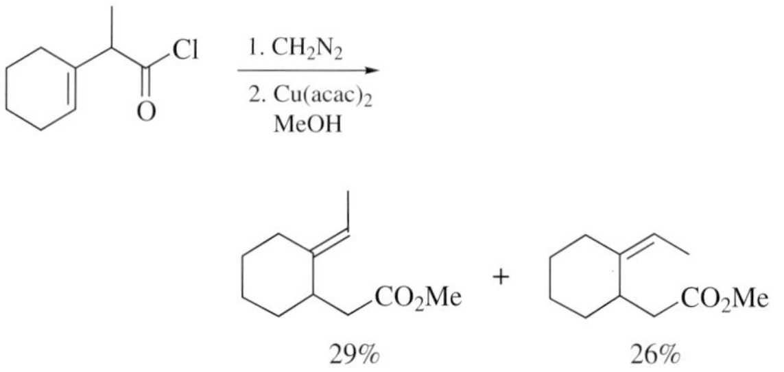

(31)

Insertions into Aldehyde C–H Bonds. The α-diazo ketones (and esters) derived from diazomethane and an acid chloride (or chloroformate) will also insert into the C–H bond of aldehydes to give 1,3-dicarbonyl derivatives.[45] The reaction is catalyzed by $SnCl_2$, but some simple Lewis acids, such as BF_3 etherate, also work. The reaction works well for aliphatic aldehydes, but gives variable results with aromatic aldehydes, at times giving none of the desired diketone (eq 32). Sterically hindered aldehydes will also participate in this reaction, as illustrated in eq 33 with the reaction of ethyl α-diazoacetate and pivaldehyde. In a related reaction, α-diazo phosphonates and sulfonates will react with aldehydes in the presence of $SnCl_2$ to give the corresponding β-keto phosphonates and sulfonates.[46] This reaction is a practical alternative to the Arbuzov reaction for the synthesis of these species.

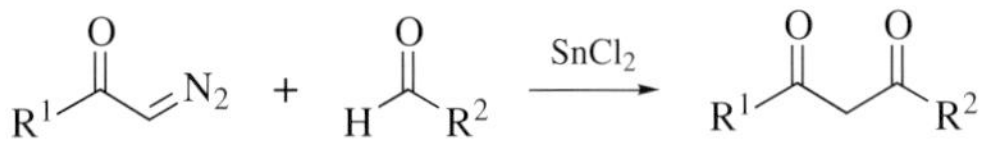

(32)

R^1	R^2	Yield (%)
Ph	H	88
Ph	$PhCH_2$	90
$PhCH_2CH_2$	Ph	0

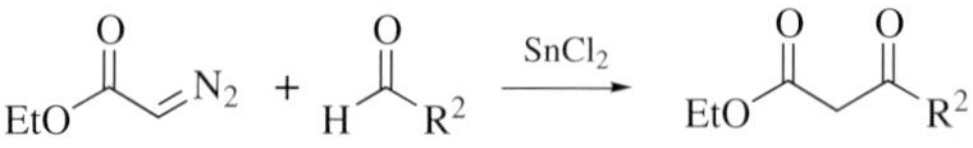

(33)

R^2	Yield (%)
t-Bu	65
Ph	50

Additions to Ketones. The addition of diazomethane to ketones[47] is also a preparatively useful method for one-carbon homologation. This reaction is a one-step alternative to the Tiffeneau–Demjanow rearrangement[48] and proceeds by the mechanism shown in eq 34. It can lead to either homologation or epoxidation depending on the substrate and reaction conditions. The addition of Lewis acids, such as BF_3 etherate, or alcoholic cosolvents tend to favor formation of the homologation products over epoxidation.

$$ \text{(34)} $$

However, the reaction is limited by the poor regioselectivity observed in the insertion when the groups R^1 and R^2 in the starting ketone are different alkyl groups. What selectivity is observed tends to favor migration of the less substituted carbon,[49] a trend which is opposite to that typically observed in rearrangements of electron-deficient species such as in the Baeyer–Villiger reaction. Furthermore, the product of the reaction is a ketone and is therefore capable of undergoing further reaction with diazomethane. Thus, ideally, the product ketone should be less reactive than the starting ketone. Strained ketones tend to react more rapidly and are therefore good substrates for this reaction (eq 35).[50] This method has also found extensive use in cyclopentane annulation reactions starting with an alkene. The overall process begins with dichloroketene addition to the alkene to produce an α-dichlorocyclobutanone. These species are ideally suited for reaction with diazomethane because the reactivity of the starting ketone is enhanced due to the strain in the cyclobutanone as well as the α-dichloro substitution. Furthermore, the presence of the α-dichloro substituents hinders migration of that group and leads to almost exclusive migration of the methylene group. Thus treatment with diazomethane and methanol leads to a rapid evolution of nitrogen, and produces the corresponding α-dichlorocyclopentanone, which can be readily dehalogenated to the hydrocarbon (eq 36).[51] Aldehydes will also react with diazomethane, but in this case homologation is not observed. Rather, the corresponding methyl ketone derived from migration of the hydrogen is produced (eq 37).

$$ \text{(35)} $$

$$ \text{(36)} $$

$$ \text{(37)} $$

Cycloadditions with Diazomethane. Diazomethane will undergo [3 + 2] dipolar cycloadditions with alkenes and alkynes to give pyrazolines and pyrazoles, respectively.[52] The reaction proceeds more rapidly with electron-deficient alkenes and strained alkenes and is controlled by FMO considerations with the HOMO of the diazomethane and the LUMO of the alkene serving as the predominant interaction[53] In the case of additions to electron-deficient alkenes, the carbon atom of the diazomethane behaves as the negatively charged end of the dipole, and therefore the regiochemistry observed is as shown in eq 38. With conjugated alkenes, such as styrene, the terminal carbon has the larger lobe in the LUMO, and as such the reaction proceeds to give the product shown in eq 39. Pyrazolines are most often used as precursors to cyclopropanes by either thermal or photochemical extrusion of N_2. In both cases the reaction may proceed by a stepwise mechanism with loss of stereospecificity. As shown in eq 40, the thermal reaction provides an almost random product distribution, while the photochemical reaction provides variable results ranging from 20:1 to stereospecific extrusion of nitrogen.[54]

$$ \text{(38)} $$

$$ \text{(39)} $$

$$ \text{(40)} $$

$$
\begin{array}{ll}
\Delta & 1.2{:}1 \\
h\nu & 20{:}1 \\
 & \text{to} >100{:}1
\end{array}
$$

Cyclopropanes can also be directly synthesized from alkenes and diazomethane, either photochemically or by using transition metal salts, usually ***Copper(II) Chloride*** or ***Palladium(II) Acetate***, as promoters. The metal-mediated reactions are more commonly used than the photochemical ones, but they are not as popular as the Simmons–Smith procedure. However, they do occasionally offer advantages. Of the two processes, the Cu-catalyzed reaction produces a more active reagent,[55] which will cyclopropanate a variety of alkenes, including enamines as shown in eq 41[56] These products can then be converted to α-methyl ketones by thermolysis. The cyclopropanation of the norbornenol derivative shown in eq 42 was problematic using the Simmons–Smith procedure and provided low yields, but occurred smoothly using the $CuCl_2$/diazomethane method.[57]

$$ \text{(41)} $$

(42)

The Pd(OAc)$_2$-mediated reaction can be used to cyclopropanate electron-deficient alkenes as well as terminal alkenes. Thus selective reaction at a monosubstituted alkene in the presence of others is readily achieved using this method (eq 43).[58] The example shown in eq 44 is one in which the Simmons–Smith procedure failed to provide any of the desired product, whereas the current method provided a 92% yield of cyclopropane.[59]

(43)

(44)

In the case of the photochemical reaction, irradiation of diazomethane in the presence of *cis*-2-butene provides *cis*-1,2-dimethylcyclopropane with no detectable amount of the *trans* isomer (eq 45).[60] This reaction is thought to proceed via a singlet carbene. However, if the same reaction is carried out via a triplet carbene, generated via triplet sensitization, then a 1.3:1 mixture of *trans* to *cis* dimethylcyclopropane is observed (eq 46).[61] The yields in the photochemical reaction are typically lower than the metal-mediated processes, and are usually accompanied by more side products.

(45)

(46)

Additions to Electron-deficient Species. Diazomethane will also add to highly electrophilic species such as sulfenes or imminium salts to give the corresponding three-membered ring heterocycles. When the reaction is performed on sulfenes, the products are episulfones which are intermediates in the Ramberg–Backlund rearrangement, and are therefore precursors for the synthesis of alkenes via chelotropic extrusion of SO$_2$. The sulfenes are typically prepared in situ by treatment of a sulfonyl chloride with a mild base, such as *Triethylamine* (eq 47).[62] Similarly, the addition of diazomethane to imminium salts has been used to methylenate carbonyls.[63] In this case, the intermediate aziridinium salt is treated with a strong base, such as *Butyllithium*, in order to induce elimination (eq 48).

(47)

(48)

Miscellaneous Reactions. Diazomethane has been shown to react with vinylsilanes derived from α,β-unsaturated esters to provide the corresponding allylsilane by insertion of CH$_2$ into the C–Si bond (eq 49).[64] The reaction has been shown to be stereospecific, with *cis*-vinylsilane providing *cis*-allylsilanes; however, the mechanism of the reaction has not been defined. Diazomethane has also been used in the preparation of trimethyloxonium salts. Treatment of a solution of dimethyl ether and trinitrobenzenesulfonic acid with diazomethane provides trimethyloxonium trinitrobenzenesulfonate, which is more stable than the fluoroborate salt.[65]

(49)

Related Reagents. 2-Diazopropane; Diphenyldiazomethane; Phenyldiazomethane; 1-Diazo-2-propene.

1. (a) Regitz, M.; Maas, G. *Diazo Compounds, Properties and Synthesis*; Academic: Orlando, 1986. (b) Black, T. H., *Aldrichim. Acta* **1983**, *16*, 3. (c) Pizey, J. S. *Synthetic Reagents*; Wiley: New York, 1974; Vol. 2, p 65.

2. Arndt, F., *Org. Synth., Coll. Vol.* **1943**, *2*, 165.

3. Moore, J. A.; Reed, D. E., *Org. Synth., Coll. Vol.* **1973**, *5*, 351. Redemann, C. E.; Rice, F. O.; Roberts, R.; Ward, H. P., *Org. Synth., Coll. Vol.* **1955**, *3*, 244. McPhee, W. D.; Klingsberg, E., *Org. Synth., Coll. Vol.* **1955**, *3*, 119.

4. De Boer, Th. J.; Backer, H. J., *Org. Synth., Coll. Vol.* **1963**, *4*, 250. Hudlicky, M., *J. Org. Chem.* **1980**, *45*, 5377. See also Aldrich Chemical Company Technical Bulletins Number AL-121 and AL-131. Note that the preparation described in *Fieser & Fieser*, **1967**, *1*, 191. is flawed and neglects to mention the addition of ethanol Failure to add ethanol can result in a buildup of diazomethane and a subsequent explosion.

5. McKay, A. F., *J. Am. Chem. Soc.* **1948**, *70*, 1974. See also Aldrich Chemical Company Technical Bulletin Number AL-132.

6. von Pechman, A., *Chem. Ber.* **1894**, *27*, 1888.

7. Ref. 2, note 3.

8. For a description of the safety hazards associated with diazomethane, see: Gutsche, C. D., *Org. React.* **1954**, *8*, 391.

9. Fujisawa, T.; Sato, T.; Itoh, T., *Chem. Lett.* **1982**, 219.

10. Nicolaou, K. C.; Paphatjis, D. P.; Claremon, D. A.; Dole, R. E., *J. Am. Chem. Soc.* **1981**, *103*, 6967.

11. Fujisawa, T.; Sato, T.; Kawashima, M.; Naruse, K.; Tamai, K., *Tetrahedron Lett.* **1982**, *23*, 3583.

12. Kozikowski, A. P.; Sugiyama, K.; Springer, J. P., *J. Org. Chem.* **1981**, *46*, 2426.

13. De, B.; Corey, E. J., *Tetrahedron Lett.* **1990**, *31*, 4831. Macomber, R. S., *Synth. Commun.* **1977**, *7*, 405.

14. Blade, R. J.; Hodge, P., *J. Chem. Soc., Chem. Commun.* **1979**, 85.

15. Kawamata, A.; Fukuzawa, Y.; Fujise, Y.; Ito, S., *Tetrahedron Lett.* **1982**, *23*, 1083.

16. Ray, J. A.; Harris, T. M., *Tetrahedron Lett.* **1982**, *23*, 1971.

17. Ogawa, H.; Chihara, T.; Taya, K., *J. Am. Chem. Soc.* **1985**, *107*, 1365.

18. Frimer, A. A.; Gilinsky-Sharon, P.; Aljadef, G., *Tetrahedron Lett.* **1982**, *23*, 1301.

19. Chavis, C.; Dumont, F.; Wightman, R. H.; Ziegler, J. C.; Imbach, J. L., *J. Org. Chem.* **1982**, *47*, 202.

20. Neeman, M.; Johnson, W. S., *Org. Synth., Coll. Vol.* **1973**, *5*, 245.

21. Ohno, K.; Nishiyama, H.; Nagase, H., *Tetrahedron Lett.* **1979**, *20*, 4405.

22. Robins, M. J.; Lee, A. S. K.; Norris, F. A., *Carbohydr. Res.* **1975**, *41*, 304.

23. Evans, D. S.; Bender, S. L.; Morris, J., *J. Am. Chem. Soc.* **1988**, *110*, 2506. For a similar example with the antibiotic lasalocid, see: Westly, J. W.; Oliveto, E. P.; Berger, J.; Evans, R. H.; Glass, R.; Stempel, A.; Toome, V.; Williams, T., *J. Med. Chem.* **1973**, *16*, 397.

24. Nishiyama, H.; Nagase, H.; Ohno, K., *Tetrahedron Lett.* **1979**, *20*, 4671.

25. Muller, v. H.; Huber-Emden, H.; Rundel, W., *Justus Liebigs Ann. Chem.* **1959**, *623*, 34.

26. Seagusa, T.; Ito, Y.; Kobayashi, S.; Hirota, K.; Shimizu, T., *Tetrahedron Lett.* **1966**, *7*, 6131.

27. In addition to acid chlorides, α-diazo ketones can be synthesized from carboxylic acid anhydrides; however, in this case one equivalent of the carboxylic acid is converted to the corresponding methyl ester. Furthermore, the anhydride can be formed in situ using DCC. See Hodson, D.; Holt, G.; Wall, D. K., *J. Chem. Soc. (C)* **1970**, 971.

28. For a review of the Wolff rearrangement, see: Meier, H.; Zeller, K.-P.; *Angew. Chem., Int. Ed. Engl.* **1975**, *14*, 32.

29. Bergmann, E. D.; Hoffmann, E., *J. Org. Chem.* **1961**, *26*, 3555.

30. Clark, R. D., *Synth. Commun.* **1979**, *9*, 325.

31. Smith, A. B.; Dorsey, M.; Visnick, M.; Maeda, T.; Malamas, M. S., *J. Am. Chem. Soc.* **1986**, *108*, 3110.

32. Smith, A. B.; Toder, B. H.; Branca, S. J.; Dieter, R. K., *J. Am. Chem. Soc.* **1981**, *103*, 1996.

33. Rosenquist, N. R.; Chapman, O. L., *J. Org. Chem.* **1976**, *41*, 3326.

34. Hudliky, T.; Sheth, J. P., *Tetrahedron Lett.* **1979**, *20*, 2667.

35. For a review of intramolecular reactions of α-diazo ketones, see: Burke, S. D.; Grieco, P. A., *Org. React.* **1979**, *26*, 361.

36. Murai, A.; Kato, K.; Masamune, T., *Tetrahedron Lett.* **1982**, *23*, 2887.

37. Iwata, C.; Fusaka, T.; Fujiwara, T.; Tomita, K.; Yamada, M., *J. Chem. Soc., Chem. Commun.* **1981**, 463.

38. For a review on the reactions of α-diazo ketones with acid, see: Smith, A. B.; Dieter, R. K., *Tetrahedron* **1981**, *37*, 2407.

39. Ackeral, J.; Franco, F.; Greenhouse, R.; Guzman, A.; Muchowski, J. M., *J. Heterocycl. Chem.* **1980**, *17*, 1081.

40. Ghatak, U. R.; Sanyal, B.; Satyanarayana, G.; Ghosh, S., *J. Chem. Soc., Perkin Trans. 1* **1981**, 1203.

41. Blair, I. A.; Mander, L. N., *Aust. J. Chem.* **1979**, *32*, 1055.

42. Smith, A. B.; Toder, B. H.; Branca, S. J.; Dieter, R. K., *J. Am. Chem. Soc.* **1981**, *103*, 1996. Smith, A. B.; Dieter, K., *J. Am. Chem. Soc.* **1981**, *103*, 2009. Smith, A. B.; Dieter, K., *J. Am. Chem. Soc.* **1981**, *103*, 2017.

43. Smith, A. B.; Toder, B. H.; Branca, S. J., *J. Am. Chem. Soc.* **1984**, *106*, 3995.

44. Smith, A. B.; Toder, B. H.; Richmond, R. E.; Branca, S. J., *J. Am. Chem. Soc.* **1984**, *106*, 4001.

45. Holmquist, C. R.; Roskamp, E. J., *J. Org. Chem.* **1989**, *54*, 3258. Padwa, A.; Hornbuckle, S. F.; Zhang, Z. Z. Zhi, L.; *J. Org. Chem.* **1990**, *55*, 5297.

46. Holmquist, C. R.; Roskamp, E. J., *Tetrahedron Lett.* **1992**, *33*, 1131.

47. For a review of this reaction, see: Gutsche, C. D., *Org. React.* **1954**, *8*, 364.

48. For a review, see: Smith, P. A. S.; Baer, D. R., *Org. React.* **1960**, *11*, 157.

49. House, H. O.; Grubbs, E. J.; Gannon, W. F., *J. Am. Chem. Soc.* **1960**, *82*, 4099.

50. Majerski, Z.; Djigas, S.; Vinkovic, V., *J. Org. Chem.* **1979**, *44*, 4064.

51. Greene, A. E.; Depres, J-P., *J. Am. Chem. Soc.* **1979**, *101*, 4003.

52. For a review, see: Regitz, M.; Heydt, H. In *1,3-Dipolar Cycloadditions Chemistry*; Padwa, A.; Ed.; Wiley: New York, 1984; p 393.

53. For a discussion of the orbital interactions that control dipolar additions of diazomethane, see: Fleming, I. *Frontier Molecular Orbitals and Organic Chemical Reactions*; Wiley: New York, 1976; p 148.

54. Van Auken, T. V.; Rienhart, K. L., *J. Am. Chem. Soc.* **1962**, *84*, 3736.

55. This is a very reactive reagent combination which will cyclopropanate benzene and other aromatic compounds. See: Vogel, E.; Wiedeman, W.; Kiefe, H.; Harrison, W. F., *Tetrahedron Lett.* **1963**, *4*, 673. Muller, E.; Kessler, H.; Kricke, H.; Suhr, H., *Tetrahedron Lett.* **1963**, *4*, 1047.

56. Kuehne, M. E.; King, J. C., *J. Org. Chem.* **1973**, *38*, 304.

57. Pincock, R. E.; Wells, J. I., *J. Org. Chem.* **1964**, *29*, 965.

58. Suda, M., *Synthesis* **1981**, 714.

59. Raduchel, B.; Mende, U.; Cleve, G.; Hoyer, G. A.; Vorbruggen, H., *Tetrahedron Lett.* **1975**, *16*, 633.

60. Doering, W. von E.; LaFlamme, P., *J. Am. Chem. Soc.* **1956**, *78*, 5447.

61. Duncan, F. J.; Cvetanovic, R. J., *J. Am. Chem. Soc.* **1962**, *84*, 3593.

62. Fischer, N.; Opitz, G., *Org. Synth., Coll. Vol.* **1973**, *5*, 877.

63. Hata, Y.; Watanabe, M., *J. Am. Chem. Soc.* **1973**, *95*, 8450.

64. Cunico, R. F.; Lee, H. M.; Herbach, J., *J. Organomet. Chem.* **1973**, *52*, C7.

65. Helmkamp, G. K.; Pettit, D. J., *Org. Synth., Coll. Vol.* **1973**, *5*, 1099.

Tarek Sammakia

University of Colorado, Boulder, CO, USA

2,3-Dichloro-5,6-dicyano-1,4-benzoquinone[1]

[84-58-2]　　　　$C_8Cl_2N_2O_2$　　　　(MW 227.01)

InChI = 1/C8Cl2N2O2/c9-5-6(10)8(14)4(2-12)3(1-11)7(5)13

InChIKey = HZNVUJQVZSTENZ-UHFFFAOYAL

(powerful oxidant, particularly useful for dehydrogenation to form aromatic[1a–d] and α,β-unsaturated carbonyl compounds;[1] oxidizes activated methylene[1a–c] and hydroxy groups[1b] to carbonyl compounds; phenols are particularly sensitive[1c])

Alternate Name: DDQ.

Physical Data: mp 213–216 °C; $E_0 \approx 1000$ mV.

Solubility: very sol ethyl acetate and THF; moderately sol dichloromethane, benzene, dioxane, and acetic acid; insol H_2O.

Form Supplied in: bright yellow solid; widely available.

Analysis of Reagent Purity: UV (λ_{max} [dioxane] 390 nm) and mp.

Purity: recrystallization from a large volume of dichloromethane.

Handling, Storage, and Precautions: indefinitely stable in a dry atmosphere, but decomposes in the presence of water with the evolution of HCN. Store under nitrogen in a sealed container.

Original Commentary

Derek R. Buckle

SmithKline Beecham Pharmaceuticals, Epsom, UK

Introduction. Quinones of high oxidation potential are powerful oxidants which perform a large number of useful reactions under relatively mild conditions. Within this class, DDQ represents one of the more versatile reagents since it combines high oxidant ability with relative stability[1] (see also *Chloranil*). Reactions with DDQ may be carried out in inert solvents such as benzene, toluene, dioxane, THF, or AcOH, but dioxane and hydrocarbon solvents are often preferred because of the low solubility of the hydroquinone byproduct. Since DDQ decomposes with the formation of hydrogen cyanide in the presence of water, most reactions with this reagent should be carried out under anhydrous conditions.[1a]

Dehydrogenation of Hydrocarbons. The mechanism by which quinones effect dehydrogenation is believed to involve an initial rate-determining transfer of hydride ion from the hydrocarbon followed by a rapid proton transfer leading to hydroquinone formation.[1d] Dehydrogenation is therefore dependent upon the degree of stabilization of the incipient carbocation and is enhanced by the presence of functionality capable of stabilizing the transition state. As a consequence, unactivated

hydrocarbons are stable to the actions of DDQ while the presence of alkenes or aromatic moieties is sufficient to initiate hydrogen transfer.[1d,2] The formation of stilbenes from suitably substituted 1,2-diarylethanes[3] and the synthesis of chromenes by dehydrogenation of the corresponding chromans (eq 1)[4] are particularly facile transformations. Similar reactions have also found considerable utility for the introduction of additional unsaturation into partially aromatized terpenes and steroids, where the ability to control the degree of unsaturation in the product is a particular feature of quinone dehydrogenations.[5] Moreover, the ability to effect exclusive dehydrogenation in the presence of sensitive substituents such as alcohols and phenols (eq 2)[5b] illustrates the mildness of the method and represents a further advantage.

$$(1)$$

$$(2)$$

DDQ is a particularly effective aromatization reagent and is frequently the reagent of choice to effect facile dehydrogenation of both simple (eq 3)[6] and complex hydroaromatic carbocyclic compounds.[1d,7] Skeletal rearrangements are relatively uncommon features of quinone-mediated dehydrogenation reactions, but 1,1-dimethyltetralin readily undergoes aromatization with a 1,2-methyl shift when subjected to the usual reaction conditions (eq 4).[8] Wagner–Meerwein rearrangements have also been observed in the aromatization of steroids (eq 5), although in this instance considerably longer reaction times are required.[9] Such reactions provide a unique method for the aromatization of cyclic systems containing quaternary carbon atoms without the loss of carbon.

$$(3)$$

$$(4)$$

$$(5)$$

DDQ is also an effective reagent for the dehydrogenation of hydroaromatic heterocycles, and pyrroles,[10] pyrazoles,[11] triazoles,[12] pyrimidines,[13] pyrazines,[14] indoles,[15] quinolines,[16] furans,[17] thiophenes,[18] and isothiazoles[19] are among the many aromatic

Essential Reactions for Organic Synthesis, First Edition. Edited by Philip L. Fuchs.
© 2016 John Wiley & Sons, Ltd. Published 2016 by John Wiley & Sons, Ltd.

compounds prepared in this manner. Rearomatization of nitrogen heterocycles following nucleophilic addition across a C=N bond (eq 6)[13] is a particularly useful application of DDQ,[13,16] and similar addition and reoxidation reactions in acyclic systems have also been reported.[20]

(6)

One particularly important use of DDQ has been in the dehydrogenation of reduced porphyrins, where the degree of aromatization of the product is highly dependent on the relative reagent:substrate stoichiometry.[1b] Under optimal conditions, excellent yields of partially or fully conjugated products may be isolated.[1b,21] The formation of porphyrins from tetrahydro precursors on reaction with 3 equiv of DDQ under very mild conditions (eq 7) typifies one of the more commonly described transformations.[21] More recently, DDQ has been used as part of a one-pot sequence for the formation of porphyrins from simple intermediates, although the overall yields in such reactions are generally comparatively low.[22]

(7)

In addition to the formation of neutral aromatic compounds, DDQ is also an effective agent for the preparation of the salts of stable aromatic cations. High yields of tropylium (eq 8) and triphenylcyclopropenyl (eq 9) cations have been isolated in the presence of acids such as perchloric, phosphoric, and picric acid,[23] and oxonium,[24] thioxonium,[23,25] and pyridinium[23,26] salts may be prepared in reasonable yields from appropriate starting materials under essentially similar conditions. The formation of the perinaphthyl radical has been reported on oxidation of perinaphthalene with DDQ under neutral conditions,[23] although such products are not usually expected.

(8)

(9)

Dehydrogenation of Carbonyl Compounds. DDQ and other high oxidation potential quinones are versatile reagents for the synthesis of α,β-unsaturated carbonyl compounds,[1e] a reaction that has found extensive application in the chemistry of 3-keto steroids.[1b] The regiochemical course of this dehydrogenation is highly dependent on the initial steroidal geometry; thus the 5α- and the 5β-series usually furnish Δ^1- and Δ^4-3-keto steroids, respectively (eq 10).[1b] The selection of one isomer over the other is likely to reflect the relative steric crowding of the C-4 hydrogen atom in the two series, but other factors may play a role in those instances where the anticipated product is not formed.[1b]

(10)

A rather more unusual situation exists during the dehydrogenation of Δ^4-3-keto steroids where the product formed is dependent on the oxidizing quinone. Thus whereas DDQ gives the $\Delta^{4,6}$ ketone, chloranil and a number of other quinones yield only the $\Delta^{1,4}$ isomer (eq 11), a result that has been rationalized on the basis of DDQ proceeding via the kinetic enolate while less reactive quinones proceed via the thermodynamic enolate.[27]

(11)

While DDQ is an effective reagent for the formation of α,β-unsaturated steroidal ketones, the dehydrogenation of cyclohexanones to the corresponding enone only proceeds well when the further reaction is blocked by *gem*-dialkyl substitution.[1b,28] Tropone, by contrast, has been prepared from 2,4-cycloheptadienone (eq 12), although the yield was somewhat low.[29] Heterocyclic enones such as flavones[30] and chromones[31] may be efficiently prepared from flavanones and chromanones, respectively, under similar conditions to those used for the dehydrogenation of steroids, and the dehydrogenation of larger ring heterocyclic ketones has been described.[32] Ketone enol ethers have also been shown to undergo facile dehydrogenation to α,β-unsaturated ketones with DDQ, although the nature of the product formed may be dependent on the presence or absence of moisture.[33] Prior formation of the silyl enol ether is a potentially more versatile procedure that has been shown to overcome the problems generally associated with the dehydrogenation of unblocked cyclohexanones (eq 13),[34] particularly when the acidic hydroquinone formed during the reaction is neutralized by the addition of *N,O-Bis(trimethylsilyl)acetamide* (BSA)[34] or a hindered base.[35] Preparation of the enone derived from either the kinetic or the thermodynamic enolate is possible in this manner.[34,35a]

$$(12)$$

$$(13)$$

Quinone dehydrogenation reactions of carbonyl compounds are mostly limited to the more readily enolized ketones, and analogous reactions on esters[36] and amides[37] require stronger conditions and are far less common unless stabilization of the incipient carbonium ion is possible. Oxidation in the presence of the silylating agent bis(trimethylsilyl)trifluoroacetamide (BSTFA) considerably improves the dehydrogenation of steroidal lactams (eq 14) by facilitating the breakdown of the intermediate quinone–lactam complex.[38] Similar dehydrogenations of carboxylic acids are rare, but reaction of the α-anion of carboxylate salts generated in the presence of HMPA has given modest yields of a number of α,β-unsaturated fatty acids.[39]

$$(14)$$

Oxidation of Alcohols. Saturated alcohols are relatively stable to the action of DDQ in the absence of light, although some hindered secondary alcohols have been oxidized in reasonable yield on heating under reflux in toluene for extended periods of time (eq 15).[40] It has been suggested that oxidation proceeds in this instance as a result of relief of steric strain.[40] Allylic and benzylic alcohols, on the other hand, are readily oxidized to the corresponding carbonyl compounds,[1b,41] and procedures have been developed which utilize catalytic amounts of the reagent in the presence of a stoichiometric amount of a second oxidant.[42] Since the rate of oxidation of allylic alcohols is greater than that for many other reactions,[43] the use of DDQ provides a selective method for the synthesis of allylic and benzylic carbonyl compounds in the presence of other oxidizable groups.

$$(15)$$

Benzylic Oxidation. The oxidation of benzylic alkyl groups proceeds rapidly in those instances in which stabilization of the

incipient carbonium ion is possible[44,45] and a number of polycyclic aromatic compounds have been oxidized in good yield to the corresponding benzylic ketones on brief treatment with DDQ in aqueous acetic acid at rt.[44] The reaction is postulated to proceed via an intermediate benzylic acetate which is hydrolyzed and further oxidized under the reaction conditions.[44] It is interesting that 1-alkylazulenes, which are cleaved by many of the more common oxidants, are cleanly oxidized following a short exposure to DDQ in aqueous acetone (eq 16), while under the same conditions no oxidation of C-2 alkyl substituents takes place.[46] As expected, the oxidation was shown to be disfavored by the presence of strongly electron-withdrawing substituents.[46]

$$(16)$$

The stabilization of benzylic carbonium ions is also a feature of arenes containing electron-donating substituents, especially those having 4-alkoxy or 4-hydroxy groups, and such compounds are particularly effective substrates for oxidation by DDQ. Thus 6-methoxytetralone has been prepared in 70% yield from 6-methoxytetralin on treatment with DDQ in methanol,[47] although it is possible to isolate intermediate benzylic acetates if the oxidation is carried out in acetic acid (eq 17).[48] An interesting variant of the oxidation in inert solvents in the presence of either *Cyanotrimethylsilane*[49] or *Azidotrimethylsilane*[50] results in the isolation of good to excellent yields of benzyl cyanides and azides, respectively.

$$(17)$$

Benzylic oxidation of alkoxybenzyl ethers is particularly facile, and since some of the more activated derivatives are cleaved under conditions which leave benzyl, various ester, and formyl groups unaffected, they have found application in the protection of primary and secondary alcohols.[51] Deprotection with DDQ in dichloromethane/water follows the order: 3,4-dimethoxy > 4-methoxy > 3,5-dimethoxy > benzyl and secondary > primary, thus allowing the selective removal of one function in the presence of another.[51] 2,6-Dimethoxybenzyl esters are readily cleaved to the corresponding acids on treatment with DDQ in wet dichloromethane at rt, whereas 4-methoxybenzyl esters are stable under these conditions.[52] Oxidative cleavage of N-linked 3,4-dimethoxybenzyl derivatives with DDQ has also been demonstrated.[53]

DDQ is a powerful oxidizing agent for phenols, and carbonium ion stabilization via the quinone methide makes benzylic oxidation of 4-alkylphenols a highly favored process.[47,54] With methanol as the solvent it is possible to isolate α-methoxybenzyl derivatives in reasonable yield.[55]

Phenolic Cyclization and Coupling Reactions. The oxidation of phenolic compounds which either do not possess benzylic hydrogen atoms, or which have an alternative reaction pathway, can result in a variety of interesting products. Cyclodehydrogenation reactions leading to oxygen heterocycles represent a particular application of phenolic oxidation by DDQ, and is common when intramolecular quenching of the intermediate phenoxyl radical is possible (eqs 18–20).[56–59] These reactions necessarily take place in nonpolar solvents and have given such products as coumarins,[56] chromenes (eq 18),[57] benzofurans (eq 19),[58] and spiro derivatives (eq 20).[59]

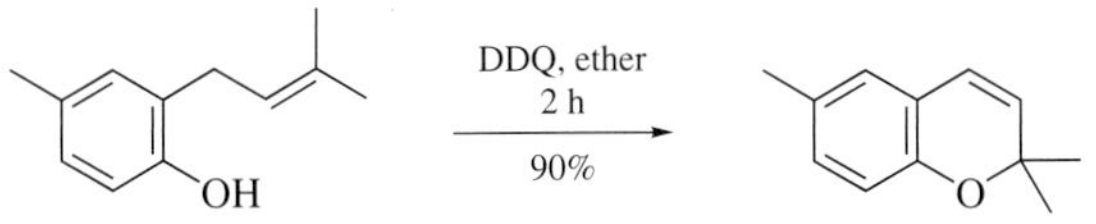

(18)

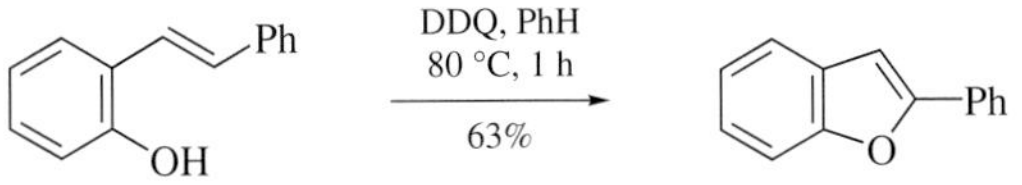

(19)

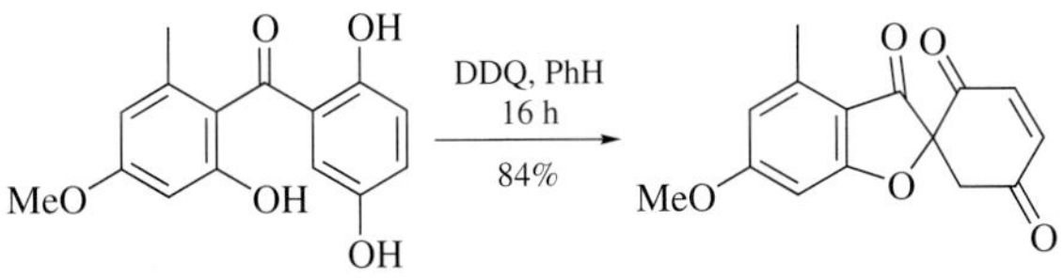

(20)

Phenols and enolizable ketones that cannot undergo α,β-dehydrogenation may afford intermolecular products arising from either C–C or C–O coupling on treatment with DDQ in methanol.[60] 2,6-Dimethoxyphenol, for example, results predominantly in oxidative dimerization (eq 21), while the hindered 2,4,6-tri-*t*-butylphenol generates the product of quinone coupling (eq 22).[60] Various other unusual products have been observed on DDQ oxidation of phenols and enolic compounds, their structure being dependent on that of the parent compound.[41,60,61]

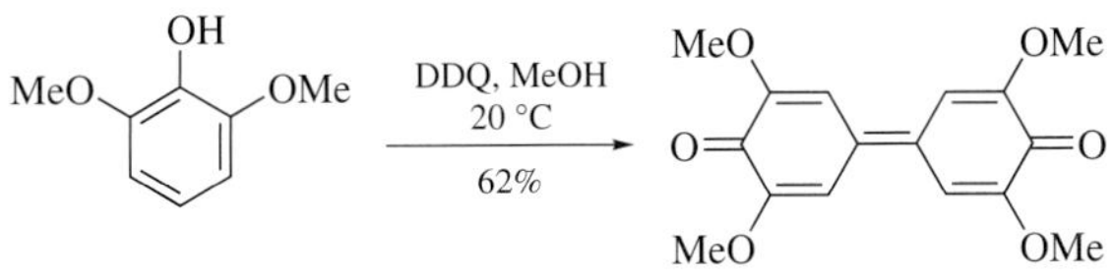

(21)

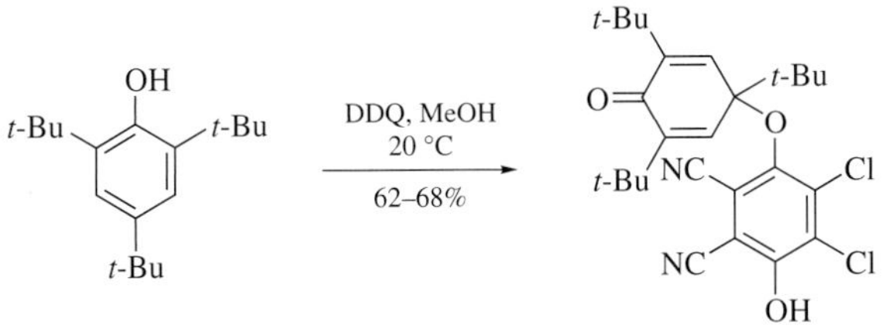

(22)

Miscellaneous Reactions. In addition to the key reactions above, DDQ has been used for the oxidative removal of chromium,[62] iron,[63] and manganese[64] from their complexes with arenes and for the oxidative formation of imidazoles and thiadiazoles from acyclic precursors.[65] Catalytic amounts of DDQ also offer a mild method for the oxidative regeneration of carbonyl compounds from acetals,[66] which contrasts with their formation from diazo compounds on treatment with DDQ and methanol in nonpolar solvents.[67] DDQ also provides effective catalysis for the tetrahydropyranylation of alcohols.[68] Furthermore, the oxidation of chiral esters or amides of arylacetic acid by DDQ in acetic acid provides a mild procedure for the synthesis of chiral α-acetoxy derivatives, although the diastereoselectivity achieved so far is only 65–67%.[69]

While quinones in general are well known dienophiles in Diels–Alder reactions, DDQ itself only rarely forms such adducts.[70] It

has, however, been shown to form 1:1 adducts with electron-rich heterocycles such as benzofurans and indoles where it forms C–O and C–C adducts, respectively.[71]

First Update

Steven J. Collier & Mark D. McLaws
Albany Molecular Research Inc., Albany, NY, USA

Dehydrogenation of Hydrocarbons. DDQ is often the oxidant of choice for the dehydrogenation of sensitive compounds. For example, acenes, known for being prone to photooxidation and photodimerization reactions, were produced in situ for cycloaddition with [60]fullerene by DDQ dehydrogenation (eq 23).[72]

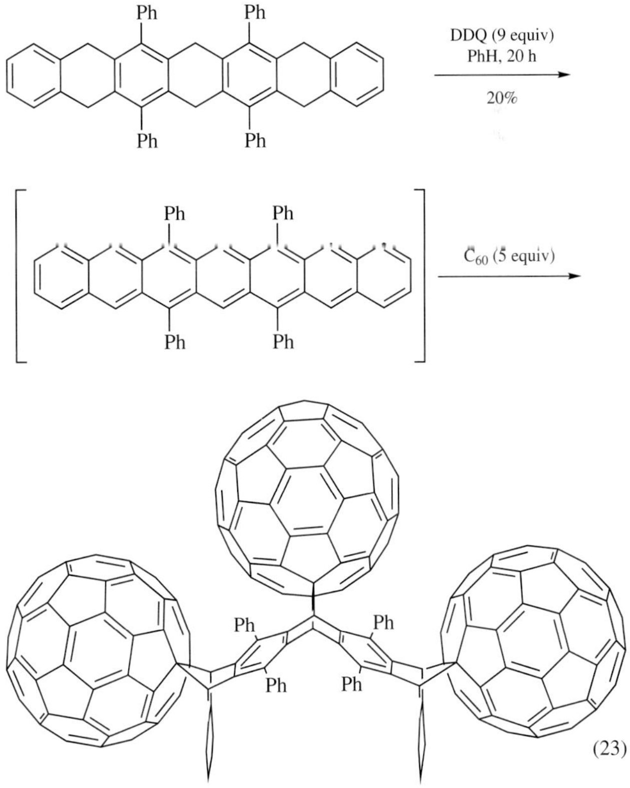

(23)

The mild conditions under which DDQ effects dehydrogenation of dihydroquinoline compounds has proved useful for activating a resin bound safety-catch linker for cleavage by aminolysis (eq 24).[73] The excess DDQ and DDQH are easily removed by filtration after the activation step, effectively purging the system of these potential impurities. Conversely, removal of DDQ and DDQH with polymer-bound scavenger resins has also proven valuable in solution phase applications.[74,75]

The aromatization of pyrrolidines to 2*H*-pyrroles occurs in good yields without sigmatropic rearrangement to give 1*H*-pyrroles, which can occur under conditions employing other oxidants such as MnO$_2$ and chloranil (eq 25).[76]

Oxidation of phenylalkylacetylenes gives rise to (*Z*)-enynes (eq 26).[77] Successful dehydrogenation is highly substrate dependant, requiring an α-substituent capable of stabilizing the charged transition state. Thus, 1-phenyl and 1-phenylthio substituted alkynes undergo oxidation while 1-unsubstituted or 1-alkyl substituted alkynes fail to react. Although the oxidation appears to proceed nonstereoselectively, the (Z)-olefin geometry is obtained due to selective removal of the (E)-enyne by DDQ through an unknown mechanism. As a result, product yields around 45–50% are generally observed.

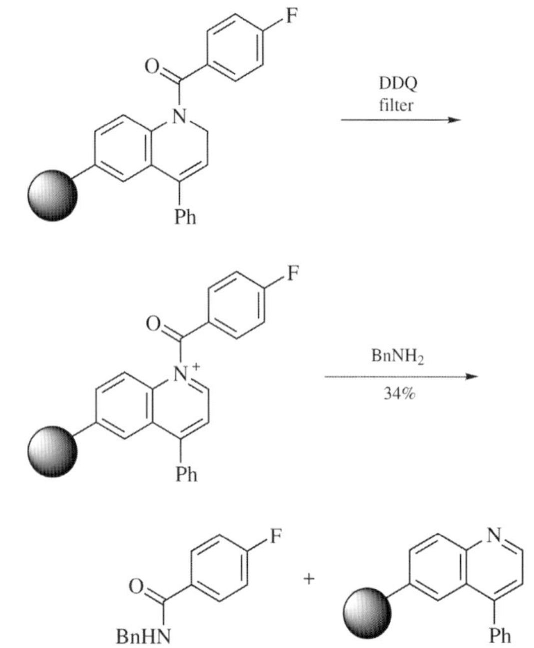

(24)

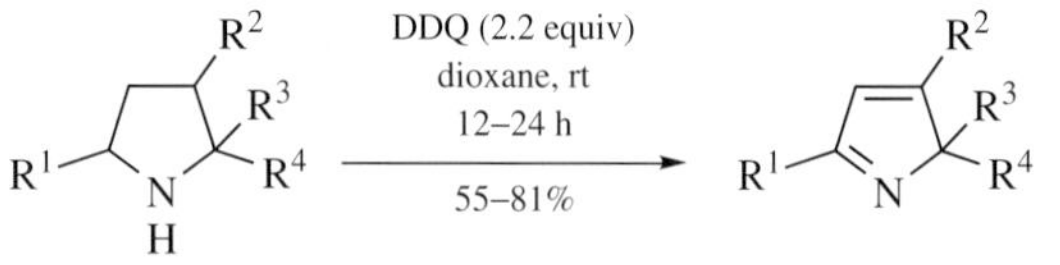

(25)

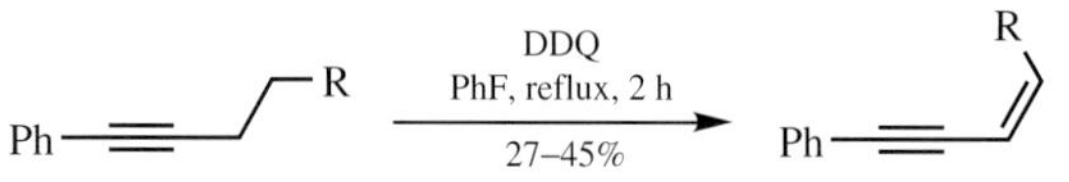

(26)

Dehydrogenation of Carbonyl Compounds. Oxidation of thioflavanones gave thioflavones in excellent yield whereas dehydrogenation with o-chloranil under like conditions failed (eq 27).[78] Inductive effects play an important role in such dehydrogenations. For example, the sulfone analog resisted dehydrogenation whereas the sulfoxide derivative gave the thioaurone product in low yield.

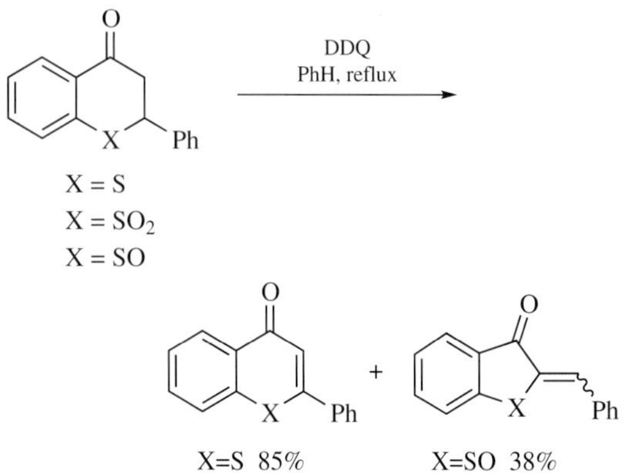

(27)

In some cases substrates with subtle structural features have allowed for alternative reaction pathways yielding compounds that are quite different than the expected dehydrogenation product.[79–81] For example, an unusual skeletal rearrangement took place when certain methanonaphthalene derivatives were dehydrogenated with DDQ (eq 28).[79] While the expected annulene was obtained when compounds with a mono-substituted bridging methyl group underwent dehydrogenation, similar substrates, with di-substituted bridging methyl groups, gave vinyl-naphthalenes through a 1,5-sigmatropic carbon shift (Berson-Willcott rearrangement).

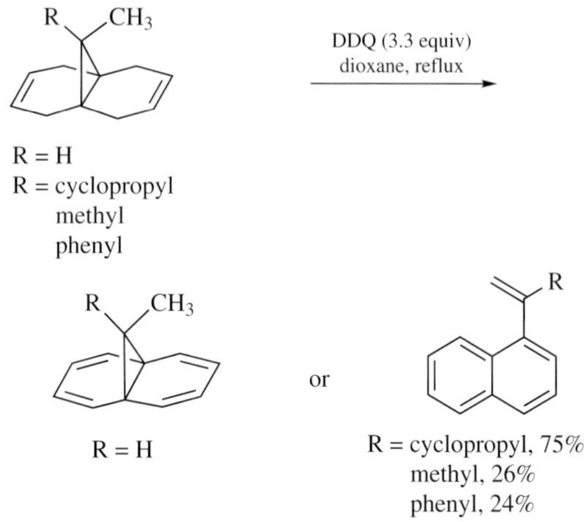

(28)

The dehydrogenation of eudesma-4-en-3-ones, like the related 3-keto steroids,[1b] are also strongly influenced by both steric and electronic factors. Although 9-hydro-eudesmaenones underwent dehydrogenation to give the corresponding dieneone in good yield, its 9-keto counterpart failed to react under comparable reaction conditions (eq 29).[81] Interestingly, dehydrogenation of similar 9-keto substrates without the 4-methyl substituent gave a rearranged aromatic product in good yield. The steric factors defined by the orientation of the angular 10-methyl group also played a major role in the success of the oxidation, with the β-epimer generally giving higher yields of dehydrogenated products.

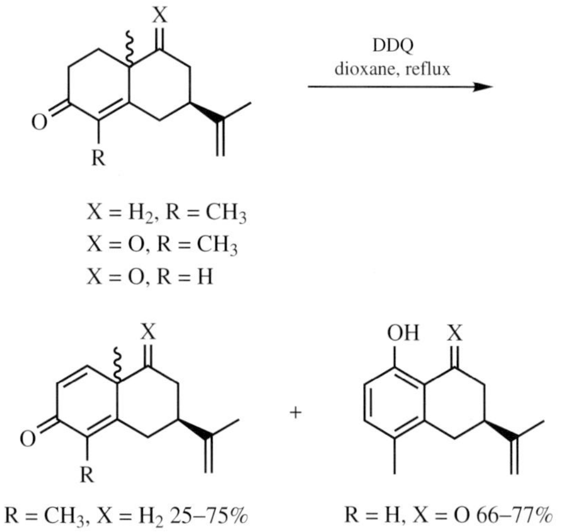

(29)

Benzylic Oxidation. During the preparation of 4-ethyl or 4-benzylpyridines via aromatization of dihydropyridine derivatives using DDQ, extended reactions lead to formation of hydroquinone ethers (eq 30). This degradation was much less significant when a Weinreb amide unit was in place of the ester function.[82]

$$\text{R = Me: 14\%; R = Ph: 47\%} \quad (30)$$

DDQ has also been known to oxidize the unreactive methyl groups of azulenes to formyl groups.[46] A range of 3-substituted Guaiazulene derivatives were oxidized using this technique, although the reaction was poorer with more electron withdrawing 3-substituents (eq 31). However, use of a large excess of DDQ gave rapid and essentially quantitative conversion to the formyl derivative.[83]

$$(31)$$

The oxidation of bisbenzyl ethers using DDQ has been studied in some detail and the reaction is mild and highly selective for the monoaldehyde products, even when excess oxidant is present (eq 32).[84] Conversion of *para*- and *meta*-bisbenzyl ethers was found to be rapid whereas *ortho*-bisbenzyl ethers gave a slower reaction. However, the rate could be increased when electron donating groups were present, without compromising the selectivity for the monoaldehyde. The reaction rate is retarded by aromatic side chain hindrance (but not side chain steric hindrance), suggesting that the reaction proceeds via the formation of an initial π-stacked complex.

$$(32)$$

The oxidation of 1-arylpropenes with DDQ in the presence of water to give cinnamaldehydes is well known.[85–88] The treatment of 3-arylpropenes under similar conditions also affords cinnamaldehydes (eq 33), although a large excess of DDQ and longer reaction times are required.[89,90] The reaction proceeds via formation of a quinol ether and is promoted by the presence of electron releasing groups on the aryl ring. Oxidation in the presence of MeOH gives esters and in the presence of acetic acid, acylals are obtained (eq 33).

$$(33)$$

Phenolic Cyclization and Coupling Reactions. DDQ has recently been successfully employed in oxidative cyclizations and coupling reactions to give naphthofuranquinones and naphthopyranquinones,[91] spirocoumaranones,[92] cyclooctadiene and spirodienone lignans,[93,94] and various porphyrin derivatives.[95,96] The ratio of naphthofuranquinone and naphthopyranquinone produced upon oxidation of alkenylnaphthoquinones was found to be temperature dependant (eq 34).[91] At elevated temperatures, the furan derivative is the preferred product, whereas the pyran derivative is favored at lower temperatures.

$$(34)$$

78% at 8 °C   70% at 60 °C

The oxidative cyclization of substituted benzophenones using DDQ can give sensitive spirocoumaranones (eq 35).[92] This phenolic cyclization is unusual in that it involves a highly constrained tetra-*ortho*-substituted benzophenone arrangement which, under other oxidative conditions, traditionally results in decomposition or further reaction to give diphenyl ether products through cleavage of the benzylic ketone.[97,98]

Oxidative cyclization of 2,3-dibenzyl-2-hydroxybutanes gives the expected lignans along with an unusual spirocyclization product which is thought to be derived from the cyclooctadiene product through oxidative formation of the bridging ether followed by a 1,2-aryl shift (eq 36).[93]

Oxidative coupling of triarylporphyrins was accomplished with DDQ-Sc(OTf)₃ (eq 37).[95] Sc(OTf)₃ is added to increase the oxidation potential of DDQ by interacting with the anion radical. Another interesting aspect of this oxidative coupling reaction is the regiochemical bias exhibited by porphyrins with certain core metals. While oxidative coupling of porphyrins with zinc cores gave triply-linked diporphyrins, porphyrins with palladium cores gave only the doubly-linked product. The influence exerted by the core metal on the relative energies of the cation radical HOMO orbital may account for the regioselectivity differences.

Deprotection Reactions. Several new alcohol protecting groups which are cleaved by DDQ have been reported. For example, *p*-phenylbenzyl ethers, which can be introduced to alcohols under either acidic (via the corresponding trichloroacetimidate and triflic acid) or basic conditions (via *p*-phenylbenzyl bromide and base), exhibit enhanced acid stability over the well-known PMB (*p*-methoxybenzyl) groups. The protecting group can be cleaved through treatment with either DDQ, or using catalytic DDQ with Mn(OAc)$_3$ as a reoxidant for the hydroquinone.[99] Similarly, 2-naphthylmethyl ethers (prepared from the corresponding bromide) are useful protecting groups for carbohydrates which can be cleaved with DDQ.[100] This protecting group is stable to the acidic conditions required for acetal cleavage, and also other conditions known to partially or completely cleave PMB groups.

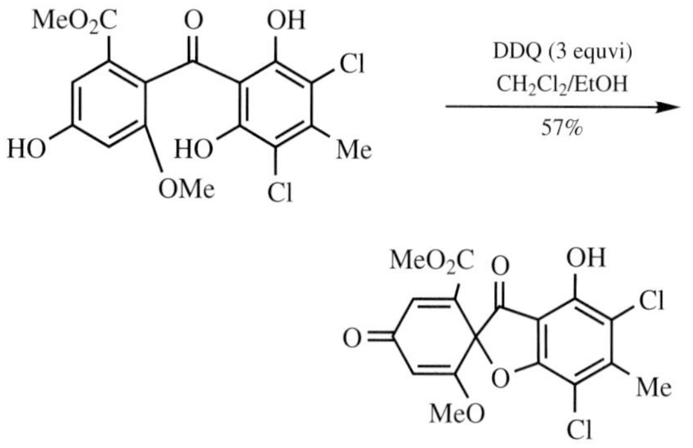

(35)

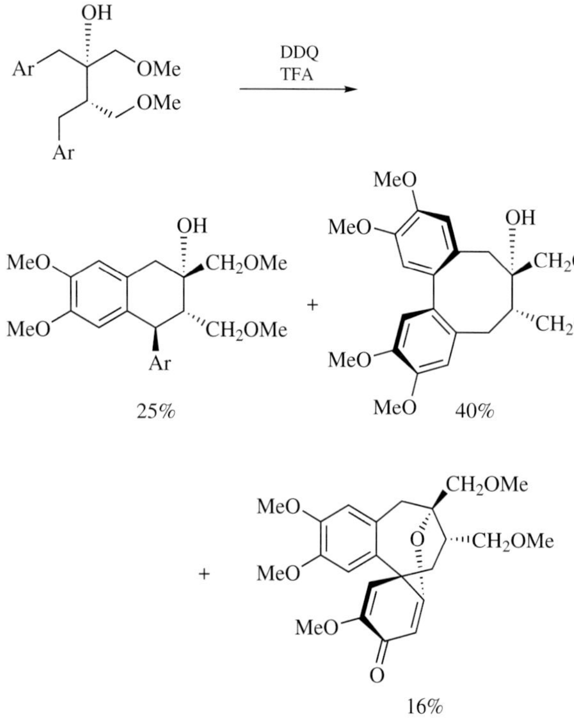

(36)

Prenyl ethers can be cleaved under mild conditions (1.2 equiv DDQ, 9:1 CH$_2$Cl$_2$:water) without cleaving other alcohol protect-

ing groups including acetals, acetates, allyl, benzyl, and TBDPS ethers (eq 38).[101] As mentioned above, the DDQ can be present in catalytic amounts if Mn(III) salts are present as a reoxidant. *O*-Allyl ethers are also cleaved using stoichiometric DDQ to give the corresponding alcohol and acrolein.[102] The reaction is selective for allyl ether in the presence of acetonides, acetates, benzyl, and TBDPS ethers.

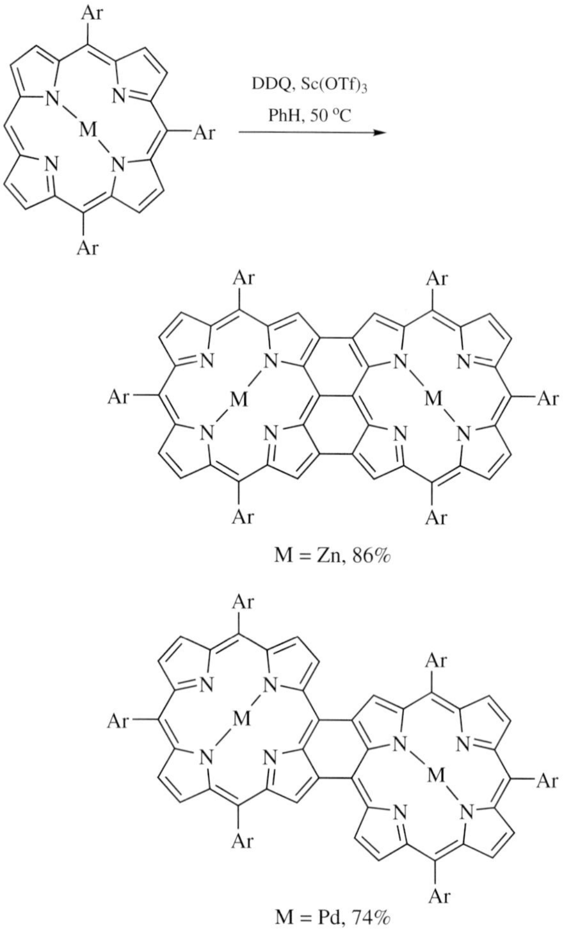

M = Zn, 86%

M = Pd, 74% (37)

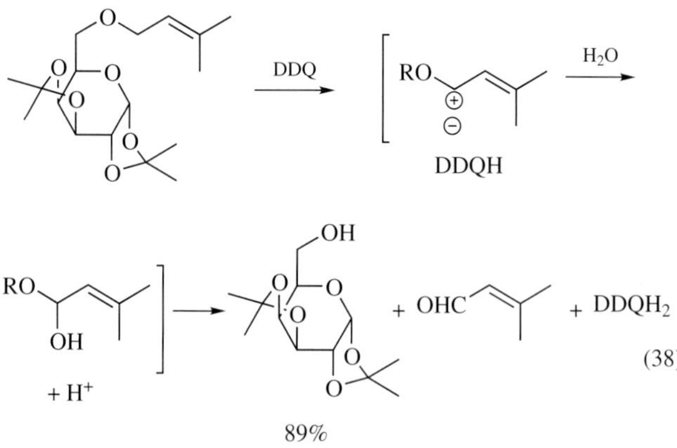

(38)

89%

Cyclic dithioketals can also be cleaved upon treatment with DDQ in the presence of water, although the outcome of the reaction can depend on the ring size of the dithioketal and upon the substituents on the carbon backbone.[103,104] For example, many 2-substituted 1,3-dithianes convert to the parent aldehyde compounds in good yield (eq 39), but competing thioester formation is observed with dithianes derived from electron-rich benzaldehydes. 1,3-Dithiolanes derived from aromatic aldehydes were converted to thioesters, whereas those derived from aliphatic and aromatic ketones were stable. Diphenyldithioacetals were also generally stable unless they were derived from electron-rich benzaldehydes. Selective cleavage of 1,3-dithianes in the presence of 1,3-dithiolanes, and diphenylthioacetals was reported.[103] 1,3-Dithianes, 1,3-dithiolanes, and diphenylthioacetals derived from cinnamaldehydes gave benzaldehydes in up to 50% yield upon treatment with DDQ in aqueous solvents (eq 40).[104] Other products were also obtained depending upon the nature of the aryl ring.

$$n = 1 \quad 1\text{–}21\% \qquad 0\text{–}75\%$$
$$n = 2 \quad 23\text{–}97\% \qquad 8\text{–}74\%$$

$$R^1 = R^2 = Ph, -(CH_2)_2-, -(CH_2)_3-$$

Cycloaddition Reactions. There are a small number of examples in which DDQ participates in cycloaddition reactions. A substituted pentacene undergoes [4 + 2] cycloaddition with DDQ (eq 41), with reaction occurring exclusively at the least hindered diene function, due to steric constraints.[105] An analogous reaction was observed with a related naphthacene derivative. DDQ also undergoes a reversible cycloaddition with N-carboxyphenyldihydropyridine derivatives (eq 42).[82] The 2-, 4-, and 6-ethyldihydropyridine derivatives were prepared as an isomeric mixture. However, only the 6-ethyl derivative underwent cycloaddition with DDQ, whereas the 4-ethyl derivative aromatized with elimination of the N-acyl function, and the 2-ethyl derivative did not react at all. Furthermore, heating of the Diels-Alder adduct resulted in quantitative regeneration of the 6-ethyl derivative and DDQ. Aromatization the 6-ethyl derivative could be achieved by using higher reaction temperatures or through a slow retro-Diels Alder/aromatization sequence of the cycloadduct.[82] [3 + 2] Cycloadditions between 4-carboxyethyl-5-alkoxyoxazoles and one of the C=O functions of DDQ have also been reported (eq 43), but the reactions only occur under high pressure conditions.[106] The absence of the ester function results in complex mixtures under a variety of conditions, including thermal, high pressure, and Lewis acid catalysis.

Mitsunobu-type Reactions. The combination of triphenylphosphine and DDQ can be used to activate alcohols and related substrates to nucleophilic displacement in a reaction which is related to the widely used Mitsunobu reaction (using triphenylphosphine and an azodicarboxylate) (eq 44). For

R = Me: 59%; R = Me(CH₂)₈: 20%

example, alcohols (including primary, secondary, allylic, benzylic alcohols), thiols, and selenols are converted in good yields to alkyl halides (chloride, bromide, iodide) when treated with tetraalkylammonium halides in the presence of the DDQ/PPh$_3$ system.[107] The method is highly selective for primary alcohols in the presence of secondary alcohols and also for primary and secondary alcohols over tertiary alcohols, thiols, and many other functional groups. Tetrabutylammonium cyanides[108] or azides[109] react with alcohols (including tertiary alcohols), thiols, and TMS ethers under similar conditions to give the corresponding alkyl cyanides or azides in high yields and the reaction exhibits selectivity between different alcohols and between alcohols, thiols, and TMS ethers. In the latter case, this method represents a safer and milder method of introducing azide groups into organic molecules when compared to many existing methods. Use of a higher amount of DDQ than PPh$_3$ prevents competitive formation of iminophosphoranes from the alkyl azide products.[109] Diethyl α-thiocyanatophosphonates have been prepared from the corresponding α-hydroxyphosphonates using ammonium thiocyanate and the DDQ/PPh$_3$ system, giving greatly improved performance over the classical Mitsunobu conditions (using DEAD).[110] The DDQ/PPh$_3$ approach was also successful for the preparation of diethyl α-bromo, iodo, and azido phosphonates from the corresponding α-hydroxy compounds and n-Bu$_4$NBr, n-Bu$_4$NI, and NaN$_3$.[111]

$$\text{RXH} + \text{R}'_4\text{NY} \xrightarrow{\text{DDQ, PPh}_3} \text{RY} \quad (44)$$

X = OH (1°, 2°), SH, SeH; Y = Cl, Br, I: 35–99%

X = OH (1°, 2°, 3°), SH, OTMS; Y = CN, N$_3$: 40–97%

X = OH ; Y = SCN: 60–90%

Nucleophilic Addition and Substitution Reactions. Aside from oxidation reactions, DDQ can undergo nucleophilic addition and substitution reactions to give adducts with interesting physical and chemical properties. The electron deficient quinone can undergo Michael-type additions with expulsion of either chloride[112] or cyano[113,114] ring substituents or both,[115] depending on the nucleophile and reaction conditions. Reaction of DDQ with ylides represents an interesting case where nucleophilic attack on the quinone can occur in both 1,2- and 1,4-fashion,[116] as well as at the cyano carbon,[116–118] depending on the nature of the ylide. An interesting cycloadduct of DDQ and bis(enamines) was formed presumably through a sequence of conjugate addition of the enamine to DDQ followed by either Diels-Alder reaction with the resulting quinone enolate or conjugate addition to the enamine and internal aldol-like cyclization (eq 45).[119]

Miscellaneous Reactions. DDQ promotes the ferrocenium tetrakis[3,5-bis(trifluoromethyl)phenyl]borate catalyzed iodination of substituted benzenes with ICl, although it does not itself act as a catalyst.[120] With toluene, a 47:53 mixture of *ortho:para* iodotoluene is obtained in a yield of 82% with 13% chlorotoluene (67:33 *ortho:para*) also observed. The role of the DDQ has not yet been elucidated, but it is assumed to effectively inhibit unfavorable radical iodination of toluene to form benzyl iodide. Other monoiodinated aromatic species were also prepared from the corresponding arenes under these conditions, including 4-

iodo-*m*-xylene (62%), iodo-*p*-xylene (56%), 4-iodoanisole (70%, 99:1 *para:ortho*) and 4-iodoaniline (54%, 99:1 *para:ortho*).

$$(45)$$

DDQ is also an effective reagent for the preparation of the highly electrophilic phenylselenium cation via oxidation of diphenyldiselenide.[121] The reaction occurs under mild conditions and is carried out in the presence of an alkene and suitable nucleophile (e.g., MeOH, water), to give the corresponding alkoxy- or hydroxy-phenylselenylation product (eq 46). The reaction tolerates functional groups including alkynes, carbonyls, nitriles, or acetoxy groups. When internal nucleophiles are present in the substrate, ring closure reactions can occur (eq 47).

$$(46)$$

R^3 = Me, H

$$(47)$$

8:2 mix of diastereomers

3-Alkoxy-2,5-diphenylfurans and 2,3,5-triphenylfuran can be oxidatively cleaved with DDQ in CH$_2$Cl$_2$ or CH$_2$Cl$_2$/DMSO to give *cis*-2-alkoxy (or 2-phenyl)-but-2-ene-1,4-diones eq (48).[122] With 2-methyl-5-pentylfuran, the *trans*-enedione was obtained, but the reaction was unsuccessful with 2,5-diphenylfuran, 2-methoxyfuran, and some other furans.

$$(48)$$

R = OMe, OEt, OPr, Oi-Pr, OBu, Oi-Bu, Ph 80–98%

As mentioned previously, the oxidation potential of DDQ can be enhanced by the addition of proton or Lewis acids. Activation by a super acid such as triflic acid can enable oxidation of adamantanes despite the extremely high oxidation potential of these systems.[123] Thus treatment of adamantanes with DDQ in triflic acid results in hydride abstraction to give the corresponding cation. Trapping of the cation with TfOH gives the alkyl triflates, which give adamantan-1-ols in good yields after an S_N1-type hydrolysis (eq 49).[123] This method gives superior results to other oxidation methods, which give mixtures of regioisomeric alcohols, plus ketones and 1,3-diols.

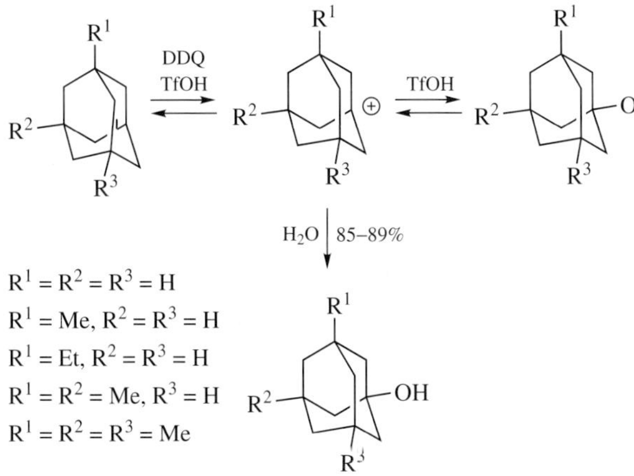

$$R^1 = R^2 = R^3 = H$$
$$R^1 = Me, R^2 = R^3 = H$$
$$R^1 = Et, R^2 = R^3 = H$$
$$R^1 = R^2 = Me, R^3 = H$$
$$R^1 = R^2 = R^3 = Me$$

A novel and interesting reaction mediated by DDQ involves the nucleophilic substitution of thiophene substituted trityl-type alcohols via gas-solid contact.[124] Co-grinding 9-thienothienylfluoren-9-ol derivatives with DDQ gives a highly colored radical cation (as suggested by ESR studies) via a charge transfer interaction (eq 50). The reaction remains in the solid state throughout the process. Collapse of the radical cation generates a proton which acts as a catalyst for the formation of a carbocation from the parent fluorenol. Upon exposure to MeOH vapor this gives the corresponding methyl ether and regenerates the proton catalyst. The reaction propagates in the solid state giving low to moderate yields of product. Methoxylation of a cephalosporin derivative was achieved using DDQ in MeOH, giving methoxylation of the β-lactam ring α to the carbonyl unit (eq 51).[125]

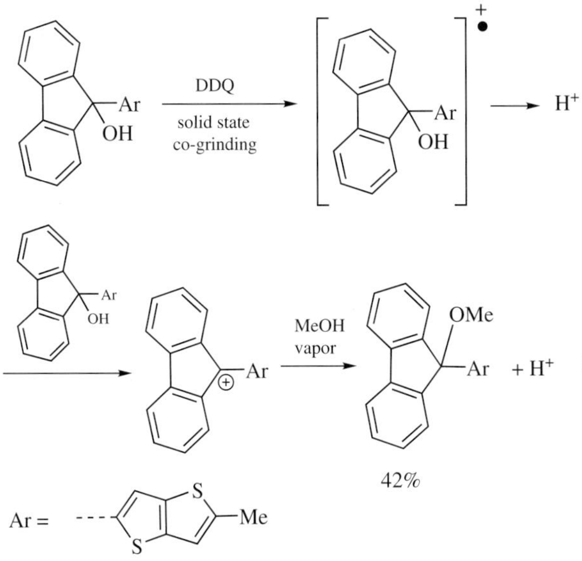

(50)

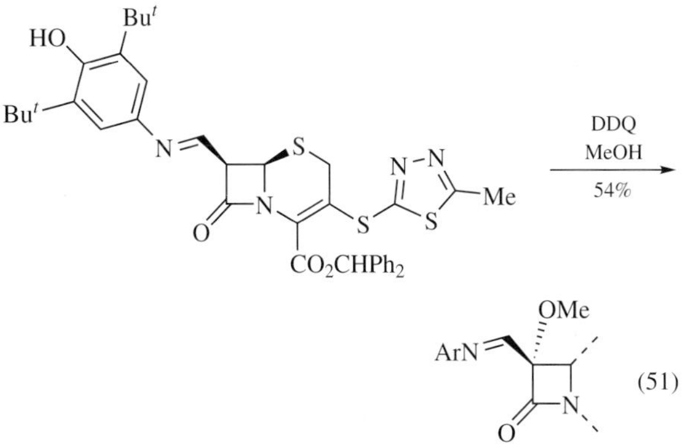

(51)

Treatment of an allylic alcohol with $Pd(OAc)_2$ and DDQ resulted in a novel rearrangement-oxidation product through an intermediate conjugate addition adduct (eq 52).[126] Subsequent dehydration and delivery of hydroxide to quench the resulting conjugated iminium ion gave a terminal allylic alcohol which underwent DDQ oxidation to the aldehyde.

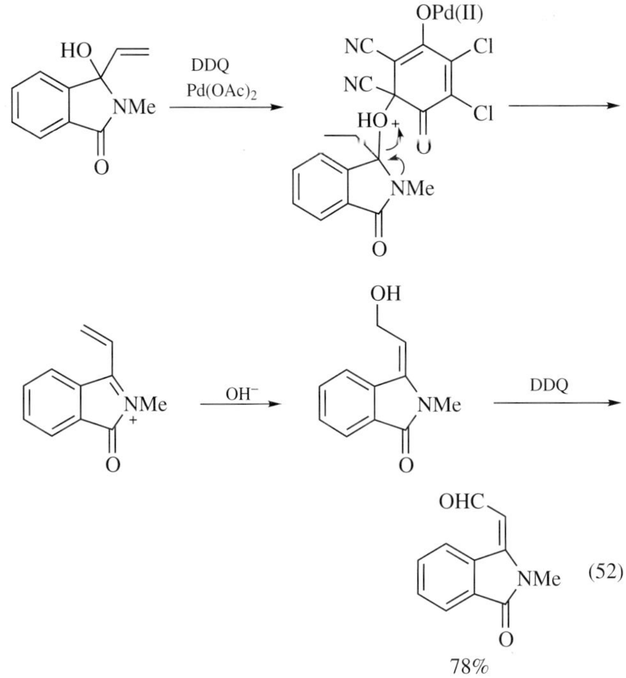

(52)

78%

1. (a) Jackman, L. M., *Adv. Org. Chem.* **1960**, *2*, 329. (b) Walker, D.; Hiebert, J. D., *Chem. Rev.* **1967**, *67*, 153. (c) Becker, H.-D. In *The Chemistry of the Quinonoid Compounds*; Patai, S., Ed.; Wiley: Chichester, 1974; Part 2, Chapter 7. (d) Fu, P. P.; Harvey, R. G., *Chem. Rev.* **1978**, *78*, 317. (e) Buckle, D. R.; Pinto, I. L., *Comprehensive Organic Synthesis* **1991**, *7*, 119.

2. Asato, A. E.; Kiefer, E. F., *J. Chem. Soc., Chem. Commun.* **1968**, 1684.

3. Findlay, J. W. A.; Turner, A. B., *Org. Synth.* **1969**, *49*, 53.

4. (a) Starratt, A. N.; Stoesl, A., *Can. J. Chem.* **1977**, *55*, 2360. (b) Ahluwalia, V. K.; Arora, K. K., *Tetrahedron* **1981**, *37*, 1437. (c) Ahluwalia, V. K.; Ghazanfari, F. A.; Arora, K. K., *Synthesis* **1981**, 526. (d) Ahluwalia, V. K.; Jolly, R. S., *Synthesis* **1982**, 74.

5. (a) Brown, W.; Turner, A. B., *J. Chem. Soc. (C)* **1971**, 2057. (b) Turner, A. B., *Chem. Ind. (London)* **1976**, 1030. (c) Fu, P. P.; Harvey, R. G., *Tetrahedron Lett.* **1977**, 2059. (d) Abad, A.; Agulló, C.; Arnó, M.; Domingo, L. R.; Zaragozá, R. J., *J. Org. Chem.* **1988**, *53*, 3761.

6. Braude, E. A.; Brook, A. G.; Linstead, R. P., *J. Chem. Soc.* **1954**, 3569.

7. (a) Muller, J. F.; Cagniant, D.; Cagniant, P., *Bull. Soc. Chem. Fr.* **1972**, 4364. (b) Diederick, F.; Staab, H. A., *Angew. Chem., Int. Ed. Engl.* **1978**, *17*, 372. (c) Stowasser, B.; Hafner, K., *Angew. Chem., Int. Ed. Engl.* **1986**, *25*, 466. (d) Funhoff, D. J. H.; Staab, H. A., *Angew. Chem., Int. Ed. Engl.* **1986**, *25*, 742. (e) Di Raddo, P.; Harvey, R. G., *Tetrahedron Lett.* **1988**, *29*, 3885.

8. Braude, E. A.; Jackman, L. M.; Linstead, R. P.; Lowe, G., *J. Chem. Soc.* **1960**, 3123.

9. Brown, W.; Turner, A. B., *J. Chem. Soc. (C)* **1971**, 2566.

10. Padwa, A.; Haffmanns, G.; Tomas, M., *J. Org. Chem.* **1984**, *49*, 3314.

11. (a) Bousquet, E. W.; Moran, M. D.; Harmon, J.; Johnson, A. L.; Summers, J. C., *J. Org. Chem.* **1975**, *40*, 2208. (b) Padwa, A.; Nahm, S.; Sato, E., *J. Org. Chem.* **1978**, *43*, 1664.

12. Gilgen, P.; Heimgartner, H.; Schmid, H., *Helv. Chim. Acta* **1974**, *57*, 1382.

13. Harden, D. B.; Mokrosz, M. J.; Strekowski, L., *J. Org. Chem.* **1988**, *53*, 4137.

14. Blake, K. W.; Porter, A. E. A.; Sammes, P. G., *J. Chem. Soc., Perkin Trans. 1* **1972**, 2494.

15. Hayakawa, K.; Yasukouchi, T.; Kanematsu, K., *Tetrahedron Lett.* **1986**, *27*, 1837.

16. Meyers, A. I.; Wettlaufer, D. G., *J. Am. Chem. Soc.* **1984**, *106*, 1135.

17. (a) Piozzi, F.; Venturella, P.; Bellino, A., *Org. Prep. Proced. Int.* **1971**, *3*, 223. (b) Stanetty, P.; Purstinger, G., *J. Chem. Res. (M)* **1991**, 581.

18. (a) Schultz, A. G.; Fu, W. Y.; Lucci, R. D.; Kurr, B. G.; Lo, K. M.; Boxer, M., *J. Am. Chem. Soc.* **1978**, *100*, 2140. (b) Moursounidis, J.; Wege, D., *Tetrahedron Lett.* **1986**, *27*, 3045. (c) Mazerolles, P.; Laurent, C., *J. Organomet. Chem.* **1991**, *35*, 402.

19. Howe, R. K.; Franz, J. E., *J. Org. Chem.* **1978**, *43*, 3742.

20. (a) Strekowski, L.; Cegla, M. T.; Kong, S.-B; Harden, D. B., *J. Heterocycl. Chem.* **1989**, *26*, 923. (b) Strekowski, L.; Cegla, M. T.; Harden, D. B.; Kong, S.-B., *J. Org. Chem.* **1989**, *54*, 2464.

21. (a) Kämpfen, U.; Eschenmoser, A., *Tetrahedron Lett.* **1985**, *26*, 5899. (b) Barnett, G. H.; Hudson, M. F.; Smith, K. M., *J. Chem. Soc., Perkin Trans. 1* **1975**, 1401.

22. (a) Hevesi, L.; Renard, M.; Proess, G., *J. Chem. Soc., Chem. Commun.* **1986**, 1725. (b) Proess, G.; Pankert, D.; Hevesi, L., *Tetrahedron Lett.* **1992**, *33*, 269.

23. Reid, D. H.; Fraser, M.; Molloy, B. B.; Payne, H. A. S.; Sutherland, R. G., *Tetrahedron Lett.* **1961**, 530.

24. Carretto, J.; Simalty, M., *Tetrahedron Lett.* **1973**, 3445.

25. Nakazumi, H.; Ueyama, T.; Endo, T.; Kitao, T., *Bull. Chem. Soc. Jpn.* **1983**, *56*, 1251.

26. Ishii, H.; Chen, I-S.; Ishikawa, T., *J. Chem. Soc., Perkin Trans. 1* **1987**, 671.

27. Turner, A. B.; Ringold, H. J., *J. Chem. Soc. (C)* **1967**, 1720.

28. (a) Kane, V. V.; Jones, M., Jr., *Org. Synth.* **1982**, *61*, 129. (b) Hagenbruch, B.; Hünig, S., *Chem. Ber.* **1983**, *116*, 3884. (c) Jeffs, P. W.; Redfearn, R.; Wolfram, J., *J. Org. Chem.* **1983**, *48*, 3861.

29. Van Tamelen, E. E.; Hildahl, G. T., *J. Am. Chem. Soc.* **1956**, *78*, 4405.

30. (a) Amemiya, T.; Yasunami, M.; Takase, K., *Chem. Lett.* **1977**, 587. (b) Matsuura, S.; Iinuma, M.; Ishikawa, K.; Kagei, K., *Chem. Pharm. Bull.* **1978**, *26*, 305.

31. Shanka, C. G.; Mallaiah, B. V.; Srimannarayana, G., *Synthesis* **1983**, 310.

32. Cliff, G. R.; Jones, G., *J. Chem. Soc. (C)* **1971**, 3418.

33. (a) Pradhan, S. K.; Ringold, H. J., *J. Org. Chem.* **1964**, *29*, 601. (b) Heathcock, C. H.; Mahaim, C.; Schlecht, M. F.; Utawanit, T., *J. Org. Chem.* **1984**, *49*, 3264.

34. Ryu, I.; Murai, S.; Hatayama, Y.; Sonoda, N., *Tetrahedron Lett.* **1978**, 3455.

35. (a) Flemming, I.; Paterson, I., *Synthesis* **1979**, 736. (b) Fevig, T. L.; Elliott, R. L.; Curran, D. P., *J. Am. Chem. Soc.* **1988**, *110*, 5064.

36. (a) Cross, A. D. (Syntex Corp.), Neth. Patent 6 503 543, **1965** (*Chem. Abstr.* **1966**, *64*, 5177). (b) Das Gupta, A. K.; Chatterje, R. M.; Paul, M., *J. Chem. Soc. (C)* **1971**, 3367.

37. Tanaka, T.; Mashimo, K.; Wagatsuma, M., *Tetrahedron Lett.* **1971**, 2803.

38. Bhattacharya, A.; DiMichele, L. M.; Dolling, U.-H.; Douglas, A. W.; Grabowski, E. J. J., *J. Am. Chem. Soc.* **1988**, *110*, 3318.

39. (a) Cainelli, G.; Cardillo, G.; Ronchi, A. U., *J. Chem. Soc., Chem. Commun.* **1973**, 94. (b) Latif, N.; Mishriki, N.; Girgis, N. S., *Chem. Ind. (London)* **1976**, 28.

40. Iwamura, J.; Hirao, N., *Tetrahedron Lett.* **1973**, 2447.

41. Becker, H.-D.; Bjork, A.; Alder, E., *J. Org. Chem.* **1980**, *45*, 1596.

42. Cacchi, S.; La Torre, F.; Paolucci, G., *Synthesis* **1978**, 848.

43. Burstein, S. H.; Ringold, H. J., *J. Am. Chem. Soc.* **1964**, *86*, 4952.

44. Lee, H.; Harvey, R. G., *J. Org. Chem.* **1988**, *53*, 4587.

45. (a) Creighton, A. M.; Jackman, L. M., *J. Chem. Soc.* **1960**, 3138. (b) Oikawa, Y.; Yonemitsu, O., *Heterocycles* **1976**, *4*, 1859.

46. Amemiya, T.; Yasunami, M.; Takase, K., *Chem. Lett.* **1977**, 587.

47. Findlay, J. W. A.; Turner, A. B., *Chem. Ind. (London)* **1970**, 158.

48. (a) Bouquet, M.; Guy, A.; Lemaire, M.; Guetté, J. P., *Synth. Commun.* **1985**, *15*, 1153. (b) Corey, E. J.; Xiang, Y. B., *Tetrahedron Lett.* **1987**, *28*, 5403.

49. Lemaire, M.; Doussot, J.; Guy, A., *Chem. Lett.* **1988**, 1581.

50. Guy, A.; Lemor, A.; Doussot, J.; Lemaire, M., *Synthesis* **1988**, 900.

51. (a) Oikawa, Y.; Yoshioka, T.; Yonemitsu, O., *Tetrahedron Lett.* **1982**, *23*, 885, 889. (b) Oikawa, Y.; Tanaka, T.; Horita, K.; Yoshioka, T.; Yonemitsu, O., *Tetrahedron Lett.* **1984**, *25*, 5393. (c) Nakajima, N.; Abe, R.; Yonemitsu, O., *Chem. Pharm. Bull.* **1988**, *36*, 4244. (d) Kozikowski, A. P.; Wu, J.-P., *Tetrahedron Lett.* **1987**, *28*, 5125.

52. Kim, C. U.; Misco, P. F., *Tetrahedron Lett.* **1985**, *26*, 2027.

53. Grunder-Klotz, E.; Ehrhardt, J.-D., *Tetrahedron Lett.* **1991**, *32*, 751.

54. (a) Becker, H.-D.; , *J. Org. Chem.* **1965**, *30*, 982. (b) Findlay, J. W. A.; Turner, A. B., *J. Chem. Soc. (C)* **1971**, 547.

55. (a) Buchan, G. M.; Findlay, J. W. A.; Turner, A. B., *J. Chem. Soc., Chem. Commun.* **1975**, 126. (b) Bouquet, M., *C.R. Hebd. Seances Acad. Sci. (II)* **1984**, *229*, 1389.

56. (a) Subba Raju, K. V.; Srimannarayana, G.; Subba Rao, N. V., *Tetrahedron Lett.* **1977**, 473. (b) Prashant, A.; Krupadanam, G. L. D.; Srimannarayana, G., *Bull. Chem. Soc. Jpn.* **1992**, *65*, 1191.

57. (a) Cardillo, C.; Cricchio, R.; Merlin, L., *Tetrahedron* **1971**, *27*, 1875. (b) Cardillo, G.; Orena, M.; Porzi, G.; Sandri, S., *J. Chem. Soc., Chem. Commun.* **1979**, 836. (c) Jain, A. C.; Khazanchi, R.; Kumar, A., *Bull. Chem. Soc. Jpn.* **1979**, *52*, 1203.

58. Imafuku, K.; Fujita, R., *Chem. Express* **1991**, *6*, 323.

59. (a) Coutts, I. G. C.; Humphreys, D. J.; Schofield, K., *J. Chem. Soc. (C)* **1969**, 1982. (b) Lewis, J. R.; Paul, J. G., *J. Chem. Soc., Perkin Trans. 1* **1981**, 770.

60. Becker, H.-D., *J. Org. Chem.* **1965**, *30*, 982, 989.

61. (a) Schmand, H. L. K.; Boldt, P., *J. Am. Chem. Soc.* **1975**, *97*, 447. (b) Barton, D. H. R.; Bergé-Lurion, R.-M.; Lusinchi, X.; Pinto, B. M., *J. Chem. Soc., Perkin Trans. 1* **1984**, 2077.

62. Semmelhack, M. F.; Bozell, J. J.; Sato, T.; Wulff, W.; Spiess, E.; Zask, A., *J. Am. Chem. Soc.* **1982**, *104*, 5850.

63. Sutherland, R. G.; Chowdhury, R. L.; Piórko, A.; Lee, C. C., *J. Org. Chem.* **1987**, *52*, 4618.

64. Miles, W. H.; Smiley, P. M.; Brinkman, H. R., *J. Chem. Soc., Chem. Commun.* **1989**, 1897.

65. (a) Begland, R. W.; Hartter, D. R.; Jones, F. N.; Sam, D. J.; Sheppard, W. A.; Webster, O. W.; Weigert, F. J., *J. Org. Chem.* **1974**, *39*, 2341. (b) Sugawara, T.; Masuya, H.; Matsuo, T.; Miki, T., *Chem. Pharm. Bull.* **1979**, *27*, 2544.

66. Tanemura, K.; Suzuki, T.; Horaguchi, T., *J. Chem. Soc., Chem. Commun.* **1992**, 979.

67. Oshima, T.; Nishioka, R.; Nagai, T., *Tetrahedron Lett.* **1980**, *21*, 3919.

68. Tanemura, K.; Horaguchi, T.; Suzuki, T., *Bull. Chem. Soc. Jpn.* **1992**, *65*, 304.

69. (a) Lemaire, M.; Guy, A.; Imbert, D.; Guetté, J.-P., *J. Chem. Soc., Chem. Commun.* **1986**, 741. (b) Guy, A.; Lemor, A.; Imbert, D.; Lemaire, M., *Tetrahedron Lett.* **1989**, *30*, 327.

70. (a) Noyori, R.; Hayashi, N.; Kato, M., *Tetrahedron Lett.* **1973**, 2983. (b) Kuroda, S.; Funamizu, M.; Kitahara, Y., *Tetrahedron Lett.* **1975**, 1973.

71. Tanemura, K.; Suzuki, T.; Haraguchi, T., *Bull. Chem. Soc. Jpn.* **1993**, *66*, 1235.

72. Miller, G. P.; Briggs, J., *Org. Lett.* **2003**, *5*, 4203.

73. Arseniyadis, S.; Wagner, A.; Mioskowski, C., *Tetrahedron Lett.* **2004**, *45*, 2251.

74. Lee, A.-L.; Ley, S. V., *Org. Biomol. Chem.* **2003**, *1*, 3957.

75. Chang, J.; Zhao, K.; Pan, S., *Tetrahedron Lett.* **2002**, *43*, 951.

76. Cheruku, S. R.; Padmanilayam, M. P.; Vennerstrom, J. L., *Tetrahedron Lett.* **2003**, *44*, 3701.

77. Montevecchi, P. C.; Navacchia, M. L., *J. Org. Chem.* **1998**, *63*, 8035.

78. Somogyi, L., *Synth. Commun.* **1999**, *29*, 1857.

79. Barasz, J. A.; Ghaffari, A. H.; Otte, D. A.; Thamattoor, D. M., *Chem. Lett.* **2002**, 64.

80. Ronan, B.; Bacqué, E.; Barrière, J.-C.; Sablé, S., *Tetrahedron* **2003**, *59*, 2929.

81. Liu, L.; Nan, F.; Xiong, Z.; Li, T.; Li, Y., *Synth. Commun.* **1996**, *26*, 551.

82. Wallace, D.; Gibb, A. D.; Cottrell, I. F.; Kennedy, D. J.; Brands, K. M. J.; Dolling, U. H., *Synthesis* **2001**, 1784.

83. Okajima, T.; Kurokawa, S., *Chem. Lett.* **1997**, 69.

84. Wang, W.; Li, T.; Attardo, G., *J. Org. Chem.* **1997**, *62*, 6598.

85. Sadler, I. H.; Stewart, J. A. G., *J. Chem. Soc., Chem. Commun.* **1969**, 773.

86. Lutz, F. E.; Kiefer, E. F., *Tetrahedron Lett.* **1970**, *11*, 4851.

87. Kiefer, E. F.; Lutz, F. E., *J. Org. Chem.* **1972**, *37*, 1519.

88. Gellerstedt, G.; Petterson, E.-L., *Acta Chem. Scand. B* **1975**, *29*, 1005.

89. Iliefski, T.; Li, S.; Lundquist, K., *Tetrahedron Lett.* **1998**, *39*, 2413.

90. Iliefski, T.; Li, S.; Lundquist, K., *Acta Chem. Scand.* **1998**, *52*, 1177.

91. Ameer, F.; Giles, R. G. F.; Green, I. R.; Nagabhushana, K. S., *Synth. Commun.* **2002**, *32*, 369.

92. Katoh, T.; Ohmori, O.; Iwasaki, K.; Inoue, M., *Tetrahedron* **2002**, *58*, 1289.

93. Venkateswarlu, R.; Kamakshi, C.; Moinuddin, S. G. A.; Subhash, P. V.; Ward, R. S.; Pelter, A.; Coles, S. J.; Hursthouse, M. B.; Light, M. E., *Tetrahedron* **2001**, *57*, 5625.

94. Ward, R. S.; Hughes, D. D., *Tetrahedron* **2001**, *57*, 5633.

95. Kamo, M.; Tsuda, A.; Nakamura, Y.; Aratani, N.; Furukawa, K.; Kato, T.; Osuka, A., *Org. Lett.* **2003**, *5*, 2079.

96. Nath, M.; Huffman, J. C.; Zaleski, J. M., *Chem. Commun.* **2003**, 858.

97. Hendrickson, J. B.; Ramsay, M. V. J.; Kelly, T. R., *J. Am. Chem. Soc.* **1972**, *94*, 6834.

98. Sala, T.; Sargent, M. V., *J. Chem. Soc., Perkin Trans. 1* **1981**, 855.

99. Sharma, G. V. M.; Rakesh, *Tetrahedron Lett.* **2001**, *42*, 5571.

100. Xia, J.; Abbas, S. A.; Locke, R. D.; Piskorz, C. F.; Alderfer, J. L.; Matta, K. L., *Tetrahedron Lett.* **2000**, *41*, 169.

101. Vatèle, J.-M., *Tetrahedron* **2002**, *58*, 5689.

102. Yadav, J. S.; Chandrasekhar, S.; Sumithra, G.; Kache, R., *Tetrahedron Lett.* **1996**, *37*, 6603.

103. Tanemura, K.; Dohya, H.; Imamura, M.; Suzuki, T.; Horaguchi, T., *J. Chem. Soc., Perkin Trans. 1* **1995**, 453.

104. Tanemura, K.; Nishida, Y.; Suzuki, T.; Satsumabayashi, K.; Horaguchi, T., *J. Heterocycl. Chem.* **1997**, *34*, 457.

105. Zhou, X.; Kitamura, M.; Shen, B.; Nakajima, K.; Takahashi, T., *Chem. Lett.* **2004**, *33*, 410.

106. Suga, H.; Shi, X.; Ibata, T.; Kakehi, A., *Heterocycles* **2001**, *55*, 1711.

107. Iranpoor, N.; Firouzabadi, H.; Aghapour, Gh.; Vaez Zadeh, A. R., *Tetrahedron* **2002**, *58*, 8689.

108. Iranpoor, N.; Firouzabadi, H.; Akhlaghinia, B.; Nowrouzi, N., *J. Org. Chem.* **2004**, *69*, 2562.

109. Iranpoor, N.; Firouzabadi, H.; Akhlaghinia, B.; Nowrouzi, N., *Tetrahedron Lett.* **2004**, *45*, 3291.

110. Firouzabadi, H.; Iranpoor, N.; Sobhani, S., *Synthesis* **2004**, 290.

111. Firouzabadi, H.; Iranpoor, N.; Sobhani, S., *Tetrahedron* **2004**, *60*, 203.

112. Alnabari, M.; Bittner, S., *Synthesis* **2000**, 1087.

113. Nakatsuji, S.; Akashi, N.; Suzuki, K.; Enoki, T.; Anzai, H., *J. Chem. Soc., Perkin Trans. 2* **1996**, 2555.

114. Aly, A. A.; El-Shaieb, K. M., *Tetrahedron* **2004**, *60*, 3797.

115. Machocho, A. K.; Win, T.; Grinberg, S.; Bittner, S., *Tetrahedron Lett.* **2003**, *44*, 5531.

116. Abdou, W. M.; Salam, M. A. E.; Sediek, A. A., *Tetrahedron* **1997**, *53*, 13945.

117. Makosza, M.; Kedziorek, M.; Ostrowski, S., *Synthesis* **2002**, 2517.

118. Xiao, Z.; Patrick, B. O.; Dolphin, D., *Chem. Commun.* **2003**, 1062.

119. Pinho e Melo, T. M. V. D.; Cabral, A. M. T. D. P. V.; Rocha Gonsalves, A. Md'A.; Beja, A. M.; Paixão, J. A.; Silva, M. R.; Alte da Veiga, L., *J. Org. Chem.* **1999**, *64*, 7229.

120. Kitagawa, H.; Shibata, T.; Matsuo, J.-I.; Makaiyama, T., *Bull. Chem. Soc. Jpn.* **2002**, *75*, 339.

121. Tiecco, M.; Testaferri, L.; Temperini, A.; Bagnoli, L.; Marini, F.; Santi, C., *Synlett* **2001**, 1767.

122. Sayama, S.; Inamura, Y., *Heterocycles* **1996**, *43*, 1371.

123. Tanemura, K.; Suzuki, T.; Nishida, Y.; Satsumabayashi, K.; Horaguchi, T., *J. Chem. Soc., Perkin Trans. 1* **2001**, 3230.

124. Tanaka, M.; Tanifuji, N.; Hatada, S.; Kobayashi, K., *J. Org. Chem.* **2001**, *66*, 803.

125. Yoshida, Y.; Matsuda, K.; Sasaki, H.; Matsumoto, Y.; Matsumoto, S.; Tawara, S.; Takasugi, H., *Bioorg. Med. Chem.* **2000**, *8*, 2317.

126. Ueki, A.; Tanaka, S.; Kumazawa, M.; Ooi, T.; Sano, S.; Nagao, Y., *Heterocycles* **2003**, *61*, 449.

Diisobutylaluminum Hydride[1]

$$i\text{-Bu}_2\text{AlH}$$

[1191-15-7]   C$_8$H$_{19}$Al   (MW 142.22)

InChI = 1/2C4H9.Al.H/c2*1-4(2)3;;/h2*4H,1H2,2-3H3;;/
    rC8H19Al/c1-7(2)5-9-6-8(3)4/h7-9H,5-6H2,1-4H3

InChIKey = AZWXAPCAJCYGIA-DFAADSFOAF

(reducing agent for many functional groups; opens epoxides;
hydroaluminates alkynes and alkenes)

Alternative Name: DIBAL; DIBAL-H; DIBAH.
Physical Data: mp −80 to −70 °C; bp 116–118 °C/1 mmHg;
    d 0.798 g cm^{-3}; fp −18 °C.
Solubility: sol pentane, hexane, heptane, cyclohexane, benzene,
    toluene, xylenes, ether, dichloromethane, THF.
Form Supplied in: can be purchased as a neat liquid or as 1.0
    and 1.5 M solutions in cyclohexane, CH$_2$Cl$_2$, heptane, hexanes,
    THF, and toluene.
Handling, Storage, and Precautions: neat liquid is pyrophoric;
    solutions react very vigorously with air and with H$_2$O, and re-
    lated compounds, giving rise to fire hazards; use in a fume hood,
    in the absence of oxygen and moisture (see ***Lithium Aluminum
    Hydride*** for additional precautions); THF solutions should only
    be used below 70 °C, as above that temperature ether cleavage
    is problematic.

Original Commentary

Paul Galatsis
University of Guelph, Guelph, Ontario, Canada

Reduction of Functional Groups.[2] Diisobutylaluminum
hydride has several acronyms: DIBAH, DIBAL-H, and DIBAL.
For the purpose of this article, DIBAL will be used.

In general, aldehydes, ketones, acids, esters, and acid chlorides
arc all reduced to the corresponding alcohols by this reagent.
Alkyl halides are unreactive towards DIBAL. Amides are
reduced to amines, while nitriles afford aldehydes upon hydrolysis of an intermediate imine. Isocyanates are also reduced to the
corresponding imines. Nitro compounds are reduced to hydroxylamines. Disulfides are reduced to thiols, while sulfides, sulfones,
and sulfonic acids are unreactive in toluene at 0 °C. Tosylates
are converted quantitatively to the corresponding alkanes. Cyclic
imides can be reduced to carbinol lactams.

Comparable stereochemistry.[2] is observed in the reduction
of ketones with DIBAL, ***Aluminum Hydride***, and ***Lithium
Aluminum Hydride***. Excellent 1,3-asymmetric induction is possible with DIBAL (eq 1); however, this is strongly solvent
dependent.[3a]

In addition to the hydroxy directing group, amines and amides
also showed excellent 1,3-asymmetric induction.[3b] In conjunction

with chiral additives, DIBAL can reduce ketones with moderate to
good enantioselectivity (eq 2).[4] The use of the Lewis
acid ***Methylaluminum Bis(2,6-di-tert-butyl-4-methylphenoxide)***
(MAD) along with DIBAL allows discrimination between carbonyl groups (eq 3).[5,6]

Reduction of α-halo ketones to the carbonyl compounds can
be accomplished with DIBAL in the presence of ***Tin(II) Chloride*** and ***N,N,N′,N′-Tetramethylethylenediamine*** (eqs 4 and 5).[7]
Under these conditions, vicinal dibromides are converted to the
corresponding alkenes.

DIBAL is the reagent of choice (see also ***Aluminum Hydride***)
for the reduction of α,β-unsaturated ketones to the corresponding
allylic alcohols (eq 6).[8a] A reagent derived from DIBAL and
Methylcopper in HMPA alters the regiochemistry such that 1,4-
reduction results (eq 7).[8b] Reductions of chiral β-keto sulfoxides
occur with high diastereoselectivity.[9] The choice of reduction conditions makes it possible to obtain both epimers at the carbinol
carbon (eqs 8 and 9).

Essential Reactions for Organic Synthesis, First Edition. Edited by Philip L. Fuchs.
© 2016 John Wiley & Sons, Ltd. Published 2016 by John Wiley & Sons, Ltd.

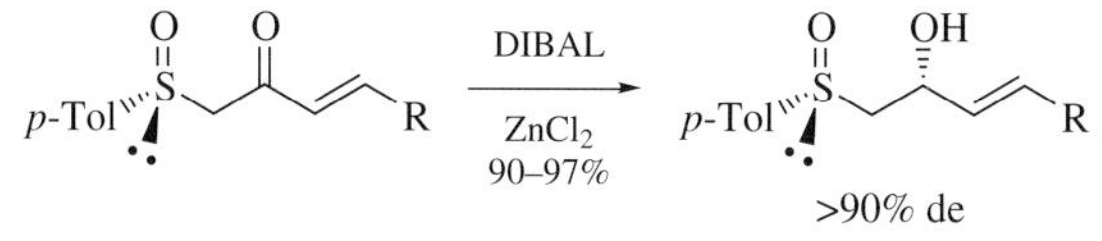

(9)

In conjuction with **Triethylaluminum**, DIBAL has been used to mediate a reductive pinacol rearrangement.[10] with enantiocontrol, as shown in eq 10.[22]

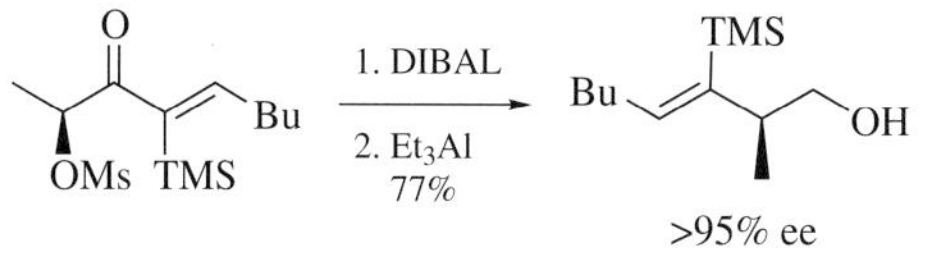

(10)

DIBAL is an excellent reagent for the reduction of α,β-unsaturated esters to allylic alcohols without complications from 1,4-addition (eq 11).[11]

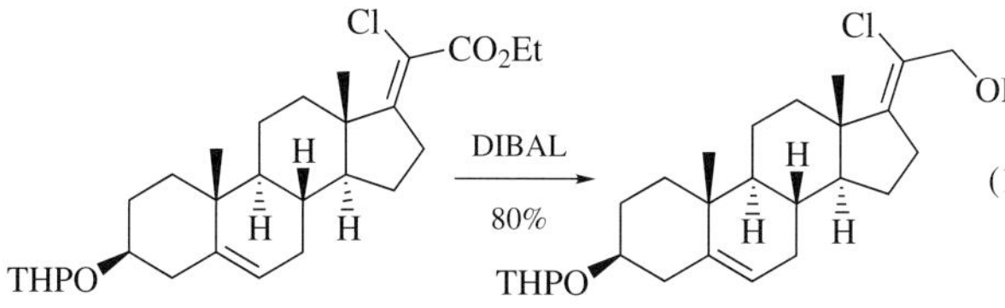

(11)

Due to its Lewis acidity, DIBAL can be used in the reductive cleavage of acetals (see also **Aluminum Hydride**). Chiral acetates are reduced with enantioselectivity (eq 12).[12] Oxidation of the intermediate alcohol followed by β-elimination gives an optically active alcohol, resulting in a net enantioselective reduction of the corresponding ketone.

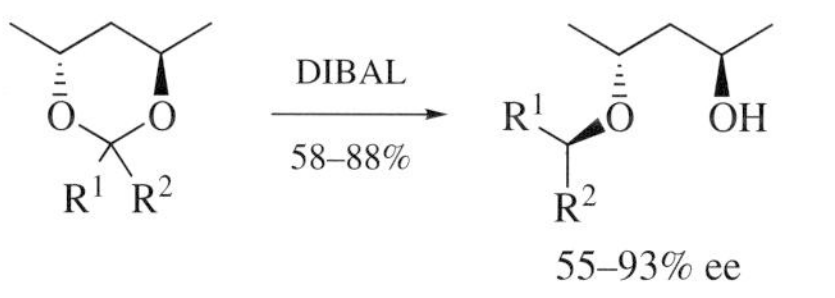

(12)

At low temperatures DIBAL converts esters to the corresponding aldehydes (eq 13)[13] and lactones to lactols (eq 14).[14] DIBAL reduction of α,β-unsaturated γ-lactones followed by an acidic work-up transforms the intermediate lactol to the furan (eq 15).[15] The reduction of nitriles can lead to aldehydes after hydrolysis of an intermediate imine (eq 16).[16] However, cyclic imines can be produced if a **Sodium Fluoride** work-up follows the reduction of halonitriles (eq 17).[17] Furthermore, imines can be reduced with excellent stereocontrol (eq 18).[18]

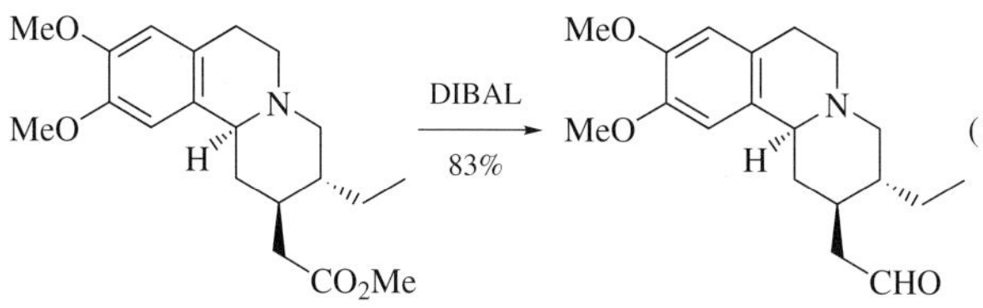

(13)

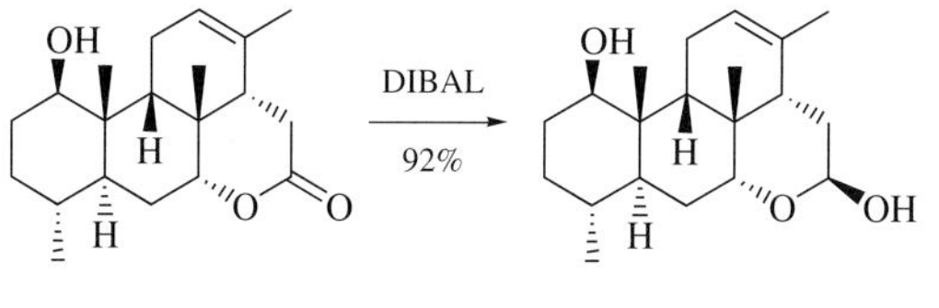

(14)

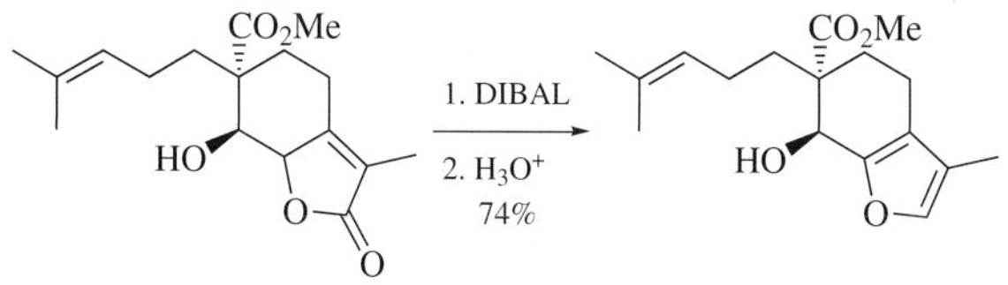

(15)

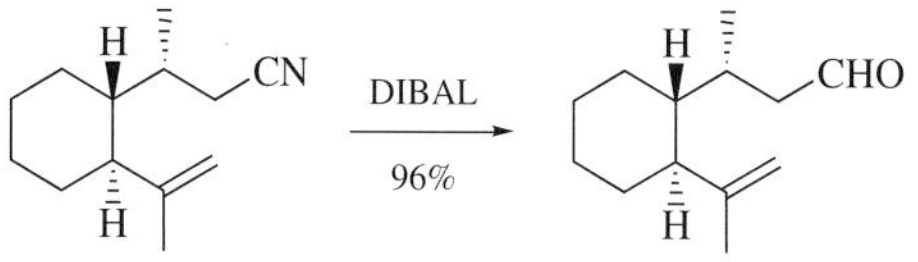

(16)

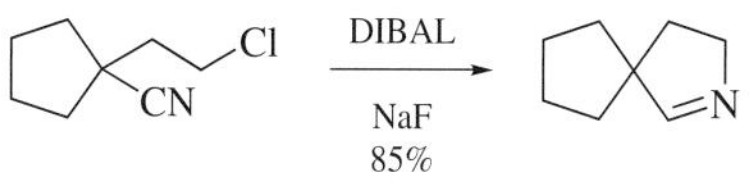

(17)

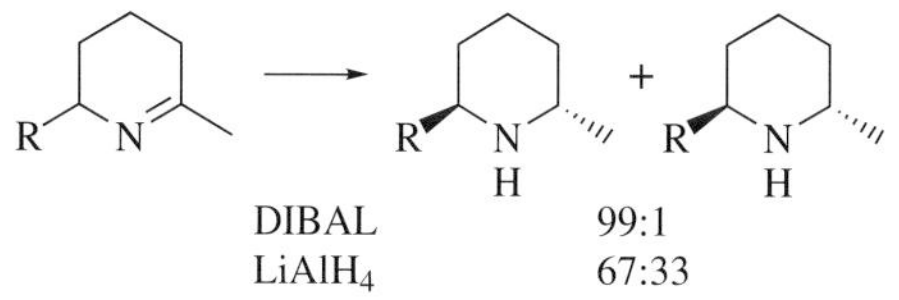

(18)

DIBAL can also be used for the reductive cleavage of cyclic aminals and amidines (eq 19).[19] Oximes can be reduced to amines. Due to the Lewis acidity of DIBAL, however, rearranged products are obtained (eq 20).[20] This chemistry was used to prepare the alkaloid pumiliotoxin C via the Beckmann rearrangement/alkylation sequence shown in eq 21.[21]

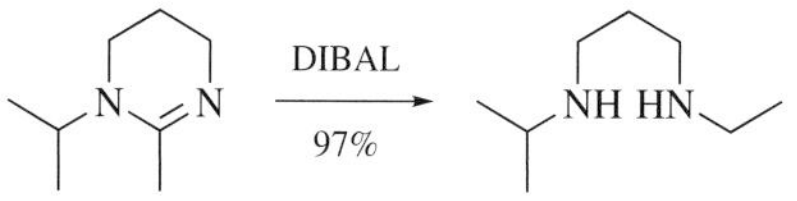

(19)

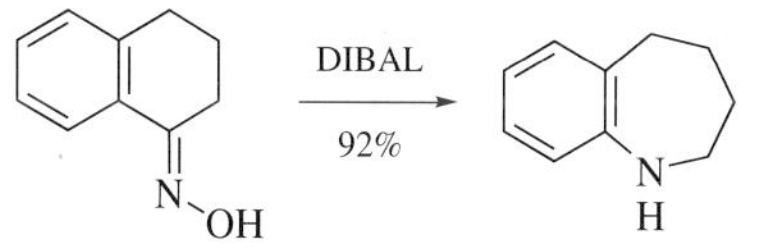

(20)

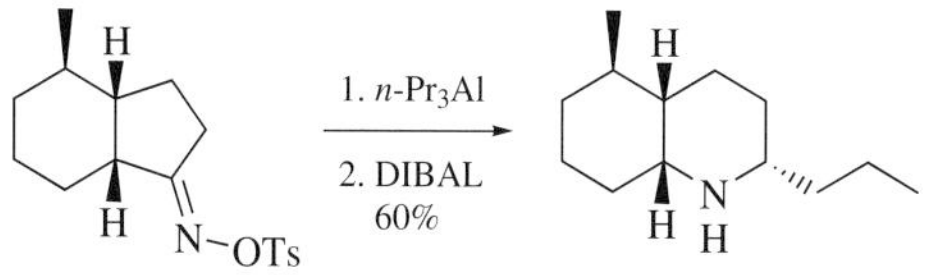

(21)

While sulfones are unreactive with DIBAL at 0 °C in toluene,[2] reduction to the corresponding sulfide has been accomplished at higher temperatures.[22] This reaction can be accomplished with **Lithium Aluminum Hydride**, but fewer equivalents are required and yields are better using DIBAL (eq 22).

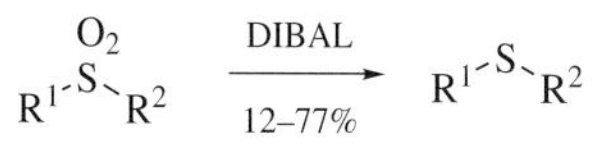

(22)

The reagent combination of DIBAL and **Butyllithium**, which is most likely lithium diisobutylbutylaluminum hydride, has also been used as a reducing agent.[23]

Epoxide Ring Opening. As a result of its Lewis acidity, several reaction pathways are followed in the reductive ring opening of epoxides by DIBAL. Attack at the more hindered carbon via carbenium ion-like intermediates (see also **Aluminum Hydride**) or S_N2' type reactions, are both known with vinyl epoxides. These modes stand in contrast to results with $LiAlH_4$ (eqs 23–25).[24]

$$\text{(23)}$$

$$\text{(24)}$$

$$\text{(25)}$$

Hydroalumination Reactions. DIBAL reacts with alkynes and alkenes to give hydroalumination products. *Syn* additions are usually observed; however, under appropriate conditions, equilibration to give a net *anti* addition is possible.[25] If the substrate has both an alkene and an alkyne group, then chemoselectivity for the alkyne is observed (eq 26).[26]

$$\text{(26)}$$

The intermediate alkenylalane can be used in several ways. If a protiolytic work-up is used, then formation of the corresponding (Z)-alkene is observed (eq 27).[27] Treatment of the alkenylalane with *Methyllithium* affords an ate complex which is nucleophilic and reacts with a variety of electrophiles, e.g. alkyl halides, CO_2, MeI, epoxides, tosylates, aldehydes, and ketones (eq 28).[1c,1h,26]

$$Me(CH_2)_nC{\equiv}CCH_2C{\equiv}C(CH_2)_mCl \xrightarrow{\text{DIBAL}}$$

$$Me(CH_2)_n \diagdown\diagup (CH_2)_mCl \quad \text{(27)}$$

n	m	
5	6	82%
3	8	94%
2	9	87%

$$\begin{array}{c} \text{1. DIBAL} \\ \text{2. MeLi} \\ \xrightarrow{\hspace{2cm}} \\ \text{3. ethylene oxide} \\ 81\% \end{array}$$

$$\text{(28)}$$

Hydroalumination of a terminal alkene followed by treatment of the intermediate alane with oxygen gives a primary alcohol (eq 29).[28]

$$\begin{array}{c} \text{1. DIBAL} \\ \xrightarrow{\hspace{2cm}} \\ \text{2. O}_2 \end{array}$$

$$\text{(29)}$$

Alkenylalanes can dimerize to afford 1,3-butadienes (eq 30),[29] and can be cyclopropanated under Simmons–Smith conditions to give cyclopropylalanes (eq 31),[30] which can be used for further chemistry.

$$Ph{-}{\equiv}{-} \xrightarrow{\text{DIBAL}} \quad \text{(30)}$$

$$Bu\diagup\diagdown Al(i\text{-}Bu)_2 \xrightarrow[\text{Zn(Cu)}]{CH_2Br_2} Bu\diagup\diagdown Al(i\text{-}Bu)_2 \quad \text{(31)}$$

Alkenylalanes can also be transmetalated (eq 32),[31] or coupled with vinyl halides via palladium catalysis (eq 33).[32]

$$\text{(32)}$$

$$M = B, Zr, Hg$$

$$Bu\diagup\diagdown Al(i\text{-}Bu)_2 + MeO_2C\diagup\diagdown Br \xrightarrow[75\%]{Pd}$$

$$Bu\diagup\diagdown\diagup CO_2Me \quad \text{(33)}$$

Addition of DIBAL to allenes has been observed to occur at the more highly substituted double bond (eq 34).[33]

$$R^1\diagdown{\bullet}\diagup R^2 \xrightarrow{\text{DIBAL}} R^1\diagdown\diagup R^2 \quad \text{(34)}$$

$$R^1 = C_7H_{15}; R^2 = H (88\%), TMS (74\%)$$

The conversion of unconjugated enynes to cyclic compounds has also been accomplished using DIBAL (eq 35).[34]

$$\xrightarrow[79\%]{\text{DIBAL}} \quad \text{(35)}$$

Related hydroalumination reactions have been reported using *Lithium Aluminum Hydride*.

First Update

Matthieu Sollogoub
Ecole Normale Supérieure, Paris, France

Pierre Sinaÿ
Université Pierre et Marie Curie, Paris, France

Reduction Reaction. DIBAL has been shown to be a practical reagent to reduce secondary phosphine oxides (eq 36).[35] In most cases, this reaction is carried out in hydrocarbon solvent at room temperature and was demonstrated to be a convenient large scale synthesis.

$$\begin{array}{c} R_1 \\ \diagup P{=}O \\ R_2 \quad H \end{array} \xrightarrow{\text{DIBAL}} \begin{array}{c} H \\ \diagup P \\ R_1 \quad R_2 \end{array} \quad \text{(36)}$$

$$R_1, R_2 = aryl, 1°, 2°, 3° alkyl \qquad 72–92\%$$

Oxidation Reactions. DIBAL is commonly known as a reducing agent, but it has also been used in oxidation reactions. Indeed, when a catalytic amount of DIBAL is introduced in presence of an aldehyde in pentane, the Tischenko reaction occurs to give the corresponding esters (eq 37).[36] Alternatively, in the case of α-hydroxy aldehydes, protected by a TBS group, the Oppenauer oxidation takes place together with a silyl migration (eq 38). Both reactions start with the reduction of the aldehyde to form an aluminum alkoxide, which is the actual substrate for those two oxidation reactions.

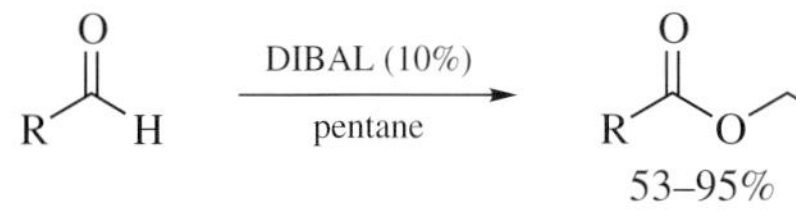

$$R \overset{O}{\underset{H}{\bigvee}} \xrightarrow[\text{pentane}]{\text{DIBAL (10\%)}} R \overset{O}{\underset{}{\bigvee}} O-R \quad (37)$$

$$53\text{–}95\%$$

$$TBSO \overset{O}{\underset{}{\bigvee}} R \xrightarrow[\text{pentane}]{\text{DIBAL (10\%)}} O \overset{OTBS}{\underset{}{\bigvee}} R \quad (38)$$

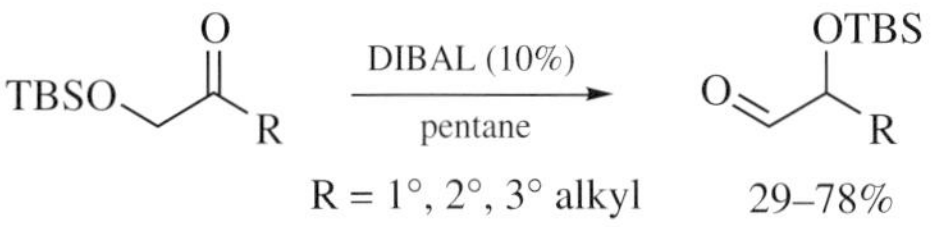

$$R = 1°, 2°, 3° \text{ alkyl} \qquad 29\text{–}78\%$$

When used in combination with TBHP, DIBAL can promote the oxidation of alcohols into the corresponding ketones via an Oppenauer type reaction with the TBHP being reduced by the aluminum alkoxide (eq 39). In the case of allylic alcohols, this combination affords the epoxidation products (eq 40).[37]

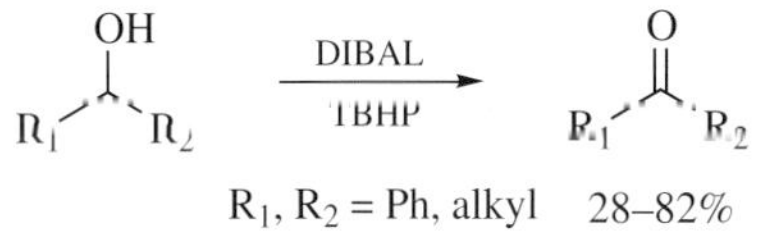

$$R_1 \overset{OH}{\underset{}{\bigvee}} R_2 \xrightarrow[\text{TBHP}]{\text{DIBAL}} R_1 \overset{O}{\underset{}{\bigvee}} R_2 \quad (39)$$

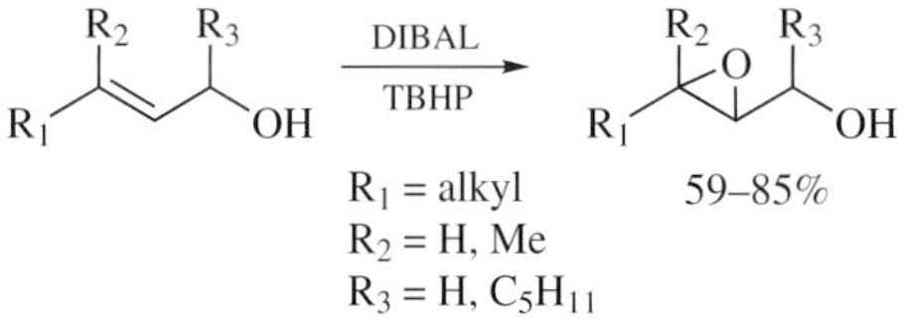

$$R_1, R_2 = Ph, \text{alkyl} \quad 28\text{–}82\%$$

$$\overset{R_2 \quad R_3}{\underset{R_1 \qquad OH}{\bigvee}} \xrightarrow[\text{TBHP}]{\text{DIBAL}} \overset{R_2 \quad R_3}{\underset{R_1 \qquad OH}{\bigvee O}} \quad (40)$$

$$R_1 = \text{alkyl} \qquad 59\text{–}85\%$$
$$R_2 = H, Me$$
$$R_3 = H, C_5H_{11}$$

Alumination-addition. While DIBAL usually reacts with alkynes to give hydroalumination products, the addition of a small quantity of triethylamine leads to the alumination of the terminal alkyne.[38] The resulting alkynylorganoaluminum reagent has then been used for diastereoselective alkynylation of oxazolidines (eq 41). Similar reactivity was observed with AlMe₃.

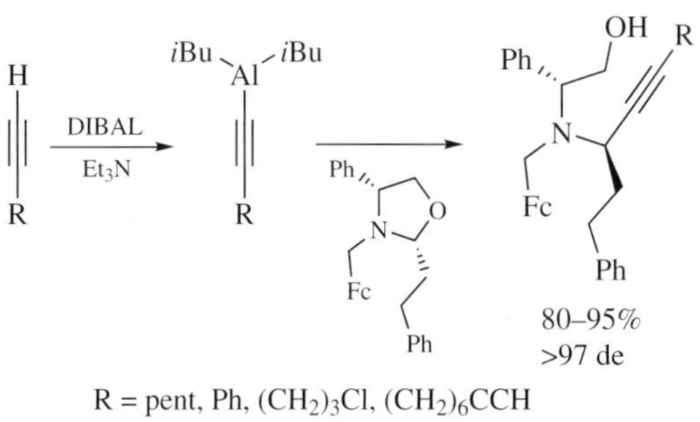

$$\text{(eq 41)}$$

$$80\text{–}95\%$$
$$>97 \text{ de}$$

$$R = \text{pent}, Ph, (CH_2)_3Cl, (CH_2)_6CCH$$

DIBAL has also been used to form aluminum-2-vinyloxy ethoxides that undergo an alkoxymetallation to form organometalic acetals, which are then added to an aldehyde.[39] The reaction yields an alcohol and a ketone (eq 42). The later is probably the product of the Oppenauer oxidation of the former due to the presence of an excess of benzaldehyde.

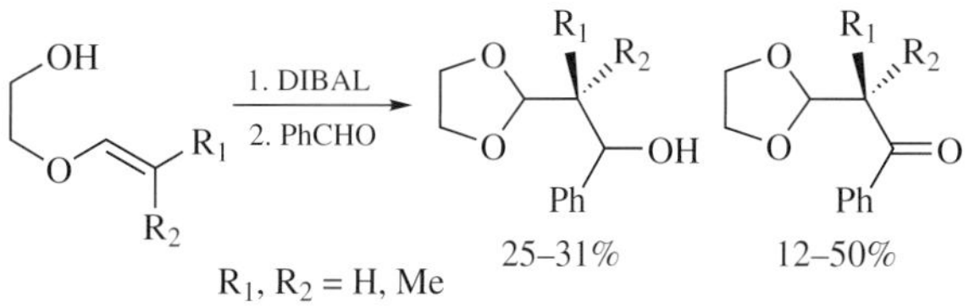

$$\text{(eq 42)}$$

$$R_1, R_2 = H, Me \qquad 25\text{–}31\% \qquad 12\text{–}50\%$$

Regioselective Reactions. An important attribute of DIBAL is its capacity to react regioselectively. Acetal cleavage is a classical example of this property, which finds its source in electronic, steric, and/or chelation effects. The reduction of aryl-1,2-ethanediol benzylidene acetals is controlled mainly by electronic effects: a change in electron donating capacity of the aryl moiety changes the regioselectivity of the reaction (eq 43).[40]

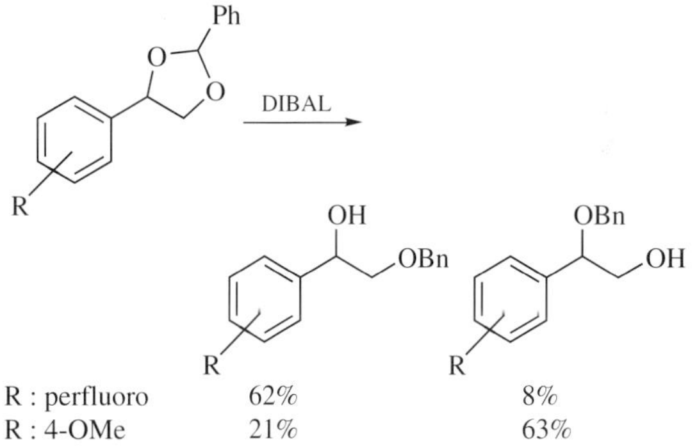

$$\text{(eq 43)}$$

$$R : \text{perfluoro} \qquad 62\% \qquad 8\%$$
$$R : 4\text{-OMe} \qquad 21\% \qquad 63\%$$

Steric control can be illustrated by the reaction of DIBAL with acetal protected spirocyclic diols (eq 44). The conformation of this system directs the outcome.[41]

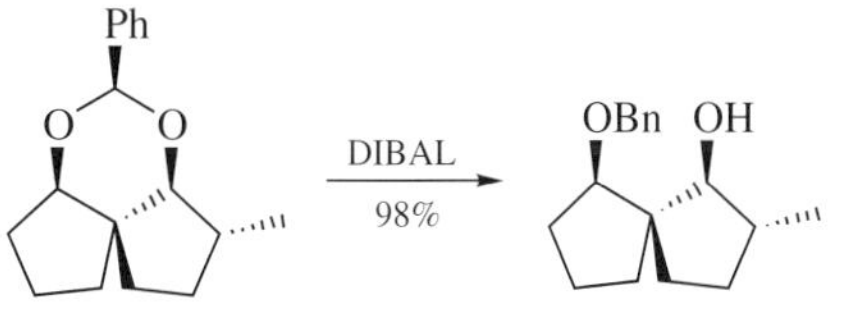

$$\text{(eq 44)}$$

$$98\%$$

The regioselectivity of acetal cleavage is also highly dependant on neighboring groups. A chelating group like an ether will direct the aluminum approach (eq 45), whereas a silyloxy group will orient the reagent towards the less hindered position (eq 46).[42]

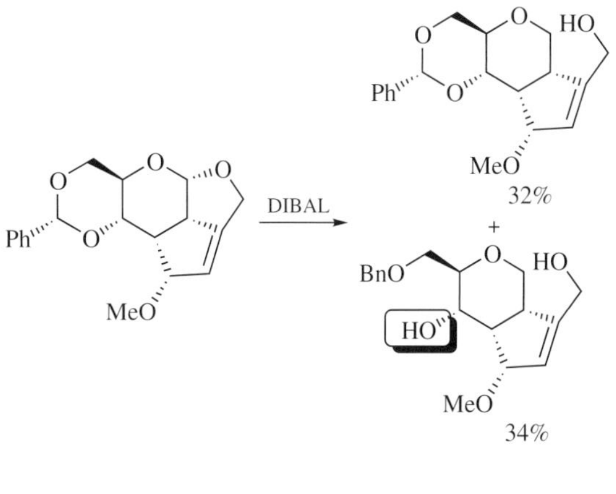

$$\text{(eq 45)}$$

$$32\%$$
$$+$$
$$34\%$$

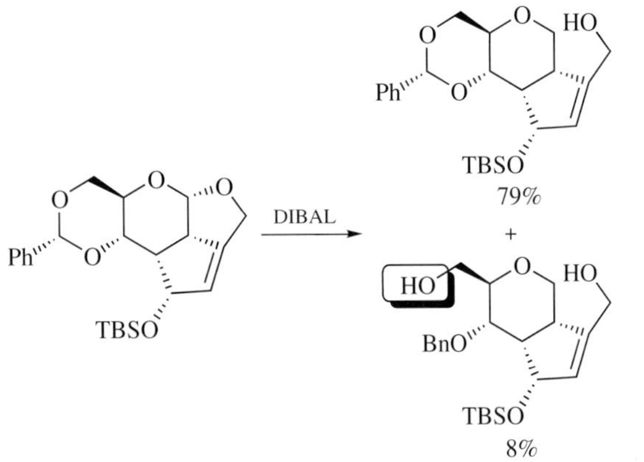

$$(46)$$

Another neighboring group assisted cleavage is spectacularly illustrated by the efficient selective ablation of two diametrically opposed benzyl groups on the primary face of perbenzylated cyclodextrins (eq 47).[43] This reaction proceeds with high dependence on steric effects, a feature exploited to duplicate debenzylation in a totally regiocontrolled manner: on a properly protected diol, only one of the two possible diols is obtained (eq 48).[44] Methyl groups can be cleaved selectively over benzyls under these conditions.[45] Triisobutylaluminum is also a good reagent for regioselective debenzylation reactions.[43]

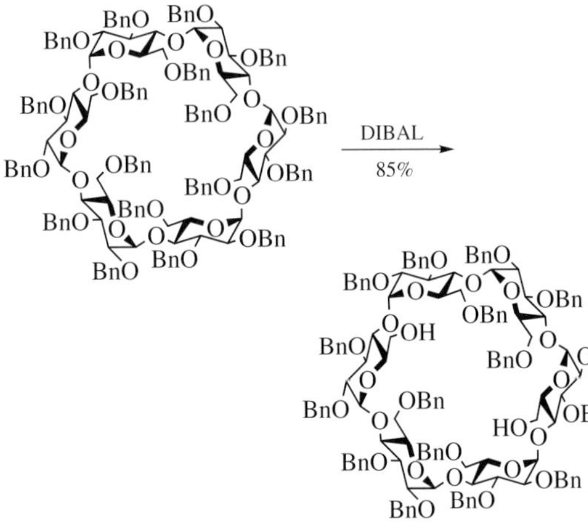

$$(47)$$

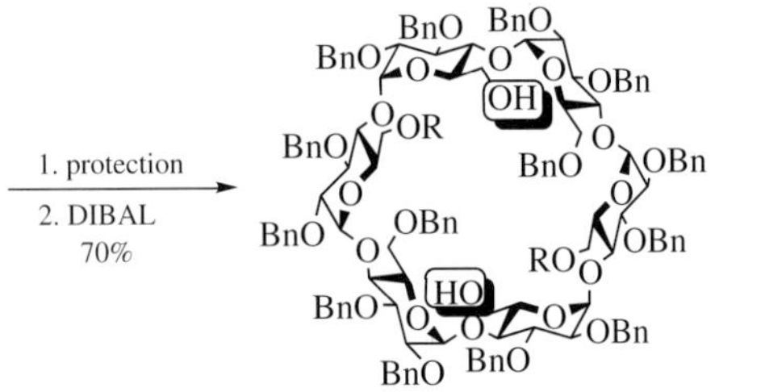

$$(48)$$

Both acetal cleavage and regioselective debenzylation reactions can be successively induced by DIBAL by reaction of a mesylated spiroketal carbohydrate derivative to form a monodebenzylated bicyclic system (eq 49).[46,47]

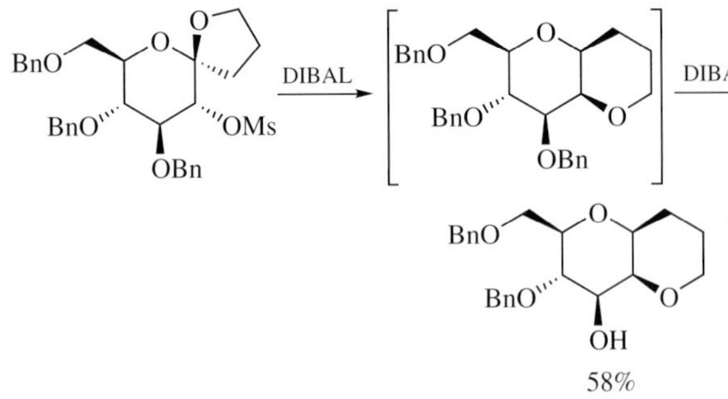

$$(49)$$

A particularly surprising neighboring group assisted reaction was discovered in the course of derivatization of myxothiazol A. In this case, DIBAL in dichloromethane was expected to reduce an ester, but preferential thiazole-ring cleavage was observed (eq 50).[48]

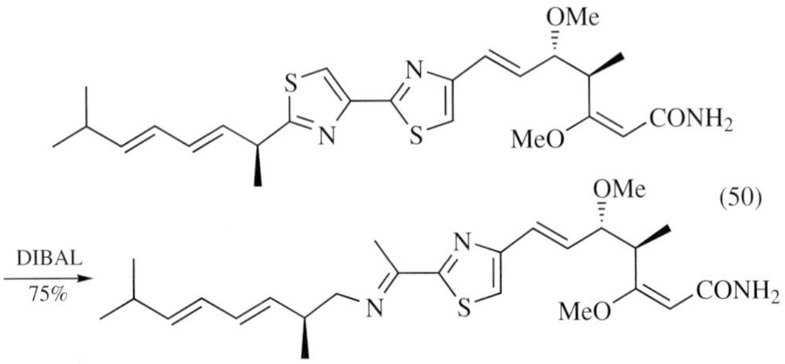

$$(50)$$

Rearrangements. The Lewis acidity of DIBAL (1.5 equiv) can be used to induce Claisen rearrangement, giving ortho products (eq 51).[49]

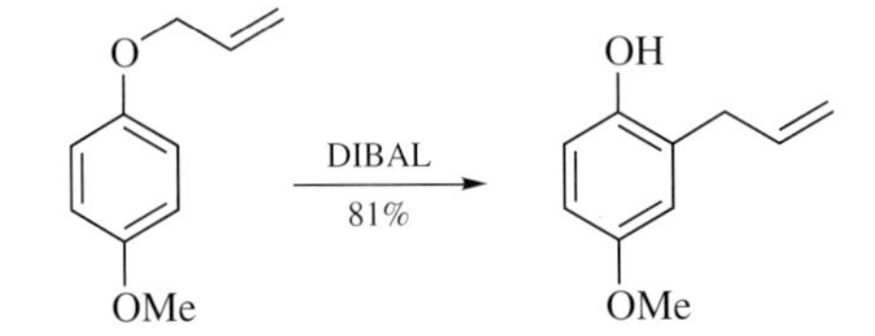

$$(51)$$

An equimolar mixture of DIBAL and THF efficiently promotes iodine transfer cyclization of hex-5-ynyl iodides to form five membered rings (eq 52).[50] The reaction proceeds via a radical pathway to afford the E and Z isomers in a 1:1 ratio, with no trace of β-fragmentation.

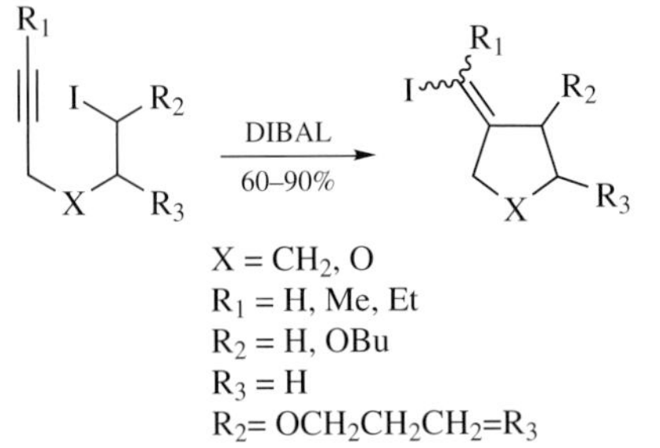

$$(52)$$

$X = CH_2, O$
$R_1 = H, Me, Et$
$R_2 = H, OBu$
$R_3 = H$
$R_2 = OCH_2CH_2CH_2 = R_3$

1. (a) Mole, T.; Jeffery, E. A. *Organoaluminum Compounds*; Elsevier: Amsterdam, 1972. (b) Winterfeldt, W., *Synthesis* **1975**, 617. (c) Zweifel, G.; Miller, J. A., *Org. React.* **1984**, *32*, 375. (d) Maruoka, K.; Yamamoto,

H., *Angew. Chem., Int. Ed. Engl.* **1985**, *24*, 668. (e) Maruoka, K.; Yamamoto, H., *Tetrahedron* **1988**, *44*, 5001. (f) Dzhemilev, V. M.; Vostrikova, O. S.; Tolstikov, G. A., *Russ. Chem. Rev. (Engl. Transl.)* **1990**, *59*, 1157. (g) Seyden-Penne, J. *Reductions by the Alumino- and Borohydrides in Organic Synthesis*; VCH: New York, 1991. (h) Eisch, J. J., *Comprehensive Organic Synthesis* **1991**, *8*, 733. (i) Eisch, J. J. In *Comprehensive Organometallic Chemistry*; Wilkinson, G., Ed.; Pergamon: Oxford, 1982; Vol. 1, p 555. (j) Zietz, J. R.; Robinson, G. C.; Lindsay, K. L. In *Comprehensive Organometallic Chemistry*; Wilkinson, G., Ed.; Pergamon: Oxford, 1982; Vol. 7, p 365.

2. Yoon, N. M.; Gyoung, Y. S., *J. Org. Chem.* **1985**, *50*, 2443.

3. (a) Kiyooka, S.; Kuroda, H.; Shimasaki, Y., *Tetrahedron Lett.* **1986**, *27*, 3009. (b) Barluenga, J.; Aguilar, E.; Fustero, S.; Olano, B.; Viado, A. L., *J. Org. Chem.* **1992**, *57*, 1219.

4. Oriyama, T.; Mukaiyama, T., *Chem. Lett.* **1984**, 2071.

5. (a) Maruoka, K.; Itoh, T.; Yamamoto, H., *J. Am. Chem. Soc.* **1985**, *107*, 4573. (b) Maruoka, K.; Sakurai, M.; Yamamoto, H., *Tetrahedron Lett.* **1985**, *26*, 3853.

6. Maruoka, K.; Araki, Y.; Yamamoto, H., *J. Am. Chem. Soc.* **1988**, *110*, 2650.

7. Oriyama, T.; Mukaiyama, T., *Chem. Lett.* **1984**, 2069.

8. (a) Wilson, K. E.; Seidner, R. T.; Masamune, S., *J. Chem. Soc., Chem. Commun.* **1970**, 213. (b) Tsuda, T.; Kawamoto, T.; Kumamoto, Y.; Saegusa, T., *Synth. Commun.* **1986**, *16*, 639.

9. Solladie, G.; Frechou, G.; Demailly, G., *Tetrahedron Lett.* **1986**, *27*, 2867.

10. Suzuki, K.; Katayama, E.; Matsumoto, T.; Tsuchihashi, G., *Tetrahedron Lett.* **1984**, *25*, 3715.

11. Daniewski, A. R.; Wojceichowska, W., *J. Org. Chem.* **1982**, *47*, 2993.

12. Mori, A.; Fujiwara, J.; Maruoka, K.; Yamamoto, H., *Tetrahedron Lett.* **1983**, *24*, 4581.

13. Szantay, C.; Toke, L.; Kolonits, P., *J. Org. Chem.* **1966**, *31*, 1447.

14. Vidari, G.; Ferrino, S.; Grieco, P. A., *J. Am. Chem. Soc.* **1984**, *106*, 3539.

15. Kido, F.; Noda, Y.; Maruyama, T.; Kabuto, C.; Yoshikoshi, A., *J. Org. Chem.* **1981**, *46*, 4264.

16. Marshall, J. A.; Andersen, N. H.; Schlicher, J. W., *J. Org. Chem.* **1970**, *35*, 858.

17. Overman, L. E.; Burk, R. M., *Tetrahedron Lett.* **1984**, *25*, 5737.

18. Matsumura, Y.; Maruoka, K.; Yamamoto, H., *Tetrahedron Lett.* **1982**, *23*, 1929.

19. Yamamoto, H.; Maruoka, K., *J. Am. Chem. Soc.* **1981**, *103*, 4186.

20. Sasatani, S.; Miyazaki, T.; Maruoka, K.; Yamamoto, H., *Tetrahedron Lett.* **1983**, *24*, 4711.

21. Hattori, K.; Matsumura, Y.; Miyazaki, T.; Maruoka, K.; Yamamoto, H., *J. Am. Chem. Soc.* **1981**, *103*, 7368.

22. Gardner, J. N.; Kaiser, S.; Krubiner, A. Lucas, H., *Can. J. Chem.* **1973**, *51*, 1419.

23. Kim, S. Ahn, K. H., *J. Org. Chem.* **1984**, *49*, 1717.

24. Lenox, R. S.; Katzenellenbogen, J. A., *J. Am. Chem. Soc.* **1973**, *95*, 957.

25. Eisch, J. J.; Foxton, M. W., *J. Org. Chem.* **1971**, *36*, 3520.

26. Utimoto, K.; Uchida, K.; Yamaya, M.; Nozaki, H., *Tetrahedron Lett.* **1977**, 3641.

27. Gensler, W. J.; Bruno, J. J., *J. Org. Chem.* **1963**, *28*, 1254.

28. Ziegler, K.; Kropp, F.; Zosel, K., *Justus Liebigs Ann. Chem.* **1960**, *629*, 241.

29. Eisch, J. J.; Kaska, W. C., *J. Am. Chem. Soc.* **1966**, *88*, 2213.

30. Zweifel, G.; Clark, G. M.; Whitney, C. C., *J. Am. Chem. Soc.* **1971**, *93*, 1305.

31. (a) Negishi, E.; Boardman, L. D., *Tetrahedron Lett.* **1982**, *23*, 3327. (b) Negishi, E.; Jadhav, K. P.; Daotien, N., *Tetrahedron Lett.* **1982**, *23*, 2085.

32. Babas, S.; Negishi, E., *J. Am. Chem. Soc.* **1976**, *98*, 6729.

33. Monturi, M.; Gore, J., *Tetrahedron Lett.* **1980**, *21*, 51.

34. Zweifel, G.; Clark, G. M.; Lynd, R., *J. Chem. Soc., Chem. Commun.* **1971**, 1593.

35. Busacca, C. A.; Lorenz, J. C.; Grinberg, N.; Haddad, N.; Hrapchak, M.; Latli, B.; Lee, H.; Sabila, P.; Saha, A.; Sarvestani, M.; Shen, S.; Varsolona, R.; Wei, X.; Senanayake, C. H., *Org. Lett.* **2005**, *7*, 4277.

36. Hon, Y.-S.; Chang, C.-P.; Wong, Y.-C., *Tetrahedron Lett.* **2004**, *45*, 3313.

37. Proto, A.; Capacchione, C.; Scettri, A.; Motta, *Appl. Catal. A* **2003**, *247*, 75.

38. Fauvrie, C.; Blanchet, J.; Bonin, M.; Micouin, L., *Org. Lett.* **2004**, *6*, 2333.

39. Maier, P.; Redlich, H., *Synlett* **2000**, 257.

40. Gauthier, D. R., Jr; Szumigala, R. H., Jr; Armstrong, J. D., III; Volante, R. P., *Tetrahedron Lett.* **2001**, *42*, 7011.

41. Cossy, J.; Gille, B.; Bellosta, V.; Duprat, A., *New J. Chem.* **2002**, *26*, 526.

42. Marco-Contelles, J.; Ruiz-Caro, J., *Carbohydr. Res.* **2001**, *335*, 63.

43. Lecourt, T.; Herault, A.; Pearce, A. J.; Sollogoub, M.; Sinaÿ, P., *Chem. Eur. J.* **2004**, *10*, 2960.

44. Bistri, O.; Sinaÿ, P.; Sollogoub, M., *Chem. Commun.* **2006**, 1112.

45. Bistri, O.; Sinaÿ, P.; Sollogoub, M., *Tetrahedron Lett.* **2005**, *46*, 7757.

46. Betancor, C.; Dorta, R. L.; Freire, R.; Prangé, T.; Suàrez, E., *J. Org. Chem.* **2000**, *65*, 8822.

47. Meng, X.; Zhang, Y.; Sollogoub, M.; Sinaÿ, P., *Tetrahedron Lett.* **2004**, *45*, 8165.

48. Söker, U.; Sasse, F.; Kunze, B.; Höfle, G., *Eur. J. Org. Chem.* **2000**, 2021.

49. Sharma, G. V. M.; Ilangovan, A.; Sreenivas, P.; Mahalingam, A. K., *Synlett* **2000**, 615.

50. Chakraborty, A.; Marek, I., *Chem. Commun.* **1999**, 2375.

4-Dimethylaminopyridine[1]

[1122-58-3] $C_7H_{10}N_2$ (MW 122.19)

InChI = 1S/C7H10N2/c1-9(2)7-3-5-8-6-4-7/h3-6H,1-2H3

InChIKey = VHYFNPMBLIVWCW-UHFFFAOYSA-N

(catalyst for acylation of alcohols or amines,[1–11] especially for acylations of tertiary or hindered alcohols or phenols[12] and for macrolactonizations;[13–15] catalyst for direct esterification of carboxylic acids and alcohols in the presence of dicyclohexylcarbodiimide (Steglich–Hassner esterification);[5] catalyst for silylation or tritylation of alcohols,[9,10] and for the Dakin–West reaction[20])

Alternate Name: DMAP.

Physical Data: colorless solid; mp 108–110 °C; pK_a 9.7.

Solubility: sol MeOH, CHCl₃, CH₂Cl₂, acetone, THF, pyridine, HOAc, EtOAc; partly sol cold hexane or water.

Form Supplied in: colorless solid; commercially available.

Preparative Method: prepared by heating 4-pyridone with HMPA at 220 °C, or from a number of 4-substituted (Cl, OPh, SO₃H, OSiMe₃) pyridines by heating with DMA.[2] Prepared commercially from the 4-pyridylpyridinium salt (obtained from pyridine and SOCl₂) by heating with DMF at 155 °C.[1,2]

Purification: can be recrystallized from EtOAc.

Handling, Storage, and Precautions: skin irritant; corrosive, toxic solid.

Original Commentary

Alfred Hassner
Bar-Ilan University, Ramat Gan, Israel

Acylation of Alcohols. Several 4-aminopyridines speed up esterification of hindered alcohols with acid anhydrides by as much as 10 000 fold; of these, DMAP is the most commonly used but 4-pyrrolidinopyridine (PPY) and 4-tetramethylguanidinopyridine are somewhat more effective.[11] DMAP is usually employed in 0.05–0.2 mol equiv amounts.

DMAP catalyzes the acetylation of hindered 11β- or 12α-hydroxy steroids. The alkynic tertiary alcohol acetal in eq 1 is acetylated at rt within 20 min in the presence of excess DMAP.[3]

$$\text{HO}\!\!-\!\!\overset{\text{OMe}}{\underset{\text{OMe}}{<}}\!\!\equiv\!\!<\quad\xrightarrow[\underset{93\%}{\text{rt}}]{\text{Ac}_2\text{O, DMAP}}\quad\text{AcO}\!\!-\!\!\overset{\text{OMe}}{\underset{\text{OMe}}{<}}\!\!\equiv\!\!< \qquad (1)$$

Esterifications mediated by *2-Chloro-1-methylpyridinium Iodide* also benefit from the presence of DMAP.[22]

DMAP acts as an efficient acyl transfer agent, so that alcohols resistant to acetylation by *Acetic Anhydride–Pyridine* usually react well in the presence of DMAP.[4a] Sterically hindered phenols can be converted into salicylaldehydes via a benzofurandione prepared by DMAP catalysis (eq 2).[4b]

Direct Esterification of Alcohols and Carboxylic Acids. Instead of using acid anhydrides for the esterification of alcohols, it is possible to carry out the reaction in one pot at rt by employing a carboxylic acid, an alcohol, *1,3-Dicyclohexylcarbodiimide*, and DMAP.[5,6] In this manner, *N*-protected amino acids and even hindered carboxylic acids can be directly esterified at rt using DCC and DMAP or 4-pyrrolidinopyridine (eq 3).[5]

$$ (2) $$

$$ (3) $$

DCC–DMAP has been used in the synthesis of depsipeptides.[6b] Macrocyclic lactones have been prepared by cyclization of hydroxy carboxylic acids with DCC–DMAP. The presence of salts of DMAP,[13a] such as its trifluoroacetate, is beneficial in such cyclizations, as shown for the synthesis of a (9S)-dihydroerythronolide.[13b] Other macrolactonizations have been achieved using *2,4,6-Trichlorobenzoyl Chloride*[1] and DMAP in *Triethylamine* at rt[14] or *Di-2-pyridyl Carbonate* (6 equiv) with 2 equiv of DMAP at 73 °C.[15]

Acylation of Amines. Acylation of amines is also faster in the presence of DMAP,[7] as is acylation of indoles,[8a] phosphorylation of amines or hydrazines,[2,8] and conversion of carboxylic acids into anilides by means of *Phenyl Isocyanate*.[1] β-Lactam formation from β-amino acids has been carried out with DCC–DMAP, but epimerization occurs.[8b]

Silylation, Tritylation, and Sulfinylation of Alcohols. Tritylation, including selective tritylation of a primary alcohol in the presence of a secondary one,[9] silylation of tertiary alcohols, selective silylation to *t*-butyldimethylsilyl ethers,[6] and sulfonylation or sulfinylation[10] of alcohols proceed more readily in the presence of DMAP. Silylation of β-hydroxy ketones with *Chlorodiisopropylsilane* in the presence of DMAP followed by treatment with a Lewis acid gives diols (eq 4).[16]

$$ (4) $$

Essential Reactions for Organic Synthesis, First Edition. Edited by Philip L. Fuchs.
© 2016 John Wiley & Sons, Ltd. Published 2016 by John Wiley & Sons, Ltd.

Miscellaneous Reactions. Alcohols, including tertiary ones, can be converted to their acetoacetates by reaction with ***Diketene*** in the presence of DMAP at rt.[17] Decarboxylation of β-keto esters has been carried out at pH 5–7 using 1 equiv of DMAP in refluxing wet toluene (eq 5).[18]

$$\text{(5)}$$

Elimination of water from a *t*-alcohol in a β-hydroxy aldehyde was carried out using an excess of ***Methanesulfonyl Chloride***–DMAP–H_2O at 25 °C.[23]

Glycosidic or allylic alcohols (even when *s*-) can be converted in a 80–95% yield to alkyl chlorides by means of ***p-Toluenesulfonyl Chloride***–DMAP. Simply primary alcohols react slower and secondary ones are converted to tosylates.[24]

Aldehydes and some ketones can be converted to enol acetates by heating in the presence of TEA, Ac_2O, and DMAP.[2] DMAP catalyzes condensation of malonic acid monoesters with unsaturated aldehydes at 60 °C to afford dienoic esters (eq 6).[19]

$$\text{(6)}$$

The conversion of α-amino acids into α-amino ketones by means of acid anhydrides (Dakin–West reaction)[20] also proceeds faster in the presence of DMAP (eq 7).

$$\text{(7)}$$

Ketoximes can be converted to nitrimines which react with Ac_2O–DMAP to provide alkynes (eq 8).[21]

$$\text{(8)}$$

For the catalysis by DMAP of the *t*-butoxylcarbonylation of alcohols, amides, carbamates, NH-pyrroles, etc., see ***Di-t-butyl Dicarbonate***.

First Update

Alison Hart and Julie A. Pigza
University of Southern Mississippi, Hattiesburg, MS, USA

Since the original article, DMAP has had further investigation into its thermal and magnetic properties,[25] as well as a thermal analysis and calorimetric study.[26] An in-depth review of the factors affecting the basicity and nucleophilicity of common nitrogen-containing catalysts, including DMAP and its derivatives, has also been presented.[27]

Esterification Reactions. While esterification of carboxylic acids and alcohols via a coupling reaction utilizing DCC and DMAP has already been covered, some new aspects will be discussed here. A DCC/DMAP esterification has been shown between heteroaromatic carboxylic acids and phenols to form functionalized (*E*)-2-arylvinyl bromides in excellent yields.[28] Other coupling agents such as diisopropylcarbodiimide (DIC) have been used with DMAP to form tyrosine esters bound to a solid support (Marshall resin).[29] The resulting amino esters were then subjected to a Pictet–Spengler reaction to form the tetrahydroisoquinoline ring. Another reagent, 1-ethyl-3-(3-dimethylaminopropyl)carbodiimide (EDC), can also be used in tangent with DMAP to perform a coupling reaction between a caffeic acid derivative and an alcohol as a precursor to the synthesis of zanamivir analogs, which display anti-inflammatory activity and are potent influenza virus neuraminidase inhibitors.[30] The combination of EDC and DMAP has also been shown to be effective for the synthesis of amino esters.[31] The benefit of EDC is that the by-product urea is water soluble and can be removed in the aqueous workup, resulting in an easier purification by chromatography.

Other methods of esterification that do not utilize a carbodiimide coupling method have also been explored. For example, the stereoselective esterification of 20-(*S*)-camptothecin, a hindered 3° alcohol, with amino acid derivatives was accomplished with the use of scandium triflate (Sc(OTf)$_3$) in high yields while retaining the integrity of the amino acid stereocenter (eq 9).[32]

$$\text{(9)}$$

Another acylating system utilizing Et_3N, DMAP, and acetic anhydride was used in the formation of mixed acetals from an *in situ*-generated hemiacetal in good yields (eq 10).[33] Similar acetate-protected acetals have also been formed from the reductive acetylation of esters, using DIBAL to reduce the ester and form the hemiacetal which is then treated with acetic anhydride in the presence of DMAP, Et_3N, and pyridine.[34]

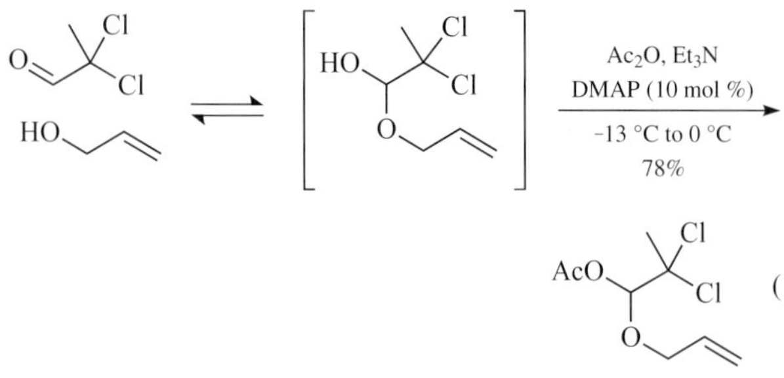

(10)

Amidation Reactions. Formation of an amide bond is also possible with the same coupling reagents as used for esterification reactions. Jeffamines ED[®], diamines containing a polyether chain, were coupled with two molecules of chelidamic acid, a heteroaromatic acid, using a DCC/DMAP catalyst system to form oligo(ether-amide)s.[35] These were used as a water-soluble chelating agent with the ability to complex with many metal ions, including Fe^{3+}. DCC and DMAP were also used in the formation of phenylethynyl-terminated bis(amide) derivatives in an effort to form heat-resistant polymers.[36] Additionally, DCC/DMAP was used in the formation of a thiazolidinethione amide as part of the synthesis of iron(III)-specific fluorescent probes (eq 11).[37]

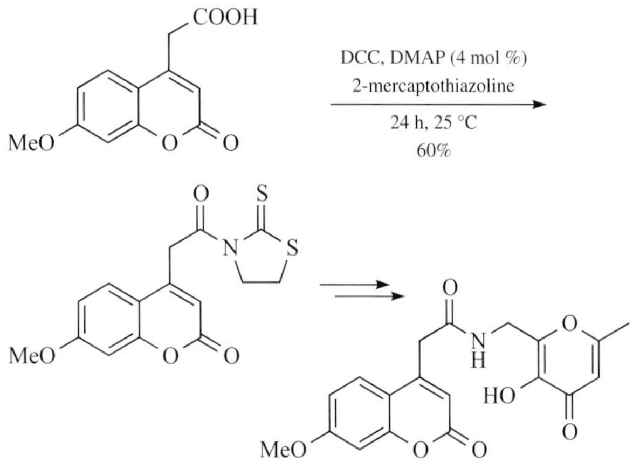

(11)

Trifluoroacetylation of primary anilines with ethyl trifluoroacetate can be accomplished in the presence of DMAP to afford good overall yields (49–98%).[38] In addition, a selectivity was observed for trifluoroacetylation of primary anilines in the presence of secondary anilines and even in the presence of hindered secondary amines.

The reaction between acid chlorides and a urea nitrogen has also been demonstrated. Hydrocinnamoyl chloride and an imidazolidinone were reacted in the presence of DMAP, leading to the formation of an imidazolidinone analog, which has attracted attention for its applicability as an HIV protease inhibitor and a

5-HT_3 receptor antagonist (eq 12).[39] In addition, this compound can also serve as a useful chiral auxiliary.

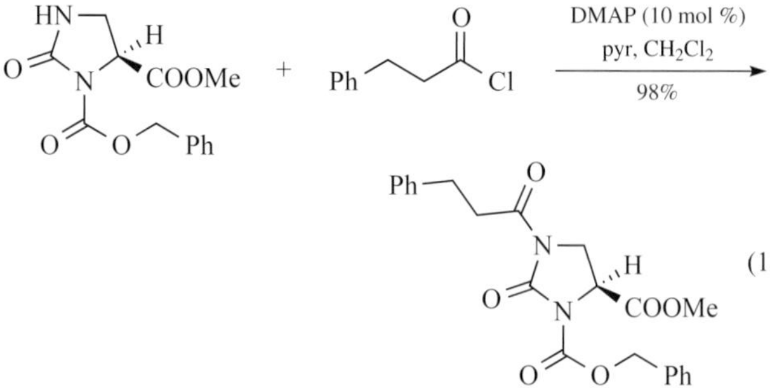

(12)

Acylation Reactions. Acylation reactions can also be accomplished using coupling reagents such as DCC/DMAP. For example, C-acylation of 1,3-diones with unactivated carboxylic acids can be accomplished with an excess of DMAP using a variety of carboxylic acids and cyclic 1,3-diones.[40] In a different example involving tetramic acids, either C- or O-acylation can be observed by changing the amount of DMAP. Substoichiometric amounts of DMAP lead to O-acylation of the ketone oxygen, while a slight excess of DMAP leads to the C-acylation product (eq 13).[41]

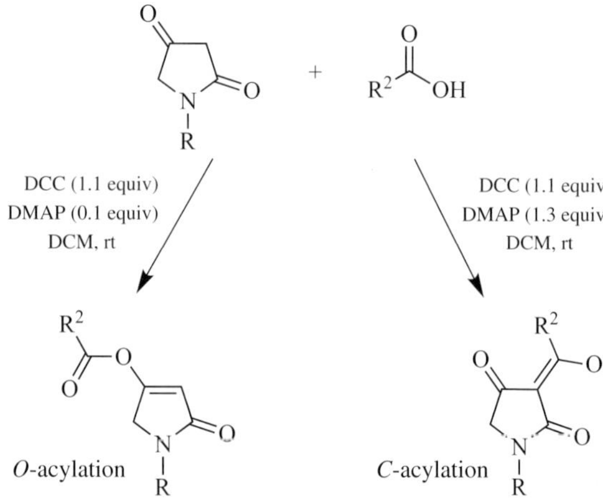

(13)

DMAP Catalysis to Form Key Substructures. DMAP has been used as a catalyst in the Morita–Baylis–Hillman (MBH) reaction. Conjugated nitroalkenes can be reacted with an amine base to form a zwitterionic species that can add to many types of ketones and aldehydes.[42] Remarkably, only DMAP and imidazole were able to effectively catalyze this reaction, likely due to their small size and ability to stabilize the intermediate. In another example, a unique catalyst system involving 1:1:1 DMAP:TMEDA:MgI_2 (each in 10 mol %) allowed the MBH reaction to proceed in fairly high yields between aldehydes and α,β-unsaturated ketones, esters, or a thioester.[43]

The catalyzed generation of a chiral enolate equivalent, formed from an enal and a triazolium-derived chiral N-heterocyclic carbene, can undergo a hetero-Diels–Alder reaction with enones with excellent enantioselectivity and diastereoselectivity (eq 14, left side).[44] Alternatively, the authors have previously shown that stronger bases, such as DBU, instead result in a homoenolate addition to the unsaturated ester, resulting in a cyclopentene ring (eq 14, right side).[45]

enolate equivalent   homoenolate equivalent

50–98% yield   50–98% yield
99% ee, >20:1 dr   99% ee, 5:1 dr

DMAP is also key in the formation of hindered, enantioenriched Michael adducts between α-branched aldehydes and nitroalkanes in conjunction with an L-threonine derivative and sulfamide (eq 15).[46] A transition state is proposed involving extensive hydrogen bonding between the chiral enamine, formed from threonine and the aldehyde, the nitroalkene, sulfamide, and protonated DMAP in which the sulfur atom acquires a distorted tetrahedral geometry. This allows one of the oxygen atoms of sulfamide to always be in a position to take part in an electrostatic interaction with the protonated nitrogen atom of DMAP, resulting in a cyclic coordination of the four components and providing 96–98% ee in all of the examples.

R = aryl, vinyl, or alkyl
70–98% yield, 96–98% ee

DMAP acts as a catalyst in the sulfinylation of diacetone glucose (DAG) with *tert*-butanesulfinyl chloride in the presence of DIPEA or Et$_3$N, resulting in the formation of the (R)-sulfinate esters in high yields ($\sim$90%, still bound to glucose).[47] Interestingly, the selectivity can be reversed to (S) by using Et$_3$N alone. The enantiopure DAG-sulfinate esters can then serve as important precursors to chiral sulfoxides, sulfinamides, or sulfinylimines via a displacement reaction with Grignards, amines, or LiHMDS, respectively.

Polysubstituted homoallylic alcohols with a trifluoromethyl group are highly sought after for their use in pharmaceutical and agrochemical applications. They can be synthesized from the allylation of trifluoropyruvate with alkenes activated by two electron-withdrawing groups in the presence of bases such as N-methylmorpholine, quinine, diethylamine, and DMAP, of which DMAP has the highest yields, using only 0.2 equiv (52–88%).[48]

In a final example, α-chloro-β-hydroxyalkanones are reacted with acetic anhydride and pyridine in the presence of catalytic DMAP to afford selectively (Z)-α-chloroenones in good yield via an elimination reaction.[49]

Cycloadditions Involving DMAP Catalysis. Bicyclic lactones containing a spirooxindole can be formed through a DMAP-catalyzed [4 + 2] Diels–Alder cycloaddition of pyrones and indole derivatives, resulting in both *endo* and *exo* products in approximately 2:1, 1:1, or 1:2 ratios in good yields. DMAP is proposed to abstract the proton on the pyrone, causing the initial cyclization to occur (eq 16).[50]

exo   *endo*

DMAP can be used to facilitate the formal [2 + 2] annulation reaction of alkynyl ketones with N-tosylimines for complete stereoselectivity of highly functionalized *trans*-substituted azetines.[51] DMAP initially adds to the ynone via a Michael-type addition, generating an anion at the 4-position via proton transfer and resonance, which then adds to the imine and can undergo a cyclization via the nitrogen. Interestingly, if Bu$_3$P is used as the catalyst instead, pyrrolidine products result (five-membered ring) from a formal [3 + 2] annulation due to an anion formation at the 2-position of the ynone. Other [3 + 2] and [4 + 2] cycloadditions incorporating C$_{60}$ fullerenes and Morita–Baylis–Hillman adducts have also been catalyzed by DMAP. When reacted in the presence of Ac$_2$O, either cyclopentene- or cyclohexene-fused C$_{60}$ fullerenes can be formed just by changing the temperature (room temperature vs. 120 °C) and conditions (DMAP/Bu$_3$P vs. DMAP alone).[52]

Multicomponent Coupling Reactions Catalyzed by DMAP. Multicomponent coupling reactions can also be facilitated with the use of a catalytic amount of DMAP. A one-pot, three-component domino coupling of β-ketodithioesters, aldehydes, and nitriles yielded 4H-thiopyrans in high yields via a cascading Knoevenagel condensation/Michael addition/cyclization sequence, forming three new bonds with high atom economy.[53] A similar strategy utilized β-diketones instead of the β-ketodithioesters resulting in pyran-annulated heterocycles.[54] Moderate to good yields are observed in these reactions as well.

DMAP-Initiated Protecting Group Strategies. DMAP has also been utilized as a key component in several protecting group strategies. The formation of cyclic acetals from 1,2-diols and alkyl propynoates in the presence of DMAP has been demonstrated.[55] These acetals are stable to acid-catalyzed hydrolysis, unlike other acetals, and methanolysis. These acetals can be deprotected by heating in neat pyrrolidine.

Demethylation of acid-sensitive aryl methyl ethers, such as 2,4-dimethoxyquinolines, can be accomplished using a trimethylsilyl iodide/DMAP reagent system, where DMAP is used as a Lewis base to prevent protonation of the alkene from the HI by-product.[56] This reaction was utilized to form the quinoline-containing alkaloid natural product, atanine, from the *Rutaceae* family.

DMAP can also be used in the removal of the thiazolidinethione moiety from *N*-acylthiazolidinethiones to the corresponding benzyl esters.[57] DMAP was found to be superior to other bases including imidazole, 2,6-lutidine, and DBU (eq 17).

$$\text{(17)}$$

Miscellaneous Non-catalytic DMAP Reactions. Iodoacetylenes, which are useful intermediates in organic synthesis and in medicinal and pharmaceutical research, can be prepared in high yields through the reaction of terminal acetylenic ketones or aryl-substituted acetylenes with molecular iodine in the presence of DMAP as a mild base.[58]

Nucleophile-promoted aldol-lactonization of ketoacids using tosyl chloride, DIPEA, and K_2CO_3 in the presence of DMAP as a Lewis base creates constrained bicyclic β-lactones with high diastereoselectivity (eq 18).[59] The β-lactone substructure is found in a numerous bioactive natural products and pharmaceutical agents.

$$\text{(18)}$$

(substituent sits equatorial)

The allylic alcohol products from Morita–Baylis–Hillman reactions were shown to participate in a DMAP-mediated Tsuji–Trost-type reaction with β-diketones or β-ketoesters, forming the C-allylation product without requiring the use of palladium. Previously, it was shown that allylic alcohols combined with β-ketoesters and DMAP afforded the transesterification products, in which the allylic alcohol displaced the ester substituent.[60] The difference between these diverging reaction pathways is likely due to the electron-withdrawing group on the allylic alcohol in the MBH adducts vs. just alkyl substituents in the latter case.

The reaction of pressurized CO_2 with substituted epoxides to yield cyclic carbonates has been demonstrated in the presence of cobalt(II) tetraphenylporphyrin chloride (CoTTP(Cl)) and DMAP

as a cocatalyst (eq 19).[61] The same author has shown that this reaction can also be carried out in the presence of chromium(III) salen complexes.[62] DMAP gave superior activity to other Lewis bases tested and resulted in high turnover numbers and short reaction times for terminal monosubstituted epoxides containing alkyl, vinyl, aryl, and even α-halo substituents. Stereochemistry on 1,2-disubstituted epoxides was retained in the product and one example of a 1,1-disubstituted epoxide was also shown.

$$\text{(19)}$$

CoTTPCl = cobalt(II) tetraphenylporphyrin

R = alkyl, aryl, α-halo
conversions >92% by NMR

DMAP has also been applied to the regioselective ring opening of epoxides and oxetanes with $POCl_3$ or PCl_3, providing α-chloroalcohols.[63] DMAP attacks the phosphorus atom and releases a chloride ion, which can then undergo a nucleophilic attack to open the ring.

Thionation or selenation of amides can be accomplished through the use of elemental sulfur or selenium, respectively, along with a hydrochlorosilane and an amine base, where DMAP gave the highest yields.[64]

Nitrilium ions are highly reactive intermediates that serve as imine synthons and are used in a multitude of synthetic named reactions; however, their lack of stability hampers their use. DMAP has been shown to stabilize nitrilium ions in the presence of trimethylsilyl triflate to create the base-stabilized triflate salts, which are stable in air with only 3% decomposition over a period of 1 month (eq 20).[65]

R^1	R^2
Mes	tBu
Ph	tBu
iPr	Ph
Cy	Ph

$$\text{(20)}$$

DMAP and alkyl bromides are used to make homologs of 1-alkyl-(*N*,*N*-dimethylamino)pyridinium bromide salts by undergoing a simple nucleophilic substitution.[66] These compounds were tested for antifungal and antibacterial capacities and show promising activity against bacterial strains such as *Mycobacterium*, *E. coli*, *Staphylococcus*, and *Salmonella*.

DMAP has been incorporated into dispirooxindole-fused heterocycles via a domino 1,4-dipolar addition and Diels–Alder reaction of DMAP, acetylenedicarboxylates, and 3-phenacylideneoxindoles.[67] The reactive intermediate is an *in situ* generated Huisgen 1,4-dipole.

DMAP has also proved useful in releasing an imine-bound borane complex formed after an alkene hydroboration reaction, thereby providing higher yields than the usual workup (56% vs. 35%).[68]

1. Hoefle, G.; Steglich, W.; Vorbrueggen, H., *Angew. Chem., Int. Ed. Engl.* **1978**, *17*, 569.

2. Scriven, E. F. V., *Chem. Soc. Rev.* **1983**, *12*, 129.

3. (a) Steglich, W.; Hoefle, G., *Angew. Chem., Int. Ed. Engl.* **1969**, *8*, 981. (b) Hoefle, G.; Steglich, W., *Synthesis* **1972**, 619.

4. (a) Salomon, R. G., Salomon, M. F.; Zagorski, M. G.; Reuter, J. M.; Coughlin, D. J., *J. Am. Chem. Soc.* **1982**, *104*, 1008. (b) Zwanenburg, D. J.; Reynen, W. A. P., *Synthesis* **1976**, 624.

5. (a) Neises, B.; Steglich, W., *Angew. Chem., Int. Ed. Engl.* **1978**, *17*, 522. (b) Hassner, A.; Alexanian, V., *Tetrahedron Lett.* **1978**, 4475.

6. (a) Ziegler, F. E.; Berger, G. D., *Synth. Commun.* **1979**, 539. (b) Gilon, C.; Klausner, Y.; Hassner, A., *Tetrahedron Lett.* **1979**, 3811.

7. Litvinenko, L. M., Kirichenko, A. C.; *Dokl. Akad. Nauk SSSR* **1967**, *176*, 97. (b) Kirichenko, A. C.; Litvinenko, L. M.; Dotsenko, I. N.; Kotenko, N. G.; Nikkel'sen, E.; Berestetskaya, V. D., *Dokl. Akad. Nauk SSSR* **1969**, *244*, 1125 (*Chem. Abstr.* **1979**, *90*, 157 601).

8. (a) Nickisch, K.; Klose, W.; Bohlmann, F., *Chem. Ber.* **1980**, *113*, 2036. (b) Kametami, T.; Nagahara, T.; Suzuki, Y.; Yokohama, S.; Huang, S.-P.; Ihara, M., *Tetrahedron* **1981**, *37*, 715.

9. (a) Chaudhary, S. K.; Hernandez, O., *Tetrahedron Lett.* **1979**, *95*, 99. (b) Hernandez, O.; Chaudhary, S. K.; Cox, R. H.; Porter, J., *Tetrahedron Lett.* **1981**, *22*, 1491.

10. Guibe-Jampel, E.; Wakselman, M.; Raulais, D., *J. Chem. Soc., Chem. Commun.* **1980**, 993.

11. Hassner, A.; Krepski, L. R.; Alexanian, V., *Tetrahedron* **1978**, *34*, 2069.

12. Vorbrueggen, H. (Schering AG) Ger. Offen. 2 517 774, **1976** (*Chem. Abstr.* **1977**, *86*, 55 293).

13. (a) Boden, E. P.; Keck, G. E., *J. Org. Chem.* **1985**, *50*, 2394. (b) Stork, G.; Rychnovsky, S. D., *J. Am. Chem. Soc.* **1987**, *109*, 1565.

14. Hikota, M.; Tone, H.; Horita, K.; Yonemitsu, O., *J. Org. Chem.* **1990**, *55*, 7.

15. (a) Kim, S.; Lee, J. I.; Ko, Y. K., *Tetrahedron Lett.* **1984**, *25*, 4943. (b) Denis, J.-N.; Greene, A. E.; Guenard, D.; Gueritte-Voegelein, F.; Mangatal, L.; Potier, P., *J. Am. Chem. Soc.* **1988**, *110*, 5917.

16. Anwar, S.; Davis, A. P., *Tetrahedron* **1988**, *44*, 3761.

17. Nudelman, A.; Kelner, R.; Broida, N.; Gottlieb, H. E., *Synthesis* **1989**, 387.

18. Taber, D. F.; Amedio, J. C., Jr.; Gulino, F., *J. Org. Chem.* **1989**, *54*, 3474.

19. Rodriguez, J.; Waegell, B., *Synthesis* **1988**, 534.

20. (a) Buchanan, G. L., *Chem. Soc. Rev.* **1988**, *17*, 91. (b) McMurry, J., *J. Org. Chem.* **1985**, *50*, 1112. (c) Hoefle, G., Steglich, W., *Chem. Ber.* **1971**, *104*, 1408.

21. Buechi, G.; Wuest, H., *J. Org. Chem.* **1979**, *44*, 4116.

22. Nicolaou, K. C.; Bunnage, M. E.; Koide, K., *J. Am. Chem. Soc.* **1994**, *116*, 8402.

23. Furukawa, J.; Morisaki, N.; Kobayashi, H.; Iwasaki, S.; Nozoe, S.; Okuda, S., *Chem. Pharm. Bull.* **1985**, *33*, 440.

24. Hwang, C. K.; Li, W. S.; Nicolaou, K. C., *Tetrahedron Lett.* **1984**, *25*, 2295.

25. Balachandran, V.; Rajeswari, S.; Lalitha, S., *Spectrochim. Acta A* **2014**, *124*, 277.

26. Shi, Q.; Tan, Z.-C.; Di, Y.-Y.; Tong, B.; Li, Y.-S.; Wang, S.-X., *J. Chem. Eng. Data* **2007**, *52*, 941.

27. De Rycke, N.; Couty, F.; David, O. R. P., *Chem. Eur. J.* **2011**, *17*, 12852.

28. Jiang, Y.; Kuang, C., *J. Chem. Sci.* **2009**, *121*, 1035.

29. Kane, T. R.; Ly, C. Q.; Kelly, D. E.; Dener, J. M., *J. Comb. Chem.* **2004**, *6*, 564.

30. Liu, K.-C.; Fang, J.-M.; Jan, J.-T.; Cheng, T.-J. R.; Wang, S.-Y.; Yang, S.-T.; Cheng, Y.-S. E.; Wong, C.-H., *J. Med. Chem.* **2012**, *55*, 8493.

31. Dhaon, M. K.; Olsen, R. K.; Ramasamy, K., *J. Org. Chem.* **1982**, *47*, 1962.

32. (a) Greenwald, R. B.; Pendri, A.; Zhao, H., *Tetrahedron: Asymmetry* **1998**, *9*, 915. (b) Liu, Y.-Q.; Tian, X.; Yang, L.; Zhan, Z.-C., *Eur. J. Med. Chem.* **2008**, *43*, 2610.

33. Roncaglia, F.; Parsons, A. F.; Bellesia, F.; Ghelfi, F., *Synth. Commun.* **2011**, *41*, 1175.

34. (a) Kopecky, D. J.; Rychnovsky, S. D., *Org. Synth.* **2003**, *80*, 177. (b) Dahanukar, V. H.; Rychnovsky, S. D., *J. Org. Chem.* **1996**, *61*, 8317. (c) Jaber, J. J.; Mitsui, K.; Rychnovsky, S. D., *J. Org. Chem.* **2001**, *66*, 4679.

35. Ignatova, M.; Sepulchre, M.; Manolova, N., *Eur. Polym. J.* **2001**, *38*, 33.

36. Wang, D. H.; Baek, J.-B.; Nishino, S. F.; Spain, J. C.; Tan, L.-S., *Polymer* **2006**, *47*, 1197.

37. Luo, W.; Ma, Y. M.; Quinn, P. J.; Hider, R. C.; Liu, Z. D., *J. Pharm. Pharmacol.* **2004**, *56*, 529.

38. Prashad, M.; Hu, B.; Har, D.; Repic, O.; Blacklock, T. J., *Tetrahedron Lett.* **2000**, *41*, 9957.

39. Doyle, M. P.; Colyer, J. T., *Tetrahedron: Asymmetry* **2003**, *14*, 3601.

40. Goncalves, S.; Nicolas, M.; Wagner, A.; Baati, R., *Tetrahedron Lett.* **2010**, *51*, 2348.

41. Jeong, Y.-C.; Moloney, M. G., *J. Org. Chem.* **2011**, *76*, 1342.

42. Deb, I.; Shanbhag, P.; Mobin, S. M.; Namboothiri, I. N. N., *Eur. J. Org. Chem.* **2009**, 4091.

43. Bugarin, A.; Connell, B. T., *J. Org. Chem.* **2009**, *74*, 4638.

44. Kaeobamrung, J.; Kozlowski, M. C.; Bode, J. W., *Proc. Natl. Acad. Sci. USA* **2010**, *107*, 20661.

45. Kaeobamrung, J.; Bode, J. W., *Org. Lett.* **2009**, *11*, 677.

46. Nugent, T. C.; Shoaib, M.; Shoaib, A., *Org. Biomol. Chem.* **2011**, *9*, 52.

47. Chelouan, A.; Recio, R.; Alcudia, A.; Khiar, N.; Fernández, I., *Eur. J. Org. Chem.* **2014**, *2014*, 6935.

48. Zhang, F.; Wang, X.-J.; Cai, C.-X.; Liu, J.-T., *Tetrahedron* **2009**, *65*, 83.

49. Concellon, J. M.; Huerta, M., *Tetrahedron* **2002**, *58*, 7775.

50. Zhou, M.-Q.; Zhao, J.-Q.; You, Y.; Xu, X.-Y.; Zhang, X.-M.; Yuan, W.-C., *Tetrahedron* **2015**, doi: 10.1016/j.tet.2015.04.032.

51. Meng, L.-G.; Cai, P.; Guo, Q.; Xue, S., *J. Org. Chem.* **2008**, *73*, 8491.

52. Yang, H.-T.; Ren, W.-L.; Miao, C.-B.; Dong, C.-P.; Yang, Y.; Xi, H.-T.; Meng, Q.; Jiang, Y.; Sun, X.-Q., *J. Org. Chem.* **2013**, *78*, 1163.

53. Khan, A. T.; Lal, M.; Ali, S.; Khan, M. M., *Tetrahedron Lett.* **2011**, *52*, 5327.

54. Verma, R. K.; Verma, G. K.; Shukla, G.; Nagaraju, A.; Singh, M. S., *ACS Comb. Sci.* **2012**, *14*, 224.

55. Ariza, X.; Costa, A. M.; Faja, M.; Pineda, O.; Vilarrasa, J., *Org. Lett.* **2000**, *2*, 2809.

56. Jones, K.; Roset, X.; Rossiter, S.; Whitfield, P., *Org. Biomol. Chem.* **2003**, *1*, 4380.

57. Wu, Y.; Sun, Y.-P.; Yang, Y.-Q.; Hu, Q.; Zhang, Q., *J. Org. Chem.* **2004**, *69*, 6141.

58. Meng, L.-G.; Cai, P.-J.; Guo, Q.-X.; Xue, S., *Synth. Commun.* **2008**, *38*, 225.

59. Liu, G.; Shirley, M. E.; Romo, D., *J. Org. Chem.* **2012**, *77*, 2496.

60. Gilbert, J. C.; Kelly, T. A., *J. Org. Chem.* **1988**, *53*, 449.

61. (a) Paddock, R. L.; Hiyama, Y.; McKay, J. M.; Nguyen, S. T., *Tetrahedron Lett.* **2004**, *45*, 2023. (b) Darensbourg, D. J.; Mackiewicz, R. M., *J. Am. Chem. Soc.* **2005**, *127*, 14026.

62. Paddock, R. L.; Nguyen, S. T., *J. Am. Chem. Soc.* **2001**, *123*, 11498.

63. Sartillo-Piscil, F.; Quintero, L.; Villegas, C.; Santacruz-Juarez, E.; Anaya de Parrodi, C., *Tetrahedron Lett.* **2002**, *43*, 15.

64. Shibahara, F.; Sugiura, R.; Murai, T., *Org. Lett.* **2009**, *11*, 3064.

65. van Dijk, T.; Bakker, M. S.; Holtrop, F.; Nieger, M.; Slootweg, J. C.; Lammertsma, K., *Org. Lett.* **2015**, *17*, 1461.

66. Sundararaman, M.; Kumar, R. R.; Venkatesan, P.; Ilangovan, A., *J. Med. Microbiol.* **2013**, *62*, 241.

67. Sun, J.; Sun, Y.; Gong, H.; Xie, Y.-J.; Yan, C.-G., *Org. Lett.* **2012**, *14*, 5172.

68. (a) Pigza, J. A.; Han, J.-S.; Chandra, A.; Mutnick, D; Pink, M.; Johnston, J. N., *J. Org. Chem.* **2013**, *78*, 822. (b) Chandra, A.; Pigza, J. A.; Han, J.-S.; Mutnick, D.; Johnston, J. N., *J. Am. Chem. Soc.* **2009**, *131*, **3470**.

Dimethyldioxirane[1]

[74087-85-7]   C₃H₆O₂   (MW 74.09)

$[74087\text{-}85\text{-}7]$ $C_3H_6O_2$ (MW 74.09)

InChI = 1/C3H6O2/c1-3(2)4-5-3/h1-2H3
InChIKey = FFHWGQQFANVOHV-UHFFFAOYAF

(selective, reactive oxidizing agent capable of epoxidation of alkenes and arenes,[11] oxyfunctionalization of alkanes,[19] and oxidation of alcohols,[23] ethers,[21] amines, imines,[32] and sulfides[35])

Alternate Name: DDO.
Physical Data: known only in the form of a dilute solution.
Solubility: soluble in acetone and CH_2Cl_2; soluble in most other organic solvents, but reacts slowly with many of them.
Form Supplied in: dilute solutions of the reagent in acetone are prepared from Oxone and acetone, as described below.
Analysis of Reagent Purity: the concentrations of the reagent can be determined by classical iodometric titration or by reaction with an excess of an organosulfide and determination of the amount of sulfoxide formed by NMR or gas chromatography.
Preparative Methods: the discovery of a convenient method for the preparation of dimethyldioxirane has stimulated important advances in oxidation technology.[1] The observation[2] that ketones enhance the decomposition of the monoperoxysulfate anion prompted mechanistic studies that implicated dioxiranes as intermediates.[3] Ultimately, these investigations led to the isolation of dilute solutions of several dioxiranes.[4] DDO is by far the most convenient of the dioxiranes to prepare and use (eq 1). Several experimental set-ups for the preparation of DDO have been described,[4–6] but reproducible generation of high concentration solutions of DDO (ca 0.1M) is aided by a well-formulated protocol.[6] The procedure involves the portionwise addition of solid Oxone (*Potassium Monoperoxysulfate*) to a vigorously stirred solution of NaHCO₃ in a mixture of reagent grade *Acetone* and distilled water at 5–10 °C. The appearance of a yellow color signals the formation of DDO, at which point the cooling bath is removed and the DDO–acetone solution is distilled into a cooled (−78 °C) receiving flask under reduced pressure (80–100 Torr). After preliminary drying over reagent grade anhydrous MgSO₄ in the cold, solutions of DDO are stored over molecular sieves in the freezer of a refrigerator at −10 to −20 °C. In instances where the concentration of DDO is crucial, analysis is typically based on reaction with an excess of an organosulfide monitored by NMR.[4,7,8]

$$\text{\Large >}\!\!=\!\!O \quad \xrightarrow[\text{5–10 °C}]{\substack{\text{Oxone} \\ \text{H}_2\text{O, NaHCO}_3}} \quad \text{\Large >}\!\!\!<\substack{O \\ | \\ O} \qquad (1)$$

Concentrated solutions of DDO in chlorinated solvents may be obtained by a simple extraction technique.

A fresh solution (50 mL) of isolated DDO (0.06–0.08 M in acetone), prepared as reported, is diluted with an equal volume of cold water (0–5 °C) and extracted in a chilled separatory funnel with four 10 mL portions of cold CH_2Cl_2, CHCl₃, or CCl₄

to yield a total volume of ca. 35 mL of extract (pale yellow DDO solution). In order to concentrate this DDO solution, the combined extracts in chlorinated solvent are washed three times in a separatory funnel at 0–5 °C with an equal volume of cold 0.01 M phosphate buffer (pH 7). The resulting solution is 0.19–0.36 M in DDO. Its concentration can be estimated by iodometry; the recovery of dioxirane from the initial acetone solution is 35–45% in most cases. ¹H NMR spectroscopy analysis reveals that initial solvent acetone is not completely eliminated.

Handling, Storage, and Precaution: solutions of the reagent can be kept in the freezer of a refrigerator (−10 to −20 °C) for as long as a week. The concentration of the reagent decreases relatively slowly, provided solutions are kept from light and traces of heavy metals. These dilute solutions are not known to decompose violently, but the usual precautions for handling peroxides should be applied, including the use of a shield. All reactions should be performed in a fume hood to avoid exposure to the volatile oxidant.

Original Commentary

Jack K. Crandall
Indiana University, Bloomington, IN, USA

Introduction. Reactions with DDO are typically performed by adding the cold reagent solution to a cold solution of a reactant in acetone or some other solvent. CH_2Cl_2 is a convenient solvent which facilitates reaction in a number of cases. After the reactant has been consumed, as monitored by TLC, etc., the solvent and excess reagent are simply removed to provide a nearly pure product. An excess of DDO is often used to facilitate conversion, provided further oxidation is not a problem. Where the product is especially sensitive to acid, the reaction can be run in the presence of solid *Potassium Carbonate* as an acid scavenger and drying agent. When it is important to minimize water content, the use of powdered molecular sieves in the reaction mixture is recommended. Reactions can be run from ambient temperatures down to −78 °C.

Dimethyldioxirane is a powerful oxidant, but shows substantial selectivity in its reactions. It has been particularly valuable for the preparation of highly reactive products, since DDO can be employed under neutral, nonnucleophilic conditions which facilitate the isolation of such species. Whereas DDO performs the general conversions of more classic reagents like *m-Chloroperbenzoic Acid*, it generates only an innocuous molecule of acetone as a byproduct. This is to be contrasted with peracids whose acidic side-products can induce rearrangements and nucleophilic attack on products. Although several other dioxiranes have been prepared, these usually offer no advantage over DDO. An important exception is *Methyl(trifluoromethyl)dioxirane*, whose greater reactivity is advantageous in situations where DDO reacts sluggishly, as in the oxyfunctionalization of alkanes.

The need to prepare DDO solutions beforehand, the low yield of the reagent based on *Potassium Monoperoxysulfate* (Oxone) (ca. 5%),[6] and the inconvenience of making DDO for large-scale reactions are drawbacks that can be avoided when the product has good stability. In these instances, an in situ method for DDO oxidations is recommended.

Essential Reactions for Organic Synthesis, First Edition. Edited by Philip L. Fuchs.
© 2016 John Wiley & Sons, Ltd. Published 2016 by John Wiley & Sons, Ltd.

Oxidation of Alkenes and Other Unsaturated Hydrocarbons. The epoxidation of double bonds has been the major area for the application of DDO methodology and a wide range of alkenes are effectively converted to epoxides by solutions of DDO.[4,7] Epoxidation is stereospecific with retention of alkene stereochemistry, as shown by the reactions of geometrical isomers; for example, (Z)-1-phenylpropene gives the *cis*-epoxide cleanly (eq 2), whereas the (E) isomer yields the corresponding *trans*-epoxide. Rate studies indicate that this reagent is electrophilic in nature and that alkyl substitution on the double bond enhances reactivity.[7] Interestingly, *cis*-disubstituted alkenes react 7–9 times faster than the *trans* isomers, an observation that has been interpreted in terms of a 'spiro' transition state.[9]

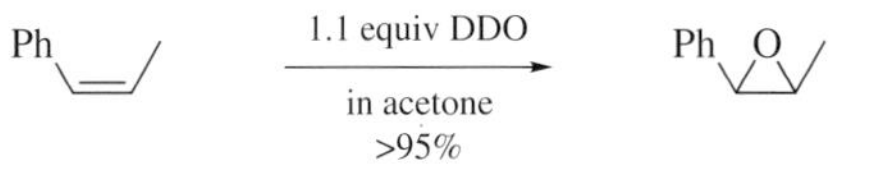

(2)

From a preparative viewpoint, the use of DDO solutions, while efficient and easy to perform, are generally not needed for simple alkenes that give stable epoxides. Rather, in situ methodology is suggested. However, the extraordinary value of isolated DDO has been amply demonstrated for the generation of unstable epoxides that would not survive most epoxidation conditions.[1] A good example of this sort of application is the epoxidation of precocenes, as exemplified in eq 3.[10] A number of impressive epoxidations have been reported for oxygen-substituted alkenes, including enol ethers, silyl enol ethers, enol carboxylates, etc.[1] Examples include a number of 1,2-anhydro derivatives of monosaccharides.[11] Steric features often result in significant stereoselection in the epoxidation, as illustrated in eq 4.[11] Conversions of alkenes with two alkoxy substituents have also been achieved (eq 5), even when the epoxides are not stable at rt.[12]

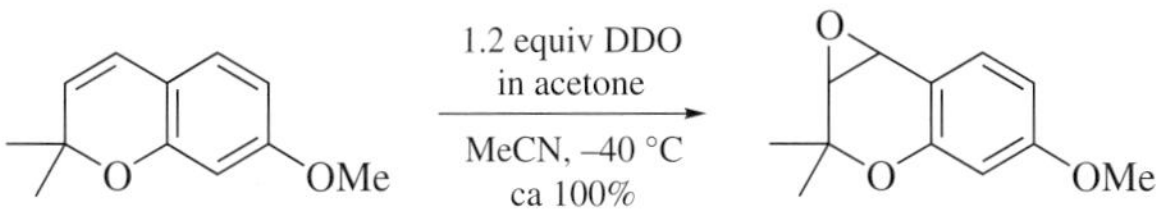

(3)

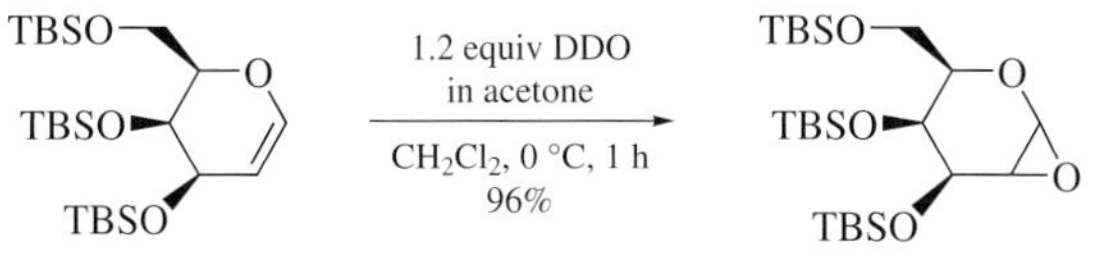

(4)

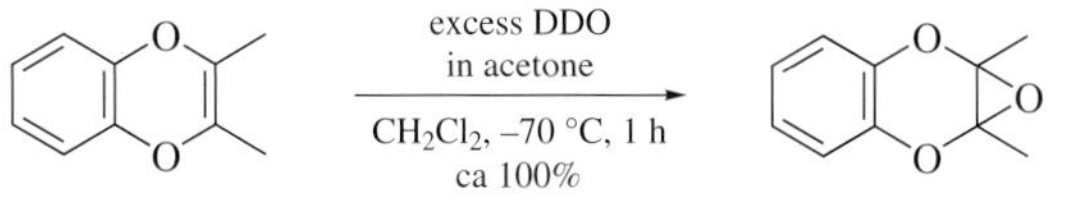

(5)

Although reactions are much slower with conjugated carbonyl compounds, DDO is still effective for the epoxidation of these electron-deficient double bonds (eq 6).[13] Alkoxy-substitution on such conjugated alkenes can also be tolerated (eq 7).[14]

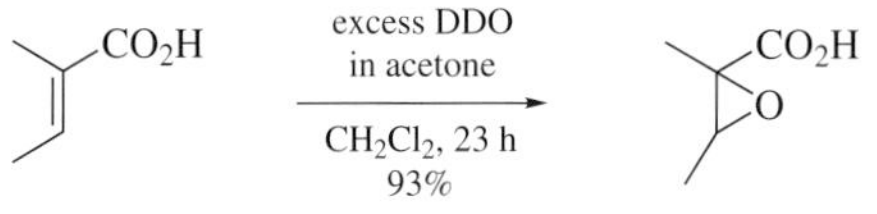

(6)

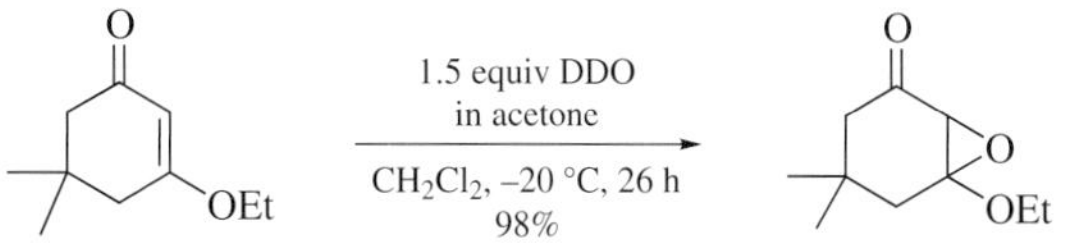

(7)

Allenes react with DDO by sequential epoxidation of the two double bonds to give the previously inaccessible, highly reactive allene diepoxides.[15] In the case of the *t*-butyl-substituted allene shown in eq 8, a single diastereomer of the diepoxide is generated, owing to steric control of the *t*-butyl group on reagent attack.

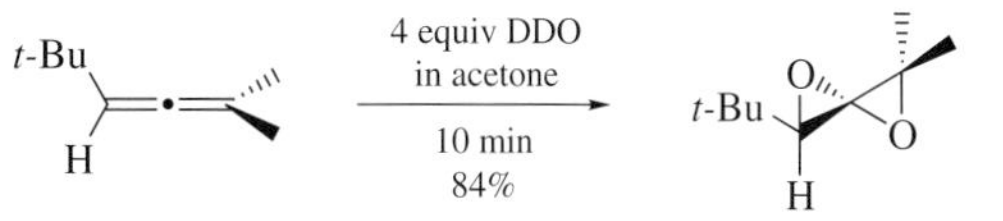

(8)

Certain polycyclic aromatic hydrocarbons can be converted to their epoxides, as typified by the reaction of phenanthrene with DDO (eq 9).[4] Aromatic heterocycles like furans and benzofurans also give epoxides, although these products are quite susceptible to rearrangement, even at subambient temperatures (eq 10).[16] The oxidation of heavily substituted phenols by DDO leads to quinones, as shown in eq 11, which illustrates the formation of an orthoquinone.[17] The corresponding hydroquinones are intermediates in these reactions, but undergo ready oxidation to the quinones.

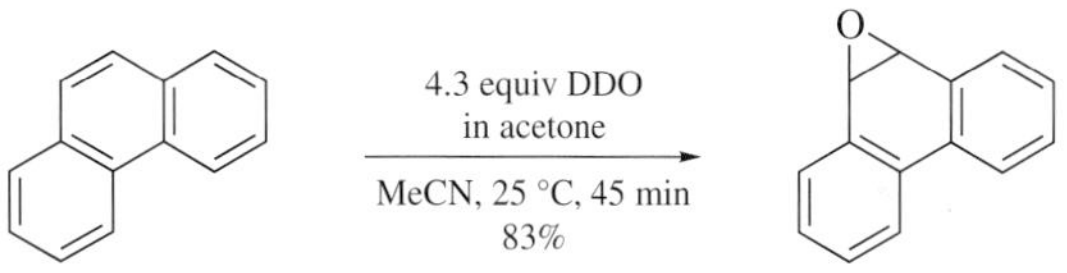

(9)

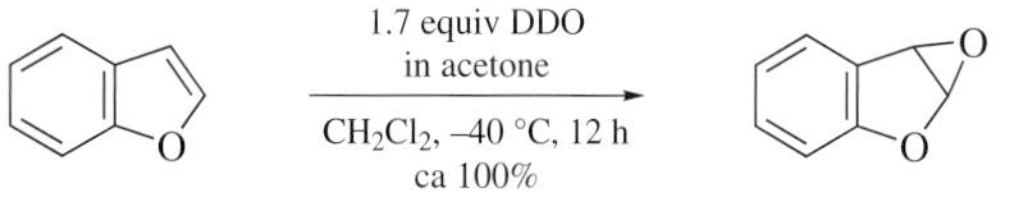

(10)

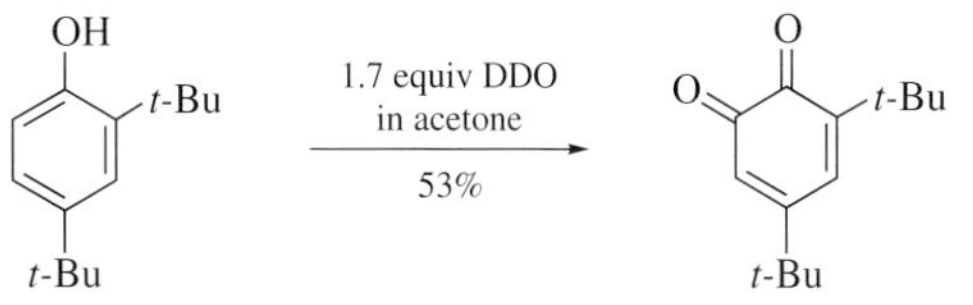

(11)

Finally, preformed lithium enolates are converted to α-hydroxy ketones by addition to a cold solution of DDO (eq 12).[18]

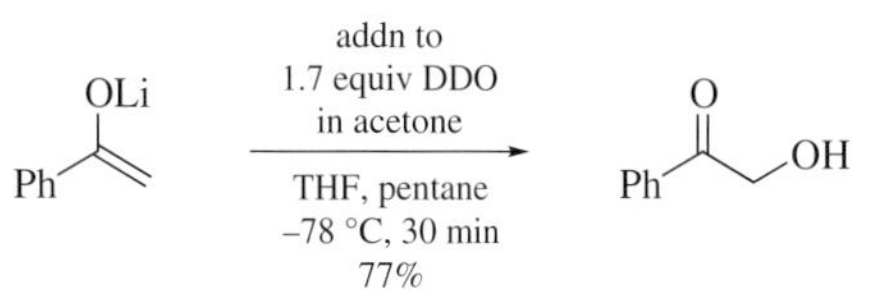

(12)

Oxidation of Saturated Hydrocarbons, Ethers, and Alcohols. Surely the most striking reaction of dioxiranes is their ability to functionalize unactivated C–H bonds by the insertion of an oxygen atom into this σ-bond. This has opened up an important new area of oxidation chemistry.[1] While DDO has been used in a number of useful transformations outlined below, the more reactive **Methyl(trifluoromethyl)dioxirane** is often a better reagent for this type of conversion, despite its greater cost and difficulty of preparation.

The discrimination of DDO for tertiary > secondary > primary C–H bonds of alkanes is more pronounced than that of the *t*-butoxide radical.[19] Good yields of tertiary alcohols can be secured in favorable cases, as in the DDO oxidation of adamantane to 1-adamantanol, which occurs with only minor reaction at C-2 (eq 13). Of major significance is the observation that these reactions are stereospecific with high retention of configuration, as illustrated by the oxidation of *cis*-dimethylcyclohexane shown in (eq 14); the *trans* isomer gives exclusively the diastereomeric alcohol. This and other data have been interpreted in terms of an 'oxenoid' mechanism for the insertion into the C–H bond. Several interesting applications in the steroid field involve significant site selectivity as well.[20] The slower reactions of DDO with hydrocarbons without tertiary hydrogens are less useful and lead to ketones owing to a rapid further oxidation of the initially formed secondary alcohol. For example, cyclododecane is converted to cyclododecanone.

(13)

(14)

Ethers and acetals are slowly converted by DDO to carbonyl compounds. This serves as a nontraditional method for deprotection of these derivatives, an example of which is shown in (eq 15).[21,22] Hemiacetals are presumed intermediates in these transformations.

(15)

While DDO has been little used for the oxidation of simple alcohols, it has found application in useful conversions of vicinal diols. The oxidation of tertiary–secondary diols to α-hydroxy ketones occurs without the usual problem of oxidative cleavage between the two functions (eq 16).[23] DDO has also been used to convert appropriate optically active diols selectively into α-hydroxy ketones of high optical purity; for example, see (eq 17).[24]

(16)

(17)

Finally, the Si–H bond of silanes suffers analogous oxidation to silanols upon reaction with DDO. This reaction takes place with retention of configuration and is, as expected, more facile than C–H oxidations.[25]

Oxidation of Nitrogen Functional Groups. Selective oxidations of nitrogen compounds are often difficult to achieve, but DDO methodology has been shown to be very useful in a number of instances. For example, one of the first applications of this reagent was in the conversion of primary amines to the corresponding nitro compounds (eq 18).[26] This process probably proceeds by successive oxidation steps via hydroxylamine and nitroso intermediates. Complications arise with unhindered primary aliphatic amines, owing to dimerization of the intermediate nitrosoalkanes and their tautomerization to oximes.[27] In oxidations of amino sugar and amino acid derivatives, it is possible to isolate the initially formed hydroxylamines (eq 19).[28]

(18)

(19)

The oxidation of secondary amines to hydroxylamines is readily achieved with 1 equiv of DDO (eq 20).[29] The use of 2 equiv of DDO results in further oxidation, the nature of which depends on the structure of the amine. Thus cyclic secondary amines which do not possess α-hydrogens are converted to nitroxides,[30] as illustrated in (eq 21). Secondary benzylamines give nitrones (eq 22).[31]

(20)

(21)

(22)

A related transformation is the oxidation of imines to nitrones by DDO (eq 23).[32] It is interesting that the isomeric oxaziridines are not produced here, given that peracids favor these heterocycles.

$$C_6Me_5CH=NMe \xrightarrow[\substack{CH_2Cl_2, \, 0\,°C, \, 2\,h \\ 71\%}]{\substack{1.1 \text{ equiv DDO} \\ \text{in acetone}}} C_6Me_5CH=N(O)Me \quad (23)$$

Reaction of α-diazo ketones with DDO leads to α-keto aldehyde hydrates (eq 24).[33] Oximes are converted to the free ketones by DDO.[34]

$$\text{(24)}$$

Oxidation of Sulfur Functional Groups. Dimethyldioxirane rapidly oxidizes sulfides to sulfoxides and converts sulfoxides to sulfones (eq 25).[4,35] The partial oxidation of sulfides to sulfoxides can be controlled by limiting the quantity of DDO. Since Oxone is one of the many reagents that can perform these reactions, the extra effort involved in preparing DDO solutions is often not warranted. An exception involves the transformation of thiophenes to the corresponding sulfones (eq 26).[36] A similar procedure gives α-oxo sulfones by DDO oxidation of thiol esters (eq 27).[37]

$$\text{PhSMe} \xrightarrow{\text{DDO}} \text{PhSOMe} \xrightarrow{\text{DDO}} \text{PhSO}_2\text{Me} \qquad \text{(25)}$$

$$\text{(26)}$$

$$\text{(27)}$$

Alkanethiols are selectively oxidized to alkanesulfinic acids by DDO (eq 28).[38] Air oxidation of an intermediate species appears to be important in this transformation.

$$\text{Me(CH}_2)_4\text{SH} \xrightarrow[\text{in acetone}]{\text{DDO}} \xrightarrow{\text{O}_2} \text{Me(CH}_2)_4\text{SO}_2\text{H} \qquad \text{(28)}$$

First Update

Ruggero Curci, Lucia D'Accolti & Caterina Fusco
Università di Bari, Bari, Italy

Direct Functionalization of C–H Bonds by Dimethyldioxirane.[1,39] The efficient oxyfunctionalization of simple, "unactivated" C–H bonds of alkanes under extremely mild conditions undoubtedly is one of the major highlights of dioxirane chemistry.[1,39]

Although less effective than methyl(trifluoromethyl)dioxirane (TFDO), oxyfunctionalization of unactivated methine C–Hs with dimethyldioxirane (DDO) is feasible for various substituted steroids related to the 5β-cholane and 5α-cholestane series to give novel mono- and dihydroxylated steroids.[40] The reactivity and site selectivity of oxyfunctionalization is affected conspicuously by the structural and steric environments of the target methine carbon atoms. This nonenzymatic procedure may be advantageously applied to selective and short-course syntheses of bioactive steroids. Thus, the major reaction product of methyl

3α-acetoxy-5α-cholan-24-oate with DDO in 36 h was the corresponding 5β-hydroxylated compound in 48% yield, while concurrent double oxyfunctionalization at the C-5 and C-17 produced the corresponding dihydroxylated derivative in comparable yield (36%) (eq 29).[40]

Both the 5β- and 17α-hydroxylation proceeded stereoselectively and the configuration of the resulting C–OH was the same as that of the original methine C–H bond. In order to accelerate the reaction of O-insertion (otherwise rather sluggish), instead of the usual DDO solution (up to 0.11 M) in acetone, the aforementioned concentrated DDO solutions (0.33–0.35 M) in CHCl$_3$ were employed.[40] These are obtained by following a procedure of extraction (outlined above) of the dioxirane into the chlorinated solvent,[41] similar to that devised to obtain ketone-free TFDO solutions.[42]

$$\text{(29)}$$

In another study where DDO was employed for *tert*-C–H hydroxylation of several di- and triacetates of (5β)-bile acid methyl esters, it was reported that derivatives bearing a 7-acetoxy group give 17α- or 14α-hydroxylated products in addition to the 5β-hydroxylated ones.[43]

For bile acid esters, the DDO oxidation of *sec*-CHOH groups of hydroxy cholate methyl esters to the corresponding carbonyls (via *gem*-diols) occurs readily; the positional order of reactivity C3 $\cong$ C7 > C6 > C12 was established.[44]

In line with the effective oxyfunctionalizations recorded for steroids, the DDO hydroxylation of cephalostatin derivatives provide another interesting example. In fact, the DDO oxidation hecogenin acetate β-hydroxyketone in (eq 30) results in O-insertion into C16–H to yield the corresponding hemiketal with amazing site-selectivity.[45a] This key transformation paves the road for synthesis of the cephalostatin North 1 hemisphere.[45]

(30)

Remarkable site-selectivity is often observed in the hydroxylation of protected N-Boc derivatives of α-amino acid esters bearing an alkyl side chain (Boc-Gly-OMe, Boc-Ala-OMe, Boc-Val-OMe, Boc-Ile-OMe, Boc-Phe-OMe); this results in different products depending on the structure.[46] Although these reactions are rather sluggish, requiring long reaction times for sizable substrate conversion, they offer a novel entry to side-chain modified α-amino acids and peptides that avoids multi-step synthetic approaches. The reactivity and site-selectivity depends on the steric environments and electron density of the target C–H bonds in the side chain; high regioselectivity for the O-insertion into the γ-C–H bond of leucine (Leu) residues with respect to the weaker α-C–H bonds is observed. Thus, Boc-Leu-OMe was found to yield the corresponding 4,4-dimethyl-4-butanolide derivative; the latter is formed by selective O-insertion into the tertiary γ-C–H bond of Leu followed by cyclization (eq 31).

(31)

A position selectivity in the oxidation of peptides containing more than one Leu residue was also reported.[46] However, it should be noted that the same transformation in eq 31 gave a markedly lower yield in γ-butanolide using a lower DDO excess.[47] The reaction occurs more rapidly (6 h, 91% conv) using the powerful TFDO (6 equiv); it gives the N-hydroxy derivative of the butanolide in 21% yield, along with the uncyclized N-hydroxy derivative of the starting Boc-Leu-OMe as the major product (57%).[47]

In general, benzhydrylic C–H bonds (only slightly more reactive than tert-C–Hs) are distinctly more reactive than benzylic C–H bonds toward dioxirane O-insertion.[39,48,49] However, special situations arise in the selective hydroxylation of complex poly-

cyclic indan hydrocarbons, i.e., the centropolyindans.[50] For instance, the rigidity of the polycyclic framework and steric factors in the angular centrotriindan 1,1'-(o-phenylene)-2,2'-spirobiindan moderate the otherwise distinct selectivity for benzhydrylic vs. benzylic C–H O-insertion by DDO; so then, spirobiindanone is formed along with the tertiary 4bα-alcohol (eq 32).[50]

(28%)   (16%)   (32)

Triptindane, another propellane-type centropolyindan, was found to react with excess DDO yielding triptindan-9-one as the major product (37% yield) at the conditions given in (eq 33).[50]

(33)

The monoketone thus obtained was fully characterized. The formation of sizeable amounts of more highly oxidized products (triptindan-9,10-dione and triptindan-9,10,11-trione) was detected by mass spectrometry.

Turning to polycyclic saturated hydrocarbons, the shown (eq 13) selective oxyfunctionalization of adamantane (*requiring 6 equiv of the powerful TFDO for exhaustive bridgehead hydroxylation*)[11] serves well to illustrate the high tertiary vs. secondary selectivities that are customarily observed. The ratio of tertiary to secondary carbons and the different reactivity towards oxidation for each type of C–H bond in Binor S renders it an attractive probe for the study of regioselectivity of oxyfunctionalization. In fact, this saturated heptacyclic hydrocarbon (the head-to-head dimer of norbornadiene) consists of two nortricyclane units, each containing one three-membered ring and three five-membered rings; it presents two symmetric methylene groups and 12 tertiary carbons ordered in four different geometries.

Treatment of Binor S in CH_2Cl_2 with aqueous (pH 7) monoperoxosulfate (caroate)/acetone (DDO in situ) afforded Binor S 1-ol in 98% yield.[51] Further oxidation of this material with isolated DDO gives the symmetrical 1,9-diol as the major product (eq 34).[51]

(34)

The examples above further demonstrate that dioxirane reactions are characterized by high selectivity and that *tert*-C–H bonds are considerably more reactive towards dioxirane *O*-insertion than their *sec*-C–H complements.[39] However, the cyclopropane moiety, if suitably oriented,[52,53] can have a marked influence in activating proximal α-C–H bonds towards dioxirane oxyfunctionalization.[52]

This is illustrated by the application of dioxiranes to polycyclic alkanes possessing a sufficiently rigid framework, such as the 2,4-didehydroadamantane case. For this substrate the reaction with DDO proceeds with 82% conversion during 12 h, yielding the expected 2,4-didehydroadamantan-7-ol, but also the precious 2,4-didehydroadamantan-10-one in comparable yield (eq 35).[54] The ketone derives from competitive dioxirane attack at the methylene positions α to the cyclopropyl moiety; the latter is encompassed in the rigid 2,4-didehydroadamantane framework to lay constrained into a "bisected"[52] orientation relative to the neighboring methylene C–H bonds.

$$ (35) $$

Hydroxylation at the bridgehead C1 and C5 does not take place because the bridgehead *tert*-C–H bonds are deactivated by the proximal cyclopropyl moiety lying perpendicularly. It is worth of note that, at variance with what is observed with other oxidants (e.g., dry ozone), no rearranged products are observed in the reaction of DDO with this target compound.[54]

It was mentioned that oxidation by DDO allows the clean conversion of secondary alcohols into carbonyls under mild conditions. In this transformation the dioxiranes rank high with respect to transition-metal oxidants because of their efficiency, superior versatility, and ease of operations. Based on kinetic data and the application of reaction probes,[55] the oxidation proceeds via a substantially concerted *O*-insertion by the dioxirane into the C–H bond "alpha" to the OH functionality generating a *gem*-diol $C(OH)_2$, hence the carbonyl. As shown by the example in eq 36, remarkable chemoselectivity is achieved in the oxidation of epoxy alcohols in that the corresponding epoxy ketones are formed in high yield, while the epoxy functionality remains untouched.[56] The epoxy ketone in eq 36 is a key intermediate in the convergent synthesis of active $1\alpha,25$-dihydroxyvitamin D_3 analogs.[57]

$$ (36) $$

Both open chain and cyclic epoxy alcohols can be neatly transformed into the corresponding epoxy ketones with high conversions and yields using just 1.1–1.5 equiv of DDO oxidant. Also, the conversion of optically active epoxy alcohols into epoxy ketones occurs selectively leaving the configuration at the chiral center(s) at the oxirane ring unaffected.[56]

Related Reagents. Potassium Monoperoxosulfate (Oxone); Potassium Monoperoxosulfate (Oxone)/Acetone (DDO in situ); Methyl(trifluoromethyl)Dioxirane.

1. (a) Adam, W.; Hadjiarapoglou, L. P.; Curci, R.; Mello, R. In *Organic Peroxides*; Ando, W., Ed.; Wiley: New York, 1992; Chapter 4, p 195. (b) Murray, R. W., *Chem. Rev.* **1989**, *89*, 1187. (c) Curci, R. In *Advances in Oxygenated Processes*; Baumstark, A., Ed; JAI: Greenwich, CT, 1990; Vol. 2, Chapter 1, p 1. (d) Adam, W.; Edwards, J. O.; Curci, R., *Acc. Chem. Res.* **1989**, *22*, 205. (e) Adam, W.; Hadjiarapoglou, L., *Top. Curr. Chem.* **1993**, *164*, 45.
2. Montgomery, R. E., *J. Am. Chem. Soc.* **1974**, *96*, 7820.
3. Edwards, J. O.; Pater, R. H.; Curci, P. R.; Di Furia, F., *Photochem. Photobiol.* **1979**, *30*, 63.
4. Murray, R. W.; Jeyaraman, R., *J. Org. Chem.* **1985**, *50*, 2847.
5. Eaton, P. E.; Wicks, G. E., *J. Org. Chem.* **1988**, *53*, 5353.
6. Adam, W.; Bialas, J.; Hadjiarapoglou, L., *Chem. Ber.* **1991**, *124*, 2377.
7. Baumstark, A. L.; Vasquez, P. C., *J. Org. Chem.* **1988**, *53*, 3437.
8. Murray, R. W.; Shiang, D. L., *J. Chem. Soc., Perkin Trans. 2* **1990**, *2*, 349.
9. Baumstark, A. L.; McCloskey, C. J., *Tetrahedron Lett.* **1987**, *28*, 3311.
10. Bujons, J.; Camps, F.; Messeguer, A., *Tetrahedron Lett.* **1990**, *31*, 5235.
11. Halcomb, R. L.; Danishefsky, S. J., *J. Am. Chem. Soc.* **1989**, *111*, 6661.
12. Adam, W.; Hadjiarapoglou, L.; Wang, X., *Tetrahedron Lett.* **1991**, *32*, 1295.
13. Adam, W.; Hadjiarapoglou, L.; Nestler, B., *Tetrahedron Lett.* **1990**, *31*, 331.
14. Adam, W.; Hadjiarapoglou, L., *Chem. Ber.* **1990**, *123*, 2077.
15. (a) Crandall, J. K.; Batal, D. J.; Sebesta, D. P.; Lin, F., *J. Org. Chem.* **1991**, *56*, 1153. (b) Crandall, J. K.; Batal, D. J.; Lin, F.; Reix, T.; Nadol, G. S.; Ng, R. A., *Tetrahedron* **1992**, *48*, 1427.
16. (a) Adger, B. M.; Barrett, C.; Brennan, J.; McGuigan, P.; McKervey, M. A.; Tarbit, B., *J. Chem. Soc., Chem. Commun.* **1993**, 1220. (b) Adger, B. M.; Barrett, C.; Brennan, J.; McKervey, M. A.; Murray, R. W., *J. Chem. Soc., Chem. Commun.* **1991**, 1553. (c) Adam, W.; Bialas, J.; Hadjiarapoglou, L.; Sauter, M., *Chem. Ber.* **1992**, *125*, 231.
17. (a) Crandall, J. K.; Zucco, M.; Kirsch, R. S.; Coppert, D. M., *Tetrahedron Lett.* **1991**, *32*, 5441. (b) Altamura, A.; Fusco, C.; D'Accolti, L.; Mello, R.; Prencipe, T.; Curci, R., *Tetrahedron Lett.* **1991**, *32*, 5445. (c) Adam, W.; Schönberger, A., *Tetrahedron Lett.* **1992**, *33*, 53.
18. Guertin, K. R.; Chan, T. H., *Tetrahedron Lett.* **1991**, *32*, 715.
19. Murray, R. W.; Jeyaraman, R.; Mohan, L., *J. Am. Chem. Soc.* **1986**, *108*, 2470.
20. (a) Bovicelli, P.; Lupattelli, P.; Mincione, E.; Prencipe, T.; Curci, R., *J. Org. Chem.* **1992**, *57*, 2182. (b) Bovicelli, P.; Lupattelli, P.; Mincione, E.; Prencipe, T.; Curci, R., *J. Org. Chem.* **1992**, *57*, 5052.
21. Curci, R.; D'Accolti, L.; Fiorentino, M.; Fusco, C.; Adam, W.; González-Nuñez, M. E.; Mello, R., *Tetrahedron Lett.* **1992**, *33*, 4225.
22. van Heerden, F. R.; Dixon, J. T.; Holzapfel, C. W., *Tetrahedron Lett.* **1992**, *33*, 7399.
23. Curci, R.; D'Accolti, L.; Detomaso, A.; Fusco, C.; Takeuchi, K.; Ohga, Y.; Eaton, P. E.; Yip, Y. C., *Tetrahedron Lett.* **1993**, *34*, 4559.
24. D'Accolti, L.; Detomaso, A.; Fusco, C.; Rosa, A.; Curci, R., *J. Org. Chem.* **1993**, *58*, 3600.
25. Adam, W.; Mello, R.; Curci, R., *Angew. Chem., Int. Ed. Engl.* **1990**, *102*, 890.
26. Murray, R. W.; Rajadhyaksha, S. N.; Mohan, L., *J. Org. Chem.* **1989**, *54*, 5783.
27. Crandall, J. K.; Reix, T., *J. Org. Chem.* **1992**, *57*, 6759.
28. Wittman, M. D.; Halcomb, R. L.; Danishefsky, S. J., *J. Org. Chem.* **1990**, *55*, 1981.

29. Murray, R. W.; Singh, M., *Synth. Commun.* **1989**, *19*, 3509.

30. Murray, R. W.; Singh, M., *Tetrahedron Lett.* **1988**, *29*, 4677.

31. Murray, R. W.; Singh, M., *J. Org. Chem.* **1990**, *55*, 2954.

32. Boyd, D. R.; Coulter, P. B.; McGuckin, M. R.; Sharma, N. D., *J. Chem. Soc., Perkin Trans. 1* **1990**, 301.

33. (a) Ihmels, H.; Maggini, M.; Prato, M.; Scorrano, G., *Tetrahedron Lett.* **1991**, *32*, 6215. (b) Darkins, P.; McCarthy, N.; McKervey, M. A.; Ye, T., *J. Chem. Soc., Chem. Commun.* **1993**, 1222.

34. Olah, G. A.; Liao, Q.; Lee, C.-S.; Prakash, G. K. S., *Synlett* **1993**, 427.

35. Murray, R. W.; Jeyaraman, R.; Pillay, M. K., *J. Org. Chem.* **1987**, *52*, 746.

36. Miyahara, Y.; Inazu, T., *Tetrahedron Lett.* **1990**, *31*, 5955.

37. Adam, W.; Hadjiarapoglou, L., *Tetrahedron Lett.* **1992**, *33*, 469.

38. Gu, D.; Harpp, D. N., *Tetrahedron Lett.* **1993**, *34*, 67.

39. Curci, R.; Dinoi, A.; Rubino, M. F., *Pure Appl. Chem.* **1995**, *67*, 811. See references therein.

40. Iida, T.; Yamaguchi, T.; Nakamori, R.; Hikosaka, M.; Mano, N.; Goto, J.; Nambara, T., *J. Chem. Soc., Perkin Trans. 1* **2001**, 2229.

41. Gilbert, M.; Ferrer, M.; Sánchez-Baeza, F.; Messeguer, M., *Tetrahedron* **1997**, *53*, 8643.

42. Adam, W.; Curci, R.; Gonzalès-Nuñez, M. E.; Mello, R., *J. Am. Chem. Soc.* **1991**, *113*, 7654.

43. Cerré, C.; Hofmann, A. F.; Schteingart, C. D., *Tetrahedron* **1997**, *53*, 435.

44. Buxton, P. C.; Marples, B. A.; Toon, R. C.; Waddington, V. L., *Tetrahedron Lett.* **1999**, *40*, 4729.

45. (a) Lee, J. S.; Fuchs, P. L., *Org. Lett.* **2003**, *5*, 2247. (b) La Cour, T. G.; Guo, C.; Boyd, M. R.; Fuchs, P. L., *Org. Lett.* **2000**, *2*, 33.

46. Saladino, R.; Mezzetti, M.; Mincione, E.; Torrini, I.; Paglialunga-Paradisi, M.; Mastropietro, G., *J. Org. Chem.* **1999**, *64*, 8468.

47. Detomaso, A.; Curci, R., *Tetrahedron Lett.* **2001**, *42*, 755.

48. Kuck, D.; Schuster, A., *Z. Naturforsch.* **1991**, *46*, 1223.

49. Fusco, C.; Fiorentino, M.; Dinoi, A.; Curci, R.; Krause, R. A.; Kuck, D., *J. Org. Chem.* **1996**, *61*, 8681.

50. Kuck, D.; Schuster, A.; Fusco, C.; Fiorentino, M.; Curci, R., *J. Am. Chem. Soc.* **1994**, *116*, 2375. See also references therein.

51. Pramod, K.; Eaton, P. E.; Gilardi, R.; Flippen-Anderson, J. L., *J. Org. Chem.* **1990**, *55*, 6105.

52. D'Accolti, L.; Dinoi, A.; Fusco, C.; Russo, A.; Curci, R., *J. Org. Chem.* **2003**, *68*, 7806.

53. Rhodes, Y. E.; DiFate, V. G., *J. Am. Chem. Soc.* **1972**, *94*, , 7582.

54. Murray, R. K.; Teager, D. S., *J. Org. Chem.* **1993**, *58*, 5548.

55. Mello, R.; Cassidei, L.; Fiorentino, M.; Fusco, C.; Hümmer, W.; Jäger, V.; Curci, R., *J. Am. Chem. Soc.* **1991**, *113*, 2205.

56. D'Accolti, L.; Fusco, C.; Annese, C.; Rella, M. R.; Turteltaub, J. S.; Williard, P. G.; Curci, R., *J. Org. Chem.* **2004**, *69*, 8510.

57. Nancy E. Lee, N. E.; Reddy, G. S.; Brown, A. J.; Williard, P. G., *Biochemistry* **1997**, *36*, 9429.

E

1-Ethyl-3-(3′-dimethylaminopropyl) carbodiimide Hydrochloride   **184**

Essential Reagents for Organic Synthesis, Edited by Philip L. Fuchs, André B. Charette, Tomislav Rovis, and Jeffrey W. Bode.
©2016 John Wiley & Sons, Ltd. Published 2016 by John Wiley & Sons, Ltd.

1-Ethyl-3-(3'-dimethylaminopropyl) carbodiimide Hydrochloride[1]

[25952-53-8] $C_8H_{18}ClN_3$ (MW 191.74)

InChI = 1/C8H17N3.ClH/c1-4-9-8-10-6-5-7-11(2)3;/h4-7H2,
   1-3H3;1H/fC8H18N3.Cl/h11H;1h/q+1;-1

InChIKey = FPQQSJJWHUJYPU-CZVSVQNTCV

(base)

[1892-57-5]

InChI = 1/C8H17N3/c1-4-9-8-10-6-5-7-11(2)3/h4-7H2,1-3H3

InChIKey = LMDZBCPBFSXMTL-UHFFFAOYAH

(.MeI)

[22572-40-3]

InChI = 1/C9H20N3.HI/c1-5-10-9-11-7-6-8-12(2,3)4;/h5-8H2,
   1-4H3;1H/q+1;/p-1/fC9H20N3.I/h;1h/qm;-1

InChIKey = AGSKWMRPXWHSPF-NFBLBWBOCN

(peptide coupling reagent;[1b,2] amide formation;[3] ester formation;[4] protein modification;[5] mild oxidations of primary alcohols[6])

Alternate Name: 1-(3-dimethylaminopropyl)-3-ethylcarbodiimide; water-soluble carbodiimide; EDC; EDCI.

Physical Data: free base is an oil, bp 47–48 °C/0.27 mmHg; HCl salt is a white powder, mp 111–113 °C; MeI salt mp 97–99 °C.

Solubility: sol H_2O, CH_2Cl_2, DMF, THF.

Form Supplied in: commercially available as HCl salt and as methiodide salt that are white solids. Reagents are >98% pure; main impurity is the urea that can form upon exposure to moisture.

Analysis of Reagent Purity: IR: 2150 cm^{-1} (N=C=N stretch); urea has C=O stretch near 1600–1700 cm^{-1}.

Purification: recrystallization from CH_2Cl_2/ether.

Handling, Storage, and Precautions: EDC is moisture-sensitive; store under N_2 in a cool dry place. It is incompatible with strong oxidizers and strong acids. EDC is a skin irritant and a contact allergen; therefore avoid exposure to skin and eyes.

Peptide Coupling Reagent. This carbodiimide (EDC) reacts very similarly to *1,3-Dicyclohexylcarbodiimide* and other carbodiimides. The advantage EDC has over DCC is that the urea produced is water soluble and, therefore, is easily extracted. Dicyclohexylurea is only sparingly soluble in many solvents and is removed by filtration which may not be as effective as extraction. A typical example of EDC for peptide coupling is shown in (eq 1).[2]

The problems associated with carbodiimide couplings are mainly from the *O*-acylisourea intermediate (**1**) having poor selectivity for specific nucleophiles. This intermediate can rearrange to an *N*-acylurea (**2**), resulting in contamination of the product and low yields. Additionally it can rearrange to 5(4*H*)-oxazolones (**3**) that tautomerize readily, resulting in racemization. Using low dielectric solvents like CH_2Cl_2 or additives to trap (**1**) minimizes

these side reactions by favoring intermolecular nucleophilic attack on (**1**).[1b]

R = carbodiimide chain; R^1 = amino acid; R^2 = amino acid side chain; R^3 = NH protecting group

When low dielectric constant solvents are used, the carboxylic acid tends to dimerize, thus promoting symmetrical anhydride formation. This intermediate is stable enough to give good yields of the desired product. High dielectric solvents such as DMF retard acylation of amino acids, and *N*-acylurea can be a major byproduct.[1b] Unfortunately, many starting materials require polar solvents for dissolution.

Addition of trapping agents such as *N-Hydroxysuccinimide*[7] and *1-Hydroxybenzotriazole* (HOBt)[8] reduce the extent of many side reactions, especially *N*-acylurea formation. Also, racemization is suppressed when these additives are present. The latter reagent eliminates the intramolecular dehydration of ω-amides of asparagine and glutamine that occurs with carbodiimides (eq 2).[8,9]

R = carbodiimide chain

With EDC the addition of *Copper(II) Chloride* suppresses racemization to <0.1% compared to 0.4% with HOBt under ideal conditions.[10] Combinations of HOBt and $CuCl_2$ are also useful.[10b] The suggested stoichiometry for the highest optical purity and yield is 2 equiv of HOBt and 0.25–0.5 equiv of $CuCl_2$ in DMF.

A practical example of EDC's utility is the solution synthesis of human epidermal growth factor, a 53-residue protein.[11] This included several couplings of fragments ranging from three to six residues in length. All couplings were performed with

Essential Reactions for Organic Synthesis, First Edition. Edited by Philip L. Fuchs.

EDC/HOBt in DMF or NMP. At the completion of the synthesis, no racemized material could be detected by HPLC.

Amide Formation. Formylated amino acids and peptides are prepared in high yields by forming the acid anhydride (eq 3).[3] These products are pure without requiring chromatography or recrystallization.

Another method for formylation of amines is with p-nitrophenyl formate, which usually gives products in high yield. However, removing the last traces of the p-nitrophenol is difficult.[12]

Treating carbon dioxide and amines with EDC gives symmetrical ureas (eq 4).[13] DCC with CO_2 at ambient pressure works equally well.

EDC facilitates the synthesis of xanthine analogs by condensing a diamine with water-soluble acids (eq 5).[14] The use of water as the solvent precludes the use of DCC in this case.

Ester Formation. Esters of N-protected amino acids are prepared in high yield with EDC and **4-Dimethylaminopyridine** (eq 6).[4] DMAP causes extensive racemization if not used in a catalytic amount.[15] However, when esterifying the α-carboxyl of β- and γ-benzyl esters of aspartyl and glutamyl derivatives, extensive racemization was observed even with DMAP present in catalytic amounts. It is postulated that the side-chain esters contribute in some fashion to the lability of the α-H.

R^1 = t-Bu, 76%; Me, 96%; CH_2CCl_3, 87%

Carbodiimides including EDC are also useful in preparing active esters such as the p-nitrophenyl, pentafluorophenyl, and N-hydroxysuccinimide esters.[1b] Numerous additional methods have been reported.[16]

Protein Modification. Carboxyl groups in proteins react with EDC, resulting in an activated group that can be trapped with a nucleophile such as glycine methyl ester.[5] The nucleophile can also be amino groups in the protein, causing cross-linking. Applications for carboxylate modification include determining which groups are buried versus exposed, and mechanistic studies for determining which carboxyl groups are essential for an enzyme's activity.

EDC is used extensively in cross-linking proteins to solid supports for affinity chromatography.[17]

Oxidation of Primary Alcohols. One example of EDC substituting for DCC in the Pfitzner–Moffatt oxidation is shown in (eq 7).[6] The use of DCC in this example resulted in a difficult purification of the product and lower yields.

Miscellaneous Reactions. EDC is a useful water scavenger as well. An example of this utility is given in (eq 8).[18] The carbodiimide is required in this reaction, but DCC does not work well in this transformation.

Another example of EDC having a marked improvement over DCC is the formation of cyanoguanidines (eq 9).[19] This was originally attempted with DCC, which gives poor yields even after extended reaction times. It is thought the positively charged nitrogen in EDC facilitates the C–S bond cleavage and since DCC lacks this atom the reaction goes slowly.

Pyrroles are formed when acetylenes undergo cycloadditions with N-acylamino acids in the presence of EDC.[20] An N-alkenyl-munchnone azomethine ylide is generated in situ and is trapped by the dipolaraphile. Loss of CO_2 yields the pyrrole (eq 10).

N-alkenylmunchnone
azomethine ylide

(10)

Related Reagents. Benzotriazol-1-yloxytris(dimethyl-amino)phosphonium Hexafluorophosphate; *N,N′*-Carbonyldiimidazole; 1-Cyclohexyl-3-(2-morpholinoethyl)carbodiimide; 1,3-Dicyclohexylcarbodiimide; Diphenyl Phosphorazidate; Di-*p*-tolylcarbodiimide; *N*-Ethyl-5-phenylisoxazolium-3′-sulfonate; 1-Hydroxybenzotriazole; Isobutene; Isobutyl Chloroformate; 1,1′-Thionylimidazole.

1. (a) Kurzer, F.; Douraghi-Zader, K., *Chem. Rev.* **1967**, *67*, 107. (b) Rich, D. H.; Singh, J., *The Peptides: Analysis, Synthesis, Biology*; Academic: New York, 1979; Vol. 1, p 241.

2. Sheehan, J. C.; Ledis, S. L., *J. Am. Chem. Soc.* **1973**, *95*, 875.

3. Chen, F. M. F.; Benoiton, N. L., *Synthesis* **1979**, 709.

4. Dhaon, M. K.; Olsen, R. K.; Ramasamy, K., *J. Org. Chem.* **1982**, *47*, 1962.

5. Carraway, K. L.; Koshland, Jr., D. E., *Methods Enzymol.* **1972**, *25*, 616.

6. Ramage, R.; MacLeod, A. M.; Rose, G. W., *Tetrahedron* **1991**, *47*, 5625.

7. Wuensch, E.; Drees, F., *Chem. Ber.* **1966**, *99*, 110.

8. Konig, W.; Geiger, R., *Chem. Ber.* **1970**, *103*, 788.

9. Gish, D. T.; Katsoyannis, P. G.; Hess, G. P.; Stedman, R. J., *J. Am. Chem. Soc.* **1956**, *78*, 5954.

10. (a) Miyazawa, T.; Otomatsu, T.; Yamada, T.; Kuwata, S., *Tetrahedron Lett.* **1984**, *25*, 771. (b) Miyazawa, T.; Otomatsu, T.; Fukui, Y.; Yamada, T.; Kuwata, S., *J. Chem. Soc., Chem. Commun.* **1988**, 419.

11. Hagiwara, D.; Neya, M.; Miyazaki, Y.; Hemmi, K.; Hashimoto, M., *J. Chem. Soc., Chem. Commun.* **1984**, 1676.

12. Okawa, K.; Hase, S., *Bull. Chem. Soc. Jpn.* **1963**, *36*, 754.

13. Ogura, H.; Takeda, K.; Tokue, R.; Kobayashi, T., *Synthesis* **1978**, 394.

14. Shamim, M. T.; Ukena, D.; Padgett, W. L.; Daly, J. W., *J. Med. Chem.* **1989**, *32*, 1231.

15. Atherton, E.; Benoiton, N. L.; Brown, E.; Sheppard, R. C.; Williams, B. J., *J. Chem. Soc., Chem. Commun.* **1981**, 336.

16. Greene, T. W., *Protective Groups in Organic Synthesis*, Wiley: New York, 1981; p 154.

17. Keeton, T. K.; Krutzsch, H.; Lovenberg, W., *Science* **1981**, *211*, 586.

18. Tam, T. F.; Thomas, E.; Kruntz, A., *Tetrahedron Lett.* **1987**, *28*, 1127.

19. Atwal, K. S.; Ahmed, S. Z.; O'Reilly, B. C., *Tetrahedron Lett.* **1989**, *30*, 7313.

20. Anderson, W. A.; Heider, A. R., *Synth. Commun.* **1986**, 357.

Richard S. Pottorf
University of Bristol, Bristol, UK

Peter Szeto
Marion Merrell Dow Research Institute, Cincinnati, OH, USA

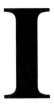

I

Essential Reagents for Organic Synthesis, Edited by Philip L. Fuchs, André B. Charette, Tomislav Rovis, and Jeffrey W. Bode.
©2016 John Wiley & Sons, Ltd. Published 2016 by John Wiley & Sons, Ltd.

N-Iodosuccinimide

[516-12-1] $C_4H_4INO_2$ (MW 224.99)

InChI = 1/C4H4INO2/c5-6-3(7)1-2-4(6)8/h1-2H2

InChIKey = LQZMLBORDGWNPD-UHFFFAOYAG

(electrophilic iodination of arenes, alkenes, and alkynes; activation of glycosyl donors)

Alternate Name: NIS; 1-iodo-2,5-pyrrolidinedione.

Physical Data: mp 193–199 °C (dec).

Solubility: soluble in dioxane, THF, MeCN; insoluble in ether, CCl_4.

Form Supplied in: white powder; widely available, but quite expensive.

Preparative Methods: reaction of *N*-silversuccinimide with I_2 in anhydrous dioxane, removal of AgI by filtration, and addition of CCl_4 to promote crystallization.[1]

Purification: recrystallization from dioxane–CCl_4 at −20 °C.

Handling, Storage, and Precautions: stored at 0 °C under nitrogen, and protected from light and moisture to avoid decomposition; an irritating solid and precautions should be taken to avoid inhalation of the powder.

Original Commentary

Scott C. Virgil

Massachusetts Institute of Technology, Cambridge, MA, USA

Introduction. *N*-Iodosuccinimide is an efficient reagent for the electrophilic iodination of organic compounds for which the reactivity of I_2 is often insufficient. A variety of functional groups are electrophilically iodinated using NIS with similar effectiveness to the corresponding brominations using *N-Bromosuccinimide* (NBS).[2] Electrophilic attack on soft nucleophiles (e.g. as in dithiane deprotection) is achieved in the presence of other sensitive groups with optimum selectivity by iodination using NIS under mild conditions (see also *1,3-Diiodo-5,5-dimethyl- hydantoin*).[3] For other iodinations a more powerful source of iodonium ion can be generated by the combination of NIS with protic acids including AcOH, TsOH, and TfOH.[4]

Reaction with Organometallic Derivatives. Similarly, the lesser reactivity of NIS compared with NBS allows it to be used for the synthesis of vinyl iodides by iodination of vinylaluminum intermediates in the presence of reactive alkenic groups.[5] NIS has also been found to be an excellent reagent for the oxidative cyclization of dipeptide dianions to β-lactams (eq 1).[6]

Iodination of Alkenic Compounds. Iodolactonization and related reactions with alkenes can be achieved using NIS in aprotic solvents. In addition, iodoesters may also be prepared by the reaction of alkenes with NIS in the presence of carboxylic

acids (eq 2),[7] and allylic trichloroacetimidates may be cyclized to 5-iododihydro-1,3-oxazines (eq 3).[8]

In combination with protic acid, NIS has been used for the preparation of α-iodo enones and β,β-dihalo enones (eqs 4 and 5; HTIB = *[Hydroxy(tosyloxy)iodo]benzene*).[9] More reactive alkenes such as enol acetates react with NIS to afford α-iodo ketones.[10] The reaction of enol ethers with NIS in the presence of an alcohol, affording iodoacetals, allows the synthesis of acetals by dehalogenation or mixed ketene acetals by elimination.[11]

Activation of Glycosyl Donors.[12] NIS is a useful reagent for the coupling of glycals to alcohols (eq 6)[13] or other carbohydrates,[14] affording 2-iodoglycosides which are readily dehalogenated to 2-deoxysugars, or can serve as precursors to 2-aminosugars via the epoxide intermediate.[15] Selective activation of glycosyl thioethers for coupling with carboxylic acids and heterocyclic compounds is achieved under mild conditions by reaction with NIS and *Trifluoromethanesulfonic Acid* (eq 7).[16] In addition, the combination of NIS and TfOH is also the reagent of choice for the activation of 4-pentenylglycosides.[17]

Essential Reactions for Organic Synthesis, First Edition. Edited by Philip L. Fuchs.
© 2016 John Wiley & Sons, Ltd. Published 2016 by John Wiley & Sons, Ltd.

$$(7)$$

Oxidation Reactions. Hypoiodite intermediates may be generated from the reaction of simple alcohols with NIS. When conducted under photochemical irradiation, the products of Barton-type or fragmentation reactions of alkoxyl radical intermediates may be obtained.[18] Aldehydes are oxidized to methyl esters via hemiacetal intermediates by reaction with NIS in methanol at rt.[19] However, such conditions are not effective for the oxidation of simple alcohols. The combination of NIS and **Tetrabutylammonium Iodide** in dichloromethane has been developed for the oxidation of a variety of alcohols to the corresponding carbonyl compounds (eq 8).[20] This reagent system is most widely used for the oxidation of lactols to lactones, in which near-quantitative yields are generally obtained under mild conditions (eq 9).[21]

$$(8)$$

$$(9)$$

First Update

Ying Zeng & Fanzuo Kong
Research Center for Eco-Environmental Sciences, Beijing, China

Reaction with Thiols. In combination with triphenylphosphine (PPh₃), NIS is used as an iodide donor to convert alkyl thiols to alkyl iodides under mild conditions (eq 10),[22] analogous to the conversion of alcohols to alkyl iodides.

$$CH_3(CH_2)_{11}SH \xrightarrow[CH_2Cl_2]{PPh_3/NIS} CH_3(CH_2)_{11}I \quad (10)$$
$$70\%$$

Reaction with Carborane Compounds. Reaction of Na₂[B₉H₉] with NIS gives the monoiodinated anion $[1\text{-}IB_9H_8]^{2-}$, while treatment of $[Me_3NH][nido\text{-}7,8\text{-}C_2B_9H_{12}]$ with NIS affords the corresponding $[Me_3NH][nido\text{-}9,11\text{-}I_2\text{-}7,8\text{-}C_2B_9H_{10}]$.[23,24] Reaction of sandwich type carborane CpCl₂Ta(2,3-Et₂C₂B₄H₄) with NIS generates B(5)-monoiodo, B(4,5)-diiodo, and in some cases B(4,5,6)-triiodo complexes.[25,26] These compounds, in turn, react with Grignard reagents to give the B-alkyl and B-aryl derivatives.

Application in Glycosylations. NIS is employed to achieve both tethering and thioglycoside activation allowing the stereoselective synthesis of 1,2-*cis*-glycosides either in a one- or two-step procedure. Thus, treatment of 2-*O*-alkenyl protected thioglycosides with NIS and a variety of alcohols in THF produces ketal intermediates, which allow the synthesis of 1,2-*cis*-glycosidic linkages (eq 11).[27] 1,2-*cis*-Glycosidic linkages are similarly obtained from 2-*O*-alkenyl protected glycosyl fluoride donors.[28]

$$(11)$$

A new method for the deprotection of benzyl ethers with NIS through a hypoiodite intermediate has been developed (eq 12).[29]

$$(12)$$

Applications in Fluorination. Iodofluorinated compounds are obtained using the so-called 'three-component reaction' [a substrate, an electrophile (NIS), and a nucleophile (fluorination agent)] from alkenes or alkynes in a regio- and stereoselective way (eq 13).[30–32]

$$(13)$$
$$80\%$$

A reagent consisting of NIS and 70% HF-pyridine converts xanthates R-OCS₂Me (R=alkyl) to R-OCF₃ (R = primary) or R-F (R = benzylic, secondary, tertiary) by an oxidative desulfurization-fluorination reaction.[33] The C-S bonds of the dithiocarbamates are directly replaced by C-F bonds using NIS and tetrabutylammonium dihydrogentrifluoride (TBAH₂F₃) (eq 14).[34]

$$(14)$$
$$60\%$$

Reactions with Alkyne and Alkene Compounds. NIS is an excellent iodinating reagent that can smoothly convert vinyl silanes and vinyl stannanes to the corresponding vinyl iodides (eqs 15 and 16).[35,36] It also enables the mild conversion of several types of alkenyl boronic acids to geometrically pure alkenyl iodides (eq 17).[37]

$$(15)$$
trans:cis = 13:1 (83%)

$$Ph\diagdown\diagup TMS \xrightarrow{NIS} Ph\diagdown\diagup I \quad (16)$$
$$90\%$$

$$Ph\diagdown\diagup B\diagup OH \xrightarrow{NIS} Ph\diagdown\diagup I \quad (17)$$
$$86\%$$

Alkenyl carboxylic acids and propiolic acids are converted to the corresponding iodoalkenes and iodoalkynes in excellent yields with NIS (eq 18).[38,39]

$$(18)$$

$$74\%$$

NIS reacts with carbonyl-conjugated alkynes in the presence of protic acid to give diiodo ketoesters, diiodo diketones (eq 19), or diiodo acetals.[40]

$$(19)$$

$$95\%$$

Applications in Retro-Mannich Reactions. Treatment of unprotected gramine with NIS results in smooth conversion to 3-iodoindole (eq 20). 3,4-Disubstituted indoles are obtained by combination of the retro-Mannich process with the directed ortho metalation reaction (DoM) or the Negishi cross-coupling protocol.[41]

$$(20)$$

$$78\%$$

Reaction with Stabilized Phosphoranes. Iodo enol lactones are obtained from reaction of the corresponding ω-carboxy keto phosphoranes with NIS (eq 21).[42]

$$(21)$$

$$95:5$$
$$88\%$$

Other Reactions. Treatment of 4-halogenocyclohexa-2,5-dienones with 4-hydroperoxycyclohexa-2,5-dienones in the presence of NIS gives substituted bis(4-oxocyclohexa-2,5-dienyl)

peroxides (eq 22), some of which have not been prepared by the classical method.[43]

$$(22)$$

$$74\%$$

NIS is used as a selective *N*-demethylating agent under mild conditions. These conditions have been successfully applied to the synthesis of metabolites of potent antibacterial agents within the macrolide and ketolide classes (eq 23).[44]

$$(23)$$

Second Update

Julie A. Pigza
Vanderbilt University, Nashville, TN, USA

Deprotection of Amines and Ethers. Catalytic NIS in MeOH has been shown to be a chemoselective catalyst for the deprotection of alkyl TBDMS ethers selectively over phenyl TBDMS ethers.[45] A method for the deprotection of dibenzylamino groups selectively to either the monobenzylamine or directly to the amine has also been developed.[46] In the case of carbohydrates containing multiple protecting groups, as in eq 24, NIS in the presence of TEMPO provided the monobenzylamine in excellent yield. Excess NIS under slightly modified conditions could be used to afford the amine in modest yield. The process is not restricted to carbohydrates but does require the presence of a nearby alcohol or alkoxy substituent to work, as for the amine in eq 25. It should also be noted that the benzyloxy group in this example remained intact.

$$(24)$$

3 equiv NIS, TEMPO   R = Bn, 98%

10 equiv NIS   R = H, 64%
TESOTf, lutidine

$$\text{Ph}\underset{\text{NBn}_2}{\overset{}{\diagdown}}\text{R} \quad \xrightarrow{\text{2 equiv NIS}} \quad \text{Ph}\underset{\text{NHBn}}{\overset{}{\diagdown}}\text{R} \quad (25)$$

R = OBn, 78%

R = CH₃, 0%

α-Iodination of Carbonyl Compounds. Recent advances in the α-iodination of carbonyl compounds has led to selectivity in formation of the monoiodinated products with little or no diiodinated material. For example, NIS in the presence of Amberlyst-15 provided the monoiodination products of β-keto esters in good yields (eq 26).[47] Cyclic ketones are selectively halogenated at the more substituted position. Iodination of aryl ketones using NIS in the presence of PTSA occurs under solvent-free microwave irradiation in high yields.[48] Both of these are an improvement over previous conditions using NIS and NaH in which diiodination can be a problem unless suitable α-substituted precursors were utilized in which only one acidic hydrogen is available for deprotonation.[49]

$$\text{Ph}\overset{\text{O}\quad\text{O}}{\diagup\!\!\diagdown}\text{OEt} \quad \xrightarrow[\substack{\text{EtOAc, rt}\\90\%}]{\substack{\text{NIS}\\\text{Amberlyst-15}}} \quad \text{Ph}\underset{\text{I}}{\overset{\text{O}\quad\text{O}}{\diagup\!\!\diagdown}}\text{OEt} \quad (26)$$

The first asymmetric direct α-iodination of aldehydes has also been described to provide products in moderate to good enantioselectivities using an organocatalyst.[50] 3-Methyl-1-butanal was reacted with NIS in the presence of a chiral pyrrolidine catalyst to provide the halogenated product in 78% yield with 89% ee (eq 27).

$$\underset{^{i}\text{Pr}}{\overset{\text{O}}{\text{H}\diagdown}} \quad \xrightarrow[\substack{\text{H}_2\text{O, CH}_2\text{Cl}_2/\text{pentane}\\\text{(20 mol \%)}}]{\text{NIS, PhCO}_2\text{H (cat.)}} \quad \underset{^{i}\text{Pr}}{\overset{\text{O}}{\text{H}\diagdown}}\text{I} \quad (27)$$

78%, 89% ee

Halogenation of Aromatic Compounds. Multiple methods have been developed utilizing NIS along with Lewis or Brønsted acids to halogenate aromatic compounds. Several involve halogenation of more electron-rich aromatic derivatives. In the presence of ZrCl₄ (5 mol%) and NIS, the iodination of activated phenyl rings was accomplished in good to excellent yields.[51] Even unprotected aniline derivatives were iodinated under the conditions. Selectivity for *para*- over *ortho*-iodination was almost exclusive. Similar conditions employing catalytic FeCl₃ (10 mol%) have been disseminated, but with lower yields and competitive *ortho*-iodination.[52] Electron-rich aromatic compounds have also been converted to their halogenated derivatives in the presence of NIS and catalytic trifluoroacetic acid (30 mol%) in high yields and with *para*-selectivity.[53] The active species for iodination is thought to be a protonated iodine trifluoroacetate, similar to that previously proposed while using trifluoromethanesulfonic acid to activate NIS.[4] Iodination of electron deficient aromatics has been described using NIS and boron trifluoride monohydrate.[54] This system has a similar acidity ($-H_0 \sim 12$) to that of trifluoromethanesulfonic acid ($-H_0 \sim 14$) and effects the iodination of pentafluorobenzene in 90% yield.

Synthesis of Heterocycles via Addition to Alkenes and Alkynes. The regioselective synthesis of heterocycles has been demonstrated using both cyclohexenyl phenols[55a] and anilines[55b] in which the products differ in both cases by either a 5-*exo-trig* or 6-*endo-trig* cyclization. For the phenolic substrate, the 6-*endo* cyclization product was formed exclusively, while under almost identical conditions, the aniline gave only the 5-*exo* cyclization (eqs 28 and 29). Similarly, hydroxycoumarin derivatives also cyclize in a 6-*endo* fashion similar to the cyclohexenylphenols.[55c]

$$(28)$$

$$\xrightarrow[\substack{0\,^{\circ}\text{C}\\85\%}]{\text{NIS, CH}_3\text{CN}}$$

$$(29)$$

$$\xrightarrow[\substack{-10\,^{\circ}\text{C}\\55\%}]{\text{NIS, CH}_3\text{CN}}$$

Addition of other nucleophiles across alkenes activated by NIS have also been described for oximes[56] via attack of the nitrogen, and for intramolecular cyclization of the carbonyl oxygen of alkylidene cyclopropyl esters.[57] Aromatic rings have also been used as nucleophiles when nearby alkenes are activated by NIS. Using NIS and Sm(OTf)₃ (10 mol%), new six-membered ring heterocycles were produced, including chromans and chromanones, as well as their nitrogen analogues in moderate yields (eq 30).[58]

$$\xrightarrow[\substack{\text{CH}_3\text{CN, rt}\\56\%}]{\text{NIS, Sm(OTf)}_3} \quad (30)$$

Trisubstituted pyridines have been synthesized from appropriate precursors via halogenation/heteroannulation in good yield with an accompanying deiodination (eq 31).[59]

$$(31)$$

$$\xrightarrow[\substack{0\,^{\circ}\text{C}\\>98\%}]{\text{NIS, EtOH}}$$

Addition of heteroatoms across alkynes activated by NIS has also been described. Aryl ketones cyclize under mild conditions in a 5-*endo-dig* fashion to afford 3-iodofuran derivatives (eq 32).[60] Normally, electrophiles such as 'I⁺' add to the 2- and 5-positions of a furan ring, rather than the 3- or 4-positions.

$$\xrightarrow[\substack{\text{acetone}\\\text{rt}\\84\%}]{\text{NIS}} \quad (32)$$

Addition of a phenolic oxygen across an alkyne has been shown using a one-pot deprotection/cyclization with BCl₃ and NIS to furnish the benzofurans in good yield, providing the vinyl iodide aptly suited for further coupling chemistry or functionalization (eq 33).[61]

$$\text{(33)}$$

Use with Ionic Liquids. Ionic liquids are recyclable alternatives to using organic solvents and are promoted in the context of green chemistry. The ionic liquid 1-butyl-3-methylimidazolium tetrafluoroborate ([bmim]BF$_4$) was successfully used to convert a series of olefins to their vic-iodohydrins using an NIS and water system (eq 34).[62] Straightforward ether extraction provided the products in good yields and with moderate regioselectivities (usually 70:20 for NIS, unless the alkene is activated towards one regioisomer as in α,β-unsaturated systems and styrene). Furthermore, the ionic liquid may be recycled by extraction with ethyl acetate; the succinimide byproduct remains in the aqueous phase. The ionic liquid [bmim]PF$_6$ has been used with NIS for the α-halogenation of β-ketoesters, β-diketones, and cyclic ketones in high yields.[63]

$$\text{(34)}$$

76%, ~20% regioisomer

Miscellaneous. Iodomethoxylation of chiral α,β-unsaturated amides has been described to afford a highly regioselective addition of NIS and methanol across an alkene in moderate diastereoselectivity using both Oppolzer's chiral sultam (eq 35) and Evans chiral oxazolidinone.[64] Subsequent cleavage of the chiral auxiliary, conversion of the iodide to the azide, and reduction to the amine results in synthetically useful α-amino acids.

$$\text{(35)}$$

Ar = 3,4-MeOC$_6$H$_3$ Ar = 3,4-MeOC$_6$H$_3$

An interesting electrophilic cleavage of a silicon–carbon bond in the presence of NIS was demonstrated with assistance by a neighboring ketone oxygen (eq 36).[65] The resulting silalactones provide interesting products and the process of selective cleavage of C(sp^3)–Si bond remains a challenging problem.

$$\text{(36)}$$

Radical mediated ring opening reactions of alcohols by CuI has been previously described to occur through a hypoiodite intermediate.[18] More recently, this process has been used to form medium-sized rings via cleavage of the C–C bond at the ring fusion. Thus, seven-membered ring formation can occur in high yield (eq 37).

$$\text{(37)}$$

Access to (Z)-trisubstituted olefins has been realized via the 1,2-metalate rearrangement[66] of α-lithiofuran derivatives using an appropriate copper source and trapping of the vinylcuprate intermediate with NIS to provide exclusively the (Z)-isomer of the vinyl iodostannane (eq 38).[67] Using (TMS)$_2$CuCNLi$_2$ instead afforded the vinyl iodosilane.

$$\text{(38)}$$

X = SnBu$_3$, 75%
X = SiMe3, 76%

NIS has been utilized minimally in the allylic iodination in the presence of catalytic TMSCl and Yb(OTf)$_3$ in which selectivity of allylic over benzylic positions was demonstrated.[68]

1. Benson, W. R.; McBee, E. T.; Rand, L., *Org. Synth., Coll. Vol.* **1973**, *5*, 663.

2. NIS is generally not effective in the range of radical halogenations possible for NBS. See also: Taneja, S. C.; Dhar, K. L.; Atal, C. K., *J. Org. Chem.* **1978**, *43*, 997.

3. Burri, K. F.; Cardone, R. A.; Chen, W. Y.; Rosen, P., *J. Am. Chem. Soc.* **1978**, *100*, 7069.

4. Olah, G. A.; Wang, Q.; Sandford, G.; Prakash, G. K. S., *J. Org. Chem.* **1993**, *58*, 3194.

5. Marshall, J. A.; Lebreton, J.; DeHoff, B. S.; Jenson, T. M., *Tetrahedron Lett.* **1987**, *28*, 723.

6. Kawabata, T.; Minami, T.; Hiyama, T., *J. Org. Chem.* **1992**, *57*, 1864.

7. (a) Adinolfi, M.; Parrilli, M.; Barone, G.; Laonigro, G.; Mangoni, L., *Tetrahedron Lett.* **1976**, 3661. (b) Haaima, G.; Weavers, R. T., *Tetrahedron Lett.* **1988**, *29*, 1085.

8. Bongini, A.; Cardillo, G.; Orena, M.; Sandri, S.; Tomasini, C., *J. Org. Chem.* **1986**, *51*, 4905.

9. (a) Iwaoka, T.; Murohashi, T.; Katagiri, N.; Sato, M.; Kaneko, C., *J. Chem. Soc., Perkin Trans. 1* **1992**, 1393. (b) Angara, G. J.; Bovonsombat, P.; McNelis, E., *Tetrahedron Lett.* **1992**, *33*, 2285.

10. Djerassi, C.; Grossman, J.; Thomas, G. H., *J. Am. Chem. Soc.* **1955**, *77*, 3826.

11. Middleton, D. S.; Simpkins, N. S., *Synth. Commun.* **1989**, *19*, 21.

12. (a) Theim, J.; Karl, H.; Schwentner, J., *Synthesis* **1978**, 696. (b) Theim, J. In *Trends in Synthetic Carbohydrate Chemistry*; Horton, D.; Hawkins, L. D.; McGarvey, G. J., Eds.; ACS: Washington, 1989; Vol. 386, Chapter 8.

13. Sebesta, D. P.; Roush, W. R., *J. Org. Chem.* **1992**, *57*, 4799.

14. Horton, D.; Priebe, W.; Sznaidman, M., *Carbohydr. Res.* **1990**, *205*, 71.

15. Friesen, R. W.; Danishefsky, S. J., *Tetrahedron* **1990**, *46*, 103.

16. (a) Veeneman, G. H.; van Leeuwen, S. H.; van Boom, J. H., *Tetrahedron Lett.* **1990**, *31*, 1331. (b) Knapp, S.; Shieh, W.-C., *Tetrahedron Lett.* **1991**, *32*, 3627. (c) Konradsson, P.; Udodong, U. E.; Fraser-Reid, B., *Tetrahedron Lett.* **1990**, *31*, 4313.

17. (a) Konradsson, P.; Mootoo, D. R.; McDevitt, R. E.; Fraser-Reid, B., *Chem. Commun.* **1990**, 270. (b) Ratcliffe, A. J.; Konradsson, P.; Fraser-Reid, B., *J. Am. Chem. Soc.* **1990**, *112*, 5665.

18. (a) McDonald, C. E.; Beebe, T. R.; Beard, M.; McMillen, D.; Selski, D., *Tetrahedron Lett.* **1989**, *30*, 4791. (b) McDonald, C. E.; Holcomb, H.; Leathers, T.; Ampadu-Nyarko, F.; Frommer, J., Jr., *Tetrahedron Lett.* **1990**, *31*, 6283.

19. McDonald, C.; Holcomb, H.; Kennedy, K.; Kirkpatrick, E.; Leathers, T.; Vanemon, P., *J. Org. Chem.* **1989**, *54*, 1213.

20. Hanessian, S.; Wong, D. H.; Therien, M., *Synthesis* **1981**, 394.

21. Hamada, Y.; Kawai, A.; Matsui, T.; Hara, O.; Shioiri, T., *Tetrahedron* **1990**, *46*, 4823.

22. Iranpoor, N.; Firouzabadi, H.; Aghapour, G., *Synlett* **2001**, 1176.

23. Siegburg, K.; Preetz, W., *Inorg. Chem.* **2000**, *39*, 3280.

24. Santos, E. C.; Pinkerton, A. B.; Kinkead, S. A.; Hurlburt, P. K.; Jasper, S. A.; Sellers, C. W.; Huffman, J. C.; Todd, L. J., *Polyhedron* **2000**, *19*, 1777.

25. Stockman, K. E.; Boring, E. A.; Sabat, M.; Finn, M. G.; Grimes, R. N., *Organometallics* **2000**, *19*, 2200.

26. Stockman, K. E.; Garrett, D. L.; Grimes, R. N., *Organometallics* **1995**, *14*, 4661.

27. Ennis, S. C.; Fairbanks, A. J.; Slinn, C. A.; Tennant-Eyles, R. J.; Yeates, H. S., *Tetrahedron* **2001**, *57*, 4221.

28. Cumpstey, I.; Fairbanks, A. J.; Redgrave, A. J., *Org. Lett.* **2001**, *3*, 2371.

29. Madsen, J.; Bols, M., *Angrew. Chem. Int. Ed.* **1998**, *37*, 3177.

30. Tamura, M.; Shibakami, M.; Sekiya, A., *Synthesis* **1995**, *5*, 515.

31. Shellhamer, D. F.; Horney, M. J.; Pettus, B. J.; Petttus, T. L.; Stringer, J. M.; Heasley, V. L.; Syvret, R. G.; Dobrolsky, J. M., *J. Org. Chem.* **1999**, *64*, 1094.

32. Shellhamer, D. F.; Jones, B. C.; Pettus, B. J.; Petttus, T. L.; Stringer, J. M.; Heasley, V. L., *J. Fluorine Chem.* **1998**, *88*, 37.

33. Kanie, K.; Tanaka, Y.; Shimizu, M.; Kuroboshi, M.; Hiyama, T., *Chem. Commun.* **1997**, 309.

34. Kanie, K.; Mizuno, K.; Kuroboshi, M.; Hiyama, T., *Bull. Chem. Soc. Jpn.* **1998**, *71*, 1973.

35. Stamos, D. P.; Taylor, A. G.; Kishi, Y., *Tetrahedron Lett.* **1996**, *37*, 8647.

36. Boden, C. D. J.; Pattenden, G.; Ye, T., *J. Chem. Soc., Perkin Trans.* **1996**, 2417.

37. Petasis, N. A.; Zavialov, L. A., *Tetrahedron Lett.* **1996**, *37*, 567.

38. Naskar, D.; Chowdhury, S.; Roy, S., *Tetrahedron Lett.* **1998**, *39*, 699.

39. Naskar, D.; Roy, S., *J. Org. Chem.* **1999**, *64*, 6896.

40. Heasley, V. L.; Shellhamer, D. F.; Chappell, A. E.; Cox, J. M.; Hill, D. J.; McGovern, S. L.; Eden, C. C.; Kissel, C. L., *J. Org. Chem.* **1998**, *63*, 4433.

41. Chauder, B.; Larkin, A.; Snieckus, V., *Org. Lett.* **2002**, *4*, 815.

42. Kayser, M. M.; Zhu, J.; Hooper, D. L., *Can. J. Chem.* **1997**, *75*, 1322.

43. Omura, K., *J. Org. Chem.* **1997**, *62*, 8790.

44. Stenmark, H. G.; Brazzale, A.; Ma, Z., *J. Org. Chem.* **2000**, *65*, 3875.

45. Karimi, B.; Zamani, A.; Zarayee, D., *Tetrahedron Lett.* **2004**, *45*, 9139.

46. Grayson, E. J.; Davis, B. G., *Org. Lett.* **2005**, *7*, 2361.

47. Meshram, H. M.; Reddy, P. N.; Sadashiv, K.; Yadav, J. S., *Tetrahedron Lett.* **2005**, *46*, 623.

48. Lee, J. C.; Bae, Y. H., *Synlett* **2003**, 507.

49. Curran, D. P.; Chang, C. T., *J. Org. Chem.* **1989**, *54*, 3140.

50. Bertelsen, S.; Halland, N.; Bachmann, S.; Marigo, M.; Braunton, A.; Jørgensen, K. A., *Chem. Commun.* **2005**, 4821.

51. Zhang, Y.; Shibatomi, K.; Yamamoto, H., *Synlett* **2005**, 2837.

52. Tanemura, K.; Suzuki, T.; Nishida, Y.; Satsumabayashi, K.; Horaguchi, T., *Chem. Lett.* **2003**, *32*, 932.

53. Castanet, A.-S.; Colobert, F.; Broutin, P.-E., *Tetrahedron Lett.* **2002**, *43*, 5047.

54. Prakash, G. K. S.; Mathew, T.; Hoole, D.; Esteves, P. M.; Wang, Q.; Rasul, G.; Olah, G. A., *J. Am. Chem. Soc.* **2004**, *126*, 15770.

55. (a) Majumdar, K. C.; Basu, P. K., *Synth. Commun.* **2002**, *32*, 3719. (b) Majumdar, K. C.; Kundu, U. K.; Das, U.; Jana, N. K.; Roy, B., *Can. J. Chem.* **2005**, *83*, 63. (c) Majumdar, K. C.; Sarkar, S., *Tetrahedron* **2002**, *58*, 8501.

56. Mernyak, E.; Benedek, G.; Schneider, G.; Wolfing, J., *Synlett* **2005**, 637.

57. Ma, S.; Lu, L., *J. Org. Chem.* **2005**, *70*, 7629.

58. Hajra, S.; Maji, B.; Karmakar, A., *Tetrahedron Lett.* **2005**, *46*, 8599.

59. Bagley, M. C.; Glover, C.; Merritt, E. A.; Xiong, X., *Synlett* **2004**, 811.

60. Sniady, A.; Wheeler, K. A.; Dembinski, R., *Org. Lett.* **2005**, *7*, 1769.

61. Colobert, F.; Castanet, A.-S.; Abillard, O., *Eur. J. Org. Chem.* **2005**, 3334.

62. Yadav, J. S.; Reddy, B. V. S.; Baishya, G.; Harshavardhan, S. J.; Chary, C. J.; Gupta, M. K., *Tetrahedron Lett.* **2005**, *46*, 3569.

63. Meshram, H. M.; Reddy, P. N.; Vishnu, P.; Sadashiv, K.; Yadav, J. S., *Tetrahedron Lett.* **2006**, *47*, 991.

64. Hajra, S.; Bhowmick, M.; Karmakar, A., *Tetrahedron Lett.* **2005**, *46*, 3073.

65. Shindo, M.; Matsumoto, K.; Shishido, K., *Angew. Chem., Int. Ed.* **2004**, *43*, 104.

66. Kocienski, P.; Wadman, S., *J. Am. Chem. Soc.* **1989**, *111*, 2363.

67. Le Menez, P.; Brion, J.-D.; Lensen, N.; Chelain, E.; Pancrazi, A.; Ardisson, J., *Synthesis* **2003**, 2530.

68. Yamanaka, M.; Arisawa, M.; Nishida, A.; Nakagawa, M., *Tetrahedron Lett.* **2002**, *43*, 2403.

Iodotrimethylsilane[1]

$$I(Me)_3Si$$

[16029-98-4] C$_3$H$_9$ISi (MW 200.11)

(a versatile reagent for the mild dealkylation of ethers, carboxylic esters, lactones, carbamates, acetals, phosphonate and phosphate esters; cleavage of epoxides, cyclopropyl ketones; conversion of vinyl phosphates to vinyl iodides; neutral nucleophilic reagent for halogen exchange reactions, carbonyl and conjugate addition reactions; use as a trimethylsilylating agent for formation of enol ethers, silyl imino esters, and *N*-silylenamines, alkyl, alkenyl and alkynyl silanes; Lewis acid catalyst for acetal formation, α-alkoxymethylation of ketones, for reactions of acetals with silyl enol ethers and allylsilanes; reducing agent for epoxides, enediones, α-ketols, sulfoxides, and sulfonyl halides; dehydrating agent for oximes)

Alternate Name: TMS-I; TMSI; trimethylsilyl iodide.
Physical Data: bp 106–109 °C; *d* 1.406 g cm^{-3}; n_D^{20} 1.4710; fp −31 °C.
Solubility: sol in CCl$_4$, CHCl$_3$, CH$_2$Cl$_2$, ClCH$_2$CH$_2$Cl, MeCN, PhMe, hexanes; reactive with THF (ethers), alcohols, and EtOAc (esters).
Form Supplied in: clear colorless liquid, packaged in ampules, stabilized with copper; widely available.
Analysis of Reagent Purity: easily characterized by ^{1}H, ^{13}C, or ^{29}Si NMR spectroscopy.
Preparative Methods: although more than 20 methods have been reported[1] for the preparation of TMS-I, only a few are summarized here. **Chlorotrimethylsilane** undergoes halogen exchange with either **Lithium Iodide**[2] in CHCl$_3$ or **Sodium Iodide**[3] in MeCN, which allows in situ reagent formation (eq 1). Alternatively, **Hexamethyldisilane** reacts with **Iodine** at 25–61 °C to afford TMS-I with no byproducts (eq 2).[4]

$$TMSCl \xrightarrow[\substack{or \\ NaI, MeCN}]{LiI, CHCl_3} TMSI + LiCl \text{ or } NaCl \quad (1)$$

$$TMS\text{–}TMS + I_2 \xrightarrow{25\text{–}61\ °C} TMSI \quad (2)$$

Several other methods for in situ generation of the reagent have been described.[5,6] It should be noted, however, that the reactivity of in situ generated reagent appears to depend upon the method of preparation.

Purity: by distillation from copper powder.
Handling, Storage, and Precautions: extremely sensitive to light, air, and moisture, it fumes in air due to hydrolysis (HI), and becomes discolored upon prolonged storage due to generation of I$_2$. It is flammable and should be stored under N$_2$ with a small piece of copper wire. It should be handled in a well ventilated fume hood and contact with eyes and skin should be avoided.

Essential Reactions for Organic Synthesis, First Edition. Edited by Philip L. Fuchs.
© 2016 John Wiley & Sons, Ltd. Published 2016 by John Wiley & Sons, Ltd.

Original Commentary

Michael E. Jung
University of California, Los Angeles, CA, USA

Michael J. Martinelli
Lilly Research Laboratories, Indianapolis, IN, USA

Use as a Nucleophilic Reagent in Bond Cleavage Reactions.

Ether Cleavage.[5,7] The first broad use of TMS-I was for dealkylation reactions of a wide variety of compounds containing oxygen–carbon bonds, as developed independently by the groups of Jung and Olah. Simple ethers initially afford the trimethylsilyl ether and the alkyl iodide, with further reaction giving the two iodides (eq 3).[7,8] This process occurs under neutral conditions, and is generally very efficient as long as precautions to avoid hydrolysis by adventitious water are taken. Since the silyl ether can be quantitatively hydrolyzed to the alcohol, this reagent permits the use of simple ethers, e.g. methyl ethers, as protective groups in synthesis. The rate of cleavage of alkyl groups is: tertiary ≈ benzylic ≈ allylic methyl > secondary > primary. Benzyl and *t*-butyl ethers are cleaved nearly instantaneously at low temperature with TMS-I. Cyclic ethers afford the iodo silyl ethers and then the diiodide, e.g. THF gives 4-iodobutyl silyl ether and then 1,4-diiodobutane in excellent yield.[7,8] Alcohols and silyl ethers are rapidly converted into the iodides as well.[8a,9] Alkynic ethers produce the trimethylsilylketene via dealkylative rearrangement.[4b] Phenolic ethers afford the phenols after workup.[5,7,10] In general, ethers are cleaved faster than esters. Selective cleavage of methyl aryl ethers in the presence of other oxygenated functionality has also been accomplished in quinoline.[11] γ-Alkoxyl enones undergo deoxygenation with excess TMS-I (2 equiv), with the first step being conjugate addition of TMS-I.[12]

$$R^1\text{–}O\text{–}R^2 \xrightarrow{TMSI} R^2I + R^1OTMS \xrightarrow{TMSI} R^1I \quad (3)$$

$$H_2O \downarrow \qquad \nearrow TMSI$$

$$R^1OH$$

Cleavage of Epoxides. Reaction of epoxides with 1 equiv of TMS-I gives the vicinal silyloxy iodide.[8e] With 2 equiv of TMS-I, however, epoxides are deoxygenated to afford the corresponding alkene (eq 4).[13a,b] However, allylic alcohols are efficiently prepared by reaction of the intermediate iodosilane with base.[13c,d] Furthermore, acyclic 2-ene-1,4-diols react with TMS-I to undergo dehydration, affording the corresponding diene.[13e]

Ester Dealkylation.[14] Among the widest uses for TMS-I involves the mild cleavage of carboxylic esters under neutral conditions. The ester is treated with TMS-I to form an initial oxonium intermediate which suffers attack by iodide (eq 5). The trimethylsilyl ester is cleaved with H_2O during workup. Although the reaction is general and efficient, it is possible to accomplish selective cleavage according to the reactivity trend: benzyl, t-butyl > methyl, ethyl, i-propyl. Neutral transesterification is also possible via the silyl ester intermediate.[15] Aryl esters are not cleaved by TMS-I, however, since the mechanism involves displacement of R^2 by I^-. Upon prolonged exposure (75 °C, 3 d) of simple esters to excess TMS-I (2.5 equiv), the corresponding acid iodides are formed.[14b,16] β-Keto esters undergo decarboalkoxylation when treated with TMS-I.[17] An interesting rearrangement reaction provides α-methylene lactones from 1-(dimethylaminomethyl)cyclopropanecarboxylates (eq 6).[18]

$$(5)$$

$$(6)$$

Lactone Cleavage.[14,19] Analogous to esters, lactones are also efficiently cleaved with TMS-I to provide ω-iodocarboxylic acids, which may be further functionalized to afford bifunctional building blocks for organic synthesis (eq 7). Diketene reacts with TMS-I to provide a new reagent for acetoacylation.[20]

$$(7)$$

Cleavage of Carbamates.[21] Since strongly acidic conditions are typically required for the deprotection of carbamates, use of TMS-I provides a very mild alternative. Benzyl and t-butyl carbamates are readily cleaved at rt,[22] whereas complete cleavage of methyl or ethyl carbamates may require higher temperatures (reflux). The intermediate silyl carbamate is decomposed by the addition of methanol or water (eq 8). Since amides are stable to TMS-I-promoted hydrolysis,[7a] this procedure can be used to deprotect carbamates of amino acids and peptides.[21d]

$$(8)$$

A recent example used TMS-I to deprotect three different protecting groups (carbamate, ester, and orthoester) in the same molecule in excellent yield (eq 9).[23]

$$(9)$$

Cleavage of Acetals.[24] Acetals can be cleaved in analogy to ethers, providing a newly functionalized product (eq 10), or simply the parent ketone (eq 11). Glycals have also been converted to the iodopyrans with TMS-I,[25] and glycosidation reactions have been conducted with this reagent.[26]

$$(10)$$

$$(11)$$

Orthoesters are converted into esters with TMS-I. The dimethyl acetal of formaldehyde, methylal, affords iodomethyl methyl ether in good yield (eq 12)[27a] (in the presence of alcohols, MOM ethers are formed).[27b] α-Acyloxy ethers also furnish the iodo ethers,[28] e.g. the protected β-acetyl ribofuranoside gave the α-iodide which was used in the synthesis of various nucleosides in good yield (eq 13).[28b] Aminals are similarly converted into immonium salts, e.g. Eschenmoser's reagent, ***Dimethyl(methylene)ammonium Iodide***, in good yield.[29]

$$(12)$$

$$(13)$$

Cleavage of Phosphonate and Phosphate Esters.[30] Phosphonate and phosphate esters are cleaved even more readily with TMS-I than carboxylic esters. The reaction of phosphonate esters proceeds via the silyl ester, which is subsequently hydrolyzed with MeOH or H_2O (eq 14).

$$(14)$$

Conversion of Vinyl Phosphates to Vinyl Iodides.[31] Ketones can be converted to the corresponding vinyl phosphates which react with TMS-I (3 equiv) at rt to afford vinyl iodides (eq 15).

$$(15)$$

Cleavage of Cyclopropyl Ketones.[32] Cyclopropyl ketones undergo ring opening with TMS-I, via the silyl enol ether (eq 16). Cyclobutanones react analogously under these conditions.[33]

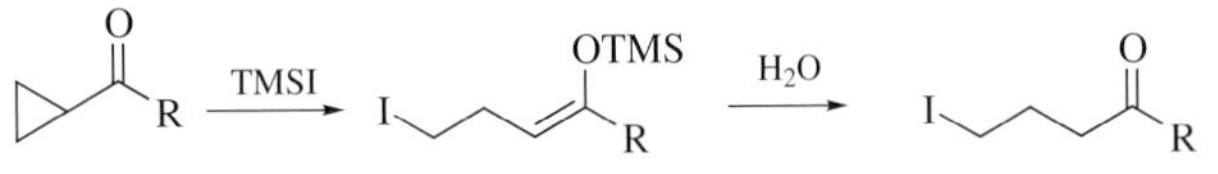

$$
\text{(16)}
$$

Halogen Exchange Reactions.[34] Halogen exchange can be accomplished with reactive alkyl halides, such as ***Benzyl Chloride*** or ***Benzyl Bromide***, and even with certain alkyl fluorides, by using TMS-I in the presence of $(n\text{-Bu})_4\text{NCl}$ as catalyst (eq 17).

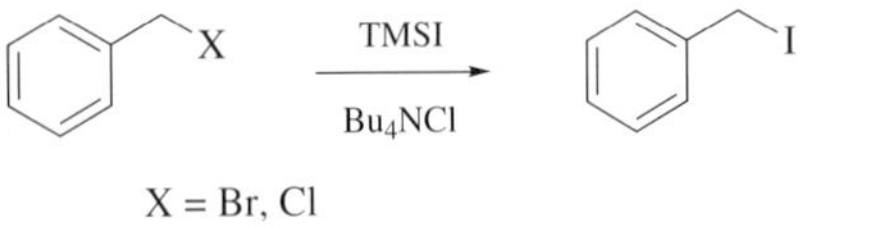

$$
\text{X = Br, Cl} \qquad \text{(17)}
$$

Use of TMS-I in Nucleophilic Addition Reactions.

Carbonyl Addition Reactions.[35] α-Iodo trimethylsilyl ethers are produced in the reaction of aldehydes and TMS-I (eq 18). These compounds may react further to provide the diiodo derivative or may be used in subsequent synthesis.

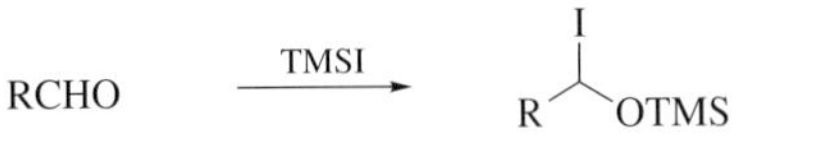

$$
\text{RCHO} \xrightarrow{\text{TMSI}} \qquad \text{(18)}
$$

An example of a reaction of an iodohydrin silyl ether with a cuprate reagent is summarized in eq 19.[36] An interesting reaction of TMS-I with phenylacetaldehydes gives a quantitative yield of the oxygen-bridged dibenzocyclooctadiene, which was then converted in a few steps to the natural product isopavine (eq 20).[35,37]

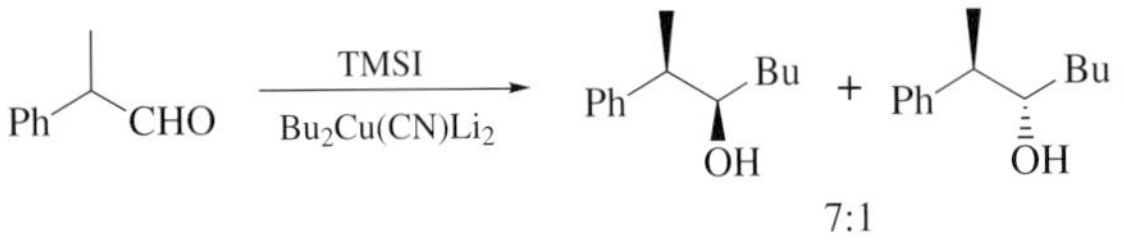

$$
\text{7:1} \qquad \text{(19)}
$$

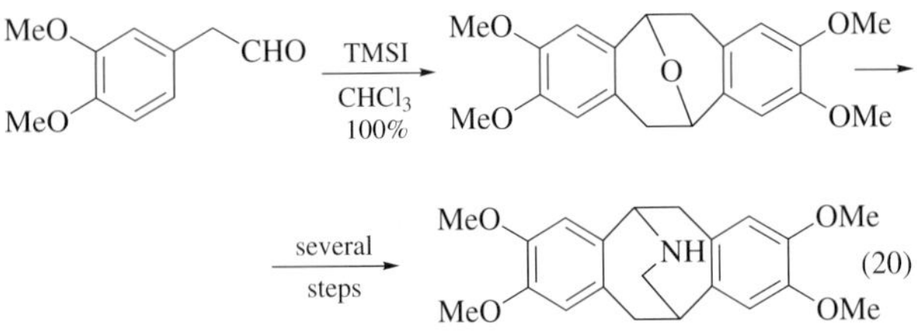

$$
\text{(20)}
$$

β-Iodo ketones have been produced from reactions of TMS-I and ketones with α-hydrogens.[38] This reaction presumably involves a TMS-I catalyzed aldol reaction followed by 1,4-addition of iodide.

Conjugate Addition Reactions.[39] α,β-Unsaturated ketones undergo conjugate addition with TMS-I to afford the β-iodo adducts in high yield (eq 21). The reaction also works well with the corresponding alkynic substrate.[40]

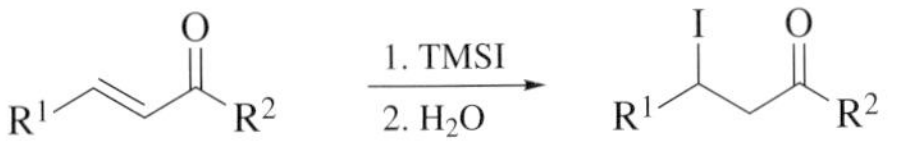

$$
\text{(21)}
$$

TMS-I has also been extensively utilized in conjunction with organocopper reagents to effect highly stereoselective conjugate additions of alkyl nucleophiles.[41]

Use of TMS-I as a Silylating Agent.

Formation of Silyl Enol Ethers.[42] TMS-I in combination with ***Triethylamine*** is a reactive silylating reagent for the formation of silyl enol ethers from ketones (eq 22). TMS-I with ***Hexamethyldisilazane*** has also been used as an effective silylation agent, affording the thermodynamic silyl enol ethers. For example, 2-methylcyclohexanone gives a 90:10 mixture in favor of the tetrasubstituted enol ether product.[42a] The reaction of TMS-I with 1,3-diketones is a convenient route to 1,3-bis(trimethylsiloxy)-1,3-dienes.[42c]

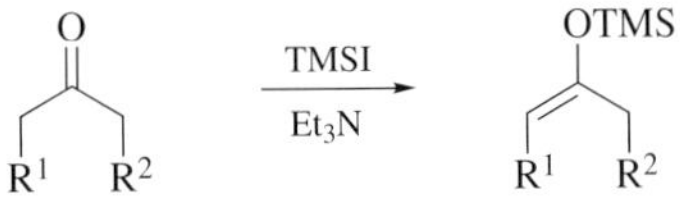

$$
\text{(22)}
$$

In an analogous process, TMS-I reacts with lactams in the presence of Et_3N to yield silyl imino ethers (eq 23).[43a]

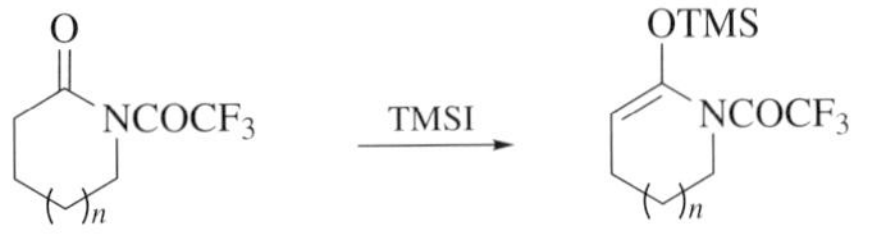

$$
\text{(23)}
$$

Halogenation of Lactams.[43b] Selective and high yielding iodination and bromination of lactams occurs with ***Iodine*** or ***Bromine***, respectively, in the presence of TMS-I and a tertiary amine base (eq 24). The proposed reaction mechanism involves intermediacy of the silyl imino ether.

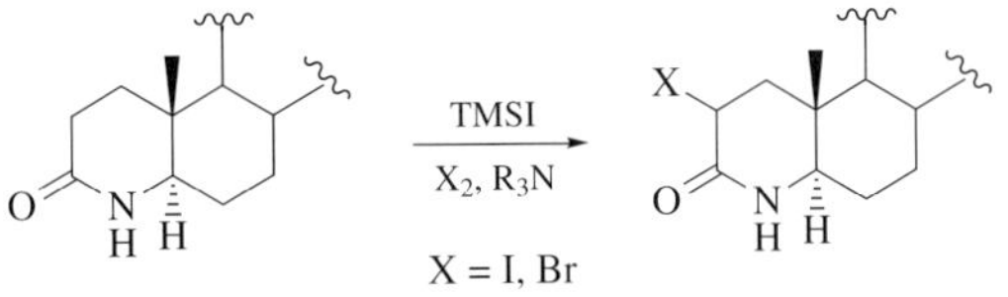

$$
\text{X = I, Br} \qquad \text{(24)}
$$

Reaction with Carbanions.[44] TMS-I has seen limited use in the silylation of carbanions, with different regioselectivity compared to other silylating reagents in the example provided in eq 25.

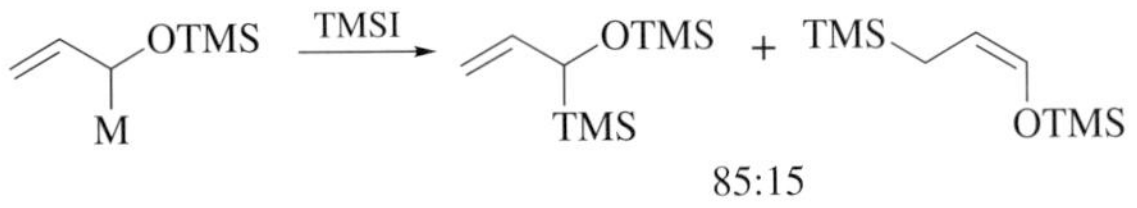

$$
\text{85:15} \qquad \text{(25)}
$$

Silylation of Alkynes and Alkenes.[45] A Heck-type reaction of TMS-I with alkenes in the presence of Pd⁰ and Et₃N affords alkenyltrimethylsilanes (eq 26).

$$Ar–CH=CH_2 \xrightarrow[Et_3N]{\underset{Pd^0}{TMSI}} Ar–CH=CH–TMS \quad (26)$$

Oxidative addition of TMS-I to alkynes can also be accomplished with a three-component coupling reaction to provide the enyne product (eq 27).

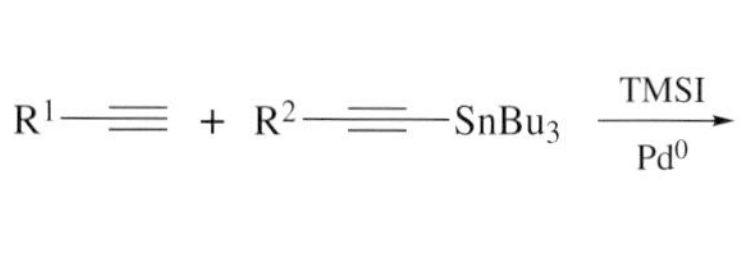

(27)

Use of TMS-I as a Lewis Acid.

Acetalization Catalyst.[46] TMS-I used in conjunction with (MeO)₄Si is an effective catalyst for acetal formation (eq 28).

$$PhCHO \xrightarrow[(MeO)_4Si]{TMSI} PhCH(OMe)_2 \quad (28)$$

Catalyst for α-Alkoxymethylation of Ketones. Silyl enol ethers react with α-chloro ethers in the presence of TMS-I to afford α-alkoxymethyl ketones (eq 29).[47]

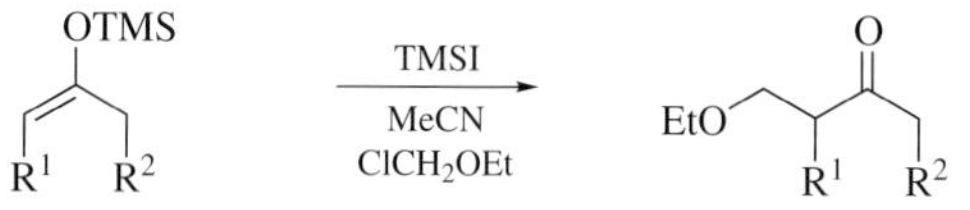

(29)

Catalyst for Reactions of Acetals with Silyl Enol Ethers and Allylsilanes. TMS-I catalyzes the condensation of silyl enol ethers with various acetals (eq 30)[48] and imines,[49] and of allylsilanes with acetals.[50]

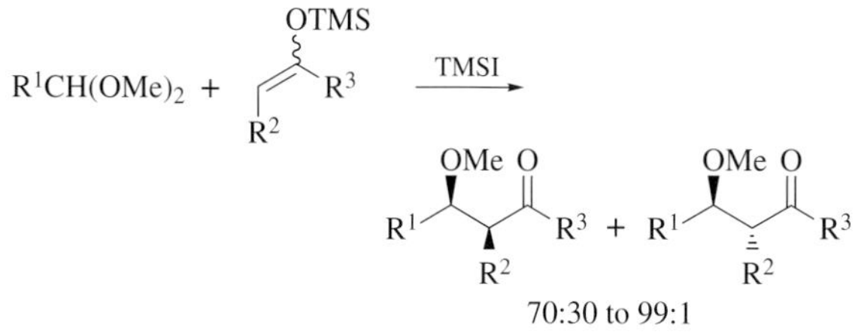

(30)
70:30 to 99:1

Use of TMS-I as a Reducing Agent.
TMS-I reduces enediones to 1,4-diketones,[51] while both epoxides and 1,2-diols are reduced to the alkenes.[13a,b,52] The Diels–Alder products of benzynes and furans are converted in high yield to the corresponding naphthalene (or higher aromatic derivative) with TMS-I (eq 31).[53]

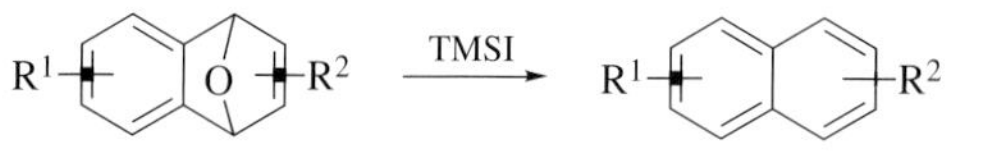

(31)

Styrenes and benzylic alcohols are reduced to the alkanes with TMS-I (presumably via formation of HI).[54] Ketones produce the symmetrical ethers when treated with trimethylsilane as a reducing agent in the presence of catalytic TMS-I.[55]

Reduction of α-Ketols.[56,57] Carbonyl compounds containing α-hydroxy, α-acetoxy, or α-halo groups react with excess TMS-I to give the parent ketone. α-Hydroxy ketone reductions proceed via the iodide, which is then reduced with iodide ion to form the parent ketone (eq 32).

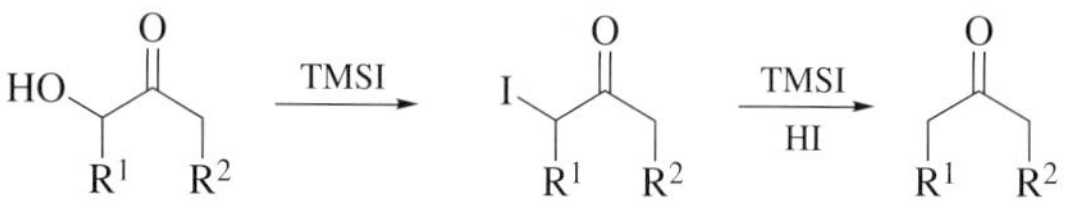

(32)

Sulfoxide Deoxygenation.[58] The reduction of sulfoxides occurs under very mild conditions with TMS-I to afford the corresponding sulfide and iodine (eq 33). Addition of I₂ to the reaction mixture accelerates the second step. The deoxygenation occurs faster in pyridine solution than the reactions with a methyl ester or alcohol.[59]

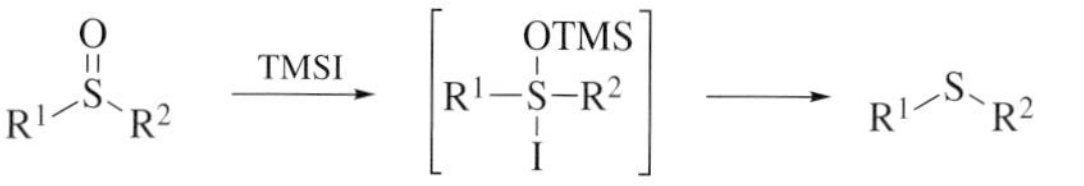

(33)

Pummerer reactions of sulfoxides can be accomplished in the presence of TMS-I and an amine base, leading to vinyl sulfides.[60] An efficient synthesis of dithioles was accomplished with TMS-I and Hünig's base (*Diisopropylethylamine*) (eq 34).[61]

(34)

Reaction with Sulfonyl Halides.[62] Arylsulfonyl halides undergo reductive dimerization to form the corresponding disulfides (eq 35). Alkylsulfonyl halides, however, undergo this process under somewhat more vigorous conditions. Although sulfones generally do not react with TMS-I, certain cyclic sulfones are cleaved in a manner analogous to lactones.[63]

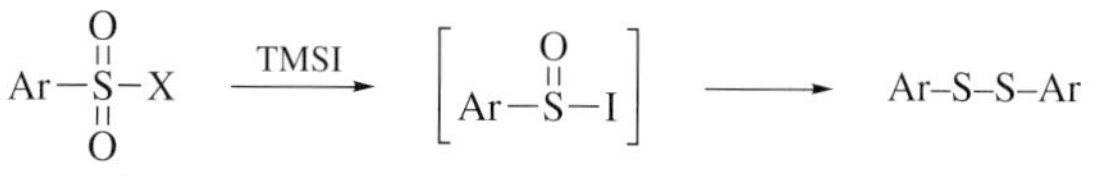

(35)

Other Reactions of TMS-I.

Reaction with Phosphine Oxides.[64] Phosphine oxides react with TMS-I to form stable adducts (eq 36). These *O*-silylated products can undergo further thermolytic reactions such as alkyl group cleavage.

$$R_3P=O \xrightarrow{TMSI} R_3\overset{+}{P}–O–TMS \; I^- \quad (36)$$

Chlorophosphines undergo halogen exchange reactions with TMS-I.[65]

Reaction with Imines. Imines react with TMS-I to form *N*-silylenamines, in a process analogous to the formation of silyl enol ethers from ketones.[66]

Reaction with Oximes.[67] Oximes are activated for dehydration (aldoximes, with hexamethylsilazane) or Beckmann rearrangement (ketoximes) with TMS-I (eq 37).

$$\text{(37)}$$

Reactions with Nitro and Nitroso Compounds.[68] Primary nitro derivatives react with TMS-I to form the oximino intermediate via deoxygenation, which then undergoes dehydration as discussed for the oximes (eq 38). Secondary nitro compounds afford the silyl oxime ethers, and tertiary nitro compounds afford the corresponding iodide. Nitroalkenes, however, react with TMS-I at 0 °C to afford the ketone as the major product (eq 39).[69]

$$\text{(38)}$$

$$\text{(39)}$$

An interesting analogy to this dehydration process is found in the reductive fragmentation of a bromoisoxazoline with TMS-I, which yields the nitrile (eq 40).[70]

$$\text{(40)}$$

Rearrangement Reactions. An interesting rearrangement occurs on treatment of a β-alkoxy ketone with TMS-I which effects dealkylation and retro-aldol reaction to give the eight-membered diketone after reductive dehalogenation (eq 41).[71] Tertiary allylic silyl ethers α to epoxides undergo a stereocontrolled rearrangement to give the β-hydroxy ketones on treatment with catalytic TMS-I (eq 42).[72]

$$\text{(41)}$$

$$\text{(42)}$$

First Update

George A. Olah, G. K. Surya Prakash, Jinbo Hu
University of Southern California, Los Angeles, CA, USA

Selective Bond Cleavage Reactions.

Ether Cleavage. Iodotrimethylsilane (TMSI) continues to be a versatile ether-cleaving agent in the past decade, and it has been widely applied to complex molecules with high chemoselectivity. Demethylation of 8-methoxy-[2,2]metacyclophanes with TMSI can be accomplished in excellent yields.[73] Treatment of isosorbide and isomannide with TMSI [in situ generated from chlorotrimethylsilane (TMSCl) and sodium iodide] in acetonitrile in the presence of acetone induces the cleavage of only one of the two rings and provides chiral trisubstituted tetrahydrofurans (eq 43).[74] The formation of cyclic sulfonium ions by TMSI-mediated intramolecular displacement of hydroxide or methoxide by sulfide has led to ring contraction reactions from thiepanes to thiolanes (eq 44).[75] The cyclization is especially favored with secondary and tertiary alcohols or ethers, and with an aliphatic more than an aromatic sulfide function.[75]

$$\text{(43)}$$

$$\text{(44)}$$

Ester and Lactone Cleavage. TMSI has been combined with triphenylphosphine (TPP) (in dichloromethane solution) as a more stable, milder, and more selective ester-cleaving agent compared with TMSI itself.[76] TPP increases the stability of TMSI as well as the selectivity by decreasing its reactivity and plays a significant role in preventing side reaction by scavenging the reactive alkyl iodides, generated from the cleavage of ester compounds, to give the corresponding phosphonium iodide salts that are inactive under the reaction conditions.[76] p-Methoxybenzyl and diphenylmethyl esters can be easily converted into the corresponding carboxylic acids using TMSI/TPP in dichloromethane at room temperature in good yields (eq 45).[76]

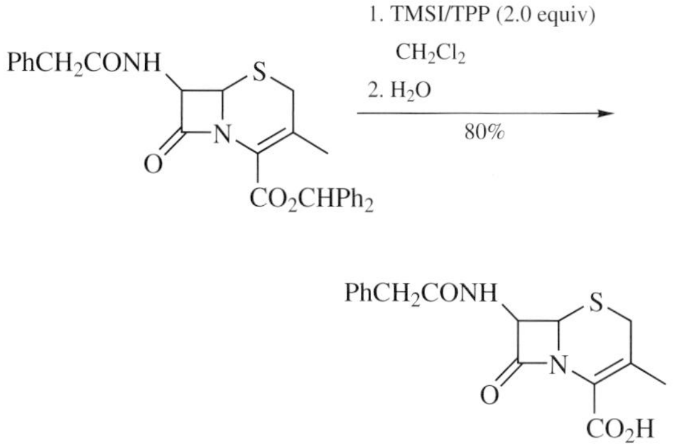

(45)

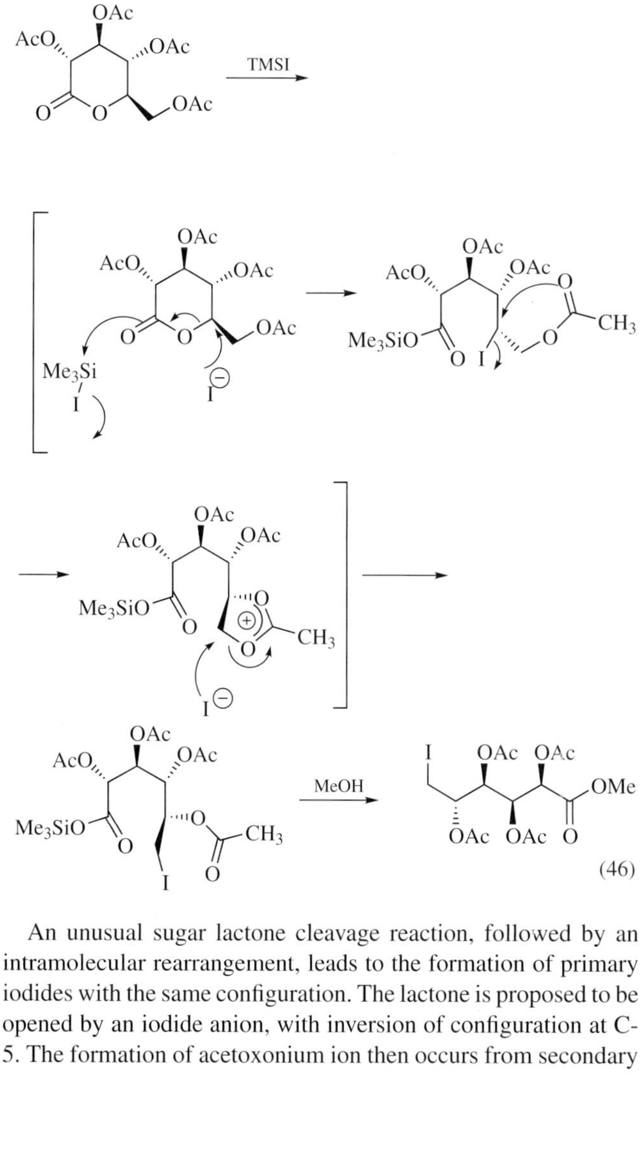

An unusual sugar lactone cleavage reaction, followed by an intramolecular rearrangement, leads to the formation of primary iodides with the same configuration. The lactone is proposed to be opened by an iodide anion, with inversion of configuration at C-5. The formation of acetoxonium ion then occurs from secondary iodide, with a second inversion at C-5. The acetoxonium ion is then opened regioselectively by an iodide ion, leading to the primary iodide (eq 46).[77]

Cleavage of Carbamates. TMSI has been used as a deblocking agent for the benzyloxycarbonyl group (Cbz) in the synthesis of the nicotinic receptor tracer 5-IA-85380 precursor (eq 47).[78] The TMSI-mediated selective carbamate cleavage can be achieved to afford amino-tin compounds without removing the stannyl moiety, which becomes the key step in the multiple-step synthesis.[78] TMSI has also been applied to the selective cleavage in chiral N-substituted 4-phenyl-2-oxazolidinones, and this method allows a more versatile use of 4-phenyl-2-oxazolidinone as a chiral auxiliary and N-protective group in the synthesis of carbacephems.[79]

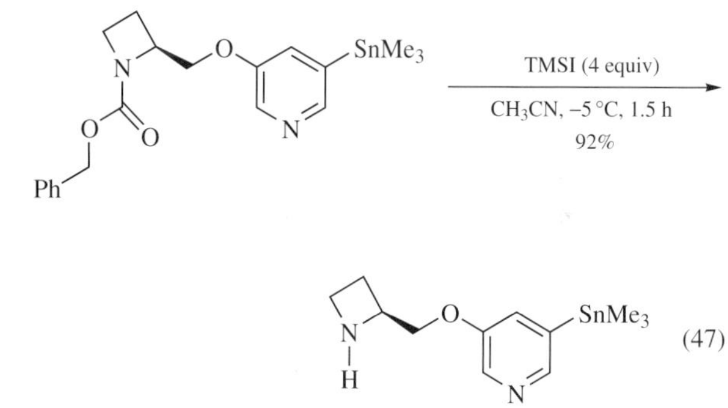

(47)

Cleavage of Acetals. The action of TMSI on 7-phenyl-6-alkynal dimethylacetals gives the oxonium ion intermediates, which undergo an intramolecular electrophilic reaction with the carbon-carbon triple bond to afford 2-(1-iodobenzylidene)cyclohexyl methyl ether (eq 48).[80] TMSI has also been used in a one-pot conversion of tetraacetal tetraoxa-cages to aza-cages in alkyl nitriles at room temperature via the ring expansion intermediates, which was interpreted to involve a Ritter-type of reaction mechanism (eq 49).[81] Interestingly, the reaction of tetraacetal tetraoxa-cages with TMSCl and NaI in nitriles at room temperature gives the amido-cages.[81]

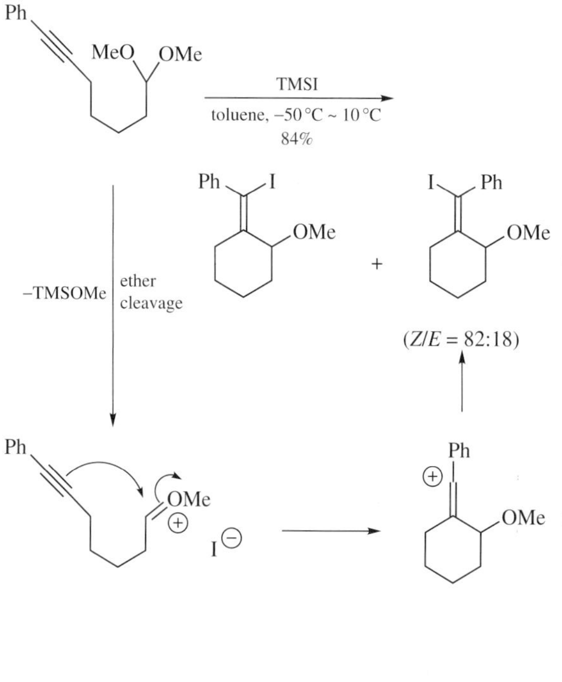

(48)

$$(49)$$

Cleavage of Phosphorous(III) Esters Leading to a Michaelis–Arbuzov Rearrangement. A new and catalytic version of the Michaelis-Arbuzov rearrangement has been reported by directly forming an alkyl halide through the action of trimethylsilyl halide (TMSX, X = I, Br) on phosphorous(III) esters (eq 50).[82] This rearrangement occurs at temperatures from 20 to 80 °C, and only a catalytic amount (5 mol%) of TMSI (or TMSBr) is needed. Unlike the usual Arbuzov rearrangement, alkyl halides are not required for this type of direct and easy-to-handle TMSX-catalyzed Michaelis-Arbuzov-like rearrangement.[82]

$$(50)$$

Use of TMSI as a Lewis-acidic Activation Agent.

For Biginelli Reaction. TMSI (in situ generated from TMSCl and NaI) is an excellent promoter for the one-pot synthesis of dihydropyrimidinones via the Biginelli reaction (eq 51), which involves the condensation of an aldehyde, a β-ketoester and urea (or thiourea).[83] The traditional Biginelli reaction commonly proceeds under strongly acidic conditions, and this protocol often

suffers from low yields particularly in case of substituted aromatic or aliphatic aldehydes. However, when TMSI is applied as a promoter, the reaction usually affords excellent yields of dihydropyrimidinones even at ambient temperature (eq 51).[83]

$$(51)$$

For the Synthesis of Hantzsch 1,4-Dihydropyridines. TMSI (in situ generated) has been used in the efficient synthesis of various substituted Hantzsch 1,4-dihydropyridines using both the classical and modified Hantzsch procedures at room temperature in acetonitrile (eq 52).[84] The usage of TMSI enables the reaction to proceed smoothly with good to excellent yields of products.[84]

$$(52)$$

For Aldol and Related Reactions. The TMSI/(TMS)$_2$NH combination can be used for the synthesis of polycyclic cyclobutane derivatives by tandem intramolecular Michael-aldol reaction.[85] TMSI-induced diastereoselective synthesis of tetrahydropyranones by a tandem Knoevenagel-Michael reaction, has also been developed.[86] More recently, the facile synthesis of α,α'-bis(substituted benzylidene)cycloalkanones has been reported, using TMSI (in situ generated) mediated cross-aldol condensations (eq 53).[87]

$$(53)$$

For the Synthesis of N-Substituted Phthalimides/Naphthalimides. N-Substituted phthalimides and naphthalimides can be synthesized in good to excellent yields, employing TMSI (in situ generated) from corresponding azides and anhydrides under mild conditions (eq 54).[88]

$$(54)$$

Use of TMSI as a Reducing Agent.

Selective Reduction of α,α-Diaryl Alcohols. TMSI has been utilized as a reducing agent for the rapid and highly selective reduction of α,α-diaryl alcohols to the corresponding alkanes.[89] The reaction proceeds particularly well for electron-rich substrates, which may be associated with the proposed intermediacy of an aryl-stabilized benzylic carbocation at the reduction site.[90] The moderately electron-deficient benzylic alcohols can also be selectively reduced to analogous toluenes, and the reaction condition tolerates other reduction-sensitive functional groups such as ketone, aldehyde, nitrile, and nitro groups (eq 55).[90] The preparation of biarylmethanes, involving benzylation via tandem Grignard reaction-TMSI-mediated reduction, has also been reported.[91]

$$(55)$$

Reduction of Azides to Amines. In situ generated TMSI has been found to be a useful reducing agent for the reduction of azides to amines (eq 56).[92] The reaction is carried out under extremely mild and neutral conditions, and a number of aryl, alkyl, and aroyl azides are suitable for this tranformation. This methodology has also been applied to the synthesis of pyrrolo[2,1-c][1,4]benzodiazepines via reductive cyclization of ω-azido carbonyl compounds.[93]

$$R{-}N_3 \xrightarrow[\substack{CH_3CN,\ rt \\ 90{\sim}98\%}]{TMSCl/NaI} R{-}NH_2 \quad (56)$$

(R= alkyl, aryl, and aroyl)

Reductive Cleavage of Phthalides and Sulfonamides. 3-Arylphthalides can be readily cleaved reductively by means of TMSI (in situ generated) to give corresponding 2-benzylbenzoic acids (eq 57) or 2-(2-thienylmethyl)benzoic acids.[94]

$$(57)$$

TMSI, in situ generated from TMSCl and NaI, is also a robust reagent for the deprotection of sulfonamides (eq 58).[95] The reductive desulfonylation usually proceeds in good yields with 1.5 equiv of TMSI in acetonitrile under reflux for 3~4 hours. The mild reaction conditions employed in this deprotection method allow the selective deprotection of sulfonamides in the presence of N-alkyl and N-benzyl groups.[95]

$$(58)$$

Reductive Cleavage of Heteroaryl C-Halogen Bonds. Regioselective reductive dehalogenation of heterocyclic antibiotic compounds, such as pyrrolnitrins, halo-uridines and pyrimidines, has been successfully accomplished (eq 59).[96a] A single-electron transfer (SET) mechanism was proposed for this type of dehalogenation (eq 59).[96a] TMSI-mediated reductive C-2 dechlorination of some 5-allyl(allenyl)-2,5-dichloro-3-dialkylamino-4,4-dimethoxy-2-cyclopentenones has also been successful.[96b]

$$(59)$$

Use of TMSI in Conjunction with Organometallic Reagents.

In Conjugate Monoorganocopper Addition to α,β-Unsaturated Carbonyl Compounds. TMSI has been demonstrated to efficiently promote the reaction of conjugate 1,4-additions of monoorganocopper compounds into a variety of α,β-unsaturated carbonyl compounds, such as cyclic and acyclic enones, β-alkoxy enones, enoates, and lactones, often at $-78\,^{\circ}C$ (eq 60).[97] The RCu(LiI)-TMSI reagent gives a good economy of group transfer with good to excellent yields of conjugate adducts. Lithium iodide, present from preparation of the organocopper compounds, increases the rate of the reaction and is a favorable component.[97] The mechanistic study of the role of TMSI in the conjugate addition of butylcopper-TMSI to α-enones shows that a direct silylation of an intermediate π-complex by TMSI is most likely (eq 60).[98] The conjugate additions of MeCu, PhCu, and n-BuCu to the chiral enoylimides in the presence of TMSI and LiI in THF give the adducts in excellent yields and high diastereoselectivity ($80 \sim 93\%\ de$).[99] In the presence of TMSI and LiI in THF, the otherwise unreactive copper acetylides can add to enones present as s-$trans$ conformers to provide good yields of the silyl enol ethers of β-acetylido carbonyl compounds.[100] Similaly, TMSI-promoted diastereoselective conjugate additions of monoorganocuprates Li[RCuI] to different α,β-unsaturated N-acyl oxazolidinones with high yields and diastereomeric ratios have also been reported.[101]

(60)

Pd-catalyzed Coupling Reactions. The reaction of terminal acetylenes with TMSI and organozinc reagents (or organostannanes) in the presence of $Pd(PPh_3)_4$ results in addition of the trimethylsilyl group of TMSI and an alkyl group of the organozinc reagent (or alkynyl group of organostannane) to the acetylenes to give vinylsilanes (eq 61).[102] This catalytic reaction involves an oxidative addition of the Si-I bond in TMSI to Pd(0) leading to a silylpalladium(II) species, and silylpalladation of an acetylene with the Si-Pd species followed by coupling with organozinc reagents (or organostannanes).[102]

(61)

TMSI as Iodination Agent in Organometallic Complexes. TMSI has been applied as an iodinating agent in organometallic complexes.[103–105] For example, reaction of the P,N-chelated dimethylplatinum complexes with TMSI stereoselectively gives the corresponding methyl iodo complexes in which only the methyl group trans to the phosphorus atom is exchanged (eq 62).[103] TMSI was also used as a halogen-exchange reagent in the thorium(IV) complex to form Th-I bond.[104] The application of TMSI as a Cl-I exchange reagent in transition metal coordination chemistry, has also been studied.[105]

(62)

Other Reactions of TMSI.

Reaction with α,β-Unsaturated Sulfoxides. The reaction of TMSI with α,β-unsaturated sulfoxides in chloroform at ambient temperature is a mild, efficient, and general method for the preparation of carbonyl compounds (eq 63).[106] The proposed reaction mechanism is shown in eq 63.[106a] Formation of a strong oxygen-silicon bond is followed by reduction of the sulfur function and oxidation of iodide to iodine, the latter precipitating in chloroform. The trimethylsiloxy anion attacks the unsaturated carbon linked to the sulfur function, which leaves the substrate, allowing the formation of the silyl enol ether species. Finally, hydrolysis converts the silyl enol ether into the carbonyl compound.[106a]

(63)

Iodination of β-Hydroxy Amino Acid Derivatives. TMSI has been used as an iodinating agent to convert β-hydroxy amino acid derivatives into corresponding β-iodo amino acid derivatives (eq 64).[107] The low yields for the reaction have been attributed in part to the sensitivity of the β-iodo products to the reflux conditions. Higher yields can be obtained when TMSCl and TMSBr are used as chlorinating and brominating agents.[107]

$$(64)$$

1. (a) Olah, G. A.; Prakash, G. K.; Krishnamurti, R., *Adv. Silicon Chem.* **1991**, *1*, 1. (b) Lee, S. D.; Chung, I. N., *Hwahak Kwa Kongop Ui Chinbo* **1984**, *24*, 735. (c) Olah, G. A.; Narang, S. C., *Tetrahedron* **1982**, *38*, 2225. (d) Hosomi, A., *Yuki Gosei Kagaku Kyokai Shi* **1982**, *40*, 545. (e) Ohnishi, S.; Yamamoto, Y., *Annu. Rep. Tohoku Coll. Pharm.* **1981**, *28*, 1. (f) Schmidt, A. H., *Aldrichim. Acta* **1981**, *14*, 31. (g) Groutas, W. C.; Felker, D., *Synthesis* **1980**, *11*, 86. (h) Schmidt, A. H., *Chem.-Ztg.* **1980**, *104* (9), 253.

2. (a) Lissel, M.; Drechsler, K., *Synthesis* **1983**, 459. (b) Machida, Y.; Nomoto, S.; Saito, I., *Synth. Commun.* **1979**, *9*, 97.

3. (a) Schmidt, A. H.; Russ, M., *Chem.-Ztg.* **1978**, *102*, 26, 65. (b) Olah, G. A.; Narang, S. C.; Gupta, B. G. B., *Synthesis* **1979**, 61. (c) Morita, T.; Okamoto, Y.; Sakurai, H., *Tetrahedron Lett.* **1978**, 2523; *J. Chem. Soc., Chem. Commun.* **1978**, 874.

4. (a) Kumada, M.; Shiiman, K.; Yamaguchi, M., *Kogyo Kagaku Zasshi* **1954**, *57*, 230. (b) Sakurai, H.; Shirahata, A.; Sasaki, K.; Hosomi, A., *Synthesis* **1979**, 740.

5. Ho, T. L.; Olah, G. A., *Synthesis* **1977**, 417.

6. (a) Jung, M. E.; Lyster, M. A., *Org. Synth., Coll. Vol.* **1988**, *6*, 353. (b) Jung, M. E.; Blumenkopf, T. A., *Tetrahedron Lett.* **1978**, 3657.

7. (a) Jung, M. E.; Lyster, M. A., *J. Org. Chem.* **1977**, *42*, 3761. (b) Voronkov, M. G.; Dubinskaya, E. I.; Pavlov, S. F.; Gorokhova, V. G., *Izv. Akad. Nauk SSSR, Ser. Khim.* **1976**, 2355.

8. (a) Olah, G. A.; Narang, S. C.; Gupta, B. G. B.; Malhotra, R., *J. Org. Chem.* **1979**, *44*, 1247. (b) Voronkov, M. G.; Dubinskaya, E. J., *J. Organomet. Chem.* **1991**, *410*, 13. (c) Voronkov, M. G.; Puzanova, V. E.; Pavlov, S. F.; Dubinskaya, E. J., *Bull. Acad. Sci. USSR, Div. Chem. Sci.* **1975**, *14*, 377. (d) Voronkov, M. G.; Dubinskaya, E. J.; Pavlov, S. F.; Gorokhova, V. G., *Bull. Acad. Sci. USSR, Div. Chem. Sci.* **1976**, *25*, 2198. (e) Voronkov, M. G.; Komarov, V. G.; Albanov, A. I.; Dubinskaya, E. J., *Bull. Acad. Sci. USSR, Div. Chem. Sci.* **1978**, *27*, 2347. (f) Hirst, G. C.; Johnson, T. O., Jr.; Overman, L. E., *J. Am. Chem. Soc.* **1993**, *115*, 2992.

9. (a) Jung, M. E.; Ornstein, P. L., *Tetrahedron Lett.* **1977**, 2659. (b) Voronkov, M. G.; Pavlov, S. F.; Dubinskaya, E. J., *Dokl. Akad. Nauk SSSR* **1976**, *227*, 607 (Eng. p. 218); *Bull. Acad. Sci. USSR, Div. Chem. Sci.* **1975**, *24*, 579.

10. (a) Casnati, A.; Arduini, A.; Ghidini, E.; Pochini, A.; Ungaro, R., *Tetrahedron* **1991**, *47*, 2221. (b) Silverman, R. B.; Radak, R. E.; Hacker, N. P., *J. Org. Chem.* **1979**, *44*, 4970. (c) Vickery, E. H.; Pahler, L. F.; Eisenbraun, E. J., *J. Org. Chem.* **1979**, *44*, 4444. (d) Brasme, B.; Fischer, J. C.; Wartel, M., *Can. J. Chem.* **1977**, *57*, 1720. (e) Rosen, B. J.; Weber, W. P., *J. Org. Chem.* **1977**, *42*, 3463.

11. Minamikawa, J.; Brossi, A., *Tetrahedron Lett.* **1978**, 3085.

12. Hartman, D. A.; Curley, R. W., Jr., *Tetrahedron Lett.* **1989**, *30*, 645.

13. (a) Denis, J. N.; Magnane, R. M.; van Eenoo, M.; Krief, A., *Nouv. J. Chim.* **1979**, *3*, 705. (b) Detty, M. R.; Seidler, M. D., *Tetrahedron Lett.* **1982**, *23*, 2543. (c) Sakurai, H.; Sasaki, K.; Hosomi, A., *Tetrahedron Lett.* **1980**, *21*, 2329. (d) Kraus, G. A.; Frazier, K., *J. Org. Chem.* **1980**, *45*, 2579. (e) Hill, R. K.; Pendalwar, S. L.; Kielbasinski, K.; Baevsky, M. F.; Nugara, P. N., *Synth. Commun.* **1990**, *20*, 1877.

14. (a) Ho, T. L.; Olah, G. A., *Angew. Chem., Int. Ed. Engl.* **1976**, *15*, 774. (b) Jung, M. E.; Lyster, M. A., *J. Am. Chem. Soc.* **1977**, *99*, 968. (c) Schmidt, A. H.; Russ, M., *Chem.-Ztg.* **1979**, *103*, 183, 285. (d) See also refs. 8a, 9a.

15. Olah, G. A.; Narang, S. C.; Salem, G. F.; Gupta, B. G. B., *Synthesis* **1981**, 142.

16. Acyl iodides are also available from acid chlorides and TMS-I: Schmidt, A. N.; Russ, M.; Grosse, D., *Synthesis* **1981**, 216.

17. (a) Ho, T. L., *Synth. Commun.* **1979**, *9*, 233. (b) Sekiguchi, A.; Kabe, Y.; Ando, W., *Tetrahedron Lett.* **1979**, 871.

18. Hiyama, T.; Saimoto, H.; Nishio, K.; Shinoda, M.; Yamamoto, H.; Nozaki, H., *Tetrahedron Lett.* **1979**, 2043.

19. Kricheldorf, H. R., *Angew. Chem., Int. Ed. Engl.* **1979**, *18*, 689.

20. Yamamoto, Y.; Ohnishi, S.; Azuma, Y., *Chem. Pharm. Bull.* **1982**, *30*, 3505.

21. (a) Jung, M. E.; Lyster, M. A., *J. Chem. Soc., Chem. Commun.* **1978**, 315. (b) Rawal, V. H.; Michoud, C.; Monestel, R. F., *J. Am. Chem. Soc.* **1993**, *115*, 3030. (c) Wender, P. A.; Schaus, J. M.; White, A. W., *J. Am. Chem. Soc.* **1980**, *102*, 6157. (d) Lott, R. S.; Chauhan, V. S.; Stammer, C. H., *J. Chem. Soc., Chem. Commun.* **1979**, 495. (e) Vogel, E.; Altenbach, H. J.; Drossard, J. M.; Schmickler, H.; Stegelmeier, H., *Angew. Chem., Int. Ed. Engl.* **1980**, *19*, 1016.

22. Olah, G. A.; Narang, S. C.; Gupta, B. G. B.; Malhotra, R., *Angew. Chem., Int. Ed. Engl.* **1979**, *18*, 612.

23. Blaskovich, M. A.; Lajoie, G. A., *J. Am. Chem. Soc.* **1993**, *115*, 5021.

24. (a) Jung, M. E.; Andrus, W. A.; Ornstein, P. L., *Tetrahedron Lett.* **1977**, 4175. (b) Bryant, J. D.; Keyser, G. E.; Barrio, J. R., *J. Org. Chem.* **1979**, *44*, 3733. (c) Muchmore, D. C.; Dahlquist, F. W., *Biochem. Biophys. Res. Commun.* **1979**, *86*, 599.

25. Chan, T. H.; Lee, S. D., *Tetrahedron Lett.* **1983**, *24*, 1225.

26. Kobylinskaya, V. I.; Dashevskaya, T. A.; Shalamai, A. S.; Levitskaya, Z. V., *Zh. Obshch. Khim.* **1992**, *62*, 1115.

27. (a) Jung, M. E.; Mazurek, M. A.; Lim, R. M., *Synthesis* **1978**, 588. (b) Olah, G. A.; Husain, A.; Narang, S. C., *Synthesis* **1983**, 896.

28. (a) Thiem, J.; Meyer, B., *Chem. Ber.* **1980**, *113*, 3075. (b) Tocik, Z.; Earl, R. A.; Beranek, J., *Nucl. Acids Res.* **1980**, *8*, 4755.

29. Bryson, T. A.; Bonitz, G. H.; Reichel, C. J.; Dardis, R. E., *J. Org. Chem.* **1980**, *45*, 524.

30. (a) Zygmunt, J.; Kafarski, P.; Mastalerz, P., *Synthesis* **1978**, 609. (b) Blackburn, G. M.; Ingleson, D., *J. Chem. Soc., Chem. Commun.* **1978**, 870. (c) Blackburn, G. M.; Ingleson, D., *J. Chem. Soc., Perkin Trans. 1* **1980**, 1150.

31. Lee, K.; Wiemer, D. F., *Tetrahedron Lett.* **1993**, *34*, 2433.

32. (a) Miller, R. D.; McKean, D. R., *J. Org. Chem.* **1981**, *46*, 2412. (b) Giacomini, E.; Loreto, M. A.; Pellacani, L.; Tardella, P. A., *J. Org. Chem.* **1980**, *45*, 519. (c) Dieter, R. K.; Pounds, S., *J. Org. Chem.* **1982**, *47*, 3174.

33. (a) Miller, R. D.; McKean, D. R., *Tetrahedron Lett.* **1980**, *21*, 2639. (b) Crimmins, M. T.; Mascarella, S. W., *J. Am. Chem. Soc.* **1986**, *108*, 3435.

34. (a) Friedrich, E. C.; Abma, C. B.; Vartanian, P. F., *J. Organomet. Chem.* **1980**, *187*, 203. (b) Friedrich, E. C.; DeLucca, G., *J. Organomet. Chem.* **1982**, *226*, 143.

35. Jung, M. E.; Mossman, A. B.; Lyster, M. A., *J. Org. Chem.* **1978**, *43*, 3698.

36. (a) Jung, M. E.; Lewis, P. K., *Synth. Commun.* **1983**, *13*, 213. (b) Lipshutz, B. H.; Ellsworth, E. L.; Siahaan, T. J.; Shirazi, A., *Tetrahedron Lett.* **1988**, *29*, 6677.

37. Jung, M. E.; Miller, S. J., *J. Am. Chem. Soc.* **1981**, *103*, 1984.

38. Schmidt, A. H.; Russ, M., *Chem.-Ztg.* **1979**, *103*, 183, 285.

39. (a) Miller, R. D.; McKean, D. R., *Tetrahedron Lett.* **1979**, 2305. (b) Larson, G. L.; Klesse, R., *J. Org. Chem.* **1985**, *50*, 3627.

40. Taniguchi, M.; Kobayashi, S.; Nakagawa, M.; Hino, T.; Kishi, Y., *Tetrahedron Lett.* **1986**, *27*, 4763.

41. (a) Corey, E. J.; Boaz, N. W., *Tetrahedron Lett.* **1985**, *26*, 6015, 6019. (b) Bergdahl, M.; Nilsson, M.; Olsson, T.; Stern, K., *Tetrahedron* **1991**, *47*, 9691, and references cited therein.

42. (a) Miller, R. D.; McKean, D. R., *Synthesis* **1979**, 730. (b) Hergott, H. H.; Simchen, G., *Liebigs Ann. Chem.* **1980**, 1718. (c) Babot, O.; Cazeau, P.; Duboudin, F., *J. Organomet. Chem.* **1987**, *326*, C57.

43. (a) Kramarova, E. P.; Shipov, A. G.; Artamkina, O. B.; Barukov, Y. I., *Zh. Obshch. Khim.* **1984**, *54*, 1921. (b) King, A. O.; Anderson, R. K.; Shuman, R. F.; Karady, S.; Abramson, N. L.; Douglas, A. W., *J. Org. Chem.* **1993**, *58*, 3384.

44. (a) Lau, P. W. K.; Chan, T. H., *J. Organomet. Chem.* **1979**, *179*, C24. (b) Wilson, S. R.; Phillips, L. R.; Natalie, K. J., Jr., *J. Am. Chem. Soc.* **1979**, *101*, 3340.

45. (a) Yamashita, H.; Kobayashi, T.; Hayashi, T.; Tanaka, M., *Chem. Lett.* **1991**, 761. (b) Chatani, N.; Amishiro, N.; Murai, S., *J. Am. Chem. Soc.* **1991**, *113*, 7778.

46. Sakurai, H.; Sasaki, K.; Hayashi, J.; Hosomi, A., *J. Org. Chem.* **1984**, *49*, 2808.

47. Hosomi, A.; Sakata, Y.; Sakurai, H., *Chem. Lett.* **1983**, 405.

48. Sakurai, H.; Sasaki, K.; Hosomi, A., *Bull. Chem. Soc. Jpn.* **1983**, *56*, 3195.

49. Mukaiyama, T.; Akamatsu, H.; Han, J. S., *Chem. Lett.* **1990**, 889.

50. Sakurai, H.; Sasaki, K.; Hosomi, A., *Tetrahedron Lett.* **1981**, *22*, 745.

51. Vankar, Y. D.; Kumaravel, G.; Mukherjee, N.; Rao, C. T., *Synth. Commun.* **1987**, *17*, 181.

52. Sarma, J. C.; Barua, N. C.; Sharma, R. P.; Barua, J. N., *Tetrahedron* **1983**, *39*, 2843.

53. Jung, K.-Y.; Koreeda, M., *J. Org. Chem.* **1989**, *54*, 5667.

54. Ghera, E.; Maurya, R.; Hassner, A., *Tetrahedron Lett.* **1989**, *30*, 4741.

55. Sassaman, M. B.; Prakash, G. K.; Olah, G. A., *Tetrahedron* **1988**, *44*, 3771.

56. (a) Ho, T.-L., *Synth. Commun.* **1979**, *9*, 665. (b) Sarma, D. N.; Sarma, J. C.; Barua, N. C.; Sharma, R. P., *J. Chem. Soc., Chem. Commun.* **1984**, 813. (c) Nagaoka, M.; Kunitama, Y.; Numazawa, M., *J. Org. Chem.* **1991**, *56*, 334. (d) Numazawa, M.; Nagaoka, M.; Kunitama, Y., *Chem. Pharm. Bull.* **1986**, *34*, 3722; *J. Chem. Soc., Chem. Commun.* **1984**, 31. (e) Hartman, D. A.; Curley, R. W., Jr., *Tetrahedron Lett.* **1989**, *30*, 645. (f) Cherbas, P.; Trainor, D. A.; Stonard, R. J.; Nakanishi, K., *J. Chem. Soc., Chem. Commun.* **1982**, 1307.

57. Olah, G. A.; Arvanaghi, M.; Vankar, Y. D., *J. Org. Chem.* **1980**, *45*, 3531.

58. (a) Olah, G. A.; Gupta, B. G. B.; Narang, S. C., *Synthesis* **1977**, 583. (b) Pitlik, J.; Sztaricskai, F., *Synth. Commun.* **1991**, *21*, 1769.

59. Nicolaou, K. C.; Barnette, W. E.; Magolda, R. L., *J. Am. Chem. Soc.* **1978**, *100*, 2567.

60. Miller, R. D.; McKean, D. R., *Tetrahedron Lett.* **1983**, *24*, 2619.

61. Schaumann, E.; Winter-Extra, S.; Kummert, K.; Scheiblich, S., *Synthesis* **1990**, 271.

62. Olah, G. A.; Narang, S. C.; Field, L. D.; Salem, G. F., *J. Org. Chem.* **1980**, *45*, 4792.

63. Shipov, A. G.; Baukov, Y. I., *Zh. Obshch. Khim.* **1984**, *54*, 1842.

64. (a) Beattie, I. R.; Parrett, F. W., *J. Chem. Soc. (A)* **1966**, 1784. (b) Livanstov, M. V.; Proskurnina, M. V.; Prischenko, A. A.; Lutsenko, I. F., *Zh. Obshch. Khim.* **1984**, *54*, 2504.

65. Kabachnik, M. M.; Prischenko, A. A.; Novikova, Z. S.; Lutsenko, I. F., *Zh. Obshch. Khim.* **1979**, *49*, 1446.

66. Kibardin, A. M.; Gryaznova, T. V.; Gryaznov, P. I.; Pudovik, A. N., *J. Gen. Chem. USSR (Engl. Transl.)* **1991**, *61*, 1969.

67. (a) Jung, M. E.; Long-Mei, Z., *Tetrahedron Lett.* **1983**, *24*, 4533. (b) Godleski, S. A.; Heacock, D. J., *J. Org. Chem.* **1982**, *47*, 4820.

68. Olah, G. A.; Narang, S. C.; Field, L. D.; Fung, A. P., *J. Org. Chem.* **1983**, *48*, 2766.

69. Singhal, G. M.; Das, N. B.; Sharma, R. P., *J. Chem. Soc., Chem. Commun.* **1989**, 1470.

70. Haber, A., *Tetrahedron Lett.* **1989**, *30*, 5537.

71. Inouye, Y.; Shirai, M.; Michino, T.; Kakisawa, H., *Bull. Chem. Soc. Jpn.* **1993**, *66*, 324.

72. Suzuki, K.; Miyazawa, M.; Tsuchihashi, G., *Tetrahedron Lett.* **1987**, *28*, 3515.

73. Yamato, T.; Matsumoto, J.; Tashiro, M., *J. Chem. Research (S)* **1994**, 246.

74. Ejjiyar, S.; Saluzzo, C.; Amouroux, R.; Massoui, M., *Tetrahedron Lett.* **1997**, *38*, 1575.

75. Cere, V.; Pollicino, S.; Fava, A., *Tetrahedron* **1996**, *52*, 5989.

76. Cha, K. H.; Kang, T. W.; Cho, D. O.; Lee, H.-W.; Shin, J.; Jin, K. Y.; Kim, K.-W.; Kim, J.-W.; Hong, C.-I., *Synth. Commun.* **1999**, 3533.

77. Heck, M. P.; Monthiller, S.; Mioskowski, C.; Guidot, J. P.; Gall, T. L., *Tetrahedron Lett.* **1994**, *35*, 5445.

78. Brenner, E.; Baldwin, R. M.; Tamagnan, G., *Tetrahedron Lett.* **2004**, *45*, 3607.

79. Fisher, J. W.; Dunigan, J. M.; Hatfield, L. D.; Hoying, R. C.; Ray, J. E.; Thomas, K. L., *Tetrahedron Lett.* **1993**, *34*, 4755.

80. Takami, K.; Yorimitsu, H.; Shinokubo, H.; Matsubara, S.; Oshima, K., *Synlett* **2001**, 293.

81. (a) Wu, H.-J.; Chern, J.-H., *Tetrahedron Lett.* **1997**, *38*, 2887. (b) Chern, J.-H.; Wu, H.-J., *Tetrahedron* **1998**, *54*, 5967.

82. Renard, P.-Y.; Vayron, P.; Mioskowski, C., *Org. Lett.* **2003**, *5*, 1661.

83. Sabitha, G.; Reddy, G. S. K. K.; Reddy, C. S.; Yadav, J. S., *Synlett* **2003**, 858.

84. Sabitha, G.; Reddy, G. S. K. K.; Reddy, C. S.; Yadav, J. S., *Tetrahedron Lett.* **2003**, 4129.

85. Ihara, M.; Taniguchi, T.; Makita, K.; Takano, M.; Ohnishi, M.; Taniguchi, N.; Fukumoto, K.; Kabuto, C., *J. Am. Chem. Soc.* **1993**, *115*, 8107.

86. Sabitha, G.; Reddy, G. S. K. K.; Reddy, Rajkumar, M.; C. S.; Yadav, J. S.; Ramakrishna, K. V. S.; Kunwar, A. C., *Tetrahedron Lett.* **2003**, 7455.

87. Sabitha, G.; Reddy, G. S.; Reddy, K. K.; Reddy, C. S.; Yadav, J. S., *Synthesis* **2004**, 263.

88. Kamal, A.; Laxman, E.; Laxman, N.; Rao, N. V., *Tetrahedron Lett.* **1998**, *39*, 8733.

89. Perry, P. J.; Pavlidis, V. H.; Coutts, I. G. C., *Synth. Commun.* **1996**, *26*, 101.

90. Cain, G. A.; Holler, E. R., *Chem. Commun.* **2001**, 1168.

91. Stoner, E. J.; Cothron, D. A.; Balmer, M. K.; Roden, B. A., *Tetrahedron* **1995**, *51*, 11043.

92. Kamal, A.; Rao, N. V.; Laxman, E., *Tetrahedron Lett.* **1997**, *38*, 6945.

93. Kamal, A.; Laxman, E.; Laxman, N.; Rao, N. V., *Bioorg. Med. Chem. Lett.* **2000**, *10*, 2311.

94. Sabitha, G.; Yadav, J. S., *Synth. Commun.* **1998**, *28*, 3065.

95. Sabitha, G.; Reddy, B. V. S.; Abraham, S.; Yadav, J. S., *Tetrahedron Lett.* **1999**, *40*, 1569.

96. (a) Sako, M.; Kihara, T.; Okada, K.; Ohtani, Y.; Kawamoto, H., *J. Org. Chem.* **2001**, *66*, 3610. (b) Akbutina, F. A.; Torosyan, S. A.; Miftakhov, M. S., *Russ. J. Org. Chem.* **2000**, *36*, 1265.

97. (a) Bergdahl, M.; Eriksson, M.; Nilsson, M.; Olsson, T., *J. Org. Chem.* **1993**, *58*, 7238. (b) Eriksson, M.; Nilsson, M.; Olsson, T., *Synlett* **1994**, 271.

98. Eriksson, M.; Johansson, A.; Nilsson, M.; Olsson, T., *J. Am. Chem. Soc.* **1996**, *118*, 10904.

99. Bergdahl, M.; Iliefski, T.; Nilsson, M.; Olsson, T., *Tetrahedron Lett.* **1995**, *36*, 3227.

100. Eriksson, M.; Iliefski, T.; Nilsson, M.; Olsson, T., *J. Org. Chem.* **1997**, *62*, 182.

101. Pollock, P.; Dambacher, J.; Anness, R.; Bergdahl, M., *Tetrahedron Lett.* **2002**, *43*, 3693.

102. (a) Chatani, N.; Amishiro, N.; Morii, T.; Yamashita, T.; Murai, S., *J. Org. Chem.* **1995**, *60*, 1834. (b) Chatani, N.; Amishiro, N.; Murai, S., *J. Am. Chem. Soc.* **1991**, *113*, 7778.

103. Pfeiffer, J.; Kickelbick, G.; Schubert, U., *Organometallics* **2000**, *19*, 62.

104. Rabinovich, D.; Bott, S. G.; Nielsen, J. B.; Abney, K. D., *Inorg. Chim. Acta* **1998**, *274*, 232.

105. Leigh, G. J.; Sanders, J. R.; Hitchcock, P. B.; Fernandes, J. S.; Togrou, M., *Inorg. Chim. Acta* **2002**, *330*, 197.

106. (a) Aversa, M.; Barattucci, A.; Bonaccorsi, P.; Giannetto, P., *Tetrahedron* **2002**, *58*, 10145. (b) Aversa, M.; Barattucci, A.; Bonaccorsi, P.; Bruno, G.; Giannetto, P.; Policicchio, M., *Tetrahedron Lett.* **2000**, *41*, 4441.

107. Choi, D.; Kohn, H., *Tetrahedron Lett.* **1995**, *39*, 7011.

2-Iodoxybenzoic Acid[1]

[61717-82-6] $C_7H_5IO_4$ (MW 280.01)

InChI = 1/C7H5IO4/c9-7-5-3-1-2-4-6(5)8(10,11)12-7/h1-4,10H

InChIKey = MEOMMSSYMAPFRJ-UHFFFAOYAV

(oxidizing agent, single-electron-transfer reagent)[1]

Alternate Name: o-iodoxybenzoic acid; 1-hydroxy-1,2-benziodo-xol-3(1H)-one-1-oxide, IBX.

Physical Data: mp 232–233 °C.

Solubility: soluble in DMSO, DMSO/THF mixtures; insoluble in H_2O and most organic solvents.

Analysis of Reagent Purity: ^{1}H-NMR (DMSO-d6) δ 8.15 (d, 1H), 8.01 (d, 1H), 7.98 (d, 1H), 7.84 (t, 1H). ^{13}C-NMR (DMSO-d6) δ167.49, 146.59, 133.39, 132.97, 131.36, 130.10, 124.99.[2]

Preparative Methods: not commercially available. IBX was first synthesized in 1893,[2] and is the penultimate precursor of the Dess-Martin periodinane reagent.[3,4] IBX is prepared by the slow addition (0.5 h) of potassium bromate (76.0 g, 0.45 mol) to a rapidly stirred sulfuric acid mixture (0.73 M, 730 mL) containing 2-iodobenzoic acid (85.2 g, 0.34 mol). The reaction temperature is maintained below 55 °C until addition is complete. The reaction mixture is heated to 65 °C for 3.6 h. The flask is cooled to 0 °C, and the solid is filtered and washed with water (1000 mL) and ethanol (2 x 50 mL) to afford IBX in 93–98% yield.[3,4] An additional ether wash (3 x 50 mL) is beneficial.[5] Other oxidants such as potassium permanganate,[2] chlorine,[6] or oxone[7] can also be employed.

Handling, Storage, and Precaution: moisture stable, explosive under excessive heating (> 200 °C) or impact.[8]

Oxidation of Alcohols to Aldehydes and Ketones. Hypervalent iodine compounds have powerful oxidizing capabilities. However, IBX possesses different properties than many similar oxidants, such as the related analogs iodoxybenzene and m-iodoxybenzoic acid.[6] Until recently, the major application of IBX was its use in DMSO for the oxidation of primary alcohols to aldehydes at room temperature, without the danger of over-oxidation to carboxylic acids. The related iodo-oxy reagents oxidize benzyl alcohols to benzaldehydes at elevated temperatures in benzene (80 °C, 5–10 h) or in acetic acid (rt, 24 h), while IBX oxidizes the same compounds in 15 min (or less) at room temperature.[9] IBX is equally effective for the oxidation of secondary alcohols to ketones under analogous conditions. Even sterically hindered alcohols are readily oxidized. Borneol undergoes smooth oxida-

(1)

tion to camphor upon treatment with a 1:1 molar ratio of IBX for 2.5 h in DMSO at room temperature (eq 1).[9]

Despite the fact that IBX is insoluble in almost all solvents except DMSO, it oxidizes alcohols in other solvent combinations. THF is the preferred cosolvent to facilitate the solubility of alcohols that are insoluble in DMSO, and allow oxidation with IBX. In some cases, the oxidation occurs entirely in the absence of DMSO. Benzoin is oxidized to benzil by treatment with IBX in THF.[10]

The mild oxidizing properties of IBX enhance its utility for the oxidation of compounds with sensitive functional groups such as thioethers and amines. Oxidation with IBX occurs cleanly in the presence of thioethers and 1,3-dithiolane protecting groups. For example, oxidation of thiochroman-4-ol, even with 10 equiv of IBX is achieved cleanly to give thiochroman-4-one with no detection of the analogous sulfoxide or sulfone by-products (eq 2).[10]

(2)

IBX is also tolerant of amine functionality, and therefore is used for the successful oxidation of amino alcohols to amino carbonyls. Frequently, this synthetic transformation requires the protection of the amino group, as a nonbasic derivative, prior to oxidation. IBX also provides an alternative method to oxidize alcohols selectively in the presence of primary, secondary, or tertiary amines, although the in situ protonation of the amine with acid is usually required (e.g., trifluoroacetic acid, 1–1.5 equiv) to avoid reduced yields.[10] Oxidation of an aminocyclohexanol occurs selectively upon treatment with both IBX and TFA (1:1 ratio), without degradation of the amine functionality, to give cyclohexanone in 91% yield (eq 3).[10]

(3)

IBX displays good tolerance for other functionalities including both isolated and conjugated double bonds,[9] carboxylic acids,[9] carboxylic esters and carboxamides,[10] as well as oxidizable heteroaromatic rings such as furans, pyridines, and indoles.[9] Phenols and anilines, however, are not tolerated and in the presence of IBX often generate side products.[10] (Aniline amides are an exception, see below.) Upon exposure to IBX, phenols (substituted with an electron-donating group) afford the corresponding o-quinone derivative. These compounds are often quite reactive, but subsequent treatment under the reducing conditions in the presence of acetic anhydride provides a useful route to catechols.[11] For example, when a tert-butyl substituted phenol is stirred in a suspension of IBX in chloroform for 53 h at ambient temperature, the corresponding o-quinone is obtained in 99% yield. Treatment of the crude sample of this quinone with potassium carbonate, acetic anhydride, and palladium on carbon, under a hydrogen atmosphere for 24 h, gives the diacetate in 87% isolated yield (eq 4).[12]

vacuum until a thick slurry remains. This slurry is filtered under a inert atmosphere and washed with ether to give DMP in 94% isolated yield (eq 6).[3,4]

Oxidation of Polyols. IBX, in common with other hypervalent iodine compounds including the Dess-Martin periodinane (DMP) and other iodoxy benzene derivatives, easily oxidizes 1,2-glycols. Unlike these reagents, however, IBX is selective and does not result in cleavage of the glycol carbon-carbon bond. The oxidation is terminated after the carbonyl is generated to afford the corresponding α-ketol or α-diketone derivatives. In addition, IBX is sufficiently mild for the selective oxidation of polyols. Primary allylic alcohols are oxidized preferentially in the presence of secondary alcohols, including *trans*-1,2-diols.[10] These results depend in part on the features of the substrates and may not be due solely to the use of IBX.

The significant bias that IBX displays towards the relative rates of oxidation of primary and secondary alcohols may also be exploited in other ways. Despite the fact that in general, even mild oxidizing agents readily oxidize secondary lactols to lactones, IBX may be successfully employed in the previously unknown one-step synthesis of γ-lactols from both 1,4-bis primary diols as well as from 1,4-primary-secondary diols.[12] Treatment of a 1,4-diol with IBX in DMSO gives the desired γ-lactol in good yield with very little over-oxidation to the corresponding lactone. Consequently this is an excellent procedure for the synthesis of γ-lactols (eq 5).[13]

Synthesis of Related IBX-type Oxidants. The desirable properties of IBX have led to the development of new derivatives based on the core structure. These derivatives display different and often advantageous properties that make them synthetically useful in their own right.

IBX is the penultimate precursor to the periodinane 1,1,1-triacetoxy-1,1-dihydro-1,2-benziodoxol-3(1H)-one [Dess-Martin periodinane (DMP)]. This is a remarkably successful reagent for the selective oxidation of alcohols to carbonyl compounds.[13] It has an extended shelf life and superior solubility in most organic solvents including chloroform, methylene chloride, acetonitrile, and to a lesser extent hexane and ether.[3,4]

The synthesis of the Dess-Martin periodinane is achieved through treatment of a stirred slurry of IBX with a solution of acetic anhydride in glacial acetic acid (11.9 M). When the solution is heated to 100 °C over a period of 40 min, the solid IBX dissolves. The solvent is removed at ambient temperature under

The desire for a more environmentally friendly oxidizing agent prompted the synthesis of water-soluble derivatives of IBX.[14] This water-soluble oxidant *m*-IBX (eq 7) contains a carboxylic acid substituent. This reagent is useful for the oxidation of a variety of alcohols to the corresponding aldehydes in both water and THF/water mixtures in fair-to-excellent yields.[14]

Conditions:

(a) (i) SOCl$_2$, heat (ii) CH$_3$OH, heat, 100%.

(b) H$_2$ (55 psi), Pd/C, CH$_3$OH, 100%.

(c) (i) NaNO$_2$, HCl, 0–5 °C (ii) KI, 91%.

(d) (i) NaOH, THF–H$_2$O (3: 1 v/v) (ii) aq HCL, 93%.

(e) KBrO$_3$, 0.73 M H$_2$SO$_4$, 55 °C, 3 h, 70%.

The synthesis of *m*-IBX employs 3-nitrophthalic acid. Esterification of the acid chloride, after treatment with methanol, gives the nitrodiester. Catalytic hydrogenation affords the corresponding aminodiester, and diazotization followed by iodination with KI provides the dimethyl iodophthalate. Saponification of the diester with NaOH, and acidic work-up give the *m*-IBX diacid precursor. Oxidation of the diacid with KBrO$_3$ in dilute acid, in a manner analogous to the synthesis of IBX, affords *m*-IBX (eq 7).[14]

Polmer-supported IBX Reagents. Two different phenoxide linked polymer-based IBX reagents have been developed. The silica supported reagent (Poly-IBX) is used in THF. The oxidation rate increases relative to DMSO as solvent. The primary alcohol in the cyclohexanol may be oxidized preferentially, and thus avoid formation of the cyclohexanone expected from the secondary alcohol (eq 8).[15] This selectively is retained even with a threefold excess of the oxidant. However, acylic ketones are generated in low yield, and 3-methyl-2-butanone is produced in only 5% yield in contrast to the conversion of 2-methylcyclohexanol to the corresponding cyclohexanone in 82% yield. This reagent is quite resilient and negligible loss of activity was observed after four cycles.

$$(8)$$

A different IBX reagent attached to Merrifield resin worked well with benzylic, allylic, primary, and secondary alcohols.[16]

Synthesis of Heterocycles. In appropriate structures IBX-mediated transformations facilitate the construction of diverse heterocyclic systems in an efficient manner.[17–19] Many of these synthetic sequences involve a single-electron-transfer (SET) mechanism.[20] For example, anilides with an appended alkene substituent undergo intramolecular radical cyclization in a 5-*exo* manner upon exposure to IBX to the corresponding heterocycles (eq 9).[20] The optimum conditions involve heating the substrate with 2 equiv of IBX in a mixed solvent system (THF/DMSO, 10:1) in a pressure tube at 90 °C for 8–12 h. Yields increase when 2 additional equiv of IBX are added and the reaction heated for a further 8 h. The best results arise when the concentration of substrate is kept constant (0.025 M).[18]

$$(9)$$

These cyclizations are quite versatile. The aryl group can tolerate a variety of functional groups, including electron-donating or electron-withdrawing substituents (with the exception of nitro groups), halides, or groups with a high degree of steric bulk at any position on the aromatic ring. Practical limitations include the necessity of an *N*-aryl moiety for a successful reaction. Displacement of the aryl group by a single carbon causes the reaction to fail. In addition, the absence of a carbonyl group or replacement of the alkene moiety with an alkyne also produces undesirable results.[18]

Amino sugars are generated in a direct sequence involving the addition of the hemiacetal to *p*-methoxyphenylisocyanate, and thermolysis with IBX in a sealed tube at 90 °C for 8 h. Subsequent exposure to CAN affords a 72% overall yield of the desired heterocycle (eq 10).[18,21]

THF is essential for these reactions in its role as the hydrogen donor and supplies the requisite hydrogen atom. This feature was established by deuterium labeling studies. However, these radical-mediated reactions fail to add to conjugated esters and cyclohexene acceptors and require an aniline substituent for successful radical-mediated ring closure. A bridged ring example is illustrated in eq 11.[20]

$$(10)$$

Conditions:

(a) *p*-MeOPhNCO. DBU (0.1 equiv), CH₂Cl₂, 25 °C, 1 h, 95%.

(b) IBX (2 equiv), THF:DMSO (10:1), 90 °C (sealed tube), 8 h, 84%.

(c) CAN (5 equiv), CH₃CN:H₂O (3:1), 30 min, 90%.

$$(11)$$

Carbamates and open-chain ureas are also useful as substrates for the synthesis of the corresponding oxazolidinones and cyclic ureas. This reaction tolerates a variety of groups on the aryl ring, and allows the formation of both quaternary centers and spirocyclic compounds. Primary allylic amines are required for the preparation of cyclic ureas, as the secondary allylic amines decompose upon heating.[20] Aryl thiocarbamates, in contrast to the other examples, do not undergo direct cyclization in the presence of IBX. Instead, a [3,3]-sigmatropic rearrangement occurs initially, followed by an IBX induced cyclization to give a thiazolidinone from the intermediate allylic thionocarbonate (eq 12).[17]

Oxidation of Carbon Skeletons with IBX. Allylic and benzylic positions are also susceptible to oxidation by IBX. These applications are not limited to the oxidation of compounds containing a pre-existing oxygen functionality but oxidize the hydrocarbon center directly to aldehydes or ketones. These oxidations also proceed via a single-electron-transfer pathway. The oxidation of aryl methyl groups to aryl aldehydes is accomplished with 3 equiv of IBX in DMSO or DMSO/fluorobenzene mixtures at 80–90 °C.[21] The first two equivalents of IBX initiate the single-electron-transfer to generate a benzylic carbocation. Subsequently, the reaction with water affords the alcohol in situ and the third equivalent of IBX completes the conversion to the desired benzaldehyde (eq 13).[21]

$$(12)$$

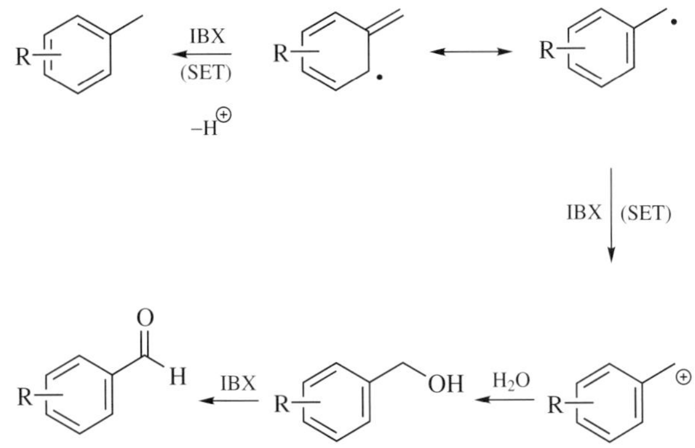

(13)

These reactions proceed rapidly in good yield and tolerate *o*-substituents and halogens, typical of the behavior of IBX with related substrates. The corresponding hydrocarbons may be oxidized directly to ketones and over-oxidation of an aldehyde to the corresponding carboxylic acids is not observed. Thus α-tetralone (**1**) and β-naphthaldehyde (**2**) are synthesized readily from the corresponding substrates. This oxidation is retarded when an electron-withdrawing group is present as illustrated by the lack of reactivity of 4-methylacetophenone (**3**). Consequently, this provides a method for the selective oxidation of polybenzylic compounds to the monocarbonyl systems (eq 14).[21]

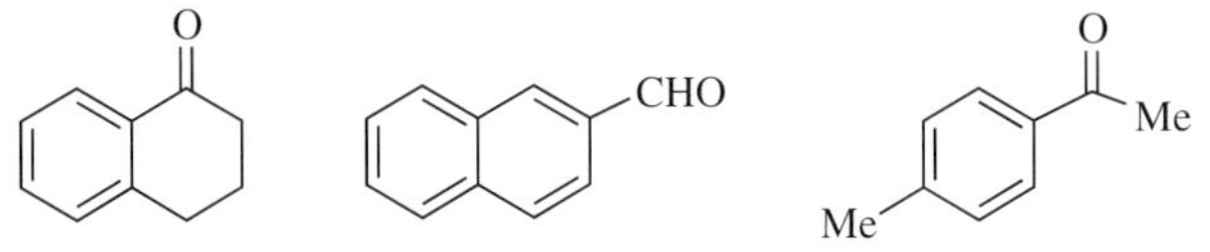

1
IBX (3 equiv)
DMSO, 12 h, 80 °C
70%

2
IBX (3 equiv)
DMSO, 20 h, 80 °C
90%

3
Unreactive

(14)

Preparation of Unsaturated Carbonyl Compounds. An attractive synthesis of α,β-unsaturated carbonyl compounds from saturated alcohol precursors and/or carbonyl compounds utilizes IBX as the oxidant.[22] Treatment of a saturated carbonyl compound with 1.5 equiv of IBX in DMSO or DMSO:fluorobenzene gives the conjugated ketone directly. This is a versatile, general reaction that tolerates a variety of functional groups. In contrast to chromium-based oxidants, this protocol is free of toxic wastes and provides an efficient route to the oxidized products. Saturated alcohols treated with an extra equivalent of IBX generate the corresponding aldehyde or ketone which may be transformed directly to the desired α,β-unsaturated carbonyl compound in a single operation. The level of unsaturation in the final product can be controlled by adjusting the amount of IBX used in the reaction (eq 15).

For example, treatment of cyclooctanol with 2.2 equiv of IBX provides cyclooctenone in 77% yield after chromatography. If the oxidant is increased to 4 equiv the corresponding cross-conjugated cyclooctadienone is generated in 80% yield (eq 15).[23] This sequence to the cyclooctadienone can be repeated in the presence of potassium carbonate and methylamine hydrochloride to provide a one-pot synthesis of a tropinone (6-methyl-8-azabicyclo[3.2.1] octan-3-one) by a double conjugate addition in 80% yield.[25] A related application is illustrated by the sequential control available in a steroid series in which the level of unsaturation in the A-ring

may be controlled with no interference from the cyclopentanone present in ring D (eq 16).[23]

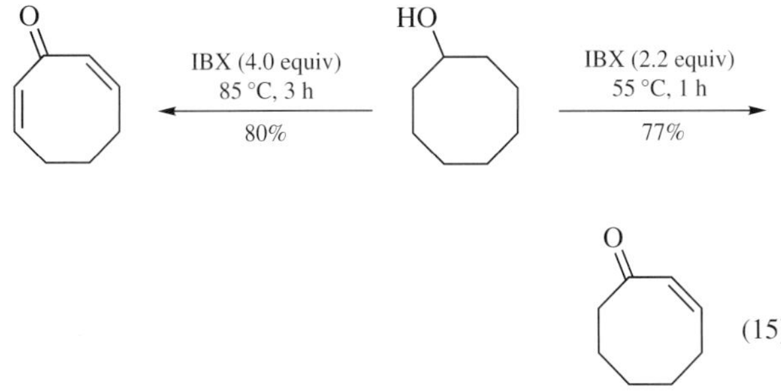

(15)

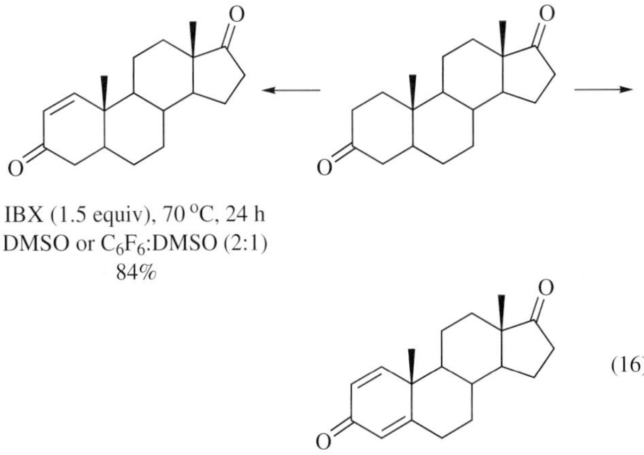

IBX (1.5 equiv), 70 °C, 24 h
DMSO or C₆F₆:DMSO (2:1)
84%

(16)

IBX (4 equiv), 85 °C, 48 h
DMSO or C₆F₆:DMSO (2:1)
72%

The reactivity of IBX may be modulated by complexation with various ligands which induce significant changes in its reactivity and increased control over competing pathways. For example, ligand complexation with 4-methoxypyridine-*N*-oxide (MPO) affords IBX–MPO in which the dehydrogenations are highly selective and may be conducted at lower temperature (25 °C).[24] If the substrate is not soluble in DMSO, aliquots of methylene chloride may be added as a cosolvent untill dissolution occurs and the oxidation rate is optimized. Thus the treatment of a saturated aldehyde in which benzylic hydrogens are also present results in the preparation of the unsaturated aldehyde exclusively in 88% yield with no concomitant benzylic oxidation (eq 17).[24]

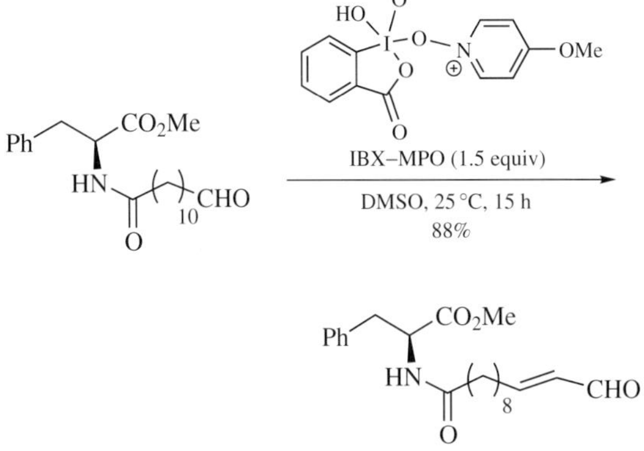

IBX–MPO (1.5 equiv)
DMSO, 25 °C, 15 h
88%

(17)

Consistent with the expectation that the dehydrogenations in eqs 15 and 16 involve IBX-mediated enolizations, silyl enol ethers also undergo a similar oxidation under mild conditions upon exposure to IBX–MPO. Thus the trimethylsilyl ether of cyclooctanone affords cyclooctenone in 96% yield after treatment with 1.1 equiv of IBX–MPO in DMSO/CH$_2$Cl$_2$ for 20 min at 25 °C.[25] This protocol is useful for challenging cyclopentenones such as indenone which is prepared in 94% yield in a 40-min reaction. The silyl enol ethers may also be generated in situ via conjugate addition to an enone, quenching with TMSCl, isolation or in situ reaction of the enol ether, and a subsequent oxidation. In this manner cyclopentenone affords 3-vinyl-2-cyclopentenone from 1,4-addition of vinyl cuprate in 90% yield (eq 18).[25]

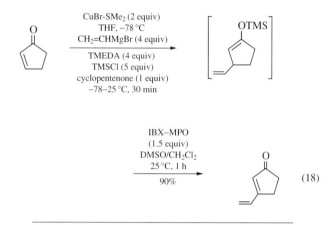

$$\text{(18)}$$

1. Wirth, K. A., *Angew. Chem. Int. Ed.* **2001**, *40*, 2812.
2. Hartmann, C.; Meyer, V., *Chem. Ber.* **1893**, *26*, 1727.
3. Dess, D. B.; Martin, J. C., *J. Org. Chem.* **1983**, *48*, 4155.
4. Boekmann, R. K.; Shao, P.; Mullins, J. J., *Org. Syn.* **1999**, *76*, 194.
5. Ireland, R. E.; Liu, L., *J. Org. Chem.* **1993**, *58*, 2899.
6. Katritzky, A. R.; Duell, B. L.; Gallos, J. K., *Org. Magn. Reson.* **1989**, *27*, 1007.
7. Frigerio, M.; Santagostino, M.; Sputore, S., *J. Org. Chem.* **1999**, *64*, 4537.
8. Barton, D. H. R.; Godfrey, C. R. A.; Morzycki, J. W.; Motherwell, W. B.; Stobie, A., *Tetrahedron Lett.* **1982**, *23*, 957.
9. Frigerio, M.; Santagostino, M., *Tetrahedron Lett.* **1994**, *35*, 8019.
10. Frigerio, M.; Santagostino, M.; Sputore, S.; Palmisano, G., *J. Org. Chem,* **1995**, *60*, 7272.
11. Pettus, T. R. R.; Magdziak, D.; Rodriguez, A. A.; Van De Water, R. W., *Org. Lett.* **2002**, *4*, 285.
12. Corey, E. J.; Palani, A., *Tetrahedron Lett.* **1995**, *36*, 3485.
13. Dess, D. B.; Martin, J. C., *J. Am. Chem. Soc.* **1991**, *113*, 7277.
14. Thottumkara, A. P.; Vinod, T. K., *Tetrahedron Lett.* **2002**, *43*, 569.
15. Mülbair, M.; Giannis, A., *Angew. Chem., Int. Ed.* **2001**, *40*, 4393.
16. Sorg, G.; Mengel, A.; Jung, G.; Radermann, J., *Angew. Chem., Int. Ed.* **2001**, *40*, 4395.
17. Nicolaou, K. C.; Baran, P. S.; Zhong, Y. -L.; Barluenga, S.; Hunt, K. W.; Kranich, R.; Vega, J. A., *J. Am. Chem. Soc.* **2002**, *124*, 2232.
18. Nicolaou, K. C.; Baran, P. S.; Zhong, Y.-L.; Vega, J. A., *Angew. Chem., Int. Ed.* **2000**, *39*, 2525.
19. Nicolaou, K. C.; Zhong, Y.-L.; Baran, P. S., *Angew. Chem., Int. Ed.* **2000**, *39*, 625.
20. Nicolaou, K. C.; Baran, P. S.; Kranich, R.; Zhong, Y.-L.; Sugita, K.; Zou, N., *Angew. Chem., Int. Ed.* **2001**, *40*, 202.
21. Nicolaou, K. C.; Baran, P. S.; Zhong, Y.-L., *J. Am. Chem. Soc.* **2001**, *123*, 3183.
22. Nicolaou, K. C.; Zhong, Y.-L.; Baran, P. S., *J. Am. Chem. Soc.* **2000**, *122*, 7596.
23. Nicolaou, K. C.; Montagon, T.; Baran, P. S.; Zhong, Y.-L., *J. Am. Chem. Soc.* **2002**, *124*, 2245.
24. Nicolaou, K. C.; Montagon, T.; Baran, P. S., *Angew. Chem., Int. Ed.* **2002**, *41*, 993.
25. Nicolaou, K. C.; Gray, D. L. F.; Montagon, T.; Harrison, S. T., *Angew. Chem., Int. Ed.* **2002**, *41*, 996.

Alex G.Fallis & Pierre E.Tessier

University of Ottawa, Ottawa, Ontario, Canada

Essential Reagents for Organic Synthesis, Edited by Philip L. Fuchs, André B. Charette, Tomislav Rovis, and Jeffrey W. Bode.
©2016 John Wiley & Sons, Ltd. Published 2016 by John Wiley & Sons, Ltd.

Lithium Aluminum Hydride[1]

[16853-85-3]　　　　AlH₄Li　　　　(MW 37.96)

InChI = 1/Al.Li.4H/q-1;+1;;;;/rAlH4.Li/h1H4;/q-1;+1
InChIKey = OCZDCIYGECBNKL-PDCCDREHAZ

(reducing agent for many functional groups;[1] can hydroaluminate double and triple bonds;[2] can function as a base[3])

Alternate Name: LAH.
Physical Data: mp 125 °C; d 0.917 g cm^{-3}.
Solubility: soluble in ether (35 g/100 mL; conc of more dil soln necessary); soluble in THF (13 g/100 mL); modestly soluble in other ethers; reacts violently with H_2O and protic solvents.
Form Supplied in: colorless or gray solid; 0.5–1 M solution in diglyme, 1,2-dimethoxyethane, ether, or tetrahydrofuran; the LiAlH₄·2THF complex is available as a 1 M solution in toluene.
Analysis of Reagent Purity: Metal Hydrides Technical Bulletin No. 401 describes an apparatus and methodology for assay by means of hydrogen evolution. See also Rickborn and Quartucci.[39a]
Handling, Storage, and Precaution: the dry solid and solutions are highly flammable and must be stored in the absence of moisture. Cans or bottles of LiAlH₄ should be flushed with N_2 and kept tightly sealed to preclude contact with oxygen and moisture. Lumps should be crushed only in a glove bag or dry box.

Original Commentary

Leo A. Paquette
The Ohio State University, Columbus, OH, USA

Functional Group Reductions. The powerful hydride transfer properties of this reagent cause ready reaction to occur with aldehydes, ketones, esters, lactones, carboxylic acids, anhydrides, and epoxides to give alcohols, and with amides, iminium ions, nitriles, and aliphatic nitro compounds to give amines. Several methods of workup for these reductions are available. A strongly recommended option[4] involves careful successive dropwise addition to the mixture containing n grams of LiAlH₄ of n mL of H_2O, n mL of 15% NaOH solution, and $3n$ mL of H_2O. These conditions provide a dry granular inorganic precipitate that is easy to rinse and filter. More simply, solid Glauber's salt ($Na_2SO_4 \cdot 10H_2O$) can be added portionwise until the salts become white.[5] In certain instances, an acidic workup (10% H_2SO_4) may prove advantageous because the inorganic salts become solubilized in the aqueous phase.[6] Should water not be compatible with the product, the use of ethyl acetate is warranted since the ethanol that is liberated usually does not interfere with the isolation.[4] Although the stoichiometry of LiAlH₄ reactions is well established,[1] excess amounts of the reagent are often employed (perhaps to make accommodation for the perceived presence of adventitious moisture). This practice is wasteful of reagent, complicates workup, and generally should be avoided.

The reduction of amides can be adjusted in order to deliver aldehydes. Acylpiperidides,[7] *N*-methylanilides,[8] aziridides,[9] imidazolides,[10] and *N,O*-dimethylhydroxylamides[11] have proven especially serviceable. All of these processes generate products that liberate the aldehyde upon hydrolytic workup. The powerful reducing ability of LiAlH₄ allows for its application in the context of other functional groups. Alkanes are often formed in good yield upon exposure of alkyl halides (I > Br > Cl; primary > secondary > tertiary)[12] and tosylates[13] to LiAlH₄ in ethereal solvents. Chloride reduction is an S_N2 process, while iodides enter principally into single electron transfer chemistry.[14] Benzylic[15] and allylic halides[16] behave comparably, although the latter can react in an S_N2' fashion as well (eq 1).[17] Select aromatic halides can be reduced under forcing conditions (e.g. diglyme, 100 °C),[18] but chemoselectivity as in eq 2 can often be achieved.[19] Vinyl,[20] bridgehead,[20] and cyclopropyl halides (eqs 3 and 4)[20,21] have all been reported to undergo reduction. The SET mechanism is also believed to operate in the latter context.[22]

$$\text{(1)}$$

$$\text{(2)}$$

$$\text{(3)}$$

$$\text{(4)}$$

LiAlH₄ is normally unreactive toward ethers.[1] Unsaturated acetals undergo reduction with double bond migration (S_N2'); in cyclic systems, the usual stereoelectronic factors often apply (eq 5).[23] Orthoesters are amenable to attack, giving acetals in good yield (eq 6).[24] The susceptibility of benzylic acetals to reduction can be enhanced by the co-addition of a Lewis acid (eq 7).[25]

$$\text{(5)}$$

$$\text{(6)}$$

$$\text{(7)}$$

Essential Reactions for Organic Synthesis, First Edition. Edited by Philip L. Fuchs.
© 2016 John Wiley & Sons, Ltd. Published 2016 by John Wiley & Sons, Ltd.

When comparison is made between LiAlH$_4$ and related reducing agents containing active Al–H and B–H bonds, LiAlH$_4$ is seen to be the most broadly effective (Table 1).[1h] Its superior reducing power is also reflected in its speed of hydride transfer.

Hydroalumination Agent. Ethylene has long been known to enter into addition with LiAlH$_4$ when the two reagents are heated under pressure at 120–140 °C; lithium tetraethylaluminate results.[2] Homogeneous hydrogenations of alkenes and alkynes to alkanes and alkenes, respectively, performed in THF or diglyme solutions under autoclave pressure, are well documented.[26] Such reductions are greatly facilitated by the presence of a transition metal halide ranging from Ti to Ni.[27] Replacement of the hydrolytic workup by the addition of appropriate halides constitutes a useful means for chain extension (eq 8).[28] 1-Chloro-1-alkynes are notably reactive toward LiAlH$_4$, addition occurring regio- and stereoselectively to give alanates that can be quenched directly (MeOH) or converted into mixed 1,1-dihaloalkenes (eq 9).[29]

A pronounced positive effect on the ease of reduction of C=C and C≡C bonds manifests itself when a neighboring hydroxyl group is present. In such cases, LiAlH$_4$ is used alone because reduction is preceded by formation of an alkoxyhydridoaluminate capable of facilitating hydride delivery (eqs 10–12).[30–32] The regio- and stereoselectivities of these reactions, where applicable, appear to be quite sensitive to the substrate structure and solvent.[33] Generally, the use of THF or dioxane results in exclusive *anti* addition. When ether is used, almost equivalent amounts of *syn* and *anti* products can result. Considerable attention has been accorded to synthetic applications of the alanate intermediates produced upon reduction of propargyl alcohols in this way. Simple heating occasions elimination of a δ-leaving group to generate homoallylic[34] and α-allenic alcohols (eqs 13 and eq 14).[35] Other variants of this chemistry have been reported.[36] The addition at −78 °C of solid iodine to alanates formed in this way greatly accelerates the elimination (eq 15).[37] Solutions of iodine in THF give rise instead to vinyl iodides.[38]

Epoxide Cleavage and Aziridine Ring Formation. Epoxides are reductively cleaved in the presence of LiAlH$_4$ with attack generally occurring at the less substituted carbon.[1g] 1,2-Epoxycyclohexanes exhibit a strong preference for axial attack (eqs 16 and 17).[39] In general, *cis* isomers are more reactive than their *trans* counterparts; ring size effects are also seen and these conform to the degree of steric inhibition to backside attack of the C–O bond.[40] Vinyl epoxides often suffer ring opening by means of the S_N' mechanism (eq 18).[41]

$$t\text{-Bu} \xrightarrow[\substack{\text{ether, 0 °C} \\ 90\%}]{\text{LiAlH}_4} t\text{-Bu} \qquad (16)$$

$$t\text{-Bu} \xrightarrow[\substack{\text{ether, 0 °C} \\ 81\%}]{\text{LiAlH}_4} t\text{-Bu} \qquad (17)$$

$$\xrightarrow[\substack{\text{Bu}_2\text{O} \\ 90\%}]{\text{LiAlD}_4} \qquad (18)$$

Two types of oximes undergo hydride reduction with ring closure to give aziridines. These are ketoximes that carry an α- or β-aryl ring and aldoximes substituted with an aromatic group at the β-carbon (eqs 19 and 20).[42]

$$\xrightarrow[\substack{\text{THF} \\ 17\%}]{\text{LiAlH}_4} \qquad (19)$$

$$\xrightarrow[\substack{\text{THF} \\ 34\%}]{\text{LiAlH}_4} \qquad (20)$$

Use as a Base. Both 1,2- and 1,3-diol monosulfonate esters react with LiAlH$_4$ functioning initially as a base and subsequently in a reducing capacity (eqs 21 and 22).[3,43]

$$\xrightarrow[\substack{\text{THF, rt} \\ 60\%}]{\text{LiAlH}_4} \qquad (21)$$

$$\xrightarrow[\substack{\text{DME, }\Delta \\ >90\%}]{\text{LiAlH}_4} \qquad (22)$$

Reduction of Sulfur Compounds. Sulfur compounds react differently with LiAlH$_4$ depending on the mode of covalent attachment of the hetero atom and its oxidation state. Reductive

desulfurization is not often encountered.[44] Arylthioalkynes undergo stereoselective *trans* reduction (eq 23)[45] and α-oxoketene dithioacetals are transformed into fully saturated *anti* alcohols under reflux conditions (eq 24).[46] While LiAlH$_4$ catalyzes the fragmentation of sulfolenes to 1,3-dienes,[47] the lithium salts of sulfolanes are smoothly ring contracted when heated with the hydride in dioxane (eq 25).[48]

$$\xrightarrow[\substack{\text{D}_2\text{O} \\ 93\%}]{\text{LiAlH}_4} \qquad (23)$$

$$\xrightarrow[\substack{\text{THF, }\Delta \\ 98\%}]{\text{LiAlH}_4} \qquad (24)$$

$$\xrightarrow[\substack{\text{2. LiAlH}_4 \\ \text{dioxane, }\Delta \\ 37\%}]{\text{1. BuLi (1.5 equiv)}} \qquad (25)$$

Stereoselective Reductions. The reduction of 4-t-butylcyclohexanone and (**1**) by LiAlH$_4$ occurs from the axial direction to the extent of 92%[49] and 85%,[50] respectively. When a polymethylene chain is affixed diaxially as in (**2**), the equatorial trajectory becomes kinetically dominant (93%).[50] Thus, although electronic factors may be an important determinant of π-facial selectivity,[51] steric demands within the ketone cannot be ignored. The stereochemical characteristics of many ketone reductions have been examined. For acyclic systems, the Felkin–Ahn model[52] has been widely touted as an important predictive tool.[53] Cram's chelation transition state proposal[54] is a useful interpretative guide for ketones substituted at C_α with a polar group. Cieplak's explanation[55] for the stereochemical course of nucleophilic additions to cyclic ketones has received considerable scrutiny.[56]

$(CH_2)_9$

(**1**)   (**2**)

Useful levels of diastereoselectivity can be realized upon reduction of selected acyclic ketones having a proximate chiral center.[57] The contrasting results in eq 26 stem from the existence of a chelated intermediate when the benzyloxy group is present and its deterrence when a large silyl protecting group is present. In the latter situation, an open transition state is involved.[58,59] In eq 27, LiAlH$_4$ alone shows no stereoselectivity, but the co-addition of **Lithium Iodide** gives rise to a *syn*-selective reducing agent as a consequence of the intervention of a Li$^+$-containing six-membered chelate.[60]

R = Bn, ether, −10 °C   98:2
R = TBDMS, THF, −20 °C   5:95   (26)

syn:anti = 95:5   (27)

The reduction of amino acids to 1,2-amino alcohols can be conveniently effected with LiAlH$_4$ in refluxing THF.[61] The higher homologous 1,3-amino alcohols have been made available starting with isoxazolines, the process being *syn* selective (eq 28).[62,63] The *syn* and *anti* O-benzyloximes of β-hydroxy ketones are reduced with good *syn* and *anti* stereoselectivity, respectively.[64] However, when NaOMe or KOMe is also added, high *syn* stereoselectivity is observed with both isomers (eq 29).[65]

88:12   (28)

syn:anti = 96:4   (29)

Addition of Chiral Ligands. If a chiral adjuvant is used to achieve asymmetric induction, it should preferentially be inexpensive, easily removed, efficiently recovered, and capable of inducing high stereoselectivity.[53] Of these, 1,3-oxathianes based on (+)-camphor[66] or (+)-pulegone,[67] and proline-derived 1,3-diamines,[68] have been accorded the greatest attention.

Extensive attempts to modify LiAlH$_4$ with chiral ligands in order to achieve the consistent and efficient asymmetric reduction of prochiral ketones has provided very few all-purpose reagents.[1f,69] Many optically active alcohols, amines, and amino alcohols have been evaluated. One of the more venerable of these reagents is that derived from DARVON alcohol.[70] The use of freshly prepared solutions normally accomplishes reasonably enantioselective conversion to alcohols.[71] As the reagent ages, its reduction stereoselectivity reverses.[70] This phenomenon is neither understood nor entirely reliable, and therefore recourse to the enantiomeric reagent[72] (from NOVRAD alcohol) is recommended.[73] The highest level of enantiofacial discrimination is usually realized with LiAlH$_4$ complexes prepared from equimolar amounts of (S)-(−)- or (R)-(+)-2,2′-dihydroxy-1,1′-binaphthyl and ethanol.[74] Often, optically pure alcohols result, irrespective of whether the ketones are aromatic,[74] alkenic,[75] or alkynic in type.[76] A useful rule of thumb is that (S)-BINAL-H generally provides (S)-carbinols and (R)-BINAL-H the (R)-antipodes when ketones of the type R$_{unsat}$–C(O)–R$_{sat}$ are involved (eq 30).[74]

For a more detailed discussion of asymmetric induction by this means, see also the following entries that deal specifically with LiAlH$_4$/additive combinations.

Related Reagents. Lithium Aluminum Hydride–(2,2′-Bipyridyl)(1,5-cyclooctadiene)nickel; Lithium Aluminum Hydride–Bis(cyclopentadienyl)nickel; Lithium Aluminum Hydride–Boron Trifluoride Etherate; Lithium Aluminum Hydride–Cerium(III) Chloride; Lithium Aluminum Hydride–2,2′-Dihydroxy-1, 1′-binaphthyl; Lithium Aluminum Hydride–Chromium(III) Chloride; Lithium Aluminum Hydride–Cobalt(II) Chloride; Lithium Aluminum Hydride–Copper(I) Iodide; Lithium Aluminum Hydride–Diphosphorus Tetraiodide; Lithium Aluminum Hydride–Nickel(II) Chloride; Lithium Aluminum Hydride–Titanium(IV) Chloride; Titanium(III) Chloride–Lithium Aluminum Hydride.

Table 1 Comparison of the reactivities of hydride reducing agents toward the More common functional groups

Reagent/functional group	Reduction products[a]						
	Aldehyde	Ketone	Acyl halide	Ester	Amide	Carboxylate salt	Iminium ion
LiAlH$_4$	Alcohol	Alcohol	Alcohol	Alcohol	Amine	Alcohol	Amine
LiAlH$_2$(OCH$_2$CH$_2$OMe)$_2$	Alcohol	Alcohol	Alcohol	Alcohol	Amine	Alcohol	–
LiAlH(O-t-Bu)3	Alcohol	Alcohol	Aldehyde	Alcohol	Aldehyde	NR	–
NaBH$_4$	Alcohol	Alcohol	–	Alcohol	NR	NR	Amine
NaBH$_3$CN	Alcohol	NR	–	NR	NR	NR	Amine
B$_2$H$_6$	Alcohol	Alcohol	–	NR	Amine	Alcohol	–
AlH$_3$	Alcohol	Alcohol	Alcohol	Alcohol	Amine	Alcohol	–
[i-PrCH(Me)]$_2$BH	Alcohol	Alcohol	–	NR	Aldehyde	NR	–
(i-Bu)$_2$AlH	Alcohol	Alcohol	–	Aldehyde	Aldehyde	Alcohol	–

[a] NR indicates that no reduction is observed.

First Update

Thierry Ollevier & Valerie Desyroy
Universite Laval, Quebec City, Quebec, Canada

Deprotection. Usually TBS ethers are not affected by LiAlH$_4$, but neighboring group participation makes the cleavage of TBS ethers possible. The presence of a group susceptible to LiAlH$_4$ reduction, such as an ester or a ketone, or an atom bearing a relatively acidic proton adjacent to the silyl ether is a prerequisite for the deprotection reaction to proceed rapidly (eq 31).[77,78] This reductive removal of TBS groups leads to unprotected ethanolamines in good yields (eq 32).[79] Due to neighboring group participation, the selective removal of one *O*-silyl group can also be achieved (eq 33).[77] This method allows the selective cleavage of one TBS ether over one that lacks the presence of the required polar group in the β-position. Compared to TBS ethers, TBDPS ethers show a similar but attenuated reactivity toward LiAlH$_4$. LiAlH$_4$ is reported to be effective for the 6-*O*-demethylation of thevinols (eq 34).[80] The selective demethylation of a tertiary alkyl methyl ether can be achieved in the presence of an aryl methyl ether. The participation of the hydroxy neighboring group is believed to explain the observed selectivity.

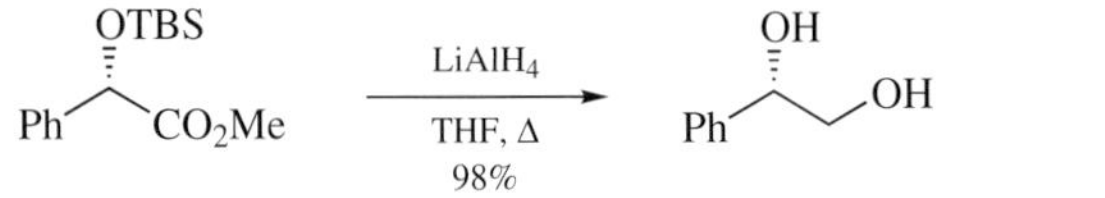

$$(31)$$

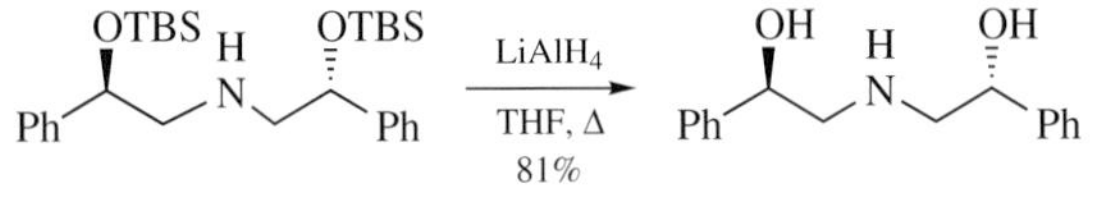

$$(32)$$

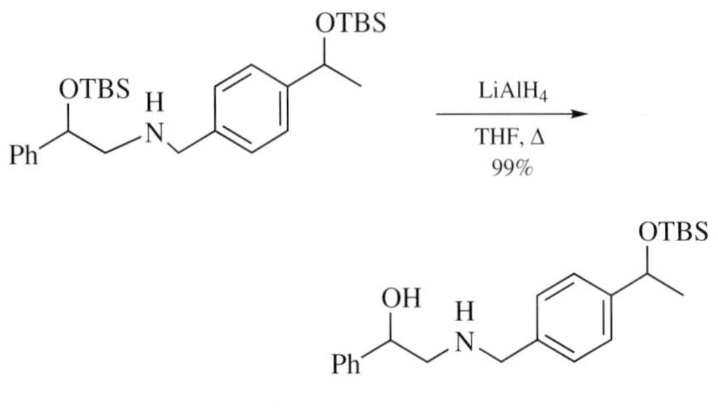

$$(33)$$

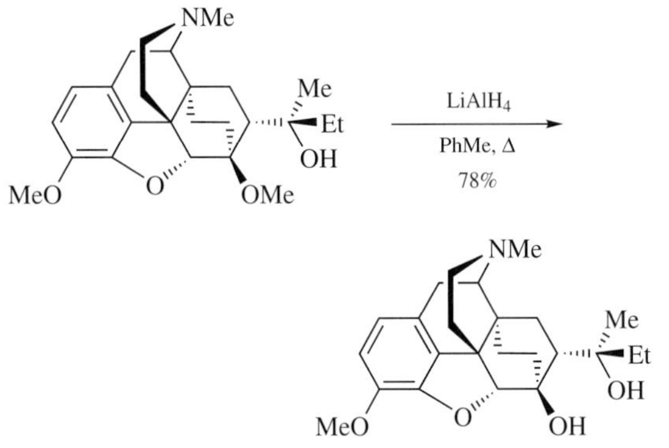

$$(34)$$

Synthesis of Heterocycles. 6-Alkoxy-1,2-oxazines are converted to aziridines by way of C=N bond reduction, N-O bond cleavage, and cyclization (eq 35).[81] A parent transformation is

described with 5-(2,2,2-trifluoromethyl)isoxazoles leading to (trifluoroethyl)aziridines in good yields (eq 36).[82] An intramolecular reductive cyclization occurs when amide esters are treated with LiAlH$_4$ in THF (eq 37).[83] Oxazole derivatives are formed under such conditions.

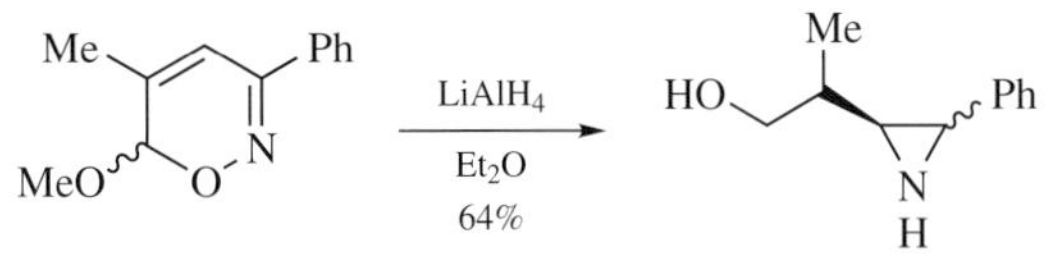

$$(35)$$

cis/trans (94/6)

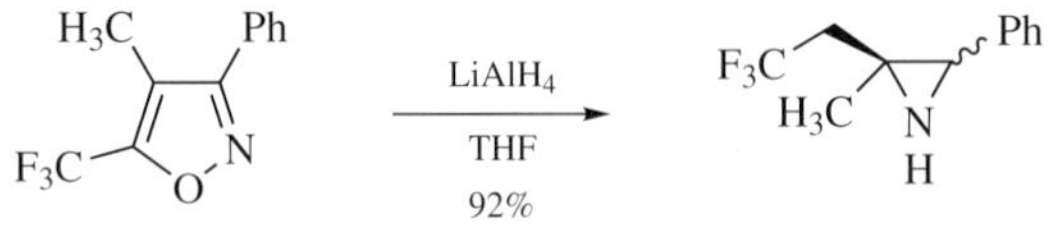

$$(36)$$

cis/trans (40/60)

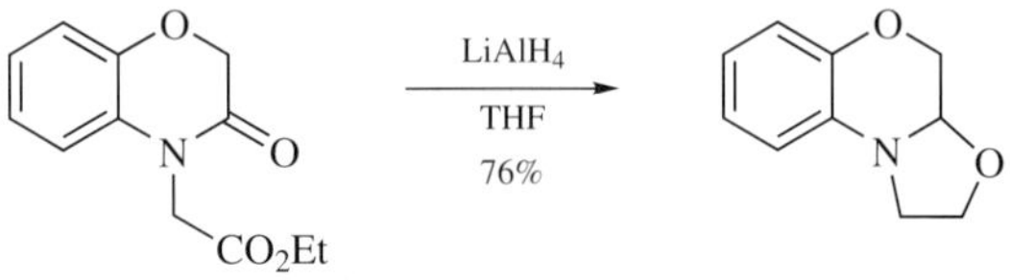

$$(37)$$

Dephosphonylations. β-Functionalized phosphonates undergo dephosphonylation (eq 38).[84–87] The reaction is effective for β-keto phosphonates or β-enamino phosphonates and occurs upon reduction of their enolate form with LiAlH$_4$. The reduction is actually performed on the β-functionalized phosphonate anions generated in the first step with sodium hydride or *n*-butyllithium (eq 38). Similarly, pyrrolidines have been obtained by the reduction of 2,4-dialkyl-5-phosphonyl pyrrolines with LiAlH$_4$ (eq 39).[88]

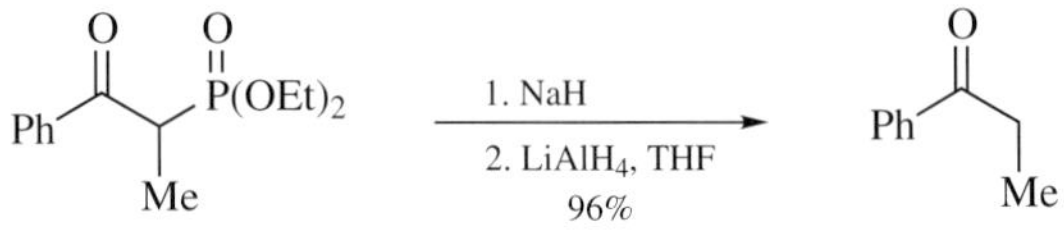

$$(38)$$

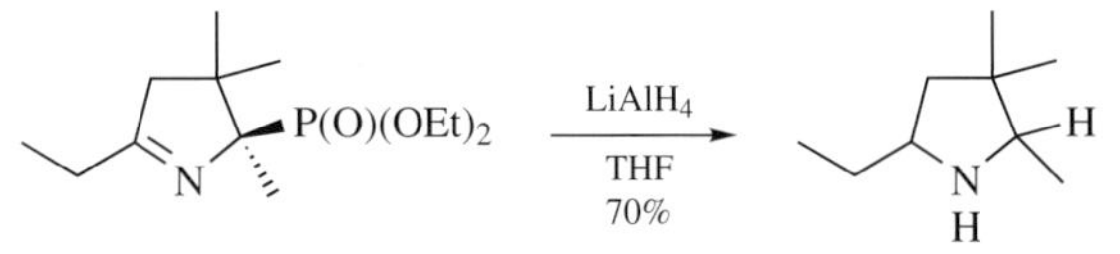

$$(39)$$

Cyclizations. Urea methyl esters afford the corresponding tetrahydro-1*H*-imidazole-2,4-diones upon cyclization with LiAlH$_4$ (eq 40).[89] Upon treatment with LiAlH$_4$, cyclization occurs rather than reduction of the methyl ester to alcohol. Treatment of 4-alkynamides with LiAlH$_4$ gives enamine pyrrolidines with an *exo* double bond in high yields.[90] 2-Benzylidene-1-methyl-pyrrolidine is obtained in good yield by treatment of 4-alkynyl-amide with LiAlH$_4$ (eq 41). The role of LiAlH$_4$ in this reaction is apparently to reduce the amide to alkynylamine and to generate a nitrogen anion which cyclizes by attack of the anion at the triple bond. Treatment of some alkynylamines with LiAlH$_4$ results in the formation of related cyclization products. The presence of a 5-phenyl group in 4-alkynylamines is necessary for a facile anionic cyclization since other substrates lacking this substituent do not cyclize into enaminopyrrolidines.[90] Ring-closure of δ-chloro aldimines is accomplished by intramolecular nucleophilic substitution using LiAlH$_4$ in ether under reflux.[91] The fast reduction of the imino bond is followed by intramolecular nucleophilic substitution (eq 42). Transannular bond formation occurs in the specific case of a 3,7-bis(methylene)-1,5-diazacyclooctane derivative reacting with LiAlH$_4$ to afford 1,5-dimethyl-3,7-diazabicyclo[3.3.0] octane in a good yield (eq 43).[92] The formation of the bicyclic system is thought to occur through hydride attack at one of the exocyclic methylene carbons followed by carbanion addition to the proximal transannular double bond. The hydridation of isopinocampheylchloroborane with LiAlH$_4$ at $-20\,°C$ in Et$_2$O generates an intermediate isopinocampheylalkylborane, which undergoes a rapid stereoselective intramolecular hydroboration to provide a cyclic trialkylborane (eq 44).[93]

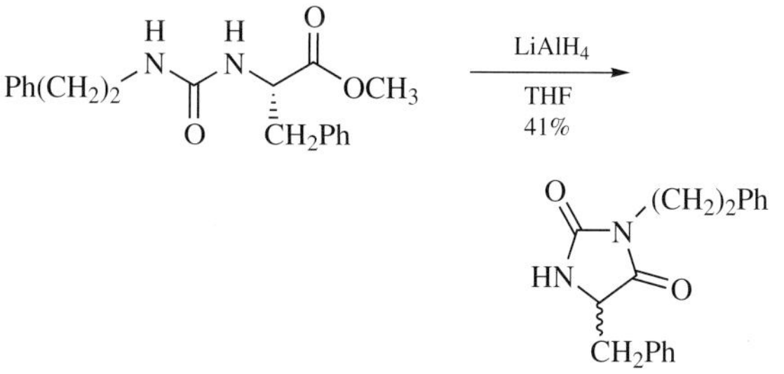

After deprotection of the trimethylsilyl ether with K$_2$CO$_3$ in MeOH, opening of the oxetane with LiAlH$_4$ in THF cleanly leads to a diol (eq 45).[94] The reaction proceeds via selective S$_N$2-type ring opening at C-4 with the 3-hydroxyl alcohol as a possible directing group. The 2-pivaloyl protected salicylaldehyde derivative undergoes S$_N$1 opening and the corresponding triol is produced in good yields (eq 46).[95] The intermediate phenol liberated in the course of the reaction acts as a directing group for the opening reaction. Phenyl analogs lacking the *ortho*-hydroxyl group react sluggishly, if at all, with LiAlH$_4$ even under reflux in THF. The LiAlH$_4$ reduction of an oxetanol bearing a hydroxyl group in the benzylic position allows the preparation of an angular hydroxy cyclobutyl carbinol in excellent yields (eq 47).[96,97]

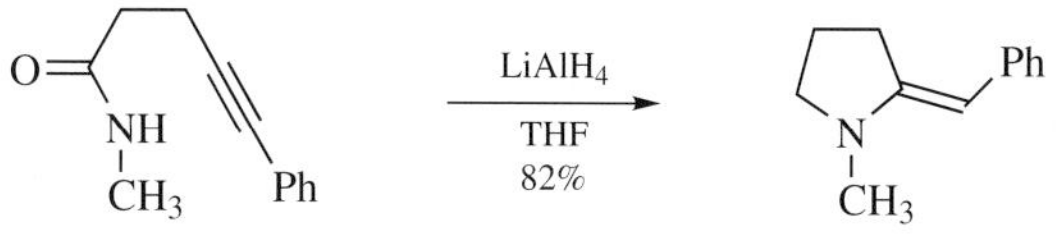

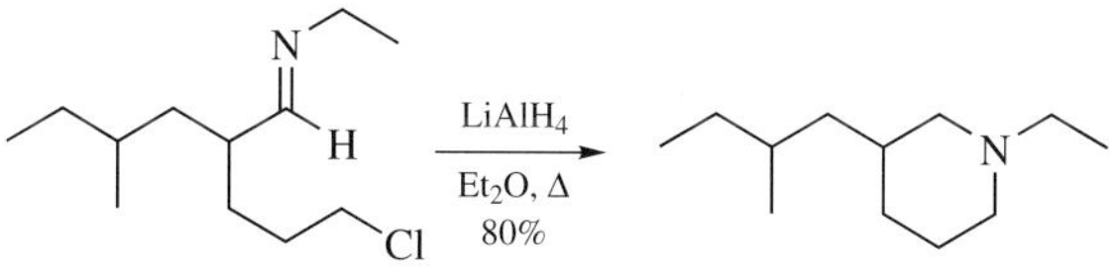

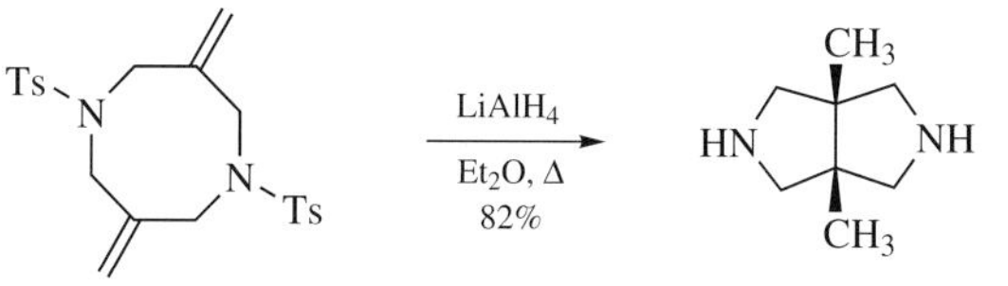

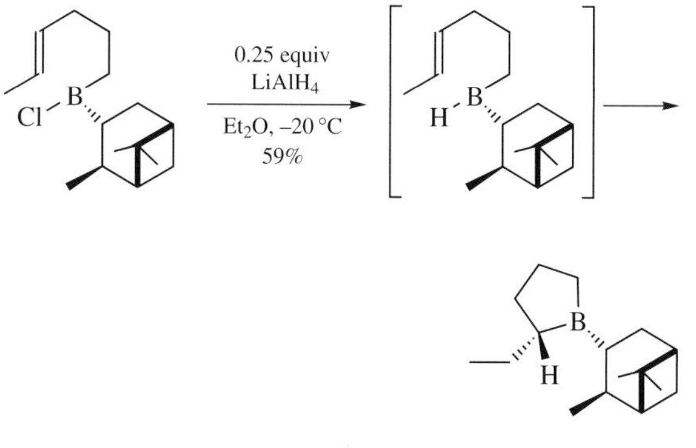

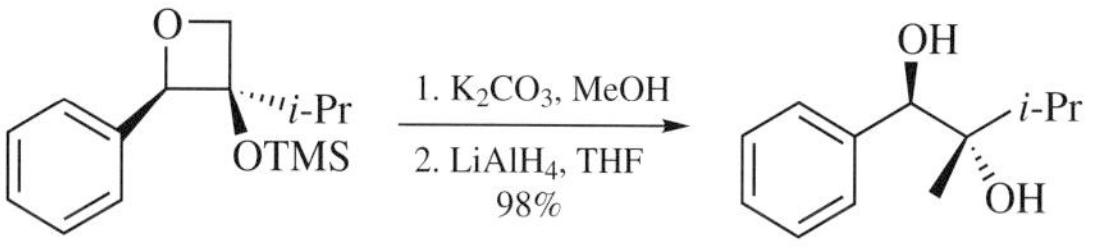

(40)

(41)

(42)

(43)

(44)

(45)

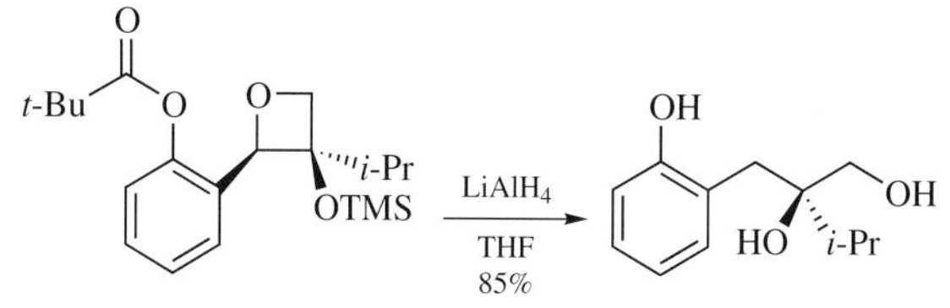

(46)

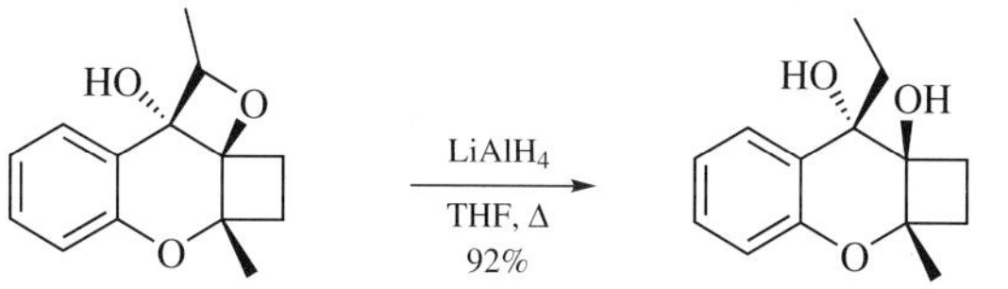

(47)

Catalyzed Reactions. Many aldehydes react with nitroalkanes in the presence of a catalytic amount of LiAlH$_4$ at $0\,°C$ to give 2-nitroalkanols in excellent yields (eq 48).[98] LiAlH$_4$ is an effective catalyst and a practical alternative to classical bases in Henry reactions. This method avoids the dehydration of 2-nitroalkanols into nitroalkenes even in the case of aromatic aldehydes. In the case of α,β-unsaturated aldehydes, no Michael addition is observed.

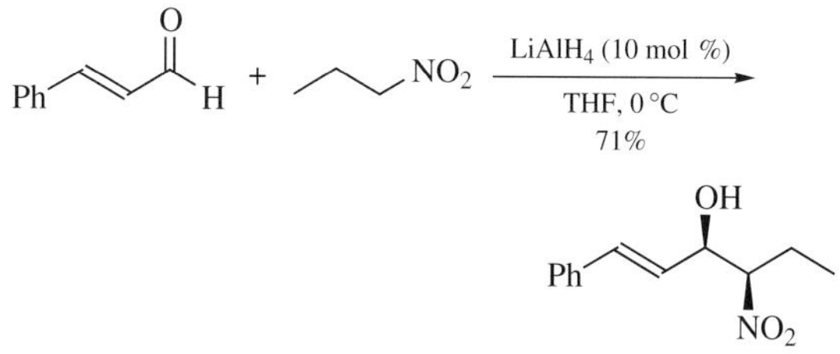

(48)

syn/anti (2/1)

Rearrangements. Benzylic methoxyamines undergo ring expansion reaction upon reduction with LiAlH$_4$ (eq 49).[99] The N-O bond of benzylic hydroxylamines is reduced with LiAlH$_4$ by a radical mechanism. This ring expansion is related to the Beckmann-type rearrangement of oximes with LiAlH$_4$. A different type of ring expansion has been observed during the LiAlH$_4$ reduction of 13-nitro oxyberberine (eq 50).[100] The ring expansion reaction results in the low yield formation of a seven-membered ring in

Oxetane Cleavages. Selective ring opening of oxetanes can be achieved by LiAlH$_4$ using appropriate directing groups.[94–97]

addition to the expected primary amine. Both nitro compounds and oximes undergo reductive rearrangement through hydroxylamine intermediates. LiAlH$_4$ treatment of 1,3-dimethyl-3-(methylthio)oxindole leads to the reduction of both the oxindole carbonyl group and the ester group (eq 51).[101] Two dethiomethylation routes compete for oxindole and indole formation. In addition, a usual rearrangement to a tetrasubstituted indole has been observed. A polycyclic structure bearing a bromo substituent at a bridgehead position undergoes a quasi-Favorskii rearrangement in nearly quantitative yield upon reaction with LiAlH$_4$.[102]

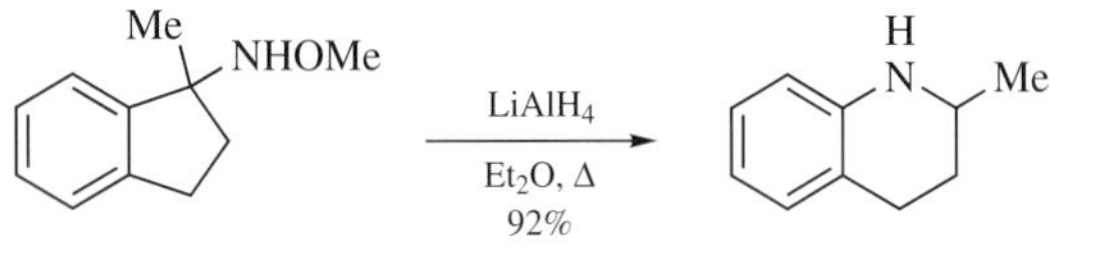

$$(49)$$

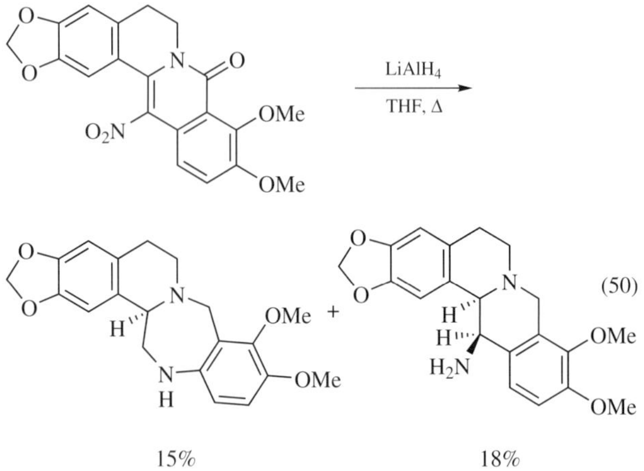

$$(50)$$

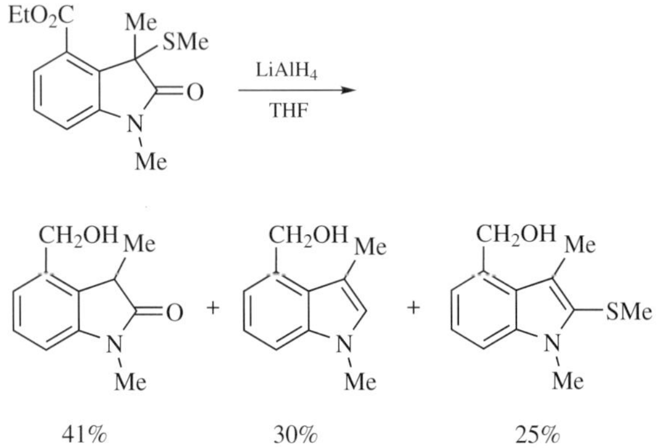

15%   18%

$$(51)$$

41%   30%   25%

Reductions of Allylic and Benzylic Substrates. LiAlH$_4$ promotes the reductive deoxygenation of hydroxybenzyl alcohols giving p-methylphenols in high yields (eq 52).[103] The mechanism proceeds by way of a benzoquinone methide intermediate that is generated in situ. δ-Fluoro and δ,δ-difluorohomoallylic alcohol derivatives can be efficiently prepared by LiAlH$_4$ reduction of δ,δ-difluoroallylic alcohols and γ-chlorodifluoromethylallylic alcohols, respectively, through a regioselective S$_N$2′ reaction, in which the presence of a free hydroxyl group is essential for the reaction to occur (eqs 53 and 54).[104] (Perfluorobutyl)anilines are selectively reduced with LiAlH$_4$ at the benzylic position (eq 55).[105] Only the *para* and the *ortho* isomers can be reduced upon treatment with LiAlH$_4$, while the *meta* isomer is inert under the same conditions.

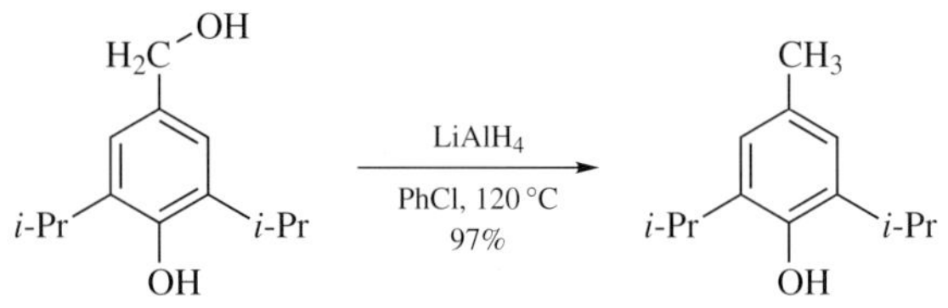

$$(52)$$

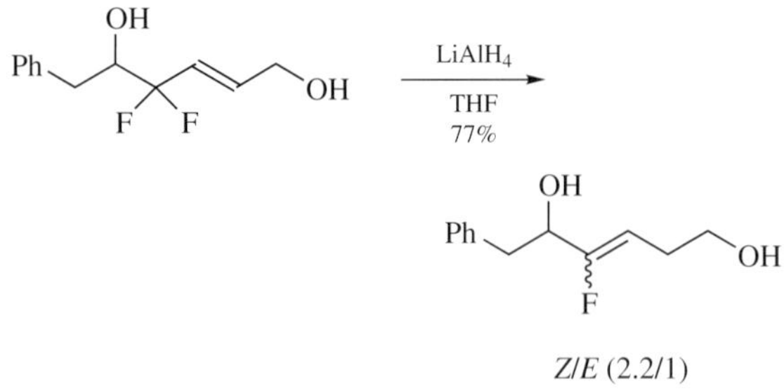

$$(53)$$

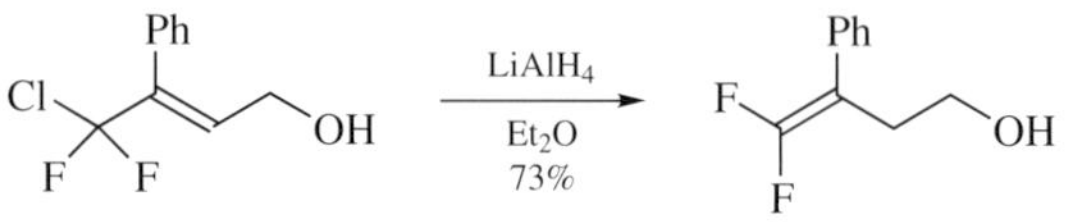

Z/E (2.2/1)

$$(54)$$

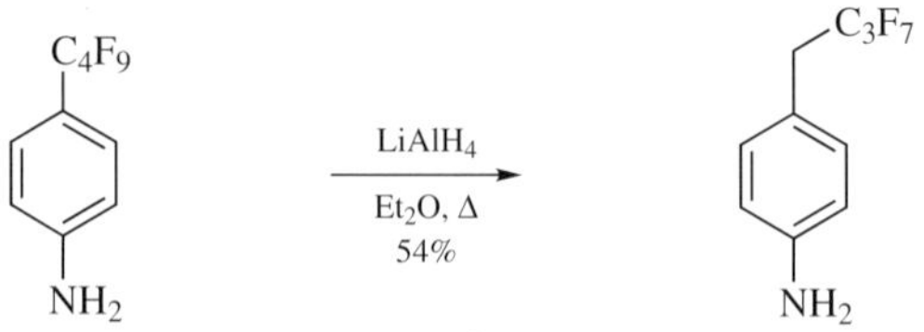

$$(55)$$

Reduction of Amides. *N*-Protected α-amino aldehydes were obtained by the LiAlH$_4$ reduction of morpholine amides (eq 56).[106] The process occurs without racemization and is applicable to most of the commonly used *N*-protected groups. This route is a useful alternative to Weinreb's method. Reduction of α,α-disubstituted selenoamides with LiAlH$_4$ successfully converts the selenocarbonyl group to a methylene group.[107] In most cases, the corresponding amines are obtained quantitatively. Reduction of tertiary *N*-aryl-*N*-benzylbenzamides generally gives secondary amines when the substituents are bulky. However, the LiAlH$_4$ reduction of similar tertiary amides on solid support (1% divinylbenzene cross-linked polystyrene resin, 100–200 mesh) affords the corresponding tertiary amines as main products (eq 57).[108] Dehalogenation of secondary *o*-halogenobenzamides at room temperature takes place before carbonyl reduction (eq 58).[109] This mild reduction is applicable to *o*-fluorobenzamides and affords defluorinated benzamides in moderate yield. However, tertiary amides do not behave similarly and undergo reduction to tertiary amines. Selective reductions of amide groups can be performed on aromatic nitro compounds using LiAlH$_4$ (eq 59).[110,111] Other functional groups are selectively reduced in the presence of the nitro group.

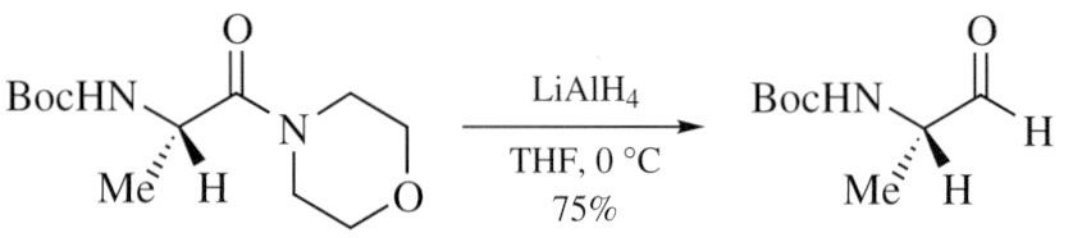

$$(56)$$

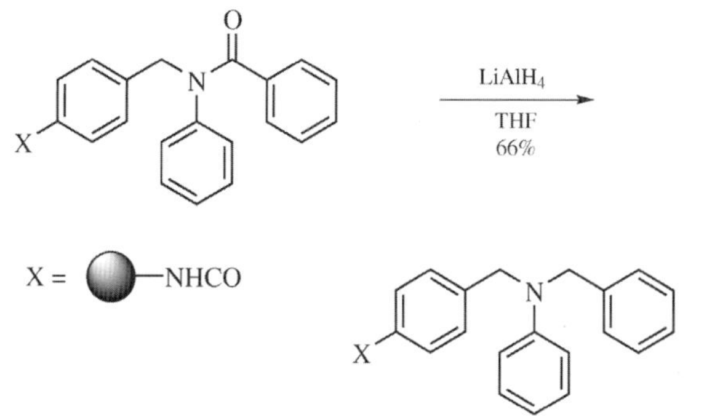

(57)

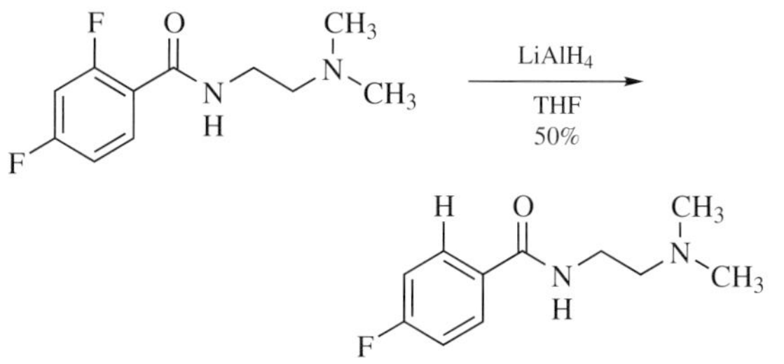

X = ⬤—NHCO

(58)

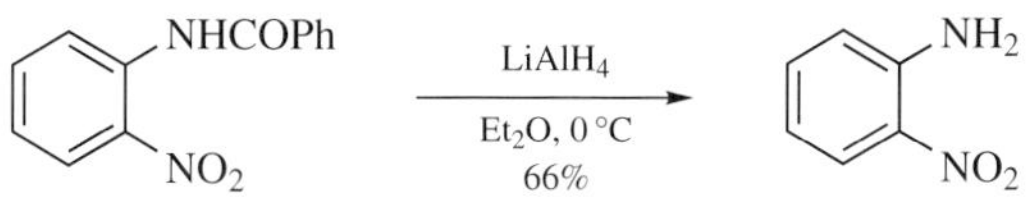

(59)

Reduction of Hydrazones. Tosylhydrazones of acetonide protected dihydroxy aldehydes undergo facile reduction with LiAlH$_4$ (eq 60).[112] The tosylhydrazones must be converted to their sodium salts with sodium hexamethyldisilazide prior to the addition of LiAlH$_4$. This combination of reagents totally blocks the usual Wolff–Kishner-type reduction in favor of a different reaction pathway. A variety of different sugar hydrazones are conveniently reduced with this method to provide the corresponding chiral allylic alcohols in good yields. Reduction of *trans*-2,3-diphenyl aziridine hydrazones leads to the corresponding alkanes (eq 61).[113] Several other hydrides have been screened but LiAlH$_4$ proved to be the reagent of choice. This procedure is a useful alternative to the existing hydrazone reductions. Yields are usually moderate but compare well with those observed for related reactions under more drastic conditions. A tandem denitration-deoxygenation is performed by reduction of (*p*-tolylsulfonyl)hydrazones of α-nitro ketones with LiAlH$_4$ in THF at 60 °C (eq 62).[114] Compared to the classical methods to effect the denitrohydrogenation of α-nitro ketones [115,116] and to the Caglioti reaction for the carbonyl to methylene conversion, this procedure has been found to provide the corresponding alkanes in good yields.

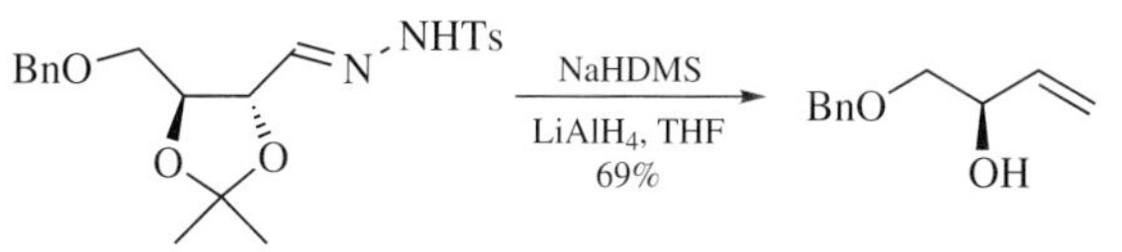

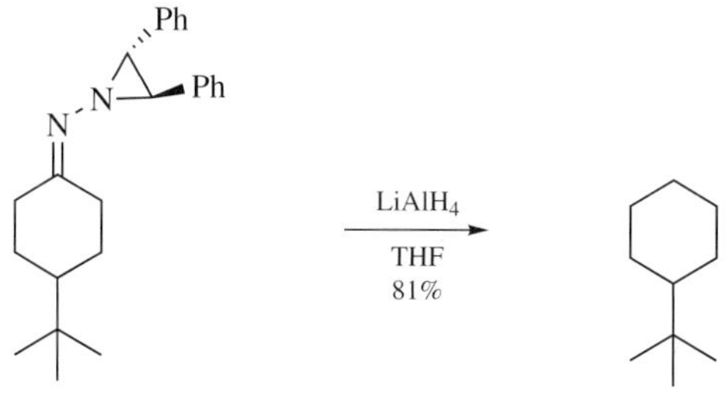

(61)

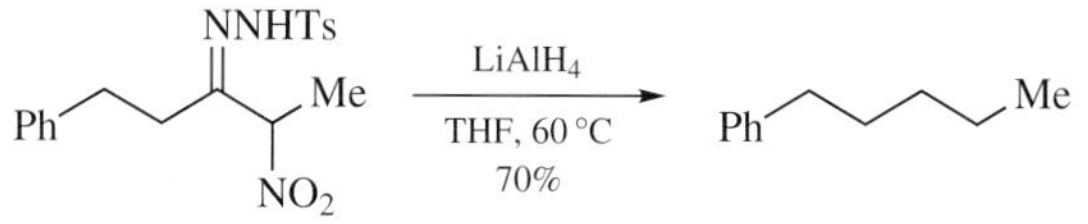

(62)

Reduction of Sulfur or Selenium Compounds. A number of sulfenates are reduced to alkenethiolate anions with LiAlH$_4$ in THF at -78 °C (eq 63).[117] Deprotonation of thiirane *S*-oxides with a base immediately followed by reduction with LiAlH$_4$ and capture with trialkylsilyl chlorides leads to the (*E*)-1-alkenylthiosilanes. This sulfenate generation/reduction sequence provides a facile route to enethiolates of thioaldehydes and selected thioketones. The reduction of propargyl *S*-thioacetate with LiAlH$_4$ constitutes an attractive preparation of unstable propargylthiol (eq 64).[118] This process is a practical alternative compared to other methods and allows the introduction of the propargylsulfenyl moiety in several types of electrophilic substrates without the need to isolate propargylthiol. The lithium propargyl thiolate resulting from the reduction of propargyl *S*-thioacetate can be directly trapped by benzyl bromide or other electrophiles. β-Chloro disulfides and β-chloro trisulfides undergo reduction with LiAlH$_4$ to afford episulfides.[119] Selenoiminium salts, which are obtained by reacting the corresponding selenoamides with methyl triflate, are converted to telluroamides upon exposure to a reagent prepared from LiAlH$_4$ and elemental tellurium (eq 65).[120] This procedure enables isolation of the first known aliphatic telluroamide in good yield. The reaction of LiAlH$_4$ with elemental selenium affords a selenating reagent that is capable of preparing a wide range of selenium-containing compounds.[121] Reduction with LiAlH$_4$ of the thiocyanate sulfate potassium salts, easily obtained by the opening of cyclic sulfates with KSCN, allows the direct synthesis of 5-deoxy-4-thio- and 6-deoxy-5-thiosugars (eq 66).[122] The process involves the in situ formation of a Li-thiolate salt which undergoes an intramolecular nucleophilic displacement of the β-sulfate group. The overall transformation makes the method attractive for converting *vic*-diols into thio derivatives in the 5-thiosugar series.

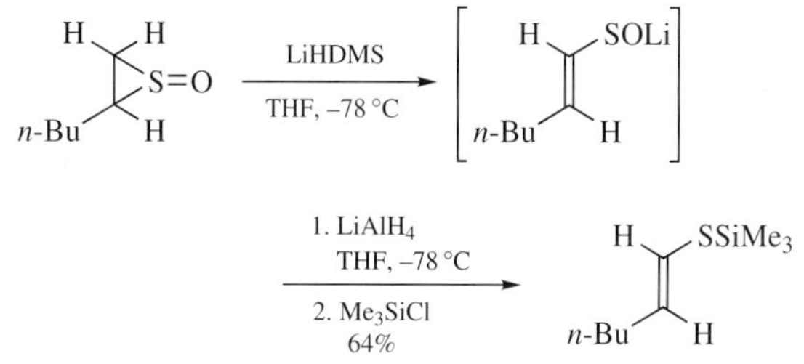

(63)

(60)

$$\text{(64)}$$

$$\text{(65)}$$

$$\text{(66)}$$

Other Reactions. A stereoselective hydroxy-directed hydroalumination is reported for the reduction of β-aryl enones with LiAlH$_4$ (eq 67).[123] The reduction occurs with a variety of aryl substituted enones with complete control of stereochemistry at the α- and/or β-position, and the *trans*-products resulting from a directed hydroalumination of the intermediate allylic alcohols are obtained. This transformation is a useful method affording *trans*-substituted cycloalkanols in good yields. The reduction of the silyl enol ether derivatives of α-ketoesters by LiAlH$_4$ is a useful method for preparing 1-hydroxy-2-alkanones (eq 68).[124] Upon treatment of the readily available silyloxy propenoates with LiAlH$_4$, the corresponding α-hydroxy ketones are exclusively obtained from reactions carried out in THF. An intermediate alkoxyhydridoaluminate is assumed to be responsible for intramolecular desilylation reaction. Improvement of the scope of this reduction is achieved by performing the reductions in Et$_2$O, in which the competing formation of α-silyloxy ketones is observed with good or no selectivity with (Z)- and (E)-substrates, respectively. 1,3-Diols are obtained by the reaction of α,β-unsaturated ketones with LiAlH$_4$ in THF under a dry oxygen atmosphere (eq 69).[125] Diols are obtained as an equimolecular mixture of *syn* and *anti* diastereoisomers. This procedure constitutes a one-pot transformation of an α,β-unsaturated ketone into a 1,3-diol in very good yields. Reduction of benzotriazole derivatives with LiAlH$_4$ leads to substitution of the benzotriazolyl moiety by a hydrogen atom (eq 70).[126,127] The reduction of 1-(α-alkoxybenzyl)benzotriazoles with LiAlH$_4$ affords the corresponding ethers. Similarly, displacement of a benzotriazole group in *N*-heterocycle derivatives with LiAlH$_4$ is a synthetic route to *N*-substituted heterocycles (eq 71).[127] The reduction of a β-aminonitroalkene with LiAlH$_4$ leads to a mixture of the corresponding *trans*-diamine derivative and a ring-cleavage product albeit in a low yield (eq 72).[128] Formation of the latter occurs through the conjugate reduction with LiAlH$_4$, followed by a retro-aldol reaction, and reduction of the intermediate. LiAlH$_4$ induces reductive coupling of 2-cyanopyrroles to provide a simple and efficient synthesis of (2-pyrrolylmethene)-(2-pyrrolylmethyl)imines (eq 73).[129] The imine is formed as the sole

product in 85% yield. A radical mechanism is reported for the reductive cyanation of an α-sulfonitrile with LiAlH$_4$ in THF.[130,131] When a bicyclic α-sulfonitrile is reduced with LiAlH$_4$, a mixture of the expected amine and the hydrocarbon is obtained in a 1:1 ratio (eq 74). Similarly, when 2,2-diphenylpropionitrile is reduced with LiAlH$_4$ in THF, a mixture of the amine and the hydrocarbon is formed (eq 75).[132] LiAlH$_4$ can be used with PCC or PDC in a 'reductive oxidation' process converting carboxylic derivatives to aldehydes.[133–136] The method involves the rapid reduction of carboxylic acids, esters, or acid chlorides with LiAlH$_4$ at 0 °C, followed by oxidation of the resulting alkoxyaluminum intermediate with PCC or PDC. Finally, various phosphine oxides are efficiently reduced by the use of a methylation reagent and LiAlH$_4$ (eq 76).[137] Optically active *P*-chirogenic phosphine oxides are also reduced with inversion of configuration at phosphorus atom by treatment with methyl triflate, followed by reduction with LiAlH$_4$.

$$\text{(67)}$$

$$\text{(68)}$$

Conditions		
1. THF, –100 °C to 20 °C, 77%	100	0
2. Et$_2$O, –100 °C to –15 °C, 83%	20	80

$$\text{(69)}$$

$$\text{(70)}$$

$$\text{(71)}$$

NH—C$_6$H$_4$—OCH$_3$

NO$_2$

LiAlH$_4$ →

NH—OCH$_3$ + NH—OCH$_3$ (72)

NH$_2$ NH$_2$

9%   12%

LiAlH$_4$, THF, 85% (73)

LiAlH$_4$, THF, 62%

exo/endo (84/16)

+ H$_2$NCH$_2$ (74)

SO$_2$*i*Pr   SO$_2$*i*Pr

exo/endo (95/5)   *exo/endo* (84/16)

Ph$_2$C—Me, CN → Ph$_2$C—Me, CH$_2$NH$_2$ + Ph$_2$C—Me, H (75)

LiAlH$_4$, THF   57%   43%

1. MeOTf, DME
2. LiAlH$_4$, –60 °C, 85% (76)

98% ee

1. (a) Brown, W. G., *Org. React.* **1951**, *6*, 469. (b) Gaylord, N. G., *Reduction with Complex Metal Hydrides*; Interscience: New York, 1956. (c) *Reduction Techniques and Applications in Organic Synthesis*; Augustine, R. L., Ed.; Dekker: New York, 1968. (d) Pizey, J. S., *Synthetic Reagents*, Wiley: New York, 1974; Vol. 1, p 101. (e) Hajos, A., *Complex Hydrides and Related Reducing Agents in Organic Synthesis*; Elsevier: New York, 1979. (f) Grandbois, E. R.; Howard, S. I.; Morrison, J. D., *Asymmetric Synthesis*; Academic: New York, 1983 Vol. 2. (g) *Comprehensive Organic Synthesis* **1991**, *8*, Chapters 1.1.8. (h) Carey, F. A.; Sundberg, R. J., *Advanced Organic Chemistry*, 3rd ed.; Plenum: New York, 1990; Part B, p 232.

2. Ziegler, K.; Bond, A. C., Jr.; Schlesinger, H. I., *J. Am. Chem. Soc.* **1947**, *69*, 1199.

3. Bates, R. B.; Büchi, G.; Matsuura, T.; Shaffer, R. R., *J. Am. Chem. Soc.* **1960**, *82*, 2327.

4. Fieser, L. F.; Fieser, M., *Fieser & Fieser* **1967**, *1*, 584.

5. (a) Paquette, L. A.; Gardlik, J. M.; McCullough, K. J.; Samodral, R.; DeLucca, G.; Ouellette, R. J., *J. Am. Chem. Soc.* **1983**, *105*, 7649. (b) Paquette, L. A.; Gardlik, J. M., *J. Am. Chem. Soc.* **1980**, *102*, 5016.

6. Sroog, C. E.; Woodburn, H. M., *Org. Synth., Coll. Vol.* **1963**, *4*, 271.

7. Mousseron, M.; Jacquier, R.; Mousseron-Canet, M.; Zagdoun, R., *Bull. Soc. Chem. Fr.* **1952**, *19*, 1042.

8. (a) Weygand, F.; Eberhardt, G., *Angew. Chem.* **1952**, *64*, 458. (b) Weygand, F.; Eberhardt, G.; Linden, H.; Schäfer, F.; Eigen, I., *Angew. Chem.* **1953**, *65*, 525.

9. Brown, H. C.; Tsukomoto, A., *J. Am. Chem. Soc.* **1961**, *83*, 2016, 4549.

10. Staab, H. A.; Bräunling, H., *Justus Liebigs Ann. Chem.* **1962**, *654*, 119.

11. Nahm, S.; Weinreb, S. M., *Tetrahedron Lett.* **1981**, *22*, 3815.

12. (a) Johnson, J. E.; Blizzard, R. H.; Carhart, H. W., *J. Am. Chem. Soc.* **1948**, *70*, 3664. (b) Krishnamurthy, S.; Brown, H. C., *J. Org. Chem.* **1982**, *47*, 276.

13. (a) Schmid, H.; Karrer, P., *Helv. Chim. Acta* **1949**, *32*, 1371. (b) Zorbach, W. W.; Tio, C. O., *J. Org. Chem.* **1961**, *26*, 3543. (c) Wang, P.-C.; Lysenko, Z.; Joullié, M. M., *Tetrahedron Lett.* **1978**, 1657.

14. (a) Ashby, E. C.; De Priest, R. N.; Goel, A. B.; Wenderoth, B.; Pham, T. N., *J. Org. Chem.* **1984**, *49*, 3545. (b) Ashby, E. C.; Pham, T. N., *J. Org. Chem.* **1986**, *51*, 3598. (c) Ashby, E. C.; Pham, T. N.; Amrollah-Madjdabadi, A., *J. Org. Chem.* **1991**, *56*, 1596.

15. (a) Trevay, L. W.; Brown, W. G., *J. Am. Chem. Soc.* **1949**, *71*, 1675. (b) Buchta, E.; Loew, G., *Justus Liebigs Ann. Chem.* **1955**, *597*, 123. (c) Parham, W. E.; Wright, C. D., *J. Org. Chem.* **1957**, *22*, 1473. (d) Ligouri, A.; Sindona, G.; Uccella, N., *Tetrahedron* **1983**, *39*, 683.

16. (a) Jefford, C. W.; Mahajan, S. N.; Gunsher, J., *Tetrahedron* **1968**, *24*, 2921. (b) Ohloff, G.; Farrow, H.; Schade, G., **1956**, *89*, 1549. (c) Fraser-Reid, B.; Tam, S. Y.-K.; Radatus, B., *Can. J. Chem.* **1975**, *53*, 2005.

17. Magid, R. M., *Tetrahedron* **1980**, *36*, 1901.

18. Karabatsos, G. J.; Shone, R. L., *J. Org. Chem.* **1968**, *33*, 619.

19. (a) Moore, L. D., *J. Chem. Eng. Data* **1964**, *9*, 251. (b) Marvel, C. S.; Wilson, B. D., *J. Org. Chem.* **1958**, *23*, 1483.

20. (a) Jefford, C. W.; Sweeney, A.; Delay, F., *Helv. Chim. Acta* **1972**, *55*, 2214. (b) Jefford, C. W.; Kirkpatrick, D.; Delay, F., *J. Am. Chem. Soc.* **1972**, *94*, 8905.

21. Jefford, C. W.; Burger, U.; Laffer, M. H.; Kabengele, T., *Tetrahedron Lett.* **1973**, 2483.

22. (a) McKinney, M. A.; Anderson, S. M.; Keyes, M.; Schmidt, R., *Tetrahedron Lett.* **1982**, *23*, 3443. (b) Hatem, J.; Meslem, J. M.; Waegell, B., *Tetrahedron Lett.* **1986**, *27*, 3723.

23. (a) Fraser-Reid, B.; Radatus, B., *J. Amm. Chem. Soc.* **1970**, *92*, 6661. (b) Radatus, B.; Yunker, M.; Fraser-Reid, B., *J. Am. Chem. Soc.* **1971**, *93*, 3086. (c) Tam, S. Y.-K.; Fraser-Reid, B., *Tetrahedron Lett.* **1973**, 4897.

24. Claus, C. J.; Morgenthau, J. L., Jr., *J. Amm. Chem. Soc.* **1951**, *73*, 5005.

25. Eliel, E. L.; Rerick, M. N., *J. Org. Chem.* **1958**, *23*, 1088.

26. (a) Slaugh, L. H., *Tetrahedron* **1966**, *22*, 1741. (b) Magoon, E. F.; Slaugh, L. H., *Tetrahedron* **1967**, *23*, 4509.

27. Pasto, D. J., *Comprehensive Organic Synthesis* **1991**, *8*, Chapter 3.3.

28. (a) Sato, F.; Kodama, H.; Sato, M., *Chem. Lett.* **1978**, 789. (b) Sato, F.; Ogura, K.; Sato, M., *Chem. Lett.* **1978**, 805.

29. Zweifel, G.; Lewis, W.; On, H. P., *J. Am. Chem. Soc.* **1979**, *101*, 5101.

30. Rossi, R.; Carpita, A., *Synthesis* **1977**, 561.

31. Solladié, G.; Berl, V., *Tetrahedron Lett.* **1992**, *33*, 3477.

32. Baudony, R.; Goré, J., *Tetrahedron Lett.* **1974**, 1593.

33. (a) Borden, W. T., *J. Am. Chem. Soc.* **1970**, *92*, 4898. (b) Grant, B.; Djerassi, C., *J. Org. Chem.* **1974**, *39*, 968.

34. Claesson, A.; Bogentoft, C., *Synthesis* **1973**, 539.

35. (a) Claesson, A., *Acta Chem. Scand.* **1975**, *B29*, 609. (b) Galantay, E.; Basco, I.; Coombs, R. V., *Synthesis* **1974**, 344. (c) Cowie, J. S.; Landor, P. D.; Landor, S. R., *J. Chem. Soc., Perkin Trans. 1* **1973**, 270. (d) Claesson, A.; Olsson, L.-I.; Bogentoft, C., *Acta Chem. Scand.* **1973**, *B27*, 2941.

36. (a) Olsson, L.-I.; Claesson, A.; Bogentoft, C., *Acta Chem. Scand.* **1974**, *B28*, 765. (b) Claesson, A., *Acta Chem. Scand.* **1974**, *B28*, 993.

37. Keck, G. E.; Webb, R. R., II, *Tetrahedron Lett.* **1982**, *23*, 3051.

38. Corey, E. J.; Katzenellenbogen, J. A.; Posner, G., *J. Am. Chem. Soc.* **1967**, *89*, 4245.

39. (a) Rickborn, B.; Quartucci, J., *J. Org. Chem.* **1984**, *29*, 3185. (b) Rickborn, B.; Lamke, W. E., II, *J. Org. Chem.* **1967**, *32*, 537. (c) Murphy, D. K.; Alumbaugh, R. L.; Rickborn, B., *J. Am. Chem. Soc.* **1969**, *91*, 2649.

40. Mihailovic, M. Lj.; Andrejevic, V.; Milovanoic, J.; Jankovic, J., *Helv. Chim. Acta* **1976**, *59*, 2305.

41. (a) Parish, E. J.; Schroepper, G. J., Jr., *Tetrahedron Lett.* **1976**, 3775. (b) Fraser-Reid, B.; Tam, S. Y.-K.; Radatus, B., *Can. J. Chem.* **1975**, *53*, 2005.

42. (a) Kotera, K.; Kitahonoki, K., *Tetrahedron Lett.* **1965**, 1059. (b) Kotera, K.; Kitahonoki, K., *Org. Synth.* **1968**, *48*, 20.

43. Kato, M.; Kurihara, H.; Yoshikoshi, A., *J. Chem. Soc., Perkin Trans. 1* **1979**, 2740.

44. Gassman, P. G.; Gilbert, D. P.; van Bergen, T. J., *J. Chem. Soc., Chem. Commun.* **1974**, 201.

45. Hojo, M.; Masuda, R.; Takagi, S., *Synthesis* **1978**, 284.

46. Gammill, R. B.; Bell, L. T.; Nash, S. A., *J. Org. Chem.* **1984**, *49*, 3039.

47. Gaoni, Y., *Tetrahedron Lett.* **1977**, 947.

48. (a) Photis, J. M.; Paquette, L. A., *J. Am. Chem. Soc.* **1974**, *96*, 4715. (b) Photis, J. M.; Paquette, L. A., *Org. Synth.* **1977**, *57*, 53.

49. Lansbury, P. T.; MacLeay, R. E., *J. Org. Chem.* **1963**, *28*, 1940.

50. Paquette, L. A.; Underiner, T. L.; Gallucci, J. C., *J. Org. Chem.* **1992**, *57*, 86.

51. (a) Wigfield, D. C., *Tetrahedron* **1979**, *35*, 449. (b) Mukherjee, D.; Wu, Y.-D.; Fronczek, R. F.; Houk, K. N., *J. Am. Chem. Soc.* **1988**, *110*, 3328. (c) Wu, Y.-D.; Tucker, J. A.; Houk, K. N., *J. Am. Chem. Soc.* **1991**, *113*, 5018.

52. (a) Chérest, M.; Felkin, H.; Prudent, N., *Tetrahedron Lett.* **1968**, 2199. (b) Ahn, N. T.; Eisenstein, O., *Nouv. J. Chim.* **1977**, *1*, 61.

53. Eliel, E. L., In *Asymmetric Synthesis*; Morrison, J. D., Ed.; Academic: New York, 1983; Vol. 2, Chapter 5.

54. (a) Cram, D. J.; Kopecky, K. R., *J. Am. Chem. Soc.* **1959**, *81*, 2748. (b) Cram, D. J.; Wilson, D. R., *J. Am. Chem. Soc.* **1963**, *85*, 1249.

55. Cieplak, A. S., *J. Am. Chem. Soc.* **1981**, *103*, 4540.

56. (a) Cheung, C. K.; Tseng, L. T.; Lin, M.-H.; Srivastava, S.; leNoble, W. J., *J. Am. Chem. Soc.* **1986**, *108*, 1598. (b) Cieplak, A. S.; Tait, B. D.; Johnson, C. R., *J. Am. Chem. Soc.* **1989**, *111*, 8447. (c) Okada, K.; Tomita, S.; Oda, M., *Bull. Chem. Soc. Jpn.* **1989**, *62*, 459. (d) Li, H.; Mehta, G.; Podma, S.; leNoble, W. J., *J. Org. Chem.* **1991**, *56*, 2006. (e) Mehta, G.; Khan, F. A., *J. Am. Chem. Soc.* **1990**, *112*, 6140.

57. Nogradi, M., *Stereoselective Synthesis*; VCH: Weinheim, 1986.

58. Overman, L. E.; McCready, R. J., *Tetrahedron Lett.* **1982**, *23*, 2355.

59. Other examples: (a) Iida, H.; Yamazaki, N.; Kibayashi, C., *J. Chem. Soc., Chem. Commun.* **1987**, 746. (b) Bloch, R.; Gilbert, L.; Girard, C., *Tetrahedron Lett.* **1988**, *29*, 1021. (c) Fukuyama, T.; Vranesic, B.; Negri, D. P.; Kishi, Y., *Tetrahedron Lett.* **1978**, 2741.

60. (a) Mori, Y.; Kuhara, M.; Takeuchi, A.; Suzuki, M., *Tetrahedron Lett.* **1988**, *29*, 5419. (b) Mori, Y.; Takeuchi, A.; Kageyama, H.; Suzuki, M., *Tetrahedron Lett.* **1988**, *29*, 5423.

61. Dickman, D. A.; Meyers, A. I.; Smith, G. A.; Gawley, R. E., *Org. Synth., Coll. Vol.* **1990**, *7*, 530.

62. (a) Jäger, V.; Buss, V., *Liebigs Ann. Chem.* **1980**, 101. (b) Jäger, V.; Buss, V.; Schwab, W., *Liebigs Ann. Chem.* **1980**, 122.

63. (a) Jäger, V.; Schwab, W.; Buss, V., *Angew. Chem., Int. Ed. Engl.* **1981**, *20*, 601. (b) Schwab, W.; Jäger, V., *Angew. Chem., Int. Ed. Engl.* **1981**, *20*, 603.

64. Narasaka, K.; Ukaji, Y., *Chem. Lett.* **1984**, 147.

65. Narasaka, K.; Yamazaki, S.; Ukaji, Y., *Bull. Chem. Soc. Jpn.* **1986**, *59*, 525.

66. Eliel, E.; Frazee, W. F., *J. Org. Chem.* **1979**, *44*, 3598.

67. Eliel, E. L.; Lynch, J. E., *Tetrahedron Lett.* **1981**, *22*, 2859.

68. Mukaiyama, T., *Tetrahedron* **1981**, *37*, 4111.

69. Nishizawa, M.; Noyori, R., *Comprehensive Organic Synthesis* **1991**, *8*, Chapter 1.7.

70. Yamaguchi, S.; Mosher, H. S.; Pohland, A., *J. Am. Chem. Soc.* **1972**, *94*, 9254.

71. (a) Reich, C. J.; Sullivan, G. R.; Mosher, H. S., *Tetrahedron Lett.* **1973**, 1505. (b) Brinkmeyer, R. S.; Kapoor, V. M., *J. Am. Chem. Soc.* **1977**, *99*, 8339. (c) Johnson, W. S.; Brinkmeyer, R. S.; Kapoor, V. M.; Yarnell, T. M., *J. Am. Chem. Soc.* **1977**, *99*, 8341. (d) Marshall, J. A.; Robinson, E. D., *Tetrahedron Lett.* **1989**, *30*, 1055.

72. Deeter, J.; Frazier, J.; Staten, G.; Staszak, M.; Weigel, L., *Tetrahedron Lett.* **1990**, *31*, 7101.

73. Paquette, L. A.; Combrink, K. D.; Elmore, S. W.; Rogers, R. D., *J. Am. Chem. Soc.* **1991**, *113*, 1335.

74. (a) Noyori, R.; Tomino, I.; Tanimoto, Y., *J. Am. Chem. Soc.* **1979**, *101*, 3129. (b) Noyori, R.; Tomino, I.; Tanimoto, Y.; Nishizawa, M., *J. Am. Chem. Soc.* **1984**, *106*, 6709.

75. (a) Noyori, R.; Tomino, I.; Nishizawa, M., *J. Am. Chem. Soc.* **1979**, *101*, 5843. (b) Beechström, P.; Björkling, F.; Högberg, H.-E.; Norin, T., *Acta Chem. Scand.* **1983**, *B37*, 1.

76. (a) Noyori, R.; Tomino, I.; Yamada, M.; Nishizawa, M., *J. Am. Chem. Soc.* **1984**, *106*, 6717. (b) Nishizawa, M.; Yamada, M.; Noyori, R., *Tetrahedron Lett.* **1981**, *22*, 247.

77. de Vries, E. F. J.; Brussee, J.; van der Gen, A., *J. Org. Chem.* **1994**, *59*, 7133.

78. Saravanan, P.; Gupta, S.; DattaGupta, A.; Gupta, S.; Singh, V. K., *Synth. Commun.* **1997**, *27*, 2695.

79. de Vries, E. F. J.; Brussee, J.; Kruse, C. G.; van der Gen, A., *Tetrahedron: Asymmetry* **1994**, *5*, 377.

80. Breeden, S. W.; Coop, A.; Husbands, S. M.; Lewis, J. W., *Helv. Chim. Acta* **1999**, *82*, 1978.

81. Zimmer, R.; Homann, K.; Reissig, H.-U., *Liebigs Ann. Chem.* **1993**, 1155.

82. Félix, C. P.; Khatimi, N.; Laurent, A. J., *J. Org. Chem.* **1995**, *60*, 3907.

83. Lohray, B. B.; Bhushan, V.; Reddy, A. S.; Rao, V. V., *Indian J. Chem., Sect. B* **2000**, *39B*, 297.

84. Hong, J. E.; Shin, W. S.; Jang, W. B.; Oh, D. Y., *J. Org. Chem.* **1996**, *61*, 2199.

85. Jang, W. B.; Shin, W. S.; Hong, J. E.; Lee, S. Y.; Oh, D. Y., *Synth. Commun.* **1997**, *27*, 3333.

86. Lee, S. Y.; Hong, J. E.; Jang, W. B.; Oh, D. Y., *Tetrahedron Lett.* **1997**, *38*, 4567.

87. Lee, S. Y.; Lee, C.-W.; Oh, D. Y., *J. Org. Chem.* **1999**, *64*, 7017.

88. Amedjkouh, M.; Grimaldi, J., *Tetrahedron Lett.* **2002**, *43*, 3761.

89. Sanders, M. L.; Donkor, I. O., *Synth. Commun.* **2002**, *32*, 1015.

90. Tokuda, M.; Fujita, H.; Nitta, M.; Suginome, H., *Helv. Chim. Acta* **1996**, *42*, 385.

91. Stevens, C.; De Kimpe, N., *J. Org. Chem.* **1993**, *58*, 132.

92. Dave, P. R.; Forohar, F.; Axenrod, T.; Das, K. K.; Qi, L.; Watnick, C.; Yazdekhasti, H., *J. Org. Chem.* **1996**, *61*, 8897.

93. Dhokte, U. P.; Pathare, P. M.; Mahindroo, V. K.; Brown, H. C., *J. Org. Chem.* **1998**, *63*, 8276.

94. Bach, T.; Kather, K., *J. Org. Chem.* **1996**, *61*, 3900.

95. Bach, T.; Lange, C., *Tetrahedron Lett.* **1996**, *37*, 4363.

96. Nath, A.; Mal, J.; Venkateswaran, R. V., *J. Org. Chem.* **1996**, *61*, 4391.

97. Mal, J.; Venkateswaran, R. V., *J. Org. Chem.* **1998**, *63*, 3855.

98. Youn, S. W.; Kim, Y. H., *Synlett* **2000**, 880.

99. Booth, S. E.; Jenkins, P. R.; Swain, C. J., *Chem. Commun.* **1993**, 147.

100. Cushman, M.; Chinnasamy, P.; Patrick, D. A.; McKenzie, A. T.; Toma, P. H., *J. Org. Chem.* **1990**, *55*, 5995.

101. Connolly, T. J.; Durst, T., *Can. J. Chem.* **1997**, *75*, 542.

102. Harmata, M.; Bohnert, G.; Kürti, L.; Barnes, C. L., *Tetrahedron Lett.* **2002**, *43*, 2347.

103. Baik, W.; Lee, H. J.; Koo, S.; Kim, B. H., *Tetrahedron Lett.* **1998**, *39*, 8125.

104. Nakamura, Y.; Okada, M.; Horikawa, H.; Taguchi, T., *J. Fluorine Chem.* **2002**, *117*, 143.

105. Strekowski, L.; Lin, S.-Y.; Lee, H.; Mason, J. C., *Tetrahedron Lett.* **1996**, *37*, 4655.

106. Douat, C.; Heitz, A.; Martinez, J.; Fehrentz, J.-A., *Tetrahedron Lett.* **2000**, *41*, 37.

107. Murai, T.; Ezaka, T.; Kanda, T.; Kato, S., *Chem. Commun.* **1996**, 1809.

108. Akamatsu, H.; Kusumoto, S.; Fukase, K., *Tetrahedron Lett.* **2002**, *43*, 8867.

109. Hendrix, J. A.; Stefany, D. W., *Tetrahedron Lett.* **1999**, *40*, 6749.

110. Joseph, M. M.; Jacob, E. D., *Indian J. Chem., Sect. B* **1996**, *35B*, 482.

111. Joseph, M. M.; Jacob, D. E., *Indian J. Chem., Sect. B* **1998**, *37B*, 945.

112. Chandrasekhar, S.; Mohapatra, S.; Takhi, M., *Synlett* **1996**, 759.

113. Leone, C. L.; Chamberlin, A. R., *Tetrahedron Lett.* **1991**, *32*, 1691.

114. Ballini, R.; Petrini, M.; Rosini, G., *J. Org. Chem.* **1990**, *55*, 5159.

115. Ono, N.; Kaji, A., *Synthesis* **1986**, 693.

116. Rosini, G.; Ballini, R., *Synthesis* **1988**, 833.

117. Schwan, A. L.; Refvik, M. D., *Synlett* **1998**, 96.

118. Castro, J.; Moyano, A.; Pericàs, M. A.; Riera, A., *Synthesis* **1997**, 518.

119. Abu-Yousef, I. A.; Harpp, D. N., *Sulfur Lett.* **2000**, *23*, 131.

120. Mutoh, Y.; Murai, T., *Org. Lett.* **2003**, *5*, 1361.

121. Ishihara, H.; Koketsu, M.; Fukuta, Y.; Nada, F., *J. Am. Chem. Soc.* **2001**, *123*, 8408.

122. Calvo-Flores, F. G.; García-Mendoza, P.; Hernández-Mateo, F.; Isac-García, J.; Santoyo-González, F., *J. Org. Chem.* **1997**, *62*, 3944.

123. Koch, K.; Smitrovich, J. H., *Tetrahedron Lett.* **1994**, *35*, 1137.

124. Dalla, V.; Catteau, J. P., *Tetrahedron* **1999**, *55*, 6497.

125. Csákÿ, A. G.; Máximo, N.; Plumet, J.; Rámila, A., *Tetrahedron Lett.* **1999**, *40*, 6485.

126. Katritzky, A. R.; Rachwal, S.; Rachwal, B.; Steel, P. J., *J. Org. Chem.* **1992**, *57*, 4925.

127. Katritzky, A. R.; Lang, H.; Lan, X., *Tetrahedron* **1993**, *49*, 2829.

128. Freeman, J. P.; Laurian, L.; Szmuszkovicz, J., *Tetrahedron Lett.* **1999**, *40*, 4493.

129. Brückner, C.; Xie, L. Y.; Dolphin, D., *Tetrahedron* **1998**, *54*, 2021.

130. Mattalia, J.-M.; Berchadsky, Y.; Péralez, E.; Négrel, J.-C.; Tordo, P.; Chanon, M., *Tetrahedron Lett.* **1996**, *37*, 4717.

131. Mattalia, J.-M.; Samat, A.; Chanon, M., *J. Chem. Soc., Perkin Trans. 1* **1991**, 1769.

132. Mattalia, J.-M.; Bodineau, N.; Négrel, J.-C.; Chanon, M., *J. Phys. Org. Chem.* **2000**, *13*, 233.

133. Cha, J. S.; Chun, J. H.; Kim, J. M.; Lee, D. Y.; Cho, S. D., *Bull. Korean Chem. Soc.* **1999**, *20*, 1373.

134. Cha, J. S.; Chun, J. H.; Kim, J. M.; Kwon, O. O.; Kwon, S. Y.; Lee, J. C., *Bull. Korean Chem. Soc.* **1999**, *20*, 400.

135. Cha, J. S.; Kim, J. M.; Chun, J. H.; Kwon, O. O.; Kwon, S. Y.; Han, S. W., *Org. Prep. Proced. Int.* **1999**, *31*, 204.

136. Cha, J. S.; Chun, J. H., *Bull. Korean Chem. Soc.* **2000**, *21*, 375.

137. Imamoto, T.; Kikuchi, S.-I.; Miura, T.; Wada, Y., *Org. Lett.* **2001**, *3*, 87.

Lithium Diisopropylamide[1,2]

[4111-54-0] $C_6H_{14}LiN$ (MW 107.15)
InChI = 1/C6H14N.Li/c1-5(2)7-6(3)4;/h5-6H,1-4H3;/q-1;+1
InChIKey = ZCSHNCUQKCANBX-UHFFFAOYAP

(hindered nonnucleophilic strong base used for carbanion generation, especially kinetic enolates,[4] α-heteroatom, allylic,[5,6] and aromatic and heteroaromatic carbanions;[7–9] unavoidable in organic synthesis)

Alternate Name: LDA.
Physical Data: powder, melts with decomposition; $pK_a = 35.7$ in THF.[3]
Solubility: soluble in Et_2O, THF, DME, HMPA; unstable above 0 °C in these solvents; stable in hexane or pentane (0.5–0.6 M) at rt for weeks, when not cooled or concentrated;[10] the complex with one molecule of THF is soluble in alkanes like cyclohexane and heptane.
Form Supplied in: solid; 2.0 M solution in heptane/THF/ethylbenzene stabilized with $Mg(i\text{-}Pr_2N)_2$; 10 wt % suspension in hexanes; LDA·THF complex, 1.5 M solution in cyclohexane and 2.0 M in heptane.
Analysis of Reagent Purity: several titration methods have been described.[120]
Preparative Methods: generally prepared directly before use from anhydrous **Diisopropylamine** and commercially available solutions of **Butyllithium**. Another process, especially useful for the preparation of large quantities, is the reaction of styrene with 2 equiv of **Lithium** and 2 equiv of $i\text{-}Pr_2NH$ in Et_2O.[11]
Handling, Storage, and Precaution: very moisture- and air-sensitive, and should always be kept in an inert atmosphere; irritating to the skin and mucous membranes; therefore contact with skin, eyes, and other tissues and organs should be avoided; proper protection is necessary. Use in a fume hood.

Original Commentary

Wouter I. Iwema Bakker, Poh Lee Wong & Victor Snieckus
University of Waterloo, Waterloo, Ontario, Canada

Enolates. LDA is undeniably the first-choice base for the quantitative formation of enolates in general and kinetic (usually less substituted) enolates in particular. It was first used for this purpose by Hamell and Levine;[12] now it is a rare day (or night) that a chemist does not use or a graduate student does not propose LDA for a carbanionic transformation. Its advantages (fast, complete, and regiospecific enolate generation, unreactivity to alkyl halides, lack of interfering products) over earlier developed bases (e.g. **Sodium Hydride**, **Triphenylmethyllithium**)[13] are no longer noted. Except for enolates of aldehydes, which are very reactive and undergo self-aldol condensation reactions,[14] the derived lithio enolates can be used for a broad spectrum of reactions including *O*- and *C*-alkylations and acylations, transmetalation to other synthetically useful metallo enolates,[15] and a multitude of carbanion-

based condensations and rearrangements. The trapping of kinetically derived lithio enolates with **Chlorotrimethylsilane** as silyl enol ethers[16] (e.g. eqs 1 and 2),[17] originally used for separation from minor regioisomers and regeneration (with **Methyllithium** or **Lithium Amide**) for the purpose of regiospecific alkylation, has been superseded in general by methods of their direct alkylation under Lewis acid catalysis.[18]

$$\text{(1)}$$

$$\text{(2)}$$
1:0.20

Consonant with its pK_a, LDA is used to derive a variety of α-stabilized carbanions, e.g. ketones (eqs 1 and 3),[19,20] α-amino ketones,[21] ω-bromo ketones (eq 4),[10] imines (eqs 6 and 7),[22–24] imino ethers (eq 8),[25] carboxylic acids (eq 9),[26] carboxylic esters,[27] lactones,[28] amides,[29] lactams,[30] imides,[31] and nitriles[32] which may be alkylated or used in aldol condensations. Enolates of methyl ketones may be *O*-alkylated with **Diethyl Phosphorochloridate** to give enol phosphates which, upon β-elimination, afford terminal alkynes in high yields (eq 5).[33]

$$\text{(3)}$$

$$\text{(4)}$$

$$\text{(5)}$$
75–95% overall

$$\text{(6)}$$

$$\text{(7)}$$

The (Z)/(E) stereoselectivity of enolate formation is dictated by the structure of the starting carbonyl compound and the base used for deprotonation. Compared to LDA, **Lithium 2,2,6,6-Tetramethylpiperidide** usually favors (E)-enolates whereas **Lithium Hexamethyldisilazide** preferentially leads to (Z)-enolates (eq 10).[34] With a *caveat* for any generalization, enolate configuration usually determines the stereochemical result in the product; for example, using a hindered ester and a bulky aldehyde combination, excellent stereoselectivities in aldol reactions are observed (eq 11).[27]

A useful reaction of ketone enolates is their oxidative coupling,[35] e.g. in the formation of a tricyclic intermediate towards the synthesis of the diterpene cerorubenic acid-III (eq 12).[36]

Alkylation of β-lactone enolates proceeds with high stereoselectivity dictated by strong stereocontrol from the C-4 R substituent (eq 13).[28] A similar result is obtained with the lithium enolates of diarylazetidinones which, on reaction with aldehydes, alkylate with high *cis* diastereoselectivity (eq 14).[30]

Sarcosine (*N*-methylglycine)-containing tripeptides and hexapeptides are poly-deprotonated in the presence of excess **Lithium Chloride** to give amide enolates which give *C*-alkylated sarcosine products with **Iodomethane**, **Allyl Bromide**, and **Benzyl Bromide** (eq 15).[29] With (S) configuration amino acids, the newly formed stereogenic center tends to have the (R) configuration. α-Nitrile carbanions are generally useful in alkylation and other reactions with electrophiles, e.g. eq 16.[32]

α,β-Unsaturated Carbonyl Compounds. Of particular synthetic value is the generation of kinetic enolates of cyclic enones which may be alkylated (e.g. eq 17)[37] or undergo more demanding processes, e.g. double Michael addition (eq 18).[38] Imines of α,β-unsaturated aldehydes and ketones with δ-acidic hydrogens undergo clean deconjugative α-alkylation (eq 19).[23,39]

Copper dienolates of α,β-unsaturated acids, prepared by **Copper(I) Iodide** transmetalation, can be selectively γ-alkylated with allylic halides; the use of nonallylic electrophiles leads mainly to α-alkylation (eq 20).[40] Copper enolates of α,β-unsaturated esters[41] and amides[42] show similar behavior.

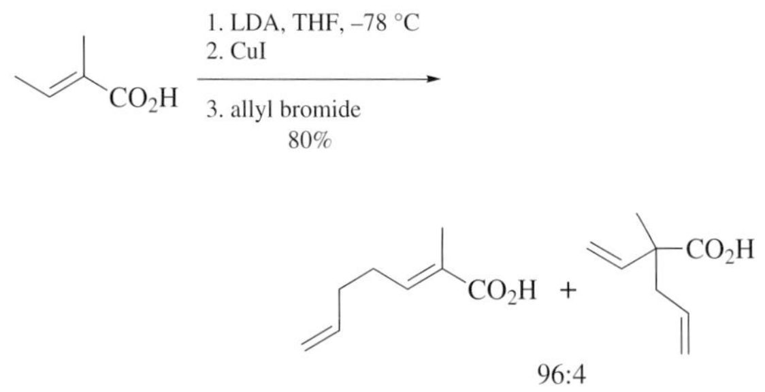

$$\text{(20)}$$

96:4

Enolates with Chiral Auxiliaries. Enantioselective alkylation of carbonyl derivatives encompassing chiral auxiliaries constitutes an important synthetic process. The anions derived from aldehydes,[43] acyclic ketones,[44,45] and cyclic ketones with **(S)-1-Amino-2-methoxymethylpyrrolidine** (SAMP)[46] are used to obtain alkylated products in good to excellent yields and high enantioselectivity (e.g. eq 21).[46]

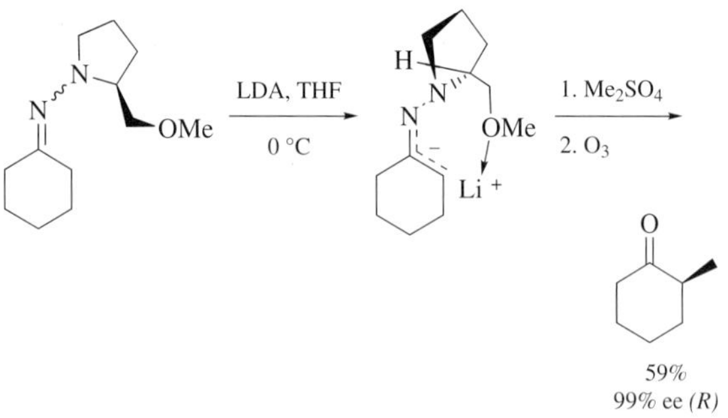

$$\text{(21)}$$

59%
99% ee (R)

Metalation[47,48] of chiral oxazolines, derived from (1S,2S)-1-phenyl-2-amino-1,3-propanediol, followed by alkylation and hydrolysis, leads to optically active dialkylacetic acids, e.g. eq 22,[49] 2-substituted butyrolactones and valerolactones,[50] β-hydroxy and β-methoxy acids,[51] 2-hydroxy carboxylic acids,[52] and 3-substituted alkanoic acids (eq 23).[53]

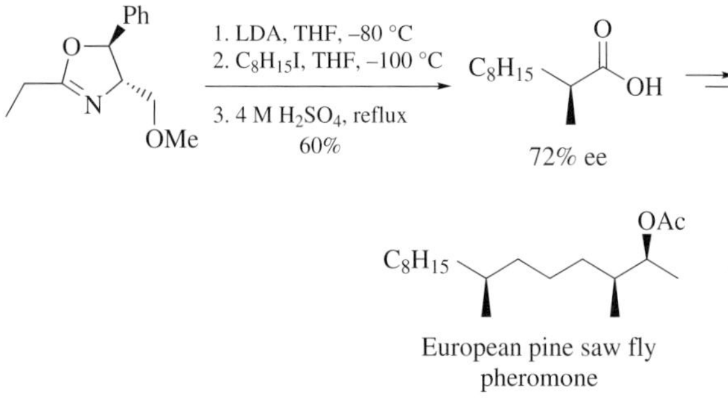

$$\text{(22)}$$

European pine saw fly pheromone

Imide enolates derived from (S)-valinol and (1S,2R)-norephedrine and obtained by either LDA or **Sodium Hexamethyldisilazide** deprotonation (eq 24)[54] exhibit complementary and highly diastereoselective alkylation properties. Mild and nondestructive removal of the chiral auxiliary to yield carboxylic acids, esters, or alcohols contributes to the significance of this protocol in small- and large-scale synthesis.[55,56]

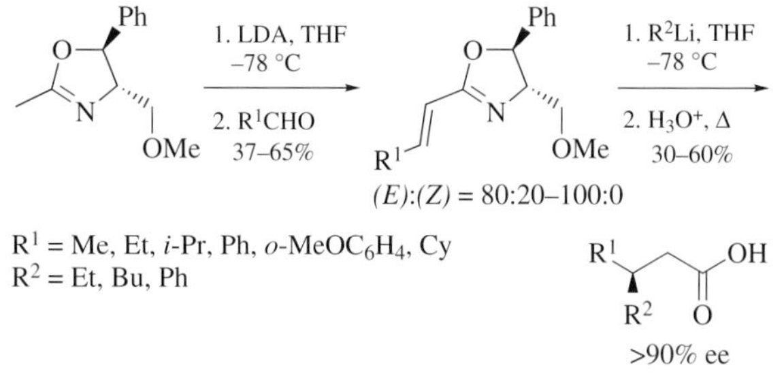

$$\text{(23)}$$

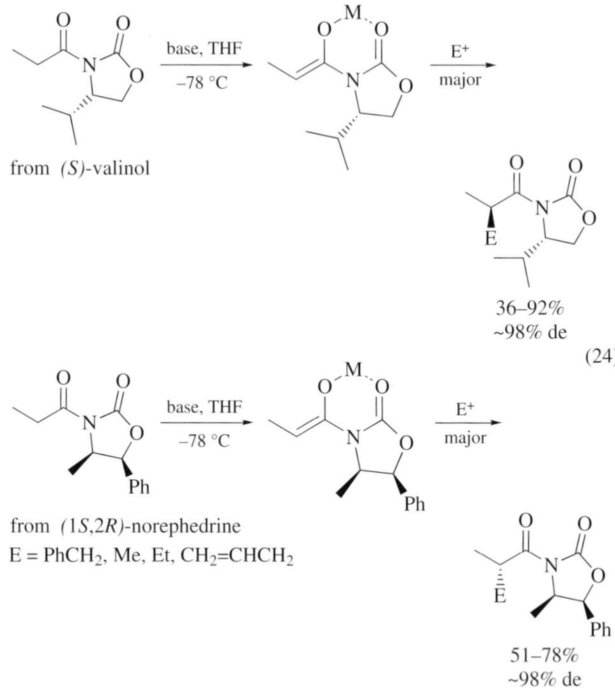

>90% ee

R^1 = Me, Et, i-Pr, Ph, o-MeOC$_6$H$_4$, Cy
R^2 = Et, Bu, Ph

from (S)-valinol

$$\text{(24)}$$

36–92%
~98% de

from (1S,2R)-norephedrine
E = PhCH$_2$, Me, Et, CH$_2$=CHCH$_2$

51–78%
~98% de

Metalated t-butyl 2-t-butyl-2,5-dihydro-4-methoxyimidazole-1-carboxylate is alkylated in good yields and with high *trans* diastereoselectivity (>99:1); hydrolysis of the resulting adducts liberates the α-amino acid methyl esters in high yields (eq 25).[57] Using this method, the (S)-t-butyl 2-t-butylimidazole derivative gave, upon isopropylation and hydrolysis, the L-valine methyl ester with 81% ee.

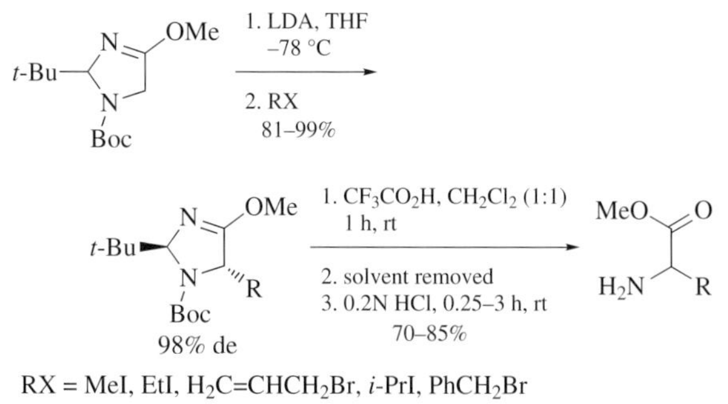

$$\text{(25)}$$

98% de

RX = MeI, EtI, H$_2$C=CHCH$_2$Br, i-PrI, PhCH$_2$Br

Alkylations and aldol condensations of aldehydes and ketones with enolates of chiral dioxolanes proceed generally with high diastereoselection, e.g. eqs 26[58,59] and 27.[60] The magnesium enolate of S-(+)-2-acetoxy-1,1,2-triphenylethanol generated by transmetalation with **Magnesium Bromide** has enjoyed considerable success in aldol condensations, e.g. eq 28.[61]

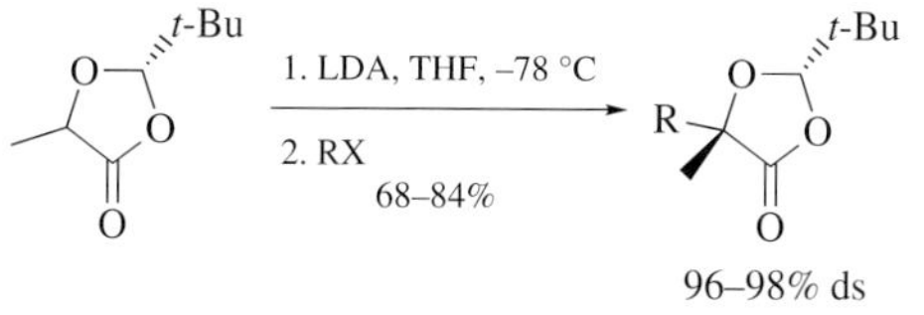

RX = EtI, PrI, BuI, $C_7H_{15}I$, H_2C=$CHCH_2Br$, $PhCH_2Br$

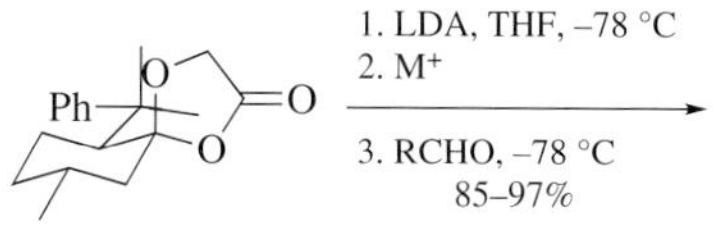

(26) 96–98% ds

R = Ph, Pr, *i*-Pr 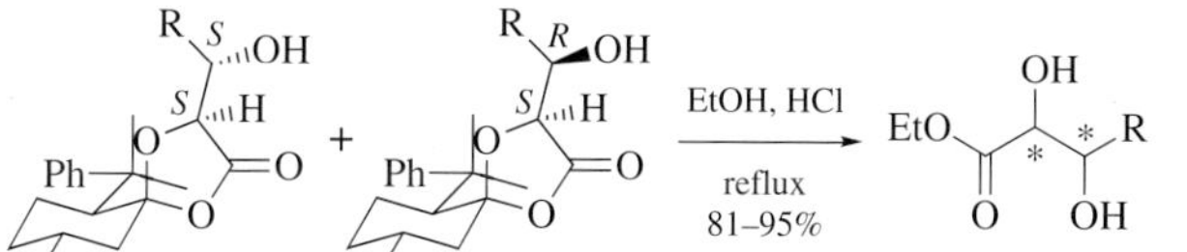 (27)

M$^+$ = Li$^+$ 68:32–96:4
M$^+$ = Mg^{2+} 73:27–90:10
M$^+$ = Zr^{4+} 45:55–19:81

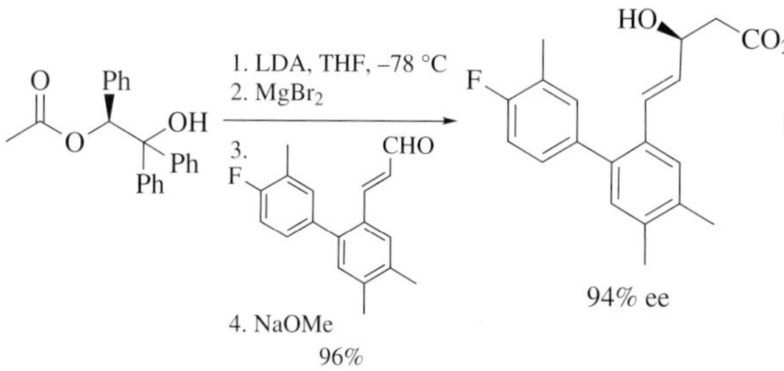

(28) 94% ee 96%

β-Lactams with high ee values[62] result from the condensation of lithium enolates of 10-diisopropylsulfonamide isobornyl esters with azadienes (eq 29).[63,64]

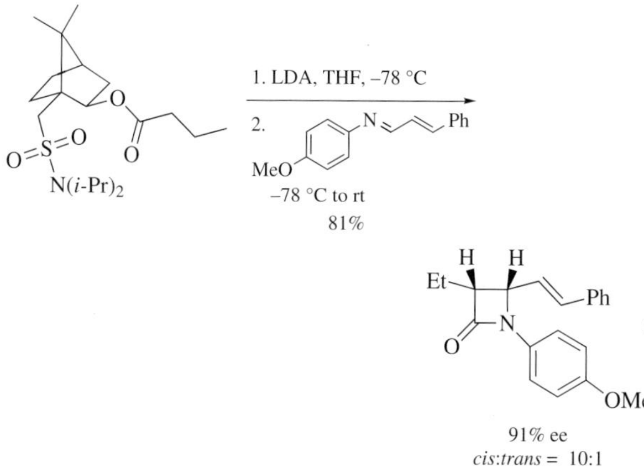

(29) 81% 91% ee *cis:trans* = 10:1

Enolate Rearrangements.

Ireland–Claisen Rearrangement.[65,66] Silyl enol ethers of allyl esters undergo highly stereoselective Ireland–Claisen rearrangement to afford 4-pentenoic acids.[67] The structure of the ester and the reaction conditions dictate the stereoselectivity. It has seen wide synthetic application, e.g. the construction of unnatural (−)-trichodiene (eq 30).[68]

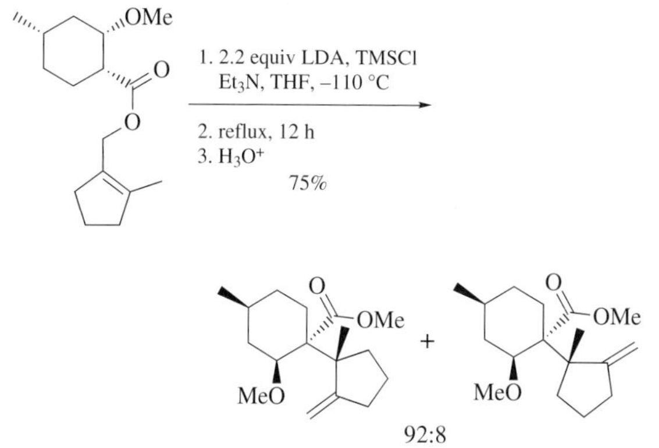

(30) 75% 92:8

Wittig Rearrangements.[69] Enolates derived from chiral propargyloxyacetic acids lead to chiral allenic esters, via a rapid diastereoselective [2,3]-Wittig rearrangement followed by esterification (eq 31).[70] Analogous rearrangements are feasible with carbohydrate derivatives[71] and steroids (eq 32).[72] In some cases, improved enantioselectivities resulted when the lithium enolate was transmetalated with ***Dichlorobis(cyclopentadienyl)zirconium***.[73] Benzylic, allylic, propargylic, and related lithio derivatives of cyanohydrin ethers undergo the [2,3]-Wittig rearrangement to afford ketones, e.g. eq 33.[74,75] Lithiated *N*-benzyl- and *N*-allylazetidinones lead, via [1,2]-Wittig rearrangement exhibiting radical character, to pyrrolidones (eq 34).[76]

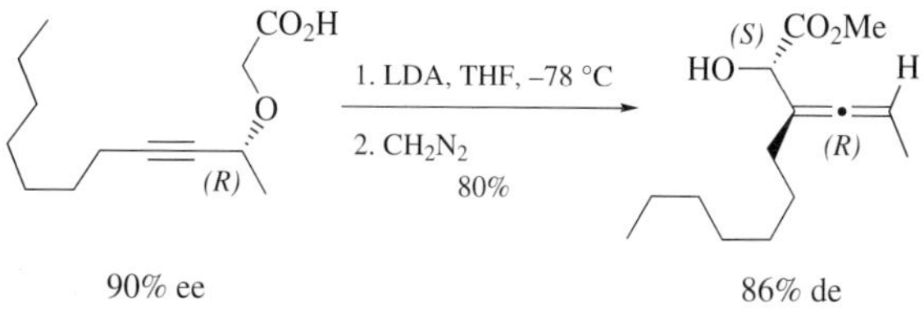

(31) 90% ee 80% 86% de

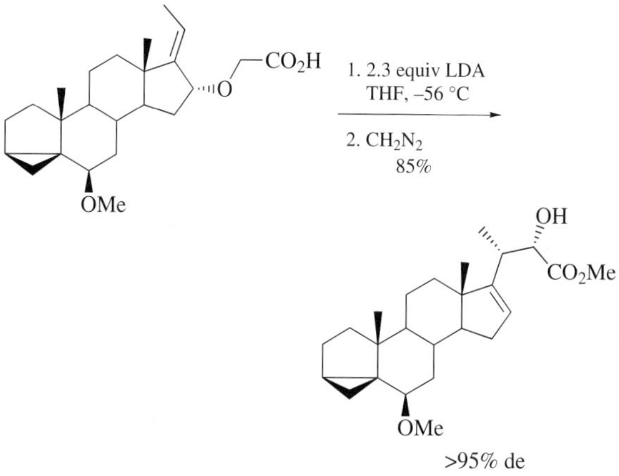

(32) 85% >95% de

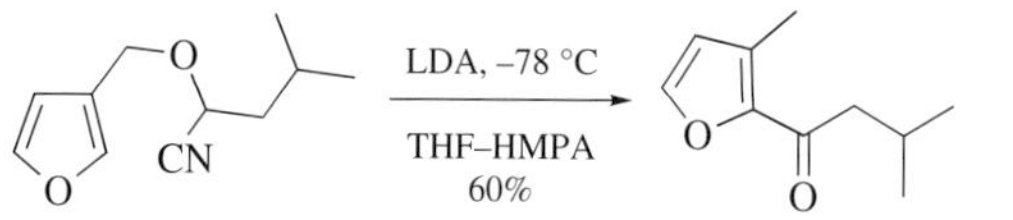

(33) 60%

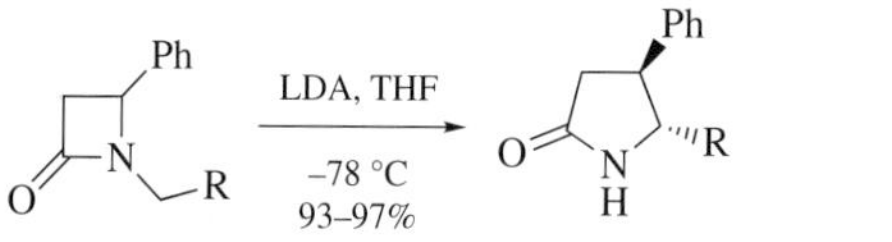

(34) 93–97%

R = Ph, CH=CH$_2$

Stevens Rearrangement. Ammonium ylides undergo a stereoselective Stevens [2,3]-rearrangement when treated with LDA to afford *N*-methyl-2,3-disubstituted piperidines (eq 35).[77]

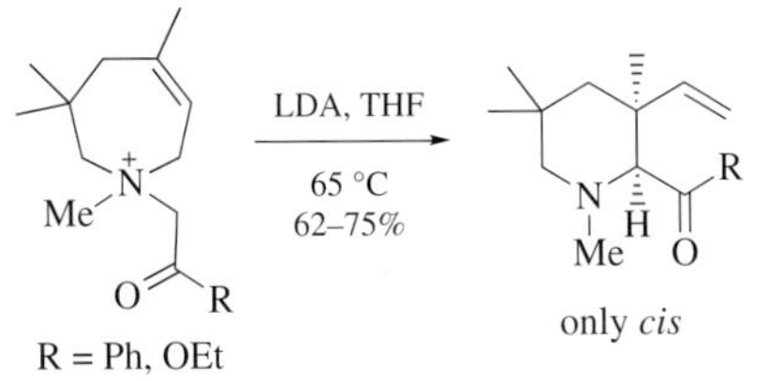

$$\text{(35)}$$

R = Ph, OEt

only *cis*

Heteroatom-stabilized Carbanions. Heteroatom-stabilized and allylic carbanions serve as homoenolate anions and acyl anion equivalents,[5,78] e.g. α-anions of protected cyanohydrins of aldehydes and α,β-unsaturated aldehydes are intermediates in general syntheses of ketones and α,β-unsaturated ketones (eq 36).[79] Allylic anions of cyanohydrin ethers may be α-alkylated (eq 37)[80,81] or, if warmed to −25 °C, may undergo 1,3-silyl migration to cyanoenolates which may be trapped with TMSCl.[82] Metalated α-aminonitriles of aldehydes are used for the synthesis of ketones and enamines (eq 38).[83] Similarly, allylic anions from 2-morpholino-3-alkenenitriles undergo predominantly α-*C*-alkylation to give, after hydrolysis, α,β-unsaturated ketones (eq 39).[84]

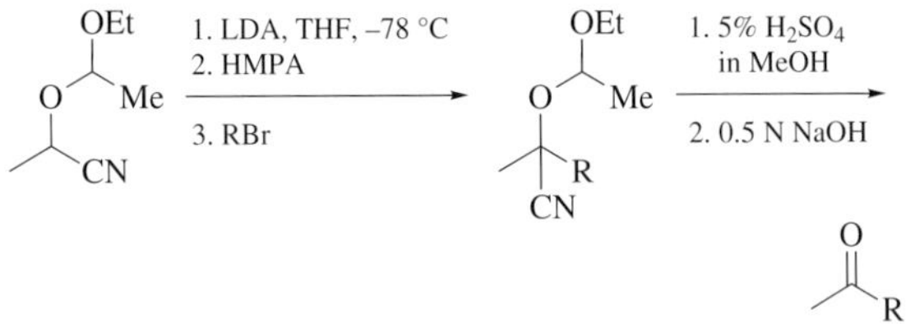

$$\text{(36)}$$

R = Bu, Hex, Dec   80–85%
R = *i*-Pr, *c*-Pent, *c*-Hex   41–80%
R = H$_2$C=CHCH$_2$ 76%

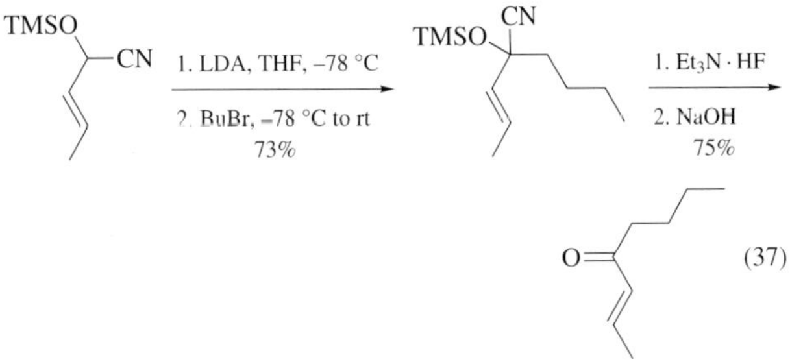

$$\text{(37)}$$

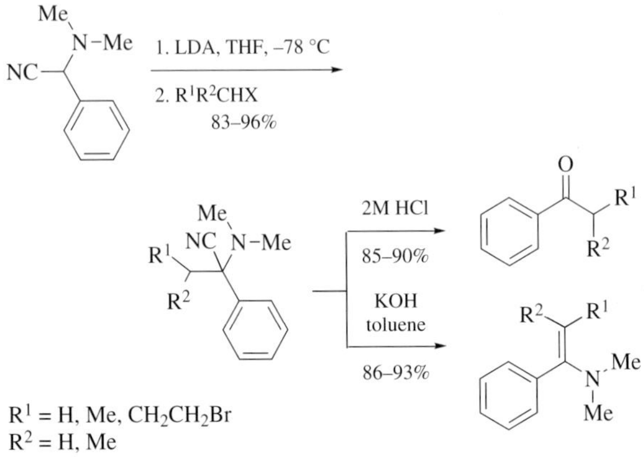

$$\text{(38)}$$

R^1 = H, Me, CH$_2$CH$_2$Br
R^2 = H, Me

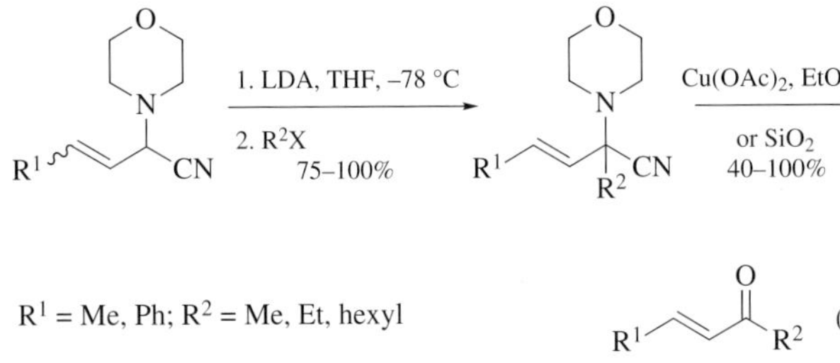

$$\text{(39)}$$

R^1 = Me, Ph; R^2 = Me, Et, hexyl

α-Sulfoxide- and sulfone-stabilized carbanions are highly useful synthetic intermediates, e.g. α-lithiated (+)-methyl *p*-tolyl (*R*)-sulfoxide reacts with α,β-unsaturated aldehydes to give, after dehydration, enantiomerically pure 1-[*p*-tolylsulfinyl]-1,3-butadienes (eq 40),[85] metalation of phenyl 3-tosyloxybutyl sulfone leads to efficient cyclopropane ring formation (eq 41),[86] and precursors for the Ramberg–Bäcklund rearrangement are prepared via α-lithiated cyclic sulfones (eq 42).[87]

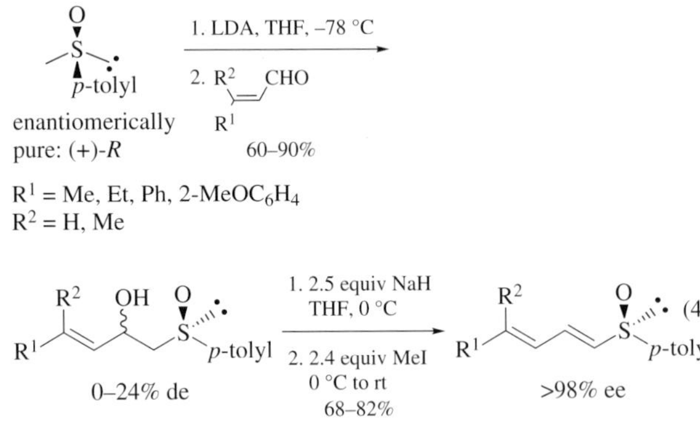

enantiomerically
pure: (+)-*R*   60–90%

R^1 = Me, Et, Ph, 2-MeOC$_6$H$_4$
R^2 = H, Me

$$\text{(40)}$$

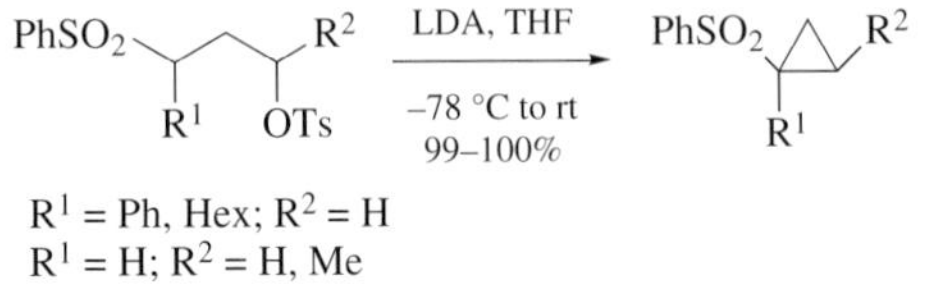

0–24% de   >98% ee

$$\text{(41)}$$

R^1 = Ph, Hex; R^2 = H
R^1 = H; R^2 = H, Me

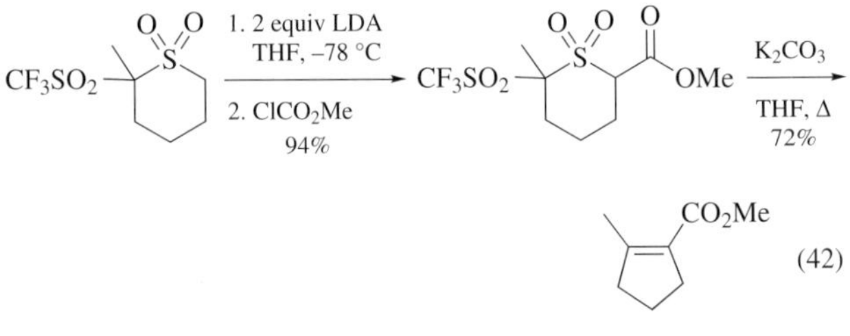

$$\text{(42)}$$

Lithiated allylic sulfoxides may be α-alkylated and the resulting products subjected to [2,3]-sigmatropic rearrangement induced by a thiophile to give allylic alcohols (eq 43).[6,88] In contrast, alkenyl aryl sulfoxides produce α-lithiated species which are alkylated with MeI or PhCHO in good yields (eq 44).[89,90] LDA has also been used to metalate allylic[91] and propargylic selenides[92,93] as well as aryl vinyl selenides.[94]

Aromatic and Heteroaromatic Metalations.

Aromatics. The utility of LDA for kinetic deprotonation of aromatic substrates is compromised by its insufficient basicity compared to the alkyllithiums.[8] Selective lateral deprotonation may be achieved which complements the *ortho*-metalation process (eq 45).[95]

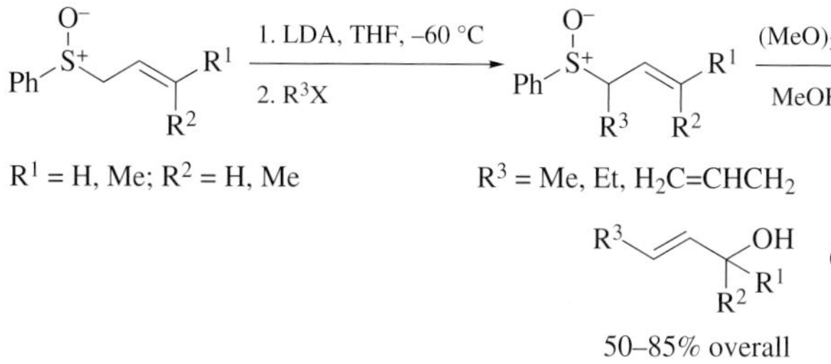

$$R^1 = H, Me; R^2 = H, Me \qquad R^3 = Me, Et, H_2C=CHCH_2$$

(43)

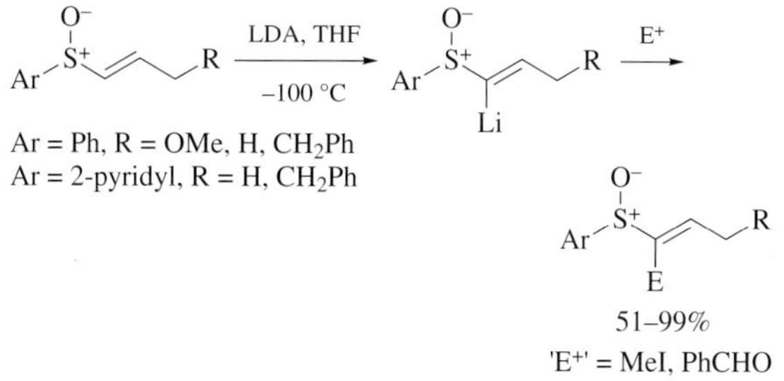

50–85% overall

Ar = Ph, R = OMe, H, CH_2Ph
Ar = 2-pyridyl, R = H, CH_2Ph

(44)

51–99%

'E⁺' = MeI, PhCHO

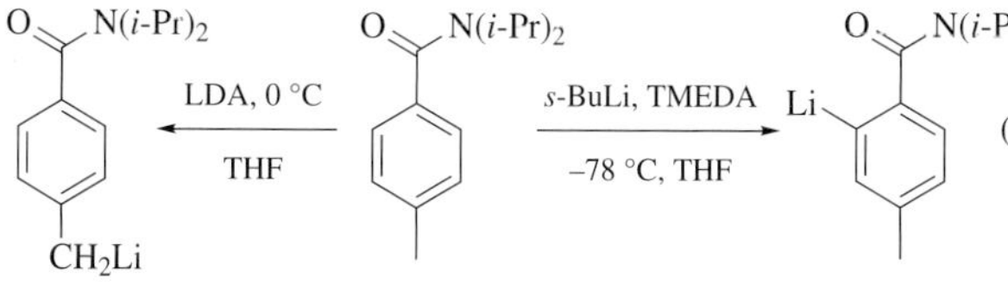

(45)

Of considerable synthetic utility is the LDA-induced conversion of bromobenzenes into benzynes. Thus low-temperature deprotonation of *m*-alkoxyaryl bromides followed by warming to rt in the presence of furan gives cycloadducts (eq 46);[96] the lithio species may also be trapped with CO_2 at −78 °C to give unusual aromatic substitution patterns. Treatment of aryl triflates with LDA leads to *N,N*-isopropylanilines in good yield (67–98%).[97]

Remote metalation of biaryls and *m*-teraryls provides a general synthesis of substituted and condensed fluorenones (eq 47).[98] Similarly, biaryl *O*-carbamates undergo remote metalation and anionic Fries rearrangement to 2-hydroxy-2'-carboxamidobiaryls, which are efficiently transformed into dibenzo[*b,d*]pyranones.[99]

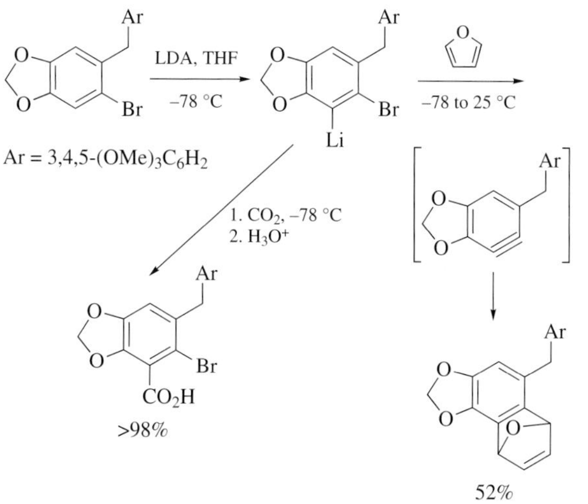

Ar = 3,4,5-(OMe)₃C₆H₂

(46)

52%

R = H, Ph

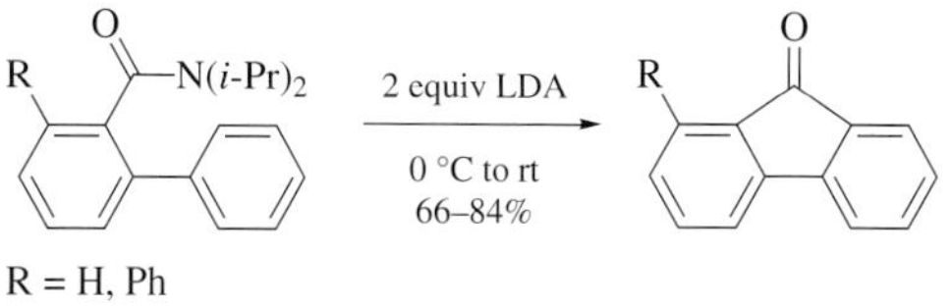

(47)

Lateral Lithiation.[7] Laterally metalated *o*-toluic acids (eq 48),[100] esters (eq 49),[101] and amides (eq 50)[102] are important intermediates for chain extension and carbo- and hetero-ring annulation. Remote lateral lithiation of 2-methyl-2'-carboxaminobiaryls constitutes a general regiospecific synthesis of 9-phenanthrols (eq 51).[103–105]

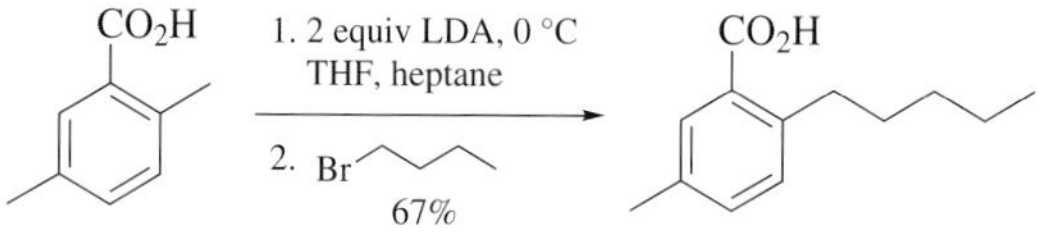

(48)

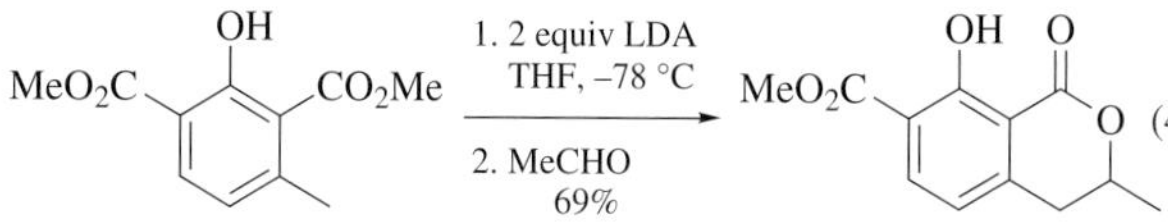

(49)

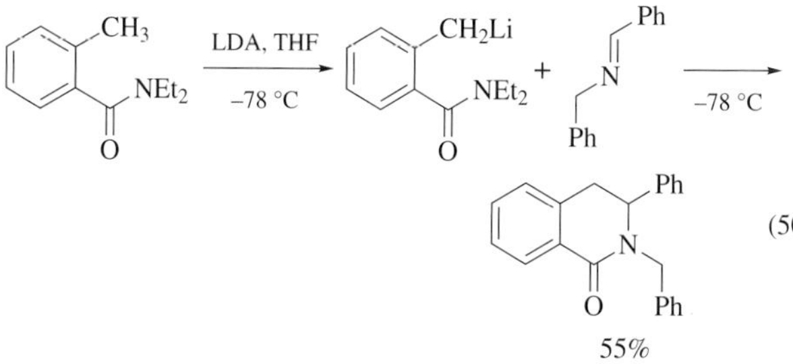

(50)

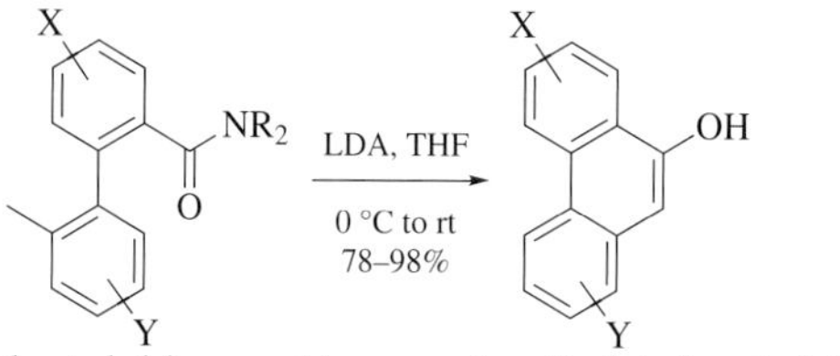

55%

(51)

α-Lithiated *o*-tolyl isocyanides may be alkylated and, following a second metalation with LiTMP, converted into 3-substituted indoles, e.g. eq 52.[106] Michael addition of α-lithiated 3-sulfonyl- and 3-cyanophthalides to α,β-unsaturated ketones and esters initiates a regioselective aromatic ring annulation process to give naphthalenes, e.g. eq 53.[107] Lithiated 3-cyanophthalides also undergo reaction with in situ-generated arynes to give anthraquinones (eq 54).[108] Similarly, aza-anthraquinones are obtained using 3-bromopyridine as the 3,4-pyridyne precursor.[109]

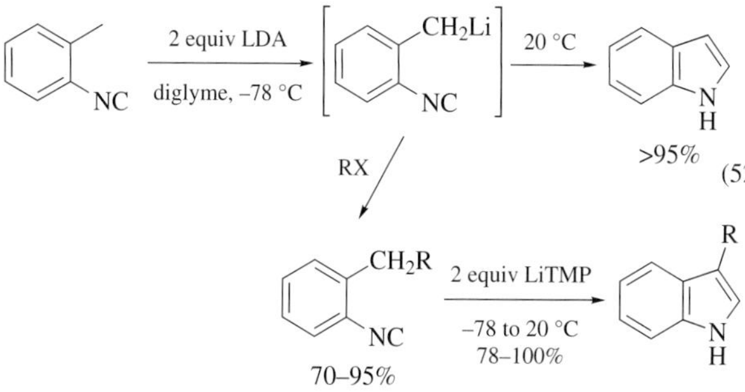

>95%

(52)

70–95%

78–100%

(53)

(54)

Heteroaromatics.[9] LDA (and LiTMP) are advantageous bases for the directed *ortho*-metalation chemistry of pyridines[110,111] and, to a much lesser extent, pyrimidines.[112] Several halopyridines can be lithiated *ortho* to the halo substituent and trapped with various electrophiles (eq 55).[113]

(55)

X = F, Cl, Br; 'E$^+$' = TMSCl, I$_2$, (PhS)$_2$, PhCHO, Ph$_2$CO

Lithiated furan- and thiophenecarboxylic acids (eq 56),[114,115] indoles (eq 57),[116] indole-3-carboxylic acids (eq 58),[117] benzofurancarboxylic acids,[118] and thiazole- and oxazolecarboxylic acids[119] serve well in heteroaromatic synthesis.

(56)

(57)

(58)

E = TMS, Me, PhCH(OH), PrCH(OH), Ph$_2$COH
PhC(OH)Me, Me$_2$COH

First Update

Jeffrey M. Warrington & Louis Barriault
University of Ottawa, Ottawa, Ontario, USA

Introduction. The past 10 years have seen a rapid development in organic chemistry, especially in the areas of asymmetric reactions and catalysis. This review attempts to highlight some of the dialkyl amide chemistry that has been examined in the past decade or so and to expand on some of the concepts introduced in the first review.

Enolates Revisited–Solution Structure. Although proven to be a ubiquitous base in organic synthesis, little was known about the structure of LDA until recently.[121] With increased importance in the enantioselectivity and stereoselectivity of aldol reactions, much of the research within the past decade has surrounded a more complete understanding of how lithium amide bases coordinate to enolate ions in solution, and what role does this play in determining the reactivity.[122]

Stereoselective aldol reactions are limited by their ability to obtain stereoisomerically pure (*E*)- or (*Z*)-enolates separately, and it has been suggested that equilibration may be occurring to erode the enolate selectivities.[17] However, it would appear that the measured rate of enolate equilibration appears to be too low to be much of an influence.[123] It was suggested by Ireland in 1976 that LDA-mediated enolizations may proceed by cyclic transition states via disolvated LDA monomers.[124] This mechanism has since been widely cited for its predictive power. Ireland proposed that the deprotonation process may be proceeding via one of two proposed transition states, where proton transfer is synchronous with metal ion transfer. Non-bonded interactions between amide alkyl groups and the enolate alkyl group cause a preference for *E*-enolate formation (Scheme 1, refs 124, 136).

R	% E	% Z
OCH$_3$	95	5
O*t*Bu	95	5
Et	77	23
i-Pr	40	60
t-Bu	0	100
Ph	0	100
NEt$_2$	0	100

Scheme 1

On the contrary, initial speculations about a THF–LDA dimer by Seebach[125] were later verified by Willard[126] with the isolation of crystalline LDA–solvent complexes. [These complexes also provide an alternative to the standard method of preparation, as the LDA–THF complex can be readily recrystalized in *n*-pentane and weighed accurately for large (>0.5 mol) scale reactions (ref 126).]

Investigations to date suggest that LDA exists as a disolvated dimer (**1**) in ethereal solvents.[127]

1

Rate studies, however, provide evidence that the reactive form of LDA is the monomer, in accordance with Ireland's initial speculation.[128] Studies by Collum and Streitwieser suggest that LDA aggregates much more readily than some of the other commonly used bases such as lithium 2,2,4,4-tetramethyl piperidine (LTMP) or lithium hexamethyldisilazide (LiHMDS).[129] Contrary to popular wisdom, it was shown by Collum and co-workers that the use of uncomplexed, solid LDA in non-polar solvents could perform enolization reactions remarkably well, forming unsolvated ketone, ester, and carboxamide enolates in reasonable yields (see ref 121b). Interestingly, X-ray crystallography of solvent-free LDA by Mulvey[130] reveals that LDA comprises an endless helical arrangement, suggesting a similar structure in non-donating solvents.

Recently Myers and co-workers demonstrated the practicality of the inexpensive amino alcohol pseudoephedrine as a highly practical chiral auxiliary for asymmetric alkylation of enolates.[131] It is believed that the high diastereoselectivities (generally >90% de) obtained with this auxiliary are a result of its highly aggregated structure (**2**). Solvation of the resulting alkoxide anions is believed to play a pivotal role in determining alkylation diastereoselectivity (ref 131). Although the initial studies were carried out with LDA as a base, it was later determined that in situ derived LiHMDS proved superior.[132] For further information on pseudoephedrine, readers should refer to the article available from this database.[133]

2

Formation of aggregates has also been shown to have an influence on enolate Claisen acylations, whereby the aggregated enolate may compete with the monomer,[134] but the exact mechanistic role of aggregates is still a subject of much debate.

Additives. It has been known for some time that additives such as hexamethylphosphoramide (HMPA), tetramethyl ethylenediamine (TMEDA), lithium chloride, magnesium bromide, or 1,3-dimethyl-3,4,5,6-tetrahydro-2(1H)-pyrimidinone (DMPU) can dramatically enhance rates in a wide array of organolithium reactions.[135] These additives are generally thought to alter the aggregation state of the reactants, but specific mechanistic reasons for these enhancements are still not fully understood. For lithium enolates, it was originally thought that increasing the donor char-

acter of the solvent results in increased contact ion separation, and thus an enhanced electron density in the enolate system (and therefore more reactive).[136] It has also been shown that the stereochemistry of enolates formed from carboxylic acid derivatives can be reversed upon the addition of HMPA[34] (Scheme 2), indicating the formation of the kinetically preferred enolate in the absence of an additive, but the thermodynamically preferred enolate is formed when aggregates are disrupted.

R_1	R_2	Additive	Ratio (Z/E)
OCH_3	Et	None	9:91
OCH_3	Et	HMPA	84:16
Ot-Bu	Et	None	5:95
Ot-Bu	Et	HMPA	77:23

Scheme 2

It was recently shown that the addition of metal salts dramatically increased the diastereoselectivity on sulfinyl ketimine enolates, which provides a rapid route to *syn*- and *anti*-1,3-amino alcohols (Scheme 3, eq 1).[137] For two other recent examples, a silver nitrate (AgNO$_3$)/LDA combination has been used in the formation of selenium containing heterocycles as potential antitumor agents (Scheme 3, eq 2),[138] while an LDA/HMPA combination was used to fragment aromatic thioacetals to form unsymmetrically substituted thiothionophthalicanhydrides (Scheme 3, eq 3).[139]

On the other hand, there are also instances reported where additives can decelerate organolithium reactions.[140] Adding complexity to the picture, Sun and Collum have explicitly demonstrated that although enolization reactions may follow different mechanistic pathways in different solvents, the rate of enolization is solvent independent. Additives such as HMPA have little influence on rates despite a significant change in the mechanism.[141] Reich and co-workers demonstrated that HMPA deaggregates phenyllithium to a monomer,[142] it is clear that this is not the case for LDA (Scheme 4).[143] HMPA instead prefers to displace THF from the LDA dimer, which may have an influence on the transition state energies for the enolization process (see ref 141). By some estimates, HMPA has an affinity for the lithium cation 300 times that of THF.[144]

Alternatively, the additives may have an effect on the aggregation of the enolate itself.[145] Mixed aggregates between the enolate and the dialkyl amide base may also be important in predicting the stereochemistry. For example, it is possible that an enolate may exhibit different reactivity at the beginning and at the end of the reaction (pure aggregates), and at the end of a reaction (mixed aggregates) (see ref 121). Similarly, the addition of large excesses of either enolate or base may change the properties of the enolate in solution. When LDA is generated by deprotonation of diisopropylamine, excess butyllithium may allow the formation of mixed aggregates,[146] but what effect this may have on typical aldol reactions is still unknown. Therefore, a predictive model for precisely how solvent addends affect enolization reactions has yet to be established.

Additive = ZnBr$_2$ or MgBr$_2$

$$dr > 90\% \quad (1)$$

$$\quad (2)$$

LDA

0.8 equiv HMPA
(no rx. with more HMPA)

$$\quad (3)$$

Scheme 3

Metal Enolates. In parallel with additives, transition metals may be added to enolates to give transmetallated species which can undergo cross-coupling chemistry. Perhaps the earliest example of metal-catalyzed enolate reactions is the Reformatsky reaction.[147] Transition metal-catalyzed enolate chemistry has been recently revived in the literature, particularly in the field of asymmetric catalysis.[148] The transition metal-catalyzed coupling reactions of aryl halides,[149] allyl epoxides,[150] and allylic esters[151] with alkyl enolates have been recently investigated. Generally the choice of base employed depends on the substrate and on the reaction performed. For enolate arylation, KHMDS seems to be the most

effective in the case of amide enolates,[152] and sodium *tert*-butoxide is the base of choice for the formation of α-aryl ketones.[153] The Pd-catalyzed asymmetric alkylation of α-tetralone enolates was shown to be most effective when LDA or LiTMP was employed as a base in the presence of trialkyltin chloride (Scheme 5, ref 151).

Terminal alkynes may be deprotonated with LDA and transmetallated with zinc to undergo palladium-catalyzed cross-couplings with aryl electrophiles (Scheme 6).[154]

5 mol %

2.5 mol % [η^3-C$_3$H$_5$PdCl$_2$]
LDA, (CH$_3$)$_3$SnCl, DME

53–99% yield
61–85% ee

Scheme 5

$$R \equiv \xrightarrow[\substack{2.\ ZnX_2 \\ 3.\ ArX,\ cat.\ PdL_n}]{1.\ LDA} R \equiv Ar$$

Scheme 6

Mannich Reaction. Enolates may undergo trapping with methylene iminium salts to give α-amino keto derivatives.[155] In these instances, generally deprotonation by KH, LiHMDS, or MeLi is preferred (ref 155) as LDA tends to give poor yields potentially due to the acidity of the Mannich products.[156] However, LDA can be used in combination with trimethylsilyl chloride to give TMS enol ethers which can be aminomethylated in high yields. More recently, Mannich-type reactions have been carried out by transmetallation using zinc chloride (Scheme 7, eq 4).[157] Badia and co-workers demonstrated the use of an asymmetric Mannich reaction on enolizable imines bearing pseudoephedrine as the chiral auxiliary (Scheme 7, eq 5).[158] An intramolecular Diels–Alder reaction was used in combination with an LDA-promoted retro-Mannich reaction in the enantioselective synthesis of (+)-luciduline (Scheme 7, eq 6).[159]

i-Pr$_2$N–Li(HMPA)

Scheme 4

R* = (–)-*trans*-2-(α-cymyl)-cyclohexyl   75%

(4)

R = Me, Et, *i*-Pr

(5)

79–86%, single diastereomer

10 equiv LDA

$(i\text{-Pr})_2\text{N}^-$

LDA

TMSCl

(6)

(+)-luciduline

Scheme 7

Single-electron Transfer: Carbene Mechanisms. The vast majority of reactions of LDA owe their mechanism to the basic character of the reagent; some recent experiments have shown that LDA may behave very differently in the presence of alkyl halides and dihalides (specifically iodides),[160] π-deficient aromatic heterocyclic compounds (Scheme 8, eq 7),[161] α-bromo imines,[162] conjugated acetylenes,[163] alkyl sulfonates,[164] or benzophenone.[165] In these instances, it has been shown that LDA may act as a single-electron transfer donor as in the case of anthracene and perylene, or via carbene mechanism in the case of benzyl halides.[166] Some substrates exhibit competition between multiple mechanisms (Scheme 8, eq 8, ref 160).

Carbenes have received some interest in the literature, with some stable *N*-heterocyclic carbenes successfully characterized by X-ray crystallography.[167] Generally these stable carbenes are generated by *C*-deprotonation of formamidinium and related cations by strong bases such as sodium hydride, potassium *tert*-butoxide, and LDA.[168] Alkylidene carbenes have been recently revived in the literature[169] for their use in a potential route to C-H insertion in a regiospecific manner. Harada and co-workers have treated terminal alkynes bearing a distal leaving group with LDA to rapidly generate bicyclo[*n*.3.0]-1-alkenes (where $n = 3$ or 4).[170]

Epoxides.[171–174] With the recent advances in technologies for asymmetric epoxidation,[175] there has been considerable activity in both the synthesis and the reactivity of epoxides. We will focus specifically upon the reactivity of epoxides with lithium dialkyl amides.

It has been known for some time that the acid-catalyzed solvolysis of epoxides can lead to products other than the anticipated 1,2-glycols,[176] but the reaction of epoxides with bases has been slower to develop. The reactivity of epoxides with strong hindered bases was originally studied by Cope,[177] Crandall,[178] and Rickborn.[179] Reactions of epoxides with dialkylamide bases take one of two general forms: (a) α-lithiation by insertion of the base between one H-CO bond or (b) β-elimination with the formation of an allylic alcohol (Scheme 9).[180]

$$+ \text{LiN}(i\text{-Pr})_2 \longrightarrow + \overset{\bullet}{\text{N}}(i\text{-Pr})_2 \quad (7)$$

LDA

Scheme 8   (8)

α-Deprotonation of epoxides results in carbenoid-type intermediates that can be trapped by an electrophile[181] or undergo rearrangement. Transannular C-H insertion results in ring-contracted saturated alkoxides.[182] Alternatively, vicinal C-H insertion results in the formation of lithium enolates.[183]

β-Deprotonations generally result in the formation of allylic alcohols, with the choice of solvent and additives playing a key role in the chemoselectivity.[184,185] This observation is most likely a manifestation of the aggregation effects discussed earlier. Morgan and Gajewski have shown via deuterium labeling studies on the ring opening of cyclopentene oxide that in the presence of polar solvents, the β-elimination pathway dominates, whereas in non-polar solvents such as benzene or ether, the α-lithiation pathway may compete (Scheme 10).[186]

Scheme 9

It has been determined that maximized yields for the allylic alcohol product can be obtained by utilizing a polar solvent such as hexamethylphosphoramide.[187] Presumably, complexation of Li^+ ion to the epoxide oxygen directs the facial selectivity via a six-atom cyclic transition state, effecting a less preferred *syn*-elimination (**3**). This coordination appears to be much less important in HMPA, with some evidence suggesting an *anti*-elimination pathway in this solvent (ref 186). Deuterium labeling experiments on *trans*-4-*tert*-butylcyclohexene oxides have shown that the β-rearrangement pathway occurs via *syn*-elimination.[188]

Solvent	α-lithiation	either pathway	β-elimination
Benzene	0.23	0.77	
Ether	0.29	0.71	–
HMPA	–	0.50	0.50

Scheme 10

3

Rearrangement of cyclooctene oxides was discovered to be an efficient route to cyclopentanoid terpenes. For example, the conversion of *cis*-cyclohexene oxide to 2-*endo*-hydroxy-*cis*-bicyclo[3.3.0] octane with an amide base was first noted by Cope, Lee, and Petree[189] and was later revisited by Whitesell and White[190] using LDA, with a marked improvement in yield and product selectivity. The protocol was further optimized with

n-butyllithium,[191] but with potassium *tert*-butoxide or Li_3PO_4 only the allylic alcohol was isolated (Scheme 11).[192]

Interestingly, the outcome of rearrangement of γ,δ-ethylenic epoxides can be dramatically changed by the solvent, with quantitative abstraction of the allylic protons being preferred in HMPA resulting in a cyclopropanyl alcohol instead of the predicted allylic alcohol (Scheme 12, ref 187).

Conditons	Allyl alcohol	Bicyclic alcohol	Reference
LDA (Ether/Hexane)	2	98	191
LDA (HMPA)	100	0	185
$LiN(Et)_2$	16	70	191
n-BuLi	0	99	192
KO-*t*Bu	Only	-	193
Li_3PO_4	Only	-	193

Scheme 11

Scheme 12

It was noted by Rickborn that the identity of the alkyl groups on the amide base has a large effect on the outcome of reaction between epoxides and amide bases.[193] Primary alkylamide bases (i.e., lithium diethylamide, lithium *n*-propylamide) give nearly quantitative yields of allylic alcohols, with the best yields obtained by cyclic primary amides such as *N*-lithiopyrrolidine. Reaction with secondary alkyl amides such as LDA proved to be more problematic, with rearrangement to the carbonyl competing successfully with proton abstraction. This observation was shown to be highly solvent dependent, as cyclooctene oxide was later converted quantitatively to 2-cyclooctenol with LDA when HMPA was used as solvent (Scheme 11).[187]

Similar to enolate chemistry, lithium dialkylamides generally prefer abstraction at the least substituted carbon, and *trans*-allylic alcohols are generally formed in preference to *cis* (Scheme 13, ref 174).

Scheme 13

Recently Schlosser and co-workers demonstrated the utility of a novel LDA/potassium *tert*-butoxide reagent (LIDAKOR[194]) in the conversion of oxiranes to allyl alcohols.[195]

Non-enantioselective rearrangements of epoxides have been exhaustively studied, but recent asymmetric versions of the rearrangement have seen considerable interest. Much research has surrounded the use of chiral diamine bases to perform the epoxide rearrangement. Specifically, bases such as 2-(*N*-pyrrolidinyl)methylpyrrolidine (Scheme 14) have been employed in stoichiometric amounts, with LDA being used to perform the initial deprotonation of the base.[196] The most probable reaction mechanism to date, proposed by Asami, involves *syn*-β-elimination, with the amine coordinated to both the leaving hydrogen and the oxygen of the epoxide.[197] The major enantiomer arises from a complex where steric interactions between cyclohexane ring and amide are minimized (Scheme 14, ref 171).

Favored   Disfavored

Scheme 14

Recently, Bertilsson, Sodergren, and Andersson have shown that these reactions can be carried out using only catalytic amounts of the a chiral diamine base because the hindered nature of LDA prevents it from competing with the chiral amide, and enantiomeric excesses of greater than 90% can readily be obtained (Scheme 15).[198]

Scheme 15

Although comparison studies were carried out with a variety of other amide bases including diethylamide (which had proved superior to LDA in the non-asymmetric epoxide rearrangement) and the more sterically hindered dicyclohexylamide, LDA proved to give the highest yield and ee by combining the proper amounts of steric hindrance and amide reactivity.[196]

Substituted Epoxides. In the presence of strong bases, β-hydroxy epoxides generally react via β-oxanol or Payne rearrangement.[199] Although classically the rearrangement is carried out with bases such as sodium sulfite,[200] potassium hydroxide,[201] in aqueous methanol, or sodium hydroxide in acetone or *tert*-butyl alcohol, LDA and other organolithiums have been employed in the Payne rearrangement in rigid tricyclic systems (Scheme 16).[202]

Scheme 16

Epoxy nitriles and esters generated in situ from chloroketones[203] can be readily transformed into the corresponding allylic alcohols with LDA, proving a rapid route to hydroxy amino acids.

While allyl and glycidyl ethers are converted into a mixture of oxetane and oxepine products with *sec*-butyllithium,[204] Mordini and co-workers[205] reported that allyl, benzyl, and propargyl epoxy ethers can be regioselectively converted into 2-vinyl, 2-phenyl, or 2-alkynyl-3-(hydroxyalkyl) oxetanes upon treatment with either Schlosser's base[206] or other mixed metal bases. Some of the best results were obtained with the LDA/potassium *tert*-butoxide mixture (LIDAKOR, ref 194). While rearrangement of propargylic or benzylic epoxide ethers formed exclusively the four-membered oxetanes, rearrangements of allyl oxiranyl ethers show a selectivity for cyclization to the seven-membered ring. Trialkylsilyl-substituted epoxide allyl ethers also show a preference for the oxepine, and mixtures are obtained as the size of the silyl substituents is increased (Scheme 17).

Aziridines. With the vast array of literature available on epoxides, it is surprising that the base-promoted elimination of aziridines is almost unknown in the literature. One of the earliest examples using sodium alkoxides in refluxing alcoholic solvents was reported in 1989 by Stamm and Speth.[207] Scheffold reported an aziridine to allylamine conversion in 1993 using a cob(I)alamin catalyst.[208] Recently, Mordini and co-workers demonstrated the use of a LDA/potassium *tert*-butoxide mixture (pentane, room temperature) to successfully rearrange *N*-substituted cyclohexyl aziridines (Scheme 18).[209]

Scheme 17

Although success was obtained with *N*-sulfonyl aziridines, the only reaction obtained with Boc and pivaloyl aziridines was removal of the protecting group. Interestingly, they also observed that *N*-trifluoroacetyl aziridines are deacetylated with LDA to initially form lithiated aziridines, followed by dimerization to give diamines. Similar to the rearrangement of cyclooctene oxide, rearrangements of 9-tosyl-9-azabicyclo[6.1.0]nonane gave formation of 1-*N*-tosylaminobicyclo[3.3.0]octane. The formation of the bicyclic compound was preferred at room temperature, while at −50 °C an almost 50/50 mixture of the bicyclic compound and allylic amine was realized.

Scheme 18

Bridgehead Enolates.[210] Enolate formation of most ketones proceeds smoothly using LDA, but there are very few instances where a bridgehead proton may be removed, as formation of a 'double bond' or enolate adjacent to a bridgehead is a violation of Bredt's rule. Such enolates are generally so unstable that they undergo self-condensation before reacting with an electrophile. There are some examples where *ent*-17-norkauran-16-one[211] underwent deuteron exchange at the bridgehead position, but this required rather forcing conditions (KO-*t*-Bu, 185 °C), and another example of a conformationally rigid system requiring milder conditions (NaOCH$_3$, 25 °C).[212] Recently Simpkins and co-workers demonstrated that certain tricyclic amides and lactones readily undergo quantitative deprotonation with LDA followed by trapping with TMS to give bridgehead silylated compounds in high yield (Scheme 19).[213]

The preference for LDA to remove the most sterically accessible proton is highlighted in the following case during synthetic studies of ledol, precapnelladiene, and compressanolide. Shea and co-workers discovered that certain 6/8-fused macrocyclic ketones undergo quantitative bridgehead enolization despite the presence of other enolizable methylene protons (Scheme 20).[214]

X = O (61%); NPh (74%)

Scheme 19

Scheme 20

Addition to Aromatics: The Halogen Dance.[215,216] In addition to *ortho*-metallation of pyridines, LDA has been employed in what has been named the halogen dance reaction, whereby the position of the halogen on the arene ring is changed. Requirements for the process are an arene ring with an exchangeable halogen (e.g., Br, I) and a directing group (carboxylic acid, ester, fluorine, etc.). After initial deprotonation with LDA, intermolecular exchange between metallated and non-metallated substrates results in the formation of dihalogenated species which can then be metallated in a manner more favored by the directing group (Scheme 21).[217]

Scheme 21

This technique was recently applied to the synthesis of natural antifungal caerulomycin C (Scheme 22).

Caerulomycin C

Scheme 22

Anionic Fries Rearrangement. Lithiated *O*-aryl carbamates constitute one of the most powerful *ortho* directors in aromatic metallation chemistry.[8] If allowed to warm in lieu of quenching with electrophiles, they can undergo anionic *ortho*-Fries rearrangement to furnish hydroxy phenyl acetamides.[218] Although generally *s*-BuLi is used for DOM of tertiary amides, Gawley and co-workers found that the use of LDA in benzylic carbamates avoids 1,2-Wittig rearrangement to the benzylic position (Scheme 23, ref 218).

Base	Yields (%)	
s-BuLi	72	0
LDA	27	61

Scheme 23

Similarly, this chemistry can be employed in rapid synthesis of the biologically relevant benzofuran-2(3*H*)-ones.[219]

Related Reagents. Lithium Amide; Lithium Hexamethyldisilazide; Lithium Diethylamide; Lithium Piperidide; Lithium Pyrrolidide; Lithium 2,2,6,6-tetramethylpiperidide; Potassium Diisopropylamide.

1. Reed, F.; Rathman, T. L., *Spec. Chem.* **1989**, *9*, 174.
2. Rathman, T. L., *Spec. Chem.* **1989**, *9*, 300.
3. Fraser, R. R.; Mansour, T. S., *J. Org. Chem.* **1984**, *49*, 3442.
4. Caine, D., *Comprehensive Organic Synthesis* **1991**, *3*, 1.
5. Yamamoto, Y., *Comprehensive Organic Synthesis* **1991**, *2*, 55.
6. Evans, D. A.; Andrews, G. C., *Acc. Chem. Res.* **1974**, *7*, 147.
7. Clark, R. D.; Jahangir, A., *Org. React.* **1995**, submitted.
8. Snieckus, V., *Chem. Rev.* **1990**, *90*, 879.
9. Queguiner, G.; Marsais, F.; Snieckus, V.; Epsztajn, J., *Adv. Heterocyclic Chem.* **1991**, *52*, 187.
10. House, H. O.; Phillips, W. V.; Sayer, T. S. B.; Yau, C. C., *J. Org. Chem.* **1978**, *43*, 700.
11. Reetz, M. T.; Maier, W. F., *Liebigs Ann. Chem.* **1980**, 1471.
12. Hamell, M.; Levine, R., *J. Org. Chem.* **1950**, *15*, 162.
13. House, H. O.; Czuba, L. J.; Gall, M.; Olmstead, H. D., *J. Org. Chem.* **1969**, *34*, 2324.
14. Mekelburger, H. B.; Wilcox, C. S., *Comprehensive Organic Synthesis* **1991**, *2*, 99.
15. Evans, D. A., In *Asymmetric Synthesis*; Morrison, J. D., Ed.; Academic: New York, 1984; Vol 3, p 1.
16. Brownbridge, P., *Synthesis* **1983**, 1.
17. Corey, E. J.; Gross, A. W., *Tetrahedron Lett.* **1984**, *25*, 495.
18. Mukaiyama, T., *Org. React.* **1982**, *28*, 203.
19. d'Angelo, J., *Tetrahedron* **1976**, *32*, 2979.
20. Stork, G.; Kraus, G. A.; Garcia, G. A., *J. Org. Chem.* **1974**, *39*, 3459.
21. Garst, M. E.; Bonfiglio, J. N.; Grudoski, D. A.; Marks, J., *J. Org. Chem.* **1980**, *45*, 2307.
22. Evans, D. A., *J. Am. Chem. Soc.* **1970**, *92*, 7593.
23. Whitesell, J. K.; Whitesell, M. A., *Synthesis* **1983**, 517.
24. Pearson, W. H.; Walters, M. A.; Oswell, K. D., *J. Am. Chem. Soc.* **1986**, *108*, 2769.
25. Ensley, H. E.; Lohr, R., *Tetrahedron Lett.* **1978**, 1415.
26. Creger, P. L., *J. Am. Chem. Soc.* **1967**, *89*, 2500.
27. Montgomery, S. H.; Pirrung, M. C.; Heathcock, C. H., *Org. Synth.* **1985**, *63*, 99; *Org. Synth., Coll. Vol.* **1990**, *7*, 190.
28. Mulzer, J.; Kerkmann, T., *J. Am. Chem. Soc.* **1980**, *102*, 3620.
29. Seebach, D.; Bossler, H.; Gründler, H.; Shoda, S., *Helv. Chim. Acta* **1991**, *74*, 197.
30. Otto, H. H.; Mayrhofer, R.; Bergmann, H. J., *Justus Liebigs Ann. Chem./Liebigs Ann. Chem.* **1983**, 1152.
31. Garratt, P. J.; Hollowood, F., *J. Org. Chem.* **1982**, *47*, 68.
32. Murata, S.; Matsuda, I., *Synthesis* **1978**, 221.
33. Negishi, E.; King, A. O.; Klima, W. L.; Patterson, W.; Silveira, A., *J. Org. Chem.* **1980**, *45*, 2526.
34. Heathcock, C. H.; Buse, C. T.; Kleschick, W. A.; Pirrung, M. C.; Sohn, J. E.; Lampe, J., *J. Org. Chem.* **1980**, *45*, 1066.
35. Frazier, R. H., Jr; Harlow, R. L., *J. Org. Chem.* **1980**, *45*, 5408.
36. Paquette, L. A.; Poupart, M.-A., *J. Org. Chem.* **1993**, *58*, 4245.
37. Kende, A. S.; Fludzinski, P., *Org. Synth.* **1986**, *64*, 68; *Org. Synth., Coll. Vol.* **1990**, *7*, 208.
38. Spitzner, D.; Engler, A., *Org. Synth.* **1988**, *66*, 37; *Org. Synth., Coll. Vol.* **1993**, *8*, 219.
39. Kieczykowski, G. R.; Schlessinger, R. H.; Sulsky, R. B., *Tetrahedron Lett.* **1976**, 597.
40. Savu, P. M.; Katzenellenbogen, J. A., *J. Org. Chem.* **1981**, *46*, 239.
41. Katzenellenbogen, J. A.; Crumrine, A. L., *J. Am. Chem. Soc.* **1974**, *96*, 5662.
42. Majewski, M.; Mpango, G. B.; Thomas, M. T.; Wu, A.; Snieckus, V., *J. Org. Chem.* **1981**, *46*, 2029.
43. Enders, D., In *Current Trends in Organic Synthesis*; Nozaki, H., Ed.; Pergamon: New York, 1983; p 151.
44. Enders, D.; Eichenauer, H., *Angew. Chem., Int. Ed. Engl.* **1979**, *18*, 397.
45. Enders, D.; Baus, U., *Liebigs Ann. Chem.* **1983**, 1439.
46. Enders, D., In *Asymmetric Synthesis*; Morrison, J. D., Ed.; Academic: New York, 1984; Vol 3, p 275.
47. Lutomski, K. A.; Meyers, A. I., In *Asymmetric Synthesis*; Morrison, J. D., Ed.; Academic: New York, 1984; Vol 3, p 213.
48. Meyers, A. I.; Knaus, G.; Kamata, K., *J. Am. Chem. Soc.* **1974**, *96*, 268.
49. Byström, S.; Högberg, H.-E.; Norin, T., *Tetrahedron* **1981**, *37*, 2249.
50. Meyers, A. I.; Yamamoto, Y.; Mihelich, E. D.; Bell, R. A., *J. Org. Chem.* **1980**, *45*, 2792.
51. Meyers, A. I.; Knaus, G., *Tetrahedron Lett.* **1974**, 1333.
52. Meyers, A. I.; Slade, J., *J. Org. Chem.* **1980**, *45*, 2785.
53. Meyers, A. I.; Smith, R. K.; Whitten, C. E., *J. Org. Chem.* **1979**, *44*, 2250.
54. Evans, D. A.; Ennis, M. D.; Mathre, D. J., *J. Am. Chem. Soc.* **1982**, *104*, 1737.
55. Evans, D. A.; Takacs, J. M.; McGee, L. R.; Ennis, M. D.; Mathre, D. J.; Bartroli, J., *Pure Appl. Chem.* **1981**, *53*, 1109.
56. Evans, D. A., *Aldrichim. Acta* **1982**, *15*, 23.
57. Blank, S.; Seebach, D., *Angew. Chem., Int. Ed. Engl.* **1993**, *32*, 1765.
58. Seebach, D.; Naef, R., *Helv. Chim. Acta* **1981**, *64*, 2704.
59. Seebach, D.; Naef, R.; Calderari, G., *Tetrahedron* **1984**, *40*, 1313.
60. Pearson, W. H.; Cheng, M.-C., *J. Org. Chem.* **1987**, *52*, 3176.

61. Lynch, J. E.; Volante, R. P.; Wattley, R. V.; Shinkai, I., *Tetrahedron Lett.* **1987**, *28*, 1385.

62. Hart, D. J.; Lee, C.-S., *J. Am. Chem. Soc.* **1986**, *108*, 6054.

63. Oppolzer, W.; Chapuis, C.; Bernardinelli, G., *Tetrahedron Lett.* **1984**, *25*, 5885.

64. Oppolzer, W.; Dudfield, P.; Stevenson, T.; Godel, T., *Helv. Chim. Acta* **1985**, *68*, 212.

65. Wipf, P., *Comprehensive Organic Synthesis* **1991**, *5*, 827.

66. Ireland, R. E.; Mueller, R. H., *J. Am. Chem. Soc.* **1972**, *94*, 5897.

67. Ireland, R. E.; Willard, A. K., *Tetrahedron Lett.* **1975**, 3975.

68. Gilbert, J. C.; Selliah, R. D., *J. Org. Chem.* **1993**, *58*, 6255.

69. Mikami, K.; Nakai, T., *Synthesis* **1991**, 594.

70. Marshall, J. A.; Wang, X., *J. Org. Chem.* **1990**, *55*, 2995.

71. Kakinuma, K.; Li, H.-Y., *Tetrahedron Lett.* **1989**, *30*, 4157.

72. Koreeda, M.; Ricca, D. J., *J. Org. Chem.* **1986**, *51*, 4090.

73. Uchikawa, M.; Katsuki, T.; Yamaguchi, M., *Tetrahedron Lett.* **1986**, *27*, 4581.

74. Cazes, B.; Julia, S., *Synth. Commun.* **1977**, *7*, 273.

75. Cazes, B.; Julia, S., *Synth. Commun.* **1977**, *7*, 113.

76. Durst, T.; Van Den Elzen, R.; LeBelle, M. J., *J. Am. Chem. Soc.* **1972**, *94*, 9261.

77. Neeson, S. J.; Stevenson, P. J., *Tetrahedron Lett.* **1988**, *29*, 3993.

78. Krief, A., *Comprehensive Organic Synthesis* **1991**, *3*, 85.

79. Stork, G.; Maldonado, L., *J. Am. Chem. Soc.* **1971**, *93*, 5286.

80. Jacobson, R. M.; Lahm, G. P.; Clader, J. W., *J. Org. Chem.* **1980**, *45*, 395.

81. Hertenstein, U.; Hünig, S.; Öller, M., *Chem. Ber.* **1980**, *113*, 3783.

82. Hertenstein, U.; Hünig, S.; Reichelt, H.; Schaller, R., *Chem. Ber.* **1982**, *115*, 261.

83. Ahlbrecht, H.; Raab, W.; Vonderheid, C., *Synthesis* **1979**, 127.

84. Takahashi, K.; Honma, A.; Ogura, K.; Iida, H., *Chem. Lett.* **1982**, 1263.

85. Solladié, G.; Ruiz, P.; Colobert, F.; Carreño, M. C.; Garcia-Ruano, J. L., *Synthesis* **1991**, 1011.

86. Chang, Y. H.; Pinnick, H. W., *J. Org. Chem.* **1978**, *43*, 373.

87. Hendrickson, J. B.; Boudreaux, G. J.; Palumbo, P. S., *J. Am. Chem. Soc.* **1986**, *108*, 2358.

88. Evans, D. A.; Andrews, G. C.; Fujimoto, T. T.; Wells, D., *Tetrahedron Lett.* **1973**, 1385.

89. Okamura, H.; Mitsuhira, Y.; Miura, M.; Takei, H., *Chem. Lett.* **1978**, 517.

90. Posner, G. H.; Tang, P.; Mallamo, J. P., *Tetrahedron Lett.* **1978**, 3995.

91. Reich, H. J.; Clark, M. C.; Willis, W. W., Jr, *J. Org. Chem.* **1982**, *47*, 1618.

92. Reich, H. J.; Shah, S. K., *J. Am. Chem. Soc.* **1977**, *99*, 263.

93. Reich, H. J.; Shah, S. K.; Gold, P. M.; Olson, R. E., *J. Am. Chem. Soc.* **1981**, *103*, 3112.

94. Reich, H. J.; Willis, W. W., Jr; Clark, P. D., *J. Org. Chem.* **1981**, *46*, 2775.

95. Beak, P.; Brown, R. A., *J. Org. Chem.* **1982**, *47*, 34.

96. Jung, M. E.; Lowen, G. T., *Tetrahedron Lett.* **1986**, *27*, 5319.

97. Wickham, P. P.; Hazen, K. H.; Guo, H.; Jones, G.; Hardee Reuter, K.; Scott, W. J., *J. Org. Chem.* **1991**, *56*, 2045.

98. Fu, J. M.; Zhao, B. P.; Sharp, M. J.; Snieckus, V., *J. Org. Chem.* **1991**, *56*, 1683.

99. Wang, W.; Snieckus, V., *J. Org. Chem.* **1992**, *57*, 424.

100. Creger, P. L., *J. Am. Chem. Soc.* **1970**, *92*, 1396.

101. Kraus, G. A., *J. Org. Chem.* **1981**, *46*, 201.

102. Clark, R. D.; Jahangir, *J. Org. Chem.* **1987**, *52*, 5378.

103. Sharp, M. J.; Cheng, W.; Snieckus, V., *Tetrahedron Lett.* **1987**, *28*, 5093.

104. Fu, J.; Sharp, M. J.; Snieckus, V., *Tetrahedron Lett.* **1988**, *29*, 5459.

105. Fu, J.; Zhao, B.; Sharp, M. J.; Snieckus, V., *J. Org. Chem.* **1991**, *56*, 1683.

106. Ito, Y.; Kobayashi, K.; Saegusa, T., *J. Am. Chem. Soc.* **1977**, *99*, 3532.

107. Hauser, F. M.; Rhee, R. P., *J. Org. Chem.* **1978**, *43*, 178.

108. Khanapure, S. P.; Reddy, R. T.; Biehl, E. R., *J. Org. Chem.* **1987**, *52*, 5685.

109. Khanapure, S. P.; Biehl, E. R., *Heterocycles* **1988**, *27*, 2643.

110. Marsais, F.; Trécourt, F.; Bréant, P.; Quéguiner, G., *J. Heterocycl. Chem.* **1988**, *25*, 81.

111. Rocca, P.; Cochennec, C.; Marsais, F.; Thomas-dit-Dumont, L.; Mallet, M.; Godard, A.; Quéguiner, G., *J. Org. Chem.* **1993**, *58*, 7832.

112. Plé, N.; Turck, A.; Martin, P.; Barbey, S.; Quéguiner, G., *Tetrahedron Lett.* **1993**, *34*, 1605.

113. Gribble, G. W.; Saulnier, M. G., *Tetrahedron Lett.* **1980**, *21*, 4137.

114. Knight, D. W.; Nott, A. P., *J. Chem. Soc., Perkin Trans. 1* **1981**, 1125.

115. Knight, D. W.; Nott, A. P., *J. Chem. Soc., Perkin Trans. 1* **1983**, 791.

116. Gribble, G. W.; Fletcher, G. L.; Ketcha, D. M.; Rajopadhye, M., *J. Org. Chem.* **1989**, *54*, 3264.

117. Buttery, C. D.; Jones, R. G.; Knight, D. W., *Synlett* **1991**, 315.

118. Buttery, C. D.; Knight, D. W.; Nott, A. P., *Tetrahedron Lett.* **1982**, *23*, 4127.

119. Cornwall, P.; Dell, C. P.; Knight, D. W., *Tetrahedron Lett.* **1987**, *28*, 3585.

120. Duhamel, L.; Plaquevent, J.-C., *J. Organomet. Chem.* **1993**, *448*, 1 and references cited therein.

121. (a) Seebach, D., *Agnew. Chem., Int. Ed. Engl.* **1988**, *27*, 1624. (b) Kim, Y. J.; Bernstein, M. P.; Roth, A. S.; Romesberg, F. E.; Willard, P. G.; Fuller, D. J.; Harrison, A. T.; Collum, D. B., *J. Org. Chem.* **1991**, *56*, 4435.

122. Mohrig, J. R.; Lee, P. K.; Stein, K. A.; Mitton, M. J.; Rosenberg, R. E., *J. Org. Chem.* **1995**, *60*, 3529.

123. Fatafah, Z. A.; Kopka, I. E.; Rathke, M. W., *J. Am. Chem. Soc.* **1980**, *102*, 3959.

124. Ireland, R. E.; Mueller, A. K., *J. Am. Chem. Soc.* **1976**, *98*, 2868.

125. Bauer, W.; Seebach, D., *Helv. Chim. Acta* **1984**, *67*, 1972.

126. Willard, P. G.; Salvino, J. M., *J. Org. Chem.* **1993**, *58*, 1.

127. Collum, D. B., *Acc. Chem. Res.* **1993**, *25*, 448.

128. Sun, X.; Kenkre, S. L.; Remenar, J. F.; Gilchrist, J. H.; Collum, D. B., *J. Am. Chem. Soc.* **1997**, *119*, 4765.

129. Streitwieser, A.; Kim, Y.-J., *Org. Lett.* **2002**, *4*, 573.

130. Barnett, N. D. R.; Mulvey, R. E.; Clegg, W.; O'Neil, P. A., *J. Am. Chem. Soc.* **1991**, *113*, 8187.

131. Myers, A. G.; Yang, B. H.; Chen, H.; McKinstry, L.; Kopecky, D. J.; Gleason, J. L., *J. Am. Chem. Soc.* **1997**, *119*, 6496 and references therein.

132. Myers, A. G.; Schnider, P.; Kwon, S.; Kung, D. W., *J. Org. Chem.* **1999**, *64*, 3322.

133. Myers, A. G.; Charest, M. G., *Electronic Encyclopedia of Reagents for Organic Synthesis*; Wiley: Chichester, 2001.

134. Streitwieser, A.; Leung, S. S. W.; Kim, Y.-J., *Org. Lett.* **1999**, *1*, 145.

135. (a) Fataftah, Z. A.; Kopka, I. E.; Rathke, M. W., *J. Am. Chem. Soc.* **1980**, *102*, 3959. (b) Fraser, R. R.; Mansour, T. S., *Tetrahedron Lett.* **1986**, *27*, 331.

136. Evans, D. A., In *Asymmetric Synthesis, Vol. 3., Stereodifferentiating Addition Reactions Part B*; Morrison, J. D., Ed.; Academic Press: London, 1984, p 2.

137. Kochi, T.; Tang, T. P.; Ellman, J. A., *J. Am. Chem. Soc.* **2002**, *124*, 6518.

138. Koketsu, M.; Yang, H. O.; Kim, Y. M.; Ichihashi, M.; Ishihara, H., *Org. Lett.* **2001**, *3*, 1705.

139. Morrison, C. F.; Burnell, J., *Org. Lett.* **2000**, *2*, 3891.

140. (a) Reich, H. J.; Green, D. P.; Phillips, N. H., *J. Am. Chem. Soc.* **1989**, *111*, 3444. (b) Reich, H. J.; Sikorski, W. H., *J. Org. Chem.* **1999**, *64*, 14.

141. Sun, X.; Collum, D. B., *J. Am. Chem. Soc.* **2000**, *122*, 2452.

142. Reich, H. J.; Borst, J. P., *J. Am. Chem. Soc.* **1991**, *113*, 1835.

143. Romesberg, F. E.; Gilchrist, J. H.; Harrison, A. T.; Fuller, D. J.; Collum, D. B., *J. Am. Chem. Soc.* **1991**, *113*, 5751.

144. Reich, H. J.; Kulicke, K. J. J., *J. Am. Chem. Soc.* **1996**, *118*, 273.

145. Leung, S. S.-W.; Streitwieser, A., *J. Org. Chem.* **1999**, *64*, 3390.

146. Pratt, L. M.; Newman, A.; St. Cyr, J.; Johnson, H.; Miles, B.; Lattier, A.; Austin, E.; Henderson, S.; Hershey, B.; Lin, M.; Balamraju, Y.; Sammonds, L.; Cheramie, J.; Karnes, J.; Woodford, B.; Carter, C., *J. Org. Chem.* **2003**, *68(16)*, 6387.

147. (a) Furstner, A., *Synthesis* **1989**, 571. (b) Conan, A.; Sibille, S.; Perichon, J., *J. Org. Chem.* **1991**, *56*, 2018. (c) Schwarz, K. H.; Kleiner, K.; Ludwig, R.; Schick, H., *J. Org. Chem.* **1992**, *57*, 4013.

148. Noyori, R., *Asymmetric Catalysis in Organic Synthesis*; Wiley-Interscience: New York, 1994.

149. Hamada, T.; Chieffi, A.; Ahman, J.; Buchwald, S. L., *J. Am. Chem. Soc.* **2002**, *7*, 1261.

150. Trost, B. M.; Jiang, C., *J. Am. Chem. Soc.* **2001**, *123*, 12907.

151. Trost, B. M.; Schroeder, G. M., *J. Am. Chem. Soc.* **1999**, *121*, 6759.

152. Shaughnessy, K. H.; Hamann, B. C.; Hartwig, J. F., *J. Org. Chem.* **1998**, 6546.

153. Fox, J. M.; Huang, X.; Chieffi, A.; Buchwald, S. L., *J. Am. Chem. Soc.* **2000**, *122*, 1360.

154. Anastasia, L.; Negishi, E.-I., *Org. Lett.* **2001**, *3*, 3111 and references therein.

155. Arend, M.; Westermann, B.; Risch, N., *Angew. Chem., Int. Ed.* **1998**, *37*, 1044.

156. Holy, N.; Fowler, R.; Burnett, E.; Lorenz, R., *Tetrahedron* **1979**, *39*, 613.

157. Comins, D. L.; Kuethe, J. T.; Hong, H.; Lakner, F. J.; Concolino, T. E.; Rheingold, A. L., *J. Am. Chem. Soc.* **1999**, *121*, 2651.

158. Vicario, J. L.; Badia, D.; Carillo, L., *J. Org. Chem.* **2001**, *66*, 9030.

159. Comins, D. L.; Brooks, C. A.; Al-awar, R. S.; Goehring, R. R., *Org. Lett.* **1999**, *1*, 229.

160. (a) Ashby, E. C.; Patil, G. S.; Gadru, K.; Gurumurthy, R., *J. Org. Chem.* **1993**, *58*, 424. (b) Ashby, E. C.; Deshpande, A. K.; Patil, G., *J. Org. Chem.* **1995**, *60*, 663.

161. Newkome, G. R.; Hager, D. C., *J. Org. Chem.* **1982**, *47*, 599.

162. Kimpe, N. D.; Yao, Z.-P.; Schamp, N., *Tetrahedron Lett.* **1986**, *27*, 1707.

163. Shen, C.; Ainsworth, C., *Tetrahedron Lett.* **1979**, *20*, 89.

164. Creary, X. J., *J. Org. Chem.* **1980**, *45*, 2419.

165. Scott, L. T.; Carlin, K. J.; Schultz, T. H., *Tetrahedron Lett.* **1978**, 4637.

166. Creary, X. J., *J. Am. Chem. Soc.* **1977**, *99*, 7632.

167. Arduengo, III, R. L.; Harlow, R. L.; Kline, M., *J. Am. Chem. Soc.* **1991**, *113*, 361.

168. Alder, R. W.; Blake, M. E.; Bortolotti, C.; Bufali, S.; Butts, C.; Linehan, E.; Oliva, J.; Orpen, A. G.; Quayle, M. J., *Chem. Commun.* **1999**, 241.

169. Stang, P. J., *Angew. Chem., Int. Ed. Engl.* **1992**, *31*, 274.

170. Harada, T.; Fujiwara, T.; Katsuhiro, I.; Oku, A., *Org. Lett.* **2000**, *2*, 1855.

171. Hodgson, D. M.; Gibbs, A. R.; Lee, G. P., *Tetrahedron* **1996**, *52*, 14361.

172. O'Brien, P. J., *J. Chem. Soc., Perkin Trans. 1* **1998**, 1439.

173. Cox, P. J.; Simpkins, N. S., *Tetrahedron: Asymmetry* **1991**, *2*, 1.

174. Smith, J. G., *Synthesis* **1984**, 629.

175. (a) Katsuki, T.; Sharpless, K. B., *J. Am. Chem. Soc.* **1980**, *102*, 5974. (b) Scott Woodard, S.; Finn, M. G.; Sharpless, B. K., *J. Am. Chem. Soc.* **1991**, *113*, 106. (c) Finn, M. G.; Sharpless, B., *J. Am. Chem. Soc.* **1991**, *113*, 113. (d) Hoshino, Y.; Yamamoto, H., *J. Am. Chem. Soc.* **2000**, *122(42)*, 10452 and references therein.

176. Cope, A. C.; Fenton, S. W.; Spencer, C. F., *J. Am. Chem. Soc.* **1952**, *74*, 5884.

177. Cope, A. C.; Heeren, J. K., *J. Am. Chem. Soc.* **1965**, *87*, 3125.

178. Crandall, J. K.; Lin, L.-H. C., *J. Org. Chem.* **1968**, *33*, 2375.

179. Thummel, R. P.; Rickborn, B., *J. Org. Chem.* **1972**, *37*, 3919.

180. Södergren, M. J.; Bertilsson, S. K.; Andersson, P. G., *J. Am. Chem. Soc.* **2000**, *122*, 6610.

181. Eisch, J. J.; Galle, J. E., *J. Org. Chem.* **1990**, *55*, 4835.

182. McDonald, R. N.; Steppel, R. N.; Cousins, R. C., *J. Org. Chem.* **1975**, *40*, 1694 and references therein.

183. (a) Yanagisawa, A.; Yasue, K.; Yamomoto, Y., *Chem. Commun.* **1994**, 2103. (b) Thies, R. W.; Chiarello, R. H., *J. Org. Chem.* **1979**, *44*, 1342.

184. Morgan, K. M.; Gronert, S., *J. Org. Chem.* **2000**, *65*, 1461.

185. Asami, M.; Suga, T.; Honda, K.; Inoue, S., *Tetrahedron Lett.* **1997**, *38*, 6425.

186. Morgan, K. M.; Gajevski, J. J., *J. Org. Chem.* **1996**, *61*, 820.

187. Apparu, M.; Barelle, M., *Tetrahedron* **1978**, *34*, 1541.

188. Thummel, R. P.; Rickborn, B., *J. Am. Chem. Soc.* **1970**, *92*, 2064.

189. Cope, A. C.; Lee, H.-H.; Petree, H. E., *J. Am. Chem. Soc.* **1958**, *80*, 2849.

190. Whitesell, J. K.; White, P. D., *Synthesis* **1975**, 602.

191. Boeckman, R. K., *Tetrahedron Lett.* **1977**, 4281.

192. Sheng, M. N., *Synthesis* **1972**, 194.

193. Kishel, C. L.; Rickborn, B., *J. Org. Chem.* **1972**, *37*, 2060.

194. Margot, C.; Schlosser, M. J., *Tetrahedron Lett.* **1985**, *26*, 1035.

195. Mordini, A.; Ben Rayana, E.; Margot, C.; Schlosser, M., *Tetrahedron* **1990**, *46*, 2401.

196. Asami, M.; Ishizaki, T.; Inoue, S., *Tetrahedron: Asymmetry* **1994**, *5*, 793.

197. Asami, M. J., *Synth. Org. Chem. Jpn.* **1996**, *54*, 188.

198. (a) Bertilsson, S. K.; Sodergren, M. J.; Andersson, P. G., *J. Org. Chem.* **2002**, *67*, 1567. (b) Andersson, P. G.; Bertilsson, S. K., *Tetrahedron Lett.* **2002**, 4665 and references therein.

199. Hanson, R. M., In *Organic Reactions*; Overman, Larry E. et al., Eds.; Wiley: New York, 2002, p 1.

200. Abou-Elzahab, M.; Adam, W.; Saha-Moller, C. R., *Liebigs Ann. Chem.* **1991**, 445.

201. Morrison, G. A.; Wilkinson, J. B., *J. Chem. Soc., Perkin Trans. 1* **1990**, 345.

202. Zwanenburg, B.; Dols, P. P. M. A.; Arnouts, E. G.; Rohaan, J.; Klunder, A. J. H., *Tetrahedron* **1994**, *50*, 3473.

203. Larcheveque, M.; Perriot, P.; Petit, Y., *Synthesis* **1983**, 297.

204. (a) Ichikawa, Y.; Niitsuma, S.; Kuniki, K.; Takita, T. J., *J. Chem. Soc., Chem. Commun.* **1988**, 625. (b) Bird, C. W.; Hormozi, N., *Tetrahedron Lett.* **1990**, *31*, 3501.

205. (a) Mordini, A.; Bindi, S.; Capperucci, A.; Nistri, D.; Reginato, G.; Valacchi, M., *J. Org. Chem.* **2001**, *66*, 3201. (b) Mordini, A.; Valcchi, M.; Nardi, C.; Bindi, S.; Poli, G.; Reginato, G., *J. Org. Chem.* **1997**, *62*, 8557. (c) Mordini, A.; Bindi, S.; Pecchi, S.; Capperucci, A.; Degl'Innocenti, A.; Reginato, G., *J. Org. Chem.* **1996**, *61*, 4466. (d) Mordini, A.; Bindi, S.; Pecchi, S.; Degl'Innocenti, A.; Reginato, G.; Serci, A., *J. Org. Chem.* **1996**, *61*, 4374.

206. Schlosser, M. J., *J. Organomet. Chem.* **1967**, *8*, 9.

207. Stamm, H.; Speth, D., *Chem. Ber.* **1989**, *122*, 1795.

208. Zhang, Z.; Scheffold, R., *Helv. Chim. Acta* **1993**, *76*, 2602.

209. Mordini, A.; Russo, F.; Valacchi, M.; Zani, L.; Degl'Innocenti, A.; Reginato, G., *Tetrahedron* **2002**, *58*, 7153.

210. Shea, K. J., *Tetrahedron* **1980**, *36*, 1683.

211. Bowen, D. H.; MacMillan, J., *Tetrahedron Lett.* **1972**, 4111.

212. Nickson, A.; Covey, D. F.; Huang, F.; Kuo, Y.-N., *J. Am. Chem. Soc.* **1975**, *97*, 904.

213. Giblin, G. M.; Kirk, D. T.; Mitchell, L.; Simpkins, N. S., *Org. Lett.* **2003**, *5*, 1673.

214. Gwaltney, II, S. L.; Sakata, S. T.; Shea, K. J., *J. Org. Chem.* **1996**, *61*, 7438.

215. Bunnett, J. F., *Acc. Chem. Res.* **1972**, *5*, 139.

216. Queguiner, G.; Rocca, P.; Cohennec, C.; Marsais, F.; Thomas-dit-Dumont, L.; Mallet, M.; Godard, A., *J. Org. Chem.* **1993**, *58*, 7832.

217. Sammakia, T.; Strangeland, E. L.; Whitcomb, M. C., *Org. Lett.* **2002**, *4*, 2385.

218. (a) Sibi, M. P.; Snieckus, V., *J. Org. Chem.* **1983**, 48, 1935. (b) Zhang, P.; Gawley, R. E., *J. Org. Chem.* **1993**, *58*, 3223.

219. Kalinin, A. V.; Chattopadhyay, M. S.; Tsukazaki, M.; Wicki, M.; Nguen, T.; Coelho, A. L.; Kerr, M.; Snieckus, V., *Synlett* **1997**, 839.

Lithium Naphthalenide[1]

[7308-67-0] $C_{10}H_8Li$ (MW 134.12)

InChI = 1S/C10H8.Li/c1-2-6-10-8-4-3-7-9(10)5-1;/h1-8H;/q-1;
+1

InChIKey = MTIJPMVWHNRZSB-UHFFFAOYSA-N

(reductive metalation reactions;[1c] reduction of metal salts;[27] initiation of polymerization reactions[28])

Alternate Name: LN.

Physical Data: no data on the isolated material; only available in solution.

Solubility: sol ether, benzene, THF; reacts with protic solvents and THF at elevated temperatures.[2]

Preparative Methods: made by addition of freshly cut **Lithium** metal to a solution of naphthalene in THF. Preparation can be accelerated by ultrasonication.[3]

Analysis of Reagent Purity: two titration methods have been described;[4] the more convenient[4b] involves conversion of 1,1-diphenylethylene by lithium naphthalenide to an intensely red-colored dianion, which is then titrated against *s*-butanol.

Handling, Storage, and Precautions: can be stored in solution up to several days; must be protected from air and moisture; can be used to ambient temperature.

Original Commentary

Kevin M. Short
Wayne State University, Detroit, MI, USA

Reductive Metalation. The powerful reductive nature of this reagent makes it an important tool for lithium–heteroatom exchange reactions. Thus, it was established early on that (phenylthio)alkanes can be converted into their requisite alkyllithium species.[5] This has become the method of choice over generation by lithium metal alone. The resultant alkyllithium species can either be quenched with a proton source (eq 1),[6] or intercepted with an electrophile. This has subsequently evolved into a powerful technique, since the reaction is general for all chalcogens (eq 2)[7] and halides (eq 3).[8]

(1)

(2)

Metal–heteroatom exchange can also be persuaded to occur with a variety of other systems, resulting in allyl-[9] and vinyllithium[10] species, as well as α-lithio ethers (eq 3),[1c,8,11] α-lithio thioethers,[1c,12] α-lithio amines,[11b] and α-lithio silanes.[1c,12b,13] The latter class provides useful intermediates for the Peterson alkenation reaction (eq 4).[13a]

(3)

(4)

In some cases, however, it may be advantageous to proceed via either **Lithium 1-(Dimethylamino)naphthalenide** (LDMAN), or **Lithium 4,4′-Di-tert-butylbiphenylide** (LDBB).[1c] With the former reagent, the byproduct formed, (dimethylamino)naphthalene, is more easily removed than is naphthalene from product mixtures. In the latter case, the greater reduction potential of di-*t*-butylbiphenyl appears to lead to more efficient halogen–lithium exchange.[8b]

Dianion Generation. Lithium naphthalenide efficiently deprotonates β-alkynyloxy[14] and carboxylate anions (eq 5).[15] In addition, the previously mentioned phenomenon of reductive metalation has been exploited to access dianions from halohydrins,[16] β-halo carboxylic acids,[17] and β-halo carboxamides,[18] and even trianions from β,ω′-dihalo alcohols.[19] A major pathway for the polyanionic species is β-elimination (eq 6);[16a,19] when such processes can be avoided, the polyanions react according to Hauser's rule (eq 7).[16a,20]

(5)

(6)

(7)

Dehalogenation Reactions. Since lithium naphthalenide is a particularly effective initiator for halogen–metal exchange, it has found widespread use for the conversion of dihalides to unsaturated species. Thus, 1,2-dichlorodisilanes have been converted to silenes (eq 8)[21] and diphosphiranes to phosphacumulenes.[22] In a related field, silicon cages[23] have been constructed from trichlorodisilanes.

Essential Reactions for Organic Synthesis, First Edition. Edited by Philip L. Fuchs.
© 2016 John Wiley & Sons, Ltd. Published 2016 by John Wiley & Sons, Ltd.

$$Ar_2Si(Cl)—Si(Cl)Ar_2 \xrightarrow[50\%]{LN, DME} Ar_2Si=SiAr_2 \quad (8)$$

$$Ar = \text{—}\!\!\!\!\begin{array}{c} i\text{-Pr} \\ \\ i\text{-Pr} \end{array}\!\!\!\!i\text{-Pr}$$

Metal Redox Reactions. Lithium naphthalenide is a convenient reducing agent for a variety of metals, and shows great promise in the synthetic area. Thus, Cu^I complexes have been reduced to Cu^0; the resultant highly reactive species adds in an oxidative fashion across the carbon–halogen bond.[24] As a consequence, the well known organocuprate addition chemistry can be carried out in one step from halocarbons, without having to initially prepare the organolithium species (eq 9). In addition to the reduction of Cu^I, lithium naphthalenide reduces Si^{IV} to Si^{II},[25] Sn^{IV} to Sn^{II} (eq 10),[26] and various lanthanide compounds.[27]

$$CuI \cdot PEt_3 \xrightarrow[\substack{3.\ PhCOCl \\ 71\%}]{\substack{1.\ LN,\ THF \\ 2.\ Br\text{—}C_6H_4\text{—}CN}} \begin{array}{c} Ph\text{—}CO\text{—}C_6H_4\text{—}CN \end{array} \quad (9)$$

$$\text{(eq 10)} \quad (10)$$

Oligomerization Reactions. Lithium naphthalenide has long been a convenient initiator for anionic 'living' polymerization reactions.[28] Thus, styrenes, acrylates, dienes, and other monomers have been polymerized using lithium naphthalenide as an anionic initiator. In some circumstances, however, oligomerization can be controlled to furnish only dimers (eq 11).[29] Also, in the presence of a secondary amine, 1,3-dienes can be persuaded to react in a formal 1,4-fashion to produce allyl amines (eq 12).[30]

$$\xrightarrow[\substack{HNEt_2,\ THF \\ 85\%}]{LN} \quad (11)$$

$$\xrightarrow[\substack{Me_2N(CH_2)_3NMe_2 \\ 72\%}]{LN,\ Et_2NH,\ PhH} \quad (12)$$

Interesting approaches toward functional polymers have recently been detailed,[31] wherein previously described chemistry involving lithium naphthalenide is conducted on suitably substituted polystyrene derivatives (eq 13).

$$\xrightarrow[\substack{2.\ ClSiMe_2H}]{1.\ LN,\ THF} \xrightarrow[\substack{H_2PtCl_6 \\ PhMe}]{} \quad (13)$$

Related Reagents. Copper(I) Iodide–Triethylphosphine–Lithium Naphthalenide; Lithium 4,4'-Di-t-butylbiphenylide; Lithium 1-(Dimethylamino)naphthalenide; Potassium Naphthalenide; Sodium Anthracenide; Sodium Naphthalenide; Sodium Phenanthrenide.

First Update

Alexander Wei
Purdue University, West Lafayette, IN, USA

Panuwat Padungros
Department of Chemistry, Faculty of Science, Chulalongkorn University, Bangkok, Thailand

Reduction and Alkylation. Organolithium species can be cleanly generated by metal–heteroatom exchange using lithium naphthalenide (LN) for conversion to unsubstituted derivatives by protonation[32–34] or reaction with electrophiles, such as alkyl halides or carbonyls.[35,36] Similarly, lithium enolates can be prepared from carbonyl species bearing α-cyano or heteroatomic functional groups[37]; subsequent reactions with appropriate electrophiles have been used in the synthesis of quaternary carbon centers (eq 14).[38,39] Notably, reductive alkylation by this approach has yielded bicyclic, *trans*-fused cyclohexenes with stereochemical outcomes that correspond to antarafacial Diels–Alder additions (eq 15).[40] The reduction of α,β-epoxyketones by LN produces β-hydroxyketones with high chemo- and regioselectivity (eq 16).[41]

$$\xrightarrow[\substack{2.\ allyl\ bromide,\ -30\ °C \\ 74\%}]{1.\ LN,\ THF,\ -30\ °C}$$

$$LN = Li^+ \quad (14)$$

$$(15)$$

$$(16)$$

Chiral phosphines are widely used as auxiliaries for various metal-catalyzed asymmetric reactions and can be prepared from stable phosphine–borane complexes. Secondary *P*-chiral phosphine–boranes can be prepared by reductive lithiation of the corresponding tertiary phosphine–borane using LN (eq 17).[42] Likewise, *P*-chiral tertiary phosphine ligands can be produced by the reductive lithiation of phosphinite–boranes followed by alkylation, both proceeding with retention of configuration (eq 18).[43,44]

$$(17)$$

$$(18)$$

Dianion Generation. A combination of *n*-BuLi and LN can be used to generate dianionic intermediates containing β-heteroatoms, without concern for rapid E1 elimination (eq 19).[45–47] Bridged bicyclic compounds can be accessed by the double reduction of aromatic diesters by LN to generate stable bis-enolates, followed by double alkylation with α,ω-electrophiles (eq 20).[48]

$$(19)$$

$$(20)$$

Reductive Cyclization. Reductive lithiation of alkyl halides by LN can trigger an intramolecular cyclization with α-aminonitriles to generate an imine anion intermediate; subsequent elimination of a neighboring *N,O*-acetal produces a polycyclic enamine in moderate yield (eq 21).[49] Intramolecular reductive cyclization of alkynes with LN followed by electrophilic trapping provides an efficient entry into siloles (eq 22)[50,51] and silaindenes.[52] Silaindenes often have useful electronic properties, such as photoluminescence, but cannot be made by traditional methods of indole synthesis due to the incompatibility of the silole with typical alkyne protecting groups. Applying this same methodology to bis(*o*-silyl)diphenylacetylenes produces fused, π-rich polycyclic compounds with silicon (eq 23)[53] and carbon[54] bridges.

$$(21)$$

$$(22)$$

$$(23)$$

$$(26)$$

$$(27)$$

Reductive Elimination. Carbohydrate-derived thioacetals undergo reductive elimination with LN to yield glycals. This elimination is compatible with pyranoside derivatives bearing various substituents and protecting groups, including benzyl ethers,[55] if the reaction conditions and temperature are carefully controlled (eq 24).[56] Lithium naphthalenide has also been applied toward the chemoselective elimination of π-allyltricarbonyliron lactone complexes, which can serve as chiral auxiliaries in the synthesis of natural products such as (−)-gloeosporone.[57] Reductive cleavage of Fe(CO)$_3$ with LN can be achieved without racemization of the allylic stereocenter (eq 25). Allylic alcohols can be prepared from α,β-epoxymesylates or epoxybenzotriazoles by reductive elimination with LN; in these cases, the alkene geometry is influenced by the steric bulk of the local substituents (eq 26).[58,59] The title reagent is also a safe and economical alternative to t-BuLi for converting α,α-dibromoesters into ynolate anions, which can form cyclic adducts with 1,5-dicarbonyls (eq 27).[60]

Deprotection. The title compound has been used for the deprotection of N-sulfonylaziridines (eq 28),[61] substituted phenyl-propargyl ethers (eq 29),[62] benzyl ethers in the presence of alkenes (eq 30),[63] and phenylsilyl ethers in the presence of acid-sensitive groups.[64]

$$(28)$$

$$(29)$$

$$(24)$$

$$(25)$$

E,E : Z,E : E,Z mixture
(5:3:2)

$$(30)$$

Organometallic Species. Treatment of lithium (2-thienyl-cyano)cuprate with LN yields a highly reactive form of copper that can oxidatively insert into alkyl halides to produce the corresponding organocuprate species, for participation in 1,4-conjugate additions, epoxide ring opening, and cross-coupling with acid chlorides (eq 31).[65,66] Organocuprates have also been prepared by reducing alkyl halides with LN in the presence of CuI–PPh$_3$,[67] but thienyl-based organocuprates avoid the unnecessary introduction of phosphines that can interfere with product purification.

The title compound is also useful for reducing the salts of Ti^{III},[68] Zn^{II},[69] Ge^{II},[70] Mg^{II},[71] and Mn^{II}[72] into reactive zero-valent metals, for subsequent generation of organometallic species.

(31)

1. (a) Wakefield, B. J., *The Chemistry of Organolithium Compounds*; Pergamon: Oxford, **1974**. (b) March, J. *Advanced Organic Chemistry*, 4th ed.; Wiley: New York, **1992**; p 729. (c) Cohen, T.; Bhupathy, M., *Acc. Chem. Res.* **1989**, *22*, 152.

2. Fujita, T.; Suga, K.; Watanabe, S., *Synthesis* **1972**, 630.

3. Azuma, T.; Yanagida, S.; Sakurai, H.; Sasa, S.; Yoshino, K., *Synth. Commun.* **1982**, *12*, 137.

4. (a) Ager, D. J., *J. Organomet. Chem.* **1983**, *241*, 139. (b) Screttas, C. G.; Micha-Screttas, M., *J. Organomet. Chem.* **1983**, *252*, 263.

5. (a) Screttas, C. G.; Micha-Screttas, M., *J. Org. Chem.* **1978**, *43*, 1064. (b) Cohen, T.; Weisenfeld, R. B., *J. Org. Chem.* **1979**, *44*, 3601.

6. Harring, S. R.; Livinghouse, T., *Tetrahedron Lett.* **1989**, *30*, 1499.

7. Agawa, T.; Ishida, M.; Ohshiro, Y., *Synthesis* **1980**, 933.

8. (a) Lesimple, P.; Beau, J.-M.; Sinaÿ, P., *Carbohydr. Res.* **1987**, *171*, 289. (b) Freeman, P. K.; Hutchinson, L. L., *J. Org. Chem.* **1980**, *45*, 1924.

9. Cohen, T.; Guo, B.-S., *Tetrahedron* **1986**, *42*, 2803.

10. Duhamel, L.; Chauvin, J.; Messier, A., *J. Chem. Res. (S)* **1982**, 48.

11. (a) Shiner, C. S.; Tsunoda, T.; Goodman, B. A.; Ingham, S.; Lee, S.; Vorndam, P. E., *J. Am. Chem. Soc.* **1989**, *111*, 1381. (b) Broka, C. A.; Shen, T., *J. Am. Chem. Soc.* **1989**, *111*, 2981. (c) Hoffmann, R.; Brückner, R., *Ber. Dtsch. Chem. Ges./Chem. Ber.* **1992**, *125*, 1957.

12. (a) McDougal, P. G.; Condon, B. D.; Laffosse, M. D., Jr.; Lauro, A. M.; Van Derveer, D., *Tetrahedron Lett.* **1988**, *29*, 2547. (b) Ager, D. J., *J. Chem. Soc., Perkin Trans. 1* **1986**, 195.

13. (a) Ager, D. J., *J. Chem. Soc., Perkin Trans. 1* **1986**, 183. (b) Mandai, T.; Kohama, M.; Sato, H.; Kawada, M.; Tsuji, J., *Tetrahedron* **1990**, *46*, 4553.

14. Watanabe, S.; Suga, K.; Suzuki, T., *Can. J. Chem.* **1969**, *47*, 2343.

15. Fujita, T.; Watanabe, S.; Suga, K., *Aust. J. Chem.* **1974**, *27*, 2205.

16. (a) Barluenga, J.; Flórez, J.; Yus, M., *J. Chem. Soc., Perkin Trans. 1* **1983**, 3019. (b) Barluenga, J.; Fernández-Simón, J. L.; Concellón, J. M.; Yus, M., *J. Chem. Soc., Perkin Trans. 1* **1988**, 3339.

17. Caine, D.; Frobese, A. S., *Tetrahedron Lett.* **1978**, 883.

18. Barluenga, J.; Foubelo, F.; Fañanás, F. J.; Yus, M., *Tetrahedron* **1989**, *45*, 2183.

19. Barluenga, J.; Fernandez, J. R.; Yus, M., *Synthesis* **1985**, 977.

20. Hauser, C. R.; Harris, T. M., *J. Am. Chem. Soc.* **1958**, *80*, 6360.

21. Watanabe, H.; Takeuchi, K.; Nakajima, K.; Nagai, Y.; Goto, M., *Chem. Lett.* **1988**, 1343.

22. Yoshifuji, M.; Toyota, K.; Yoshimura, H., *Chem. Lett.* **1991**, 491.

23. Kabe, Y.; Kawase, T.; Okada, J.; Yamashita, O.; Goto, M.; Masamune, S., *Angew. Chem., Int. Ed. Engl.* **1990**, *29*, 794.

24. Rieke, R. D.; Dawson, B. T.; Stack, D. E.; Stinn, D. E., *Synth. Commun.* **1990**, *20*, 2711.

25. Jutzi, P.; Holtmann, U.; Kanne, D.; Krüger, C.; Blom, R.; Gleiter, R.; Hyla-Kryspin, I., *Ber. Dtsch. Chem. Ges./Chem. Ber.* **1989**, *122*, 1629.

26. Jutzi, P.; Hielscher, B., *Organometallics* **1986**, *5*, 1201.

27. (a) Arnaudet, L.; Ban, B., *Nouv. J. Chim.* **1988**, *12*, 201. (b) Bochkarev, M. N.; Trifonov, A. A.; Fedorova, E. A.; Emelyanova, N. S.; Basalgina, T. A.; Kalinina, G. S.; Razuvaev, G. A., *J. Organomet. Chem.* **1989**, *372*, 217.

28. Ishizone, T.; Wakabayashi, S.; Hirao, A.; Nakahama, S., *Macromolecules* **1991**, *24*, 5015.

29. (a) Takabe, K.; Ohkawa, S.; Katagiri, T., *Synthesis* **1981**, 358. (b) Fujita, T.; Watanabe, S.; Suga, K.; Sugahara, K.; Tsuchimoto, K., *Chem. Ind. (London)* **1983**, 167.

30. Sugahara, K.; Fujita, T.; Watanabe, S.; Hashimoto, H., *J. Chem. Technol. Biotechnol.* **1987**, *37*, 95.

31. (a) O'Brien, R. A.; Rieke, R. D., *J. Org. Chem.* **1990**, *55*, 788. (b) Itsuno, S.; Shimizu, K.; Kamahori, K.; Ito, K., *Tetrahedron Lett.* **1992**, *33*, 6339.

32. Azzena, U.; Pittalis, M., *Tetrahedron* **2011**, *67*, 3360.

33. Yu, J.; Cho, H. S.; Falck, J. R., *J. Org. Chem.* **1993**, *58*, 5892.

34. (a) Amancha, P. K.; Liu, H.-J.; Ly, T. W.; Shia, K.-S., *Eur. J. Org. Chem.* **2010**, *2010*, 3473. (b) Liu, H.-J.; Shang, X., *Tetrahedron Lett.* **1998**, *39*, 367.

35. (a) Yus, M.; Herrera, R. P.; Guijarro, A., *Tetrahedron Lett.* **2001**, *42*, 3455. (b) Yus, M.; Herrera, R. P.; Guijarro, A., *Chem. Eur. J.* **2002**, *8*, 2574. (c) Kondo, Y.; Murata, N.; Sakamoto, T., *Heterocycles* **1994**, *37*, 1467.

36. (a) Foubelo, F.; Moreno, B.; Soler, T.; Yus, M., *Tetrahedron* **2005**, *61*, 90 82. (b) Tsao, J.-P.; Tsai, T.-Y.; Chen, I.-C.; Liu, H.-J.; Zhu, J.-L.; Tsao, S.-W., *Synthesis* **2010**, 4242.

37. (a) Zhu, J.-L.; Ko, Y.-C.; Kuo, C.-W.; Shia, K.-S., *Synlett* **2007**, 1274. (b) Ko, Y.-C.; Zhu, J.-L., *Synthesis* **2007**, 3659.

38. Liao, C.; Zhu, J.-L., *J. Org. Chem.* **2009**, *74*, 7873.

39. (a) Zhu, J.-L.; Shia, K.-S.; Liu, H.-J., *Chem. Commun.* **2000**, 1599. (b) Amancha, P. K.; Lai, Y.-C.; Chen, I.-C.; Liu, H.-J.; Zhu, J.-L., *Tetrahedron* **2010**, *66*, 871. (c) Liu, H.-J.; Zhu, J.-L.; Shia, K.-S., *Tetrahedron Lett.* **1998**, *39*, 4183.

40. Lee, J.-H.; Zhang, Y.; Danishefsky, S. J., *J. Am. Chem. Soc.* **2010**, *132*, 14330.

41. Jankowska, R.; Mhehe, G. L.; Liu, H. J., *Chem. Commun.* **1999**, 1581.

42. (a) Wolfe, B.; Livinghouse, T., *J. Org. Chem.* **2001**, *66*, 1514. (b) Miura, T.; Yamada, H.; Kikuchi, S.; Imamoto, T., *J. Org. Chem.* **2000**, *65*, 1877.

43. Takahashi, Y.; Yamamoto, Y.; Katagiri, K.; Danjo, H.; Yamaguchi, K.; Imamoto, T., *J. Org. Chem.* **2005**, *70*, 9009.

44. Vedejs, E.; Donde, Y., *J. Org. Chem.* **2000**, *65*, 2337.

45. (a) Foubelo, F.; Yus, M., *Tetrahedron Asymmetry* **1996**, 2911. (b) Barfoot, C. W.; Harvey, J. E.; Kenworthy, M. N.; Kilburn, J. P.; Ahmed, M.; Taylor, R. J. K., *Tetrahedron* **2005**, *61*, 3403.

46. Schäfer, A.; Thiem, J., *J. Org. Chem.* **1999**, *65*, 24.

47. (a) Barluenga, J.; Montserrat, J. M.; Florez, J., *J. Org. Chem.* **1993**, *58*, 5976. (b) Jung, M. E.; Vu, B. T., *J. Org. Chem.* **1996**, *61*, 4427.

48. Lobato, R.; Veiga, A. X.; Pérez-Vázquez, J.; Fernández-Nieto, F.; Paleo, M. R.; Sardina, F. J., *Org. Lett.* **2013**, *15*, 4090.

49. Roulland, E.; Cecchin, F.; Husson, H.-P., *J. Org. Chem.* **2005**, *70*, 4474.

50. (a) Morra, N.; Pagenkopf, B., *Org. Synth.* **2008**, *85*, 53. (b) Boydston, A. J.; Yin, Y.; Pagenkopf, B. L., *J. Am. Chem. Soc.* **2004**, *126*, 3724. (c) Tamao, K.; Yamaguchi, S.; Shiro. M., *J. Am. Chem. Soc.* **1994**, *116*, 11715.

51. Horst, S.; Evans, N. R.; Bronstein, H. A.; Williams, C. K., *J. Polym. Sci. A* **2009**, *47*, 5116.

52. Xu, C.; Wakamiya, A.; Yamaguchi, S., *Org. Lett.* **2004**, *6*, 3707.

53. Yamaguchi, S.; Xu, C.; Tamao, K., *J. Am. Chem. Soc.* **2003**, *125*, 13662.

54. Zhang, H.; Karasawa, T.; Yamada, H.; Wakamiya, A.; Yamaguchi, S., *Org. Lett.* **2009**, *11*, 3076.

55. (a) Boulineau, F. P.; Wei, A., *Carbohydr. Res.* **2001**, *334*, 271. (b) Boulineau, F. P.; Wei, A., *J. Org. Chem.* **2004**, *69*, 3391.

56. Alberch, L.; Cheng, G.; Seo, S.; Li, X.; Boulineau, F. P.; Wei, A., *J. Org. Chem.* **2011**, *76*, 2532.

57. (a) Ley, S.; Cleator, E.; Harter, J.; Hollowood, C. J., *Org. Biomol. Chem.* **2003**, *22*, 3263. (b) Hollowood, C. J.; Ley, S., *Chem. Commun.* **2002**, *18*, 2130.

58. Kang, Y. H.; Lee, C. J.; Kim, K., *J. Org. Chem.* **2001**, *66*, 2149.

59. Wu, Y.-K.; Liu, H.-J.; Zhu, J.-L., *Synlett* **2008**, *2008*, 621.

60. Shindo, M.; Koretsune, R.; Yokota, W.; Itoh, K.; Shishido, K., *Tetrahedron Lett.* **2001**, *42*, 8357.

61. (a) Alonso, A. D.; Andersson, P. G., *J. Org. Chem.* **1998**, *63*, 9455. (b) Concellon, J.; Rodriguez-Solla, H.; Simal, C., *Org. Lett.* **2008**, *20*, 4457. (c) Shu, C.; Liu, M.-Q.; Wang, S.-S.; Li, L.; Ye, L.-W., *J. Org. Chem.* **2013**, *78*, 3292.

62. Crich, D.; Karatholuvhu, M., *J. Org. Chem.* **2008**, *73*, 5173.

63. (a) Liu, H.-J.; Yip, J.; Shia, K. S., *Tetrahedron Lett.* **1997**, *38*, 2253. (b) Huo, H.-H.; Xia, X.-E.; Zhang, H.-K.; Huang, P.-Q., *J. Org. Chem.* **2013**, *78*, 455. (c) Pabbaraja, S.; Satyanarayana, K.; Ganganna, B.; Yadav, J. S., *J. Org. Chem.* **2011**, *76*, 1922. (d) Poigny, S.; Nouri, S.; Chiaroni, A.; Guyot, M.; Samadi, M., *J. Org. Chem.* **2001**, *66*, 7263.

64. Behoul, C.; Guijarro, D.; Yus, M., *Tetrahedron* **2005**, *61*, 6908.

65. Rieke, R. D.; Klein, W. R.; Wu, T.-C., *J. Org. Chem.* **1993**, *58*, 2492.

66. Ebert, G. W.; Juda, W. L.; Kosakowski, R. H.; Ma, B.; Dong, L.; Cummings, K. E.; Phelps, M. V. B.; Mostafa, A. E.; Luo, J., *J. Org. Chem.* **2005**, *70*, 4314.

67. Rieke, R. D.; Stack, D. E.; Dawson, B. T.; Wu, T. C., *J. Org. Chem.* **1993**, *58*, 2483.

68. Rele, S.; Talukdar, S.; Banerji, A.; Chattopadhyay, S., *J. Org. Chem.* **2001**, *66*, 2990.

69. (a) Zhu, L.; Wehmeyer, R. M.; Rieke, R. D., *J. Org. Chem.* **1991**, *56*, 1445. (b) Rieke, R. D.; Hanson, M. V.; Brown, J. D.; Niu, Q. J., *J. Org. Chem.* **1996**, *61*, 2726.

70. Kagoshima, H.; Hashimoto, Y.; Oguro, D.; Saigo, K., *J. Org. Chem.* **1998**, *63*, 691.

71. Rieke, R. D.; Xiong, H., *J. Org. Chem.* **1992**, *57*, 6560.

72. (a) Rieke, R. D.; Kim, S.-H.; Wu, X., *J. Org. Chem.* **1997**, *62*, 6921. (b) Rieke, R. D.; Kim, S.-H., *J. Org. Chem.* **1998**, *63*, 5235. (c) Kim, S.-H.; Rieke, R. D., *J. Org. Chem.* **2000**, *65*, 2322.

M

Manganese Dioxide   **248**

Essential Reagents for Organic Synthesis, Edited by Philip L. Fuchs, André B. Charette, Tomislav Rovis, and Jeffrey W. Bode.
©2016 John Wiley & Sons, Ltd. Published 2016 by John Wiley & Sons, Ltd.

Manganese Dioxide[1]

[1313-13-9] MnO$_2$ (MW 86.94)

InChI = 1/Mn.2O/rMnO2/c2-1-3

InChIKey = NUJOXMJBOLGQSY-MQJJFXMEAL

(useful selective oxidizing reagent for organic synthesis; oxidation of allylic alcohols to α,β-ethylenic aldehydes or ketones;[2] conversion of allylic alcohols to α,β-ethylenic esters or amides;[3] oxidation of propargylic alcohols,[4] benzylic or heterocyclic alcohols,[5] and saturated alcohols;[6] oxidative cleavage of 1,2-diols;[7] hydration of nitriles to amides;[8] dehydrogenation and aromatization reactions;[9] oxidation of amines to aldehydes, imines, amides, and diazo compounds[10])

Alternate Names: manganese oxide; manganese(IV) oxide.

Physical Data: mp 535 °C (dec.); *d* 5.03 g cm^{-3}.

Solubility: insoluble in H$_2$O and organic solvents.

Form Supplied in: dark brown powder, widely available. The commercial 'active' MnO$_2$ used as oxidizing reagent for organic synthesis is a synthetic nonstoichiometric hydrated material. The main natural source of MnO$_2$ is the mineral pyrolusite, a poor oxidizing reagent.

Original Commentary

Gérard Cahiez & Mouâd Alami

Université Pierre & Marie Curie, Paris, France

Structure of Active Manganese Dioxide.[1,11,12] The structure and the reactivity of active manganese dioxides used as oxidizing reagents in organic synthesis closely depends on their method of preparation (see below). Active manganese oxides are nonstoichiometric materials (generally MnO$_x$; $1.93 < x < 2$) and magnetic measurements reveal the presence of lower valency Mn species, probably Mn^{II} and Mn^{III} oxides and hydroxides. Thermogravimetric analysis experiments show the existence of bonded and nonbonded water molecules (hydrated MnO$_2$). On the basis of ESR studies and other experiments, a locked-water associated structure **1** has been proposed for the apomorphous precipitated MnO$_2$.[12]

$$\left[\begin{array}{ccc} & O{\cdot\cdot}H{-}O & O{-}H{\cdot\cdot}O \\ Mn & Mn & Mn \\ & O{\cdot\cdot}H{-}O & O{-}H{\cdot\cdot}O \end{array} \right]$$

(1)

In addition, variable amounts of alkaline and alkaline earth metal derivatives are detected by atomic adsorption analysis. X-ray studies have shown that the structures of active MnO$_2$ are quite complex; they are either amorphous or of moderate crystallinity (variable proportions of β- and γ-MnO$_2$).

Preparation of Active MnO$_2$, Oxidizing Power, and Reproducibility of the Results.[1,13] Pyrolusite (natural MnO$_2$) and pure synthetic crystalline MnO$_2$ are poor oxidants.[1] The oxidation of organic compounds requires an active, specially prepared MnO$_2$ and several procedures have been reported.[1,13] According to the method of preparation, the structure, the composition, and therefore the reactivity of active MnO$_2$ are variable. On this account, the choice of a procedure is of considerable importance to obtain the desired oxidation power and the reaction conditions must be carefully controlled to obtain a consistent activity. The active manganese dioxides described in the literature are generally prepared either by mixing aqueous solutions of KMnO$_4$ and a Mn^{II} salt (MnSO$_4$, MnCl$_2$) between 0 and 70 °C under acid, neutral, or basic conditions[13a–e] or by pyrolysis of a Mn^{II} salt (carbonate, oxalate, nitrate) at 250–300 °C.[13d,f] In this case the activity of the resulting material can be increased by washing with dilute nitric acid.[13f] A similar treatment has also been used to activate pyrolusite (natural MnO$_2$).[13h] On the other hand, it has been reported that the efficiency of an active MnO$_2$ depends on the percentage of the γ-form present in the material.[13i] Indeed, active γ-MnO$_2$ is sometimes clearly superior to the classical active MnO$_2$ prepared according to Attenburrow.[13j]

It is worthy of note that the percentage water content strongly influences both the oxidizing power and the selectivity (oxidation of multifunctional molecules) of active MnO$_2$. Thus it is well known that the wet material (40–60% H$_2$O) obtained after filtration must be activated by drying[1,13] (heating to 100–130 °C for 12–24 h[13a–d] or, better, at 125 °C for 52 h).[13e] Indeed, an excess of water decreases the oxidation power[1e,13k] since, according to the triphasic mechanism generally postulated,[12] it would prevent the adsorption of the substrate to the oxidatively active polar site on the surface of MnO$_2$.[1a,b,c] On the other hand, it is very important not to go past the point of complete activation since the presence of hydrated MnO$_2$ species is essential to obtain an active reagent. For this reason, the drying conditions must be carefully controlled.[1,13a,d,g,k] Alternatively, the wet material can be activated by azeotropic distillation since this mild procedure preserves the active hydrated species.[13l] Thus azeotropic distillation has been used to remove the water produced during the oxidation reaction to follow the rate of MnO$_2$ oxidations.[13e] Finally, the active MnO$_2$ mentioned above contains various metallic salts as impurities. According to their nature, which depends on the method of preparation, they can also influence the oxidizing power of the reagent (for instance, permanganate).[13d] Finally, it should be noted that the preparation of active MnO$_2$ on carbon[13m] or on silica gel[13n] as well as the activation of nonactive MnO$_2$ (pure crystalline MnO$_2$) by ultrasonic irradiation[13o] has also been reported. Some typical procedures to prepare active MnO$_2$ are reported below.

Preparation of Active MnO$_2$ from KMnO$_4$ Under Basic Conditions (Attenburrow).[13a] A solution of MnSO$_4\cdot$4H$_2$O (110 g) in H$_2$O (1.5 L) and a solution of NaOH (40%; 1.17 L) were added simultaneously during 1 h to a hot stirred solution of KMnO$_4$ (960 g) in H$_2$O (6 L). MnO$_2$ precipitated soon after as a fine brown solid. Stirring was continued for an additional hour and the solid was then collected with a centrifuge and washed with water until the washings were colorless. The solid was dried in an oven at 100–120 °C and ground to a fine powder (960 g) before use.

Preparation of Active MnO₂ from KMnO₄ Under Acidic Conditions.[13k] Active MnO_2 was made by mixing hot solutions of $MnSO_4$ and $KMnO_4$, maintaining a slight excess of the latter for several hours, washing the product thoroughly with water and drying at 110–120 °C. Its activity was unchanged after storage for many months, but it was deactivated by H_2O, MeOH, thiols, or excessive heat (500 °C). MnO_2 was less active when prepared in the presence of alkali and ineffective when precipitated from hot solutions containing a large excess of $MnSO_4$.

Preparation of Highly Active MnO₂ from KMnO₄.[1a] A solution of $MnCl_2 \cdot 4H_2O$ (200 g) in H_2O (2 L) at 70 °C was gradually added during 10 min, with stirring, to a solution of $KMnO_4$ (160 g) in H_2O (2 L) at 60 °C in a hood. A vigorous reaction ensued with evolution of chlorine; the suspension was stirred for 2 h and kept overnight at rt. The precipitate was filtered off, washed thoroughly with H_2O (4 L) until pH 6.5–7 and the washing gave a negligible chloride test. The filter cake was then dried at 120–130 °C for 18 h; this gave a chocolate-brown, amorphous powder; yield 195–200 g. Alternatively, the wet cake was mixed with benzene (1.2 L) and H_2O was removed by azeotropic distillation giving a chocolate-brown, amorphous powder; yield 195 g. The last procedure gave a slightly less active material.

Preparation of Active MnO₂ by Pyrolysis of MnCO₃.[13f] Powdered $MnCO_3$ was spread in a one-inch thick layer in a Pyrex glass and heated at 220–280 °C for about 18 h in an oven in which air circulated by convection. The initially tan powder turned darker at about 180 °C, and black when maintained at over 220 °C. No attempt was made to determine lower temperature or time limits, nor the upper limit of temperature. The MnO_2 prepared as above was stirred with about 1 L of a solution made up of 15% HNO_3 in H_2O. The slurry was filtered with suction, the solid was washed on the Buchner funnel with distilled water until the washes were about pH 5, and finally was dried at 220–250 °C. The caked, black solid was readily crushed to a powder which retained its oxidizing ability even after having been stored for several months in a loosely stoppered container.

Preparation of γ-MnO₂.[1a] To a solution of $MnSO_4$ (151 g) in H_2O (2.87 L) at 60 °C was added, with stirring, a solution of $KMnO_4$ (105 g) in H_2O (2 L), and the suspension was stirred at 60 °C for 1 h, filtered, and the precipitate washed with water until free of sulfate ions. The precipitate was dried to constant weight at 60 °C; yield 120 g (dark-brown, amorphous powder).

Preparation of Active MnO₂ on Silica Gel.[13n] $KMnO_4$ (3.79 g) was dissolved in water (60 mL) at rt. Chromatographic grade silica gel (Merck, 70–230 mesh, 60 g) was added with stirring, and the flask connected to a rotary evaporator to strip off the water at 60 °C. The purple solid was ground to fine powder and then added with vigorous stirring to a solution of $MnSO_4 \cdot H_2O$ (9.3 g) in H_2O (100 mL). The resulting brown precipitate was filtered with water until no more Mn^{II} ion could be detected in the wash water by adding ammonia. After being dried at 100 °C for 2 h, each gram of this supported reagent contained 0.83 mmol of MnO_2.

As shown above, a wide range of products of various activities are called active MnO_2 and the results described in the literature are sometimes difficult to reproduce since the nature of the MnO_2 which was used is not always well defined. Now, the commercial materials give reproducible results but they are not always convenient to perform all the oxidation reactions described in the literature. In addition, their origin (method of preparation) is not often indicated and comparison with the active MnO_2 described in the literature is sometimes difficult. For all these reasons, the use of activated MnO_2 has been somewhat restricted in spite of the efficiency and selectivity of its reactions since only an empirical approach and a careful examination of the literature allow selection of the suitable activity of MnO_2 and the optimum reaction conditions for a defined substrate.

Oxidation of Organic Compounds with MnO₂: Reaction Conditions.[1,13]

Solvent. Oxidation of organic compounds with MnO_2 has been performed in many solvents. The choice of the solvent is important; thus primary or secondary alcohols (or water) are unsatisfactory since they can compete with the substrate being adsorbed on the MnO_2 surface and they have a strong deactivating effect.[13k] A similar but less pronounced influence has also been observed with various polar solvents such as acetone, ethyl acetate, DMF, and DMSO. However, these polar solvents, including water,[13p] acetic acid, and pyridine, can be used successfully at higher temperatures. This deactivating influence due to the polarity of the solvent can be used to control the reactivity of active MnO_2 and sometimes to avoid side reactions or to improve the selectivity (for instance, allylic alcohol vs. saturated alcohol). Most of the reactions described in the literature were carried out in aliphatic or aromatic hydrocarbons, chlorinated hydrocarbons, diethyl ether, THF, ethyl acetate, acetone, and acetonitrile (caution: MeCN can react with highly active MnO_2 or with classically activated MnO_2 on prolonged treatment). In the case of the oxidation of benzylic[13f] and allylic[13d] alcohols, the best results have been obtained using diethyl ether (diethyl ether > petroleum ether > benzene). Caution: a spontaneous ignition has been observed when highly active MnO_2 was used in this solvent.[13f]

Temperature and Reaction Time. At rt, the reaction time can vary from 10 min to several days according to the nature of the substrate and the activity of the MnO_2.[13a,d,e,f,l,p] The reaction times are shortened by heating[13d] but the selectivity is very often much lower.[13q]

Ratio MnO₂:Substrate. The amount of active MnO_2 required to perform the oxidation of an organic substrate depends on the type of MnO_2, on the substrate, and on the particle size of the MnO_2.[13d,f,g] With a classical material (100–200 mesh), the ratio varies from 5:1 to 50:1 by weight.

Oxidation of Allylic Alcohols.[2] MnO_2 was used as oxidizing reagent for the first time by Ball et al. to prepare retinal from vitamin A_1 (eq 1).[2a] Since that report, the use of MnO_2 for the conversion of allylic alcohols to α,β-ethylenic aldehydes has been extensively utilized. Interestingly, the configuration of the double bond is conserved during the reaction (eqs 2–4).[2b,c] In some cases a significant rate difference between axial and equatorial alcohols has been observed[2d] (eq 5).[2e]

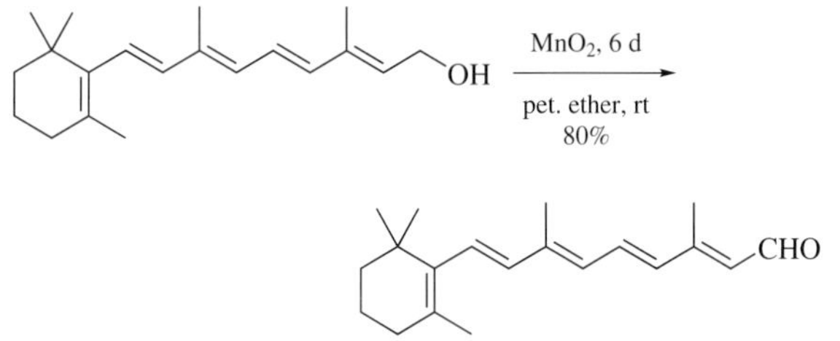

MnO$_2$, 6 d

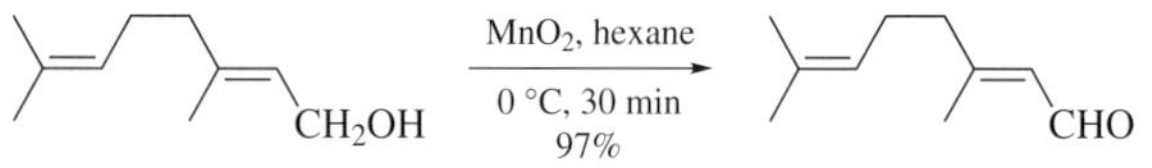

pet. ether, rt
80%

(1)

MnO$_2$, hexane

0 °C, 30 min
97%

(2)

MnO$_2$, acetone

6 h, rt
75%

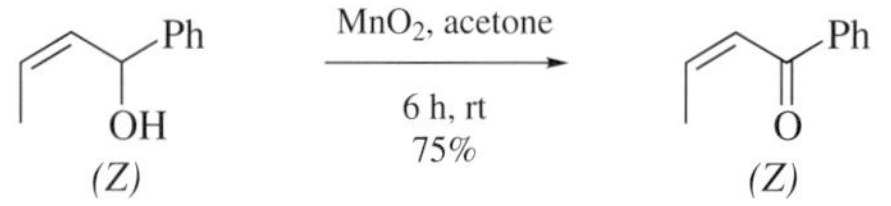

(3)

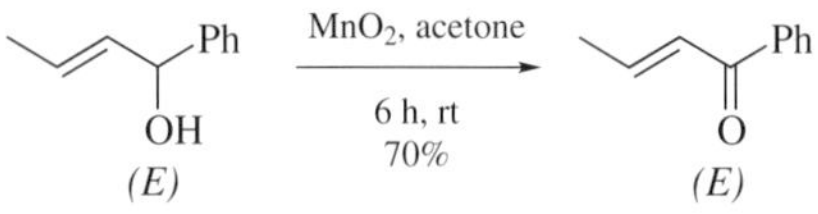

(Z) → *(Z)*

MnO$_2$, acetone

6 h, rt
70%

(4)

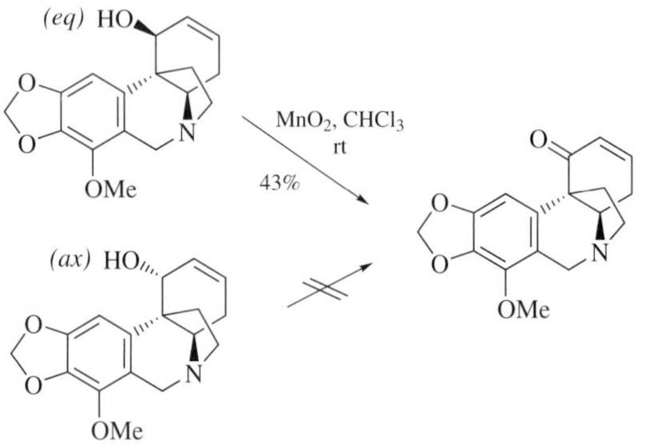

(E) → *(E)*

(eq) HO

MnO$_2$, CHCl$_3$
rt
43%

(5)

(ax) HO

MnO$_2$ has been frequently used for the preparation of very sensitive polyunsaturated aldehydes or ketones (eq 6).[2f]

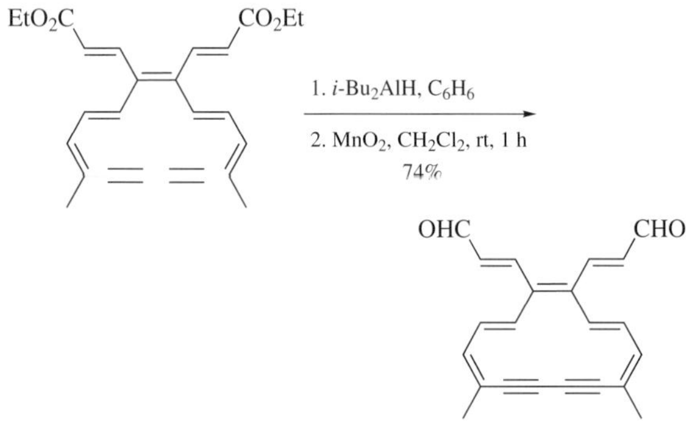

1. *i*-Bu$_2$AlH, C$_6$H$_6$
2. MnO$_2$, CH$_2$Cl$_2$, rt, 1 h
74%

(6)

Numerous functionalized α,β-ethylenic aldehydes are readily obtained by chemoselective oxidation of the corresponding allylic alcohols (eqs 7–11).[2g–k]

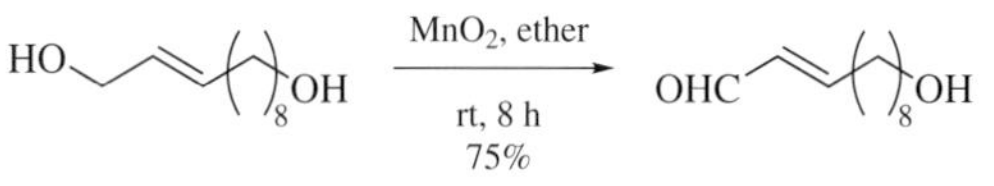

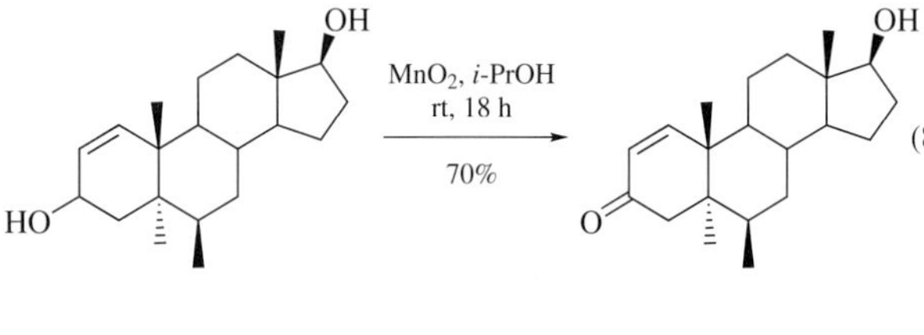

MnO$_2$, ether

rt, 8 h
75%

(7)

MnO$_2$, *i*-PrOH

rt, 18 h
70%

(8)

MnO$_2$, CHCl$_3$

rt, 96 h
79%

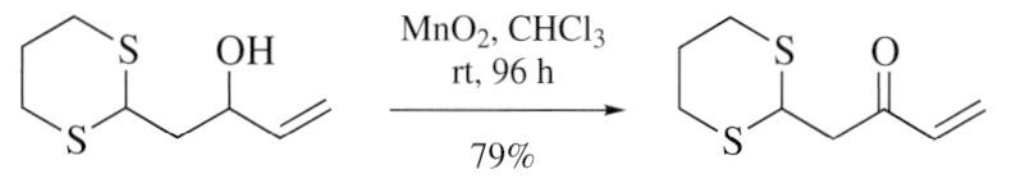

(9)

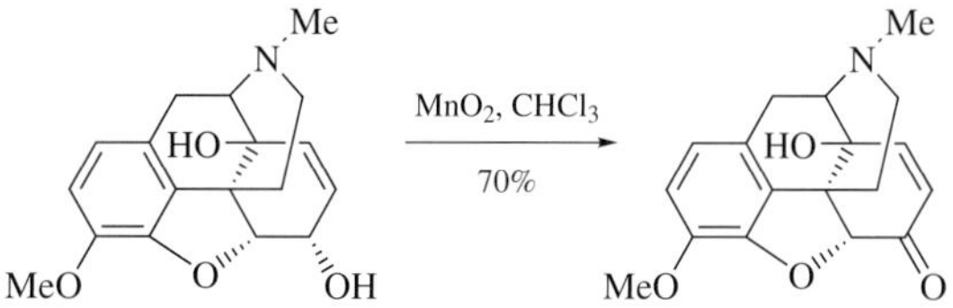

MnO$_2$, CHCl$_3$

70%

(10)

MnO$_2$, CH$_2$Cl$_2$

18 d, rt

90%

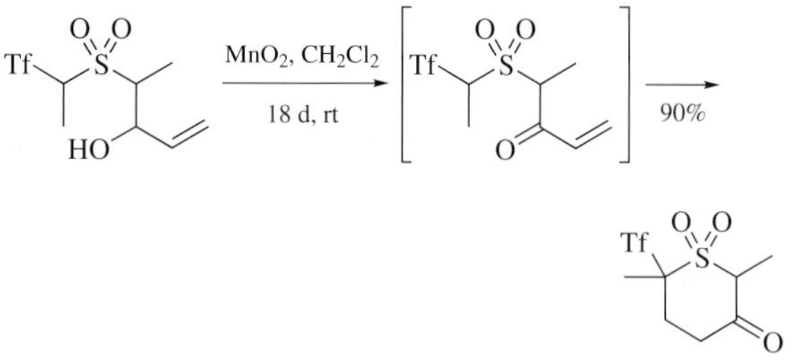

(11)

The α,β-ethylenic ketone obtained by treatment of an allylic alcohol with MnO$_2$ can undergo an in situ Michael addition (eqs 11 and 12).[2l]

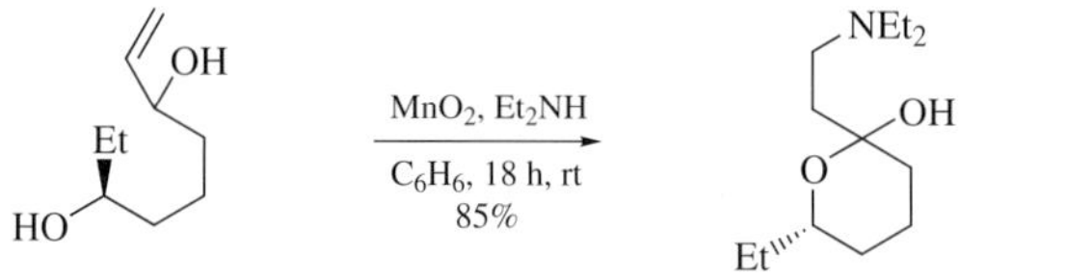

MnO$_2$, Et$_2$NH

C$_6$H$_6$, 18 h, rt
85%

(12)

Conversion of Allylic Alcohols to α,β-Ethylenic Esters and Amides.[2b,3] This procedure was first described by Corey.[2b,3] The key step is the sequential formation and oxidation of a cyanohydrin. In the presence of an alcohol or an amine the resulting acyl cyanide leads by alcoholysis or aminolysis to the corresponding α,β-ethylenic ester[2b] or amide[3] (eqs 13–15).

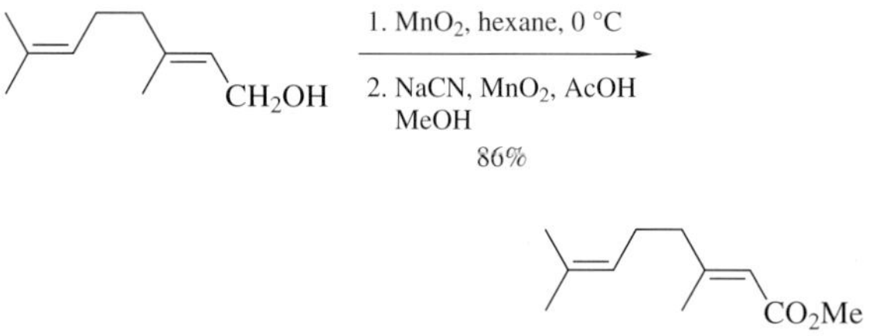

1. MnO$_2$, hexane, 0 °C
2. NaCN, MnO$_2$, AcOH
   MeOH
   86%

(13)

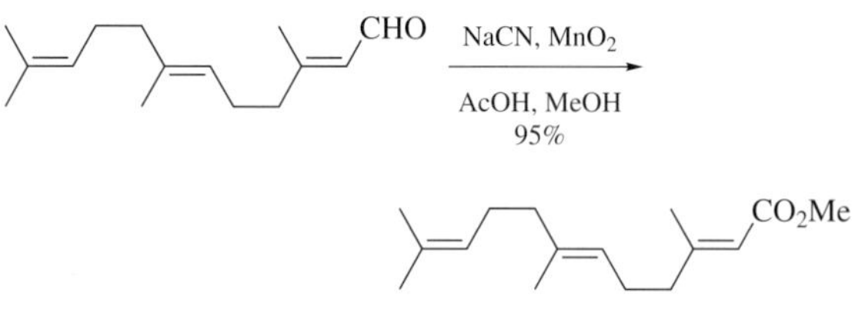

NaCN, MnO$_2$

AcOH, MeOH
95%

(14)

NaCN, MnO$_2$

i-PrOH, NH$_3$
100%

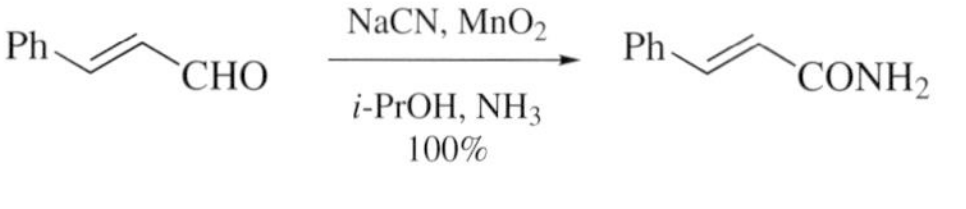

(15)

Oxidation of Propargylic Alcohols.[4] Propargylic alcohols are easily oxidized by MnO$_2$ to alkynic aldehydes and ketones (eqs 16–18).[4a–c] In the example in eq 19 the unstable propargyl aldehyde is trapped as a Michael adduct.[4d]

$$\text{HO}_2\text{C} \quad \xrightarrow[\substack{\text{rt, 24 h} \\ 74\%}]{\text{MnO}_2,\ \text{CH}_2\text{Cl}_2} \quad \text{HO}_2\text{C} \qquad (16)$$

$$\xrightarrow[\substack{0\ °\text{C} \\ 76\%}]{\text{MnO}_2,\ \text{CH}_2\text{Cl}_2} \qquad (17)$$

$$\xrightarrow[\substack{\text{rt, 2 h} \\ 88\%}]{\text{MnO}_2,\ \text{CH}_2\text{Cl}_2} \qquad (18)$$

$$\xrightarrow[\substack{\text{piperidine, rt, 12 h} \\ 86\%}]{\text{MnO}_2,\ \text{C}_6\text{H}_6} \qquad (19)$$

Oxidation of Benzylic and Heterocyclic Alcohols.[2b,5] Conjugated aromatic aldehydes or ketones can be efficiently prepared by treatment of benzylic alcohols with MnO$_2$ (eq 20).[5a] Numerous functional groups are tolerated (eqs 21–24).[5a–c]

$$\xrightarrow[\substack{\text{rt, 23 h} \\ 89\%}]{\text{MnO}_2,\ \text{CHCl}_3} \qquad (20)$$

$$\xrightarrow[\substack{\text{rt, 24 h} \\ 76\%}]{\text{MnO}_2,\ \text{CHCl}_3} \qquad (21)$$

$$\xrightarrow[\substack{\text{8 h, 60 °C} \\ 80\%}]{\text{MnO}_2,\ \text{dioxane}} \qquad (22)$$

$$\xrightarrow[\substack{\text{rt, 45 min} \\ 76\%}]{\text{MnO}_2,\ \text{acetone}} \qquad (23)$$

$$\xrightarrow[\substack{\text{rt, 4 d} \\ 93\%}]{\text{MnO}_2,\ \text{acetone}} \qquad (24)$$

Benzyl allyl and benzyl propargyl alcohols have been oxidized successfully to ketones (eqs 25 and 26).[5d,e]

$$\xrightarrow[\substack{\text{ether, hexane} \\ 86\%}]{\text{MnO}_2} \qquad (25)$$

$$\xrightarrow[\substack{\text{1 h} \\ 80\%}]{\text{MnO}_2,\ \text{CHCl}_3} \qquad (26)$$

The oxidation reaction can be extended to heterocyclic alcohols (eq 27)[5f] and the Corey procedure gives the expected esters (eq 28).[2b]

$$\xrightarrow[\substack{\text{24 h, rt} \\ 81\%}]{\text{MnO}_2,\ \text{C}_6\text{H}_6} \qquad (27)$$

$$\xrightarrow[\substack{\text{AcOH, MeOH} \\ 95\%}]{\text{NaCN, MnO}_2} \qquad (28)$$

Oxidation of Saturated Alcohols.[6] Cyclic and acyclic saturated alcohols react with MnO$_2$ to give the saturated aldehydes or ketones in good yields (eqs 29–32).[6a,b]

$$\xrightarrow[\substack{\text{16 h, rt} \\ 62\%}]{\text{MnO}_2,\ \text{pet. ether}} \qquad (29)$$

trans → *trans*

$$\xrightarrow[\substack{\text{3 d} \\ 71\%}]{\text{MnO}_2,\ \text{MeCN}} \qquad (30)$$

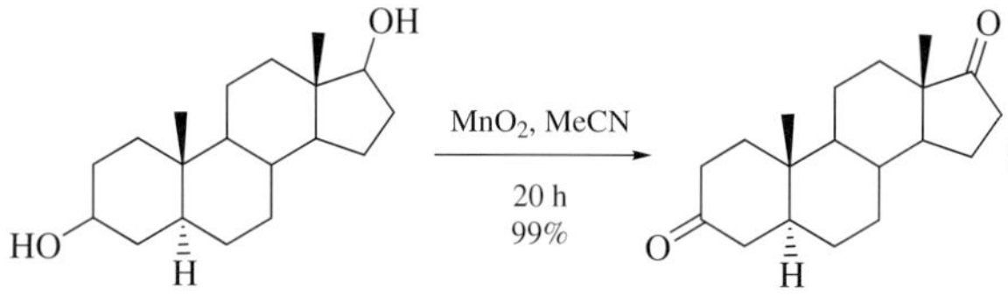

(31)

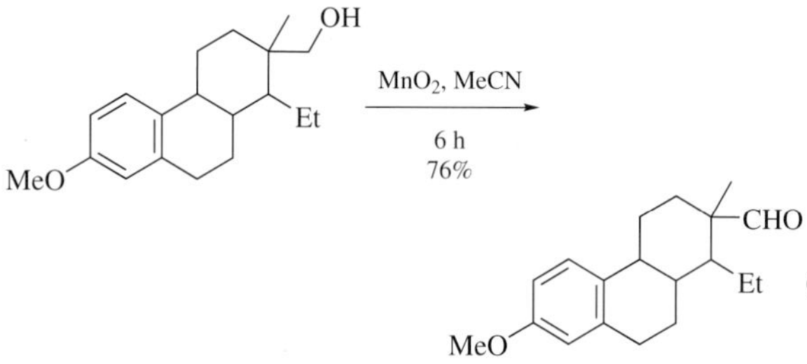

(32)

Oxidative Cleavage of 1,2-Diols.[7] 1,2-Diols are cleaved by MnO_2 to aldehydes or ketones. With cyclic 1,2-diols, the reaction leads to dialdehydes or diketones (eqs 33 and 34)[7] and the course of the reaction depends on the configuration of the starting material (eqs 34 and 35).[7]

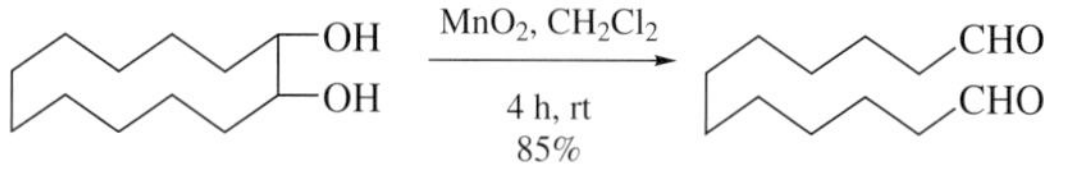

(33)

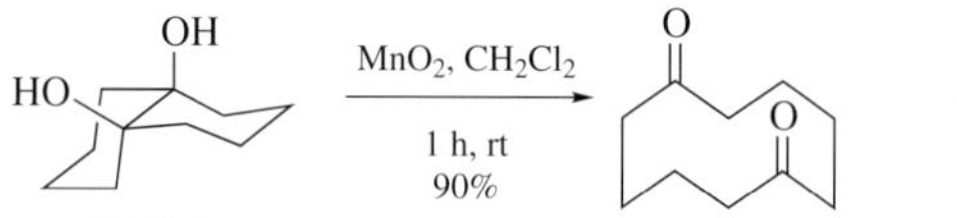

cis-diol

(34)

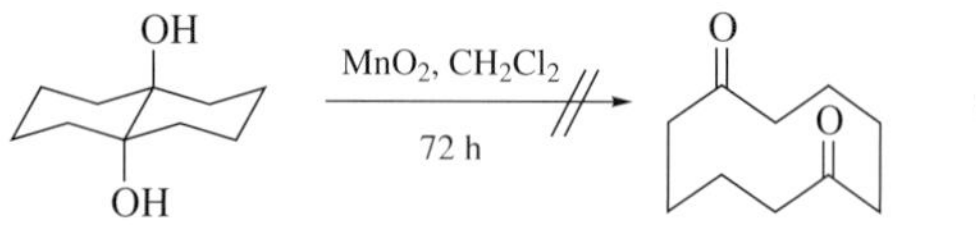

trans-diol

(35)

Hydration of Nitriles to Amides.[8] By treatment with MnO_2, nitriles are readily converted to amides. MnO_2 on silica gel is especially efficient to perform this reaction (eqs 36 and 37).[8a,b]

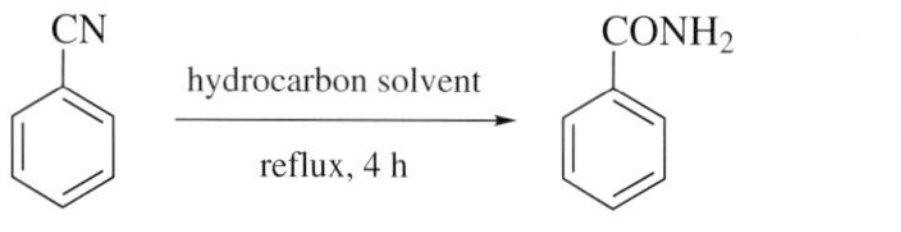

active MnO_2	30%
MnO_2, SiO_2	100%

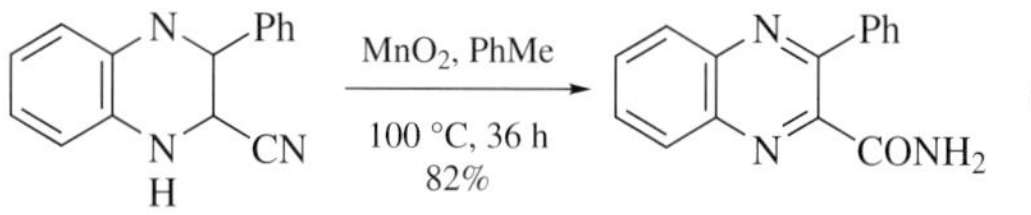

(36)

(37)

Dehydrogenation and Aromatization Reactions.[9,13j,q] MnO_2 has been widely used to carry out various dehydrogenation and aromatization reactions (eqs 38–41).[9a–c] In some cases the dehydrogenation can occur as a side reaction during, for instance, the hydration of nitriles (eq 37)[8b] or the oxidation of allylic alcohols.[13q]

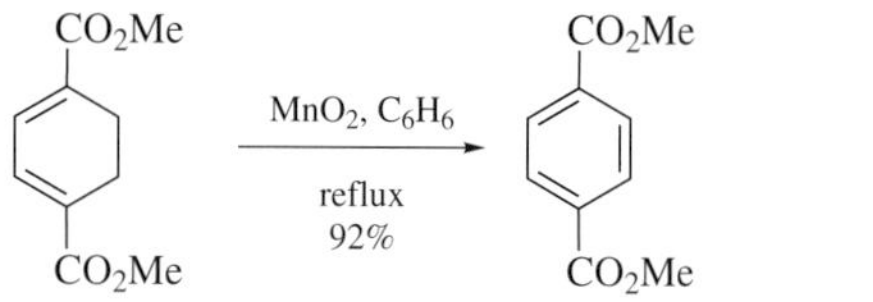

(38)

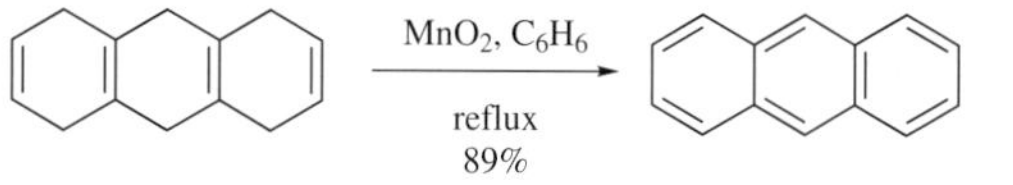

(39)

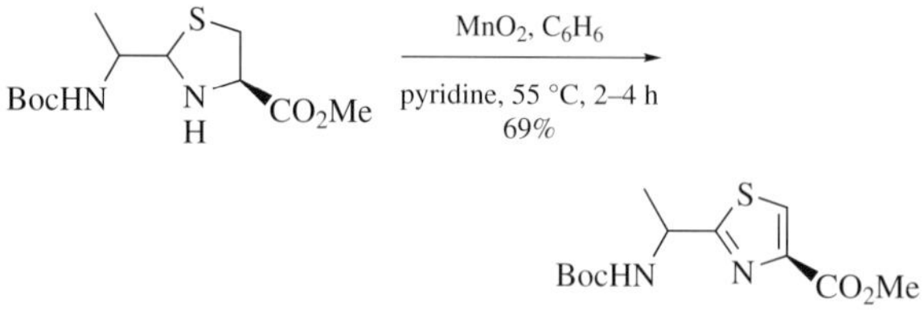

(40)

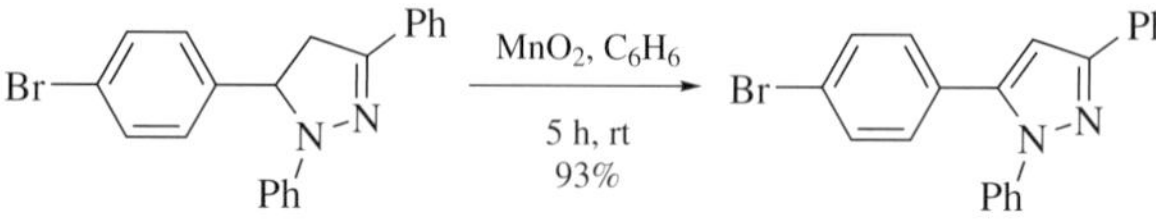

(41)

It is interesting to note that the use of γ-MnO_2 is essential to achieve the following dehydrogenation reactions (eqs 42 and 43).[13j]

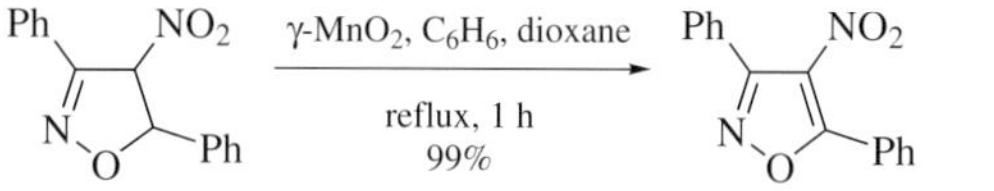

(42)

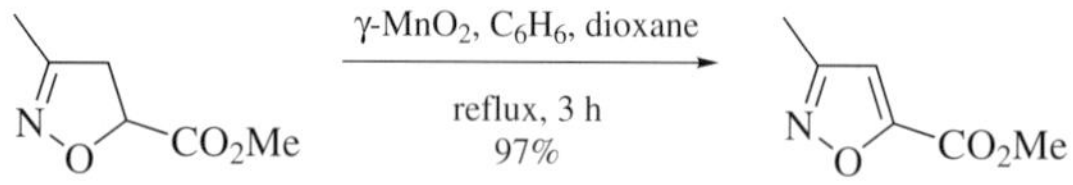

(43)

Oxidation of Amines to Aldehydes, Imines, Amides, and Diazo Compounds.[10,13g,p] The oxidation of amines by MnO_2 can lead to various products according to the structure of the starting material. Thus the formation of imines (eq 44),[10] formamides (eqs 45 and 46),[13g] and diazo compounds (eq 47)[13p] have all been described.

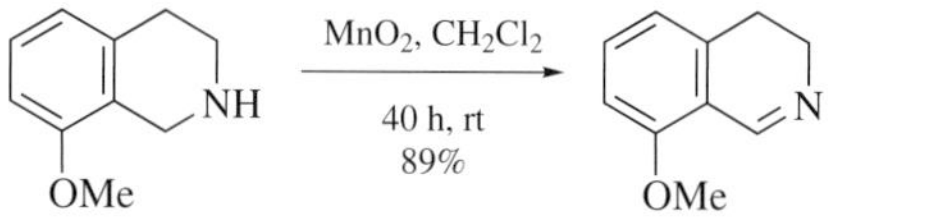

(44)

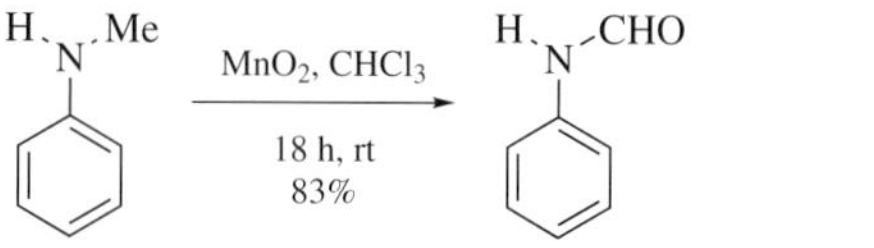

(45)

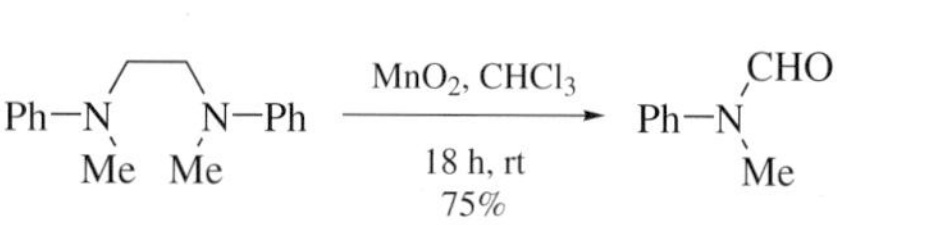

(46)

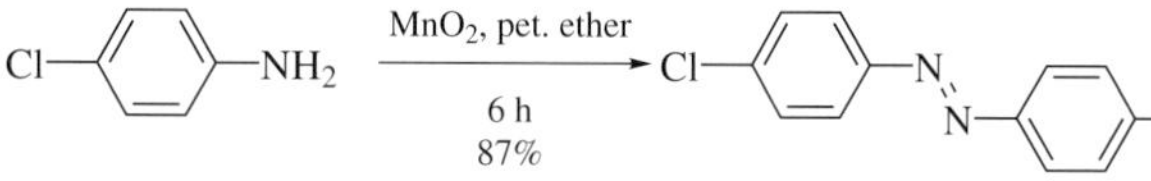

(47)

Miscellaneous Reactions.[13p,14] MnO_2 has also been used to perform various oxidation reactions: the oxidative cleavage of

α-hydroxy acids (eq 48),[13p] the oxidative dimerization of diarylmethanes (eq 49),[14a] or their conversion to diaryl ketones (eq 50),[14a] the oxidation of aldehydes to carboxylic acids,[13p] the preparation of disulfides from thiols,[14b] of phosphine oxides from phosphines,[13p] or of ketones from amines.[14c]

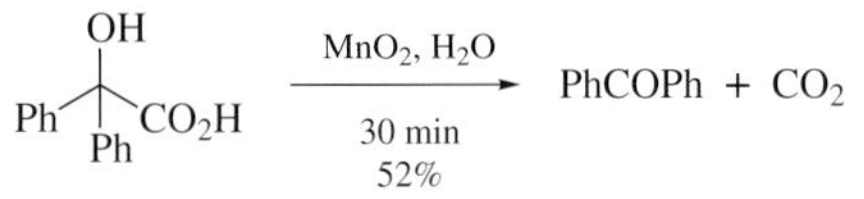

$$\text{Ph}\overset{\overset{\displaystyle OH}{|}}{\underset{\underset{\displaystyle Ph}{|}}{C}}\text{CO}_2\text{H} \xrightarrow[\substack{30 \text{ min} \\ 52\%}]{\text{MnO}_2, \text{H}_2\text{O}} \text{PhCOPh} + \text{CO}_2 \qquad (48)$$

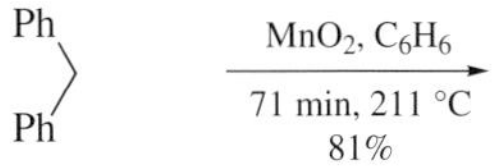

(49)

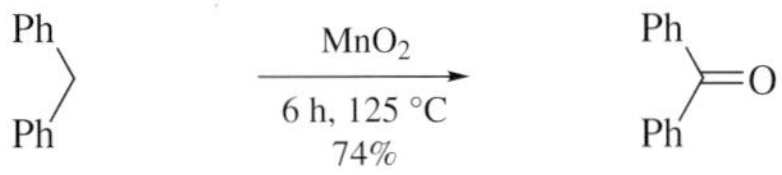

71 min, 211 °C
81%

(50)

6 h, 125 °C
74%

First Update

Richard J. K. Taylor, Mark Reid & Jonathan S. Foot
University of York, York, UK

Alternative Reaction Conditions For Alcohol Oxidation.[15] Various solvent-free oxidations using MnO_2 have been developed in recent years.[15a–f] Oxidation of activated and non-activated alcohols has been accomplished by merely adding the substrate to MnO_2 and stirring at room temperature (eqs 51 and 52).[15a] These reactions can be accelerated considerably by supporting the oxidant on silica and exposing the mixture to microwave irradiation (eq 51).[15b]

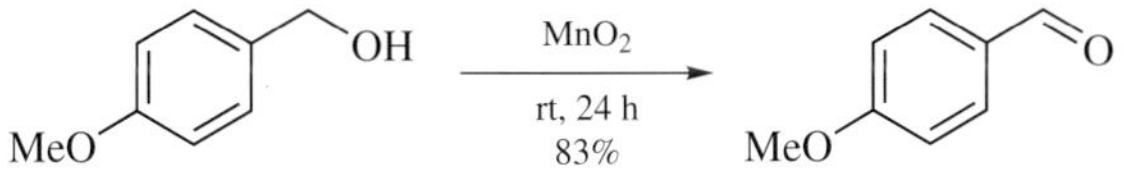

(51)

rt, 24 h
83%

Under microwave conditions: 30 sec, 83%

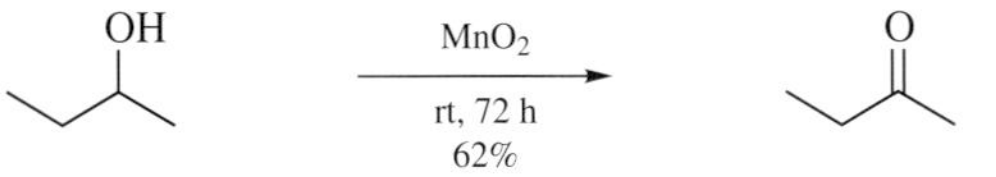

(52)

rt, 72 h
62%

Oxidation of Non-activated Alcohols.[16,17] The oxidation of non-activated cyclic and acyclic alcohols using virtually stoichiometric MnO_2 has been achieved through the addition of a catalytic co-oxidant **2** and ruthenium catalyst **3** to THF (eqs 53 and 54).[16]

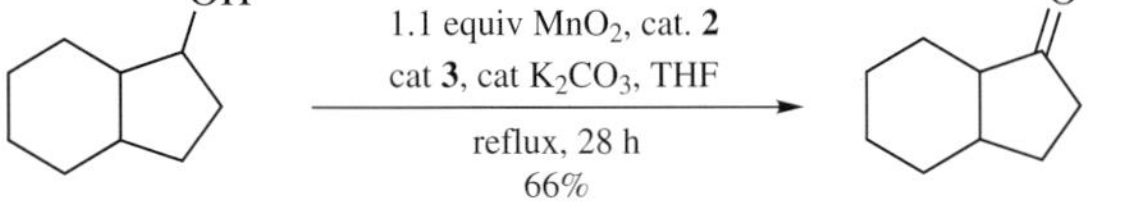

1.1 equiv MnO₂, cat. **2**
cat **3**, cat K₂CO₃, THF
reflux, 28 h
66%

(53)

1.1 equiv MnO₂, cat **2**
cat **3**, cat K₂CO₃, THF
reflux, 17 h
80%

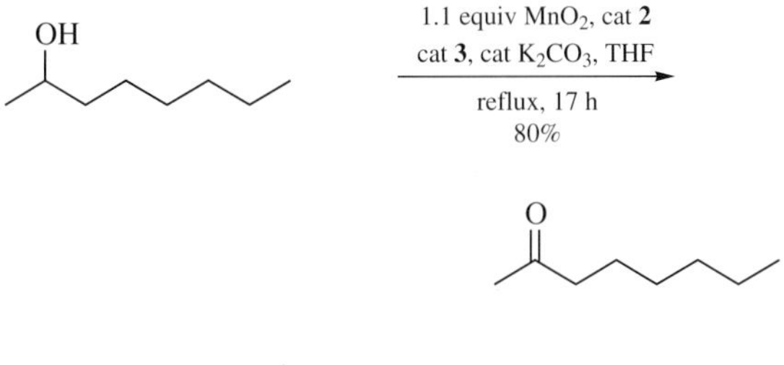

(54)

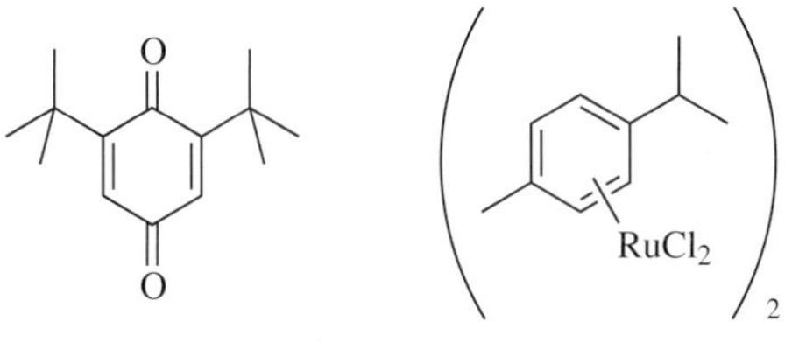

2   **3**

MnO_2 has been found to facilitate the one-pot conversion of both TMS- and TBDMS-protected allylic alcohols into their respective aldehydes and ketones (eq 55).[17]

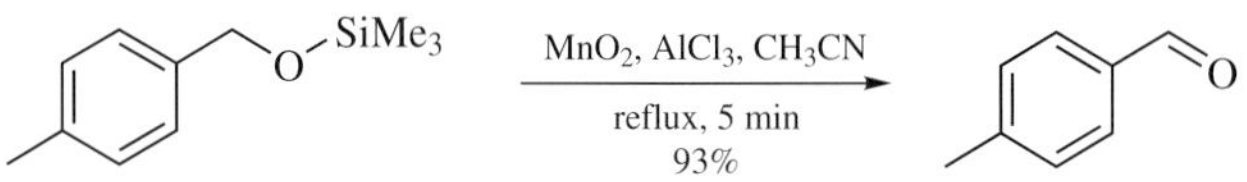

$$\xrightarrow[\substack{\text{reflux, 5 min} \\ 93\%}]{\text{MnO}_2, \text{AlCl}_3, \text{CH}_3\text{CN}}$$

(55)

Oxidation of Sulfur Compounds.[18,19] Treatment of thiols with MnO_2 in hexane at reflux provides the symmetrical disulfides in good yield (eq 56).[18a] This procedure has also been carried out in solvent-free conditions at room temperature, affording the oxidatively coupled products in shorter reaction times (eq 57).[18b]

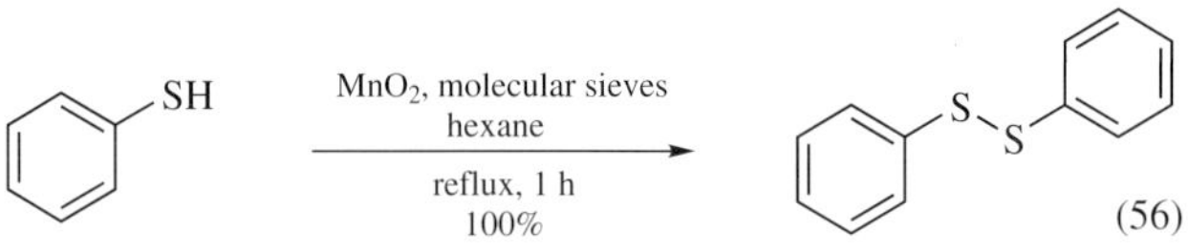

MnO₂, molecular sieves
hexane
reflux, 1 h
100%

(56)

MnO₂
rt, 5 min
92%

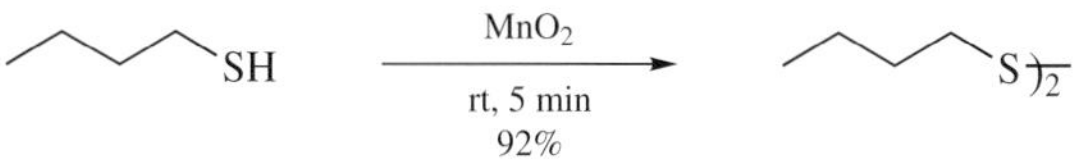

(57)

MnO_2 has also been used to selectively oxidize thioethers to their corresponding sulfoxides under acidic conditions. The reaction can be carried out either in methanolic HCl (eq 58)[19a] or using silica gel-supported sulfuric acid under solvent-free conditions (eq 59),[19b] and gives the desired sulfoxides only.

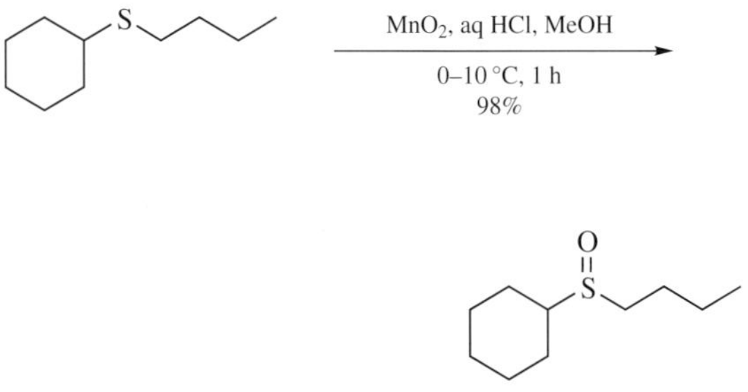

$$\xrightarrow[\substack{0-10\,°\text{C, 1 h} \\ 98\%}]{\text{MnO}_2, \text{aq HCl, MeOH}}$$

(58)

$$\xrightarrow[\substack{35-40\,°\text{C, 1 h} \\ 85\%}]{\text{MnO}_2, \text{H}_2\text{SO}_4/\text{SiO}_2}$$

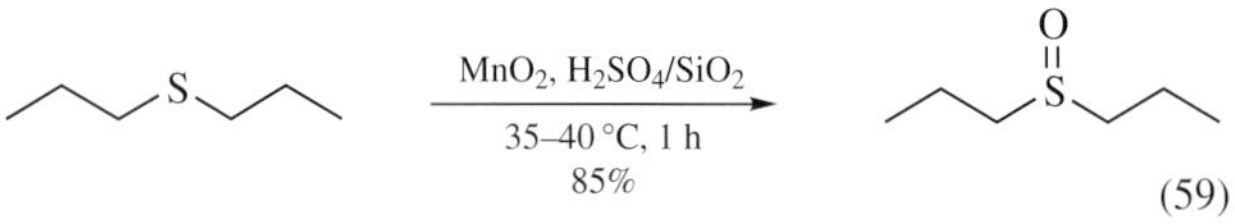

(59)

Oxidation of N–O Compounds.[20] The preparation of disubstituted oxazolines from aldoximes has been accomplished via in situ nitrile oxide formation using MnO_2 (eq 60).[20a] Oxidation of hydroxylamines and oximes to nitrones[20b] and aldehydes/ketones,[20c] respectively, has been found to occur upon treatment with MnO_2 (eqs 61 and 62).

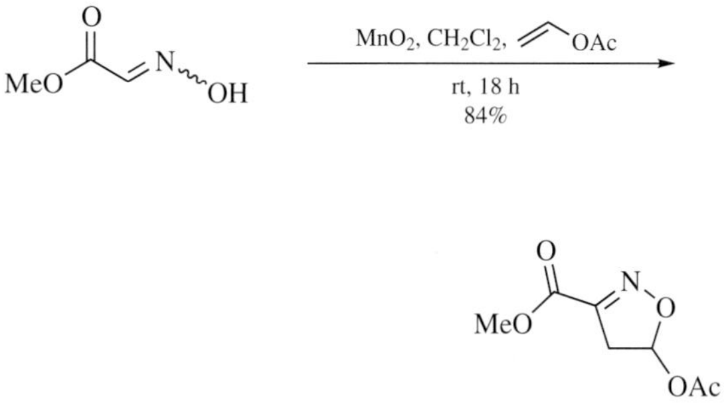

MnO₂, CH₂Cl₂, ⟍⟍OAc
rt, 18 h
84%

$$
\text{MeO}_2\text{C}\underset{\text{isoxazoline}}{\diagup}\quad (60)
$$

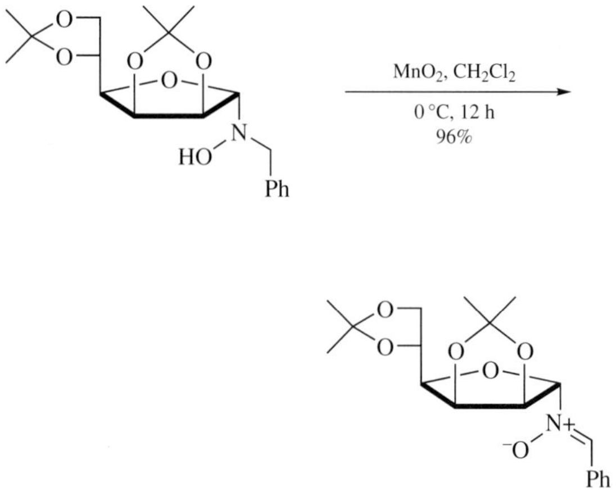

MnO₂, CH₂Cl₂
0 °C, 12 h
96%

(61)

MnO₂, hexane
rt, 15 min
91%

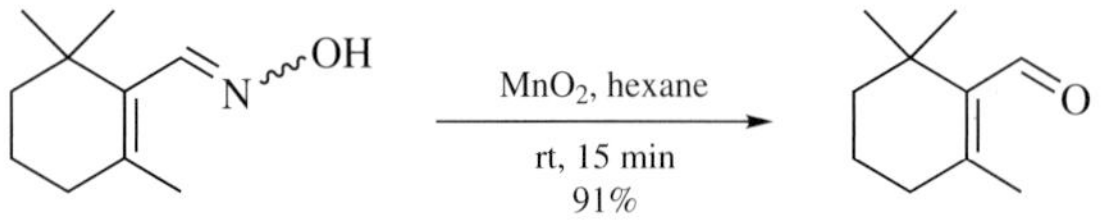

(62)

α-Acylazo compounds (eq 63)[15e] and amides (eq 64)[15f] have also been synthesized via MnO₂ oxidation under solvent-free conditions with silica-supported reagents.

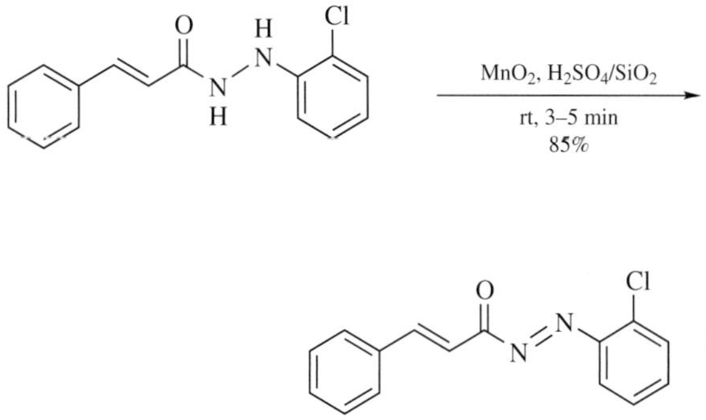

MnO₂, H₂SO₄/SiO₂
rt, 3–5 min
85%

(63)

MnO₂/SiO₂
microwave conditions, 4 min
99%

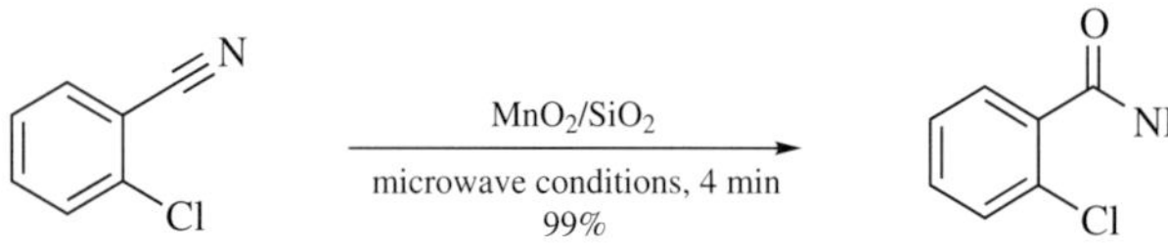

(64)

Miscellaneous MnO₂-mediated Reactions.[21] Treatment of a quinoxalinium salt with various alkenes in the presence of MnO₂ has been found to effect a novel 1,3-dipolar cycloaddition reaction to give the corresponding pyrrolo[1,2-a]quinoxalines (eq 65).[21]

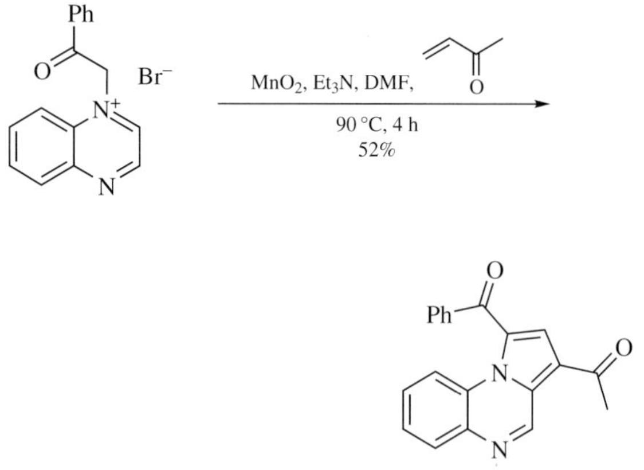

MnO₂, Et₃N, DMF,
90 °C, 4 h
52%

(65)

Use of MnO₂ for In Situ Tandem Oxidation Processes (TOP's).

Using Stabilized Ylides.[22–34] The use of MnO₂ for in situ tandem oxidation processes (TOP's) was first reported by Wei and Taylor for the synthesis of bromodienoate esters (eqs 66 and 67),[22,23] thus avoiding isolation of the toxic and unstable bromoenal intermediates **4**. Since this initial publication, MnO₂ oxidation–nucleophilic trapping reactions have been extensively developed to encompass a range of functional group transformations.

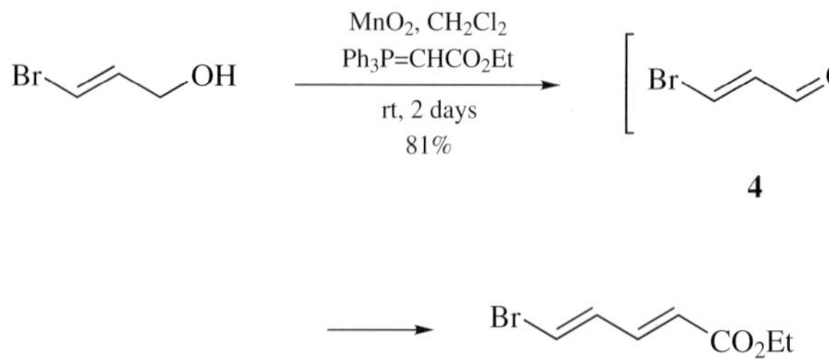

Br⟍⟍OH
MnO₂, CH₂Cl₂
Ph₃P=CHCO₂Et
rt, 2 days
81%
[Br⟍⟍⟍=O]
4

⟶ Br⟍⟍⟍CO₂Et (66)

E,E:Z,E = 6.8:1

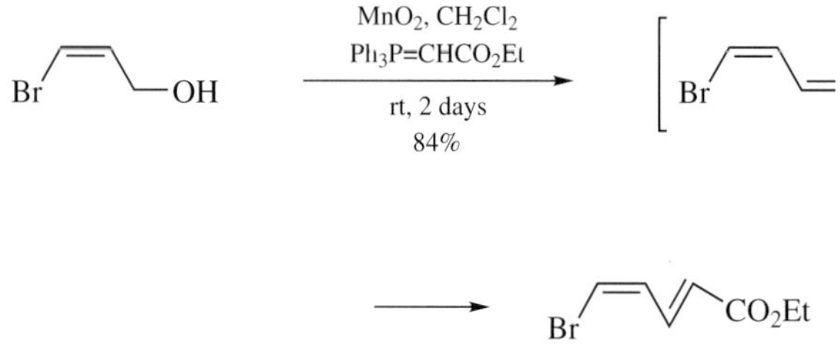

Br⟍⟍OH
MnO₂, CH₂Cl₂
Ph₃P=CHCO₂Et
rt, 2 days
84%
[Br⟍⟍=O]

⟶ Br⟍⟍CO₂Et (67)

Z,E:Z,Z = ca. 4:1

Good to excellent yields were also obtained using substrates bearing diverse functionalities (eq 68), and also with activated diols (eq 69).

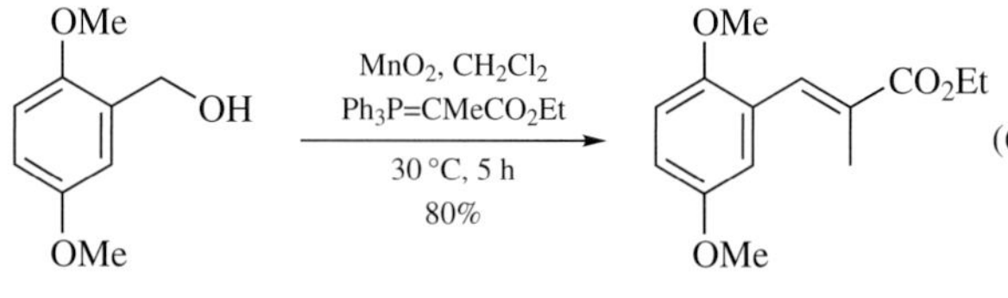

MnO₂, CH₂Cl₂
Ph₃P=CMeCO₂Et
30 °C, 5 h
80%

(68)

>98% E

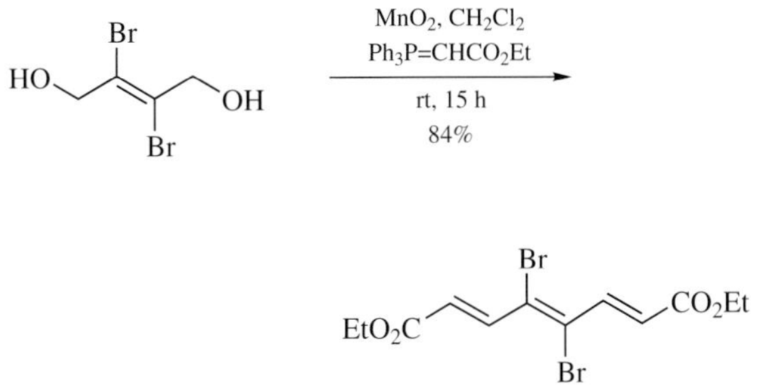

$$EtO_2C \quad \text{(69)}$$

>98% E,E,E

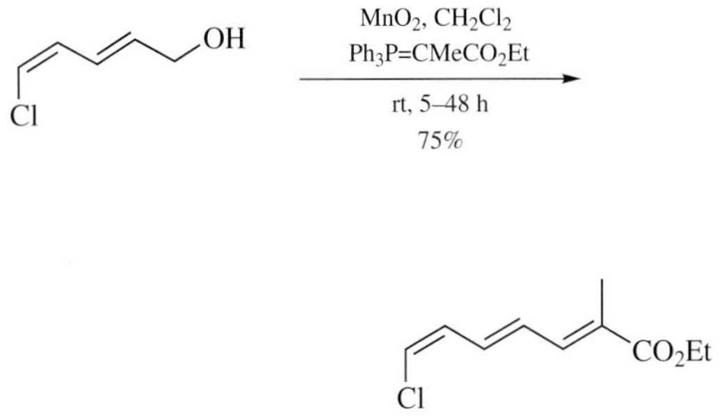

$$\text{(73)}$$

E,E,Z:Z,E,Z = 87:13

Along with carbonyl stabilized Wittig reagents, other stabilized ylides have also been used to prepare unsaturated nitrile[23] (eq 70) and Weinreb amide[24] (eq 71) adducts.

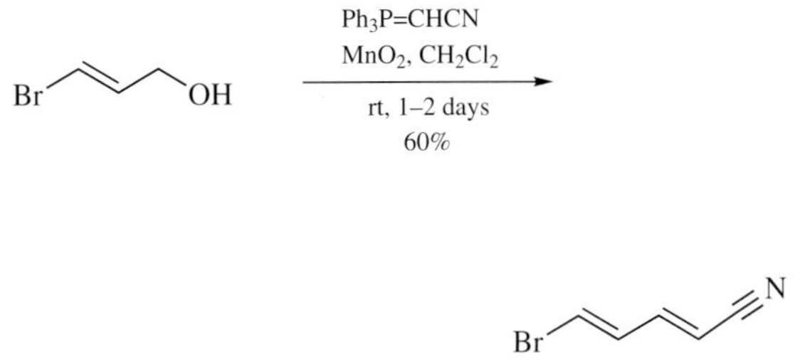

$$\text{(70)}$$

E:Z = 3:1

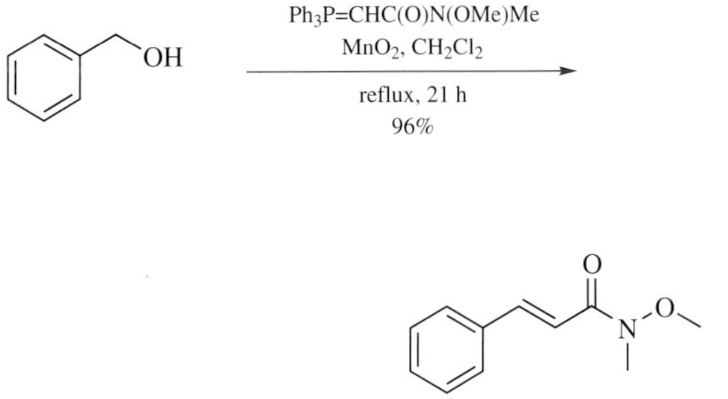

$$\text{(71)}$$

> 95% E

This methodology was later applied by Nicolaou's group to the synthesis of a bromodieneoate intermediate required in the total synthesis of Apoptolidin (eq 72),[25] and by Mladenova et al. for the synthesis of chloromethyltrienoic esters (eq 73).[26]

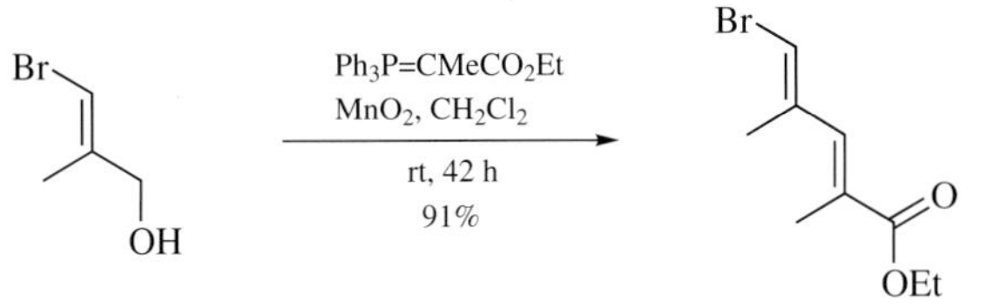

$$\text{(72)}$$

Along with activated alcohols, so-called semi-activated alcohols have also been used in this methodology (eqs 74 and 75) whereby proximal heteroatoms activate the alcohol—possibly through inductive effects or by providing a coordination site for the activated MnO_2.[27,28]

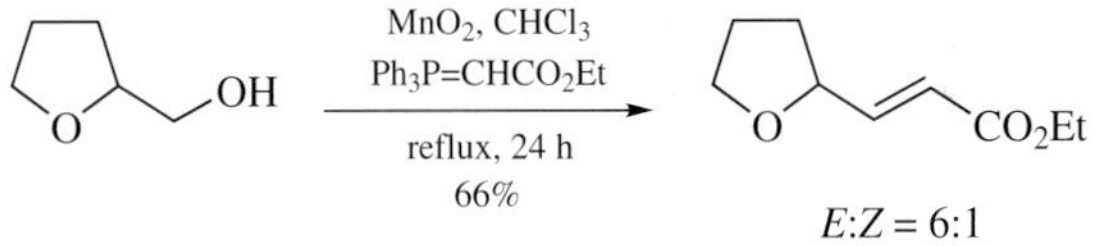

$$\text{(74)}$$

E:Z = 6:1

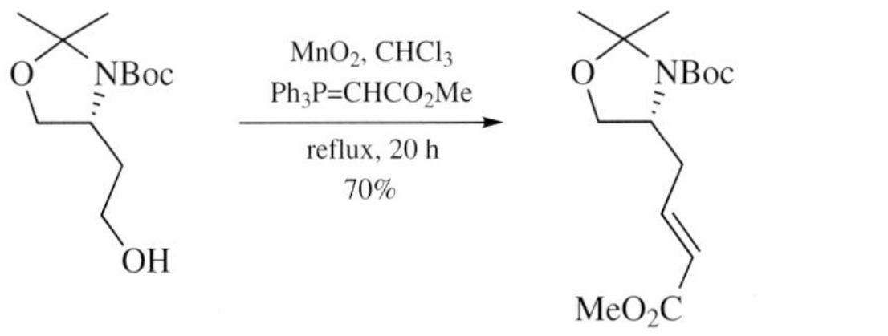

$$\text{(75)}$$

>95% E

McKervey and co-workers have also used MnO_2 for the in situ oxidation–Wittig reaction of non-racemic, protected amino acids to good effect, without any epimerization at the stereogenic center (eq 76).[28] These unsaturated ester products were then further elaborated to afford various alkaloids.

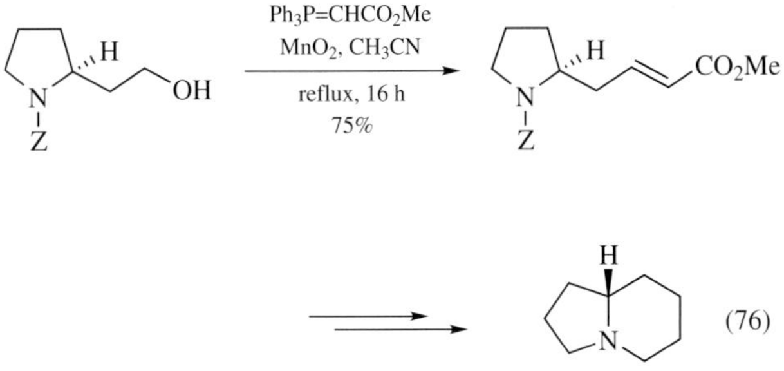

$$\text{(76)}$$

(S)-(−) coniceine

Furthermore, in the presence of stabilized phosphorane reagents, MnO_2 has been shown to be an efficient oxidant of unactivated alcohols to afford unsaturated esters in good yield (eq 77).[27]

$$CH_3(CH_2)_8CH_2OH \xrightarrow[\substack{reflux, 24\ h \\ 80\%}]{\substack{Ph_3P=CHCO_2Et \\ MnO_2,\ toluene}}$$

$$CH_3(CH_2)_8 \diagup\hspace{-0.3em}\diagdown CO_2Et \quad (77)$$

$$>95\%\ E$$

Other sequential and one-pot oxidation–Wittig reactions have been reported using Swern,[29] Dess-Martin,[30] and barium permanganate[31] oxidation.

Despite the use of different oxidizing agents in one-pot reaction sequences, MnO$_2$ has been established as the reagent of choice. This is due, in part, to several practical considerations including low toxicity, low cost, ease of handling and work-up (simple filtration), recycling potential, and the commercial availability of activated MnO$_2$. These factors make MnO$_2$ an ideal oxidant for use in TOP sequences.

Another class of activated alcohols employed in the tandem MnO$_2$ reaction are α-hydroxyketones. The synthetic utility of the intermediate α-ketoaldehydes had previously been limited by the hyperreactivity of the aldehyde group.[29] However, by employing the TOP MnO$_2$ protocol, good to excellent yields of the resulting Wittig products may be obtained (eqs 78 and 79).[32]

$$\underset{H_3C}{\overset{O}{\diagdown}}\diagup OH \xrightarrow[\substack{rt, 24\ h \\ 91\%}]{\substack{Ph_3P=CMeCO_2Et \\ MnO_2,\ CH_2Cl_2}} \underset{H_3C}{\overset{O}{\diagdown}}\diagup\hspace{-0.3em}\underset{CO_2Et}{\diagup}$$

$$E\ only \quad (78)$$

$$\xrightarrow[\substack{rt, 5\ min \\ 62\%}]{\substack{Ph_3P=CHCO_2Et \\ MnO_2,\ CH_2Cl_2}}$$

$$\quad (79)$$

$$E{:}Z = 94{:}6$$

These α-hydroxyketones have now also been used as efficient substrates in the synthesis of various N-containing heterocycles (vide infra).

Aldehyde formation–trapping reaction sequences have also been applied to vicinal diol systems employing MnO$_2$ to carry out oxidative cleavage.[33] The intermediate aldehydes have then been trapped using stabilized Wittig reagents (eq 80). Moderate yields, however, were generally obtained using MnO$_2$ and the reaction appears to be very substrate dependent. Alternatively, the use of silica-supported sodium periodate resulted in improved yields and shortened reaction times (eq 81).[34]

$$\xrightarrow[\substack{rt, 24\ h \\ 64\%}]{\substack{Ph_3P=CHCO_2Et \\ MnO_2,\ CH_2Cl_2}}$$

$$\quad (80)$$

$$E,E{:}E,Z = 4.5{:}1$$

$$\xrightarrow[\substack{rt, 1-3\ h \\ 73\%}]{\substack{Ph_3P=CHCO_2Et \\ NaIO_4,\ Et_2O}}$$

$$\quad (81)$$

$$E,E{:}E,Z = 7{:}1$$

Use of Non-stabilized Ylides.[35–37] TOP sequences using non-stabilized Wittig reactions have also been carried out to great effect, generating the ylide in situ from the corresponding phosphonium salt and the bicyclic-guanidine base, MTBD[35] (1-methyl-1,5,7-triazabicyclo[4.4.0]dec-5-ene) (**5**) (eqs 82 and 83).[36]

MTBD, **5**

$$\xrightarrow[\substack{reflux, 4\ h \\ 91\%}]{\substack{Ph_3PCH_3Br \\ MnO_2,\ 5 \\ THF,\ 4\text{Å MS}}}$$

$$\quad (82)$$

$$C_6H_{13}\equiv\diagup OH \xrightarrow[\substack{reflux, 4\ h \\ 62\%}]{\substack{Ph_3PCH_3Br \\ MnO_2,\ 5 \\ THF,\ 4\text{Å MS}}}$$

$$C_6H_{13}\equiv\diagup\hspace{-0.3em}\diagdown Ph \quad (83)$$

$$Z{:}E = 1{:}2$$

This work was later extended to provide a concise synthesis of unsaturated aldehydes through the intermediacy of unsaturated dioxolanes (eq 84).[37]

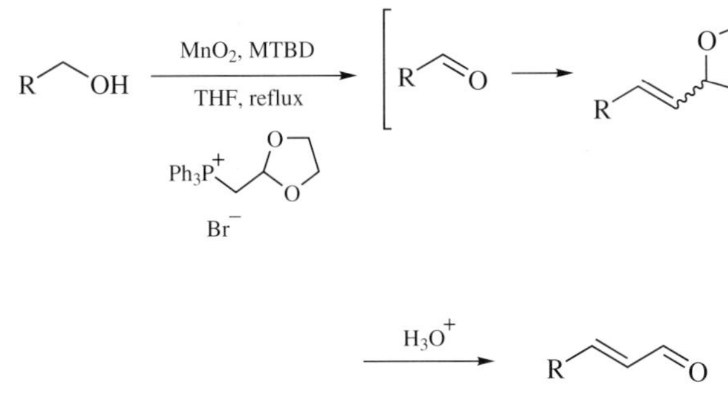

R⟶OH $\xrightarrow[\text{THF, reflux}]{\text{MnO}_2,\ \text{MTBD}}$ [R⟶O → dioxolane]

Ph₃P⁺ (dioxolane) Br⁻

$\xrightarrow{\text{H}_3\text{O}^+}$ R⟶⟶⟶O (84)

R = 2-furyl, 76%
R = 3-pyridyl, 79%

Good to excellent yields were obtained over a range of substrates without the need to isolate either the intermediate aldehyde or dioxolane.

Nitrogen Based Nucleophiles.[38–41] Apart from Wittig reagents, several other nucleophiles have been used in the in situ trapping of intermediate aldehydes including amines to afford imines (eq 85).[38]

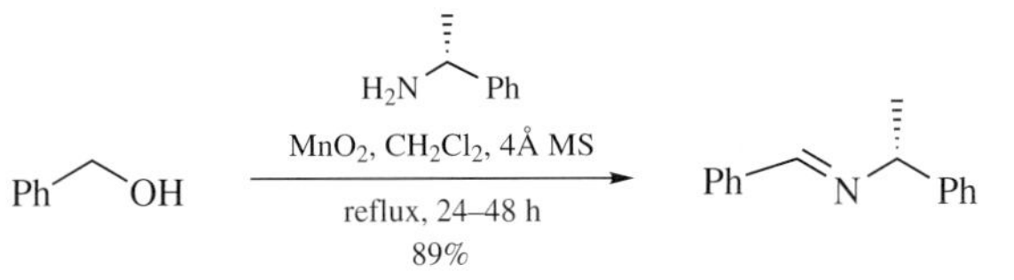

Ph⟶OH $\xrightarrow[\text{reflux, 24–48 h} \atop 89\%]{\text{MnO}_2,\ \text{CH}_2\text{Cl}_2,\ \text{4Å MS}}$ Ph⟶N⟶Ph (85)

The TOP sequence can also be carried out in the presence of a heterogeneous reductant to effect an oxidation–imination–reduction process leading from activated alcohols to amines (eqs 86 and 87). Polymer-supported cyanoborohydride[38] (PSCBH) or sodium borohydride[39] can be used as the reductant. The use of an oxidant and reductant in the same one-pot procedure is noteworthy.

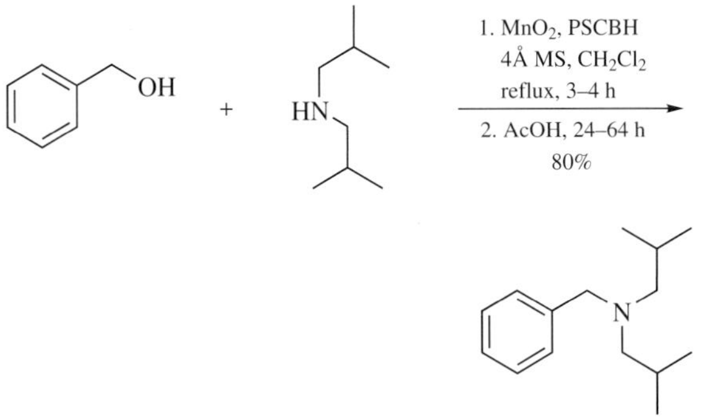

(benzyl alcohol) + HN(iBu)₂ $\xrightarrow[\substack{2.\ \text{AcOH, 24–64 h}\\80\%}]{\substack{1.\ \text{MnO}_2,\ \text{PSCBH}\\ \text{4Å MS, CH}_2\text{Cl}_2\\ \text{reflux, 3–4 h}}}$

(amine product) (86)

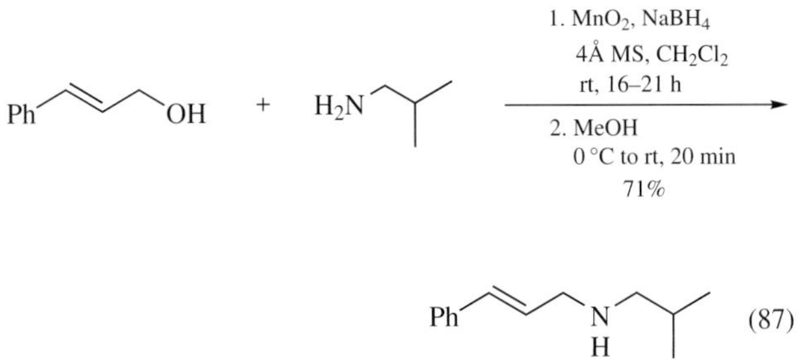

Ph⟶OH + H₂N⟶ $\xrightarrow[\substack{2.\ \text{MeOH}\\ 0\,°\text{C to rt, 20 min}\\71\%}]{\substack{1.\ \text{MnO}_2,\ \text{NaBH}_4\\ \text{4Å MS, CH}_2\text{Cl}_2\\ \text{rt, 16–21 h}}}$

Ph⟶⟶N(H)⟶ (87)

These methods are complementary with the less expensive sodium borohydride affording good to excellent yields with a range of primary amines, but with the more sterically hindered

primary and secondary amines, PSCBH is generally higher yielding.

Along with imines and amines, *O*-alkyl oximes have also been prepared in a TOP sequence (eq 88).[40]

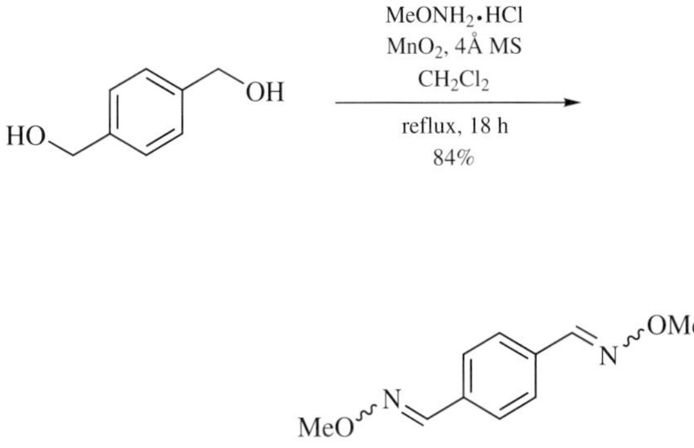

HO⟶(p-xylylene)⟶OH $\xrightarrow[\substack{\text{CH}_2\text{Cl}_2\\ \text{reflux, 18 h}\\84\%}]{\substack{\text{MeONH}_2\cdot\text{HCl}\\ \text{MnO}_2,\ \text{4Å MS}}}$

MeO⟶N⟶(aryl)⟶N⟶OMe (88)

As alluded to above, α-hydroxyketones have also been used in TOP sequences with 1,2-diamines to afford a diverse range of nitrogen-containing heterocycles.[41] Thus using MnO₂ oxidation and trapping in situ with 1,2-diamines, quinoxalines (eq 89) and dihydropyrazines (eq 90) can be accessed in good to excellent yield.

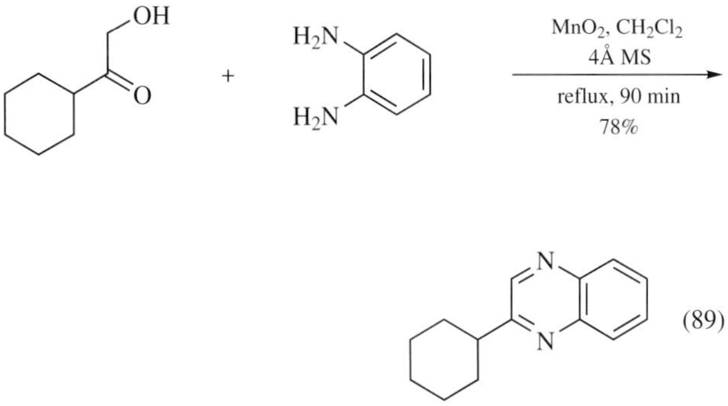

(cyclohexyl hydroxyketone) + (o-phenylenediamine) $\xrightarrow[\substack{\text{4Å MS}\\ \text{reflux, 90 min}\\78\%}]{\text{MnO}_2,\ \text{CH}_2\text{Cl}_2}$

(2-cyclohexylquinoxaline) (89)

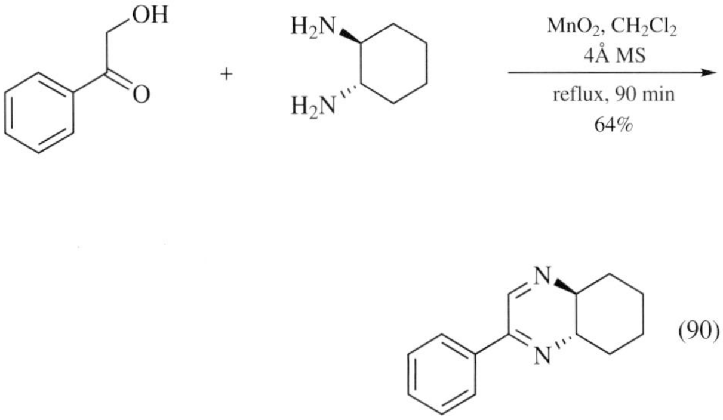

(phenyl hydroxyketone) + (trans-cyclohexanediamine) $\xrightarrow[\substack{\text{4Å MS}\\ \text{reflux, 90 min}\\64\%}]{\text{MnO}_2,\ \text{CH}_2\text{Cl}_2}$

(phenyl dihydropyrazine fused cyclohexane) (90)

However, by applying the sodium borohydride conditions described by Kanno and Taylor (vide supra),[39] a rapid and efficient route to substituted piperazines was achieved in a TOP sequence (eq 91).

$$(91)$$

Finally, by the addition of methanolic potassium hydroxide after the formation of the dihydropyrazines, aromatization was achieved to afford the pyrazine adducts in fair to good yields (eqs 92 and 93).

Functional Group Interconversions.[42–46] Lai's group showed that aromatic aldehydes could be converted into the corresponding nitriles by treatment with ammonia in isopropanol (IPA) and THF containing MnO_2, and magnesium sulfate at room temperature (eq 94).[42]

$$(92)$$

$$(93)$$

$$(94)$$

This work was later extended by Taylor's group to the direct conversion of activated primary alcohols into nitriles (eqs 95 and 96).[43] Again good to excellent yields were obtained with a wide range of activated alcohols.

$$(95)$$

$$(96)$$

A TOP synthesis of methyl esters and amides has also been reported (extending on the Corey and Gilman methodologies, see eqs 13–15),[2b,3] employing sodium cyanide (eq 97).[44,45]

$$(97)$$

Attempts to modify the conditions to prepare ethyl esters were met with limited success but the synthesis of a wide range of primary, secondary, and tertiary amide products was possible in good to excellent yields (eq 98).

$$(98)$$

Combination of MnO_2 with triazidochlorosilane ($SiCl_4/NaN_3$) affords a reagent capable of transforming activated aldehydes to their appropriate acyl azides (eq 99).[46]

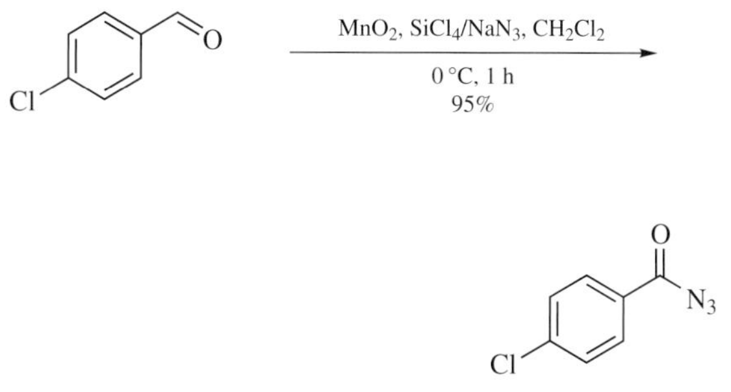

(99)

Second Update

Lee Fader

Boehringer-Ingelheim (Canada) Ltd., Laval, Québec, Canada

Manganese dioxide has, in recent years, continued to be an import reagent in synthetic organic chemistry. New advances and applications of the reagent since the last update are dominated by tandem oxidation processes, originally developed with this reagent by Taylor and coworkers, although a few unrelated transformations have also been described. The low toxicity, low cost, availability and ease of handling of the reagent ensures that it will continue to find application in place of more complex and sensitive oxidants or for substrates found to be delicate in nature.

Tandem Oxidation Processes (TOP).[47] Tandem oxidation processes that employ manganese dioxide as the oxidant have recently been reviewed.[47] Oxidation-Wittig tandem sequences, as described above, are the most explored process of this type. Although recent applications have appeared in the literature, only a few constitute improvements or enhancements to what is known. Notable among these is the conversion of activated alcohols into 1,1-dibromoalkenes, a very mild process analogous to a Corey–Fuchs approach to this type of compound (eq 100).[48]

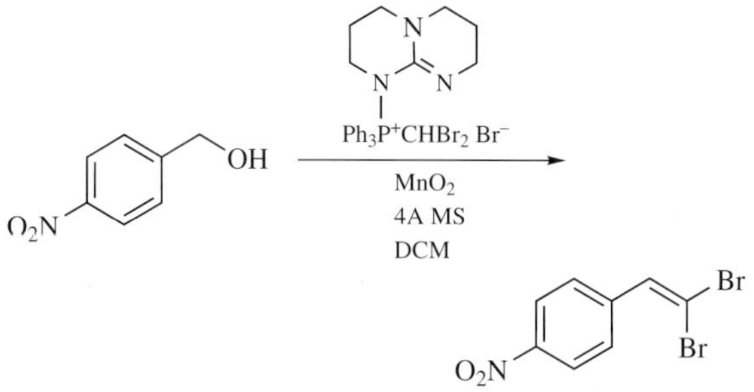

(100)

In recent years, a number of other practical tandem sequences involving MnO_2 oxidations have emerged.[47,49] For example, when alcohols are treated with MnO_2 and a stabilized sulfur ylide, the intermediate enone undergoes cyclopropanation to form the corresponding ketocyclopropanes in good yields (eq 101). Note that the terminal substituents of the olefin must be either hydrogen or aryl for the in situ cyclopropanation to be successful.

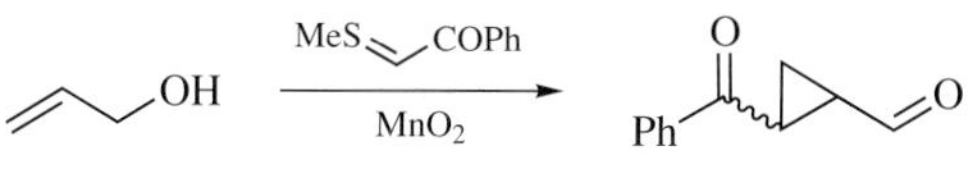

(101)

77%, *trans/cis* 3.1:1

If an activated alcohol is treated with MnO_2, a phosphorous ylide and a stabilized sulfur ylide in a one-pot procedure, a tandem oxidation–olefination–cyclopropanation sequence ensues giving the cyclopropane products in modest to good yields (eq 102).[47,50] At the current level of development, poor stereocontrol of the cyclopropanation step is observed and only α-hydroxyketones have been employed as substrates.

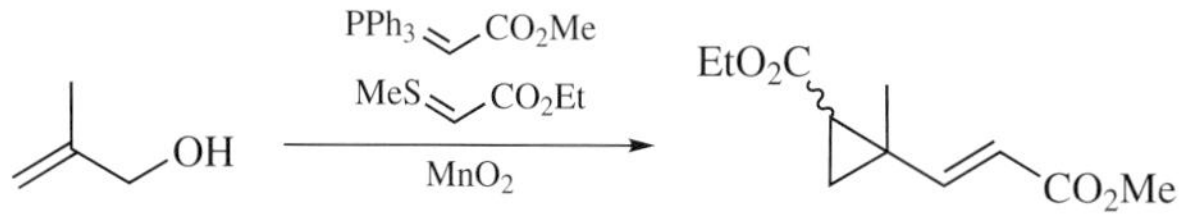

81%, *trans/cis* 3.5:1

(102)

Another attractive TOP sequence involves use of a combination of the Bestmann–Ohira reagent and MnO_2.[47,51] In these examples, a range of activated alcohols are oxidized to the corresponding aldehydes, which in turn are efficiently trapped by the subsequently added alkynylating reagent (eq 103).

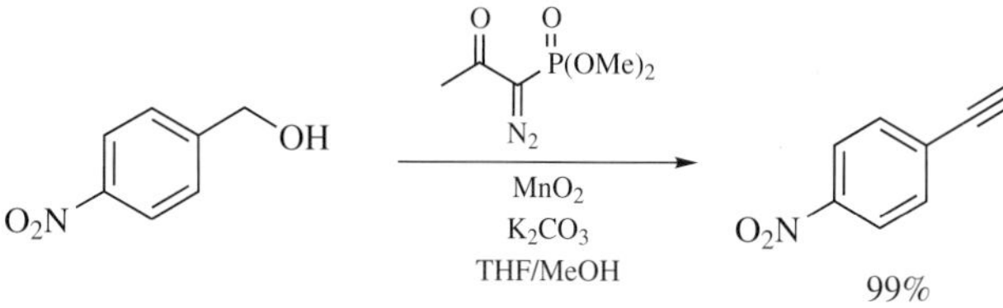

(103)

The tandem oxidation processes employing MnO_2 continue to be applied to functional group interconversions. A new entry in this area is the direct conversion of activated alcohols to nitriles via an intermediate aldehyde followed by its corresponding imine (eq 104).[47,52]

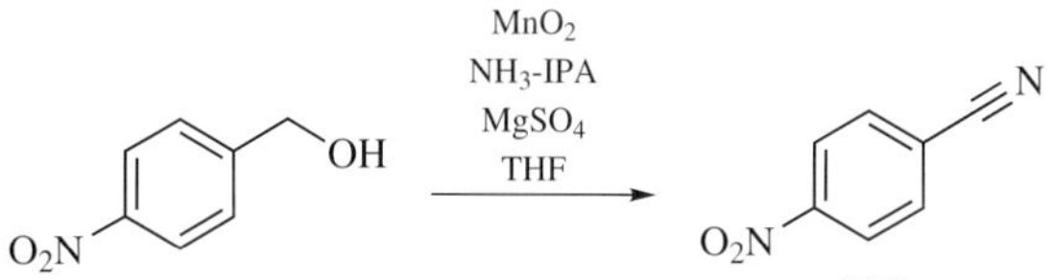

(104)

Tandem oxidation processes have also been applied to five- and six-membered heterocycle synthesis.[47] New examples of six-membered rings include pyrimides and pyridines.[47] In one approach, an α-arylpropargyl alcohol is oxidized to the corresponding ynone, which then reacts with ethyl α-aminocrotonate or an amidine to afford pyridines or pyrimidines, respectively (eqs 105 and 106).[53] In both cases, the reactions were limited in scope, but the former was subsequently improved by implementation of β-keto esters and ammonium acetate in place of the α-amino crotonate synthon (eq 107).[53]

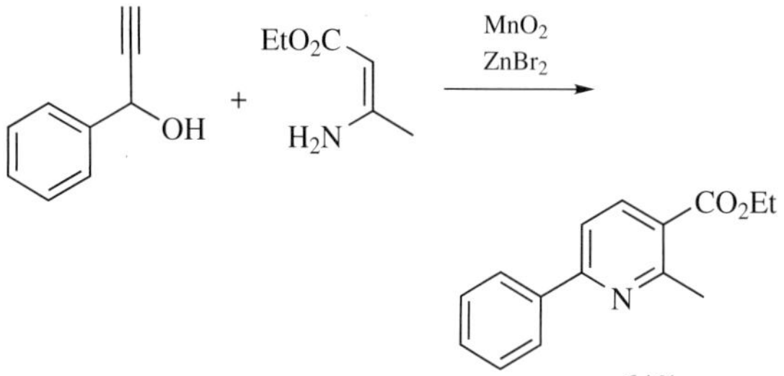

(105)

64%

(106)

84%

(107)

96%

Synthesis of five-membered heterocycles via a TOP sequence from activated alcohols affords 4,5-benzo-fused-2-substituted imidazoles, oxazoles and thiazoles in synthetically useful yields over the presumed oxidation–condensation–oxidation sequence (eq 108).[47,54]

(108)

X = NMe   90%
X = O   73%
(plus HCl for X = NMe)   X = S   66%

Related Reagents. Barium Manganate; Nickel(II) Peroxide.

1. (a) Fatiadi, A. J., *Synthesis* **1976**, 65 and 133. (b) Pickering, W. F., *Rev. Pure Appl. Chem.* **1966**, *16*, 185. (c) Evans, R. M., *Q. Rev., Chem. Soc.* **1959**, *13*, 61. (d) Hudlicky, M., *Oxidations in Organic Chemistry*; American Chemical Society: Washington, 1990. (e) Fatiadi, A. J., In *Organic Synthesis by Oxidation With Metal Compounds*; Plenum: New York, 1986; Chapter 3.

2. (a) Ball, S.; Goodwin, T. W.; Morton, R. A., *Biochem. J.* **1948**, *42*, 516. (b) Corey, E. J.; Gilman, N. W.; Ganem, B. E., *J. Am. Chem. Soc.* **1968**, *90*, 5616. (c) Bharucha, K. R., *J. Chem. Soc.* **1956**, 2446. (d) Nickon, A.; Bagli, J. F., *J. Am. Chem. Soc.* **1961**, *83*, 1498. (e) Fales, H. M.; Wildman, W. C., *J. Org. Chem.* **1961**, *26*, 881. (f) Cresp, T. M.; Sondheimer, F., *J. Am. Chem. Soc.* **1975**, *97*, 4412. (g) Babler, J. H.; Martin, M. J., *J. Org. Chem.* **1977**, *42*, 1799. (h) Counsell, R. E.; Klimstra, P. D.; Colton, F. B., *J. Org. Chem.* **1962**, *27*, 248. (i) Trost, B. M.; Kunz, R. A., *J. Am. Chem. Soc.* **1975**, *97*, 7152. (j) Sargent, L. J.; Weiss, U., *J. Org. Chem.* **1960**, *25*, 987. (k) Hendrickson, J. B.; Palumbo, P. S., *J. Org. Chem.* **1985**, *50*, 2110. (l) Saucy, G.; Borer, R., *Helv. Chim. Acta* **1971**, *54*, 2121.

3. Gilman, N. W., *Chem. Commun.* **1971**, 733.

4. (a) Struve, G.; Seltzer, S., *J. Org. Chem.* **1982**, *47*, 2109. (b) Van Amsterdam, L. J. P.; Lugtenburg, J., *Chem. Commun.* **1982**, 946. (c) Bentley, R. K.; Jones, E. R. H.; Thaller, V., *J. Chem. Soc. (C)* **1969**, 1096. (d) Makin, S. M.; Ismail, A. A.; Yastrebov, V. V.; Petrv, K. I., *Zh. Org. Khim.* **1971**, *7*, 2120 (*Chem. Abstr.* **1972**, *76*, 13 712).

5. (a) Highet, R. J.; Wildman, W. C., *J. Am. Chem. Soc.* **1955**, *77*, 4399. (b) Hänsel, R.; Su, T. L.; Schulz, J., *Ber. Dtsch. Chem. Ges.* **1977**, *110*, 3664. (c) Trost, B. M.; Caldwell, C. G.; Murayama, E.; Heissler, D., *J. Org. Chem.* **1983**, *48*, 3252. (d) König, B. M.; Friedrichsen, W., *Tetrahedron Lett.* **1987**, *28*, 4279. (e) Barrelle, M.; Glenat, R., *Bull. Soc. Chem. Fr. Part 2* **1967**, 453. (f) Loozen, H. J. J.; Godefroi, E. F., *J. Org. Chem.* **1973**, *38*, 3495.

6. (a) Crombie, L.; Crossley, J., *J. Chem. Soc.* **1963**, 4983. (b) Harrison, I. T., *Proc. Chem. Soc. London* **1964**, 110.

7. Ohloff, G.; Giersch, W., *Angew. Chem.* **1973**, *85*, 401.

8. (a) Liu, K. T.; Shih, M. H.; Huang, H. W.; Hu, C. J., *Synthesis* **1988**, 715. (b) Taylor, E. C.; Maryanoff, C. A.; Skotnicki, J. S., *J. Org. Chem.* **1980**, *45*, 2512.

9. (a) Mashraqui, S.; Keehn, P., *Synth. Commun.* **1982**, *12*, 637. (b) Hamada, Y.; Shibata, M.; Sugiura, T.; Kato, S.; Shioiri, T., *J. Org. Chem.* **1987**, *52*, 1252. (c) Bhatnagar, I.; George, M. V., *Tetrahedron* **1968**, *24*, 1293.

10. Kashdan, D. S.; Schwartz, J. A.; Rapoport, H., *J. Org. Chem.* **1982**, *47*, 2638.

11. *Comprehensive Inorganic Chemistry*; Bailar, J. C.; Trotman-Dickenson, A. F., Eds.; Pergamon: Oxford, 1973; Vol 3, p 801.

12. Fatiadi, A. J., *J. Chem. Soc. (B)* **1971**, 889.

13. (a) Attenburrow, J.; Cameron, A. F. B.; Chapman, J. H.; Evans, R. M.; Hems, B. A.; Jansen, A. B. A.; Walker, T., *J. Chem. Soc.* **1952**, 1094. (b) Mancera, O.; Rosenkranz, G.; Sondheimer, F., *J. Chem. Soc.* **1953**, 2189. (c) Henbest, H. B.; Jones, E. R. H.; Owen, T. C., *J. Chem. Soc.* **1957**, 4909. (d) Gritter, R. J.; Wallace, T. J., *J. Org. Chem.* **1959**, *24*, 1051. (e) Pratt, E. F.; Van de Castle, J. F., *J. Org. Chem.* **1961**, *26*, 2973. (f) Harfenist, M.; Bavley, A.; Lazier, W. A., *J. Org. Chem.* **1954**, *19*, 1608. (g) Henbest, H. B.; Thomas, A., *J. Chem. Soc.* **1957**, 3032. (h) Cohen, N.; Banner, B. L.; Blount, J. F.; Tsai, M.; Saucy, G., *J. Org. Chem.* **1973**, *38*, 3229. (i) Vereshchagin, L. I.; Gainulina, S. R.; Podskrebysheva, L. A.; Gaivoronskii, L. A.; Okhapkina, L. L.; Vorob'eva, V. G.; Latyshev, V. P., *J. Org. Chem. USSR (Engl. Transl.)* **1972**, *8*, 1143. (j) Barco, A.; Benetti, S.; Pollini, G. P.; Baraldi, P. G., *Synthesis* **1977**, 837. (k) Mattocks, A. R., *J. Chem. Res. (S)* **1977**, 40. (l) Goldman, I. M., *J. Org. Chem.* **1969**, *34*, 1979. (m) Carpino, L. A., *J. Org. Chem.* **1970**, *35*, 3971. (n) Liu, K. T.; Shih, M. H.; Huang, H. W.; Hu, C. J., *Synthesis* **1988**, 715. (o) Kimura, T.; Fujita, M.; Ando, T., *Chem. Lett.* **1988**, 1387. (p) Barakat, M. Z.; Abdel-Wahab, M. F.; El-Sadr, M. M., *J. Chem. Soc.* **1956**, 4685. (q) Sondheimer, F.; Amendolla, C.; Rozenkranz, G., *J. Am. Chem. Soc.* **1953**, *75*, 5930 and 5932.

14. (a) Pratt, E. F.; Suskind, S. P., *J. Org. Chem.* **1963**, *28*, 638. (b) Papadopoulos, E. P.; Jarrar, A.; Issidorides, C. H., *J. Org. Chem.* **1966**, *31*, 615. (c) Curragh, E. F.; Henbest, H. B.; Thomas, A., *J. Chem. Soc.* **1960**, 3559.

15. (a) Lou, J.-D.; Xu, Z.-N., *Tetrahedron Lett.* **2002**, *43*, 6149. (b) Varma, R. S.; Saini, R. K.; Dahiya, R., *Tetrahedron Lett.* **1997**, *38*, 7823. (c) Khandekar, A. C.; Paul, A. M.; Khadilkar, B. M., *Synth. Commun.* **2002**, *32*, 2931. (d) Stavrescu, R.; Kimura, T.; Fujita, M.; Vinatoru, M.; Ando, T., *Synth. Commun.* **1999**, *29*, 1719. (e) Wang, Y.-l.; Zhao, Y.-W.; Ma, D.-I., *J. Chem. Res.* **2002**, 158. (f) Khadilkar, B. M.; Madyar, V. R., *Synth. Commun.* **2002**, *32*, 1731.

16. Karlsson, U.; Wang, G.-Z.; Bäckvall, J.-E., *J. Org. Chem.* **1994**, *59*, 1196.

17. Firouzabadi, H.; Etemadi, S.; Karimi, B.; Jarrahpour, A. A., *Synth. Commun.* **1999**, *29*, 4333.

18. (a) Hirano, M.; Yakabe, S.; Chikamori, H.; Clark, J. H.; Morimto, T., *J. Chem. Res.* **1998**, 310. (b) Firouzabadi, H.; Abbassi, M.; Karimi, B., *Synth. Commun.* **1999**, *29*, 2527.

19. (a) Fabretti, A.; Ghelfi, F.; Grandi, R.; Pagnoni, U. M., *Synth. Commun.* **1994**, *24*, 2393. (b) Firouzabadi, H.; Abbasi, M., *Synth. Commun.* **1999**, *29*, 1485.

20. (a) Kiegiel, J.; Poplawska, M.; Jóźwik, J.; Koisor, M.; Jurczak, J., *Tetrahedron Lett.* **1999**, *40*, 5605. (b) Cicchi, S.; Marradi, M.; Goti, A.;

Brandi, A., *Tetrahedron Lett.* **2001**, *42*, 6503. (c) Shinada, T.; Oshihara, K., *Tetrahedron Lett.* **1995**, *36*, 6701.

21. Zhou, J.; Zhang, L.; Hu, Y.; Hu, H., *J. Chem. Res.* **1999**, 552.

22. (a) Wei, X.; Taylor, R. J. K., *Tetrahedron Lett.* **1998**, *39*, 3815. (b) MacDonald, G.; Alcaraz, L.; Wei, X.; Lewis, N. J.; Taylor, R. J. K., *Tetrahedron* **1998**, *54*, 9823.

23. Wei, X.; Taylor, R. J. K., *J. Org. Chem.* **2000**, *65*, 616.

24. Blackburn, L.; Kanno, H.; Taylor, R. J. K., *Tetrahedron Lett.* **2003**, *44*, 115.

25. Nicolaou, K. C.; Li, Y.; Fylaktakidou, K. C.; Mitchell, H. J.; Wei, H.-X.; Weyershausen, B., *Angew. Chem., Int. Ed.* **2001**, 40, 3849.

26. Mladenova, M.; Tavlinova, M.; Momchev, M.; Alami, M.; Ourévitch, M.; Brion, J. D., *Eur. J. Org. Chem.* **2003**, 2713.

27. Blackburn, L.; Wei, X.; Taylor, R. J. K., *Chem. Commun.* **1999**, 1337.

28. Davies, S. B.; McKervey, M. A., *Tetrahedron Lett.* **1999**, *40*, 1229.

29. Ireland, R. E.; Norbeck, D. W., *J. Org. Chem.* **1985**, *50*, 2198.

30. (a) Barrett, A. G. M.; Hamprecht, D.; Ohkubo, M., *J. Org. Chem.* **1997**, *62*, 9376. (b) Huang, C. C., *J. Labelled Compd. Radiopharm.* **1987**, *24*, 675. (c) Crich, D.; Mo, X., *Synlett* **1999**, 67.

31. Shuto, S.; Niizuma, S.; Matsuda, A., *J. Org. Chem.* **1998**, *63*, 4489.

32. Runcie, K. A.; Taylor, R. J. K., *Chem. Commun.* **2002**, 974.

33. Outram, H. S.; Raw, S. A.; Taylor, R. J. K., *Tetrahedron Lett.* **2002**, *43*, 6185.

34. Dunlap, N. K.; Wosenu, M.; Jones, J. M.; Carrick, J. D., *Tetrahedron Lett.* **2002**, *43*, 3923.

35. Simoni, D.; Rossi, M.; Rondanin, R.; Mazzani, A.; Baruchello, C.; Malagutti, C.; Roberti, M.; Indiviata, F. P., *Org. Lett.* **2000**, *2*, 3766.

36. Blackburn, L.; Pei, C.; Taylor, R. J. K., *Synlett* **2002**, *2*, 215.

37. Reid, M.; Rowe, D.; Taylor, R. J. K., *Chem. Commun.* **2003**, 2284.

38. Blackburn, L.; Taylor, R. J. K., *Org. Lett.* **2001**, *3*, 1637.

39. Kanno, H.; Taylor, R. J. K., *Tetrahedron Lett.* **2002**, *43*, 7337.

40. Kanno, H.; Taylor, R. J. K., *Synlett* **2002**, 1287.

41. Raw, S. A.; Wilfred, C. D.; Taylor, R. J. K., *Chem. Commun.* **2003**, 2286.

42. Lai, G.; Bhamare, N. K.; Anderson, W. K., *Synlett* **2001**, 230.

43. McAllister, G. D.; Wilfred, C. D.; Taylor, R. J. K., *Synlett* **2002**, 1291.

44. Foot, J. S.; Kanno, H.; Giblin, G. M. P.; Taylor, R. J. K., *Synlett* **2002**, 1293.

45. Foot, J. S.; Kanno, H.; Giblin, G. M. P.; Taylor, R. J. K., *Synthesis* **2003**, 1055.

46. Elmorsy, S. S., *Tetrahedron Lett.* **1995**, *36*, 1341.

47. Taylor, J. K.; Reid, M.; Foot, J.; Raw, S. A., *Acc. Chem. Res.* **2005**, *38*, 851.

48. Raw, S. A.; Reid, M.; Roman, E.; Taylor, R. J. K., *Synlett* **2004**, 819.

49. Oswald, M. F.; Raw, S. A.; Taylor, R. J. K., *Org. Lett.* **2004**, *6*, 3997.

50. Oswald, M. F.; Raw, S. A.; Taylor, R. J. K., *Chem. Commun.* **2005**, 2253.

51. Quesada, E.; Taylor, R. J. K., *Tetrathedron Lett.* **2005**, *46*, 6473.

52. McAllister, G. D.; Wilfred, C. D.; Taylor, R. J. K., *Synlett* **2002**, 1291.

53. Bagley, M. C.; Hughes, D. D.; Sabo, H. M.; Taylor, P. H.; Xiong, X., *Synlett* **2003**, 1443.

54. Wilfred, C. D.; Taylor, R. J. K., *Synlett* **2004**, 1628.

Osmium Tetroxide[1]

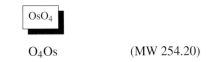

[20816-12-0]   O$_4$Os   (MW 254.20)
InChI = 1S/4O.Os
InChIKey = VUVGYHUDAICLFK-UHFFFAOYSA-N

(*cis* dihydroxylation of alkenes; osmylation; asymmetric and diastereoselective dihydroxylation; oxyamination of alkenes)

Physical Data: mp 39.5–41 °C; bp 130 °C; *d* 4.906 g cm^{-3}; chlorine- or ozone-like odor.

Solubility: soluble in water (5.3% at 0 °C, 7.24% at 25 °C); soluble in many organic solvents (toluene, *t*-BuOH, CCl$_4$, acetone, methyl *t*-butyl ether).

Form Supplied in: pale yellow solid in glass ampule, as 4 wt% solution in water, and as 2.5 wt% in *t*-BuOH.

Handling, Storage, and Precautions: vapor is toxic, causing damage to the eyes, respiratory tract, and skin; may cause temporary blindness; LD$_{50}$ 14 mg/kg for the rat, 162 mg/kg for the mouse. Because of its high toxicity and high vapor pressure, it should be handled with extreme care in a chemical fume hood; chemical-resistant gloves, safety goggles, and other protective clothing should be worn; the solid reagent and its solutions should be stored in a refrigerator.

Original Commentary

Yun Gao

Sepracor, Marlborough, MA, USA

Dihydroxylation of Alkenes. The *cis* dihydroxylation (osmylation) of alkenes by osmium tetroxide to form *cis*-1,2-diols (*vic*-glycols) is one of the most reliable synthetic transformations (eq 1).[1]

The reaction has been proposed to proceed through a [3 + 2] or [2 + 2] pathway to give the common intermediate osmium(VI) monoglycolate ester (osmate ester), which is then hydrolyzed reductively or oxidatively to give the *cis*-1,2-diol (eq 2). The *cis* dihydroxylation of alkenes is accelerated by tertiary amines such as *Pyridine*, quinuclidine, and derivatives of dihydroquinidine (DHQD) or dihydroquinine (DHQ) (eq 3).

Due to the electrophilic nature of osmium tetroxide, electron-withdrawing groups connected to the alkene double bond retard the dihydroxylation.[2] This is in contrast to the oxidation of alkenes by *Potassium Permanganate*, which preferentially attacks electron-deficient double bonds. However, in the presence of a tertiary amine such as pyridine, even the most electron-deficient alkenes can be osmylated by osmium tetroxide (eq 4).[3] The more highly substituted double bonds are preferentially oxidized (eq 5).

Under stoichiometric and common catalytic osmylation conditions, alkene double bonds are hydroxylated by osmium tetroxide without affecting other functional groups such as hydroxyl groups, aldehyde and ketone carbonyl groups, acetals, triple bonds, and sulfides (see also *Osmium Tetroxide–N-Methylmorpholine N-Oxide*).

The *cis* dihydroxylation can be performed either stoichiometrically, if the alkene is precious, or more economically and conveniently with a catalytic amount of osmium tetroxide (or its precursors such as osmium chloride or potassium osmate) in conjunction with a cooxidant. In the stoichiometric dihydroxylation, the diol product is usually obtained by the reductive hydrolysis of the osmate ester with a reducing agent such as *Lithium Aluminum Hydride*, *Hydrogen Sulfide*, K$_2$SO$_3$ or Na$_2$SO$_3$, and KHSO$_3$ or NaHSO$_3$. The reduced osmium species is normally removed by filtration. Osmium can be recovered as osmium tetroxide by oxidation of low-valent osmium compounds with hydrogen peroxide.[4] In the catalytic dihydroxylation, the osmate ester is usually hydrolyzed under basic aqueous conditions to produce the diol and osmium(VI) compounds, which are then reoxidized by the cooxidant to osmium tetroxide to continue the catalytic cycle. Normally 0.01% to 2% equiv of osmium tetroxide or precursors are used in the catalytic dihydroxylation. Common cooxidants are metal chlorates, *N-Methylmorpholine N-Oxide* (NMO), *Trimethylamine N-Oxide*, *Hydrogen Peroxide*, *t-Butyl Hydroperoxide*, and *Potassium Ferricyanide*. *Oxygen* has also been used as cooxidant in dihydroxylation of certain alkenes.[5] Excess cooxidant and osmium tetroxide are reduced with a reducing agent such as those mentioned above during the workup. The stoichiometric dihydroxylation can be carried out in almost any inert organic solvent, including most commonly MTBE, toluene, and *t*-BuOH. In the catalytic dihydroxylation, in order to dissolve the inorganic cooxidant and other additives, a mixture of water and an organic solvent are often used. The most common solvent combinations in this case are acetone–water and *t*-BuOH–water. Because of the high cost and toxicity of osmium tetroxide, the stoichiometric dihydroxylation has been mostly replaced by the catalytic version in preparative organic chemistry (see also **Osmium Tetroxide–tert-Butyl Hydroperoxide, Osmium Tetroxide–N-Methylmorpholine N-Oxide**, and **Osmium Tetroxide–Potassium Ferricyanide**).

Diastereoselective Dihydroxylation. Dihydroxylation of acyclic alkenes containing an allylic, oxygen-bearing stereocenter proceeds with predictable stereochemistry. In general, regardless of the double-bond substitution pattern and geometry, the relative stereochemistry between the pre-existing hydroxyl or alkoxyl group and the adjacent newly formed hydroxyl group of the major diastereomer will be *erythro* (i.e. *anti* if the carbon chain is drawn in the zig-zag convention) (eq 6).[6,7]

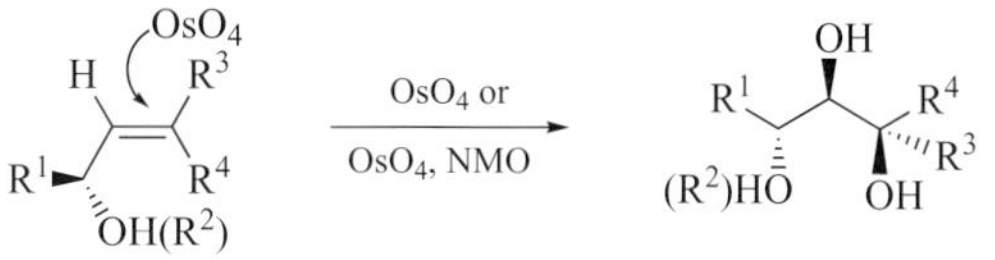

$$(6)$$

In the osmylation of 1,2-disubstituted allylic alcohols and derivatives, *cis*-alkenes provide higher diastereoselectivity than the corresponding *trans*-alkenes (eqs 7 and 8).[6] Opposite selectivities have been observed in the osmylation of (*Z*)-enoate and (*E*)-enoate esters (eqs 9 and 10).[8] High selectivity has also been observed in the osmylation of 1,1-disubstituted[9] and (*E*)-trisubstituted allylic alcohols and derivatives[10] and bis-allylic compounds[11] (eqs 11–13).

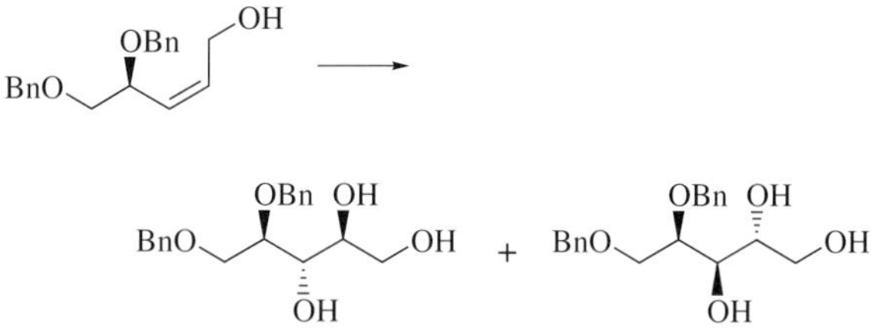

$$(7)$$

OsO$_4$ 8.0:1.0
OsO$_4$, NMO   7.0:1.0

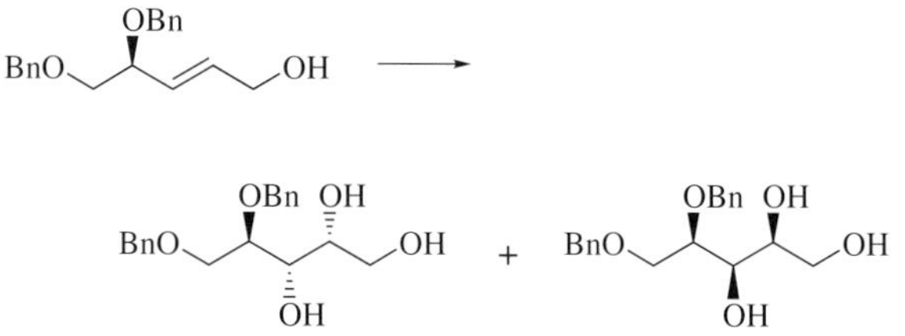

$$(8)$$

OsO$_4$ 4.2:1.0
OsO$_4$, NMO   3.1:1.0

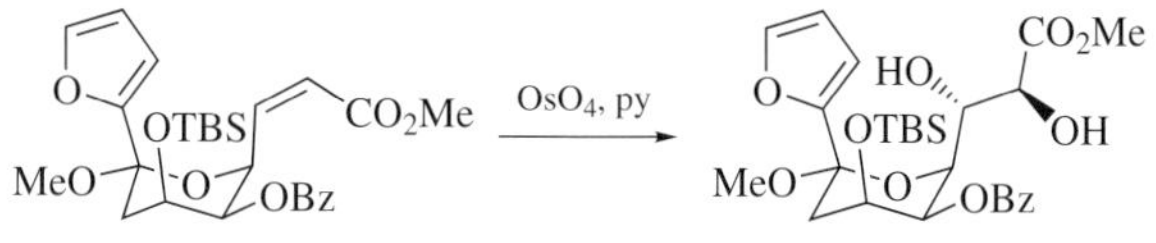

$$(9)$$

de > 18:1

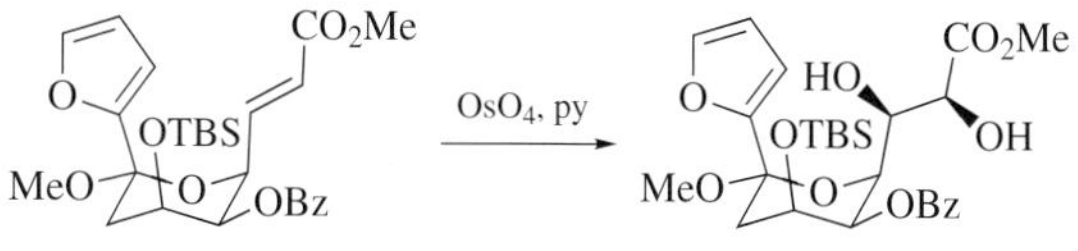

$$(10)$$

de > 25:1

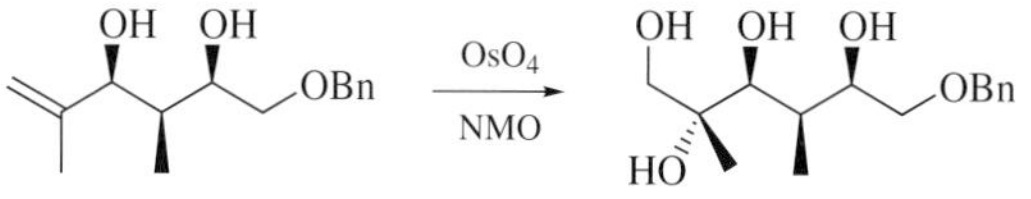

$$(11)$$

de = 35:1

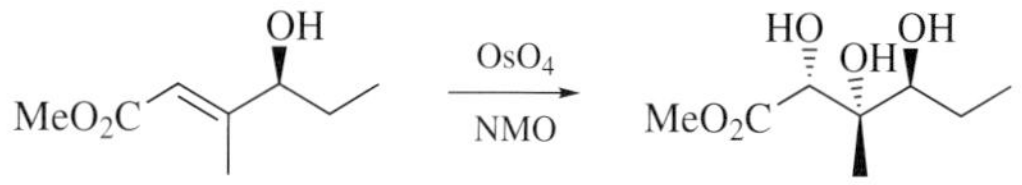

$$(12)$$

de > 100:1

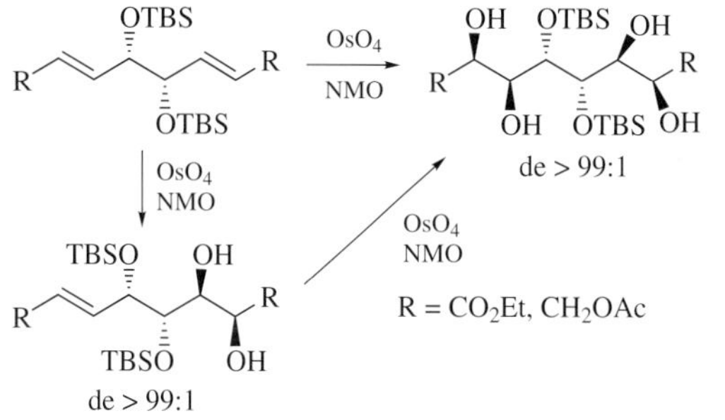

$$(13)$$

de > 99:1

R = CO$_2$Et, CH$_2$OAc

de > 99:1

The diastereoselective osmylation has been extended to oxygen-substituted allylic silane systems, and the general rule observed for the allylic alcohol system also applies (eq 14).[12] High selectivity is also observed in the osmylation of allylsilanes where the substituent on the chiral center bearing the silyl group is larger than a methyl group (eq 15).[13] These diastereoselectivities have been achieved in both stoichiometric and catalytic dihydroxylations. Slightly higher selectivity has been observed in the stoichiometric reaction than in the catalytic reaction; this may be due to less selective bis-osmate ester formation in the catalytic reaction using NMO as the cooxidant. Use of K$_3$Fe(CN)$_6$ may solve this discrepancy. Several rationales have been proposed for the observed selectivity.[14] The conclusion appears to be that the osmylation of these systems is controlled by steric bias, rather than by the electronic nature of the allylic system, and osmylation will occur from the sterically more accessible face. The high diastereoselectivity of osmium tetroxide in the dihydroxylation of chiral unsaturated compounds has been applied widely in organic synthesis.[8,15]

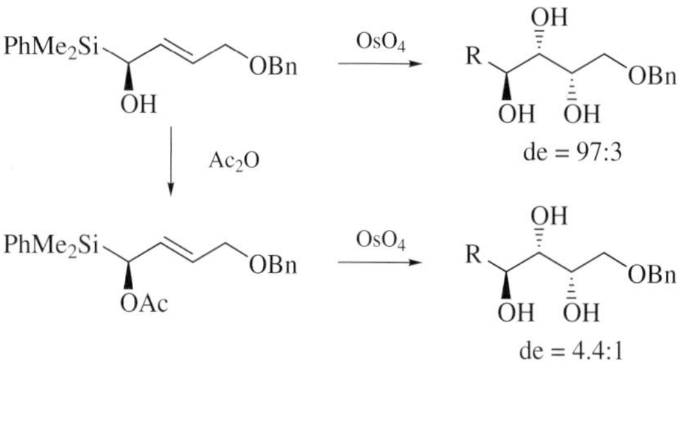

$$(14)$$

de = 97:3

de = 4.4:1

$$R = Me \quad 34{:}65$$
$$R = i\text{-}Pr \quad 67{:}33$$
$$R = Ph \quad 92{:}8$$

Sulfoxide groups direct the dihydroxylation of a remote double bond in an acyclic system perhaps by prior complexation of the sulfoxide oxygen with osmium tetroxide (eqs 16 and 17).[16] Chiral sulfoximine-directed diastereoselective osmylation of cycloalkenes has been used for the synthesis of optically pure dihydroxycycloalkanones (eq 18).[17] Nitro groups also direct the osmylation of certain cycloalkenes, resulting in dihydroxylation from the more hindered side of the ring. In contrast, without the nitro group the dihydroxylation proceeds from the less hindered side (eq 19).[18]

Enantioselective Dihydroxylation. The acceleration of osmylation by tertiary amines brought about the use of chiral amines as chiral ligands for the asymmetric dihydroxylation (AD).[1] The AD can be classified into two types: (a) noncatalytic reaction, where stoichiometric amounts of ligand and osmium tetroxide are used, and (b) catalytic reaction, where catalytic amounts of ligand and osmium tetroxide are employed in conjunction with stoichiometric amounts of cooxidant. Generally, in the stoichiometric AD systems, chiral chelating diamines are used as chiral auxiliaries with osmium tetroxide for the introduction of asymmetry to the diol products.[1,19] Although high asymmetric inductions have been achieved in these systems, the stoichiometric ADs have limited use in practical organic synthesis because of the cost of both ligand and osmium tetroxide. The discovery of the ligand-accelerated catalysis in AD made the transition from stoichiometric to catalytic AD possible.[1c,1d,20] In the most effective catalytic system, osmium tetroxide or its precursors and chiral ligands derived from cinchona alkaloids, dihydroquinidine (DHQD), or dihydroquinine (DHQ), are used catalytically in the presence of a stoichiometric amount of cooxidant such as NMO or $K_3Fe(CN)_6$. Besides alkaloid-derived ligands, other types of ligand have been designed and used in the catalytic AD with moderate success.[21] (see also *Osmium Tetroxide–N-Methylmorpholine N-Oxide* and *Osmium Tetroxide–Potassium Ferricyanide*).

Double Diastereoselective Dihydroxylation. AD of homochiral alkenes gives matched and mismatched diastereoselectivities due to the steric interaction of the chiral osmium tetroxide–ligand complex with the chiral center in the vicinity of the alkene double bond. For example, in the noncatalytic osmylation of the monothioacetal (eq 20),[22] the ratio of (2S,3R) to (2R,3S) diastereomers is 2.5:1 with the achiral quinuclidine as ligand, 40:1 with *Dihydroquinidine Acetate* (DHQD-OAc) as ligand in the matched case, and 1:16 with *Dihydroquinine Acetate* (DHQ-OAc) as ligand in the mismatched case.

Quinuclidine 2.5:1
DHQD-OAc 40:1
DHQ-OAc 1:16

Diastereoselectivities in the catalytic AD with OsO_4–NMO of several α,β-unsaturated uronic acid derivatives are significantly enhanced when the alkenes are matched with the chiral ligands DHQ-CLB (dihydroquinine *p*-chlorobenzoate) and DHQD-CLB (dihydroquinidine *p*-chlorobenzoate) (eq 21).[23] Using the chiral ligands (DHQD)$_2$-PHAL and (DHQ)$_2$-PHAL, the double diastereoselective dihydroxylation of a chiral unsaturated ester has been tested using *Osmium Tetroxide–Potassium Ferricyanide* (eq 22).[1c] These results show that enhanced diastereoselectivity in the dihydroxylation can be achieved by matching of alkene diastereoselectivity with catalyst enantioselectivity. Double diastereoselective dihydroxylation with a bidentate ligand has also been reported.[24] Kinetic resolutions of racemic alkenes with OsO_4

in the presence of a chiral ligand have been demonstrated.[25] An elegant example is the kinetic resolution of the enantiomers of C_{76}, the smallest chiral fullerene, by asymmetric osmylation in the presence of DHQD- and DHQ-derived ligands.[26]

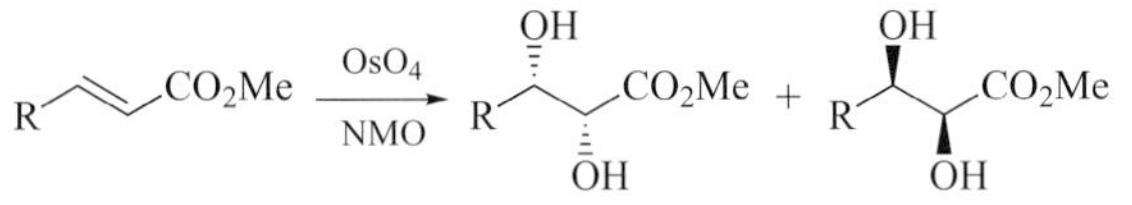

R	Reagent	Ratio
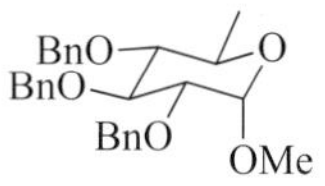 BnO (sugar, OMe)	OsO$_4$ only	10.3:1
	DHQD-CLB	1.3:1
	DHQ-CLB	20.5:1
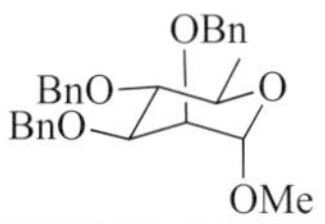 BnO (sugar, OMe)	OsO$_4$ only	7.4:1
	DHQD-CLB	3.4:1
	DHQ-CLB	15.9:1

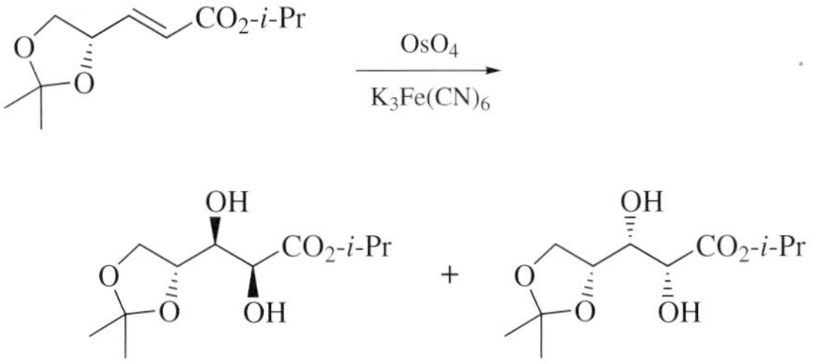

no ligand	2.8:1
(DHQD)$_2$-PHAL	39:1
(DHQ)$_2$-PHAL	1:1.3

Oxyamination of Alkenes and Oxidation of Other Functional Groups. Osmium tetroxide catalyzes the vicinal oxyamination of alkenes to give *cis*-vicinal hydroxyamides with *Chloramine-T* (eq 23)[27] and alkyl *N*-chloro-*N*-argentocarbamate, generated in situ by the reaction of alkyl *N*-chlorosodiocarbamate (such as ethyl or *t*-butyl *N*-chlorosodiocarbamate) with *Silver(I) Nitrate* (eq 24).[28]

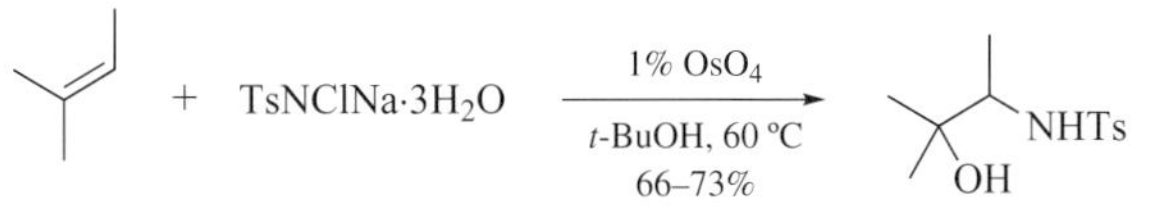

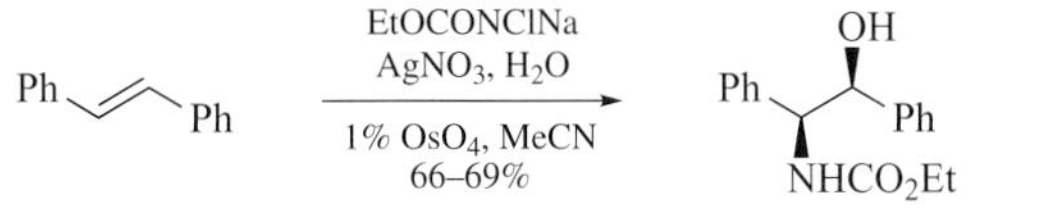

Since chloramine-T is readily available, the former method offers a practical and direct method for introducing a vicinal hydroxyl group and a tolylsulfonamide to a double bond. While the sulfonamide protecting group is difficult to remove (and undesirable in some cases), the *N*-chloro-*N*-argentocarbamate system provides an alternative method, since the carbamate group can be easily removed to give free amine. This latter system is also more regioselective and reactive towards electron-deficient alkenes such as **Dimethyl Fumarate** and (*E*)-stilbene than the procedures based on chloramine-T. In all of these oxyamination reactions, monosubstituted alkenes react more rapidly than di- or trisubstituted alkenes. In the presence of tetraethylammonium acetate, trisubstituted alkenes can be oxyaminated with catalytic OsO$_4$ and *N*-chloro-*N*-metallocarbamates. So far, attempts to effect catalytic asymmetric oxyamination have not been successful.

Osmium tetroxide also catalyzes the oxidation of organic sulfides to sulfones with NMO or trimethylamine *N*-oxide (see **Osmium Tetroxide–N-Methylmorpholine N-Oxide**). In contrast, most sulfides are not oxidized with stoichiometric amounts of OsO$_4$. Oxidations of alkynes and alcohols with OsO$_4$ without and in the presence of cooxidants have also been reported.[1a, 1b] However, these reactions have not found wide synthetic applications because of the availability of other methods.

First Update

Timothy J. Donohoe, Robert M. Harris & Majid J. Chughtai
Dyson Perrins Laboratory, Oxford, UK

Asymmetric Dihydroxylation Reactions. A substantial amount of work has been reported on the development of the asymmetric dihydroxylation (AD) reaction as originally described by Sharpless.[29] A greater understanding has emerged of the functional group tolerance of the AD reaction and also its applicability towards differing alkene substitution patterns. The mechanism of the AD reaction has been the subject of intense debate especially with respect to the question of whether a [2 + 2] or [3 + 2] pathway is followed, and some insightful mechanistic studies have followed from this discussion.[30]

Two representative examples of the AD process are shown below: the first (eq 25)[31] helps to define the relative reactivity of the two alkene units within geraniol with respect to asymmetric dihydroxylation (the electron rich C6,7 alkene reacts first), and the second (eq 26)[32] illustrates the compatibility of the AD process with easily oxidized sulfides.

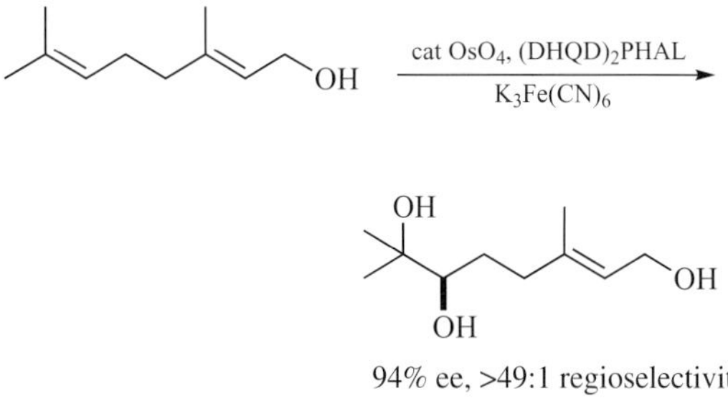

94% ee, >49:1 regioselectivity

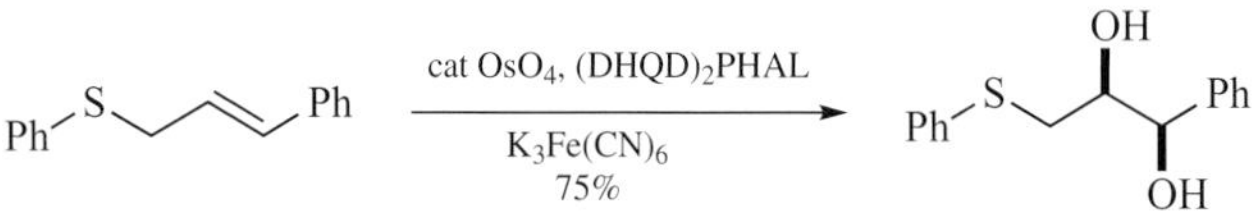

98% ee

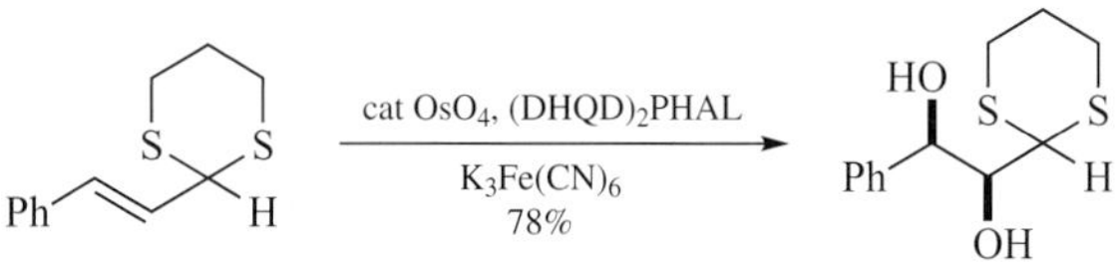

97% ee

Moreover, Sharpless has recently defined a series of bidentate ligands that help to promote a reaction through the 'second catalytic cycle.'[33] Addition of citric acid to a dihydroxylation reaction gives excellent yields of diols from electron deficient alkenes (eq 27). In addition to lowering the pH of the reaction, the citrate additive is presumed to chelate to the transition metal throughout the reaction so favoring turnover via the second cycle. Dihydroxylation through this mechanism is commonly believed to destroy any enantioselectivity imparted by the cinchona alkaloid ligands, and clearly one consequence of the addition of citric acid to a dihydroxylation reaction is, therefore, the formation of racemic products when chiral amine additives are added.

$$(27)$$

Conditions	Yield (%)
1. water/acetone/*t*BuOH (5:2:1)	10
2. water/*t*BuOH (1:1), citric acid (25 mol %)	76

One possible solution to the lack of enantioselectivity observed with citric acid additive is the addition of a chiral amino alcohol (eq 28),[34] which performs the same chelating role as the citrate and locks the reaction into the second cycle. However, the chiral backbone of the ligand is then capable of imparting enantioselectivity during a subsequent dihydroxylation reaction. While this area clearly has much potential for development, the enantioselectivities obtained are not yet competitive with those commonly realized with the conventional AD process.

$$(28)$$

Ligand R	TsHN (phenyl COOH OH)	HO (NHTs OH)
Me	48% (ee)	51% (ee)
Et	50% (ee)	48% (ee)

Hydrogen Bonding Control During Dihydroxylation. The reagent combination of OsO$_4$ and TMEDA produces a complex that acts as a hydrogen bond acceptor and dihydroxylates allylic alcohols and amides with control of both stereo- and regiochemistry

(eq 29).[35,36] The selectivity observed during this oxidation is opposite to that normally observed under dihydroxylation with more standard (i.e., Upjohn) conditions and is difficult to accomplish by other means. For example, compare the oxidation of geraniol under the action of OsO$_4$/TMEDA (eq 30,[35] oxidation of the C-2,3 alkene ensues) with that observed under the action of an AD mix (eq 25).

$$(29)$$

$$(30)$$

There is evidence (NMR, IR, X-ray) that TMEDA forms a (reactive) bidentate complex with the osmium tetroxide whereby the diamine ligand increases the electron density on both the osmium center and also the outlying oxo ligands: this in turn makes them more electron rich and better hydrogen bond acceptors and so leads to the selectivity observed.[36]

The chelating nature of the diamine precludes osmate ester hydroysis in situ and means that the oxidation of an alkene requires 1 equiv of osmium tetroxide, followed by a step to release the glycol. However, the use of catalytic amounts of transition metal in the directed dihydroxylation of allylic amides was shown to be possible (eq 31)[37] by using a monodentate amine (quinuclidine, which is released in situ from quinuclidine-*N*-oxide). These conditions are catalytic in osmium but generally exhibit a less powerful directing effect than stoichiometric osmium tetroxide and TMEDA.

$$(31)$$

In Situ Protection of Diols from Dihydroxylation. Sharpless recently examined the applicability of Narasaka's dihydroxylation reaction conditions (catalytic osmium tetroxide and NMO in the presence of phenylboronic acid). The boronic acid reagent replaces the water which is normally present in dihydroxylation reactions to hydrolyze the osmate ester and liberate the diol.[38] In this case, however, the phenyl boronic acid promotes this reaction forming a cyclic boronate ester in the process (eq 32). Sharpless discovered that these modified conditions could sometimes give different stereoselectivities to that observed under UpJohn conditions and also that conjugated dienes could be dihydroxylated only once (something that is not possible with the UpJohn reaction because overoxidation ensues). The cyclic boronate esters so produced are useful protecting groups for the glycol unit and can be easily cleaved with aqueous hydrogen peroxide.

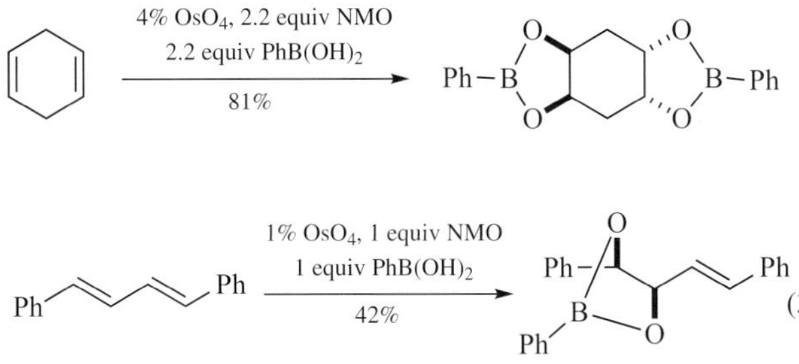

$$ (32) $$

Oxidative Cyclization of 1,5-Dienes. The oxidative cyclization of 1,5-dienes to produce *cis*-tetrahydrofurans has previously been reported only with strong oxidants such as $KMnO_4$ and RuO_4; yields are not usually above 50% probably because of overoxidation of the products. This reaction is an interesting one because addition of oxygen is (*syn*) stereospecific across both alkenes and also stereoselective for the formation of 2,5-*cis*-tetrahydrofuran; it is possible to make up to four new chiral centers and one heterocyclic ring in one step. Recently, there have been reports of osmium tetroxide promoting the same type of oxidative cyclization, using either periodate[39] or amine-oxides as reoxidants (eq 33).[40] The yields for such reactions (especially using amine-*N*-oxides in conjunction with acidic conditions) are very high, and this method promises to become a useful way of preparing heavily substituted heterocycles in one step. The role of stereodirecting groups on the 1,5-diene backbone was also probed.[40] Studies showed clearly that lowering the pH of the reaction was essential in promoting oxidative cyclization at the expense of simple dihydroxylation.

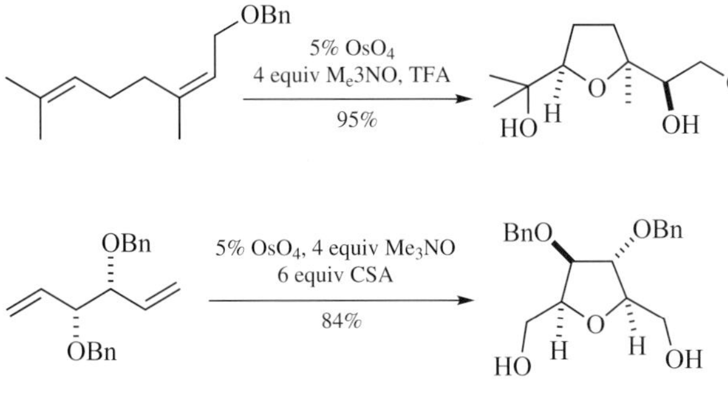

$$ (33) $$

New Reoxidants for Dihydroxylation. Bäckvall has introduced some well thought out reoxidants for the dihydroxylation and asymmetric dihydroxylation processes. Hydrogen peroxide is the terminal oxidant that oxidizes osmium(VI) back to osmium(VIII) by first oxidizing flavin (the product flavin-OOH in turn oxidizes an amine to its *N*-oxide in situ; this *N*-oxide then acts conventionally in the dihydroxylation reaction). Originally, *N*-methylmorpholine was the amine of choice and this additive could be added in catalytic amounts. Recent developments have

seen the use of even cheaper amine additives such as triethylamine (eq 34).[41] This 'triple catalytic' system is also compatible with Sharpless' chiral ligands and allows oxidation of a range of alkenes with high enantioselectivity.

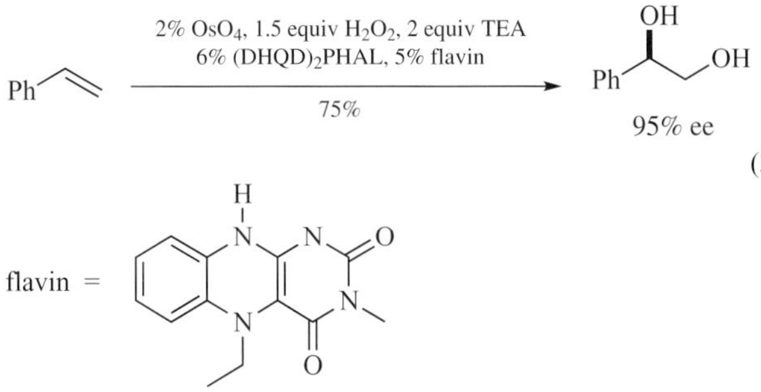

$$ (34) $$

In another development, Krief has shown that selenoxides (especially aryl/alkyl selenoxides) are capable of replacing amine-oxides as the reoxidant in the dihydroxylation reaction. An asymmetric process based on $(DHQD)_2PHAL$ (named SeOAD) was also developed and this gave good to excellent ee's with a range of alkenes.[42]

Aminohydroxylation of Alkenes. Asymmetric aminohydroxylation (AA) reaction is a powerful method for preparing vicinal aminoalcohols in a stereospecific reaction from alkenes using osmium catalysis; the use of cinchona based ligands enables the products to be formed with high levels of enantioselectivity.[43] It should be noted, however, that Sharpless' aminohydroxylation reaction does not usually involve the addition of osmium tetroxide, rather potassium osmate $[K_2(OH)_4OsO_2]$ is employed as the catalyst. There are several reasons for this, such as the ease of handling and minimization of the diol side-product. Therefore, this review will concentrate only on the recently published examples that use osmium tetroxide as the transition metal additive. The first example shown below is unusual in that oxidation of a silyl enol-ether means that one of the newly formed chiral centers is lost as an amino-ketone is formed in situ (eq 35);[44] however, good levels of enantioselectivity were observed within these products.

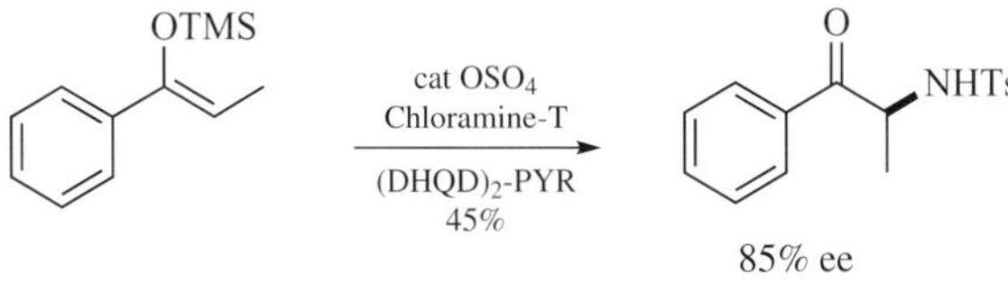

$$ (35) $$

Another example shown regards the regio- and stereoselective hydroxyamination of chiral alkenes using a similar reagent combination (eq 36);[45] in this case the nitrogen was placed at the least hindered end of the alkene (this is the commonly expected outcome of the AA process, although exceptions are known). There is no need for a chiral ligand here because the facial selectivity is set by the steric bias of the bridged ring system.

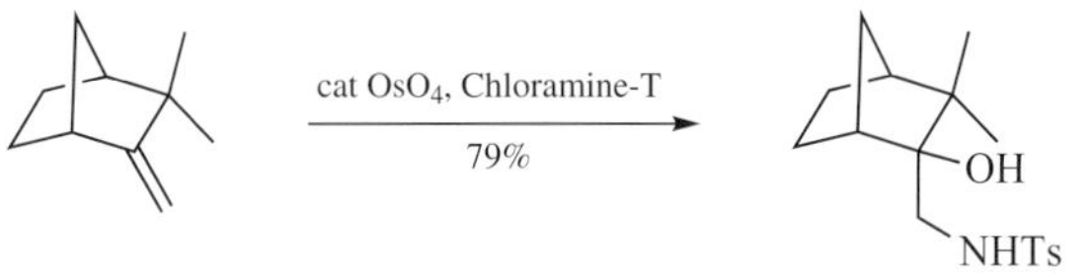

$$ (36) $$

Oxidative Cleavage of Alkenes. An alternative to the oxidative cleavage of alkenes using ozone or the Lemieux–Johnson protocol has been reported recently. Under the action of catalytic osmium tetroxide, with oxone as a reoxidant, a variety of substituted alkenes were cleaved efficiently to furnish carbonyl compounds (eq 37).[46] Any of the aldehydes that are produced via this sequence are immediately oxidized in situ to give the corresponding acid; clearly this does not happen for any ketones so produced. Even electron deficient alkenes such as α,β-unsaturated carbonyl compounds could be conveniently oxidized, although the products then underwent a decarboxylation reaction to produce the corresponding diacid.

The authors postulate that vicinal diols are not produced as intermediates during this reaction sequence, primarily because the reintroduction of such diols does not lead to the formation of cleaved products but leaves them unaltered.

Photoinduced Charge Transfer Osmylation. The reaction of osmium tetroxide with benzenoid derivatives can only be accomplished by irradiation of the mixture. This reaction had previously been shown to work (with stoichiometric osmium tetroxide) by promotion of charge transfer between the two reactants to form an ion-pair; this can then collapse to form an osmate ester of benzene diol. Subsequent dihydroxylation of this intermediate appears to follow a more conventional (and stereoselective) course. The use of catalytic osmium tetroxide (in conjunction with barium perchlorate as a reoxidant) for this reaction is noteworthy, as is the formation of both inositol and conduritol derivatives in one-pot (eq 38).[47] The photoinduced osmylation of mono-substituted arenes was possible although the yields were lower and the amount of cyclitol-type products reduced.

Immobilized Forms of Osmium Tetroxide. Several groups have been actively searching for immobilized forms of osmium

tetroxide in order to overcome the problems of toxicity and volatility associated with this reagent. Ley has encapsulated osmium tetroxide in a polyurea matrix forming a stable catalyst that is completely insoluble in water or organic solvents (eq 39).[48] This catalyst is then capable of promoting the dihydroxylation reaction of a range of alkenes using NMO as the reoxidant. The microcapsules (called Os EnCat) could be stored for long periods of time without special precautions and could also be reused five times without any loss in activity.

Kobayashi has recently encapsulated osmium tetroxide into a polymer derived from acrylonitrile, butadiene, and polystyrene. This catalyst (called ABS-MC OsO_4) has also been used to give geminal diols from alkenes using NMO as a reoxidant; it can be easily recovered and then reused (eq 40).[49] In addition, Kobayashi has shown that this immobilized catalyst is compatible with conditions for the AD reaction and gives diols with good levels of enantioselectivity when $(DHQD)_2PHAL$ ligands are included in the dihydroxylation reaction.

Other approaches to immobilization have included the use of macroporous resins and functionalized silica solids that contain residual vinyl groups which can be dihydroxylated as a means of anchoring the transition metal to the solid support: the resulting osmium(VI) complexes are then reoxidized in situ.[50] The AD reaction has also been investigated with polymeric versions of Sharpless' chiral cinchona based ligands.[51]

Oxidation of Alcohols and Amines using Osmium Tetroxide. Recent reports have shown that benzylic amines can be oxidized to yield nitriles using an oxidizing system that includes catalytic osmium tetroxide. Unfortunately, only benzylic amines could be successfully oxidized, with aliphatic derivatives giving rise to complex reaction mixtures (eq 41).[52]

There have also been reports of the oxidation of alcohols into aldehydes and ketones using oxygen and catalytic osmium tetroxide in the presence of copper (eq 42).[53] Remarkably, allylic alcohols are not dihydroxylated at all and the reagent does not lead to any overoxidation to carboxylic acids. This reagent combination will only oxidize activated alcohols (such as benzylic or allylic) and is, therefore, a useful alternative to manganese dioxide.

$$(42)$$

Related Reagents. Sodium Periodate–Osmium Tetroxide.

Second Update

Jung Woon Yang, Sun Min Kim, Joong Suk Oh & Choong Eui Song
Department of Chemistry and Department of Energy Science, Institute of Basic Science, Sungkyunkwan University, Suwon, Korea

Dihydroxylation of Alkenes: Synthetic Applications.

Using Upjohn condition (OsO4-NMO). The utility of the Upjohn protocol for the dihydroxylation was recently demonstrated in the synthesis of bicyclic analogues of pentopyranoses,[54a] (–)-anisomycin,[54b] trisubstituted γ-butyrolactone,[54c] 6-bromo-4-(1,2-dihydroxyethyl)-7-hydroxycoumarine (Bhc-diol) as a photoremovable protecting group,[54d] 3,4-dihydroxy-2-(3-methylbut-2-enyl)-3,4-dihydronaphthalen-1(2*H*)-one,[54e] benzo-[*c*]pyrano[3,2-*h*]acridin-7-ones,[54f] both enantiomers of conduritol C tetraacetate and of *meso*-conduritol-D-tetraacetate.[54g]

Recently, Richarson and coworkers developed a convenient method for in situ generation of NMO from NMM (*N*-methylmorpholine) using CO_2 and H_2O_2. The NMO is quantitatively formed in situ without isolation, and undergoes asymmetric dihydroxylation of alkenes to yield the resulting diols with high enantioselectivities.[55]

Using OsO4-TMEDA condition. Dihydroxylation of homoallylic alcohol using stoichiometric OsO4 with TMEDA (tetramethylethylene diamine) as a ligand predominantly provides *syn* product. The enhanced *syn* selectivity can be ascribed to the coordination of TMEDA to OsO4, thereby enhancing the electronegativity of the *oxo*-ligand on osmium and favoring hydrogen bonding to the homoallylic OH-group. This methodology was efficiently applied to an enantioselective formal synthesis of (–)-Secosyrin 1, a natural product isolated from *Pseudomonas syringae*, showing an unusual response in resistant soybean plants (eq 43) .[56]

Donohoe et al. also studied the directed dihydroxylation of homoallylic trihaloacetamide under OsO4–TMEDA conditions. In particular, replacement of trichloroacetamide with trifluoroacetamide as a directing group produced *syn* selectivity that is remarkably enhanced under OsO4-TMEDA conditions (eq 44).[57]

The Fuchs and coworkers approach to apoptolidin commenced with a hydrogen-bonding directed dihydroxylation of a single (*Z*)-allylic hemiacetal by employing Donohoe's OsO4-TMEDA condition to give Apoptolidin diols in quantitative yield with high *syn* selectivity, whereas (*E*)-allylic acetal (if R = CH_3) produced the opposite selectivity (eq 45).[58]

$$(43)$$

syn:anti = 99:1

secosyrin 1

$$(44)$$

X = CCl$_3$; 1. OsO4-TMEDA, 1.2:1 (*syn:anti*), 92% yield
2. UpJohn, 1:1 (*syn:anti*), 99% yield

X = CF$_3$; 1. OsO4-TMEDA, >20:1 (*syn:anti*), 98% yield
2. UpJohn, 1:1 (*syn:anti*), 98% yield

Conversion of a spirodiene to the desired diol was also achieved in 60% yield as a single diastereomer by employing Donohoe's OsO4-TMEDA complex (eq 46).[59]

New Cooxidants for Dihydroxylation. In recent studies, several oxidants, O_2,[60a] air,[60b] $NaClO_2$ (sodium chlorite),[60c] NaOCl (sodium hypochlorite),[60d] PhSe(O)Bn,[60e] Ba(ClO3)3,[60f] and Me3NO[60g–i] were successfully used as cooxidants in Os-catalyzed dihydroxylation of olefins. For instance, Donohoe reported on the use of Me3NO as a cooxidant in the dihydroxylation process, which allows the synthesis of (±)-1-epiaustraline in 14% overall yield from *N*-Boc-pyrrole 2,5-methyl diester (eq 47).[61]

The construction of *syn*-diol moiety is employed by means of OsO4-Ba(ClO3)3 oxidation (eq 48).[62]

Allylic hydroperoxides can be directly converted to the corresponding triols upon treatment with a tiny amount of OsO4 (0.2 mol %) in aqueous acetone. Hydroperoxide acts as both a directing group and an internal cooxidant (eq 49).[63]

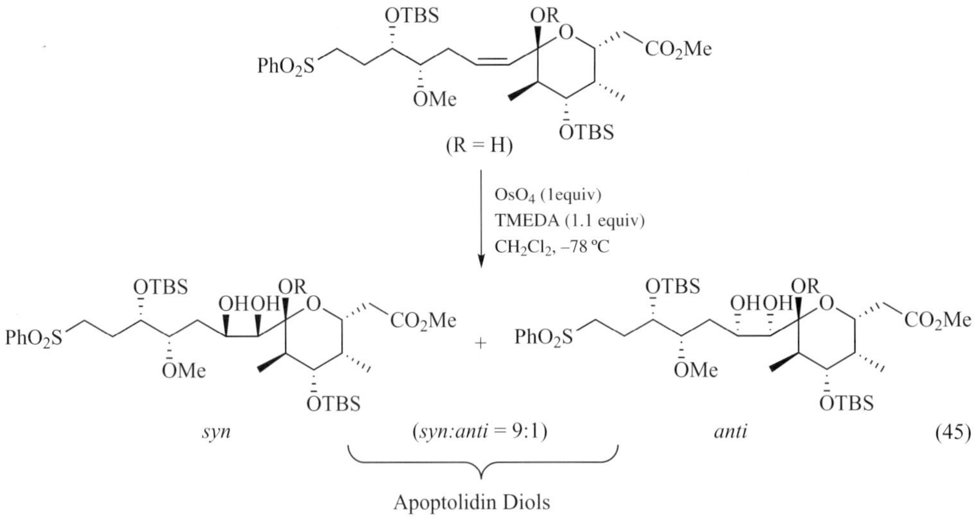

(45)

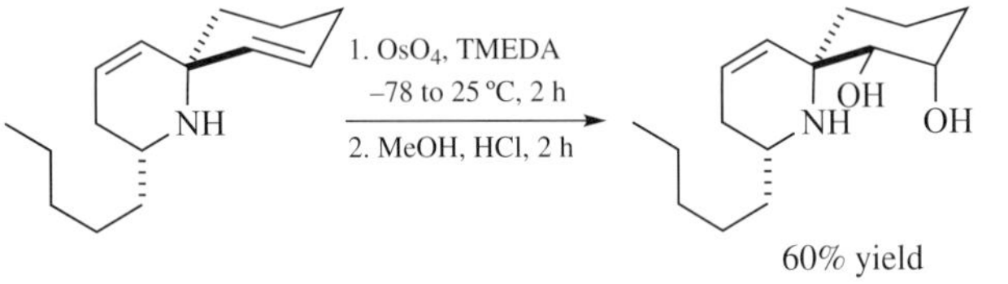

syn   (*syn:anti* = 9:1)   *anti*

Apoptolidin Diols

(46)

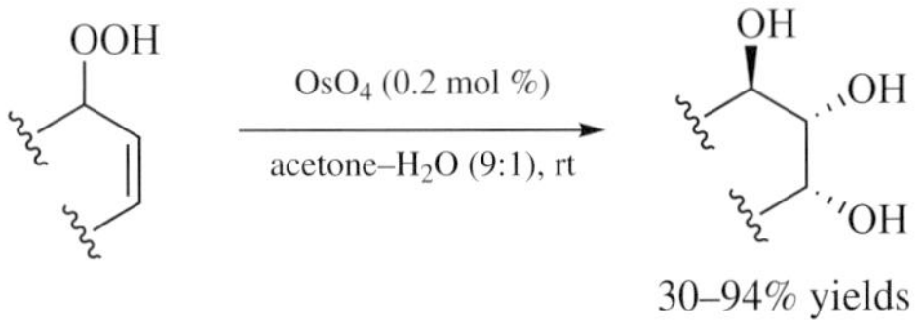

60% yield

(49)

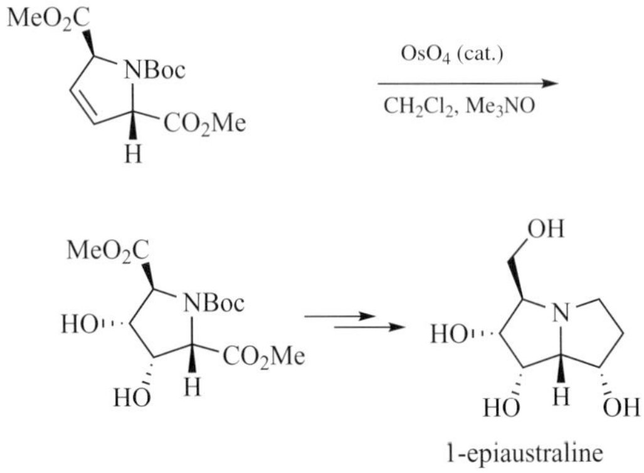

30–94% yields

(47)

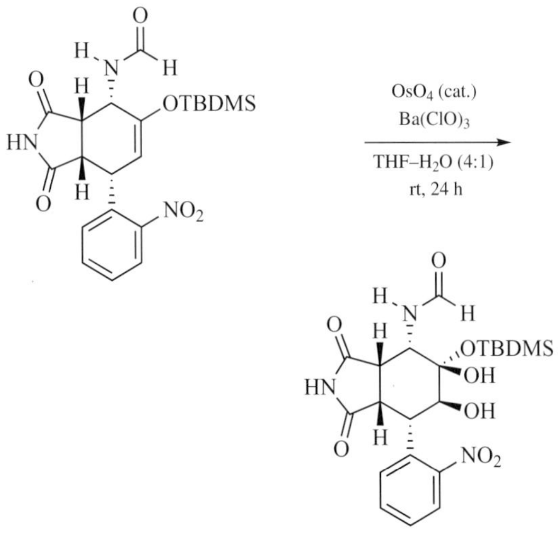

1-epiaustraline

Deprotection by Using OsO₄/NMO. The selective cleavage of propargyl ethers and tosyl groups can be achieved under OsO₄-catalyzed dihydroxylation conditions. For example, deprotection of the propargyl ethers was achieved by isomerization to the corresponding allenes with potassium *tert*-butoxide, followed by cleavage with catalytic amounts of OsO₄ and NMO to release the parent alcohols (eq 50).[64a,b]

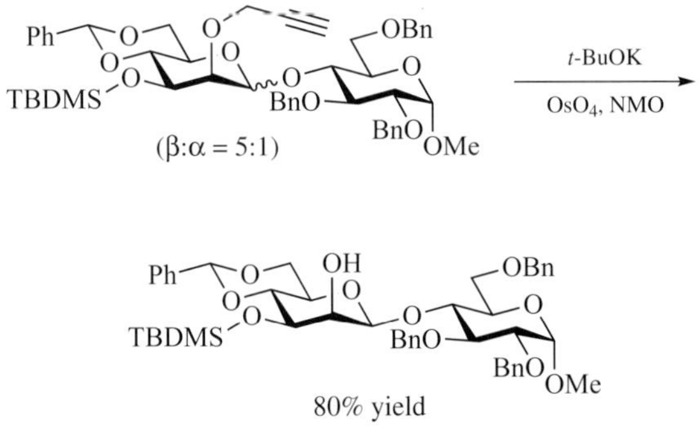

(50)

80% yield

Deprotection of tosyl groups is highly dependent on the double bond position during the course of dihydroxylation of olefins. For instance, the tosyl group was deprotected to produce alcohol if the numbers of methylene units are less than 2, while the tosyl group was tolerated if the number of methylene units is greater than 2 (eq 51).[65]

(48)

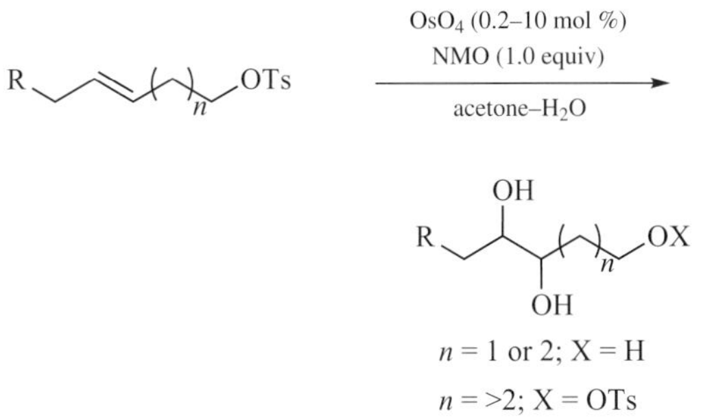

$$n = 1 \text{ or } 2; X = H$$
$$n = >2; X = OTs$$
(51)

Asymmetric Dihydroxylation: Synthetic Applications. Recently, the synthesis of D-xylose-protected alkanediol,[66a] (S)-oxybutynin,[66b] natural (−) and unnatural (+) glyceollin I,[66c] L-DOPA, and (R)-Selegiline[66d] was realized by using Sharpless asymmetric dihydroxylation as a key step.

Stereoselective Oxidative Cyclization of 1,5-dienes Catalyzed by OsO$_4$. There have been reports of osmium tetroxide promoting the nonasymmetric oxidative cyclization of 1,5-dienes to produce *cis*-tetrahydrofurans, using either periodate[67] or amineoxides[68] as reoxidants. Stereoselective oxidative cyclization of a 1,5-diene to furnish enantioenriched *cis*-tetrahydrofuran was also reported by Donohoe et al.[69] Os-catalyzed asymmetric dihydroxylation followed by oxidative cyclization of 1,5-diene was accomplished with Me$_3$NO (TMO) as an oxidant under acidic conditions using catalytic OsO$_4$. It was postulated that the 1,2-diol was tightly bound to Os prior to oxidation of the pendant alkene, leading to cyclization of the conformationally rigid vicinal diol with a high level of stereocontrol. This protocol was successfully applied to a formal synthesis of (+)-*cis*-solamin (eq 52).[69]

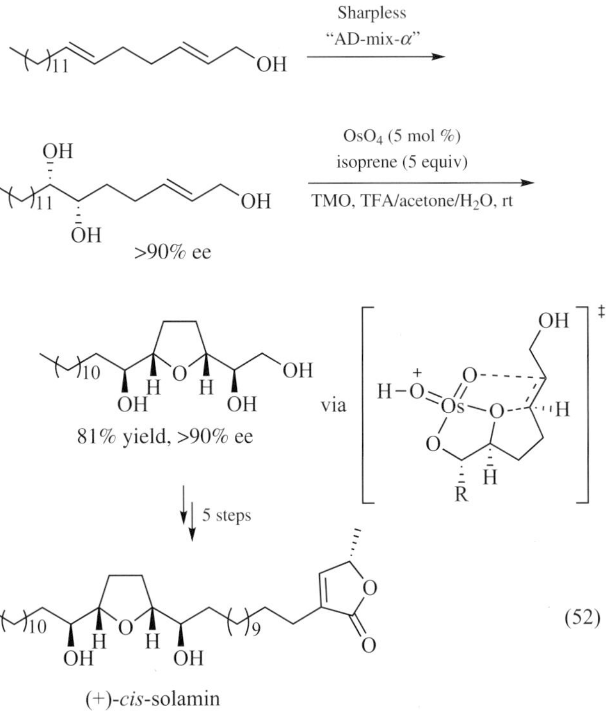

(52)

The same group further extended this methodology to the enantioselective total synthesis of the potent antitumor agent (+)-*cis*-

sylvaticin through double oxidative cyclization using OsO$_4$ to establish the bis-THF core. Synthesis of the bis-THF fragment began with commercially available tetradecatetraene. Dihydroxylation of the tetraene followed by in situ protection selectively yielded terminal diene. Mono-dihydroxylation of the terminal diene using the Sharpless AD-mix-β produced a diol. The aldehyde was then generated using sodium periodate, followed by Wittig reaction to produce a new diene that contained a long chain alkyl group. The resulting diene underwent double oxidative cyclization in the presence of OsO$_4$ to produce a single diastereomer of the key bis-THF fragment (eq 53).[70]

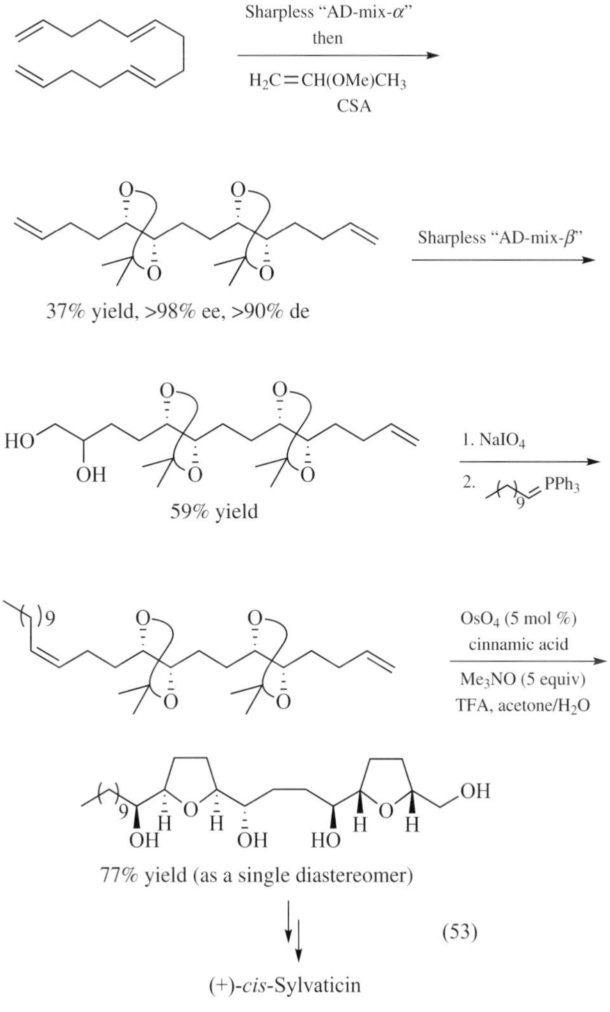

(53)

The same group also described a highly stereoselective synthesis of pyrrolidines utilizing a OsO$_4$-catalyzed oxidative cyclization strategy. Preferential mode of cyclization is most likely due to the robust intramolecular coordination of osmium with nitrogen and oxygen atoms of the N-tosylated aminoalcohol (eq 54).[71]

Aminohydroxylation of Alkenes. Sharpless' asymmetric aminohydroxylation (AA) allows for the catalytic and enantioselective synthesis of protected vicinal aminoalcohols in a single step.[72a] This reaction is significant as it applies to the synthesis of a wide variety of biologically active agents and natural products. For example, new monoterpene β-amino alcohols can effectively be synthesized from (+)-2-carene, (+)-3-carene, (−)-β-pinene, and

(−)-camphene using commercially available chloramine-T as the nitrogen source.[72b]

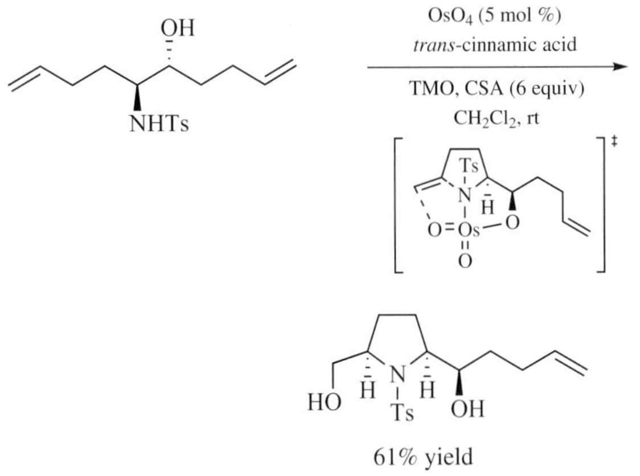

$$OsO_4 \ (5 \ mol \ \%)$$
trans-cinnamic acid

TMO, CSA (6 equiv)
CH_2Cl_2, rt

(54)

61% yield

Aminohydroxylation Reaction using a New Nitrogen Source.

Luxenburger and coworkers reported a base-free, intermolecular asymmetric aminohydroxylation (AA) reaction of olefins with alkyl 4-chlorobenzoyloxycarbamates as a nitrogen source that is readily prepared.[73] Generally, the reoxidant for an AA reaction is typically carbamate salt prepared in situ by treatment of NaOH with *t*-BuOCl. The addition of a base was mandatory for the promotion of the AA reaction. However, Luxenburger's reagent is found to be effective under base-free reaction conditions. The reaction proceeds under neutral conditions and various base-sensitive functional substituents were not affected (eq 55).

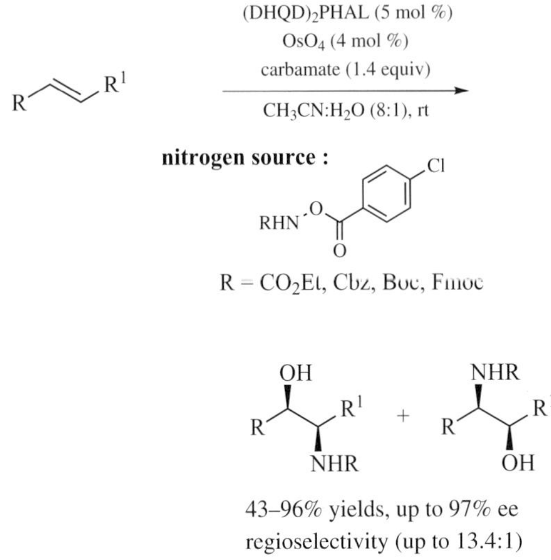

$$(DHQD)_2PHAL \ (5 \ mol \ \%)$$
$OsO_4 \ (4 \ mol \ \%)$
carbamate (1.4 equiv)

$CH_3CN:H_2O$ (8:1), rt

nitrogen source :

R = CO$_2$Et, Cbz, Boc, Fmoc

(55)

43–96% yields, up to 97% ee
regioselectivity (up to 13.4:1)

Applications of Imido-Osmium (VIII) Complex to Diamination and Aminohydroxylation.

Imidoosmium(VIII) compounds are versatile reagents that can be used for the transformation of olefins to vicinal diamines or vicinal aminoalcohols. Since the pioneering work of Sharpless in the development of the diamination of olefins,[74] Muñiz et al. reported the first example of asymmetric diamination of alkene bearing an (−)-8-phenylmenthol ester as a chiral auxiliary with bisimidoosmium complex at −15°C to give the corresponding osmaimidazolidine as a 94:6 dr (diastereomeric ratio). The stabilized osmaimidazolidine compound

can be purified by column chromatography, followed by reduction with LiAlH$_4$ to give an enantiomerically pure vicinal diamine (eq 56). They also investigated diamination of electron-deficient olefins including amides, ketones, and aldehydes as the functional groups with bis(imido) complex. As a result, the corresponding osmaimidazolidines were obtained as a single product in high chemical yields (84–92%).[73a−c]

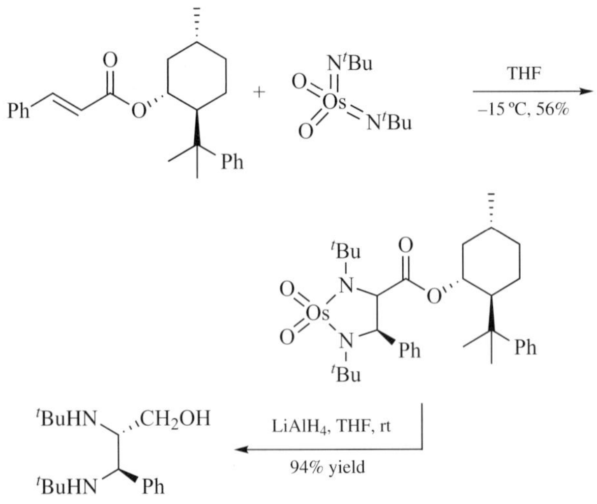

$$\xrightarrow{THF} {-15\,°C, \ 56\%}$$

(56)

$$\xleftarrow{LiAlH_4, \ THF, \ rt} {94\% \ yield}$$

Interestingly, the use of monoimidoosmium(VIII) complex in combination with dimethyl fumarate provided complicated products such as diols and aminoalcohols. On the other hand, bis- and tris(imido)osmium complexes displayed complete chemoselectivity and underwent diamination exclusively (eq 57).[76]

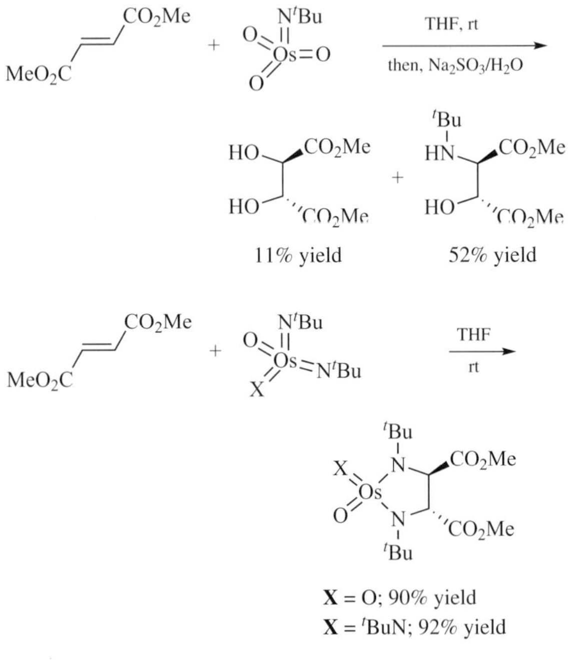

$$\xrightarrow{THF, \ rt} {then, \ Na_2SO_3/H_2O}$$

11% yield   52% yield

$$\xrightarrow{THF} {rt}$$

(57)

X = O; 90% yield
X = ^{t}BuN; 92% yield

In particular, this protocol showed a highly chemoselective diamination of vinyl acrylate. For example, the diaminated product was the sole product with neither diol nor aminoalcohol being detected. Diamination of the acrylate C=C double bond was exclusive. (eq 58).[76]

(58)

Recycling of Osmium by Immobilization of Osmium Tetroxide.

Several groups have been actively searching for immobilized forms of osmium tetroxide in order to overcome the problems of toxicity and volatility associated with this reagent. The following are some representative examples of immobilization methods for a catalytic system (OsO_4 and/or cinchona alkaloid ligand).

Immobilization of Osmium Tetroxide onto Macroporous Resins.

Song and coworkers reported simple immobilization of OsO_4 onto macroporous resins such as Amberlite XAD-4 (divinylbenzene and styrene copolymer) or XAD-7 (divinylbenzene and acrylate copolymer) bearing residual vinyl groups via osmylation. The osmylated resins are air-stable, nonvolatile, and much easier to handle than their homogeneous counterpart. Moreover, the resin-bound OsO_4 exhibited excellent catalytic activity in the asymmetric dihydroxylation (AD) of olefins. After reaction, the recovered osmylated resins could be reused for five consecutive reactions without any significant decrease in product yield. The osmylated resin in particular maintained its catalytic efficiency for almost three cycles even if the catalyst loading of Amberlite XAD-7 was reduced to 0.2 mol % (eq 59).[77]

(59)

Although this approach is highly efficient for recycling the osmium component, an additional acid-base work-up procedure is required to recover the cinchona alkaloid-derived chiral ligand. For this reason, the same group attempted to develop an immobilization method for simultaneous recovery and reuse of both catalytic components (osmium and cinchona alkaloid-derived chiral ligand).[78] Addition of OsO_4 to macroporous cinchona alkaloid resin bearing residual vinyl groups resulted in the formation of rigid Os-ligand complexes and partially osmylated product onto resins. The catalytic system exhibited excellent activity and enantioselectivity (up to 99% ee) in heterogeneous Os-catalyzed AD reactions (eq 60).

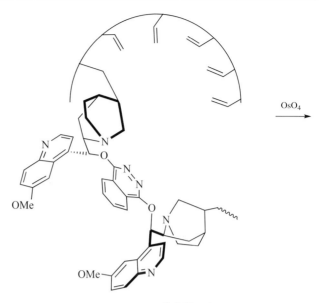

macroporous alkaloid resin
bearing residual vinyl groups
(white-colored resin)

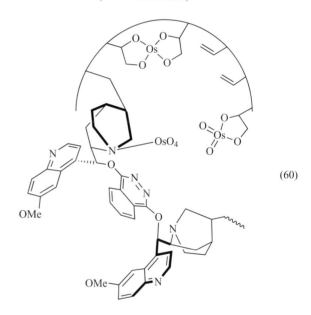

(60)

Os-complex of macroporous alkaloid resin
(black-colored resin)

Application of Osmylated Macroporous Resins for Asymmetric Aminohydroxylation.

Song and coworkers also examined the utility of osmylated macroporous resins for the asymmetric aminohydroxylation (AA) of olefins with AcNHBr/LiOH as the oxidant/nitrogen source. Most of the reactions proceeded with excellent enantioselectivity (>99% ee for *trans*-cinnamate derivatives) and extremely high regioselectivity (>20:1). After reaction, the osmylated resins can be quantitatively recovered and reused in three consecutive reactions without any significant loss of catalytic efficiency (eq 61).[79]

Macroencapsulation Method for Immobilization of Osmium Tetroxide.

Macroencapsulation is a technique for entrapping precious metals in a polymeric coating of microspheres. Kobayashi and coworkers introduced a phenoxyethoxymethyl-polystyrene microencapsulation (PEM-MC) method for the immobilization of OsO_4. To examine the efficiency of the PEM-MC OsO_4, the AD of olefin was performed using $(DHQD)_2PHAL$ as a chiral ligand and $K_3Fe(CN)_6$ as a cooxidant in acetone/H_2O. The microencapsulated OsO_4 catalyst can be recovered quantitatively by simple filtration and reused without loss of activity. However, microencapsulated OsO_4 exhibited somewhat inferior catalytic activity and enantioselectivity compared with its homogeneous analogue. Interestingly, when using PEM-MC OsO_4 and Triton® X-405 as a nonionic phase transfer catalyst, the reaction proceeds smoothly and produces the desired product with water as the sole solvent.[80]

$$>99\% \text{ ee}, >20:1 \text{ (regioselectivity)},$$
$$93\% \text{ yield (XAD-4·}OsO_4\text{)}, 90\% \text{ yield (XAD-16·}OsO_4\text{)}$$

Ley and Malla Reddy independently reported OsO_4 microencapsulated in polymer matrices such as polyurea or polysulfone, respectively. Polyurea-microencapsulated OsO_4 has proven to be extremely robust and long-term storable. High chemical yield and excellent enantioselectivity was also achieved using polysulfone-microencapsulated OsO_4 even without the need for slow addition of olefins.[81a,b]

Immobilization of Osmium Tetroxide into Ionic Liquids. Room temperature ionic liquids (RTILs) have emerged as powerful media for the immobilization of catalysts. Several research groups have pursued AD reaction by immobilizing OsO_4 into ionic liquids. In 2002, Yao initially studied the Os-catalyzed dihydroxylation under standard Upjohn conditions (OsO_4/NMO in t-BuOH/H_2O) in the presence of RTILs such as 1-butyl-3-methylimidazolium hexafluorophosphate ($[Bmim]PF_6$). Both the catalyst and the RTIL can be recycled and reused in the AD reaction, but with a decrease of catalytic activity due to the Os leaching. 4-dimethylaminopyridine (4-DMAP) as a ligand was added to the reaction mixture for the creation of amine-OsO_4 complex in order to prevent Os metal leaching. The resulting complex strongly binds to cationic imidazolium moieties of RTIL, which enhances its immobilizability in polar ionic liquids (eq 62).[82]

Yanada et al. also described the immobilization of OsO_4 in ionic liquid $[Emim][BF_4]$ applied to several substrates including mono-, di-, and trisubstituted aliphatic olefins as well as aromatic olefins.[83]

Song et al. employed 1,4-bis(9-O-quininyl)phthalazine $[(QN)_2PHAL]$ as a chiral ligand for an AD reaction using NMO in mixtures of $[Bmim][PF_6]$ and acetone–H_2O. $(QN)_2PHAL$ can also be converted into a new ligand bearing highly polar residues (four hydroxy groups) during AD reactions. This highly polar ligand can be strongly immobilized in the ionic liquid phases and thus minimize Os leaching during the extraction of product. The recovered ionic liquid phase containing Os and chiral ligand are capable to reuse it three times even at a low catalyst loading of Os (0.1 mol %). A distinctive feature of this methodology is that it requires neither structural modification of the ligand nor high catalyst loading of OsO_4 (eq 63).[84]

On the other hand, Afonso and coworkers described an AD reaction in a biphasic solvent system ($[Bmim][PF_6]$/H_2O using OsO_4 (0.5 mol %), $(DHQD)_2PHAL$ (1.0 mol %), and $K_3Fe(CN)_6$ (3 equiv) as a cooxidant at room temperature.[85]

Immobilization of Osmium Tetroxide onto Sugar. Table sugar (sucrose), a disaccharide of glucose and fructose, contains eight free hydroxyl functional groups that can be used as a new media for convenient immobilization methods. Song and coworkers demonstrated that the combination of sucrose and highly polarized alkaloid ligands generated in situ from $(QN)_2PHAL$ during the reaction provided a highly simple and efficient approach for recycling both catalytic components (OsO_4 and chiral cinchona alkaloid ligand). The chiral catalyst is strongly bounded into sugar moiety in aqueous phase and thus isolated chiral diol from organic layers by extraction with Et_2O. 0.1 mol % of Os-catalyst loading was sufficient to reach excellent enantioselectivity (>99% ee) (eq 64).[86]

Immobilization of Osmium Tetroxide onto Ionic Polymer Support. The incorporation of OsO_4 onto short-length PEGylated (PEG: Poly(ethylene glycol)) ionic polymers was achieved by Janda and coworkers. Immobilized OsO_4 exhibited excellent catalytic performance and recyclability in the AD reaction. The high recyclability could be attributed to the interaction between the induced dipole of OsO_4 and the ions of the polymeric scaffolds in a manner similar to ionic liquids. Moreover, Os-bounded

$$(64)$$

polymer is also able to effectively immobilize a significant amount of chiral ligand (eq 65).[87]

$$(65)$$

P: poly(4-vinylpyridine/styrene)
MsO-PEG3-OMe: tri(ethylene glycol) monomesylate monomethyl ether

Immobilization of Osmium Tetroxide onto PEG Support.
PEG [Poly(ethylene glycol), molecular weight: 400] can be used as an alternative recyclable medium for the AD reaction. Using (DHQD)$_2$ PHAL ligand, OsO$_4$, and NMO in PEG, it was possible to reuse the catalytic system for up to five cycles, producing the desired diols in high yields with excellent enantioselectivities without any loss of catalytic activity.[88]

Immobilization of Osmium Tetroxide onto Other Supports.
Good performance in terms of chemical yield (up to 90%), enantioselectivity (up to 86% ee), and recyclability of both components (OsO$_4$ and chiral ligand) was achieved by employing silica-supported chitosan-OsO$_4$[89a] and wool-OsO$_4$ complex in the AD reaction of olefins.[89b,c] Fullerene was also shown to serve as a support for efficient immobilization of OsO$_4$, which was successfully recovered and recycled several times.[90]

Recycling of Osmium from Homogeneous Systems. Conceptually new methodology for recycling of osmium species from homogeneous systems was developed by Donohoe et al. For example, the reaction of bis(imino)osmium(VI) complex, which is synthesized by mixing *trans*-stilbene with OsO$_4$, and *trans*-stilbene with *tert*-butyl hydroperoxide (^{t}BuOOH) in CH$_2$Cl$_2$ produced a corresponding osmium(VII) intermediate. This intermediate was isolated by column chromatography, which was subjected to hydrolysis in acetone:H$_2$O (4:1) at 60 °C to produce a stilbene diol (87% yield) and liberate bis(imino)osmium(VI) complex for reuse (eq 66).[91]

$$(66)$$

Oxidative Cleavage of Olefins. Oxidative cleavage of olefin is used widely in organic synthesis. Most oxidative cleavage of olefin reactions can be classified into two categories: direct cleavage via ozonolysis; and dihydroxylation of olefin followed by diol cleavage. Recently, Borhan and coworkers demonstrated that the combination of OsO_4/Oxone is an effective system for the oxidative cleavage of alkenes, where Oxone plays three distinct roles in this system: (i) reoxidation of Os(VI) to form Os(VIII) species; (ii) promotion of oxidative cleavage to intermediate aldehyde beginning with an attack of a peroxomonosulfate ion derived from Oxone into Os(VIII) species; and (iii) oxidation of aldehyde into carboxylic acid. The reaction can be considered as an alternative to ozonolysis because the reaction proceeds without the intermediacy of 1,2-diols (eq 67).[92]

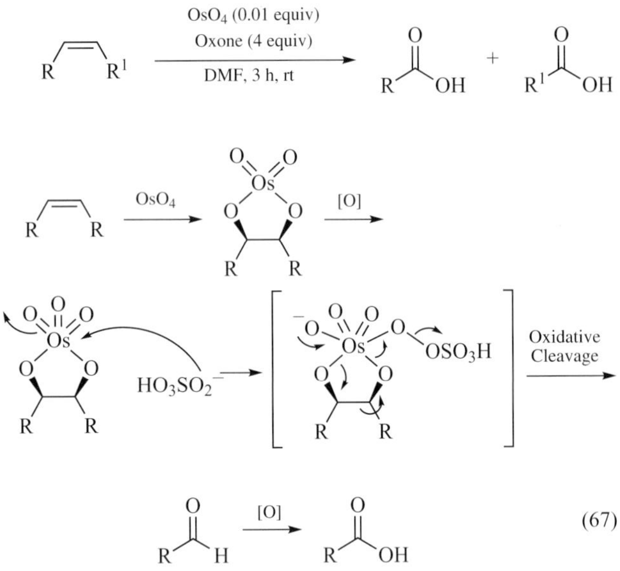

$$(67)$$

OsO_4-mediated oxidative cleavage/oxidative lactonization of alkenols with Oxone as the cooxidant in alcoholic solvents produced the corresponding lactones via the formation of aldehydes as a key intermediate (eq 68).[93]

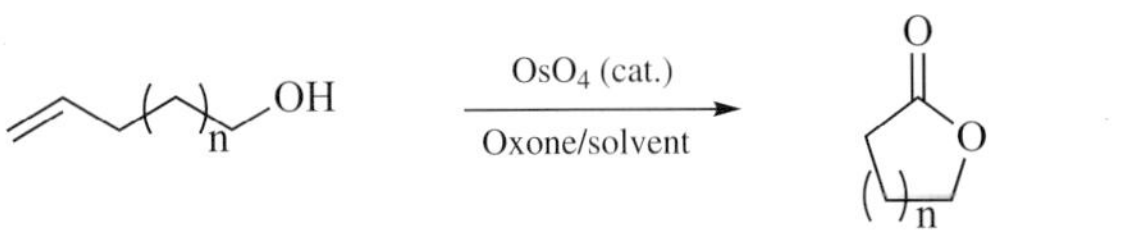

$$(68)$$

Borhan and coworkers described the stereoselective total synthesis of (+)-tanikolide, a natural product isolated from the marine cyanobacteria *Lyngbya majuscula*, using O_sO_4-mediated tandem oxidative cleavage–lactonization of the alkenol in the presence of soluble Oxone (n-Bu$_4$NHSO$_5$) to produce the lactone (eq 69).[94]

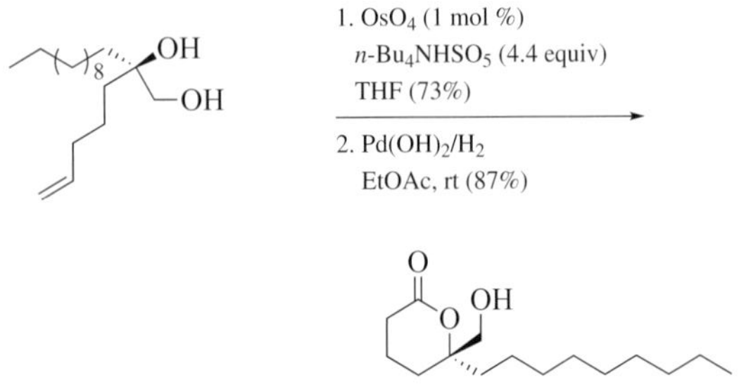

$$(69)$$

(+)-tanikolide

Similarly, the importance of the tandem oxidative cleavage–lactonization reaction using OsO_4–Oxone was demonstrated by

Boland and coworkers during a short and efficient total synthesis of several alkanolides, alkenolides, and dihydroisocoumarins. For instance, micromolide was synthesized that is potent in vitro activity against *Mycobacterium tuberculosis* (H37Rv) (eq 70).[95]

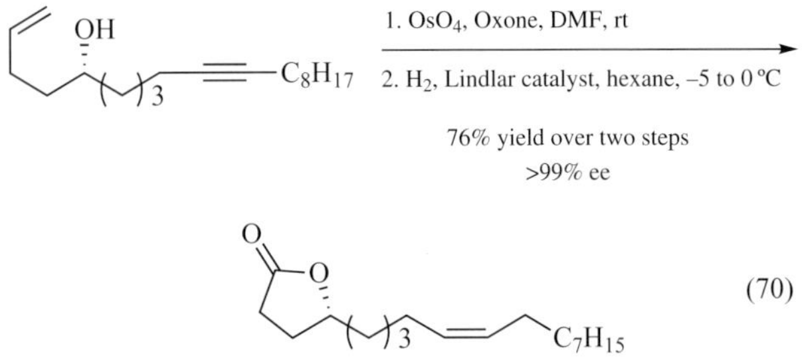

$$(70)$$

Micromolide

Due to the high volatility and toxicity of OsO_4, the same group applied OsEnCat^{TM}40 (microencapsulation of OsO_4 in polyurea) to oxidative cleavage of olefins. The OsEnCat^{TM}40 catalyst was recovered and reused three times but with a sacrifice of catalytic activity (eq 71).[96]

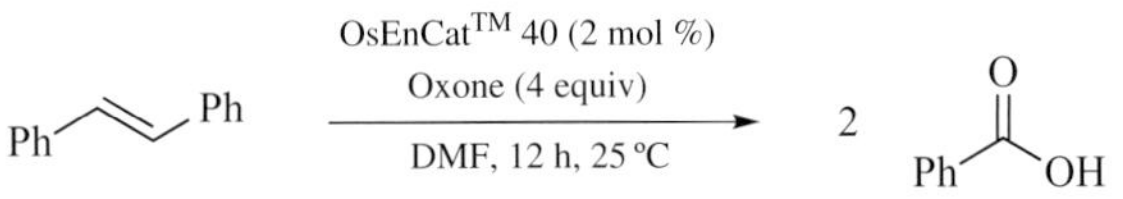

$$(71)$$

A catalytic amount of OsO_4 in combination with H_2O_2 in DMF has also been shown to efficiently cleave alkenes, producing aldehydes rather than carboxylic acids. Aryl olefins are cleaved in good yield, whereas alkyl olefins cleaved moderate yield for di- and tri-substituted alkenes (eq 72).[97]

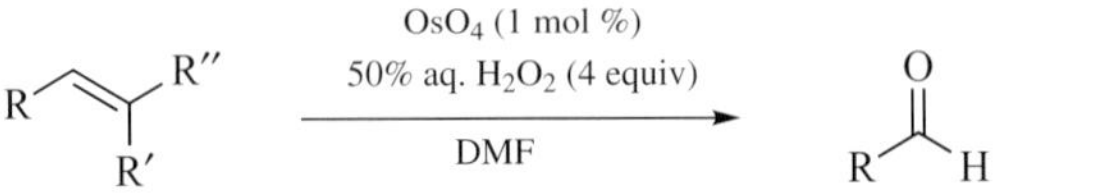

$$(72)$$

R = Ph, alkyl
R', R'' = alkyl or H

Dias et al. reported that the oxidation of *N*-Boc-protected homoallylic alcohols with OsO_4 as a catalyst in the presence of NaIO$_4$ efficiently occurred in Et$_2$O:H$_2$O (1:1) to provide 4-*N*-Boc-amino-3-hydroxy ketones in high yields (eq 73).[98]

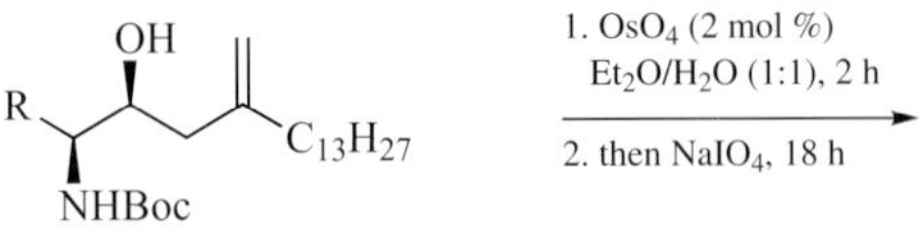

R = Me, *i*-Pr, Bn, *i*-Bu, CH$_2$OBn

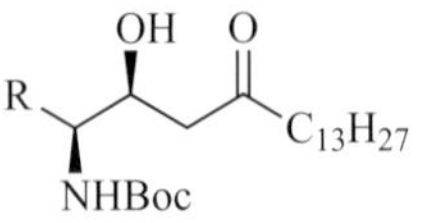

$$(73)$$

80–94% yield

Conventional oxidative cleavage of olefin can be achieved by a two-step sequencing reaction of OsO_4-mediated dihydroxylation of olefin, followed by periodate cleavage of diol. The resulting aldehyde can be reduced to the desired alcohol with sodium borohydride in a quantitative yield (eq 74).[99]

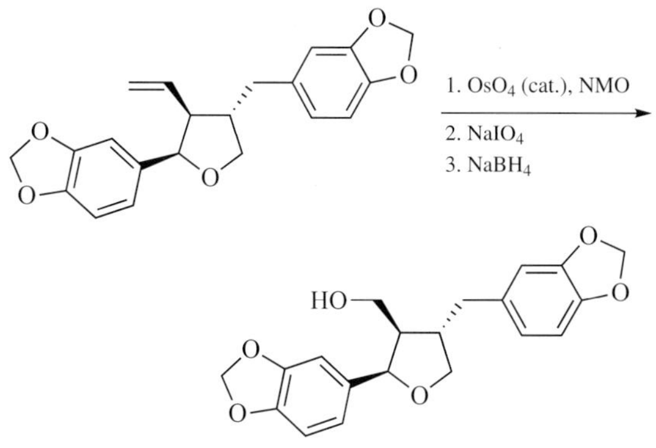

1. OsO_4 (cat.), NMO
2. $NaIO_4$
3. $NaBH_4$
HO

furanofuran lignan

(74)

The main drawback of the oxidative cleavage of olefins by OsO_4-$NaIO_4$ is the formation of side product such as α-hydroxy ketone derivatives presumably formed via the overoxidation of the diol intermediate. Jin and coworkers found that the use of 2,6-lutidine in dioxane-H_2O enables higher chemical yield of the corresponding aldehydes by suppressing unwanted products (eq 75).[100]

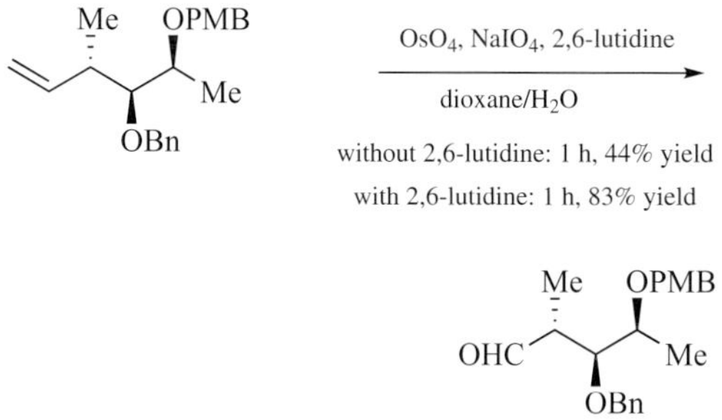

Me
OPMB
Me
OBn
OsO_4, $NaIO_4$, 2,6-lutidine
dioxane/H_2O
without 2,6-lutidine: 1 h, 44% yield
with 2,6-lutidine: 1 h, 83% yield
OHC

(75)

An efficient method for the oxidative cleavage of solid-supported peptide olefins into aldehydes using a combination of OsO_4/$NaIO_4$/DABCO system was described by Meldal and coworkers. The resulting aldehydes are subjected to intermolecular *N*-acylium Pictet–Spengler reactions to furnish pyrroloisoquinoline derivatives in high purity (eq 76).[101]

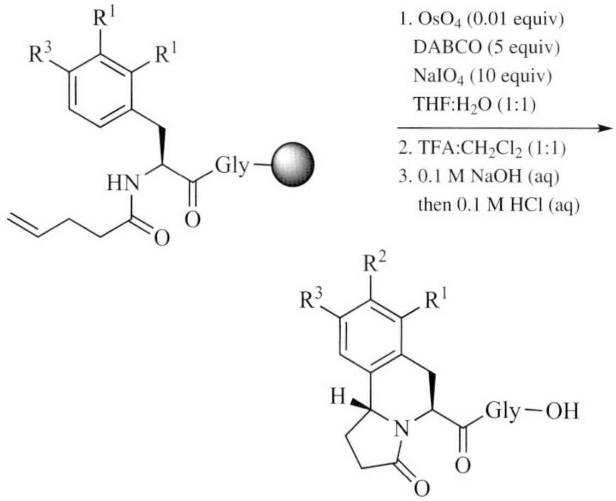

1. OsO_4 (0.01 equiv)
DABCO (5 equiv)
$NaIO_4$ (10 equiv)
THF:H_2O (1:1)
2. TFA:CH_2Cl_2 (1:1)
3. 0.1 M NaOH (aq)
then 0.1 M HCl (aq)
Gly
Gly—OH

(76)

The 1-azabicyclo[3.1.0]hexane moiety is present relatively often in biologically active compounds. A new synthetic route of 1-azabicyclo[3.1.0]hexane is reported via dihydroxylation/oxidative cleavage/reduction (eq 77).[102]

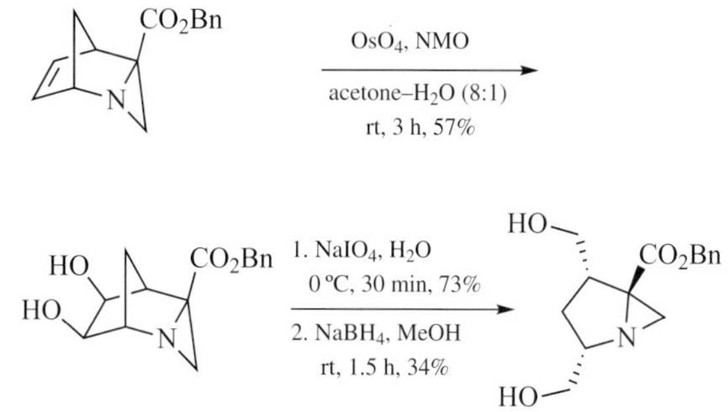

CO_2Bn
OsO_4, NMO
acetone–H_2O (8:1)
rt, 3 h, 57%
1. $NaIO_4$, H_2O
0 °C, 30 min, 73%
2. $NaBH_4$, MeOH
rt, 1.5 h, 34%

(77)

L-Hamamelose, which is used as a starting material for the synthesis of potential glycosidase inhibitors, was also synthesized via dihydroxylation/oxidative cleavage/reduction (eq 78).[103]

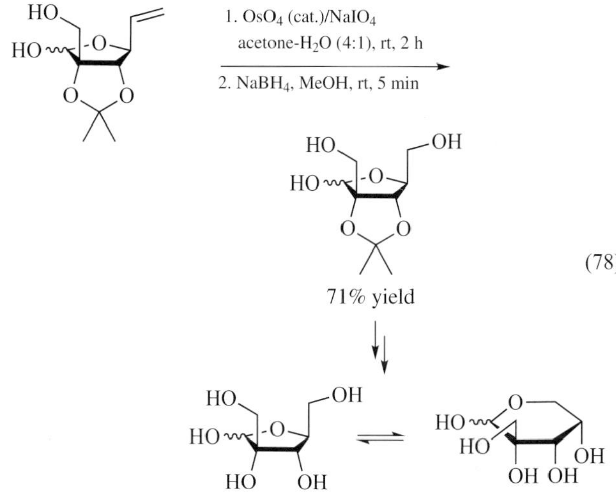

1. OsO_4 (cat.)/$NaIO_4$
acetone-H_2O (4:1), rt, 2 h
2. $NaBH_4$, MeOH, rt, 5 min
71% yield

L-Hamamelose

(78)

C–H and Si–H Bond Oxidation. Mayer and coworkers reported selective C–H bond oxidation of unactivated alkanes, including primary, secondary, and tertiary C–H bonds, using stoichiometric amounts of OsO_4 in aqueous pH = 12.1 solution at 85 °C. For example, isobutane can be oxidized to *tert*-butanol (eq 79). The catalytic version of this reaction using $NaIO_4$ as a terminal oxidant exhibited unsatisfactory results. The turnover number of OsO_4 was only ca. 4.[104]

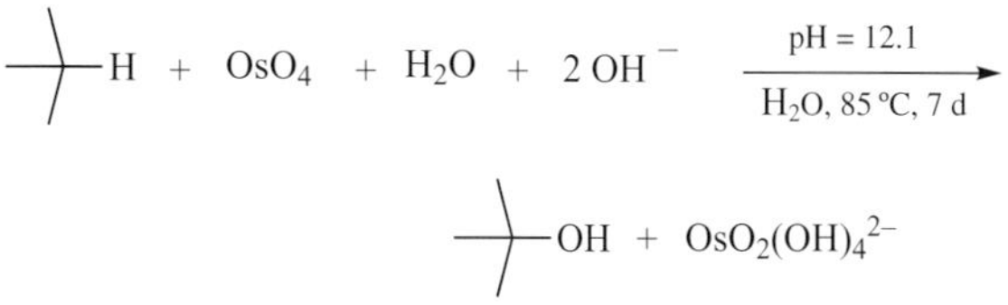

H + OsO_4 + H_2O + 2 OH^-
pH = 12.1
H_2O, 85 °C, 7 d
OH + $OsO_2(OH)_4^{2-}$

(79)

The author proposed that the reaction presumably proceeded via [3+2] mechanism, which is quite similar to [3+2] mechanisms for olefin dihydroxylation. Insertion of OH^- into OsO_4 at pH = 12.1 led to the formation of $OsO_4(OH)^-$ species. Addition of C–H bonds into two oxo-groups of $OsO_4(OH)^-$ provided an intermediate, which can be hydrolyzed by aqueous base, and liberated a corresponding alcohol and reduced osmate [$OsO_2(OH)_4^{2-}$] (eq 80).

$$\left[\begin{matrix} R \diagdown \underset{H}{\overset{}{O}} \\ O \diagup \overset{\|}{\underset{\|}{Os}}=O \\ O \diagdown \underset{OH}{} \end{matrix}\right]^{-} \longrightarrow \left[\begin{matrix} R' \cdots \overset{H}{\overset{}{O}} \\ O \diagup \overset{\|}{Os}=O \\ O \diagdown \underset{OH}{} \end{matrix}\right]^{\ddagger} \longrightarrow \left[\begin{matrix} RO \diagdown \overset{OH^-}{\overset{}{Os}}=O \\ O \diagup \overset{\|}{Os}=O \\ O \diagdown \underset{OH}{} \end{matrix}\right] \begin{matrix} OH^- \\ H_2O \end{matrix}$$

$$ROH + \left[\begin{matrix} HO \diagdown \overset{O}{\overset{\|}{Os}} \diagup OH \\ HO \diagup \overset{\|}{\underset{O}{Os}} \diagdown OH \end{matrix}\right]^{2-} \quad (80)$$

Methane could also be oxidized to methanol at $50\,°C$ by utilizing OsO_4 and $NaIO_4$ in D_2O. No methane oxidation product was detected using either $NaIO_4$ without OsO_4 or OsO_4 without $NaIO_4$. It was observed that the presence of methane noticeably inhibited the further oxidation of methanol, which is a competitor for methane oxidation. The overoxidation of methanol was ca.1000 times slower in methane than in argon (eq 81).[105]

$$CH_4 \xrightarrow[\text{D_2O, 50 °C}]{\text{$OsO_4/NaIO_4$}} CH_3OH + CH_2(OH)_2 + CO_2 \quad (81)$$

Exposure of silanes (Et_3SiH, i-Pr_3SiH, Ph_3SiH, or $PhMe_2SiH$) to OsO_4 in the presence of an excess of pyridine ligand furnished the corresponding silanols. In the cases of Et_3SiH and $PhMe_2SiH$, OsO_4(^{t}Bupy) complex derived from OsO_4 and 4-$tert$-butylpyridine (^{t}Bupy) gave superior yields (ca. 100%) in comparison to OsO_4(py) complex (40–60% yield). The relatively lower yield of the OsO_4(py) system could be attributed to the coprecipitation of osmium(VIII) with $[Os(O)_2py_2]_2(\mu\text{-}O)_2$ (eq 82).[106]

$$\underset{L}{\overset{O}{\underset{\|}{O=Os=O}}} + L + \underset{R'}{\overset{H}{\underset{\|}{R\diagup Si\diagdown R''}}} \longrightarrow$$

$$\underset{R'}{\overset{OH}{\underset{|}{R\diagup Si\diagdown R''}}} + 1/2 \left[\begin{matrix} L \overset{O}{\underset{\|}{Os}} O \overset{O}{\underset{\|}{Os}} L \\ L \overset{\|}{\underset{O}{}} \overset{\|}{\underset{O}{}} L \end{matrix}\right] \quad (82)$$

$$L \text{ (ligand)} = \underset{(^t\text{Bupy})}{N\diagup\diagdown} , \underset{(\text{py})}{N\diagup\diagdown}$$

$tert$-Amine and Sulfur Oxidation with OsO_4. Oxidation of $tert$-amines to lactams, hydroxylactams, and ketolactams[107a] and selective oxidation of optically pure 2-acyl-2-alkyl-1,3-dithiolane-1-oxides to 2-acyl-2-alkyl-1,3-dithiolane-1,1-dioxides[107b] were also performed by using conventional Sharpless dihydroxylation protocol.

Osmylation on Carbon Nanotubes. The covalent linkage of OsO_4 to the sidewalls of single-wall carbon nanotubes (SWCNTs) was achieved via base-catalyzed [2+3] cycloaddition.[108] Osmylation on the SWCNT sidewall can also be achieved with OsO_4 from the gas phase under UV photoirradiation. The resulting osmate ester adduct led to an increase in the electrical resistance with decreasing the π-electron density in the nanotubes. However, the cycloaddition product can be cleaved by UV light in a vacuum or under oxygen atmosphere due to the reversibility of cycloaddition reactions, thereby restoring the original electronic properties.[109]

1. (a) Schröder, M., *Chem. Rev.* **1980**, *80*, 187. (b) Singh, H. S., In *Organic Synthesis by Oxidation with Metal Compounds*; Mijs, W. J.; De Jonge, C. R. H. I., Eds.; Plenum: New York, **1986**; Chapter 12. (c) Johnson, R. A.; Sharpless, K. B., In *Catalytic Asymmetric Synthesis*; Ojima, I., Ed.; VCH: New York, **1993**. (d) Lohray, B. B., *Tetrahedron: Asymmetry* **1992**, *3*, 1317. (e) Haines, A. H., *Comprehensive Organic Synthesis* **1991**, *7*, 437.

2. Henbest, H. B.; Jackson, W. R.; Robb, B. C. G., *J. Chem. Soc. (B)* **1966**, 803.

3. Herrmann, W. A.; Eder, S.; Scherer, W., *Angew. Chem., Int. Ed. Engl.* **1992**, *31*, 1345.

4. Rüegger, U. P.; Tassera, J., *Swiss Chem.* **1986**, *8*, 43.

5. Austin, R. G.; Michaelson, R. C.; Myers, R. S., In *Catalysis of Organic Reactions*; Augustine, R. L., Ed.; Dekker: New York, **1985**; p 269.

6. (a) Cha, J. K.; Christ, W. J.; Kishi, Y., *Tetrahedron* **1984**, *40*, 2247. (b) Christ, W. J.; Cha, J. K.; Kishi, Y., *Tetrahedron Lett.* **1983**, *24*, 3943 and 3947.

7. Brimacombe, J. S.; Hanna, R.; Kabir, A. K. M. S.; Bennett, F.; Taylor, I. D., *J. Chem. Soc., Perkin Trans. 1* **1986**, 815.

8. DeNinno, M. P.; Danishefsky, S. J.; Schulte, G., *J. Am. Chem. Soc.* **1988**, *110*, 3925.

9. Evans, D. A.; Kaldor, S. W., *J. Org. Chem.* **1990**, *55*, 1698.

10. Stork, G.; Kahn, M., *Tetrahedron Lett.* **1983**, *24*, 3951.

11. Saito, S.; Morikawa, Y.; Moriwake, T., *J. Org. Chem.* **1990**, *55*, 5424.

12. Panek, J. S.; Cirillo, P. F., *J. Am. Chem. Soc.* **1990**, *112*, 4873.

13. Fleming, I.; Sarker, A. K.; Thomas, A. P., *Chem. Commun.* **1987**, 157.

14. (a) Vedejs, E.; McClure, C. K., *J. Am. Chem. Soc.* **1986**, *108*, 1094. (b) Vedejs, E.; Dent, W. H., III, *J. Am. Chem. Soc.* **1989**, *111*, 6861.

15. (a) Ikemoto, N.; Schreiber, S. L., *J. Am. Chem. Soc.* **1990**, *112*, 9657. (b) Hanselmann, R.; Benn, M., *Tetrahedron Lett.* **1993**, *34*, 3511.

16. (a) Hauser, F. M.; Ellenberger, S. R.; Clardy, J. C.; Bass, L. S., *J. Am. Chem. Soc.* **1984**, *106*, 2458. (b) Solladié, G.; Fréchou, C.; Demailly, G., *Tetrahedron Lett.* **1986**, *27*, 2867. (c) Solladié, G.; Fréchou, C.; Hutt, J.; Demailly, G., *Bull. Soc. Chem. Fr.* **1987**, 827.

17. (a) Johnson, C. R.; Barbachyn, M. R., *J. Am. Chem. Soc.* **1984**, *106*, 2459. (b) Johnson, C. R., *Pure Appl. Chem.* **1987**, *59*, 969.

18. (a) Trost, B. M.; Kuo, G.-H.; Benneche, T., *J. Am. Chem. Soc.* **1988**, *110*, 621. (b) Poli, G., *Tetrahedron Lett.* **1989**, *30*, 7385.

19. Hanessian, S.; Meffre, P.; Girard, M.; Beaudoin, S.; Sancéau, J.-Y.; Bennani, Y., *J. Org. Chem.* **1993**, *58*, 1991 and references therein.

20. Anderson, P. G.; Sharpless, K. B., *J. Am. Chem. Soc.* **1993**, *115*, 7047.

21. (a) Oishi, T.; Hirama, M., *Tetrahedron Lett.* **1992**, *33*, 639. (b) Imada, Y.; Saito, T.; Kawakami, T.; Murahashi, S.-I., *Tetrahedron Lett.* **1992**, *33*, 5081.

22. Annuziata, R.; Cinquini, M.; Cozzi, F.; Raimondi, L.; Stefanelli, S., *Tetrahedron Lett.* **1987**, *28*, 3139.

23. Brimacombe, J. S.; McDonald, G.; Rahman, M. A., *Carbohydr. Res.* **1990**, *205*, 422.

24. Oishi, T.; Iida, K.-I.; Hirama, M., *Tetrahedron Lett.* **1993**, *34*, 3573.

25. (a) Ward, R. A.; Procter, G., *Tetrahedron Lett.* **1992**, *33*, 3363. (b) Lohray, B. B.; Bhushan, V., *Tetrahedron Lett.* **1993**, *34*, 3911.

26. Hawkins, J. M.; Meyer, A., *Science* **1993**, *260*, 1918.

27. Herranz, E.; Sharpless, K. B., *Org. Synth., Coll. Vol.* **1990**, *7*, 375.

28. Herranz, E.; Sharpless, K. B., *Org. Synth., Coll. Vol.* **1990**, *7*, 223.

29. Kolb, H. C.; VanNieuwenhze, M. S.; Sharpless, K. B., *Chem. Rev.* **1994**, *94*, 2483.

30. (a) Corey, E. J.; Noe, M. C.; Grogan, M. J., *Tetrahedron Lett.* **1996**, *37*, 4899. (b) Delmonte, A. J.; Haller, J.; Houk, K. N.; Sharpless, K. B.; Singleton, D. A.; Strassner, T.; Thomas, A. A., *J. Am. Chem. Soc.* **1997**, *119*, 9907.

31. Xu, D.; Park, C. K.; Sharpless, K. B., *Tetrahedron Lett.* **1994**, *35*, 2495.

32. Walsh, P. J.; Ho, P. T.; King, S. B.; Sharpless, K. B., *Tetrahedron Lett.* **1994**, *35*, 5129.

33. Dupau, P.; Epple, R.; Thomas, A. A.; Fokin, V. V.; Sharpless, K. B., *Adv. Synth., Catal.* **2002**, *344*, 421.

34. Anderson, M. A.; Epple, R.; Fokin, V. V.; Sharpless, K. B., *Angew. Chem., Int. Ed. Engl.* **2002**, *41*, 472.

35. Donohoe, T. J.; Blades, K; Helliwell, M.; Waring, M. J.; Newcombe, N. J., *Tetrahedron Lett.* **1998**, *39*, 8755.

36. Donohoe, T. J.; Blades, K.; Moore, P. R.; Waring, M. J.; Winter, J. J. G.; Helliwell, M.; Newcombe, N. J.; Stemp, G., *J. Org. Chem.* **2002**, *67*, 7946.

37. Blades, K.; Donohoe, T. J.; Winter, J. J. G.; Stemp, G., *Tetrahedron Lett.* **2000**, *41*, 4701.

38. Gypser, A.; Michel, D.; Nirschi, D. S.; Sharpless, K. B., *J. Org. Chem.* **1998**, *63*, 7322.

39. de Champdoré, M.; Lasalvia, M.; Piccialli, V., *Tetrahedron Lett.* **1998**, *39*, 9781.

40. Donohoe, T. J.; Butterworth, S., *Angew. Chem., Int. Ed. Engl.* **2003**, *42*, 948.

41. Jonsson, S. Y.; Adolfsson, H.; Bäckvall, J., *Org. Lett.* **2001**, *3*, 3463.

42. Krief, A.; Castillo-Colaux, C., *Synlett* **2001**, 501.

43. Kolb H. C.; Sharpless, K. B., In *Transition Metals For Organic Synthesis*; Beller, M.; Bolm, C., Eds; Wiley: New York, **1998**, Vol. 2, p 243.

44. Phukan, P.; Sudalai, A., *Tetrahedron: Asymmetry* **1998**, *9*, 1001.

45. Pinheiro, S.; Pedraza, S. F.; Farias, F. M. C.; Gonçalves, A. S.; Costa, P. R. R., *Tetrahedron: Asymmetry* **2000**, *11*, 3845.

46. Travis, B. R.; Narayan, R. S.; Borhan, B., *J. Am. Chem. Soc.* **2002**, *124*, 3824.

47. Motherwell, W. B.; Williams, A. S., *Angew. Chem., Int. Ed. Engl.* **1995**, *34*, 2031.

48. Ley, S. V.; Ramarao, C.; Lee, A.-L.; Østergaard, N.; Smith, S. C.; Shirley, I. M., *Org. Lett.* **2003**, 5, 185.

49. Kobayashi, S.; Endo, M.; Nagayama, S., *Angew. Chem., Int. Ed.* **1999**, *121*, 11229.

50. (a) Severeyns, A.; De Vos, D. E.; Fiermans, L.; Verpoot, F.; Grobet, P. J.; Jacobs, P. A., *Angew. Chem., Int. Ed.* **2001**, *40*, 586. (b) Yang, J. W.; Han, H.; Roh, E. J.; Lee, S.; Song, C. E., *Org. Lett.* **2002**, 4, 4685.

51. Choudary, B. M.; Chowdari, N. S.; Jyothi, K.; Kumar, S.; Kantam, M. L., *Chem. Commun.* **2001**, 586.

52. Gao, S.; Herzig, D.; Wang, B., *Synthesis* **2001**, *4*, 544.

53. (a) Coleman, K. S.; Coppe, M.; Thomas, C.; Osborn, J. A., *Tetrahedron Lett.* **1999**, *40*, 3723. (b) Muldoon, J.; Brown, S. N., *Org. Lett.* **2002**, 4, 1043.

54. (a) Miles, J. A. L.; Mitchell, L.; Percy, J. M.; Singh, K.; Uneyama, E., *J. Org. Chem.* **2007**, *72*, 1575. (b) Ono, M.; Tanikawa, S.; Suzuki, K.; Akita, H., *Tetrahedron* **2004**, *60*, 10187. (c) Dias, L. C.; de Castro, I. B. D.; Steil, L. J.; Augusto, T., *Tetrahedron Lett.* **2006**, *47*, 213. (d) Lu, M.; Fedoryak, O. D.; Moister, B. R.; Dore, T. M., *Org. Lett.* **2003**, *5*, 2119. (e) Suresh, V.; Selvam, J. J. P.; Rajesh, K.; Shekhar, V.; Babu, D. C.; Venkateswarlu, Y., *Synthesis* **2010**, 1763. (f) Bongui, J.-B.; Elomri, A.; Cahard, D.; Tillequin, F.; Pfeiffer, B.; Pierré, A.; Seguin, E., *Chem. Pharm. Bull.* **2005**, *53*, 1540. (g) Lang, M.; Ziegler, T., *Eur. J. Org. Chem.* **2007**, 768.

55. Balagam, B.; Mitra, R.; Richardson, D. E., *Tetrahedron Lett.* **2008**, *49*, 1071.

56. Donohoe, T. J.; Fisher, J. W.; Edwards, P. J., *Org. Lett.* **2004**, *6*, 465.

57. Donohoe, T. J.; Mitchell, L.; Waring, M. J.; Helliwell, M.; Bell, A.; Newcombe, N. J., *Org. Biomol. Chem.* **2003**, *1*, 2173.

58. Kim, Y.; Fuchs, P. L., *Org. Lett.* **2007**, *9*, 2445.

59. Wybrow, R. A. J.; Edwards, A. S.; Stevenson, N. G.; Adams, H.; Johnstone, C.; Harrity, J. P. A., *Tetrahedron* **2004**, *60*, 8869.

60. (a) Döbler, C.; Mehltretter, G. M.; Sundermeier, U.; Beller, M., *J. Am. Chem. Soc.* **2000**, *122*, 10289. (b) Döbler, C.; Mehltretter, G. M.; Sundermeier, U.; Beller, M., *J. Organomet. Chem.* **2001**, *621*, 70.

(c) Junttila, M. H.; Hormi, O. E., *J. Org. Chem.* **2004**, *69*, 4816. (d) Mehltretter, G. M.; Bhor, S.; Klawonn, M.; Döbler, C.; Sundermeier, U.; Eckert, M.; Militzer, H.-C.; Beller, M., *Synthesis* **2003**, 295. (e) Krief, A.; Castillo-Colaux, C., *Synlett* **2001**, 501. (f) Alonso, D.; Caballero, E.; Medarde, M.; Tomé, F., *Tetrahedron Lett.* **2007**, *48*, 907. (g) Bacherikov, V. A.; Tsai, T.-J.; Chang, J.-Y.; Chou, T.-C.; Lee, R.-Z.; Su, T.-L., *Eur. J. Org. Chem.* **2006**, 4490. (h) Donohoe, T. J.; Sintim, H. O., *Org. Lett.* **2004**, *6*, 2003. (i) Fernández de la Pradilla, R.; Lwoff, N.; Viso, A., *Eur. J. Org. Chem.* **2009**, 2312.

61. Donohoe, T. J.; Sintim, H. O., *Org. Lett.* **2004**, *6*, 2003.

62. Alonso, D.; Caballero, E.; Medarde, M.; Tomé, F., *Tetrahedron Lett.* **2007**, *48*, 907.

63. Alp, C.; Atmaca, U.; Çelik, M.; Gültekin, M. S., *Synlett* **2009**, 2765.

64. (a) Crich, D.; Jayalath, P., *Org. Lett.* **2005**, *7*, 2277. (b) Crich, D.; Jayalath, P.; Hutton, T. K., *J. Org. Chem.* **2006**, *71*, 3064.

65. Batt, F.; Piva, O.; Fache, F., *Tetrahedron Lett.* **2008**, *49*, 566.

66. (a) Len, C.; Sélouane, A.; Weiling, A.; Coicou, F.; Postel, D., *Tetrahedron Lett.* **2003**, *44*, 663. (b) Gupta, P.; Fernandes, R. A.; Kumar, P., *Tetrahedron Lett.* **2003**, *44*, 4231. (c) Khupse, R. S.; Erhardt, P. W., *Org. Lett.* **2008**, *10*, 5007. (d) Sayyed, I. A.; Sudalai, A., *Tetrahedron: Asymmetry* **2004**, *15*, 3111

67. de Champdoré, M.; Lasalvia, M.; Piccialli, V., *Tetrahedron Lett.* **1998**, *39*, 9781.

68. Donohoe, T. J.; Butterworth, S., *Angew. Chem., Int. Ed.* **2003**, *42*, 948.

69. Donohoe, T. J.; Butterworth, S., *Angew. Chem., Int. Ed.* **2005**, *44*, 4766.

70. Donohoe, T. J.; Harris, R. M.; Williams, O.; Hargaden, G. C.; Burrows, J.; Parker, J., *J. Am. Chem. Soc.* **2009**, *131*, 12854.

71. Donohoe, T. J.; Churchill, G. H.; Wheelhouse, K. M. P.; Glossop, P. A., *Angew. Chem., Int. Ed.* **2006**, *45*, 8025.

72. (a) Li, G.; Chang, H.-T.; Sharpless, K. B., *Angew. Chem., Int. Ed.* **1996**, *35*, 451. (b) Łączkowski, K. Z.; Kmieciak, A.; Kozakiewicz, A., *Tetrahedron: Asymmetry* **2009**, *20*, 1487.

73. Harris, L.; Mee, S. P. H.; Furneaux, R. H.; Gainsford, G. J.; Luxenburger, A., *J. Org. Chem.* **2011**, *76*, 358.

74. Chong, A. O.; Oshima, K.; Sharpless, K. B., *J. Am. Chem. Soc.* **1977**, *99*, 3420.

75. (a) Muñiz, K.; Nieger, M., *Synlett* **2003**, 211. (b) Muñiz, K.; Iesato, A.; Nieger, M., *Chem. Eur. J.* **2003**, *9*, 5581. (c) Muñiz, K., *Chem. Soc. Rev.* **2004**, *33*, 166.

76. Muñiz, K., *Eur. J. Org. Chem.* **2004**, 2243.

77. Yang, J. W.; Han, H.; Roh, E. J.; Lee, S.-g.; Song, C. E., *Org. Lett.* **2002**, *4*, 4685.

78. Park, Y. S.; Jo, C. H.; Choi, H. Y.; Kwon, E. K.; Song, C. E., *Bull. Korean Chem. Soc.* **2004**, *25*, 1671.

79. Jo, C. H.; Han, S.-H.; Yang, J. W.; Roh, E. J.; Shin, U.-S.; Song, C. E., *Chem. Commun.* **2003**, 1312.

80. Ishida, T.; Akiyama, R.; Kobayashi, S., *Adv. Synth. Catal.* **2003**, *345*, 576.

81. (a) Ley, S. V.; Ramarao, C.; Lee, A.-L.; Østergaard, N.; Smith, S. C.; Shirley, I. M., *Org. Lett.* **2003**, *5*, 185. (b) Malla Reddy, S.; Srinivasulu, M.; Venkat Reddy, Y.; Narasimhulu, M.; Venkateswarlu, Y., *Tetrahedron Lett.* **2006**, *47*, 5285.

82. Yao, Q., *Org. Lett.* **2002**, *4*, 2197.

83. Yanada, R.; Takemoto, Y., *Tetrahedron Lett.* **2002**, *43*, 6849.

84. Song, C. E.; Jung, D.-u.; Roh, E. J.; Lee, S.-g.; Chi, D. Y., *Chem. Commun.* **2002**, 3038.

85. Branco, L. C.; Afonso, C. A. M., *J. Org. Chem.* **2004**, *69*, 4381.

86. Kwon, E. K.; Choi, D. S.; Choi, H. Y.; Lee, Y. J.; Jo, C. H.; Hwang, S. H.; Park, Y. S.; Song, C. E., *Bull. Korean Chem. Soc.* **2005**, *26*, 1839.

87. Lee, B. S.; Mahajan, S.; Janda, K. D., *Tetrahedron Lett.* **2005**, *46*, 4491.

88. Chandrasekhar, S.; Narsihmulu, C.; Sultana, S. S.; Reddy, N. R., *Chem. Commun.* **2003**, 1716.

89. (a) Huang, K.; Liu, H.-W.; Dou, X.; Huang, M.-Y.; Jiang, Y.-Y., *Polym. Adv. Technol.* **2003**, *14*, 364. (b) Miao, J.-H.; Yang, J.-H.; Chen, L.-Y.; Tu, B.-X.; Huang, M.-Y.; Jiang, Y.-Y., *Polym. Adv. Technol.* **2004**, *15*, 221. (c) Miao, J. H.; Yang, J. H.; Chen, L. Y.; Huang, M. Y.; Jiang, Y. Y., *Chin. Chem. Lett.* **2003**, *14*, 1008.

90. Lazarus, L. L.; Brutchey, R. L., *Dalton Trans.* **2010**, *39*, 7888.

91. Donohoe, T. J.; Harris, R. M.; Butterworth, S.; Burrows, J. N.; Cowley, A.; Parker, J. S., *J. Org. Chem.* **2006**, *71*, 4481.

92. Travis, B. R.; Narayan, R. S.; Borhan, B., *J. Am. Chem. Soc.* **2002**, *124*, 3824.

93. Schomaker, J. M.; Travis, B. R.; Borhan, B., *Org. Lett.* **2003**, *5*, 3089.

94. Schomaker, J. M.; Borhan, B., *Org. Biomol. Chem.* **2004**, *2*, 621.

95. Habel, A.; Boland, W., *Org. Biomol. Chem.* **2008**, *6*, 1601.

96. Whitehead, D. C.; Travis, B. R.; Borhan, B., *Tetrahedron Lett.* **2006**, *47*, 3797.

97. Hart, S. R.; Whitehead, D. C.; Travis, B. R.; Borhan, B., *Org. Biomol. Chem.* **2011**, *9*, 4741.

98. Dias, L. C.; Fattori, J.; Perez, C. C.; de Oliveira, V. M.; Aguilar, A. M., *Tetrahedron* **2008**, *64*, 5891.

99. Miles, S. M.; Marsden, S. P.; Leatherbarrow, R. J.; Coates, W. J., *J. Org. Chem.* **2004**, *69*, 6874.

100. Yu, W.; Mei, Y.; Kang, Y.; Hua, Z.; Jin, Z., *Org. Lett.* **2004**, *6*, 3217.

101. Nielsen, T. E.; Meldal, M., *Org. Lett.* **2005**, *7*, 2695.

102. Rodriguez-Borges, J. E.; Vale, M. L. C.; Aguiar, F. R.; Alves, M. J.; García-Mera, X., *Synthesis* **2008**, 971.

103. Kim, W. H.; Kang, J.-A.; Lee, H.-R.; Park, A.-Y.; Chun, P.; Lee, B.; Kim, J.; Kim, J.-A.; Jeong, L. S.; Moon, H. R., *Carbohydr. Res.* **2009**, *344*, 2317.

104. Bales, B. C.; Brown, P.; Dehestani, A.; Mayer, J. M., *J. Am. Chem. Soc.* **2005**, *127*, 2832.

105. Osako, T.; Watson, E. J.; Dehestani, A.; Bales, B. C.; Mayer, J. M., *Angew. Chem., Int. Ed.* **2006**, *45*, 7433.

106. Valliant-Saunders, K.; Gunn, E.; Shelton, G. R.; Hrovat, D. A.; Borden, W. T.; Mayer, J. M., *Inorg. Chem.* **2007**, *46*, 5212.

107. (a) Fujii, H.; Ogawa, R.; Jinbo, E.; Tsumura, S.; Nemoto, T.; Nagase, H., *Synlett* **2009**, 2341. (b) Barros, M. T.; Henriques, A. S.; Leitão, A. J.; Maycock, C. D., *Helv. Chim. Acta* **2002**, *85*, 4079.

108. Lu, X.; Tian, F.; Feng, Y.; Xu, X.; Wang, N.; Zhang, Q., *Nano Lett.* **2002**, *2*, 1325.

109. Cui, J.; Burghard, M.; Kern, K., *Nano Lett.* **2003**, *3*, 613.

Oxalyl Chloride

[79-37-8]   C₂Cl₂O₂   (MW 126.92)

$[79\text{-}37\text{-}8]$ $C_2Cl_2O_2$ (MW 126.92)

InChI = 1/C2Cl2O2/c3-1(5)2(4)6

InChIKey = CTSLXHKWHWQRSH-UHFFFAOYAG

(versatile agent for preparation of carboxylic acid chlorides;[1] phosphonic acid dichlorides;[2] alkyl chlorides;[3] β-chloro enones;[4] acyl isocyanates[5])

Physical Data: mp $-12\,°C$; bp $63\text{–}64\,°C/763\,mmHg$; $d\ 1.48\,g\,cm^{-3}$; $n_D^{20}\ 1.4305$.

Solubility: sol hexane, benzene, diethyl ether, halogenated solvents, e.g. dichloromethane and chloroform, acetonitrile.

Form Supplied in: colorless, fuming liquid; widely available; 2 M soln in dichloromethane.

Handling, Storage, and Precautions: liquid and solution are toxic, corrosive, and severely irritating to the eyes, skin, and respiratory tract. Use in a fume hood and wear protective gloves, goggles, and clothing. Bottles should be stored in a cool, dry place and kept tightly sealed to preclude contact with moisture. Decomposes violently with water, giving toxic fumes of CO, CO_2, HCl.

Original Commentary

Roger Salmon
Zeneca Agrochemicals, Bracknell, UK

Preparation of Carboxylic Acid Chlorides (and Anhydrides). Oxalyl chloride has found general application for the preparation of carboxylic acid chlorides since the reagent was introduced by Adams and Ulich.[1] Acid chlorides produced by this means have subsequently featured in the synthesis of acyl azides,[6] bromoalkenes,[7] carboxamides,[8] cinnolines,[9] diazo ketones,[10] (thio)esters,[11] lactones,[12] ketenes for cycloaddition reactions,[13] intramolecular Friedel–Crafts acylation reactions,[14] and the synthesis of pyridyl thioethers.[11]

Like *Thionyl Chloride*, oxalyl chloride gives gaseous byproducts with acids and the chlorides can be readily isolated in a pure form by evaporation of the solvent and any excess reagent, or used in situ for further elaboration (eq 1).

$$R\underset{O}{\overset{}{-}}Cl + CO_2 + CO + HCl \quad (1)$$

Prior formation of an amine or alkali metal salt, with or without pyridine,[1] has been used to advantage with substrates that are sensitive to strong acids or are bases (see also *Oxalyl Chloride–Dimethylformamide* for a procedure conducted under neutral conditions using silyl esters). By adjusting the molar proportions of oxalyl chloride to substrate, anhydrides can also be prepared using these methods (eq 2).[15] N-Carboxy-α-amino acid anhydrides can also be made this way.[16]

The use of nonpolar solvents such as hexane or toluene allows for the removal of inorganic or amine salts which may otherwise interfere with subsequent reactions.

Under the mild conditions employed (eqs 3 and 4),[17] racemization of stereogenic centers, skeletal rearrangement, or byproduct formation, seen with other reagents such as thionyl chloride/pyridine,[18] are seldom observed.

Conversion of β-bromoacrylic acid to the acid chloride using thionyl chloride/DMF, *Phosphorus(III) Chloride*, or benzotrichloride/zinc chloride also resulted in bromine for chlorine exchange. Use of oxalyl chloride with the preformed ammonium salt provided a mild, general method to β-bromoacryloyl chlorides (eq 5)[19] without halogen exchange or (E/Z) equilibration. β-Fluoro- and iodoacrylic acids have been cleanly converted to the acid chlorides without prior salt formation.

As well as forming acid chlorides, α-tertiary amino acids can react with oxalyl chloride and undergo an oxidative decarboxylation to give iminium salts, or ring expansion, depending on the substituents and their stereochemistry (eq 6).[20]

R^1	R^2	R^3	% Yield (A)	% Yield (B)
Me	H	*t*-Bu	0	69
H	CO_2Me	*c*-Hex	0	59
H	Me	Me	30	0

Essential Reactions for Organic Synthesis, First Edition. Edited by Philip L. Fuchs.

Preparation of Phosphonic Acid Chlorides. Phosphonic acid dichlorides have been obtained in high yield (determined by ^{31}P NMR) at low temperature from the corresponding acids using oxalyl chloride and *Pyridine* (eq 7).[2]

$$\text{(7)}$$

R = Et, PhCH$_2$, CF$_2$H, arabinomethyl, phthalidyl

Similarly, monoalkyl methylphosphonochloridates (eq 8)[21] can be made from dialkyl esters; thionate acid chlorides could not be made by this method. Thionyl chloride and PCl$_5$ were also used to make this type of compound (see also *Oxalyl Chloride–Dimethylformamide*).

$$\text{(8)}$$

R = Me, Et, Pr, *i*-Pr, Bu

Numerous other reagents such as PCl$_3$, PCl$_5$, POCl$_3$, and Ph$_3$P/CCl$_4$ are available for the preparation of acid chlorides and anhydrides but may not be as convenient as the byproducts are not so easily removed, or the reactions require more vigorous conditions.

Direct Introduction of the Chlorocarbonyl Group (Halocarbonylation). Alkanes or cycloalkanes react with oxalyl chloride under radical conditions; typically, mixtures are produced.[22] However, bicyclo[2.2.1]heptane undergoes regio- and stereospecific chlorocarbonylation, giving the ester on subsequent methanolysis (eq 9).[23]

$$\text{(9)}$$

Certain alkenes such as 1-methylcyclohexene and styrene react with oxalyl chloride, under ionic conditions without added catalyst, to give alkenoic acid chlorides in variable yields. Alkenes such as octene and stilbene did not react under these conditions.[24]

Reactions of aromatic compounds with oxalyl chloride/Lewis acid catalysts have been reviewed.[25] Anthracene is unusual as it undergoes substitution without added catalyst (eq 10).[26]

$$\text{(10)}$$

Preparation of Chloroalkanes. Alcohols react with oxalyl chloride to give oxalyl monoalkyl esters, which if heated in the presence of pyridine give the alkyl chloride (eq 11).[3]

$$\text{ROH} \xrightarrow[\text{benzene, rt}]{(COCl)_2} \xrightarrow[\text{120–125 °C}]{\text{pyridine}} \text{RCl} + \text{CO} + \text{CO}_2 \quad \text{(11)}$$

Tertiary alcohols have been converted to tertiary chlorides in a Barton–Hunsdiecker type radical process using hydroxamate esters (eq 12).[27]

$$\text{(12)}$$

e.g. R = Me(CH$_2$)$_{16}$CMe$_2$–

Chlorination of Alkenes. A novel stereospecific dichlorination of electron rich alkenes has been reported using a manganese reagent generated from *Benzyltriethylammonium Chloride* and oxalyl chloride (eqs 13–17).[28] No oxygenation byproducts are observed.

$$\text{PhCH}_2\text{NEt}_3^+ \; \text{MnO}_4^- + (COCl)_2 \longrightarrow [\text{Mn}] \quad \text{(13)}$$

$$[\text{Mn}] \xrightarrow{98\%} \quad \text{(14)}$$

$$[\text{Mn}] \xrightarrow{69\%} \quad \text{(15)}$$

$$[\text{Mn}] \xrightarrow{80\%} \quad \text{(16)}$$

$$[\text{Mn}] \xrightarrow{96\%} \quad \text{(17)}$$

Reactions with Carbonyl Groups. Unsaturated 3-keto steroids give the corresponding 3-chloro derivatives with oxalyl chloride (eq 18).[4] Prolonged heating can give rise to aromatization.[4] Tropone gives the chlorotropylium chloride in high yield.[4] In a related reaction, 1,2-dithiol-3-ones and -3-thiones give dithiolium salts when heated in toluene or chloroform with the reagent.[4] A range of β-chloro enones has been prepared from diketones. Dimedone gives the β-chloro enone in high yield (eq 19).[29] Keto esters did not react to give β-chloro esters.

$$\text{(18)}$$

(19)

β-Keto aldehydes give a single regio- and stereospecific isomer, the chlorine being *cis* to the carbonyl group (eq 20).

(20)

Certain triketones give 3-chlorides with excess oxalyl chloride, in good yield (eq 21).[30]

(21)

Preparation of Acyl Isocyanates and Aryl Isocyanates. Certain primary carboxamides can be converted to acyl isocyanates in yields from 36–97% with the reagent (eq 22);[5] *Phosgene* gives nitriles under similar conditions. Oxalyl chloride has found limited application for the preparation of triazine and quinone isocyanates.[5]

(22)

$R = ClCH_2, CCl_3, PhCH_2, 3,4\text{-}Cl_2C_6H_3, Ph_2CH$

Miscellaneous Applications. Oxalyl chloride has been used in the preparation of 2,3-furandiones from alkenyloxysilanes,[31] *o*-aminophenols from *N*-aryl nitrones,[32] dihydroquinolines via a modified Bischler–Napieralski ring closure,[33] 2,3-β-furoquinoxalines from quinoxazolones,[34] sterically hindered salicylaldehydes from phenoxyoxalyl chlorides,[35] and in mild cleavage of 7-carboxamido groups in cephalosporin natural products, without cleavage of the lactam ring or disruption of optical centers.[36]

First Update

Ivan V. Efremov

Pfizer Inc., Groton, CT, USA

Preparation of Carboxylic Acid Chlorides. As described in the original article, oxalyl chloride is widely need for the synthesis of carboxylic acid chlorides. This general approach has found use in new chemistry fields such as combinatorial chemistry[37] and dendrimer synthesis.[38–40] An interesting downstream application was formation of macrocyclic diamides without resorting to high dilution.[41]

Source of Other Oxalyl Derivatives. A variety of other oxalyl derivatives is known—oxalyl diimidazole can serve as an

example (see also *Oxalyl Bromide* and *Diethyl Oxalate*). Recently, the preparation of new reagents starting from oxalyl chloride has been reported. Thus, reaction of 1*H*-benzotriazole with oxalyl chloride led to formation of the corresponding dibenzotriazole derivative which, in turn, was shown to be an excellent tool for the preparation of unsymmetrical tetrasubstituted oxamides.[42] Drawing upon the utility of Weinreb amides, *N, N'*-dimethoxy-*N, N'*-dimethylethanediamide was prepared from oxalyl chloride and demonstrated to have utility for the synthesis of α-keto amides and 1,2-diketones.[43]

Formation of Chloroiminium Salts. Oxalyl chloride reacts readily with amides or lactams to afford chloroiminium salts that have many synthetic applications (eq 23) (see also *Oxalyl Chloride-Dimethylformamide*).

(23)

For example, efficient syntheses of thioamides and thiolactams are based on this methodology.[44,45] Certain types of chloroiminium salts can serve as precursors for high-energy synthetic intermediates. Thus, azomethine ylides were obtained by treatment of β-acylamino carboxylic esters with oxalyl chloride (eq 24).[46]

(24)

Chloroiminium salts prepared from formaldehyde led to aminochlorocarbenes when treated with base (eq 25).[47,48]

(25)

Oxalyl Chloride as a C2 Synthon. Oxalyl chloride has found widespread use as a C2 building block in organic synthesis. Applications of this reagent in such a fashion for the synthesis of heterocycles have been reviewed.[49] A particular area of interest involves the synthesis and utility of oxazolidine-4,5-diones.[50]

A general synthesis of butenolides taking advantage of one-pot cyclization of silyl enol ethers with oxalyl chloride was developed by Langer and applied to a number of synthetic problems. This useful methodology has recently been reviewed.[51]

Additional examples of the utilization of oxalyl chloride as a C2 synthon include preparation of maleic anhydrides,[52,53] 1,4-dioxane-2,3-diones[54,55] and 1,2-diketones.[56–58] A useful in situ reduction of the latter was developed to allow for a streamlined synthesis of vicinal diamines (eq 26).[59]

$$(26)$$

Oxalyl Chloride as a C1 Synthon. Although oxalyl chloride is mainly employed as a C2 equivalent in organic synthesis, there are successful examples of using this reagent for the introduction of a single carbon functionality. Thus, in addition to the expected formation of 1,4-dioxane-2,3-diones when reacted with 1,2-diols, the reaction can also lead to the preparation of cyclic carbonates.[55,60] Also, oxalyl chloride can react with ambident dianions as a C1 synthon (eq 27).[61]

$$(27)$$

Miscellaneous Applications. Oxalyl chloride has been used as a reagent for stereospecific synthesis of 2-azetidinones from aziridine-2-carboxylates (eq 28).[62]

$$(28)$$

Another interesting stereospecific transformation is the conversion of enantiomerically pure α-Li alkyl sulfoxides to vicinal chloroamines (eq 29).[63] The "nonoxidative" chloro-Pummerer rearrangement was proposed as the mechanism. The final products can be converted to the corresponding aziridines by treatment with sodium borohydride followed by sodium hydride.

$$(29)$$

Treatment of serine-containing peptides with oxalyl chloride resulted in mild dehydration to afford the corresponding α-Ala derivatives (eq 30).[64]

$$(30)$$

N-Formylimidazole was shown to be a convenient formylating agent for a variety of amines and could be prepared in situ from oxalyl chloride, formic acid, and imidazole. The reaction occurs through initial formation of formyl chloride.[65] Symmetrical tetrasubstituted oxamides could be prepared from *N*-alkyl cyclic amines. The postulated first step is the formation of an *N*-acyltrialkylammonium salt followed by selective loss of the alkyl substituent most capable of forming a stable carbocation. The observed substituent effects were consistent with the proposed mechanism.[66] Aryl isocyanates could be obtained directly from aniline hydrochlorides by treatment with oxalyl chloride. The final products result from thermal decomposition of initially formed oxamic chlorides.[67]

Related Reagents. Dimethyl Sulfoxide–Oxalyl Chloride; Oxalyl Chloride–Aluminum Chloride.

1. Adams, R.; Ulich, L. H., *J. Am. Chem. Soc.* **1920**, *42*, 599.

2. Stowell, M. H. B.; Ueland, J. M.; McClard, R. W., *Tetrahedron Lett.* **1990**, *31*, 3261.

3. Rhoads, S. J.; Michel, R. E., *J. Am. Chem. Soc.* **1963**, *85*, 585.

4. (a) Moersch, G. W.; Neuklis, W. A.; Culbertson, T. P.; Morrow, D. F.; Butler, M. E., *J. Org. Chem.* **1964**, *29*, 2495. (b) Haug, E.; Fohlisch, B., *Z. Naturforsch., Tell B* **1969**, *24*, 1353 (*Chem. Abstr.* **1970**, *72*, 43 079m). (c) Bader, J., *Helv. Chim. Acta* **1968**, *51*, 1409. (d) Faust, J.; Mayer, R., *Justus Liebigs Ann. Chem.* **1965**, *688*, 150.

5. (a) Speziale, A. J.; Smith, L. R., *J. Org. Chem.* **1962**, *27*, 3742. (b) von Gizycki, U., *Angew. Chem., Int. Ed. Engl.* **1971**, *10*, 403.

6. (a) van Reijendam, J. W.; Baardman, F., *Synthesis* **1973**, 413. (b) Lemmens, J. M.; Blommerde, W. W. J. M.; Thijs, L.; Zwanenburg, B., *J. Org. Chem.* **1984**, *49*, 2231.

7. Paquette, L. A.; Dahnke, K.; Doyon, J.; He, W.; Wyant, K.; Friendrich, D., *J. Org. Chem.* **1991**, *56*, 6199.

8. Keller-Schierlein, W.; Muller, A.; Hagmann, L.; Schneider, U.; Zähner, H., *Helv. Chim. Acta* **1985**, *68*, 559.

9. Hutchings, M. G.; Devonald, D. P., *Tetrahedron Lett.* **1989**, *30*, 3715.

10. (a) Wilds, A. L.; Shunk, C. H., *J. Am. Chem. Soc.* **1948**, *70*, 2427. (b) Hudlicky, T; Kutchan, T., *Tetrahedron Lett.* **1980**, *21*, 691. (c) Duddeck, H.; Ferguson, G.; Kaitner, B.; Kennedy, M., *J. Chem. Soc., Perkin Trans. 1* **1990**, 1055.

11. (a) Szmuszkovicz, J., *J. Org. Chem.* **1964**, *29*, 843. (b) Hatanaka, M.; Yamamoto, Y.; Nitta, H.; Ishimaru, T., *Tetrahedron Lett.* **1981**, *22*, 3883. (c) Cochrane, E. J.; Lazer, S. W.; Pinhey, J. T.; Whitby, J. D., *Tetrahedron Lett.* **1989**, *30*, 7111.

12. Bacardit, R.; Cervelló, J.; de March, P.; Marquet, J.; Moreno-Mañas, M.; Roca, J. L., *J. Heterocycl. Chem.* **1989**, *26*, 1205.

13. (a) Brady, W. T.; Giang, Y. F., *J. Org. Chem.* **1985**, *50*, 5177. (b) Snider, B. B.; Allentoff, A. J., *J. Org. Chem.* **1991**, *56*, 321.

14. (a) Burke, S. D.; Murtiashaw, C. W.; Dike, M. S.; Smith-Strickland, S. M.; Saunders, J. O., *J. Org. Chem.* **1981**, *46*, 2400. (b) Satake, K.; Imai, T.; Kimura, M.; Morosawa, S., *Heterocycles* **1981**, *16*, 1271.

15. (a) Wingfield, H. N.; Harlan, W. R.; Hanmer, H. R., *J. Am. Chem. Soc.* **1953**, *75*, 4364. (b) Schrecker, A. W.; Maury, P. B., *J. Am. Chem. Soc.* **1954**, *76*, 5803.

16. Konopinska, D.; Siemion, I. Z., *Angew. Chem., Int. Ed. Engl.* **1967**, *6*, 248.

17. (a) Ihara, M.; Yasui, K.; Takahashi, M.; Taniguchi, N.; Fukumoto, K., *J. Chem. Soc., Perkin Trans. 1* **1990**, 1469. (b) Nordlander, J. E.; Njoroge, F. G.; Payne, M. J.; Warman, D., *J. Org. Chem.* **1985**, *50*, 3481.

18. (a) Simon, M. S.; Rogers, J. B.; Saenger, W.; Gououtas, J. Z., *J. Am. Chem. Soc.* **1967**, *89*, 5838. (b) Krubsack, A. J.; Higa, T., *Tetrahedron Lett.* **1968**, 5149.

19. (a) Stack, D. P.; Coates, R. M., *Synthesis* **1984**, 434. (b) Gillet, J. P.; Sauvêtre, R.; Normant, J. F., *Synthesis* **1982**, 297. (c) Wilson, R. M.; Commons, T. J., *J. Org. Chem.* **1975**, *40*, 2891.

20. (a) Wasserman, H. H.; Han, W. T.; Schaus, J. M.; Faller, J. W., *Tetrahedron Lett.* **1984**, *25*, 3111. (b) Clough, S. C.; Deyrup, J. A., *J. Org. Chem.* **1974**, *39*, 902. (c) Sardina, F. J.; Howard, M. H.; Koskinen, A. M. P.; Rapoport, H. J., *J. Org. Chem.* **1989**, *54*, 4654.

21. Pelchowicz, Z., *J. Chem. Soc.* **1961**, 238.

22. (a) Kharasch, M. S.; Brown, H. C., *J. Am. Chem. Soc.* **1942**, *64*, 329. (b) Tabushi, I.; Hamuro, J.; Oda, R., *J. Org. Chem.* **1968**, *33*, 2108.

23. Tabushi, I.; Okada, T.; Oda, R., *Tetrahedron Lett.* **1969**, 1605.

24. (a) Kharasch, M. S.; Kane, S. S.; Brown, H. C., *J. Am. Chem. Soc.* **1942**, *64*, 333. (b) Bergmann, F.; Weizmann, M.; Dimant, E.; Patai, J; Szmuskowicz, J., *J. Am. Chem. Soc.* **1948**, *70*, 1612.

25. Olah, G. A.; Olah, J. A. In *The Friedel–Crafts Reaction*; Olah, G. A., Ed.; Interscience: New York, 1964; Vol. 3, p. 1257.

26. Latham, H. G.; May, E. L.; Mosettig, E., *J. Am. Chem. Soc.* **1948**, *70*, 1079.

27. (a) Barton, D. H. R.; Crich, D., *J. Chem. Soc., Chem. Commun.* **1984**, 774. (b) Crich, D.; Forth, S. M., *Synthesis* **1987**, 35.

28. Markó, I. E.; Richardson, P. F., *Tetrahedron Lett.* **1991**, *32*, 1831.

29. (a) Clark, R. D.; Heathcock, C. H., *J. Org. Chem.* **1976**, *41*, 636. (b) Pellicciari, R; Fringuelli, F; Sisani, E.; Curini, M., *J. Chem. Soc., Perkin Trans. 1* **1981**, 2566. (c) Büchi, G.; Carlson, J. A., *J. Am. Chem. Soc.* **1969**, *91*, 6470.

30. Lakhvich, F. A.; Khlebnicova, T. S.; Ahren, A. A., *Synthesis* **1985**, 784.

31. Murai, S.; Hasegawa, K.; Sonoda, N., *Angew. Chem., Int. Ed. Engl.* **1975**, *14*, 636.

32. Liotta, D.; Baker, A. D.; Goldman, N. L.; Engel, R., *J. Org. Chem.* **1974**, *39*, 1975.

33. Larsen, R. D.; Reamer, R. A.; Corley, E. G.; Davis, P.; Grabowski, E. J. J.; Reider, P. J.; Shinkai, I., *J. Org. Chem.* **1991**, *56*, 6034.

34. Kollenz, G., *Justus Liebigs Ann. Chem.* **1972**, *762*, 13.

35. Zwanenburg, D. J.; Reynen, W. A. P., *Synthesis* **1976**, 624.

36. Shiozaki, M.; Ishida, N.; Iino, K.; Hiraoka, T., *Tetrahedron Lett.* **1977**, 4059.

37. Georgiadis, T. M.; Baindur, N.; Player, M. R., *J. Comb. Chem.* **2004**, *6*, 224.

38. Zimmerman, S. C.; Zharov, I.; Wendland, M. S.; Rakow, N. A.; Suslick, K. S., *J. Am. Chem. Soc.* **2003**, *125*, 13504.

39. Yamaguchi, N.; Wang, J.-S.; Hewitt, J. M.; Lenhart, W. C.; Mourey, T. H., *J. Polym. Sci., Part A* **2002**, *40*, 2855.

40. Chaumette, J.-L.; Laufersweiler, M. J.; Parquette, J. R., *J. Org. Chem.* **1998**, *63*, 9399.

41. Sharghi, H.; Niknam, K.; Massah, A. R., *J. Heterocycl. Chem.* **1999**, *36*, 601.

42. Katritzki, A. R.; Levell, J. R.; Pleynet, D. P. M., *Synthesis* **1998**, 153.

43. Sibi, M. P.; Marvin, M.; Sharma, R., *J. Org. Chem.* 1995; *60*, 5016.

44. Ilankumaran, P.; Ramesha, A. R.; Chandrasekaran, S., *Tetrahedron Lett.* **1995**, *36*, 8311.

45. Smith, D. C.; Lee, S. W.; Fuchs, P. L., *J. Org. Chem.* **1994**, *59*, 348.

46. Anderson, R. J.; Batsanov, A. S.; Belskaia, N.; Groundwater, P. W.; Meth-Cohn, O.; Zaytsev, A., *Tetrahedron Lett.* **2004**, *45*, 943.

47. Cheng, Y.; Yang, H.; Meth-Cohn, O., *Org. Biomol. Chem.* **2003**, *1*, 3605.

48. Cheng, Y.; Yang, H.; Meth-Cohn, O., *Chemical Commun.* **2003**, 90.

49. Beckert, R.; Mayer, R., *Wissenschaftliche Zeitschrift der Technischen Universitaet Dresden* **1987**, *36*, 87 (CAN 108:167337).

50. Nekrasov, D. D., *Izbrannye Metody Sinteza i Modifikatsii Geterotsiklov* **2003**, *2*, 396 (CAN 141:295900).

51. Langer, P., *Synlett* **2006**, 3369.

52. Roy, S.; Eastman, A.; Gribble, G. W., *Tetrahedron* **2006**, *62*, 7838.

53. Davis, P. D.; Bit, R. A.; Hurst, S. A., *Tetrahedron Lett.* **1990**, *31*, 2353.

54. Itaya, T.; Iida, T., *J. Chem. Soc., Perkin Trans. 1* **1994**, 1671.

55. Iida, T.; Itaya, T., *Tetrahedron* **1993**, *49*, 10511.

56. Faust, R.; Weber, C.; Fiandanese, V.; Marchese, G.; Punzi, A., *Tetrahedron* **1997**, *53*, 14655.

57. Malanga, C.; Aronica, L. A.; Lardicci, L., *Tetrahedron Lett.* **1995**, *36*, 9185.

58. Babudri, F.; Fiandanese, V.; Marchese, G.; Punzi, A., *Tetrahedron Lett.* **1995**, *36*, 7305.

59. Nutaitis, C. F., *Synth. Commun.* **1992**, *22*, 1081.

60. Itaya, T.; Iida, T.; Eguchi, H., *Chem. Pharm. Bull.* **1993**, *41*, 408.

61. Langer, P.; Doering, M.; Seyferth, D., *Synlett* **1999**, 135.

62. Sharma, S. D.; Kanwar, S.; Rajpoot, S., *J. Heterocycl. Chem.* **2006**, *43*, 11.

63. Volonterio, A.; Bravo, P.; Panzeri, W.; Pesenti, C.; Zanda, M., *Eur. J. Org. Chem.* **2002**, 3336.

64. Ranganathan, D.; Shah, K.; Vaish, N., *J. Chem. Soc., Chem. Commun.* **1992**, 1145.

65. Kitagawa, T.; Ito, J.; Tsutsui, C., *Chem. Pharm. Bull.* **1994**, *42*, 1931.

66. Paritosh, R. D.; Kumar, K. A.; Duddu, R.; Axenrod, T.; Dai, R.; Das, K. K.; Guan, X.-P.; Sun, J.; Trivedi, N. J., *J. Org. Chem.* **2000**, *65*, 1207.

67. Oh, L. M.; Spoors, P. G.; Goodman, R. M., *Tetrahedron Lett.* **2004**, *45*, 4769.

Oxalyl Chloride–Dimethylformamide

(COCl)$_2$
[79-37-8]   C$_2$Cl$_2$O$_2$ (MW 126.92)
InChI = 1/C2Cl2O2/c3-1(5)2(4)6
InChIKey = CTSLXHKWHWQRSH-UHFFFAOYAG
(DMF)
[68-12-2]   C$_3$H$_7$NO   (MW 73.11)
InChI = 1/C3H7NO/c1-4(2)3-5/h3H,1-2H3
InChIKey = ZMXDDKWLCZADIW-UHFFFAOYAS

(versatile agent combination for in situ generation of Vilsmeier reagent used in preparation of carboxylic acid chlorides and other acid derived products,[1] phosphonic acid chlorides,[2] β-chloro enones,[3] ketoximes[4] and formyl derivatives[5])

Physical Data: see entries for **Oxalyl Chloride** and **N,N-Dimethylformamide**.
Form Supplied in: generated in situ under anhydrous conditions from oxalyl chloride and either catalytic or stoichiometric dimethylformamide, typically in halogenated solvents such as dichloromethane, acetonitrile, or hexane. The intermediate dimethylforminium chloride (DMFCl) can be isolated as a hygroscopic solid, mp 140 °C (see also **Thionyl Chloride, Phosphorus Oxychloride**).
Handling, Storage, and Precautions: (see also **Oxalyl Chloride**); reagent generated in situ under anhydrous conditions; corrosive, toxic. Oxalyl chloride should be used in the fume hood.

Synthesis of Carboxylic Acid Chlorides and Anhydrides.
Carboxylic acids, including amino acids,[6] readily react with DMFCl generated in situ to produce acid chlorides on heating (eq 1).[1] The procedure has been shown to be general, and may provide acid chlorides not available with oxalyl chloride alone.[7,8]

The mild reaction conditions employed enables the preparation of chlorides without disrupting chiral or other sensitive functionality. The use of hexane has been found particularly suitable for microscale preparation of high purity Mosher's acid chloride as it enables residual DMFCl to be filtered from solution (eq 2).[9]

DMFCl has been used to prepare carboxylic esters (eq 3),[10] N-hydroxycarboxamides and O,N-dimethylcarboxamides,[11] and α-hydroxycarboxamides (eq 4),[12] via a one pot process.

A procedure using silyl esters, developed to allow synthesis of acid chlorides under neutral conditions, has been applied to a range of substrates (eq 5),[13] including amino acid anhydrides.[1]

Reduction of the intermediate generated from a carboxylic acid and DMFCl provides aldehydes with **Lithium Tri-tert-butoxyaluminum Hydride**,[14] and alcohols with **Sodium Borohydride**,[15] both in high yield and chemoselectivity.

Preparation of Phosphonic Acid Chlorides. Disilyl esters of phosphonic acids react with oxalyl chloride–DMF to give the phosphonyl chlorides and silyl chlorides under mild conditions (eq 6).[2] Prior treatment of an acid-sensitive phosphonate monoester with **Trimethylsilyldiethylamine** was used to minimize exposure to HCl (eq 7).[16]

A general process suitable for large scale synthesis of phosphonic acid dichlorides, which avoids the need for preforming silyl esters, has been reported (eq 8).[17]

β-Chloro (or Bromo) Enones. Cyclic diketones in solution with DMF and dichloromethane react with oxalyl chloride (or bromide) to give β-halo enones in high yield (eq 9).[3] The reported procedure reduces overall reaction time and avoids solubility problems.

$$(9)$$

Formylation. Vilsmeier formylation reactions have been reviewed.[1] Oxalyl chloride–dimethylformamide has found occasional use; one application has been preparation of a piperidine substituted dialdehyde (eq 10).[5]

$$(10)$$

Alkyl Chlorides. *N,N*-Diphenylformamide is reported to be an improvement on DMF when used with oxalyl chloride to prepare primary and secondary alkyl chlorides from alcohols.[18]

Ketoximes. Lithium salts of nitroalkanes react with DMFCl, preformed from oxalyl chloride and DMF, to give intermediates which on reaction with Grignard reagents give ketoximes on workup (eq 11).[4]

$$(11)$$

1. (a) Standler, P. A., *Helv. Chim. Acta* **1978**, *61*, 1675. (b) Marson, C. M., *Tetrahedron* **1992**, *48*, 3659.

2. Bhongle, N. N.; Notter, R. N.; Turcotte, J. G., *Synth. Commun.* **1987**, *17*, 1071.

3. Mewshaw, R. E., *Tetrahedron Lett.* **1989**, *30*, 3753.

4. Fujisawa, T.; Kurita, Y.; Sata, T., *Chem. Lett.* **1983**, 1537.

5. Reichardt, C.; Schagerer, K., *Angew. Chem., Int. Ed. Engl.* **1973**, *12*, 323.

6. Ookawa, A.; Soai, K.; *J. Chem. Soc., Perkin Trans. 1* **1987**, 1465.

7. (a) Bosshard, H. H.; Mory, R.; Schmid, M.; Zollinger, H., *Helv. Chim. Acta* **1959**, *42*, 1653. (b) Borer, B. C.; Balogh, D. W., *Tetrahedron Lett.* **1991**, *32*, 1039. (c) Buschmann, H.; Scharf, H-D., *Synthesis* **1988**, 827.

8. Ireland, R. E.; Anderson, R. C.; Badoud, R.; Fitzsimmons, B. J.; McGarvey, G. J.; Thaisrivongs, S.; Wilcox, C. S., *J. Am. Chem. Soc.* **1983**, *105*, 1988.

9. Ward, D. E.; Rhee, C. K., *Tetrahedron Lett.* **1991**, *32*, 7165.

10. Stetter, H.; Kogelnik, H-J., *Synthesis* **1986**, 140.

11. (a) Nakonieczna, L.; Milewska, M.; Kolasa, T.; Chimiak, A., *Synthesis* **1985**, 929 (b) Ward, J. S.; Merritt, L., *J. Heterocycl. Chem.* **1990**, *27*, 1709.

12. Kelly, S. E.; LaCour, T. G., *Synth. Commun.* **1992**, *22* 859.

13. (a) Wissner, A.; Grudzinskas, C. V., *J. Org. Chem.* **1978**, *43*, 3972. (b) Hausler, J.; Schmidt, U., *Chem. Ber.* **1974**, *107*, 145. (c) Mobashery, S; Johnston, M., *J. Org. Chem.* **1985**, *50*, 2200.

14. Fujisawa, T.; Mori, T.; Sato, T., *Chem. Lett.* **1983**, 835.

15. Fujisawa T.; Toshiki, M.; Tsuge, S.; Sato, T., *Tetrahedron Lett.* **1983**, *24*, 1543.

16. Biller, S. A.; Forster, C.; Gordon, E. M.; Harrity, T.; Scott, W.; Ciosek, C. P., *J. Med. Chem.* **1988**, *31*, 1869.

17. Rogers, R. S., *Tetrahedron Lett.* **1992**, *33*, 7473.

18. Fujisawa, T.; Tida, S; Sato, T., *Chem. Lett.* **1984**, 1173.

Roger Salmon
Zeneca Agrochemicals, Bracknell, UK

Ozone[1]

[10028-15-6] O_3 (MW 48.00)

InChI = 1/O3/c1-3-2

InChIKey = CBENFWSGALASAD-UHFFFAOYAY

(powerful oxidant; capable of oxidizing many electron rich functional groups;[1c,d] most widely used to cleave alkenes, affording a variety of derivatives depending on workup conditions[1b,2])

Physical Data: mp $-193\,°C$; bp $-111.9\,°C$; d ($0\,°C$, gas) $2.14\,g\,L^{-1}$.

Solubility: 0.1–0.3% by weight in hydrocarbon solvents at -80 to $-100\,°C$.[3]

Preparative Methods: ozone is a colorless to faint blue gas which is usually generated in the laboratory by passing dry air or oxygen through two electrodes connected to an alternating current source of several thousand volts. From air, ozone is typically generated at concentrations of 1–2%; from oxygen, concentrations are typically 3–4%. Several laboratory scale generators are commercially available. In addition, it is possible to generate ozone by reacting O_2^+ salts with aq hydrogen fluoride at low temperature.[4] This method allows for the incorporation and control of isotopically labeled ozone depending on whether labeled oxygen or water is employed in the reagent formation.

Analysis of Reagent Purity: the amount of ozone generated can be determined based on the liberation of iodine from potassium iodide solution followed by thiosulfate titration to determine the amount of iodine produced.[5] Photometric detectors are available which can determine the concentration of ozone in a metered gas stream. In this manner, exact amounts of ozone introduced into a reaction can be determined.

Handling, Storage, and Precautions: ozone is irritating to all mucous membranes and is highly toxic in concentrations greater than 0.1 ppm by volume. It has a characteristic odor which can be detected at levels as low as 0.01 ppm. All operations with ozone should be carried out in an efficient fume hood and scrubbing systems employing thiosulfate solutions can be used to destroy excess ozone. Liquefied ozone poses a severe explosion hazard.

Original Commentary

Richard A. Berglund

Eli Lilly and Company, Lafayette, IN, USA

Ozonolysis of Alkenes. Ozone has been most widely used for cleavage of carbon–carbon double bonds to produce carbonyl compounds or alcohols, depending on workup conditions. These reactions usually are performed by passing a stream of ozone in air or oxygen through a solution of the substrate in an inert solvent at low temperature (-25 to $-78\,°C$). Useful solvents include pentane, hexane, ethyl ether, CCl_4, $CHCl_3$, CH_2Cl_2, EtOAc, DMF, MeOH, EtOH, H_2O, or HOAc. The solvents most commonly used are CH_2Cl_2 and MeOH or a combination of the two. Reaction

endpoint can be determined using photometric monitors to detect ozone in the exit gas stream, or by the appearance of a blue color in the reaction medium, which indicates excess ozone in solution. A wide variety of alkenes undergo ozonolysis, and those in which the double bond is connected to electron-donating groups react substantially faster than alkenes substituted with electron-withdrawing groups.[1b,6] With haloalkenes, the rate of ozone attack is decelerated and a greater variety of products are obtained, although double bond cleavage is still prevalent.[7]

The reaction mechanism for ozonolysis has been studied extensively and is thought to involve a 1,3-dipolar cycloaddition to afford an initial 1,2,3-trioxolane or primary ozonide, which cleaves to a carbonyl compound and a carbonyl oxide (eq 1).[8] The carbonyl oxide generally forms at the fragment containing the more electron donating group. Recombination of the fragments affords a 1,2,4-trioxolane or ozonide which is sometimes isolated, but due to the danger of explosion is usually directly converted to carbonyl compounds via either a reductive or oxidative procedure. If alcoholic solvents are used, trapping of the carbonyl oxide can occur to afford an α-alkoxy hydroperoxide.

$$R_2C=O + R_2C=\overset{+}{O}-\overset{-}{O} \longrightarrow R_2C\overset{O}{\underset{O-O}{\diagdown}}CR_2 \qquad (1)$$

Reductive workup procedures afford aldehydes, ketones, or alcohols. An extensive number of reducing agents have been used including catalytic hydrogenation, sulfite ion, bisulfite ion, iodide, phosphine, phosphite, tetracyanoethylene (TCNE), Zn–HOAc, BH_3, $SnCl_2$, Me_2S, thiourea, or, to obtain alcohols, $LiAlH_4$ or $NaBH_4$.[1a,b,9] **Dimethyl Sulfide** offers several advantages. It rapidly reduces peroxidic ozonolysis products to carbonyl compounds, it operates under neutral conditions, excess sulfide is easily removed by evaporation, and the oxidation product is DMSO.[10] In cases where the odor of Me_2S is a problem, **Thiourea** is a convenient substitute: results are comparable to those obtained with Me_2S, and thiourea *S,S*-dioxide separates out from the reaction mixture.[11] A polymer-based diphenylphosphine system also has been developed which offers the advantage of a simple filtration and evaporative workup and eliminates potential product contamination by PPh_3 or its oxide.[12] A comparison of these workup options is shown in eq 2.

$$Ph\diagup\diagdown \xrightarrow[\text{2. [H]}]{\text{1. } O_3} PhCHO \qquad (2)$$

Me_2S, 89%
$(NH_2)_2CS$, 81%
poly-PPh_2, 80%

Oxidative workup procedures convert peroxidic ozonolysis products to ketones or carboxylic acids. Typical oxidative reagents include peroxy acids, silver oxide, chromic acid, permanganate, molecular oxygen, and the most widely used reagent, **Hydrogen Peroxide**.[1a,b,9]

Additional terminal functionalization can be accomplished by several methods. Schreiber has developed general ozonolysis and workup procedures which enable a variety of products to be prepared from cycloalkenes (eq 3).[13] Also, iron or copper salts can be

Essential Reactions for Organic Synthesis, First Edition. Edited by Philip L. Fuchs.
© 2016 John Wiley & Sons, Ltd. Published 2016 by John Wiley & Sons, Ltd.

used to convert ozonides or α-alkoxy hydroperoxides to chlorides or alkenes with one less carbon atom than in the original alkene (eqs 4 and 5),[14] and treatment of ozonides with hydrogen and an amine in the presence of a catalyst provides a direct route for the production of amines from alkenes (eq 6).[15] In a more specific case, stilbenes can be converted to alkyl benzoates by treatment of the intermediate α-alkoxybenzyl hydroperoxides with amines or DMSO.[16]

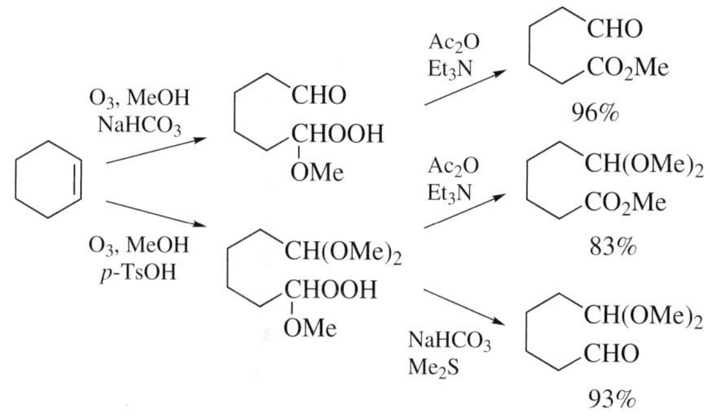

(3)

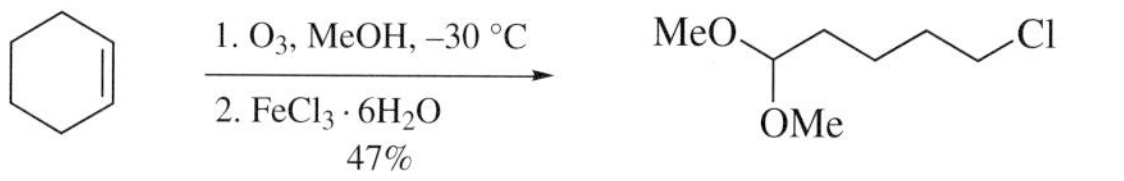

(4)

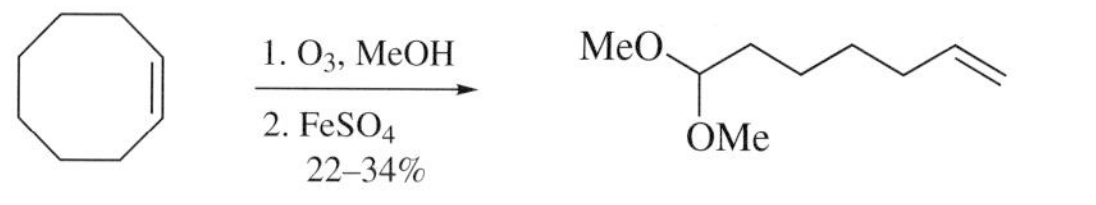

(5)

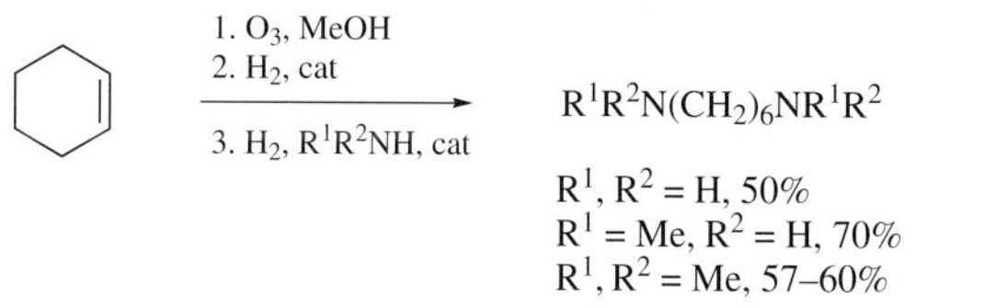

$$R^1R^2N(CH_2)_6NR^1R^2 \quad (6)$$

$R^1, R^2 = H, 50\%$
$R^1 = Me, R^2 = H, 70\%$
$R^1, R^2 = Me, 57\text{--}60\%$

Vinyl ethers are more reactive toward ozone than alkenes due to the electron-donating oxygen substituent and double bond cleavage products are often obtained.[17] Notably, the ozonolysis of cyclic vinyl ethers provides a path to aldol and homoaldol type products.[18] In addition, silyloxyalkenes undergo clean oxidative cleavage with ozone to afford diacids by using oxidative workups (eq 7),[19] or hydroxyl or oxo derivatives by using reductive workups (eqs 8–10).[20] Overall, the two-step process of forming and cleaving a silyloxyalkene provides a method for regiospecific cleavage of an unsymmetrical ketone. Eq 10 shows that the silyloxyalkene double bond is sufficiently nucleophilic to allow for selective oxidative cleavage of this bond in the presence of less activated double bonds; with appropriate workup conditions, the method can thus complement Baeyer–Villiger oxidation.

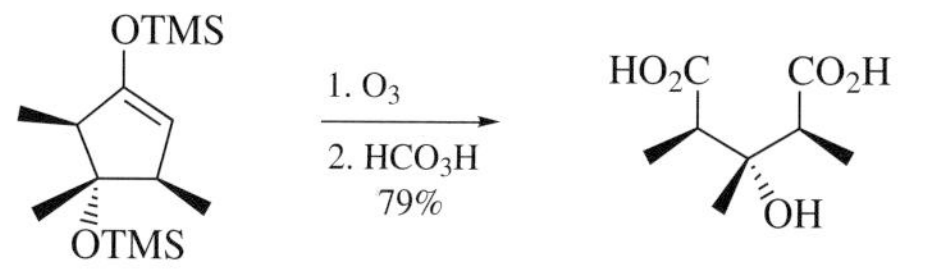

(7)

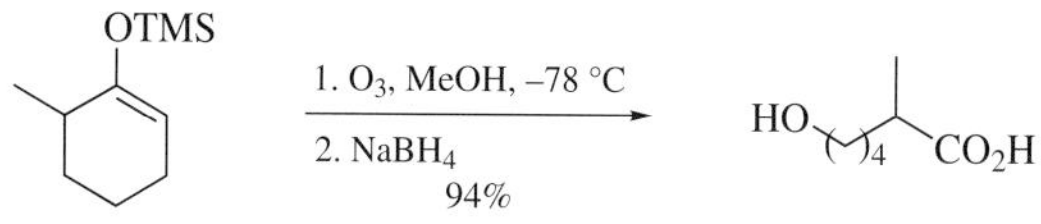

(8)

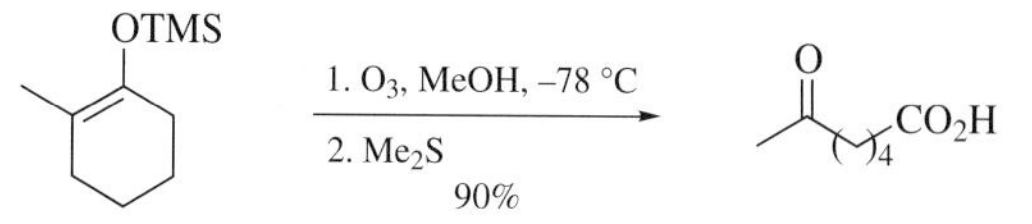

(9)

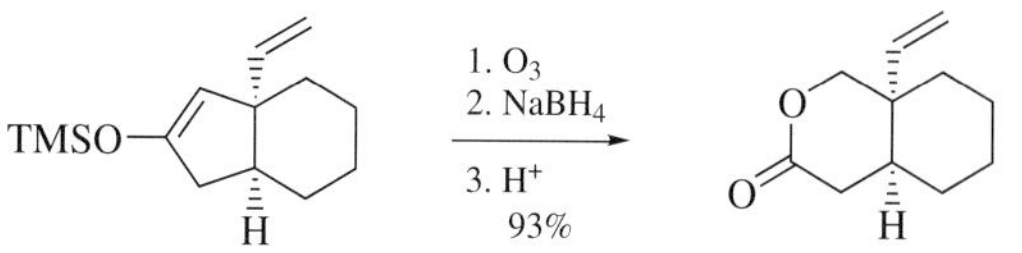

(10)

While ozone generally reacts with vinyl sulfides and enamines to provide both the expected products of double bond cleavage and anomalous products,[21,22] ketene dithioacetals have been efficiently cleaved to ketones using ozone (eq 11).[23]

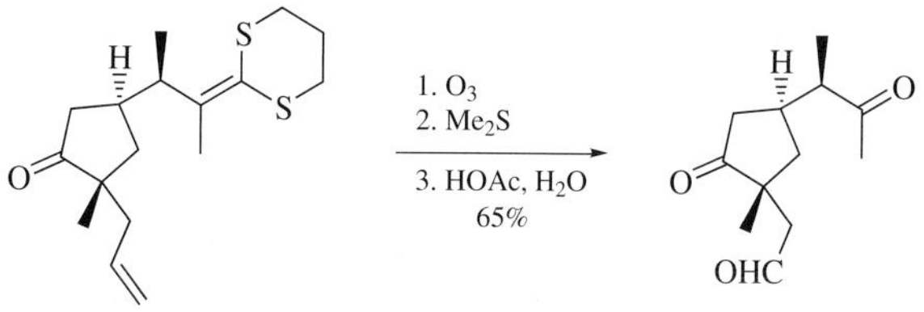

(11)

The reaction of α,β-unsaturated ketones with ozone usually affords keto acids containing one less carbon than in the original molecule (eq 12).[24]

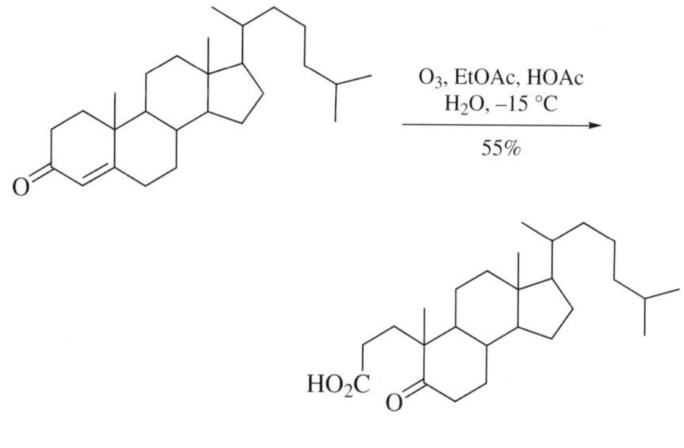

(12)

In the ozonolysis of 1,3-dienes, one double bond often can be cleaved selectively, and in 1,3-cyclodienes, the regioselectivity in fragmentation of the primary ozonide depends upon the size of the rings (eq 13).[25] Eq 13 shows that as the ring size contracts from cyclooctadiene to cyclohexadiene, the α,β-unsaturated ester becomes favored over the enal.

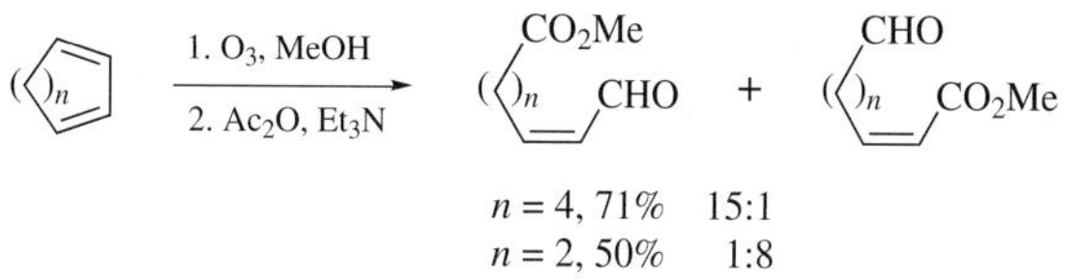

(13)

$n = 4, 71\% \quad 15{:}1$
$n = 2, 50\% \quad 1{:}8$

Hindered alkenes often afford epoxides upon ozonation due to difficulty in forming the primary ozonide by a cycloaddition process.[26] Examples are displayed in eqs 14 and 15.[27,28]

$$\text{(14)}$$

$$\text{(15)}$$

Overall, ozone compares favorably with other approaches for oxidative alkene cleavage involving **Osmium Tetroxide**, **Potassium Permanganate**, **Ruthenium(VIII) Oxide**, **Sodium Periodate**, or chromyl carboxylates which are costly, toxic, involve metal wastes, and may require detailed workup procedures.

Ozonation of Alkynes. Reactions of alkynes with ozone afford either carboxylic acids or, if reductive procedures are used, α-dicarbonyl compounds.[1c] For the production of carboxylic acids, MeOH has been shown to be superior to CH_2Cl_2 as reaction solvent.[29] As with alkenes, a number of reducing agents can be used to produce α-dicarbonyl compounds. An easy option which results in high yields of α-dicarbonyl compounds involves the addition of **Tetracyanoethylene** directly to an ozonation reaction mixture as an in situ reducing agent (eq 16).[30]

$$\text{(16)}$$

R = Ph, 92%; H, 60%; Pr, 71%

Alkynes react slower with ozone than do alkenes and selective reaction of alkenes can be achieved in the presence of alkynes (eq 17).[31] Conversely, the example of eq 18 shows that alkynes are more reactive toward ozone than aromatic rings.[32]

$$\text{(17)}$$

$$\text{(18)}$$

The ozonation of terminal alkynes to afford α-oxoaldehydes is significant (see eq 16). While reagents such as $KMnO_4$, RuO_4, OsO_4, and **Thallium(III) Nitrate** can be used to convert internal alkynes to α-dicarbonyl compounds, terminal alkynes are generally cleaved to carboxylic acids. Only **Mercury(II) Acetate**-catalyzed oxidations using a molybdenum peroxide complex afford α-oxoaldehydes.[33] However, high catalyst loads are required, the oxidant is not commercially available, and metal wastes are generated.

Oxygenated alkynes also can be ozonated. The reaction of ozone with alkynyl ethers followed by reductive workup provides a convenient method for the production of α-keto esters in moderate yields (eq 19).[34]

$$\text{(19)}$$

R = Me, R' = Et, 25%
R = Pr, R' = Et, 30%

Ozonation of Aromatic Systems. Aromatic compounds are less reactive toward ozone than either alkenes or alkynes. As a consequence, more forcing conditions are required to ozonize aromatic systems. These conditions typically involve the use of acetic acid as solvent, excess ozone, and oxidative decomposition often using H_2O_2. Electron-withdrawing groups deactivate aromatic systems toward electrophilic ozone attack, while electron-donating groups activate aromatic systems toward ozone attack. An example of this is provided in Woodward's strychnine synthesis where methoxy substitution allows for selective oxidation of one aromatic ring over two others to afford an often difficult to prepare, terminally functionalized, conjugated (Z,Z)-diene (eq 20).[35]

$$\text{(20)}$$

With polycyclic aromatic hydrocarbons, the site of ozone attack may be dependent upon substrate structure and reaction solvent (eq 21).[36]

$$\text{(21)}$$

Synthetically useful ozonolyses of heteroaromatic systems include the preparation of pyridine derivatives from quinolines (eq 22),[37] the preparation of versatile N-acyl amides by the ozonolysis of imidazoles (eq 23),[38] and the unmasking of a latent carboxylic acid function by the ozonolysis of a furan system (eq 24).[39]

$$\text{(22)}$$

$$\text{(23)}$$

$$\text{(24)}$$

Ozonation of Heteroatoms. Phosphines are converted to phosphine oxides and phosphites to phosphates by ozone.[40,41] These reactions are quite general and a wide range of substitutents can be tolerated. Phosphine oxides also can be produced by the ozonation of alkylidenetriphenylphosphoranes or of thio- or selenophosphoranes.[42,43] Organic sulfides are converted to sulfoxides and sulfones by ozonation.[41,44] Tertiary amines are converted to amine oxides, while nitro compounds can be produced in modest yields by ozonation of primary amines.[44,45] This preparation of nitroalkanes compares well with alternate approaches using peroxides, peroxy acids, permanganate, or *Monoperoxysulfuric Acid*, but ozonation on silica gel has proven to be superior (see *Ozone–Silica Gel*). Selenides are converted to selenoxides by ozone and this reaction is often used to achieve overall production of unsaturated carbonyl compounds. An example is shown in eq 25.[46]

$$\text{(25)}$$

Modification of Ozone Reactivity. The reactivity of ozone toward various unsaturated moieties can be moderated by the addition of either Lewis acids or pyridine to the ozonations. Enhanced electrophilic ozone reactivity toward aromatic substrates is observed when the Lewis acids *Aluminum Chloride* or *Boron Trifluoride* are added to reaction mixtures.[47] Conversely, an apparent decrease in ozone reactivity and a concurrent increase in the regioselectivity of ozone attack can be achieved by adding small amounts of pyridine to ozonolyses (eq 26).[48] It is thought that coordination of either the Lewis acid or basic pyridine to ozone results in the modified reactivity.

$$\text{(26)}$$

70% without pyridine

In a related procedure, ozonizable dyes have been used as endpoint indicators for selective ozonation of substrates containing multiple unsaturated linkages.[49] The dye affords colored solutions and the ozonation is carried out just until the color is discharged. If the dye is of suitable reactivity such that the most reactive substrate

unsaturated linkage reacts first, and the dye second, the reaction can be stopped before further oxidation of the substrate occurs.

Interestingly, addition of BF$_3$ etherate to the ozonolysis of o-dimethoxybenzene derivatives results in increased yields of (Z,Z)-dienes (eq 27, compare to eq 20).[50] In this case, it is thought that coordination of the Lewis acid to the diene reduces its electron density and suppresses further attack by ozone. Also, the fact that the BF$_3$ is already coordinated to ether may limit its ability to coordinate to ozone and increase its electrophilic reactivity.

$$\text{(27)}$$

20% without BF$_3$·OEt$_2$

Ozonation of Acetals. Ozone reacts very efficiently with acetals to afford the corresponding esters (eqs 28 and 29).[51] The aldehyde and alcohol components of the acetal function can be varied and yields are excellent. Cyclic acetals react much faster than acyclic acetals as a result of conformational effects.

$$C_6H_{13}CH(OR)_2 \xrightarrow[-78\,°C]{O_3,\ EtOAc} C_6H_{13}CO_2R \qquad \text{(28)}$$

R = Me, 15 h, 91%
R = Et, 8 h, 94%

$$\text{(29)}$$

n = 2, 10 min, 98%
n = 3, 2 h, 97%

Miscellaneous Ozonations. Ozonation offers a simple neutral alternative for oxidation of secondary alcohols to ketones (eq 30).[52]

$$\text{(30)}$$

R^1	R^2	%
Me	Me	83
Me	Me	72
-(CH$_2$)$_4$-		53
-(CH$_2$)$_5$-		65

Upon reaction of allene with one equivalent of ozone, trisection occurs to provide carbon monoxide derived from the central carbon atom and carbonyl compounds from the remaining carbon atoms.[53] In the example of eq 31, allene ozonolysis is used to prepare a versatile protected α-hydroxyaldehyde.[54]

$$\text{(31)}$$

R = C$_5$H$_{11}$, Ph, t-Bu, 89–99%

Ozonation of benzyl ethers affords high yields of benzoate esters (eq 32).[55] Coupled with deacylation by NaOMe, this reaction offers a mild alternative for removal of benzyl ether protecting groups (eq 33).[56] However, due to the higher reactivity of alkenes, selective oxidative cleavage of carbon–carbon double bonds can be accomplished in the presence of benzyl ethers (eq 34).[57]

$$
\text{(32)}
$$

$$
\text{(33)}
$$

yield for β-anomer, 75%

$$
\text{(34)}
$$

Ozone has been used to cleave nitronate anions, resulting in the high yield production of either aldehydes or ketones.[58] An example of this reaction is shown in eq 35.[58a] This is a very general method and has advantages over the Nef reaction which requires strong acid conditions, and other procedures utilizing permanganate or *Titanium(III) Chloride*.

$$
\text{(35)}
$$

Aldehydes can be converted to peroxy acids via ozonation in methyl or ethyl acetate,[59] or to methyl esters via ozonation in 10% methanolic KOH (eq 36).[60] Ethyl esters can be produced analogously, but the use of higher alcohols results in low KOH solubility and poor conversion. This problem can be overcome by adding the aldehyde to a solution of lithium alkoxide in THF at −78 °C and treating this mixture with ozone (eq 37). Additionally, the direct preparation of methyl esters can be accomplished via alkene ozonolysis in methanolic NaOH or by addition of NaOMe to a MeOH–CH$_2$Cl$_2$ ozonolysis solvent system.[61]

$$
R^1CHO \xrightarrow[\text{O}_3,\ -78\ °C]{R^2OH,\ KOH} R^1CO_2R^2
$$

R^1	R^2	%	
Cy	Me	58	(36)
	Et	60	
Ph	Me	66	
	Et	60	
3-Oxobisnor-	Me	85	
4-cholenyl	Et	87	

$$
\text{(37)}
$$

$$
CO_2CHMe_2
$$

First Update

Matthew M. Kreilein

University of North Carolina, Chapel Hill, NC, USA

Additional Ozonolysis Quenching Reagents. The reduction of the intermediate ozonide can be achieved via several methods to deliver various products as reviewed previously.[62] Dimethyl sulfide and triphenylphosphine are very commonly employed to deliver aldehydes from the olefins due to their availability and cost; however, these two and other methods can sometimes require long reaction times, substantial amounts of reagent, and the need to remove the by-products formed during the quench. Two newer methods allow for inexpensive and more rapid quenching of ozonide intermediates and form little to no by-products that need to be removed from the reaction medium. Triethylamine can be used to decompose the peroxide intermediates formed in ozonolysis reactions to deliver aldehyde products in less time using less equivalents than required for the same workup with dimethyl sulfide (eq 38).[63] The cleavage is achieved through deprotonation of the intermediate ozonide rather than through attack of one of the peroxidic oxygens. The isolation of carboxylic acid derivatives in the cyclic alkenes used shows that no oxygen is being incorporated into the quenching agent. This method of quenching by deprotonation was also observed when Me$_2$S was used, as some examples gave the aldehydo-acid products in minor yield.

Since an oxygen atom is not bonded to the reducing agent, the use of triethylamine seems best suited to acyclic olefins as the resulting carboxylic acid fragment generated can be removed from the reaction mixture if the keto- or aldehydo-acid is not desired (eq 39). The time required to quench the intermediate is cut up to twenty-fold in certain cases, and only 2 equiv of Et$_3$N were necessary to quench the reaction instead of the 10 equiv used in the comparison experiment with Me$_2$S. In addition, the yields for several reactions were significantly higher and there was no noticeable epimerization of chiral centers positioned α to the aldehyde.

$$
\text{(38)}
$$

Starting Alkene $\xrightarrow[\text{then 2 equiv Et}_3\text{N or 10 equiv Me}_2\text{S}]{\text{O}_3,\ \text{CH}_2\text{Cl}_2,\ -78\ °\text{C}}$ Carbonyl Compounds

$$(39)$$

Starting Material	Product	Et$_3$N Treatment	Me$_2$S Treatment
(cyclopentene-Ph)	(COPh / COR chain)	12 h R = H 9% R = OH 73%	48 h R = H 83% R = OH 11%
(cyclohexanone allyl)	(cyclohexanone aldehyde)	3 h 88%	72 h 54% aldehyde 20% ozonide
(Boc oxazolidine vinyl, MeO$_2$C)	(Boc oxazolidine aldehyde, MeO$_2$C)	3 h 78%	24 h 78%

Clean cleavage of the intermediate ozonide from ozonolysis in MeOH was also achieved in high yield using 3,3′-thiodipropionic acid or its mono- or disodium salts. The desired aldehydes were isolated without quenching reagent by-products and required only solvent evaporation and extraction.[64] For several systems, the time required to quench the ozonide was shortened as well as the equivalents necessary to achieve the transformation. The sodium salts proved to be more useful than the parent diacid as they provided the desired aldehydes that were isolated as their dimethyl acetal derivative when the acidic diacid or Me$_2$S was used (eq 40).

Deprotection of Aziridines. Previous work by Ito showed that ozone cleaved the benzyl groups of dibenzylaniline. In addition, it was observed that cleavage of the nitrogen-benzyl bonds in substituted aziridines was thwarted and that cleavage of the nitrogen substituent resulted without fragmentation of the aziridine ring (eq 41).[65] This was also observed, albeit in lower yield, for *N*-alkylated aziridines.

Starting Alkene $\xrightarrow[\text{then}\ \substack{\text{XO}_2\text{C}\diagdown\diagup\text{S}\diagdown\diagup\text{CO}_2\text{X}\\ \text{X = H, H or Na, H or Na, Na}}]{\text{O}_3,\ \text{MeOH},\ -78\ °\text{C}}$ Aldehyde

$$(40)$$

Starting Material	Product	Quench	Time	Yield
(cyclopentene-CO$_2$Et)	(OHC / CHO with CO$_2$Et)	2 equiv Me$_2$S	1.5 h	60%
		2 equiv with X = H, H	18 h	62%
		2 equiv with X = Na, Na	<1 h	93%
1-undecene	decanal	Me$_2$S		only dimethyl acetal
		2 equiv with X = H, H		only dimethyl acetal
		2 equiv with X = Na, Na		95%

$$\text{Ph}\diagup\text{N(CH}_2\text{Ph)}_2 \xrightarrow{\text{O}_3} \underset{78\%}{\text{PhCHO}} + \underset{20\%}{\text{PhCH}_2\text{NH}\diagup\text{Ph}}$$

(aziridine, R = Ph, *n*-Bu) $\xrightarrow{\text{O}_3}$ (NH aziridine)

$$(41)$$

This effect seems to be due to the lessened reactivity of the benzylic carbons when incorporated adjacent to the aziridine ring. In similar systems, use of Pearlman's catalyst, palladium(II) hydroxide on carbon, to remove the useful benzhydryl protecting group led to fragmentation of the aziridine ring in certain cases (eq 42).

(aziridine with CHPh$_2$, R, CO$_2$Et) $\xrightarrow[\text{H}_2\ (1\ \text{atm})\\ \text{MeOH, rt, 3 h}]{10\%\ \text{Pd(OH)}_2/\text{C}}$ (NH aziridine, R, CO$_2$Et) or (Ph, H$_2$N, CO$_2$Et)

$$(42)$$

R	Aziridine	Amine
n-Pr	83%	N/A
Cy	93%	N/A
t-Bu	100%	N/A
Ph	N/A	79%

Borrowing from Ito's original work, smooth removal of the benzhydryl group was achieved in moderate yield (40–60%) after ozonolysis followed by workup with 10 equiv of sodium borohydride in MeOH at low temperature (eq 43).[66] Since there is no aziridine bond breaking in the reaction sequence, the deprotected aziridines are recovered without loss of enantiopurity.

(aziridine with CHPh$_2$, R, CO$_2$Et) $\xrightarrow[\text{2. NaBH}_4\ (10\ \text{equiv), MeOH, }-78\ °\text{C}]{\text{1. O}_3,\ \text{CH}_2\text{Cl}_2,\ -78\ °\text{C, 3 h}}$ (NH aziridine, R, CO$_2$Et)

$$(43)$$

R = Cy, *t*-Bu, -(CH$_2$)$_5$-, 4-Me-C$_6$H$_4$, 4-Ph-C$_6$H$_4$, 4-Br-C$_6$H$_4$

Use in Solid Phase Synthesis. Solid phase synthesis offers the ability to perform transformations that involve little to no workup and purification. Ozone, possessing the same properties in certain cases, seems well suited to solid phase synthesis. Undec-10-enoic acid was coupled to merrifield resin under standard conditions using 1,3-dicyclohexylcarbodiimide (DCC).[67] The terminal olefin was then subjected to ozonolysis at low temperature in CH$_2$Cl$_2$. The intermediate secondary ozonide could be converted into an alcohol, aldehyde, or carboxylic acid depending on the workup conditions.

When treated with sodium borohydride in *i*-PrOH, the terminal alcohol was obtained after the substrate was released from the resin via hydrolysis (eq 44). Reductive workup using triphenylphosphine yielded the aldehyde product after release from the resin using chlorotrimethylsilane in MeOH/*i*-PrOH. This removal condition proved very mild and useful as it did not interfere with the aldehyde generated in the reaction. If the ozonolysis was conducted with acetic acid in the medium and was followed by stirring in an oxygen atmosphere overnight, the carboxylic acid can be obtained after release from the resin using the aforementioned TMSCl-mediated removal condition (eq 45).

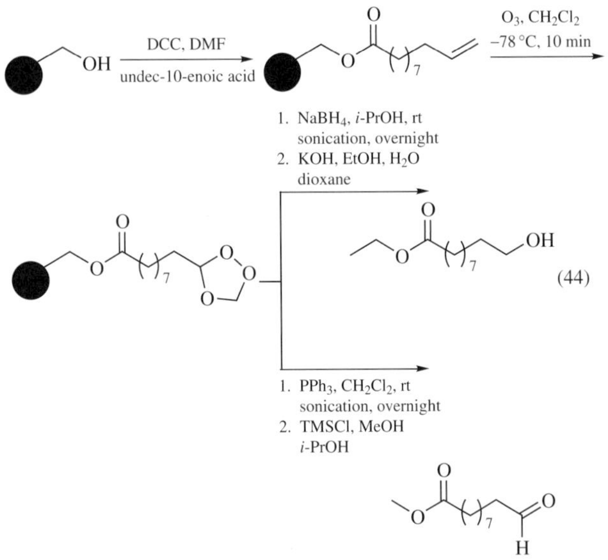

(44)

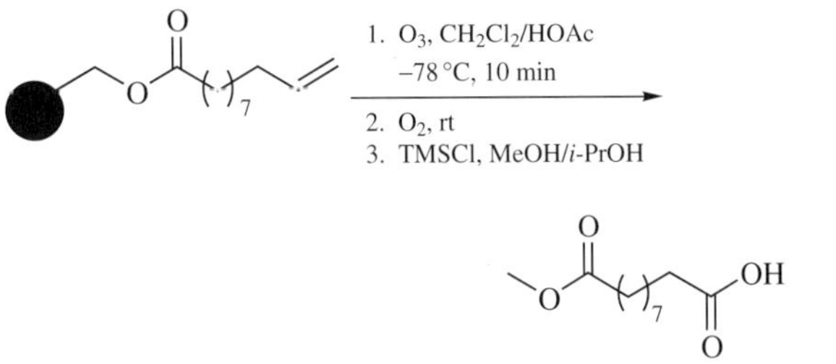

(45)

Synthesis of Substituted Methyl Esters. Cleavage of double bonds using ozone with MeOH as a solvent is known to give methyl esters. A useful application of the formation of methyl esters uses ozonolysis with a workup employing either sodium hydroxide and MeOH or sodium methoxide.[68,69] This condition offered the controlled synthesis of α- and β-amino and α- and β-oxygenated methyl esters. All of the products were obtained from aldehydes in four to six steps in good overall yield. The α-oxygenated esters were obtained from protected allylic alcohols and no loss of the protecting group was observed, including an acetate ester (eq 46). In addition, the reaction conditions did not disturb chiral centers present in the substrates (eq 47).

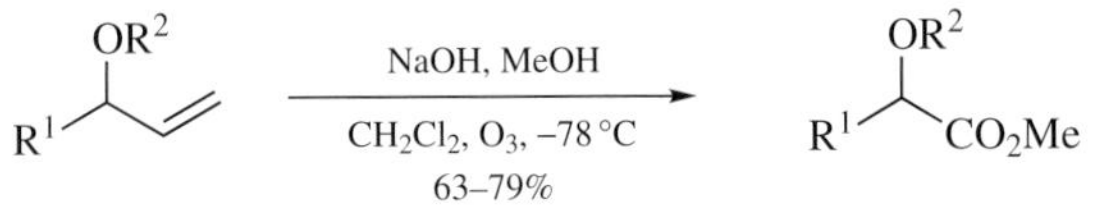

(46)

$$R^1 = n\text{-}C_6H_{13}, \ n\text{-}C_{10}H_{21}, \ c\text{-}C_6H_{11}$$
$$R^2 = Me, \ Bn, \ TBS, \ Ac$$

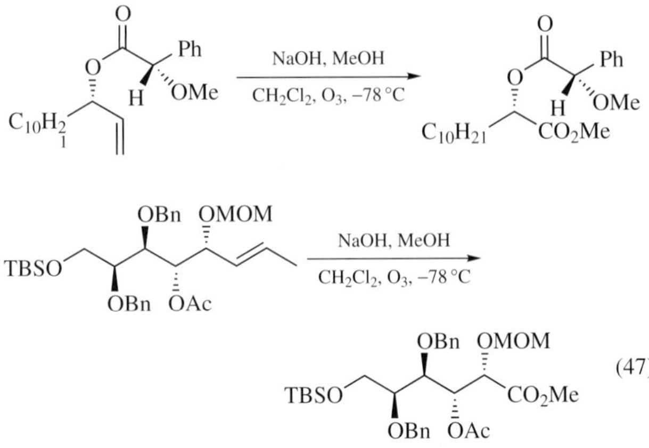

(47)

When simple aldehydes were converted to allylic alcohols, they could be transformed to the corresponding trichloroacetimidates and subjected to thermal rearrangement to provide allyl amines. The allyl amine functionality was also accessible from enantioselective reduction of propargyl ketones followed by conversion to the phthalamide derivative and reduction of the triple bond. Ozonolysis of these substrates provided the corresponding α-amino methyl esters in good yield (eq 48). As with the allyl alcohols, optically active substrates were treated with ozone without loss of enantiopurity of the chiral center.

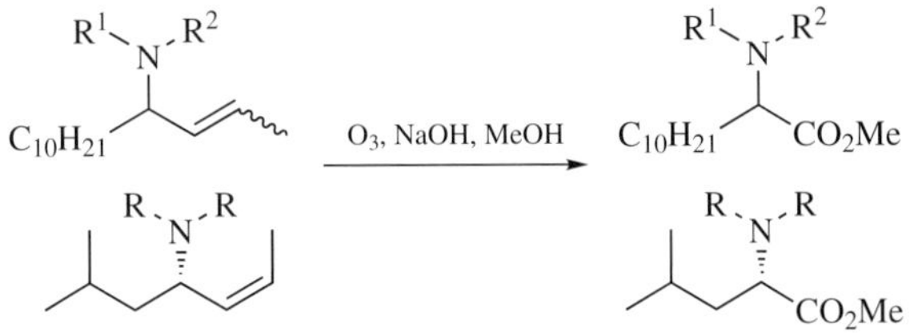

(48)

Synthesis of the β-amino methyl esters was accomplished by Grignard addition to aldehydes or epoxides to deliver the homoallylic alcohols. After conversion to the homoamino derivative as prescribed in the allylic alcohol to allylic amine conversion, ozonolysis delivered the targets in good yield, without loss of optical activity or nitrogen protection (eq 49). Numerous additional examples as well as synthetic possibilities exist for the synthesis of natural product fragments utilizing this methodology.

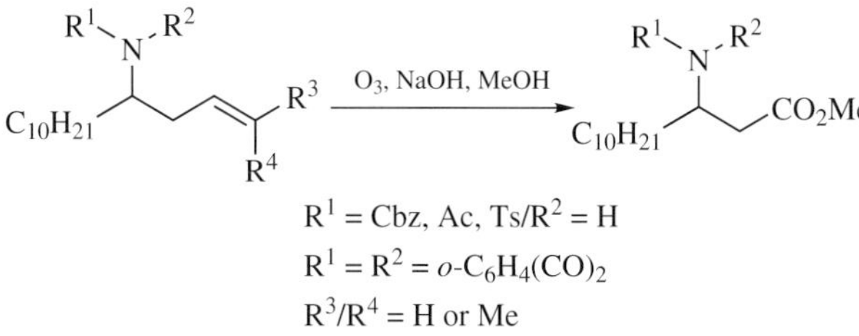

$$R^1 = Cbz, Ac, Ts/R^2 = H$$
$$R^1 = R^2 = o\text{-}C_6H_4(CO)_2$$
$$R^3/R^4 = H \text{ or } Me$$

Preparation of Tertiary Amines from Alkenes and Secondary Amines. A useful preparation of tertiary amines from alkenes can be achieved when a secondary ozonide is treated with a secondary amine (eq 50).[70] The reaction is quite versatile and provides tertiary amines when the reaction is carried out at reflux after addition of the amine. When the reaction medium was kept at room temperature, isolation of the enamine was observed, making this a clean, four-step, one-pot preparation of morpholino enamines and tertiary amines for use in synthesis. Overall, the reaction performed best with morpholine; some problems were encountered with methylenecyclohexane and piperidine, as it seems the enamine intermediate is difficult to form in such a hindered system.

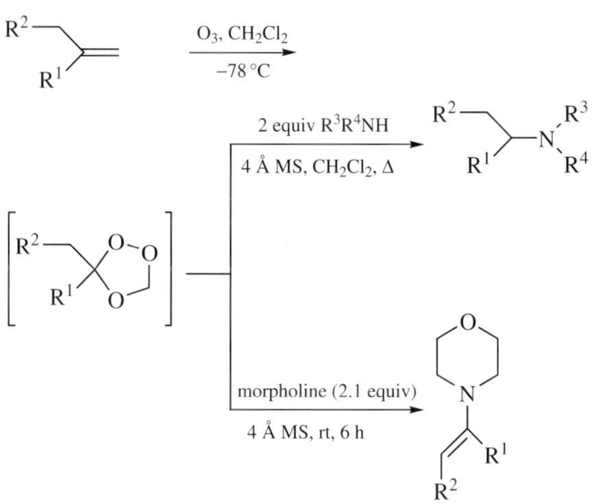

Mechanistically, the first step mimics the quench with Et_3N proceeding through deprotonation of the intermediate ozonide (eq 51). The second equivalent of the amine then forms the enamine intermediate observed if the reaction is not kept at the reflux point. The enamine intermediate undergoes a modified Wallach reduction with piperidinium formate that is generated during the initial deprotonation step of the reaction to give the target amine.

Ozonolysis of Carbon–Heteroatom Bonds and Heterocyclic Compounds. The ozonolysis of nitrogen, phosphorus, sulfur, and selenium to form N-oxides, phosphine oxides and phosphonates, sulfoxides, and selenoxides that can be used to functionalize various substrates has been well documented.[62,71,72] A useful application of the power of ozonolysis is to cleave carbon heteroatom bonds for conversion into carbonyl compounds. A useful

conversion of silicon- and tin-containing molecules has been reported to provide a variety of carbonyl compounds (eq 52).[73,74] Treatment of (α-hydroxyalkyl)trialkylsilanes with ozone provided the corresponding carboxylic acids. Ozonolysis of the silicon component was more rapid than direct ozonolysis of other functional groups such as sulfur. In addition, a library of (α-alkoxyalkyl)trialkylstannanes were prepared and treated with ozone to arrive at the corresponding esters. The cleavage of the carbon tin bond was also possible without the α-alkoxyalkyl group to arrive at carbonyl compounds and tertiary alcohols.

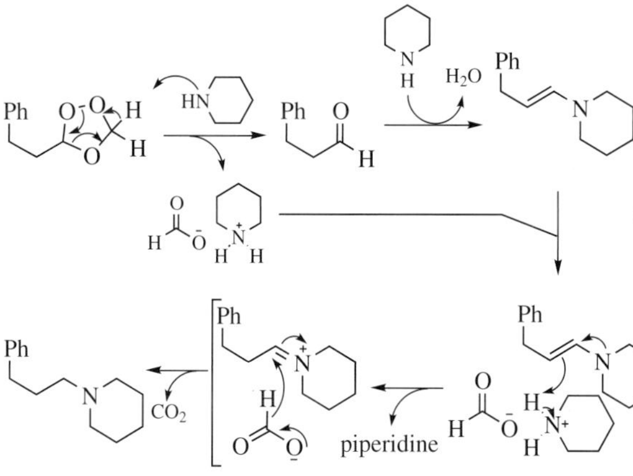

(51)

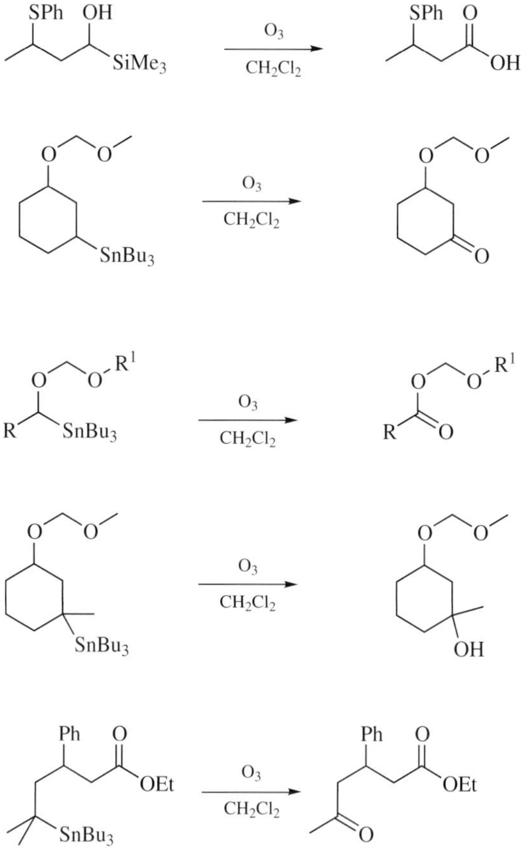

(52)

In addition to tin and silicon, carbon-nitrogen double bonds as hydrazones, oximes, and nitrones can be treated with ozone to form carbonyl compounds (eq 53).[75–77] While the transformation can be accomplished on several types of substrates, ketone-derived starting materials (e.g., ketoximes) perform better in the reaction, as the aldehyde-derived starting materials have the potential for overoxidation to the corresponding carboxylic acid product.

Phosphorus-containing compounds provide useful substrates for ozonolysis reactions as well and can provide several products depending on the reaction workup. Several biological uses exist for β-amino-α-hydroxy phosphonic acid derivatives and they can be readily prepared by ozonolysis of N-(ethoxycarbonyl)-β-amino-α-methylene phosphonic esters after reductive workup with sodium borohydride (eq 54).[78] When the reaction mixture is treated with sodium hydroxide in MeOH, an anomalous ozonolysis reaction occurs and cleavage of the methylene as well as the carbon-phosphorus bond occurs to yield N-(ethoxycarbonyl)-α-amino methyl carboxylic esters.

(Cyanomethylene)phosphoranes also provide useful and easy to obtain substrates for ozonolysis reactions. Acid chlorides and carboxylic acids can be reacted with (cyanomethylene)triphenylphosphorane in the presence of N,O-bis(trimethylsilyl)acetamide (BSA) or 1-ethyl-3-(3-dimethylaminopropyl)carbodimide (EDC or EDCI) and N,N-dimethylaminopyridine (DMAP), respectively, to yield the starting phosphoranes (eq 55).[79]

The resulting (cyanomethylene)phosphoranes can be treated with ozone followed by nucleophiles to yield an array of carbonyl compounds (eq 56). Addition of alcohols to the reaction medium yields α-keto esters while addition of water leads to α-keto acids. Addition of amines leads to α-keto amides thereby offering another method for the synthesis of amides, a common structural motif in natural products.

$$(57)$$

1, 2, 3, 4, 6: R = OH, $R^1 = R^2 =$ -CH=CH-CH=CH-

7, 8, 9, 10: R = $R^1 = R^2 =$ H

11, 12, 13, 14: **a** R = OH, $R^1 = n$-C_8H_{17}, $R^2 = CH_3$; **b** R = OH, $R^1 = i$-C_4H_9, $R^2 = CH_3$

c R = OH, $R^1 = R^2 =$ -CH$_2$CH$_2$CH$_2$CH$_2$-; **d** R = OH, $R^1 =$ H, $R^2 = CH_3$

Method A: HOAc, rt, 0.5 h; Method B: HOAc:H$_2$O (1:1), rt, 1 h;

Method C: CH$_2$Cl$_2$:EtOH (1:1), rt, 1 h; Method D: dry CH$_2$Cl$_2$, rt, 0.5 h

Numerous heterocyclic compounds can be oxidized with ozone to deliver derivatives useful in synthesis and the reaction products can be tuned according to the additives and conditions in the reaction medium. When treated with ozone, pyrimidine-2-thiones and 2-thiouracils react to give several pyrimidine derivatives (eq 57).[80] Use of aprotic solvents or ozonolysis without an active nucleophile yields dimerization products, whereas ozonolysis in protic solvents or in the presence of a nucleophile leads to the sulfinic acid derivative that can then be converted to several products depending on workup conditions. The parent pyrimidine can be isolated when acid is introduced into the medium, and with an equal volume of water the pyrimidinone product is isolated. While a protic solvent, EtOH in the medium acts as a nucleophile to deliver the 2-ethoxypyrimidine.

In a similar manner, 2-thiouracils can be converted to numerous products when treated with ozone in the presence of dimethyldioxirane (DMDO or DDO). Again, a sulfinic acid intermediate is observed and the products of the reaction are dependent on the additives in the system. This methodology allows for the arrival at functionalized uracils in a manner akin to the reactivity pattern observed with pyrimidine-2-thiones (eq 58).[81]

A novel fragmentation of N-arylidene- or N-(alkylideneamino)-β-lactams can be induced by ozone to lead to various enol ethers after a reductive workup with sodium borohydride.[82] The starting β-lactams can be prepared via [2 + 2] cycloaddition of alkoxy ketenes and an azine and upon treatment with ozone at low temperature, yield the expected secondary ozonides (eq 59). Reduction of the ozonide leads to the corresponding N-nitroso intermediate, which is susceptible to fragmentation of the C4–N1 bond to give a zwitterion intermediate that rearranges to yield the product enol ethers. In the reaction sequence, *trans*-β-lactams yield predominantly the *E*-enol ether while the *cis*-β-lactams preferentially form the *Z*-configured enol ethers.

α-Amino-β-hydroxy esters are another set of useful building blocks that can be obtained from the ozonolysis of heterocyclic compounds.[83] Addition of lithiated alkoxyallenes to carbohydrate-derived aldonitrones yields 1,2-oxazines, which are treated with ozone in MeOH and fragmented across the enol ether resident in the molecule to yield the target esters after treatment with acetic anhydride and triethylamine (eq 60). In the series of oxazines studied, only the *cis*-oriented oxazines yielded useful products. The *trans*-series yielded products from addition of a methoxy substituent across the enol ether double bond. The target esters can also be functionalized into useful amino triols after N-benzyl deprotection with concomitant N-protection via hydrogenation in the presence of palladium on carbon and Boc$_2$O, ester reduction with lithium aluminum hydride, selective protection of the resulting primary alcohol with *t*-butylchlorodiphenylsilane (TBDPSCl), and removal of the isopropylidene protecting group with Amberlist-15 acidic resin.

Nitration of Aromatic Rings. Ozone can serve as a coreagent for the nitration of aromatic rings in the *kyodai*-nitration protocol, which is achieved by treating an aromatic substrate with nitrogen dioxide. This method is especially useful as the use of ozone with NO$_2$ allows for the nitration of deactivated aromatic systems, even a low temperature (eq 61).[84–92] A wide range of product ratios has been observed insofar as the amount of *mono-* and *di*-nitro products. In addition, varying degrees of *ortho-* and *para*-substitution patterns have emerged and can be substrate, solvent, or concentration dependent.

The *kyodai*-nitration has some very useful advantages over traditional conditions that require the use of high heat or strong protic acids to achieve nitration. In addition to mildly deactivated systems such as acetanilides and phenolic esters, the *kyodai*-nitration protocol is "forcing" enough to induce nitration of many deactivated aromatic systems such as halogenobenzenes, polyhalogenobenzenes, aromatic carbonyls, aryl chlorides, aryl carboxylic acid salts, nitrobenzenes, dinitrobenzenes, nitrophenols, and polyaromatic systems. While this heightened reactivity is observed, the conditions are gentle enough to be used for nitration of more sensitive aromatic systems such as aromatic acetals

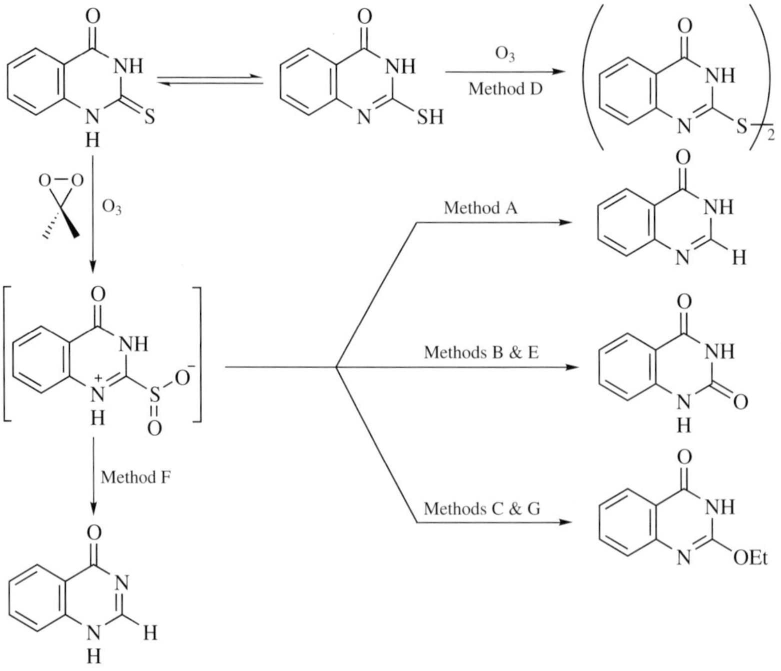

(58)

Method A: HOAc, rt, 0.5 h; Method B: HOAc/H$_2$O (1:1), rt, 1 h
Method C: CH$_2$Cl$_2$/EtOH (1:1), rt; Method D: CH$_2$Cl$_2$, rt, 0.5 h
Method E: DMDO, CH$_2$Cl$_2$, rt; Method F: DMDO, water, rt
Method G: DMDO, CH$_2$Cl$_2$/EtOH (1:1), rt

and acylals. Aryl sulfides can be subjected to *kyodai*-nitration to deliver nitrated aryl sulfoxides. In some highly deactivated systems, the addition of Lewis acids such as boron trifluoride etherate or aluminum trichloride or protic acids such as methanesulfonic acid is necessary to achieve successful nitration.

Miscellaneous and Anomalous Ozonolysis. Numerous cases of nearby alcohols and other functional groups participating in novel fragmentation of the ozonide can be found; however, their general use in synthesis is somewhat limited. In addition, "anomalous" ozonolysis, usually defined as an ozonolysis reaction where the double bond as well as an adjacent single bond is cleaved, can be found throughout the literature. Most of these miscellaneous

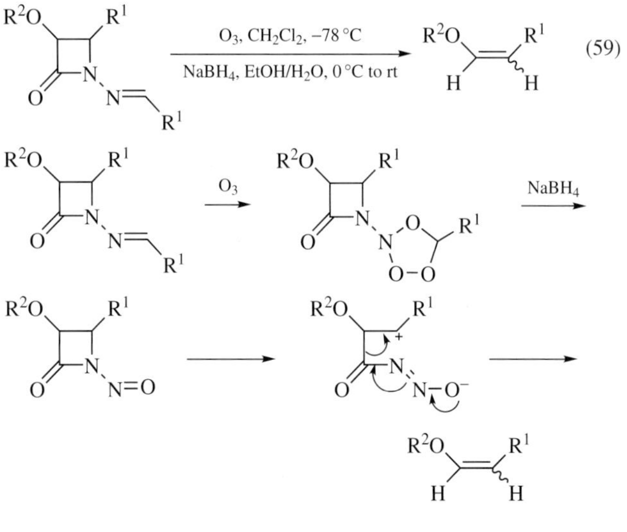

and anomalous ozonolysis reactions do not fit into a general class of reactivity. There are, however, several cases where general reactivity patterns can be found and these reactions can be used to form useful albeit specific structural motifs. Cyclic allylic alcohols

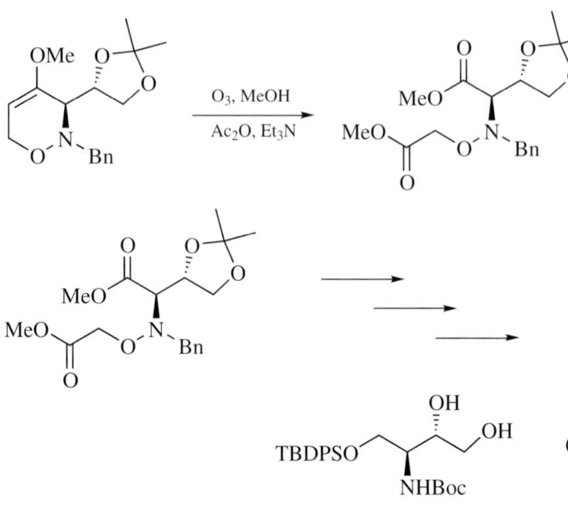

(59)

(60)

can be treated with ozone in the presence of NaOH to form the intermediate ozonide, which is fragmented after deprotonation of the allylic alcohol. Migration of the group on the alcohol carbon can be achieved when the group is larger than methyl to give a variety of products resulting in oxidation of the alcohol to the ketone (eq 62).[93]

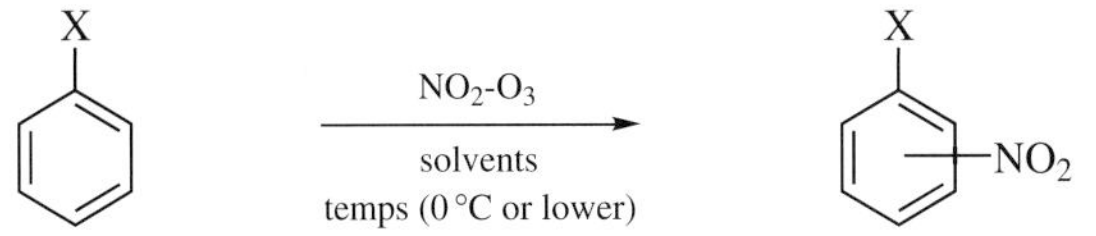

$$\text{(61)}$$

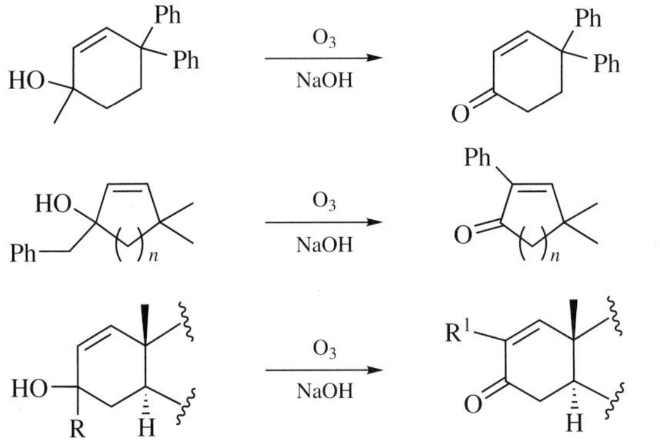

$$\text{(62)}$$

R = Me, R^1 = H; R = Et, R^1 = Me

R = n-Pr, R^1 = Et; R = i-Bu, R^1 = i-Pr

R = Bn; R^1 = Ph

During the course of a synthesis of the steroid ouabain, a methylene cyclobutane was synthesized and treated with ozone followed by reductive workup with Me$_2$S. In the event, the expected ketone was not obtained. Instead, the primary ozonide was fragmented via a Grob-like fragmentation initiated by the neighboring alcohol (eq 63).[94]

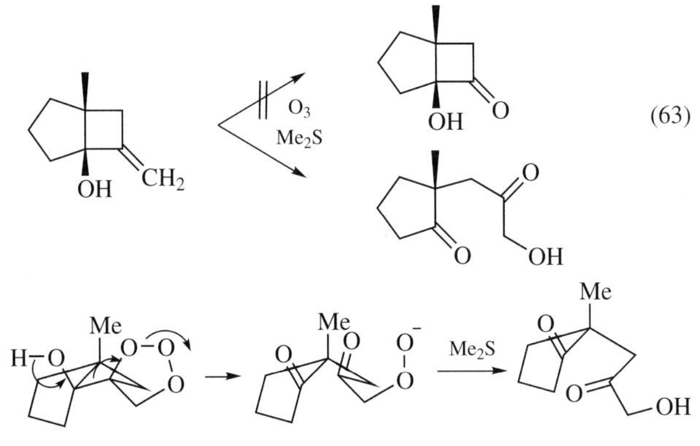

$$\text{(63)}$$

The resulting α-hydroxy ketones were obtained in good yield and could be useful building blocks for synthesis. The ozonolysis of a similarly functionalized camphor-derived starting material provides additional proof for the suggested mechanism (eq 64). In addition, it was possible to control the stereochemistry of the alcohol.

When a Z-configured methylenecyclobutene was subjected to ozonolysis at low temperature, only ozonolysis from the more accessible exo-face of the bicycle occurred thereby allowing for stereoselective installation of the hydroxyl group. Interestingly,

the corresponding E-configured starting material provided the expected bicyclic ketone in good yield and no explanation for the difference of reactivity was offered. The participation of the neighboring hydroxyl group was also verified by ozonolysis of the starting materials with the neighboring amine and alcohol protected. In these cases, the neighboring group could not participate in the fragmentation due to protection and only the ketone resulting from normal ozonolysis was obtained.

Starting Olefin $\xrightarrow[\text{Me}_2\text{S}]{\overset{\text{O}_3}{\text{CH}_2\text{Cl}_2, -78\,°\text{C}}}$ Product $\quad$ (64)

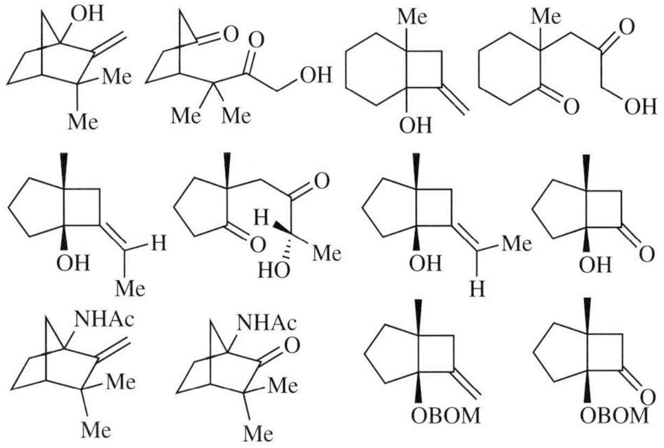

While the ozonolysis of the camphor-derived substrate gave insight into the probable mechanism for formation of the α-hydroxy carbonyls, the course of the reaction cannot be considered general since the ozonolysis of a slightly different bicyclo-alkan-1-ol led to a 4:1 mixture of products favoring the pinacolic Wagner-Meerwein rearrangement over the Grob-like fragmentation pathway (eq 65).[95]

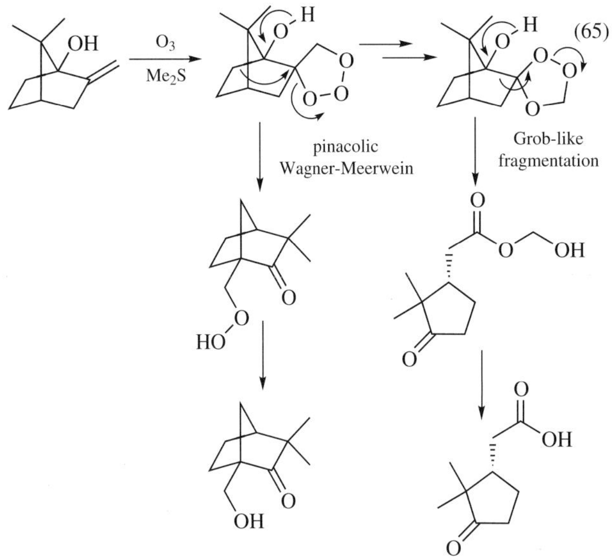

$$\text{(65)}$$

pinacolic Wagner-Meerwein

Grob-like fragmentation

Some biologically active natural products contain the 1,2-dioxolane and 1,2-dioxane structural motif (eq 66). In studies aimed at the formation of this functional group, Dussault was able to afford 1,2-dioxolanes in a single step in good yield via the spontaneous 5-exo-cyclization of hydroperoxyacetals after their treatment with ozone at low temperature (eq 66).[96] Use of this methodology for the synthesis of the 1,2-dioxanes was not as successful as incorporation of a methoxy substituent allowed for oxocarbenium ion formation and then slow cyclization to mixtures of isomers of several products.

core of Plakinic acid A   core of Peroxyplakoric acid A_1 core of Plakortide F

(66)

As discussed previously, the ozonolysis of selenium in organic molecules leads to formation of the selenoxide, which is typically eliminated to form carbon-carbon double bonds. It is possible to use ozone in systems containing a selenium atom provided a more reactive functional group is present. Several examples of the successful ozonolysis of a carbon-carbon double bond in systems containing a phenylseleno moiety allow for functionalization of double bonds while preserving the synthetic handle resident in the starting olefin (eq 67).[97]

Seleno Olefins $\xrightarrow[\text{PPh}_3]{\substack{\text{O}_3 \\ \text{CH}_2\text{Cl}_2, \, -78\,^\circ\text{C}}}$ Seleno Carbonyls (67)

1. (a) Bailey, P. S., *Chem. Rev.* **1958**, *58*, 925. (b) Bailey, P. S. *Ozonation in Organic Chemistry*; Academic: New York, 1978; Vol. 1. (c) Bailey, P. S. *Ozonation in Organic Chemistry*; Academic: San Diego CA, 1982; Vol. 2. (d) Razumovskii, S. D.; Zaikov, G. E. *Ozone and Its Reactions With Organic Compounds*; Elsevier: Amsterdam, 1984.

2. Odinokov, V. N.; Tolstikov, G. A., *Russ. Chem. Rev. (Engl. Transl.)* **1981**, *50*, 636.

3. Varkony, H.; Pass, S.; Mazur, Y., *J. Chem. Soc., Chem. Commun.* **1974**, 437.

4. Dmitrov, A.; Seppelt, K.; Schleffler, D.; Willner, H., *J. Am. Chem. Soc.* **1998**, *120*, 8711.

5. Dietz, R. N.; Pruzansky, J.; Smith, J. D., *Angew. Chem.* **1973**, *45*, 402.

6. (a) Pryor, W. A.; Giamalva, D.; Church, D. F., *J. Am. Chem. Soc.* **1983**, *105*, 6858. (b) Fleet, G. W. J., *Org. React. Mech.* **1984**, 179.

7. Gilles, C. W.; Kuczkowski, R. L., *Isr. J. Chem.* **1983**, *24*, 446.

8. (a) Murray, R. W., *Acc. Chem. Res.* **1968**, *1*, 313. (b) Criegee, R., *Angew. Chem., Int. Ed. Engl.* **1975**, *14*, 745. (c) Razumovskii, S. D.; Zaikov, G. E., *Russ. Chem. Rev. (Engl. Transl.)* **1980**, *49*, 1163. (d) Kuczkowski, R. L., *Acc. Chem. Res.* **1983**, *16*, 42.

9. For further discussion of reductive or oxidative reagents, see: (a) Belew, J. S. In *Oxidation*; Augustine, R. L., Ed.; Dekker: New York, 1969; Vol. 1, p 259. (b) Hudlicky, M. *Oxidation in Organic Chemistry*; American Chemical Society: Washington, 1990.

10. Pappas, J. J.; Keaveney, W. P.; Gancher, E.; Berger, M., *Tetrahedron Lett.* **1966**, 4273.

11. Gupta, D.; Soman, R.; Dev, S. K., *Tetrahedron* **1982**, *38*, 3013.

12. Ferraboschi, P.; Gambero, C.; Azadani, M. N.; Santaniello, E., *Synth. Commun.* **1986**, *16*, 667.

13. Schreiber, S. L.; Claus, R. E.; Reagan, J., *Tetrahedron Lett.* **1982**, *23*, 3867.

14. (a) Cardinale, G.; Grimmelikhuysen, J. C.; Laan, J. A. M.; Ward, J. P., *Tetrahedron* **1984**, *40*, 1881. (b) Cardinale, G.; Laan, J. A. M.; Ward, J. P., *Tetrahedron* **1985**, *41*, 2899.

15. (a) Benton, F. L.; Kiess, A. A., *J. Org. Chem.* **1960**, *25*, 470. (b) Diaper, D. G. M.; Mitchell, D. L., *Can. J. Chem.* **1962**, *40*, 1189. (c) Pollart, K. A.; Miller, R. E., *J. Org. Chem.* **1962**, *27*, 2392. (d) White, R. W.; King, S. W.; O'Brien, J. L., *Tetrahedron Lett.* **1971**, 3591.

16. Ellam, R. M.; Padbury, J. M., *J. Chem. Soc., Chem. Commun.* **1972**, 1086.

17. (a) Corey, E. J.; Katzenellenbogen, J. A.; Gilman, N. W.; Roman, S. A.; Erickson, B. W., *J. Am. Chem. Soc.* **1968**, *90*, 5618. (b) Effenberger, F., *Angew. Chem., Int. Ed. Engl.* **1969**, *8*, 295. (c) Keul, H.; Choi, H.-S.; Kuczkowski, R. L., *J. Org. Chem.* **1985**, *50*, 3365. (d) Wojciechowski, B. J.; Pearson, W. H.; Kuczkowski, R. L., *J. Org. Chem.* **1989**, *54*, 115. (e) Wojciechowski, B. J.; Chiang, C.-Y. Kuczkowski, R. L., *J. Org. Chem.* **1990**, *55*, 1120. (f) Griesbaum, K.; Kim, W.-S.; Nakamura, N.; Mori, M.; Nojima, M.; Kusabayashi, S., *J. Org. Chem.* **1990**, *55*, 6153. (g) Kuczkowski, R. L. *Advances in Oxygenated Processes*; JAI Greenwich, CT, 1991.

18. (a) Danishefsky, S.; Kato, N.; Askin, D.; Kerwin, J. F., Jr., *J. Am. Chem. Soc.* **1982**, *104*, 360. (b) Hillers, S.; Niklaus, A.; Reiser, O., *J. Org. Chem.* **1993**, *58*, 3169.

19. Vedejs, E.; Larsen, S. D., *J. Am. Chem. Soc.* **1984**, *106*, 3031.

20. (a) Clark, R. D.; Heathcock, C. H., *Tetrahedron Lett.* **1974**, 2027. (b) Clark, R. D.; Heathcock, C. H., *J. Org. Chem.* **1976**, *41*, 1396.

21. (a) Chaussin, R.; Leriverend, P.; Paquer, D., *J. Chem. Soc., Chem. Commun.* **1978**, 1032. (b) Strobel, M.-P.; Morin, L.; Paquer, D., *Tetrahedron Lett.* **1980**, 523. (c) Barillier, D. Strobel, M. P., *Nouv. J. Chim.* **1982**, *6*, 201. (d) Barillier, D.; Vazeux, M., *J. Org. Chem.* **1986**, *51*, 2276.

22. Witkop, B., *J. Am. Chem. Soc.* **1956**, *78*, 2873.

23. Ziegler, F. E.; Fang, J.-M., *J. Org. Chem.* **1981**, *46*, 825.

24. Dauben, W. G.; Wight, H. G.; Boswell, G. A., *J. Org. Chem.* **1958**, *23*, 1787.

25. Wang, Z.; Zvlichovsky, G., *Tetrahedron Lett.* **1990**, *31*, 5579.

26. (a) Bailey, P. S.; Lane, A. G., *J. Am. Chem. Soc.* **1967**, *89*, 4473. (b) Bailey, P. S.; Ward, J. W.; Hornish, R. E.; Potts, F. E., III, *Adv. Chem. Ser.* **1972**, *112*, 1. (c) Griesbaum, K.; Zwick, G., *Chem. Ber.* **1985**, *118*, 3041.

27. Bailey, P. S.; Hwang, H. H.; Chiang, C.-Y., *J. Org. Chem.* **1985**, *50*, 231.

28. Hochstetler, A. R., *J. Org. Chem.* **1975**, *40*, 1536.

29. Silbert, L. S.; Foglia, T. A., *Angew. Chem.* **1985**, *57*, 1404.

30. Yang, N. C.; Libman, J., *J. Org. Chem.* **1974**, *39*, 1782.

31. McCurry, P. M., Jr.; Abe, K., *Tetrahedron Lett.* **1974**, 1387.

32. Cannon, J. G.; Darko, L. L., *J. Org. Chem.* **1964**, *29*, 3419.

33. Ballistreri, F. P.; Failla, S.; Tomaselli, G. A.; Curci, R., *Tetrahedron Lett.* **1986**, *27*, 5139.

34. Wisaksono, W. W.; Arens, J. F., *Recl. Trav. Chim. Pays-Bas* **1961**, *80*, 846.

35. Woodward, R. B.; Cava, M. P.; Ollis, W. D.; Hunger, A.; Daeniker, H. U.; Schenker, K., *Tetrahedron* **1963**, *19*, 247.

36. Dobinson, F.; Bailey, P. S., *Tetrahedron Lett.* **1960**, (13), 14.

37. O'Murchu, C., *Synthesis* **1989**, 880.

38. Kashima, C.; Harada, K.; Hosomi, A., *Heterocycles* **1992**, *33*, 385.

39. Schmid, G.; Fukuyama, T.; Akasaka, K.; Kishi, Y., *J. Am. Chem. Soc.* **1979**, *101*, 259.

40. Caminade, A.; El Khatib, F.; Baceiredo, A.; Koenig, M., *Phosphorus Sulfur Silicon* **1987**, *29*, 365.

41. Thompson, Q. E., *J. Am. Chem. Soc.* **1961**, *83*, 845.

42. Caminade, A. M.; El Khatib, F.; Koening, M., *Phosphorus Sulfur Silicon* **1983**, *14*, 381.

43. Skowronska, A.; Krawczyk, E., *Synthesis* **1983**, 509.

44. Horner, L.; Schaefer, H.; Ludwig, W., *Chem. Ber.* **1958**, *91*, 75.

45. Bachman, G. B.; Strawn, K. G., *J. Org. Chem.* **1968**, *33*, 313.

46. Grese, T. A.; Hutchinson, K. D.; Overman, L. E., *J. Org. Chem.* **1993**, *58*, 2468.

47. (a) Wilbaut, J. P.; Sixma, F. L. J.; Kampschmidt, L. W. F.; Boer, H., *Recl. Trav. Chim. Pays-Bas* **1950**, *69*, 1355. (b) Sixma, F. L. J.; Boer, H.; Wilbaut, J. P.; Pel, H. J.; de Bruyn, J., *Recl. Trav. Chim. Pays-Bas* **1951**, *70*, 1005. (c) Wilbaut, J. P.; Boer, H., *Recl. Trav. Chim. Pays-Bas* **1955**, *74*, 241.

48. (a) Shepherd, D. A.; Donia, R. A.; Campbell, J. A.; Johnson, B. A.; Holysz, R. P.; Slomp, G., Jr.; Stafford, J. E.; Pederson, R. L.; Ott, A. C., *J. Am. Chem. Soc.* **1955**, *77*, 1212. (b) Slomp, G., Jr., *J. Org. Chem.* **1957**, *22*, 1277. (c) Slomp, G. Jr.; Johnson, J. L., *J. Am. Chem. Soc.* **1958**, *80*, 915. (d) Boddy, I. K.; Boniface, P. J.; Cambie, R. C.; Craw, P. A.; Huang, Z.-D.; Larsen, D. S.; McDonald, H.; Rutledge, P. S.; Woodgate, P. D., *Angew. Chem., Int. Ed. Engl.* **1984**, *37*, 1511. (e) Haag, T.; Luu, B.; Hetru, C., *J. Chem. Soc., Perkin Trans. 1* **1988**, 2353.

49. Veysoglu, T.; Mitscher, L. A.; Swayze, J. K., *Synthesis* **1980**, 807.

50. Isobe, K.; Mohri, K.; Tokoro, K.; Fukushima, C.; Higuchi, F.; Taga, J.-I.; Tsuda, Y., *Chem. Pharm. Bull.* **1988**, *36*, 1275.

51. (a) Deslongchamps, P.; Moreau, C., *Can. J. Chem.* **1971**, *49*, 2465. (b) Deslongchamps, P.; Moreau, C.; Fréhel, D.; Atlani, P., *Can. J. Chem.* **1972**, *50*, 3402. (c) Deslongchamps, P.; Atlani, P.; Fréhel, D.; Malaval, A.; Moreau, C., *Can. J. Chem.* **1974**, *52*, 3651. (d) Deslongchamps, P.; Moreau, C.; Fréhel, D.; Chênevert, R., *Can. J. Chem.* **1975**, *53*, 1204. (e) Deslongchamps, P., *Tetrahedron* **1975**, *31*, 2463.

52. Waters, W. L.; Rollin, A. J.; Bardwell, C. M.; Schneider, J. A.; Aanerud, T. W., *J. Org. Chem.* **1976**, *41*, 889.

53. Kolsaker, P.; Teige, B., *Acta Chem. Scand.* **1970**, *24*, 2101.

54. Corey, E. J.; Jones, G. B., *Tetrahedron Lett.* **1991**, *32*, 5713.

55. Hirama, M.; Shimizu, M., *Synth. Commun.* **1983**, *13*, 781.

56. Angibeaud, P.; Defaye, J.; Gadelle, A.; Utille, J.-P., *Synthesis* **1985**, 1123.

57. Hirama, M.; Uei, M., *J. Am. Chem. Soc.* **1982**, *104*, 4251.

58. (a) McMurry, J. E.; Melton, J.; Padgett, H., *J. Org. Chem.* **1974**, *39*, 259. (b) Crossley, M. J.; Crumbie, R. L.; Fung, Y. M.; Potter, J. J.; Pegler, M. A., *Tetrahedron Lett.* **1987**, *28*, 2883. (c) Aizpurua, J. M.; Oiarbide, M.; Palomo, C., *Tetrahedron Lett.* **1987**, *28*, 5365.

59. Dick, C. R.; Hanna, R. F., *J. Org. Chem.* **1964**, *29*, 1218.

60. Sundararaman, P.; Walker, E. C.; Djerassi, C., *Tetrahedron Lett.* **1978**, 1627.

61. Marshall, J. A.; Garofalo, A. W., *J. Org. Chem.* **1993**, *58*, 3675.

62. Ozone; Berglund, R. A., Ed. In *Encyclopedia of Reagents for Organic Synthesis*; Paquette, L. A., Ed. in Chief; Wiley: West Sussex, UK, 1995.

63. Hon, Y.-S.; Lin, S.-W.; Chen, Y.-J., *Synth. Commun.* **1993**, *23*, 1543.

64. Appell, R. B.; Tomlinson, I. A.; I, H., *Synth. Commun.* **1995**, *25*, 3589.

65. Ito, Y.; Ida, H.; Matsuura, T., *Tetrahedron Lett.* **1978**, *19*, 3119.

66. Patawardhan, A. P.; Lu, Z.; Pulgam, V. R.; Wulff, W. D., *Org. Lett.* **2005**, *7*, 2201.

67. Sylvain, C.; Wagner, A.; Mioskowski, C., *Tetrahedron Lett.* **1997**, *38*, 1043.

68. Marshall, J. A.; Garofalo, A. W.; Sedrani, R. C., *Synlett* **1992**, 643.

69. Marshall, J. A.; Garofalo, A. W., *J. Org. Chem.* **1993**, *58*, 3675.

70. Hon, Y.-S.; Lu, L., *Tetrahedron Lett.* **1993**, *34*, 5309.

71. Noecker, L.; Giuliano, R. M.; Cooney, M.; Boyko, W.; Zajac, W. W., Jr., *J. Carbohydr. Chem.* **2002**, *21*, 539.

72. Zhang, M.-X.; Eaton, P. E.; Gilardi, R., *Angew. Chem. Int. Ed.* **2000**, *39*, 401.

73. Linderman, R. J.; Chen, K., *Tetrahedron Lett.* **1992**, *33*, 6767.

74. Linderman, R. J.; Jaber, M., *Tetrahedron Lett.* **1994**, *35*, 5993.

75. Yang, Y.-T.; Li, T.-S.; Li, Y.-L., *Synth. Commun.* **1993**, *23*, 1121.

76. Kraft, P.; Eichenberger, W.; Frech, D., *Eur. J. Org. Chem.* **2005**, 3233.

77. Gagnon, J. L.; Walters, T. R.; Zajac, Jr., W. W.; Buzby, J. H., *J. Org. Chem.* **1993**, *58*, 6712.

78. Francavilla, M.; Gasperi, T.; Loreto, M. A.; Tardella, P. A.; Bassetti, M., *Tetrahedron Lett.* **2002**, *43*, 7913.

79. Wasserman, H. A.; Ho, W.-B., *J. Org. Chem.* **1994**, *59*, 4364.

80. Crestini, C.; Saladino, R.; Nicoletti, R., *Tetrahedron Lett.* **1993**, *34*, 1631.

81. Crestini, C.; Mincione, E.; Saladino, R.; Nicoletti, R., *Tetrahedron* **1994**, *50*, 3259.

82. Alcaide, B.; Miranda, M.; Pérez-Castells, J.; Sierra, M. A., *J. Org. Chem.* **1993**, *58*, 297.

83. Helms, M.; Reißig, H.-U., *Eur. J. Org. Chem.* **2005**, 998.

84. Mori, T.; Suzuki, H., *Synlett* **1995**, 383.

85. Suzuki, H.; Mori, T.; Maeda, K., *J. Chem. Soc., Chem. Commun.* **1993**, 1335.

86. Suzuki, H.; Murashima, T.; Kozai, I.; Mori, T., *J. Chem. Soc., Perkin Trans. 1* **1993**, 1591.

87. Suzuki, H.; Yonezawa, S.; Mori, T.; Maeda, K., *J. Chem. Soc., Perkin Trans. 1* **1994**, 1367.

88. Suzuki, H.; Tomaru, J.; Murashima, T., *J. Chem. Soc., Perkin Trans. 1* **1994**, 2413.

89. Suzuki, H.; Murashima, T., *J. Chem. Soc., Perkin Trans. 1* **1994**, 903.

90. Suzuki, H.; Mori, T., *J. Chem. Soc., Perkin Trans. 2* **1995**, 41.

91. Suzuki, H.; Mori, T.; Maeda, K., *Synthesis* **1994**.

92. Nose, M.; Suzuki, H., *Synthesis* **2002**, 1065.

93. DeNinno, M. P., *J. Am. Chem. Soc.* **1995**, *117*, 9927.

94. Jung, M. E.; Davidov, P., *Org. Lett.* **2001**, *3*, 627.

95. Martínez, A. G.; Vilar, E. T.; Fraile, A. G.; de la Moya Cerero, S.; Maroto, B. L., *Tetrahedron Lett.* **2005**, *46*, 5157.

96. Dai, P.; Dussault, P. H., *Org. Lett.* **2005**, *7*, 4333.

97. Clive, D. L. J.; Postema, M. H. D., *J. Chem. Soc., Chem. Commun.* **1994**, 235.

P

Pinacolborane

[25015-63-8] C$_6$H$_{13}$BO$_2$ (MW 127.98)

InChI = 1S/C6H13BO2/c1-5(2)6(3,4)9-7-8-5/h7H,1-4H3

InChIKey = UCFSYHMCKWNKAH-UHFFFAOYSA-N

(monofunctional hydroborating agent, also used in transition metal catalyzed aryl couplings)

Alternate Name: 4,4,5,5-tetramethyl-1,3,2-dioxaborole.

Physical Data: bp 42–43 °C, 50 mm Hg; *fp* 5 °C; *d* 0.882 g cm^{-3}.

Solubility: soluble in ether, THF, CH$_2$Cl$_2$, and other organic solvents.

Form Supplied in: neat colorless liquid and also as 1.0 M solution in THF.

Analysis of Reagent Purity: ^{1}H NMR, ^{11}B NMR (δ 28.0, d).

Preparative Methods: reaction of borane-methyl sulfide with pinacol at 0 °C in CH$_2$Cl$_2$ furnishes pinacolborane.[1] This solution can be used directly for further reactions or can be distilled to obtain a colorless liquid (*bp* 42–43 °C, 50 mm Hg) (eq 1).

Alternatively, the reaction of H$_3$B·N(Ph)Et$_2$ with pinacol at room temperature and subsequent distillation produces the title compound in 75% yield (75 mmol scale).[26]

Purity: distillation.

Handling, Storage, and Precautions: packaged under nitrogen. Store and handle under a nitrogen atmosphere and refrigerate.

Original Commentary

P. Veeraraghavan Ramachandran & J. Subash Chandra

Purdue University, West Lafayette, IN, USA

Hydroboration. Pinacolborane is a stable, easily prepared and stored hydroborating agent. Unlike catecholborane[2] which requires harsh reaction conditions for hydroboration of alkenes (100 °C) and alkynes (70 °C), hydroboration with pinacolborane proceeds under mild conditions furnishing the boronates. Knochel and co-workers[1] observed an excellent level of regioselectivity for hydroboration of alkynes with pinacolborane at room temperature (eq 2). Alkenes, however, react slowly with pinacolborane and often require heating for 2–3 days to furnish the terminal pinacolboronates as the major regioisomer (>98%) (eq 3).

Metal-catalyzed Hydroboration. Pereira and Srebnik discovered that HZrCp$_2$Cl is an excellent catalyst for hydroboration of alkynes[3] (eq 4) and terminal alkenes[4] (eq 5) with pinacolborane. However, HZrCp$_2$Cl is not compatible with many functional groups and hence the effect of other catalysts, such as Rh(PPh$_3$)$_3$Cl and Rh(PPh$_3$)$_2$(CO)Cl was studied. Wilkinson's catalyst [Rh(PPh$_3$)$_3$Cl] hydroborates alkynes with very poor regioselectivity (eq 6), however, terminal alkenes undergo facile hydroboration (eq 7). Hydroboration of internal alkenes with pinacolborane in the presence of HZrCp$_2$Cl or Rh(PPh$_3$)$_3$Cl leads to isomerization furnishing the terminal pinacolboronates[4] (eq 8). Changing the catalyst system to Rh(PPh$_3$)$_2$(CO)Cl[4] overcomes this problem and the expected internal boronate is obtained as the major product (eq 9). This catalyst also dramatically increases the regioselectivity in the hydroboration of alkynes[5] (eq 10). Recently Pt(dba)$_2$/P(2,4,6-MeO-C$_6$H$_2$)$_3$ has also been reported as an efficient catalyst for hydroboration of alkynes with pinacolborane under mild reaction conditions and in good yields.[6,7]

71%   29%

Essential Reactions for Organic Synthesis, First Edition. Edited by Philip L. Fuchs.

© 2016 John Wiley & Sons, Ltd. Published 2016 by John Wiley & Sons, Ltd.

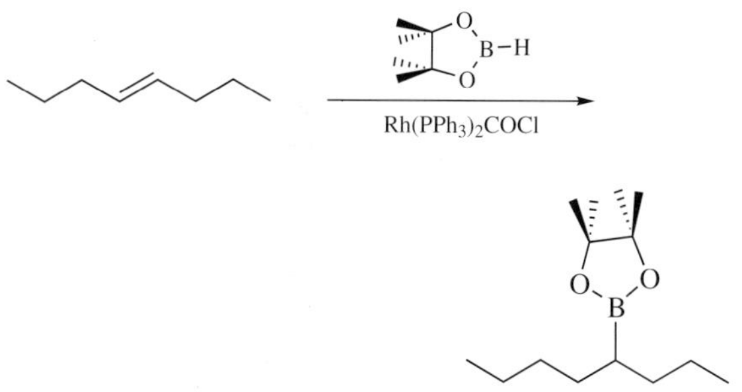

(9)

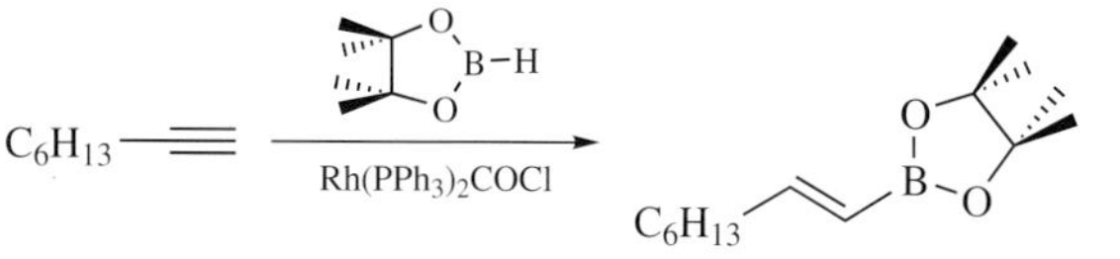

(10)

Vinylic ethers, acetals, and esters, also undergo catalytic hydroboration with pinacolborane without difficulty.

Vinyl bromides, however, do not provide the expected hydroboration product under these conditions. Instead, initial hydroboration occurs β- to the bromine atom and is followed by a fast *syn*-elimination to furnish the terminal alkene. This then undergoes hydroboration to provide the debrominated boronate. The intermediate *B*-bromopinacolborane cleaves the ether C–O bond in the solvent THF to provide 4-bromobutanol upon oxidation[4,8] (eq 11).

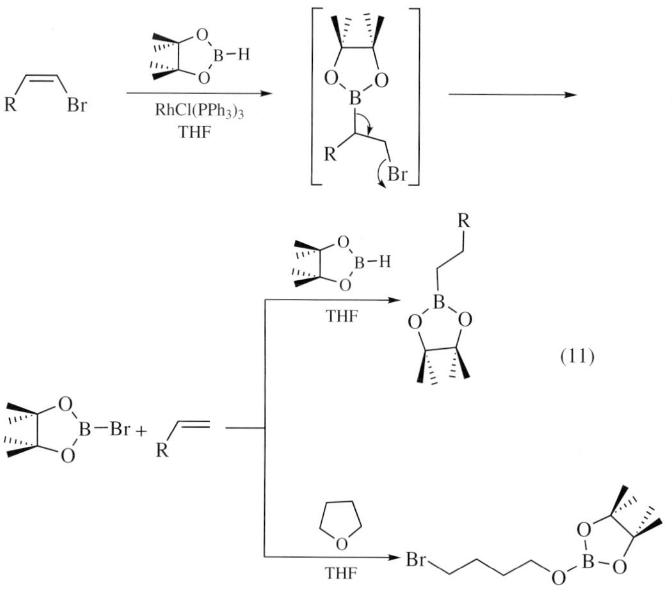

(11)

Hydroboration of allenes with pinacolborane provides the allylboronate or vinylboronates regioselectively depending on the bulk and basicity of the supporting phosphine ligand[6] (eq 12).

Recently Miyaura and co-workers[9] have reported a *trans*-hydroboration of terminal alkynes using [Rh(COD)Cl]$_2$[P(^{i}Pr)$_3$]$_4$ or [Ir(COD)Cl]$_2$[P(^{i}Pr)$_3$]$_4$ (eq 13). Mechanistic studies via deuterium labeling show that after the oxidative addition of the alkyne to the metal, the acetylenic deuterium undergoes migration to the β-carbon resulting in the formation of a vinylidene metal complex. Oxidative addition of borane to the metal complex and 1,2-

boryl migration to the α-carbon results in the stereospecific formation of thermodynamically stable alkenylmetal complex. This subsequently undergoes reductive elimination to provide the *Z*-vinylboronate as the sole product (eq 14).

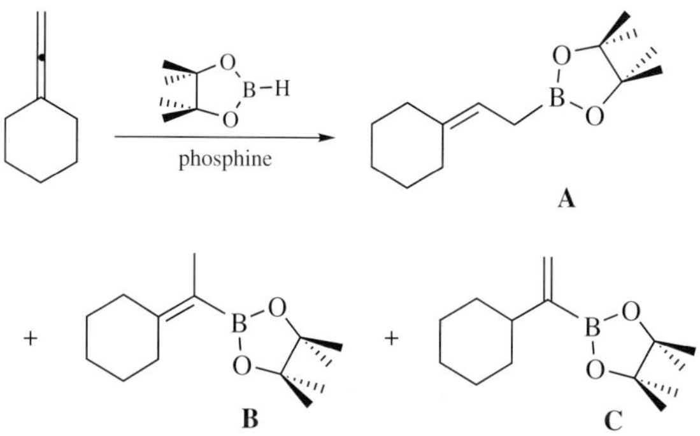

(12)

	A	**B**	**C**
Pd(C$_6$H$_{11}$)$_3$	56	0	44
P[2,4,6-(OCH$_3$)$_3$C$_6$H$_2$]$_3$	1	0	99
P[C(CH$_3$)$_3$]$_3$	0	100	0

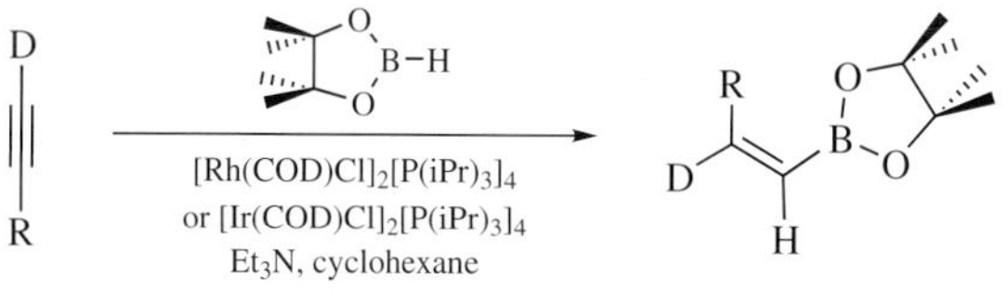

(13)

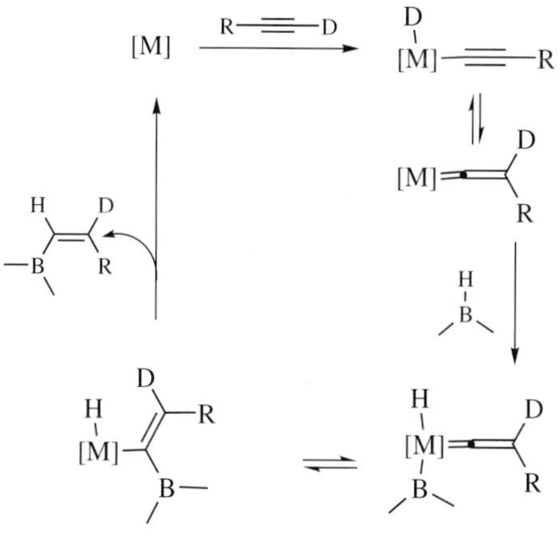

(14)

Generation of Boron Enolates. Mukaiyama and co-workers[10] have reported the formation of boron enolates from α-iodoketones by reaction with pinacolborane in the presence of a base (pyridine or Et$_3$N). The resulting enolborates upon reaction with aldehydes furnished the aldol products in moderate yield and diastereoselectivity (eq 15). Cyclic α-iodoketones provided higher diastereoselectivities (>98% *syn*) than the acyclic α-iodoketones (*syn:anti* 83:17). The moderate yields were ascribed to the decomposition of the highly labile boron enolates during their isolation or during their reaction with aldehydes.

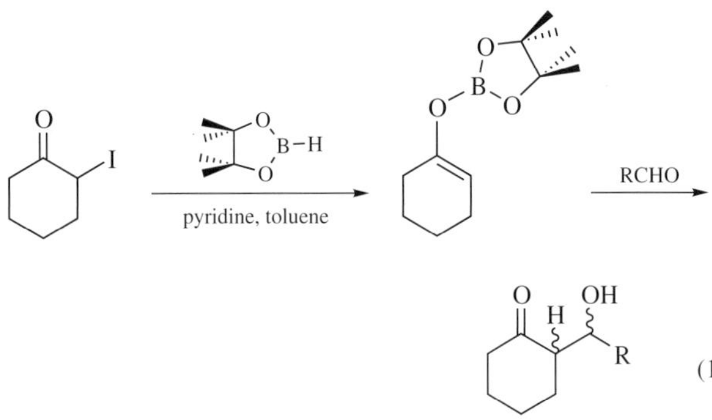

(15)

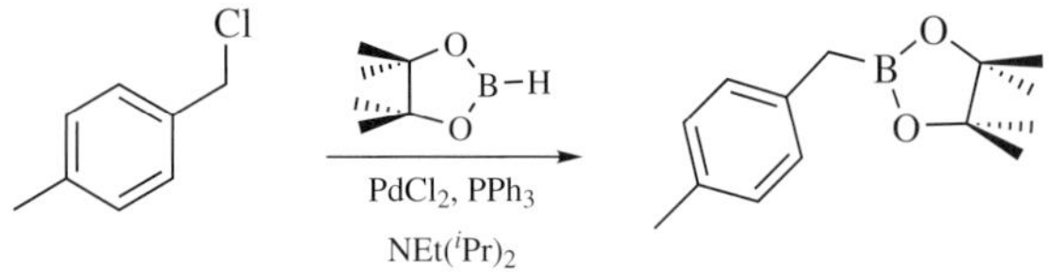

syn:anti >98:<2

(18)

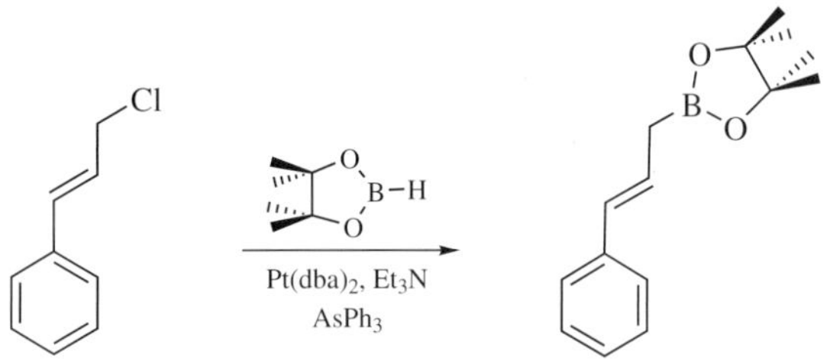

(19)

Suzuki–Miyaura Cross Coupling. Arylboronates are valuable reagents in organic synthesis owing to their widespread use in Suzuki-Miyaura cross coupling reactions.[11,12] Pinacolborane is extensively used in the borylation of aryl halides in the presence of a base (typically pyridine, Et$_3$N, or KOAc) and a catalytic amount of PdCl$_2$(dppf) affording arylboronates[13,14] (eq 16).

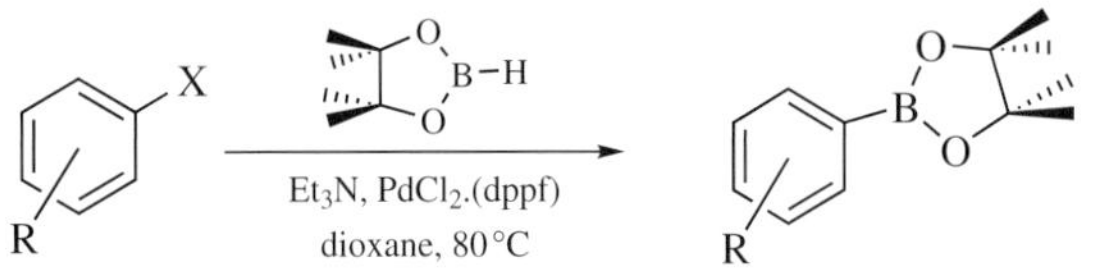

(16)

X = Br, I, OTf
R = Alkyl, Aryl, nitrile, ester, ketone, ether, amine, etc.

Pinacolborane is tolerant towards several functional groups including esters, ketones, ethers, tertiary amines, nitriles, etc. The resulting pinacolboronates are stable to air and moisture and can be purified by column chromatography. Hence the reaction has broad scope and can be used on a variety of substrates. Aryl iodides react faster than bromides or triflates. The typical solvents are dioxane, toluene, acetonitrile, and 1,2 dichloroethane. Pinacolborane reacts with polar solvents such as DMF, decomposing to pinacolatodiboron. Electron donating groups such as -NMe$_2$ increase the reactivity of the aryl halides in this reaction.

Aryl chlorides typically do not react under these conditions. However, Miyaura[15] has been able to extend the scope of this reaction to include aryl chlorides by changing the catalyst system to Pd(dba)$_2$ and PCy$_3$ and replacing pinacolborane with bis(pinacolato)diboron. Vinyl iodides and triflates undergo borylation with pinacolborane under similar conditions in the presence of triphenylarsine (AsPh$_3$)[16] (eq 17). Benzylic halides react with pinacolborane in the presence of PdCl$_2$, PPh$_3$, and *N,N*-diisopropylethyl amine[17] (eq 18). Borylation of allylic halides in the presence of Pt(dba)$_2$, AsPh$_3$, and Et$_3$N leads to highly regio- and stereoselective allylboronates[18] (eq 19).

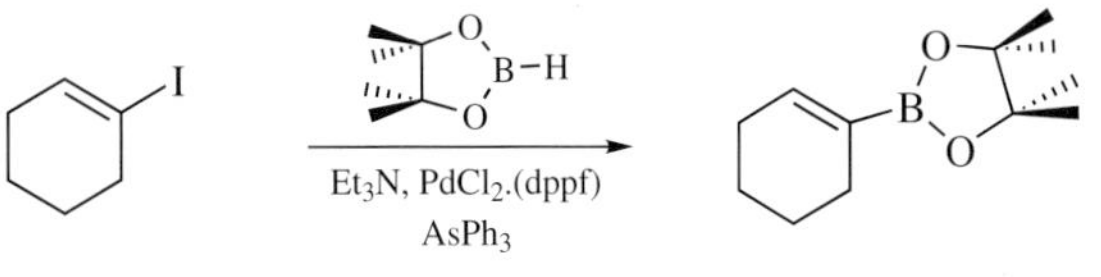

Dehydrogenative Borylation. Murata and co-workers[19,20] observed that the reaction of pinacolborane with olefins in the presence of bis(chloro-1,5-cyclooctadienylrhodium) at room temperature provides vinyl pinacolboronate. It is interesting to note that dehydrogenative borylation occurs in the presence of phosphine-free rhodium catalyst whereas olefin hydroboration is the predominant reaction with phosphine-containing rhodium catalysts such as Rh(PPh$_3$)$_2$COCl and Wilkinson's catalyst (eq 20). However, Westcott and co-workers[21] were successful in achieving dehydrogenative borylation of vinyl ethers under refluxing conditions in the presence of Wilkinson's catalyst (eq 21).

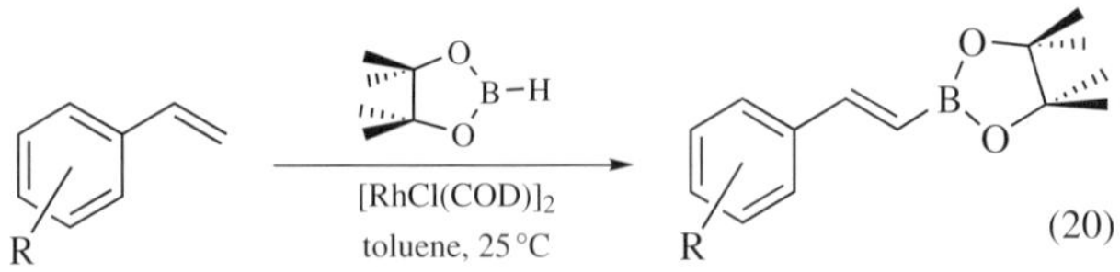

(20)

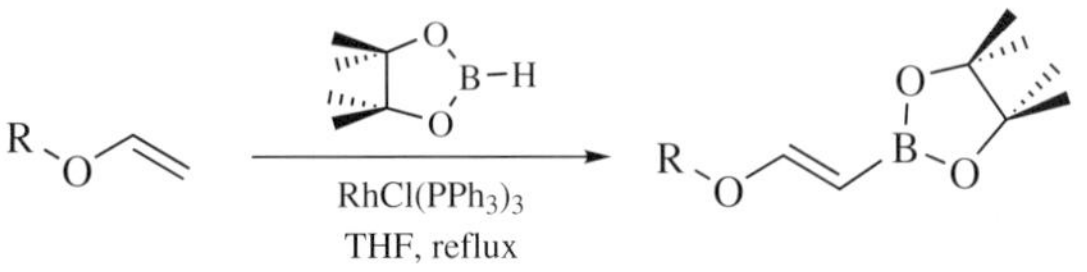

(21)

Smith and Marder reported the dehydrogenative borylation of arenes, yielding arylboronates, with pinacolborane in the presence of rhodium and iridium catalysts such as Cp*Rh(η^4-C$_6$Me$_6$),[22,23] CpIrPMe$_3$,[24] and [RhClP(^{i}Pr)$_3$]$_2$N$_2$[25] (eq 22). Toluene and other methyl substituted arenes react with pinacolborane in the presence of [RhClP(^{i}Pr)$_3$]$_2$N$_2$ and furnish benzylboronates via benzylic C–H activation and dehydrogenative borylation[25] (eq 23).

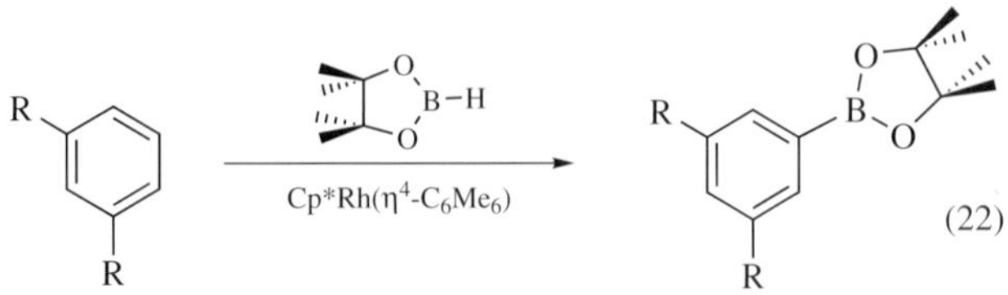

(22)

R = Alkyl, OR, NR$_2$ etc.

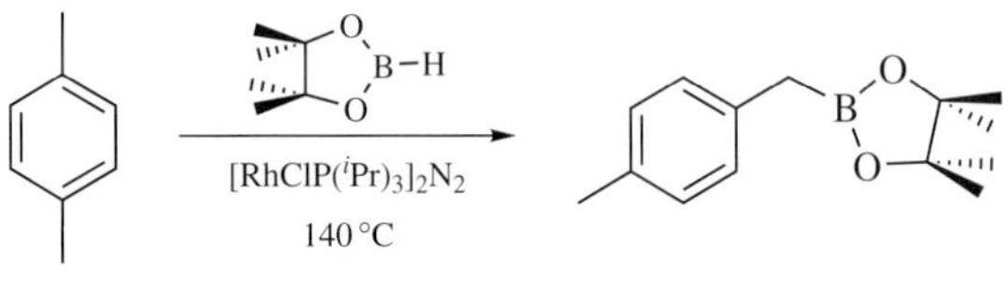

(23)

(17)

Related Reagents. Catecholborane (benzo-1,3,2-dioxaborole); bis(pinacolato)diboron.

First Update

Abel Ros, Rosario Fernández & José M. Lassaletta

Instituto de Investigaciones Químicas (IIQ), Sevilla, Spain

Transition-metal-catalyzed Hydroboration of Alkynes. It has been observed that alkynes can be hydroborated at room temperature using HBpin (2 equiv) to afford the corresponding *anti*-Markovnikov products with high regio- and *E*-stereoselectivity. *Z*-Vinylboronates could be obtained with complete regio- and stereoselectivity via hydroboration of terminal alkynes catalyzed by a nonclassical ruthenium hydride (eq 24).[27] Geometrically defined *Z*-vinylboronates bearing a synthetically useful α-alkoxycarbonyl group have been obtained by a CuH-catalyzed 1,2-addition/transmetalation sequence from acetylenic esters (eq 25).[28]

R = Ph, *n*-hex, *i*-Pent
CH₂OPh, CH₂NMeBn

Cat. (0.1–0.2 mol %)
Toluene, –15 °C

Z selectivity >95%
52–95% yield

(24)

R = Ph, Et, *n*-Pr
n-C₅H₁₁, TMS
R′ = Me, Et

[(PPh₃)CuH] (2 mol %)
PPh₃ (3 mol %)
THF, rt

Z/E 5:1 to >25:1
80–95% yield

(25)

Hydroboration of Alkenes. Pinacolborane is able to hydroborate a wide variety of alkenes in the presence of rhodium catalysts, with the major product being the linear adduct (eq 26).[29] Linear regioselectivity is attributable to the greater steric demand of the pinacol group relative to the catechol analog. Pinacolborane is also able to hydroborate cyclopropenes in the presence of Wilkinson's catalyst, affording cyclopropylboronates with good *trans/cis* selectivities.[30] The introduction of a coordinating alkoxycarbonyl or alkoxymethyl groups in the cyclopropene substrate in presence of [Rh(COD)Cl]₂/(R)-BINAP or related chiral catalysts lead to the corresponding *cis*-cyclopropylboronates as single diastereomers in nearly quantitative yields and with excellent enantioselectivities (eq 27). Although rhodium-catalyzed hydroboration of alkenes have demonstrated that the linear

pinacolboronate is formed as the major product,[29] it has been[31] demonstrated that cationic rhodium complexes bearing chelating phosphines catalyze the formation of branched pinacolboronates as the major isomers. Hydroboration of styrene derivatives catalyzed by the system (R,S)-Josiphos/[Rh(COD)₂]⁺BF₄⁻ (1.2:1) at 25 °C afforded the corresponding branched pinacolboronates in moderate to excellent yields and with high levels of regio- and enantioselectivity (eq 28). Interestingly, a reversal in the sense of enantioselection was observed relative to catecholborane. It has also been observed that iridium-based catalysts such us [Ir(COD)Cl]₂/dppb promoted the hydroboration of vinylarenes with HBpin to give the linear isomer as the only observed product (eq 29). The change in regioselectivity was attributed to a change in mechanism from Rh–H insertion to Ir–B insertion. Based on this observation and exploiting that hydroboration in the β-position (linear product) of 1,1-disubstituted olefins generates chiral boronates,[32] an iridium-catalyzed asymmetric hydroboration of terminal olefins using the system [Ir(OMe)COD]₂/phosphinooxazoline ligand as the catalyst was developed (eq 30). Hydroboration reactions occurred with complete regioselectivity at β-position, yielding the desired pinacolboronates in excellent yields and with enantioselectivities up to 92%.

R = Ar, alkyl
CH₂OPh, CH₂Si(OEt)₃
OCOCH₃

Rh(PPh₃)₃Cl (1 mol %)
CH₂Cl₂, 25 °C

BPin

linear product >90%
>80% yield

(26)

R = Ph, Me, TMS, CO₂Me
R¹ = CO₂Me, CO₂Et, CH₂OMe

[Rh(COD)Cl]₂ (3 mol %)
(R)-BINAP (6 mol %)
THF, rt, 20 min

cis/trans >99:1, ee 87–98%
92–99% yield

(27)

Ar = Ph, Tol, 2-Naph
p-OMe-C₆H₄
p-Cl-C₆H₄, *p*-Br-C₆H₄

[Rh(COD)₂]BF₄ (5 mol %)
(R,S)-Josiphos (6.5 mol %)
DCE, rt

BPin

linear:branched 80:20
ee 80–88%
40–90% yield

(28)

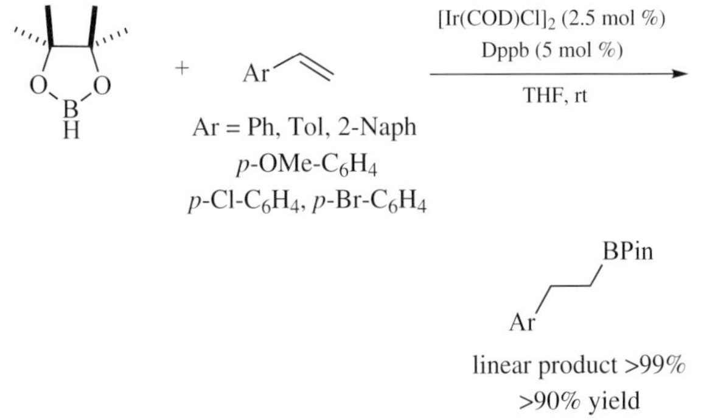

Ar = Ph, Tol, 2-Naph
p-OMe-C$_6$H$_4$
p-Cl-C$_6$H$_4$, p-Br-C$_6$H$_4$

[Ir(COD)Cl]$_2$ (2.5 mol %)
Dppb (5 mol %)
THF, rt

$$\text{(29)}$$

linear product >99%
>90% yield

Ar = Ph, Tol, p-OMe-C$_6$H$_4$, p-F-
C$_6$H$_4$ p-Cl-C$_6$H$_4$, p-Br-C$_6$H$_4$
R = Me, Et, Cy

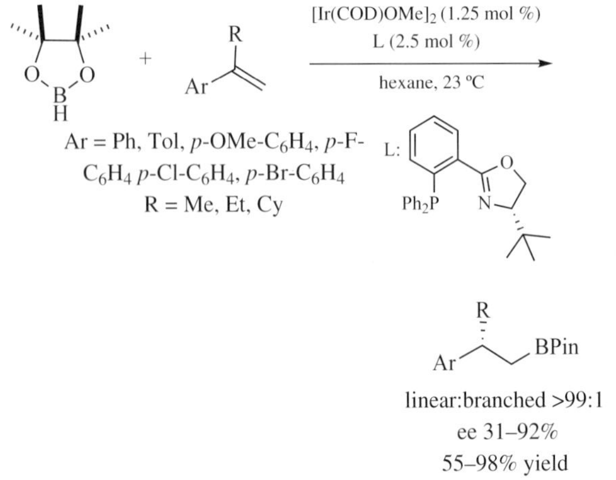

[Ir(COD)OMe]$_2$ (1.25 mol %)
L (2.5 mol %)
hexane, 23 °C

$$\text{(30)}$$

linear:branched >99:1
ee 31–92%
55–98% yield

1,4-Hydroboration of 1,3-Dienes. Allylboranes are versatile intermediates in organic synthesis employed mainly in the synthesis of allylic alcohols after oxidation, allylation of aldehydes and imines to give homoallylic alcohols and amines, and Suzuki–Miyaura cross-coupling reactions. The hydroboration of 1,3-dienes constitutes an elegant and efficient methodology for the synthesis allylboranes. The introduction of the pinacolborane moiety provides air and water stability, and allylpinacolboronates can be conveniently purified by chromatography on silica gel. An iron-catalyzed methodology for the 1,4-hydroboration of 1,3-dienes has been developed that was reported to be chemo-, regio-, and stereoselective.[33] The linear (E)-γ-disubstituted allyl pinacolboronates were produced in 66–92% yields, with linear:branched ratios of 90:10 to 99:1 and >99:1 E:Z ratios (eq 31).

The combination of a nitrogen-based ligand, FeCl$_2$, and a catalytic amount of activated Mg as reducing agent was necessary for catalysis. In certain cases, the 1,4-hydroboration to the linear or branched product was controlled by modulation of R^2 substituent of the ligand. The 1,4-hydroboration reaction tolerates electrophilic functionalities, such as esters and acetals in the 1,3-diene substrates, and can also be used in a one-pot hydroboration–allylation reaction. Though the reported methodology is efficient, it is limited to the hydroboration of 2-substituted 1,3-dienes and is less general for terminally substituted substrates. As such,[34] a complementary methodology was developed for the 1,4-hydroboration of 1-substituted dienes with HBpin catalyzed by nickel complexes. The system Ni(COD)$_2$/PCy$_3$ (1:1) catalyzes, at room temperature in a few hours, the 1,4-hydroboration of a wide family of 1-substituted dienes containing different functional groups to give, after oxidation, the corresponding Z-allylic alco-

hols in 29–93% yields (eq 32). This methodology was extended to 1,4-substituted 1,3-dienes, and the borylation took place with high regioselectivity at the less hindered carbon.

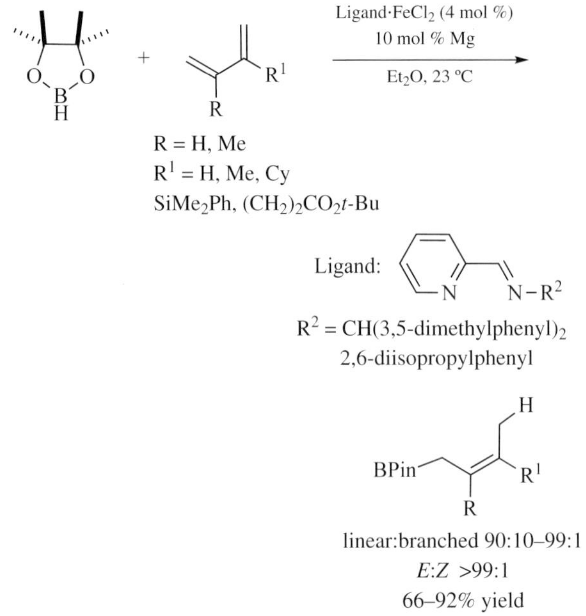

R = H, Me
R^1 = H, Me, Cy
SiMe$_2$Ph, (CH$_2$)$_2$CO$_2$$t$-Bu

Ligand·FeCl$_2$ (4 mol %)
10 mol % Mg
Et$_2$O, 23 °C

Ligand:

R^2 = CH(3,5-dimethylphenyl)$_2$
2,6-diisopropylphenyl

$$\text{(31)}$$

linear:branched 90:10–99:1
E:Z >99:1
66–92% yield

R = Ph, n-hex, n-pent
CH$_2$OTBDPS, (CH$_2$)$_2$CO$_2$Et
(CH$_2$)$_2$OH
R^1 = H, Me; R^2 = H, Me

(1) Ni(COD)$_2$ (2.5 mol %)
PCy$_3$ (2.5 mol %)
Toluene, rt
(2) H$_2$O$_2$, NaOH
THF

$$\text{(32)}$$

Z:E >99:1
29–93% yield

Complete stereoselectivity was obtained, and following oxidation the corresponding substituted Z-allylic alcohols were obtained in 54–91% yields (eq 33).

R= Ph, n-hex, Cy
R^1 = n-pent, CH$_2$OTBS
CH$_2$OH, Me

(1) Ni(COD)$_2$ (2.5 mol %)
PCy$_3$ (2.5 mol %)
Toluene, rt
(2) H$_2$O$_2$, NaOH
THF

$$\text{(33)}$$

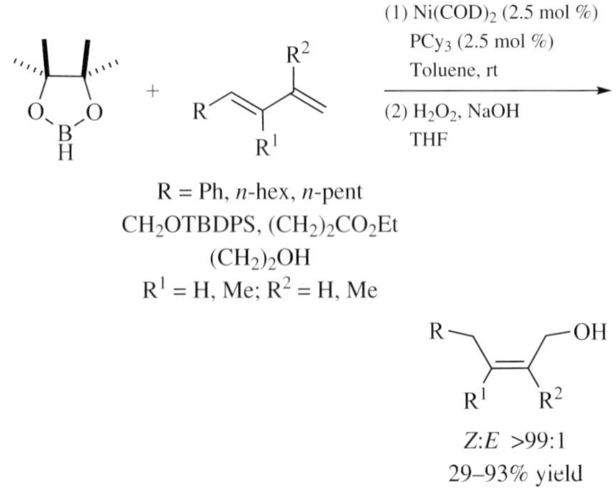

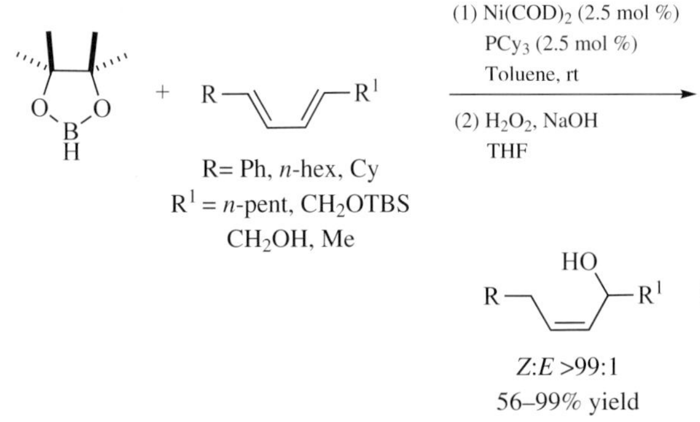

Z:E >99:1
56–99% yield

In a similar manner to the Fe-catalyzed 1,4-hydroboration,[33] a one-pot hydroboration–allylation procedure was carried out to give the corresponding homoallylic alcohol with a quaternary stereocenter in good yield and with high diastereoselectivity.

C–H Borylation on Arenes. The transformation of a C–H bond to a C–B bond is an elegant and useful methodology, accessing valuable boronic esters in an atom-economical manner.[35,36] Various direct C–H borylation of arenes and heteroarenes have been reported using HBpin as the boron source and the 1:2 $[Ir(OMe)COD]_2$/dtbpy (dtbpy = 4,4′-di-*tert*-butyl bipyridine) system as the catalyst (eq 34). The corresponding borylation reactions were performed at room temperature using 1.1 equiv of HBpin to afford the desired pinacol boronic esters in 22–99% yield. The regioselectivity was governed by steric factors, furnishing a single regioisomer only for substrates with C–H bonds with well-distinguished levels of steric hindrance.

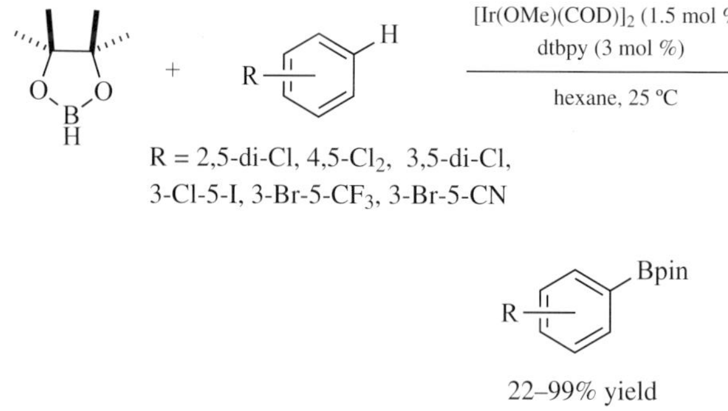
[Ir(OMe)(COD)]2 (1.5 mol %)
dtbpy (3 mol %)
hexane, 25 °C
R = 2,5-di-Cl, 4,5-Cl2, 3,5-di-Cl,
3-Cl-5-I, 3-Br-5-CF3, 3-Br-5-CN
Bpin
22–99% yield

(34)

A catalytic system based on the use on hemilabile nitrogen ligand was designed able to catalyze the nitrogen-directed *ortho*-borylation of aromatic *N,N*-dimethylhydrazones (eq 35).[37] The system (1:2) $[Ir(OMe)COD]_2$/pyridinohydrazone catalyzed the borylation of a broad family of aromatic *N,N*-dimethylhydrazones giving the *ortho*-borylated products in good to excellent conversions. A partial decomposition of these products was regularly observed during chromatographic purification on silica gel or alumina, but a one-pot borylation/Suzuki–Miyaura coupling was developed to afford functionalized biaryl products in moderate to excellent yields (56–99%). The overall procedure tolerates both electron donating and withdrawing groups in *ortho*, *meta*, and *para* positions of the phenyl ring, as well as disubstituted derivatives.

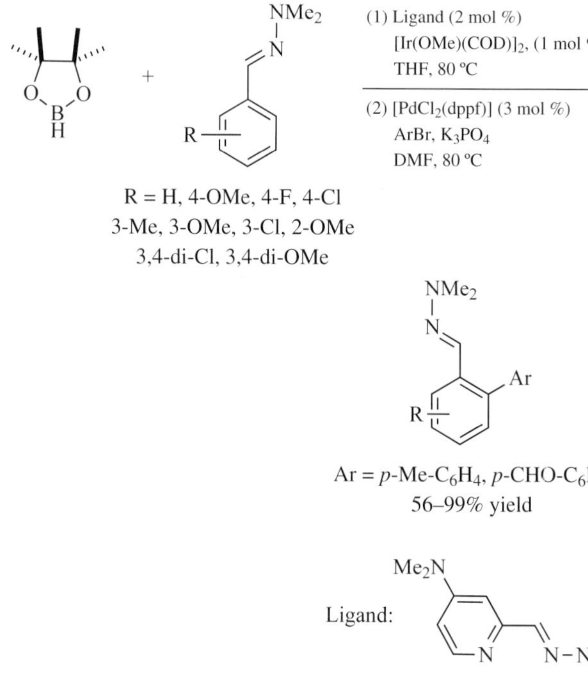
(1) Ligand (2 mol %)
[Ir(OMe)(COD)]2, (1 mol %)
THF, 80 °C
(2) [PdCl2(dppf)] (3 mol %)
ArBr, K3PO4
DMF, 80 °C
R = H, 4-OMe, 4-F, 4-Cl
3-Me, 3-OMe, 3-Cl, 2-OMe
3,4-di-Cl, 3,4-di-OMe
Ar = p-Me-C6H4, p-CHO-C6H4
56–99% yield
Ligand:

(35)

Pinacolborane as Reducing Agent. Although pinacolborane has been extensively used as borylating agent, it has been also successfully employed as a mild hydride source. A borylated Ru complex analog to Shvo's catalyst was applied for the hydroboration of aldehydes, imines, and ketones.[38] It was reported that a Ru-dimer (2 mol %) in the presence of 1.5 equiv of pinacolborane generates a catalytically active species for the hydroboration of aldehydes in a very efficient manner (eq 36). The hydroboration reaction was applicable to a variety of aromatic and aliphatic aldehydes to afford, after workup, the corresponding alcohols in moderate to good yields. The methodology was also extended to the hydroboration of imines and ketones although with a limited scope.

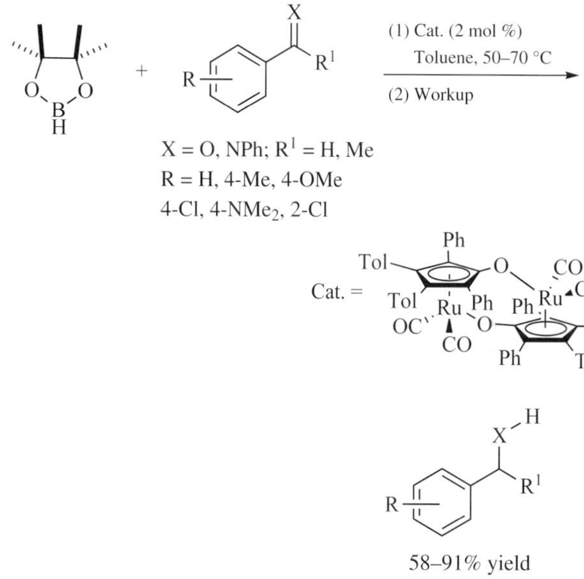
(1) Cat. (2 mol %)
Toluene, 50–70 °C
(2) Workup
X = O, NPh; R1 = H, Me
R = H, 4-Me, 4-OMe
4-Cl, 4-NMe2, 2-Cl
Cat. =
58–91% yield

(36)

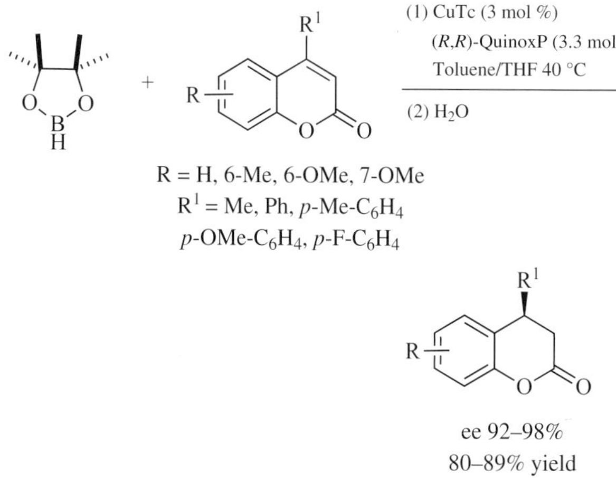
(1) CuTc (3 mol %)
(R,R)-QuinoxP (3.3 mol %)
Toluene/THF 40 °C
(2) H2O
R = H, 6-Me, 6-OMe, 7-OMe
R1 = Me, Ph, p-Me-C6H4
p-OMe-C6H4, p-F-C6H4
ee 92–98%
80–89% yield

(37)

A Cu-catalyzed asymmetric 1,4-reduction of coumarins was developed employing pinacolborane as the reducing agent.[39] Screening led to the identification of the 1:1.1 copper(I)

$$\text{(38)}$$

Cat: (R)-DTBM-SEGPHOS (5 mol %)
PCy$_3$ (5 mol %), CuOAc (2.5 mol %)
γ/α 3:1–25:1, ee 84–99%, 70–96% yield

$$\text{(39)}$$

Cat: Taniaphos (5 mol %)/CuF·3PPh$_3$·2EtOH (2.5 mol %)
only α product, syn/anti 6:1–10:1, ee 66–84%, 86–91% yield

thiophene-2-carboxylate/(R,R)-QuinoxP combination as the most active and selective catalyst to afford 4-aryl- and 4-alkyl-substituted dihydrocoumarins with full conversions and excellent ee values (92–98%) (eq 37). The role of pinacolborane as reducing agent was crucial, as reactions did not take place when catecholborane, diethylborane, or 9-BBN was used as the reagents. It was also delineated that the chiral boron enolate could be trapped with another electrophile instead of water, and a 3,4-substituted dihydrocoumarin as single diastereoisomer can be obtained.

An asymmetric reductive aldol reaction of allenic esters with ketones using pinacolborane has also been reported.[40] Employing a chiral copper(I) complex as the catalyst, the reaction proceeds with complete chemoselectivity (allenic ester vs. ketone in the initial reduction step), but the α,γ-regioselectivity in the vinylogous aldol reaction step was highly dependent on the structure of chiral diphosphine ligands used. The combination CuOAc/(R)-DTBM-SEGPHOS/PCy$_3$ (1:2:2) afforded the $\gamma - cis$-adducts in excellent yields, good γ:α regioselectivities (3:1–25:1), and excellent enantioselectivities (84–99%) (eq 38). The regioselectivity could be switched to α:$\gamma >$ 99:1 when Taniaphos-type ligands were used. After a ligand screening, the system CuF·3PPh$_3$·2EtOH/Taniaphos (1:2), where the Taniaphos ligand contains a bulky di-(3,5-xylyl)phosphine and a morpholine unit, showed to be the most active and selective catalyst affording the α-adducts in >86% yields with complete regioselectivity, good syn/anti diastereoselectivities (6:1–10:1), and moderate to good enantioselectivities (eq 39).

1. Tucker, C. E.; Davidson, J.; Knochel, P., *J. Org. Chem.* **1992**, *57*, 3482.

2. Kabalka, G. W., *Org. Prep. Proced. Int.* **1977**, *9*, 131.

3. Pereira, S.; Srebnik, M., *Organometallics* **1995**, *14*, 3127.

4. Pereira, S.; Srebnik, M., *J. Am. Chem. Soc.* **1996**, *118*, 909.

5. Pereira, S.; Srebnik, M., *Tetrahedron Lett.* **1996**, *37*, 3283.

6. Yamamoto, Y.; Fujikawa, R.; Yamada, A.; Miyaura, N., *Chem. Lett.* **1999**, 1069.

7. Yamamoto, Y.; Kurihara, K.; Yamada, A.; Takahashi, M.; Takahashi, Y.; Miyaura, N., *Tetrahedron* **2003**, *59*, 537.

8. Colin, S.; Vaysse-Ludot, L.; Lecouve, J.-P.; Maddaluno, J., *J. Chem. Soc., Perkin Trans. 1* **2000**, 4505.

9. Ohmura, T.; Yamamoto, Y.; Miyaura, N., *J. Am. Chem. Soc.* **2000**, *122*, 4990.

10. Mukaiyama, T.; Takuwa, T.; Yamane, K.; Imachi, S., *Bull. Chem Soc. Jpn.* **2003**, *76*, 813.

11. Suzuki, A., *J. Organomet. Chem.* **1999**, *576*, 147.

12. Miyaura, N.; Suzuki, A., *Chem. Rev.* **1995**, *95*, 2457.

13. Murata, M.; Oyama, T.; Watanabe, S.; Masuda, Y., *J. Org. Chem.* **2000**, *65*, 164.

14. Murata, M.; Watanabe, S.; Masuda, Y., *J. Org. Chem.* **1997**, *62*, 6458.

15. Ishiyama, T.; Ishida, K.; Miyaura, N., *Tetrahedron* **2001**, *57*, 9813.

16. Murata, M.; Oyama, T.; Watanabe, S.; Masuda, Y., *Synthesis* **2000**, *6*, 778.

17. Murata, M.; Oyama, T.; Watanabe, S.; Masuda, Y., *Synth. Commun.* **2002**, *32*, 2513.

18. Murata, M.; Watanabe, S.; Masuda, Y., *Tetrahedron Lett.* **2000**, *41*, 5877.

19. Murata, M.; Kawakita, K.; Asana, T.; Watanabe, S.; Masuda, Y., *Bull. Chem. Soc. Jpn.* **2002**, *75*, 825.

20. Murata, M.; Watanabe, S.; Masuda, Y., *Tetrahedron Lett.* **1999**, *40*, 2585.

21. Vogels, C. M.; Hayes, P. G.; Shaver, M. P.; Westcott, S. A., *Chem. Commun.* **2000**, 51.

22. Tse, M. K.; Cho, J.-Y.; Smith, M. R., *Org. Lett.* **2001**, *3*, 2831.

23. Cho, J.-Y.; Iverson, C. N.; Smith, M. R., *J. Am. Chem. Soc.* **2000**, *122*, 12868.

24. Iverson, C. N.; Smith, M. R., *J. Am. Chem. Soc.* **1999**, *121*, 7696.

25. Shimada, S.; Batsanov, A. S.; Howard, J. A. K.; Marder, T. B., *Angew. Chem., Int. Ed.* **2001**, *40*, 2168.

26. Kikuchi, T.; Nobuta, Y.; Umeda, J.; Yamamoto, Y.; Ishiyama, T.; Miyaura, N., *Tetrahedron* **2008**, *64*, 4967.

27. Gunanathan, C.; Hölscher, M.; Pan, F.; Leitner, W., *J. Am. Chem. Soc.* **2012**, *134*, 14349.

28. Lipshutz, B. H.; Bošković, Ž. V.; Aue, D. H., *Angew. Chem., Int. Ed.* **2008**, *47*, 10183.

29. Crudden, C. M.; Edwards, D., *Eur. J. Org. Chem.* **2003**, 4695.

30. Rubina, M.; Rubin, M.; Gevorgyan, V., *J. Am. Chem. Soc.* **2003**, *125*, 7198.

31. Crudden, C. M.; Hleba, Y. B.; Chen, A. C., *J. Am. Chem. Soc.* **2004**, *126*, 9200.

32. Mazet, C.; Gérard, D., *Chem. Commun.* **2011**, *47*, 298.

33. Wu, J. Y.; Moreau, B.; Ritter, T., *J. Am. Chem. Soc.* **2009**, *131*, 12915.

34. Ely, R. J.; Morken, J. P., *J. Am. Chem. Soc.* **2010**, *132*, 2534.

35. Mkhalid, I. A. I.; Barnard, J. H.; Marder, T. B.; Murphy, J. M.; Hartwig, J. F., *Chem. Rev.* **2010**, *110*, 890.

36. Ishiyama, T.; Nobuta, Y.; Hartwig, J. F.; Miyaura, N., *Chem. Commun.* **2003**, 2924.

37. López-Rodríguez, R.; Ros, A.; Fernández, R.; Lassaletta, J. M., *J. Org. Chem.* **2012**, *77*, 9915.

38. Koren-Selfridge, L.; Londino, H. N.; Vellucci, J. K.; Simmons, B. J.; Casey, C. P.; Clark, T. B., *Organometallics* **2009**, *28*, 2085.

39. Kim, H.; Yun, J., *Adv. Synth. Catal.* **2010**, *352*, 1881.

40. Zhao, D.; Oisaki, K.; Kanai, M.; Shibasaki, M., *J. Am. Chem. Soc.* **2006**, *128*, 14440.

Potassium Hexamethyldisilazide

$$KN(SiMe_3)_2$$

[40949-94-8] $C_6H_{18}KNSi_2$ (MW 199.53)

InChI = 1/C6H18NSi2.K/c1-8(2,3)7-9(4,5)6;/h1-6H3;/q-1;+1

InChIKey = IUBQJLUDMLPAGT-UHFFFAOYAI

(sterically hindered base)

Alternate Names: KHMDS; potassium bis(trimethylsilyl)amide.

Solubility: soluble in THF, ether, benzene, toluene.[1]

Form Supplied in: commercially available as moisture-sensitive, tan powder, 95% pure, and 0.5 M solution in toluene.

Analysis of Reagent Purity: solid state structures of $[KN(SiMe_3)_2]_2$[3] and $[KN(SiMe_3)_2 \cdot 2 \text{ toluene}]_2$[4] have been determined by X-ray diffraction; solutions may be titrated using fluorene,[2] 2,2'-bipyridine,[5] and 4-phenylbenzylidene benzylamine[6] as indicators.

Preparative Methods: prepared and isolated by the procedure of Wannagat and Niederpruem.[1] A more convenient in situ generation from **Potassium Hydride** and **Hexamethyldisilane** is described by Brown.[2]

Handling, Storage, and Precautions: the dry solid and solutions are inflammable and must be stored in the absence of moisture. These should be handled and stored under a nitrogen atmosphere. Use in a fume hood.

Original Commentary

Brett T. Watson

Bristol-Myers Squibb Pharmaceutical Research Institute, Wallingford, CT, USA

Use as a Sterically Hindered Base for Enolate Generation.
Potassium bis(trimethysilyl)amide, $KN(TMS)_2$, has been shown to be a good base for the formation of kinetic enolates from carbonyl groups bearing α-hydrogens.[7] For example, treatment of 2-methylcyclohexanone with $KN(TMS)_2$ at low temperature followed by trapping with **Triethylborane** and **Iodomethane** gave good selectivity for 2,6-dimethylcyclohexanone (eq 1). In comparison, the use of **Potassium Hydride** for this transformation gave good selectivity for 2,2-dimethylcyclohexanone, which is the product derived from the thermodynamic enolate (eq 2).[8]

This reagent has been shown to be a good base for the generation of highly reactive potassium enolates;[9] for example, treatment of various ketones and esters bearing α-hydrogens with $KN(TMS)_2$ followed by 2 equiv of N-F-saccharinsultam allowed isolation of the difluorinated product (eq 3).

(3)

mono-:difluorination = 2:98

In a study on the electrophilic azide transfer to chiral enolates, Evans[10] found that the use of potassium bis(trimethylsilyl)amide was crucial for this process. The $KN(TMS)_2$ played a dual role in the reaction; as a base, it was used for the stereoselective generation of the (Z)-enolate (**1**). Reaction of this enolate with trisyl azide gave an intermediate triazene species (**2**) (eq 4). The potassium counterion from the $KN(TMS)_2$ used for enolate formation was important for the decomposition of the triazene to the desired azide. Use of other hindered bases such as **Lithium Hexamethyldisilazide** allowed preparation of the intermediate triazene; however, the lithium ion did not catalyze the decomposition of the triazene to the azide.[10a] This methodology has been utilized in the synthesis of cyclic tripeptides.[10b]

Treatment of carbonyl species bearing acidic α-hydrogens with potassium bis(trimethylsilyl)amide has also been shown to generate anions which, due to the larger, less coordinating potassium cation, allow the negative charge to be stabilized by other features in the molecule rather than as the potassium enolate. Treatment of 9-acetyl-*cis,cis,cis,cis*-cyclonona-1,3,5,7-tetraene with this reagent gave an anionic species which was characterized by spectroscopic methods to be more like the [9]-annulene anion than the nonafulvene enolate. In this case the negative charge is more fully stabilized by delocalization into the ring to form the aromatic species rather than as the potassium enolate. Use of the bis(trimethylsilyl)amide bearing the more strongly coordinating lithium cation led to an intermediate which appeared to be lithium nonafulvene enolate. Addition of **Chlorotrimethylsilane** to each of these intermediates gave the same nonafulvenesilyl enol ether (eq 5).[11]

(4)

93% 2,6-
86% yield

90% 2,2-
79% yield

Essential Reactions for Organic Synthesis, First Edition. Edited by Philip L. Fuchs.
© 2016 John Wiley & Sons, Ltd. Published 2016 by John Wiley & Sons, Ltd.

$$(5)$$

Selective Formation of Linear Conjugated Dienolates. Potassium bis(trimethylsilyl)amide has been shown to be an efficient base for the selective generation of linear-conjugated dienolates from α,β-unsaturated ketones.[12] As shown in eqs 6 and 7, treatment of both cyclic and acyclic α,β-unsaturated enones with KN(TMS)$_2$ in a solvent mixture of DMF/THF (2:1) followed by quenching with **Methyl Chloroformate** gave excellent selectivities for the products derived from the linear dienolate anion. In comparison, the use of lithium bases for this reaction gave products derived from the cross-conjugated dienolate anions. This methodology, however, did not work for 1-cyclohexenyl methyl ketone, in which case the product from the cross-conjugated dienolate anion was isolated exclusively (eq 8).

$$(6)$$

Base	Linear	Cross	Yield (%)
KN(TMS)$_2$	>99	–	34
LiN(TMS)$_2$	75	25	68
LDA	–	99	44

$$(7)$$

Base	Linear	Cross	Yield (%)
KN(TMS)$_2$	99	–	34
LDA	16	84	50

$$(8)$$

not detected   >99

Stereoselective Generation of Alkyl (Z)-3-Alkenoates. Deconjugative isomerization of 2-alkenoates to 3-alkenoates occurs via γ-deprotonation of the α,β-unsaturated ester to form an intermediate dienolate anion. In most cases, the α-carbon is more reactive to protonation[13] and allows for the isolation of the 3-alkenoate. If the C-4 position bears a methyl group, this transformation is usually stereospecific, leading to the (Z)-3-alkenoate; however, when groups larger than a methyl occupy the C-4 position, the reaction becomes increasingly stereorandom.[13d] Potassium bis(trimethylsilyl)amide, however, was shown to be a good base for the stereoselective isomerization of 2,4-dimethyl-3-pentyl (E)-2-dodecenoate, which bears a long C-4 substituent, to the corresponding (Z)-3-dodecenoate (eq 9).[14a]

$$(9)$$

Base	(Z):(E)	Yield (%)
LDA/HMPA	84:16	85
KN(TMS)$_2$	97:3	64

This reagent has been used to stereoselectively prepare (Z)-3-alkenoate moieties for use in the syntheses of insect pheromones.[14]

Generation of α-Keto Acid Equivalents (Dianions of Glycolic Acid Thioacetals). Potassium bis(trimethylsilyl)amide was found to be the optimal reagent for the generation of the dianion of glycolic acid thioacetals. This reagent may be used to effect a nucleophilic α-keto acid homologation. Treatment of the starting bis(ethylthio)acetic acid with KN(TMS)$_2$ proceeded to give the corresponding soluble dianionic species. This underwent alkylation with a variety of halides and tosylates (eq 10) and subsequent hydrolysis allowed isolation of the desired α-keto acids.[15]

$$(10)$$

RX	Yield (%)
MeI	100
EtOTs	100
i-PrOTs	72
CyOTs	64
PhCH$_2$Cl	100

The dianion was also shown to undergo ring-opening reactions with epoxides and aziridines (eq 11).

Generation of Ylides and Phosphonate Anions.

Ylides. In the Wittig reaction, lithium salt-free conditions have been shown to improve (Z/E) ratios of the alkenes which are prepared;[16] **Sodium Hexamethyldisilazide** has been shown to be a good base for generating these conditions. In a Wittig-based synthesis of (Z)-trisubstituted allylic alcohols, potassium bis(trimethylsilyl)amide was shown to be the reagent of choice for preparing the starting ylides.[17] These were allowed to react with protected α-hydroxy ketones and depending upon the substitution pattern of the ylide and/or the ketone, stereoselectivities ranging from good to excellent were achieved (eqs 12–14).

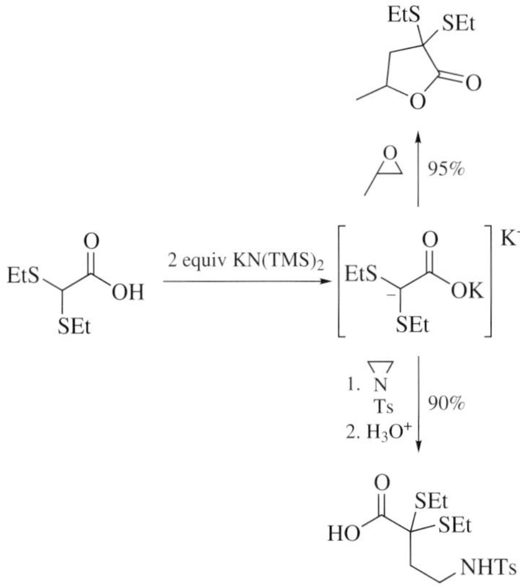

$$\text{(11)}$$

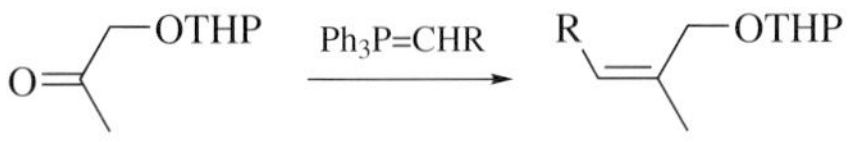

95%

2 equiv KN(TMS)$_2$

1. N–Ts | 90%
2. H$_3$O$^+$

$$\text{(12)}$$

Ph$_3$P=CHR

R	(Z):(E)	Yield (%)
n-Pr	60:1	87
i-Pr	6:1	45

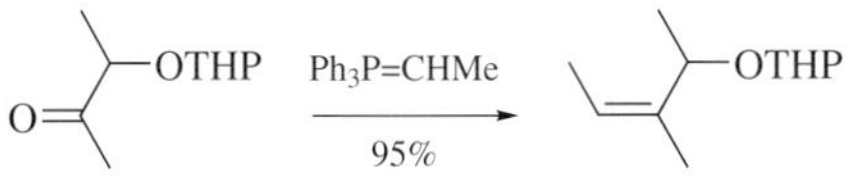

$$\text{(13)}$$

Ph$_3$P=CHMe
95%

(Z):(E) = 200:1

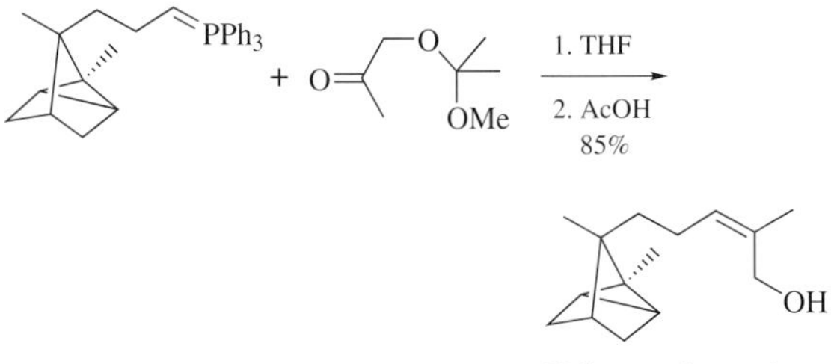

$$\text{(14)}$$

1. THF
2. AcOH
85%

>99% stereoisomeric purity

Phosphonates. In a Horner–Emmons-based synthesis of di- and trisubstituted (Z)-α,β-unsaturated esters, the strongly dissociated base system of potassium bis(trimethylsilyl)amide/**18-Crown-6** was used to prepare the desired phosphonate anions. This base system, coupled with highly electrophilic bis(trifluoroethyl)phosphono esters, gave phosphonate anions which, when allowed to react with aldehydes, gave excellent selectivity for the (Z)-α,β-unsaturated esters (eq 15).[18]

Intramolecular Cyclizations.

Haloacetal Cyclizations. Intramolecular closure of a carbanion onto an α-haloacetal has been shown to be a valuable method for the formation of carbocycles.[19a] Potassium bis(trimethylsilyl) amide was found to be the most useful base for the formation of the necessary carbanions. This methodology may be used for the formation of single carbocycles (eq 16), for annulation onto existing ring systems (eq 17), and for the formation of multiple ring systems in a single step (eq 18). In the case of annulations forming decalin or hydrindan systems, this ring closure proceeded to give

largely the *cis*-fused bicycles (eq 17). For the reaction shown in eq 18, in which two rings are being formed, the stereochemistry of the ring closure was found to be dependent upon the counter ion of the bis(trimethylsilyl)amide; use of the potassium base allowed isolation of the *cis*-decalin system as the major product (95%), whereas use of lithium bis(trimethylsilyl)amide led to the isolation of the *trans*-decalin (95%).[19]

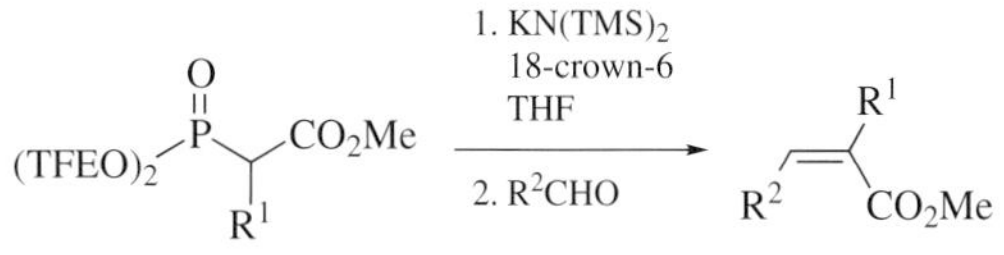

1. KN(TMS)$_2$
   18-crown-6
   THF
2. R^{2}CHO

$$\text{(15)}$$

R^{2}CHO	R^1	(Z):(E)	Yield (%)
Me(CH$_2$)$_6$CHO	H	12:1	90
Me(CH$_2$)$_6$CHO	Me	46:1	88
Me(CH$_2$)$_2$CH=CHCHO	H	>50:1	87
Me(CH$_2$)$_2$CH=CHCHO	Me	>50:1	79
CyCHO	H	4:1	71
CyCHO	Me	>50:1	80
PhCHO	H	>50:1	95
PhCHO	Me	30:1	95

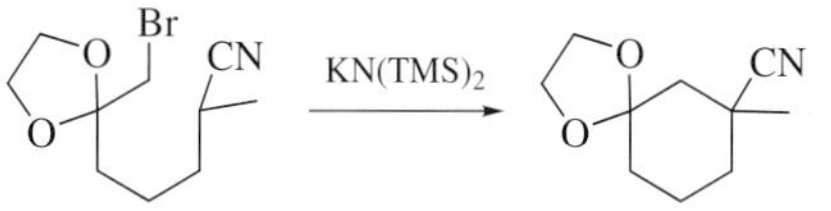

KN(TMS)$_2$

$$\text{(16)}$$

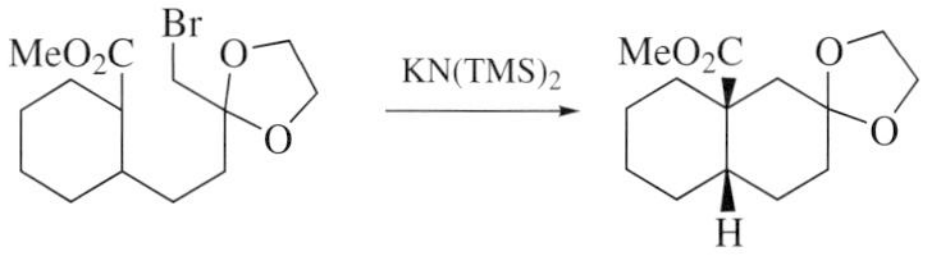

KN(TMS)$_2$

$$\text{(17)}$$

KN(TMS)$_2$

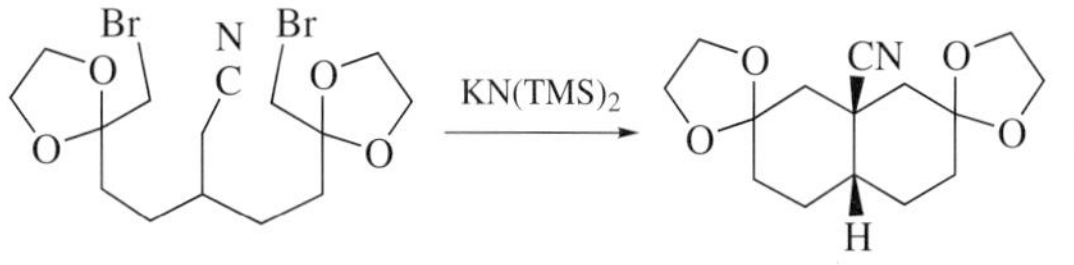

$$\text{(18)}$$

95% cis + 5% trans

Intramolecular Lactonization. In a general method for the formation of 14- and 16-membered lactones via intramolecular alkylation,[20] potassium bis(trimethylsilyl) amide was shown to be a useful base for this transformation (eq 19).

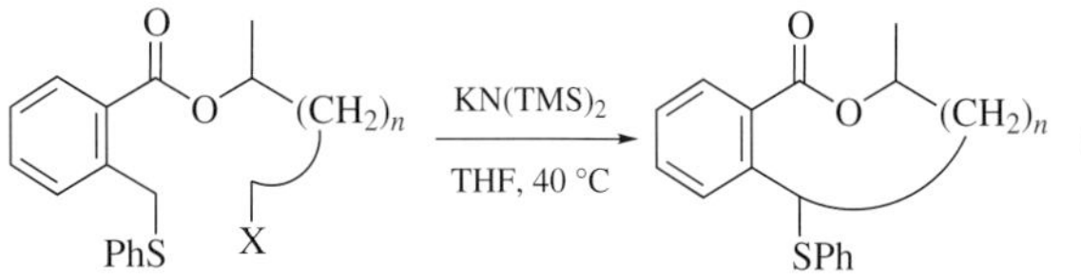

KN(TMS)$_2$
THF, 40 °C

$$\text{(19)}$$

n = 5, 0%; 7, 75%; 8, 71%

Intramolecular Rearrangement. Potassium bis(trimethylsilyl)amide was shown to be a good base for the generation of a diallylic anion which underwent a biogenetically inspired intramolecular cyclization, forming ($\pm$)-dictopterene B (eq 20).[21]

Synthesis of Vinyl Fluorides. Addition of potassium bis(trimethylsilyl)amide to β-fluoro-β-silyl alcohols was shown to selectively effect a Peterson-type alkenation reaction to form vinyl fluorides (eq 21).[22] Treatment of a primary β-fluoro-β-silyl alcohol with KN(TMS)$_2$ led cleanly to the terminal alkene. Use of a *syn*-substituted secondary alcohol led to the stereoselective formation of the (Z)-substituted alkene (eq 22); reaction of the *anti*-isomer, however, demonstrated no (Z:E) selectivity.

Oxyanionic Cope Rearrangement. Potassium bis(trimethylsilyl)amide/18-crown-6 was shown to be a convenient alternative to potassium hydride for the generation of anions for oxyanionic Cope rearrangements (eq 23).[23]

Stereoselective Synthesis of Functionalized Cyclopentenes. Potassium bis(trimethylsilyl)amide was shown to be an effective base for the base-induced ring contraction of thiocarbonyl Diels–Alder adducts (eq 24).[24] *Lithium Diisopropylamide* has also been shown to be equally effective for this transformation.

First Update

Hélène Lebel

Université de Montréal, Montréal, Quebec, Canada

Nitrogen Source: Synthesis of Amines. KHMDS has been exploited as a nitrogen source in a few systems. For instance, the ring opening of (trifluoromethyl)oxirane with KHMDS, followed by acidic hydrolysis produced the desired amino alcohol in 61% yield without any variation in the enantiomeric purity (eq 25).[25]

KHMDS was shown to be as effective as LHMDS[26] in the boron-assisted displacement of secondary chloride which led to the corresponding protected amine with inversion of configuration (eq 26).[27] The chemoselective displacement of the chloride is greatly facilitated by the α-boro substituent, which is known to cause a rate acceleration on the attack of nucleophiles of approximately two orders of magnitude relative to the primary halide.[28]

KHMDS was also used in the synthesis of aminocinnolines from trifluoromethylhydrazones (eq 27).[29] The deprotonation of the hydrazone is followed by the spontaneous loss of the fluoride anion to produce the corresponding difluoroalkene. Subsequent nucleophilic attack of KHMDS and cyclization with loss of the second fluoride leads to an intermediate that upon aromatization produces the desired product.

Both LHMDS and KHMDS have been successfully utilized as nitrogen sources in palladium(0) catalyzed aminations of allyl chlorides (eq 28). Up to 90% conversion and 55% isolated yield were obtained for the synthesis of allyl hexamethyldisilazide and no cyclization via deprotonation or significant hydrolysis during the work-up was observed.[30]

$$\text{(28)} \quad 55\%$$

Deprotonation of Imidazolium and Pyrazolium Salts: Synthesis of N-Heterocyclic Carbene Ligands. Potassium bases, such as potassium t-butoxide[31] and KHMDS, have been selected for the deprotonation of imidazolium and pyrazolium salts to generate N-heterocyclic carbene ligands, which have emerged over the last decade as an important part of transition metal-catalyzed homogeneous catalysis.[32,33] For instance, 1,1′-methylene-3,3′-di-tert-butyldiimidazole-2,2′-diylidene and 1,2-ethylene-3,3′-di-tert-butyldiimidazole-2,2′-diylidene were prepared in good yields (ca. 60%) by deprotonation of the corresponding imidazolium dibromides with KHMDS in THF (eq 29).[34] It is important to note that this base leaves methylene and ethylene bridges intact.

$$\text{(29)} \quad n = 1, 2$$

KHMDS was also used in the preparation of pyridine- and phosphine-functionalized N-heterocyclic carbenes (eqs 30 and 31),[35] as well as for the synthesis of N-functionalized pincer biscarbene ligands (eq 32).[36]

$$\text{(30)} \quad 75\text{–}80\%$$

A series of chiral triazolium salts have been reacted with a base to form the corresponding chiral carbenes, which was shown to catalyze the Stetter reaction efficiently and to provide 1,4-dicarbonyl products in high yields and enantioselectivities (eq 33).[37] A survey of common bases identified KHMDS as providing an optimal balance between the yield and selectivity in this reaction. The reaction is sensitive to the nature of the Michael acceptor: while electron deficient E-alkenes provided the desired product in good yields and enantioselectivities, no reaction was observed in the case of Z-alkenes.[38]

$$\text{(31)} \quad 60\text{–}70\%$$

$$\text{(32)} \quad 70\text{–}80\%$$

$$\text{(33)} \quad 63\text{–}94\% \quad 82\text{–}96\% \text{ ee}$$

Palladium-catalyzed α-Arylation of Carbonyl Compounds. A variety of bases have been used in the palladium-catalyzed α-arylation of carbonyl derivatives.[39,40] The pKa of the carbonyl moiety determines the choice of the base. The preferred bases for the α-arylation with ester derivatives are either NaHMDS (t-butyl propionate) or LiHMDS (t-butyl acetate); as KHMDS was reported to lead to lower yield because of competing hydrodehalogenation.[41–43] More sensitive substrates such as α-imino esters,[44] malonates, or cyanoesters[45] required the use of a milder base, as decomposition was observed with HMDS bases.

Both sodium t-butoxide and potassium hexamethyldisilazide were used in palladium-catalyzed ketone arylations.[46,47] While reactions involving electron-neutral or electron-rich aryl halides were more selective for monoarylation when KHMDS was used, sodium t-butoxide gave good selectivity with electron-poor aryl halides, without direct decomposition of the aryl halide. Bis(diphenylphosphino)ferrocene-ligated palladium complexes were active catalysts and led to the formation of a variety of substituted ketones in excellent yields (eq 34).

$$R^1 = H, Me$$
$$R^2 = Ph, t\text{-}Bu,$$

Ar = Ph, 2-MePh, 4-MeOPh

More recently, the use of potassium phosphate has also been reported for the α-arylation of ketones with chloroarenes.[48]

KHMDS was the most effective base for the intermolecular arylation of N,N-dialkylamides (eq 35).[49] Higher yields were observed with KHMDS compared to LiTMP. Use of NaOt-Bu resulted in low conversion together with a high level of undesired side products, while LDA appeared to deactivate the system. Coupling of unfunctionalized and electron-rich aryl bromides with N,N-dimethylacetamide afforded α-aryl amides in moderate to good yields when the reaction was conducted with at least 2 equiv of KHMDS base. Diarylation of acetamides as well as hydro-dehalogenation of the aryl halides were side reactions that limited this process.

Ar = 4-MePh, 2-MePh,
4-MeOPh, 2-naphthyl

Alkylation of Nitrile Derivatives. Tertiary benzylic nitriles were prepared from aryl fluorides and secondary nitrile anion.[50] In the presence of 4 equiv of nitrile and 1.5 equiv of a base, the nucleophilic aromatic substitution of fluoroarenes led to tertiary benzylic nitriles (eq 36). KHMDS was the best base for this reaction, as LiHMDS and NaHMDS provided lower yields. The desired product was not observed when Cs_2CO_3, LDA, or t-BuOK were used. With KHMDS, the reaction proceeded in high yields with a variety of substrates.

Treatment of readily available arylacetonitriles with KHMDS and subsequent alkylation with α,ω-dibromo or dichloroalkanes produces cycloalkyl adducts in good yields and short reaction time (eq 37).[51] In this process, the nitrile moiety serves as a masked aldehyde, which could be revealed upon reduction with DIBAL-H.

R^1 = 2-OMe, 3-OMe, 4-OMe, 3,5-OMe, 2-Cl, 4-Cl, H, 3-Me,
4-CN, 4-CF$_3$

Synthesis of Pyridines. The annulation of α-aryl carbonyl derivatives with vinamidinium hexafluorophosphate salts in the presence of a base gave access to the corresponding 3-arylpyridine.[52] While potassium t-butoxide was used with ketones, the annulation of aldehydes was performed with KHMDS (eq 38).

Deprotonation of Oxazolines. New chiral oxazoline-sulfoxide ligands have recently been prepared and utilized as chiral ligands in copper(II)-catalyzed enantioselective Diels–Alder reactions.[53] These new ligands were prepared via sulfinylation of chiral 1,3-oxazoline to lead to unsubstituted oxazoline sulfoxide derivatives. While the first methylation was conducted using LDA and methyl iodide, the introduction of the second methyl was carried out using KHMDS and methyl iodide to afford the desired chiral ligand (eq 39).

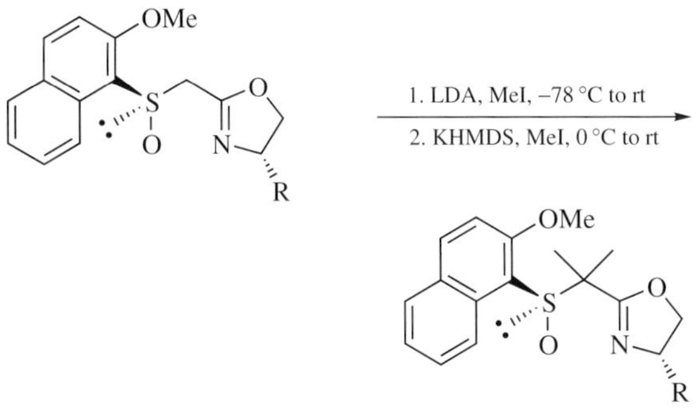

(39)

R = *t*-Bu, Ph, Bn

Deprotonation of Alkyne and Propargylic Derivatives.
KHMDS is a strong enough base to deprotonate terminal alkynes.
It has been particularly useful for the intramolecular condensation
of alkyne anions with aldehydes. For instance, the final cyclization in the synthesis of a new bicyclic tetrahydropyridine system
was carried out in the presence of KHMDS (eq 40).[54] The same
reaction failed when using LDA in THF.

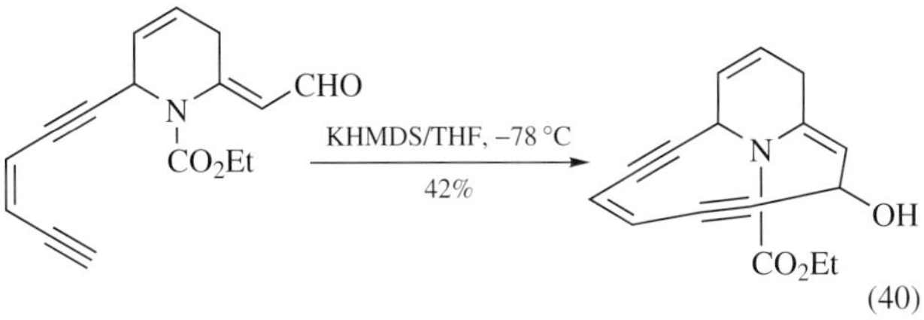

(40)

In a tandem cyclization of propynyloxyethyl derivatives,
KHMDS is as efficient as LDA or NaHMDS (eq 41).[55] This cyclization proceeded through the formation of an alkynyl metal,
followed by a cyclization which led to a carbenic species that
underwent an intramolecular C–H insertion.

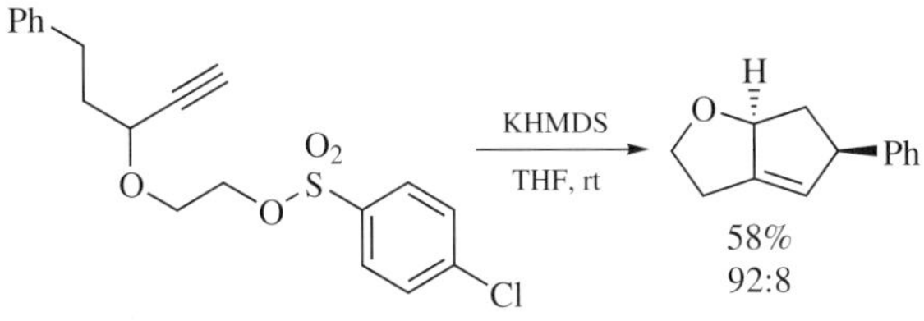

(41)

58%
92:8

A convenient method for the preparation of 1-phenylthio-3-
alken-1-ynes from 4-tetrahydropyranyloxy-1-phenylthio-2-alky-
nes in the presence of base was reported (eq 42).[56]

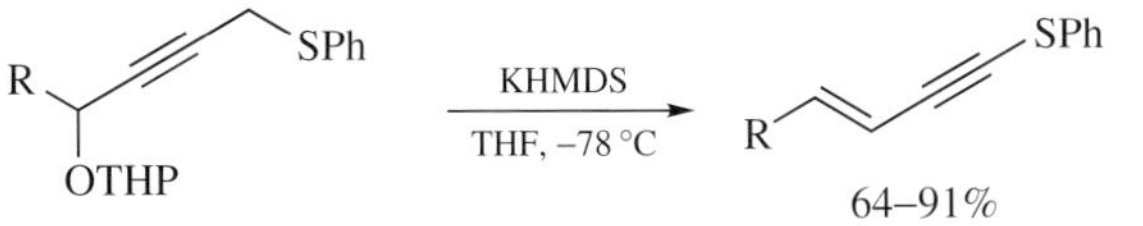

(42)

64–91%
1:1.3 to 17:1 (*E:Z*)

R = Ph, Me, PhCH$_2$CH$_2$, *c*-C$_6$H$_{11}$,
1-naphthyl, 9-anthryl

The use of KHMDS led mainly to the formation of the
E-isomer, whereas the *Z*-isomer was formed as the major product
in the presence of MeLi. The presence of the masked hydroxyl
group at the 4-position of the substrate is essential, as treatment of
corresponding 4-hydroxy-1-phenylthio-2-alkynes with KHMDS
produced 4-hydroxy-1-phenylthio-1,2-alkadienes (eq 43).

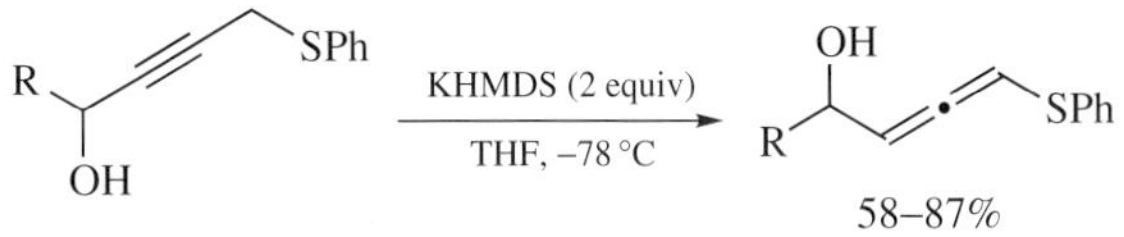

(43)

58–87%

R = Ph, Me, PhCH$_2$CH$_2$

Cyclization of 1,1-Disubstituted Alkenes to Cyclopentenes.
A general method for the cyclization of an unactivated 1,1-disubs-
tituted alkene to the corresponding cyclopentene has been described.[57] Bromination followed by the addition of KHMDS in
the same pot gave the vinyl bromide intermediate, which reacted
further in situ to give the alkylidene carbene. This is then inserted
in a 1,5-fashion into a C–H bond to yield the corresponding cyclopentene (eq 44). KHMDS proved to be superior in this process as
compared to LiHMDS and NaHMDS.

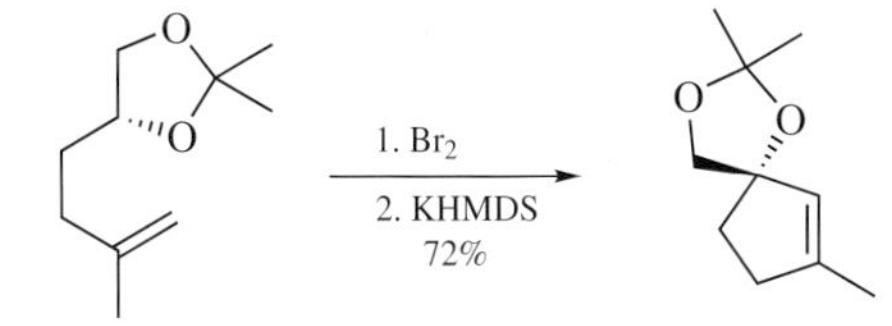

(44)

Darzens Reaction. KHMDS was used to deprotonate bro-
momethylketones to generate the corresponding bromoenolate
that upon reaction with various carbonyl derivatives produced a
variety of epoxides. This method has been particularly useful to
synthesize epoxides derived from amino acids with high yields and
selectivity (eq 45).[58] Both lithium and sodium enolate produced
lower yields and diastereoselectivities.

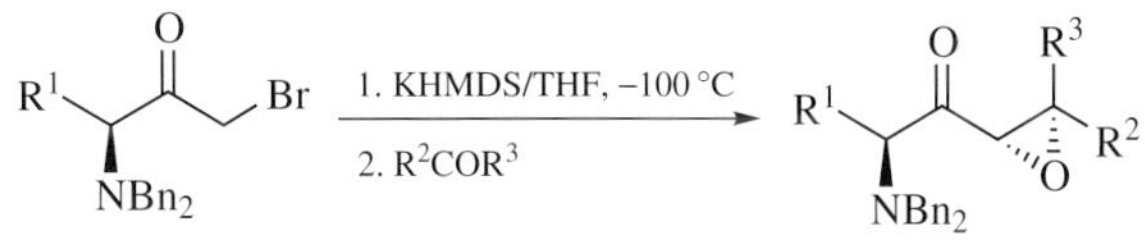

(45)

68–87%
90 to >95% de

Deprotonation of Amino Acid Derivatives. The direct asymmetric α-methylation of α-amino acid derivatives was recently
reported to proceed with good efficiency and enantioselectivity
when using KHMDS in toluene (eq 46).[59] It was established that
the chiral induction was based on the dynamic chirality of enolates, as the di-boc derivative and the cyclic acetal gave racemic
products.[60]

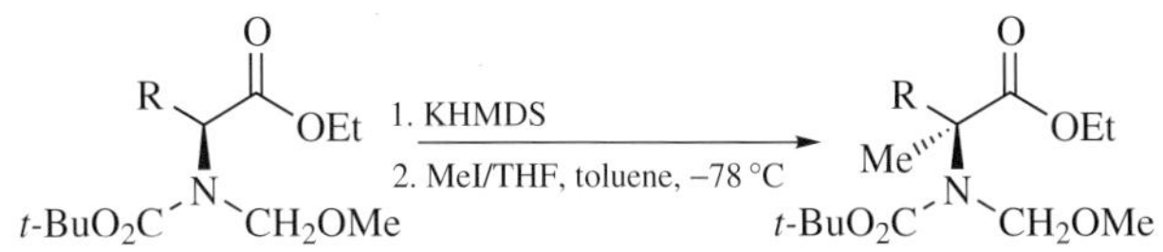

(46)

83–96%
76–93% ee

An intramolecular version of this process has been also reported, for which it was also shown that KHMDS provided superior results than LHDMS.[61] The corresponding cyclic amino acids
could be recovered in 61–95% yield and with 94–98% enantiomeric excess.

Epoxidation with Alkylhydroperoxide. The epoxidation of quinone monoacetals in the presence of tritylhydroperoxide using KHMDS was shown to produce the desired epoxides in good yields (eq 47).[62,63]

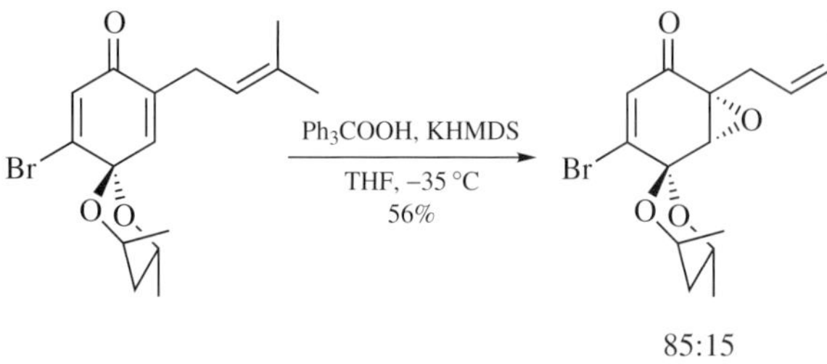

$$85:15 \quad (47)$$

Alkylation of Phosphorus Derivatives. The formation of diethyl phosphite anions has been achieved successfully with KHMDS and LiHMDS.[64] The following reaction with acid chlorides is strongly influenced by the counter-ion, as only the rearrangement product was observed with the lithium base, whereas the desired hydroxybisphosphate was the major product when using the potassium base (eq 48).

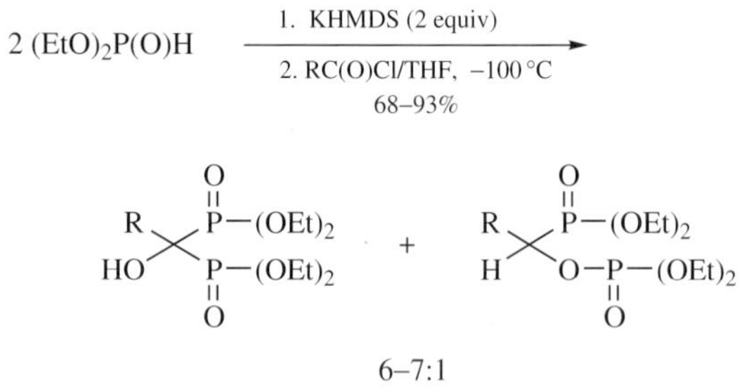

$$6–7:1 \quad (48)$$

A series of new phosphinooxazoline ligands[65] have been recently prepared and tested in the asymmetric Heck reaction.[66,67] Synthesis of the ligands involved the aromatic nucleophilic substitution of aryl fluorides with phosphide nucleophile generated from the corresponding phosphine and KHMDS (eq 49). The reaction proceeded in good yields, but proved to be more sluggish with electron-rich aryl fluorides and failed completely when the addition of electron-deficient phosphines was attempted.

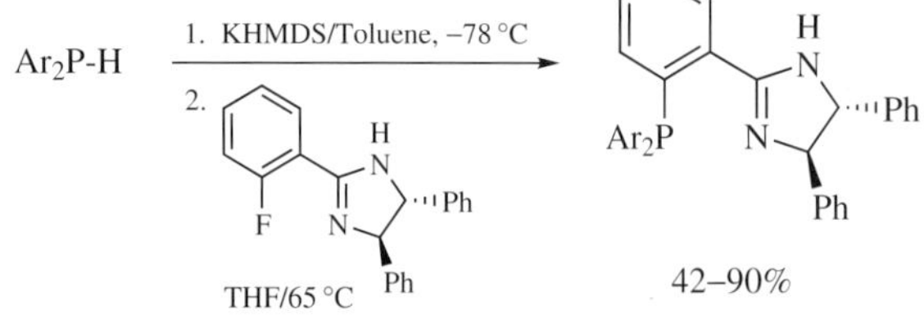

Acetal Fission in Desymmetrization of 1,2-Diols. The asymmetric desymmetrization of cyclic *meso*-1,2-diols was accomplished via diastereoselective acetal fission. After acetalization of *meso*-1,2-diols with the C_2-symmetric bis-sulfoxide ketone, the resulting acetal was subjected to base-promoted acetal fission, followed by acetylation or benzylation to give the desymmetrized diol derivatives. Interestingly, the counter-cation of the base had a remarkable effect on the diastereoselectivity of the reaction. While LHMDS produce the desired compound with low diastereoselectivity, 90% and >96% diastereomeric excesses were obtained

with NaHMDS and KHMDS, respectively. The best results were obtained using 3 equiv of KHMDS and 18-crown-6 in THF, which led to the formation of the desired acetate in 90% yield and 96% de (eq 50).[68]

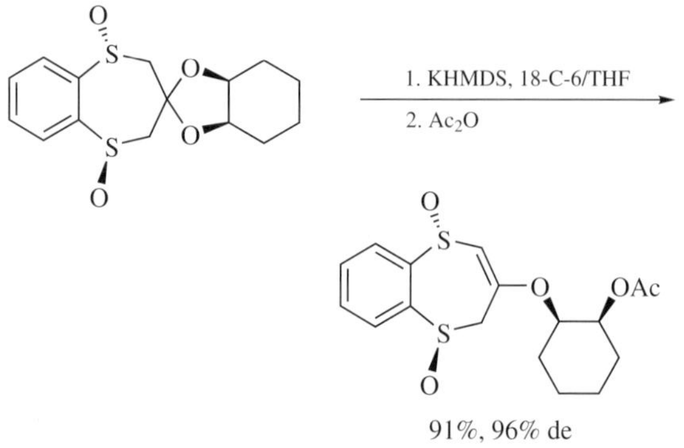

$$91\%, 96\% \text{ de} \quad (50)$$

Generation of Sulfonium Ylides. Sulfonium ylides could be generated very efficiently by deprotonation of the corresponding sulfonium salt with a strong base, such as KHMDS. The formation of vinyloxiranes from sulfonium ylides and aldehydes has been recently investigated. It was shown that in the presence of lithium bromide and silylated diphenylsulfonium allylide, produced from the corresponding sulfonium salt and KHMDS, aldehydes led to the corresponding vinyl epoxide with high *cis*-selectivity (eq 51).[69]

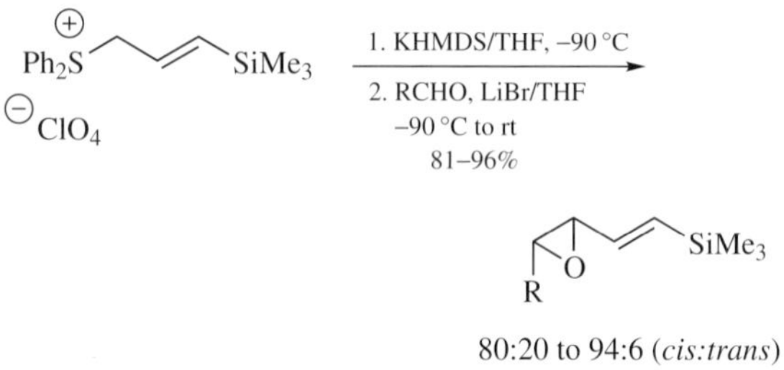

$$80:20 \text{ to } 94:6 \ (cis:trans)$$

More recently, an asymmetric version has been reported based on the use of chiral sulfonium salts.[70] In the presence of a base, the chiral sulfonium salt was reacted with various carbonyl derivatives to provide the corresponding *trans*-epoxide with excellent diastereomeric ratios and enantiomeric excesses. Potassium hydroxide and phosphazene bases are usually utilized for this process, although it has been shown that KHMDS is indeed equally effective (eq 52).

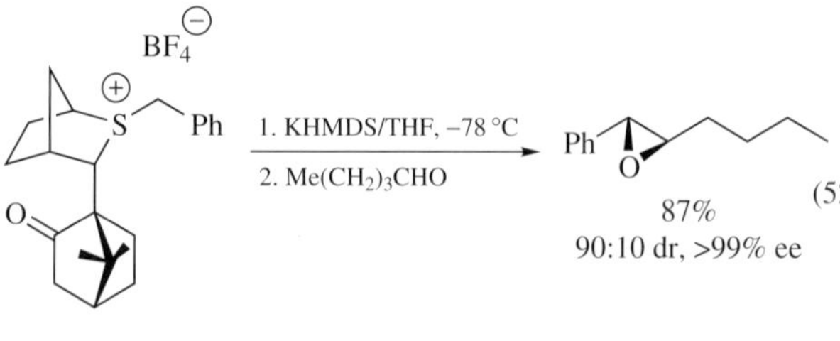

$$87\% \quad (52)$$
$$90:10 \text{ dr}, >99\% \text{ ee}$$

The formation of metal carbenes from sulfonium ylides has been also recently disclosed.[71] Treatment of the sulfonium salt with KHMDS followed by the addition of tris(triphenylphosphine) ruthenium dichloride and phosphine exchange with tricyclohexylphosphine led to the formation of the Grubb's metathesis catalyst in 98% yield (eq 53). This process has been shown to be quite general and to work with various transition metals.

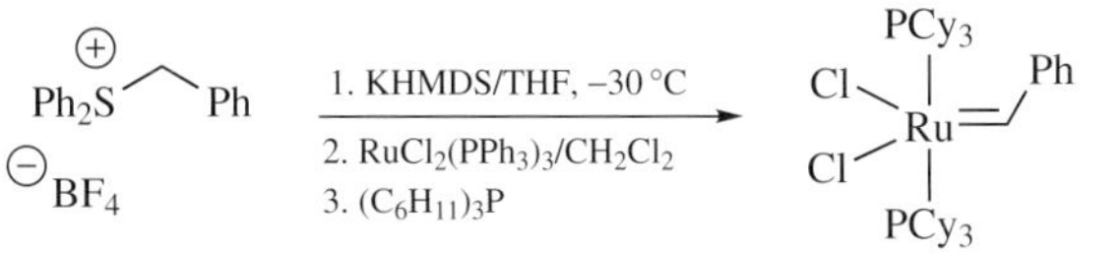

(53)

98%

Generation of Sulfur Ylides: Julia Olefination and Related Processes. Recently, the 'modified' Julia olefination, which employed certain heteroarylsulfones instead of the traditional phenylsulfones, has emerged as a powerful tool for alkene synthesis.[72] Although the reaction was first reported with LDA, bases such as LHMDS, NaHMDS, and KHMDS are now commonly used.[73] In addition, solvent as well as base counter-cation have been shown to markedly affect the stereochemical outcome of the olefination reaction. For instance, KHMDS was less selective than NaHMDS for the coupling between benzothiazoylsulfone (**1**) and cyclopropane carboxaldehyde (**2**) in toluene, furnishing a 3.7:1 ratio compared to a 10:1 ratio favoring the Z-isomer. However, both bases provided a 1:1 mixture of isomers when the reaction was run in DMF (eq 54).[74]

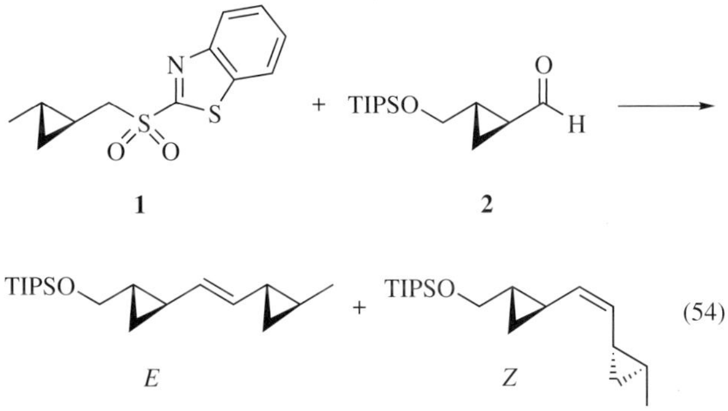

(54)

Conditions	E:Z
NaHMDS, THF	1.1:1
KHMDS, THF	1.2:1
NaHMDS, Toluene	1:10
KHMDS, Toluene	1:3.7

In a different system, KHMDS proved to be the most Z-selective for the formation of a triene from an allylic sulfone and a conjugated aldehyde (eq 55).

1-Phenyl-1H-tetrazol-5-yl sulfones were introduced as another alternative to phenylsulfones; these lead to alkenes in a one-pot procedure from aldehydes. Usually the highest selectivities are obtained when using KHMDS.[75] For instance, alkene **3** was obtained in 100% yield and an 84:16 E:Z ratio with NaHMDS, whereas the use of KHMDS led to the same product in 59% yield and a 99:1 E:Z ratio (eq 56).

Many examples of the use of this strategy with KHMDS has been reported to construct double bonds in the total synthesis of natural products. For instance, the E,E,E-triene of thiazinotrienomycin E was constructed according to this process in 85% yield and with 10:1 selectivity (eq 57).[76]

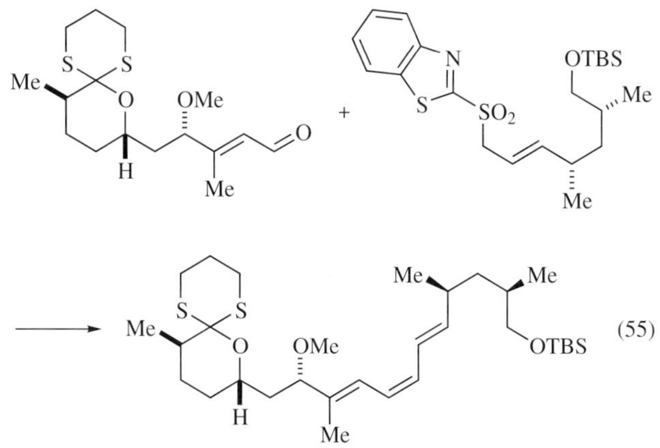

(55)

Conditions	E:Z
LiHMDS	1:2.4
NaHMDS	1:1.3
KHMDS	1:4.6

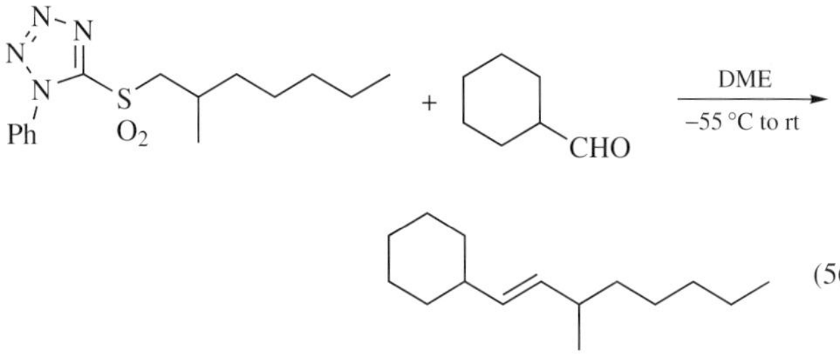

(56)

3

NaHMDS: 100%; 84:16 (E:Z)
KHMDS: 59%; 99:1 (E:Z)

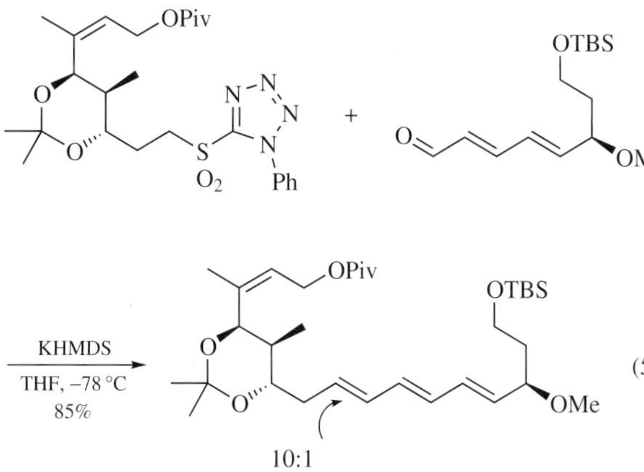

(57)

10:1

Disubstituted E-alkenes were also prepared according to this strategy as exemplified in the total synthesis of brefeldin A,[77] amphidinolide A,[78] and leucascandrolide A.[79]

High Z-selectivities were observed with benzylic and allylic 1-tert-butyl-1H-tetrazol-5-yl sulfones when reacted with aliphatic aldehydes in the presence of KHMDS (eq 58).[80]

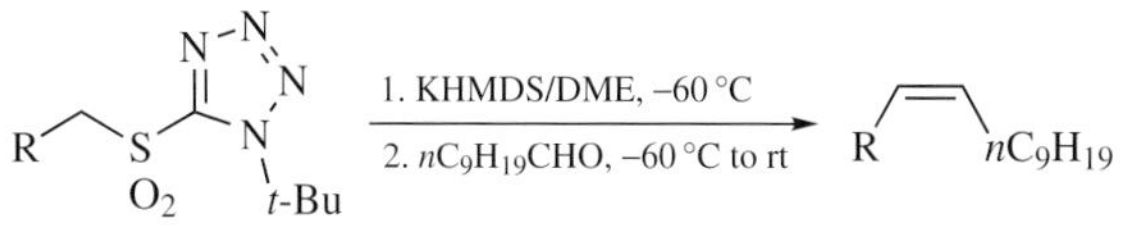

(58)

R = Ph, 95%; >99:1 (E:Z)
R = CH=CH$_2$, 60%; 96:4 (E:Z)

It was shown that E,Z-dienes could be synthesized from α,β-unsaturated aldehydes and 2-pyridylsulfones in good yields and selectivities.[81,82] NaHMDS and KHMDS were equally effective in promoting this condensation (eq 59).

$$(59)$$

67%, 91:9 (E,Z:E,E)

Generation of Other Heteroatom Ylides. KHMDS has been also used in the generation of other heteroatom ylides, such as tellurium[83] and arsonium ylides.[84]

Second Update

Helena C. Malinakova & John C. Hershberger
University of Kansas, Lawrence, KS, USA

Base for α-Deprotonation of Carbonyl Compounds. Potassium hexamethyldisilazide, KN(TMS)$_2$ (KHMDS) has been identified as the optimum base for an efficient epimerization of the stereogenic C-2 carbon in the esters of phenyl kainic acid and in *cis*-prolinoglutamic esters.[85] Bases included in an initial survey, e.g., NaH, DBU, LHMDS, MeONa, and KOH afforded low yields of the epimerized products. Although promising results were observed with KH activated by the 18-crown-6 ether, the protocol suffered from poor reproducibility. In contrast, the treatment of *N*-Cbz-protected methyl *cis*-3-prolinoglutamate with solid KHMDS (2.5 equiv, 0 °C, THF, 6 h) resulted in 80% epimerization without the occurrence of undesired degradation processes (eq 60). The application of KHMDS (5 equiv, THF, 0 °C–rt, 16 h) to epimerization of *endo* (*cis*(C2–C3), *cis*(C3–C4)) *N*-Cbz-protected 4-phenyl methyl kainate afforded 95% epimerization accompanied by significant saponification, and subsequent treatment with diazomethane was required to recover the epimerized ester in a good yield (60–75%) (eq 61). In contrast, the *exo* (*cis*(C2–C3), *trans*(C3–C4)), 3-phenyl methyl kainate ester proved to be much more resistant to epimerization, which only proceeded to the extent of 40%.

$$(60)$$

80% epimerization

In the synthesis of a second-generation anticancer taxoid ortataxel, deprotonation of 13-oxobaccatin III by KHMDS (THF–HMPA, 83:17) followed by electrophilic oxidation with (1*R*)-(10-camphorsulfonyl)oxaziridine afforded the C14-hydroxylated

taxoid in 70% yield (eq 62).[86] An analogous oxidation mediated by *t*-BuOK (THF–DMPU, 83:17) afforded the product in 83% yield (eq 62). The 14β-OH epimer was formed exclusively, in accord with the approach of the oxidant from the less sterically hindered β-face. The performance of several bases was compared, indicating the order of reactivity *t*-BuOK > LDEA (lithium diethylamide) > KHMDS > NaHMDS, in THF/HMPA or DMPU solvent mixtures.

$$(61)$$

95% epimerization

Oxalactims, for example 5*H*-methyl-2-phenyl-oxazol-4-one, were investigated as synthetically powerful synthons for the construction of α-hydroxy acids via α-deprotonation followed by molybdenum-catalyzed asymmetric allylation and a subsequent hydrolysis of the oxalactim rings. Among the series of the HMDS bases used in the Mo-catalyzed asymmetric allylation, LHMDS was found to be the base of choice providing better yields, regio- (branched/linear), diastereo- and enantioselectivity than the corresponding allylation reactions mediated by KHMDS (eq 63).[87]

$$(62)$$

70% (KHMDS)
83% (*t*-BuOK)

Potassium hexamethyldisilazide, KHMDS, has been used successfully in the past for α-deprotonation of sulfoxides.[53,68] Aiming to probe the nature of the transition state in the reaction of α-metalated (*R*)-methyl *p*-tolyl sulfoxide with *N*-(benzylidene) aniline under kinetic conditions, the effect of the choice of the bases for the α-deprotonation on the reactivity and diastereoselectivity was studied, using LDA, LHMDS, NaHMDS, and KHMDS.[88] Best yields and diastereoselectivities (90%, 84:16) were achieved with LDA, whereas LHMDS showed diminished reactivity (50%) without a loss of diastereoselectivity (87:13). Improved yields and lower diastereoselectivities were recorded in reactions mediated by NaHMDS and KHMDS, providing the amine in 60% (dr 81:19) and 77% (dr 69:31) yields and diastereoselectivities, respectively (eq 64).

Base	Yield (%)	Regioselectivity branched/linear	dr	ee (%)
LHMDS	91	99 : 1	11.5 : 1	99
KHMDS	88	6.7 : 1	1.4 : 1	80

77% (69:31)

These results, along with quantum mechanical calculations, supported the conclusion that a six-membered chelated "flat-chair-like" transition state operated under the kinetic conditions. The diastereoselectivity of reactions between Li-salts of the (R)-methyl p-tolyl sulfoxide formed by α-deprotonation with n-BuLi, could be further enhanced by the addition of external C_2-symmetrical bidentate ligand (R,R)-1,2-N,N-bis(trifluoromethanesulfonyla-mino)cyclohexane in the form of the lithium N,N-dianion providing the amine in 80% yield with a diastereoselectivity of 99:1, apparently achieving a "match" between the chirality of the additive and the chirality of the sulfoxide reagent.

KHMDS has been used to effect α-deprotonation of O-silyl protected cyanohydrins derived from 2-p-tolylsulfinyl benzaldehyde followed by trapping of the C-nucleophile with diverse C-electrophiles, providing a powerful alternative approach to cyanohydrins of ketones.[89] The remote 1,4-asymmetric induction was equally effective for either epimer (diastereomer) of the O-TIPS protected cyanohydrin, and an equimolar mixture of the two epimers was employed. Both KHMDS and LHMDS bases provided the substituted cyanohydrins from reactions with highly reactive electrophiles (ClCOOMe and ClCOMe) in excellent yields and diastereoselectivities (dr > 98:2) (eq 65). The deprotonation induced by KHMDS led to more reactive nucleophiles, shortening the reaction times. Notably, in alkylations of Eschenmoser's salt, and benzyl and allyl bromides, the application of LHMDS instead of KHMDS improved the diastereoselectivity. The stereoselectivity of the alkylations mediated by KHMDS could be increased by the inclusion of the 18-crown-6 ether

additive. To rationalize the observed trends, the deprotonation with KHMDS was proposed to give rise to pyramidal potassium C-bonded enolates, in contrast to planar lithium N-bonded enolates arising from the deprotonation with LHMDS. The steric discrimination of the enantiotopic faces would be achieved more efficiently with the planar N-enolates considering the different directions of the approach by the electrophiles. Thus, the 18-crown-6 ether additive serves to minimize the pyramidal shape of the potassium enolates, improving diastereoselectivity. The increase in the diastereoselectivity observed with highly reactive electrophiles (e.g., ClCOOMe) presumably reflects the involvement of an early transition state resembling the structures of the planar N-lithium enolates, or "planarized" potassium enolates complexed with the 18-crown-6 additive, effectively discriminating against one of the two enantiotopic pathways for the approach of the electrophiles.

The deprotonation of a stereogenic carbon in chiral non-racemic substrates with KHMDS has been used as a first step of sequences relying on the memory of chirality for asymmetric synthesis of organic products.[90,91] The treatment of N-Boc-protected amino esters featuring a Michael acceptor group with KHMDS in DMF–THF (1:1) at −78 °C for 30 min provided trisubstituted pyrrolidines, piperidines and tetrahydroisoquinolines, and in good yields (65–74%) and diastereoselectivities (4:1 for pyrrolidine, and 1:0 for piperidines and tetrahydroisoquinoline) and excellent enantiomeric excesses (91–98% ee) (eqs 66 and 67).[90]

The choice of the N-protecting group (N-Boc) proved to be critical for achieving a high enantiomeric excess of the cyclization reaction. In contrast to KHMDS, lithium amide bases (LHMDS or LiTMP) did not afford detectable quantities of the anticipated heterocycles. The mechanism of asymmetry transfer was proposed to rely on the formation of axially chiral nonracemic enolate (eq 66) with a chiral C–N axis, the racemization barrier for which was found to be 16.0 kcal mol^{-1}.[90]

The deprotonation of C-3 substituted 1,4-benzodiazepin-2-ones with KHMDS afforded an enantiopure, conformationally chiral enolate with a relatively bulky N −1 substituent, which provided a sufficient barrier to enolate racemization.[91] In contrast, LHMDS proved to be ineffective in this transformation, and KHMDS served as an operationally simpler alternative to a mixed base consisting of LDA/n-BuLi. At $-100\,°C$ a *simultaneous* addition of KHMDS and an excess of benzyl bromide to the substrate provided the alkylated product in 75% ee (78% yield). However, the simultaneous addition could not be carried out with a more reactive benzyl iodide. Ultimately, the best result was achieved in a stepwise addition mode, when the deprotonation with KHMDS (20 min) at $-109\,°C$ (THF–HMPA) was followed by the addition of an excess of benzyl iodide affording the 3,3-disubstituted benzodiazepine in 97% ee (93% yield) (eq 68).[91]

$$\text{(66)}$$

65%, 4:1 dr
(91% ee major diastereomer)

$$\text{(67)}$$

$n = 2$

66% (97% ee)

93% (97% ee)

The utility of a series of different inorganic bases, including LHMDS, NaHMDS, KHMDS, t-BuONa, t-BuOK, t-BuOLi, LTMP, LDA, PhOLi, MeONa, and EtONa for the alkylation of cyclohexanone with benzyl bromide in a microreactor was studied.[92] The microreactor employed a field-induced electrokinetic flow acting as a pump. To achieve a sufficient electrokinetic mobilization, the inorganic bases had to be solubilized by the addition of stoichiometric quantities of the appropriate crown ethers. With a single exception (LHMDS with 12-crown-4 ether), the desired electrokinetic mobility was reached with all other bases under these conditions.

α-Deprotonation of aliphatic nitriles with NaHMDS or KHMDS bases operated as a key step in a mild and transition metal-free α-arylation of primary and secondary aliphatic nitriles with activated heteroaryl halides.[93] The α-nitrile carbanion had to be generated in the presence of the heteroaryl halide to minimize competing decomposition reactions. Notably, monoarylation occurred selectively with primary carbonitriles. A correlation between the choice of the metal ion in the silylamide bases and the leaving group in the heteroaryl halide was observed. The S_NAr reaction performed best when NaHMDS was used with bromide and chloride leaving groups, and KHMDS with substrates featuring a fluoride leaving group (eq 69).[93]

$$\text{(69)}$$

X = Cl, M = Na; yield: 61%
X = F, M = K; yield: 88%

Carbon nucleophiles can be used to construct carbon-transition metal bonds in stable organometallic complexes via an S_N2 displacement at the metal center. For example, deprotonation at the α-position to the ester groups in the chiral nonracemic (ethoxycarbonylmethyl)-N-(trifluoromethanesulfonyl)-2-(aminophenyl)iodopalladium(II) complex followed by a rapid irreversible displacement of the iodide afforded new C(sp^3)–Pd bonds, yielding a diastereomerically enriched palladacycle with a Pd-bonded stereogenic carbon.[94] The ratio of atropisomers arising from a hindered rotation about the aryl–Pd bond in the aryliodopalladium complex proved to play an important role, and the results shown in eq 70 were achieved with an aza-aryliodopalladium complex featuring a 98:2 ratio of atropisomers. Apparently, the ring-closure reaction was faster than the interconversion of the atropisomers. Deprotonation with KHMDS followed by the displacement of the iodide afforded excellent yields (89% and 93%) of the palladacycle in good diastereoselectivities (60% and 72%) favoring the diastereomer with the (S) configuration at the Pd-bonded stereogenic carbon. The application of sterically hindered bases with tightly chelating cations proved to be important for achieving high diastereoselectivites. Consequently, deprotonation with LDA (THF, rt) afforded the palladacycle in 61% yield and 80% de. The optimum diastereoselectiviy was realized following the replacement of the iodide counterion with acetate utilizing AgOAc additive and t-BuOK base (THF, $-78\,°C$) yielding the palladacycle in 99% yield and 92% de (eq 70).[94]

98:2 atropisomer ratio

(70)

base: KHMDS; 93% yield (72% de)
base: AgOAc/t-BuOK; 99% yield (92% de)

Base for Deprotonation of Heteroatom–Hydrogen (X–H) Bonds. The deprotonation of heteroatom–hydrogen bonds with KHMDS constitutes a key step in diverse methods for the preparation of heterocycles.[95,96] A cyclization of 3,4-dihydro-2-pyridones derived from (S)-phenylglycinol featuring a β-enamino ester functionality and an internal hydroxyl group via an intramolecular Michael addition was achieved though the deprotonation of the hydroxyl group with KHMDS, NaHMDS, or LHMDS in THF at 0 °C.[95] In all cases, cis-diastereomers of the bicyclic lactams were formed, and only the configuration of the carbon adjacent to the ester group was variable. Reactions performed under kinetic conditions with 0.1 equiv of LHMDS (2 h) gave the best diastereoselectivity (>98:2 dr). However, the same diastereoselectivity (>98:2 dr) was also realized with KHMDS (0.1 equiv) when the reaction was stopped after just 1 h (eq 71).[95] In contrast, the application of 0.5 equiv of a strong base (KHMDS) induced the epimerization of the acidic proton providing almost 1:1 ratios of cis/trans diastereomers.

(71)

>80% (>98:2 dr)

KHMDS was used to promote the condensation of an imidazole precursor with commercially available carbamates providing 1-substituted 7-PMB-protected xanthines in a single step.[96] KHMDS served to deprotonate the N–H bonds in both substrates, yielding a metal amide and an isocyanate via fragmentation of the metalated carbamate, which participated in a condensation reaction providing potassium salts of the xanthines. In most cases, the potassium salt precipitated directly from the crude reaction mixtures, and was easily N-allylated in a subsequent step under mild conditions. The choice of the base for the condensation reaction was optimized. Sodium bases (EtONa, MeONa, t-BuONa), and unhindered potassium bases (EtOK) gave low to moderate yields of the corresponding N-metal salts, whereas lithium bases (LHMDS, LDA) gave modest yields and required longer reaction

times. A problematic base-induced degradation of the imidazole substrate could be minimized by using strong sterically hindered potassium bases (e.g., KHMDS, t-AmOK, t-BuOK) and by adding a solution of the base in diglyme slowly to the preheated (70–80 °C) and stirred diglyme solutions of substrates (eq 72).

(72)

In the preparation of a series of phenylalkylphosphonamidate derivatives of glutamic acid, KHMDS was employed to deprotonate the H–P bond in the dibenzyl phosphite, generating a P-nucleophile that reacted with phenylalkyl bromides to provide targeted phosphonates in good yields (eq 73).[97]

(73)

n = 1, 3 and 5

2.5 h reflux
THF

The deprotonation of N–H bonds in diverse oxazolidin-2-ones with KHMDS as the base followed by the treatment of the crude reaction mixtures with trimethylsilylethynyl iodonium triflate electrophiles afforded trimethylsilyl-terminated N-ethynyl oxazolidinones in 50–60% yields (eq 74).[98] Desilylation could be realized on the purified products or the crude reaction mixtures, and the alkynyl oxazolidinones were elaborated into novel stannyl enamines in the subsequent steps. In contrast, protocols employing n-BuLi in toluene, or Cs_2CO_3 in DMF gave yields lower than 20%. The procedure could be successfully applied to chiral oxazolidinones, since substitution at the C4 position of the oxazolidinones did not have a detrimental effect on reactivity.

(74)

(50–60%)

Pyrrole carbinols were identified as new base-labile protecting groups for aldehydes.[99] The optimal bases for the preparation of a pyrrole carbinol via the reaction of metal pyrrolates with iso-butyraldehyde were sought. Alkali metal pyrrolates performed better than their alkaline earth counterparts. Lithium and sodium pyrrolates prepared via the deprotonation of pyrrole with n-BuLi, LHMDS, and NaHMDS afforded the corresponding pyrrole carbinol in 90%, 85% and 85% yields, respectively, whereas potassium pyrrolate obtained using KHMDS afforded the pyrrole carbinol in a lower yield (45%) (eq 75).[99]

KHMDS could also be used for the deprotonation of an N–H bond present in a ligand of an aminoiridium complex, providing an amidoiridium complex required in a study elucidating the kinetic and thermodynamic barriers to interconversions of late metal amido hydride complexes and late metal amine complexes.[100] The treatment of an ammonia complex (PCP)Ir(H)(Cl)(NH$_3$) with KHMDS (2 equiv) in THF ($-78\,^{\circ}$C) afforded the targeted amido complex (PCP)Ir(H)(NH$_2$) in a quantitative yield accompanied by HMDS and KCl (eq 76).[100] The product was stable below $-10\,^{\circ}$C, and could be characterized by NMR techniques.

KHMDS has been used as the key reagent in different approaches to a stereocontrolled glycosidation and functionalization of carbohydrates.[101,102] The choice of the base and solvent for the deprotonation of the glycosidic O–H group in reactions of 2,3,4,6-tetra-*O*-benzyl-D-glucopyranose or 2,3,4,6-tetra-*O*-benzyl-D-mannopyranose with 2-chloro-6-nitrobenzothiazole was found to be a critical factor in controlling the α/β selectivity.[101] In the synthesis of 2,3,4,6-tetra-*O*-benzyl-D-mannopyranoside both stereoisomers (α or β) could be prepared selectively. The α-isomer was obtained by using KHMDS (THF, rt, 0.5 h) giving 90% yield of a 73:27 ratio of α:β stereoisomers (eq 77), whereas a preferential formation of the β-stereoisomer was realized using NaHMDS (THF, rt, 0.5 h) giving 92% yield of a 34:66 ratio of α:β stereoisomers (eq 78).

In contrast, a stereoselective preparation of the α-isomer of the 2,3,4,6-tetra-*O*-benzyl-D-glucopyranoside employed LHMDS base (THF–DMF, $0\,^{\circ}$C, 48 h) and gave a 92% yield of an 88:12 ratio of α:β stereoisomers.[101]

A conversion of bromohydrins featuring a glycosidic hydroxyl group of cyclic hemiacetals into epoxides mediated by different bases, for example KHMDS, followed by the epoxide opening by nucleophiles represents a useful strategy for a stereocontrolled functionalization of carbohydrates.[102,103] For example, an α-stereoselective coupling of *N*-acetyllactosamine-derived bromohydrins to sialic acid-derived thiols or thioacetates was realized via an *in situ* formation of an epoxide, which was effectively mediated by KHMDS at low temperatures (eqs 79 and 80).[103] The reaction temperature proved to be critical for achieving a good diastereoselectivity.

Deprotonation of the N–H Bond in Sulfonamides: Catalyst in an Anionic Ring-opening Polymerization. The addition of 5 mol % each of *N*-benzyl methanesulfonamide and KHMDS to a solution of 2-*n*-decyl-*N*-mesylaziridine in DMF initiated a ring-opening polymerization providing a very low polydispersity polyamine polymer (eq 81).[104]

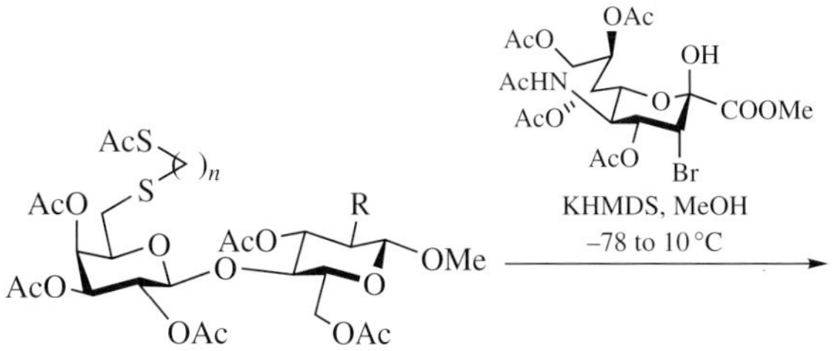

(80)

R = NHAc or OAc

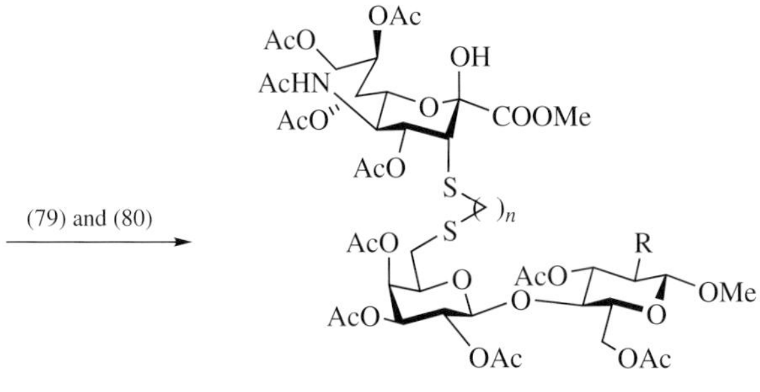

(79) and (80) →

21–54% (R = NHAc, n = 1–5)
22–68% (R = OAc, n = 1–5)

Polymerization of racemic aziridines afforded soluble polymers, whereas the use of enantiopure aziridines provided sparingly soluble polymers. KHMDS serves to deprotonate the primary sulfonamide initiator, providing an amide anion, which then functions as a nucleophile to promote the ring-opening polymerization of the aziridine. The polymerization kinetics were shown to feature a first-order dependence on both the aziridine substrate and the anionic initiator.

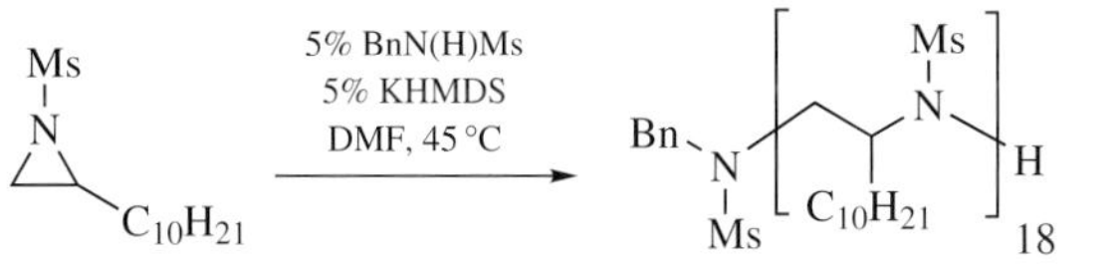

(81)

racemic

$M_n = 4800$, PDA = 1.06

Nucleophilic Nitrogen Source in Polymer Synthesis.
KHMDS was used as a stoichiometric N-nucleophile to initiate an anionic polymerization of ethylene oxide, providing the α-amino-ω-hydroxyl-poly(ethyleneglycol) (PEG) (THF, 50 °C, 15 h) polymers with different molecular weights (2100, 4400, 7200) (eq 82).[105] The N-silyl groups were removed under acidic conditions, and terminating functionalities with structures of known pharmaceutical agents (sulfadiazine and chlorambucil) were attached to the amino and hydroxyl groups via carbamate or ester groups. The anticancer activity of the resulting polymers was studied.

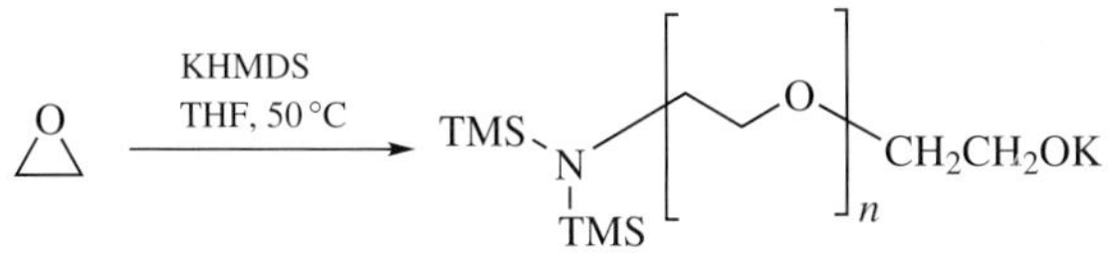

(82)

A method for terminal functionalization of polystyrene chains relying on platinum-catalyzed hydrosilylation of ω-silyl hydride

functionalized polymers with bis(trimethylsilyl)allylamine was developed. KHMDS has been treated with allyl bromide in HMDS (0 °C, 1 h) to provide the requisite TMS-protected allylamine.[106]

Base-mediated Rearrangements of the Carbon Skeleton.
KHMDS has been utilized to induce the [3,3] sigmatropic anionic oxy-Cope rearrangement with subsequent C-methylation, providing the ring-expanded α-methylated ketone in an excellent yield (81%) in a one step reaction (eq 83).[107]

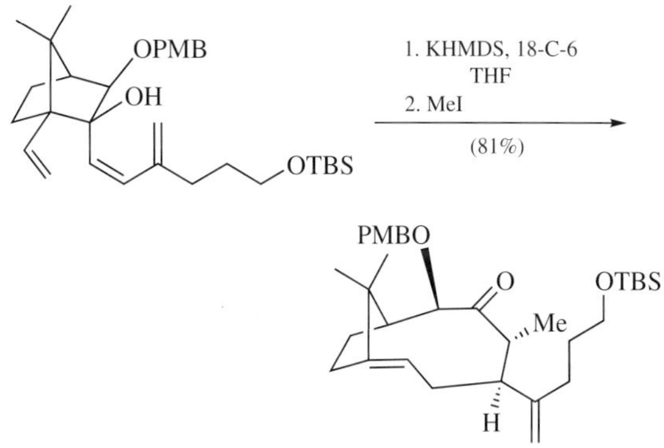

(83)

An attempted KHMDS-mediated oxygenation reaction of cis-C9–C10 taxol precursor with an excess (10 equiv) of KHMDS (THF, −78 °C) utilizing an excess of 18-crown-6 ether (30 equiv) afforded a hydroxylated product featuring a rearranged carbon skeleton corresponding to the ring system found in taxol. Apparently, the cis-C9–C10 ring system was prone to a facile α-ketol rearrangement, and the α-oxygenation occurred only after the skeletal rearrangement took place (eq 84).[107]

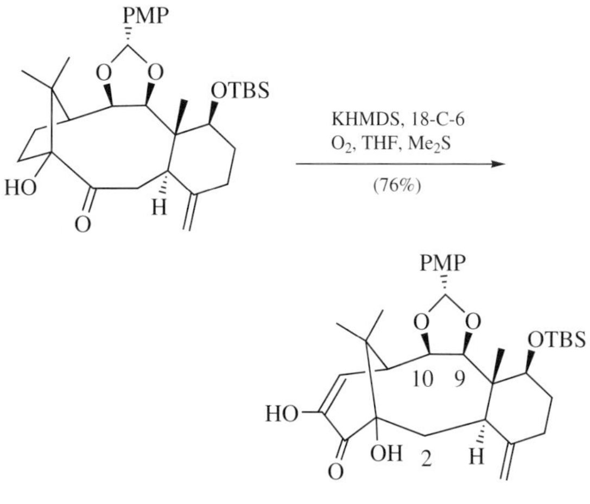

(84)

In contrast, an analogous KHMDS-mediated oxygenation reaction applied to the trans-C9–C10 PMP acetal-protected taxol precursor afforded the C-2 oxygenated product in a good yield. The oxidation was not accompanied by the α-ketol rearrangement of the trans-C9–C10 ring system. Apparently, the trans-locked PMP-acetal introduced a significant steric strain preventing the operation of the α-ketol rearrangement.

KHMDS has been studied as a reagent capable of inducing base-mediated rearrangements of 2-benzyloxycyclooctanone and its Δ^5-unsaturated congener.[108] Exploring the feasibility of an O → C 1,2-shift, 2-benzyloxycyclooctanone was treated with bases including NaH in DMF, LDA in THF, and KHMDS in

THF. However, the O → C 1,2-shift failed to occur. In contrast, the skeletal rearrangement took place in 10 min at room temperature when the reactivity of KHMDS was increased by the addition of the 18-crown-6 ether. Under these conditions, both 2-benzyloxycyclooctanone and 2-benzyl-2-hydroxycyclooctanone afforded a ring-contracted 1-benzylcarbonyl-1-hydroxycycloheptanone in an identical excellent yield (91%) (eqs 85 and eq 86).[108]

The rearrangement pathway was rationalized by a sequential O → C 1,2-shift (in the case of the 2-benzyloxycyclooctanone) followed by the α-ketol rearrangement. In contrast, analogous rearrangements of the Δ⁵-unsaturated analogs under the same conditions required a minimum of 8 h, providing only modest yields of the rearranged product (39–41%). Notably, the α-ketol rearrangement is a thermodynamically controlled process implying greater thermodynamic stability of the seven-membered isomers.

KHMDS in the presence or absence of the 18-crown-6 ether were used to induce an acyloin rearrangement of 1,3-di-*tert*-butyl-dimethylsilyloxybicyclo[2.2.2]oct-5-en-2-ones to the 1,8-*tert*-butyldimethylsilyloxybicyclo[3.2.1]oct-3-en-2-ones.[109] The bicyclo[2.2.2]octenone bearing a formyl (CHO) group afforded good yields of the rearranged bicyclo[3.2.1]octenone with *exo* stereochemistry at C-7 in reactions mediated by KHMDS (80%) or KH (66%). Interestingly, the addition of the 18-crown-6 to the reaction mediated by KHMDS gave only a low yield (35%) and a 1:1 ratio of C-7 *endo/exo* rearranged products (eq 87).[109] The bicyclo[2.2.2]octenone bearing a cyano (CN) group failed to proceed with KH base, whereas the reactions mediated by KHMDS without or with the addition of 18-crown-6 ether afforded 77% and 62% combined yields of C-7 *endo/exo* epimers, respectively (eq 88).[109]

In contrast, treatment of the bicyclo[2.2.2]octenone bearing a nitro (NO₂) group with KHMDS led only to the epimerization of the C-7 stereocenter in the substrate, whereas KH in THF or NaH in DMF induced the acyloin rearrangement in good yields.

The application of KHMDS was essential to realizing the Ireland–Claisen rearrangement of esters of allylic cyclohexenols.[110] Under the optimized conditions (KHMDS, toluene, −78 °C to reflux, TMSCl), the rearranged cyclohexene product was obtained in a high yield (76%, dr 3:2) (eq 89).[110] Other bases, including LDA, LHMDS or NaHMDS, gave lower yields, and the application of TIPSOTf and Hunig's base resulted in significant decom-

position. However, the presence of Me or Br substituents at the C-2 carbon in the cyclohexenol ring of the substrates reduced the yields to less than 50%.

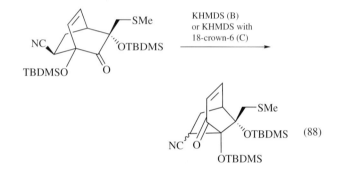

Base-mediated (Formal) Cycloadditions. KHMDS, as well as other bases, have been used in the synthesis of pyrroles via a formal [3 + 2] cycloaddition of metalated isocyanides with ethyl cyclopropylpropiolate.[111] The effect of the nature of the base on the ratio of the desired pyrrole product to the undesired ketone byproduct was investigated. The proposed mechanism consists of a formal cycloaddition of the metalated isocyanides across the triple bond, followed by a 1,5-hydrogen shift and protonation. The reaction between benzyl isocyanide and *tert*-butyl cyclopropyl-propiolate mediated by *t*-BuOK (2 h, rt) afforded a mixture of both the products in 96% combined yield and 1:1 ratio (eq 90).[111] Similar results were achieved in reactions mediated by KHMDS both at rt and −78 °C (93% and 86% combined yields of 1:1 mixtures) (eq 90). The replacement of KHMDS with NaHMDS and LiHMDS led to a nearly exclusive formation of the undesired ketone product. The content of the undesired ketone product appeared to increase when harder cations were used. Accordingly, the application of *t*-BuOCs (2 h, rt) afforded 91% combined yield of the two products in a 96:4 ratio favoring the heterocycle (eq 90).

Base	Combined yield (%)	Pyrrole:ketone ratio
t-BuOK	96	51:49
KHMDS	93	47:53
NaHMDS	98	6:94
LHMDS	98	2:98
t-BuOCs	91	96:4

Several examples of the formal cycloaddition reaction mediated effectively by KHMDS were described (eq 91).

The effect of the base on the yield and diastereoselectivity of the 1,3-dipolar cycloaddition of azomethine ylides derived from iminoesters with (S)-2-*p*-tolylsulfinyl-2-cyclopentenone was studied.[112] With a variety of bases (e.g., AgOAc/DBU, AgOAc/TEA) the cycloaddition proceeded with complete regioselectivity and endo selectivity, but low diastereoselectivity (facial selectivity). The diastereoselectivity was dependent on the cation of the base, and the comparison of the performance of LHMDS, NaHMDS and KHMDS indicated that the best yields and diastereoselectivities could be achieved with the lithium cation, giving 91% yield of the diastereomer A and a 98/2 ratio of A/B diastereomers (eq 92). The lower rate and stereoselectivity of reactions involving sodium and potassium cations (eq 92) was rationalized by the operation of a stepwise addition/ring-closure mechanism, in which the cation acts as a tether between the ester oxygen and the oxygen of the *p*-tolylsulfinyl groups in the two substrates. In contrast to the lithium cation, sodium or potassium cations were unable to fulfill this role with comparable efficiency.

Base	Yield of **A** (%)	**A:B** ratio
LHMDS	91	98:2
NaHMDS	48	68:32
KHMDS	54	75:25

Base-mediated Olefination Reactions. A Peterson olefination reagent (*t*-BuO)Ph$_2$SiCH$_2$CN, which has shown good selectivity for the formation of (Z)-β-monosubstituted α,β-unsaturated cyanides was developed, and the effect of the cations in the bases and the substrate structures on the reaction yield and the (Z)-stereoselectivity was investigated.[113] The results of the olefinations of various *aromatic aldehydes* mediated by *n*-BuLi, KHMDS and NaHMDS revealed only minimal effects of the cation in the base. In general, (Z)-selectivity exceeded 9:1, and good yields were obtained. For example, olefination of 2-pyridylcarboxaldehyde gave optimum results (88% yield, 92:8, Z:E ratio) when mediated by KHMDS (eq 93). In contrast, the olefinations of diverse *aliphatic aldehydes* with (*t*-BuO)Ph$_2$SiCH$_2$CN mediated by KHMDS proved to be sensitive to the steric bulk of substituents at the α-carbon of the aldehyde, giving a lower (Z)-selectivity. The (Z)-selectivity increased to good levels with only minimal decrease in the overall yields when KHMDS was replaced with NaHMDS and *n*-BuLi (eq 93).

KHMDS has been used as a base in Wittig-type olefinations, which served as the key steps in the synthesis of all eight geometric isomers of methyl 2,4,6-decatrienoate desired as precursors for several pheromones.[114] Dienyl acetals (eq 94 and eq 95) were converted into the corresponding trienyl esters with *cis*-stereoselectivity under the modified Horner–Wadsworth–Emmons olefination conditions utilizing electrophilic phosphonate (CF$_3$CH$_2$O)$_2$P(O)CH$_2$COOCH$_3$ and a strongly dissociated base system. Thus, reactions mediated by KHMDS along with the 18-crown-6 ether provided the trienyl esters in 87% and 88% yields, and 95:5 and 93:7 Z:E ratios, respectively, after flash column chromatography (eqs 94 and 95). A partial isomerization of the 6-Z bond in the ester product (eq 95) occurred in the presence of the excess of the 18-crown-6 ether due to the increased polarity of the medium, accounting for the lower selectivity (93:7).

Generation of Ylides. A cyclopropanation of alkylidene and arylidene malonates with arsonium ylides was developed, providing access to *trans*-2,3-disubstituted cyclopropane 1,1-dicarboxylic acids bearing a vinyl substituent.[115] The choice of bases for the reactions between ethyl propylidene malonate and arsonium cinnamylide, generated in situ by deprotonation of a corresponding arsonium bromide, was optimized. Experiments with LHMDS, NaHMDS and KHMDS revealed that the best yields and diastereoselectivities could be achieved with KHMDS (88% yield, *trans:cis* = 16:1) (eq 96). The application of LHMDS led to a decrease in both the yields and diastereoselectivities (30%, 1.5:1). The transformation proved to be generally applicable, allowing an efficient cyclopropanation of arylidene malonates (eq 97).

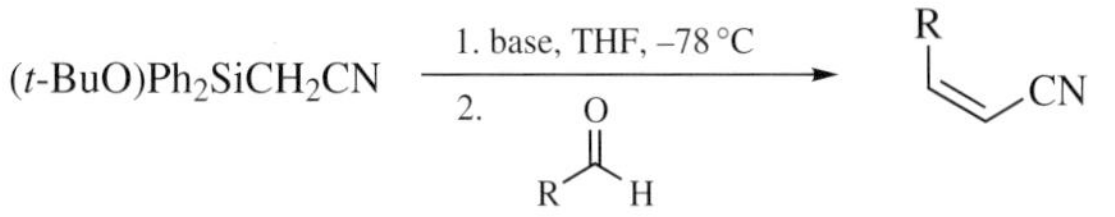

$$(t\text{-BuO})Ph_2SiCH_2CN \xrightarrow[\quad 2.\quad]{1.\ \text{base, THF, } -78\,°C} \quad (93)$$

R	Base	Yield	*Z:E* ratio
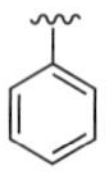 phenyl	KHMDS	74	96:4
	NaHMDS	79	98:2
	n-BuLi	93	99:1
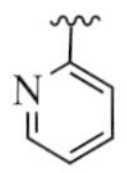 pyridyl	KHMDS	88	92:8
	NaHMDS	75	90:10
	n-BuLi/TMEDA	71	91:9
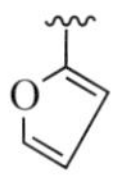 furyl	KHMDS	62	98:2
	NaHMDS	85	94:6
	n-BuLi	90	97:3

R	Base	Yield	*Z:E* ratio
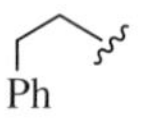 Ph	KHMDS	99	96:4
	NaHMDS	88	98:2
	n-BuLi	99	96:4
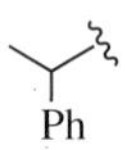 Ph	KHMDS	20	98:2
	NaHMDS	55	96:4
	n-BuLi/TMEDA	60	98:2
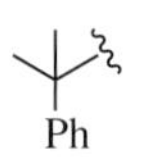 Ph	KHMDS	93	86:14
	NaHMDS	85	95:5
	n-BuLi	87	95:5

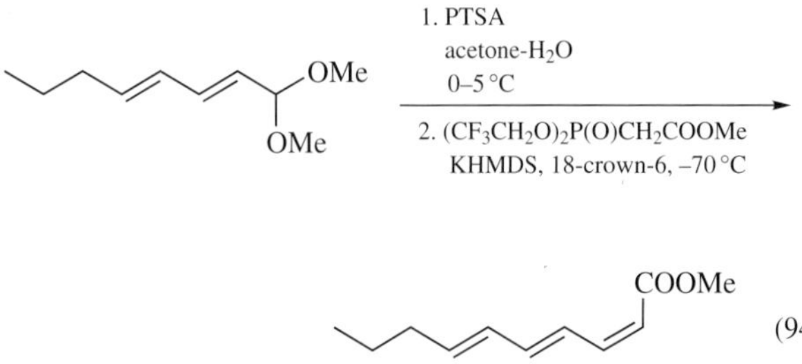

$$(94)$$

95:5 (2Z), 87% yield

KHMDS was used as a base for the deprotonation of benzyltrimethylphosphonium iodide (KHMDS, 0.5 M in toluene, 4 h, 0 °C to rt) to afford the corresponding phosphorus ylide, which was reacted *in situ* with phenylnitrile or 2-furylnitrile, providing *N*-vinylic phosphazenes (eq 98).[116] A mechanistic pathway involving the construction and subsequent ring-opening of an azaphosphetane (eq 98) was proposed to rationalize the reac-

tion course. The *N*-vinylicphosphazenes were reacted with alkynyl esters to afford unsaturated amines or imines.

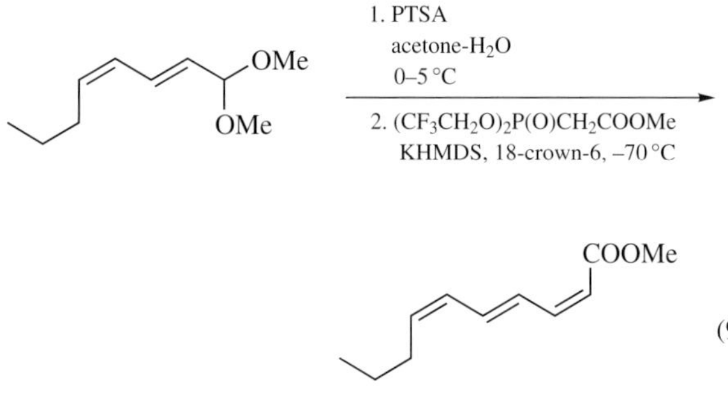

$$(95)$$

93:7 (2Z), 88% yield

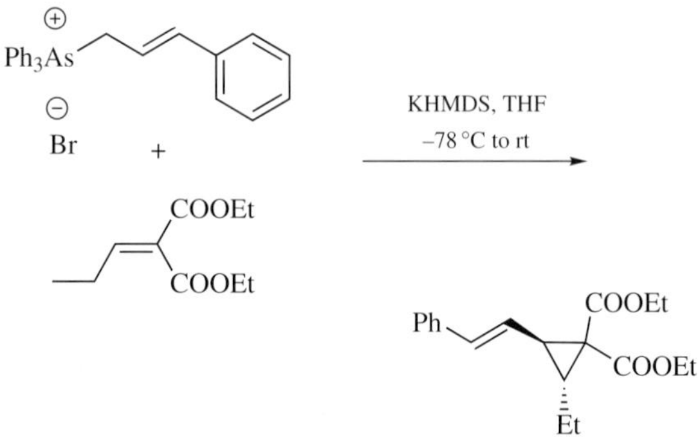

$$(96)$$

88% (*trans/cis* 16/1)

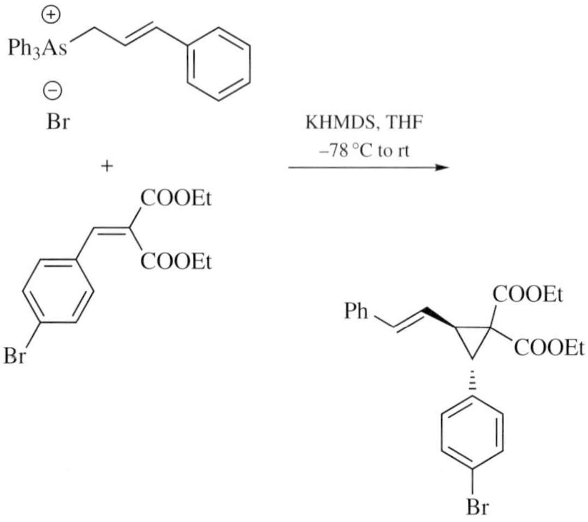

$$(97)$$

94% (*trans/cis* 27/1)

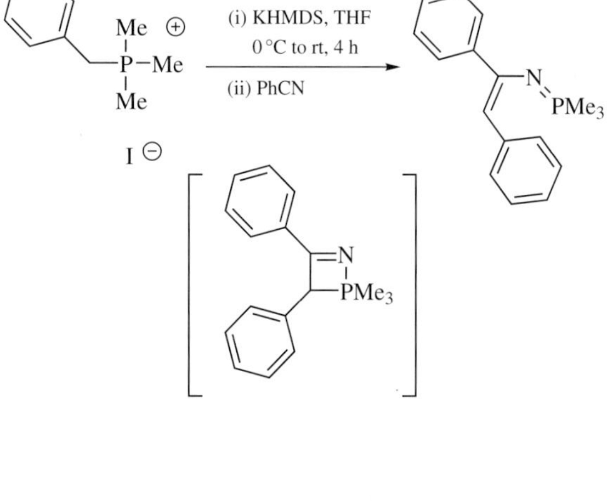

$$(98)$$

Base in Pd-catalyzed C–N Bond-forming Reactions. Suitable bases for successful cross-coupling of aromatic bromides bearing reactive functionalities (OH, AcNH, CH$_3$CO, NH$_2$CO) with amines via a catalytic system employing a new bicyclic triaminophosphine ligand along with Pd(OAc)$_2$ catalyst (toluene, 80 °C) were sought.[117] The cross-coupling reactions, which failed using weak bases, could be efficiently realized employing t-BuONa/LDA mixtures, LHMDS, KHMDS, or NaHMDS bases. The results for coupling of 3-hydroxyphenyl bromide to morpholine (toluene, 80 °C, 22 h) are shown in eq 99.

The reasons for the effectiveness of silylated amide bases might consist in an internal protection of the reactive functionalities, either via the formation of tightly associated lithium ion pairs following the deprotonation, or via the formation of silylated alcohols or silylated amides operating as protecting groups. Both pathways would prevent the alkoxide or amide coordination to palladium resulting in catalyst deactivation.

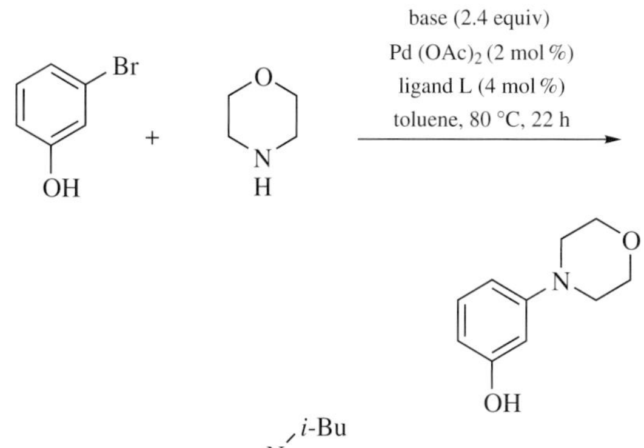

(99)

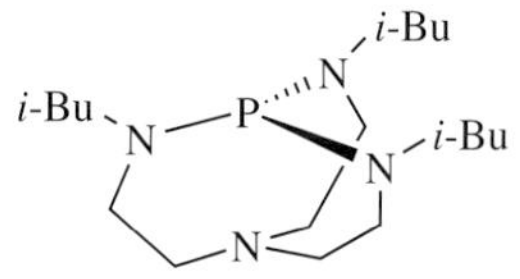

Base	Yield (%)	Base	Yield (%)
t-BuONa	Trace	LiHMDS	92
Cs$_2$CO$_3$	NR	KHMDS	85
t-BuONa/LDA	80	NaHMDS	88

Base for C–H or N–H Deprotonation of Precursors to Carbene or Amide Ligands in the Preparation of Metal Complexes. KHMDS has been used as a base for the deprotonation of N–H or B–H bonds in imine-based ligands or trispyrazolyl borate-derived ligands to synthesize calcium complexes, which were used as initiators of the ring-opening polymerization of lactides.[118] A first example of a two-coordinate, neutral In^I singlet carbene complex has been prepared via an in situ deprotonation of the imine ligand precursor with KHMDS in the presence of In^{I}I to afford the In(I) complex in 36% unoptimized yield as pale yellow, thermally stable crystals.[119]

A bidentate ferrocenyl-N-heterocyclic carbene ligand precursor in the form of the bis-imidazolinium salt was deprotonated in situ with KHMDS and reacted with [PdCl$_2$(cod)] to provide a bis(carbene)palladium(II) dichloride complex, which was structurally characterized (eq 100).[120]

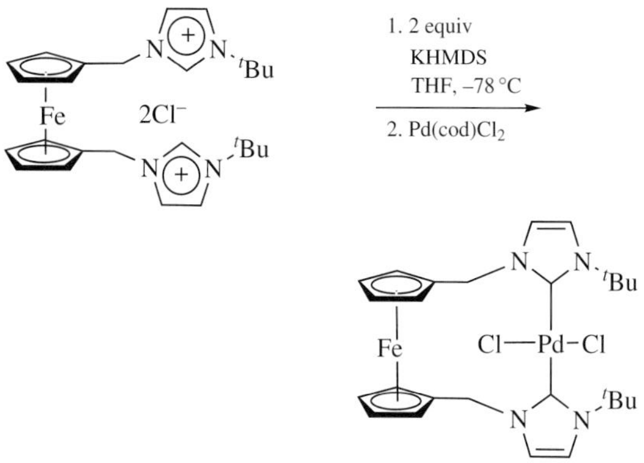

(100)

An in situ treatment of the iminium salt shown in eq 101 with KHMDS afforded a four-membered N-heterocyclic carbene ligand, which reacted with the ruthenium complex (PPh$_3$)Cl$_2$Ru=CH-o-O-i-PrC$_6$H$_5$ (60 °C, overnight) providing the ruthenium carbene complex in 30% yield as an air stable brown solid (eq 101), which was characterized by NMR and tested for its reactivity in ring closing metathesis.[121]

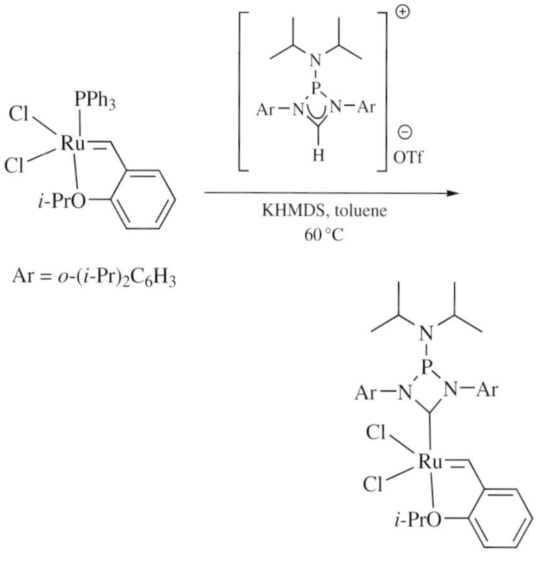

(101)

Synergistic Effects in Metalation of Unactivated C–H Bonds. The metalation of toluene has been achieved using a bimetallic base KZn(HMDS)$_3$, providing the metalated benzyl product [KZn(HMDS)$_2$CH$_2$Ph], whereas an analogous metalation could not be achieved either with KHMDS or Zn(HMDS)$_2$.[122] A mixture of KHMDS and Zn(HMDS)$_2$ (10 mmol each) in toluene (20 mL) was stirred for 15 min at room temperature followed by vigorous heating for 5 min to afford the metalated benzyl product as a solid [KZn(HMDS)$_2$CH$_2$Ph]. Under more strenuous conditions the metalation could not be realized with either KHMDS or Zn(HMDS)$_2$, clearly demonstrating that the described deprotonative metalation is synergic by its origin. The synergic effect did not operate when Zn was replaced with Mg, and the synergy-driven deprotonation could be extended to m-xylene and mesitylene.

1. Wannagat, U.; Niederpruem, H., *Ber. Dtsch. Chem. Ges.* **1961**, *94*, 1540.

2. (a) Brown, C. A., *Synthesis* **1974**, 427. (b) Brown, C. A., *J. Org. Chem.* **1974**, *39*, 3913.

3. Tesh, K. F.; Hanusa, T. P.; Huffman, J. C., *Inorg. Chem.* **1990**, *29*, 1584.

4. Williard, P. G., *Acta Crystallogr.* **1988**, *C44*, 270.

5. Ireland, R. E.; Meissner, R. S., *J. Org. Chem.* **1991**, *56*, 4566.

6. Duhamel, L.; Plaquevent, J.-C., *J. Organomet. Chem.* **1993**, *448*, 1.

7. (a) Evans, D. A. in *Asymmetric Synthesis*; Morrison, J. D., Ed.; Academic: New York, 1984; Vol. 3, p 1. (b) Brown, C. A., *J. Org. Chem.* **1974**, *39*, 3913.

8. Negishi, E.; Chatterjee, S., *Tetrahedron Lett.* **1983**, *24*, 1341.

9. Differding, E.; Rueegg, G.; Lang, R. W., *Tetrahedron Lett.* **1991**, *32*, 1779.

10. (a) Evans, D. A.; Britton, T. C., *J. Am. Chem. Soc.* **1987**, *109*, 6881. (b) Evans, D. A.; Ellman, J. A., *J. Am. Chem. Soc.* **1989**, *111*, 1063.

11. (a) Boche, G.; Heidenhain, F., *Angew. Chem., Int. Ed. Engl.* **1978**, *17*, 283. (b) Boche, G.; Heidenhain, F.; Thiel, W.; Eiben, R., *Ber. Dtsch. Chem. Ges.* **1982**, *115*, 3167.

12. Kawanisi, M.; Itoh, Y.; Hieda, T.; Kozima, S.; Hitomi, T.; Kobayashi, K., *Chem. Lett.* **1985**, 647.

13. (a) Rathke, M. W.; Sullivan, D., *Tetrahedron Lett.* **1972**, 4249. (b) Herrman, J. L.; Kieczykowski, G. R.; Schlessinger, R. H., *Tetrahedron Lett.* **1973**, 2433. (c) Krebs, E. P., *Helv. Chim. Acta* **1981**, *64*, 1023. (d) Kende, A. S.; Toder, B. H., *J. Org. Chem.* **1982**, *47*, 163. (e) Ikeda, Y.; Yamamoto, H., *Tetrahedron Lett.* **1984**, *25*, 5181.

14. (a) Ikeda, Y.; Ukai, J.; Ikeda, N.; Yamamoto, H., *Tetrahedron* **1987**, *43*, 743. (b) Chattopadhyay, A.; Mamdapur, V. R., *Synth. Commun.* **1990**, *20*, 2225.

15. (a) Bates, G. S., *Chem. Commun.* **1979**, 161. (b) Bates, G. S.; Ramaswamy, S., *Can. J. Chem.* **1980**, *58*, 716.

16. (a) Schlosser, M.; Christmann, K. F., *Justus Liebigs Ann. Chem.* **1967**, *708*, 1. (b) Schlosser, M., *Top. Stereochem.* **1970**, *5*, 1. (c) Schlosser, M.; Schaub, B.; de Oliveira-Neto, J.; Jeganathan, S., *Chimia* **1986**, *40*, 244. (d) Schaub, B.; Jeganathan, S.; Schlosser, M., *Chimia* **1986**, *40*, 246.

17. Sreekumar, C.; Darst, K. P.; Still, W. C., *J. Org. Chem.* **1980**, *45*, 4260.

18. Still, W. C.; Gennari, C., *Tetrahedron Lett.* **1983**, *24*, 4405.

19. (a) Stork, G.; Gardner, J. O.; Boeckman, R. K., Jr.; Parker, K. A., *J. Am. Chem. Soc.* **1973**, *95*, 2014. (b) Stork, G.; Boeckman, R. K., Jr., *J. Am. Chem. Soc.* **1973**, *95*, 2016.

20. Takahashi, T.; Kazuyuki, K.; Tsuji, J., *Tetrahedron Lett.* **1978**, 4917.

21. Abraham, W. D.; Cohen, T., *J. Am. Chem. Soc.* **1991**, *113*, 2313.

22. Shimizu, M.; Yoshioka, H., *Tetrahedron Lett.* **1989**, *30*, 967.

23. Paquette, L. A.; Pegg, N. A.; Toops, D.; Maynard, G. D.; Rogers, R. D., *J. Am. Chem. Soc.* **1990**, *112*, 277.

24. Larsen, S. D., *J. Am. Chem. Soc.* **1988**, *110*, 5932.

25. Brown, H. C.; Ramachandran, P. V.; Gong, B. Q.; Brown, H. C., *J. Org. Chem.* **1995**, *60*, 41.

26. Mantri, P.; Duffy, D. E.; Kettner, C. A., *J. Org. Chem.* **1996**, *61*, 5690.

27. Wityak, J.; Earl, R. A.; Abelman, M. M.; Bethel, Y. B.; Fisher, B. N.; Kauffman, G. S.; Kettner, C. A.; Ma, P.; McMillan, J. L.; Mersinger, L. J.; Pesti, J.; Pierce, M. E.; Rankin, F. W.; Chorvat, R. J.; Confalone, P. N., *J. Org. Chem.* **1995**, *60*, 3717.

28. Matteson, D. S.; Schaumberg, G. D., *J. Org. Chem.* **1966**, *31*, 726.

29. Kiselyov, A. S., *Tetrahedron Lett.* **1995**, *36*, 1383.

30. Bruning, J., *Tetrahedron Lett.* **1997**, *38*, 3187.

31. See for example: Grasa, G. A.; Guveli, T.; Singh, R.; Nolan, S. P., *J. Org. Chem.* **2003**, *68*, 2812.

32. Herrmann, W. A., *Angew. Chem. Int. Ed.* **2002**, *41*, 1290.

33. Yong, B. S.; Nolan, S. P., *Chemtracts: Org. Chem.* **2003**, *16*, 205.

34. Douthwaite, R. E.; Haussinger, D.; Green, M. L. H.; Silcock, P. J.; Gomes, P. T.; Martins, A. M.; Danopoulos, A. A., *Organometallics* **1999**, *18*, 4584.

35. Danopoulos, A. A.; Winston, S.; Gelbrich, T.; Hursthouse, M. B.; Tooze, R. P., *Chem. Commun.* **2002**, 482.

36. Danopoulos, A. A.; Winston, S.; Motherwell, W. B., *Chem. Commun.* **2002**, 1376.

37. Kerr, M. S.; de Alaniz, J. R.; Rovis, T., *J. Am. Chem. Soc.* **2002**, *124*, 10298.

38. Kerr, M. S.; Rovis, T., *Synlett* **2003**, 1934.

39. Culkin, D. A.; Hartwig, J. F., *Acc. Chem. Res.* **2003**, *36*, 234.

40. Miura, M.; Nomura, M., *Top. Curr. Chem.* **2002**, *219*, 211.

41. Lloyd-Jones, G. C., *Angew. Chem. Int. Ed.* **2002**, *41*, 953.

42. Lee, S.; Beare, N. A.; Hartwig, J. F., *J. Am. Chem. Soc.* **2001**, *123*, 8410.

43. Jorgensen, M.; Lee, S.; Liu, X. X.; Wolkowski, J. P.; Hartwig, J. F., *J. Am. Chem. Soc.* **2002**, *124*, 12557.

44. Gaertzen, O.; Buchwald, S. L., *J. Org. Chem.* **2002**, *67*, 465.

45. Beare, N. A.; Hartwig, J. F., *J. Org. Chem.* **2002**, *67*, 541.

46. Hamann, B. C.; Hartwig, J. F., *J. Am. Chem. Soc.* **1997**, *119*, 12382.

47. Fox, J. M.; Huang, X. H.; Chieffi, A.; Buchwald, S. L., *J. Am. Chem. Soc.* **2000**, *122*, 1360.

48. Ehrentraut, A.; Zapf, A.; Beller, M., *Adv. Synth. Catal.* **2002**, *344*, 209.

49. Shaughnessy, K. H.; Hamann, B. C.; Hartwig, J. F., *J. Org. Chem.* **1998**, *63*, 6546.

50. Caron, S.; Vazquez, E.; Wojcik, J. M., *J. Am. Chem. Soc.* **2000**, *122*, 712.

51. Papahatjis, D. P.; Nikas, S.; Tsotinis, A.; Vlachou, M.; Makriyannis, A., *Chem. Lett.* **2001**, 192.

52. Marcoux, J. F.; Corley, E. G.; Rossen, K.; Pye, P.; Wu, J.; Robbins, M. A.; Davies, I. W.; Larsen, R. D.; Reider, P. J., *Org. Lett.* **2000**, *2*, 2339.

53. Watanabe, K.; Hirasawa, T.; Hiroi, K., *Heterocycles* **2002**, *58*, 93.

54. Brana, M. F.; Moran, M.; Devega, M. J. P.; Pitaromero, I., *Tetrahedron Lett.* **1994**, *35*, 8655.

55. Harada, T.; Fujiwara, T.; Iwazaki, K.; Oku, A., *Org. Lett.* **2000**, *2*, 1855.

56. Ogawa, A.; Sakagami, K.; Shima, A.; Suzuki, H.; Komiya, S.; Katano, Y.; Mitsunobu, O., *Tetrahedron Lett.* **2002**, *43*, 6387.

57. Taber, D. F.; Christos, T. E.; Neubert, T. D.; Batra, D., *J. Org. Chem.* **1999**, *64*, 9673.

58. Barluenga, J.; Baragana, B.; Concellon, J. M.; Pinera-Nicolas, A.; Diaz, M. R.; Garcia-Granda, S., *J. Org. Chem.* **1999**, *64*, 5048.

59. Kawabata, T.; Suzuki, H.; Nagae, Y.; Fuji, K., *Angew. Chem. Int. Ed.* **2000**, *39*, 2155.

60. Kawabata, T.; Fuji, K., *Top. Stereo. Chem.* **2003**, *23*, 175.

61. Kawabata, T.; Kawakami, S.; Majumdar, S., *J. Am. Chem. Soc.* **2003**, *125*, 13012.

62. Corey, E. J.; Wu, L. I., *J. Am. Chem. Soc.* **1993**, *115*, 9327.

63. Li, C. M.; Johnson, R. P.; Porco, J. A., *J. Am. Chem. Soc.* **2003**, *125*, 5095.

64. Ruel, R.; Bouvier, J. P.; Young, R. N., *J. Org. Chem.* **1995**, *60*, 5209.

65. Helmchen, G.; Pfaltz, A., *Acc. Chem. Res.* **2000**, *33*, 336.

66. Busacca, C. A.; Grossbach, D.; So, R. C.; O'Brien, E. M.; Spinelli, E. M., *Org. Lett.* **2003**, *5*, 595.

67. Erratum: Busacca, C. A.; Grossbach, D.; So, R. C.; O'Brien, E. M.; Spinelli, E. M., *Org. Lett.* **2003**, *5*, 1595.

68. Maezaki, N.; Sakamoto, A.; Nagahashi, N.; Soejima, M.; Li, Y. X.; Imamura, T.; Kojima, N.; Ohishi, H.; Sakaguchi, K.; Iwata, C.; Tanaka, T., *J. Org. Chem.* **2000**, *65*, 3284.

69. Zhou, Y. G.; Li, A. H.; Hou, X. L.; Dai, L. X., *Chem. Commun.* **1996**, 1353.

70. Aggarwal, V. K.; Bae, I.; Lee, H. Y.; Richardson, J.; Williams, D. T., *Angew. Chem. Int. Ed.* **2003**, *42*, 3274.

71. Gandelman, M.; Rybtchinski, B.; Ashkenazi, N.; Gauvin, R. M.; Milstein, D., *J. Am. Chem. Soc.* **2001**, *123*, 5372.

72. Blakemore, P. R., *J. Chem. Soc., Perkin Trans. 1* **2002**, 2563.

73. Baudin, J. B.; Hareau, G.; Julia, S. A.; Lorne, R.; Ruel, O., *Bull. Soc. Chim. Fr.* **1993**, *130*, 856.

74. Charette, A. B.; Lebel, H., *J. Am. Chem. Soc.* **1996**, *118*, 10327.

75. Blakemore, P. R.; Cole, W. J.; Kocienski, P. J.; Morley, A., *Synlett* **1998**, 26.

76. Smith, A. B.; Wan, Z. H., *J. Org. Chem.* **2000**, *65*, 3738.

77. Trost, B. M.; Crawley, M. L., *J. Am. Chem. Soc.* **2002**, *124*, 9328.

78. Trost, B. M.; Chisholm, J. D.; Wrobleski, S. T.; Jung, M., *J. Am. Chem. Soc.* **2002**, *124*, 12420.

79. Fettes, A.; Carreira, E. M., *Angew. Chem. Int. Ed.* **2002**, *41*, 4098.

80. Kocienski, P. J.; Bell, A.; Blakemore, P. R., *Synlett* **2000**, 365.

81. Charette, A. B.; Berthelette, C.; St-Martin, D., *Tetrahedron Lett.* **2001**, *42*, 5149.

82. Erratum: Charette, A. B.; Berthelette, C.; St-Martin, D., *Tetrahedron Lett.* **2001**, *42*, 6619.

83. Tang, Y.; Ye, S.; Huang, Z. Z.; Huang, Y. Z., *Heteroatom Chem.* **2002**, *13*, 463.

84. Dai, W.-M.; Wu, A.; Wu, H., *Tetrahedron: Asymmetry* **2002**, *13*, 2187.

85. Klotz, P.; Mann, A., *Tetrahedron Lett.* **2003**, *44*, 1927.

86. Baldelli, E.; Battaglia, A.; Bombardelli, E.; Carenzi, G.; Fontana, G.; Gambini, A.; Gelmi, M. L.; Guerrini, A.; Pocar, D., *J. Org. Chem.* **2003**, *68*, 9773.

87. Trost, B. M.; Dogra, K.; Franzini, M., *J. Am. Chem. Soc.* **2004**, *126*, 1944.

88. Pedersen, B.; Rein, T.; Sotofte, I.; Norrby, P. O.; Tanner, D., *Collect. Czech. Chem. Commun.* **2003**, *68*, 885.

89. García Ruano, J. L.; Martín-Castro, A. M.; Tato, F.; Pastor, C. J., *J. Org. Chem.* **2005**, *70*, 7346.

90. Kawabata, T.; Majumdar, S.; Tsubaki, K.; Monguchi, D., *Org. Biomol. Chem.* **2005**, *3*, 1609.

91. Carlier, P. R.; Lam, P. C. H.; DeGuzman, J. C.; Zhao, H., *Tetrahedron: Asymmetry* **2005**, *16*, 2998.

92. Wiles, C.; Watts, P.; Haswell, S. J.; Pombo-Villar, E., *Tetrahedron* **2005**, *61*, 10757.

93. Klapars, A.; Waldman, J. H.; Campos, K. R.; Jensen, M. S.; McLaughlin, M.; Chung, J. Y. L.; Cvetovich, R. J.; Chen, C., *J. Org. Chem.* **2005**, *70*, 10186.

94. Lu, G.; Malinakova, H. C., *J. Org. Chem.* **2004**, *69*, 4701.

95. Agami, C.; Dechoux, L.; Hebbe, S., *Tetrahedron Lett.* **2003**, *44*, 5311.

96. Zavialov, I. A.; Dahanukar, V. H.; Nguyen, H.; Orr, C.; Andrews, D. R., *Org. Lett.* **2004**, *6*, 2237.

97. Maung, J.; Mallari, J. P.; Girtsman, T. A.; Wu, L. Y.; Rowley, J. A.; Santiago, N. M.; Brunelle, A. N.; Berkman, C. E., *Bioorg. Med. Chem.* **2004**, *12*, 4969.

98. Naud, S.; Cintrat, J. C., *Synthesis* **2003**, 1391.

99. Dixon, D. J.; Scott, M. S.; Luckhurst, C. A., *Synlett* **2003**, 2317.

100. Kanzelberger, M.; Zhang, X.; Emge, T. J.; Goldman, A. S.; Zhao, J.; Incarvito, C.; Hartwig, J. F., *J. Am. Chem. Soc.* **2003**, *125*, 13644.

101. Hashihayata, T.; Mandai, H.; Mukaiyama, T., *Bull. Chem. Soc. Jpn.* **2004**, *77*, 169.

102. Kobayashi, S.; Takahashi, Y.; Komano, K.; Alizadeh, B. H.; Kawada, Y.; Oishi, T.; Tanaka, S. I.; Ogasawara, Y.; Sasaki, S. Y.; Hirama, M., *Tetrahedron* **2004**, *60*, 8375.

103. Hinou, H.; Sun, X. S.; Ito, Y., *J. Org. Chem.* **2003**, *68*, 5602.

104. Stewart, I. C.; Lee, C. C.; Bergman, R. G.; Toste, D. F., *J. Am. Chem. Soc.* **2005**, *127*, 17616.

105. Jia, Z.; Zhang, H.; Huang, J., *Bioorg. Med. Chem. Lett.* **2003**, *13*, 2531.

106. Quirk, R. P.; Kim, H.; Polce, M. J.; Wesdemiotis, C., *Macromolecules* **2005**, *38*, 7895.

107. Paquette, L. A.; Hofferberth, J. E., *J. Org. Chem.* **2003**, *68*, 2266.

108. Vilotijevic, I.; Yang, J.; Hilmey, D.; Paquette, L. A., *Synthesis* **2003**, 1872.

109. Katayama, S.; Yamauchi, M., *Chem. Pharm. Bull.* **2005**, *53*, 666.

110. Beaulieu, P.; Ogilvie, W. W., *Tetrahedron Lett.* **2003**, *44*, 8883.

111. Larionov, O. V.; de Meijere, A., *Angew. Chem. Int. Ed.* **2005**, *44*, 5664.

112. García Ruano, J. L.; Tito, A.; Peromingo, M. T., *J. Org. Chem.* **2003**, *68*, 10013.

113. Kojima, S.; Fukuzaki, T.; Yamakawa, A.; Murai, Y., *Org. Lett.* **2004**, *6*, 3917.

114. Khrimian, A., *Tetrahedron* **2005**, *61*, 3651.

115. Jiang, H.; Deng, X.; Sun, X.; Tang, Y.; Dai, L. X., *J. Org. Chem.* **2005**, *70*, 10202.

116. Palacios, F.; Alonso, C.; Pagalday, J.; Ochoa de Retana, A. M.; Rubiales, G., *Org. Biomol. Chem.* **2003**, *1*, 1112.

117. Urgaonkar, S.; Verkade, J. G., *Adv. Synth. Catal.* **2004**, *346*, 611.

118. Chisholm, M. H.; Gallucci, J. C.; Phomphrai, K., *Inorg. Chem.* **2004**, *43*, 6717.

119. Hill, M. S.; Hitchcock, P. B., *Chem. Commun.* **2004**, 1818.

120. Coleman, K. S.; Turberville, S.; Pascu, S. I.; Green, M. L. H., *J. Organomet. Chem.* **2005**, *690*, 653.

121. Despagnet-Ayoub, E.; Grubbs, R. H., *Organometallics* **2005**, *24*, 338.

122. Clegg, W.; Forbes, G. C.; Kennedy, A. R.; Mulvey, R. E.; Liddle, S. T., *Chem. Commun.* **2003**, 406.

Potassium Monoperoxysulfate

$$2KHSO_5 \cdot KHSO_4 \cdot K_2SO_4$$

[37222-66-5]　　　　$H_3K_5O_{18}S_4$　　　(MW 614.81)

InChI = 1S/5K.2H2O5S.2H2O4S/c;;;;;2*1-5-6(2,3)4;2*1-5(2,3)
　　4/h;;;;;2*1H,(H,2,3,4);2*(H2,1,2,3,4)/q5*+1;;;;/p-5

InChIKey = HJKYXKSLRZKNSI-UHFFFAOYSA-I

(oxidizing agent for a number of functional groups, including alkenes,[16] arenes,[17] amines,[26] imines,[30] sulfides;[37] used for the preparation of dioxiranes[5])

Alternative Names: potassium caroate, potassium hydrogen persulfate, Oxone®, potassium peroxymonosulfate.

Physical Data: mp dec; d 1.12–1.20 g cm^{-3}.

Solubility: sol water (25.6 g 100 g, 20 °C), aqueous methanol, ethanol, acetic acid; insol common organic solvents.

Form Supplied in: white, granular, free flowing solid. Available as Oxone® and as Curox® and Caroat®.

Analysis of Reagent Purity: iodometric titration, as described in the Du Pont data sheet for Oxone®.

Handling, Storage, and Precautions: the Oxone triple salt $2KHSO_5 \cdot KHSO_4 \cdot K_2SO_4$ is a relatively stable, water-soluble form of potassium monopersulfate that is convenient to handle and store. Oxone has a low order of toxicity, but is irritating to the eyes, skin, nose, and throat. It should be used with adequate ventilation and exposure to its dust should be minimized. Traces of heavy metal salts catalyze the decomposition of Oxone. For additional handling instructions, see the Du Pont data sheet.

Original Commentary

Jack K. Crandall
Indiana University, Bloomington, IN, USA

Oxidation Methodology. Oxone ($2KHSO_5 \cdot KHSO_4 \cdot K_2SO_4$) is a convenient, stable source of potassium monopersulfate (caroate), which serves as a stoichiometric oxidizing agent under a variety of conditions. Thus aqueous solutions of Oxone can be used to perform oxidations in homogeneous solution and in biphasic systems using an immiscible cosolvent and a phase-transfer catalyst. Recently, solid–liquid processes using supported Oxone reagents have been developed. Other oxidation methods involve the generation and reaction of a secondary reagent under the reaction conditions, as with the widely employed aqueous Oxone–ketone procedures, which undoubtedly involve dioxirane intermediates.[1–4] In other instances, oxaziridine derivatives and metal oxo complexes appear to be the functional oxidants formed in situ from Oxone. Synthetically useful examples of these oxidations are grouped below according to the functional groups being oxidized.

Ketones and Other Oxygen Functions. Various ketones can be converted to the corresponding dioxiranes by treatment with buffered aqueous solutions of Oxone (eq 1). Of particular interest are dimethyldioxirane[5] ($R^1 = R^2 = Me$) and methyl(trifluoromethyl)dioxirane[6] ($R^1 = Me, R^2 = CF_3$) derived

from acetone and 1,1,1-trifluoro-2-propanone, respectively. The discovery of a method for the isolation of dilute solutions of these volatile dioxiranes in the parent ketone by codistillation from the reaction mixture has opened an exciting new area of oxidation chemistry (see **Dimethyldioxirane** and **Methyl(trifluoromethyl)-dioxirane**). Solutions of dioxiranes derived from higher molecular weight ketones have also been prepared.[5,7]

$$(1)$$

Interestingly, the reaction of a solid slurry of Oxone and wet **Alumina** with solutions of cyclic ketones in CH_2Cl_2 provokes Baeyer–Villiger oxidation to give the corresponding lactones (eq 2).[8] The same wet alumina–Oxone reagent can be used to oxidize secondary alcohols to ketones (eq 3).[9] Aldehydes are oxidized to acids by aqueous Oxone.[5,10]

$$(2)$$

$$(3)$$

Alkenes, Arenes, and Alkanes. Aqueous solutions of Oxone can epoxidize alkenes which are soluble under the reaction conditions; for example, sorbic acid (eq 4).[11] (the high selectivity for epoxidation of the 4,5-double bond here is noteworthy). Alternatively, the use of a cosolvent to provide homogeneous solutions promotes epoxidation (eq 5).[11] Control of the pH to near neutrality is usually necessary to prevent hydrolysis of the epoxide. Rapidly stirred heterogeneous mixtures of liquid alkenes and aqueous Oxone solutions buffered with $NaHCO_3$ also produce epoxides, as shown in eq 6.[12]

$$(4)$$

$$(5)$$

$$(6)$$

An in situ method for epoxidations with dimethyldioxirane using buffered aqueous acetone solutions of Oxone has been widely applied.[1–4] The epoxidation of 1-dodecene is particularly impressive in view of the difficulty generally encountered in the epoxidation of relatively unreactive terminal alkenes (eq 7).[13] A biphasic procedure using benzene as a cosolvent and a phase-transfer

Essential Reactions for Organic Synthesis, First Edition. Edited by Philip L. Fuchs.
© 2016 John Wiley & Sons, Ltd. Published 2016 by John Wiley & Sons, Ltd.

agent was utilized in this case. Equally remarkable is the epoxidation of the methylenecyclopropane derivatives indicated in eq 8, given the propensity of the products to rearrange to the isomeric cyclobutanones.[14]

$$\text{(7)}$$

$$\text{(8)}$$

The epoxidation of conjugated double bonds also proceeds smoothly with the Oxone–acetone system, as illustrated by eq 9.[15] The conversion of water-insoluble enones can be accomplished with this method using CH_2Cl_2 as a cosolvent and a quaternary ammonium salt as a phase-transfer catalyst. However, a more convenient procedure utilizes 2-butanone both as a dioxirane precursor and as an immiscible cosolvent (eq 10).[16] No phase-transfer agent is required in this case.

$$\text{(9)}$$

$$\text{(10)}$$

The epoxides of several polycyclic aromatic hydrocarbons have been prepared by the use of a large excess of oxidant in a biphasic Oxone–ketone system under neutral conditions, as shown for the oxidation of phenanthrene (eq 11).[17] However, the use of isolated dioxirane solutions is more efficient for the synthesis of reactive epoxides, since hydrolysis of the product is avoided.[5,18] A number of unstable epoxides of various types have been produced in a similar manner, as discussed for *Dimethyldioxirane* and *Methyl(trifluoromethyl)dioxirane*.

$$\text{(11)}$$

Epoxidations have also been performed with other oxidizing agents generated in situ from Oxone. An intriguing method uses a catalytic amount of an immonium salt to facilitate alkene epoxidation in a process which apparently involves an intermediate oxaziridium species as the active oxidant.[19] This procedure is carried out by adding solid Oxone and $NaHCO_3$ to a solution of the alkene and catalyst in MeCN containing a very limited quantity of

water (eq 12). Finally, Oxone is the stoichiometric oxidant in interesting modifications of the widely studied metal porphyrin oxidations, where it has obvious advantages over some of the other oxidants commonly used.[20] The potential of this method is illustrated by the epoxidation reaction in eq 13.[21] In this conversion, only 1.4 mol% of the robust catalyst tetrakis(pentafluorophenyl)-porphyrinatomanganese chloride (TFPPMnCl) is required. The catalytic hydroxylation of unactivated hydrocarbons is also possible (eq 14).[22] Other metal complexes promote similar oxidations.[23–25]

$$\text{(12)}$$

$$\text{(13)}$$

$$\text{(14)}$$

Nitrogen Compounds. The aqueous Oxone–acetone combination has been developed for the transformation of certain anilines to the corresponding nitrobenzene derivatives, as exemplified in eq 15.[26] This process involves sequential oxidation steps proceeding by way of an intermediate nitroso compound. In the case of primary aliphatic amines, other reactions of the nitrosoalkane species compete with the second oxidation step (for example, dimerization and tautomerization to the isomeric oxime), thereby limiting the synthetic generality of these oxidations.[27] An overwhelming excess of aqueous Oxone has been used to convert cyclohexylamine to nitrocyclohexane (eq 16).[27]

$$\text{(15)}$$

$$\text{(16)}$$

Pyridine is efficiently converted to its *N*-oxide by the Oxone–acetone oxidant.[5] Cytosine and several of its derivatives give the *N*3-oxides selectively upon reaction with buffered Oxone (eq 17).[28] A similar transformation of adenosine 5′-monophosphate yields the *N*1-oxide.[29]

$$\text{(17)}$$

The very useful *N*-sulfonyloxaziridines are conveniently prepared by treating *N*-sulfonylimines with Oxone in a biphasic solvent system (eq 18).[30,31] Either bicarbonate or carbonate can be used to buffer this reaction, but reaction is much faster with carbonate, suggesting that the monopersulfate dianion is the oxidizing species (for illustrations of the remarkable chemistry of these oxaziridines, see ***N-(Phenylsulfonyl)(3,3-dichlorocamphoryl)-oxaziridine***).

$$\text{PhCH=NSO}_2\text{Ph} \xrightarrow[\substack{\text{KHCO}_3\text{, 2 h or} \\ \text{K}_2\text{CO}_3\text{, 15 min} \\ 95\%}]{\substack{1.2 \text{ equiv Oxone} \\ \text{H}_2\text{O, toluene}}} \text{Ph}\diagup\!\!\diagdown\text{NSO}_2\text{Ph} \quad \text{(18)}$$

The Oxone–acetone system has also been employed for the synthesis of simple oxaziridines from *N*-alkylaldimines (eq 19).[32] Interestingly, the *N*-phenyl analogs produce the isomeric nitrones rather than the oxaziridines (eq 20). It is noteworthy that MeCN can replace acetone as the solvent in this procedure.

$$\text{PhCH=N-}t\text{-Bu} \xrightarrow[\substack{\text{acetone} \\ 98\%}]{\substack{1.2 \text{ equiv Oxone} \\ \text{aq. KHCO}_3}} \text{Ph}\diagup\!\!\diagdown\text{N-}t\text{-Bu} \quad \text{(19)}$$

$$\text{PhCH=NPh} \xrightarrow[\substack{\text{acetone} \\ 98\%}]{\substack{1.2 \text{ equiv Oxone} \\ \text{aq. KHCO}_3}} \text{PhCH=N(O)Ph} \quad \text{(20)}$$

Finally, the chlorination of aldoximines gives the corresponding hydroximoyl chlorides, as shown in eq 21.[33] The combination of Oxone and anhydrous HCl in DMF serves as a convenient source of hypochlorous acid, the active halogenating agent.

$$\text{(21)}$$

Sulfur Compounds. Some of the earliest applications of Oxone in organic synthesis involved the facile oxidation of sulfur functions. For example, aqueous Oxone selectively oxidizes sulfides to sulfones even in highly functionalized molecules, as illustrated in eq 22.[34] Sulfones can also be prepared by a convenient two-phase system consisting of a mixture of solid Oxone, 'wet' ***Montmorillonite K10*** clay, and a solution of the sulfide in an inert solvent.[35]

$$\xrightarrow[\substack{0 \,°\text{C, 4 h} \\ 77\%}]{\substack{3 \text{ equiv Oxone} \\ \text{H}_2\text{O, MeOH}}} \quad \text{(22)}$$

The partial oxidation of sulfides to sulfoxides has been accomplished in a few cases by careful control of the reaction stoichiometry and conditions.[34] A biphasic procedure for sulfoxide formation from diaryl sulfides is shown in eq 23.[36] However, a more attractive and versatile procedure uses a solid Oxone–wet alumina reagent with a solution of the sulfide.[37] This method permits control of the reaction to form either the sulfoxide or the sulfone simply by adjusting the amount of oxidant and the reaction temperature, as illustrated in eq 24. These oxidations are compatible with other functionality.

$$(p\text{-MeOC}_6\text{H}_4)_2\text{S} \xrightarrow[\substack{\text{H}_2\text{O, CH}_2\text{Cl}_2\text{, 18 h} \\ 92\%}]{\substack{2 \text{ equiv Oxone} \\ \text{Bu}_4\text{NBr}}} (p\text{-MeOC}_6\text{H}_4)_2\text{S=O} \quad \text{(23)}$$

$$\text{PhSCH}_2\text{CH}_2\text{OH} \begin{cases} \xrightarrow[\substack{\text{CH}_2\text{Cl}_2\text{, reflux} \\ 84\%}]{\substack{1 \text{ equiv Oxone} \\ \text{wet alumina}}} \text{PhSOCH}_2\text{CH}_2\text{OH} \\ \\ \xrightarrow[\substack{\text{CHCl}_3\text{, reflux} \\ 89\%}]{\substack{3 \text{ equiv Oxone} \\ \text{wet alumina}}} \text{PhSO}_2\text{CH}_2\text{CH}_2\text{OH} \end{cases} \quad \text{(24)}$$

Another intriguing method for the selective oxidation of sulfides to sulfoxides (eq 25) uses buffered Oxone in a biphasic solvent mixture containing a catalytic amount of an *N*-phenylsulfonylimine as the precursor of the actual oxidizing agent, the corresponding *N*-sulfonyloxaziridine.[38] The oxaziridine is smoothly and rapidly formed by reaction of the imine with buffered Oxone and regenerates the imine upon oxygen transfer to the sulfide. The greater reactivity of the sulfide relative to the sulfoxide accounts for the preference for monooxidation in this procedure. The biphasic nature of this reaction prevents direct oxidation by Oxone, which would be less selective.

$$\text{PhSCH=CH}_2 \xrightarrow[\substack{\text{H}_2\text{O, CH}_2\text{Cl}_2 \\ \text{K}_2\text{CO}_3\text{, 0.5 h} \\ 90\%}]{\substack{4.5 \text{ equiv Oxone} \\ \text{PhSO}_2\text{N=CHC}_6\text{H}_4\text{NO}_2\text{-}p}} \text{PhSO}_2\text{CH=CH}_2 \quad \text{(25)}$$

Oxone sulfoxidations can show appreciable diastereoselectivity in appropriate cases, as demonstrated in eq 26.[39] Enantioselective oxidations of sulfides to sulfoxides have been achieved by buffered aqueous Oxone solutions containing bovine serum albumin (BSA) as a chiral mediator (eq 27).[40] As little as 0.05 equiv of BSA is required and its presence discourages further oxidation of the sulfoxide to the sulfone. Oxone can be the active oxidant or reaction can be performed in the presence of acetone, trifluoroacetone, or other ketones, in which case an intermediate dioxirane is probably the actual oxidizing agent. The level of optical induction depends on structure of the sulfide and that of any added ketone. Sulfoxide products show ee values ranging from 1% to 89%, but in most examples the ee is greater than 50%.

$$trans{:}cis = 10{:}1 \qquad (26)$$

$$(27)$$
$$89\% \ ee$$

1-Decanethiol is efficiently oxidized to decanesulfonic acid (97% yield) by aqueous Oxone.[10] In a similar manner an acylthio function was converted into the potassium sulfonate salt, as shown in eq 28.[41]

$$AcS(CH_2)_{10}CO_2Me \xrightarrow[\substack{K_2CO_3 \\ 92\%}]{\substack{5 \text{ equiv Oxone} \\ H_2O, \text{ MeOH}}} KO_3S(CH_2)_{10}CO_2Me \qquad (28)$$

Finally, certain relatively stable thioketones can be transformed into the corresponding thione S-oxides by the aqueous Oxone–acetone reagent (eq 29).[42]

$$(p\text{-MeOC}_6H_4)_2C{=}S \xrightarrow[\substack{\text{aq. KHCO}_3 \\ 18\text{-crown-6, 6 h} \\ 97\%}]{\substack{2 \text{ equiv Oxone} \\ \text{acetone, benzene}}} (p\text{-MeOC}_6H_4)_2C{=}S{=}O \qquad (29)$$

First Update

Yian Shi & Christopher P. Burke
Colorado State University, Fort Collins, CO, USA

Modification of Oxone. Since Oxone is a triple salt (2KHSO$_5$·KHSO$_4$·K$_2$SO$_4$) only about 50% per mole is active oxidant. A convenient method for the preparation of purified KHSO$_5$·H$_2$O on a large scale has been developed which allows for significant reduction in the amount of oxidizing agent needed for a reaction.[43] One of the main drawbacks of using Oxone is that aqueous/alcoholic or at least biphasic reaction conditions are usually necessary. To circumvent this problem several organic-soluble forms of Oxone have been developed including tetra-*n*-butylammonium peroxymonosulfate (*n*-Bu$_4$NHSO$_5$),[44] tetraphenylphosphonium peroxymonosulfate (Ph$_4$PHSO$_5$),[45,46] and benzyltriphenylphosphonium peroxymonosulfate (BnPh$_3$PHSO$_5$).[47] These oxidants can be used under anhydrous conditions and in many cases show similar reactivity to Oxone. A study has been done comparing the activity of these

different oxidants in the oxidation of benzaldehyde, *trans*-stilbene, triphenylphosphine, thioanisole, and phenylboronic acid.[43]

Ketones, Aldehydes, and Alcohols. The well-known Baeyer-Villiger oxidation of ketones by Oxone (eq 2) has been exploited in a variety of reactions. This protocol has been used with KHSO$_5$ for cleavage of α- and β-dicarbonyl compounds to esters or acids (eqs 30 and 31).[48,49] This process is simpler, cheaper, and milder than the commonly used haloform reaction.

$$(30)$$

$$(31)$$

α-Hydroxy- and α-nitroketones are oxidatively cleaved by Oxone in a similar manner to yield the corresponding esters and acids (eqs 32 and 33).[48–50] α-Nitroketones can be cleaved to dicarboxylic acids or dicarboxylic acid monomethyl esters depending on reaction conditions. It is proposed that in the case of α-hydroxy ketones Bayer-Villiger oxidation is followed by oxidation of the resulting aldehyde to give the diacid/ester.

$$(32)$$

$$(33)$$

Oxone has also been recently more thoroughly studied as reagent for the oxidation of aldehydes.[51,52] Aryl- and aliphatic aldehydes can be efficiently converted directly to acids or esters depending on the choice of solvent (eq 34).[52] They can also be converted to nitriles in one pot by reaction with hydroxylamine on alumina with microwave irradiation (eq 35).[53]

$$(34)$$

$$(35)$$

Oxone oxidizes metal complexes including tris[(2-oxazolinyl)-phenolato] manganese(III) which, in conjunction with n-Bu_4NBr, is an effective oxidant for aromatic and primary and secondary aliphatic alcohols.[54]

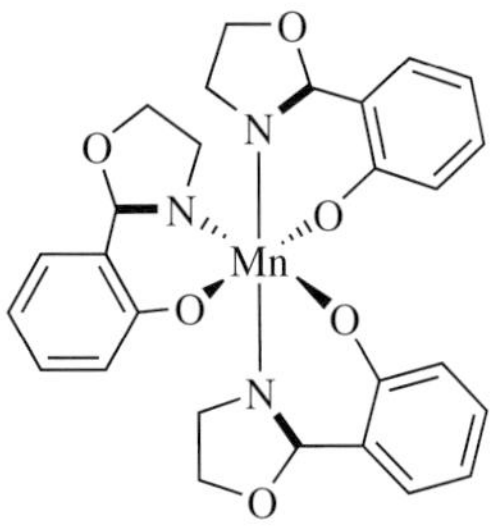

tris[(2-oxazolinyl)phenolato] manganese(III)

Additionally, Oxone is a suitable stoichiometric oxidant for alcohol oxidations with TEMPO and n-Bu_4NBr even in aprotic solvents (eq 36).[55] Aliphatic and electron-rich benzylic alcohols give lower yields than electron-neutral benzylic alcohols in this case. A simple combination of Oxone and NaBr can oxidize benzylic alcohols to aldehydes and ketones.[56] Once again, electron-rich benzylic alcohols gave lower yields; in this case it is due to competing halogenation of the aromatic ring. $BnPh_3PHSO_5$ has also been used with $AlCl_3$ to oxidize benzylic and allylic alcohols under aprotic solvent-free conditions (eq 37).[57] This protocol gives high yields for both electron-poor and electron-rich primary and secondary benzylic alcohols.

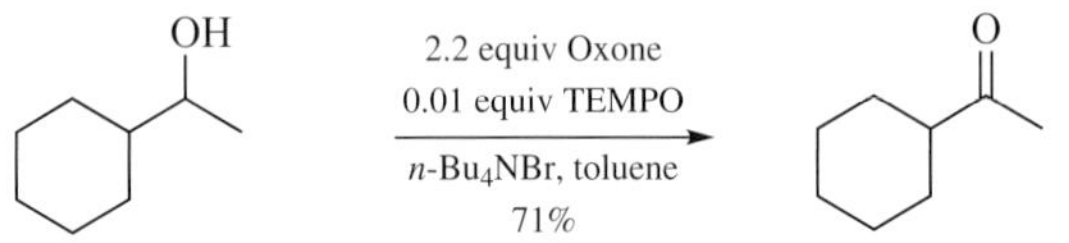

OH
2.2 equiv Oxone
0.01 equiv TEMPO
n-Bu4NBr, toluene
71%
O

(36)

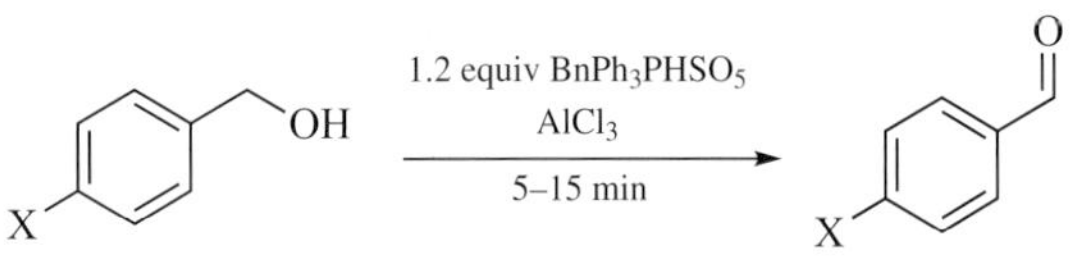

OH
X
1.2 equiv BnPh3PHSO5
AlCl3
5–15 min
O
X

(37)

$X = NO_2$, 100%
X = OMe, 80%

Alkenes, Alkynes, Arenes, and Alkanes. One of the most common applications of Oxone in organic synthesis is the in situ formation of dioxiranes from ketones (eq 1). Dioxirane chemistry has grown significantly in recent years, particularly in the area of enantioselective epoxidation, and a wide variety of chiral ketones have been designed for this purpose.[58–69] Notably, ketones (**5** and **6**) derived from fructose and glucose, respectively, have been shown to be effective catalysts for enantioselective epoxidations of a variety of *trans*-, trisubstituted, *cis*-, and terminal olefins with Oxone as primary oxidant (eqs 38 and 39).[70–72]

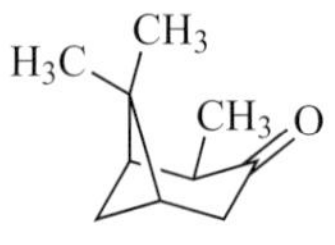

1

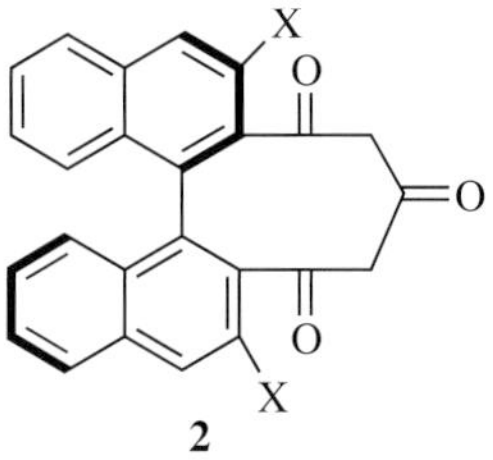

2

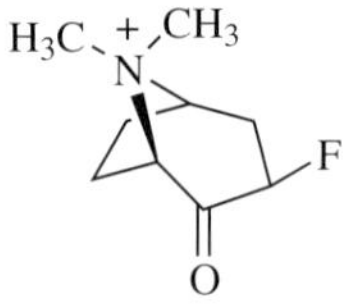

3

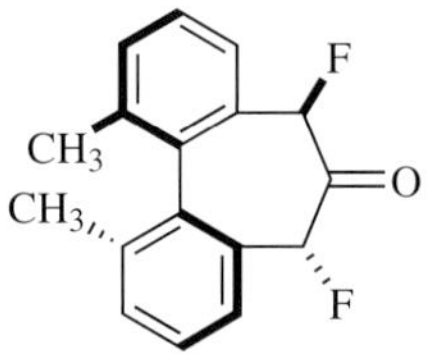

4

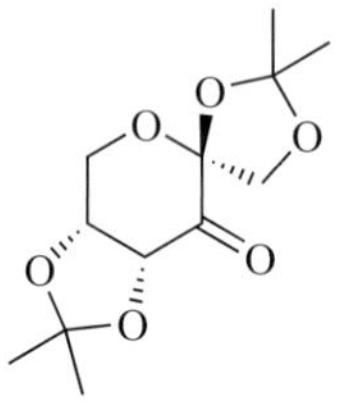

5

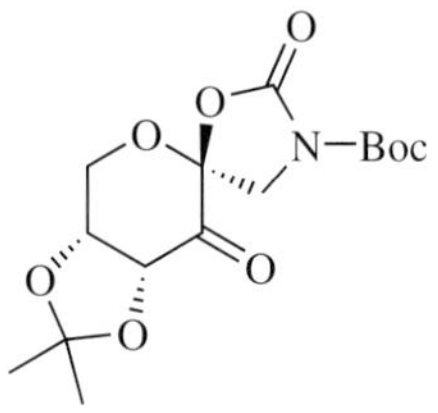

6

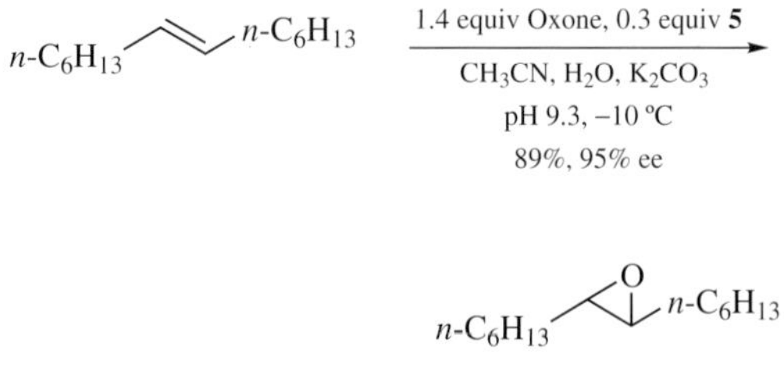

n-C6H13
n-C6H13
1.4 equiv Oxone, 0.3 equiv 5
CH3CN, H2O, K2CO3
pH 9.3, −10 °C
89%, 95% ee
O
n-C6H13
n-C6H13

(38)

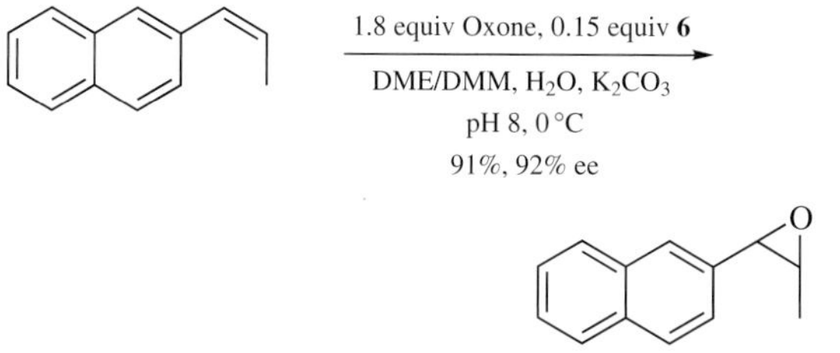

1.8 equiv Oxone, 0.15 equiv **6**

DME/DMM, H$_2$O, K$_2$CO$_3$
pH 8, 0 °C
91%, 92% ee

(39)

Oxone and its derivatives have also been used with chiral iminium salts and amines to form enantiomerically enriched epoxides. The scope and enantioselectivity of epoxidation with chiral iminium salts with Oxone and Ph$_4$PHSO$_5$ have made progress during recent years.[46,73–77] Chiral iminium salts (**7** and **8**) have been particularly successful for various olefins (eqs 40 and 41).[78,79]

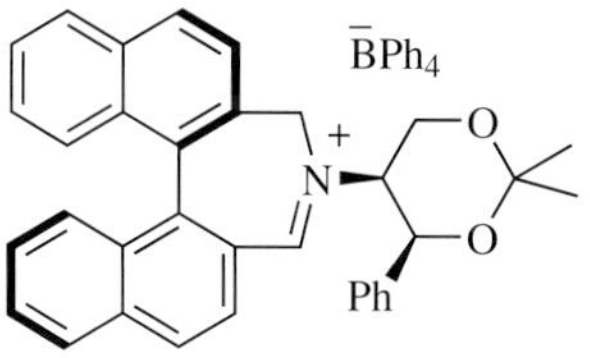

$\bar{B}Ph_4$

7

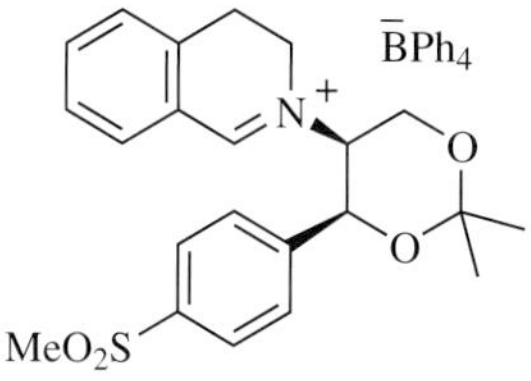

$\bar{B}Ph_4$

MeO$_2$S

8

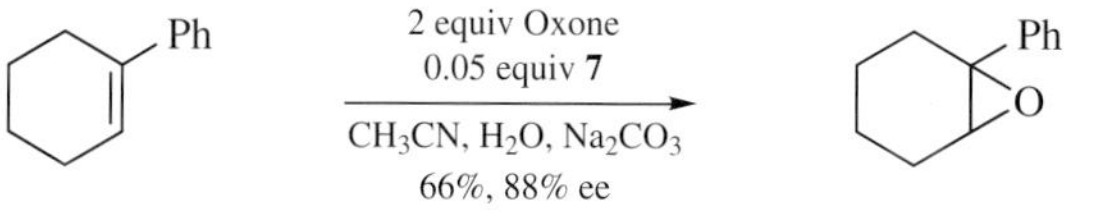

2 equiv Oxone
0.05 equiv **7**

CH$_3$CN, H$_2$O, Na$_2$CO$_3$
66%, 88% ee

(40)

2 equiv Ph$_4$PHSO$_5$
0.1 equiv **8**

CHCl$_3$, –40 °C
59%, 97% ee

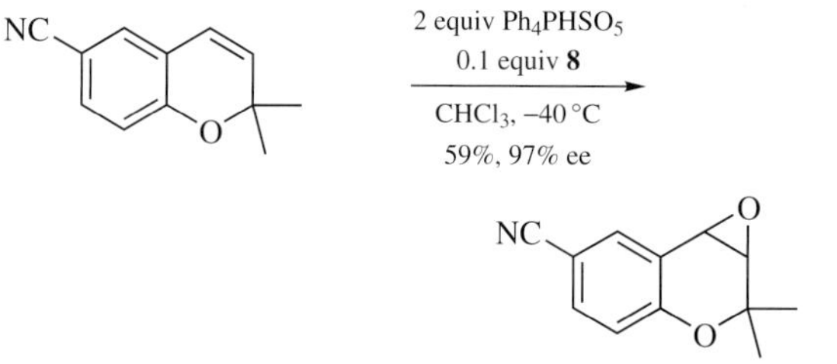

(41)

Amine-catalyzed epoxidation is a relatively new area, and the active species is thought to be an ammonium peroxymonosulfate salt which acts as a phase transfer catalyst and undergoes electrophilic attack by an olefin.[80–82] Use of chiral amines has given rise to enantiomerically enriched epoxides (eq 42).

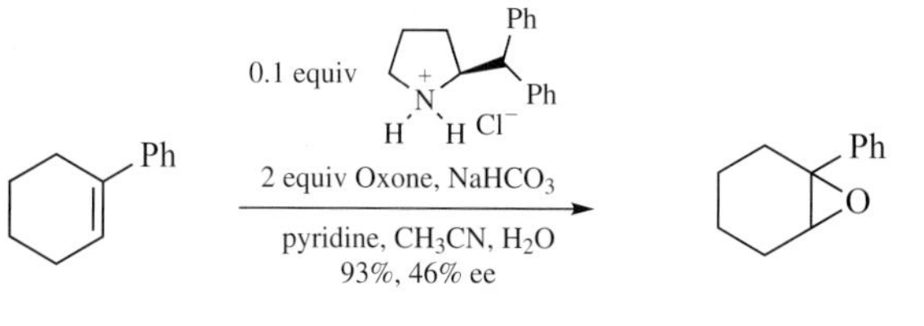

0.1 equiv

2 equiv Oxone, NaHCO$_3$

pyridine, CH$_3$CN, H$_2$O
93%, 46% ee

(42)

Oxone in conjunction with OsO$_4$ cleaves alkenes to ketones or carboxylic acids (eq 43).[83] This protocol has the advantage over traditional methods in that there is no need for intermediate 1,2-diols. This method has been exploited in the direct synthesis of lactones from alkenols (eq 44) and tetrahydrofuran-diols from 1,4-dienes as well.[84,85]

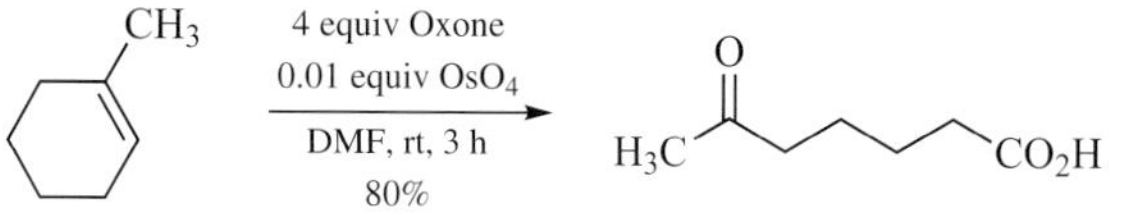

4 equiv Oxone
0.01 equiv OsO$_4$

DMF, rt, 3 h
80%

(43)

4 equiv Oxone
0.01 equiv OsO$_4$

DMF, rt
73%

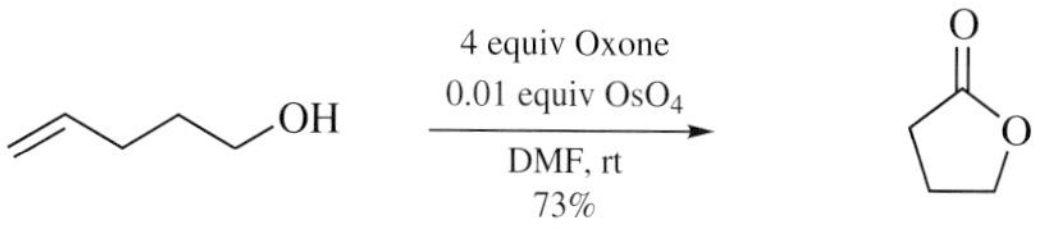

(44)

When used in conjunction with RuCl$_3$, Oxone cleaves alkenes to aldehydes in high yields (eq 45).[86] This method is less effective for aliphatic olefins, and NaIO$_4$ is suggested as an alternate oxidant in these cases. The same combination with more Oxone and less water oxidizes alkenes to α-hydroxy ketones (eq 46).[87,88] The reaction is fairly regioselective depending on the electronic properties of the substrate, with the hydroxy group preferentially ending up next to the more electron-withdrawing substituent. 1,2-Diols, which are the possible intermediates/by-products in the above keto-hydroxylation, are also oxidized under the same conditions to α-hydroxy ketones, and enantiopurity of the starting materials is preserved during the reaction (eq 47).[89] Oxone in aqueous acetone has been shown to dihydroxylate various 1,2-glycals in one step in moderate to good yields.[90]

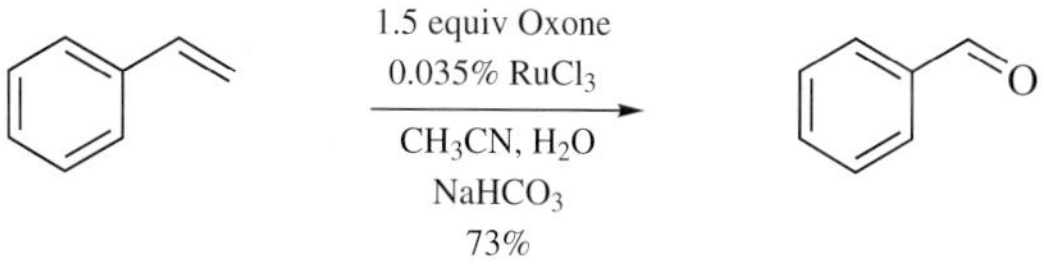

1.5 equiv Oxone
0.035% RuCl$_3$

CH$_3$CN, H$_2$O
NaHCO$_3$
73%

(45)

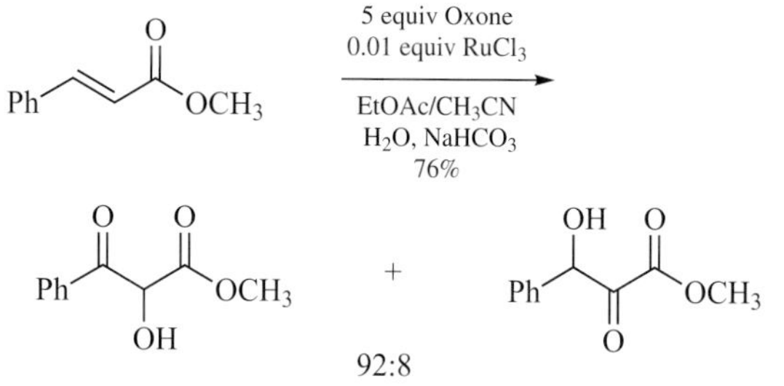

5 equiv Oxone
0.01 equiv RuCl$_3$

EtOAc/CH$_3$CN
H$_2$O, NaHCO$_3$
76%

+

92:8

(46)

5 equiv Oxone
0.01 equiv RuCl$_3$

EtOAc/CH$_3$CN
H$_2$O, NaHCO$_3$
87%

99% ee   99% ee

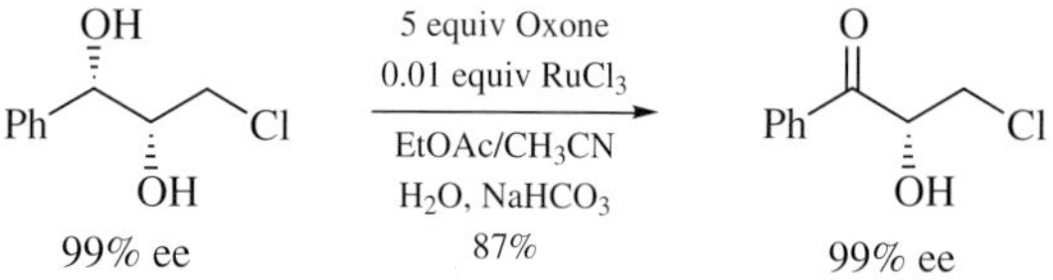

(47)

Oxone, in conjunction with RuO$_2$, cleaves alkynes to carboxylic acids.[91] Both internal and terminal alkynes are cleaved in short reacton times with high yields (eq 48). Intermediate 1,2-diketones are proposed.

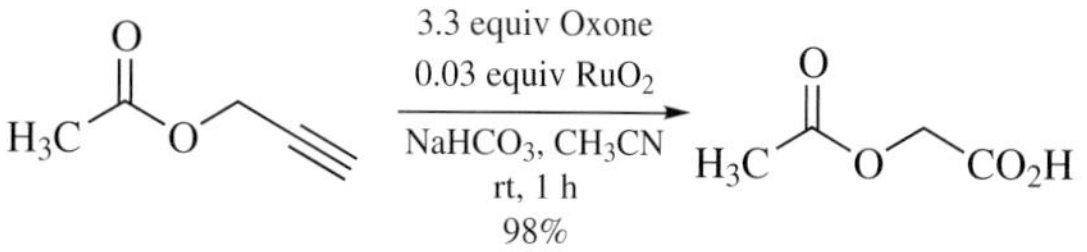

3.3 equiv Oxone
0.03 equiv RuO$_2$

NaHCO$_3$, CH$_3$CN
rt, 1 h
98%

(48)

Dioxiranes generated from Oxone have recently been shown to undergo C–H insertion reactions with activated and unactivated

C–H bonds. This strategy has been used in an intramolecular fashion for the oxidation of hydrocarbons (eq 49) and steroids.[92,93] Fructose-derived ketone **5** has also been used for this purpose in an intermolecular reaction for the desymmetrization and kinetic resolution of 1,2-diols to α-hydroxy ketones (eq 50).[94,95] There has also been a report of the direct oxidation of hydrocarbons to ketones and lactones by Mn-porphyrin complexes with Oxone.[96]

$$(49)$$

$$(50)$$

Nitrogen Compounds. Previous methods for oxidizing anilines to nitro compounds with Oxone/acetone were ineffective with carboxylic acid and alcohol-containing systems.[26] A new method using Oxone and acetone under totally aqueous conditions allows oxidation of carboxylic and alcoholic anilines as well (eq 51).[97] In this case the reaction occurred in the absence of acetone but yields were significantly lower. The combination of Oxone and ozone has been used to oxidize amino sugars to nitro sugars as well (eq 52).[98]

$$(51)$$

$$(52)$$

Oxone adsorbed on silica gel or alumina is a very effective oxidant for the selective oxidation of primary and secondary amines to hydroxylamines without overoxidation. These reactions can even be accomplished under solvent-free conditions and with very short reaction times with heating or microwave irradiation (eq 53).[99] Pyridine and trialkylamines were also readily oxidized to their N-oxides. It is suggested that the hydroxylamines are protected from overoxidation because of their strong adsorption to the silica gel or alumina surface.

$$(53)$$

The Oxone/acetone system is also very effective for the formation of nitroxides from secondary amines without α-hydrogens (eq 54).[100]

$$(54)$$

Because of Oxone's acidic nature, N-nitrosation of secondary amines is possible with the use of sodium nitrite in the presence of wet SiO_2 (eq 55).[101] Nitrophenols can be obtained via nitrosation-oxidation of phenols under similar conditions (eq 56).[102] Although acidic, the use of Oxone for these reactions eliminates the need for strong acids to generate NO^+ unlike traditional methods. Nitrosoarenes can also be prepared by oxidation of anilines with Oxone (eq 57).[103]

$$(55)$$

$$(56)$$

$$(57)$$

Oximes can be converted to their corresponding nitro compounds with Oxone and refluxing acetonitrile (eq 58).[104] They can also be cleaved to their parent carbonyl compounds by Oxone in conjunction with glacial acetic acid, or silica gel/alumina and microwave irradiation (eq 58).[105–107] Ketoximes and aldoximes are both converted to carbonyl compounds in high yields using the microwave and alumina procedure. Several of the above transformations are highlighted in the oxidative decarboxylation of α-amino acids to form ketones and carboxylic acids.[108]

$$(58)$$

Aldoximes can be converted to isothiocyanates in a convenient one-pot procedure with Oxone and HCl based on a procedure for synthesizing hydroximoyl chlorides.[33,109] The reaction presumably proceeds through an oxathiazoline that decomposes to give the isothiocyanate (eq 59).

$$Ph{-\!\!=}NOH \xrightarrow[\text{0.5 N HCl, DMF}]{\text{1.1 equiv Oxone}} \left[\begin{array}{c} Ph \\ | \\ Cl \end{array}{=}NOH \right] \xrightarrow[\substack{\text{NEt}_3, \text{ THF} \\ 92\%}]{\text{thiourea}}$$

$$\left[\begin{array}{c} Ph \\ \diagdown \\ S \end{array} \begin{array}{c} N \\ \diagup \diagdown \\ NH_2 \\ NH_2 \end{array} O \right] \longrightarrow Ph{-}N{=}C{=}S \quad (59)$$

An interesting method for the protection of carboxylic acids as diphenylmethyl esters in high yield using Oxone and benzophenone hydrazone has been reported (eq 60).[110] Various aromatic tosylhydrazones can also be cleaved to carbonyl compounds by Oxone/acetone (eq 61).[111] It is proposed that cleavage occurs via collapse of an oxaziridine intermediate. $BnPh_3PHSO_5$ has been used with $BiCl_3$ to regenerate carbonyl compounds from oximes, phenylhydrazones, 2,4-dinitrophenylhydrazones, and semicarbazones under nonaqueous conditions.[112] Yields are generally very good, and it is reported that it is also possible to oxidize alcohols to ketones under these conditions without affecting any of the above mentioned carbonyl derivatives.

$$\text{(cycloheptyl)}{-}CO_2H \xrightarrow[\substack{I_2, \text{ alumina, CH}_2Cl_2 \\ 70\%}]{\substack{\text{1.5 equiv Oxone} \\ \text{benzophenone hydrazone}}}$$

(60)

$$\text{(cycloheptyl-C(=O)-O-CHPh}_2)$$

$$\underset{\text{NNHTs}}{\text{(tetralin-NNHTs)}} \xrightarrow[\substack{\text{phosphate buffer (pH 6)} \\ 98\%}]{\text{6 equiv Oxone, acetone}} \underset{O}{\text{(tetralone)}} \quad (61)$$

Likewise, Oxone is useful for the conversion of nitro groups into carbonyl compounds (Nef reaction) in the presence of aqueous base (eq 62).[113]

$$\underset{\substack{| \\ OAc}}{\overset{NO_2}{|}} \xrightarrow[\substack{\text{MeOH, phosphate buffer} \\ 93\%}]{\text{1 equiv Oxone, 1 N NaOH}} \underset{\substack{| \\ OAc}}{\overset{O}{||}} \quad (62)$$

Benzimidazoles can be synthesized in one step by condensing 1,2-phenylenediamines and aldehydes in the presence of Oxone (eq 63).[114] The reaction gives good selectivity and tolerates electron-rich and electron-poor phenylenediamines as well as a wide variety of aromatic and aliphatic aldehydes.

$$\underset{\substack{\\ H}}{\text{(EtO}_2C\text{-aryl-NH}_2\text{, NHcyclohexyl)}} + \underset{N}{\text{(pyridine-2-CHO)}} \xrightarrow[\substack{\text{aq. DMF, rt, 1 h} \\ 85\%}]{\text{0.65 equiv Oxone}}$$

$$\text{(EtO}_2C\text{-benzimidazole-2-pyridyl, N-cyclohexyl)} \quad (63)$$

Urazoles are oxidized to triazolinediones when subjected to Oxone and $NaNO_2$ (eq 64).[115] These compounds, which have typically been difficult to synthesize and purify, are relatively easily made in high yields and purity by this procedure.

$$\underset{\substack{| \\ CH_3}}{\overset{\text{HN}-\text{NH}}{O{=}\diagdown\diagup{=}O \atop N}} \xrightarrow[\substack{\text{SiO}_2, \text{ CH}_2Cl_2, \text{ rt, 2 h} \\ 100\%}]{\text{1.5 equiv Oxone, NaNO}_2} \underset{\substack{| \\ CH_3}}{\overset{\text{N}=\text{N}}{O{=}\diagdown\diagup{=}O \atop N}} \quad (64)$$

Oxaziridines, which have previously been produced from reaction of imines with Oxone,[30,31] have recently been made in excellent yields with n-Bu_4NHSO_5 in acetonitrile (eq 65).[116] The reactions are generally E-selective. However, as the size of the group on nitrogen decreases more Z-isomer is produced. The effects of solvent and Lewis acids on the E/Z selectivity and rate of reaction were studied as well.

$$\underset{\substack{| \\ H}}{\text{Ph-C=N-}t\text{-Bu}} \xrightarrow[\substack{\text{CH}_3\text{CN, rt, 50 min} \\ 100\% (E)}]{\text{1 equiv }n\text{-Bu}_4\text{NHSO}_5} \underset{\substack{| \\ H}}{\text{Ph-C-N-}t\text{-Bu (oxaziridine)}} \quad (65)$$

Acylhydrazides are oxidized to N,N'-diacylhydrazines in high yields with aqueous Oxone (eq 66), although only aromatic hydrazides were effective.[117] The oxidation of alkylcyanohydrazines to azo-bis nitriles can be carried out with an Oxone/KBr system (eq 67).[118] Nitriles can also be hydrolyzed to amides in moderate to good yields with Oxone/acetone (eq 68).[119]

$$\underset{\substack{| \\ H}}{\overset{O}{\underset{Bn}{||}}\text{-N-NH}_2} \xrightarrow[\substack{\text{H}_2\text{O, rt, 30 min} \\ 93\%}]{\text{1 equiv Oxone}} \underset{\substack{| \\ H}}{\text{Bn-C(=O)-N-N-C(=O)-Bn}} \quad (66)$$

$$\underset{\substack{CH_3 \quad CH_3}}{\overset{\text{CN} \quad H \quad H \quad CN}{H_3C{-}C{-}N{-}N{-}C{-}CH_3}} \xrightarrow[\substack{\text{H}_2\text{O, 3.5 h} \\ 66\%}]{\text{1 equiv Oxone, KBr}}$$

$$\underset{\substack{CH_3 \quad CH_3}}{\overset{\text{CN} \quad\quad \text{CN}}{H_3C{-}C{-}N{=}N{-}C{-}CH_3}} \quad (67)$$

$$\text{(Ph-CN)} \xrightarrow[\substack{\text{phosphate buffer} \\ \text{NaHCO}_3, \text{ pH 7.5} \\ 83\%}]{\text{5 equiv Oxone, acetone}} \text{(Ph-C(=O)-NH}_2) \quad (68)$$

Sulfur Compounds. A convenient procedure for high yielding oxidation of sulfides to either sulfoxides or sulfones using aqueous acetone and Oxone has been reported (eq 69) and has been used to produce SK&F [107310] in kilogram quantities.[120] Selectivity is attained by controlling stoichiometry and reaction temperature. These same transformations have also been accomplished with good yield and selectivity on silica gel and alumina. The role of the surfaces has been investigated, and it was found that Oxone is activated by being dispersed on the surface of the adsorbent allowing greater contact with the substrate.[121] Mechanistic studies of sulfide oxidation and thioester hydrolysis by Oxone have also been performed.[122] In addition, oxidation of glycosyl sulfides to glycosyl sulfoxides has been accomplished with Oxone on silica gel.[123]

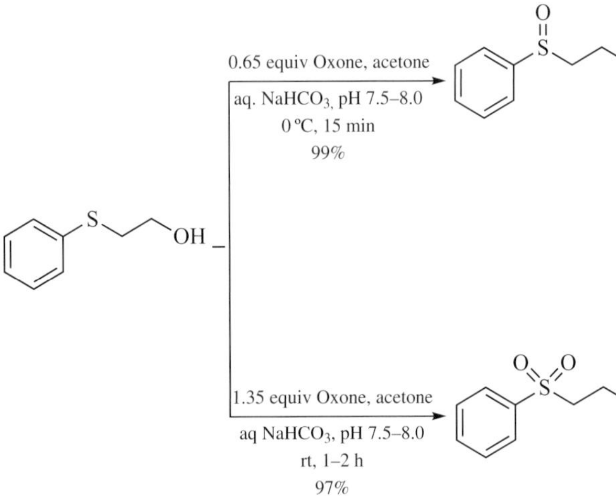

(69)

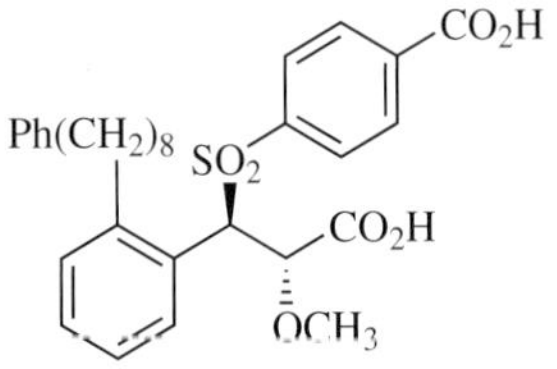

SK&F [107310]

Episulfides can be oxidized to episulfones with Oxone and 1,1,1-trifluoroacetone without considerable episulfoxide formation in many cases (eq 70).[124] Sulfides and thiols can be selectively oxidized to sulfoxides and disulfides, respectively, with the use of BnPh$_3$PHSO$_5$ in anhydrous aprotic solvents or solvent-free conditions (eq 71).[125,126]

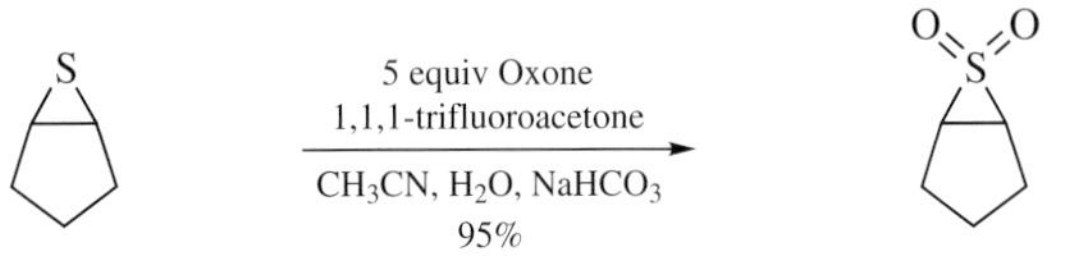

(70)

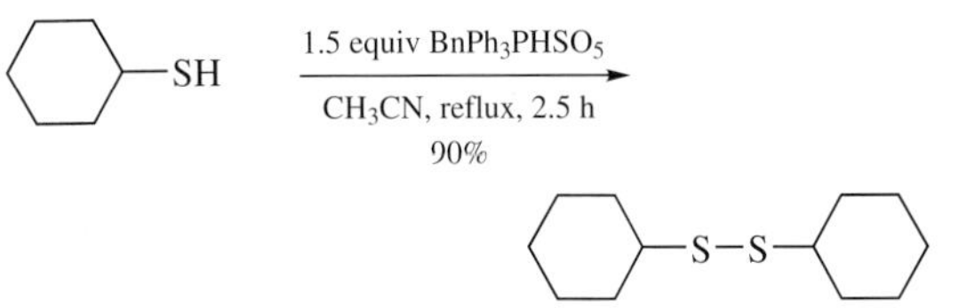

(71)

A variety of thiocarbonyl compounds including thioamides, thioureas, and thioesters are converted to their corresponding carbonyl compounds in good yield by simply grinding them with solid Oxone with a mortar and pestle (eq 72).[127] Thioketones remained unchanged under the reaction conditions.

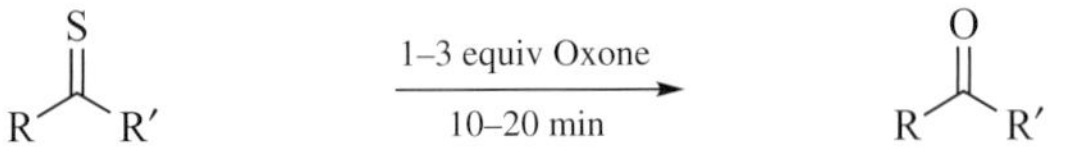

(72)

R	R′	Yield (%)
Me	NMePh	95
NH$_2$	NHPh	85
Ph	OEt	80

Boron, Phosphorus, and Selenium Compounds. Oxone has been used to oxidize carbon-boron bonds during the work-up of hydroboration reactions to obtain high yields of the resultant alcohols (eq 73).[128] Aqueous Oxone/acetone oxidizes electron-poor and electron-rich aromatic and aliphatic boronic acids and esters to the corresponding alcohols rapidly and efficiently (eq 74).[129] A one-pot procedure for the synthesis of *meta*-substituted phenols from benzenes has been developed, and a similar strategy has been devised for the synthesis of Boc-oxindoles from Boc-indoles.[130,131]

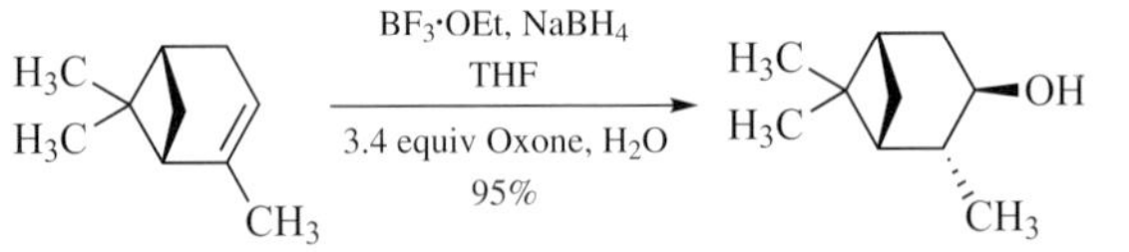

(73)

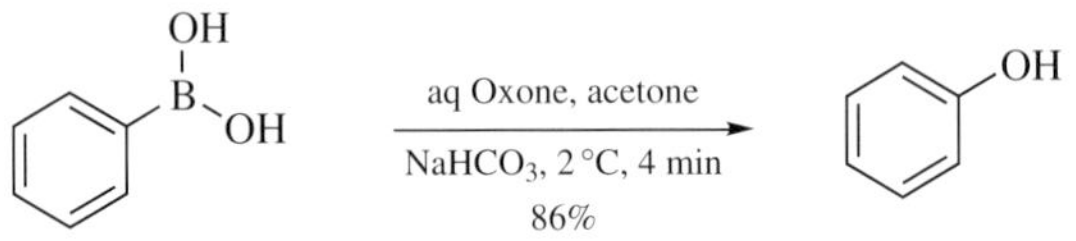

(74)

Phosphorus(III), phosphothio-, and phosphoseleno-compounds are oxidized by Oxone in THF/MeOH to produce phosphono-, phosphonothio-, and phosphonoseleno- compounds, respectively, with predominant retention of configuration at phosphorus (eq 75).[132] Thioalkyl or amino groups attached to phosphorus are unaffected.

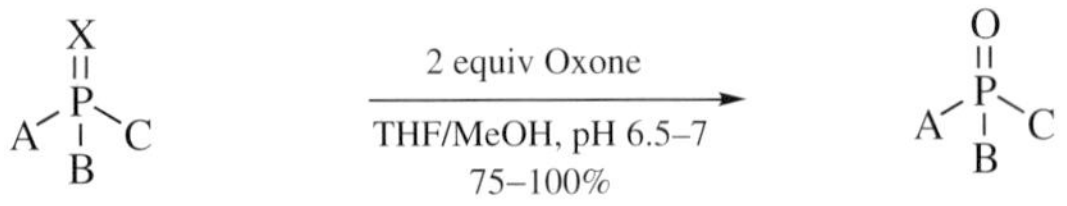

(75)

X = lone pair, S, Se
A, B, C, = R, OR, SR, NR$_2$

Selenides can be oxidized directly to selenones with methanolic buffered Oxone solutions under mild conditions. Selenones can be isolated directly, or if a nucleophile is present, they are displaced giving the substitution product (eq 76).[133] Some α-sulfonyl selenides can be oxidized to selenol esters with Oxone in MeOH or THF.[134]

$$\text{(76)}$$

Halides. Oxone oxidizes halides to form electrophilic halogens in situ for a variety of halogenation reactions. Use of in situ generated halogens is advantageous because it obviates the need for storage and handling of dangerous chlorine/bromine and keeps these compounds at low concentrations during reaction. Electron-rich aromatic compounds undergo predominantly *para*-halogenation with Oxone and KX or NH_4I (eq 77).[135–139] Significant amounts of *ortho*-halogenated products are sometimes observed. Phenols can also be iodinated with high yields and selectivity with $BnPh_3PHSO_5$ and KI.[140]

$$\text{(77)}$$

Conjugated enones and simple alkenes also undergo halogenation with chlorine or bromine generated from Oxone and the corresponding sodium halide or hydrohalic acid.[141,142] In the case of enones, the addition product can be treated with base to give conjugated vinyl halides (eq 78).

$$\text{(78)}$$

A variety of alkenes undergo azidoiodination with sodium azide, potassium iodide, and Oxone on wet alumina to give azido-iodo compounds regioselectively in high yield (eq 79).[143] These compounds are useful precursors to vinyl azides, amines, and aziridines and are typically synthesized with more expensive and exotic reagents. Similar methods have been used in the iodolactonization and iodoetherification of unsaturated carboxylic acids and alcohols to make five- and six-membered lactones, tetrahydrofurans, and tetrahydropyrans (eq 80).[144]

$$\text{(79)}$$

$$\text{(80)}$$

gem-Halo-nitro compounds can be prepared in one step from oximes with Oxone supported on wet basic alumina and NaCl or KBr (eq 81).[145,146] Use of basic alumina is essential in this case due to oxidative deprotection of the oximes to ketones when neutral alumina is used.

$$\text{(81)}$$

Halodecarboxylation of aromatic α,β-unsaturated carboxylic acids (Hunsdiecker reaction) to make β-halostyrenes has been accomplished with Oxone and sodium halide (eq 82).[147] Bromodecarbonylation and bromodecarboxylation of electron-rich benzaldehydes and benzoic acids has also been observed.[148]

$$\text{(82)}$$

Aromatic methyl ketones can be halogenated at the α-position with Oxone and sodium halide, however, competing halogenation of the aromatic ring is significant.[149] α,α-Dichloroketones can be synthesized from alkynes by reaction with Oxone in HCl/DMF (eq 83).[150] Oxone consistently gave better results than MCPBA for this transformation.

$$\text{(83)}$$

Reaction of amides and carbamates with Oxone on wet alumina and NaCl gives *N*-chlorinated products in high yields (eq 84).[151]

$$\text{(84)}$$

Protecting Group Removal. Several types of alcohol and carbonyl protecting groups can be removed with solutions of Oxone. Deprotection with Oxone offers a mild alternative to traditional methods which often require harshly acidic or basic conditions. Primary alkyl and phenolic TBS ethers are cleaved with aqueous methanolic Oxone (eq 85).[152] Primary alkyl TBS ethers are much more labile and thus can be cleaved in the presence of phenolic TBS ethers by limiting the reaction time. Secondary and tertiary TBS ethers and TBDPS ethers are unaffected. When the reactions were carried out in the absence of Oxone with solutions of HCl and HF adjusted to the same pH as the Oxone solution, no cleavage was observed for any type of TBS ether after 2.5–3 h, suggesting that Oxone's deprotective ability is not due solely to its acidic nature.

$$\text{(85)}$$

Oxone in refluxing acetonitrile cleaves TMS and THP ethers to alcohols and acetals to carbonyls.[153] In contrast, BnPh₃PHSO₅ has been used to oxidatively cleave TMS and THP ethers and ethylene acetals to carbonyl compounds under microwave irradiation with BiCl₃ (eqs 86 and 87).[154] No overoxidation products were observed with this method. Oxone on alumina has also been used to cleave ketals to diols and carbonyl compounds under solvent-free conditions with microwave irradiation.[155]

$$Ph\diagdown OR \xrightarrow[\text{CH}_2\text{Cl}_2,\ \text{microwave, 3.5 min}]{\substack{\text{1 equiv BnPh}_3\text{PHSO}_5 \\ \text{0.4 equiv BiCl}_3}} Ph\diagdown O \quad (86)$$

R = TMS, 90%
R = THP, 90%

$$\xrightarrow[\text{CH}_2\text{Cl}_2,\ \text{microwave, 6 min} \\ 80\%]{\substack{\text{1 equiv BnPh}_3\text{PHSO}_5 \\ \text{0.4 equiv BiCl}_3}} \quad (87)$$

Cleavage of acetals to esters (eq 88) as well as cleavage of THP ethers with Oxone on wet alumina has also been reported.[156] THP ethers gave mainly the deprotected alcohols along with significant amounts of esterified products.

$$\xrightarrow[\text{CH}_2\text{Cl}_2,\ 50\,°\text{C, 10 h} \\ 98\%]{\text{3 equiv Oxone, alumina}} \quad (88)$$

1,3-Dithiolanes and 1,3-dithianes can also be cleaved by Oxone on wet alumina to give the parent carbonyl compounds in high yields (eq 89).[157] The combination of BnPh₃PHSO₅ and BiCl₃ has also been applied successfully to the deprotection of 1,3-dithiolanes and 1,3-dithianes under nonaqueous conditions.[158]

$$\xrightarrow[\text{CH}_2\text{Cl}_2,\ \text{reflux} \\ n = 1,\ 77\% \\ n = 2,\ 97\%]{\text{5 equiv Oxone, alumina}} \quad (89)$$

Miscellaneous. Ring opening of a variety of epoxides and aziridines with NaN₃ in the presence of Oxone in high yields has been accomplished (eq 90).[159] The specific role of Oxone is unclear, however, no ring opening takes place in its absence. It is suggested that the results are due to Oxone's acidic nature.

$$\text{Cl}\diagdown\!\!\triangle\!\!O \xrightarrow[\text{CH}_3\text{CN, H}_2\text{O} \\ 95\%]{\text{0.5 equiv Oxone, NaN}_3} \text{Cl}\diagdown\!\!\diagup^{\text{OH}}\!\!\diagdown N_3 \quad (90)$$

Oxone has been used with triarylbismuth and copper salts to effect aryl transfer reactions.[160] Aryl groups can be transferred to alcohols to make phenyl ethers (eq 91). However, two coordination sites on the substrate are necessary.

$$\text{H}_3\text{CO}\diagup\!\!\overset{O}{\underset{O}{\diagdown}}\!\!\overset{\text{OH}}{\diagup}\!\!\overset{}{\diagdown}\text{OCH}_3 \xrightarrow[\substack{\text{acetone} \\ \text{Cu(II) pivalate} \\ 65\%}]{\substack{\text{0.5 equiv Oxone} \\ \text{1 equiv BiPh}_3}} \text{H}_3\text{CO}\diagup\!\!\overset{O}{\underset{O}{\diagdown}}\!\!\overset{\text{OPh}}{\diagup}\!\!\overset{}{\diagdown}\text{OCH}_3 \quad (91)$$

Second Update

Benjamin R. Buckley
Loughborough University, Loughborough, UK

Ketones, Aldehydes, and Alcohols. The use of Oxone as a stoichiometric reoxidant for the in situ generation of IBX has been reported by several groups. The use of a substoichiometric amount of 2-iodobenzoic acid and Oxone in acetonitrile/water under Vinod's conditions affords a range of carboxylic acids (eq 92) or ketones (eq 93) from the corresponding alcohols, respectively.[161] Application of Giannis' conditions, which differ slightly by using an ethyl acetate/water system and an additional phase transfer catalyst, can be used to afford aldehydes rather than carboxylic acids (eq 94).[162] An in-depth study in this area has been conducted by Ishihara and coworkers who found that 2-iodoxybenzene sulfonic acid is an extremely active precatalyst for the oxidation of alcohols to aldehydes, ketones, carboxylic acids, and enones when Oxone is employed as the terminal oxidant.[163] A comparison between in situ generated IBX and its sulfonic acid analog is shown is eq 95. The sulfonic acid analog is far more reactive in acetonitrile and ethyl acetate compared to the in situ generated IBX; it is important to note that the sulfonic acid derivative has been prepared and isolated previously, but studies on its use as an oxidant were not achieved due to its poor stability profile.[164] Several related reports employing alternative terminal oxidants have appeared; for example, tetraphenylphosphonium monoperoxysulfate, which has the advantage that it is soluble in organic solvents, can be used to generate IBX in situ, and primary alcohols are selectively oxidized to the corresponding aldehydes in excellent yields when employing acetonitrile as solvent.[165] A recyclable fluorous IBX analog has also been reported; again, this fluorous derivative is generated using Oxone and can be recycled up to five times without loss of activity (eq 96).[166]

$$\diagup\!\!\diagdown\!\!\diagup\!\!\diagdown\!\!\diagup\text{OH} \xrightarrow[\substack{\text{1.3 equiv Oxone} \\ \text{MeCN/H}_2\text{O (2:1)} \\ 70\,°\text{C, 6 h} \\ 94\%}]{\substack{\text{0.2 equiv}}} \diagup\!\!\diagdown\!\!\diagup\!\!\overset{O}{\diagdown}\text{OH} \quad (92)$$

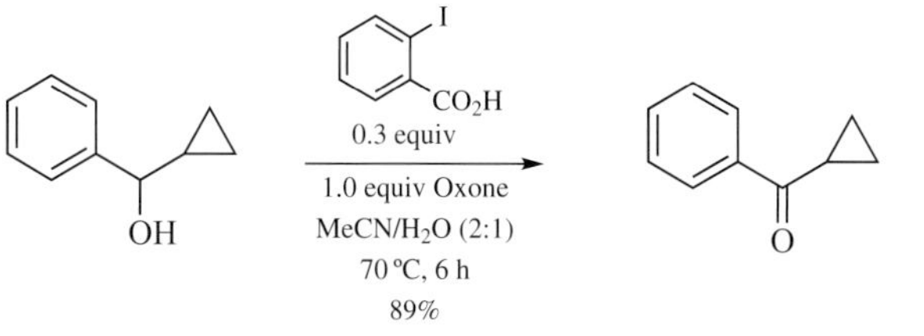

$$(93)$$

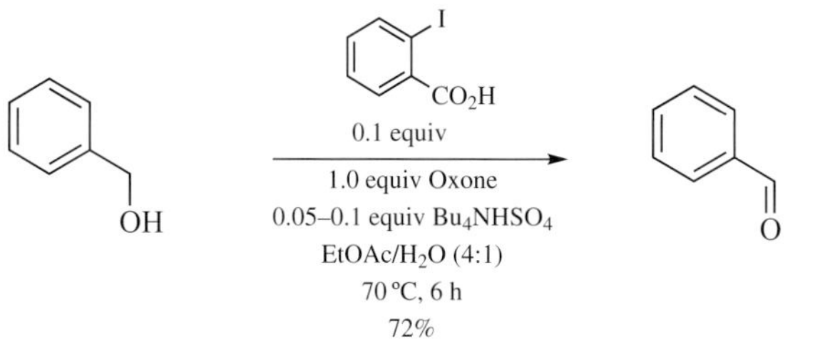

$$(94)$$

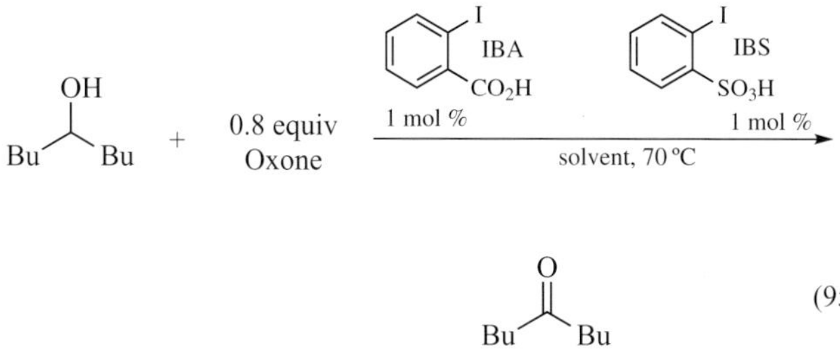

$$(95)$$

	IBA	IBS
MeCN	24 h, <5%	1.6 h, >99%
EtOAc	24 h, <5%	10 h, >99%

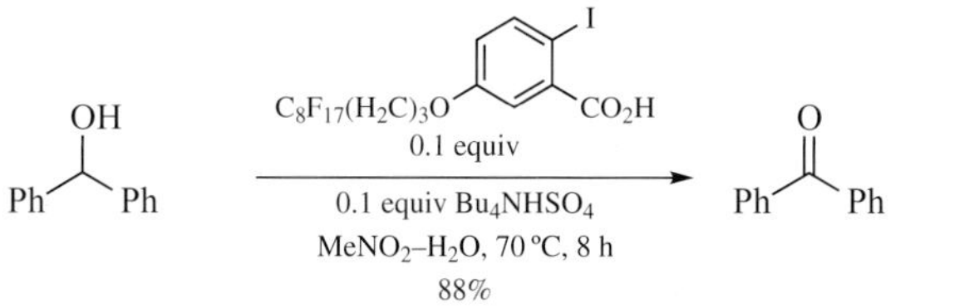

$$(96)$$

Inoue and coworkers have reported the manganese-catalyzed oxidation of alcohols to ketones using persulfate as the terminal oxidant.[167] Recrystallization of Oxone to give purified $KHSO_5$ was found to be beneficial for improving the yield of the reaction (eq 97).

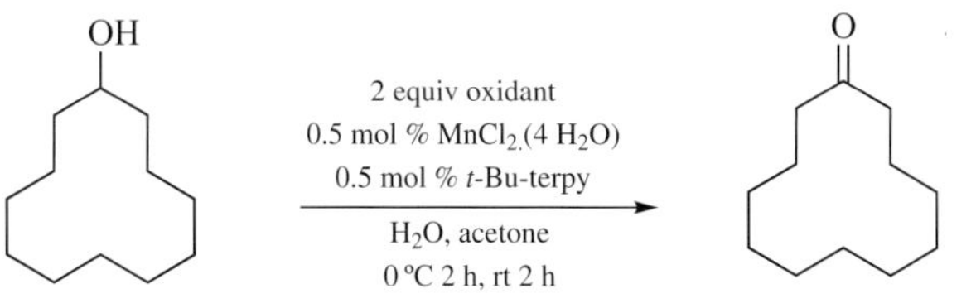

oxidant =	Oxone:	61%
	$KHSO_5$	83%

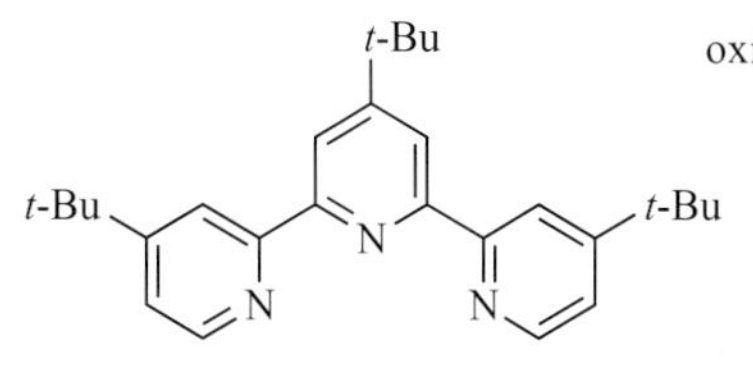

Alkenes, Arenes, and Alkanes. The catalytic *cis*-dihydroxylation of alkenes with iron complexes and Oxone as terminal oxidant has been reported by Che and coworkers.[168] A wide range of electron-rich, electron-poor, aliphatic, aromatic, terminal, and α,β-unsaturated alkenes were tolerated under the optimized reaction conditions. For example, methyl cinnamate could be transformed into the diol **9** on a 30 mmol scale with traces of undesired products **10** and **11** being isolated as side products (eq 98). Oxone has also been employed in the diastereoselective dihydroxylation of alkenes in a catalytic approach to the Prevost–Woodward reaction.[169] Up to 77% yield was obtained in the dihydroxylation of styrene; however, $NaIO_4$ was found to be a superior oxidant affording the diol in 87% yield (eq 99).

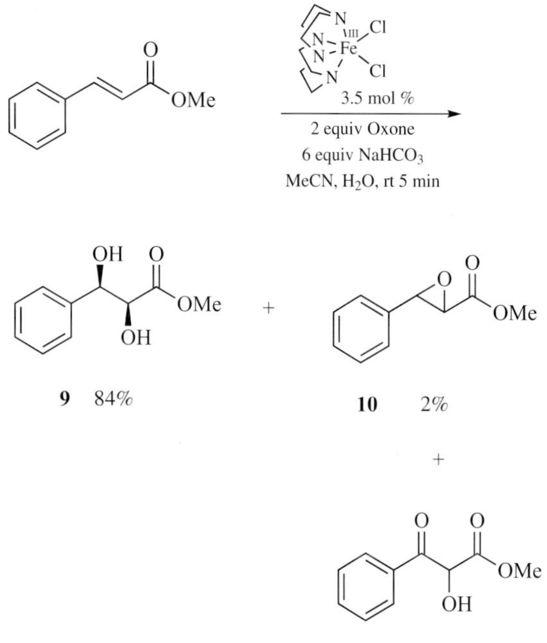

9 84%   **10** 2%

$$(98)$$

11 3%

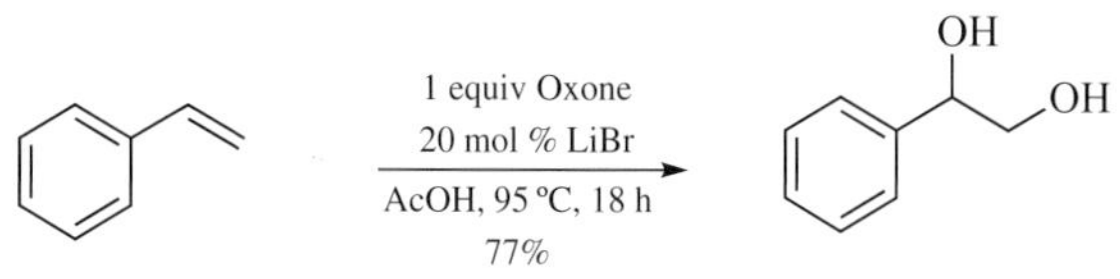

$$(99)$$

Oxone continues to be the terminal oxidant of choice for organocatalyzed epoxidation of alkenes, when employing dioxirane or oxaziridine catalysis. Shi and workers have reported several investigations into their fructose-derived ketone precatalysts, with substructure alterations having dramatic effects on the enantioselectivity of the process (eq 100).[60,170] Indeed, Singleton and Wang have thoroughly investigated isotope effects on the enantioselectivity of various Shi catalysts.[171] The Shi epoxidation process has also been applied to the pilot plant scale synthesis of the chiral lactone **12**, in which over 100 kg was produced (eq 101).[172] Oxone was employed as the terminal oxidant and the epoxidation proceeded in 63% yield, with 97% purity and 88% ee.

$$(100)$$

$$(102)$$

$$(103)$$

$$(101)$$

12

$$(104)$$

Several other groups have reported effective dioxirane systems employing Oxone as the terminal oxidant. For example, Armstrong et al. have developed a spirocyclic *N*-carbethoxy-azabicyclo[3.2.1]octanone precatalyst, which affords up to 91.5% ee in the epoxidation of stilbenes (eq 102).[173] Shing et al. have developed an arabinose-derived ketone and employed this in the enantioselective synthesis of the Taxol side chain; however, enantioselectivities for the epoxidation were only up to 68% (eq 103).[174] Bortolini et al. have also described the epoxidation of alkenes with the stoichiometric keto bile acid–Oxone system, a range of ee values were observed over several substrate types but up to 98% was observed for the epoxidation of *trans*-stilbene, although the yield was only 50% (eq 104).[175]

Page et al. have reported several oxaziridinium systems for highly enantioselective epoxidation of unfunctionalized alkenes. For example, the simple binaphthalene-derived iminium salt **13** affords up to 84% ee in the epoxidation of 1-phenylcyclohexene employing Oxone as the stoichiometric oxidant.[176] They also report that the Oxone conditions employed have a drastic effect on the product ratio when certain sensitive alkene substrates are exposed to the reaction conditions; for example, precatalyst **13** affords 91% yield of the diol **14** when using 2 equiv of Oxone and 5 equiv of sodium hydrogen carbonate in water:acetonitrile (1:10; known as Yang's conditions[63]), but when 2 equiv of Oxone and 4 equiv of sodium carbonate in water:acetonitrile (1:1) is employed, 90% conversion to the epoxide **15** is observed with 83% ee (eq 105). Page et al. have also reported the in situ generation of persulfate using boron-doped diamond electrodes and applied this technology to the asymmetric epoxidation of unfunctionalized alkenes.[177] Enantioselectivities of up to 64% were obtained when employing precatalyst **16** (eq 106). The alternative oxidant tetraphenylphosphonium monoperoxysulfate derived from Oxone has also been successfully employed in asymmetric epoxidation using oxaziridinium salts.[46] Page and coworkers have used the epoxidation conditions outlined in eq 41 followed by subsequent ring opening to afford the asymmetric syntheses of (−)-levcromakalim,[79] (−)-(3′S)-lomatin,[178] (+)-(3′S, 4′R)-*trans*-khellactone,[178] and (+)-scuteflorin.[179]

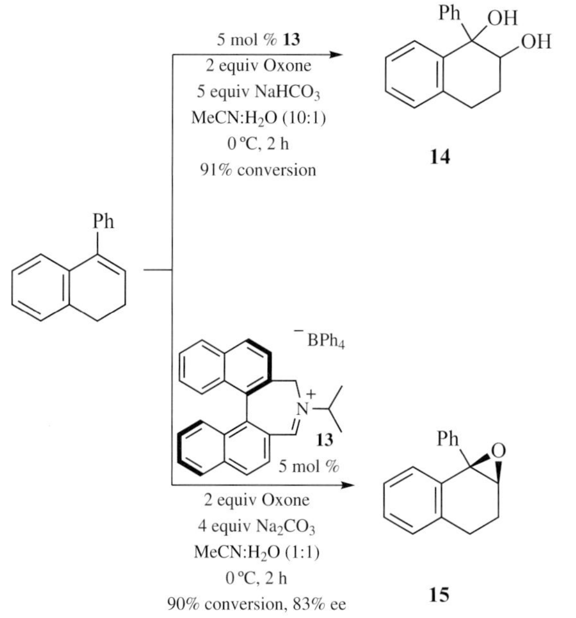

5 mol % **13**
2 equiv Oxone
5 equiv NaHCO$_3$
MeCN:H$_2$O (10:1)
0 °C, 2 h
91% conversion

14

(105)

13
5 mol %

2 equiv Oxone
4 equiv Na$_2$CO$_3$
MeCN:H$_2$O (1:1)
0 °C, 2 h
90% conversion, 83% ee

15

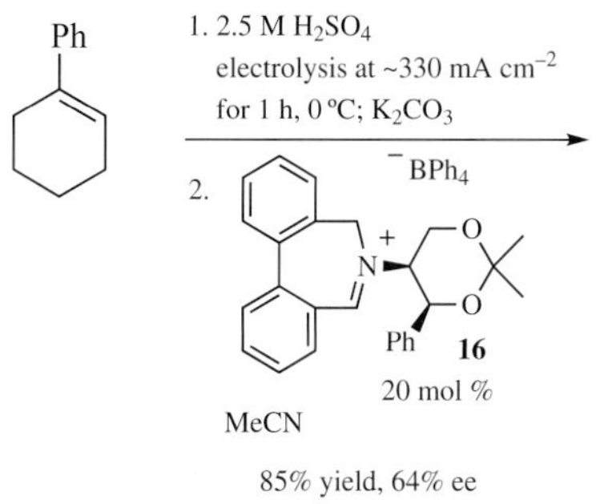

1. 2.5 M H$_2$SO$_4$
electrolysis at ~330 mA cm^{-2}
for 1 h, 0 °C; K$_2$CO$_3$

2.
16
20 mol %

MeCN

85% yield, 64% ee

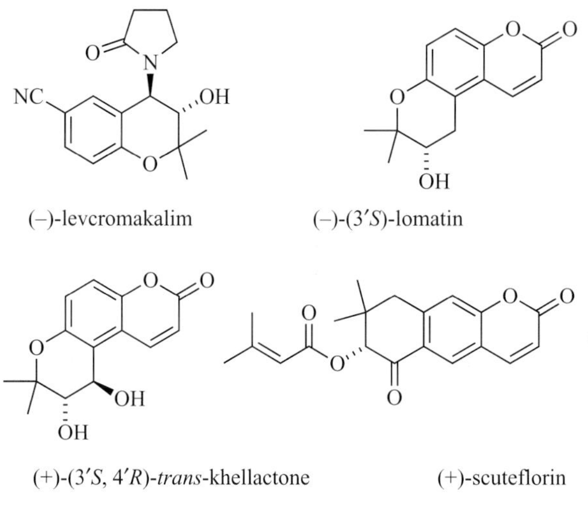

(−)-levcromakalim

(−)-(3′S)-lomatin

(106)

(+)-(3′S, 4′R)-trans-khellactone

(+)-scuteflorin

Lacour and coworkers have also reported several iminium ion precatalysts, for example, **17–19**.[180] Originally, Lacour and coworkers reported the use of the chiral counterion TRISPHAT to investigate the effect on the ee; however, no increase was observed.[181] The nature of the anion can have a bearing on the enantioselectivity of the epoxidation reaction; degradation of the anion tetraphenylborate to biphenyl under Oxone oxidation can occur, thus higher loadings of catalyst affording lower ee values due to biphenyl altering the partition ability of the iminium/oxaziridinium species in the biphasic reaction. Since this report

they have adopted, the SbF$_6$ counterion and excellent ee values were obtained with precatalyst **20** for the epoxidation of hindered trisubstituted allylic alcohols using Oxone as the terminal oxidant (eq 107).[182]

Very few reports have emerged on the amine-catalyzed epoxidation protocol employing Oxone reported independently by Aggarwal et al. in 2003[81] and Yang and coworkers in 2005,[82] although both Page and Lacour have shown that amines such as **21** catalyze the epoxidation of simple alkenes.[183–185] The reactions, unlike those reported by Aggarwal and Yang that are believed to enjoy hydrogen bonding to persulfate, here employing Oxone initially converts the amine to the iminium ion and subsequently to the oxaziridinium ion intermediate, thus enabling epoxidation (eq 108).

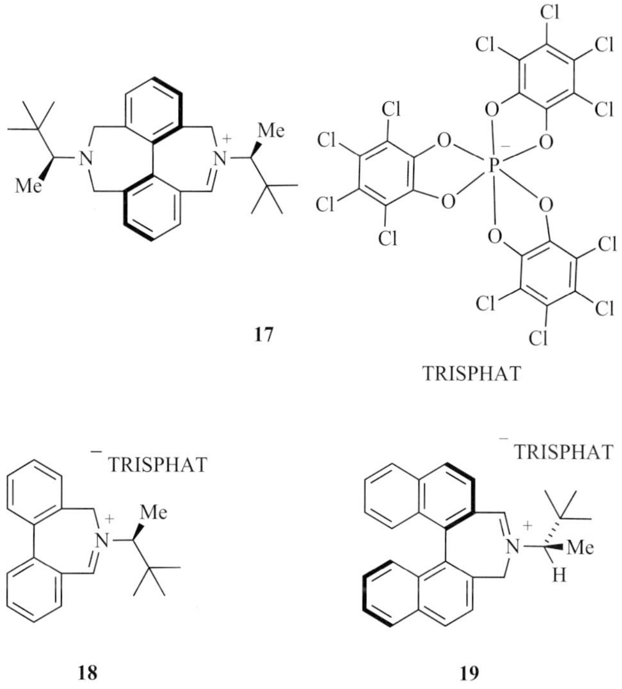

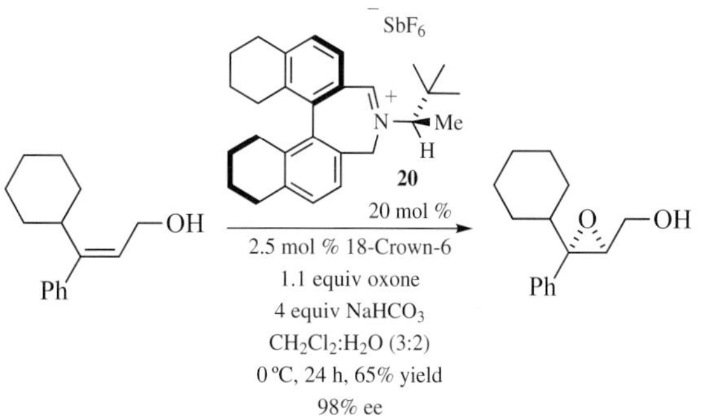

17

TRISPHAT

TRISPHAT

18

TRISPHAT

19

20
20 mol %
2.5 mol % 18-Crown-6
1.1 equiv oxone
4 equiv NaHCO$_3$
CH$_2$Cl$_2$:H$_2$O (3:2)
0 °C, 24 h, 65% yield
98% ee

(107)

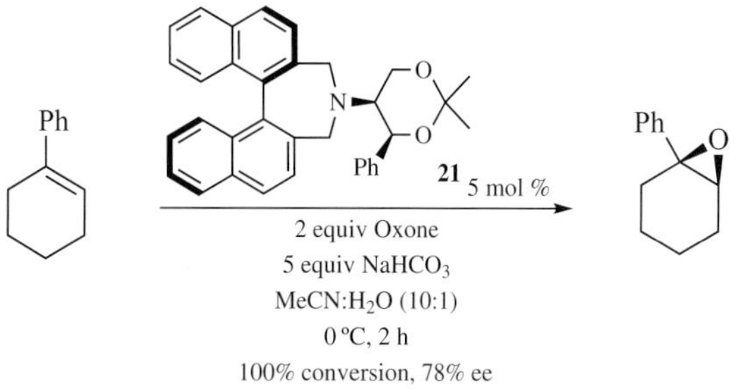

21 5 mol %
2 equiv Oxone
5 equiv NaHCO$_3$
MeCN:H$_2$O (10:1)
0 °C, 2 h
100% conversion, 78% ee

(108)

Oxone has also been used in the metalloporphyrin-catalyzed diastereoselective epoxidation of allyl-substituted alkenes (eq 109)[186] and in the manganese, salen–polystyrene-bound imidazole catalyzed deoxidation of alkenes (eq 110).[187]

Sanford and coworkers have reported the use of Oxone in combination with palladium acetate for the oxime–ether directed C–H oxygenation of arenes (eq 111).[188] A range of substituted arenes are tolerated under the reaction conditions and the use of Oxone rather than periodate-based oxidants afforded N-oxidized products in some cases.

$$(109)$$

$$(110)$$

$$(111)$$

Yakura et al. have shown that Oxone in combination with a catalytic quantity of 4-iodophenoxy acetic acid is an effective system for the hypervalent iodine oxidation of phenols and aryl ethers (eq 112).[189]

$$(112)$$

The selective oxygenation of saturated C–H bonds using a Mn(μ-O)$_2$Mn reactive center and a ligand based on Kemp's triacid has been reported.[190] The ligand **22** is believed to direct a –COOH group attached to the substrate and thus aid the selective C–H oxidation of substrates such as ibuprofen (eq 113). In comparison to ligand **23**, which does not bear the proposed directing group, poor selectivity in the oxidation process is observed. The same group has also reported an in-depth study on the use of manganese catalysts for C–H oxygenation, and desaturation of a range of substrates has been reported using Oxone as a stoichiometric oxidant. For example, desaturation of **24** could be achieved exclusively when employing the manganese catalyst **25**, Oxone, and N-methylmorpholine as an additive (eq 114).[191]

$$(113)$$

| Using **23** | 77% | 23% |
| **22** | 97.5% | 2.5% |

$$(114)$$

Ekkati and Kodanko has reported the cleavage of an amino acid backbone through persulfate and iron-catalyzed C–H oxidation (eq 115).[192]

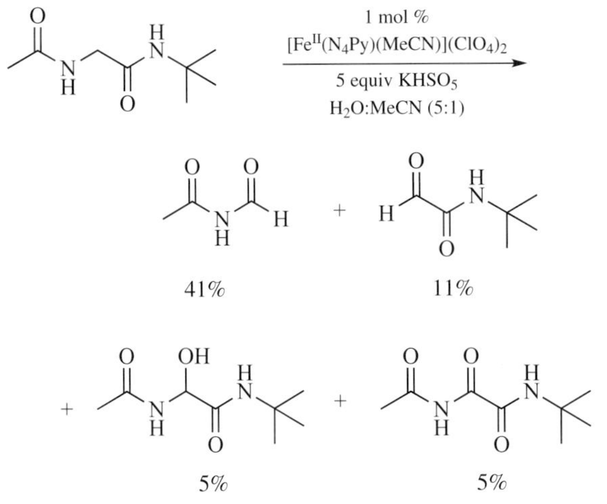

(115)

(118)

(119)

Nitrogen Compounds. Zolfigol et al. have again reported the catalytic oxidation of urazoles to the corresponding triazolinediones, but this time using in situ generation of Br$^+$ using Oxone/KBr.[193] Good yields of the products could be obtained when using 20 mol % KBr and 2 equiv of Oxone (eq 116); however, improved yields could be obtained if sodium periodate was employed as the terminal oxidant.

Related Reagents. Dimethyldioxirane; methyl-(trifluoromethyl)dioxirane.

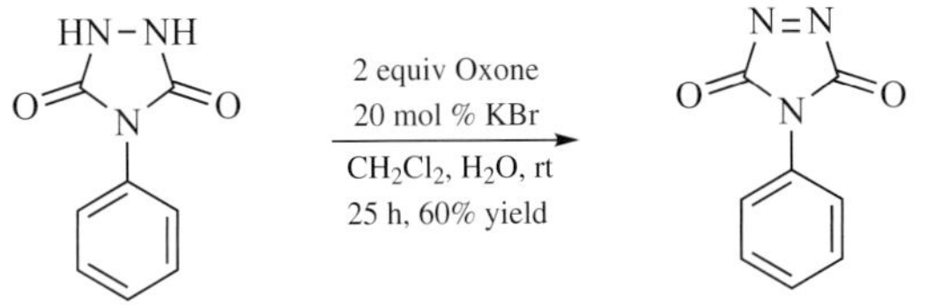

(116)

Priewisch and Rück-Braun have reported the efficient preparation of nitrosoarenes employing Oxone in dichloromethane/water (eq 117).[194] Excellent yields are obtained for a range of aryl substrates; the authors show that these nitrosoarenes can then be efficiently converted to the corresponding azobenzene derivatives.

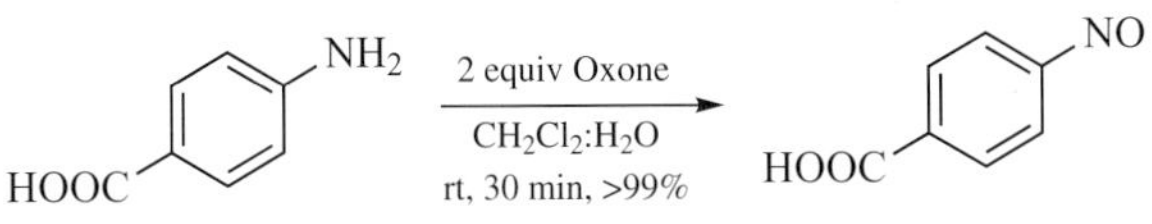

(117)

Sulfur Compounds. The Shi series of catalysts in conjunction with Oxone have been employed by Khiar et al. in the asymmetric monooxidation of disulfides (eq 118).[195] The authors observed up to 96% ee and found this system to provide superior ee values to the corresponding reactions employing Schiff base–vanadium complexes and hydrogen peroxide as stoichiometric oxidant. Colonna et al. reported similar conditions earlier in 2005; however, this study only gave a maximum of 75% ee.[196] The selective oxidation of sulfides to sulfoxides using a titanium-based catalyst, Ti$_4$[(OCH$_2$)$_3$CMe]$_2$(i-PrO)$_{10}$, and Oxone has been reported by Reddy and Verkade (eq 119), although their standard conditions employed hydrogen peroxide as the terminal oxidant.[197]

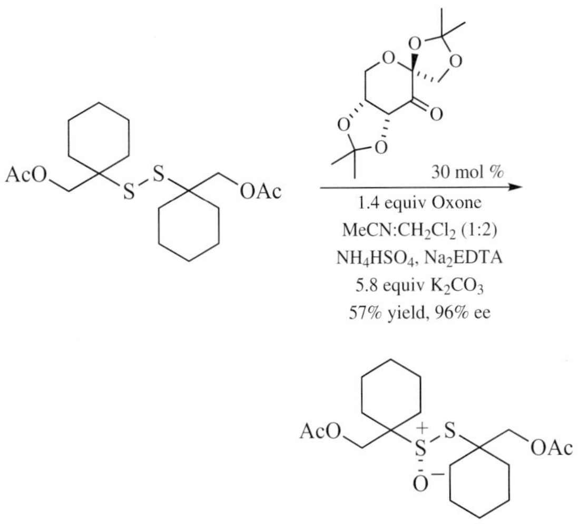

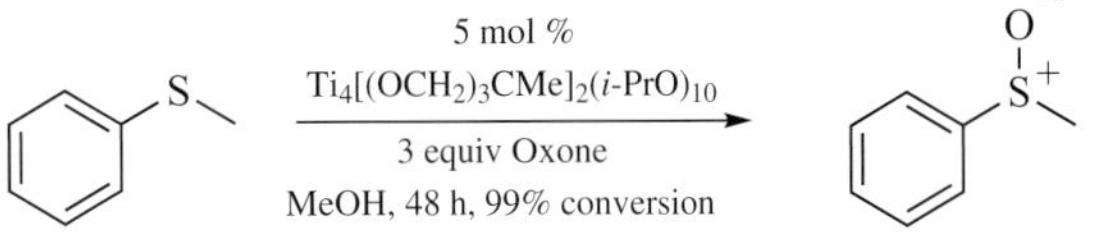

1. Adam, W.; Hadjiarapoglou, L. P.; Curci, R.; Mello, R., In *Organic Peroxides*; Ando, W.; Ed.; Wiley: New York, 1992; Chapter 4, p 195.

2. Curci, R., In *Advances in Oxygenated Processes*; Baumstark A., Ed.; JAI: Greenwich, CT, 1990; Vol. 2; Chapter 1, p 1.

3. Murray, R. W., *Chem. Rev.* **1989**, *89*, 1187.

4. Adam, W.; Edwards, J. O.; Curci, R., *Acc. Chem. Res.* **1989**, *22*, 205.

5. Murray, R. W.; Jeyaraman, R., *J. Org. Chem.* **1985**, *50*, 2847.

6. Mello, R.; Fiorentino, M.; Sciacovelli, O.; Curci, R., *J. Org. Chem.* **1988**, *53*, 3890.

7. Murray, R. W.; Singh, M.; Jeyaraman, R., *J. Am. Chem. Soc.* **1992**, *114*, 1346.

8. Hirano, M.; Oose, M.; Morimoto, T., *Chem. Lett.* **1991**, 331.

9. Hirano, M.; Oose, M.; Morimoto, T., *Bull. Chem. Soc. Jpn.* **1991**, *64*, 1046.

10. Kennedy, R. J.; Stock, A., *J. Org. Chem.* **1960**, *25*, 1901.

11. Bloch, R.; Abecassis, J.; Hassan, D., *J. Org. Chem.* **1985**, *50*, 1544.

12. Zhu, W.; Ford, W. T., *J. Org. Chem.* **1991**, *56*, 7022.

13. Curci, R.; Fiorentino, M.; Troisi, L.; Edwards, J. O.; Pater, R. H., *J. Org. Chem.* **1980**, *45*, 4758.

14. Hofland, A.; Steinberg, H.; De Boer, T. J., *Recl. Trav. Chim. Pays-Bas* **1985**, *104*, 350.

15. Corey, P. F.; Ward, F. E., *J. Org. Chem.* **1986**, *51*, 1925.

16. Adam, W.; Hadjiarapoglou, L.; Smerz, A., *Chem. Ber.* **1991**, *124*, 227.

17. Jeyaraman, R.; Murray, R. W., *J. Am. Chem. Soc.* **1984**, *106*, 2462.

18. Mello, R.; Ciminale, F.; Fiorentino, M.; Fusco, C.; Prencipe, T.; Curci, R., *Tetrahedron Lett.* **1990**, *31*, 6097.

19. Hanquet, G.; Lusinchi, X.; Milliet, P., *C. R. Hebd. Seances Acad. Sci., Ser. C* **1991**, *313*, 625.

20. Meunier, B., *Nouv. J. Chim.* **1992**, *16*, 203.

21. De Poorter, B.; Meunier, B., *Nouv. J. Chim.* **1985**, *9*, 393.

22. De Poorter, B.; Ricci, M.; Meunier, B., *Tetrahedron Lett.* **1985**, *26*, 4459.

23. Neumann, R.; Abu-Gnim, C., *J. Chem. Soc., Chem. Commun.* **1989**, 1324.

24. Strukul, G.; Sinigalia, R.; Zanardo, A.; Pinna, F.; Michelin, R., *Inorg. Chem.* **1989**, *28*, 554.

25. Khan, M. M. T.; Chetterjee, D.; Merchant, R. R.; Bhatt, A., *J. Mol. Catal.* **1990**, *63*, 147.

26. Zabrowski, D. L.; Moormann, A. E.; Beck, K. R. J., *Tetrahedron Lett.* **1988**, *29*, 4501.

27. Crandall, J. K.; Reix, T., *J. Org. Chem.* **1992**, *57*, 6759.

28. Itahara, T., *Chem. Lett.* **1991**, 1591.

29. Kettani, A. E.; Bernadou, J.; Meunier, B., *J. Org. Chem.* **1989**, *54*, 3213.

30. Davis, F. A.; Chattopadhyay, S.; Towson, J. C.; Lal, S.; Reddy, T., *J. Org. Chem.* **1988**, *53*, 2087.

31. Davis, F. A.; Weismiller, M. C.; Murphy, C. K.; Reddy, R. T.; Chen, B. C., *J. Org. Chem.* **1992**, *57*, 7274.

32. Hajipour, A. R.; Pyne, S. G., *J. Chem. Res. (S)* **1992**, 388.

33. Kim, J. N.; Ryu, E. K., *J. Org. Chem.* **1992**, *57*, 6649.

34. Trost, B. M.; Curran, D. P., *Tetrahedron Lett.* **1981**, *22*, 1287.

35. Hirano, M.; Tomaru, J.; Morimoto, T., *Bull. Chem. Soc. Jpn.* **1991**, *64*, 3752.

36. Evans, T. L.; Grade, M. M., *Synth. Commun.* **1986**, *16*, 1207.

37. Greenhalgh, R. P., *Synlett* **1992**, 235.

38. Davis, F. A.; Lal, S. G.; Durst, H. D., *J. Org. Chem.* **1988**, *53*, 5004.

39. Quallich, G. J.; Lackey, J. W., *Tetrahedron Lett.* **1990**, *31*, 3685.

40. Colonna, S.; Gaggero, N.; Leone, M.; Pasta, P., *Tetrahedron* **1991**, *47*, 8385.

41. Reddey, R. N., *Synth. Commun.* **1987**, *17*, 1129.

42. Tabuchi, T.; Nojima, M.; Kusabayashi, S., *J. Chem. Soc., Perkin Trans. 1* **1991**, 3043.

43. Travis, B. R.; Ciaramitaro, B. P.; Borhan, B., *Eur. J. Org. Chem.* **2002**, 3429.

44. Trost, B. M.; Braslau, R., *J. Org. Chem.* **1988**, *53*, 532.

45. Campestrini, S.; Di Furia, F.; Labat, G.; Novello, F., *J. Chem. Soc., Perkin Trans. 2* **1994**, 2175.

46. Page, P. C. B.; Barros, D.; Buckley, B. R.; Ardakani, A.; Marples, B. A., *J. Org. Chem.* **2004**, *69*, 3595.

47. Hajipour, A. R.; Mallakpour, S. E.; Adibi, H., *Chem. Lett.* **2000**, 460.

48. Yan, J.; Travis, B. R.; Borhan, B., *J. Org. Chem.* **2004**, *69*, 9299.

49. Ashford, S. W.; Grega, K. C., *J. Org. Chem.* **2001**, *66*, 1523.

50. Ballini, R.; Curini, M.; Epifano, F.; Marcotullio, M. C.; Rosati, O., *Synlett* **1998**, 1049.

51. Webb, K. S.; Ruszkay, J. S., *Tetrahedron* **1998**, *54*, 401.

52. Travis, B. R.; Sivakumar, M.; Hollist, G. O.; Borhan, B., *Org. Lett.* **2003**, *5*, 1031.

53. Bose, D. S.; Narsaiah, A. V., *Tetrahedron Lett.* **1998**, *39*, 6533.

54. Bagherzadeh, M., *Tetrahedron Lett.* **2003**, *44*, 8943.

55. Bolm, C.; Magnus, A. S.; Hildebrand, J. P., *Org. Lett.* **2000**, *2*, 1173.

56. Koo, B.-S.; Lee, C. K.; Lee, K.-J., *Synth. Commun.* **2002**, *32*, 2115.

57. Hajipour, A. R.; Mallakpour, S.; Adibi, H., *Chem. Lett.* **2000**, 460.

58. Denmark, S. E.; Wu, Z., *Synlett* **1999**, 847.

59. Frohn, M.; Shi, Y., *Synthesis* **2000**, 1979.

60. Shi, Y., *Acc. Chem. Res.* **2004**, *37*, 488.

61. Yang, D., *Acc. Chem. Res.* **2004**, *37*, 497.

62. Curci, R.; Fiorentino, M.; Serio, M. R., *J. Chem. Soc., Chem. Commun.* **1984**, 155.

63. Yang, D.; Tang, Y.-C.; Wong, M.-K.; Zheng, J.-H.; Cheung, K.-K., *J. Am. Chem. Soc.* **1996**, *118*, 491.

64. Yang, D.; Wang, X.-C.; Wong, M.-K.; Yip, Y.-C.; Tang, M.-W., *J. Am. Chem. Soc.* **1996**, *118*, 11311.

65. Yang, D.; Wong, M.-K.; Yip, Y.-C.; Wang, X.-C.; Tang, M.-W.; Zheng, J.-H.; Cheung, K.-K., *J. Am. Chem. Soc.* **1998**, *120*, 5943.

66. Denmark, S. E.; Wu, Z.; Crudden, C. M.; Matsuhashi, H., *J. Org. Chem.* **1997**, *62*, 8288.

67. Denmark, S. E.; Matsuhashi, H., *J. Org. Chem.* **2002**, *67*, 3479.

68. Stearman, C. J.; Behar, V., *Tetrahedron Lett.* **2002**, *43*,1943.

69. Denmark, S. E.; Forbes, D. C.; Hays, D. S.; De Pue, J. S.; Wilde, R. G., *J. Org. Chem.* **1995**, *60*, 1391.

70. Tu, Y.; Wang, Z.-X.; Shi, Y., *J. Am. Chem. Soc.* **1996**, *118*, 9806.

71. Wang, Z.-X.; Tu, Y.; Frohn, M.; Zhang, J.-R.; Shi, Y., *J. Am. Chem. Soc.* **1997**, *119*, 11224.

72. Shu, L.; Shi, Y., *Tetrahedron Lett.* **2004**, *45*, 8115.

73. Aggarwal, V. K.; Wang, M. F., *J. Chem. Soc., Chem. Commun.* **1996**, 191.

74. Page, P. C. B.; Rassias, G. A.; Bethell, D.; Schilling, M. B., *J. Org. Chem.* **1998**, *63*, 2774.

75. Minakata, S.; Takemiya, A.; Nakamura, K.; Ryu, I.; Komatsu, M., *Synlett* **2000**, *12*, 1810.

76. Wong, M.-K.; Ho, L.-M.; Zheng, Y.-S.; Ho, C.-Y.; Yang, D., *Org. Lett.* **2001**, *16*, 2587.

77. Armstrong, A.; Ahmed, G.; Garnett, I.; Goacolou, K.; Wailes, J. S., *Tetrahedron* **1999**, *55*, 2341.

78. Page, P. C. B.; Buckley, B. R.; Blacker, A. J., *Org. Lett.* **2004**, *6*, 1543.

79. Page, P. C. B.; Buckley, B. R.; Heaney, H.; Blacker, A. J., *Org. Lett.* **2005**, *7*, 375.

80. Armstrong, A., *Angew. Chem. Int. Ed.* **2004**, *43*, 1460.

81. Aggarwal, V. K.; Lopin, C.; Sandrinelli, F., *J. Am. Chem. Soc.* **2003**, *125*, 7596.

82. Ho, C.-Y.; Chen, Y.-C.; Wong, M.-K.; Yang, D., *J. Org. Chem.* **2005**, *70*, 898.

83. Travis, B. R.; Narayan, R. S.; Borhan, B., *J. Am. Chem. Soc.* **2002**, *124*, 3824.

84. Schomaker, J. M.; Travis, B. R.; Borhan, B., *Org. Lett.* **2003**, *5*, 3089.

85. Travis, B.; Borhan, B., *Tetrahedron Lett.* **2001**, *42*, 7741.

86. Yang, D.; Zhang, C., *J. Org. Chem.* **2001**, *66*, 4814.

87. Plietker, B., *J. Org. Chem.* **2003**, *68*, 7123.

88. Plietker, B., *J. Org. Chem.* **2004**, *69*, 8287.

89. Plietker, B., *Org. Lett.* **2004**, *6*, 289.

90. Rani, S.; Vankar, Y. D., *Tetrahedron Lett.* **2003**, *44*, 907.

91. Yang, D.; Chen, F.; Dong, Z.-M.; Zhang, D.-W., *J. Org. Chem.* **2004**, *69*, 2221.

92. Yang, D.; Wong, M.-K.; Wang, X.-C.; Tang, Y.-C., *J. Am. Chem. Soc.* **1998**, *120*, 6611.

93. Wong, M.-K.; Chung, N.-W.; He, L.; Wang, X.-C.; Yan, Z.; Tang, Y-C.; Yang, D., *J. Org. Chem.* **2003**, *68*, 6321.

94. Adam, W.; SahaMoller, C. R.; Zhao, C.-G., *Tetrahedron: Asymmetry* **1998**, *9*, 4117.

95. Adam, W.; Saha-Moller, C. R.; Zhao, C.-G., *J. Org. Chem.* **1999**, *64*, 7492.

96. Cammarota, L.; Campestrini, S.; Carrieri, M.; Di Furia, F.; Ghiotti, P., *J. Mol. Catal. A* **1999**, *137*, 155.

97. Webb, K. S.; Seneviratne, V., *Tetrahedron Lett.* **1995**, *36*, 2377.

98. Noecker, L.; Giuliano, R. M.; Cooney, M.; Boyko, W.; Zajac, Jr., W. W., *J. Carbohydr. Chem.* **2002**, *21*, 539.

99. Fields, J. D.; Kropp, P. J., *J. Org. Chem.* **2000**, *65*, 5937.

100. Brik, M. E., *Tetrahedron Lett.* **1995**, *36*, 5519.

101. Zolfigol, M. A.; Bagherzadeh, M.; Choghamarani, A. G.; Keypour, H.; Salehzadeh, S., *Synth. Commun.* **2001**, *31*, 1161.

102. Zolfigol, M. A.; Bagherzadeh, M.; Madrakian, E.; Ghaemi, E.; Taquian-Nasab, A., *J. Chem. Res. (S).* **2001**, 140.

103. Priewisch, B.; Rück-Braun, K., *J. Org. Chem.* **2005**, *70*, 2350.

104. Bose, D. S.; Vanajatha, G., *Synth. Commun.* **1998**, *28*, 4531.

105. Bose, D. S.; Srinivas, P., *Synth. Commun.* **1997**, *27*, 3835.

106. Bose, D. S.; Narsaiah, A. V.; Lakshminarayana, V., *Synth. Commun.* **2000**, *30*, 3121.

107. Bigdeli, M. A.; Nikje, M. M. A.; Heravi, M. M., *Phosphorus, Sulfur, Silicon* **2002**, *177*, 15.

108. Paradkar, V. M.; Latham, T. B.; Demko, D. M., *Synlett* **1993**, 1059.

109. Kim, J. N.; Jung, K. S.; Lee, H. J.; Son, J. S., *Tetrahedron Lett.* **1997**, *38*, 1597.

110. Curini, M.; Rosati, O.; Pisani, E., *Tetrahedron Lett.* **1997**, *38*, 1239.

111. Jung, J. C.; Kim, K. S.; Kim, Y. H., *Synth. Commun.* **1992**, *22*, 1583.

112. Hajipour, A. R.; Mallakpour, S. E.; Baltork, I. M.; Adibi, H., *Synth. Commun.* **2001**, *31*, 3401.

113. Ceccherelli, P.; Curini, M.; Marcotullio, M. C.; Epifano, F.; Rosati, O., *Synth. Commun.* **1998**, *28*, 3057.

114. Beaulieu, P. L.; Haché, B.; von Moos, E., *Synthesis* **2003**, 1683.

115. Zolfigol, M. A.; Bagherzadeh, M.; Chehardoli, G.; Mallakpour, S. E., *Synth. Commun.* **2001**, *31*, 1149.

116. Mohajer, D.; Iranpoor, N.; Rezaeifard, A., *Tetrahedron Lett.* **2004**, *45*, 631.

117. Kulkarni, P. P.; Kadam, A. J.; Desai, U. V.; Mane, R. B.; Wadgaonkar, P. P., *J. Chem. Res. (S).* **2000**, 184.

118. Tamhankar, B. V.; Desai, U. V.; Mane, R. B.; Kulkarni, P. P.; Wadgaonkar, P. P., *Synth. Commun.* **2002**, *32*, 3643.

119. Bose, D. S.; Baquer, S. M., *Synth. Commun.* **1997**, *27*, 3119.

120. Webb, K. S., *Tetrahedron Lett.* **1994**, *35*, 3457.

121. Kropp, P. J.; Breton, G. W.; Fields, J. D.; Tung, J. C.; Loomis, B. R., *J. Am. Chem. Soc.* **2000**, *122*, 4280.

122. Bunton, C. A.; Foroudian, H. J.; Kumar, A., *J. Chem. Soc., Perkin Trans. 2* **1995**, 33.

123. Chen, M.-Y.; Patkar, L. N.; Chen, H.-T.; Lin, C.-C., *Carbohydr. Res.* **2003**, *338*, 1327.

124. Johnson, P.; Taylor, R. J. K., *Tetrahedron Lett.* **1997**, *38*, 5873.

125. Hajipour, A. R.; Mallakpour, S. E.; Adibi, H., *J. Org. Chem.* **2002**, *67*, 8666.

126. Hajipour, A. R.; Mallakpour, S. E.; Adibi, H., *Phosphorus, Sulfur, Silicon* **2002**, *177*, 2277.

127. Baltork, I. M.; Sadeghi, M. M.; Esmayilpour, K., *Synth. Commun.* **2003**, *33*, 953.

128. Ripin, D. H. B.; Cai, W.; Brenek, S. J., *Tetrahedron Lett.* **2000**, *41*, 5817.

129. Webb, K. S.; Levy, D., *Tetrahedron Lett.* **1995**, *36*, 5117.

130. Maleczka, Jr. R. E.; Shi, F.; Holmes, D.; Smith, III, M. R., *J. Am. Chem. Soc.* **2003**, *125*, 7792.

131. Vazquez, E.; Payack, J. F., *Tetrahedron Lett.* **2004**, *45*, 6549.

132. Woźniak, L. A.; Stec, W. J., *Tetrahedron Lett.* **1999**, *40*, 2637.

133. Ceccherelli, P.; Curini, M.; Epifano, F.; Marcotullio, M. C.; Rosati, O., *J. Org. Chem.* **1995**, *60*, 8412.

134. Yi, J. S.; Kim, K., *J. Chem. Soc., Perkin Trans. 1* **1999**, 71.

135. Tamhankar, B. V.; Desai, U. V.; Mane, R. B.; Wadgaonkar, P. P.; Bedekar, A. V., *Synth. Commun.* **2001**, *31*, 2021.

136. Narender, N.; Srinivasu, P.; Kulkarni, S. J.; Raghavan, K. V., *Synth. Commun.* **2002**, *32*, 279.

137. Narender, N.; Srinivasu, P.; Prasad, M. R.; Kulkarni, S. J.; Raghavan, K. V., *Synth. Commun.* **2002**, *32*, 2313.

138. Narender, N.; Srinivasu, P.; Kulkarni, S. J.; Raghavan, K. V., *Synth. Commun.* **2002**, *32*, 2319.

139. Mohan, K. V. V. K.; Narender, N.; Kulkarni, S. J., *Tetrahedron Lett.* **2004**, *45*, 8015.

140. Hajipour, A. R.; Adibi, H., *J. Chem. Res.* **2004**, 294.

141. Dieter, R. K.; Nice, L. E.; Velu, S. E., *Tetrahedron Lett.* **1996**, *37*, 2377.

142. Kim, K.-M.; Park, I.-H., *Synthesis* **2004**, 2641.

143. Curini, M.; Epifano, F.; Marcotullio, M. C.; Rosati, O., *Tetrahedron Lett.* **2002**, *43*, 1201.

144. Curini, M.; Epifano, F.; Marcotullio, M. C.; Montanari, F., *Synlett* **2004**, 368.

145. Ceccherelli, P.; Curini, M.; Epifano, F.; Marcotullio, M. C.; Rosati, O., *Tetrahedron Lett.* **1998**, *39*, 4385.

146. Curini, M.; Epifano, F.; Marcotullio, M. C.; Rosati, O.; Rossi, M., *Tetrahedron* **1999**, *55*, 6211.

147. You, H.-W.; Lee, K.-J., *Synlett* **2001**, 105.

148. Koo, B.-S.; Kim, E.-H.; Lee, K.-J., *Synth. Commun.* **2002**, *32*, 2275.

149. Kim, E.-H.; Koo, B.-S.; Song, C. E.; Lee, K. J., *Synth. Commun.* **2001**, *31*, 3627.

150. Kim, K. K.; Kim, J. N.; Kim, K. M.; Kim, H. R.; Ryu, E. K., *Chem. Lett.* **1992**, 603.

151. Curini, M.; Epifano, F.; Marcotullio, M. C.; Rosati, O.; Tsadjout, A., *Synlett* **2000**, 813.

152. Sabitha, G.; Syamala, M.; Yadav, J. S., *Org. Lett.* **1999**, *1*, 1701.

153. Baltork, I. M.; Amini, M. K.; Farshidipoor, S., *Bull. Chem. Soc. Jpn.* **2000**, *73*, 2775.

154. Hajipour, A. R.; Mallakpour, S. E.; Baltork, I. M.; Adibi, H., *Synth. Commun.* **2001**, *31*, 1625.

155. Bose, D. S.; Jayalakshmi, B.; Narsaiah, A. V., *Synthesis* **2000**, 67.

156. Curini, M.; Epifano, F.; Marcotullio, M. C.; Rosati, O., *Synlett* **1999**, 777.

157. Ceccherelli, P.; Curini, M.; Marcotullio, M. C.; Epifano, F.; Rosati, O., *Synlett* **1996**, 767.

158. Hajipour, A. R.; Mallakpour, S. E.; Baltork, I. M.; Adibi, H., *Phosphorus, Sulfur, Silicon* **2002**, *177*, 2805.

159. Sabitha, G.; Babu, R. S.; Reddy, M. S. K.; Yadav, J. S., *Synthesis* **2002**, 2254.

160. Sheppard, G. S., *Synlett* **2002**, 1207.

161. Thottumkara, A. P.; Bowsher, M. S.; Vinod, T. K., *Org. Lett.* **2005**, *7*, 2933.

162. Schulze, A.; Giannis, A, *Synthesis,* **2006**, 257.

163. Uyanik, M.; Akakura, M.; Ishihara, K., *J. Am. Chem. Soc.* **2009**, *131*, 251.

164. Koposov, A. Y.; Litvinov, D. N.; Zhdankin, V. V.; Ferguson, M. J.; McDonald, R.; Tykwinski, R. R., *Eur. J. Org. Chem.* **2006**, 4791.

165. Page, P. C. B.; Appleby, L. F.; Buckley, B. R.; Allin, S. M.; McKenzie, M. J., Synlett **2007**, 1565.

166. Miura, T.; Nakashima, K.; Tada, N.; Itoh, A., *Chem. Commun.* **2011**, 1875.

167. Kamijo, S.; Amaoka, Y.; Inoue, M., *Synthesis* **2010**, 2475.

168. Chow, T. W.-S.; Wong, E. L.-M.; Guo, Z.; Liu, Y.; Huang, J.-S.; Che, C.-M., *J. Am. Chem. Soc.* **2010**, *132*, 13229.

169. Emmanuvel, L.; Shaikh, T. M. A.; Sudalai, A., *Org. Lett.* **2005**, *7*, 5071.

170. Wong, O. A.; Wang, B.; Zhao, M.-X.; Shi, Y., *J. Org. Chem.* **2009**, *74*, 6335.

171. Singleton, D. A.; Wang, Z., *J. Am. Chem. Soc.* **2005**, *127*, 6679.

172. Ager, D. J.; Anderson, K.; Oblinger, E.; Shi, Y.; VanderRoest J., *Org. Process Res. Dev.* **2007**, *11*, 44.

173. Armstrong, A.; Bettati, M.; White, A. J. P., *Tetrahedron* **2010**, *66*, 6309.

174. Shing, T. K. M.; Luk, T.; Lee, C. M., *Tetrahedron* **2006**, *62*, 6621.

175. Bortolini, O.; Fantin, G.; Fogagnolo, M.; Mari, L., **2006**, *62*, 4482.

176. Page, P. C. B.; Farah, M. M.; Buckley, B. R.; Blacker, A. J., *J. Org. Chem.* **2007**, *72*, 4424.

177. Page, P. C. B.; Marken, F.; Williamson, C.; Chan, Y.; Buckley, B. R.; Bethell, D., *Adv. Synth. Catal.* **2008**, *350*, 1149.

178. Page, P. C. B.; Appleby, L. F.; Day, D.; Chan, Y.; Buckley, B. R.; Allin, S. M.; Mckenzie, M. J., *Org. Lett.* **2009**, *11*, 1991.

179. Bartlett, C. J.; Day, D. P.; Chan, Y.; Allin, S. M.; Mckenzie, M. J.; Slawin, A. M. Z.; Page, P. C. B., *J. Org. Chem.* **2012**, *77*, 772.

180. Novikov, R.; Bernardinelli, G.; Lacour, J., *Adv. Synth. Catal.* **2009**, *351*, 596.

181. Vachon, J.; Pérollier, C.; Monchaud, D.; Marsol, C.; Ditrich, K.; Lacour, J., *J. Org. Chem.* **2005**, *70*, 5903.

182. Novikov, R.; Lacour, J., *Tetrahedron: Asymmetry* **2010**, *21*, 1611.

183. Gonçlaves, M.-H.; Martinez, A.; Grass, S.; Page, P. C. B.; Lacour, J., *Tetrahedron Lett.* **2006**, *47*, 5297.

184. Page, P. C. B.; Farah, M. M.; Buckley, B. R.; Blacker, A. J.; Lacour, J., *Synlett* **2008**, 1381.

185. Vachon, J.; Lauper, C.; Ditrich, K.; Lacour, J., *Tetrahedron: Asymmetry* **2006**, *17*, 2334.

186. Chan, W.-K.; Wong, M.-K.; Che, C.-M., *J. Org. Chem.* **2005**, *70*, 4226.

187. Mirkhani, V.; Moghadam, M.; Tangestaninejad, S.; Bahramian, B., *App. Catal. A* **2006**, *311*, 43.

188. Desai, L. V.; Malik, H. A.; Sanford, M. S., *Org. Lett.* **2006**, *8*, 1141.

189. Yakura, T.; Omoto, M.; Yamauchi, Y.; Tian, Y.; Ozono, A., *Tetrahedron* **2010**, *66*, 5833.

190. Das, S.; Incarvito, C. D.; Crabtree, R. H.; Brudvig, G. W., *Science* **2006**, *312*, 194.

191. Hull, J. F.; Balcells, D.; Sauer, E. L. O.; Raynaud, C.; Brudvig, G. W.; Crabtree, R. H.; Eisenstein, O., *J. Am. Chem. Soc.* **2010**, *132*, 7605.

192. Ekkati, A. R.; Kodanko, J. J., *J. Am. Chem. Soc.* **2007**, *129*, 12390.

193. Zolfigol, M. A.; Bagherzadeh, B.; Mallakpour, S.; Chehardoli, G.; Ghorbani-Choghamarani, A.; Koukabi, N.; Dehghanian, M.; Doroudgar, M., *J. Mol. Catal. A* **2007**, *270*, 219.

194. Priewisch, B.; Rück-Braun, K., *J. Org. Chem.* **2005**, *70*, 2350.

195. Khiar, N.; Mallouk, S.; Valdivia, V.; Bougrin, K.; Soufiaoui, M.; Fernández, I., *Org. Lett.* **2007**, *9*, 1255.

196. Colonna, S.; Pironti, V.; Drabowicz, B. F.; Fensterbank, L.; Malacria, M., *Eur. J. Org. Chem.* **2005**, 1727.

197. Reddy, C. V.; Verkade, J. G., *J. Mol. Catal. A* **2007**, *272*, 233.

Potassium *tert*-Butoxide

$$\boxed{t\text{-BuOK}}$$

[865-47-4]   C$_4$H$_9$KO   (MW 112.23)

InChI = 1/C4H9O.K/c1-4(2,3)5;/h1-3H3;/q-1;+1

InChIKey = LPNYRYFBWFDTMA-UHFFFAOYAU

(strong alkoxide base capable of deprotonating many carbon and other Brφnsted acids; relatively poor nucleophile[1])

Physical Data: mp 256–258 °C (dec).

Solubility: sol/100g solvent at 25–26 °C: hexane 0.27 g, toluene 2.27 g, ether 4.34 g, *t*-BuOH 17.80 g, THF 25.00 g.

Form Supplied in: white, hygroscopic powder; widely available commercially; also available as a 1.0 M solution in THF.

Preparative Methods: sublimation (220 °C/1 mmHg; 180 °C/0.05 mmHg) of the commercial material prior to use is recommended. For critical experiments, the reagent should be freshly prepared prior to use. Solutions of the reagent in *t*-BuOH may be prepared by reaction of the anhydrous alcohol with **Potassium** under nitrogen,[2] or the solvent may be removed (finally at 150 °C/0.1–1.0 mmHg for 2 h) and solutions prepared using other solvents.[3]

Handling, Storage, and Precaution: do not breathe dust; avoid contact with eyes, skin, and clothing. Reacts with water, oxygen, and carbon dioxide, and may ignite on exposure to air at elevated temperatures. Handle in an inert atmosphere box or bag and conduct reactions under an inert atmosphere in a fume hood. Store in small lots in sealed containers under nitrogen.

Original Commentary

Drury Caine

The University of Alabama, Tuscaloosa, AL, USA

Introduction. Potassium *t*-butoxide is intermediate in power among the bases which are commonly employed in modern organic synthesis. It is a stronger base than the alkali metal hydroxides and primary and secondary alkali metal alkoxides,[1] but it is a weaker base than the alkali metal amides and their alkyl derivatives, e.g. the versatile strong base **Lithium Diisopropylamide**.[4] The continued popularity of *t*-BuOK results from its commercial availability and the fact that its base strength is highly dependent on the choice of reaction solvent. It is strongly basic in DMSO, where it exists primarily as ligand-separated ion pairs and dissociated ions, but its strength is significantly decreased in solvents such as benzene, THF, and DME, where its state of aggregation is largely tetrameric. DMSO is able to enhance the basicity of the *t*-butoxide anion by selectively complexing with the potassium cation. Other additives, such as the dipolar aprotic solvent **Hexamethylphosphoric Triamide** (HMPA) and **18-Crown-6**, have a similar effect. This section will provide cursory coverage of the reactions of *t*-BuOK in *t*-BuOH and in relatively nonpolar aprotic solvents. The unique features of **Potassium tert-Butoxide–Dimethyl Sulfoxide**, **Potassium tert-Butoxide–Hexamethylphosphoric Triamide**, and **Potassium tert-Butoxide–18-Crown-6**, as well as those of the **Potassium tert-**

Essential Reactions for Organic Synthesis, First Edition. Edited by Philip L. Fuchs.
© 2016 John Wiley & Sons, Ltd. Published 2016 by John Wiley & Sons, Ltd.

Butoxide– *tert*-Butyl Alcohol Complex (1:1), are described elsewhere in this encyclopedia.

Alkylations. Many bases which are weaker than *t*-BuOK are capable of essentially quantitative conversion of active methylene compounds into the corresponding enolates or other anions.[5] However, the alkylation of diethyl malonate with a bicyclic secondary tosylate (eq 1)[6] and the alkylation of ethyl *n*-butylacetoacetate with *n*-BuI (eq 2)[7] provide examples of cases where the use of *t*-BuOK in *t*-BuOH is very effective. In the latter reaction, cleavage of the product via a retro-Claisen reaction is minimized with the sterically hindered base and yields obtained are higher than when **Sodium Ethoxide** or EtOK in EtOH, **Sodium** in dioxane or toluene, or **Sodium Hydride** in toluene are used for the enolate formation.

$$(1)$$

$$(2)$$

Potassium *t*-butoxide in *t*-BuOH or ethereal solvents is not capable of effecting quantitative formation of enolates of unactivated saturated ketones;[8] also, because potassium enolates are subject to rapid proton transfer reactions, their intermolecular alkylations are complicated by equilibration of structurally isomeric enolates, polyalkylation, and aldol condensation reactions.[9] Thus reactions of preformed lithium enolates (generated by deprotonation of ketones with LDA or by indirect procedures) with alkylating agents provide the method of choice for regioselective alkylation of unsymmetrical ketones.[9,10] However, as illustrated in eqs 3 and 4,[11,12] ketones capable of forming only a single enolate, such as symmetrical cyclic ketones and those containing α-methylene blocking groups, are readily alkylated via *t*-BuOK-promoted reactions. Also, *t*-BuOK in *t*-BuOH or benzene frequently has been employed for the alkylation of α,α-disubstituted aldehydes.[9]

$$(3)$$

$$(4)$$

As illustrated in eq 5, α,β-unsaturated ketones undergo α,α-dialkylation when treated with excess *t*-BuOK/*t*-BuOH and an alkylating agent.[13] The α,α-dimethylated β,γ-unsaturated ketone is formed by conversion of the initially produced α-methylated β,γ-unsaturated ketone to its dienolate, which undergoes a second methylation faster than the β,γ-double bond is isomerized to the α,β-position.[14]

In contrast to intermolecular processes, intramolecular alkylations are frequently performed with *t*-BuOK in various solvents.[19] In eqs 6 and 7 are shown examples of *endo*- and *exo*-cycloalkylations of cyclic saturated ketones which lead to new five-membered rings.[15,16]

An interesting example of how a change in the base can influence the course of a cycloalkylation reaction is shown in eq 8.[17] Since the reaction with *t*-BuOK involves equilibrating conditions, *exo*-cycloalkylation occurs via the more-substituted enolate, which is more thermodynamically stable. On the other hand, when LDA is used as the base, *endo*-cycloalkylation occurs via the kinetically formed terminal enolate.

Intramolecular alkylations of α,β-unsaturated ketones may occur at the α-, α'-, or γ-positions depending upon the nature of the base, the leaving group, and other structural features. A recent example involving α-cycloalkylation using *t*-BuOK is shown in eq 9.[18]

The reaction of α-halo esters, ketones, nitriles, and related compounds with appropriate organoboranes in the presence of *t*-BuOK can lead to replacement of the halogen with an alkyl or an aryl group.[19] An example of this reaction using an α-bromo ester

is shown in eq 10.[20] THF is a more effective solvent than *t*-BuOH for alkylations of α-bromo ketones using this methodology.[21] Potassium 2,6-di-*t*-butylphenoxide, a mild, sterically hindered base, is much more effective than *t*-BuOK for the alkylation of highly reactive α-halo ketones, e.g. bromoacetone, and α-halo nitriles.[22]

Condensation Reactions. Traditionally, intermolecular aldol condensation reactions have been performed under equilibrating conditions using weaker bases than *t*-BuOK in protic solvents.[23] Since the mid-1970s, new methodology has focused on directed aldol condensations which involve the use of preformed *Lithium* and Group 2 enolates,[24] Group 13 enolates,[25] and transition metal enolates.[26] Although examples of the use of *t*-BuOK in intramolecular aldol condensations are limited, complex diketones (eq 11)[27] and keto aldehydes (eq 12)[28] have been cyclized with this base.

t-BuOK is the most commonly used base for the Darzens condensation of an α-halo ester with a ketone or an aromatic aldehyde to yield an α,β-epoxy or glycidic ester.[29] In the example shown in eq 13,[30] the aldol step is reversible and the epoxide ring closure step is rate limiting. This leads to the product with the ester group and the bulky β-substituent *trans*. However, in other systems the opposite stereochemical result often occurs because the aldol condensation step is rate limiting. In addition to esters, a wide variety of α-halo compounds that contain electron-withdrawing groups may participate in these types of reaction.[29c]

Ph—CHO + Ph—C(Cl)(CO₂Et) $\xrightarrow[75\%]{\substack{t\text{-BuOK} \\ t\text{-BuOH}}}$ epoxide (Ph, Ph, CO₂Et) (13)

The Dieckmann cyclization of diesters and related reactions has found an enormous amount of use in the synthesis of five-, six-, seven-, and even larger-membered rings.[31] In unsymmetrical systems, steric effects and the stability of the product enolates determine the regiochemistry of the reaction. As shown in eqs 14 and 15,[32,33] *t*-BuOK is an effective base for these reactions when used in *t*-BuOH or other solvents such as benzene. In the former example, 40% of unchanged starting material is recovered when MeONa/PhH is used to effect the cyclization.

(eq 14) steroid diester $\xrightarrow[\text{PhH}]{\substack{t\text{-BuOK} \\ t\text{-BuOH}}}$ products 84% + 3% (14)

(eq 15) $\xrightarrow[90\%]{\substack{t\text{-BuOK, PhH} \\ \text{reflux}}}$ products 8:1 (15)

The nature of the base can profoundly influence the regiochemistry of the reaction. *t*-BuOK favors kinetic control in the reaction shown in eq 16 and the product derived from cyclization of the enolate having a β-amino group is obtained. However, when EtONa/EtOH is employed, the more stable β-keto ester enolate resulting from thermodynamic control is obtained.[34] In addition to diesters, dinitriles, ε-keto esters, ε-cyano esters, ε-sulfinyl esters, and ε-phosphonium esters may participate in these reactions.[31]

t-BuOK is a better base than EtONa for the Stobbe condensation (eq 17) because yields are higher, reaction times are shorter, and frequent side reactions of ketone or aldehyde reduction are avoided.[35]

Although the use of *t*-BuOK for the generation of unstabilized ylides from phosphonium salts has been rare, it is the base of choice for the generation of **Methylenetriphenylphosphorane** for the Wittig reaction of hindered ketones (eq 18).[36] No 1,1-di-*t*-butylethylene is obtained from di-*t*-butyl ketone when NaH/DMSO is used to generate the ylide.[37]

(eq 16) $\xrightarrow[70\%]{\substack{t\text{-BuOK, PhH} \\ \text{heat}}}$ and $\xrightarrow[80\%]{\substack{\text{EtONa, EtOH} \\ \text{heat}}}$ (16)

(eq 17)

Base, solvent	Yield (%)
EtONa (2 equiv), EtOH, reflux, 6 h	79
t-BuOK (1.1 equiv), *t*-BuOH, reflux, 40 min	89–94

+ Ph₃PMe Br⁻ $\xrightarrow[\text{reflux}]{t\text{-BuOK, PhH}}$ $\xrightarrow[\substack{120–130 \,^\circ\text{C} \\ 48\,\text{h}}]{t\text{-BuCO-}t\text{-Bu}}$ *t*-Bu₂C=CH₂ 96% (18)

Elimination Reactions. *t*-BuOK is a widely used base for both α- and β-elimination reactions. It is the most effective base in the conventional alkoxide–haloform reaction for the generation of dihalocarbenes.[38] This procedure still finds general use (eq 19),[39] but since it requires anhydrous conditions, it has been replaced to a degree by use of phase-transfer catalysts.[40] Vinylidene carbenes have also been produced from the reaction of α-halo allenes with *t*-BuOK.[41]

(eq 19) $\xrightarrow[56.5–63.5\%]{\substack{\text{CHBr}_3, \ t\text{-BuOK} \\ 0 \text{ to } 25\,^\circ\text{C}}}$ (19)

Substrates containing a host of leaving groups, such as alkyl chlorides (eqs 20 and 21),[42,43] bromides (eqs 22–24),[44–46] tosylates, mesylates (eq 25),[47] and even sulfenates and sulfinates (eq 26),[48] undergo β-eliminations with *t*-BuOK in the solid phase or in *t*-BuOH or various nonpolar solvents.

(eq 20) $\xrightarrow[80–88\%]{\substack{t\text{-BuOK, THF} \\ -78 \text{ to } 25\,^\circ\text{C}}}$ (20)

(eq 21) $\xrightarrow[80\%]{t\text{-BuOK, THF}}$ (21)

$$ (22) $$

anti-elimination $$ (23) $$

$$ (24) $$

X = O, 32%
X = S, 65% $$ (25) $$

$$ (26) $$

The regiochemical and stereochemical results of β-eliminations of 2-substituted acyclic alkanes with *t*-BuOK in various solvents have been extensively investigated.[49] Space does not permit the details of these investigations to be presented, but a few brief generalizations may be noted. (1) *t*-BuOK normally gives more of the terminal alkene than primary alkoxide bases but less than extremely bulky bases such as potassium tricyclohexylmethoxide. (2) A greater proportion of the terminal alkene is produced in solvents such as *t*-BuOH, where the base is highly aggregated, than in DMSO where it is substantially monomeric and partially dissociated. (3) The lower the state of aggregation of the base, the higher the *trans:cis* ratio of the disubstituted 2-alkene. (4) As base aggregation increases, the *syn:anti* elimination ratio increases. As far as the *trans:cis* ratio is concerned, cyclic halides give the opposite results from open-chain systems, i.e. this ratio is higher when base aggregation is greater (eq 27).[50]

$$ (27) $$

Solvent	trans (%)	cis (%)
PhH	83	1.5
PhH, dicyclo-hexyl-18-crown-6	9	76
DMF	7	72

Fragmentation Reactions. Because it is a relatively strong base and a relatively weak nucleophile, *t*-BuOK has been the most popular base for effecting Grob-type fragmentations of cyclic 1,3-diol derivatives.[51] In *t*-BuOH or nonpolar solvents, sufficiently high concentrations of alkoxide ions of 1, 3-diol derivatives are produced for fragmentations to proceed smoothly if relatively good leaving groups are present (eq 28).[52] When relatively poor leaving groups, e.g. sulfinates, are involved, it is necessary to increase the strength of *t*-BuOK by the addition of a dipolar aprotic solvent (eq 29).[53] Other base–solvent combinations which may offer advantages over *t*-BuOK/*t*-BuOH for certain fragmentation substrates include NaH/THF, DMSO⁻Na⁺/DMSO, and LAH/ether.[51]

$$ (28) $$

$$ (29) $$

Isomerizations of Unsaturated Compounds. *t*-BuOK is an effective base for bringing about migrations of double bonds in alkenes and alkynes via carbanion intermediates,[1] but since the base promotes these reactions most effectively in DMSO, they will be described in more detail under **Potassium *tert*-Butoxide–Dimethyl Sulfoxide**. Important examples of enone deconjugations with *t*-BuOK/*t*-BuOH which proceed via di- and trienolate intermediates are shown in eqs 30 and 31.[54,55] Potassium *t*-pentoxide is effective in promoting the latter reaction, but various lithium amide bases are not, apparently because they deprotonate the enone at the α'-position regioselectively. The isomerization of α,β-unsaturated imines to alkenyl imines (eq 32) is an important step in an alternative method for reduction–alkylation of α,β-unsaturated ketones.[56]

$$ (30) $$

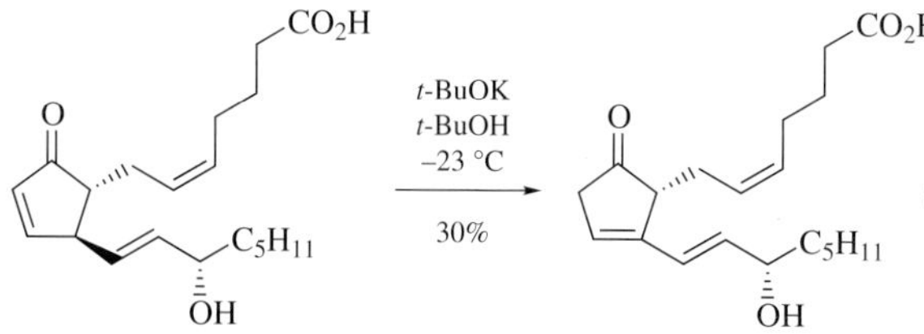

(31)

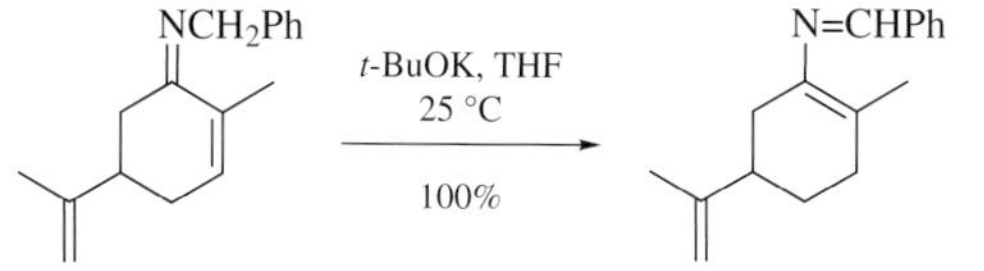

(32)

(37)

(38)

Ketone Cleavage Reactions. *t*-BuOK in ether containing water is the medium of choice for the cleavage of nonenolizable ketones such as nortricyclanone (eq 33).[57] Optically active tertiary α-phenyl phenyl ketones are cleaved in fair yields with high retention of configuration with the base in *t*-BuOH or PhH (eq 34).[58] *t*-BuOK is more effective than *t*-BuONa or *Lithium tert-Butoxide* in these reactions.

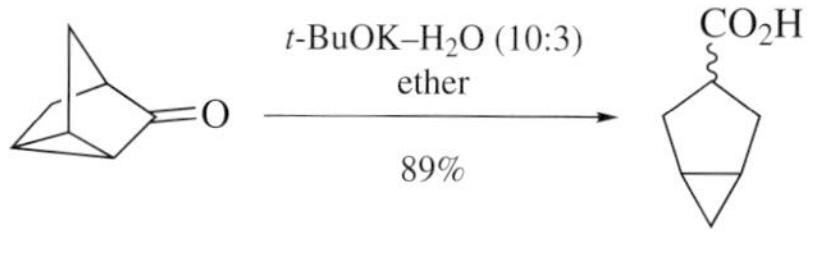

(33)

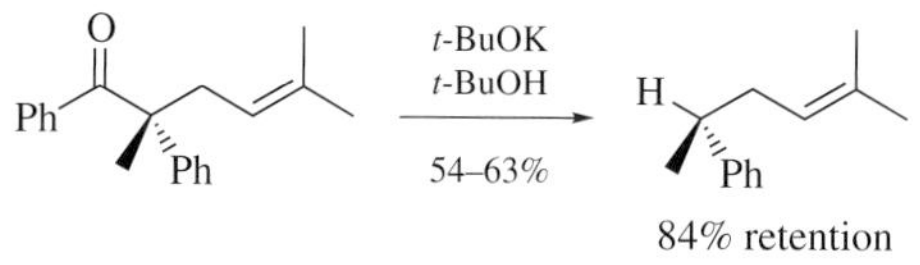

(34)

84% retention

Michael Additions. *t*-BuOK is one of an arsenal of bases that can be used in Michael addition reactions.[59] A catalyst prepared by impregnation of xonotlite with the base promotes the Michael addition of the β-diketone dimedone to 2 mol of MVK (eq 35).[60] Unsymmetrical ketones like 2-methylcyclohexanone yield Michael adducts primarily at the more substituted α-carbon atom (eq 36).[8a] Active methylene compounds undergo double Michael additions to enynes in the presence of *t*-BuOK (eq 37).[61] This type of reaction has been used in the total synthesis of griseofulvin. A tricyclo[5.3.1.0^{1,5}]undecanedione, a precursor of (±)-cedrene, is available by a highly regioselective intramolecular Michael addition using *t*-BuOK (eq 38).[62]

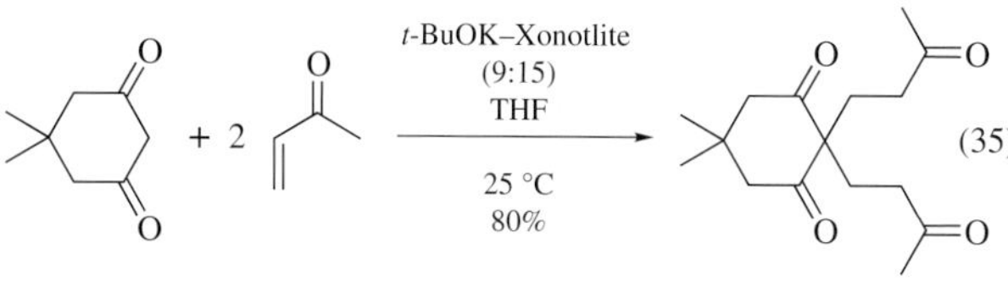

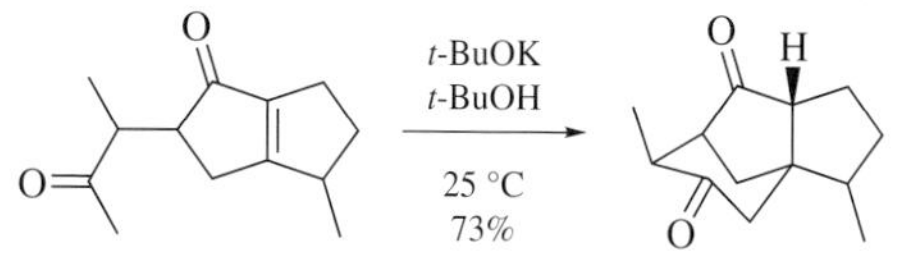

(35)

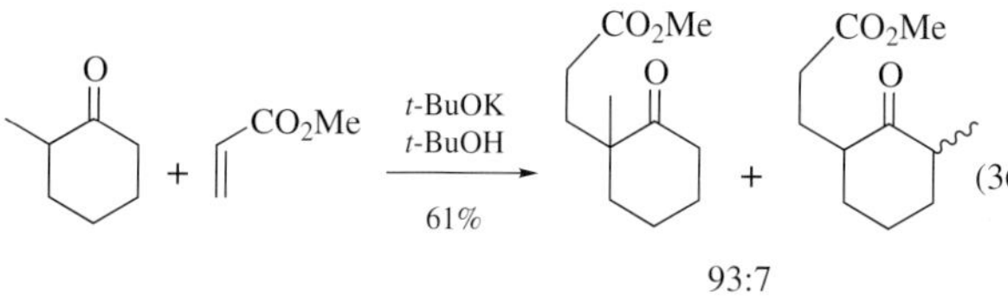

(36)

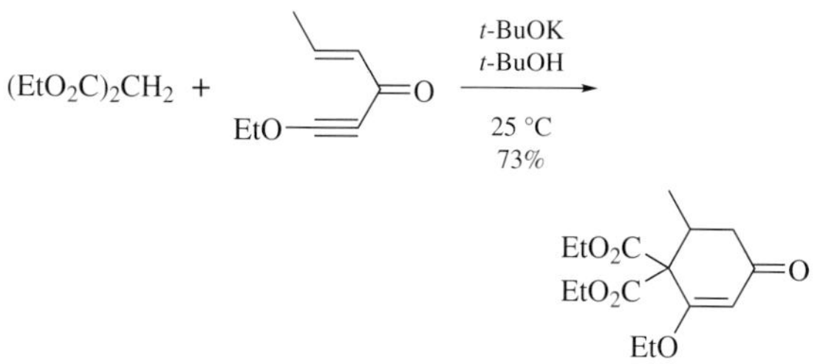

93:7

Oxidation Reactions. The use of *t*-BuOK to convert organic substrates to carbanionic species which react with molecular oxygen via a radical process has been reviewed.[1] Ketones and esters are the most common substrates for these reactions. Oxidations of unsymmetrical ketones occur via the more thermodynamically stable potassium enolates; tetrasubstituted enolates yield stable hydroperoxides, while hydroperoxides derived from trisubstituted enolates are further oxidized to α-diketones or their enol forms.[9a]

Rearrangement Reactions. Benzil is converted into the ester of benzylic acid in high yield upon treatment with *t*-BuOK in *t*-BuOH/PhH (eq 39).[63] Lower yields are obtained if the individual solvents alone are used or if the base is replaced by MeONa or EtONa.

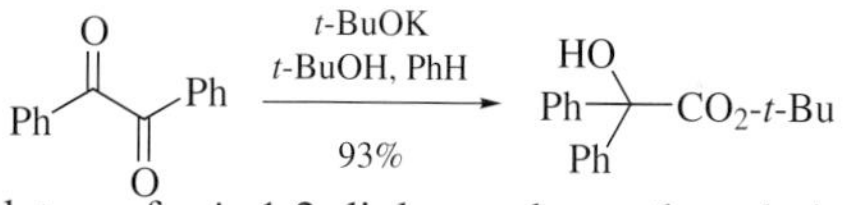

(39)

Monotosylates of *cis*-1,2-diols, such as that derived from α-pinene, undergo pinacol-type rearrangements upon treatment with *t*-BuOK (eq 40).[64] The *trans* isomer is converted to the epoxide under the same conditions (eq 41).[64] 4-Benzoyloxycyclohexanone undergoes a mechanistically interesting rearrangement to benzoylcyclopropanepropionic acid when treated with *t*-BuOK/*t*-BuOH (eq 42).[65]

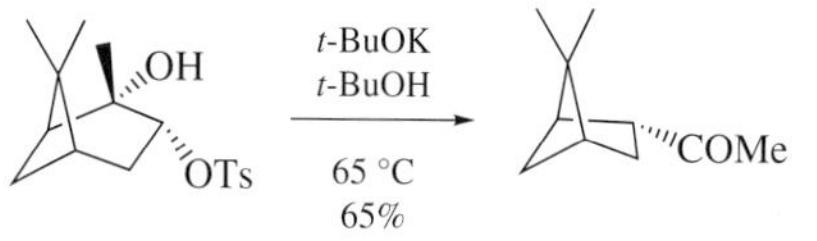

(40)

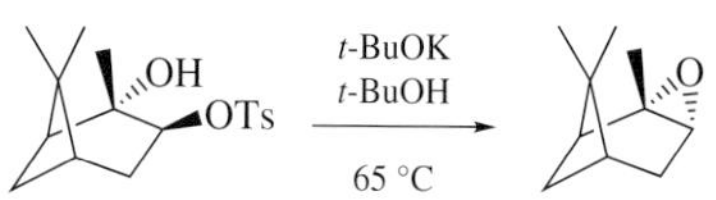

(41)

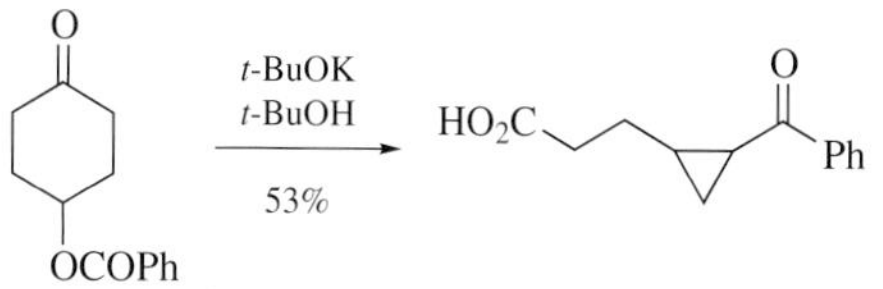

(42)

The Ramberg–Bäcklund rearrangement of α-halo sulfones is frequently carried out with *t*-BuOK.[66] This reaction provides a useful route to deuterium-labeled alkenes (eq 43).[67] *t*-BuOK is presumably the active base in a modification which involves the direct conversion of a sulfone into an alkene with KOH and a mixture of *t*-BuOH/CCl₄.[68] The base converts the sulfone to the α-sulfonyl carbanion, which undergoes chlorination with CCl_4 by a single-electron transfer process. Proton abstraction from the α'-position of the α-chloro ketone by the base yields a thiirane 1,1-dioxide which loses SO_2 to yield the alkene. *t*-BuOK is frequently used directly in this modified procedure (eq 44).[69] Cyclic enediynes are available from the corresponding α-chloro sulfones using *t*-BuOK to effect the Ramberg–Bäcklund rearrangement.[70]

Bromomethylenecyclobutanes undergo ring enlargement reactions to 1-bromocyclopentenes when heated in the presence of solid *t*-BuOK (eq 45).[71]

Reduction Reactions. A modification of the Wolff–Kishner reduction, which is particularly useful for the reduction of α, β-unsaturated ketones, involves the reaction of carbonyl hydrazones and semicarbazones with *t*-BuOK in PhMe at reflux.[72] The reduction of (±)-3-oxo-α-cadinol to (±)-α-cadinol and its isomer (eq 46) provides an example of this method.[73]

First Update

Drury Caine
The University of Alabama, Tuscaloosa, AL, USA

Introduction. *t*-BuOK is a widely used base whose basicity is strongly dependent upon the nature of the medium. It has lower basicity in nonpolar, protic, or aprotic solvents, e.g., *t*-BuOH, THF,

and benzene, where it exists primarily in the form of aggregated ion pairs, but its basicity is considerably enhanced in the presence of dipolar, aprotic solvents such as DMSO, HMPA, and DMF, where solvation of the potassium cation produces ligand-separated and free anions. Because it strongly complexes the potassium cation, addition of 18-crown-6 to solutions of the base in nonpolar aprotic solvents have enhanced basicity; but, more importantly, significantly enhanced nucleophilicity.

Addition Reactions. *t*-BuOK is capable of deprotonating a variety of weakly acidic carbon and heteroatomic acids to produce anions, either quantitatively or in low equilibrium concentrations, which add to a variety of unsaturated compounds. Catalytic, equimolar, or excess quantities of the reagent are employed, depending upon the process involved.

The *t*-BuOK-catalyzed reaction of a terminal alkyne with cyclohexanone in DMSO to give a tertiary alcohol in 91% yield (eq 47) provides a straightforward illustration of an addition to a carbonyl compound.[74] The same type of addition takes place in the nonpolar solvent benzene but the rate is slower and the yield lower. Treatment of cyclohexanone with ethynylbenzene under the same reaction conditions yields 1-(phenylethynyl)cyclohexanol in 83% yield; when the reaction is carried out using 1.0 equiv of *t*-BuOK in the absence of solvent the yield of the tertiary alcohol is 93%.[75] Other aliphatic and aromatic ketones give similar results. Ketones with relatively acidic α hydrogens are capable of undergoing intermolecular aldol additions in the presence of the base; but, apparently, the reversibility of this reaction allows the irreversible addition of the acetylide anion to compete favorably.[74]

Enolate anions of active methylene compounds (such as ethyl acetoacetate produced with a catalytic amount of *t*-BuOK) undergo Michael additions to conjugated enones and enals to produce adducts, which then undergo sequential intramolecular aldol additions, bicyclic lactone formation, and base-catalyzed decarboxylation in refluxing *t*-BuOH.[76] Equation 48 illustrates the preparation of 5-methylcyclohexenone by this method. Michael-Claisen [3+3] reactions involving symmetrical or unsymmetrical ketones and conjugated enoates provide a useful route to 1,3-cyclohexanediones.[77] For example, 4-*n*-pentylcyclohexan-1,3-dione is formed in 88% yield by treating 2-octanone with ethyl acrylate in the presence of 1.2 equiv of *t*-BuOK in THF at room temperature (eq 49). The structures of the products formed from a variety of ketone-enoate combinations show that the addition step involves nucleophilic attack by the more hindered enolate of an unsymmetrical ketone, followed by cyclization via the less hindered enolate of the keto moiety and the acyl carbon of the enoate.

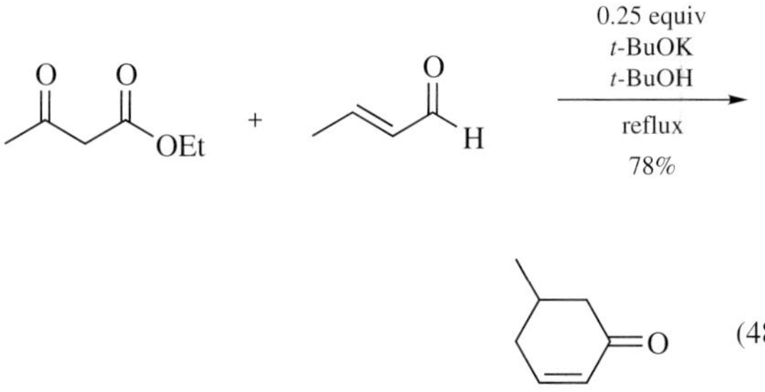

(48)

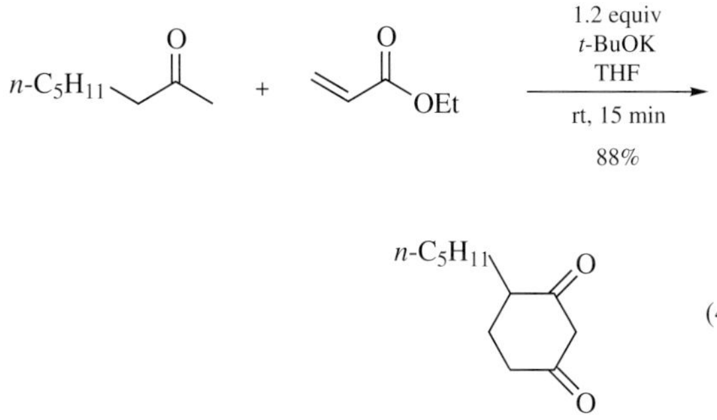

(49)

A one-pot synthesis of (*E*)-4-cyano-5-arylpent-4-enenitriles is accomplished by treating diethyl(cyanomethyl)phosphonate with *t*-BuOK in THF containing 2.0 equiv of HMPA with acrylonitrile, followed by the addition of an aromatic aldehyde (eq 50).[78] The sequence involves anion formation, Michael addition, and proton transfer followed by a Horner-Wadsworth-Emmons reaction. Nitriles add to vinylic silanes, phosphines, and thio derivatives in the presence of a catalytic amount of *t*-BuOK in DMSO (eq 51).[79] The medium is also effective for promoting addition of ketones to vinylphosphines. Similarly, Michael adducts are also obtained in good to excellent yields by reacting ketones or imines with styrenes in the presence of 20–30 mol % *t*-BuOK in DMSO with warming to 40 °C. DMSO and *N*-methylpyrrolidinone (NMP) are useful solvents when nitrile anions are involved as nucleophiles.[80]

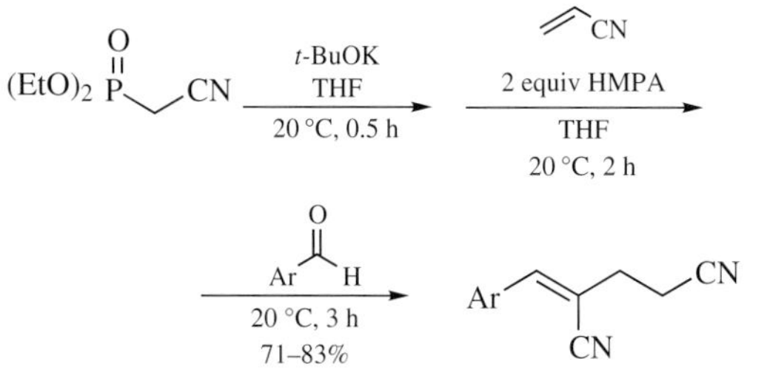

(50)

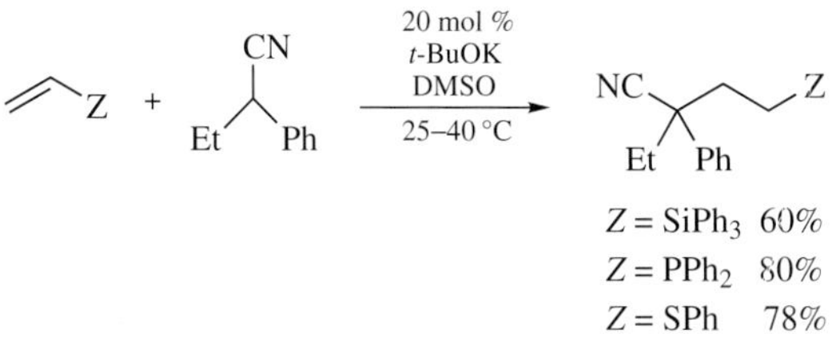

(51)

Z = SiPh$_3$ 60%
Z = PPh$_2$ 80%
Z = SPh   78%

Chloromethyl methyl sulfone undergoes deprotonation and addition to carbonyl compounds faster than degradation via the Ramberg-Baecklund reaction when treated with *t*-BuOK in THF

at low temperature.[81] Chlorohydrins are isolated if the reaction mixture is quenched with an acid at low temperature, or oxiranes are formed if the temperature is allowed to rise to 0 °C prior to quenching (eq 52). When the carbonyl component is an aldehyde the chlorohydrin or oxirane product undergoes the Ramberg-Baecklund reaction to give a vinyl alcohol upon treatment with an excess of the base.

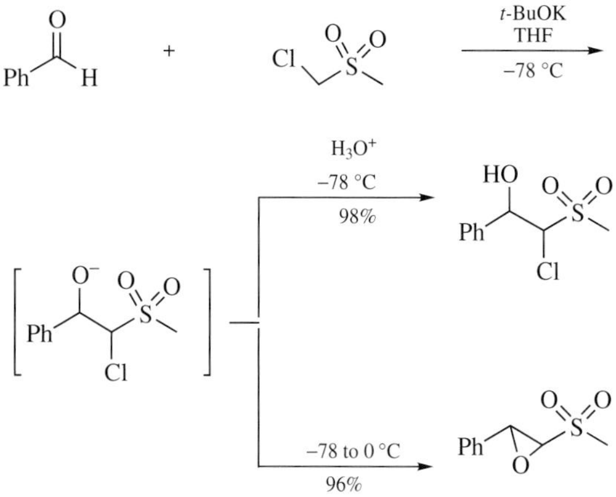

(52)

Dimethyl malonate derivatives containing an allenyl sulfone substituent at the γ- or δ-position undergo *endo*-mode ring closure to give cyclopentene or cyclohexene derivatives upon treatment with 1.5 equiv of *t*-BuOK in *t*-BuOH at rt.[82] The intermediate unsaturated cycloadducts undergo demethoxycarbonylation and double bond isomerization under the reaction conditions (eq 53). Other active methine compounds behave similarly. A threefold excess of *t*-BuOK is more effective than certain amine bases for the conversion of phenylsulfonyl methyl derivatives of aromatic or heteroaromatic compounds into the corresponding dithio esters upon reaction with an excess of elemental sulfur.[83]

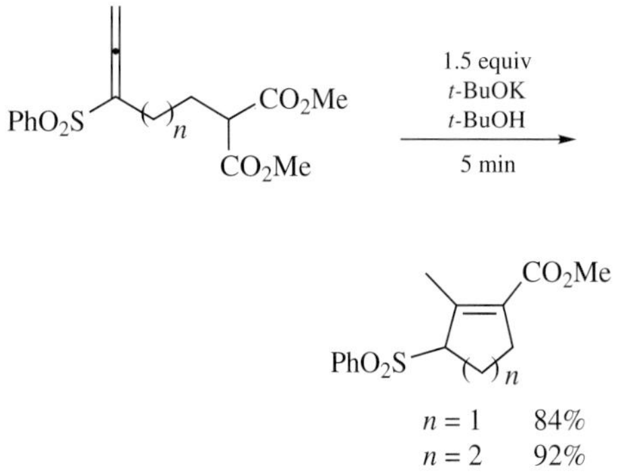

(53)

n = 1 84%
n = 2 92%

Nucleophilic attack of *t*-BuO anion on carbon, sulfur, or silicon compounds such as TMSCF$_3$,[84] secondary trifluoroamides and trifluoroacetates,[85] secondary trifluoromethanesulfinamides and trifluoromethanesulfinates,[86] and trifluoromethyl sulfoxides and sulfones[87] allows the transfer of a trifluoromethyl carbanion to various electrophilic acceptors, such as nonenolizable ketones and aldehydes (eq 54).[87]

$$Ph\overset{O}{\underset{O}{S}}CF_3 \;+\; \underset{3\ equiv}{Ph\overset{O}{\underset{H}{\parallel}}} \xrightarrow[\substack{-50\ ^\circ C\ to\ rt \\ 91\%}]{\substack{2\ equiv \\ t\text{-BuOK} \\ DMF}} Ph\overset{OH}{\underset{CF_3}{\vert}} \qquad (54)$$

In addition to various carbon acids, *t*-BuOK is capable of converting many compounds containing O–H, S–H, and N–H bonds into the corresponding conjugate bases, which may participate in addition reactions with appropriate acceptors. For example, the base promotes th sequential addition of BnSH to an α-bromo-α,β-unsaturated sulfone followed by a Ramberg-Baeklund rearrangement of the adduct to give an (E,Z) mixture of allylic sulfides (eq 55).[88] As illustrated in eq 56, tosyl amides undergo addition-elimination to give (Z)-β-trifluoromethyl enamines in good yield when treated with 2-bromo-3,3,3-trifluoropropene in the presence of an excess of *t*-BuOK in DMSO/*t*-BuOH.[89]

$$Ph\diagdown SH \;+\; \overset{O\ O}{\underset{Br}{S}}Ph \xrightarrow[\substack{rt,\ 15\ min \\ 77\%}]{\substack{t\text{-BuOK} \\ t\text{-BuOH, CH}_2Cl_2}}$$

$$Ph\diagdown S\diagdown\diagdown Ph \qquad (55)$$

$$\begin{array}{c} E{:}Z \\ 93{:}7 \end{array}$$

$$\overset{Ts}{\underset{H}{N}} \;+\; \overset{CF_3}{\underset{Br}{=}} \xrightarrow[\substack{80\ ^\circ C,\ 4\ h \\ 82\%}]{\substack{1.4\ equiv\ t\text{-BuOK} \\ t\text{-BuOH, DMSO}}}$$

$$\diagdown\diagdown\overset{Ts}{\underset{N}{}}\diagdown\overset{CF_3}{=} \qquad (56)$$

t-BuOK in THF is more effective than K_2CO_3, *n*-BuLi, or *t*-BuONa for the hydroamination of styrenes with anilines.[90,91] The use of 2- or 3-chlorostyrenes in the reaction provides an interesting route to *N*-arylindolines via an addition-intramolecular substitution (aryne intermediate) sequence (eq 57).[91]

$$\underset{Cl}{\diagdown} \;+\; H_2N{-}Ph \xrightarrow[\substack{135\ ^\circ C \\ (sealed\ tube) \\ 53\%}]{\substack{3\ equiv \\ t\text{-BuOK} \\ toluene}}$$

$$\underset{\underset{Ph}{\vert}}{N} \qquad (57)$$

The potassium salts of (R)- or (S)-4-phenyl-2-oxazolidinone, produced from the conjugated acids using *t*-BuOK in THF in the presence of 18-crown-6, undergo highly stereoselective additions to monosubstituted nitroalkenes.[92] Equation 58 illustrates the addition of the (R)-nucleophile to a typical acceptor.

$$\underset{\underset{H}{\vert}}{\overset{O}{\underset{N}{\diagup}}}Ph \xrightarrow[\substack{THF \\ 0\ ^\circ C,\ 15\ min}]{\substack{t\text{-BuOK} \\ 18\text{-Crown-6}}}$$

$$K^+ \overset{O}{\underset{N}{\diagup}}Ph \xrightarrow[\substack{-78\ ^\circ C,\ 15\ min \\ 87\%}]{\substack{t\text{-Bu}\diagup\diagdown NO_2}} \overset{O}{\underset{\underset{t\text{-Bu}}{\diagup}\diagdown NO_2}{N}}Ph \qquad (58)$$

$$>98\%\ de$$

Oxa- and aza-Michael additions followed by intramolecular cyclizations occur when propargyl alcohols are treated with α,β-unsaturated nitroalkenes in the presence of *t*-BuOK in THF.[93] Although *exo*-3-methylene tetrahydrofurans are usually the major products of these reactions, 3,4-dihydropyrans are also obtained in some cases, depending upon the substituents present within the donor and acceptor. Equation 59 shows the results of the reactions of propargyl alcohol with four nitroalkenes. Similar reactions of methyl propargylamine with these same acceptors give exclusively 3-methylenepyrrolidines.[93]

$$\underset{R_2}{\overset{R_1}{\diagdown}}{=}NO_2 \;+\; HO\diagdown{=} \xrightarrow{\substack{t\text{-BuOK} \\ THF}}$$

$$\underset{R_2}{\overset{R_1}{\underset{\underset{H}{}}{\overset{NO_2}{\diagup}}}}O \;+\; \underset{R_2}{\overset{R_1}{\underset{\underset{H}{}}{\overset{NO_2}{\diagup}}}}O \qquad (59)$$

	Yield (%)	5-*exo*/6-*endo*
$R_1,R_2 = -(CH_2)_4-$	78	1.7/1
$R_1,R_2 = -(CH_2)_3-$	84	100/1
$R_1 = Me, R_2 = C_5H_{11}$	57	20/1
$R_1 = Me, R_2 = Ph$	73	100/1

A variety of heterocyclic compounds are available by intramolecular additions of oxygen or nitrogen anions to alkynyl groups. Cyclizations of 4-pentynones to furans occur in good yields by 5-*exo-trig* addition of their potassium enolates to the alkyne or the isomeric allene, followed by migration of the double bond into the ring when the substrate is treated with *t*-BuOK in DMF at 60 °C.[94] *t*-BuOK-18-crown-6 promoted 5-*endo-trig* cyclizations of 2-hydromethyl-[95] and 2-thiomethyl 1,3-enynes[96] to provide the corresponding furans and thiofurans in good yields (eq 60). *syn*-Phenylselenoalkynyl alcohols cyclize to *trans*-(Z)-2-(phenylselenomethylene)tetrahydrofurans stereoselectively upon treatment with excess *t*-BuOK in *t*-BuOH containing 18-crown-6 (eq 61).[97] *Anti* substrates give the corresponding *cis* isomers and the corresponding thio compounds also undergo a similar reaction.

Five- to eight-membered ring oxacycles are readily prepared by treating allene sulfones[98] and allene phosphonates[99] bearing ω-hydroxyl side chains with *t*-BuOK in *t*-BuOH at slightly elevated temperatures. The yields obtained with the hydroxyphosphonates are shown in eq 62.[99]

t-BuOK-18-crown-6 promoted cyclization and subsequent dehydroethoxylation of δ-phenylseleno-γ-alkynyl amides leads to (Z)-5-(phenylselenomethylene)pyrrol-2-ones stereoselectively (eq 63).[100] A number of 2-(trimethylsilylethynyl)phenylcarbamates cyclize to various substituted indoles upon reaction with *t*-BuOK in *t*-BuOH (eq 64).[101] Other heterocyclic compounds of synthetic interest which are obtained by *t*-BuOK-promoted inter- or intramolecular reactions include 4-aminoquinazolines,[102] 2-acyl-2-alkyl-3-nitro-2*H*-chromenes,[103] iodoaziridines,[104] and imidazoles.[105]

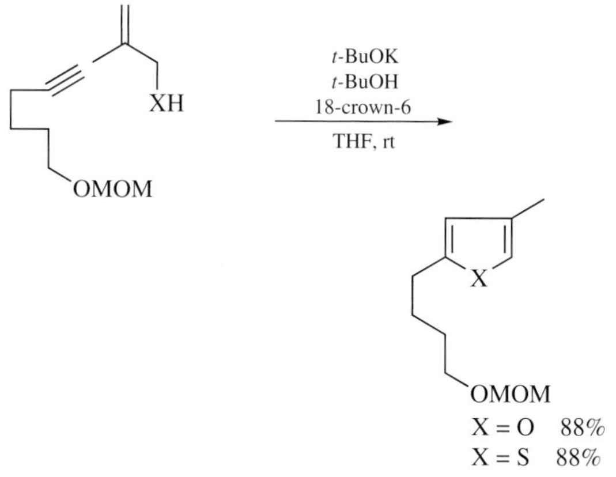

$$X = O \quad 88\%$$
$$X = S \quad 88\%$$

(60)

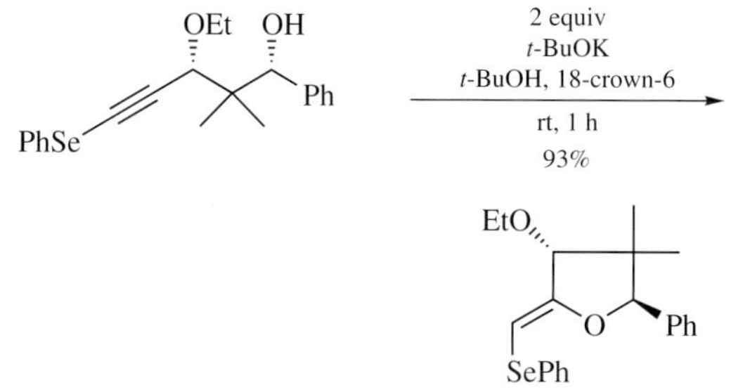

(61)

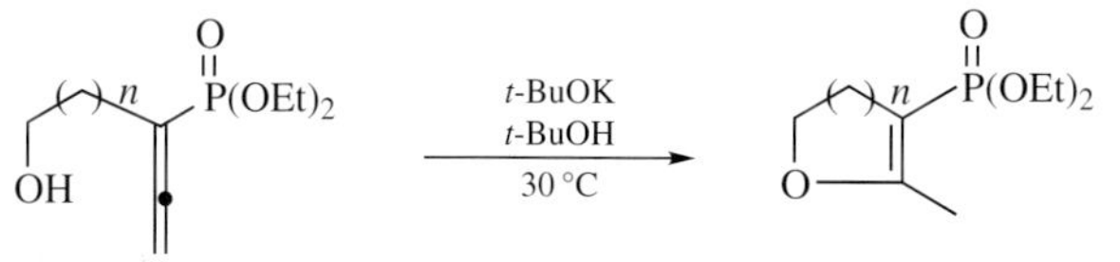

n	Time	Yield (%)
1	30 min	98
2	30 min	71
3	30 min	94
4	6 h	75

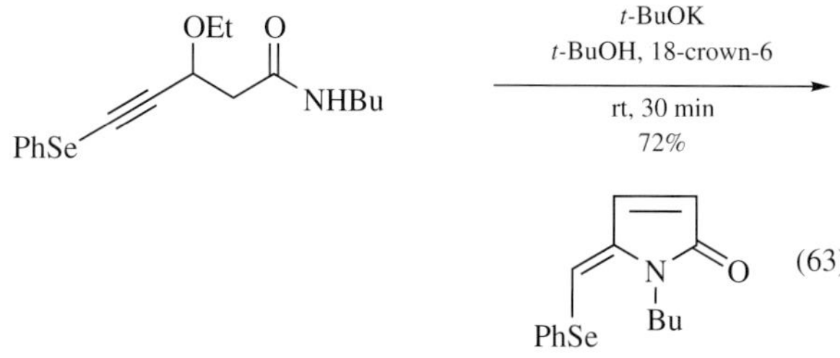

(62)

(63)

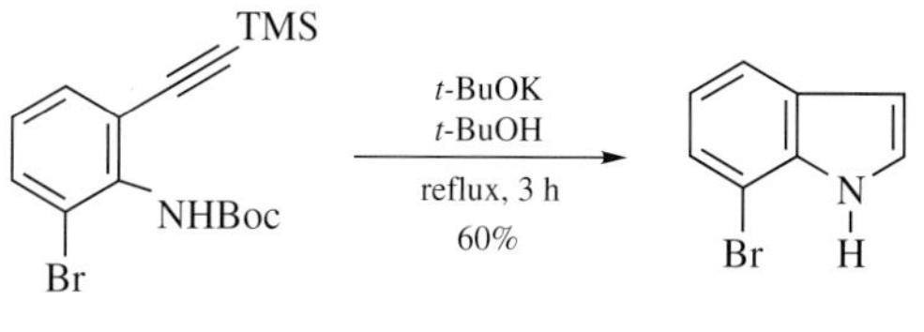

(64)

Alkylation Reactions. 2-(*N*-Methylanilino)-2-phenylsulfanylacetonitrile undergoes alkylation and elimination of benzenethiol upon treatment with *t*-BuOK in THF, followed by the addition of alkyl, allyl, or benzyl halides to give (*E*)-amino-α,β-unsaturated nitriles (eq 65).[106] Potassium enolates of nitriles containing electron withdrawing substituents at the α-position are produced when their conjugate acids are treated with *t*-BuOK in DMSO or HMPA.[107] These species undergo alkylation via an $S_{RN}1$ mechanism when treated with tertiary α-haloketones or nitriles (eq 66). *t*-BuOK in THF promotes the alkylation of BetMIC with methyl iodide or benzyl chloride (eq 67).[105]

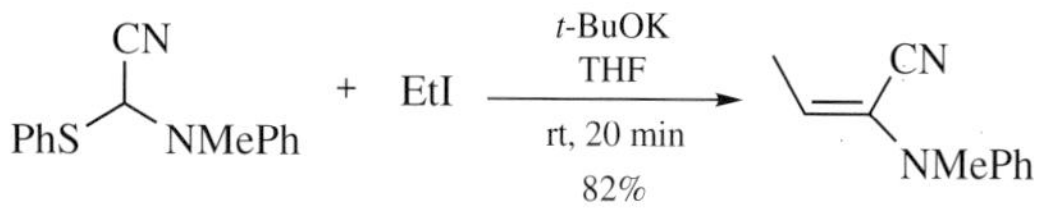

(65)

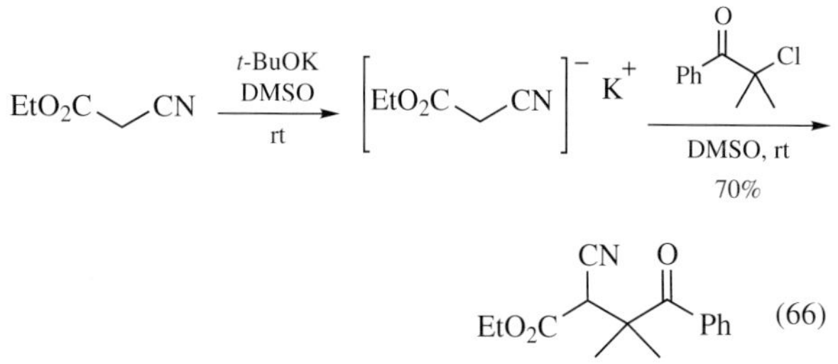

(66)

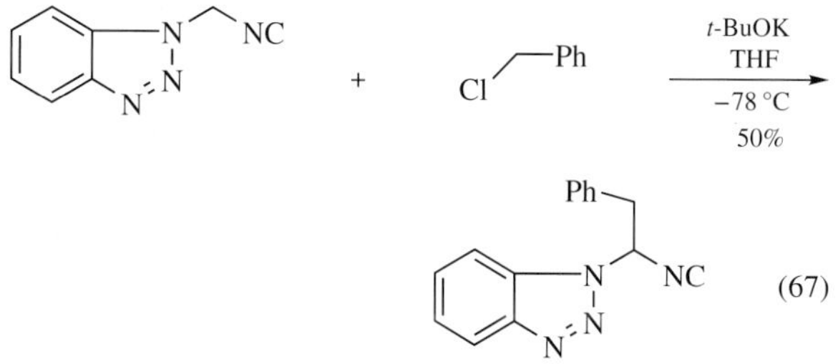

(67)

t-BuOK in DMSO or DMF converts alanine ester imines into their potassium enolates, which undergo alkylation with 2-fluoro-alkyl bromides or iodides to give (after mild hydrolysis) γ-fluoro-α-methyl-α-amino esters.[108] 1,1-Difluoro-1-alkenes are available from difluoromethyl phenyl sulfone by deprotonation with 2 equiv of *t*-BuOK in DMF, alkylation with primary iodides at −50 °C and elimination of the benzenesulfonyl group at −20 °C to rt with the same base in THF.[109] Treatment of β-phenylselenyl silyl enol ethers with *t*-BuOK in THF leads to α-phenylseleno potassium enolates which undergo monoallylation and benzylation in reasonable yields. Problems of transfer of the phenylseleno group and diallylation are encountered when parent ketones are directly deprotonated with various bases prior to alkylation.[110]

An acetal of 1-bromo-3-chloro-3-phenylacetone undergoes deprotonation at the 3-position, intramolecular cyclization, and dehydrochlorination to give a cyclopropene acetal in 70% yield upon treatment with 2 equiv of *t*-BuOK in THF containing 1 equiv of DMI ((1,3-dimethyl) imidazolidinone) (eq 68).[111] In DMSO, *t*-BuOK is basic enough to form 2,2-dimethyl-3-phenylcyclo-propane by deprotonation of 1-iodo-2,2-dimethyl-;3-phenylpro-pane, followed by intramolecular displacement of iodide.[112] An interesting route to optically active cyclopropyl ketones involves cleavage of a silylated cyclic hemiacetal by nucleophilic attack of *t*-BuOK on silicon, ring opening, transfer of a phosphonyl group from carbon to oxygen, and stereocontrolled intramolecular ring closure (eq 69).[113] Anions of *N*-protected ethyl oxamates are produced with *t*-BuOK in DMF,[114] and *O*-protected hydroxamic acids using *t*-BuOK with no solvent under phase transfer conditions[115] are readily alkylated.

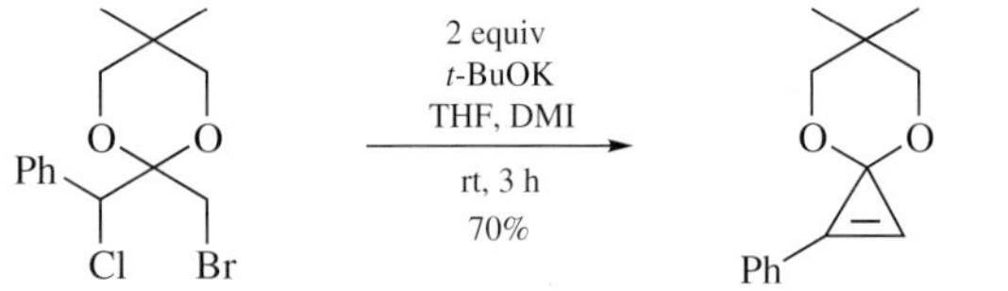

$$(68)$$

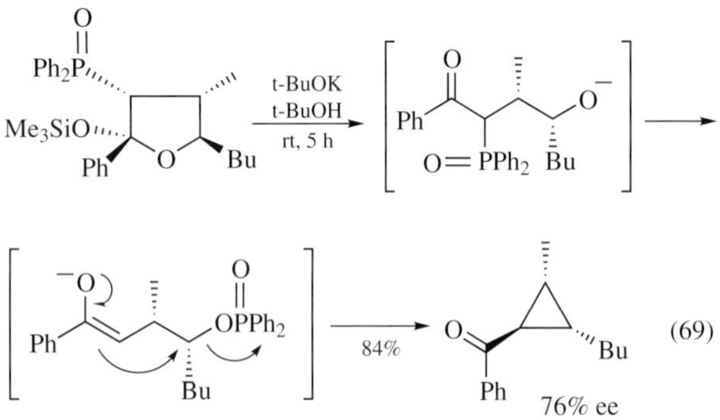

$$(69)$$

γ- and *δ*-Chloroalkaneamides undergo cycloalkylation to the corresponding *γ*- and *δ*-lactams in excellent yields upon treatment with *t*-BuOK in THF at 0 °C (eq 70); the corresponding *β*-homologs exclusively undergo elimination and dimerization or trimerization under the same conditions.[116] *N*-Phenylthiocaprolactam is an excellent reagent for phenylsulfenylation of potassium enolates of acyclic or cyclic ketones produced by deprotonation of the substrates with *t*-BuOK in DMSO.[117]

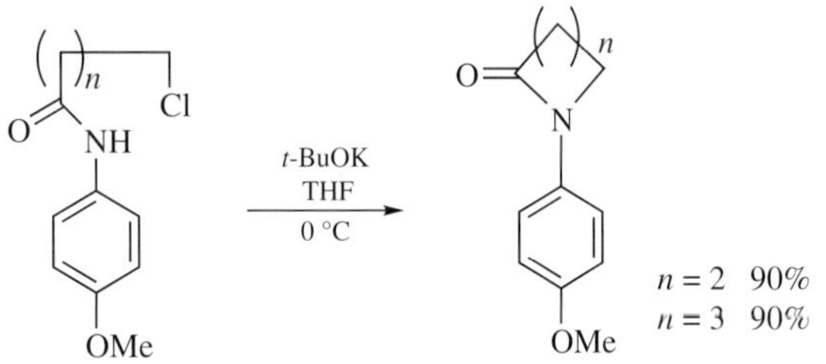

$$(70)$$

Condensation Reactions. The Darzens condensation of symmetrical ketones with (−)-8-phenylmenthyl-*α*-chloroacetate gives mixtures of glycidic esters with diastereoselectivities in the 77–96% range upon treatment with *t*-BuOK in CH$_2$Cl$_2$ (eq 71).[118] When an unsymmetrical ketone such as acetophenone is employed, the major diastereomer has a (*Z*,2*R*,3*R*) configuration, which is consistent with the enolate anion adopting a conformation with its π-bond interacting with the phenyl ring and attacking the ketone via an open transition state.

Treatment of esters with powdered *t*-BuOK in the absence of a solvent provides *α*-substituted *β*-keto esters in moderate yields, whereas related Claisen reactions in solution, e.g., *t*-BuOK/toluene, fail.[119] The formation of *N*-alkyl-(*E*)-1-alkenesulfonamides by condensation of *N*-Boc-methanesulfonamides with aliphatic or aromatic aldehydes in the presence of *t*-BuOK proceeds by formation of the anion of the sulfonamide, addition to the aldehyde, transfer of the Boc group from nitrogen to oxygen, and elimination of the OBoc group (eq

72).[120] *α*,*β*-Unsaturated nitriles containing five- or six-membered rings are readily formed when nitriles containing appropriately disposed carbonyl groups are treated with *t*-BuOK in *t*-BuOH at rt (eq 73).[121] Although the *α*-protons of carbonyl compounds are much more acidic than those of nitriles, the success of the reaction depends on the fact that cyano groups are much less electrophilic than carbonyl groups. Not only carbonyl groups of tethered saturated ketones, but also those of enones, lactams, and imides act as acceptors in these reactions.

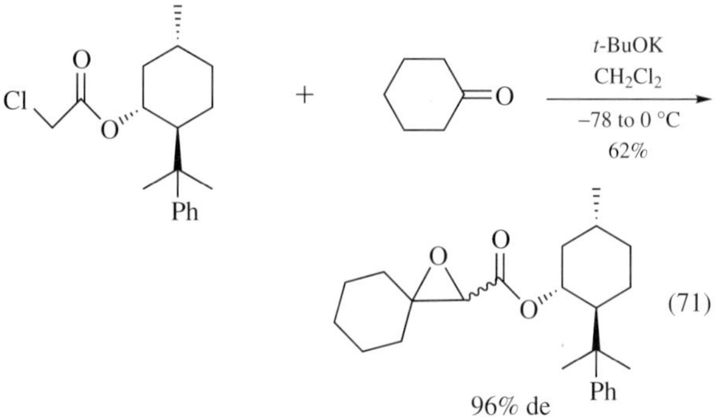

$$(71)$$

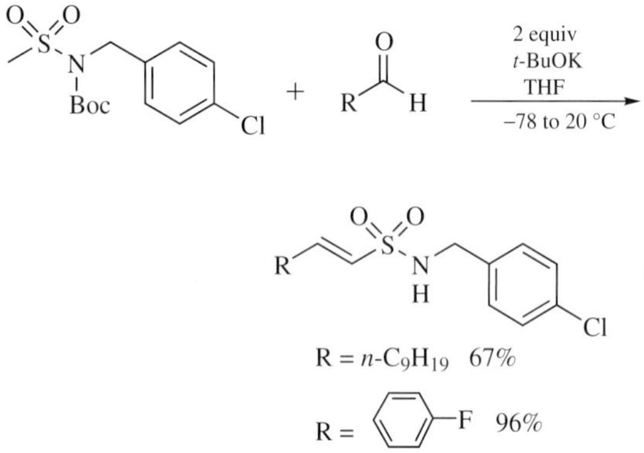

96% de

$$(72)$$

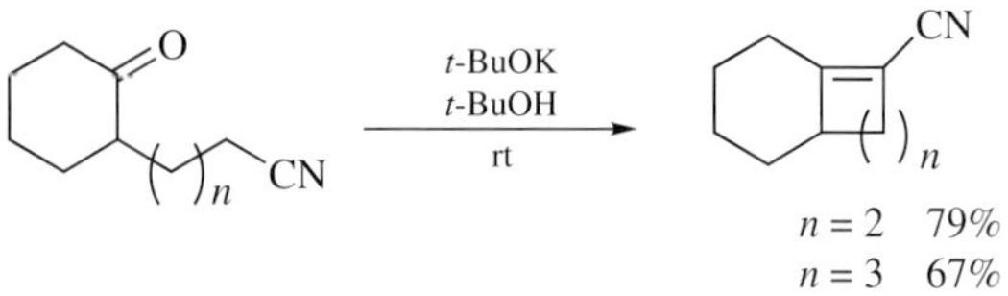

R = *n*-C$_9$H$_{19}$ 67%

R = ⟨benzene⟩–F 96%

$$(73)$$

n = 2 79%
n = 3 67%

4-Arylfuran-3-ols, produced by condensation of dimethyl diglycolate with aryl glycolates in the presence of *t*-BuOK in *t*-BuOH-toluene (2:1), contain a 2-methoxycarbonyl and a 5-*tert*-butoxycarbonyl group (eq 74).[122] In the intermediate dimethoxy-carbonyl compound, the 2-methoxycarbonyl group is protected from exchange with a *t*-butoxy group by resonance stabilization involving the anion of the furan-3-ol.

t-BuOK in various solvents is a useful base for the preparation of a variety of nitrogen heterocycles, including maleimides,[123] 3-arylalkylidene-2,5-piperazinediones,[124] 3-cyano-2-pyridones,[125] 3-substituted 4-hydroxy-1,8-naphthyridin-2-(1*H*)-ones,[126] and *δ*-lactams,[127] which contain complex functionality. A synthesis of xanthines in which *t*-BuOK is more effective than other lithium, sodium, or potassium bases provides a recent example in this area (eq 75).[128]

$$ (74) $$

$$ (75) $$

$$ (78) $$

$$ (79) $$

$$ (80) $$

$$ (81) $$

Elimination Reactions. *t*-BuOK continues to enjoy wide-spread use as a base for promoting β-elimination reactions. 2-Iodomethyl tetrahydrofurans give furan derivatives (eq 76)[129] or open-chain unsaturated ketones (eq 77),[130] depending upon the substitution pattern, when treated with excess of *t*-BuOK in THF. A CF$_3$ anion serves as a leaving group in the synthesis of an unsaturated nitrile (eq 78).[131] β-Elimination and substitution are competitive when 1-substituted 2-bromomethylaziridines are treated with *t*-BuOK in THF (eq 79),[132] but other base/solvent combinations give much poorer results. β-Amino-α,β-unsaturated ketones[133] and nitro compounds[134] are obtained when the corresponding β-methoxyamino compounds are treated with 2 equiv of *t*-BuOK in DMF. Microwave irradiation of 2-ethoxyanisole in the presence of *t*-BuOK-18-crown-6 gives exclusively the de-methylation product by β-elimination, while demethylation by substitution is favored when ethylene glycol is added.[135] Indoles containing 2-phenylsulfonyl protecting groups on nitrogen are readily deprotected using *t*-BuOK in DMF (eq 80).[136] Electrophilic acetone equivalents such as 1-chloro-2-alkoxy-2-propenes are obtained in excellent yields when 1,3-dichloro-2-alkoxypropanes are treated with *t*-BuOK in THF.[137] 1-Alkynylphosphonates are available by β-elimination of enol phosphates with *t*-BuOK in THF (eq 81).[138]

Isomerization Reactions. A number of propargylic compounds such as aryl and alkyl propargyl ethers,[139] and *N*-propargyllactams, 2-oxazolidinones, and *N*-methyl-2-imidazolidinones[140] undergo isomerization to the corresponding allenic derivative upon treatment with *t*-BuOK under various conditions. Chiral *N*-progargyl urethanes also produce allenic urethanes, but the related amides are converted into chiral ynamides under similar conditions (eq 82).[141] Interestingly, 2-halobenzyl 1-alkynyl sulfides undergo cyclization to dihydrothiophenes in moderate yields upon treatment with 2 equiv of *t*-BuOK in acetonitrile (eq 83).[142] The presence of the 2-halo substituent appears to be essential for the success of the reaction. The reaction mechanism presumably involves a 5-*endo-dig* cyclization of a benzylic anion of a propargylic or allenic sulfide produced by base-promoted isomerization of the substrate. *N*-Benzylaldimines undergo transamination to the corresponding *N*-benzylidene derivatives upon treatment with an excess of *t*-BuOK in THF (eq 84),[143] while the corresponding *N*-diphenylmethylimines undergo a similar isomerization with a catalytic amount of the base.[144]

$$ (76) $$

$$ (77) $$

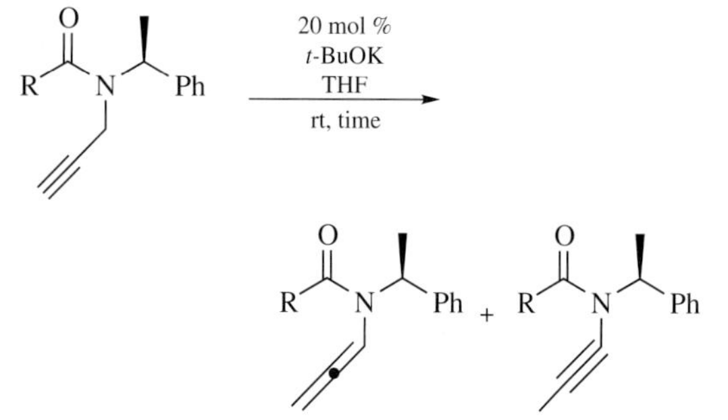

(82)

Time (h)	R	Yield (%)	
72	OMe	60	0
8	Me	0	83

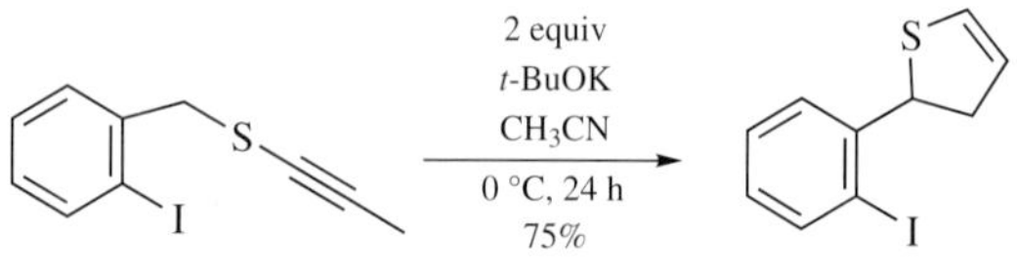

(83)

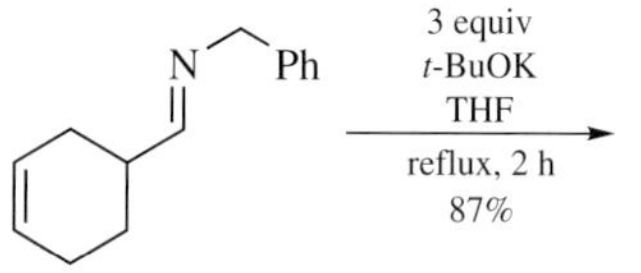

(84)

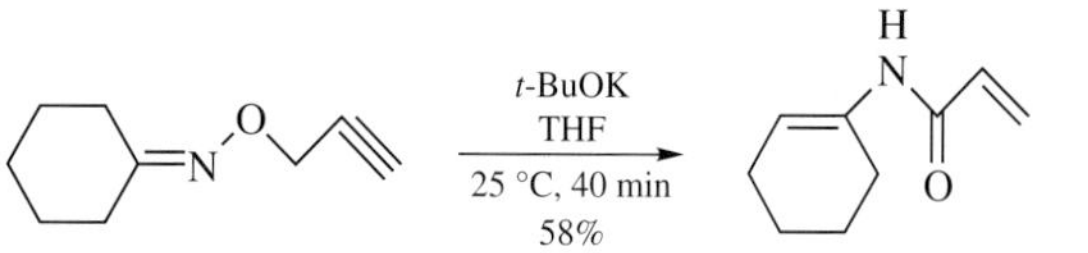

(85)

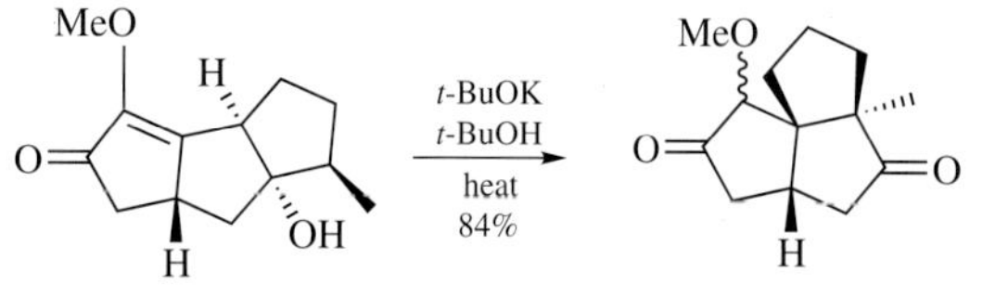

(86)

β-MeO/α-MeO = 2/1

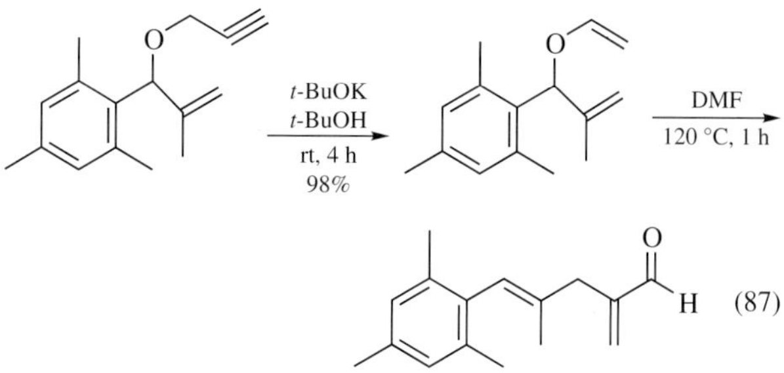

(87)

Several medium and large ring chlorosulfone azacycles undergo Ramberg-Baeklund rearrangements to azacycloalkenes when treated with *t*-BuOK in DMSO (eq 88).[148] α′-Chloro-α,β-unsaturated sulfones undergo a Diels-Alder reaction with conjugated dienes followed by a *t*-BuOK-induced Ramberg-Baeklund rearrangement to give exomethylene cyclohexane derivatives.[149]

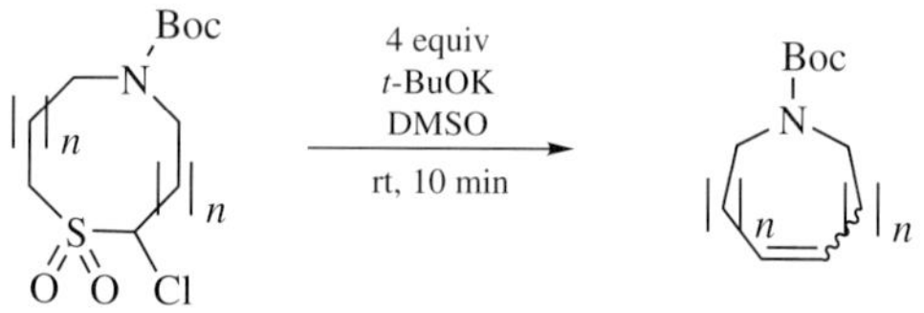

(88)

n	Yield (%)	trans/cis
1	66	0/100
2	100	94/6
3	97	98/2

Substitution Reactions. Although there are exceptions (cf. eq 79), the high basicity and low nucleophilicity of *t*-BuOK makes it much more likely to participate in E2-type elimination rather than S$_N$2 reactions in aliphatic systems. On the other hand, the *t*-BuO anion is known to participate as a nucleophile in S$_N$Ar reactions in appropriate arenes.[150,151] Such reactions occur much more rapidly using microwave irradiation than conventional heating. For example, 4-fluorobenzaldehyde is converted into 4-*tert*-butoxybenzaldehyde 12 times more rapidly under microwave irradiation than by reflux in DMSO without irradiation.[150] *t*-BuOK is frequently employed as a base to generate anions which participate in S$_N$Ar reactions. The conversion of 4-fluorobenzonitrile into 4-hydroxy benzonitrile by treatment with 2-butyne-1-ol and the base in DMSO (again under microwave irradiation) involves formation of the propargylic alkoxide, S$_N$Ar substitution of fluoride, and isomerization of the propargyl ether product into an allenyl ether, followed by mild acid hydrolysis (eq 89).[152] *t*-BuOK in DMSO under microwave irradiation also promotes the amination of electron-rich aryl halides via an elimination-addition (benzyne) mechanism.[153] Aryne intermediate formation promoted by *t*-BuOK is also involved in the synthesis of indolines from halostyrenes and amines discussed above (eq 57).[91]

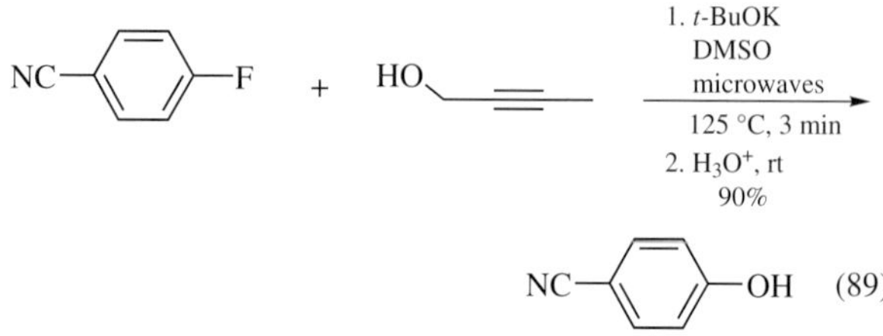

Rearrangement Reactions. *O*-Propargyl ketoximes undergo rearrangement into *N*-(1-alkenyl) acrylamides in low to moderate yields upon treatment with *t*-BuOK in THF (eq 85).[145] The shift of a carbanionic species from oxygen to nitrogen bears a resemblance to the Wittig rearrangement of ether α-carbanions. Treatment of linear δ-hydroxy triquinane enones with *t*-BuOK in *t*-BuOH leads to angular triquinane δ-diketones in moderate yields by a retroaldol ring opening, proton transfer, and an intramolecular Michael addition pathway (eq 86).[146] Allyl propargyl ethers isomerize to the corresponding allenyl ethers, which undergo a Claisen rearrangement to substituted α,β-unsaturated aldehydes upon treatment with *t*-BuOK in THF followed by heating (eq 87).[147]

(89)

Nitrobenzene and many of its 2-, 3-, and 4-substituted derivatives are converted into nitroaniline derivatives by treatment with sulfenamides in the presence of *t*-BuOK (eq 90).[154] In this conversion, termed vicarious nucleophilic substitution (VNS), the base presumably promotes both the formation of the nucleophilic sulfenamide anion and the β-elimination of the thiocarbamoyl group from the σ-adduct. Numerous other examples of *t*-BuOK-promoted VNS reactions of nitrobenzenes have appeared in recent years.[155–157] An interesting example of this process involves the synthesis of dithianylated nitrobenzenes which are hydrolyzable to aldehydes (eq 91).[157] The treatment of mixtures of *m*-nitroaniline and enolizable ketones with *t*-BuOK in DMSO leads to nitroindoles by oxidative nucleophilic substitution of hydrogen (eq 92).[158] The proposed mechanism for this transformation involves attack of the potassium enolate of the ketone on the ring, spontaneous oxidation of the σ-adduct, and imine formation and tautomerization.

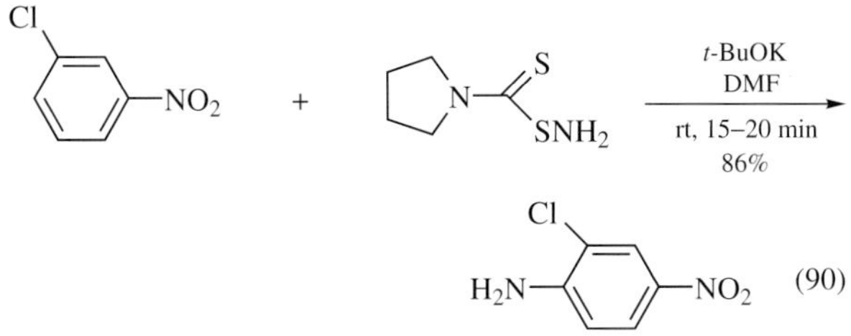

(90)

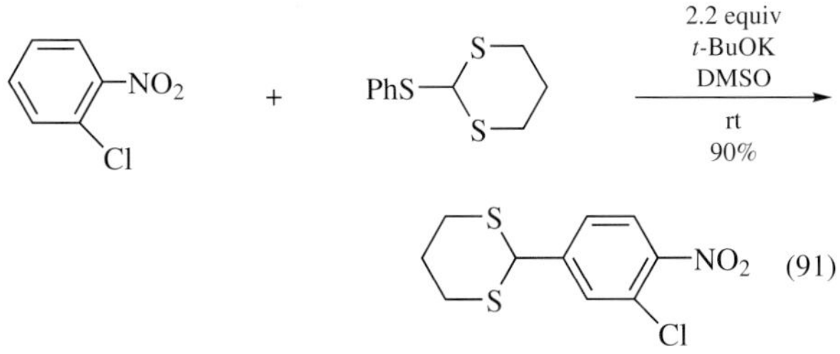

(91)

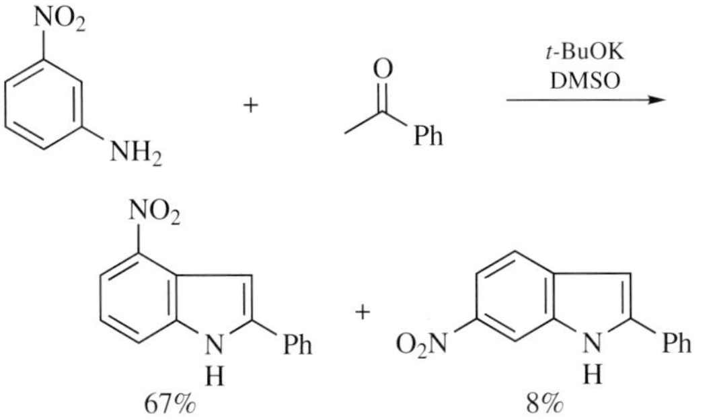

(92)

Synthesis and Reactions of Carboxylic Derivatives. *N*-Substituted amides are readily prepared by treating mixtures of nonenolizable[159] or enolizable esters[160] and amines with solid *t*-BuOK under microwave irradiation. Primary amides are obtained with formamide.[160] Methyl or ethyl esters are converted to the corresponding *tert*-butyl esters by ester interchange using an excess of *tert*-butyl acetate and a catalytic amount of *t*-BuOK,[161] or with 1–1.2 equiv of the base in diethyl ether.[162] *t*-BuOK in wet THF allows the cleavage of simple aliphatic esters in molecules containing enolizable dialkyl malonate moieties.[163] A 50:50 mixture of epimeric aziridine containing a *tert*-butyl ester group and a chiral group capable of chelating the potassium cation in a transient

tetrahedral intermediate undergoes kinetic resolution upon treatment with excess *t*-BuOK in THF at low temperature (eq 93).[164]

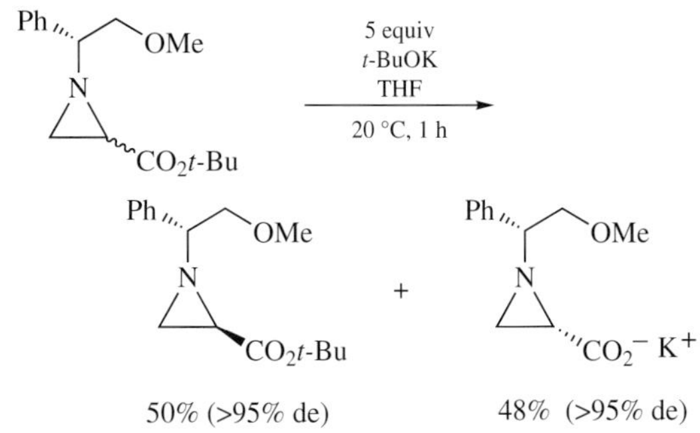

(93)

50% (>95% de)   48% (>95% de)

Ylide and Related Reactions. Allyloxymethylenetriphenylphosphonium chlorides yield allyl vinyl ethers by conversion into ylides with *t*-BuOK and treatment with aryl and enolizable acyclic aldehydes and ketones.[165] α,β-Unsaturated nitriles are obtained with high *Z*-stereoselectivity when diphenylcyanomethylenephosphonate is converted into its potassium salt by treatment with *t*-BuOK in THF, followed by addition of aldehydes with bulky substituents.[166] α-Fluoro-α,β-unsaturated thiazolines are obtained with slight *E*-stereoselectivity when mixtures of fluorophosphonomethylthioamides, or the corresponding thiazolines (eq 94), and aldehydes are treated with *t*-BuOK in THF at 0 °C; higher *E*-selectivity is usually observed when BuLi in THF is used as the base at −78 °C.[167]

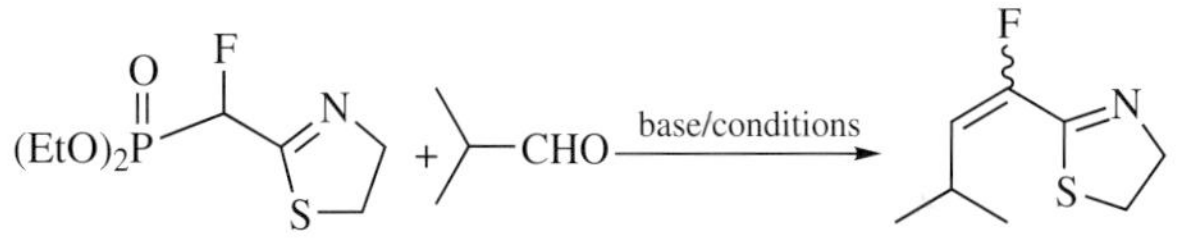

(94)

base/conditions	Yield (%)	*E/Z*
1.1 equiv *t*-BuOK, THF 0 °C, 4 h	88	75/25
1.1 equiv *n*-BuLi, THF −78 °C, 4 h	92	92/8

Vinyl sulfides, selenides, and tellurides are prepared in a one-pot procedure by treating mixtures of the chloromethyl phenyl chalcogenide and triphenylphosphine in THF with *t*-BuOK (to form the ylide), followed by the addition of a carbonyl compound (eq 95).[168] The yields are good to poor and the *Z* isomer is usually favored.

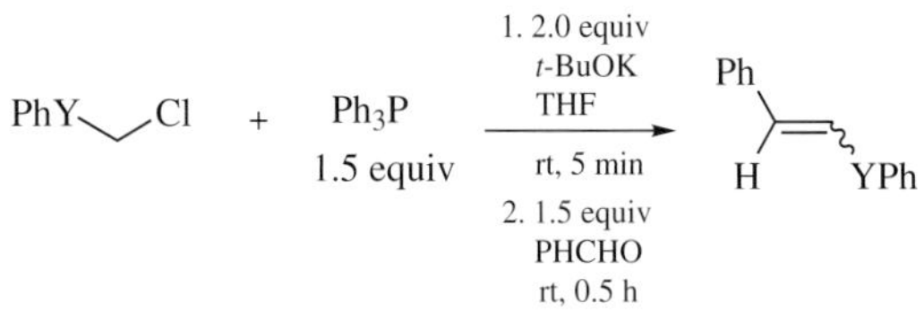

(95)

Y	Yield (%)	*E/Z*
S	84	30/70
Se	75	36/64
Te	45	<2/>98

1-Alkyl-1-aryloxiranes are obtained by heating a mixture of trimethylsulphonium iodide, *t*-BuOK, and an alkyl aryl ketone in the absence of a solvent.[169] Semistabilized allylic telluronium or arsonium ylides are obtained by treating their precursor salts with *t*-BuOK as well as other bases in THF.[170] Additions of these ylides to α,β-unsaturated ketones yield 2-vinyl-*trans*-3-substituted cyclopropyl ketones (eq 96); telluronium ylides yield primarily *cis*-2-vinyl isomers, while arsonium ylides produce mainly *trans*-2-vinyl isomers. Similarly, allylic arsonium ylides react with alkyl 2-pyran-5-carboxylates to give substituted vinyl-cyclopropanecarboxylates alone, or as a mixture with the corresponding vinyldihydrofurans.[171] Acyl bismuthonium ylides, prepared by deprotonation of 2-oxoalkylbismuthonium salts with *t*-BuOK in THF, yield 2,3-diacyloxiranes when treated with 1,2-dicarbonyl compounds such as ethyl pyruvate.[172]

(96)

R	Y	Yield (%)	Ratio
i-Bu	Te	92	>99/<1
Ph	As	84	6/94

Related Reagents. *n*-Butyllithium–potassium *t*-butoxide; Potassium Amide; Potassium Hexamethyldisilazide; Potassium *t*-butoxide–benzophenone; Potassium *t*-butoxide–*t*-butyl Alcohol Complex; *n*-butyllithium–Potassium ıt t-butoxide; Potassium *t*-butoxide–18-crown-6; Potassium *t*-butoxide–dimethyl Sulfoxide; Potassium *t*-butoxide–hexamethylphosphoric Triamide; Potassium Diisopropylamide; Potassium *t*-heptoxide; Potassium Hydroxide Potassium 2-methyl-2-butoxide.

1. Pearson, D. E.; Buehler, C. A., *Chem. Rev.* **1974**, *74*, 45.

2. Johnson, W. S.; Schneider, W. P., *Org. Synth., Coll. Vol.* **1963**, *4*, 132.

3. (a) Skattebøl, L.; Solomon, S., *Org. Synth.* **1969**, *49*, 35. (b) Skattebøl, L.; Solomon, S., *Org. Synth., Coll. Vol.* **1973**, *5*, 306.

4. House, H. O.; Czuba, L. J.; Gall, M.; Olmstead, H. D., *J. Org. Chem.* **1969**, *34*, 2324.

5. (a) Carey, F. A.; Sundberg, R. J. *Advanced Organic Chemistry*, 3rd ed.; Plenum: New York, 1990; Part B, p 3. (b) House, H. O. *Modern Synthetic Reactions*, 2nd ed.; Benjamin: Menlo Par, CA, 1972; p 510.

6. Marshall, J. A.; Carroll, R. D., *J. Org. Chem.* **1965**, *30*, 2748.

7. Renfrow, W. B.; Renfrow, A., *J. Am. Chem. Soc.* **1946**, *68*, 1801.

8. (a) House, H. O.; Roelofs, W. L.; Trost, B. M., *J. Org. Chem.* **1966**, *31*, 646. (b) Malhotra, S. K.; Johnson, F., *J. Am. Chem. Soc.* **1965**, *87*, 5513. (c) Brown, C. A., *J. Chem. Soc., Chem. Commun.* **1974**, 680.

9. (a) Caine, D. In *Carbon–Carbon Bond Formation*; Augustine, R. L., Ed.; Dekker: New York, 1979; Vol. 1, Chapter 2. (b) Caine, D., *Comprehensive Organic Synthesis* **1991**, *3*, 1.

10. Evans, D. A. In *Asymmetric Synthesis*; Morrison, J. D., Ed.; Academic: New York, 1984; Vol. 3, Chapter 1.

11. Ruber, S. M.; Ronald, R. C., *Tetrahedron Lett.* **1984**, *25*, 5501.

12. Piers, E.; Britton, R. W.; deWaal, W., *Can. J. Chem.* **1969**, *47*, 831.

13. Dauben, W. G.; Ashcraft, A. C., *J. Am. Chem. Soc.* **1963**, *85*, 3673.

14. Ringold, H. J.; Malhotra, S. K., *J. Am. Chem. Soc.* **1962**, *84*, 3402.

15. Cargill, R. L.; Bushey, D. F.; Ellis, P. D.; Wolff, S.; Agosta, W. C., *J. Org. Chem.* **1974**, *39*, 573.

16. Christol, H.; Mousserson, M.; Plenat, M. F., *Bull. Soc. Chem. Fr.* **1959**, 543.

17. House, H. O.; Sayer, T. S. B.; Yau, C. C., *J. Org. Chem.* **1978**, *43*, 2153.

18. Srikrishna, A.; Hemamalini, P.; Sharma, G. V. R., *J. Org. Chem.* **1993**, *58*, 2509.

19. (a) Brown, H. C.; Rogić, M. M., *Organomet. Chem. Synth.* **1972**, *1*, 305. (b) Rogić, M. M., *Intra-Sci. Chem. Rep.* **1973**, *7*, 155.

20. (a) Brown, H. C.; Rogić, M. M.; Rathke, M. W.; Kabalka, G. W., *J. Am. Chem. Soc.* **1968**, *90*, 818, 1911; (b) Brown, H. C.; Rogić, M. M.; Rathke, M. W.; Kabalka, G. W., *J. Am. Chem. Soc.* **1969**, *91*, 2150.

21. Brown, H. C.; Rogić, M. M.; Rathke, M. W., *J. Am. Chem. Soc.* **1968**, *90*, 6218.

22. Brown, H. C.; Nambu, H.; Rogić, M. M., *J. Am. Chem. Soc.* **1969**, *91*, 6852, 6854, 6855.

23. Heathcock, C. H., *Comprehensive Organic Synthesis* **1991**, *2*, 133.

24. Heathcock, C. H., *Comprehensive Organic Synthesis* **1991**, *2*, 181.

25. Kim, B. M.; Williams, S. F.; Masamune, S., *Comprehensive Organic Synthesis* **1991**, *2*, 239.

26. Paterson, I., *Comprehensive Organic Synthesis* **1991**, *2*, 301.

27. Trost, B. M.; Shuey, C. D.; DiNinno, F., Jr.; McElvain, S. S., *J. Am. Chem. Soc.* **1979**, *101*, 1284.

28. Murai, A.; Tanimoto, N.; Sakamoto, N.; Masamune, T., *J. Am. Chem. Soc.* **1988**, *110*, 1985.

29. (a) Newman, M. S.; Magerlein, B. J., *Org. React.* **1949**, *5*, 413. (b) Ballester, M., *Chem. Rev.* **1955**, *55*, 283. (c) Rosen, T., *Comprehensive Organic Synthesis* **1991**, *2*, 409.

30. Zimmerman, H. E.; Ahramjiam, L., *J. Am. Chem. Soc.* **1960**, *82*, 5459.

31. (a) Schaefer, J. P.; Bloomfield, J. J., *Org. React.* **1967**, *15*, 1. (b) Davis, B. R.; Garratt, P. J., *Comprehensive Organic Synthesis* **1991**, *2*, 795.

32. Nace, H. R.; Smith, A. H., *J. Org. Chem.* **1973**, *38*, 1941.

33. Georges, M.; Tam, T.-F.; Fraser-Reid, B., *J. Org. Chem.* **1985**, *50*, 5747.

34. Blasko, G.; Kardos, J.; Baitz-Gács, E.; Simonyi, M.; Szantay, C., *Heterocycles* **1986**, *24*, 2887.

35. Johnson, W. S.; Daub, G. H., *Org. React.* **1951**, *6*, 1.

36. Fitjer, L.; Quabeck, U., *Synth. Commun.* **1985**, *15*, 855.

37. Abruscato, G. J.; Binder, R. G.; Tidwell, T. T., *J. Org. Chem.* **1972**, *37*, 1787.

38. Parham, W. E.; Schweizer, E. E., *Org. React.* **1963**, *13*, 55.

39. Taylor, R. T.; Paquette, L. A., *Org. Synth., Coll. Vol.* **1990**, *7*, 200.

40. (a) Starks, C. M.; Liotta, C. *Phase Transfer Catalysis: Principles and Techniques*; Academic: New York, 1978; Chapter 6. (b) Starks, C. M.; Liotta, C. L.; Halpern, M. *Phase-Transfer Catalysis Fundamentals, Applications, and Industrial Perspectives*; Chapman & Hall: New York, 1994; Chapter 8.

41. Hartzler, H. D., *J. Org. Chem.* **1964**, *29*, 1311.

42. Pulwer, M. J.; Blacklock, T. J., *Org. Synth., Coll. Vol.* **1990**, *7*, 203.

43. Halton, B.; Milsom, P. J., *J. Chem. Soc., Chem. Commun.* **1971**, 814.

44. Johnson, W. S.; Johns, W. F., *J. Am. Chem. Soc.* **1957**, *79*, 2005.

45. Tremelling, M. J.; Hopper, S. P.; Mendelowitz, P. C., *J. Org. Chem.* **1978**, *43*, 3076.

46. (a) Paquette, L. A.; Barrett, J. H., *Org. Synth.* **1969**, *49*, 62. (b) Paquette, L. A.; Barrett, J. H., *Org. Synth., Coll. Vol.* **1973**, *5*, 467.

47. Quinn, C. B.; Wiseman, J. R., *J. Am. Chem. Soc.* **1973**, *95*, 1342, 6120.

48. Mandai, T.; Hara, K.; Nakajima, T.; Kawada, M.; Otera, J., *Tetrahedron Lett.* **1983**, *24*, 4993.

49. (a) Bartsch, R. A.; Zavada, J., *Chem. Rev.* **1980**, *80*, 453. (b) Krebs, A.; Swienty-Busch, J., *Comprehensive Organic Synthesis* **1991**, *6*, 949.

50. Swoboda, M.; Hapala, J.; Zavada, J., *Tetrahedron Lett.* **1972**, 265.

51. (a) Becker, K. B.; Grob, C. A. In *The Chemistry of Double-Bonded Functional Groups*; Patai, S., Ed.; Wiley: New York, 1977; Part 2, p 653. (b) Caine, D., *Org. Prep. Proced. Int.* **1988**, *20*, 3. (c) Weyerstahl, P.; Marschall, H., *Comprehensive Organic Synthesis* **1991**, *6*, 1041.

52. Heathcock, C. H.; Badger, R. A., *J. Org. Chem.* **1972**, *37*, 234.

53. Fischli, A.; Branca, Q.; Daly, J., *Helv. Chim. Acta* **1976**, *59*, 2443.

54. Ringold, H. J.; Malhotra, S. K., *Tetrahedron Lett.* **1962**, 669.

55. Corey, E. J.; Cyr, C. R., *Tetrahedron Lett.* **1974**, 1761.

56. Wender, P. A.; Eissenstat, M. A., *J. Am. Chem. Soc.* **1978**, *100*, 292.

57. (a) Gassman, P. G.; Lumb, J. T.; Zalar, F. V., *J. Am. Chem. Soc.* **1967**, *89*, 946. (b) Gassman, P. G.; Zalar, F. V., *J. Am. Chem. Soc.* **1966**, *88*, 2252.

58. (a) Paquette, L. A.; Gilday, J. P., *J. Org. Chem.* **1988**, *53*, 4972. (b) Paquette, L. A.; Ra, C. S., *J. Org. Chem.* **1988**, *53*, 4978.

59. Jung, M. E., *Comprehensive Organic Synthesis* **1991**, *4*, 1.

60. Houbreckts, Y.; Laszlo, P.; Pennetreau, P., *Tetrahedron Lett.* **1986**, *27*, 705.

61. Stork, G.; Tomasz, M., *J. Am. Chem. Soc.* **1964**, *86*, 471.

62. Horton, M.; Pattenden, G., *Tetrahedron Lett.* **1983**, *24*, 2125.

63. von Doering, W. E.; Urban, R. S., *J. Am. Chem. Soc.* **1956**, *78*, 5938.

64. Carlson, R. G.; Pierce, J. K., *Tetrahedron Lett.* **1968**, 6213.

65. Yates, P.; Anderson, C. D., *J. Am. Chem. Soc.* **1963**, *85*, 2937.

66. (a) Paquette, L. A., *Org. React.* 1977, *25*, 1. (b) Clough, J. M., *Comprehensive Organic Synthesis* **1991**, *3*, 861.

67. Paquette, L. A.; Wingard, R. E., Jr.; Photis, J. M., *J. Am. Chem. Soc.* **1974**, *96*, 5801.

68. (a) Meyers, C. Y.; Matthews, W. S.; Ho, L. L.; Kolb, V. M.; Parady, T. E. In *Catalysis in Organic Synthesis*; Smith, G. V., Ed.; Academic: New York, 1977; p 197. (b) Meyers, C. Y. In *Topics in Organic Sulfur Chemistry*; Tisler, M., Ed.; University Press: Ljubljana, 1978; p 207.

69. Matsuyama, H.; Miyazawa, Y.; Takei, Y.; Kobayashi, M., *J. Org. Chem.* **1987**, *52*, 1703.

70. Nicolaou, K. C.; Zuccarello, G.; Ogawa, Y.; Schweiger, E. J.; Kumazawa, T., *J. Am. Chem. Soc.* **1988**, *110*, 4866.

71. Erickson, K. L., *J. Org. Chem.* **1971**, *36*, 1031.

72. Grundon, M. F.; Henbest, H. B.; Scott, M. D., *J. Chem. Soc.* **1963**, 1855.

73. Caine, D.; Frobese, A. S., *Tetrahedron Lett.* **1977**, 3107.

74. Babler, J. H.; Liptak, V. P.; Phan, N., *J. Org. Chem.* **1996**, *61*, 416.

75. Miyamoto, H.; Yasaka, S.; Tanaka, K., *Bull. Chem. Soc. Jpn.* **2001**, *74*, 185.

76. Chong, B. -D.; Ji, Y. -I.; Oh, S. -S.; Yang, J. -D.; Baik, W.; Koo, S., *J. Org. Chem.* **1997**, *62*, 9323.

77. Ishikara, T.; Kadoya, R.; Arai, M.; Takahashi, H.; Kaisi, Y.; Mizuta, T.; Yoshikai, K.; Saito, S., *J. Org. Chem.* **2001**, *66*, 8000.

78. Shen, Y.; Zhang, Z., *Synth. Commun.* **2000**, *30*, 445.

79. Bunlaksananusorn, T.; Rodriguez, A. L.; Knochel, P., *J. Chem. Soc., Chem. Commun.* **2001**; 745.

80. Rodriguez, A. L.; Bunlaksananusorn, T.; Knochel, P., *Org. Lett.* **2000**, *2*, 3285.

81. Makosza, M.; Urbanska, N.; Chesnokov, A. A., *Tetrahedron Lett.* **2003**, *44*, 1473.

82. Mukai, C.; Ukon, R.; Kuroda, N., *Tetrahedron Lett.* **2003**, *44*, 1583.

83. Abrunhosa, I.; Gulea, M.; Masson, S., *Synthesis* **2004**; 928.

84. Nelson, D. W.; Easley, R. A.; Pintea, B. N. V., *Tetrahedron Lett.* **1999**; *40*, 25.

85. Jablonski, L.; Joubert, J.; Billard, T.; Langlois, B. R., *Synlett* **2003**; 230.

86. Inschauspe, D.; Sortais J. -B; Billard, T.; Langlois B. R., *Synlett* **2003**; 233.

87. Prakash, G. K. S; Hu, J.; Olah, G. A., *Org. Lett.* **2003**, *5*, 3253.

88. Evans, P.; Taylor, R. J. K., *Synlett* **1997**, 1043.

89. Jiang, B.; Zhang, F.; Xiong, W., *Tetrahedron* **2002**, *58*, 265.

90. Beller, M.; Breindl, C.; Riermeier, T. H.; Eichberger, M.; Trauthwein, H., *Angew. Chem. Int. Ed. Eng.* **1998**, *37*, 3389.

91. Beller, M.; Breindl, C.; Riermeier, T. H.; Tillack, A., *J. Org. Chem.* **2001**, *66*, 1403.

92. Lucet, D.; Toupet, L.; Le Gall, T.; Mioskowski, C., *J. Org. Chem.* **1997**, *62*, 2682.

93. Dumez, E.; Rodriguez, J.; Dulcere J. P., *J. Chem. Soc., Chem. Commun.* **1997**, 1831.

94. Arcadi, A.; Marinelli, F.; Pini, E.; Rossi, E., *Tetrahedron Lett.* **1996**, *37*, 3387.

95. Marshall, J. A.; DuBay W. J., *J. Org. Chem.* **1993**, *58*, 3435.

96. Marshall, J. A.; DuBay W. J., *Synlett* **1993**, 209.

97. Yoshimatsu, M.; Naito, M.; Shimizu, H.; Muraoka, O.; Tanabe, G.; Kataoka, T., *J. Org. Chem.* **1996**, *61*, 8200.

98. Mukai, C.; Yamashita, H.; Hanaoka M., *Org. Lett.* **2001**, *3*, 3385.

99. Mukai, C.; Ohta, M.; Yamashita, H.; Kitagaki, S., *J. Org. Chem.* **2004**, *69*, 6867.

100. Yoshimatsu, M.; Machida, K.; Fuseya, T.; Shimizu, H.; Kataoka, T., *J. Chem. Soc., Perkin Trans. 1* **1996**, 1839.

101. Kondo, Y.; Kojima, S.; , Sakamoto, T., *Heterocycles* **1996**, *43*, 2741; *J. Org. Chem.* **1997** *62*, 6507.

102. Kitagawa, O.; Suzuki, T.; Taguchi, T., *J. Org. Chem.* **1998**, *63*, 4842.

103. Seijas, J. A.; Vazquez-Tato, M. P.; Matinez, M. M., *Tetrahedron Lett.* **2000**, *41*, 2215.

104. Ahrach, M.; Gerardin, P.; Loubinoux B., *Synth. Commun.* **1997**, *27*, 1877.

105. Katritzky, A. R.; Cheng, D.; Musgrave R. P., *Heterocycles* **1997**, *44*, 67.

106. Cheng, C.-C.; Chen, S.-T.; Chuang, T.-H.; Fang, J.-M., *J. Chem. Soc., Perkin Trans. 1* **1994**, 2217.

107. Ros, F.; de la Rosa, J.; Enfedaque, J., *J. Org. Chem.* **1995**, *60*, 5419.

108. Haufe, G.; Laue, K. W.; Triller, M. U.; Takeuchi, Y.; Shibata, N., *Tetrahedron* **1998**, *54*, 5929.

109. Prakash, G. K. S.; Hu, J.; Wang, Y.; Olah, G. A., *Angew. Chem. Int. Ed.* **2004**, *43*, 5203.

110. Ponthieux, S.; Outurquin, F.; Paulmier, C., *Tetrahedron* **1997**, *53*, 6365.

111. Ando, R.; Sakaki, T.; Jikihara, T., *J. Org. Chem.* **2001**, *66*, 3617.

112. Arguello, J. E.; Penenory, A. B.; Rossi, R. A., *J. Org. Chem.* **1999**, *64*, 6115.

113. Nelson, A.; Warren, S., *J. Chem. Soc., Perkin Trans. 1* **1999**, 3425.

114. Berree, F.; Bazureau, J.-P.; Michelot, G.; Le Corre, M., *Synth. Commun.* **1999**, *29*, 2685.

115. Hoffmann, P.; Doucet, J.-B.; Li, W.; Vergnes, L.; Labidalle, S., *J. Chem. Res. (S)* **1997**, 218.

116. Wang, E. C.; Lin, H.-J., *Heterocycles* **1998**, *48*, 481.

117. Foray, G.; Penenory, A. B.; Rossi, R. A., *Tetrahedron Lett.* **1997**, *38*, 2035.

118. Ohkata, K.; Kimura, J.; Shinohara, Y.; Takagi, R.; Hiraga, Y., *J. Chem. Soc., Chem. Commun.* **1996**, 2411.

119. Yoshizawa, K.; Toyota, S.; Toda, F., *Tetrahedron Lett* **2001**, *42*, 7983.

120. Tozer, M. J.; Woolford, A. J. A.; Linney, I. D., *Synlett* **1998**, 186.

121. Fleming, F. F.; Funk, L. A.; Altundas, R.; Sharief, V., *J. Org. Chem.* **2002**, *67*, 9414.

122. Tse, B.; Jones, A. B., *Tetrahedron Lett.* **2001**, *42*, 6429.

123. Faul, M. M.; Winneroski, L. L.; Krumrich, C. A., *Tetrahedron Lett.* **1999**, *40*, 1109.

124. Gonzalez, J. F.; de la Cuesta, E.; Avendano, C., *Synth. Commun.* **2004**, *34*, 1589.

125. Jain, R.; Roschangar, F.; Ciufolini, M. A., *Tetrahedron Lett.* **1995**, *36*, 3307.

126. Delieza, V.; Detsi, A.; Bardakos, V.; Igglessi-Markopoulou, O., *J. Chem. Soc., Perkin Trans. 1* **1997**, 1487.

127. Nemes, P.; Balazs, B.; Toth, G.; Scheiber P., *Synlett* **2000**, 1327.

128. Zavialov, I. A.; Dahanukar, V. H.; Nguyen, H.; Orr, C.; Andrews, D. R., *Org. Lett.* **2004**, *6*, 2237.

129. Jung, J. H.; Lee, J. W.; Oh, D. Y., *Tetrahedron Lett.* **1995**, *36*, 923.

130. Lee, C.-W.; Hong, J. E.; Oh, D. Y., *J. Org. Chem.* **1995**, *60*, 7027.

131. Kende, A. S.; Liu, K., *Tetrahedron Lett.* **1995**, *36*, 4035.

132. De Kimpe, N.; De Smaele, D.; Sakonyi, Z., *J. Org. Chem.* **1997**, *62*, 2448.

133. Seko, S.; Tani, N., *Tetrahedron Lett.* **1998**, *39*, 8117.

134. Seko, S.; Komoto, I., *J. Chem. Soc., Perkin Trans. 1* **1998**, 2975.

135. Oussaid, A.; Thach, L. N.; Loupy, A., *Tetrahedron Lett.* **1997**, *38*, 2451.

136. Bashford, K. E.; Cooper, A. L.; Kane, P. D.; Moody, C. J., *Tetrahedron Lett.* **2002**, *43*, 135.

137. Janicki, S. Z.; Fairgrieve, J. M.; Petillo, P. A., *J. Org. Chem.* **1998**, *63*, 3694.

138. Hong, J. E.; Lee, C.-W.; Kwon, Y.; Oh, D. Y., *Synth. Commun.* **1996**, *26*, 1563.

139. Moghaddam, F. M.; Emami, R., *Synth. Commun.* **1997**, *27*, 4073.

140. Wei, L.-L.; Mulder, J. A.; Xiong, H.; Zificsak, C. A.; Douglas, C. J.; Hsung, R. P., *Tetrahedron* **2001**, *57*, 459.

141. Huang, J.; Xiong, H.; Hsung, R. P.; Rameshkumar, C.; Mulder, J. A.; Grebe, T. P., *Org. Lett.* **2002**, *4*, 2417.

142. McConachie, L. K.; Schwan, A. L., *Tetrahedron Lett.* **2000**, *41*, 5637.

143. DeKimpe, N.; De Smaele, D.; Hofkens, A.; Dejaegher, Y.; Kesteleyn, B., *Tetrahedron* **1997**, *53*, 10803.

144. Cainelli, G.; Giacomini, D.; Trere, A.; Boyl, P. P., *J. Org. Chem.* **1996**, *61*, 5134.

145. Trofimov, B. A.; Tarasova, O. A.; Sigalov, M. V.; Mikhaleva, A. I., *Tetrahedron Lett.* **1995**, *36*, 9181.

146. MacDougall, J. M.; Moore, H. W., *J. Org. Chem.* **1997**, *62*, 4554.

147. Parsons, P. J.; Thomson, P.; Taylor, A.; Sparks, T., *Org. Lett.* **2000**, *2*, 571.

148. MaGee, D. I.; Beck, E. J., *J. Org. Chem.* **2000**, *65*, 8367

149. Block, E.; Jeon, H. R.; Zhang, S.-Z.; Dikarev, E. V., *Org. Lett.* **2004**, *6*, 437.

150. Salmoria, G. V.; Dall'Oglio, E.; Zucco, C., *Tetrahedron Lett.* **1998**, *39*, 2471.

151. Woiwode, T. F.; Rose, C.; Wandless, T. J., *J. Org. Chem.* **1998**, *63*, 9594.

152. Levin, J. I.; Du, M. T., *Synth. Commun.* **2002**, *32*, 1401.

153. Shi, L.; Wang, M.; Fan, C.-A.; Zhang, F.-M.; Tu, Y.-Q., *Org. Lett.* **2003**, *5*, 3515.

154. Makosza, M.; Bialecki, M., *J. Org. Chem.* **1998**, *63*, 4878.

155. Florio, S.; Lorusso, P.; Luisi, R.; Granito, C.; Ronzini, L.; Troisi, L., *Eur. J. Org. Chem.* **2004**, 2118.

156. Seko, S.; Miyake, K., *J. Chem. Soc., Chem. Commun.* **1998**, 1519.

157. Kim, W.-K.; Paik, S.-C.; Lee, H.; Cho, C. -G., *Tetrahedron Lett.* **2000**, *41*, 5111.

158. Moskalev, N.; Barbasiewicz, M.; Makosza, M., *Tetrahedron* **2004**, *60*, 347.

159. Varma, R. S.; Naicker, K. P., *Tetrahedron Lett.* **1999**, *40*, 6177.

160. Zradni, F.-Z; Hamelin, J.; Derdour, A., *Synth. Commun.* **2002**, *32*, 3525.

161. Stanton, M. G.; Gagne, M. R., *J. Org. Chem.* **1997**, *62*, 8240.

162. Vasin, V. A.; Razin, V. V., *Synlett* **2001**, 658.

163. Wilk, B. K., *Synth. Commun.* **1996**, *26*, 3859.

164. Alezra, V.; Bouchet, C.; Micouin, L.; Bonin, M.; Husson, H.-P., *Tetrahedron Lett.* **2000**, *41*, 655.

165. Kulkarni, M. G.; Pendharkar, D. S.; Rasne, R. M., *Tetrahedron Lett.* **1997** *38*, 1459.

166. Zhang, T. Y.; O'Toole, J. C.; Dunigan, J. M., *Tetrahedron Lett.* **1998**, *39*, 1461.

167. Pfund, E.; Masson, S.; Vazeux, M.; Lequeux, T., *J. Org. Chem.* **2004**, *69*, 4670.

168. Silveira, C. C.; Begnini, M. L.; Boeck, P.; Braga, A. L., *Synthesis* **1997**, 221.

169. Toda, F.; Kanemoto, K., *Heterocycles* **1997**, *46*, 185.

170. Tang, Y.; Huang, Y.-Z; Dai, L.-X; Sun, J.; Xia, W., *J. Org. Chem.* **1997**, *62*, 954.

171. Moorhoff, C. M., *Tetrahedron Lett.* **1996**, *37*, 9349.

172. Matano, Y.; Suzuki, H., *J. Chem. Soc., Chem. Commun.* **1996**, 2697.

R

Ruthenium(II), Tris(2,2'-bipyridine-κN1,κN1')-, (OC-6-11)-

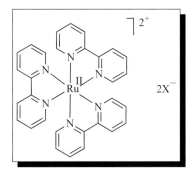

[14323-06-9]

Ruthenium(II), tris(2,2'-bipyridine-κN1, κN1'),[1] chloride(1:2), (OC-6-11)-

$C_{30}H_{24}N_6RuCl_2$ (MW 640.53)

InChI = 1S/3C10H8N2.2ClH.Ru/c3*1-3-7-11-9(5-1)10-6-2-4-8-12-10;;;/h3*1-8H;2*1H;/q;;;;;;+2/p-2

InChIKey = SJFYGUKHUNLZTK-UHFFFAOYSA-L

(cation)

[15158-62-0]

InChI = 1S/3C10H8N2.Ru/c3*1-3-7-11-9(5-1)10-6-2-4-8-12-10;/h3*1-8H;/q;;;+2

InChIKey = HNVRWFFXWFXICS-UHFFFAOYSA-N

(PF$_6$ complex)

[60804-74-2]

InChI = S/3C10H8N2.2F6P.Ru/c3*1-3-7-11-9(5-1)10-6-2-4-8-12-10;2*1-7(2,3,4,5)6;/h3*1-8H;;;/q;;;2*-1;+2

InChIKey = KLDYQWXVZLHTKT-UHFFFAOYSA-N

(chloride hydrate)

[50525-27-4]

InChI = 1S/3C10H8N2.2ClH.6H2O.Ru/c3*1-3-7-11-9(5-1)10-6-2-4-8-12-10;;;;;;;;;/h3*1-8H;2*1H;6*1H2;/q;;;;;;;;;;;+2/p-2

InChIKey = WHELTKFSBJNBMQ-UHFFFAOYSA-L

(perchlorate)

[15635-95-7]

InChI = S/3C10H8N2.2ClHO4.Ru/c3*1-3-7-11-9(5-1)10-6-2-4-8-12-10;2*2-1(3,4)5;/h3*1-8H;2*(H,2,3,4,5);/q;;;;;+2/p-2

InChIKey = BXKPAPTYLLPPEO-UHFFFAOYSA-L

(BF$_4$ complex)

[63950-81-2]

InChI = 1S/3C10H8N2.2BF4.Ru/c3*1-3-7-11-9(5-1)10-6-2-4-8-12-10;2*2-1(3,4)5;/h3*1-8H;;;/q;;;2*-1;+2

InChIKey = GWPUNVSVEJLECO-UHFFFAOYSA-N

(visible light-active catalyst used in single-electron transfer reductions and oxidations[1])

Alternate Names: [Ru(bpy)$_3$]$^{2+}$, Ruthenium-tris(2,2'-bipyridyl).
Physical Data: λ_{max} = 452 nm, redox potentials[2] (vs. SCE, saturated calomel electrode, as 0 V in CH$_3$CN); -0.86 V (Ru^{III}/Ru^{II}*), 1.29 V (Ru^{III}/Ru^{II}), 0.84 V (Ru^{II}*/Ru^{I}), and -1.33 (Ru^{II}/Ru^{I}).

Solubility: for X = Cl, sol in H$_2$O, CH$_3$OH, CH$_3$CN, DMF, and DMSO; insoluble in nonpolar, aprotic solvents; for X = PF$_6$, sol in CH$_3$CN, acetone, DMF, and DMSO; insoluble in H$_2$O.
Form Supplied in: the dichloride is commercially available from Strem Chemicals, Inc. and Sigma-Aldrich as the hexahydrate. [Ru(bpy)$_3$]$^{2+}$ complexes appear as red to orange crystals or powder.
Preparation Methods: for X = PF$_6$, the catalyst is prepared from RuCl$_3$, 2,2'-bipyridine, and NH$_4$PF$_6$ in EtOH.[3]
Purification: recrystallized in boiling H$_2$O.
Handling, Storage, and Precautions: containers should be stored under nitrogen atmosphere in the dark. Keep away from strong oxidizing agents.

Upon visible light excitation of [Ru(bpy)$_3$]$^{2+}$ (bpy = 2,2'-bipyridine), the resulting metal-to-ligand charge transfer (MLCT) excited state species, *[Ru(bpy)$_3$]$^{2+}$, can be reductively or oxidatively quenched with the appropriate single-electron donor or acceptor, respectively. Reductive quenching of the excited state forms [Ru(bpy)$_3$]$^{+}$, which then performs a single-electron reduction to regenerate [Ru(bpy)$_3$]$^{2+}$. Alternatively, oxidative quenching forms [Ru(bpy)$_3$]$^{3+}$, which then oxidizes a suitable donor to regenerate ground state [Ru(bpy)$_3$]$^{2+}$.

Reductions of Carbon–Sulfur, Carbon–Phosphorus, and Carbon–Nitrogen Bonds. Suitably activated carbon–heteroatom bonds may be reduced by *[Ru(bpy)$_3$]$^{2+}$ resulting in C–H bond formation.[4–7] Phenylacylsulfonium salts have been reduced in the presence of [Ru(bpy)$_3$]Cl$_2$ and a 1,4-dihydropyridine derivative under visible light irradiation to produce the corresponding acetophenone (eq 1).[8] Similarly, ammonium and phosphonium salts undergo efficient reduction under these conditions.[4] Reduction of aryl diazonium salts has historically found usage in radical aromatic substitution reactions via decomposition into N$_2$ and aryl radical species, a transformation classically promoted by copper(I) salts. A milder version of this reductive decomposition pathway has been demonstrated in a photocatalytic Pschorr reaction that uses *[Ru(bpy)$_3$]$^{2+}$ to reduce the C–N bond (eq 2).[5] Excellent yields are reported and no additional oxidants are necessary. Recently, reductive ring openings of chalcone-derived tosyl aziridines have been disclosed using Hantzsch ester (HE) to produce β-amino ketones (eq 3).[6]

$$\text{(1)}$$

quantitative yields

Essential Reactions for Organic Synthesis, First Edition. Edited by Philip L. Fuchs.
© 2016 John Wiley & Sons, Ltd. Published 2016 by John Wiley & Sons, Ltd.

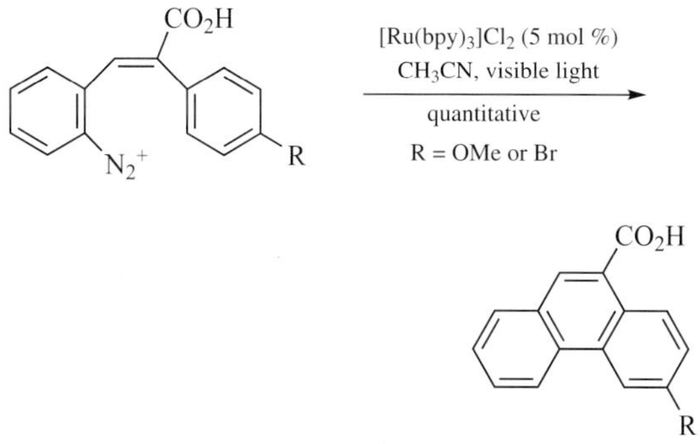

(2)

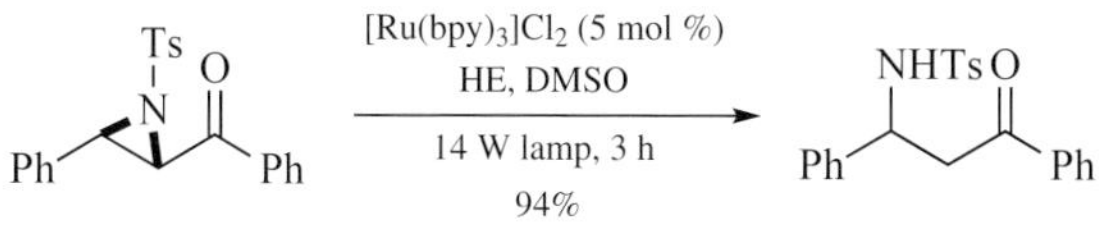

(3)

(5)

Reductions of Carbon–Oxygen Bonds. Aryl carbonyls can be transformed into alcohols through single-electron reduction of the carbon–oxygen double bond by [Ru(bpy)$_3$]$^+$, which is generated in situ using 1-benzyl-1,4-dihydronicotinimide (BNAH)[7] or Et$_3$N[9] as the reductive quencher of excited state *[Ru(bpy)$_3$]$^{2+}$. The radical HO–C$^\bullet$R$_2$ intermediate formed in the *[Ru(bpy)$_3$]$^{2+}$/ BNAH photocatalytic cycle can be intercepted by the BNA$^\bullet$ species to form secondary benzylic alcohols in good yields (eq 4).[10] [Ru(bpy)$_3$](PF$_6$)$_2$ has been used in combination with catalytic viologen species under oxidative quenching conditions of *[Ru(bpy)$_3$]$^{2+}$ to promote Meerwein–Ponndorf–Verley-type reductions of aryl ketones.[11] Under reductive quenching conditions, radical decarboxylation of N-(acyloxy)phthalimides proceeds in the presence of [Ru(bpy)$_3$]Cl$_2$, BNAH, and light to generate alkyl radicals that react with α,β-unsaturated ketones to form β-alkylated products (eq 5).[12] Cleavage of carbon–oxygen single bonds can also be achieved via the photoreductive ring opening of β-epoxy ketones to form β-hydroxy ketones[6] using [Ru(bpy)$_3$]Cl$_2$ and a dimethylbenzimidazoline (eq 6).[13]

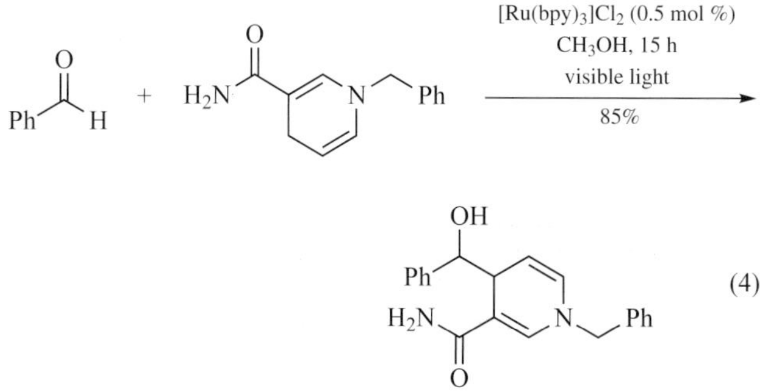

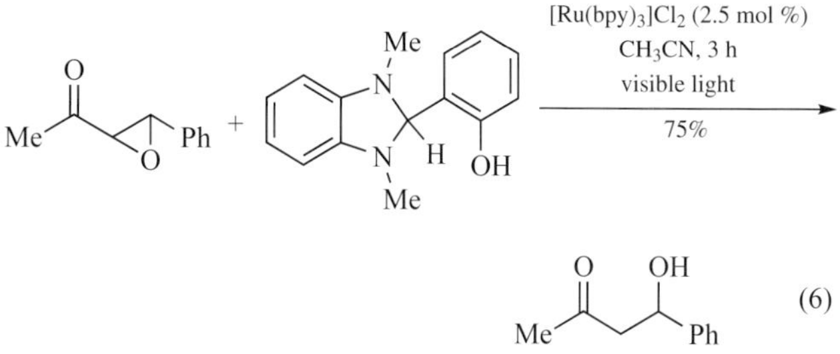

(6)

the absence and presence of perchloric acid, respectively, using [Ru(bpy)$_3$]Cl$_2$, N-methyldihydroacridine and light (eq 7).[15] Modern development of this transformation has led to an expansion of the substrate scope to include bromopyrroloindolines and α-bromo- and α-chloroamides using [Ru(bpy)$_3$]Cl$_2$ in combination with iPr$_2$NEt and either HE or formic acid using a fluorescent light bulb as a light source (eq 8).[16] This study also demonstrated the chemoselective conversion of activated halides over aryl halides as well as high functional group tolerance. Dehalogenations of vicinal dibromides is known to produce stilbenes under oxidative quenching conditions utilizing light, [Ru(bpy)$_3$]Cl$_2$, substoichiometric viologen, and (NH$_4$)$_3$EDTA in a biphasic EtOAc-H$_2$O system (eq 9).[17] Reductive quenching conditions for similar transformations can be accessed using a [Ru(bpy)$_3$]Cl$_2$/Et$_3$N[9] or [Ru(bpy)$_3$]Cl$_2$/1,5-dimethoxynaphthalene/ascorbic acid[18] catalytic cycle.

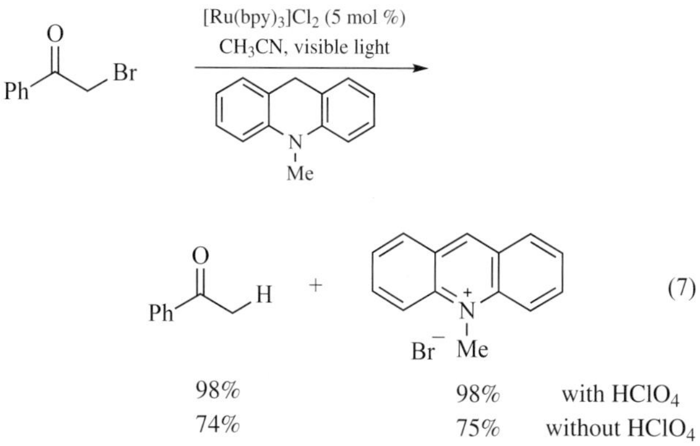

(4)

Reductions of Carbon–Halogen Bonds. Reductive dehalogenation of substrates bearing an α-electron withdrawing group have been well studied by several groups in the field of photocatalysis. [Ru(bpy)$_3$]Cl$_2$ has been used with 3-methyl-2,3-dihydrobenzothiazole and a fluorescent lamp to carry out the reduction of various organohalides such as α-chloride ketones and α-chloride esters, α-bromo esters, and bromomalonates.[14] Both the reductive and oxidative quenching pathways of *[Ru(bpy)$_3$]$^{2+}$ have been utilized for the reduction of α-bromoacetophenone in

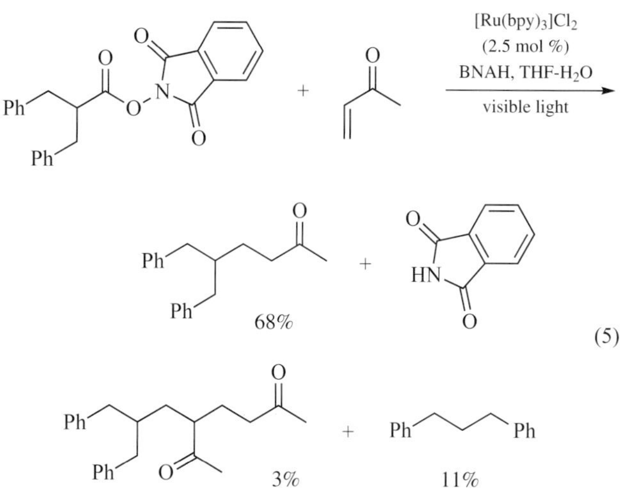

(7)

$$(8)$$

$$(10)$$

$$(9)$$

$$(11)$$

Oxidation of Carbon–Hydrogen Bonds. Alkyl radicals produced from halide reductions can be exploited for additions onto various π-systems through modification of the $[Ru(bpy)_3]Cl_2$ cycle. Enantioselective alkylations of aldehydes have been carried out using dual photocatalytic/organocatalytic cycles (eq 10).[19] Excellent enantioselectivities and yields are obtained using this system for the asymmetric α-trifluoromethylation[20] and α-benzylation[21] of aldehydes. Additions of TEMPO onto morpholine-derived enamines have been described using the reductive quenching cycle of $[Ru(bpy)_3](PF_6)_2$.[22] Silyl enol ethers of ketones, amides, and aldehydes can undergo α-perfluoroalkylations in the presence of $[Ru(bpy)_3]Cl_2$ and light in a one-pot process from the corresponding carbonyl compounds (eq 11).[23] The $[Ru(bpy)_3]Cl_2$/Et$_3$N reductive quenching cycle has also been shown to mediate the intramolecular addition of activated bromides onto heterocycles[24] and electron-rich olefins (eq 12).[25] The analogous intermolecular bromomalonate additions onto heterocycles require the use of a non-hydrogen atom donor triarylamine as the reductive quencher.[26] Bromopyrroloindolines can be coupled with indoles under similar reductive quenching conditions to obtain complex bisindole structures, which have been recently demonstrated in the synthesis of gliocladin C (eq 13).[27] Related iridium-based photocatalysts are also competent in many of these reactions, including atom transfer reactions.[28] Glycosyl halides can be reduced using the reductive quenching cycle of $[Ru(bpy)_3](BF_4)_2$ and the resultant radical can add onto electron-deficient alkenes to obtain exclusive α-substituted products in high yields (eq 14).[29] In addition to C–H oxidation via reduction of organohalides, $[Ru(bpy)_3]Cl_2$ is known to perform direct oxidations of phenols to generate cross-linked proteins. Nearby tyrosine residues oxidize in the presence of $[Ru(bpy)_3]^{3+}$ through oxidative quenching of *$[Ru(bpy)_3]^{2+}$ with persulfate ion and combine rapidly ($\sim$0.5 s) to form the bis(phenol)-linked protein in good yield (eq 15).[30]

$$(12)$$

$$(13)$$

$$dr = 1.5:1 \tag{14}$$

$$ \tag{15}$$

$$ \tag{17}$$

Oxidations of Carbon–Heteroatom Bonds.

Iminium ions are produced by the single-electron oxidation of tertiary amines by *[Ru(bpy)$_3$]$^{2+}$, followed by hydrogen atom abstraction and can be trapped by nucleophiles to obtain diverse scaffolds. N-Aryl tetrahydroisoquinolines have been oxidized in the presence of [Ru(bpy)$_3$]Cl$_2$, visible light, and nitroalkanes to promote aza-Henry reactions under aerobic conditions in high yields (eq 16).[31] Azomethine ylides can be formed under similar conditions to undergo efficient [3+2] cycloadditions with various dipolarophiles, generating highly functionalized pyrrolo[2,1-a]isoquinolines.[32] Mannich reactions between acetone and N-aryl tetrahydroisoquinolines have been afforded using [Ru(bpy)$_3$](PF$_6$)$_2$, a fluorescent lamp, and a proline catalyst.[33] For transformations involving tetrahydroisoquinolines, the iridium-based photocatalysts have also been successful. The use of [Ru(bpy)$_3$]Cl$_2$, visible light, and base in the presence of oxygen has been shown to promote intramolecular cyclizations of amines onto in situ oxidized benzylamines to form tetrahydroimidazoles.[34] In addition to amine oxidations, when [Ru(bpy)$_3$]Cl$_2$ is combined with CBr$_4$ or CHI$_3$ in the presence of DMF, an aminal radical intermediate is formed, which is then oxidized by [Ru(bpy)$_3$]$^{3+}$ to form a Vilsmeier–Haack reagent that activates primary or secondary alcohols toward displacement by bromide or iodide to give high yields of halogenated products (eq 17).[35]

$$ \tag{16}$$

Cycloadditions.

Bis-enones are well known to undergo single-electron transfer-mediated cyclizations under a variety of conditions.[36] The reduction of aryl enones to the radical anion species by [Ru(bpy)$_3$]$^+$ has been reported under reductive quenching conditions of *[Ru(bpy)$_3$]$^{2+}$ with [Ru(bpy)$_3$]Cl$_2$, iPr$_2$NEt, and LiBr as a Lewis acid using a flood lamp or under ambient sunlight (eq 18).[37] This process was shown to initiate both intra- and intermolecular formal [2+2] cycloadditions of symmetrical and unsymmetrical[38] enones in high diastereomeric ratio (dr) and yield. The intramolecular formal [3+2] cycloaddition of aryl cyclopropyl ketones and alkenes likewise forms fused cyclopentane structures with good dr and in high yield in the presence of [Ru(bpy)$_3$]Cl$_2$, TMEDA, La(OTf)$_3$, and light (eq 19).[39] These conditions have considerable advantages over other metal-mediated or electrochemical-promoted processes with respect to yield and chemoselectivity. One of the most common types of cycloadditions is the Diels–Alder [4+2] reaction, which can establish six-membered cycles with excellent regio- and stereocontrol. Accordingly, single-electron reduction of bis-enones initiates intramolecular hetero [4+2] cycloadditions in the presence of [Ru(bpy)$_3$]Cl$_2$, iPr$_2$NEt, LiBr, and light in H$_2$O–CH$_3$CN mixtures to give products in >10:1 dr (eq 20).[40] Notably, both reactive partners are electron deficient and the observed regioselectivity is opposite to that obtained using methods involving a two-electron pathway. The oxidative quenching cycle of *[Ru(bpy)$_3$]$^{2+}$ can alternatively be accessed for [2+2] cycloadditions of electron-rich styrenes using a [Ru(bpy)$_3$](PF$_6$)$_2$/methyl viologen (MV) system under sunlight via single-electron oxidation of the cyclopropyl ketone (eq 21).[41]

$$ \tag{18}$$

Miscellaneous.

The reduction of alkyl azides and biomolecule-containing azides to the corresponding primary amine has been recently accomplished with [Ru(bpy)$_3$]Cl$_2$ and visible light using an appropriate reductive quencher in organic as

$$\xrightarrow[\substack{CH_3CN, \ MgSO_4, \ 29 \ h \\ 23 \ W \ \text{fluorescent bulb}}]{\substack{[Ru(bpy)_3]Cl_2 \ (2.5 \ mol \ \%) \\ TMEDA \ (5 \ equiv) \\ La(OTf)_3 \ (1 \ equiv)}}$$

63%
10:1 dr (19)

$$\xrightarrow[\substack{H_2O \ (10 \ equiv), \ CH_3CN \\ 200 \ W \ \text{floodlight}, \ 1 \ h}]{\substack{[Ru(bpy)_3]Cl_2 \ (5 \ mol \ \%) \\ iPr_2NEt \ (3 \ equiv) \\ LiBr \ (2 \ equiv)}}$$

86%
>10:1 dr (20)

$$\xrightarrow[\substack{MeNO_2, \ MgSO_4, \ 3.5 \ h \\ \text{visible light} \\ 88\%}]{\substack{[Ru(bpy)_3](PF_6)_2 \ (5 \ mol \ \%) \\ MV(PF_6)_2 \ (15 \ mol \ \%)}}$$

(21)

well as aqueous solvents (eq 22).[42] This transformation exhibits high levels of chemoselectivity and functional group tolerance, with alkyl azides reduced in the presence of sensitive moieties such as disulfides, aldehydes, alcohols, and alkyl halides. Aryl radicals have also been generated from aryl iodonium salts using [Ru(bpy)₃]Cl₂ and silane additives in the presence of visible light to promote the ring-opening radical polymerization of epoxides.[43] Irradiation with a green fluorescent light bulb was shown to improve the efficiency of the transformation.

$$\xrightarrow[\substack{(HOCH_2)_3CNH, \ pH \ 7.4 \\ \text{visible light}, \ 10 \ min, \ 25 \ ^\circ C \\ >95\%}]{\substack{[Ru(bpy)_3]Cl_2 \ (1 \ mM) \\ \text{ascorbate} \ (50 \ mM)}}$$

DNA
oligonucleotide

(22)

Related Reagents. [Bis(2-phenylpyridine)(4,4′-*tert*-butyl-2,2′-bipyridine)iridium(III)]hexafluorophosphate, [Ir(ppy)₂-(dtbbpy)]PF₆; [4,4′-bis(1,1-dimethylethyl)-2,2′-bipyridine]-bis[3,5-difluoro-2-[5-(trifluoromethyl)-2-pyridinyl-κN]phenyl]-iridium(III)hexafluorophosphate, [Ir(dF(CF₃)ppy)₂(dtbbpy)]PF₆.

1. (a) Yoon, T. P.; Ischay, M. A.; Du, J., *Nat. Chem.* **2010**, *2*, 527. (b) Narayanam, J. M. R.; Stephenson, C. R. J., *Chem. Soc. Rev.* **2011**, *40*, 102. (c) Teplý, F., *Collect. Czech. Chem. Commun.* **2011**, *76*, 859.

2. (a) Kalyanasundaram, K., *Coord. Chem. Rev.* **1982**, *46*, 159. (b) Juris, A.; Balzani, V.; Barigelletti, F.; Campagna, S.; Belser, P.; von Zelewsky, A., *Coord. Chem. Rev.* **1988**, *84*, 85. (c) Balzani, V.; Bergamini, G.; Marchioni, F.; Ceroni, P., *Coord. Chem. Rev.* **2006**, *250*, 1254.

3. Mabrouk, P. A.; Wrighton, M. S., *Inorg. Chem.* **1986**, *25*, 526.

4. Van Bergen, T. J.; Hedstrand, D. M.; Kruizinga, W. H.; Kellog, R. M., *J. Org. Chem.* **1979**, *44*, 4953.

5. Cano-Yelo, H.; Deronzier, A., *J. Chem. Soc., Perkin Trans. 2.* **1984**, 1093.

6. Larraufie, M. H.; Pellet, R.; Fensterbank l L, Goddard, J. P.; Lacôte, E.; Malacria, M.; Ollivier, C., *Angew. Chem., Int. Ed.* **2011**, *50*, 4463.

7. Ishitani, O.; Pac, C.; Sakurai, H., *J. Org. Chem.* **1983**, *48*, 2941.

8. Hedstrand, D. M.; Kruizinga, W. H.; Kellog, R. M., *Tetrahedron Lett.* **1978**, *19*, 1255.

9. Willner, I.; Tsfania, T.; Eichen, Y., *J. Org. Chem.* **1990**, *55*, 2656.

10. Ishitani, O.; Yanagida, S.; Takamuku, S.; Pac, C., *J. Org. Chem.* **1987**, *52*, 2790.

11. Herance, J. R.; Ferrer, B.; Bourdelande, J. L.; Marquet, J.; Garcia, H., *Chem. Eur. J.* **2006**, *12*, 3890.

12. Okada, K.; Okamoto, K.; Morita, N.; Okubo, K.; Oda, M., *J. Am. Chem. Soc.* **1991**, *113*, 9401.

13. Hasegawa, E.; Takizawa, S.; Seida, T.; Yamaguchi, A.; Yamaguchi, N.; Chiba, N.; Takahashi, T.; Ikeda, H.; Akiyama, K., *Tetrahedron* **2006**, *62*, 6581.

14. Mashraqui, S. H.; Kellog, R. M., *Tetrahedron Lett.* **1985**, *26*, 1453.

15. Fukuzumi, S.; Mochizuki, S.; Tanaka, T., *J. Phys. Chem.* **1990**, *94*, 722.

16. Narayanam, J. M. R.; Tucker, J. W.; Stephenson, C. R. J., *J. Am. Chem. Soc.* **2009**, *131*, 8756.

17. (a) Maidan, R.; Goren, Z.; Becker, J. Y.; Willner, I., *J. Am. Chem. Soc.* **1984**, *106*, 6217. (b) Goren, Z.; Willner, I., *J. Am. Chem. Soc.* **1983**, *105*, 7764.

18. Maji, T.; Karmakar, A.; Reiser, O., *J. Org. Chem.* **2011**, *76*, 736.

19. Nicewicz, D. A.; MacMillan, D. W. C., *Science* **2008**, *322*, 77.

20. Nagib, D. A.; Scott, M. E.; MacMillan, D. W. C., *J. Am. Chem. Soc.* **2009**, *131*, 10875.

21. Shih, H.-W.; Vander Wal, M. N.; Grange, R. L.; MacMillan, D. W. C., *J. Am. Chem. Soc.* **2010**, *132*, 13600.

22. Koike, T.; Akita, M., *Chem. Lett.* **2009**, *38*, 166.

23. Pham, P. V.; Nagib, D. A.; MacMillan, D. W. C., *Angew. Chem., Int. Ed.* **2011**, *50*, 6119.

24. Tucker, J. W.; Narayanam, J. M. R.; Krabbe, S. W.; Stephenson, C. R. J., *Org. Lett.* **2010**, *12*, 368.

25. Tucker, J. W.; Nguyen, J. D.; Narayanam, J. M. R.; Krabbe, S. W.; Stephenson, C. R. J., *Chem., Commun.* **2010**, *46*, 4985.

26. Furst, L.; Matsuura, B. S.; Narayanam, J. M. R.; Tucker, J. W.; Stephenson, C. R. J., *Org. Lett.* **2010**, *12*, 3104.

27. Furst, L.; Narayanam, J. M. R.; Stephenson, C. R. J., *Angew. Chem., Int. Ed.* **2011**, DOI: 10.1002/anie.201103145.

28. Nguyen, J. D.; Tucker, J. W.; Konieczynska, M.; Stephenson, C. R. J., *J. Am. Chem. Soc.* **2011**, *133*, 4160.

29. Andrews, R. S.; Becker, J. J.; Gagné, M. R., *Angew. Chem., Int. Ed.* **2010**, *49*, 7274.

30. Fancy, D. A.; Kodadek, T., *Proc. Natl. Acad. Sci. USA* **1999**, *96*, 6020.

31. Condie, A. G.; González-Gómez, J.-C., Stephenson, C. R. J., *J. Am. Chem. Soc.* **2010**, *132*, 1464.

32. Zou, Y.-Q.; Lu, L.-Q.; Fu, L.; Chang, N.-J.; Rong, J.; Chen, J.-R.; Xiao, W.-J., *Angew. Chem., Int. Ed.* **2011**, *50*, 7171.

33. Rueping, M.; Vila, C.; Koenigs, R. M.; Poscharny, K.; Fabry, D. C., *Chem. Commun.* **2011**, *47*, 2360.

34. Xuan, J.; Cheng, Y.; An, J.; Lu, L.-Q.; Zhang, X.-X.; Xiao, W.-J., *Chem. Commun.* **2011**, *47*, 8337.

35. Dai, C.; Narayanam, J. M. R.; Stephenson, C. R. J., *Nature Chem.* **2011**, *3*, 140.

36. (a) Yang, J.; Felton, G. A. N.; Bauld, N. L.; Krische, M. J., *J. Am. Chem. Soc.* **2004**, *126*, 1634. (b) Wang, L.-C.; Jang, H.-Y.; Lynch, V.; Krische, M. J., *J. Am. Chem. Soc.* **2002**, *124*, 9448. (c) Baik, T. G.; Wang, L.-C.; Luiz, A.-L., Krische, M. J., *J. Am. Chem. Soc.* **2001**, *123*, 6716.

37. Ischay, M. A.; Anzovino, M. E.; Du, J.; Yoon, T. P., *J. Am. Chem. Soc.* **2008**, *130*, 12886

38. Du, J.; Yoon, T. P., *J Am. Chem. Soc.* **2009**, *131*, 14604.

39. Lu, Z.; Shen, M.; Yoon, T. P., *J. Am. Chem. Soc.* **2011**, *133*, 1162.

40. Hurtley, A. E.; Cismesia, M. A.; Ischay, M. A.; Yoon, T. P., *Tetrahedron* **2011**, *67*, 4442.

41. Ischay, M. A.; Lu, Z.; Yoon, T. P., *J. Am. Chem. Soc.* **2010**, *132*, 8572.

42. Chen, Y.; Kamlet, A. S.; Steinman, J. B.; Liu, D. R., *Nat. Chem.* **2011**, *3*, 146.

43. Lalevee, J.; Blanchard, N.; Tehfe, M. A.; Morlet-Savary, F.; Fouassier, J. P., *Macromolecules* **2010**, *43*, 10191.

Laura Furst & Corey R. J. Stephenson
Boston University, Boston, MA, USA

S

Essential Reagents for Organic Synthesis, Edited by Philip L. Fuchs, André B. Charette, Tomislav Rovis, and Jeffrey W. Bode.
©2016 John Wiley & Sons, Ltd. Published 2016 by John Wiley & Sons, Ltd.

Samarium(II) Iodide[1]

[32248-43-4] I$_2$Sm (MW 404.16)

InChI = 1/2HI.Sm/h2*1H;/q;;+2/p-2/f2I.Sm/h2*1h;/q2*-1;m

InChIKey = UAWABSHMGXMCRK-ZFDXCKRNCZ

(one-electron reducing agent possessing excellent chemoselectivity in reduction of carbonyl, alkyl halide, and α-heterosubstituted carbonyl substrates;[1] promotes Barbier-type coupling reactions, ketyl–alkene coupling reactions, and radical cyclizations[1])

Physical Data: mp 527 °C; bp 1580 °C; d 0.922 g cm^{-3}.

Solubility: soluble 0.1M in THF.

Form Supplied in: commercially available as a 0.10 M solution in THF.

Preparative Methods: typically prepared in situ for synthetic purposes. SmI$_2$ is conveniently prepared by oxidation of ***Samarium(0)*** metal with organic dihalides.[2]

Handling, Storage, and Precautions: is air sensitive and should be handled under an inert atmosphere. SmI$_2$ may be stored over THF for long periods when it is kept over a small amount of samarium metal.

Original Commentary

Gary A. Molander & Christina R. Harris
University of Colorado, Boulder, CO, USA

Reduction of Organic Halides and Related Substrates.
Alkyl halides are readily reduced to the corresponding hydrocarbon by SmI$_2$ in the presence of a proton source. The ease with which halides are reduced by SmI$_2$ follows the order I > Br > Cl. The reduction is highly solvent dependent. In THF solvent, only primary alkyl iodides and bromides are effectively reduced;[2] however, addition of HMPA effects the reduction of aryl, alkenyl, primary, secondary, and tertiary halides (eq 1)[3,4] Tosylates are also reduced to hydrocarbons by SmI$_2$. Presumably, under these reaction conditions the tosylate is converted to the corresponding iodide which is subsequently reduced.[4,5]

$$\text{(1)}$$

Samarium(II) iodide provides a means to reduce substrates in which the halide is resistant to reduction by hydride reducing agents (eq 2).

$$\text{(2)}$$

Samarium(II) iodide has been utilized as the reductant in the Boord alkene-type synthesis involving ring scission of 3-halotetrahydrofurans (eq 3).[6] SmI$_2$ provides an alternative to the sodium-induced reduction which typically affords mixtures of stereoiso-

meric alkenes and overreduction in these transformations. When SmI$_2$ is employed as the reductant, isomeric purities are generally >97% and overreduction products comprise <3% of the reaction mixture.

$$\text{(3)}$$

Reduction of α-Heterosubstituted Carbonyl Compounds.
Samarium(II) iodide provides a route for the reduction of α-heterosubstituted carbonyl substrates. A wide range of α-heterosubstituted ketones is rapidly reduced to the corresponding unsubstituted ketone under mild conditions (eq 4).[7] The reaction is highly selective and may be performed in the presence of isolated iodides as well as isolated ketones.[7]

$$\text{(4)}$$

Y	Isolated yield (%)
Cl	100
SPh	76
S(O)Ph	64
SO$_2$Ph	88

Samarium(II) iodide-induced reductive cleavage of α-hydroxy ketones provides a useful entry to unsubstituted ketones (eq 5).[8]

$$\text{(5)}$$

Samarium(II) iodide promotes the reductive cleavage of α-alkoxy ketones. Pratt and Hopkins have utilized this protocol in synthetic studies en route to betaenone B (eq 6).[9]

$$\text{(6)}$$

Likewise, this procedure provides a route for the reduction of α,β-epoxy ketones and α,β-epoxy esters to generate the corresponding β-hydroxy carbonyl compounds (eqs 7 and 8).[3,10] The epoxy ketone substrates may be derived from Sharpless asymmetric epoxidation. Consequently, this procedure provides a means to prepare a variety of chiral, nonracemic β-hydroxy carbonyl compounds that are difficult to acquire by more traditional procedures.

$$\text{(7)}$$

$$ (8) $$

$$ >98\% \text{ ee} $$

Vinyloxiranes undergo reductive epoxide ring opening with samarium(II) iodide to provide (E)-allylic alcohols (eq 9).[3,10b,11] These reaction conditions are tolerant of ketone, ester, and nitrile functional groups. Again, Sharpless asymmetric epoxidation chemistry may be utilized to gain entry to the desired non-racemic substrates, thereby providing a useful entry to highly functionalized, enantiomerically enriched allylic alcohols.

$$ (9) $$

Y	Isolated yield (%)
COSEt	80
SO$_2$Ph	82
P(O)(OEt)$_2$	84
H	69
Me	42
SPh	54

A useful method for preparation of β-hydroxy esters is accomplished by SmI$_2$-promoted deoxygenation of an α-hydroxy ester followed by condensation with a ketone (eq 10).[12] In some instances, excellent diastereoselectivities are achieved, although this appears to be somewhat substrate dependent.

$$ (10) $$

A useful reaction sequence for transforming carbonyl compounds to one-carbon homologated nitriles has evolved from the ability of SmI$_2$ to deoxygenate cyanohydrin O,O'-diethyl phosphates (eq 11).[13] The procedure is tolerant of a number of functional groups including alcohols, esters, amides, sulfonamides, acetals, alkenes, alkynes, and amines. Furthermore, it provides a distinct advantage over other previously developed procedures for similar one-carbon homologations.

$$ (11) $$

Deoxygenation Reactions. Sulfoxides are reduced to sulfides by SmI$_2$ (eq 12).[2,3,14] This process is rapid enough that reduction of isolated ketones is not a competitive process. Likewise, aryl sulfones are reduced to the corresponding sulfides by SmI$_2$ (eq 13).[3,12]

$$ (12) $$

$$ (13) $$

Barbier-type Reactions. Samarium(II) iodide is quite useful in promoting Barbier-type reactions between aldehydes or ketones and a variety of organic halides. The efficiency of SmI$_2$ promoted Barbier-type coupling processes is governed by the substrate under consideration in addition to the reaction conditions employed. In general, alkyl iodides are most reactive while alkyl chlorides are virtually inert. Typically, catalytic **Iron(III) Chloride** or **Hexamethylphosphoric Triamide** can be added to SmI$_2$ to reduce reaction times or temperatures and enhance yields. Kagan and co-workers have recently applied an intermolecular SmI$_2$-promoted Barbier reaction towards the synthesis of hindered steroidal alcohols. An intermolecular Barbier-type reaction between the hindered ketone and **Iodomethane** produced a 97:3 mixture of diastereomers in excellent yield (eq 14).[15]

$$ (14) $$

Samarium(II) iodide-promoted intramolecular Barbier-type reactions have also been employed to produce a multitude of cyclic and bicyclic systems.[1] Molander and McKie have employed an intramolecular Barbier-type reductive coupling reaction to promote the formation of bicyclo[m.n.1]alkan-1-ols from the corresponding iodo ketone substrates in good yield (eq 15).[16]

$$ (15) $$

Annulation of five- and six-membered rings proceeds with excellent diastereoselectivity via an intramolecular Barbier-type process (eq 16).[17] The Barbier-type coupling scheme provides a reliable and convenient alternative to other such methods for preparing fused bicyclic systems.

$$ (16) $$

only diastereomer

The SmI$_2$-promoted Barbier-type reaction has also been utilized in the synthesis of polyquinanes. Cook and Lannoye have employed this method to effect a bis-annulation of an appropriately substituted diketone (eq 17).[18]

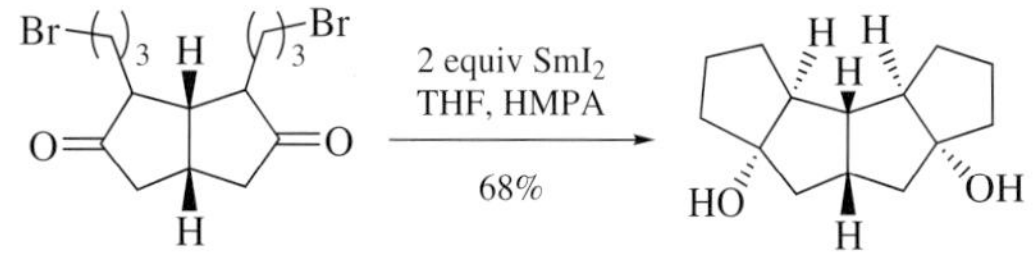

(17)

Substituted β-keto esters also provide excellent substrates for the intramolecular Barbier cyclization (eq 18).[19] Diastereoselectivities are typically quite good but are highly dependent on substituent and solvent effects.

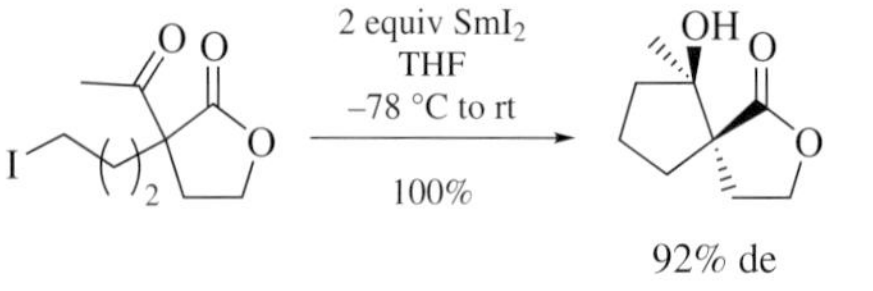

(18)

Nucleophilic Acyl Substitutions. Samarium(II) iodide facilitates the highly selective intramolecular nucleophilic acyl substitution of halo esters (eqs 19 and 20).[20]

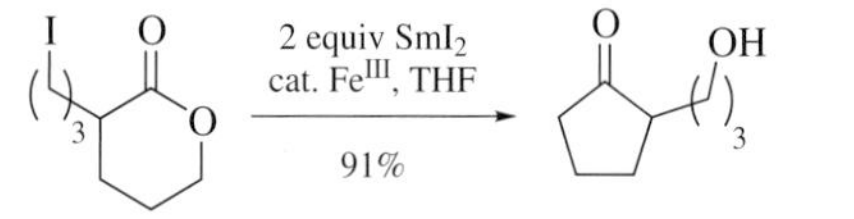

(19)

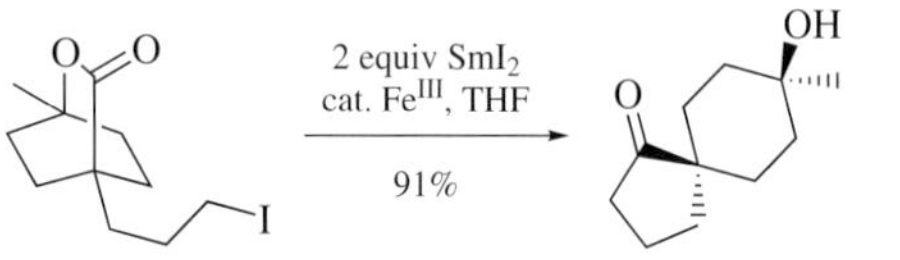

(20)

Unlike organolithium or organomagnesium reagents, SmI$_2$-promoted nucleophilic substitution does not proceed with double addition to the carbonyl, nor are any products resulting from reduction of the final product observed. With suitably functionalized substrates, this procedure provides a strategy for the formation of eight-membered rings (eq 21).

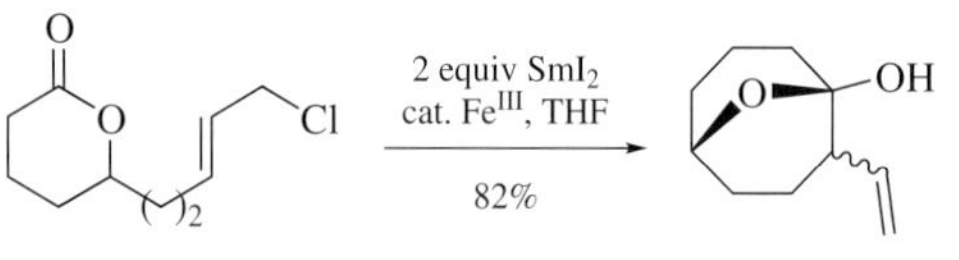

(21)

Ketone–Alkene Coupling Reactions. Ketyl radicals derived from reduction of ketones or aldehydes with SmI$_2$ may be coupled both inter- and intramolecularly to a variety of alkenic species. Excellent diastereoselectivities are achieved with intramolecular coupling of the ketyl radical with α,β-unsaturated esters.[21] In the following example, ketone–alkene cyclization took place in a stereocontrolled manner established by chelation of the resulting Sm(III) species with the hydroxyl group incorporated in the substrate (eq 22).[21b]

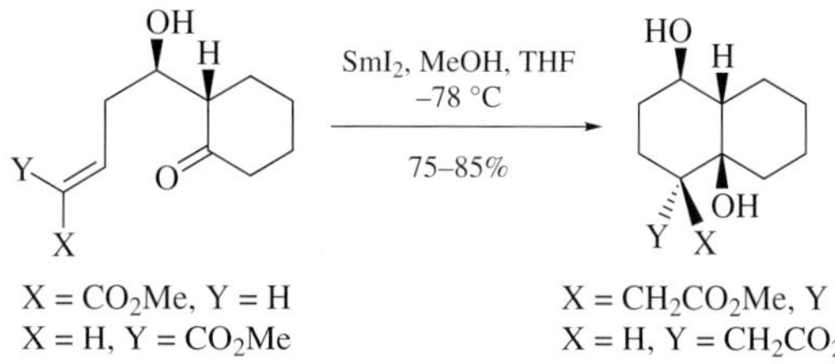

(22)

A similar strategy utilizing β-keto esters provided very high diastereoselectivities in the ketyl–alkene coupling process. In these examples, chelation control about the developing hydroxyl and carboxylate stereocenters was the source of the high diastereoselectivity achieved (eq 23).[22]

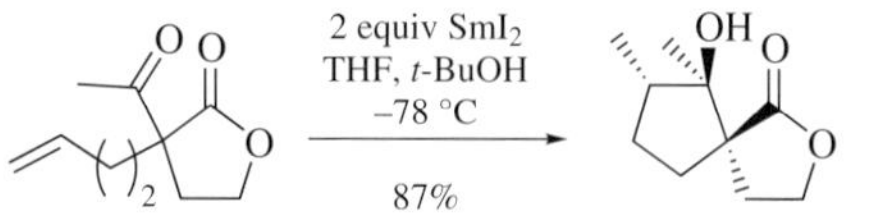

(23)

Alkynic aldehydes likewise undergo intramolecular coupling to generate five- and six-membered ring carbocycles. This protocol has been utilized as a key step in the synthesis of isocarbacyclin (eq 24).[23] SmI$_2$ was found to be superior to several other reagents in this conversion.

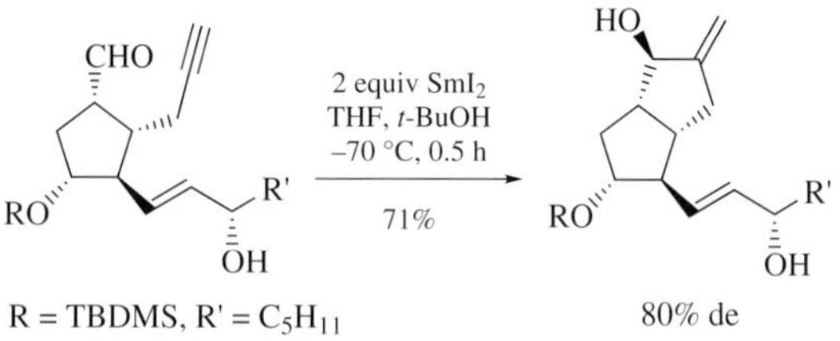

(24)

R = TBDMS, R' = C$_5$H$_{11}$ 80% de

Samarium(II) iodide in the presence of HMPA effectively promotes the intramolecular coupling of unactivated alkenic ketones by a reductive ketyl–alkene radical cyclization process (eq 25). This protocol provides a means to generate rather elaborate carbocycles through a sequencing process in which the resulting organosamarium species is trapped with various electrophiles to afford the cyclized product in high yield.[24]

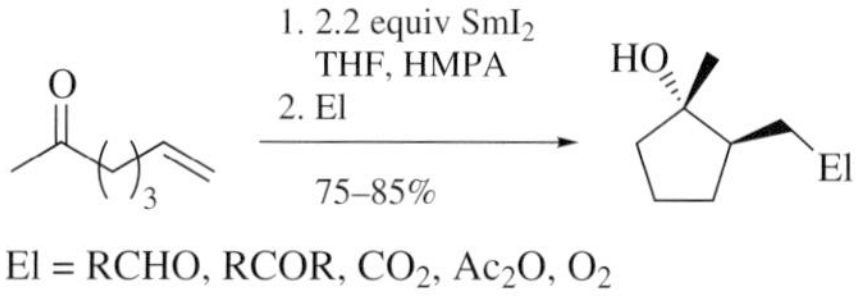

(25)

El = RCHO, RCOR, CO$_2$, Ac$_2$O, O$_2$

Pinacolic Coupling Reactions. In the absence of a proton source, both aldehydes and ketones are cleanly coupled in the presence of SmI$_2$ to the corresponding pinacol.[25] Considerable diastereoselectivity has been achieved in the coupling of aliphatic 1,5- and 1,6-dialdehydes, providing near exclusive formation of the cis-diols (eq 26).[26]

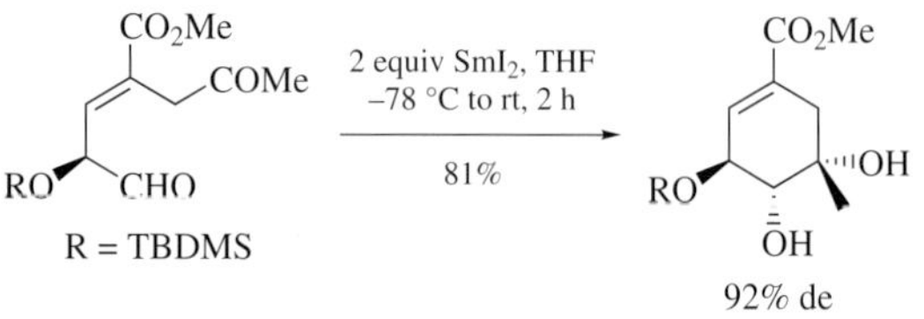

(26)

R = TBDMS   92% de

Intramolecular cross coupling of aldehydes and ketones proceeds with excellent diastereoselectivity and high yield in suitably functionalized systems wherein chelation control by the resulting Sm III species directs formation of the newly formed stereocenters (eq 27).[22a, 27] A similar strategy has been utilized with a β-keto amide substrate to provide a chiral, nonracemic oxazolidinone species. This strategy permits entry to highly functionalized, enantiomerically pure dihydroxycyclopentanecarboxylate derivatives (eq 28).

$$ (27) $$

$$ (28) $$

Radical Addition to Alkenes and Alkynes. Samarium(II) iodide has proven effective for initiation of various radical addition reactions to alkenes and alkynes. Typically, tin reagents are used in the initiation of these radical cyclization reactions; however, the SmI$_2$ protocol often provides significant advantages over these more traditional routes.

Samarium(II) iodide-mediated cyclization of aryl radicals onto alkene and alkyne acceptors provides an excellent route to nitrogen- and oxygen-based heterocycles (eq 29).[28]

$$ (29) $$

The SmI$_2$ reagent is unique in that it provides the ability to construct more highly functionalized frameworks through a sequential radical cyclization/intermolecular carbonyl addition reaction.[29] Thus the intermediate radical formed after initial cyclization may be further reduced by SmI$_2$, forming an organosamarium intermediate which may be trapped by various electrophiles, affording highly functionalized products (eq 30).

$$ (30) $$

Samarium(II) iodide further mediates the cyclization reactions of alkynyl halides (eq 31).[30] When treated with SmI$_2$, the alkynyl halides are converted to the cyclized product in good yield. Addition of DMPU as cosolvent provides slightly higher yields in some instances.

$$ (31) $$

Highly functionalized bicyclic and spirocyclic products are obtained in good yield and high diastereoselectivity by a tandem reductive cleavage–cyclization strategy (eq 32).[31] Radical ring opening of cyclopropyl ketones mediated by samarium(II) iodide-induced electron transfer permits the elaboration of a tandem ring opening–cyclization strategy wherein the resultant samarium enolate may be trapped by either oxygen or carbon electrophiles.

$$ (32) $$

First Update

André Charette
Université de Montréal, Montréal, Québec, Canada

Reductive Cross-coupling of Imines and Aldehydes. Samarium iodide-mediated intramolecular reductive cross-coupling of aldehydes or ketones with oximes (eq 33),[32] hydrazones (eq 34),[33] and imines (eq 35)[34] is well-documented.

$$ (33) $$

$$ (34) $$

$$ (35) $$

The intermolecular coupling of imine derivatives and aldehydes can be achieved using samarium iodide. For example,

N-tosylimines and aldehydes gave *syn-β*-amino alcohol derivatives in good yields and diastereoselectivities (eq 36).[35] Access to enantiopure *syn-β*-amino alcohols can be achieved if chiral chromium complex of the aldehyde is used (eq 37).

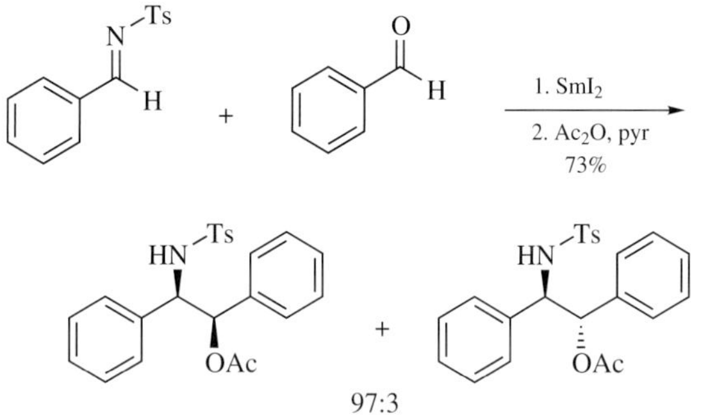

(36)

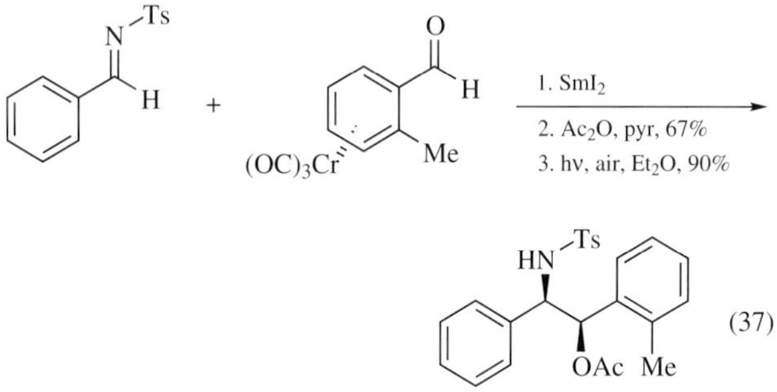

97:3

(37)

It is also possible to couple planar chiral ferrocenecarboxaldehydes with imines with excellent diastereocontrol.[36] Oximes can be coupled with aldehydes in good to excellent yields. However, the level of diastereocontrol is usually quite modest (eq 38).

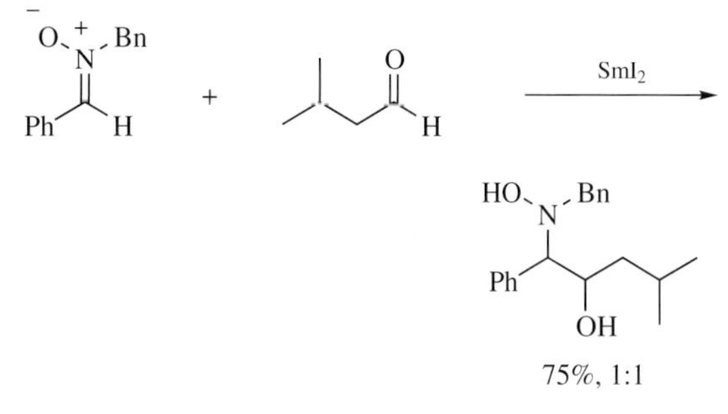

(38)

75%, 1:1

It is possible to use samarium iodide catalytically in several reactions if a cheap alloy of the light lanthanides (La, Ce, Nd, Pr, Sm) called Mischmetall is used.[37]

Synthesis of Alkenes by Reductive Elimination. The treatment of 2-halo-3-hydroxy esters and amides with samarium iodide leads to the corresponding di- or trisubstituted (*E*)-*α,β*-unsaturated derivatives in high yields and diastereoselectivities (eqs 39 and 40).[38] The precursors are readily accessible by condensation of the lithium enolate of *α*-haloesters or amides. If the substrate contains *γ,δ*-unsaturation, the *β,γ*-unsaturated ester is generated in the process (eq 41).

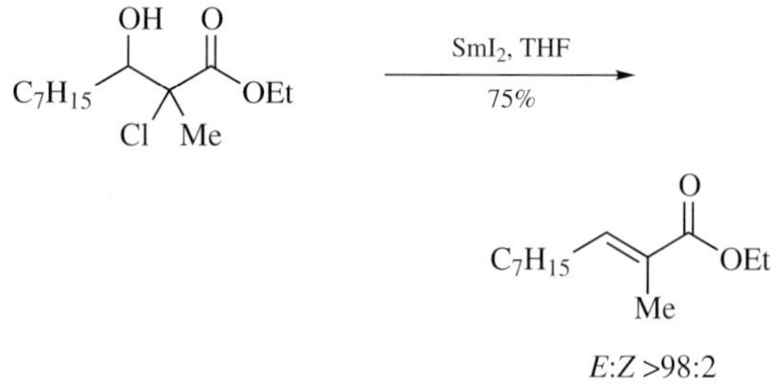

(39)

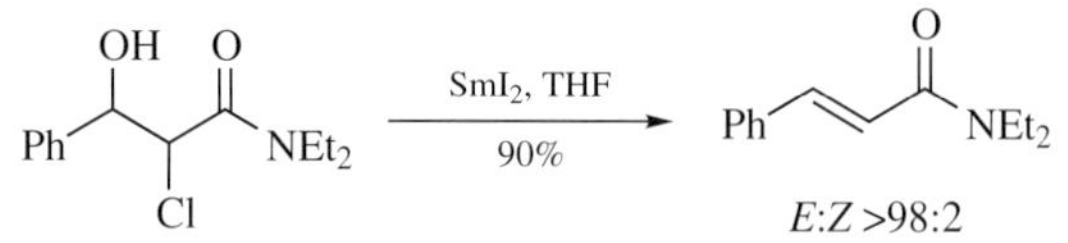

E:Z >98:2

(40)

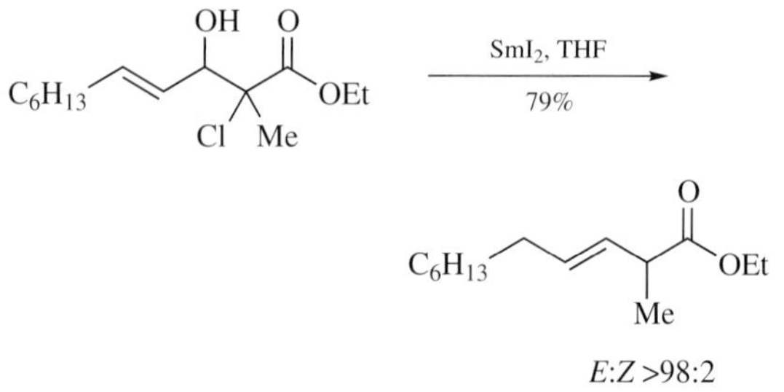

E:Z >98:2

(41)

E:Z >98:2

The stereoselective reduction of *α,α*-dichloro-*β*-hydroxy esters using samarium iodide yields (*Z*)-*α*-chloro-*α,β*-unsaturated esters (eq 42).[39]

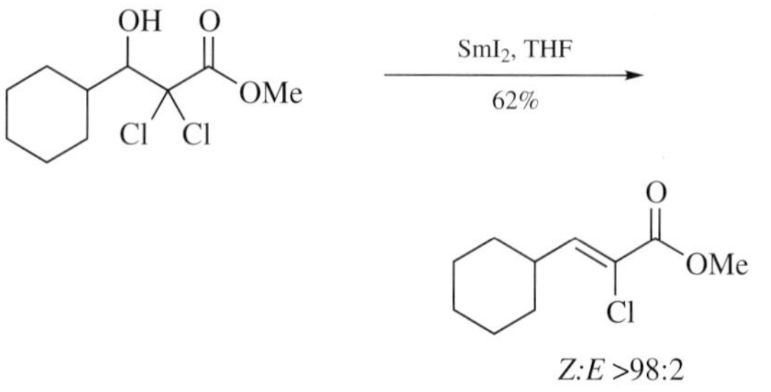

(42)

Z:E >98:2

Similarly, *γ*-acetoxy-*α,β*-enoates are reduced by samarium diiodide to generate dienolates which are kinetically trapped at the *α*-position by electrophiles (proton, aldehydes, or ketones).[40]

(*Z*)-Alkenylsilanes are obtained in high diastereoselectivities if *O*-acetyl-1-chloro-1-trimethylsilylalkan-2-ols are treated with samarium iodide (eq 43). The stereochemical outcome is independent from the relative stereochemistry of the starting material.[41]

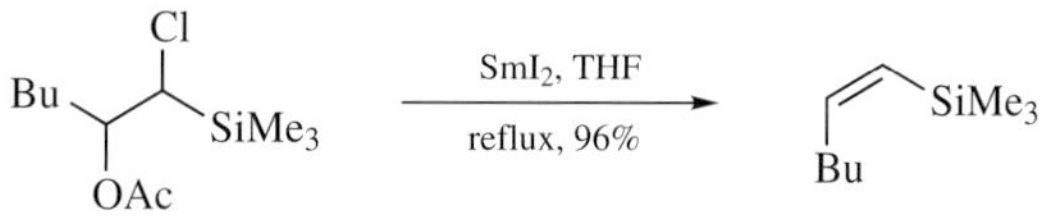

(43)

Samarium iodide can also be used as an alternative to sodium/mercury amalgam for the reductive elimination of 1,2-acetoxy-sulfones in the Julia-Lythgoe olefination.[42] The alkene is generated in a two-step process that first involves DBU or LDA treatment to generate a vinyl sulfone that is then reductively cleaved with samarium iodide (eq 44). The diastereoselectivity of both transformations is usually quite good and the method is compatible with the synthesis of monoalkenes as well as dienes and trienes.

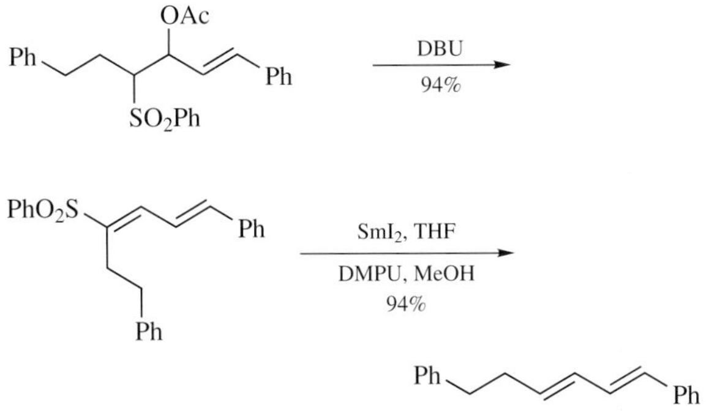

OAc

Ph ⟋⟍⟋ Ph
SO₂Ph

→ DBU / 94%

PhO₂S ⟍ Ph

Ph

→ SmI₂, THF / DMPU, MeOH / 94%

Ph ⟍⟋⟍⟋ Ph   (44)

Synthesis of α-Heteroalkyl Samarium. Samarium iodide is the reagent of choice to generate α-alkoxyalkylsamarium species from suitable precursors. For example, the anomeric position of glycosides can be functionalized by treating a pyridylsulfone precursor with samarium iodide (eq 45). A subsequent quench with an aldehyde generates the corresponding C-glycoside via the Barbier reaction with outstanding diastereoselectivity.[43] It is also possible to generate similar reactive intermediates from the corresponding glycosyl phenylsulfones[44] or glycosyl phosphates.[45]

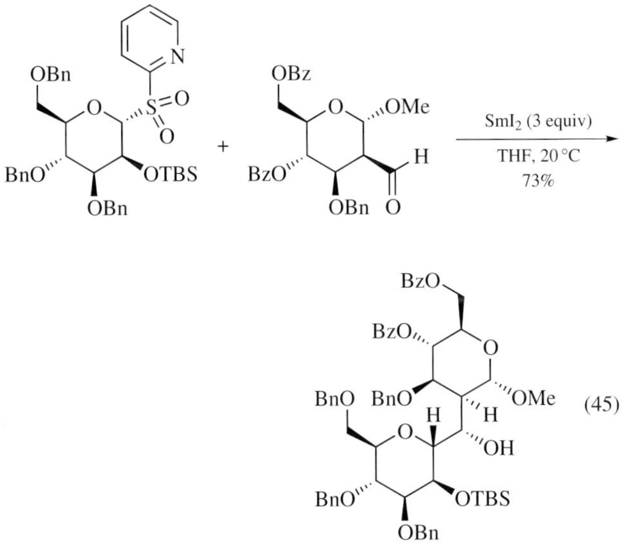

(45)

An alternative but related approach involves the coupling of epoxides and carbonyl compounds (eq 46).[46] In this reaction, the addition of a catalytic amount of nickel(II) iodide[47] produced slightly higher yields of the C-glycoside.

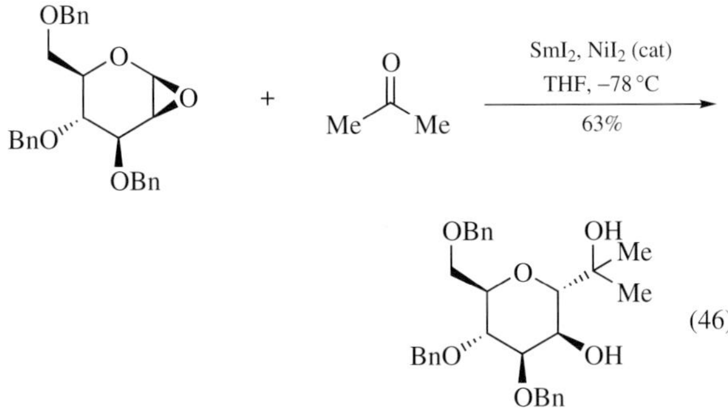

SmI₂, NiI₂ (cat)
THF, −78 °C
63%

(46)

In a related fashion, benzyloxymethyl 2-pyridylsulfone can be used as a hydroxymethylation equivalent to provide a convenient approach for the one-carbon homologation of carbonyl compounds (eq 47).[48] The pyridylsulfone derivative is a superior precursor than the corresponding chloride.

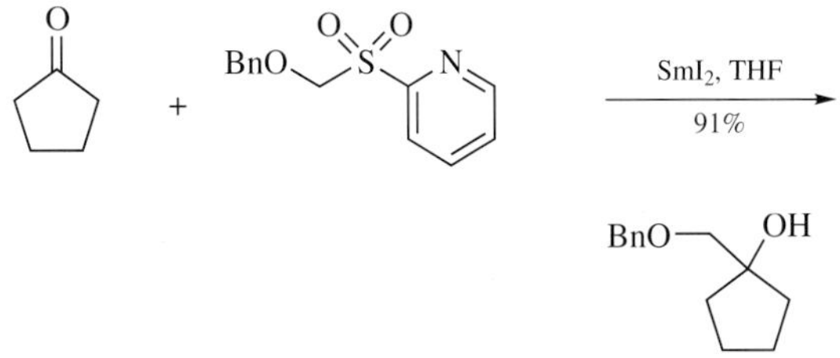

SmI₂, THF
91%

(47)

Another hydroxymethyl equivalent is the silylmethyl group. Tamao oxidation of the product obtained from the samarium iodide-promoted intramolecular reductive cyclization of bromosilyloxy derivatives leads to the hydroxymethyl group (eq 48).[49]

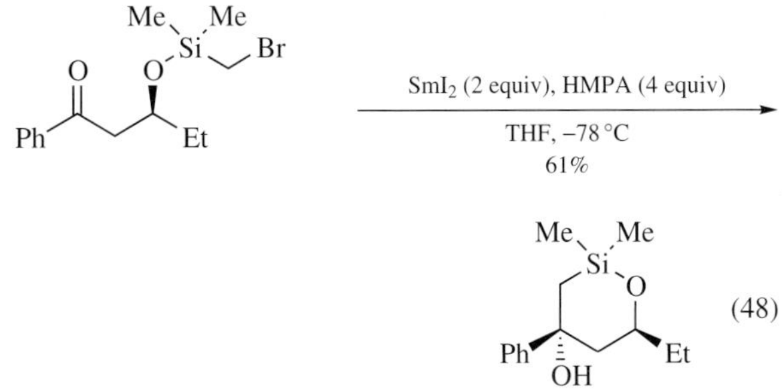

SmI₂ (2 equiv), HMPA (4 equiv)
THF, −78 °C
61%

(48)

The diiodomethylation of carbonyl compounds is also possible if samarium iodide is used in conjunction with iodoform.[50] The products are synthetically useful since they are easily converted into α-hydroxyacids or α-iodoaldehydes upon basic treatment (eq 49).

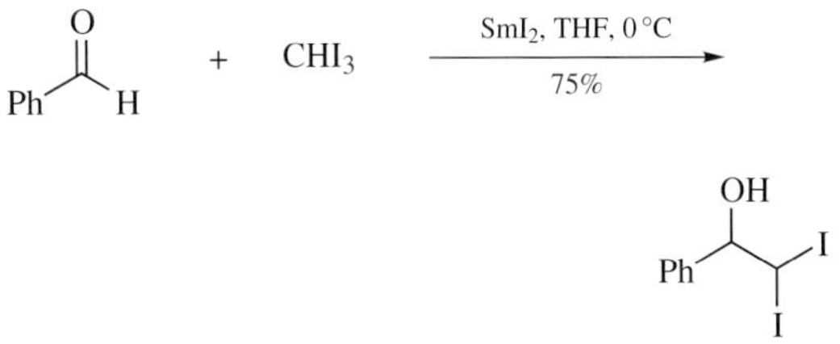

SmI₂, THF, 0 °C
75%

(49)

It is possible to generate an α-heteroalkyl radical by a 1,5-hydrogen atom transfer from the radical obtained from an *o*-iodobenzyl protected amine (eq 50). It can then be subjected to several reactions such as condensation with a ketone.[51]

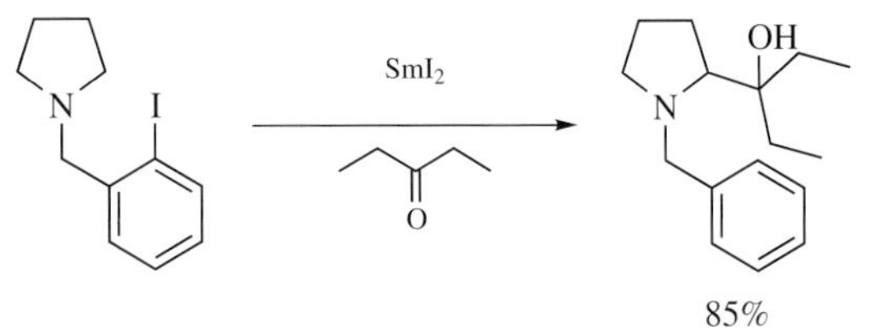

SmI₂

85%

(50)

Alternatively, an α-amino radical can be generated from an α-benzotriazolylamine precursor (eq 51).

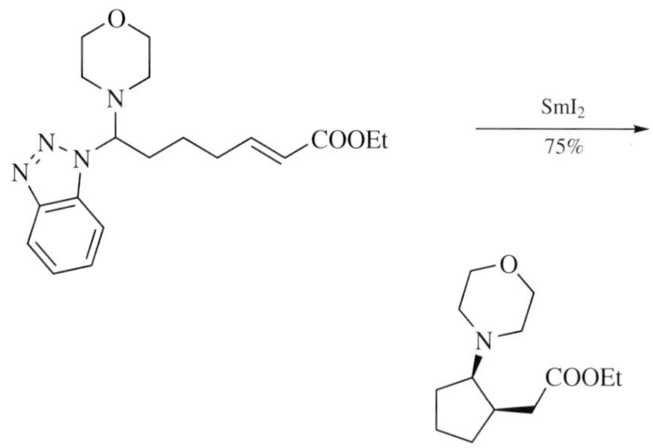

(51)

Opening of α,β-Epoxy Esters and Amides. Treatment of aromatic α,β-epoxyamides with samarium iodide leads to the highly stereoselective synthesis of α,β-unsaturated amides with high diastereocontrol (eq 52).[52] If the reaction is run on a substrate that contains γ-protons, then a base-promoted reaction produces the (E)-α-hydroxy-β,γ-unsaturated amide (eq 53).[53]

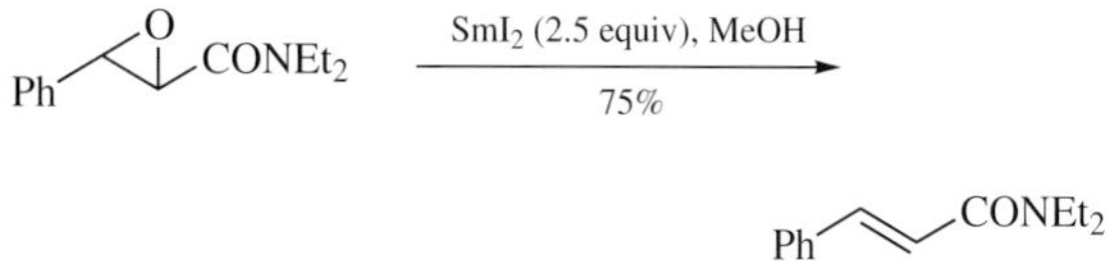

(52)

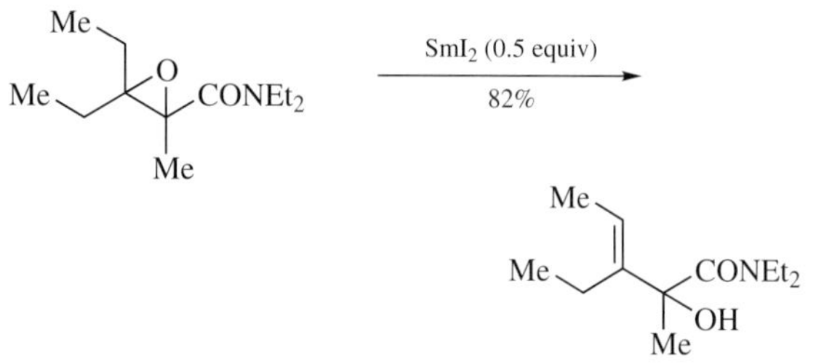

(53)

Analogous reactions with the α,β-unsaturated ester generates the saturated ester derivative (eq 54).[54]

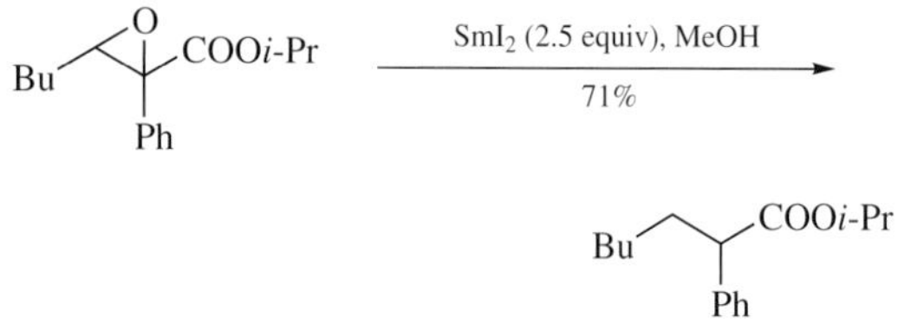

(54)

Addition of Vinylsamarium to Aldehydes. Treatment of (Z)-α-chloro-α,β-unsaturated ketones with samarium iodide leads to the vinylsamarium reagent that can be trapped with aldehydes or ketones to produce Baylis-Hillman type adducts with inversion of stereochemistry at the alkene (eq 55).[55]

Coupling of N-Acyl Lactams with Aldehydes or Ketones. Treatment of N-acyl lactams with samarium iodide leads to an acylsamarium species that is trapped by ketones or aldehydes (eq 56).[56]

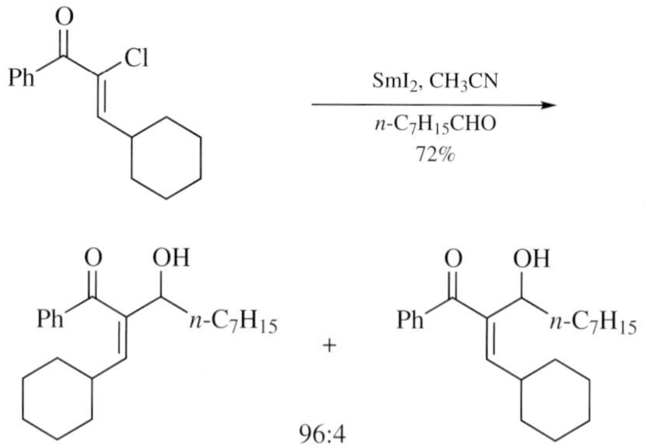

(55)

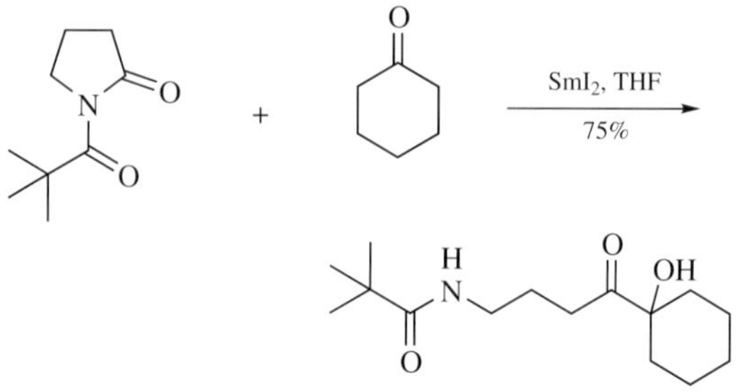

(56)

It is also possible to couple imides with alkyl halides both interintramolecularly[57] and intramolecularly.[58] Alternative precursors to generate acylsamarium species also include acyl chlorides[59] and amides.[60]

Synthesis of 1,2-Dicarbonyl by Coupling Reactions. It is possible to generate 1,2-diketones easily by treating an appropriate precursor with samarium iodide. For example, the transformation of N-acylbenzotriazoles into 1,2-diketones can be achieved in good to excellent yields (eq 57).[61]

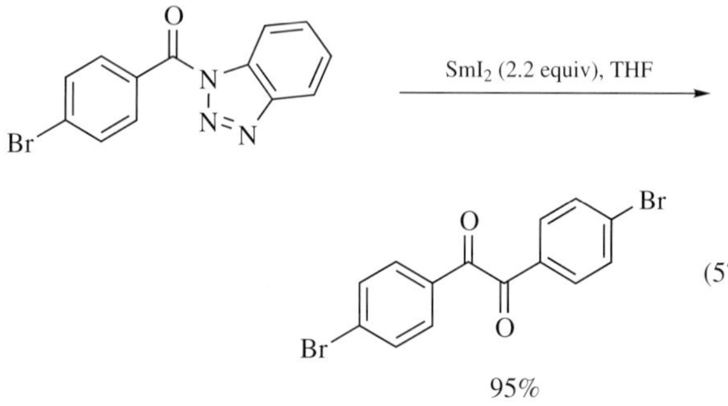

(57)

Synthesis of Homoenolate Equivalent. The samarium iodide-induced coupling of carbonyl derivatives with methoxyallene provides 4-hydroxy 1-enol ethers in high yields (eq 58).[62] An almost equimolar mixture of the two enol ethers are usually observed but acid hydrolysis leads to the aldehyde.

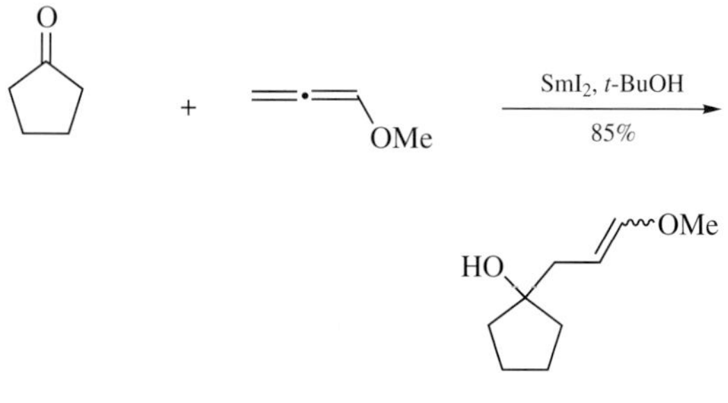

(58)

Related examples include the coupling of ketones with indole (eq 59)[63] and alkynyl moieties (eq 60).[64] In the latter case, tetrakis(triphenylphosphine)palladium must also be added to generate the electrophilic component.

$$(59)$$

$$(60)$$

Synthesis of Amidines from Amines and Nitriles. An efficient one-step preparation of N,N'-disubstituted amidines is possible by direct nucleophilic addition of an amine to a nitrile using catalytic amounts of samarium iodide (eq 61).[65] Alternatively, an azide can be used instead of an amine.[66]

$$(61)$$

Chemoselective Reduction of Carboxylic Acids. The facile chemoselective reduction of carboxylic acids in the presence of an aldehyde proceeds smoothly with samarium iodide in combination with lanthanide triflate and methanol (eq 62).

$$(62)$$

Reduction of Azides. Reduction of alkyl, aryl, and aroyl-azides to the corresponding primary amine or amide occurs

in good yield upon treatment with excess samarium iodide in THF.[67,68]

Reductive Cleavage of N–O Bonds. An efficient process for the reductive cleavage of N–O bonds using samarium iodide that is compatible when base sensitive substrate is available (eqs 63 and 64).[69] This reagent is sometimes superior to aluminum amalgam or sodium amalgam. Furthermore, the direct quenching of the reduction mixture with acylating agents provides high yields of the corresponding protected amine.

$$(63)$$

$$(64)$$

Cleavage of Haloethyl Derived Protecting Groups. Samarium diiode is a mild and effective reagent for the deprotection of 2-bromoethyl and 2-iodoethyl esters[70] and (2,2,2-trichloroethoxy) methoxy ethers.[71]

Cleavage of N-Tosyl Protecting Groups. The deprotection of N-benzenesulfonamides or N-p-toluenesulfonamides of the parent primary or secondary amines occurs in good yield upon heating with excess samarium iodide in a mixture of THF and DMPU (eq 65).[72] The method has also been used in the epimerization-free deprotection of protected α-chiral amines.[73]

$$(65)$$

It is also possible to deprotect N-sulfonylated amides under similar conditions.[74]

Tishchenko Reduction of Carbonyl Derivatives. The samarium iodide-catalyzed Tishchenko reaction has been used quite extensively in synthesis. Interesting examples include the diastereoselective synthesis of *anti*-1,3-diols (eq 66)[75] and δ-lactones (eq 67).[76]

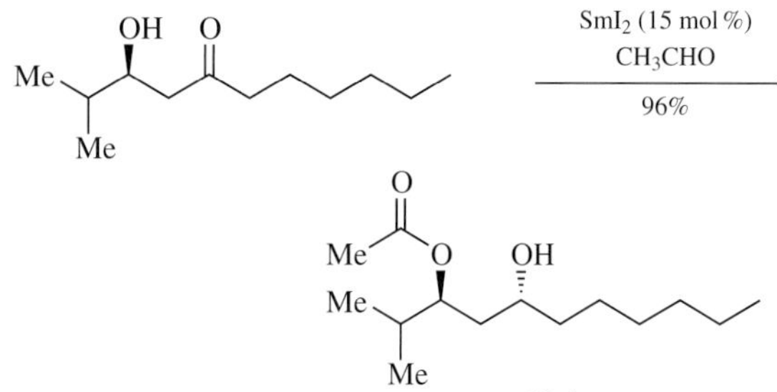

(66)

>99:1

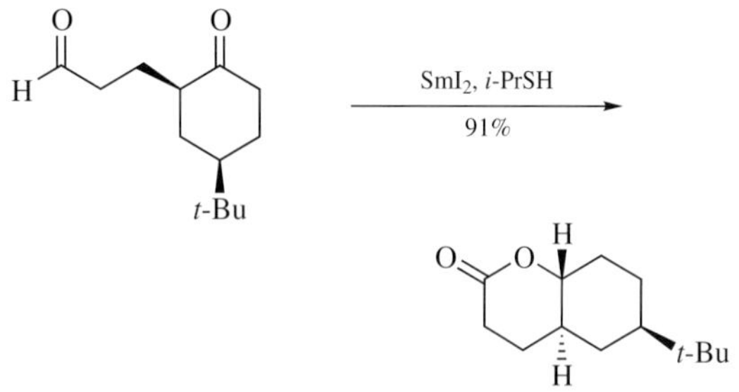

(67)

trans:cis 81:19

A mechanistically different stereoselective reduction of β-hydroxy ketones leading to *anti*-1,3-diol using stoichiometric amounts of samarium iodide has been reported.[77]

Preparation of Silyl Enol Ethers. Ketones and α-substituted aldehydes are converted into their corresponding silyl enol ethers by the reaction with trimethylsilyl ketene acetal derived from methyl isobutyrate in the presence of a catalytic amount of samarium iodide (eqs 68 and 69).[78] Mixtures are usually obtained with unsymmetrical ketones.

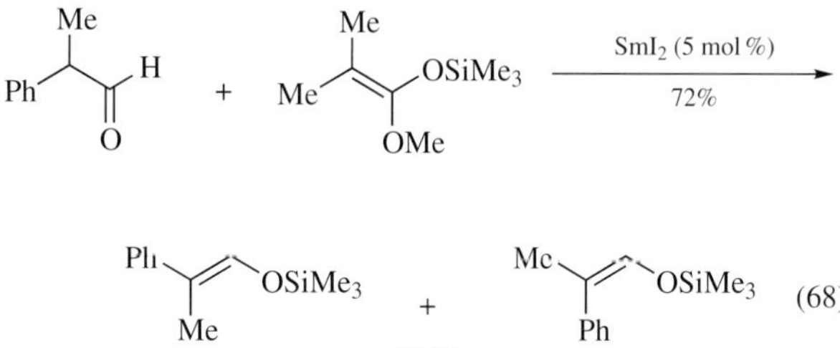

(68)

90:10

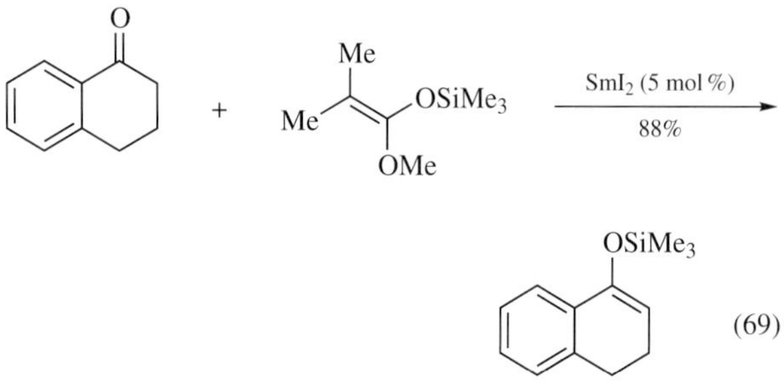

(69)

Lewis Acid Catalyzed Reactions. Samarium iodide catalyzes several transformations by presumably acting as a Lewis acid. For example, it is an efficient catalyst for the imino-Diels-Alder (eq 70) and for imino-aldol reactions.[79] Tandem Mukaiyama-Michael-aldol (eq 71)[80] and Michael imino-aldol processes have also been reported.[81]

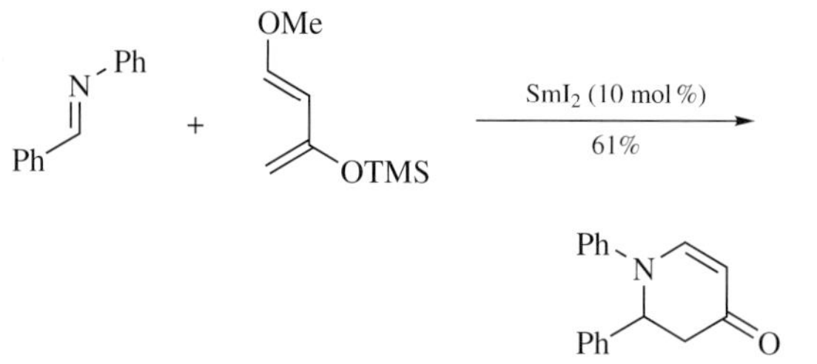

(70)

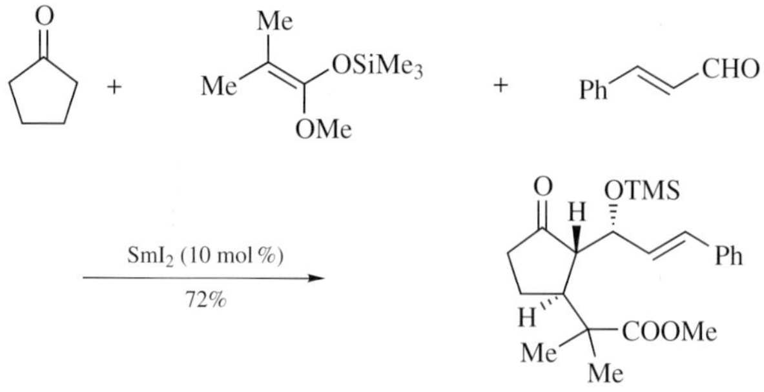

(71)

Three-component α-Amino Phosphonate Synthesis. A simple and efficient synthesis of α-amino phosphonates is possible under relatively mild conditions by the reaction of aldehydes, amines, and a dialkylphosphite using samarium iodide in catalytic amounts (eq 72).[82]

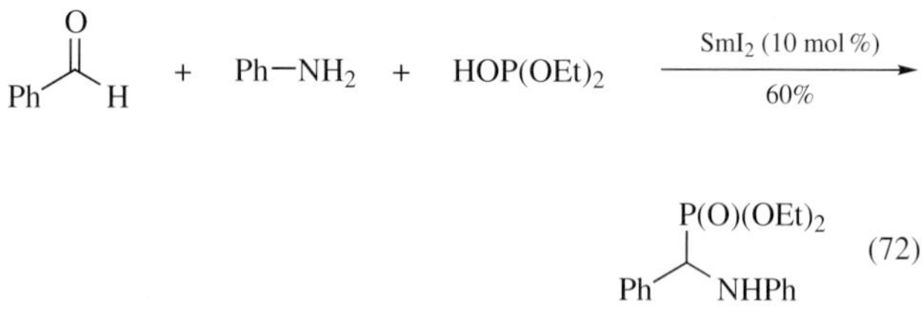

(72)

Related Reagents. Samarium(II) Iodide–1,3-Dioxolane.

1. (a) Molander, G. A., *Chem. Rev.* **1992**, *92*, 29. (b) Molander, G. A. In *The Chemistry of the Metal–Carbon Bond*; Hartley, F. R., Ed.; Wiley: Chichester, 1989; Vol. 5, Chapter. 8. (c) Kagan, H. B., *Nouv. J. Chim.* **1990**, *14*, 453. (d) Soderquist, J. A., *Aldrichim. Acta* **1991**, *24*, 15. (e) Molander, G. A., *Comprehensive Organic Synthesis* **1991**, *1*, Chapter 1.9.

2. (a) Girard, P.; Namy, J. L.; Kagan, H. B., *J. Am. Chem. Soc.* **1980**, *102*, 2693. (b) Namy, J. L.; Girard, P.; Kagan, H. B., *Nouv. J. Chim.* **1977**, *1*, 5. (c) Namy, J. L.; Girard, P.; Kagan, H. B., *Nouv. J. Chim.* **1981**, *5*, 479.

3. Inanaga, J., *Heteroatom Chem.* **1990**, *3*, 75.

4. Inanaga, J.; Ishikawa, M.; Yamaguchi, M., *Chem. Lett.* **1987**, 1485.

5. Kagan, H. B.; Namy, J. L.; Girard, P., *Tetrahedron* **1981**, *37*, 175, Suppl. 1.

6. Crombie, L.; Rainbow, L. J., *Tetrahedron Lett.* **1988**, *29*, 6517.

7. (a) Molander, G. A.; Hahn, G., *J. Org. Chem.* **1986**, *51*, 1135. (b) Smith, A. B., III; Dunlap, N. K.; Sulikowski, G. A., *Tetrahedron Lett.* **1988**, *29*, 439. (c) Castro, J.; Sörensen, H.; Riera, A.; Morin, C.; Moyano, A.; Pericàs, M. A.; Greene, A. E., *J. Am. Chem. Soc.* **1990**, *112*, 9388.

8. (a) White, J. D.; Somers, T. C., *J. Am. Chem. Soc.* **1987**, *109*, 4424. (b) Holton, R. A.; Williams, A. D., *J. Org. Chem.* **1988**, *53*, 5981.

9. Pratt, D. V.; Hopkins, P. B., *Tetrahedron Lett.* **1987**, *28*, 3065.

10. (a) Molander, G. A.; Hahn, G., *J. Org. Chem.* **1986**, *51*, 2596. (b) Otsubo, K.; Inanaga, J.; Yamaguchi, M., *Tetrahedron Lett.* **1987**, *28*, 4437.

11. Molander, G. A.; La Belle, B. E.; Hahn, G., *J. Org. Chem.* **1986**, *51*, 5259.

12. Enholm, E. J.; Jiang, S., *Tetrahedron Lett.* **1992**, *33*, 313.

13. (a) Yoneda, R.; Harusawa, S.; Kurihara, T., *Tetrahedron Lett.* **1989**, *30*, 3681. (b) Yoneda, R.; Harusawa, S.; Kurihara, T., *J. Org. Chem.* **1991**, *56*, 1827.

14. Handa, Y.; Inanaga, J.; Yamaguchi, M., *J. Chem. Soc., Chem. Commun.* **1989**, 298.

15. Sasaki, M.; Collin, J.; Kagan, H. B., *Nouv. J. Chim.* **1992**, *16*, 89.

16. Molander, G. A.; McKie, J. A., *J. Org. Chem.* **1991**, *56*, 4112.

17. (a) Molander, G. A.; Etter, J. B., *J. Org. Chem.* **1986**, *51*, 1778. (b) Zoretic, P. A.; Yu, B. C.; Caspar, M. L., *Synth. Commun.* **1989**, *19*, 1859. (c) Daniewski, A. R.; Uskokovic, M. R., *Tetrahedron Lett.* **1990**, *31*, 5599.

18. (a) Lannoye, G.; Cook, J. M., *Tetrahedron Lett.* **1988**, *29*, 171. (b) Lannoye, G.; Sambasivarao, K.; Wehrli, S.; Cook, J. M.; Weiss, U., *J. Org. Chem.* **1988**, *53*, 2327.

19. Molander, G. A.; Etter, J. B.; Zinke, P. W., *J. Am. Chem. Soc.* **1987**, *109*, 453.

20. Molander, G. A.; McKie, J. A., *J. Org. Chem.* **1993**, *58*, 7216.

21. (a) Hon, Y.-S.; Lu, L.; Chu, , K.-P., *Synth. Commun.* **1991**, *21*, 1981. (b) Kito, M.; Sakai, T.; Yamada, K.; Matsuda, F.; Shirahama, H., *Synlett* **1993**, 158. (c) Fukuzawa, S.; Iida, M.; Nakanishi, A.; Fujinami, T.; Sakai, S., *J. Chem. Soc., Chem. Commun.* **1987**, 920. (d) Fukuzawa, S.; Nakanishi, A.; Fujinami, T.; Sakai, S., *J. Chem. Soc., Perkin Trans. 1* **1988**, 1669. (e) Enholm, E. J.; Trivellas, A., *Tetrahedron Lett.* **1989**, *30*, 1063. (f) Enholm, E. J.; Satici, H.; Trivellas, A., *J. Org. Chem.* **1989**, *54*, 5841. (g) Enholm, E. J.; Trivellas, A., *J. Am. Chem. Soc.* **1989**, *111*, 6463.

22. (a) Molander, G. A.; Kenny, C., *J. Am. Chem. Soc.* **1989**, *111*, 8236. (b) Molander, G. A.; Kenny, C., *Tetrahedron Lett.* **1987**, *28*, 4367.

23. (a) Shim, S. C.; Hwang, J.-T.; Kang, H.-Y.; Chang, M. H., *Tetrahedron Lett.* **1990**, *31*, 4765. (b) Bannai, K.; Tanaka, T.; Okamura, N.; Hazato, A.; Sugiura, S.; Manabe, K.; Tomimori, K.; Kato, Y.; Kurozumi, S.; Noyori, R., *Tetrahedron* **1990**, *46*, 6689.

24. Molander, G. A.; McKie, J. A., *J. Org. Chem.* **1992**, *57*, 3132.

25. (a) Namy, J. L.; Souppe, J.; Kagan, H. B., *Tetrahedron Lett.* **1983**, *24*, 765. (b) Fürstner, A.; Csuk, R.; Rohrer, C.; Weidmann, H., *J. Chem. Soc., Perkin Trans. 1* **1988**, 1729.

26. Chiara, J. L.; Cabri, W.; Hanessian, S., *Tetrahedron Lett.* **1991**, *32*, 1125.

27. Molander, G. A.; Kenny, C., *J. Am. Chem. Soc.* **1989**, *111*, 8236.

28. Inanaga, J.; Ujikawa, O.; Yamaguchi, M., *Tetrahedron Lett.* **1991**, *32*, 1737.

29. (a) Molander, G. A.; Harring, L. S., *J. Org. Chem.* **1990**, *55*, 6171. (b) Curran, D. P.; Totleben, M. J., *J. Am. Chem. Soc.* **1992**, *114*, 6050.

30. Bennett, S. M.; Larouche, D., *Synlett* **1991**, 805.

31. Batey, R. A.; Motherwell, W. B., *Tetrahedron Lett.* **1991**, *32*, 6649.

32. Boiron, A.; Zillig, P.; Faber, D.; Giese, B., *J. Org. Chem.* **1998**, *63*, 5877–5882.

33. Sturino, C. F.; Fallis, A. G., *J. Am. Chem. Soc.* **1994**, *116*, 7447.

34. Taniguchi, N.; Hata, T.; Uemura, M., *Angew. Chem. Int. Ed.* **1999**, *38*, 1232.

35. Tanaka, Y.; Taniguchi, N.; Kimura, T.; Uemura, M., *J. Org. Chem.* **2002**, *67*, 9227.

36. Taniguchi, N.; Uemura, M., *J. Am. Chem. Soc.* **2000**, *122*, 8301.

37. Helion, F.; Namy, J. L., *J. Org. Chem.* **1999**, *64*, 2944.

38. Concellon, J. M.; Perez-Andres, J. A.; Rodriguez-Solla, H., *Angew. Chem., Int. Ed.* **2000**, *39*, 2773.

39. Concellon, J. M.; Huerta, M.; Llavona, R., *Tetrahedron Lett.* **2004**, *45*, 4665.

40. Otaka, A.; Yukimasa, A.; Watanabe, J.; Sasaki, Y.; Oishi, S.; Tamamura, H.; Fujii, N., *Chem. Commun.* **2003**, 1834.

41. Concellon, J. M.; Bernad, P. L.; Bardales, E., *Org. Lett.* **2001**, *3*, 937.

42. Keck, G. E.; Savin, K. A.; Weglarz, M. A., *J. Org. Chem.* **1995**, *60*, 3194.

43. Jarreton, O.; Skrydstrup, T.; Espinosa, J. F.; Jimenez-Barbero, J.; Beau, J. M., *Chem. Eur. J.* **1999**, *5*, 430.

44. Urban, D.; Skrydstrup, T.; Riche, C.; Chiaroni, A.; Beau, J. M., *Chem. Commun.* **1996**, 1883.

45. Hung, S. C.; Wong, C. H., *Angew. Chem., Int. Ed.* **1996**, *35*, 2671.

46. Chiara, J. L.; Sesmilo, E., *Angew. Chem., Int. Ed.* **2002**, *41*, 3242.

47. Machrouhi, F.; Namy, J. L., *Tetrahedron Lett.* **1999**, *40*, 1315.

48. Skrydstrup, T.; Jespersen, T.; Beau, J. M.; Bols, M., *Chem. Commun.* **1996**, 515.

49. Park, H. S.; Lee, I. S.; Kwon, D. W.; Kim, Y. H., *Chem. Commun.* **1998**, 2745.

50. Concellon, J. M.; Bernad, P. L.; Perez-Andres, J. A., *Tetrahedron Lett.* **1998**, *39*, 1409.

51. Booth, S. E.; Benneche, T.; Undheim, K., *Tetrahedron* **1995**, *51*, 3665.

52. Concellon, J. M.; Bardales, E., *J. Org. Chem.* **2003**, *68*, 9492.

53. Concellon, J. M.; Bernad, P. L.; Bardales, E., *Chem. Eur. J.* **2004**, *10*, 2445.

54. Concellon, J. M.; Bardales, E.; Llavona, R., *J. Org. Chem.* **2003**, *68*, 1585.

55. Concellon, J. M.; Bernad, P. L.; Huerta, M.; Garcia-Granda, S.; Diaz, M. R., *Chem. Eur. J.* **2003**, *9*, 5343.

56. Farcas, S.; Namy, J. L., *Tetrahedron Lett.* **2000**, *41*, 7299.

57. Farcas, S.; Namy, J. L., *Tetrahedron Lett.* **2001**, *42*, 879.

58. Ha, D. C.; Yun, C. S.; Lee, Y., *J. Org. Chem.* **2000**, *65*, 621.

59. Namy, J. L.; Colomb, M.; Kagan, H. B., *Tetrahedron Lett.* **1994**, *35*, 1723.

60. McDonald, C. E.; Galka, A. M.; Green, A. I.; Keane, J. M.; Kowalchick, J. E.; Micklitsch, C. M.; Wisnoski, D. D., *Tetrahedron Lett.* **2001**, *42*, 163.

61. Wang, X. X.; Zhang, Y. M., *Tetrahedron Lett.* **2002**, *43*, 5431.

62. Holemann, A.; Reissig, H. U., *Org. Lett.* **2003**, *5*, 1463.

63. Gross, S.; Reissig, H. U., *Org. Lett.* **2003**, *5*, 4305.

64. Aurrecoechea, J. M.; Anton, R. F. S., *J. Org. Chem.* **1994**, *59*, 702.

65. Xu, F.; Sun, J. H.; Shen, Q., *Tetrahedron Lett.* **2002**, *43*, 1867.

66. Su, W.; Li, Y. S.; Zhang, Y. M., *J. Chem. Res. (S)* **2001**, 32.

67. Goulaouic-Dubois, C.; Hesse, M., *Tetrahedron Lett.* **1995**, *36*, 7427.

68. Benati, L.; Montevecchi, P. C.; Nanni, D.; Spagnolo, P.; Volta, M., *Tetrahedron Lett.* **1995**, *36*, 7313.

69. Keck, G. E.; Wager, T. T.; McHardy, S. F., *Tetrahedron* **1999**, *55*, 11755.

70. Pearson, A. J.; Lee, K., *J. Org. Chem.* **1994**, *59*, 2257.

71. Evans, D. A.; Kaldor, S. W.; Jones, T. K.; Clardy, J.; Stout, T. J., *J. Am. Chem. Soc.* **1990**, *112*, 7001.

72. Vedejs, E.; Lin, S., *J. Org. Chem.* **1994**, *59*, 1602.

73. Fujihara, H.; Nagai, K.; Tomioka, K., *J. Am. Chem. Soc.* **2000**, *122*, 12055.

74. Knowles, H. S.; Parsons, A. F.; Pettifer, R. M.; Rickling, S., *Tetrahedron* **2000**, *56*, 979.

75. Evans, D. A.; Hoveyda, A. H., *J. Am. Chem. Soc.* **1990**, *112*, 6447.

76. Hsu, J. L.; Chen, C. T.; Fang, J. M., *Org. Lett.* **1999**, *1*, 1989.

77. Keck, G. E.; Wager, C. A.; Sell, T.; Wager, T. T., *J. Org. Chem.* **1999**, *64*, 2172.

78. Hydrio, J.; VandeWeghe, P.; Collin, J., *Synthesis* **1997**, 68.

79. Collin, J.; Jaber, N.; Lannou, M. I., *Tetrahedron Lett.* **2001**, *42*, 7405.

80. Giuseppone, N.; Collin, J., *Tetrahedron* **2001**, *57*, 8989.

81. Jaber, N.; Assie, M.; Fiaud, J. C.; Collin, J., *Tetrahedron* **2004**, *60*, 3075.

82. Xu, F.; Luo, Y. Q.; Deng, M. Y.; Shen, Q., *Eur. J. Org. Chem.* **2003**, 4728.

Scandium Trifluoromethanesulfonate

[144026-79-9] C₃F₉O₉S₃Sc (MW 492.16)

$C_3F_9O_9S_3Sc$ (MW 492.16)

InChI = 1S/3CHF3O3S.Sc/c3*2-1(3,4)8(5,6)7;/h3*(H,5,6,7);/
q;;;+3/p-3
InChIKey = HZXJVDYQRYYYOR-UHFFFAOYSA-K

(a Lewis acid catalyst)

Physical Data: mp >300 °C.
Solubility: soluble in H_2O, alcohol, acetonitrile, and most polar organic solvents.
Form Supplied in: colorless solid.
Preparative Methods: [1]scandium triflate is commercially available. On the other hand, it can also be prepared from the corresponding oxide (Sc_2O_3) and aqueous trifluoromethanesulfonic acid (TfOH). After filtration and concentration of the clear aqueous solution in vacuo, the resulting hydrated salt is dried in vacuo (<1 mmHg) at 200 °C for 40 h to afford the anhydrous triflate, which is stored over P_2O_5.

$$Sc_2O_3 \;+\; 6\,TfOH \quad \xrightarrow{100\,°C,\,1\,h} \quad 2Sc(OTf)_3 \;+\; 3\,H_2O$$
$$(1{:}1\;TfOH{:}H_2O)$$

Handling, Storage, and Precautions: the anhydrous triflate is fairly hygroscopic (not decomposed, forms a hydrate) and must be kept in a desiccator over P_2O_5 and freshly dried in vacuo (<1 mmHg) at 200 °C for 1 h before using.

Original Commentary

Masaharu Sugiura & Shu Kobayashi
The University of Tokyo, Tokyo, Japan

Scandium triflate [$Sc(OTf)_3$] is a new type of Lewis acid that is different from typical Lewis acids such as $AlCl_3$, BF_3, $SnCl_4$, etc.[2] While most Lewis acids are decomposed or deactivated in the presence of water, $Sc(OTf)_3$ is stable and works as a Lewis acid catalyst in water solution. Many reactions proceed smoothly when this reagent is used in catalytic quantities, while stoichiometric amounts of conventional Lewis acids are needed in the same reactions. Moreover, many nitrogen-containing compounds such as imines and hydrazones are also successfully activated by a catalytic amount of $Sc(OTf)_3$ in both organic and aqueous solvents. $Sc(OTf)_3$ can be recovered almost quantitatively after reactions are complete and can be reused. While lanthanide triflates [$Ln(OTf)_3$] have similar properties, the catalytic activity of $Sc(OTf)_3$ is higher than that of $Ln(OTf)_3$ in several cases.

Aldol-type Reactions. $Sc(OTf)_3$ is an effective catalyst in aldol-type reactions of silyl enol ethers with aldehydes. The activities of typical rare-earth triflates [Sc, Y, Yb(OTf)₃] were evaluated in the reaction of 1-trimethylsiloxycyclohexene with benzalde-

hyde in dichloromethane (eq 1).[3] While the reaction proceeds sluggishly at −78 °C in the presence of $Yb(OTf)_3$ or $Y(OTf)_3$, the aldol-type adduct is obtained in 81% yield in the presence of $Sc(OTf)_3$. Obviously, $Sc(OTf)_3$ is more active than $Y(OTf)_3$ or $Yb(OTf)_3$ in this case.

$Sc(OTf)_3$ also catalyzes aldol-type reactions of silyl enolates with acetals. For example, the reaction of 3-phenylpropionaldehyde dimethyl acetal with the ketene silyl acetal of methyl isobutyrate proceeds at 0 °C to room temperature to give the desired adduct in 97% yield (eq 2).[2]

$Sc(OTf)_3$ is effective in the aldol-type reaction of silyl enolates with aldehydes in aqueous media (H_2O–THF) without any significant decomposition of the water-sensitive silyl enolates. Thus, aldehydes available in aqueous-solution such as formaldehyde and chloroacetaldehyde can be directly used to afford the corresponding aldol adduct in high yield (eq 3).[2]

The $Sc(OTf)_3$-catalyzed aldol-type reactions of silyl enol ethers with aldehydes can be performed in micellar systems using a catalytic amount of a surfactant such as sodium dodecylsulfate (SDS).[4,5] In these systems, reactions proceed smoothly in water without using any organic solvent.

$Sc(OTf)_3$ is more soluble in water than in organic solvents such as dichloromethane. After the reaction is complete, the catalyst can be recovered almost quantitatively from the aqueous layer. The recovered catalyst is also effective in a second reaction, and the yield of the subsequent run is comparable to that of the first experiment (eq 4).[2]

Essential Reactions for Organic Synthesis, First Edition. Edited by Philip L. Fuchs.
© 2016 John Wiley & Sons, Ltd. Published 2016 by John Wiley & Sons, Ltd.

Aldol-type reactions of polymer-supported silyl enol ethers with aldehydes are also catalyzed by Sc(OTf)$_3$.[6] A 2-silyloxypyrrole and aldehydes undergo aldol-type reactions in the presence of 5 mol % Sc(OTf)$_3$.[7] The aldol-type reaction of silyl enol ethers with chiral η^6-(benzaldehyde)chromium complexes has been reported.[8] Ketones unusually show higher reactivity than aldehydes in the presence of a Lewis acid such as Sc(OTf)$_3$ or (C$_6$F$_5$)$_2$SnBr$_2$ due to the differentiated recognition of ketone and aldehyde carbonyls.[9] The cross-aldol reactions between aldehydes and scandium enolates prepared from ketones, Sc(OTf)$_3$, and diisopropylethylamine have been reported,[10] while the Sc(OTf)$_3$/triphenylphosphine reagent is recognized to promote Reformatsky-type reactions between α-bromo carboxylic acid derivatives and aldehydes.[11]

Related Sc(OTf)$_3$-catalyzed reactions, i.e. alkylation of silyl enol ethers with sulfur dioxide adduct of 1-methoxybutadiene,[12] that of lithium enolates with epoxides[13], and benzopyran formation from o-hydroxybenzaldehydes and dimethoxypropane[14] have also been reported.

Michael Reactions. The Michael reactions of silyl enol ethers or ketene silyl acetals with α,β-unsaturated carbonyl compounds are catalyzed by Sc(OTf)$_3$ to give the corresponding 1,5-dicarbonyl compounds in high yields after acid work-up (eq 5).[2] When the crude adducts were worked up without acid, the synthetically valuable silyl enol ethers could be isolated. The catalyst can be recovered almost quantitatively and reused. Sc(OTf)$_3$ also catalyzes 1,4-addition of PhMe$_2$Si-ZnMe$_2$Li to enones in the presence of 3 mol % of Me$_2$Cu(CN)Li$_2$.[15]

$$(5)$$

Mannich-type Reactions. The reactions of imines with ketene silyl acetals proceed smoothly in the presence of Sc(OTf)$_3$ to afford the corresponding β-amino ester derivative in moderate yield (eq 6).[16] Sc(OTf)$_3$ shows higher activity than Yb(OTf)$_3$ does in this case. The catalyst can be recovered after the reaction is complete and reused. A Mannich-type reaction of N-(β-aminoalkyl)benzotriazoles with silyl enolates has also been developed.[17] Mannich-type reactions of polymer-supported silyl enol ethers with imines[18] or of polymer-supported α-iminoacetates with silyl enolates[19] are also catalyzed by Sc(OTf)$_3$.

$$(6)$$

80% (M = Sc)
65% (M = Yb)

Four-component (silyl enolates, α,β-unsaturated thioesters, amines, and aldehydes) coupling reactions are catalyzed by Sc(OTf)$_3$ to produce the corresponding amino thioester and γ-acyl-δ-lactam derivatives stereoselectively in high yields (eq 7).[20]

$$(7)$$

Mannich-type reactions of aldehydes, amines, and vinyl ethers proceed smoothly in the presence of a catalytic amount of Sc(OTf)$_3$ in aqueous media (eq 8).[21] Interestingly, dehydration accompanied by imine formation and successive addition of a vinyl ether proceed smoothly in aqueous solution.

$$(8)$$

The Sc(OTf)$_3$-catalyzed three-component reactions of silyl enol ethers with aldehydes and aromatic amines can be performed in micellar systems using SDS without using any organic solvent (eq 9).[22]

$$(9)$$

Polymer-supported silyl enol ethers (PSSEEs) can be employed in Sc(OTf)$_3$-catalyzed reactions with aldehydes and aromatic amines.[23] This process provides a convenient method for the construction of a β-amino alcohol library.

In the presence of a catalytic amount of Sc(OTf)$_3$, benzoylhydrazones react with ketene silyl acetals to afford the corresponding adducts, β-N-benzoylhydrazino esters, in high yield (eq 10).[24] Several benzoylhydrazones including those derived from aromatic, aliphatic, α,β-unsaturated aldehydes, and glyoxylate work well. The reactions of polymer-supported hydrazones with silyl enolates are also catalyzed by Sc(OTf)$_3$ to produce pyrazolone derivatives after base treatment.[25]

$$\text{(10)}$$

Allylation Reactions. The allylation reactions of carbonyl compounds with tetraallyltin proceed smoothly under the influence of a catalytic amount of Sc(OTf)$_3$. The reaction can be carried out in aqueous media as exemplified by unprotected sugars that react directly to give the adducts in high yield (eq 11).[26]

$$\text{(11)}$$

Related Sc(OTf)$_3$-catalyzed allylation of aldehydes with allyltributyltin in nitromethane,[27] allylation of aldehydes with allylgermanes,[28] allylation of acylsilanes,[29] and intramolecular allenylation of propargyl silanes[30] have also been reported. The allylation of aldehydes with tetraallyltin proceeds smoothly in micellar systems without using any organic solvents.[31] Novel allylation of aldehydes with alkeneylepoxides is catalyzed by Sc(OTf)$_3$ to produce δ-hydroxy-α,β-unsaturated aldehydes.[32]

While allylation of imines is also catalyzed by Sc(OTf)$_3$ in dichloromethane to give homoallylic amines in moderate yield,[33] three-component reactions of aldehydes, amines, and allyltributyltin proceed in micellar systems to afford the corresponding homoallylic amines in good to high yield (eq 12).[34] It is suggested that imine formation from aldehydes and amines is very fast under these conditions, and that the selective activation of imines rather than aldehydes is achieved. Allylation of N-benzoylhydrazone or three-component reaction of aldehydes, benzoylhydrazine, and tetraallyltin are also catalyzed by Sc(OTf)$_3$.[35]

$$\text{(12)}$$

Friedel–Crafts Acylation, Alkylation, and Related Reactions. While a stoichiometric amount of AlCl$_3$ is needed in Friedel–Crafts acylations, a small amount of Sc(OTf)$_3$ smoothly catalyzes the same reaction.[36] In the acetylation of thioanisole and o- or m-dimethoxybenzene, a single acetylated product is formed in an excellent yield. In the benzoylation of anisole, both benzoic anhydride and benzoyl chloride are effective, while benzoic anhydride gives a slightly higher yield. Addition of lithium perchlorate (LiClO$_4$) as a cocatalyst improves the yield dramatically (eq 13).[37]

$$\text{(13)}$$

The Fries rearrangement of acyloxybenzene or naphthalene derivatives proceeds smoothly in the presence of a catalytic amount of Sc(OTf)$_3$ (eq 14).[38] The direct acylation of phenol or naphthol derivatives with acid chlorides is also catalyzed to give the 2-acylated product in high yields.

$$\text{(14)}$$

Friedel–Crafts alkylations of arenes with mesylates,[39] benzyl or allyl alcohols,[40] aldehyde/diol combinations (reductive alkylation),[36b] 1,3-dienes,[41] or alkenes in an ionic liquid[42] are also effectively catalyzed by Sc(OTf)$_3$. Sc(OTf)$_3$ works as an efficient catalyst for the condensation reaction of trimethylhydroquinone with isophytol to afford α-tocopherol.[43] 2-Aminoalkylation of phenols with α-iminoacetates (or glyoxylate/amine) is catalyzed by Sc(OTf)$_3$ to produce amino acid derivatives.[44] The Sc(OTf)$_3$-catalyzed alkylations of indoles with α-hydroxy esters,[45] aziridines,[46] acetals,[47] and aldehydes[48] have been utilized as key steps of total syntheses as exemplified in eq 15.[48]

$$\longrightarrow \quad (-)\text{-Penitrem D} \qquad (15)$$

Sc(OTf)$_3$ shows high catalytic activity for the nitration of arenes with nitric acid.[49] Sequential Claisen rearrangement/cyclization of allyl phenyl ethers is efficiently catalyzed by Sc(OTf)$_3$ at high temperature in an ionic liquid to give 2,3-dihydrobenzofurans.[50]

Diels–Alder Reactions and Related Cycloadditions. In the Diels–Alder reaction of methyl vinyl ketone (MVK) with isoprene, the adduct is obtained in 91% yield in the presence of 10 mol % Sc(OTf)$_3$, while 10 mol % Y(OTf)$_3$ or Yb(OTf)$_3$ gives only a trace amount of the adduct.[1c, 51] Sc(OTf)$_3$-catalyzed Diels–Alder reactions generally provide high yields with high endo selectivities. The present Diels–Alder reaction even proceeds in aqueous media. Thus, naphthoquinone reacts with cyclopentadiene in H$_2$O–THF (9:1) at room temperature to give the corresponding adduct in high yield (100% endo) (eq 16). Sc(OTf)$_3$ also serves as an effective catalyst for Diels–Alder reactions in supercritical carbon dioxide (sc CO$_2$).[52]

In the presence of 10 mol % Sc(OTf)$_3$, N-benzylideneaniline reacts with 2-trans-1-methoxy-3-trimethylsiloxy-1,3-butadiene (Danishefsky's diene) to afford the corresponding aza Diels–Alder adduct, a tetrahydropyridine derivative, quantitatively (eq 17).[16] On the other hand, in the reaction of N-benzylideneaniline with

cyclopentadiene under the same conditions, the tetrahydroquinoline derivative is obtained (eq 18).[16]

Sc(OTf)$_3$-catalyzed three-component coupling reactions of aldehydes, amines, and dienes have also been developed.[53] In the presence of 10 mol % Sc(OTf)$_3$ and magnesium sulfate, the reaction of benzaldehyde with aniline and Danishefsky's diene produces the tetrahydropyridine derivative in 83% yield, while the reaction with cyclopentadiene instead of Danishefsky's diene produces the tetrahydroquinoline derivative. Various combinations of aldehydes, amines, and alkenes are possible in these reactions to produce diverse tetrahydroquinoline derivatives in high yield. Moreover, the three-component coupling reactions proceed smoothly in aqueous solution, and commercial formaldehyde-water solution can be used directly (eq 19). Sc(OTf)$_3$-catalyzed three-component aza Diels–Alder cycloadditions also proceed smoothly in an ionic liquid.[54]

The three-component coupling reaction of benzaldehyde, N-benzylhydroxylamine, and N-phenylmaleimide proceeds smoothly in the presence of a catalytic amount of Sc(OTf)$_3$, to afford the corresponding isoxazolidine derivative in good yield and with high diastereoselectivity (eq 20).[55]

In the presence of a catalytic amount of Sc(OTf)$_3$, imines and alkynyl sulfides undergo [2 + 2] cycloaddition and successive fragmentation to afford α,β-unsaturated thioimidates (eq 21).[56] Alkynyl selenides[57] or alkynyl silyl ethers[58] undergo similar reactions in the presence of Sc(OTf)$_3$.

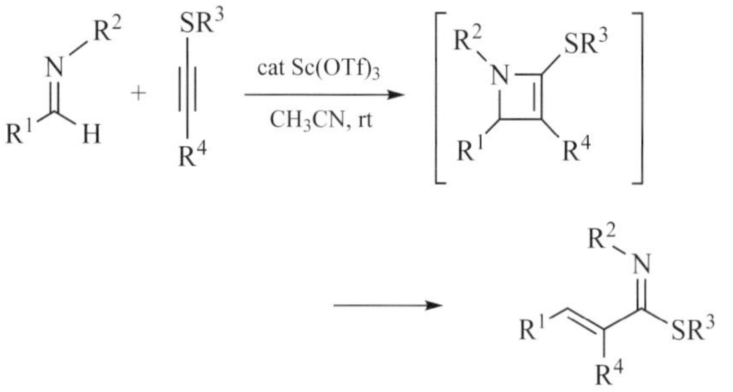

$$(21)$$

Sc(OTf)$_3$-catalyzed [5 + 2] cycloaddition of a methyleneoxindole and a η^3-pyridinylmolybdenum complex[59] and [4 + 3] cycloaddition of 2-silyloxyacrolein and furan[60] have also been reported.

Asymmetric Catalysts Using Sc(OTf)$_3$. The chiral Sc catalyst prepared from Sc(OTf)$_3$, (R)-BINOL, and a tertiary amine in dichloromethane (eq 22) serves as an asymmetric catalyst for the Diels–Alder reactions of an acrylic acid derivative with dienes (eq 23).[61] The highest enantioselectivities are observed when cis-1,2,6-trimethylpiperidine is employed as the amine. Even 3 mol % of the catalyst is enough to complete the reaction yielding the endo adduct in 92% ee. The structure of the chiral Sc catalyst is indicated by ^{13}C-NMR and IR spectra.[62] A similar Sc(OTf)$_3$/(R)-BINOL/diisopropylethylamine complex serves as an effective catyst for the Diels–Alder reaction of N-benzyloxy-carbonyl-1-aminobutadiene with 3-acyl-1,3-oxazolidin-2-one.[63]

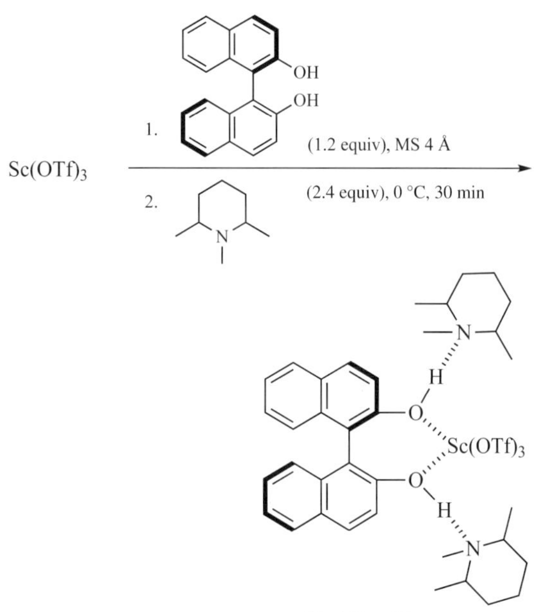

$$(22)$$

chiral Sc triflate

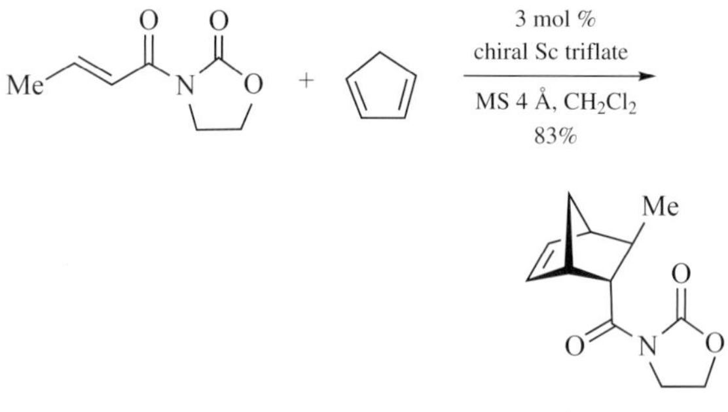

$$(23)$$

endo/exo = 87/13

92% ee (endo)

A chiral scandium catalyst prepared from Sc(OTf)$_3$, (R)-BINOL, and DBU is effective in enantioselective aza Diels–Alder reactions (eq 24).[64] The reaction of N-alkylidene- or N-arylidene-2-hydroxyaniline with cyclopentadiene proceeds in the presence of the chiral catalyst and 2,6-di-tert-butyl-4-methylpyridine (DTBMP) to afford the corresponding 8-hydroxyquinoline derivatives in good to high yields with good to excellent diastereo selectivity and enantioselectivity.

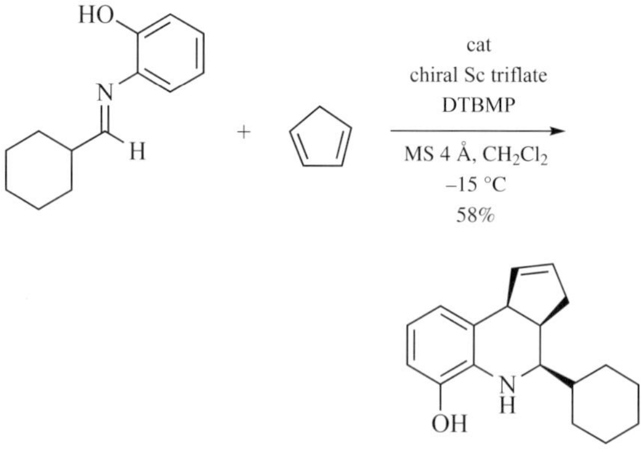

$$(24)$$

cis/trans = >99/1

73% ee (cis)

In the presence of the chiral scandium catalyst, the 1,3-dipolar cycloaddition of benzylbenzylideneamine N-oxide with 3-(2-butenoyl)-1,3-oxazolidin-2-one proceeds to yield the endo adduct (endo/exo 99/1) in 69% ee (eq 25).[65] Chiral scandium complexes prepared from Sc(OTf)$_3$ and 2,2'-bis(oxazolyl)-BINOL derivatives also serve as efficient catalysts.[66]

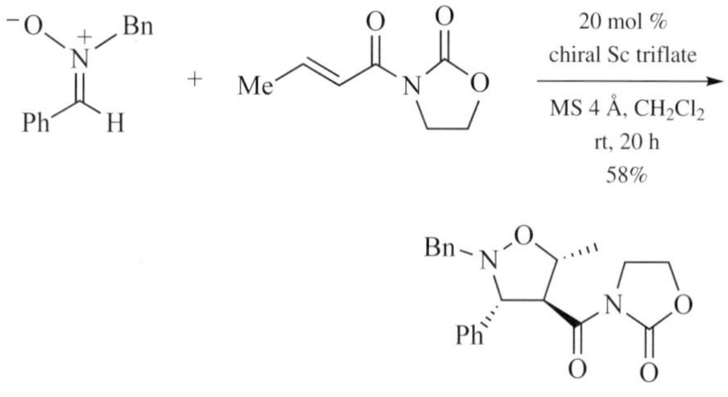

$$(25)$$

69% ee

Sc(OTf)$_3$/3,3'-bis(aminomethyl)-BINOL complexes work as asymmetric catalysts for Michael-type reaction of 2-(trimethylsilyloxy)furan to 3-(2-butenoyl)-1,3-oxazolidin-2-one.[67]

Sc(OTf)$_3$/^{i}Pr-pybox complexes are found to catalyze asymmetric glyoxylate-ene reaction, though enantioselectivity is low.[68]

Miscellaneous Reactions. Sc(OTf)$_3$ also catalyzes Meerwein–Ponndrof–Verley reductions,[1d] Tishchenko reductions,[69] Baeyer–Villiger oxidations,[70] acetalization reactions,[71] acylal formation,[72] β-selective glycosilation reactions with thioglycosides,[73] acylation reactions of alcohols,[74] chemoselective deprotection of silyl alkyl ethers,[75] silyl ether protection of alcohols with allyl silanes,[76] deprotection of benzylic poly(ethyleneglycol) ethers,[77] guanidium formation reactions of carbodiimide with benzylamine,[78] stereoselective radical reactions,[79] reactions of α-diazocarbonyl compounds,[80] decarbonylation of aromatic aldehydes,[81] dehydration reactions of aldoximes to nitriles,[82] rearrangements of epoxides,[83] ring expansions of cyclic ethers,[84] Ferrier rearrangements,[85] Prins-type cyclizations,[86] additions of 1-trimethylsilyl nitropropanoate to imines,[87] sequential carbonyl ene reaction/acetylations,[88] three-component reactions of aldehydes/amines/tributyltin cyanide,[89] Biginelli reactions,[90] and a new three-component condensation of 2-aminopyridine/aldehydes/isonitrile.[91] Interestingly, multiple reactions in one-pot (Diels–Alder reaction, allylation of aldehydes, and acetylation of alcohols) proceed smoothly in the presence of Sc(OTf)$_3$ (eq 26).[92]

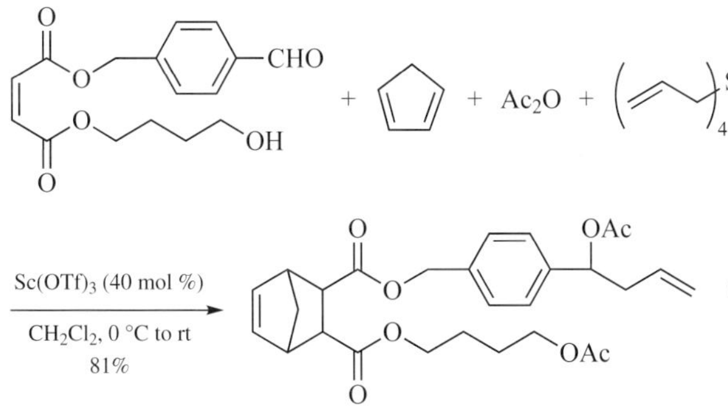

Related Reagents. Related polymer-supported scandium triflates, i.e. Nafion-Sc,[93] MC Sc(OTf)$_3$,[94] PA-Sc-TAD,[95] and a polymer-supported scandium that works efficiently in water.[96] A Lewis acid-surfactant combined scandium catalyst, scandium tris(dodecylsulfate).[97]

First Update

Thierry Ollevier

Université Laval, Québec, QC, Canada

Intramolecular Redox Reaction. Polycyclic tetrahydroquinolines were prepared by a scandium triflate-catalyzed 1,5-hydride shift/ring-closure sequence (eq 27).[98] While the title compound readily catalyzes this intramolecular neutral C–H bond functionalization, Gd(OTf)$_3$ proved to be more efficient.

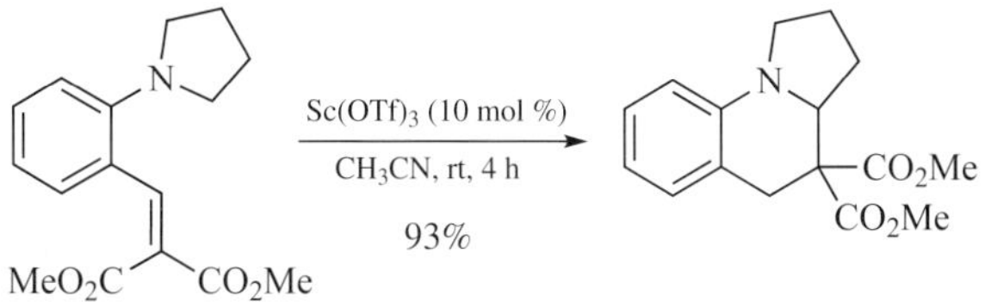

Scandium triflate catalyzed the intramolecular redox reaction from yne-enones to ring-fused tetrahydroquinolines (eq 28).[99] This Lewis acid-catalyzed intramolecular redox domino reaction occurs via domino 1,5-hydride shift and cyclization to afford tetrahydroquinolines in moderate to excellent yields and with high diastereoselectivity. A similar Sc(OTf)$_3$-catalyzed tandem 1,5-hydride transfer cyclization process was applied in the construction of 3-amino-3-carboxy-tetrahydroquinoline derivatives (eq 29).[100]

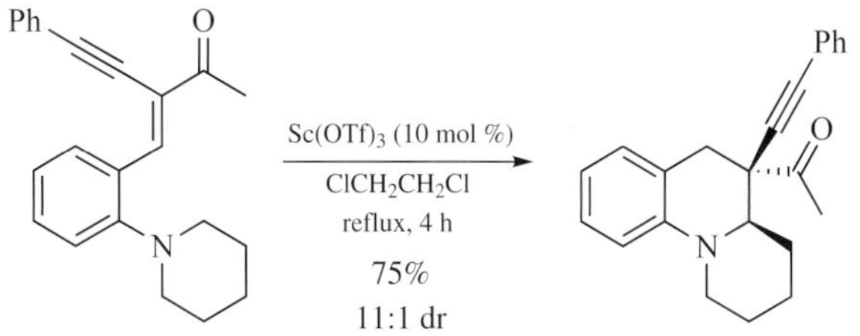

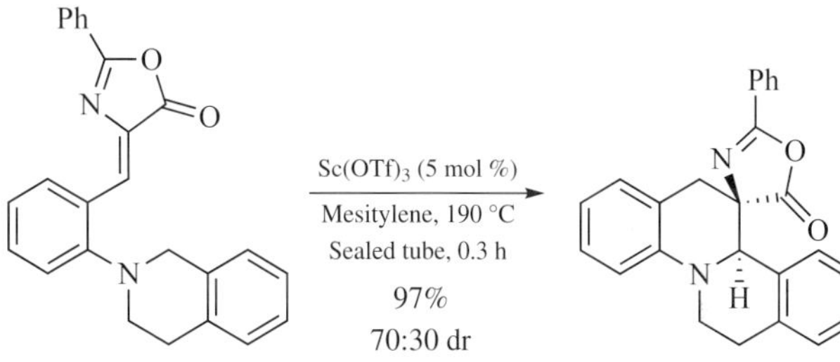

Functionalization of Indole Derivatives. Fully substituted 1,2-dihydro-β-carbolines could be efficiently synthesized via a tandem process involving α-indolyl propargylic alcohols and nitrones using 30 mol % of Sc(OTf)$_3$ (eq 30).[101]

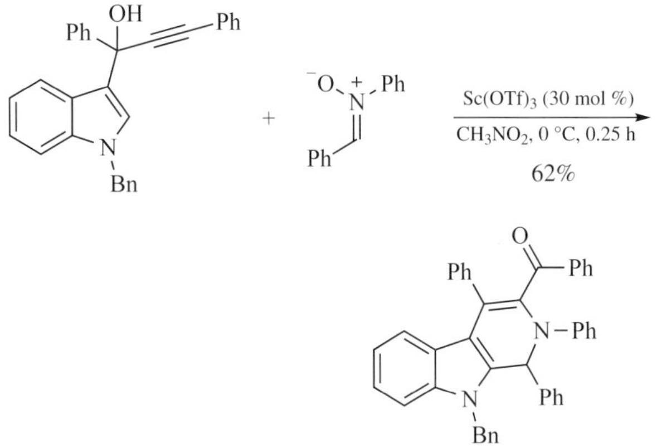

The enantioselective Friedel–Crafts reaction of α,β-unsaturated 2-acyl N-methylimidazoles with electron-rich heterocycles, such as indole derivatives, 2-methoxyfuran, and pyrrole, was catalyzed by Sc(OTf)$_3$ conjointly used with a chiral bis(oxazolinyl)pyridine ligand (eq 31).[102] The reaction afforded good enantioselectivities (>90% ee) for a broad range of substrates.

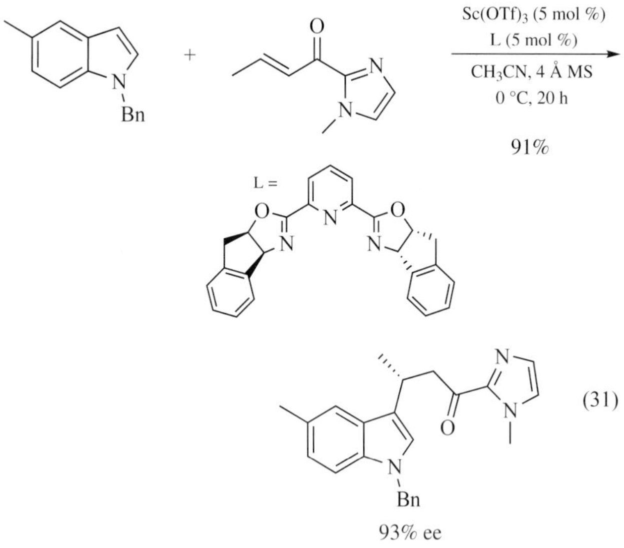

$$(31)$$

93% ee

$$(33)$$

98% ee

The title compound, when used in partner with a chiral *N,N'*-dioxide, efficiently catalyzes a highly enantioselective Friedel–Crafts reaction of indoles and pyrroles with chalcone derivatives (eq 32).[103a] A series of β-heteroaryl-substituted dihydrochalcones was obtained with high enantioselectivities and in moderate to excellent yields. The enantioselective Friedel–Crafts alkylation of indoles with alkylidene malonates has also been developed by the same authors using chiral *N,N'*-dioxide–Sc(OTf)₃ complexes as catalysts.[103b] Asymmetric vinylogous Michael reaction of α-angelica lactone and its derivatives to α,β-unsaturated γ-keto esters, affording the corresponding γ,γ-disubstituted butenolide products in moderate to good yields (up to 93%), was also reported using the same type of Sc(OTf)₃-derived catalyst.[103c]

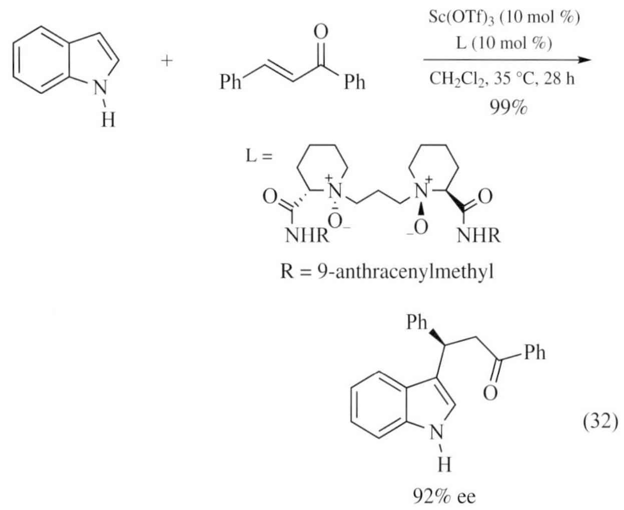

Scandium triflate is an efficient catalyst for the reaction of indolylmethyl Meldrum's acids with a variety of nucleophiles (eq 34).[105] The reaction proceeds through the nucleophilic displacement of Meldrum's acid moiety via a gramine-type fragmentation. This method allowed access to unsymmetric diindolemethanes in moderate yields.

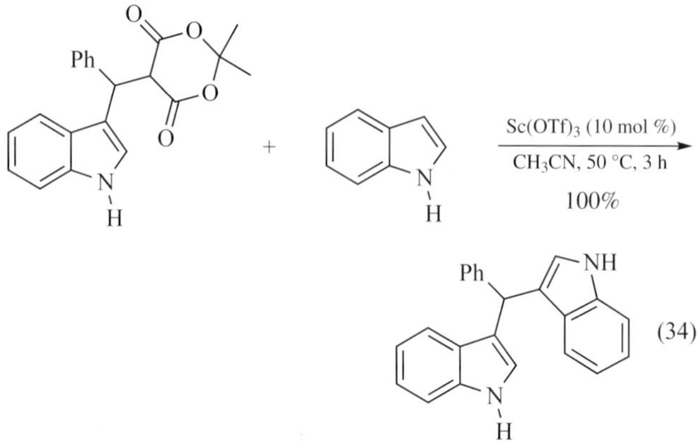

$$(34)$$

$$(32)$$

92% ee

The direct C₆ functionalization reaction of 2,3-disubstituted indoles with various N–Ts aziridines was catalyzed by Sc(OTf)₃ (eq 35).[106] The reaction proceeds in mild conditions using various 2,3-disubstituted indoles and aziridines, leading to the C₆-functionalized indole as the major regioisomer.

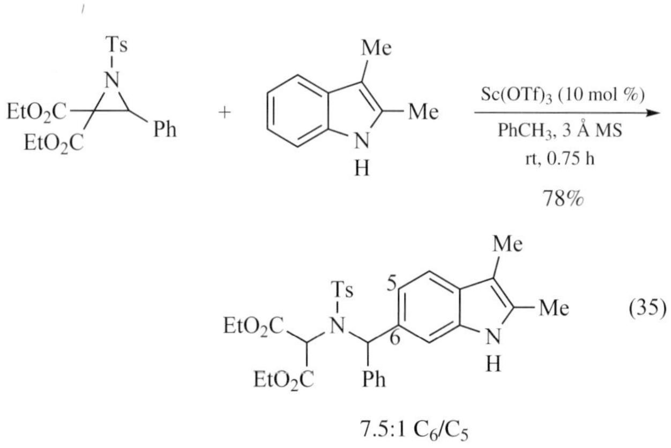

A direct catalytic asymmetric aldol-type reaction of 3-substituted-2-oxindoles with glyoxal derivatives was catalyzed by a similar chiral *N,N'*-dioxide–Sc(OTf)₃ complex (eq 33).[104] The complex efficiently catalyzed the aldol reaction affording the 3-(α-hydroxy-β-carbonyl) oxindoles with vicinal quaternary–tertiary stereocenters, in up to 93% yield, 99:1 dr, and >99% ee under mild conditions.

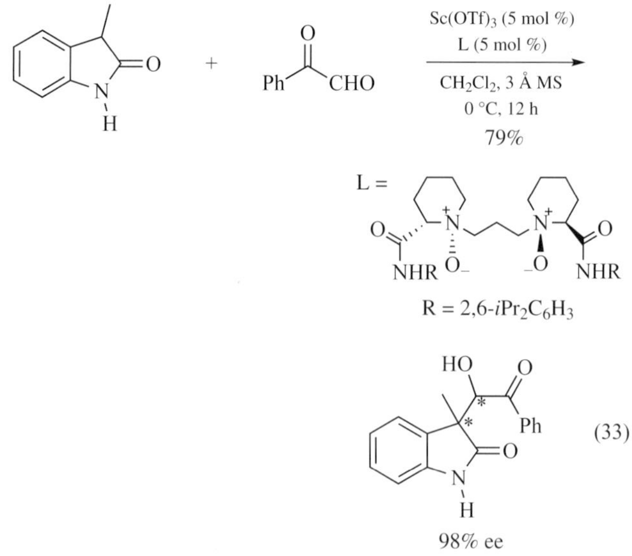

$$(35)$$

7.5:1 C₆/C₅

Lewis Acid-catalyzed [3+2] Cycloaddition. The title compound efficiently catalyzed an efficient and highly regioselective 1,3-dipolar cycloaddition reaction of alkynes and *N*-tosylazomethine ylides (eq 36).[107] Cleavage of the C–C bond of *N*-tosyl aziridines under mild conditions was catalyzed by Sc(OTf)$_3$, affording the *N*-tosylazomethine ylides. The reaction proceeds smoothly with various internal alkynes to afford the desired cycloadducts in moderate to good yields.

Alkylation Using Alkanes. Direct alkylation of pyridines and quinolines using simple alkanes and *tert*-butyl peroxide as oxidant was developed.[108] This C–C bond forming reaction was carried out by using Sc(OTf)$_3$ as a Lewis acid to increase the reactivity of pyridine and quinoline derivatives (eq 37). Scandium triflate demonstrated the best catalytic activity among the Lewis acids tested. While bis-alkylation product was obtained using quinoline, only mono-alkylation products were afforded when isoquinoline was used. Cycloheptane, cyclohexane, and norbornane were determined to be suitable reaction partners.

1. (a) Massaux, J.; Duyckarts, G., *Anal. Chim. Acta.* **1974**, *73*, 416. (b) Roberts, J. E.; Bykowsky, J. S., *Thermochimica Acta.* **1978**, *25*, 233. (c) Kobayashi, S.; Hachiya, I.; Araki, M.; Ishitani, H., *Tetrahedron Lett.* **1993**, *34*, 3755. (d) Castellani, C. B.; Carugo, O.; Perotti, A.; Sacchi, D.; Invernizzi, A. G.; Vidari, G., *J. Mol. Cat.* **1993**, *85*, 65.

2. (a) Kobayashi, S., *Chem. Lett.* **1991**, 2187. (b) Kobayashi, S., *Synlett* **1994**, 689. (c) Marshman, R. W., *Aldrichimica Acta* **1995**, *28*, 77.

3. Kobayashi, S.; Hachiya, I.; Ishitani, H.; Araki, M., *Synlett* **1993**, 472.

4. Kobayashi, S.; Wakabayashi, T.; Nagayama, S.; Oyamada, H., *Tetrahedron Lett.* **1997**, *38*, 4559.

5. Tian, H.-Y.; Chen, Y.-J.; Wang, D.; Zenf, C.-C.; Li, C.-J., *Tetrahedron Lett.* **2000**, *41*, 2529.

6. Kobayashi, S.; Hachiya, I., Yasuda, M., *Tetrahedron Lett.* **1996**, *37*, 4559.

7. Uno, H.; Nishihara, Y.; Mizobe, N.; Ono, N., *Bull. Chem. Soc. Jpn.* **1999**, *72*, 1533.

8. Swamy, V. M.; Bhadbhade, M. M.; Puranik, V. G.; Sarkar, A., *Tetrahedron Lett.* **2000**, *41*, 6137.

9. Chen, J.-X.; Sakamoto, A.; Orita, A.; Otera, J., *J. Org. Chem.* **1998**, *63*, 6739.

10. Fukuzawa, S.; Tsuchimoto, T.; Kanai, T., *Bull. Chem. Soc. Jpn.* **1994**, *67*, 2227.

11. Kagoshima, H., Hashimoto, K., Saigo, K., *Tetrahedron Lett.* **1998**, *39*, 8465.

12. Deguin, B.; Roulet, J.-M.; Vogel, P., *Tetrahedron Lett.* **1997**, *38*, 6197.

13. Crotti, P.; Bussolo, V. D.; Favero, L.; Pineschi, M.; Pasero, M., *J. Org. Chem.* **1996**, *61*, 9548.

14. Yadav, J. S.; Reddy, B. V. S.; Rao, T. P., *Tetrahedron Lett.* **2000**, *41*, 7943.

15. Lipshtz, B. H.; Sclafani, J. A.; Takanami, T., *J. Am. Chem. Soc.* **1998**, *120*, 4021.

16. Kobayashi, S.; Araki, M.; Ishitani, H.; Nagayama, S.; Hachiya, I., *Synlett* **1995**, 233.

17. Kobayashi, S.; Ishitani, H.; Komiyama, S.; Oniciuv, D. C.; Katritzky, A. R., *Tetrahedron Lett.* **1996**, *37*, 3731.

18. Kobayashi, S.; Hachiya, I.; Suzuki, S.; Moriwaki, M., *Tetrahedron Lett.* **1996**, *37*, 2809.

19. (a) Kobayashi, S.; Akiyama, R.; Kitagawa, H., *J. Comb. Chem.* **2000**, *2*, 438. (b) Kobayashi, S.; Akiyama, R.; Kitagawa, H., *J. Comb. Chem.* **2001**, *3*, 196.

20. Kobayashi, S.; Akiyama, R.; Moriwaki, M., *Tetrahedron Lett.* **1997**, *38*, 4819.

21. Kobayashi, S.; Ishitani, H., *J. Chem. Soc., Chem. Commun.* **1995**, 1379.

22. Kobayashi, S.; Busujima, T.; Nagayama, S., *Synlett* **1999**, 545.

23. Kobayashi, S.; Akiyama, R.; Moriwaki, M.; Suzuki, S.; Hachiya, I., *Tetrahedron Lett.* **1996**, *37*, 7783.

24. (a) Oyamada, H.; Kobayashi, S., *Synlett* **1998**, 249. (b) Okitsu, O.; Oyamada, H.; Furuta, T.; Kobayashi, S., *Heterocycles* **2000**, *52*, 1143.

25. Kobayashi, S.; Furuta, T.; Sugita, K.; Okitsu, O.; Oyamada, H., *Tetrahedron Lett.* **1999**, *40*, 1341.

26. Hachiya, I.; Kobayashi, S., *J. Org. Chem.* **1993**, *58*, 6958.

27. (a) Aggarwal, V. K.; Vennall, G. P., *Tetrahedron Lett.* **1996**, *37*, 3745. (b) Aggarwal, V. K.; Vennall, G. P., *Synthesis* **1998**, 1822.

28. (a) Akiyama, T.; Iwai, J., *Tetrahedron Lett.* **1997**, *38*, 853. (b) Akiyama, T.; Iwai, J.; Sugano, M., *Tetrahedron* **1999**, *55*, 7499.

29. Bonini, B. F.; Comes-Franchini, M.; Fochi, M.; Mazzanti, G.; Nanni, C.; Ricci, A., *Tetrahedron Lett.* **1998**, *39*, 6737.

30. Clive, D. L. J.; He, X.; Postema, M. H. D.; Mashimbye, M. J., *J. Org. Chem.* **1999**, *64*, 4397.

31. Kobayashi, S.; Wakabayashi, T.; Oyamada, H., *Chem. Lett.* **1997**, 831.

32. Lautens, M.; Ouellet, S. G.; Raeppel, S., *Angew. Chem., Int. Ed., Engl.* **2000**, *39*, 4079.

33. Bellucci, C.; Cozzi, P. G.; Umani-Ronchi, A., *Tetrahedron Lett.* **1995**, *36*, 7289.

34. Kobayashi, S.; Busujima, T.; Nagayama, S., *Chem. Commun.* **1998**, 19.

35. Kobayashi, S.; Sugita, K.; Oyamada, H., *Synlett* **1999**, 138.

36. Kawada, A.; Mitamura, S.; Kobayashi, S., *Synlett* **1994**, 545.

37. (a) Kawada, A.; Mitamura, S.; Kobayashi, S., *Chem. Commun.* **1996**, 183. (b) Kawada, A.; Mitamura, S.; Matuso, J.; Tsuchiya, T.; Kobayashi, S., *Bull. Chem. Soc. Jpn.* **2000**, *73*, 2325.

38. (a) Kobayashi, S.; Moriwaki, M.; Hachiya, I., *J.Chem. Soc., Chem. Commun.* **1995**, 1527. (b) Kobayashi, S.; Moriwaki, M.; Hachiya, I., *Synlett* **1995**, 1153. (c) Kobayashi, S.; Moriwaki, M.; Hachiya, I., *Bull. Chem. Soc. Jpn.* **1997**, *70*, 267.

39. (a) Kostuki, H.; Ohishi, T.; Inoue, M., *Synlett* **1998**, 255. (b) Kostuki, H.; Ohishi, T.; Inoue, M.; Kojima, T., *Synthesis* **1999**, 603.

40. (a) Tsuchimoto, T.; Tobita, K.; Hiyama, T.; Fukuzawa, S., *Synlett* **1996**, 557. (b) Tsuchimoto, T.; Hiyama, T.; Fukuzawa, S., *Chem. Chommun.* **1996**, 2345. (c) Tsuchimoto, T.; Tobita, K.; Hiyama, T.; Fukuzawa, S., *J. Org. Chem.* **1997**, *62*, 557.

41. Matsui, M.; Yamamoto, H., *Bull. Chem. Soc. Jpn.* **1995**, *68*, 2663.

42. Song, C. E.; Shim, W. H.; Roh, E. J.; Choi, J. H., *Chem. Commun.* **2000**, 1695.

43. Martsui, M.; Karibe, N.; Hayashi, K.; Yamamoto, H., *Bull. Chem. Soc. Jpn.* **1995**, *68*, 3569.

44. Huang, T.; Li, C.-J., *Tetrahedron Lett.* **2000**, *41*, 6715.

45. El Gihani, M. T.; Heaney, H.; Shuhaibar, K. F., *Synlett* **1996**, 871.

46. (a) Bennani, Y. L.; Zhu, G.-D.; Freeman, J. C., *Synlett* **1998**, 754. (b) Nakagawa, M.; Kawahara, M., *Org. Lett.* **2000**, *2*, 953.

47. Heaney, H.; Simcox, M.; Slawin, A. M. Z.; Giles, R. G., *Synlett* **1998**, 640.

48. Smith, A. B., III; Kanoh, N.; Ishiyama, H.; Hartz, R. A., *J. Am. Chem. Soc.* **2000**, *122*, 11254.

49. Waller, F. J.; Barrett, A. G. M.; Braddock, D. C.; Ramprasad, D., *Chem. Commun.* **1997**, 613.

50. Zulfigar, F.; Kitazume, T., *Green Chem.* **2000**, *2*, 296.

51. (a) Arseniyadis, S.; Rodriguez, R.; Yashunsky, D. V.; Camara, J.; Ourisson, G., *Tetrahedron Lett.* **1994**, *35*, 4843. (b) Ishihara, K.; Kubota, M.; Kurihara, H.; Yamamoto, H., *J. Am. Chem. Soc.* **1995**, *11*, 4413. (c) Sammakia, T.; Berliner, M. A., *J. Org. Chem.* **1995**, *60*, 6652.

52. Oakes, R. S.; Heppenstall, T. J.; Shezad, N.; Clifford, A. A.; Rayner, C. M., *Chem. Commun.* **1999**, 1459.

53. (a) Kobayashi, S.; Ishitani, H.; Nagayama, S., *Chem. Lett.* **1995**, 423. (b) Kobayashi, S.; Ishitani, H.; Nagayama, S., *Synthesis* **1995**, 1195.

54. Zulfigar, F.; Kitazume, T., *Green Chem.* **2000**, *2*, 137.

55. Kobayashi, S.; Akiyama, R.; Kawamura, M.; Ishitani, H., *Chem. Lett.* **1997**, 1039.

56. Ishitani, H.; Nagayama, S.; Kobayashi, S., *J. Org. Chem.* **1996**, *61*, 1902.

57. Ma, Y.; Qian, C., *Tetrahedron Lett.* **2000**, *41*, 945.

58. Shindo, M.; Oya, S.; Sato, Y.; Shishido, K., *Heterocycles* **2000**, *52*, 545.

59. Malinakova, H. C.; Liebeskind, L. S., *Org. Lett.* **2000**, *2*, 4083.

60. Harwata, M.; Sharma, U., *Org. Lett.* **2000**, *2*, 2703.

61. Kobayashi, S.; Araki, M.; Hachiya, I., *J. Org. Chem.* **1994**, *59*, 3758.

62. Kobayashi, S.; Ishitani, H.; Araki, M.; Hachiya, I., *Tetrahedron Lett.* **1994**, *35*, 6325.

63. Wipf, P.; Wang, X., *Tetrahedron Lett.* **2000**, *41*, 8747.

64. Ishitani, H.; Kobayashi, S., *Tetrahedron Lett.* **1996**, *37*, 7357.

65. Kobayashi, S.; Kawamura, M., *J. Am. Chem. Soc.* **1998**, *120*, 5840.

66. Kodama, H.; Ito, J.; Hori, K.; Ohta, T.; Furukawa, I., *J. Organomet. Chem.* **2000**, *603*, 6.

67. (a) Kitajima, H.; Katsuki, T., *Synlett* **1997**, 568. (b) Kitajima, H.; Ito, K.; Katsuki, T., *Tetrahedron* **1997**, *53*, 17015.

68. Qian, C.; Wang, L., *Tetrahedron: Asymmetry* **2000**, *11*, 2347.

69. Gillespie, K. M.; Munslow, I. J.; Scott, P., *Tetrahedron Lett.* **1999**, *40*, 9371.

70. Kotsuki, H.; Arimura, K.; Araki, T.; Shinohara, T., *Synlett* **1999**, 462.

71. (a) Ishihara, K.; Karumi, Y.; Kubota, M.; Yamamoto, H., *Synlett* **1996**, 839. (b) Fukuzawa, S.; Tsucihmoto, T.; Hotaka, T.; Hiyama, T., *Synlett* **1995**, 1077. (c) Pozsgay, V., *Tetrahedron: Asymmetry* **2000**, *11*, 151.

72. Aggarwal, V. K.; Fonquerna, S.; Vennall, G. P., *Synlett* **1998**, 849.

73. Fukase, K.; Kinoshita, I.; Kanoh, T.; Nakai, Y.; Hasuoka, A.; Kusumoto, S., *Tetrahedron* **1996**, *52*, 3897.

74. (a) Ishihara, K.; Kubota, M.; Kurihara, H.; Yamamoto, H., *J. Org. Chem.* **1996**, *61*, 4560. (b) Barrett, A. G. M.; Braddock, D. C., *J. Chem. Soc. Chem. Commun.* **1997**, 351. (c) Kajiro, H.; Mitamura, S.; Mori, A.; Hiyama, T., *Synlett* **1998**, 51. (d) Kajiro, H.; Mori, A.; Nishihara, Y.; Hiyama, T., *Chem. Lett.* **1999**, 459. (e) Kajiro, H.; Mitamura, S.; Mori, A.; Hiyama, T., *Tetrahedron Lett.* **1999**, *40*, 1689. (f) Kajiro, H.; Mitamura, S.; Mori, A.; Hiyama, T., *Bull. Chem. Soc. Jpn.* **1998**, *72*, 1553. (g) Koiwa, M.; Hareau, G. P. J.; Sato, F., *Tetrahedron Lett.* **2000**, *41*, 2389. (h) Mehta, S.; Whitfield, D. M., *Tetrahedron* **2000**, *56*, 6415. (i) Zhao, H.; Pendri, A.; Greenwald, R. B., *J. Org. Chem.* **1998**, *63*, 7559. (j) Greenwald, R. B.; Pendari, A.; Zhao, H., *Tetrahedron: Asymmetry* **1998**, *9*, 915.

75. Oriyama, T.; Kobayashi, Y.; Noda, K., *Synlett* **1998**, 1047.

76. Suzuki, T.; Watahiki, T.; Oriyama, T., *Tetrahedron Lett.* **2000**, *41*, 8903.

77. Mehta, S.; Whitfield, D., *Tetrahedron Lett.* **1998**, *39*, 5907.

78. Yamamoto, N.; Isobe, M., *Chem. Lett.* **1994**, 2299.

79. (a) Sibi, M. P.; Ji, J., *Angew. Chem., Int. Ed. Engl.* **1996**, *35*, 190. (b) Sibi, M. P.; Ji, J., *Angew. Chem., Int. Ed. Engl.* **1997**, *36*, 274. (c) Sibi, M. P.; Jasperse, C. P.; Ji, J., *J. Am Chem. Soc.* **1995**, *117*, 10779. (d) Sibi, M. P.; Ji, J.; Sausker, J. B.; Jasperse, C. P., *J. Am. Chem. Soc.* **1999**, *121*, 7517. (e) Mero, C. L.; Porter, N. A., *J. Am. Chem. Soc.* **1999**, *121*, 5155. (f) Sibi, M. P.; Porter, N. A., *Acc. Chem. Res.* **1999**, *32*, 163.

80. (a) Burgess, K.; Lim, H.-J.; Porte, A. M.; Sulikowski, G. A., *Angew. Chem., Int. Ed., Engl.* **1996**, *35*, 220. (b) Xie, W.; Fang, J.; Li, J.; Wang, P. G., *Tetrahedron* **1999**, *55*, 12929. (c) Pansare, S. V.; Jain, R. P.; Bhattacharya, A., *Tetrahedron Lett.* **1999**, *40*, 5255.

81. Castellani, C. B.; Carugo, O.; Giusti, M.; Leopizzi, C.; Perotti, A.; Invernizzi, A. G.; Vidari, G., *Tetrahedron* **1996**, *52*, 11045.

82. Fukuzawa, S.; Yamaishi, Y.; Furuya, H.; Terao, K.; Iwasaki, F., *Tetrahedron Lett.* **1997**, *38*, 7203.

83. (a) Prein, M.; Padwa, A., *Tetrahedron Lett.* **1996**, *37*, 6981. (b) Bazin, H. G.; Kerns, R. J.; Linhardt, R. J., *Tetrahedron Lett.* **1997**, *38*, 923. (c) Prein, M.; Manley, P. J.; Padwa, A., *Tetrahedron* **1997**, *53*, 7777.

84. Hori, N.; Nagasawa, K.; Shimizu, T.; Nakata, T., *Tetrahedron Lett.* **1999**, *40*, 2145.

85. Yadav, J. S.; Reddy, B. V. S.; Murthy, C. V. S.; Kumar, G. M., *Synlett* **2000**, 1450.

86. (a) Zhang, W.-C.; Viswanathan, G. S.; Li, C.-J., *Chem. Commun.* **1999**, 291. (b) Zhang, W.-C.; Li, C.-J., *Tetrahedron* **2000**, *56*, 2403.

87. Anderson, J. C.; Peace, S.; Pih, S., *Synlett* **2000**, 850.

88. Aggarwal, V. K.; Vennall, G. P.; Davey, P. N.; Newman, C., *Tetrahedron Lett.* **1998**, *39*, 1997.

89. Kobayashi, S.; Busujima, T.; Nagayama, S., *Chem. Commun.* **1998**, 981.

90. Ma, Y.; Qian, C.; Wang, L.; Tang, M., *J. Org. Chem.* **2000**, *65*, 3864.

91. Blackburn, C.; Guan, B.; Fleming, P.; Shiosaki, K.; Tsai, S., *Tetrahedron Lett.* **1998**, *39*, 7203.

92. Orita, A.; Nagano, Y.; Nakazawa, K.; Otera, J., *Synlett* **2000**, 599.

93. Kobayashi, S.; Nagayama, S., *J. Org. Chem.* **1996**, *61*, 2256.

94. Kobayashi, S.; Nagayama, S., *J. Am. Chem. Soc.* **1998**, *120*, 2985.

95. Kobayashi, S.; Nagayama, S., *J. Am. Chem. Soc.* **1996**, *118*, 8977.

96. Nagayama, S.; Kobayashi, S., *Angew. Chem., Int. Ed., Engl.* **2000**, *39*, 567.

97. (a) Kobayashi, S.; Wakabayashi, T., *Tetrahedron Lett.* **1998**, *39*, 5389. (b) Manabe, K.; Mori, Y.; Wakabayashi, T.; Nagayama, S.; Kobayashi, S., *J. Am. Chem. Soc.* **2000**, *122*, 7202.

98. Murarka, S.; Zhang, C.; Konieczynska, M. D.; Seidel, D., *Org. Lett.* **2009**, *11*, 129.

99. Zhou, G.; Zhang, J., *Chem. Commun.* **2010**, *46*, 6593.

100. Han, W.-Y.; Zuo, J.; Wu, Z.-J.; Zhang, X.-M.; Yuan, W.-C., *Tetrahedron* **2013**, *69*, 7019.

101. Wang, L.; Xie, X.; Liu, Y., *Org. Lett.* **2012**, *14*, 5848.

102. Evans, D. A.; Fandrick, K. R.; Song, H.-J., *J. Am. Chem. Soc.* **2005**, *127*, 8942.

103. (a) Wang, W.; Liu, X.; Cao, W.; Wang, J.; Lin, L.; Feng, X., *Chem. Eur. J.* **2010**, *16*, 1664. (b) Liu, Y.; Zhou, X.; Shang, D.; Liu, X.; Feng, X., *Tetrahedron* **2010**, *66*, 1447. (c) Ji, J.; Lin, L.; Zhou, L.; Zhang, Y.; Liu, Y.; Liu, X.; Feng, X., *Adv. Synth. Catal.* **2013**, *355*, 2764.

104. (a) Shen, K.; Liu, X.; Zheng, K.; Li, W.; Hu, X.; Lin, L.; Feng, X., *Chem. Eur. J.* **2010**, *16*, 3736. (b) Hui, Y.; Zhang, Q.; Jiang, J.; Lin, L.; Liu, X.; Feng, X., *J. Org. Chem.* **2009**, *74*, 6878.

105. Armstrong, E. L.; Grover, H. K.; Kerr, M. A., *J. Org. Chem.* **2013**, *78*, 10534.

106. Liu, H.; Zheng, C.; You, S.-L., *J. Org. Chem.* **2014**, *79*, 1047.

107. Li, L.; Zhang, J., *Org. Lett.* **2011**, *13*, 5940.

108. Deng, G.; Li, C.-J., *Org. Lett.* **2009**, *11*, 1171.

Sodium Azide[1]

[26628-22-8] N₃Na (MW 65.02)

N_3Na

InChI = 1/N3.Na/c1-3-2;/q-1;+1
InChIKey = PXIPVTKHYLBLMZ-UHFFFAOYAH

(nucleophilic azide source for organoazide preparation;[2] precursor to reagents such as hydrazoic acid,[3] halogen azides,[4] trimethylsilyl azide,[5] tosyl azide,[6] and diphenyl phosphorazidate[7])

Physical Data: dec ca. 300 °C; d 1.850 g cm^{-3}.
Solubility: sol water (39 g/100 g at 0 °C, 55 g/100 g 100 °C); slightly sol alcohol; insol ether.
Form Supplied in: white solid; widely available.
Handling, Storage, and Precautions: while relatively insensitive to impact, the solid can decompose explosively above its melting point. It forms highly explosive azides with metals such as Cu, Pb, Hg, Ag, Au, their alloys and compounds, and reacts with acids to form hydrazoic acid (HN₃) which is a toxic, spontaneously explosive gas. Explosive *gem*-diazides can be formed in CH₂Cl₂ or other chlorinated solvents and shock or heat sensitive metal azidothioformates in CS₂. All work with NaN₃ and other azides should be conducted on a very small scale behind a shield, in a fume hood. Excess NaN₃ on flasks, paper, etc. can be destroyed in a fume hood by soaking with acidified *Sodium Nitrite* or by oxidation with *Cerium(IV) Ammonium Nitrate*.[8]

Original Commentary

Kenneth Turnbull
Wright State University, Dayton, OH, USA

Introduction. The reaction of NaN₃ with I₂ (releasing N₂) is catalyzed by thiols and thiones and this has been used as a spot test for such compounds.[9] NaN₃ has been used to assess the interactions between charged sites in myoglobin.[10]

Preparation of Organic Azides. Organic azides can be reduced readily to amines, utilized for amine, azide or diazo transfer, act as nitrene or nitrenium precursors, and undergo Curtius and Schmidt rearrangements, cycloadditions and Staudinger reactions.[1b] They are prepared most often by nucleophilic displacement of a leaving group by azide ion (commonly NaN₃) (eq 1). Various leaving groups have been used, including halides, sulfonates (mainly OTs, OMs, or OTf, although brosylates[11] and nosylates[12] have been employed), sulfites,[13] and anhydrides.[14] Displacement of allylic acetates (and related species),[15] with *Tetrakis(triphenylphosphine)palladium(0)* as catalyst, and groups such as nitro,[16] phosphine sulfides (from thiaphosphonium species),[17] and phenylseleninates[18] has been reported.

$$R–X + NaN_3 \longrightarrow R–N_3 \qquad (1)$$

Eliminative azidation to form α-azidovinyl ketones occurs

on NaN₃ treatment of some dibromo ketones (eq 2).[19] This approach also works well for the preparation of α-azidostyrenes from styrene.[20] Usually, *gem*-dihalo compounds react with NaN₃ to give *gem*-diazides;[21] however, an unusual nitrile formation has been reported (eq 3) under these conditions.[22] An interesting, stereospecific, solvent-dependent, azide-induced ring opening reaction of a dioxaphospholane has been observed (eq 4).[23]

$$ (2) $$

$$ (3) $$

$$ (4) $$

Displacements can occur at the carbon atom of alkyl (primary, or secondary; tertiary requires *Zinc Chloride* or *Zinc Iodide* catalysis[24]), allyl, benzyl, acyl,[25] activated vinyl,[26] aryl,[27] or heteroaryl[28] species, and aryl diazonium salts (ArN₂⁺, from ArNH₂/HNO₂),[29] or at the heteroatom of, amongst others, organosulfonyl, silyl and phosphoryl halides. Of the latter, *p-Toluenesulfonyl Azide* (PTSN₃),[6] *Azidotrimethylsilane* (TMSN₃),[5] and *Diphenyl Phosphorazidate*, (PhO)₂P(O)N₃ (DPPA),[7] are the most common and the last two are commercially available.

Nucleophilic azide ion displacements are enhanced by polar, aprotic solvents (e.g. DMSO) with which high yield, aryl halide displacement to form even mononitrophenyl azides can occur.[27] Phase-transfer catalysis[30] (permitting the use of less polar solvents) or ultrasonication (for activated primary halides)[31] has also been used. Under such conditions, S_N2 inversion of configuration occurs and this has been observed also for alcohols under Mitsunobu conditions (*Triphenylphosphine*, *Diethyl Azodicarboxylate*, HN₃).[32] Retention is possible where a neighboring group is present.[33]

Tertiary alcohols are converted directly to azides using NaN₃/ *Sulfuric Acid* or HN₃/*Boron Trifluoride* or *Titanium(IV) Chloride* (eq 5),[34] and the carboxylic acid to acyl azide transformation (often en route to Curtius rearrangements to isocyanates) occurs with DPPA[7,1b] or via activation with DMF/*Thionyl Chloride*.[35]

$$ (5) $$

NaN₃ reacts with epoxides at 25–30 °C (pH 6–7) to give azido alcohols.[36] Usually, inversion of stereochemistry takes place and attack at the least hindered site is preferred. The regio- and stereoselectivity of the reaction can often be enhanced by using $TMSN_3$ with a Lewis acid.[37] High selectivity was shown by NaN_3 on a calcium cation-exchanged Y-type zeolite (CaY) (eq 6),[38] but less so with NaN_3 on silica or alumina or the NaN_3/NH_4Cl system. $NaN_3/ZnCl_2$ gave lower yields than $TMSN_3/BF_3·OEt_2$ for the ring opening of 1,2-epoxysilanes;[39a] selective azide opening at the site of silyl substitution has been reported.[39b] With a $PhSO_2$ group attached to the epoxide, azidation–elimination occurs to form the corresponding azidoaldehydes.[40] Reaction of the epoxy ester (**1**) with NaN_3 under more vigorous conditions gave (**2**) in 60% yield (eq 6).[41]

(6)

Hydrazoic acid (HN_3; NaN_3/H^+) reacts with alkenes to form azidoalkanes.[42] Alkenes bearing a phenyl group or two geminal alkyl groups require a Lewis acid ($TiCl_4$ is best). Mono- or 1,2-dialkyl alkenes do not react and Michael additions occur with α,β-unsaturated alkenes.[26b] Enol ethers and silylenol ethers give azido ethers[42] and a similar process occurs with *Trifluoroacetic Acid* catalysis[43] or from acetals[44] or aldehydes with $TMSN_3$.[43] Interestingly, $TiCl_4$-catalyzed HN_3 addition to silyl enol ethers in the presence of primary or secondary alcohols gives the azido ethers shown (eq 7).[42] Recently, it has been found that (**3**) reacts with NaN_3/CAN to form the α-azido ketone (**4**) (eq 8).[45]

(7)

(8)

Other oxidative double bond azidations have been reported. Thus an azidohydrin was formed from pregnenolone acetate and chromyl azide (NaN_3, *Chromium(VI) Oxide*)[46] and steroidal dienones reacted with $TMSN_3$/*Lead(IV) Acetate*[47] to give diazido compounds. Vicinal diazides also result from alkenes and Fe^{III},[48] *Manganese(III) Acetate* (eq 9),[49] or *Iodosylbenzene* and NaN_3.[50] Anti-Markovnikov selenoazido products were prepared from the reaction of azide ion with alkenes and *(Diacetoxyiodo)benzene/Diphenyl Diselenide* (eq 10);[51] α-keto azides (with $TMSN_3$) are formed without PhSeSePh.[52]

(9)

(10)

Azide ion (or congener) attack upon nonconjugated alkenes is aided by the use of *Dimethyl(methylthio)sulfonium Tetrafluoroborate*.[53] *Trans* products are obtained and, in general, the amount of anti-Markovnikov product increases with increased azide nucleophilicity, and vice versa. Monosubstituted alkenes favor anti-Markovnikov addition, whereas the opposite occurs with trisubstitution. 1,1-Disubstituted alkenes can give either orientation.

Schmidt Reactions.[54] This term is used for several transformations, general examples of which are shown in eqs 11 and 12. The former is used infrequently due to the drastic conditions required compared to the analogous Curtius and Hoffmann rearrangements and the discovery that DPPA effects the transformation under mild conditions.[7,1b] $TMSN_3$ has been used frequently.

$$RCO_2H \xrightarrow[H^+]{NaN_3} RNH_2 \qquad (11)$$

$$RCOR \xrightarrow[H^+]{NaN_3} RCONHR \qquad (12)$$

The ketone to amide transformation (eq 12) is still of considerable utility (with the provisos regarding the hazards associated with HN_3) and various acids have been employed, including H_2SO_4 (the most common), *Polyphosphoric Acid*, and *Methanesulfonic Acid*. With an unsymmetrical ketone a mixture of amide products can result although preferential migration of an aryl group (over alkyl) has been reported. In one case, the amount of 'aryl migration' product (**6**) (R = H, 75%) was greater (80%) starting from the 7-nitroketone (**5**) (R = NO_2) and lower (70%) from the 7-amino species (**5**) (R = NH_2) (eq 13).[55] Aldehydes usually give nitriles under Schmidt conditions.[56]

(13)

Curtius Reaction.[57] The Curtius reaction involves conversion of an acid chloride (or anhydride) to an isocyanate (eq 14). Trapping of the isocyanate is possible in the presence of a nucleophile. Some cyclic anhydrides react to give isocyanates which can cyclize subsequently.

$$RCOCl \xrightarrow[2. \Delta]{1. NaN_3 \text{ or } TMSN_3} RNCO \qquad (14)$$

Preparation of Heterocycles. As mentioned, heterocycles can be obtained via Schmidt or Curtius reactions. In addition, organic azides react with alkenes to form triazolines (triazoles from alkynes), aziridines, or other heterocycles.[58] In situ triazoline generation and subsequent cleavage can lead to other heterocycles (see eq 15).[59] Reaction of NaN_3 with other α,β-unsaturated alkenes (or alkynes) provides different heterocycles dependent on the substituents. Such reactions are too numerous to mention in detail and only selected examples are shown in eqs 16–18.[60–62]

$$\text{MsO} \quad \overset{\text{Bu}}{\diagup} \quad \text{CO}_2\text{Me} \xrightarrow[\text{rt}]{\text{NaN}_3, \text{DMF}} \quad \overset{\text{H}}{\underset{\text{N}}{\bigcirc}}\text{-Bu} \quad (15)$$

$$\overset{\text{EtO}}{\underset{\text{EtO}_2\text{C}}{\diagup}}\text{CO}_2\text{Et} \xrightarrow[67\%]{\text{NaN}_3, \text{TFA}} \quad \overset{\text{N}}{\underset{\text{EtO}}{\bigcirc}}\text{CO}_2\text{Et} \quad (16)$$

$$\text{Ph}\overset{\text{O}}{\diagdown}\text{≡} \xrightarrow{\text{NaN}_3}_{\text{DMF}} \quad \overset{\text{PhOC}}{\underset{\text{N}\,\text{N}}{\bigcirc}}\text{NH} \quad (17)$$

$$\overset{\text{O}_2\text{N}}{\underset{\text{NO}_2}{\diagup}} \xrightarrow{\text{NaN}_3} \quad \overset{\text{-O}\,\text{N}^+\,\text{O}^-\,\text{N}}{\underset{}{\bigcirc}} \quad (18)$$

A useful tetrazole preparation is the addition of NaN_3 (under acidic conditions) to nitriles.[63] Similar processes occur with **Tri-n-butyltin Azide** or $TMSN_3$.[64]

First Update

B. Narsaiah, J. S. Yadav, T. Yakaiah & B. P. V. Lingaiah
*Indian Institute of Chemical Technology,
Hyderabad, India*

Sodium azide is a nucleophilic reagent used in the synthesis of ring systems such as tetrazole and triazole derivatives. It is also used in the ring cleavage of epoxides and aziridines, and also for the conversion of alkyl halides, alcohols, amines, esters, alkenes, alkynes, cyclic ketones, and nitro compounds to the respective azides. It is prepared[65] in high yield by the reaction of alkyl nitrites with 30–50% NaOH solution containing 0.8–1.5 equiv of hydrazine hydrate. The details of the reactions and syntheses of specific ring systems using sodium azide are discussed below.

Tetrazole Derivatives. Tetrazole-containing organic compounds are considered potent HIV-1 protease inhibitors, a target in many ongoing medicinal chemistry programs.[66,67] The tetrazole derivatives are mainly synthesized from nitriles, amides, acid nitriles, acid azides, aldehydes, dienones, isocyanates by the reaction with sodium azide using various catalysts. More specifically, aryl nitriles[68–71] with tri(n-butyl)tin chloride[72] or aliphatic nitriles using $AlCl_3$,[73,74] triethylammonium chloride,[75,76] tetrabutylammonium salts,[77] magnesium salts,[78]

zinc chloride (Demko Sharpless tetrazole synthesis),[79] zinc bromide,[80] tetrachlorosilane,[81,82] and others[83–87] were reacted with sodium azide to obtain tetrazole derivatives. Typical examples are outlined in eqs 19–26.

$$\text{PhCH=CH-CN} + \text{NaN}_3 \xrightarrow[20\,\text{W}]{\text{MW}} \text{tetrazole} \quad (19)$$
60%

$$\text{MeO-C}_6\text{H}_4\text{-CN} + \text{NaN}_3 \xrightarrow[20\,\text{W}]{\text{MW}} \text{tetrazole} \quad (20)$$
96%

$$\text{CH}_3\text{CN} + \text{NaN}_3 \xrightarrow[\Delta]{\text{AlCl}_3} \text{tetrazole} \quad (21)$$

$$\text{NC-CN} + \text{NaN}_3 \xrightarrow{\text{AlCl}_3} \text{bis-tetrazole} \quad (22)$$
75.6%

$$\text{2-F-C}_6\text{H}_4\text{-CN} + \text{NaN}_3 \xrightarrow{\text{Et}_3\text{N}\cdot\text{HCl}} \text{tetrazole} \quad (23)$$

$$(\text{NO}_2)_3\text{CCN} + \text{NaN}_3 \xrightarrow[\text{CCl}_4\text{-CHCl}_3\text{-HOAc}]{(\text{Bu})_4\text{N}^+\text{X}^-} \text{tetrazole} \quad (24)$$

$$\text{biphenyl-CN} + \text{NaN}_3 \xrightarrow[\text{DMF}]{\text{MgCl}_2} \text{tetrazole} \quad (25)$$

$$\text{CH}_2(\text{CN})_2 + \text{ZnCl}_2 + \text{NaN}_3 + \text{bipyridine} + \text{H}_2\text{O} \longrightarrow$$

$$\left[\text{Zn complex} \right]^{+2} \cdot \text{H}_2\text{O} \quad (26)$$

Similarly, amides such as quinone amide using triflic anhydride,[88] *N*-cyclohexyl-5-hydroxy pentamide using PCl_5,[89] dicyandiamide,[90,91] thiourea,[92] and thiocyanates,[93] on reaction with sodium azide, resulted in the formation of tetrazole derivatives. Some examples are outlined in eqs 27–31.

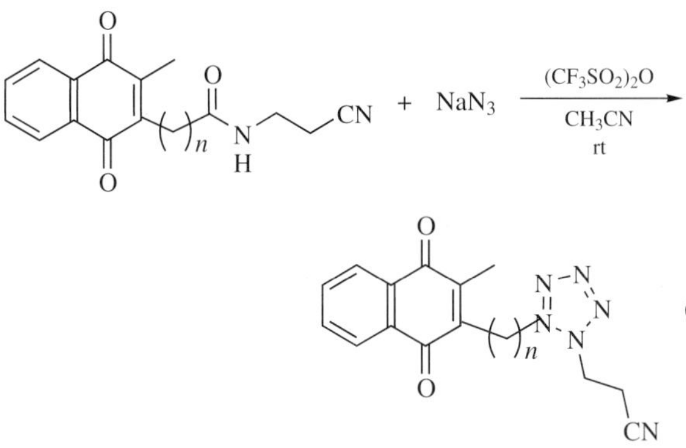

(27)

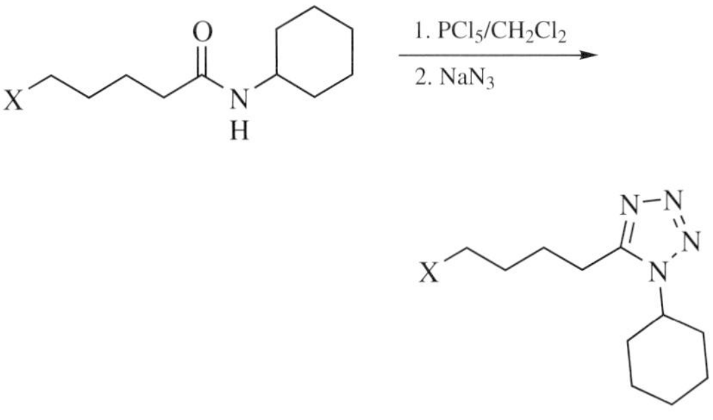

$X = Cl$ 92%

(28)

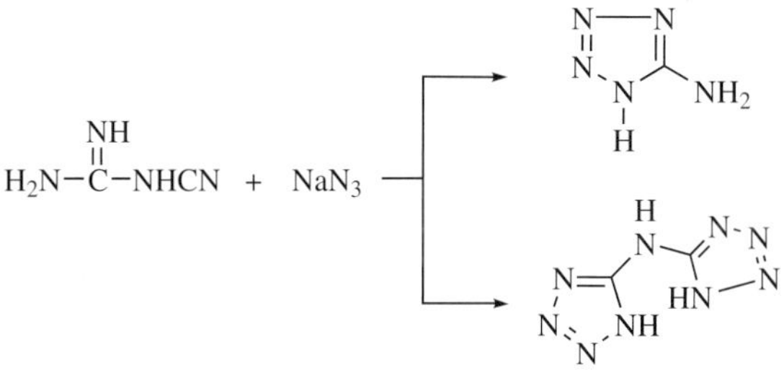

(29)

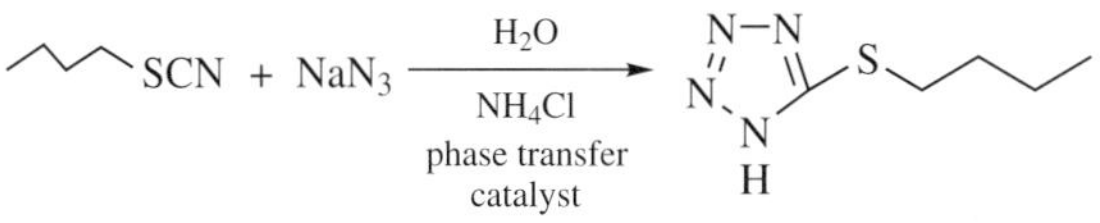

(30)

$$RCHO \xrightarrow[\substack{CH_3CN \\ \Delta}]{IN_3} [RNCO] \longrightarrow RNHCON_3 \longrightarrow Tetrazoles$$

(31)

Reaction of carboxylic acids[96,97] with sodium azide formed intermediate acid azides for peptide synthesis, and these are further cyclized to tetrazole derivatives. Acid nitriles,[98] aldehydes,[99,100] and dienones[101] also reacted with sodium azide to form tetrazoles. Representative examples are displayed in eqs 32–34.

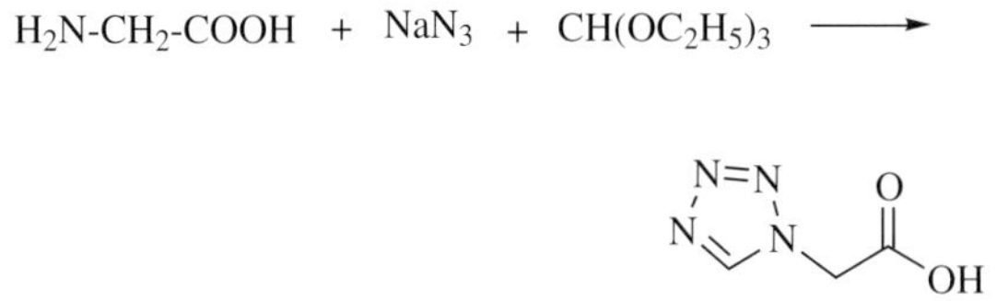

$$H_2N-CH_2-COOH + NaN_3 + CH(OC_2H_5)_3 \longrightarrow$$

(32)

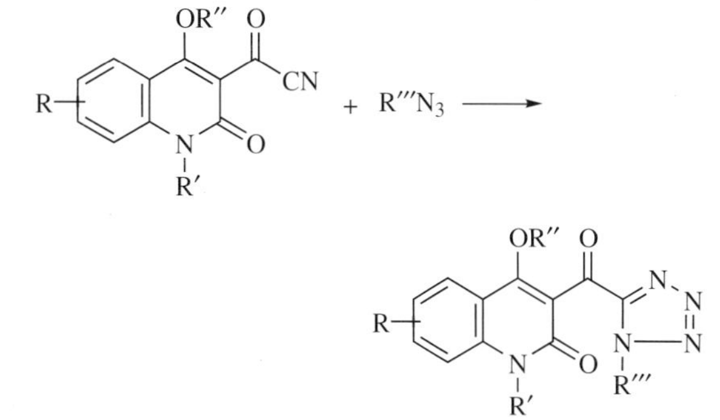

(33)

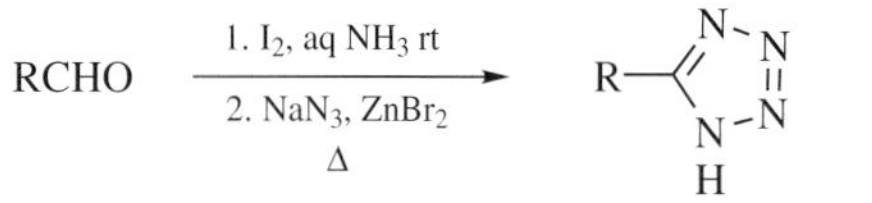

(34)

Reaction of sodium azide with alcohols using DEAD/PPh$_3$,[102] PPh$_3$,[103] BF$_3$·Et$_2$O,[104] or Zeolites[105] can be used to form azides. Similarly, alkyl halides in ionic liquids[106,107] or under other conditions,[108–114] chlorocyclodextrin,[115] and certain amines[116] can be reacted with sodium azide to form organic azides as outlined in eqs 35–40.

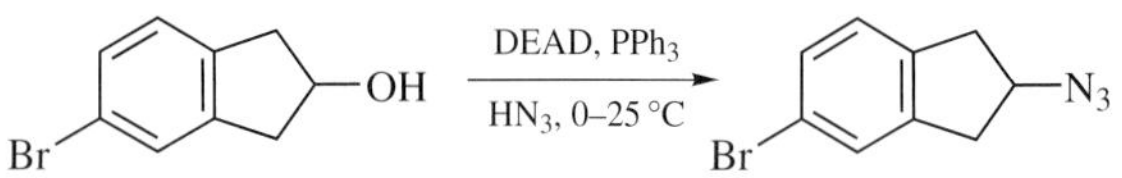

(35)

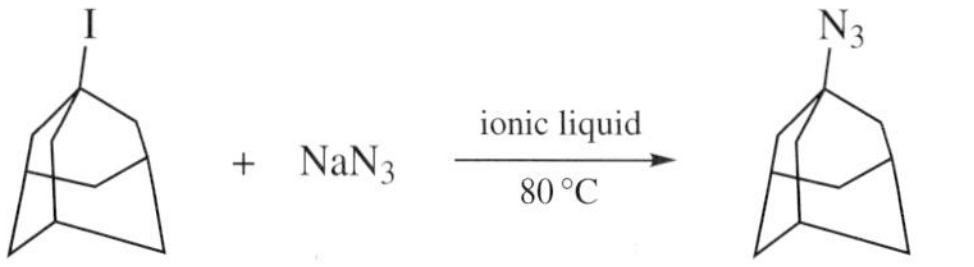

(36)

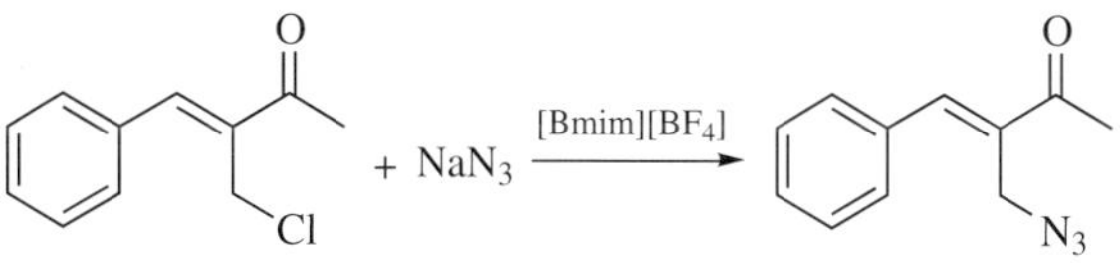

(37)

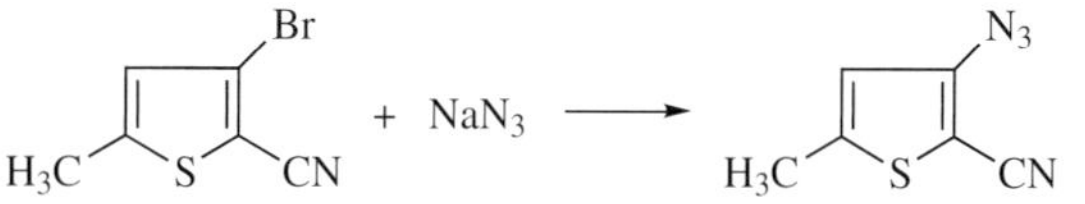

(38)

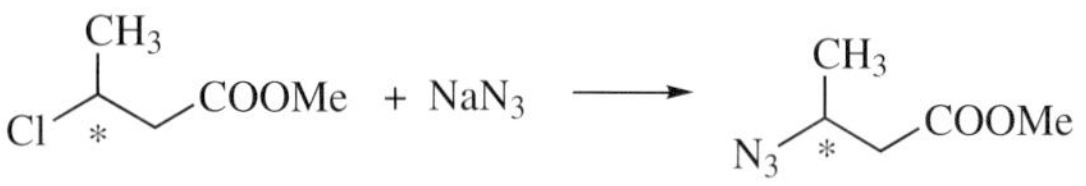

(39)

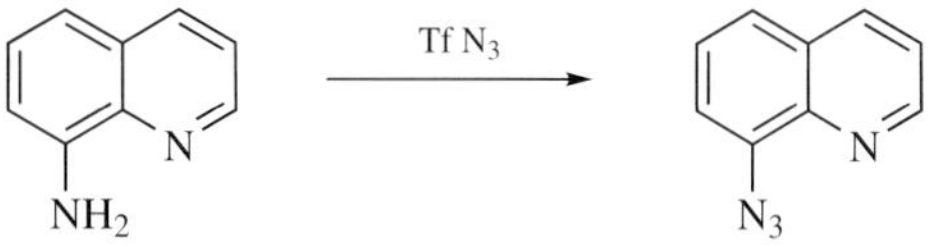

(40)

Similarly, reaction of carboxylic acid halides with sodium azide led to the formation of the respective acyl azides, which are stable.[117–119] By contrast, reaction of carboxylic acid chlorides[96] with sodium azide in toluene at 70–100 °C transpires via a Curtius rearrangement to generate isocyanates. The details of these transformations are outlined in eqs 41–43.

$$Fmoc-NHCHRCOCl + NaN_3 \xrightarrow[0\,°C]{acetone} Fmoc-NHCHRCON_3$$

(41)

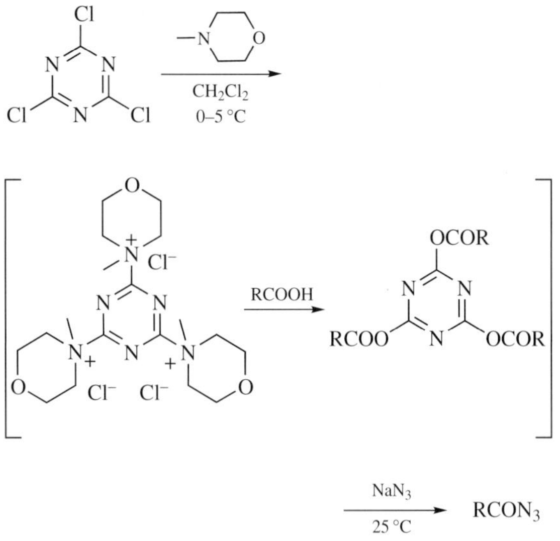

$$\xrightarrow{\underset{25\,°C}{NaN_3}} RCON_3 \quad (42)$$

$$RCOCl + NaN_3 \longrightarrow [RCON_3] \longrightarrow RNCO \quad (43)$$

Triazole Derivatives. Triazole derivatives are known to possess tumor necrosis factor-α (TNF-α) production inhibitor activity. The synthesis of triazole derivatives can be achieved from alkynes[120] or diynes[121] by a tandem cascade reaction involving 1,3-dipolar cycloaddition, anionic cyclization and sigmatropic rearrangement on reaction with sodium azide. Some of the benzoyl triazole derivatives were considered to be potent local anaesthetics and are comparable with Lidocaine. The triazoles can also be prepared from benzoyl acetylenes,[122] triazoloquinazoline derivatives,[123] 2-trifluoromethyl chromones,[124] aliphatic alkynes,[125] 2-nitroazobenzenes,[126] ring opening of [1,2,4]triazolo[5,1-c][2,4]benzothiazepin-10(5H)-one,[127] alkenyl esters[128] and dendrimers.[129] A number of these reactions are outlined in eqs 44–48.

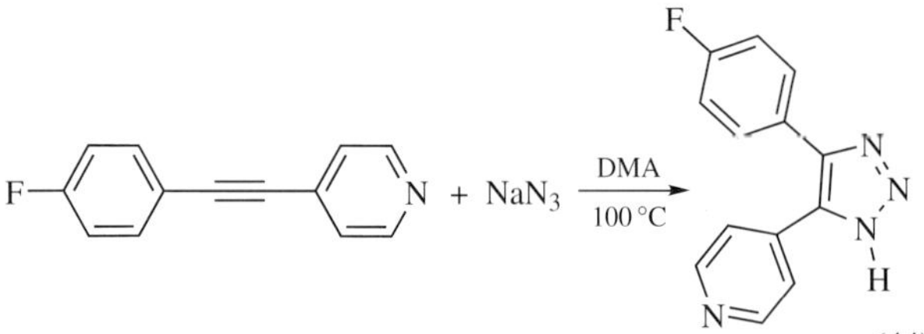

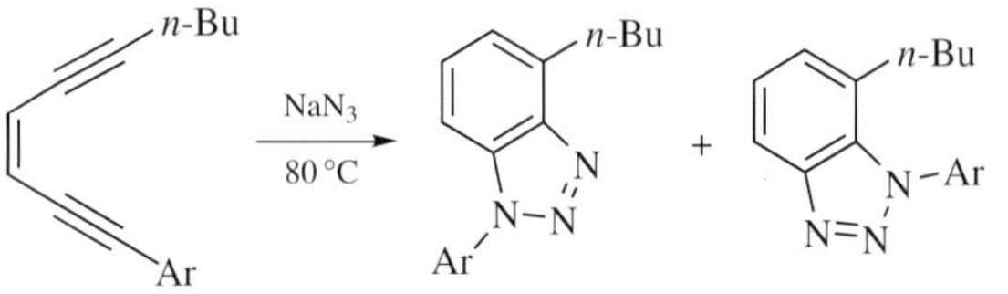

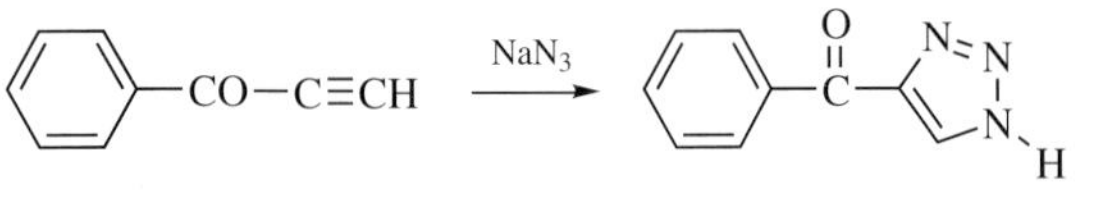

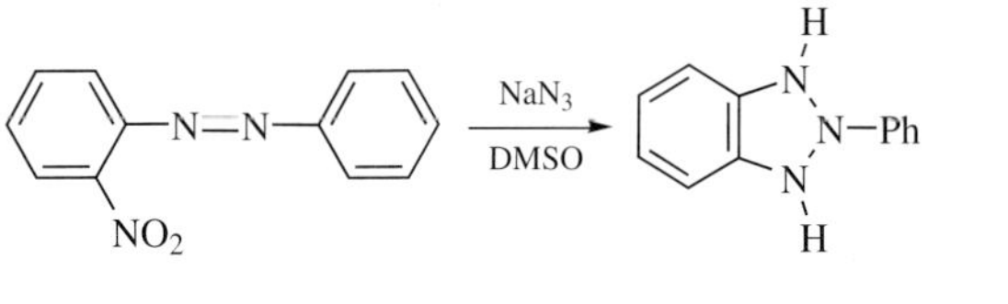

$$R'COCH{=}CHCO_2R'' + NaN_3 \xrightarrow[DMF]{FeCl_3} \text{...} \quad (48)$$

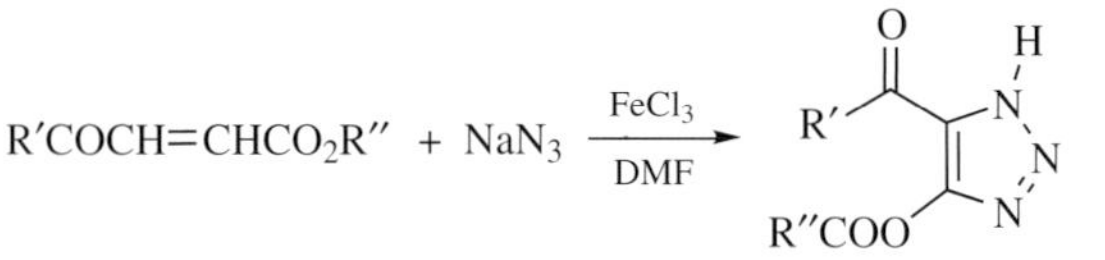

Ring Cleavage of Aziridines and Epoxides. Several activated aziridines were cleaved by sodium azide in aqueous acetonitrile and led to the formation of chiral 1,2-azidoamines in the absence of Lewis acid. The rate of reaction can be increased in unactivated aziridines by employing 50 mol % CuCl$_2$·2H$_2$O.[130] Azidolysis of epoxides in the presence of water[131] or polyacrylamide[132] resulted in the regioselective formation of azidohydrins. Similarly,[133,134] a wide variety of epoxides and aziridines were also converted to the respective β-azido alcohols and β-azido amines with sodium azide using Oxone as a catalyst. The highly efficient sodium-azide-mediated endocyclic cleavage of N-acyloxazolidinones resulted in the formation of N-acyl-β-amino alcohols.[135] Details of representative examples are outlined in eqs 49–54.

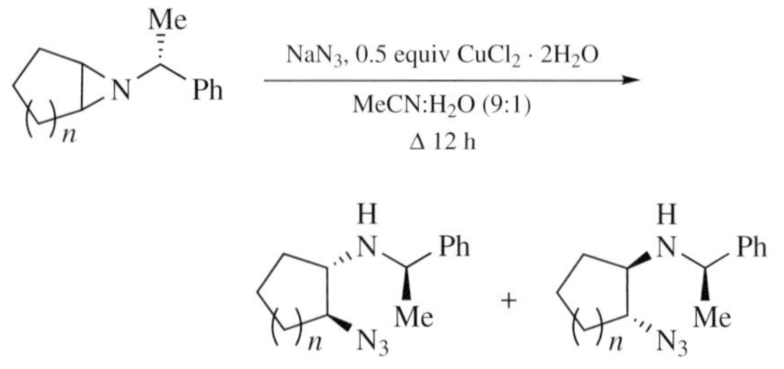

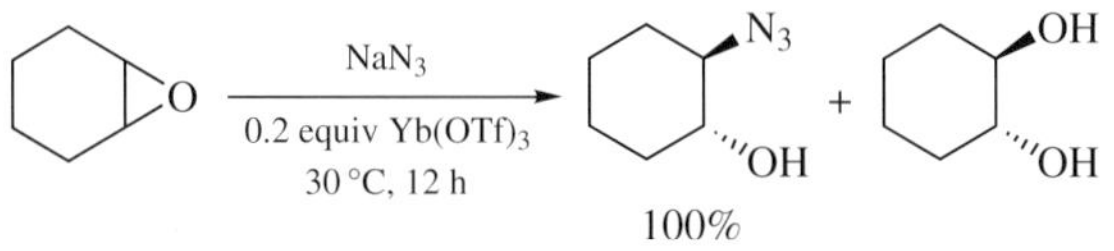

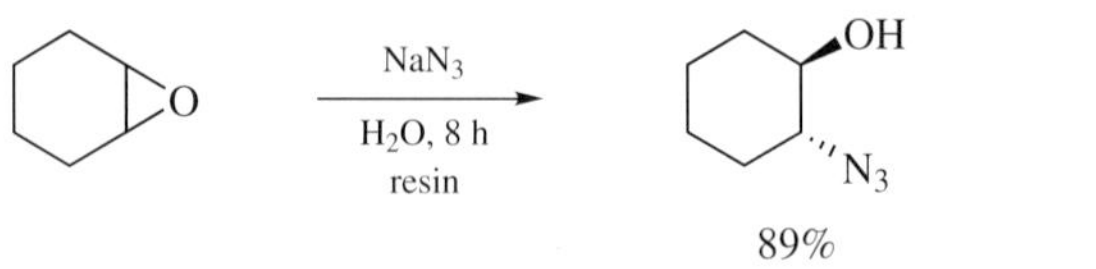

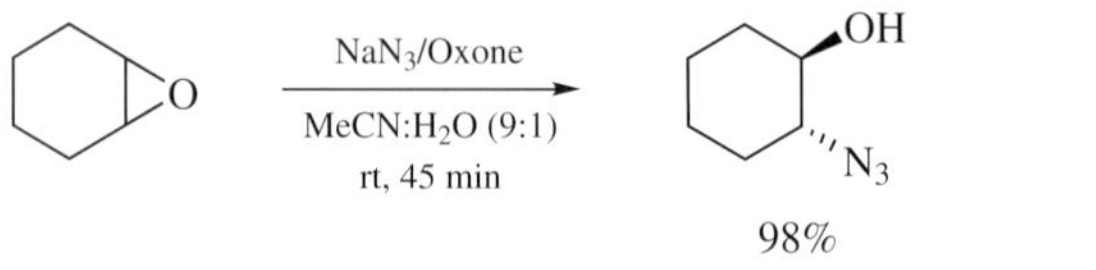

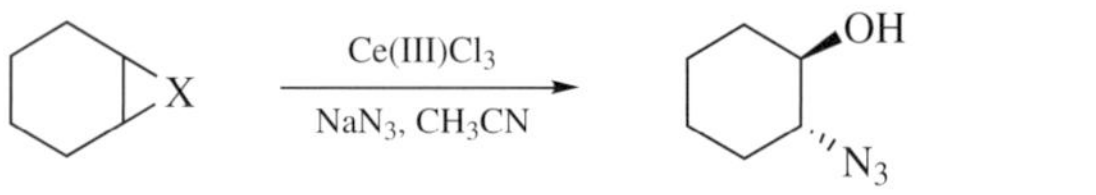

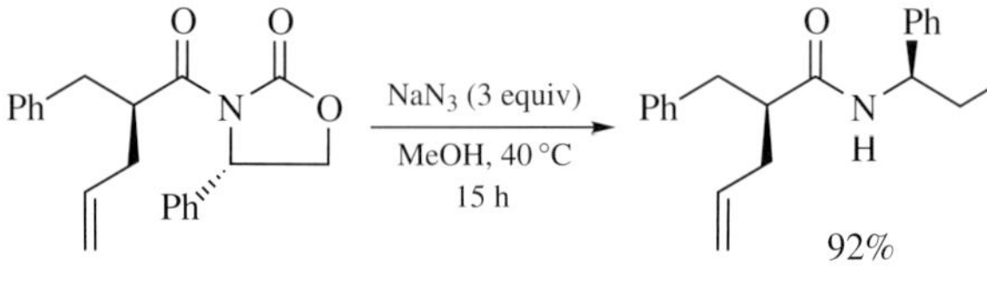

Reaction with Cyclic Enones. Conjugate addition of azide ion to cyclic enones in water using sodium azide in the presence of Lewis base[136] resulted in the formation of β-azido carbonyl compounds (eq 55). The Schmidt reaction[137] of benzopyranones with sodium azide led to pyrano[3,2-b]azepines in reasonable yields (eq 56).

$$\text{(55)}$$

$$\text{(56)}$$

3.2:1 (96%)

Miscellaneous Reactions. Regiospecific substitution of a 4-NO_2 group with sodium azide in benzothiophene led to the formation of 4-azidobenzothiophenes (eq 57).[138] Similarly, nitroalkenes react with sodium azide to afford the corresponding azido derivatives, which subsequently cyclize to generate triazoles (eq 58).[139]

$$\text{(57)}$$

60%

$$\text{(58)}$$

Finally, electrophilic amination of methylbenzenes[140] or exposing NaN$_3$ to N_2O_4 can be used to produce a surface film of nitrogen.[141]

Related Reagents. Azidotrimethylsilane; N-Bromosuccinimide–Sodium Azide; Hydrazoic Acid.

1. (a) Fieser, L. F.; Fieser, M., *Fieser & Fieser* **1967**, *1*, 1041. (b) Fieser, L. F.; Fieser, M., *Fieser & Fieser* **1969**, *2*, 376. (c) Fieser, L. F.; Fieser,

M., *Fieser & Fieser* **1972**, *3*, 259. (d) Fieser, L. F.; Fieser, M., *Fieser & Fieser* **1974**, *4*, 440. (e) Fieser, L. F.; Fieser, M., *Fieser & Fieser* **1975**, *5*, 593. (f) For synthesis and reactions of organic azides, see: Scriven, E. F. V.; Turnbull, K., *Chem. Rev.* **1988**, *88*, 297 and references therein.

2. Biffin, M. E. C.; Miller, J.; Paul, D. B. In *The Chemistry of the Azido Group*; Patai, S., Ed.; Wiley: New York, 1971; p 57.

3. (a) Fieser, L. F.; Fieser, M., *Fieser & Fieser* **1967**, *1*, 446. (b) Fieser, L. F.; Fieser, M., *Fieser & Fieser* **1969**, *2*, 211. (c) Fieser, L. F.; Fieser, M., *Fieser & Fieser* **1975**, *5*, 329.

4. Dehnicke, K., *Adv. Inorg. Chem. Radiochem.* **1983**, *26*, 169.

5. Groutas, W. C.; Felker, D., *Synthesis* **1980**, 861.

6. (a) Fieser, L. F.; Fieser, M., *Fieser & Fieser* **1967**, *1*, 1178. (b) Fieser, L. F.; Fieser, M., *Fieser & Fieser* **1969**, *2*, 415. (c) Fieser, L. F.; Fieser, M., *Fieser & Fieser* **1972**, *3*, 291. (d) Fieser, L. F.; Fieser, M., *Fieser & Fieser* **1974**, *4*, 510. (e) Fieser, L. F.; Fieser, M., *Fieser & Fieser* **1977**, *6*, 597. (f) Fieser, M., *Fieser & Fieser* **1981**, *9*, 472.

7. Shioiri, T.; Yamada, S., *Yuki Gosei Kagaku Kyokai Shi* **1973**, *31*, 666 (*Chem. Abstr.* **1974**, *80*, 60 160p).

8. (a) Bretherick, L. *Handbook of Reactive Chemical Hazards*, 4th ed.; Butterworths: London, 1990; p 1360. (b) Military Specification MIL-S-20552A, *Sodium Azide, Technical*, 1952, July 24. (c) Armour, M.-A. *Waste Disposal in Academic Institutions*, Kaufman, J. A., Ed.; Lewis: Chelsea, MI, 1990; p 122.

9. Feigl, F. *Spot Tests*; Elsevier: Amsterdam, 1954; Vol. 2, p 164.

10. Friend, S. H.; March, K. L.; Hanania, G. I. H.; Gurd, F. R. N., *Biochemistry* **1980**, *19*, 3039.

11. Banert, K.; Kirmse, W., *J. Am. Chem. Soc.* **1982**, *104*, 3766.

12. Fleming, P. R.; Sharpless, K. B., *J. Org. Chem.* **1991**, *56*, 2869.

13. (a) Lohray, B. B.; Ahuja, J. R., *J. Chem. Soc., Chem. Commun.* **1991**, 95. (b) Dubois, L.; Dodd, R. H., *Tetrahedron* **1993**, *49*, 901.

14. Kaiser, C.; Weinstock, J., *Org. Synth.* **1971**, *51*, 48.

15. (a) Safi, M.; Sinou, D., *Tetrahedron Lett.* **1991**, *32*, 2025. (b) Murahashi, S.-I.; Taniguchi, Y.; Imada, Y.; Tanigawa, Y., *J. Org. Chem.* **1989**, *54*, 3292. (c) Murahashi, S.-I.; Tanigawa, Y.; Imada, Y.; Taniguchi, Y., *Tetrahedron Lett.* **1986**, *27*, 227.

16. Norris, R. K.; Smyth-King, R. J., *Tetrahedron* **1982**, *38*, 1051.

17. Krafft, G. A.; Siddall, T. L., *Tetrahedron Lett.* **1985**, *26*, 4867.

18. Krief, A.; Dumont, W.; Denis, J.-N., *J. Chem. Soc., Chem. Commun.* **1985**, 571.

19. Kakimoto, M.; Kai, M.; Kondo, K., *Chem. Lett.* **1982**, 525.

20. Hortmann, A. G.; Robertson, D. A.; Gillard, B. K., *J. Org. Chem.* **1972**, *37*, 322.

21. (a) Ogilvie, W.; Rank, W., *Can. J. Chem.* **1987**, *65*, 166. (b) Landen, G.; Moore, H. W., *Tetrahedron Lett.* **1976**, 2513.

22. Kappe, C. O., *Liebigs Ann. Chem.* **1990**, 505.

23. Pautard-Cooper, A.; Evans, S. A. Jr., *Tetrahedron* **1991**, *47*, 1603.

24. (a) Ravindranath, B.; Srinivas, P., *Indian J. Chem., Sect. B* **1985**, *24*, 1178. (b) Quast, H.; Seiferling, B., *Liebigs Ann. Chem.* **1982**, 1566.

25. Lwowski, W. *Azides and Nitrenes*, Scriven, E. F. V., Ed.; Academic: Orlando, 1984; p 205.

26. (a) Beltrame, P.; Favini, G.; Cattaria, M. G.; Guella, F., *Gazz. Chim. Ital.* **1968**, *98*, 380. (b) Boyer, J. H., *J. Am. Chem. Soc.* **1951**, *73*, 5248.

27. Grieco, P. A.; Mason, J. P., *J. Chem. Eng. Data* **1967**, *12*, 623.

28. (a) Castillón, S.; Meléndez, E.; Pascual, C.; Vilarrasa, J., *J. Org. Chem.* **1982**, *47*, 3886. (b) Choi, P.; Rees, C. W.; Smith, E. H., *Tetrahedron Lett.* **1982**, *23*, 121.

29. Sheradsky, T. In *The Chemistry of the Azido Group*, Patai, S., Ed.; Wiley: New York, 1971; p 331.

30. Marti, M. J.; Rico, I.; Ader, J. C.; de Savignac, A.; Lattes, A., *Tetrahedron Lett.* **1989**, *30*, 1245.

31. Priebe, H., *Acta Chem. Scand.* **1984**, *B38*, 895.

32. Mitsunobu, O., *Synthesis* **1981**, 1.

33. Dieter, R. K.; Deo, N.; Lagu, B.; Dieter, J. W., *J. Org. Chem.* **1992**, *57*, 1663.

34. Khuong-Huu, Q.; Pancrazi, A.; Kabore, I., *Tetrahedron* **1974**, *30*, 2579.

35. Arrieta, A.; Aizpurua, J. M.; Palomo, C., *Tetrahedron Lett.* **1984**, *25*, 3365.

36. (a) Swift, G.; Swern, D., *J. Org. Chem.* **1966**, *31*, 4226. (b) VanderWerf, C. A.; Heisler, R. Y.; McEwen, W. E., *J. Am. Chem. Soc.* **1954**, *76*, 1231.

37. Sutowardoyo, K. I.; Emziane, M.; Lhoste, P.; Sinou, D., *Tetrahedron* **1991**, *47*, 1435 and references therein.

38. (a) Onaka, M.; Sugita, K.; Izumi, Y., *Chem. Lett.* **1986**, 1327. (b) Onaka, M.; Sugita, K.; Izumi, Y., *J. Org. Chem.* **1989**, *54*, 1116.

39. (a) Tomoda, S.; Matsumoto, Y.; Takeuchi, Y.; Nomura, Y., *Chem. Lett.* **1986**, 1193. (b) Chakraborty, T. K.; Reddy, G. V., *Tetrahedron Lett.* **1991**, *32*, 679.

40. Barone, A. D.; Snitman, D. L.; Watt, D. S., *J. Org. Chem.* **1978**, *43*, 2066.

41. Orr, D. E., *Synthesis* **1984**, 618.

42. Hassner, A.; Fibiger, R.; Andisik, D., *J. Org. Chem.* **1984**, *49*, 4237.

43. Kyba, E. P.; John, A. M., *Tetrahedron Lett.* **1977**, 2737.

44. Kirchmeyer, S.; Mertens, A.; Olah, G. A., *Synthesis* **1983**, 500.

45. Magnus, P.; Barth, L., *Tetrahedron Lett.* **1992**, *33*, 2777.

46. Draper, R. W., *J. Chem. Soc., Perkin Trans. 1* **1983**, 2781.

47. Draper, R. W., *J. Chem. Soc., Perkin Trans. 1* **1983**, 2787.

48. (a) Minisci, F.; Galli, R., *Tetrahedron Lett.* **1962**, 533. (b) Minisci, F.; Galli, R., *Tetrahedron Lett.* **1963**, 357. (c) Minisci, F.; Galli, R.; Cecere, M., *Gazz. Chim. Ital.* **1964**, *94*, 67.

49. Fristad, W. E.; Brandvold, T. A.; Peterson, J. R.; Thompson, S. R., *J. Org. Chem.* **1985**, *50*, 3647.

50. Moriarty, R. M.; Khosrowshahi, J. S., *Tetrahedron Lett.* **1986**, *27*, 2809 and references therein.

51. Tingoli, M.; Tiecco, M.; Chianelli, D.; Balducci, R.; Temperini, A., *J. Org. Chem.* **1991**, *56*, 6809.

52. Ehrenfreund, J.; Zbiral, E., *Justus Liebigs Ann. Chem.* **1973**, 290.

53. Trost, B. M.; Shibata, T., *J. Am. Chem. Soc.* **1982**, *104*, 3225.

54. (a) Wolff, H., *Org. React.* **1946**, *3*, 307. (b) Krow, G. R., *Tetrahedron* **1981**, *37*, 1283.

55. Escale, R.; El Khayat, A.; Vidal, J.-P.; Girard, J.-P.; Rossi, J.-C., *J. Heterocycl. Chem.* **1984**, *21*, 1033.

56. (a) Yeh, M. Y.; Tien, H. J.; Huang, L. Y.; Chen, M. H., *J. Chin. Chem. Soc. Taipei* **1983**, *30*, 29. (b) Nishiyama, K.; Watanabe, A., *Chem. Lett.* **1984**, 773.

57. Smith, P. A. S., *Org. React.* **1946**, *3*, 337.

58. Lwowski, W. *1,3-Dipolar Cycloaddition Chemistry*; Padwa, A., Ed.; Wiley: New York, 1984; Vol. 1, p 559.

59. (a) Sundberg, R. J.; Pearce, B. C., *J. Org. Chem.* **1982**, *47*, 725. (b) See also: Choi, J.-R.; Han, S.; Cha, J. K., *Tetrahedron Lett.* **1991**, *32*, 6469.

60. (a) Donkor, A.; Prager, R. H.; Thompson, M. J., *Angew. Chem., Int. Ed. Engl.* **1992**, *45*, 1571. (b) Similarly: Rybinskaya, M. I.; Nesmeyanov, A. N.; Kochetkov, N. K., *Russ. Chem. Rev. (Engl. Transl.)* **1969**, *38*, 433.

61. (a) Türck, U.; Behringer, H., *Chem. Ber.* **1965**, *98*, 3020. (b) Similarly: Meek, J. S.; Fowler, J. S.; *J. Am. Chem. Soc.* **1967**, *89*, 1967. (c) Meek, J. S.; Fowler, J. S., *J. Org. Chem.* **1968**, *33*, 985. (d) Zefirov, N. S.; Chapovskaya, N. K.; Kolesnikov, V. V., *J. Chem. Soc., Chem. Commun.* **1971**, 1001.

62. (a) Emmons, W. D.; Freeman, J. P., *J. Org. Chem.* **1957**, *22*, 456. (b) Stevens, T. E.; Emmons, W. D., *J. Am. Chem. Soc.* **1958**, *80*, 338. (c) Similarly: Smith, P. A. S.; Boyer, J. H., *Org. Synth., Coll. Vol.* **1963**, *4*, 75. (d) Boulton, A. J.; Ghosh, P. B.; Katritzky, A. R., *Tetrahedron Lett.* **1966**, 2887.

63. Bernstein, P. R.; Vacek, E. P., *Synthesis* **1987**, 1133.

64. (a) Duncia, J. V.; Pierce, M. E.; Santella, J. B., III *J. Org. Chem.* **1991**, *56*, 2395. (b) Birkofer, L.; Wegner, P., *Chem. Ber.* **1966**, *99*, 2512.

65. Takahito, T.; Takaaki, H.; Jun, M.; Ryozo, O. JP 0912,309, 1997 (*Chem. Abstr.* **1997**, *126*, 173818w).

66. Alteman, M.; Bjorsne, M.; Muhlman, A.; Classon, B.; Kvarnstrom, I.; Danielson, H.; Markgren, P.-O.; Nillroth, U.; Unge, T.; Hallberg, A.; Samuelsson, B., *J. Med. Chem.* **1998**, *41*, 3782.

67. Alterman, M.; Andersson, H. O.; Garg, N.; Ahlsen, G.; Lovgren, S.; Classon, B.; Danielson, U. H.; Kvarnstrom, I.; Vrang, L.; Unge, T.; Samuelsson, B.; Hallberg, A., *J. Med. Chem.* **1999**, *42*, 3835.

68. Alterman, M.; Hallberg, A., *J. Org. Chem.* **2000**, *65*, 7984.

69. Le, V. D.; Rees, C. W.; Sivadasan, S., *J. Chem. Soc., Perkin Trans 1* **2002**, 1543.

70. Germano, C.; David, M. C.; Avellana Jaime, P.; Maria Germano, G. P. WO 97,34,885, 1997 (*Chem. Abstr.* **1997**, *127*, 318881u).

71. Kiyoto, K.; Toshikaza, O.; Norihito, T.; Sunao, M.; Ryozo, O.Eur. Pat. 796,852, 1997 (*Chem. Abstr.* **1997**, *127*, 307389 p).

72. Sadao, I.; Fujjo, S.; Hidekazu, M.; Keita, K. WO 98,56,757, 1998 (*Chem. Abstr.* **1999**, *130*, 52423n).

73. Magano, E.; Magaki, N.; Iscuo, M. JP 1,0218,868, 1998 (*Chem. Abstr.* **1998**, *129*, 161564j).

74. Shigeru, T.; Magami, T.; Yoshtaka, M.; Yuki, K.; Takashi, O. JP 281,662,2000 (*Chem. Abstr.* **2000**, *133*, 252440q).

75. Pardhasaradhi, M.; Srinivas, K.; Snehalatha, N. C. K. S.U. S. Pat. 6,326,498, 2001 (*Chem. Abstr.* **2002**, *136*, 5996r)

76. Pardhasaradhi, M.; Srinivas, K.; Snehalatha, N. C. K. S. JP 284,770, 2002 (*Chem. Abstr.* **2002**, *137*, 263040t).

77. Shastin, A. V.; Godovikova, T. I.; Korsunskii, B. L., *Chem. Heterocycl. Compd.* **1998**, *34*, 383.

78. Kiyoshi, S.; Tadashi, K.; Nobushige, I. JP 1, 1171,873, 1999 (*Chem. Abstr.* **1999**, *131*, 73659e).

79. Wang, L.; Wang, X.; Li, Y.; Bai, Z.; Xiong, R.; Xiong, M.; Li, G., *Wuji Huaxue Xuebao* **2002**, *18*, 1191 (*Chem. Abstr.* **2003**, *138*, 280327e).

80. Demko, Z. P.; Sharpless, K. B., *Org. Lett.* **2002**, *4*, 2525.

81. Esikov, K. A.; Morozova, S. E.; Mahin, A. A.; Ostrovskii, V. A., *Russ. J. Org. Chem.* **2002**, *38*, 1370.

82. Morozova, S. E.; Esikov, K. A.; Dmitrieva, T. N.; Malin, A. A.; Ostrovskii, V. A., *Russ. J. Org. Chem.* **2004**, *40*, 443.

83. Tran, T., *Hoa Chat.* **1998**, 3 (*Chem. Abstr.* **1999**, *131*, 45064p).

84. Phoebe, C. Y.-C.; Eric, A. G. JP 226,359, 2001 (*Chem. Abstr.* **2001**, *135*, 180771q).

85. Hideo, H.; Hideaki, M. WO, 03,14,112, 2003 (*Chem. Abstr.* **2003**, *138*, 170242r).

86. Magatoshi, O.; Koichi, N.; Yasuhiro, S. WO, 03,16,290, 2003 (*Chem. Abstr.* **2003**, *138*, 205061k).

87. Yukiyoshi, W.; Jun, M.; Akihiko, Y.; Yuichi, O.; Katsuhiko, S.; Akira, E. WO, 106,436, 2003 (*Chem. Abstr.* **2004**, *140*, 59643z).

88. Biot, C.; Bauer, H.; Schirmer, R. H.; Davioud-Charvet, E., *J. Med. Chem.* **2004**, *47*, 5972.

89. Suker, L. B.; Jisun, Y. JP 229,953, 2000 (*Chem. Abstr.* **2000**, *133*, 150564r).

90. Ogawa, H.; Tanaka, H.; Oonishi, A. JP 08,333,354, 1996 (*Chem. Abstr.* **1997**, *126*, 157505m).

91. Li, N. D.; Hiskey Michael, A. U. S. Pat. 60,634, 2003 (*Chem. Abstr.* **2003**, *138*, 255236s)

92. Batey, R. A.; Powell, D. A., *Org. Lett.* **2000**, *2*, 3237.

93. Blaise, W. L.; Branko, S. J., *Synth. Commun.* **1998**, *28*, 3591.

94. Aatsushi, T.; Shingo, S. JP 181,265, 2001 (*Chem. Abstr.* **2001**, *135*, 78538g).

95. Cho, J. H.; Lee, G. S.; Lee, I. U.; Oh, G. H.; Oh, C. S., *Repub. Korean Kongkae Taheho Cangbu KR*, **2002**, *59*, 043 (*Chem. Abstr.* **2005**, *142*, 113617a).

96. Vasanthakumar, G. R.; Ananda, K.; Babu, V. V. S. B., *Ind. J. Chem.* **2002**, *41B*, 1733 (*Chem. Abstr.* **2003**, *138*, 900 36m).

97. Xie, C.; Yu, S.; Liu, F., *Kexueban* **2003**, *24*, 304 (*Chem. Abstr.* **2003**, *139*, 397153g).

98. Wolfgang, G.; Hermani, B.; Christopher, R.; Thomas, K. DE 19,727,410, 1999 (*Chem. Abstr.* **1999**, *130*, 97667q).

99. Shie, J. J.; Fang, J.-M., *J. Org. Chem.* **2003**, *68*, 1158.

100. Marinescu, L.; Thinggaard, J.; Thomsen, I. B.; Bois, M., *J. Org. Chem.* **2003**, *68*, 9453.

101. Salama, T. A.; El-Ahl, A. S.; Khalil, A.-G. M.; Girges, M. M.; Lackner, B.; Steindl, C.; Elmorsy, S. S., *Monatsh. Chem.* **2003**, *134*, 1241.

102. Göksu, S.; Secen, H., *Tetrahedron* **2005**, *61*, 6801.

103. Reddy, G. V. S.; Rao, G. V.; Subramanyam, R. V. K.; Iyenger, D. S., *Synth. Commun.* **2000**, *30*, 2233.

104. Sampath Kumar, H. M.; Subba Reddy, B. V.; Anjaneyulu, S.; Yadav, J. S., *Tetrahedron Lett.* **1998**, *39*, 7385.

105. Sreekumar, R.; Padmakumar, R.; Rugmini, P., *Chem. Commun.* **1997**, 1133.

106. Ciappe, C.; Pieraccini, D.; Saullo, P., *J. Org. Chem.* **2003**, *68*, 6710.

107. Saibabu Kotti, S. R. S.; Xu, X.; Li, G.; Headley, A. D., *Tetrahedron Lett.* **2004**, *45*, 1427.

108. Martins, F.; Duarte, M. F.; Fernandez, M. T.; Langley, G. J.; Rodrigues, P.; Barros, M. T.; Costa, M. L., *Rapid Commun. Mass Spectrom.* **2004**, *18*, 363.

109. Chapyshew, S. V., *Chem. Heterocycl. Compd.* **2001**, *37*, 968.

110. Yoon, S.-K.; Kim, B.-C.; Jung, W.-H.; Lee, J.-C.; Lee, K.; Park, C.-W. WO, 0281,465, 2002 (*Chem. Abstr.* **2002**, *137*, 310807k).

111. Ma, Y., *Heteroat. Chem.* **2002**, *13*, 307.

112. Park, S., *Bull. Korean Chem. Soc.* **2003**, *24*, 253.

113. Akira, A.; Daisuke, I.; Takashi. M. Eur. Pat. 1,344,763, 2003.

114. Smirnov, O. Y.; Churakov, A. M.; Tyurin, A. Y.; Strelenko, Y. A.; Ioffe, S. L.; Tartakovsky, V. A., *Russ. Chem. Bull.* **2002**, *51*, 1849 (*Chem. Abstr.* **2003**, *138*, 320773f).

115. Takayuki, K. JP 08,269,105, 1996 (*Chem. Abstr.* **1997**, *126*, 61801d).

116. Liu, Q.; Tor, Y., *Org. Lett.* **2003**, *5*, 2571.

117. Bandgar, B. P.; Pandit, S. S., *Tetrahedron Lett.* **2002**, *43*, 3413.

118. Yazaki, A.; Yoshidaziro; Niino, Y. WO 9738,971, 1997 (*Chem. Abstr.* **1997**, *127*, 318778r).

119. Yu, Y.; Yu, Z.; Zhang, J.; Zhu, M.; Hu, X.; *Qingdao Yixyeyuan Xhebao*, **1996**, *32*, 207 (*Chem. Abstr.* **1997**, *126*, 89064q).

120. Tullis, J. S.; VanRens, J. C.; Natchus, M. G.; Clark, M. P.; De, B.; Hsieh, L.; Janusz, M. J., *Bioorg. Med. Chem. Lett.* **2003**, *13*, 1665.

121. Chen, Z.-Y.; Wu, M.-J., *Org. Lett.* **2005**, *7*, 475.

122. Caliendo, G.; Fiorino, F.; Grieco, P.; Perissutti, E.; Santagada, V.; Meli, R.; Raso, G. M.; Zanesco, A.; De, N. G., *Eur. J. Med. Chem.* **1999**, *34*, 1043 (*Chem. Abstr.* **2000**, *132*, 207830k)

123. Mori, N.; Kaneko, M.; Torii, Y.; Takahashi, T.; Imaoka, T.; Tanida, K. WO 0034,278, 1998 (*Chem. Abstr.* **2000**, *133*, 43529m).

124. Sosnovskikh, V. Y.; Usachev, B. I., *Mendelev Commun.* **2002**, *2*, 75 (*Chem. Abstr.* **2003**, *138*, 170152m).

125. Setsu, F.; Umemura, E.; Sagaki, K.; Tadauchi, K.; Okutomsi, T.; Ohtsuka, K.; Takahata, S. WO 0342,188, 2003 (*Chem. Abstr.* **2003**, *138*, 401733v).

126. Fischer, W.; Fritzsche, K.; Wolf, W.; Bore, L. WO 0224,668, 2001 (*Chem. Abstr.* **2002**, *136*, 279458a).

127. Bakavoli, M.; Davoodnia, A.; Rahimizadeh, M.; Heravi, M. M., *Phosphorus, Sulfur Silicon Relat. Elem.* **2002**, *177*, 2303.

128. Yasuda, S.; Imura, K.; Okada, Y.; Tsujiyama, S. WO 0364,400 2002 (*Chem. Abstr.* **2003**, *139*, 164793 a).

129. Joralemon, M. J.; Nugent, A. K.; Matson, J. B.; O'Reilly, R. K.; Hawaker, C. J.; Wooley, K. L., *Polym. Mater. Sci. Eng.* **2004**, *91*, 195.

130. Bisai, A.; Pandey, G.; Pandey, M. K.; Singh, V. K., *Tetrahedron Lett.* **2003**, *44*, 5839.

131. Fringuelli, F.; Piermatti, O.; Pizzo, F.; Vaccaro, L., *J. Org. Chem.* **1999**, *64*, 6094.

132. Tamami, B.; Mahdavi, H., *Tetrahedron Lett.* **2001**, *42*, 8721.

133. Sabitha, G.; Babu, R.; Satheesh, M.; Reddy, S. K.; Yadav, J. S., *Synthesis* **2002**, *15*, 2254.

134. Sabitha, G.; Babu, R. S.; Rajkumar, M.; Yadav, J. S., *Org. Lett.* **2002**, *4*, 343.

135. Bouzide, A.; Sauvé, G., *Tetrahedron Lett.* **2002**, *43*, 1961.

136. Xu, L.-W.; Xia, C.-G.; Li, J.-W.; Zhou, S.-L., *Synlett* **2003**, *14*, 2246.

137. Pozgan, F.; Polanc, S.; Kocevar, M., *Heterocycles* **2002**, *56*, 379.

138. Shevelev, S. A.; Dalinger, I. L.; Cherkasova, T. I., *Tetrahedron Lett.* **2001**, *42*, 8539.

139. Bakhareva, S. V.; Bereshvitskaya, V. M.; Aboskalova, N. I., *Russ. J. Gen. Chem.* **2001**, *71*, 1493 (*Chem. Abstr.* **2002**, *137*, 20338s).

140. Borodkin, G. I.; Elanov, I. R.; Popav, S. A.; Pokrovski, L. M.; Shubin, V. G., *Russ. J. Org. Chem.* **2003**, *39*, 672 (*Chem. Abstr.* **2004**, *140*, 217141r).

141. Vaulin, S. D.; Fevifilaktov, V. I., *Advances in confined Detonations* **2002**, 186, (*Chem. Abstr.* **2004**, *140*, 238044b).

Sodium Borohydride

NaBH₄

[16940-66-2]   BH₄Na   (MW 37.84)

InChI = 1S/BH4.Na/h1H4;/q-1;+1

InChIKey = YOQDYZUWIQVZSF-UHFFFAOYSA-N

(reducing agent for aldehydes and ketones, and many other functional groups in the presence of additives[1])

Physical Data: mp 400 °C; d 1.0740 g cm^{-3}.

Solubility: sol H₂O (stable at pH 14, rapidly decomposes at neutral or acidic pH); sol MeOH (13 g/100 mL)[1b], and EtOH (3.16 g/100 mL),[1b] but decomposes to borates; sol polyethylene glycol (PEG),[2a] sol and stable in *i*-PrOH (0.37 g/100 mL)[3] and diglyme (5.15 g/100 mL);[1b] insol ether;[1b] slightly sol THF.[1c]

Form Supplied in: colorless solid in powder or pellets; supported on silica gel or on basic alumina; 0.5 M solution in diglyme; 2.0 M solution in triglyme; 12 wt % solution in 14 M aqueous NaOH. Typical impurities are sodium methoxide and sodium hydroxide.

Analysis of Reagent Purity: can be assessed by hydrogen evolution.[4]

Purification: crystallize from diglyme[3] or isopropylamine.[4]

Handling, Storage, and Precautions: harmful if inhaled or absorbed through skin. It is decomposed rapidly and exothermically by water, especially if acid solutions are used. This decomposition forms toxic diborane gas and flammable/explosive hydrogen gas, and thus must be carried out under a hood. Solutions in DMF can undergo runaway thermal reactions, resulting in violent decompositions.[5] The addition of supported noble metal catalysts to solutions of NaBH₄ can result in ignition of liberated hydrogen gas.[5]

Original Commentary

Luca Banfi, Enrica Narisano & Renata Riva
Università, di Genova, Italy

Reduction of Aldehydes and Ketones. Sodium borohydride is a mild and chemoselective reducing agent for the carbonyl function. At 25 °C in hydroxylic solvents it rapidly reduces aldehydes and ketones, but it is essentially inert to other functional groups such as epoxides, esters, lactones, carboxylic acid salts, nitriles, and nitro groups. Acyl halides, of course, react with the solvent.[1a] The simplicity of use, the low cost, and the high chemoselectivity make it one of the best reagents for this reaction. Ethanol and methanol are usually employed as solvents, the former having the advantage of permitting reductions in homogeneous solutions with relatively little loss of reagent through the side reaction with the solvent.[1a] Aprotic solvents such as diglyme greatly decrease the reaction rates.[1a] On the other hand, NaBH₄ in polyethylene glycol (PEG) shows a reactivity similar to that observed in EtOH.[2a] Although the full details of the mechanism of ketone reduction by NaBH₄ remain to be established,[6] it has been demonstrated that all four hydrogen atoms can be transferred. Moreover, the rate of reduction was shown to slightly increase when the

hydrogens on boron are replaced by alkoxy groups.[1a,c,d] However, especially when NaBH₄ is used in MeOH, an excess of reagent has to be used in order to circumvent the competitive borate formation by reaction with the solvent. Ketone reduction has been accelerated under phase-transfer conditions[7] or in the presence of HMPA supported on a polystyrene-type resin.[8]

The isolation of products is usually accomplished by diluting the reaction mixture with water, making it slightly acidic to destroy any excess hydride, and then extracting the organic product from the aqueous solution containing boric acid and its salts.

Kinetic examination of the reduction of benzaldehyde and acetophenone in isopropyl alcohol indicated a rate ratio of 400:1.[1a] Thus it is in principle possible to reduce an aldehyde in the presence of a ketone.[9a] Best results (>95% chemoselectivity) have been obtained using a mixed solvent system (EtOH–CH₂Cl₂ 3:7) and performing the reduction at −78 °C,[9a] or by employing an anionic exchange resin in borohydride form.[10] This reagent can also discriminate between aromatic and aliphatic aldehydes. On the other hand, reduction of ketones in the presence of aldehydes can be performed by NaBH₄–***Cerium(III) Chloride***. NaBH₄ in MeOH–CH₂Cl₂ (1:1) at −78 °C reduces ketones in the presence of conjugated enones and aldehydes in the presence of conjugated enals.[9]

Conjugate Reductions. NaBH₄ usually tends to reduce α,β-unsaturated ketones in the 1,4-sense,[1d] affording mixtures of saturated alcohol and ketone. In alcoholic solvents, saturated β-alkoxy alcohols can be formed as byproducts via conjugate addition of the solvent.[11] The selectivity is not always high. For example, while cyclopentenone is reduced only in the conjugate fashion, cyclohexenone affords a 59:41 ratio of allylic alcohol and saturated alcohol.[1d] Increasing steric hindrance on the enone increases 1,2-attack.[11] Aldehydes undergo more 1,2-reduction than the corresponding ketones.[1c,1d] The use of pyridine as solvent may be advantageous in increasing the selectivity for 1,4-reduction, as exemplified (eq 1) by the reduction of (R)-carvone to dihydrocarveols and (in minor amounts) dihydrocarvone.[12]

$$(1)$$

Trialkyl borohydrides such as ***Lithium Tri-s-butylborohydride*** and ***Potassium Tri-s-butylborohydride*** are superior reagents for the chemoselective 1,4-reduction of enones. On the other hand, 1,2-reduction can be obtained by using NaBH₄ in the mixed solvent MeOH–THF (1:9),[13] or with NaBH₄ in combination with CeCl₃ or other lanthanide salts.[14]

NaBH₄ in alcoholic solvents has been used for the conjugate reduction of α,β-unsaturated esters,[15] including cinnamates and alkylidenemalonates, without affecting the alkoxycarbonyl group. Conjugate nitroalkenes have been reduced to the corresponding nitroalkanes.[16] Saturated hydroxylamines are obtained by reducing nitroalkenes with the ***Borane–Tetrahydrofuran*** complex in the presence of catalytic amounts of NaBH₄, or by using a combination of NaBH₄ and ***Boron Trifluoride Etherate*** in 1:1.5 molar ratio.[17] Extended reaction can lead also to the saturated amines.[17]

Essential Reactions for Organic Synthesis, First Edition. Edited by Philip L. Fuchs.
© 2016 John Wiley & Sons, Ltd. Published 2016 by John Wiley & Sons, Ltd.

Reduction of Carboxylic Acid Derivatives. The reduction of carboxylic esters[1c, 1d] by NaBH$_4$ is usually slow, but can be performed by the use of excess reagent in methanol or ethanol[18] at room temperature or higher. The solvent must correspond to the ester group, since NaBH$_4$ catalyzes ester interchange. This transformation can also be achieved at 65–80 °C in *t*-BuOH[19] or polyethylene glycol.[2b] Although the slow rate and the need to use excess reagent makes other stronger complex hydrides such as *Lithium Borohydride* or *Lithium Aluminum Hydride* best suited for this reaction, in particular cases the use of NaBH$_4$ allows interesting selectivity: see, for example, the reduction of eq 2,[20] where the β-lactam remains unaffected, or of eq 3,[21] where the epoxide and the cyano group do not react.

$$(2)$$

$$(3)$$

Borohydrides cannot be used for the reduction of α,β-unsaturated esters to allylic alcohols since the conjugate reduction is faster.[18b] The reactivity of NaBH$_4$ toward esters has been enhanced with various additives. For example, the system NaBH$_4$–CaCl$_2$ (2:1) shows a reactivity similar to LiBH$_4$.[18b] Esters have also been reduced with NaBH$_4$–*Zinc Chloride* in the presence of a tertiary amine,[22] or with NaBH$_4$–*Copper(II) Sulfate*. The latter system reduces selectively aliphatic esters in the presence of aromatic esters of amides.[23] Finally, esters have also been reduced with NaBH$_4$–*Iodine*.[24a] In this case the reaction seems to proceed through diborane formation, and so it cannot be used for substrates containing an alkenic double bond. A related methodology, employing *Borane–Dimethyl Sulfide* in the presence of catalytic NaBH$_4$,[25] is particularly useful for the regioselective reduction of α-hydroxy esters, as exemplified by the conversion of (*S*)-diethyl malate into the vicinal diol (eq 4).

$$(4)$$

regioisomer ratio = 200:1

Lactones are only slowly reduced by NaBH$_4$ in alcohol solvents at 25 °C, unless the carbonyl is flanked by an α-heteroatom functionality.[1d] Sugar lactones are reduced to the diol when the reduction is carried out in water at neutral pH, or to the lactol when the reaction is performed at lower (∼3) pH.[26] Thiol esters are more reactive and are reduced to primary alcohols with NaBH$_4$ in EtOH, without reduction of ester substituents.[27]

Carboxylic acids are not reduced by NaBH$_4$. The conversion into primary alcohols can be achieved by using NaBH$_4$ in combination with powerful Lewis acids,[1k, 28] *Sulfuric Acid*,[28] *Catechol*,[24b] *Trifluoroacetic Acid*,[24b] or I$_2$.[24a] In these cases the actual reacting species is a borane, and thus hydroboration of double bonds present in the substrate can be a serious side reaction. Alternatively, the carboxylic acids can be transformed into activated derivatives,[29] such as carboxymethyleneiminium salts[29a] or

mixed anhydrides,[29b] followed by reduction with NaBH$_4$ at low temperature. These methodologies tolerate the presence of double bonds, even if conjugated to the carboxyl.[29a]

Nitriles are, with few exceptions,[21] not reduced by NaBH$_4$.[1k] Sulfurated NaBH$_4$,[30] prepared by the reaction of sodium borohydride with sulfur in THF, is somewhat more reactive than NaBH$_4$, and reduces aromatic nitriles (but not aliphatic ones) to amines in refluxing THF. Further activation has been realized by using the *Cobalt Boride* system, (NaBH$_4$–CoCl$_2$) which appears to be one of the best methods for the reduction of nitriles to primary amines. More recently it has been found that *Zirconium(IV) Chloride*,[31] Et$_2$SeBr$_2$,[32] CuSO$_4$,[23] *Chlorotrimethylsilane*,[33] and I$_2$[24a] are also efficient activators for this transformation. The NaBH$_4$–Et$_2$SeBr$_2$ reagent allows the selective reduction of nitriles in the presence of esters or nitro groups, which are readily reduced by NaBH$_4$–CoCl$_2$.

NaBH$_4$ in alcoholic solvents does not reduce amides.[1a, 1c–d] However, under more forcing conditions (NaBH$_4$ in pyridine at reflux), reduction of tertiary amides to the corresponding amines can be achieved.[32] Secondary amides are inert, while primary amides are dehydrated to give nitriles. Also, NaBH$_4$–Et$_2$SeBr$_2$ is specific for tertiary amides.[32] Reagent combinations which show enhanced reactivity, and which are thus employable for all three types of amides, are NaBH$_4$–CoCl$_2$, NaBH$_4$ in the presence of strong acids[34] (e.g. *Methanesulfonic Acid* or *Titanium(IV) Chloride*) in DMF or DME, NaBH$_4$–Me$_3$SiCl,[33] and NaBH$_4$–I$_2$.[24a]

An indirect method for the reduction of amides to amines by NaBH$_4$ (applicable only to tertiary amides) involves conversion into a Vilsmeier complex [(R$_2$N=C(Cl)R)$^+$Cl$^-$], by treatment with *Phosphorus Oxychloride*, followed by its reduction.[35] In a related methodology, primary or secondary (also cyclic) amides are first converted into ethyl imidates by the action of *Triethyloxonium Tetrafluoroborate*, and the latter reduced to amines with NaBH$_4$ in EtOH or, better, with NaBH$_4$–*Tin(IV) Chloride* in Et$_2$O.[36]

In addition to the above-quoted methods, tertiary δ-lactams have been reduced to the corresponding cyclic amines by dropwise addition of MeOH to the refluxing mixture of NaBH$_4$ and substrate in *t*-BuOH,[37] or by using trifluoroethanol as solvent.[38] This reaction was applied during a synthesis of indolizidine alkaloid swainsonine for the reduction of lactam (**1**) to amine (**2**) (eq 5).[38]

$$(5)$$

Acyl chlorides can be reduced to primary alcohols by reduction in aprotic solvents such as PEG,[2a] or using NaBH$_4$–*Alumina* in Et$_2$O.[39] More synthetically useful is the partial reduction to the aldehydic stage, which can be achieved by using a stoichiometric amount of the reagent at −70 °C in DMF–THF,[40] with the system NaBH$_4$–*Cadmium Chloride*–DMF,[41] or with *Bis(triphenylphosphine)copper(I) Borohydride*.

Alternative methodologies for the indirect reduction of carboxylic derivatives employ as intermediates 2-substituted 1,3-benzoxathiolium tetrafluoborates (prepared from carboxylic acids, acyl chlorides, anhydrides, or esters)[42] and dihydro-1,3-

thiazines or dihydro-1,3-oxazines (best prepared from nitriles).[43] These compounds are smoothly reduced by $NaBH_4$, to give acetal-like adducts, easily transformable into the corresponding aldehydes by acidic hydrolysis. Conversion of primary amides into the N-acylpyrrole derivative by reaction with 1,4-dichloro-1,4-dimethoxybutane in the presence of a cationic exchange resin, followed by $NaBH_4$ reduction, furnished the corresponding aldehydes.[44]

Cyclic anhydrides are reduced by $NaBH_4$ to lactones in moderate to good yields. Hydride attack occurs principally at the carbonyl group adjacent to the more highly substituted carbon atom.[45] Cyclic imides are more reactive than amides and can be reduced to the corresponding α'-hydroxylactams by using methanolic or ethanolic $NaBH_4$ in the presence of HCl as buffering agent.[1c] These products are important as precursors for N-acyliminium salts. The carbonyl adjacent to the most substituted carbon is usually preferentially reduced[46] (see also **Cobalt Boride**). N-Alkylphthalimides may be reduced with $NaBH_4$ in 2-propanol to give an open-chain hydroxy-amide which, upon treatment with AcOH, cyclizes to give phthalide (a lactone) and the free amine. This method represents a convenient procedure for releasing amines from phthalimides under nonbasic conditions.[47]

Reduction of C=N Double Bonds. The C=N double bond of imines is generally less reactive than the carbonyl C=O toward reduction with complex hydrides. However, imines may be reduced by $NaBH_4$ in alcoholic solvents under neutral conditions at temperatures ranging from $0\,°C$ to that of the refluxing solvent.[1c,1d,48] Protonation or complexation with a Lewis acid of the imino nitrogen dramatically increases the rate of reduction.[1i] Thus $NaBH_4$ in AcOH (see **Sodium Triacetoxyborohydride**) or in other carboxylic acids is an efficient reagent for this transformation (although the reagent of choice is probably **Sodium Cyanoborohydride**). Imines are also reduced by **Cobalt Boride**, $NaBH_4$–**Nickel(II) Chloride**, and $NaBH_4$–$ZrCl_4$.[31] Imine formation, followed by in situ reduction, has been used as a method for synthesis of unsymmetrical secondary amines.[48] Once again, $Na(CN)BH_3$ represents the best reagent.[1c,1d,48] However, this transformation was realized also with $NaBH_4$,[48,49] either by treating the amine with excess aqueous formaldehyde followed by $NaBH_4$ in MeOH, or $NaBH_4$–CF_3CO_2H, or through direct reaction of the amine with the $NaBH_4$–carboxylic acid system. In the latter case, part of the acid is first reduced in situ to the aldehyde, which then forms an imine. The real reagent involved is $NaB(OCOR)_3H$ (see **Sodium Triacetoxyborohydride**). Reaction of an amine with glutaric aldehyde and $NaBH_4$ in the presence of H_2SO_4 represents a good method for the synthesis of N-substituted piperidines.[49c] Like protonated imines, iminium salts are readily reduced by $NaBH_4$ in alcoholic media.[1c,50] N-Silylimines are more reactive than N-alkylimines. Thus α-amino esters can be obtained by reduction of N-silylimino esters.[51] α,β-Unsaturated imines are reduced by $NaBH_4$ in alcoholic solvents in the 1,2-mode to give allylic amines.[52] Enamines are transformed into saturated amines by reduction with $NaBH_4$ in alcoholic media.[48,53]

The reduction of oximes and oxime ethers is considerably more difficult and cannot be realized with $NaBH_4$ alone. Effective reagent combinations for the reduction of oximes include sulfurated $NaBH_4$,[30] $NaBH_4$–$NiCl_2$, $NaBH_4$–$ZrCl_4$,[31] $NaBH_4$–MoO_3,[54] $NaBH_4$–$TiCl_4$,[55] and $NaBH_4$–**Titanium(III) Chloride**.[56] In all cases the main product is the corresponding

primary amine. $NaBH_4$–$ZrCl_4$ is efficient also for the reduction of oxime ethers. $NaBH_4$–MoO_3 reduces oximes without affecting double bonds, while $NaBH_4$–$NiCl_2$ reduces both functional groups. The reduction with $NaBH_4$–$TiCl_3$ in buffered (pH 7) aqueous media has been used for the chemoselective reduction of α-oximino esters to give α-amino esters (eq 6).[56]

$$\underset{\substack{Ph}}{\overset{\substack{N\diagup OH}}{\diagdown}}\!\!\!\!\!\underset{O}{\overset{}{\diagdown}}\!\!\!OMe \quad \xrightarrow[\substack{2.\ HCl\\82\%}]{\substack{1.\ NaBH_4,\ TiCl_3\\ \text{L-tartaric acid, pH 7}\\ MeOH-H_2O}} \quad \underset{\substack{Ph}}{\overset{\substack{NH_2\cdot HCl}}{\diagdown}}\!\!\!\!\!\underset{O}{\overset{}{\diagdown}}\!\!\!OMe \qquad (6)$$

$NaBH_4$ reduces hydrazones only when they are N,N-dialkyl substituted. The reaction is slow and yields are not usually satisfactory.[57] More synthetically useful is the reduction of N-p-tosylhydrazones to give hydrocarbons,[1c,1d,58] which has been carried out with $NaBH_4$ in refluxing MeOH, dioxane, or THF.[58] Since N-p-tosylhydrazones are easily prepared from aldehydes or ketones, the overall sequence represents a mild method for carbonyl deoxygenation. α,β-Unsaturated tosylhydrazones show a different behavior yielding, in MeOH, the allylic (or benzylic) methyl ethers.[58c] The reduction of tosylhydrazones with $NaBH_4$ is not compatible with ester groups, which are readily reduced under these conditions. More selective reagents for this reduction are $NaBH(OAc)_3$ and $NaCNBH_3$.

Reduction of Halides, Sulfonates, and Epoxides. The reduction of alkyl halides or sulfonates by $NaBH_4$ is not an easy reaction.[1d] It is best performed in polar aprotic solvents[59] such as DMSO, sulfolane, HMPA, DMF, diglyme, or PEG (polyethylene glycol),[2a] at temperatures between $60\,°C$ and $100\,°C$ (unless for highly reactive substrates), or under phase-transfer conditions.[60a] The mechanism is believed to be S_N2 (I > Br > Cl and primary > secondary). Although the more nucleophilic **Lithium Triethylborohydride** seems better suited for these reductions,[59b] the lower cost of $NaBH_4$ and the higher chemoselectivity (for example esters, nitriles, and sulfones can survive)[59a] makes it a useful alternative. Also, some secondary and tertiary alkyl halides, capable of forming relatively stable carbocations, for example benzhydryl chloride, may be reduced by $NaBH_4$. In this case the mechanism is different (via a carbocation) and the reaction is accelerated by water.[59a,b] Primary, secondary, and even aryl iodides and bromides[1d] have been reduced in good yields by $NaBH_4$ under the catalysis of soluble polyethylene- or polystyrene-bound tin halides (PE–$Sn(Bu)_2Cl$ or PS–$Sn(Bu)_2Cl$).[61] Aryl bromides and iodides have also been reduced with $NaBH_4$–**Copper(I) Chloride** in MeOH.[62]

$NaBH_4$ reduces epoxides only sluggishly.[1d] Aryl-substituted and terminal epoxides can be reduced by slow addition of MeOH to a refluxing mixture of epoxide and $NaBH_4$ in t-BuOH,[63] or by $NaBH_4$ in polyethylene glycol.[2b] The reaction is regioselective (attack takes place on the less substituted carbon), and chemoselective (nitriles, carboxylic acids, and nitro groups are left intact).[63] The opposite regioselectivity was realized by the $NaBH_4$-catalyzed reduction with diborane.[64]

Other Reductions. Aromatic and aliphatic nitro compounds are not reduced to amines by $NaBH_4$ in the absence of an activator.[1d] The $NaBH_4$–$NiCl_2$ system (see **Nickel Boride**) is a good reagent combination for this reaction, being effective also for primary and secondary aliphatic compounds. Other additives that permit $NaBH_4$ reduction are $SnCl_2$,[65] Me_3SiCl,[33] $CoCl_2$ (see *Cobalt Boride*), and MoO_3 (only for aromatic compounds),[66] Cu^{2+} salts (for aromatic and tertiary aliphatic),[23,67] and *Palladium on Carbon* (good for both aromatic and aliphatic).[68] Also, sulfurated $NaBH_4$[30] is an effective and mild reducing agent for aromatic nitro groups. In the presence of catalytic selenium or tellurium, $NaBH_4$ reduces nitroarenes to the corresponding *N*-arylhydroxylamines.[69]

The reduction of azides to amines proceeds in low yield under usual conditions, but it can be performed efficiently under phase-transfer conditions,[60b] using $NaBH_4$ supported on an ion-exchange resin,[70] or using a THF–MeOH mixed solvent (this last method is well suited only for aromatic azides).[71]

Tertiary alcohols or other carbinols capable of forming a stable carbocation have been deoxygenated by treatment with $NaBH_4$ and CF_3CO_2H or $NaBH_4$–CF_3SO_3H.[72] Under the same conditions,[72] or with $NaBH_4$–*Aluminum Chloride*,[73] diaryl ketones have also been deoxygenated.

Cyano groups α to a nitrogen atom can be replaced smoothly by hydrogen upon reaction with $NaBH_4$.[74] Since α-cyano derivatives of trisubstituted amines can be easily alkylated with electrophilic agents, the α-aminonitrile functionality can be used as a latent α-amino anion,[74a] as exemplified by eq 7 which shows the synthesis of ephedrine from a protected aminonitrile. The reduction, proceeding with concurrent benzoyl group removal, is only moderately stereoselective (77:23).

Primary amines have been deaminated in good yields through reduction of the corresponding bis(sulfonimides) with $NaBH_4$ in HMPA at 150–175 °C.[75] $NaBH_4$ reduction of ozonides is rapid at -78 °C and allows the one-pot degradation of double bonds to alcohols[1b] (see also *Ozone*). The reduction of organomercury(II) halides (see also *Mercury(II) Acetate*) is an important step in the functionalization of double bonds via oxymercuration– or amidomercuration–reduction. This reduction, which proceeds through a radical mechanism, is not stereospecific, but it can be in some cases diastereoselective.[76] In the presence of *Rhodium(III) Chloride* in EtOH, $NaBH_4$ completely saturates arenes.[77] $NaBH_4$ has also been employed for the reduction of quinones,[78] sulfoxides (in combination with *Aluminum Iodide*[79] or Me_3SiCl[33]), and sulfones (with Me_3SiCl),[33] although it does not appear to be the reagent of choice for these reductions. Finally, $NaBH_4$ was used for the reduction of various heterocyclic systems (pyridines, pyridinium salts, indoles, benzofurans, oxazolines, and so on).[1c,1d,48,80] The discussion of these reductions is beyond the scope of this article.

Diastereoselective Reductions. $NaBH_4$, like other small complex hydrides (LiBH$_4$ and LiAlH$_4$), shows an intrinsic preference for axial attack on cyclohexanones,[1c,1d,81] as exemplified by the reduction of 4-*t*-butylcyclohexanone (eq 8).[81a] This preference, which is due to stereoelectronic reasons,[82] can be counterbalanced by steric biases. For example, in 3,3,5-

trimethylcyclohexanone, where a β-axial substituent is present, the stereoselectivity is nearly completely lost (eq 9).[81a]

Also, in 2-methylcyclopentanone[81c] the attack takes place from the more hindered side, forming the *trans* isomer (dr = 74:26). In norcamphor,[81a] both stereoelectronic and steric effects favor *exo* attack, forming the *endo* alcohol in 84:16 diastereoisomeric ratio. In camphor, however, the steric bias given by one of the two methyls on the bridge brings about an inversion of stereoselectivity toward the *exo* alcohol.[81a]

The stereoselectivity for equatorial alcohols has been enhanced by using the system $NaBH_4$–*Cerium(III) Chloride*, which has an even higher propensity for attack from the more hindered side,[83] or by precomplexing the ketone on *Montmorillonite K10* clay.[84] On the other hand, bulky trialkylborohydrides (see *Lithium Tri-s-butylborohydride*) are best suited for synthesis of the axial alcohol through attack from the less hindered face.

$NaBH_4$ does not seem to be the best reagent for the stereoselective reduction of chiral unfunctionalized acyclic ketones. Bulky complex hydrides such as $Li(s\text{-}Bu)_3BH$ usually afford better results.[1c,1d] When a heteroatom is present in the α- or β-position, the stereochemical course of the reduction depends also on the possible intervention of a cyclic chelated transition state. Also, in this case other complex hydrides are often better suited for favoring chelation (see *Zinc Borohydride*). Nevertheless, cases are known[85] where excellent degrees of stereoselection have been achieved with the simpler and less expensive $NaBH_4$. Some examples are shown in eqs 10–15.

$$ (14) $$

$$ (15) $$

The stereoselective formation of *anti* adduct (**4**) in the reduction of ketone (**3**) was explained through the intervention of a chelate involving the methoxy group,[85a] although there is some debate on what the acidic species is that is coordinated (probably Na$^+$). A chelated transition state is probably the cause of the stereoselective formation of anti product (**6**) from (**5**).[85b] Methylation of the NH group indeed provokes a decrease of stereoselection. On the other hand, when appropriate protecting groups that disfavor chelation are placed on the heteroatom, the reduction proceeds by way of the Felkin model where the heteroatomic substituent plays the role of 'large' group, and *syn* adducts are formed preferentially. This is the case of α-dibenzylamino ketones (eqs 12 and 13)[85c,d] and of the α-silyloxy ketone of eq 14.[85e] Finally, the sulfonium salt of eq 15 gives, with excellent stereocontrol, the *anti* alcohol.[85f] This result was explained by a transition state where the S$^+$ and carbonyl oxygen are close due to a charge attraction.

The reduction of a diastereomeric mixture of enantiomerically pure β-keto sulfoxides (**7**) furnished one of the four possible isomers with good overall stereoselectivity (90%), when carried out under conditions which favor epimerization of the α chiral center (eq 16). This outcome derives from a chelation-controlled reduction (involving the sulfoxide oxygen) coupled with a kinetic resolution of the two diastereoisomers of (**7**).[86]

$$ (16) $$

The reduction of cyclic imines and oximes follows a trend similar to that of corresponding ketones. However, the tendency for attack from the most hindered side is in these cases attenuated.[1c,1d,57,87] In the case of oximes, while NaBH$_4$–MoO$_3$ attacks from the axial side, NaBH$_4$–NiCl$_2$ attacks from the equatorial side.[88] An example of diastereoselective reduction of acyclic chiral imines is represented by the one-pot transformation of α-alkoxy or α,β-epoxynitriles into *anti* vicinal amino alcohols (eq 17) or epoxyamines. The outcome of these reductions was explained on the basis of a cyclic chelated transition state.[89]

$$ (17) $$

$$ 80{:}20 < dr < 98{:}2 $$

Enantioselective Reductions. NaBH$_4$ has been employed with less success than LiAlH$_4$ or BH$_3$ in enantioselective ketone reductions.[1d,90,91] Low to moderate ee values have been obtained in the asymmetric reduction of ketones with chiral phase-transfer catalysts, chiral crown ethers,[91a] β-cyclodextrin,[91b] and bovine serum albumin.[91c] On the other hand, good results have been realized in the reduction of propiophenone with NaBH$_4$ in the presence of isobutyric acid and of diisopropylidene-D-glucofuranose (ee = 85%),[91d] or in the reduction of α-keto esters and β-keto esters with NaBH$_4$–L-tartaric acid (ee >86%).[91e]

Very high ee values have been obtained in the asymmetric conjugate reduction of α,β-unsaturated esters and amides with NaBH$_4$ in the presence of a chiral semicorrin (a bidentate nitrogen ligand) cobalt catalyst.[92] Good to excellent ee values were realized in the reduction of oxime ethers with NaBH$_4$–ZrCl$_4$ in the presence of a chiral 1,2-amino alcohol.[93]

First Update

Nikola Stiasni & Martin Hiersemann
Technische Universität Dortmund, Dortmund, Germany

Conjugate Reductions. A wide range of activated conjugated alkenes such as α,α-dicyano olefins and α,β-unsaturated nitriles underwent a highly chemoselective reduction of the carbon-carbon double bond in the presence of NaBH$_4$ and a catalytic amount of InCl$_3$ (10–15 mol%) to give saturated products in good to high yields (57–96%).[94] Activated $\alpha,\beta,\gamma,\delta$-unsaturated alkenes could be reduced chemoselectively to afford exclusively γ,δ-unsaturated derivatives in 77–97% yields.[95] 1,4-Reduction of α,β-unsaturated ketones, aldehydes, carboxylic acids, and esters could be achieved chemoselectively by NaBH$_4$–NiCl$_2$ to give saturated products in 61–96% yields.[96,97]

Reduction of Carboxylic Acid Derivatives. Carboxylic acids were reduced to the corresponding alcohols by activation of the carboxy function with cyanuric chloride and N-methylmorpholine, and subsequent reduction by NaBH$_4$ in 73–98% yields.[98] The method was shown to be particularly useful for the reduction of N-protected amino acids containing Boc, Cbz, or Fmoc protective groups. Enantiopure amino acids were reduced without any observable racemization. NaBH$_4$ could also reduce carboxylic acids in the presence of a catalytic amount of 3,4,5-trifluorophenylboronic acid (1 mol%) to the corresponding alcohols in 78–99% yields.[99] It was shown that functional groups such as halogeno, cyano, nitro, and azido have not been attacked under these conditions. α-Aza *gem*-dicarboxylic esters underwent chemo- and stereoselective decarboxylative reduction, when treated with NaBH$_4$ and a catalytic amount of InCl$_3$ (20 mol% based on diester), to afford monoalcohols in 77–84% yields (eq 18).[100,101] The proposed mechanism includes radical decarboxylation by the in situ formed indium hydride, Cl$_2$InH, followed by chemoselective reduction of the ester group. The presence of the N-atom alpha to the dicarboxylic esters proved to be essential for the observed chemo- and stereoselectivity. Lactams were not reduced under these conditions and could be chemoselectively reduced in the presence of *gem*-dicarboxylic esters by NaBH$_4$–I$_2$.[102]

$$\text{(18)}$$

$NaBH_4$ modified with equimolar amounts of Co^{II} salts (see *Cobalt Boride*) or Ni^{II} salts (see *Nickel Boride*) reduces nitriles to amines. It was found that nitriles could be reduced with $NaBH_4$ in the presence of a catalytic amount (10 mol %) of Ni^{II} salts.[103] To circumvent dimerization of formed amines, Boc_2O was added in a one-pot reaction to furnish the corresponding Boc-protected amines in 45–96% yields. Pure products could be obtained after removal of the metal catalyst with diethylenetriamine and extractive workup. Upon treatment with $NaBH_4$ in ethanol, δ-hydroxynitriles afforded pyranosylamines in moderate to good yields (50–70%).[104] The reductive cyclization resulted in the preferential formation of the β-anomer. The method is, however, limited in substrate scope.

Primary amides could be reduced to alcohols in a two step procedure by initially converting them into di-*tert*-butyl acylimidodicarbonates with 2 equiv of Boc_2O and a catalytic amount of DMAP, followed by subsequent reduction with $NaBH_4$ in ethanol (54–97% yields).[105]

Reduction of Halides, Sulfonates, and Epoxides. $NaBH_4$ in the presence of a catalytic amount of $InCl_3$ (10–20 mol %) forms indium hydride Cl_2InH capable of reducing alkyl halides under mild conditions (eq 19).[106,107] This reagent can serve as an alternative to toxic *Tri-n-butylstannane*, however, the disadvantage is that the BH_3, generated during the transmetalation causes side reactions with substrates containing double bonds.[108] Another alternative to minimize the use of toxic Bu_3SnH is to employ catalytic amounts of Bu_3SnCl (5–20 mol %), together with an excess of $NaBH_4$ in the presence of a radical initiator.[109–111]

$$\text{(19)}$$

α,β-epoxy ketones and α,β-epoxy esters were successfully reduced with a $NaBH_4$–PhSeSePh system (see also *Diphenyl Diselenide*).[112] $Na[PhSeB(OEt)_3]$ formed in situ opens epoxides regioselectively α to ester or carbonyl groups to form the corresponding β-hydroxy compounds in good to excellent yields (80–99%). In conjunction with subsequent $NaBH_4$–Et_2B-(OMe) reduction (see *Diastereoselective Reductions*), this method allowed access to *syn* 1,3-diols.[113] 2,3-Epoxy bromides have been reduced to allylic alcohols with $NaBH_4$ and a catalytic amount of $InCl_3$ (20 mol %) in 50–85% yields (eq 20).[114] Trisubstituted epoxides failed to undergo this reaction.

$$\text{(20)}$$

Other Reductions. The reductive removal of 2-oxazolidinone chiral auxiliaries with $NaBH_4$ is a mild and chemoselective method that provided the corresponding alcohols in good to excellent yields (75–95%), with complete preservation of stereochem-

istry (>99% ee).[115,116] As exemplified by eq 21, the removal of 2-oxazolidinone proceeded in quantitative yield and without any observable racemization.[117] *Lithium Borohydride*, on the contrary, led to considerable racemization, whereas *Lithium Aluminum Hydride* was not usable due to the presence of nitrile functional group in the substrate.

$$\text{(21)}$$

The conversion of the azide (**8**) to the amine (**9**) could be achieved with $NaBH_4$ in the presence of a catalytic amount of $CoCl_2$ (2 mol %) in 93% yield (eq 22).[118] Under catalytic hydrogenation conditions debromination was a significant side reaction.

$$\text{(22)}$$

$NaBH_4$–$CoCl_2$ reduction of azides could also be carried out in water to give products in 91–98% yields. In the case of hydrophobic azides a phase-transfer catalyst was added.[119] $NaBH_4$ with a catalytic amount of *1,3-Propanedithiol* (1 mol %) is a chemoselective and efficient reagent for the reduction of azides to amines.[120] α-Hydroxy-β-azido carboxylic acids were reduced to the corresponding α-hydroxy-β-amino acids in 85–96% yields by employing $NaBH_4$ with a catalytic amount of $Cu(NO_3)_2$ (10 mol %).[121] γ-Azido-α,β-unsaturated esters underwent reductive cyclization when treated with $NaBH_4$ in the presence of a catalytic amount of $CoCl_2$ (1 mol %) to afford the corresponding lactams in 77–93% yields. By adding an appropriate chiral ligand, the products could be obtained in moderate to high ee's (51–98% ee, eq 23)[122] (see also *Enantioselective Reductions*).

$$\text{(23)}$$

α,β-Alkynoic-γ-hydroxy esters were reduced with $NaBH_4$ in methanol to form α,β-alkenoic-γ-hydroxy esters in 60–86% yields and with excellent (*E*)-selectivity (eq 24).[123] It was found that the free γ-hydroxy group is involved in *E/Z* stereocontrol and crucial for the reaction to proceed.

$$\text{(24)}$$

$NaB(OMe)_4$, easily prepared from $NaBH_4$ and methanol, was found to be a mild catalyst for various Michael additions at room temperature. The reaction was applicable to a wide variety of nucleophiles and Michael acceptors, providing products in good to excellent yields (69–100%).[124] Baylis–Hillman adducts could be reduced by $NaBH_4$ with a catalytic amount of $InCl_3$ (13 mol %) to give the corresponding trisubstituted (*E*)-alkenones chemo-

and diastereoselectively in 54–84% yields (eq 25).[125] It could be shown that the presence of acetonitrile was necessary to obtain good yields.

$$R^1 \xrightarrow{\begin{array}{c} \text{NaBH}_4, \text{InCl}_3 \ (13 \text{ mol \%}) \\ \text{MeCN} \\ \hline 54-84\% \end{array}} R^1 \quad (25)$$

Diastereoselective Reductions. The diastereoselective reductions of β-hydroxy ketones could be accomplished with $NaBH_4$ in the presence of $Et_2B(OMe)$ to form 1,3-*syn* diols in 68–99% yields and excellent stereochemical purities (dr>98:2) (eq 26).[126,127] $Et_2B(OMe)$ could be generated in situ from **Triethylborane** and methanol. The method has been successfully utilized in several natural product syntheses[128–130] and complements the *anti*-selective reductions performed with **Tetramethylammonium Triacetoxyborohydride** (see also **Zinc Borohydride**).

$$R \xrightarrow{\begin{array}{c} \text{Et}_2\text{B(OMe)}, \text{NaBH}_4 \\ \text{THF}, \text{MeOH} \\ \hline 68-99\%, \text{dr}>98:2 \end{array}} R \quad (26)$$

In the synthesis of (*R*)-salmeterol, a phenylglycinyl ketone derivative (**10**) was diastereoselectively reduced by $NaBH_4$ in the presence of a stoichiometric amount of $CaCl_2$ to afford the alcohol (**11**) in 76% yield and good diastereoselectivity (dr 10:1).[131] It was shown that the presence of free amino- and hydroxy-groups were essential for the observed diastereoselectivity. This selectivity was explained to arise by chelate formation between the calcium ion, the carbonyl oxygen atom, and the primary hydroxy group resulting in a complex, which is then attacked by the borohydride from the less hindered face.

(10)

(11)

A new synthetic route to *anti*-α,β-dihydroxy esters was developed by performing a stereoselective reduction of α-keto-β-hydroxy esters with $NaBH_4$.[132] The best results were achieved in CH_2Cl_2 at low temperatures to give the corresponding products in 60–86% yields. Reductions of the α-keto-β-hydroxy esters with β-aryl substituents proceeded with high diastereoselectivities (>95:5 *anti/syn*) (eq 28), whereas the corresponding substrates with β-alkyl substituents yielded products with lower diastereoselectivities (85:15–93:7 *anti/syn*).

$$\text{Ar} \xrightarrow{\begin{array}{c} \text{NaBH}_4, \text{CH}_2\text{Cl}_2 \\ \hline 60-86\%, \text{dr}>95/5 \end{array}} \text{Ar} \quad (28)$$

This method has been extended to α-hydroxy-β-(*N*-tosyl)-amino esters to provide *anti*-α-hydroxy-β-amino esters in 40–95% yields (eq 29).[133] The observed diastereoselectivity (>95:5 *anti/syn*) was shown to be highly dependent on the presence of the amino-group hydrogen atom.

$$\text{Ar} \xrightarrow{\begin{array}{c} \text{NaBH}_4, \text{THF} \\ \hline 40-95\%, \text{dr}>95:5 \end{array}} \text{Ar} \quad (29)$$

The corresponding reductions under catalytic hydrogenation conditions (Pd/C) led exclusively to the *syn*-configured derivatives (>95:5 *syn/anti*).[133] Anti-configured α-hydroxy β-amino esters could be efficiently converted into *syn*-configured isomers.[134] Reduction of α-amino-γ-oxo acids by $NaBH_4$ in the presence of a catalytic amount of $MnCl_2$ (20 mol %) afforded *syn*-α-amino-γ-hydroxy acids in 60–80% yields and high diastereoselectivities (>95:5 dr) (eq 30).[135–137]

$$\text{Ar} \xrightarrow{\begin{array}{c} \text{NaBH}_4 \\ \text{MnCl}_2 \ (20 \text{ mol \%}) \\ \text{MeOH} \\ \hline 60-80\%, \text{dr}>95/5 \end{array}} \text{Ar} \quad (30)$$

Enantioselective Reductions. Enantioselective reductions utilizing $NaBH_4$ premodified with alcohols such as tetrahydrofurfuryl alcohol, methanol, or ethanol in the presence of catalytic amounts (1–5 mol %) of enantiopure β-ketoiminato Co^{II} complexes[138,139] could be applied to aromatic ketones[140,141] (eq 31) and *N*-diphenylphosphinyl imines[142] to give the corresponding products in good yields (>81%) and high ee's (>90% ee). The reduction of α,β-unsaturated amides[143] proceeded in good to high yields (69–99%) and moderate to high ee's (49–91% ee).

$$\text{MeO} \xrightarrow{\begin{array}{c} \text{NaBH}_4, \text{EtOH}, \\ \text{tetrahydrofurfuryl alcohol} \\ \text{CHCl}_3, \text{Co}^{II} \text{ catalyst } (1 \text{ mol \%}) \\ \hline 98\%, 94\% \text{ ee} \end{array}} \quad (31)$$

The method could be applied to reduce 1,3-diaryl-1,3-diketones to 1,3-diaryl-1,3-propanediols in 93–100% yields and high ee's (>97% ee). The products were obtained as a mixture of diastereomers (81:19 – 90:10) and could be purified by recrystallization.[144] Symmetrical 2-substituted-1,3-diaryl-1,3-diketones underwent reductive desymmetrization to give rise to β-hydroxyketones (45–97% yields) in high ee's (>91% ee) and diastereoselectivities (dr>99:1) (eq 32).[145] Unsymmetrical 2-alkyl-3-aryl-1,3-diketones were reduced chemoselectively at the aryl-substituted carbonyl group to yield 2-alkyl-3-aryl-3-hydroxyketones (41–48% yields) in high ee's (>95%) and diastereoselectivities (dr>97:3).[146] The yields could be further improved by the addition of a stoichiometric amount of sodium methoxide in a one-pot reaction.[147] 1,4-Diaryl-1,4-butanediones were reduced to enantiopure 1,4-diaryl-1,4-butanediols (60–100% yields), which could be elaborated in two steps to enantiopure 2,5-disubstituted pyrrolidines (>99% ee).[148] It was found that the use of an appropriate alcohol and the addition of a catalytic amount of chloroform (10–25 mol% based on the Co^{II} catalyst) were indispensable for achieving good yields and high ee's.[138,149] Both enantiomers of the Co^{II} catalyst are accessible, thus allowing generation of both enantiomers of the desired product.

Excellent results were obtained using $NaBH_4$ in combination with the tartaric acid-derived boronic ester (TarB-NO_2) to reduce aromatic ketones in good yields (65–90%) and high ee's (93–98% ee).[150,151] In reduction of aliphatic, sterically hindered ketones, TarB-NO_2 performed comparable to *Tetrahydro-1-methyl-3,3-diphenyl-1H,3H-pyrrolo[1,2-c]-[1,3,2]oxazaborole*.[152] Conjugate reduction of α,β-unsaturated esters and amides could be achieved with $NaBH_4$ in the presence of catalytic amounts of chiral aza-bis(oxazoline) ligand (2.8 mol%) and $CoCl_2$, giving rise to products in 85–89% yields and high ee's (92–96% ee).[153]

Related Reagents. Cerium(III) Chloride; Nickel Boride; Potassium Triisopropoxyborohydride; Sodium Cyanoborohydride; Sodium Triacetoxyborohydride; Cobalt Boride; Lithium Borohydride; Lithium Aluminium Hydride; Zinc Borohydride; Tetramethylammonium Triacetoxyborohydride.

Second Update

Tohru Yamada & Tatsuyuki Tsubo
Keio University, Yokohama, Japan

Conjugate Reduction. The carbon–carbon double bonds of various α,β-unsaturated ketones and related compounds can be selectively reduced with a heterogeneous palladium catalyst.[154] When nonpolar solvents such as toluene were employed, 1,4-reduction products of α,β-unsaturated ketones, esters, amides, and nitriles were obtained in high yields (87–100%) with high chemoselectivities in the presence of a catalytic amount of Pd/C(2.5–5 mol%) and acetic acid. Highly-polar solvents enhanced 1,2-reduction to afford alcohols as a result of overreduction.

Reduction of Carboxylic Acid Derivatives. Reduction of carboxylic esters has been enhanced with various activators.

$CoCl_2$ effects the reduction of carboxylic esters with appropriate additives. While $CoCl_2 \cdot 6H_2O$-catalyzed reduction using $NaBH_4$ does not react with alkyl esters, phenyl esters can be reduced to alcohols.[155] When ethyl cinnamate was used as substrate, the carbon–carbon double bond was reduced to the allylic alcohol. In contrast, when phenyl cinnamate was used as substrate, both the carbon–carbon double bonds and the ester carbonyl functions were reduced to give the saturated alcohol. Various saturated and α,β-unsaturated aryl esters were converted into the corresponding saturated alcohols with $CoCl_2 \cdot 6H_2O$ (1 mol%) and $NaBH_4$ in EtOH under mild conditions in high yields (83–99%) (eq 33). Synthesis of (R)-tolterodine, a muscarinic receptor antagonist, was achieved with this methodology. Cobalt-catalyzed reduction of carboxylic esters to alcohols is employed with alkyl esters with amine additives.[156] The $CoCl_2 \cdot 6H_2O$–$NaBH_4$ system reduces ethyl cinnamate to give only saturated ethyl ester, ethyl 3-phenylpropanoate, in 95% yield. However, using the $CoCl_2 \cdot 6H_2O$–$NaBH_4$ system with a catalytic amount of iPr_2NH smoothly reduces ethyl cinnamate to give the saturated alcohol, 3-phenyl-1-propanol, in 85% yield (eq 34). The combination of $CoCl_2 \cdot 6H_2O$ (10 mol%) and iPr_2NH (20 mol%) was an effective catalytic system for the reduction of α,β-unsaturated, aliphatic and aromatic esters and lactones affording the corresponding saturated alcohols in good to high yields (79–85%).

The combination of $NaBH_4$ and $CeCl_3$ used for chemoselective or diastereoselective reduction of ketones (see Original Commentary and First Update) could also be employed for the reduction of carboxylic esters.[157] In the presence of a catalytic amount of $CeCl_3 \cdot 7H_2O$ (1 mol%), various methyl carboxylates were reduced to alcohols with $NaBH_4$ in EtOH at ambient temperature in good to high yields (75–95%).

Reduction of Halides, Sulfonates, and Epoxides. Aryl tosylates could be reduced by Ni catalyst and $NaBH_4$ under mild conditions.[158] In the presence of a catalytic amount of $Ni(PPh_3)_4Cl_2$ (7 mol%) and an electron-rich monodentate phosphine ligand such as tricyclohexylphosphine (PCy_3) (28 mol%), hydrogenolysis of various aryl tosylates was achieved in high yields (47–100%) (eq 35).

Regioselective reduction of aryl-substituted epoxides with $NaBH_4$ is catalyzed by $PdCl_2$ in the presence of moist alumina (H_2O content, 19 wt%) in hexane.[159] The selectivity in reduction of aryl-substituted epoxides was the same as with $NaBH_4$ and diborane.[64] In addition, the alumina can be easily recovered and reused without further treatment.

Other Reductions. The (porphinato)irons could realize the reduction of alkenes and alkynes with $NaBH_4$.[160,161] Various unsaturated carbon–carbon bonds were saturated by *meso*-tetraphenylporphinatoiron chloride (TPPFe^{III}Cl) derivatives (up to 81% yield). Ruthenium(III) complexes also pair with $NaBH_4$ in the reduction of unsaturated carbon–carbon bonds (as does cobalt boride). In the presence of a catalytic amount of $Ru(PPh_3)_4H_2$ (0.5–1 mol %) and $NaBH_4$, unsaturated carbon–carbon bonds in a wide variety of alkenes and alkynes were saturated in toluene at 100 °C.[162] Addition of water was required to provide a proton source. Similar systems with $RuCl_3$ in aqueous solution reduce unsaturated bonds under milder conditions. Various unactivated mono- or disubstituted olefins and activated trisubstituted olefins were reduced with $RuCl_3$ (10 mol %) and $NaBH_4$ in THF–H_2O at 0 °C to room temperature (eq 36).[163] When the $RuCl_3$-catalyzed reductions of olefins were carried out in aqueous amide solution, unactivated trisubstituted olefins were also hydrogenated.[164,165]

$$
\begin{array}{c}
\text{NaBH}_4,\ \text{H}_2\text{O} \\
\text{Ru(PPh}_3)_4\text{H}_2\ (0.5\ \text{mol \%}) \\
\text{toluene, 100 °C, 2.5 h} \\
\hline
\text{71\% yield} \\
\text{(100\% conversion)}
\end{array}
\quad (36)
$$

Deoxygenation of sulfoxides by $NaBH_4$ is effected by the combination of $CoCl_2 \cdot 6H_2O$ and moist alumina in hexane (eq 37).[166] The reactivity is a function of steric hindrance. When the combination of TPPFe^{III}Cl and $NaBH_4$ was used, sulfilimines could be as smoothly reduced as the sulfoxides.[167]

$$
\begin{array}{c}
\text{O} \\
\parallel \\
R^1 \diagdown S \diagup R^2
\end{array}
\xrightarrow[\substack{\text{hexane} \\ 12-95\%}]{\substack{\text{NaBH}_4,\ \text{almina-H}_2\text{O} \\ \text{CoCl}_2 \cdot 6\text{H}_2\text{O (catal.)}}}
R^1 \diagdown S \diagup R^2
\quad (37)
$$

The TPPFe^{III}Cl–$NaBH_4$ system also achieves the reduction of nitrobenzene to aniline.[168] Although catalysts such as TPPCo^{II} or TPPMn^{III}Cl can also effect nitrobenzene reduction, the reactivities were lower.

Enantioselective Reductions. Enantioselective borohydride reduction of aromatic ketones catalyzed by optically active β-ketoiminato Co^{II} complexes[138] could be applied to a continuous-flow microreactor and the corresponding alcohols were obtained in gram scale maintaining enantioselectivity as high as the batch reaction.[169] The β-ketoiminato Co^{II}-catalyzed reduction system enables reduction of aliphatic as well as aromatic ketones. By changing a precursor of the axial ligand on the cobalt complex from chloroform to 1,1,1-trichloroethane, the enantioselectivity was improved and various *tert*-alkyl ketones were reduced with 80–90% ee (eq 38).[170,171] In addition, the chlorovinyl-Co^{III} derived from the original Co^{II} and 1,1,1-trichloroethane in the presence of $NaBH_4$ could be recovered via silica gel column chromatography and the enantioselective catalytic reduction performed several times without loss of reactivity or enantioselectivity.[172]

$$
\begin{array}{c}
\text{NaBH}_4,\ \text{MeOH} \\
\text{Co}^{II}\ \text{catal. (5 mol \%)} \\
\text{CH}_3\text{CCl}_3,\ \text{THF, }-20\ °\text{C, 16 h} \\
\hline
\text{85\% yield, 88\% ee}
\end{array}
\quad (38)
$$

By modification of the reaction condition, biaryl lactones could be efficiently reduced to the corresponding biaryl products by this cobalt-catalyzed system.[173] Various axially chiral biaryl compounds were obtained with high ee values (80–93% ee) by the *atropo*-enantioselective borohydride reduction with the dynamic kinetic resolution of biaryl lactones in the presence of EtOH and 1-(2-pyridinyl)ethanol (eq 39).

$$
\begin{array}{c}
\text{NaBH}_4,\ \text{EtOH} \\
\text{1-(2-pyridinyl)ethanol} \\
\text{CHCl}_3,\ \text{Co}^{II}\ \text{catal. (5 mol \%)} \\
\hline
\text{64–96\% yield, 80–93\% ee}
\end{array}
\quad (39)
$$

Reduction of cycloalkenone with α-substituents using $NaBH_4$ and TarB-NO_2 could proceed with high enantioselectivity and regioselectivity.[174] Carbonyl groups of cycloalkenones substituted with alkyl or aryl halides in the α-position were successfully reduced to the corresponding allylic alcohols with high ee values (80–99% ee) (eq 40). Asymmetric 1,2-reductions of enones by borohydride were also achieved with a *N,N*-dioxide-Sc^{III} catalyst.[175] $NaBH_4$ reduces (*E*)-4-phenylbut-3-en-2-one to the corresponding allylic alcohol in excellent yield and with good enantioselectivity (99% yield and 77% ee). In this reaction, however, $NaBH_4$ does not seem to be the best hydride source for achieving high enantioselectivity. When KBH_4 was used as a reductant, the product derived from (*E*)-4-phenylbut-3-en-2-one was produced with higher enantioselectivity (90% ee) and the other substrates were also reduced with high ee values (85–95% ee).

$$X = Br, I, alkyl$$
$$R = H, Me, Cl$$
$$n = 1, 2$$

(40)

Asymmetric synthesis of β-functionalized optically active secondary alcohols was realized by TarB-NO$_2$-catalyzed enantioselective reduction[150–152] of α-halo ketones to an intermediate terminal epoxide and sequential ring opening with various nucleophiles.[176] Optically active styrene oxide was prepared from α-bromoacetophenone with NaBH$_4$ and TarB-NO$_2$ in high yield and with high enantioselectivity (98% yield and 94% ee). β-Functionalized secondary alcohols could be obtained from the epoxides by nucleophilic attack under appropriate conditions (eq 41).

(41)

$$R = NEt_2, OiPr, SC_4H_9$$
$$CN, CH_2CN$$

Functionalization of Olefins. Oxygenation of olefins by molecular oxygen was performed in the presence of NaBH$_4$ and metal catalysts such as Mn^{III},[177,178] Co^{II},[179] and Fe^{III} (eq 42).[180,181]

(42)

Various functional groups could also be introduced to unactivated alkenes by excess NaBH$_4$ and Fe^{III}.[182–184] NaBH$_4$ and iron complexes coordinated by appropriate ligands mediated the Markovnikov olefin addition of radical traps such as sodium azide, potassium thiocyanate, air (O$_2$), N-acetylsulfanilyl chloride, potassium cyanate, tosyl cyanide, and TEMPO (eq 43). Hydrofluorination of unactivated alkenes were also realized by this system with Selectfluor$^{®}$.[185]

(43)

$$X = N_3, OH, SCN, Cl, CN$$
$$NHCONH_2, TMPO, NO$$

Synthesis of oximes from aryl-conjugated ethylenes could be achieved by cobalt or iron catalysts using borohydride. The system of Co^{II} and ethyl nitrite and ammonium borohydride rather than sodium borohydride was more optimal with most substrates.[186] The reaction with Fe^{II} and *tert*-butyl nitrite gave the oxime products in moderate to high yield using sodium borohydride (eq 44).[187]

(44)

1. (a) Brown, H. C.; Krishnamurthy, S., *Tetrahedron* **1979**, *35*, 567. (b) *Fieser & Fieser* **1967**, *1*, 1049. (c) Seyden-Penne, J. *Reductions by the Alumino- and Borohydrides in Organic Synthesis*; VCH–Lavoisier: Paris, 1991. (d) *Comprehensive Organic Synthesis* **1991**, *8*, Chapters 1.1, 1.2, 1.7, 1.10, 1.11, 1.14, 2.1, 2.3, 3.3, 3.5, 4.1, 4.4, 4.7.

2. (a) Santaniello, E.; Fiecchi, A.; Manzocchi, A.; Ferraboschi, P., *J. Org. Chem.* **1983**, *48*, 3074. (b) Santaniello, E.; Ferraboschi, P.; Fiecchi, A.; Grisenti, P.; Manzocchi, A., *J. Org. Chem.* **1987**, *52*, 671.

3. Brown, H. C.; Mead, E. J.; Subba Rao, B. C., *J. Am. Chem. Soc.* **1955**, *77*, 6209.

4. Stockmayer, W. H.; Rice, D. W.; Stephenson, C. C., *J. Am. Chem. Soc.* **1955**, *77*, 1980.

5. *The Sigma-Aldrich Library of Chemical Safety Data*, Sigma-Aldrich: Milwaukee, 1988.

6. Wigfield, D. C., *Tetrahedron* **1979**, *35*, 449.

7. Bunton, C. A.; Robinson, L.; Stam, M. F., *Tetrahedron Lett.* **1971**, *121*. Subba Rao, Y. V.; Choudary, B. M., *Synth. Commun.* **1992**, *22*, 2711.

8. Tomoi, M.; Hasegawa, T.; Ikeda, M.; Kakiuchi, H., *Bull. Chem. Soc. Jpn.* **1979**, *52*, 1653.

9. (a) Ward, D. E.; Rhee, C. K., *Can. J. Chem.* **1989**, *67*, 1206. (b) Ward, D. E.; Rhee, C. K.; Zoghaib, W. M., *Tetrahedron Lett.* **1988**, *29*, 517.

10. Yoon, N. M.; Park, K. B.; Gyoung, Y. S., *Tetrahedron Lett.* **1983**, *24*, 5367.

11. Johnson, M. R.; Rickborn, B., *J. Org. Chem.* **1970**, *35*, 1041.

12. Raucher, S.; Hwang, K.-J., *Synth. Commun.* **1980**, *10*, 133.

13. Varma, R. S.; Kabalka, G. W., *Synth. Commun.* **1985**, *15*, 985.

14. Komiya, S.; Tsutsumi, O., *Bull. Chem. Soc. Jpn.* **1987**, *60*, 3423.

15. Schauble, J. H.; Walter, G. J.; Morin, J. G., *J. Org. Chem.* **1974**, *39*, 755. Salomon, R. G.; Sachinvala, N. D.; Raychaudhuri, S. R.; Miller, D. B., *J. Am. Chem. Soc.* **1984**, *106*, 2211.

16. Hassner, J.; Heathcock, C. H., *J. Org. Chem.* **1964**, *29*, 1350.

17. Varma, R. S.; Kabalka, G. W., *Synth. Commun.* **1985**, *15*, 843.

18. (a) Olsson, T.; Stern, K.; Sundell, S., *J. Org. Chem.* **1988**, *53*, 2468. (b) Brown, H. C.; Narasimhan, S.; Choi, Y. M., *J. Org. Chem.* **1982**, *47*, 4702.

19. Soai, K.; Oyamada, H.; Takase, M.; Ookawa, A., *Bull. Chem. Soc. Jpn.* **1984**, *57*, 1948.

20. Kawabata, T.; Minami, T.; Hiyama, T., *J. Org. Chem.* **1992**, *57*, 1864.

21. Mauger, J.; Robert, A., *J. Chem. Soc., Chem. Commun.* **1986**, 395.

22. Yamakawa, T.; Masaki, M.; Nohira, H., *Bull. Chem. Soc. Jpn.* **1991**, *64*, 2730.

23. Yoo, S.; Lee, S., *Synlett* **1990**, *419*.

24. (a) Prasad, A. S. B.; Kanth, J. V. B.; Periasamy, M., *Tetrahedron* **1992**, *48*, 4623. (b) Suseela, Y.; Periasamy, M., *Tetrahedron* **1992**, *48*, 371.

25. Saito, S.; Ishikawa, T.; Kuroda, A.; Koga, K.; Moriwake, T., *Tetrahedron* **1992**, *48*, 4067.

26. Wolfrom, M. L.; Anno, K., *J. Am. Chem. Soc.* **1952**, *74*, 5583. Attwood, S. V.; Barrett, A. G. M., *J. Chem. Soc., Perkin Trans. 1* **1984**, *1315*.

27. Liu, H.-J.; Bukownik, R. R.; Pednekar, P. R., *Synth. Commun.* **1981**, *11*, 599.

28. Abiko, A.; Masamune, S., *Tetrahedron Lett.* **1992**, *33*, 5517.

29. (a) Fujisawa, T.; Mori, T.; Sato, T., *Chem. Lett.* **1983**, *835*. (b) Rodriguez, M.; Llinares, M.; Doulut, S.; Heitz, A.; Martinez, J., *Tetrahedron Lett.* **1991**, *32*, 923.

30. Lalancette, J. M.; Freche, A.; Brindle, J. R.; Laliberté, M., *Synthesis* **1972**, *526*.

31. Itsuno, S.; Sakurai, Y.; Ito, K., *Synthesis* **1988**, *995*.

32. Akabori, S.; Takanohashi, Y., *J. Chem. Soc., Perkin Trans. 1* **1991**, *479*.

33. Giannis, A.; Sandhoff, K., *Angew. Chem., Int. Ed. Engl.* **1989**, *28*, 218.

34. (a) Wann, S. R.; Thorsen, P. T.; Kreevoy, M. M., *J. Org. Chem.* **1981**, *46*, 2579. (b) Kano, S.; Tanaka, Y.; Sugino, E.; Hibino, S., *Synthesis* **1980**, *695*.

35. Rahman, A.; Basha, A.; Waheed, N.; Ahmed, S., *Tetrahedron Lett.* **1976**, *219*.

36. Tsuda, Y.; Sano, T.; Watanabe, H., *Synthesis* **1977**, *652*.

37. Mandal, S. B.; Giri, V. S.; Sabeena, M. S.; Pakrashi, S. C., *J. Org. Chem.* **1988**, *53*, 4236.

38. Setoi, H.; Takeno, H.; Hashimoto, M., *J. Org. Chem.* **1985**, *50*, 3948.

39. Santaniello, E.; Farachi, C.; Manzocchi, A., *Synthesis* **1979**, *912*.

40. Babler, J. H.; Invergo, B. J., *Tetrahedron Lett.* **1981**, *22*, 11.

41. Entwistle, I. D.; Boehm, P.; Johnstone, R. A. W.; Telford, R. P., *J. Chem. Soc., Perkin Trans. 1* **1980**, *27*.

42. Barbero, M.; Cadamuro, S.; Degani, I.; Fochi, R.; Gatti, A.; Regondi, V., *Synthesis* **1986**, *1074*.

43. Meyers, A. I.; Nabeya, A.; Adickes, H. W.; Politzer, I. R.; Malone, G. R.; Kovelesky, A.; Nolan, R. L.; Portnoy, R. C., *J. Org. Chem.* **1973**, *38*, 36. Politzer, I. R.; Meyers, A. I., *Org. Synth., Coll. Vol.* **1988**, *6*, 905.

44. Lee, S. D.; Brook, M. A.; Chan, T. H., *Tetrahedron Lett.* **1983**, *24*, 1569.

45. Takano, S.; Ogasawara, K., *Synthesis* **1974**, *42*.

46. Goto, T.; Konno, M.; Saito, M.; Sato, R., *Bull. Chem. Soc. Jpn.* **1989**, *62*, 1205.

47. Osby, J. O.; Martin, M. G.; Ganem, B., *Tetrahedron Lett.* **1984**, *25*, 2093.

48. Gribble, G. W.; Nutaitis, C. F., *Org. Prep. Proced. Int.* **1985**, *17*, 317.

49. (a) Sondengam, B. L.; Hentchoya Hémo, J.; Charles, G., *Tetrahedron Lett.* **1973**, *261*. (b) Gribble, G. W.; Nutaitis, C. F., *Synthesis* **1987**, *709*. (c) Verardo, G.; Giumanini, A. G.; Favret, G.; Strazzolini, P., *Synthesis* **1991**, *447*.

50. Guerrier, L.; Royer, J.; Grierson, D. S.; Husson, H.-P., *J. Am. Chem. Soc.* **1983**, *105*, 7754; Polniaszek, R. P.; Kaufman, C. R., *J. Am. Chem. Soc.* **1989**, *111*, 4859.

51. Matsuda, Y.; Tanimoto, S.; Okamoto, T.; Ali, S. M., *J. Chem. Soc., Perkin Trans. 1* **1989**, *279*.

52. De Kimpe, N.; Stanoeva, E.; Verhé, R.; Schamp, N., *Synthesis* **1988**, *587*.

53. Borch, R. F.; Bernstein, M. D.; Durst, H. D., *J. Am. Chem. Soc.* **1971**, *93*, 2897.

54. Mundy, B. P.; Bjorklund, M., *Tetrahedron Lett.* **1985**, *26*, 3899.

55. Spreitzer, H.; Buchbauer, G.; Püringer, C., *Tetrahedron* **1989**, *45*, 6999.

56. Hoffman, C.; Tanke, R. S.; Miller, M. J., *J. Org. Chem.* **1989**, *54*, 3750.

57. Walker, G. N.; Moore, M. A.; Weaver, B. N., *J. Org. Chem.* **1961**, *26*, 2740.

58. (a) Caglioti, L., *Org. Synth., Coll. Vol.* **1988**, *6*, 62. (b) Rosini, G.; Baccolini, G.; Cacchi, S., *Synthesis* **1975**, *44*. (c) Grandi, R.; Marchesini, A.; Pagnoni, U. M.; Trave, R., *J. Org. Chem.* **1976**, *41*, 1755.

59. (a) Hutchins, R. O.; Kandasamy, D.; Dux III, F.; Maryanoff, C. A.; Rotstein, D.; Goldsmith, B.; Burgoyne, W.; Cistone, F.; Dalessandro, J.; Puglis, J., *J. Org. Chem.* **1978**, *43*, 2259. (b) Krishnamurthy, S.; Brown, H. C., *J. Org. Chem.* **1980**, *45*, 849. (c) Kociensky, P.; Street, S. D. A., *Synth. Commun.* **1984**, *14*, 1087.

60. (a) Rolla, F., *J. Org. Chem.* **1981**, *46*, 3909. (b) Rolla, F., *J. Org. Chem.* **1982**, *47*, 4327.

61. Bergbreiter, D. E.; Walker, S. A., *J. Org. Chem.* **1989**, *54*, 5138.

62. Narisada, M.; Horibe, I.; Watanabe, F.; Takeda, K., *J. Org. Chem.* **1989**, *54*, 5308.

63. Ookawa, A.; Hiratsuka, H.; Soai, K., *Bull. Chem. Soc. Jpn.* **1987**, *60*, 1813.

64. Brown, H. C.; Yoon, N. M., *J. Am. Chem. Soc.* **1968**, *90*, 2686.

65. Satoh, T.; Mitsuo, N.; Nishiki, M.; Inoue, Y.; Ooi, Y., *Chem. Pharm. Bull.* **1981**, *29*, 1443.

66. Yanada, K.; Yanada, R.; Meguri, H., *Tetrahedron Lett.* **1992**, *33*, 1463.

67. Cowan, J. A., *Tetrahedron Lett.* **1986**, *27*, 1205.

68. Neilson, T.; Wood, H. C. S.; Wylie, A. G., *J. Chem. Soc* **1962**, *371*. Petrini, M.; Ballini, R.; Rosini, G., *Synthesis* **1987**, 713.

69. Uchida, S.; Yanada, K.; Yamaguchi, H.; Meguri, H., *Chem. Lett.* **1986**, 1069; *J. Chem. Soc., Chem. Commun.* **1986**, 1655.

70. Kabalka, G. W.; Wadgaonkar, P. P.; Chatla, N., *Synth. Commun.* **1990**, *20*, 293.

71. Soai, K.; Yokoyama, S.; Ookawa, A., *Synthesis* **1987**, 48.

72. Olah, G. A.; Wu, A.; Farooq, O., *J. Org. Chem.* **1988**, *53*, 5143.

73. Ono, A.; Suzuki, N.; Kamimura, J., *Synthesis* **1987**, 736.

74. (a) Stork, G.; Jacobson, R. M.; Levitz, R., *Tetrahedron Lett.* **1979**, 771; (b) Santoyo-Gonzalez, F.; Hernandez-Mateo, F.; Vargas-Berenguel, A., *Tetrahedron Lett.* **1991**, *32*, 1371.

75. Hutchins, R. O.; Cistone, F.; Goldsmith, B.; Heuman, P., *J. Org. Chem.* **1975**, *40*, 2018.

76. Gouzoules, F. H.; Whitney, R. A., *J. Org. Chem.* **1986**, *51*, 2024. Takahata, H.; Bandoh, H.; Hanayama, M.; Momose, T., *Tetrahedron: Asymmetry* **1992**, *3*, 607 and refs. therein.

77. Nishiki, M.; Miyataka, H.; Niino, Y.; Mitsuo, N.; Satoh, T., *Tetrahedron Lett.* **1982**, *23*, 193.

78. Cho, H.; Harvey, R. G., *J. Chem. Soc., Perkin Trans. 1* **1976**, *836*.

79. Babu, J. R.; Bhatt, M. V., *Tetrahedron Lett.* **1986**, *27*, 1073.

80. *Comprehensive Organic Synthesis* **1991**, *8*, Chapters 3.6–3.8, pp 579–666.

81. (a) Boone, J. R.; Ashby, E. C., *Top. Stereochem.* **1979**, *11*, 53. (b) Ref. 6. (c) Caro, B.; Boyer, B.; Lamaty, G.; Jaouen, G., *Bull. Soc. Claim. Fr., Part 2* **1983**, 281.

82. Wong, S. S.; Paddon-Row, M. N., *J. Chem. Soc., Chem. Commun.* **1990**, *456* and refs. therein.

83. Krief, A.; Surleraux, D.; Ropson, N., *Tetrahedron: Asymmetry* **1993**, *4*, 289.

84. Sarkar, A.; Rao, B. R.; Konar, M. M., *Synth. Commun.* **1989**, *19*, 2313.

85. (a) Glass, R. S.; Deardorff, D. R.; Henegar, K., *Tetrahedron Lett.* **1980**, *21*, 2467. (b) Maugras, I.; Poncet, J.; Jouin, P., *Tetrahedron* **1990**, *46*, 2807. (c) Guanti, G.; Banfi, L.; Narisano, E.; Scolastico, C., *Tetrahedron* **1988**, *44*, 3671. (d) Reetz, M. T.; Drewes, M. W.; Lennick, K.; Schmitz, A.; Holdgrün, X., *Tetrahedron: Asymmetry* **1990**, *1*, 375. (e) Saito, S.; Harunari, T.; Shimamura, N.; Asahara, M.; Moriwake, T., *Synlett* **1992**, *325*. (f) Shimagaki, M.; Matsuzaki, Y.; Hori, I.; Nakata, T.; Oishi, T., *Tetrahedron Lett.* **1984**, *25*, 4779. (g) Morizawa, Y.; Yasuda, A.; Uchida, K., *Tetrahedron Lett.* **1986**, *27*, 1833. (h) Fujii, H.; Oshima, K.;

Utimoto, K., *Chem. Lett.* **1992**, *967*. (i) Fujii, H.; Oshima, K.; Utimoto, K., *Tetrahedron Lett.* **1991**, *32*, 6147. (j) Kobayashi, Y.; Uchiyama, H.; Kanbara, H.; Sato, F., *J. Am. Chem. Soc.* **1985**, *107*, 5541. (k) Oppolzer, W.; Tamura, O.; Sundarababu, G.; Signer, M., *J. Am. Chem. Soc.* **1992**, *114*, 5900. (l) Elliott, J.; Hall, D.; Warren, S., *Tetrahedron Lett.* **1989**, *30*, 601.

86. Guanti, G.; Narisano, E.; Pero, F.; Banfi, L.; Scolastico, C., *J. Chem. Soc., Perkin Trans. 1* **1984**, *189*.

87. Hutchins, R. O.; Su, W.-Y.; Sivakumar, R.; Cistone, F.; Stercho, Y. P., *J. Org. Chem.* **1983**, *48*, 3412.

88. Ipaktschi, J., *Chem. Ber.* **1984**, *117*, 856.

89. (a) Brussee, J.; Van der Gen, A., *Recl. Trav. Chim. Pays-Bas* **1991**, *110*, 25. (b) Urabe, H.; Aoyama, Y.; Sato, F., *J. Org. Chem.* **1992**, *57*, 5056 and refs. therein.

90. Brown, H. C.; Park, W. S.; Cho, B. T.; Ramachandran, P. V., *J. Org. Chem.* **1987**, *52*, 5406.

91. (a) Takahashi, I.; Odashima, K.; Koga, K., *Chem. Pharm. Bull.* **1985**, *33*, 3571. (b) Fornasier, R.; Reniero, F.; Scrimin, P.; Tonellato, U., *J. Org. Chem.* **1985**, *50*, 3209. (c) Utaka, M.; Watabu, H.; Takeda, A., *J. Org. Chem.* **1986**, *51*, 5423. (d) Hirao, A.; Itsuno, S.; Owa, M.; Nagami, S.; Mochizuki, H.; Zoorov, H. H. A.; Niakahama, S.; Yamazaki, N., *J. Chem. Soc., Perkin Trans. 1* **1981**, *900*. (e) Yatagai, M.; Ohnuki, T., *J. Chem. Soc., Perkin Trans. 1* **1990**, *1826*.

92. von Matt, P.; Pfaltz, A., *Tetrahedron: Asymmetry* **1991**, *2*, 691.

93. Itsuno, S.; Sakurai, Y.; Shimizu, K.; Ito, K., *J. Chem. Soc., Perkin Trans. 1* **1990**, *1859*.

94. Ranu, B. C.; Samanta, S., *Tetrahedron Lett.* **2002**, *43*, 7405.

95. Ranu, B. C.; Samanta, S., *J. Org. Chem.* **2003**, *68*, 7130.

96. Khurana, J. M.; Sharma, P., *Bull. Chem. Soc. Jpn.* **2004**, *77*, 549.

97. Kangani, C. O.; Brückner, A. M.; Curran, D. P., *Org. Lett.* **2005**, *7*, 379.

98. Falorni, M.; Porcheddu, A.; Taddei, M., *Tetrahedron Lett.* **1999**, *40*, 4395.

99. Tale, R. H.; Patil, K. M.; Dapurkar, S. E., *Tetrahedron Lett.* **2003**, *44*, 3427.

100. Haldar, P.; Ray, J. K., *Org. Lett.* **2005**, *7*, 4341.

101. Seayad, J.; Seayad, A. M.; List, B., *J. Am. Chem. Soc.* **2006**, *128*, 1086.

102. Haldar, P.; Ray, J. K., *Tetrahedron Lett.* **2003**, *44*, 8229.

103. Caddick, S.; Judd, D. B.; Lewis, A. K. de K.; Reich, M. T.; Williams, M. R. V., *Tetrahedron* **2003**, *59*, 5417.

104. Dorsey, A. D.; Barbarow, J. E.; Trauner, D., *Org. Lett.* **2003**, *5*, 3237.

105. Ragnarsson, U.; Grehn, L.; Monteiro, L. S.; Maia, H. L. S., *Synlett* **2003**, *15*, 2386.

106. Inoue, K.; Sawada, A.; Shibata, I.; Baba, A., *J. Am. Chem. Soc.* **2002**, *124*, 906.

107. Donohoe, T. J.; Chiu, J. Y. Y.; Thomas, R. E., *Org. Lett.* **2007**, *9*, 421.

108. Hayashi, N.; Shibata, I.; Baba, A., *Org. Lett.* **2004**, *6*, 4981.

109. Corey, E. J.; Suggs, J. W., *J. Org. Chem.* **1975**, *40*, 2554.

110. Stork, G.; Sher, P. M., *J. Am. Chem. Soc.* **1986**, *108*, 303.

111. Attrill, R. P.; Blower, M. A.; Mulholland, K. R.; Roberts, J. K.; Richardson, J. E.; Teasdale, M. J.; Wanders, A., *Org. Process Res. Dev.* **2000**, *4*, 98.

112. Miyashita, M.; Suzuki, T.; Hoshino, M.; Yoshikoshi, A., *Tetrahedron* **1997**, *53*, 12469.

113. Tosaki, S.; Horiuchi, Y.; Nemoto, T.; Ohshima, T.; Shibasaki, M., *Chem. Eur. J.* **2004**, *10*, 1527.

114. Ranu, B. C.; Banerjee, S.; Das, A., *Tetrahedron Lett.* **2004**, *45*, 8579.

115. Prashad, M.; Har, D.; Kim, H.-Y.; Repič, O., *Tetrahedron Lett.* **1998**, *39*, 7067.

116. Prashad, M.; Kim, H.-Y.; Lu, Y.; Liu, Y.; Har, D.; Repič, O.; Blacklock, T. J.; Giannousis, P., *J. Org. Chem.* **1999**, *64*, 1750.

117. Prashad, M.; Har, D.; Chen, L.; Kim, H.-Y; Repič, O.; Blacklock, T. J., *J. Org. Chem.* **2002**, *67*, 6612.

118. Tschaen, D. M.; Abramson, L.; Cai, D.; Desmond, R.; Dolling, U.-H.; Frey, L.; Karady, S.; Shi, Y.-S.; Verhoeven, T. R., *J. Org. Chem.* **1995**, *60*, 4324.

119. Fringuelli, F.; Pizzo, F.; Vaccaro, L., *Synthesis* **2000**, *5*, 646.

120. Pei, Y.; Wickham, B. O. S., *Tetrahedron Lett.* **1993**, *34*, 7509.

121. Fringuelli, F.; Pizzo, F.; Rucci, M.; Vaccaro, L., *J. Org. Chem.* **2003**, *68*, 7041.

122. Paraskar, A. S.; Sudalai, A., *Tetrahedron* **2006**, *62*, 4907.

123. Meta, C. T.; Koide, K., *Org. Lett.* **2004**, *6*, 1785.

124. Campaña, A. G.; Fuentes, N.; Gómez-Bengoa, E.; Mateo, C.; Oltra, J. E.; Echavarren, A. M.; Cuerva, J. M., *J. Org. Chem.* **2007**, *72*, 8127.

125. Das, B.; Banerjee, J.; Chowdhury, N.; Majhi, A.; Holla, H., *Synlett* **2006**, *12*, 1879.

126. Chen, K.-M.; Hardtmann, G. E.; Prasad, K.; Repič, O.; Shapiro, M. J., *Tetrahedron Lett.* **1987**, *28*, 155.

127. Chen, K.-M.; Gunderson, K. G.; Hardtmann, G. E.; Prasad, K.; Repič, O.; Shapiro, M. J., *Chem. Lett.* **1987**, *1923*.

128. Hosaka, M.; Hayakawa, H.; Miyashita, M., *J. Chem. Soc., Perkin Trans. 1* **2000**, *4227*.

129. Aubele, D. L.; Wan, S.; Floreancig, P. E., *Angew. Chem., Int. Ed.* **2005**, *44*, 3485.

130. Snider, B. B.; Song, F., *Org. Lett.* **2001**, *3*, 1817.

131. Bream, R. N.; Ley, S. V.; McDermott, B.; Procopiou, P. A., *J. Chem. Soc., Perkin Trans. 1* **2002**, *2237*.

132. Liao, M.; Yao, W.; Wang, J., *Synthesis* **2004**, *16*, 2633.

133. Zhao, Y.; Jiang, N.; Chen, S.; Peng, C.; Zhang, X.; Zou, Y.; Zhang, S.; Wang, J., *Tetrahedron* **2005**, *61*, 6546.

134. Lee, J.-M.; Lim, H.-S.; Seo, K.-C.; Chung, S.-K., *Tetrahedron: Asymmetry* **2003**, *14*, 3639.

135. Berkeš, D.; Kolarovič, A.; Považanec, F., *Tetrahedron Lett.* **2000**, *41*, 5257.

136. Berkeš, D.; Kolarovič, A.; Manduch, R.; Baran, P.; Považanec, F., *Tetrahedron: Asymmetry* **2005**, *16*, 1927.

137. Jakubec, P.; Berkeš, D.; Šiška, R.; Gardianova, M.; Považanec, F., *Tetrahedron: Asymmetry* **2006**, *17*, 1629.

138. Yamada, T.; Nagata, T.; Sugi, K. D.; Yorozu, K.; Ikeno, T.; Ohtsuka, Y.; Miyazaki, D.; Mukaiyama, T., *Chem. Eur. J.* **2003**, *9*, 4485.

139. Yamada, T.; Nagata, T.; Ikeno, T.; Ohtsuka, Y.; Sagara, A.; Mukaiyama, T., *Inorg. Chim. Acta* **1999**, *296*, 86.

140. Sugi, K. D.; Nagata, T.; Yamada, T.; Mukaiyama, T., *Chem. Lett.* **1996**, *25*, 737.

141. Sugi, K. D.; Nagata, T.; Yamada, T.; Mukaiyama, T., *Chem. Lett.* **1996**, *25*, 1081.

142. Sugi, K. D.; Nagata, T.; Yamada, T.; Mukaiyama, T., *Chem. Lett.* **1997**, *26*, 493.

143. Yamada, T.; Ohtsuka, Y.; Ikeno, T., *Chem. Lett.* **1998**, *27*, 1129.

144. Ohtsuka, Y.; Kubota, T.; Ikeno, T.; Nagata, T.; Yamada, T., *Synlett* **2000**, *4*, 535.

145. Ohtsuka, Y.; Koyasu, K.; Ikeno, T.; Yamada, T., *Org. Lett.* **2001**, *3*, 2543.

146. Ohtsuka, Y.; Koyasu, K.; Miyazaki, D.; Ikeno, T.; Yamada, T., *Org. Lett.* **2001**, *3*, 3421.

147. Ohtsuka, Y.; Miyazaki, D.; Ikeno, T.; Yamada, T., *Chem. Lett.* **2002**, *24*.

148. Sato, M.; Gunji, Y.; Ikeno, T.; Yamada, T., *Synthesis* **2004**, *9*, 1434.

149. Kokura, A.; Tanaka, S.; Teraoka, H.; Shibahara, A.; Ikeno, T.; Nagata, T.; Yamada, T., *Chem. Lett.* **2007**, *36*, 26.

150. Cordes, D. B.; Nguyen, T. M.; Kwong, T. J.; Suri, J. T.; Luibrand, R. T.; Singaram, B., *Eur. J. Org. Chem.* **2005**, *2005*(24), 5289.

151. Kim, J.; Suri, J. T.; Cordes, D. B.; Singaram, B., *Org. Process Res. Dev.* **2006**, *10*, 949.

152. Eagon, S.; Kim, J.; Yan, K.; Haddenham, D.; Singaram, B., *Tetrahedron Lett.* **2007**, *48*, 9025.

153. Geiger, C.; Kreitmeier, P.; Reiser, O., *Adv. Synth. Catal.* **2005**, *347*, 249.

154. Russo, A. T.; Amezcua, K. L.; Huynh, V. A.; Rousslang, Z. M.; Cordes, D. B., *Tetrahedron Lett.* **2011**, *52*, 6823.

155. Jagdale, A. R.; Sudalai, A., *Tetrahedron Lett.* **2008**, *49*, 3790.

156. Sudalai, A.; Jagdale, A.; Paraskar, A., *Synthesis* **2009**, *2009*, 660.

157. Xu, Y.; Wei, Y., *Synth. Commun.* **2010**, *40*, 3423.

158. Kogan, V., *Tetrahedron Lett.* **2006**, *47*, 7515.

159. Yakabe, S., *Synth. Commun.* **2010**, *40*, 1339.

160. Kano, K.; Takeuchi, M.; Hashimoto, S.; Yoshida, Z.-i., *J. Chem. Soc., Chem. Commun.* **1991**, 1728.

161. Takeuchi, M.; Kano, K., *Organometallics* **1993**, *12*, 2059.

162. Adair, G. R. A.; Kapoor, K. K.; Scolan, A. L. B.; Williams, J. M. J., *Tetrahedron Lett.* **2006**, *47*, 8943.

163. Sharma, P. K.; Kumar, S.; Kumar, P.; Nielsen, P., *Tetrahedron Lett.* **2007**, *48*, 8704.

164. Babler, J. H.; White, N. A., *Tetrahedron Lett.* **2010**, *51*, 439.

165. Babler, J. H.; Ziemke, D. W.; Hamer, R. M., *Tetrahedron Lett.* **2013**, *54*, 1754.

166. Yakabe, S.; Hirano, M.; Morimoto, T., *Synth. Commun.* **2011**, *41*, 2251.

167. Nagata, T.; Fujimori, K.; Yoshimura, T.; Furukawa, N.; Oae, S., *J. Chem. Soc., Perkin Trans. 1* **1989**, 1431.

168. Sakaki, S.; Mitarai, S.; Ohkubo, K., *Chem. Lett.* **1991**, *20*, 195.

169. Hayashi, T.; Kikuchi, S.; Asano, Y.; Endo, Y.; Yamada, T., *Org. Process Res. Dev.* **2012**, *16*, 1235.

170. Tsubo, T.; Chen, H.-H.; Yokomori, M.; Fukui, K.; Kikuchi, S.; Yamada, T., *Chem. Lett.* **2012**, *41*, 780.

171. Tsubo, T.; Yokomori, M.; Chen, H.-H.; Komori-Orisaku, K.; Kikuchi, S.; Koide, Y.; Yamada, T., *Chem. Lett.* **2012**, *41*, 783.

172. Tsubo, T.; Chen, H.-H.; Yokomori, M.; Kikuchi, S.; Yamada, T., *Bull. Chem. Soc. Jpn.* **2013**, 983.

173. Ashizawa, T.; Tanaka, S.; Yamada, T., *Org. Lett.* **2008**, *10*, 2521.

174. Kim, J.; Bruning, J.; Park, K. E.; Lee, D. J.; Singaram, B., *Org. Lett.* **2009**, *11*, 4358.

175. He, P.; Liu, X.; Zheng, H.; Li, W.; Lin, L.; Feng, X., *Org. Lett.* **2012**, *14*, 5134.

176. Eagon, S.; Ball-Jones, N.; Haddenham, D.; Saavedra, J.; DeLieto, C.; Buckman, M.; Singaram, B., *Tetrahedron Lett.* **2010**, *51*, 6418.

177. Tabushi, I.; Koga, N., *J. Am. Chem. Soc.* **1979**, *101*, 6456.

178. Perree-Fauvet, M.; Gaudemer, A., *J. Chem. Soc., Chem. Commun.* **1981**, 874.

179. Okamoto, T.; Oka, S., *J. Org. Chem.* **1984**, *49*, 1589.

180. Santa, T.; Mori, T.; Hirobe, M., *Chem. Pharm. Bull.* **1985**, *33*, 2175.

181. Kano, K.; Takagi, H.; Takeuchi, M.; Hashimoto, S.; Yoshida, Z.-i., *Chem. Lett.* **1991**, *20*, 519.

182. Ishikawa, H.; Colby, D. A.; Boger, D. L., *J. Am. Chem. Soc.* **2007**, *130*, 420.

183. Ishikawa, H.; Colby, D. A.; Seto, S.; Va, P.; Tam, A.; Kakei, H.; Rayl, T. J.; Hwang, I.; Boger, D. L., *J. Am. Chem. Soc.* **2009**, *131*, 4904.

184. Leggans, E. K.; Barker, T. J.; Duncan, K. K.; Boger, D. L., *Org. Lett.* **2012**, *14*, 1428.

185. Barker, T. J.; Boger, D. L., *J. Am. Chem. Soc.* **2012**, *134*, 13588.

186. Okamoto, T.; Kobayashi, K.; Oka, S.; Tanimoto, S., *J. Org. Chem.* **1988**, *53*, 4897.

187. Prateeptongkum, S.; Jovel, I.; Jackstell, R.; Vogl, N.; Weckbecker, C.; Beller, M., *Chem. Commun.* **2009**, 1990.

Sodium Cyanoborohydride[1]

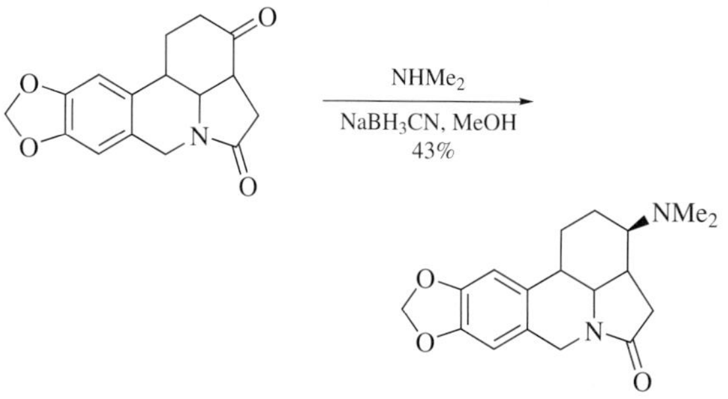

$$\boxed{\text{NaBH}_3\text{CN}}$$

[25895-60-7]   CH$_3$BNNa   (MW 62.85)

InChI = 1/CH3BN.Na/c2-1-3;/h2H3;/q-1;+1

InChIKey = CVDUGUOQTVTBJH-UHFFFAOYAX

(selective, mild reducing reagent for reductive aminations of aldehydes and ketones, reductions of imines, iminium ions, oximes and oxime derivatives, hydrazones, enamines; reductive deoxygenation of carbonyls via sulfonyl hydrazones, reductions of aldehydes and ketones, polarized alkenes, alkyl halides, epoxides, acetals, and allylic ester groups)

Physical Data: white, hygroscopic solid, mp 240–242 °C (dec).

Solubility: sol most polar solvents (e.g. MeOH, EtOH, H$_2$O, carboxylic acids) and polar aprotic solvents (e.g. HMPA, DMSO, DMF, sulfolane, THF, diglyme); insol nonpolar solvents (e.g. ether, CH$_2$Cl$_2$, benzene, hexane).

Form Supplied in: widely available; the corresponding deuterated (or tritiated) reagent is available via acid-catalyzed exchange with D$_2$O (or T$_2$O).[1a,2]

Handling, Storage, and Precautions: store under dry N$_2$ or Ar.

Original Commentary

Robert O. Hutchins
Drexel University, Philadelphia, PA, USA

MaryGail K. Hutchins
LNP Engineering Plastics, Exton, PA, USA

Functional Group Reductions: General. The chemoselectivity available with NaBH$_3$CN is remarkably dependent on solvent and pH. Under neutral or slightly acidic conditions (pH > 5), only iminium ions are reduced in protic and ether (e.g. THF) solvents.[2] Most other functional groups including aldehydes, ketones, esters, lactones, amides, nitro groups, halides, and epoxides are inert under these conditions.

Reductive Aminations. The relative inertness of aldehydes and ketones toward NaBH$_3$CN at pH > 5 allows reductive aminations with amine and amine derivatives (usually in MeOH) via in situ generation of iminium ions which are then reduced to

amines (eqs 1 and 2).[2–4] This protocol is compatible with most other functional groups, can be used to prepare N-heterocycles with stereochemical control (eqs 3 and 4),[5,6] and serves as a methylation process using CH$_2$O as the aldehyde (eq 5).[7,8] For difficult cases (e.g. aromatic amines, hindered and trifluoromethyl ketones), yields may be greatly improved by prior treatment of the carbonyl and amine with *Titanium(IV) Chloride* (eq 6)[9] or *Titanium Tetraisopropoxide*[10a] A reagent system prepared from NaBH$_3$CN and *Zinc Chloride* also is also effective for reductive aminations[10b] (see also *Sodium Triacetoxyborohydride* and *Tetrabutylammonium Cyanoborohydride*).

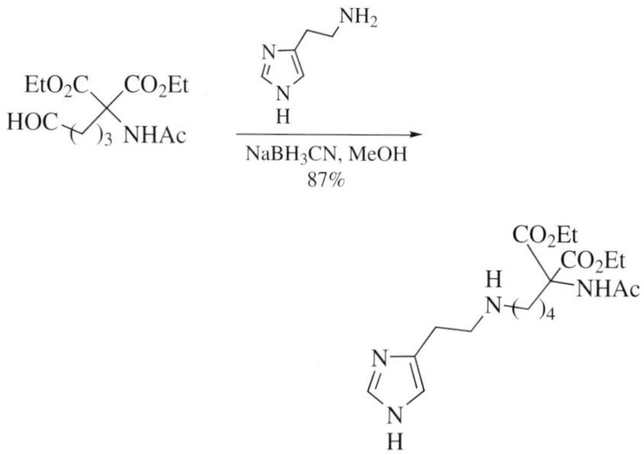

(2)

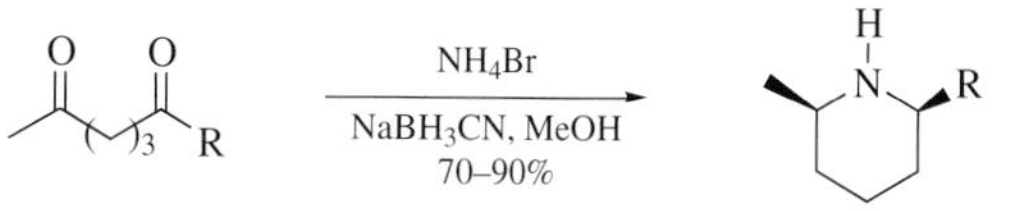

(3)

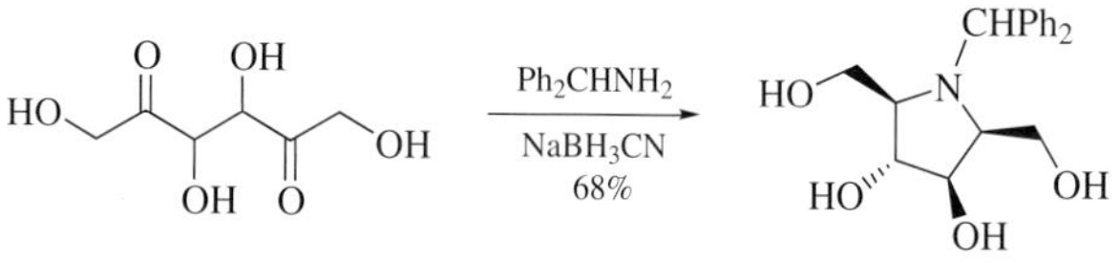

(4)

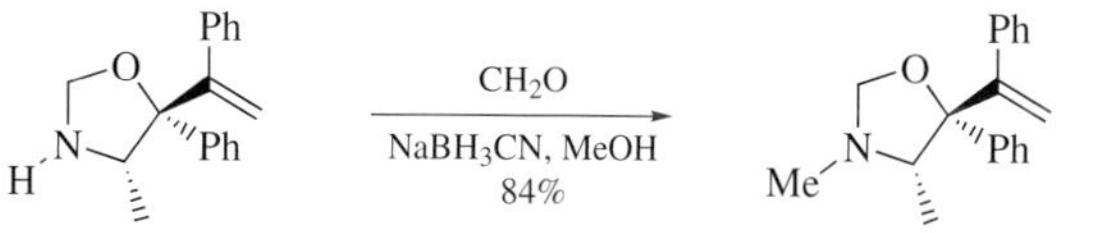

(5)

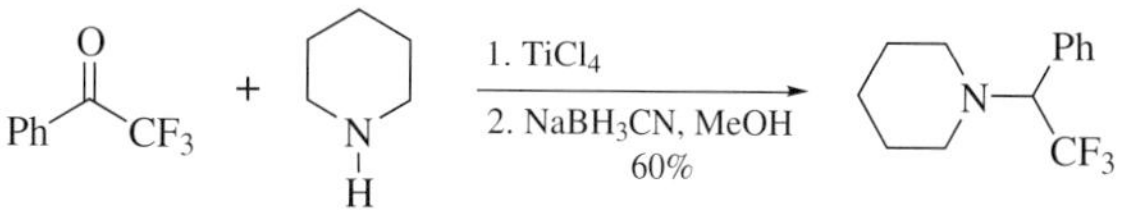

(6)

Reductions of Imines and Derivatives. Preformed imines (eq 7),[2,11,12] iminium ions (eq 8),[2,11,13] oximes (eq 9),[2,14] oxime derivatives (eqs 10 and 11),[15,16] hydrazones (eq 12),[17] and other N-heterosubstituted imines (eqs 13 and 14)[18,19] are reduced to the corresponding amine derivatives by NaBH$_3$CN, usually in acidic media (see also *Lithium Aluminum Hydride* and *Sodium Borohydride*).

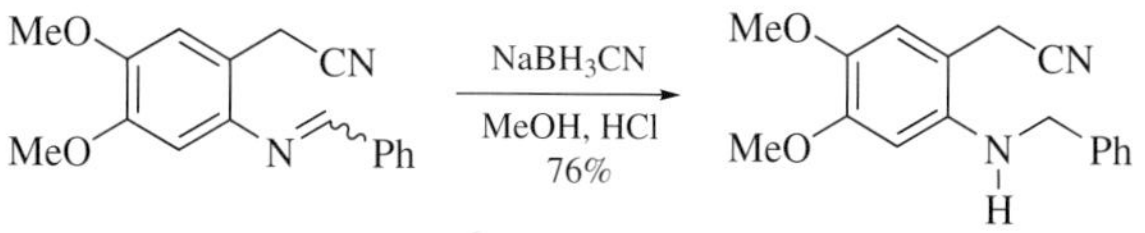

(7)

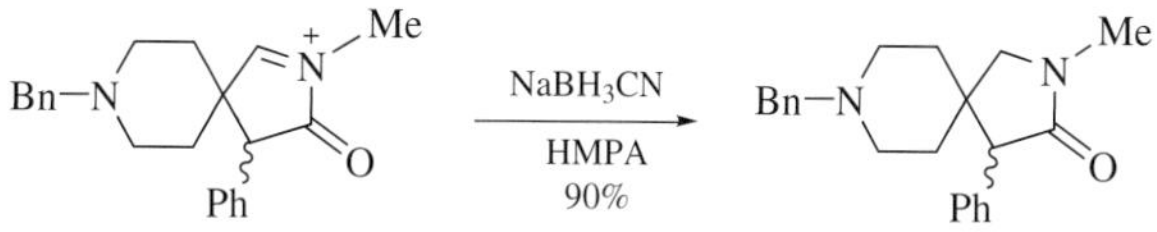

(8)

(1)

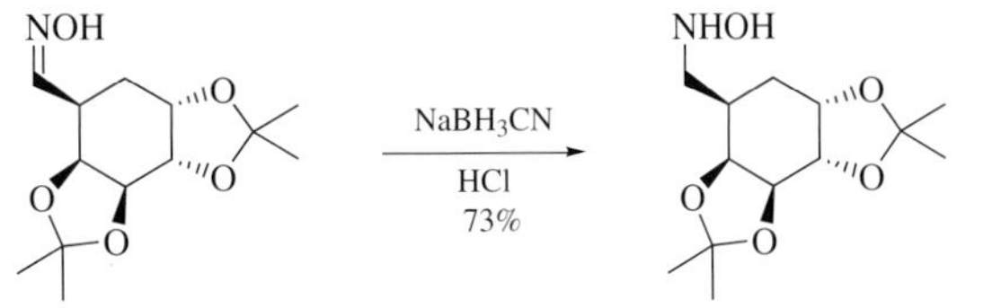

(9)

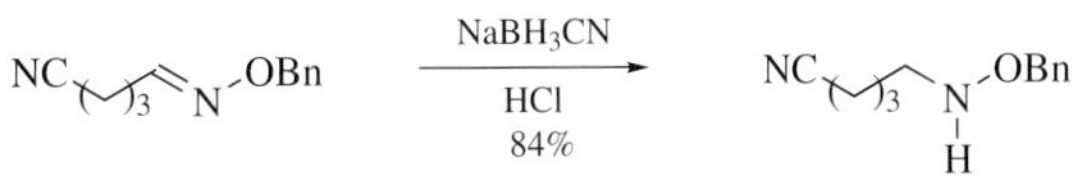

(10)

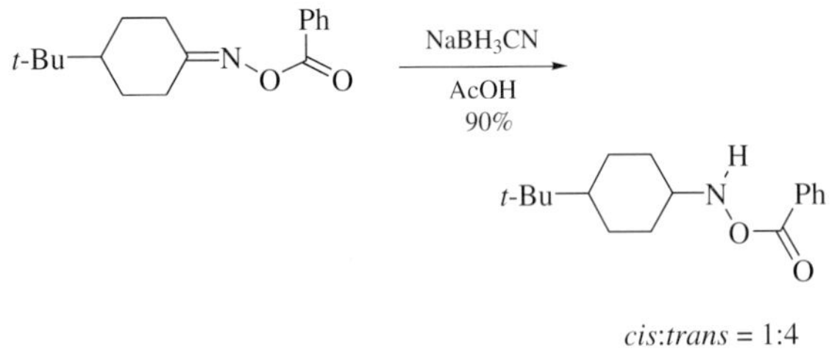

(11)

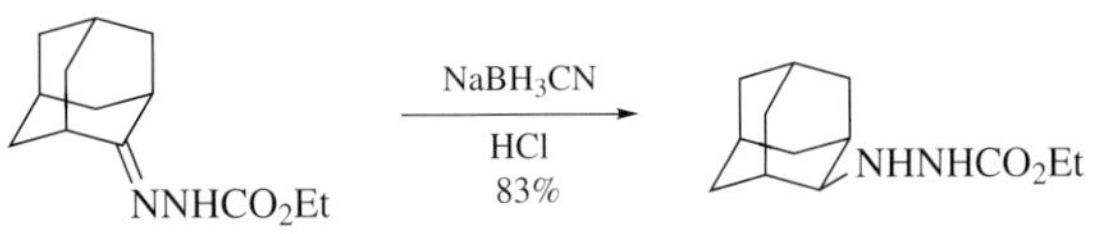

cis:trans = 1:4

(12)

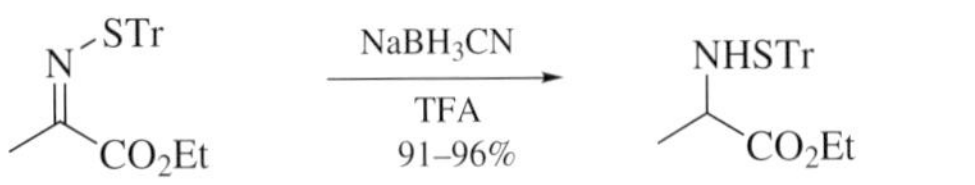

(13)

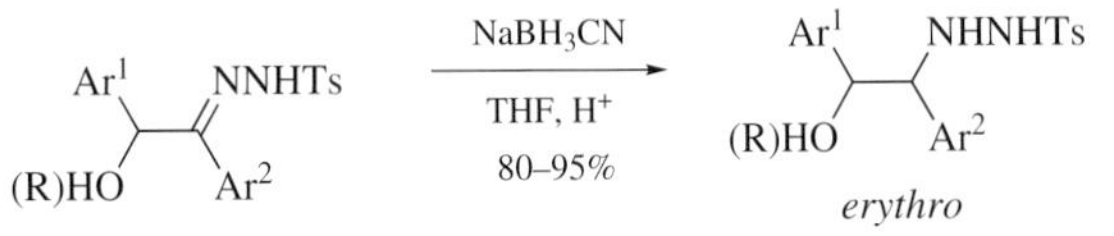

(14)

erythro

Also under acidic conditions, enamines are reduced to amines via iminium ions by NaBH$_3$CN (eq 15).[2,20] This type of conversion is also effected with NaBH$_3$CN/ZnCl$_2$.[10b] Pyridines and related nitrogen heterocycles are reduced by NaBH$_3$CN/H$^+$ to di- or tetrahydro derivatives (eq 16).[1e,21] Likewise, pyridinium and related salts (e.g. quinolinium, isoquinolinium) are reduced. With 4-substituted derivatives, 1,2,5,6-tetrahydropyridine products are produced (eq 17).[22]

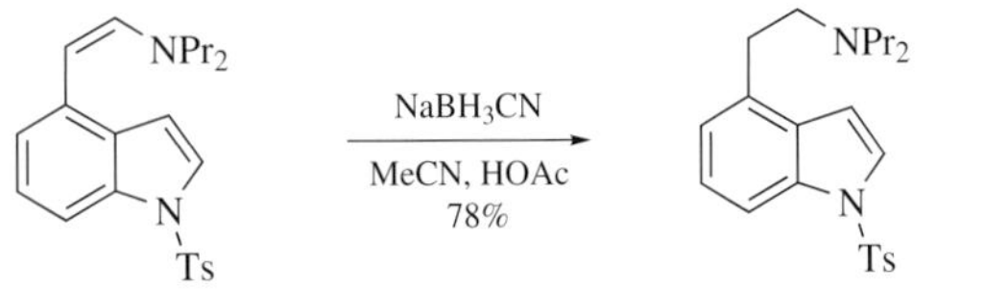

(15)

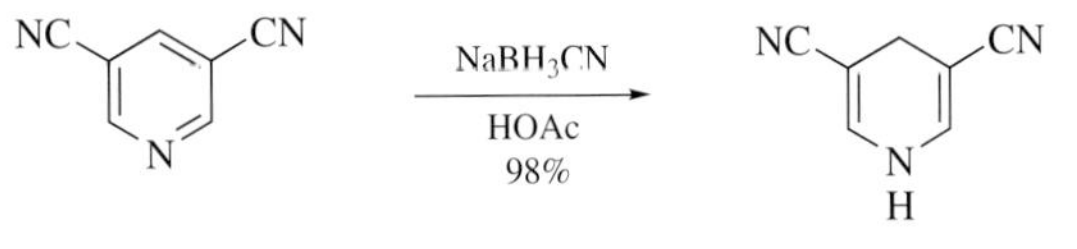

(16)

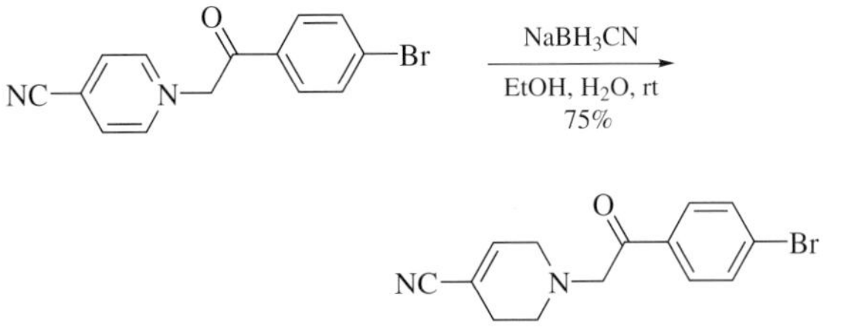

(17)

Reductive Deoxygenation of Aldehydes and Ketones.[23,24]

p-Toluenesulfonylhydrazones (tosylhydrazones), generated in situ from unhindered aliphatic aldehydes and ketones and tosylhydrazine, are reduced by NaBH$_3$CN in slightly acidic DMF/sulfolane (ca. 100–110 °C) to hydrocarbons via diazene intermediates (eq 18).[24] With hindered examples the tosylhydrazones must be preformed and large excesses (5–10×) of NaBH$_3$CN used in more acidic media (e.g. pH < 4) (eq 19).[24,25] Likewise, aryl tosylhydrazones are nearly inert to the reagent, but exceptions are known.[26] Reduction of tosylhydrazones to hydrocarbons also occurs with NaBH$_3$CN/ZnCl$_2$ in refluxing MeOH. This combination also gives poor yields with aryl systems.[10b]

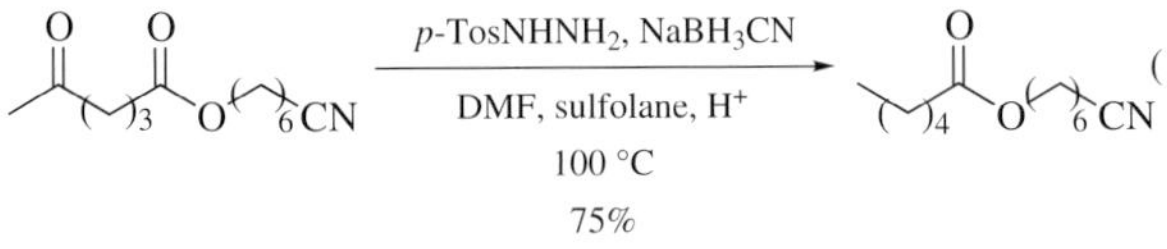

(18)

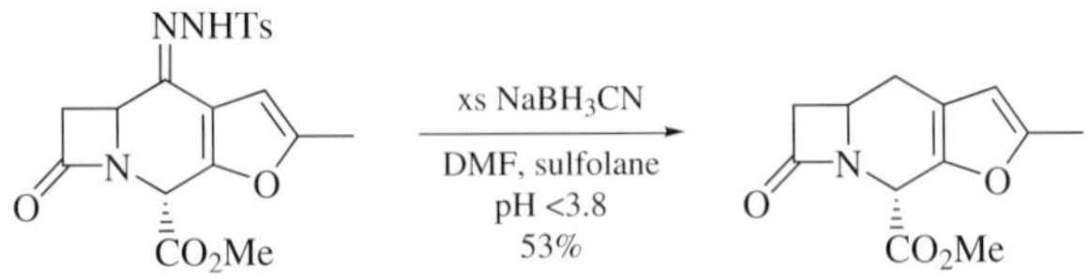

(19)

Reductive deoxygenation of most α,β-unsaturated tosylhydrazones with NaBH$_3$CN cleanly affords alkenes in which the double bond migrates to the former tosylhydrazone carbon (eq 20).[24,27,28] However, the process gives mixtures of alkenes and alkanes with cyclohexenones[27] (see also ***Bis(triphenylphosphine)copper(I) Borohydride***, ***Catecholborane*** and ***Sodium Borohydride***).

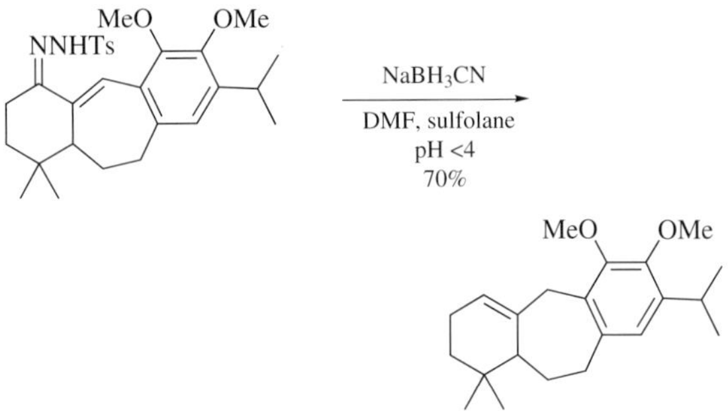

(20)

Reduction of Other π-Bonded Functional Groups.

In acidic media (i.e. pH < 4), aldehydes and ketones are selectively reduced to alcohols (eq 21).[2a,28] α,β-Unsaturated ketones are reduced primarily to allylic alcohols (eq 22)[29] except cyclohexenones, which give mixtures of allylic and saturated alcohols. Allylic ethers are also produced concomitantly with substrates further conjugated with aryl rings.[29]

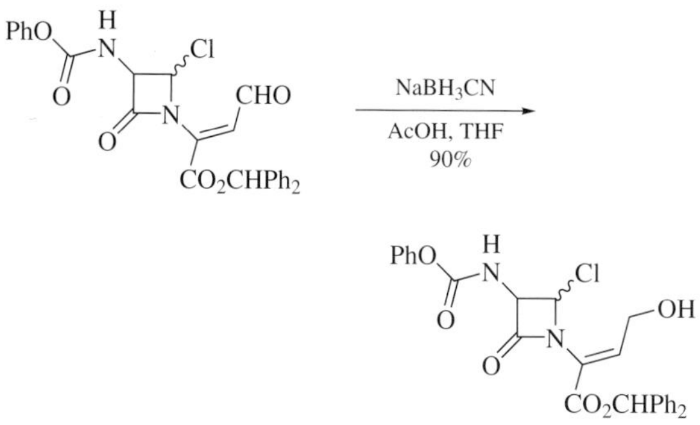

(21)

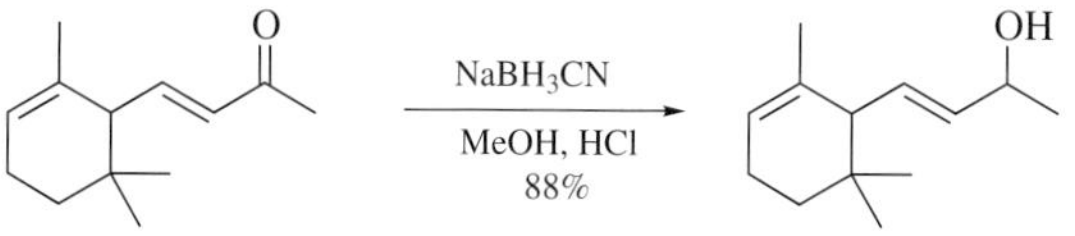

(22)

The combination of NaBH₃CN/ZnCl₂ in ether also reduces aldehydes and ketones to alcohols.[10b] With 5% H₂O present, cyclohexanones are selectively reduced in the presence of aliphatic derivatives.[30a] With NaBH₃CN/**Zinc Iodide**, however, aryl aldehydes and ketones are converted to aryl alkanes.[30b]

While isolated alkenes are inert toward NaBH₃CN, highly polarized double bonds (i.e. containing an attached nitro or two other electron-withdrawing groups) are reduced to hydrocarbons in acidic EtOH (eq 23).[31,32] Reductions of iron carbonyl–alkene complexes to the corresponding alkyl complexes also occurs readily with NaBH₃CN in MeCN.[33]

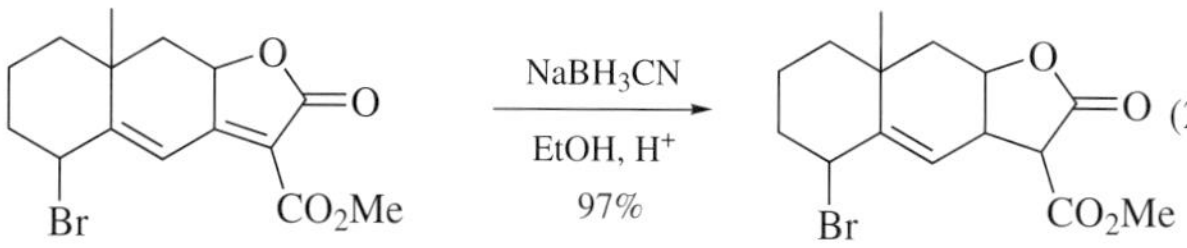

(23)

Nitriles are inert toward NaBH₃CN even under strongly acidic conditions. However, methylation with Me₂Br⁺SbF₆⁻ and subsequent reduction with NaBH₃CN affords the corresponding methylamine (eq 24).[34]

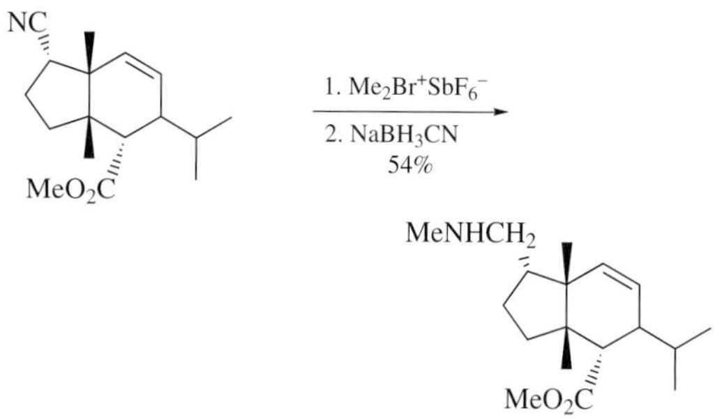

(24)

order of reactivity is I > Br, RSO₃ > Cl.[35a] In addition, primary alcohols are reduced to hydrocarbons via in situ conversion to iodides with **Methyltriphenoxyphosphonium Iodide** (MTPI) and subsequent reduction (eq 27).[35a,37] On the other hand, the combinations NaBH₃CN/**Tin(II) Chloride**[38a] or NaBH₃CN/ZnCl₂[38b] reduce tertiary, allylic, and benzylic halides but are inert toward primary, secondary, and aryl derivatives (eq 28)[38] (see also **Lithium Aluminum Hydride**, **Lithium Tri-sec-butylborohydride** and **Sodium Borohydride**).

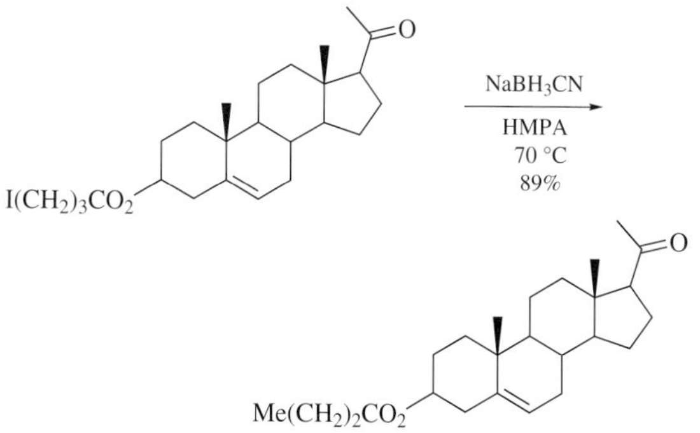

(25)

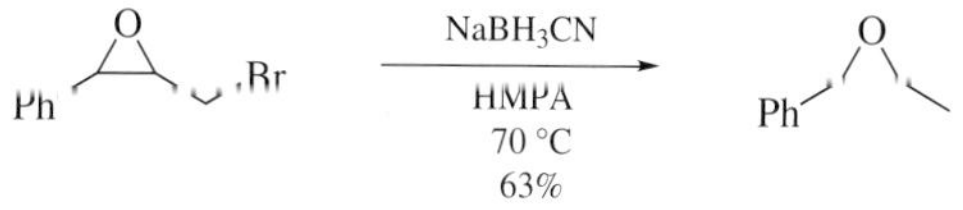

(26)

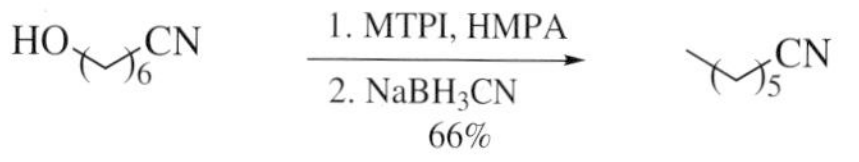

(27)

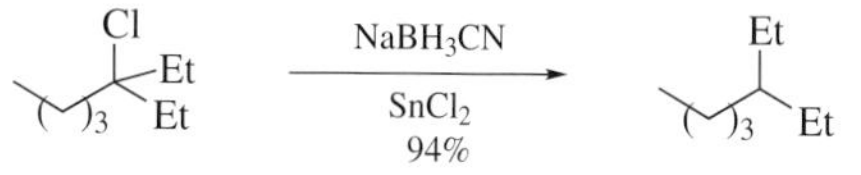

(28)

In the presence of **Boron Trifluoride Etherate**, NaBH₃CN reduces epoxides to alcohols[39] with attack of hydride at the site best able to accommodate a carbocation. Epoxide opening occurs primarily *anti* (eq 29).[39] Acetals are also reduced to ethers by NaBH₃CN in acetic media (eq 30).[40]

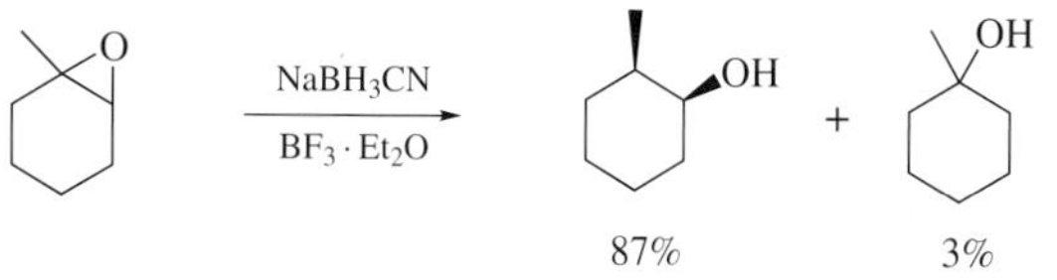

(29)

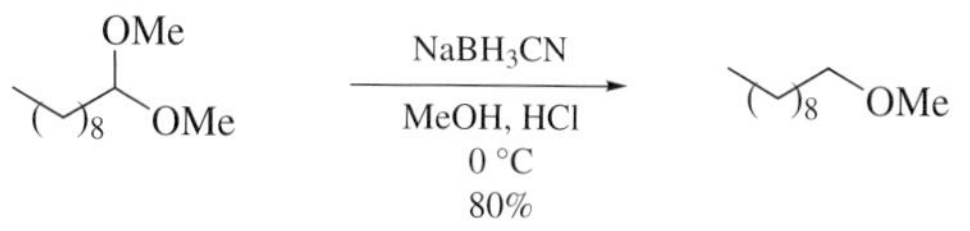

(30)

Reduction of σ-Bonded Functional Groups. In S$_N$2 rate enhancing polar aprotic solvents (e.g. HMPA, DMSO), primary and secondary alkyl, benzylic and allylic halides, sulfonate esters (eqs 25 and 26),[35a] and quaternary ammonium salts[36] are reduced to hydrocarbons. As expected for an S$_N$2-type process, the

Allylic groups that are normally not displaced by hydrides (e.g. carboxylates, ethers) are effectively activated via Pd⁰ complexation to give π-allyl complexes which are reduced by NaBH₃CN to alkenes (eq 31).[41]

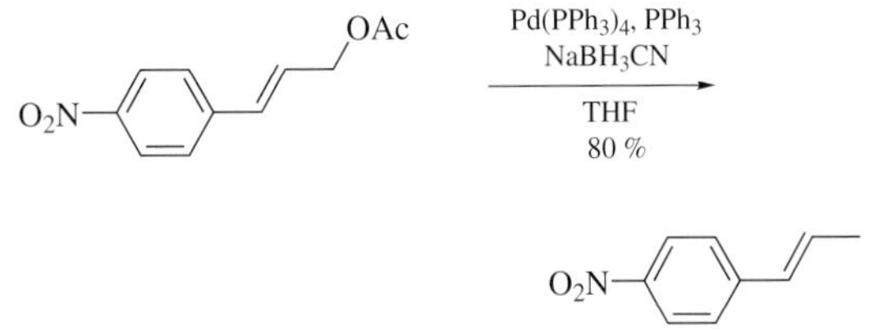

$$O_2N-\text{(structure)}-OAc \xrightarrow[\substack{\text{THF} \\ 80\%}]{\substack{Pd(PPh_3)_4, PPh_3 \\ NaBH_3CN}}$$

$$O_2N-\text{(structure)} \quad (31)$$

First Update

Matthew L. Crawley

Wyeth Research, Collegeville, PA, USA

Expanded Utility of Sodium Cyanoborohydride. Sodium cyanoborohydride is a highly chemoselective reducing agent with reactivity that can be controlled and directed by subtle changes in reaction conditions. Its primary uses have historically been reductive aminations,[1–10] reduction of imines and derivatives,[11–22] reductive deoxygenation,[23–27] and reductive cleavage of σ-bonded functional groups.[35–41] While these applications continue to be reported, the use of NaBH$_3$CN has grown to include an expanded scope of reducible σ-bonded functional groups, protocols for direct deoxygenation of aldehydes, ketones, and esters under mild conditions, participation in radical cyclization reactions, and polymer supported applications. Additionally, in recent years, utilization of sodium cyanoborohydride in reductive aminations and imine reductions has focused on stereoselective methods, work with compounds containing highly sensitive functional groups, and testing of biological hypotheses in enzymatic systems.

Reductive Aminations. The selective reactivity profile of sodium cyanoborohydride for iminium ions over aldehydes continues to drive the use of this reagent for reductive amination transformations in the synthesis of drug targets (eq 32).[42] As shown in this reaction, reductive aminations with NaBH$_3$CN are sensitive to substitution on the amine and compatible with an array of functional groups.

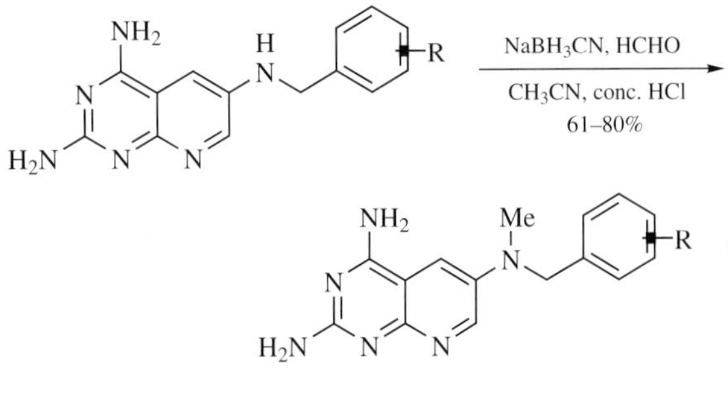

$$(32)$$

Synthesis involving sugar chemistry continues to take advantage of the mildness and selectivity of sodium cyanoborohydride as well, demonstrated in the synthesis of glucuronides (eq 33).[43]

A large portion of recent articles featuring NaBH$_3$CN have reported stereoselective chemistry, such as in the report of diastereoselective reductive amination of 3-hydroxy ketones (eq 34).[44] In some cases, essentially complete control of diastereoselectivity could be obtained. In another example, where a galac-

tose sugar was used as a chiral auxiliary, the synthesis of a chiral pyrrolidine was achieved (eq 35).[45]

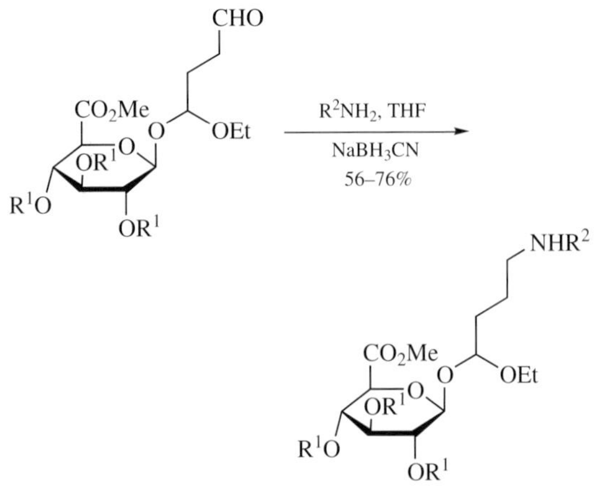

$$(33)$$

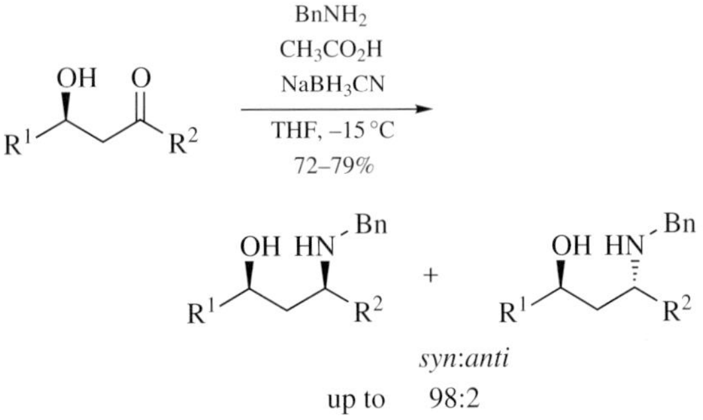

syn:anti
up to   98:2

$$(34)$$

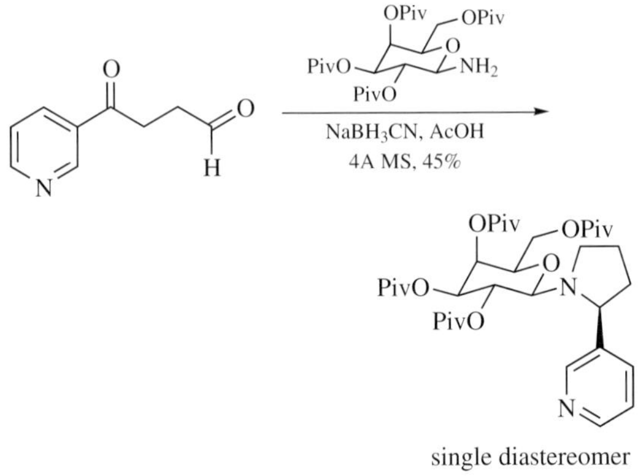

$$(35)$$

single diastereomer

Highly conjugated systems, such as a hydrophilic porphyrin, can be appended to a β-cyclodextrin conjugate system utilizing excess NaBH$_3$CN as the reducing agent (eq 36).[46]

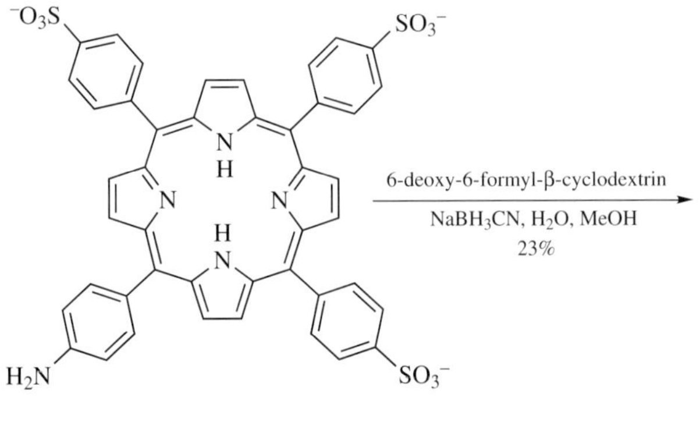

porphyrin–NH–CH$_2$–cyclodextrin   (36)

Applications with sodium cyanoborohydride now include the synthesis of large molecular entities used in biological testing. Conjugation of lactose to human serum albumin by reductive amination (eq 37)[47] and carbohydrate conjugation with PEG amines[48] have been added to the array of transformations reported for $NaBH_3CN$.

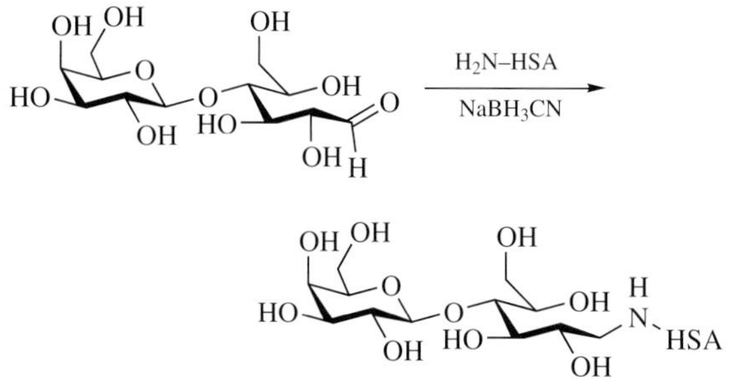

(37)

Sodium cyanoborohydride is selective and stable under physiological conditions of biological assays and its use has been reported in work including enzyme mechanism studies[49] and in biology involving substrate deactivation of synthases.[50]

Polymer-supported sodium cyanoborohydride (PSCBH),[51] now available commercially, has become more widely used. It is stable and effective even in the presence of oxidants, such as manganese dioxide, shown in this one pot oxidation–reduction amination protocol (eq 38).[52]

$$R^1CH_2OH \xrightarrow[\text{(b) AcOH, 41–80\%}]{\substack{\text{(a) } MnO_2, R^2NH_2, \text{PSCBH} \\ \text{DCM, 4A MS, reflux}}} R^1CH_2NHR^2 \quad (38)$$

Reduction of Imines and Derivatives. Imines are often reduced with $NaBH_3CN$ using Bronsted acids to form the more reactive iminium ion.[11,12] However, since the first report of utilizing zinc chloride with sodium borohydride,[10b] the use of Lewis acids has become common. A hydroamination/reduction series, where the imine was preformed followed by the addition of $NaBH_3CN$, highlighted this combination of reagents (eq 39).[53]

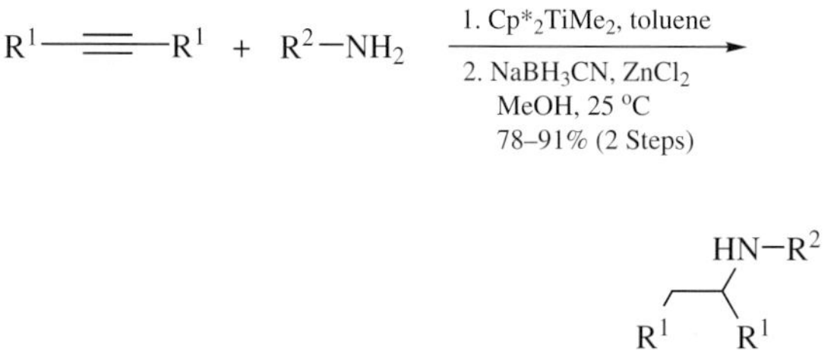

(39)

Sodium cyanoborohydride has been utilized in C_{60} chemistry, illustrated in the dual reduction of an imine and cyclopropane ring appended to a [60]fullerene (eq 40).[54]

$NaBH_3CN$ reductions often proceed with complete control of diastereoselectivity if the adjacent chiral groups are sufficiently large (eq 41)[55] or coordinating (eq 42).[56] In both reactions high yield of a single diastereomer was obtained as product.

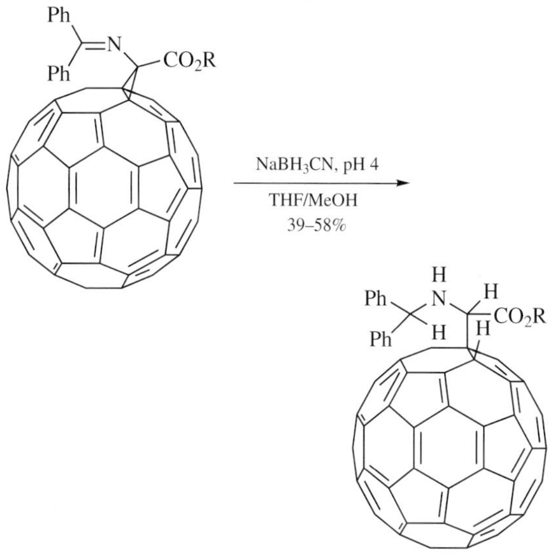

(40)

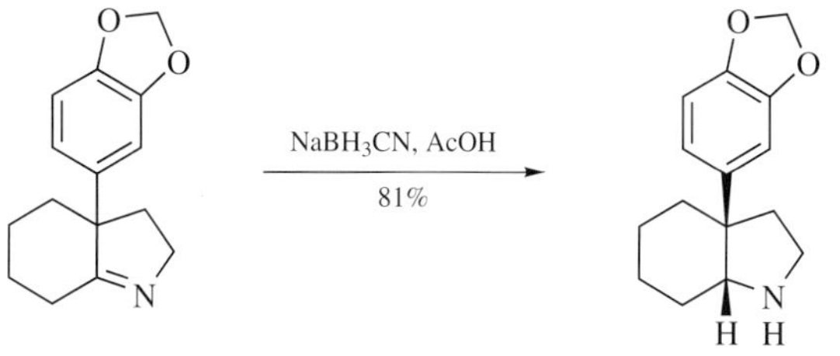

(41)

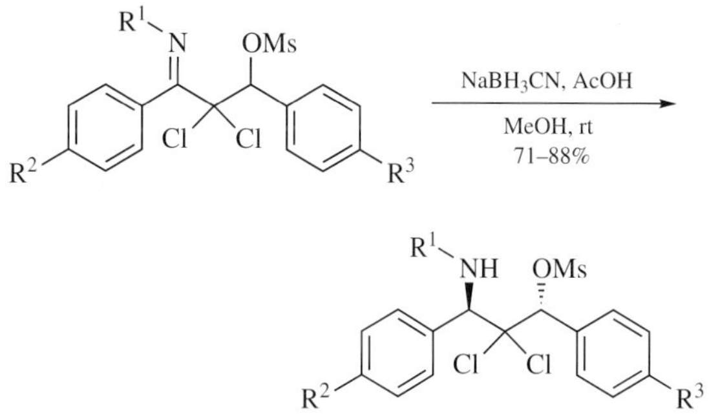

(42)

Nitrogen containing heterocycles can be reduced by $NaBH_3CN$ under acidic conditions.[1e] Two recent examples involved the stereoselective reduction of pyrazines to piperazines (eq 43)[57] and pyrazinones to piperazinones (eq 44).[58] The effectiveness of the ongoing use of sodium cyanoborohydride in reduction of cyclic imine derivatives was demonstrated by these reactions. In both cases only a single diastereomer of product was detected and isolated.

Reduction of Enamines. The reduction of enamines with $NaBH_3CN$ under acidic conditions proceeds via an iminium ion, similar to that seen in the reduction of imines.[2] Examples, such as the reduction of an enamine-fulvene to the corresponding amine target (eq 45)[59] and the reduction of the enamino-butenolide to the expected lactone (eq 46),[60] are abundant in the current literature.

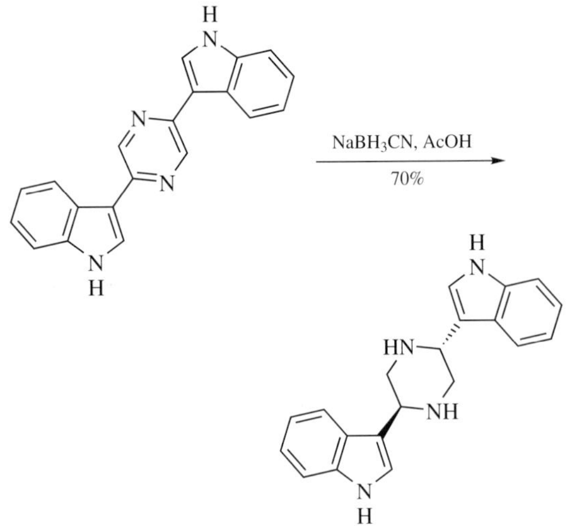

(43)

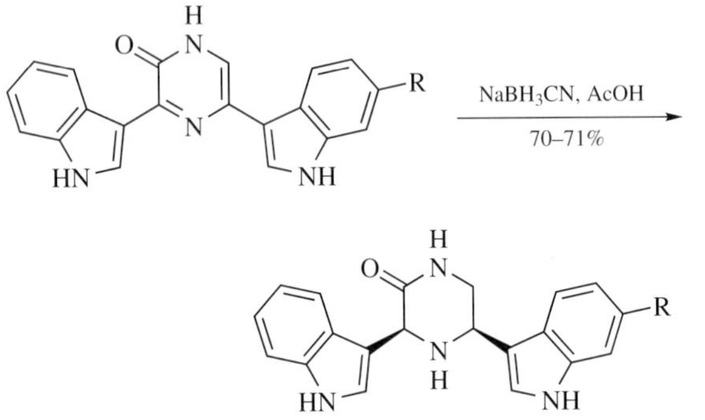

(44)

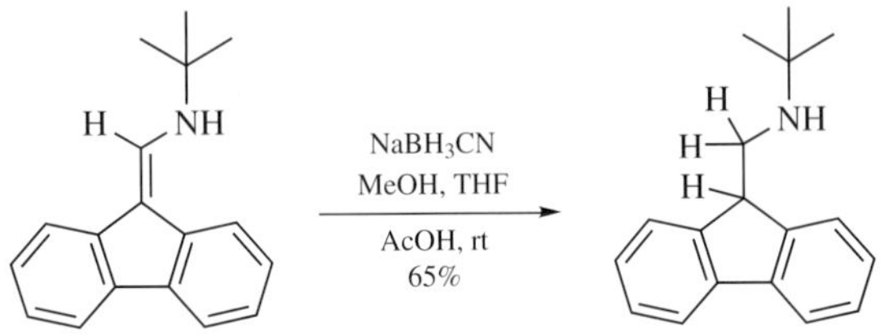

(45)

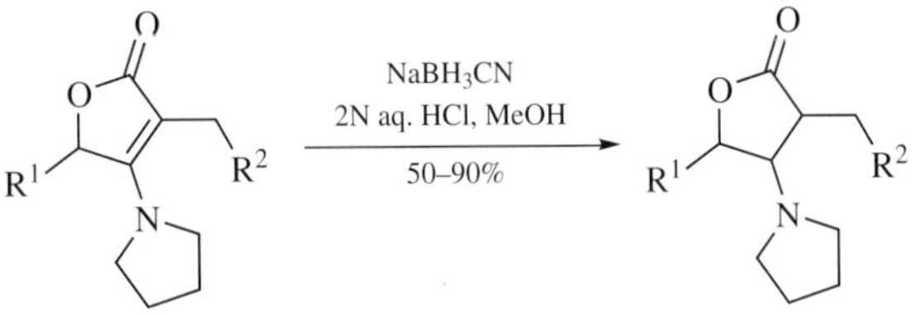

(46)

As with reductive aminations and imine reductions that employ sodium cyanoborohydride, a significant number of enamine reductions now are reported to afford products stereoselectively. Two representative examples include a reduction and subsequent transformation of a bicyclic enamine (eq 47)[61] and the reduction of a highly functionalized enamine (eq 48).[62] In both instances the reduction proceeded with high (>9:1 ratio) diastereoselectivity.

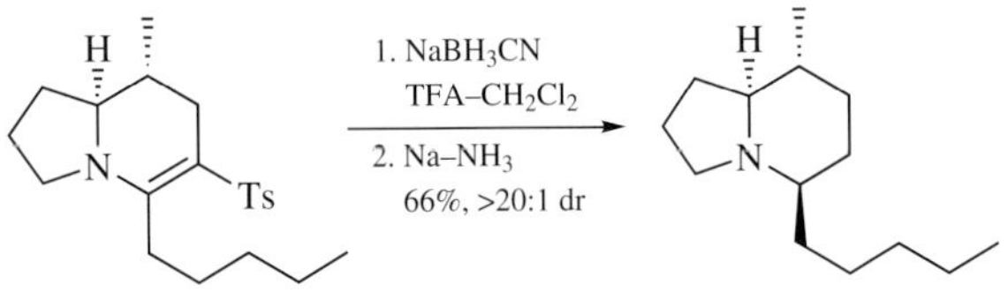

(47)

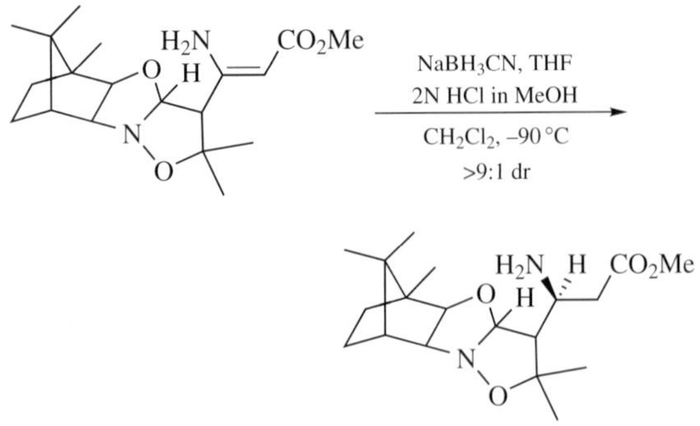

(48)

Reductive Deoxygenation of Aldehydes and Ketones. The most common procedure to deoxygenate an aldehyde or ketone with sodium cyanoborohydride has involved in situ generation and reduction of a tosylhydrazone under acidic conditions.[24] While the targets for deoxygenation have not significantly changed, several direct deoxygenation protocols have emerged, including the combination of NaBH$_3$CN/BF$_3$-OEt$_2$ (eqs 49 and 50)[63,64] and the combination of NaBH$_3$CN/TMSCl (eq 51).[65] These reactions, particularly when TMSCl is used, are highly selective over what are typically viewed as Lewis acid sensitive functional groups. While these deoxygenations are high yielding for aldehydes and ketones, little or no reduction is observed with esters or acids.

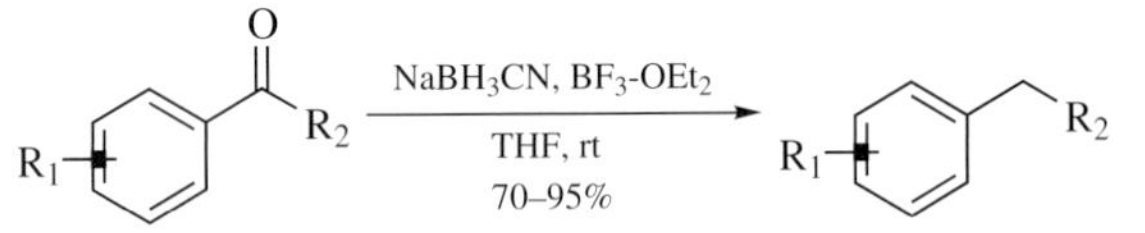

(49)

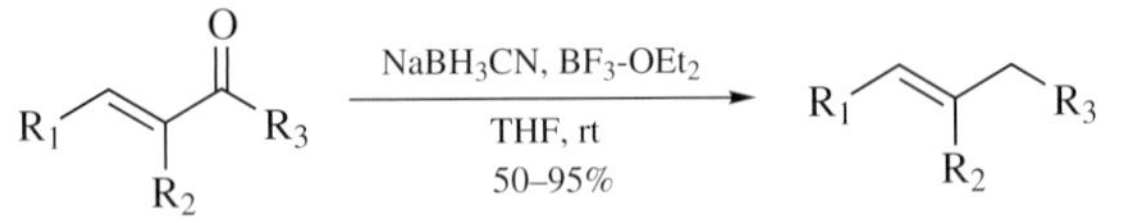

(50)

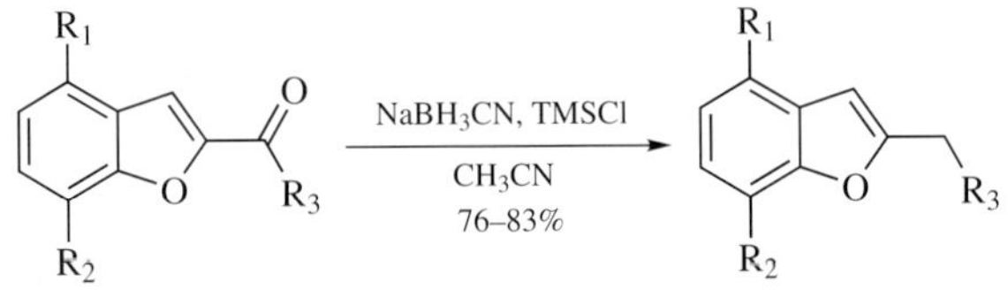

(51)

Lactones, previously not reported as substrates for direct or indirect deoxygenation reactions, can be deoxygenated by a NaBH$_3$CN/Bu$_3$SnCl/AIBN combination originally utilized for the purpose of a radical cyclization reaction (eq 52).[66] The authors reported all three reagents in combination are needed to observe the deoxygenation of the lactone carbonyl.

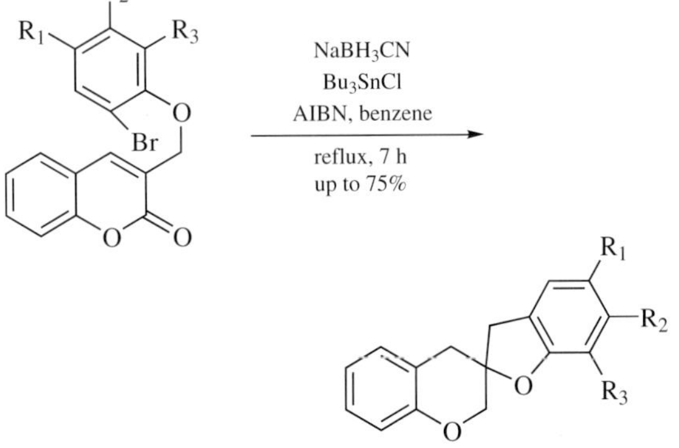

(52)

Radical Cyclization Reactions. Since the first report of a catalytic tin system used in combination with sodium cyanoborohydride for trapping radicals from cyclization reactions (eq 53),[67] its use has seen steady growth. Recent examples, including the one discussed previously where unexpected lactone reduction occurred (eq 52),[66] have focused on regioselective (eq 54)[68] and stereoselective (eq 55)[69] cyclizations. It has been postulated that the role of NaBH$_3$CN in these reactions is to maintain a low concentration of tin hydride and thus maintain the turnover of the catalytic cycle.

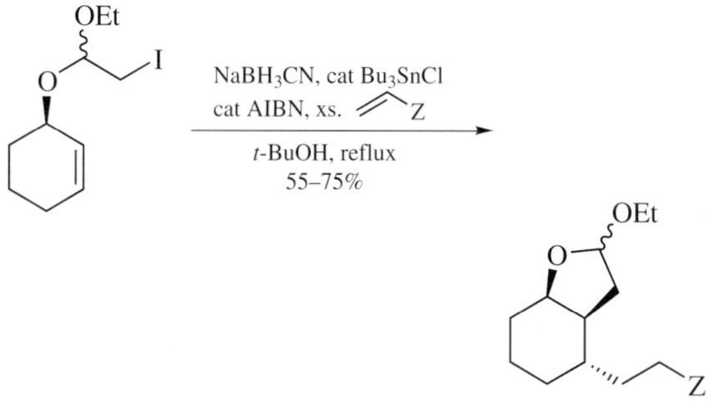

(53)

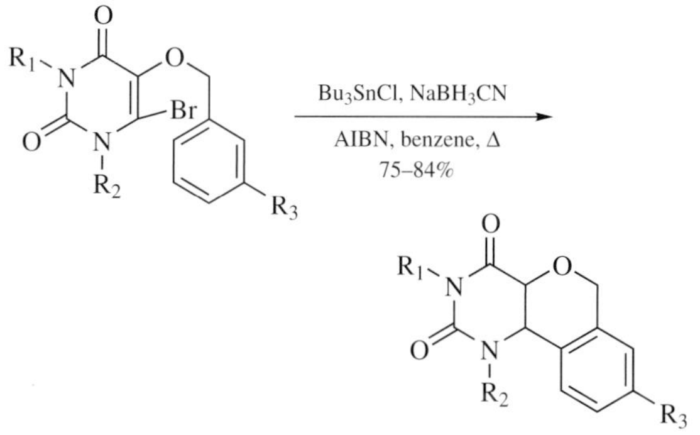

(54)

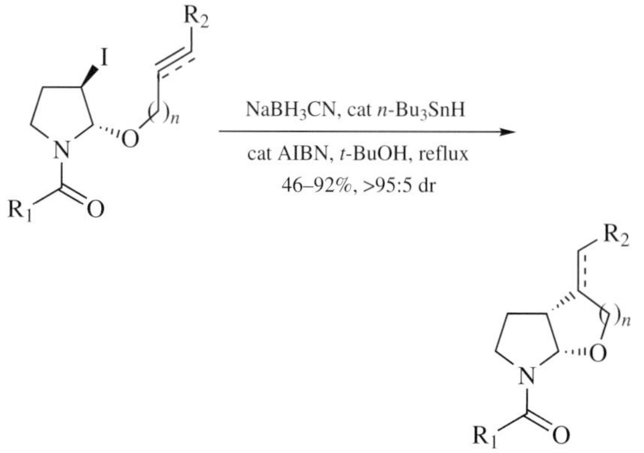

(55)

1,4-Reductions. Reduction of other π-bonded functional groups, including 1,4-type reductions of olefins activated by ketones and esters, continue to be reported with NaBH$_3$CN.[32] The new applications push the limit of sodium cyanoborohydride in terms of functional group compatibility and selectivity; a representative example of this type of process was reported in the stereoselective reduction of a highly functionalized natural product derivative (eq 56).[70]

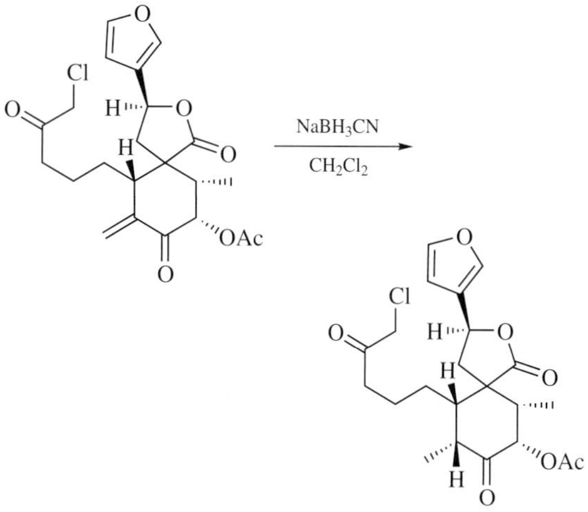

(56)

Reduction and Replacement of σ-Bonded Functional Groups. More than three decades have elapsed since the use of NaBH$_3$CN in combination with Me$_3$SnCl was reported for the selective catalytic dehalogenation of alkyl iodides (eq 57).[71] Since that time the application of sodium cyanoborohydride to reduce σ-bonded functional groups, including other halides, alcohols, and epoxides in an S$_N$2-type process has been widely reported.[35–41] The scope of functional groups that could be reduced by this method was recently expanded by the report of reductive decyanation of aminonitriles with a NaBH$_3$CN/Hg(O$_2$CCF$_3$)$_2$ combination (eq 58).[72]

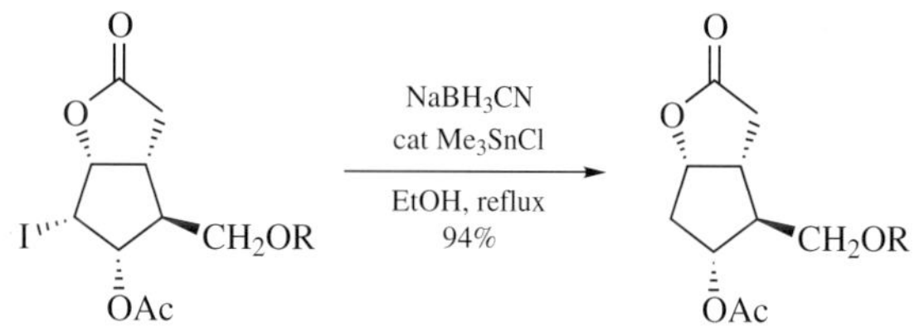

(57)

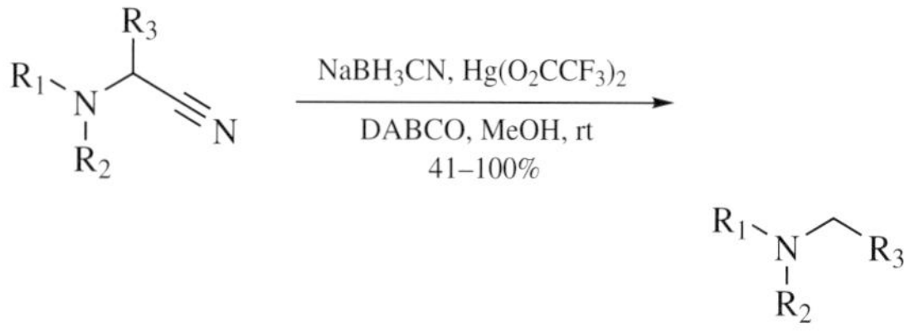

(58)

There have been a significant number of reports of the reductive removal of alcohols and ethers from hemiacetals (eq 59),[73] hemicyliminals (eq 60),[74] and acetals (eqs 61 and 62).[75,76] These reactions employ a variety of Lewis acids that are best suited for the activation of the σ-bonded leaving group. In many cases the processes are both high yielding and diastereoselective. Either the ether/amide products (eqs 59, 60 and 62) or the alcohol products (eq 61) can be isolated from these reactions.

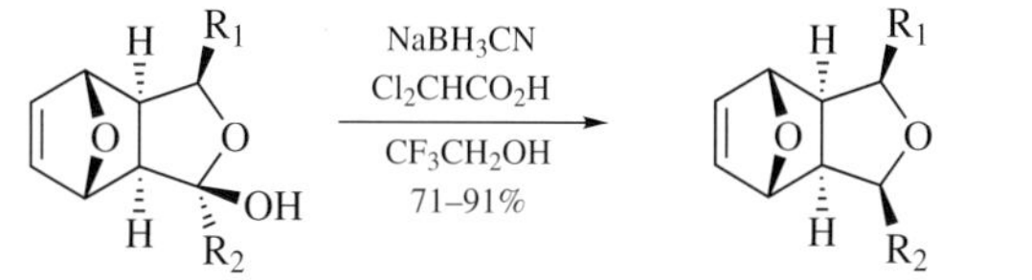

(59)

single diastereomer

(60)

(61)

(62)

Another example of acetal cleavage that utilized a mild Lewis acid is the cleavage of a chiral acetal to a differentially protected diol product (eq 63).[77] In this example the ether and the alcohol generated are tethered together and are thus both isolated. Additionally, the silyl chloride serves the dual role as the acid catalyst and a trapping group for the resulting alcohol. A similar tethered system gave high yield in the reductive opening of iminocyclitol derivatives catalyzed by acetic acid (eq 64).[78]

(63)

(64)

A new method to reduce 2,3-epoxy alcohols and vinylic epoxides in a regioselective and stereoselective manner was reported using a combination of NaBH$_3$CN and a zeolite (H-ZSM-5) catalyst (eq 65).[79] This represents a new and significantly more mild method of reductive epoxide opening than the combination of NaBH$_3$CN/BF$_3$-OEt$_2$[39] or NaBH$_3$CN/MeOH–HCl[40] as the active reducing agent.

(65)

1. (a) Hutchins, R. O.; Natale, N. R., *Org. Prep. Proced. Int.* **1979**, *11*, 201. (b) Lane, C. F., *Synthesis* **1975**, 135. (c) *Comprehensive Organic Synthesis* **1991**, *8*, Chapters 1.2, 1.14, 3.5, 4.1, 4.2. (d) Seyden-Penne, J. *Reductions by the Alumino- and Borohydrides in Organic Synthesis*; VCH: New York, 1991. (e) Gribble, G. W.; Nutaitis, C. F., *Org. Prep. Proced. Int.* **1985**, *17*, 317.

2. (a) Borch, R. F.; Bernstein, M. D.; Durst, H. D., *J. Am. Chem. Soc.* **1971**, *93*, 2897. (b) Hutchins, R. O.; Hutchins M. K., *Comprehensive Organic Synthesis* **1991**, *8*, 25.

3. Mori, K.; Sugai, T.; Maeda, Y.; Okazaki, T.; Noguchi, T.; Naito, H., *Tetrahedron* **1985**, *41*, 5307.

4. Umezawa, B.; Hoshino, O.; Sawaki, S.; Sashida, H.; Mori, K.; Hamada, Y.; Kotera, K.; Iitaka, Y., *Tetrahedron* **1984**, *40*, 1783.

5. (a) Abe, K.; Okumura, H.; Tsugoshi, T.; Nakamura, N., *Synthesis* **1984**, 597. (b) Abe, K.; Tsugoshi, T.; Nakamura, N., *Bull. Chem. Soc. Jpn.* **1984**, *57*, 3351.

6. Reitz, A. B.; Baxter, E. W., *Tetrahedron Lett.* **1990**, *31*, 6777.

7. Borch, R. F.; Hassid, A. I., *J. Org. Chem.* **1972**, *37*, 1673.

8. Jacobsen, E. J.; Levin, J.; Overman, L. E., *J. Am. Chem. Soc.* **1988**, *110*, 4329.

9. Barney, C. L.; Huber, E. W.; McCarthy, J. R., *Tetrahedron Lett.* **1990**, *31*, 5547.

10. (a) Mattson, R. J.; Pham, K. M.; Leuck, D. J.; Cowen, K. A., *J. Org. Chem.* **1990**, *55*, 2552. (b) Kim, S.; Oh, C. H.; Ko, J. S.; Ahn, K. H.; Kim, Y. J., *J. Org. Chem.* **1985**, *50*, 1927.

11. Hutchins, R. O.; Su, W.-Y.; Sivakumar, R.; Cistone, F.; Stercho, Y. P., *J. Org. Chem.* **1983**, *48*, 3412.

12. Orlemans, E. O.; Schreuder, A. H.; Conti, P. G. M.; Verboom, W.; Reinhoudt, D. N., *Tetrahedron* **1987**, *43*, 3817.

13. Van Parys, M.; Vandewalle, M., *Bull. Soc. Chim. Belg.* **1981**, *90*, 757.

14. Reonchet, J. M. J.; Zosimo-Landolfo, G.; Bizzozero, N.; Cabrini, D.; Habaschi, F.; Jean, E.; Geoffroy, M., *J. Carbohydr. Chem.* **1988**, *7*, 169.

15. Bergeron, R. J.; Pegram, J. J., *J. Org. Chem.* **1988**, *53*, 3131.

16. Sternbach, D. D.; Jamison, W. C. L., *Tetrahedron Lett.* **1981**, *22*, 3331.

17. Zinner, G.; Blass, H.; Kilwing, W.; Geister, B., *Arch. Pharm. (Weinheim, Ger.)* **1984**, *317*, 1024.

18. Branchaud, B. P., *J. Org. Chem.* **1983**, *48*, 3531.

19. Rosini, G.; Medici, A.; Soverini, M., *Synthesis* **1979**, 789.

20. Cannon, J. G.; Lee, T.; Ilhan, M.; Koons, J.; Long, J. P., *J. Med. Chem.* **1984**, *27*, 386.

21. Booker, E.; Eisner, U., *J. Chem. Soc., Perkin Trans. 1* **1975**, 929.

22. Hutchins, R. O.; Natale, N. R., *Synthesis* **1979**, 281.

23. Hutchins, R. O.; Hutchins, M. K., *Comprehensive Organic Synthesis* **1991**, *8*, 327.

24. Hutchins, R. O.; Milewski, C. A.; Maryanoff, B. E., *J. Am. Chem. Soc.* **1973**, *95*, 3662.

25. Sato, A.; Hirata, T.; Nakamizo, N., *Agric. Biol. Chem.* **1983**, *47*, 799.

26. Schultz, A. G.; Lucci, R. D.; Fu, W. Y.; Berger, M. H.; Erhardt, J.; Hagmann, W. K., *J. Am. Chem. Soc.* **1978**, *100*, 2150.

27. Hutchins, R. O.; Kacher, M.; Rua, L., *J. Org. Chem.* **1975**, *40*, 923.

28. Koft, E. R., *Tetrahedron* **1987**, *43*, 5775.

29. Hutchins, R. O.; Kandasamy, D., *J. Org. Chem.* **1975**, *40*, 2530.

30. (a) Kim, S.; Kim, Y. J.; Oh, C. H.; Ahn, K. H., *Bull. Korean Chem. Soc.* **1984**, *5*, 202. (b) Lau, C. K.; Dufresne, C.; Bélanger, P. C.; Piétré, S.; Scheigetz, J., *J. Org. Chem.* **1986**, *51*, 3038.

31. Hutchins, R. O.; Rotstein, D.; Natale, N.; Fanelli, J.; Dimmel, D., *J. Org. Chem.* **1976**, *41*, 3328.

32. Schultz, A. G.; Godfrey, J. D.; Arnold, E. V.; Clardy, J., *J. Am. Chem. Soc.* **1979**, *101*, 1276.

33. (a) Florio, S. M.; Nicholas, K. M., *J. Organomet. Chem.* **1978**, *144*, 321. (b) Whitesides, T. H.; Neilan, J. P., *J. Am. Chem. Soc.* **1976**, *98*, 63.

34. (a) Borch, R. F.; Evans, A. J.; Wade, J. J., *J. Am. Chem. Soc.* **1975**, *97*, 6282. (b) Borch, R. F.; Evans, A. J.; Wade, J. J., *J. Am. Chem. Soc.* **1977**, *99*, 1612.

35. (a) Hutchins, R. O.; Kandasamy, D.; Maryanoff, C. A.; Masilamani, D.; Maryanoff, B. E., *J. Org. Chem.* **1977**, *42*, 82. (b) Hutchins, R. O.; Milewski, C. A.; Maryanoff, B. E., *Org. Synth., Coll. Vol.* **1988**, *6*, 376.

36. Yamada, K.; Itoh, N.; Iwakuma, T., *J. Chem. Soc., Chem. Commun.* **1978**, 1089.

37. (a) Okada, K.; Kelley, J. A.; Driscoll, J. S., *J. Org. Chem.* **1977**, *42*, 2594. (b) Borchers, F.; Levsen, K.; Schwarz, H.; Wesdemiotis, C.; Winkler, H. U., *J. Am. Chem. Soc.* **1977**, *99*, 6359.

38. (a) Kim, S.; Ko, J. S., *Synth. Commun.* **1985**, *15*, 603. (b) Kim, S.; Kim, Y. J.; Ahn, K. H., *Tetrahedron Lett.* **1983**, *24*, 3369.

39. Hutchins, R. O.; Taffer, I. M.; Burgoyne, W., *J. Org. Chem.* **1981**, *46*, 5214.

40. Horne, D. A.; Jordan, A., *Tetrahedron Lett.* **1978**, 1357.

41. (a) Hutchins, R. O.; Learn, K.; Fulton, R. P., *Tetrahedron Lett.* **1980**, *21*, 27. (b) Hutchins, R. O.; Learn, K., *J. Org. Chem.* **1982**, *47*, 4380.

42. Gangjee, A.; Vasudevan, A.; Queener, S. F.; Kisliuk, R. L., *J. Med. Chem.* **1996**, *39*, 1438.

43. Bakina, E.; Wu, Z.; Rosenblum, M.; Farquhar, D., *J. Med. Chem.* **1997**, *40*, 4013.

44. Haddad, M.; Dorbais, J.; Larcheveque, M., *Tetrahedron Lett.* **1997**, *38*, 5981.

45. Loh, T. P.; Zhou, J. R.; Li, X. R.; Sim, K. Y., *Tetrahedron Lett.* **1999**, *40*, 7847.

46. Carofiglio, T.; Fornasier, R.; Lucchini, V.; Simonato, L.; Tonellato, U., *J. Org. Chem.* **2000**, *65*, 9013.

47. Jeong, J. M.; Hong, M. K.; Lee, J.; Son, M.; So, Y.; Lee, D. S.; Chung, J. K.; Lee, M. C., *Bioconjugate Chem.* **2004**, *15*, 850.

48. Larson, R. S.; Menard, V.; Jacobs, H.; Kim, S. W., *Bioconjugate Chem.* **2001**, *12*, 861.

49. Rasche, M. E.; White, R. H., *Biochemistry* **1998**, *37*, 11343.

50. Parker, E. J.; Bulloch, E. M. M.; Jameson, G. B.; Abell, C., *Biochemistry* **2001**, *40*, 14821.

51. Hutchins, R. O.; Natale, N. R.; Taffer, I. M., *Chem. Commun.* **1978**, 1088.

52. Blackburn, L.; Taylor, R. J. K., *Org. Lett.* **2001**, *3*, 1637.

53. Heutling, A.; Doye, S., *J. Org. Chem.* **2002**, *67*, 1961.

54. Burley, G. A.; Keller, P. A.; Pyne, S. G.; Ball, G. E., *J. Org. Chem.* **2002**, *67*, 8316.

55. Schkeryantz, J. M.; Pearson, W. H., *Tetrahedron* **1996**, *52*, 3107.

56. Aelterman, W.; De Kimpe, N., *J. Org. Chem.* **1998**, *63*, 6.

57. Miyake, F. Y.; Yakushijin, K.; Horne, D. A., *Org. Lett.* **2000**, *2*, 3185.

58. Miyake, F. Y.; Yakushijin, K.; Horne, D. A., *Org. Lett.* **2002**, *4*, 941.

59. Koenemann, M.; Erker, G.; Froelich, R.; Wuerthwein, E. U., *J. Am. Chem. Soc.* **1997**, *119*, 11155.

60. Pashkovsky, F. S.; Katok, Y. M.; Khlebnicova, T. S.; Lakhvich, F. A., *Tetrahedron Lett.* **2001**, *42*, 3657.

61. Back, T. G.; Nakajima, K., *J. Org. Chem.* **2000**, *65*, 4543.

62. Mauduit, M.; Kouklovsky, C.; Langlois, Y.; Riche, C., *Org. Lett.* **2000**, *2*, 1053.

63. Srikrishna, A.; Sattigeri, J. A.; Viswajanani, R.; Yelamaggad, C. V., *Synlett* **1995**, 93.

64. Srikrishna, A.; Viswajanani, R.; Sattigeri, J. A.; Yelamaggad, C. V., *Tetrahedron Lett.* **1995**, *36*, 2347.

65. Box, V. G. S.; Meleties, P. C., *Tetrahedron Lett.* **1998**, *39*, 7059.

66. Majumdar, K. C.; Chattopadhyay, S. K., *Tetrahedron Lett.* **2004**, *45*, 6871.

67. Stork, G.; Sher, P. M., *J. Am. Chem. Soc.* **1986**, *108*, 303.

68. Majumdar, K. C.; Basu, P. K.; Mukhopadhyay, P. P.; Sarkar, S.; Ghosh, S. K.; Biswas, P., *Tetrahedron* **2003**, *59*, 2151.

69. Matos, M. R. P. N.; Afonso, C. A. M.; McGarvey, T.; Lee, P.; Batey, R. A., *Tetrahedron Lett.* **1999**, *40*, 9189.

70. Moessner, E.; de la Torre, M. C.; Rodriguez, B., *J. Nat. Prod.* **1996**, *59*, 367.

71. Corey, E. J.; Suggs, J. W., *J. Org. Chem.* **1975**, *40*, 2554.

72. Sassaman, M. B., *Tetrahedron* **1996**, *52*, 10835.

73. Shi, H.; Liu, H.; Bloch, R.; Mandville, G., *Tetrahedron* **2001**, *57*, 9335.

74. Wang, E. C.; Chen, H. F.; Feng, P. K.; Lin, Y. L.; Hsu, M. K., *Tetrahedron Lett.* **2002**, *43*, 9163.

75. Srikrishna, A.; Sattigeri, J. A.; Viswajanani, R.; Yelamaggad, C. V., *J. Org. Chem.* **1995**, *60*, 2260.

76. Srikrishna, A.; Viswajanani, R.; Yelamaggad, C. V., *Tetrahedron* **1997**, *53*, 10479.

77. Gustafsson, T.; Schou, M.; Almqvist, F.; Kihlberg, J., *J. Org. Chem.* **2004**, *69*, 8694.

78. Fuentes, J.; Gasch, C.; Olano, D.; Pradera, M. A.; Repetto, G.; Sayago, F. J., *Tetrahedron: Asymmetry* **2002**, *13*, 1743.

79. Gupta, A.; Vankar, Y. D., *Tetrahedron Lett.* **1999**, *40*, 1369.

Sodium Hexamethyldisilazide[1]

$$NaN(SiMe_3)_2$$

[1070-89-9] $C_6H_{18}NNaSi_2$ (MW 183.42)

InChI = 1/C6H18NSi2.Na/c1-8(2,3)7-9(4,5)6;/h1-6H3;/q-1;+1/
     rC6H18NNaSi2/c1-9(2,3)7(8)10(4,5)6/h1-6H3

InChIKey = WRIKHQLVHPKCJU-JSJAVMDOAQ

(useful as a sterically hindered base and as a nucleophile)

Alternate Names: NaHMDS; sodium bis(trimethylsilyl)amide.
Physical Data: mp 171–175 °C; bp 170 °C/2 mmHg.
Solubility: soluble in THF, ether, benzene, toluene.[1]
Form Supplied in: (a) off-white powder (95%); (b) solution in THF (1.0 M); (c) solution in toluene (0.6 M).
Analysis of Reagent Purity: THF solutions of the reagent may be titrated using 4-phenylbenzylidenebenzylamine as an indicator.[2]
Handling, Storage, and Precautions: the dry solid and solutions are flammable and must be stored in the absence of moisture. These should be handled and stored under a nitrogen atmosphere. Use in a fume hood.

Original Commentary

Brett T. Watson

Bristol-Myers Squibb Pharmaceutical Research Institute, Wallingford, CT, USA

Introduction. Sodium bis(trimethylsilyl)amide is a synthetically useful reagent in that it combines both high basicity[3] and nucleophilicity,[4] each of which may be exploited for useful organic transformations such as selective formation of enolates,[5] preparation of Wittig reagents,[6] formation of acyl anion equivalents,[7] and the generation of carbenoid species.[8] As a nucleophile, it has been used as a nitrogen source for the preparation of primary amines.[9,10]

Sterically Hindered Base for Enolate Formation. Like other metal dialkylamide bases, sodium bis(trimethylsilyl)amide is sufficiently basic to deprotonate carbonyl-activated carbon acids[5] and is sterically hindered, allowing good initial kinetic vs. thermodynamic deprotonation ratios.[11] The presence of the sodium counterion also allows for subsequent equilibration to the thermodynamically more stable enolate.[5f] More recently, this base has been used in the stereoselective generation of enolates for subsequent alkylation or oxidation in asymmetric syntheses.[12] As shown in eq 1, NaHMDS was used to selectively generate a (*Z*)-enolate; alkylation with **Iodomethane** proceeded with excellent diastereoselectivity.[12a] In this case, use of the sodium enolate was preferred as it was more reactive than the corresponding lithium enolate at lower temperatures.

The reagent has been used for the enolization of carbonyl compounds in a number of syntheses.[13] For ketones and aldehydes which do not have enolizable protons, NaHMDS may be used to prepare the corresponding TMS-imine.[14]

Generation of Ylides for Wittig Reactions. In the Wittig reaction, salt-free conditions have been shown to improve (*Z*):(*E*) ratios of the alkenes which are prepared.[15] NaHMDS has been shown to be a good base for generating ylides under lithium-salt-free conditions.[6] It has been used in a number of syntheses to selectively prepare (*Z*)-alkenes.[16] Ylides generated under these conditions have been shown to undergo other ylide reactions such as *C*-acylations of thiolesters and inter- and intramolecular cyclization.[6] Although Wittig-based syntheses of vinyl halides exist,[17] NaHMDS has been shown to be the base of choice for the generation of iodomethylenetriphenylphosphorane for the stereoselective synthesis of (*Z*)-1-iodoalkenes from aldehydes and ketones (eq 2).[18]

79%
99:1 diastereoselectivity (1)

96%   61%
(*Z*):(*E*) = 62:1   (2)

NaHMDS has been shown to be the necessary base for the generation of the ylide anion of sodium cyanotriphenylphosphoranylidenemethanide, which may be alkylated with various electrophiles and in turn used as an ylide to react with carbonyl compounds.[19] NaHMDS was used as the base of choice in a Horner–Emmons–Wadsworth-based synthesis of terminal conjugated enynes.[20]

Intramolecular Alkylation via Protected Cyanohydrins (Acyl Anion Equivalents). Although NaHMDS was not the base of choice for the generation of protected cyanohydrin acyl carbanion equivalents in the original references,[21] it has been shown to be an important reagent for intramolecular alkylation using this strategy (eqs 3 and 4).[7,22] The advantages of this reagent are (a) that it allows high yields of intramolecularly cyclized products with little intermolecular alkylation and (b) the carbanion produced in this manner acts only as a nucleophile without isomerization of double bonds α,β to the anion or other existing double bonds in the molecule. Small and medium rings as well as macrocycles[22a] have been reported using this methodology (eqs 3 and 4).

Essential Reactions for Organic Synthesis, First Edition. Edited by Philip L. Fuchs.
© 2016 John Wiley & Sons, Ltd. Published 2016 by John Wiley & Sons, Ltd.

(3)

(4)

Generation of Carbenoid Species. Metal bis(trimethylsilyl) amides may be used to effect α-eliminations.[23] It is proposed that these nucleophilic agents undergo a hydrogen–metal exchange reaction with polyhalomethanes to give stable carbenoid species.[23b] NaHMDS has been used to generate carbenoid species which have been used in a one-step synthesis of monobromocyclopropanes (eqs 5 and 6).[23c,d] NaHMDS has been shown to give better yields than the corresponding lithium or potassium amides in this reaction.

cis:trans = 1.5:1

(5)

(6)

A similar study which evaluated the use of NaHMDS versus **Butyllithium** for the generation of the active carbenoid species from 1,1-dichloroethane and subsequent reaction with alkenes, forming 1-chloro-1-methylcyclopropanes, suggested that the amide gave very similar results to those with *n*-butyllithium.[24]

In an initial report, the carbenoid species formed by the treatment of diiodomethane with NaHMDS was shown to react as a nucleophile, displacing primary halides and leading to a synthesis of 1,1-diiodoalkanes; this is formally a 1,1-diiodomethylene homologation (eq 7).[25] This methodology is limited in that electrophiles which contain functionality that allows facile E2 elimination (i.e. allyl) form a mixture of the desired 1,1-diiodo compound and the iododiene. In the case of **Allyl Bromide**, addition of 2 equiv of the sodium reagent allows isolation of the iododiene as the major product.

(*E*):(*Z*) = 40:60

(7)

Synthesis of Primary Amines. The nucleophilic properties of this reagent may be utilized in the S_N2 displacement of primary alkyl bromides, iodides, and tosylates to form bis(trimethylsilyl) amines (**1**) (eq 8).[9a] HCl hydrolysis of (**1**) allows isolation of the corresponding hydrochloride salt of the amine, which may be readily separated from the byproduct, bis(trimethylsilyl) ether. In one example a secondary allylic bromide also underwent the conversion with good yield.

$$RX + NaHMDS \longrightarrow R-N(TMS)_2 \longrightarrow RNH_3Cl + (TMS)_2O \quad (8)$$

R	X	(1)	(2)
Me	I	75%	99%
Et	OTs	77%	98%
Br	Br	73%	100%
	Br	66%	97%

Aminomethylation. NaHMDS may be used as the nitrogen source in a general method for the addition of an aminomethyl group (eq 9).[10] The reagent is allowed to react with chloromethyl methyl ether, forming the intermediate aminoether. Addition of Grignard reagents to this compound allows the displacement of the methoxy group, leaving the bis(trimethylsilyl)-protected amines. Acidic hydrolysis of these allows isolation of the hydrochloride salt of the corresponding amine in good yields.

(9)

R = Me, 89%; allyl, 76%; Cy, 75%; Ph, 78%; propargyl, 66%

First Update

Professor Hélène Lebel
Université de Montréal, Montréal, Quebec, Canada

Synthesis of Sulfinamides. The nucleophilic substitution of chiral enantiopure sulfinate esters with NaHMDS led to S-O bond breakage with inversion of configuration at the S atom, giving quantitative conversion to chiral sulfinamides (eq 10). Lithium amide in liquid ammonia at −78 °C could be also used, although in some cases such as in the synthesis of (*R*)-(+)-*p*-toluenesulfinamide higher ee's were observed when using NaHMDS.[26]

N-Alkylation of Aryl Amines. Various aryl amines and aminopyridines have been reacted with di-*t*-butyldicarbonate in the presence of 2 equiv of NaHMDS in THF to lead to the corresponding Boc-protected amines in high yields.[27] A series of anilines were treated with 1.5 to 2.0 equiv of NaHMDS and reacted with the appropriate esters or lactones to give the corresponding N-aryl amides in 88–99% yields (eq 11).[28] No diminution of the reaction rate was observed with sterically hindered esters and the method could be applied to the preparation of N-aryl amides containing functional groups that are incompatible with strong bases such as n-BuLi. The reaction with alkyl amines was not efficient as the mechanism involved the formation of the corresponding sodium amide, and NaHMDS is strong enough to deprotonate aryl amines, but not alkyl amines.

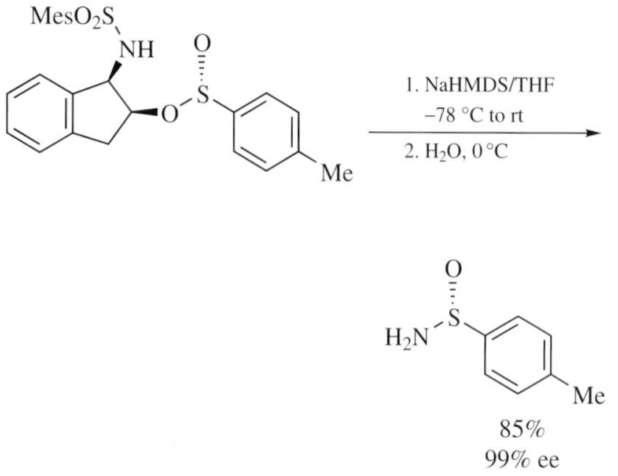

salts, α-quaternization is achieved with high enantioselectivities and moderate to excellent yields.[36]

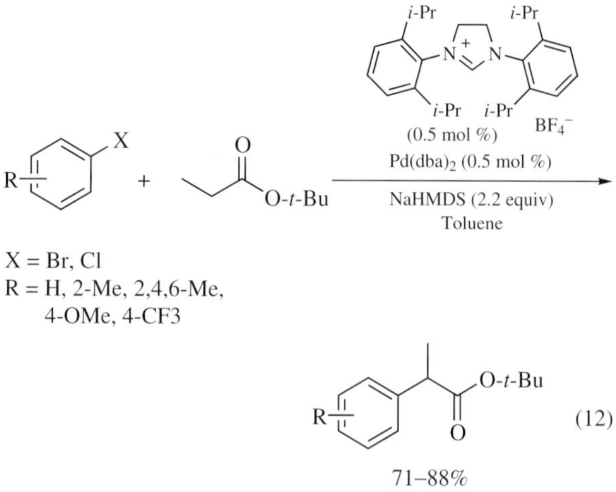

(10)

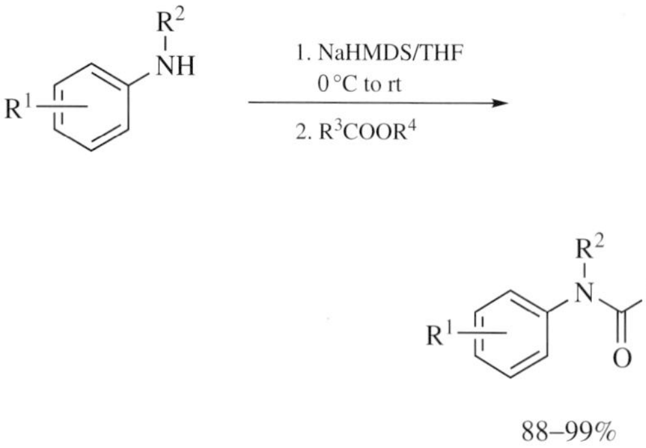

85%
99% ee

(11)

88–99%

R^1 = H, F, Br, I, OMe, CF$_3$, PhCO

R^2 = H, Me

R^3 = H, alkyl, aryl

R^4 = Me, Et, t-Bu

Palladium-catalyzed α-Arylation of Carbonyl Compounds and Nitriles. A variety of bases have been used in the palladium-catalyzed α-arylation of carbonyl derivatives.[29,30] The pKa of the carbonyl moiety determines the choice of the base. For instance, α-arylation with ester derivatives requires the use of strong bases such as NaHMDS.[31] The best yields for the arylation of t-butyl acetate were reported with LHMDS, whereas the arylation of t-butyl propionate occurred in higher yields in the presence of NaHMDS (eq 12).[32]

It was shown later that LiNCy$_2$ could also be very effective for this reaction.[33] More sensitive substrates such as α-imino esters,[34] malonates, or cyanoesters[35] required the use of a milder base, as decomposition was observed with NaHMDS.

However, NaHMDS was used with success in nickel–BINAP-catalyzed enantioselective α-arylation of α-substituted γ-butyro-lactones (eq 13). Coupled with an accelerating effect of Zn(II)

(12)

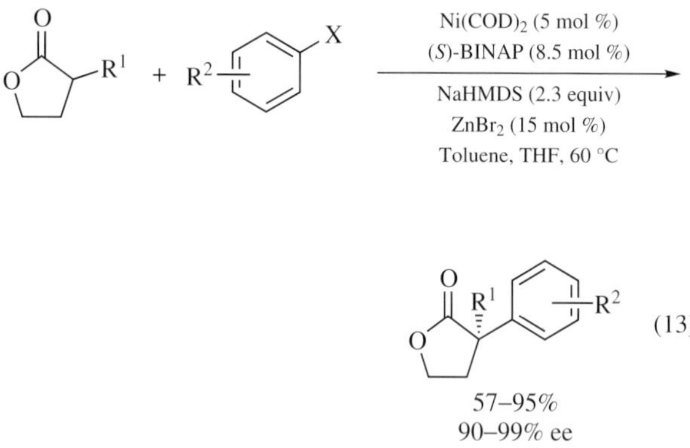

71–88%

(13)

57–95%
90–99% ee

R^1 = Me, Bn, Allyl, n-Pr

X = Br, Cl

R^2 = H, 3-NMe$_2$, 4-OTBS, 4-t-Bu, 3-OMe, 4-OMe, 4-CF$_3$, 2-napthyl, 3-CO$_2t$-Bu, 4-CO$_2t$-Bu

Sodium t-butoxide is usually used in palladium-catalyzed ketone arylations. However, this base was incompatible with aryl halides substituted with an electron-withdrawing group. Replacement by NaHMDS led to α-aryl ketones in good yields (eq 14).[37]

Finally, the α-arylation of nitrile derivatives was also reported using NaHMDS (eq 15).[38]

Transition Metal-catalyzed Allylic Alkylation. Simple ketone enolates were found to be suitable nucleophiles for palladium-catalyzed allylic alkylations.[39] The palladium-catalyzed asymmetric allylic alkylation of α-aryl ketones will take place in the presence of NaHMDS and allyl acetate to produce the desired α,α-disubstituted ketone derivative in high yields and ee (eq 16).[40]

Synthesis of 1-Bromoalkynes. The dehydrobromination of 1,1-dibromoalkenes leading to 1-bromoalkynes was usually performed with NaHMDS (eq 17).[41] In comparison, stronger bases such as butyllithium led to the corresponding acetylide anion via a metal-halogen process.[42]

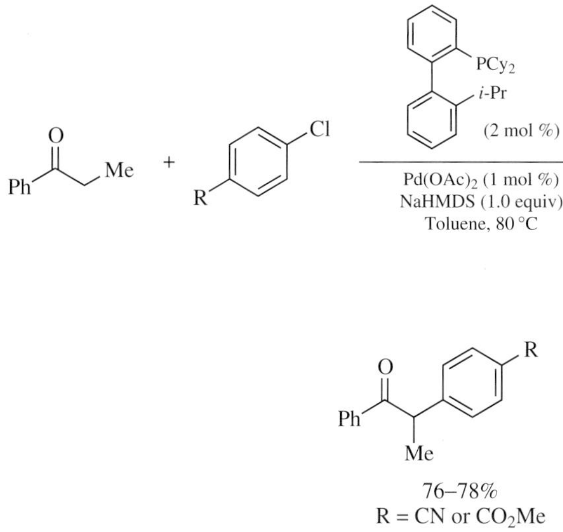

$$\text{(14)}$$

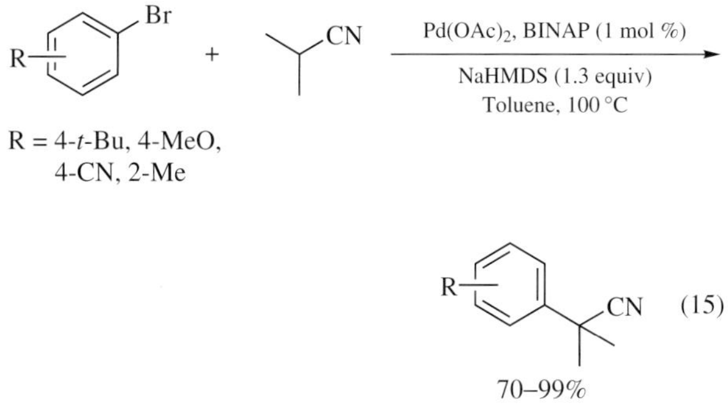

76–78%

R = CN or CO$_2$Me

R = 4-t-Bu, 4-MeO,
4-CN, 2-Me

$$\text{(15)}$$

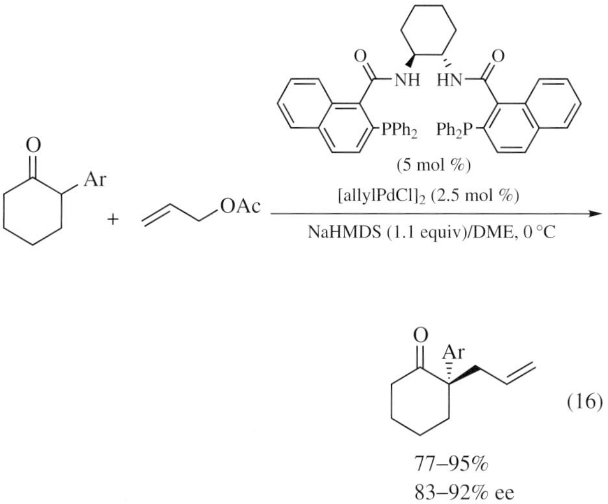

70–99%

$$\text{(16)}$$

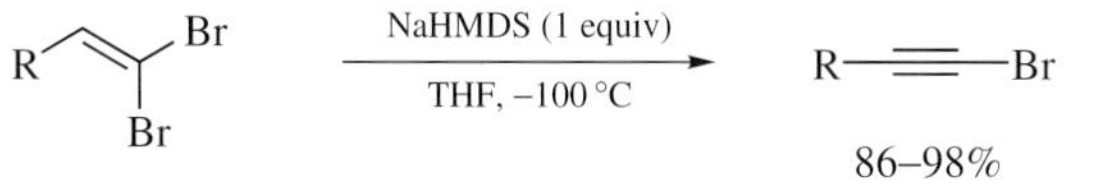

77–95%
83–92% ee

$$\text{(17)}$$

86–98%

Synthesis of Cyclopropene. When allyl chloride was dropped into a solution of NaHMDS in toluene at 110 °C, cyclopropene could be isolated in a trap/ampoule at −80 °C in 40% yield and >95% purity (eq 18).[43]

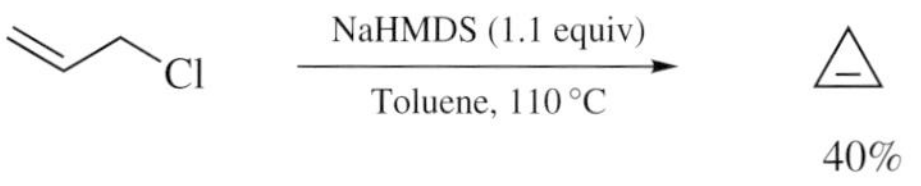

$$\text{(18)}$$

40%

Deprotection of Alkyl Aryl Ether. Both NaHMDS and LDA can act as efficient nucleophiles in O-demethylation reactions (eq 19).[44] The reaction is run in a sealed tube at 185 °C in a mixture of THF and 1,3-dimethyl-2-imidazolidinone (DMEU). This synthetic strategy is applicable to methoxyarenes, including benzene, naphthalene, anthracene, biphenyl, and pyridine, which bear either an electron-withdrawing or an electron-donating group, in addition to being efficient with benzodioxoles that lead to the corresponding catechols in excellent yields. As the activity of NaHMDS is lower than that of LDA, mono-O-demethylation of o-dimethoxybenzenes could be accomplished only with the former.

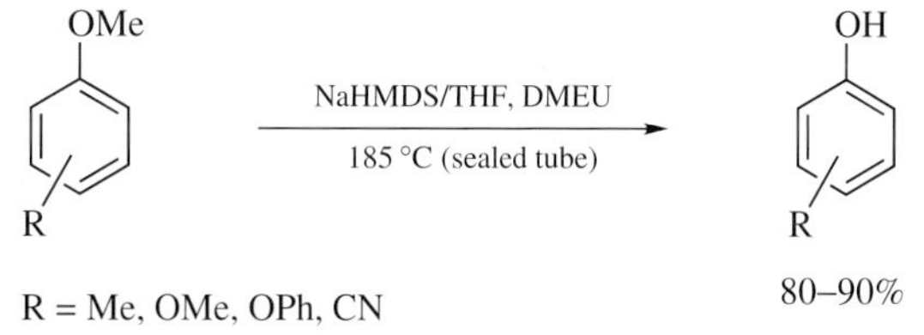

$$\text{(19)}$$

R = Me, OMe, OPh, CN

80–90%

Synthesis of Aromatic Nitriles from Esters. A one-flask method has been developed for the conversion of aromatic esters to the corresponding nitriles by use of NaHMDS in a sealed tube at 185 °C in a mixture of THF and 1,3-dimethyl-2-imidazolidinone (DMEU) (eq 20).[45] The transformation proceeded with good to excellent yields. The synthetic strategy is only applicable to aromatic esters that bear an electron-donating substituent such as hydroxy or methoxy. In the latter case, competitive O-demethylation is observed, thus leading to a mixture of nitrile products. The reaction has been also applied to indole-3-carboxylate. However, simple unsubstituted methyl benzoate failed to give the desired product.

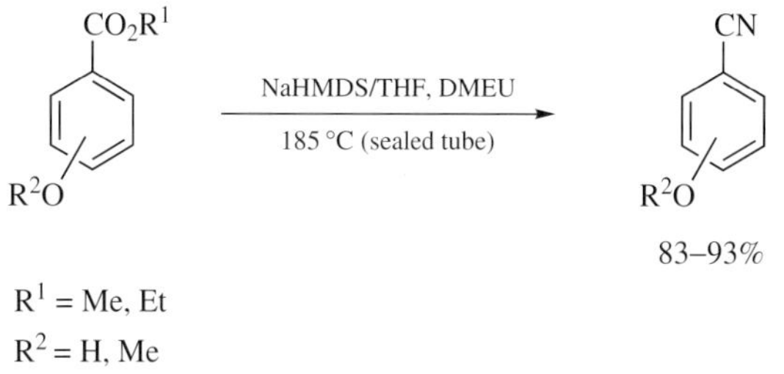

$$\text{(20)}$$

R^1 = Me, Et
R^2 = H, Me

83–93%

Deprotonation of Alkynes. NaHMDS is a base strong enough to deprotonate terminal alkynes. It has been particularly useful for the condensation of alkyne anions with aldehydes, both intermolecularly (eq 21)[46] and intramolecularly (eq 22).[47]

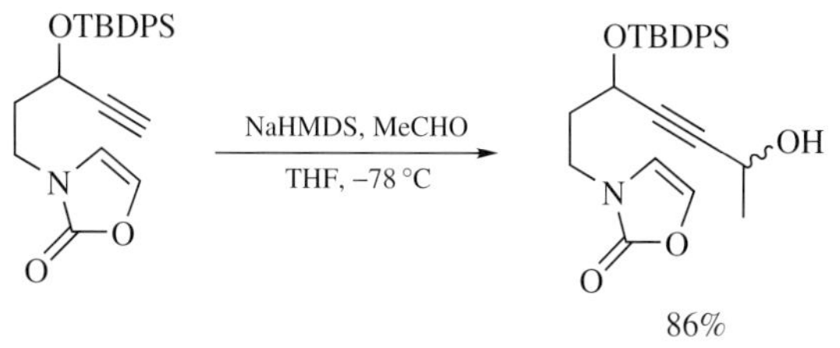

$$\text{(21)}$$

86%

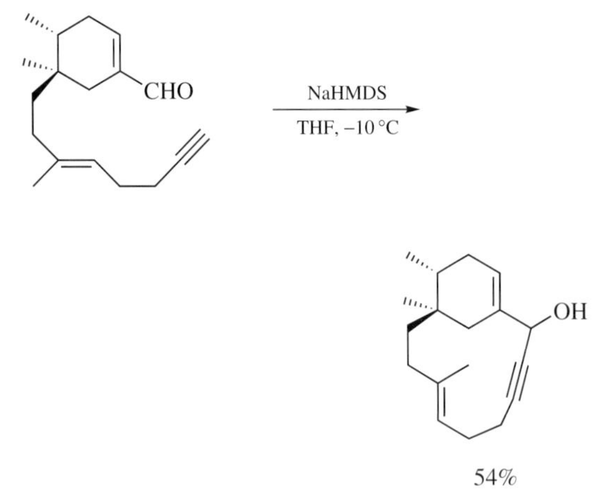

$$(22)$$

54%

In a tandem cyclization of propynyloxyethyl derivatives, NaHMDS is as efficient as LDA or KHMDS (eq 23).[48] This cyclization proceeded through the formation of an alkyne metal, followed by a cyclization which led to a carbenic species that underwent an intramolecular C-H insertion.

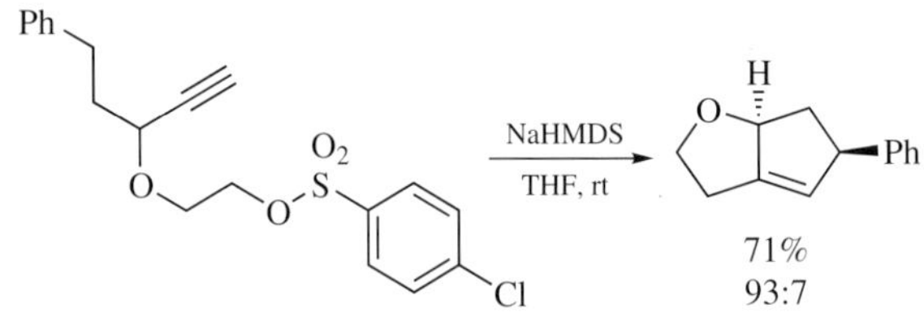

$$(23)$$

71%
93:7

Enantioselective Oxidation with (Camphorsulfonyl) Oxaziridines. The synthesis of 2-substituted-2-hydroxy-1-tetralones from 2-substituted-1-tetralone sodium enolates, obtained from the corresponding ketones and NaHMDS, proceeded with high enantioselectivities in the presence of (+)-(2R,8aR*)-[(8,8-dichlorocamphoryl)sulfonyl]oxaziridine (eq 24).[49]

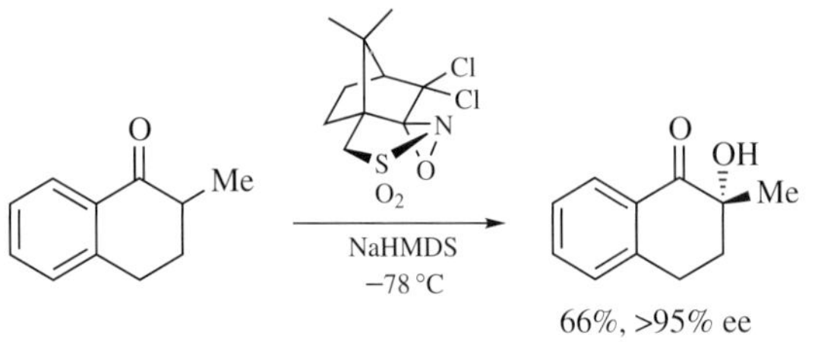

$$(24)$$

66%, >95% ee

A similar strategy was used to prepare α-hydroxy phosphonates and the highest selectivity was observed when NaHMDS was used as a base (eq 25).[50,51]

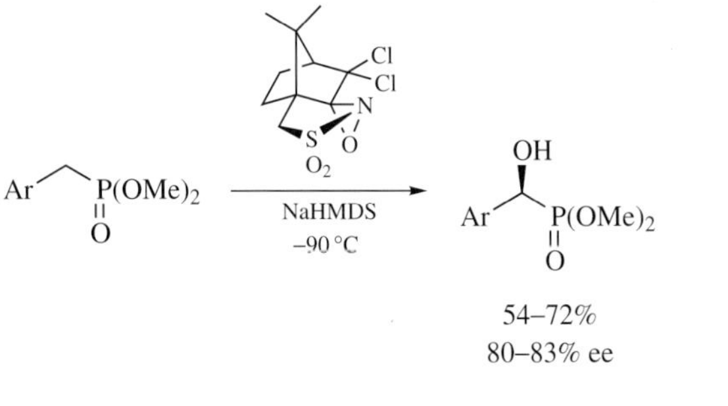

$$(25)$$

54–72%
80–83% ee

Deprotonation of Trimethylsilyldiazomethane. Treatment of trimethylsilyldiazomethane with NaHMDS led efficiently to the formation of the corresponding sodium ion. Ring opening of N-carboxylated β-lactams with this anion followed by photolytic Wolff rearrangement provided γ-lactams in very good yields (eq 26).[52]

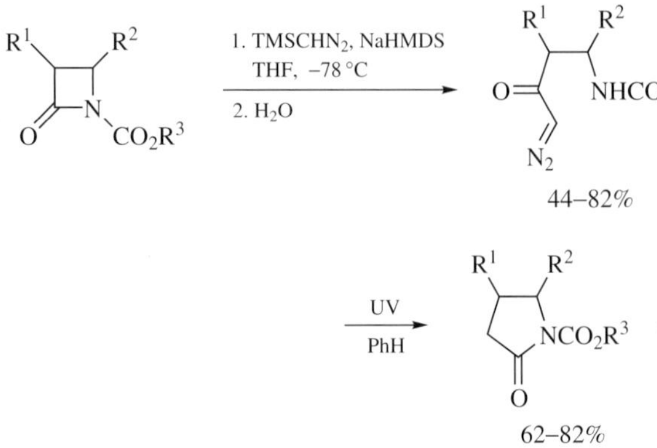

44–82%

$$(26)$$

62–82%

Deprotonation of Tosylhydrazones. The deprotonation of tosylhydrazones with NaHMDS provides the corresponding sodium salts very efficiently. The formation of allylic alcohols from sugar hydrazones was accomplished when NaHMDS was combined with LiAlH$_4$ (eq 27).[53]

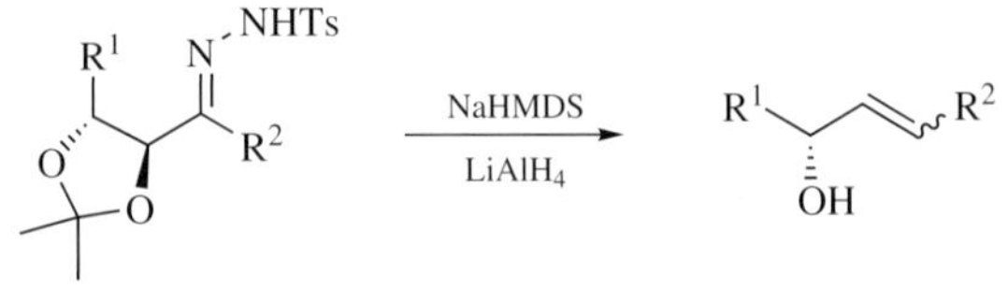

$$(27)$$

Tosylhydrazone salts could also be decomposed into the corresponding diazo compounds in the presence of a phase transfer catalyst. The addition of late transition metal complexes leads to the formation of metal carbenoid species which undergo various reactions such as cyclopropanation, aziridination, epoxidation, and C-H insertion.[54,55] While both cyclopropanation and aziridination work equally well in the presence of the sodium and lithium tosylhydrazone salts, it was established that the sodium salt provided higher yields and selectivities in the reaction with aldehydes which led to the formation of epoxides (eq 28).[56,57]

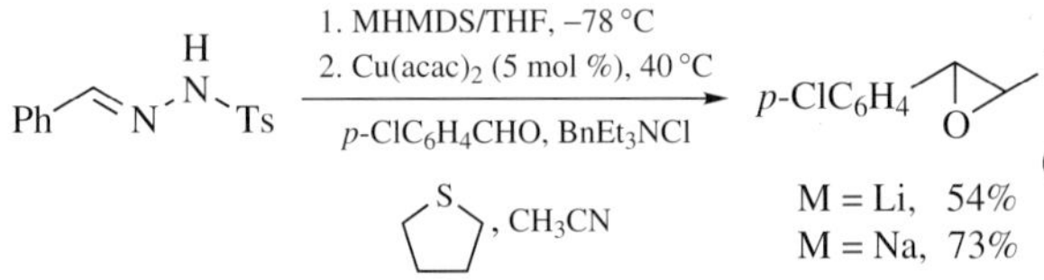

M = Li, 54%
M = Na, 73%

$$(28)$$

Most tosylhydrazone sodium salts could be usually isolated from the reaction of the corresponding tosylhydrazones with sodium methoxide. However, a number of more functionalized hydrazone sodium salts, such as those derived from alkenyl aryl sulfonylhydrazones, ketone sulfonylhydrazones, and trimethylsilylacrolein sulfonylhydrazones, could not be isolated and an in situ salt generation procedure using NaHMDS was developed. A mixture of the desired hydrazones and NaHMDS was stirred at −78 °C then concentrated down before the addition of rhodium acetate, tetrahydrothiophene, $BnEt_3NCl$, and the aldehyde to produce the required epoxide in high yields (eq 29).

$$(29)$$

76–90%
2:1–1:1 (*trans:cis*)

Generation of Phosphorus Ylides and Phosphonate Anions. NaHMDS is the most utilized base for the deprotonation of a variety of phosphonium salts to generate the corresponding ylides, which then undergo Wittig reaction with a carbonyl compound. More recently, it was shown that such a base is compatible with a variety of other systems. For instance, it was shown that allenes and dienes could be prepared, respectively, from aromatic and alicyclic aldehydes when reacted with $(Me_2N)_3P=CH_2$ in the presence of 4 equiv of NaHMDS and titanium trichloride isopropoxide (eqs 30 and 31).[58,59]

$$(30)$$

40–69%

$$(31)$$

85%
E:Z = 7.9:1

This strategy has also been applied to the one-pot double deoxygenation of simple alkyl- and polyether-tethered aromatic dialdehydes to give macrocyclic allenes in high yield without the need for slow-addition techniques.[60]

NaHMDS was also used for the generation of chloroallyl phosphonamide anion, which could further be added to oximes to yield a variety of *cis*-disubstituted *N*-alkoxy aziridines in enantiomerically pure form (eq 32).[61] Oxidative cleavage of the chiral auxiliary followed by derivatization of products led to the formation of enantiopure *N*-alkoxy aziridines.

The preparation of *Z*-α,β-unsaturated amides by using Horner–Wadsworth–Emmons reagents, (diphenylphosphono)acetamides, has been recently reported.[62] High *Z:E* ratios were observed with aromatic aldehydes and potassium *t*-butoxide, whereas the best *Z*-selectivities for aliphatic aldehydes were obtained when NaHMDS was used as base (eq 33).

$$(32)$$

74–94%

$$(PhO)_2P(O)CH_2CONR^1R^2 \xrightarrow[\text{2. } R^3CHO]{\text{1. NaHMDS/THF}}$$

$$(33)$$

90–94%
Z:E = 85:15 to 94:6

The electrophilic fluorination of benzylic phosphonates has been reported and the optimal reaction conditions involved the use of 2 equiv of NaHMDS and 2.5 equiv of *N*-fluorobenzenesulfonimide.[63]

New semi-stabilized ylides such as $PhCH=P(MeNCH_2-CH_2)_3N$ have been prepared from $RCH_2P(MeNCH_2CH_2)_3N$ employing various bases: NaHMDS, KHMDS, LDA, *t*-BuOK.[64] The reaction of such ylides with aldehydes provided alkenes in high yield with quantitative *E*-selectivity (eq 34). Although the *E*-selectivity is maintained despite the change in the metal ion of the ionic base, NaHMDS furnished usually the highest selectivities.

$$(34)$$

86–94%
>99:1, (*E:Z*)

Generation of Sulfur Ylides: Julia Olefination and Related Reactions. Deprotonation at a position α to a sulfone group is effected using a strong lithium base, typically *n*-butyllithium, LDA, or LiHMDS, to give an anion that then reacts with an aldehyde or ketone leading to the corresponding β-hydroxy sulfone. There are, however, few examples that used NaHMDS as the base. For instance, in the synthesis of curacin A, condensation of the requisite sulfone and aldehyde in the presence of NaHMDS led to the corresponding hydroxy sulfone, which was further elaborated to the desired diene (eq 35).[65]

The β-hydroxy imidazolyl sulfone derivatives were readily prepared from imidazolyl sulfones and aldehydes in the presence of NaHMDS (eq 36).[66]

The reductive elimination reaction of the β-hydroxy imidazolyl sulfone derivatives with sodium amalgam or samarium diiodide provided mainly the desired *E*-alkenes in good yields.

More recently, the 'modified' Julia olefination, which employed certain heteroarylsulfones instead of the traditional phenylsulfones, has emerged as a powerful tool for alkene synthesis.[67] Although the reaction was first reported with LDA, bases such as LHMDS, NaHMDS, and KHMDS are now commonly used.[68] In addition, solvent as well as base countercation have been shown to markedly affect the stereochemical outcome of the olefination reaction. For instance, it was shown that the *E*-isomer was major when benzothiazoylsulfone (**1**) and cyclopropane carboxaldehyde (**2**) were reacted with NaHMDS in DMF, whereas the *Z*-isomer was obtained in the presence of NaHMDS in either toluene or dichloromethane (eq 37).[69]

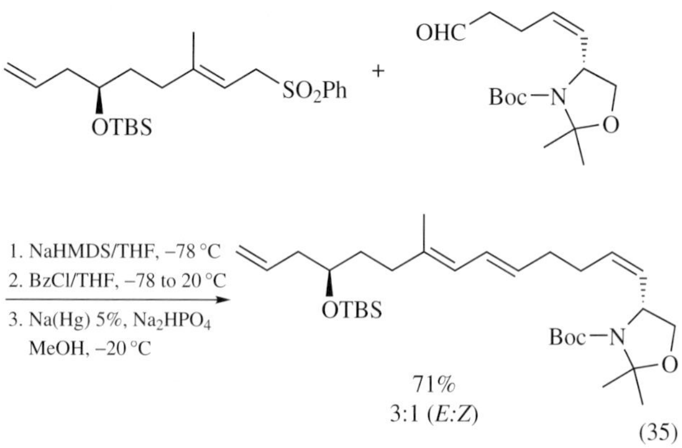

(35)

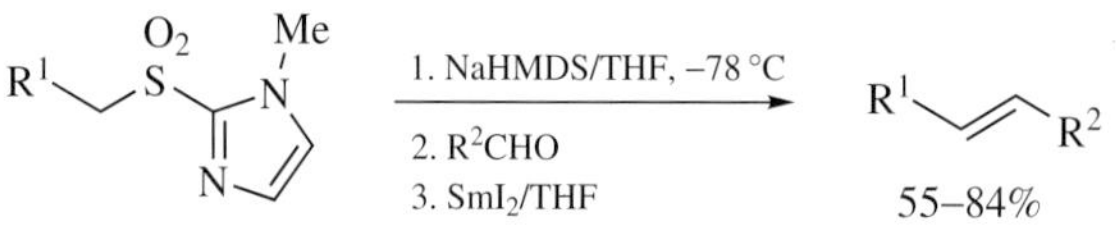

(36)

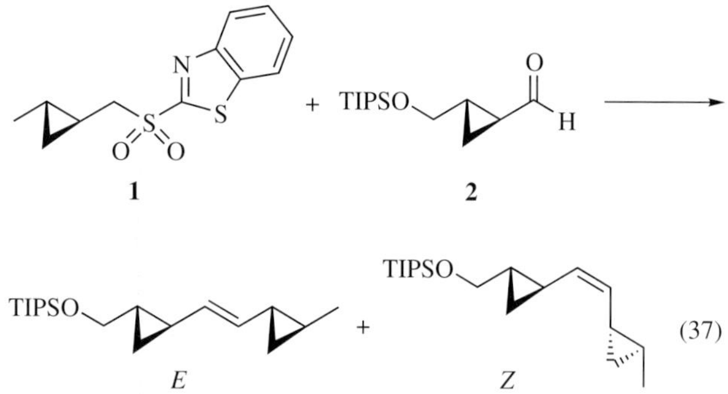

The former reaction conditions were used for the synthesis of *E*-polycyclopropane alkene (**3**), a precursor to the natural product U-106305 (eq 38).

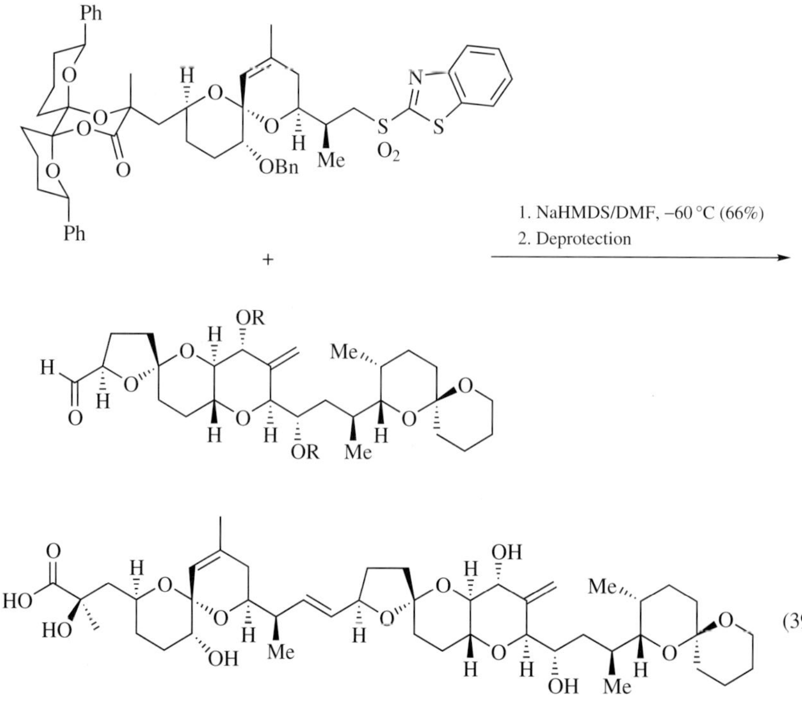

(37)

Conditions	*E:Z*
NaHMDS, DMF	3.5:1
NaHMDS, Toluene	1:10
NaHMDS, CH₂Cl₂	1:10

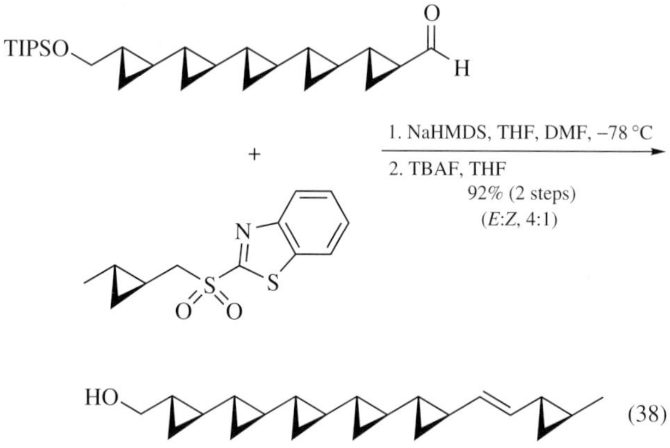

1. NaHMDS, THF, DMF, −78 °C
2. TBAF, THF
   92% (2 steps)
   (*E:Z*, 4:1)

(38)

(*E*)-**3**

1. NaHMDS/DMF, −60 °C (66%)
2. Deprotection

(39)

A similar strategy was used for the final coupling in a synthesis of okadaic acid, which involves an olefination reaction (eq 39).[70]

More recently, the same reaction conditions have been used for the preparation of another cyclopropyl alkene fragment in the total synthesis of ambruticin S (eq 40).[71]

The synthesis of fluoroalkene derivatives has recently been disclosed using 2-(1-fluoroethyl)sulfonyl-1,3-benzothiazole in the presence of various aldehydes and ketones.[72,73] Either NaHMDS or t-BuOK could be used as a base.

1-Phenyl-1H-tetrazol-5-yl sulfones were introduced as another alternative to phenyl sulfones, which led to alkenes in a one-pot procedure from aldehydes. Usually the highest selectivities are obtained while using KHMDS, although higher yields are observed with sodium hexamethyldisilazide.[74] For instance, alkene **4** was obtained in 100% yield and an 84:16 E:Z ratio with NaHMDS, whereas the use of KHMDS led to the same product with 59% yield and a 99:1 E:Z ratio (eq 41).

More recently, this strategy has been used to construct the double bond of strictifolione.[75] The use of NaHMDS led to formation of a mixture (4:1) of isomeric E- and Z-alkenes in 34% yield (eq 42).

It was shown that E,Z-dienes could be synthesized from α,β-unsaturated aldehydes and 2-pyridyl sulfones with good yields and selectivities.[76,77] NaHMDS and KHMDS were equally effective in promoting this condensation (eq 43).

Finally, sulfonium ylides could be generated from the corresponding salt and NaHMDS. When reacted with imines, sulfonium ylides furnished the corresponding aziridines in high yields and selectivities (eq 44).[78]

Generation of Bismuthonium Ylides. The use of NaHMDS as base to deprotonate bismuthonium salts has been reported. Higher yields for the subsequent condensation were usually obtained when using NaHMDS compared to potassium t-butoxide (eq 45).[79]

Related Reagents. Lithium Hexamethyldisilazide; Potassium Hexamethyldisilazide.

1. Wannagat, U.; Niederpruem, H., *Chem. Ber.* **1961**, *94*, 1540.

2. Duhamel, L.; Plaquevent, J. C., *J. Organomet. Chem.* **1993**, *448*, 1.

3. Barletta, G.; Chung, A. C.; Rios, C. B.; Jordan, F.; Schlegel, J. M., *J. Am. Chem. Soc.* **1990**, *112*, 8144.

4. (a) Capozzi, G.; Gori, L.; Menichetti, S., *Tetrahedron Lett.* **1990**, *31*, 6213. (b) Capozzi, G.; Gori, L.; Menichetti, S.; Nativi, C., *J. Chem. Soc., Perkin Trans. 1* **1992**, 1923.

5. (a) Evans, D. A., In *Asymmetric Synthesis*; Morrison, J. D., Ed.; Academic: New York, 1984; Vol 3, p 1. (b) Tanabe, M.; Crowe, D. F., *Chem. Common.* **1969**, 1498. (c) Barton, D. H. R.; Hesse, R. H.; Pechet, M. M.; Wiltshire, C., *Chem. Common.* **1972**, 1017. (d) Krüger, C. R.; Rochow, E., *J. Organomet. Chem.* **1964**, *1*, 476. (e) Krüger, C. R.; Rochow, E. G., *Angew. Chem., Int. Ed. Engl.* **1963**, *2*, 617. (f) Gaudemar, M.; Bellassoued, M., *Tetrahedron Lett.* **1989**, *30*, 2779.

6. Bestmann, H. J.; Stransky, W.; Vostrowsky, O., *Ber. Dtsch. Chem. Ges.* **1976**, *109*, 1694.

7. Stork, G.; Depezay, J. C.; d'Angelo, J., *Tetrahedron Lett.* **1975**, 389.

8. Martel, B.; Hiriart, J. M., *Synthesis* **1972**, 201.

9. (a) Bestmann, H. J.; Woelfel, G., *Ber. Dtsch. Chem. Ges.* **1984**, *117*, 1250. (b) Anteunis, M. J. O.; Callens, R. De Witte, M.; Reyniers, M. F.; Spiessens, L., *Bull. Soc. Chim. Belg.* **1987**, *96*, 545.

10. Bestmann, H. J.; Woelfel, G.; Mederer, K., *Synthesis* **1987**, 848.

11. Barton, D. H. R.; Hesse, R. H.; Tarzia, G.; Pechet, M. M., *Chem. Commun.* **1969**, 1497.

12. (a) Evans, D. A.; Ennis, M. D.; Mathre, D. J., *J. Am. Chem. Soc.* **1982**, *104*, 1737. (b) Evans, D. A.; Morrissey, M. M.; Dorow, R. L., *J. Am. Chem. Soc.* **1985**, *107*, 4346. (c) Davis, F. A.; Haque, M. S. Przeslawski, R. M., *J. Org. Chem.* **1989**, *54*, 2021.

13. (a) Schmidt, U.; Riedl, B., *Chem. Commun.* **1992**, 1186. (b) Glazer, E. A.; Koss, D. A.; Olson, J. A.; Ricketts, A. P.; Schaaf, T. K.; Wiscount, R. J., Jr., *J. Med. Chem.* **1992**, *35*, 1839.

14. Krueger, C.; Rochow, E. G.; Wannagat, U., *Ber. Dtsch. Chem. Ges.* **1963**, *96*, 2132.

15. (a) Schlosser, M.; Christmann, K. F., *Justus Liebigs Ann. Chem.* **1967**, *708*, 1. (b) Schlosser, M., *Top. Stereochem.* **1970**, *5*, 1. (c) Schlosser, M.; Schaub, B.; de Oliveira-Neto, J.; Jeganathan, S., *Chimia* **1986**, *40*, 244. (d) Schaub, B.; Jeganathan, S.; Schlosser, M., *Chimia* **1986**, *40*, 246.

16. (a) Corey, E. J.; Su, W., *Tetrahedron Lett.* **1990**, *31*, 3833. (b) Niwa, H.; Inagaki, H.; Yamada, K., *Tetrahedron Lett.* **1991**, *32*, 5127. (c) Chattopadhyay, A.; Mamdapur, V. R., *Synth. Commun.* **1990**, *20*, 2225. (d) Mueller, S.; Schmidt, R. R., *Helv. Chim. Acta* **1993**, *76*, 616.

17. (a) Miyano, S.; Izumi, Y.; Fuji, K.; Ohno, Y.; Hashimoto, H., *Bull. Chem. Soc. Jpn.* **1979**, *52*, 1197. (b) Smithers, R. H., *J. Org. Chem.* **1978**, *43*, 2833.

18. Stork, G.; Zhao, K., *Tetrahedron Lett.* **1989**, *30*, 2173.

19. Bestmann, H. J.; Schmidt, M., *Angew. Chem., Int. Ed. Engl.* **1987**, *26*, 79.

20. Gibson, A. W.; Humphrey, G. R.; Kennedy, D. J.; Wright, S. H. B., *Synthesis* **1991**, 414.

21. (a) Stork, G.; Maldonado, L., *J. Am. Chem. Soc.* **1971**, *93*, 5286. (b) Stork, G.; Maldonado, L., *J. Am. Chem. Soc.* **1974**, *96*, 5272.

22. (a) Takahashi, T.; Nagashima, T. Tsuji, J., *Tetrahedron Lett.* **1981**, 1359. (b) Takahashi, T.; Nemoto, H.; Tsuji, J., *Tetrahedron Lett.* **1983**, 2005.

23. (a) Martel, B.; Aly, E., *J. Organomet. Chem.* **1971**, *29*, 61. (b) Martel, B.; Hiriart, J. M., *Tetrahedron Lett.* **1971**, 2737. (c) Martel, B.; Hiriart, J. M., *Synthesis* **1972**, 201. (d) Martel, B.; Hiriart, J. M., *Angew. Chem., Int. Ed. Engl.* **1972**, *11*, 326.

24. Arora, S.; Binger, P., *Synthesis* **1974**, 801.

25. Charreau, P.; Julia, M.; Verpeaux, J. N., *Bull. Soc., Chem. Fr. Part 2* **1990**, *127*, 275.

26. Han, Z. X.; Krishnamurthy, D.; Grover, P.; Fang, Q. K.; Senanayake, C. H., *J. Am. Chem. Soc.* **2002**, *124*, 7880.

27. Kelly, T. A.; McNeil, D. W., *Tetrahedron Lett.* **1994**, *35*, 9003.

28. Wang, J. J.; Rosingana, M.; Discordia, R. P.; Soundararajan, N.; Polniaszek, R., *Synlett* **2001**, 1485.

29. Culkin, D. A.; Hartwig, J. F., *Acc. Chem. Res.* **2003**, *36*, 234.

30. Miura, M.; Nomura, M., *Top. Curr. Chem.* **2002**, *219*, 211.

31. Lloyd-Jones, G. C., *Angew. Chem., Int. Ed.* **2002**, *41*, 953.

32. Lee, S.; Beare, N. A.; Hartwig, J. F., *J. Am. Chem. Soc.* **2001**, *123*, 8410.

33. Jorgensen, M.; Lee, S.; Liu, X. X.; Wolkowski, J. P.; Hartwig, J. F., *J. Am. Chem. Soc.* **2002**, *124*, 12557.

34. Gaertzen, O.; Buchwald, S. L., *J. Org. Chem.* **2002**, *67*, 465.

35. Beare, N. A.; Hartwig, J. F., *J. Org. Chem.* **2002**, *67*, 541.

36. Spielvogel, D. J.; Buchwald, S. L., *J. Am. Chem. Soc.* **2002**, *124*, 3500.

37. Fox, J. M.; Huang, X. H.; Chieffi, A.; Buchwald, S. L., *J. Am. Chem. Soc.* **2000**, *122*, 1360.

38. Culkin, D. A.; Hartwig, J. F., *J. Am. Chem. Soc.* **2002**, *124*, 9330.

39. Kazmaier, U., *Curr. Org. Chem.* **2003**, *7*, 317.

40. Trost, B. M.; Schroeder, G. M.; Kristensen, J., *Angew. Chem. Int. Ed.* **2002**, *41*, 3492.

41. Grandjean, D.; Pale, P.; Chuche, J., *Tetrahedron Lett.* **1994**, *35*, 3529.

42. Corey, E. J.; Fuchs, P. L., *Tetrahedron Lett.* **1972**, 3769.

43. Binger, P.; Wedemann, P.; Brinker, U. H., *Org. Synth.* **2000**, *77*, 254.

44. Hwu, J. R.; Wong, F. F.; Huang, J. J.; Tsay, S. C., *J. Org. Chem.* **1997**, *62*, 4097.

45. Hwu, J. R.; Hsu, C. H.; Wong, F. F.; Chung, C. S.; Hakimelahi, G. H., *Synthesis* **1998**, 329.

46. Nomura, L.; Mukai, C., *Org. Lett.* **2002**, *4*, 4301.

47. Houghton, T. J.; Choi, S.; Rawal, V. H., *Org. Lett.* **2001**, *3*, 3615.

48. Harada, T.; Fujiwara, T.; Iwazaki, K.; Oku, A., *Org. Lett.* **2000**, *2*, 1855.

49. Chen, B. C.; Murphy, C. K.; Kumar, A.; Reddy, R. T.; Clark, C.; Zhou, P.; Lewis, B. M.; Gala, D.; Mergelsberg, I.; Scherer, D.; Buckley, J.; DiBenedetto, D.; Davis, F. A., *Org. Synth.* **1996**, *73*, 159.

50. Pogatchnik, D. M.; Wiemer, D. F., *Tetrahedron Lett.* **1997**, *38*, 3495.

51. Skropeta, D.; Schmidt, R. R., *Tetrahedron: Asymmetry* **2003**, *14*, 265.

52. Ha, D. C.; Kang, S.; Chung, C. M.; Lim, H. K., *Tetrahedron Lett.* **1998**, *39*, 7541.

53. Chandrasekhar, S.; Mohapatra, S.; Takhi, M., *Synlett* **1996**, 759.

54. Aggarwal, V. K.; Alonso, E.; Fang, G. Y.; Ferrar, M.; Hynd, G.; Porcelloni, M., *Angew. Chem. Int. Ed.* **2001**, *40*, 1433.

55. Aggarwal, V. K.; de Vicente, J.; Bonnert, R. V., *Org. Lett.* **2001**, *3*, 2785.

56. Aggarwal, V. K.; Alonso, E.; Fang, G. Y.; Ferrar, M.; Hynd, G.; Porcelloni, M., *Angew. Chem. Int. Ed.* **2001**, *40*, 1430.

57. Aggarwal, V. K.; Alonso, E.; Bae, I.; Hynd, G.; Lydon, K. M.; Palmer, M. J.; Patel, M.; Porcelloni, M.; Richardson, J.; Stenson, R. A.; Studley, J. R.; Vasse, J.-L.; Winn, C. L., *J. Am. Chem. Soc.* **2003**, *125*, 10926.

58. Reynolds, K. A.; Dopico, P. G.; Sundermann, M. J.; Hughes, K. A.; Finn, M. G., *J. Org. Chem.* **1993**, *58*, 1298.

59. Reynolds, K. A.; Finn, M. G., *J. Org. Chem.* **1997**, *62*, 2574.

60. Brody, M. S.; Williams, R. M.; Finn, M. G., *J. Am. Chem. Soc.* **1997**, *119*, 3429.

61. Hanessian, S.; Cantin, L. D., *Tetrahedron Lett.* **2000**, *41*, 787.

62. Ando, K., *Synlett* **2001**, 1272.

63. Taylor, S. D.; Dinaut, A. N.; Thadani, A. N.; Huang, Z., *Tetrahedron Lett.* **1996**, *37*, 8089.

64. Wang, Z. G.; Zhang, G. T.; Guzei, I.; Verkade, J. G., *J. Org. Chem.* **2001**, *66*, 3521.

65. Aube, J.; Hoemann, M. Z.; Agrios, K. A.; Aube, J., *Tetrahedron* **1997**, *53*, 11087.

66. Kende, A. S.; Mendoza, J. S., *Tetrahedron Lett.* **1990**, *31*, 7105.

67. Blakemore, P. R., *J. Chem. Soc., Perkin Trans. 1* **2002**, 2563.

68. Baudin, J. B.; Hareau, G.; Julia, S. A.; Lorne, R.; Ruel, O., *Bull. Soc. Chim. Fr.* **1993**, *130*, 856.

69. Charette, A. B.; Lebel, H., *J. Am. Chem. Soc.* **1996**, *118*, 10327.

70. Ley, S. V.; Humphries, A. C.; Eick, H.; Downham, R.; Ross, A. R.; Boyce, R. J.; Pavey, J. B. J.; Pietruszka, J., *J. Chem. Soc., Perkin Trans. 1* **1998**, 3907.

71. Kirkland, T. A.; Colucci, J.; Geraci, L. S.; Marx, M. A.; Schneider, M.; Kaelin, D. E.; Martin, S. F., *J. Am. Chem. Soc.* **2001**, *123*, 12432.

72. EP Pat. 1209154 A (to Pazenok, S.; Demoute, J.-P.; Zard, S.; Lequeux, T.) (2002).

73. Chevrie, D.; Lequeux, T.; Demoute, J. P.; Pazenok, S., *Tetrahedron Lett.* **2003**, *44*, 8127.

74. Blakemore, P. R.; Cole, W. J.; Kocienski, P. J.; Morley, A., *Synlett* **1998**, 26.

75. Juliawaty, L. D.; Watanabe, Y.; Kitajima, M.; Achmad, S. A.; Takayama, H.; Aimi, N., *Tetrahedron Lett.* **2002**, *43*, 8657.

76. Charette, A. B.; Berthelette, C.; St-Martin, D., *Tetrahedron Lett.* **2001**, *42*, 5149.

77. Erratum: Charette, A. B.; Berthelette, C.; St-Martin, D., *Tetrahedron Lett.* **2001**, *42*, 6619.

78. Yang, X. F.; Mang, M. J.; Hou, X. L.; Dai, L. X., *J. Org. Chem.* **2002**, *67*, 8097.

79. Rahman, M. M.; Matano, Y.; Suzuki, H., *Synthesis* **1999**, 395.

Sodium Hydride[1]

$$\boxed{\text{NaH}}$$

[7646-69-7]   HNa   (MW 24.00)

InChI = 1/Na.H/rHNa/h1H

InChIKey = MPMYQQHEHYDOCL-RVEWWXDUAC

(used as a base for the deprotonation of alcohols, phenols, amides (NH), ketones, esters, and stannanes; used as a reducing agent for disulfides, disilanes, azides, and isoquinolines)

Physical Data: mp 800 °C (dec); d 1.396 g cm^{-3}.

Solubility: decomposes in water; insol all organic solvents; insol liq NH$_3$; sol molten sodium.

Form Supplied in: free-flowing gray powder (95% dry hydride); gray powder dispersed in mineral oil.

Handling, Storage, and Precautions: the dispersion is a solid and may be handled in the air. The mineral oil may be removed from the dispersion by stirring with pentane, then allowing the hydride to settle. The pentane/mineral oil supernatant may be pipetted off, but care should be exercised to quench carefully any hydride in the supernatant with a small amount of an alcohol before disposal. The dry powder should only be handled in an inert atmosphere.

Sodium hydride dust is a severe irritant and all operations should be done in a fume hood, under a dry atmosphere. Sodium hydride is stable in dry air at temperatures of up to 230 °C before ignition occurs; in moist air, however, the hydride rapidly decomposes, and if the material is a very fine powder, spontaneous ignition can occur as a result of the heat evolved from the hydrolysis reaction. Sodium hydride reacts more violently with water than sodium metal (eq 1); the heat of reaction usually causes hydrogen ignition.

$$\text{NaH} + \text{H}_2\text{O} \longrightarrow \text{NaOH} + \text{H}_2 \tag{1}$$

Original Commentary

Robert E. Gawley

University of Miami, Coral Gables, FL, USA

Introduction. The following is arranged by reaction type: NaH acting as a base on oxygen, nitrogen, germanium/silicon, and carbon acids, and as a reducing agent.

Oxygen Acids (Alcohol Deprotonation). Sodium hydride may be used as a base in the Williamson ether synthesis in neat benzyl chloride,[2] in DMSO,[3] or in THF (eq 2).[4] Phenols may also be deprotonated and alkylated in THF.[4b]

$$\text{Ph}_3\text{COH} + \text{NaH} + \text{MeI} \xrightarrow[85\%]{\text{THF}} \text{Ph}_3\text{COMe} \tag{2}$$

Curiously, tertiary propargylic alcohols may be alkylated in preference to either axial or equatorial secondary alcohols, using sodium hydride in DMF (eq 3).[5]

C-5β (C-3 axial): 91%
C-5α (C-3 equatorial): 93%

Sodium hydride in DMF is also used to deprotonate carbohydrate derivatives, for methylation or benzylation (eq 4).[6]

Unstable benzyl tosylates may be made by deprotonation of benzyl alcohols and acylation with ***p*-Toluenesulfonyl Chloride** (eq 5).[7]

$$\text{PhCH}_2\text{OH} \xrightarrow[\substack{2.\ p\text{-TsCl} \\ -70\ \text{to}\ 25\ °\text{C} \\ 80\%}]{\substack{1.\ \text{NaH, ether} \\ \text{reflux, 15 h}}} \text{PhCH}_2\text{OTs} \tag{5}$$

An interesting conformational effect is seen when *p-t*-butylcalix[4]arene is tetraethylated. When ***Potassium Hydride*** is used as the base, the partial cone conformation predominates (i.e. one of the aryl groups is inverted), whereas with sodium hydride, the cone is produced exclusively (eq 6).[8]

Deprotonation of vinylsilane-allylic alcohols using sodium hydride in HMPA is followed by an 'essentially quantitative' C → O silicon migration (eq 7).[9]

Nitrogen Acids. Sodium hydride in DMSO, HMPA, NMP, or DMA assists the transamination of esters (eq 8).[10]

Acyl amino acids and peptides may be alkylated on nitrogen using sodium hydride, with no racemization (eq 9).[11] A slight change of reaction conditions allows simultaneous esterification.[12]

Essential Reactions for Organic Synthesis, First Edition. Edited by Philip L. Fuchs.
© 2016 John Wiley & Sons, Ltd. Published 2016 by John Wiley & Sons, Ltd.

$$(9)$$

A similar intramolecular alkylation has been used to make β-lactams (eq 10).[13]

$$(10)$$

Germanium/Silicon Acids. Germanium–hydrogen and silicon–hydrogen bonds are quantitatively cleaved with sodium hydride in ethereal solvents (eq 11).[14]

$$R_3XH + NaH \xrightarrow[>95\%]{DME} R_3XNa \qquad (11)$$

$$R = Bu, Ph; X = Si, Ge$$

Carbon Acids. Active methylene compounds, such as malonates and β-keto esters, can be deprotonated with sodium hydride and alkylated on carbon (eqs 12 and 13).[15] Alkylation of Reissert anions is also facile with sodium hydride (eq 14).[16]

$$(12)$$

$$(13)$$

$$(14)$$

Normally, sodium enolates of ketones alkylate on oxygen. A 'superactive' form of sodium hydride is formed when butylsodium is reduced with hydrogen; superactive sodium hydride is an excellent base for essentially quantitative deprotonation of ketones, trapped as their silyl ethers.[17] For example, cyclododecanone is converted to its enol ether (containing a 'hyperstable' double bond) in 92% yield (eq 15).

$$(15)$$

More commonly, sodium hydride is used as a base for carbonyl condensation reactions. For example, Claisen condensations of ethyl acetate[18] and ethyl isovalerate[19] are effected by sodium hydride. Condensations of cyclohexanone with methyl benzoate[19] and ethyl formate (eq 16)[20] are also facile.[21] Sodium hydride can also doubly deprotonate β-diketones, allowing acylation at the less acidic site (eq 17).[22]

$$(16)$$

$$(17)$$

An undergraduate experiment using sodium hydride involves the crossed condensation of ethyl acetate and dimethyl phthalate (eq 18).[23] Sodium hydride is also effective as a base in the Stobbe condensation[24] and the Darzens condensation.[25] It is also effective in a stereoselective intramolecular Michael reaction (eq 19).[26]

$$(18)$$

$$(19)$$

The Dieckmann condensation of esters[27] and thioesters[28] is mediated by sodium hydride (eq 20). Conditions for the latter are significantly more mild than for the former, and the yields are higher.

$$(20)$$

Sodium hydride may be used to cleave formate esters and formanilides (eqs 21 and 22).[29] The mechanism apparently involves removal of the formyl proton and loss of carbon monoxide.

$$BuOCHO + NaH \xrightarrow[68\%]{DME} BuONa + CO + H_2 \qquad (21)$$

$$Ph(Me)NCHO + NaH \xrightarrow[70\%]{DME} Ph(Me)NNa + CO + H_2 \qquad (22)$$

Dehydrohalogenation with sodium hydride is a means of making methylenecyclopropanes (eq 23).[30]

$$(23)$$

Enolate formation apparently accelerates the Diels–Alder cycloaddition/cycloreversion, shown in eq 24, which occurs at room temperature.[31]

$$(24)$$

An unusual cyclization of *N*-allyl-α,β-unsaturated amides is mediated by sodium hydride in refluxing xylene (eq 25).[32] The reaction is thought to proceed by intramolecular 1,4-addition of the dianion shown.

$$\text{(25)}$$

Reductions. The sodium salt of trimethylsilane is produced quantitatively by reduction of hexamethyldisilane with sodium hydride (eq 26).[33]

$$\text{TMS–TMS} + \text{NaH} \xrightarrow[>95\%]{} \text{TMSNa} + \text{TMSH} \quad \text{(26)}$$

Sodium hydride in DMSO is an effective medium for the reduction of disulfide bonds in proteins under aprotic conditions.[34] When the molar ratio of hydride to 1/2 cystine residues exceeds 2:1, essentially complete reduction of the disulfide bonds of bovine serum albumin is achieved.

Azides are reduced to amines by sodium hydride, although the yields are moderate (eq 27).[35] Sodium hydride also reduces isoquinoline to 1,2-dihydroisoquinoline in good yield (eq 28).[36]

$$\text{BuN}_3 + \text{NaH} \xrightarrow[39\%]{} \text{BuNH}_2 \quad \text{(27)}$$

$$\text{(28)}$$

First Update

D. David Hennings
Array BioPharma, Boulder, CO, USA

Introduction. This update is arranged by reaction type: NaH acting as a base on oxygen, nitrogen, sulfur/selenium, and carbon acids, followed by reductions, deprotection, and debromination.

Oxygen Acids. Alkylations using sodium hydride can be attenuated depending upon the conditions used. Diaryloxymethanes have been prepared using dichloromethane as the methylenation agent under harsh conditions (>120 °C). These same substrates can be conveniently prepared in very good yields using sodium hydride with NMP as the solvent at 40 °C (eq 29).[37]

$$\text{(29)}$$

R = Cl, Me, OMe   92–99% yield
(*m*- and *p*-)

The formation of trichloroacetimidates in the presence of an Fmoc-protected hydroxyl group was achieved using catalytic

sodium hydride and trichloroacetonitrile without cleavage of the base-labile Fmoc group (eq 30).[38]

$$\text{(30)}$$

Conversion of optically active 1-(benzothiazol-2-ylsulfanyl)-alkanols to thiiranes can be achieved without racemization by treatment with sodium hydride in THF at ambient temperature (eq 31).[39]

$$\text{(31)}$$

Addition of alkoxides generated from sodium hydride to iminodithiazoles give 4-alkoxyquinazoline-2-carbonitriles (eq 32).[40] This reaction generally requires 40 h in refluxing alcoholic solvents. The reaction times can be reduced to a couple of hours using microwaves, giving similar yields of the desired products.

$$\text{(32)}$$

Several examples of S_NAr reactions using sodium hydride to generate the necessary alkoxide have been reported. An interesting example of an intramolecular variation was reported using β-lactams to give the 4-oxa-7-azabicyclo[4.2.0]octane skeleton (eq 33).[41]

Selectivity can be achieved using sodium hydride to generate various alkoxides. In the case of *myo*-inositol orthoesters, sulfonylation using sodium hydride as the base led to 4,6-di-*O*-sulfonylated products whereas the use of pyridine or triethylamine gave 2,4-di-*O*-sulfonylated products.[42] Coordination of an alkoxide to metals can also be used to direct facial selectivity in various types of organometallic processes. In the preparation of the C(29–40) fragment of pectenotoxin-2, the key step was alkoxide-directed hydrogenation using a cationic rhodium catalyst (eq 34).[43] The use of sodium hydride not only promoted the selective hydrogenation, it also prevented dehydration of the substrate to the corresponding furan.

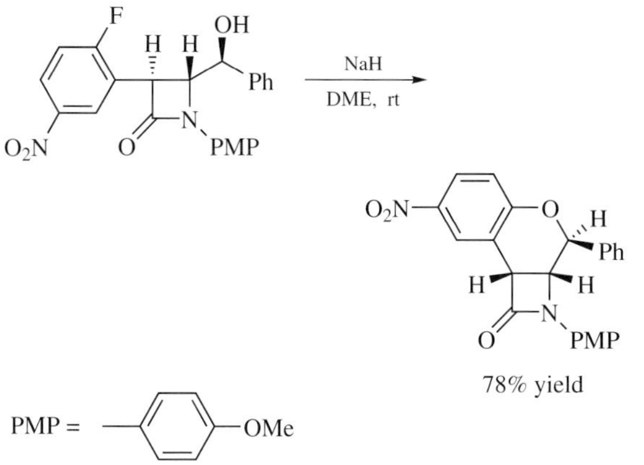

(33)

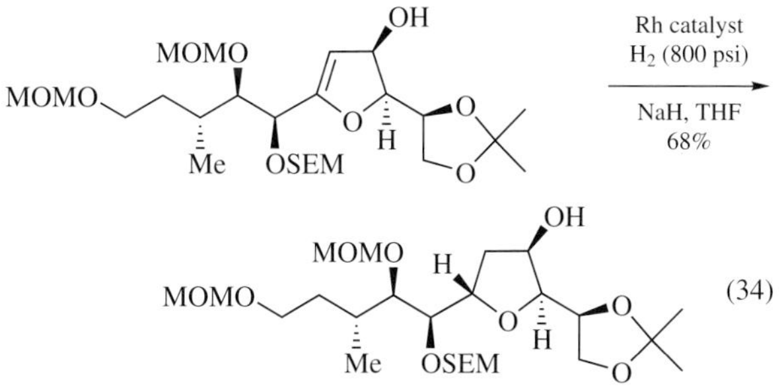

78% yield

PMP = ⟨4-methoxyphenyl structure⟩ OMe

(34)

An interesting selectivity between sodium hydride and potassium hydride was observed in the preparation of the macrocyclic dilactone core of Macroviracin A.[44] In both cases, the carboxylic acid was activated using 2-chloro-1,3-dimethylimidazolinium chloride followed by addition of the base. Using potassium hydride as the base led exclusively to monomer cyclization, whereas sodium hydride gave a 44% yield of the desired dilactone. Sodium hydride has also been useful in manipulating the paclitaxel core. Epimerization and subsequent conversion of the C-7 hydroxyl group to the corresponding xanthate using NaH/CS$_2$/MeI ultimately provided a route to prepare 7-deoxygenated taxolanalogs.[45] Deoxygenation of the C-2 hydroxyl group was also achieved using similar chemistry (eq 35).[46] In this example, a novel NaH-promoted benzoyl group migration occurs followed by formation of the C-2 xanthate, which was ultimately deoxygenated under standard conditions.

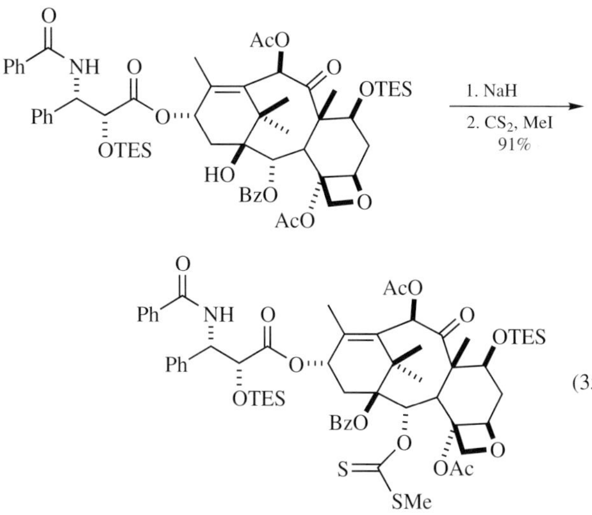

(35)

Nitrogen Acids. Sodium hydride can be used to achieve selective *N*-alkylation of 2-hydroxycarbazole (eq 36).[47]

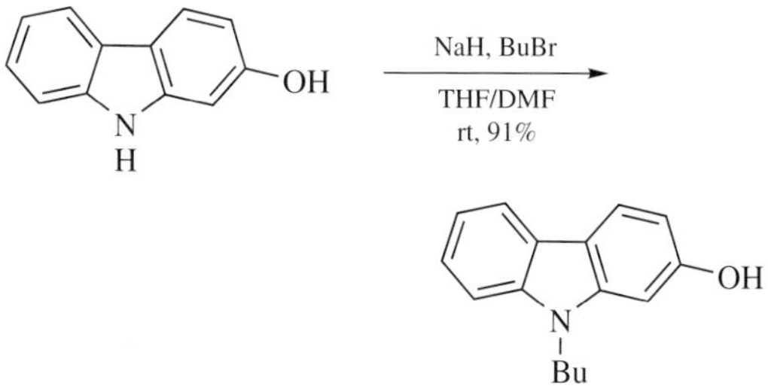

(36)

S$_N$Ar reactions have been performed using sodium hydride to deprotonate various aromatic primary amines and diamines.[48] In the case of 2-aminobenzenethiol, the S$_N$Ar pathway can be minimized giving good yields of fluorinated 2-substituted phenylbenzothiazoles (eq 37).[49] The choice of solvent is critical to achieving addition of the aniline anion to the benzonitrile.

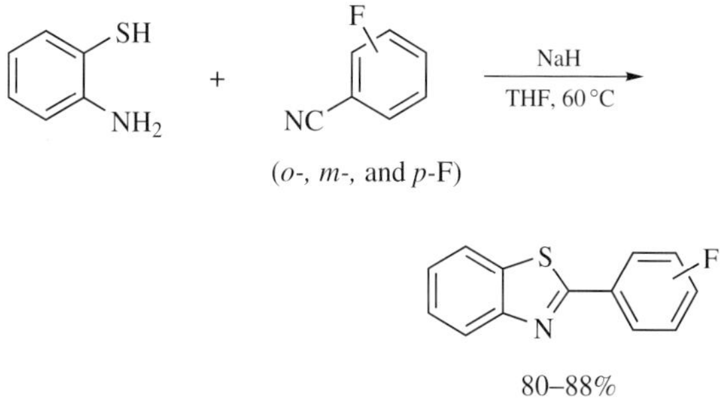

(*o-*, *m-*, and *p-*F)

(37)

80–88%

The synthesis of *α*-lactams from *α*-haloamides can be accomplished using *t*-BuOK or KOH/18-crown-6. Alternatively, using sodium hydride with 15-crown-5 ether provided the products in superior yields while reducing the necessary reaction time (eq 38).[50]

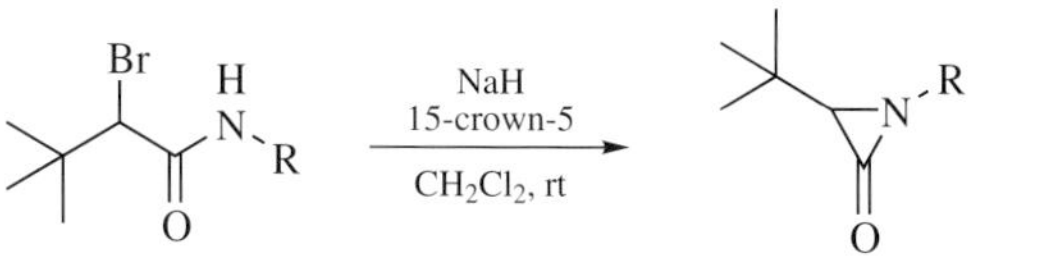

(38)

R = adamantyl (2 h)
R = trityl (1 h)

Aziridines can readily be prepared by intramolecular cyclization of protected amines with the appropriate electrophilic functionality. The method of cyclization, the nature of the electrophilic component, and the configuration of the substrate can lead to different stereochemical outcomes. Using allylic mesylates it is possible to prepare 2,3-*trans*-2-alkenyl-3-alkylaziridines (eq 39).[51] The cyclization of sulfonyl-protected amino-bromoallenes using sodium hydride gives 2,3-*cis*-2-ethynylazidiridnes with good selectivity (eq 40).[52] It is interesting to note that the *cis* selectivity is only observed with sulfonyl-protected amines. The use of Boc-protected amines gives little to no selectivity.

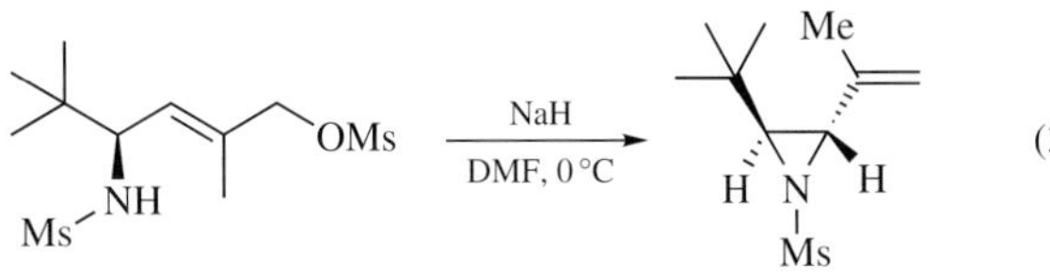

(39)

93% yield
100:0 *trans* selectivity

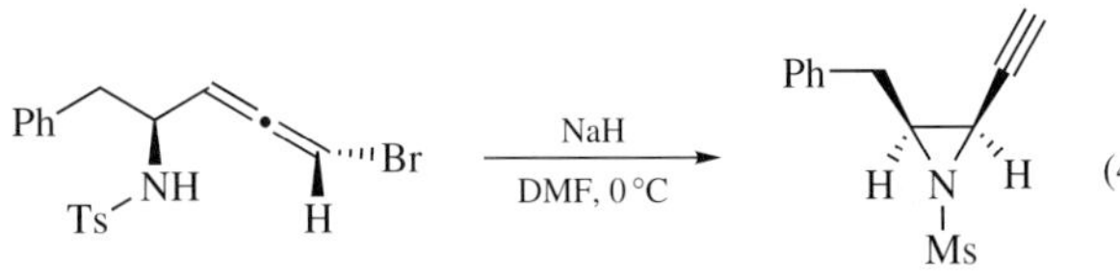

(40)

88% yield
89:11 *cis* selectivity

Aziridination of unsaturated ketones can be achieved directly using sodium hydride in conjunction with *N,N'*-diamino-1,4-diazoniabicyclo[2.2.2]octane dinitrate (eq 41).[53] This reaction proceeds through a nitrogen-nitrogen ylide. The ylide undergoes Michael addition followed by cyclization of the resulting enolate and expulsion of the tertiary amine.

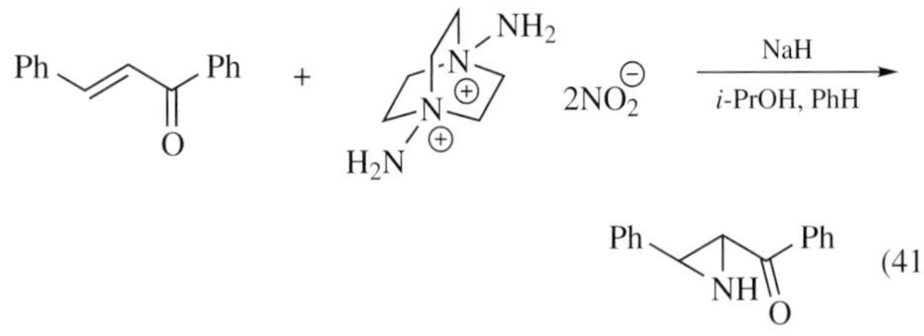

(41)

95%

Urethanes can be deprotonated with sodium hydride, and with the assistance of silver triflate, can perform intramolecular allylic displacements to give bicyclic structures (eq 42). This method was employed in the stereoselective formal synthesis of palustrine.[54]

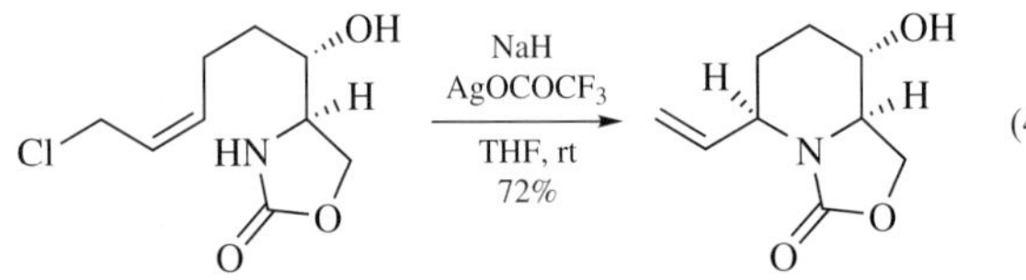

(42)

Conjugate addition of amines can be useful in cascade sequences. The addition of secondary propargylic amines have been used in conjunction with palladium catalysis to generate highly functionalized pyrrolidines (eq 43).[55] The sequence involves conjugate addition followed by carbopalladation using an aryl halide. This method has also been used with proargylic alcohols to generate 2,3-disubstituted furans.[56]

Electrophilic amination can be achieved using hydroxylamine-based ammonia equivalents. The preferred conditions for this transformation utilize sodium hydride in dioxane. Addition of the anion to *O*-(*p*-nitrobenzoyl)hydroxylamine (NbzONH$_2$) generates the *N*-amino-2-oxazolidinone (eq 44).[57] Although the amino compound can be isolated, typically the product is converted to the more stable hydrazone.

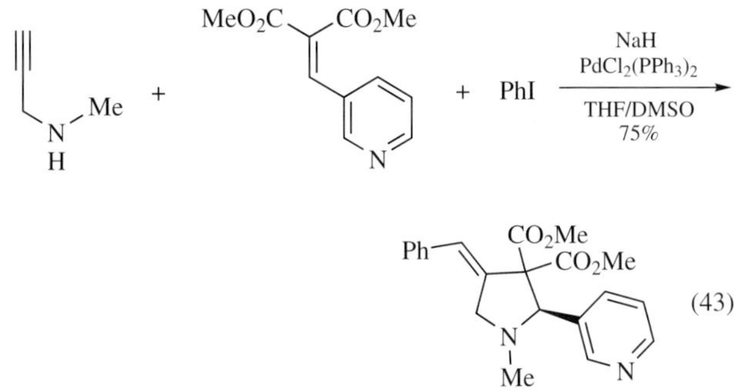

(43)

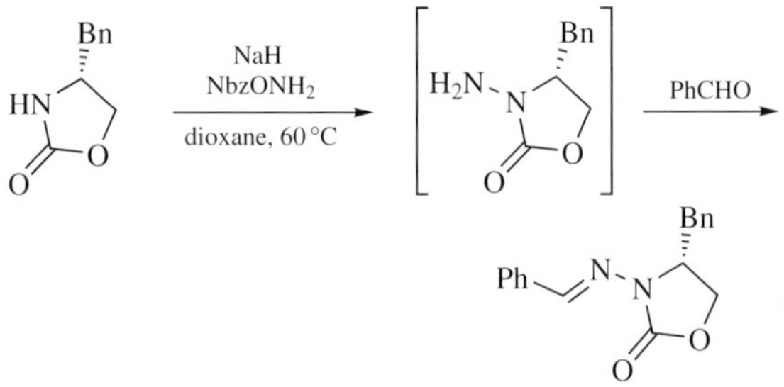

(44)

80%

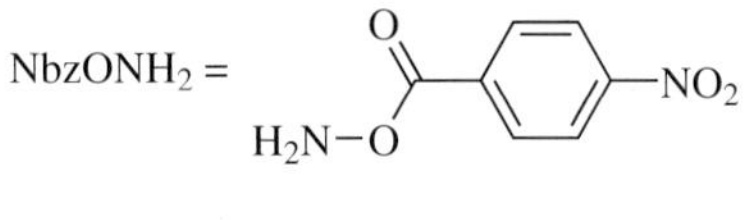

$$NbzONH_2 =$$

Selenium/Sulfur Acids. A strong base such as sodium hydride is not normally needed to facilitate alkylation of thiols. However, it was necessary to use sodium hydride for the alkylation of 5-bromo-D-pentano-1,4-lactones (eq 45).[58] The base typically employed for this transformation (NaOMe) is not compatible with the lactone functionality.

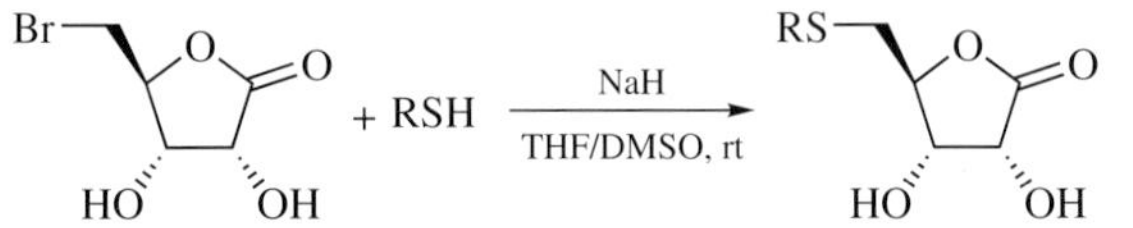

(45)

R = hexyl, octyl, decyl, and dodecyl

82–95%

The method used to prepare phenyl selenide can have a pronounced effect on the selectivity of the reagent. It was observed that Michael addition or acetate displacement could be selectively achieved by changing the method of preparation of phenyl selenide.[59] In the illustrated case, the reagent prepared using sodium hydride gave the conjugate addition product (eq 46).

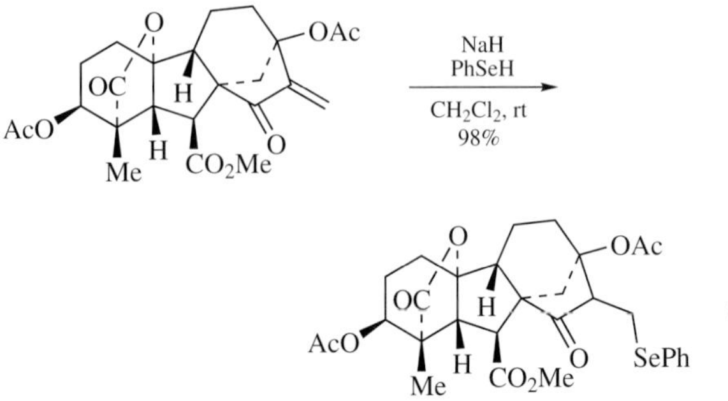

(46)

Carbon Acids. One of the more common uses of sodium hydride is the deprotonation of activated methylene compounds to generate highly reactive carbanions. A shuttle-deprotonation system has been described in which sodium hydride is the stoichiometric base and is used in conjunction with a crown ether cocatalyst to generate ketenes used in asymmetric catalysis.[60] Enaminones can be converted to naphthyridinones using sodium hydride in THF (eq 47).[61]

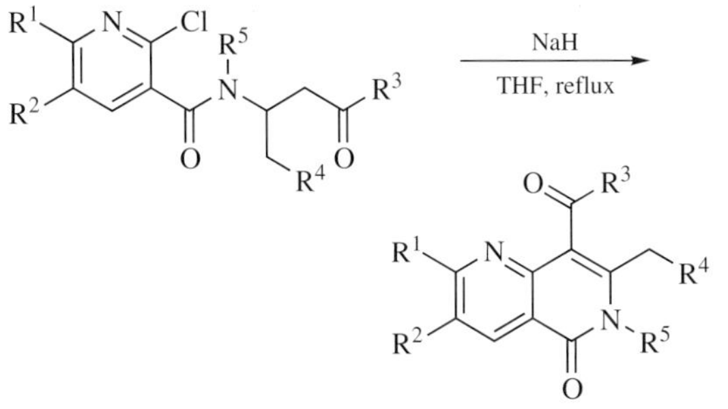

$$R^1 = H; R^2 = Cl, F; R^3 = Me, Ph$$
$$R^4 = H, Ph; R^5 = H, alkyl, Ph$$

30–80%

Lactams[62] and ketones[63] have been enolized using sodium hydride and have subsequently undergone intramolecular alkylation and acylation on the oxygen atom (eqs 48 and 49).

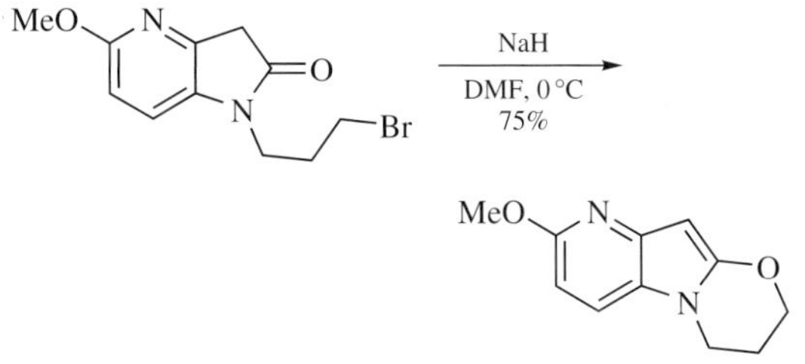

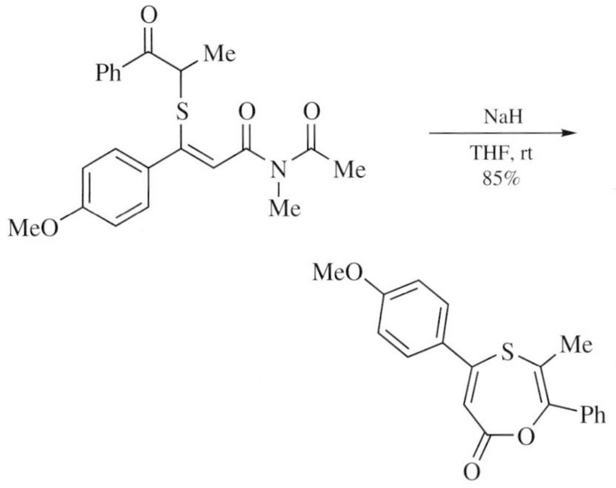

A general route to pyrroles has been developed using isocyanides with sodium hydride in the presence of Michael acceptors. The reagents tosylbenzyl isocyanide (TosBIC)[64] and tosylmethyl isocyanide (TosMIC)[65] have been the preferred reagents in these transformations (eq 50). Pyrroles have also been prepared using sodium hydride in conjuction with β-carbonyl-O-methyloximes via alkylation followed by intramolecular Michael addition.[66]

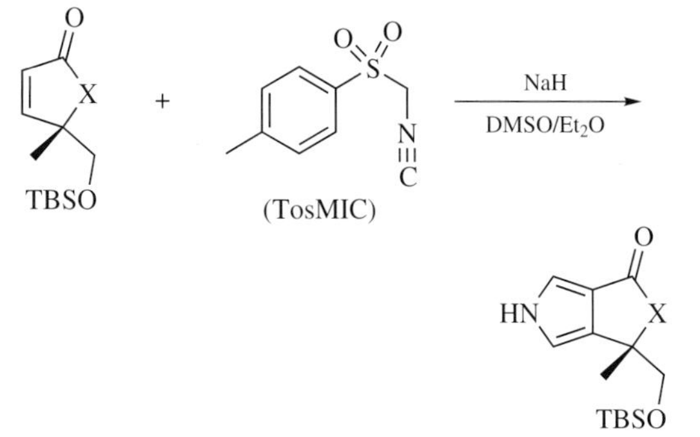

X = O (87%)
X = NBoc (97%)

A novel approach to aziridines employs the use of sodium hydride in DMSO with dichloroazetidines (eq 51).[67] The choice of DMSO is critical in obtaining the desired product. This transformation is proposed to go through a highly strained 2-azetine.

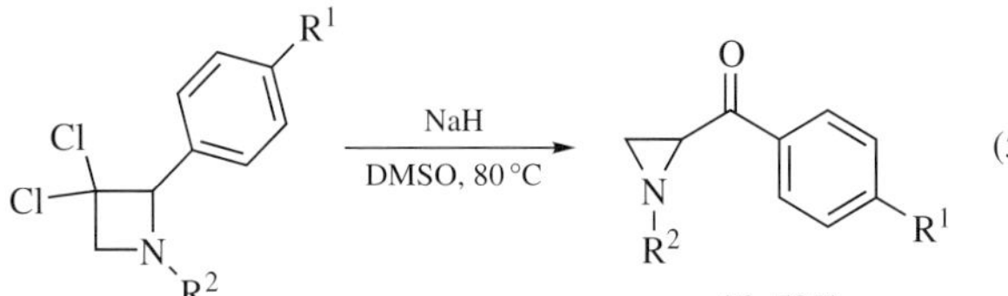

$$R^1 = H, Me, OMe, F$$
$$R^2 = i\text{-}Pr, Cy$$

46–59%

It is possible to oxidize the carbon center of enolates generated using sodium hydride. This transformation is assumed to occur by addition of the enolate to traces of O_2 present in the solvent.[68] The oxidation of a carboxylated tetrahydroisoquinoline can be achieved in quantitative yield using NaH in DMF (eq 52).[69] This transformation is presumed to occur by addition of the enolate to oxygen, followed by oxidative decarboxylation.

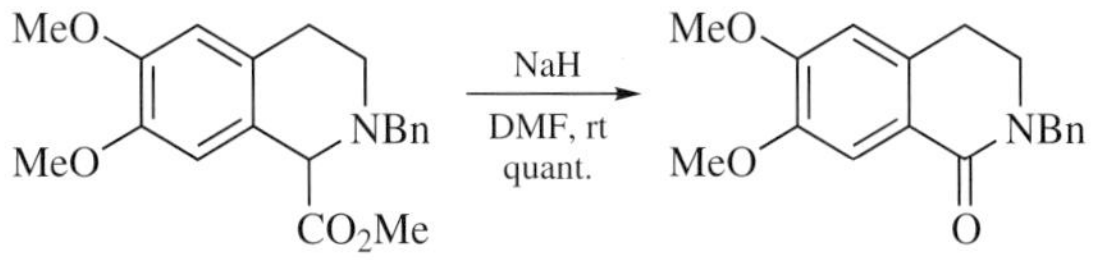

Sodium hydride can also be used to deprotonate certain sp^2 carbon atoms. Deprotonation of acyldiazomethanes using sodium hydride followed by treatment with N-sulfonylamines gave N-tosyl diazoketamines in good yields (eq 53).[70]

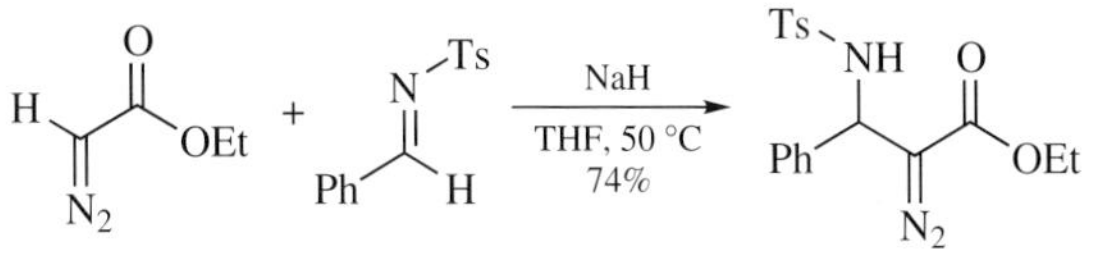

The carbene of 1,3-dimethylimidazolium iodide can be generated using sodium hydride. In the presence of a benzaldehyde and an electron deficient aryl fluoride, this carbene can promote acylation of the aryl fluoride with the aldehyde (eq 54).[71]

(54)

Lastly, sodium hydride can be used with allenyl ketones and dialkyl phosphites to give β-alkynyl-enol phosphates (eq 55).[72] The use of sodium hydride made the process highly stereoselctive and nearly pure Z-isomer was isolated.

(55)

R = various aryl groups

61–74% yields
>95:5 Z selective

Reductions. Elemental selenium can be reduced using sodium hydride. The resulting sodium diselenide (Na_2Se_2) can be used to prepare dialkyl selenides (eq 55).[73] Similarly, elemental sulfur can be reduced to diatomic sulfur by sodium hydride in the presence of a phase transfer catalyst.[74]

(56)

R = alkyl

60–88%

Sodium hydride has been used in combination with tellurium to reduce carbon-selenium bonds in α-phenylseleno carbonyl compounds using DMF as the solvent without reducing the carbonyl functionality.[75] Changing the solvent NMP enhances the reducing ability of the telluride anion.[76] Sodium telluride (Na_2Te) is generated by heating tellurium and sodium hydride in NMP at 100 °C. Aromatic aldehydes can be reduced by Na_2Te at 80 °C. Surprisingly, 7-deazapurines can be obtained by treatment of benzonitriles with sodium hydride/tellurium in NMP (eq 57).

(57)

Ar = Ph, p-tolyl, m-tolyl   15–28%

Reduction of α-chloro boronic esters using sodium hydride and DMSO has been reported to give clean conversion to the dechlorinated product.[77] The combination of lanthanide chloride catalysts

with sodium hydride effected reductive dehalogentaion of aryl chlorides and fluorides.[78] The reduction of aldehydes and ketones has been demonstrated using sodium hydride with dialkylzincs. Dialkylzinc hydride "ate" complexes are useful in the reduction of aliphatic aldehydes without promoting aldol reactions.[79] Sodium hydride is also useful in converting aryl halides into the corresponding biaryls when used with other metal cocatalysts. It is proposed that sodium hydride reduces $Ni(OAc)_2$ and $Al(acac)_3$ to produce subnanometrical Ni-Al clusters that are responsible for the coupling reaction (eq 58).[80] A similar system using nickel and zinc with sodium hydride to promote homocoupling of aryl and vinyl halides has also been reported.[81]

(58)

X = Cl or Br   34–99%

Deprotection. Deprotection of 2-(trimethylsilyl)ethyl esters using sodium hydride in DMF occurred cleanly in the presence of other silyl ethers (eq 59).[82] The reactive intermediate is presumed to be traces of "anhydrous" sodium hydroxide generated from adventitious water in the solvent.

(59)

R = TIPS (82%)
R = SEM (82%)

Debromination. Treatment of 1,1,2-tribromocyclopropanes with sodium hydride and dialkyl phosphates provides an efficient route to 1-bromo-2-alkylcyclopropenes (eq 60).[83]

(60)

R = pentyl (64%)
R = octyl (96%)

Related Reagents. Calcium Hydride; Iron(III) Chloride–Sodium Hydride; Lithium Aluminum Hydride; Potassium Hydride; Potassium Hydride–s-Butyllithium–N,N,N',N'-Tetramethylethylenediamine; Potassium Hydride–Hexamethylphosphoric Triamide; Sodium Borohydride; Sodium Hydride–copper(II) Acetate–Sodium t-Pentoxide; Sodium Hydride–nickel(II) Acetate–Sodium t-Pentoxide; Sodium Hydride–palladium(II) Acetate–Sodium t-Pentoxide; Tris(cyclopentadienyl)lanthanum–Sodium Hydride; Lithium Hydride; Sodium Telluride.

1. (a) Mackay, K. M. *Hydrogen Compounds of the Metallic Elements*; Spon: London, 1966. (b) Hurd, D. T. *An Introduction to the Chemistry of*

the Hydrides; Wiley: New York, 1952. (c) Wiberg, E.; Amberger, E. *Hydrides of the Elements of Main Groups I–IV*; Elsevier: New York, 1971.

2. Tate, M. E.; Bishop, C. T., *Can. J. Chem.* **1963**, *41*, 1801.

3. Doornbos, T.; Strating, J., *Synth. Commun.* **1971**, *1*, 175.

4. (a) Stoochnoff, B. A.; Benoiton, N. L., *Tetrahedron Lett.* **1973**, 21. (b) Brown, C. A.; Barton, D.; Sivaram, S., *Synthesis* **1974**, 434.

5. Hajos, Z. G.; Duncan, G. R., *Can. J. Chem.* **1975**, *53*, 2971.

6. (a) Brimacombe, J. S.; Jones, B. D.; Stacey, M.; Willard, J. J., *Carbohydr. Res.* **1966**, *2*, 167. (b) Brimacombe, J. S.; Ching, O. A.; Stacey, M., *J. Chem. Soc. (C)* **1969**, 197.

7. Kochi, J. K.; Hammond, G. S., *J. Am. Chem. Soc.* **1953**, *75*, 3443.

8. Groenen, L. C.; Ruël, B. H. M.; Casnati, A.; Timmerman, P.; Verboom, W.; Harkema, S.; Pochini, A.; Ungaro, R.; Reinhoudt, D. N., *Tetrahedron Lett.* **1991**, *32*, 2675.

9. Sato, F.; Tanaka, Y.; Sato, M., *J. Chem. Soc., Chem. Commun.* **1983**, 165.

10. Singh, B., *Tetrahedron Lett.* **1971**, 321.

11. Coggins, J. R.; Benoiton, N. L., *Can. J. Chem.* **1971**, *49*, 1968.

12. McDermott, J. R.; Benoiton, N. L., *Can. J. Chem.* **1973**, *51*, 1915.

13. (a) Baldwin, J. E.; Christie, M. A.; Haber, S. B.; Kruse, L. I., *J. Am. Chem. Soc.* **1976**, *98*, 3045. (b) Wasserman, H. H.; Hlasta, D. J.; Tremper, A. W.; Wu, J. S., *Tetrahedron Lett.* **1979**, 549. (c) Wasserman, H. H.; Hlasta, D. J., *J. Am. Chem. Soc.* **1978**, *100*, 6780.

14. Corriu, R. J. P.; Guerin, C., *J. Organomet. Chem.* **1980**, *197*, C19.

15. Zaugg, H. E.; Dunnigan, D. A.; Michaels, R. J.; Swett, L. R.; Wang, T. S.; Sommers, A. H.; DeNet, R. W., *J. Org. Chem.* **1961**, *26*, 644.

16. (a) Kershaw, J. R.; Uff, B. C., *J. Chem. Soc., Chem. Commun.* **1966**, 331. (b) Uff, B. C.; Kershaw, J. R., *J. Chem. Soc. (C)* **1969**, 666.

17. Pi, R.; Friedl, T.; Schleyer, P. v. R.; Klusener, P.; Brandsma, L., *J. Org. Chem.* **1987**, *52*, 4299.

18. Hinckley, A. A. Sodium Hydride Dispersions; Metal Hydrides, Inc., 1964 (quoted in: *Fieser & Fieser* **1967**, *1*, 1075) .

19. Swamer, F. W.; Hauser, C. R., *J. Am. Chem. Soc.* **1946**, *68*, 2647.

20. Ainsworth, C., *Org. Synth., Coll. Vol.* **1963**, *4*, 536.

21. (a) Bloomfield, J. J., *J. Org. Chem.* **1961**, *26*, 4112. (b) Bloomfield, J. J., *J. Org. Chem.* **1962**, *27*, 2742. (c) Anselme, J. P., *J. Org. Chem.* **1967**, *32*, 3716.

22. Miles, M. L.; Harris, T. M.; Hauser, C. R., *J. Org. Chem.* **1965**, *30*, 1007.

23. Gruen, H.; Norcross, B. E., *J. Chem. Educ.* **1965**, *42*, 268.

24. (a) Ref. 18. (b) Daub, G. H.; Johnson, W. S., *J. Am. Chem. Soc.* **1948**, *70*, 418. (c) Daub, G. H.; Johnson, W. S., *J. Am. Chem. Soc.* **1950**, *72*, 501.

25. (a) Ref. 18. (b) Della Pergola, R.; DiBattista, P., *Synth. Commun.* **1984**, *14*, 121.

26. Stork, G.; Winkler, J. D.; Saccomano, N. A., *Tetrahedron Lett.* **1983**, *24*, 465.

27. Pinkney, P. S., *Org. Synth., Coll. Vol.* **1943**, *2*, 116.

28. Liu, H.-J.; Lai, H. K., *Tetrahedron Lett.* **1979**, 1193.

29. Powers, J. C.; Seidner, R.; Parsons, T. G., *Tetrahedron Lett.* **1965**, 1713.

30. Carbon, J. A.; Martin, W. B.; Swett, L. R., *J. Am. Chem. Soc.* **1958**, *80*, 1002.

31. Tamura, Y.; Sasho, M.; Nakagawa, K.; Tsugoshi, T.; Kita, Y., *J. Org. Chem.* **1984**, *49*, 473.

32. Bortolussi, M.; Bloch, R.; Conia, J. M., *Tetrahedron Lett.* **1977**, 2289.

33. Corriu, R. J. P.; Guérin, C., *J. Chem. Soc., Chem. Commun.* **1980**, 168.

34. Krull, L. H.; Friedman, M., *Biochem. Biophys. Res. Commun.* **1967**, *29*, 373.

35. Lee, Y.-J.; Closson, W. D., *Tetrahedron Lett.* **1974**, 381.

36. Natsume, M.; Kumadaki, S.; Kanda, Y.; Kiuchi, K., *Tetrahedron Lett.* **1973**, 2335.

37. Liu, W.; Szewczyk, J.; Waykole, L.; Repic, O.; Blacklock, T. J., *Synth. Commun.* **2003**, *33*, 2719.

38. Roussel, F.; Knerr, L.; Grathwohl, M.; Schmidt, R. R., *Org. Lett.* **2000**, *2*, 3043.

39. Di Nunno, L.; Franchini, C.; Nacci, A.; Scilimati, A.; Sinicropi, M. S., *Tetrahedron: Asymmetry* **1999**, *10*, 1913.

40. Besson, T.; Rees, C. W., *J. Chem. Soc., Perkin Trans. 1* **1996**, 2857.

41. Buttero, P. D.; Molteni, G.; Papagni, A.; Pilati, T., *Tetrahedron* **2003**, *59*, 5259.

42. Sureshan, K. M.; Shashidhar, M. S.; Praveen, T.; Gonnade, R. G.; Bhadbhade, M. M., *Carbohydr. Res.* **2002**, *337*, 2399.

43. Peng, X.; Bondar, D.; Paquette, L. A., *Tetrahedron* **2004**, *60*, 9589.

44. Takahashi, S.; Souma, K.; Hashimoto, R.; Koshino, H.; Nakata, T., *J. Org. Chem.* **2004**, *69*, 4509.

45. Chaudhary, A. G.; Chordia, M. D.; Kingston, D. G., I., *J. Org. Chem.* **1995**, *60*, 3260.

46. Chaudhary, A. G.; Rimoldi, J. M.; Kingston, D. G., I., *J. Org. Chem.* **1993**, *58*, 3798.

47. Albanese, D.; Landini, D.; Penso, M.; Spano, G.; Trebicka, A., *Tetrahedron* **1995**, *51*, 5691.

48. Subrayan, R. P.; Rasmussen, P. G., *Tetrahedron* **1995**, *51*, 6167.

49. Mettery, Y.; Michaud, S.; Vierfond, J. M., *Heterocycles* **1994**, *38*, 1001.

50. Cesare, V.; Lyons, T. M.; Lengyel, I., *Synthesis* **2002**, 1716.

51. Ohno, H.; Toda, A.; Fujii, N.; Miwa, Y.; Taga, T.; Yamaoka, Y.; Osawa, E.; Ibuka, T., *Tetrahedron Lett.* **1999**, *40*, 1331.

52. Ohno, H.; Hamaguchi, H.; Tanaka, T., *Org. Lett.* **2001**, *3*, 2269.

53. Xu, J.; Jiao, P., *J. Chem. Soc., Perkin Trans. 1* **2002**, 1491.

54. Hirai, Y.; Watanabe, J.; Nozaki, T.; Yokoyama, H.; Yamaguchi, S., *J. Org. Chem.* **1997**, *62*, 776.

55. Azoulay, S.; Monteiro, N.; Balme, G., *Tetrahedron Lett.* **2002**, *43*, 9311.

56. Garcon, S.; Vassiliou, S.; Cavicchioli, M.; Hartmann, B.; Monteiro, N.; Balme, G., *J. Org. Chem.* **2001**, *66*, 4069.

57. Shen, Y.; Friestad, G. K., *J. Org. Chem.* **2002**, *67*, 6236.

58. Lalot, J.; Manier, G.; Stasik, I.; Demailly, G.; Bequpere, D., *Carbohydr. Res.* **2001**, *335*, 55.

59. Harrison, P. A.; Murtagh, L.; Willis, C. L., *J. Chem. Soc., Perkin Trans. 1* **1993**, 3047.

60. Taggi, A. E.; Wack, H.; Hafez, A. M.; France, S.; Lectka, T., *Org. Lett.* **2002**, *4*, 627.

61. Vales, M.; Lokshin, V.; Pepe, G.; Guglielmetti, R.; Samat, A., *Tetrahedron* **2002**, *58*, 8543.

62. Van de Poel, H.; Guillaumet, F.; Viaud-Massuard, M. C., *Tetrahedron Lett.* **2002**, *43*, 1205.

63. Kim, B. S.; Kim, K., *Tetrahedron Lett.* **2001**, *42*, 4637.

64. Di Santo, R.; Costi, R.; Massa, S.; Artico, M., *Synth. Commun.* **1995**, *25*, 795.

65. Frieman, B. A.; Bock, C. W.; Bhat, K. L., *Heterocycles* **2001**, *55*, 2099.

66. Song, Z.; Reiner, J.; Zhao, K., *Tetrahedron Lett.* **2004**, *45*, 3959.

67. Dejaegher, Y.; Mangelinckx, S.; De Kimpe, N., *J. Org. Chem.* **2002**, *67*, 2075.

68. Stefanic, P.; Breznik, M.; Lah, N.; Leban, I.; Plavec, J.; Kikelj, D., *Tetrahedron Lett.* **2001**, *42*, 5295.

69. Bois-Choussy, M.; De Palois, M.; Zhu, J., *Tetrahedron Lett.* **2001**, *42*, 3427.

70. Jiang, N.; Ma, Z.; Qu, Z.; Xing, X.; Xie, L.; Wang, J., *J. Org. Chem.* **2003**, *68*, 893.

71. Suzuki, Y.; Toyota, T.; Imada, F.; Sato, M.; Miyashita, A., *Chem. Commun.* **2003**, 1314.

72. Ding, Y.; Huang, X., *Synth. Commun.* **2001**, *31*, 449.

73. Krief, A.; Derock, M., *Tetrahedron Lett.* **2002**, *43*, 3083.

74. Okuma, K.; Kuge, S.; Koga, Y.; Shioji, K.; Wakita, H.; Machiguchi, T., *Heterocycles* **1998**, *48*, 1519.

75. Silveira, C. C.; Lenardao, E. J.; Comasseto, J. V., *Synth. Commun.* **1994**, *24*, 575.

76. Suzuki, H.; Nakamura, T., *J. Org. Chem.* **1993**, *58*, 241.

77. Matteson, D. S.; Soundararajan, R.; Ho, O. C.; Gatzweiler, W., *Organometallics* **1996**, *15*, 152.

78. Zhang, Y.; Liao, S.; Xu, Y.; Yu, D., *Synth. Commun.* **1997**, *27*, 4327.

79. Uchiyama, M.; Furumoto, S.; Saito, M.; Kondo, Y.; Sakamoto, T., *J. Am. Chem. Soc.* **1997**, *119*, 11425.

80. Massicot, F.; Schneider, R.; Fort, Y.; Illy-Cherry, S.; Tillement, O., *Tetrahedron* **2001**, *57*, 531.

81. Lin, G.; Hong, R., *J. Org. Chem.* **2001**, *66*, 2877.

82. Serrano-Wu, M. H.; Regueiro-Ren, A.; St. Laurent, D. R.; Carroll, T. M.; Balasubramanian, B. N., *Tetrahedron Lett.* **2001**, *42*, 8593.

83. Al Dulayymi, A. R.; Baird, M. S., *J. Chem. Soc., Perkin Trans. 1* **1994**, 1547 and 3047.

Sodium Periodate[1]

[7790-28-5]   INaO$_4$ (MW 213.89)

InChI = 1/HIO4.Na/c2-1(3,4)5;/h(H,2,3,4,5);/q;+1/p-1/fIO4.Na/
q-1;m

InChIKey = JQWHASGSAFIOCM-QMNIUIOECX

(oxidative cleavage of 1,2-diols;[2] oxidation of sulfides,[3] selenides,[4] phenols,[5] indoles,[6] and tetrahydro-β-carbolines[7])

Alternate Name: sodium metaperiodate.
Physical Data: mp 300 °C (dec); specific gravity 3.865.
Solubility: sol H$_2$O (14.4 g/100 mL H$_2$O at 25 °C; 38.9 g/100 mL at 51.5 °C), H$_2$SO$_4$, HNO$_3$, acetic acid; insol organic solvents.
Form Supplied in: colorless to white tetragonal, efflorescent crystals; readily available.
Handling, Storage, and Precautions: irritant; gloves and safety goggles should be worn when handling this oxidant; avoid inhalation of dust and avoid contact of oxidant with combustible matter.

Original Commentary

Andrew G. Wee & Jason Slobodian
University of Regina, Regina, Saskatchewan, Canada

Introduction. Sodium periodate is widely used for the oxidation of a variety of organic substrates and as a cooxidant in other oxidation reactions (see **Sodium Periodate–Osmium Tetroxide** and **Sodium Periodate–Potassium Permanganate**).[8] The NaIO$_4$ oxidation is usually conducted in water; however, for organic substrates that are insoluble in water, an organic cosolvent (e.g. MeOH, 95% EtOH, 1,4-dioxane, acetone, MeCN) is used. Alternatively, the oxidation can be conducted either with phase-transfer catalysis (PTC) using quaternary ammonium[5] or phosphonium[9] salts in a two-phase system, or in an organic solvent if the oxidant is first coated on an inert support.[10]

Oxidative Cleavage of 1,2-Diols. NaIO$_4$ is widely used for the oxidative cleavage[2] of a variety of 1,2-diols to yield aldehydes or ketones (eq 1). In this respect, it complements the **Lead(IV) Acetate** method for oxidation. 1,2-Diols have been shown to be chemoselectively cleaved by NaIO$_4$ in the presence of a sulfide group.[11] NaIO$_4$ coated on wet silica gel efficiently oxidizes 1,2-diols to the aldehydes (eq 2).[10a] This method is particularly useful for the preparation of aldehydes which readily form hydrates, and it is also convenient to conduct because isolation of the product involves simple filtration of the reaction mixture and evaporation.

$$\text{(1)}$$

Oxidation of Sulfides to Sulfoxides. The selective oxidation of sulfides to sulfoxides is an important transformation because sulfoxides are useful intermediates in synthesis.[12] The reaction is conducted using an equimolar amount of NaIO$_4$ in aqueous methanol at 0 °C (eq 3).[3] Higher reaction temperatures or the use of an excess of NaIO$_4$ result in overoxidation to give sulfones. NaIO$_4$ supported on acidic alumina (eq 4)[10b,c] or silica gel[10d] is effective for the selective oxidation of sulfides, at ambient temperature, to afford good yields of sulfoxides. Phase transfer-catalyzed NaIO$_4$ oxidation of sulfides also results in the selective formation of sulfoxides.[9]

$$\text{(2)}$$

$$\text{(3)}$$

$$\text{(4)}$$

α-Phosphoryl sulfoxides, useful for the preparation of vinylic sulfoxides,[13] are prepared in high yields by the oxidation of α-phosphoryl sulfides using NaIO$_4$.[14] Vinylic sulfoxides can also be prepared in good yields by the oxidation of vinylic sulfides using NaIO$_4$ (eq 5).[15] Poor yields of sulfoxides are obtained in the NaIO$_4$ oxidation of acetylenic sulfides.[15a]

$$\text{(5)}$$

2-Substituted 1,3-dithianes are stereoselectively oxidized to the *trans*-1-oxide by NaIO$_4$ at low temperatures.[16] Dimethyl dithioacetals of aldehydes and ketones suffer NaIO$_4$-mediated hydrolysis to give carbonyl compounds.[17] This method could be useful for the deprotection of dimethyl dithioacetals. Oxidation of dithioethers such as 1,4-dithiacycloheptane using NaIO$_4$ at 0 °C furnishes the 1-oxide in modest yield.[18] The use of **m-Chloroperbenzoic Acid** for this oxidation leads to an appreciable amount of the 1,4-dioxide. Oxidation of a naphtho-1,5-dithiocin using an excess of NaIO$_4$ at rt results in a high yield of the *cis*-1,5-dioxide.[19] The sulfide unit in thiosulfoxides is selectively oxidized to the S,S-dioxide in good yields using an equimolar amount of NaIO$_4$ at 0 °C.[20] Unsymmetrical thiosulfinic S-esters are efficiently converted to the thiosulfonic S-esters, without concomitant cleavage of the S–S bond, by NaIO$_4$ oxidation.[21] NaIO$_4$ is effective for the selective oxidation of the sulfide moiety in (**1**) to the sulfoxide in the presence of a disulfide linkage (eq 6).[22] Other oxidants such as CrO$_3$ in acetic acid, H$_2$O$_2$, and *m*-CPBA, which are useful for the oxidation of simple sulfides, only cause the decomposition of (**1**).

$$\text{PhS}_{\text{S}}\underset{\text{H}}{\overset{\text{O}}{\diagup\hspace{-0.3em}\text{N}}}\diagdown_{\text{S}}\diagdown\text{CHPh}_2 \quad\xrightarrow[\text{0 °C}]{\text{NaIO}_4,\ \text{MeOH}}$$

(1)

$$\text{PhS}_{\text{S}}\diagdown\diagdown\underset{\text{H}}{\overset{\text{O}}{\diagup\hspace{-0.3em}\text{N}}}\diagdown\diagdown\overset{\text{O}}{\underset{\text{Ph}}{\overset{\|}{\text{S}}}}\diagdown\text{Ph} \qquad (6)$$

Oxidation of Selenides to Selenoxides. Diaryl, dialkyl, and aryl alkyl selenides are oxidized[4] to the corresponding selenoxides in high yields using a slight excess of $NaIO_4$ at 0 °C (eq 7).[4a] The presence of an electron-withdrawing substituent in diaryl selenides inhibits the oxidation of the selenium center. Vinylic selenides can be oxidized[23] with $NaIO_4$ to give high yields of vinylic selenoxides (eq 8).[23a] In contrast, oxidation with **Hydrogen Peroxide** results in the cleavage of the double bond to give carboxylic acids. The oxidation of organoselenides possessing β-hydrogens results in the formation of highly unstable organoselenoxides that undergo facile *syn* elimination, often at room temperature, to give alkenes (eq 9).[24,25] Such a process constitutes a useful method for the introduction of a double bond into organic molecules.

$$\text{Ph}\diagdown^{\text{Se}}\diagup\text{Me} \quad\xrightarrow[\substack{\text{H}_2\text{O, 0 °C}\\79\%}]{\text{NaIO}_4,\ \text{MeOH}}\quad \text{Ph}\diagdown^{\overset{\text{O}}{\overset{\|}{\text{Se}}}}\diagup\text{Me} \qquad (7)$$

$$\underset{\text{C}_5\text{H}_{11}}{\overset{\text{Ph}}{\diagdown}}\diagup^{\text{SePh}} \quad\xrightarrow[\substack{\text{H}_2\text{O, rt}\\90\%}]{\text{NaIO}_4,\ \text{MeOH}}\quad \underset{\text{C}_5\text{H}_{11}}{\overset{\text{Ph}}{\diagdown}}\diagup^{\overset{\text{O}}{\overset{\|}{\text{SePh}}}} \qquad (8)$$

$$\underset{\text{SePh}}{\overset{\text{O}}{\text{Ph}\diagdown\hspace{-0.3em}\diagup\diagdown}} \quad\xrightarrow[\substack{\text{H}_2\text{O, NaHCO}_3,\ \text{rt}\\89\%}]{\text{NaIO}_4,\ \text{MeOH}}\quad \text{Ph}\diagdown\hspace{-0.3em}\overset{\text{O}}{\diagup}\hspace{-0.3em}\diagdown \qquad (9)$$

Oxidation of Phenols and Its Derivatives. Dihydroxybenzenes are oxidized to give high yields of the corresponding quinones using $NaIO_4$ supported on silica gel (eq 10)[10a] or under PTC (see also **Tetra-butylammonium Periodate**).[5] Treatment of *p*-hydroxybenzyl alcohol with $NaIO_4$ in aqueous acetic acid results in the formation of *p*-benzoquinone, albeit in low yield (23%).[26] On the other hand, *o*-(hydroxymethyl)phenols possessing at least one bulky group at the C-4 position are efficiently oxidized to give spiroepoxy-2,4-cyclohexadienones (eq 11).[27] In the absence of a bulky group, self-dimerization of the spiroepoxycyclohexadienone via Diels–Alder reaction occurs. In the case of *o*-(hydroxymethyl)phenols that are substituted with one or two aryl groups at the benzylic carbon, a novel oxidative rearrangement occurs to yield benzylidene protected catechols in modest yields (eq 12).[28] However, this oxidative rearrangement is only successful if substituents are present in the C-2 and C-4 positions of the phenol unit, because oxidation of α-(2-hydroxyphenyl) benzyl alcohol only results in the formation of a dimer.

$$\underset{\text{OH}}{\overset{\text{OH}}{\diagup\hspace{-0.3em}\bigcirc\hspace{-0.3em}\diagdown}} \quad\xrightarrow[\substack{\text{CH}_2\text{Cl}_2,\ \text{H}_2\text{O, rt}\\100\%}]{\text{NaIO}_4,\ \text{SiO}_2}\quad \underset{\text{O}}{\overset{\text{O}}{\diagup\hspace{-0.3em}\bigcirc\hspace{-0.3em}\diagdown}} \qquad (10)$$

$$\text{(11)}$$

$$\text{(12)}$$

Oxidation of Indoles and Tetrahydro-β-carbolines. The indolic double bond in 2,3-dialkyl- and 3-alkylindoles is readily oxidized by 2 mole equiv of $NaIO_4$ at room temperature to give *o*-amidoacetophenone derivatives in good yields (eq 13).[6] However, the oxidation of 2,3-diphenylindole under the same conditions results in a lower yield of the oxidative cleavage product.[29] Interestingly, the oxidation of 2-alkylindoles results in the formation of a mixture of products comprised of indoxyl derivatives.[29] Tetrahydrocarbazoles are also efficiently oxidized by $NaIO_4$ to afford benzocyclononene-2,7-dione derivatives.[6] Tetrahydro-β-carbolines have also been subjected to $NaIO_4$ oxidation.[7] Thus, in the oxidation of the tetrahydro-β-carboline-3-carboxylates, the type of product that is formed depends upon the degree of substitution at C-1 of the starting material (eqs 14 and 15).[7a]

$$\text{(13)}$$

$$\text{(14)}$$

$$\text{(15)}$$

Other Applications. 1,3-Cyclohexanedione and its 3-substituted derivatives are oxidized with $NaIO_4$, with concomitant loss of the C-2 carbon unit, to give glutaric acid in good yields.[30] 1,3-Cyclopentanediones react more slowly under the same conditions, and aromatic diketones such as 1,3-indandione give poor yields of the dicarboxylic acid product. α-Hydroxy carboxylic acids undergo oxidative decarboxylation to give aldehydes (eq 16)[31] upon treatment with aqueous $NaIO_4$; however, long reaction times are required. The use of PTC[9] or Bu_4NIO_4 allows for shorter reaction times without adversely affecting the yield. The oxidation of hydrazine with $NaIO_4$ in the presence of trace amounts of aqueous **Copper(II) Sulfate** and **Acetic Acid** results in the formation of **Diimide**. The in situ generation of diimide by this method has been successfully applied to a one-pot procedure for the reduction of alkenes (eq 17).[32]

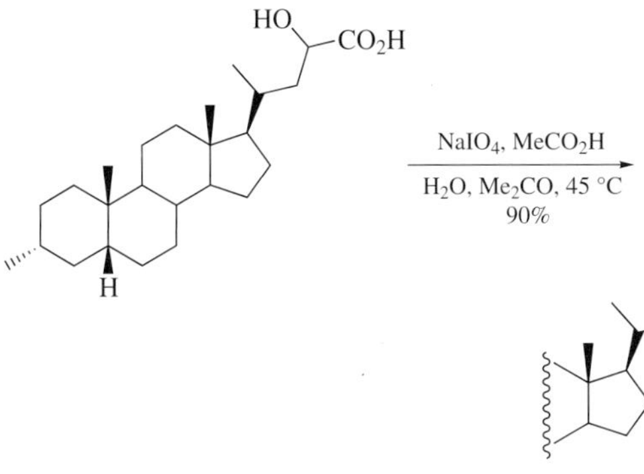

(16)

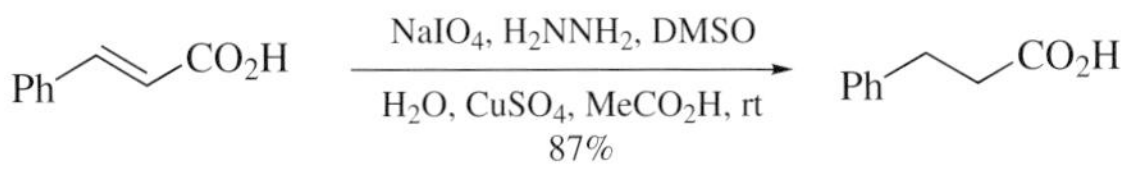

(17)

A secondary amide is obtained by selective oxidation of a tertiary carbon center in adamantane with $NaIO_4$ in the presence of iron(III) perchlorate in acetonitrile (eq 18).[33] Dimethylhydrazones undergo periodate induced hydrolysis, at pH 7, to give carbonyl compounds in high yields (eq 19).[34] However, these conditions are unsuitable for the hydrolysis of dimethylhydrazones derived from aromatic or α,β-unsaturated aldehydes because mixtures of aldehydes and nitriles are formed.

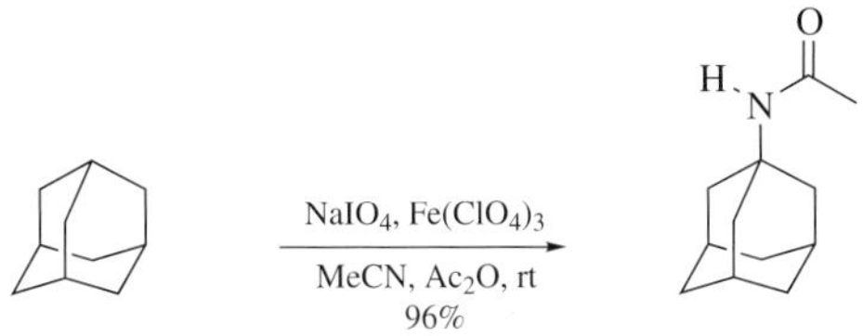

(18)

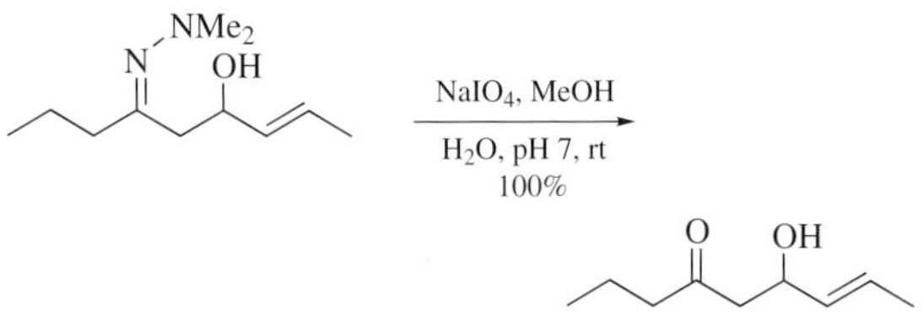

(19)

Acylphosphoranes are oxidized to α,β-dicarbonyl compounds in fair yields using aqueous $NaIO_4$ (eq 20).[35] This method complements other methods such as the **Potassium Permanganate**[36a] or **Ruthenium(VIII) Oxide** oxidation[36b] of alkynes. $NaIO_4$ is also used for the oxidation of hydroxamic acids and N-hydroxycarbamic esters at pH 6 to generate highly reactive nitroso compounds.[37] The oxidations are usually conducted in the presence of conjugated dienes so that the nitroso intermediates are trapped as their Diels–Alder cycloadducts (eq 21).

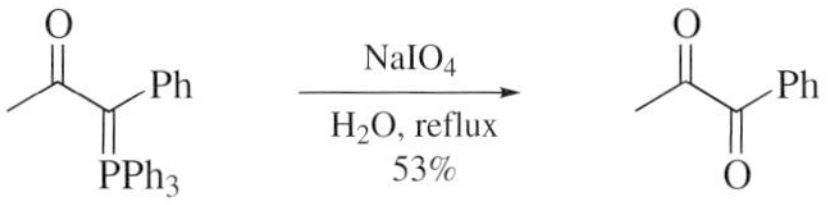

(20)

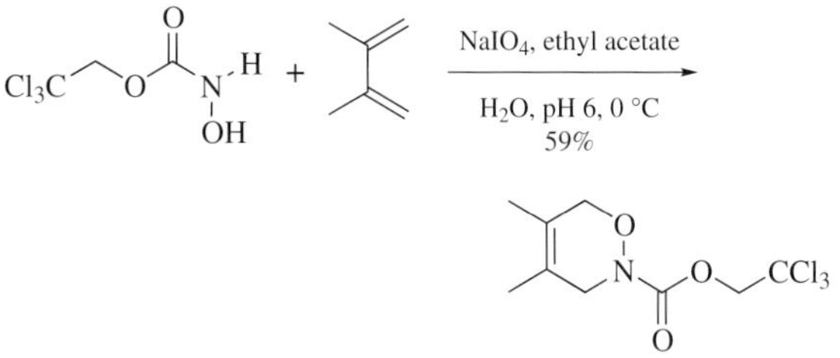

(21)

First Update

Manuel A. Fernández-Rodríguez & Enrique Aguilar
Universidad de Oviedo, Asturias, Spain

Oxidative Cleavage of 1,2-Diols. The oxidative cleavage of vicinal diols and related functional groups to afford aldehydes or ketones remains as one of the most useful reactions of sodium periodate. In this regard, an improved method using silica gel-supported $NaIO_4$ reagent in powder form has been developed.[38] The reaction is performed by stirring a suspension of the oxidant and the glycol in dichloromethane at room temperature or below, to afford the resultant aldehyde in high yield. This new protocol displays the advantages of sodium metaperiodate coated on wet silica gel[10a] but, on the other hand, the drawbacks that this reagent occasionally presents, such as colloid formation and the need for column chromatography purification, are suppressed. A one-pot sequential oxidative cleavage/Wittig olefination can be achieved by using an excess of the improved $NaIO_4$ reagent on silica gel in the presence of stabilized ylides under anhydrous conditions (eq 22).[39] Manganese dioxide also promotes this one-pot transformation albeit in lower yields.[39a]

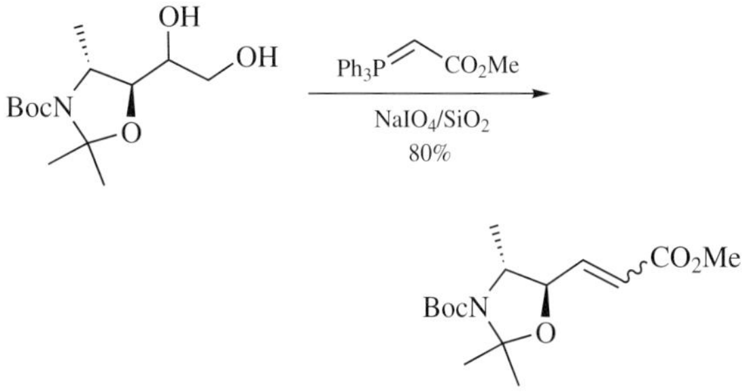

(22)

The oxidative cleavage of both saturated and unsaturated α-ketols can also be carried out by using sodium periodate in aq THF. The initially formed carboxylic acids are transformed into their methyl esters by treating the crude product in situ with diazomethane (eq 23).[40] In contrast, other oxidants give lower yields (e.g., lead tetraacetate), or fail with unsaturated substrates (e.g., sodium bismuthate).

Oxidation of Sulfides and Selenides. Sulfoxides and sulfones are valuable intermediates in organic synthesis, and their preparation mainly relies on the selective oxidation of sulfides. In this regard, sulfoxide formation can be accomplished, in 3 min or less, using 1.7 equiv of wet silica-supported sodium periodate under microwave thermolysis conditions.[41] Furthermore, an excess of the reagent (3 equiv) gives the corresponding sulfones at similar rates. A wide range of both alkyl- and aryl sulfides can be oxidized to the corresponding sulfoxide or sulfone derivatives by using the appropriate amount of the oxidant. Some of the features of this reaction are (1) the presence of some moisture is necessary for the reaction to be complete, (2) the microwave irradiation is crucial to accelerate the formation of the sulfone but not for the sulfoxides, as the transformation of benzyl phenyl sulfide into the corresponding sulfoxide takes place in just 5 min under purely thermal conditions (oil bath, 140 °C).

The oxidation of 3-sulfanyl alcohols with an excess of sodium metaperiodate in acetonitrile-water at room temperature leads either to disulfides, sultines, or a mixture of both types of compounds depending on the amount of oxidant reagent added (eq 24).[42]

Dihydroselenopyrans with an electron-withdrawing group in the α-position undergo an oxidative ring contraction with sodium periodate at room temperature in MeOH/H_2O to afford selenophenes in low yields (eq 25).[43] m-CPBA is also a useful reagent for this transformation, but a high excess of this oxidant is required to obtain the selenophene products in comparable yields to when $NaIO_4$ is employed. In contrast, the use of 1.5 equiv of m-CPBA produces a mixture of benzoates and selenopyrans in moderate to good combined yields and only traces of the selenophenes are detected.

Oxidation of Thioureas, Selenoureas, and Related Systems. The addition of sodium periodate to 1,3-disubstituted thioureas in water at ambient temperature leads to the corresponding ureas in good yields; on the other hand, a highly polar DMF-water medium is needed when trisubstituted thioureas are employed.[44] Nevertheless, other oxidants, such as sodium chlorite and ammonium persulfate, provide similar results although longer reaction times and higher temperatures are required. Moreover, 1,3-disubstituted aryl- or alkyl thioureas are transformed into guanidines in high yields in DMF-H_2O media at room temperature, or at 80–85 °C in certain cases, using $NaIO_4$ as oxidant in the presence of ammonia or of a wide array of amines (eq 26).[45] A catalytic amount of triethylamine accelerates the reaction for the aliphatic analogs. Once more, sodium chlorite is also a useful oxidative agent in the guanidine formation, although higher temperatures are occasionally required. In contrast, the oxidation of trisubstituted thioureas is unsuccessful with both reagents.

The addition of sodium periodate to selenoureas in refluxing DMF leads to carbodiimides in moderate to high yields, depending on the steric hindrance of the substituents (eq 27).[46] However, a mixture of the carbodiimide and the corresponding urea derivative are provided in low combined yield if the reaction is performed either at room temperature or during longer reaction times. Also, only moderate yields of the urea are obtained when other oxidants such as $NaClO_4$, $KMnO_4$, and Na_2CrO_4 are employed. Interestingly, neither the dimerization of the seleneourea nor the dimerization or trimerization of the carbodiimide is observed, whereas those processes are the major pathways with other oxidants such as HCl and ferricyanide or H_2O_2, respectively.

Dimethylthiocarbamate has been suggested as a potential protecting group for alcohols due to its high compatibility with a wide range of reaction conditions, functional groups, and cleavage protocols used for conventional alcohol protecting groups.[47] In fact, it can be easily removed by heating at 45 °C in MeOH/H_2O in the presence of an excess of sodium periodate followed by hydrolysis under basic conditions to provide the corresponding alcohol derivative in high yield (eq 28). Alternatively, the cleavage can be performed by using hydrogen peroxide as oxidant under basic conditions. Significantly, both the addition and the cleavage of the dimethylthiocarbamate group are widely compatible with protecting groups such as silyl ethers, aryloxy derivatives, acetates, and MOM and MEM ethers, among others.

$$TBSO \overbrace{}_{5} O \overset{S}{\underset{\underset{|}{N}}{||}} \xrightarrow[\substack{MeOH/H_2O \\ 95\%}]{NaIO_4} TBSO \overbrace{}_{5} OH \quad (28)$$

Oxidation of Alkenes. Sodium periodate is efficiently used as the oxygen source for the catalyzed epoxidation of olefins with manganese tetraphenylporphyrin chloride in the presence of imidazole as a donor ligand.[48] The reaction is conducted in a dichloromethane-water media using tetrabutylammonium bromide as a phase transfer catalyst to afford the resultant epoxides in short reaction times and high yields. Alternatively, tetrabutylammonium periodate is efficiently employed as an oxidizing agent with the same catalytic system in a single phase medium of dichloromethane.[49] Various supported Mn(III)-porphyrin catalyst systems have been developed for alkene epoxidation with comparable activities to their homogeneous counterparts in terms of the yields of the transformation.[50] Additionally, the supported catalyst can be recycled with minimum loss of activity. Furthermore, ultrasonic irradiation accelerates the heterogeneous olefin epoxidation described above and, more interestingly, the side reactions, such as double bond cleavage and allylic oxidation, are minimized.[50c]

On the other hand, an asymmetric version of the epoxidation can be achieved using Salen-manganese complexes (**2**) instead of porphyrins (eq 29).[51] Nevertheless, modest yields and selectivities are obtained when $NaIO_4$ is used, whereas the previously described single phase process with n-Bu_4NIO_4 leads to higher asymmetric inductions.

$$Ph \diagup\!\!\!\diagdown \xrightarrow[\substack{NaIO_4/imidazole \\ rt}]{catalyst\ \mathbf{2}} \overset{O}{\triangle}\!\!\!\diagdown^{'''}Ph \quad (29)$$

60% yield
46% ee

2

The transformation of olefins to iodohydrins can be achieved using iodohydroxylation agents such as hypoiodous acid (IOH). That reagent is generated in situ by treating periodic acid or sodium periodate with an appropriate reducing agent such as sodium bisulfate ($NaHSO_3$).[52] The reaction takes place at room temperature in CH_3CN/H_2O solvent system, providing the iodohydrin derivative in good yield when $H_5IO_6/NaHSO_3$ is employed. If a combination of $NaIO_4/NaHSO_3$ is used, the reaction remains incomplete unless it is conducted in strong acid media (pH = 1).

In contrast, the addition of the former reagent combination to *trans*-homoallylic alcohols leads to small amounts of the corresponding iodohydrins whereas the main product is iodo-substi-

tuted tetrahydrofuran (**3**) (eq 30).[53] Both compounds probably arise from an iodonium intermediate that can evolve either by an intramolecular stereoselective cyclization to afford the furan derivatives or by nucleophilic addition of water to furnish the iodohydrins. However, the reaction is highly dependent on the alkene employed: *cis*-disubstituted homoallylic alcohols and homoallylic alcohols having a terminal double bond lead to a mixture of hydroxy-substituted tetrahydrofurans and iodohydrins, due to the preference of the initially formed iodonium intermediate to undergo nucleophilic attack by H_2O rather than the intramolecular cyclization.

$$(30)$$

3   61%   3%

On the other hand, sodium periodate has been shown to oxidize alkali metal halides in aqueous media and promotes the halogenation of olefins in a regio- and diastereoselective fashion.[54] Thus, $NaIO_4$ (25 mol %) in the presence of sodium- or lithium halides under acidic conditions (usually 30% aqueous sulfuric acid) may selectively furnish three different halogenated compounds, depending on the reaction conditions used, in high yield (eq 31). Therefore, the corresponding bromohydrin (**4**) is obtained when a mixture of CH_3CN/H_2O is used as solvent; a $MeOH/H_2O$ mixture causes the formation of brominated methoxy compounds (**5**) and, finally, dibromides (**6**) are favored in acetic acid solution. Significantly, an asymmetric version of the bromohydroxylation, followed by an epoxidation reaction, has been developed; thus, the enantioselective synthesis of phenyl epoxides is achieved through the formation of a β-cyclodextrin complex of the corresponding styrenes, albeit with chiral inductions up to 55%.

$$(31)$$

Aromatic Halogenation. Sodium periodate has been described as a versatile oxidant in the iodination of arenes. Therefore, both activated and deactivated arenes are synthesized using strongly electrophilic I^+ reagent prepared from diiodine in the presence of sodium periodate as oxidant (eq 32).[55] The reaction proceeds at room temperature under anhydrous acidic conditions or in concentrated sulfuric acid to afford mono- and diiodinated products selectively depending on the amount of the arene employed. In addition, other oxidants such as $CrCO_3$, $KMnO_4$, active MnO_2, HIO_3, or $NaIO_3$ provide similar results.

$$ (32) $$

The sodium periodate-mediated oxidation of alkali halides is a useful method to accomplish the halogenation of aromatic compounds. Accordingly, NaCl is successfully employed as a chlorine source for the chlorination of various aromatic compounds, although mixtures of regioisomers are sometimes obtained; on the other hand, the related sodium periodate-mediated bromination reaction, which uses lithium bromide or sodium bromide as halogen source, affords the expected brominated products in a regioselective fashion (eq 33).[54]

$$ (33) $$

Oxidation of Halocompounds. Sodium periodate has been presented as a useful oxidant for the conversion of iodoarenes to the corresponding iodylarenes $(ArIO_2)$.[56] The reaction takes place in boiling aq acetic acid solutions giving the iodylarene products in good yields after 3–6 h of reaction. However, these reaction conditions fail to transform 2-iodobenzoic acid into 2-iodylbenzoic acid (IBX); instead, the reaction product, in nearly quantitative yield, is the stable tautomeric form of 2-iodosylbenzoic acid (eq 34).

$$ (34) $$

Primary and secondary halides are oxidized to the corresponding aldehydes or ketones, respectively, in good yields and short reaction times using sodium periodate in DMF at reflux (eq 35).[57] In an interesting result, a one carbon moiety is lost when phenacyl bromide furnishes benzaldehyde as a single isolated product un-

der such reaction conditions. This result may be explained considering that the initial oxidation does not stop at the expected phenyl glyoxal; instead, it proceeds to form phenylglyoxalic acid, which finally undergoes a decarboxylation to give the observed benzaldehyde.

$$ (35) $$

Formation of Stable Radicals. The oxidation of N,N'-dihydroxyimidazolidines (**7**) with sodium periodate or silver(I) oxide affords several stable oligopyridine-based nitronyl-nitroxide biradicals (**8**) in moderate to good yields, under phase transfer conditions (dichloromethane-water medium) at 0 °C (eq 36).[58] Also, the related bis-N-hydroxyimidazolidines react in a similar way furnishing the corresponding imino-nitroxide biradicals albeit in modest yields.[59]

$$ (36) $$

Furthermore, optically active and racemic α-nitroyl nitroxides having a stereogenic center at the 4-position are prepared by treating the 1,3-dihydroxyimidazolidine precursors with aq $NaIO_4$ in chloroform at room temperature.[60] Additionally, N-tert-butylhydroxylamine derivatives (**9**) are suitable precursors for the high yielding synthesis of the corresponding 2'-deoxyribofuranyl-purines (**10**), which bear an N-tert-butylaminoxyl radical, by reaction with 1 equiv of sodium periodate in a biphasic medium at ambient temperature (eq 37).[61]

(37)

Similarly, sodium periodate is efficiently employed in the preparation of 1,5-dimethyl-6-oxoverdazyl radicals (**11**) in modest to high yields at room temperature using a MeOH/H$_2$O solvent (eq 38). Other oxidizing reagents such as ferricyanide, lead(IV) oxide, or silver oxide seem to generate the radical structures although all the isolation attempts were unsuccessful.[62]

(38)

On the other hand, the addition of sodium periodate supported on silica gel, which had been previously used for the oxidation of quinols to quinones,[10a] to 6-methoxybenzofuran-5-ol derivative (**12**) in dichloromethane promotes a single-electron oxidation process which leads to a stable semiquinone radical (**13**) instead of the expected o-quinone (eq 39).[63] Other oxidants including AgO, CAN, DDQ, or nonsupported NaIO$_4$ give complex mixtures of compounds.

(39)

Other Applications. Activated aryl N,N-dimethylenamines are transformed into aldehydes at room temperature in aq THF by oxidative cleavage using an excess of NaIO$_4$ alone (eq 40).[64] It has been previously described that such scission occurs with a wide range of reagents including ozone, singlet oxygen, sodium dichromate, RuO$_4$/NaIO$_4$, or OsO$_4$/NaIO$_4$. However, the sodium periodate method cited above is very convenient because the oxidative rupture can be achieved without a transition metal catalyst, the major drawback being its failure with unactivated derivatives. Indeed, the method has been efficiently employed to prepare nitroquinolinecarbaldehydes as part of the synthesis of substituted phenanthrolines.[64a]

(40)

β-Amino acids are synthesized by an oxidative cleavage of 2,3-dihydropyridones using sodium periodate.[65] Other oxidants employed such as KMnO$_4$, KMnO$_4$/NaIO$_4$, RuCl$_3$/NaIO$_4$, or OsO$_4$/NaIO$_4$ require harsh conditions and lead to complex mixtures of products where the desired compound is obtained in low yield. The reaction with sodium periodate gives the β-amino acid together with variable amounts of the N-formyl amino acid which is formed in a competitive reaction pathway. However, an improved one-pot procedure that consists on the addition at room temperature of an excess of the oxidant and subsequent alkaline hydrolysis of the crude mixture, affords the β-amino acid as a single product in good yield, which is isolated, following an ion-exchange chromatography, as its hydrochloride derivative (eq 41). Importantly, the method is useful for unprotected 2,3-dihydropyridones and for the synthesis of β-amino acids bearing functional groups sensitive to other oxidizing agents. On the other hand, the addition of Fmoc chloride to the crude product affords the N-protected amino acid in high yield, which can even be improved if, previously to the Fmoc protection, the crude product is purified through cation exchange chromatography.

(41)

Chiral and achiral acylnitroso dienophiles can be generated by oxidation of N-acyl hydroxylamines with sodium periodate; such intermediates are trapped in situ with cyclopentadiene to undergo a hetero-Diels-Alder reaction. Good yields, although moderate diastereoselectivities, are achieved when this transformation is performed from sulfonamide-containing hydroxamic acids (**14**) to give bicyclic cycloadducts (**15**) (eq 42); in fact, little variation in diastereoselectivity was observed upon the use of the sulfonamides relative to other α-amino protective groups. Other oxidants such as Dess-Martin periodinane, Bu$_4$NIO$_4$, Et$_4$NIO$_4$, or Swern conditions, were also tested, without significant improvements in terms of yields and diastereoselectivities.[66]

$$(42)$$

Sodium periodate supported on wet silica gel can be used for the conversion of ketoximes to the corresponding ketones under microwave irradiation.[67] The reaction needs 2 equiv of the oxidant and less than 2 min to be complete (eq 43); on the other hand, the alternative heating protocol affords the products after 36 h at 110 °C.

$$(43)$$

The treatment of trifluoroacetyl-substituted hydrazones with sodium periodate in THF/H$_2$O (in a 2/3 ratio), usually at room temperature, produces their cyclization to trifluoromethylated oxadiazine derivatives (**16**) in good yields (eq 44).[68] The reaction is highly solvent dependent and considerable amounts of oxadiazines (**17**) are observed when the solvent ratio is changed to 2/1. The cyclization to oxadiazines (**17**) may be accomplished in high yield using other acid catalysts such as silica gel, trifluoroacetic acid, or hot acetic acid.

$$(44)$$

A mild and efficient method for the deprotection of silyl ethers has been reported using sodium periodate.[69] The authors pointed out that NaIO$_4$ is the authentic cleaving reagent rather than the residual periodic acid that can be found in the commercial source of NaIO$_4$, because even at neutral or slightly basic conditions the reaction still takes place. Therefore, TBS, TIPS, TMS, TES, TIBS (triisobutylsilyl), and TPS (triphenylsilyl) groups are effectively removed by reaction with an excess of sodium periodate in THF at room temperature affording the corresponding alcohols in high

yields (eq 45). However, the TBDPS group provides only low yields even at higher temperatures or after longer reaction times.

$$(45)$$

The oxidation of either homoallylic and homopropargylic secondary alcohols to the corresponding ketones can be accomplished using Na$_2$Cr$_2$O$_7$/HNO$_3$ or Na$_2$Cr$_2$O$_7$/H$_2$SO$_4$ as a catalyst system and sodium periodate as the stoichiometric oxidant at or below room temperature.[70] As expected, carboxylic acids are obtained when such conditions are applied to the analogous primary alcohols. Remarkably, the mild conditions employed allow the oxidation of homopropargylic alcohols in high yield, preventing the rearrangement to the allenic isomer (eq 46). Moreover, homoallylic alcohols can be oxidized without isomerization to the conjugated carbonyl compounds.

$$(46)$$

1. (a) Shing, T. K. M., *Comprehensive Organic Synthesis* **1991**, *7*, 703. (b) Sklarz, B., *Q. Rev., Chem. Soc.* **1967**, *21*, 3. (c) House, H. O. *Modern Synthetic Reactions*, 2nd ed.; Benjamin/Cummings: Menlo Park, CA, 1972.

2. (a) Kovar, J.; Baer, H. H., *Can. J. Chem.* **1971**, *49*, 3238. (b) Torii, S ; Uneyama, K.; Ueda, K., *J. Org. Chem.* **1984**, *49*, 1830. (c) Schmid, C. R.; Bryant, J. D.; Dowlatzedah, M.; Phillips, J. L.; Prather, D. E.; Renee, D. S.; Sear, N. L.; Vianco, C. S., *J. Org. Chem.* **1991**, *56*, 4056. (d) Jackson, D. Y., *Synth. Commun.* **1988**, *18*, 337.

3. (a) Leonard, N. J.; Johnson, C. R., *J. Org. Chem.* **1962**, *27*, 282. (b) Johnson, C. R.; Keiser, J. E., *Org. Synth.* **1966**, *46*, 78. (c) Lee, J. B.; Yergatian, S. Y.; Crowther, B. C.; Downie, I. M., *Org. Prep. Proced. Int.* **1990**, *22*, 544.

4. (a) Cinquini, M.; Colonna, S.; Giovini, R., *Chem. Ind. (London)* **1969**, 1737. (b) Entwistle, I. D.; Johnstone, R. A. W.; Varley, J. H., *J. Chem. Soc., Chem. Commun.* **1976**, 61. (c) Masuyama, Y.; Ueno, Y.; Okawara, M., *Chem. Lett.* **1977**, 835.

5. Takata, T.; Tajima, R.; Ando, W., *J. Org. Chem.* **1983**, *48*, 4764.

6. (a) Dolby, L. J.; Booth, D. L., *J. Am. Chem. Soc.* **1966**, *88*, 1049. (b) Rivett, D. E.; Wilshire, J. F. K., *Aust. J. Chem.* **1971**, *24*, 2717.

7. (a) Gatta, F.; Misiti, D., *J. Heterocycl. Chem.* **1989**, *26*, 537. (b) Akimoto, H.; Okamura, K.; Yui, M.; Shiori, T.; Kuramoto, M.; Kikugawa, Y.; Yamada, S.-I., *Chem. Pharm. Bull.* **1974**, *22*, 2614. (c) Hutchinson, C. R.; O'Loughlin, G. J.; Brown, R. T.; Fraser, S. B., *J. Chem. Soc., Chem. Commun.* **1974**, 928.

8. Carlsen, P. H. J.; Katsuki, T.; Martin, V. S.; Sharpless, K. B., *J. Org. Chem.* **1981**, *46*, 3936.

9. Ferraboschi, P.; Azadani, M. N.; Santaniello, E.; Trave, S., *Synth. Commun.* **1986**, *16*, 43.

10. (a) Daumas, M.; Vo-Quang, Y.; Vo-Quang, L.; Le Goffic, F., *Synthesis* **1989**, 64. (b) Liu, K.-T.; Tong, Y.-C., *J. Org. Chem.* **1978**, *43*, 2717. (c) Liu, K.-T.; Tong, Y.-C., *J. Chem. Res. (S)* **1979**, 276. (d) Gupta, D. N.; Hodge, P.; Davies, J. E., *J. Chem. Soc., Perkin Trans. 1* **1981**, 2970.

11. (a) Fleet, G. W. J.; Shing, T. K. M., *J. Chem. Soc., Chem. Commun.* **1984**, 835. (b) Wolfrom, M. L.; Yosizawa, Z., *J. Am. Chem. Soc.* **1959**, *81*, 3477.

12. (a) Trost, B. M.; Salzmann, T. N., *J. Am. Chem. Soc.* **1973**, *95*, 6840. (b) Trost, B. M.; Salzmann, T. N., *J. Org. Chem.* **1975**, *40*, 148.

13. Mikolajczyk, M.; Grzejszczak, S.; Zatorski, A., *J. Org. Chem.* **1975**, *40*, 1979.

14. Mikolajczyk, M.; Zatorski, A., *Synthesis* **1973**, 669.

15. (a) Russel, G. A.; Ochrymowycz, L. A., *J. Org. Chem.* **1970**, *35*, 2106. (b) Evans, D. A.; Bryan, C. A.; Sims, C. L., *J. Am. Chem. Soc.* **1972**, *94*, 2891.

16. (a) Carey, F. A.; Dailey, O. D., Jr.; Hernandez, O.; Tucker, J. R., *J. Org. Chem.* **1976**, *41*, 3975. (b) Carey, F. A.; Dailey, O. D., Jr.; Fromuth, T. E., *Phosphorus Sulfur Silicon* **1981**, *10*, 163.

17. Nieuwenhuyse, H.; Louw, R., *Tetrahedron Lett.* **1971**, 4141.

18. Roush, P. B.; Musker, W. K., *J. Org. Chem.* **1978**, *43*, 4295.

19. Glass, R. S.; Broeker, J. L., *Tetrahedron* **1991**, *47*, 5077.

20. Ogura, K.; Suzuki, M.; Tsuchihashi, G.-I., *Bull. Chem. Soc. Jpn.* **1980**, *53*, 1414.

21. (a) Takata, T.; Kim, Y. H.; Oae, S., *Bull. Chem. Soc. Jpn.* **1981**, *54*, 1443. (b) Kim, Y. H.; Takata, T.; Oae, S., *Tetrahedron Lett.* **1978**, 2305.

22. Hiskey, R. G.; Harpold, M. A., *J. Org. Chem.* **1967**, *32*, 3191.

23. (a) Sevrin, M.; Dumont, W.; Krief, A., *Tetrahedron Lett.* **1977**, 3835. (b) Harirchian, B.; Magnus, P., *J. Chem. Soc., Chem. Commun.* **1977**, 522.

24. Reich, H. J.; Reich, I. L.; Renga, J. M., *J. Am. Chem. Soc.* **1973**, *95*, 5813.

25. Clive, D. L. J., *J. Chem. Soc., Chem. Commun.* **1973**, 695.

26. Adler, E.; Holmberg, K.; Ryrfors, L.-O., *Acta Chem. Scand.* **1974**, *B28*, 883.

27. Becker, H.-D.; Bremholt, T.; Adler, E., *Tetrahedron Lett.* **1972**, 4205.

28. Becker, H.-D.; Bremholt, T., *Tetrahedron Lett.* **1973**, 197.

29. Dolby, L. J.; Rodia, R. M., *J. Org. Chem.* **1970**, *35*, 1493.

30. Wolfrom, M. L.; Bobbitt, J. M., *J. Am. Chem. Soc.* **1956**, *78*, 2489.

31. Yanuka, Y.; Katz, R.; Sarel, S., *Tetrahedron Lett.* **1968**, 1725.

32. Hoffman, J. M., Jr.; Schlessinger, R. H., *J. Chem. Soc., Chem. Commun.* **1971**, 1245.

33. Kotani, E.; Kobayashi, S.; Ishii, Y.; Tobinaga, S., *Chem. Pharm. Bull.* **1985**, *33*, 4680.

34. Corey, E. J.; Enders, D., *Tetrahedron Lett.* **1976**, 11.

35. Bestmann, H.-J.; Armsen, R.; Wagner, H., *Chem. Ber.* **1969**, *102*, 2259.

36. (a) See **Potassium Permanganate**. (b) Zibuck, R.; Seebach, D., *Helv. Chim. Acta* **1988**, *71*, 237.

37. (a) Kirby, G. W.; McLean, D., *J. Chem. Soc., Perkin Trans. 1* **1985**, 1443. (b) Sklarz, B.; Al-Sayyab, A. F., *J. Chem. Soc.* **1964**, 1318. (c) See **Tetraethylammonium Periodate**.

38. Zhong, Y.-L.; Shing, T. K. M., *J. Org. Chem.* **1997**, *62*, 2622.

39. (a) Outram, H. S.; Raw, S. A.; Taylor, R. J. K., *Tetrahedron Lett.* **2002**, *43*, 6185. (b) Dunlap, N. K.; Mergo, W.; Jones, J. M.; Carrick, J. D., *Tetrahedron Lett.* **2002**, *43*, 3923.

40. Floresca, R.; Kurihara, M.; Watt, D. S., *J. Org. Chem.* **1993**, *58*, 2196.

41. Varma, R. S.; Saini, R. K.; Meshram, H. M., *Tetrahedron Lett.* **1997**, *38*, 6525.

42. Yolka, S.; Fellous, R.; Lizzani-Cuvelier, L.; Loiseau, M., *Tetrahedron Lett.* **1998**, *39*, 991.

43. Kataoka, T.; Ohe, Y.; Umeda, A.; Iwamura, T.; Yoshimatsu, M.; Shimizu, H., *J. Chem. Soc., Chem. Commun.* **1993**, 577.

44. Ramadas, K.; Janarthanan, N., *Synth. Commun.* **1997**, *27*, 2357.

45. Ramadas, K.; Janarthanan, N.; Pritha, R., *Synlett* **1997**, 1053.

46. Koketsu, M.; Suzuki, N.; Ishihara, H., *J. Org. Chem.* **1999**, *64*, 6473.

47. Barma, D. K.; Bandyopadhyay, A.; Capdevila, J. H.; Falck, J. R., *Org. Lett.* **2003**, *5*, 4755.

48. Mohajer, D.; Tangestaninejad, S., *J. Chem. Soc., Chem. Commun.* **1993**, 240.

49. Mohajer, D.; Tangestaninejad, S., *Tetrahedron Lett.* **1994**, *35*, 945.

50. (a) Brulé, E.; de Miguel, Y. R., *Tetrahedron Lett.* **2002**, *43*, 8555. (b) Tangestaninejad, S.; Habibi, M. H.; Mirkhani, V.; Moghadam, M., *Synth. Commun.* **2002**, *32*, 3331. (c) Tangestaninejad, S.; Mirkhani, V., *Chem. Lett.* **1998**, 1265.

51. Pietikäinen, P., *Tetrahedron Lett.* **1995**, *36*, 319.

52. Masuda, H.; Takase, K.; Nishio, M.; Hasegawa, A.; Nishiyama, Y.; Ishii, Y., *J. Org. Chem.* **1994**, *59*, 5550.

53. Okimoto, Y.; Kikuchi, D.; Sakaguchi, S.; Ishii, Y., *Tetrahedron Lett.* **2000**, *41*, 10223.

54. Dewkar, G. K.; Srinivasarao, V. N.; Sudalai, A., *Org. Lett.* **2003**, *5*, 4501.

55. (a) Kraszkiewicz, L.; Sosnowski, M.; Skulski, L., *Tetrahedron* **2004**, *60*, 9113. (b) Lulinski, P.; Skulski, L., *Bull. Chem. Soc. Jpn.* **2000**, *73*, 951.

56. (a) Kraszkiewicz, L.; Skulski, L., *Arkivoc* **2003**, *6*, 120. (b) Kraszkiewicz, P.; Skulski, L.; Kraszkiewicz, L., *Molecules* **2001**, *6*, 881.

57. Das, S.; Panigrahi, A. K.; Maikap, G. C., *Tetrahedron Lett.* **2003**, *44*, 1375.

58. Ulrich, G.; Ziessel, R.; Luneau, D.; Rey, P., *Tetrahedron Lett.* **1994**, *35*, 1211.

59. Ulrich, G.; Ziessel, R., *Tetrahedron Lett.* **1994**, *35*, 1215.

60. Shimono, S.; Tamura, R.; Ikuma, N.; Takimoto, T.; Kawame, N.; Tamada, O.; Sakai, N.; Matsuura, H.; Yamauchi, J., *J. Org. Chem.* **2004**, *69*, 475.

61. Kaneko, T.; Aso, M.; Koga, N.; Suemune, H., *Org. Lett.* **2005**, *7*, 303.

62. Barr, C. L.; Chase, P. A.; Hicks, R. G.; Lemaire, M. T.; Stevens, C. L., *J. Org. Chem.* **1999**, *64*, 8893.

63. Green, M. P.; Pichlmair, S.; Marques, M. M. B.; Martin, H. J.; Diwald, O.; Berger, T.; Mulzer, J., *Org. Lett.* **2004**, *6*, 3131.

64. (a) Riesgo, E. C.; Jin, X.; Thummel, R. P., *J. Org. Chem.* **1996**, *61*, 3017. (b) Vetelino, M. G.; Coe, J. W., *Tetrahedron Lett.* **1994**, *35*, 219.

65. Ege, M.; Wanner, K. T., *Org. Lett.* **2004**, *6*, 3553.

66. (a) Vogt, P. F.; Miller, M. J., *Tetrahedron* **1998**, *54*, 1317. (b) Mulvihill, M. J.; Gage, J. L.; Miller, M. J., *J. Org. Chem.* **1998**, *63*, 3357.

67. Varma, R. S.; Dahiya, R.; Saini, R. K., *Tetrahedron Lett.* **1997**, *38*, 8819.

68. Kamitori, Y.; Hojo, M.; Masuda, R.; Fujitani, T.; Sukegawa, K., *Heterocycles* **1991**, *32*, 1693.

69. Wang, M.; Li, C.; Yin, D.; Liang, X.-T., *Tetrahedron Lett.* **2002**, *43*, 8727.

70. Schmieder-van de Vondervoort, L.; Bouttemy, S.; Padrón, J. M.; Le Bras, J.; Muzart, J.; Alsters, P. L., *Synlett* **2002**, 243.

T

Tetrabutylammonium Fluoride[1]

$$\boxed{n\text{-Bu}_4\text{NF}}$$

(TBAF)

[429-41-4] $C_{16}H_{36}FN$ (MW 261.53)

InChI = 1/C16H36N.FH/c1-5-9-13-17(14-10-6-2,15-11-7-3)16-12-8-4;/h5-16H2,1-4H3;1H/q+1;/p-1/fC16H36N.F/h;1h/qm;-1

InChIKey = FPGGTKZVZWFYPV-FKCYTGJCCB

(TBAF·3H$_2$O)

[87749-50-6] $C_{16}H_{42}FNO_3$ (MW 315.59)

InChI = 1/C16H36N.FH.3H2O/c1-5-9-13-17(14-10-6-2,15-11-7-3)16-12-8-4;;;;/h5-16H2,1-4H3;1H;3*1H2/q+1;;;;/p-1/fC16H36N.F.3H2O/h;1h;;;/qm;-1;;;

InChIKey = VEPTXBCIDSFGBF-MCKFDJAWCR

(TBAF·xH$_2$O)

[22206-57-1]

(can be used for most fluoride-assisted reactions; deprotection of silyl groups;[1e] desilylation;[2,3] fluorination;[4] used as a base[5,6])

Alternate Name: TBAF.

Physical Data: TBAF·xH$_2$O: mp 62–63 °C.

Solubility: sol H$_2$O, THF, MeCN, DMSO, anhydrous TBAF is incompatible with halogenated solvents such as CH$_2$Cl$_2$ and CHCl$_3$.

Form Supplied in: trihydrate, 1.0 M solution in THF, and 75 wt % solution in water.

Preparative Method: aqueous **Hydrofluoric Acid** is passed through an Amberlite IRA 410 OH column, followed by an aqueous solution of **Tetrabutylammonium Bromide**. After the resin is washed with water, the combined water fractions are repeatedly evaporated until no water is present. Tetrabutylammonium fluoride is collected as an oil in quantitative yield. TBAF·xH$_2$O: aqueous hydrofluoric acid is passed through an Amberlite IRA 410 OH column, followed by an aqueous solution of tetra-*n*-butylammonium bromide. After the resin is washed with water, the combined water fractions are repeatedly evaporated until no water is present. Tetrabutylammonium fluoride is collected as an oil in quantitative yield. The hydrofluoric acid method has been described most recently by Kumar.[87] TBAF$_{anh}$: under nitrogen, hexafluorobenzene is added to a cold (−50 °C) THF solution of tetra-*n*-butylammonium cyanide. The highly colored solution is stirred for 4 h at −15 °C, cooled to −60 °C, and filtered. The resulting colorless solid is washed with cold THF and the residual solvent is removed in vacuo to yield anhydrous TBAF (60%).[88]

Handling, Storage, and Precautions: use a fume hood for TBAF·xH$_2$O; TBAF$_{anh}$ is hydroscopic; it should be handled and stored under dry argon or nitrogen, and stored in a freezer.

Original Commentary

Hui-Yin Li

DuPont Merck Pharmaceutical Company, Wilmington, DE, USA

Deprotection of Silyl Groups. Tetrabutylammonium fluoride has been used widely as a reagent for the efficient cleav-

age of various silyl protecting groups such as *O*-silyls of nucleosides,[7,8] pyrophosphate,[9] *N*-silyls,[10,11] CO_2-silyl, and *S*-silyl derivatives.[1e] These reactions are often carried out under very mild conditions in excellent yields. Thus it has been used in the synthesis of base-sensitive chlorohydrins (eq 1)[12] and β-lactams.[10,13] 2-(Trimethylsilyl)ethoxymethyl groups can also be effectively removed from various substrates (eq 2).[14–18] Silyl ethers can be converted to esters in one pot when they are treated with TBAF, followed by exposure to acyl chlorides[19,20] or anhydride[21] in the presence of base (eq 3). Treatment of triisopropylsilyl enol ethers with **Iodosylbenzene/Azidotrimethylsilane**, followed by desilylation and elimination with TBAF, gives good yields of the α,β-unsaturated ketones (eq 4).[22,23]

Cyclobutanone alkyl silyl acetals, obtained from [2 + 2] cycloadditions, can be deprotected with 1 equiv of TBAF in THF to give the open-chain cyano esters in excellent yields (eq 5).[24] When 4-chloro-2-cyanocyclobutane alkyl silyl acetals are used as substrates for this reaction, (*E/Z*) mixtures of 2-cyanocyclopropanecarboxylates are obtained by an intramolecular cyclization (eq 6).

11-Membered pyrrolizidine dilactones have been synthesized by treating a trimethylsilylethyl ester with TBAF in MeCN to form an anion, which then undergoes cyclization by displacement of the mesylate.

Desilylation Reagent. Cleavage of carbon–silicon bonds with fluoride has been studied very extensively. TBAF is a very powerful reagent for desilylation of a wide range of silicon-containing

Essential Reactions for Organic Synthesis, First Edition. Edited by Philip L. Fuchs.

compounds, such as vinylsilanes,[2,25,26] alkynylsilanes,[23,27] arylsilanes,[28,29] acylsilanes,[30] β-silyl sulfones,[31–33] and other silane derivatives.[3,34–37] It appears that cleavage of sp-C–Si bonds is more facile than that of sp²-C–Si and sp³-C–Si bonds and that substituted groups, such as phenyl and alkoxyl, can often facilitate cleavage. A dimethylphenylsilyl group can be removed from a vinyl carbon by TBAF with retention of the alkene stereochemistry (eq 7).[2] This method has been applied to the synthesis of terminal conjugated trienes (eqs 8 and 9).[38] The five-membered siloxanes can be desilylated with 3 equiv of TBAF in DMF and this protodesilylation is very sensitive to subtle structure changes (eq 10).[39]

The anions generated in situ by desilylation of silylacetylenes,[40,41] allylsilanes,[42–44] propargylsilanes,[45] α-silyloxetanones,[46] bis(trimethylsilylmethyl) sulfides,[47] and other silane derivatives,[48–51] can undergo nucleophilic addition to ketones and aldehydes (eq 11).[52] N-(C,C-bis(trimethylsilyl)methyl) amido derivatives can add to aldehydes followed by Peterson alkenation to form acyl enamines.[48,53] Treatment of 2-trimethylsilyl-1,3-dithianes can generate dithianyl anions, which are capable of carbocyclization via direct addition to carbonyl or Michael addition (eq 12). The fluoride-catalyzed Michael additions are more general than Lewis acid-catalyzed reactions and proceed well even for those compounds with enolizable protons and/or severe steric hindrance (eq 13).[54,55]

Direct fluoride-induced trifluoromethylation of α-keto esters (eq 14),[56] ketones,[57] aldehydes,[58,59] and sulfoxides[59] have been reported using *Trifluoromethyltrimethylsilane* with TBAF in THF.

Desilylation of some compounds can generate very reactive species such as benzynes,[60] pyridynes,[61] xylylenes,[62,63] and benzofuran-2,3-xylylenes.[64] 1,4-Elimination of o-(α-trimethylsilylalkyl)benzyltrimethylammonium halides with TBAF in acetonitrile generates o-xylylenes, which undergo intermolecular and intramolecular cycloadditions (eq 15).[62–64] Treatment of α-silyl disulfides with *Cesium Fluoride* or TBAF forms thioaldehydes, which have been trapped by cycloaddition with cyclopentadiene (eq 16).[65]

exo:endo = 1:7

Use as a Base. TBAF has been widely used for a variety of base-catalyzed reactions such as alkylation,[66] elimination,[67] halogenation,[68] Michael addition,[69–71] aldol condensation, and intramolecular cyclizations.[5,72–74] It is especially useful when other inorganic bases face solubility problems in organic solvents. The reactions are usually carried out below 100 °C due to the low thermal stability of TBAF.[1e]

TBAF is very useful for alkylation of nucleic acid derivatives. Methylation[75] or benzylation[66] of uracil gives almost quantitative yields of alkylated product when using alkyl bromides, dialkyl sulfates (eq 17), trialkyl phosphates, or alkyl chlorides with TBAF. Alkylation of the thiol anions generated from deprotection by 1,2-dibromoethane produces interesting tetrachalcogenofulvalenes.[14] Under phase-transfer conditions, selective mono- and dialkylations of malononitrile have been achieved by using neat TBAF with *Potassium Carbonate* or *Potassium tert-Butoxide* and controlling the amount of alkyl bromides or iodides used (eqs 18 and 19).[76]

$$\text{(17)}$$

$$\text{(18)}$$

$$\text{(19)}$$

Enol silyl ethers react with aldehydes with a catalytic amount of TBAF to give the aldol silyl ethers in good yields. These reactions generally proceed under very mild conditions and within shorter periods of time than conventional strong acidic or basic conditions. The products from 4-t-butyl-1-methyl-2-(trimethylsilyloxy) cyclohexene and benzaldehyde show very good axial selectivity and a little *anti–syn* selectivity (eq 20).[77] The aldol condensation of ketones and aldehydes can be achieved in one pot when ethyl (trimethylsilyl)acetate is used as a silylation agent with TBAF (eq 21).

$$\text{(20)}$$

$$\text{(21)}$$

Silyl nitronates undergo aldol condensation with aldehydes in the presence of a catalytic amount of anhydrous TBAF to form highly diastereoselective *erythro* products, which can be elaborated to give synthetically useful 1,2-amino alcohols (eq 22).[6,78] A one-pot procedure has been developed for direct aldol condensation of nitroalkanes with aldehydes by using TBAF trihydrate with **Triethylamine** and **tert-Butyldimethylchlorosilane**.[79] It appears that silyl nitronates are not reactive intermediates in this case, and the reactions proceed by a different mechanism.

$$\text{(22)}$$

Miscellaneous. Fluoride ion from anhydrous TBAF undergoes nucleophilic displacement of tosylates,[4,80] halides,[80] and aryl nitro compounds[81] to give fluorinated products. When used with **N-Bromosuccinimide**, bromofluorination products are obtained.[82]

Several important peptide-protecting groups such as 9-fluorenylmethyloxycarbonyl,[83] benzyl,[84] 4-nitrobenzyl,[85] 2,2,2-trichloroethyl,[85] and acetonyl (eq 23)[86] can be removed by TBAF under mild conditions.

$$\text{(23)}$$

First Update

Haoran Sun & Stephen G. DiMagno

University of Nebraska-Lincoln, Lincoln, NE, USA

TBAF$_{anh}$.[88] Quaternary ammonium fluoride salts have been prepared by neutralization of quaternary ammonium hydroxides with aqueous HF,[89] liberation of cations from an ion-exchange column with aqueous HF, reaction of AgF with tetraalkylammonium halides,[90] and anion exchange with KF.[91] These aqueous preparative methods have limited the synthetic chemistry of TBAF to hydrates of the salt. The reason for this limitation is clear; attempts to remove the strongly hydrogen bonded water molecules from TBAF·xH$_2$O require conditions that lead to decomposition of the cation. Formation of butene, tributylamine, and bifluoride ion is observed during drying TBAF·3H$_2$O under dynamic vacuum at moderate (40 °C) temperature.[92] These observations led Sharma and Fry to conclude in 1983 that "it is very unlikely that pure, anhydrous tetraalkylammonium fluoride salts have ever, in fact, been produced in the case of ammonium ions susceptible to E2 eliminations".[92] Recently, low-temperature nucleophilic aromatic substitution reactions performed in anhydrous polar aprotic solvents (eq 24) have been shown to yield anhydrous tetra-n-butylammonium fluoride (TBAF$_{anh}$) for the first time.[88] Although the chemistry of this new form of the reagent has yet to be explored systematically, TBAF$_{anh}$ has shown enhanced reactivity (relative to TBAF·xH$_2$O) in nucleophilic fluorination reactions.

$$\text{(24)}$$

Nucleophilic Fluorination with TBAF. TBAF$_{anh}$ fluorinates primary alkyl halides, tosylates, and mesylates within minutes at or below room temperature (eq 25).[88] The pronounced nucleophilicity of the anhydrous reagent is also exhibited in nucleophilic aromatic substitution (S_NAr) reactions. In polar aprotic solvents such as acetonitrile and DMSO, aromatic halides undergo halogen exchange (Halex) reactions with $TBAF_{anh}$ to form the corresponding fluoroaromatics (eq 26), and activated nitroaromatic compounds are fluorodenitrated in high yield (eq 27).[93] Various functional groups, including aryl esters, aldehydes, ketones, and N-benzyl protecting groups, are compatible with these fluorination conditions.[93]

$$RX + TBAF \xrightarrow[\text{$-40\,^\circ$C to rt}]{\text{polar aprotic solvents}} RF + TBAX \qquad (25)$$

R = primary (secondary) alkyl; X = Cl, Br, I, OTs, OMs

yield 95% (26)

yield 95% (27)

The nucleophilicity of commercial $TBAF \cdot xH_2O$ in S_NAr reactions is attenuated significantly compared to that of $TBAF_{anh}$. Substrates are limited to strongly activated nitropyridines and nitroarenes. At rt or at elevated temperature in DMF, fluorodenitration is relatively efficient, although hydrolysis is a competing side reaction.[94] Another possible example of a $TBAF \cdot xH_2O$ assisted S_NAr denitration reaction was reported recently. 2-Aryl-5-substituted-2,3-dihydrobenzofurans were synthesized in good yields (eq 28)[95] by a TBAF-promoted displacement of an aromatic nitro group by an alkoxide. A fluoroaromatic compound may be an intermediate in this reaction sequence.

yield 93% (28)

TBAF-promoted C–C Bond Formation. The repertoire of fluoride-promoted, transition metal ion-catalyzed coupling reactions has expanded significantly of late. Metal ions that form strong bonds to fluoride and carbon are key to fluoride promoted activation strategies. For example phenyl boronic acids,[96] and organotin compounds (eq 29) are among the stable organometallic species used in fluoride-promoted C–C bond forming reactions.[97,98] With tin compounds, a hypervalent organotin fluoride is assumed to be the active species.[98] The fluorophilicity of silicon, tin, and boron (Si–F BDE = 144.7 kcal mol^{-1}, Sn–F BDE = 146.1 kcal mol^{-1}, and B–F BDE = 171.5 kcal mol^{-1})[99,100] provides the driving force for organic group transfer.

Organosilicon compounds are the most commonly used, stable surrogates for carbanion donors in fluoride-promoted coupling reactions; transmetalation of alkyl, alkenyl, alkynyl, or aryl groups can occur under relatively mild reaction conditions. Organosilicon cross-coupling precursors include siletanes, silanols,[101] silyl ethers, siloxanes, polysiloxanes, halosilanes, and silyl hydrides. Transfer of strongly basic alkyl groups generally requires an activated organosilicon compound possessing at least one Si–heteroatom bond. Fluoride promoted coupling reactions have been used to form alkyl–aryl, aryl–aryl, alkyl–vinyl, and alkyne–alkyne bonds.

Lee and Fu reported that the TBAF-promoted, Pd-catalyzed coupling of alkyl halides with arylsilanes gave fairly good yields (65–81%) of the alkylated arenes at room temperature. The catalyst was formed *in situ* from the electron-rich phosphine P(t-Bu)$_2$Me and PdBr$_2$ (eq 30).[102] TBAF was used in excess.

yield 81% (30)

TBAF can promote biaryl formation from triallyl(aryl)silanes and a wide variety of aryl halide substrates. For example, chlorobenzenes undergo smooth cross-coupling with triallyl(aryl)silanes to give biaryls in good to excellent yields (eq 31).[103] Surprisingly, an appreciable amount of added water is required to achieve high yields; however, the role of added water is not yet clear.[103] In a recent variant of the Hiyama reaction, TBAF and palladium-coated nickel nanoclusters were employed to prepare biaryl compounds from aryl bromides and phenyltrimethoxysilane.[104]

$$\text{(31)}$$

$$\text{yield 100\%}$$

$$\text{(34)}$$

$$\text{yield 83\%}$$

An inexpensive vinylpolysiloxane, 1,3,5,7-tetramethyl-1,3,5,7-tetravinylcyclotetrasiloxane, (D_4V) furnished substituted styrenes from haloarenes by palladium-catalyzed coupling in the presence of TBAF as an activator (eq 32).[105] Simple phenyldimethylvinylsilanes have also been used directly in TBAF-promoted alkene-coupling reactions (eq 33).[106]

$$\text{(35)}$$

$$\text{yield 63\%} \qquad \text{18\%}$$

Initial forays by the Denmark group investigated the use of silanols in fluoride-promoted coupling reactions that demonstrated high yielding, stereospecific vinyl–aryl coupling reactions.[110] Later, unsymmetrical 1,4-disubstituted 1,3-butadienes were successfully prepared in high yields by similar TBAF promoted cross coupling reactions (eq 36).[111] The mechanism of these reactions was elucidated through spectroscopic and kinetic analyses.[110,112] With silanols, the potent activator TBAF is not always necessary, since these coupling reactions can be catalyzed by base; when TBAOH was used in place of TBAF, similar coupling efficiencies were achieved (eq 37).[113]

$$\text{(32)}$$

$$\text{yield 86\%}$$

$D_4V = 1,3,5,7$ tetramethyl-1,3,5,7-tetravinylcyclotetrasiloxane

$$\text{(36)}$$

$$\text{yield 86\%}$$

$$\text{(33)}$$

$$\text{yield 85\% (90\% trans)}$$

In the presence of TBAF and electron-rich Pd phosphine complexes, aryl chlorides react smoothly with arylchlorosilanes and alkenylchlorosilanes to form substituted arenes. The starting (E or Z) geometry of C–C double bond of alkenylchlorosilanes was maintained in the products (eq 34).[107] Trimethoxyvinylsilane was also used in arene vinylation (eq 35), though homocoupling side products were observed with this reagent.[108,109]

$$\text{(37)}$$

$$\text{yield 81\%}$$

Aryl and alkyl-substituted diynes and tetraynes have been synthesized in good yields (82–99%) by TBAF-promoted desilylation and Cu-catalyzed oxidative dimerization of triisopropylsilyl (TIPS)-protected acetylenes (eq 38). Copper acetate was used as oxidant in this reaction.[114] Aryl- and alkenyl alkynes were made under similar conditions (eq 39).[115] Pd/C with TBAF was used in ligand- and copper-free, one-pot, 'domino' Halex–Sonogashira reactions.[116] Similarly, TBAF promoted the synthesis of 2-substituted indoles by a tandem Sonogashira/cyclization reaction of 2-iodoanilines and terminal alkynes.[117]

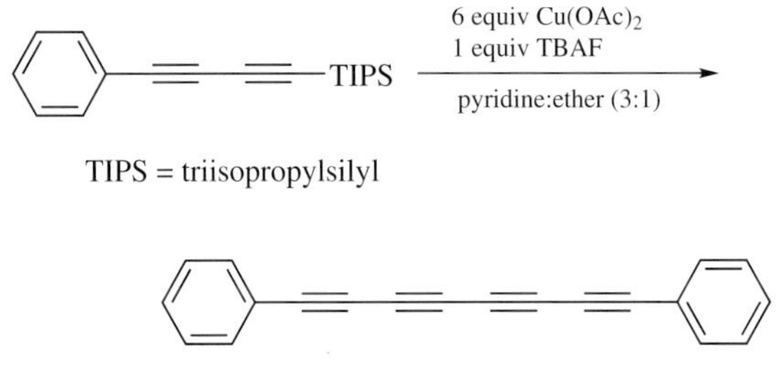

TIPS = triisopropylsilyl

(38)

yield 100%

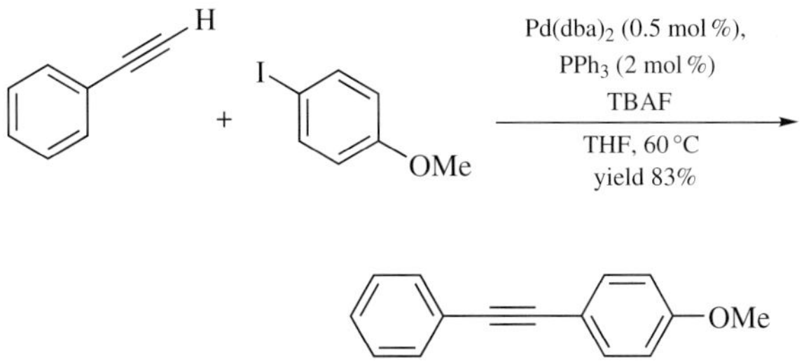

(39)

Catalytic amounts of TBAF promote the addition of organosilanes to carbonyl compounds. Examples of this approach include allylation of aldehydes to give allylic alcohols (eq 40)[118] and the preparation of α-fluoroallylic alcohols from benzaldehyde (eq 41).[119]

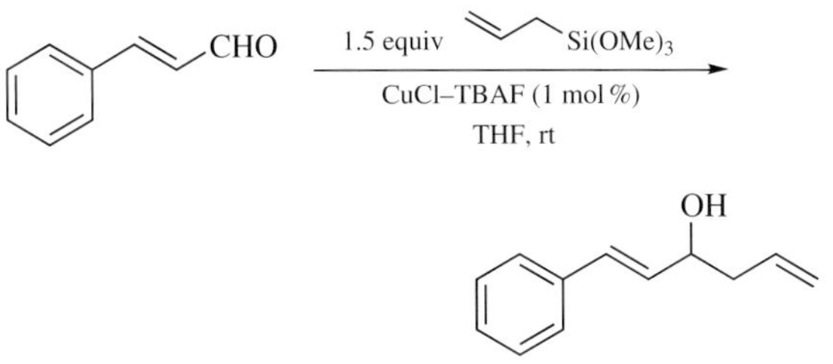

(40)

yield 92%

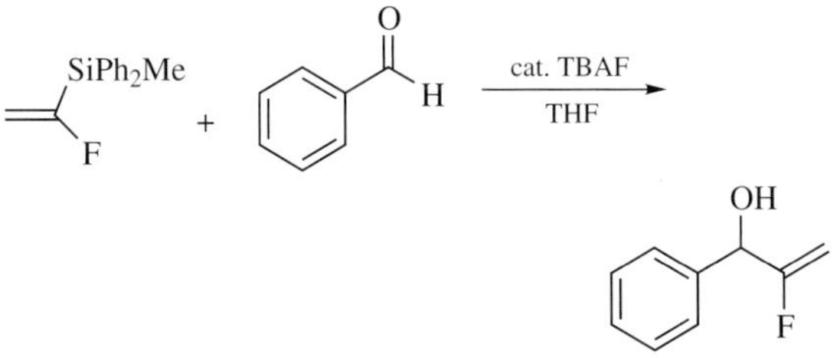

(41)

yield 61%

TBAF-promoted C–N and C–S Bond Formation. Fluoride can induce C–N and C–S bond formation in reactions of organosilanes with nitrogen- or sulfur-containing electrophiles such as imines, nitriles, nitro compounds, or sulfines, or by oxidative coupling of organosilanes with nitrogen or sulfur nucleophiles. Copper-promoted N-arylation and N-vinylation gives good yield of N-phenylbenzimidazole and N-vinylbenzimidazole (eq 42).[120] Allyl and benzyl sulfoxides were prepared from sulfines by fluoride-induced C–S coupling (eq 43).[121] More interestingly, a combination of chiral palladium catalyst and TBAF promoted asymmetric allylation of imines with tetraallylsilane yielded chiral benzylic amines in good yield and high enantiomeric excess (eq 44).[122]

TBAF efficiently unmasks nucleophilic azide for C–N bond formation. TBAF stimulated the reaction of benzonitrile with $TMSN_3$ to form 5-substituted 1H-tetrazoles (eq 45).[123] Under similar conditions, 4-aryl-1H-1,2,3-triazoles can be synthesized by TBAF-catalyzed cycloadditions of electron poor nitroolefins (eq 46).[124,125]

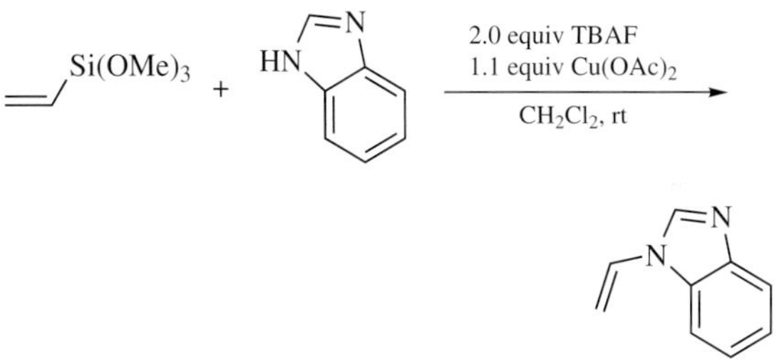

(42)

yield 88%

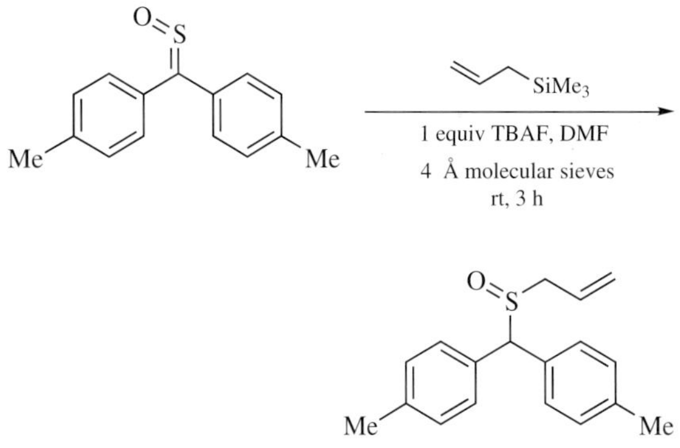

(43)

yield 68%

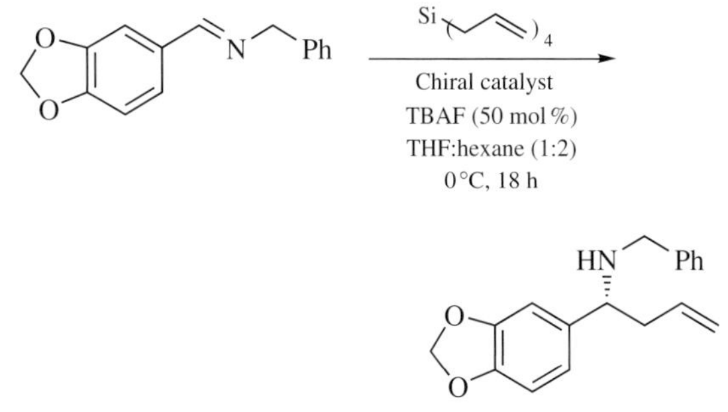

(44)

yield 80%, 89% ee

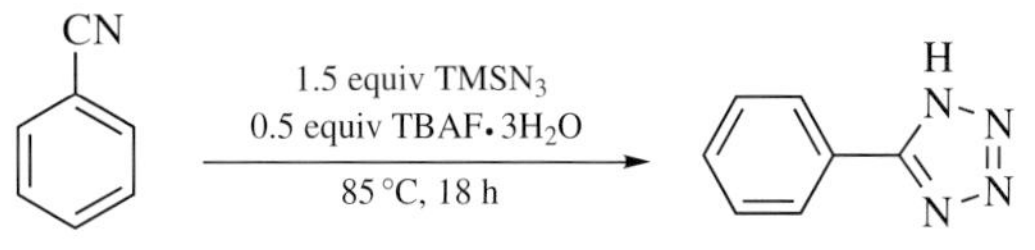

(45)

yield 86%

$$\text{(46)}$$

TBAF-assisted Trifluoromethylation, Perfluoroalkylation, and Polyfluoroalkoxylation.[126] TBAF combines with TMSCF$_3$ to form nucleophilic trifluoromethyl anion equivalents. The precise structure of this nucleophile is unknown, although it is likely that a pentacoordinate silicon intermediate is involved. The TBAF/TMSCF$_3$ combination leads to smooth trifluoromethylation of aldehydes,[127,128] ketones,[127,128] esters,[129] imines,[130] nitroso compounds, α-ketoesters, α-ketoamides (eq 47),[131] and sulfur-based electrophiles (eq 48)[127] in fairly good to excellent yields. A catalytic amount of TBAF suffices to promote trifluoromethylation of aldehydes and ketones with TMSCF$_3$, as the initially formed α-trifluoromethyl alkoxide is adequately nucleophilic to unmask TMSCF$_3$ (eq 49).[128,132] Trifluoromethyl ketones, potential inhibitors of hydrolytic enzymes, are readily prepared by reaction of TMSCF$_3$ with esters in presence of 2.5 mol % molecular sieve-dried TBAF (eq 50).[129] The use of TBAF$_{anh}$ in trifluoromethylation reactions has not yet been reported.

$$\text{(47)}$$

yield 91%

$$\text{(48)}$$

$$\text{(49)}$$

yield 78%

$$\text{(50)}$$

yield 95%

Polyfluoroalkoxylation of aromatic rings has been achieved by S$_N$Ar reactions of dinitrobenzenes with polyfluoro alcohols in the presence of excess TBAF (eq 51).[133] Finally, TBAF-assisted Horner–Wadsworth–Emmons reactions provide a convenient entry into trifluoromethylated alkenes (eq 52).[134]

$$\text{(51)}$$

yield 98%

$$\text{(52)}$$

yield 98%

Miscellaneous. The uses of TBAF as a silyl group deprotection reagent, a desilylation reagent, and a base in organic synthesis are legion. Recently, highly specialized, mild, and selective deprotection methods featuring TBAF have been optimized for various synthetic applications.[135–141] Under appropriate conditions, even unactivated C$_{(sp3)}$–SiMe$_2$Ph bonds can be cleaved efficiently by TBAF.[142] A relatively recent development is the use of catalytic amounts of TBAF to silylate primary and secondary alcohols with hydrosilanes, disilanes, and silazanes.[143,144] TBAF was also used as a decomplexation reagent to unmask 1-cyclodecene-3,9-diyne cobalt complexes.[145] TBAF promoted the functionalization of o-carborane with carbonyl compounds to give the corresponding carbinols.[146–148] Finally, TBAF has been touted as an efficient base for rapid N-alkylation of purines.[149]

Concluding Remarks. New uses for TBAF, a highly soluble source of nucleophilic and basic fluoride ion, continue to be developed. The ability to tune the reactivity of TBAF using solvent effects is a key to its exploitation in organic synthesis. The recent synthesis of anhydrous variants of this reagent should further expand the synthetic utility of this reagent.

1. (a) Corey, E. J.; Snider, B. B., *J. Am. Chem. Soc.* **1972**, *94*, 2549. (b) Hudlicky, M. *Chemistry of Organic Fluorine Compounds*, 2nd ed.; Horwood: New York, 1992. (c) Umemoto, T., *Yuki Gosei Kagaku Kyokaishi* **1992**, *50*, 338. (d) Clark, J. H., *Chem. Rev.* **1980**, *80*, 429. (e) Greene, T. W.; Wuts, P. G. M. *Protective Groups in Organic Synthesis*, 2nd ed.; Wiley: New York, 1991. (f) Sharma, R. K.; Fry, J. L., *J. Org. Chem.* **1983**, *48*, 2112. (g) Cox, D. P.; Terpinski, J.; Lawrynowicz, W., *J. Org. Chem.* **1984**, *49*, 3216.

2. Oda, H.; Sato, M.; Morizawa, Y.; Oshima, K.; Nozaki, H., *Tetrahedron* **1985**, *41*, 3257.

3. Dhar, R. K.; Clawson, D. K.; Fronczek, F. R.; Rabideau, P. W., *J. Org. Chem.* **1992**, *57*, 2917.

4. Gerdes, J. M.; Bishop, J. E.; Mathis, C. A., *J. Fluorine Chem.* **1991**, *51*, 149.

5. Pless, J., *J. Org. Chem.* **1974**, *39*, 2644.

6. Seebach, D.; Beck, A. K.; Mukhopadhyay, T.; Thomas, E., *Helv. Chim. Acta* **1982**, *65*, 1101.

7. Krawczyk, S. H.; Townsend, L. B., *Tetrahedron Lett.* **1991**, *32*, 5693.

8. Meier, C.; Tam, H.-D., *Synlett* **1991**, 227.

9. Valentijn, A. R. P. M.; van der Marel, G. A.; Cohen, L. H.; van Boom, J., *Synlett* **1991**, 663.

10. Hanessian, S.; Sumi, K.; Vanasse, B., *Synlett* **1992**, 33.

11. Kita, Y.; Shibata, N.; Tamura, O.; Miki, T., *Chem. Pharm. Bull.* **1991**, *39*, 2225.

12. Solladié-Cavallo, A.; Quazzotti, S.; Fischer, J.; DeCian, A., *J. Org. Chem.* **1992**, *57*, 174.

13. Konosu, T.; Oida, S., *Chem. Pharm. Bull.* **1991**, *39*, 2212.

14. Zambounis, J. S.; Mayer, C. W., *Tetrahedron Lett.* **1991**, *32*, 2737.

15. Kita, H.; Tohma, H.; Inagaki, M.; Hatanaka, K., *Heterocycles* **1992**, *33*, 503.

16. Stephenson, G. R.; Owen, D. A.; Finch, H.; Swanson, S., *Tetrahedron Lett.* **1991**, *32*, 1291.

17. Fugina, N.; Holzer, W.; Wasicky, M., *Heterocycles* **1992**, *34*, 303.

18. Shakya, S.; Durst, T., *Heterocycles* **1992**, *34*, 67.

19. Beaucage, S. L.; Ogilvie, K. K., *Tetrahedron Lett.* **1977**, 1691.

20. Ma, C.; Miller, M. J., *Tetrahedron Lett.* **1991**, *32*, 2577.

21. Mandai, T.; Murakami, T.; Kawada, M.; Tsuji, J., *Tetrahedron Lett.* **1991**, *32*, 3399.

22. Magnus, P.; Evans, A.; Lacour, J., *Tetrahedron Lett.* **1992**, *33*, 2933.

23. Ihara, M.; Suzuki, S.; Taniguchi, N.; Fukumoto, K.; Kabuto, C., *J. Chem. Soc., Chem. Commun.* **1991**, 1168.

24. Rousseau, G.; Quendo, A., *Tetrahedron* **1992**, *48*, 6361.

25. Fleming, I.; Newton, T. W.; Sabin, V.; Zammattio, F., *Tetrahedron* **1992**, *48*, 7793.

26. Ito, T.; Okamoto, S.; Sato, F., *Tetrahedron Lett.* **1990**, *31*, 6399.

27. Lopp, M.; Kanger, T.; Müraus, A.; Pehk, T.; Lille, Ü., *Tetrahedron: Asymmetry* **1991**, *2*, 943.

28. Yu, S.; Keay, B. A., *J. Chem. Soc., Perkin Trans. 1* **1991**, 2600.

29. Mukai, C.; Kim, I. J.; Hanaoka, M., *Tetrahedron: Asymmetry* **1992**, *3*, 1007.

30. Degl'Innocenti, A.; Stucchi, E.; Capperucci, A.; Mordini, A.; Reginato, G.; Ricci, A., *Synlett* **1992**, 329.

31. Kocienski, P. J., *Tetrahedron Lett.* **1979**, 2649.

32. Kocienski, P. J., *J. Org. Chem.* **1980**, *45*, 2037.

33. Hsiao, C. N.; Hannick, S. M., *Tetrahedron Lett.* **1990**, *31*, 6609.

34. Bonini, B. F.; Masiero, S.; Mazzanti, G.; Zani, P., *Tetrahedron Lett.* **1991**, *32*, 2971.

35. Nativi, C.; Palio, G.; Taddei, M., *Tetrahedron Lett.* **1991**, *32*, 1583.

36. Okamoto, S.; Yoshino, T.; Tsujiyama, H.; Sato, F., *Tetrahedron Lett.* **1991**, *32*, 5793.

37. Kobayashi, Y.; Ito, T.; Yamakawa, I.; Urabe, H.; Sato, F., *Synlett* **1991**, 813.

38. Kishi, N.; Maeda, T.; Mikami, K.; Nakai, T., *Tetrahedron* **1992**, *48*, 4087.

39. Hale, M. R.; Hoveyda, A. H., *J. Org. Chem.* **1992**, *57*, 1643.

40. Nakamura, E.; Kuwajima, I., *Angew. Chem., Int. Ed. Engl.* **1976**, *15*, 498.

41. Mohr, P., *Tetrahedron Lett.* **1991**, *32*, 2223.

42. Furuta, K.; Mouri, M.; Yamamoto, H., *Synlett* **1991**, 561.

43. Hosomi, A.; Shirahata, A.; Sakurai, H., *Tetrahedron Lett.* **1978**, 3043.

44. Nakamura, H.; Oya, T.; Murai, A., *Bull. Chem. Soc. Jpn.* **1992**, *65*, 929.

45. Pornet, J., *Tetrahedron Lett.* **1981**, *22*, 455.

46. Mead, K. T.; Park, M., *J. Org. Chem.* **1992**, *57*, 2511.

47. Hosomi, A.; Ogata, K.; Ohkuma, M.; Hojo, M., *Synlett* **1991**, 557.

48. Lasarte, J.; Palomo, C.; Picard, J. P.; Dunogues, J.; Aizpurua, J. M., *J. Chem. Soc., Chem. Commun.* **1989**, 72.

49. Watanabe, Y.; Takeda, T.; Anbo, K.; Ueno, Y.; Toru, T., *Chem. Lett.* **1992**, 159.

50. Paquette, L. A.; Blankenship, C.; Wells, G. J., *J. Am. Chem. Soc.* **1984**, *106*, 6442.

51. Seitz, D. E.; Milius, R. A.; Quick, J., *Tetrahedron Lett.* **1982**, *23*, 1439.

52. Grotjahn, D. B.; Andersen, N. H., *J. Chem. Soc., Chem. Commun.* **1981**, 306.

53. Palomo, C.; Aizpurua, J. M.; Legido, M.; Picard, J. P.; Dunogues, J.; Constantieux, T., *Tetrahedron Lett.* **1992**, *33*, 3903.

54. Majetich, G.; Casares, A.; Chapman, D.; Behnke, M., *J. Org. Chem.* **1986**, *51*, 1745.

55. Majetich, G.; Desmond, R. W.; Soria, J. J., *J. Org. Chem.* **1986**, *51*, 1753.

56. Ramaiah, P.; Prakash, G. K. S., *Synlett* **1991**, 643.

57. Coombs, M. M.; Zepik, H. H., *J. Chem. Soc., Chem. Commun.* **1992**, 1376.

58. Bansal, R. C.; Dean, B.; Hakomori, S.; Toyokuni, T., *J. Chem. Soc., Chem. Commun.* **1991**, 796.

59. Patel, N. R.; Kirchmeier, R. L., *Inorg. Chem.* **1992**, *31*, 2537.

60. Himeshima, Y.; Sonoda, T.; Kobayashi, H., *Chem. Lett.* **1983**, 1211.

61. Tsukazaki, M.; Snieckus, V., *Heterocycles* **1992**, *33*, 533.

62. Ito, Y.; Nakatsuka, M.; Saegusa, T., *J. Am. Chem. Soc.* **1980**, *102*, 863.

63. Ito, Y.; Miyata, S.; Nakatsuka, M.; Saegusa, T., *J. Org. Chem.* **1981**, *46*, 1043.

64. Bedford, S. B.; Begley, M. J.; Cornwall, P.; Knight, D. W., *Synlett* **1991**, 627.

65. Krafft, G. A.; Meinke, P. T., *Tetrahedron Lett.* **1985**, *26*, 1947.

66. Botta, M.; Summa, V.; Saladino, R.; Nicoletti, R., *Synth. Commun.* **1991**, *21*, 2181.

67. Ben Ayed, T.; Amri, H.; El Gaied, M. M., *Tetrahedron* **1991**, *47*, 9621.

68. Sasson, Y.; Webster, O. W., *J. Chem. Soc., Chem. Commun.* **1992**, 1200.

69. Kuwajima, I.; Murofushi, T.; Nakamura, E., *Synthesis* **1976**, 602.

70. Yamamoto, Y.; Okano, H.; Yamada, J., *Tetrahedron Lett.* **1991**, *32*, 4749.

71. Arya, P.; Wayner, D. D. M., *Tetrahedron Lett.* **1991**, *32*, 6265.

72. Taguchi, T.; Suda, Y.; Hamochi, M.; Fujino, Y.; Iitaka, Y., *Chem. Lett.* **1991**, 1425.

73. Ley, S. V.; Smith, S. C.; Woodward, P. R., *Tetrahedron* **1992**, *48*, 1145, 3203.

74. White, J. D.; Ohira, S., *J. Org. Chem.* **1986**, *51*, 5492.

75. Ogilvie, K. K.; Beaucage, S. L.; Gillen, M. F., *Tetrahedron Lett.* **1978**, 1663.

76. Díez-Barra, E.; De La Hoz, A.; Moreno, A.; Sánchez-Verdú, P., *J. Chem. Soc., Perkin Trans. 1* **1991**, 2589.

77. Nakamura, E.; Shimizu, M.; Kuwajima, I.; Sakata, J.; Yokoyama, K.; Noyori, R., *J. Org. Chem.* **1983**, *48*, 932.

78. Colvin, E. W.; Seebach, D., *J. Chem. Soc., Chem. Commun.* **1978**, 689.

79. Fernández, R.; Gasch, C.; Gómez-Sánchez, A.; Vílchez, J. E., *Tetrahedron Lett.* **1991**, *32*, 3225.

80. Cox, D. P.; Terpinski, J.; Lawrynowicz, W., *J. Org. Chem.* **1984**, *49*, 3216.

81. Clark, J. H.; Smith, D. K., *Tetrahedron Lett.* **1985**, *26*, 2233.

82. Maeda, M.; Abe, M.; Kojima, M., *J. Fluorine Chem.* **1987**, *34*, 337.

83. Ueki, M.; Amemiya, M., *Tetrahedron Lett.* **1987**, *28*, 6617.

84. Ueki, M.; Aoki, H.; Katoh, T., *Tetrahedron Lett.* **1993**, *34*, 2783.

85. Namikoshi, M.; Kundu, B.; Rinehart, K. L., *J. Org. Chem.* **1991**, *56*, 5464.

86. Kundu, B., *Tetrahedron Lett.* **1992**, *33*, 3193.

87. Kumar, M. B., *Synlett* **2002**, 2125.

88. Sun, H.; DiMagno, S. G., *J. Am. Chem. Soc.* **2005**, *127*, 2050,

89. Harmon, K. M.; Gennick, I., *Inorg. Chem.* **1975**, *14*, 1840.

90. Hayami, J.; Ono, N.; Kaji, A., *Tetrahedron Lett.* **1968**, 1385.

91. Dermeik, S.; Sasson, Y., *J. Org. Chem.* **1989**, *54*, 4827.

92. Sharma, R. K.; Fry, J. L., *J. Org. Chem.* **1983**, *48*, 2112.

93. Sun, H.; DiMagno, S. G., *Angew Chem., Int. Ed.* **2006**, *45*, 2720.

94. Kuduk, S. D.; DiPardo, R. M.; Bock, M. G., *Org. Lett.* **2005**, *7*, 577.

95. Kuethe, J. T.; Wong, A.; Journet, M.; Davies, I. W., *J. Org. Chem.* **2005**, *70*, 3727.

96. Lee, C. K. Y.; Holmes, A. B.; Ley, S. V.; McConvey, I. F.; Al-Duri, B.; Leeke, G. A.; Santos, R. C. D.; Seville, J. P. K., *Chem. Commun.* **2005**, 2175.

97. Grasa, G. A.; Nolan, S. P., *Org. Lett.* **2001**, *3*, 119.

98. Mee, S. P. H.; Lee, V.; Baldwin, J. E., *Chem. Eur. J.* **2005**, *11*, 3294.

99. Leroy, G.; Temsamani, D. R.; Wilante, C., *Theochem.* **1994**, *112*, 21.

100. Poon, C.; Mayer, P. M., *Can. J. Chem.* **2002**, *80*, 25.

101. Denmark, S. E.; Ober, M. H., *Adv. Synth. Catal.* **2004**, *346*, 1703.

102. Lee, J.-Y.; Fu, G. C., *J. Am. Chem. Soc.* **2003**, *125*, 5616.

103. Sahoo, A. K.; Oda, T.; Nakao, Y.; Hiyama, T., *Adv. Synth. Catal.* **2004**, *346*, 1715.

104. Duran Pachon, L.; Thathagar, M. B.; Hartl, F.; Rothenberg, G., *Phys. Chem. Chem. Phys.* **2006**, *8*, 151.

105. Denmark, S. E.; Butler, C. R., *Org. Lett.* **2006**, *8*, 63.

106. Anderson, J. C.; Anguille, S.; Bailey, R., *Chem. Commun.* **2002**, 2018.

107. Gouda, K.-I.; Hagiwara, E.; Hatanaka, Y.; Hiyama, T., *J. Org. Chem.* **1996**, *61*, 7232.

108. Soli, E. D.; DeShong, P., *J. Org. Chem.* **1999**, *64*, 9724.

109. Mowery, M. E.; DeShong, P., *J. Org. Chem.* **1999**, *64*, 1684.

110. Denmark, S. E.; Wehrli, D.; Choi, J. Y., *Org. Lett.* **2000**, *2*, 2491.

111. Denmark, S. E.; Tymonko, S. A., *J. Am. Chem. Soc.* **2005**, *127*, 8004.

112. Denmark, S. E.; Sweis, R. F.; Wehrli, D., *J. Am. Chem. Soc.* **2004**, *126*, 4865.

113. Denmark, S. E.; Neuville, L., *Org. Lett.* **2000**, *2*, 3221.

114. Heuft, M. A.; Collins, S. K.; Yap, G. P. A.; Fallis, A. G., *Org. Lett.* **2001**, *3*, 2883.

115. Mori, A.; Kawashima, J.; Shimada, T.; Suguro, M.; Hirabayashi, K.; Nishihara, Y., *Org. Lett.* **2000**, *2*, 2935.

116. Thathagar, M. B.; Rothenberg, G., *Org. Biomol. Chem.* **2006**, *4*, 111.

117. Suzuki, N.; Yasaki, S.; Yasuhara, A.; Sakamoto, T., *Chem. Pharm. Bull.* **2003**, *51*, 1170.

118. Yamasaki, S.; Fujii, K.; Wada, R.; Kanai, M.; Shibasaki, M., *J. Am. Chem. Soc.* **2002**, *124*, 6536.

119. Hanamoto, T.; Harada, S.; Shindo, K.; Kondo, M., *Chem. Commun.* **1999**, 2397.

120. Lam, P. Y. S.; Deudon, S.; Averill, K. M.; Li, R.; He, M. Y.; DeShong, P.; Clark, C. G., *J. Am. Chem. Soc.* **2000**, *122*, 7600.

121. Capperucci, A.; Degl'Innocenti, A.; Leriverend, C.; Metzner, P., *J. Org. Chem.* **1996**, *61*, 7174.

122. Fernandes, R. A.; Yamamoto, Y., *J. Org. Chem.* **2004**, *69*, 735.

123. Amantini, D.; Beleggia, R.; Fringuelli, F.; Pizzo, F.; Vaccaro, L., *J. Org. Chem.* **2004**, *69*, 2896.

124. D'Ambrosio, G.; Fringuelli, F.; Pizzo, F.; Vaccaro, L., *Green Chem.* **2005**, *7*, 874.

125. Amantini, D.; Fringuelli, F.; Piermatti, O.; Pizzo, F.; Zunino, E.; Vaccaro, L., *J. Org. Chem.* **2005**, *70*, 6526.

126. Prakash, G. K. S.; Yudin, A. K., *Chem. Rev.* **1997**, *97*, 757.

127. Patel, N. R.; Kirchmeier, R. L., *Inorg. Chem.* **1992**, *31*, 2537.

128. Krishnamurti, R.; Bellew, D. R.; Prakash, G. K. S., *J. Org. Chem.* **1991**, *56*, 984.

129. Wiedemann, J.; Heiner, T.; Mloston, G.; Prakash, G. K. S.; Olah, G. A., *Angew. Chem., Int. Ed.* **1998**, *37*, 820.

130. Prakash, G. K. S.; Mandal, M.; Olah, G. A., *Angew. Chem., Int. Ed.* **2001**, *40*, 589.

131. Singh, R. P.; Kirchmeier, R. L.; Shreeve, J. N. M., *J. Org. Chem.* **1999**, *64*, 2579.

132. Prakash, G. K. S.; Krishnamurti, R.; Olah, G. A., *J. Am. Chem. Soc.* **1989**, *111*, 393.

133. Tejero, I.; Huertas, I.; Gonzalez-Lafont, A.; Lluch, J. M.; Marquet, J., *J. Org. Chem.* **2005**, *70*, 1718.

134. Kobayashi, T.; Eda, T.; Tamura, O.; Ishibashi, H., *J. Org. Chem.* **2002**, *67*, 3156.

135. Ruzza, P.; Calderan, A.; Borin, G., *Tetrahedron Lett.* **1996**, *37*, 5191.

136. Gevorgyan, V.; Yamamoto, Y., *Tetrahedron Lett.* **1995**, *36*, 7765.

137. Routier, S.; Sauge, L.; Ayerbe, N.; Coudert, G.; Merour, J.-Y., *Tetrahedron Lett.* **2002**, *43*, 589.

138. Jacquemard, U.; Beneteau, V.; Lefoix, M.; Routier, S.; Merour, J.-Y.; Coudert, G., *Tetrahedron* **2004**, *60*, 10039.

139. Aronica, L. A.; Raffa, P.; Valentini, G.; Caporusso, A. M.; Salvadori, P., *Tetrahedron Lett.* **2006**, *47*, 527.

140. Fustero, S.; Garcia Sancho, A.; Chiva, G.; Sanz-Cervera, J. F.; Del Pozo, C.; Acena, J. L., *J. Org. Chem.* **2006**, *71*, 3299.

141. Smith, E. D.; Fevrier, F. C.; Comins, D. L., *Org. Lett.* **2006**, *8*, 179.

142. Heitzman, C. L.; Lambert, W. T.; Mertz, E.; Shotwell, J. B.; Tinsley, J. M.; Va, P.; Roush, W. R., *Org. Lett.* **2005**, *7*, 2405.

143. Tanabe, Y.; Okumura, H.; Maeda, A.; Murakami, M., *Tetrahedron Lett.* **1994**, *35*, 8413.

144. Tanabe, Y.; Murakami, M.; Kitaichi, K.; Yoshida, Y., *Tetrahedron Lett.* **1994**, *35*, 8409.

145. Jones, G. B.; Wright, J. M.; Rush, T. M.; Plourde, G. W. II.; Kelton, T. F.; Mathews, J. E.; Huber, R. S.; Davidson, J. P., *J. Org. Chem.* **1997**, *62*, 9379.

146. Cai, J.; Nemoto, H.; Nakamura, H.; Singaram, B.; Yamamoto, Y., *Spec. Publ.- R. Soc. Chem.* **1997**, *201*, 112.

147. Nakamura, H.; Aoyagi, K.; Yamamoto, Y., *J. Am. Chem. Soc.* **1998**, *120*, 1167.

148. Cai, J.; Nemoto, H.; Nakamura, H.; Singaram, B.; Yamamoto, Y., *Chem. Lett.* **1996**, 791.

149. Brik, A.; Wu, C.-Y.; Best, M. D.; Wong, C.-H., *Bioorg. Med. Chem.* **2005**, *13*, 4622.

Tetrakis(triphenylphosphine)-palladium(0)[1]

[14221-01-3] $C_{72}H_{60}P_4Pd$ (MW 1155.62)

InChI = 1/4C18H15P.Pd/c4*1-4-10-16(11-5-1)19(17-12-6-2-7-13-17)18-14-8-3-9-15-18;/h4*1-15H;

InChIKey = NFHFRUOZVGFOOS-UHFFFAOYAV

(catalyzes carbon–carbon bond formation of organometallics with a wide variety of electrophiles;[2] in combination with other reagents, catalyzes the reduction of a variety of functional groups;[3] catalyzes carbon–metal (Sn, Si) bond formation;[4] catalyzes deprotection of the allyloxycarbonyl group[5])

Physical Data: mp has been reported to vary between 100–116 °C (dec) and is not a good indication of purity.

Solubility: insol saturated hydrocarbons; moderately sol many other organic solvents including CHCl$_3$, DME, THF, DMF, PhMe, benzene.

Form Supplied in: yellow, crystalline solid from various sources. The quality of batches from the same source have been noted to be highly variable and can dramatically alter the expected reactivity.

Preparative Methods: readily prepared by the reduction of PdCl$_2$ (Ph$_3$P)$_2$[6] with **Hydrazine** or by the reaction of **Tris(dibenzylideneacetone)dipalladium–Chloroform** with **Triphenylphosphine**.[7]

Handling, Storage, and Precautions: is air and light sensitive and should be stored in an inert atmosphere in the absence of light. It can be handled for short periods quickly in the air but best results are achieved by handling in a glove box or glove bag under argon or nitrogen.

Direct Carbon–Carbon Bond Formation. One of the most attractive features of Pd(Ph$_3$P)$_4$ is its ability to catalyze carbon–carbon bond formation under mild conditions by the cross-coupling of organometallic (typically organoaluminum,[2e] -boron,[2f] -copper, -magnesium, -tin,[2b–d] or -zinc[2e] reagents) and unsaturated electrophilic partners (halides or sulfonates such as trifluoromethanesulfonates[2d] (triflates)). Although Pd(Ph$_3$P)$_4$ is the catalyst of choice in many of these reactions, numerous other Pd^0 and Pd^{II} catalysts have been used successfully.

Symmetrical or unsymmetrical biaryls are efficiently produced by the Pd(Ph$_3$P)$_4$ catalyzed cross-coupling of aryl halides or aryl triflates (I, Br > OTf in terms of rate of reaction[8a]) with a variety of metalated aromatics such as arylboronic acids,[8] arylstannanes,[9] aryl Grignards[10] and arylzincs[11] (eqs 1 and 2). Reactions employing ArB(OH)$_2$ are carried out in aqueous base (2M Na$_2$CO$_3$ or K$_3$PO$_4$) while the remainder are conducted under anhydrous conditions. A recent report documents the reaction of ArB(OR)$_2$ with ArBr under nonaqueous conditions in the presence of **Thallium(I) Carbonate**.[8c] Reactions employing ArOTf and ArSnR$_3$ as coupling partners require greater than stoichiometric amounts of **Lithium Chloride**.[9a,b] Acceleration of the reaction rate has been noted in the coupling of ArSnMe$_3$ and either ArOTf or ArBr/I by the addition of catalytic **Copper(I)**

Bromide[9c] or stoichiometric **Silver(I) Oxide**,[9f] respectively. In many cases, one or both of the aromatic species can be heteroaromatic such as pyridine, furan, thiophene, quinoline, oxazole, thiazole, or indole.[8e,9f,g,11] Symmetrical biaryls have been prepared in excellent yields by the Pd(Ph$_3$P)$_4$-catalyzed homocoupling of ArBr/I under phase transfer conditions.[12]

$$\text{Ph}-X + M-\text{Ph} \xrightarrow{\text{Pd(Ph}_3\text{P)}_4} \text{Ph}-\text{Ph} \quad (1)$$

M = MgBr, X = I 70%
M = ZnCl, X = OSO$_2$F 95%

$$ (2) $$

92%

In a similar manner, Pd(Ph$_3$P)$_4$ catalyzes the cross-coupling of metalated alkenes, such as vinylaluminum reagents,[13] vinylboronates or -boronic acids,[8d,14] vinylstannanes,[15] vinylsilanes,[16] vinylzincs or -zirconates,[17] vinylcuprates[18] or -copper reagents[19] and vinyl Grignards[20] with vinyl halides (I, Br), triflates, or phosphates to form 1,3-dienes (eq 3). Vinylallenes,[17f] dienyl sulfides,[14h] and dienyl ethers[17d] have also been prepared using this strategy. In most instances the dienes are formed with retention of double bond geometry in both reacting partners.[14g] However, the formation of (E,Z)- and (Z,Z)-diene combinations has been documented to suffer from poor reaction yields or scrambling of alkene geometry in some cases (eq 4).[14a,b,18–20] A dramatic rate enhancement has been noted in the coupling of vinyl iodides with vinylboronic acids by replacing the aqueous bases that are normally used (NaOEt or NaOH) with **Thallium(I) Hydroxide**[14f,i] or **Thallium(I) Carbonate**.[21] LiCl[15a,c] is required when vinyl triflates are used in coupling reactions with vinylstannanes. Intramolecular versions of the vinylstannane–vinyl triflate coupling have been reported[22] and the necessity of adding LiCl in these reactions has been debated.[22a] 1,3-Dienes have been prepared via the Pd(Ph$_3$P)$_4$-catalyzed reaction between allylic alcohols and aldehydes in the presence of **Phenyl Isocyanate** and **Tributylphosphine** (eq 5).[23] (E/E):(E/Z) ratios range from 1:1 to 4:1.

$$R^1\diagdown\diagup X + M\diagdown\diagup R^2 \xrightarrow{\text{Pd(Ph}_3\text{P)}_4} R^1\diagdown\diagup\diagdown\diagup R^2 \quad (3)$$

R^1 = Hex, X = Br; M = B(DOB), R^2 = Bu 86%
R^1 = Bu, X = I; M = ZrCp$_2$Cl, R^2 = Hex 93%

$$ (4) $$

R^1 = Bu, X = Br; M = B(sia)$_2$, R^2 = Hex 49%
R^1 = t-Bu, X = I; M = Cu•MgCl$_2$, R^2 = Pent 53%

$$\text{Ph}\diagdown\diagup\diagdown\text{OH} + \text{C}_7\text{H}_{15}\text{CHO} \xrightarrow[\text{PhNCO, Bu}_3\text{P}]{\text{Pd(Ph}_3\text{P)}_4} $$

82%

$$\text{Ph}\diagdown\diagup\diagdown\diagup\text{C}_7\text{H}_{15} \quad (5)$$

4:1

Similar cross-coupling procedures have been used to prepare styrenes by the reaction of metalated aromatics with vinyl halides/

Essential Reactions for Organic Synthesis, First Edition. Edited by Philip L. Fuchs.
© 2016 John Wiley & Sons, Ltd. Published 2016 by John Wiley & Sons, Ltd.

triflates[11d,24] or, conversely, metalated alkenes with aromatic halides/triflates[9b,16,25] in the presence of Pd(Ph$_3$P)$_4$ (eq 6). Typically, ArCl are poor substrates in Pd(PPh$_3$)$_4$-catalyzed coupling reactions. However, by forming the chromium tricarbonyl complex of the aryl chloride, a facile coupling reaction with vinylstannanes can be achieved (eq 7).[26]

$$
\text{MeO}\!-\!\!\bigcirc\!\!-\!\text{Br} + \overset{\displaystyle \diagup}{}\text{SnBu}_3 \xrightarrow[\substack{\text{PhMe} \\ 76\%}]{\text{Pd(Ph}_3\text{P)}_4} \text{MeO}\!-\!\!\bigcirc\!\!-\!\!\overset{\displaystyle \diagup}{} \quad (6)
$$

$$
\text{(7)} \qquad 68\%
$$

Enynes and arenynes are available from the Pd(Ph$_3$P)$_4$-catalyzed coupling of metalated alkynes (Mg,[20] Al,[13,27] Zn,[17f,28] Sn[15a,b,29]) with vinyl or aryl halides, triflates or phosphates (eq 8). Alternatively, 1-haloalkynes and metalated alkenes (B[14c] or Zn[11a,17a]) can be utilized in similar procedures.

$$
\text{Bu}\diagdown\!\!=\!\!\diagup\text{I} + \text{M}\!-\!\!\equiv \xrightarrow{\text{Pd(Ph}_3\text{P)}_4} \text{Bu}\diagdown\!\!=\!\!\diagup\!-\!\!\equiv \quad (8)
$$

M = SnMe$_3$ 50%
M = ZnCl   71%

Enynes and arenynes can also be prepared by the Pd(Ph$_3$P)$_4$-catalyzed reaction between vinyl halides (I, Br, or Cl) and 1-alkynes in the presence of **Copper(I) Iodide** and an amine base such as RNH$_2$ (R = Bu, Pr), Et$_2$NH, or Et$_3$N.[30] A modified procedure employing aqueous base under phase transfer conditions has also been described (eq 9).[31] The arenyne products derived from such coupling reactions provide ready access to substituted indoles (eq 10).[29c,30d]

$$
\text{Bu}\diagdown\!\!=\!\!\diagup\text{I} + \text{C}_5\text{H}_{11}\!-\!\!\equiv \xrightarrow[\substack{\text{R}_4\text{NCl, PhH} \\ 50\%}]{\substack{\text{Pd(Ph}_3\text{P)}_4 \\ \text{CuI, aq NaOH}}} \text{Bu}\diagdown\!\!=\!\!\diagup\!-\!\!\equiv\!-\!\text{C}_5\text{H}_{11} \quad (9)
$$

$$
\text{(10)}
$$

Pd(Ph$_3$P)$_4$ catalyzes the coupling of simple alkyl metals and vinyl halides (Br, I) or triflates to form substituted alkenes (eq 11). Alkylboron,[32] alkyl Grignard,[20,33] alkylzinc,[34] and alkylaluminum[13] reagents have been particularly useful in this regard. A variety of functional groups on either reacting partner are tolerated and the reaction proceeds with retention of alkene geometry (usually >98%), providing stereochemically pure, highly substituted alkenes. For example, allylsilanes (eq 12)[35] and vinylcyclopropanes (eq 13)[36] have been prepared employing **Trimethylsilylmethylmagnesium Chloride** and cyclopropylzinc chloride, respectively, as the organometallic partner.

$$
\text{Hex}\diagdown\!\!=\!\!\diagup\text{I} \xrightarrow[\text{Pd(Ph}_3\text{P)}_4]{\text{EtMgBr}} \text{Hex}\diagdown\!\!=\!\!\diagup\text{Et} \quad (11)
$$

87% (E)
85% (Z)

$$
\text{Pr}\diagdown\!\!=\!\!\diagup\text{I} \xrightarrow[\substack{\text{TMSCH}_2\text{MgCl, THF} \\ 84\%}]{\text{Pd(Ph}_3\text{P)}_4,\ \text{rt}} \text{Pr}\diagdown\!\!=\!\!\diagup\text{TMS} \quad (12)
$$

$$
\text{(13)} \qquad \text{P = TBDMS}, 82\%
$$

Ketones are obtained by the Pd(Ph$_3$P)$_4$-catalyzed coupling of acid chlorides with organometallic reagents (eq 14). Organozinc[37] and organocopper[38] reagents have been used most successfully, while other reports document the utility of R$_4$Pb (R = Bu, Et),[39] R$_4$Sn (R = Me, Bu, Ph),[40] and R$_2$Hg (R = Et, Ph)[41] reagents in this reaction. For those cases in which the organometallic reagent is alkenic, complete retention of alkene geometry is observed (see eq 14). In addition, the formation of tertiary alcohols is not observed under the conditions employed. A modified procedure that substitutes alkyl chloroformates for acid chlorides leads to an efficient preparation of esters (eq 15).[37b,38] In a related reaction, Pd(Ph$_3$P)$_4$ catalyzes the coupling reaction between substituted aryl- or alkylsulfonyl chlorides and vinyl- or allylstannanes, providing a general route to sulfones (eq 16).[42]

$$
\text{(14)} \qquad 85\%
$$

$$
\text{(15)} \qquad 73\%
$$

$$
\text{Ph}\diagdown\!\!=\!\!\diagup\text{SnBu}_3 \xrightarrow[\substack{\text{THF, 70 °C} \\ 90\%}]{\substack{\text{MeSO}_2\text{Cl} \\ \text{Pd(Ph}_3\text{P)}_4}} \text{Ph}\diagdown\!\!=\!\!\diagup\text{SO}_2\text{Me} \quad (16)
$$

The Pd(Ph$_3$P)$_4$-catalyzed reaction of various allylic electrophiles with carbon-based nucleophiles is a very useful method for the formation of C–C bonds under relatively mild conditions (eq 17) and has been extensively reviewed elsewhere.[2a,43] The most commonly used electrophilic substrates for Pd(Ph$_3$P)$_4$-catalyzed allylic substitution reactions are allylic esters, carbonates, phosphates, carbamates, halides, sulfones, and epoxides. More recently, allylic alcohols themselves have been demonstrated to be useful substrates.[44] Commonly employed nucleophiles include soft stabilized carbanions such as malonate and, to a lesser extent, a variety of organometallic reagents (Al, Grignards, Sn, Zn, Zr). A few selected examples begin to illustrate the scope of this reaction in terms of the general patterns of reactivity with respect to regioselectivity (eqs 18 and 19) and stereoselectivity (eqs 17 and 18). It should be noted that a variety of other Pd catalysts (including **Palladium(II) Acetate–1,2-Bis(diphenylphosphino)ethane** and **Bis(dibenzylideneacetone)palladium(0)**) have been shown to be useful in these alkylations. In addition, certain nitrogen-, sulfur-, and oxygen-based reagents are suitable nucleophilic substrates. The utility of some of these latter reagents are covered in subsequent sections.

(17)

(18)

$(E):(Z) = 98:2$

(19)

94%   6%

Aldehydes and α-bromo ketones or esters are efficiently coupled in an aldol reaction in the presence of *Diethylaluminum Chloride–Tributylstannyllithium* (or *Tin(II) Chloride*) and catalytic Pd(Ph$_3$P)$_4$, providing β-hydroxy ketones or esters (eq 20).[45]

(20)

R = Ph, 70%; EtO, 75%

Carbonylative Carbon–Carbon Bond Formation. A general, mild (50 °C), and high yielding conversion of halides and triflates into aldehydes via Pd(Ph$_3$P)$_4$-catalyzed carbonylation (1–3 atm CO) in the presence of *Tributylstannane* has been described (eq 21).[46] The range of usable substrates is extensive and includes ArI, benzyl and allyl halides, and vinyl iodides and triflates. The reaction has been extended to include ArBr by carrying out the carbonylation at 80 °C under pressure (50 atm CO), using poly(methylhydrosiloxane) (PMHS) instead of tin hydride.[47]

(21)

Pd(Ph$_3$P)$_4$ catalyzes the carbonylation of benzyl[48a] and vinyl[48b] bromides under phase transfer conditions in the presence of hydroxide to form the corresponding carboxylic acids. A wide variety of substitution is tolerated and the products are formed in moderate to excellent yield at room temperature and at normal pressure (1 atm CO). Extension of the reaction to the formation of esters from aryl, alkyl, and vinyl bromides has been described.[49] These transformations usually require a co-catalyst system of Pd(Ph$_3$P)$_4$ and [(1,5-cyclohexadiene)RhCl]$_2$ in the presence of either M(OR)$_4$ (M = Ti, Zr) or M(OR)$_3$ (M = B, Al) (eq 22).

(22)

Ti(OBu)$_4$ (no Rh cat)   85%
Al(OEt)$_3$ 66%

Vinyl triflates serve as substrates for Pd(Ph$_3$P)$_4$-catalyzed carbonylation and have been converted into the corresponding esters[50] or ketones[15c,51] (eq 23).

(23)

R = CHCH$_2$, 76%; Ph, 93%

Aromatic and Vinyl Nitriles. Aromatic halides (Br, I) have been converted into nitriles in excellent yields by Pd(Ph$_3$P)$_4$ catalysis in the presence of *Sodium Cyanide/Alumina*,[52] *Potassium Cyanide*,[53] or *Cyanotrimethylsilane*[54] (eq 24). While the latter two procedures require the use of ArI as substrates, a more extensive range of substituents are tolerated than the alternative method employing ArBr. A Pd(Ph$_3$P)$_4$-catalyzed extrusion of CO from aromatic and heteroaromatic acyl cyanides (readily available from cyanohydrins) at 120 °C provides aryl nitriles in excellent yields (eq 25).[55]

(24)

(25)

Similarly, vinyl halides (Br, Cl) provide vinyl nitriles upon treatment with Pd(Ph$_3$P)$_4$/*Potassium Cyanide/18-Crown-6*.[56]

Carbon–Heteroatom (N, S, O, Sn, Si, Se, P) Bond Formation. Primary and secondary amines (but not ammonia) undergo reaction with allylic acetates,[57] halides,[58] phosphates,[59] and nitro compounds[60] in the presence of Pd(Ph$_3$P)$_4$ to provide the corresponding allylic amines (eq 26). A variety of ammonia equivalents have been demonstrated to be useful in this Pd(Ph$_3$P)$_4$-catalyzed alkylation, including 4,4′-dimethoxybenzhydrylamine,[57c] NaNHTs,[58] and NaN$_3$[61] (eq 26). Both allylic phosphates and chlorides react faster than the corresponding acetates[58,59a] and (Z)-alkenes are isomerized to the (E)-isomers.[57a,59a] The use of primary amines as nucleophiles in the synthesis of secondary allyl amines is sometimes problematic since the amine that is formed undergoes further alkylation to form the tertiary amine. Thus hydroxylamines have been shown to be useful primary amine equivalents (eq 27) since the reaction products are easily reduced to secondary amines.[59b]

(26)

X = PO(OEt)$_2$, Nu = NEt$_2$ 68%
X = Ac, Nu = N$_3$ 88%

$$\text{(27)}$$

Allyl sulfones can be obtained by the Pd(Ph$_3$P)$_4$-catalyzed reaction of allylic acetates[62] and allylic nitro compounds[63,64] with NaSO$_2$Ar (eq 28). Pd(Ph$_3$P)$_4$ also catalyzes the addition of HOAc to vinyl epoxides, providing a facile entry into 1,4-hydroxy acetates.[65]

$$\text{(28)}$$

Aryl halides (Br, I) have been converted in good yields into the corresponding arylstannanes or -silanes by treatment with R$_6$Sn$_2$ (R = Bu, Me)[4,66] or *Hexamethyldisilane*,[67] respectively, in the presence of catalytic Pd(Ph$_3$P)$_4$ (eq 29). Aryl[9a,b] and vinyl[15b,68] triflates produce aryl- and vinylstannanes under similar conditions, provided *Hexamethyldistannane* and LiCl are used as co-reactants. In some cases, the presence of additional Ph$_3$P has been observed to improve yields.[68] By using the (Ph$_3$Sn)$_2$Zn–TMEDA complex, vinyl halides can also be converted into vinylstannanes.[69]

$$\text{(29)}$$

Acyltrimethylstannanes can be prepared in moderate yields by the treatment of acid chlorides with Me$_6$Sn$_2$ and catalytic Pd(Ph$_3$P)$_4$ in refluxing THF.[70c] However, Pd(Ph$_3$P)$_2$Cl$_2$ is a superior catalyst for this transformation when using sterically bulky or electron-poor acyl halides.

Pd(Ph$_3$P)$_4$ catalyzes the addition of R$_6$Sn$_2$,[70] R$_6$Si$_2$,[71] Ph$_2$S$_2$, and *Diphenyl Diselenide*[72] across the triple bond of 1-alkynes to provide the respective (Z)-1,2-addition products (eq 30). The (Z)-distannanes can be partially isomerized to the (E)-isomers by photolysis.[70] The reaction cannot be extended to include internal alkynes containing alkyl substituents, but allenes do undergo 1,2-addition of disilane[73] and ditin.[74] In a similar fashion, Pd(PPh$_3$)$_4$ catalyzes the regio- and stereospecific addition of R$_3$Sn–SiR$_3$ to 1-alkynes, the (Z)-1-silyl-2-stannylalkene isomers being the sole products (eq 30).[75]

$$\text{(30)}$$

M = N = SPh	79%
M = N = SnBu$_3$	59%
M = SnMe$_3$, N = TMS	51%

α,β-Acetylenic esters react with R$_6$Sn$_2$ in the presence of Pd(Ph$_3$P)$_4$ at room temperature to provide only the (Z)-2,3-distannylalkenoates (eq 31).[76] When heated to 75–95 °C, clean isomerization to the (E)-isomers is observed. The corresponding amides are also useful substrates but provide either the (E)-isomers directly or (E/Z)-distannane mixtures.[76] Surprisingly, under similar reaction conditions, α,β-alkynic aldehydes and ketones form (Z)-β-stannyl enals and enones in excellent yields (eq 32).[77]

$$\text{(31)}$$

R = Me, TBDMSOCH$_2$

$$\text{(32)}$$

P = TBDMS   X = H, 87%; Me, 90%

The Pd(Ph$_3$P)$_4$-catalyzed hydrostannylation of 1-alkynes with R$_3$SnH provides mixtures of vinylstannane regio- and stereoisomers, the ratios depending upon the nature of the alkyne substituents and the R group of the tin reagent.[78] In general, when using alkyl-substituted 1-alkynes, *Triphenylstannane* provides (E)-1-stannylalkenes[78a] as the major products, while 2-stannylalkenes are obtained predominantly with (Bu$_3$Sn)$_2$Zn[78c] (eq 33). The Pd(Ph$_3$P)$_4$ mediated *cis* addition of R$_3$Sn–H across symmetrical internal alkynes has also been demonstrated to be generally high yielding.[78b] The hydrostannylation of α,β-unsaturated nitriles with Bu$_3$SnH/Pd(Ph$_3$P)$_4$ is regioselective, providing α-stannyl nitriles as the sole products.[79]

$$\text{(33)}$$

Ph$_3$SnH,	77%	89:11
(Bu$_3$Sn)$_2$Zn,	70%	<5:>95

The Pd(Ph$_3$P)$_4$-mediated preparation of allyl- and benzylstannanes has been achieved by treatment of allylic acetates with Et$_2$AlSnBu$_3$[80] or by reacting benzyl halides (Br, Cl) with R$_6$Sn$_2$.[66c] In a similar fashion, allyl- and benzylsilanes are prepared by the Pd(Ph$_3$P)$_4$-catalyzed reaction of R$_6$Si$_2$ with allylic halides[81] and benzyl halides,[4] respectively. Allylstannanes have also been prepared by the addition of Bu$_3$SnH to 1,3-dienes in the presence of Pd(Ph$_3$P)$_4$ (eq 34).[82]

$$\text{(34)}$$

Dialkyl aryl- and vinylphosphonates (RPO(OR')$_2$) are readily prepared in excellent yields by the reaction of aryl or vinyl bromides, respectively, with dialkyl phosphite (HPO(OR')$_2$) in the presence of catalytic Pd(Ph$_3$P)$_4$ and Et$_3$N.[83] In a related process, unsymmetrical alkyl diarylphosphinates (ArPhPO(OR)) are obtained in good yields, regardless of aromatic substitution, by the Pd(Ph$_3$P)$_4$-catalyzed coupling reaction between aryl bromides and alkyl benzenephosphonites (HPhPO(OR)).[84]

Oxidation Reactions. α-Bromo ketones are dehydrobrominated to produce enones in low to good yields, especially when the products are phenolic, by treatment with stoichiometric Pd(Ph$_3$P)$_4$ in hot benzene.[85] Primary and secondary alcohols are oxidized in the presence of PhBr, base (NaH or K$_2$CO$_3$) and Pd(Ph$_3$P)$_4$ as catalyst to the corresponding aldehydes or ketones.[86] The practical advantages of these methods to alternative strategies have yet to be demonstrated.

Reduction Reactions. At elevated temperatures (100–110 °C), ArBr and ArI are reduced to ArH in the presence of catalytic Pd(Ph$_3$P)$_4$ and reducing agents such as HCO$_2$Na (eq 35),[3,87] NaOMe,[88] and PMHS/Bu$_3$N.[3] Aldehydes, ketones, esters, acids, and nitro substituents are unaffected. ArCl are poor substrates unless the aromatic nucleus is substituted with NO$_2$.[88] ArOTf are reduced in poorer yield under similar conditions; Pd(OAc)$_2$ and Pd(Ph$_3$P)$_2$Cl$_2$ are superior catalysts with these substrates.[89] A limited number of examples of the Pd(PPh$_3$)$_4$-catalyzed reduction of vinyl bromides and triflates to alkenes in the presence of HCO$_2$Na[3] and Bu$_3$SnH,[15a,b] respectively, have been described.

$$X = CHO, 80\%; COMe, 84\% \qquad (35)$$

Pd(Ph$_3$P)$_4$ catalyzes the reductive displacement of a variety of allylic substituents with hydride transfer reagents. Allylic acetates are reduced to simple alkenes in the presence of **Sodium Cyanoborohydride**,[90] PMHS,[91] **Samarium(II) Iodide**/i-PrOH,[92] or Bu$_3$SnH[93] (eq 36). Bu$_3$SnH/Pd(Ph$_3$P)$_4$ also reduces allylic amines[93] and thiocarbamates.[94] All of these reductive procedures are accompanied by positional and/or geometrical isomerization of the alkene to an extent dependent upon the substrate structure. Interestingly, the Pd(Ph$_3$P)$_4$-catalyzed reduction of allylic sulfones with **Sodium Borohydride** is high yielding with no double bond positional isomerization observed.[95] By using a more bulky reducing agent, **Lithium Triethylborohydride**, a range of allylic groups, such as methyl, phenyl, and silyl ethers, sulfides, sulfones, selenides, and chlorides, are reduced to alkenes in the presence of Pd(Ph$_3$P)$_4$ and excess Ph$_3$P with little or no loss of alkene regio- or stereochemistry (eq 37).[96] An interesting Pd(Ph$_3$P)$_4$-catalyzed reduction of allylic acetates, tosylates, and chlorides to the corresponding alkenes has been described that uses n-BuZnCl as the hydride source.[97] This procedure proceeds with high levels of regio- and stereospecificity, but the scope has yet to be explored.

$$[H] = NaBH_3CN, \quad 68\% \ 42:58$$
$$[H] = SmI_2, \ i\text{-PrOH}, \quad 92\% \ 93:7$$
$$(36)$$

$$X = SPh, \quad 90\% \ (E):(Z) = 94:5$$
$$X = OTBDMS, \quad 80\% \ (E):(Z) = 98:1$$
$$(37)$$

Acid chlorides[98] and acyl selenides[99] are efficiently reduced in the presence of Bu$_3$SnH under Pd(Ph$_3$P)$_4$ catalysis to provide the corresponding aldehydes in good to excellent yields without the formation of ester or alcohol byproducts (eq 38). A wide variety of substrate substituents, such as alkenes, nitriles, bromides, and nitro groups, are tolerated.

$$R = Hex, X = Cl \qquad 77\%$$
$$R = Oct, X = SePh \qquad 69\%$$
$$(38)$$

The conjugate reduction of α,β-unsaturated ketones and aldehydes to the saturated analogs can be accomplished in the presence of Pd(Ph$_3$P)$_4$ and hydride transfer reagents such as Bu$_3$SnH[100] or mixed systems of Bu$_3$SnH/HOAc or Bu$_3$SnH/ZnCl$_2$.[79] A potentially more versatile, general, and selective reduction procedure involves a three-component system of Pd(Ph$_3$P)$_4$/Ph$_2$SiH$_2$/ZnCl$_2$ (eq 39).[101] α,β-Unsaturated esters and nitriles are untouched using this latter method.

$$(39)$$

α-Bromo ketones are reductively debrominated to the parent ketones in the presence of Pd(Ph$_3$P)$_4$ and Ph$_2$SiH$_2$/K$_2$CO$_3$ at room temperature.[102] However, **Hexacarbonylmolybdenum** appears to be a superior catalyst for this conversion. Reductive debromination of α-bromo ketones, acids, and nitriles can also be accomplished by Pd(Ph$_3$P)$_4$ catalysis using PMHS/Bn$_3$N,[87b] HCO$_2$Na,[87b] or Me$_6$Si$_2$[103] as hydrogen donors but under much more drastic conditions (110–170 °C).

Removal of Allyloxycarbonyl (Aloc) and Allyl Protecting Groups. The allyloxycarbonyl protecting group[5] has been used extensively for the protection of a variety of alcohols[104] and amines,[105] including the amines of nucleotide bases,[106] N-terminal amines of amino acids and peptides,[107] and amino sugars.[108] Pd(Ph$_3$P)$_4$ mediates the high yielding removal of the Aloc group in the presence of nucleophilic allyl scavengers (eq 40) such as Bu$_3$SnH, 2-ethylhexanoic acid, dimedone, malonate, **Ammonium Formate**, **N-Hydroxysuccinimide**, and various amines including **Morpholine** and **Pyrrolidine**. The deprotections are effected at or below room temperature, often in the presence of excess Ph$_3$P. The mild conditions used in these procedures leave most of the common N or O protecting groups intact, including Boc, TBDMS, MMT, DMT, and carbonates.

$$(40)$$

The Pd(Ph$_3$P)$_4$-catalyzed removal of the O-allyl protecting group has been described for a number of systems.[5] For example, allyl esters are efficiently cleaved to the parent acid in chemically sensitive systems such as penicillins[105,109] and glycopeptides.[110] The internucleotide phosphate linkage, protected as the allyl phospho(III)triester, remains intact upon deprotection under Pd(Ph$_3$P)$_4$ catalysis.[111] Allyl ethers that protect the anomeric hydroxy in carbohydrates (mono- and disaccharides) are efficiently removed under Pd(Ph$_3$P)$_4$ catalysis in hot (80 °C) HOAc.[112]

Rearrangements, Isomerizations and Eliminations. Pd(Ph$_3$P)$_4$ catalyzes several [3,3]-sigmatropic rearrangements including those of *O*-allyl phosphoro- and phosphonothionates to the corresponding *S*-allyl thiolates (eq 41)[113] and allylic *N*-phenylformimidates to *N*-allyl-*N*-phenylformamides (eq 42).[114] The 3-aza-Cope rearrangement of *N*-allylenamines to γ,δ-unsaturated imines (eq 43)[115] and a Claisen rearrangement but with no allyl inversion[116] have also been described. In general, these Pd(Ph$_3$P)$_4$-catalyzed rearrangements provide compounds that are either not accessible via thermal reactions or are produced only under much more forcing reaction conditions. For those cases in which the allyl moiety contains substituents that may lead to regioisomers upon rearrangement, the less substituted isomer is usually favoured[113,115] (see eq 41), although exceptions are known.[114]

(EtO)$_2$P(=S)O⟶ $\xrightarrow[\text{93\%}]{\text{Pd(Ph$_3$P)$_4$}}{\text{DME, 80 °C}}$ (EtO)$_2$P(=O)S⟶ (41)

$\xrightarrow[\text{100\%}]{\text{Pd(Ph$_3$P)$_4$}}{\text{THF, reflux}}$ (42)

$\xrightarrow[\text{82\%}]{\text{Pd(Ph$_3$P)$_4$, TFAA}}{\text{PhH, 50 °C}}$ (43)

Pd(Ph$_3$P)$_4$ catalyzes stereoselective, intramolecular metalloene reactions of acetoxydienes in HOAc, efficiently generating a range of cyclic 1,4-dienes (eq 44).[117] The reactions proceed in good to excellent yields and have been extended to include the preparation of pyrrolidines, piperidines, and tetrahydrofurans by incorporating N and O atoms into the bridge that tethers the reactive alkenes.

$\xrightarrow[\text{HOAc, 75 °C}]{\text{Pd(Ph$_3$P)$_4$}}$ (44)

X = C(CO$_2$Me)$_2$, 52% 10:90
X = NCOCF$_3$, 67% 28:72

The isomerization of allylic acetates is a useful method for allylic oxygen interconversion. Although these reactions are typically carried out in the presence of Pd^{II} catalysts such as Pd(OAc)$_2$, Pd(Ph$_3$P)$_4$ has proven to be useful in the 1,3-rearrangement of α-cyanoallylic acetates to γ-acetoxy-α,β-unsaturated nitriles (eq 45).[118] These compounds are conveniently transformed into furans.

$\xrightarrow[\text{91\%}]{\text{Pd(Ph$_3$P)$_4$}}{\text{THF, rt}}$ $\xrightarrow[\text{2. DIBAL}]{\text{1. aq NaOH}}{\text{74\%}}$ (45)

Several 1,3-diene syntheses involving elimination reactions that are catalyzed by Pd(Ph$_3$P)$_4$ have been reported. The first involves the Et$_3$N mediated elimination of HOAc from allylic acetates in refluxing THF.[119] A complementary procedure involves the Pd(Ph$_3$P)$_4$ catalyzed decarboxylative elimination of β-acetoxycarboxylic acids (eq 46).[120] The substrates are easily prepared by the condensation of enals and carboxylate enolates; irrespective of the diastereomeric mixture, (*E*)-alkenes are formed in a highly stereocontrolled manner. The geometry of the double bond present in the enal precursor remains unaffected in the elimination and the reaction is applicable to the formation of 1,3-cyclohexadienes.

$\xrightarrow[\text{90\%}]{\text{Pd(Ph$_3$P)$_4$, Et$_3$N}}{\text{PhMe, 85 °C}}$ (46)

+ 10% (*Z,Z*) isomer

Monoepoxides of simple cyclic 1,3-dienes are smoothly converted in good yield to β,γ-unsaturated ketones in the presence of Pd(Ph$_3$P)$_4$ catalyst (eq 47).[121] Other vinyl epoxides, such as those in open chains or in cyclic systems in which the double bond is not in the ring, are converted under similar conditions into dienols.

$\xrightarrow[\text{63\%}]{\text{Pd(Ph$_3$P)$_4$, Ph$_3$P}}{\text{PhH, rt}}$ (47)

A mixture of Pd(Ph$_3$P)$_4$/dppe catalyzes the transformation of α,β-epoxy ketones, readily available via several methods, into β-diketones (eq 48).[122] The reaction is applicable to both cyclic and acyclic compounds, although epoxy ketones bearing an α-alkyl group are poor substrates.

$\xrightarrow[\text{94\%}]{\text{Pd(Ph$_3$P)$_4$, dppe}}{\text{PhMe, 80 °C}}$ (48)

Related Reagents. Diphenylsilane–Tetrakis(triphenylphosphine)palladium(0)–Zinc Chloride.

1. (a) Heck, R. F. *Palladium Reagents in Organic Syntheses*; Academic: London, 1985. (b) Tsuji, J. *Organic Synthesis with Palladium Compounds*; Springer: Berlin, 1980.

2. (a) Frost, C. G.; Howarth, J.; Williams, J. M. J., *Tetrahedron: Asymmetry* **1991**, *3*, 1089. (b) Stille, J. K., *Pure Appl. Chem.* **1985**, *57*, 1771. (c) Stille, J. K., *Angew. Chem., Int. Ed. Engl.* **1986**, *25*, 508. (d) Scott, W. J.; Stille, J. K., *Acc. Chem. Res.* **1988**, *21*, 47. (e) Negishi, E., *Acc. Chem. Res.* **1982**, *15*, 340. (f) Suzuki, A., *Pure Appl. Chem.* **1985**, *57*, 1749. (g) Beletskaya, I. P., *J. Organomet. Chem.* **1983**, *250*, 551. (h) Mitchell, T. N., *J. Organomet. Chem.* **1986**, *304*, 1.

3. Pri-Bar, I.; Buchman, O., *J. Org. Chem.* **1986**, *51*, 734.

4. Azarian, D.; Dua, S. S.; Eaborn, C.; Walton, D. R. M., *J. Organomet. Chem.* **1976**, *117*, C55.

5. Greene, T. W.; Wuts, P. G. M. *Protective Groups in Organic Synthesis*, 2nd ed.; Wiley: New York, 1991.

6. Coulson, D., *Inorg. Synth* **1972**, *13*, 121.

7. Ito, T.; Hasegawa, S.; Takahashi, Y.; Ishii, Y., *J. Organomet. Chem.* **1974**, *73*, 401.

8. (a) Fu, J.; Snieckus, V., *Tetrahedron Lett.* **1990**, *31*, 1665. (b) Miyaura, N.; Yanagi, T.; Suzuki, A., *Synth. Commun.* **1981**, *11*, 513. (c) Sato, M.; Miyaura, N.; Suzuki, A., *Chem. Lett.* **1989**, 1405. (d) Takayuki, O.; Miyaura, N.; Suzuki, A., *J. Org. Chem.* **1993**, *58*, 2201. (e) Sharp, M. J.; Snieckus, V., *Tetrahedron Lett.* **1985**, *26*, 5997. (f) Sharp, M. J.; Cheng, W.; Snieckus, V., *Tetrahedron Lett.* **1987**, *28*, 5093. (g) Miller, R. B.; Dugar, S., *Organometallics* **1984**, *3*, 1261. (h) Thompson, W. J.; Gaudino, J., *J. Org. Chem.* **1984**, *49*, 5237. (i) Huth, A.; Beetz, I.; Schumann, I., *Tetrahedron* **1989**, *45*, 6679.

9. (a) Echavarren, A. M.; Stille, J. K., *J. Am. Chem. Soc.* **1987**, *109*, 5478. (b) Echavarren, A. M.; Stille, J. K., *J. Am. Chem. Soc.* **1988**, *110*, 4051. (c) Gómez-Bengoa, E.; Echavarren, A. M., *J. Org. Chem.* **1991**, *56*, 3497. (d) Dondoni, A.; Fantin, G.; Fogagnolo, M.; Medici, A.; Pedrini, P., *Synthesis* **1987**, 693. (e) Clough, J. M.; Mann, I. S.; Widdowson, D. A., *Tetrahedron Lett.* **1987**, *28*, 2645. (f) Malm, J.; Björk, P.; Gronowitz, S.; Hörnfeldt, A.-B., *Tetrahedron Lett.* **1992**, *33*, 2199. (g) Achab, S.; Guyot, M.; Potier, P., *Tetrahedron Lett.* **1993**, *34*, 2127.

10. Widdowson, D. A.; Zhang, Y.-Z., *Tetrahedron* **1986**, *42*, 2111.

11. (a) Negishi, E.; Luo, F.-T.; Frisbee, R.; Matsushita, H., *Heterocycles* **1982**, *18*, 117. (b) Negishi, E.; Takahashi, T.; King, A. O., *Org. Synth.* **1988**, *66*, 67. (c) Roth, G. P.; Fuller, C. E., *J. Org. Chem.* **1991**, *56*, 3493. (d) Arcadi, A.; Burini, A.; Cacchi, S.; Delmastro, M.; Marinelli, F.; Pietroni, B., *Synlett* **1990**, 47. (e) Pelter, A.; Rowlands, M.; Clements, G., *Synthesis* **1987**, 51. (f) Pelter, A.; Rowlands, M.; Jenkins, I. H., *Tetrahedron Lett.* **1987**, *28*, 5213.

12. Torii, S.; Tanaka, H.; Morisaki, K., *Tetrahedron Lett.* **1985**, *26*, 1655.

13. Takai, K.; Sato, M.; Oshima, K.; Nozaki, H., *Bull. Chem. Soc. Jpn.* **1984**, *57*, 108.

14. (a) Miyaura, N.; Yamada, K.; Suzuki, A., *Tetrahedron Lett.* **1979**, 3437. (b) Miyaura, N.; Suginome, H.; Suzuki, A., *Tetrahedron Lett.* **1981**, *22*, 127. (c) Miyaura, N.; Yamada, K.; Suginome, H.; Suzuki, A., *J. Am. Chem. Soc.* **1985**, *107*, 972. (d) Miyaura, N.; Satoh, M.; Suzuki, A., *Tetrahedron Lett.* **1986**, *27*, 3745. (e) Satoh, M.; Miyaura, N.; Suzuki, A., *Chem. Lett.* **1986**, 1329. (f) Uenishi, J.; Beau, J.-M.; Armstrong, R. W.; Kishi, Y., *J. Am. Chem. Soc.* **1987**, *109*, 4756. (g) Miyaura, N.; Suginome, H.; Suzuki, A., *Tetrahedron* **1983**, *39*, 3271. (h) Ishiyama, T.; Miyaura, N.; Suzuki, A., *Chem. Lett.* **1987**, 25. (i) Roush, W. R.; Moriarty, K. J.; Brown, B. B., *Tetrahedron Lett.* **1990**, *31*, 6509.

15. (a) Scott, W. J.; Crisp, G. T.; Stille, J. K., *J. Am. Chem. Soc.* **1984**, *106*, 4630. (b) Scott, W. J.; Stille, J. K., *J. Am. Chem. Soc.* **1986**, *108*, 3033. (c) Scott, W. J.; Crisp, G. T.; Stille, J. K., *Org. Synth.* **1990**, *68*, 116. (d) Stille, J. K.; Groh, B. L., *J. Am. Chem. Soc.* **1987**, *109*, 813.

16. Hatanaka, Y.; Hiyama, T., *Tetrahedron Lett.* **1990**, *31*, 2719.

17. (a) Negishi, E.; Okukado, N.; King, A. O.; Van Horn, D. E.; Spiegel, B. I., *J. Am. Chem. Soc.* **1978**, *100*, 2254. (b) Negishi, E.; Takahashi, T.; Baba, S., *Org. Synth.* **1988**, *66*, 60. (c) Negishi, E.; Luo, F.-T., *J. Org. Chem.* **1983**, *48*, 1562. (d) Negishi, E.; Takahashi, T.; Baba, S.; Van Horn, D. E.; Okukado, N., *J. Am. Chem. Soc.* **1987**, *109*, 2393. (e) Okukado, N.; Van Horn, D. E.; Klima, W. L.; Negishi, E., *Tetrahedron Lett.* **1978**, 1027. (f) Ruitenberg, K.; Kleijn, H.; Elsevier, C. J.; Meijer, J.; Vermeer, P., *Tetrahedron Lett.* **1981**, *22*, 1451.

18. (a) Jabri, N.; Alexakis, A.; Normant, J. F., *Tetrahedron Lett.* **1981**, *22*, 959. (b) Jabri, N.; Alexakis, A.; Normant, J. F., *Bull. Soc. Chem. Fr., Part 2* **1983**, 321.

19. (a) Jabri, N.; Alexakis, A.; Normant, J. F., *Tetrahedron Lett.* **1982**, *23*, 1589. (b) Jabri, N.; Alexakis, A.; Normant, J. F., *Bull. Soc. Chem. Fr., Part 2* **1983**, 332.

20. Dang, H. P.; Linstrumelle, G., *Tetrahedron Lett.* **1978**, 191.

21. Hoshino, Y.; Miyaura, N.; Suzuki, A., *Bull. Chem. Soc. Jpn.* **1988**, *61*, 3008.

22. (a) Piers, E.; Friesen, R. W.; Keay, B. A., *J. Chem. Soc., Chem. Commun.* **1985**, 809. (b) Stille, J. K.; Tanaka, M., *J. Am. Chem. Soc.* **1987**, *109*, 3785.

23. Okukado, N.; Uchikawa, O.; Nakamura, Y., *Chem. Lett.* **1988**, 1449.

24. (a) Miller, R. B.; Al-Hassan, M., *J. Org. Chem.* **1985**, *50*, 2121. (b) McCague, R., *Tetrahedron Lett.* **1987**, *28*, 701.

25. (a) McKean, D. R.; Parrinello, G.; Renaldo, A. F.; Stille, J. K., *J. Org. Chem.* **1987**, *52*, 422. (b) Miyaura, N.; Suzuki, A., *J. Chem. Soc., Chem. Commun.* **1979**, 866. (c) Miyaura, N.; Maeda, K.; Suginome, H.; Suzuki, A., *J. Org. Chem.* **1982**, *47*, 2117.

26. Scott, W. J., *J. Chem. Soc., Chem. Commun.* **1987**, 1755.

27. (a) Takai, K.; Oshima, K.; Nozaki, H., *Tetrahedron Lett.* **1980**, *21*, 2531. (b) Sato, M.; Takai, K.; Oshima, K.; Nozaki, H., *Tetrahedron Lett.* **1981**, *22*, 1609.

28. (a) King, A. O.; Okukado, N.; Negishi, E., *J. Chem. Soc., Chem. Commun.* **1977**, 683. (b) King, A. O.; Negishi, E., *J. Org. Chem.* **1978**, *43*, 358. (c) Negishi, E.; Okukado, N.; Lovich, S. F.; Luo, F.-T., *J. Org. Chem.* **1984**, *49*, 2629. (d) Carpita, A.; Rossi, R., *Tetrahedron Lett.* **1986**, *27*, 4351. (e) Chen, Q.-Y.; He, Y.-B., *Tetrahedron Lett.* **1987**, *28*, 2387.

29. (a) Castedo, L.; Mouriño, A.; Sarandeses, L. A., *Tetrahedron Lett.* **1986**, *27*, 1523. (b) Stille, J. K.; Simpson, J. H., *J. Am. Chem. Soc.* **1987**, *109*, 2138. (c) Rudisill, D. E.; Stille, J. K., *J. Org. Chem.* **1989**, *54*, 5856.

30. (a) Ratovelomanana, V.; Linstrumelle, G., *Tetrahedron Lett.* **1981**, *22*, 315. (b) Ratovelomanana, V.; Linstrumelle, G., *Synth. Commun.* **1981**, *11*, 917. (c) Jeffery-Luong, T.; Linstrumelle, G., *Synthesis* **1983**, 32. (d) Arcadi, A.; Cacchi, S.; Marinelli, F., *Tetrahedron Lett.* **1989**, *30*, 2581. (e) Scott, W. J.; Peña, M. R.; Swärd, K.; Stoessel, S. J.; Stille, J. K., *J. Org. Chem.* **1985**, *50*, 2302. (f) Mandai, T.; Nakata, T.; Murayama, H.; Yamaoki, H.; Ogawa, M.; Kawada, M.; Tsuji, J., *Tetrahedron Lett.* **1990**, *31*, 7179.

31. Rossi, R.; Carpita, A.; Quirici, M. G.; Gaudenzi, M. L., *Tetrahedron* **1982**, *38*, 631.

32. Hoshino, Y.; Ishiyama, T.; Miyaura, N.; Suzuki, A., *Tetrahedron Lett.* **1988**, *29*, 3983.

33. (a) Murahashi, S.-I.; Yamamura, M.; Yanagisawa, K.; Mita, N.; Kondo, K., *J. Org. Chem.* **1979**, *44*, 2408. (b) Huynh, C.; Linstrumelle, G., *Tetrahedron Lett.* **1979**, 1073.

34. (a) Negishi, E.; Valente, L. F.; Kobayashi, M., *J. Am. Chem. Soc.* **1980**, *102*, 3298. (b) Negishi, E.; Zhang, Y.; Cederbaum, F. E.; Webb, M. B., *J. Org. Chem.* **1986**, *51*, 4080. (c) Tamaru, Y.; Ochiai, H.; Nakamura, T.; Yoshida, Z., *Tetrahedron Lett.* **1986**, *27*, 955.

35. Negishi, E.; Luo, F.-T.; Rand, C. L., *Tetrahedron Lett.* **1982**, *23*, 27.

36. Piers, E.; Jean, M.; Marrs, P. S., *Tetrahedron Lett.* **1987**, *28*, 5075.

37. (a) Sato, T.; Itoh, T.; Fujisawa, T., *Chem. Lett.* **1982**, 1559. (b) Negishi, E.; Bagheri, V.; Chatterjee, S.; Luo, F.-T.; Miller, J. A.; Stoll, A. T., *Tetrahedron Lett.* **1983**, *24*, 5181. (c) Tamaru, Y.; Ochiai, H.; Nakamura, T.; Yoshida, Z., *Org. Synth.* **1989**, *67*, 98.

38. Jabri, N.; Alexakis, A.; Normant, J. F., *Tetrahedron* **1986**, *42*, 1369.

39. Yamada, J.; Yamamoto, Y., *J. Chem. Soc., Chem. Commun.* **1987**, 1302.

40. Kosugi, M.; Shimizu, Y.; Migita, T., *Chem. Lett.* **1977**, 1423.

41. Takagi, K.; Okamoto, T.; Sakakibara, Y.; Ohno, A.; Oka, S.; Hayama, N., *Chem. Lett.* **1975**, 951.

42. Labadie, S. S., *J. Org. Chem.* **1989**, *54*, 2496.

43. (a) Trost, B. M., *Angew. Chem., Int. Ed. Engl.* **1989**, *28*, 1173. (b) Tsuji, J.; Minami, I., *Acc. Chem. Res.* **1987**, *20*, 140. (c) Trost, B. M., *Acc. Chem. Res.* **1980**, *13*, 385. (d) Trost, B. M., *Tetrahedron* **1977**, *33*, 2615.

44. Starý, I.; Stará, I. G.; Kocovský, P., *Tetrahedron Lett.* **1993**, *34*, 179.

45. Matsubara, S.; Tsuboniwa, N.; Morizawa, Y.; Oshima, K.; Nozaki, H., *Bull. Chem. Soc. Jpn.* **1984**, *57*, 3242.

46. (a) Baillargeon, V. P.; Stille, J. K., *J. Am. Chem. Soc.* **1983**, *105*, 7175. (b) Baillargeon, V. P.; Stille, J. K., *J. Am. Chem. Soc.* **1986**, *108*, 452.

47. Pri-Bar, I.; Buchman, O., *J. Org. Chem.* **1984**, *49*, 4009.

48. (a) Alper, H.; Hashem, K.; Heveling, J., *Organometallics* **1982**, *1*, 775. (b) Galamb, V.; Alper, H., *Tetrahedron Lett.* **1983**, *24*, 2965.

49. (a) Woell, J. B.; Fergusson, S. B.; Alper, H., *J. Org. Chem.* **1985**, *50*, 2134. (b) Hashem, K. E.; Woell, J. B.; Alper, H., *Tetrahedron Lett.* **1984**, *25*, 4879. (c) Alper, H.; Antebi, S.; Woell, J. B., *Angew. Chem., Int. Ed. Engl.* **1984**, *23*, 732.

50. Hashimoto, H.; Furuichi, K.; Miwa, T., *J. Chem. Soc., Chem. Commun.* **1987**, 1002.

51. Crisp, G. T.; Scott, W. J.; Stille, J. K., *J. Am. Chem. Soc.* **1984**, *106*, 7500.

52. Dalton, J. R.; Regen, S. L., *J. Org. Chem.* **1979**, *44*, 4443.

53. Sekiya, A.; Ishikawa, N., *Chem. Lett.* **1975**, 277.

54. Chatani, N.; Hanafusa, T., *J. Org. Chem.* **1986**, *51*, 4714.

55. Murahashi, S.-I.; Naota, T.; Nakajima, N., *J. Org. Chem.* **1986**, *51*, 898.

56. Yamamura, K.; Murahashi, S.-I., *Tetrahedron Lett.* **1977**, 4429.

57. (a) Genêt, J. P.; Balabane, M.; Bäckvall, J. E.; Nyström, J. E., *Tetrahedron Lett.* **1983**, *24*, 2745. (b) Nyström, J. E.; Rein, T.; Bäckvall, J. E., *Org. Synth.* **1989**, *67*, 105. (c) Trost, B. M.; Keinan, E., *J. Org. Chem.* **1979**, *44*, 3451.

58. Byström, S. E.; Aslanian, R.; Bäckvall, J. E., *Tetrahedron Lett.* **1985**, *26*, 1749.

59. (a) Tanigawa, Y.; Nishimura, K.; Kawasaki, A.; Murahashi, S.-I., *Tetrahedron Lett.* **1982**, *23*, 5549. (b) Murahashi, S.-I.; Imada, Y.; Taniguchi, Y.; Kodera, Y., *Tetrahedron Lett.* **1988**, *29*, 2973.

60. Tamura, R.; Hegedus, L. S., *J. Am. Chem. Soc.* **1982**, *104*, 3727.

61. Murahashi, S.-I.; Tanigawa, Y.; Imada, Y.; Taniguchi, Y., *Tetrahedron Lett.* **1986**, *27*, 227.

62. Inomata, K.; Yamamoto, T.; Kotake, H., *Chem. Lett.* **1981**, 1357.

63. (a) Tamura, R.; Hayashi, K.; Kakihana, M.; Tsuji, M.; Oda, D., *Tetrahedron Lett.* **1985**, *26*, 851. (b) Ono, N.; Hamamoto, I.; Yanai, T.; Kaji, A., *J. Chem. Soc., Chem. Commun.* **1985**, 523.

64. Tamura, R.; Hayashi, K.; Kakihana, M.; Tsuji, M.; Oda, D., *Chem. Lett.* **1985**, 229.

65. Deardorff, D. R.; Myles, D. C., *Org. Synth.* **1989**, *67*, 114.

66. (a) Kosugi, M.; Shimizu, K.; Ohtani, A.; Migita, T., *Chem. Lett.* **1981**, 829. (b) Kosugi, M.; Ohya, T.; Migita, T., *Bull. Chem. Soc. Jpn.* **1983**, *56*, 3855. (c) Azizian, H.; Eaborn, C.; Pidcock, A., *J. Organomet. Chem.* **1981**, *215*, 49.

67. (a) Matsumoto, H.; Nagashima, S.; Yoshihiro, K.; Nagai, Y., *J. Organomet. Chem.* **1975**, *85*, C1. (b) Matsumoto, H.; Yoshihiro, K.; Nagashima, S.; Watanabe, H.; Nagai, Y., *J. Organomet. Chem.* **1977**, *128*, 409.

68. Wulff, W. D.; Peterson, G. A.; Bauta, W. E.; Chan, K.-S.; Faron, K. L.; Gilbertson, S. R.; Kaesler, R. W.; Yang, D. C.; Murray, C. K., *J. Org. Chem.* **1986**, *51*, 277.

69. Nonaka, T.; Okuda, Y.; Matsubara, S.; Oshima, K.; Utimoto, K.; Nozaki, H., *J. Org. Chem.* **1986**, *51*, 4716.

70. (a) Mitchell, T. N.; Amamria, A.; Killing, H.; Rutschow, D., *J. Organomet. Chem.* **1983**, *241*, C45. (b) Mitchell, T. N.; Amamria, A.; Killing, H.; Rutschow, D., *J. Organomet. Chem.* **1986**, *304*, 257. (c) Mitchell, T. N.; Kwetkat, K., *J. Organomet. Chem.* **1992**, *439*, 127.

71. Watanabe, H.; Kobayashi, M.; Saito, M.; Nagai, Y., *J. Organomet. Chem.* **1981**, *216*, 149.

72. Kuniyasu, H.; Ogawa, A.; Miyazaki, S.-I.; Ryu, I.; Kambe, N.; Sonoda, N., *J. Am. Chem. Soc.* **1991**, *113*, 9796.

73. Watanabe, H.; Saito, M.; Sutou, N.; Kishimoto, K.; Inose, J.; Nagai, Y., *J. Organomet. Chem.* **1982**, *225*, 343.

74. Killing, H.; Mitchell, T. N., *Organometallics* **1984**, *3*, 1318.

75. (a) Chenard, B. L.; Laganis, E. D.; Davidson, F.; RajanBabu, T. V., *J. Org. Chem.* **1985**, *50*, 3666. (b) Chenard, B. L.; Van Zyl, C. M., *J. Org. Chem.* **1986**, *51*, 3561. (c) Mitchell, T. N.; Killing, H.; Dicke, R.; Wickenkamp, R., *J. Chem. Soc., Chem. Commun.* **1985**, 354. (d) Mitchell, T. N.; Wickenkamp, R.; Amamria, A.; Dicke, R.; Schneider, U., *J. Org. Chem.* **1987**, *52*, 4868.

76. Piers, E.; Skerlj, R. T., *J. Chem. Soc., Chem. Commun.* **1986**, 626.

77. Piers, E.; Tillyer, R. D., *J. Chem. Soc., Perkin Trans. 1* **1989**, 2124.

78. (a) Ichinose, Y.; Oda, H.; Oshima, K.; Utimoto, K., *Bull. Chem. Soc. Jpn.* **1987**, *60*, 3468. (b) Miyake, H.; Yamamura, K., *Chem. Lett.* **1989**, 981. (c) Matsubara, S.; Hibino, J.-I.; Morizawa, Y.; Oshima, K.; Nozaki, H., *J. Organomet. Chem.* **1985**, *285*, 163.

79. Four, P.; Guibe, F., *Tetrahedron Lett.* **1982**, *23*, 1825.

80. Trost, B. M.; Herndon, J. W., *J. Am. Chem. Soc.* **1984**, *106*, 6835.

81. Matsumoto, H.; Yako, T.; Nagashima, S.; Motegi, T.; Nagai, Y., *J. Organomet. Chem.* **1978**, *148*, 97.

82. Miyake, H.; Yamamura, K., *Chem. Lett.* **1992**, 507.

83. (a) Hirao, T.; Masunaga, T.; Ohshiro, Y.; Agawa, T., *Tetrahedron Lett.* **1980**, *21*, 3595. (b) Hirao, T.; Masunaga, T.; Yamada, N.; Ohshiro, Y.; Agawa, T., *Bull. Chem. Soc. Jpn.* **1982**, *55*, 909.

84. Xu, Y.; Li, Z.; Xia, J.; Guo, H.; Huang, Y., *Synthesis* **1983**, 377.

85. Townsend, J. M.; Reingold, I. D.; Kendall, M. C. R.; Spencer, T. A., *J. Org. Chem.* **1975**, *40*, 2976.

86. Tamaru, Y.; Yamada, Y.; Inoue, K.; Yamamoto, Y.; Yoshida, Z., *J. Org. Chem.* **1983**, *48*, 1286.

87. Helquist, P., *Tetrahedron Lett.* **1978**, 1913.

88. Zask, A.; Helquist, P., *J. Org. Chem.* **1978**, *43*, 1619.

89. (a) Chen, Q.-Y.; He, Y.-B.; Yang, Z.-Y., *J. Chem. Soc., Chem. Commun.* **1986**, 1452. (b) Peterson, G. A.; Kunng, F.-A.; McCallum, J. S.; Wulff, W. D., *Tetrahedron Lett.* **1987**, *28*, 1381.

90. Hutchins, R. O.; Learn, K.; Fulton, R. P., *Tetrahedron Lett.* **1980**, *21*, 27.

91. Keinan, E.; Greenspoon, N., *J. Org. Chem.* **1983**, *48*, 3545.

92. Tabuchi, T.; Inanaga, J.; Yamaguchi, M., *Tetrahedron Lett.* **1986**, *27*, 601.

93. Keinan, E.; Greenspoon, N., *Tetrahedron Lett.* **1982**, *23*, 241.

94. Yamamoto, Y.; Hori, A.; Hutchinson, C. R., *J. Am. Chem. Soc.* **1985**, *107*, 2471.

95. Kotake, H.; Yamamoto, T.; Kinoshita, H., *Chem. Lett.* **1982**, 1331.

96. Hutchins, R, O.; Learn, K., *J. Org. Chem.* **1982**, *47*, 4380.

97. Matsushita, H.; Negishi, E., *J. Org. Chem.* **1982**, *47*, 4161.

98. (a) Guibe, F.; Four, P.; Riviere, H., *J. Chem. Soc., Chem. Commun.* **1980**, 432. (b) Four, P.; Guibe, F., *J. Org. Chem.* **1981**, *46*, 4439.

99. Kuniyasu, H.; Ogawa, A.; Higaki, K.; Sonoda, N., *Organometallics* **1992**, *11*, 3937.

100. Keinan, E.; Gleize, P. A., *Tetrahedron Lett.* **1982**, *23*, 477.

101. (a) Keinan, E.; Greenspoon, N., *Tetrahedron Lett.* **1985**, *26*, 1353. (b) Keinan, E.; Greenspoon, N., *J. Am. Chem. Soc.* **1986**, *108*, 7314.

102. Perez, D.; Greenspoon, N.; Keinan, E., *J. Org. Chem.* **1987**, *52*, 5570.

103. Urata, H.; Suzuki, H.; Moro-Oka, Y.; Ikawa, T., *J. Organomet. Chem.* **1982**, *234*, 367.

104. Guibe, F.; Saint M'Leux, Y., *Tetrahedron Lett.* **1981**, *22*, 3591.

105. Jeffrey, P. D.; McCombie, S. W., *J. Org. Chem.* **1982**, *47*, 587.

106. Hayakawa, Y.; Kato, H.; Uchiyama, M.; Kajino, H.; Noyori, R., *J. Org. Chem.* **1986**, *51*, 2400.

107. (a) Kunz, H.; Unverzagt, C., *Angew. Chem., Int. Ed. Engl.* **1984**, *23*, 436. (b) Kinoshita, H.; Inomata, K.; Kameda, T.; Kotake, H., *Chem. Lett.* **1985**, 515.

108. Boullanger, P.; Descotes, G., *Tetrahedron Lett.* **1986**, *27*, 2599.

109. Deziel, R., *Tetrahedron Lett.* **1987**, *28*, 4371.

110. (a) Kunz, H.; Waldmann, H., *Angew. Chem., Int. Ed. Engl.* **1984**, *23*, 71. (b) Kunz, H.; Waldmann, H., *Helv. Chim. Acta* **1985**, *68*, 618. (c) Friedrich-Bochnitschek, S.; Waldmann, H.; Kunz, H., *J. Org. Chem.* **1989**, *54*, 751.

111. Hayakawa, Y.; Uchiyama, M.; Kato, H.; Noyori, R., *Tetrahedron Lett.* **1985**, *26*, 6505.

112. Nakayama, K.; Uoto, K.; Higashi, K.; Soga, T.; Kusama, T., *Chem. Pharm. Bull.* **1992**, *40*, 1718.

113. (a) Tamaru, Y.; Yoshida, Z.; Yamada, Y.; Mukai, K.; Yoshioka, H., *J. Org. Chem.* **1983**, *48*, 1293. (b) Yamada, Y.; Mukai, K.; Yoshioka, H.; Tamaru, Y.; Yoshida, Z., *Tetrahedron Lett.* **1979**, 5015.

114. Ikariya, T.; Ishikawa, Y.; Hirai, K.; Yoshikawa, S., *Chem. Lett.* **1982**, 1815.

115. Murahashi, S.-I.; Makabe, Y.; Kunita, K., *J. Org. Chem.* **1988**, *53*, 4489.

116. Trost, B. M.; Runge, T. A.; Jungheim, L. N., *J. Am. Chem. Soc.* **1980**, *102*, 2840.

117. Oppolzer, W., *Angew. Chem., Int. Ed. Engl.* **1989**, *28*, 38.

118. Mandai, T.; Hashio, S.; Goto, J.; Kawada, M., *Tetrahedron Lett.* **1981**, *22*, 2187.

119. Trost, B. M.; Verhoeven, T. R.; Fortunak, J. M., *Tetrahedron Lett.* **1979**, 2301.

120. Trost, B. M.; Fortunak, J. M., *J. Am. Chem. Soc.* **1980**, *102*, 2843.

121. Suzuki, M.; Oda, Y.; Noyori, R., *J. Am. Chem. Soc.* **1979**, *101*, 1623.

122. Suzuki, M.; Watanabe, A.; Noyori, R., *J. Am. Chem. Soc.* **1980**, *102*, 2095.

Richard W. Friesen
Merck Frosst Centre for Therapeutic Research, Kirkland,
Quebec, Canada

Tetra-*n*-propylammonium Perruthenate[1]

$$Pr_4N^+RuO_4^-$$

[114615-82-6] $C_{12}H_{28}NO_4Ru$ (MW 351.48)

InChI = 1S/C12H28N.4O.Ru/c1-5-9-13(10-6-2,11-7-3)12-8-4;;;;;/h5-12H2,1-4H3;;;;;/q+1;;;;-1;

InChIKey = NQSIKKSFBQCBSI-UHFFFAOYSA-N

(mild oxidant for conversion of multifunctionalized alcohols to aldehydes and ketones;[1] can selectively oxidize primary–secondary diols to lactones;[2] can cleave carbon–carbon bonds of 1,2-diols[3])

Alternate Name: TPAP.

Physical Data: mp 165 °C (dec).

Solubility: sol CH_2Cl_2, and MeCN; partially sol C_6H_6.

Form Supplied in: dark green solid; commercially available.

Analysis of Reagent Purity: microanalysis.

Handling, Storage, and Precautions: stable at room temperature and may be stored for long periods of time without significant decomposition, especially if kept refrigerated in the dark. The reagent should not be heated neat, as small quantities decompose with flame at 150–160 °C in air.

Original Commentary

Steven V. Ley & Joanne Norman
University of Cambridge, UK

General Procedures. TPAP is a convenient, mild, neutral and selective oxidant of primary alcohols to aldehydes and secondary alcohols to ketones.[1] These reactions are carried out with catalytic TPAP at rt in the presence of stoichiometric or excess *N-Methylmorpholine N-Oxide* (NMO) as a cooxidant. A kinetic study with 2-propanol as the substrate has shown that these oxidations are strongly autocatalytic.[4] Turnovers of up to 250 are obtainable if activated powdered molecular sieves are introduced to remove both the water formed during oxidations and the water of crystallization of the NMO. Solvents commonly employed are dichloromethane and acetonitrile, or combinations of these. This reagent also works well on a small scale where other methods, such as those employing activated *Dimethyl Sulfoxide* reagents, are inconvenient.[1] For large scale oxidations it is necessary to moderate these reactions by cooling and by slow portionwise addition of the TPAP. To achieve full conversion on a large scale, the NMO should be predried (by first treating an organic solution of NMO with anhydrous magnesium sulfate) and the use of 10% acetonitrile–dichloromethane as solvent is recommended.[1]

Oxidation of Primary Alcohols. Primary alcohols can be oxidized in the presence of a variety of functional groups, including tetrahydropyranyl ethers (eq 1),[1,5] epoxides (eq 2),[1,6] acetals (eq 3),[1,7] silyl ethers,[1,8] peroxides,[9] lactones,[1,10] alkenes,[1,11] alkynes,[1,12] esters,[1,13] amides,[1,14] sulfones,[1] and indoles.[10] Oxidation of substrates with labile α-centers proceeds without epimerization.[1]

Oxidation of Secondary Alcohols. In a similar fashion, multifunctional secondary alcohols are oxidized to ketones (eqs 4 and 5)[9,15] in good yields.[1,5–14]

A particularly hindered secondary alcohol (an intermediate in the latter stages of the synthesis of tetronolide) resisted oxidation with activated DMSO, *Pyridinium Chlorochromate*, and activated *Manganese Dioxide*, yet stoichiometric TPAP afforded the ketone in 81% yield.[16]

Allylic Alcohols. These alcohols are successfully oxidized to the corresponding enones and enals (eq 6).[1,17]

Oxidation of a primary allylic alcohol over a secondary allylic alcohol has been achieved.[18] Oxidation of the homoallylic alcohol of cholesterol by TPAP and NMO under ultrasonication conditions gives the dienone cholest-4-ene-3,6-dione in 80% yield.[19] This oxidation was subsequently carried out in the presence of a labile TBDMS enol ether group which remained intact, while with both PCC and activated DMSO this protecting group did not survive.[19] Oxidation of homopropargylic alcohols leads to allenones, as with other common oxidants.[20]

Lactols. Oxidations of lactols to lactones are facile and high yielding; several examples have been reported in the literature (eq 7).[1,21]

$$(7)$$

Selective Oxidations. The selective oxidation of 1,4 and 1,5 primary–secondary diols to lactones is a valuable application of this reagent.[2] Few general mild reagents for the chemoselective oxidation to the hydroxy aldehyde are available.[22] The most widely known reagents are Pt and O_2, and *Dihydridotetrakis-(triphenylphosphine)ruthenium(II)*.[22] Hydroxy aldehydes, in their lactol form, are then oxidized further to lactones. The use of TPAP is advantageous in that it is commercially available, employs mild catalytic reaction conditions, and reacts with high selectivity in unsymmetrical cases (eq 8).[2] Lactones have also been formed from primary–tertiary diols.[23]

$$(8)$$

Functional Group Compatibility. The neutral conditions of these oxidations have been utilized to provide improved yields with acid sensitive substrates compared to the well established Swern method (eqs 9–10).[1,16]

$$(9)$$

TPAP 73%
Swern 30%

$$(10)$$

TPAP 73%
Swern 0%

Highly sensitive alcohols have also been oxidized, albeit in low yield, when most other conventional methods, such as

Dess–Martin periodinane (*1,1,1-Triacetoxy-1,1-dihydro-1,2-benziodoxol-3(1H)-one*),[24] PCC, *Pyridinium Dichromate*, and DMSO activated with *Sulfur Trioxide–Pyridine* have failed (eq 10).[25]

Oxidative Cleavage Reactions. Among the numerous methods for 1,2-diol cleavage there exist only a few that involve catalytic ruthenium reagents, for example *Ruthenium(III) Chloride* with *Sodium Periodate*.[22] Attempted selective monooxidation of a 1,2-diol to the hydroxy aldehyde with catalytic TPAP and NMO resulted in carbon–carbon bond cleavage to provide the aldehyde (eq 11).[3] Furthermore, attempted oxidation of an anomeric α-hydroxy ester failed; instead, in this case decarboxylation/decarbonylation and formation of the lactone was observed (eq 12). However, *Dimethyl Sulfoxide–Acetic Anhydride* provided the required α-dicarbonyl unit.[26] Retro-aldol fragmentations can also be a problem.[27]

$$(11)$$

$$(12)$$

Heteroatom Oxidation. Thus far, TPAP has only been used to oxidize sulfur. Oxidation of an oxothiazolidine *S*-oxide to the corresponding *S,S*-dioxide gave only poor results in comparison to the standard RuO_2 with $NaIO_4$ conditions.[28] However, oxidation of sulfides to sulfones in the presence of isolated double bonds has been investigated. The yields with TPAP/NMO range from 61–99% which are higher than to those obtained with *m-Chloroperbenzoic Acid*, *Potassium Monoperoxysulfate* (Oxone®), or *Hydrogen Peroxide–Acetic Acid*.[29]

First Update

Anthony J. Wilson
University of Nottingham, Nottingham, UK

Additional Co-oxidants. Previous literature has exclusively described the use of tetrapropylammonium perruthenate (TPAP) with *N*-methylmorpholine *N*-oxide (NMO) as a co-oxidant. More recent work has investigated the use of several alternative co-oxidants that further increase the versatility of TPAP. The use of molecular oxygen in particular as the co-oxidant[30–33] greatly reduces both the cost and environmental impact of this reagent system.

Polymer-supported *N*-methylmorpholine *N*-oxide has been developed[34] as has the use of sodium hypochlorite.[35] Additional work has been completed on the use of this reagent with ionic liquids,[36] which can facilitate catalyst recovery.[37]

Oxidation of Secondary Nitro Compounds. Secondary nitro compounds have been selectively oxidized to the corresponding keto compounds,[38] and a variety of conditions for this transformation have been detailed. Initial investigations used stoichiometric TPAP in the presence of NMO and 4 Å molecular sieves (eq 13); however, the incorporation of a variety of silver salts into the reaction mixture has allowed the use of catalytic quantities of TPAP.

$$\text{(13)}$$

81%

Isomerization of Allylic Alcohols. An unusual isomerization of allylic alcohols has been reported, in which treatment of allylic alcohols with 5 mol % of TPAP in the presence of molecular oxygen and 4 Å molecular sieves gives the expected aldehyde in good yield. However, when the reaction is run without molecular sieves under an argon atmosphere, the inclusion of a high molecular weight alcohol such as 1-decanol or 2-undecanol leads exclusively to the formation of the saturated aldehyde or ketone in excellent yield (eq 14).[39] A range of examples are given and a mechanism is proposed.

$$\text{(14)}$$

90%

Cyclization of Triazenes to Triazoles. TPAP/NMO has been used to cyclize 1,2,4-triazines to the corresponding 1,2,4-triazoles.[40] Such compounds are of interest to medicinal chemists due to their wide range of biological properties. Classical methods of synthesizing such compounds typically employ harsh conditions such as the use of H_2O_2 and thus have limited functional group compatibility. The use of alternative reagents such as NaClO, $Ca(ClO)_2$, Dess–Martin periodinane, and TPAP/NMO has broadened the range of functional groups that can be incorporated into 1,2,4-triazoles by this method (eq 15).

$$\text{(15)}$$

R = Ph, $CH_2{=}CH_2{-}$, 3-Pyr   R = Ph (79%), $CH_2{=}CH_2{-}$ (60%),
$MeO_2CCH_2CH_2{-}$ 3-Pyr (45%), $MeO_2CCH_2CH_2{-}$ (75%)

Oxidation of Pyrrolines. Oxidation of 3-pyrrolines to the corresponding imines has been conducted with TPAP/NMO in MeCN (eq 16).[41,42]

$$\text{(16)}$$

92%

Sequential/One-pot TPAP–Wittig Oxidation Olefination. Another interesting development is the possibility of performing sequential oxidation and olefination reactions without the requirement for workup and isolation of sensitive intermediate aldehyde.[43] Generation of the aldehyde is accomplished by treatment of the starting alcohol with TPAP/NMO in CH_2Cl_2 over 4 Å molecular sieves, after which the crude reaction mixture was added slowly to a $-78\,°C$ solution of triphenylphosphonium bromide in anhydrous THF that had been pretreated with *n*-BuLi for 1 h. Aqueous workup then gave the desired products in good yield (eq 17).

$$\text{(17)}$$

R' = H, Me, Cl, Br, CO_2Et
X = Cl, Br   44–94% yield

Synthesis of Nitronyl Nitroxide Radicals. Nitronyl nitroxide radicals are conventionally synthesized by oxidation of dihydroxyimidazolidines with reagents such as $NaIO_4$, SeO_2, and PdO_2. However, the use of TPAP/NMO in CH_2Cl_2 confers the numerous established advantages of this mild and selective oxidizing agent in yields comparable to the traditional techniques (eq 18).[44]

$$\text{(18)}$$

44 90%

Related and Modified Reagents. In addition to the standard reagent, a polymer-supported perruthenate (PSP) compound has been described.[45,46] Recent work has also examined the doping of organically modified silicas (ormosils) with TPAP via a sol–gel process,[47–53] which enhances the general versatility and reusability of TPAP catalysts. The use of supercritical carbon dioxide as a solvent has also been investigated.[54] TPAP has also found use as a convenient source of ruthenium in ruthenium-catalyzed hypochlorite oxidations.[55]

1. (a) Griffith, W. P.; Ley, S. V.; Whitcombe, G. P.; White, A. D., *J. Chem. Soc., Chem. Commun.* **1987**, 1625. (b) Griffith, W. P.; Ley, S. V., *Aldrichim. Acta* **1990**, *23*, 13. (c) Ley, S. V.; Norman, J.; Griffith, W. P.; Marsden, S. P., *Synthesis* **1994**, 639.

2. Bloch, R.; Brillet, C., *Synlett* **1991**, 829.

3. Queneau, Y.; Krol, W. J.; Bornmann, W. G.; Danishefsky, S. J., *J. Org. Chem.* **1992**, *57*, 4043.

4. Lee, D. G.; Wang, Z.; Chandler, W. D., *J. Org. Chem.* **1992**, *57*, 3276.

5. Guanti, G.; Banfi, L.; Ghiron, C.; Narisano, E., *Tetrahedron Lett.* **1991**, *32*, 267.

6. (a) Stürmer, R.; Ritter, K.; Hoffmann, R. W., *Angew. Chem., Int. Ed. Engl.* **1993**, *32*, 101. (b) Kim, G.; Chu-Moyer, M. Y.; Danishefsky, S. J., *J. Am. Chem. Soc.* **1990**, *112*, 2003.

7. (a) Anthony, N. J.; Armstrong, A.; Ley, S. V.; Madin, A., *Tetrahedron Lett.* **1989**, *30*, 3209. (b) Romeyke, Y.; Keller, M.; Kluge, H.; Grabley, S.; Hammann, P., *Tetrahedron* **1991**, *47*, 3335.

8. (a) Ley, S. V.; Maw, G. N.; Trudell, M. L., *Tetrahedron Lett.* **1990**, *31*, 5521. (b) Rosini, G.; Marotta, E.; Raimondi, A.; Righi, P., *Tetrahedron: Asymmetry* **1991**, *2*, 123.

9. (a) Hu, Y.; Ziffer, H., *J. Labelled Comp. Radiopharm.* **1991**, *29*, 1293.

10. Linz, G.; Weetman, J.; Hady, A. F. A.; Helmchen, G., *Tetrahedron Lett.* **1989**, *30*, 5599.

11. (a) Cole, P. A.; Bean, J. M.; Robinson, C. H., *Proc. Natl. Acad. Sci. USA* **1990**, *87*, 2999. (b) Piers, E.; Roberge, J. Y., *Tetrahedron Lett.* **1991**, *32*, 5219.

12. Desmaële, D.; Champion, N., *Tetrahedron Lett.* **1992**, *33*, 4447.

13. Hori, K.; Hikage, N.; Inagaki, A.; Mori, S.; Nomura, K.; Yoshii, E., *J. Org. Chem.* **1992**, *57*, 2888.

14. Guanti, G.; Banfi, L.; Narisano, E.; Thea, S., *Synlett* **1992**, 311.

15. Sulikowski, M. M.; Ellis Davies, G. E. R.; Smith, A. B., III, *J. Chem. Soc., Perkin Trans. 1* **1992**, 979.

16. Takeda, K.; Kawanishi, E.; Nakamura, H.; Yoshii, E., *Tetrahedron Lett.* **1991**, *32*, 4925.

17. (a) Ninan, A.; Sainsbury, M., *Tetrahedron* **1992**, *48*, 6709. (b) Rychnovsky, S. D.; Rodriguez, C., *J. Org. Chem.* **1992**, *57*, 4793. (c) Kang, H.-J. Ra, C. S.; Paquette, L. A., *J. Am. Chem. Soc.* **1991**, *113*, 9384. (d) Schreiber, S. L.; Kiessling, L. L., *Tetrahedron Lett.* **1989**, *30*, 433.

18. Hitchcock, S. A.; Pattenden, G., *Tetrahedron Lett.* **1992**, *33*, 4843.

19. Moreno, M. J. S. M.; Melo, M. L. S.; Neves, A. S. C., *Tetrahedron Lett.* **1991**, *32*, 3201.

20. Marshall, J. A.; Robinson, E. D.; Lebreton, J., *J. Org. Chem.* **1990**, *55*, 227.

21. Paquette, L. A.; Kang, H-J.; Ra, C. S., *J. Am. Chem. Soc.* **1992**, *114*, 7387.

22. *Comprehensive Organic Synthesis* **1991**, 7.

23. (a) Mehta, G.; Karra, S. R., *Tetrahedron Lett.* **1991**, *32*, 3215. (b) Mehta, G.; Karra, S. R., *J. Chem. Soc., Chem. Commun.* **1991**, 1367.

24. Yamashita, D. S.; Rocco, V. P.; Danishefsky, S. J., *Tetrahedron Lett.* **1991**, *32*, 6667.

25. Tokoroyama, T.; Kotsuji, Y.; Matsuyama, H.; Shimura, T.; Yokotani, K.; Fukuyama, Y., *J. Chem. Soc., Perkin Trans. 1* **1990**, 1745.

26. Watanabe, T.; Nishiyama, S.; Yamamura, S.; Kato, K.; Nagai, M.; Takita, T., *Tetrahedron Lett.* **1991**, *32*, 2399.

27. Shih, T. L.; Mrozik, H.; Holmes, M. A.; Arison, B. H.; Doss, G. A.; Waksmunski, F.; Fisher, M. H., *Tetrahedron Lett.* **1992**, *33*, 1709.

28. White, G. J.; Garst, M. E., *J. Org. Chem.* **1991**, *56*, 3177.

29. Guertin, K. R.; Kende, A. S., *Tetrahedron Lett.* **1993**, *34*, 5369.

30. Lenz, R.; Ley, S. V., *J. Chem. Soc., Perkin Trans. 1* **1997**, 3291.

31. Marko, I. E.; Giles, P. R.; Tsukazaki, M.; Chelle-Regnaut, I.; Urch, C. J.; Brown, S. M., *J. Am. Chem. Soc.* **1997**, *119*, 12661.

32. Coleman, K. S.; Lorber, C. Y.; Osborn, J. A., *Eur. J. Inorg. Chem.* **1998**, 1673.

33. Hasan, M.; Musawir, M.; Davey, P. N.; Kozhevnikov, I. V., *J. Mol. Catal. A* **2002**, *180*, 77.

34. Brown, D. S.; Kerr, W. J.; Lindsay, D. M.; Pike, K. G.; Ratcliffe, P. D., *Synlett* **2001**, 1257.

35. Gonsalvi, L.; Arends, I. W. C. E.; Moilanen, P.; Sheldon, R. A., *Adv. Synth. Catal.* **2003**, *345*, 1321.

36. Farmer, V.; Welton, T., *Green Chem.* **2002**, *4*, 97.

37. Ley, S. V.; Ramarao, C.; Smith, M. D., *Chem. Commun.* **2001**, 2278.

38. Tokunaga, Y.; Ihara, M.; Fukumoto, K., *J. Chem. Soc., Perkin Trans. 1* **1997**, 207.

39. Marko, I. E.; Gautier, A.; Tsukazaki, M.; Llobet, A.; Plantalech-Mir, E.; Urch, C. J.; Brown, S. M., *Angew. Chem., Int. Ed.* **1999**, *38*, 1960.

40. Paulvannan, K.; Hale, R.; Sedehi, D.; Chen, T., *Tetrahedron* **2001**, *57*, 9677.

41. Green, M. P.; Prodger, J. C.; Hayes, C. J., *Tetrahedron Lett.* **2002**, *43*, 2649.

42. Goti, A.; Romani, M., *Tetrahedron Lett.* **1994**, *35*, 6567.

43. MacCoss, R. N.; Balskus, E. P.; Ley, S. V., *Tetrahedron Lett.* **2003**, *44*, 7779.

44. Gorini, L.; Caneschi, A.; Menichetti, S., *Synlett* **2006**, 948.

45. Langer, P., *J. Prakt. Chem.* **2000**, *342*, 728.

46. Hinzen, B.; Ley, S. V., *J. Chem. Soc., Perkin Trans. 1* **1997**, 1907.

47. Pagliaro, M.; Ciriminna, R., *Tetrahedron Lett.* **2001**, *42*, 4511.

48. Ciriminna, R.; Campestrini, S.; Pagliaro, M., *Adv. Synth. Catal.* **2003**, *345*, 1261.

49. Ciriminna, R.; Pagliaro, M., *Chem. Eur. J.* **2003**, *9*, 5067.

50. Campestrini, S.; Carraro, M.; Ciriminna, R.; Pagliaro, M.; Tonellato, U., *Tetrahedron Lett.* **2004**, *45*, 7283.

51. Campestrini, S.; Carraro, M.; Ciriminna, R.; Pagliaro, M.; Tonellato, U., *Adv. Synth. Catal.* **2005**, *347*, 825.

52. Campestrini, S.; Carraro, M.; Franco, L.; Ciriminna, R.; Pagliaro, M., *Tetrahedron Lett.* **2008**, *49*, 419.

53. Ciriminna, R.; Campestrini, S.; Pagliaro, M., *Adv. Synth. Catal.* **2004**, *346*, 231.

54. Fidalgo, A.; Ciriminna, R.; Ilharco Laura, M.; Campestrini, S.; Carraro, M.; Pagliaro, M., *Phys. Chem. Chem. Phys.* **2008**, *10*, 2026.

55. Gonsalvi, L.; Arends, I. W. C. E.; Sheldon, R. A., *Org. Lett.* **2002**, *4*, 1659.

p-Toluenesulfonyl Chloride[1]

[98-59-9] $C_7H_7ClO_2S$ (MW 190.66)

InChI = 1/C7H7ClO2S/c1-6-2-4-7(5-3-6)11(8,9)10/h2-5H,1H3

InChIKey = YYROPELSRYBVMQ-UHFFFAOYAN

(sulfonyl transfer reagent; O-sulfonylation of alcohols[2] for conversion to chlorides[3] or intermolecular[4] and intramolecular[5] displacements, vicinal diols for epoxidation,[6] 1,3-diols for oxetane formation,[7] carboxylic acids for esterification[8] or decarboxylation,[9] oximes for Beckmann rearrangements[10] or fragmentations[11] and Neber rearrangements,[12] hydroxamic acids for Lossen rearrangements,[13] nitrones for rearrangements,[14] conversion of N-cyclopropylhydroxylamines to β-lactams;[15] N-sulfonylation of aliphatic amines[16] for subsequent deamination[17] or displacement,[18] aromatic amines for protection;[19] C-sulfonylation of alkenes[20] and silylalkynes;[21] dehydration of ureas,[22] formamides,[23] and amides[24])

Alternative Name: tosyl chloride.

Physical Data: mp 67–69 °C; bp 146 °C/15 mmHg.

Solubility: insol H_2O; freely sol ethanol, benzene, chloroform, ether.

Form Supplied in: white solid, widely available.

Purification: upon prolonged standing the material develops impurities of p-toluenesulfonic acid and HCl. Tosyl chloride is purified by dissolving 10 g in a minimum volume of $CHCl_3$ (ca. 25 mL), filtering, and diluting with five volumes (ca. 125 mL) of petroleum ether (bp 30–60 °C) to precipitate impurities (mostly tosic acid, mp 101–104 °C). The solution is filtered, clarified with charcoal, and concentrated to ca. 40 mL by evaporation. Further evaporation to a very small volume gives 7 g of pure white crystals (mp 67.5–68.5 °C).[25]

Tosyl chloride may also be recrystallized from petroleum ether, from benzene, or from toluene/petroleum ether (bp 40–60 °C) in the cold. Tosyl chloride in diethyl ether can be washed with aqueous 10% NaOH until colorless, then dried with Na_2SO_4 and crystallized by cooling in powdered dry ice. It can also be purified by dissolving in benzene, washing with aqueous 5% NaOH, drying with K_2CO_3 or $MgSO_4$, and distilling under reduced pressure.[26]

Handling, Storage, and Precautions: freshly purified tosyl chloride should be used for best results. Tosyl chloride is a moisture-sensitive, corrosive lachrymator.

Original Commentary

D. Todd Whitaker, K. Sinclair Whitaker & Carl R. Johnson
Wayne State University, Detroit, MI, USA

General Discussion. The tosylation of alcohols is one of the most prevalent reactions in organic chemistry.[1] Optimized conditions for this reaction include the use of a 1:1.5:2 ratio of alcohol/tosyl chloride/pyridine in chloroform (eq 1).[2a] This procedure avoids formation of unwanted pyridinium salts inherent to reactions where higher relative quantities of pyridine have been employed.[2b]

Good yields (81–88%) for tosylation have also been observed under biphasic conditions where **Benzyltriethylammonium Chloride** is employed as a phase-transfer catalyst between benzene and aqueous sodium hydroxide solution.[2c]

It has long been known that it is possible to selectively tosylate primary over secondary alcohols (eq 2).[2d]

It is also possible to regioselectively mono-O-tosylate various nonprotected hydroxyl functionalities, as illustrated in Scheme 1.[2e] This method employs a preliminary activation of a glycopyranoside with **Dibutyltin Oxide** and usually requires the use of a basic catalyst such as **4-Dimethylaminopyridine** (DMAP) in conjunction with tosyl chloride. Regioselectivity differs markedly from acylation reactions and is thought to be a function of changes

Scheme 1

Essential Reactions for Organic Synthesis, First Edition. Edited by Philip L. Fuchs.
© 2016 John Wiley & Sons, Ltd. Published 2016 by John Wiley & Sons, Ltd.

in the kinetics of the reactions with the various tin intermediates which are in equilibrium.

Regioselective mono- and ditosylation of aldonolactones has also been reported.[2f] A few of these products are shown in Scheme 2. In no instances was the β-hydroxy function tosylated.

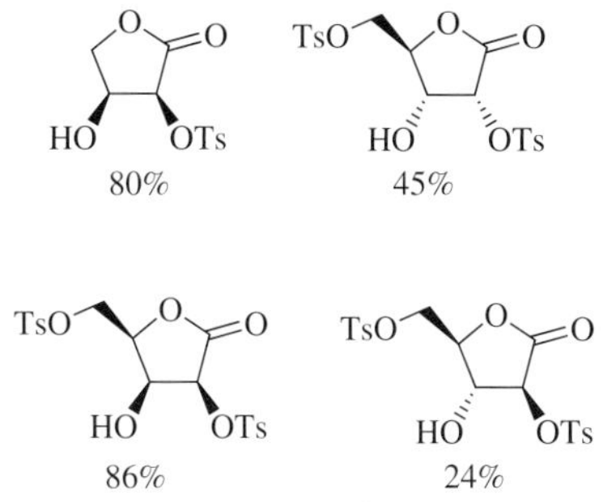

Scheme 2

The regioselective tosylation of various cyclodextrins has been reported. The two major products resulting from the tosylation of β-cyclodextrin are heptakis(6-*O*-(*p*-tosyl))-β-cyclodextrin (**1**) and heptakis(6-*O*-(*p*-tosyl))-2-*O*-(*p*-tosyl))-β-cyclodextrin (**2**).[2g,h] Yamamura has also prepared hexakis(6-*O*-(*p*-tosyl))-α-cyclodexdtrins[2i] as well as polytosylated γ-cyclodextrins.[2j]

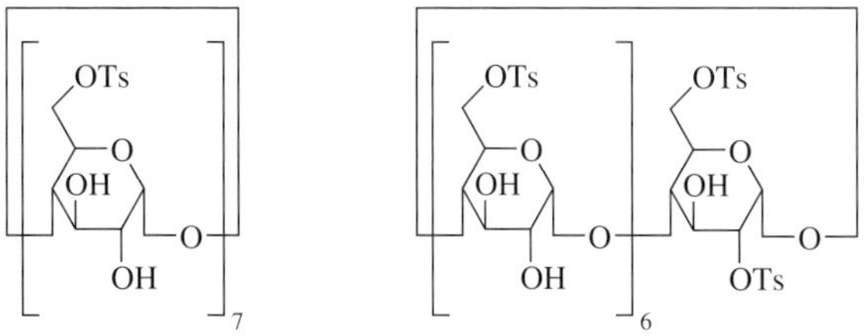

(1) 24%   **(2)** 15%

By proper choice of base it is possible to selectively *O*-tosylate in the presence of a free amine or *N*-tosylate in the presence of a free hydroxyl, with yields in excess of 94% (eq 3). The probable explanation for selective *O*-tosylation in the presence of the stronger base is formation of an adequate amount of phenoxide to allow the anion to act as the nucleophile.[2k]

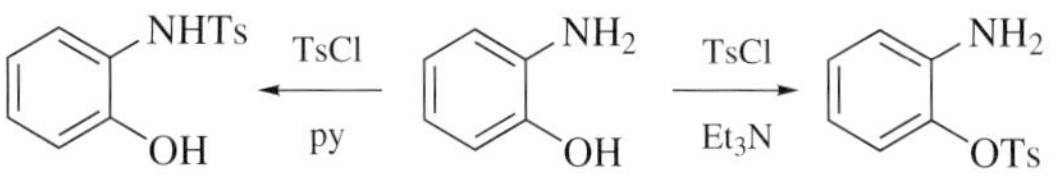

Tosyl chloride has been used in the preparation of allylic chlorides from their respective alcohols, leaving in place sensitive groups and with no rearrangement of the allylic substrate (eq 4).[21]

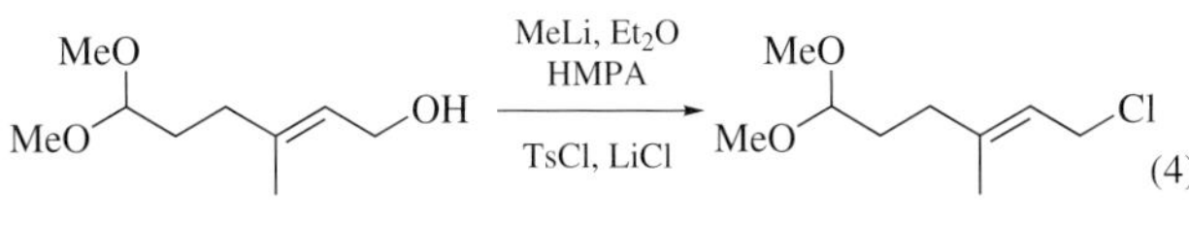

Studies on reactions of various types of alcohols with the related tosyl chloride/dimethylaminopyridine (TsCl/DMAP) system have led to the following conclusions: allylic, propargylic, and glycosidic hydroxyls quickly react to form the corresponding chlorides, 2,3-epoxy and selected primary alcohols yield chlorides

but at a slower rate, and reactions of aliphatic secondary alcohols stop at the tosylate stage. Example reactions are illustrated in eqs 5–10.[3a]

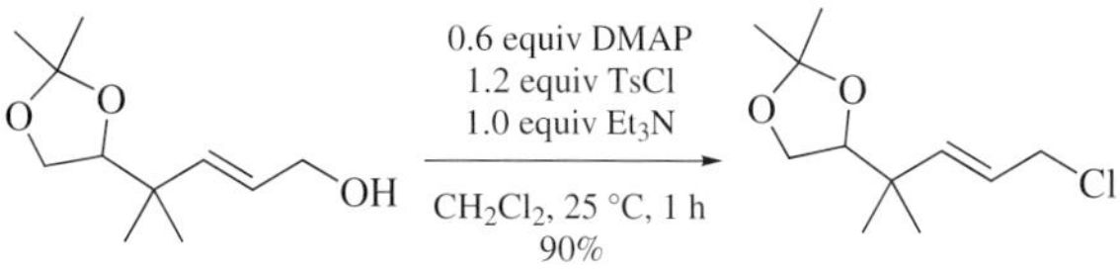

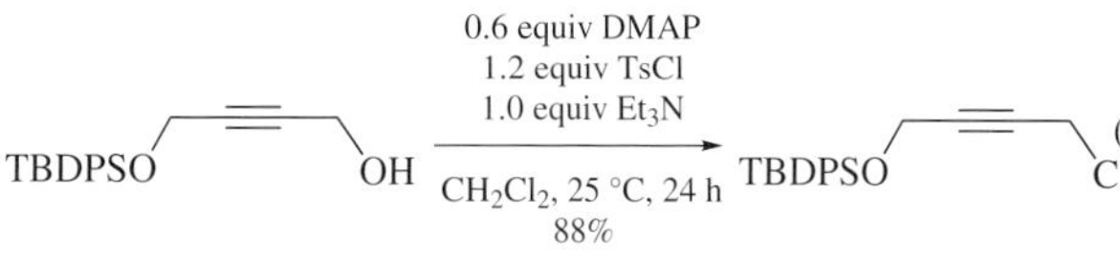

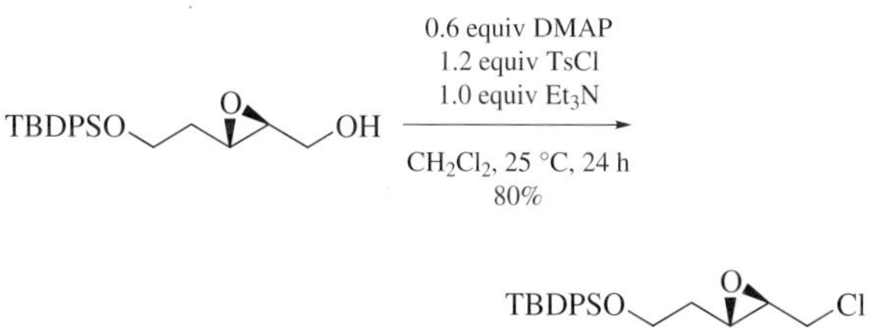

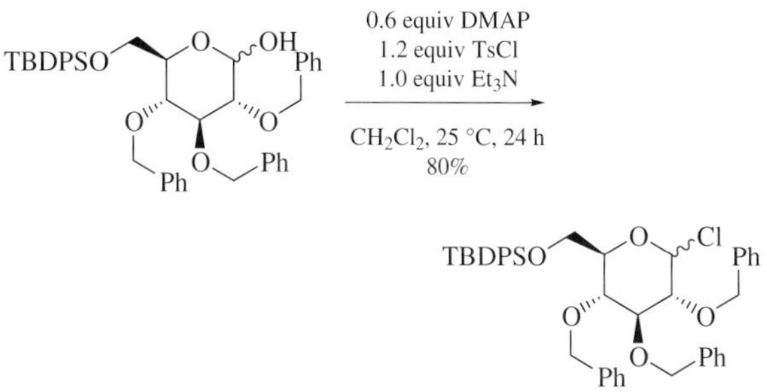

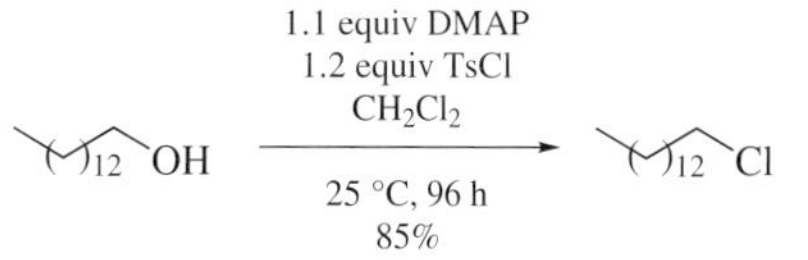

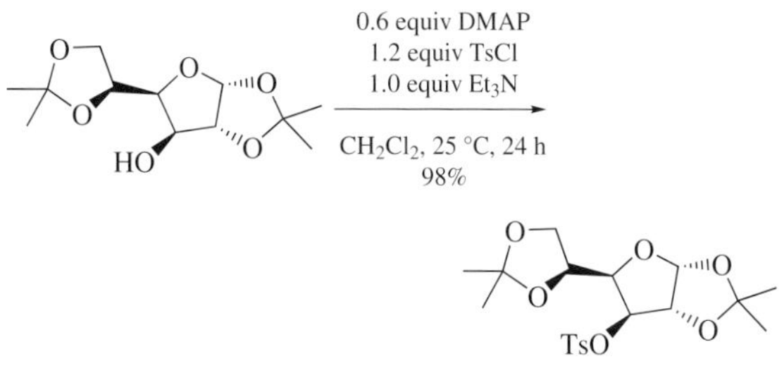

Attempted tosylation of 3β-methoxy-21-hydroxy-5α-pregnan-20-one afforded its α-chloro derivative.[3b] This reinforces the rule that as hybridization α to the alcohol increases in s character, displacement of the intermediate tosylate is facilitated (eq 11).

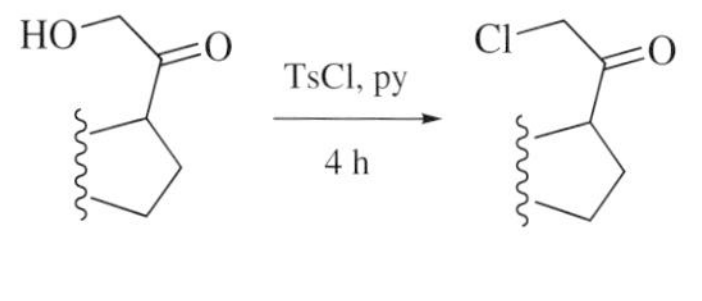

The α-chlorination of sulfoxides can also be achieved using tosyl chloride and pyridine, albeit in poor yield (32%), as shown in eq 12.[3c]

$$R \overset{O}{\underset{}{\overset{\parallel}{S}}} R^1 \xrightarrow[\text{py}]{\text{TsCl}} R \overset{O}{\underset{Cl}{\overset{\parallel}{S}}} R^1 \quad (12)$$

Alcohols treated with tosyl chloride are transformed into their corresponding sulfonate esters without manipulation of existing stereochemistry in the substrate. The *p*-toluenesulfonyloxy moiety subsequently serves as a good leaving group in intermolecular nucleophilic substitution or elimination reactions.[4] Treatment of multifunctional alcohols with tosyl chloride can result in the formation of intermediate *O*-sulfonylated species, which can undergo intramolecular displacement of the tosylate group. For example, treatment of *trans*-2-hydroxycyclohexaneacetamide with tosyl chloride in pyridine is the key step in the conversion of the lactone of *trans*-2-hydroxycyclohexaneacetic acid to its *cis* isomer (eq 13).[5a]

$$\text{(eq 13)}$$

Alcohols treated with tosyl chloride can also serve as alkylating agents for thioamides to make thiazolines (eq 14).[5b]

$$\text{(eq 14)}$$

Several useful one-pot procedures employing the reaction of tosyl chloride with vicinal diols have been used to form epoxides. Treatment of variously substituted diols with 1 equiv of tosyl chloride and sodium hydroxide in monoglyme provides access to a variety of 1-alkynyloxiranes (R[1] and/or R[3] being terminal or substituted alkynes) in moderate to good yield (eq 15).[6a]

$$R^1 \overset{R^2 \quad R^3}{\underset{HO \quad OH}{\bigwedge}} \xrightarrow[\text{monoglyme}]{\text{TsCl, NaOH}} R^1 \overset{R^2 \quad R^3}{\underset{O}{\triangle}} \quad (15)$$
$$55–85\%$$

Another method entails treatment of the diol in THF with 2.2 equiv of **Sodium Hydride** followed by reaction with a slight molar excess of tosyl chloride. This fast reaction, illustrated in eq 16, can be used to prepare enantiopure epoxides in good to excellent yields.[6b]

$$HO \overset{OH}{\underset{}{\bigvee}} R \xrightarrow[\text{THF, 0 °C}]{\substack{\text{2.2 equiv NaH} \\ \text{1.2 equiv TsCl}}} \overset{O}{\underset{R}{\triangle}} H \quad (16)$$
$$61–88\%$$

Phase-transfer catalysis has also been successfully employed to achieve epoxidation. A variety of cyclic *trans*-substituted diols in dichloromethane were treated with a 50% aqueous solution of sodium hydroxide in the presence of the phase-transfer catalyst

benzyltriethylammonium bromide. Consistently good yields were achieved for these as well as glycosidic and acyclic substrates (eq 17).[6c]

$$\text{(eq 17)}$$
$$n = 1, 2, 3$$

A one-pot conversion of a variety of 1,3-diols to oxetanes has also been reported.[7] The procedure entailed alcohol deprotonation with one equivalent of **Butyllithium** followed by treatment with tosyl chloride and then a second equivalent of base (eq 18).

$$\underset{R^2 \quad R^3 \quad R^4}{\overset{HO \quad OH}{R^1 \bigvee R^6}} \xrightarrow[\substack{3. \text{ BuLi, } 0\,°C \\ 68–84\%}]{\substack{1. \text{ BuLi, THF, } 0\,°C \\ 2. \text{ TsCl, } 60\,°C}} \underset{R^2 \quad R^3 \quad R^4}{\overset{R^1 \quad O \quad R^6}{\square R^5}} \quad (18)$$

When a solution of a carboxylic acid and an alcohol in pyridine is treated with tosyl chloride, an ester is formed rapidly in excellent yield. This procedure is useful especially in the esterification of tertiary alcohols. The combination of a carboxylic acid and tosyl chloride serves as a convenient method of in situ preparation of symmetrical acid anhydrides for further formation of esters and amides (eq 19). The novelty of this protocol is that the acid can be recycled through the anhydride stage in the presence of the alcohol, thereby resulting in complete conversion to the ester (eq 20).[8a] Reactivity is determined by the strength of the acid: strong acids facilitate the esterifications.[8b]

$$R^1 \overset{O}{\underset{}{\overset{\parallel}{\diagup}}} OH \xrightarrow[\text{py}]{\text{TsCl}} R^1 \overset{O \quad O}{\underset{}{\diagup O \diagdown}} R^1 \quad (19)$$

$$\text{(eq 20)}$$
$$78–96\%$$

Similarly, a two-step procedure employing treatment of a mixture of tosic acid and various amino acids with alcoholic tosyl chloride results in the isolation of the esters of the amino acids as their *p*-toluenesulfonate salts in excellent yield. The tosic acid used in the esterification is added to make the amino acids more soluble and to prevent *N*-tosylation (eq 21).[8c]

$$ROH \xrightarrow{\text{TsCl}} [ROTs] \xrightarrow{\text{TsOH·H}_2\text{NCHR'CO}_2\text{H}}$$
a. ethyl alcohol   amino acids: Gly, Leu, Tyr, Trp
b. benzyl alcohol

$$\text{TsOH·H}_2\text{NCHR'CO}_2\text{R} \quad (21)$$
a. 90–97%
b. 81–85%

A novel formal decarboxylation of the amino acid α-anilino-α,α-diphenylacetic acid has been observed upon treatment with tosyl chloride in pyridine.[9] It has been proposed that a mixed anhydride of *p*-toluenesulfonic acid has undergone an elimination via *N*-deprotonation and synchronous extrusion of carbon monoxide in this reaction. Interestingly, no *N*-tosylation occurs (eq 22).

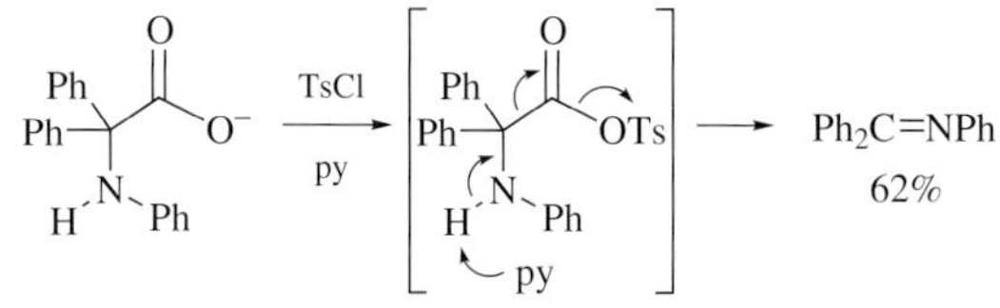

$$Ph_2C=NPh \quad (22)$$
$$62\%$$

Spontaneous Beckmann rearrangement has been observed upon *O*-tosylation of oximes. Lactams can therefore be conveniently prepared from cyclic oximes, as shown in eq 23.[10a] The rearrangement has long been known to proceed with retention of configuration of the migrating group.[10b,c] This is complementary to the reaction of oximes with *Sulfuric Acid* (eq 24).[10b]

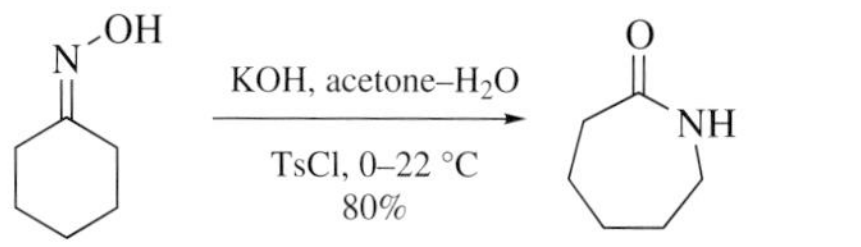

$$(23)$$

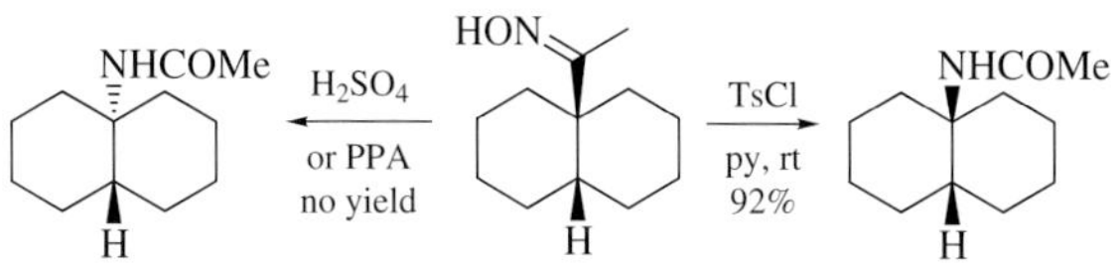

$$(24)$$

A Beckmann fragmentation of oximes using tosyl chloride in a basic ethanol–water system has also been observed.[11] Formation of the cyclic product shown in eq 25 was consistent with a base-induced opening of an intermediate lactone followed by rearrangement and incipient extrusion of benzoic acid.

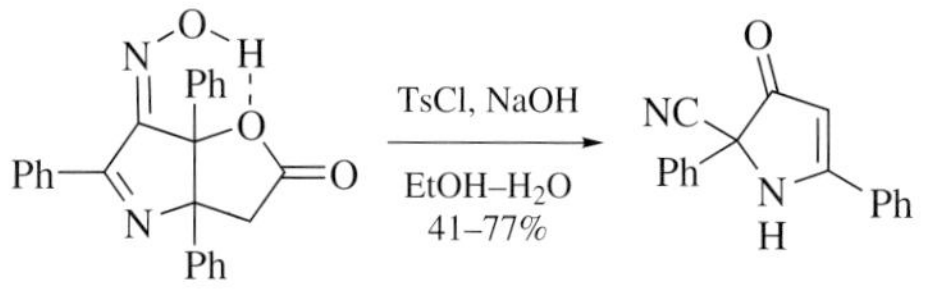

$$(25)$$

Oximes which are treated with tosyl chloride can also be used as substrates in the Neber rearrangement to *α*-amino ketones.[12a] The mixture shown in eq 26 was further subjected to reductive amination to yield 14% of the CNS-active *cis-N,N*-dimethylated *α*-amino alcohol.[12b]

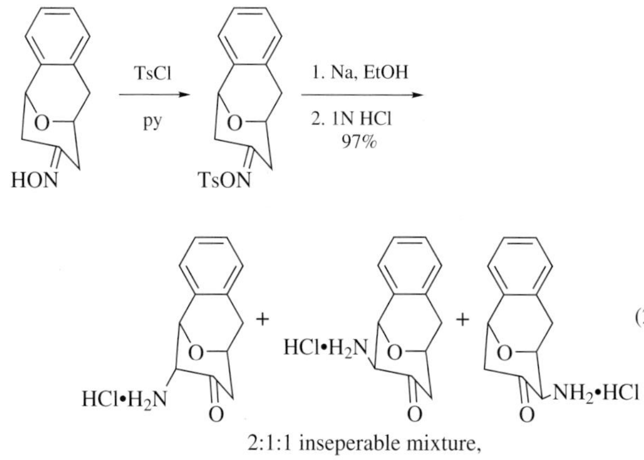

$$(26)$$

2:1:1 inseperable mixture,

Rearrangements of hydroxamic acids using sulfonyl chlorides have been accomplished, albeit without reported yields, the net result being a unique variation of the Lossen rearrangement (eq 27).[13]

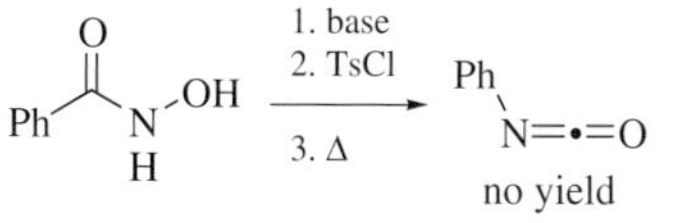

$$(27)$$

no yield

As an alternative to the Beckmann rearrangement, ketonic nitrones can be treated with tosyl chloride in pyridine in the presence of water, as illustrated in eq 28.[14]

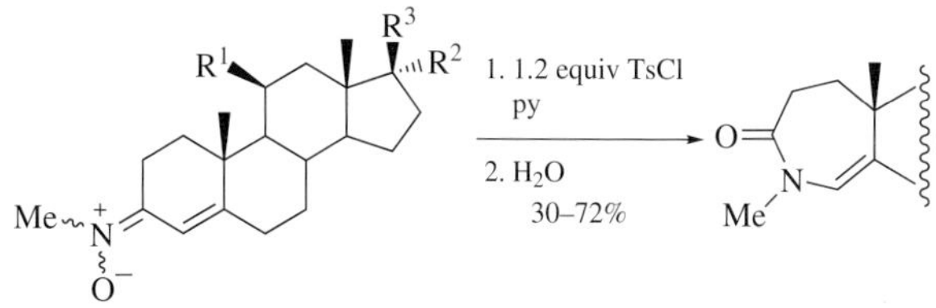

$$(28)$$

Rearrangement reactions which utilize tosyl chloride seem to progress in a two-step process: the displacement of chlorine from the sulfonyl halide resulting in the formation of an oxygen–sulfur bond, followed by the migration, elimination, or intramolecular displacement of the sulfonylate anion.[8a] When the series of carbinolamines depicted in eq 29 were treated with tosyl chloride, they decomposed into the corresponding electron-deficient nitrogen species, which subsequently triggered ring enlargement to the *β*-lactams in moderate yields.[15]

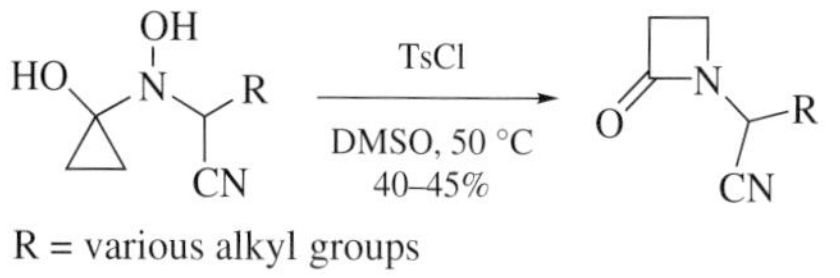

$$(29)$$

R = various alkyl groups

N-Tosylation is a facile procedure. For inexpensive amines a useful procedure entails treating 2 equiv of the amine with tosyl chloride. This is the first step in the convenient preparation of Diazald (*N-Methyl-N-nitroso-p-toluenesulfonamide*), a *Diazomethane* precursor (eq 30).[16a]

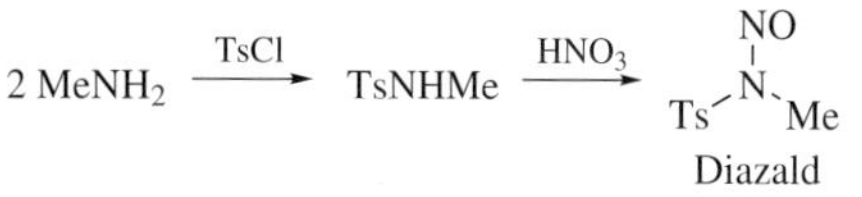

$$2 \text{ MeNH}_2 \xrightarrow{\text{TsCl}} \text{TsNHMe} \xrightarrow{\text{HNO}_3} \quad (30)$$

Diazald

Protection of 3-amino-3-pyrazoline sulfate, a key intermediate in the preparation of 3(5)-aminopyrazole, was achieved in the presence of an excess of sodium bicarbonate in good yields (eq 31).[16b]

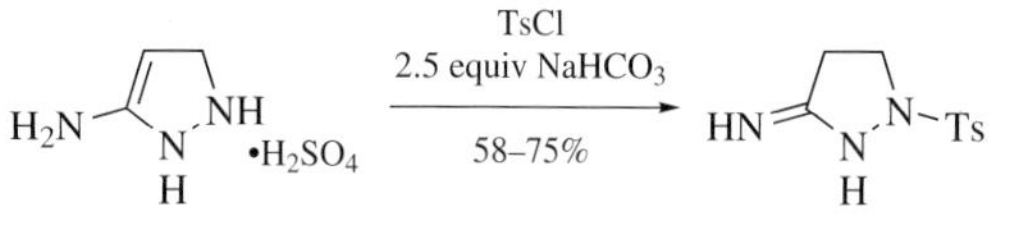

$$(31)$$

Sodium Carbonate can also be used as the base in the tosylation of amines, as shown in the reaction of anthranilic acid with tosyl chloride. There was no competing nucleophilic attack at sulfur by the resonance-stabilized carboxylate group (eq 32).[16c]

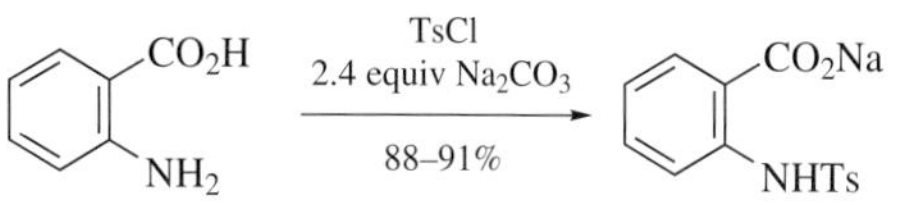

$$(32)$$

Primary, benzyl, and unhindered secondary amines can be ditosylated and deaminated in good to excellent yields, using *Sodium*

Borohydride as the nucleophile, to afford the corresponding alkanes; only highly congested substrates experience competing attack at nitrogen.[17] Similarly, primary aliphatic amines, when ditosylated and treated with iodide ion in DMF at 90–120 °C yielded, as their major products, the corresponding alkyl iodides, with some competition arising from elimination reactions (eqs 33 and 34).[18]

$$H_2NCH_2R \xrightarrow[\text{NaOH}]{\underset{\text{DMF–H}_2\text{O}}{\text{TsCl}} \quad \underset{\text{DMF, NaH}}{\text{TsCl}}} Ts_2NCH_2R \quad (33)$$

$$ICH_2R \xleftarrow[110–120\,°\text{C}]{2\text{ equiv KI, DMF}} Ts_2NCH_2R \xrightarrow[175\,°\text{C}]{2\text{ equiv NaBH}_4} MeR \quad (34)$$

Tosyl chloride has been used in the protection of the imidazole residue of N^α-acylhistidine peptides[19a,b] and the guanidino residue of arginine peptides.[19c] The resulting nitrogen-protecting group is stable to most conditions but is easily removable in anhydrous hydrogen fluoride at 0 °C.

The tosylation of carbon can be accomplished using electron transfer conditions. Treatment of styrene[20a] and analogs[20b] with **Copper(II) Chloride** and tosyl chloride or **Benzenesulfonyl Chloride** results in a formal replacement of the vinyl proton by the sulfonyl moiety (eq 35). The intermediacy of a *trans-β-*chloro sulfone has been demonstrated by ^{1}H NMR. Treatment with base induced the elimination of HCl. A variety of other sulfonyl transfer reagents can be employed in the synthesis of isolated β-chloro sulfones, with good results (60–97% yield) for a variety of alkenes (ethylene, 1-butene, 2-butene, 1-octene, acrylonitrile, methyl acrylate, and 1,3-butadiene).[20a]

$$\text{(35)}$$

Although it was found that tosyl chloride does not react with 3-sulfolene at temperatures below 110 °C, upon further warming to 140 °C addition to the diene afforded 1-chloro-4-tosyl-2-butene in a convenient procedure that did not require the use of a sealed tube or bomb (eq 36).[20c] Cycloheptatriene, 1,4-norbornadiene, and phenylacetylene are a few examples of a variety of compounds which are reactive substrates in this protocol.

$$\text{(36)}$$

Radical addition of tosyl chloride to norbornene and aldrin without skeletal rearrangement can be achieved using **Dibenzoyl Peroxide** as an initiator (eq 37).[20a,d] Reaction with 1,4-norbornadiene gives rise to a rearranged addition product.

$$\text{(37)}$$

Treatment of bis(trimethylsilyl)acetylene with tosyl chloride and a slight excess of anydrous **Aluminum Chloride** affords

p-tolyl (trimethylsilyl)ethynyl sulfone, a precursor to the reactive Michael acceptor ethynyl *p*-tolyl sulfone (eq 38).[21]

$$TsCl + TMS\text{——}\equiv\text{——}TMS \xrightarrow[\substack{\text{CH}_2\text{Cl}_2 \\ 80\%}]{\text{AlCl}_3} Ts\text{——}\equiv\text{——}TMS \quad (38)$$

The dehydration of ureas employing tosyl chloride provides access to substituted carbodiimides, as depicted in eq 39. The urea made from treatment of ethyl isocyanate with *N,N*-dimethyl-1,3-propanediamine was treated in situ with 1.1 equiv of tosyl chloride and a large excess of **Triethylamine** to afford the corresponding carbodiimide in high yield.[22a,b]

$$\text{(39)}$$

Various ureas upon treatment with tosyl or benzenesulfonyl chloride in the presence of a phase transfer catalyst, benzyltriethylammonium chloride, results in moderate to excellent yields of carbodiimides (eq 40).[22c] The polymeric carbodiimide in eq 41 offers the advantages of cleaner workup and recyclability if used to prepare aldehydes under Moffatt oxidation conditions.[22d]

$$\text{(40)}$$

$$\text{(41)}$$

(P) = styrene–divinylbenzene copolymer

Isocyanides can be made from treatment of formamides with tosyl chloride in pyridine, as illustrated in eq 42.[23a,b] Similarly, nitriles can be synthesized in moderate to good yields by dehydration of primary amides using tosyl chloride in pyridine[23a,b] or quinoline (eq 43).[23c]

$$\underset{R}{\overset{\overset{\text{O}}{\|}}{\diagup}}_{NH_2} \xrightarrow[30–82\%]{\text{TsCl, py}} R\text{—CN} \quad (42)$$

$$\underset{\underset{H}{|}}{R}\underset{}{N}\overset{\overset{\text{O}}{\|}}{\diagdown}\text{II} \xrightarrow[\substack{2\text{ equiv py or} \\ 4\text{ equiv quinoline} \\ 50–74\%}]{1.5–2\text{ equiv TsCl}} R\text{—NC} \quad (43)$$

First Update

Julia Haas

Array BioPharma, Boulder, CO, USA

Tosylation of Alcohols. Various new reaction conditions for the tosylation of alcohols have recently been reported. Many of them are superior to the traditionally used pyridine in chloroform in that they are less toxic, environmentally friendly, and more economical. In addition, they may need shorter reaction times, have easier workup procedures, and circumvent undesired chloride formation.

Et_3N with catalytic $Me_3N \cdot HCl$ has been used to tosylate a variety of alcohols in good yields (87–98%).[27] The reaction times are generally short: most reactions are complete within 1 h, and even unreactive alcohols such as 3,3-dimethylbutan-2-ol have been shown to react within 5 h. Various solvents can be used in this reaction including halogenated solvents, toluene, and acetonitrile. It is interesting to note that in the absence of $Me_3N \cdot HCl$, less than 10% yield of the desired tosylates are obtained. The formation of chloride side products can be almost completely circumvented under these conditions; even 2-alkenyl- and 2-alkynyl alcohols can be sulfonylated on large scale in good yields. For example, 2-methallyl alcohol was sulfonylated in 94% yield (eq 44).

$$94\% \qquad\qquad \text{trace}$$
$$\text{trace} \qquad\qquad >90\%$$

Diamines of the general structure $Me_2N(CH_2)_nNMe_2$ promote efficient tosylations (87–95% yield) in short reaction times.[28] The diamine can be used as the stoichiometric base or in a catalytic fashion together with Et_3N. Toluene or acetonitrile are used as solvents.

Inorganic bases such as K_2CO_3 or metal alkoxides in the presence of tertiary amines[27,28] or phase-transfer catalysts[29] are an efficient and environmentally friendly option for the tosylation of alcohols. The reaction conditions are generally mild and avoid side reactions. Even 2-alkenyl- and 2-alkynyl alcohols can be sulfonylated on large scale in good yields (59–91%) using $Me_3N \cdot HCl$ and K_2CO_3 in methyl isobutyl ketone (MIBK).[30] No substantial formation of the corresponding chlorides is observed under these conditions. Avoiding the formation of 2-propynyl halides is very important on large scale because these substances can possess explosive characteristics.

K_2CO_3 in combination with microwave irradiation can be used for the synthesis of aryl tosylates under solvent-free conditions.[31] Excellent yields (92–99%) of the desired aryl tosylates are obtained in 3–5 min. The products can be isolated by stirring the reaction mixture with water followed by filtration.

DABCO (1,4-diazabicyclo[2.2.2]octane) is an efficient base for the sulfonylation of alcohols in good yields.[32] DABCO is easy to handle and less toxic than pyridine, and more environmentally friendly solvents such as MTBE can be used. The procedure is amendable to large-scale syntheses, and the product can be isolated by filtration.

Grinding alcohols with tosyl chloride and DABCO in a mortar also affords the corresponding tosylates in 80–100% yield.[33] Under these conditions, amino alcohols can be selectively *O*-tosylated. For example, 2-amino-1-propanol was ground with tosyl chloride and DABCO for 10 min to afford the corresponding sulfonate ester in 90% yield (see also eq 47).

Primary and secondary alcohols can be sulfonylated in the presence of Ag(I) oxide and KI in CH_2Cl_2.[34] The sulfonylations occur under mild, neutral reaction conditions, and good yields (66–99%) have been obtained for a variety of substrates. The products can be isolated by filtration, and in situ replacement of the formed tosylate with sodium azide is also possible.

Selective Tosylation of Diols and Polyols. Dialkylstannylene acetals are often used for the monofunctionalization of 2-substituted 1,2-diols. To form the corresponding tin acetal, the diol is typically treated with R_2SnX (where X = O or $(OMe)_2$), with azeotropic or desiccant removal of water or MeOH. When di-*n*-butyl acetals are used, the primary alcohol functionality can be tosylated with high regioselectivity by treatment with tosyl chloride.[35] An interesting reversal of this regioselectivity has been observed for hexamethylenestannylene acetals, which react preferentially at the secondary carbon atom for most substrates.[36] The regislectivities are very substrate dependent and range from 4:96 in favor of the secondary isomer for a glucose-derived substrate to 92:8 in favor of the undesired isomer for an allose derivative. Other dialkylstannylene acetals gave varying regioselectivities. A dibutylstannylene acetal has also been used for the selective tosylation of the *β*-alcohol functionality in *syn*-2,3-dihydroxy esters.[37]

In a catalytic version, the regioselective sulfonylation of *α*-chelatable alcohols can be accomplished with tosyl chloride and 1 equiv of an amine in the presence of 2 mol % of Bu_2SnO.[38] The reactions are generally high yielding, regioselective, and exclusive monotosylation is observed in many cases. For unsymmetrical diols, the less hindered alcohol is tosylated. The reaction rate is dependent on the solvent and the pKa of the amine.[39] This reaction was also used for the selective tosylation of various carbohydrates and for the selective monosulfonylation of internal 1,2-diols.[40]

Selective monotosylations of symmetrical diols have been accomplished in good yields (75–93%) with a stoichiometric amount of tosyl chloride in the presence of Ag(I) oxide and a catalytic amount of potassium iodide.[41] When excess Ag(I) oxide is used, polysubstituted cyclic ethers are formed in 61–88% yield. This method was also used for the synthesis of substituted crown ethers.

Zinc bromide has been reported to promote low temperature tosylations in pyridine. At temperatures below −25 °C, this reaction is selective for primary hydroxyl groups and has been used to tosylate the six primary hydroxyl groups in *β*-cyclodextrin simultaneously.[42]

The regioselective 2′-*O*-tosylation of adenosine with tosyl chloride has been accomplished using 10–20 mol % of dibutyltin dichloride as a catalyst (eq 45).[43] Dibutylchlorotin hydroxide, dibutyltin oxide, bis(dibutylchlorotin) oxide, and tributyltin chloride are also effective catalysts for this reaction and the method has also been used for the ditosylation of several methyl glycosides.

(45)

Sodium hydride in DMF followed by tosyl chloride has been used for the regioselective tosylation of *myo*-inositol orthoesters.[44] Very selective mono- and ditosylation could be achieved depending on the reaction conditions. The tosyl groups were used as protecting groups and later removed using Mg(0) in methanol.

Formation of Esters, Amides, and Anhydrides. Tosyl chloride and *N*-methylimidazole in acetonitrile can be used to promote esterifications, thioesterifications, and amide bond formations in excellent yields (eq 46).[45] Various functionalities are tolerated in this reaction and enolizable substrates, including *N*-Cbz amino acids, can be esterified without racemization.

1. TsCl (1.2 equiv), *N*-methylimidazole (3.0 equiv) CH₃CN, 0–5 °C, 30 min
2. R₂XH (1.0 equiv), CH₃CN, 0–5 °C, 2 h 82–95% (X = O, S, N)

(46)

Similarly, amides of CBz- and Boc-protected amino acids can be synthesized in 81–93% yield with minimal racemization of the α-stereocenter using tosyl chloride and pyridine in THF followed by treatment of the intermediate anhydrides with ammonia solution.[46]

A facile one-pot procedure for the synthesis of symmetric carboxylic acid anhydrides utilizes tosyl chloride (0.5–0.6 equiv) in K_2CO_3 media under solvent-free conditions. The desired anhydrides are formed by grinding the reaction mixture for 15–45 min in a mortar. The products were isolated by stirring the reaction mixture with CH_2Cl_2 or $CHCl_3$ followed by filtration and concentration.[47]

Formation of Sulfonamides. The tosyl group is a popular protecting group for alcohols, amines, and amides. It can be cleaved by various methods, including reductive cleavage with Mg, Zn, or Li/NH₃.[48] Amines are generally tosyl-protected by stirring the amine with tosyl chloride in an organic solvent in the presence of an acid scavenger.[48] In a solvent-free process, grinding amines with tosyl chloride in a mortar affords the corresponding tosyl amides in 90–100% yield.[33] Most aliphatic amines react without the addition of a catalyst. For aromatic amines and less reactive aliphatic amines such as amino acids, DABCO can be added to promote the reaction. The solvent-free synthesis of aryl sulfonamides has also been reported using microwave irradiation.[49]

The selective *N*-tosylation of amino alcohols can be accomplished using tosyl chloride in the presence of a metal oxide (MgO, CuO, Ag₂O) in aq THF solution (eq 47).[50] This reaction

complements the solid-state reaction mentioned above that leads to selective *O*-tosylation.[33]

(47)

The methyl esters of L-tyrosine and D-(4-hydroxyphenyl) glycine have been *N*-tosylated without protection of the phenolic hydroxyl group and without racemization of the stereogenic carbon centers using a THF/DMF mixture as the solvent (eq 48).[51] The DMF (2.6 mol/mol amino ester) is believed to solvate the oxygen atom of the formed *N*,*O*-dianion specifically, reducing its nucleophilicity and dramatically increasing the chemoselectivity of the *N*-substitution. In the absence of DMF, mixtures of the *N*-, *O*-, and disulfonyl esters were produced.

(48)

The selective *N*-tosylation of pyrrole has been reported in ionic liquids.[52] Treatment of pyrrole with KOH and tosyl chloride in 1-butyl-3-methylimidazolium hexafluorophosphate ([bmim] [PF₆]) afforded the corresponding *N*-tosylate in 98% yield. Azides can be directly converted to the corresponding sulfonamides in good yields using Fe and NH₄Cl in MeOH (eq 49). Ultrasonication was found to enhance the reaction rate greatly and improve the yields in this reaction.[53]

(49)

α-Chlorination of Ketones. Treatment of ketones with LDA followed by tosyl chloride affords the corresponding α-chloro ketones in good yields.[54] Various ketones have been chlorinated by this method as illustrated in eq 50. The regioselectivity of this reaction appears to depend on the selectivity of the enolate formation. Resin-bound tosyl chloride can also be used to effect this transformation.

(50)

Dehydration of Oximes. Tosyl chloride was used for the conversion of aromatic aldoximes to nitriles under microwave irradiation as shown in eq 51.[55] The reagents were adsorbed on alumina and irradiated to provide the desired nitriles in moderate yields. For certain ketoximes, up to 40% yields of Beckmann rearranged products could be obtained.

$$\text{(51)}$$

Synthesis of Aziridines. Reaction of *N*-aryl-β-amino alcohols with tosyl chloride under phase-transfer condition affords the corresponding *N*-aryl aziridines in good yields (eq 52).[56] Under the same conditions, L-ephedrine, an *N*-alkyl-β-amino alcohol, gave the corresponding *N*-tosyl derivative as the major product (78%), along with the expected *N*-alkyl aziridine in low yield (20%).

$$\text{(52)}$$

The direct transformation of 2-amino alcohols to *N*-tosyl aziridines can be achieved in moderate to good yields using potassium hydroxide and tosyl chloride in water/dichloromethane.[57] This procedure works well for small amino alcohols. Higher substituted amino alcohols give better yields using a two-step procedure: formation of the *N,O*-ditosylate followed by cyclization with potassium carbonate in acetonitrile.

Synthesis of β-Amino Acids. β-Amino acids can be prepared in an Arndt-Eistert-type reaction employing tosyl chloride for the activation of the carboxyl group of *N*-protected amino acids.[58] Formation of the corresponding diazomethane derivatives proceeds in good yield and without racemization of the α-stereogenic center (eq 53). Wolff rearrangement then affords the corresponding β-amino acids.

$$\text{(53)}$$

Synthesis of *N*-Sulfonyl Imines. Nonenolizable aldehydes and ketones can be converted to the corresponding *N*-tosyl imines in a two-step procedure.[59] Addition of LiHMDS to the carbonyl compound leads to the corresponding *N*-(trimethylsilyl) imines, which can be treated with tosyl chloride to afford the corresponding *N*-tosyl imines in excellent yields. The reaction produces the thermodynamically favored (*E*)-*N*-sulfonyl imines exclusively (eq 54).

$$\text{(54)}$$

Synthesis of Heterocycles. The convenient and efficient synthesis of a variety of 2-amino-substituted 1-aza-3-(aza, oxa, or thia) heterocycles has been described using tosyl chloride/NaOH (eq 55).[60] Different substitution patterns are tolerated and five- or six- membered rings can be formed in moderate to good yields. The application of polymer-supported tosyl chloride is possible in this reaction: it facilitates the work-up and renders the reaction conditions very suitable for parallel or automated synthesis.

$$\text{(55)}$$

Formation of 2-Chloromethylpyridines. A combination of tosyl chloride and NEt$_3$ can be used to convert substituted 2-picoline-*N*-oxides to the corresponding 2-chloromethylpyridines.[61] This reaction has been exploited for the synthesis of a variety of 2-[2-(pyridylmethyl)-thio]-1*H*-benzimidazoles, key intermediates in the manufacturing of H$^+$/K$^+$-ATPase inhibitors, in a one-pot procedure (eq 56).

$$\text{(56)}$$

1. *Fieser & Fieser* **1967**, *1*, 1179; *Fieser & Fieser* **1972**, *3*, 292; *Fieser & Fieser* **1974**, *4*, 510; *Fieser & Fieser* **1975**, *5*, 676; *Fieser & Fieser* **1977**, *6*, 598; *Fieser & Fieser* **1980**, *8*, 489; *Fieser & Fieser* **1981**, *9*, 472; *Fieser & Fieser* **1984**, *11*, 536; *Fieser & Fieser* **1988**, *13*, 313.

2. (a) Kabalka, G. W.; Varma, M.; Varma, R. S., *J. Org. Chem.* **1986**, *51*, 2386. (b) Marvel, C. S.; Sekera, V. C., *Org. Synth., Coll. Vol.* **1955**, *3*, 366. (c) Szeja, W., *Synthesis* **1979**, 822. (d) Johnson, W. S.; Collins, J. C., Jr.; Pappo, R.; Rubin, M. B.; Kropp, P. J.; Johns, W. F.; Pike, J. E.; Bartmann, W., *J. Am. Chem. Soc.* **1963**, *85*, 1409. (e) Tsuda, Y.; Nishimura, M.; Kobayashi, T.; Sato, Y.; Kanemitsu, K., *Chem. Pharm. Bull.* **1991**, *39*, 2883. (f) Lundt, I.; Madsen, R., *Synthesis* **1992**, 1129. (g) Yamamura, H.; Fujita, K., *Chem. Pharm. Bull.* **1991**, *39*, 2505. (h) Ashton, P. R.; Ellwood, P.; Staton, I.; Stoddart, J. F., *J. Org. Chem.* **1991**, *56*, 7274. (i) Fujita, K. E.; Ohta, K.; Masunari, K.; Obe, K.; Yamamura, H., *Tetrahedron Lett.* **1992**, *33*, 5519. (j) Yamamura, H.; Kawase, Y.; Kawai, M.; Butsugan, Y., *Bull. Chem. Soc. Jpn.* **1993**, *66*, 585. (k) Kurita, K., *Chem. Ind. (London)* **1974**, 345. (l) Stork, G.; Grieco, P. A.; Gregson, M., *Org. Synth., Coll. Vol.* **1988**, *6*, 638.

3. (a) Hwang, C. K.; Li, W. S.; Nicolaou, K. C., *Tetrahedron Lett.* **1984**, *25*, 2295. (b) Revelli, G. A.; Gros, E. G., *Synth. Commun.* **1993**, *23*, 1111. (c) Hojo, M.; Yoshita, Z.-i., *J. Am. Chem. Soc.* **1968**, *90*, 4496.

4. (a) For a general discussion of S$_N$2 displacement reactions see: Carey, F. A.; Sundberg, R. J., *Advanced Organic Chemistry Part A: Structure and Mechanisms*, 3rd ed.; Plenum: New York, 1990; p 261. (b) For a general discussion of leaving group effects see: *ibid.*, p 290.

5. (a) Brewster, J. H.; Kucera, C. H., *J. Am. Chem. Soc.* **1955**, *77*, 4564. (b) Eberle, M. K.; Nuninger, F., *J. Org. Chem.* **1993**, *58*, 673.

6. (a) Holand, S.; Epsztein, R., *Synthesis* **1977**, 706. (b) Murthy, V. S.; Gaitonde, A. S.; Rao, S. P., *Synth. Commun.* **1993**, 285. (c) Sjeja, W., *Synthesis* **1985**, 983.

7. Picard, P.; Leclercq, D.; Bats, J.-P.; Moulines, J., *Synthesis* **1981**, 550.

8. (a) Brewster, J. H.; Ciotti, C. J., Jr., *J. Am. Chem. Soc.* **1955**, *77*, 6214. (b) Hennion, G. F.; Barrett, S. O., *J. Am. Chem. Soc.* **1957**, *79*, 2146. (c) Arai, I.; Muramatsu, I., *J. Org. Chem.* **1983**, *48*, 121.

9. Sheehan, J. C.; Frankenfeld, J. W., *J. Org. Chem.* **1962**, *27*, 628.

10. (a) Oxley, P.; Short, W. F., *J. Chem. Soc.* **1948**, 1514. (b) Hill, R. K.; Chortyk, O. T., *J. Am. Chem. Soc.* **1962**, *84*, 1064. (c) Wheland, G. W., *Advanced Organic Chemistry*, 3rd ed., Wiley: New York, 1960; p 597.

11. Mataka, S.; Suzuki, H.; Sawada, T.; Tashiro, M., *Bull. Chem. Soc. Jpn.* **1993**, *66*, 1301.

12. (a) O'Brien, C., *Chem. Rev.* **1964**, *64*, 81. (b) Wünsch, B.; Zott, M.; Höfner, G., *Liebigs Ann. Chem.* **1992**, 1225.

13. Hurd, C. D.; Bauer, L., *J. Am. Chem. Soc.* **1954**, *76*, 2791.

14. (a) Barton, D. H. R.; Day, M. J.; Hesse, R. H.; Pechet, M. M., *J. Chem. Soc., Perkin Trans. 1* **1975**, 1764. (b) Barton, D. H. R.; Day, M. J.; Hesse, R. H.; Pechet, M. M., *J. Chem. Soc., Chem. Commun.* **1971**, 945.

15. Wasserman, H. H.; Glazer, E. A.; Hearn, M. J., *Tetrahedron Lett.* **1973**, 4855.

16. (a) De Boer, T. J.; Backer, H. J., *Org. Synth., Coll. Vol.* **1963**, *4*, 943. (b) Dorn, H.; Zubek, A., *Org. Synth., Coll. Vol.* **1973**, *5*, 39. (c) Scheifele, H. J., Jr.; De Tar, D. F., *Org. Synth., Coll. Vol.* **1963**, *4*, 35.

17. Hutchins, R. O.; Cistone, F.; Goldsmith, B; Heuman, P., *J. Org. Chem.* **1975**, *40*, 2018.

18. DeChristopher, P. J.; Adamek, J. P.; Lyon, G. D.; Galante, J. J.; Haffner, H. E.; Boggio, R. J.; Baumgarten, R. J., *J. Am. Chem. Soc.* **1969**, *91*, 2384.

19. (a) Sakakibari, S.; Fujii, T., *Bull. Chem. Soc. Jpn.* **1969**, *42*, 1466. (b) Fujii, T.; Sakakibari, S., *Bull. Chem. Soc. Jpn.* **1974**, *47*, 3146. (c) Mazur, R. H.; Plume, G., *Experientia* **1968**, *24*, 661.

20. (a) Asscher, M.; Vofsi, D., *J. Chem. Soc.* **1964**, 4962. (b) Truce, W. E.; Goralski, C. T., *J. Org. Chem.* **1970**, *35*, 4220. (c) Truce, W. E.; Goralski, C. T.; Christensen, L. W.; Bavry, R. H., *J. Org. Chem.* **1970**, *35*, 4217. (d) Cristol, S. J.; Reeder, J. A., *J. Org. Chem.* **1961**, *26*, 2182.

21. Waykole, L.; Paquette, L. A., *Org. Synth.* **1989**, *67*, 149.

22. (a) Sheehan, J. C.; Cruickshank, P. A., *Org. Synth., Coll. Vol.* **1973**, *5*, 555. (b) Sheehan, J. C.; Cruickshank, P. A.; Boshart, G. L., *J. Org. Chem.* **1961**, *26*, 2525. (c) Jászay, Z. M.; Petneházy, I.; Töke, L.; Szajáni, B., *Synthesis* **1987**, 520. (d) Weinshenker, N. M.; Shen, C. M.; Wong, J. Y., *Org. Synth., Coll. Vol.* **1988**, *6*, 951.

23. (a) Stephens, C. R.; Bianco, E. J.; Pilgrim, F. J., *J. Am. Chem. Soc.* **1955**, *77*, 1701. (b) Stephens, C. R.; Conover, L. H.; Pasternack, R.; Hochstein, F. A.; Moreland, W. T.; Regna, P. P.; Pilgrim, F. J.; Pilgrim, F. J.; Brunings, K. J.; Woodward, R. B., *J. Am. Chem. Soc.* **1954**, *76*, 3568.

24. (a) Hertler, W. R.; Corey, E. J., *J. Org. Chem.* **1958**, *23*, 1221. (b) Corey, E. J.; Hertler, W. R., *J. Am. Chem. Soc.* **1959**, *81*, 5209. (c) Schuster, R. E.; Scott, J. E.; Casanova, J., Jr., *Org. Synth., Coll. Vol.* **1973**, *5*, 772.

25. Pelletier, S. W., *Chem. Ind. (London)* **1953**, 1034.

26. Perrin, D. D.; Armarego, W. L. F., *Purification of Laboratory Chemicals*, 3rd ed.; Pergamon: Oxford, 1988; p 291.

27. Yoshida, Y.; Sakakura, Y.; Aso, N.; Okada, S.; Tanabe, Y., *Tetrahedron* **1999**, *55*, 2183.

28. Yoshida, Y.; Shimonishi, K.; Sakakura, Y.; Okada, S.; Aso, N.; Tanabe, Y., *Synthesis* **1999**, 1633.

29. Jaszay, Z. M.; Petnehazy, I.; Toke, L., *Heteroatom Chem.* **2004**, *15*, 447.

30. Tanabe, Y.; Yamamoto, H.; Yoshida, Y.; Miyawaki, T.; Utsumi, N., *Bull. Chem. Soc. Jpn.* **1995**, *68*, 297.

31. Xu, L.; Xia, C., *Synth. Commun.* **2004**, *34*, 1199.

32. Hartung, J.; Huenig, S.; Kneuer, R.; Schwarz, M.; Wenner, H., *Synthesis* **1997**, 1433.

33. Hajipour, A. R.; Mallakpour, S. E.; Najafi, A. R.; Mazlumi, G., *Sulfur Letters* **2000**, *24*, 137.

34. Bouzide, A.; LeBerre, N.; Sauve, G., *Tetrahedron Lett.* **2001**, *42*, 8781.

35. David, S.; Hannessian, S., *Tetrahedron* **1985**, *41*, 643.

36. Kong, X.; Grindley, B., *Can. J. Chem.* **1994**, *72*, 2396.

37. Park, J. N.; Ko, S. Y., *Bull. Kor. Chem. Soc.* **2002**, *23*, 507.

38. Martinelli, M. J.; Nayyar, N. K.; Moher, E. D.; Dhokte, U. P.; Pawlak, J. M.; Vaidyanathan, R., *Org. Lett.* **1999**, *1*, 447.

39. Martinelli, M. J.; Vaidyanathan, R.; Pawlak, J. M.; Nayyar, N. K.; Dhokte, U. P.; Doecke, C. W.; Zollars, L. M. H.; Moher, E. D.; Van Khau, V.; Kosmrlj, B., *J. Am. Chem. Soc.* **2002**, *124*, 3578.

40. Martinelli, M. J.; Vaidyanathan, R.; Van Khau, V., *Tetrahedron Lett.* **2000**, *41*, 3773.

41. Bouzide, A.; Sauve, G., *Org. Lett.* **2002**, *4*, 2329.

42. Yamamura, H.; Kawasaki, J.; Saito, H.; Araki, S.; Kawai, M., *Chem. Lett.* **2001**, 706.

43. Kawana, M.; Tsujimoto, M.; Takahashi, S., *J. Carbohydrate Chem.* **2000**, *19*, 67.

44. Sureshan, K. M.; Shashidhar, M. S., *Tetrahedron Lett.* **2001**, *42*, 3037.

45. Wakasugi, K.; Iida, A.; Misaki, T.; Nishii, Y.; Tanabe, Y., *Advanced Synthesis & Catalysis* **2003**, *345*, 1209.

46. Ananda, K.; Vasanthakumar, G.; Babu, V. V. S., *Protein and Peptide Letters* **2001**, *8*, 45.

47. Kazemi, F.; Sharghi, H.; Nasseri, M. A., *Synthesis* **2004**, 205.

48. See *Protective Groups in Organic Synthesis*, Greene, T. W.; Wuts, P. G. M. (Eds). Wiley Sons, In: New York, 1999. Protection of alcohols p 199, phenols p 285. amines p 616. and amides p 644.

49. Sharma, A. K.; Das, S. K., *Synth. Commun.* **2004**, *34*, 3807.

50. Kang, H. H.; Rho, H. S.; Kim, D. H.; Oh, S. G., *Tetrahedron Lett.* **2003**, *44*, 7225.

51. Penso, M.; Albanese, D.; Landini, D.; Lupi, V.; Tricarico, G., *Euro. J. Org. Chem.* **2003**, 4513.

52. Le, Z.-G.; Chen, Z.-C.; Hu, Y.; Zheng, Q., *Synthesis* **2004**, 1951.

53. Chandrasekhar, S.; Narsihmulu, C., *Tetrahedron Lett.* **2000**, *11*, 7969.

54. Brummond, K. M.; Gesenberg, K. D., *Tetrahedron Lett.* **1999**, *40*, 2231.

55. Ghiaci, M.; Bakhtiari, K., *Synth. Commun.* **2001**, *31*, 1803.

56. Sriraghavan, K.; Ramakrishnan, V. T., *Synth. Commun.* **2001**, *31*, 1105.

57. Bieber, L. W.; de Araujo, M. C. F., *Molecules* **2002**, *7*, 902.

58. Vasanthakumar G.-R.; Babu V. V. S., *Synth. Commun.* **2002**, *32*, 651.

59. Georg, G..; Harriman, G. C. B.; Peterson, S. A., *J. Org. Chem.* **1995**, *60*, 7366.

60. Heinelt, U.; Schultheis, D.; Jaeger, S.; Lindenmaier, M.; Pollex, A.; Beckmann, H. S. G., *Tetrahedron* **2004**, *60*, 9883.

61. Rane, R. A.; Pathak, R. K.; Kaushik, C. P.; Rao, K. V. V.; Prasad, K. A., *Synth. Commun.* **2002**, *32*, 1211.

Triethylsilane[1]

$$Et_3SiH$$

[617-86-7] $C_6H_{16}Si$ (MW 116.31)

InChI = 1/C6H16Si/c1-4-7(5-2)6-3/h7H,4-6H2,1-3H3
InChIKey = AQRLNPVMDITEJU-UHFFFAOYAL

(mild reducing agent for many functional groups)

Physical Data: mp $-156.9\,°C$; bp $107.7\,°C$; d $0.7309\ \mathrm{g\,cm^{-3}}$.
Solubility: insol H_2O; sol hydrocarbons, halocarbons, ethers.
Form Supplied in: colorless liquid; widely available.
Purification: simple distillation, if needed.
Handling, Storage, and Precautions: triethylsilane is physically very similar to comparable hydrocarbons. It is a flammable, but not pyrophoric, liquid. As with all organosilicon hydrides, it is capable of releasing hydrogen gas upon storage, particularly in the presence of acids, bases, or fluoride-releasing salts. Proper precautions should be taken to vent possible hydrogen buildup when opening vessels in which triethylsilane is stored.

Original Commentary

James L. Fry
The University of Toledo, Toledo, OH, USA

Introduction. Triethylsilane serves as an exemplar for organosilicon hydride behavior as a mild reducing agent. It is frequently chosen as a synthetic reagent because of its availability, convenient physical properties, and economy relative to other organosilicon hydrides which might otherwise be suitable for effecting specific chemical transformations.

Hydrosilylations. Addition of triethylsilane across multiple bonds occurs under the influence of a large number of metal catalysts.[2] Terminal alkynes undergo hydrosilylations easily with triethylsilane in the presence of platinum,[3] rhodium,[3a,4] ruthenium,[5] osmium,[6] or iridium[4] catalysts. For example, phenylacetylene can form three possible isomeric hydrosilylation products with triethylsilane; the (Z)-β-, the (E)-β-, and the α-products (eq 1). The (Z)-β-isomer is formed exclusively or preferentially with ruthenium[5] and some rhodium[4] catalysts, whereas the (E)-β-isomer is the major product formed with platinum[3] or iridium[4] catalysts. In the presence of a catalyst and carbon monoxide, terminal alkynes undergo silylcarbonylation reactions with triethylsilane to give (Z)- and (E)-β-silylacrylaldehydes.[7] Phenylacetylene gives an 82% yield of a mixture of the (Z)- and (E)-isomers in a 10:1 ratio when 0.3 mol % of ***Dirhodium(II) Tetrakis (perfluorobutyrate)*** catalyst is used under atmospheric pressure at $0\,°C$ in dichloromethane (eq 2).[7d] Terminal alkenes react with triethylsilane in the presence of this catalyst to form either 'normal' anti-Markovnikov hydrosilylation products or allyl- or vinylsilanes, depending on whether the alkene is added to the silane or vice versa.[8] A mixture of 1-hexene and triethylsilane in the presence of 2 mol % of an iridium catalyst ($[IrCl(CO)_3]_n$) reacts under 50 atm of carbon monoxide to give a 50% yield of a mixture of the (Z)- and (E)-enol silyl ether isomers in a 1:2 ratio (eq 3).[9] Hydrolysis yields the derived acylsilane quantitatively.[9]

$$Ph\!-\!\!\equiv\ +\ Et_3SiH\ \xrightarrow{\ cat\ }$$

$$
\begin{array}{ccc}
\overset{H}{\underset{Ph}{\diagdown}}\!\!=\!\!\overset{H}{\underset{SiEt_3}{\diagup}} &
\overset{H}{\underset{Ph}{\diagdown}}\!\!=\!\!\overset{SiEt_3}{\underset{H}{\diagup}} &
\overset{Et_3Si}{\underset{Ph}{\diagdown}}\!\!=\!\!\overset{H}{\underset{H}{\diagup}}
\end{array}\quad(1)
$$

$$Ph\!-\!\!\equiv\ +\ Et_3SiH\ +\ CO\ \xrightarrow{\ cat\ }$$

$$
\overset{OHC}{\underset{Ph}{\diagdown}}\!\!=\!\!\overset{SiEt_3}{\underset{H}{\diagup}}\ +\ \overset{OHC}{\underset{Ph}{\diagdown}}\!\!=\!\!\overset{H}{\underset{SiEt_3}{\diagup}}\quad(2)
$$

$$BuCH=CH_2\ +\ Et_3SiH\ +\ CO\ \xrightarrow{\ cat\ }$$

$$
\overset{Bu}{\underset{H}{\diagdown}}\!\!=\!\!\overset{OSiEt_3}{\underset{SiEt_3}{\diagup}}\ +\ \overset{Bu}{\underset{H}{\diagdown}}\!\!=\!\!\overset{SiEt_3}{\underset{OSiEt_3}{\diagup}}\quad(3)
$$

A number of metal complexes catalyze the hydrosilylation of various carbonyl compounds by triethylsilane.[10] Stereoselectivity is observed in the hydrosilylation of ketones[11] as in the reactions of 4-*t*-butylcyclohexanone and triethylsilane catalyzed by ruthenium,[12] chromium,[13] and rhodium[12,14] metal complexes (eq 4). Triethylsilane and ***Chlorotris(triphenylphosphine)rhodium(I)*** catalyst effect the regioselective 1,4-hydrosilylation of α,β-unsaturated ketones and aldehydes.[15,16] Reduction of mesityl oxide in this manner results in a 95% yield of product that consists of 1,4- and 1,2-hydrosilylation isomers in a 99:1 ratio (eq 5). This is an exact complement to the use of phenylsilane, where the ratio of respective isomers is reversed to 1:99.[16]

$$
t\text{-Bu}\!\!-\!\!\bigcirc\!\!=\!\!O\ \xrightarrow[\ cat\ ]{\ Et_3SiH\ }
$$

(Ph$_3$P)$_3$RuCl$_2$, AgTFA, PhMe, Δ	5:95
Et$_4$N$^+$ [HCr$_2$(CO)$_{10}$]$^-$, DME, Δ	10:90
(Ph$_3$P)$_3$RhCl, PhMe, Δ	11:89
[Rh(η^3-C$_3$H$_5$){P(OMe)$_3$}$_3$], PhH	29:71

(4)

(5)   99:1

Silane Alcoholysis. Triethylsilane reacts with alcohols in the presence of metal catalysts to give triethylsilyl ethers.[17] The use of dirhodium(II) perfluorobutyrate as a catalyst enables regioselective formation of monosilyl ethers from diols (eq 6).[17a]

$$
\begin{array}{c}
\underset{Bu}{\overset{Bu}{\diagdown}}\!\!C\!\!\diagup\!\!\diagdown\!\!\diagup\!\!OH \\
\end{array}
\ \xrightarrow[\underset{92\%}{Rh_2(pfb)_4}]{\ Et_3SiH\ }\
\begin{array}{c}
\underset{Bu}{\overset{Bu}{\diagdown}}\!\!C\!\!\diagup\!\!\diagdown\!\!\diagup\!\!OSiEt_3\ +\ H_2
\end{array}\quad(6)
$$

Essential Reactions for Organic Synthesis, First Edition. Edited by Philip L. Fuchs.

Formation of Singlet Oxygen. Triethylsilane reacts with ozone at −78 °C in inert solvents to form triethylsilyl hydrotrioxide, which decomposes at slightly elevated temperatures to produce triethylsilanol and *Singlet Oxygen*. This is a convenient way to generate this species for use in organic synthesis.[18]

Reduction of Acyl Derivatives to Aldehydes. Aroyl chlorides and bromides give modest yields of aryl aldehydes when refluxed in diethyl ether with triethylsilane and *Aluminum Chloride*.[19] Better yields of both alkyl and aryl aldehydes are obtained from mixtures of acyl chlorides or bromides and triethylsilane by using a small amount of 10% *Palladium on Carbon* catalyst (eq 7).[20] This same combination of triethylsilane and catalyst can effect the reduction of ethyl thiol esters to aldehydes, even in sensitive polyfunctional compounds (eq 8).[21]

$$C_7H_{15}COCl + Et_3SiH \xrightarrow[83\%]{10\% \text{ Pd/C}} C_7H_{15}CHO \quad (7)$$

Radical Chain Reductions. Triethylsilane can replace toxic and difficult to remove organotin reagents for synthetic reductions under radical chain conditions. Although it is not as reactive as *Tributylstannane*,[22] careful choice of initiator, solvent, and additives leads to effective reductions of alkyl halides,[23,24] alkyl sulfides,[23] and alcohol derivatives such as O-alkyl S-methyl dithiocarbonate (xanthate) and thionocarbonate esters.[22,23,25,26] Portionwise addition of 0.6 equiv of *Dibenzoyl Peroxide* to a refluxing triethylsilane solution of O-cholestan-3β-yl O′-(4-fluorophenyl) thionocarbonate gives a 93% yield of cholestane (eq 9).[22] The same method converts bis-xanthates of *vic*-diols into alkenes (eq 10).[22] Addition of a small amount of thiol such as *t*-dodecanethiol to serve as a 'polarity reversal catalyst'[24] with strong radical initiators in nonaromatic solvents also gives good results.[23,25] Treatment of ethyl 4-bromobutanoate with four equiv of triethylsilane, two equiv of dilauroyl peroxide (DLP), and 2 mol% of *t*-dodecanethiol in refluxing cyclohexane for 1 hour yields ethyl butanoate in 97% yield (eq 11).[23]

Ionic Hydrogenations and Reductive Substitutions. The polar nature of the Si–H bond enables triethylsilane to act as a hydride donor to electron-deficient centers. Combined with Brønsted or Lewis acids this forms the basis for many useful synthetic transformations.[27] Use of *Trifluoromethanesulfonic Acid* (triflic acid) at low temperatures enables even simple alkenes to be reduced to alkanes in high yields (eq 12).[28] *Boron Trifluoride* monohydrate is effective in promoting the reduction of polycyclic aromatic compounds (eq 13).[29] Combined with thiols,

it enables sulfides to be prepared directly from aldehydes and ketones (eq 14).[30] Combinations of triethylsilane with either *Trifluoroacetic Acid*/ammonium fluoride or *Pyridinium Poly (hydrogen fluoride)* (PPHF) are effective for the reductions of alkenes, alcohols, and ketones (eq 15).[31] Immobilized strong acids such as iron- or copper-exchanged *Montmorillonite K10*[32] or the superacid *Nafion-H*[33] facilitate reductions of aldehydes and ketones[32] or of acetals[33] by increasing the ease of product separation (eq 16). Boron trifluoride and triethylsilane are an effective combination for the reduction of alcohols, aldehydes, ketones (eq 17),[34] and epoxides.[35] *Boron Trifluoride Etherate* sometimes may be substituted for the free gas.[36]

$$(16)$$

$$(17)$$

Triethylsilane in 3M ethereal **Lithium Perchlorate** solution effects the reduction of secondary allylic alcohols and acetates (eq 18).[37] The combination of triethylsilane and **Titanium(IV) Chloride** is a particularly effective reagent pair for the selective reduction of acetals.[38] Treatment of (±)-frontalin with this pair gives an 82% yield of tetrahydropyran products with a *cis:trans* ratio of 99:1 (eq 19).[38b] This exactly complements the 1:99 product ratio of the same products obtained with **Diisobutylaluminum Hydride**.[38b]

$$(18)$$

$$(19)$$

Triethylsilane and trityl salts.[39] or **Trimethylsilyl Trifluoromethanesulfonate**[40] are effective for the reduction of various ketones and acetals, as are combinations of **Chlorotrimethylsilane** and indium(III) chloride[41] and **Tin(II) Bromide** and **Acetyl Bromide**.[42] Isophthaldehyde undergoes reductive polycondensation to a polyether when treated with triethylsilane and **Triphenylmethyl Perchlorate**.[43]

Triethylsilane reduces nitrilium ions to aldimines,[44] diazonium ions to hydrocarbons,[45] and aids in the deprotection of amino acids.[46] With aluminum halides, it reduces alkyl halides to hydrocarbons.[47]

First Update

Ronald J. Rahaim Jr & Robert E. Maleczka Jr
Michigan State University, East Lansing, MI, USA

Additional Hydrosilylations. Hydrosilylations of terminal alkynes with triethylsilane (eq 1) have been improved in terms of their regio- and stereocontrol as well as in other aspects of their operation. Through the employment of Pt(DVDS),[48] Pt-catalyzed hydrosilylations of 1-alkynes[3] can now be performed at room temperature and in water with very high selectivity for the (*E*)-β-vinylsilanes (eq 20). It has also been shown that PtO$_2$ catalyzes the internal hydrosilylation of aryl alkynes under *ortho*-substituent regiocontrol (eq 21).[49] Strong preference for the (*E*)-β-vinylsilanes during the hydrosilylation of 1-alkynes has

also been observed with cationic Ru-catalysts[50] and [RhCl(nbd)]$_2$/dppp,[51] the latter of which can also be employed in water (eq 20). Hydrosilylations afford α-vinylsilanes when catalyzed by [CpRu (MeCN)$_3$]PF$_6$[52] (eq 20). (*Z*)-β-Vinylsilanes are similarly made under [RuCl$_2$(*p*-cymene)]$_2$ catalysis[53] or by the *trans* hydrosilylation of 1-alkynes under Lewis acid (AlCl$_3$)[54] catalysis.

$$(20)$$

R = C$_{10}$H$_{21}$	[RhCl(nbd)]$_2$ + dppp	93:3:4
R = Ph	[Cp*Rh(BINAP)](SbF$_6$)$_2$	97:0:3
R = C$_4$H$_9$	Pt(DVDS)EP	100:0:0
R = C$_4$H$_9$	[RuCl$_2$(*p*-cymene)]$_2$	4:96:0
R = PhCH$_2$	AlCl$_3$	0:100:0
R = C$_6$H$_{12}$CO$_2$H	[CpRu(MeCN)$_3$]PF$_6$	2.5:2.5:95

$$(21)$$

AlCl$_3$ can also promote the hydrosilylation of allenes and alkenes.[54] With regard to the hydrosilylation of alkenes; Rh-catalyzed reactions of Et$_3$SiH and methylenecyclopropanes provide a convenient route to homoallylic silanes (eq 22).[55]

$$(22)$$

The hydrosilylation of carbonyl compounds with Et$_3$SiH (eq 4) has also been the subject of additional research. Owing to these efforts, carbonyls can now be directly converted to their triethylsilyl (TES) ethers with copper catalysts in the company of a bidentate phosphine[56] or *N*-heterocyclic carbene[57] ligand. Triethylsilyl ethers can also be made from carbonyl compounds and Et$_3$SiH in the presence of rhenium(V) oxo-complexes.[58]

Additional Silane Alcoholysis. The direct silylation of alcohols with triethylsilane (eq 26)[17] continues to be an interesting, if somewhat underused, method to TES protect alcohols. Recent works have demonstrated that this process is promoted by a number of catalysts including PdCl$_2$,[59] a Au(I) catalyst,[60] and the Lewis acid B(C$_6$F$_6$)$_3$[61] (eq 23).

$$R-OH \xrightarrow[\text{catalyst}]{Et_3SiH} R-OSiEt_3 \qquad (23)$$

R =	Catalyst	% Yield
1° and 2° aliphatic	PdCl$_2$	78–98
1°, 2°, 3° aliphatic or Ar	Au(I)	80–100
2°, 3° aliphatic or Ar	B(C$_6$F$_5$)$_3$	95–100

Additional Ionic Hydrogenation and Reductive Substitutions.

Nitrogen Containing Functional Group Reductions. As previously discussed, triethylsilane can donate its hydride to carbonyls and other functional groups (eqs 12–19)[27–43] A variety of transition metals have recently emerged as promoters of such reactions, especially for reductions of nitrogen containing moieties. For example, organic azides are efficiently transformed to their Boc-protected amines with catalytic palladium in the presence of di-*tert*-butyl dicarbonate (eq 24).[62]

$$Ar-N\overset{H}{\underset{Me}{}} \quad \xrightarrow[\text{SnCl}_4]{\text{Et}_3\text{SiH}} \quad R-N_3 \quad \xrightarrow[\text{cat Pd, Boc}_2\text{O}]{\text{Et}_3\text{SiH}} \quad R-N\overset{Boc}{\underset{H}{}} \quad (24)$$
$$\underset{(0–91\%)}{} \qquad \underset{(64–98\%)}{}$$

$$R = ArCH_2 \qquad R = \text{aryl, } 1°, 2° \text{ aliphatics}$$

Alternatively, if such azides bear a 1°-benzylic group they can be converted to *N*-methylanilines by reaction with Et$_3$SiH and SnCl$_4$.[63] Wilkinson's catalyst and Et$_3$SiH reduce aromatic nitro groups to their amines in moderate to good yields,[64] while the combination of Pd(OAc)$_2$ and Et$_3$SiH in a THF–water mixture reduces aliphatic nitro groups to the *N*-hydroxylamines (eq 25).[65]

$$R-NO_2 \xrightarrow[\text{cat}]{\text{Et}_3\text{SiH}} R-N\overset{H}{\underset{Y}{}} \quad \begin{array}{l} R = \text{aryl; cat} = \text{Rh(PPh}_3)_3\text{Cl}; \\ Y = H \ (0–90\%) \\ \\ R = \text{aliphatic; cat} = \text{Pd(OAc)}_2; \\ Y = OH \ (31–89\%) \end{array} \quad (25)$$

Imines are reduced by triethylsilane to their amines when the proper Ir[66,67] or Ni[68] catalysts are employed. Non-metal-mediated reductions of C=N groups by Et$_3$SiH are also possible. Among these, the trifluorosulfonic acid promoted reductive amidation of aliphatic and aromatic aldehydes with Et$_3$SiH is an excellent way to mono *N*-alkylate aliphatic and aromatic amides, thioamides, carbamates, and ureas (eq 26).[69] It is also worth noting that trifluorosulfonic acid/Et$_3$SiH reduces acyl- and tosylhydrazones to hydrazines[70,71] and 2-aminopyrimidines to 2-amino-dihydro- or 2-aminotetrahydropyrimidines (eq 27).[72]

$$\underset{R^1}{\overset{X}{\|}}\!-\!NH_2 + R^2CHO \xrightarrow[\text{TFA}]{\text{Et}_3\text{SiH}} \underset{R^1}{\overset{X}{\|}}\!-\!N\underset{H}{}\!R^2 \quad 63–97\% \quad (26)$$

X = O or S; R^1 = aliphatic, aryl, OR, NR'R''; R^2 = aliphatic, aryl

$$R^1\underset{R^2}{\overset{}{N}}\!\!\overset{N}{\underset{N}{\bigcirc}}\!\!R^3 \xrightarrow[\text{TFA}]{\text{Et}_3\text{SiH}} \quad (60–90\%)$$

$$R^1\underset{R^2}{\overset{}{N}}\!\!\overset{N}{\underset{N}{\bigcirc}}\!\!R^3 \;+\; R^1\underset{R^2}{\overset{}{N}}\!\!\overset{N}{\underset{N}{\bigcirc}}\!\!R^3 \quad (27)$$

R^1 = H, CH$_2$, Me, Ph, SO$_2$Ar; R^2 = H, CH$_2$, Me;
R^3 = H, Me, C(O)Me, Ar, vinyl, Br

Reductive Etherifications and Acetal Reductions. Additional applications of triethylsilane in the reduction of C–O bonds also continue to surface. The Kusanov–Parnes dehydrative reduction[27] of hemiacetals and acetals with trifluorosulfonic acid/Et$_3$SiH has proven especially valuable. Under such conditions, 4,6-*O*-benzylidene acetal glucose derivatives can be asymmetrically deprotected to 6-*O*-benzyl-4-hydroxy derivatives (eq 28)[73] and thioketone derivatives can be converted to *syn*-2,3-bisaryl (or heteroaryl) dihydrobenzoxanthins with excellent stereo- and chemoselectivity (eq 29).[74] Triethylsilane is also useful in a number of related acetal reductions, including those used for the formation of *C*-glycosides. For example, Et$_3$SiH reductively opens 1,3-dioxolan-4-ones to 2-alkoxy carboxylic acids when catalyzed by TiCl$_4$.[75] Furthermore, functionalized tetrahydrofurans are generated in good yield from 1,2-*O*-isopropylidenefuranose derivatives with boron trifluoride etherate and Et$_3$SiH (eq 30).[76] These same conditions lead to 1,4- or 1,5-anhydroalditols when applied to methyl furanosides or pyranosides.[77]

$$\text{(structure)} \xrightarrow[\substack{\text{TFA} \\ (80–95\%)}]{\text{Et}_3\text{SiH}}$$

R^1 = Ac, Bn; R^2 = OAc, OBn, NHAc

$$\text{(structure)} \quad (28)$$

$$\text{(structure)} \xrightarrow[\substack{\text{TFA} \\ \text{CH}_2\text{Cl}_2 \\ (30–95\%)}]{\text{Et}_3\text{SiH}}$$

$$\text{(structure)} \quad (29)$$

$$\text{(structure)} \xrightarrow[\text{BF}_3\cdot\text{Et}_2\text{O}]{\text{Et}_3\text{SiH}} \text{(structure)} \quad 72–95\% \quad (30)$$

Novel syntheses of amino acids have also employed triethylsilane C–O bond cleavages of *N,O*-acetals. In this way, *N*-methylamino acid derivatives are isolated in high yields from the Fmoc or Cbz protected 5-oxazolidinone precursors, using TFA and Et$_3$SiH,[78,79] or with the Lewis acid AlCl$_3$ and Et$_3$SiH (eq 31).[80] A one-pot preparation of *N*-methyl-α-amino acid dipeptides can be accomplished from an oxazolidinone, amino acid, TFA, and Et$_3$SiH combination.[81]

$$P = Fmoc \text{ or } Cbz \tag{31}$$

Triethylsilane can also facilitate the high yielding reductive formation of dialkyl ethers from carbonyls and silyl ethers. For example, the combination of 4-bromobenzaldehyde, trimethylsilyl protected benzyl alcohol, and Et_3SiH in the presence of catalytic amounts of $FeCl_3$ will result in the reduction *and* benzylation of the carbonyl group (eq 32).[82] Similarly, $Cu(OTf)_2$ has been shown to aid Et_3SiH in the reductive etherification of variety of carbonyl compounds with *n*-octyl trimethylsilyl ether to give the alkyl ethers in moderate to good yields.[83] Likewise, TMSOTf catalyzes the conversion of tetrahydropyranyl ethers to benzyl ethers with Et_3SiH and benzaldehyde, and diphenylmethyl ethers with Et_3SiH and diphenylmethyl formate.[84] Symmetrical and unsymmetrical ethers are afforded in good yield from carbonyl compounds with silyl ethers (or alcohols) and Et_3SiH catalyzed by bismuth trihalide salts.[85] An intramolecular version of this procedure has been nicely applied to the construction of *cis*-2,6-di- and trisubstituted tetrahydropyrans.[86]

In a related process, triethylsilane plus $SnCl_4$ can expediently convert appropriately protected aldol products to fully protected 1,3-diols. Moreover, the synthesis of *syn*-1,3-ethylidene acetals from 1-(2-methoxyethoxy)ethyl-protected β-hydroxy ketones with $SnCl_4$ and Et_3SiH can occur with very high levels of diastereocontrol (eq 33).[87]

syn/anti: >200/1

Ether Cleavages. Triethylsilane and $B(C_6F_5)_3$ can also be used for the general cleavage of ether bonds to their corresponding triethylsilyl ether and hydrocarbon.[61b] This chemistry can selectively cleave differently substituted ethers (e.g., primary alkyl ethers cleave preferentially over secondary, tertiary, or aryl ether groups), but it should be noted that only a limited number of such examples have been reported. Furthermore, chemoselectivity can be an issue as $Et_3SiH/B(C_6F_5)_3$ can deoxygenate primary alcohols and acetals, as well as perform the aforementioned silane alcoholyses. Nonetheless, Et_3SiH and TFA are well suited for taking triphenylmethyl (trityl, Tr) protective groups off hydroxyls (eq 34),[73] aziridines,[88] or peptides[89] even when other acid-sensitive functional groups are present. Triethylsilane has also been employed in the deprotection of triphenylmethyl-protected nucleotides, but with dichloroacetic acid in dichloromethane.[90]

Ester Reductions and Miscellaneous Reductive Substitutions. Triethylsilane can react with esters in a number of ways. Aliphatic esters and lactones are reduced to acyclic and cyclic ethers when treated with $TiCl_4$, TMSOTf, and Et_3SiH (eq 35).[91] Propargylic acetates, on the other hand, will undergo reductive cleavage of their C–O bonds when treated with catalytic amounts of indium(III) bromide and Et_3SiH.[92] Aryl and enol triflates are reduced when exposed to Et_3SiH and a Pd–phosphine complex[93] (eq 36), whereas aromatic and aliphatic iodides, bromides, and chlorides are dehalogenated with Et_3SiH and catalytic $PdCl_2$ (also see eq 11[23]). Curiously, Et_3SiH and $PdCl_2$ can also be used to make C–X bonds, as alcohols are converted to the corresponding halide with $PdCl_2$, Et_3SiH, and iodomethane, dibromomethane, or hexachloroethane (eq 37).[94] Likewise, lactones will undergo a ring-opening halosilylation with $PdCl_2$, Et_3SiH, and iodomethane, or allyl bromide, producing the triethylsilyl ω-iodo- or ω-bromoalkanoates.[95]

Reductive Couplings and Cyclizations. As previously discussed, triethylsilane can react with both activated (eq 5)[15,16] and non-activated olefins (eq 12[28]). Recent developments in this area include the saturation of alkenes by Et_3SiH under catalysis by Grubb's 1st generation catalyst. A particularly elegant application of this chemistry is possible when ring closing metathesis (RCM) is kinetically favored. In such cases one can effect a one-pot ring closure/alkene reduction in good overall yield (eq 38).[96]

Alkenes, along with alkynes, allenes, or dienes, can also participate in triethylsilane promoted reductive couplings. Aldehydes, in particular, are good at coupling with the intermediates of nickel-catalyzed additions of Et_3SiH across alkenes, allenes, dienes, or alkynes (eq 39).[97] These reactions tend to be highly regioselective; as are the indium(III) bromide catalyzed reductive syn aldol between aldehydes, enones, and Et_3SiH (eq 40).[98] Finally, in the presence of ethylaluminum sesquichloride and Et_3SiH, alkylchloroformates participate in what have been termed Friedel–Crafts alkylations of alkenes.[99]

$$PhCHO + Me\!\!=\!\!=\!\!C_4H_9 \xrightarrow[\substack{(84\%)}]{Et_3SiH \\ cat\ Ni}$$

$$(39)$$

$$(40)$$

syn/anti >99/1

Imines can serve as electrophiles in similar processes. For example, tetrahydropyran or tetrahydrofuran containing amino acids are synthesized in good yield from a $TiCl_4$ catalyzed coupling of cyclic enol ethers, *N*-tosyl imino ester, and triethylsilane (eq 41).[100] Triethylsilane and a palladium-catalyst can prompt the cyclization–hydrosilylation of 1,6- and 1,7-dienes with good yields and moderate to high stereoselectivity (eq 42).[101] Cationic rhodium catalyzes a cyclization–hydrosilylation of 1,6-enynes,[102] whereas palladium catalyzes a regio- and stereoselective cycloreduction.[103] This reaction has also been applied to haloenynes[104] and bimetallic cobalt/rhodium nanoparticles in an atmosphere of carbon monoxide to effect a carbonylative silylcarbocyclization (eq 43).[105] 1,6-Diynes with cationic platinum, or cationic rhodium, in conjunction with Et_3SiH undergo a cyclization–hydrosilylation.[106,107] The combination of Et_3SiH, rhodium, carbon monoxide, and allenyl–carbonyl compounds yields *cis*-2-triethylsilylvinyl-cyclopentanols and cyclohexanols (eq 44).[108] For this reaction the investigators mention that Et_3SiH is superior to Ph_3SiH, Me_2PhSiH, and $(EtO)_3SiH$. This contrasts most other reports of hydrosilylations with Et_3SiH, where no particular advantage is either attributed or demonstrated for Et_3SiH over other silanes. Finally, reductive Nazarov cyclizations can also take place with Et_3SiH and a Lewis acid.[109]

When reacting alkenes with triethylsilane it is necessary to keep in mind that the $PdCl_2/Et_3SiH$ combination also promotes the double bond isomerization of monosubstituted aliphatic olefins[110] and α-alkylidene cyclic carbonyl compounds are isomerized to α,β-unsaturated cyclic carbonyls with tris(triphenylphosphine) rhodium chloride.[111]

$$(41)$$

$n = 1$ (71%)
$n = 2$ (98%)

R^1 = ester
R^2 = H, Me, Ph, $$(42)$$
ester, CN,
SO_2Me

$$(43)$$

R^1	R^2	R^3	Catalyst	R^4	R^5	% Yield
CO_2Et	H	Ph	$Pd(dppe)Cl_2$	H	Ph	85
Me	Me	H	$[Rh(COD)_2]SbF_6$	$SiEt_3$	H	65
CO_2Me	H	H	$Co_2Rh_2 + CO$	$SiEt_3$	CHO	93

$$(44)$$

$X = C(CO_2Et)_2$, NTs, O; R = H, Me, Et; $n = 1, 2$

Aromatic Silylations. Aryltriethylsilanes are synthesized in moderate to good yield from electron-rich *meta-* and *para*-substituted aryl iodides, by $Pd(P-tBu)_3$ in the presence of K_3PO_4 and triethylsilane.[112] Platinum oxide in conjunction with sodium acetate and Et_3SiH silylates *meta* and *para* substituted aryl iodides and bromides that contain electron-withdrawing groups.[113] *ortho*-Triethylsilyl aromatics are accessed with $Ru_3(CO)_{12}$ using azoles, imines, pyridines, amides, and esters as directing groups; the system tolerates electron-donating and withdrawing groups (eq 45).[114] This method has also been applied to the silylation of benzylic C–H bonds.[115]

R = azole, imine,
ester, amide, $$(45)$$
pyridine,

Generation of Other Triethylsilyl Reagents, etc. Triethylsilane is also used in the synthesis of various other reagents for organic synthesis. Triethylsilyl cyanide, which is used for the silylcyanation of aldehydes and ketones, can be prepared from

Et$_3$SiH and acetonitrile in the presence of catalytic amounts of Cp(CO)$_2$FeMe.[116] Bromotriethylsilane is prepared when Et$_3$SiH reacts with copper(II) bromide and catalytic amounts of copper(I) iodide[117] or with PdCl$_2$ and allyl bromide.[118] Et$_3$SiH can also reduce Bu$_3$SnCl to Bu$_3$SnH, which when carried out in the presence of alkynes, allenes, or alkenes can undergo Lewis acid promoted hydrostannation reactions (eq 46).[119] This represents the first example of Lewis acid catalyzed hydrostannations with in situ generated tributyltin hydride. Significantly, Et$_3$SiH succeeded in this reaction where hydrosiloxanes failed. Lastly, Et$_3$SiH reacts with indium(III) chloride to generate dichloroindium hydride.[119]

$$R\!\!\equiv\!\!R^1 \xrightarrow[\substack{\text{toluene, 0 °C to rt} \\ (70–90\%)}]{\substack{\text{Et}_3\text{SiH, Bu}_3\text{SnCl} \\ 10\ \text{mol}\%\ \text{B(C}_6\text{F}_3)_3}} \quad \begin{array}{c} R \\ | \\ H \end{array}\!\!\!=\!\!\!\begin{array}{c} \text{SnBu}_3 \\ | \\ R^1 \end{array} \quad (46)$$

Z:E ~ 90:10

Related Reagents. Phenylsilane–Cesium Fluoride; Tri-*n*-butylstannane; Tricarbonylchloroiridium–Diethyl(methyl) Silane–Carbon Monoxide; Triethylsilane–Trifluoroacetic Acid.

1. (a) Fleming, I. In *Comprehensive Organic Chemistry*; Barton, D.; Ollis, W. D., Eds.; Pergamon: New York, 1979; Vol. 3, p 541. (b) Colvin, E. *Silicon in Organic Synthesis*; Butterworths: Boston, 1981. (c) Weber, W. P. *Silicon Reagents for Organic Synthesis*; Springer: New York, 1983. (d) *The Chemistry of Organic Silicon Compounds*; Patai, S.; Rappoport, Z., Eds.; Wiley: New York, 1989. (e) Corey, J. Y. In *Advances in Silicon Chemistry*; Larson, G. L., Ed.; JAI: Greenwich, CT, 1991; Vol. 1, p 327.

2. (a) Lukevics, E., *Russ. Chem. Rev. (Engl. Transl.)* **1977**, *46*, 264. (b) Speier, J. L., *Adv. Organomet. Chem.* **1979**, *17*, 407. (c) Keinan, E., *Pure Appl. Chem.* **1989**, *61*, 1737.

3. (a) Doyle, M. P.; High, K. G.; Nesloney, C. L.; Clayton, T. W., Jr.; Lin, J., *Organometallics* **1991**, *10*, 1225. (b) Lewis, L. N.; Sy, K. G.; Bryant, G. L., Jr.; Donahue, P. E., *Organometallics* **1991**, *10*, 3750.

4. Kopylova, L. I.; Pukhnarevich, V. B.; Voronkov, M. G., *Zh. Obshch. Khim.* **1991**, *61*, 2418.

5. Esteruelas, M. A.; Herrero, J.; Oro, L. A., *Organometallics* **1993**, *12*, 2377.

6. Esteruelas, M. A.; Oro, L. A.; Valero, C., *Organometallics* **1991**, *10*, 462.

7. (a) Murai, S.; Sonada, N., *Angew. Chem., Int. Ed. Engl.* **1979**, *18*, 837. (b) Matsuda, I.; Ogiso, A.; Sato, S.; Izumi, Y., *J. Am. Chem. Soc.* **1989**, *111*, 2332. (c) Ojima, I.; Ingallina, P.; Donovan, R. J.; Clos, N., *Organometallics* **1991**, *10*, 38. (d) Doyle, M. P.; Shanklin, M. S., *Organometallics* **1993**, *12*, 11.

8. Doyle, M. P.; Devora, G. A.; Nefedov, A. O.; High, K. G., *Organometallics* **1992**, *11*, 549.

9. Chatani, N.; Ikeda, S.; Ohe, K.; Murai, S., *J. Am. Chem. Soc.* **1992**, *114*, 9710.

10. (a) Eaborn, C.; Odell, K.; Pidcock, A., *J. Organomet. Chem.* **1973**, *63*, 93. (b) Corriu, R. J. P.; Moreau, J. J. E., *J. Chem. Soc., Chem. Commun.* **1973**, 38. (c) Ojima, I.; Nihonyanagi, M.; Kogure, T.; Kumagai, M.; Horiuchi, S.; Nakatsugawa, K.; Nagai, Y., *J. Organomet. Chem.* **1975**, *94*, 449.

11. Ojima, I.; Nihonyanagi, M.; Nagai, Y., *Bull. Chem. Soc. Jpn.* **1972**, *45*, 3722.

12. Semmelhack, M. F.; Misra, R. N., *J. Org. Chem.* **1982**, *47*, 2469.

13. Fuchikami, T.; Ubukata, Y.; Tanaka, Y., *Tetrahedron Lett.* **1991**, *32*, 1199.

14. Bottrill, M.; Green, M., *J. Organomet. Chem.* **1976**, *111*, C6.

15. Ojima, I.; Kogure, T.; Nihonyanagi, M.; Nagai, Y., *Bull. Chem. Soc. Jpn.* **1972**, *45*, 3506.

16. Ojima, I.; Kogure, T., *Organometallics* **1982**, *1*, 1390.

17. (a) Doyle, M. P.; High, K. G.; Bagheri, V.; Pieters, R. J.; Lewis, P. J.; Pearson, M. M., *J. Org. Chem.* **1990**, *55*, 6082. (b) Zakharkin, L. I.; Zhigareva, G. G., *Izv. Akad. Nauk SSSR, Ser. Khim.* **1992**, 1284. (c) Barton, D. H. R.; Kelly, M. J., *Tetrahedron Lett.* **1992**, *33*, 5041.

18. Corey, E. J.; Mehrota, M. M.; Khan, A. U., *J. Am. Chem. Soc.* **1986**, *108*, 2472.

19. Jenkins, J. W.; Post, H. W., *J. Org. Chem.* **1950**, *15*, 556.

20. Citron, J. D., *J. Org. Chem.* **1969**, *34*, 1977.

21. Fukuyama, T.; Lin, S.-C.; Li, L., *J. Am. Chem. Soc.* **1990**, *112*, 7050.

22. Barton, D. H. R.; Jang, D. O.; Jaszberenyi, J. C., *Tetrahedron Lett.* **1991**, *32*, 7187: *Tetrahedron* **1993**, *49*, 2793.

23. Cole, S. J.; Kirwan, J. N.; Roberts, B. P.; Willis, C. R., *J. Chem. Soc., Perkin Trans. 1* **1991**, 103.

24. Allen, R. P.; Roberts, B. P.; Willis, C. R., *J. Chem. Soc., Chem. Commun.* **1989**, 1387.

25. Kirwin, J. N.; Roberts, B. P.; Willis, C. R., *Tetrahedron Lett.* **1990**, *31*, 5093.

26. Cf. Chatgilialoglu, C.; Ferreri, C.; Lucarini, M., *J. Org. Chem.* **1993**, *58*, 249.

27. (a) Kursanov, D. N.; Parnes, Z. N., *Russ. Chem. Rev. (Engl. Transl.)* **1969**, *38*, 812. (b) Kursanov, D. N.; Parnes, Z. N.; Loim, N. M., *Synthesis* **1974**, 633. (c) Nagai, Y., *Org. Prep. Proced. Int.* **1980**, *12*, 13. (d) Kursanov, D. N.; Parnes, Z. N.; Kalinkin, M. I.; Loim, N. M. *Ionic Hydrogenation and Related Reactions*; Harwood: Chur, Switzerland, 1985.

28. Bullock, R. M.; Rappoli, B. J., *J. Chem. Soc., Chem. Commun.* **1989**, 1447.

29. (a) Larsen, J. W.; Chang, L. W., *J. Org. Chem.* **1979**, *44*, 1168. (b) Eckert-Maksic, M.; Margetic, D., *Energy Fuels* **1991**, *5*, 327. (c) Eckert-Maksic, M.; Margetic, D., *Energy Fuels* **1993**, *7*, 315.

30. Olah, G. A.; Wang, Q.; Trivedi, N. J.; Prakash, G. K. S., *Synthesis* **1992**, 465.

31. Olah, G. A.; Wang, Q.; Prakash, G. K. S., *Synlett* **1992**, 647.

32. Izumi, Y.; Nanami, H.; Higuchi, K.; Onaka, M., *Tetrahedron Lett.* **1991**, *32*, 4741.

33. Olah, G. A.; Yamato, T.; Iyer, P. S.; Prakash, G. K. S., *J. Org. Chem.* **1986**, *51*, 2826.

34. (a) Adlington, M. G.; Orfanopoulos, M.; Fry, J. L., *Tetrahedron Lett.* **1976**, 2955. (b) Fry, J. L.; Orfanopoulos, M.; Adlington, M. G.; Dittman, W. R., Jr.; Silverman, S. B., *J. Org. Chem.* **1978**, *43*, 374. (c) Fry, J. L.; Silverman, S. B.; Orfanopoulos, M., *Org. Synth.* **1981**, *60*, 108.

35. Fry, J. L.; Mraz, T. J., *Tetrahedron Lett.* **1979**, 849.

36. (a) Doyle, M. P.; West, C. T.; Donnelly, S. J.; McOsker, C. C., *J. Organomet. Chem.* **1976**, *117*, 129. (b) Dailey, O. D., Jr., *J. Org. Chem.* **1987**, *52*, 1984. (c) Krause, G. A.; Molina, M. T., *J. Org. Chem.* **1988**, *53*, 752. (d) Gil, J. F.; Ramón, D. J.; Yus, M., *Tetrahedron* **1993**, *49*, 4923.

37. Wustrow, D. J.; Smith, W. J., III; Wise, L. D., *Tetrahedron Lett.* **1994**, *35*, 61.

38. (a) Kotsuki, H.; Ushio, Y.; Kadota, I.; Ochi, M., *Chem. Lett.* **1988**, 927. (b) Ishihara, K.; Mori, A.; Yamamoto, H., *Tetrahedron* **1990**, *46*, 4595.

39. (a) Tsunoda, T.; Suzuki, M.; Noyori, R., *Tetrahedron Lett.* **1979**, 4679. (b) Kato, J.; Iwasawa, N.; Mukaiyama, T., *Chem. Lett.* **1985**, *6*, 743. (c) Kira, M.; Hino, T.; Sakurai, H., *Chem. Lett.* **1992**, 555.

40. (a) Bennek, J. A.; Gray, G. R., *J. Org. Chem.* **1987**, *52*, 892. (b) Sassaman, M. B.; Kotian, K. D.; Prakash, G. K. S.; Olah, G. A., *J. Org. Chem.* **1987**, *52*, 4314.

41. (a) Mukaiyama, T.; Ohno, T.; Nishimura, T.; Han, J. S.; Kobayashi, S., *Chem. Lett.* **1990**, 2239. (b) Mukaiyama, T.; Ohno, T.; Nishimura, T.; Han, J. S.; Kobayashi, S., *Bull. Chem. Soc. Jpn.* **1991**, *64*, 2524.

42. (a) Oriyama, T.; Iwanami, K.; Tsukamoto, K.; Ichimura, Y.; Koga, G., *Bull. Chem. Soc. Jpn.* **1991**, *64*, 1410. (b) Oriyama, T.; Ichimura, Y.; Koga, G., *Bull. Chem. Soc. Jpn.* **1991**, *64*, 2581.

43. Yokozawa, T.; Nakamura, F., *Makromol. Chem., Rapid Commun.* **1993**, *14*, 167.

44. (a) Fry, J. L., *J. Chem. Soc., Chem. Commun.* **1974**, 45. (b) Fry, J. L.; Ott, R. A., *J. Org. Chem.* **1981**, *46*, 602.

45. Nakayama, J.; Yoshida, M.; Simamura, O., *Tetrahedron* **1970**, *26*, 4609.

46. Mehta, A.; Jaouhari, R.; Benson, T. J.; Douglas, K. T., *Tetrahedron Lett.* **1992**, *33*, 5441.

47. (a) Doyle, M. P.; McOsker, C. C.; West, C. T., *J. Org. Chem.* **1976**, *41*, 1393. (b) Parnes, Z. N.; Romanova, V. S.; Vol'pin, M. E., *J. Org. Chem. USSR (Engl. Transl.)* **1988**, *24*, 254.

48. (a) Aneetha, H.; Wu, W.; Verkada, J. G., *Organometallics* **2005**, *24*, 2590. (b) Wu, W.; Li, C.-J., *Chem. Commun.* **2003**, 1668.

49. Hamze, A.; Provot, O.; Alami, M.; Brion, J.-D., *Org. Lett.* **2005**, *7*, 5625.

50. (a) Takeuchi, R.; Nitta, S.; Watanabe, D., *J. Chem. Soc. Chem. Commun.* **1994**, 1777. (b) Faller, J. W.; D'Alliessi, D. G., *Organometallics* **2002**, *21*, 1743.

51. Sato, A.; Kinoshita, H.; Shinokubo, H.; Oshima, K., *Org. Lett.* **2004**, *6*, 2217.

52. Trost, B. M.; Ball, Z. T., *J. Am. Chem. Soc.* **2005**, *127*, 17644.

53. Na, Y.; Chang, S., *Org. Lett.* **2000**, *2*, 1887.

54. (a) Sudo, T.; Asao, N.; Gevorgyan, V.; Yamamoto, Y., *J. Org. Chem.* **1999**, *64*, 2494. (b) Song, Y.-S.; Yoo, B. R.; Lee, G.-H.; Jung, I. N., *Organometallics* **1999**, *18*, 3109.

55. Bessmertnykh, A. G.; Blinov, K. A.; Grishin, Y. K.; Donskaya, N. A.; Tveritinova, E. V.; Yur'eva, N. M.; Beletskaya, I. P., *J. Org. Chem.* **1997**, *62*, 6069.

56. Lipshutz, B. H.; Caires, C. C.; Kuipers, P.; Chrisman, W., *Org. Lett.* **2003**, *5*, 3085.

57. Díez-González, S.; Kaur, H.; Zinn, F. K.; Stevens, E. D.; Nolan, S. P., *J. Org. Chem.* **2005**, *70*, 4784.

58. Ison, E. A.; Trivedi, E. R.; Corbin, R. A.; Abu-Omar, M. M., *J. Am. Chem. Soc.* **2005**, *127*, 15374.

59. Mirza-Aghayan, M.; Boukherroub, R.; Bolourtchian, M., *J. Organomet. Chem.* **2005**, *690*, 2372.

60. Ito, H.; Takagi, K.; Miyahara, T.; Sawamura, M., *Org. Lett.* **2005**, *7*, 3001.

61. (a) Blackwell, J. M.; Foster, K. L.; Beck, V. H.; Piers, W. E., *J. Org. Chem.* **1999**, *64*, 4887. (b) Gevorgyan, V.; Rubin, M.; Benson, S.; Liu, J.-X.; Yamamoto, Y., *J. Org. Chem.* **2000**, *65*, 6179.

62. Kotsuki, H.; Ohishi, T.; Araki, T., *Tetrahedron Lett.* **1997**, *38*, 2129.

63. Lopez, F. J.; Nitzan, D., *Tetrahedron Lett.* **1999**, *40*, 2071.

64. Brinkman, H. R.; Miles, W. H.; Hilborn, M. D.; Smith, M. C., *Synth. Commun.* **1996**, *26*, 973.

65. Rahaim, R. J., Jr.; Maleczka, R. E., Jr. *Org. Lett.* **2005**, *7*, 5087.

66. Field, L. D.; Messerle, B. A.; Rumble, S. L., *Eur. J. Org. Chem.* **2005**, 2881.

67. In-situ formed imine: Mizuta, T.; Sakaguchi, S.; Ishii, Y., *J. Org. Chem.* **2005**, *70*, 2195.

68. Vetter, A. H.; Berkessel, A., *Synthesis* **1995**, 419.

69. Dubé, D.; Scholte, A. A., *Tetrahedron Lett.* **1999**, *40*, 2295.

70. Wu, P.-L.; Peng, S.-Y.; Magrath, J., *Synthesis* **1995**, 435.

71. Wu, P.-L.; Peng, S.-Y.; Magrath, J., *Synthesis* **1996**, 249.

72. Baskaran, S.; Hanan, E.; Byun, D.; Shen, W., *Tetrahedron Lett.* **2004**, *45*, 2107.

73. Imagawa, H.; Tsuchihashi, T.; Singh, R. K.; Yamamoto, H.; Sugihara, T.; Nishizawa, M., *Org. Lett.* **2003**, *5*, 153.

74. Kim, S.; Wu, J. Y.; Chen, H. Y.; DiNinno, F., *Org. Lett.* **2003**, *5*, 685.

75. Winneroski, L. L.; Xu, Y., *J. Org. Chem.* **2004**, *69*, 4948.

76. Ewing, G. J.; Robins, M. J., *Org. Lett.* **1999**, *1*, 635.

77. (a) Rolf, D.; Gray, G. R., *J. Am. Chem. Soc.* **1982**, *104*, 3539. (b) Rolf, D.; Bennek, J. A.; Gray, G. R., *J. Carbohydr. Chem.* **1983**, *2*, 373. (c) Bennek, J. A.; Gray, G. R., *J. Org. Chem.* **1987**, *52*, 892.

78. Luke, R. W. A.; Boyce, P. G. T.; Dorling, E. K., *Tetrahedron Lett.* **1996**, *37*, 263.

79. Aurelio, L.; Brownlee, R. T. C.; Hughes, A. B., *Org. Lett.* **2002**, *4*, 3767.

80. Zhang, S.; Govender, T.; Norstrom, T.; Arvidsson, P. I., *J. Org. Chem.* **2005**, *70*, 6918.

81. Dorow, R. L.; Gingrich, D. E., *Tetrahedron Lett.* **1999**, *40*, 467.

82. (a) Iwanami, K.; Seo, H.; Tobita, Y.; Oriyama, T., *Synthesis* **2005**, 183. (b) Iwanami, K.; Yano, K.; Oriyama, T., *Synthesis* **2005**, 2669.

83. Yang, W.-C.; Lu, X.-A.; Kulkarni, S. S.; Hung, S.-C., *Tetrahedron Lett.* **2003**, *44*, 7837.

84. Suzuki, T.; Kobayashi, K.; Noda, K.; Oriyama, T., *Synth. Commun.* **2001**, *31*, 2761.

85. (a) Wada, M.; Nagayama, S.; Mizutani, K.; Hiroi, R.; Miyoshi, N., *Chem. Lett.* **2002**, 248. (b) Komatsu, N.; Ishida, J.; Suzuki, H., *Tetrahedron Lett.* **1997**, *38*, 7219. (c) Bajwa, J. S.; Jiang, X.; Slade, J.; Prasad, K.; Repic, O.; Blacklock, T. J., *Tetrahedron Lett.* **2002**, *43*, 6709.

86. Evans, P. A.; Cui, J.; Gharpure, S. J.; Hinkle, R. J., *J. Am. Chem. Soc.* **2003**, *125*, 11456.

87. Cullen, A. J.; Sammakia, T., *Org. Lett.* **2004**, *6*, 3143.

88. Vedejs, E.; Klapars, A.; Warner, D. L.; Weiss, A. H., *J. Org. Chem.* **2001**, *66*, 7542.

89. Kadereit, D.; Deck, P.; Heinemann, I.; Waldmann, H., *Chem. Eur. J.* **2001**, *7*, 1184.

90. Ravikumar, V. T.; Krotz, A. H.; Cole, D. L., *Tetrahedron Lett.* **1995**, *36*, 6587.

91. Yato, M.; Homma, K.; Ishida, A., *Tetrahedron* **2001**, *57*, 5353.

92. Sakai, N.; Hirasawa, M.; Konakahara, T., *Tetrahedron Lett.* **2005**, *46*, 6407.

93. Kotsuki, H.; Datta, P. K.; Hayakawa, H.; Suenaga, H., *Synthesis* **1995**, 1348.

94. Ferreri, C.; Costantino, C.; Chatgilialoglu, C.; Boukherroub, R.; Manuel, G., *J. Organomet. Chem.* **1998**, *554*, 135.

95. Iwata, A.; Ohshita, J.; Tang, H.; Kunai, A.; Yamamoto, Y.; Matui, C., *J. Org. Chem.* **2002**, *67*, 3927.

96. Menozzi, C.; Dalko, P. I.; Cossy, J., *Synlett* **2005**, 2449.

97. (a) Montgomery, J., *Angew. Chem., Int. Ed.* **2004**, *43*, 3890. (b) Mahandru, G. M.; Liu, G.; Montgomery, J., *J. Am. Chem. Soc.* **2004**, *126*, 3698. (c) Ng, S.-S.; Jamison, T. F., *J. Am. Chem. Soc.* **2005**, *127*, 7320. (d) Knapp-Reed, B.; Mahandru, G. M.; Montgomery, J., *J. Am. Chem. Soc.* **2005**, *127*, 13156.

98. Shibata, I.; Kato, H.; Ishida, T.; Yasuda, M.; Baba, A., *Angew. Chem. Int. Ed.* **2004**, *43*, 711.

99. Biermann, U.; Metzger, J. O., *Angew. Chem. Int. Ed.* **1999**, *38*, 3675.

100. Ghosh, A. K.; Xu, C.-X.; Kulkarni, S. S.; Wink, D., *Org. Lett.* **2005**, *7*, 7.

101. (a) Widenhoefer, R. A.; Stengone, C. N., *J. Org. Chem.* **1999**, *64*, 8681. (b) Widenhoefer, R. A.; Perch, N. S., *Org. Lett.* **1999**, *1*, 1103. (c) Perch, N. S.; Pei, T.; Widenhoefer, R. A., *J. Org. Chem.* **2000**, *65*, 3836. (d) Wang, X.; Chakrapani, H.; Stengone, C. N.; Widenhoefer, R. A., *J. Org. Chem.* **2001**, *66*, 1755.

102. Chakrapani, H.; Liu, C.; Widenhoefer, R. A., *Org. Lett.* **2003**, *5*, 157.

103. Oh, C. H.; Jung, H. H.; Sung, H. R.; Kim, J. D., *Tetrahedron* **2001**, *57*, 1723.

104. Oh, C. H.; Park, S. J., *Tetrahedron Lett.* **2003**, *44*, 3785.

105. Park, K. H.; Jung, I. G.; Kim, S. Y.; Chung, Y. K., *Org. Lett.* **2003**, *5*, 4967.

106. Wang, X.; Chakrapani, H.; Madine, J. W.; Keyerleber, M. A.; Widenhoefer, R. A., *J. Org. Chem.* **2002**, *67*, 2778.

107. Liu, C.; Widenhoefer, R. A., *Organometallics* **2002**, *21*, 5666.

108. Kang, S.-K.; Hong, Y.-T.; Leen, J.-H.; Kim, W.-Y.; Lee, I.; Yu, C.-M., *Org. Lett.* **2003**, *5*, 2813.

109. (a) Giese, S.; West, F. G., *Tetrahedron Lett.* **1998**, *39*, 8393. (b) Giese, S.; West F. G., *Tetrahedron* **2000**, *56*, 10221.

110. Miraz-Aghayan, M.; Boukherroub, R.; Bolourtchian, M.; Hoseini, M.; Tabar-Hydar, K., *J. Organomet. Chem.* **2003**, *678*, 1.

111. Tanaka, M.; Mitsuhashi, H.; Maruno, M.; Wakamatsu, T., *Chem. Lett.* **1994**, 1455.

112. Yamanoi, Y., *J. Org. Chem.* **2005**, *70*, 9607.

113. Hamze, A.; Provot, O.; Alami, M.; Brion, J.-D., *Org. Lett.* **2006**, *8*, 931.

114. Kakiuchi, F.; Matsumoto, M.; Tsuchiya, K.; Igi, K.; Hayamizu, T.; Chatani, N.; Murai, S., *J. Organomet. Chem.* **2003**, *686*, 134.

115. Kakiuchi, F.; Tsuchiya, K.; Matsumoto, M.; Mizushima, E.; Chatani, N., *J. Am. Chem. Soc.* **2004**, *126*, 12792.

116. Itazaki, M.; Nakazawa, H., *Chem. Lett.* **2005**, *34*, 1054.

117. Kunai, A.; Ochi, T.; Iwata, A.; Ohshita, J., *Chem. Lett.* **2001**, 1228.

118. Gevorgyan, V.; Liu, J.-X.; Yamamoto, Y., *Chem. Commun.* **1998**, 37.

119. Hayashi, N.; Shibata, I.; Baba, A., *Org. Lett.* **2004**, *6*, 4981.

Trifluoromethanesulfonic Acid[1]

[1493-13-6] CHF_3O_3S (MW 150.09)

InChI = 1/CHF3O3S/c2-1(3,4)8(5,6)7/h(H,5,6,7)/f/h5H

InChIKey = ITMCEJHCFYSIIV-JSWHHWTPCD

(one of the strongest organic acids; catalyst for oligomerization/polymerization of alkenes and ethers; precursor for triflic anhydride and several metal triflates; acid catalyst in various reactions)

Alternate Name: triflic acid.

Physical Data: bp 162 °C/760 mmHg, 84 °C/43 mmHg, 54 °C/8 mmHg; d 1.696 g cm^{-3}.

Solubility: sol water and in many polar organic solvents such as DMF, sulfolane, DMSO, dimethyl sulfone, acetonitrile; sol alcohols, ketones, ethers, and esters, but these generally are not suitable inert solvents (see below).

Analysis of Reagent Purity: IR;[2] ^{19}F NMR.[3]

Preparative Methods: best prepared by basic hydrolysis of CF_3SO_2F followed by acidification.[2]

Purification: distilled with a small amount of Tf$_2$O.[4]

Handling, Storage, and Precautions: is a stable, hygroscopic liquid which fumes copiously on exposure to moist air. Transfer under dry nitrogen is recommended. Contact with cork, rubber, and plasticized materials will cause rapid discoloration of the acid and deterioration of the materials. Samples are best stored in sealed glass ampules or glass bottles with Kel-F^{TM} or PTFE plastic screw cap linings. Use in a fume hood.

Original Commentary

Lakshminarayanapuram R. Subramanian, Antonio García Martínez & Michael Hanack

Universität Tübingen, Tübingen, Germany

Reaction with P$_2$O$_5$. Trifluoromethanesulfonic acid (TfOH) reacts with an excess of **Phosphorus(V) Oxide** to give **Trifluoromethanesulfonic Anhydride** (eq 1),[5] while treatment with a smaller amount of P$_2$O$_5$ (TfOH:P$_2$O$_5$ = 6:1) and slower distillation leads to trifluoromethyl triflate (eq 2).[6]

$$CF_3SO_3H + P_2O_5 \text{ (excess)} \xrightarrow[-H_2O]{\Delta} (CF_3SO_2)_2O \quad (1)$$

$$6\,CF_3SO_3H + P_2O_5 \xrightarrow[70\%]{\Delta}$$
$$3\,CF_3SO_2OCF_3 + 3\,SO_2 + 2\,H_3PO_4 \quad (2)$$

The synthetic utility of trifluoromethyl triflate as a trifluoromethanesulfonylating agent is severely limited, because the reagent is rapidly destroyed by a fluoride-ion chain reaction in the presence of other nucleophiles.[7]

Dehydration of a 2:1 mixture of CF$_3$CO$_2$H and TfOH with P$_2$O$_5$ affords trifluoroacetyl triflate (eq 3),[8] which is a very reactive agent for trifluoroacetylations at O, N, C, or halogen centers (eq 3).[8a]

$$CF_3SO_3H \xrightarrow[P_2O_5, \Delta]{CF_3CO_2H} CF_3CO_2OSO_2CF_3 \xrightarrow[-MeOTf]{anisole, 65\,°C, 28\,h}$$
$$75\% \qquad\qquad\qquad 86\%$$

$$\text{(3)}$$

Protonation and Related Reactions. TfOH is one of the strongest monoprotic organic acids known. The acid, and its conjugate base (CF$_3$SO$_3^-$), have extreme thermal stability, are resistant to oxidation and reduction, and are not a source of fluoride ions, even in the presence of strong nucleophiles. They do not lead to sulfonation as do **Sulfuric Acid**, **Fluorosulfuric Acid**, and **Chlorosulfonic Acid** in some reactions. TfOH is therefore effectively employed in protonation reactions.

The strong protonating property of TfOH is used to generate allyl cations from suitable precursors in low-temperature ionic Diels–Alder reactions. 3,3-Diethoxypropene and 2-vinyl-1,3-dioxolane add to cyclohexa-1,3-diene in the presence of TfOH to give the corresponding Diels–Alder adducts, the latter in high yield (eq 4).[9]

$$\text{(4)}$$

An intramolecular Diels–Alder reaction with high stereoselectivity occurs involving allyl cations by protonation of allyl alcohols (eq 5).[10]

$$\text{(5)}$$

Alkynes and allenes are protonated with TfOH to give vinyl triflates (eqs 6 and 7),[11] which are precursors to vinyl cations.

$$\text{(6)}$$
$$65:35$$

$$\text{(7)}$$
$$60–80\%$$

A convenient synthesis of pyrimidines is developed by protonation of alkynes with TfOH in the presence of nitriles (eq 8).[12]

$$R^1 \underline{\quad\quad} + R^2CN \xrightarrow[\text{68–94\%}]{\text{CF}_3\text{SO}_3\text{H} \atop 0\,°\text{C to rt, 24 h}} \quad (8)$$

$$R^1 = R^2 = \text{alkyl, aryl}$$

Triflic acid catalyzes the transformation of α-hydroxy carbonyl compounds to ketones (eq 9).[13]

$$\xrightarrow[\substack{-\text{PhCHO} \\ 97\%}]{\substack{\text{CF}_3\text{SO}_3\text{H} \\ \text{CH}_2\text{Cl}_2,\ 0\,°\text{C to rt}}} \quad (9)$$

Oximes undergo Beckmann rearrangement with TfOH in the presence of Bu_4NReO_4 to give amides in high yield (eq 10).[14]

$$\xrightarrow[\substack{\text{MeNO}_2,\ 20\ \text{min} \\ 98\%}]{\text{CF}_3\text{SO}_3\text{H},\ \text{Bu}_4\text{NReO}_4} \quad (10)$$

TfOH protonates nitroalkenes, even nitroethylene, to give N,N-dihydroxyiminium carbenium ions, which react with arenes to give arylated oximes. This overall process provides a route to α-aryl methyl ketones from 2-nitropropene (eq 11)[15] and constitutes a versatile synthetic method for the preparation of α-arylated ketones, otherwise difficult to synthesize by the conventional Friedel–Crafts reaction.

$$\xrightarrow{\text{CF}_3\text{SO}_3\text{H}}$$

$$\left[\quad \xrightarrow{\text{PhH}} \quad \xrightarrow[85\%]{\text{H}_2\text{O}} \right]$$

$$\xrightarrow{} \quad (11)$$

TfOH catalyzes the removal of N-t-butyl groups from N-substituted N-t-butylcarbamates to give carbamate-protected primary amines (eq 12).[16]

$$\xrightarrow[95\%]{1\ \text{mol \%}\ \text{CF}_3\text{SO}_3\text{H}}$$

$$\quad (12)$$

The methyl group attached to the phenolic oxygen of tyrosine is smoothly cleaved by TfOH in the presence of **Thioanisole** (eq 13).[17] This deblocking method was successfully applied to the synthesis of a new potent enkephalin derivative.

$$\xrightarrow[\sim100\%]{\substack{\text{CF}_3\text{SO}_3\text{H},\ \text{PhSMe} \\ \text{CF}_3\text{CO}_2\text{H},\ 25\,°\text{C, 50 min}}}$$

$$\quad + \ \text{Ph}\overset{+}{\text{S}}\text{Me}_2 \quad (13)$$

1,3,4-Oxadiazoles are prepared in good yields from silylated diacylhydrazines (formed in situ) by acid-catalyzed cyclization using TfOH (eq 14).[18]

$$\xrightarrow[\substack{\text{MeCN, 0\,°C, 24 h} \\ 60\%}]{\text{CF}_3\text{SO}_3\text{H},\ \text{Me}_2\text{SiCl}_2} \quad (14)$$

TfOH protonates naphthalene at room temperature to give a complex mixture of products.[19] TfOH promotes aldol reaction of silyl enol ethers with aldehydes and acetals, leading to new C–C bond formation (eq 15).[20] TfOH competes well with other reagents employed for the aldol reaction, while **Methanesulfonic Acid** does not afford any product.

$$\xrightarrow[91\%]{\substack{\text{CF}_3\text{SO}_3\text{H},\ \text{PhCHO} \\ \text{CH}_2\text{Cl}_2,\ -95\,°\text{C}}} \quad (15)$$

Cyclization of 3- and 4-arylalkanoic acids to bicyclic ketones is effected by TfOH via the corresponding acid chlorides (eq 16).[21]

$$\xrightarrow[\substack{\text{CH}_2\text{Cl}_2,\ -78\,°\text{C to rt} \\ 87\%}]{\text{CF}_3\text{SO}_3\text{H}} \quad (16)$$

Allylic O-methylisoureas are cyclized with TfOH containing **Benzeneselenenyl Trifluoromethanesulfonate** to 5,6-dihydro-1,3-oxazines (eq 17).[22]

$$\xrightarrow[66\%]{\substack{\text{CF}_3\text{SO}_3\text{H} \\ \text{PhSeOTf}}} \quad (17)$$

Tscherniac amidomethylation of aromatics with N-hydroxymethylphthalimide in TfOH proceeds smoothly at room temperature to give the corresponding α-amido-methylated products (eq 18).[23]

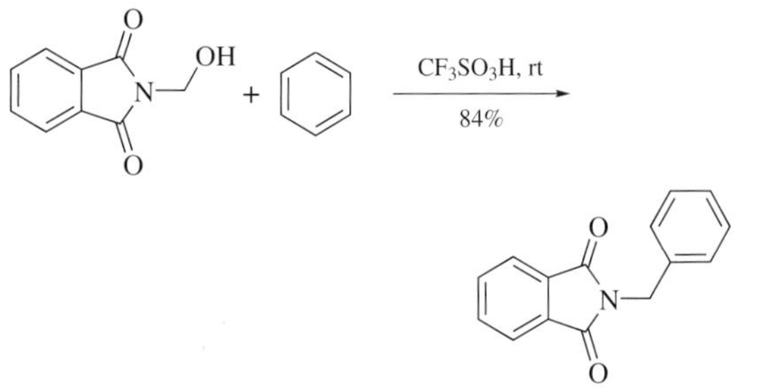

(18)

TfOH catalyzes the amination[24] and phenylamination[25] of aromatics via the corresponding aminodiazonium ion generated from *Azidotrimethylsilane* and *Phenyl Azide* respectively (eq 19).

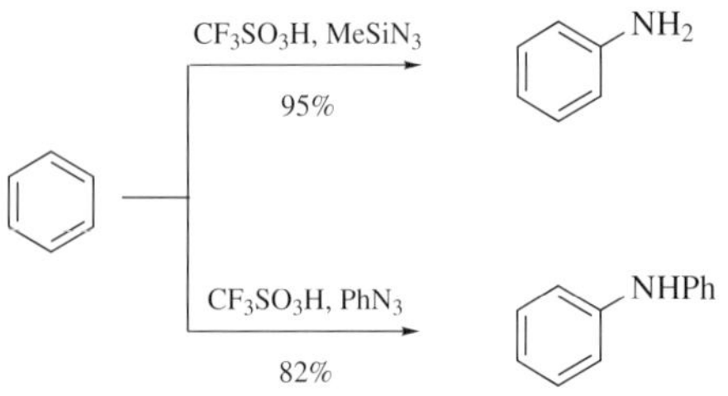

(19)

Electrophilic hydroxylation of aromatics is carried out by protonation of *Bis(trimethylsilyl) Peroxide* with TfOH in the presence of the substrate (eq 20).[26]

(20)

Phenol and 2,3,5,6-tetramethylphenol are protonated with TfOH under irradiation to afford rearranged products (eqs 21 and 22).[27]

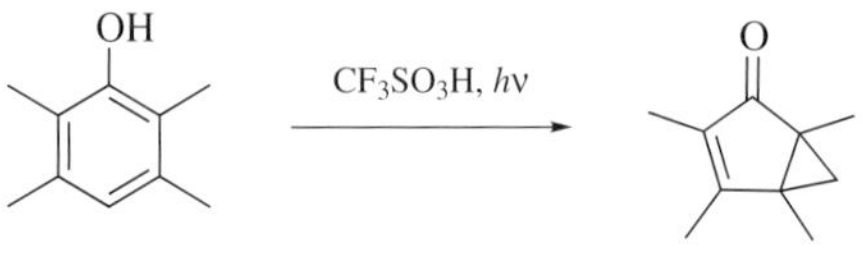

(21)

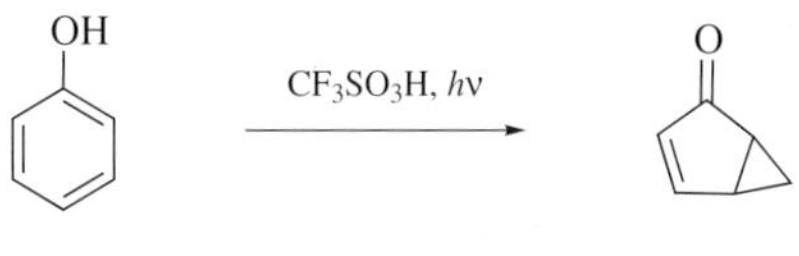

(22)

Other Applications. TfOH is the starting material for the preparation of the electrophilic reagent *Trimethylsilyl Trifluoro-methanesulfonate*. The latter is prepared by reacting TfOH with *Chlorotrimethylsilane*[28] or more conveniently with Me$_4$Si (eq 23).[29]

$$CF_3SO_3H \xrightarrow[12-99\%]{TMSCl \ or \ Me_4Si} CF_3SO_3TMS \quad (23)$$

Functionalized silyl triflates can also be prepared using TfOH (eq 24).[30]

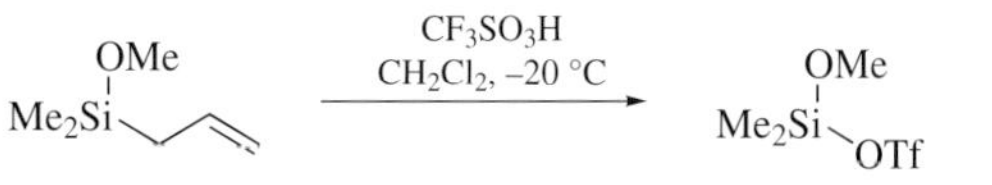

(24)

Reaction of aromatic compounds with *Bis(pyridine) iodonium(I) Tetrafluoroborate* in the presence of TfOH

is an effective method to form the monoiodo compounds regioselectively (eq 25).[31]

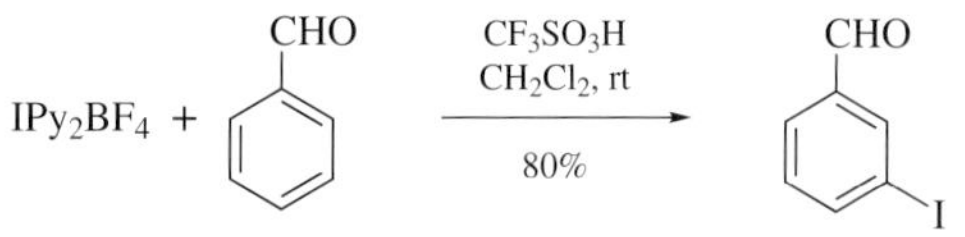

(25)

Ionic hydrogenation of alkenes with trialkylsilanes is possible in the presence of the strong acid TfOH, even at −75 °C (eq 26).[32]

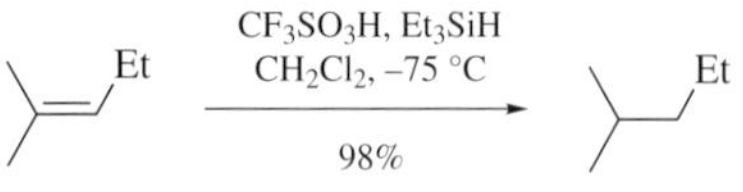

(26)

Hydroxycarbonyl compounds can be selectively reduced to carbonyl compounds by means of TfOH in the presence of tri-alkylboranes (eq 27).[33]

(27)

The triphenylmethyl cation is nitrated with *Nitronium Tetra-fluoroborate* in the presence of TfOH (eq 28).[34]

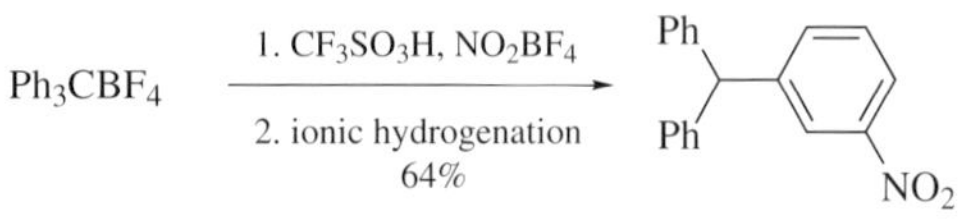

(28)

Sterically hindered azidophenyltriazines decompose in TfOH at 0 °C to give isomeric triflates (eq 29).[35]

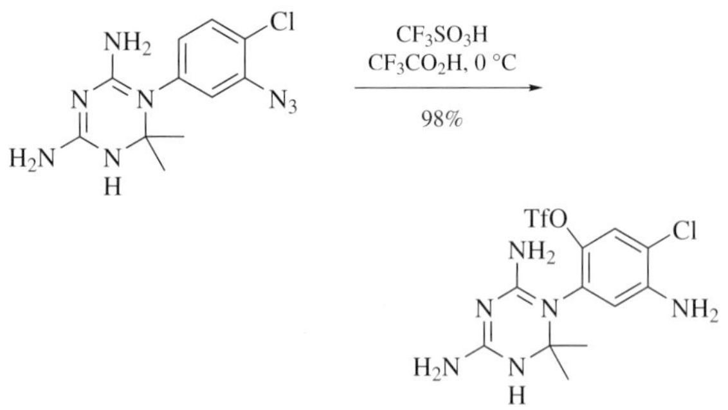

(29)

Benzoyl triflate prepared from TfOH and *Benzoyl Chloride* is a mild and effective benzoylating agent for sterically hindered alcohols[36] and acylative ring expansion reactions.[37] The applications of TfOH in Koch–Haaf carboxylation,[38] Fries rearrangement,[39] and sequential chain extension in carbohydrates[40] are also documented. Recent applications of TfOH in cyclization reactions have been published.[41–43]

First Update

G. K. Surya Prakash
University of Southern California, Los Angeles, CA, USA

Jinbo Hu
Shanghai Institute of Organic Chemistry, Shanghai, China

Superelectrophilic Activation or Superelectrophilic Solvation. Trifluoromethanesulfonic acid (triflic acid, TfOH) has been extensively employed as a superacid ($H_o = -14.1$) in superelectrophilic activation (or superelectrophilic solvation), both concepts advanced by Olah.[44,45] Superelectrophilic activations may occur when a cationic electrophile reacts with a Brønsted or Lewis acid to give a dicationic (doubly electron-deficient) superelectrophile. However, it should be recognized that the activation may proceed through superelectrophilic solvation without necessarily forming limiting dicationic intermediates. The frequently used depiction of protosolvated species as their limiting dications is just for simplicity.[45]

Carboxonium ions are highly stabilized by strong oxygen participation and therefore are much less reactive compared to alkyl cations. However, under the superelectrophilic solvation by triflic acid, the Friedel–Crafts-type reactions still can occur via a protosolvated reactive intermediate. For example, 1-phenyl-2-propen-1-ones can be readily transformed into 1-indanones in good yields through triflic acid-catalyzed reaction (eq 30).[46]

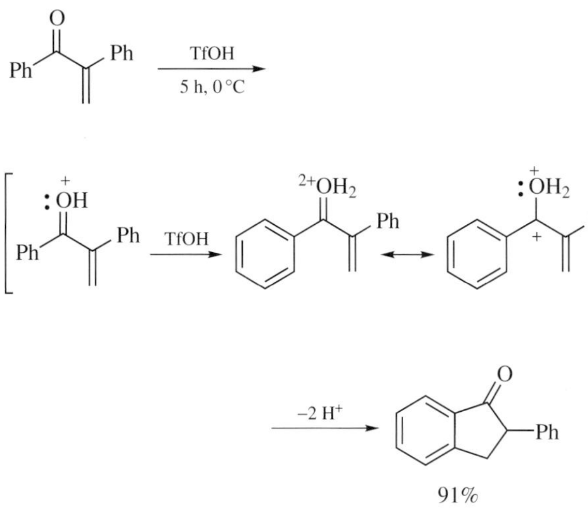

(30)
91%

A one-pot synthesis of 1-indanones and 1-tetralones in good to excellent yields have been developed by reacting a series of alkenyl carboxylic acid derivatives with arenes in TfOH medium. The reaction involves dicationic intermediates involving intermolecular alkylation followed by intramolecular acylation (eq 31).[47] These reactions have been further investigated.[48]

Dicarboxylic acids can also form a variety of distonic superelectrophilic intermediates by TfOH-mediated protonation of the carboxylic acid group and ionization of adjacent functional groups. α-Keto dicarboxylic acids in strongly acidic medium generate reactive multiply charged electrophilic species capable of condensing with arenes in high yields (eq 32).[49] The cascade of reactions also involve the loss of carbon monoxide.

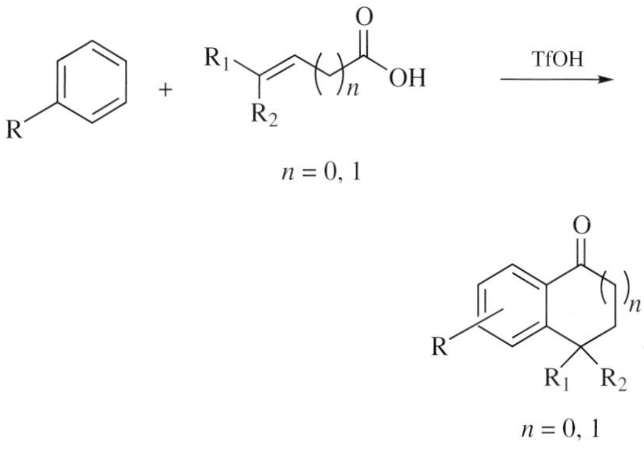

(31)

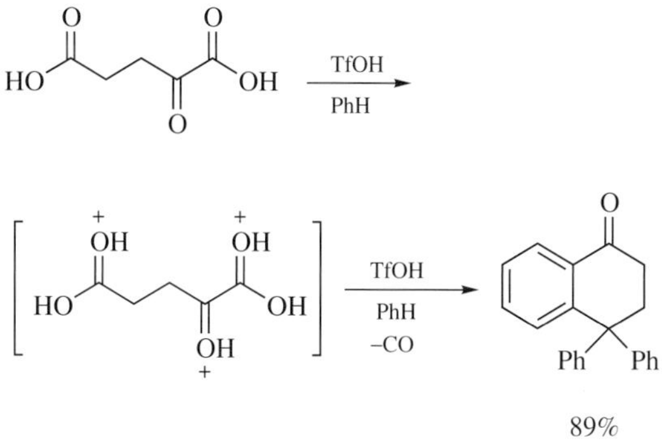

$n = 0, 1$

(32)
89%

Triflic acid-catalyzed Friedel–Crafts acylation reactions of aromatics with methyl benzoate give benzophenone products in good to excellent yields (eq 33).[50] To explain the high level of electrophilic reactivity of this system, protosolvated species are proposed as possible intermediates (eq 32). In the triflic acid-catalyzed cyclization of some ethylene dications, protonation of the ester group is thought to be a key activation step. Reaction of α-(methoxycarbonyl)diphenylmethanol with TfOH gives the fluorene product in 94% yield (eq 34).[51]

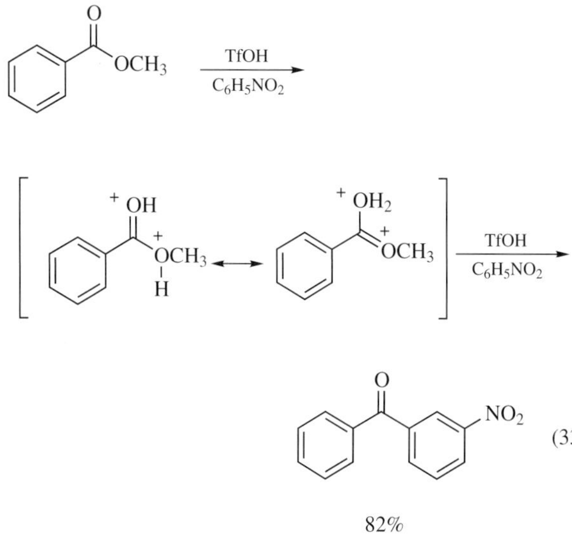

(33)
82%

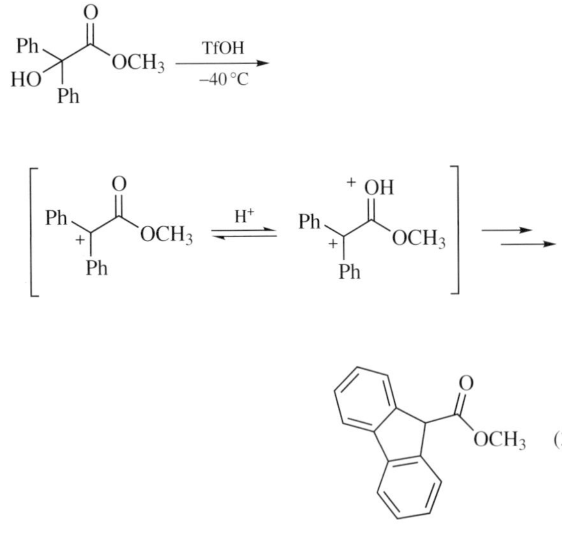

(34)

94%

(37)

80%

(38)

95%

The same concept was applied in the synthesis of aryl-substituted piperidines by the TfOH-catalyzed reaction of piperidones with benzene (eq 35).[52] In the TfOH-catalyzed reactions, acetyl-substituted heteroaromatic compounds, such as pyridines, thiazoles, quinolines, and pyrazines can condense with benzene in good yields via the dicationic intermediates (eq 36).[53] Amino alcohols have also been found to ionize cleanly to the dicationic intermediates, which were directly observed by low-temperature ^{13}C NMR.[54] Amino alcohols can react with benzene in triflic acid by electrophilic aromatic substitution with 70~99% yields (eq 37).[54] Similarly, amino acetals can react with benzene in triflic acid medium to give 1-(3,3-diphenylpropyl)amines or 1-(2,2-diphenylethyl)amines in 50~99% yield (eq 38).[55]

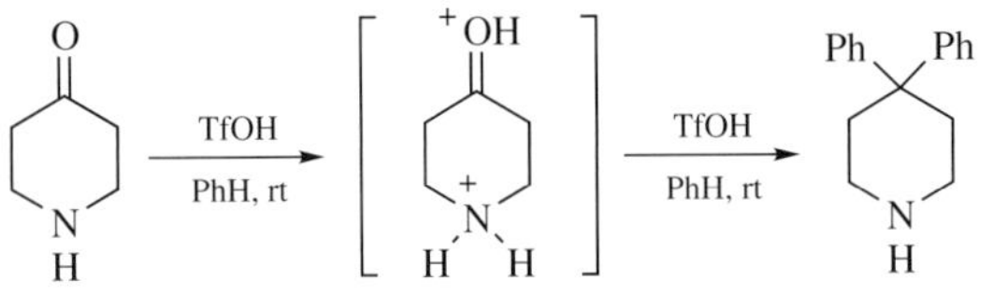

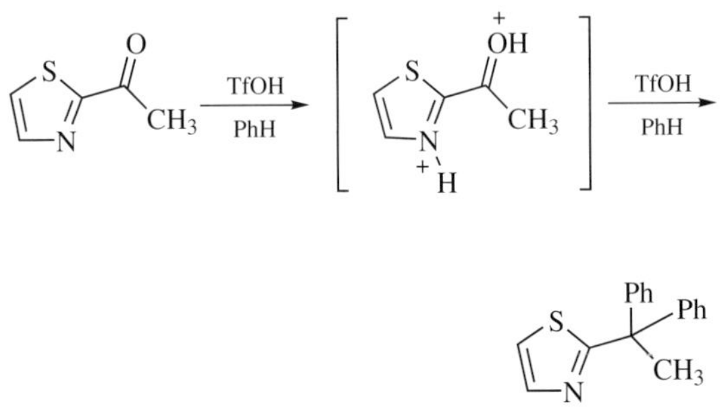

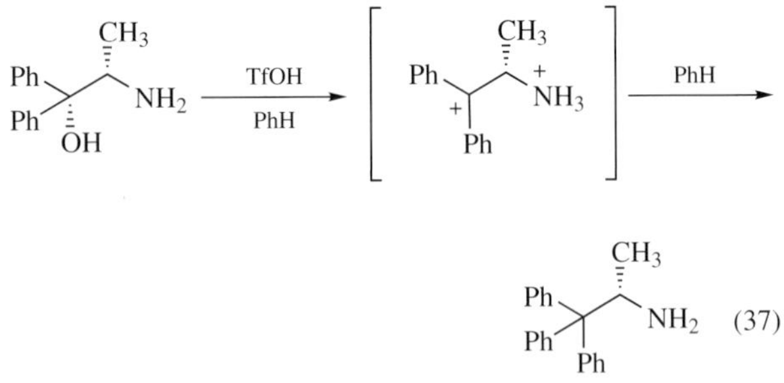

Klumpp and co-workers reported the triflic acid-catalyzed reactions of olefinic amines with benzene to give addition products in 75~99% yields.[56] Remarkably, the chemistry was also used to conveniently prepare functionalized polystyrene beads having pendant amine groups (eq 39).[56] In the triflic acid medium, amides are also able to form reactive, dicationic electrophiles.[57] It has been shown that protonated amide increases the reactivity of an adjacent electrophilic group (eq 40), and the protonated amide itself shows enhanced reactivity for Friedel-Crafts acylation arising from an adjacent cationic charge (eq 41).[57] Similar types of TfOH-catalyzed Friedel-Crafts acylation of aromatics with β-lactams have been reported.[58] TfOH-mediated activation of α,β-unsaturated amides for condensation with arenes have been disclosed by Koltunov co-workers.[59] Klumpp and co-workers have also demonstrated the triflic acid-catalyzed superelectrophilic reactions of 2-oxazolines with benzene to give the corresponding amide products.[60] When aminoalkynes and related heterocycles reacted with benzene in triflic acid, diarylated products were obtained in generally good yields (69~99%) via dicationic intermediates.[61] Triflic acid also promotes reactions of pyrazolecarboxaldehydes with arenes.[62]

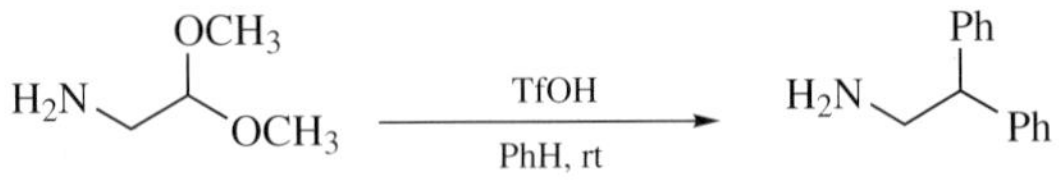

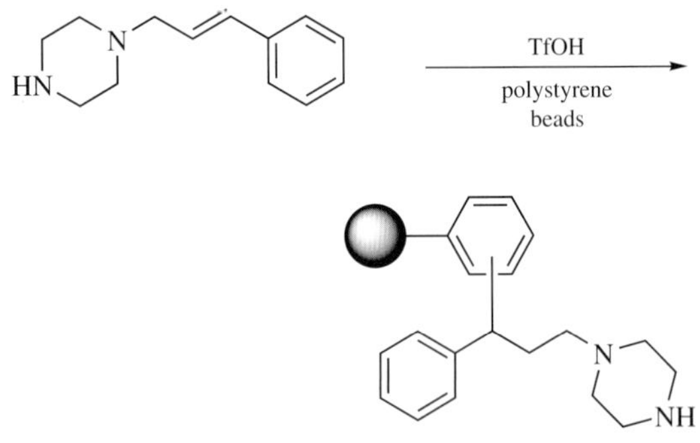

(35)

99%

(36)

75%

(39)

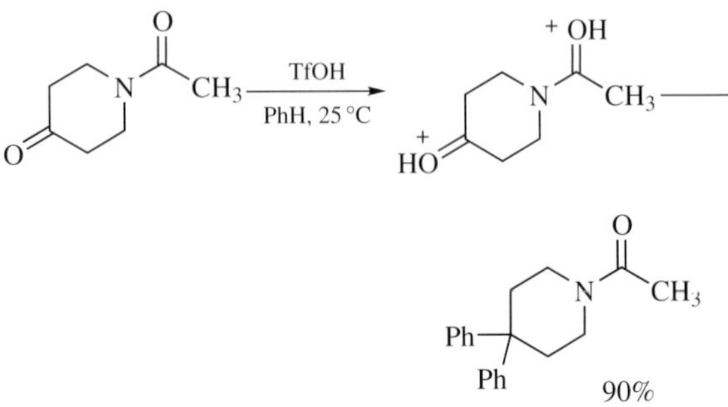

(40)

90%

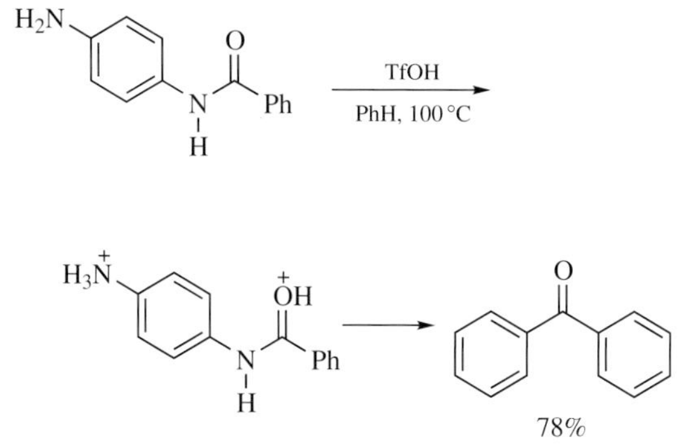

$$(41)$$

78%

In TfOH medium, vinylpyrazine undergoes anti-Markownikow addition involving superelectrophilic intermediates. Arylation of such an electrophile with benzene gave 2-phenylethylpyrazine in high yield (eq 42).[63]

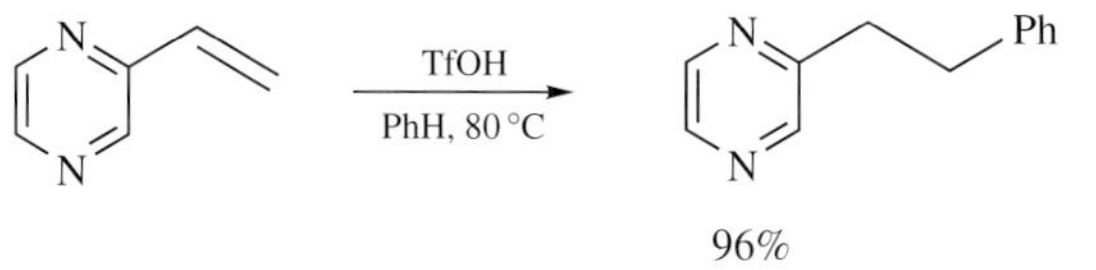

$$(42)$$

96%

Ethyl trifluoropyruvate has been activated in TfOH medium for the hydroxyalkylation of arenes to give valuable Mosher's acid derivatives in good to excellent yields.[64] Even Selectfluor® has been activated in TfOH to effect electrophilic fluorination of arenes including fluorobenzene and chlorobenzene.[65]

A novel, mild method for the preparation of diaryl sulfoxides from arenes and thionyl chloride has been developed in TfOH medium. Under the nonoxidative reaction conditions only sulfoxides are produced without any contamination from the corresponding sulfones (eq 43).[66]

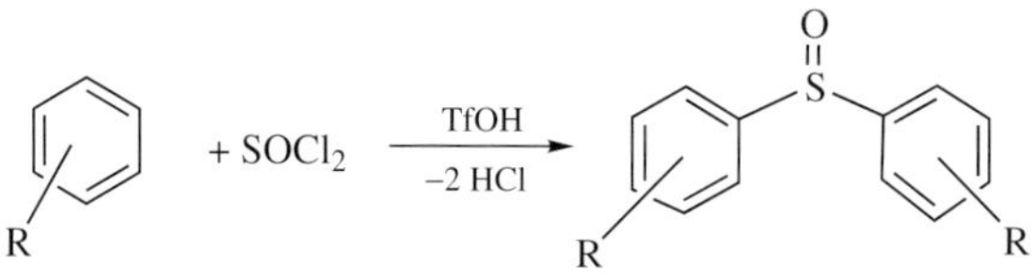

$$(43)$$

Other Protonation and Acid-catalyzed Reactions. The catalytic activity of triflic acid can be dramatically increased by the addition of a catalytic amount of bismuth(III) chloride. For example, triflic acid or $BiCl_3$ by itself poorly catalyzes the sulfonylation of arenes using arenesulfonyl chlorides. However, the $BiCl_3$-triflic acid combination catalysts can efficiently catalyze the sulfonylation reactions (eq 44).[67] Similar synergistic effects between TfOH and bismuth(III) or antimony(III) chlorides have been observed in methanesulfonylation of arenes.[68]

Corey et al. have shown the asymmetric Diels-Alder reactions catalyzed by a triflic acid activated chiral oxazaborolidine (eq 45).[69] Triflic acid has also been found to be an efficient catalyst (1 mol %) for the hetero-Diels–Alder reaction between aromatic aldehydes and unactivated dienes.[70]

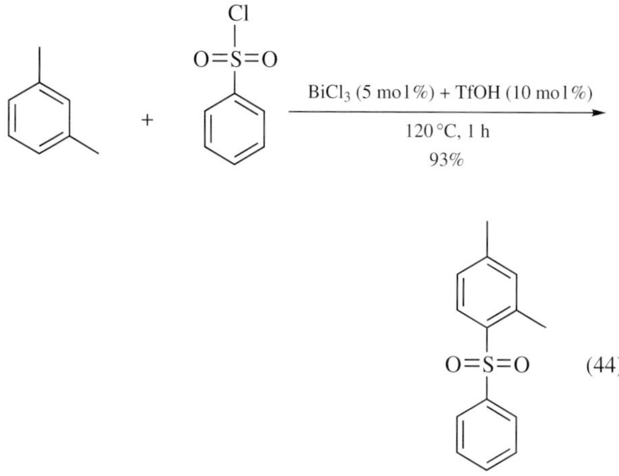

$$(44)$$

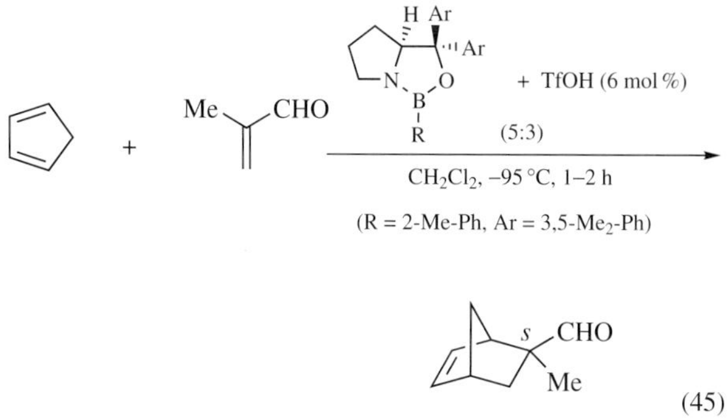

$$(45)$$

99% yield, ee = 96%
(*exo:endo* = 89:11)

The synthetic scope of the Thiele-Winter reaction of quinines with acetic anhydride can be increased by the use of triflic acid (eq 46).[71] Reaction of cyclopropylacylsilanes with triflic acid in aprotic solvent affords the corresponding cyclobutanone or 2-silyl-4,5-dihydrofuran derivatives.[72] Triflic acid can react with *o*-iodosylbenzoic acid to form a hypervalent iodine reagent, which reacts with 1-trimethylsilylalkynes to afford alkynyliodonium triflates bearing a carboxy group in high yields (eq 47).[73] Reaction of (diacetoxyiodo)benzene [PhI(OAc)$_2$] with excess triflic acid results in oligomerization of PhI(OAc)$_2$.[74]

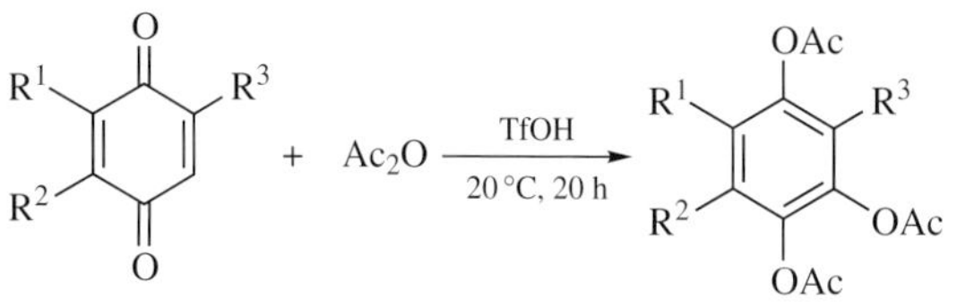

$$(46)$$

(R = H, CH$_3$, Br, OEt, etc.)   92–42%

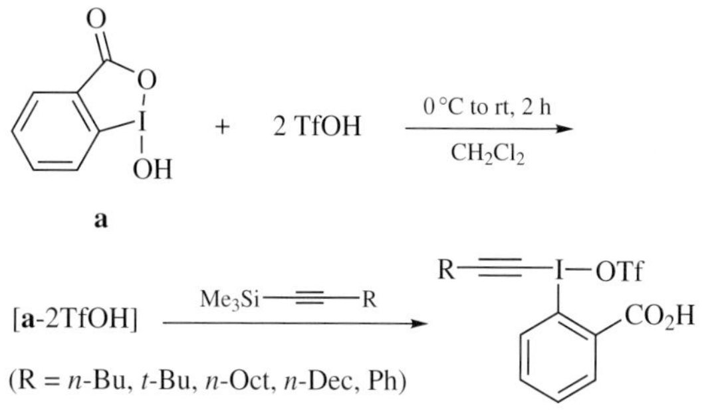

$$(47)$$

(R = *n*-Bu, *t*-Bu, *n*-Oct, *n*-Dec, Ph)

Olah et al. reported the triflic acid-catalyzed isobutene-iso-butylene alkylation, modified with trifluoroacetic acid (TFA) or water. They found that the best alkylation conditions were at an acid strength of about $H_o = -10.7$, giving a calculated research octane number (RON) of 89.1 (TfOH/TFA) and 91.3 (TfOH/H$_2$O).[75] Triflic acid-modified zeolites can be used for the gas phase synthesis of methyl *tert*-butyl ether (MTBE), and the mechanism of activity enhancement by triflic acid modification appears to be related to the formation of extra-lattice Al rather than the direct presence of triflic acid.[76] A thermally stable solid catalyst prepared from amorphous silica gel and triflic acid has also been reported.[77,78] The obtained material was found to be an active catalyst in the alkylation of isobutylene with *n*-butenes to yield high-octane gasoline components.[77] A similar study has been carried out with triflic acid-functionalized mesoporous Zr-TMS catalysts.[79,80] Triflic acid-catalyzed carbonylation,[81] direct coupling reactions,[82] and formylation[83] of toluene have also been reported. Triflic acid also promotes transalkylation[84] and adamantylation of arenes in ionic liquids.[85] Triflic acid-mediated reactions of methylenecyclopropanes with nitriles have also been investigated to provide [3 + 2] cycloaddition products as well as Ritter products.[86] Triflic acid also catalyzes cyclization of unsaturated alcohols to cyclic ethers.[87]

Loh et al. found a triflic acid-catalyzed 2-oxonia Cope rearrangement, which was used in the stereocontrolled synthesis of linear 22*R*-homoallylic sterols (eq 48).[88] Interestingly, poor stereoselectivity was observed when In(OTf)$_3$ was employed as the catalyst for this reaction. Stereoselective Mannich-type reaction of chiral aldimines with 2-silyloxybutadienes in the presence of triflic acid gives the corresponding products with 70–92% de in 62–74% chemical yield, which are not obtained by general Lewis acid-promoted methods (eq 49).[89]

82% yield
22*R*/22*S* = 98:2 (48)

Triflic acid was also used in the synthesis of dixanthones and poly(dixanthones) by cyclization of 2-aryloxybenzonitriles at room temperature.[90] Addition of dialkyl disulfides to terminal alkynes is catalyzed by a rhodium-phosphine complex and triflic acid giving (*Z*)-bis(alkylthio)olefins stereoselectively (eq 50).[91]

R^1 = Tol, R^2 = Ph, R^3 = H   70% yield, 90% de   (49)

95%   (50)

Marko and co-workers have reported the role of triflic acid in the metal triflate-catalyzed acylation of alcohols with carboxylic anhydrides.[92] Their mechanistic insights demonstrate that triflic acid is generated under the acylation reaction conditions, and that two competing catalytic cycles are operating at the same time: a rapid one involving triflic acid and a slower one involving the metal triflates. A straightforward synthesis of aziridines is reported by treating electron rich alkyl- or aryl azide with electron deficient olefin and TfOH in cold acetonitrile.[93]

1. (a) Howells, R. D.; McCown, J. D., *Chem. Rev.* **1977**, *77*, 69. (b) Stang, P. J.; White, M. R., *Aldrichim. Acta* **1983**, *16*, 15.

2. (a) Haszeldine, R. N.; Kidd, J. M., *J. Chem. Soc.* **1954**, 4228. (b) Burdon, J.; Farazmand, I.; Stacey, M.; Tatlow, J. C., *J. Chem. Soc.* **1957**, 2574.

3. Matjaszewski, K.; Sigwalt, P., *Monatsh. Chem.* **1986**, *187*, 2299 (*Chem. Abstr.* **1987**, *106*, 5495).

4. (a) Sagl, D.; Martin, J. C., *J. Am. Chem. Soc.* **1988**, *110*, 5827. (b) Saito, S.; Sato, Y.; Ohwada, T.; Shudo, K., *Chem. Pharm. Bull.* **1991**, *39*, 2718.

5. Stang, P. J.; Hanack, M.; Subramanian, L. R., *Synthesis* **1982**, 85.

6. Hassani, M. O.; Germain, A.; Brunel, D.; Commeyras, A., *Tetrahedron Lett.* **1981**, *22*, 65.

7. Taylor, S. L.; Martin, J. C., *J. Org. Chem.* **1987**, *52*, 4147.

8. (a) Forbus, T. R., Jr.; Taylor, S. L.; Martin, J. C., *J. Org. Chem.* **1987**, *52*, 4156. (b) Taylor, S. L.; Forbus, T. R., Jr.; Martin, J. C., *Org. Synth., Coll. Vol.* **1990**, *7*, 506.

9. Gassman, P. G.; Singleton, D. A.; Wilwerding, J. J.; Chavan, S. P., *J. Am. Chem. Soc.* **1987**, *109*, 2182.

10. (a) Gassman, P. G.; Singleton, D. A., *J. Org. Chem.* **1986**, *51*, 3075. (b) Gorman, D. B.; Gassman, P. G., *J. Org. Chem.* **1995**, *60*, 977.

11. (a) Stang, P. J.; Summerville, R. H., *J. Am. Chem. Soc.* **1969**, *91*, 4600. (b) Summerville, R. H.; Senkler, C. A.; Schleyer, P. v. R.; Dueber, T. E.; Stang, P. J., *J. Am. Chem. Soc.* **1974**, *96*, 1100.

12. García Martínez, A.; Herrera Fernandez Martínez, A.; Alvarez, R.; Silva Losada, M. C.; Molero Vilchez, D.; Subramanian, L. R.; Hanack, M., *Synthesis* **1990**, 881.

13. Olah, G. A.; Wu, A., *J. Org. Chem.* **1991**, *56*, 2531.

14. Narasaka, K.; Kusama, H.; Yamashita, Y.; Sato, H., *Chem. Lett.* **1993**, 489.

15. Okabe, K.; Ohwada, T.; Ohta, T.; Shudo, K., *J. Org. Chem.* **1989**, *54*, 733.

16. Earle, M. J.; Fairhurst, R. A.; Heaney, H.; Papageorgiou, G., *Synlett* **1990**, 621.

17. Kiso, Y.; Nakamura, S.; Ito, K.; Ukawa, K.; Kitagawa, K.; Akita, T.; Moritoki, H., *J. Chem. Soc., Chem. Commun.* **1979**, 971.

18. Rigo, B.; Cauliez, P.; Fasseur, D.; Couturier, D., *Synth. Commun.* **1988**, *18*, 1247.

19. Launikonis, A.; Sasse, W. H. F.; Willing, I. R., *Aust. J. Chem.* **1993**, *46*, 427.

20. Kawai, M.; Onaka, M.; Izumi, Y., *Bull. Chem. Soc. Jpn.* **1988**, *61*, 1237.

21. Hulin, B.; Koreeda, M., *J. Org. Chem.* **1984**, *49*, 207.

22. Freire, R.; León, E. Z.; Salazar, J. A.; Suárez, E., *J. Chem. Soc., Chem. Commun.* **1989**, 452.

23. Olah, G. A.; Wang, Q.; Sandford, G.; Oxyzoglou, A. B.; Prakash, G. K. S., *Synthesis* **1993**, 1077.

24. Olah, G. A.; Ernst, T. D., *J. Org. Chem.* **1989**, *54*, 1203.

25. Olah, G. A.; Ramaiah, P.; Wang, Q.; Prakash, G. S. K., *J. Org. Chem.* **1993**, *58*, 6900.

26. Olah, G. A.; Ernst, T. D., *J. Org. Chem.* **1989**, *54*, 1204.

27. Childs, R. F.; Shaw, G. S.; Varadarajan, A., *Synthesis* **1982**, 198.

28. Marsmann, H. C.; Horn, H. G., *Z. Naturforsch., Tell B* **1972**, *27*, 1448.

29. Demuth, M.; Mikhail, G., *Synthesis* **1982**, 827.

30. Uhlig, W., *J. Organomet. Chem.* **1993**, *452*, 29.

31. Barluenga, J.; González, J. M.; García-Martín, M. A.; Campos, P. J.; Asensio, G., *J. Org. Chem.* **1993**, *58*, 2058.

32. Bullock, R. M.; Rappoli, B. J., *J. Chem. Soc., Chem. Commun.* **1989**, 1447.

33. Olah, G. A.; Wu, A.-H., *Synthesis* **1991**, 407.

34. Olah, G. A.; Wang, Q.; Orlinkov, A.; Ramaiah, P., *J. Org. Chem.* **1993**, *58*, 5017.

35. Stevens, M. F. G.; Chui, W. K.; Castro, M. A., *J. Heterocycl. Chem.* **1993**, *30*, 849.

36. Brown, L.; Koreeda, M., *J. Org. Chem.* **1984**, *49*, 3875.

37. Takeuchi, K.; Ohga, Y.; Munakata, M.; Kitagawa, T.; Kinoshita, T., *Tetrahedron Lett.* **1992**, *33*, 3335.

38. Booth, B. L.; El-Fekky, T. A., *J. Chem. Soc., Perkin Trans. 1* **1979**, 2441.

39. Effenberger, F.; Klenk, H.; Reiter, P. L., *Angew. Chem., Int. Ed. Engl.* **1973**, *12*, 775.

40. Auzanneau, F.-I.; Bundle, D. R., *Can. J. Chem.* **1993**, *71*, 534.

41. Marson, C. M.; Fallah, A., *Tetrahedron Lett.* **1994**, *35*, 293.

42. Saito, S.; Sato, Y.; Ohwada, T.; Shudo, K., *J. Am. Chem. Soc.* **1994**, *116*, 2312.

43. Pearson, W. H.; Fang, W.; Kamp, J. W., *J. Org. Chem.* **1994**, *59*, 2682.

44. Olah, G. A., *Angew. Chem. Int. Ed. Engl.* **1993**, *32*, 767.

45. Olah, G. A.; Klumpp, D. A., *Acc. Chem. Res.* **2004**, *37*, 211.

46. Suzuki, T.; Ohwada, T.; Shudo, K., *J. Am. Chem. Soc.* **1997**, *119*, 6774.

47. Prakash, G. K. S.; Yan, P.; Török, B.; Olah, G. A., *Catal. Lett.* **2003**, *87*, 109; *89*, 159.

48. Rendy, R.; Zhang, Y.; McElrea, A.; Gomez, A.; Klumpp, D. A., *J. Org. Chem.* **2004**, *69*, 2340.

49. Klumpp, D. A.; Garza, M.; Lau, S.; Shick, B.; Kantardjieff, K., *J. Org. Chem.* **1999**, *64*, 7635.

50. Hwang, J. P.; Prakash, G. K. S.; Olah, G. A., *Tetrahedron* **2000**, *56*, 7199.

51. Ohwada, T.; Suzuki, T.; Shudo, K., *J. Am. Chem. Soc.* **1998**, *120*, 4629.

52. Klumpp, D. A.; Garza, M.; Jones, A.; Mendoza, S., *J. Org. Chem.* **1999**, *64*, 6702.

53. Klumpp, D. A.; Garza, M.; Sanchez, G. V., Jr.; Lau, S.; de Leon, S., *J. Org. Chem.* **2000**, *65*, 8997.

54. Klumpp, D. A.; Aguirre, S. L.; Sanchez, G. V., Jr.; de Leon, S., *J. Org. Lett.* **2001**, *3*, 2781.

55. Klumpp, D. A.; Sanchez, G. V., Jr.; Aguirre, S. L.; Zhang, Y.; de Leon, S., *J. Org. Chem.* **2002**, *67*, 5028.

56. Zhang, Y.; McElrea, A.; Sanchez, G. V., Jr.; Do, D.; Gomez, A.; Aguirre, S. L.; Rendy, R.; Klumpp, D. A., *J. Org. Chem.* **2003**, *68*, 5119.

57. Klumpp, D. A.; Rendy, R.; Zhang, Y.; Gomez, A.; McElrea, A., *Org. Lett.* **2004**, *6*, 1789.

58. Anderson, K. W.; Tepe, J. J., *Tetrahedron* **2002**, *58*, 8475.

59. Koltunov, K. Y.; Walspurger, S.; Sommer, J., *Eur. J. Org. Chem.* **2004**, *19*, 4039.

60. Klumpp, D. A.; Rendy, R.; McElrea, A., *Tetrahedron Lett.* **2004**, *45*, 7959.

61. Klumpp, D. A.; Rendy, R.; Zhang, Y.; McElrea, A.; Gomez, A.; Dang, H., *J. Org. Chem.* **2004**, *69*, 8108.

62. Klumpp, D. A.; Kindelin, P. J.; Li, A., *Tetrahedron Lett.* **2005**, *46*, 2931.

63. Zhang, Y.; Briski, J.; Zhang, Y.; Rendy, R.; Klumpp, D. A., *Org. Lett.* **2005**, *7*, 2505.

64. Prakash, G. K. S.; Yan, P.; Török, B.; Olah, G. A., *Synlett* **2003**, 527.

65. Shamma, T.; Buchholz, H.; Prakash, G. K. S.; Olah, G. A., *Isr. J. Chem.* **1999**, *39*, 207.

66. Olah, G. A.; Marinez, E. R.; Prakash, G. K. S., *Synlett* **1999**, 1397.

67. Repichet, S.; Le Roux, C.; Dubac, J., *Tetrahedron Lett.* **1999**, *40*, 9233.

68. Peyronneau, M.; Boisdon, M-T.; Roques, N.; Mazieres, S.; Le Roux, C., *Eur. J. Org. Chem.* **2004**, *22*, 4636.

69. Corey, E. J.; Shibata, T.; Lee, T. W., *J. Am. Chem. Soc.* **2002**, *124*, 3808.

70. Aggarwal, V. K.; Vennall, G. P.; Davey, P. N.; Newman, C., *Tetrahedron Lett.* **1997**, *38*, 2569.

71. Villemin, D.; Bar, N.; Hammadi, M., *Tetrahedron Lett.* **1997**, *38*, 4777.

72. Nakajima, T.; Segi, M.; Mituoka, T.; Fukute, Y.; Honda, M.; Naitou, K., *Tetrahedron Lett.* **1995**, *36*, 1667.

73. Kitamura, T.; Nagata, K.; Taniguchi, H., *Tetrahedron Lett.* **1995**, *36*, 1081.

74. Kitamura, T.; Inoue, D.; Wakimoto, I.; Nakamura, T.; Katsuno, R.; Fujiwara, Y., *Tetrahedron* **2004**, *60*, 8855.

75. Olah, G. A.; Batamack, P.; Deffieux, D.; Torok, B.; Wang, Q.; Molnar, A.; Prakash, G. K. S., *Appl. Catalysis A: General* **1996**, *146*, 107.

76. Nikolopoulos, A. A.; Kogelbauer, A.; Goodwin, J. G., Jr.; Marcelin, G., *J. Catalysis* **1996**, *158*, 76.

77. de Angelis, A.; Flego, C.; Ingallina, P.; Montanari, L.; Clerici, M. G.; Carati, C.; Perego, C., *Catalysis Today* **2001**, *65*, 363.

78. Marziano, N. C.; Ronchin, L.; Tortato, C.; Zingales, A.; Sheikh-Osman, A. A., *J. Molecular Catalysis A: Chemical* **2001**, *174*, 265.

79. Chidambaram, M.; Curulla-Ferre, D.; Singh, A. P.; Anderson, B. G., *J. Catalysis* **2003**, *220*, 442.

80. Landge, S. M.; Chidambaram, M.; Singh, A. P., *J. Molecular Catalysis A: Chemical* **2004**, *213*, 257.

81. Xu, B.-Q.; Sood, D. S.; Gelbaum, L. T.; White, M. G., *J. Catalysis* **1999**, *186*, 345.

82. Xu, B.-Q.; Sood, D. S.; Iretskii, A. V.; White, M. G., *J. Catalysis* **1999**, *187*, 358.

83. Sood, D. S.; Sherman, S. C.; Iretskii, A. V.; Kenvin, J. C.; Schiraldi, D. A.; White, M. G., *J. Catalysis* **2001**, *199*, 149.

84. Al-kinany, M. C.; Jibril, B. Y.; Al-Khowaiter, S. H.; Al-Dosari, M. A.; Al-Megren, H. A.; Al-Zahrani, S. M.; Al-Humaizi, K., *Chem. Eng. & Proc.* **2005**, *44*, 841.

85. Laali, K. K.; Sarca, V. D.; Okazaki, T.; Brock, A.; Der, P., *Org. & Biomol. Chem.* **2005**, *3*, 1034.

86. Huang, J.-W.; Shi, M., *Synlett* **2004**, 2343.

87. Coulombel, L.; Dunach, E., *Green Chem.* **2004**, *6*, 499.

88. Loh, T.-P.; Hu, Q.-Y.; Ma, L.-T., *Org. Lett.* **2002**, *4*, 2389.

89. Ishimaru, K.; Kojima, T., *Tetrahedron Lett.* **2003**, *44*, 5441.

90. Colquhoun, H. M.; Lewis, D. F.; Williams, D. J., *Org. Lett.* **2001**, *3*, 2337.

91. Arisawa, M.; Yamaguchi, M., *Org. Lett.* **2001**, *3*, 763.

92. Dumeunier, R.; Marko, I. E., *Tetrahedron Lett.* **2004**, *45*, 825.

93. Mahoney, J. M.; Smith, C. R.; Johnston, J. N., *J. Am. Chem. Soc.* **2005**, *127*, 1354.

Trifluoromethanesulfonic Anhydride

$$(CF_3SO_2)_2O$$

[358-23-6] $C_2F_6O_5S_2$ (MW 282.16)

InChI = 1S/C2F6O5S2/c3-1(4,5)14(9,10)13-15(11,12)2(6,7)8

InChIKey = WJKHJLXJJJATHN-UHFFFAOYSA-N

(preparation of triflates;[1] mild dehydrating reagent; promoter for coupling reactions in carbohydrates[2])

Alternate Name: triflic anhydride.

Physical Data: bp 81–83 °C/745 mmHg; *d* 1.677 g cm^{-3}; n_D^{20} 1.3210.

Solubility: soluble in dichloromethane; insoluble in hydrocarbons.

Form Supplied in: colorless liquid in ampules. Once opened it should be immediately used.

Analysis of Reagent Purity: IR, NMR.

Preparative Method: by distillation of *Trifluoromethanesulfonic Acid* with an excess of *Phosphorus (V) Oxide.*[1] Purity: by redistillation with a small amount of P_2O_5. It is advisable to freshly distill the reagent from a small quantity of P_2O_5 before use.

Handling, Storage, and Precautions: the pure reagent is a colorless liquid that does not fume in air and is stable for a long period. It is not soluble in water and hydrolyzes only very slowly to triflic acid over several days at room temperature. Preferably stored under N_2 in a stoppered flask. Dangerously exothermic reactions have been reported when attempting to triflate hindered alcohols.[67]

Original Commentary

Antonio García Martínez,

Lakshminarayanapuram R. Subramanian & Michael Hanack

Universität Tübingen, Tübingen, Germany

Reaction with Alcohols and Phenols. The reaction of alcohols and phenols with triflic anhydride (Tf$_2$O) at ∼0 °C in the presence of a base (usually *Pyridine*) in an inert solvent (usually dichloromethane) for 2–24 h affords the corresponding reactive trifluoromethanesulfonate esters (triflates).[1] When triflic anhydride and pyridine are combined, the pyridinium salt forms immediately and normally precipitates out from the reaction mixture. Nevertheless, the salt is an effective esterifying agent, reacting with the added alcohol to give triflates in high yields (eq 1).[3]

(1)

R = alkyl, aryl

Pyridine can become involved in nucleophilic substitution when very reactive triflates are being synthesized.[2,3] One approach to minimize this disadvantage is to replace it with sterically hindered bases, such as 2,6-di-*t*-butyl-4-methylpyridine,[3,4] 2,4,6-trisubstituted pyrimidines,[5] or nonnucleophilic aliphatic

Essential Reactions for Organic Synthesis, First Edition. Edited by Philip L. Fuchs.
© 2016 John Wiley & Sons, Ltd. Published 2016 by John Wiley & Sons, Ltd.

amines (usually *N,N*-diisobutyl-2,4-dimethyl-3-pentylamine). No salt formation appears to take place under these conditions. The triflic anhydride seems to be the direct triflating agent and the base only neutralizes the triflic acid formed. Numerous alkyl triflates have been prepared in the literature[1b] by the above method. Some recent examples of triflates prepared from alcohols are illustrated in eqs 2 and 3.[6,7] As an exception, 2,6-dinitrobenzyl alcohol does not react with Tf$_2$O although similar sulfonyl esters could be prepared.[8]

(2)

(3)

Alkyl triflates have come to be recognized as useful intermediates for the functionalization of organic substrates by nucleophilic substitution, e.g. in carbohydrate chemistry.[9] Triflate is the best leaving group known[1b] next to the nonaflate and hence a large number of triflates, obtained in good yields by reaction of the corresponding alcohols (or alkoxides) with Tf$_2$O, have been used to generate unstable or destabilized carbocations under solvolytic conditions.[1b] Some new typical examples are shown in [1–4].[10–13]

Alkyl triflates are known to be powerful reagents for the alkylation of aromatic compounds.[1b,14] However, the reaction of alkyl triflates with heterocycles affords *N*-alkylation products.[15]

In an improved modification of the Ritter reaction, primary and secondary alcohols react with Tf$_2$O in CH$_2$Cl$_2$ in the presence of a 2:1 excess of nitriles to give the corresponding amides in good yields (eq 4).[16]

(4)

Aryl triflates are prepared from phenols at 0 °C using pyridine as solvent.[1b] Sometimes it is useful to conduct the reaction in CH$_2$Cl$_2$ at −77 °C, as in the preparation of 3,5-di-*t*-butyl-4-hydroxyphenyl triflate (eq 5).[17] Aryl triflates are synthetically transformed into several products of interest and applications in organic chemistry, by cross-coupling reactions with organometallics (eqs 5 and 6).[18]

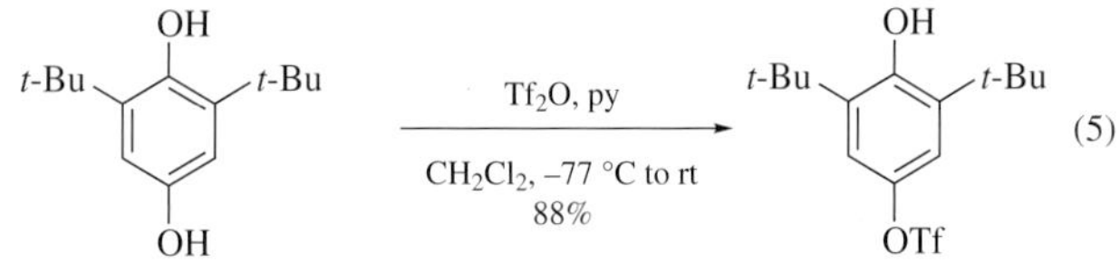

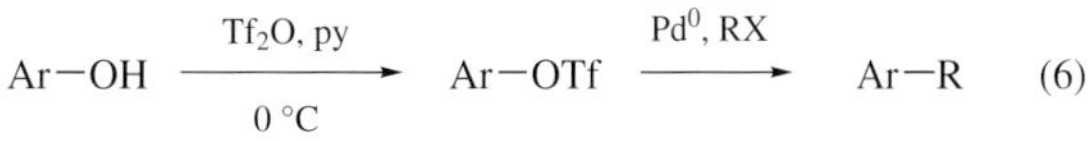

$$ \text{(5)} $$

$$ Ar{-}OH \xrightarrow[0\,°C]{Tf_2O,\ py} Ar{-}OTf \xrightarrow{Pd^0,\ RX} Ar{-}R \quad (6) $$

Reaction of Tf$_2$O with Amines. The reaction of 1 equiv of Tf$_2$O in CH$_2$Cl$_2$ and Et$_3$N with amines (or their salts) affords trifluoromethanesulfonamides (triflamides) in good yields.[19,20] If 2 equiv of Tf$_2$O are used, triflimides are formed. The triflamides are soluble in alkali and readily alkylated to triflimides (eq 7).[19,20]

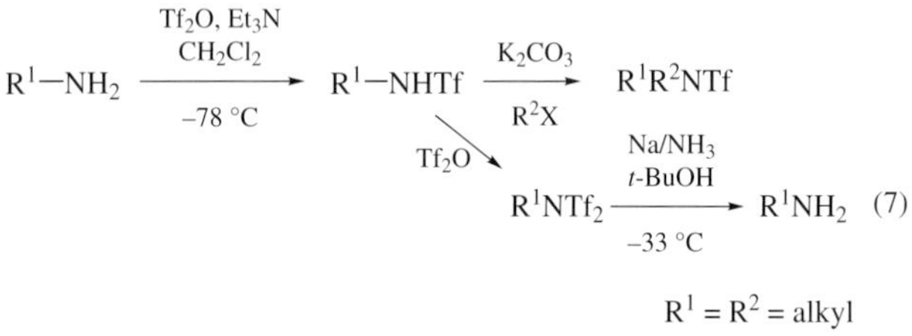

$$ \text{(7)} $$

$$ R^1 = R^2 = alkyl $$

Triflamides can be deprotected reductively (**Sodium–Ammonia**) to yield the corresponding amines.[21] This protocol has been employed in the facile two-step synthesis of aza macrocycles starting from trifluoromethanesulfonyl derivatives of linear tetramines (eq 8).[22]

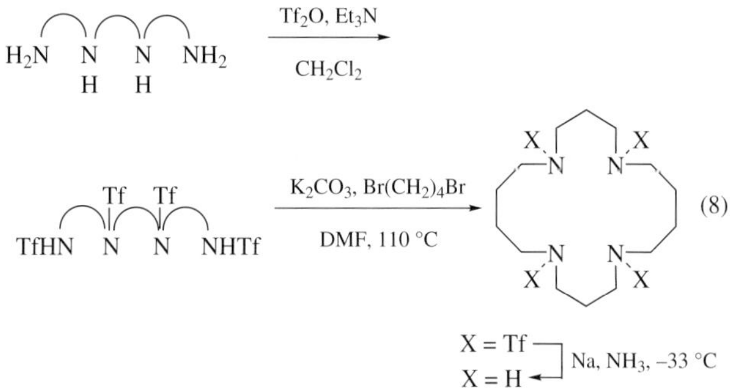

$$ \text{(8)} $$

$$ X = Tf $$
$$ X = H \quad \xleftarrow{Na,\ NH_3,\ -33\,°C} $$

Several triflamides (**5–8**)[23] and *O*-triflylammonium salts[24] have been used for the formation of vinyl triflates from regiospecifically generated metalloenolates or for preparing triflates from alcohols.

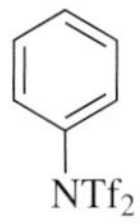

(5)

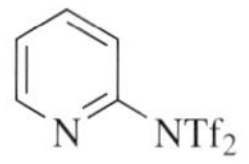

(6)

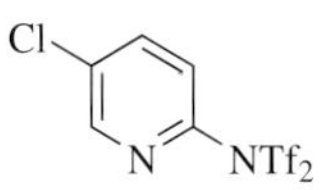

(7)

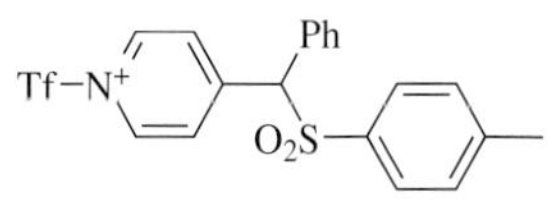

(8)

Reaction of Tf$_2$O with Carbonyl Compounds. The reaction of Tf$_2$O with carbonyl compounds consists of the electrophilic attack of the anhydride on the carboxylic oxygen, resulting in the formation of triflyloxycarbenium ions as intermediates (eq 9). According to the nature of the carbonyl compound, the triflyloxy-carbenium cations can eliminate a proton giving a vinyl triflate, undergo a rearrangement, or be trapped by the gegenion yielding *gem*-bistriflates (eq 9).

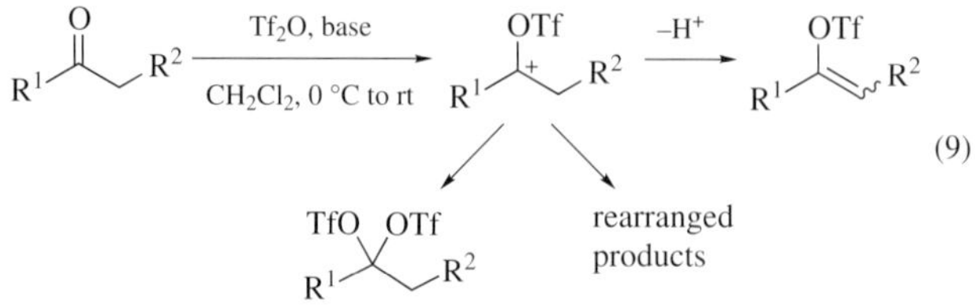

$$ \text{(9)} $$

In the case of acyclic and monocyclic ketones, the reaction with Tf$_2$O affords vinyl triflates in good yields. Several methods exist to realize this reaction.[1b,25] For example, the reaction is carried out at room temperature in CH$_2$Cl$_2$ (or pentane) in the presence of 2,4-di-*t*-butyl-4-methylpyridine (DTBMP) (eq 10).[4]

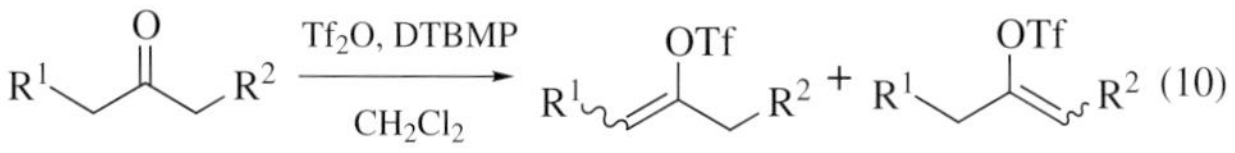

$$ \text{(10)} $$

Other bases such as pyridine,[1b] lutidine,[1b] Et$_3$N,[1b] polymer-bound 2,6-di-*t*-butyl-4-methylpyridine,[26] and 2,4,6-trialkyl-substituted pyrimidines[27] were also used. The commercially available *N,N*-diisobutyl-2,4-dimethyl-3-pentylamine is a very convenient base to prepare the vinyl triflates.[28] In the case of nonfunctionalized ketones, anhydrous Na$_2$CO$_3$ has been proved to be very successful.[1b,25]

The reaction of ketones with Tf$_2$O is governed by Markovnikov's rule and results in the formation of the more substituted triflate as the major product. When the reaction of ketones[27] and α-halo ketones[29] with Tf$_2$O is carried out in the presence of a nitrile, the intermediate trifloxy cation (eq 9) can be trapped, forming pyrimidines in good yields (eq 11).

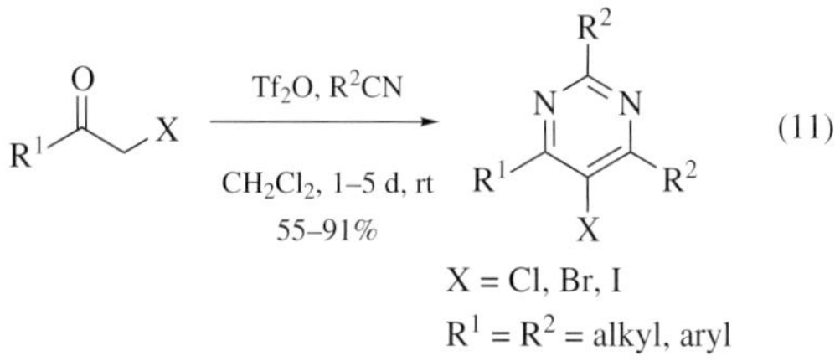

$$ \text{(11)} $$

$$ X = Cl, Br, I $$
$$ R^1 = R^2 = alkyl, aryl $$

The reaction of Tf$_2$O with strained bicyclic ketones such as 2-norbornanone and nopinone takes place with Wagner–Meerwein rearrangement of the corresponding triflyloxy cations, forming bridgehead triflates in good yields (eq 12).[30] These triflates are key compounds in the preparation of other bridgehead derivatives by substitution[31] and of substituted cyclopentanes by fragmentation.[32]

In the reaction of Tf_2O with norcaranones and spiro[2.5]octan-4-one, the cyclopropane ring undergoes fragmentation to give vinyl triflates (eqs 13 and 14).[33]

However, a cyclopropane ring is formed in the reaction of 5-methylnorborn-5-en-2-one with Tf_2O under the same conditions (eq 15).[34]

When the ketone can accomplish neither the stereoelectronic conditions for the elimination of TfOH nor for a rearrangement, the reaction of ketones with Tf_2O results in the formation of a *gem*-bistriflate (eqs 16 and 17).[35]

Sensitive ketones such as 3-pentyn-2-one also afford the corresponding vinyl triflate on treatment with Tf_2O in the presence of a base (eq 18).[36]

Substituted cyclopropenones and tropones react with Tf_2O with the formation of the corresponding dication ether salts (eq 19).[37]

Treatment of trifluoroacetyl ylides with Tf_2O results in the formation of *gem*-bistriflates (eq 20).[38]

Reaction of Tf_2O with Aldehydes. The reaction of aliphatic aldehydes with Tf_2O in the presence of 2,6-di-*t*-butyl-4-methylpyridine (DTBMP) in refluxing CH_2Cl_2 or $ClCH_2CH_2Cl$ for 2 h affords the corresponding vinyl triflates as a mixture of (*Z*)- and (*E*)-isomers.[4,39] When the reaction is carried out at 0 °C, *gem*-bistriflates are formed as products (eq 21).[40] The *gem*-bistriflates result due to the trapping of the intermediate triflyloxycarbenium ion by the triflate anion. Primary vinyl triflates have been used extensively in the generation of alkylidene carbenes,[41] and *gem*-bistriflates are interesting precursors for *gem*-dihaloalkanes[42,43] and (*E*)-iodoalkenes.[44]

Reaction with Dicarbonyl Compounds. 1,3-Diketones can be reacted with an equimolar amount of Tf_2O or in excess to furnish the corresponding vinyl triflates or dienyl triflates (eq 22).[45] These triflates are transferred into monoketones, monoalcohols, alkanes, and unsaturated ketones by means of various reducing reagents.[45]

The reaction of 3-methylcyclopentane-1,2-dione with Tf$_2$O/Et$_3$N affords the vinyl triflate in 53% yield (eq 23).[46] The reaction takes place probably through the enol form. The product was coupled with alkenylzinc compounds in the presence of a palladium catalyst.[46]

$$\text{(23)}$$

The reaction of β-keto esters[47] with Tf$_2$O in the presence of a base results in the formation of 2-carboxyvinyl triflates (eq 24). These substrates undergo nucleophilic substitution of the TfO-group (eq 24)[47] and also coupling reactions.[48]

$$\text{(24)}$$

PNB = p-nitrobenzyl carbamate

Reaction with Carboxylic Acids and Esters. The reaction of carboxylic acids and esters with Tf$_2$O takes place according to the scheme shown in eq 25.[49]

$$R^1-CO_2Tf + TfOR^2(H) \quad \text{(25)}$$

R^1 = alkyl, aryl; R^2 = alkyl, H

The trifluoromethanesulfonic carboxylic anhydrides are highly effective acylation agents, which react without catalysts even with deactivated aromatics to yield aryl ketones (eq 26).[50]

$$\text{(26)}$$

Alkyl arylacetates react with Tf$_2$O to give a cation which in the presence of a nitrile affords isoquinoline derivatives via cyclization of the intermediate nitrilium cation (eq 27).[50]

$$\text{(27)}$$

R^1 = H, 6-Me, 7–Cl, 5–NO$_2$, 6,7-(OMe)$_2$; R^2 = Et, Me

Reaction of Tf$_2$O with Amides. The reaction of a 2-oxo-1,2-dihydroquinoline with Tf$_2$O in the presence of pyridine affords the corresponding 2-quinoline triflate (eq 28).[51]

$$\text{(28)}$$

The reaction of tertiary amides with Tf$_2$O gives a mixture of O-sulfonylated (major) and N-sulfonylated (minor) products. In the presence of collidine and an alkene, [2 + 2] cycloadducts are formed which hydrolyze to give cyclobutanones (eq 29).[52]

$$\text{(29)}$$

R^1, R^2 = H, Me, Ph

Treatment of DMF with Tf$_2$O results in the formation of an imminium triflate, which formylates less active aromatics. It is a convenient variation of the Vilsmeier–Haack reaction (eq 30).[53]

$$\text{(30)}$$

The reaction of N-methylpyridone and substituted urea systems with Tf$_2$O gives heteroatom-stabilized dicarbonium salts (eqs 31 and 32).[37,54]

$$\text{2OTf}^- \quad \text{(31)}$$

$$\text{2OTf}^- \quad \text{(32)}$$

Secondary amides can be converted to tetrazoles with Tf_2O in the presence of **Sodium-Azide** (eq 33).[55]

$$R^1 \overset{O}{\underset{H}{\overset{\|}{C}}} \overset{R^2}{N} \xrightarrow[\substack{20\ °C \\ 0-72\%}]{\substack{Tf_2O,\ NaN_3 \\ CH_2Cl_2\ or\ MeCN}} \overset{N-N}{\underset{N-N}{\|}} R^1 \quad (33)$$

R^1 = alkyl, Ph; R^2 = alkyl, $(CH_2)_2OAc$, $(CH_2)_2OTBDMS$

Other Applications. Activated arenes can be converted to aryl triflones by Friedel–Crafts reaction with Tf_2O using **Aluminum Chloride** as catalyst (eq 34).[56]

$$Ar-H \xrightarrow[\substack{18\ h,\ rt \\ 10-73\%}]{Tf_2O,\ AlCl_3} ArSO_2CF_3 \quad (34)$$

The reaction of Tf_2O with Ph_3PO in CH_2Cl_2 at $0\,°C$ affords triphenylphosphine ditriflate, which can be used as an oxygen activator, and then to a diphosphonium salt (eq 35).[57]

$$Ph_3\overset{+}{P}-O^- \xrightarrow[0\,°C]{Tf_2O} Ph_3\overset{+}{P}-OTf\ OTf^- \xrightarrow{Ph_3PO}$$

$$Ph_3\overset{+}{P}-O-\overset{+}{P}Ph_3\ 2OTf^- \quad (35)$$

The less stable dimethyl sulfide ditriflate, obtained from Tf_2O and DMSO, has been used to oxidize alcohols (eq 36).[58]

$$Me_2S=O \xrightarrow[-78\,°C]{Tf_2O,\ CH_2Cl_2} Me_2\overset{+}{S}-OTf\ OTf^- \xrightarrow{\substack{R^1 \\ R^2}\overset{|}{C}H-OH}$$

$$\overset{R^1}{\underset{R^2}{C}}=O \quad (36)$$

Tetrahydropyran is not a suitable solvent in reactions involving Tf_2O because it is cleaved, affording 1,5-bistrifloxypentane (eq 37).[59]

$$\underset{O}{\bigcirc} \xrightarrow[20\,°C]{Tf_2O} TfO \diagdown\diagup\diagdown\diagup OTf \quad (37)$$

Diols react with Tf_2O to yield the corresponding ditriflates; however, the reaction of 1,1,2,2-tetraphenyl-1,2-ethanediol with Tf_2O takes place with rearrangement (eq 38).[59]

$$\underset{HO}{\overset{Ph}{\underset{|}{\overset{|}{Ph}}}}\overset{Ph}{\underset{|}{\underset{OH}{\overset{|}{Ph}}}} \xrightarrow[20\,°C]{Tf_2O,\ py} \underset{Ph}{\overset{Ph}{\underset{|}{Ph}}}\overset{O}{\underset{Ph}{\overset{\|}{C}}} \quad (38)$$

Vinylene 1,2-bistriflates are formed by the reaction of azobenzils with Tf_2O (eq 39).[60]

$$\underset{O}{\overset{Ph}{\underset{\|}{}}}\overset{Ph}{\underset{N_2}{}} \xrightarrow{Tf_2O} \underset{TfO}{\overset{Ph}{=}}\overset{Ph}{\underset{OTf}{}} + \underset{TfO}{\overset{Ph}{=}}\overset{OTf}{\underset{Ph}{}} \quad (39)$$

$$83\%3\%$$

The reaction of enolates, prepared from silyl enol ethers and , with Tf_2O affords vinyl triflates (eq 40).[61]

$$\underset{}{\overset{}{\bigtriangleup}}=O \xrightarrow[DMF,\ Et_3N]{TMSCl} \overset{}{\bigtriangleup}-OTMS \xrightarrow[2.\ Tf_2O]{1.\ MeLi}$$

$$\overset{}{\bigtriangleup}-OTf \quad (40)$$

The combination of equimolecular quantities of **Iodosylbenzene** and Tf_2O generates $PhI(OTf)_2$, a compound also formed by treatment of Zefiro's reagent with Tf_2O. As shown in eq 41, this compound can be used to prepare *para*-disubstituted benzene derivatives in good yields.[62]

$$PhIO \xrightarrow{Tf_2O} PhI(OTf)_2 \xrightarrow{ArH}$$

$$\underset{TfO}{\overset{Ph}{\underset{|}{I}}}\diagdown\diagup\underset{OTf}{\overset{Ar}{\underset{|}{I}}} \xrightarrow{PPh_3}$$

$$Ph_3\overset{+}{P}-\diagdown\diagup-\overset{+}{P}Ph_3\ 2OTf^- \quad (41)$$

Tf_2O is a suitable promoter for the stereoselective glucosidation of glycosyl acceptors using sulfoxides as donors.[63]

The reaction of Tf_2O with a catalytic amount of **Antimony(V)Fluoride** at $25\,°C$ produces trifluoromethyl triflate in 94% yield (eq 42).[64]

$$(CF_3SO_2)_2O \xrightarrow[80\%]{SbF_5} CF_3OSO_2CF_3 \quad (42)$$

Useful application of Tf_2O as dehydrating reagent is accounted by the synthesis of isocyanides from formamides and vinylformamides (eq 43).[65]

$$\underset{O\ NHCHO}{\overset{O}{\diagup}}\text{''STol} \xrightarrow[86\%]{\substack{Tf_2O,\ i\text{-}Pr_2NEt \\ CH_2Cl_2,\ -78\,°C}} \underset{O\ NC}{\overset{O}{\diagup}}\text{''STol} \quad (43)$$

Reaction of enaminones with Tf_2O in a 1:1 molar ratio affords 3-trifloxypropeniminium triflates by *O*-sulfonylation. From a cyclic enaminone, by using a 2:1 molar ratio, the corresponding bis(3-amino-2-propenylio) bistriflate is obtained (eq 44).[66]

$$\underset{BuHN}{\overset{}{\bigcirc}}O \xrightarrow[81\%]{Tf_2O} \underset{BuHN}{\overset{}{\bigcirc}}\overset{OTf^-}{\underset{+}{}} \quad (44)$$

First Update

Spencer J. Williams
University of Melbourne, Parkville, Victoria, Australia

Reactions with Alcohols, Phenols, and Thiols. Chemical transformations effected by triflic anhydride have been comprehensively reviewed.[68] Formation of alkyl triflates from alcohols is typically performed using triflic anhydride (Tf$_2$O) and pyridine or hindered pyridines such as 2,6-di-*tert*-butyl-4-methylpyridine (DTBMP) according to the methods already outlined. Purification of the resultant triflate from the excess base can sometimes prove problematic, and polymeric pyridine equivalents such as poly(4-vinylpyridine) or poly(2,6-di-*tert*-butylpyridine) have been successfully used allowing the removal of excess base and its triflate salt by simple filtration.[69] Alternatively, phenolic triflates can be prepared from Tf$_2$O under Schotten–Baumann conditions using toluene and 30% aqueous K$_3$PO$_3$.[70] Work-up is effected by simple separation of the organic phase and evaporation of the solvent. When attempting to triflate alcohols of low reactivity with Tf$_2$O, alternative redox processes can occur resulting in the formation of sulfinates as the major products. Such undesired redox processes are favored at higher temperatures and particularly when using triethylamine as base (eq 45).[67]

$$85:15$$
$$70\%$$

Methyl substituted pyridines such as 2,6-lutidine and 2,4,6-collidine can also afford undesirable by-products. These pyridines can react with Tf$_2$O to generate sulfinate- and trifluoromethyl-substituted derivatives (eq 46).[71] In such cases, the use of 2,6-di-*tert*-butyl-substituted pyridines is recommended.

Triflates formed from hept-1-enitols spontaneously cyclize with debenzylation to afford vinyl C-glycosides (eq 47).[72,73]

$$90\%$$

Dimethyl substituted homoallylic alcohols undergo smooth cyclization to *trans*-cyclopropanes upon treatment with Tf$_2$O and a base (eq 48).[74,75] Various functionalities can be introduced adjacent to the newly formed cyclopropane ring by altering the quenching conditions.[74]

$$87\%$$

Eleven-membered carbocycles are formed by way of a five-carbon ring expansion in a homo-Cope rearrangement upon treatment of β-(hydroxymethyl)allylsilanes with Tf$_2$O (eq 49).[76]

$$75\%$$

Reaction with Amines. Reaction of Tf$_2$O with hexamethyldisilazane affords the reactive silylating reagent *N,N*-bis(trimethylsilyl)trifluoromethylsulfonamide.[77,78] This compound exists in two tautomeric forms in solution, related by trimethylsilyl shifts between oxygen and nitrogen (eq 50).[77]

$$85\%$$

Reaction with Carbonyl and Thiocarbonyl Compounds. Treatment of ketones with α-protons with Tf$_2$O and magnesium bromide or Grignard reagents results in α-bromination (eq 51).[79]

$$89\%$$

An alternative to the Arndt–Eistert reaction for the homologation of carboxylic acids uses *N*-(acylmethyl)benzotriazoles prepared from acyl chlorides and *N*-(trimethylsilyl)-methylbenzotriazole. Upon treatment with Tf$_2$O these compounds are converted to the corresponding vinyl triflate, and then to the homologous ester by sequential treatment with sodium methoxide in acetonitrile and then HCl in methanol (eq 52).[80]

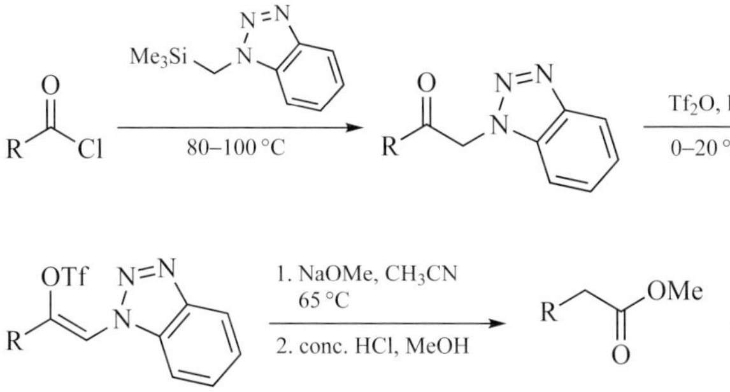

$$(52)$$

50–70% overall

Tf$_2$O reacts with thioketones to generate triflylthiocarbenium ions that undergo Wagner–Meerwein rearrangement to bridgehead thiotriflates (eq 53).[81]

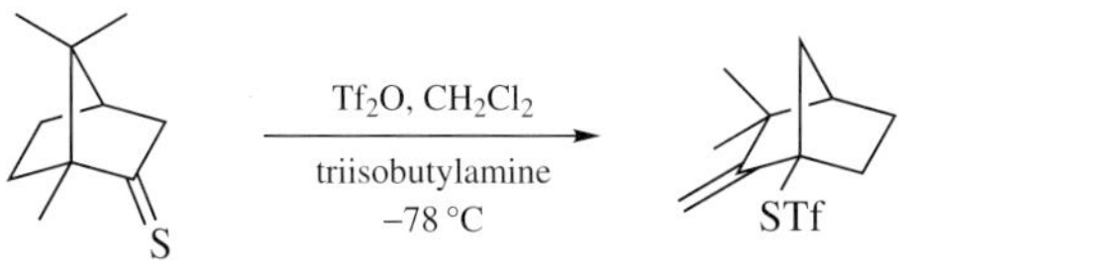

$$(53)$$

79%

Ketones react with methylthiocyanate in the presence of Tf$_2$O to afford 2,4-bis(methylthio)pyrimidines (eq 54).[82] Methylthio substituted pyrimidines are useful synthetic intermediates as the methylthio groups can be substituted by other functionalities such as alkoxy and amino groups. Substitution occurs faster at the 4-position, allowing the selective introduction of two different nucleophiles.[68,82]

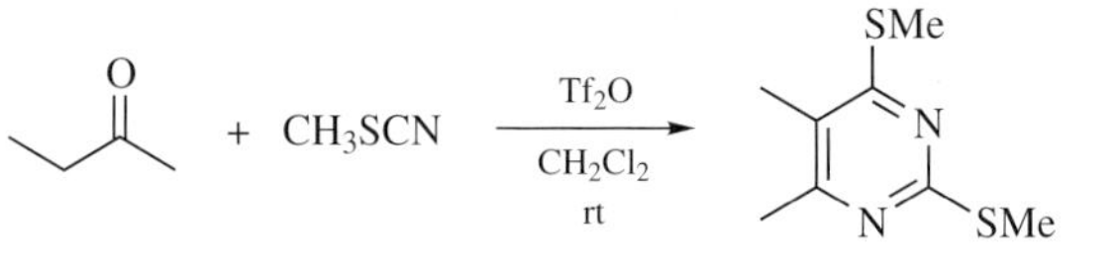

$$(54)$$

75%

When the reaction is carried out with an ester and a nitrile, good yields of 4-alkoxypyrimidines are obtained (eq 55).[83]

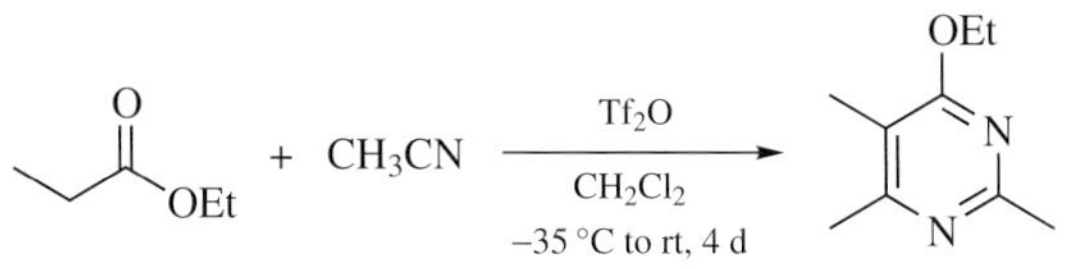

$$(55)$$

75%

Ketones and nitriles when treated with Tf$_2$O react to give pyrimidines as outlined earlier (eq 11). In the case of benzyl nitriles bearing electron-donating substitutents, naphthalene derivatives are obtained instead (eq 56).[84]

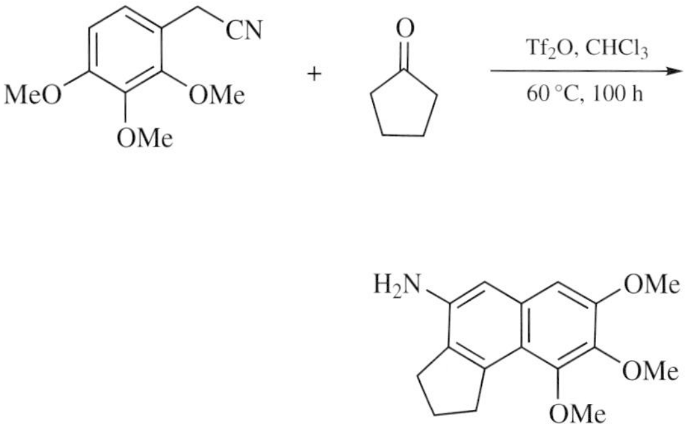

$$(56)$$

23%

Reaction with Amides. It is well known that treatment of secondary or tertiary amides with Tf$_2$O affords imino and iminium triflates, respectively (eq 57).

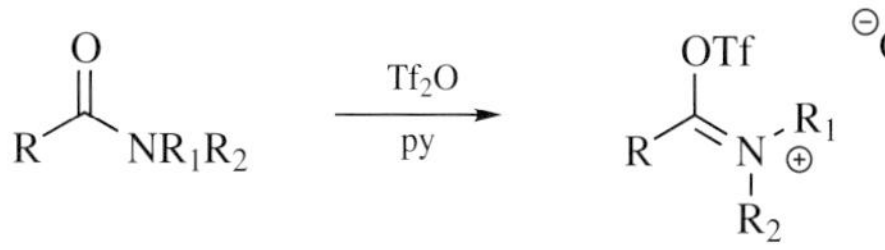

$$(57)$$

These salts can react with a variety of nucleophiles including amines (to afford amidines),[85–87] alcohols (to afford esters),[85,88] hydrogen sulfide (to afford thioamides),[89] and isotopically labeled water (to afford ^{18}O-labeled amides).[89] Iminium triflates are also excellent intermediates for the preparation of assorted heterocycles. Iminium triflates react with azides (to afford tetrazoles),[55] β-mercaptoamines (to afford thiazolines),[90] and triols (to afford orthoesters) (eq 58).[91] The reaction of lactams with Tf$_2$O and sodium azide affords tetrazolo-fused bicyclics.[92]

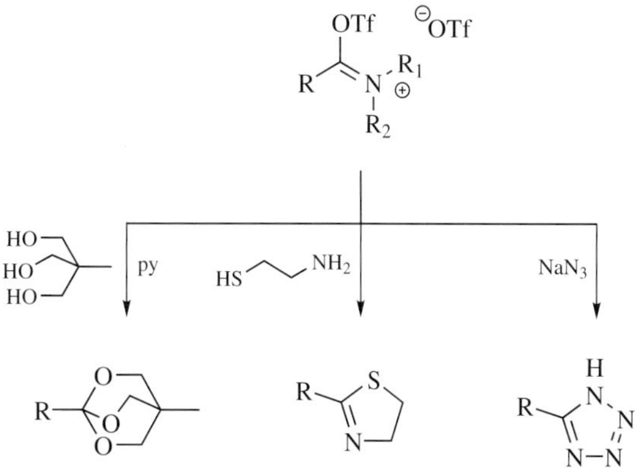

$$(58)$$

Treatment of unsaturated amides with Tf$_2$O generates intermediate iminium triflates that cyclize under radical conditions (eq 59).[93]

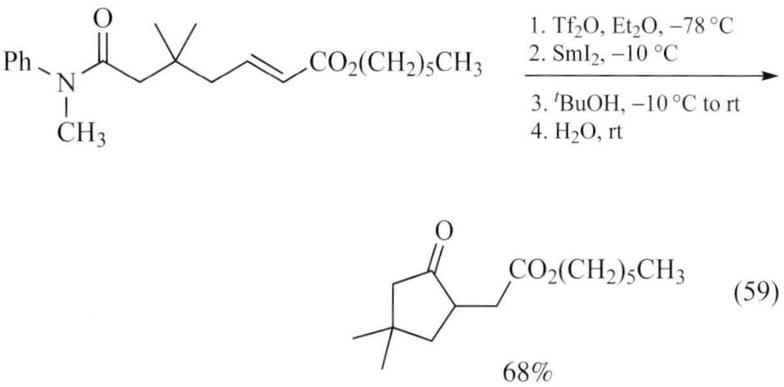

$$(59)$$

68%

Treatment of bis-hydrazides with Tf$_2$O and pyridine provides a mild method for the formation of 1,3,4-oxadiazoles (eq 60).[94]

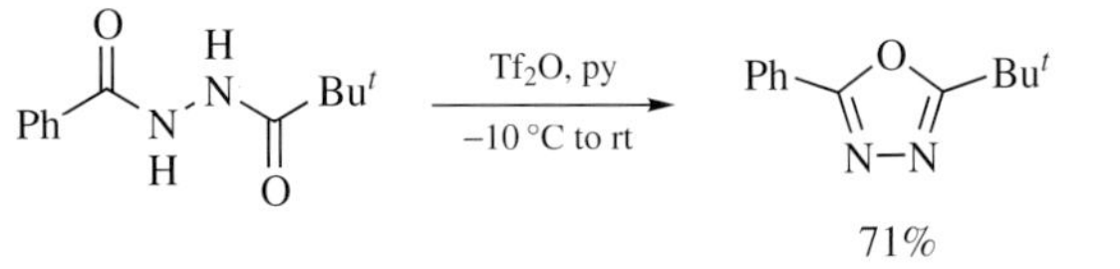

$$\text{(60)}$$

71%

Carbinolamides (derived from keto amides) afford 2-sulfonamido-substituted furans when treated with Tf$_2$O and pyridine (eq 61).[95]

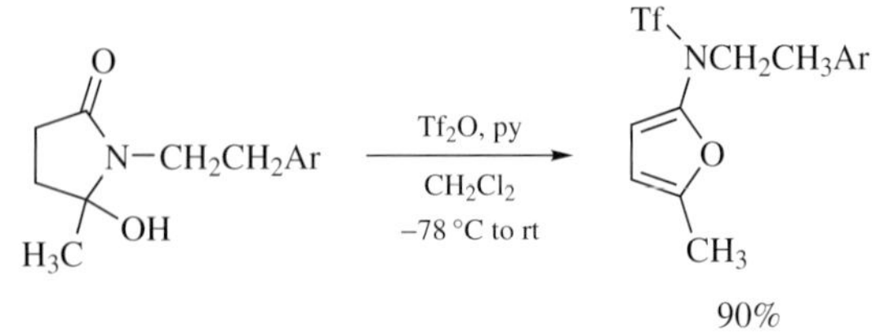

$$\text{(61)}$$

90%

Indolin-2-ones can participate in Vilsmeier reactions with reactive aromatics under the agency of Tf$_2$O. In the reaction with indoles and benzofurans, Tf$_2$O was found to be the reagent of choice; no reaction took place when phosphorous oxychloride was used (eq 62).[96]

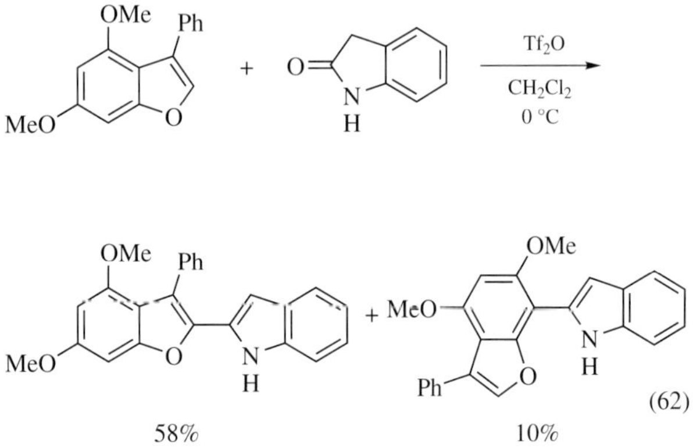

$$\text{(62)}$$

58%   10%

Tf$_2$O and a base can effect dehydration of primary amides to nitriles.[97] This procedure is particularly mild, with no epimerization observed at the stereogenic α-carbon of several substrates (eq 63).[97]

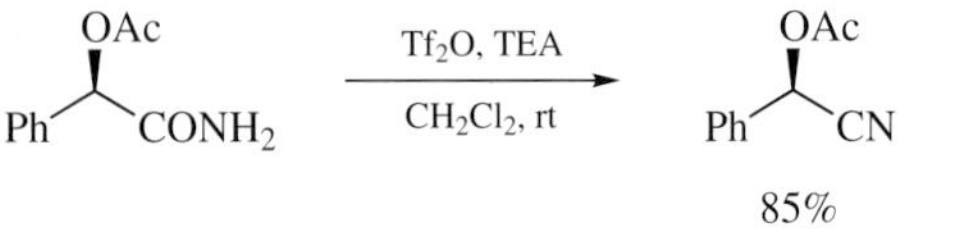

$$\text{(63)}$$

85%

Oxidations. Tf$_2$O reacts with sulfides affording a trifluoromethylsulfonylsulfonium triflate, which can be used to oxidize alcohols (eq 64).[98]

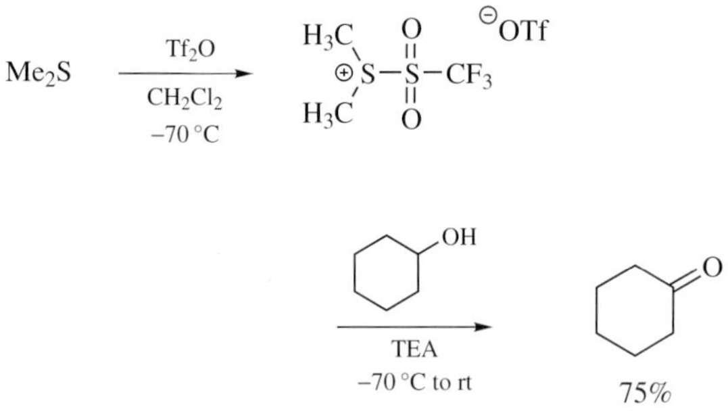

$$\text{(64)}$$

75%

Alternatively, the intermediate sulfonium salts prepared in the same manner can be treated with aqueous sodium acetate to afford the sulfoxide; none of the corresponding sulfones were observed (eq 65).[98]

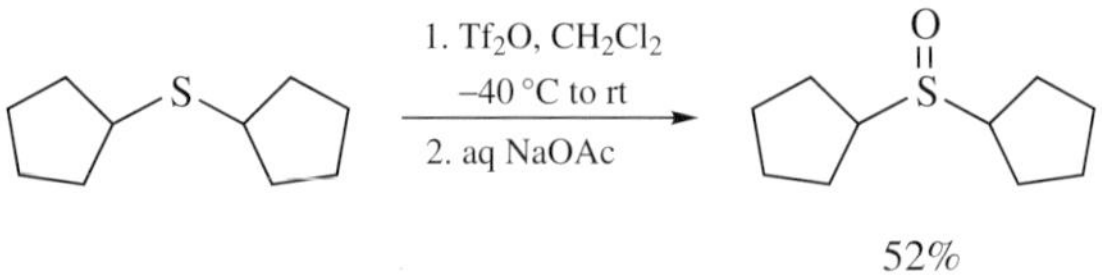

$$\text{(65)}$$

52%

Tf$_2$O reacts with Grignard reagents to afford the self-coupled alkanes (eq 66).[99]

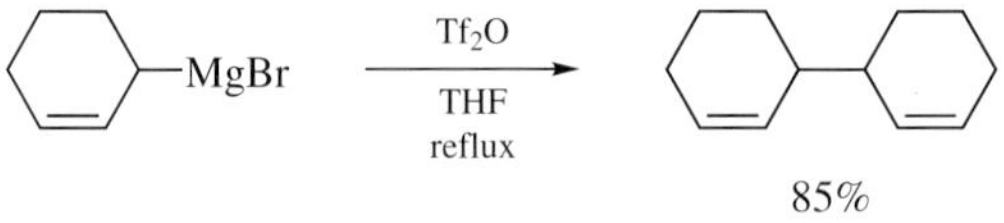

$$\text{(66)}$$

85%

Addition of Tf$_2$O to 85% hydrogen peroxide affords the powerful oxidant trifluoromethanepersulfonic acid. This oxidant will convert unreactive thioethers to the corresponding sulfones (eq 67).[100]

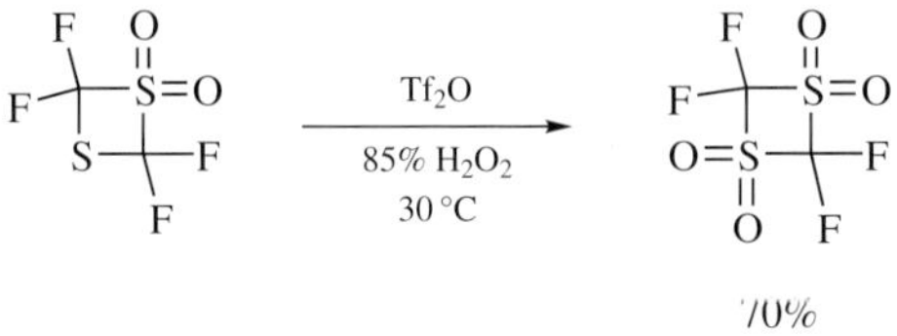

$$\text{(67)}$$

70%

Owing to the electrophilicity of the divalent sulfur of a thiosulfinate, thiols do not react with Tf$_2$O to afford thiosulfonates; disulfides are obtained instead (eq 68).[101] In the presence of a base, only diphenyl disulfide is observed.[101]

$$\text{PhSH} + \text{Tf}_2\text{O} \xrightarrow[\text{rt}]{\text{CH}_2\text{Cl}_2} \text{PhSSPh} + \text{CF}_3\text{SSPh} \quad \text{(68)}$$

Glycosylations and Acetalations. The combination of Tf$_2$O and a hindered pyridine base such as 2,6-di-*tert*-butylpyridine promotes glycosylation of oxygen and nitrogen nucleophiles by thioglycoside sulfoxides.[102] This reaction has a broad scope, enabling the glycosylation of unreactive nucleophiles under mild conditions, and has been extended to solid phase glycosylations.[103] The anomeric stereochemistry of the product is determined by the nature of the protecting groups ('P'), with esters affording 1,2-*trans* products and ethers affording an anomeric stereochemistry that is sensitive to the reaction conditions (eq 69).[104]

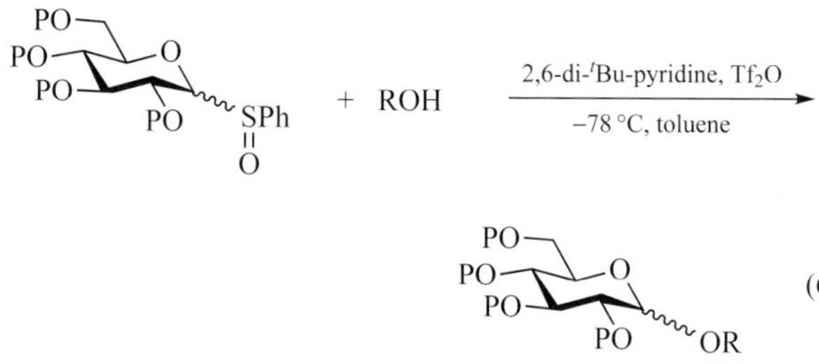

(69)

R = steroids, phenols, carbohydrates

An important modification of the original sulfoxide glycosylation uses 4,6-*O*-benzylidene-protected thioglycoside sulfoxides and a modified protocol for the addition of reagents, requiring the addition of the nucleophile after activation of the sulfoxide with Tf$_2$O.[105] This approach is effective in the production of the challenging β-manno 1,2-*cis*-glycosidic linkage, and there is strong evidence that the reaction proceeds via an intermediate glycosyl triflate (eq 70).[106,107]

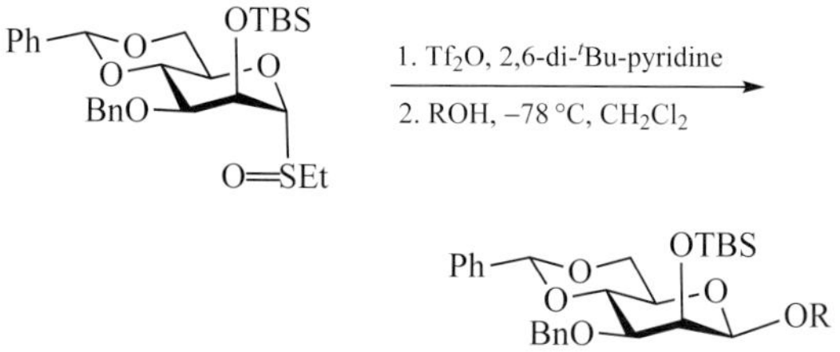

(70)

In some cases the addition of electrophilic alkenes, such as 4-allyl-1,2-dimethoxybenzene, to the reaction mixture can improve the outcome of the reaction, particularly when run in the presence of thioglycosides, by scavenging a by-product of the reaction, phenylsulfenyl triflate.[108] Non-carbohydrate thioacetal sulfoxides also undergo similar acetalation reactions when treated with Tf$_2$O (eq 71).[109]

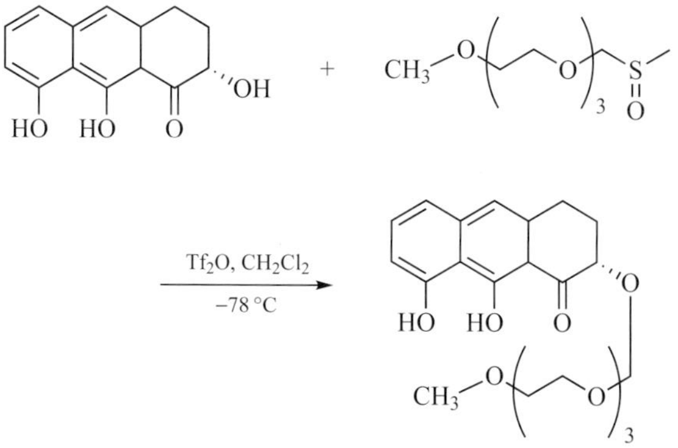

(71)

30%

Thioglycosides can be activated towards glycosylation using a variety of reagents in combination with Tf$_2$O. In particular, the reagent combination of 1-benzenesulfinylpiperidine (BSP) (**9**), Tf$_2$O, and a hindered base, such as 2,6-di-*tert*-butylpyridine or 2,4,6-tri-*tert*-butylpyrimidine (TTBP) (**10**), is very effective.[110] The combination of BSP and TTBP is especially noteworthy as these reagents are crystalline and shelf-stable and may be stored premixed. The combination of diphenyl sulfoxide, Tf$_2$O, and TTBP has been shown to be a more powerful promoter than BSP/TTBP/Tf$_2$O for the activation of thioglycosides.[111] Again, for all of these reagent combinations, the use of 4,6-*O*-benzylidene

acetals or 4,6-*O*-phenylboronates[112] enables the stereoselective formation of β-mannosides. The method has been used with moderate success on solid phase.[112] In the case of rhamnosides, where it is not possible to use a 4,6-*O*-benzylidene acetal, judicious choice of a 2-*O*-(3-trifluoromethylphenylsulfonyl) protecting group allows the preparation of the corresponding β-rhamnosides.[113]

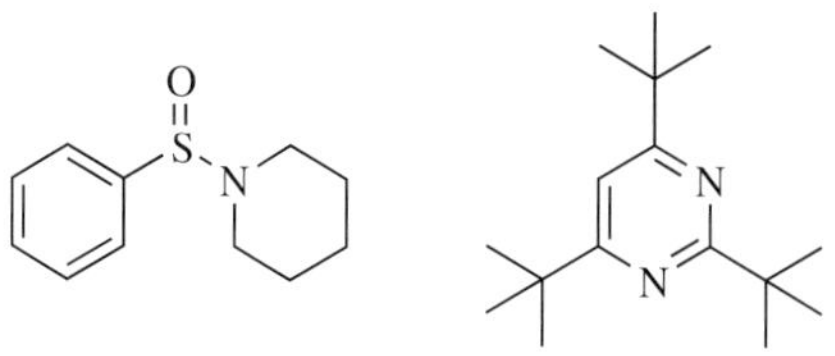

(**9**)   (**10**)

Tf$_2$O in combination with sulfoxides lacking α-protons, such as diphenylsulfoxide or dibenzothiophene sulfoxide, generates a sulfoxonium triflate, which is capable of effecting the formation of glycosides from carbohydrate hemiacetals and oxygen nucleophiles.[114] This is a dehydrative glycosylation, and appears to proceed by way of a sulfoxonium intermediate (eq 72).[115]

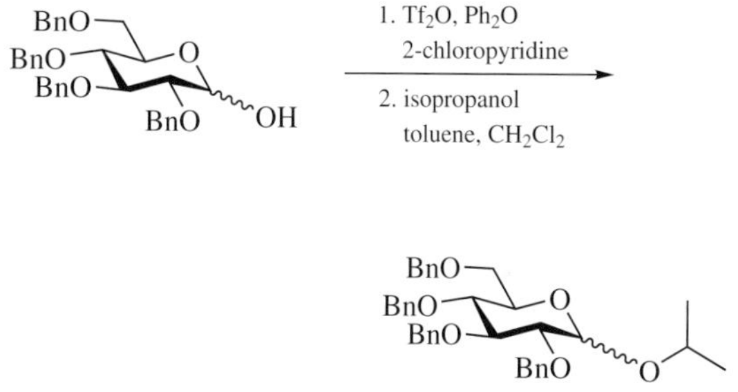

(72)

89%

Carbohydrate glycals undergo oxidative glycosylation when treated first with Tf$_2$O and diphenyl sulfoxide (or tributylphosphine oxide)[116] and then with a nucleophile in the presence of a Lewis acid.[117] This reaction proceeds via oxygen transfer to generate a 1,2-anhydrosugar intermediate which is opened by nucleophiles in the presence of a Lewis acid, affording 2-hydroxy β-glucosides (eq 73). If a more hindered sulfoxide (dibenzothiophene sulfoxide) is used, the reaction occurs with the complimentary stereochemistry to give 2-hydroxy α-mannosides.[118]

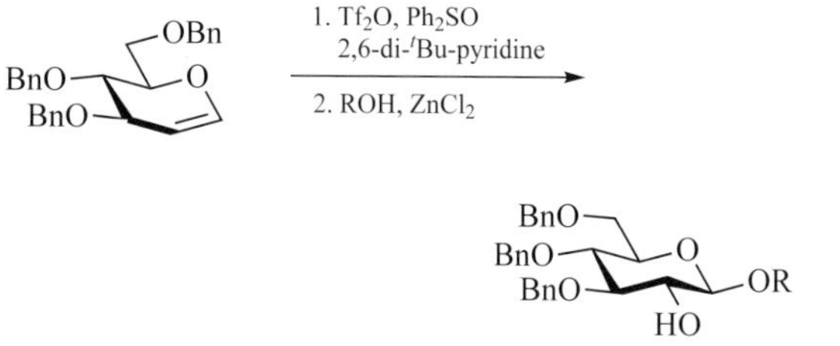

(73)

A variant of this reaction is termed as the C2-acetamidoglycosylation. Treatment of glycals with Tf$_2$O and thianthrene-5-oxide, followed by trimethylsilylacetamide and a base, and then finally with an alcohol nucleophile and an acid catalyst affords 2-acetamido glycosides (eq 74).[119]

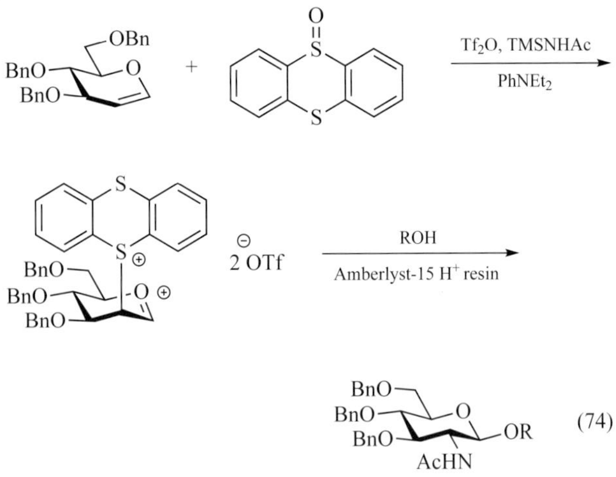

(74)

R = carbohydrates, amino acids, and steroids

Tf$_2$O acts as a promoter in glycosylations of alcohols by 2′-carboxylbenzyl glycosides (eq 75).[120] This reaction most likely proceeds via the mixed anhydride.

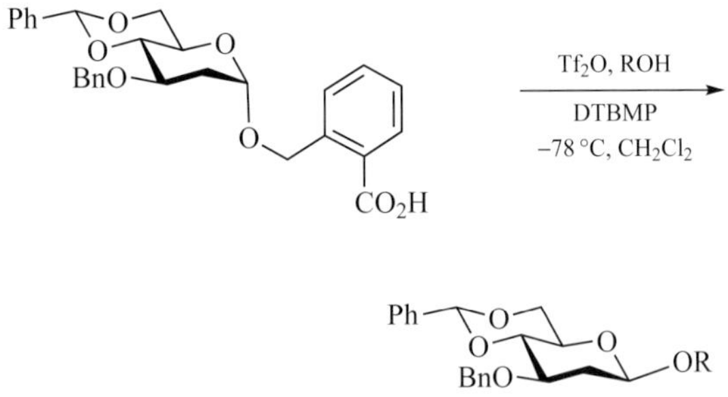

(75)

Tf$_2$O has been used as a promoter in glycosylations of carbohydrate alcohols by glycosyl fluorides.[121]

Aminals. Allylic and propargylic aminals when treated with Tf$_2$O afford iminium salts, which react with Grignard reagents to afford allylic and propargylic amines (eq 76).[122]

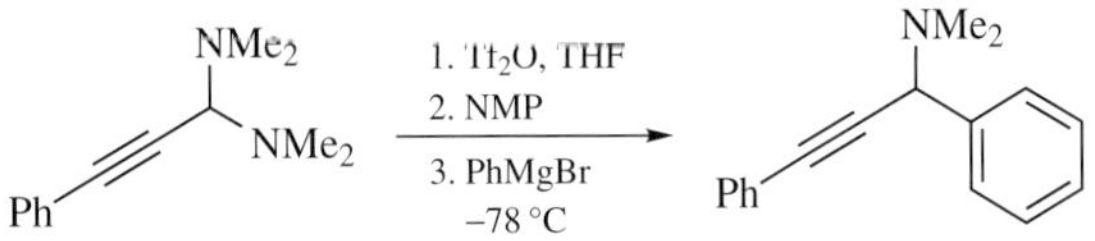

(76)

67%

1,1-Difluorovinyl methyl ethers and aminal bis(carbamates) or bis(sulfonamides) react in the presence of Tf$_2$O to generate β-amino-α,α-difluoroketones (eq 77).[123]

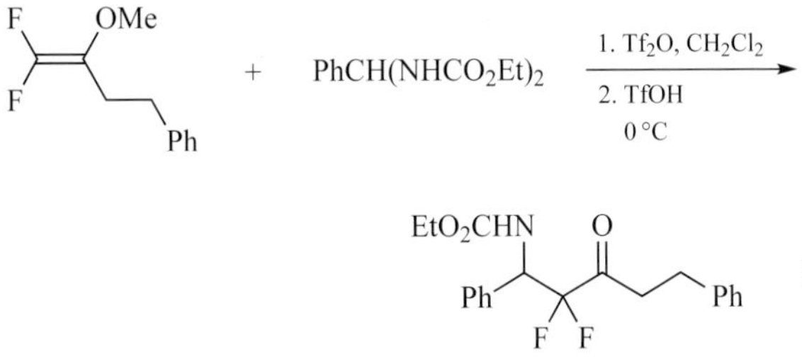

(77)

84%

An azaallenium salt has been prepared by treatment of a bis(iminomethane) compound with Tf$_2$O (eq 78).[124]

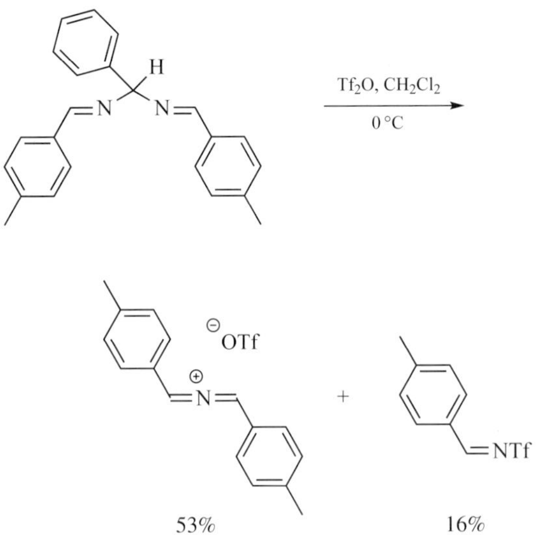

(78)

53%   16%

Other Applications. O-Silylated hydroxamic acids when treated with Tf$_2$O and a base yield nitrile oxides.[125] In the presence of olefins, these undergo [3 + 2] dipolar cycloadditions affording isoxazolines (eq 79).[125]

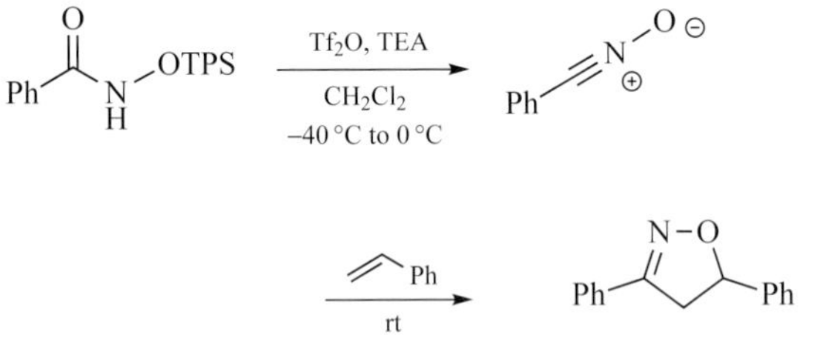

(79)

88%

Triphenylphosphine oxide can be converted to the corresponding unsubstituted phosphinimines by repeated sequential treatment with Tf$_2$O and then ammonia (eq 80).[126]

$$\text{Ph}_3\text{P=O} \quad \xrightarrow[\text{2. NH}_3]{\text{1. Tf}_2\text{O, CH}_2\text{Cl}_2} \quad \text{Ph}_3\text{P=NH}$$

repeat seven times   82%   (80)

The combination of tetraalkylammonium nitrates and Tf$_2$O generates nitronium triflate, which acts as a convenient anhydrous nitrating reagent.[127,128] This reagent system is effective for nitration of homo and heteroaromatic systems (eq 81),[127] and for N-nitration of saturated nitrogen heterocycles (eq 82).[128]

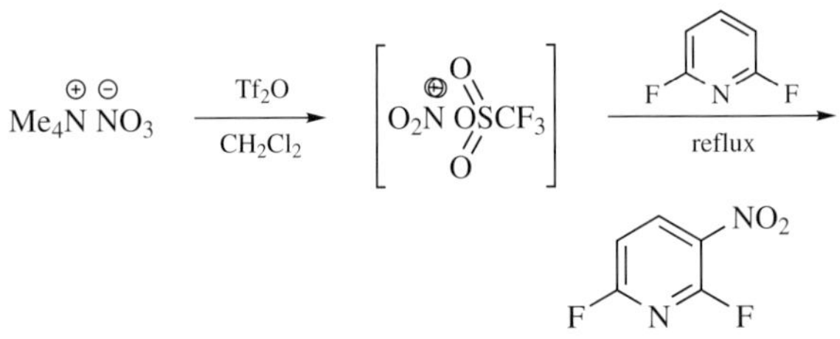

(81)

98%

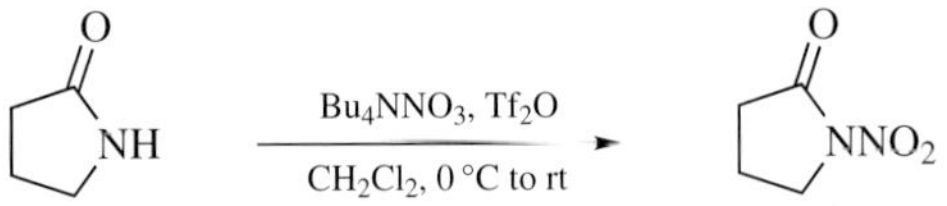

(82)

56%

Treatment of sulfoxide-substituted electron-rich heterocycles with Tf$_2$O promotes substitution at sulfur, resulting in the formation of large-ring N,S-heterocycles (eq 83).[129]

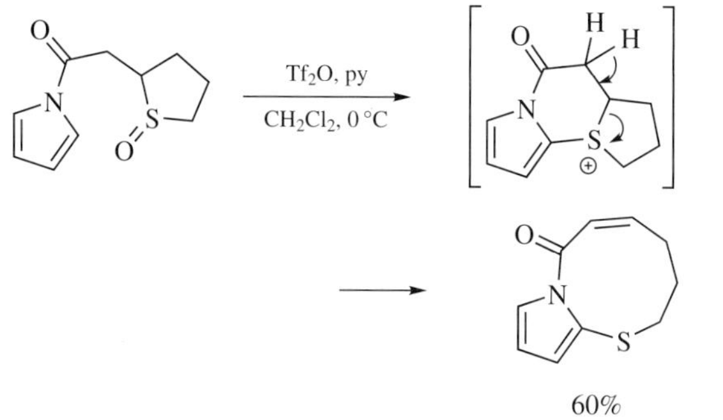

(83)

60%

Dimethyl sulfide ditriflate, obtained from Tf$_2$O and DMSO, has been used to generate o-quinone methides (eq 84).[130]

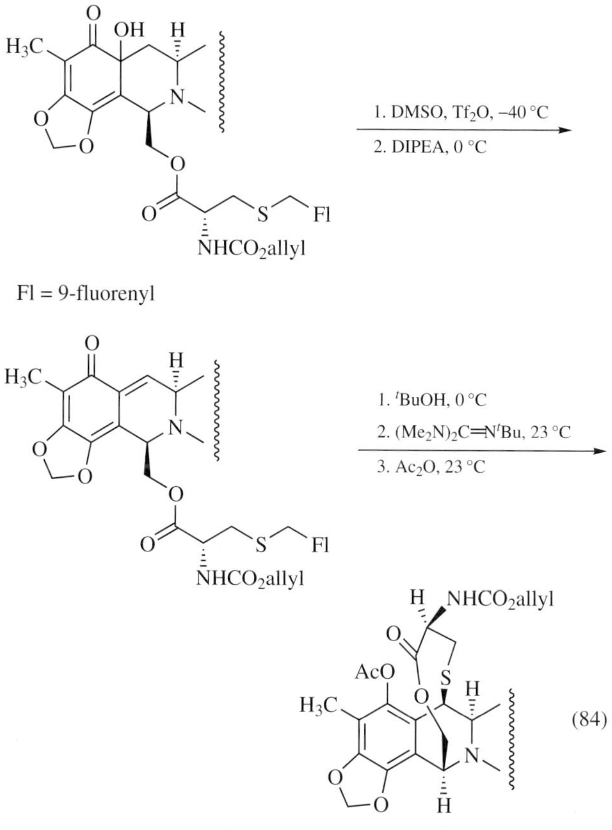

Fl = 9-fluorenyl

(84)

Bis- and tris-chalcogen monoxides when treated with Tf$_2$O afford di- or tri-chalcogona dications, respectively; these salts can be considered 2-center, 2-electron or 3-center, 4-electron bonded species (eq 85).[131,132] These compounds are susceptible to nucleophilic attack at carbon or at the chalcogen and can act as alkylating agents under very mild conditions.[132]

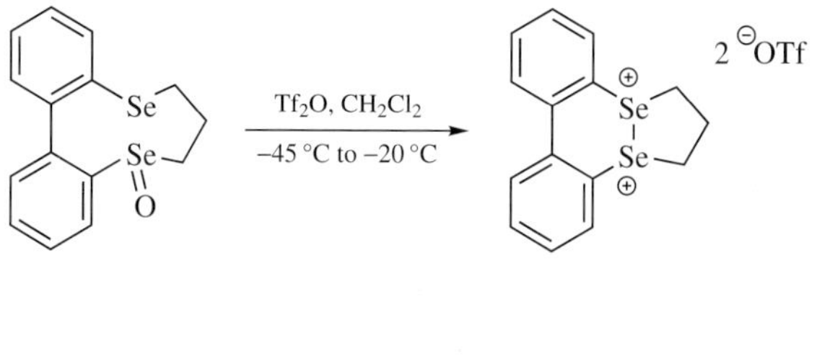

(85)

A one-pot procedure for the synthesis of thioamides relies on the successive reaction of Grignard reagents with carbon disulfide and then with Tf$_2$O to generate a mixed anhydride intermediate. This intermediate reacts with amines to generate thioamides (eq 86).[133]

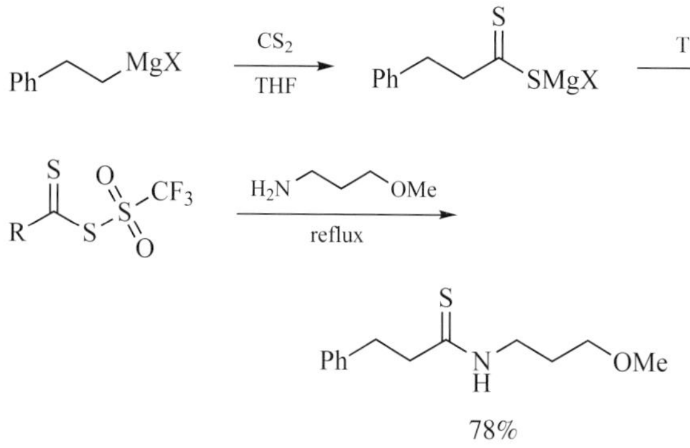

(86)

78%

Tf$_2$O reacts with dimethylcyanamide to afford an O,N-ditriflylisourea.[134] Reaction of this isourea with alcohols, thiols, or amines affords isoureas, thioisoureas, or guanidines, respectively (eq 87).[134]

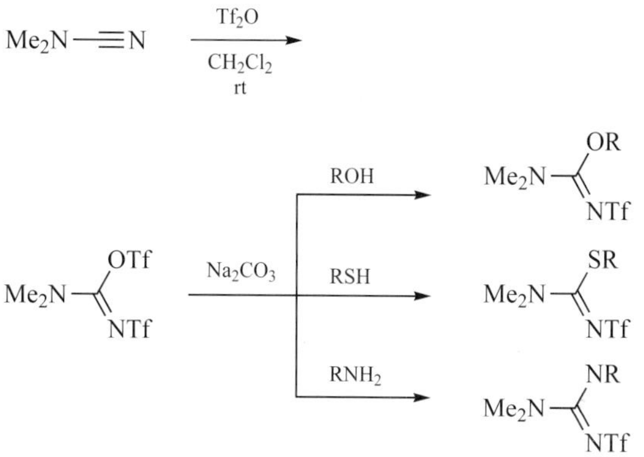

(87)

Trimethylsilylalkynes can be converted to alkynyl(phenyl) iodonium triflates by treatment with PhI(OAc)$_2$ and Tf$_2$O (eq 88).[135] Alkynyl(phenyl)iodonium triflates are useful 'electrophilic acetylene' equivalents and can act as Michael acceptors, 1,3-dipolariphiles and alkynyl cation equivalents.[136]

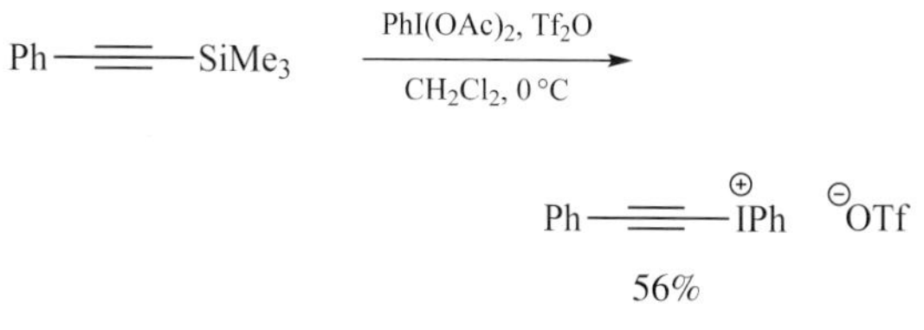

(88)

56%

The reaction of Tf$_2$O with phenylseleninic anhydride and diphenyl diselenide directly affords phenylselenenyl triflate (eq 89).[137] This procedure is a convenient alternative to the preparation of phenylselenenyl triflate from light sensitive AgOTf and moisture sensitive PhSeCl.

$$(PhSeO)_2O + Ph_2Se_2 \xrightarrow[CH_2Cl_2]{Tf_2O} PhSeOTf \qquad (89)$$

Electrophilic trifluoromethylation reagents can be generated from the treatment of trifluoromethyl biphenyl sulfoxides with Tf$_2$O.[138,139] These reagents will react with a wide range of nucleophiles including carbanions, aromatics, silyl enol ethers, thiolates, and phosphines (eq 90).[139]

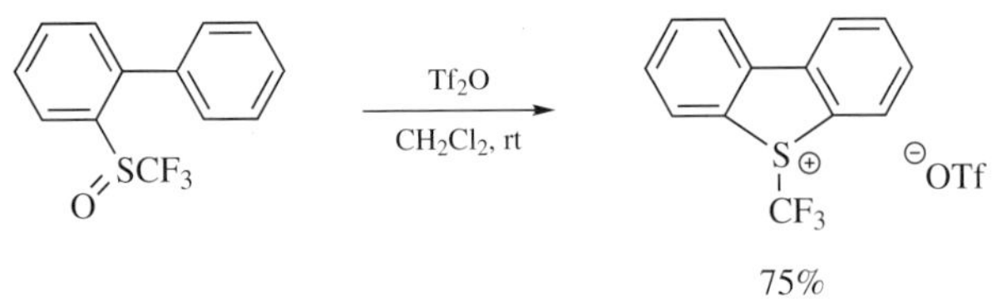

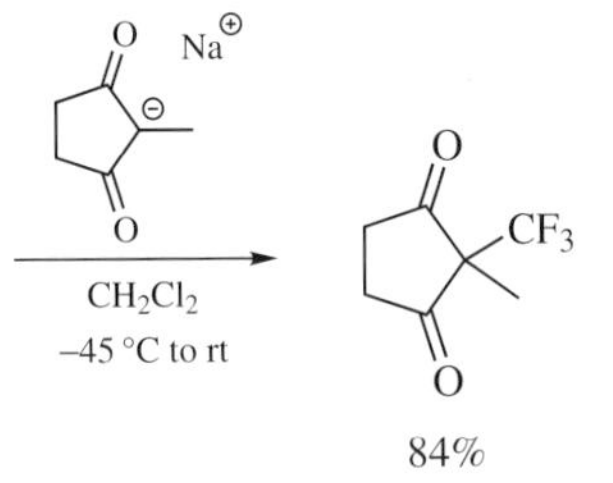

75%

(93)

92%
>19:1 dr

(90)

84%

Second Update

Sophie Régnier
Université de Montréal, Montréal, Québec, Canada

Reaction with Alcohols and Phenols. Reaction of alcohols with triflic anhydride in the presence of a base affords the corresponding alkyl triflate, a versatile intermediate that can be used to generate an allylic carbocation in a [4+3] cycloaddition (91).[140]

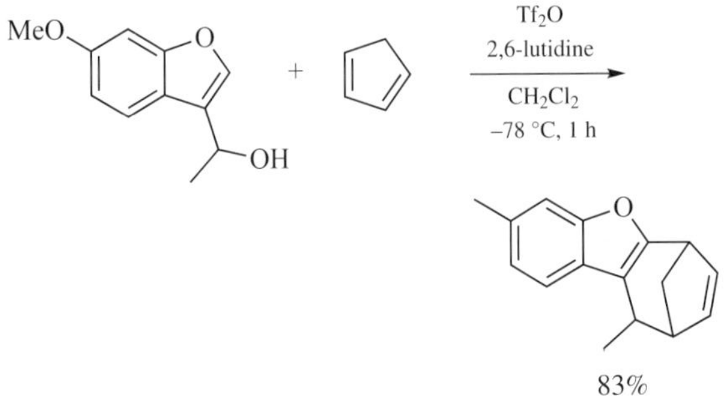

83%

Triflates are also known to be excellent leaving groups and, therefore, can participate in substitution reactions to generate saturated heterocycles, such as azetidines[141,142] and aziridines (92).[143]

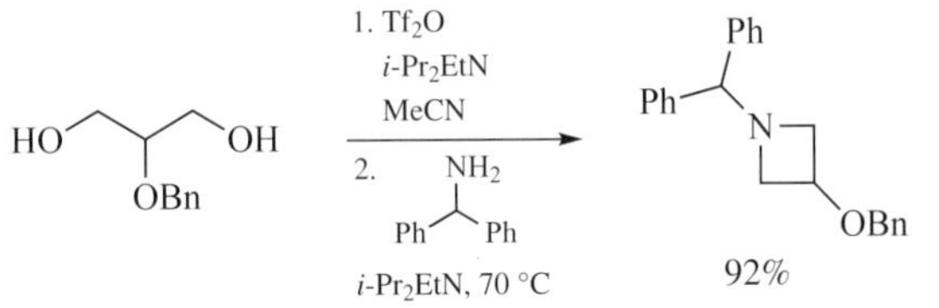

(92)

92%

Alkyl triflates can also be used as efficient leaving groups for intramolecular cyclization in the process of generating heterocycles, such as thioindenes,[144] cyclic enones,[145,146] *N*-heterocyclic carbene precursors,[147,148] or tetrahydropyridines via a Grob fragmentation process (93).[149] They can also be reduced to the corresponding alkane in the presence of a copper catalyst.[150]

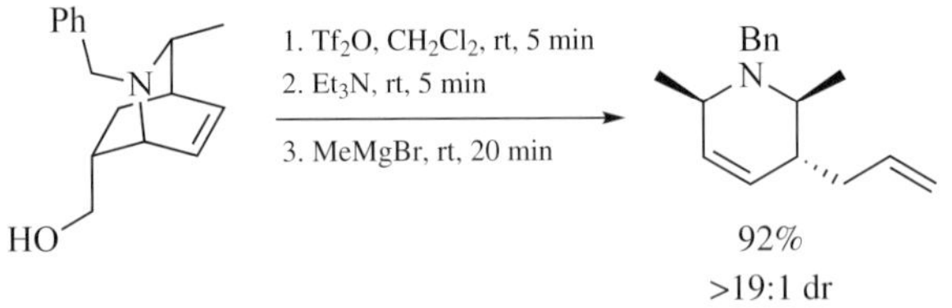

The reaction of *o*-trimethylsilylphenoxytrimethylsilane with excess butyllithium followed by treatment with triflic anhydride affords *o*-trimethylsilylphenyl triflate,[151] which, in the presence of a fluoride source, can generate the benzyne and engage in various reactions.

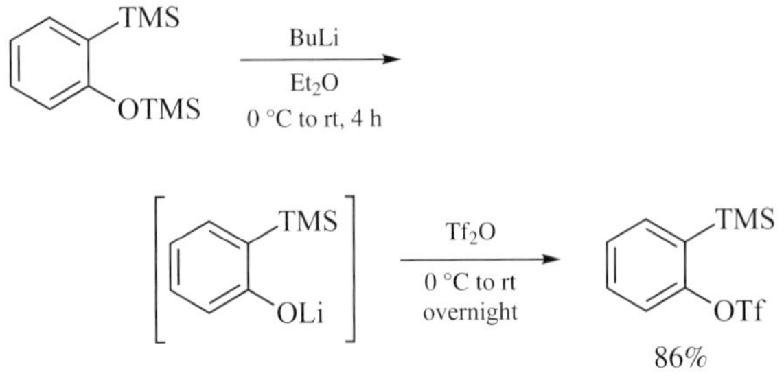

(94)

86%

Reaction with Amides. Iminium triflates, readily obtained by the reaction of amides with triflic anhydride, can be reduced to the corresponding amine in the presence of a reducing agent, such as Hantzsch ester (95)[152] or sodium borohydride.[153]

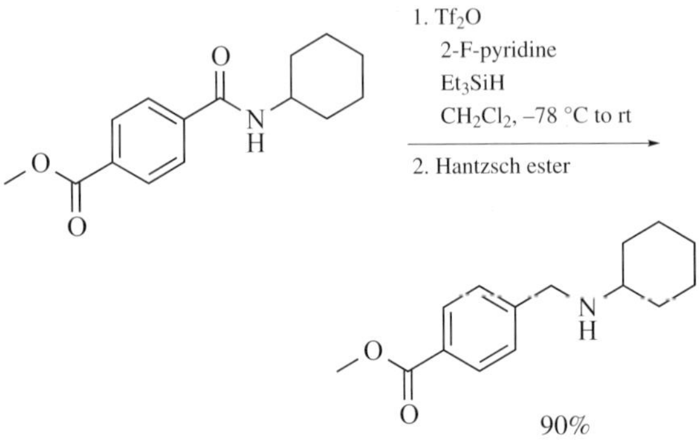

(95)

90%

Iminium triflates can also undergo nucleophilic addition with a broad spectrum of nucleophiles. For example, the activated amide can react with malonates in a Knoevenagel-type condensation reaction (96).[154] Other nucleophiles include, but are not limited to, Grignard reagents to form ketones[155,156] or alkylamines,[157–159] phosphites,[160] amino acids,[161] pyridines,[162,163] and pyridine oxydes[164,165] or alcohols to afford lactones via intramolecular cyclization (97).[166]

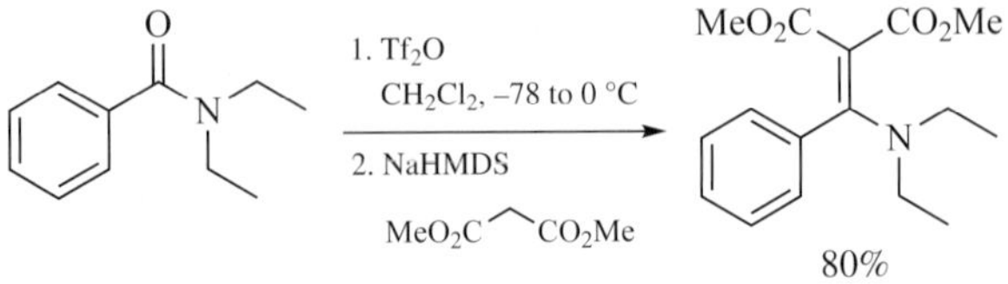

(96)

80%

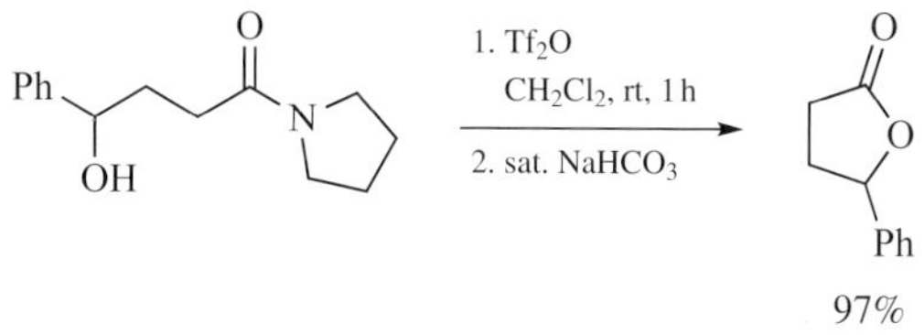

(97)

97%

Iminium triflates can also react with diazonium compounds to form new C–C bonds.[167] Condensation with nitriles or hydrazides will lead to the formation of various heterocycles, such as substituted pyridines,[168,169] pyrimidines,[170,171] triazoles,[172] and substituted indazoles.[173]

Reaction of amides with triflic anhydride can also lead to interesting spirocyclic moieties through a Bischler–Napieralski-type cyclization (98).[174]

Reaction of amides with diphenyl sulfoxide in the presence of triflic anhydride affords the α-arylation product (99).[175]

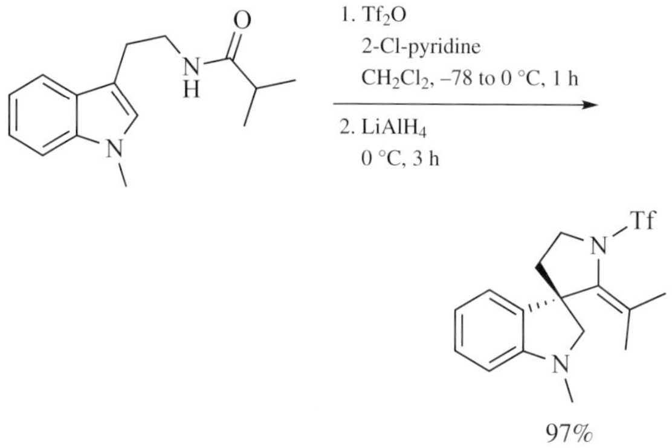

(98)

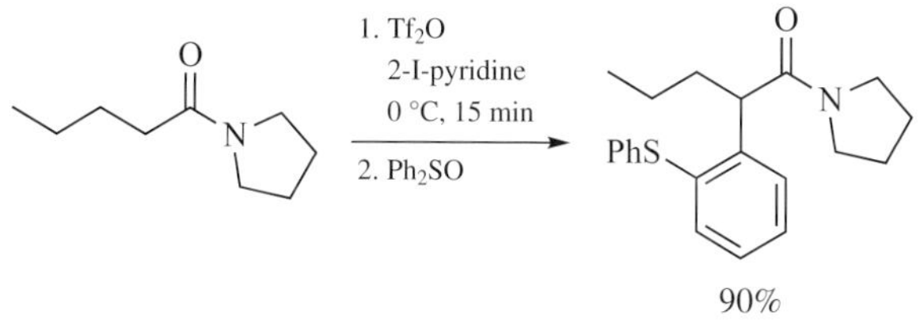

97%

(99)

90%

Amino-substituted naphthalenes can be obtained by reacting phenyl vinyl acetamides with triflic anhydride in the presence of a base via the ketene-iminium intermediate (100).[176]

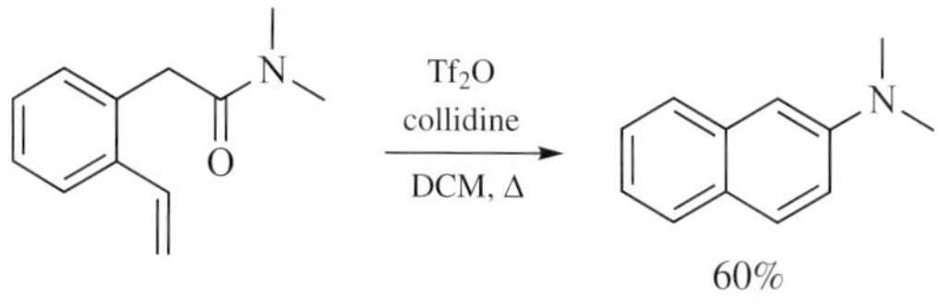

(100)

60%

Reaction with Sulfoxides and Phosphine Oxides. Sulfoxides and phosphine oxides can be reduced to the corresponding sulfide[177,178] and phosphine[179] in the presence of triflic anhydride and a mild reducing agent.

Sulfinimines can be obtained by activating the corresponding sulfoxide with triflic anhydride, followed by a nitrile attack and hydrolysis.[180,181]

Sulfoxide derivatives can be activated by triflic anhydride and participate in a Pummerer-type rearrangement to afford benzofurans (101),[182] benzothiophenes,[183,184] or spirocyclic indolenines.[185]

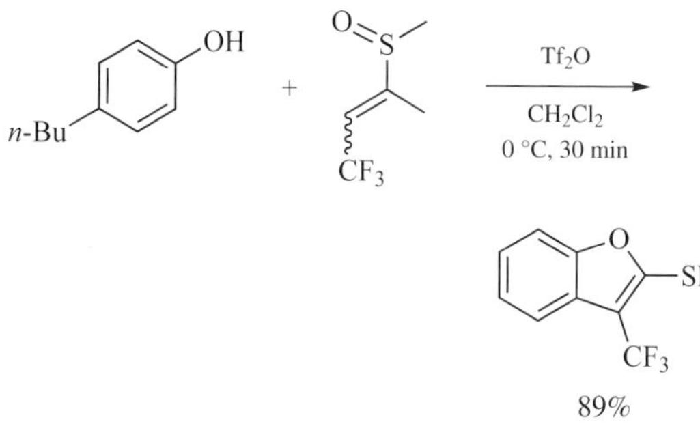

(101)

89%

Reaction of diphenylsulfoxide with triflic anhydride affords, as precedently mentioned, the corresponding sulfoxonium triflate, a powerful reagent for numerous reactions, such as α-arylation of carbonyls,[186] enolate additions,[187,188] or nucleophilic substitution on alkenes (102).[189,190]

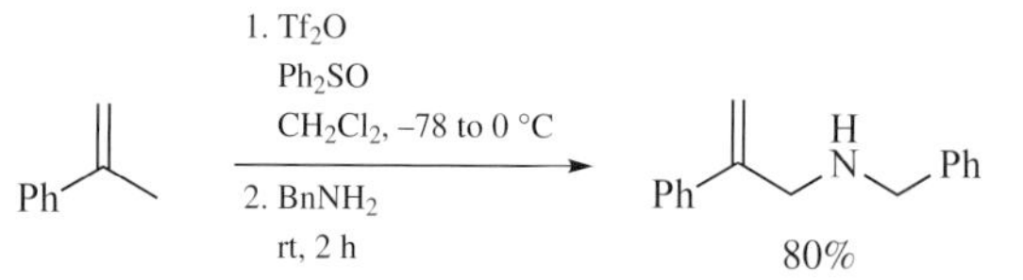

(102)

80%

Similarly, reaction of benzenesulfinylpiperidine (BSP) with triflic anhydride affords a sulfoxonium triflate useful to perform the hydrolysis of dithioketals (103).[191]

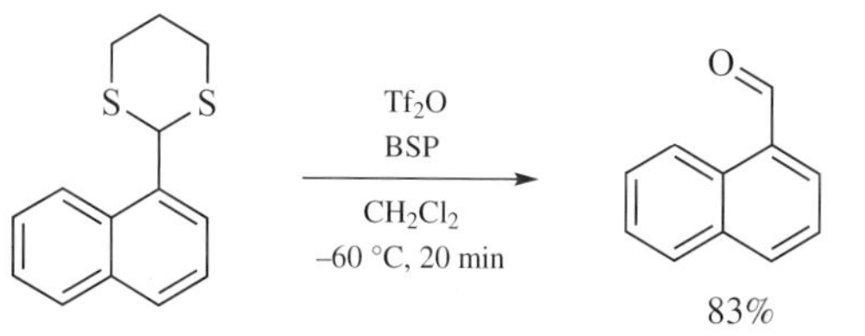

(103)

83%

Reactions with Carbonyls. Reaction of 1-(methylthio) acetone with nitriles in the presence of triflic anhydride can afford the substituted oxazole (104).[192]

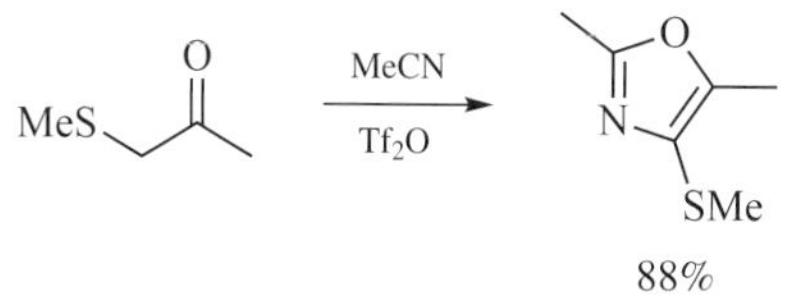

(104)

88%

Enones can be activated by triflic anhydride and undergo an intramolecular or intermolecular conjugate addition of an arene ring (105).[193]

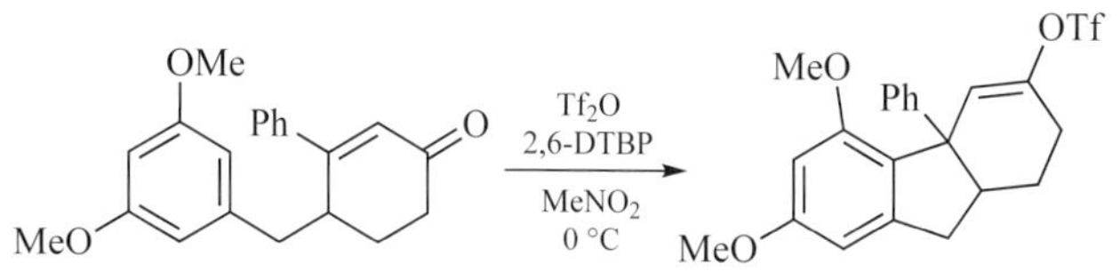

72% (105)

Reactions with Pyridines. It is well known that the reaction of triflic anhydride with pyridine and pyridine derivatives affords the pyridinium triflate, thus activating the *para* position for nucleophilic attacks, such as ketene acetals (106),[194–197] activated arenes (107),[198,199] azulenes,[200,201] or N-alkenes.[202]

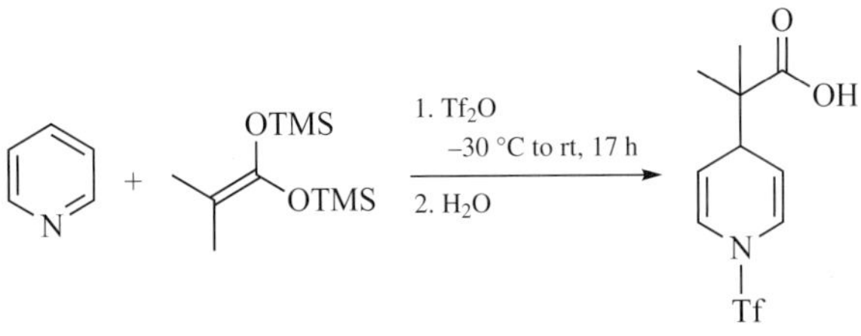

(106)

97%

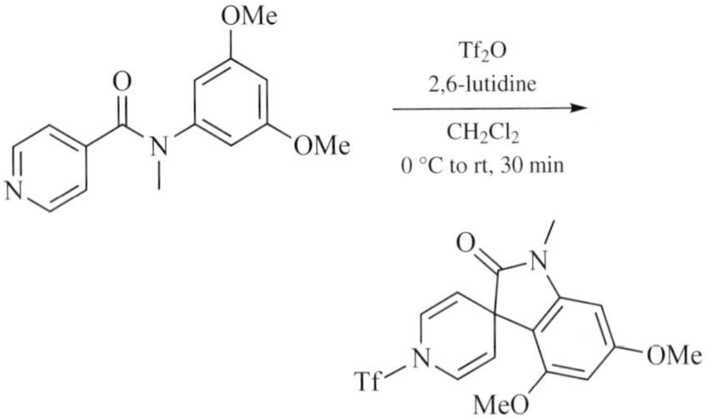

(107)

92%

Electron-deficient pyridines and pyridines derivatives can also be oxidized to the corresponding *N*-oxide by a mixture of triflic anhydride and sodium carbonate (108).[203]

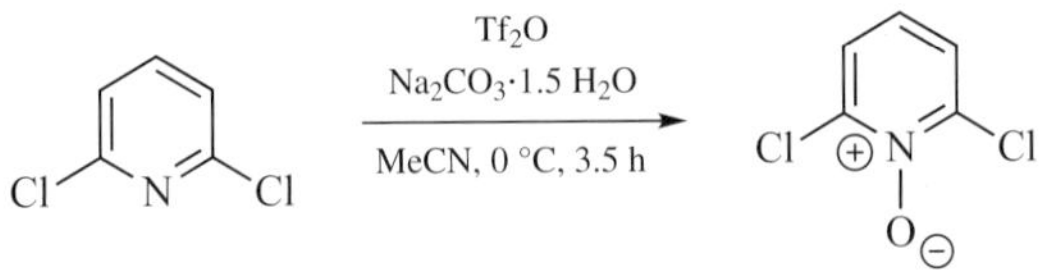

(108)

70%

Other Applications. Boc-protected amines react with triflic anhydride in the presence of a base to form the corresponding isocyanate. This intermediate can then undergo a nucleophilic attack by an amine or an activated arene to form ureas (109)[204] or tetrahydroisoquinolinones.[205]

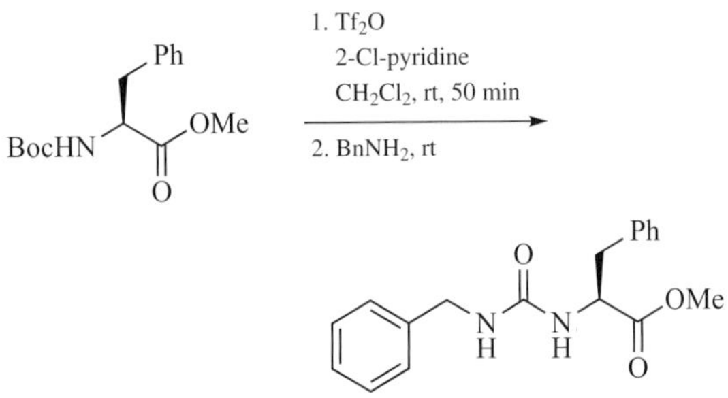

(109)

96%

Carboxilic acids can react with triflic anhydride and activated aromatic compounds to form the acylated product in a Friedel–Crafts-type acylation reaction (110).[206,207] This type of Friedel–Crafts acylation can also be applied to ureas to obtain the corresponding aryl amide.[208]

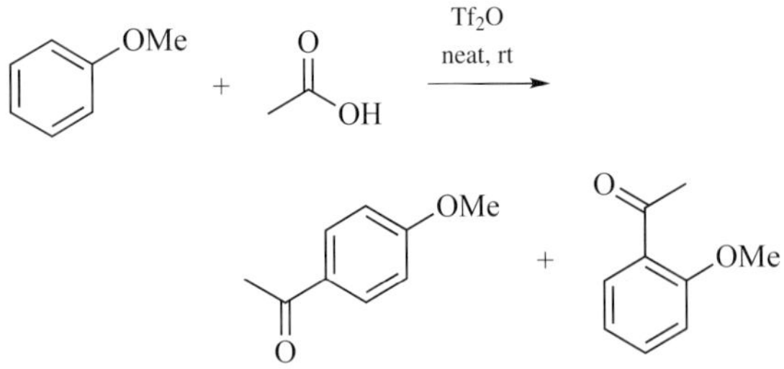

(110)

98%
p:o = 93:7

Oximes can react with triflic anhydride to afford amidines or enamines after a nucleophilic capture of the iminocarbocation formed (111).[209] The activated oxime can also eliminate to form the corresponding nitrile in the presence of imidazole[210] or be converted into benzisoxazoles through an intramolecular cyclization process.[211]

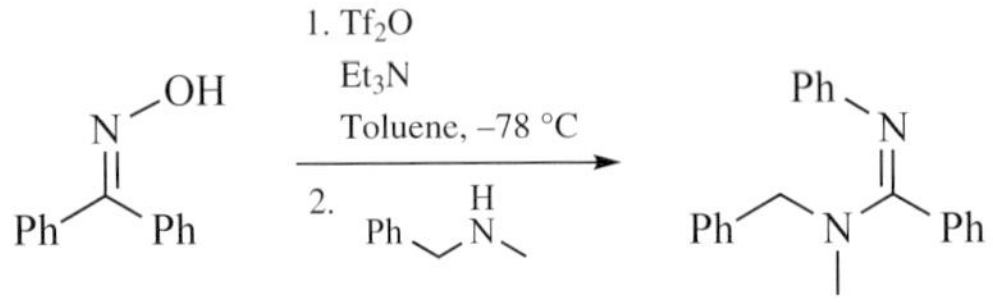

(111)

1. (a) Gramstad, T.; Haszeldine, R. N., *J. Chem. Soc.* **1957**, 4069. (b) Stang, P. J.; Hanack, M.; Subramanian, L. R., *Synthesis* **1982**, 85. (c) Stang, P. J.; White, M. R., *Aldrichim. Acta* **1983**, *16*, 15.

2. Binkley, R. W.; Ambrose, M. G., *J. Carbohydr. Chem.* **1984**, *3*, 1.

3. Ambrose, M. G.; Binkley, R. W., *J. Org. Chem.* **1983**, *48*, 674.

4. Stang, P. J.; Treptow, W., *Synthesis* **1980**, 283.

5. García Martínez, A.; Herrera Fernandez, A.; Martínez Alvarez, R.; Silva Losada, M. C.; Molero Vilchez, D.; Subramanian, L. R.; Hanack, M., *Synthesis* **1990**, 881.

6. Yoshida, M.; Takeuchi, K., *J. Org. Chem.* **1993**, *58*, 2566.

7. Jeanneret, V.; Gasparini, F.; Péchy, P.; Vogel, P., *Tetrahedron* **1992**, *48*, 10637.

8. Neenan, T. X.; Houlihan, F. M.; Reichmanis, E.; Kometani, J. M.; Bachman, B. J., Thompson, L. F., *Macromolecules* **1990**, *23*, 145.

9. (a) Sato, K.; Hoshi, T.; Kajihara, Y., *Chem. Lett.* **1992**, 1469. (b) Izawa, T.; Nakayama, K.; Nishiyama, S.; Yamamura, S.; Kato, K.; Takita, T., *J. Chem. Soc., Perkin Trans. 1* **1992**, 3003. (c) Knapps, S.; Naughton, A. B. J.; Jaramillo, C.; Pipik, B., *J. Org. Chem.* **1992**, *57*, 7328.

10. Takeuchi, K.; Kitagawa, T.; Ohga, Y.; Yoshida, M.; Akiyama, F.; Tsugeno, A., *J. Org. Chem.* **1992**, *57*, 280.

11. Eaton, P. E.; Zhou, J. P., *J. Am. Chem. Soc.* **1992**, *114*, 3118.

12. Spitz, U. P., *J. Am. Chem. Soc.* **1993**, *115*, 10174.

13. Zheng, C. Y.; Slebocka-Tilk, H.; Nagorski, R. W.; Alvarado, L.; Brown, R. S., *J. Org. Chem.* **1993**, *58*, 2122.

14. Effenberger, F.; Weber, Th., *Ber. Dtsch. Chem. Ges./Chem. Ber.* **1988**, *121*, 421.

15. (a) Rubinsztajn, S.; Fife, W. K.; Zeldin, M., *Tetrahedron Lett.* **1992**, *33*, 1821. (b) Dodd, R. H.; Poissonnet, G.; Potier, P., *Heterocycles* **1989**, *29*, 365.

16. García Martínez, A.; Martínez Alvarez, R.; Teso Vilar, E.; García Fraile, A.; Hanack, M.; Subramanian, L. R., *Tetrahedron Lett.* **1989**, *30*, 581.

17. Sonoda, T.; García Martínez, A.; Hanack, M.; Subramanian, L. R., *Croat. Chim. Acta* **1992**, *65*, 585 (*Chem. Abstr.* **1993**, *118*, 168491).

18. Ritter, K., *Synthesis* **1993**, 735.

19. Hendrickson, J. B.; Bergeron, R., *Tetrahedron Lett.* **1973**, 4607.

20. Hendrickson, J. B.; Bergeron, R.; Giga, A.; Sternbach, D. D., *J. Am. Chem. Soc.* **1973**, *95*, 3412.

21. Edwards, M. L.; Stemerick, D.; McCarthy, J. R., *Tetrahedron Lett.* **1990**, *31*, 3417.

22. Panetta, V.; Yaouanc, J. J.; Handel, H., *Tetrahedron Lett.* **1992**, *33*, 5505.

23. (a) McMurry, J. E.; Scott, W. J., *Tetrahedron Lett.* **1983**, *24*, 979. (b) Comins, J. E.; Dehghani, A., *Tetrahedron Lett.* **1992**, *33*, 6299. (c) Crisp, G. T.; Flynn, B. L., *Tetrahedron* **1993**, *49*, 5873.

24. Anders, E.; Stankowiak, A., *Synthesis* **1984**, 1039.

25. García Martínez, A.; Herrera Fernandez, A.; Alvarez, R. M.; Sánchez García, J. M., *An. Quim., Ser. C* **1981**, *77c*, 28 (*Chem. Abstr.* **1982**, *97*, 5840).

26. (a) Wright, M. E.; Pulley, S. R., *J. Org. Chem.* **1987**, *52*, 5036. (b) Dolle, R. E.; Schmidt, S. J.; Erhard, K. F.; Kruse, L. I., *J. Am. Chem. Soc.* **1989**, *111*, 278.

27. García Martínez, A.; Herrera Fernandez, A.; Moreno Jiménez, F.; García Fraile, A.; Subramanian, L. R.; Hanack, M., *J. Org. Chem.* **1992**, *57*, 1627.

28. Stang, P. J.; Kowalski, M. H.; Schiavelli, M. D.; Longford, R., *J. Am. Chem. Soc.* **1989**, *111*, 3347.

29. García Martínez, A.; Herrera Fernández, A.; Molero Vilchez, D.; Hanack, M.; Subramanian, L. R., *Synthesis* **1992**, 1053.

30. (a) Bentz, H.; Subramanian, L. R.; Hanack, M.; García Martínez, A.; Gómez Marin, M.; Perez-Ossorio, R., *Tetrahedron Lett.* **1977**, 9. (b) Kraus, W.; Zartner, G., *Tetrahedron Lett.* **1977**, 13. (c) García Martínez, A.; García Fraile, A.; Sánchez García, J. M., *Ber. Dtsch. Chem. Ges./Chem. Ber.* **1983**, *116*, 815. (d) García Martínez, A.; Teso Vilar, E.; Gómez Marin, M.; Ruano Franco, C., *Ber. Dtsch. Chem. Ges./Chem. Ber.* **1985**, *118*, 1282.

31. (a) García Martínez, A.; Teso Vilar, E.; López, J. C.; Manrique Alonso, J.; Hanack, M.; Subramanian, L. R., *Synthesis* **1991**, 353. (b) García Martínez, A.; Teso Vilar, E.; García Fraile, A.; Ruano Franco, C.; Soto Salvador, J.; Subramanian, L. R.; Hanack, M., *Synthesis* **1987**, 321.

32. (a) García Martínez, A.; Teso Vilar, E.; García Fraile, A.; Osío Barcina, J.; Hanack, M.; Subramanian, L. R., *Tetrahedron Lett.* **1989**, *30*, 1503. (b) García Martínez, A.; Teso Vilar, E.; Osío Barcina, J.; Manrique Alonso, J.; Rodríguez Herrero, E.; Hanack, M.; Subramanian, L. R., *Tetrahedron Lett.* **1992**, *33*, 607.

33. García Martínez, A.; Herrera Fernandez, A.; Sánchez García, J. M., *An. Quim.* **1979**, *75*, 723 (*Chem. Abstr.* **1980**, *93*, 45 725).

34. García Martínez, A.; Espada Rios, I.; Osío Barcina, J.; Teso Vilar, E., *An. Quim., Ser. C* **1982**, *78c*, 299 (*Chem. Abstr.* **1983**, *98*, 159 896).

35. (a) García Martínez, A.; Espada Rios, I.; Teso Vilar, E., *Synthesis* **1979**, 382. (b) García Martínez, A.; Espada Rios, I.; Osío Barcina, J.; Montero Hernando, M., *Ber. Dtsch. Chem. Ges./Chem. Ber.* **1984**, *117*, 982.

36. Hanack, M.; Hassdenteufel, J. R., *Ber. Dtsch. Chem. Ges./Chem. Ber.* **1982**, *115*, 764.

37. Stang, P. J.; Maas, G.; Smith, D. L.; McCloskey, J. A., *J. Am. Chem. Soc.* **1981**, *103*, 4837.

38. Wittmann, H.; Ziegler, E.; Sterk, H., *Monatsh. Chem.* **1987**, *118*, 531.

39. Wright, M. E.; Pulley, S. R., *J. Org. Chem.* **1989**, *54*, 2886.

40. García Martínez, A.; Martínez Alvarez, R.; García Fraile, A.; Subramanian, L. R.; Hanack, M., *Synthesis* **1987**, 49.

41. Stang, P. J., *Chem. Rev.* **1978**, *78*, 383.

42. García Martínez, A.; Herrera Fernandez, A.; Martínez Alvarez, R.; García Fraile, A.; Calderón Bueno, J.; Osío Barcina, J.; Hanack, M.; Subramanian, L. R., *Synthesis* **1986**, 1076.

43. García Martínez, A.; Osío Barcina, J.; Rys, A. Z.; Subramanian, L. R., *Tetrahedron Lett.* **1992**, *33*, 7787.

44. García Martínez, A.; Martínez Alvarez, R.; Martínez Gonzalez, S.; Subramanian, L. R.; Conrad, M., *Tetrahedron Lett.* **1992**, *33*, 2043.

45. García Martínez, A.; Martínez Alvarez, R.; Madueño Casado, M.; Subramanian, L. R.; Hanack, M., *Tetrahedron* **1987**, *43*, 275.

46. Negishi, E.; Owczarczyk, Z.; Swanson, D. R., *Tetrahedron Lett.* **1991**, *32*, 4453.

47. Evans, D. A.; Sjogren, E. B., *Tetrahedron Lett.* **1985**, *26*, 3787.

48. (a) Houpis, I. N., *Tetrahedron Lett.* **1991**, *32*, 6675. (b) Cook, G. K.; Hornback, W. J.; Jordan, C. L.; McDonald III, J. H.; Munroe, J. E., *J. Org. Chem.* **1989**, *54*, 5828.

49. (a) Effenberger, F.; Sohn, E.; Epple, G., *Ber. Dtsch. Chem. Ges./Chem. Ber.* **1983**, *116*, 1195. (b) Effenberger, F., *Angew. Chem.* **1980**, *92*, 147; *Angew. Chem., Int. Ed. Engl.* **1980**, *19*, 151.

50. García Martínez, A.; Herrera Fernández, A.; Molero Vilchez, D.; Laorden Gutiérrez, L.; Subramanian, L. R., *Synlett* **1993**, 229.

51. Robl, J. A., *Synthesis* **1991**, 56.

52. Falmagne, J.-B.; Escudero, J.; Taleb-Sahraoui, S.; Ghosez, L., *Angew. Chem.* **1981**, *93*, 926; *Angew. Chem., Int. Ed. Engl.* **1981**, *20*, 879.

53. García Martínez, A.; Martínez Alvarez, R.; Osío Barcina, J.; de la Moya Cerero, S.; Teso Vilar, E.; García Fraile, A.; Hanack, M.; Subramanian, L. R., *Chem. Commun./J. Chem. Soc., Chem. Commun.* **1990**, 1571.

54. Gramstad, T.; Husebye, S.; Saebo, J., *Tetrahedron Lett.* **1983**, *24*, 3919.

55. Thomas, E. W., *Synthesis* **1993**, 767.

56. Hendrickson, J. B.; Bair, K. W., *J. Org. Chem.* **1977**, *42*, 3875.

57. Hendrickson, J. B.; Schwartzman, S. M., *Tetrahedron Lett.* **1975**, 277.

58. Hendrickson, J. B.; Schwartzman, S. M., *Tetrahedron Lett.* **1975**, 273.

59. Lindner, E.; v. Au, G.; Eberle, H.-J., *Ber. Dtsch. Chem. Ges./Chem. Ber.* **1981**, *114*, 810.

60. Maas, G.; Lorenz, W., *J. Org. Chem.* **1984**, *49*, 2273.

61. (a) Stang, P. J.; Magnum, M. G.; Fox, D. P.; Haak, P., *J. Am. Chem. Soc.* **1974**, *96*, 4562. (b) Hanack, M.; Märkl, R.; García Martínez, A., *Ber. Dtsch. Chem. Ges./Chem. Ber.* **1982**, *115*, 772.

62. (a) Kitamura, T.; Furuki, R.; Nagata, K.; Taniguchi, H.; Stang, P. J., *J. Org. Chem.* **1992**, *57*, 6810. (b) Stang, P. J.; Zhdankin, V. V.; Tykwinski, R.; Zefirov, N., *Tetrahedron Lett.* **1992**, *33*, 1419.

63. Raghavan, S.; Kahne, D., *J. Am. Chem. Soc.* **1993**, *115*, 1580.

64. Taylor, S. L.; Martin, J. C., *J. Org. Chem.* **1987**, *52*, 4147.

65. Baldwin, J. E.; O'Neil, I. A., *Synlett* **1990**, 603.

66. Singer, B.; Maas, G., *Ber. Dtsch. Chem. Ges./Chem. Ber.* **1987**, *120*, 485.

67. Netscher, T.; Bohrer, P., *Tetrahedron Lett.* **1996**, *37*, 8359–8362.

68. Baraznenok, I. L.; Nenajdenko, V. G.; Balenkova, E. S., *Tetrahedron* **2000**, *56*, 3077–3119.

69. Ross, A. S.; Pitie, M.; Meunier, B., *J. Chem. Soc., Perkin Trans. 1* **2000**, 571–574.

70. Frantz, D. E.; Weaver, D. G.; Carey, J. P.; Kress, M. H.; Dolling, U. H., *Org. Lett.* **2002**, *4*, 4717–4718.

71. Binkley, R. W.; Ambrose, M. G., *J. Org. Chem.* **1983**, *48*, 1776–1777.

72. Martin, O. R.; Yang, F.; Xie, F., *Tetrahedron Lett.* **1995**, *36*, 47–50.

73. Yang, B.-H.; Jiang, J.-Q.; Ma, K.; Wu, H.-M., *Tetrahedron Lett.* **1995**, *36*, 2831–2834.

74. Nagasawa, T.; Handa, Y.; Onoguchi, Y.; Ohba, S.; Suzuki, K., *Synlett* **1995**, 739–741.

75. Lebel, H.; Marcoux, J. F.; Molinaro, C.; Charette, A. B., *Chem. Rev.* **2003**, *103*, 977–1050.

76. Suzuki, H.; Kuroda, C., *Tetrahedron* **2003**, *59*, 3157–3174.

77. Jonas, S.; Westerhausen, M.; Simchen, G., *J. Organometallic Chem.* **1997**, *548*, 131–137.

78. Simchen, G.; Jonas, S., *J. Prakt. Chem./Chem.-Ztg.* **1998**, *340*, 506–512.

79. Nishiyama, T.; Ono, Y.; Kurokawa, S.; Kimura, S., *Chem. Pharm. Bull.* **2000**, *48*, 1999–2002.

80. Katritzky, A. R.; Zhang, S.; Fang, Y., *Org. Lett.* **2000**, *2*, 3789–3791.

81. Martinez, A. G.; Vilar, E. T.; Jimenez, F. M.; Bilbao, C. M., *Tetrahedron: Asymmetry* **1997**, *8*, 3031–3034.

82. Martinez, A. G.; Fernandez, A. H.; Moreno-Jimenez, F.; Fraile, M. J. L.; Subramanian, L. R., *Synlett* **1994**, 559–560.

83. Martinez, A. G.; Fernandez, A. H.; Alvarez, R. M.; Vilchez, M. D. M.; Gutierrez, M. L. L.; Subramanian, L. R., *Tetrahedron* **1999**, *55*, 4825–4830.

84. Herrera, A.; Martinez-Alvarez, R.; Ramiro, P.; Chioua, M.; Torres, R., *Tetrahedron* **2002**, *58*, 3755–3764.

85. Sforza, S.; Dossena, A.; Corradini, R.; Virgili, E.; Marchelli, R., *Tetrahedron Lett.* **1998**, *39*, 711–714.

86. Charette, A. B.; Grenon, M., *Tetrahedron Lett.* **2000**, *41*, 1677–1680.

87. Alder, R. W.; Blake, M. E.; Bufali, S.; Butts, C. P.; Orpen, A. G.; Schutz, J.; Williams, S. J., *J. Chem. Soc., Perkin Trans. 1* **2001**, 1586–1593.

88. Charette, A. B.; Chua, P., *Synlett* **1998**, 163–165.

89. Charette, A. B.; Chua, P., *Tetrahedron Lett.* **1998**, *39*, 245–248.

90. Charette, A. B.; Chua, P., *J. Org. Chem.* **1998**, *63*, 908–909.

91. Charette, A. B.; Chua, P., *Tetrahedron Lett.* **1997**, *38*, 8499–8502.

92. Vonhoff, S.; Vasella, A., *Synth. Commun.* **1999**, *29*, 551–560.

93. McDonald, C. E.; Galka, A. M.; Green, A. I.; Keane, J. M.; Kowalchick, J. E.; Micklitsch, C. M.; Wisnoski, D. D., *Tetrahedron Lett.* **2001**, *42*, 163–166.

94. Liras, S.; Allen, M. P.; Segelstein, B. E., *Synth. Commun.* **2000**, *30*, 437–443.

95. Padwa, A.; Crawford, K. R.; Rashatasakhon, P.; Rose, M., *J. Org. Chem.* **2003**, *68*, 2609–2617.

96. Black, D. S. C.; Rezaie, R., *Tetrahedron Lett.* **1999**, *40*, 4251–4254.

97. Bose, D. S.; Jayalakshmi, B., *Synthesis* **1999**, 64–65.

98. Nenajdenko, V. G.; Vertelezkij, P. V.; Koldobskij, A. B.; Alabugin, I. V.; Balenkova, E. S., *J. Org. Chem.* **1997**, *62*, 2483–2486.

99. Nishiyama, T.; Seshita, T.; Shodai, H.; Aoki, K.; Kameyama, H.; Komura, K., *Chem. Lett.* **1996**, 549–550.

100. Eschwey, M.; Sundermeyer, W.; Stephenson, D. S., *Chem. Ber.* **1983**, *116*, 1623–1630.

101. Billard, T.; Langlois, B. R.; Large, S.; Anker, D.; Roidot, N.; Roure, P., *J. Org. Chem.* **1996**, *61*, 7545–7550.

102. Kahne, D.; Walker, S.; Cheng, Y.; Van Engen, D., *J. Am. Chem. Soc.* **1989**, *111*, 6881–6882.

103. Liang, R.; Yan, L.; Loebach, J.; Ge, M.; Uozumi, Y.; Sekanina, K.; Horan, N.; Gildersleeve, J.; Thompson, C.; Smith, A.; Biswas, K.; Still, W. C.; Kahne, D., *Science* **1996**, *274*, 1520–1522.

104. Yan, L.; Kahne, D., *J. Am. Chem. Soc.* **1996**, *118*, 9239–9248.

105. Crich, D.; Sun, S., *J. Org. Chem.* **1997**, *62*, 1198–1199.

106. Crich, D.; Sun, S., *J. Am. Chem. Soc.* **1997**, *119*, 11217–11223.

107. Crich, D., *J. Carbohydr. Chem.* **2002**, *21*, 667–690.

108. Gildersleeve, J.; Smith, A.; Sakurai, K.; Raghavan, S.; Kahne, D., *J. Am. Chem. Soc.* **1999**, *121*, 6176–6182.

109. Silva, D. J.; Kahne, D.; Kraml, C. M., *J. Am. Chem. Soc.* **1994**, *116*, 2641–2642.

110. Crich, D.; Smith, M., *J. Am. Chem. Soc.* **2001**, *123*, 9015–9020.

111. Codee, J. D. C.; Litjens, R. E. J. N.; den Heeten, R.; Overkleeft, H. S.; van Boom, J. H.; van der Marel, G. A., *Org. Lett.* **2003**, *5*, 1519–1522.

112. Crich, D.; Smith, M., *J. Am. Chem. Soc.* **2002**, *124*, 8867–8869.

113. Crich, D.; Picione, J., *Org. Lett.* **2003**, *5*, 781–784.

114. Gin, D., *J. Carbohydr. Chem.* **2002**, *21*, 645–665.

115. Garcia, B. A.; Poole, J. L.; Gin, D. Y., *J. Am. Chem. Soc.* **1997**, *119*, 7597–7598.

116. Mukaiyama, T.; Suda, S., *Chem. Lett.* **1990**, 1143–1146.

117. Di Bussolo, V.; Kim, Y.-J.; Gin, D. Y., *J. Am. Chem. Soc.* **1998**, *120*, 13515–13516.

118. Kim, J.-Y.; Di Bussolo, V.; Gin, D. Y., *Org. Lett.* **2001**, *3*, 303–306.

119. Liu, J.; Gin, D. Y., *J. Am. Chem. Soc.* **2002**, *124*, 9789–9797.

120. Kim, K. S.; Park, J.; Lee, Y. J.; Seo, Y. S., *Angew. Chem. Int. Ed.* **2003**, *42*, 459–462.

121. Toshima, K., *Carbohydr. Res.* **2000**, *327*, 15–26.

122. Gommermann, N.; Koradin, C.; Knochel, P., *Synthesis* **2002**, 2143–2149.

123. Kodama, Y.; Okumura, M.; Yanabu, N.; Taguchi, T., *Tetrahedron Lett.* **1996**, *37*, 1061–1064.

124. Bottger, G.; Geisler, A.; Frohlich, R.; Wurthwein, E.-U., *J. Org. Chem.* **1997**, *62*, 6407–6411.

125. Muri, D.; Bode, J. W.; Carreira, E. M., *Org. Lett.* **2000**, *2*, 539–541.

126. Hendrickson, J. B.; Sommer, T. J.; Singer, M., *Synthesis* **1995**, 1496–1496.

127. Shackelford, S. A.; Anderson, M. B.; Christie, L. C.; Goetzen, T.; Guzman, M. C.; Hananel, M. A.; Kornreich, W. D.; Li, H.; Pathak, V. P.; Rabinovich, A. K.; Rajapakse, R. J.; Truesdale, L. K.; Tsank, S. M.; Vazir, H. N., *J. Org. Chem.* **2003**, *68*, 267–275.

128. Adams, C. M.; Sharts, C. M.; Shackelford, S. A., *Tetrahedron Lett.* **1993**, *34*, 6669–6672.

129. Bates, D. K.; Xia, M., *J. Org. Chem.* **1998**, *63*, 9190–9196.

130. Corey, E. J.; Gin, D. Y.; Kania, R. S., *J. Am. Chem. Soc.* **1996**, *118*, 9202–9203.

131. Furukawa, N.; Kobayashi, K.; Sato, S., *J. Organometallic Chem.* **2000**, *611*, 116–126.

132. Nenajdenko, V. G.; Shevchenko, N. E.; Balenkova, E. S.; Alabugin, I. V., *Chem. Rev.* **2003**, *103*, 229–282.

133. Katritzky, A. R.; Moutou, J.-L.; Yang, Z., *Synthesis* **1995**, 1497–1505.

134. Martinez, A. G.; Fernandez, A. H.; Jimenez, F. M.; Ruiz, P. M.; Subramanian, L. R., *Synlett* **1995**, 161–162.

135. Kitamura, T.; Kotani, M.; Fujiwara, Y., *Synthesis* **1998**, 1416–1418.

136. Stang, P. J.; Zhdankin, V. V., *Chem. Rev.* **1996**, *96*, 1123–1178.

137. Kutateladze, A. G.; Kice, J. L.; Kutateladze, T. G.; Zefirov, N. S.; Zyk, N. V., *Tetrahedron Lett.* **1992**, *33*, 1949–1952.

138. Umemoto, T.; Ishihara, S., *Tetrahedron Lett.* **1990**, *31*, 3579–3582.

139. Umemoto, T.; Ishihara, S., *J. Am. Chem. Soc.* **1993**, *115*, 2156–2164.

140. Gong, W.; Liu, Y.; Zhang, J.; Jiao, Y.; Xue, J.; Li, Y., *Chem. Asian J.* **2013**, *8*, 546.

141. Hillier, M. C.; Chen, C.-Y., *J. Org. Chem.* **2006**, *71*, 7885.

142. Miller, R. A.; Lang, F.; Marcune, B.; Zewge, D.; Song, Z. J.; Karady, S., *Synth. Commun.* **2003**, *33*, 3347.

143. Wan, B.; Li, H.; Huang, Y.; Jin, W.; Xue, F.; *Tetrahedron Lett.* **2008**, *49*, 1686.

144. Zhou, H.; Xie, Y.; Ren, L.; Wang, K., *Adv. Synth. Catal.* **2009**, *351*, 1289.

145. Colombo, M. I.; Justribó, V., *Tetrahedron Lett.* **2003**, *44*, 8023.

146. Colombo, M. I.; Justribó, V.; Pellegrinet, S. C., *J. Org. Chem.* **2007**, *72*, 3702.

147. Bertrand, G.; Martin, D.; Lassauque, N.; Donnadieu, B., *Angew. Chem., Int. Ed.* **2012**, *51*, 6172.

148. Zhang, J.; Su, X.; Fu, J.; Qin, X.; Zhao, M.; Shi, M., *Chem. Commun.* **2012**, *48*, 9192.

149. Charette, A. B.; Lemonner, G., *J. Org. Chem.* **2010**, *75*, 7465.

150. Lalic, G.; Dang, H.; Cox, N., *Angew. Chem., Int. Ed.* **2014**, *53*, 752.

151. Kobayashi, H.; Himeshima, Y.; Sonoda, T., *Chem. Lett.* **1983**, *12*, 1211.

152. Charette, A. B.; Pelletier, G.; Bechara, W. S., *J. Am. Chem. Soc.* **2010**, *132*, 12817.

153. Huang, P.-Q.; Geng, H., *Org. Chem. Front.* **2015**, *2*, 150.

154. Huang, P.-Q.; Ou, W.; Xiao, K.-J.; Wang, A. E., *Chem. Commun.* **2014**, *50*, 8761.

155. Charette, A. B.; Bechara, W.; Pelletier, G., *Nat. Chem.* **2012**, *4*, 228.

156. Huang, P.-Q.; Wang, Y.; Xiao, K.-J.; Huang, Y.-H., *Tetrahedron* **2015**, *71*, 4248.

157. Huang, P.-Q.; Xiao, K.-J.; Luo, J.-M.; Xia, X.-E.; Wang, Y., *Chem. Eur. J.* **2013**, *19*, 13075.

158. Huang, P.-Q.; Xiao, K.-J.; Luo, J.-M.; Ye, K.-Y.; Wang, Y., *Angew. Chem., Int. Ed.* **2010**, *49*, 3037.

159. Huang, P.-Q.; Huang, Y.-H.; Xiao, K.-J.; Wang, Y.; Xia, X.-E., *J. Org. Chem.* **2015**, *80*, 2861.

160. Huang, P.-Q.; Wang, A.-E.; Chang, Z.; Sun, W.-T., *Org. Lett.* **2015**, *17*, 732.

161. Kuhnert, N.; Clemens, I.; Walsh, R., *Org. Biomol. Chem.* **2005**, *3*, 1694.

162. Charette, A. B.; Pelletier, G., *Org. Lett.* **2013**, *15*, 2290.

163. Charette, A. B.; Mathieu, S.; Martel, J., *Org. Lett.* **2005**, *7*, 5401.

164. Movassaghi, M.; Medley, J. W., *J. Org. Chem.* **2009**, *74*, 1341.

165. Chen, Z.; Li, Y.; Gao, L.; Zhu, H.; Li, G., *Org. Biomol. Chem.* **2014**, *12*, 6982.

166. Maulide, N.; Valerio, V.; Petkova, D.; Madelaine, C., *Chem. Eur. J.* **2013**, *19*, 2606.

167. Wang, Y.-G.; Cui, S.-L.; Wang, J., *J. Am. Chem. Soc.* **2008**, *130*, 13526.

168. Movassaghi, M.; Hill, M. D.; Ahmad, O. K., *J. Am. Chem. Soc.* **2007**, *129*, 10096.

169. Shi, M.; Zhang, J.; Fu, J.; Su, X.; Wang, X.; Song, S., *Chem. Asian J.* **2013**, *8*, 552.

170. Movassaghi, M.; Hill, M. D., *J. Am. Chem. Soc.* **2006**, *128*, 14254.

171. Movassaghi, M.; Ahmad, O. K.; Hill, M. D., *J. Org. Chem.* **2009**, *74*, 8460.

172. Charette, A. B.; Bechara, W. S.; Khazhieva, I. S.; Rodriguez, E., *Org. Lett.* **2015**, *17*, 1184.

173. Charette, A. B.; Cyr, P.; Régnier, S.; Bechara, W. S., *Org. Lett.* **2015**, *17*, 3386.

174. Movassaghi, M.; Medley, J. W., *Org. Lett.* **2013**, *15*, 3614.

175. Maulide, N.; Peng, B.; Geerdink, D. Farés, C., *Angew. Chem., Int. Ed.* **2014**, *53*, 5462.

176. De Mesmaeker, A.; Villedieu-Percheron, E.; Catak, S.; Zurwerra, D.; Staiger, R.; Lachia, M., *Tetrahedron Lett.* **2014**, *55*, 2446.

177. Bahrami, K.; Khodaei, M. M.; Karimi, A., *Synthesis* **2008**, 2543.

178. Blazejewski, J.-C.; Macé, Y.; Pradet, C.; Magnier, E., *Eur. J. Org. Chem.* **2010**, 5772.

179. Jenkins, I. D.; Petersson, M. J.; Loughlin, W. A., *Chem. Commun.* **2008**, 4493.

180. Magnier, E.; Macé, Y.; Urban, C.; Pradet, C.; Blazejewski, J.-C., *Eur. J. Org. Chem.* **2009**, 5313.

181. Magnier, E.; Macé, Y.; Urban, C.; Pradet, C.; Marrot, J.; Blazejewski, J.-C., *Eur. J. Org. Chem.* **2009**, 3150.

182. Oshima, K.; Kobatake, T.; Fujino, D.; Yoshida, S.; Yorimitsu, H., *J. Am. Chem. Soc.* **2010**, *132*, 11838.

183. Oshima, K.; Yoshida, S.; Yorimitsu, H., *Org. Lett.* **2007**, *9*, 5573.

184. Oshima, K.; Yoshida, S.; Yorimitsu, H., *Chem. Lett.* **2008**, *47*, 786.

185. Feldman, K.; Vidulova, D. B., *Org. Lett.* **2004**, *6*, 1869.

186. Maulide, N.; Huang, X., *J. Am. Chem. Soc.* **2011**, *133*, 8510.

187. Mukaiyama, T.; Takuwa, T.; Onishi, J. Y.; Matsuo, J.-I., *Chem. Lett.* **2004**, *33*, 8.

188. Mukaiyama, T.; Takuwa, T.; Minowa, T.; Fujisawa, H., *Chem. Pharm. Bull.* **2005**, *53*, 476.

189. Mukaiyama, T.; Yamanaka, H.; Matsuo, J.-I.; Kawana, A., *Chem. Lett.* **2003**, *32*, 626.

190. Mukaiyama, T.; Matsuo, J.-I.; Yamanaka, H.; Kawana, A., *Chem. Lett.* **2003**, *32*, 392.

191. Crich, D.; Picione, J., *Synlett* **2003**, 1257.

192. Martinez-Alvarez, R.; Herrera, A.; Ramiro, P.; Molero, D.; Almys, J., *J. Org. Chem.* **2006**, *71*, 3026.

193. Trauner, D.; Grundl, M. A.; Kaster, A.; Beaulieu, E. D., *Org. Lett.* **2006**, *8*, 5429.

194. Rudler, H.; Parlier, A.; Sandoval-Chavez, C.; Herson, P.; Daran, J.-C., *Angew. Chem., Int. Ed.* **2008**, *47*, 6843.

195. Álvarez-Toledano, C.; Gualo-Soberanes, N.; Ortega-Alfaro, M. C.; López-Cortés, J. G.; Toscano, R. A.; Rudler, H., *Tetrahedron Lett.* **2010**, 3186.

196. Rudler, H.; Parlier, A.; Denneval, C.; Herson, P., *J. Fluorine Chem.* **2010**, *131*, 738.

197. Rivera-Hernández, A.; Alvarez-Toledano, C.; Chans, G. M.; Rudler, H.; López Cortés, J. G.; Toscano, R. A., *Tetrahedron* **2014**, *70*, 1861.

198. Clayden, J.; Arnott, G.; Brice, H.; Blaney, E., *Org. Lett.* **2008**, *10*, 3089.

199. Corey, E. J.; Tian, Y., *Org. Lett.* **2005**, *7*, 5535.

200. Shoji, T.; Ito, S.; Toyota, K.; Morita, N., *Eur. J. Org. Chem.* **2010**, 1059.

201. Shoji, T.; Ito, S.; Okujima, T.; Higashi, J.; Yokoyama, R.; Toyota, K.; Yasunami, M.; Morita, N., *Eur. J. Org. Chem.* **2009**, 1554.

202. Clayden, J.; Senczyszyn, J., *Org. Lett.* **2013**, *15*, 1922.

203. Zhu, X.; Kreutter, K. D.; Hu, H.; Player, M. R.; Gaul, M. D., *Tetrahedron Lett.* **2008**, 832.

204. Kokotos, C. G.; Spyropoulot, C., *J. Org. Chem.* **2014**, *79*, 4477.

205. Kim, S.; In, J.; Hwang, S.; Kim, C.; Seo, J. H., *Eur. J. Org. Chem.* **2013**, 965.

206. Khodaei, M. M.; Alizadeh, A.; Nazari, E., *Tetrahedron Lett.* **2007**, *48*, 4199.

207. Comegna, D.; DellaGreca, M.; Iesce, M. R.; Previtera, L.; Zarrelli, A.; Zuppolini, S., *Org. Biomol. Chem.* **2012**, *10*, 1219.

208. Herndon, J. W.; Ying, W.; Gamage, L. S. R.; Lovro, L. R.; Hendron, III, J. W.; Jenkins, N. W., *Tetrahedron Lett.* **2014**, *55*, 4541.

209. Mukaiyama, T.; Takuwa, T.; Minowa, T.; Onishi, J. Y., *Chem. Lett.* **2004**, *33*, 322.

210. Laali, K. K.; Kalkhambkar, R. G.; Bunge, S. D., *Tetrahedron Lett.* **2011**, *52*, 5184.

211. Kalkhambkar, R. G.; Yuvaraj, H., *Synth. Commun.* **2014**, *44*, 547.

Trimethylsilyl Trifluoromethane sulfonate[1]

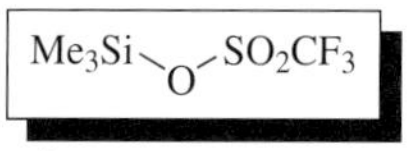

$$Me_3Si \diagdown_O \diagup SO_2CF_3$$

[27607-77-8] $C_4H_9F_3O_3SSi$ (MW 222.29)

InChI = 1/C4H9F3O3SSi/c1-12(2,3)10-11(8,9)4(5,6)7/h1-3H3

InChIKey = FTVLMFQEYACZNP-UHFFFAOYAR

Alternate Name: TMSOTf.

Physical Data: bp 45–47 °C/17 mmHg, 39–40 °C/12 mmHg; d 1.225 g cm^{-3}.

Solubility: sol aliphatic and aromatic hydrocarbons, haloalkanes, ethers.

Form Supplied in: colorless liquid; commercially available.

Preparative Method: may be prepared by a variety of methods.[2]

Handling, Storage, and Precaution: flammable; corrosive; very hygroscopic.

Original Commentary

Joseph Sweeney & Gemma Perkins
University of Bristol, Bristol, UK

Silylation. TMSOTf is widely used in the conversion of carbonyl compounds to their enol ethers. The conversion is some 10^9 faster with TMSOTf/*Triethylamine* than with *Chlorotrimethylsilane* (eqs 1–3).[3–5]

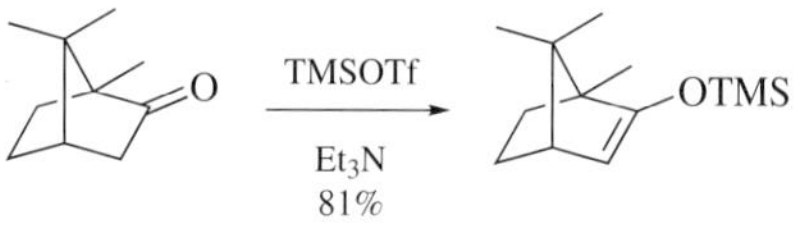

(1)

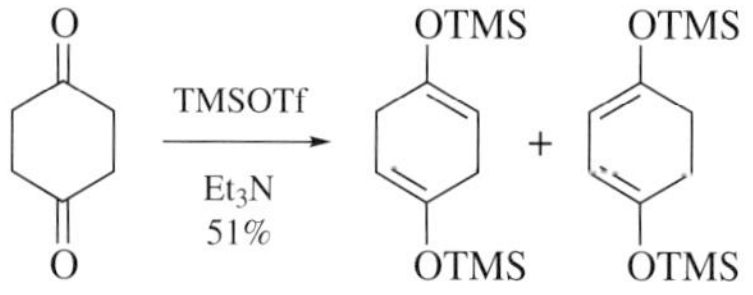

(2)

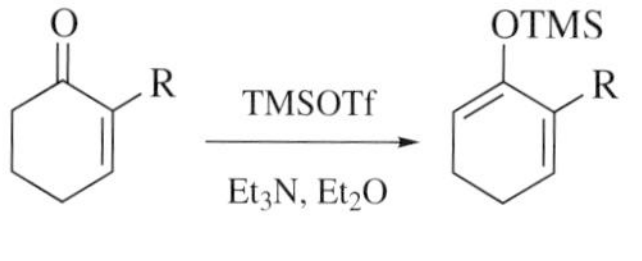

(3)

Dicarbonyl compounds are converted to the corresponding bis-enol ethers; this method is an improvement over the previous two-step method (eq 4).[6]

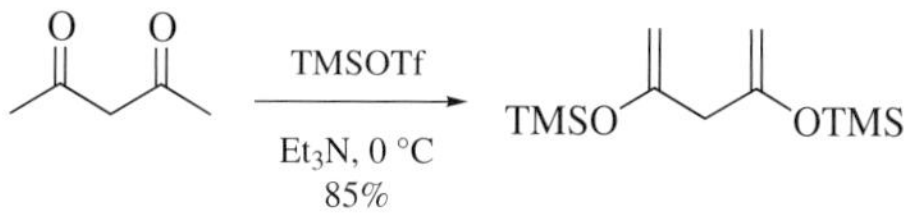

(4)

In general, TMSOTf has a tendency to *C*-silylation which is seen most clearly in the reaction of esters, where *C*-silylation dominates over *O*-silylation. The exact ratio of products obtained depends on the ester structure[7] (eq 5).[8] Nitriles undergo *C*-silylation; primary nitriles may undergo *C*,*C*-disilylation.[9]

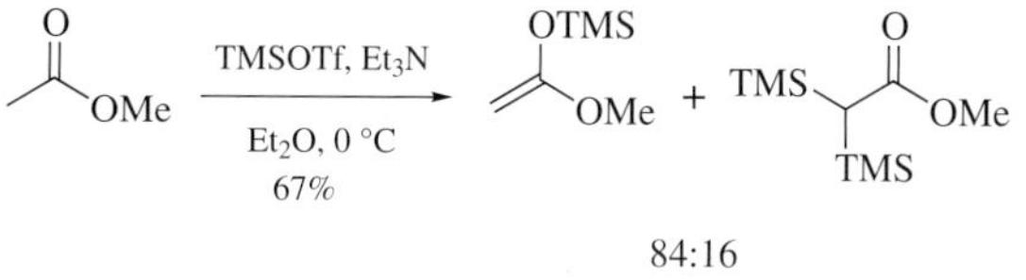

(5)

84:16

TMS enol ethers may be prepared by rearrangement of α-ketosilanes in the presence of catalytic TMSOTf (eq 6).[10,11]

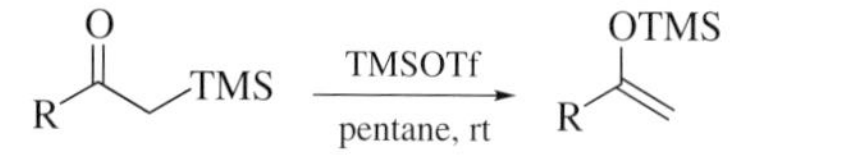

(6)

Enhanced regioselectivity is obtained when trimethylsilyl enol ethers are prepared by treatment of α-trimethylsilyl ketones with catalytic TMSOTf (eq 7).[12]

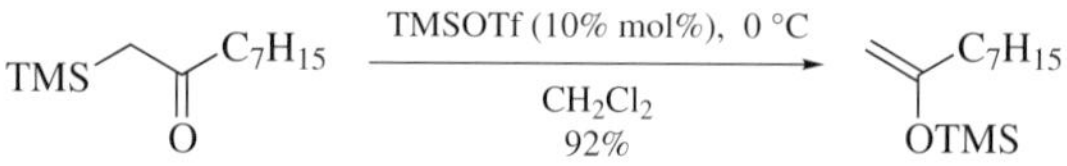

(7)

The reaction of imines with TMSOTf in the presence of Et$_3$N gives *N*-silylenamines.[13]

Ethers do not react, but epoxides are cleaved to give silyl ethers of allylic alcohols in the presence of TMSOTf and *1,8-Diazabicyclo[5.4.0]undec-7-ene*; The regiochemistry of the reaction is dependent on the structure of the epoxide (eq 8).[14]

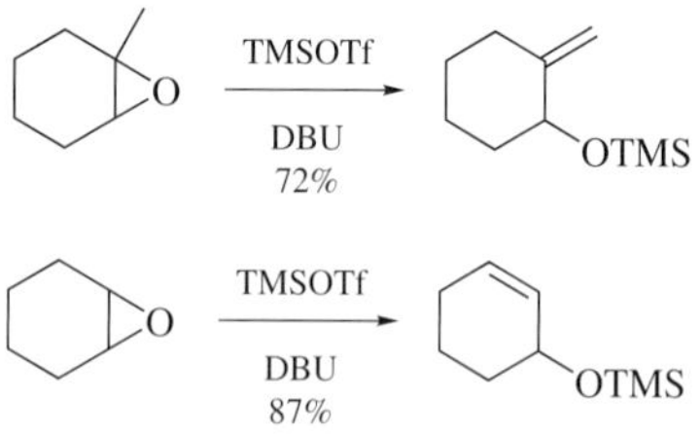

(8)

Indoles and pyrroles undergo efficient *C*-silylation with TMSOTf (eq 9).[15]

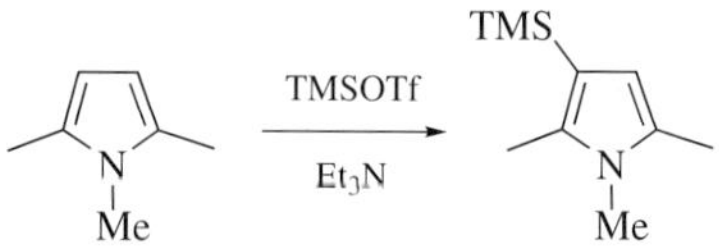

(9)

t-Butyl esters are dealkylatively silylated to give TMS esters by TMSOTf; benzyl esters are inert under the same conditions.[16]

Imines formed from unsaturated amines and α-carbonyl esters undergo ene reactions in the presence of TMSOTf to form cyclic amino acids.[17]

Carbonyl Activation. 1,3-Dioxolanation of conjugated enals is facilitated by TMSOTf in the presence of 1,2-bis(trimethylsilyloxy)ethane. In particular, highly selective protection of sterically differentiated ketones is possible (eq 10).[18] Selective protection of ketones in the presence of enals is also facilitated (eq 11).[19]

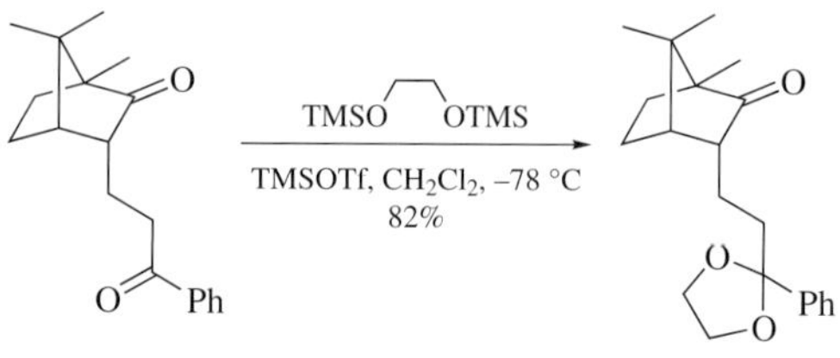

(10)

Essential Reactions for Organic Synthesis, First Edition. Edited by Philip L. Fuchs.
© 2016 John Wiley & Sons, Ltd. Published 2016 by John Wiley & Sons, Ltd.

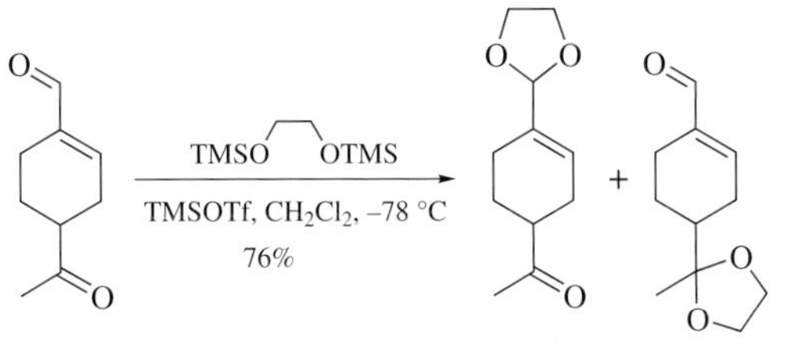

(11)

1:27

The similar reaction of 2-alkyl-1,3-disilyloxypropanes with chiral ketones is highly selective and has been used to prepare spiroacetal starting materials for an asymmetric synthesis of α-tocopherol subunits (eq 12).[20]

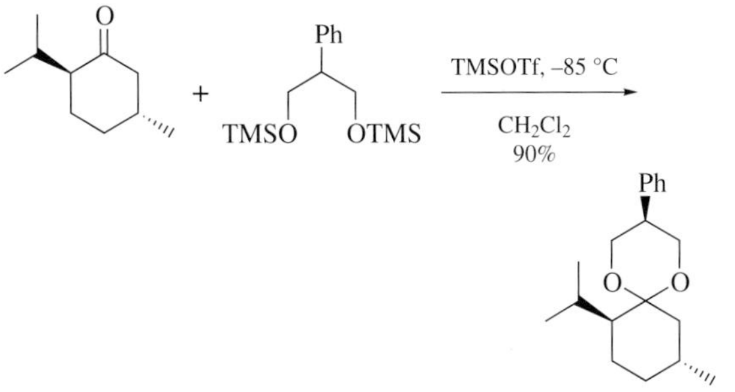

(12)

The preparation of spiro-fused dioxolanes (useful as chiral glycolic enolate equivalents) also employs TMSOTf (eq 13).[21]

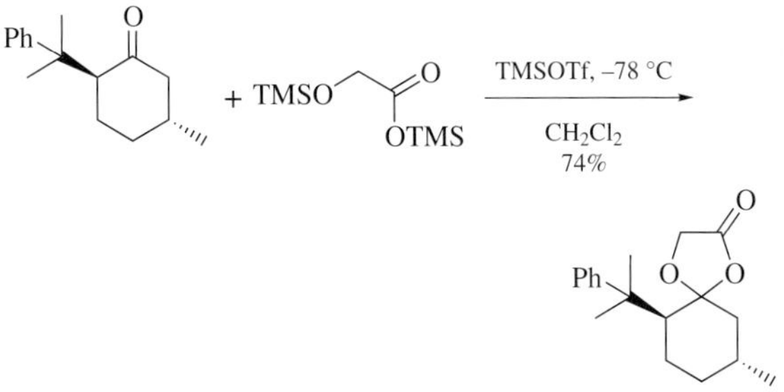

(13)

~ 1:1 mixture

TMSOTf mediates a stereoselective aldol-type condensation of silyl enol ethers and acetals (or orthoesters). The nonbasic reaction conditions are extremely mild. TMSOTf catalyzes many aldol-type reactions; in particular, the reaction of relatively non-nucleophilic enol derivatives with carbonyl compounds is facile in the presence of the silyl triflate. The activation of acetals was first reported by Noyori and has since been widely employed (eq 14).[22,23]

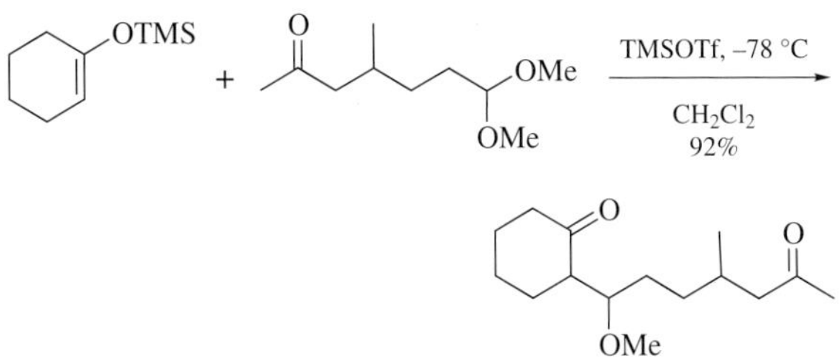

(14)

In an extension to this work, TMSOTf catalyzes the first step of a [3 + 2] annulation sequence which allows facile synthesis of fused cyclopentanes possessing bridgehead hydroxy groups (eq 15).[24]

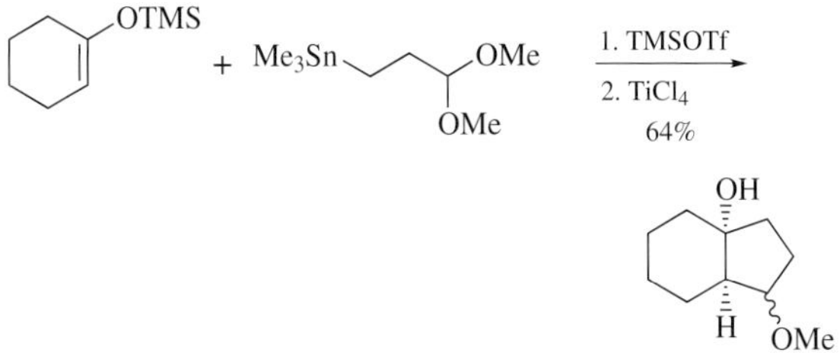

(15)

The use of TMSOTf in aldol reactions of silyl enol ethers and ketene acetals with aldehydes is ubiquitous. Many refinements of the basic reaction have appeared. An example is shown in eq 16.[25]

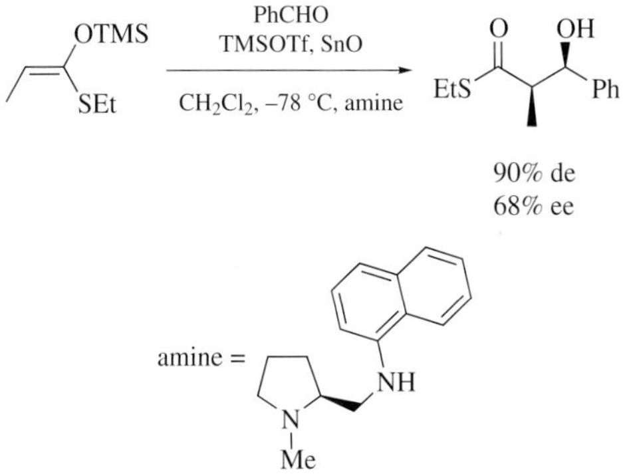

(16)

90% de
68% ee

amine =

The use of TMSOTf in the reaction of silyl ketene acetals with imines offers an improvement over other methods (such as Ti^{IV}- or Zn^{II}-mediated processes) in that truly catalytic amounts of activator may be used (eq 17);[26] this reaction may be used as the crucial step in a general synthesis of 3-(1'-hydroxyethyl)-2-azetidinones (eq 18).[27]

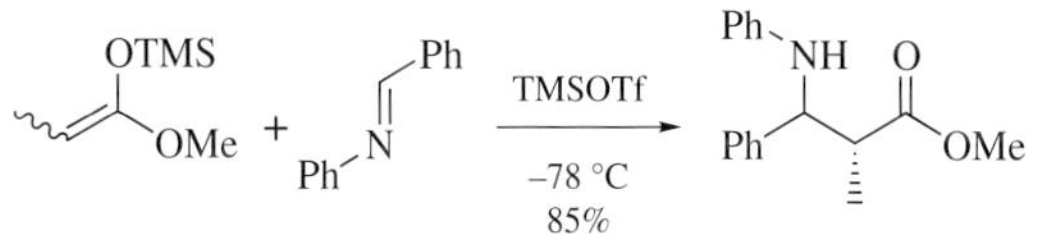

(17)

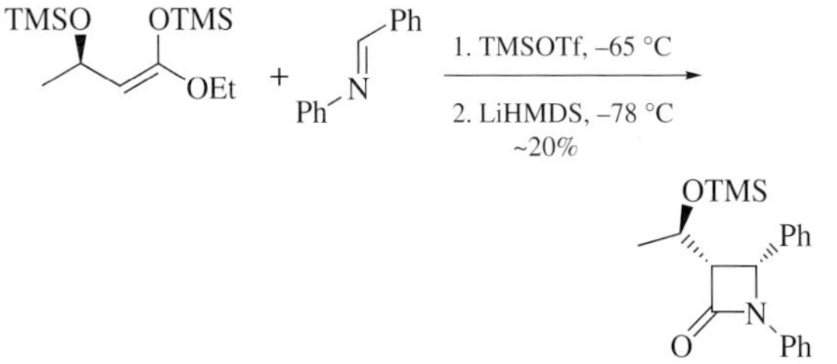

(18)

Stereoselective cyclization of α,β-unsaturated enamide esters is induced by TMSOTf and has been used as a route to quinolizidines and indolizidines (eq 19).[28]

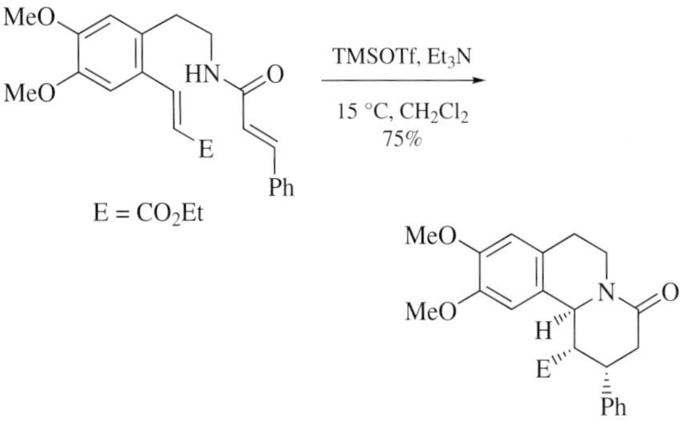

E = CO$_2$Et

(19)

The formation of nitrones by reaction of aldehydes and ketones with **N-Methyl-N,O-bis(trimethylsilyl)hydroxylamine** is accelerated when TMSOTf is used as a catalyst; the acceleration is particularly pronounced when the carbonyl group is under a strong electronic influence (eq 20).[29]

$$ (20) $$

β-Stannylcyclohexanones undergo a stereoselective ring contraction when treated with TMSOTf at low temperature. When other Lewis acids were employed, a mixture of ring-contracted and protiodestannylated products was obtained (eq 21).[30]

$$ (21) $$

The often difficult conjugate addition of alkynyl organometallic reagents to enones is greatly facilitated by TMSOTf. In particular, alkynyl zinc reagents (normally unreactive with α,β-unsaturated carbonyl compounds) add in good yield (eq 22).[31] The proportion of 1,4-addition depends on the substitution pattern of the substrate.

$$ (22) $$

The 1,4-addition of phosphines to enones in the presence of TMSOTf gives β-phosphonium silyl enol ethers, which may be deprotonated and alkylated in situ (eq 23).[32]

$$ (23) $$

Miscellaneous. Methyl glucopyranosides and glycopyranosyl chlorides undergo allylation with allylsilanes under TMSOTf catalysis to give predominantly α-allylated carbohydrate analogs (eq 24).[33]

$$ (24) $$

Glycosidation is a reaction of massive importance and widespread employment. TMSOTf activates many selective glycosidation reactions (eq 25).[34]

$$ (25) $$

TMSOTf activation for coupling of 1-O-acylated glycosyl donors has been employed in a synthesis of avermectin disaccharides (eq 26).[35]

$$ (26) $$

Similar activation is efficient in couplings with trichloroimidates[36] and O-silylated sugars.[37,38]

2-Substituted Δ^3-piperidines may be prepared by the reaction of 4-hydroxy-1,2,3,4-tetrahydropyridines with a variety of carbon and heteronucleophiles in the presence of TMSOTf (eqs 27 and 28).[39]

$$ (27) $$

$$ (28) $$

Iodolactamization is facilitated by the sequential reaction of unsaturated amides with TMSOTf and **Iodine** (eq 29).[40]

$$ (29) $$

By use of a silicon-directed Beckmann fragmentation, cyclic (E)-β-trimethylsilylketoxime acetates are cleaved in high yield in the presence of catalytic TMSOTf to give the corresponding unsaturated nitriles. Regio- and stereocontrol are complete (eq 30).[41]

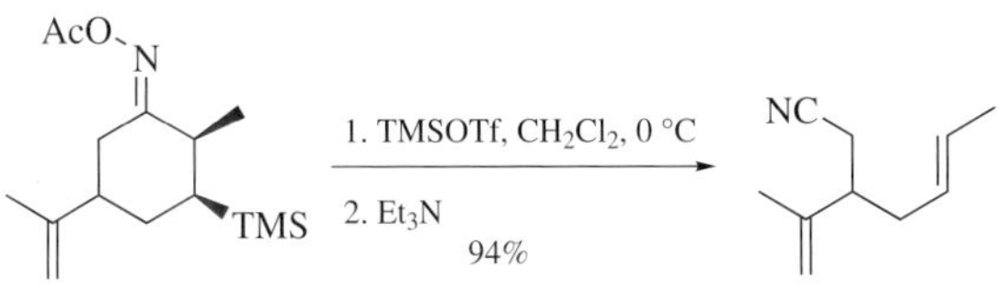

(30)

A general route to enol ethers is provided by the reaction of acetals with TMSOTf in the presence of a hindered base (eq 31).[42] The method is efficient for dioxolanes and noncyclic acetals.

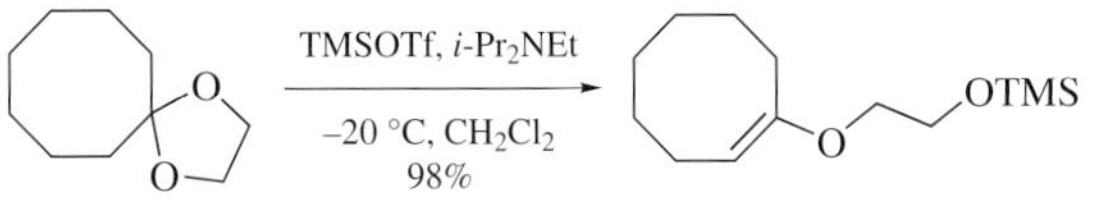

(31)

α-Halo sulfoxides are converted to α-halovinyl sulfides by reaction with excess TMSOTf (eq 32),[43] while α-cyano- and α-alkoxycarbonyl sulfoxides undergo a similar reaction (eq 33).[44] TMSOTf is reported as much superior to *Iodotrimethylsilane* in these reactions.

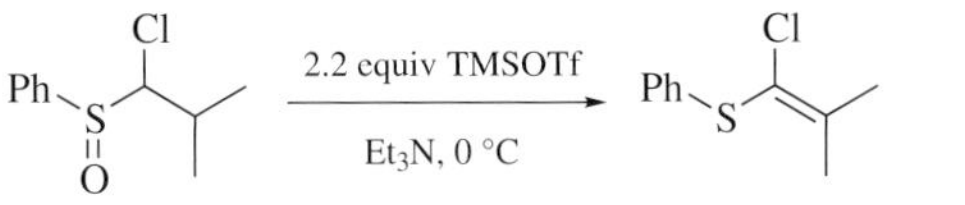

(32)

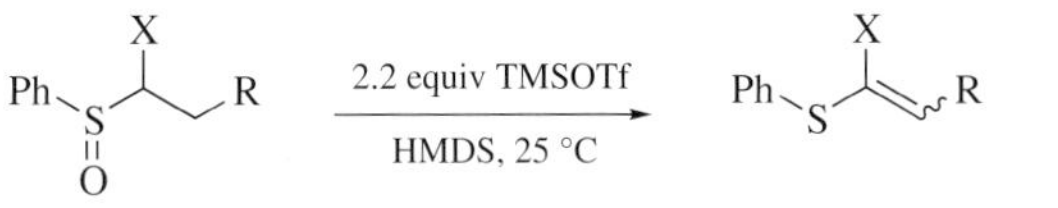

(33)

$$X = CN \text{ or } CO_2R'$$

First Update

Enrique Aguilar & Manuel A. Fernández-Rodríguez
Universidad de Oviedo, Oviedo, Spain

The introduction of TMSOTf as a highly electrophilic silylating reagent led to its widespread use and there are a very large number of reactions which employ the reagent in either stoichiometric or catalytic amounts.

O-Silylation. The formation of TMS ethers can be achieved by reacting the requisite alcohol with TMSOTf and an amine (triethylamine, pyridine, or 2,6-lutidine) in dichloromethane (eq 34).[45]

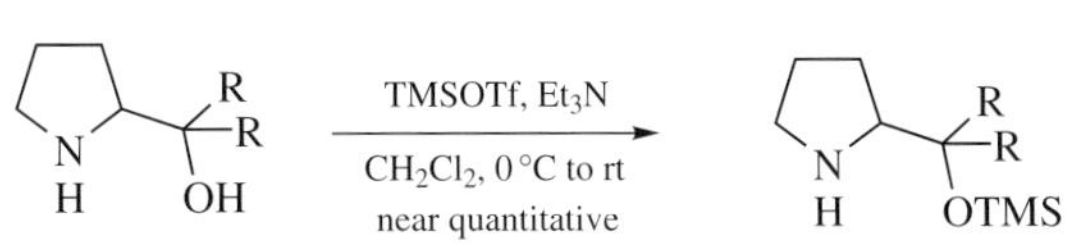

(34)

The combination of TMSOTf and Et$_3$N in dichloromethane (DCM) allows the direct conversion of *p*-methoxybenzyl ethers into silyl-protected alcohols, thus affording an expedient way to replace the benzyl ether-type protective group with the silyl ether-type one (eq 35).[46]

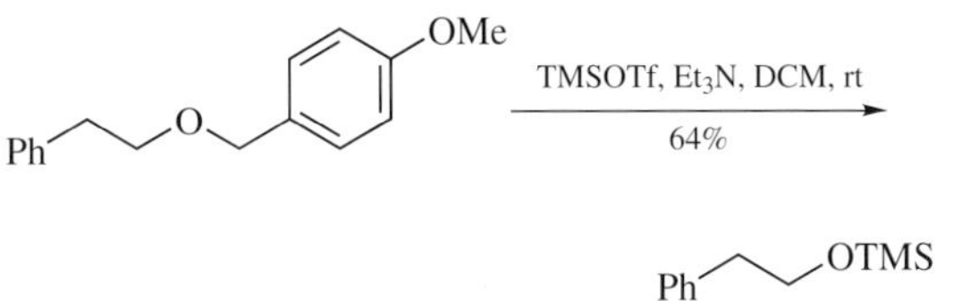

(35)

The preparation of silyl enol ethers from carbonyl compounds represents one of the major uses of TMSOTf. Recently, the stereochemistry and regiospecificity of such transformation has been addressed for aldehydes[47] and α-(*N*-alkoxycarbonylamino) ketones,[48] respectively. On the other hand, enantiopure silyl enol ethers can be formed by addition of TMSOTf to zinc enolates, which are obtained from the copper-catalyzed enantioselective conjugate addition of dialkylzinc reagents to cyclic (eq 36) and acyclic enones.[49]

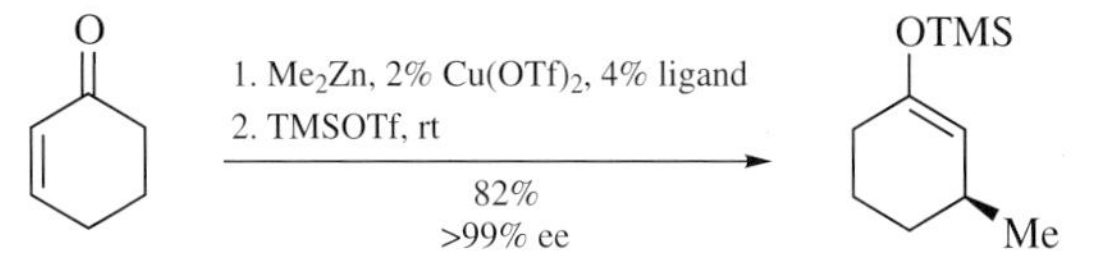

(36)

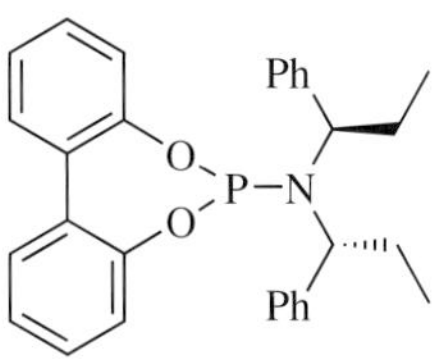

The addition of dimethyl sulphide to α,β-unsaturated carbonyl compounds in the presence of TMSOTf generates highly reactive 3-trialkylsilyloxyalk-2-enylenesulfonium salts, which permits the introduction of a wide variety of nucleophiles at the β-position as well as α-functionalization.[50]

Bis-silylation of imides in both oxygen atoms results in the formation of cyclic 2-azadienes (eq 37).[51]

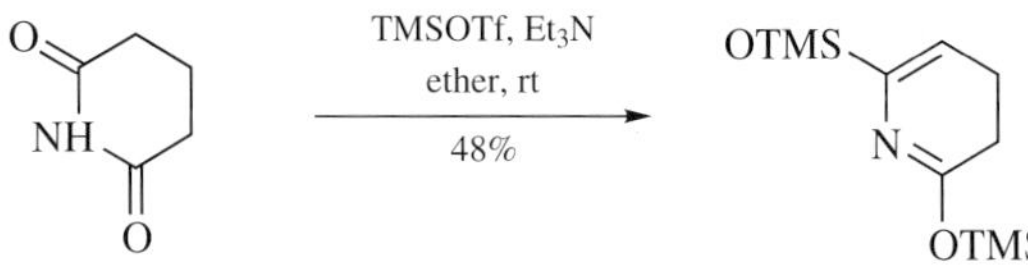

(37)

Silylated 2-oxidienes and bis-silylated 2,3-dioxidienes are prepared from α,β-unsaturated ketones[52] and 1,2-dicarbonyl compounds,[53] respectively; in the latter case, the reagent employed for the first silylation is trimethylsilyl bromide while TMSOTf is required for the second silylation. 1,3-Bis-silyloxydienes are prepared by direct silylation of 1,3-dicarbonyl compounds with TMSOTf/Et$_3$N.[54] They also can be obtained as mixtures of *Z/E*-isomers by lithium reductive cleavage of isoxazoles followed by the slow addition of an excess of TMSOTf/Et$_3$N (eq 38).[55]

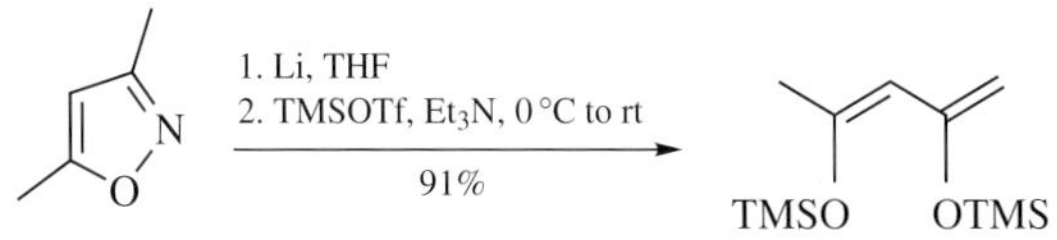

(38)

The formation of stable *N,O*-acetal TMS ethers, which are excellent precursors of *N*-acyliminium ions, is easily achieved by DIBAL reduction of *N*-acylamides followed by in situ protection with TMSOTf/pyridine (eq 39).[56] 2,6-Lutidine has also been used as base.[57] The DIBAL reduction-TMSOTf/pyridine silylation sequence has also been applied to the formation of monosilyl *O,O*-acetal from esters.[58]

$$(39)$$

Silylated α,β-unsaturated oximes, which can be easily desilylated to enoximes, can be readily prepared by treatment of aliphatic nitro compounds with an excess of TMSOTf in the presence of Et$_3$N, provided that an electron-withdrawing group is located at the β-position (eq 40).[59]

$$(40)$$

The in situ formation of silyl enol ether intermediates appears to be a valuable strategy to accomplish further transformations. That is what happens for the palladium(II) acetate-mediated preparation of indenones from indanones (eq 41),[60] or for diastereoselective enol silane coupling reactions (eq 42).[61]

$$(41)$$

$$(42)$$

A Claisen rearrangement of allylic esters is also effected by TMSOTf in the presence of triethylamine; when chiral esters are employed the transfer of chirality is, however, quite low. Higher degrees of 1,4-asymmetric transmission are reached if bulkier silyl triflates and bulkier amines are employed (eq 43).[62]

$$(43)$$

C-Silylation. Depending on the reaction conditions, secondary amides can be either C-silylated or N-silylated. Kinetic C-silylation of N-methylacetamide can be quantitatively achieved by treatment with excess TMSOTf (3 equiv) and Et$_3$N (3 equiv) at 0 °C for 2 min, followed by aqueous work-up (eq 44).[63]

$$(44)$$

The silylation of aromatic[64] or heteroaromatic compounds can be performed via magnesiated or lithiated intermediates, which can be accessed by hydrogen-metal or halogen-metal atom exchange. Thus, C-4 silylation of imidazoles is achieved from 4-iodo imidazoles (eq 45),[65] while C-2 silylated oxazoles are prepared by addition of TMSOTf to the lithiated oxazole (eq 46); interestingly, for this reaction, the employment of TMSCl as electrophile resulted in the formation of the O-silylation product.[66] Pyrazole N-oxides and 1,2,3-triazole N-oxides are also C-silylated on the ring and at exocyclic α-positions in high yields by a one-pot procedure, which is initiated by O-silylation followed by deprotonation, and subsequently terminated by silylation of the generated anion.[67]

$$(45)$$

$$(46)$$

The enantioselective C-silylation of allylic substrates such as N-(tert-butoxycarbonyl)-N-(p-methoxyphenyl)allylamines[68] or 1,3-diphenylpropene[69] is accomplished with butyllithium in the presence of (−)-sparteine, followed by the addition of TMSOTf (eq 47). The same procedure allows the asymmetric deprotonation-substitution of arenetricarbonyl(0) complexes,[70] while chiral bis(oxazolines) have been the ligands of choice to perform such transformation with aryl benzyl sulfides;[71] in these reactions, different yields and enantioselectivities are reached if trimethylsilyl chloride is used as silylating reagent, although there is a substrate dependence and no definite rules can be established.

$$(47)$$

1,1-Dibromo-4-methyl-3-(trimethylsilyloxy)pentane is cleanly silylated with TMSOTf through a preformed carbenoid species (eq 48). On the other hand, silylation can be carried out with TMS-imidazole albeit in lower yields, while the treatment with trimethylsilyl chloride is unsuccessful.[72]

C-Silylated cyclohexyl diazo esters can be prepared from the appropriate diazo acetates by treatment with TMSOTf and ethyl diisopropylamine in ether at −78 °C (eq 49).[73]

N-Silylation. The *N*-bis-silylation of α-amino acids with TMSOTf is only effective for glycine (eq 50); for other α-amino acids *N*-mono-silylation prevails because the larger size of the carbon chain at the α-position hinders bis-silylation.[74]

TMSOTf is the reagent of choice to transform imines into *N*-silyl iminium salts, which have been postulated to be the reactive intermediates in several reactions (eq 51).[75]

Secondary amides[76] and *O*-aryl *N*-isopropylcarbamates[77] can be converted in situ into their *N*-silylated derivatives which, although not isolated, are stable for further transformations (eq 52).

C,O-Bis-silylation. Bis-silylation of α,β-unsaturated carbonyl compounds can be achieved by palladium-TMSOTf-catalyzed addition of disilanes to enones, enals, or aromatic aldehydes via an η^3-silyloxyallylpalladium intermediate.[78] The scope of the reaction is wide and it generally leads to 1,4-bis-silylated compounds,

although for cinnamaldehyde, it gives a mixture of 1,4- and 1,2-addition products, and 1,2-bis-silylated adducts are obtained for aromatic aldehydes (eq 53).

Si-Si = PhMe$_2$Si-SiMe$_2$Ph or Me$_3$Si-SiMe$_3$

Carbonyl Activation. TMSOTf frequently acts as a Lewis acid and it is able to activate several functional groups (the carbonyl group, the acetal unit, the nitrone moiety,...) thus facilitating different kinds of reactions.

In this regard, it has promoted the direct conversion of carbonyl compounds into *O,S*-acetals by reaction of aldehydes with 1 equiv each of Me$_3$SSiPh and a silyl ether in CH$_2$Cl$_2$ at low temperature; the mentioned compounds are isolated as major reaction products (eq 54).[79]

1,3-Dioxolanation is a very common strategy for the protection of carbonyl groups. In this regard, an efficient dioxolanation of ketones in the presence of aldehydes is achieved by treating the dicarbonyl compound with TMSOTf and dimethyl sulphide in dichloromethane (eq 55).[80] On the other hand, both aldehydes and ketones can be protected as 4-trimethylsilyl-1,3-dioxolanes by using catalytic amounts of TMSOTf; this protective group can be selectively cleaved to regenerate the carbonyl compound in the presence of a 1,3-dioxolane with either LiBF$_4$ or HF in acetonitrile.[81]

The role of the ligand has been found to be crucial in the silyl Lewis acid Mukaiyama aldol reaction, which opens interesting applications for synthetic organic chemistry. When TMSOTf induces the reaction, the silyl group of TMSOTf remains in the product and that of the silyl enol ether becomes the catalyst for the next catalytic cycle; however, if the reaction is promoted by TMSNTf$_2$, the silyl group of the catalyst is not released from

-NTf$_2$ and that of the silyl enol ether intermolecularly transfers to the product (eq 56).[82]

$$(56)$$

catalyst			
TMSOTf	24%	99	1
TMSNTf$_2$	>99%	1	>99

The direct preparation of esters from aldehydes has been reported but different conditions are required depending on the nature of the aldehyde. The aldehydes are treated with the corresponding silyl ether followed by addition of AIBN and NBS in CCl$_4$; for aromatic aldehydes, the presence of TMSOTf is required to complete the radical oxidation to the ester; for aliphatic aldehydes, the acetals are isolated under the above-mentioned reaction conditions (eq 57), while the radical oxidation to the esters only takes place in the absence of TMSOTf.[83]

$$(57)$$

The condensation of a carbonyl compound with an isonitrile usually takes place in the presence of a carboxylic acid to give carboxamides (Passerini reaction) and plays a central role in combinatorial chemistry. Activation of the carbonyl group can be achieved by TMSOTf (eq 58), although higher yields can be obtained with the system Zn(OTf)$_2$/TMSCl.[84]

$$(58)$$

A previously known thermal ring-contraction, which follows an initial 1,3-dipolar cycloaddition between azides and cyclic enones, can be smoothly promoted by TMSOTf to give enaminones (eq 59).[85]

$$(59)$$

$$Z/E \text{ ratio} = 1{:}1 \text{ to } 2{:}1$$

Also, 3-stannyl cyclohexanones experience a ring contraction to a 2-methylcyclopentanone upon treatment with TMSOTf in dichloromethane (eq 60). This transformation has been used as a key step for the synthesis of (+)-β-cuparenone.[86]

$$(60)$$

The addition of bis(trimethylsilyl)formamide to methyl pyruvate is readily accomplished in the presence of a catalytic amount of TMSOTf (eq 61).[87]

$$(61)$$

Similarly, pyridine addition to aldehydes is also promoted by TMSOTf in a three-component reaction to form [1-(trimethylsiloxy)alkyl]pyridinium salts (eq 62), which may act as group transfer reagents or as precursors for the analogous phosphonium salts.[88]

$$(62)$$

The allylation of aldehydes can be achieved in the presence of ketones in a highly chemoselective fashion by preferential in situ conversion of aldehydes into 1-silyloxysulfonium salts, by treatment with dimethyl sulphide and TMSOTf and subsequent displacement with allyltributylstannane (eq 63).[89]

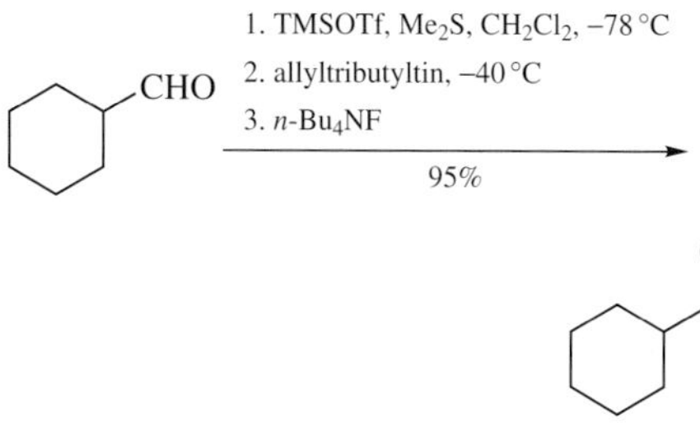

1. TMSOTf, Me$_2$S, CH$_2$Cl$_2$, −78 °C
2. allyltributyltin, −40 °C
3. *n*-Bu$_4$NF

95%

(63)

TMSOTf (1 equiv)

ClCH$_2$CH$_2$Cl, reflux

88%

(66)

Z:E = 24:76

TMSOTf also promotes an intramolecular addition of allyl-silanes to aldehydes (Sakurai reaction), when used with 2,6-di-*tert*-butyl-4-methylpyridine[90] (2,6-DTBMP) (eq 64), and ketones.[91]

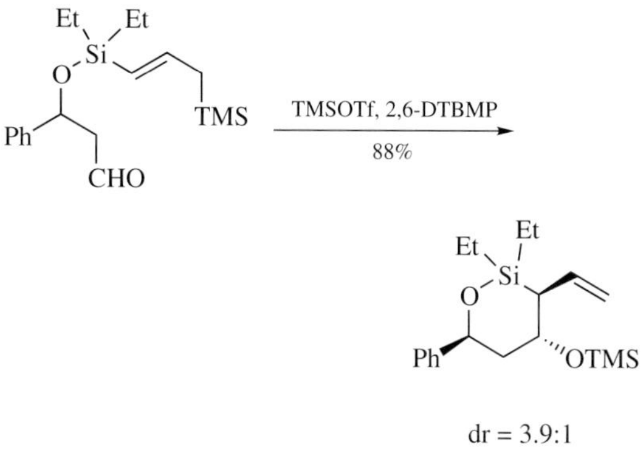

An improving effect of TMSOTf (and other trialkylsilyl triflates) in the enantioselectivity has been disclosed for the Cu-catalyzed 1,4-addition of Me$_3$Al to a 4,4-disubstituted cyclo-hexa-2,5-dienone, using a chiral oxazoline as ligand.[95] The presence of TMSOTf also helps to increase the yield for the conjugate addition of copper acetylides to α,β-unsaturated ketones and aldehydes; however, trimethylsilyl iodide has been more effective while other trimethylsilyl halides or BF$_3$ inhibit the addition.[96]

The conjugate addition of organozinc reagents to α,β-enones is readily mediated by TMSOTf (eq 67), thus avoiding previous requirements of transmetallation to organocopper compounds or the use of nickel complexes as catalysts.[97]

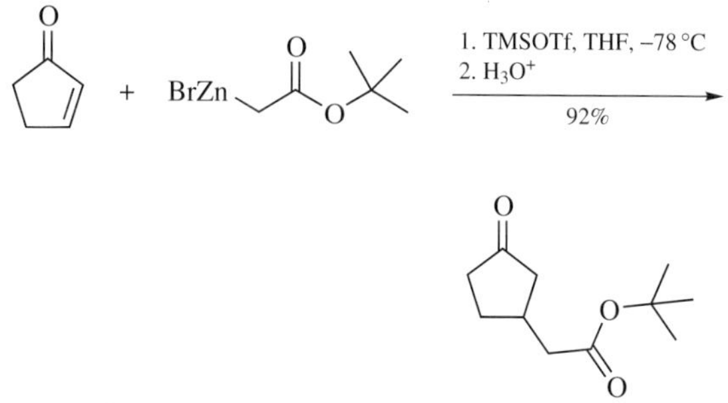

TMSOTf, 2,6-DTBMP

88%

(64)

dr = 3.9:1

An intermolecular version of the Sakurai reaction has been developed. It proceeds at temperatures as low as −78 °C, using substoichiometric amounts of the Lewis acid, to form five- or six-membered oxygenated heterocycles from cyclic allylsiloxanes and aldehydes through a chairlike transition state (eq 65). Acetals may also act as electrophiles in this kind of reaction.[92] β-Borylallylsilanes also undergo nucleophilic allylation to lead to the preparation of functionalized alkenylboranes, which may participate in further transformations.[93]

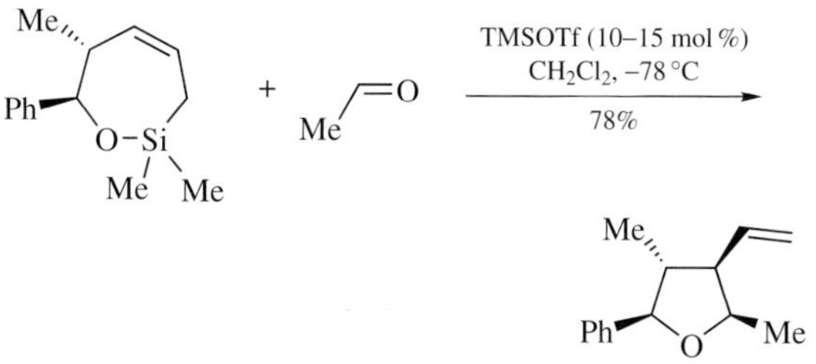

1. TMSOTf, THF, −78 °C
2. H$_3$O$^+$

92%

(67)

TMSOTf (10–15 mol %)
CH$_2$Cl$_2$, −78 °C

78%

(65)

The Lewis acid-mediated ring-opening reaction of arylcarbonyl activated cyclopropanes with arylaldehydes is successfully performed with 1 equiv of TMSOTf in refluxing 1,2-dichloroethane to form α-substituted α,β-enones in good yields (eq 66).[94]

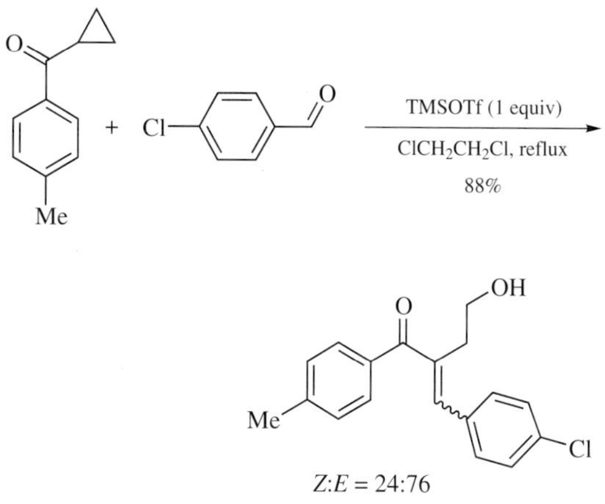

An asymmetric intramolecular Michael-aldol reaction which leads to nonracemic tricyclic cyclobutanes is performed by using TMSOTf and bis[(*R*)-1-phenylethyl]amine as chiral amine, but only moderate enantioselectivities are reached (eq 68).[98] A similar reaction sequence can also be carried out with TMSOTf and HMDS as base, with (−)-8-phenylmenthol as the chiral auxiliary; however, the iodotrimethylsilane-HMDS system is more efficient in terms of yield and diastereoselectivity.[99] The combination Et$_3$N/TMSOTf (or some other trialkylsilyl triflates) has been used to accomplish an intramolecular Michael reaction, which was the key step for the synthesis of sesquiterpene (±)-ricciocarpin A.[100]

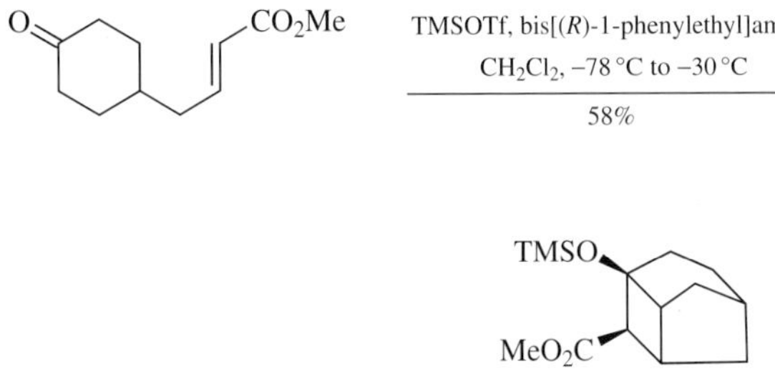

$$\text{(68)}$$

ee = 23%

On the other hand, β-functionalization of α,β-unsaturated lactones and esters is readily achieved through a phosphoniosilylation process: the addition of an aldehyde to the generated ylide followed by the fast addition of TMSOTf gave optimum results for the hydroxyalkylation reaction (eq 69). In the absence of TMSOTf alkylated products are obtained instead of the hydroxyalkylated ones.[101]

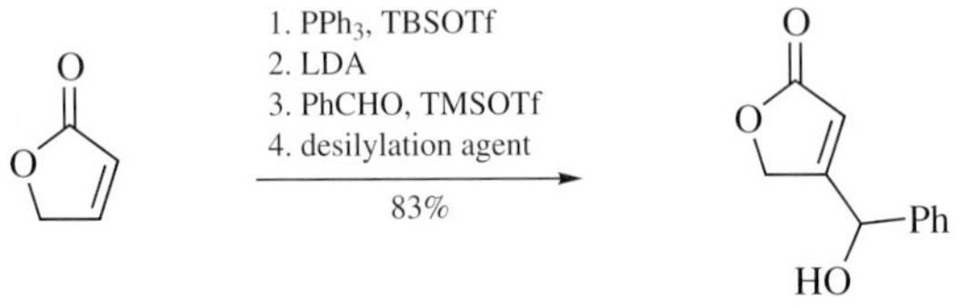

1. PPh₃, TBSOTf
2. LDA
3. PhCHO, TMSOTf
4. desilylation agent

$$\text{(69)}$$

83%

The combination of Me₃Al and TMSOTf allows the conversion of a carbonyl into a geminal dimethyl functionality (eq 70).[102] Other methylating reagents such as MeTiCl₃, Me₂TiCl₂, Me₂Zn, or Me₃Al itself were not successful for this transformation.

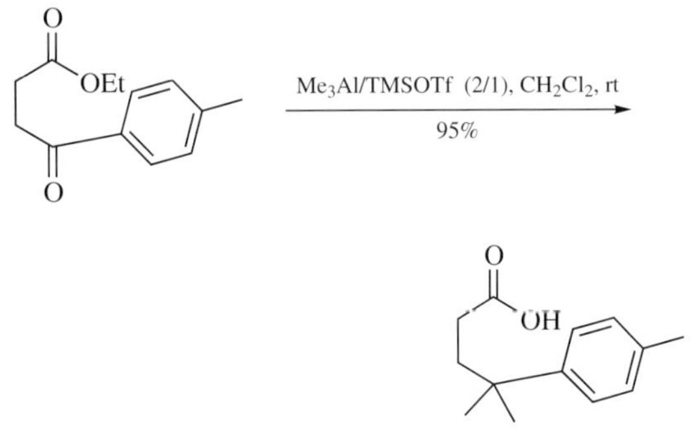

Me₃Al/TMSOTf (2/1), CH₂Cl₂, rt

95%

$$\text{(70)}$$

The coupling reaction between an electron-deficient alkene and an aldehyde (Baylis–Hillman reaction) usually requires a catalyst/catalytic system (typically, a tertiary amine and a Lewis acid) to be successful. The base catalyst is not necessary when pyridine-2-carboxaldehyde is employed as electrophile; it is enough with the activation effected by stoichiometric amounts of TMSOTf for the reaction to proceed to give indolizidine derivatives (eq 71).[103]

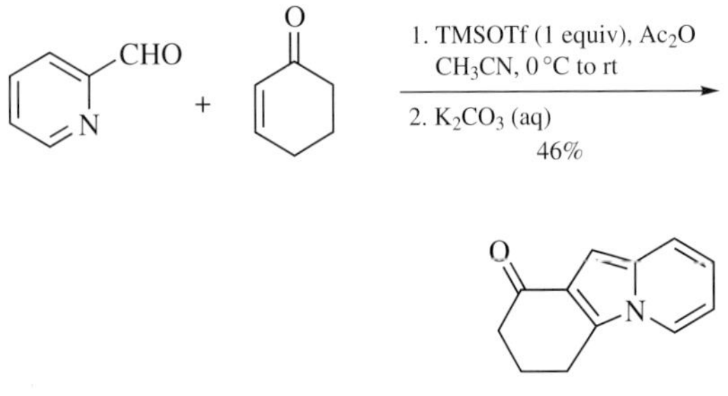

1. TMSOTf (1 equiv), Ac₂O
   CH₃CN, 0 °C to rt
2. K₂CO₃ (aq)
   46%

$$\text{(71)}$$

This reaction can be promoted by the use of dimethyl sulphide as a catalyst (Chalcogeno-Baylis-Hillman reaction) in the presence of stoichiometric amounts of TMSOTf; in the case of the reaction between methyl vinyl ketone and p-nitrobenzaldehyde a 2:1 adduct was isolated (eq 72), with incorporation of two units of ketone into the final product.[104]

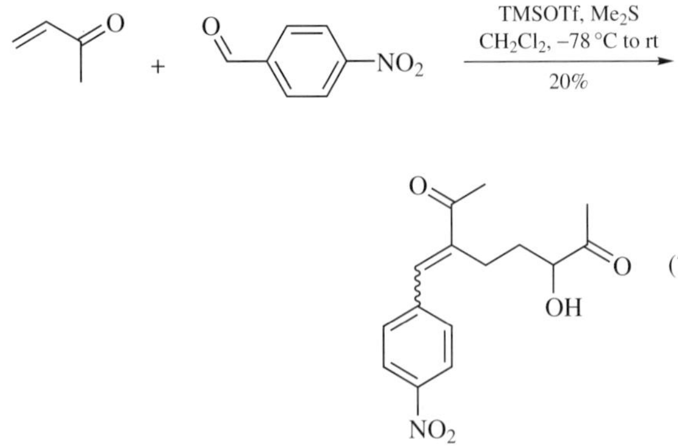

TMSOTf, Me₂S
CH₂Cl₂, −78 °C to rt

20%

$$\text{(72)}$$

A synthesis of cross-conjugated 2-cyclopenten-1-ones from dialkenyl ketones is readily induced by TMSOTf (eq 73). A strong fluorine-directing effect has been observed for such Nazarov-type cyclization, as mixtures of products have been observed for nonfluorinated dialkenyl ketones. The addition of 1,1,1,3,3,3-hexafluoro-2-propanol (HFIP) as a cosolvent dramatically accelerates the cyclization. Other acids such as BF₃·OEt₂, FeCl₃, polyphosphoric acid, or TiOH are less effective while neither TMSI nor TMSOMe promote this cyclization at all in CH₂Cl₂.[105] 3-Ethoxycarbonyltetrahydro-γ-pyrones also undergo such Nazarov-type cyclization.[106]

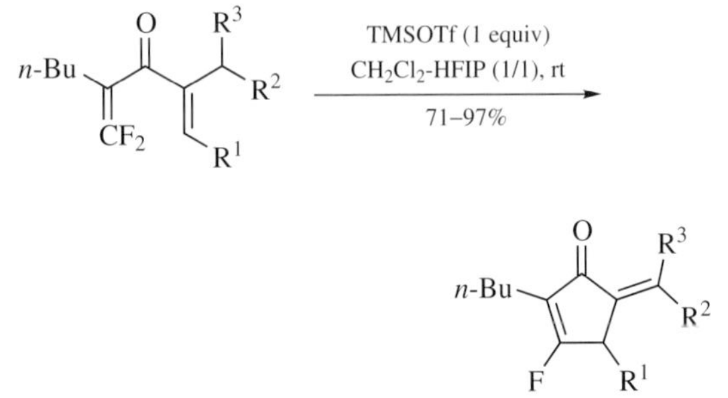

TMSOTf (1 equiv)
CH₂Cl₂-HFIP (1/1), rt

71–97%

$$\text{(73)}$$

A catalytic amount of TMSOTf is enough to catalyze the [4 + 3] cycloaddition between 1,4-dicarbonyl compounds and bis(trimethylsilyl) enol ethers; the process is highly chemo- and regioselective (eq 74).[107] When 1,5-diketones are employed, the title compound is regarded as the ideal catalyst to achieve [5 + 3] cyclizations with the bis(trimethylsilyl)enol ether of methyl acetoacetate.[108]

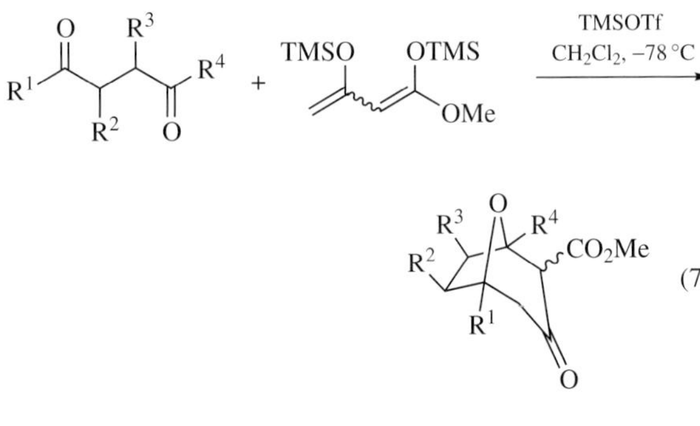

TMSOTf
CH₂Cl₂, −78 °C

$$\text{(74)}$$

In a highly polar medium such as 3.0 M lithium perchlorate-ethyl acetate, TMSOTf is an effective reagent for promoting intermolecular[109] or intramolecular[110] all-carbon cationic [5 + 2] cycloaddition reactions (eq 75).

$$(75)$$

TMSOTf also interacts with carbonyl[111] or imino[112] groups promoting intermolecular hetero-Diels-Alder reactions with efficient control of the stereochemistry (eq 76).

$$(76)$$

Acetal Activation. TMSOTf acts as a catalyst for the addition of several nucleophiles (allylsilanes, allylstannanes, silyl enol ethers, trimethylsilyl cyanide) toward *N,O*-acetals, which were obtained following a DIBAL reduction-silylation sequence from acyl amides, by promoting the formation of the corresponding acyliminium ions (eq 77).[56a] In such reactions, boron trifluoride etherate gives usually slightly higher yields. TMSOTf also acts as a catalyst for the analogous addition of silyl enol ethers to *O,O*-acetals.[58]

$$(77)$$

In other reactions that proceed via an acyliminium ion, *O*-vinyl *N,O*-acetals rearrange smoothly to β-(*N*-acylamino)aldehydes at 0 °C in CH_2Cl_2 in the presence of TMSOTf with moderate to high diastereoselectivities (eq 78).[113] However, TMSOTf failed to promote aprotic alkyne-iminium cyclizations, which are readily enhanced by TMSCl, TMSBr, or $SiCl_4$.[114] On the other hand, TMSOTf assists in the addition of enols to heteroaromatic imines or hydroxyaminal intermediates.[115]

$$(78)$$

dr = 20–96%

The tropane alkaloid skeleton can be accessed in one pot via domino ene-type reactions of acetone silyl enol ether, the first one of them being intermolecular, with catalytic use of TMSOTf (eq 79).[116] Alternatively, asymmetric tropinones can be reached by cyclization of 1,3-bis-silyl enol ethers with acyliminium triflates.[117]

$$(79)$$

TMSOTf is effective in both stoichiometric and catalytic modes to promote the [4 + 3] cycloaddition of dienes to 1,1-dimethoxy-acetone-derived silyl enol ethers (eq 80). $SnBr_4$, $SnCl_4$, $TiCl_4$, $SbCl_5$, BCl_3, and $Ph_3C^+BF_4^-$ gave satisfactory results when employed in stoichiometric amounts, while the cycloaddition does not take place for $Sn(OAc)_4$, $SnCl_2$, $ZnBr_2$, or $AgBF_4$.[118]

$$(80)$$

A previously described procedure, reported to be efficient for dioxolanes and noncyclic acetals,[42] has been generalized for both cyclic and acyclic acetals of ketones and aldehydes,[119] and it has been successfully applied to the opening of 1,3-dioxanes to enol ethers (eq 81).[120] However, aromatization has occurred for 4-(*N*-benzoylamino)-4-phenyl-2,5-cyclohexadienone dimethyl ketal (eq 82).[121] This method has also allowed the development of 1-methylcyclopropyl ether as a protecting group for the hydroxyl moiety which is prepared regioselectively from terminal acetonides.[122]

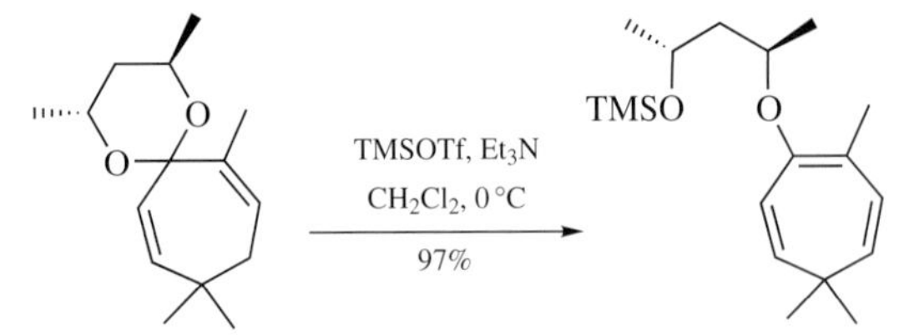

(81)

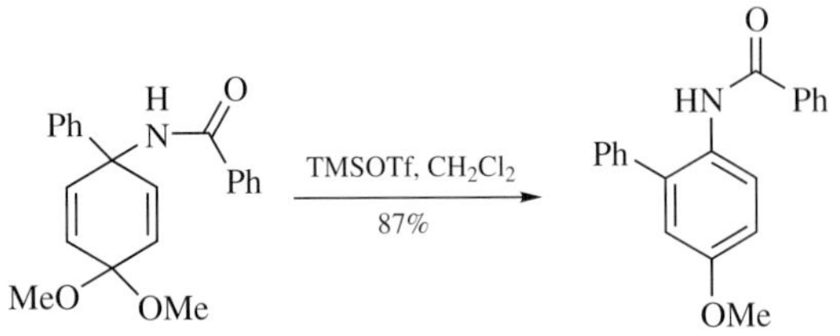

(82)

The allylation of acetals is also promoted by TMSOTf. In fact, TMSOTf is the only viable Lewis acid for the intermolecular allylation which is achieved employing allylborates, generated in situ from triallylborane, as the allyl source, and butyllithium in THF at $-78\,^{\circ}$C (eq 83).[123] The intramolecular allylation of mixed silyl-substituted acetals reaches high diastereoselectivities when promoted by TMSOTf (eq 84), although the diastereoselectivities are even higher and very good yields are reached when the intramolecular allylation reaction is promoted by boron trifluoride etherate without prior synthesis of the mixed acetal.[124] However, depending on the structure of the mixed acetal, TiCl$_4$ provides higher diastereoselectivities while the intramolecular allylation with TMSOTf as Lewis acid proceeds without stereoselectivity.[125] Thioacetals are suitable substrates for the TMSOTf-promoted addition of allylmetals.[126] On the other hand TiCl$_4$, SnCl$_4$, or Et$_2$AlCl provide higher yields although slightly lower diastereoselectivities than TMSOTf for the addition of other nucleophiles such as silyl enol ethers to trialkylsilyl-substituted acyclic-mixed acetals.[127]

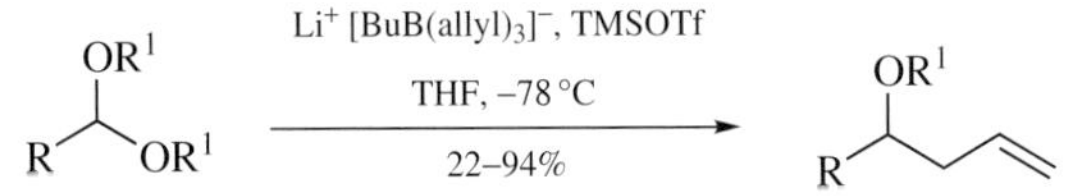

(83)

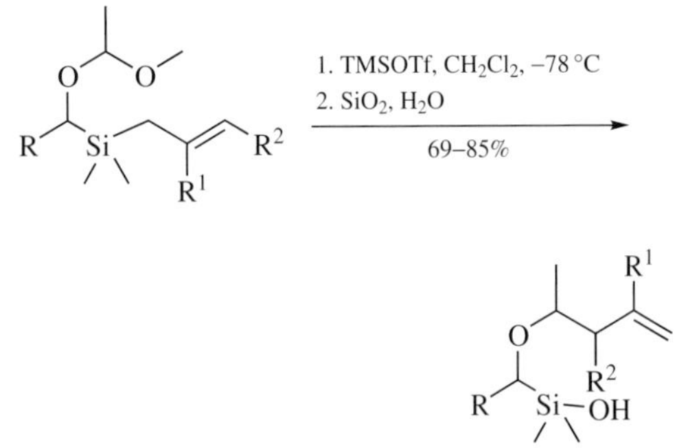

(84)

diastereoselectivity = 1.6–34:1

The reduction of ketals to protected secondary alcohols is readily accomplished in high yield with borane-dimethyl sulphide upon activation with TMSOTf at $-78\,^{\circ}$C (eq 85). Other Lewis acids require higher temperatures, and 1 equiv of TMSOTf is essential for complete conversion of the ketal.[128] The solvent has

a decisive role for controlling the site selectivity in the case of unsymmetrical diols.[129]

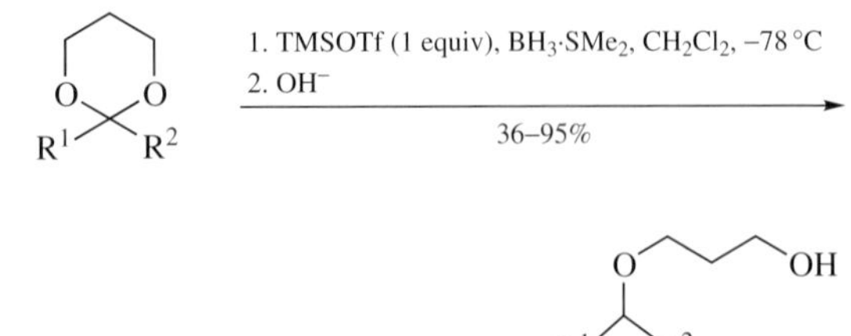

(85)

TMSOTf also plays an important role in sugar chemistry. In fact it often mediates in the formation of the glycosidic bond, either alone[130] or in conjunction with trimethylsilyl chloride,[131] *N*-bromosuccinimide (NBS),[132] or *N*-iodosuccinimide (NIS).[133] In such reactions, different moieties have acted as leaving groups[134] in the glycosyl donor and *O*-, *N*-, or *C*-nucleophiles[135] have been employed (eq 86). The addition of hydroxyl unprotected sugar units to the *exo*-anomeric double bond of ketene dithioacetals, which act as glycosyl donors, is also promoted by TMSOTf.[136]

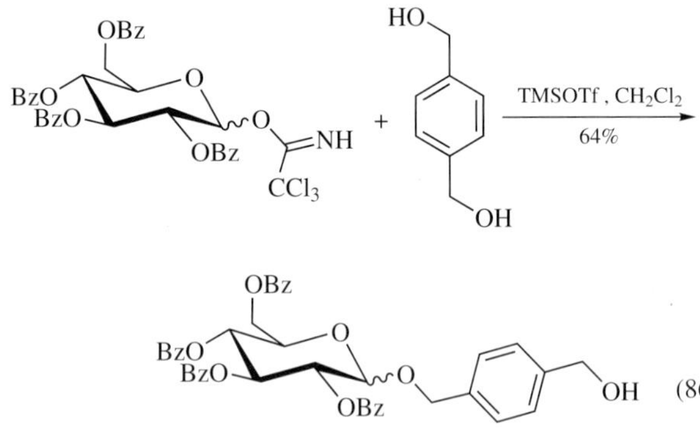

(86)

A TMSOTf-induced intramolecular cyclization of silyl enol ethers also takes place with orthoesters (eq 87).[137]

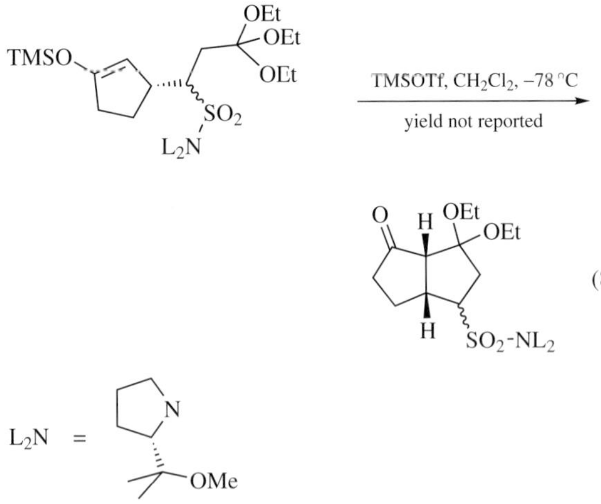

(87)

Nitrone Activation. The nucleophilic addition to aldonitrones depends on the nature of the metal involved and the presence/absence of an activator. Thus, allylsilanes add to aromatic nitrones to give homoallylhydroxylamines in excellent yields (eq 88);[138] other Lewis acids were unsuccessful with TiCl$_4$ as the only exception, although it produced just a 32% yield.

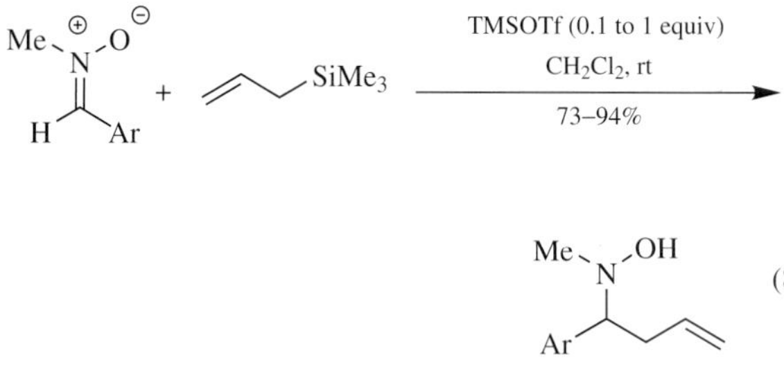

TMSOTf (0.1 to 1 equiv)
CH$_2$Cl$_2$, rt
73–94%

(88)

The [3 + 2] dipolar cycloaddition of nonaromatic nitrones with allyltrimethylsilane, which requires temperatures >100 °C without a catalyst, is carried out at temperatures ≤20 °C in the presence of TMSOTf (eq 89), with moderate to good yields and low to moderate diastereoselectivities.[139]

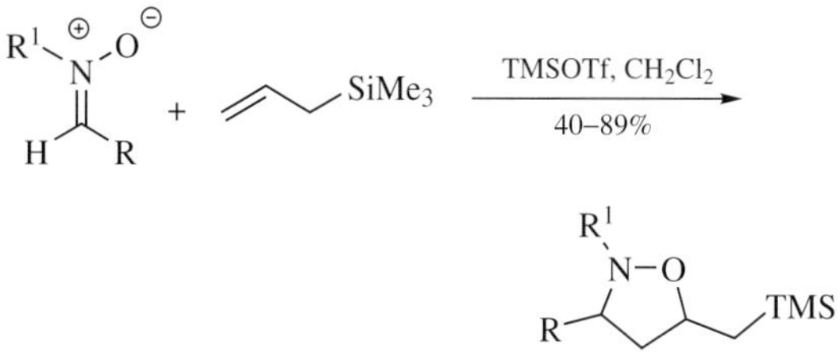

TMSOTf, CH$_2$Cl$_2$
40–89%

(89)

cis/trans = 61/39 to 15/85

On the other hand, allylstannanes add to nitrones in the presence of TMSOTf to give *O*-silylated homoallylic hydroxylamines in high yields; when the crude reaction mixture is quenched with NIS, 5-iodomethylisoxazolines are formed in excellent yields (eq 90).[140]

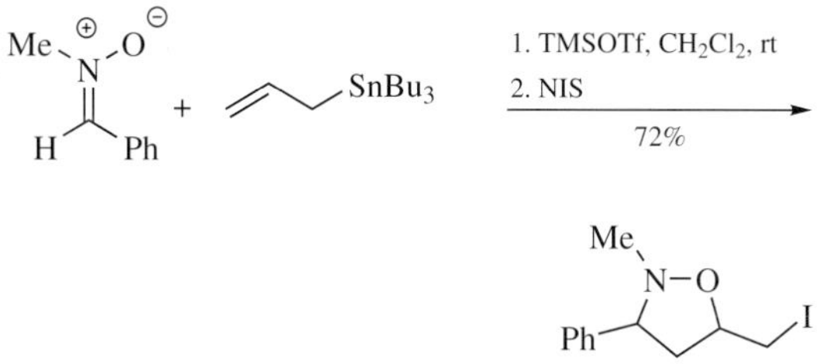

1. TMSOTf, CH$_2$Cl$_2$, rt
2. NIS
72%

(90)

cis/trans = 55/45

The addition of silyl ketene acetals to chiral aldonitrones requires the use of Lewis acids as activating reagents. Whereas activation with TMSOTf followed by a treatment with hydrofluoric acid-pyridine leads to the *syn*-adducts of isoxazolidin-5-ones (eq 91), the use of diethylaluminium chloride or boron trifluoride etherate leads to *anti*-compounds.[141]

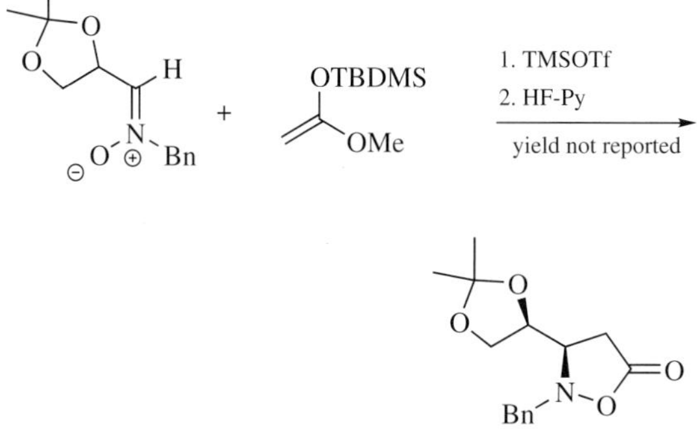

1. TMSOTf
2. HF-Py
yield not reported

(91)

Epoxide Ring Opening. One-pot alkylation-*O*-silylation reactions of epoxides take place in excellent yields when the epoxides are treated with trialkylaluminum-TMSOTf and Et$_3$N (1.5 equiv of each one) in methylene chloride at −50 °C (eq 92). Other trialkylsilyl triflates also undergo this reaction, which is stereospecific: *anti*-compounds are obtained from *trans*-epoxides while *cis*-epoxides yield *syn*-adducts.[142]

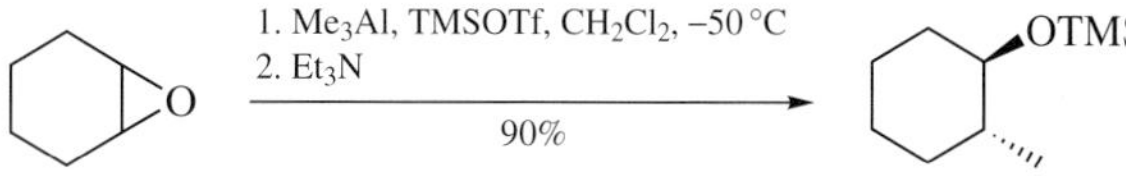

1. Me$_3$Al, TMSOTf, CH$_2$Cl$_2$, −50 °C
2. Et$_3$N
90%

(92)

Chiral aryl epoxides are converted into hydroxy-protected chiral α-hydroxy aryl ketones with complete retention of the chiral center and good regioselectivity by treatment with a combination of DMSO and TMSOTf (eq 93). Other reagent combinations such as DMSO/BF$_3$·OEt$_2$ or DMSO/TfOH lead to lower yields of the desired alcohols.[143]

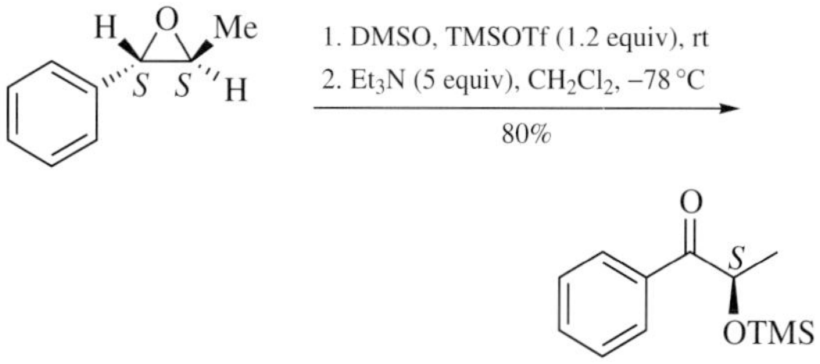

1. DMSO, TMSOTf (1.2 equiv), rt
2. Et$_3$N (5 equiv), CH$_2$Cl$_2$, −78 °C
80%

(93)

ee > 96%

The TMSOTf-induced intramolecular reaction of allylstannanes with epoxides causes a regio- and stereoselective cyclization to five- or six-membered carbocycles (eq 94), depending on the substituents of the starting material. TiCl$_4$ also gives high yields and slightly higher selectivities; catalytic amounts of Lewis acids are insufficient for completion of the reactions.[144]

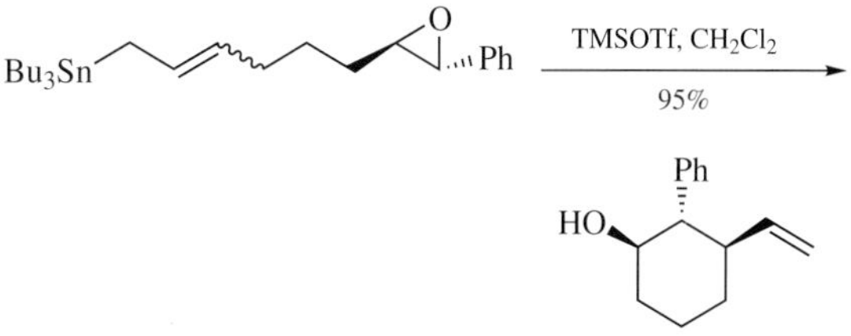

TMSOTf, CH$_2$Cl$_2$
95%

(94)

Silylated aldol adducts can be reached by using a nonaldol rearrangement promoted by treatment of bulky epoxy bis-silyl ethers with TMSOTf/*i*-Pr$_2$NEt in methylene chloride at −50 °C (eq 95).[145] Bulky mesylated epoxy silyl ethers also undergo this transformation; however, a silyl triflate-promoted Payne rearrangement was observed as a side reaction depending on the stereochemistry of the starting epoxide.[146]

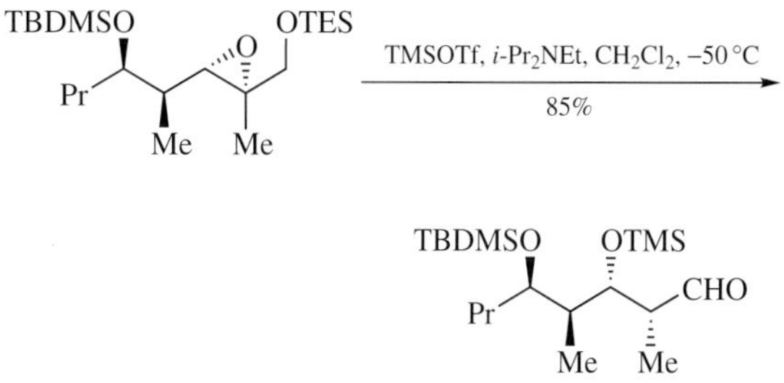

TMSOTf, *i*-Pr$_2$NEt, CH$_2$Cl$_2$, −50 °C
85%

(95)

Cleavage of Protecting Groups. THP ethers of primary, secondary, and phenolic alcohols can be conveniently deprotected at room temperature by treatment with 1.2 equiv of TMSOTf in methylene chloride (eq 96).[147] Deprotection of *tert*-butyldimethylsilyl ethers in the presence of a *tert*-butyldiphenylsilyl ether has been smoothly accomplished using TMSOTf at $-78\,°C$, as one of the steps of the total synthesis of marine macrolide ulapualide A.[148]

$$Ph\text{—}\text{OTHP} \xrightarrow[99\%]{\text{TMSOTf, CH}_2\text{Cl}_2} Ph\text{—}\text{OH} \quad (96)$$

TMSOTf has been reported to be a selective and efficient regent for the cleavage of *tert*-butyl esters in the presence of *tert*-butyl ethers, under virtually neutral conditions (eq 97), while iodotrimethysilane effects the deprotection of both species. No racemization was observed when this procedure was applied to the synthesis of *tert*-butoxy amino acids.[149]

$$R\overset{O}{\underset{}{\text{—}}}O\text{—} \xrightarrow[53-90\%]{\substack{\text{1. TMSOTf, Et}_3\text{N, dioxane, reflux} \\ \text{2. H}_2\text{O}}} R\overset{O}{\underset{}{\text{—}}}\text{OH} \quad (97)$$

6-Substituted isopropyl-protected guaiacols undergo transprotection in one step when treated with TMSOTf in anhydrous acetonitrile (eq 98); very high yields are obtained for this transformation if the substituent at position 6 is bulky and electron-withdrawing. However, mono- and disubstituted isopropyl phenyl ethers undergo Friedel-Crafts acetylation rather than transprotection.[150]

$$\xrightarrow[62\%]{\text{TMSOTf, Ac}_2\text{O, CH}_3\text{CN, rt}} \quad (98)$$

The system TMSOTf/2,6-lutidine has been proved to be highly effective for the deprotection of cyclic and acyclic acetals from aldehydes (eq 99); on the other hand, high chemoselectivity has been achieved when using triethylsilyl triflate/2,6-lutidine, which can deprotect acetals in the presence of ketals.[151]

$$\xrightarrow[83\%]{\substack{\text{1. TMSOTf, 2,6-lutidine, CH}_2\text{Cl}_2, 0\,°C \\ \text{2. H}_2\text{O}}} \quad (99)$$

The same combination of reagents (TMSOTf/2,6-lutidine) has been employed to deprotect *N-tert*-butoxycarbonyl groups from substrates in the solid phase synthesis of several peptides, without cleaving the substrates from the support as would occur in the case of using TFA with TFA-sensitive resins (such as Rink's amide resin) (eq 100). This method has great potential for the solid-phase synthesis of small molecule libraries.[152] Such a reagent combination had been previously developed for solution phase reactions of nonpeptidic substrates.[153]

$$\xrightarrow[\substack{\text{2. MeOH}}]{\substack{\text{1. TMSOTf, 2,6-lutidine} \\ \text{CH}_2\text{Cl}_2, \text{rt}}} \quad (100)$$

A combination of thioanisole and TMSOTf was used to nucleophilically release phenols from a polystyrene resin (eq 101).[154]

$$\xrightarrow[\substack{10\% \text{ TFA:CH}_2\text{Cl}_2, \text{rt}}]{\text{TMSOTf, anisole (10 equiv)}} \quad (101)$$

Hypervalent Iodine Chemistry. The formation of hypervalent iodine complexes is often promoted by TMSOTf. Thus, TMS-alkynes can be transformed with iodosobenzene and TMSOTf into phenyl iodonium triflates in moderate to good yields, which may be subsequently converted into alkyneamides (eq 102).[155]

$$\xrightarrow[64\%]{\text{TMSOTf, CH}_2\text{Cl}_2, -20\,°C}$$

$$\xrightarrow[46\%]{\substack{\text{BuLi, toluene, 0}\,°C}} \quad (102)$$

The TMSOTf-mediated reaction of benziodoxoles with aromatic nitrogenated heterocycles[156] or azides[157] permits the preparation of hypervalent iodine complexes bearing nitrogen ligands (eq 103) as well as several other unsymmetrical tricoordinate iodinanes.[158] TMSOTf has found widespread use as a reagent for the preparation of fluorinated hypervalent iodine compounds.[159]

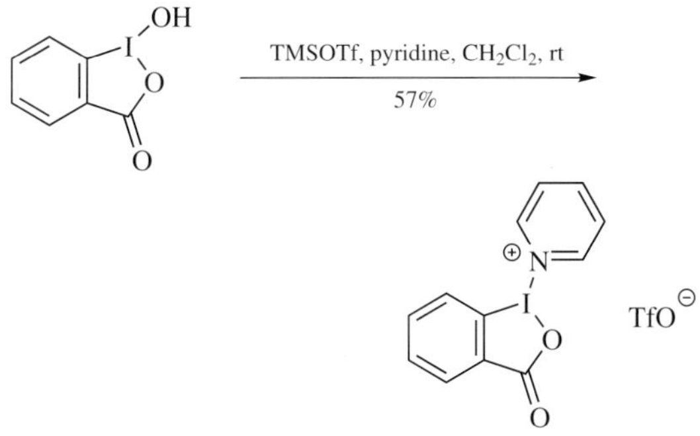

(103)

1-Alkynylbenziodoxoles are selectively formed in high preparative yield by the sequential addition of trimethylsilylacetylenes and pyridine to benzoiodoxole triflates,[160] which are prepared in situ from 2-iodosylbenzoic acid and TMSOTf (eq 104).

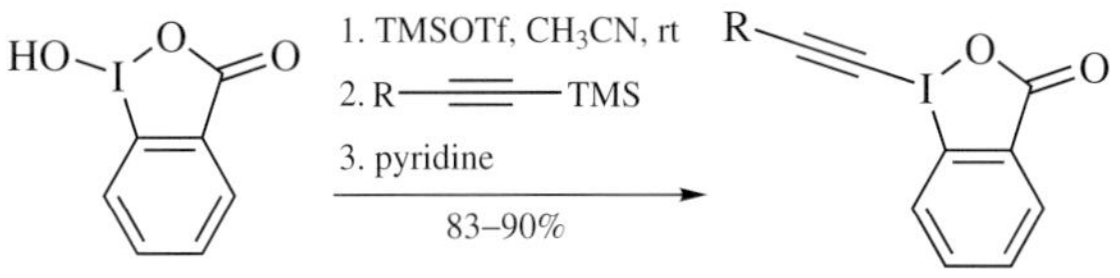

(104)

On the other hand, TMSOTf activates acetoxybenziodazoles to afford 3-iminobenziodoxoles upon reaction with amides and alcohols via an acid-catalyzed rearrangement (eq 105).[161]

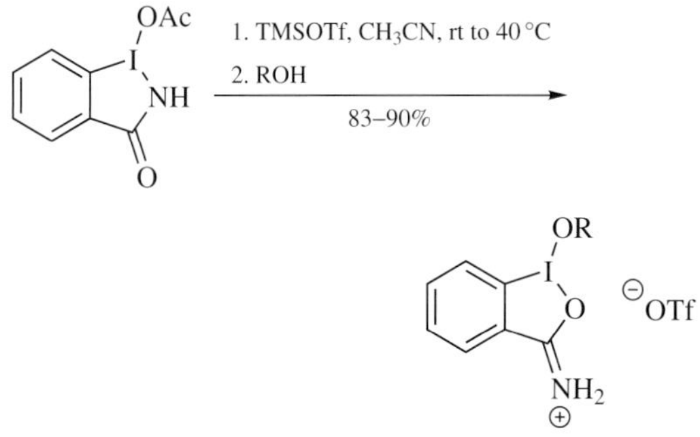

(105)

The formation of cationic iodonium macrocycles such as rhomboids, squares, or pentagons can be carried out in high yields by reaction of hypervalent iodoarenes with TMS-substituted aromatic compounds in the presence of TMSOTf (eq 106).[162]

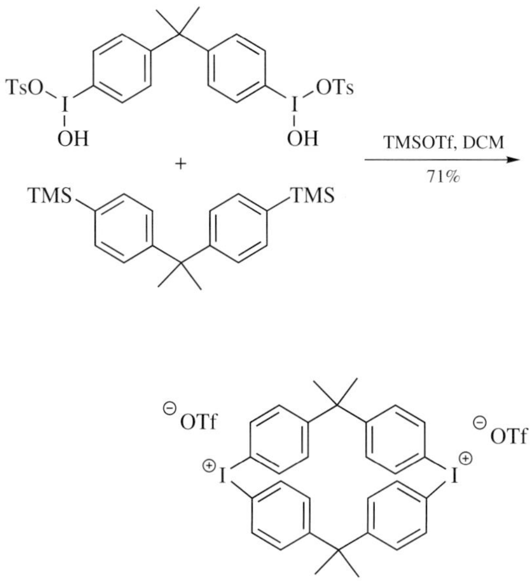

(106)

A combination of a hypervalent iodine(III) reagent [phenyl-iodine(III) bis(trifluoroacetate)] and TMSOTf has shown great efficiency for the synthesis of pyrroloiminoquinones from 3-(azidoethyl)indole derivatives (eq 107).[163]

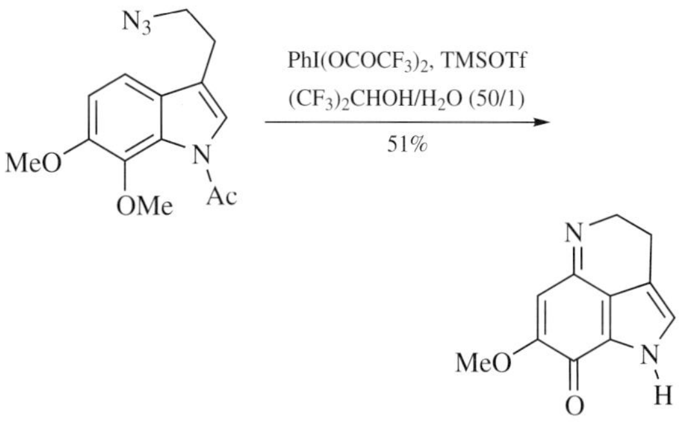

(107)

Miscellaneous. TMSOTf has been also the Lewis acid of choice (2 equiv) to promote the olefination reaction to give α-cyano α,β-unsaturated aldehydes from 2-cyano-3-ethoxy-2-en-1-ols (eq 108), which takes place at $-78\,^{\circ}\mathrm{C}$.[164]

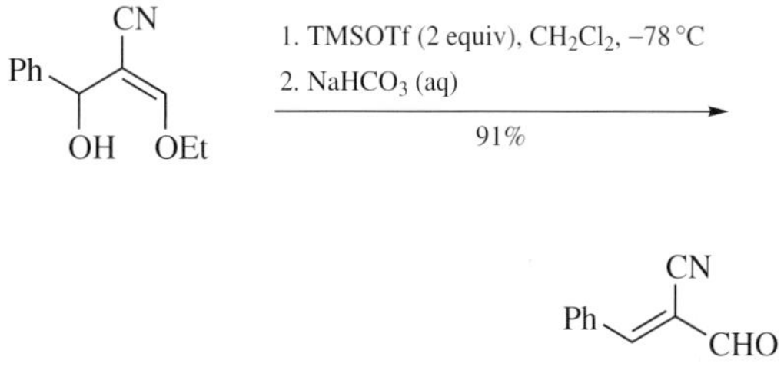

(108)

The reaction of pyridine with chiral acyl chlorides mediated by TMSOTf results in the formation of chiral N-acylpyridinium ions which can be trapped with organometallic reagents to form N-acyldihydropyridines or N-acyldihydropyridones in a diastereoselective manner (eq 109).[165] Catalytic TMSOTf also partakes in the acylation of cyclic bis(trimethylsiloxy)-1,3-dienes, which are synthons of 1,3-dicarbonyl dianions.[54]

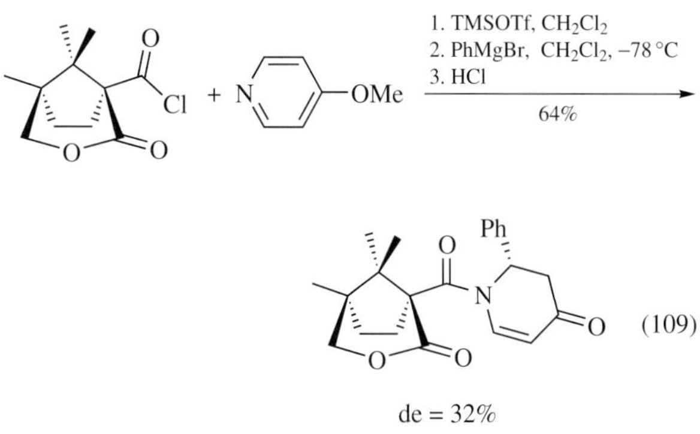

(109)

de = 32%

The reaction of 1,1-bis(trimethylsiloxy)ketene acetals with oxalyl chloride results in a straightforward one-pot synthesis of a variety of 3-hydroxymaleic anhydrides (eq 110).[166]

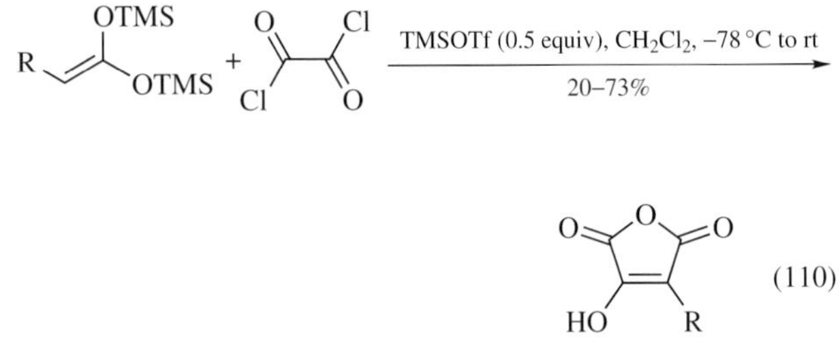

(110)

TMSOTf gives the best yields, among the tested Lewis acids ($ZnBr_2$, $TiCl_4$, $SnCl_4$), for the synthesis of functionalized 1,2,3,4-β-tetrahydrocarbolines from enamido ketones,[167] through the formation of N-acyliminium ions as intermediates (eq 111).

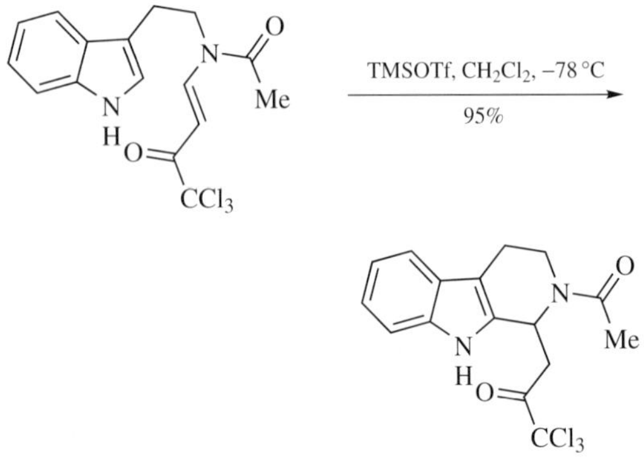

(111)

4-Trimethylsilyloxy-1-benzothiopyrylium triflate and 4-trimethylsilyloxyquinolinium triflate, which can be easily generated by TMSOTf addition to 4H-1-benzothiopyran-4-one[168] and 4-quinolone,[169] respectively, behave as Michael acceptors as they undergo the in situ TMSOTf-promoted addition of C-nucleophiles such as allylstannanes and silyl enol ethers, respectively. In a similar manner, 4-trimethylsilyloxypyrylium triflate, easily accessible in situ from pyran-4-one, undergoes a domino reaction sequence with 2 equiv of a silyloxybuta-1,3-diene leading to the formation of tetrahydro-2H-chromenes as sole products with high yields (eq 112).[170]

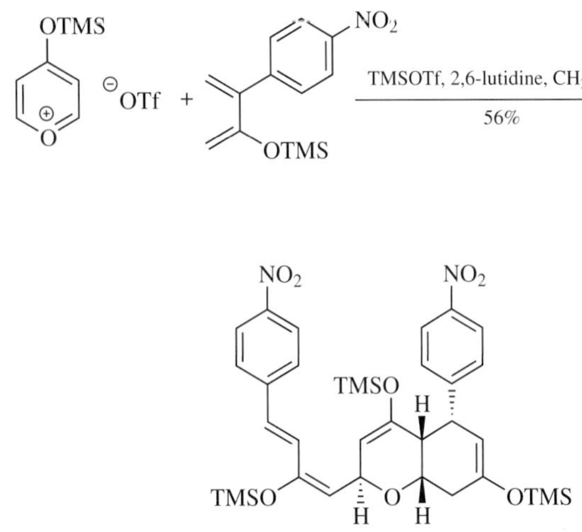

(112)

TMSOTf promotes the Diels-Alder cycloaddition of p-tolyl vinyl sulfoxide with furan or cyclopentadiene;[171] the reaction is slower and less stereoselective than when induced by ethyl Meerwein's reagent (Et_3OBF_4). It also promotes the [4 + 2] cycloaddition between 2,2-dimethoxyethyl acrylate and several

dienes with high yields and stereoselectivities (eq 113);[172] $TiCl_4$ and Ph_3CClO_4 also give satisfactory results in such processes.

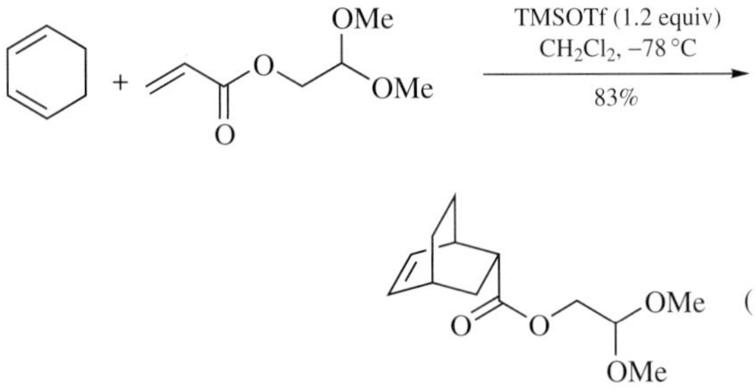

(113)

endo:exo = >99:1

Just catalytic amounts of the title reagent are required to efficiently mediate in the conversion of Diels-Alder adducts of 1-methoxy-3-[(trimethylsilyl)oxy]-1,3-butadiene (Danishefsky's diene) to cyclohexenones (eq 114).[173] Dilute HCl or concentrated acid lead to a greater percentage of side products.

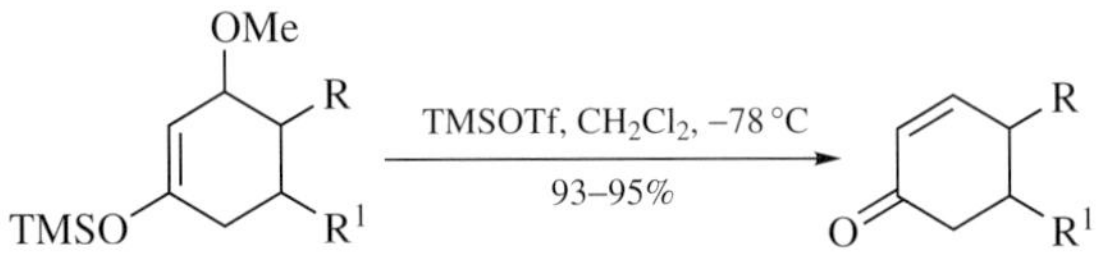

(114)

The addition of Lewis acids also improves the yield for a Beckmann fragmentation followed by carbon-carbon bond formation through reaction of oxime acetates with organoaluminum reagents (eq 115). $ZnCl_2$ is generally more effective than TMSOTf for this transformation.[174]

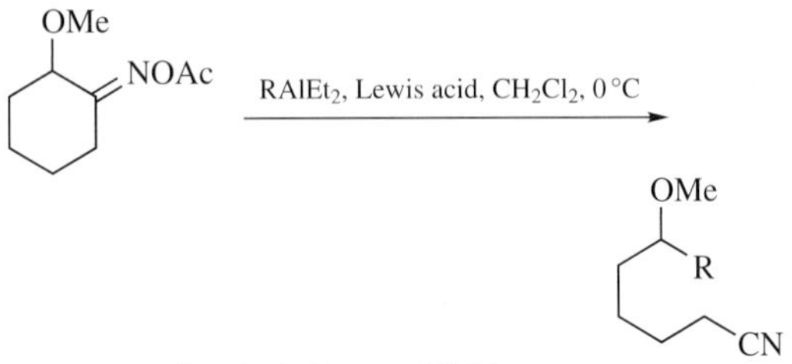

(115)

Lewis Acid	Yields
TMSOTf	39–62%
$ZnCl_2$	35–91%

The mixture of 1,3-dibromo-5,5-dimethylhydantoin (DBDMH)-TMSOTf in methylene chloride has proved to be more reactive than DBDMH alone for the aromatic bromination of arenes.[175]

On the contrary, debromination of α-bromo carboxylic acid derivatives can be achieved in high yield by the combination of triphenylphosphine and TMSOTf (eq 116). Other Lewis acids such as germanium(IV) chloride proved to be even more efficient for such transformation.[176]

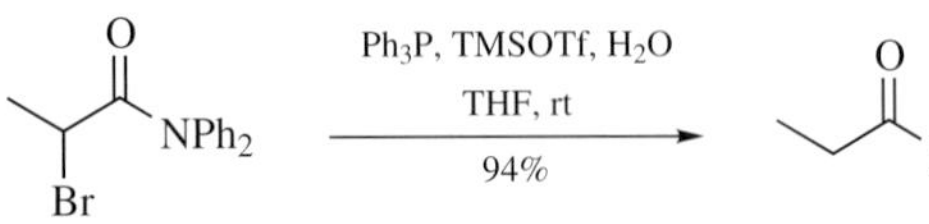

(116)

The generation of thionium ions from sulfoxides bearing an α-hydrogen can be carried out by using TMSOTf-Et_3N as initiator to afford Pummerer reaction products (eq 117).[177]

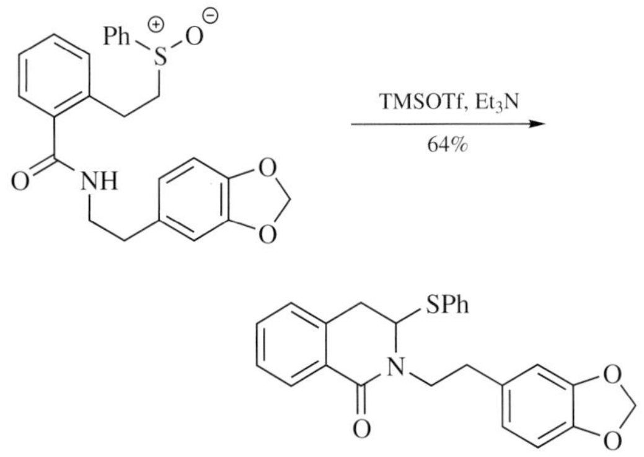

(117)

Several coupling reactions have been activated by the addition of TMSOTf.

It has been pointed out that the triflate ion increases the electrophilic character of *N*-acyl quaternary salts of imidazoles and accelerates its coupling reaction to silyl enol ethers.[178] In fact, in the absence of this Lewis acid, the yield of the coupling product diminishes (eq 118). Thiazoles, benzothiazoles, and benzoimidazoles react in a similar manner.[179]

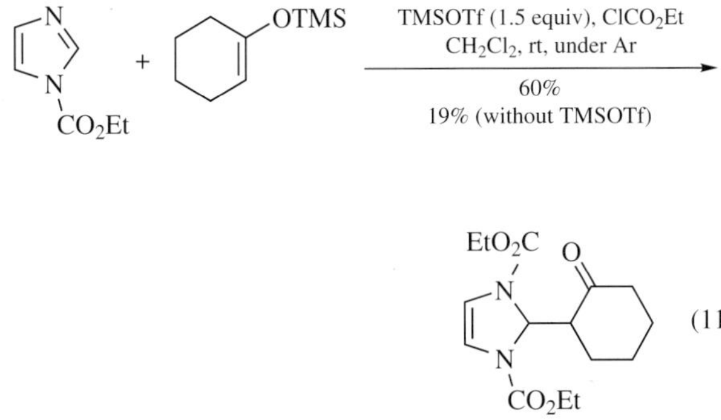

(118)

The nucleophilic attack of silyl enol glycine derivative to a dienyl acetate organometallic complex proceeds readily and regiospecifically at −78 °C in the presence of TMSOTf to afford a dienyl glycine derivative in excellent yield (eq 119).[180]

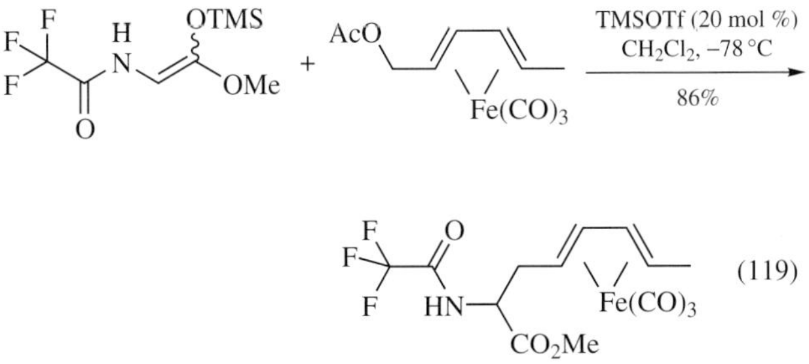

(119)

The nucleophilic substitution of α-acetoxyhydrazones by silyl enol ethers takes place with high chemical yields and moderate diastereomeric excesses. Among the several Lewis acids tested (AsCl$_3$, AlCl$_3$, TiCl$_4$, ZnCl$_2$,...), only BF$_3$·OEt$_2$ provides a slightly higher diastereoselectivity than TMSOTf.[181]

α-Propargyl carbocations, which can be generated by treatment of (arene)Cr(CO)$_3$-substituted propargyl acetates with TMSOTf,[182] react with a variety of *C*-, *S*-, or *N*-nucleophiles to give good yields of the corresponding propargyl derivatives. If the α-propargyl carbocations are prepared from arene-substituted propargyl silyl ethers,[183] the nature of the products (allenyl or propargyl compounds) depends mainly on the substituents of

the starting material, although the nucleophile also may exert some influence; for example, diphenyl-substituted propargyl silyl ethers usually lead to the formation of allenyl derivatives. α,β-Unsaturated carbonyl compounds are obtained when these substituted propargyl silyl ethers are treated with TMSOTf followed by the addition of water (eq 120).[183]

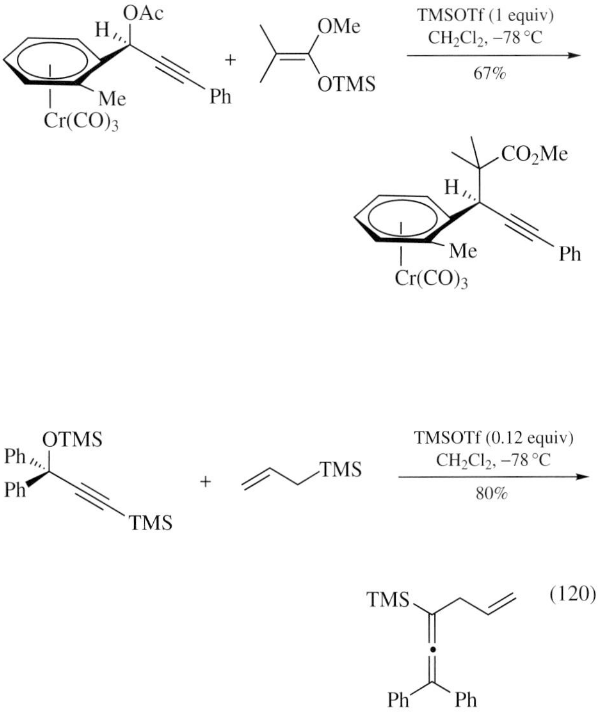

(120)

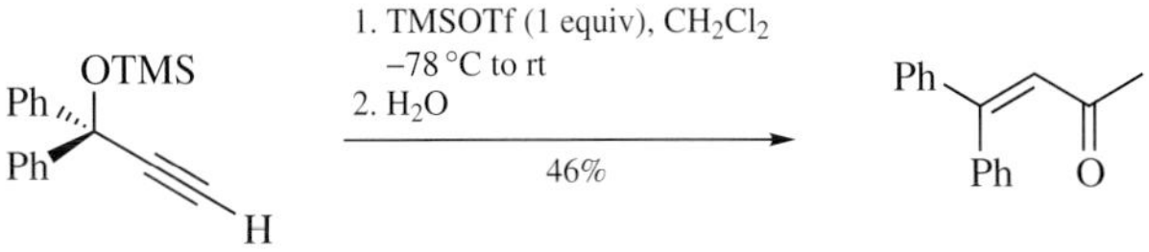

A TMSOTf-induced stereospecific cationic *syn*-[1,2] silyl shift occurs with retention of the stereochemistry at the migrating terminus (eq 121).[184]

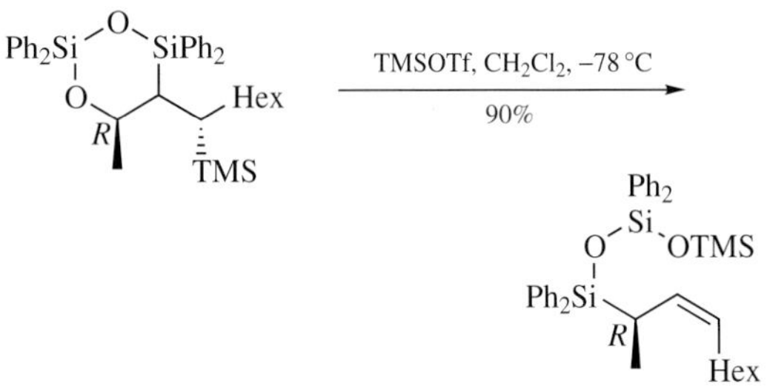

(121)

Iron carbene complexes bearing chirality at the carbene ligand can be generated from optically pure bimetallic (chromium-iron) complexes by the addition of 1 equiv of TMSOTf in the presence of an olefin, which in situ undergoes an asymmetric cyclopropanation. Excellent ee's are obtained when the reaction is carried out with *gem*-disubstituted olefins (eq 122).[185]

$$(122)$$

ee > 95%

Benzylic silanes bearing an electron-donating group experience an oxidative carbon-silicon bond cleavage selectively with oxovanadium(V) compounds to permit an intermolecular carbon-carbon bond formation; the addition of TMSOTf resulted in a more facile coupling (eq 123).[186]

$$(123)$$

TMSOTf may act as a Lewis acid promoter for the chiral oxovanadium complex-catalyzed oxidative coupling of 2-naphthols. In the enantioselective version, chlorotrimethylsilane affords higher enantiomeric excesses (eq 124).[187]

$$(124)$$

ee = 42%
ee = 48% (with TMSCl)

catalyst =

1. Reviews: (a) Emde, H.; Domsch, D.; Feger, H.; Frick, U.; Götz, H. H.; Hofmann, K.; Kober, W.; Krägeloh, K.; Oesterle, T.; Steppan, W.; West, W.; Simchen, G., *Synthesis* **1982**, 1. (b) Noyori, R.; Murata, S.; Suzuki, M., *Tetrahedron* **1981**, *37*, 3899. (c) Stang, P. J.; White, M. R., *Aldrichim. Acta* **1983**, *16*, 15. Preparation: (d) Olah, G. H.; Husain, A.; Gupta, B. G. B.; Salem, G. F.; Narang, S. C., *J. Org. Chem.* **1981**, *46*, 5212. (e) Morita, T.; Okamoto, Y.; Sakurai, H., *Synthesis* **1981**,

745. (f) Demuth, M.; Mikhail, G., *Synthesis* **1982**, 827. (g) Ballester, M.; Palomo, A. L., *Synthesis* **1983**, 571. (h) Demuth, M.; Mikhail, G., *Tetrahedron* **1983**, *39*, 991. (i) Aizpurua, J. M.; Palomo, C., *Synthesis* **1985**, 206.

2. Simchen, G.; Kober, W., *Synthesis* **1976**, 259.

3. Hergott, H. H.; Simchen, G., *Liebigs Ann. Chem.* **1980**, 1718.

4. Simchen, G.; Kober, W., *Synthesis* **1976**, 259.

5. Emde, H.; Götz, A.; Hofmann, K.; Simchen, G., *Liebigs Ann. Chem.* **1981**, 1643.

6. Krägeloh, K.; Simchen, G., *Synthesis* **1981**, 30.

7. Emde, H.; Simchen, G., *Liebigs Ann. Chem.* **1983**, 816.

8. Emde, H.; Simchen, G., *Synthesis* **1977**, 636.

9. Emde, H.; Simchen, G., *Synthesis* **1977**, 867.

10. Yamamoto, Y.; Ohdoi, K.; Nakatani, M.; Akiba, K., *Chem. Lett.* **1984**, 1967.

11. Emde, H.; Götz, A.; Hofmann, K.; Simchen, G., *Liebigs Ann. Chem.* **1981**, 1643.

12. Matsuda, I.; Sato, S.; Hattori, M.; Izumi, Y., *Tetrahedron Lett.* **1985**, *26*, 3215.

13. Ahlbrecht, H.; Düber, E. O., *Synthesis* **1980**, 630.

14. Murata, S.; Suzuki, M.; Noyori, R., *J. Am. Chem. Soc.* **1980**, *102*, 2738.

15. Frick, U.; Simchen, G., *Synthesis* **1984**, 929.

16. Borgulya, J.; Bernauer, K., *Synthesis* **1980**, 545.

17. Tietze, L. F.; Bratz, M., *Synthesis* **1989**, 439.

18. Hwu, J. R.; Wetzel, J. M., *J. Org. Chem.* **1985**, *50*, 3946.

19. Hwu, J. R.; Robl, J. A., *J. Org. Chem.* **1987**, *52*, 188.

20. Harada, T.; Hayashiya, T.; Wada, I.; Iwa-ake, N.; Oku, A., *J. Am. Chem. Soc.* **1987**, *109*, 527.

21. Pearson, W. H.; Cheng, M-C., *J. Am. Chem. Soc.* **1986**, *51*, 3746.

22. Murata, S.; Suzuki, M.; Noyori, R., *J. Am. Chem. Soc.* **1980**, *102*, 3248.

23. Murata, S.; Suzuki, M.; Noyori, R., *Tetrahedron* **1988**, *44*, 4259.

24. Lee, T. V.; Richardson, K. A., *Tetrahedron Lett.* **1985**, *26*, 3629.

25. Mukaiyama, T.; Uchiro, H.; Kobayashi, S., *Chem. Lett.* **1990**, 1147.

26. Guanti, G.; Narisano, E.; Banfi, L., *Tetrahedron Lett.* **1987**, *28*, 4331.

27. Guanti, G.; Narisano, E.; Banfi, L., *Tetrahedron Lett.* **1987**, *28*, 4335.

28. Ihara, M.; Tsuruta, M.; Fukumoto, K.; Kametani, T., *J. Chem. Soc., Chem. Commun.* **1985**, 1159.

29. Robl, J. A.; Hwu, J. R., *J. Org. Chem.* **1985**, *50*, 5913.

30. Sato, T.; Watanabe, T.; Hayata, T.; Tsukui, T., *J. Chem. Soc., Chem. Commun.* **1989**, 153.

31. Kim, S.; Lee, J. M., *Tetrahedron Lett.* **1990**, *31*, 7627.

32. Kim, S.; Lee, P. H., *Tetrahedron Lett.* **1988**, *29*, 5413.

33. Hosomi, A.; Sakata, Y.; Sakurai, H., *Tetrahedron Lett.* **1984**, *25*, 2383.

34. Yamada, H.; Nishizawa, M., *Tetrahedron* **1992**, 3021.

35. Rainer, H.; Scharf, H.-D.; Runsink, J., *Liebigs Ann. Chem.* **1992**, 103.

36. Schmidt, R. R., *Angew. Chem., Int. Ed. Engl.* **1986**, *25*, 212.

37. Tietze, L.-F.; Fischer, R.; Guder, H.-J., *Tetrahedron Lett.* **1982**, *23*, 4661.

38. Mukaiyama, T.; Matsubara, K., *Chem. Lett.* **1992**, 1041.

39. Kozikowski, A. P.; Park, P., *J. Org. Chem.* **1984**, *49*, 1674.

40. Knapp, S.; Rodriques, K. E., *Tetrahedron Lett.* **1985**, *26*, 1803.

41. Nishiyama, H.; Sakuta, K.; Osaka, N.; Itoh, K., *Tetrahedron Lett.* **1983**, *24*, 4021.

42. Gassman, P. G.; Burns, S. J., *J. Org. Chem.* **1988**, *53*, 5574.

43. Miller, R. D.; Hässig, R., *Synth. Commun.* **1984**, *14*, 1285.

44. Miller, R. D.; Hässig, R., *Tetrahedron Lett.* **1985**, *26*, 2395.

45. (a) Marigo, M.; Wabnitz, T. C.; Fielenbach, D.; Jorgensen, K. A., *Angew. Chem., Int. Ed.* **2005**, *44*, 794. (b) Lilly, M. J.; Sherburn, M. S., *Chem. Commun.* **1997**, 967.

46. Oriyama, T.; Yatabe, K.; Kawada, Y.; Koga, G., *Synlett* **1995**, 45.

47. Guha, S. K.; Shibayama, A.; Abe, D.; Sakaguchi, M.; Ukaji, Y.; Inomata, K., *Bull. Chem. Soc. Jpn.* **2004**, *77*, 2147.

48. Rossi, L.; Pecunioso, A., *Tetrahedron Lett.* **1994**, *35*, 5285.

49. Knopf, O.; Alexakis, A., *Org. Lett.* **2002**, *4*, 3835.

50. Kim, S.; Park, J. H.; Kim, Y. G.; Lee, J. M., *J. Chem. Soc., Chem. Commun.* **1993**, 1188.

51. Ghosez, L.; Bayard, P.; Nshimyumukiza, P.; Gouverneur, V.; Sainte, F.; Beaudegnies, R.; Rivera, M.; Frisque-Hesbain, A.-M.; Wynants, C., *Tetrahedron* **1995**, *51*, 11021.

52. (a) Plenio, H.; Aberle, C., *Chem. Commun.* **1996**, 2123. (b) Trost, B. M.; Urabe, H., *Tetrahedron Lett.* **1990**, *31*, 615.

53. Rivera, J. M.; Rebek, J., Jr., *J. Am. Chem. Soc.* **2000**, *122*, 7811.

54. Langer, P.; Schneider, T., *Synlett* **2000**, 497.

55. Barbero, A.; Pulido, F. J., *Synthesis* **2004**, 401.

56. Suh, Y.-G.; Shin, D.-Y.; Jung, J.-K.; Kim, S.-H., *Chem. Commun.* **2002**, 1064.

57. Wybrow, R. A. J.; Edwards, A. S.; Stevenson, N. G.; Adams, H.; Johnstone, C.; Harrity, J. P. A., *Tetrahedron* **2004**, *60*, 8869.

58. (a) Kanwar, S.; Trehan, S., *Tetrahedron Lett.* **2005**, *46*, 1329. (b) Kiyooka, S.-i.; Shirouchi, M.; Kaneko, Y., *Tetrahedron Lett.* **1993**, *34*, 1491.

59. Danilenko, V.; Tishkov, A. A.; Ioffe, S. L.; Lyapkalo, I. M.; Strelenko, Y. A.; Tartakovsky, V. A., *Synthesis* **2002**, 635.

60. Hauser, F. M.; Zhou, M.; Sun, Y., *Synth. Commun.* **2001**, *31*, 77.

61. Miller, S. J.; Bayne, C. D., *J. Org. Chem.* **1997**, *62*, 5680.

62. Kobayashi, M.; Masumoto, K.; Nakai, E.-i; Nakai, T., *Tetrahedron Lett.* **1996**, *37*, 3005.

63. Werner, R. M.; Barwick, M.; Davis, J. T., *Tetrahedron Lett.* **1995**, *36*, 7395.

64. (a) Slagt, M. Q.; Rodríguez, G.; Grutters, M. M. P.; Gebbink, R. J. M. K.; Klopper, W.; Jenneskens, L. W.; Lutz, M.; Spek, A. L.; van Koten, G., *Chem. Eur. J.* **2004**, *10*, 1331. (b) Motoyama, Y.; Koga, Y.; Kobayashi, K.; Aoki, K.; Nishiyama, H., *Chem. Eur. J.* **2002**, *8*, 2968.

65. Turner, R. M.; Ley, S. V.; Lindell, S. D., *Synlett* **1993**, 748.

66. Miller, R. A.; Smith, R. M.; Karady, S.; Reamer, R. A., *Tetrahedron Lett.* **2002**, *43*, 935.

67. Begtrup, M.; Vedsø, P., *J. Chem. Soc., Perkin Trans. 1* **1993**, 625.

68. Weisenburger, G. A.; Faibish, N. C.; Pippel, D. J.; Beak, P., *J. Am. Chem. Soc.* **1999**, *121*, 9522 and references cited therein.

69. Marr, F.; Fröhlich, R.; Hoppe, D., *Tetrahedron: Asymmetry* **2002**, *13*, 2587.

70. Wilhelm, R.; Sebhat, I. K.; White, A. J. P.; Williams, D. J.; Widdowson, D. A., *Tetrahedron: Asymmetry* **2000**, *11*, 5003.

71. Nakamura, S.; Nakagawa, R.; Watanabe, Y.; Toru, T., *J. Am. Chem. Soc.* **2000**, *122*, 11340.

72. Hoffmann, R. W.; Bewersdorf, M.; Krüger, M.; Mikolaiski, W.; Stürmer, R., *Chem. Ber.* **1991**, *124*, 1243.

73. Müller, P.; Lacrampe, F.; Bernardinelli, G., *Tetrahedron: Asymmetry* **2003**, *14*, 1503.

74. Cavelier-Frontin, F.; Jacquier, R.; Paladino, J.; Verducci, J., *Tetrahedron* **1991**, *47*, 9807.

75. Grumbach, H.-J.; Arend, M.; Risch, N., *Synthesis* **1996**, 883.

76. Esker, J. L.; Newcomb, M., *Tetrahedron Lett.* **1992**, *33*, 5913.

77. Kauch, M.; Hoppe, D., *Can. J. Chem.* **2001**, *79*, 1736.

78. Ogoshi, S.; Tomiyasu, S.; Morita, M.; Kurosawa, H., *J. Am. Chem. Soc.* **2002**, *124*, 11598.

79. (a) Hoffmann, R.; Brückner, R., *Chem. Ber.* **1992**, *125*, 1471. (b) Kusche, A.; Hoffmann, R.; Münster, I.; Keiner, P.; Brückner, R., *Tetrahedron Lett.* **1991**, *32*, 467.

80. Kim, S.; Kim, Y. G.; Kim, D.-i., *Tetrahedron Lett.* **1992**, *33*, 2565.

81. Lillie, B. M.; Avery, M. A., *Tetrahedron Lett.* **1994**, *35*, 969.

82. Ishihara, K.; Hiraiwa, Y.; Yamamoto, H., *Chem. Commun.* **2002**, 1564.

83. Markó, I. E.; Mekhalfia, A.; Ollis, W. D., *Synlett* **1990**, 347.

84. Xia, Q.; Ganem, B., *Org. Lett.* **2002**, *4*, 1631.

85. Reddy, D. S.; Judd, W. R.; Aubé, J., *Org. Lett.* **2003**, *5*, 3899.

86. Sato, T.; Hayashi, M.; Hayata, T., *Tetrahedron* **1992**, *48*, 4099.

87. Roos, E. C.; López, M. C.; Brook, M. A.; Hiemstra, H.; Speckamp, W. N.; Kaptein, B.; Kamphuis, J.; Schoemaker, H. E., *J. Org. Chem.* **1993**, *58*, 3259.

88. Anders, E.; Hertlein, K.; Stankowiak, A.; Irmer, E., *Synthesis* **1992**, 577.

89. Kim, S.; Kim, S. H., *Tetrahedron Lett.* **1995**, *36*, 3723.

90. Beignet, J.; Cox, L. R., *Org. Lett.* **2003**, *5*, 4231.

91. Roux, M.-C.; Wartski, L.; Nierlich, M.; Lance, M., *Tetrahedron* **1996**, *52*, 10083.

92. (a) Miles, S. M.; Marsden, S. P.; Leatherbarrow, R. J.; Coates, W. J., *J. Org. Chem.* **2004**, *69*, 6874. (b) Meyer, C.; Cossy, J., *Tetrahedron Lett.* **1997**, *38*, 7861.

93. Suginome, M.; Ohmori, Y.; Ito, Y., *J. Am. Chem. Soc.* **2001**, *123*, 4601.

94. Shi, M.; Yang, Y.-H.; Xu, B., *Tetrahedron* **2005**, *61*, 1983.

95. Takemoto, Y.; Kuraoka, S.; Hamaue, N.; Iwata, C., *Tetrahedron: Asymmetry* **1996**, *7*, 993.

96. Eriksson, M.; Iliefski, T.; Nilsson, M.; Olsson, T., *J. Org. Chem.* **1997**, *62*, 182.

97. Kim, S.; Han, W. S.; Lee, J. M., *Bull. Korean Chem. Soc.* **1992**, *13*, 466.

98. Takasu, K.; Misawa, K.; Yamada, M.; Furuta, Y.; Taniguchi, T.; Ihara, M., *Chem. Commun.* **2000**, 1739.

99. Takasu, K.; Ueno, M.; Ihara, M., *J. Org. Chem.* **2001**, *66*, 4667.

100. Ihara, M.; Suzuki, S.; Taniguchi, N.; Fukumoto, K., *J. Chem. Soc., Chem. Commun.* **1993**, 755.

101. Jung, S. H.; Kim, J. H., *Bull. Korean Chem. Soc.* **2002**, *23*, 1375.

102. Kim, C. U.; Misco, P. F.; Luh, B. Y.; Mansuri, M. M., *Tetrahedron Lett.* **1994**, *35*, 3017.

103. Basavaiah, D.; Rao, A. J., *Chem. Commun.* **2003**, 604.

104. Kataoka, T.; Iwama, T.; Tsujiyama, S.-i., Iwamura, T.; Watanabe, S.-i., *Tetrahedron* **1998**, *54*, 11813.

105. Ichikawa, J.; Miyazaki, S.; Fujiwara, M.; Minami, T., *J. Org. Chem.* **1995**, *60*, 2320.

106. Andrews, J. F. P.; Regan, A. C., *Tetrahedron Lett.* **1991**, *32*, 7731.

107. Molander, G. A.; Cameron, K. O., *J. Org. Chem.* **1991**, *56*, 2617.

108. (a) Molander, G. A.; Cameron, K. O., *J. Org. Chem.* **1993**, *58*, 5931. (b) Molander, G. A.; Cameron, K. O., *J. Am. Chem. Soc.* **1993**, *115*, 830.

109. Collins, J. L.; Grieco, P. A.; Walker, J. K., *Tetrahedron Lett.* **1997**, *38*, 1321.

110. Grieco, P. A.; Walker, J. K., *Tetrahedron* **1997**, *53*, 8975.

111. Tietze, L. F.; Schneider, C., *Synlett* **1992**, 755.

112. Akiba, K.-y.; Motoshima, T.; Ishimaru, K.; Yabuta, K.; Hirota, H.; Yamamoto, Y., *Synlett* **1993**, 657.

113. Arenz, T.; Frauenrath, H.; Raabe, G.; Zorn, M., *Liebigs Ann. Chem.* **1994**, 931.

114. Murata, Y.; Overman, L. E., *Heterocycles* **1996**, *42*, 549.

115. Moutou, J. L.; Schmitt, M.; Wermuth, C. G.; Bourguignon, J. J., *Tetrahedron Lett.* **1994**, *35*, 6883.

116. Mikami, K.; Ohmura, H., *Chem. Commun.* **2002**, 2626.

117. Albrecht, U.; Armbrust, H.; Langer, P., *Synlett* **2004**, 143.

118. Murray, D. H.; Albizati, K. F., *Tetrahedron Lett.* **1990**, *29*, 4109.

119. Gassman, P. G.; Burns, S. J.; Pfister, K. B., *J. Org. Chem.* **1993**, *58*, 1449.

120. Futagawa, T.; Nishiyama, N.; Tai, A.; Okuyama, T.; Sugimura, T., *Tetrahedron* **2002**, *58*, 9279.

121. Swenton, J. S.; Bonke, B. R.; Clark, W. M.; Chen, C.-P.; Martin, K. V., *J. Org. Chem.* **1990**, *55*, 2027.

122. Rychnovsky, S. D.; Kim, J., *Tetrahedron Lett.* **1991**, *32*, 7219.

123. Hunter, R.; Michael, J. P.; Tomlinson, G. D., *Tetrahedron* **1994**, *50*, 871.

124. Linderman, R. J.; Chen, K., *J. Org. Chem.* **1996**, *61*, 2441.

125. Fujita, K.; Inoue, A.; Shinokubo, H.; Oshima, K., *Org. Lett.* **1999**, *1*, 917.

126. Sato, T.; Otera, J.; Nozaki, H., *J. Org. Chem.* **1990**, *55*, 6116.

127. Linderman, R. J.; Anklekar, T. V., *J. Org. Chem.* **1992**, *57*, 5078.

128. Hunter, R.; Bartels, B.; Michael, J. P., *Tetrahedron Lett.* **1991**, *32*, 1095.

129. Hunter, R.; Bartels, B., *J. Chem. Soc., Perkin Trans. 1* **1991**, 2887.

130. (a) Roush, W. R.; Bennett, C. E.; Roberts, S. E., *J. Org. Chem.* **2001**, *66*, 6389. (b) Kovensky, J.; Sinaÿ, P., *Eur. J. Org. Chem.* **2000**, 3523. (c) Iimori, T.; Kobayashi, H.; Hashimoto, S.-i.; Ikegami, S., *Heterocycles* **1996**, *42*, 485.

131. Yadav, V.; Chu, C. K.; Rais, R. H.; Al Safarjalani, O. N.; Guarcello, V.; Naguib, F. N. M.; el Kouni, M. H., *J. Med. Chem.* **2004**, *47*, 1987.

132. Qin, Z.-H.; Li, H.; Cai, M.-S.; Li, Z.-J., *Carbohydr. Res.* **2002**, *337*, 31.

133. Crich, D.; de la Mora, M.; Vinod, A. U., *J. Org. Chem.* **2003**, *68*, 8142.

134. (a) Tanimori, S.; Ohta, T.; Kirihata, M., *Bioorg. Med. Chem. Lett.* **2002**, *12*, 1135. (b) Jenkins, D. J.; Marwood, R. D.; Potter, B. V. L., *Chem. Commun.* **1997**, 449.

135. (a) Pradeepkumar, P. I.; Amirkhanov, N. V.; Chattopadhyaya, J., *Org. Biomol. Chem.* **2003**, *1*, 81. (b) Ali, I. A. I.; Abdel-Rahman, A. A.-H.; El Ashry, E. S. H.; Schmidt, R. R., *Synthesis* **2003**, 1065.

136. Mlynarski, J.; Banaszek, A., *Tetrahedron: Asymmetry* **2000**, *11*, 3737.

137. Huart, C.; Ghosez, L., *Angew. Chem., Int. Ed. Engl.* **1997**, *36*, 634.

138. Wuts, P. G. M.; Jung, Y.-W., *J. Org. Chem.* **1988**, *53*, 1957.

139. Dhavale, D. D.; Trombini, C., *Heterocycles* **1992**, *34*, 2253.

140. Gianotti, M.; Lombardo, M.; Trombini, C., *Tetrahedron Lett.* **1998**, *39*, 1643.

141. Merino, P.; Mates, J. A., *Arkivoc* **2000**, *xi*, 12 and references cited therein.

142. Shanmugam, P.; Miyashita, M., *Org. Lett.* **2003**, *5*, 3265.

143. (a) Gala, D.; DiBenedetto, D. J., *Tetrahedron Lett.* **1994**, *35*, 8299. (b) Raubo, P.; Wicha, J., *J. Org. Chem.* **1994**, *59*, 4355.

144. Yoshitake, M.; Yamamoto, M.; Kohmoto, S.; Yamada, K., *J. Chem. Soc., Perkin Trans. 1* **1990**, 1226.

145. Jung, M. E.; Hoffmann, B.; Rausch, B.; Contreras, J.-M., *Org. Lett.* **2003**, *5*, 3159 and references cited therein.

146. Jung, M. E.; van der Heuvel, A., *Tetrahedron Lett.* **2002**, *43*, 8169.

147. Oriyama, T.; Yatabe, K.; Sugawara, S.; Machiguchi, Y.; Koga, G., *Synlett* **1996**, 523.

148. Chattopadhyay, S. K.; Pattenden, G., *Tetrahedron Lett.* **1998**, *39*, 6095.

149. Trzeciak, A.; Bannwarth, W., *Synthesis* **1996**, 1433.

150. Williams, C. M.; Mander, L. N., *Tetrahedron* **2004**, *45*, 667.

151. Fujioka, H.; Sawama, Y.; Murata, N.; Okitsu, T.; Kubo, O.; Matsuda, S.; Kita, Y., *J. Am. Chem. Soc.* **2004**, *126*, 11800.

152. Zhang, A.; Russell, D. H.; Zhu, J.; Burgess, K., *Tetrahedron Lett.* **1998**, *39*, 7439.

153. Sakaitani, M.; Ohfune, Y., *J. Org. Chem.* **1990**, *55*, 870.

154. Todd, M. H.; Abell, C., *Tetrahedron Lett.* **2000**, *41*, 8183.

155. Klein, M.; König, B., *Tetrahedron* **2004**, *60*, 1087.

156. Zhdankin, V. V.; Koposov, A. Y.; Yashin, N. V., *Tetrahedron Lett.* **2002**, *43*, 5735.

157. Zhdankin, V. V.; Kuehl, C. J.; Krasutsky, A. P.; Formaneck, M. S.; Bolz, J. T., *Tetrahedron Lett.* **1994**, *35*, 9677.

158. Zhdankin, V. V.; Crittell, C. M.; Stang, P. J.; Zefirov, N. S., *Tetrahedron Lett.* **1990**, *31*, 4821.

159. Zhdankin, V. V.; Kuehl, C. J.; Simonsen, A. J., *J. Org. Chem.* **1996**, *61*, 8272.

160. Zhdankin, V. V.; Kuehl, C. J.; Krasutsky, A. P.; Bolz, J. T.; Simonsen, A. J., *J. Org. Chem.* **1996**, *61*, 6547.

161. Zhdankin, V. V.; Arbit, R. M.; McSherry, M.; Mismash, B.; Young, V. G., Jr., *J. Am. Chem. Soc.* **1997**, *119*, 7408.

162. Radhakrishnan, U.; Stang, P. J., *J. Org. Chem.* **2003**, *68*, 9209.

163. Kita, Y.; Watanabe, H.; Egi, M.; Saiki, T.; Fukuoka, Y.; Tohma, H., *J. Chem. Soc., Perkin Trans. 1* **1998**, 635 and references cited therein.

164. Yoshimatsu, M.; Yamaguchi, S.; Matsubara, Y., *J. Chem. Soc., Perkin Trans. 1* **2001**, 2560.

165. (a) Hoesl, C. E.; Maurus, M.; Pabel, J.; Polborn. K.; Wanner, K. T., *Tetrahedron* **2002**, *58*, 6757. (b) Pabel, J.; Hösl, C. E.; Maurus, M.; Ege, M.; Wanner, K. T., *J. Org. Chem.* **2000**, *65*, 9272.

166. Ullah, E.; Langer, P., *Synlett* **2004**, 2782.

167. Tietze, L. F.; Wichmann, J., *Liebigs Ann. Chem.* **1992**, 1063.

168. (a) Beifuss, U.; Tietze, M.; Gehm, H., *Synlett* **1996**, 182. (b) Beifuss, U.; Gehm, H.; Noltemeyer, M.; Schmidt, H.-G., *Angew. Chem., Int. Ed.* **1995**, *34*, 647.

169. Beifuss, U.; Schniske, U.; Feder, G., *Tetrahedron* **2001**, *57*, 1005.

170. Beifuss, U.; Goldenstein, K.; Döring, F.; Lehmann, C.; Noltemeyer, M., *Angew. Chem., Int. Ed.* **2001**, *40*, 568.

171. Ronan, B.; Kagan, H. B., *Tetrahedron: Asymmetry* **1991**, *2*, 75.

172. Hashimoto, Y.; Saigo, K.; Machida, S.; Hasegawa, M., *Tetrahedron Lett.* **1990**, *39*, 5625.

173. Vorndam, P. E., *J. Org. Chem.* **1990**, *55*, 3693.

174. Fujioka, H.; Yamanaka, T.; Takuma, K.; Miyazaki, M.; Kita, Y., *J. Chem. Soc., Chem. Commun.* **1991**, 533.

175. Chassaing, C.; Haudrechy, A.; Langlois, Y., *Tetrahedron Lett.* **1997**, *38*, 4415.

176. Kagoshima, H.; Hashimoto, Y.; Oguro, D.; Kutsuna, T.; Saigo, K., *Tetrahedron Lett.* **1998**, *39*, 1203.

177. (a) Padwa, A.; Waterson, A. G., *Tetrahedron Lett.* **1998**, *39*, 8585. (b) Padwa, A.; Hennig, R.; Kappe, C. O.; Reger, T. S., *J. Org. Chem.* **1998**, *63*, 1144. (c) Padwa, A.; Gunn, D. E., Jr.; Osterhout, M. H., *Synthesis* **1997**, 1353.

178. Itoh, T.; Miyazaki, M.; Nagata, K.; Ohsawa, A., *Tetrahedron* **2000**, *56*, 4383.

179. Itoh, T.; Miyazaki, M.; Nagata, K.; Matsuya, Y.; Ohsawa, A., *Heterocycles* **1999**, *50*, 667.

180. Kalita, B.; Nicholas, K. M., *Tetrahedron* **2004**, *60*, 10771.

181. Enders, D.; Maaßen, R.; Runsink, J., *Tetrahedron: Asymmetry* **1998**, *9*, 2155.

182. Müller, T. J. J.; Netz, A., *Tetrahedron Lett.* **1999**, *40*, 3145 and references cited therein.

183. Ishikawa, T.; Okano, M.; Aikawa, T.; Saito, S., *J. Org. Chem.* **2001**, *66*, 4635.

184. Suginome, M.; Takama, A.; Ito, Y., *J. Am. Chem. Soc.* **1998**, *120*, 1930.

185. (a) Wang, Q.; Mayer, M. F.; Brennan, C.; Yang, F.; Hossain, M. M.; Grubisha, D. S.; Bennet, D., *Tetrahedron* **2000**, *56*, 4881. (b) Vargas, R. M.; Theys, R. D.; Hossain, M. M., *J. Am. Chem. Soc.* **1992**, *114*, 777.

186. Hirao, T.; Fujii, T.; Ohshiro, Y., *Tetrahedron Lett.* **1994**, *35*, 8005.

187. Chu, C.-Y.; Hwang, D.-R.; Wang, S.-K.; Uang, B.-J., *Chem. Commun.* **2001**, 980.

Trimethylsilyldiazomethane[1]

[18107-18-1] $C_4H_{10}N_2Si$ (MW 114.25)

InChI = 1/C4H10N2Si/c1-7(2,3)4-6-5/h4H,1-3H3

InChIKey = ONDSBJMLAHVLMI-UHFFFAOYAN

(one-carbon homologation reagent; stable, safe substitute for diazomethane; [C–N–N] 1,3-dipole for the preparation of azoles[1])

Physical Data: bp 96 °C/775 mmHg; n_D^{25} 1.4362.[2]

Solubility: sol most organic solvents; insol H_2O.

Form Supplied in: commercially available as 2 M and 10 w/w% solutions in hexane, and 10 w/w% solution in CH_2Cl_2; also available as 2 M solutions in diethyl ether or hexanes.

Analysis of Reagent Purity: concentration in hexane is determined by ^1H NMR.[3]

Preparative Methods: prepared by the diazo-transfer reaction of **Trimethylsilylmethylmagnesium Chloride** with **Diphenyl Phosphorazidate** (DPPA) (eq 1).[3]

$$\text{TMS} \diagup \text{Cl} \xrightarrow{\text{Mg}} \text{TMS} \diagup \text{MgCl} \xrightarrow{(\text{PhO})_2\text{P(O)N}_3} \text{TMS} \diagup_{N_2} \quad (1)$$

Handling, Storage, and Precautions: should be protected from light.

Original Commentary

Takayuki Shioiri & Toyohiko Aoyama
Nagoya City University, Nagoya, Japan

One-carbon Homologation. Along with its lithium salt, which is easily prepared by lithiation of trimethylsilyldiazomethane (TMSCHN₂) with **Butyllithium**, TMSCHN₂ behaves in a similar way to **Diazomethane** as a one-carbon homologation reagent. TMSCHN₂ is acylated with aromatic acid chlorides in the presence of **Triethylamine** to give α-trimethylsilyl diazo ketones. In the acylation with aliphatic acid chlorides, the use of 2 equiv of TMSCHN₂ without triethylamine is recommended. The crude diazo ketones undergo thermal Wolff rearrangement to give the homologated carboxylic acid derivatives (eqs 2 and 3).[4]

$$\text{(eq 2): } \underset{\text{COCl}}{\text{naphthyl}} \xrightarrow[\substack{2.\ \text{PhNH}_2,\ 180\ °C \\ 2,4,6\text{-trimethylpyridine} \\ 80\%}]{1.\ \text{TMSCHN}_2,\ \text{Et}_3\text{N}} \underset{\text{CONHPh}}{\text{naphthyl}} \quad (2)$$

$$\text{(eq 3): } \underset{\substack{\text{N} \\ \text{CO}_2\text{Bn}}}{\text{COCl}} \xrightarrow[\substack{2.\ \text{PhCH}_2\text{OH},\ 180\ °C \\ 2,4,6\text{-trimethylpyridine} \\ 77\%}]{1.\ 2\ \text{equiv TMSCHN}_2} \underset{\substack{\text{N} \\ \text{CO}_2\text{Bn}}}{\text{CH}_2\text{CO}_2\text{Bn}} \quad (3)$$

Various ketones react with TMSCHN₂ in the presence of **Boron Trifluoride Etherate** to give the chain or ring homologated ketones (eqs 4–6).[5] The bulky trimethylsilyl group of TMSCHN₂ allows for regioselective methylene insertion (eq 5). Homologation of aliphatic and alicyclic aldehydes with TMSCHN₂ in the presence of **Magnesium Bromide** smoothly gives methyl ketones after acidic hydrolysis of the initially formed β-keto silanes (eq 7).[6]

$$\text{PhCOCH}_2\text{Ph} \xrightarrow[\substack{\text{CH}_2\text{Cl}_2,\ -15\ °C,\ 1\ h \\ 74\%}]{\text{TMSCHN}_2,\ \text{BF}_3\cdot\text{Et}_2\text{O}} \text{PhCOCH}_2\text{CH}_2\text{Ph} \quad (4)$$

$$\text{(eq 5, 2-methylcyclohexanone} \to \text{2-methylcycloheptanone)} \xrightarrow[\substack{\text{CH}_2\text{Cl}_2,\ -15\ °C,\ 4\ h \\ 69\%}]{\text{TMSCHN}_2,\ \text{BF}_3\cdot\text{Et}_2\text{O}} \quad (5)$$

$$\text{(eq 6, fluorenone} \to \text{phenanthren-9-ol)} \xrightarrow[\substack{\text{CH}_2\text{Cl}_2,\ -15\ \text{to}\ -10\ °C,\ 3\ h \\ 80\%}]{\text{TMSCHN}_2,\ \text{BF}_3\cdot\text{Et}_2\text{O}} \quad (6)$$

$$t\text{-BuCHO} \xrightarrow[\substack{2.\ 10\%\ \text{aq HCl} \\ 89\%}]{1.\ \text{TMSCHN}_2,\ \text{MgBr}_2} t\text{-BuCOMe} \quad (7)$$

O-Methylation of carboxylic acids, phenols, enols, and alcohols can be accomplished with TMSCHN₂ under different reaction conditions. TMSCHN₂ instantaneously reacts with carboxylic acids in benzene in the presence of methanol at room temperature to give methyl esters in nearly quantitative yields (eq 8).[7] This method is useful for quantitative gas chromatographic analysis of fatty acids. Similarly, *O*-methylation of phenols and enols with TMSCHN₂ can be accomplished, but requires the use of **Diisopropylethylamine** (eqs 9 and 10).[8] Although methanol is recommended in these *O*-methylation reactions, methanol is not the methylating agent. Various alcohols also undergo *O*-methylation with TMSCHN₂ in the presence of 42% aq. **Tetrafluoroboric Acid**, smoothly giving methyl ethers (eq 11).[9]

$$\text{HO-C}_6\text{H}_4\text{-CO}_2\text{H} \xrightarrow[\substack{\text{rt, 30 min}}]{\substack{\text{TMSCHN}_2 \\ \text{MeOH–benzene}}} \text{HO-C}_6\text{H}_4\text{-CO}_2\text{Me} \quad (8)$$

quantitative

$$\underset{\text{OH}}{\text{C}_6\text{H}_4\text{-CO}_2\text{Me}} \xrightarrow[\substack{\text{rt, 15 h} \\ 78\%}]{\substack{\text{TMSCHN}_2,\ i\text{-Pr}_2\text{NEt} \\ \text{MeOH–MeCN}}} \underset{\text{OMe}}{\text{C}_6\text{H}_4\text{-CO}_2\text{Me}} \quad (9)$$

$$\underset{O}{\text{PhCO-CH}_2\text{-CO}_2\text{Et}} \xrightarrow[\substack{\text{rt, 15 h} \\ 89\%}]{\substack{\text{TMSCHN}_2,\ i\text{-Pr}_2\text{NEt} \\ \text{MeOH–MeCN}}} \underset{\text{MeO}}{\text{Ph-C=CH-CO}_2\text{Et}} \quad (10)$$

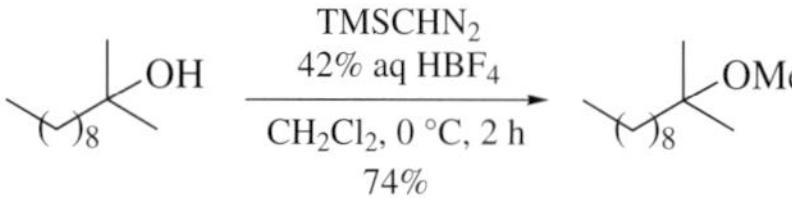

$$(11)$$

Alkylation of the lithium salt of TMSCHN$_2$ (TMSC(Li)N$_2$) gives α-trimethylsilyl diazoalkanes which are useful for the preparation of vinylsilanes and acylsilanes. Decomposition of α-trimethylsilyl diazoalkanes in the presence of a catalytic amount of **Copper(I) Chloride** gives mainly (E)-vinylsilanes (eq 12),[10] while replacement of CuCl with rhodium(II) pivalate affords (Z)-vinylsilanes as the major products (eq 12).[11] Oxidation of α-trimethylsilyl diazoalkanes with **m-Chloroperbenzoic Acid** in a two-phase system of benzene and phosphate buffer (pH 7.6) affords acylsilanes (α-keto silanes) (eq 12).[12]

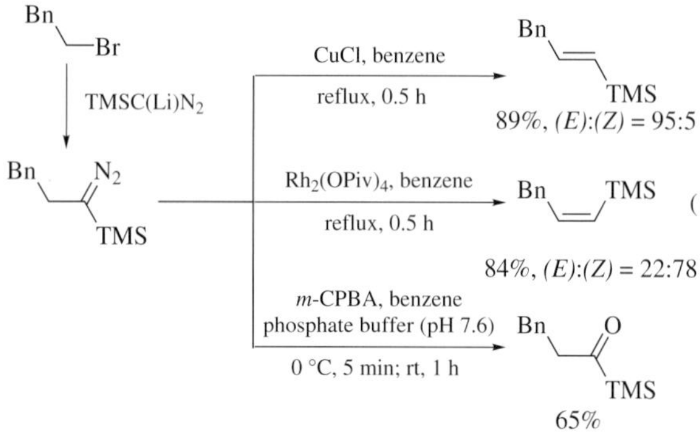

$$(12)$$

(E)-β-Trimethylsilylstyrenes are formed by reaction of alkanesulfonyl chlorides with TMSCHN$_2$ in the presence of triethylamine (eq 13).[13] TMSC(Li)N$_2$ reacts with carbonyl compounds to give α-diazo-β-hydroxy silanes which readily decompose to give α,β-epoxy silanes (eq 14).[14] However, benzophenone gives diphenylacetylene under similar reaction conditions (eq 15).[15]

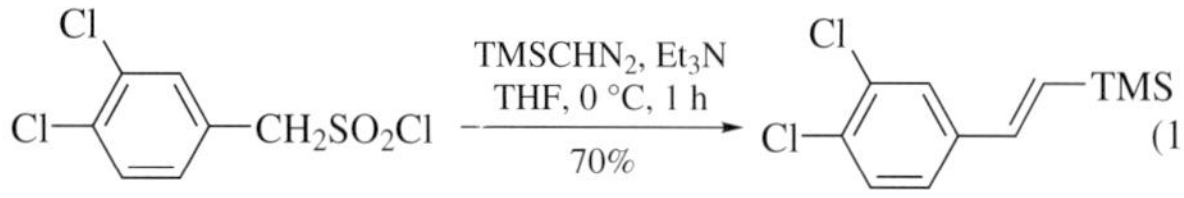

$$(13)$$

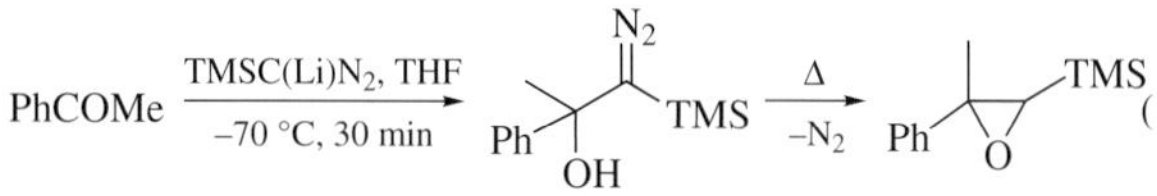

$$(14)$$

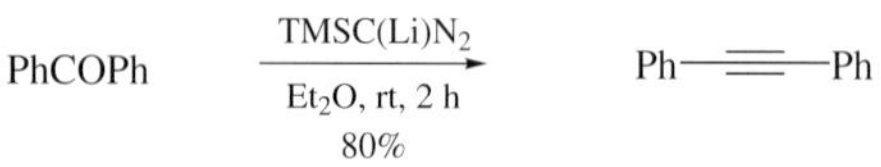

$$(15)$$

Silylcyclopropanes are formed by reaction of alkenes with TMSCHN$_2$ in the presence of either **Palladium(II) Chloride** or CuCl depending upon the substrate (eqs 16 and 17).[16] Silylcyclopropanones are also formed by reaction with trialkylsilyl and germyl ketenes (eq 18).[17]

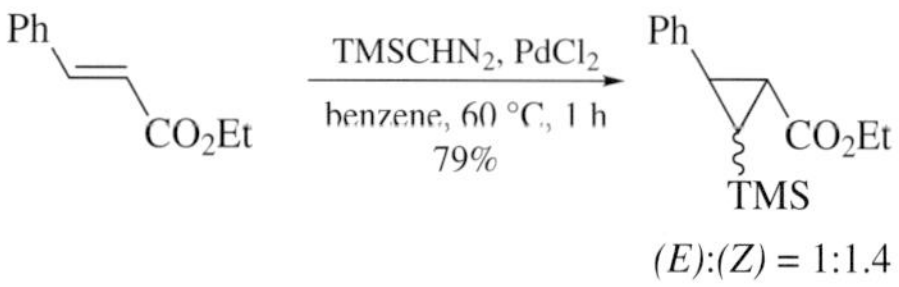

$$(16)$$

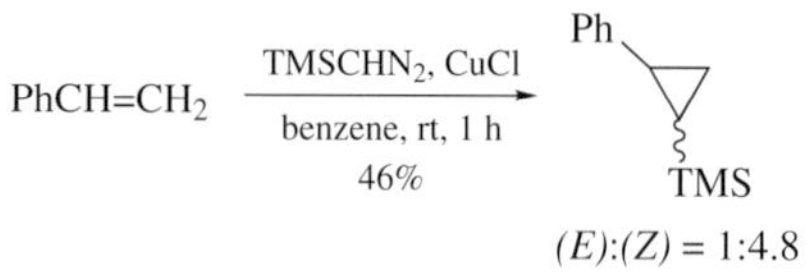

(E):(Z) = 1:1.4

$$(17)$$

(E):(Z) = 1:4.8

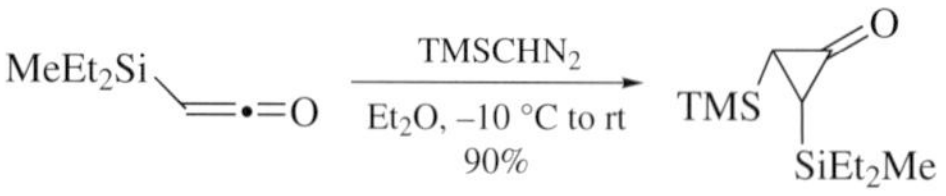

$$(18)$$

[C–N–N] Azole Synthon. TMSCHN$_2$, mainly as its lithium salt, TMSC(Li)N$_2$, behaves like a 1,3-dipole for the preparation of [C–N–N] azoles. The reaction mode is similar to that of diazomethane but not in the same fashion. TMSC(Li)N$_2$ (2 equiv) reacts with carboxylic esters to give 2-substituted 5-trimethylsilyltetrazoles (eq 19).[18] Treatment of thiono and dithio esters with TMSC(Li)N$_2$ followed by direct workup with aqueous methanol gives 5-substituted 1,2,3-thiadiazoles (eq 20).[19] While reaction of di-t-butyl thioketone with TMSCHN$_2$ produces the episulfide with evolution of nitrogen (eq 21),[20] its reaction with TMSC(Li)N$_2$ leads to removal of one t-butyl group to give the 1,2,3-thiadiazole (eq 21).[20]

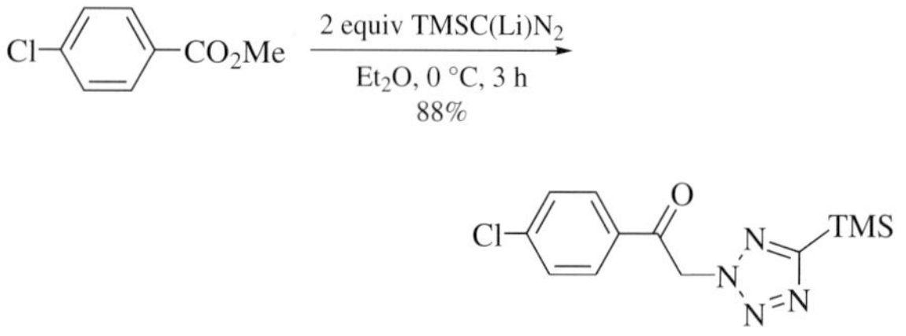

$$(19)$$

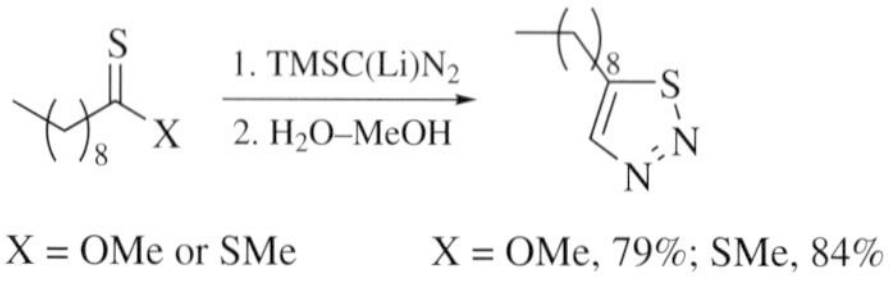

$$(20)$$

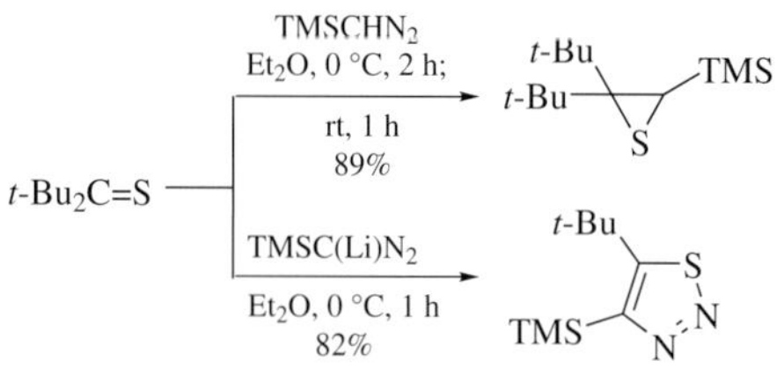

X = OMe or SMe   X = OMe, 79%; SMe, 84%

$$(21)$$

TMSCHN$_2$ reacts with activated nitriles only, such as cyanogen halides, to give 1,2,3-triazoles.[21] In contrast with this, TMSC(Li)N$_2$ smoothly reacts with various nitriles including aromatic, heteroaromatic, and aliphatic nitriles, giving 4-substituted 5-trimethylsilyl-1,2,3-triazoles (eq 22).[22] However, reaction of α,β-unsaturated nitriles with TMSC(Li)N$_2$ in Et$_2$O affords 3(or 5)-trimethylsilylpyrazoles, in which the nitrile group acts as a leaving group (eq 23).[23] Although α,β-unsaturated nitriles bearing bulky substituents at the α- and/or β-positions of the nitrile group undergo reaction with TMSC(Li)N$_2$ to give pyrazoles, significant amounts of 1,2,3-triazoles are also formed. Changing the reaction solvent from Et$_2$O to THF allows for predominant formation of pyrazoles (eq 24).[23] Complete exclusion of the formation of 1,2,3-triazoles can be achieved when the nitrile group is replaced by a phenylsulfonyl species.[24] Thus reaction of α,β-unsaturated sulfones with TMSC(Li)N$_2$ affords pyrazoles in excellent yields (eq 25). The geometry of the double bond of α,β-unsaturated sulfones is not critical in the reaction. When both a cyano and a sulfonyl group are present as a leaving group, elimination of the sulfonyl group occurs preferentially (eq 26).[24] The trimethylsilyl group attached to the heteroaromatic products is easily removed with 10% aq. KOH in EtOH or HCl–KF.

$$(22)$$

$$(23)$$

$$(24)$$

$$(25)$$

$$(26)$$

Various 1,2,3-triazoles can be prepared by reaction of TMSC-(Li)N$_2$ with various heterocumulenes. Reaction of isocyanates with TMSC(Li)N$_2$ gives 5-hydroxy-1,2,3-triazoles (eq 27).[25] It has been clearly demonstrated that the reaction proceeds by a stepwise process and not by a concerted 1,3-dipolar cycloaddition mechanism. Isothiocyanates also react with TMSC(Li)N$_2$ in THF to give lithium 1,2,3-triazole-5-thiolates which are treated in situ with alkyl halides to furnish 1-substituted 4-trimethylsilyl-5-alkylthio-1,2,3-triazoles in excellent yields (eq 28).[26] However, changing the reaction solvent from THF to Et$_2$O causes a dramatic solvent effect. Thus treatment of isothiocyanates with TMSC(Li)N$_2$ in Et$_2$O affords 2-amino-1,3,4-thiadiazoles in good yields (eq 28).[27] Reaction of ketenimines with TMSC(Li)N$_2$ smoothly proceeds to give 1,5-disubstituted 4-trimethylsilyl-1,2,3-triazoles in high yields (eq 29).[28] Ketenimines bearing an electron-withdrawing group at one position of the carbon–carbon

double bond react with TMSC(Li)N$_2$ to give 4-aminopyrazoles as the major products (eq 30).[29]

$$(27)$$

$$(28)$$

$$(29)$$

$$(30)$$

Pyrazoles are formed by reaction of TMSCHN$_2$ or TMSC(Li)N$_2$ with some alkynes (eqs 31 and 32)[24,30] and quinones (eq 33).[31] Some miscellaneous examples of the reactivity of TMSCHN$_2$ or its lithium salt are shown in eqs 34–36.[20,31,32]

$$(31)$$

$$(32)$$

$$(33)$$

$$(34)$$

$$\text{(35)}$$

$$\text{(36)}$$

$$\text{(38)}$$

R = Cbz (66%)
R = CO$_2$Et (66%)

First Update

Timothy Snowden
The University of Alabama, Tuscaloosa, AL, USA

Trimethylsilyldiazomethane has been shown to have a nucleophilicity between that of silyl enol ethers and enamines in dichloromethane. Trimethylsilyldiazomethane is 1.5 orders of magnitude less nucleophilic than diazomethane and 4 orders of magnitude more nucleophilic than ethyl diazoacetate.[33] A crystal structure of lithium trimethylsilyldiazomethane, TMSC(Li)N$_2$, has also been obtained.[34]

One-carbon Homologation. When acid chlorides are unavailable or problematic, the Arndt-Eistert-type homologation of carboxylic acid derivatives is possible by adding TMSCHN$_2$ to a mixed anhydride or to a solution of carboxylic acid and DCC, followed by Wolff rearrangement of the intermediate diazoketone (eq 37).[35] The use of DCC as the activating agent limits the reaction to a maximum 50% yield because of the intermediate symmetric anhydride formed. A two-step ring expansion of N-carboxylated-β-lactams to the corresponding γ-lactams is affected by treating the former with NaHMDS and TMSCHN$_2$, followed by photo-Wolff rearrangement of the ring-opened diazoketone. The reaction occurs in moderate overall yield without epimerization of resident stereocenters (eq 38).[36]

One-carbon homologation of aldehydes and ketones using trimethylaluminum or methylaluminum bis(2,6-di-*tert*-butyl-4-methylphenoxide) (MAD) has been reported to be more regioselective and higher yielding than homologation using boron trifluoride etherate in some cases (eqs 39 and 40).[37] Organoaluminum reagents also promote the direct conversion of aliphatic, alicyclic, and aromatic aldehydes to homologous methyl ketones using TMSCHN$_2$ (eq 41).[38] The complementary conversion of dialkyl ketones to homologated aldehydes is possible by reacting TMSC(Li)N$_2$ and lithium diisopropylamine with the ketone (eq 42). This is a particularly singular transformation, although yields (16–84%) are highly substrate dependent and conditions are unsuitable for base-sensitive substrates.[39]

$$\text{(37)}$$

77%

$$\text{(39)}$$

Conditions	Yield
Me$_3$Al, CH$_2$Cl$_2$, −20 °C	68% (96:2)
BF$_3$·Et$_2$O, CH$_2$Cl$_2$, −20 °C	35% (64:23)

$$\text{(40)}$$

Conditions	Yield
MAD, CH$_2$Cl$_2$, −78 °C	75% (85:15:0)
Me$_3$Al, CH$_2$Cl$_2$, −20 °C	92% (31:36:33)
BF$_3$·Et$_2$O, CH$_2$Cl$_2$, −20 °C	87% (51:40:9)

$$\text{(41)}$$

$$\text{(42)}$$

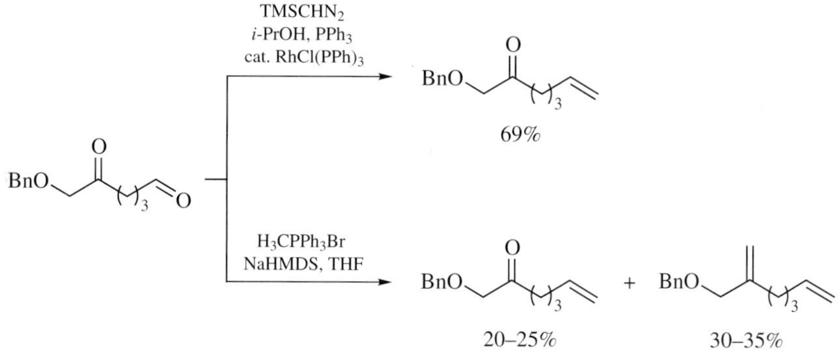

(46)

Alkenylation. Trimethylsilyldiazomethane is a useful reagent for the methylenation of carbonyl compounds in the presence of catalytic ruthenium[40] or rhodium complexes. Wilkinson's catalyst along with stoichiometric TMSCHN$_2$, 2-propanol, and triphenylphosphine rapidly methylenates a variety of sensitive ketones and aldehydes under nonbasic conditions in high yield (eqs 43–45).[41] The use of 2-trimethylsilanylethyl 4-diphenylphosphanylbenzoate (DPPBE)[42] in place of triphenylphosphine simplifies alkene purification without sacrificing overall yield.[41b] The TMSCHN$_2$ and Wilkinson's catalyst system is superior to standard Wittig conditions not only for methylenation of enolizable substrates but also for enhanced chemoselectivity in the methylenation of keto-aldehydes (eq 46).[41a,43]

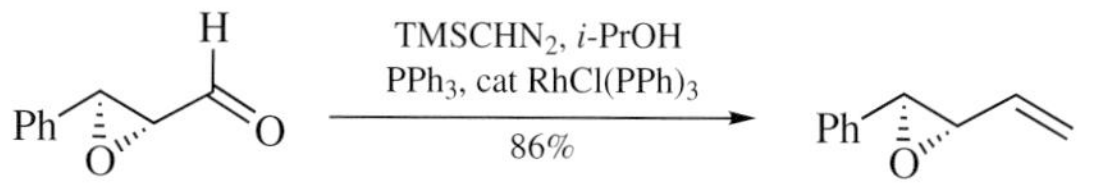

(43)

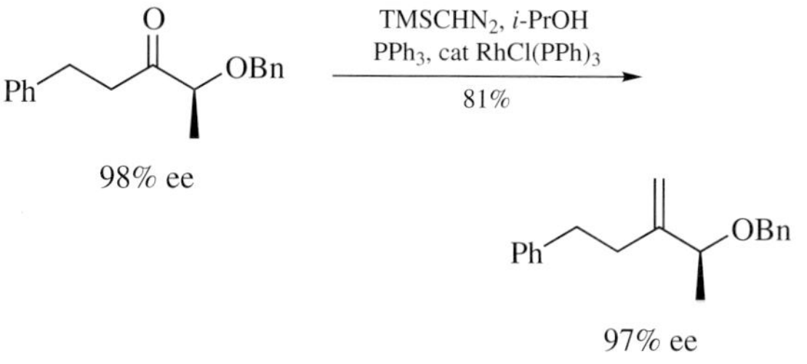

(44)

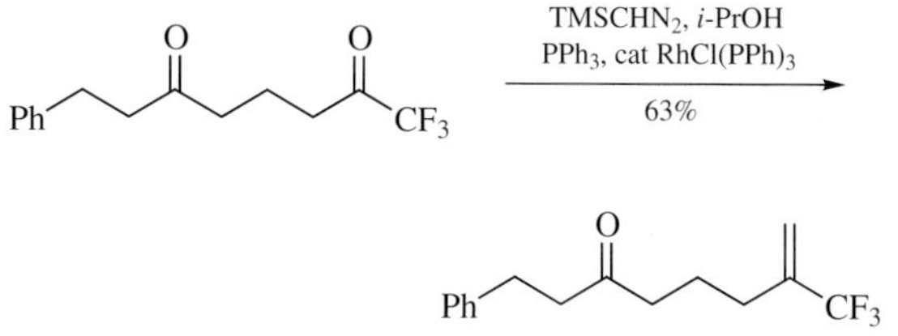

(45)

Substrates suitable for the Colvin rearrangement[15] have been extended to enolizable aryl alkyl ketones and both aromatic and aliphatic aldehydes using TMSC(Li)N$_2$ (eq 47). These conditions are reported to be superior to those employing dimethyl diazomethylphosphonate (DAMP) with regard to reaction times and range of permissible substrates. However, the reaction is not suitable for dialkyl ketones (see eq 42).[44]

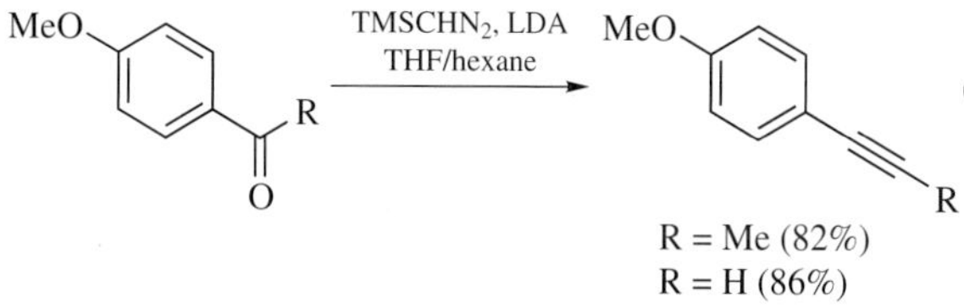

(47)

R = Me (82%)
R = H (86%)

The reaction of 2 equiv of TMSCHN$_2$ with an alkyne in the presence of catalytic RuCl(cod)Cp* affords 1,4-bis(trimethylsilyl)buta-1,3-dienes in moderate to excellent yield (eq 48).[45] The resulting terminal vinylsilanes may be readily converted to a variety of functionalities. This ruthenium catalyst also facilitates enyne metathesis with TMSCHN$_2$ to generate bicyclo[3.1.0] hexane derivatives bearing a heteroatom in the cyclopentane ring (eq 49). The reactivity of the enynes is greater with propargyl sulfonamides than with the corresponding ethers.[46] A related cascade metathesis reaction involving the Ni(cod)$_2$-catalyzed [4 + 2 + 1] cycloaddition of TMSCHN$_2$ and a dienyne offers an efficient approach to unsaturated bicyclo[5.3.0]decane systems (eq 50).[47]

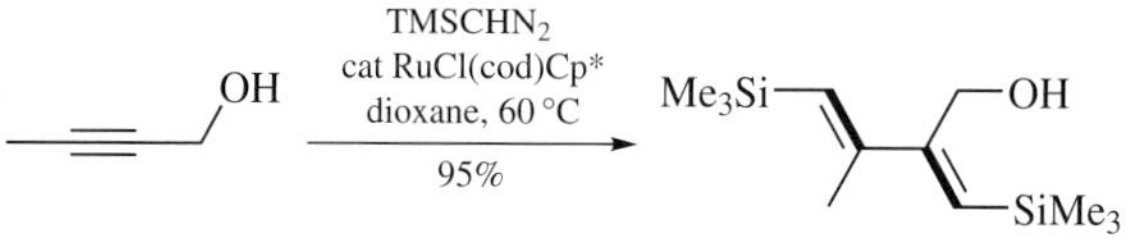

(48)

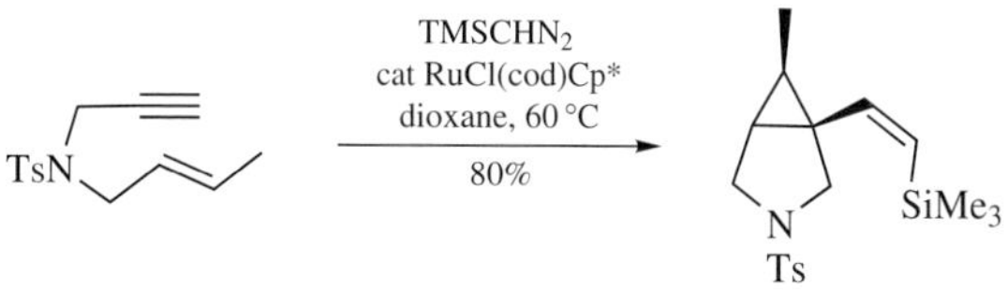

(49)

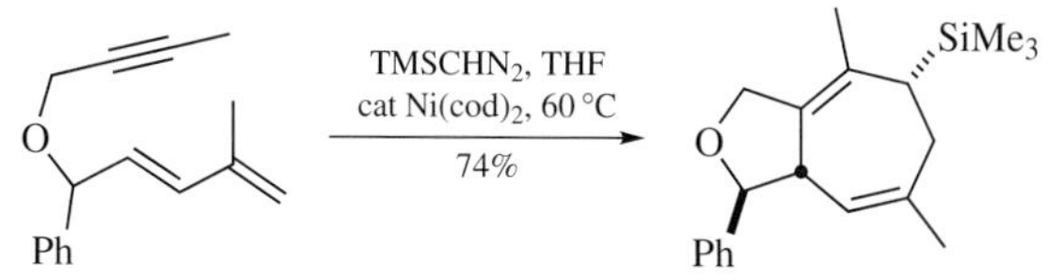

(50)

>95:5 dr

Cyclopropane and Aziridine Formation. Aliphatic ketones react with TMSC(Li)N$_2$ in DME to generate an intermediate singlet alkylidene carbene. The carbene reacts with a large excess of alkene to afford alkylidenyl cyclopropanes in yields ranging from <30% to 69% (eq 51).[48]

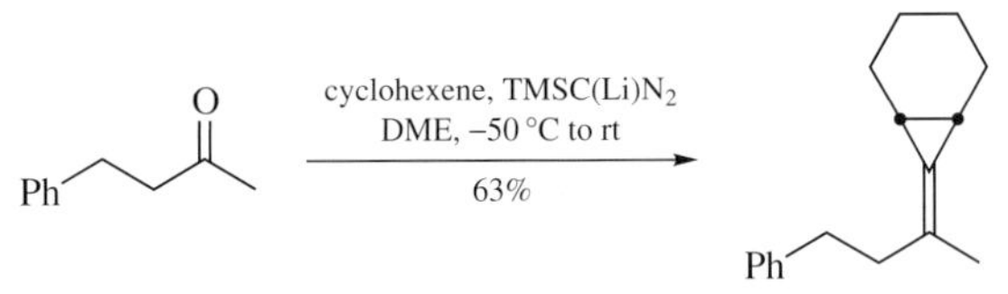

$$(51)$$

Improvements in the stereoselective preparation of silylcyclopropanes[16] (eqs 16 and 17) from alkenes and TMSCHN$_2$ have been reported based upon the judicious choice of catalytic metal and ligand. The *meso*-tetra-*p*-tolylporphyrin iron(II) complex, (TTP)Fe, and TMSCHN$_2$ convert styrene to the corresponding *trans*-1-phenyl-2-trimethylsilylcyclopropane in 89% yield with 13:1 diastereoselectivity.[49] The selection of [Cu(CH$_3$CN)$_4$]PF$_6$ and an appropriate bis(oxazoline) ligand provides highly diastereo- and enantioselective cyclopropanation of styrene and its derivatives, albeit with compromised yield compared to other protocols.[50]

Substituted *N*-sulfonyltrimethylsilylaziridines are formed when *N*-sulfonylaldimines are treated with TMSCHN$_2$. The reaction is highly diastereoselective, preferentially affording *cis*-products in good yields,[51] although electron releasing substituents attached to the *N*-sulfonylaldimine decrease the efficiency (eq 52).[52] The *C*-silylaziridines are useful precursors to a variety of products created by desilylation and trapping with electrophiles. This process proceeds with complete retention of stereochemistry. Alternatively, nucleophilic ring opening proceeds with excellent regiocontrol, occurring only at the silicon-bearing carbon atom (eq 53).[53] Substituted *N*-sulfonyltrimethylsilylaziridines are also formed with moderate enantioselectivity by treating *N*-tosyl α-imino esters with TMSCHN$_2$ and a catalytic BINAP-copper(I) complex to give the corresponding *cis*-aziridine or a bis(oxazoline)-copper(I) complex to create the analogous *trans*-diastereomer.[54]

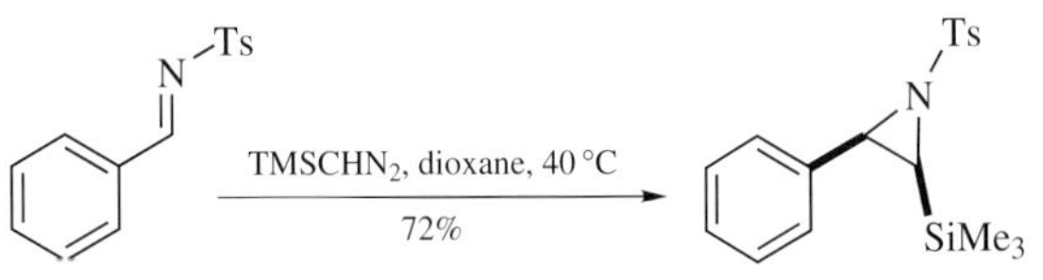

$$(52)$$

95:5 dr

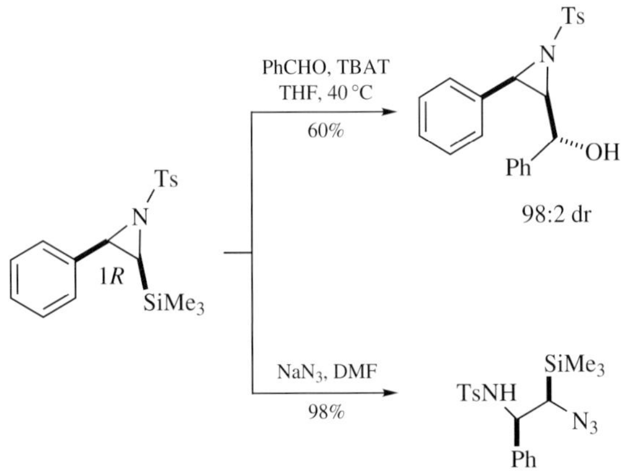

$$(53)$$

TBAT = tetrabutylammonium
triphenyldifluorosilicate

Insertion Reactions. Trimethylsilyldiazomethane undergoes net insertion between the B–C bond of borinate and boronate esters. Thus, olefin hydroboration, followed by treatment with TMSCHN$_2$, oxidation, and desilylation offers a method for hydroxymethylation of alkenes in fair to moderate yield (eq 54).[55] Stable, chiral allenylboranes (**1R** and **1S**) are prepared by reaction of TMSCHN$_2$ with *B*-MeO-9-BBN followed by resolution with pseudoephedrine and reaction with allenyl magnesium bromide (eq 55). Compounds **1R** and **1S** are useful for the asymmetric allenylboration of aldehydes with predictable absolute stereochemistry. The precursors to **1** are then readily recovered during work-up.[56]

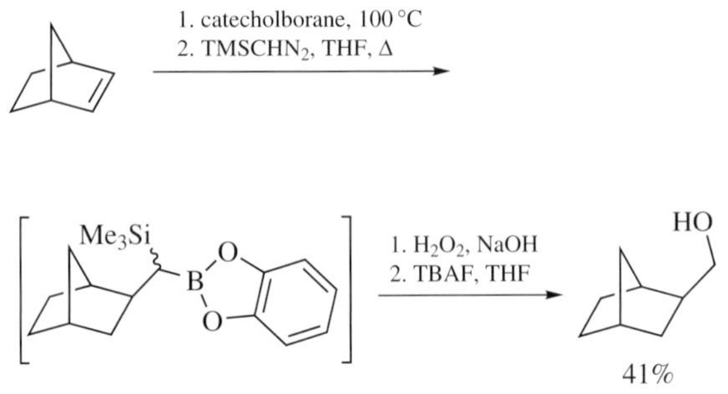

$$(54)$$

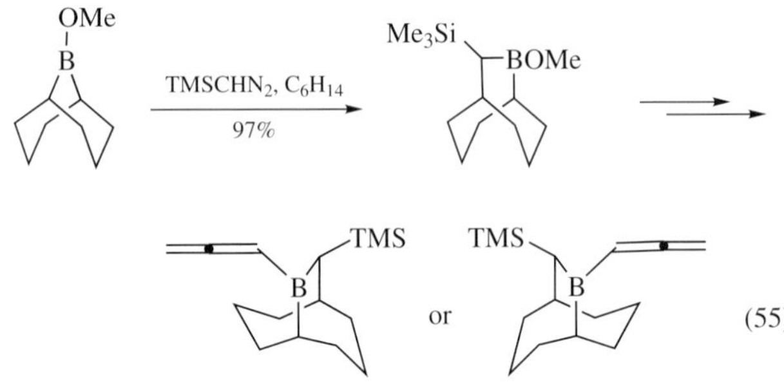

41%

$$(55)$$

1R   **1S**

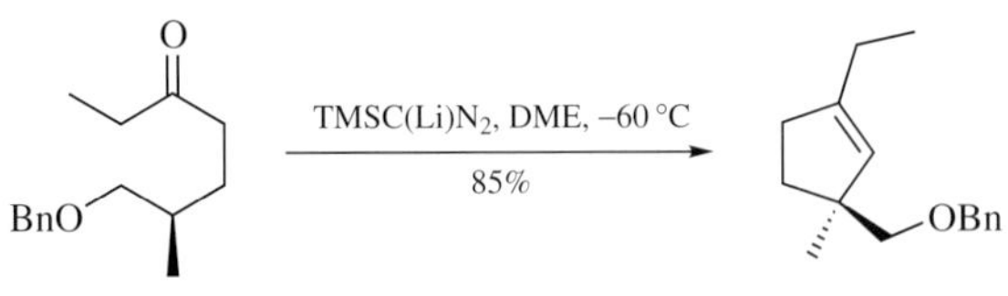

$$(56)$$

Lithium trimethylsilyldiazomethane has proved particularly useful in the conversion of ketones into alkylidene carbenes, vide supra, that readily undergo 1,5 C–H insertion reactions to afford cyclopentenes (eq 56).[57] Yields are generally good and the chemoselectivity of C–H insertion is predictable. The C–H insertion of the singlet carbene into heteroatom-bearing stereocenters proceeds with retention of stereochemistry (eqs 57 and 58).[58] Reaction with acetals affords spiroketals (eq 59)[59] or 2-cyclopentenones after acetal hydrolysis (eq 60).[60]

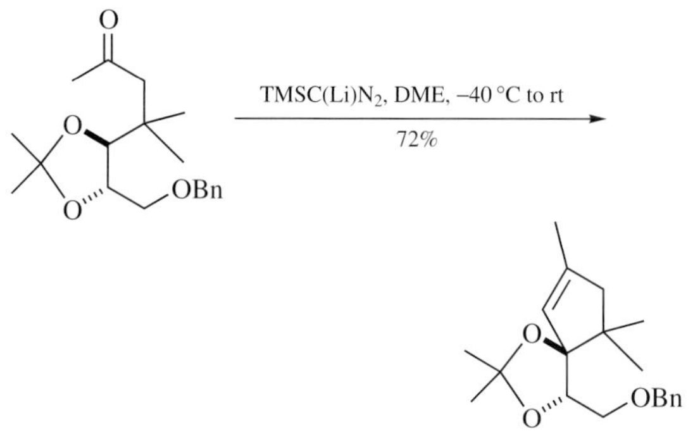

(57)

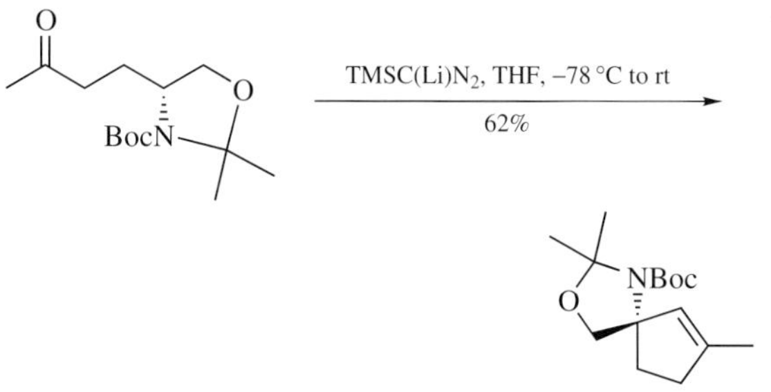

(58)

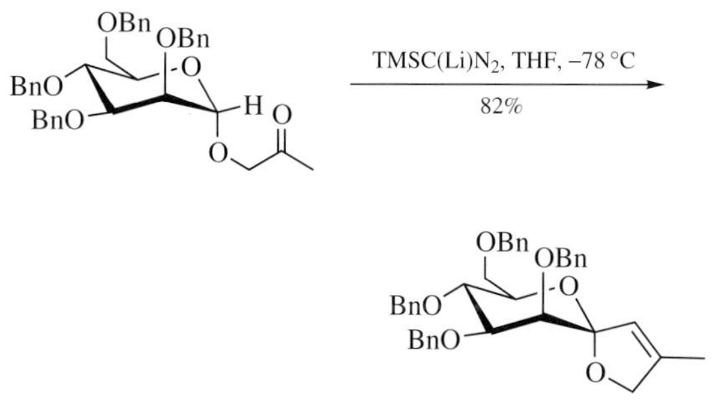

(59)

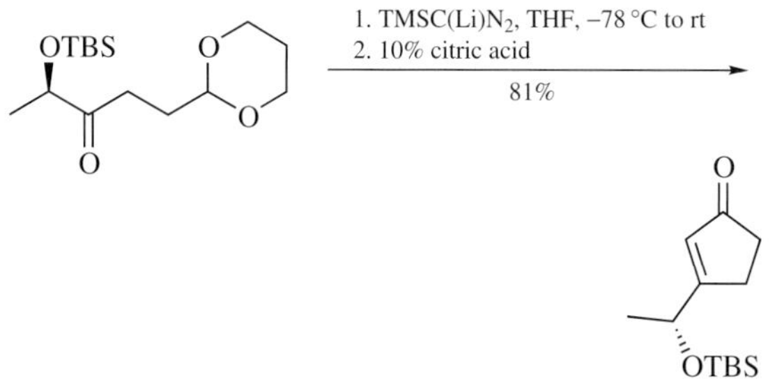

(60)

>99% ee

The intermediate carbenes generated from ketones and TMSC-(Li)N$_2$ have also been shown to undergo 1,5 O–Si and N–H insertion to produce 5-trimethylsilyl-2,3-dihydrofurans[61] and 2-pyrrolines or 3-substituted indoles, respectively (eqs 61 and 62).[62] Azulene 1-carboxylic esters may be generated in moderate yield by the reaction of TMSC(Li)N$_2$ with alkyl 4-aryl-2-oxobutanoates, followed by oxidation with manganese dioxide.[63]

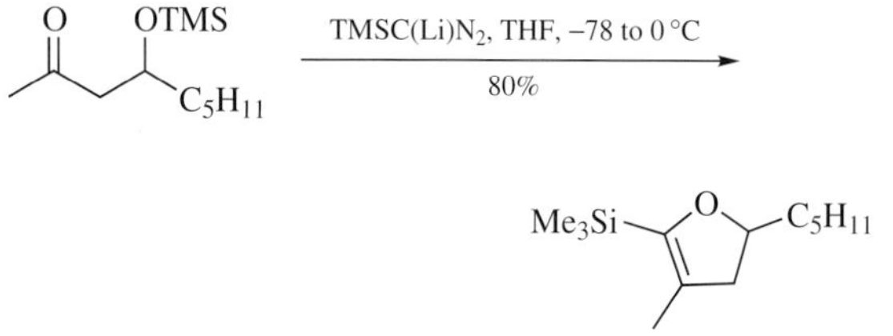

(61)

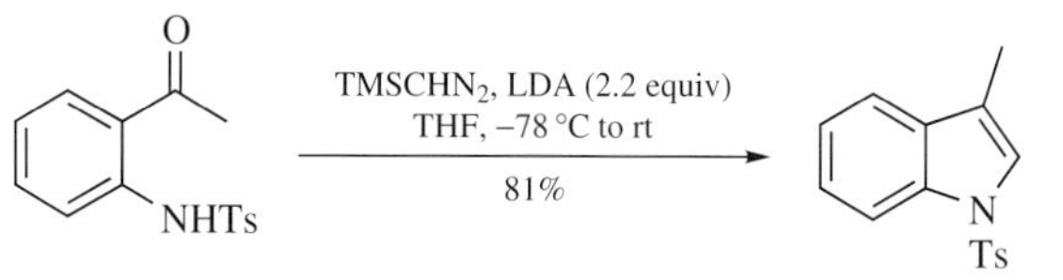

(62)

Ketenylation Reactions. Lithium silylethynolates are conveniently prepared by introducing carbon monoxide to TMSC(Li)N$_2$. The resulting ynolate, when mixed with trimethylaluminum, adds to epoxides to give 2-trimethylsilyl-γ-lactones with excellent regioselectivity and stereospecificity (eq 63).[64] Silylethynolates also add to aziridines, without Lewis acid activation, leading to 2-trimethylsilyl-γ-lactams. Introduction of excess aldehyde during lactam preparation affords α-alkylidene lactams in good yields via Peterson olefination (eq 64).[65]

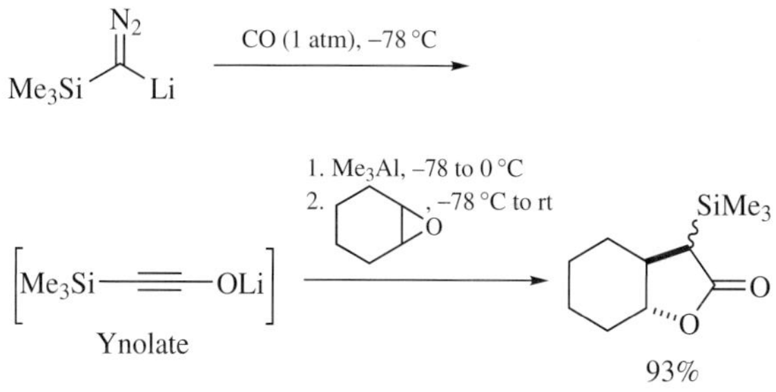

(63)

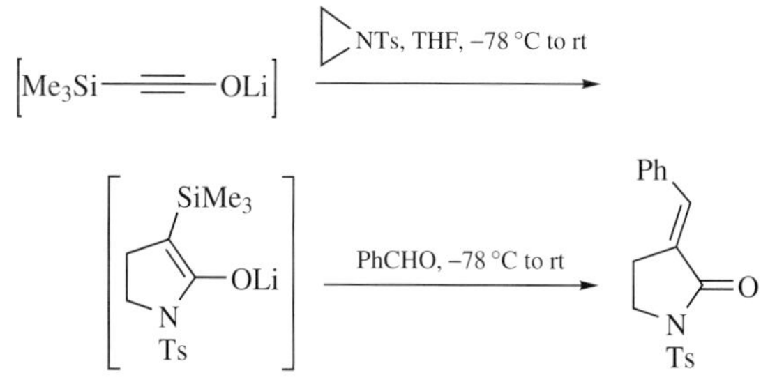

93%

(64)

61% (96:4 E/Z)

Preparation of Homoallylic Sulfides. When allyl sulfides are treated with TMSCHN$_2$ and a transition metal catalyst, an allylsulfonium ylide is formed that rapidly undergoes a diastereoselective [2,3] sigmatropic rearrangement in high yield (eq 65). Trimethylsilyldiazomethane is superior to ethyl diazoacetate for the transformation and rhodium(II) acetate is the catalyst of choice.[66] The homoallylic α-silyl sulfides thus obtained are suitable substrates for subsequent Peterson olefination or conversion to homoallylic aldehydes. Propargyl sulfides are also suitable reactants with TMSCHN$_2$ and catalytic iron(II) to generate allenyl α-silyl sulfides.[67]

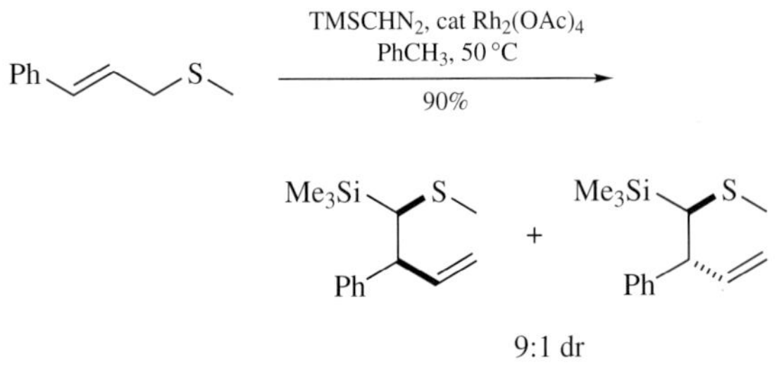

(65)

9:1 dr

(69)

Pericyclic Reactions. Trimethylsilyldiazomethane reacts with chiral acrylates to create optically active Δ^1-pyrazolines via regioselective asymmetric [3+2] cycloaddition. Subsequent protodesilylation affords Δ^2-pyrazolines in good yields (eq 66). This procedure offers a convenient route to azaprolines.[68]

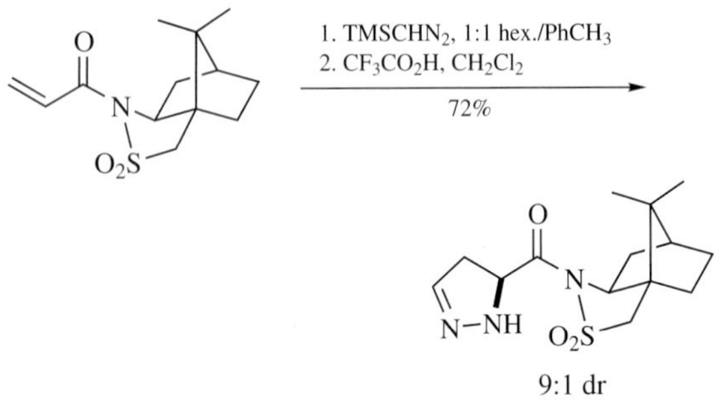

(66)

9:1 dr

Di- and trisubstituted furans may also be prepared by reacting TMSCHN$_2$ with an acyl isocyanate. The intermediate 4-trimethylsilyloxy oxazole readily undergoes Diels-Alder cycloaddition in situ with a suitable dienophile to construct the product (eq 67).[69]

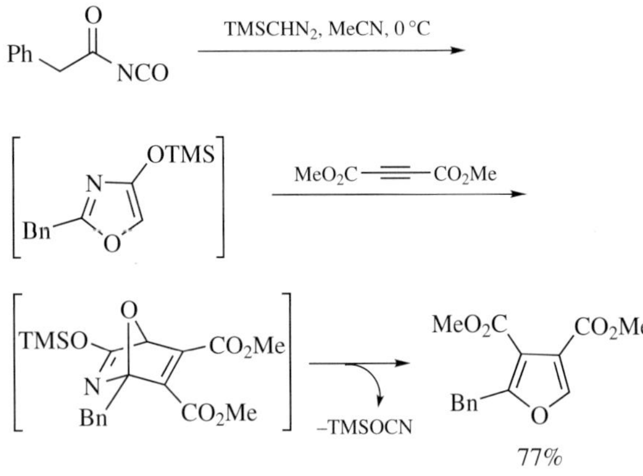

(67)

77%

Trimethylsilyldiazomethane smoothly reacts with (trialkylsilyl)vinylketenes, ultimately prepared from diazoketones, in a net [4 + 1] annulation process to afford 2-trialkylsilylcyclopentenones in good to excellent yields (eq 68).[70] The products are readily desilylated by treatment with methanesulfonic acid in methanol. An analogous reaction with (trialkylsilyl)arylketenes gives trialkylsilyl-2-indanone derivatives (eq 69).[71]

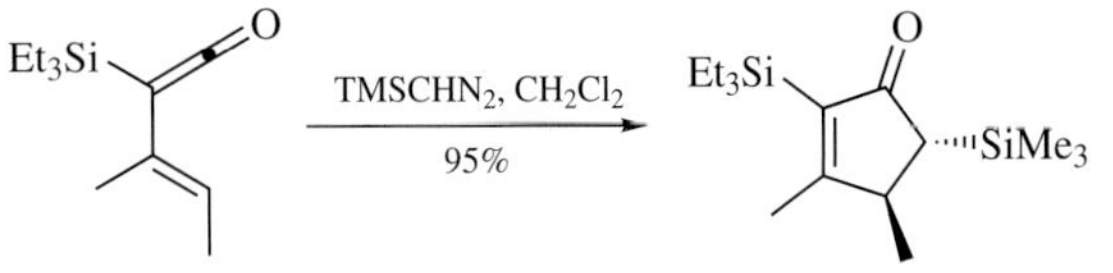

(68)

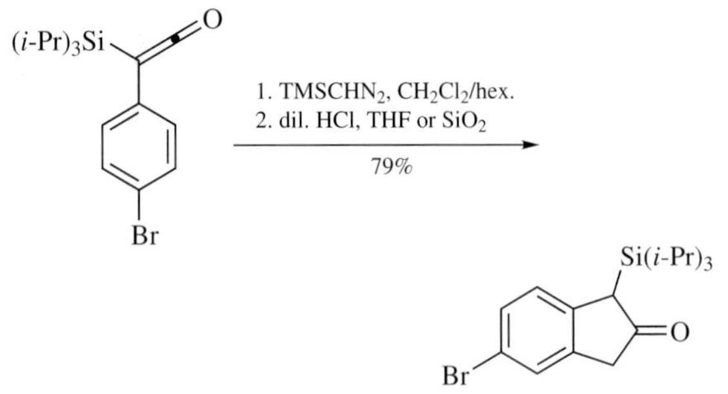

Miscellaneous Reactions. Trimethylsilyldiazomethane converts acid- and base-sensitive maleic anhydride derivatives into the corresponding bis(methyl esters) (eq 70).[72] Terminal silyl enol ethers are conveniently prepared from aldehydes by first treating the carbonyl compound with TMSC(Li)N$_2$, followed sequentially by methanol and Rh$_2$(OAc)$_4$ (eq 71). The method works well with base-sensitive substrates and is superior to the attempted regioselective deprotonation/O-silylation of the corresponding methyl ketone.[73]

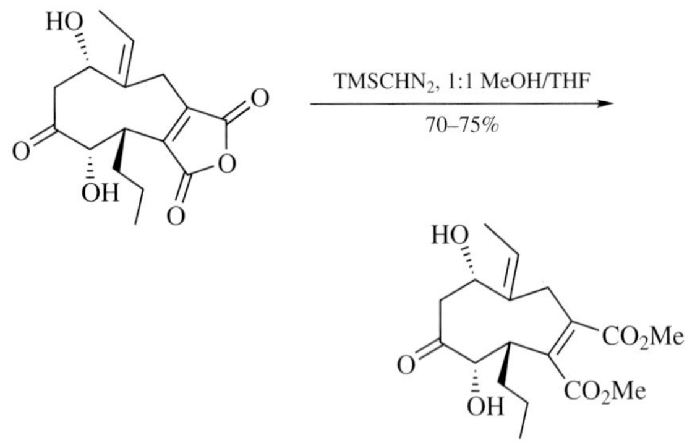

(70)

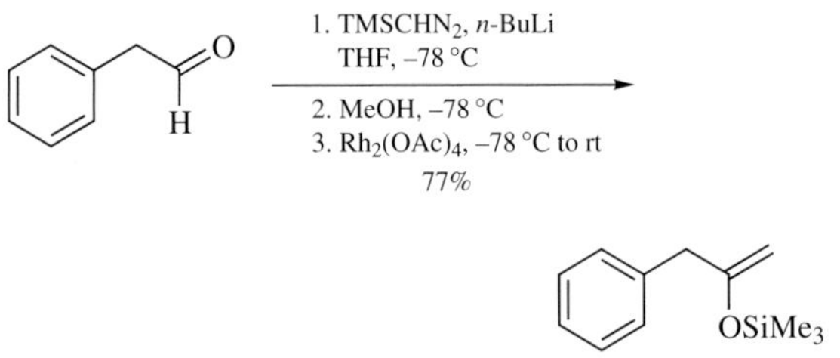

(71)

1. (a) Shioiri, T.; Aoyama, T., *J. Synth. Org. Chem. Jpn* **1986**, *44*, 149 (*Chem. Abstr.* **1986**, *104*, 168 525q). (b) Aoyama, T., *Yakugaku Zasshi* **1991**, *111*, 570 (*Chem. Abstr.* **1992**, *116*, 58 332q). (c) Anderson, R.; Anderson, S. B. In *Advances in Silicon Chemistry*; Larson, G. L., Ed.; JAI: Greenwich, CT, 1991; Vol. 1, p 303. (d) Shioiri, T.; Aoyama, T. In *Advances in the Use of Synthons in Organic Chemistry*; Dondoni, A., Ed.; JAI: London, 1993; Vol. 1, p 51.

2. Seyferth, D.; Menzel, H.; Dow, A. W.; Flood, T. C., *J. Organomet. Chem.* **1972**, *44*, 279.

3. Shioiri, T.; Aoyama, T.; Mori, S., *Org. Synth.* **1990**, *68*, 1.

4. Aoyama, T.; Shioiri, T., *Chem. Pharm. Bull.* **1981**, *29*, 3249.

5. Hashimoto, N.; Aoyama, T.; Shioiri, T., *Chem. Pharm. Bull.* **1982**, *30*, 119.

6. Aoyama, T.; Shioiri, T., *Synthesis* **1988**, 228.

7. Hashimoto, N.; Aoyama, T.; Shioiri, T., *Chem. Pharm. Bull.* **1981**, *29*, 1475.

8. Aoyama, T.; Terasawa, S.; Sudo, K.; Shioiri, T., *Chem. Pharm. Bull.* **1984**, *32*, 3759.

9. Aoyama, T.; Shioiri, T., *Tetrahedron Lett.* **1990**, *31*, 5507.

10. Aoyama, T.; Shioiri, T., *Tetrahedron Lett.* **1988**, *29*, 6295.

11. Aoyama, T.; Shioiri, T., *Chem. Pharm. Bull.* **1989**, *37*, 2261.

12. Aoyama, T.; Shioiri, T., *Tetrahedron Lett.* **1986**, *27*, 2005.

13. Aoyama, T.; Toyama, S.; Tamaki, N.; Shioiri, T., *Chem. Pharm. Bull.* **1983**, *31*, 2957.

14. Schöllkopf, U.; Scholz, H.-U., *Synthesis* **1976**, 271.

15. Colvin, E. W.; Hamill, B. J., *J. Chem. Soc., Perkin Trans. 1* **1977**, 869.

16. Aoyama, T.; Iwamoto, Y.; Nishigaki, S.; Shioiri, T., *Chem. Pharm. Bull.* **1989**, *37*, 253.

17. Zaitseva, G. S.; Lutsenko, I. F.; Kisin, A. V.; Baukov, Y. I.; Lorberth, J., *J. Organomet. Chem.* **1988**, *345*, 253.

18. Aoyama, T.; Shioiri, T., *Chem. Pharm. Bull.* **1982**, *30*, 3450.

19. Aoyama, T.; Iwamoto, Y.; Shioiri, T., *Heterocycles* **1986**, *24*, 589.

20. Shioiri, T.; Iwamoto, Y.; Aoyama, T., *Heterocycles* **1987**, *26*, 1467.

21. Crossman, J. M.; Haszeldine, R. N.; Tipping, A. E., *J. Chem. Soc., Dalton Trans.* **1973**, 483.

22. Aoyama, T.; Sudo, K.; Shioiri, T., *Chem. Pharm. Bull.* **1982**, *30*, 3849.

23. Aoyama, T; Inoue, S.; Shioiri, T., *Tetrahedron Lett.* **1984**, *25*, 433.

24. Asaki, T.; Aoyama, T.; Shioiri, T., *Heterocycles* **1988**, *27*, 343.

25. Aoyama, T.; Kabeya, M.; Fukushima, A.; Shioiri, T., *Heterocycles* **1985**, *23*, 2363.

26. Aoyama, T.; Kabeya, M.; Shioiri, T., *Heterocycles* **1985**, *23*, 2371.

27. Aoyama, T.; Kabeya, M.; Fukushima, A.; Shioiri, T., *Heterocycles* **1985**, *23*, 2367.

28. Aoyama, T.; Katsuta, S.; Shioiri, T., *Heterocycles* **1989**, *28*, 133.

29. Aoyama, T.; Nakano, T.; Marumo, K.; Uno, Y.; Shioiri, T., *Synthesis* **1991**, 1163.

30. Chan, K. S.; Wulff, W. D., *J. Am. Chem. Soc.* **1986**, *108*, 5229.

31. Aoyama, T.; Nakano, T.; Nishigaki, S.; Shioiri, T., *Heterocycles* **1990**, *30*, 375.

32. Rösch, W.; Hees, U.; Regitz, M., *Chem. Ber.* **1987**, *120*, 1645.

33. Bug, T.; Hartnagel, M.; Schlierf, C.; Mayr, H., *Chem. Eur. J.* **2003**, *9*, 4068.

34. Feeder, N.; Hendy, M. A.; Raithby, P. R.; Snaith, R.; Wheatley, A. E. H., *Eur. J. Org. Chem.* **1998**, 861.

35. Cesar, J.; Dolenc, M. S., *Tetrahedron Lett.* **2001**, *42*, 7099.

36. Ha, D.-C.; Kang, S.; Chung, C.-M.; Lim, H.-K., *Tetrahedron Lett.* **1998**, *39*, 7541.

37. (a) Maruoka, K.; Concepcion, A. B.; Yamamoto, H., *J. Org. Chem.* **1994**, *59*, 4725. (b) Maruoka, K.; Concepcion, A. B.; Yamamoto, H., *Synthesis* **1994**, 1283.

38. Maruoka, K.; Concepcion, A. B.; Yamamoto, H., *Synlett* **1994**, 521.

39. Miwa, K.; Aoyama, T.; Shioiri, T., *Synlett* **1994**, 109.

40. Lebel, H.; Paquet, V., *Organometallics* **2004**, *23*, 1187.

41. (a) Lebel, H.; Guay, D.; Paquet, V.; Huard, K., *Org. Lett.* **2004**, *6*, 3047. (b) Lebel, H.; Paquet, V., *J. Am. Chem. Soc.* **2004**, *126*, 320.

42. Yoakim, C.; Guse, I.; O'Meara, J. A.; Thavonekham, B., *Synlett* **2003**, 473.

43. Lebel, H.; Paquet, V.; Proulx, C., *Angew. Chem., Int. Ed.* **2001**, *40*, 2887.

44. Miwa, K.; Aoyama, T.; Shioiri, T., *Synlett* **1994**, 107.

45. Le Paih, J.; Derien, S.; Oezdemir, I.; Dixneuf, P. H., *J. Am. Chem. Soc.* **2000**, *122*, 7400.

46. Monnier, F.; Castillo, D.; Derien, S.; Toupet, L.; Dixneuf, P. H., *Angew. Chem., Int. Ed.* **2003**, *42*, 5474.

47. Yike, N.; Montgomery, J., *J. Am. Chem. Soc.* **2004**, *126*, 11162.

48. Sakai, A.; Aoyama, T.; Shioiri, T., *Tetrahedron* **1999**, *55*, 3687.

49. Hamaker, C. G.; Mirafzal, G. A.; Woo, L. K., *Organometallics* **2001**, *20*, 5171.

50. France, M. B.; Milojevich, A. K.; Stitt, T. A.; Kim, A. J., *Tetrahedron Lett.* **2003**, *44*, 9287.

51. Hori, R.; Aoyama, T.; Shioiri, T., *Tetrahedron Lett.* **2000**, *41*, 9455.

52. Aggarwal, V. K.; Ferrara, M., *Org. Lett.* **2000**, *2*, 4107.

53. Aggarwal, V. K.; Alonso, E.; Ferrara, M.; Spey, S. E., *J. Org. Chem.* **2002**, *67*, 2335.

54. Juhl, K.; Hazell, R. G.; Jorgensen, K. A., *J. Chem. Soc., Perkin Trans. 1* **1999**, 2293.

55. Goddard, J.-P.; Le Gall, T.; Mioskowski, C., *Org. Lett.* **2000**, *2*, 1455.

56. Lai, C.; Soderquist, J. A., *Org. Lett.* **2005**, *7*, 799.

57. (a) Ohira, S.; Moritani, T., *J. Chem. Soc., Chem. Commun.* **1992**, 721. (b) Taber, D. F.; Meagley, R. P., *Tetrahedron Lett.* **1994**, *35*, 7909.

58. (a) Taber, D. F.; Walter, R.; Meagley, R. P., *J. Org. Chem.* **1994**, *59*, 6014. (b) Gabaitsekgosi, R.; Hayes, C. J., *Tetrahedron Lett.* **1999**, *40*, 7713.

59. Wardrop, D. J.; Zhang, W.; Fritz, J., *Org. Lett.* **2002**, *4*, 489.

60. Sakai, A.; Aoyama, T.; Shioiri, T., *Tetrahedron Lett.* **2000**, *41*, 6859.

61. Miwa, K.; Aoyama, T.; Shioiri, T., *Synlett* **1994**, 461.

62. (a) Yagi, T.; Aoyama, T.; Shioiri, T., *Synlett* **1997**, 1063. (b) Miyagi, T.; Hari, Y.; Aoyama, T., *Tetrahedron Lett.* **2004**, *45*, 6303.

63. Hari, Y.; Tanaka, S.; Takuma, Y.; Aoyama, T., *Synlett* **2003**, 2151.

64. Kai, K.; Iwamoto, K.; Chatani, N.; Murai, S., *J. Am. Chem. Soc.* **1996**, *118*, 7634.

65. Iwamoto, K.; Kojima, M.; Chatani, N.; Murai, S., *J. Org. Chem.* **2001**, *66*, 169.

66. (a) Carter, D. S.; Van Vranken, D. L., *Tetrahedron Lett.* **1999**, *40*, 1617. (b) Aggarwal, V. K.; Ferrara, M.; Hainz, R.; Spey, S. E., *Tetrahedron Lett.* **1999**, *40*, 8923.

67. Prabharasuth, R.; Van Vranken, D. L., *J. Org. Chem.* **2001**, *66*, 5256.

68. Mish, M. R.; Guerra, F. M.; Carreira, E. M., *J. Am. Chem. Soc.* **1997**, *119*, 8379.

69. Hari, Y.; Iguchi, T.; Aoyama, T., *Synthesis* **2004**, 1359.

70. Loebach, J. L.; Bennett, D. M.; Danheiser, R. L., *J. Am. Chem. Soc.* **1998**, *120*, 9690.

71. Dalton, A. M.; Zhang, Y.; Davie, C. P.; Danheiser, R. L., *Org. Lett.* **2002**, *4*, 2465.

72. Fields, S. C.; Dent, W. H., III; Green, F. R., III; Tromiczak, E. G., *Tetrahedron Lett.* **1996**, *37*, 1967.

73. Aggarwal, V. K.; Sheldon, C. G.; MacDonald, G. J.; Martin, W. P., *J. Am. Chem. Soc.* **2002**, *124*, 10300.

Zinc–Acetic Acid 554

Zinc–Acetic Acid[1]

$$\boxed{\text{Zn–AcOH}}$$

(Zn)

[7440-66-6]   Zn   (MW 65.39)

InChI = 1/Zn

InChIKey = HCHKCACWOHOZIP-UHFFFAOYAS

(AcOH)

[64-19-7] $C_2H_4O_2$ (MW 60.06)

InChI = 1/C2H4O2/c1-2(3)4/h1H3,(H,3,4)/f/h3H

InChIKey = QTBSBXVTEAMEQO-TULZNQERCK

(reducing agent; causes reductive elimination of vicinal hetero-atoms;[2–15] cleaves heteroatom–heteroatom bonds;[16–27] reduces allylic, benzylic, or α-carbonyl-substituted heteroatoms,[28–34] activated carbonyls,[35–37] and alkenes[38–40])

Physical Data: see entries for **Zinc** and **Acetic Acid**.

Form Supplied in: although zinc is available in a variety of forms, the overwhelming majority of zinc–acetic acid reductions use zinc powder.

Purification: acid washing is not uncommon, but not always vital.

Introduction. Zinc in acetic acid is capable of a wide range of reduction reactions. Although many of these can also be performed by a great number of other reagents, this reagent is of particular value in that good chemoselectivities can often be achieved. Some such instances are noted in the text and equations below; many of the references have also been chosen to demonstrate selective reduction in sensitive, polyfunctional molecules.

Reductive Elimination of Vicinal Heteroatoms.[2–15] A great variety of combinations of heteroatom substituents has been successfully reductively eliminated by Zn/AcOH (eq 1 and Table 1). Cosolvents such as ether, THF, CH_2Cl_2, *i*-PrOH, or water have all been used. Reaction temperatures vary from case to case, but yields reported are typically good to excellent.

$$(1)$$

A set of protecting groups, based upon Zn/AcOH elimination of the 2,2,2-trichloroethoxy group, has been developed, and is discussed below as a special case. Table 1 records several of the combinations of heteroatoms successfully eliminated under these

Table 1 Reductive eliminations of vicinal heterosubstituents (eq 1)

X	Y	Yield (%)
Cl	OR	see next section
Cl	Cl	96[2]
Cl	SO_2R	87[3]
Cl	NO_2	79[4]
Br	OR	88[5]
Br	Br	94[6]
OR	OR	92[7]

conditions. Perhaps surprisingly, the relative stereochemistry of the two carbon centers does not have to permit *trans* coplanarity of the heteroatoms.[4,7]

Heterocyclic rings may be cleaved by these reactions. Thus, effective reversal of iodolactonization (eq 2)[8] or of epoxidation (after iodide ring opening) (eq 3)[9] may be achieved.

$$(2)$$

$$(3)$$

Halogenated oximes (**1**) eliminate to give the nitrile in variable, but often good to excellent, yields (eq 4).[10]

$$(4)$$

2,2,2-Trichloroethoxy-based Protecting Groups.[11–15] Valuable protecting groups for alcohols,[11] phenols,[11c] amines,[11c] and carboxylic acids,[12] and an introduction/protection reagent for thiols,[13] all dependent upon the lability of this group to Zn/AcOH reductive elimination, have been developed. They are summarized in Table 2.

Table 2 Protecting groups based on OCH_2CCl_3 elimination

Functionality	Protected form
ROH[11a]	$ROCH_2CCl_3$
ROH[11b]	$ROCH_2OCH_2CCl_3$
ROH[11c,d]	$ROC(O)OCH_2CCl_3$
ROH[11e]	$ROC(O)OCMe_2CCl_3$
ArOH[11c]	$ArOC(O)OCH_2CCl_3$
R_2NH[11c]	$R_2NC(O)OCH_2CCl_3$
RCO_2H[12]	$RC(O)OCH_2CCl_3$
RSH[13]	$RSC(O)OCH_2CCl_3$

The trichloroacetylidene acetal has also been proposed as a potential protection for diols, similarly deprotectable.[14] Also, along similar lines, the 2-iodoethyl carbamate protection for amines, deprotectable by Zn dust alone in MeOH, has been proposed,[15] but has found relatively little use.

Heteroatom–Heteroatom Cleavage.[16–27] N=N double bonds can be cleaved[16] by Zn/AcOH; less frequently, hydrazines[17] may be obtained from the reduction (eqs 5 and 6).[17] Hydrazones may also be reduced to amines,[18] as used[18b] in a variant of the classical Knorr pyrrole synthesis. Diazo ketone (**2**) has been successfully reduced, despite the apparent potential for adverse side reactions (eq 7).[19]

$$PhN=NAr \xrightarrow[\text{90–100\%}]{\text{Zn, AcOH}} PhNH\text{-}NHAr \qquad (5)$$

$$\text{Ar}^1\underset{\overset{\Vert}{N}\text{Ar}^2}{\overset{+}{N}}\text{O}^- \xrightarrow[89\%]{\text{Zn, AcOH}} \text{Ar}^1\text{NH-NHAr}^2 \qquad (6)$$

$$\text{(2)} \qquad \xrightarrow[\substack{0\,°C, THF \\ 50\%}]{\text{Zn, AcOH}} \qquad (7)$$

Aromatic nitro groups[20a] (aliphatic nitro groups can yield oximes,[20b] even though not all such would be stable under all reaction conditions), hydroxylamines,[21] oximes[22] (once again, finding use[22b] in a Knorr pyrrole synthesis), N-nitro-[23a] and N-nitrosoamines,[23b] aromatic N-oxides,[24] and aromatic N–S bonds[25] have all been reduced to amines with Zn/AcOH. These references include ample evidence of the ability of this reagent to perform the desired reaction, while leaving intact, for example, bromides,[22] carbon-bonded sulfur atoms,[20,23b] and isolated C=C double bonds.[21] It will, of course, be understood that the newly liberated amines often undergo spontaneous intramolecular reactions. Sulfonamides[26] can also be reduced to the thiols.

Zn/AcOH can act as a useful alternative reagent for reductive workup of ozonolysis reactions,[27] which can be considered, at least formally, to involve O–O bond cleavage (see *Ozone*).

Curiously, the vigor of the conditions reported to have been employed for these disparate reactions does not seem to follow any consistent pattern; reaction conditions, therefore, may need individual determination in many cases.

Carbon–Heteroatom Cleavage.[28–34] Substitution of N,[28] O,[29] S,[30] or halogen,[31] for example, at an allylic, benzylic, or α-carbonyl-substituted carbon atom renders the heteroatom liable to cleavage with Zn/AcOH. In allylic systems, double bond migration usually occurs. If conjugated to a carbonyl group, migration still occurs, giving the β,γ-product (eq 8)[32a] but these can be easily reconjugated (eq 9).[32b] Homoallylic reduction[33] has also been reported in a constrained system (eq 10).

$$ (8) $$

$$ (9) $$

$$ (10) $$

Cleavage α to a carbonyl group has been exploited in the use of phenacyl protecting groups.[34] The reduction of compounds such as (3) (for their formation, via *Dichloroketene* see *Trichloroacetyl Chloride*) retains the strained four-membered ring, and often gives excellent yields (eq 11).[31c]

$$ (11) $$

Carbonyl Reduction.[35–37] Quinones are reduced to hydroquinones by Zn/AcOH at reflux.[35a] Incorporation of Ac_2O into the reaction mixture gives the respective diacetate.[35a] Under milder conditions (rt), the intermediate γ-hydroxycyclohexenone may be intercepted in surprisingly high yield (72–90%).[35b]

Diaryl ketones may be reduced to the alcohols,[36a] but a competing dimerization has also been reported[36b] in some unusual cases. Reduction of phthalimide (4) proceeds well and, notably, with regiospecificity (eq 12).[37]

$$ (12) $$

Reduction of Activated Alkenes.[38–40] Carbonyl (mono-[38a] or di-[38b]) substitution renders a C=C double bond liable to reduction by Zn/AcOH. α,β,γ,δ-Dienones may yield either α,β-[38a] or β,γ-enones[38c] in equally high yields. A recent modification,[39] at lower temperature (rt), and with much shorter reaction times, uses ultrasonication; excellent yields were achieved.

α,β-Unsaturated nitro compounds can also be reduced.[40] Under mild conditions an oxime can be obtained (eq 13)[40a] (cf. *Sodium Borohydride*, which reduces the C=C double bond); more vigorous reaction leads to the corresponding ketone.[40b]

$$ (13) $$

Aza-heterocycle Ring Contraction. A variety of polyaza six-ring heterocycles undergo contraction with formal excision of N, e.g. eq 14.[41a] Further heteroatoms in the ring are tolerated: 1,2,3-triazines yield pyrazoles[41b] and 1,2,4-triazines yield imidazoles,[41c] with the latter usually requiring reflux temperature. The utility of indole syntheses from cinnolines[41d] by this method should be noted. Adjacency of heteroatoms is not required: conversions of pyrimidine to pyrrole[42a,b] and 1,3,5-triazine to imidazole[42c] have been recorded. The mechanism of these conversions is unclear, but ring dihydro derivatives[41a,42b] are thought to be involved.

$$ (14) $$

Reduction of Aryl Substituents. Although, as mentioned above, many potentially labile functionalities are stable to

Zn/AcOH, aryl iodides may be dehalogenated.[43] Substitution of Cl or Br at the 2-position of heteroaromatics also renders them liable to reduction, often with superb selectivity (eqs 15 and 16).[44]

Related Reagents. Iron; Tin; Zinc Amalgam; Zinc–Zinc Chloride.

1. *Fieser & Fieser* **1967**, *1*, 1276.

2. Attenburrow, J.; Connett, J. E.; Graham, W.; Oughton, J. F.; Ritchie, A. C.; Wilkinson, P. A., *J. Chem. Soc.* **1961**, 4547.

3. Kay, I. T.; Punja, N., *J. Chem. Soc. (C)* **1968**, 3011.

4. Komeichi, Y.; Osawa, Y.; Duax, W. L.; Cooper, A., *Steroids* **1970**, *15*, 619.

5. Goodman, L.; Winstein, S.; Boschan, R., *J. Am. Chem. Soc.* **1958**, *80*, 4312.

6. Martin, J. D.; Pérez, C.; Ravelo, J. L., *J. Am. Chem. Soc.* **1985**, *107*, 516.

7. Goto, G., *Bull. Chem. Soc. Jpn.* **1977**, *50*, 186.

8. Hamanaka, N.; Seko, T.; Miyazaki, T.; Naka, M.; Furuta, K.; Yamamoto, H., *Tetrahedron Lett.* **1989**, *30*, 2399.

9. Hatakeyama, S.; Numata, H.; Takano, S., *Tetrahedron Lett.* **1984**, *25*, 3617.

10. Sakamoto, T.; Mori, H.; Takizawa, M.; Kikugawa, Y., *Synthesis* **1991**, 750.

11. (a) Lemieux, R. U.; Driguez, H., *J. Am. Chem. Soc.* **1975**, *97*, 4069. (b) Jacobson, R. M.; Clader, J. W., *Synth. Commun.* **1979**, *9*, 57. (c) Windholz, T. B.; Johnston, D. B. R., *Tetrahedron Lett.* **1967**, 2555. (d) Imoto, M.; Kusunose, N.; Kusumoto, S.; Shiba, T., *Tetrahedron Lett.* **1988**, *29*, 2227. (e) Eckert, H.; Listl, M.; Ugi, I., *Angew. Chem., Int. Ed. Engl.* **1978**, *17*, 361.

12. Woodward, R. B.; Heusler, K.; Gosteli, J.; Naegeli, P.; Oppolzer, W.; Ramage, R.; Rangamathan, S.; Vorbrüggen, H., *J. Am. Chem. Soc.* **1966**, *88*, 852.

13. Sheehan, J. C.; Commons, T. J., *J. Org. Chem.* **1978**, *43*, 2203.

14. Greene, T. W.; Wuts, P. G. M. *Protective Groups in Organic Synthesis*, 2nd ed.; Wiley: New York, 1991; p 122.

15. Grimshaw, J., *J. Chem. Soc.* **1965**, 7136.

16. Barton, D. H. R.; Lamotte, G.; Motherwell, W. B.; Narang, S. C., *J. Chem. Soc., Perkin Trans. 1* **1979**, 2030.

17. (a) Ruggli, P.; Hölzle, K., *Helv. Chim. Acta* **1943**, *26*, 814 (*Chem. Abstr.* **1944**, *38*, 2640). (b) Budziarek, R.; Drain, D. J.; Macrae, F. J.; McLean, J.; Newbold, G. T.; Seymour, D. E.; Spring, F. S.; Stansfield, M., *J. Chem. Soc.* **1955**, 3158.

18. (a) Goodwin, R. C.; Bailey, J. R., *J. Am. Chem. Soc.* **1925**, *47*, 167. (b) Treibs, A.; Schmidt, R.; Zinsmeister, R., *Chem. Ber.* **1957**, *90*, 79 (*Chem. Abstr.* **1957**, *51*, 10 480).

19. Vedejs, E., *J. Chem. Soc., Chem. Commun.* **1971**, 536.

20. (a) Saupe, T.; Krieger, C.; Staab, H. A., *Angew. Chem., Int. Ed. Engl.* **1986**, *25*, 451. (b) Johnson, K.; Degering, E. F., *J. Am. Chem. Soc.* **1939**, *61*, 3194.

21. Howell, A. R.; Pattenden, G., *J. Chem. Soc., Chem. Commun.* **1990**, 103.

22. (a) Prasitpan, N.; Johnson, M. E.; Currie, B. L., *Synth. Commun.* **1990**, *20*, 3459. (b) Fischer, H., *Org. Synth., Coll. Vol.* **1955**, *3*, 513.

23. (a) Shriner, R. L.; Neumann, F. W., *Org. Synth., Coll. Vol.* **1955**, *3*, 73. (b) Allen, C. F. H.; Vanallan, J. A., *J. Org. Chem.* **1948**, *13*, 603.

24. Freeman, J. P.; Gannon, J. J.; Surbey, D. L., *J. Org. Chem.* **1969**, *34*, 187.

25. Shealy, Y. F.; O'Dell, C. A., *J. Org. Chem.* **1964**, *29*, 2135.

26. Hendrickson, B.; Bergeron, R., *Tetrahedron Lett.* **1970**, 345.

27. Callant, P.; Ongena, R.; Vandewalle, M., *Tetrahedron* **1981**, *37*, 2085.

28. Gaskell, A. J.; Joule, J. A., *Tetrahedron* **1968**, *24*, 5115.

29. Cope, A. C.; Barthel, J. W.; Smith, R. D., *Org. Synth., Coll. Vol.* **1963**, *4*, 218.

30. Gotthardt, H.; Nieberl, S.; Doenecke, J., *Liebigs Ann. Chem.* **1980**, 873 (*Chem. Abstr.* **1980**, *93*, 239 355).

31. (a) Grieco, P. A., *J. Org. Chem.* **1972**, *37*, 2363. (b) Danheiser, R. L.; Savariar, S., *Tetrahedron Lett.* **1987**, *28*, 3299. (c) Danheiser, R. L.; Savariar, S.; Cha, D. D., *Org. Synth., Coll. Vol.* **1993**, *8*, 82.

32. (a) Moppett, C. E.; Sutherland, J. K., *J. Chem. Soc. (C)* **1968**, 3040. (b) Roth, B. D.; Roark, W. H., *Tetrahedron Lett.* **1988**, *29*, 1255.

33. Rakhit, S.; Gut, M., *J. Am. Chem. Soc.* **1964**, *86*, 1432.

34. Greene, T. W.; Wuts, P. G. M. *Protective Groups in Organic Synthesis*, 2nd ed.; Wiley: New York, 1991; pp 153, 238.

35. (a) Crawford, H. M.; Lumpkin, M.; Mcdonald, M., *J. Am. Chem. Soc.* **1952**, *74*, 4087. (b) Speziale, A. J.; Stephens, J. A.; Thompson, Q. E., *J. Am. Chem. Soc.* **1954**, *76*, 5011.

36. (a) Gross, M. E.; Lankelman, H. P., *J. Am. Chem. Soc.* **1951**, *73*, 3439. (b) Agranat, I.; Tapuhi, Y., *J. Am. Chem. Soc.* **1979**, *101*, 665.

37. Brewster, J. H.; Fusco, A. M., *J. Org. Chem.* **1963**, *28*, 501.

38. (a) Corey, E. J.; Watt, D. S., *J. Am. Chem. Soc.* **1973**, *95*, 2303. (b) Büchi, G.; Foulkes, D. M.; Kurono, M.; Mitchell, G. F.; Schneider, R. S., *J. Am. Chem. Soc.* **1967**, *89*, 6745. (c) Fieser, L. F.; Rajagopalan, S.; Wilson, E.; Tishler, M., *J. Am. Chem. Soc.* **1951**, *73*, 4133.

39. Marchand, A. P.; Reddy, G. M., *Synthesis* **1991**, 198.

40. (a) Baer, H. H.; Rank, W., *Can. J. Chem.* **1972**, *50*, 1292. (b) Anagnostopoulos, C. E.; Fieser, L. F., *J. Am. Chem. Soc.* **1954**, *76*, 532.

41. (a) Boger, D. L.; Coleman, R. S.; Panek, J. S.; Yohannes, D., *J. Org. Chem.* **1984**, *49*, 4405. (b) Chandross, E. A.; Smolinsky, G., *Tetrahedron Lett.* **1960** (13), 19. (c) Laakso, P. V.; Robinson, R.; Vandrewala, H. P., *Tetrahedron* **1957**, *1*, 103. (d) Besford, L. S.; Bruce, J. M., *J. Chem. Soc.* **1964**, 4037.

42. (a) Patterson, J. M., *Synthesis* **1976**, 281. (b) Longridge, J. L.; Thompson, T. W., *J. Chem. Soc. (C)* **1970**, 1658. (c) Cook, A. H.; Jones, D. G., *J. Chem. Soc.* **1941**, 278.

43. Giza, C. A.; Hinman, R. L., *J. Org. Chem.* **1964**, *29*, 1453.

44. (a) Marais, J. L. C.; Backeburg, O. G., *J. Chem. Soc.* **1950**, 2207. (b) Gronowitz, S.; Raznikiewicz, T., *Org. Synth., Coll. Vol.* **1973**, *5*, 149.

Peter Ham
SmithKline Beecham Pharmaceuticals, Harlow, UK

List of Contributors

Enrique Aguilar   *Universidad de Oviedo, Oviedo, Spain*
- Sodium Periodate   447
- Trimethylsilyl Trifluoromethanesulfonate   524

Mouâd Alami   *Université Pierre & Marie Curie, Paris, France*
- Manganese Dioxide   248

Toyohiko Aoyama   *Nagoya City University, Nagoya, Japan*
- Trimethylsilyldiazomethane   543

Alan Armstrong   *University of Bath, Bath, UK*
- N,N'-Carbonyl Diimidazole   72

Luca Banfi   *Universitá, di Genova, Italy*
- Sodium Borohydride   406

Louis Barriault   *University of Ottawa, Ottawa, Ontario, USA*
- Lithium Diisopropylamide   224

Richard A. Berglund   *Eli Lilly Company, Lafayette, IN, USA*
- Ozone   290

Justin Du Bois   *Stanford University, Stanford, CA, USA*
- (Diacetoxyiodo)benzene   136

Derek R. Buckle   *SmithKline Beecham Pharmaceuticals, Epsom, UK*
- 2,3-Dichloro-5,6-dicyano-1,4-benzoquinone   152

Benjamin R. Buckley   *Loughborough University, Loughborough, UK*
- Potassium Monoperoxysulfate   334

Kevin Burgess   *Texas A&M University, College Station, TX, USA*
- Chlorotris(triphenylphosphine)-rhodium(I)   121

Christopher P. Burke   *Colorado State University, Fort Collins, CO, USA*
- Potassium Monoperoxysulfate   334

Gérard Cahiez   *Université Pierre & Marie Curie, Paris, France*
- Manganese Dioxide   248

Drury Caine   *The University of Alabama, Tuscaloosa, AL, USA*
- Potassium *tert*-Butoxide   353

J. Subash Chandra   *Purdue University, West Lafayette, IN, USA*
- Pinacolborane   306

Calvin J. Chany II   *University of Illinois at Chicago, Chicago, IL, USA*
- (Diacetoxyiodo)benzene   136

André Charette   *Université de Montréal, Montréal, Québec, Canada*
- *m*-Chloroperbenzoic Acid   87
- Samarium(II) Iodide   378

Ke Chen   *Bristol-Myers Squibb, New Brunswick, NJ, USA*
- Bis(dibenzylideneacetone)palladium(0)   2

Paul Galatsis	*University of Guelph, Guelph, Ontario, Canada*	
	• Diisobutylaluminum Hydride	164
Yun Gao	*Sepracor, Marlborough, MA, USA*	
	• Osmium Tetroxide	264
M. D. García Romero	*University of Leicester, Leicester, UK*	
	• *N*-Bromosuccinimide	43
Robert E. Gawley	*University of Miami, Coral Gables, FL, USA*	
	• Sodium Hydride	438
Julia Haas	*Array BioPharma, Boulder, CO, USA*	
	• *p*-Toluenesulfonyl Chloride	480
Peter Ham	*SmithKline Beecham Pharmaceuticals, Harlow, UK*	
	• ZincAcetic Acid	554
Michael Hanack	*Universität Tübingen, Tübingen, Germany*	
	• Trifluoromethanesulfonic Acid	498
	• Trifluoromethanesulfonic Anhydride	507
Christina R. Harris	*University of Colorado, Boulder, CO, USA*	
	• Samarium(II) Iodide	378
Robert M. Harris	*Dyson Perrins Laboratory, Oxford, UK*	
	• Osmium Tetroxide	264
Alison Hart	*University of Southern Mississippi, Hattiesburg, MS, USA*	
	• 4-Dimethylaminopyridine	170
Alfred Hassner	*Bar-Ilan University, Ramat Gan, Israel*	
	• 4-Dimethylaminopyridine	170
D. David Hennings	*Array BioPharma, Boulder, CO, USA*	
	• Sodium Hydride	438
Jung-Nyoung Heo	*Korea Research Institute for Chemical Technology, Daejeon, Korea*	
	• 9-Borabicyclo[3.3.1]nonane Dimer	17
John C. Hershberger	*University of Kansas, Lawrence, KS, USA*	
	• Potassium Hexamethyldisilazide	313
Martin Hiersemann	*Technische Universität Dortmund, Dortmund, Germany*	
	• Sodium Borohydride	406
Tse-Lok Ho	*National Chiao-Tung University, Hsinchu, Taiwan, Republic of China*	
	• Cerium (IV) Ammonium Nitrate	80
Jinbo Hu	*University of Southern California, Los Angeles, CA, USA*	
	• Iodotrimethylsilane	194
	Shanghai Institute of Organic Chemistry, Shanghai, China	
	• Trifluoromethanesulfonic Acid	498
Terry V. Hughes	*J&JPRD, Raritan, NJ, USA*	
	• *N*-Chlorosuccinimide	98
MaryGail K. Hutchins	*LNP Engineering Plastics, Exton, PA, USA*	
	• Sodium Cyanoborohydride	419
Robert O. Hutchins	*Drexel University, Philadelphia, PA, USA*	
	• Sodium Cyanoborohydride	419
Wouter I. Iwema Bakker	*University of Waterloo, Waterloo, Ontario, Canada*	
	• Lithium Diisopropylamide	224
P. R. Jenkins	*University of Leicester, Leicester, UK*	
	• *N*-Bromosuccinimide	43

Subject Index